The Martin-Gay Student Success Program

Each Martin-Gay product is motivated by Elayn's firm belief that every student can succeed. This Student Success Program is designed to help students review and retain basic algebra concepts, and gain the knowledge of basic Euclidean geometry and study skills necessary for success in all levels of mathematics!

Options to support a variety of classroom environments!

MyMathLab®

The Video Organizer was written to help students achieve organized course notes and examples. This organizer contains an Outline presented in the same order as the Video Lecture Series along with all of the video Examples.

New Student Success Tips
My new 3–4 minute video segments on study skills are daily reminders to renew organizational and study habits.

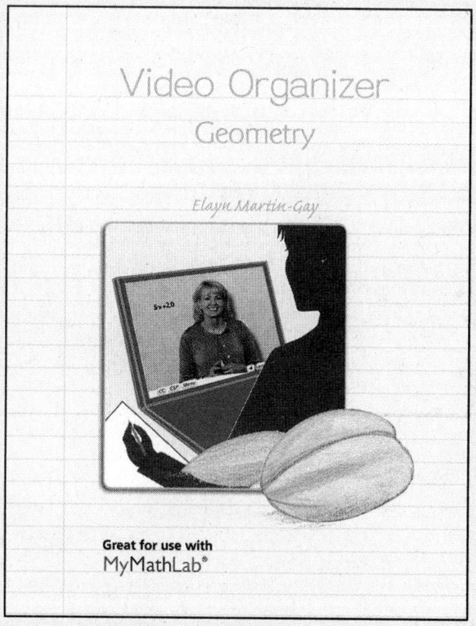

Martin-Gay Interactive Lecture Series Contains:

- Video Lecture Series
- Student Success Tips Videos
- Chapter Test Prep Videos

Complete Interactive Video Notes

Geometry

Elayn Martin-Gay

PEARSON

Boston Columbus Hoboken Indianapolis New York San Francisco
Amsterdam Cape Town Dubai London Madrid Milan Munich Paris Montréal Toronto
Delhi Mexico City São Paulo Sydney Hong Kong Seoul Singapore Taipei Tokyo

Editorial Director, Mathematics: *Christine Hoag*
Editor-in-Chief: *Michael Hirsch*
Acquisitions Editor: *Mary Beckwith*
Project Manager: *Christine Whitlock*
Project Team Lead: *Peter Silvia*
Assistant Editor: *Matthew Summers*
Editorial Assistant: *Megan Tripp*
Executive Development Editor: *Dawn Nuttall*
Program Team Lead: *Karen Wernholm*
Program Manager: *Patty Bergin*
Cover and Illustration Design: *Tamara Newnam*
Program Design Lead: *Heather Scott*
Director, Course Production: *Ruth Berry*
Executive Content Manager, MathXL: *Rebecca Williams*
Senior Content Developer, TestGen: *John Flanagan*
Media Producer: *Shana Siegmund/Marielle Guiney*
Executive Marketing Manager: *Jacquelyn Flynn*
Director of Marketing, Mathematics: *Roxanne McCarley*
Senior Marketing Manager: *Rachel Ross*
Marketing Assistant: *Kelly Cross*
Senior Author Support/Technology Specialist: *Joe Vetere*
Senior Procurement Specialist: *Carol Melville*
Interior Design, Production Management, Answer Art, and Composition: *Integra Software Services Pvt. Ltd.*
Text Art: *Scientific Illustrators*

Acknowledgments of third party content appear on pages P1–P2, which constitutes an extension of this copyright page.

Martin-Gay, K. Elayn
Geometry/Elayn Martin-Gay, University of New Orleans.—1st edition.
pages cm
Audience: College.
Summary: Euclidean geometry, an introduction. To be used with Algebra 1 & 2.—Publisher's note.
ISBN 0-13-417365-1 (alk. paper)
1. Geometry—Textbooks. 2. Geometry—Study and teaching (Higher) 3. Geometry—Study and teaching (Graduate) I. Title.
QA445.M285 2015
516.2—dc23

2014049493

1 2 3 4 5 6 7 8 9 10—CRK—19 18 17 16 15

PEARSON

www.pearsonhighered.com

ISBN-10: 0-13-417365-1 (Student Edition)
ISBN-13: 978-0-13-417365-8

Contents

CHAPTER 1 A Beginning of Geometry 1

1.1 Tips for Success in Mathematics 2
1.2 Geometry—A Mathematical System 7
1.3 Points, Lines, and Planes 12
1.4 Segments and Their Measure 19
1.5 Angles and Their Measure 25
1.6 Angle Pairs and Their Relationships 32
1.7 Coordinate Geometry—Midpoint and Distance Formulas 39
1.8 Constructions—Basic Geometry Constructions 44

CHAPTER 2 Introduction to Reasoning and Proofs 53

2.1 Perimeter, Circumference, and Area 54
2.2 Patterns and Inductive Reasoning 60
2.3 Conditional Statements 67
2.4 Biconditional Statements and Definitions 73
2.5 Deductive Reasoning 78
2.6 Reviewing Properties of Equality and Writing Two-Column Proofs 84
2.7 Proving Theorems About Angles 90

CHAPTER 3 Parallel and Perpendicular Lines 101

3.1 Lines and Angles 102
3.2 Proving Lines Are Parallel 108
3.3 Parallel Lines and Angles Formed by Transversals 115
3.4 Proving Theorems About Parallel and Perpendicular Lines 122
3.5 Constructions—Parallel and Perpendicular Lines 127
3.6 Coordinate Geometry—The Slope of a Line 131
3.7 Coordinate Geometry—Equations of Lines 139

CHAPTER 4 Triangles and Congruence 156

4.1 Types of Triangles 157
4.2 Congruent Figures 165
4.3 Congruent Triangles by SSS and SAS 170
4.4 Congruent Triangles by ASA and AAS 176
4.5 Proofs Using Congruent Triangles 182
4.6 Isosceles, Equilateral, and Right Triangles 190

CHAPTER 5 Special Properties of Triangles 204

5.1 Perpendicular and Angle Bisectors 205
5.2 Bisectors of a Triangle 214
5.3 Medians and Altitudes of a Triangle 222
5.4 Midsegments of Triangles 227
5.5 Indirect Proofs and Inequalities in One Triangle 234
5.6 Inequalities in Two Triangles 242

CHAPTER 6 Quadrilaterals 253

6.1 Polygons 254
6.2 Parallelograms 260
6.3 Proving that a Quadrilateral Is a Parallelogram 266
6.4 Rhombuses, Rectangles, and Squares 274
6.5 Trapezoids and Kites 282

CHAPTER 7 Similarity 293

7.1 Ratios and Proportions 294
7.2 Proportion Properties and Problem Solving 300
7.3 Similar Polygons 307
7.4 Proving Triangles Are Similar 312
7.5 Geometric Mean and Similarity in Right Triangles 320
7.6 Additional Proportions in Triangles 326

CHAPTER 8 Transformations 336

8.1 Rigid Transformations 337
8.2 Translations 341
8.3 Reflections 348

8.4 Rotations 353
8.5 Dilations 360
8.6 Compositions of Reflections 365
Extension—Frieze Patterns 371

CHAPTER 9 Right Triangles and Trigonometry 376

9.1 The Pythagorean Theorem and Its Converse 377
9.2 Special Right Triangles 383
9.3 Trigonometric Ratios 389
9.4 Solving Right Triangles 395
9.5 Vectors 401
Extension—Law of Sines 408
Extension—Law of Cosines 416

CHAPTER 10 Area 426

10.1 Angle Measures of Polygons and Regular Polygon Tessellations 427
10.2 Areas of Triangles and Quadrilaterals with a Review of Perimeter 436
10.3 Areas of Regular Polygons 445
10.4 Perimeters and Areas of Similar Figures 452
10.5 Arc Measures, Circumferences, and Arc Lengths of Circles 457
10.6 Areas of Circles and Sectors 464
10.7 Geometric Probability 471

CHAPTER 11 Surface Area and Volume 482

11.1 Solids and Cross Sections 483
11.2 Surface Areas of Prisms and Cylinders 492
11.3 Surface Areas of Pyramids and Cones 500
11.4 Volumes of Prisms and Cylinders and Cavalieri's Principle 506
11.5 Volumes of Pyramids and Cones 513
11.6 Surface Areas and Volumes of Spheres 518
11.7 Areas and Volumes of Similar Solids 525

CHAPTER 12 Circles and Other Conic Sections 534

12.1 Circle Review and Tangent Lines 535
12.2 Chords and Arcs 544
12.3 Inscribed Angles 551
12.4 Additional Angle Measures and Segment Lengths 558
12.5 Coordinate Plane—Circles 564
12.6 Locus 570
Extension—Parabolas 574

Student Success Resource Section 580

A Review of Basic Concepts 581

A.1 Measurement Conversions 581
A.2 Probability 582
A.3 Exponents, Order of Operations, and Variable Expressions 583
A.4 Operations on Real Numbers 584
A.5 Simplifying Expressions 586
A.6 Solving Linear Equations 587
A.7 Solving Linear Inequalities 588
A.8 Solving Formulas for a Variable 589
A.9 The Coordinate Plane 590
A.10 Graphing Linear Equations 591
A.11 Solving Systems of Linear Equations in Two Variables 592
A.12 Exponents 594
A.13 Multiplying Polynomials 595
A.14 Simplifying Radical Expressions 596
A.15 Solving Quadratic Equations by Factoring 597
A.16 Solving Quadratic Equations by the Square Root Property 598
A.17 Solving Quadratic Equations by the Quadratic Formula 599

Tables 600

1. Math Symbols 600
2. Formulas 601
3. Measures 603
4. Properties of Real Numbers 604

Postulates, Theorems, and Additional Proofs 605

B Additional Lessons 615

B.1 Ellipses and Hyperbolas 615

B.2 Measurement, Rounding Error, and Reasonableness 624

B.3 The Effect of Measurement Errors on Calculations 625

Answers to Selected Exercises A1

Index I1

Photo Credits P1

Preface

Geometry was written to provide a solid foundation in Euclidean geometry for students who might not have previous experience in geometry. Specific care was taken to make sure students have the most up-to-date relevant text preparation for their next mathematics course or for nonmathematical courses that require an understanding of geometry or logical reasoning. I have tried to achieve this by writing a user-friendly text that is keyed to objectives and contains many worked-out examples and illustrations. As suggested by AMATYC and the NCTM Standards (plus Addenda), real-life and real-data applications, data interpretation, conceptual understanding, problem solving, writing, cooperative learning, appropriate use of technology, mental mathematics, number sense, estimation, critical thinking, and geometric concepts are emphasized and integrated throughout the book.

What's in this text?

- **The Martin-Gay Program** and MyMathLab® actively encourage students to use the text, video program, and Video Organizer as an integrated learning system.
- **The Video Organizer** is designed to help students take notes and work practice exercises while watching the Interactive Lecture Series videos (available in MyMathLab). All content in the Video Organizer is presented in the same order as it is presented in the videos, making it easy for students to create a course notebook and build good study habits.
    - Covers all of the video examples in order.
    - Provides ample space for students to write down key definitions and properties.
    - Includes "Play" and "Pause" button icons to prompt students to follow along with the author for some exercises while they try others on their own.

The Video Organizer is available in a loose-leaf, notebook-ready format. It is also available for download in MyMathLab.

- **Student Success Tips Videos** are 3- to 5-minute video segments designed to be daily reminders to students to continue practicing and maintaining good organizational and study habits. They are organized in three categories and are available in MyMathLab and the Interactive Lecture Series. The categories are:

  1. Success Tips that apply to any course in college in general, such as Time Management.
  2. Success Tips that apply to any mathematics course. One example is based on understanding that mathematics is a course that requires homework to be completed in a timely fashion.
  3. Section- or Content-specific Success Tips to help students avoid common mistakes or to better understand concepts that often prove challenging. One example of this type of tip is how to apply the order of operations to simplify an expression.

- **Interactive Lecture Series**, featuring your text author (Elayn Martin-Gay), provides students with active learning at their own pace. The videos are available in MyMathLab and offer the following resources and more:

  A complete lecture for each section of the text highlights key examples and exercises from the text. "Pop-ups" reinforce key terms, definitions, and concepts.

  An interface with menu navigation features allows students to quickly find and focus on the examples and exercises they need to review.

  Student Success Tips Videos.

- **The Interactive Lecture Series** also includes the following resources for test prep:

  The Chapter Test Prep Videos help students during their most teachable moment—when they are preparing for a test. This innovation provides step-by-step solutions for the exercises found in each Chapter Test. The chapter test prep videos are also available on YouTube™. The videos are captioned in English and Spanish.
- **The Martin-Gay MyMathLab** course includes extensive exercise coverage and a comprehensive video program. There are section lecture videos for every section, which students can also access at the specific objective level; Student Success Tips Videos; and watch clips at the exercise level to help students while doing homework in MathXL.

Key Pedagogical Features

Chapters Chapters are divided into Sections. Below is an overview of a Section, then an Exercise Set, then the End-of-Chapter features.

Sections Each section begins with a list of Objectives. These objectives are also repeated at the place of discussion within the section. When applicable, under the list of objectives there is a list of new Vocabulary words. Throughout the section, each new vocabulary word is highlighted at place of definition.

Examples Detailed, step-by-step examples are available throughout each section. Many examples reflect real life and include illustrations. Additional instructional support is provided in the annotated examples.

Practice Exercises Throughout the text, each worked-out example has a parallel Practice exercise. These invite students to be actively involved in the learning process. Students should try each Practice exercise after finishing the corresponding example. Learning by doing will help students grasp ideas before moving on to other concepts. All answers to the Practice exercises are provided at the back of the text.

Helpful Hints Helpful Hints contain practical advice on applying mathematical concepts. Strategically placed where students are most likely to need immediate reinforcement, Helpful Hints help students avoid common trouble areas and mistakes.

Exercise Sets The exercise sets have been carefully written with a special focus on making sure that even- and odd-numbered exercises are paired and that they contain real-life applications and illustrations. In addition, many types of exercises were included to help students obtain a full conceptual knowledge of the section's topics. These types of exercises are labeled and include: Multiple Choice, Complete the Table, Multiple Steps, Sketch, Construction, Fill in the Blank, Complete the Proof, Proof, Coordinate Geometry, and Find the Error.

Overall, the exercises in an exercise set are written starting with less difficult ones and then increasing in difficulty. This allows students to gain confidence while working the earlier exercises. To help achieve this, the exercises at the beginning of a section are keyed to previously worked examples. If applicable, a section of Mixed Practice exercises are included. The odd answers to these exercises are found at the end of this text.

Vocabulary and Readiness Check These questions are immediately prior to a section's exercise set. These exercises quickly check a student's understanding of new vocabulary words. Also, the readiness exercises center on a student's understanding of a concept that is necessary in order to continue to the exercise set. The odd answers to these exercises are in the back of this text.

Applications Real-world and real-data application exercises occur in almost every exercise set and show the relevance of mathematics and geometry and help students gradually and continuously develop their problem-solving skills.

Concept Extensions These exercises are found toward the end of every exercise set, but before the Review and Preview exercises (described below). Concept Extension exercises require students to take the concepts from that section a step further by combining them with concepts learned in previous sections or by combining several concepts from the current section.

Writing Exercises These exercises occur in almost every exercise set and require students to provide a written response to explain concepts or justify their thinking.

Review and Preview Exercises These exercises occur at the end of each exercise set (except in Chapter 1) and are keyed to earlier sections. They review concepts learned earlier in the text that will be needed in the next section or chapter.

Exercise Set Resource Icons Located at the opening of each exercise set, these icons remind students of the resources available for extra practice and support:

MyMathLab®

See Student Resources descriptions on page xiii for details on the individual resources available.

End-of-Chapter The following features can be found at the end of each chapter. They are meant to give students an overall view of the chapter and thus help them have an understanding of how the concepts of a chapter fit together. All answers to these features below are found at the end of this text.

Mixed Practice Exercises In the section exercise sets, these exercises require students to determine the problem type and strategy needed to solve it just as they would need to do on a test.

Vocabulary Check This feature provides an opportunity for students to become more familiar with the use of mathematical terms as they strengthen their verbal skills. These appear at the end of each chapter before the Chapter Review.

Chapter Review The end of every chapter contains a comprehensive review of topics introduced in the chapter. The Chapter Review offers exercises keyed to every section in the chapter, as well as Mixed Review exercises that are not keyed to sections.

Chapter Test and Chapter Test Prep Videos The Chapter Test is structured to include those problems that involve common student errors. The **Chapter Test Prep Videos** give students instant access to a step-by-step video solution of each exercise in the Chapter Test.

Chapter Standardized Test After each Chapter Test, there is a standardized test. These chapter standardized tests are written to help students prepare for standardized tests in the future. They are multiple choice tests and cover the material presented in the associated chapter.

Student and Instructor Resources

STUDENT RESOURCES

Interactive Lecture Series Videos	Video Organizer	Student Solutions Manual
Provides students with active learning at their pace. The videos offer: • A complete lecture for each text section. The interface allows easy navigation to examples and exercises students need to review. • Interactive Concept Check exercises • Student Success Tips Videos • Chapter Test Prep Videos	Designed to help students take notes and work practice exercises while watching the Interactive Lecture Series Videos. • Covers all of the video examples in order. • Provides ample space for students to write down key definitions and rules. • Includes "Play" and "Pause" button icons to prompt students to follow along with the author for some exercises while they try others on their own. Available in loose-leaf, notebook-ready format and in MyMathLab.	Provides completely worked-out solutions to the odd-numbered section exercises; all exercises in the Integrated Reviews, Chapter Reviews, Chapter Tests, and Cumulative Reviews

INSTRUCTOR RESOURCES

Instructor's Solutions Manual (Available for download from the IRC) **TestGen®** (Available for download from the IRC)	**MyMathLab®** (access code required) **Instructor-to-Instructor Videos**—available in the Instructor Resources section of the MyMathLab course. **MathXL®** (access code required)

ABOUT THE AUTHOR

Elayn Martin-Gay has taught mathematics at the University of New Orleans for more than 25 years. Her numerous teaching awards include the local University Alumni Association's Award for Excellence in Teaching, and Outstanding Developmental Educator at University of New Orleans, presented by the Louisiana Association of Developmental Educators.

Prior to writing textbooks, Elayn Martin-Gay developed an acclaimed series of lecture videos to support developmental mathematics students in their quest for success. These highly successful videos originally served as the foundation material for her texts. Today, the videos are specific to each book in the Martin-Gay series. The author has also created Chapter Test Prep Videos to help students during their most "teachable moment"—as they prepare for a test—along with Instructor-to-Instructor videos that provide teaching tips, hints, and suggestions for each developmental mathematics course, including basic mathematics, prealgebra, beginning algebra, and intermediate algebra.

Elayn is the author of 12 published textbooks as well as multimedia, interactive mathematics, all specializing in developmental mathematics courses. She has also published series in Algebra 1 and Algebra 2. She has participated as an author across the broadest range of educational materials: textbooks, videos, tutorial software, and courseware. This provides an opportunity of various combinations for an integrated teaching and learning package offering great consistency for the student.

CHAPTER

1 A Beginning of Geometry

1.1 Tips for Success in Mathematics

1.2 Geometry—A Mathematical System

1.3 Points, Lines, and Planes

1.4 Segments and Their Measure

1.5 Angles and Their Measure

1.6 Angle Pairs and Their Relationships

1.7 Coordinate Geometry—Midpoint and Distance Formulas

1.8 Constructions—Basic Geometry Constructions

Below is an attempt to show a beginning of Euclidean Geometry. I hope you see that we are only scratching the surface in this text.

Geometry arose independently in many early cultures as a practical need for consistent surveying and measuring. The Greek philosopher Thales (6th century BC) is given credit for starting to formalize this topic.

I'll call it the Parallel Postulate.

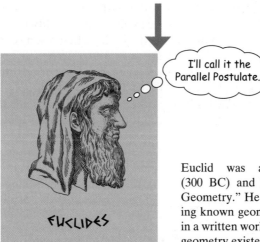

Euclid was a Greek mathematician (300 BC) and known as the "Father of Geometry." He is given credit for organizing known geometry (and number theory) in a written work called *Elements*. No other geometry existed for over 2000 years!

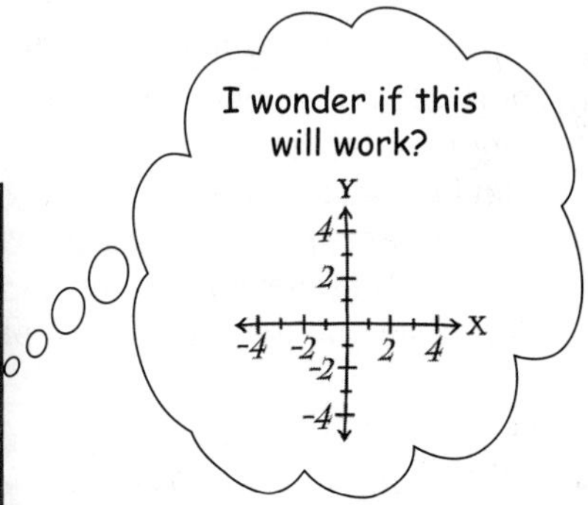

During the reign of Euclid's geometry, the 17th century mathematician René Descartes merged geometry and algebra. He "invented" the coordinate system we still use, called the plane coordinate system or the Cartesian coordinate system.

What is the geometry of our universe? See the Historical Note at the end of this chapter.

?

1.1 TIPS FOR SUCCESS IN MATHEMATICS

OBJECTIVES

1 Get Ready for This Course.

2 Understand Some General Tips for Success.

3 Understand How to Use the Resources Provided in MyMathLab and Math XL.

4 Get Help as Soon as You Need It.

5 Learn How to Prepare for and Take an Exam.

6 Develop Good Time Management.

Before reading this section, remember that your instructor (or teacher) is your best source for information. Please see your instructor (or teacher) for any additional help or information.

OBJECTIVE 1 ▶ Getting Ready for This Course. Now that you have decided to take this course, remember that a *positive attitude* will make all the difference in the world. Your belief that you can succeed is just as important as your commitment to this course. Make sure you are ready for this course by having the time and positive attitude that it takes to succeed.

Next, make sure you have scheduled your math course at a time that will give you the best chance for success. For example, if you are also working, you may want to check with your employer to make sure that your work hours will not conflict with your course schedule. Also, schedule your class during a time of day when you are more attentive and do your best work.

This online geometry course is different from traditional math courses that you have taken in the past. You will work exercises and complete homework online. Because of this, it is your responsibility to keep a written notebook or journal of your work. You will need this documentation of your work when it comes to studying for a quiz, exam, or test.

On the day of your first class period, double-check your schedule and allow yourself plenty of time to arrive. Make sure you bring a notebook or binder, paper, and a pencil or some other writing instrument. If you are required to have a lab manual, graph paper, calculator, or some other supply, bring these with you also.

OBJECTIVE 2 ▶ General Tips for Success. Below are some general tips that will increase your chance for success in a mathematics class. Many of these tips will also help you in other courses you may be taking.

Note: Many tips have to do with the specifics of this online course and will be listed in Objective 3.

Exchange names and phone numbers with at least one other person in class. This contact person can be a great help if you miss an assignment or want to discuss math concepts or exercises that you find difficult.

Choose to attend all class periods and be on time. If possible, sit near the front of the classroom. This way, you will see and hear the presentation better. It may also be easier for you to participate in classroom activities.

Do your homework. You've probably heard the phrase "practice makes perfect" in relation to music and sports. It also applies to mathematics. You will find that the more time you spend solving math exercises, the easier the process becomes. In this online course, homework can be submitted as many times as you like. This means you can work and rework those exercises that you struggle with until you master them. It is a good idea to work through all homework exercises twice before the submission deadline. Also, be sure to schedule enough time to complete your assignments before the due date assigned by your teacher.

Check your work. Checking work is the same for an online course as it is for a traditional course. This is why it is imperative that you work each exercise on paper before submitting the answer. If it's on paper, you can go back, check your work, and follow your steps to ensure the answer is correct, or find any mistakes and correct them. If you can't find your mistake or if you have any questions, make sure you talk to your teacher.

Learn from your mistakes and be patient with yourself. Everyone, even your instructor, makes mistakes. (That definitely includes me–Elayn Martin-Gay.) Use your errors to learn and to become a better math student. The key is finding and understanding your errors.

Was your mistake a careless one, or did you make it because you can't read your own math writing? If so, try to work more slowly or write more neatly and make a conscious effort to carefully check your work.

Did you make a mistake because you don't understand a concept? Take the time to review the concept or ask questions to better understand it.

Did you skip too many steps? Skipping steps or trying to do too many steps mentally may lead to preventable mistakes.

Know how to get help if you need it. It's always a good idea to ask for help whenever there is something that you don't understand. One great advantage about doing homework in MyMathLab is that there is built-in "help" whenever you need it. Should you get a wrong answer, a box will appear and offer you hints for working the exercise correctly. Again, this is why it is so important to keep a record of all your work on paper so you can go back and follow the suggestions. You will have three attempts to get each exercise correct before it is marked wrong. **Remember:** Even though you are working online, your teacher is your most valuable resource for answering questions. Having a written journal of neatly worked exercises helps your teacher identify mistakes on an exercise or about a concept in general.

Organize your class materials, including homework assignments, graded quizzes and tests, and notes from your class or lab. All of these items will be valuable references throughout your course, especially when studying for upcoming tests or your final exam. Make sure you can locate these materials when you need them. An excellent way to do this is by using the Organizer, which is reviewed in Objective 3.

Read your ebook or watch the section lecture videos before class. Your course ebook is available through MyMathLab. Use this online text just as you would a printed textbook. Read the assigned section(s), and then write down any questions you may have. You will then be prepared to ask any questions in class the next day. Also, familiarizing yourself with the material before class will help you understand it much more readily when it is presented. There is also a reading assessment homework if assigned by your teacher.

Lecture videos, approximately 20 minutes in length, are available for every section of your ebook. These videos are specific to the material in the ebook and are presented by the ebook author, Elayn Martin-Gay. Watching a section video before class is another way to familiarize yourself with the material. Write down any questions you may have so that you can ask them in class. Watching a section video after class is also an excellent way to review concepts that are difficult for you.

Don't be afraid to ask questions. Teachers are not mind readers. Many times they do not know a concept is unclear until a student asks a question. You are not the only person in class with questions. Other students are normally grateful that someone has spoken up.

Turn in assignments on time. Always be aware of the schedule of assignments and due dates set by your teacher. Do not wait until the last minute to submit your work online. It is a good idea to submit your assignments 6–8 hours before the submission deadline to ensure some "cushion" time in case you have technology trouble.

When assignments are turned in online, it is extremely important for you to keep a copy of your written work. You will find it helpful to organize this work in a 3-ring binder. This way, you can refer to your written work to ask questions and can use it later to study for tests. (See the Video Organizer in Objective 3.)

OBJECTIVE 3 ▶ Understanding How to Use the Resources Provided in MathXL and MyMathLab. There are many helpful resources available to you through MathXL and MyMathLab. It is important that you understand these resources and know when to use them. Let's start with the resources that are available within MathXL to help you successfully complete and master the exercises in your assigned homework. When working your homework assignments, you will find the following buttons listed on the right-hand side of the screen.

- **Help Me Solve This**—Select this resource to get guided, step-by-step help for the exercise you are working. Once you have reached the correct answer (through the help feature), you must work an additional exercise of the same type before you receive credit for having worked it correctly.
- **View an Example**—Select this resource to view a correctly worked example similar to the exercise you are working on. After viewing the example, you can go back to your original exercise and complete it on your own.
- **Textbook**—Select this resource to go to the section of the ebook where you can find exercises similar to the one you are working on.
- **Video**—Select this resource to view a video clip of Elayn Martin-Gay (your ebook author) working an exercise similar to the one you need help with. **Not all exercises have an accompanying video clip. This button will not be listed if no video clip is available.

Let's now take a moment to go over a few of the features available in MyMathLab to help you prepare for class, review outside class, organize, improve your study skills, and succeed.

- **Ebook and Videos**—You can choose to read the ebook and/or watch the videos for every section of the text. The ebook includes worked examples, helpful hints, practice exercises, and section exercises for every text section. Read the material actively, and make a note of any questions you have so that you can then ask them in class.

  There are lecture **videos,** approximately 20 minutes in length, for every ebook section. Watch these videos to prepare for class, to review after class, or to help you catch up if you miss class. The videos are presented by your ebook author, Elayn Martin-Gay, so all material covered in the videos is consistent with the coverage in your ebook. Make a note of any questions you have after watching the videos so that you can ask your instructor. Your instructor may assign watching the videos as homework to prepare for class.
- **Video Organizer**—The **Video Organizer** is designed to help you take notes and work practice exercises while watching Elayn Martin-Gay's lecture series (available in MyMathLab and on DVD). All content in the Video Organizer is presented in the same order as it is presented in the videos, making it easy for students to create a course notebook and build good study habits!

  - Covers all of the video examples in order
  - Provides sample space for student to write down key definitions and rules
  - Includes "Play" and "Pause" button icons to prompt students to follow along with Elayn for some exercises while they try others on their own

  The Video Organizer is available in a loose-leaf, notebook-ready format. It is also available for download in MyMathLab.

OBJECTIVE 4 ▸ Getting Help. If you have trouble completing assignments or understanding the mathematics, get help as soon as you need it! This tip is presented as an objective on its own because it is so important. In mathematics, usually the material presented in one section builds on your understanding of the previous section. This means that if you don't understand the concepts covered during a class period, there is a good chance that you will not understand the concepts covered during the next class period. If this happens to you, get help as soon as you can.

Where can you get help? Many suggestions have been made in the section on where to get help, and now it is up to you to do it. Try your instructor, a tutoring center, or a math lab, or you may want to form a study group with fellow classmates. If you do decide to see your instructor or go to a tutoring center, make sure that you have a neat notebook and are ready with your questions.

OBJECTIVE 5 ▶ Preparing for and Take an Exam. Make sure that you allow yourself plenty of time to prepare for a test. If you think that you are a little "math anxious," it may be that you are not preparing for a test in a way that will ensure success. The way that you prepare for a test in mathematics is important. To prepare for a test,

1. Review your previous homework assignments. You may also want to rework some of them.
2. Review any notes from class and section-level quizzes you have taken. (If this is a final exam, also review chapter tests you have taken.)
3. Practice working out exercises by completing the Chapter Review found at the end of each chapter.
4. Since homework exercises are online, you may easily work new homework exercises. If you open an already submitted homework assignment, you can get new exercises by clicking "similar exercise." This will generate new exercises similar to the homework exercises you have already submitted. You can then work and rework exercises until you fully understand them. *Don't stop here!*
5. It is important that you place yourself in conditions similar to test conditions to find out how you will perform. In other words, as soon as you feel that you know the material, try taking some sample tests.
   In your ebook, there are two forms of chapter tests at the end of each chapter. One form is an open response test form and the second is a standardized test form. You can use these two tests as practice tests. Do not use your notes or any other help when completing these tests. Check your answers by using the answer section in the ebook. There are also exact video clip solutions to the open response practice test form. Finally, identify any concepts that you do not understand and consult your teacher.
6. Get a good night's sleep before the exam.
7. On the day of the actual test, allow yourself plenty of time to arrive at your exam location.

When taking your test,

1. Read the directions on the test carefully.
2. Read each problem carefully as you take the test. Make sure that you answer the question asked.
3. Pace yourself by first completing the problems you are most confident with. Then work toward the problems you are least confident with. Watch your time so you do not spend too much time on one particular problem.
4. Do not turn your test in early. If you have extra time, spend it double-checking your work and answers.

OBJECTIVE 6 ▶ Managing Your Time. As a student, you know the demands that classes, homework, work, and family place on your time. Some days you probably wonder how you'll ever get everything done. One key to managing your time is developing a schedule. Here are some hints for making a schedule:

1. Make a list of all of your weekly commitments for the term. Include classes, work, regular meetings, extracurricular activities, etc.
2. Next, estimate the time needed for each item on the list. Also make a note of how often you will need to do each item. Don't forget to include time estimates for the reading, studying, and homework you do outside of your classes. You may want to ask your instructor for help estimating the time needed.
3. In the following exercise set, you are asked to block out a typical week on the schedule grid given. Start with items with fixed time slots like classes and work.
4. Next, include the items on your list with flexible time slots. Think carefully about how best to schedule items such as study time.
5. Don't fill up every time slot on the schedule. Remember that you need to allow time for eating, sleeping, and relaxing! You should also allow a little extra time in case some items take longer than planned.
6. If you find that your weekly schedule is too full for you to handle, you may need to make some changes in your workload, classload, or in other areas of your life. If you work, you may want to talk to your advisor, manager or supervisor, or someone in your school's counseling center for help with such decisions.

1.1 EXERCISE SET MyMathLab®

1. How many times is it suggested that you work through homework exercises before the submission deadline?
2. How does the "Help Me Solve This" feature work?
3. Why is it important that you write your step-by-step solutions to homework exercises and keep a hard copy of all work submitted online?
4. How many times are you allowed to submit homework online?
5. How can the lecture videos for each section help you in this course? When is the best time to use them?
6. In the homework assignments, how many attempts do you get to correct an exercise before it is marked incorrect?
7. If the "View an Example" feature is used, is it necessary to work an additional exercise before continuing the assignment?
8. How does reading the ebook section before class help you prepare for class?
9. Do all homework exercises in MyMathLab come with an accompanying video clip solution?
10. How can you use MathXL to contact your teacher about an exercise you don't understand?
11. When are your homework assignments due?
12. How much "cushion" time is recommended before your deadline when submitting homework online?
13. Is it still OK to ask your teacher for help even though this is an online course?
14. Name two ways you can prepare for any chapter tests.
15. If you are absent, name two ways you can review the material you missed.
16. List the resources available in MyMathLab to help you. Which of these resources do you think will be most helpful to you?
17. Are you allowed to use a calculator in this class?
18. Review Objective 6 and fill in the schedule grid that follows.
19. Study your completed grid from Exercise 18. Decide whether you have the time necessary to successfully complete this course and any others you have scheduled.

	Monday	*Tuesday*	*Wednesday*	*Thursday*	*Friday*	*Saturday*	*Sunday*
4:00 a.m.							
5:00 a.m.							
6:00 a.m.							
7:00 a.m.							
8:00 a.m.							
9:00 a.m.							
10:00 a.m.							
11:00 a.m.							
12:00 p.m.							
1:00 p.m.							
2:00 p.m.							
3:00 p.m.							
4:00 p.m.							
5:00 p.m.							
6:00 p.m.							
7:00 p.m.							
8:00 p.m.							
9:00 p.m.							
10:00 p.m.							
11:00 p.m.							
Midnight							
1:00 a.m.							
2:00 a.m.							
3:00 a.m.							

1.2 GEOMETRY—A MATHEMATICAL SYSTEM

OBJECTIVES

1. Use Logic to Recognize Patterns.
2. Understand How a Mathematical System, Like Geometry, Is Formed.

OBJECTIVE 1 ▶ Using Logic to Recognize Patterns. The word **geometry** comes from the Greek language and means to *measure the earth*. (For more on the origin of geometry, see the Chapter 1 Opener, page 1.)

▶ Helpful Hint

To be more specific, "geo" means earth and "metron" means measurement.

We use geometry in many different situations. It is still needed for measuring the perimeter of a backyard for the purpose of constructing a fence, or for calculating the area of a backyard for the purpose of ordering grass seed. Nowadays, the use of geometry is all around you, and some applications can easily be seen, especially in architecture and construction. Look at the angles of a building—the windows, the roof, any brick or stone placement, etc. The next time you approach a bridge, notice the engineering and geometry involved in constructing the bridge. More recently, geometry is greatly used in the construction of cell phones and cell towers, GPS systems, and in the development of 3-D movies.

VOCABULARY

- geometry
- logic
- undefined term
- defined term
- postulate
- axiom
- theorem

Helpful Hint

The number of mathematical systems has become so large, that there is a Mathematics Classification System. It is used by authors of research papers to clearly classify the subject of their paper.

Below, we take great liberty and diagram a few of the major divisions that you may be familiar with.

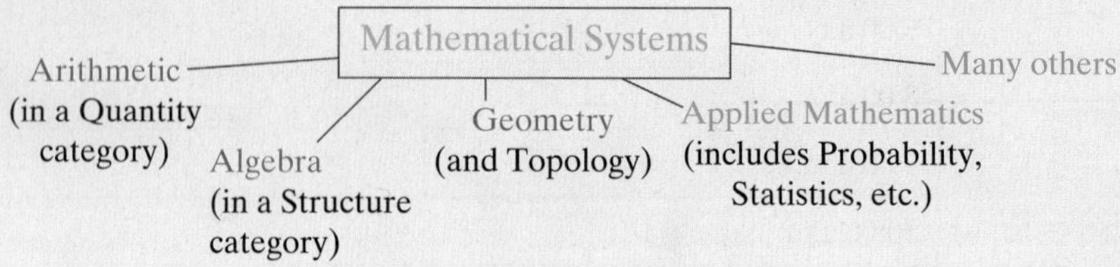

Perimeter needed for fencing.
Area of yard needed for grass seed.

Notice the angles in this building.

Ravenel Bridge near Charleston, SC

Before we talk more about geometry and mathematical systems, let's check our reasoning skills, or **logic.** We need to learn how to use reason and logic in order to recognize patterns, make conjectures, and develop the laws of this system. Let's start to examine the mathematical system of geometry by first recognizing patterns and making conjectures. Later in this text, we will discuss reasoning further, including different types of reasoning.

We can use logic to look for patterns in figures, as shown in Example 1.

EXAMPLE 1 Using Logic to Recognize Patterns

Sketch the next figure in this pattern.

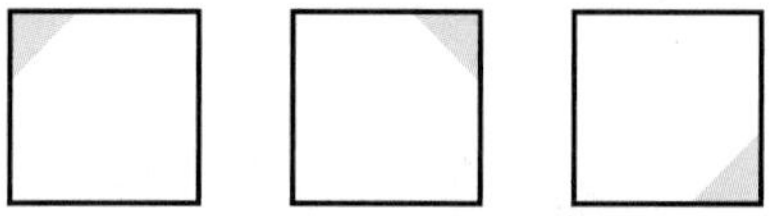

Solution From left to right, each figure is a square, with a triangle shaded in a different corner. From left to right, the shading is moving clockwise. Thus, we predict that the next figure is

.

□

PRACTICE 1 Sketch the next figure in the pattern.

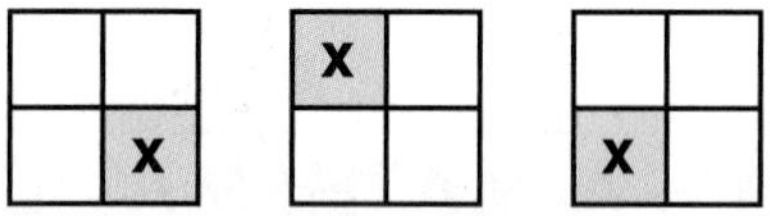

Patterns can also appear in lists of numbers.

EXAMPLE 2 Using Logic to Recognize Patterns

Look for a pattern and predict the next number.

a. 2, 10, 50, 250, ... **b.** −3, 0, 3, 6, 9, ...

Solution **a.** 2, 10, 50, 250
×5 ×5 ×5

Each number is 5 times the previous number. We predict the next number to be 1250 because

$$250 \times 5 = 1250$$

b. Each number is 3 more than the previous number. We predict the next number to be 12 because

$$9 + 3 = 12$$

□

PRACTICE 2 Look for a pattern and predict the next number.

a. $-1, 1, -2, 2, -3, \ldots$

b. $\frac{1}{3}, \frac{1}{9}, \frac{1}{27}, \ldots$

OBJECTIVE 2 ▶ Understanding How a Mathematical System Is Formed. Now we can use our logic and reasoning skills to develop the mathematical system of geometry. How do we begin?

We begin with **undefined terms,** which we first describe.

Then we use these undefined terms to formally define terms. From these, we use our **defined terms** to write statements that we do not prove, but instead agree and accept them to be true. These statements are called **postulates** or **axioms.** Our mathematical system grows by using terms, postulates, and axioms to prove **theorems.**

It continues to grow by using theorems to prove other theorems.

▶ Helpful Hint

What is the difference between a postulate and an axiom? Although we may use these words interchangeably, the table below may help:

	Axiom	*Postulate*
Similar	Do *not* prove these statements	
Possible Differences	Thought of as a global, self-evident truth or common notion	A truth that is more specific to geometry

A Mathematical System consists of

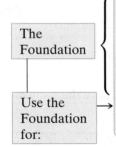

The Foundation:

Undefined terms are the most basic of terms that we do not formally define, but have a meaning or description that we agree upon.

To form **definitions,** we use undefined terms or already defined terms to formally define them.

Postulates (or Axioms) are statements that we accept as true and do not try to prove.

Use the Foundation for: →

Theorems are statements that we prove using logic, the foundation above, and other previously proved theorems.

Some examples of undefined terms are *point*, *line*, *plane*, and the idea of *between*. For example, what does it mean for a point on a line to be *between* two other points?

In the next section, we begin to develop our mathematical system called geometry.

Sometimes in geometry we may be faced with unmarked figures. If this occurs, we must be careful when making assumptions about these figures. For example, let's look at the following figures and learn the assumptions that we may or may not make.

Figure Assumptions

Be careful when assuming anything from a marked or unmarked figure.
Examples: Can you assume that …

- This is a line?

Yes

- This is a right angle (measures 90°)?

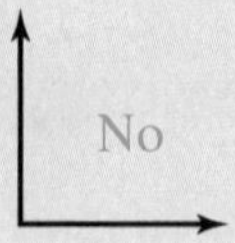

No

- This is a right angle?

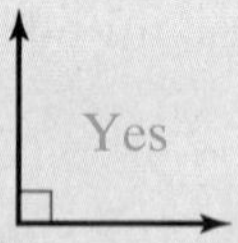

Yes

- These angles are supplementary (sum of 180°)?

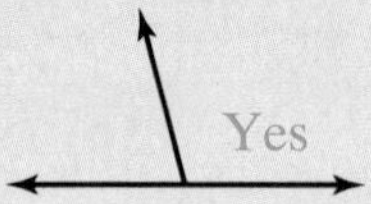

Yes

- These angles have a common side?

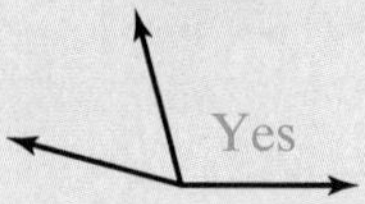

Yes

- These two lines intersect to form angles?

Yes

- These angles are complementary (sum of 90°)?

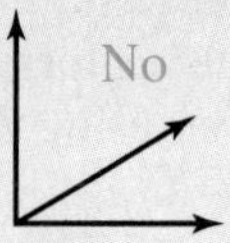

No

- These angles are complementary (sum of 90°)?

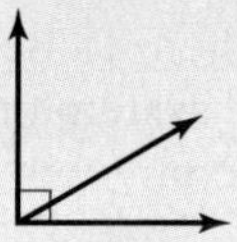

Yes

- These segments have equal length?

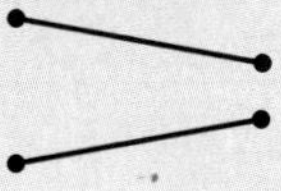

No

- These segments have equal length?

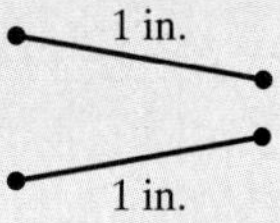

Yes

In Example 3 and associated exercises, we assume a basic knowledge of squares, triangles, rectangles, and some angle measures.

EXAMPLE 3 Given the figure, can you conclude the statement?

This is a square.

Solution No, we cannot assume that the figure is a square. A square has 4 right angles and 4 sides of equal length. Correctly marked,

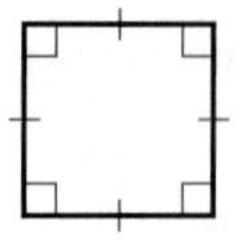

This is a square.

Helpful Hint
The same number of tick marks on sides indicates equal length.

PRACTICE 3 Given the figure, can you conclude the statement?

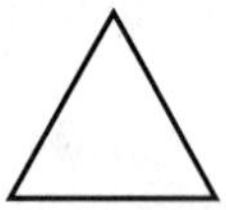

The angles in this triangle have the same measure.

VOCABULARY & READINESS CHECK

Word Bank *Use the choices to fill in each blank. Some choices may be used more than once and some not at all.*

undefined term	axiom	theorem
postulate	defined term	

1. Which word or phrase above describes a word that has been defined? ____________
2. Which word or phrase above describes a word that is described, but not defined? ____________
3. A statement that we prove is called a(n) ____________.
4. A statement that we accept as true but do not prove is called a(n) ____________ or a(n) ____________.

1.2 EXERCISE SET MyMathLab®

Look for a pattern and sketch the next figure. See Example 1.

1. 

2. 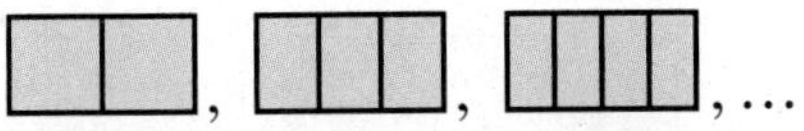

3. 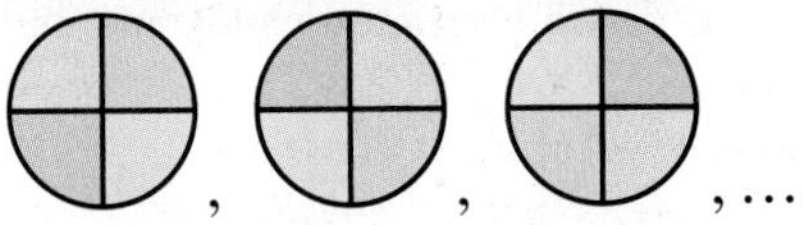

4. 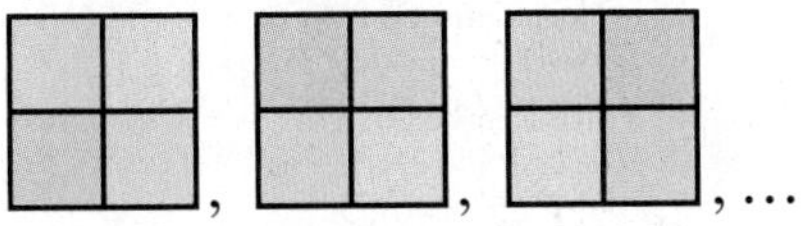

5. 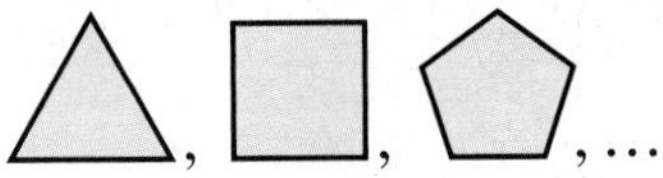

6. 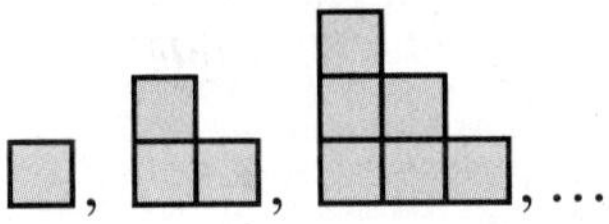

7. 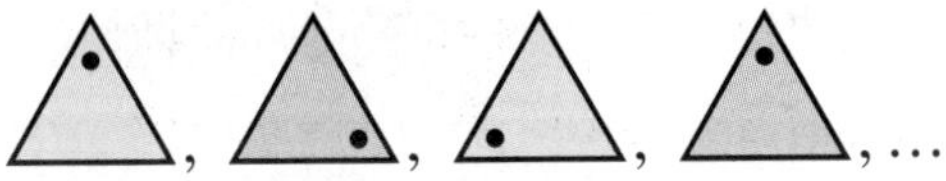

8. 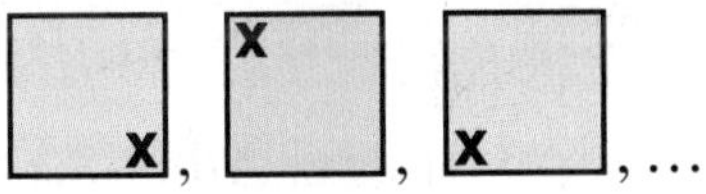

9. 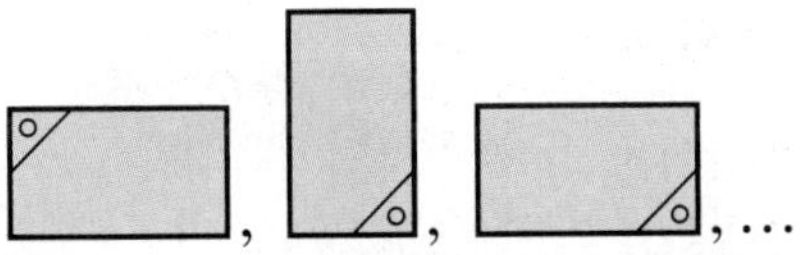

10. 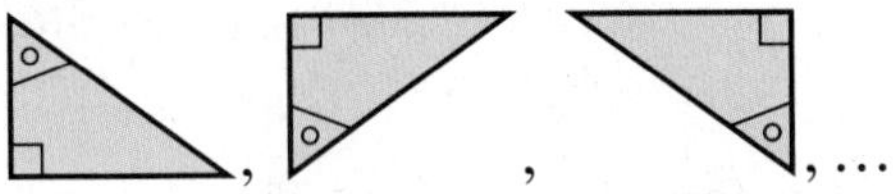

Look for a pattern and predict the next two numbers. See Example 2.

11. $-10, 0, 10, 20, \ldots$

12. $-6, -4, -2, 0, \ldots$

13. $1, \frac{1}{2}, \frac{1}{4}, \frac{1}{8}, \ldots$

14. $1, \frac{1}{2}, \frac{1}{3}, \frac{1}{4}, \ldots$

15. $1, 4, 16, 64, \ldots$

16. $2, 6, 18, 54, \ldots$

17. $\frac{1}{2}, \frac{1}{10}, \frac{1}{50}, \frac{1}{250}, \ldots$

18. $\frac{1}{5}, \frac{1}{10}, \frac{1}{20}, \ldots$

19. $-7, 0, 7, 0, -7, \ldots$

20. $-1, 0, 1, 0, -1, \ldots$

Given each figure, conclude whether or not each statement is true. Answer yes or no. See the Figure Assumptions in this section and Example 3.

21. 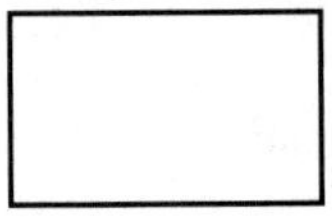

This is a rectangle. (A rectangle has 4 right angles.)

22. 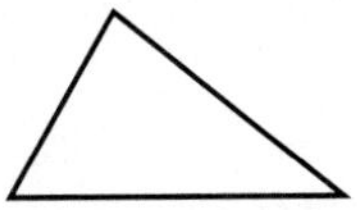

This is a triangle.

23. 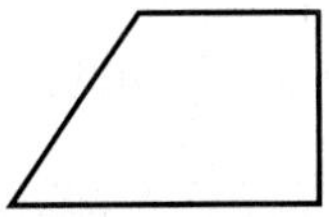

This is a quadrilateral. (A quadrilateral has 4 sides.)

24. 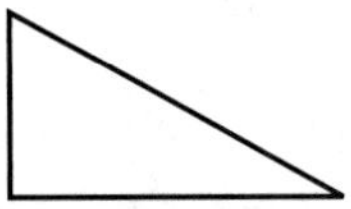

This is a right triangle (a triangle with a right angle).

25. 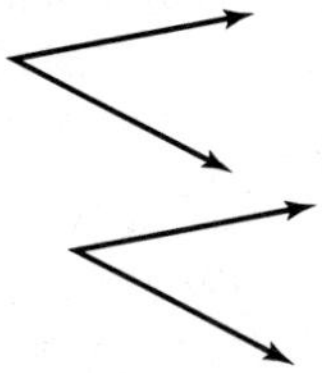

These angles have the same measures.

26. 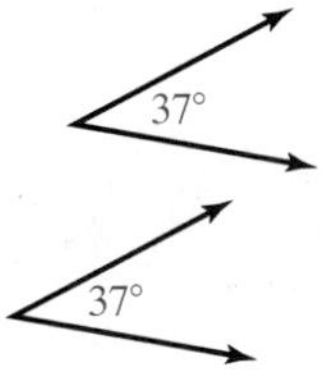

These angles have the same measures.

27. 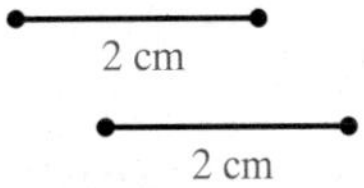

These segments have the same length.

28. 

These segments have the same length.

29. 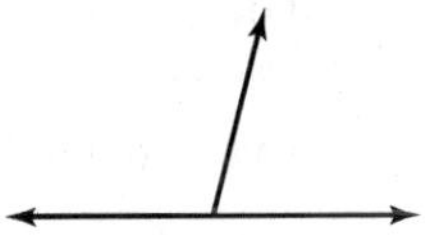

These angles are supplementary.

30. 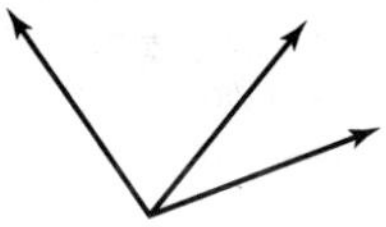

These angles have a common side.

31. 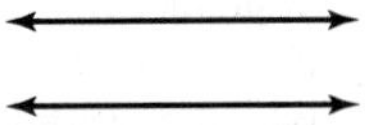

These lines are parallel. (If extended, they never touch.)

32. 

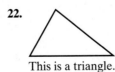

This is a parallelogram. (A parallelogram has opposite sides parallel.)

CONCEPT EXTENSIONS

Find the Error *A fellow student shows you his/her next number or figure in a list. Decide whether this next number or figure is correct. If not correct, tell why and provide the correct one.*

33. $\frac{1}{2}, \frac{1}{4}, \frac{1}{8}, \ldots$? $\frac{1}{64}$ Square the denominator to get the next denominator.

34. $\frac{1}{2}, \frac{1}{4}, \frac{1}{6}, \ldots$? $\frac{1}{12}$ Double the denominator to get the next denominator.

35. 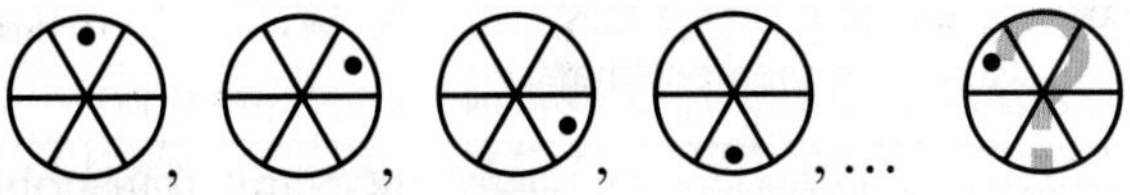

36. 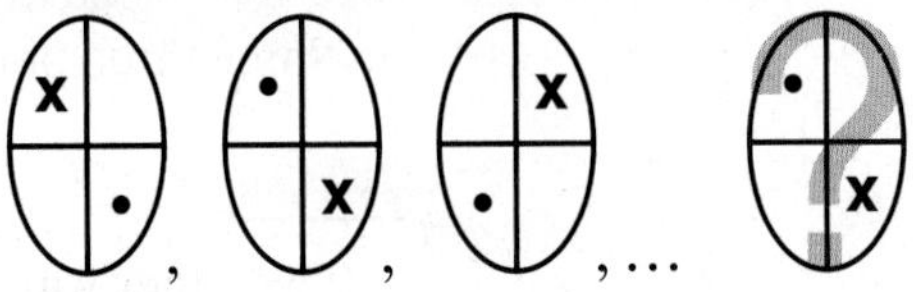

37. Describe a difference between a postulate and a theorem.

38. Describe a similarity between a postulate and an axiom.

1.3 POINTS, LINES, AND PLANES

OBJECTIVE

1 Learn the Basic Terms and Postulates of Geometry.

VOCABULARY

- point
- line
- plane
- lie on
- collinear
- coplanar
- space
- geometric figure
- between
- segment or line segment
- ray
- opposite rays
- intersection

OBJECTIVE 1 ▶ Learning Terms and Postulates of Geometry. Let's begin to examine the mathematical system of geometry as we should—with the undefined terms of point, line, and plane. These are described after the Helpful Hint.

▶ Helpful Hint

This is the beginning of the language of geometry. It is very important to study and understand ALL of the new vocabulary words and phrases.

You may want to start your own index card listing of vocabulary words.

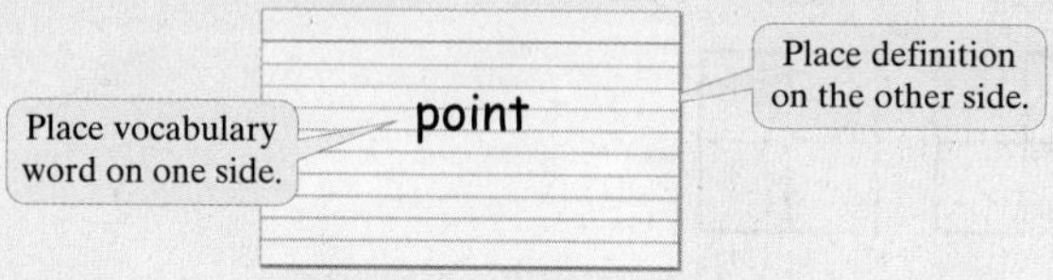

Undefined Terms

Description	*How to Name It*	*Example*
A **point** has no dimension (no length, width, or height). It does have a location, or position.	Name a point by a single capital letter.	*A* • or Point *A*
A **line** extends in opposite directions without end and has one dimension—length.	Name a line by a single lowercase letter or by any two points on the line.	 Line ℓ or $\overleftrightarrow{AB}$ or $\overleftrightarrow{BA}$
A **plane** extends in two dimensions without end. The two dimensions are length and width, but no thickness. We represent a plane by a flat surface.	Name a plane by a single capital letter or by any three points on the plane (that do not lie on the same line).	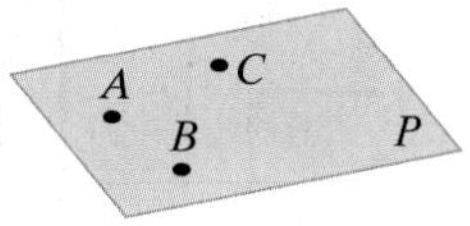 Plane *P* or plane *ABC*

Now we use our undefined terms to write a few formal definitions. The phrase **lie on** is used in both definitions below. This is an undefined concept that is assumed to be commonly understood.

Collinear points are points that lie on the same *line*.

Coplanar points are points that lie on the same *plane*.

▶ Helpful Hint

In this text:

- A line is a straight line.
- A line contains an infinite number of points.
- A plane contains an infinite number of lines.
- Our drawings of planes appear to end and have edges, but we must remember that planes extend without end.

EXAMPLE 1 **Naming Points, Lines, and Planes**

a. Write two other ways to name $\overleftrightarrow{QT}$.

b. Write two other ways to name plane *P*.

c. Name three points that are collinear.

d. Name four points that are coplanar.

e. Name three points that are not collinear.

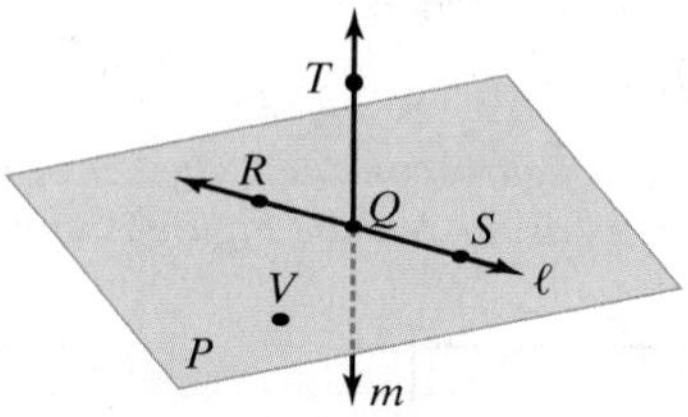

Solution

a. Two other ways to name $\overleftrightarrow{QT}$ are $\overleftrightarrow{TQ}$ and line *m*.

b. There are many correct answers. Two other ways to name plane *P* are plane *RQV* and plane *RSV*.

c. Points *R*, *Q*, and *S* lie on the same line, so they are collinear.

> **Helpful Hint**
> Do *not* use collinear points to name a plane since many planes exist that go through a single line.

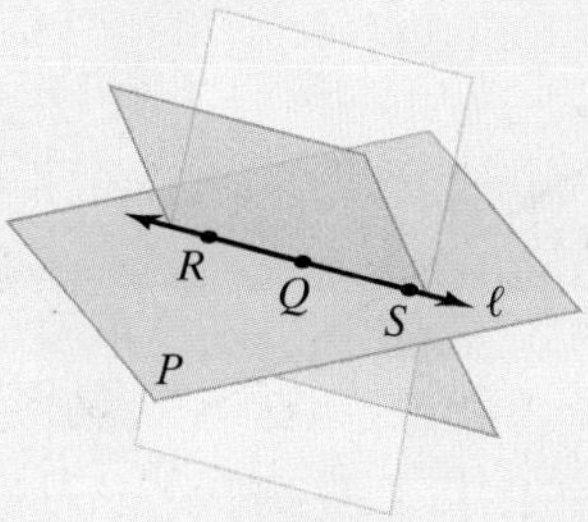

d. Points R, Q, S, and V lie on the same plane, so they are coplanar.

e. There are many correct answers. For instance, points T, Q, and V do not lie on the same line. □

PRACTICE 1

a. Write two other ways to name $\overleftrightarrow{EH}$.

b. Write two other ways to name plane M.

c. Name three points that are collinear.

d. Name four points that are coplanar.

e. Name three points that are not collinear.

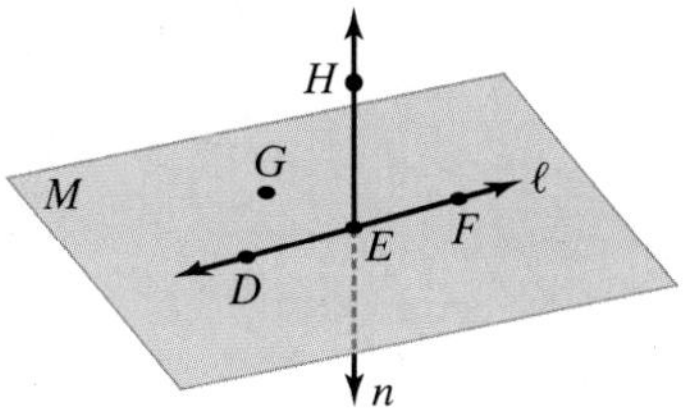

> **Helpful Hint**
> Notice that by our description on the previous page, the simplest geometric figure is a point.

Let's continue to use our undefined terms to define another term.

Space is the set (a collection of objects) of all points in three dimensions.
A **geometric figure** is any nonempty subset of space (means that a geometric figure contains points).

As mentioned earlier, the idea of between is another undefined term or concept. We have a basic idea of what it means for a point on a line to be between two other points on a line.

Use the term **between** when points are on the same line (or line segment).

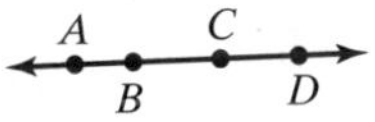

Point B is between points A and D.
Point B is between points A and C.

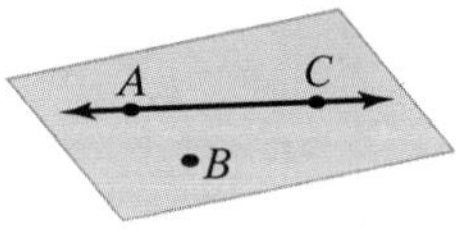

Point B is NOT between points A and C.

Defined Terms

Definition	*How to Name It*	*Example*
A **line segment** or simply **segment** is part of a line. It consists of two endpoints and all the points between them.	Name a segment by its end points: $\overline{AB}$ (segment AB) or $\overline{BA}$ (segment BA)	segment A B $\overline{AB}$ or $\overline{BA}$
A **ray** is part of a line. It consists of an endpoint and all points of a line on one side of the endpoint.	Name a ray by its endpoint and any other point on the ray. Here, the order of points is important—list the endpoint first. $\overrightarrow{AB}$ (ray AB)	ray A B $\overrightarrow{AB}$ ray A B $\overrightarrow{BA}$
Opposite rays are two rays that share the same endpoint and form a line.	Name each opposite ray as you would name a ray.	opposite rays A C B $\overrightarrow{CB}$ and $\overrightarrow{CA}$

> **Helpful Hint**
> - Rays: Take care when naming rays. For example, $\overrightarrow{AB}$ and $\overrightarrow{BA}$ are *not* the same ray, $\overrightarrow{AB}$ has endpoint A and $\overrightarrow{BA}$ has endpoint B.
> - Lines: $\overleftrightarrow{AB}$ and $\overleftrightarrow{BA}$ name the same line.
> - Segments: $\overline{AB}$ and $\overline{BA}$ name the same segment.

EXAMPLE 2 Naming Segments and Rays

Use the given figure to find the following.

a. Name the segments in the figure.

b. Name the rays in the figure.

c. Which of the rays in part **b** are opposite rays?

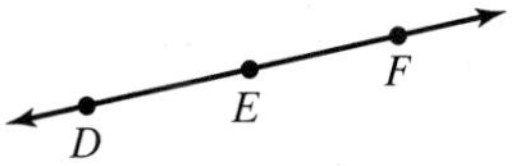

Solution

a. The three segments are $\overline{DE}$ or $\overline{ED}$, $\overline{EF}$ or $\overline{FE}$, and $\overline{DF}$ or $\overline{FD}$.

b. The four rays are $\overrightarrow{DE}$ or $\overrightarrow{DF}$, $\overrightarrow{ED}$, $\overrightarrow{EF}$, and $\overrightarrow{FD}$ or $\overrightarrow{FE}$.

c. The opposite rays are $\overrightarrow{ED}$ and $\overrightarrow{EF}$. □

PRACTICE 2 Use the given figure to find the following.

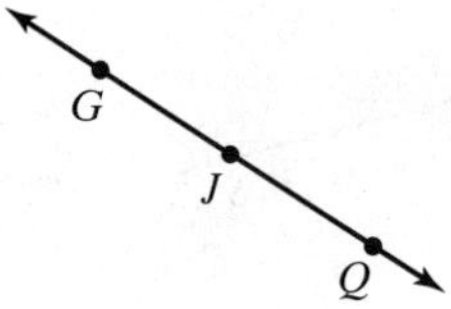

a. Name the segments.

b. Name the rays.

c. Which of the rays in part b are opposite rays?

Once we know how to name terms, we can draw them and use our proper labeling, as in Example 3.

EXAMPLE 3 Drawing Lines, Segments, and Rays

Perform parts **a** through **d** in order to create a single figure.

a. Draw three noncollinear points, M, N, and L.

b. Draw $\overleftrightarrow{MN}$.

c. Draw $\overline{NL}$.

d. Draw $\overrightarrow{LM}$.

Solution

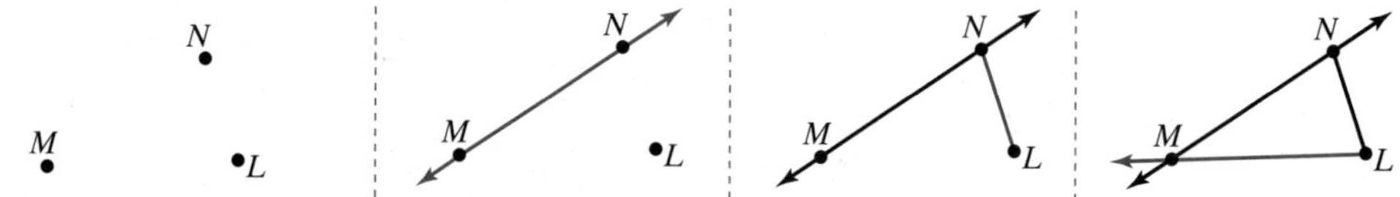

a. Draw M, N, and L. **b.** Draw $\overleftrightarrow{MN}$. **c.** Draw $\overline{NL}$. **d.** Draw $\overrightarrow{LM}$. □

PRACTICE 3 Perform parts **a** through **d** in order to create a single figure.

a. Draw three noncollinear points, A, B, and C.

b. Draw $\overrightarrow{BC}$.

c. Draw $\overleftrightarrow{CA}$.

d. Draw $\overline{AB}$.

Recall that a **postulate** or **axiom** is a statement accepted without proof.

We use some of the following geometry postulates in algebra. For example, Postulate 1.3-1 is used to graph equations such as $y = 2x + 8$. We graph two points that are solutions to the equation and draw the line through the points.

Postulate 1.3-1 Two Points Determine a Line

Through any two points there is exactly one line.

Line t passes through points A and B. Line t is the only line that passes through both points.

When we have two or more geometric figures, their **intersection** is the set of points that the figures have in common.

Postulate 1.3-2 Intersection of Lines

If two distinct lines intersect, then they intersect at exactly one point.

$\overleftrightarrow{AE}$ and $\overleftrightarrow{DB}$ intersect at point C.

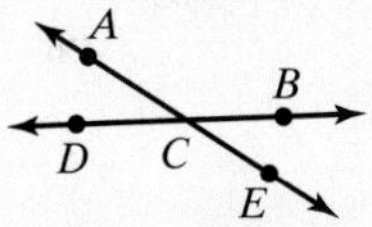

Here is an example of the point of intersection of two lines on the coordinate plane.

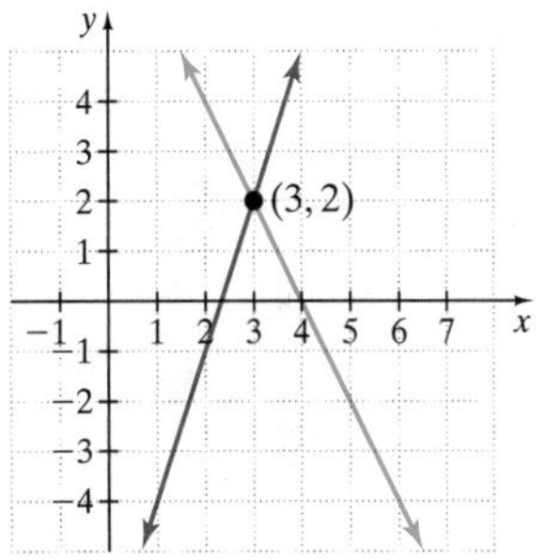

Let's make sure we can name opposite rays.

EXAMPLE 4 Opposite Rays

Name two pairs of opposite rays.

Solution

Points M, N, and X are collinear and X is between M and N. So, $\overrightarrow{XM}$ and $\overrightarrow{XN}$ are opposite rays.

Points P, Q, and X are collinear and X is between P and Q. So, $\overrightarrow{XP}$ and $\overrightarrow{XQ}$ are opposite rays.

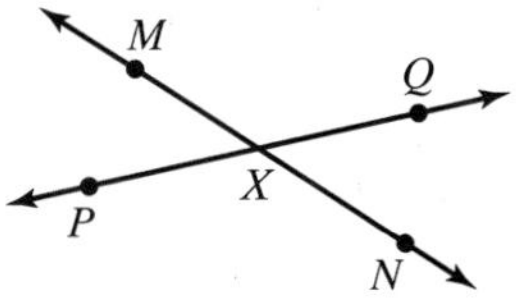

PRACTICE **4** Name two pairs of opposite rays.

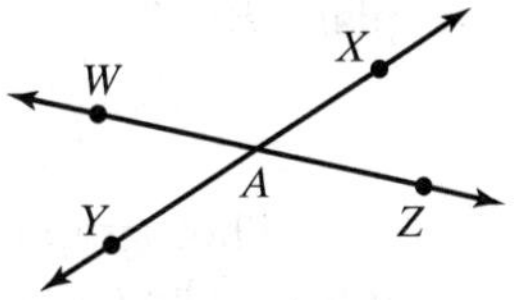

Postulate 1.3-2 in this section has to do with the intersection of lines. Now let's introduce a postulate that has to do with the intersection of planes.

Postulate 1.3-3 Intersection of Planes

If two distinct planes intersect, then they intersect in exactly one line.

Plane RST and plane WST intersect in $\overleftrightarrow{ST}$.

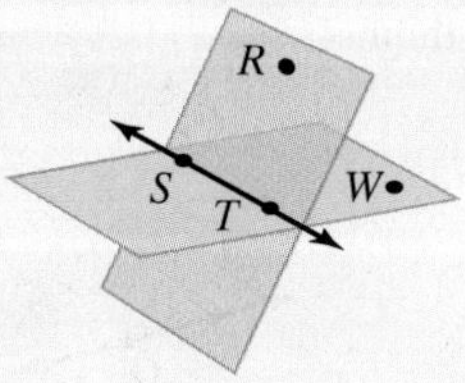

EXAMPLE 5 Finding the Intersection of Two Planes

Each surface of the box shown represents part of a plane. What is the intersection of plane *ADC* and plane *BFG*?

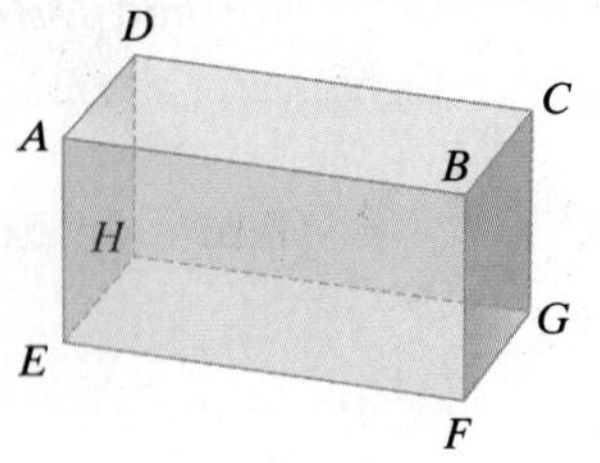

Solution

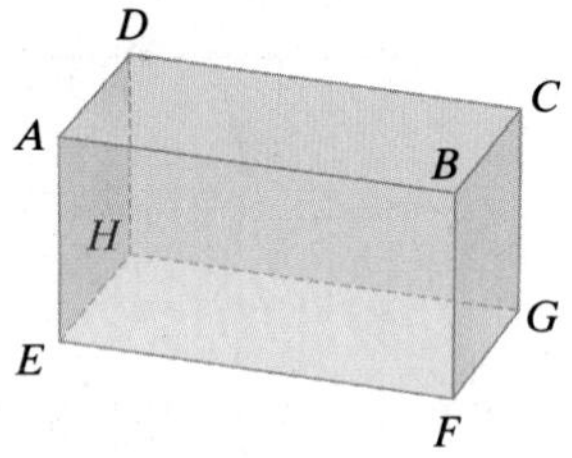

Shade plane *ADC* and plane *BFG* to see where they intersect. Remember from Postulate 1.3-3 that two distinct planes intersect in a line.

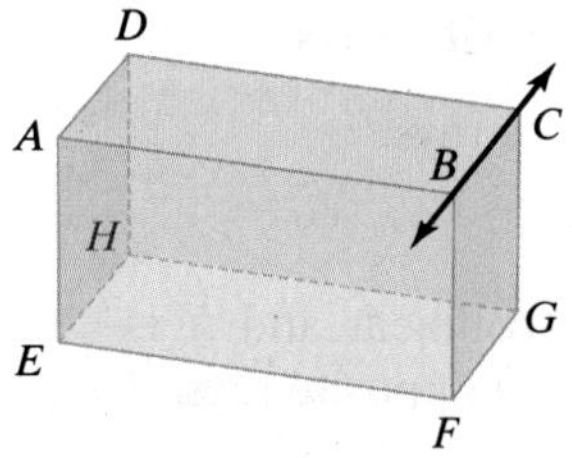

We can see that both planes contain point *B* and point *C*. These points define a distinct line.

The planes intersect in $\overleftrightarrow{BC}$. □

PRACTICE 5 Use the figure for Example 5. What is the intersection of plane *ADH* and plane *EFG*?

Photographers use three-legged tripods to make sure that a camera is steady. The feet of the tripod all touch the floor at the same time. You can think of the feet as points and the floor as a plane. As long as the feet do not all lie in one line, they will lie in exactly one plane.

This illustrates Postulate 1.3-4.

Helpful Hint

Don't forget that the points *must* be noncollinear. If they are collinear, many planes pass through them.

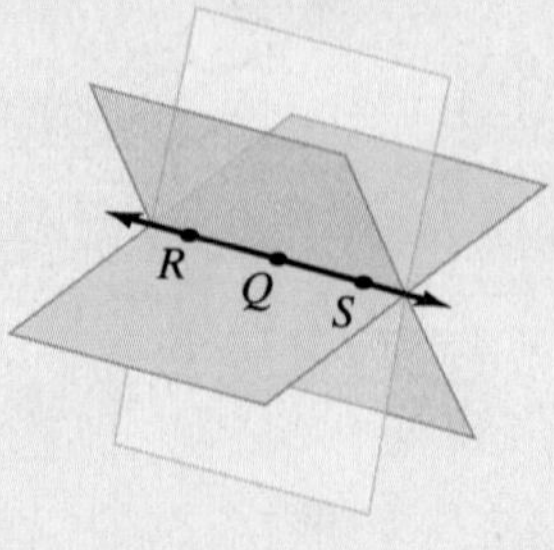

Postulate 1.3-4 Three Noncollinear Points Determine a Plane

Through any three noncollinear points there is exactly one plane.

Points *Q*, *R*, and *S* are noncollinear. Plane *P* is the only plane that contains them.

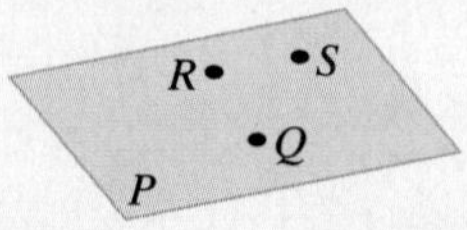

EXAMPLE 6 **Using Postulate 1.3-4**

Use the figure shown to answer the question.
What plane contains points N, P, and Q? Shade the plane.

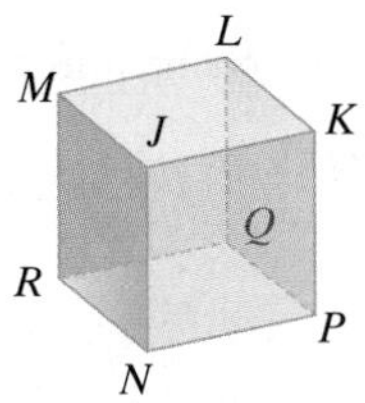

Solution

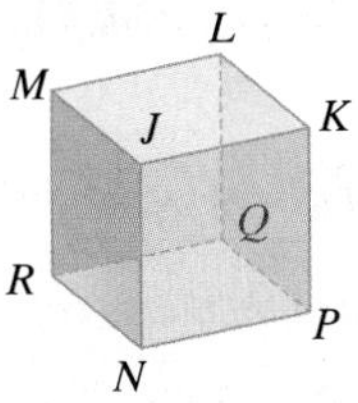

The plane on the bottom of the figure contains points N, P, and Q.

PRACTICE
6 What plane contains points J, M, and Q? Shade the plane. (Hint: The plane does not have to be already drawn in the figure.)

VOCABULARY & READINESS CHECK

Matching *Match each symbol with its meaning.*

1. $\overline{RS}$ **2.** $\overleftrightarrow{RS}$ **3.** $\overrightarrow{RS}$ **4.** $\overrightarrow{SR}$

A. line RS **B.** ray SR **C.** segment RS **D.** ray RS

Word Bank *Use the choices to fill in each blank.*

point	collinear	segment	opposite rays
line	coplanar	ray	geometric figure
plane	space		

5. A(n) ____________ extends in opposite directions without end.
6. A(n) ____________ extends in two dimensions without end.
7. A(n) ____________ consists of an endpoint and all points of a line on one side of the endpoint.
8. A(n) ____________ consists of two endpoints and all the points between them.
9. ____________ points lie on the same plane.
10. ____________ points lie on the same line.
11. A(n) ____________ is any nonempty subset of space.
12. ____________ are two rays that share an endpoint and form a line.
13. A(n) ____________ has a location but no dimension.
14. ____________ is the set of all points.

1.3 EXERCISE SET MyMathLab®

Use the figures at the right for Exercises 1–8. See Example 1.

1. What are two other ways to name $\overleftrightarrow{EF}$?
2. What are two other ways to name $\overleftrightarrow{RT}$?
3. What are two other ways to name plane C?
4. What are two other ways to name plane Q?
5. Name three collinear points on plane C.
6. Name three collinear points on plane Q.

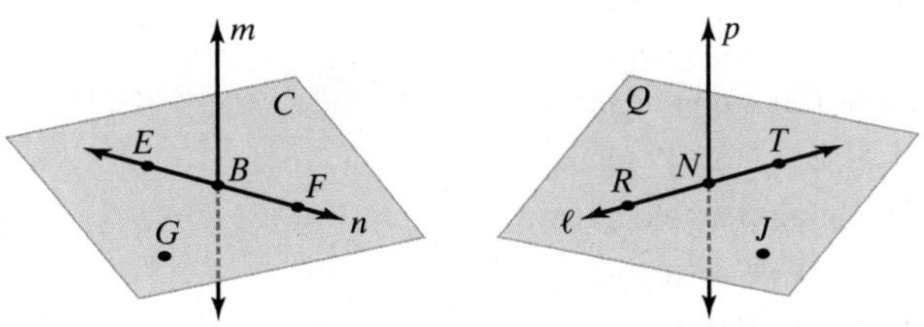

7. Name four coplanar points on plane C.

8. Name four coplanar points on plane Q.

Use the figures below for Exercises 9–14. See Example 2.

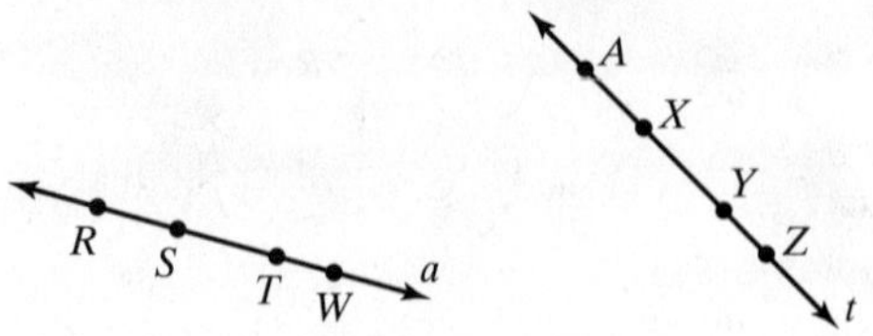

9. Name the segments on line a.

10. Name the segments on line t.

11. Name the rays on line a.

12. Name the rays on line t.

13. Name the pair of opposite rays with endpoint T.

14. Name the pair of opposite rays with endpoint X.

Use the figure at the right for Exercises 15–18. See Examples 2 and 3.

15. Name three points.

16. Name the line(s).

17. Name the line segment(s).

18. Name the ray(s).

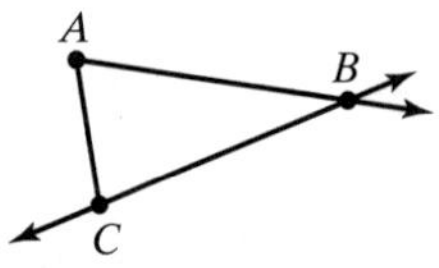

True or False *Use the figure at the right for Exercises 19–22. See Example 4.*

19. $\overrightarrow{XL}$ and $\overrightarrow{LX}$ are opposite rays.

20. $\overrightarrow{XP}$ and $\overrightarrow{XN}$ are opposite rays.

21. $\overrightarrow{XM}$ and $\overrightarrow{XP}$ are opposite rays.

22. $\overrightarrow{XL}$ and $\overrightarrow{XN}$ are opposite rays.

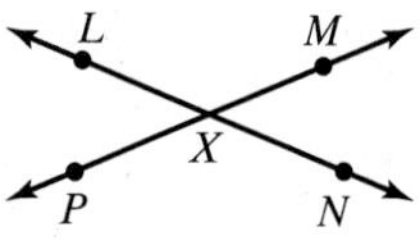

Mixed Practice *See Examples 1–4. For Exercises 23–26, name three points in each diagram that are not collinear.*

23.

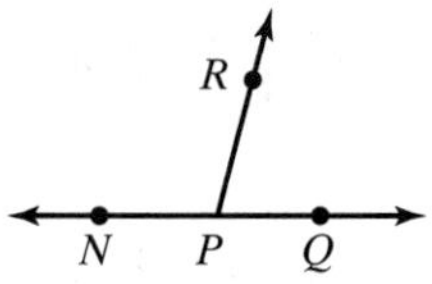

24.

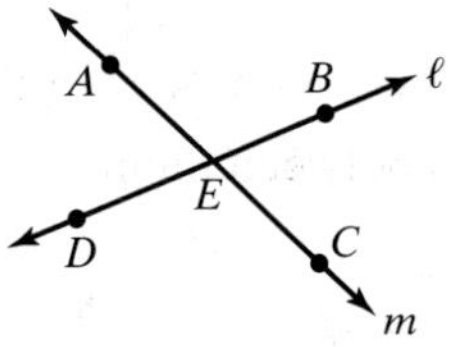

25.

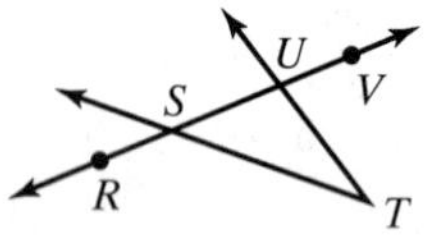

26.

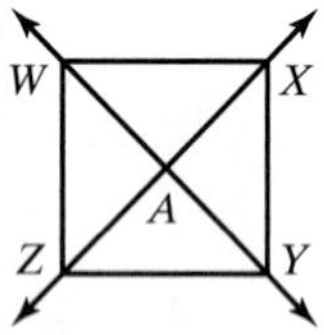

True or False *Use the figure for Exercise **24** to answer each question either true or false.*

27. Point A lies on line l.

28. Point D lies on line m.

29. Point B lies on line l.

30. Point C lies on line m.

31. A, B, and C are coplanar.

32. D, E, and B are collinear.

33. A, B, and C are collinear.

34. D, E, and B together lie on a single plane only.

Use the figure at the right for Exercises 35–42. See Example 5. Name the intersection of each pair of planes.

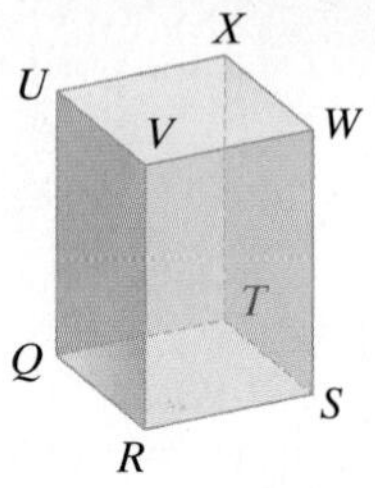

35. planes QRS and RSW

36. planes UXV and WVS

37. planes XWV and UVR

38. planes TXW and TQU

Name two planes that intersect in the given line.

39. $\overleftrightarrow{QU}$

40. $\overleftrightarrow{TS}$

41. $\overleftrightarrow{XT}$

42. $\overleftrightarrow{VW}$

Name a point that is coplanar with the given points. See Example 6.

43. Q, R, and S

44. T, S, and F

45. G, Q, and T

46. E, F, and G

47. Q, R, and H

48. R, S, and F

49. Q, R, and F

50. R, S, and G

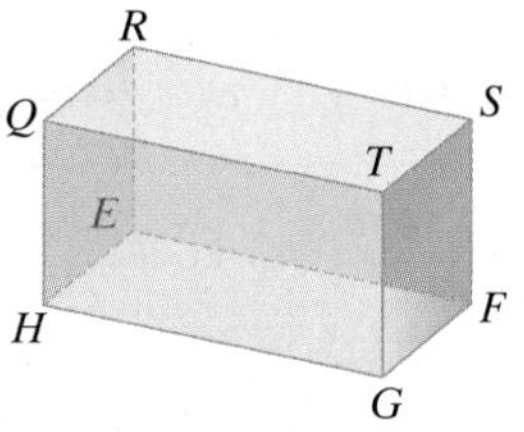

Postulate 1.3-4 states that any three noncollinear points lie in exactly one plane. Find the plane that contains the first three points listed. Then determine whether the fourth point is in that plane. Write "coplanar" or "noncoplanar" to describe the points. See Example 6.

51. Z, S, Y, C

52. S, U, V, Y

53. X, Y, Z, U

54. X, S, V, U

55. X, Z, S, V

56. S, V, C, Y

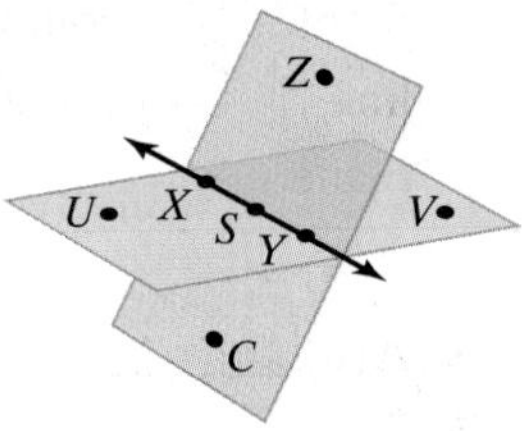

Sketch *Sketch the figure described. Assume that each figure lies on a plane.*

57. Two lines that do not intersect.

58. Three lines that do not intersect.

59. Three lines that intersect at a point.

60. Four lines that intersect at a point.

61. Two lines that intersect at point A and another line that intersects both lines, but not at point A.

62. Three lines that intersect at point A and another line that intersects all three lines, but not at point A.

CONCEPT EXTENSIONS

For Exercises 63–68, determine whether each statement is "always," "sometimes," or "never" true.

63. $\overleftrightarrow{TQ}$ and $\overleftrightarrow{QT}$ are the same line.

64. $\overrightarrow{JK}$ and $\overrightarrow{JL}$ are the same ray.

65. Two distinct intersecting lines are coplanar.

66. Four points are coplanar.

67. A plane containing two points of a line contains the entire line.

68. Two distinct lines intersect in more than one point.

69. A cell phone tower at point A receives a cell phone signal from the southeast. A cell phone tower at point B receives a signal from the same cell phone from due west. Trace the diagram below and find the location of the cell phone. Describe how Postulates 1.3-1 and 1.3-2 help you locate the phone.

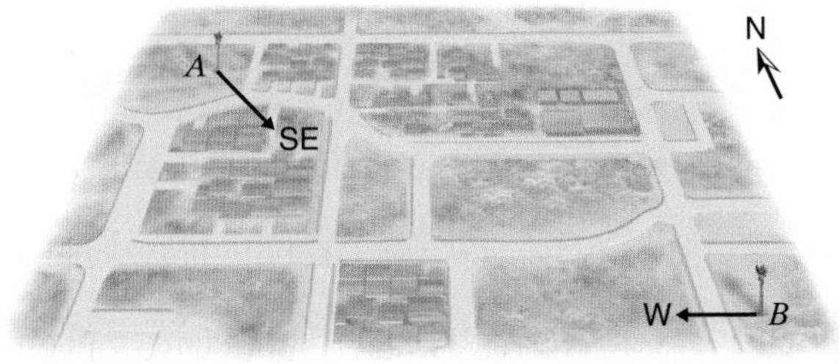

70. You can represent the hands on a clock at 6:00 as opposite rays. Estimate the other 11 times on a clock that you can represent as opposite rays. (It may help to sketch the face of a clock.)

Coordinate Geometry *Graph the points on a coordinate plane and state whether they appear to be collinear.*

71. (1, 1), (4, 4), (−3, −3)

72. (2, 4), (4, 6), (0, 2)

73. (0, 0), (−5, 1), (6, −2)

74. (0, 0), (8, 10), (4, 6)

75. (0, 0), (0, 3), (0, −10)

76. (−2, −6), (1, −2), (4, 1)

77. How many planes contain the same three collinear points? Explain.

78. How many planes contain a given line? Explain.

79. Suppose two points are in plane P. Explain why the line containing the points is also in plane P.

80. Suppose two lines intersect. How many planes do you think contain both lines? Use the diagram at the right and your answer to Exercise **79** to explain your answer.

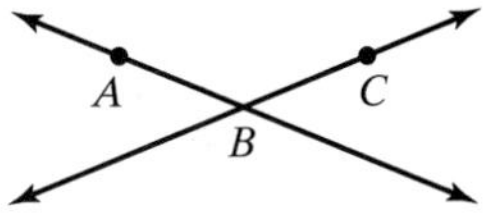

1.4 SEGMENTS AND THEIR MEASURE

OBJECTIVES

1 Understand the Measure of Segments.

2 Use Segment Postulates and Algebra to Find Segment Lengths.

VOCABULARY

- coordinate
- distance
- congruent segments
- midpoint
- bisect
- segment bisector

Helpful Hint

Be very careful with notation.

$\overline{AB}$ means line segment AB.

AB means the length of $\overline{AB}$.

OBJECTIVE 1 ▶ Understanding Segment Measures. In the previous section, we defined many terms and stated a few postulates. Now, let's expand our geometry system by measuring segments. To measure a segment, let's learn how to find the distance between two points on a segment.

Postulate 1.4-1 Ruler Postulate

The points on a line can be paired, one-to-one, with a real number. The real number that corresponds to a point is called the **coordinate** of the point.

The above allows us to calculate distance. The **distance** between points A and B is the absolute value of the difference of their coordinates.

We use the notation AB for distance or the length of $\overline{AB}$.

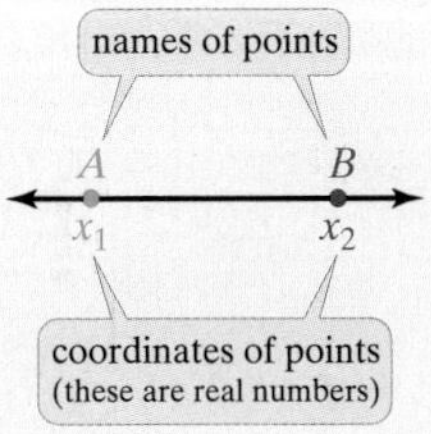

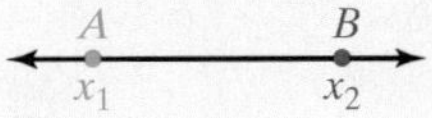

$AB = |x_2 - x_1|$

In Example 1, we find the distance between two points on a number line.

EXAMPLE 1 **Measuring Lengths of Segments**

Find MP.

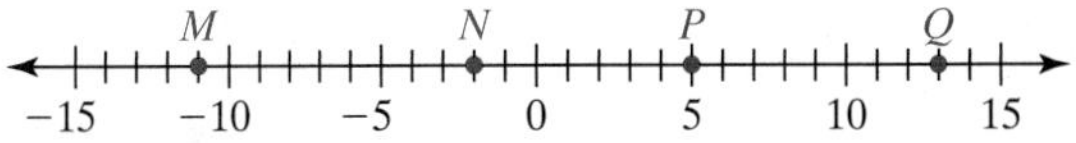

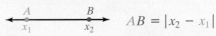

Solution Recall that MP is the distance between points M and P. Use the Ruler Postulate to find MP.

The coordinate of M is −11. The coordinate of P is 5.

$$MP = |5 - (-11)| = |16| = 16 \text{ or } MP = |-11 - 5| = |-16| = 16$$

Thus, the distance is 16 units. □

PRACTICE

1 Find NQ.

> **Helpful Hint**
> Segments are congruent, $\cong$.
> Lengths are equal, $=$.

Now that we can measure segments, we can talk about the notation used for segments of equal ($=$) length. Two segments that have the same length are called **congruent segments.** The symbol $\cong$ means congruent.

Congruent Segments vs. Equal Lengths

If $\overline{AB} \cong \overline{CD}$, then $AB = CD$.
Also, if $AB = CD$, then $\overline{AB} \cong \overline{CD}$.

We mark congruent segments in a figure with exactly the same number of tick marks.

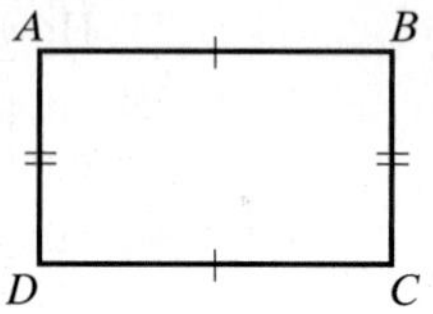

Note: $\overline{AB}$ and $\overline{DC}$ both have 1 tick mark, so $\overline{AB} \cong \overline{DC}$.
$\overline{AD}$ and $\overline{BC}$ both have 2 tick marks, so $\overline{AD} \cong \overline{BC}$.

EXAMPLE 2 Measuring Congruent Segments

a. Use the figure above to write the congruent segments and the equal distances.
b. If $BC = 2$ feet, find AD.
c. If $DC = 6$ feet, find AB.

Solution **a.** $\overline{AD}$ and $\overline{BC}$ are marked the same, so $\overline{AD} \cong \overline{BC}$ and $AB = DC$.

Segments | lengths

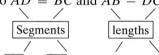

Also, $\overline{AB}$ and $\overline{DC}$ are marked the same, so $\overline{AB} \cong \overline{DC}$ and $AB = DC$.

b. $BC = AD$, so if $BC = 2$ feet, then $AD = 2$ feet.
c. $DC = AB$, so if $DC = 6$ feet, then $AB = 6$ feet. □

PRACTICE 2 **a.** Use the figures below to write the congruent segments and the equal distances.
b. If $AC = 10$ cm, find DF.
c. If $CB = 25$ cm, find FE.

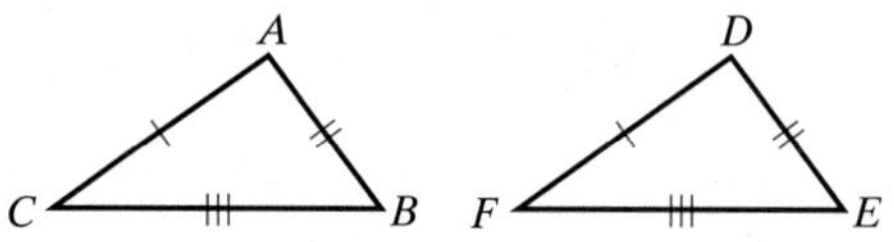

EXAMPLE 3 Determining if Segments Are Congruent

Given the figure, is $\overline{BG} \cong \overline{AC}$?

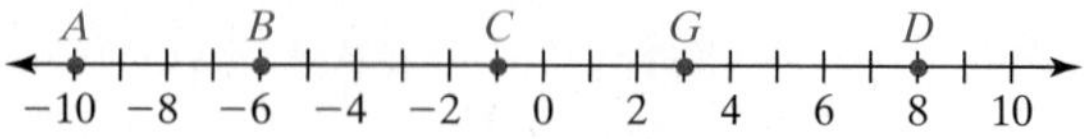

Solution Remember: If $BG = AC$, then $\overline{BG} \cong \overline{AC}$.

$$BG = |3 - (-6)| = |9| = 9,$$
$$AC = |-1 - (-10)| = |-1 + 10| = |9| = 9$$

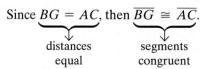

Since $BG = AC$, then $\overline{BG} \cong \overline{AC}$.
($BG = AC$: distances equal; $\overline{BG} \cong \overline{AC}$: segments congruent) □

PRACTICE 3 Using the figure for Example 3, is $\overline{BC} \cong \overline{GD}$?

Answer yes or no.

Remember that **between** is an undefined term. If we have three distinct points on a line, then one of the points will be between the other two.

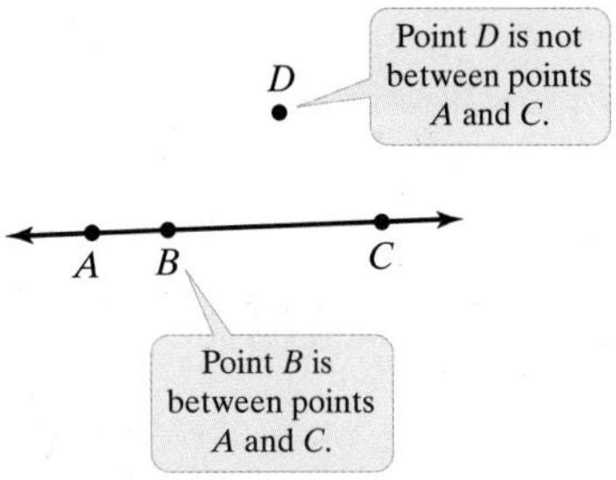

> **Helpful Hint**
> To use the concept of betweenness, the points *must* lie on the same line (be collinear).

We can use the concept of betweenness to add the lengths of segments.

> **Helpful Hint**
> Since *AB*, *BC*, and *AC* are real numbers, then segment subtraction is possible also. We review this in Section 2.6.

Postulate 1.4-2 Segment Addition Postulate

If point B is between points A and C, then $AB + BC = AC$. Also, if $AB + BC = AC$, then point B is between points A and C.

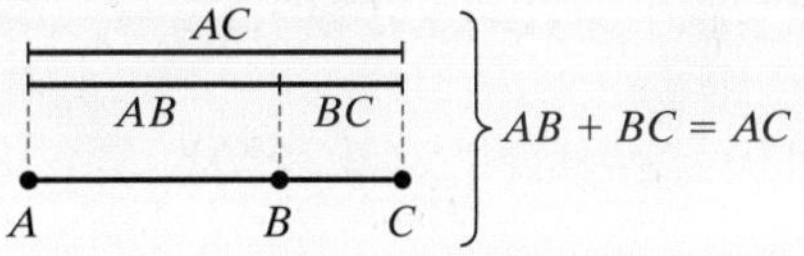

The **midpoint** of a segment is a point that divides, or **bisects,** a segment into two congruent segments.

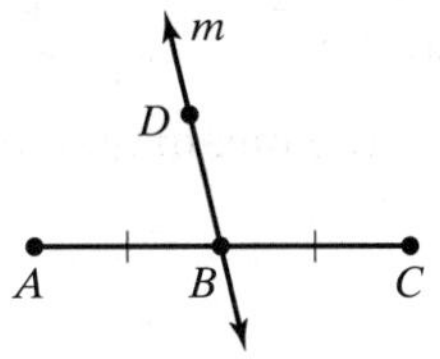

True statements:

- B is the midpoint of $\overline{AC}$.
- Line m bisects $\overline{AC}$.
- $\overleftrightarrow{BD}$ bisects $\overline{AC}$.
- $\overrightarrow{BD}$ bisects $\overline{AC}$.
- $\overline{AB} \cong \overline{BC}$.
- $AB = BC$.

> **Helpful Hint**
> A segment bisector contains the midpoint of a segment, but no other point of the segment.

A line, ray, segment, or plane that intersects a segment at its midpoint is called a **segment bisector.**

OBJECTIVE 2 ▶ Using Algebra to Find Segment Lengths.

EXAMPLE 4 Using the Segment Addition Postulate

If $MP = 47$ units, find MN and NP.

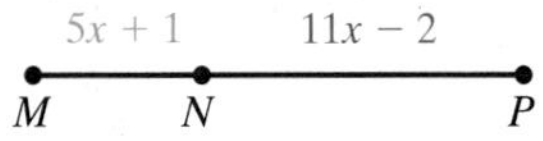

Solution Since we know MP, let's use the Segment Addition Postulate and algebra to first find x.

$$MN + NP = MP \quad \text{Segment Addition Postulate}$$

$$(5x + 1) + (11x - 2) = 47 \quad \text{Substitution}$$

$$16x - 1 = 47 \quad \text{Combine like terms.}$$

$$16x = 48 \quad \text{Add 1 to both sides.}$$

$$\frac{16x}{16} = \frac{48}{16} \quad \text{Divide both sides by 16.}$$

$$x = 3 \quad \text{Simplify.}$$

Now use the value of x to find MN and NP.

$$\left.\begin{aligned} MN &= 5x + 1 = 5(3) + 1 = 16 \\ NP &= 11x - 2 = 11(3) - 2 = 31 \end{aligned}\right\} \text{Let } x = 3.$$

Check: See that $MP = 47$.

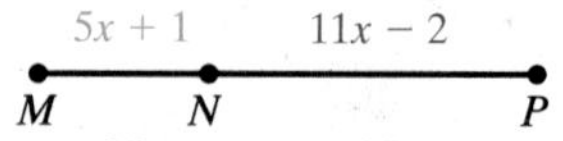

or or

16 + 31 = 47 Check. □

PRACTICE 4 If $QS = 79$, find QR and RS.

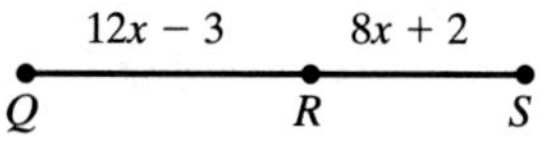

EXAMPLE 5 Finding the Midpoint

Point C is the midpoint of $\overline{AB}$. Find AC, CB, and AB.

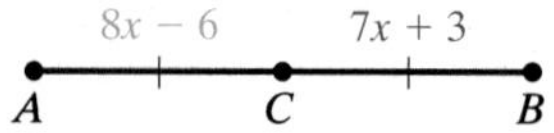

Solution Let's use the meaning of midpoint to find the value of x.

$$AC = CB \quad \text{Definition of midpoint}$$

$$8x - 6 = 7x + 3 \quad \text{Substitute.}$$

$$x - 6 = 3 \quad \text{Subtract } 7x \text{ from both sides.}$$

$$x = 9 \quad \text{Add 6 to both sides.}$$

Now let's use the value of x to find AC and CB, which should be the same.

$$\begin{aligned} AC &= 8x - 6 & & & CB &= 7x + 3 \\ &= 8(9) - 6 & & \text{Substitute 9 for } x. & &= 7(9) + 3 \\ &= 72 - 6 & & \text{Multiply.} & &= 63 + 3 \\ &= 66 & & \text{Simplify.} & &= 66 \end{aligned}$$

Notice that $AC = CB = 66$, as expected.
To find AB, we add.

$$\begin{aligned} AB &= AC + CB & & \text{Use the Segment Addition Postulate.} \\ &= 66 + 66 & & \text{Substitute.} \\ &= 132 & & \text{Add.} \end{aligned}$$

□

PRACTICE 5 Point P is the midpoint of QR. Find QP, PR, and QR.

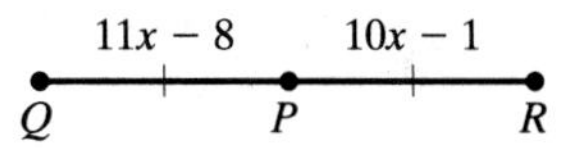

VOCABULARY & READINESS CHECK

Word Bank *Use the choices to fill in each blank.*

bisect	distance	between	segment bisector
midpoint	coordinate	equal	congruent

1. Two segments that have the same length are called ____________ segments.
2. The real number that corresponds to a point is called the ____________ of the point.
3. Given three distinct points on a line, one of the points will always be ____________ the other two.
4. A point that divides a segment into two congruent segments is called a(n) ____________.
5. If two segments are congruent, then their lengths are ____________.
6. The ____________ between two points is the absolute value of the difference of their coordinates.
7. A line, ray, segment, or plane that intersects a segment at its midpoint is called a(n) ____________.
8. In geometry, to ____________ a segment means to divide it into two equal parts.

1.4 EXERCISE SET MyMathLab®

Use the figure shown to answer each question as yes or no.

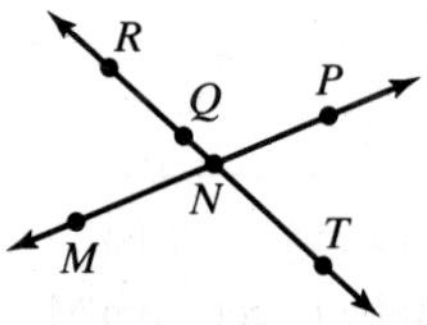

1. Is N between M and P?
2. Is N between Q and T?
3. Is Q between N and T?
4. Is M between N and P?
5. Is T between M and P?
6. Is P between R and T?
7. Is R between Q and T?
8. Is N between R and Q?
9. Is Q between R and N?
10. Is Q between R and T?

Find the length of each segment. See Example 1.

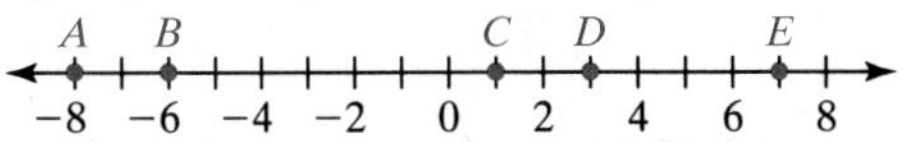

11. $\overline{BD}$ **12.** $\overline{AB}$
13. $\overline{CE}$ **14.** $\overline{AD}$

Each figure is marked separately, showing their congruent segments. Use the figures to find each measure in Exercises 15–20. See Example 2.

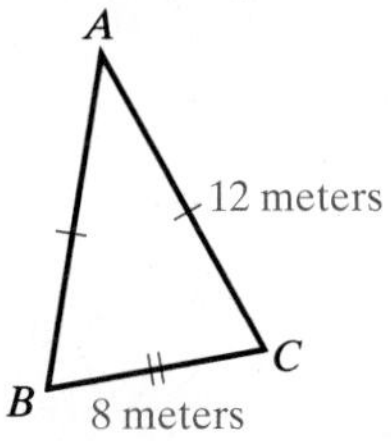

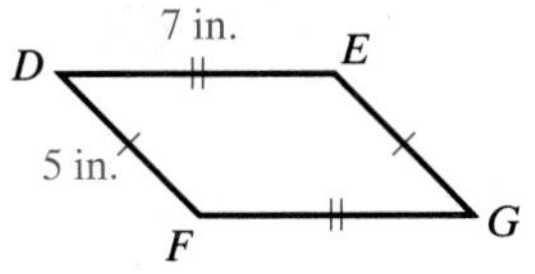

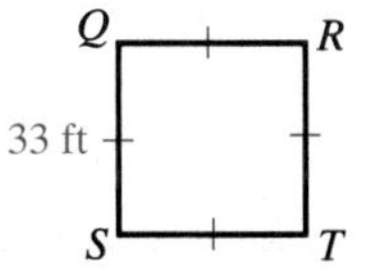

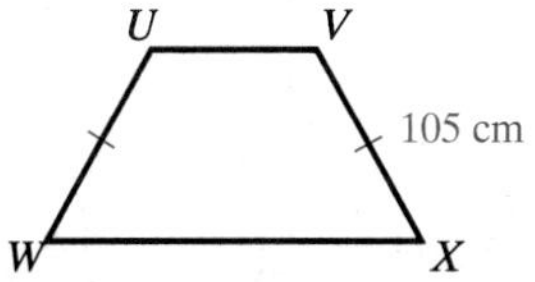

15. AC **16.** UW
17. EG **18.** FG
19. QR **20.** RT

Use the number line below for Exercises 21–24. Tell whether the segments are congruent. See Example 3.

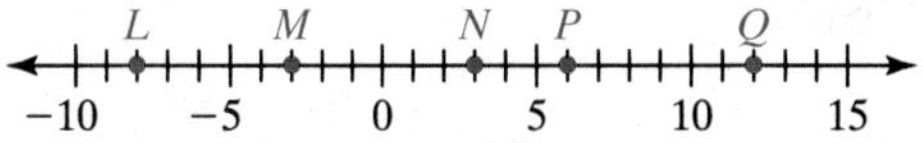

21. $\overline{LN}$ and $\overline{MQ}$ **22.** $\overline{MP}$ and $\overline{NQ}$
23. $\overline{MN}$ and $\overline{PQ}$ **24.** $\overline{LP}$ and $\overline{MQ}$

Use the number line below for Exercises 25–28. See Example 4.

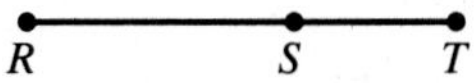

25. If $RS = 15$ and $ST = 9$, then $RT =$ ____________.
26. If $ST = 15$ and $RT = 40$, then $RS =$ ____________.
27. **Multiple Steps** $RS = 8y + 4$, $ST = 4y + 8$, and $RT = 96$.
  a. What is the value of y?
  b. Find RS and ST.
28. **Multiple Steps** $RS = 14x - 8$, $ST = 9x + 10$, and $RT = 232$.
  a. What is the value of x?
  b. Find RS and ST.

Multiple Steps *For Exercises 29 and 30, see Example 5.*

29. A is the midpoint of $\overline{XY}$.

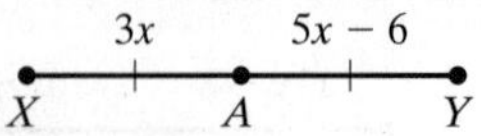

a. Find the value of x.

b. Find XA, AY, and XY.

30. C is the midpoint of $\overline{QR}$.

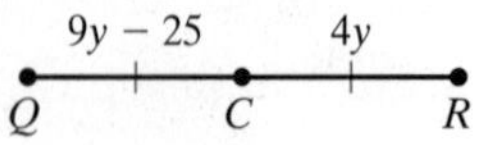

a. Find the value of y.

b. Find QC, CR, and QR.

For Exercises 31–34, use the figure below. Find the value of PT. See Example 5.

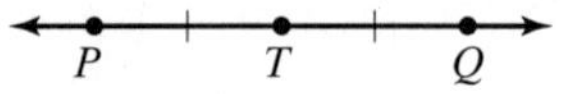

31. $PT = 5x + 3$ and $TQ = 7x - 9$

32. $PT = 6x - 35$ and $TQ = 2x - 3$

33. $PT = 4x - 6$ and $TQ = 3x + 4$

34. $PT = 7x - 24$ and $TQ = 6x - 2$

On a number line, the coordinates of X, Y, Z, and W are −7, −3, 1, and 5, respectively. Find the lengths of the two segments. Then tell whether they are congruent.

35. $\overline{XY}$ and $\overline{ZW}$ **36.** $\overline{ZX}$ and $\overline{WY}$

Suppose the coordinate of A is 0, AR = 5 units, and AT = 7 units. What are the possible coordinates of the midpoint of the given segment?

37. $\overline{AR}$

38. $\overline{AT}$

39. Suppose point E has a coordinate of 3 and $EG = 5$ units. What are the possible coordinates of point G?

40. Suppose point P has a coordinate of 10 and $PG = 6$ units. What are the possible coordinates of point G?

Use the diagram below for Exercises 41 and 42.

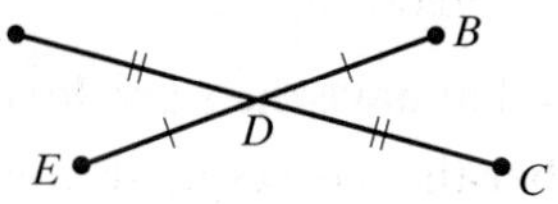

41. If $AD = 12$ and $AC = 4y - 36$, find the value of y. Then find AC and DC.

42. If $ED = x + 4$ and $DB = 3x - 8$, find ED, DB, and EB.

CONCEPT EXTENSIONS

The numbers labeled on the map of Florida are mile markers. Assume that Route 10 between Quincy and Jacksonville is straight.

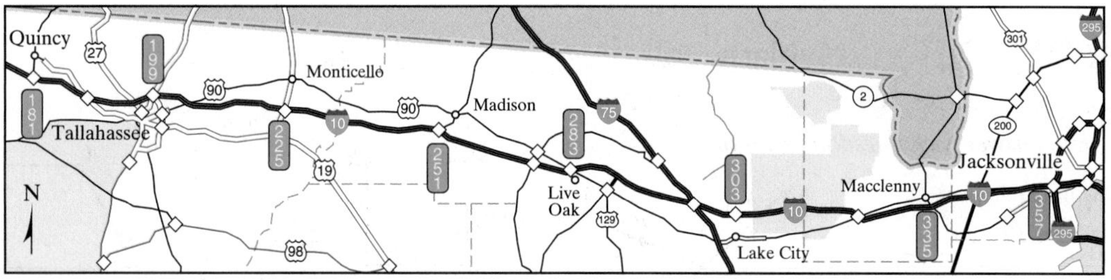

43. Suppose you drive at an average speed of 55 mph. How long will it take to get from Live Oak to Jacksonville?

44. Suppose you drive at an average speed of 58 mph. How long will it take to get from Macclenny to Tallahassee?

Find the Error *Use the highway sign for Exercises 45 and 46.*

45. A driver reads the highway sign and says, "It's 145 miles from Mitchell to Watertown." What error did the driver make? Explain.

46. Your friend reads the highway sign and says, "It's 71 miles to Watertown." Is your friend correct? Explain.

47. C is the midpoint of $\overline{AB}$, D is the midpoint of $\overline{AC}$, E is the midpoint of $\overline{AD}$, F is the midpoint of $\overline{ED}$, G is the midpoint of $\overline{EF}$, and H is the midpoint of $\overline{DB}$. If $DC = 16$, what is GH?

48. Multiple Steps

a. Use the diagram below. What algebraic expression represents GK?

b. If $GK = 30$, what are GH and JK?

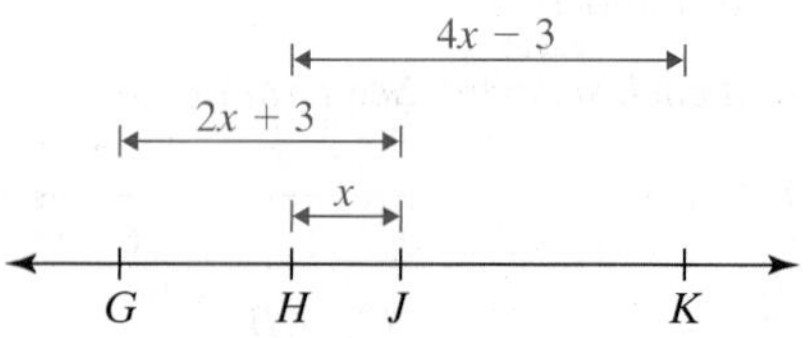

49. Suppose you know PQ and QR. Can you use the Segment Addition Postulate to find PR? Explain.

50. Use the map above for Exercises **43** and **44** and explain how to use mile markers to find distances between points.

1.5 ANGLES AND THEIR MEASURE

OBJECTIVES

1 Understand the Measure of Angles.

2 Use Algebra and the Angle Addition Postulate to Solve Applications and Find Angle Measures.

VOCABULARY

- angle
- sides of an angle
- vertex
- interior of an angle
- exterior of an angle
- protractor
- degrees
- acute angle
- right angle
- obtuse angle
- straight angle
- congruent angles

OBJECTIVE 1 ▶ Understanding Angle Measures. In Section 1.3 we defined rays. What can we call a figure formed by two rays that share a common endpoint? Can you see that this type of thinking leads to new figures in our geometry mathematical system? Let's use the terms we've learned thus far to define new terms and figures.

Defined Terms

Definition	*How to Name It*	*Example*
An **angle** consists of two different rays with a common endpoint. The rays are the **sides** of the angle. The common endpoint is the **vertex** of the angle.	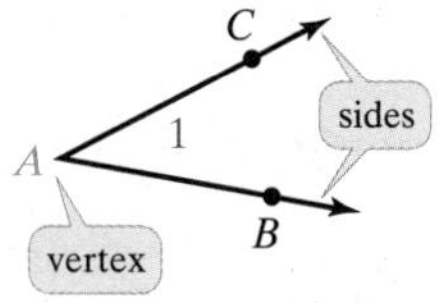 Point A is the vertex. The sides are rays $\overrightarrow{AC}$ and $\overrightarrow{AB}$.	Ways to name this angle: $\angle A$, $\angle 1$, $\angle CAB$, $\angle BAC$

(*Note:* The number 1 is inserted to show a numerical way to name this angle.)

▶ Helpful Hint
The vertex must be in the middle when naming an angle using three points.

The **interior** of an angle contains all points between the two sides of the angle.

The **exterior** of an angle contains all points that are not in the interior of the angle and are not on the angle.

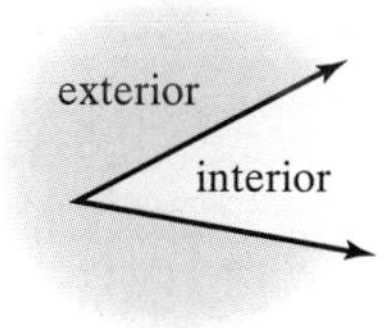

▶ Helpful Hint
Given an angle in a plane, every point on the plane

- is in the interior of the angle,
- is in the exterior of the angle, or
- lies on the angle

EXAMPLE 1 **Identifying and Naming Angles**

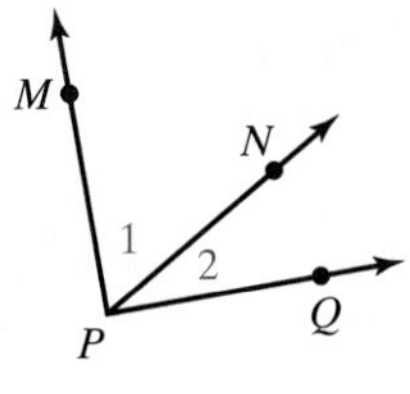

a. How many different angles are in the diagram?

b. Write two other ways to name $\angle 1$.

Solution

a. There are three different angles in the diagram: $\angle 1$, $\angle 2$, and $\angle MPQ$.

b. $\angle 1$ can also be named as $\angle MPN$ or $\angle NPM$. □

PRACTICE 1

a. Write two other ways to name $\angle 2$.

b. Write one other way to name $\angle MPQ$.

▶ Helpful Hint
In the diagram for Example 1, no angle should be named $\angle P$ because all three angles have point P as their vertex. $\angle P$ does not clearly tell us what angle we are referring to.

The instrument to the right is called a **protractor.** It can be used to measure angles in units called **degrees** (°). For example, the measure of $\angle A$ (also denoted by $m\angle A$) is 30°.

A Protractor

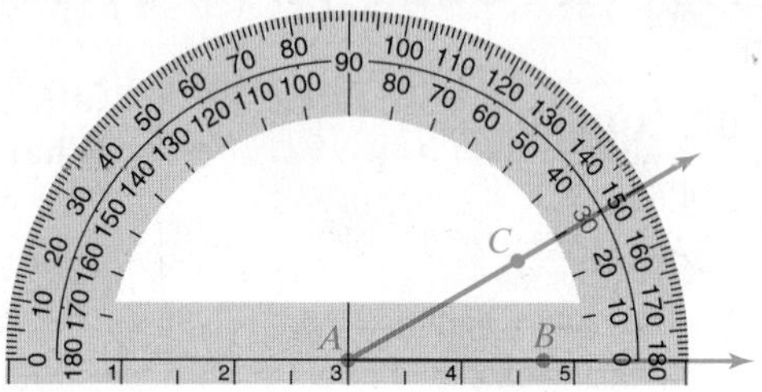

$m\angle A = 30°$

How do we understand the measure of one degree, or 1°? We start with a circle. A circle has 360°; thus 1° is $\frac{1}{360}$ of the rotation to form a circle.

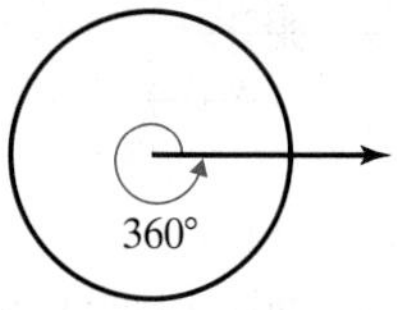

Helpful Hint

Notice that a protractor reminds us of half a circle, and shows half of 360°, or 180°.

Postulate 1.5-1 Protractor Postulate

Suppose we have $\overrightarrow{AB}$ as shown and point C on one side of $\overrightarrow{AB}$. Every ray, for example, $\overrightarrow{AC}$ can be paired one-to-one with a real number from 0 to 180. The measure of $\angle CAB$ (in degrees) equals the absolute value of the difference between the real numbers on the protractor for $\overrightarrow{AC}$ and for $\overrightarrow{AB}$.

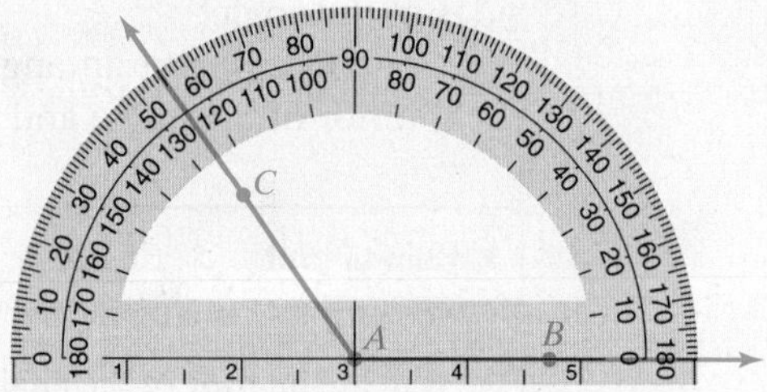

$$m\angle CAB = |125 - 0|° = 125°$$

Helpful Hint

Learn how to use your protractor. Notice that there are two sets of numbering about the protractor, both from 0 to 180. There is lower numbering and upper numbering. Be consistent and use lower numbering for both rays or upper numbering for both rays.

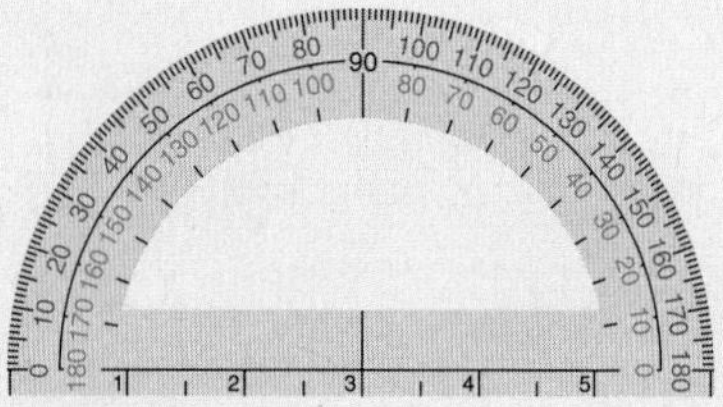

Now we can use the measure of an angle to classify it.

Classifying Angles

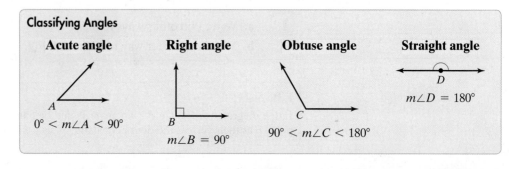

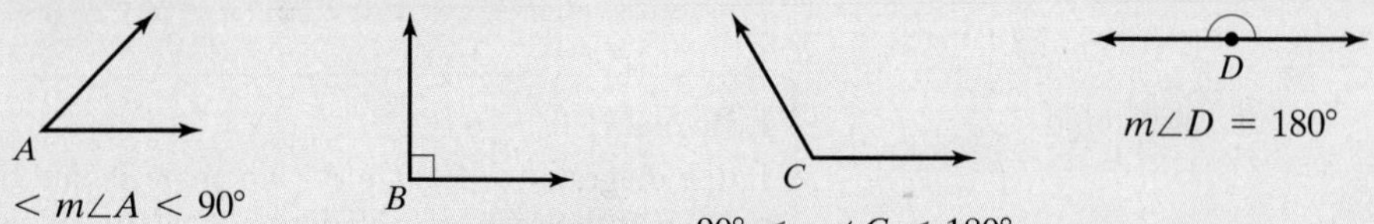

$0° < m\angle A < 90°$

$m\angle B = 90°$

$90° < m\angle C < 180°$

$m\angle D = 180°$

> **Helpful Hint**
> The symbol ⊾ indicates a right angle.

EXAMPLE 2 **Measuring and Classifying Angles**

Find $m\angle RQM$, $m\angle RQS$, and $m\angle RQN$. Then classify each angle as acute, right, obtuse, or straight.

Solution

$$m\angle RQM = |45 - 0|^\circ = 45^\circ\text{; acute angle}$$
$$m\angle RQS = |90 - 0|^\circ = 90^\circ\text{; right angle}$$
$$m\angle RQN = |165 - 0|^\circ = 165^\circ\text{; obtuse angle}$$

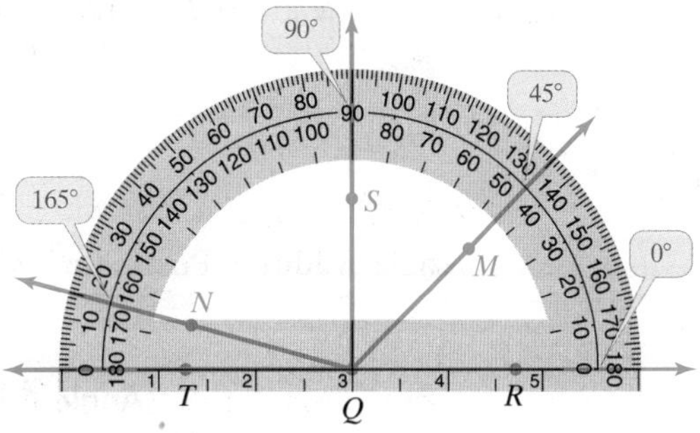

PRACTICE
2 Find $m\angle TQN$, $m\angle TQM$, and $m\angle TQR$. Then classify each angle as acute, right, obtuse, or straight.

Just as we did for segments, we can use the same notation for angles of equal (=) measure.

Two angles that have the same measure are called **congruent angles.** Recall that the symbol ≅ means congruent.

> **Helpful Hint**
> Angles are congruent, ≅. Angle measures are equal, =.

Congruent Angles vs. Equal Angle Measures

If $\angle A \cong \angle B$, then $m\angle A = m\angle B$.
Also, if $m\angle A = m\angle B$, then $\angle A \cong \angle B$.

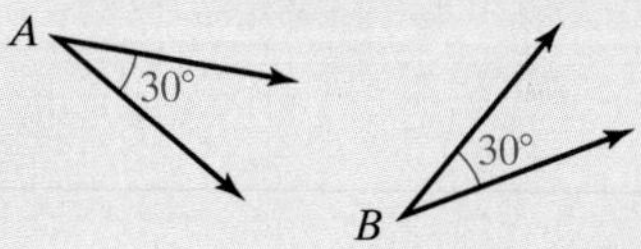

> **Helpful Hint**
> There may be many angles and figures considered at one time. If so, make sure each set of congruent angles is marked with the same number of arcs.

We mark congruent angles with exactly the same number of arcs, as shown in the figure below.

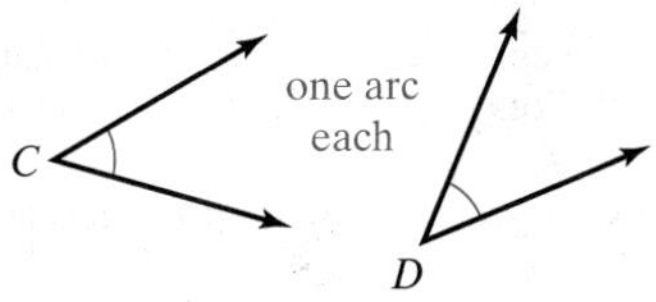

$\angle C \cong \angle D$

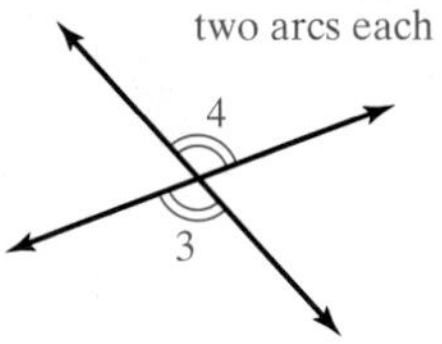

$\angle 3 \cong \angle 4$

EXAMPLE 3 **Measuring Congruent Angles**

Use the figures on the previous page to answer each question.

a. If $m\angle C = 45°$, find $m\angle D$.

b. If $m\angle 3 = 113°$, find $m\angle 4$.

Solution

a. $\angle C \cong \angle D$, so $m\angle C = m\angle D$. Since $m\angle C = 45°$, then $m\angle D = 45°$.

b. $\angle 3 \cong \angle 4$, so $m\angle 3 = m\angle 4$. Since $m\angle 3 = 113°$, then $m\angle 4 = 113°$. □

PRACTICE 3

a. Use the figures at the right to write the congruent angles and equal angle measures.

b. If $m\angle A = 25°$, find $m\angle D$.

c. If $m\angle F = 120°$, find $m\angle C$.

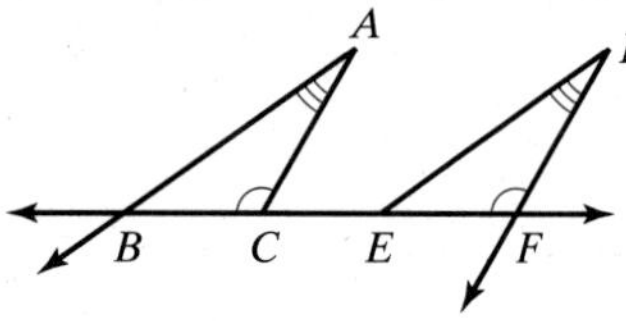

OBJECTIVE 2 ▶ Using Algebra and the Angle Addition Postulate. Notice that the Angle Addition Postulate is similar to the Segment Addition Postulate.

Postulate 1.5-2 Angle Addition Postulate

If P is in the interior of $\angle ABC$, then

$m\angle ABP + m\angle PBC = m\angle ABC$.

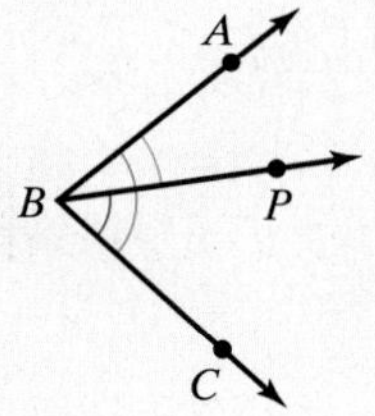

> **Helpful Hint**
> Just as for segment lengths, since the angle measures are real numbers (in degrees), then angle subtraction is also possible.

We can use our knowledge that a circle contains 360° and the Angle Addition Postulate to solve applications.

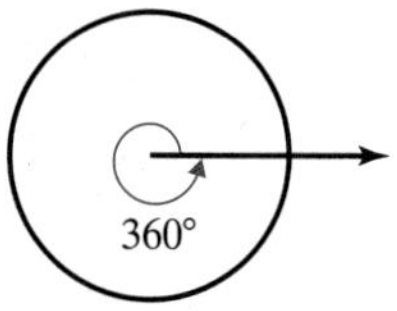

EXAMPLE 4 **Dividing a Pizza Application**

A circular pizza can easily be cut into 8 slices because of the ease of using a straight knife, as shown below.

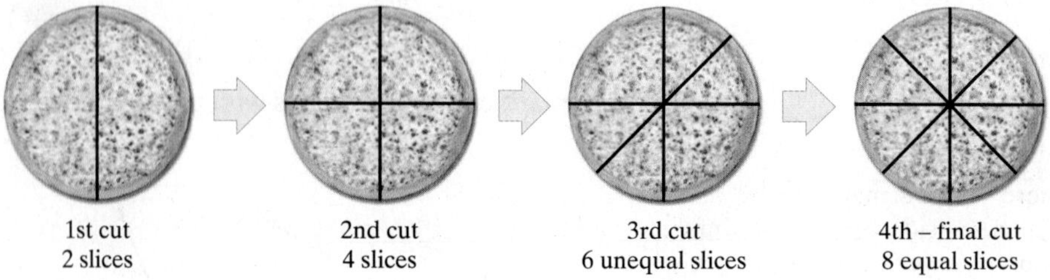

Find the angle measure of a slice of pizza cut into 8 equal-size slices, as shown.

Solution Recall that a circle measures 360°. Since each of the 8 slices has equal measure, we divide by 8.

$$\text{angle measure of a slice} = \frac{360°}{8}$$
$$= 45°$$

If a circular pizza is cut into 8 slices, then the angle measure of each slice is 45°.

Check: To check, use the Angle Addition Postulate. Add 45° eight times, or use the definition of multiplication and see that $8 \cdot 45° = 360°$.

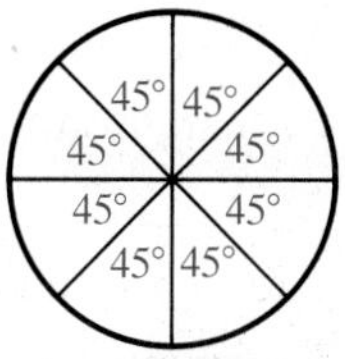

PRACTICE 4 A circular pizza is cut into 6 equal-size slices, as shown. Find x, the angle measure of each slice.

EXAMPLE 5 Using the Angle Addition Postulate

If $\angle DEG$ is a right angle, find $m\angle DEF$ and $m\angle FEG$.

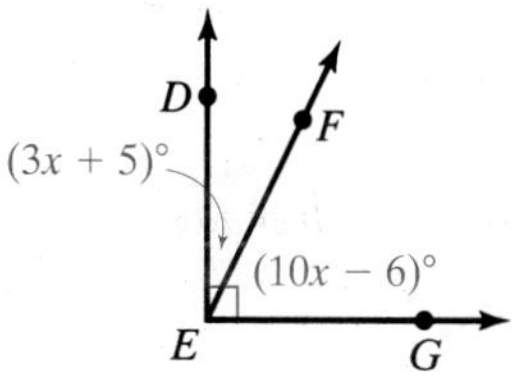

Solution Since $\angle DEG$ is a right angle, then $m\angle DEG = 90°$. Thus,

$m\angle DEF + m\angle FEG$	$= m\angle DEG$	Angle Addition Postulate
$(3x + 5) + (10x - 6)$	$= 90$	Substitution
$13x - 1$	$= 90$	Combine like terms.
$13x$	$= 91$	Add 1 to both sides.
x	$= 7$	Divide both sides by 13.

Use the value of x to find angle measures.

$$\left.\begin{aligned} m\angle DEF &= 3x + 5 = 3(7) + 5 = 26° \\ m\angle FEG &= 10x - 6 = 10(7) - 6 = 64° \end{aligned}\right\} \text{Let } x = 7.$$

Check: $m\angle DEF + m\angle FEG = 26° + 64° = 90°$, a right angle.

PRACTICE 5 If $\angle PQS$ is a straight angle, find $m\angle PQR$ and $m\angle RQS$.

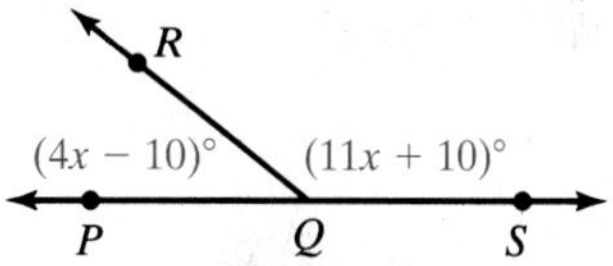

VOCABULARY & READINESS CHECK

Word Bank *Use the choices to fill in each blank.*

vertex	sides	congruent
angle	equal	protractor

1. An instrument used to measure angles in degrees is called a(n) ____________.
2. Two angles that have the same measure are called ____________ angles.
3. A(n) ____________ consists of two different rays with a common endpoint.
4. The rays of an angle are also called the ____________ of the angle.
5. The common endpoint of the rays of an angle is called the ____________ of the angle.
6. If two angles are congruent, then their measures are ____________.

Matching *Match each angle with the correct degree measure.*

7. obtuse angle	**A.** 180°
8. acute angle	**B.** between 90° and 180°
9. straight angle	**C.** between 0° and 90°
10. right angle	**D.** 90°

1.5 EXERCISE SET MyMathLab®

Name each shaded angle in three different ways. See Example 1.

1.

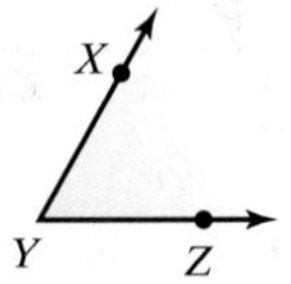

2.

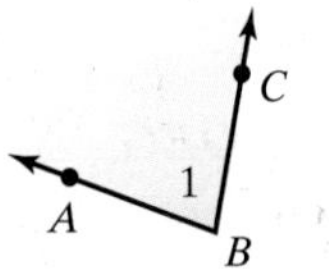

3.

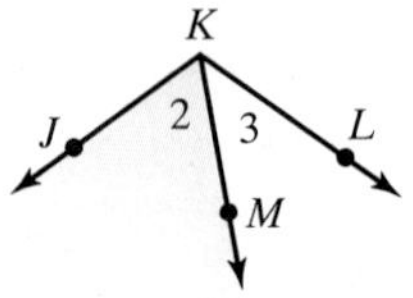

4.

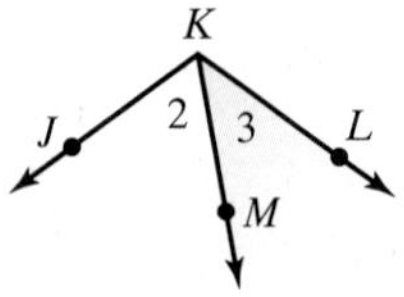

Use a Protractor *Use the diagram below. Find the measure of each angle. Then classify the angle as "acute," "right," "obtuse," or "straight." See Example 2.*

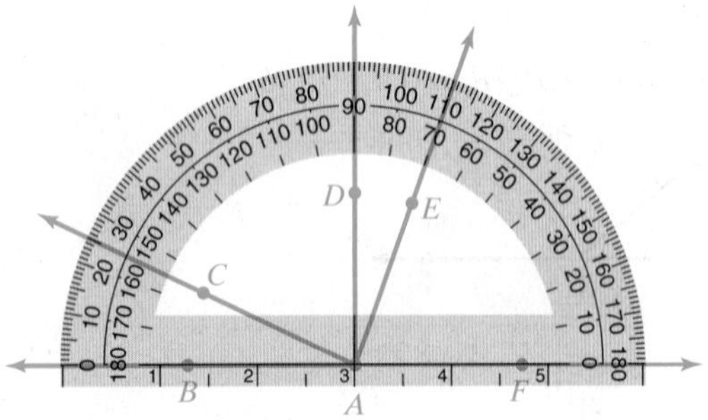

5. $\angle EAF$
6. $\angle DAF$
7. $\angle BAE$
8. $\angle BAC$
9. $\angle CAE$
10. $\angle DAE$

Sketch *Draw a figure that fits each description. See Example 3.*

11. an obtuse angle, $\angle RST$
12. an acute angle, $\angle GHJ$
13. a straight angle, $\angle KLM$
14. a right angle, $\angle QRS$

Use the diagram below. Use the arc marks to complete each statement. See Example 3.

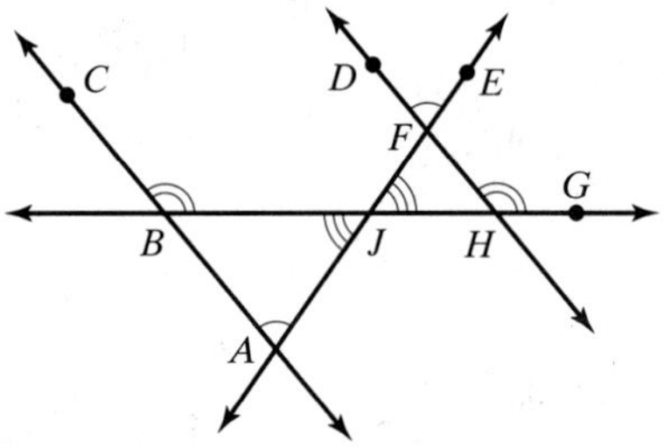

15. $\angle CBJ \cong$ ____________
16. $\angle FJH \cong$ ____________
17. If $m\angle EFD = 75°$, then $m\angle JAB =$ ____________.
18. If $m\angle GHF = 130°$, then $m\angle JBC =$ ____________.

Solve. See Example 4.

Use the diagram at the right to find the angle measure of the hands of a clock for each given time. (Each angle measure answer is ≤180°.)

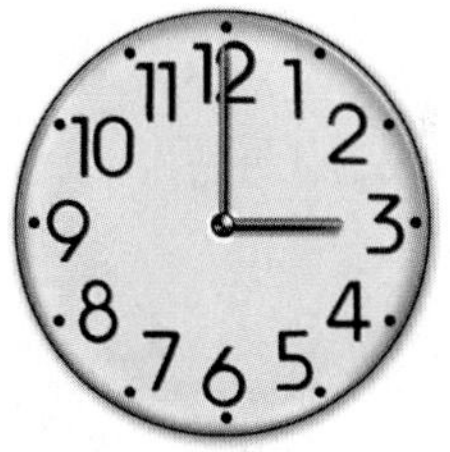

19. 3:00
20. 5:00
21. 10:00
22. 11:00

Solve. See Example 4.

23. An orange normally has 10 segments, as shown in the cross section below. If each segment has an equal angle measure, find x, the measure of the angle of one segment.

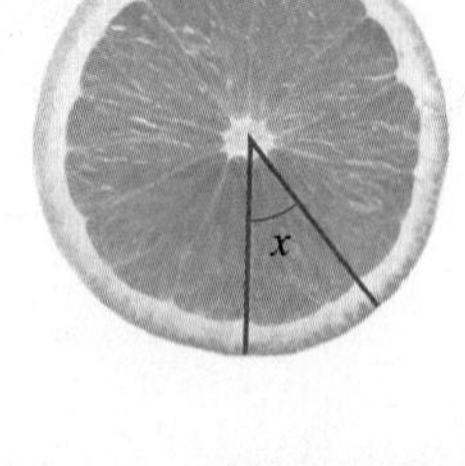

24. An apple normally has 5 chambers (containing one or two seeds), as shown by the cross section below. If the chambers are equally placed, find the angle measure, y.

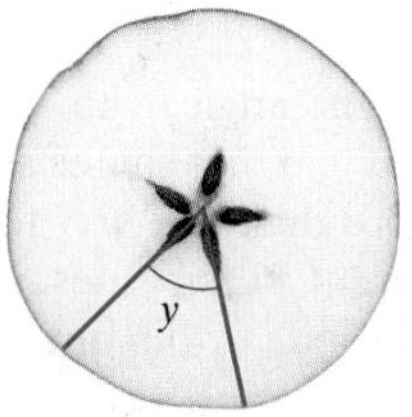

25. This rose window below is divided into 16 equal sections. Find z, the angle measure of a section.

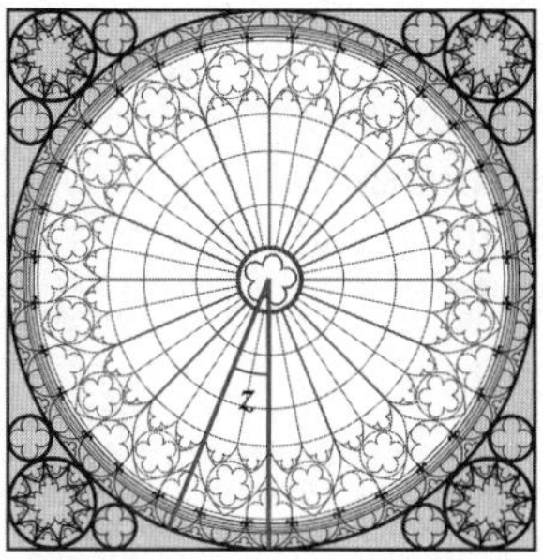

26. The wagon wheel below has 12 spokes dividing the wheel into 12 equal sections. Find x, the angle measure of a section.

Solve Exercises 27–32. See Example 5.

27. If $m\angle ABD = 79°$, what are $m\angle ABC$ and $m\angle DBC$?

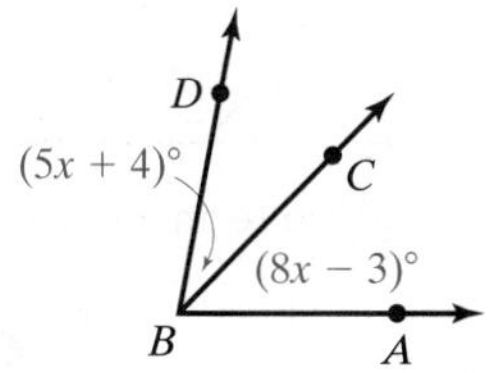

28. If $m\angle YNM = 45°$, what are $m\angle YNZ$ and $m\angle ZNM$?

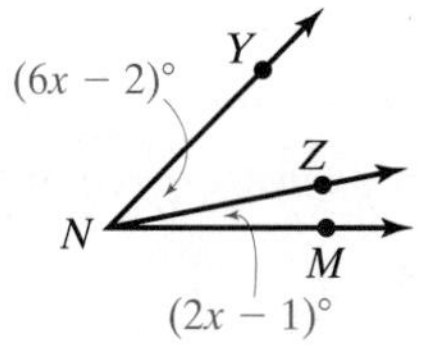

29. $\angle RQT$ is a straight angle. What are $m\angle RQS$ and $m\angle TQS$?

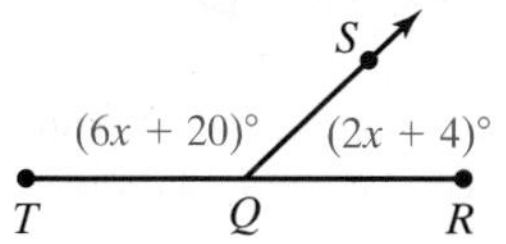

30. $\angle ABC$ is a straight angle. What are $m\angle ABD$ and $m\angle DBC$?

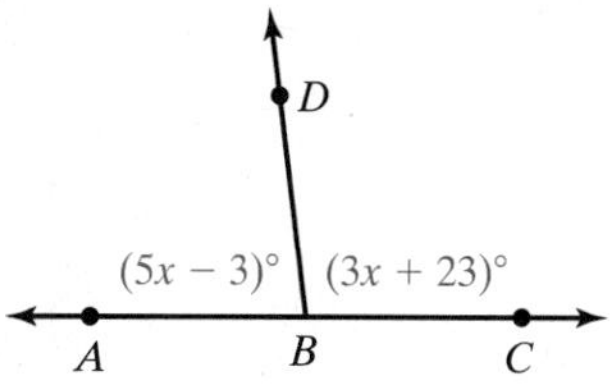

31. What are $m\angle RST$ and $m\angle TSU$?

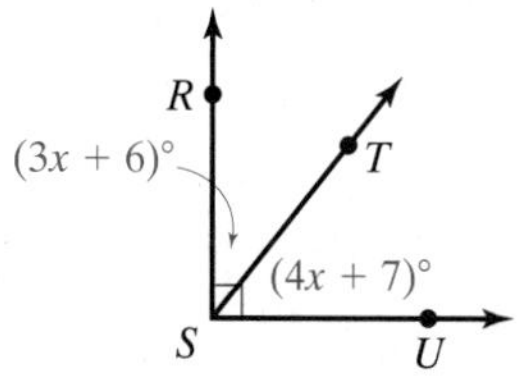

32. What are $m\angle MRJ$ and $m\angle JRK$?

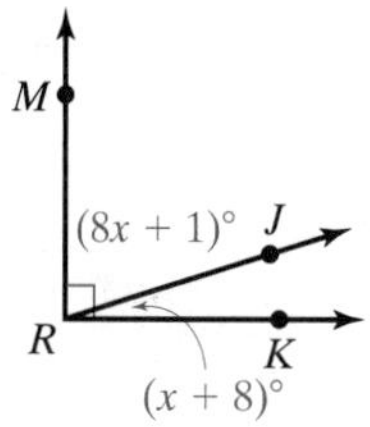

Use a Protractor *Measure and classify each angle. See Example 2.*

33.

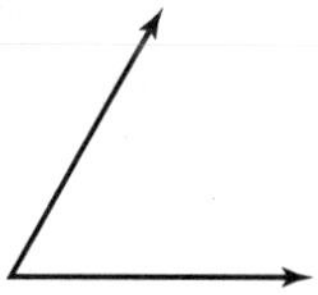

34.

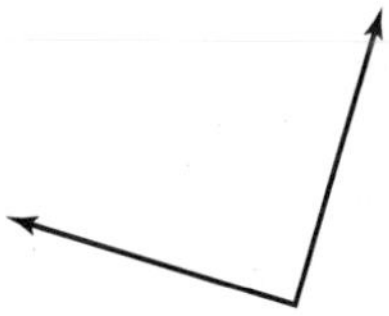

35.

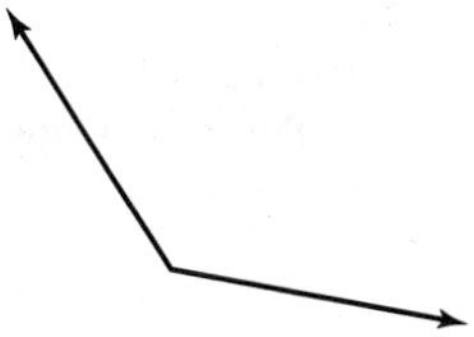

36.

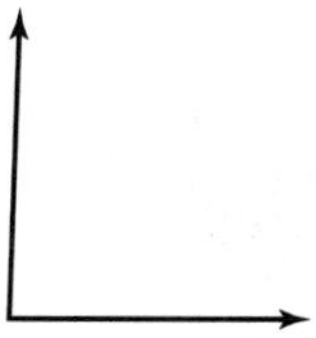

CONCEPT EXTENSIONS

Use the diagram shown for Exercises 37 and 38. Solve for x. Find the angle measures to check your work. See Example 5.

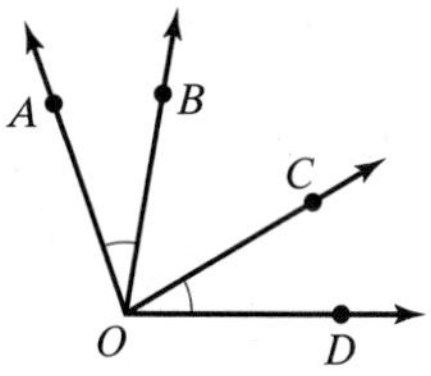

37. $m\angle AOB = (4x - 2)°, m\angle BOC = (5x + 10)°,$
$m\angle COD = (2x + 14)°$

38. $m\angle AOB = 28°, m\angle BOC = (3x - 2)°, m\angle AOD = 6x°$

Use the diagram at the right to find the angle measure of the hands of a clock at each time. (Each angle measure is ≤180°.)

39. 1:20 **40.** 8:40

41. 3:30 **42.** 4:30

43. Find the Error Your classmate concludes from the diagram to the right that $\angle JKL \cong \angle LKM$. Is your classmate correct? Explain.

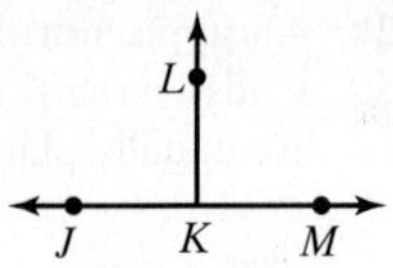

44. Sketch Sketch a right angle with vertex V. Name it $\angle 1$. Then sketch a 135° angle that shares a side with $\angle 1$. Name it $\angle PVB$. Is there more than one way to sketch $\angle PVB$? If so, sketch all the different possibilities. (*Hint:* Two angles are the same if you can rotate or flip one to match the other.)

1.6 ANGLE PAIRS AND THEIR RELATIONSHIPS

OBJECTIVES

1. Learn Special Relationships Between Pairs of Angles.
2. Use Algebra to Find Angle Measures.

VOCABULARY

- adjacent angles
- vertical angles
- linear pair
- complementary angles
- complement
- supplementary angles
- supplement
- angle bisector

OBJECTIVE 1 ▶ Learning Special Relationships Between Angle Pairs. Pairs of angles have special relationships that we give special names. Before we continue, recall the different parts of an angle, shown to the right.

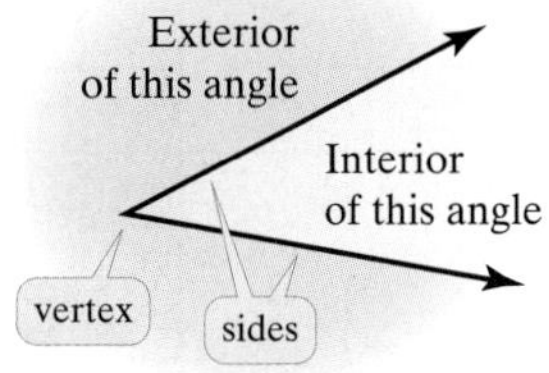

Special Angle Pairs

Definition	*Examples*
Two angles are called **adjacent angles** if they share a common side and a common vertex, but have no interior points in common.	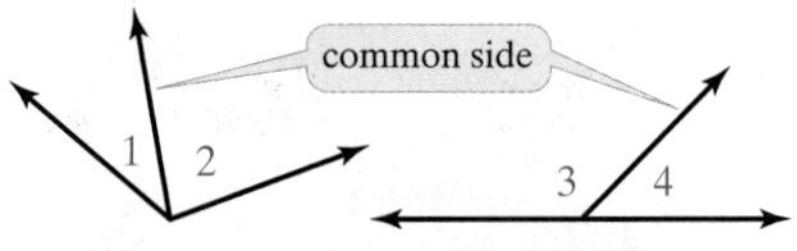 $\angle 1$ and $\angle 2$ are adjacent angles. $\angle 3$ and $\angle 4$ are adjacent angles.
Two angles are called **vertical angles** if their sides form opposite rays.	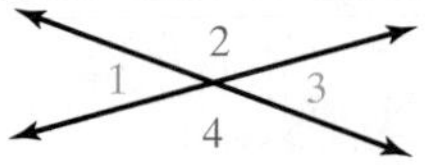 $\angle 1$ and $\angle 3$ are vertical angles. $\angle 2$ and $\angle 4$ are vertical angles.
Two adjacent angles form a **linear pair** if their noncommon sides are opposite rays.	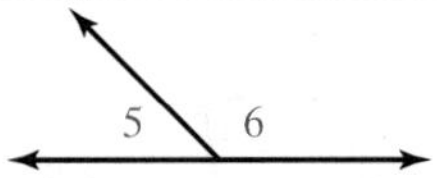 $\angle 5$ and $\angle 6$ form a linear pair.
Two angles are **complementary** if their measures have a sum of 90°. Each angle is called the **complement** of the other.	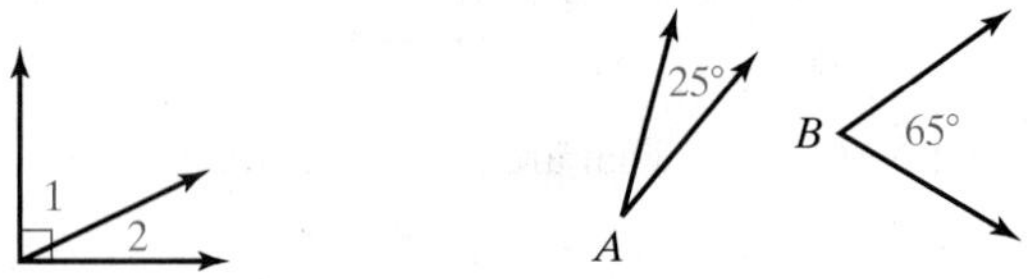 $\angle 1$ and $\angle 2$ are complementary angles. $\angle A$ and $\angle B$ are complementary angles.
Two angles are **supplementary** if their measures have a sum of 180°. Each angle is called the **supplement** of the other.	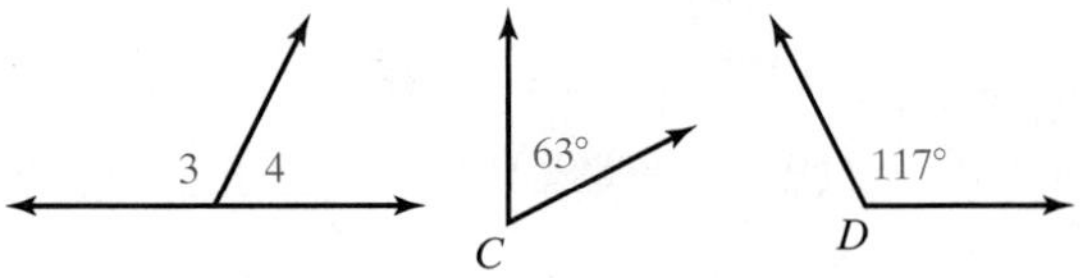 $\angle 3$ and $\angle 4$ are supplementary angles. (Notice that these angles form a linear pair.) $\angle C$ and $\angle D$ are supplementary angles.

EXAMPLE 1 **Identifying Vertical Angles and Linear Pairs**

Use the figure to answer each statement as true or false.

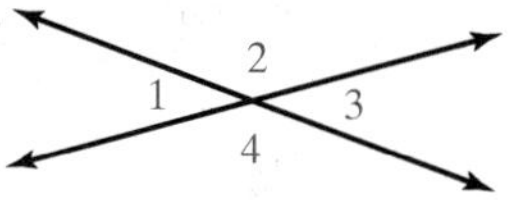

a. ∠2 and ∠3 are vertical angles.

b. ∠2 and ∠4 are vertical angles

c. ∠3 and ∠4 form a linear pair.

d. ∠3 and ∠1 form a linear pair.

Solution

a. False. The sides of the angles do not form opposite rays.

b. True. The sides of the angles form opposite rays.

c. True. They are adjacent angles and their noncommon sides form opposite rays.

d. False. They are not adjacent angles. □

PRACTICE 1 Use the figure to answer each statement as true or false.

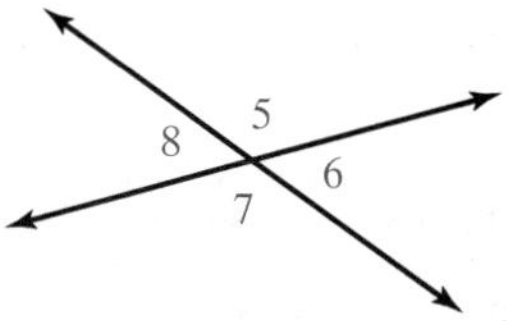

a. ∠8 and ∠5 form a linear pair.

b. ∠8 and ∠6 are vertical angles.

c. ∠7 and ∠5 form a linear pair.

d. ∠7 and ∠6 are vertical angles.

EXAMPLE 2 **Identifying Complementary Angles**

Use the figure to identify each pair of complementary angles.

Solution

∠1 and ∠2 Because $35° + 55° = 90°$
∠2 and ∠3 Because $55° + 35° = 90°$
∠3 and ∠4 Because $35° + 55° = 90°$
∠1 and ∠4 Because $35° + 55° = 90°$

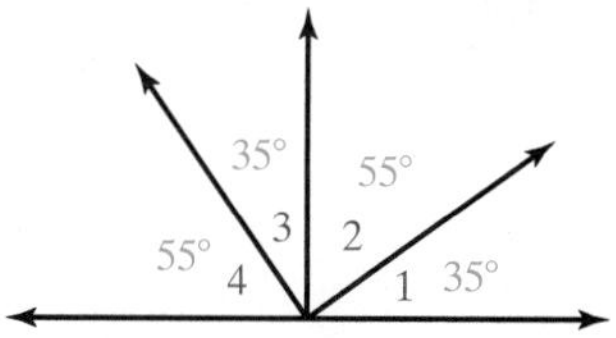

□

PRACTICE 2 Use the figure to identify each pair of complementary angles.

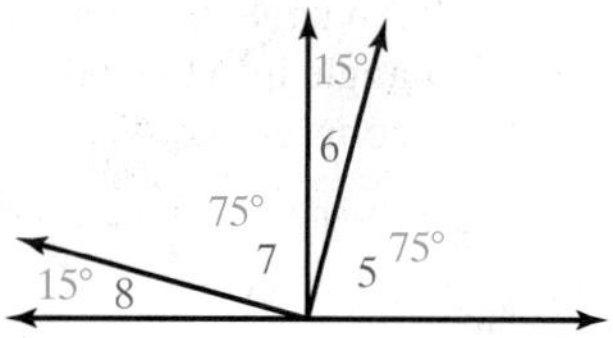

EXAMPLE 3 **Identifying Supplementary Angles**

Use the figure to identify each pair of supplementary angles.

Solution

∠1 and ∠2 Because $33° + 147° = 180°$
∠2 and ∠3 Because $147° + 33° = 180°$
∠3 and ∠4 Because $33° + 147° = 180°$
∠4 and ∠1 Because $147° + 33° = 180°$

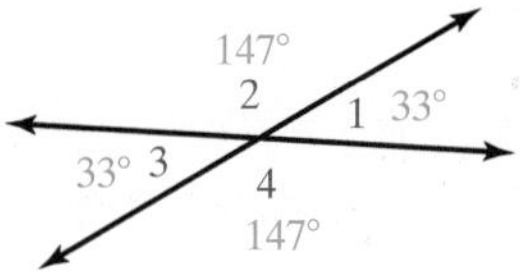

□

PRACTICE **3** Use the figure to identify each pair of supplementary angles.

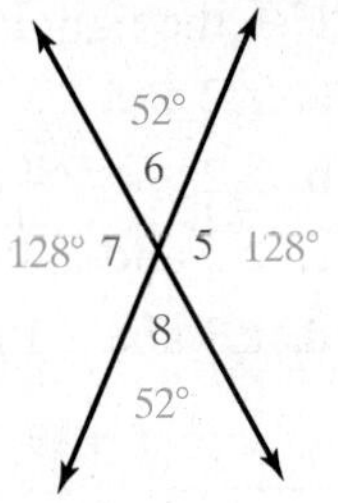

EXAMPLE 4 **Finding Measures of Complementary and Supplementary Angles**

Given that $m\angle P = 73°$:

a. If $\angle A$ and $\angle P$ are supplementary angles, find $m\angle A$.

b. If $\angle B$ and $\angle P$ are complementary angles, find $m\angle B$.

Solution

a. Since $m\angle A + m\angle P = 180°$, then $m\angle A = 180° - m\angle P = 180° - 73° = 107°$

b. Since $m\angle B + m\angle P = 90°$, then $m\angle B = 90° - m\angle P = 90° - 73° = 17°$ □

PRACTICE **4** Given that $m\angle C = 16°$:

a. If $\angle M$ and $\angle C$ are supplementary angles, find $m\angle M$.

b. If $\angle N$ and $\angle C$ are complementary angles, find $m\angle N$.

Just as a segment has a bisector (or midpoint), an angle has a bisector, called an angle bisector. An **angle bisector** is a ray that divides an angle into two adjacent angles that are congruent.

Defined Term

Definition	*Example*
An **angle bisector** is a ray that divides an angle into two adjacent angles that are congruent.	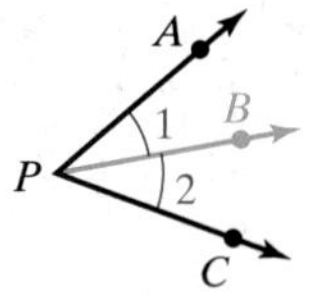 $m\angle 1 = m\angle 2$ — $\overrightarrow{PB}$ is the angle bisector of $\angle APC$.

EXAMPLE 5 **Using Angle Bisectors to Find Angle Measures**

Use the figure shown to find the measure of each unknown angle.

a. Find $m\angle BAC$.

b. Find $m\angle BAE$.

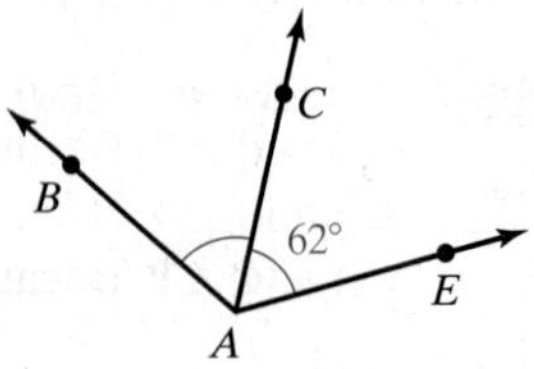

Solution From the given notation, $\angle BAC \cong \angle CAE$; thus $\overrightarrow{AC}$ is the angle bisector of $\angle BAE$. Then,

a. $m\angle CAE = m\angle BAC = 62°$.

b. $m\angle BAE = m\angle BAC + m\angle CAE = 62° + 62° = 124°$. □

PRACTICE 5 Use the figure shown to find the measure of each unknown angle.

a. Find $m\angle LKM$.

b. Find $m\angle MKN$.

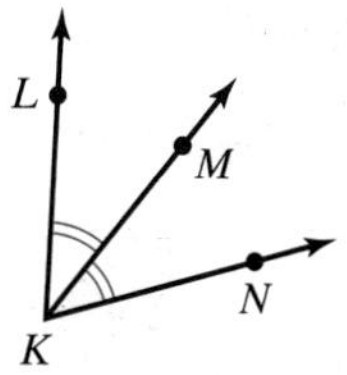

$m\angle LKN = 72°$

▸ Helpful Hint

You may have noticed by now that vertical angles are congruent. (This is *not* a postulate, and we will prove this in Chapter 2.)

Vertical Angles:

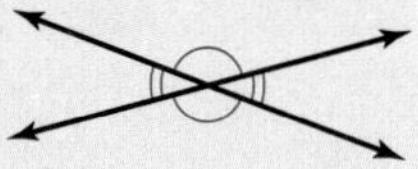

OBJECTIVE 2 ▸ Using Algebra to Find Angle Measures.

EXAMPLE 6 Finding Angle Measures

In the figure, $\overrightarrow{QY}$ bisects $\angle RQZ$.
Find the value of x; then find $m\angle RQY$ and $m\angle YQZ$.

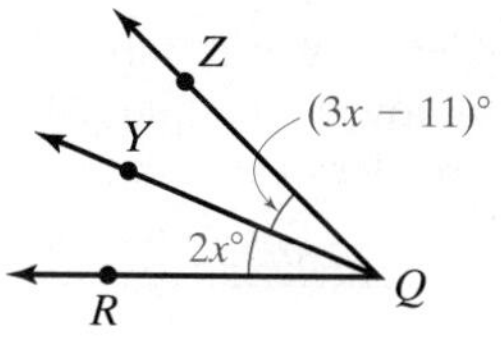

Solution

$$\begin{aligned} m\angle RQY &= m\angle YQZ && \text{Congruent angles have the same measure.} \\ 2x° &= (3x - 11)° && \text{Substitution} \\ 0 &= x - 11 && \text{Subtract } 2x \text{ from both sides.} \\ 11 &= x && \text{Add 11 to both sides.} \end{aligned}$$

$$\left.\begin{aligned} m\angle RQY &= 2x° = 2(11)° = 22° \\ m\angle YQZ &= (3x - 11)° = (3 \cdot 11 - 11)° = 22° \end{aligned}\right\} \text{The values are the same, so the work is correct.}$$

Thus, $x = 11$, and $m\angle RQY = m\angle YQZ = 22°$.

PRACTICE 6 In the figure, $\overrightarrow{HL}$ bisects $\angle KHJ$.

Find the value of x; then find $m\angle KHL$ and $m\angle LHJ$.

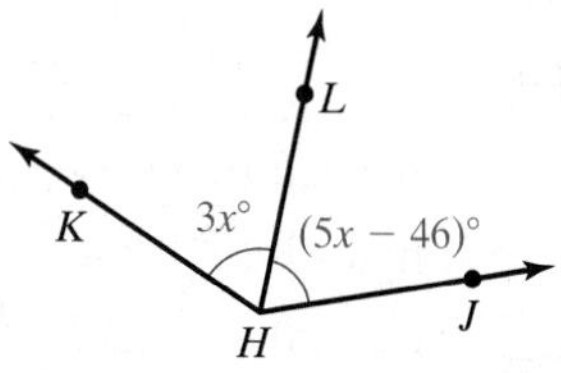

EXAMPLE 7 Finding Measures of Angles

Solve for x and y. Then find the measure of each angle.

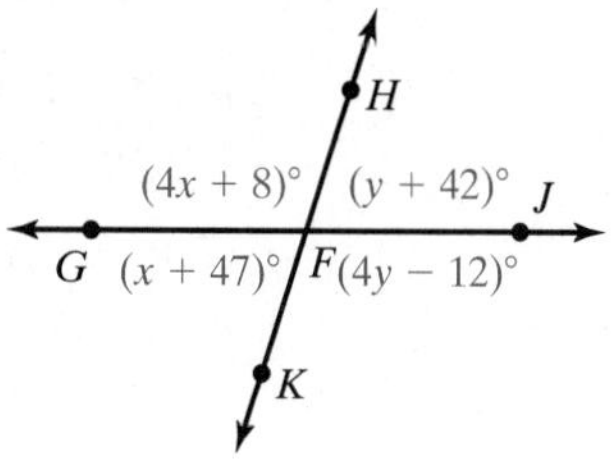

Solution To solve for x (or y), we will choose the two angles in x (or y). These two angles form a linear pair; thus, their sum is 180°. (Angles that form a linear pair are supplementary.)

$$m\angle GFK + m\angle GFH = 180^\circ \qquad m\angle KFJ + m\angle JFH = 180^\circ$$
$$(x + 47)^\circ + (4x + 8)^\circ = 180^\circ \qquad (4y - 12)^\circ + (y + 42)^\circ = 180^\circ$$
$$5x + 55 = 180 \qquad 5y + 30 = 180$$
$$5x = 125 \qquad 5y = 150$$
$$x = 25 \qquad y = 30$$

Thus, $x = 25$ and $y = 30$.
Let's use substitution to find the measure of each angle.

$$m\angle GFK = (x + 47)^\circ = (25 + 47)^\circ = 72^\circ$$
$$m\angle GFH = (4x + 8)^\circ = (4 \cdot 25 + 8)^\circ = 108^\circ$$
$$m\angle KFJ = (4y - 12)^\circ = (4 \cdot 30 - 12)^\circ = 108^\circ$$
$$m\angle JFH = (y + 42)^\circ = (30 + 42)^\circ = 72^\circ$$

Check: Notice the angle measures of 72°, 108°, 108°, and 72°. The pairs of angles with the same measure form vertical angles, so the pairs should be congruent, which they are. Also, 72° + 108° = 180° so the proper angles form linear pairs. □

> **Helpful Hint**
> We have not formally stated nor proved anything about vertical angles. They are indeed congruent, so it is a nice way to check.

PRACTICE 7 Solve for x and y. Then find the measure of each angle.

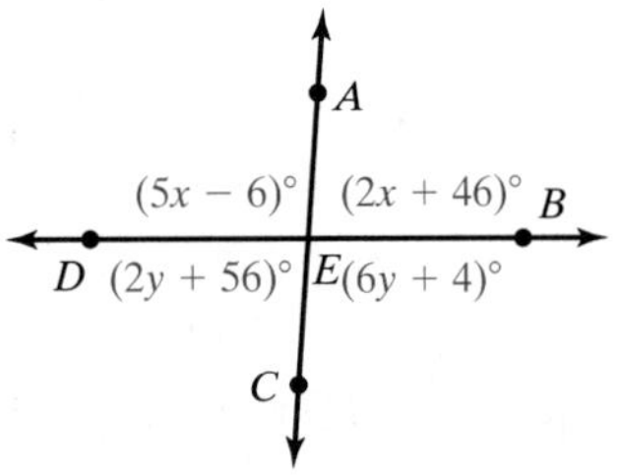

VOCABULARY & READINESS CHECK

Word Bank *Use the choices to fill in each blank.*

vertical	angle bisector	supplementary
adjacent	linear pair	complementary

1. Two angles are ____________ angles if their measures have a sum of 90°.
2. Two angles are ____________ angles if their measures have a sum of 180°.
3. Two angles are ____________ angles if their sides form opposite rays.
4. Two angles are ____________ angles if they share a common side and a common vertex, but have no interior points in common.
5. Two adjacent angles form a(n) ____________ if their noncommon sides are opposite rays.
6. A ray that divides an angle into two adjacent congruent angles is called a(n) ____________.

1.6 EXERCISE SET MyMathLab®

MIXED PRACTICE

Fill in the Blank *Use the diagram for Exercises 1–6. See Example 1.*

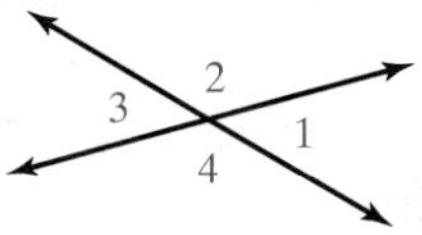

1. ∠2 and ____________ are vertical angles.

2. ∠3 and ____________ are vertical angles.

True or False *Answer each statement true or false.*

3. ∠4 and ∠1 form a linear pair.

4. ∠2 and ∠1 form a linear pair.

5. ∠3 and ∠1 form a linear pair.

6. ∠4 and ∠2 form a linear pair.

Use the diagram below. Is each statement true? Explain. See Examples 1–3.

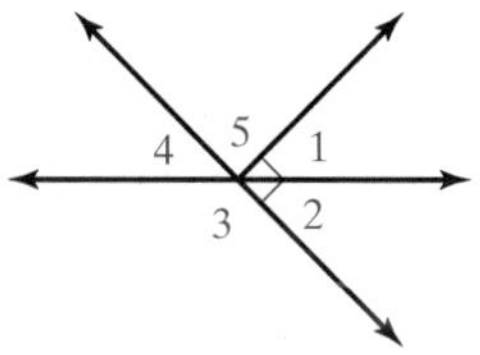

7. ∠1 and ∠5 are adjacent angles.

8. ∠3 and ∠5 are vertical angles.

9. ∠3 and ∠4 are complementary.

10. ∠1 and ∠2 are supplementary.

Name an angle or angles in the diagram described by each of the following. See Examples 1–3.

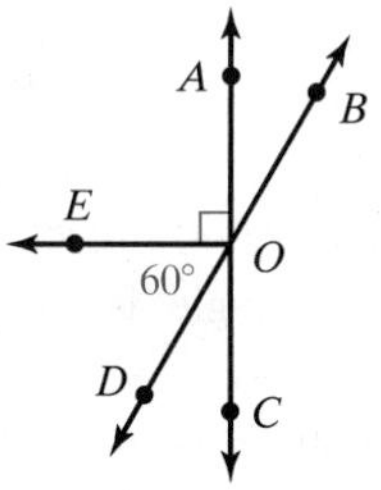

11. supplementary to ∠*AOD*

12. adjacent and congruent to ∠*AOE*

13. supplementary to ∠*EOA*

14. complementary to ∠*EOD*

15. an angle that is vertical to ∠*AOB*

16. an angle that is vertical to ∠*BOC*

True or False *Use the diagram to answer each statement true or false. See Example 2.*

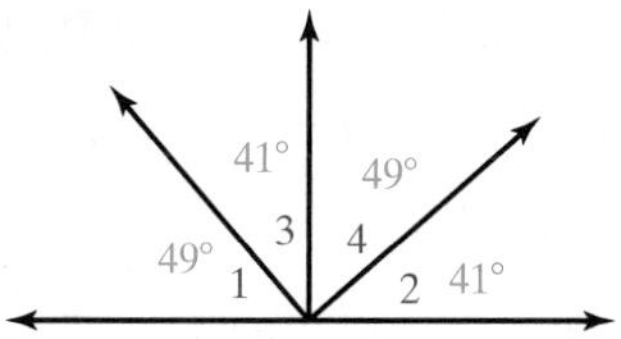

17. ∠2 and ∠3 are complementary angles.

18. ∠1 and ∠4 are complementary angles.

19. ∠2 and ∠4 are complementary angles.

20. ∠3 and ∠1 are complementary angles.

21. ∠1 and ∠2 are complementary angles.

22. ∠3 and ∠4 are complementary angles.

True or False *Use the diagram to answer each statement true or false. See Example 3.*

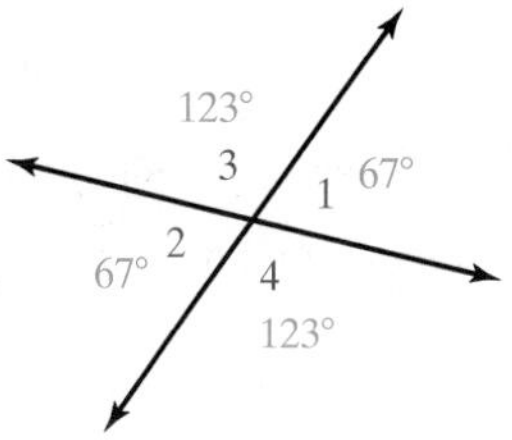

23. ∠1 and ∠2 are supplementary angles.

24. ∠3 and ∠4 are supplementary angles.

25. ∠3 and ∠2 are supplementary angles.

26. ∠4 and ∠1 are supplementary angles.

For Exercises 27–34, can you make the given conclusion from the information in the diagram? Explain. See Examples 1–3.

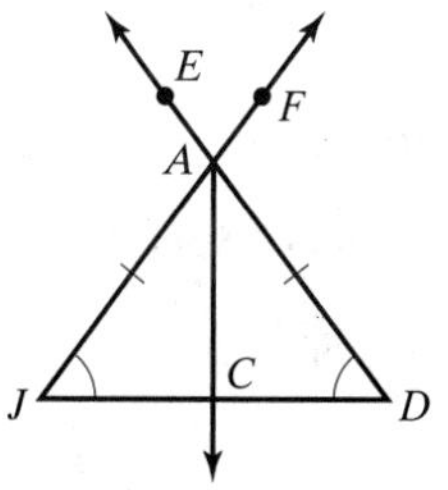

27. $\angle J \cong \angle D$

28. $\angle JAC \cong \angle DAC$

29. $m\angle FAD + m\angle DAJ = 180°$

30. $m\angle JCA + m\angle ACD = 180°$

31. $\overline{AJ} \cong \overline{AD}$

32. *C* is the midpoint of $\overline{JD}$.

33. ∠*JAE* and ∠*EAF* are vertical angles.

34. ∠*EAF* and ∠*JAD* are vertical angles.

In the diagram below, $m\angle ACB = 65°$. Find each of the following. See Examples 1–4.

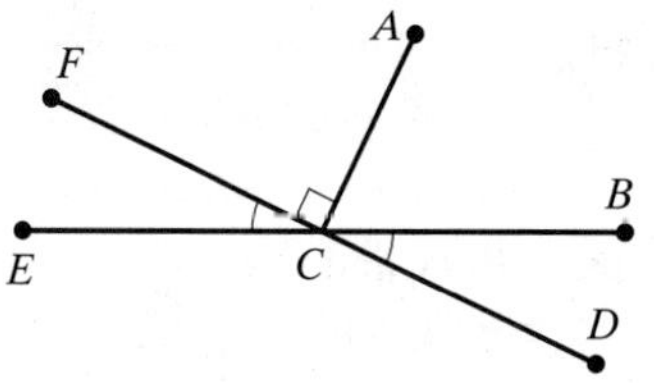

35. $m\angle ACD$

36. $m\angle BCD$

37. $m\angle ECD$

38. $m\angle ACE$

Fill in the Chart *The first row has been completed for you. See Example 4.*

	Angle Measure	Measure of Its Supplement	Measure of Its Complement
	30°	150°	60°
39.	10°		
40.	70°		
41.	88°		
42.	1°		
43.	123°		
44.	175°		

Use the notation on each diagram to find each angle measure. See Example 5.

45. $m\angle DJG$

46. $m\angle DJH$

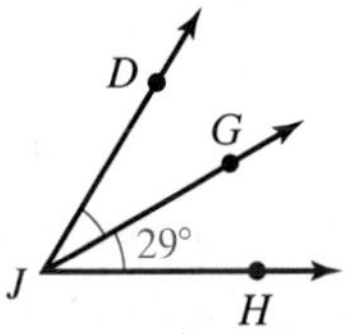

47. $m\angle RVT$

48. $m\angle SVT$

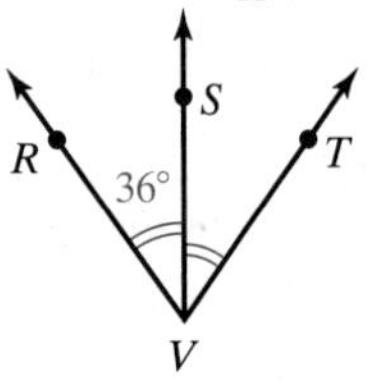

Multiple Steps *Solve. See Example 6.*

49. In the diagram, $\overrightarrow{GH}$ bisects $\angle FGI$.

a. Solve for x.

b. Find $m\angle FGH$.

c. Find $m\angle HGI$.

d. Find $m\angle FGI$.

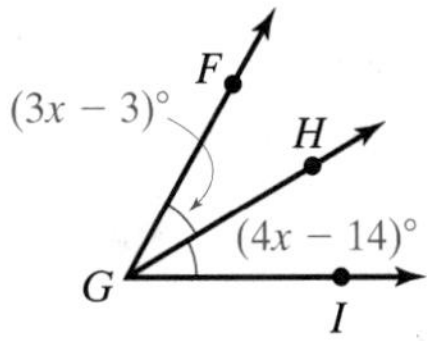

50. In the diagram, $\overrightarrow{QB}$ bisects $\angle AQG$.

a. Solve for x.

b. Find $m\angle AQB$.

c. Find $m\angle BQG$.

d. Find $m\angle AQG$.

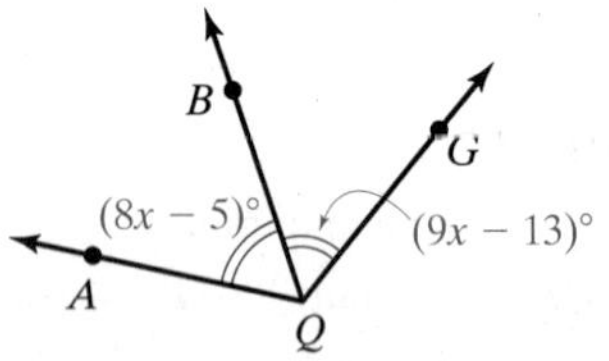

For Exercises 51–54, $\overrightarrow{BD}$ bisects $\angle ABC$. Solve for x and find $m\angle ABC$. See Example 6.

51. $m\angle ABD = 5x°, m\angle DBC = (3x + 10)°$

52. $m\angle ABC = (4x - 12)°, m\angle ABD = 24°$

53. $m\angle ABD = (4x - 16)°, m\angle CBD = (2x + 6)°$

54. $m\angle ABD = (3x + 20)°, m\angle CBD = (6x - 16)°$

55. Multiple Steps $\angle RQS$ and $\angle TQS$ are a linear pair where $m\angle RQS = (2x + 4)°$ and $m\angle TQS = (6x + 20)°$. See Example 6.

a. Solve for x.

b. Find $m\angle RQS$ and $m\angle TQS$.

c. Show how you can check your answer.

56. Multiple Steps $\angle EFG$ and $\angle GFH$ are a linear pair where $m\angle EFG = (2n + 21)°$, and $m\angle GFH = (4n + 15)°$.

a. Solve for n.

b. Find $m\angle EFG$ and $m\angle GFH$.

c. Show how you can check your answer.

Solve for x and y. Then find the measure of each angle. See Example 7.

57.

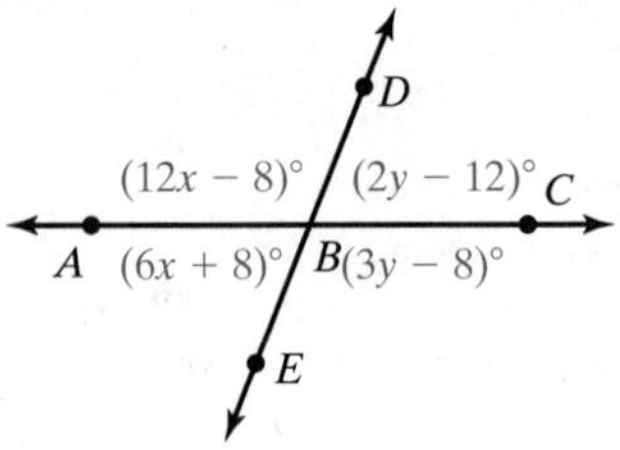

58.

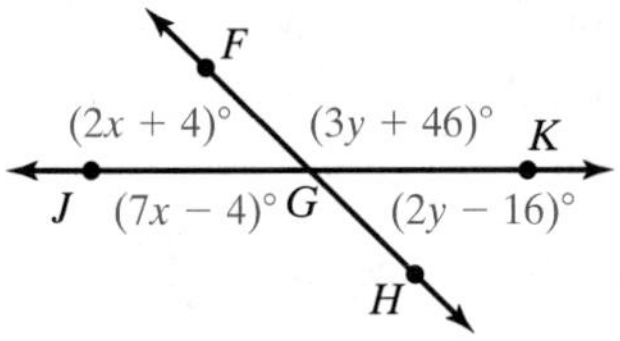

59.

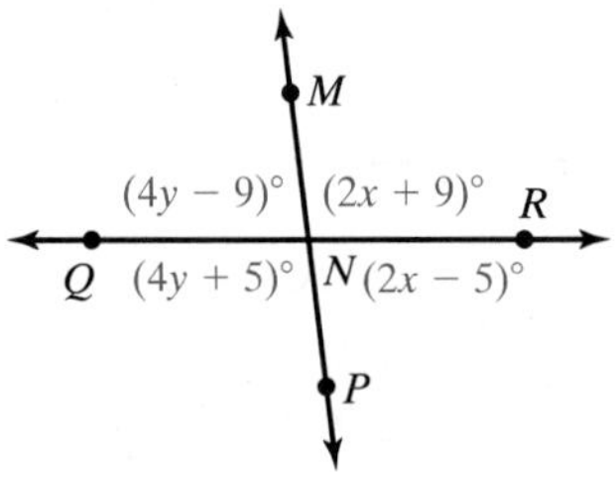

60.

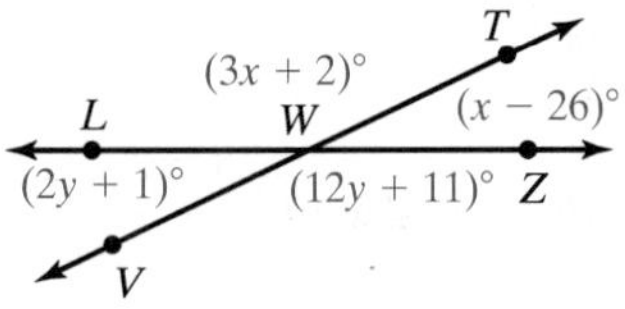

CONCEPT EXTENSIONS

For Exercises 61 and 62, find the measure of each angle in the angle pair described. Start by drawing a diagram.

61. The measure of one angle is twice the measure of its supplement.

62. The measure of one angle is 20 less than the measure of its complement.

63. In the diagram below, are $\angle 1$ and $\angle 2$ adjacent? Justify your reasoning.

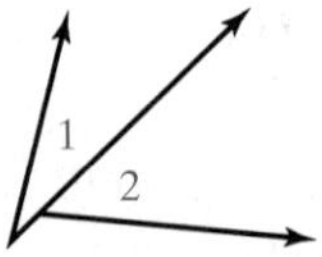

64. When $\overrightarrow{BX}$ bisects $\angle ABC$, $\angle ABX \cong \angle CBX$. One student claims there is always a related equation $m\angle ABX = \frac{1}{2}m\angle ABC$. Another student claims the related equation is $2m\angle ABX = m\angle ABC$. Who is correct? Explain.

65. **Multiple Steps** A beam of light and a mirror can be used to study the behavior of light. Light that strikes the mirror is reflected so that the angle of reflection and the angle of incidence are congruent. In the diagram, $\angle ABC$ has a measure of 41°.

a. Name the angle of reflection and find its measure.

b. Find $m\angle ABD$.

c. Find $m\angle ABE$ and $m\angle DBF$.

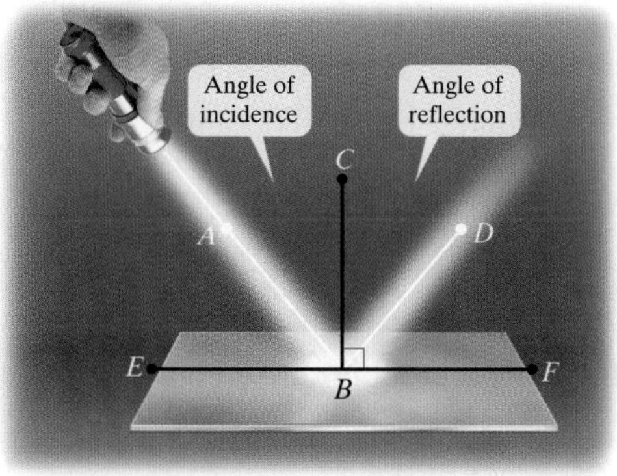

66. Describe all situations where vertical angles are also supplementary.

For Exercises 67–70, name all of the angle(s) in the diagram that are described by the following.

67. adjacent and congruent to $\angle KMQ$

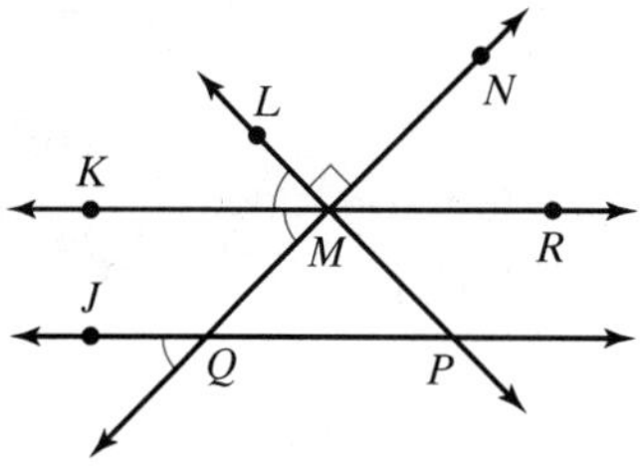

68. supplementary to $\angle JQM$

69. complementary to $\angle NMR$

70. a linear pair with $\angle LMQ$

1.7 COORDINATE GEOMETRY—MIDPOINT AND DISTANCE FORMULAS

OBJECTIVES

1 Find the Midpoint of a Segment.

2 Find the Distance Between Two Points on the Coordinate Plane.

VOCABULARY

- Midpoint Formula
- Distance Formula

OBJECTIVE 1 ▶ Finding the Midpoint of a Segment. Recall from Section 1.4 that a **midpoint** of a segment is a point that divides, or bisects, a segment into two congruent segments.

Now we study the midpoint of a segment that is either on a number line or on the coordinate plane (also called the rectangular coordinate system).

Midpoint Formulas

	Formula	*Diagram*
On a Number Line The coordinate of the midpoint is the *average* or *mean* of the coordinates of the endpoints.	Given $\overline{AB}$ on a number line: The coordinate of the midpoint M of $\overline{AB}$ is $\frac{a+b}{2}$.	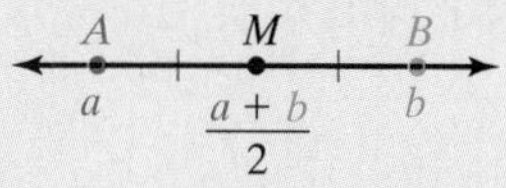
On the Coordinate Plane The coordinates of the midpoint are the average of the x-coordinates and the average of the y-coordinates of the endpoints.	Given $\overline{AB}$ where $A(x_1, y_1)$ and $B(x_2, y_2)$: The coordinates of the midpoint of $\overline{AB}$ are $\left(\frac{x_1+x_2}{2}, \frac{y_1+y_2}{2}\right)$.	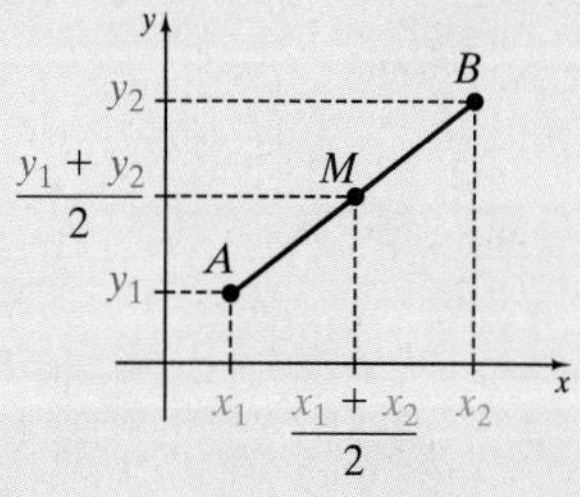

EXAMPLE 1 **Finding the Midpoint on a Number Line**

Use the diagram and find the coordinate of the midpoint, M, of $\overline{PQ}$.

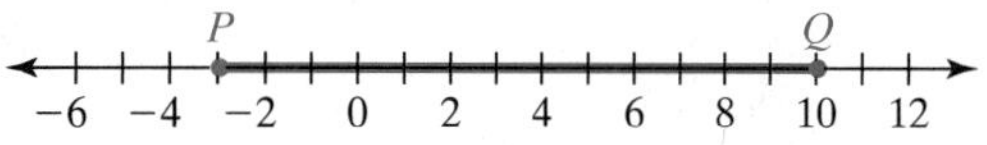

Solution The coordinate of P is –3. The coordinate of Q is 10. Let $a = -3$ and $b = 10$, or simply find the average of –3 and 10.

$$M = \frac{a+b}{2} = \frac{-3+10}{2} = \frac{7}{2} = 3.5$$

The coordinate, M, of the midpoint of $\overline{PQ}$ is 3.5. □

To **check**, graph point M on the number line and see that $PM = MQ$.

PRACTICE 1 Use the diagram and find the coordinate of the midpoint, M, of $\overline{RS}$.

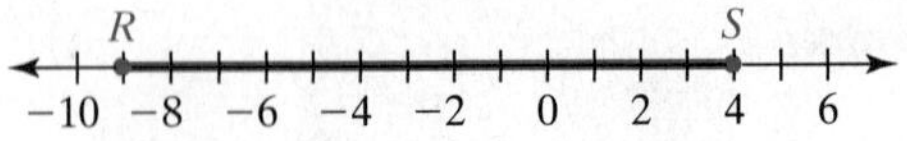

EXAMPLE 2 Finding the Midpoint in the Coordinate Plane

Find the midpoint of the line segment, $\overline{PQ}$, that joins points $P(-3, 3)$ and $Q(1, 0)$.

Solution Use the Midpoint Formula. It makes no difference which point we call (x_1, y_1) and which point we call (x_2, y_2). Let $(x_1, y_1) = P(-3, 3)$ and $(x_2, y_2) = Q(1, 0)$.

$$x\text{-coordinate of } M = \frac{x_1 + x_2}{2} = \frac{-3+1}{2} = \frac{-2}{2} = -1$$

$$y\text{-coordinate of } M = \frac{y_1 + y_2}{2} = \frac{3+0}{2} = \frac{3}{2} = 1.5$$

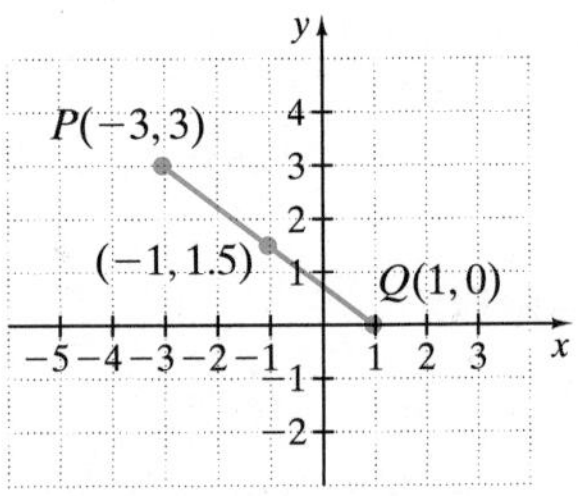

The midpoint of segment $\overline{PQ}$ is $M(-1, 1.5)$, as shown on the graph. □

PRACTICE 2 Find the midpoint of the line segment that joins points $P(5, -2)$ and $Q(8, -6)$.

So far, if we know the coordinates of the endpoints of a segment, we can use the **Midpoint Formula** to find the midpoint of the segment.

If we have the midpoint and an endpoint of a segment, we can also use the Midpoint Formula to find the other endpoint.

EXAMPLE 3 Finding an Endpoint

The midpoint of $\overline{CD}$ is $M(-2, 1)$. One endpoint is $C(-5, 7)$. What are the coordinates of the other endpoint, D?

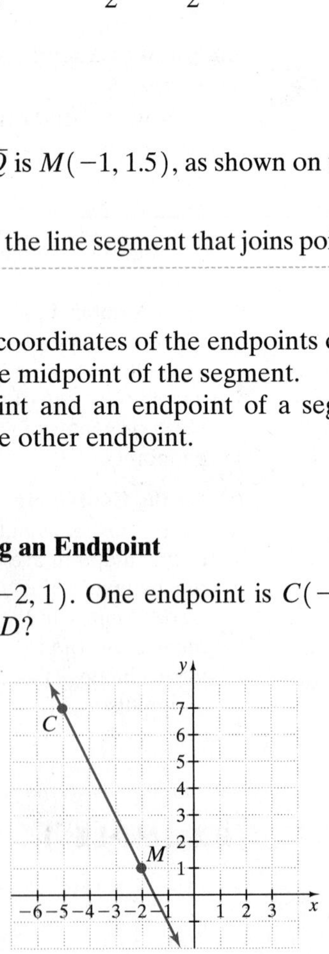

Solution Let $M(-2, 1)$ be (x, y) and $C(-5, 7)$ be (x_1, y_1). Let the coordinates of D be (x_2, y_2).

$$(-2, 1) = \left(\frac{-5 + x_2}{2}, \frac{7 + y_2}{2}\right)$$

x		y
$-2 = \dfrac{-5 + x_2}{2}$	Use the Midpoint Formula.	$1 = \dfrac{7 + y_2}{2}$
$-4 = -5 + x_2$	Multiply each side by 2.	$2 = 7 + y_2$
$1 = x_2$	Solve for x_1 and y_2.	$-5 = y_2$

The coordinates of D are $(1, -5)$.

To **check**, use the Midpoint Formula and see that the midpoint of segment $\overline{CD}$ is $M(-2, 1)$. □

PRACTICE **3** The midpoint of $\overline{AB}$ has coordinates $(4, -9)$. Endpoint A has coordinates $(-3, -5)$. What are the coordinates of B?

OBJECTIVE 2 ▶ Finding the Distance Between Two Points on the Coordinate Plane. In Section 1.4, we learned how to find the distance between two points on a number line. To find the distance between two points on a coordinate plane, we can use the Distance Formula.

Distance Formula

The distance between two points $A(x_1, y_1)$ and $B(x_2, y_2)$ is

$$d = \sqrt{(x_2 - x_1)^2 + (y_2 - y_1)^2}.$$

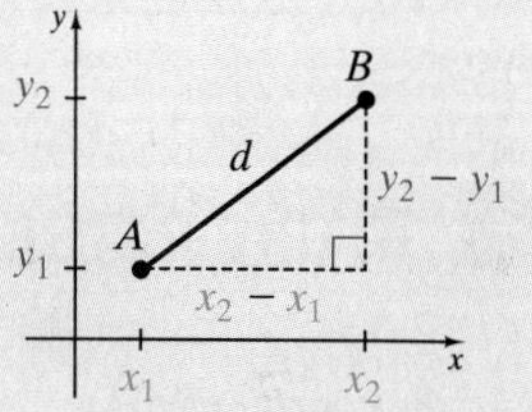

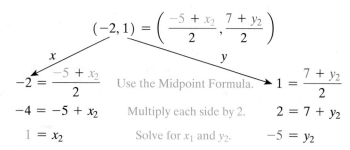

The **Distance Formula** is based on the *Pythagorean Theorem*, which we will study later in this book. When we use the Distance Formula, we are really finding the length of the hypotenuse of a right triangle.

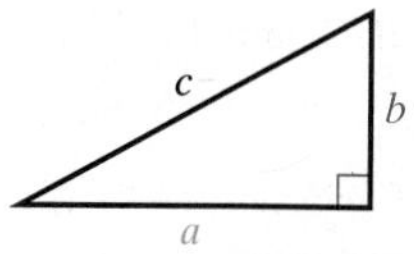

$a^2 + b^2 = c^2$

EXAMPLE 4 Find the distance between $A(2, -5)$ and $B(1, -4)$. Give an exact distance and a one-decimal-place approximation.

Solution To use the Distance Formula, it makes no difference which point we call (x_1, y_1) and which point we call (x_2, y_2). We will let $(x_1, y_1) = (2, -5)$ and $(x_2, y_2) = (1, -4)$.

$$\begin{aligned} d &= \sqrt{(x_2 - x_1)^2 + (y_2 - y_1)^2} \\ &= \sqrt{(1 - 2)^2 + [-4 - (-5)]^2} \\ &= \sqrt{(-1)^2 + (1)^2} \\ &= \sqrt{1 + 1} \\ &= \sqrt{2} \approx 1.4 \end{aligned}$$

The distance between the two points is exactly $\sqrt{2}$ units, or approximately 1.4 units. □

PRACTICE 4 Find the distance between $P(-3, 7)$ and $Q(-2, 3)$. Give an exact distance and a one-decimal-place approximation.

Helpful Hint

The distance between two points is a distance.

The midpoint of a line segment is the point halfway between the endpoints of the segment.

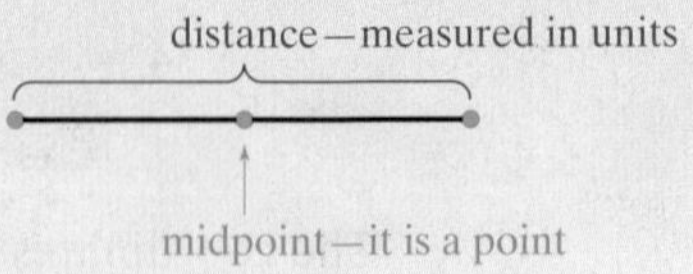

VOCABULARY & READINESS CHECK

Word Bank *Use the choices below to fill in each blank. Some choices may be used more than once.*

distance   midpoint   point

1. The ____________ of a line segment is a ____________ exactly halfway between the two endpoints of the line segment.
2. The ____________ Formula is $d = \sqrt{(x_2 - x_1)^2 + (y_2 - y_1)^2}$.
3. The ____________ Formula for a segment on the coordinate plane is $M = \left(\frac{x_1 + x_2}{2}, \frac{y_1 + y_2}{2}\right)$.
4. The ____________ Formula for a segment on a number line is $M = \frac{a + b}{2}$.

1.7 EXERCISE SET MyMathLab®

Find the coordinate of the midpoint of the segment with the given endpoints. See Example 1.

1. 2 and 4
2. −9 and 6
3. 2 and −5
4. −8 and −12

For Exercises 5–12, find the midpoint of the line segment whose endpoints are given. See Example 2.

5. $(6, -8), (2, 4)$
6. $(3, 9), (7, 11)$
7. $(-2, -1), (-8, 6)$
8. $(-3, -4), (6, -8)$
9. $(7, 3), (-1, -3)$
10. $(-2, 5), (2, 15)$
11. $\left(\frac{1}{2}, \frac{3}{8}\right), \left(-\frac{3}{2}, \frac{5}{8}\right)$
12. $\left(-\frac{2}{5}, \frac{7}{15}\right), \left(-\frac{2}{5}, -\frac{4}{15}\right)$

The coordinates of point T are given. The midpoint of $\overline{ST}$ is $(5, -8)$. Find the coordinates of point S. See Example 3.

13. $T(0, 4)$
14. $T(5, -15)$
15. $T(10, 18)$
16. $T(-2, 8)$

Find the distance between each pair of points. For Exercises 23 and 24, give an exact answer and a one-decimal-place approximation. See Example 4.

$-1), K(2, 5)$
18. $L(10, 14), M(-8, 14)$
$1, -11), P(-1, -3)$
20. $A(0, 3), B(0, 12)$
21. $E(6, -2), F(-2, 4)$
22. $Q(12, -12), T(5, 12)$
23. $R(0, 5), S(12, 3)$
24. $C(12, 6), D(-8, 18)$

For Exercises 25–28, use the map shown. Find the distance between the cities to the nearest tenth.

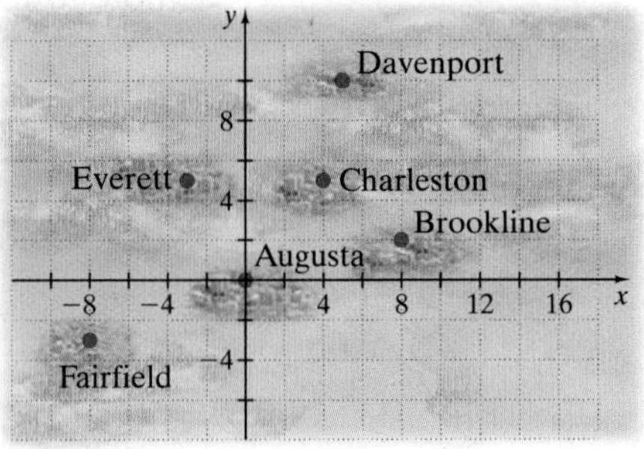

25. Augusta and Brookline
26. Brookline and Charleston
27. Brookline and Davenport
28. Everett and Fairfield

On a zip-line course, you are harnessed to a cable that travels through the treetops. You start at platform A and zip to each of the other platforms. Each grid unit represents 5 m. Round each answer to the nearest tenth.

29. How far do you travel from Platform B to Platform C?

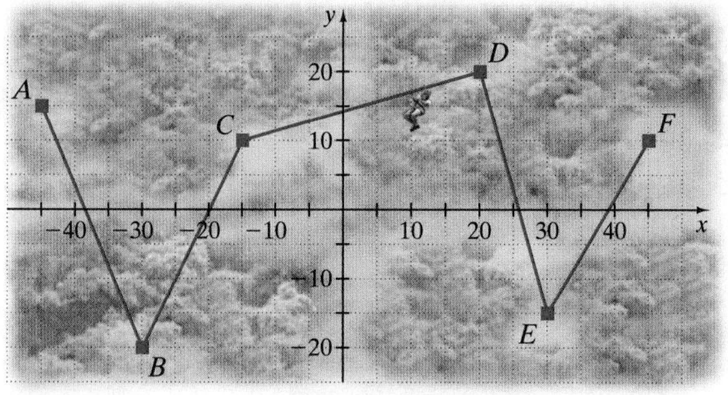

30. How far do you travel from Platform D to Platform E?

MIXED PRACTICE

Multiple Steps *For Exercises 31–36, find (**a**) PQ to the nearest tenth and (**b**) the coordinates of the midpoint of* $\overline{PQ}$.

31. $P(3, 2), Q(6, 6)$

32. $P(-3, -1), Q(5, -7)$

33. $P(0, -2), Q(3, 3)$

34. $P(-4, -2), Q(1, 3)$

35. $P(2, 3), Q(4, -2)$

36. $P(-5, -3), Q(-3, -5)$

MIXED PRACTICE

Multiple Steps *For each graph, find (**a**) AB to the nearest tenth and (**b**) the coordinates of the midpoint of* $\overline{AB}$.

37.

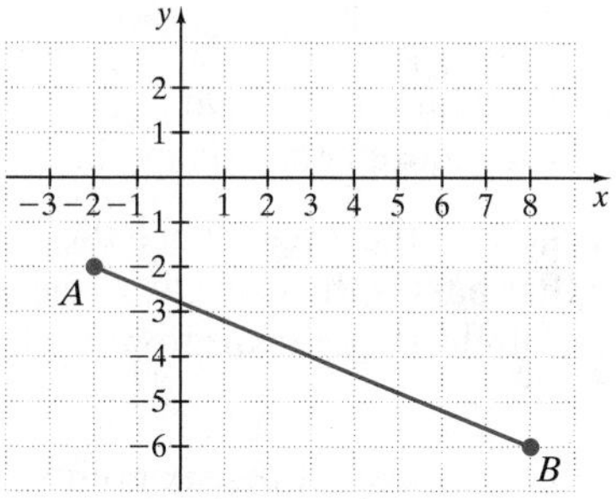

38.

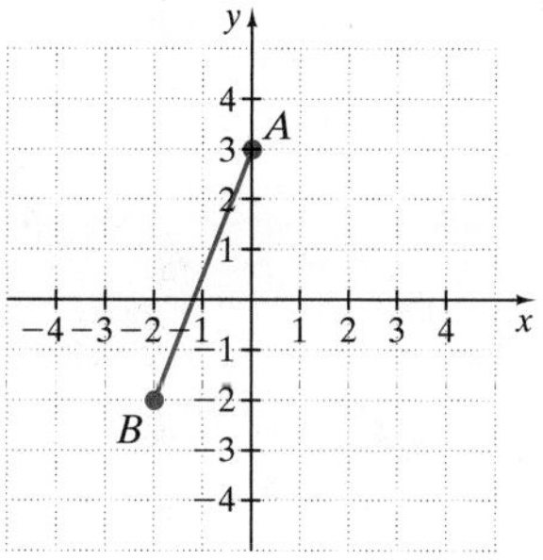

The units of the subway map below are in miles. Suppose the routes between stations are straight. Find the distance you would travel between each pair of stations to the nearest tenth of a mile.

39. Oak Station and Jackson Station

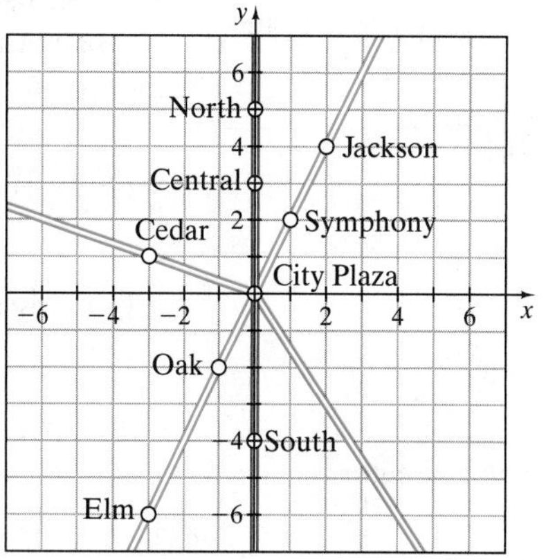

40. Central Station and South Station

41. Elm Station and Symphony Station

42. Cedar Station and City Plaza Station

43. Maple Station (not shown) is located 6 mi west and 2 mi north of City Plaza. What is the distance between Cedar Station and Maple Station?

44. Sycamore Station (not shown) is located 4 mi east and 6 mi south of City Plaza. What is the distance between Cedar Station and Sycamore Station?

CONCEPT EXTENSIONS

45. **Multiple Steps** Point H (2, 2) is the midpoint of many segments.

a. Find the coordinates of the endpoints of four noncollinear segments that have point H as their midpoint.

b. You know that a segment with midpoint H has length 8. How many possible noncollinear segments match this description? Explain.

46. **Multiple Steps** Points $P(-4, 6)$, $Q(2, 4)$, and R are collinear. One of the points is the midpoint of the segment formed by the other two points.

a. What are the possible coordinates of R?

b. $RQ = \sqrt{160}$. Does this information affect your answer to part **a**? Explain.

47. The midpoint of $\overline{TS}$ is the origin. Point T is located in quadrant II. What quadrant contains point S? Explain.

48. How does the Distance Formula ensure that the distance between two different points is positive?

49. Do you use the Midpoint Formula or the Distance Formula to find the following?

a. Given points K and P, find the distance from K to the midpoint of $\overline{KP}$.

b. Given point K and the midpoint of $\overline{KP}$, find KP.

50. **Find the Error** Your friend calculates the distance between points Q (1, 5) and R (3, 8). What is his error?

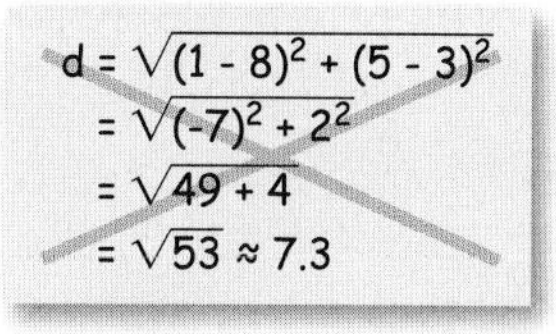

1.8 CONSTRUCTIONS—BASIC GEOMETRY CONSTRUCTIONS

OBJECTIVE

1 Make Basic Constructions Using a Straight Edge and a Compass.

VOCABULARY

- straight edge
- compass
- construction
- perpendicular lines
- perpendicular bisector

Helpful Hint
A ruler makes a great straight edge—just don't use the measurement markings on it.

OBJECTIVE 1 ▶ Make Basic Constructions Using a Straight Edge and a Compass. We can use special geometric tools to make a figure that is congruent to an original figure without measuring. This method is more accurate than sketching, drawing, or measuring.

A **straight edge** is a ruler with no markings on it. A **compass** is a geometric tool used to draw circles and parts of circles called *arcs*. A **construction** is a geometric figure drawn using a straight edge and a compass.

A compass is also used as a symbol of precision. Look carefully and you can see it in many icons, such as a mason symbol and many computer icons, as shown below.

Mason Symbol

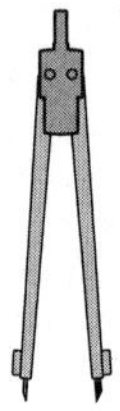

East German flag

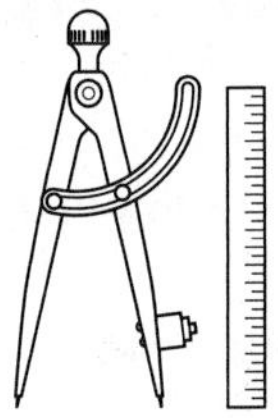

It is very important to note that we use compass and straight edge constructions to illustrate the theory of constructions. In theory, the ideal compass can draw a perfect circle, but in reality, such an instrument does not exist. To see this for yourself, use your compass and try to draw a perfect circle.

To practice using your compass, you may want to try the activity below.

TRY THE ACTIVITY BELOW

Activity—A Compass Design

STEP 1. Open your compass to about 2 in. Make a circle and mark the point at the center of the circle. Keep the opening of your compass fixed. Place the compass point on the circle. With the pencil end, make a small arc to intersect the circle.

STEP 2. Place the compass point on the circle at the arc. Mark another arc. Continue around the circle this way to draw four more arcs—six in all.

STEP 3. Place your compass point on an arc you marked on the circle. Place the pencil end at the next arc. Draw a large arc that passes through the circle's center and continues to another point on the circle. (See illustration on the next page.)

STEP 4. Draw six large arcs in this manner, each centered at one of the six points marked on the circle. The end result is shown in color.

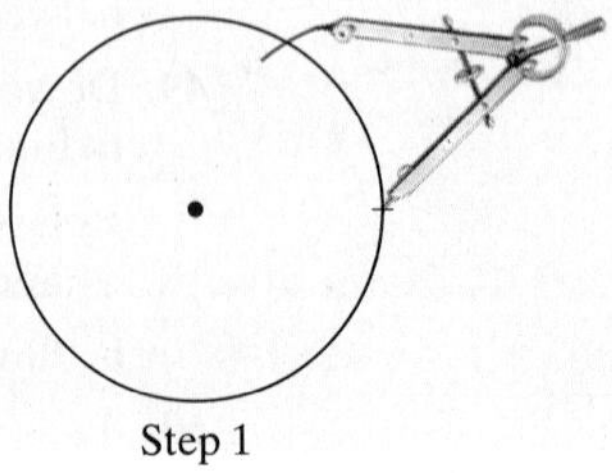

Step 1

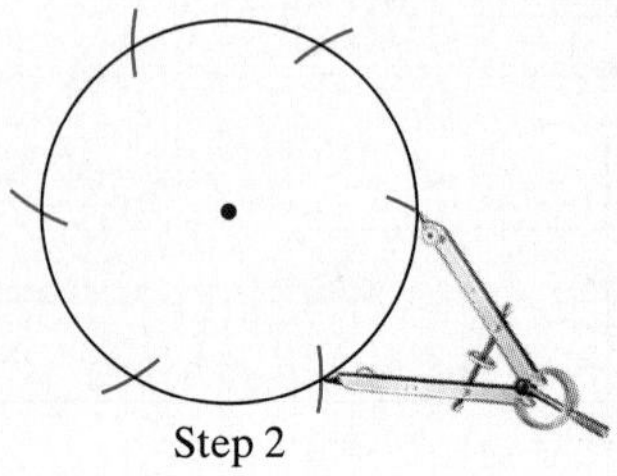

Step 2

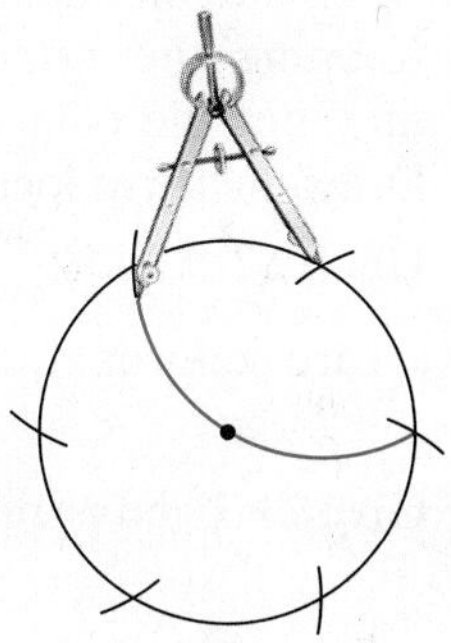

Step 3

Step 4

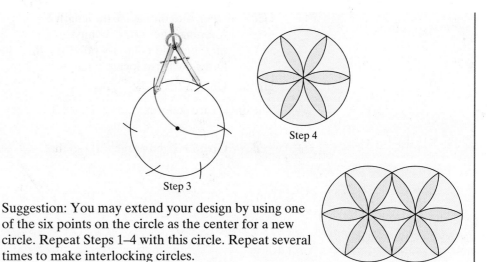

Suggestion: You may extend your design by using one of the six points on the circle as the center for a new circle. Repeat Steps 1–4 with this circle. Repeat several times to make interlocking circles.

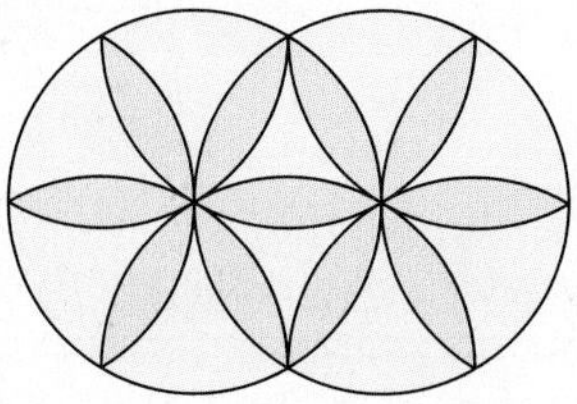

EXAMPLE 1 Constructing Congruent Segments

Construct a segment congruent to a given segment.

Given: $\overline{AB}$
Construct: $\overline{CD}$ so that $\overline{CD} \cong \overline{AB}$

Solution

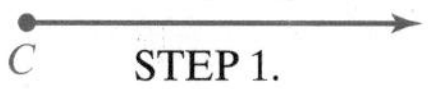

STEP 1.

> **Helpful Hint**
> Using the same compass setting keeps segments congruent. It guarantees that the lengths of $\overline{AB}$ and $\overline{CD}$ are exactly the same.

STEP 1. Draw a ray with endpoint C.

STEP 2. Open the compass to the length of $\overline{AB}$.

STEP 3. With the same compass setting, put the compass point on point C. Draw an arc that intersects the ray. Label the point of intersection D.

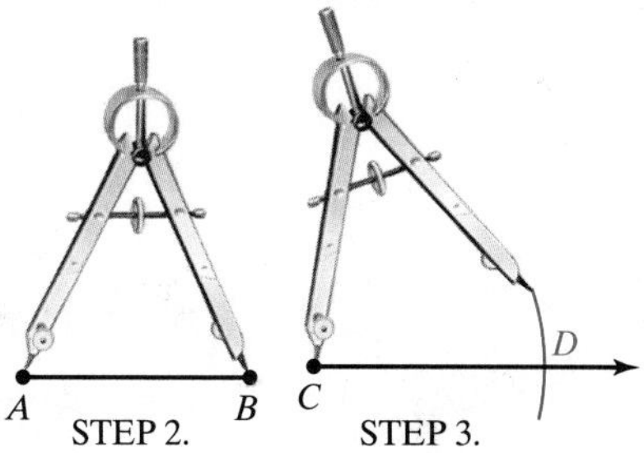

STEP 2. STEP 3.

The segments are congruent, or $\overline{CD} \cong \overline{AB}$. □

PRACTICE 1 Given $\overline{XY}$, construct $\overline{RS}$ so that $RS = 2XY$.

EXAMPLE 2 Constructing Congruent Angles

Construct an angle congruent to a given angle.

Given: $\angle A$

Construct: $\angle S$ so that $\angle S \cong \angle A$

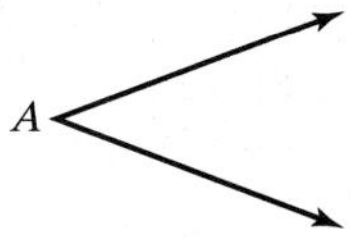

Solution

STEP 1.

STEP 1. Draw a ray with endpoint S.

STEP 2. With the compass point on vertex A, draw an arc that intersects the sides of $\angle A$. Label the points of intersection B and C.

STEP 3. With the same compass setting, put the compass point on point S. Draw an arc and label its point of intersection with the ray as R.

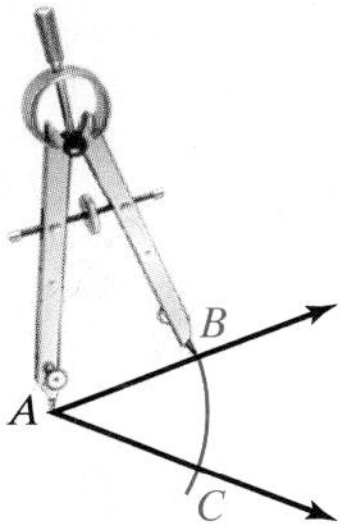

STEP 2.

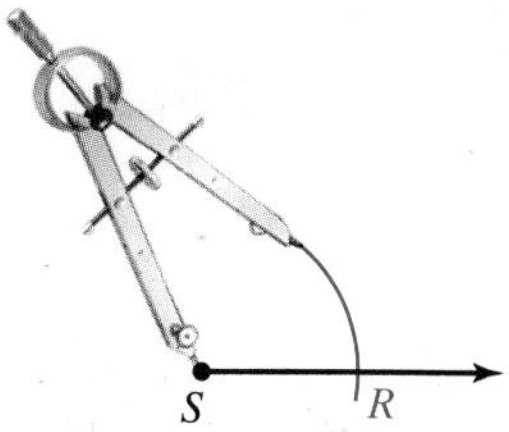

STEP 3.

STEP 4. Open the compass to the length BC. Keeping the same compass setting, put the compass point on R. Draw an arc to locate point T.

STEP 5. Draw $\overrightarrow{ST}$.

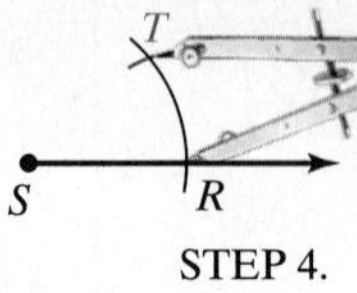

STEP 4.

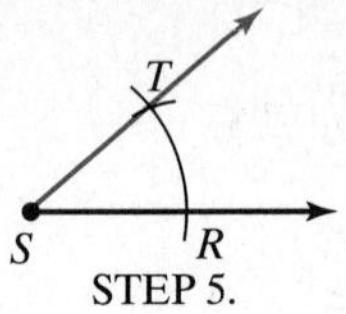

STEP 5.

The angles are congruent, or $\angle S \cong \angle A$. □

PRACTICE 2 Given $\angle B$, construct $\angle Q$ so that $\angle Q \cong \angle B$.

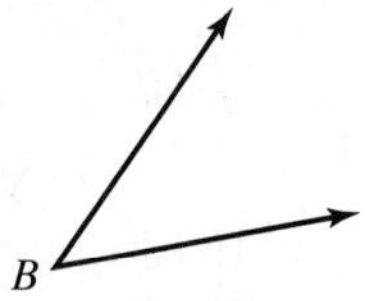

Perpendicular lines are two lines that intersect to form right angles. The symbol $\perp$ means "is perpendicular to." In the diagram at the right, $\overleftrightarrow{AB} \perp \overleftrightarrow{CD}$ and $\overleftrightarrow{CD} \perp \overleftrightarrow{AB}$.

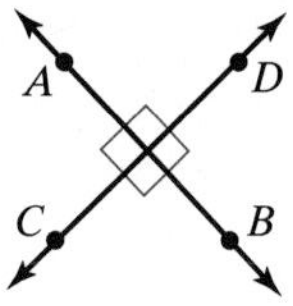

A **perpendicular bisector** of a segment is a line, segment, ray, or even a plane that is perpendicular to the segment at its midpoint. In the diagram at the right, $\overleftrightarrow{EF}$ is the perpendicular bisector of $\overline{GH}$. The perpendicular bisector bisects the segment into two congruent segments. This construction is demonstrated in Example 3. We will justify the steps for this construction in Chapter 5, as well as for the other constructions in this section.

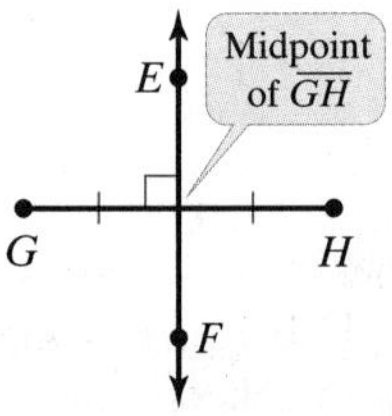

EXAMPLE 3 Constructing the Perpendicular Bisector

Construct the perpendicular bisector of a segment.

Given: $\overline{AB}$

Construct: $\overleftrightarrow{XY}$ so that $\overleftrightarrow{XY}$ is the perpendicular bisector of $\overline{AB}$.

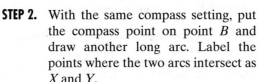

Solution

STEP 1. Put the compass point on point A and draw a long arc as shown. Be sure the opening is greater than $\frac{1}{2}AB$.

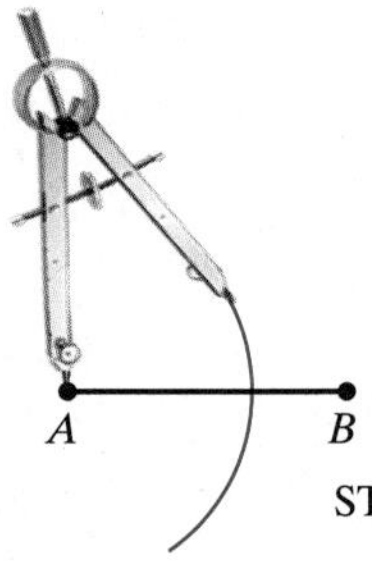

STEP 1.

> **Helpful Hint**
> Why must the compass opening be greater than $\frac{1}{2}AB$? If the opening is less than $\frac{1}{2}AB$, the two arcs will not intersect in Step 2.

STEP 2. With the same compass setting, put the compass point on point B and draw another long arc. Label the points where the two arcs intersect as X and Y.

STEP 3. Draw $\overleftrightarrow{XY}$. Label the point of intersection of $\overline{AB}$ and $\overleftrightarrow{XY}$ as M, the midpoint of $\overline{AB}$.

$\overleftrightarrow{XY} \perp \overline{AB}$ at midpoint M, so $\overleftrightarrow{XY}$ is the perpendicular bisector of $\overline{AB}$. □

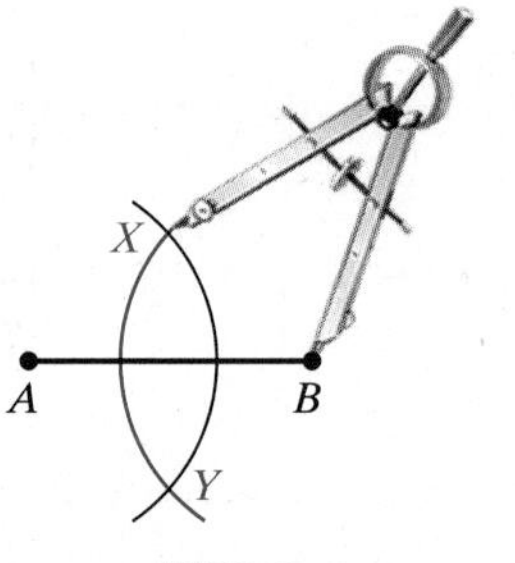

STEP 2.

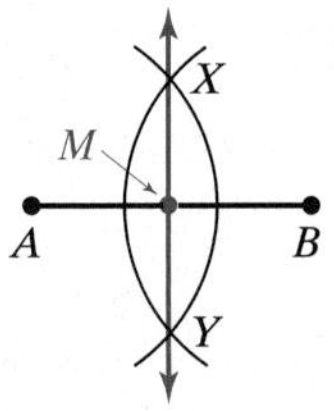

STEP 3.

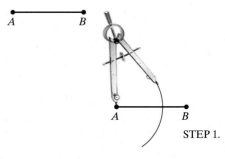

PRACTICE 3 Draw $\overline{ST}$. Construct its perpendicular bisector.

Recall that in Section 1.6, we first introduced the angle bisector.

EXAMPLE 4 **Constructing the Angle Bisector**

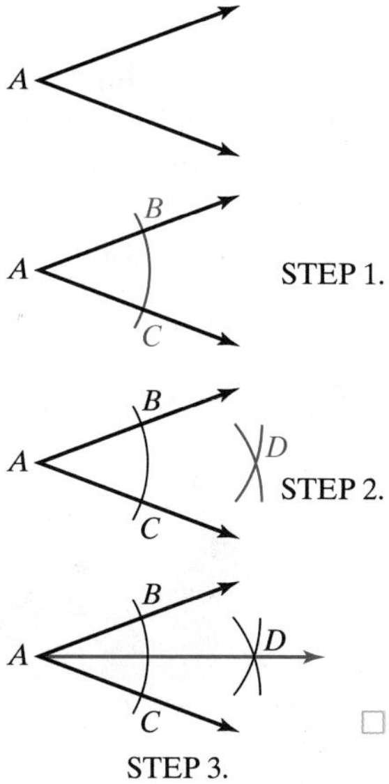

Construct the bisector of an angle.

Given: $\angle A$

Construct: $\overrightarrow{AD}$, the bisector of $\angle A$

Solution

STEP 1. Put the compass point on vertex A. Draw an arc that intersects the sides of $\angle A$. Label the points of intersection B and C.

STEP 2. Put the compass point on point C and draw an arc. With the same compass setting, draw an arc using point B. Be sure the arcs intersect. Label the point where the two arcs intersect as D.

STEP 3. Draw $\overrightarrow{AD}$.

$\overrightarrow{AD}$ is the angle bisector of $\angle CAB$. □

PRACTICE 4 Draw obtuse $\angle XYZ$. Then construct its bisector $\overrightarrow{YP}$.

1.8 EXERCISE SET MyMathLab®

MIXED PRACTICE

Construction *When needed, for Exercises 1–8, draw diagrams similar to the given ones. Then do the construction. Check your work with a ruler or a protractor. See Examples 1–4.*

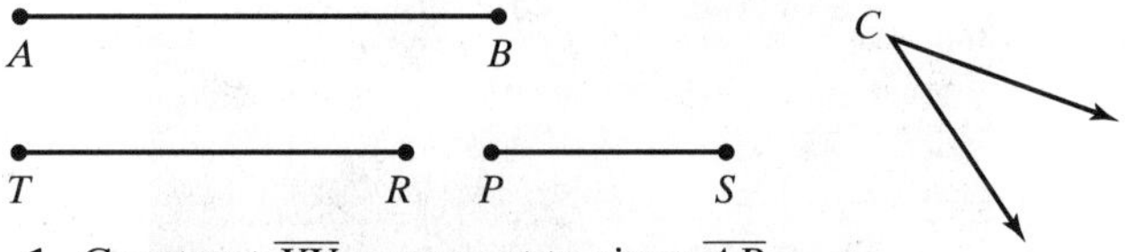

1. Construct $\overline{XY}$ congruent to given $\overline{AB}$.
2. Construct $\overline{VW}$ so that $VW = 2AB$.
3. Construct $\overline{DE}$ so that $DE = TR + PS$.
4. Construct $\overline{QJ}$ so that $QJ = TR - PS$.
5. Construct $\angle D$ so that $\angle D \cong \angle C$.
6. Construct $\angle F$ so that $m\angle F = 2m\angle C$.
7. Construct the perpendicular bisector of $\overline{AB}$.
8. Construct the perpendicular bisector of $\overline{TR}$.
9. Draw acute $\angle PQR$. Then construct its bisector.
10. Draw obtuse $\angle XQZ$. Then construct its bisector.

Construction *Sketch the figure described. Explain how to construct it. Then do the construction. See Examples 1–4.*

11. $\overleftrightarrow{XY} \perp \overleftrightarrow{YZ}$
12. $\overrightarrow{ST}$ bisects right $\angle PSQ$.

CONCEPT EXTENSIONS

13. How is constructing an angle bisector similar to constructing a perpendicular bisector?
14. Given an angle, $\angle A$, how can you construct an angle whose measure is $\frac{1}{4}m\angle A$?

Answer the questions about a segment in a plane. Explain each answer.

15. How many midpoints does the segment have?
16. How many bisectors does it have?
17. How many lines in the plane are its perpendicular bisectors?
18. How many lines in space are its perpendicular bisectors?

Construction *For Exercises 19–20, copy $\angle 1$ and $\angle 2$. Construct each angle described.*

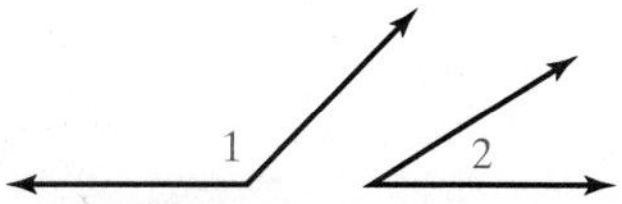

19. $\angle B$; $m\angle B = m\angle 1 + m\angle 2$
20. $\angle C$; $m\angle C = m\angle 1 - m\angle 2$
21. Given $\overline{PQ}$, explain how to divide $\overline{PQ}$ into four congruent segments with a compass and straight edge.
22. Draw a large triangle with three acute angles. Construct the bisectors of the three angles. What appears to be true about the three angle bisectors?

Construction *Use a ruler to draw segments of 2 cm, 4 cm, and 5 cm. Then construct each triangle with the given side measures, if possible. If it is not possible, explain why not.*

23. 4 cm, 4 cm, and 5 cm
24. 2 cm, 5 cm, and 5 cm
25. 2 cm, 2 cm, and 5 cm
26. 2 cm, 2 cm, and 4 cm
27. 5 cm, 5 cm, and 5 cm
28. 4 cm, 4 cm, and 4 cm
29. Describe how to construct a 45° angle using what you know.
30. Which step best describes how to construct the pattern at the right?

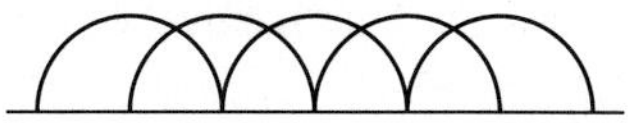

   A. Use a straight edge to draw the segment and then a compass to draw five half circles.

   B. Use a straight edge to draw the segment and then a compass to draw six half circles.

   C. Use a compass to draw five half circles and then a straight edge to join their ends.

   D. Use a compass to draw six half circles and then a straight edge to join their ends.

Historical Note—The Current Status of Geometry

In about 1830, mathematicians János Bolyai (Hungarian, 1802–1860) and Nikolai Lobachevsky (Russian, 1792–1856) separately published work on non-Euclidean geometry. Their work is based on Euclid's Fifth Postulate (also called the Parallel Postulate) not being valid. (See Chapter 3 of this text.)

This led, along with mathematician Georg Riemann (German 1826–1866), to two popular Non-Euclidean Geometries summarized below.

	Non-Euclidean Geometries	
Euclidean Geometry	***Spherical or Elliptic Geometry***	***Hyperbolic Geometry***
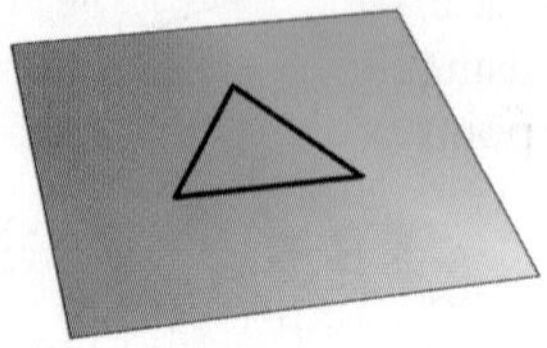 Triangle Angle Sum is 180°	 Triangle Angle Sum is greater than 180°	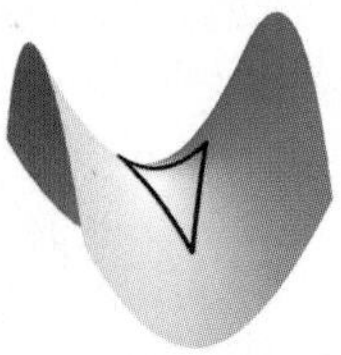 Triangle Angle Sum is less than 180°
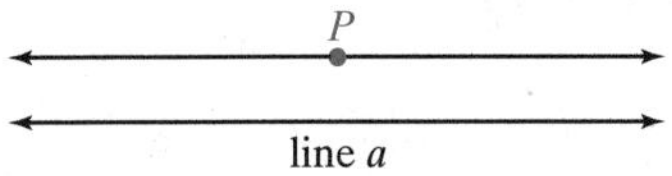 1 parallel line through point P to line *a*.	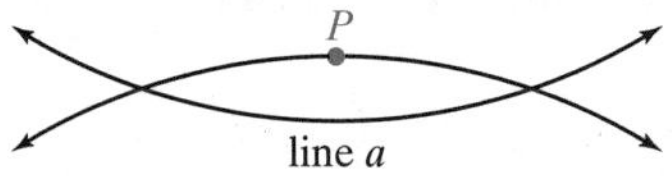 No parallel lines through point P to line *a*.	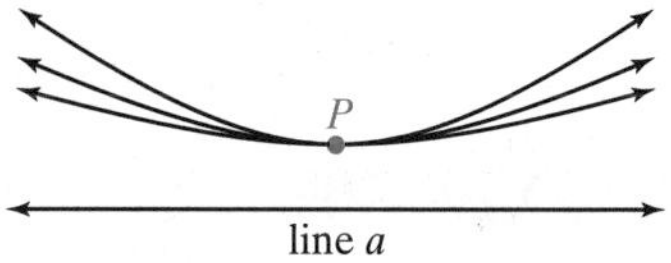 Infinite number of parallel lines through point P to line *a*.

In the 20th century, Einstein's theory of relativity shows that Euclidean geometry is not the geometry of space-time. Basically Einstein predicted that gravity would cause deviations from Euclidean geometry, such as the Earth's gravity causing a slight bending of a line.

In 1919, the slight bending of a line was verified by the bending of starlight during a solar eclipse. This knowledge is now incorporated in appropriate applications such as the software for GPS systems.

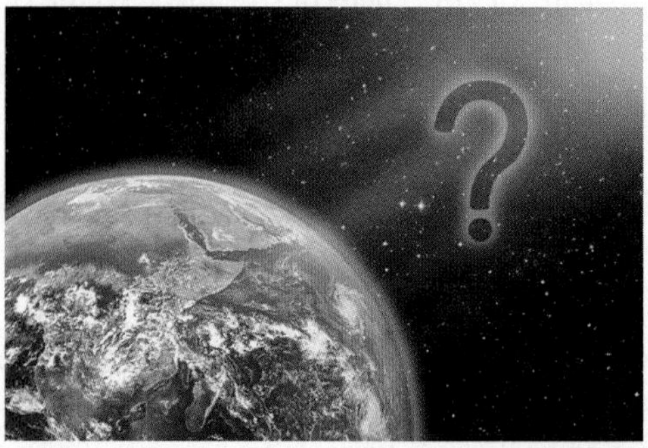

Unfortunately, we don't know whether light rays are proper models of Euclid's geometry. Also, at this time, there is no test to do so.

Thus, the question still remains....

What is the geometry of our universe—Euclidean or some non-Euclidean?

CHAPTER 1 REVIEW

***(1.2)** Look for a pattern. If a figure pattern, sketch the next figure. If a number pattern, predict the next two numbers.*

1. 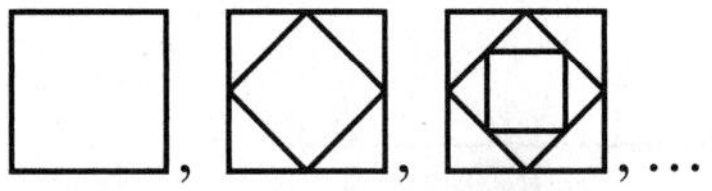

2. 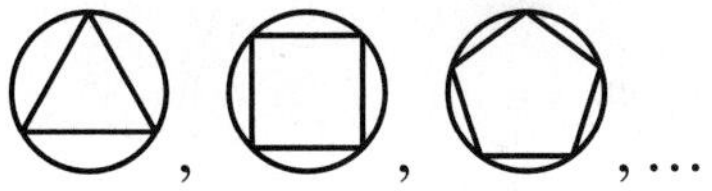

3. $\frac{1}{3}, \frac{1}{6}, \frac{1}{12}, \frac{1}{24}, \ldots$

4. 40, 35, 30, 25, ...

***(1.3)** Use the figure below for Exercises 5–8.*

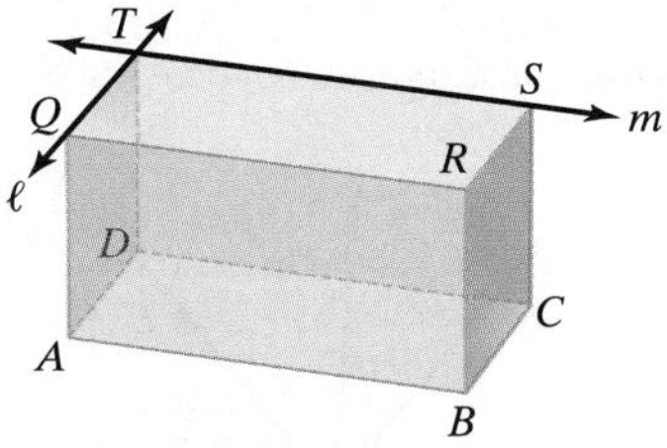

5. Name two intersecting lines.

6. Name the intersection of planes *QRBA* and *TSRQ*.

7. Name three noncollinear points.

8. Name the intersection of lines *l* and *m*.

Determine whether the statement is true or false. Explain your reasoning.

9. Two points are always collinear.

10. $\overrightarrow{LM}$ and $\overrightarrow{ML}$ are the same ray.

***(1.4)** For Exercises 11 and 12, use the number line below.*

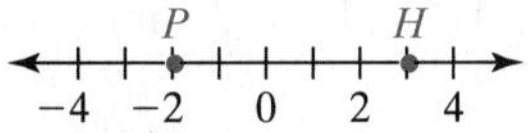

11. Find two possible coordinates of *Q* such that $PQ = 5$.

12. Find the coordinate of the midpoint of $\overline{PH}$.

13. Find the value of *m*.

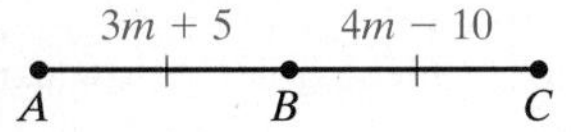

14. If $XZ = 50$, what are *XY* and *YZ*?

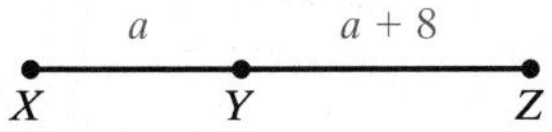

***(1.5)** Classify each angle as "acute," "right," "obtuse," or "straight."*

15. 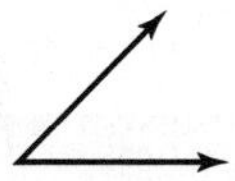

16. 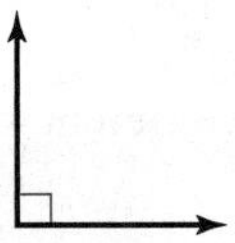

Use the diagram below for Exercises 17 and 18.

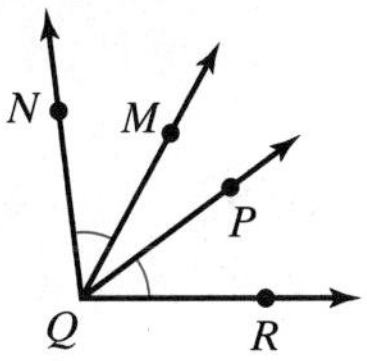

17. If $m\angle MQR = 61°$ and $m\angle MQP = 25°$, find $m\angle PQR$.

18. If $m\angle NQM = (2x + 8)°$ and $m\angle PQR = (x + 22)°$, find the value of *x*.

***(1.6)** Name a pair of each of the following.*

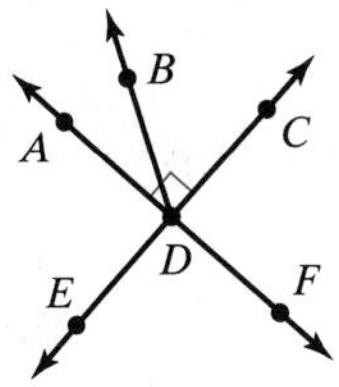

19. complementary angles

20. supplementary angles

21. vertical angles

22. linear pair

Find the value of x.

23. 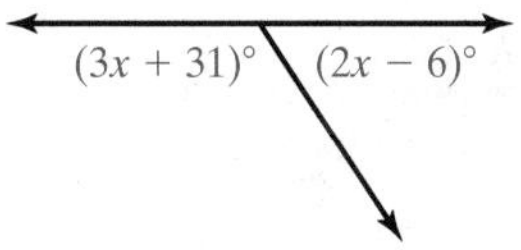

24. 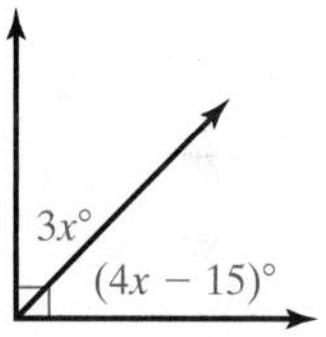

***(1.7)** Find the distance between the points to the nearest tenth.*

25. $A(-1, 5), B(0, 4)$

26. $C(-1, -1), D(6, 2)$

$\overline{AB}$ has endpoints $A(-3, 2)$ and $B(3, -2)$.

27. Find the coordinates of the midpoint of $\overline{AB}$.

28. Find *AB* to the nearest tenth.

M is the midpoint of $\overline{JK}$. Find the coordinates of K.

29. $J(-8, 4), M(-1, 1)$

30. $J(9, -5), M(5, -2)$

(1.8)

31. Use a protractor to draw a 73° angle. Then construct an angle congruent to it.

32. Use a protractor to draw a 60° angle. Then construct the bisector of the angle.

33. Sketch $\overline{LM}$ on paper. Construct a line segment congruent to $\overline{LM}$. Then construct the perpendicular bisector of your line segment.

34. **a.** Sketch $\angle B$ on paper. Construct an angle congruent to $\angle B$.
b. Construct the bisector of your angle from part a.

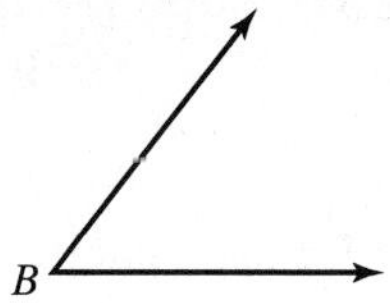

MIXED REVIEW

Determine whether the given points are coplanar. If yes, name the plane. If no, explain.

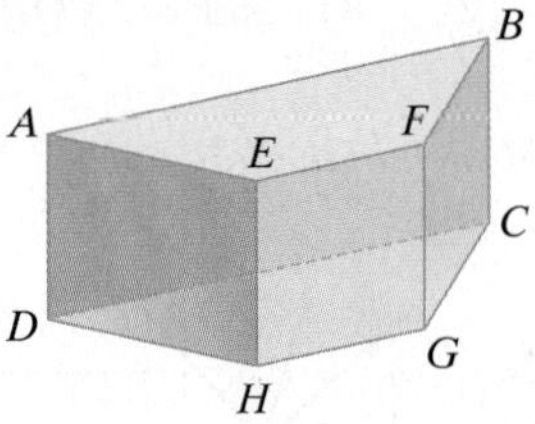

35. A, E, F, and B

36. D, C, E, and F

37. H, G, F, and B

38. A, E, B, and C

39. Use the figure from Exercises **35–38**. Name the intersection of each pair of planes.
a. plane $AEFB$ and plane $CBFG$
b. plane $EFGH$ and plane $AEHD$

Use the figure below for Exercises 40–47.

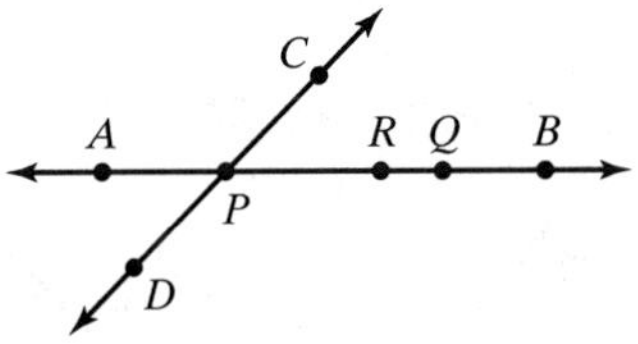

40. Give two other names for $\overleftrightarrow{AB}$.

41. Give two other names for $\overrightarrow{PR}$.

42. Give two other names for $\angle CPR$.

43. Name three collinear points.

44. Name two opposite rays.

45. Name three segments.

46. Name two angles that form a linear pair.

47. Name a pair of vertical angles.

48. a. Find the value of x in the diagram below.
b. Classify $\angle ABC$ and $\angle CBD$ as "acute," "right," or "obtuse."

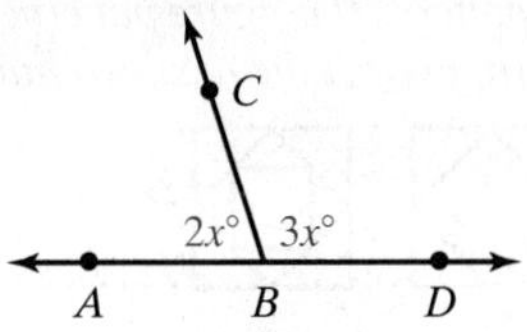

Find the length of each segment.

49. $\overline{PQ}$

50. $\overline{RS}$

51. $\overline{ST}$

52. $\overline{QT}$

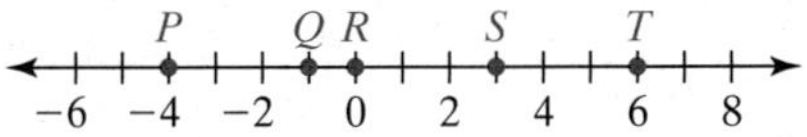

Use the figure below for Exercises 53–55.

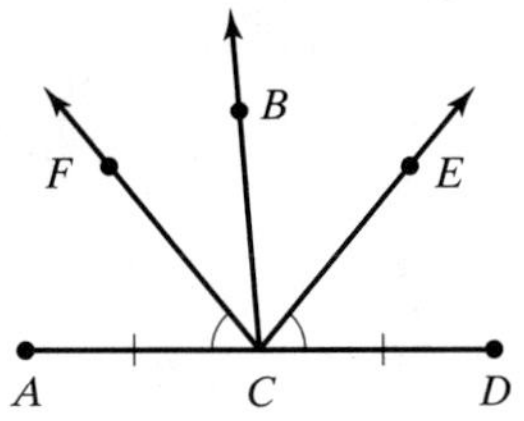

53. If $AC = (4x + 5)$ and $DC = (3x + 8)$, find AD.

54. If $m\angle FCD = 130°$ and $m\angle BCD = 95°$, find $m\angle FCB$.

55. If $m\angle FCA = 50°$, find $m\angle FCE$.

56. Find the Error Suppose $PQ = QR$. Your friend says that Q is always the midpoint of $\overline{PR}$. Is he correct? Explain.

57. Determine whether the following situation is possible. Explain your reasoning. Include a sketch.
Collinear points C, F, and G lie in plane M. $\overleftrightarrow{AB}$ intersects plane M at C. $\overleftrightarrow{AB}$ and $\overleftrightarrow{GF}$ do not intersect.

CHAPTER 1 TEST

TEST PREP VIDEO The fully worked-out solutions to any exercises you want to review are available in MyMathLab.

1. Look for a pattern and predict the next two numbers.
7, 13, 19, 25, . . .

2. Look for a pattern and predict the next figure.

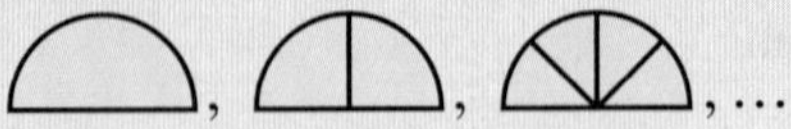

Multiple Choice *Fill in each blank with the letter of the correct choice.*

3. Statements that we prove true, using logic, are called ________.
a. postulates
b. theorems
c. axioms

4. If two distinct planes intersect, their intersection is a ________.
a. point
b. plane
c. line
d. ray

5. Given the figure, which choice below cannot be used to uniquely name $\angle 1$?
a. $\angle SRT$
b. $\angle TRS$
c. $\angle R$

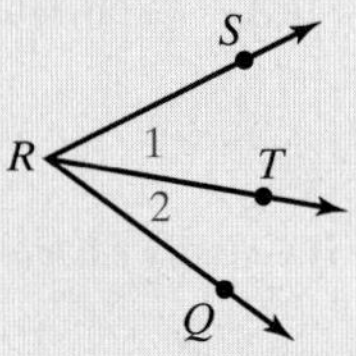

6. How many different angles are in the figure above?
a. 1
b. 2
c. 3
d. 4

7. $\overrightarrow{MN}$ bisects $\angle EMG$. If $m\angle EMN = 78°$, what is $m\angle EMG$?
a. 156°
b. 78°
c. 39°
d. 168°

Fill in the Blank *For Exercises 8–10, use the figure below.*

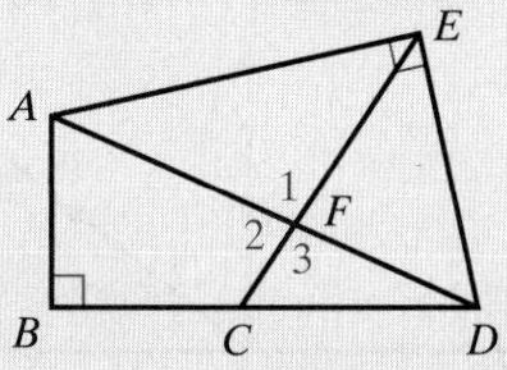

8. $\angle EFD$ and ____________ are vertical angles.

9. $\angle EFD$ and ____________ form a linear pair.

10. The complementary angles are ____________

True or False *Answer Exercises 11–15 true or false.*

11. Angles that form a linear pair are supplementary.

12. Complementary angles are always congruent.

13. Vertical angles are always congruent.

14. Right angles are always congruent.

15. Obtuse angles are always congruent.

Label each angle below as "acute," "right," "obtuse," or "straight."

16.

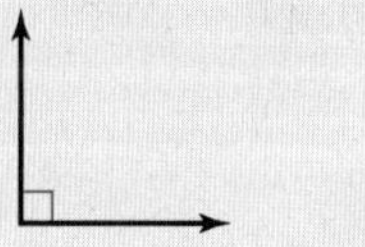

____

17.

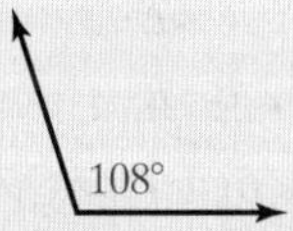

____

18.

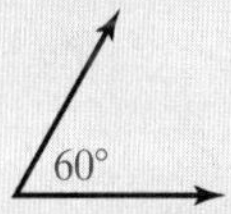

____

19.

____

Use the number line for Exercises 20–21.

20. Find the length of $\overline{AD}$.

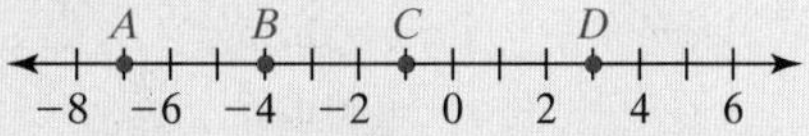

21. Decide whether segments $\overline{AC}$ and $\overline{BD}$ are congruent.

22. M is the midpoint of $\overline{XY}$.

a. Find the value of x.

b. Find XM, MY, and XY.

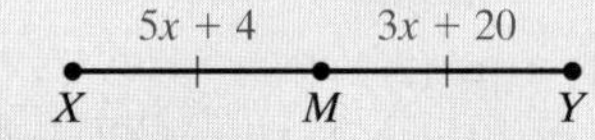

23. Construction Draw an obtuse $\angle ABC$. Use a compass and a straight edge to bisect the angle.

Use the figure for Exercises 24–27.

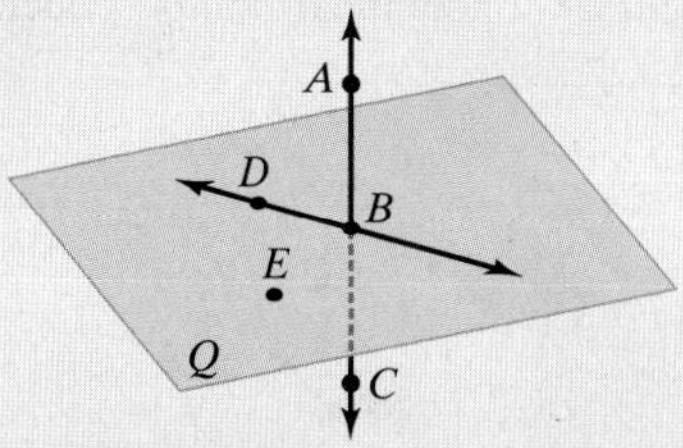

24. Name three collinear points.

25. What is the intersection of $\overrightarrow{AC}$ and plane Q?

26. How many planes contain the given line and point?

a. $\overleftrightarrow{DB}$ and point A

b. $\overleftrightarrow{AC}$ and point D

27. Name four coplanar points.

28. If $JK = 48$, find the value of x.

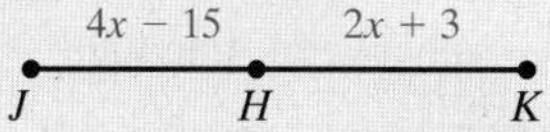

Fill in the Blank *Use the figure to complete each statement.*

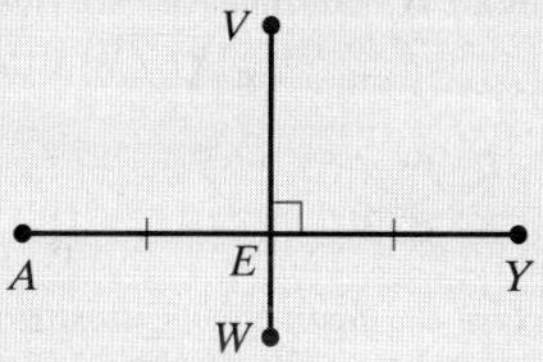

29. $\overline{VW}$ is the ____________ of $\overline{AY}$.

30. If $EY = 3.5$, then $AY =$ ____________.

31. ____________ is the midpoint of ____________.

32. In the diagram, $m\angle BDJ = (7y + 2)°$, and $m\angle JDR = (2y + 7)°$. Find the value of y.

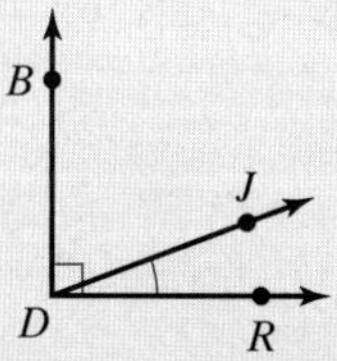

33. $\overrightarrow{PB}$ bisects $\angle RPT$ so that $m\angle RPB = (x + 2)°$ and $m\angle TPB = (2x - 6)°$. Draw a figure; then find $m\angle RPT$.

34. $\overline{RS}$ has endpoints $R(2, 4)$ and $S(-1, 7)$. What are the coordinates of its midpoint, M?

35. The midpoint of $\overline{BC}$ is $(5, -2)$. One endpoint is $B(3, 4)$. What are the coordinates of endpoint C?

36. What is the distance between points $K(-9, 8)$ and $L(-6, 0)$?

CHAPTER 1 STANDARDIZED TEST

Look for a pattern and predict the next two numbers.

1. 4, 5, 7, 10, ...

a. 14, 19 **b.** 11, 13

c. 14, 18 **d.** 15, 21

2. $\frac{1}{2}, \frac{1}{6}, \frac{1}{18}, \frac{1}{54}, \ldots$

a. $\frac{1}{162}, \frac{1}{486}$ **b.** $\frac{1}{108}, \frac{1}{216}$

c. $\frac{1}{108}, \frac{1}{324}$ **d.** $\frac{1}{162}, \frac{1}{324}$

Choose the best word or phrase to complete each statement.

3. A word that is described but not defined is a(n) ____________.

a. theorem **b.** axiom

c. undefined term **d.** defined term

4. The number of noncollinear points needed to describe a plane is ____________.

a. one **b.** two
c. three **d.** four

Use this figure for Questions 5–7.

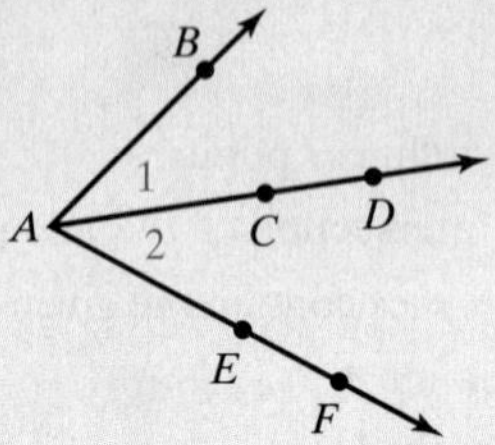

5. Another way to describe $\angle BAC$ is ____________.

a. $\angle BAE$ **b.** $\angle BAD$
c. $\angle 2$ **d.** $\angle BAF$

6. ____________ is between A and D.

a. $\overrightarrow{AC}$ **b.** point C
c. point B **d.** $\overrightarrow{AF}$

7. $\overrightarrow{AD}$ bisects $\angle BAE$. If $m\angle 1 = (3x - 2)°$ and $m\angle 2 = (4x - 7)°$, then $m\angle BAE =$ ____________.

a. 5° **b.** 13°
c. 30° **d.** 26°

Use the figure for Questions 8–11.

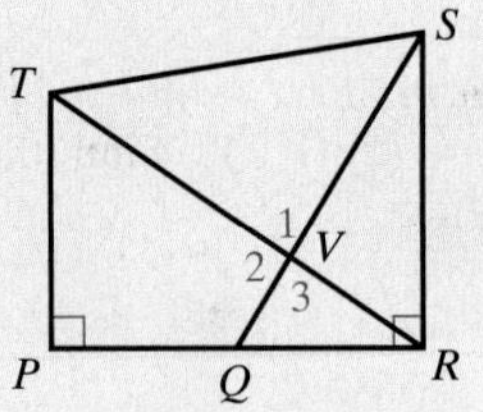

8. A pair of vertical angles is ____________.

a. $\angle 1$ and $\angle 2$ **b.** $\angle SRQ$
c. $\angle SRV$ and $\angle VRQ$ **d.** $\angle 1$ and $\angle 3$

9. A pair of complementary angles is ____________.

a. $\angle 1$ and $\angle 2$ **b.** $\angle SRQ$ and $\angle TPQ$
c. $\angle SRV$ and $\angle VRQ$ **d.** $\angle 1$ and $\angle 3$

10. ____________ form a linear pair.

a. $\angle 1$ and $\angle 2$ **b.** $\angle SRQ$ and $\angle TPQ$
c. $\angle SRV$ and $\angle VRQ$ **d.** $\angle 1$ and $\angle 3$

11. ____________ are congruent angles.

a. $\angle 1$ and $\angle 2$ **b.** $\angle SRQ$ and $\angle TPQ$
c. $\angle SRV$ and $\angle VRQ$ **d.** $\angle STP$ and $\angle SRQ$

Use the figure below for Questions 12–15.

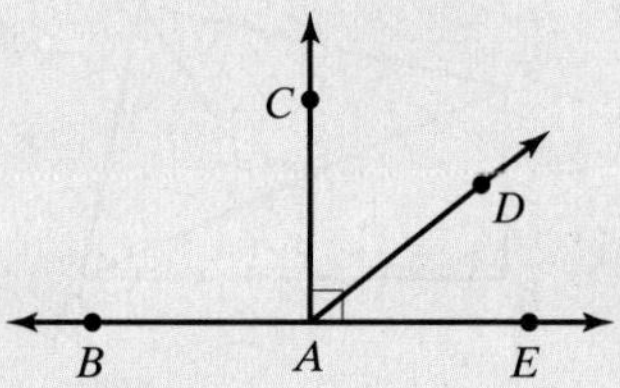

12. ____________ is an acute angle.

a. $\angle CAE$ **b.** $\angle DAE$
c. $\angle BAE$ **d.** $\angle BAD$

13. ____________ is an obtuse angle.

a. $\angle CAE$ **b.** $\angle DAE$
c. $\angle BAE$ **d.** $\angle BAD$

14. ____________ is a right angle.

a. $\angle CAE$ **b.** $\angle DAE$
c. $\angle BAE$ **d.** $\angle BAD$

15. ____________ is a straight angle.

a. $\angle CAE$ **b.** $\angle DAE$
c. $\angle BAE$ **d.** $\angle BAD$

Use this figure for Questions 16 and 17.

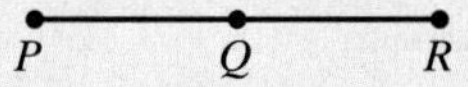

16. If $PQ = 3x + 7$, $QR = 4x - 2$, and $PR = 61$, the value of x is ____________.

a. 8 **b.** 31
c. 30 **d.** 7

17. If $PQ = 5x + 6$, $QR = 7x - 4$, and $\overline{PQ} \cong \overline{QR}$, then $PR =$ ____________.

a. 62 **b.** 5
c. 31 **d.** 7

18. $\overline{AB}$ has endpoints $A(7, -8)$ and $B(-3, 10)$. The coordinates of the midpoint of $\overline{AB}$ are ____________.

a. $(5, -9)$ **b.** $(4, 2)$
c. $(2, 1)$ **d.** $(10, -18)$

19. The midpoint of $\overline{CD}$ is $(7, -2)$. One endpoint is $C(9, -8)$. The other endpoint is ____________.

a. $D(11, -14)$ **b.** $D(-2, 6)$
c. $D(5, 4)$ **d.** $D(8, -5)$

20. The distance between points $G(-2, 3)$ and $H(4, -5)$ is ____________.

a. $2\sqrt{2}$ **b.** 100
c. $\sqrt{10}$ **d.** 10

CHAPTER

2 Introduction to Reasoning and Proofs

5	3			7				
6			1	9	5			
	9	8					6	
8				6				3
4			8		3			1
7				2				6
	6					2	8	
			4	1	9			5
				8			7	9

A Sudoku puzzle

5	3	4	6	7	8	9	1	2
6	7	2	1	9	5	3	4	8
1	9	8	3	4	2	5	6	7
8	5	9	7	6	1	4	2	3
4	2	6	8	5	3	7	9	1
7	1	3	9	2	4	8	5	6
9	6	1	5	3	7	2	8	4
2	8	7	4	1	9	6	3	5
3	4	5	2	8	6	1	7	9

The same puzzle with solution numbers marked in red

2.1 Perimeter, Circumference, and Area

2.2 Patterns and Inductive Reasoning

2.3 Conditional Statements

2.4 Biconditional Statements and Definitions

2.5 Deductive Reasoning

2.6 Reviewing Properties of Equality and Writing Two-Column Proofs

2.7 Proving Theorems About Angles

Some Variations of a Sudoku Puzzle

3								4
		2		6		1		
	1		9		8		2	
		5				6		
	2						1	
		9				6		
	8		3		4		6	
		4		1		9		
5								7

A Jigsaw Sudoku puzzle

	P		K		R	I		D
D			B					R
	B		E			P	A	
P				K	W	A		B
						R	K	
	A	D						
B				E				P
A						E		
E	R		P		K	B		

ABDEIKPRW

A Wordoku puzzle

							1	
		2					3	4
				5	1			
					6	5		
	7		3				8	
		3						
				8				
5	8					9		
6	9							

A Hypersudoku puzzle

You may be familiar with a popular puzzle named Sudoku. The word *Sudoku* is Japanese and means *single number*.

Notice that the first Sudoku puzzle shown is a 9-by-9 grid with nine 3-by-3 sub-grids. Solve this puzzle by filling in each blank space with a digit from 1 to 9 so that the same digit does not appear twice in any row or column, or any of the nine 3-by-3 sub-grids.

Believe it or not, knowledge of operations on numbers is not needed to solve a typical puzzle. Although operations on numbers are not needed to solve a puzzle, logic and reasoning are definitely needed, and that is the main focus of this chapter. (A few variations of this puzzle are shown.)

2.1 PERIMETER, CIRCUMFERENCE, AND AREA

OBJECTIVES

1. Find the Perimeter or Circumference of Basic Shapes
2. Find the Area of Basic Shapes

VOCABULARY

- perimeter
- circumference
- area

Although the main focus of this chapter is logic and reasoning, we will first review some basic shapes and ways to measure these shapes. We may then use these concepts reviewed for examples throughout this chapter. Perimeter and area are two different ways of measuring geometric figures.

The **perimeter** P of a geometric figure is the distance around the figure. The distance around a circle is given a special name, called its **circumference** C. The **area** A of a geometric figure is the number of square units it encloses. In this section, we concentrate on the known figures of squares, rectangles, triangles, and circles. We may use the formulas below for calculating perimeter, circumference, and area.

Perimeter, Circumference, and Area Formulas

Square

side length s

$P = 4s$

$A = s^2$

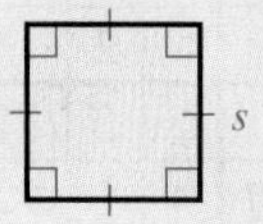

Triangle

side lengths a, b, and c, base b, and height h

$P = a + b + c$

$A = \frac{1}{2}bh$

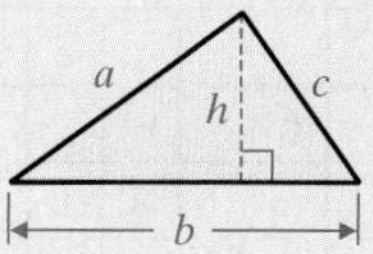

Rectangle

length l and width w

$P = 2l + 2w$ or $2(l + w)$

$A = lw$

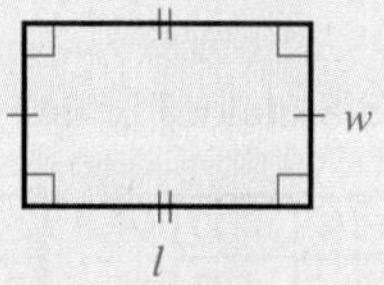

Circle

radius r and diameter d

$C = \pi d$, or $C = 2\pi r$

$A = \pi r^2$

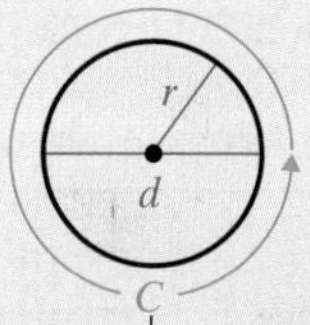

Helpful Hint

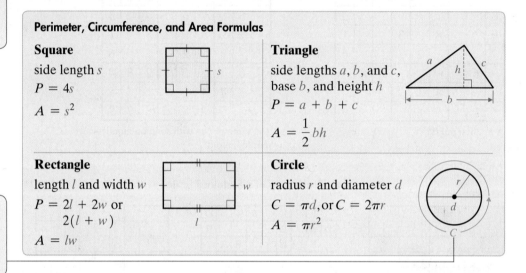

The diameter d of a circle is twice its radius r.

$$d = 2 \cdot r$$

Perimeter and circumference are measured in units, such as inches, feet, yards, miles, centimeters, and meters. When measuring area, use square units such as square inches (in.^{2}), square feet (ft^2), square yards (yd^2), square miles (mi^2), square centimeters (cm^2), and square meters (m^2).

OBJECTIVE 1 ▶ Finding the Perimeter or Circumference of Basic Geometric Figures.

EXAMPLE 1 **Finding the Perimeter of a Rectangle**

The botany club members are designing a rectangular garden for the courtyard of your school as shown to the right. They plan to place edging on the outside of the path. How much edging material will they need?

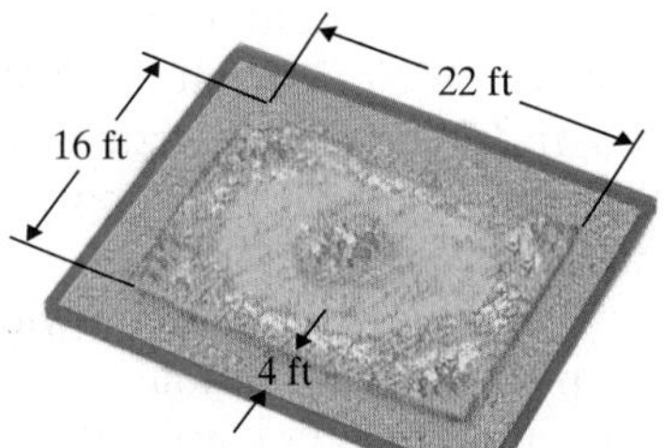

Solution The red line about the outside of the path is where edging is to be placed. Since this is the distance around a rectangle, first find the dimensions of this rectangle. Then calculate the perimeter P.

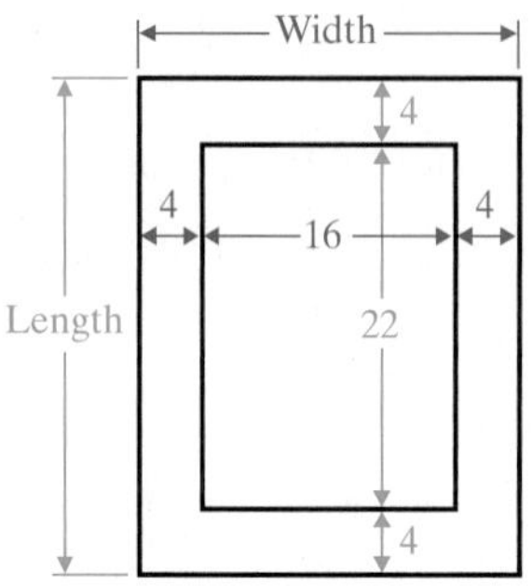

Find the dimensions of the rectangle formed by the edging.

Width of the garden and path

$= 4 + 16 + 4 = 24$

Length of the garden and path

$= 4 + 22 + 4 = 30$

Next, we calculate the perimeter of the rectangle formed by the edging.

$P = 2l + 2w$	Use the formula for the perimeter of a rectangle.
$= 2(30) + 2(24)$	Substitute 30 for l and 24 for w.
$= 60 + 48$	Multiply.
$= 108$	Add.

We need 108 feet of edging material. □

PRACTICE
1 We want to frame a picture that is 5 in. by 7 in. with a 1-in.-wide frame.

a. What is the perimeter of the picture?

b. What is the perimeter of the outside edge of the frame?

We will name a circle with the symbol ⊙. For example, the circle with center A is written $\odot A$.

The formulas for a circle involve the special number *pi* (π). Pi is the ratio of any circle's circumference to its diameter. Since π is an irrational number,

$$\pi = 3.1415926\ldots,$$

we cannot write it as a terminating decimal. For an approximate answer, we will use

$$\pi \approx 3.14 \text{ or } \pi \approx \frac{22}{7}$$

We can also use the π key on a calculator to get a rounded decimal for π. For an exact answer, we leave the result in terms of π.

> **Helpful Hint**
> Circle formulas for area and circumference contain the irrational number, π.
> - For exact answers, leave the result in terms of π.
> - For approximate answers, use
>
> $\pi \approx 3.14$,
>
> $\pi \approx \frac{22}{7}$ or the
>
> π Key on your calculator.

To better understand circumference and π (pi), try the following experiment. Take any can and measure its circumference and its diameter.

The can in the figure above has a circumference of 23.5 centimeters and a diameter of 7.5 centimeters. Now divide the circumference by the diameter.

$$\frac{\text{circumference}}{\text{diameter}} = \frac{23.5 \text{ cm}}{7.5 \text{ cm}} \approx 3.13$$

Try this with other sizes of cylinders and circles—you should always get a number close to 3.1. The exact ratio of circumference to diameter is π. (Recall that $\pi \approx 3.14$ or $\approx \frac{22}{7}$.)

EXAMPLE 2 **Finding Circumference**

What is the circumference of each circle in terms of π? What is the circumference of the circle to the nearest tenth? (Recall that the circumference of a circle is the distance around the circle.)

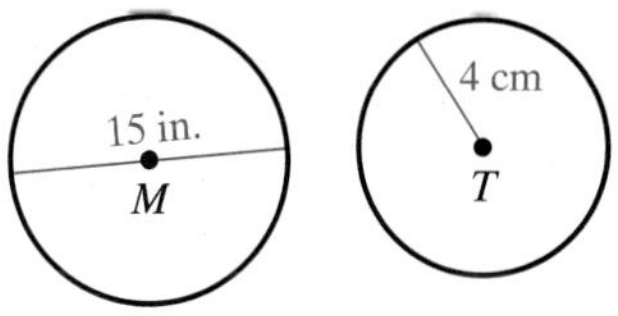

a. $\odot M$

b. $\odot T$

> **Helpful Hint**
> - In part **a.**, we are given the diameter d of the $\odot$, so we choose to use $C = \pi d$.
> - In part **b.**, we are given the radius of the circle, so we choose the formula $C = 2\pi r$.
> - Remember: We may always use either formula since we know $d = 2r$.

Solution

a. For $\odot M$, we are given the circle's diameter $d = 15$ in., so we choose the formula $C = \pi d$.

$C = \pi d$ Choose a formula for circumference of a circle.

$= \pi(15)$ Replace d with 15.

$= 15\pi$ This is the exact answer.

≈ 47.1238898 Use a calculator approximation for π.

The circumference of $\odot M$ is exactly 15π in., or about 47.1 in., rounded to the nearest tenth.

b. For $\odot T$, we are given radius $r = 4$ cm, so we use the formula $C = 2\pi r$.

$C = 2\pi r$ Choose a formula for circumference of a circle.

$= 2\pi(4)$ Replace r with 4.

$= 8\pi$ This is the exact answer.

≈ 25.13274123 Use a calculator.

The circumference of $\odot T$ is exactly 8π cm, or about 25.1 cm, rounded to the nearest tenth. □

PRACTICE 2

a. What is the exact circumference of a circle with radius 24 m?

b. What is the circumference of a circle with diameter 3 m to the nearest tenth?

EXAMPLE 3 Finding Perimeter in the Coordinate Plane

Coordinate Geometry What is the perimeter of ΔEFG?

Solution To find the perimeter of a triangle, first find the measure of each side.

$EF = |6 - (-2)| = 8$
$FG = |3 - (-3)| = 6$ } Use the Ruler Postulate.

$EG = \sqrt{(3 - (-3))^2 + (6 - (-2))^2}$ Use the Distance Formula.

$= \sqrt{6^2 + 8^2}$ Simplify within the parentheses.

$= \sqrt{36 + 64}$ Simplify.

$= \sqrt{100}$

$= 10$

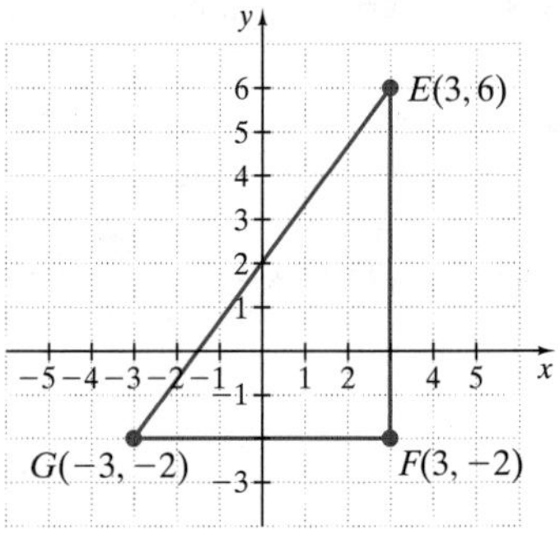

Next, we add the side lengths of the triangle to find the perimeter.

$$EF + FG + EG = 8 + 6 + 10 = 24$$

The perimeter of ΔEFG is 24 units. □

PRACTICE 3 Graph quadrilateral $JKLM$ with vertices $J(-3, -3)$, $K(1, -3)$, $L(1, 4)$, and $M(-3, 1)$. What is the perimeter of $JKLM$? (Recall that perimeter means the distance around, so no formula is needed.)

OBJECTIVE 2 ▶ Finding the Area of Basic Geometric Figures. To use the formulas for perimeter and area, always check to see that the measurement units are all the same.

EXAMPLE 4 Finding the Area of a Rectangle

Banners We want to make a rectangular banner similar to the one at the right. The banner shown is $2\frac{1}{2}$ ft in width and 5 ft in length. If we round up to the nearest square yard, how much material do we need?

Solution Since we are asked for the area in square yards, let's first convert the dimensions of the banner to yards. To do so, we will use the conversion factor $\frac{1 \text{ yd}}{3 \text{ ft}}$.

$$\text{Width: } \frac{5}{2}\cancel{\text{ft}} \cdot \frac{1 \text{ yd}}{3 \cancel{\text{ft}}} = \frac{5}{6}\text{ yd} \quad \text{Recall that } 2\frac{1}{2} = \frac{5}{2}.$$

$$\text{Length: } 5\cancel{\text{ft}} \cdot \frac{1 \text{ yd}}{3 \cancel{\text{ft}}} = \frac{5}{3}\text{ yd}$$

Now let's find the area of the banner.

$$A = lw \quad \text{Use the formula for area of a rectangle.}$$
$$= \frac{5}{3} \cdot \frac{5}{6} \quad \text{Substitute } \frac{5}{3} \text{ for } l \text{ and } \frac{5}{6} \text{ for } w.$$
$$= \frac{25}{18} \quad \text{Multiply.}$$

The area of the banner is $\frac{25}{18}$, or $1\frac{7}{18}$ square yards (yd^2). Rounding up, we need 2 square yards of material. □

PRACTICE 4 You are designing a poster that will be 3 yd wide and 8 ft high. How much paper do you need to make the poster? Give your answer in square feet.

EXAMPLE 5 Finding the Area of a Circle

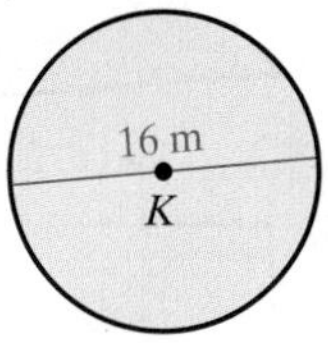

Find the area of $\odot K$ in terms of π.

Solution The formula for the area of a circle is $A = \pi r^2$. In $\odot K$, we are given diameter, so we first find the radius of $\odot K$.

$$r = \frac{16}{2}, \text{ or } 8 \quad \text{The radius is half the diameter.}$$

Next, let's use the radius to find the area.

$$A = \pi r^2 \quad \text{Use the formula for area of a circle.}$$
$$= \pi(8)^2 \quad \text{Substitute 8 for } r.$$
$$= 64\pi \quad \text{Simplify.}$$

The area of $\odot K$ is exactly 64π square meters. □

PRACTICE 5 The diameter of a circle is 14 ft.

a. What is the area of the circle in terms of π?

b. What is the area of the circle using the approximation 3.14 for π?

VOCABULARY & READINESS CHECK

Word Bank *Use the choices below to fill in each blank. Some choices may not be used.*

circumference	radius	π	$\frac{22}{7}$
diameter	perimeter	3.14	

1. The ____________ of a geometric figure is the distance around the figure.
2. The distance around a circle is called the ____________.
3. The exact ratio of circumference to diameter is ____________.
4. The diameter of a circle is double its ____________.
5. Both ____________ and ____________ are approximations for π.
6. The ____________ of a geometric figure is the number of square units it encloses.

2.1 EXERCISE SET MyMathLab®

Mixed Practice

Find the perimeter of each figure. See Examples 1 and 3.

1.
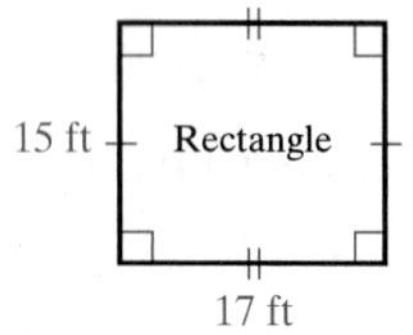

2.
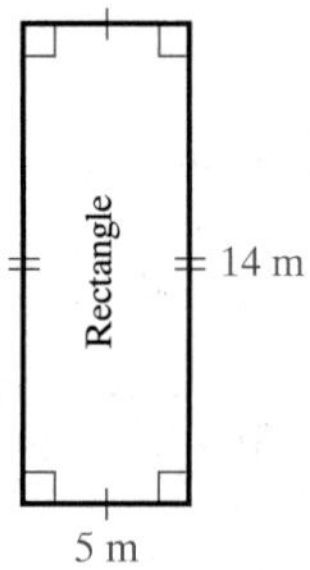

3.
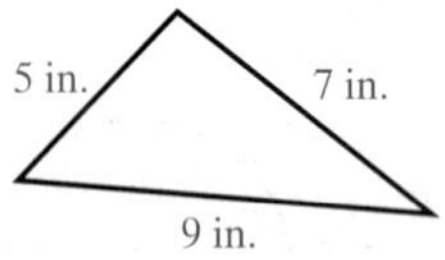

4.
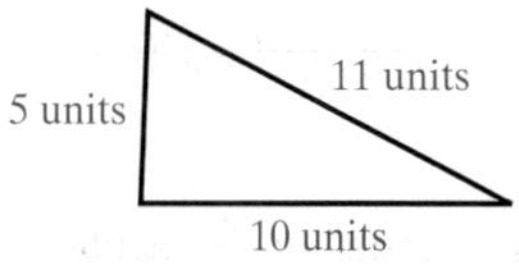

5.
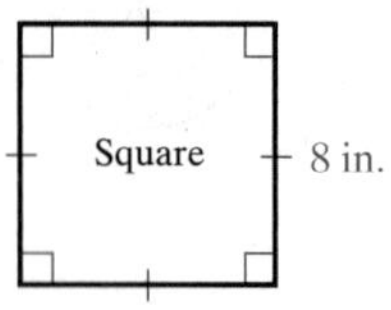

6.
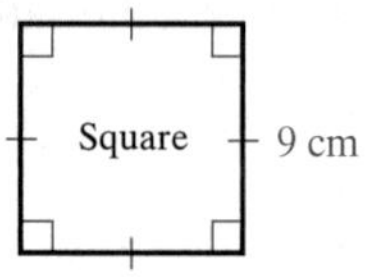

Find the perimeter of each regular figure. (The sides of each figure have the same length.) See Examples 1 and 3.

7.
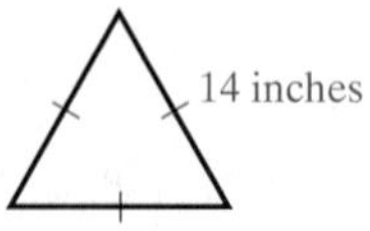

8.
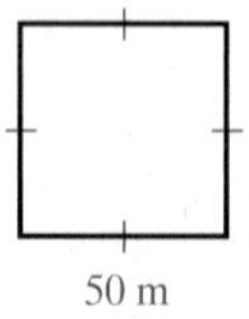

9.
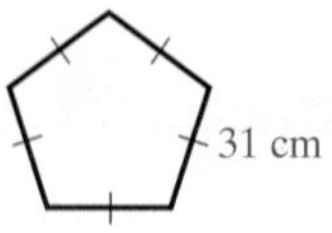

10.
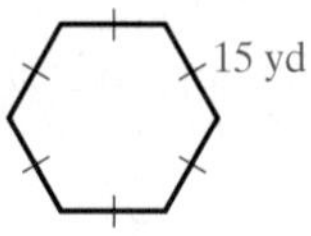

Find the circumference of each circle. Give the exact circumference and then an approximation. Use $\pi \approx 3.14$. See Example 2.

11.
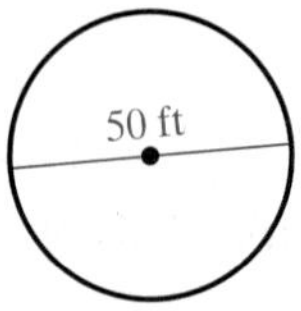

12.
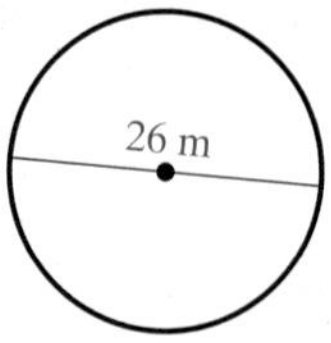

13.

14.
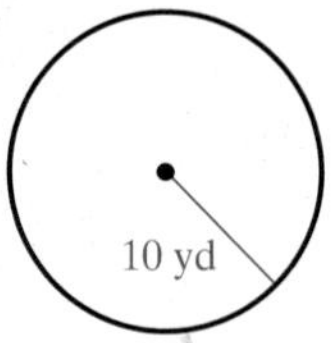

15.
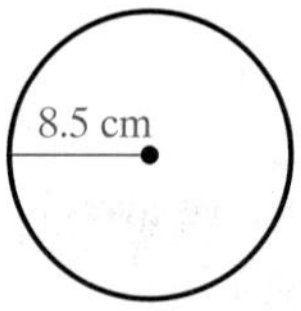

16.

Find the area of each geometric figure. If the figure is a circle, give an exact area and then use the approximation 3.14 for π to approximate the area. See Examples 4 and 5.

17.
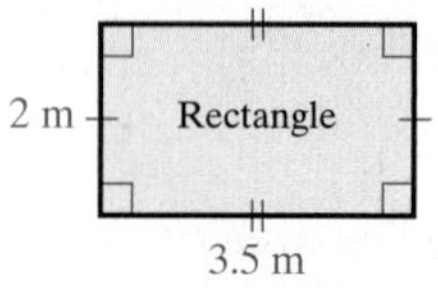

18.
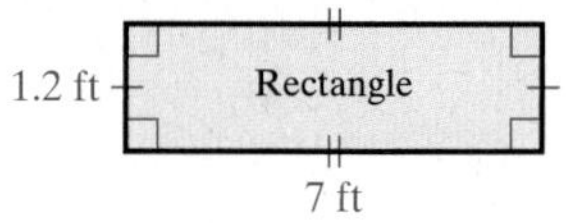

19.
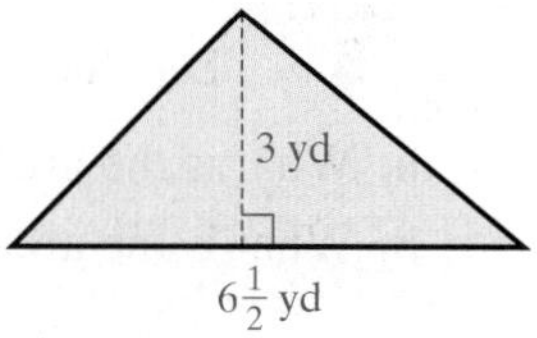

20.
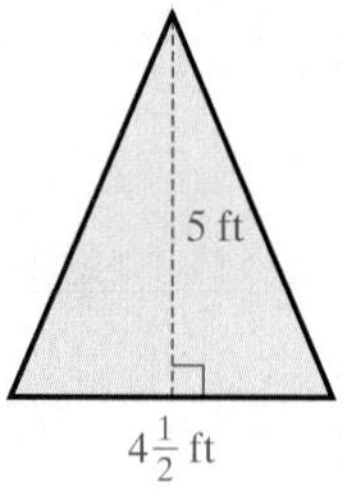

21.

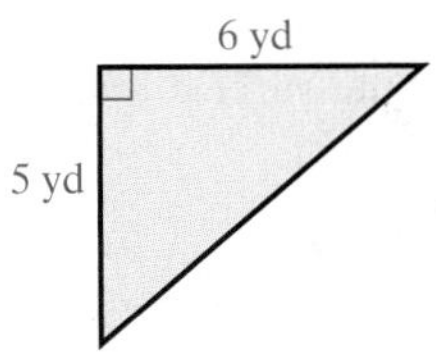

22.

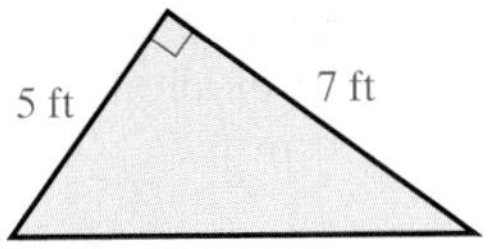

23.

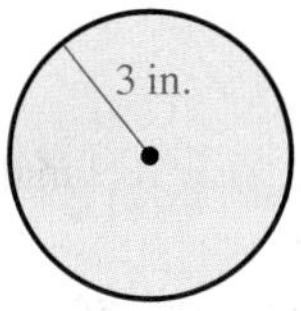

24.

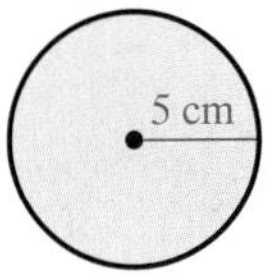

25.

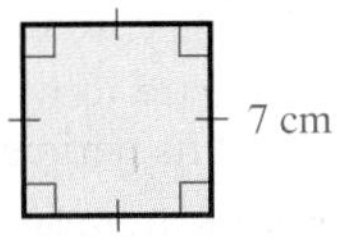

26.

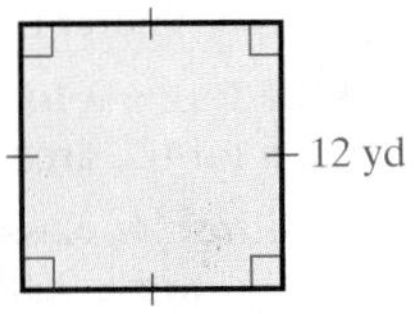

27.

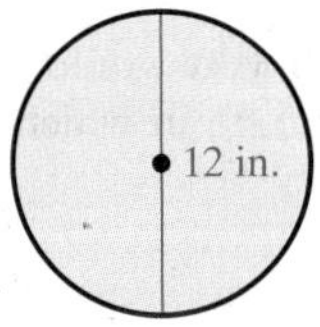

28.

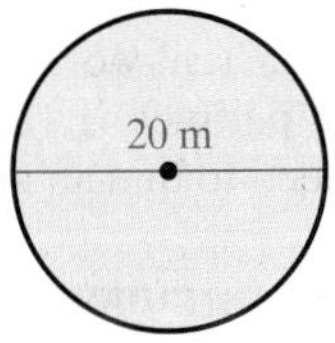

Coordinate Geometry *Graph each figure with the given vertices in the coordinate plane. Find each perimeter. See Example 3.*

29. $X(0, 2), Y(4, -1), Z(-2, -1)$

30. $A(-4, -1), B(4, 5), C(4, -2)$

31. $L(0, 1), M(3, 5), N(5, 5), P(5, 1)$

32. $S(-5, 3), T(7, -2), U(7, -6), V(-5, -6)$

Mixed Practice *Solve. See Examples 1 through 5.*

33. What is the area of a section of pavement that is 20 ft wide and 100 yd long? Give your answer in square feet.

34. A drapery panel measures 6 ft by 7 ft. Find how many square feet of material are needed for *four* panels.

35. The largest American flag measures 505 feet by 225 feet. It's the U.S. "Super flag" owned by "Ski" Demski of Long Beach, California. Find its area. (*Source: Guinness World Records*)

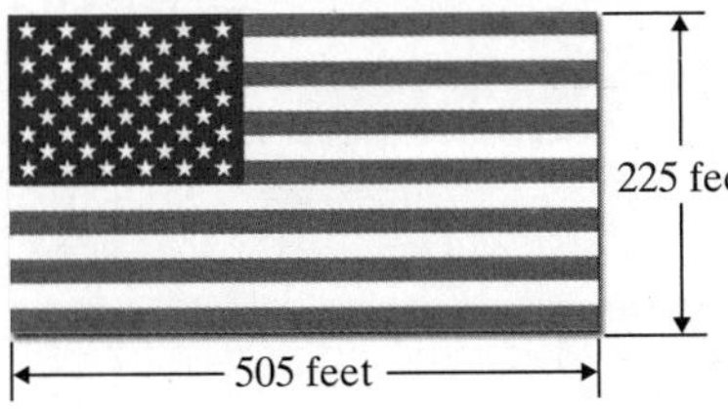

36. The longest illuminated sign is in Ramat Gan, Israel, and measures 197 feet by 66 feet. Find its area. (*Source: The Guinness Book of World Records*)

37. A pool measures 15 feet by 25 feet and it is surrounded by a 6-foot deck. How much fencing is needed to fence the outside of the decking?

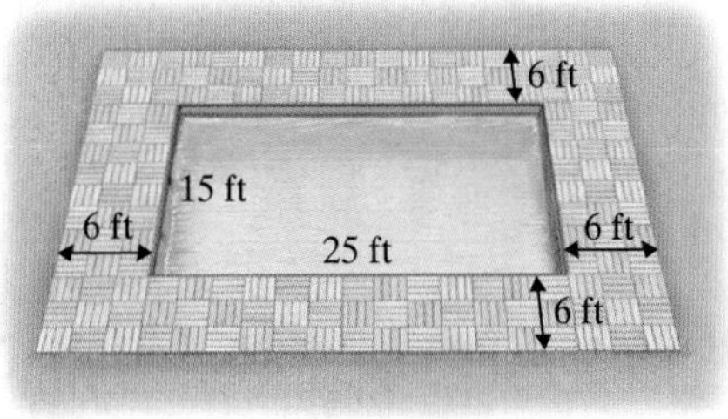

38. A 4 in. by 6 in. picture is surrounded by matting that is 3 inches wide all around. Find the perimeter of the outside of the matting.

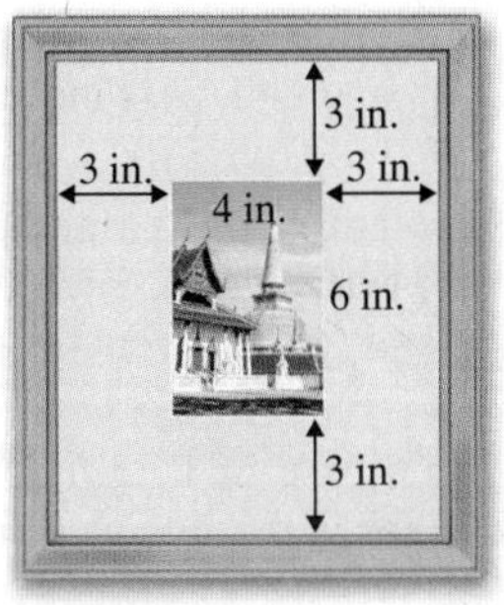

For Exercises 39 and 40, give an exact answer and a one-decimal place approximation.

39. Wyley Robinson just bought a trampoline for his children to use. The trampoline has a diameter of 15 feet. If Wyley wishes to buy netting to go around the outside of the trampoline, how many feet of netting does he need?

40. The largest round barn in the world is located at the Marshfield Fairgrounds in Wisconsin. The barn has a diameter of 150 ft. What is the circumference of the barn? (*Source: The Milwaukee Journal Sentinel*)

CONCEPT EXTENSIONS

Given the following situations, tell whether you are more likely to be concerned with area or perimeter.

41. buying carpet to install in a room

42. buying gutters to install on a house

43. ordering paint to paint a wall

44. ordering baseboards to install in a room

45. buying a wallpaper border to go on the walls around a room

46. buying fertilizer for your yard

47. ordering fencing to fence a yard

48. ordering grass seed to plant in a yard

Find the area of each described figure.

49. drafting triangle

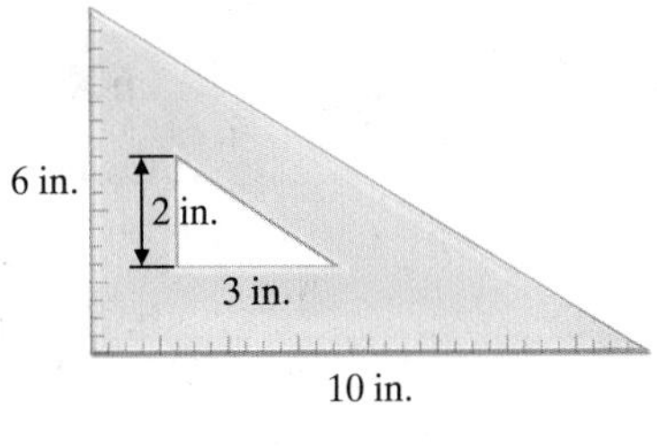

50. picture frame

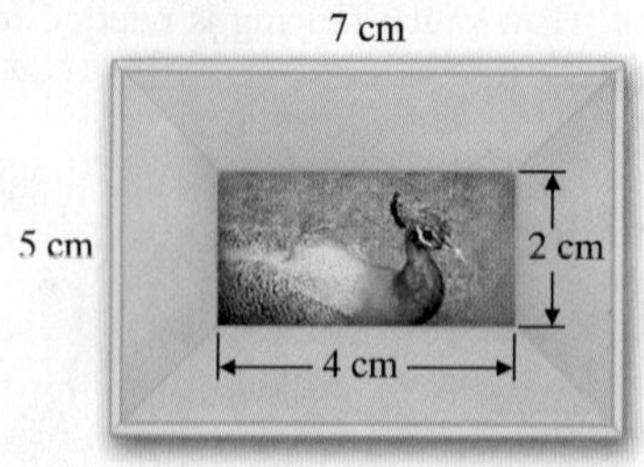

Coordinate Geometry *Graph each rectangle in the coordinate plane. Find its perimeter and area.*

51. $A(-3, 2), B(-2, 2), C(-2, -2), D(-3, -2)$

52. $A(-2, -6), B(-2, -3), C(3, -3), D(3, -6)$

53. **Coordinate Geometry** The endpoints of a diameter of a circle are $A(2, 1)$ and $B(5, 5)$. Find the area of the circle in terms of π.

54. **Coordinate Geometry** On graph paper, draw polygon $ABCDEFG$ with vertices $A(1, 1)$, $B(10, 1)$, $C(10, 8)$, $D(7, 5)$, $E(4, 5)$, $F(4, 8)$, and $G(1, 8)$. Find the perimeter and the area of the polygon.

55. A rectangle has a base of x units. The area is $(4x^2 - 2x)$ square units. What is the height of the rectangle in terms of x?

a. $(4 - x)$ units
b. $(x - 2)$ units
c. $(4x^3 - 2x^2)$ units
d. $(4x - 2)$ units

56. The surface area of a three-dimensional figure is the sum of the areas of all of its surfaces. Find the surface area of the figure shown.

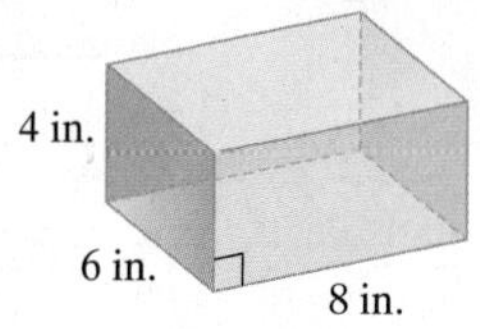

57. Can you use the formula for the perimeter of a rectangle to find the perimeter of any square? Explain.

58. a. Can you use the formula for the perimeter of a square to find the perimeter of any rectangle? Explain.

b. Use the formula for the perimeter of a square to write a formula for the area of a square in terms of its perimeter.

59. **Find the Error** A classmate finds the area of a circle with radius 30 in. to be 900 in.^{2}. What error did your classmate make?

60. Describe a real-world situation in which you would need to find a perimeter. Then describe a situation in which you would need to find an area.

REVIEW AND PREVIEW

Evaluate. See an appendix on Evaluating Expressions.

61. 5^2 **62.** 7^2 **63.** 3^2 **64.** 20^2

65. $1^2 + 2^2$ **66.** $5^2 + 3^2$ **67.** $4^2 + 2^2$ **68.** $1^2 + 6^2$

2.2 PATTERNS AND INDUCTIVE REASONING

OBJECTIVES

1 Use Logic to Understand Patterns.

2 Understand and Use Inductive Reasoning.

3 Form Conjectures and Find Counterexamples.

VOCABULARY

- inductive reasoning
- induction
- conjecture
- counterexample

OBJECTIVE 1 ▶ Using Logic. In Section 1.2, we learned how a mathematical system, such as geometry, grows. To start the growth of geometry, we introduced terms, postulates, and axioms in Chapter 1. We now continue to increase our knowledge of geometry by using these terms, postulates, and axioms to prove theorems.

To prove theorems, we need a basic study of types of logic (or reasoning) and types of statements.

We begin with a further study of predicting the next number or figure in a pattern. Then we give this logical process a name.

▶ Helpful Hint

Recall the names of statements that we prove and those that we do not prove, but assume to be true.

postulates—do not prove postulates

axioms—do not prove axioms

theorems—do prove theorems

EXAMPLE 1 **Using Logic**

Look for a pattern in each list. Then use this pattern to predict the next number.

a. 3, 4, 6, 9, 13, 18, ____________

b. 3, 6, 18, 36, 108, 216, ____________

Solution

a. Because 3, 4, 6, 9, 13, 18, ______ is increasing relatively slowly, let's try addition as a possible basis for this pattern.

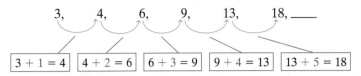

3, 4, 6, 9, 13, 18, ____

| $3 + 1 = 4$ | $4 + 2 = 6$ | $6 + 3 = 9$ | $9 + 4 = 13$ | $13 + 5 = 18$ |

From these observations, we conclude that each number after the first is obtained by adding a counting number to the previous number. The additions begin with 1 and continue through each successive counting number. Using this pattern, the next number is $18 + 6$, or 24.

b. Because 3, 6, 18, 36, 108, 216, ______ is increasing relatively rapidly, let's try multiplication as a possible basis for this pattern.

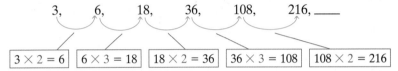

3, 6, 18, 36, 108, 216, ____

| $3 \times 2 = 6$ | $6 \times 3 = 18$ | $18 \times 2 = 36$ | $36 \times 3 = 108$ | $108 \times 2 = 216$ |

From these observations, we conclude that each number after the first is obtained by multiplying the previous number by 2 or by 3. The multiplications begin with 2 and then alternate, multiplying by 2, then 3, then 2, then 3, and so on. Using this pattern, the next number is 216×3, or 648. □

PRACTICE 1 Look for a pattern in each list. Then use this pattern to predict the next number.

a. 3, 6, 18, 72, 144, 432, 1728, ______ **b.** 1, 9, 17, 3, 11, 19, 5, 13, 21, ______

OBJECTIVE 2 ▶ Using Inductive Reasoning. We learned in Chapter 1 that geometry involves the study of patterns. In everyday life, we frequently rely on patterns and routines to draw conclusions. Here is an example:

> The last six times I went to the beach, the traffic was light on Wednesdays and heavy on Sundays. My conclusion is that weekdays have lighter traffic than weekends.

This type of reasoning process is referred to as **inductive reasoning,** or **induction.**

Inductive Reasoning

Inductive reasoning is the process of arriving at a general conclusion based on observing patterns or observing specific examples.

Although inductive reasoning is a powerful method of drawing conclusions, we can never be absolutely certain that these conclusions are true. For this reason, the conclusions are called **conjectures** or educated guesses. Inductive reasoning does not guarantee the truth of the conjecture (conclusion), but rather provides strong support for the conjecture (conclusion).

▶ Helpful Hint

Studying a pattern and using inductive reasoning can result in more than one educated guess or conjecture for the next number in a list.

Example: 1, 2, 4, ______

Possible Pattern: Each number after the first is obtained by multiplying the previous number by 2. The missing number is 4×2, or 8.

Possible Pattern: Each number after the first is obtained by adding successive counting numbers, starting with 1, to the previous number. The second number is $1 + 1$, or 2. The third number is $2 + 2$, or 4. The missing number is $4 + 3$, or 7.

In our next example, the patterns are a bit more complex than the additions and multiplications in Example 1 and Practice 1.

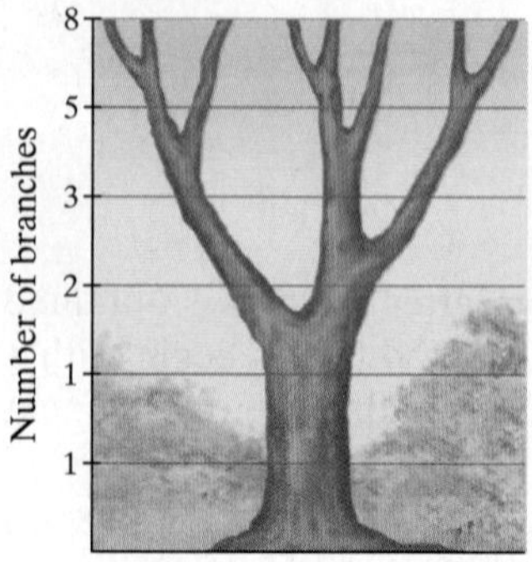

As this tree branches, the number of branches forms the Fibonacci sequence.

EXAMPLE 2 Using Inductive Reasoning

Look for a pattern in each list. Then use this pattern to predict the next number.

a. 1, 1, 2, 3, 5, 8, 13, 21, ____________

b. 23, 54, 95, 146, 117, 98, ____________

Solution

a. Starting with the third number in the list, let's form our observations by comparing each number with the two numbers that immediately precede it.

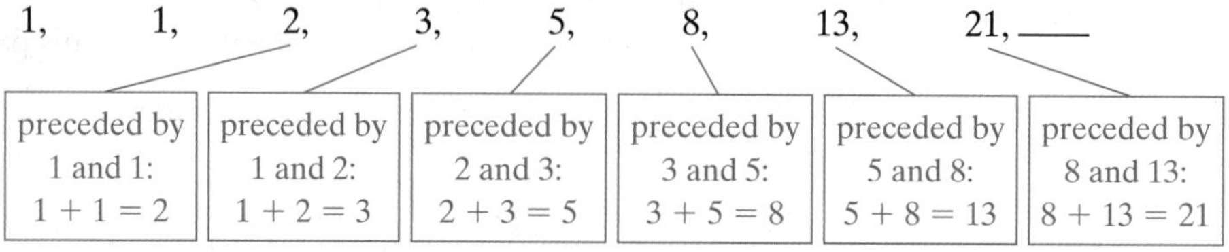

Notice that the first two numbers are 1. Each number thereafter is the sum of the two preceding numbers. Using this pattern, the next number is $13 + 21$, or 34. (The numbers 1, 1, 2, 3, 5, 8, 13, 21, and 34 are the first nine terms of a list of numbers called the *Fibonacci sequence.*)

b. Let's study the digits that form each number. First, focus on the sum of the digits.

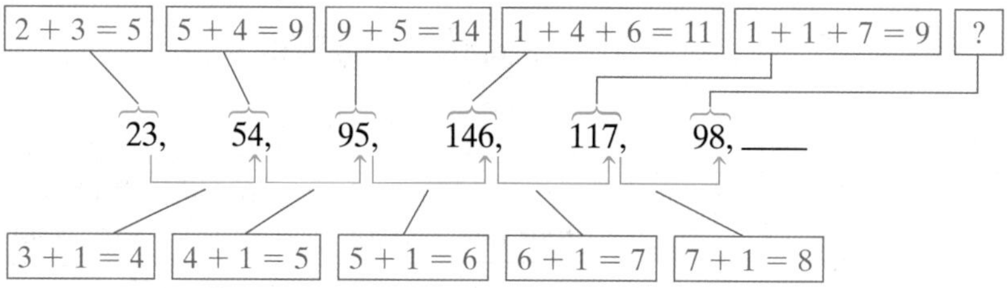

Notice that starting with the second number, we obtain the first digit of the number by adding the digits of the previous number. We obtain the last digit of the number by adding 1 to the final digit of the preceding number. Applying this pattern to the number that follows 98, the first part of the number is $9 + 8$, or 17. The last digit is $8 + 1$, or 9. Thus, the next number in the list is 179. □

PRACTICE 2 Look for a pattern in each list. Then use this pattern to predict the next number.

a. 1, 3, 4, 7, 11, 18, 29, 47, ____________

b. 2, 3, 5, 9, 17, 33, 65, 129, ____________

(*Hint:* Study the numbers added to each previous number to have a sum of the next number.)

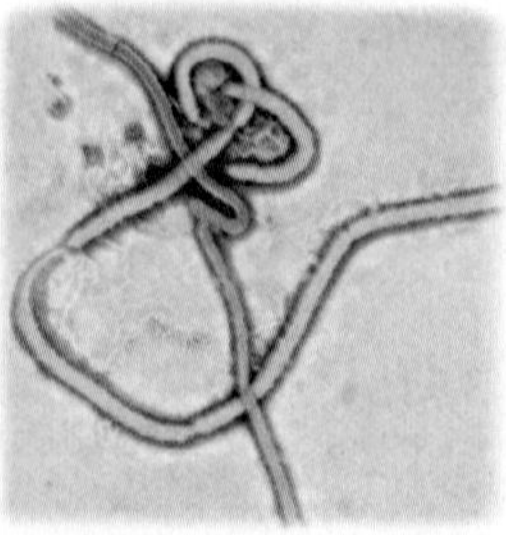
This electron microscope photograph shows the knotty shape of the Ebola virus.

In addition to number patterns, we also have visual patterns. These visual patterns have real-life applications. For example, visual patterns can be used to illustrate the knotty shapes and patterns of viruses. Scientists can then use these patterns to form weapons against viruses. This is based on recognizing visual patterns that illustrate the possible ways that knots can be tied.

For our next example, we practice recognizing visual patterns.

EXAMPLE 3 **Using Inductive Reasoning**

Notice two patterns in this sequence of figures. Use the patterns to draw the next figure in the sequence.

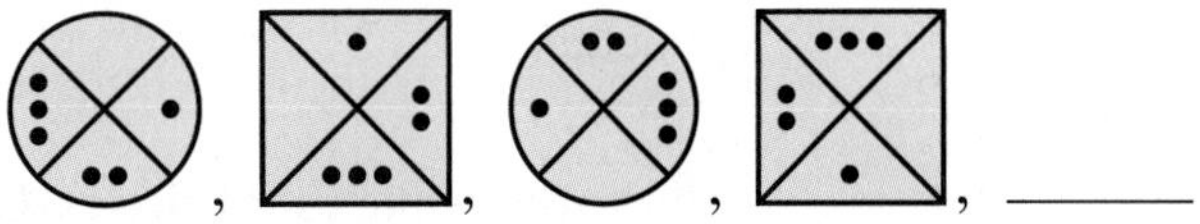

Solution First, the figures alternate between circles and squares. Thus, the next figure will be a circle. The second pattern in the four regions is the dot pattern. This dot pattern remains the same except that it rotates counterclockwise as we follow the figures from left to right.

This means that the next figure should be a circle with a single dot in the right-hand region, two dots in a bottom region, three dots in the left-hand region, and no dots in the top region. This figure is drawn to the right. □

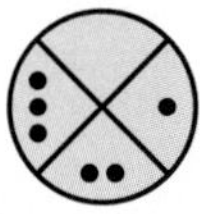

PRACTICE 3 Notice two patterns in this sequence of figures. Use the patterns to draw the next figure in the sequence.

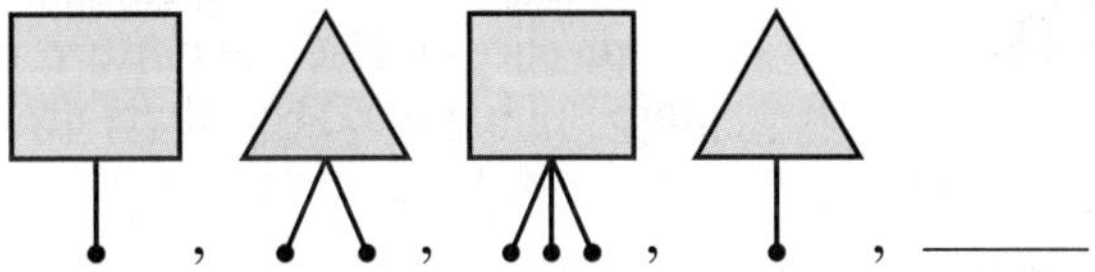

OBJECTIVE 3 ▶ Forming Conjectures and Finding Counterexamples. Remember that a conjecture is a conclusion or educated guess based on observations. Thus, a conjecture is a general conclusion we make based on inductive reasoning.

Let's continue to study patterns and use inductive reasoning to make conjectures. In these examples, we use our conjecture to find multiple future numbers or figures—not just the next number or figure in the list.

EXAMPLE 4 **Forming Conjectures**

Study the list of circles. Use the pattern to answer the questions.

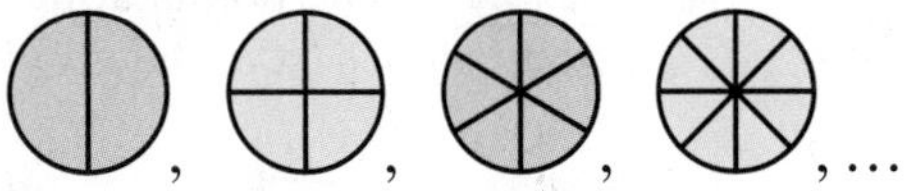

a. Make a conjecture about the color of the 11th circle.

b. Make a conjecture about the number of regions in the 11th circle.

c. Make a conjecture about the appearance of the 11th circle.

d. Make a conjecture about the appearance of the 30th circle.

Solution Let's study the list of circles.

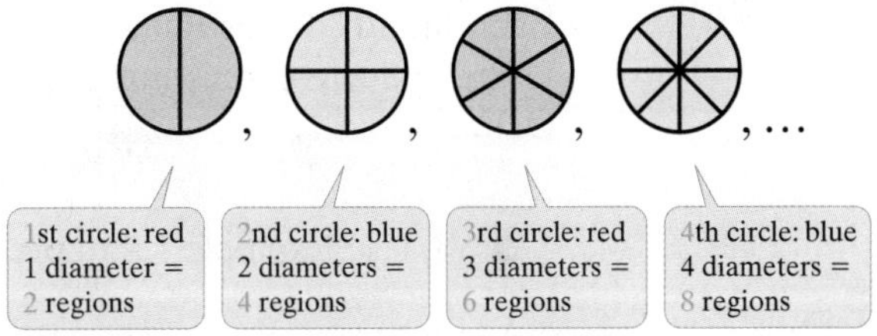

a. Odd-numbered circles are red; even-numbered circles are blue. Since 11 is an odd number, we conjecture that the 11th circle is red.

b. If we number the circles in the list, each numbered circle has twice as many regions as its number in the list. Thus, the 11th circle has $11 \cdot 2$ or 22 regions formed by diameters.

c. The appearance of the 11th circle is red with 22 regions formed by diameters.

d. The 30th circle is blue, since 30 is an even number.

The 30th circle has $30 \cdot 2$ or 60 regions formed by diameters.

Thus, the appearance of the 30th circle is a blue circle with 60 regions formed by diameters. □

PRACTICE 4 Study the list of circles. Use the pattern to answer each question.

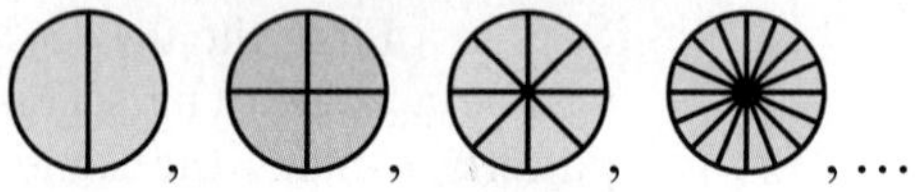

a.–d. Answer the same questions as for Example 4.

Conjectures can be made from graphs of data.

EXAMPLE 5 Forming a Conjecture

The number of hybrid cars sold at a certain dealership is increasing according to the graph shown. Use the pattern to predict the number of hybrid cars sold by this dealership in October. (Numbers are rounded.)

A hybrid car is a car that has a gasoline engine and an electric motor.

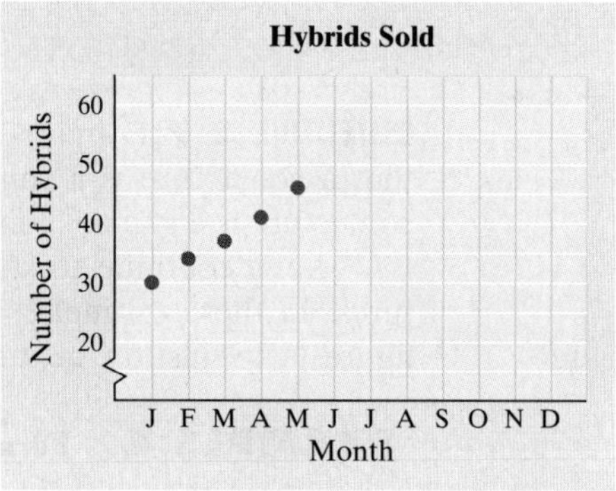

Solution The points appear to lie on a line that is increasing. Below, we list the data point values from the graph.

Month	*Number of Hybrid Cars Sold*
Jan.	30
	+4
Feb.	34
	+3
March	37
	+4
April	41
	+5
May	46

Each month, we estimate that the number of hybrid cars sold increases by about 4. Using inductive reasoning, we predict that the dealership sells 66 hybrid cars in October. □

PRACTICE 5 Use Example 5 and predict the number of hybrid cars sold in December.

- To prove that a conjecture is *true*, we need to prove it true in all cases.
- To prove that a conjecture is *false*, we need to give only a single counterexample.

A **counterexample** is an example that shows that a conjecture is false, or incorrect.

EXAMPLE 6 **Finding a Counterexample**

Find a counterexample to show that each conjecture is false.

a. Conjecture: The product of two numbers is always greater than either number.

b. Conjecture: All apples are red.

Solution

a. Sometimes this conjecture is true; for example, $3 \cdot 4 = 12$. For the product 12, it is true that $12 > 4$ and $12 > 3$.

Counterexample: $-2 \cdot 4 = -8$. Notice that $-8 \not> 4$. Thus, the conjecture is false because it is not always true.

b. Counterexample: Simply find one apple that is not red. For example, Granny Smith apples are green. (See photo.)

PRACTICE 6 Find a counterexample to show that each conjecture is false.

a. Conjecture: The product of two positive numbers is always greater than either number.

b. Conjecture: The area of a square is always an even number.

VOCABULARY & READINESS CHECK

Word Bank *Fill in the blanks.*

counterexample   conjecture   inductive reasoning

1. Another word for an educated guess is a(n) ____________.
2. A(n) ____________ is an example that shows that a conjecture is false.
3. The process of arriving at a general conclusion based on observing patterns or specific examples is called ____________.

2.2 EXERCISE SET MyMathLab®

Look for a pattern in each list. Then use this pattern to predict the next number. See Example 1.

1. 5, 10, 20, 40, . . .
2. 5, 20, 80, 320, . . .
3. 20, 17, 14, 11, . . .
4. 20, 18, 16, 14, . . .
5. 2, −2, 3, −3, 4, . . .
6. 5, −5, 0, 4, −4, 0, 3, . . .
7. 1, 4, 9, 16, 25, . . .
8. 1, 3, 7, 13, 21, . . .
9. 1, 2, 6, 24, 120, . . .
10. 2, 6, 12, 36, 72, . . .

Use the same directions as above. See Example 2.

11. 51, 42, 33, 24, . . .
12. 79, 68, 77, 66, . . .
13. 2, 6, 24, 72, 288, . . .
14. 2, 4, 12, 24, 72, . . .
15. 0.1, 0.01, 0.001, . . .
16. 0.1, 0.11, 0.111, . . .
17. $0, \frac{1}{2}, \frac{3}{4}, \frac{7}{8}, \frac{15}{16}, \ldots$
18. $\frac{1}{3}, \frac{4}{9}, \frac{13}{27}, \frac{40}{81}, \ldots$

Study the list of figures. Use the pattern to draw the next figure in each list. See Example 3.

19. 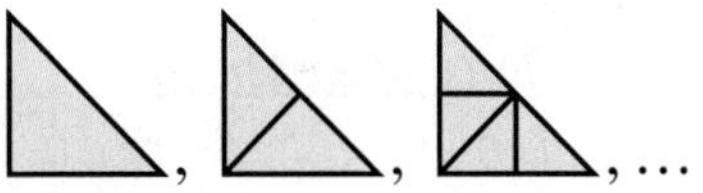

20. 

21. 

22. 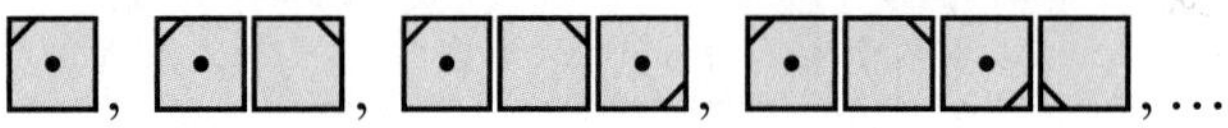

Use the figures and inductive reasoning to make a conjecture. See Example 4.

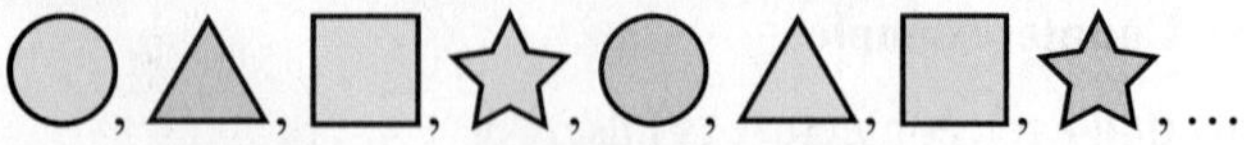

23. What is the color of the 15th figure?

24. What is the color of the 40th figure?

25. What is the shape of the 12th figure?

26. What is the shape of the 50th figure?

27. Multiple Steps

a. What is the color of the 17th figure?

b. What is the shape of the 17th figure?

c. Describe the appearance of the 17th figure.

28. Multiple Steps

a. What is the color of the 20th figure?

b. What is the shape of the 20th figure?

c. Describe the appearance of the 20th figure.

Make a conjecture for each scenario. Try a few examples, if necessary. See Example 4.

29. the sum of two odd numbers

30. the sum of an even and an odd number

31. the product of two even numbers

32. the product of two odd numbers

33. the sum of a number and its opposite

34. the product of a number and its reciprocal

Multiple Steps *For Exercises 35–36, answer parts a–d. See Example 4.*

a. Make a conjecture about the color of the 13th figure.

b. Make a conjecture about the number of regions in the 13th figure.

c. Make a conjecture about the appearance of the 13th figure.

d. Make a conjecture about the appearance of the 29th figure.

35. 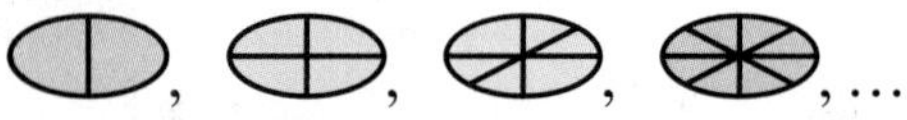

36. 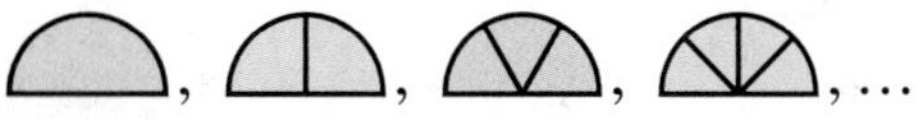

Multiple Steps *For Exercises 37–38, answer parts a–d. See Example 4.*

a. Make a conjecture about the color(s) of the 12th figure.

b. Make a conjecture about the shape of the 12th figure.

c. Sketch the 13th figure. Label the color(s).

d. Sketch the 26th figure. Label the color(s).

37. 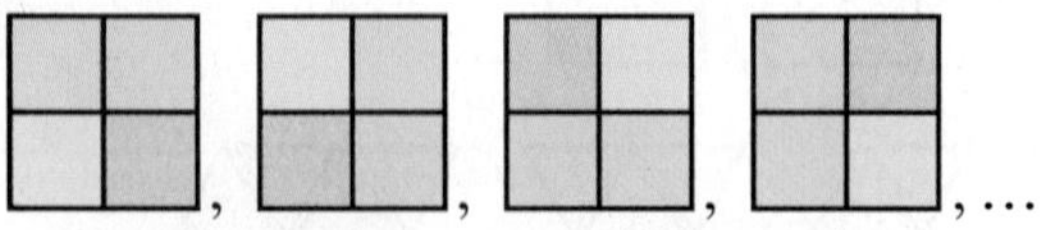

38. 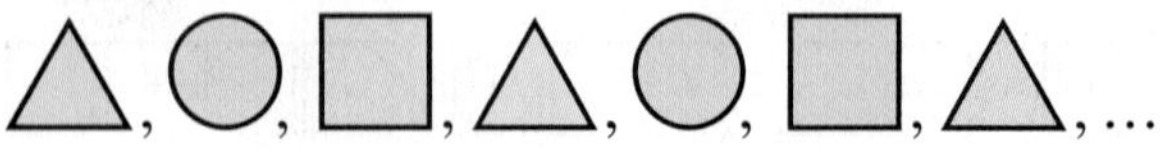

Use inductive reasoning to make a conjecture about the weather. See Example 5.

39. Light travels much faster than sound, so you see lightning before you hear thunder. If you count 5 seconds between the lightning and thunder, how far away is the storm? See the graph below.

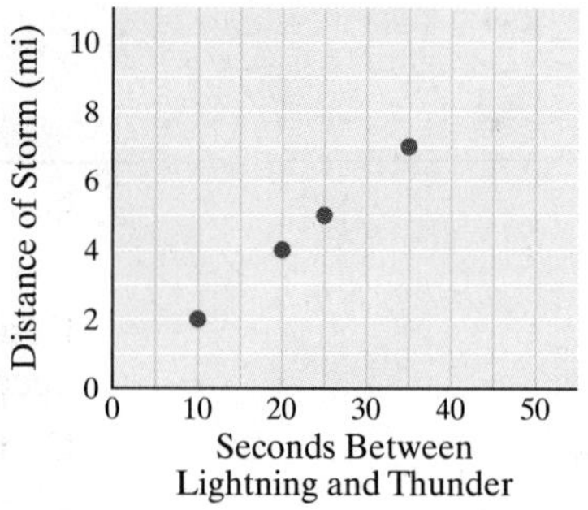

40. The speed at which a cricket chirps is affected by the temperature. If you hear a cricket chirp 20 times in 14 seconds, what is the temperature?

Number of Chirps per 14 Seconds	*Temperature (°F)*
5	45
10	55
15	65

Find one counterexample to show that each conjecture is false. See Example 6.

41. $\angle 1$ and $\angle 2$ are supplementary, so one of the angles is acute.

42. ΔABC is a right triangle, so $\angle A$ measures $90°$.

43. The sum of two numbers is always greater than either number.

44. The product of two negative numbers is always less than either number.

45. If a month of the year starts with the letter A, then it must be April.

46. The numerical value of the area of a rectangle is always greater than its length and its width.

CONCEPT EXTENSIONS

Predict the next term in each sequence. Use your calculator to verify your answer. (Commas are not present to aid in finding the pattern.)

47.
$1 \times 1 = 1$
$11 \times 11 = 121$
$111 \times 111 = 12321$
$1111 \times 1111 = 1234321$
$11111 \times 11111 =$ ________

48.
$12345679 \times 9 = 111111111$
$12345679 \times 18 = 222222222$
$12345679 \times 27 = 333333333$
$12345679 \times 36 = 444444444$
$12345679 \times 45 =$ ________

Solve.

49. Find the perimeter (distance around the figure) when 100 triangles are put together in the pattern shown. Assume that all triangle sides are 1 cm long.

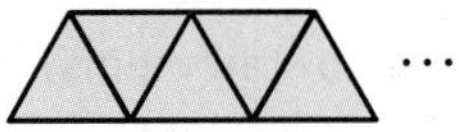

50. The small squares on a chessboard can be combined to form larger squares. For example, there are sixty-four 1×1 squares and one 8×8 square. Use inductive reasoning to determine how many 2×2 squares, 3×3 squares, and so on, are on a chessboard. What is the total number of squares on a chessboard?

Coordinate Geometry *For Exercises 51 and 52, plot the points on a coordinate system.*

$A(6, -2), B(6, 5), C(8, 0), D(8, 7), E(10, 2), F(10, 6), G(11, 4), H(12, 3), I(4, 0), J(7, 6), K(5, 6), L(4, 7), M(2, 2), N(1, 4), O(2, 6)$

51. Most of the points fit a pattern. Which do not?

52. What pattern do the majority of the points fit?

53. Write two different lists of numbers that begin with the same two numbers.

54. Describe a real-life situation in which you recently used inductive reasoning.

During bird migration, volunteers get up early on Bird Day to record the number of bird species they observe in their community during a 24-hour period. Results are posted online to help scientists and students track the migration.

Bird Count

Year	*Number of Species*
2008	70
2009	83
2010	80
2011	85
2012	90

55. Make a graph of the data.

56. Use the graph and inductive reasoning to make a conjecture about the number of bird species the volunteers in this community will observe in 2019.

57. Find the Error For each of the past four years, Paulo has grown 2 in. every year. He is now 16 years old and is 5 ft 10 in. tall. He figures that when he is 22 years old, he will be 6 ft 10 in. tall. What would you tell Paulo about his conjecture?

58. Clay thinks the next term in the squence 2, 4, . . . is 6. Given the same pattern, Ott thinks the next term is 8, and Stacie thinks the next term is 7. What conjecture is each person making? Is there enough information to decide who is correct?

59. Multiple Steps

a. Write the first six terms of the sequence that starts with 1, and for which the difference between consecutive terms is first 2, and then 3, 4, 5, and 6.

b. Evaluate $\dfrac{n^2 + n}{2}$ for $n = 1, 2, 3, 4, 5,$ and 6. Compare the sequence you get with your answer for part **a.**

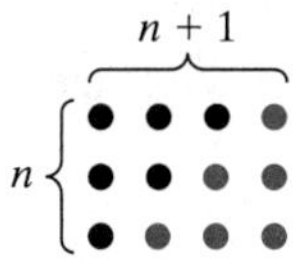

c. Examine the diagram at the right and explain how it illustrates a value of $\dfrac{n^2 + n}{2}$.

d. Draw a similar diagram to represent $\dfrac{n^2 + n}{2}$ for $n = 5$.

60. Multiple Steps When he was in the third grade, German mathematician Karl Gauss (1777–1855) took ten seconds to sum the integers from 1 to 100.

a. See if you can find a fast way to sum the integers from 1 to 100.

b. Find a fast way to sum the integers from 1 to n.

(*Hint*: Write a few of the integers to be added. Under each integer, write a few of the integers to be added but in reverse order. Find the sum of each pair of integers that you formed.)

REVIEW AND PREVIEW

Solve. See Section 1.4.

61. Solve for x if B is the midpoint of $\overline{AC}$.

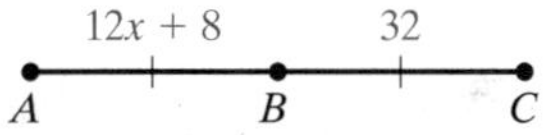

62. Use the value for x in Exercise **61** to find AB.

Tell whether each conjecture is true or false. Explain. See this section.

63. The sum of two even numbers is even.

64. The sum of three odd numbers is odd.

2.3 CONDITIONAL STATEMENTS

OBJECTIVES

1. Recognize Conditional Statements and Their Parts.
2. Write Converses, Inverses, and Contrapositives of Conditional Statements.

OBJECTIVE 1 ▶ Recognizing Conditional Statements. Now that we know about inductive reasoning, let's study a special type of statement.

A **conditional statement** is a statement that is written, or that can be written, in **"if-then" form.**

When a statement is in "if-then" form, the phrase that follows "if" is called the **hypothesis,** and the phrase that follows "then" is called the **conclusion.**

If a figure is a pentagon, then it has five sides.
(Hypothesis: a figure is a pentagon; Conclusion: it has five sides)

If it is snowing, then it is cloudy.
(Hypothesis: it is snowing; Conclusion: it is cloudy)

VOCABULARY

- conditional statement
- "if-then" form
- hypothesis
- conclusion
- negation
- converse
- inverse
- contrapositive
- equivalent statements

Symbols

When working with conditional statements, it is often handy to use shortcut notations. You may see any of the following:

Hypothesis		Conclusion
if p	then	q
p	implies	q
p	$\rightarrow$	q

Helpful Hint

Notice from the symbols above that we usually use the letter p to represent the hypothesis and the letter q to represent the conclusion.

EXAMPLE 1 Identifying the Hypothesis and the Conclusion

Identify the hypothesis (p) and the conclusion (q).

a. If an animal is a turtle, then the animal is a reptile.

b. If a number is even, then the number is not odd.

Solution

a. Hypothesis (p): an animal is a turtle
Conclusion (q): the animal is a reptile

b. Hypothesis (p): a number is even
Conclusion (q): the number is not odd

PRACTICE 1 Identify the hypothesis (p) and the conclusion (q).

a. If an animal is a pig, then the animal has 44 teeth.

b. If $x = 7$, then $x^2 = 49$.

Let's practice writing conditional statements.

EXAMPLE 2 Writing a Conditional Statement

A hypothesis (p) and a conclusion (q) are given. Use them to write a conditional statement, $p \rightarrow q$.

a. p: a figure is a square $\quad$ q: the figure is not a triangle

b. p: 9 is a perfect square $\quad$ q: 9 is not a prime number

Solution

a. If a figure is a square, then the figure is not a triangle.

b. If 9 is a perfect square, then 9 is not a prime number.

PRACTICE 2 A hypothesis (p) and a conclusion (q) are given. Use them to write a conditional statement, $p \rightarrow q$.

a. p: an angle measures 36° $\quad$ q: the angle is an acute angle

b. p: you live in Texas $\quad$ q: you live in the continental U.S.

> **Helpful Hint**
> It takes practice and reasoning skills to be able to recognize some conditional statements. Learn to take your time and be patient.

Some conditional statements may be written without using the words "if" and "then." For example, the Example 2, part **a** statement on the previous page about the square may be written as

A square is not a triangle.

EXAMPLE 3 **Writing a Conditional Statement**

Write the following statement in "if-then" form.
Acute angles measure less than 90°.

Solution If an angle is acute, then it measures less than 90°. □

PRACTICE 3 Write the following statement in "if-then" form.
Fractions are real numbers.

A conditional statement may be either *true* or *false*.

- A conditional statement is true if every time the hypothesis is true, then the conclusion is also true.
- A conditional statement is false if there is a *counterexample* in which the hypothesis is true, but the conclusion is false.

EXAMPLE 4 **Is a Conditional Statement True or False?**

Determine whether each conditional statement is true or false.

a. If an angle measures 92°, then it is an obtuse angle.

b. If a month begins with the letter J, then the month has 31 days.

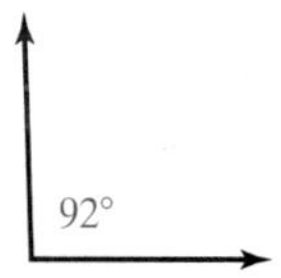

Solution

a. This conditional statement is true. All angles that measure 92° are obtuse angles.

b. This conditional statement is false. The month June starts with a J but has 30 days. □

PRACTICE 4 Determine whether each conditional statement is true or false.

a. If two angles are complementary, then they are adjacent angles.

b. If a month begins with the letter M, then the month has 31 days.

OBJECTIVE 2 ▸ Writing Converses, Inverses, and Contrapositives. Now that we know about conditional statements ($p \rightarrow q$), we will study a few related conditional statements. Before we do so, let's introduce the negation of a statement. The **negation** of a statement is formed by writing the negative of the statement. (*Note:* The notation for negation is **~**, so $\sim p$ is read "not *p*.")

Statement	Negation
The computer cover is red.	The computer cover is *not* red.
The rarest blood group for humans is group AB.	The rarest blood group for humans is *not* group AB.

Now let's use a given p statement for the hypothesis and q statement for the conclusion to write the conditional statement $p \rightarrow q$ and some related conditional statements.

Related Conditional Statements

Name	*How to Write It (in words)*	*How to Write It (in symbols)*	*Statement*	
Conditional	Given: p: $m\angle B = 160°$; q: $\angle B$ is obtuse	$p \rightarrow q$ (if p then q)	If $m\angle B = 160°$, then $\angle B$ is obtuse.	True
Converse	Switch the hypothesis and the conclusion.	$q \rightarrow p$ (if q then p)	If $\angle B$ is obtuse, then $m\angle B = 160°$.	False
Inverse	Negate the hypothesis and the conclusion of the conditional statement.	$\sim p \rightarrow \sim q$ (if not p then not q)	If $m\angle B \neq 160°$, then $\angle B$ is not obtuse.	False
Contrapositive	Negate the hypothesis and the conclusion of the converse statement.	$\sim q \rightarrow \sim p$ (if not q then not p)	If $\angle B$ is not obtuse, then $m\angle B \neq 160°$.	True

Helpful Hint

Remember which conditional statements are equivalent statements (both always true or both always false):

conditional and contrapositive statements ($p \rightarrow q$) and ($\sim q \rightarrow \sim p$)
converse and inverse statements ($q \rightarrow p$) and ($\sim p \rightarrow \sim q$)

If two statements are both always true or both always false, we call them **equivalent statements.**

- The conditional statement and the contrapositive statement in the table above are both true and are examples of equivalent statements.
- The converse statement and the inverse statement are both false and are examples of equivalent statements.

Why is this important? When writing proofs or using logic in general, we may always substitute one conditional statement for an equivalent conditional statement.

EXAMPLE 5 Writing Related Conditional Statements

Write the (**a**) converse, (**b**) inverse, and (**c**) contrapositive of the given conditional statement.

If it is raining, then it is cloudy.
(p: it is raining; q: it is cloudy)

Solution

a. Converse: If it is cloudy, then it is raining.
(q: it is cloudy; p: it is raining)

b. Inverse: If it is not raining, then it is not cloudy.
($\sim p$: it is not raining; $\sim q$: it is not cloudy)

c. Contrapositive: If it is not cloudy, then it is not raining.
($\sim q$: it is not cloudy; $\sim p$: it is not raining) □

PRACTICE

5 Write the (**a**) converse, (**b**) inverse, and (**c**) contrapositive of the given conditional statement.

If the figure is a triangle, then the figure has three sides.
(p: the figure is a triangle; q: the figure has three sides)

VOCABULARY & READINESS CHECK

Word Bank *Use the choices to fill in each blank. Some choices may be used more than once.*

equivalent	inverse	contrapositive	hypothesis
negation	converse	conclusion	conditional
$p \rightarrow q$	$\sim p$		

1. In symbols, an "if-then" statement can be written as ____________.

2. In symbols, the negation of p is written as ____________.

3. For an "if-then" statement, the part following "then" is called the ____________.

4. For an "if-then" statement, the part following "if" is called the ____________.

For a conditional statement $p \to q$, complete the table.

	Symbols	Statement Name
	$p \to q$	conditional
5.	$\sim p \to \sim q$	____________
6.	$\sim q \to \sim p$	____________
7.	$q \to p$	____________
8.	$\sim p$	____________

9. Two statements that are both always true or both always false are called ____________ statements.

10. Conditional and contrapositive statements are ____________ statements.

2.3 EXERCISE SET MyMathLab®

Identify the hypothesis and conclusion of each conditional. See Example 1.

1. If you are an American citizen, then you have the right to vote.

2. If you want to be healthy, then you should eat vegetables.

3. If a figure is a rectangle, then it has four sides.

4. If a figure is a pentagon, then it has five sides.

Use the given hypothesis (p) and conclusion (q) to write a conditional statement. See Example 2.

5. p: the animal is an alligator
q: the animal is a reptile

6. p: the animal is a frog
q: the animal is an amphibian

7. p: $x = 9$
q: $x^2 = 81$

8. p: $x = 9$
q: $\sqrt{x} = 3$

Write each statement in "if-then" form. See Example 3.

9. Straight angles measure 180°.

10. Right angles measure 90°.

11. Pianists are musicians.

12. Whales are mammals.

13. The equation $3x - 7 = 14$ implies that $3x = 21$.

14. The equation $2x + 3 = 11$ implies that $2x = 8$.

15. A line segment measures 1 foot if the segment's measure is 12 inches.

16. A side of a square measures 2 yards if the side measures 6 feet.

Determine if the conditional is true or false. If it is false, find a counterexample. See Example 4.

17. If you live in a country that borders the United States, then you live in Canada.

18. If you play a sport with a ball and a bat, then you play baseball.

19. If an angle measures 80°, then it is acute.

20. If a polygon has eight sides, then it is an octagon.

Multiple Choice *Choose the negation of each statement. See Example 5.*

21. The car is blue.

a. The car is red.
b. There is no car.
c. The car is not red.
d. The car is not blue.

22. A rectangle has 3 sides.

a. A rectangle has no sides.
b. A rectangle has 4 sides.
c. A rectangle does not have 3 sides.
d. A triangle has 3 sides.

23. An acute angle measures 108°.

a. An acute angle measures $< 90°$.
b. An acute angle does not measure 108°.
c. An obtuse angle measures 108°.
d. An obtuse angle does not measure 108°.

24. The avacado is soft.

a. The avacado is not soft.
b. The avacado is green.
c. The avacado is stiff.
d. The avacado is ripe.

Write the converse, inverse, and contrapositive of the given conditional statement. See Example 5.

25. If you are a quarterback, then you play football.

26. If you are a goalkeeper, then you play soccer.

27. If $4x + 8 = 28$, then $x = 5$.

28. If $5x - 3 = 32$, then $x = 7$.

Complete the Table *A true conditional statement ($p \rightarrow q$) is given. Complete the table by identifying each given statement as converse ($q \rightarrow p$), inverse ($\sim p \rightarrow \sim q$), contrapositive ($\sim q \rightarrow \sim p$), or none of these. See Example 5.*

	If m∠C = 14°, then ∠C is acute.	***Conditional***
29.	If $\angle C$ is not acute, then $m\angle C \neq 14°$.	
30.	If $m\angle C = 14°$, then $\angle C$ is not acute.	
31.	If $m\angle C \neq 14°$, then $\angle C$ is not acute.	
32.	If $\angle C$ is acute, then $m\angle C = 14°$.	

	If a figure is a hexagon, then it has 6 sides.	***Conditional***
33.	If a figure is not a hexagon, then it has 6 sides.	
34.	If a figure is not a hexagon, then it does not have 6 sides.	
35.	If a figure has 6 sides, then it is a hexagon.	
36.	If a figure does not have 6 sides, then it is not a hexagon.	

Recall that a conditional statement and its related contrapositive statement are equivalent statements. Write the related contrapositive statement for each true conditional statement. In your own words, explain why the contrapositive statement is also true. See Examples 4 and 5.

37. If a number is odd, then it is not divisible by 2.

38. If a number is even, then it is divisible by 2.

39. If I live in Canada, then I live in North America.

40. If I live in Brazil, then I live in South America.

41. If $m\angle C = 105°$, then $\angle C$ is obtuse.

42. If $m\angle B = 15°$, then $\angle B$ is acute.

Write each postulate as a conditional statement. See Example 3.

43. Two distinct intersecting lines meet in exactly one point.

44. Through any two points there is exactly one line.

CONCEPT EXTENSIONS

If the given statement is not in "if-then" form, rewrite it. Write the converse, inverse, and contrapositive of the given conditional statement. Determine if each statement is true or false. If a statement is false, give a counterexample.

45. Odd natural numbers less than 8 are prime.

46. Two lines that lie in the same plane are coplanar.

Write the converse of each statement. If the converse is true, write "true." If it is not true, provide a counterexample.

47. If $x = -6$, then $|x| = 6$.

48. If y is negative, then $-y$ is positive.

49. If $x < 0$, then $x^3 < 0$.

50. If $x < 0$, then $x^2 > 0$.

For Exercises 51–54, write each statement as a conditional.

51. An event with probability 1 is certain to occur.

52. An event with probability 0 is certain not to occur.

53. "We're half the people; we should be half the Congress." —Jeanette Rankin, former U.S. congresswoman, calling for more women in office

54. "Anyone who has never made a mistake has never tried anything new." —Albert Einstein

55. **Find the Error** A given conditional is true. Natalie claims its contrapositive is also true. Sean claims its contrapositive is false. Who is correct and how do you know?

56. Write a true conditional that has a true converse, and write a true conditional that has a false converse.

Your classmate claims that the conditional and contrapositive of the following statement are both true. If $x = 2$, then $x^2 = 4$.

57. Is he correct? Explain.

58. Can you find a counterexample of the conditional?

Advertisements often suggest conditional statements. What conditional does each ad imply?

59. IHOP: Come hungry. Leave happy.

60. Subway, eat fresh.

61. Trix Are for Kids!

62. Burger King: Have it your way

Let a *represent an integer. Consider the three statements* r, t, *and* u.
r: a is even. t: $2a$ is even. u: $2a$ is odd.

63. How many conditional statements (true or false) of the form $p \rightarrow q$ can you make from these statements?

64. Decide which statements from Exercise **63** are true, and provide a counterexample if they are false.

REVIEW AND PREVIEW

Find a counterexample to show that each statement is false. See Section 2.2.

65. You can connect any four points to form a rectangle.

66. The square of a number is always greater than the number.

Find the perimeter of each rectangle with the given width and length. See Section 2.1.

67. 6 in., 12 in.

68. 3.5 cm, 7 cm

69. $1\frac{3}{4}$ yd, 18 in.

70. 11 yd, 60 ft

2.4 BICONDITIONAL STATEMENTS AND DEFINITIONS

OBJECTIVES

1 Write and Understand Biconditional Statements.

2 Identify and Understand Good Definitions.

VOCABULARY

- biconditional statement
- good definition

OBJECTIVE 1 ▶ Writing Biconditional Statements. Recall from the previous section that

- A *conditional statement* $(p \rightarrow q)$ may be true or false.
- The *converse* of a *conditional statement* is written by switching the hypothesis and the conclusion $(q \rightarrow p)$.
- The *converse* of a *conditional statement* may be true or false.

A **biconditional statement** is equivalent to writing a conditional statement and its converse. For a biconditional statement, we use the phrase "if and only if."

EXAMPLE 1 **Rewriting a Biconditional Statement**

Write the given biconditional statement as a conditional statement and its converse.

Two angles are supplementary *if and only if* the sum of their measures is 180°.

Solution

Conditional Statement: If two angles are supplementary, then the sum of their measures is 180°.

Converse: If the sum of the measures of two angles is 180°, then the angles are supplementary. □

PRACTICE 1 Write the biconditional statement as a conditional statement and its converse.

Two angles are complementary *if and only if* the sum of their measures is 90°.

We will use the notation $\leftrightarrow$ for "if and only if." Thus,

"$p \leftrightarrow q$" is read as "p if and only if q."

Notice from Example 1 that a biconditional statement combines a conditional statement $(p \rightarrow q)$ and its converse $(q \rightarrow p)$. The result is the biconditional $(p \leftrightarrow q)$.

Below, we show an example of combining $p \rightarrow q$ and $q \rightarrow p$ to form $p \leftrightarrow q$.

Statement Name	*Notation*	*Example*
Conditional	$p \rightarrow q$	If $2x = 10$, then $x = 5$.
Converse	$q \rightarrow p$	If $x = 5$, then $2x = 10$.
Biconditional	$p \leftrightarrow q$	$2x = 10$ if and only if $x = 5$.

Just as with any other statement, a biconditional statement may be true or false. A biconditional statement is true if both the conditional statement and its converse are true.

▶ Helpful Hint Don't forget when each statement is true or false.

	True	False
Conditional Statement $(p \rightarrow q)$	If p true, then q true.	If p true, but q false. (Find counterexample.)
Biconditional Statement $(p \leftrightarrow q)$	Both $p \rightarrow q$ true and $q \rightarrow p$ true.	

EXAMPLE 2 **Writing a True Biconditional Statement**

A true conditional statement is given:
If a closed figure is a triangle, then it has three sides.

a. Write the converse of the conditional statement.

b. Decide whether the converse statement is true or false.

c. If the converse statement is true, write a true biconditional statement. If the converse statement is false, give a counterexample.

Solution

a. Converse: If a closed figure has 3 sides, then it is a triangle.

b. The converse statement is true, as the sketch to the right reminds us.

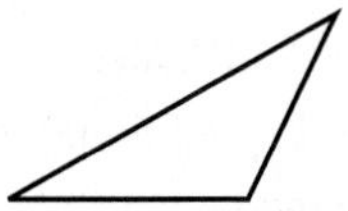

A triangle.

c. Biconditional: A closed figure is a triangle if and only if it has 3 sides. □

PRACTICE

2 Use the same directions as for Example 2.
True Conditional Statement:
If point B is between points A and C, then $AB + BC = AC$.

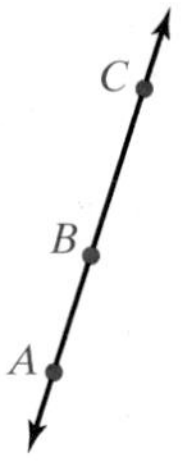

EXAMPLE 3 **Writing a True Biconditional Statement**

Use the same directions as for Example 2.
True Conditional Statement: If $x = 5$, then $x^2 = 25$.

Solution

a. Converse: If $x^2 = 25$, then $x = 5$.

b. The converse statement is false.

c. For a counterexample, notice that if $x^2 = 25$, then x may also be -5, since $(-5)^2 = 25$. □

PRACTICE

3 Use the same directions as for Example 2.
True Conditional Statement: If $x = 7$, then $|x| = 7$.

OBJECTIVE 2 ▶ Identifying and Understanding Definitions. We are studying biconditional statements to help us identify and understand definitions. Identifying and understanding definitions helps us correctly use statements in proving theorems and in basic reasoning in general. Before we continue, make sure you understand the following.

- A **good definition** can be written as a true biconditional. In other words, the conditional $(p \rightarrow q)$ is true and the converse $(q \rightarrow p)$ is true.
- Also, if you have a true biconditional statement, then it can be used "forward" (the conditional) or "backward" (the converse) to help us with logical reasoning or to verify a part of a proof.

One way to show that a statement is *not* a good definition is to find a counterexample.

EXAMPLE 4 Identifying Good Definitions

Multiple Choice Which of the following is a good definition?

a. A fish is an animal that swims.
b. Rectangles have four corners.
c. Giraffes are animals with very long necks.
d. A penny is a coin worth one cent.

Solution

Choice **a** is not reversible. A whale is a counterexample. A whale is an animal that swims, but it is a mammal, not a fish.

Choice **b:** "Corners" is not clearly defined. Thus, a corner does not necessarily measure 90°, as needed for a figure to be a rectangle.

Choice **c:** "Very long" is not precise. Also, Choice **c** is not reversible because ostriches also have very long necks.

Choice **d** is a good definition. It is reversible, and all of the terms in the definition are clearly defined and precise.

The answer is **d**. □

PRACTICE 4 **Multiple Choice** Which of the following is a good definition?

a. A square is a figure with four right angles.
b. January is a month that begins with the letter J.
c. A minute is a measure of time that equals 60 seconds.
d. The month of February has 29 days.

Can you see how important it is to carefully check your definitions? For example, let's look at a definition that is not precise at all.

Not a precise definition: A black widow spider is an arachnid with 8 legs.

Can you see that this definition will not help you identify a black widow spider? If you check, the converse is false. (Converse: An arachnid with 8 legs is a black widow spider. A counterexample is any spider that is not a black widow.) Thus, this is not a good or precise definition.

EXAMPLE 5 Writing a Definition as a True Biconditional

Determine whether or not the following definition is a good one. To do so, attempt to write it as a true biconditional statement.

Definition: A right angle is an angle that measures 90°.

Solution Let's see if we can write this definition as a true biconditional.

Is the conditional true? If an angle is a right angle, then it measures 90°. TRUE

Is the converse true? If an angle measures 90°, then it is a right angle. TRUE

Since the conditional and the converse are true, we can write a true biconditional. } An angle is a right angle if and only if it measures 90°.

This means that the original definition is a good definition and can be used both "forward" and "backward." □

PRACTICE 5 Determine whether or not the following definition is a good one. To do so, attempt to write it as a true biconditional statement.

Definition: A yard measures 36 inches.

VOCABULARY & READINESS CHECK

Word Bank *Use the choices to fill in each blank. Some choices may not be used, and some may be used more than once.*

$p \leftrightarrow q$	$\sim p \rightarrow \sim q$	if-then	conditional	$\leftrightarrow$
$q \rightarrow p$	$\sim q \rightarrow \sim p$	if and only if	biconditional	$p \rightarrow q$

1. A(n) ____________ statement is written using a conditional statement and its converse.
2. A biconditional statement may be written using the phrase ____________.
3. The converse of the statement $p \rightarrow q$ is ____________.
4. The symbol for "if and only if" is ____________.
5. A good definition may be rewritten as a(n) ____________ statement.
6. A true biconditional statement $p \leftrightarrow q$ may be used "forward" and "backward." In symbols, forward and backward mean, respectively, ____________ and ____________.

2.4 EXERCISE SET MyMathLab®

Write the two conditional statements that form each biconditional. See Example 1.

1. A line bisects a segment if and only if the line intersects the segment only at its midpoint.
2. A ray bisects a segment if and only if the ray intersects the segment only at its midpoint.
3. An integer is divisible by 100 if and only if its last two digits are zeros. *Hint:* Recall that the integers are $\{\ldots -3, -2, -1, 0, 1, 2, 3, \ldots\}$.
4. An integer is divisible by 5 if and only if its last digit is 0 or 5. See the hint for Exercise **3**.
5. You live in Washington, D.C., if and only if you live in the capital of the United States.
6. You live in Baton Rouge, Louisiana, if and only if you live in the capital of the state of Louisiana.
7. $x^2 = 144$ if and only if $x = 12$ or $x = -12$.
8. $|x| = 13$ if and only if $x = 13$ or $x = -13$.

Multiple Steps *For Exercises 9–18, each conditional statement is true.*

a. Write its converse.
b. Determine whether the converse statement is true or false.
c. If the converse is also true, combine the statements as a biconditional. If the converse is false, give a counterexample. See Examples 2 and 3.

9. If two segments have the same length, then they are congruent.
10. If two angles have the same degree measure, then they are congruent.
11. If $x = 12$, then $2x - 5 = 19$.
12. If $x = 0$, then $8x + 14 = 14$.
13. If a number is divisible by 20, then it is an even number.
14. If a number is divisible by 6, then it is a multiple of 3.
15. In the United States, if it is July 4, then it is Independence Day.
16. In the United States, if it is the 4th Thursday in November, then it is Thanksgiving.
17. If $x = 3$, then $|x| = 3$.
18. If $x = -5$, then $|x| = 5$.

Multiple Steps *For Exercises 19–22, perform the following parts. See Examples 2 and 3.*

a. Write each statement as a conditional statement.
b. Determine whether the conditional statement is true or false.
c. If part b is true, write the converse of the conditional. (If part b is false, give a counterexample for the conditional and do not continue to part d and part e.)
d. Determine whether the converse statement is true or false.
e. If part d is true, write a true biconditional statement. If part d is false, give a counterexample for the converse.

19. Two angles that form a linear pair are adjacent.
20. Two angles that are vertical angles are congruent.

21. An acute angle, A, measures $0° < m\angle A < 90°$.

22. An obtuse angle, B, measures $90° < m\angle B < 180°$.

Is each statement below a good definition? If so, rewrite it as a true biconditional statement. If not, explain. See Examples 4 and 5.

23. A cat is an animal with whiskers.

24. A dolphin is a mammal.

25. A segment is part of a line.

26. A compass is a geometric tool.

27. Collinear points are points that lie on the same line.

28. Perpendicular lines are two lines that intersect to form right angles.

Multiple Steps *For Exercises 29 and 30, which conditional and its converse form a true biconditional? See Examples 2 and 3.*

29. **a.** If $x > 0$, then $|x| > 0$.

b. If $x = 3$, then $x^2 = 9$.

c. If $x^3 = 5$, then $x = 125$.

d. If $x = 19$, then $2x - 3 = 35$.

30. **a.** If $x = -3$, then $x^2 = 9$.

b. If $x < 0$, then $|x| > 0$.

c. If $x = 10$, then $5x + 5 = 55$.

d. If $x = -2$, then $|x| = 2$.

Mixed Practice *Write each statement as a biconditional.*

31. Points in Quadrant III have two negative coordinates.

32. Points in Quadrant I have two positive coordinates.

33. A hexagon is a six-sided polygon.

34. A triangle is a three-sided polygon.

35. When the sum of the digits of an integer is divisible by 9, the integer is divisible by 9 and vice versa.

36. When the last digit of an integer is even, the number is divisible by 2 and vice versa.

CONCEPT EXTENSIONS

*For Exercises 37 and 38, is the following a good definition? Explain by answering parts **a** and **b**.*

a. *Can you write the statement as two true conditionals?*

b. *Are the two true conditionals converses of each other?*

37. A ligament is a band of tough tissue connecting bones or holding organs in place.

38. An obtuse angle is an angle with measure greater than 90°.

39. **Find the Error** Your friend defines a right angle as an angle with a greater measure than an acute angle. Use a biconditional to show that this is not a good definition.

40. **Find the Error** Why is the following statement a poor definition?

Elephants are gigantic animals.

41. Explain how the term *biconditional* is fitting for a statement composed of *two* conditionals.

42. Which of the following statements is a better definition of a linear pair? Explain.

A linear pair is a pair of supplementary angles.

A linear pair is a pair of adjacent angles with noncommon sides that are opposite rays.

Let statements p, q, r, and s be as follows:

p: $\angle A$ and $\angle B$ are a linear pair.

q: $\angle A$ and $\angle B$ are supplementary angles.

r: $\angle A$ and $\angle B$ are adjacent angles.

s: $\angle A$ and $\angle B$ are adjacent and supplementary angles.

Substitute for p, q, r, and s, and write each statement the way you would read it.

43. $p \rightarrow q$

44. $p \rightarrow r$

45. $p \rightarrow s$

46. $p \leftrightarrow s$

For Exercises 47–50, use the chart below. Decide whether the description of each letter is a good definition. If not, provide a counterexample by giving another letter that could fit the description.

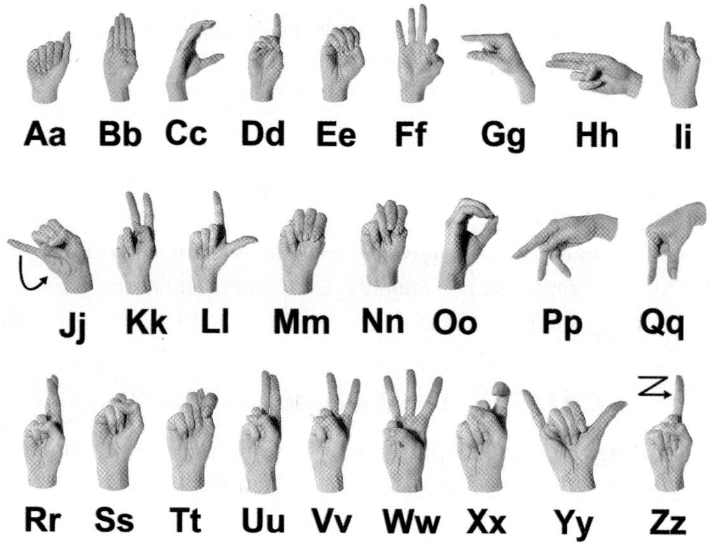

47. The letter D is formed by pointing straight up with the finger beside the thumb and folding the other fingers and the thumb so that they all touch.

48. The letter K is formed by making a V with the two fingers beside the thumb.

49. You have formed the letter I if and only if the smallest finger is sticking up and the other fingers are folded into the palm of your hand with your thumb folded over them and your hand is held still.

50. You form the letter B by holding all four fingers tightly together and pointing them straight up while your thumb is folded into the palm of your hand.

REVIEW AND PREVIEW

Solve. For Exercises 51 and 52, write the converse of each statement. See Section 2.3.

51. If you do not sleep enough, then your grades suffer.

52. If you are in the school chorus, then you have a good voice.

Multiple Steps *For Exercises 53 and 54, answer*

a. *true,* ***b.*** *false, or* ***c.*** *sometimes true, sometimes false.*

53. If a conditional is true, its contrapositive is ____________.

54. If a conditional is true, its converse is ____________.

What are the next two terms in each list? See Section 2.2.

55. 100, 90, 80, 70, …

56. 2500, 500, 100, 20, …

57. 1, 2, 0, 3, −1, …

58. 1, 3, 2, 4, 3, …

2.5 DEDUCTIVE REASONING

OBJECTIVES

1 Review Conditional, Converse, Inverse, and Contrapositive Statements.

2 Understand and Use Two Laws of Deductive Reasoning: the Law of Detachment and the Law of Syllogism.

VOCABULARY

- deductive reasoning
- Law of Detachment
- Law of Syllogism

OBJECTIVE 1 ▶ Reviewing Related Conditional Statements. Before we introduce another type of reasoning, deductive reasoning, let's review three statements that can be formed from a conditional statement. We will also review the symbols we use.

Related Conditional Statements

Name	*How to Write It (in words)*	*How to Write It (in symbols)*	*How to Say It (in words)*
Conditional	Given: hypothesis p: it is raining conclusion q: it is cloudy	$p \rightarrow q$ (if p then q)	If it is raining, then it is cloudy.
Converse	Switch the hypothesis and the conclusion.	$q \rightarrow p$ (if q then p)	If it is cloudy, then it is raining.
Inverse	Negate the hypothesis and the conclusion of the conditional statement.	$\sim p \rightarrow \sim q$ (if not p then not q)	If it is not raining, then it is not cloudy.
Contrapositive	Negate the hypothesis and the conclusion of the converse statement.	$\sim q \rightarrow \sim p$ (if not q then not p)	If it is not cloudy, then it is not raining.

▶ Helpful Hint

Don't forget the equivalent statements (both always true or both always false):

Equivalent Statements: Conditional and Contrapositive Statements

Equivalent Statements: Converse and Inverse Statements

EXAMPLE 1 **Writing Related Conditional Statements**

Use the given conditional statement to write its **(a)** converse, **(b)** inverse, and **(c)** contrapositive statements. Then write each statement in symbols, as done for the given statement:

If a number is a whole number, then it is an integer. ($p \rightarrow q$)

Solution

a. Converse: If a number is an integer, then it is a whole number. $(q \rightarrow p)$

b. Inverse: If a number is not a whole number, then it is not an integer. $(\sim p \rightarrow \sim q)$

c. Contrapositive: If a number is not an integer, then it is not a whole number. $(\sim q \rightarrow \sim p)$ □

PRACTICE

1 Use the given conditional statement to write its (a) converse, (b) inverse, and (c) contrapositive statements. Then write each statement in symbols.

If a number is an integer, then it is a real number.

OBJECTIVE 2 ▶ Using Two Laws of Deductive Reasoning: The Law of Detachment and the Law of Syllogism. Recall that for inductive reasoning, we can never be sure that our conclusion (conjecture) is true because we base our conclusion on observing a few specific examples. This is not true for deductive reasoning. A conclusion based on deductive reasoning is always true.

Deductive Reasoning

Deductive reasoning is the process of proving a specific conclusion from one or more general statements. A conclusion that is proved true by deductive reasoning is called a theorem.

Let's try to understand the difference between inductive and deductive reasoning.

Inductive Reasoning—Observing patterns or examples, and then using them to form a general conclusion that may or may not be true.

Deductive Reasoning—Using known information to write a logical argument that arrives at a specific conclusion that is true. (This is how we prove theorems.)

We will study two laws of deductive reasoning. They are the **Law of Detachment** and the Law of Syllogism.

Law of Detachment

If $p \rightarrow q$ is a true conditional statement and p is true, then q is true.

How do we use this law? We first identify the hypothesis (p) and the conclusion (q) of a true conditional statement. Then, if we are given the same true hypothesis (p), we use this Law of Detachment to conclude that the same conclusion (q) is true. For example:

Given: If it is raining, then it is cloudy. ($p \rightarrow q$ is true.)
It is raining. (same true p)

Conclusion: It is cloudy. (q is then true.)

EXAMPLE 2 Using the Law of Detachment

Use the Law of Detachment to make a true conclusion. Assume that the first statement $p \rightarrow q$ is true.

$\underbrace{\text{If you shop at Save Market,}}_{p}$ $\underbrace{\text{you will save money.}}_{q}$

$\underbrace{\text{You shop at Save Market.}}_{p}$

Solution **Conclusion:** $\underbrace{\text{You save money.}}_{q}$ □

PRACTICE 2 Use the Law of Detachment to make a true conclusion. (Assume that the first statement $p \rightarrow q$ is true.)

If you live and drive legally in California, you have a license plate on the front and the back of your car.

You live and drive legally in California.

EXAMPLE 3 Using the Law of Detachment

Determine whether each reasoning is valid using the Law of Detachment. (Assume that the first statement $p \rightarrow q$ is true.)

a. If any student gets an A on a final exam, then the student will pass the course.
Leo got an A on his final exam, so he will pass the course.

b. If two angles are adjacent, then they share a common vertex.
$\angle 1$ and $\angle 2$ share a common vertex, so the angles are adjacent.

Solution

a. In symbols, the first sentence is a true conditional ($p \rightarrow q$). The first part of the second sentence is a true hypothesis (p); thus q, the second part of the second sentence, is true.

This argument is valid.

b. In symbols, the first sentence is a true conditional ($p \rightarrow q$). The first part of the second sentence is a true conclusion (q), *not* a true hypothesis (p).

Thus, this argument is not valid. □

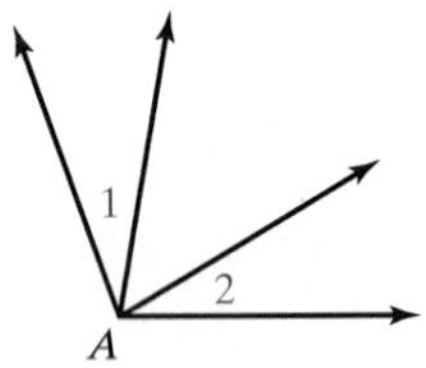

Angles 1 and 2 share a common vertex but are not adjacent.

PRACTICE 3 Determine whether each reasoning is valid using the Law of Detachment.

a. If there is lightning, then it is not safe to be out in the open. Marla sees lightning from the soccer field. She concludes it is not safe to be out on the field.

b. If a figure is a square, then its sides have equal length. Figure $ABCD$ has sides of equal length, so the figure is a square.

Another law of deductive reasoning is the **Law of Syllogism.**

Law of Syllogism
If $p \rightarrow q$ is true and $q \rightarrow r$ is true, then $p \rightarrow r$ is a true conditional statement.

EXAMPLE 4 **Using the Law of Syllogism**

Use the Law of Syllogism to form a true conclusion ($p \rightarrow r$).

Given: If a figure is a square, then the figure is a rectangle.
If a figure is a rectangle, then the figure has four sides.

Solution

Given: If $\underbrace{\text{a figure is a square}}_{p}$, then $\underbrace{\text{the figure is a rectangle}}_{q}$. $p \rightarrow q$ is true.

If $\underbrace{\text{a figure is a rectangle}}_{q}$, then $\underbrace{\text{the figure has four sides}}_{r}$. $q \rightarrow r$ is true.

Conclusion: $p \rightarrow r$ Thus, $p \rightarrow r$ is true.
If a figure is a square, then the figure has four sides. $p \rightarrow r$ in words □

PRACTICE 4 If a natural number ends in 0, then it is divisible by 10.
If a natural number is divisible by 10, then it is divisible by 5.

EXAMPLE 5 **Using the Laws of Deductive Reasoning**

Decide what you can conclude from the true conditional statements given, and note whether your reasoning involves the Law of Detachment or the Law of Syllogism.

a. Given: All elephants are mammals.
All mammals have hair.

b. Given: If a number is prime, then it has exactly two factors.
The number 11 is prime.

Solution

a. Here, the given statements are in the form

$p \rightarrow q$ (given to be true)

$q \rightarrow r$ (given to be true)

Thus, we conclude: All elephants have hair. ($p \rightarrow r$ is then true.) The Law of Syllogism is used.

b. The given statements are in the form

$p \rightarrow q$ (given to be true)

p (true)

Thus, we conclude: The number 11 has exactly two factors. (q is then true.)
The Law of Detachment is used. □

PRACTICE 5 Decide what you can conclude from the true conditional statements given, and note whether your reasoning involves the Law of Detachment or the Law of Syllogism.

a. Given: All triangles have three sides.
The figure is a triangle.

b. Given: All squares are rectangles.
If the figure is a rectangle, then the diagonals bisect each other.

VOCABULARY & READINESS CHECK

Word Bank *Use the choices to fill in each blank.*

inductive	Law of Syllogism
deductive	Law of Detachment

1. The ____________ says that if $p \to q$ is true and p is true, then q is true.
2. The ____________ says that if $p \to q$ is true and $q \to r$ is true, then $p \to r$ is true.
3. ____________ reasoning is used to prove theorems.
4. With ____________ reasoning, your conclusion may or may not be true.

2.5 EXERCISE SET MyMathLab®

Given p and q, write each statement. See Example 1.

p: $m\angle A = m\angle B$
q: $\angle A$ is congruent to $\angle B$

1. $\sim p$ **2.** $\sim q$ **3.** $p \to q$ **4.** $q \to p$
5. $\sim p \to q$ **6.** $p \to \sim q$ **7.** $\sim p \to \sim q$ **8.** $\sim q \to \sim p$

Some of the statements above have special names. Give the name for each exercise given below. (Here, $p \to q$ is our original conditional statement.) See Example 1.

9. Exercise **3** **10.** Exercise **4**
11. Exercise **7** **12.** Exercise **8**

Use symbols to write the statement that is equivalent to each statement below. See Example 1.

13. $p \to q$ **14.** $q \to p$

Use the hypothesis (p) and conclusion (q) given to write each conditional statement. See Example 1.

p: a figure has seven sides
q: the figure is a heptagon

A Heptagon

15. converse **16.** conditional
17. inverse **18.** contrapositive

If possible, use the Law of Detachment to make a conclusion. If it is not possible to make a conclusion, tell why. (Assume that the first statement $p \to q$ is true.) See Examples 2 and 3.

19. If a doctor suspects her patient has a broken bone, then she should take an X-ray.
Dr. Ngemba suspects Lilly has a broken arm.

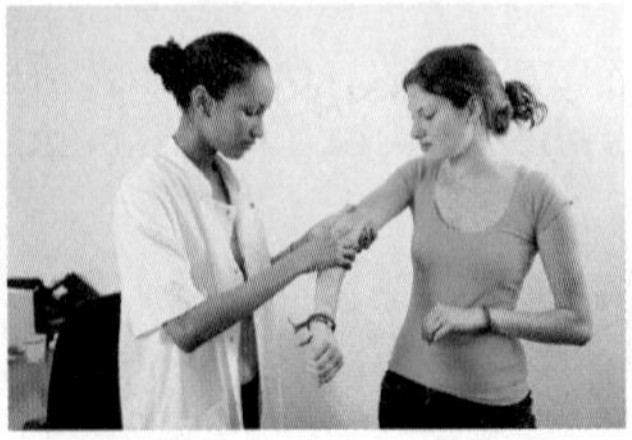

20. If a student wants to go to college, then the student must study hard.
Rashid wants to go to Portland State University.

21. If a rectangle has side lengths 3 cm and 4 cm, then it has area 12 cm^2.
Rectangle $ABCD$ has area 12 cm^2.

22. If an angle is obtuse, then it is not acute.
$\angle XYZ$ is not obtuse.

23. If three points are on the same line, then they are collinear.
Points X, Y, and Z are on line m.

24. If three points are on the same plane, then they are coplanar.
Points X, Y, and Z are on plane P.

If possible, use the Law of Syllogism to make a conclusion. If it is not possible to make a conclusion, tell why. Assume these statements are true. See Example 4.

25. If an animal is a Florida panther, then its scientific name is *Puma concolor coryi*.
If an animal is a *Puma concolor coryi*, then it is endangered.

26. If a line intersects a segment at its midpoint, then the line bisects the segment.
If a line bisects a segment, then it divides the segment into two congruent segments.

27. If a whole number ends in 6, then it is divisible by 2.
If a whole number ends in 4, then it is divisible by 2.

28. If you improve your vocabulary, then you will improve your score on a standardized test.
If you read often, then you will improve your vocabulary.

Multiple Steps *For Exercises 29–32, complete the following parts. See Example 5.*

a. *Make conclusions from the following statements.*

b. *Note which law you used to make a conclusion, the Law of Detachment or the Law of Syllogism.*

29. If you live in the Bronx, then you live in New York.
Tracy lives in the Bronx.

30. If you ride a bicycle, then you are exercising.
You ride a bicycle.

31. If you are studying botany, then you are studying biology.
If you are studying biology, then you are studying a science.

32. If an Alaskan mountain is more than 20,300 ft high, then it is the highest in Alaska.
If a mountain is the highest in Alaska, then it is the highest in the United States.

CONCEPT EXTENSIONS

33. **Complete the Table** Consider the following procedure:

Select a number. Multiply the number by 6. Add 8 to the product. Divide this sum by 2. Subtract 4 from the quotient.

a. Inductive Reasoning: Complete the chart to follow this procedure for four different numbers. The first number has been done for you. Then write a conjecture that relates the result of this process to the original number selected.

Select a number.	4	7	11	100
Multiply the number by 6.	$4 \times 6 = 24$			
Add 8 to the product.	$24 + 8 = 32$			
Divide this sum by 2.	$\frac{32}{2} = 16$			
Subtract 4 from the quotient.	$16 - 4 = 12$			

b. Deductive Reasoning: Now go through the same steps, but represent the original number by the variable n and use deductive reasoning to prove the conjecture in part **a**.

34. Answer the same parts as for Exercise **33**, but with the following procedure: Choose an integer. Multiply the integer by 3. Add 6 to the product. Divide the sum by 3.

Write the first statement as a conditional. If possible, use the Law of Detachment to make a conclusion. If it is not possible to make a conclusion, tell why. (Assume that the first statement $p \rightarrow q$ is true.)

35. All national parks are interesting.
Mammoth Cave is a national park.

36. All squares are rectangles.
ABCD is a square.

37. The temperature is always above 32°F in Key West, Florida.
The temperature is 62°F.

38. Every high school student likes art.
Ling likes art.

Kira Julie Curtis Maria

For Exercises 39–44, assume that the following statements are true.

a. If Maria is drinking juice, then it is breakfast time.

b. If it is lunchtime, then Kira is drinking milk and nothing else.

c. If it is mealtime, then Curtis is drinking water and nothing else.

d. If it is breakfast time, then Julie is drinking juice and nothing else.

e. Maria is drinking juice.

Multiple Choice *Use only the information given above. For each statement, write "must be true," "may be true," or "is not true." Explain your reasoning.*

39. Julie is drinking juice.
40. Curtis is drinking water.
41. Kira is drinking milk.
42. Curtis is drinking juice.
43. Maria is drinking water.
44. Julie is drinking milk.
45. Give an example of a rule used in your school that could be written as a conditional. Explain how the Law of Detachment is used in applying that rule.
46. Consider the following given statements and conclusion.

Given: If an animal is a fish, then it has gills.
A turtle does not have gills.

You conclude: A turtle is not a fish.

a. Make a Venn diagram to illustrate the given information.

b. Use the Venn diagram to help explain why the argument uses good reasoning.

REVIEW AND PREVIEW

Use the figure at the right. See Sections 1.5 and 1.6.

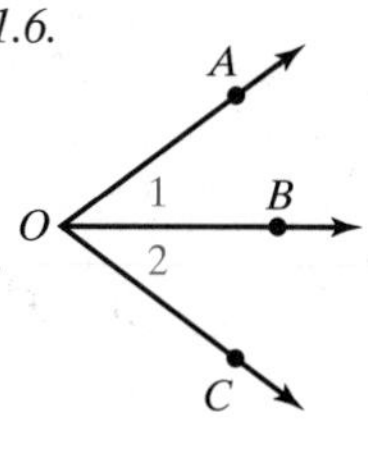

47. Name ∠1 in two other ways.
48. Name ∠2 in two other ways.
49. If $\angle 1 \cong \angle 2$, name the bisector of $\angle AOC$.
50. If $m\angle 1 = 43°$ and $m\angle 2 = 43°$, classify $\angle AOC$ as acute, right, or obtuse.

2.6 REVIEWING PROPERTIES OF EQUALITY AND WRITING TWO-COLUMN PROOFS

OBJECTIVES

1 Use Properties of Equality to Justify Reasons for Steps.

2 Write a Two-Column Proof.

VOCABULARY

- Reflexive Property
- Symmetric Property
- Transitive Property
- Substitution Property
- proof
- two-column proof

OBJECTIVE 1 ▶ Justifying Reasons for Steps. As we prepare to prove geometric theorems, let's review some properties that we learned in algebra. We used these properties often as we simplified expressions and solved equations. These same properties are also needed to prove geometric theorems.

Algebra Properties of Equality

Let a, b, and c be any real numbers.

Addition Property	If $a = b$, then $a + c = b + c$.
Subtraction Property	If $a = b$, then $a - c = b - c$.
Multiplication Property	If $a = b$, then $a \cdot c = b \cdot c$.
Division Property	If $a = b$ and $c \neq 0$, then $\frac{a}{c} = \frac{b}{c}$.
Reflexive Property	$a = a$
Symmetric Property	If $a = b$, then $b = a$.
Transitive Property	If $a = b$ and $b = c$, then $a = c$.
Substitution Property	If $a = b$, then b can replace a in any expression.

Also, don't forget the Distributive Property.

The Distributive Property

Use multiplication to distribute a to each term of the sum or difference within the parentheses.

$a(b + c) = ab + ac$ $\qquad$ $a(b - c) = ab - ac$

We use this property to multiply, but we also use the Distributive Property to combine like terms. To minimize any confusion, see the examples of possible justifications below.

$5(2x - 7)$	Multiply	$7x + 12x$	Combine Like Terms
$= 10x - 35$	or	$= 19x$	or
	Distributive Property		Simplify

EXAMPLE 1 **Giving Reasons for Statements**

Solve $5x + 12 = 47$. Give a reason to justify each statement.

Solution

Statements	Reasons
$5x + 12 = 47$	Given
$5x = 35$	Subtraction Property of Equality
$x = 7$	Division Property of Equality

□

▶ Helpful Hint

Remember to concentrate on the reasons to justify each algebraic statement. This will help as we move to geometric proofs.

PRACTICE 1 Solve $9x - 45 = 45$. Give a reason to justify each statement.

EXAMPLE 2 Giving Reasons for Statements

Solve $7x - 10(5 + 3x) = 2x$. Give a reason to justify each statement.

Solution

Statements	Reasons
$7x - 10(5 + 3x) = 2x$	Given
$7x - 50 - 30x = 2x$	Multiply or Distributive Property
$-23x - 50 = 2x$	Simplify or Combine like terms.
$-50 = 25x$	Addition Property of Equality
$-2 = x$	Division Property of Equality

□

PRACTICE 2 Solve $20x - 8(3 + 2x) = 28x$. Give a reason to justify each statement.

Let's now continue to review some properties of equality that will be stated in terms of segment lengths and angle measures.

Properties of Equality

	Angle Measures	*Segment Lengths*
Reflexive Property	$m\angle A = m\angle A$	$AB = AB$
Symmetric Property	If $m\angle A = m\angle B$, then $m\angle B = m\angle A$.	If $AB = CD$, then $CD = AB$.
Transitive Property	If $m\angle A = m\angle B$ and $m\angle B = m\angle C$, then $m\angle A = m\angle C$.	If $AB = CD$ and $CD = EF$, then $AB = EF$.

Let's practice recognizing what property is used.

EXAMPLE 3 Stating Properties

Fill in each blank with the reason to justify the statement.

Statements	Reasons
a. $CG = MN$ $MN = CG$	Given ________
b. $m\angle P = m\angle Q$ $m\angle Q = m\angle B$ $m\angle P = m\angle B$	Given Given ________
c. $m\angle 1 + m\angle 2 + m\angle 3 = 50°$ $m\angle 1 + m\angle 2 = 43°$ $43° + m\angle 3 = 50°$	Given Given ________

Solution

a. Symmetric Property of Equality
b. Transitive Property of Equality
c. Substitution Property

□

PRACTICE 3 Fill in each blank with the reason to justify the statement.

a.

Statements	Reasons
$m\angle B = m\angle E$	Given
$m\angle E = m\angle B$	________

b.

Statements	Reasons
$AD = NF$	Given
$NF = PG$	Given
$AD = PG$	________

c.

Statements	**Reasons**
$m\angle 2 + m\angle 4 + m\angle 6 = 75°$	Given
$m\angle 4 + m\angle 6 = 60°$	Given
$m\angle 2 + 60° = 75°$	____________

OBJECTIVE 2 ▶ Writing a Two-Column Proof. We have studied logic in this chapter, and we reviewed properties of equality. We are now ready to begin our study of proofs. A **proof** is an argument that uses logic to establish the truth of a statement. There are many formats for proofs, but for now, we will use a two-column proof.

A **two-column proof** lists numbered statements on the left and corresponding numbered reasons or justifications on the right. These statements show the logical order of the proof.

EXAMPLE 4 **Writing a Two-Column Proof**

Write a two-column proof.

Given: $m\angle 1 = m\angle 3$

Prove: $m\angle BAG = m\angle EAC$

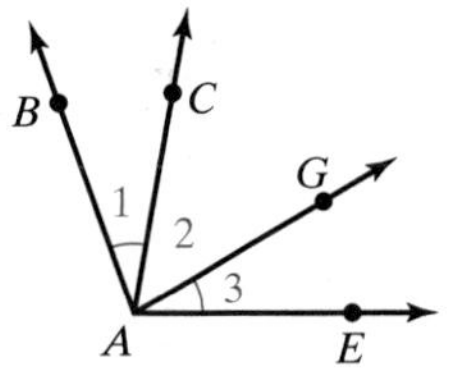

Solution Take a moment to study the figure and make sure that it is marked with the given information. Next, study what you want to prove and form a plan.

For example, since $m\angle 1 = m\angle 3$, we can add $m\angle 2$ to both $m\angle 1$ and $m\angle 3$. The resulting angles are the angles we are interested in and will have equal measure.

Statements	**Reasons**
1. $m\angle 1 = m\angle 3$	**1.** Given
2. $m\angle 2 = m\angle 2$	**2.** Reflexive Property of Equality
3. $m\angle 1 + m\angle 2 = m\angle 3 + m\angle 2$	**3.** Addition Property of Equality
4. $m\angle 1 + m\angle 2 = m\angle BAG$ $m\angle 3 + m\angle 2 = m\angle EAC$	**4.** Angle Addition Postulate
5. $m\angle BAG = m\angle EAC$	**5.** Substitution Property (Steps 3, 4)

Helpful Hint
Notice the reason for Step 5 contains the steps used for substitution. This helps organize the proof.

PRACTICE 4 Write a two-column proof.

Given: $AB = CD$

Prove: $AC = BD$

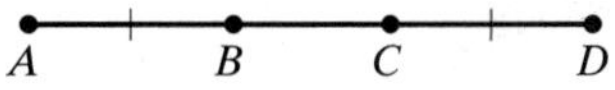

Recall from Chapter 1 that if segment lengths are equal, then the segments are congruent. For example, if $AB = CD$, then $\overline{AB} \cong \overline{CD}$. Also, if angle measures are equal, then the angles are congruent. For example, if $m\angle A = m\angle B$, then $\angle A \cong \angle B$.

This means that the Reflexive, Symmetric, and Transitive Properties of Equality are also true for congruence.

Properties of Congruence

	Segment Congruence	*Angle Congruence*
Reflexive Property	$\overline{AB} \cong \overline{AB}$	$\angle A \cong \angle A$
Symmetric Property	If $\overline{AB} \cong \overline{CD}$, then $\overline{CD} \cong \overline{AB}$.	If $\angle A \cong \angle B$, then $\angle B \cong \angle A$.
Transitive Property	If $\overline{AB} \cong \overline{CD}$, and $\overline{CD} \cong \overline{EF}$, then $\overline{AB} \cong \overline{EF}$.	If $\angle A \cong \angle B$ and $\angle B \cong \angle C$, then $\angle A \cong \angle C$.

EXAMPLE 5 **Writing Two-Column Proof Involving Congruence**

Write a two-column proof.

Given: $AB = 5$ units, $CD = 5$ units

Prove: $\overline{AB} \cong \overline{CD}$

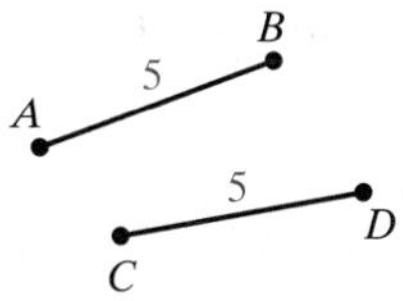

Helpful Hint
Remember: We are given lengths, so we can show that since the segment lengths are equal, the segments are congruent.

Solution If we show that $AB = CD$, we will then have that $\overline{AB} \cong \overline{CD}$.

Helpful Hint
Both reasons are correct depending on our thoughts.

Statements	Reasons
1. $AB = 5$ units	**1.** Given
2. $CD = 5$ units	**2.** Given
3. 5 units $= CD$	**3.** Symmetric Property of Equality
4. $AB = CD$	**4.** Transitive Property of Equality or Substitution Property (Steps 1, 3)
5. $\overline{AB} \cong \overline{CD}$	**5.** Definition of congruent segments

Helpful Hint
As we become more proficient at writing proofs, we may be allowed to assume Example 5, step 3, the Symmetric Property of Equality, and not need to write it as a step.

PRACTICE
5 Write a two-column proof.

Given: $m\angle A = 32°$, $m\angle B = 32°$

Prove: $\angle A \cong \angle B$

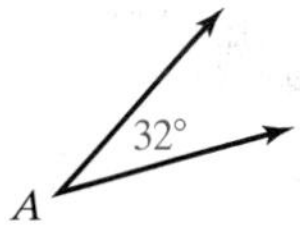

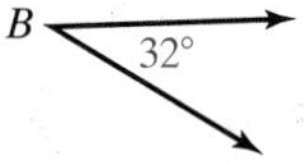

VOCABULARY & READINESS CHECK

Word Bank *Use the choices to fill in each blank.*

Reflexive	Transitive
Symmetric	proof

1. The statement "If $m\angle 1 = m\angle 2$, then $m\angle 2 = m\angle 1$" is an example of the ____________ Property.
2. The statement "$AB = AB$" is an example of the ____________ Property.
3. The statement "If $m\angle C = m\angle G$ and $m\angle G = m\angle Q$, then $m\angle C = m\angle Q$" is an example of the ____________ Property.
4. A ____________ is a logical argument used to establish the truth of a statement.

2.6 EXERCISE SET MyMathLab®

Proof *Fill in each blank with a reason that justifies the statement. See Examples 1 and 2.*

1.

Statements	Reasons
$3x - 10 = 65$	Given
$3x = 75$	**a.** ?
$x = 25$	**b.** ?

2.

Statements	Reasons
$11x + 14 = 80$	Given
$11x = 66$	**a.** ?
$x = 6$	**b.** ?

3.

Statements	Reasons
$9x - 4 = 7x + 44$	Given
$2x - 4 = 44$	**a.** ?
$2x = 48$	**b.** ?
$x = 24$	**c.** ?

4.

Statements	Reasons
$3x + 6 = -2x + 31$	Given
$5x + 6 = 31$	**a.** ?
$5x = 25$	**b.** ?
$x = 5$	**c.** ?

5.

Statements	Reasons
$\frac{1}{2}x - 5 = 10$	Given
$2\left(\frac{1}{2}x - 5\right) = 20$	**a.** ?
$x - 10 = 20$	**b.** ?
$x = 30$	**c.** ?

6.

Statements	Reasons
$5(x + 3) = -4$	Given
$5x + 15 = -4$	**a.** ?
$5x = -19$	**b.** ?
$x = -\frac{19}{5}$	**c.** ?

7.

Statements	Reasons
$15 + 7(3x + 4) = -20$	Given
$15 + 21x + 28 = -20$	**a.** ?
$21x + 43 = -20$	**b.** ?
$21x = -63$	**c.** ?
$x = -3$	**d.** ?

8.

Statements	Reasons
$-20 + 2(11 - 2x) = 26$	Given
$-20 + 22 - 4x = 26$	**a.** ?
$2 - 4x = 26$	**b.** ?
$-4x = 24$	**c.** ?
$x = -6$	**d.** ?

9. Given: $\angle CDE$ and $\angle EDF$ are supplementary.

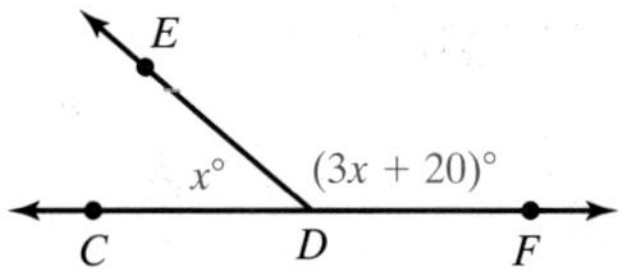

Statements	Reasons
$m\angle CDE + m\angle EDF = 180°$	**a.** ?
$x° + (3x + 20)° = 180°$	**b.** ?
$4x° + 20° = 180°$	**c.** ?
$4x° = 160°$	**d.** ?
$x° = 40°$	**e.** ?

10. Given: $XY = 42$

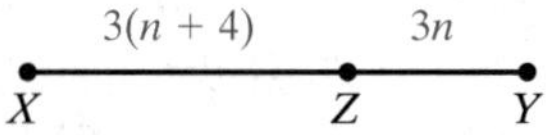

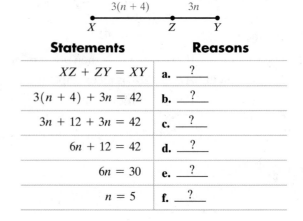

Statements	Reasons
$XZ + ZY = XY$	**a.** ?
$3(n + 4) + 3n = 42$	**b.** ?
$3n + 12 + 3n = 42$	**c.** ?
$6n + 12 = 42$	**d.** ?
$6n = 30$	**e.** ?
$n = 5$	**f.** ?

Name the property of equality or congruence that justifies going from the first statement to the second statement. See Examples 3 and 5.

11. $2x + 1 = 7$
$2x = 6$

12. $5x = 20$
$x = 4$

13. $\overline{ST} \cong \overline{QR}$
$\overline{QR} \cong \overline{ST}$

14. $AB - BC = 12$
$AB = 12 + BC$

Matching *Match each statement with the appropriate property that it describes.*

15. If $\angle M \cong \angle N$, then $\angle N \cong \angle M$.

16. If $CD = PQ$ and $PQ = RF$, then $CD = RF$.

17. If $RT = 5$, then $9(RT) = 45$.

18. If $3x - 12 = 15$, then $3x = 27$.

19. If $m\angle A = 27°$ and $m\angle C = m\angle A + 10°$, then $m\angle C = 37°$.

20. If $-2(x - 7) = 5$, then $-2x + 14 = 5$.

A. Multiplication Property

B. Substitution

C. Addition Property

D. Transitive Property

E. Distributive Property

F. Symmetric Property

Fill in the Blank *Use the given property to complete each statement.*

21. Symmetric Property of Congruence
If $\angle H \cong \angle K$, then __?__ $\cong \angle H$.

22. Symmetric Property of Equality
If $AB = YU$, then __?__.

23. Distributive Property
$3(x - 1) = 3x -$ __?__

24. Reflexive Property of Congruence
$\angle POR \cong$ __?__

25. Transitive Property of Congruence
If $\angle XYZ \cong \angle AOB$ and $\angle AOB \cong \angle WYT$, then __?__.

26. Substitution Property
If $LM = 7$ and $EF + LM = NP$, then __?__ $= NP$.

Proof *For Exercises 27 and 28, fill in the missing statements or reasons for the two-column proof. See Examples 4 and 5.*

27. Given: C is the midpoint of $\overline{AD}$.
Prove: $x = 6$

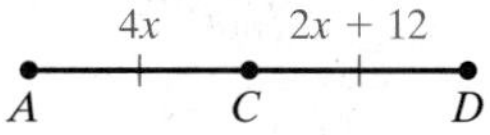

Statements	Reasons
1. C is the midpoint of $\overline{AD}$.	1. a. __?__
2. $\overline{AC} \cong \overline{CD}$	2. b. __?__
3. $AC = CD$	3. $\cong$ segments have equal length.
4. $4x = 2x + 12$	4. c. __?__
5. d. __?__	5. Subtraction Property of Equality
6. $x = 6$	6. e. __?__

28. Given: $\overrightarrow{AC}$ is the angle bisector of $\angle DAG$.
Prove: $x = 3$

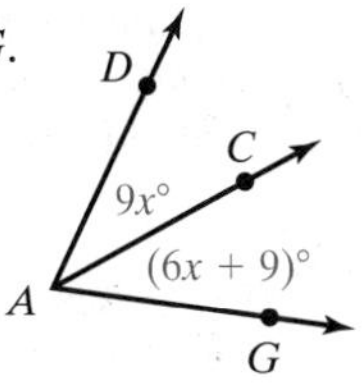

Statements	Reasons
1. $\overrightarrow{AC}$ is the bisector of $\angle DAG$.	1. a. __?__
2. $\angle DAC \cong \angle CAG$	2. b. __?__
3. $m\angle DAC = m\angle CAG$	3. $\cong$ angles have equal measure.
4. $9x = 6x + 9$	4. c. __?__
5. d. __?__	5. Subtraction Property of Equality
6. $x = 3$	6. e. __?__

CONCEPT EXTENSIONS

Proof *For Exercises 29 and 30, write a two-column proof.*

29. Given: $m\angle GFI = 128°$
Prove: $m\angle EFI = 40°$

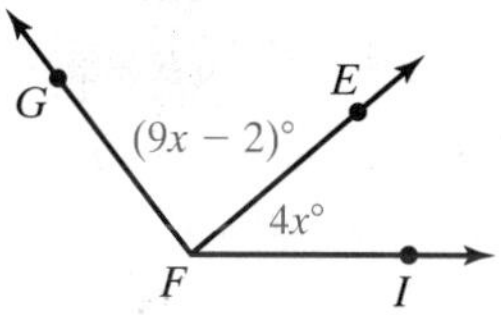

30. Given: $KM = 35$
Prove: $KL = 15$

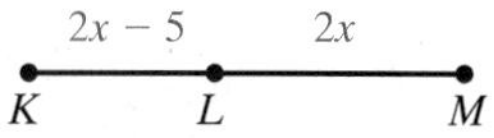

31. Explain why the statements $\overline{LR} \cong \overline{RL}$ and $\angle CBA \cong \angle ABC$ are both true by the Reflexive Property of Congruence.

32. Complete the following statement. Describe the reasoning that supports your answer.
The Transitive Property of Falling Dominoes: If Domino A causes Domino B to fall, and Domino B causes Domino C to fall, then Domino A causes Domino __?__ to fall.

A very important part in writing proofs is analyzing the diagram for key information. Use this for Exercises 33 and 34.

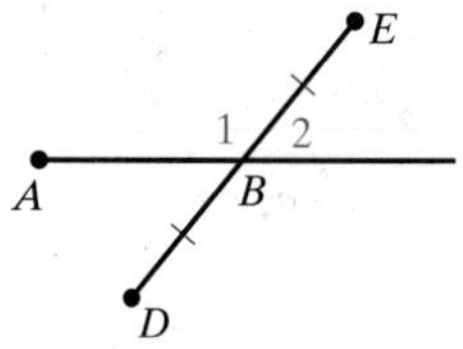

33. What true statement(s) can you make about segment lengths based on the diagram above?

34. What true statement(s) can you make about angle measures based on the diagram above?

Consider the following relationships among people. Tell whether each relationship is "reflexive," "symmetric," "transitive," or "none of these." Explain.

Sample: The relationship "is younger than" is not reflexive because Sue is not younger than herself. It is not symmetric because if Sue is younger than Fred, then Fred is not younger than Sue. It is transitive because if Sue is younger than Fred and Fred is younger than Alana, then Sue is younger than Alana. (*Hint*: These names are made up to help with reasoning.)

35. has the same birthday as

36. lives in a different state than

37. is taller than

38. is shorter than

39. Find the Error The statements below "show" that $1 = 2$. Describe the error.

Statements	Reasons
$a = b$	Given
$ab = b^2$	Multiplication Property of Equality
$ab - a^2 = b^2 - a^2$	Subtraction Property of Equality
$a(b - a) = (b + a)(b - a)$	Distributive Property
$a = b + a$	Division Property of Equality
$a = a + a$	Substitution Property
$a = 2a$	Simplify.
$1 = 2$	Division Property of Equality

40. Find the Error The statements below "show" that $0 = 3$. Describe the error.

Statements	Reasons
$x = -3$	Given
$-3x = 9$	Multiplication Property of Equality
$-3x - x^2 = 9 - x^2$	Subtraction Property of Equality
$-x(3 + x) = (3 - x)(3 + x)$	Distributive Property
$-x = 3 - x$	Division Property of Equality
$0 = 3$	Addition Property of Equality

REVIEW AND PREVIEW

Use the diagram at the right. Find each measure. See Section 1.5.

41. $m\angle AOC$

42. $m\angle DOB$

43. $m\angle AOD$

44. $m\angle BOE$

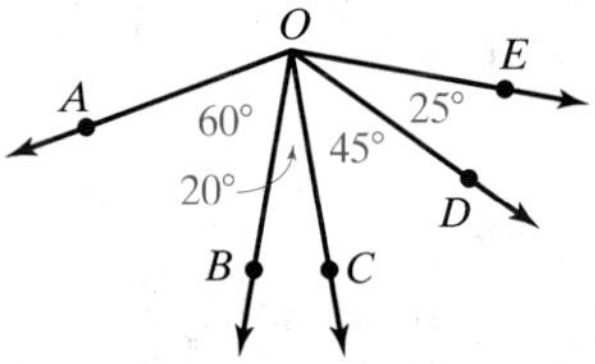

Find the value of each variable. See Section 1.6.

45.

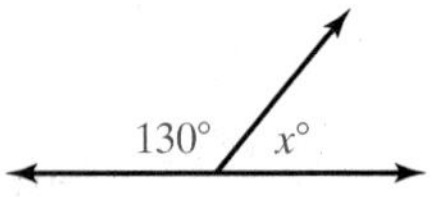

46.

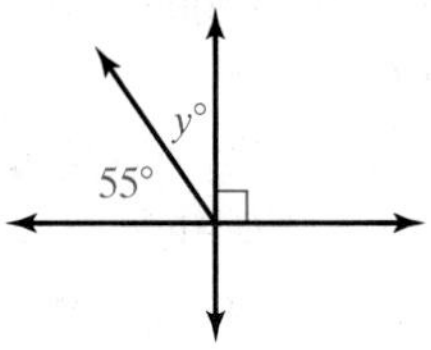

2.7 PROVING THEOREMS ABOUT ANGLES

OBJECTIVE

1 Prove and Use Theorems About Angles.

VOCABULARY

- paragraph proof

OBJECTIVE 1 ▶ Proving and Using Theorems About Angles. Let's continue to learn how to use our properties from Section 2.6 and earlier to prove theorems. Remember, once we prove a theorem, it is then a true statement and can be used to prove other theorems. That is how our mathematical system of geometry continues to grow.

In this section, we will concentrate on theorems involving angles.

When writing a proof, it is often helpful to take a moment to write a statement in "if-then" form. This helps identify the given information and what we want to prove.

The given is the hypothesis, following the word "if."

What we want to prove is the conclusion, following the word "then."

Let's begin our proofs with a few theorems about complementary and supplementary angles.

Theorem 2.7-1 Equal Complements Theorem

Complements of the same angle (or of equal angles) are equal in measure.

If $m\angle 1 + m\angle 2 = 90°$ and $m\angle 3 + m\angle 2 = 90°$, then $m\angle 1 = m\angle 3$.

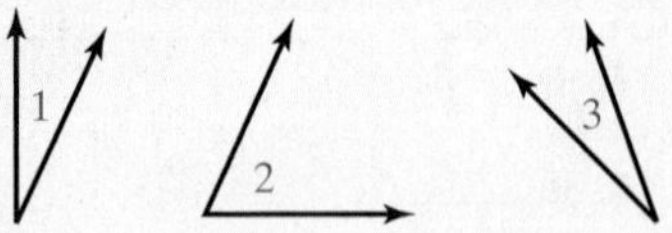

We prove this theorem in Example 1.

Theorem 2.7-2 Equal Supplements Theorem

Supplements of the same angle (or of equal angles) are equal in measure.

If $m\angle 1 + m\angle 2 = 180°$ and $m\angle 3 + m\angle 2 = 180°$, then $m\angle 1 = m\angle 3$.

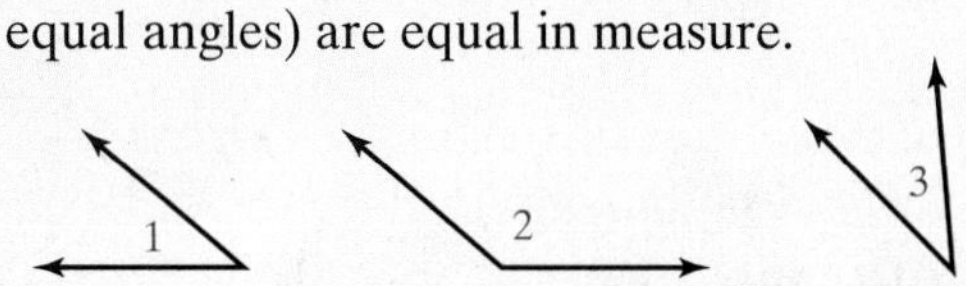

We prove this theorem in Exercise 25 of Exercise Set 2.7.

> **Helpful Hint**
> Remember to write theorems in "if-then" form to help identify the "given" and "what to prove."

The "if-then" form of the Equal Complements Theorem is: If two angles are complementary to the same angle (or to equal angles), then they are equal in measure.

EXAMPLE 1 Proving the Equal Complements Theorem

Given: $\angle A$ and $\angle B$ are complementary.
$\angle C$ and $\angle D$ are complementary.
$m\angle A = m\angle C$

Prove: $m\angle B = m\angle D$

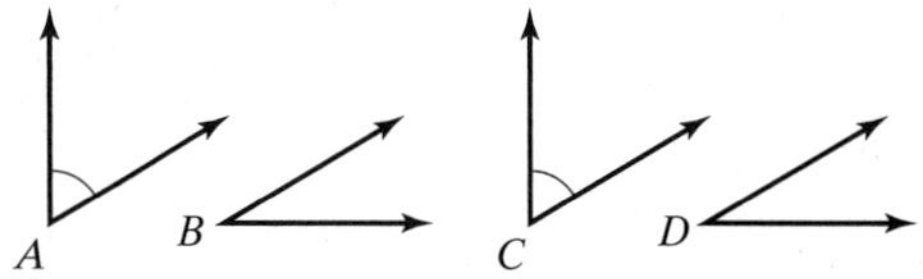

Solution

Statements	Reasons
1. $\angle A$ and $\angle B$ are complements. $\angle C$ and $\angle D$ are complements.	**1.** Given
2. $m\angle A = m\angle C$	**2.** Given
3. $m\angle A + m\angle B = 90°$ $m\angle C + m\angle D = 90°$	**3.** Definition of complementary $\angle$'s
4. $m\angle A + m\angle B = m\angle C + m\angle D$	**4.** Transitive Property (Some might use substitution.) (Step 3)
5. $m\angle A + m\angle B = m\angle A + m\angle D$	**5.** Substitution (Steps 2 and 4)
6. $m\angle B = m\angle D$	**6.** Subtraction Property of Equality (Step 5)

> **Helpful Hint**
> Write your proofs so that someone else reading them can follow them. For example, in the Example 1 proof:
> - Step 5 Reason—It is easier to follow the Step 5 statement if you've cited the steps used for substitution.

PRACTICE 1 Write the Equal Supplements Theorem in "if-then" form. We will prove this theorem in the exercise set.

Now that we know more about complementary and supplementary angles, let's prove a theorem about angles forming a linear pair being supplements of each other.

Theorem 2.7-3 Linear Pair Theorem

If two angles form a linear pair, then the angles are supplementary.

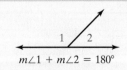

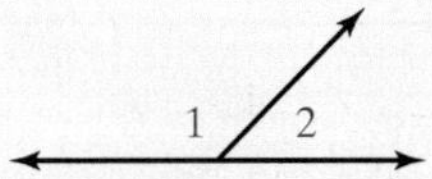

$m\angle 1 + m\angle 2 = 180°$

We prove this theorem in Example 2.

EXAMPLE 2 Proving the Linear Pair Theorem

Given: $\angle 1$ and $\angle 2$ form a linear pair.

Prove: $\angle 1$ and $\angle 2$ are supplementary angles.

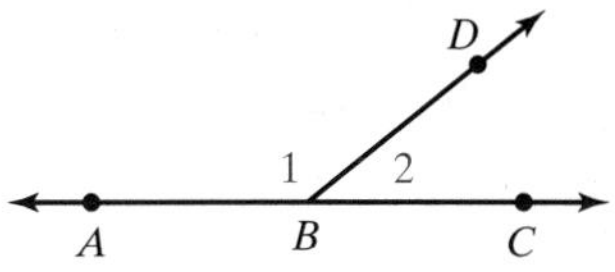

Solution

Statements	Reasons
1. $\angle 1$ and $\angle 2$ form a linear pair.	**1.** Given
2. $\overrightarrow{BA}$ and $\overrightarrow{BC}$ are opposite rays.	**2.** Definition of linear pair
3. A, B, and C lie on the same line, with B between A and C.	**3.** Definition of opposite rays
4. $\angle ABC$ is a straight angle.	**4.** Definition of straight angle
5. $m\angle ABC = 180°$	**5.** Definition of straight angle
6. $m\angle 1 + m\angle 2 = m\angle ABC$	**6.** Angle Addition Postulate
7. $m\angle 1 + m\angle 2 = 180°$	**7.** Transitive Property (Steps 5, 6)
8. $\angle 1$ and $\angle 2$ are supplementary.	**8.** Definition of supplementary angles (Step 7)

Helpful Hint

It is always a good idea to review definitions of terms in a proof. For Example 2, it might be necessary to review the definition of "linear pair."

PRACTICE 2 Write the Linear Pair Theorem in a form other than the "if-then" form.

Let's continue writing and proving statements about angles.

Theorem 2.7-4 Vertical Angles Theorem

Vertical angles are congruent.

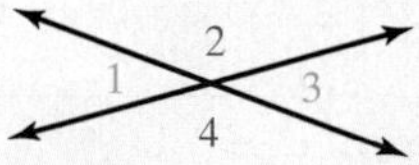

$\angle 1 \cong \angle 3$
$\angle 2 \cong \angle 4$

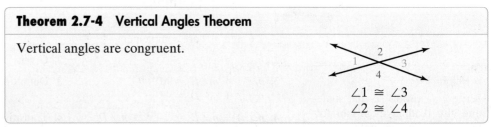

We prove this theorem in Example 3.

This proof is an excellent example of how important it is to know and use any theorems we already proved.

EXAMPLE 3 **Proving the Vertical Angles Theorem**

Given: $\angle 1$ and $\angle 3$ are vertical angles.

Prove: $\angle 1 \cong \angle 3$

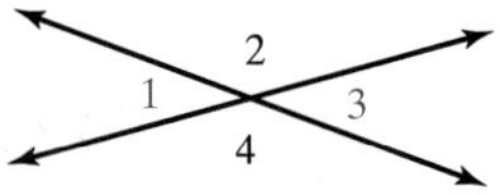

Solution

Statements	Reasons
1. $\angle 1$ and $\angle 3$ are vertical angles.	**1.** Given
2. $\angle 1$ and $\angle 2$ form a linear pair. $\angle 3$ and $\angle 2$ form a linear pair.	**2.** Definition of linear pair angles
3. $\angle 1$ and $\angle 2$ are supplementary. $\angle 3$ and $\angle 2$ are supplementary.	**3.** Linear Pair Theorem (Step 2)
4. $\angle 1 \cong \angle 3$	**4.** Equal Supplements Theorem (Step 3)

PRACTICE 3 Write the Vertical Angles Theorem in "if-then" form.

As we mentioned earlier, there are many forms of proofs. We have been using the two-column form.

The **paragraph proof** is a form where the proof is written as a paragraph.

Theorem 2.7-5 Right Angles Congruent Theorem

All right angles are congruent.

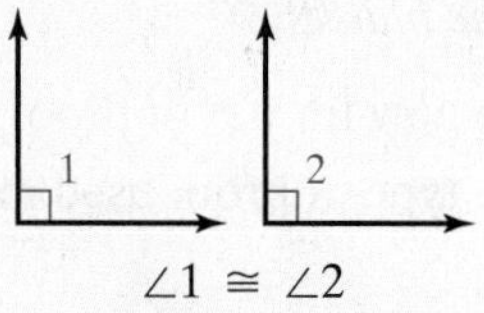

We prove this theorem in Example 4, paragraph form, and in Practice 4, two-column form.

EXAMPLE 4 Proving the Right Angles Congruent Theorem

Given: $\angle 1$ and $\angle 2$ are right angles.

Prove: $\angle 1 \cong \angle 2$

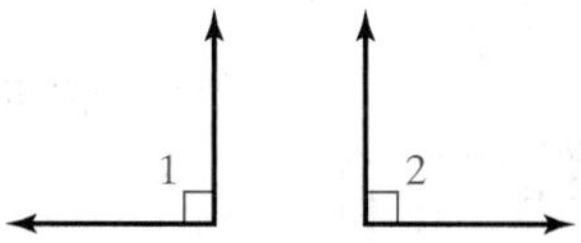

Solution We write this proof in paragraph form.

Proof: $\angle 1$ and $\angle 2$ being right angles is given. By the definition of right angles, $m\angle 1 = 90°$ and $m\angle 2 = 90°$. By the Transitive Property, since both angles equal 90°, it is true that $m\angle 1 = m\angle 2$. Then $\angle 1 \cong \angle 2$ by the definition of congruent angles. □

PRACTICE 4 Prove the Right Angles Congruent Theorem using a two-column proof format.

We prove our last theorem in the exercise set.

Theorem 2.7-6 Equal Supplementary Angles Theorem

Two equal supplementary angles are right angles.
If $m\angle 1 = m\angle 2$ and $m\angle 1 + m\angle 2 = 180°$,
then $\angle 1$ and $\angle 2$ are right angles.

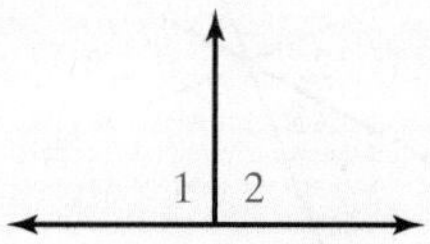

We prove this theorem in Exercise 26 of Exercise Set 2.7.

Let's now apply our knowledge of algebra and these theorems to solve problems.

EXAMPLE 5 Using the Vertical Angles Theorem

Find the value of x.

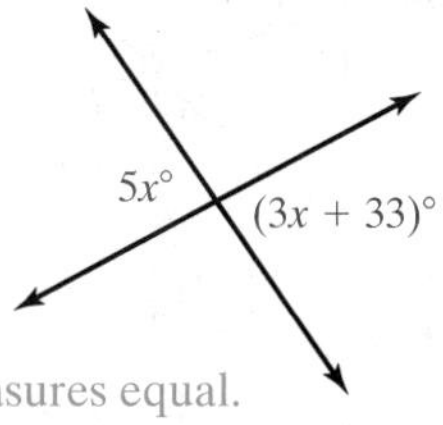

Solution The angles are vertical angles and are thus equal in measure.

$5x = 3x + 33$ Set the vertical angles measures equal.

$2x = 33$ Subtract $3x$ from both sides.

$x = \frac{33}{2}$ or $x = 16\frac{1}{2}$ or $x = 16.5$ Divide both sides by 2.

Thus, $x = \frac{33}{2}$ or $16\frac{1}{2}$ or 16.5

To **check**, let x be 16.5 in each angle expression, and simplify. They should be and are equal in measure, 82.5°. □

PRACTICE 5 Find the value of x.

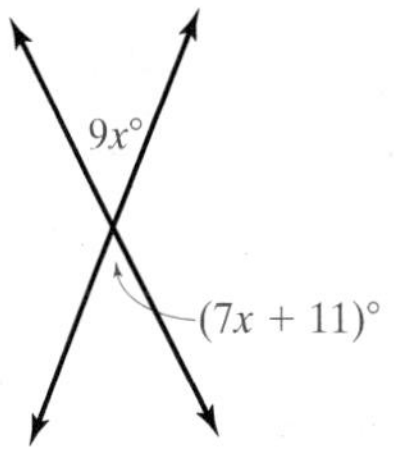

VOCABULARY & READINESS CHECK

Fill in the blanks.

1. The new format of proof shown in this section is the ____________ proof.

2. The type of proof used most often thus far is the ____________ proof.

2.7 EXERCISE SET MyMathLab®

Mixed Practice

Use the theorems in this section to find the measures of the angles. See Examples 1–4.

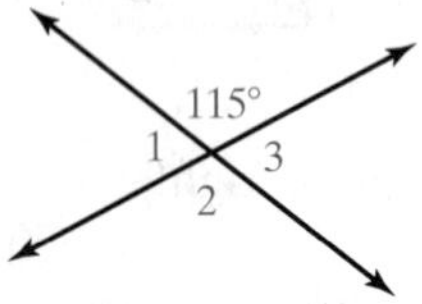

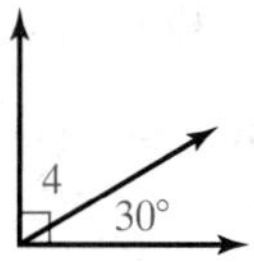

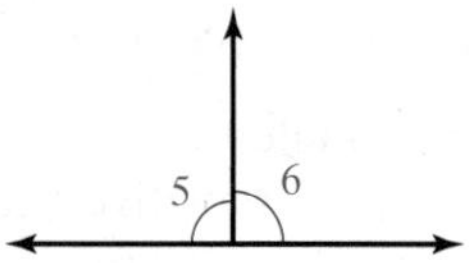

1. $m\angle 1$ **2.** $m\angle 2$ **3.** $m\angle 3$

4. $m\angle 4$ **5.** $m\angle 5$ **6.** $m\angle 6$

For each figure, find the measures of $\angle 1$, $\angle 2$, and $\angle 3$. See Examples 1–4.

7.

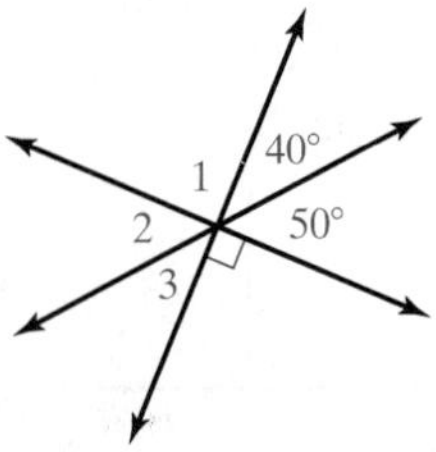

8.

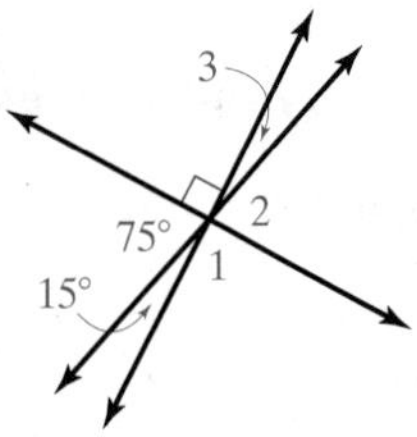

Find the value of each variable. See Example 5.

9.

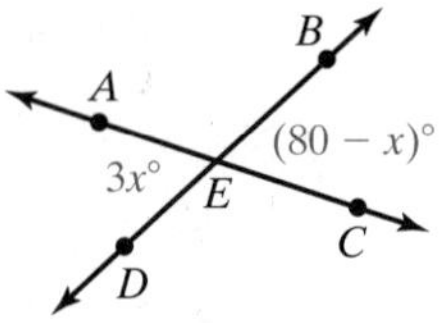

10.

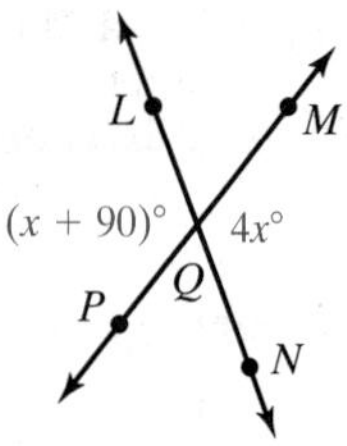

11.

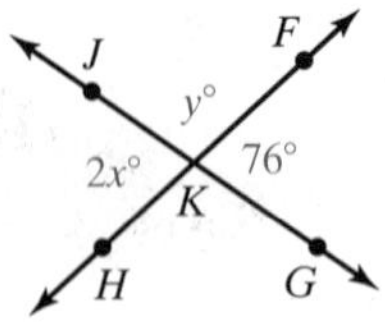

12.

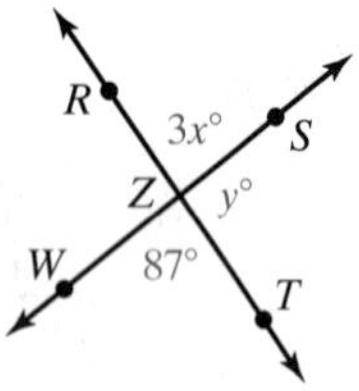

Find the measures of all four angles in each exercise.

13. Exercise 9 **14.** Exercise 10

15. Exercise 11 **16.** Exercise 12

17. Proof Complete the following proof by filling in the blanks.

Given: $\angle 1 \cong \angle 3$

Prove: $\angle 6 \cong \angle 4$

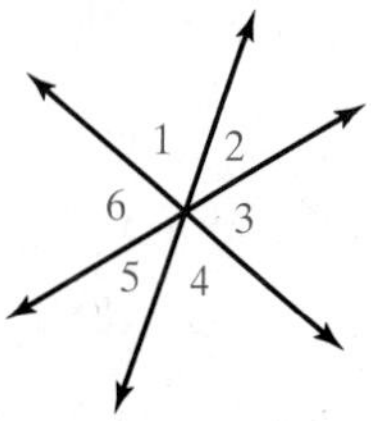

Statements	Reasons
1. $\angle 1 \cong \angle 3$	**1.** Given
2. $\angle 3 \cong \angle 6$	**2. a.** ___?___
3. b. ___?___	**3.** Transitive Property
4. $\angle 1 \cong \angle 4$	**4. c.** ___?___
5. $\angle 6 \cong \angle 4$	**5. d.** ___?___

18. Proof Use a 2-column proof and the given figure:

Given: $m\angle 1 \cong m\angle 2$

Prove: $m\angle 3 \cong m\angle 4$

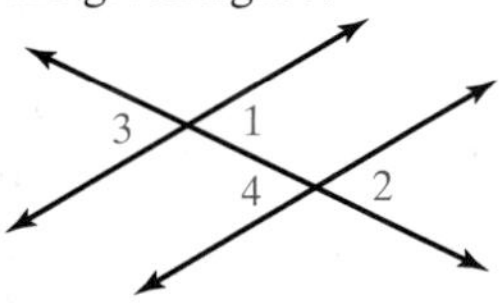

Name two pairs of congruent angles in each figure. Justify your answers by naming the theorem.

19.

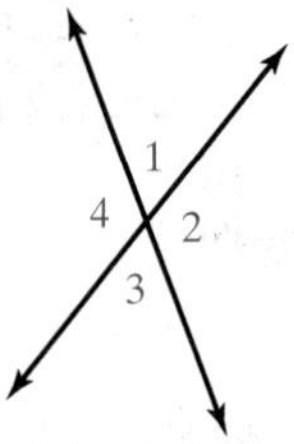

20.

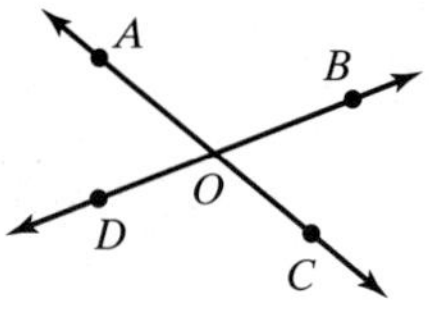

21.

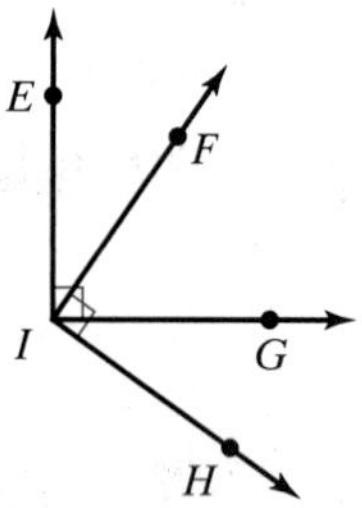

22.

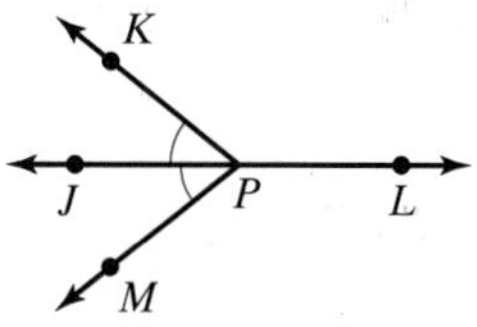

23. In the photograph, the legs of the table are constructed so that $\angle 1 \cong \angle 2$. What theorem can you use to justify the statement that $\angle 3 \cong \angle 4$?

24. Explain why this statement is true: If $m\angle ABC + m\angle XYZ = 180°$ and $\angle ABC \cong \angle XYZ$, then $\angle ABC$ and $\angle XYZ$ are right angles.

25. Proof Prove Theorem 2.7-2: Supplements of the same angle (or of equal angles) are equal in measure. (You may follow Example 1 as a guide.)

Given: $\angle A$ and $\angle B$ are supplementary.

$\angle C$ and $\angle D$ are supplementary.

$m\angle A = m\angle C$

Prove: $m\angle B = m\angle D$

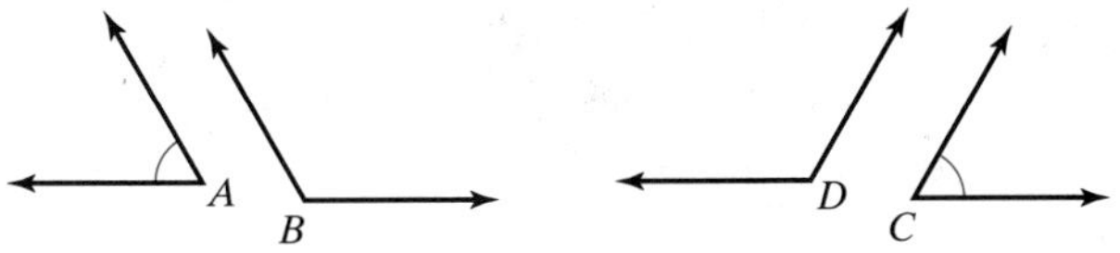

26. Proof Prove Theorem 2.7-6: Two equal supplementary angles are right angles.

Given: $m\angle 1 = m\angle 2$

$m\angle 1 + m\angle 2 = 180°$

Prove: $\angle 1$ and $\angle 2$ are right angles.

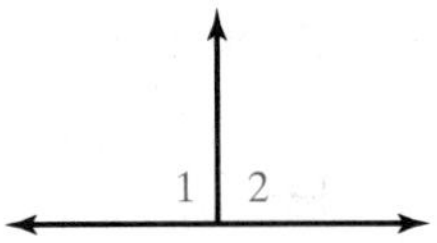

27. Multiple Steps

a. What is the measure of the angle formed by Main St. and 116th St.?

b. What is the measure of the angle formed by Park St. and 116th St.?

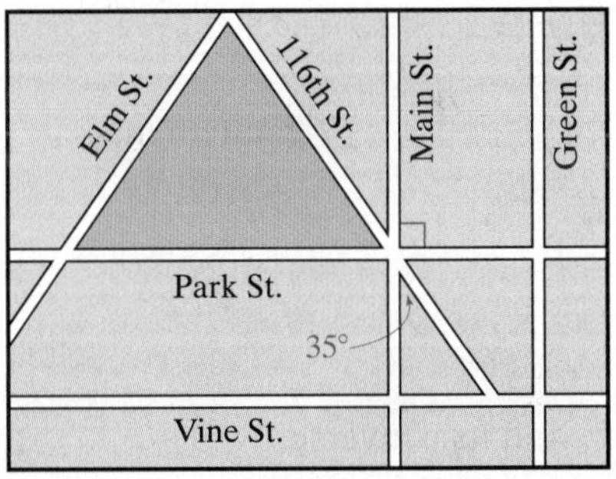

28. In the game of miniature golf, the ball bounces off the wall at the same angle it hit the wall. (This is the angle formed by the path of the ball and the line perpendicular to the wall at the point of contact.) In the diagram, the ball hits the wall at a 40° angle. Using Theorem 2.7-1, what are the values of x and y?

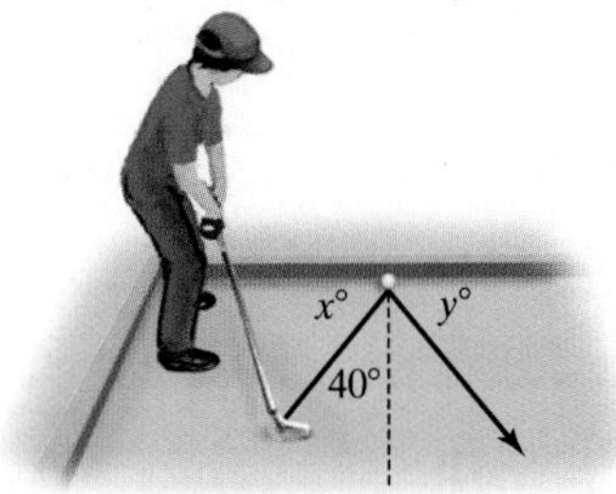

Find the value of each variable and the measure of each angle. See Example 5.

29.

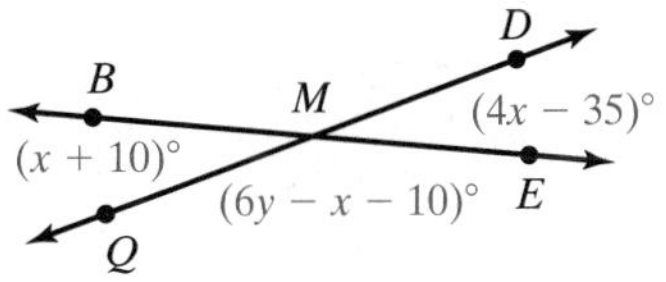

30.

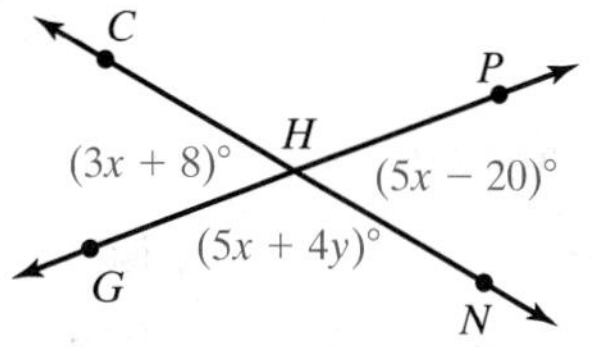

Find the measure of each angle.

31. $\angle A$ is twice as large as its complement, $\angle B$.

32. $\angle A$ is half as large as its complement, $\angle B$.

33. $\angle A$ is twice as large as its supplement, $\angle B$.

34. $\angle A$ is half as large as twice its supplement, $\angle B$.

Find the value of each variable and the measure of each angle.

35.

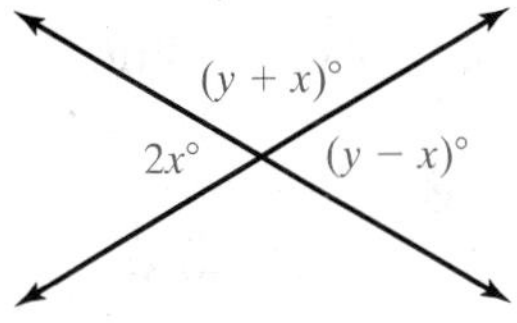

36.

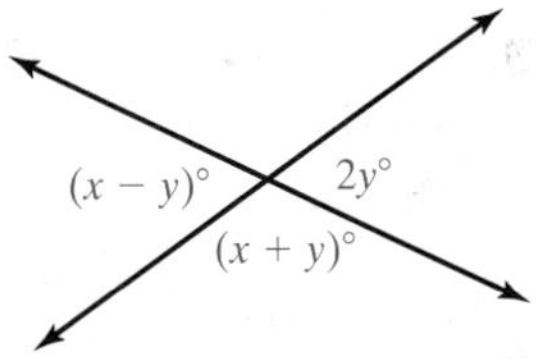

37.

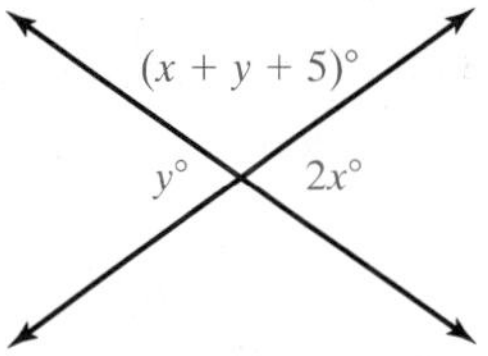

38.

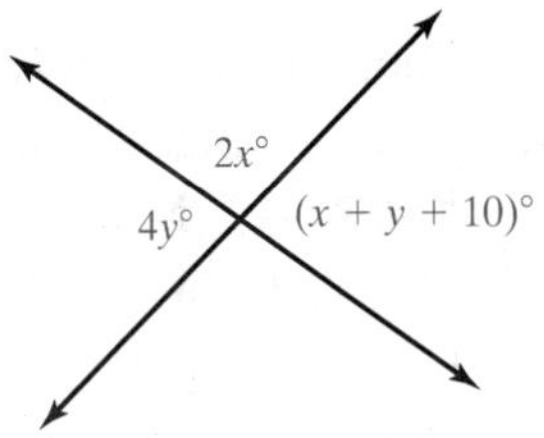

CONCEPT EXTENSIONS

39. Give an example of vertical angles in your home or classroom.

40. How are angles that form a right angle related?

Coordinate Geometry *$\angle AOX$ contains points $A(1, 3)$, $O(0, 0)$, and $X(4, 0)$.*

41. Find the coordinates of a point B so that $\angle BOA$ and $\angle AOX$ are adjacent complementary angles.

42. Find the coordinates of a point C so that $\overrightarrow{OC}$ is a side of a different angle that is adjacent and complementary to $\angle AOX$.

REVIEW AND PREVIEW

Which property of equality or congruence justifies going from the first statement to the second? See Section 2.6.

43. $3x + 7 = 19$

$3x = 12$

44. $4x = 20$

$x = 5$

45. $\angle 1 \cong \angle 2$ and $\angle 3 \cong \angle 2$

$\angle 1 \cong \angle 3$

46. $m\angle Q = m\angle R$

$m\angle R = m\angle Q$

For Exercises 47–52, use the figure at the right. See Section 1.3.

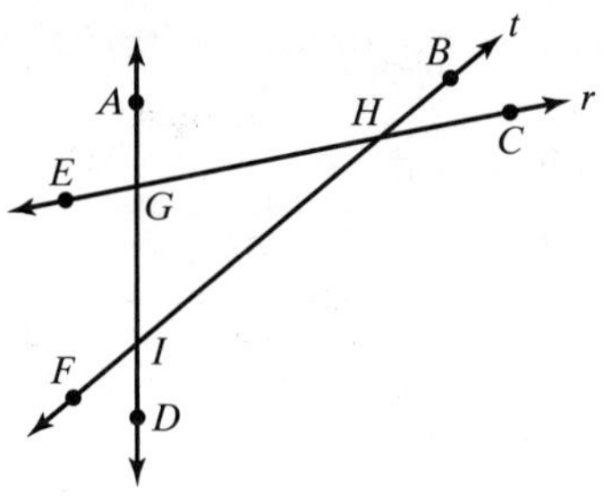

47. Name four points on line t.

48. Are points G, A, and B collinear?

49. Are points F, I, and H collinear?

50. Name the line on which point E lies.

51. Name line t in three other ways.

52. Name the point at which lines t and r intersect.

CHAPTER 2 VOCABULARY CHECK

Word Bank *Use the choices to fill in each blank. Some choices may not be used, and some may be used more than once.*

deductive	$q \rightarrow p$	Law of Detachment	if and only if	perimeter
inductive	Transitive	Law of Syllogism	Reflexive	area
inverse	equivalent	contrapositive	Symmetric	circumference
converse	negation	conclusion	conditional	$\sim p$
counterexample	$\sim p \rightarrow \sim q$	conditional	biconditional	$\leftrightarrow$
$p \rightarrow q$	$\sim q \rightarrow \sim p$	if-then	hypothesis	

1. The distance around a circle is given a special name, called ____________.
2. A(n) ____________ is an example that shows that a conjecture is false.
3. The process of arriving at a general conclusion based on observing specific examples is called ____________ reasoning.
4. The distance around a geometric figure is called____________.
5. The number of square units a geometric figure encloses is called____________.
6. The statement "If $m\angle 1 = m\angle 2$, then $m\angle 2 = m\angle 1$" is an example of the ____________ Property.
7. The statement "$AB = AB$" is an example of the ____________ Property.
8. The statement "If $m\angle C = m\angle G$ and $m\angle G = m\angle Q$, then $m\angle C = m\angle Q$" is an example of the ____________ Property.
9. For an "if-then" statement, the part following "then" is called the ____________.
10. For an "if-then" statement, the part following "if" is called the ____________.
11. Two statements that are both always true or both always false are called ____________ statements.
12. In symbols, the negation of p is written as ____________.
13. In symbols, an "if p then q" statement can be written as ____________.
14. The converse of the statement $p \rightarrow q$ is ____________.
15. The symbol for "if and only if" is ____________.
16. A(n) ____________ statement is written using a conditional statement and its converse.
17. A biconditional statement may be written using the phrase ____________.
18. A good definition may be rewritten as a(n) ____________ statement.
19. A true biconditional statement $p \leftrightarrow q$ may be used "forward" and "backward." In symbols, forward and backward mean ____________ and ____________, respectively.
20. The ____________ says if $p \rightarrow q$ is true and p is true, then q is true.
21. The ____________ says if $p \rightarrow q$ is true and $q \rightarrow r$ is true, then $p \rightarrow r$ is true.
22. ____________ reasoning is used to prove theorems.
23. With ____________ reasoning, your conclusion may or may not be true.

For a conditional statement $p \rightarrow q$, complete the table.

	Symbols	Statement Name
24.	$p \rightarrow q$	____________
25.	$\sim p \rightarrow \sim q$	____________
26.	$\sim q \rightarrow \sim p$	____________
27.	$q \rightarrow p$	____________
28.	$\sim p$	____________

CHAPTER 2 REVIEW

(2.1) Find the perimeter and area of each figure.

1. 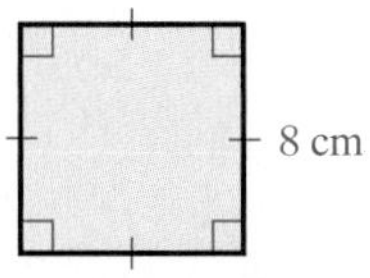

2. Find the perimeter and area of a rectangle with $l = 12$ m and $w = 8$ m.

Find the circumference and the area for each circle in terms of π.

3. $r = 3$ in. **4.** $d = 15$ m

(2.2) Find a pattern for each list of numbers. Describe the pattern and use it to show the next two terms.

5. 1000, 100, 10, ... **6.** 5, −5, 5, −5, ...

7. 34, 27, 20, 13, ... **8.** 6, 24, 96, 384, ...

Find a counterexample to show that each conjecture is false.

9. The product of any integer and 2 is greater than 2.

10. The city of Portland is in Oregon.

(2.3) Rewrite each sentence as a conditional statement.

11. All motorcyclists wear helmets.

12. Two nonparallel lines intersect in one point.

13. Angles that form a linear pair are supplementary.

14. School is closed on certain holidays.

Write the converse, inverse, and contrapositive of the given conditional. Then determine whether each statement is true or false.

15. If an angle is obtuse, then its measure is greater than 90° and less than 180°.

16. If a figure is a square, then it has four sides.

17. If you play the tuba, then you play an instrument.

18. If you are the manager of a restaurant, then you are busy on Saturday night.

(2.4) For Exercises 19–22, determine whether each statement is a good definition. If not, explain.

19. A bird has feathers.

20. A newspaper has articles you read.

21. A linear pair is a pair of adjacent angles whose noncommon sides are opposite rays.

22. An angle is a geometric figure.

23. Write the following definition as a biconditional.

An oxymoron is a phrase that contains contradictory terms.

24. Write the following biconditional as two statements, a conditional and its converse.

Two angles are complementary if and only if the sum of their measures is 90°.

(2.5) Use the Law of Detachment to make a conclusion.

25. If you practice tennis every day, then you will become a better player. Colin practices tennis every day.

26. ∠1 and ∠2 are supplementary. If two angles are supplementary, then the sum of their measures is 180°.

Use the Law of Syllogism to make a conclusion.

27. If two angles are vertical, then they are congruent. If two angles are congruent, then their measures are equal.

28. If your father buys new gardening gloves, then he will work in his garden. If he works in his garden, then he will plant tomatoes.

(2.6) Fill in the reason that justifies each step.

Given: $QS = 42$

Prove: $x = 13$

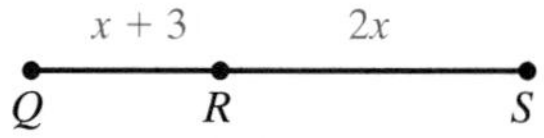

Statements	Reasons
1. $QS = 42$	1. _?_ (Exercise 29)
2. $QR + RS = QS$	2. _?_ (Exercise 30)
3. $(x + 3) + 2x = 42$	3. _?_ (Exercise 31)
4. $3x + 3 = 42$	4. _?_ (Exercise 32)
5. $3x = 39$	5. _?_ (Exercise 33)
6. $x = 13$	6. _?_ (Exercise 34)

Use the given property to complete the statement.

35. Division Property of Equality

If $2(AX) = 2(BY)$, then $AX =$ _?_.

36. Distributive Property: $3p - 6q = 3($ _?_ $)$

(2.7) Use the diagram for Exercises 37–40.

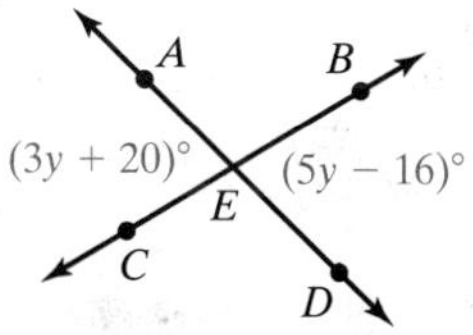

37. Find the value of y.

38. Find $m\angle AEC$.

39. Find $m\angle BED$.

40. Find $m\angle AEB$.

41. Given: ∠1 and ∠2 are complementary. ∠3 and ∠4 are complementary. ∠2 ≅ ∠4

Prove: ∠1 ≅ ∠3

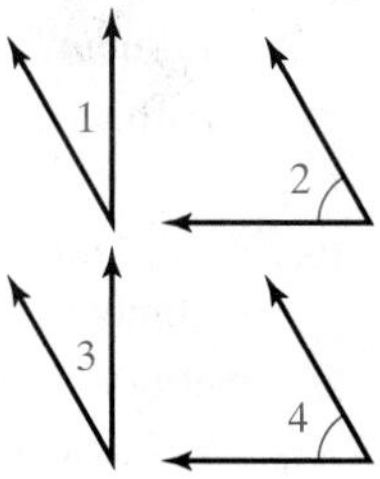

MIXED PRACTICE

42. Find a pattern for the list. What are the next two terms in the list?

1, −3, 9, −27, . . .

43. What is the converse of the conditional statement below? Is the converse true or false?

If you are a teenager, then you are younger than 20.

44. What is the name of the property that justifies going from the first line to the second line?

∠A ≅ ∠B and ∠B ≅ ∠C

∠A ≅ ∠C

45. Proof Write a two-column proof.

Given: ∠1 ≅ ∠4

Prove: ∠2 ≅ ∠3

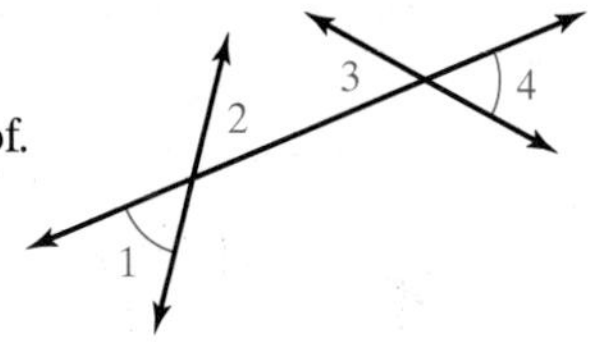

46. Is the following definition reversible? If yes, write it as a true biconditional.

A rectangle is a geometric figure with exactly four sides.

47. What can you conclude from the given information?
Given: If you play hockey, then you are on the team. If you are on the team, then you are a varsity athlete.

CHAPTER 2 TEST

TEST PREP VIDEO The fully worked-out solutions to any exercises you want to review are available in MyMathLab.

Find the perimeter and area of the figure.

1.

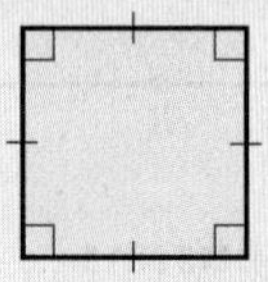

2. A garden that is 5 ft by 6 ft has a walkway 2 ft wide around it. What is the amount of fencing needed to surround the outside of the walkway?

Find the circumference and area of $\odot C$ in terms of π. For Exercise 4, also approximate the circumference of the circle by using $\pi \approx 3.14$.

3.

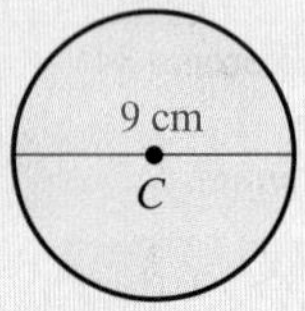

4.

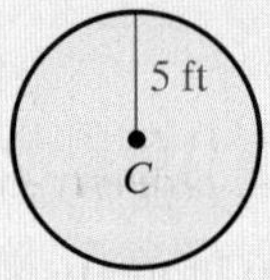

Use inductive reasoning to describe each pattern and find the next two terms of each list.

5. $-16, 8, -4, 2, \ldots$

6. $1, 4, 9, 16, 25, \ldots$

For Exercises 7 and 8, find a counterexample.

7. All snakes are poisonous.

8. If two angles are complementary, then they are not congruent.

9. Identify the hypothesis and conclusion:
If $x + 9 = 11$, then $x = 2$.

10. Write "quadrilaterals have four sides" as a conditional statement.

11. **Multiple Choice** Choose the negation of the statement: An obtuse angle measures 79°.

a. An obtuse angle measures between 90° and 180°.

b. An acute angle measures 79°.

c. An obtuse angle does not measure 79°.

d. An acute angle does not measure 79°.

For Exercises 12 and 13 write the converse, inverse, and contrapositive for each statement.

12. If a figure is a square, then it has at least two right angles.

13. If a square has side length 3 meters, then its perimeter is 12 meters.

14. Rewrite this biconditional as two conditionals.

A fish is a bluegill if and only if it is a bluish, freshwater sunfish.

15. **Multiple Steps** The following conditional statement is true. If A, B, and C are collinear, then they lie on the same line.

a. Write its converse.

b. Determine whether the converse is true.

c. If the converse is true, write the statement as a biconditional.

16. **Multiple Choice** Which of the following is a good definition?

a. Grass is a green growing plant.

b. Dinosaurs are extinct.

c. A pound weighs less than a ton.

d. A yard is a unit of measure exactly 3 feet long.

Explain why the statement is not a good definition.

17. Supplementary angles are angles that form a straight line.

Name the property that justifies each statement.

18. If $UV = KL$ and $KL = 6$, then $UV = 6$.

19. If $m\angle 1 + m\angle 2 = m\angle 4 + m\angle 2$, then $m\angle 1 = m\angle 4$.

20. $\angle B \cong \angle B$

21. If $\angle 1 \cong \angle 2$, then $\angle 2 \cong \angle 1$.

For each diagram, state two pairs of angles that are congruent. Justify your answers.

22.

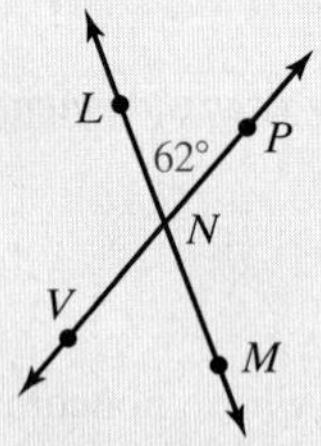

23.

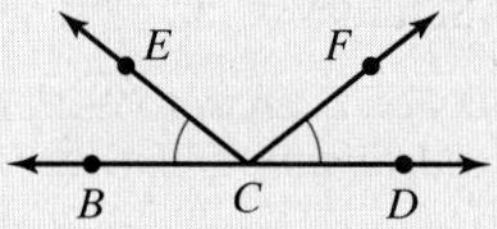

24. Find the value of the variable and the measures of all four angles.

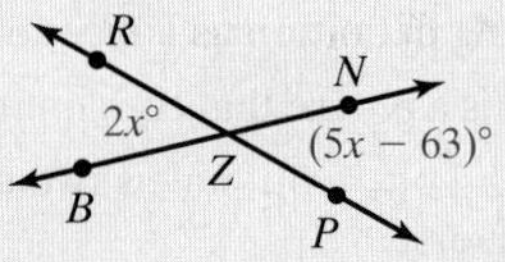

For Exercises 25 and 26, use the Law of Detachment and the Law of Syllogism to make any possible conclusion. Write "not possible" if you cannot make any conclusion.

25. If a student wants to be a chemical engineer, then that student must graduate from college. James wants to graduate from college.

26. If the traffic light is red, then you must stop.
If you must stop, then you must apply your brakes.

27. **Proof** Complete the proof.
Given: B is the midpoint of $\overline{AC}$.
Prove: $AB = \frac{AC}{2}$

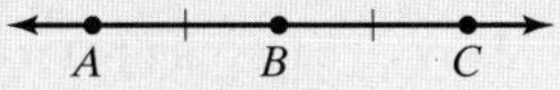

Proof:

Statements	Reasons
1. __?__ (a.)	1. Given
2. $AB = BC$	2. __?__ (b.)
3. $AB + BC = AC$	3. __?__ (c.)
4. $AB + AB = AC$	4. __?__ (d.)
5. __?__ (e.)	5. Simplify.
6. $AB = \frac{AC}{2}$	6. __?__ (f.)

28. **Proof** Complete this proof by filling in the blanks.
Given: $\angle FED$ and $\angle DEW$ are complementary.
Prove: $\angle FEW$ is a right angle.

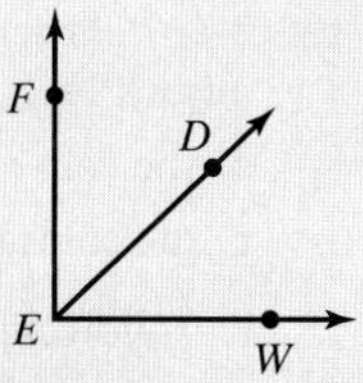

$\angle FED$ and $\angle DEW$ are complementary because it is given. By the definition of complementary angles, $m\angle FED + m\angle DEW =$ **a.** __?__. $m\angle FED + m\angle DEW = m\angle FEW$ by the **b.** __?__. $90° = m\angle FEW$ by the **c.** __?__ Property of Equality. Then $\angle FEW$ is a right angle by the **d.** __?__.

CHAPTER 2 STANDARDIZED TEST

Choose the best answer choice.

1. The perimeter of a square that measures 5 inches on a side is:

a. 25 in. **b.** 10 in.
c. 20 in. **d.** 25 sq. in.

2. The area of a rectangle that measures 5 yards in length and 2 feet in width is:

a. 10 sq. ft **b.** 10 sq. yd
c. 34 sq. ft **d.** 30 sq. ft

3. The exact circumference of a circle with a diameter of 10 meters is:

a. 10π m **b.** 5π m
c. 100π m **d.** 25π m

Give the next two terms in each list.

4. 1, 2, 4, 7, 11, . . .

a. 18, 29 **b.** 16, 22
c. 22, 44 **d.** 32, 64

5. 1, −2, 4, −8, 16, . . .

a. 32, −64 **b.** −32, 64
c. −48, 96 **d.** −64, 128

Identify the hypothesis and conclusion of each statement.

6. Pianists are musicians.

a. Hypothesis: A person is a musician.
Conclusion: A person is a pianist.

b. Hypothesis: A person is a pianist.
Conclusion: A person is a musician.

c. None of these

7. Dogs are retrievers.

a. Hypothesis: An animal is a dog.
Conclusion: An animal is a retriever.

b. Hypothesis: An animal is a retriever.
Conclusion: An animal is a dog

c. None of these

Use the conditional statement, "pine trees have needles."

8. "If a tree is not a pine tree, then it does not have needles" is ____________.

a. the converse **b.** the inverse
c. the contrapositive **d.** None of these

9. "If a tree does not have needles, then it is not a pine tree" is ____________.

a. the converse **b.** the inverse
c. the contrapositive **d.** None of these

10. "If a tree has needles, then it is not a pine tree" is ____________.

a. the converse **b.** the inverse
c. the contrapositive **d.** None of these

Use the conditional statement, "trucks have four-wheel drive."

11. "If a vehicle has four-wheel drive, then it is a truck" is ____________.

a. the converse **b.** the inverse
c. the contrapositive **d.** None of these

12. "If a vehicle is not a truck, then it does not have four-wheel drive" is ____________.

a. the converse **b.** the inverse
c. the contrapositive **d.** None of these

13. "If a vehicle does not have four-wheel drive, then it is not a truck" is ____________.

a. the converse **b.** the inverse

c. the contrapositive **d.** None of these

Use the following statements for Questions 11–12.

All good skiers wear helmets.
Emily is a good skier.
If you wear a helmet, you concerned about safety.
Erick wears a helmet.

14. What conclusion can be drawn using only the Law of Detachment?

a. Erick is a good skier.

b. Emily is a snowboarder and a skier.

c. Emily wears a helmet.

d. Erick is a snowboarder that wears a helmet.

15. What conclusion can be drawn using only the Law of Syllogism?

a. Erick is a good skier.

b. Emily is a snowboarder.

c. If you wear a helmet, you are a good skier.

d. All good skiers are concerned about safety.

16. Choose the negation of the statement: An acute angle measures 135°.

a. An acute angle does not measure 135°.

b. An obtuse angle measures 135°.

c. An obtuse angle does not measure 135°.

d. An acute angle measures between 0° and 90°.

17. Which of the following is a good definition?

a. Tables have exactly four legs.

b. Computers have screens.

c. A liter is close to a quart.

d. An inch is a unit of measure that is exactly 2.54 centimeters long.

Name the property that justifies each statement.

18. If $6x = 72$, then $x = 12$.

a. Addition Property of Equality

b. Symmetric Property of Equality

c. Division Property of Equality

d. Transitive Property of Equality

19. If $AB = DE$, then $AB + CD = CD + DE$.

a. Addition Property of Equality

b. Multiplication Property of Equality

c. Division Property of Equality

d. Transitive Property of Equality

20. If $2x - 5 = 27$, then $2x = 32$.

a. Addition Property of Equality

b. Multiplication Property of Equality

c. Division Property of Equality

d. Transitive Property of Equality

21. If $AB = CD$ and $CD = 15$, then $AB = 15$.

a. Addition Property of Equality

b. Multiplication Property of Equality

c. Division Property of Equality

d. Transitive Property of Equality

CHAPTER

3 Parallel and Perpendicular Lines

3.1 Lines and Angles

3.2 Proving Lines Are Parallel

3.3 Parallel Lines and Angles Formed by Transversals

3.4 Proving Theorems About Parallel and Perpendicular Lines

3.5 Constructions—Parallel and Perpendicular Lines

3.6 Coordinate Geometry—The Slope of a Line

3.7 Coordinate Geometry—Equations of Lines

An aircraft marshal w/ parallel signal lights

An aircraft marshal w/ perpendicular signal lights

It is simply not possible to name all of the real-life examples of parallel and perpendicular lines. For example, most rooms contain multiple examples of these special pairs of lines. Trees are usually parallel to each other and perpendicular to the ground. Lined paper contains many parallel lines, and so on. Start opening your mind so that you notice these special lines all around you.

Above are just two examples of the concepts of parallel and perpendicular lines that you may not have thought of. Do you know what these signals mean? I can assure you that pilots know and depend on them.

Rebecca A. Beavers

3.1 LINES AND ANGLES

OBJECTIVES

1 Identify Relationships Between Lines and Planes That Do Not Intersect.

2 Learn the Names of Angles Formed by Lines and a Transversal.

VOCABULARY

- parallel lines
- skew lines
- parallel planes
- parallel segments
- transversal
- interior angles
- exterior angles
- alternate interior angles
- same-side interior angles
- consecutive interior angles
- corresponding angles
- alternate exterior angles

OBJECTIVE 1 ▶ Identifying Lines and Planes That Do Not Intersect. In Chapter 1, we learned about lines and planes that intersect. In this chapter, we study lines and planes that do not intersect.

Defined Terms—Parallel and Skew

Definition	*Diagram*	*How to Name It*
Parallel lines are coplanar lines that do not intersect. The symbol ∥ means "is parallel to."	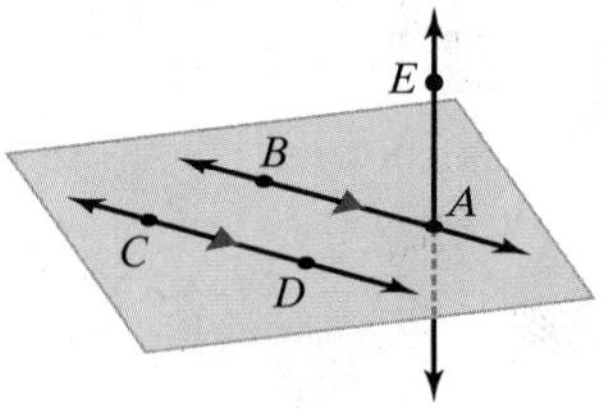	$\overleftrightarrow{AB}$ and $\overleftrightarrow{CD}$ are parallel lines. Symbols: $\overleftrightarrow{AB} \parallel \overleftrightarrow{CD}$
Skew lines are noncoplanar; they are not parallel and do not intersect.		$\overleftrightarrow{CD}$ and $\overleftrightarrow{EA}$ are skew lines.
Parallel planes are planes that do not intersect.	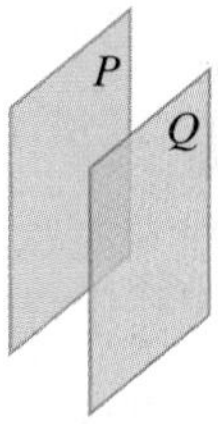	Planes P and Q are parallel planes. Symbols: Plane $P \parallel$ Plane Q

There are many examples of parallel lines, skew lines, and parallel planes in a simple box, as shown in the next example.

▶ Helpful Hint

If you have trouble visualizing a three-dimensional box drawn on flat paper, study any box that is handy such as a shoe box or a box containing food such as cereal, crackers, etc.

EXAMPLE 1 **Identifying Lines and Planes That Do Not Intersect**

Recall that each segment in the figure shown is part of a line, as shown for $\overleftrightarrow{GH}$. Answer the questions based on the appearance of the figure.

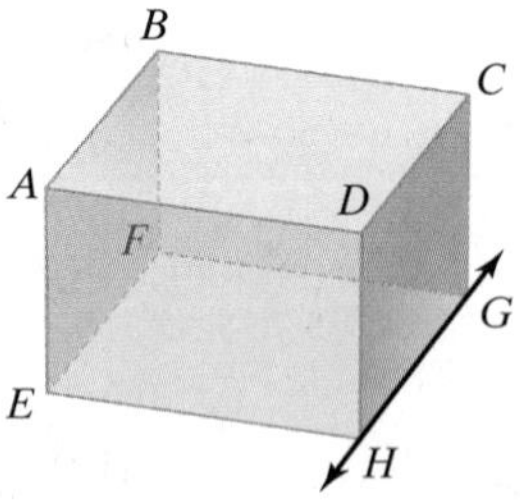

a. Which line(s) are parallel to $\overleftrightarrow{GH}$?

b. Which line(s) are skew to $\overleftrightarrow{GH}$ and also pass through point A?

c. Name any line(s) perpendicular to $\overleftrightarrow{GH}$.

d. Name any plane(s) parallel to plane $CDHG$.

Solution

a. $\overleftrightarrow{DC}$, $\overleftrightarrow{AB}$, and $\overleftrightarrow{EF}$ are parallel to $\overleftrightarrow{GH}$.

b. $\overleftrightarrow{AD}$ and $\overleftrightarrow{AE}$ are skew to $\overleftrightarrow{GH}$ and pass through point A.

c. $\overleftrightarrow{DH}$, $\overleftrightarrow{CG}$, $\overleftrightarrow{EH}$, and $\overleftrightarrow{FG}$ are perpendicular to $\overleftrightarrow{GH}$.

d. plane $ABFE \parallel$ plane $CDHG$. □

PRACTICE

1 Use the same figure and information from Example 1 to answer each question.

a. Which line(s) are parallel to $\overleftrightarrow{AD}$?

b. Which line(s) are skew to $\overleftrightarrow{AD}$ and also pass through point F?

c. Name any line(s) perpendicular to $\overleftrightarrow{AD}$.

d. Name any plane(s) parallel to plane $ABCD$.

We may sometimes hear or read the phrase **parallel segments.** To make sure there is no confusion, let's define that phrase.

Two segments are parallel if the lines containing them are parallel.

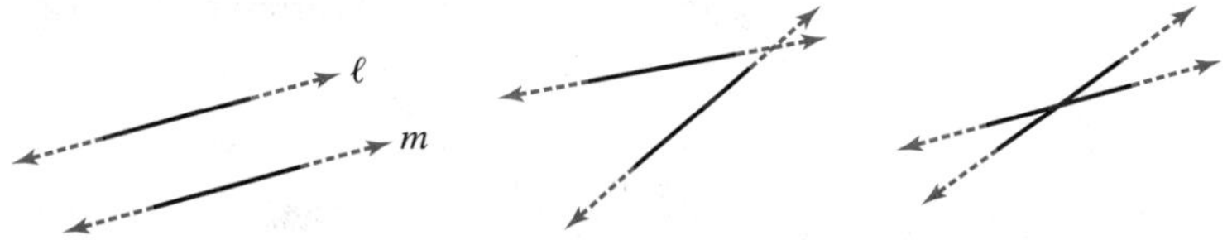

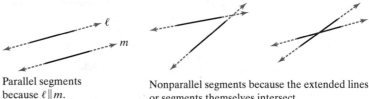

Parallel segments because $\ell \parallel m$.

Nonparallel segments because the extended lines or segments themselves intersect.

OBJECTIVE 2 ▶ Learning the Names of Angles Formed by Lines and a Transversal. Now that we know the definitions of parallel, skew, and perpendicular lines, let's define some additional terms that will help us analyze the relationships among these lines and the resulting angles that form. (Notice that by doing so, we are, once again, expanding our mathematical system called geometry.)

A **transversal** is a line that intersects two or more coplanar lines at different points. The figure below shows the eight angles formed by a transversal t and two lines ℓ and m.

> **▶ Helpful Hint**
> Notice that our definition of transversal has nothing to do with whether lines ℓ and m are parallel or not.

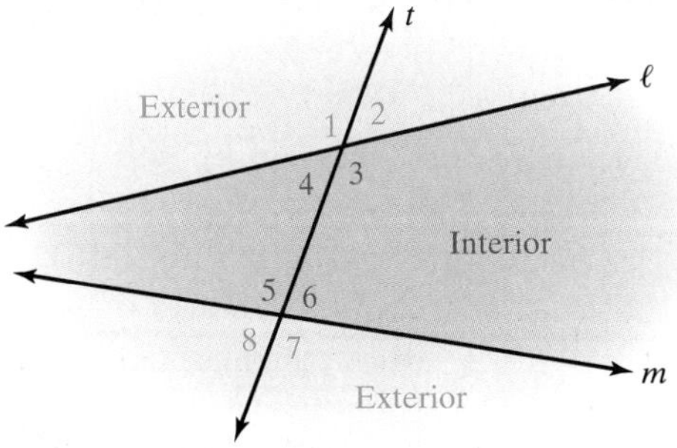

Angles 3, 4, 5, and 6 lie between ℓ and m. They are **interior angles.**
Angles 1, 2, 7, and 8 lie outside of ℓ and m. They are **exterior angles.**

Pairs of the eight angles have special names, as suggested by their positions.

Defined Terms—Angle Pairs Formed by Transversals

Definition	*Example and Diagram*	
Alternate interior angles are nonadjacent interior angles that lie on opposite sides of the transversal.	$\angle 4$ and $\angle 6$ $\angle 3$ and $\angle 5$	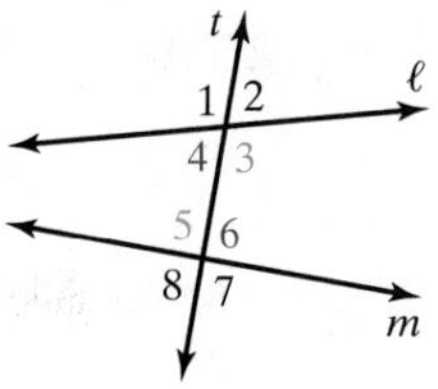
Same-side interior angles are interior angles that lie on the same side of the transversal (sometimes called **consecutive interior angles**).	$\angle 4$ and $\angle 5$ $\angle 3$ and $\angle 6$	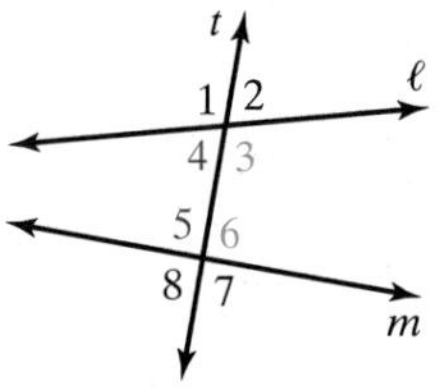
Corresponding angles lie on the same side of the transversal and in corresponding positions.	$\angle 1$ and $\angle 5$ $\angle 4$ and $\angle 8$ $\angle 2$ and $\angle 6$ $\angle 3$ and $\angle 7$	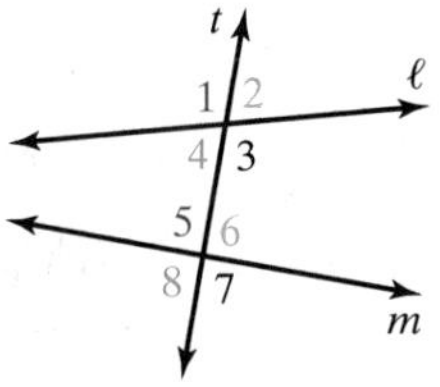
Alternate exterior angles are nonadjacent exterior angles that lie on opposite sides of the transversal.	$\angle 1$ and $\angle 7$ $\angle 2$ and $\angle 8$	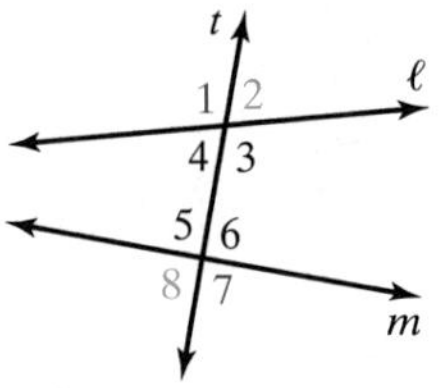

▶ **Helpful Hint**

The names of these defined angle pairs are the same whether lines ℓ and m are parallel or not.

▶ **Helpful Hint**

To help identify angle pairs, make sure we read their names carefully. For example:

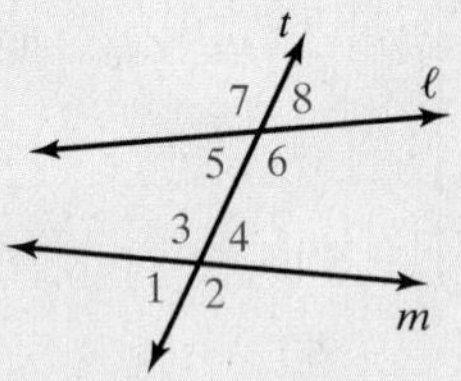

alternate interior angles

↗ lie on alternate sides of tranversal t

↖ lie in the interior of lines ℓ and m

Examples: $\angle 5$ and $\angle 4$
$\angle 3$ and $\angle 6$

corresponding angles

↑ lie in corresponding (or same) positions with respect to the transversal t, except one angle is formed with line ℓ and one angle is formed with line m

Examples: $\angle 8$ and $\angle 4$; $\angle 7$ and $\angle 3$
$\angle 6$ and $\angle 2$; $\angle 5$ and $\angle 1$

EXAMPLE 2 **Identifying Angle Pairs**

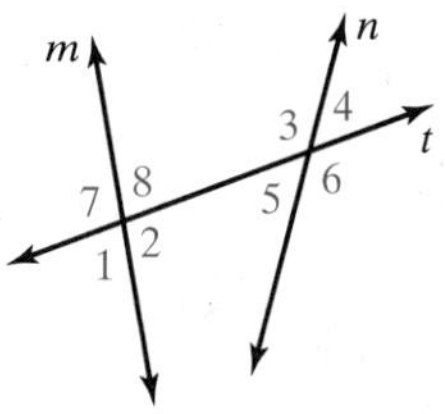

List all angle pairs in the figure.

a. alternate interior

b. corresponding

Solution

a. $\angle 8$ and $\angle 5$
$\angle 2$ and $\angle 3$

b. $\angle 7$ and $\angle 3$
$\angle 1$ and $\angle 5$
$\angle 8$ and $\angle 4$
$\angle 2$ and $\angle 6$ □

PRACTICE 2 List all angle pairs in the figure for Example 2.

a. alternate exterior **b.** same-side interior

EXAMPLE 3 **Classifying an Angle Pair**

The photo shows the Hearst Building in New York City. The new tower (showing many triangles) was completed in 2006.

Fill in the blank.

a. $\angle 1$ and $\angle 5$ are ____________ angles.

b. $\angle 2$ and $\angle 7$ are ____________ angles.

Solution

a. $\angle 1$ and $\angle 5$ are alternate exterior angles.

b. $\angle 2$ and $\angle 7$ are same-side interior angles. □

PRACTICE 3 Fill in the blanks.

a. $\angle 3$ and $\angle 7$ are ____________ angles.

b. $\angle 4$ and $\angle 6$ are ____________ angles.

VOCABULARY & READINESS CHECK

Word Bank *Use the choices to fill in each blank. Some choices may be used more than once and some not at all.*

skew lines	transversal	same-side interior
parallel planes	interior	corresponding
alternate exterior	exterior	
parallel lines	alternate interior	

1. Planes that do not intersect are called ____________.
2. Coplanar lines that do not intersect are called ____________.
3. A line that intersects two or more coplanar lines at different points is called a(n) ____________.
4. ____________ are not coplanar; they are not parallel and do not intersect.

Use the given figure and the Word Bank to fill in each blank.

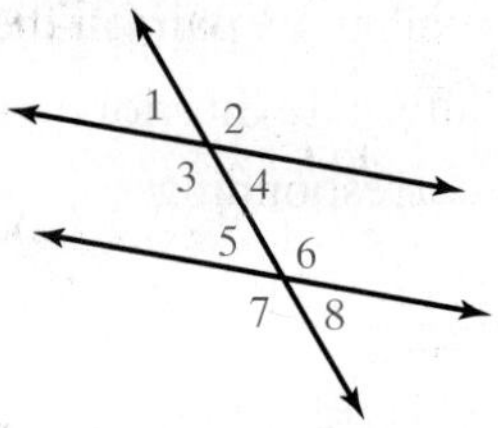

5. Angles 3, 4, 5, and 6 are called ____________ angles.
6. Angles 1, 2, 7, and 8 are called ____________ angles.
7. Angles 2 and 6 are called ____________ angles.
8. Angles 4 and 5 are called ____________ angles.
9. Angles 1 and 8 are called ____________ angles.
10. Angles 3 and 5 are called ____________ angles.

3.1 EXERCISE SET MyMathLab®

Fill in the Blank *Using the given figure, assume lines (extended segments) and planes that appear to be parallel are parallel. Fill in the blank with "parallel," "perpendicular," or "skew." See Example 1.*

1. $\overleftrightarrow{EH}$ and $\overleftrightarrow{HD}$ are ____________.

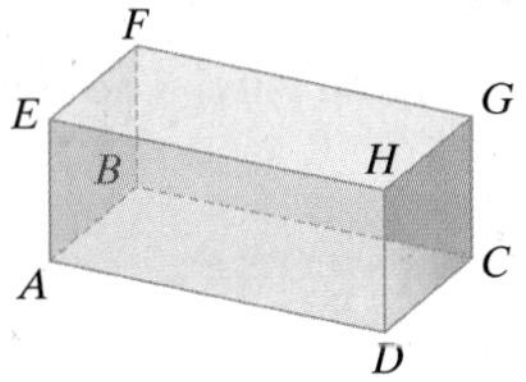

2. $\overleftrightarrow{EH}$ and $\overleftrightarrow{EA}$ are ____________.
3. $\overleftrightarrow{EH}$ and $\overleftrightarrow{GC}$ are ____________.
4. $\overleftrightarrow{HD}$ and $\overleftrightarrow{BC}$ are ____________.
5. $\overleftrightarrow{BC}$ and $\overleftrightarrow{AD}$ are ____________.
6. $\overleftrightarrow{FB}$ and $\overleftrightarrow{GC}$ are ____________.
7. Plane $DHGC$ and plane $AEFB$ are ____________.
8. Plane $ABCD$ and plane $EFGH$ are ____________.

Use the figure to name each of the following. Assume lines and planes that appear to be parallel are parallel. See Example 1.

9. a plane parallel to plane HDG

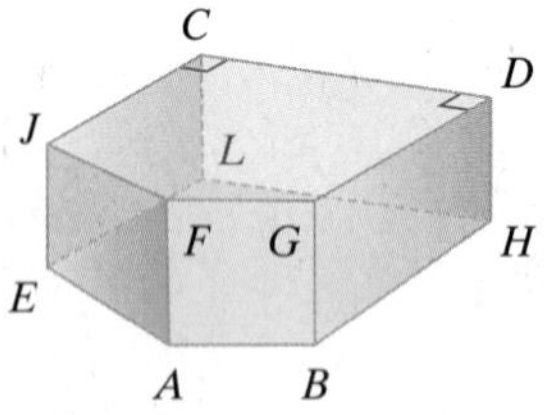

10. a plane parallel to plane JCD
11. a line that is parallel to $\overleftrightarrow{AB}$
12. a line that is parallel to $\overleftrightarrow{DH}$
13. a line that is skew to $\overleftrightarrow{EJ}$
14. a line that is skew to $\overleftrightarrow{BH}$

Multiple Choice *Are the angles labeled in the same color* **A.** *alternate interior angles,* **B.** *same-side interior angles,* **C.** *corresponding angles,* **D.** *alternate exterior angles? See Example 2.*

15.

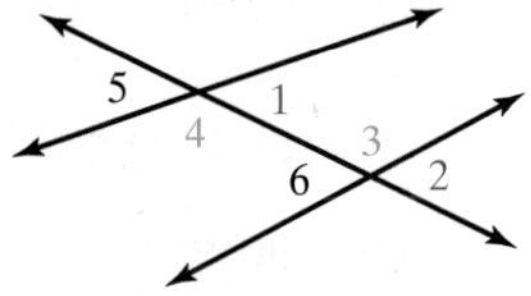

Blue angles are ____________.
Red angles are ____________.

16.

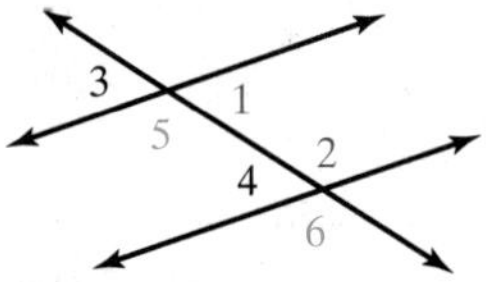

Blue angles are ____________.
Red angles are ____________.

17.

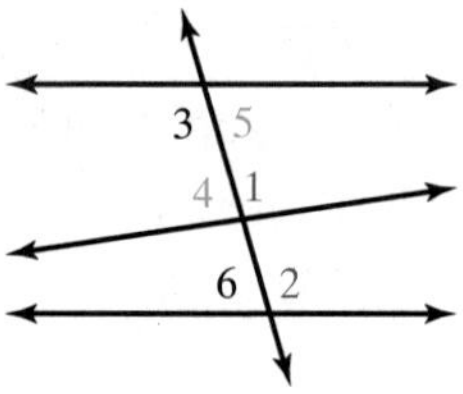

Blue angles are ____________.
Red angles are ____________.

18.

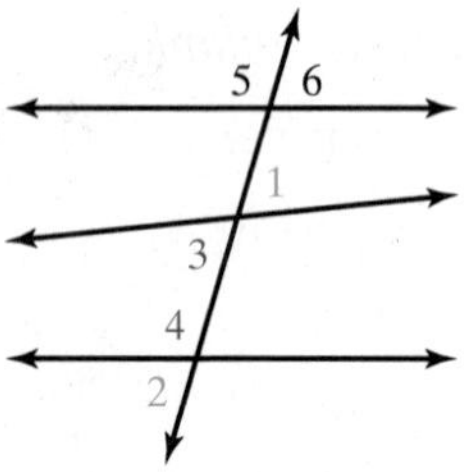

Blue angles are ____________.
Red angles are ____________.

The photo shows an overhead view of airport runways. Complete each statement with "alternate interior angles," "same-side interior angles," "corresponding angles," or "alternate exterior angles." See Examples 2 and 3.

19. ∠1 and ∠2 are ____________.

20. ∠2 and ∠3 are ____________.

Identify all pairs of each type of angles in the figure for the two lines and the transversal named. (Hint: If it helps, you may want to make your own drawing of the two lines, the transversal, and the numbered angles as shown.) See Example 2.

For Exercises 21–24, lines d and e with transversal line a.

21. corresponding angles

22. alternate interior angles

23. same-side interior angles

24. alternate exterior angles

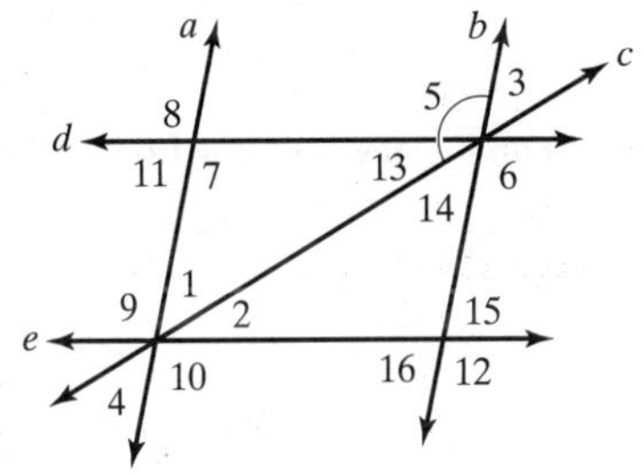

For Exercises 25–28, lines a and b with transversal line c.

25. alternate interior angles

26. corresponding angles

27. alternate exterior angles

28. same-side interior angles

Fill in the Blank *For Exercises 29–32, use the numbered angles in the figure to name the angle pairs. See Examples 2 and 3.*

29. ∠3 and ∠5 are ____________ angles.

30. ∠4 and ∠8 are ____________ angles.

31. ∠2 and ∠5 are ____________ angles.

32. ∠3 and ∠7 are ____________ angles.

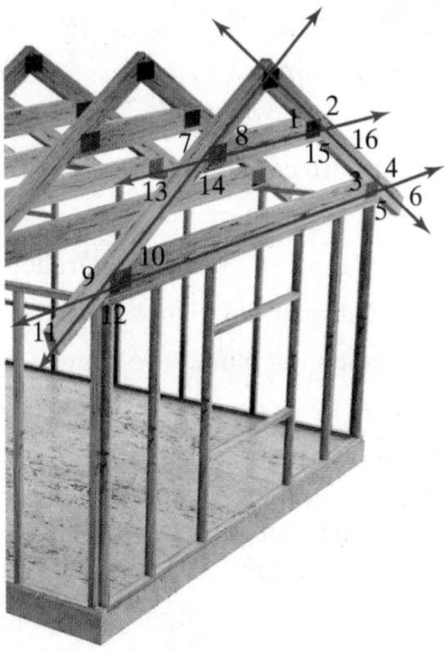

Fill in the Blank *Use the numbered angles in the construction drawing to name the angle pairs. See Examples 2 and 3.*

33. ∠2 and ∠4 are ____________ angles.

34. ∠13 and ∠11 are ____________ angles.

35. ∠13 and ∠10 are ____________ angles.

36. ∠15 and ∠4 are ____________ angles.

37. ∠1 and ∠6 are ____________ angles.

38. ∠7 and ∠12 are ____________ angles.

In Exercises 39–44, use the figure shown and describe the statement as true or false. If false, explain. Assume that lines and planes that appear to be parallel are parallel. See Example 1.

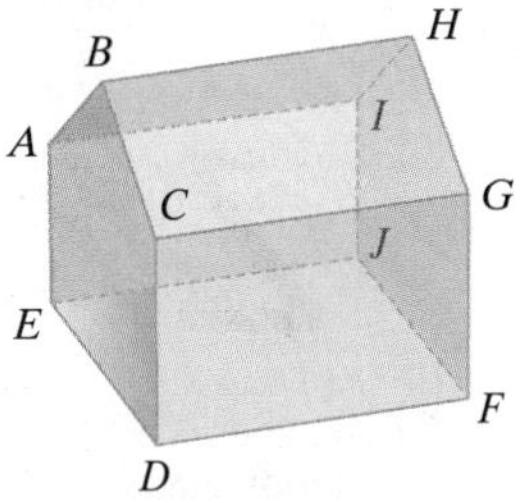

39. $\overleftrightarrow{CB} \parallel \overleftrightarrow{HG}$

40. $\overleftrightarrow{ED} \parallel \overleftrightarrow{HG}$

41. plane $AED \parallel$ plane FGH

42. plane $ABH \parallel$ plane CDF

43. $\overleftrightarrow{AB}$ and $\overleftrightarrow{HG}$ are skew lines.

44. $\overleftrightarrow{AE}$ and $\overleftrightarrow{BC}$ are skew lines.

Sketch two lines and a transversal. Use your sketch to count how many pairs of each type of angles are formed. See Example 2.

45. alternate interior angles

46. corresponding angles

47. alternate exterior angles

48. vertical angles

CONCEPT EXTENSIONS

In Exercises 49–54, determine whether each statement is always, sometimes, or never true. See Example 1.

49. Two parallel lines are coplanar.

50. Two skew lines are coplanar.

51. Two planes that do not intersect are parallel.

52. Two lines that lie in parallel planes are parallel.

53. Two lines in intersecting planes are skew.

54. A line and a plane that do not intersect are skew.

55. You and a friend are driving go-karts on two different tracks. As you drive on a straight section heading east, your friend passes above you on a straight section heading south. Are these sections of the two tracks parallel, skew, or neither? Explain.

56. A rectangular rug covers the floor in a living room. One of the walls in the same living room is painted blue. Are the rug and the blue wall parallel? Explain.

57. Multiple Steps

a. Describe the three ways in which two lines may be related.

b. Give examples from the real world to illustrate each of the relationships you described in part a.

58. Multiple Steps

a. Suppose two parallel planes A and B are each intersected by a third plane, C. Make a conjecture about the intersection of planes A and C and the intersection of planes B and C.

b. Find examples in your classroom to illustrate your conjecture in part a.

59. Sketch The letter Z illustrates alternate interior angles, as shown in the figure. Sketch at least two other letters that illustrate pairs of angles presented in this lesson. Then mark and describe the angles.

$\angle 1$ and $\angle 2$ are alternate interior angles

60. Sketch A transversal r intersects lines ℓ and m. If ℓ and r form $\angle 1$ and $\angle 2$ and m and r form $\angle 3$ and $\angle 4$, sketch a diagram that meets all of the following conditions:

- $\angle 1 \cong \angle 2$
- $\angle 3$ is an interior angle.
- $\angle 4$ is an exterior angle.
- $\angle 3$ and $\angle 4$ are supplementary.
- $\angle 2$ and $\angle 4$ lie on opposite sides of r.

Use the figure for Exercises 61 and 62.

61. Do planes A and B have lines in common other than $\overleftrightarrow{CD}$? Explain.

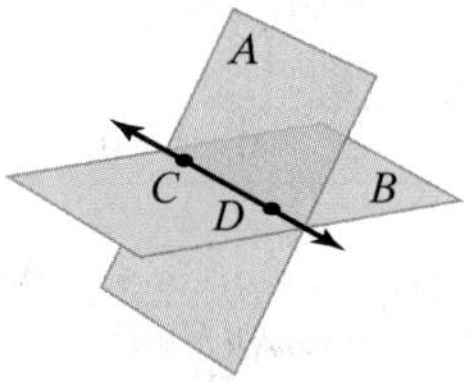

62. Sketch Are there planes that intersect planes A and B in lines parallel to $\overleftrightarrow{CD}$? Draw a sketch to support your answer.

REVIEW AND PREVIEW

If $m\angle YDF = 121°$ and $\overrightarrow{DR}$ bisects $\angle FDI$, find the measure of each angle. See Sections 1.6 and 2.6.

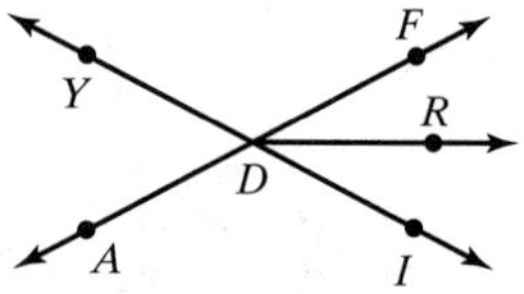

63. $\angle IDA$ **64.** $\angle YDA$ **65.** $\angle RDI$

66. $\angle FDR$ **67.** $\angle YDR$ **68.** $\angle ADR$

Classify each pair of angles. See this section.

69. $\angle 4$ and $\angle 2$

70. $\angle 6$ and $\angle 3$

71. $\angle 4$ and $\angle 5$

72. $\angle 6$ and $\angle 7$

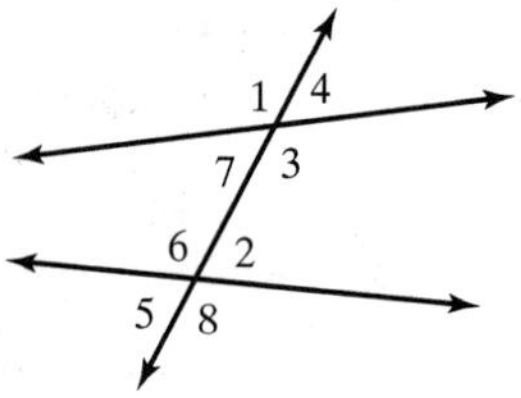

3.2 PROVING LINES ARE PARALLEL

OBJECTIVES

1 Use Theorems to Prove that Two Lines are Parallel.

2 Use Algebra to Find the Measures of Angles Needed so that Lines are Parallel.

OBJECTIVE 1 ▶ Using Theorems to Prove that Two Lines Are Parallel. Examples of parallel lines and transversals are all around us in everyday situations, as shown in the margins on the following page.

In Section 3.1, we learned the names of angles formed by two lines and a transversal. In this section, we use these names of angles and learn how to prove whether two lines are parallel or not.

First, let's introduce the most important postulate in our geometry on a plane—the Parallel Postulate.

VOCABULARY
- flow proof

Postulate 3.2-1 Parallel Postulate

Through a point not on a line, there is one and only one line parallel to the given line.

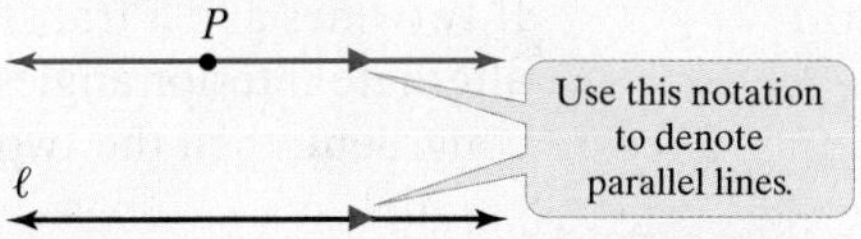

There is exactly one line through P parallel to ℓ.

Real-Life Examples of Parallel Lines and Transversals

Railroad tracks

Handicap ramp with railings

Parking lot striping

Recall that we agreed to accept postulates as true. For about 2000 years, mathematicians did try to prove this Parallel Postulate, but failed to do so. In fact, there are options other than this postulate that lead to additional valid geometries. (See the History Note at the end of this section.) You may be wondering why we introduce the Parallel Postulate now—in this section on proving lines are parallel. The reason is amazing. The proofs of the theorems we introduce in this section use and depend on the existence of the Parallel Postulate, and also the following postulate and theorem.

Postulate 3.2-2 Perpendicular Postulate

Through a point not on a line, there is one and only one line perpendicular to the given line.

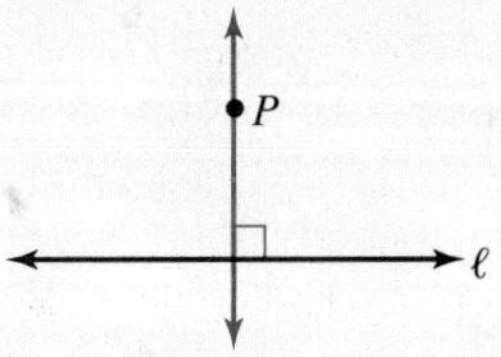

Remember, we accept postulates as true and do not prove them.

Theorem 3.2-3 Two Lines Perpendicular to a Third Line

Theorem	If . . .	Then . . .
In a plane, if two lines are perpendicular to the same line, then they are parallel to each other.	$m \perp t$ and $n \perp t$ 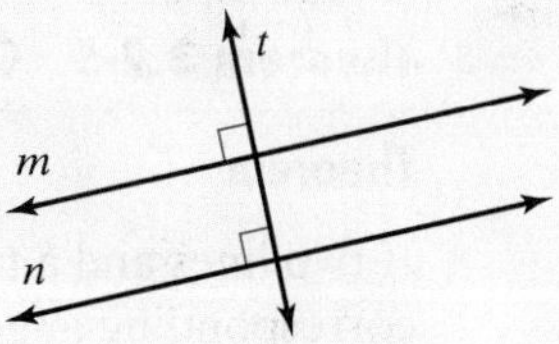	$m \parallel n$ 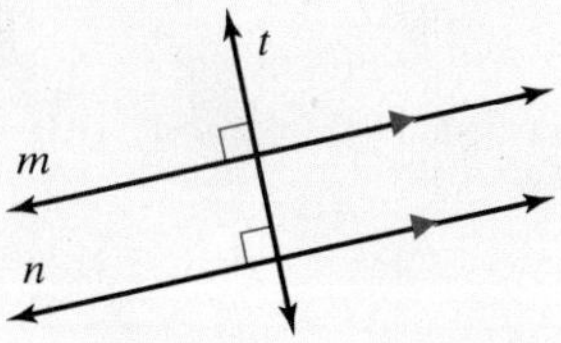

We prove this theorem in the back of this text.

Helpful Hint

Make sure you read and understand this important information about a valid geometry.

It is VERY IMPORTANT to understand that in order to develop a valid geometry, as we are doing, we may only prove a theorem using earlier proved theorems. In this text, I thought it best to cluster material on parallel and perpendicular lines, and next do the same with triangles. The end result is that the proof of a few theorems in this chapter may not be shown until later, when a needed postulate is stated or a needed type of proof is shown. Given the above, please be assured that the proof of theorems is done in an order so that our geometry is valid.

Theorem 3.2-4 Alternate Interior Angles Theorem

Theorem	If . . .	Then . . .
If two lines and a transversal form alternate interior angles that are congruent, then the two lines are parallel.	$\angle 4 \cong \angle 6$ 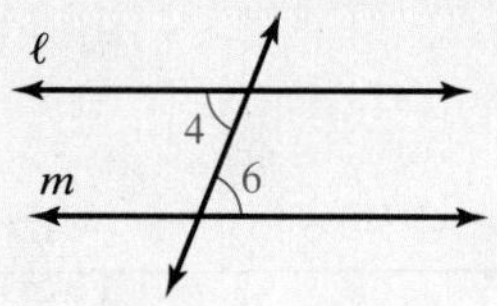	$\ell \parallel m$

We prove this theorem in Section 4.5, but we use no "new" theorems to do so.

For now, let's use the Alternate Interior Angles Theorem to show that two lines are parallel.

EXAMPLE 1 Identifying Parallel Lines

Which lines are parallel if $\angle 1 \cong \angle 7$? Justify your answer.

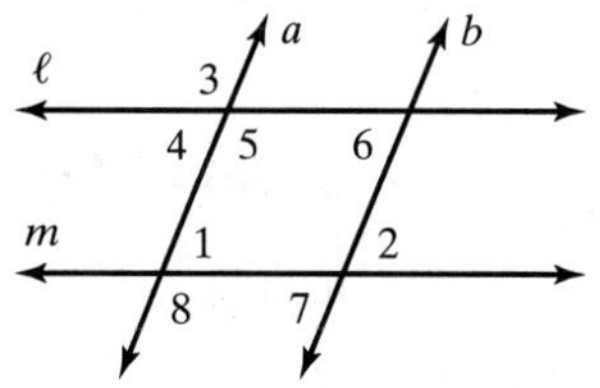

Solution $\angle 1$ and $\angle 7$ are *not* formed by line ℓ, so we concentrate on line a and line b with transversal m.

$\angle 1$ and $\angle 7$ are alternate interior angles. If $\angle 1 \cong \angle 7$, then $a \parallel b$ by the Alternate Interior Angles Theorem. □

PRACTICE 1 Which lines are parallel if $\angle 6 \cong \angle 2$? Justify your answer.

Let's continue with other theorems we may use to prove that two lines are parallel.

Theorem 3.2-5 Corresponding Angles Theorem

Theorem	If . . .	Then . . .
If two lines and a transversal form corresponding angles that are congruent, then the lines are parallel.	$\angle 2 \cong \angle 6$ 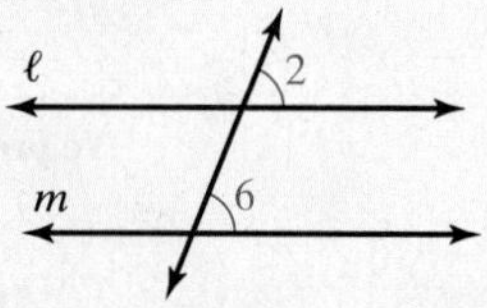	$\ell \parallel m$

Theorem 3.2-6 Same-Side Interior Angles Theorem

Theorem	If . . .	Then . . .
If two lines and a transversal form same-side interior angles that are supplementary, then the two lines are parallel.	$m\angle 3 + m\angle 6 = 180°$ 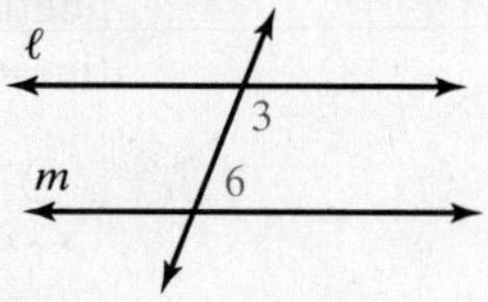	$\ell \parallel m$

Theorem 3.2-7 Alternate Exterior Angles Theorem

Theorem	If . . .	Then . . .
If two lines and a transversal form alternate exterior angles that are congruent, then the two lines are parallel.	$\angle 1 \cong \angle 7$ 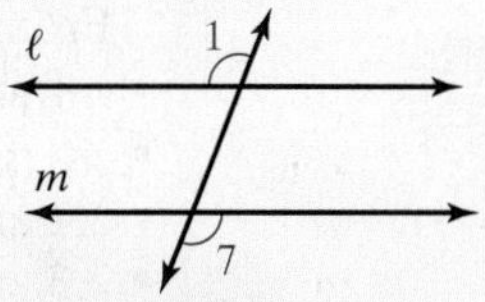	$\ell \parallel m$

We prove Theorem 3.2-5 next. We prove Theorems 3.2-6 and 3.2-7 in Practice 2 and Example 2 of this section.

We now use the Alternate Interior Angles Theorem to prove the Corresponding Angles Theorem.

Proof Proof of the Corresponding Angles Theorem (Theorem 3.2-5)

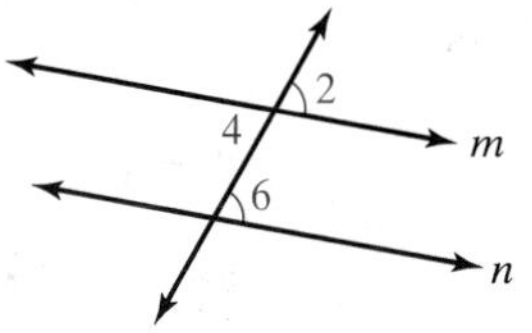

Given: $\angle 2 \cong \angle 6$

Prove: $m \parallel n$

Here, we use a two-column, statement/reason proof.

Statements	Reasons
1. $\angle 2 \cong \angle 6$	**1.** Given
2. $\angle 2 \cong \angle 4$	**2.** Vertical Angles Theorem
3. $\angle 4 \cong \angle 6$	**3.** Substitution (or Transitive Property) (Steps 1, 2)
4. $m \parallel n$	**4.** Alternate Interior Angles Theorem (Step 3)

Thus far, we have learned two forms of proof—paragraph and two-column (statement/reason). In a third form, called **flow proof,** arrows show the logical connections between the statements. Reasons are written below the statements. Here we show the same proof as above, but in flow-proof form.

Proof Flow Proof of the Corresponding Angles Theorem (Theorem 3.2-5)

Given: $\angle 2 \cong \angle 6$

Prove: $m \parallel n$

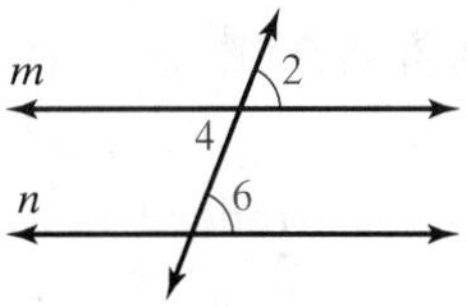

Flow Proof:

$\angle 2 \cong \angle 6$
Given
→

$\angle 2 \cong \angle 4$
Vertical angles are $\cong$.
→

$\angle 4 \cong \angle 6$
Substitution (or Transitive Property)
→

$m \parallel n$
If alternate interior angles are $\cong$, then the lines are parallel.

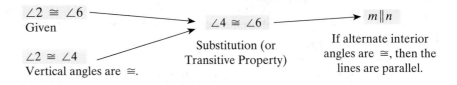

EXAMPLE 2 **Writing a Flow Proof of the Alternate Exterior Angles Theorem (Theorem 3.2-7)**

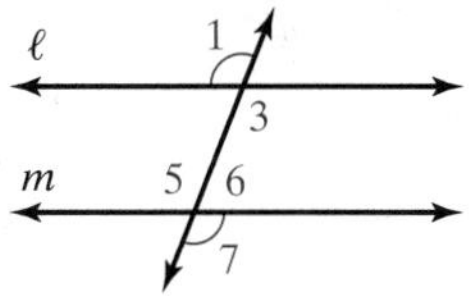

Given: $\angle 1 \cong \angle 7$
Prove: $\ell \parallel m$

Solution It may help to organize a plan.

What We Know:	What We Need:	What to Do:
• $\angle 1 \cong \angle 7$ From the diagram we know • $\angle 1$ and $\angle 3$ are vertical • $\angle 5$ and $\angle 7$ are vertical • $\angle 1$ and $\angle 5$ are corresponding • $\angle 3$ and $\angle 7$ are corresponding • $\angle 3$ and $\angle 5$ are alternate interior	One pair of corresponding or alternate interior angles congruent to prove $\ell \parallel m$	Use a pair of congruent vertical angles to relate either $\angle 1$ or $\angle 7$ to its corresponding angle.

Flow Proof:

$\angle 1 \cong \angle 7$
Given

$\angle 3 \cong \angle 1$
Vertical angles are $\cong$.

→ $\angle 3 \cong \angle 7$
Substitution (or Transitive Property)

→ $\ell \parallel m$
If corresponding angles are $\cong$, then the lines are parallel.

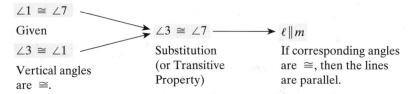

PRACTICE 2 Use the same figure from Example 2 to prove that $\ell \parallel m$ given that $\angle 3$ and $\angle 6$ are supplementary. Write a flow proof of the Same-Side Interior Angles Theorem 3.2-6.

We now have five ways to prove that two lines *m* and *n* are parallel.

1. m and n are both $\perp$ to transversal t. (Theorem 3.2-3)
2. Alternate interior angles are equal. (Theorem 3.2-4)
3. Corresponding angles are equal. (Theorem 3.2-5)
4. Interior angles on the same side of a transversal are supplementary. (Theorem 3.2-6)
5. Alternate exterior angles are equal. (Theorem 3.2-7)

Now let's practice what we have learned by solving an application.

EXAMPLE 3 Determining Whether Lines Are Parallel

The fence gate at the right is made up of pieces of wood arranged in various directions. Suppose $\angle 1 \cong \angle 2$. Are lines r and s parallel? Explain.

Solution Yes, $r \parallel s$. $\angle 1$ and $\angle 2$ are alternate exterior angles. If two lines and a transversal form congruent alternate exterior angles, then the lines are parallel (Alternate Exterior Angles Theorem).

PRACTICE 3 In Example 3, what is another way to explain why $r \parallel s$? Justify your answer.

OBJECTIVE 2 ▶ Using Algebra to Find the Angle Measures Needed so that Lines Are Parallel. We can use algebra along with the theorems from this section to solve problems involving parallel lines.

EXAMPLE 4 **Using Algebra to Prove Lines Are Parallel**

What is the value of x that makes $a \| b$?

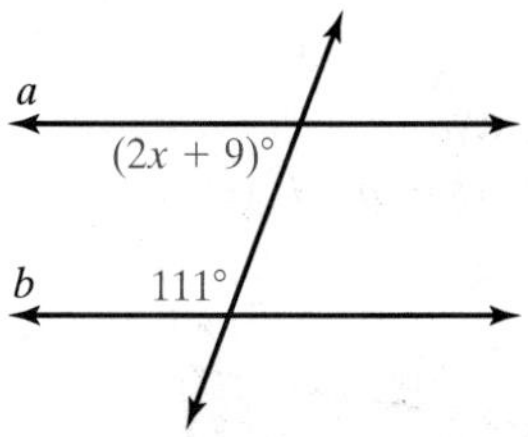

Solution The two angles are same-side interior angles. By the Same-Side Interior Angles Theorem, $a \| b$ if the angles are supplementary.

$(2x + 9) + 111 = 180$ Same-Side Interior Angles Theorem

$2x + 120 = 180$ Simplify.

$2x = 60$ Subtract 120 from each side.

$x = 30$ Divide each side by 2.

Thus, if $x = 30$, then $a \| b$. □

PRACTICE 4 What is the value of w that makes $c \| d$?

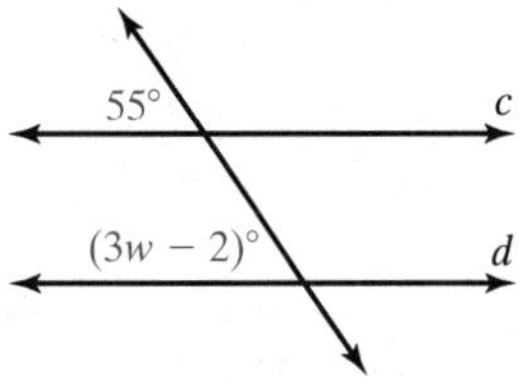

VOCABULARY & READINESS CHECK

Fill in the Blank *Refer to the figure to fill in each blank.*

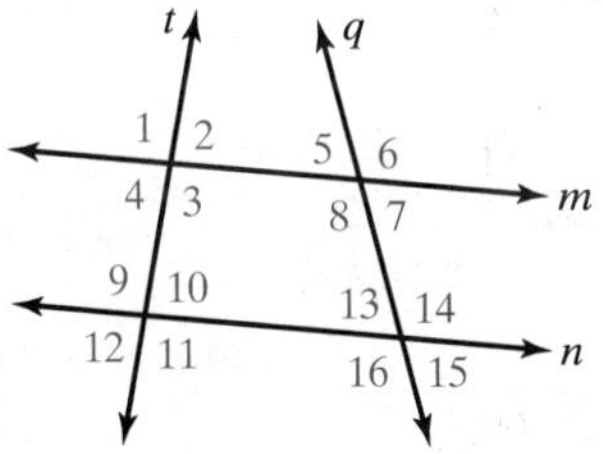

1. ∠5 and ____________ are vertical angles.
2. ∠10 and ____________ are vertical angles.

Use lines m and n with transversal t.

3. ∠4 and ____________ are alternate interior angles.
4. ∠2 and ____________ are alternate exterior angles.
5. ∠10 and ____________ are same-side interior angles.
6. ∠1 and ____________ are corresponding angles.

Use lines m and n with transversal q.

7. ∠5 and ____________ are alternate exterior angles.
8. ∠8 and ____________ are alternate interior angles.
9. ∠14 and ____________ are corresponding angles.
10. ∠7 and ____________ are same-side interior angles.

3.2 EXERCISE SET MyMathLab®

With the given information, can we determine that $m \parallel n$? Write yes or no. If yes, state the theorem used. See Examples 1 and 3.

1.

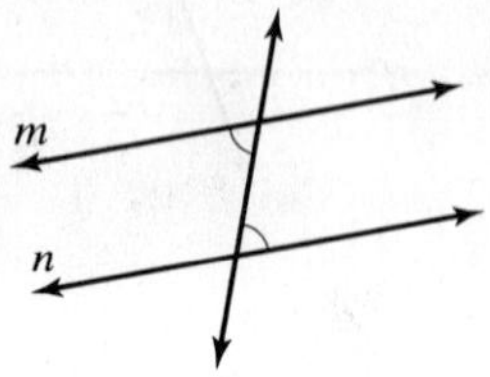

2.

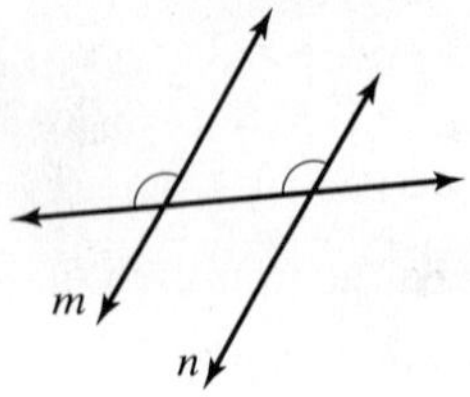

3.

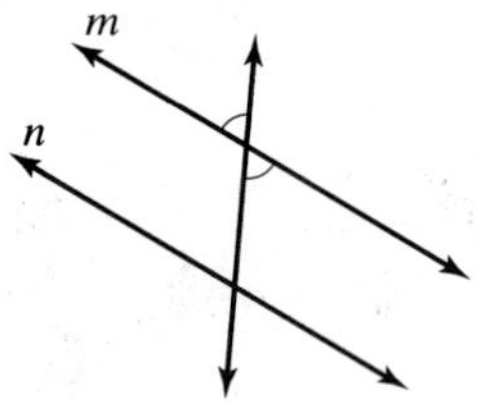

4.

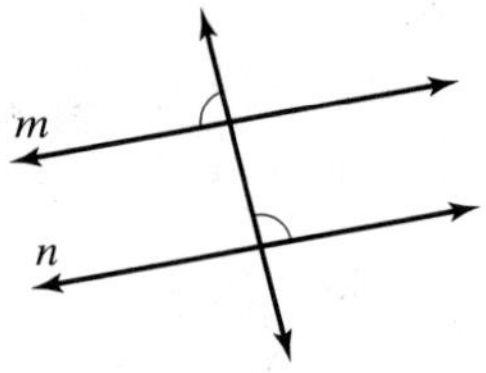

5.

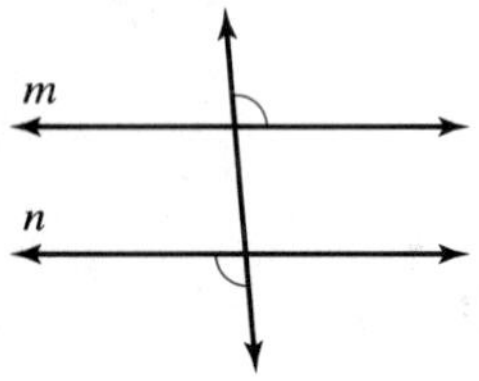

6.

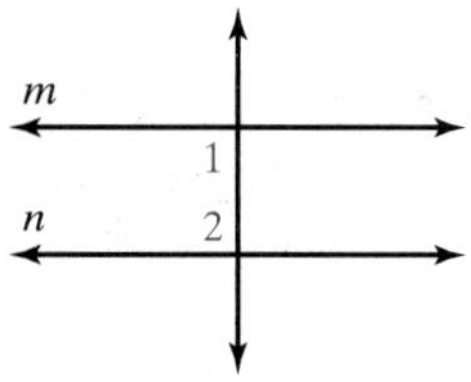

$m\angle 1 + m\angle 2 = 180°$

7.

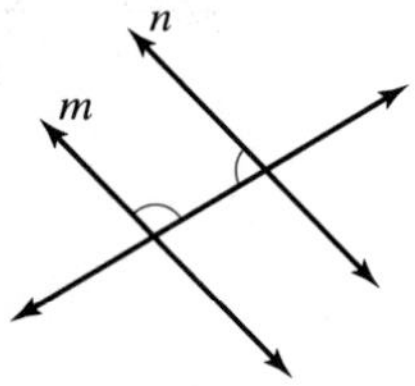

8.

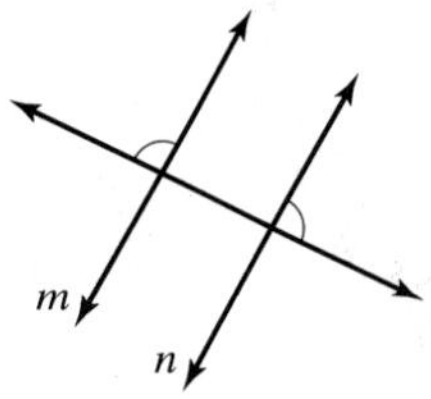

Which lines or segments are parallel? Justify your answer. See Examples 1 and 3.

9.

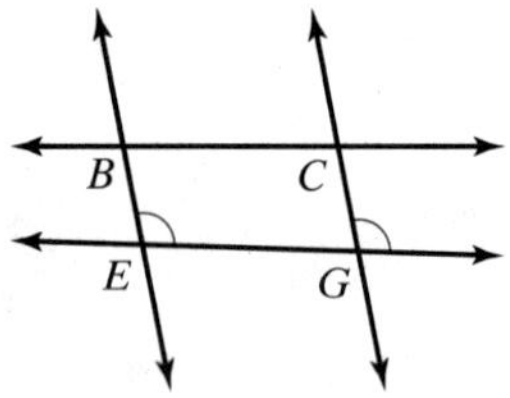

10.

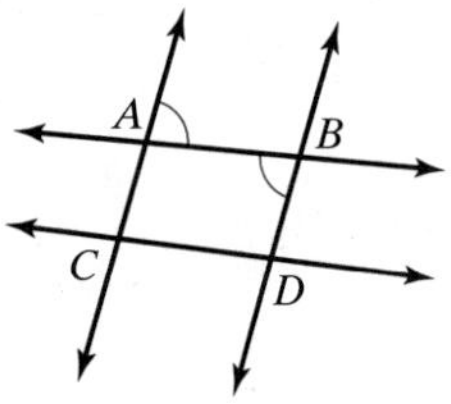

11.

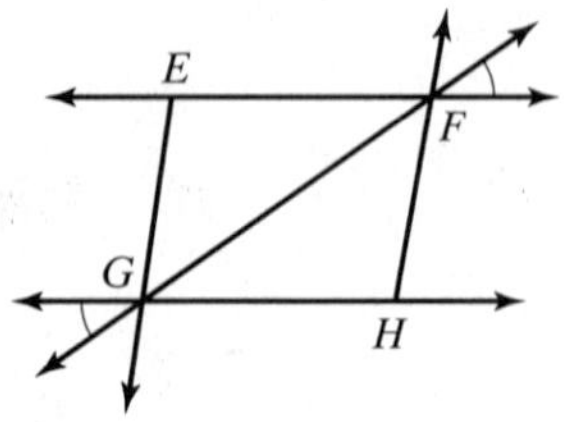

12.

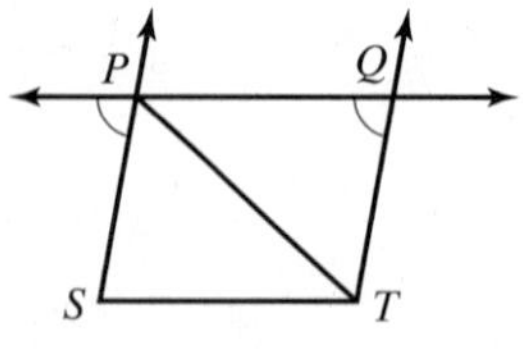

13.

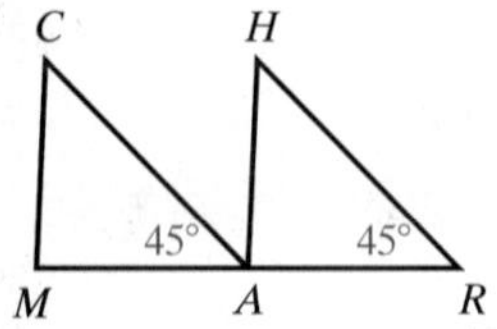

14.

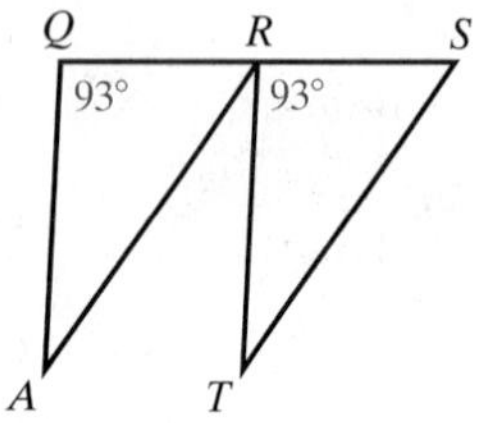

15.

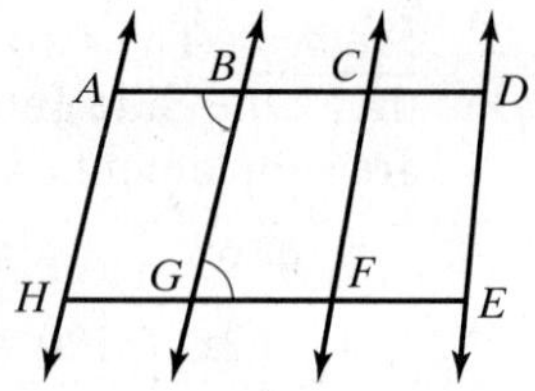

16.

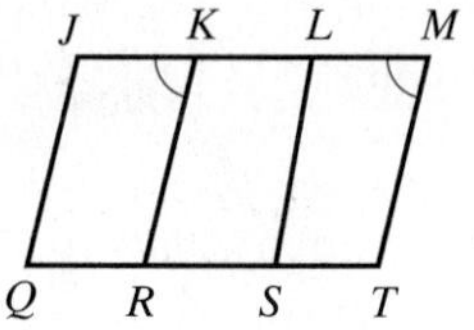

Which lines, if any, are parallel? If none of the lines are parallel, write "no." Explain the theorem used.

17.

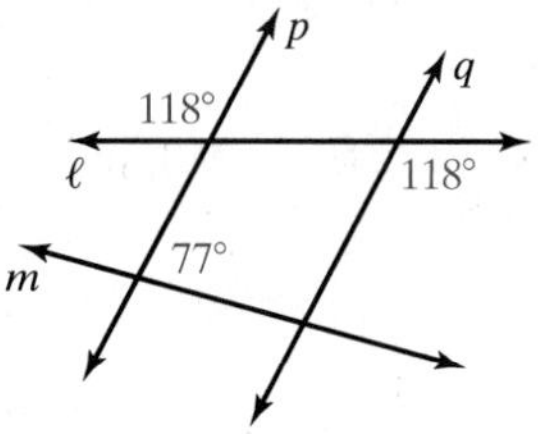

18.

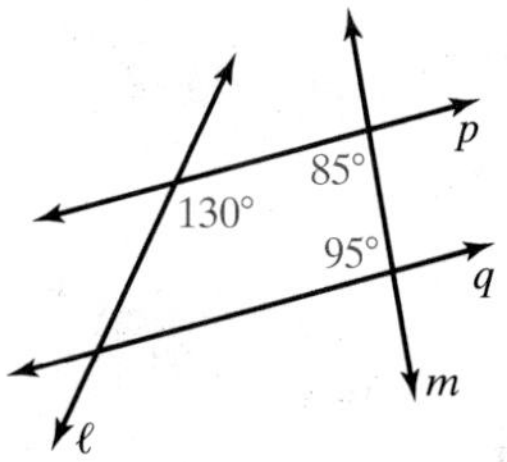

19.

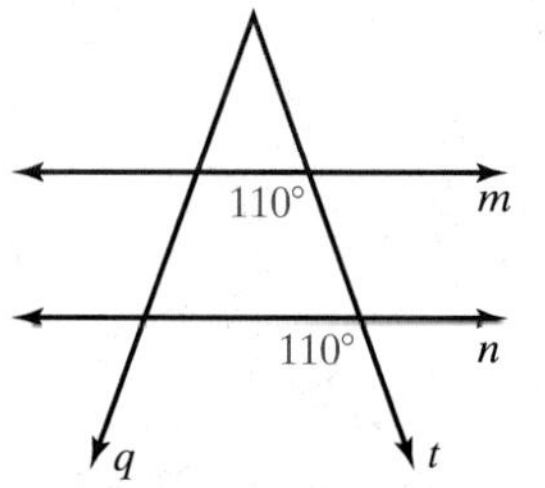

20.

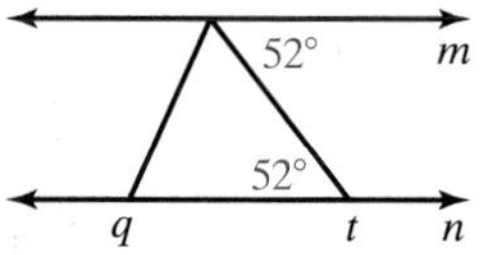

21.

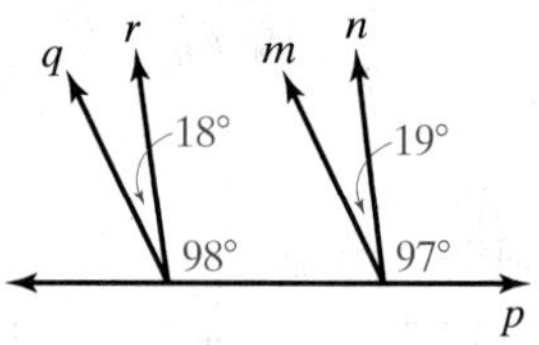

22.

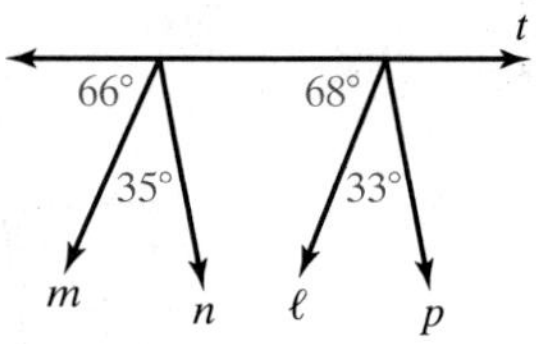

23.

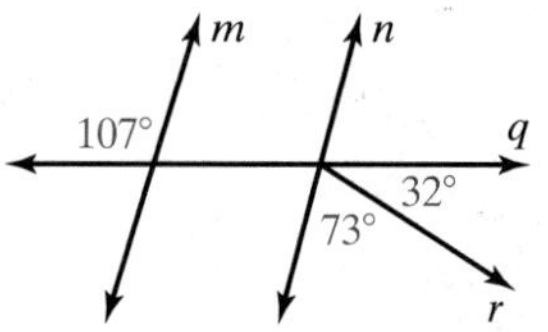

24.

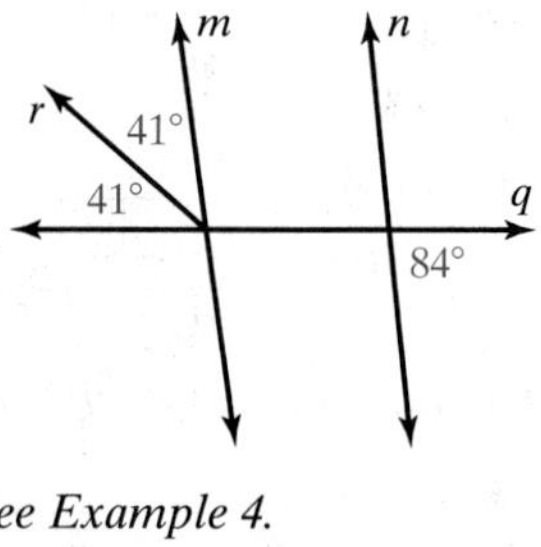

Find the value of x that makes $\ell \parallel m$. See Example 4.

25.

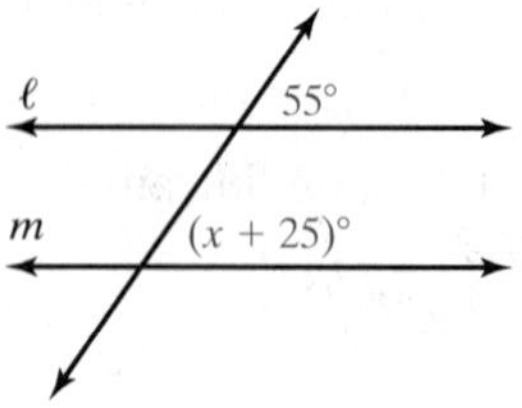

26.

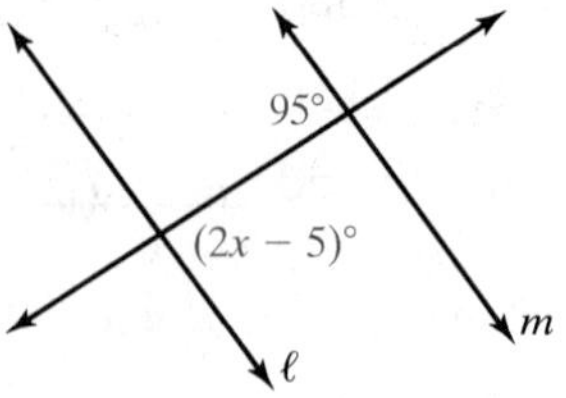

27.

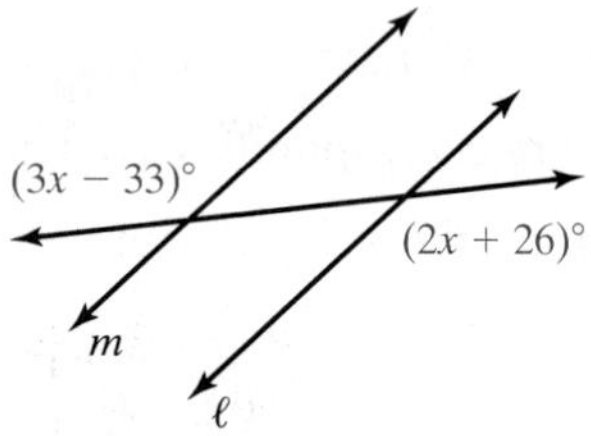

28.

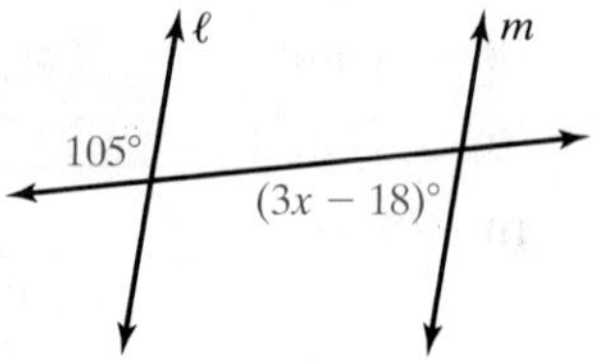

Solve. See Example 3.

29. Two workers paint lines for angled parking spaces. One worker paints a line so that $m\angle 1 = 65°$. The other worker paints a line so that $m\angle 2 = 65°$. Are their lines parallel? Explain.

30. From Exercise **29**, suppose one worker paints a line so that $m\angle 1 = 65°$, and the other worker paints a line so that $m\angle 3 = 65°$. Are their lines parallel? Explain.

Determine the value of x that makes $r \parallel s$. Then find $m\angle 1$ and $m\angle 2$. See Example 4.

31. $m\angle 1 = 80 - x, m\angle 2 = 90 - 2x$

32. $m\angle 1 = 60 - 2x, m\angle 2 = 70 - 4x$

33. $m\angle 1 = 40 - 4x, m\angle 2 = 50 - 8x$

34. $m\angle 1 = 20 - 8x, m\angle 2 = 30 - 16x$

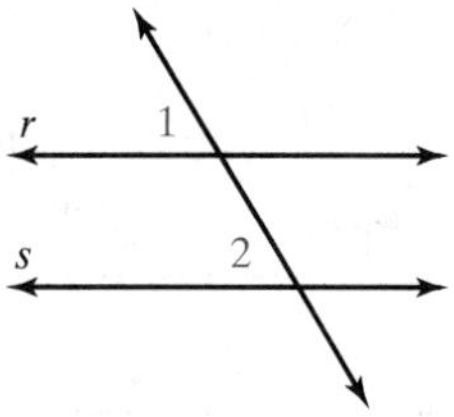

Use the diagram below for Exercises 35–38.

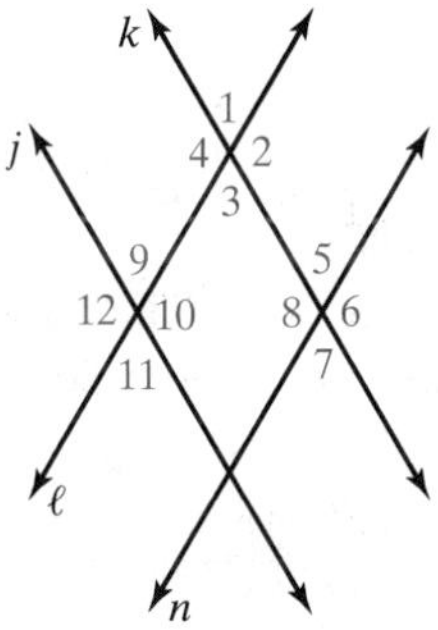

35. If $\angle 3 \cong \angle 9$, what theorem can you use to show that $j \parallel k$?

36. If $\angle 1 \cong \angle 7$, what theorem can you use to show that $\ell \parallel n$?

Proofs *Write a flow proof. See Example 3.*

37. Given: $\angle 12 \cong \angle 8, \angle 8 \cong \angle 4$
Prove: $j \parallel k$ and $\ell \parallel n$

38. Given: $m\angle 8 + m\angle 9 = 180°, m\angle 8 + m\angle 3 = 180°$
Prove: $\ell \parallel n$ and $j \parallel k$

CONCEPT EXTENSIONS

Name a Theorem *Use the given information to determine which lines, if any, are parallel. Justify each conclusion with a theorem.*

39. $\angle 2$ is supplementary to $\angle 3$.

40. $\angle 6$ is supplementary to $\angle 7$.

41. $\angle 1 \cong \angle 8$

42. $\angle 11 \cong \angle 7$

43. $\angle 1 \cong \angle 3$

44. $\angle 2 \cong \angle 10$

45. $\angle 8 \cong \angle 6$

46. $\angle 5 \cong \angle 10$

47. $\angle 9 \cong \angle 12$

48. $\angle 5 \cong \angle 2$

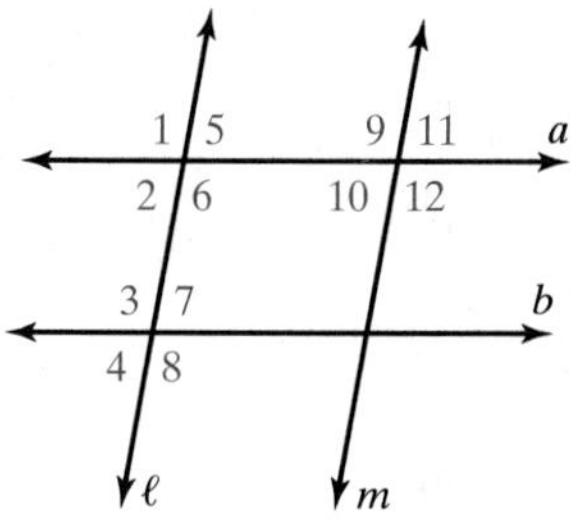

For Exercises 49 and 50, find the value of x that makes $\ell \parallel m$.

49.

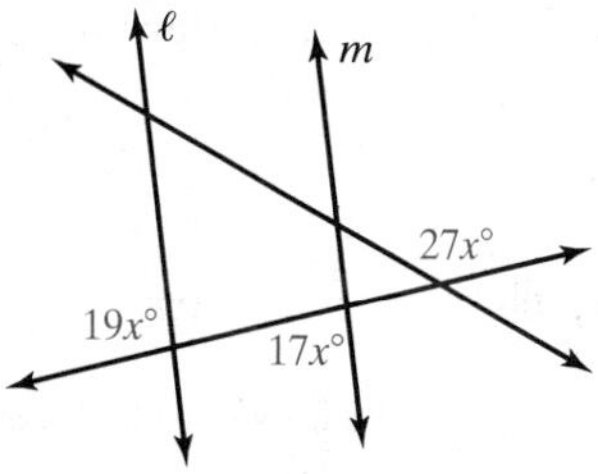

50.

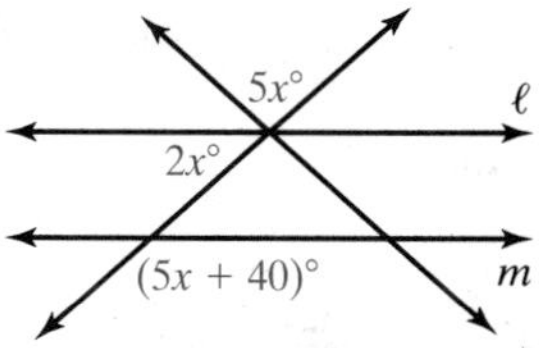

51. **Find the Error** A classmate says that $\overleftrightarrow{AB} \parallel \overleftrightarrow{DC}$ based on the diagram at the right. Explain your classmate's error.

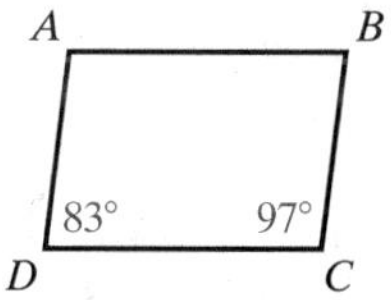

52. How are flow proofs and two-column proofs alike? How are they different?

REVIEW AND PREVIEW

Determine whether each statement is always, sometimes, or never true. See Section 3.1.

53. Skew lines are coplanar.

54. Skew lines intersect.

55. Parallel planes intersect.

56. Rays are parallel.

3.3 PARALLEL LINES AND ANGLES FORMED BY TRANSVERSALS

OBJECTIVES

1 Prove and Use Theorems About Parallel Lines Cut by a Transversal.

2 Use Algebra to Find Measures of Angles Formed by Parallel Lines Cut by a Transversal.

OBJECTIVE 1 ▶ Proving and Using Theorems About Parallel Lines and Transversals. In this section, we see if the *converses* of the theorems proved in Section 3.2 are also true. If so, we can use these converse theorems to prove that angles formed by parallel lines cut by a transversal are congruent, or supplementary, or both.

Recall what we know about converses:

Helpful Hint

- If $p \rightarrow q$, then $q \rightarrow p$
  ($p \rightarrow q$ is the conditional; $q \rightarrow p$ is the converse.)
- Also, if $p \rightarrow q$ is a true statement, then $q \rightarrow p$ might be true or false.

This Helpful Hint reminds us that we cannot use the converses of the theorems in Section 3.2 unless we prove the converses to be true.

Let's start with the Converse of the Alternate Interior Angles Theorem.

Theorem 3.3-1 Alternate Interior Angles Converse (Converse of Theorem 3.2-4)

Theorem	If...	Then...
If two parallel lines are cut by a transversal, then alternate interior angles are congruent.	$\ell \parallel m$ 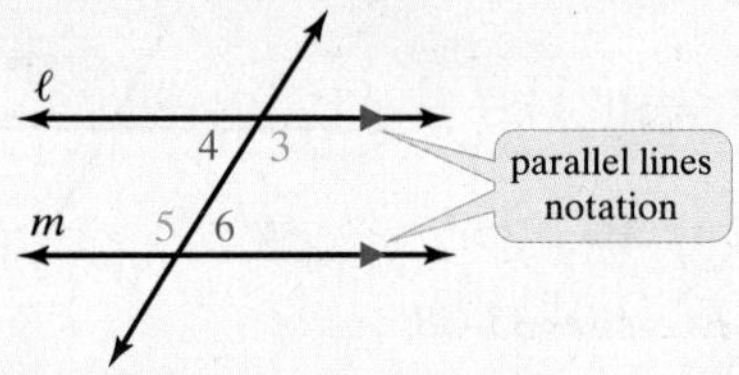	$\angle 4 \cong \angle 6$ $\angle 3 \cong \angle 5$

We prove this theorem in the back of this text.

Helpful Hint

As stated in Section 3.2, we may prove a theorem using only earlier proved theorems. This is VERY IMPORTANT to ensure that our mathematical system is valid. To prove this theorem (3.3-1) now, we only need the concept of Indirect Proofs (Section 5.5). Just know that our geometry is a valid system.

Let's use the postulates and theorems we already know to find angle measures.

EXAMPLE 1 Finding Angle Measures

Using the figure shown and given that $m\angle 3 = 55°$, find the measure of each angle. Tell what theorem or postulate you used. (Recall that red arrow head notation means $\parallel$ lines.)

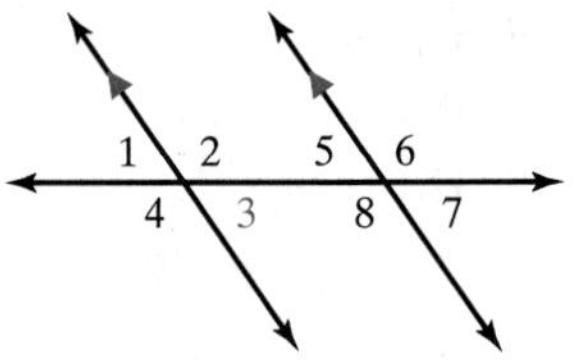

a. $m\angle 5$ **b.** $m\angle 7$

c. $m\angle 4$ **d.** $m\angle 2$

Solution

a. $m\angle 5 = 55°$ Alternate Interior Angles Converse (Theorem 3.3-1)

b. $m\angle 7 = m\angle 5 = 55°$ Vertical Angles Theorem (Theorem 2.7-4)

c. $m\angle 4 = 180° - m\angle 3 = 180° - 55° = 125°$ Linear Pair Theorem (Theorem 2.7-3)

d. $m\angle 2 = m\angle 4 = 125°$ Vertical Angles Theorem (Theorem 2.7-4) □

PRACTICE 1 Using the figure for Example 1 and given that $m\angle 3 = 48°$, find the measure of each angle.

a. $m\angle 1$ **b.** $m\angle 2$ **c.** $m\angle 8$ **d.** $m\angle 6$

We can use the Alternate Interior Angles Converse to prove similar theorems about parallel lines. Let's introduce a few of these theorems about parallel lines and a transversal.

Theorem 3.3-2 Corresponding Angles Converse (Converse of Theorem 3.2-5)

Theorem

If two parallel lines are cut by a transversal, then corresponding angles are congruent.

If...

$\ell \parallel m$

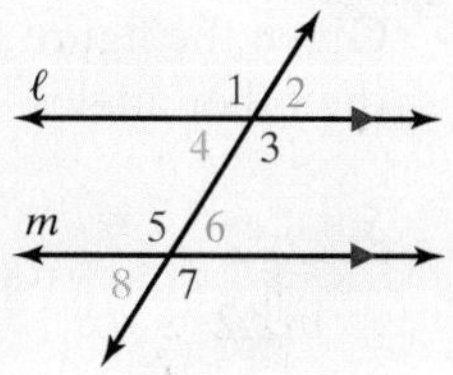

Then...

$\angle 1 \cong \angle 5$
$\angle 2 \cong \angle 6$
$\angle 3 \cong \angle 7$
$\angle 4 \cong \angle 8$

Theorem 3.3-3 Same-Side Interior (or Consecutive) Angles Converse (Converse of Theorem 3.2-6)

Theorem

If two parallel lines are cut by a transversal, then same-side interior angles are supplementary.

If...

$\ell \parallel m$

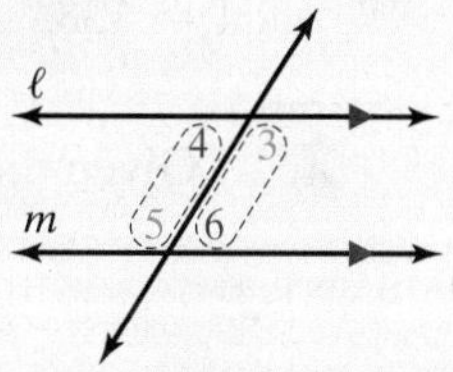

Then...

$m\angle 4 + m\angle 5 = 180°$
$m\angle 3 + m\angle 6 = 180°$

Theorem 3.3-4 Alternate Exterior Angles Converse (Converse of Theorem 3.2-7)

Theorem

If two parallel lines are cut by a transversal, then alternate exterior angles are congruent.

If...

$\ell \parallel m$

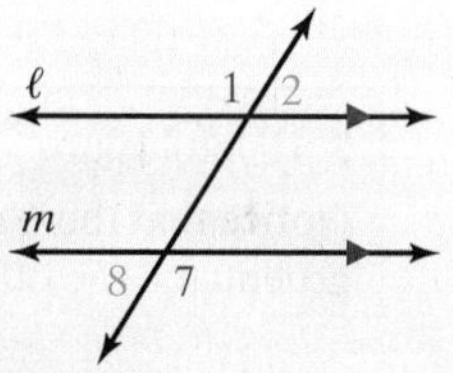

Then...

$\angle 1 \cong \angle 7$
$\angle 2 \cong \angle 8$

We prove Theorem 3.3-2 below. We prove Theorems 3.3-3 and 3.3-4 in Exercises 59 and 60, Exercise Set 3.3.

We can use the Alternate Interior Angles Converse to prove the Corresponding Angles Converse.

Proof of Theorem 3.3-2: Corresponding Angles Converse

Given: $\ell \parallel m$

Prove: $\angle 2 \cong \angle 6$

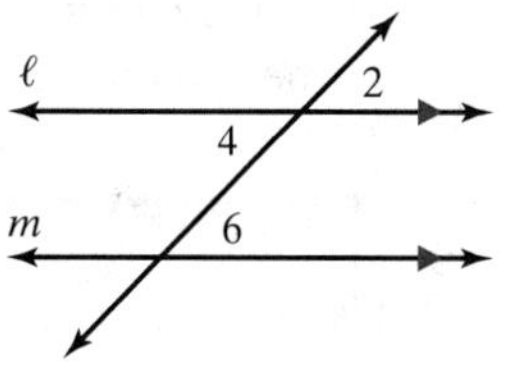

Statements	Reasons
1. $\ell \parallel m$	**1.** Given
2. $\angle 4 \cong \angle 6$	**2.** If lines are parallel, then alternate interior angles are $\cong$.
3. $\angle 4 \cong \angle 2$	**3.** Vertical angles are $\cong$.
4. $\angle 2 \cong \angle 6$	**4.** Substitution (or Transitive Property of $\cong$) (Steps 2, 3)

(Note: We prove one pair of corresponding angles are congruent. The remaining corresponding angles are proved congruent in a similar manner.)

If we know the measure of one of the angles formed by two parallel lines and a transversal, then we can use theorems to find the measures of the other angles formed.

EXAMPLE 2 **Finding Angle Measures**

Given the figure shown and $m\angle 4 = 42°$, find the measures of the other angles.

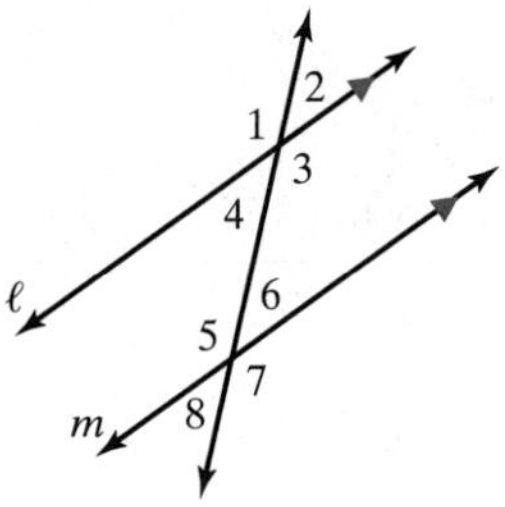

Solution

$m\angle 2 = 42°$ Vertical Angles Theorem ($\angle 2 \cong \angle 4$)

$m\angle 6 = 42°$ Alternate Interior Angles Converse ($\angle 4 \cong \angle 6$)

$m\angle 8 = 42°$ Vertical Angles Theorem ($\angle 6 \cong \angle 8$)

$m\angle 7 = 138°$ Linear Pair Theorem ($\angle 8$ and $\angle 7$)

$m\angle 5 = 138°$ Vertical Angles Theorem ($\angle 7 \cong \angle 5$)

$m\angle 3 = 138°$ Alternate Interior Angles Converse ($\angle 5 \cong \angle 3$)

$m\angle 1 = 138°$ Vertical Angles Theorem ($\angle 3 \cong \angle 1$) □

PRACTICE 2 Given the figure shown and $m\angle 1 = 97°$, find the measures of the other angles.

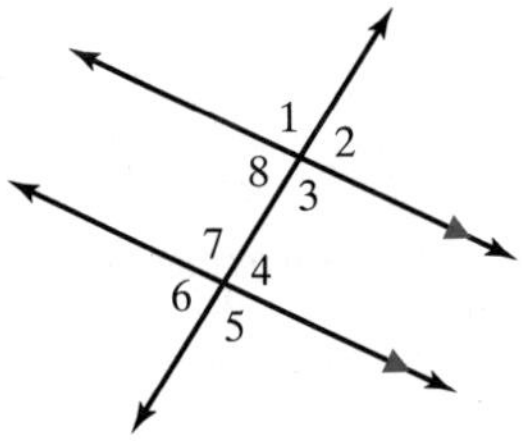

Helpful Hint

Notice that there are 8 angles formed by 2 parallel lines and a transversal. Four angles are congruent and the other 4 angles are congruent and supplementary to the first 4 congruent angles.

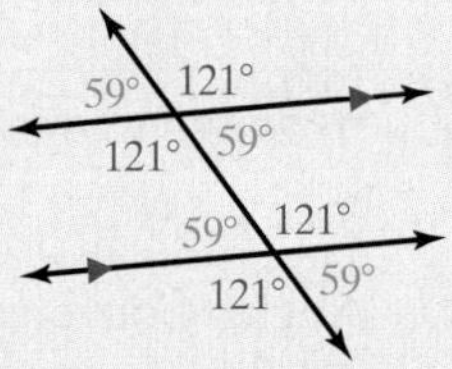

It is important to be able to identify parallel lines, transversals, and the angles formed in diagrams that are more complicated, as shown in the next example.

EXAMPLE 3 **Finding Measures of Angles**

Using the figure shown, what are the measures of $\angle 3$ and $\angle 4$? Which theorem justifies each answer?

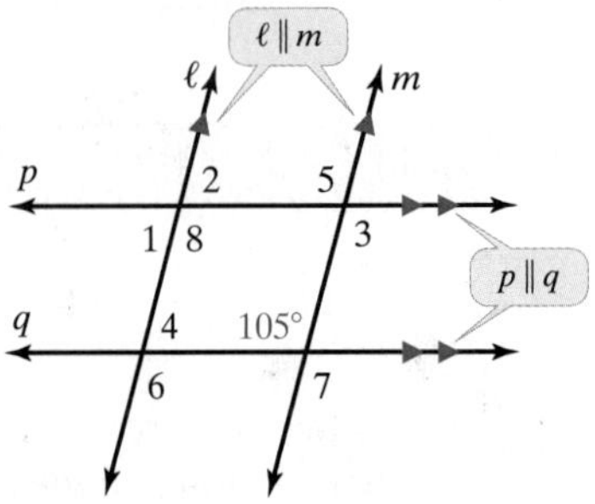

Solution Since $p \parallel q$, $m\angle 3 = 105°$ by the Alternate Interior Angles Converse.

Since $\ell \parallel m$, $m\angle 4 + 105° = 180°$ by the Same-Side Interior Angles Converse. (They are supplementary.)

So, $m\angle 4 = 180° - 105° = 75°$. □

PRACTICE 3 Use the figure in Example 3 to find the measure of each angle. Justify each answer by listing the appropriate theorem.

a. $\angle 1$ **b.** $\angle 2$

c. $\angle 5$ **d.** $\angle 6$

e. $\angle 7$ **f.** $\angle 8$

OBJECTIVE 2 ▶ Using Algebra to Find Angle Measures Formed by Parallel Lines and a Transversal. Let's practice using our algebra skills and our knowledge of angle measures formed by parallel lines and a transversal.

EXAMPLE 4 **Using Algebra and Parallel Line Theorems to Find Angle Measures**

Given the figure and $\ell \parallel m$, find the value of x.

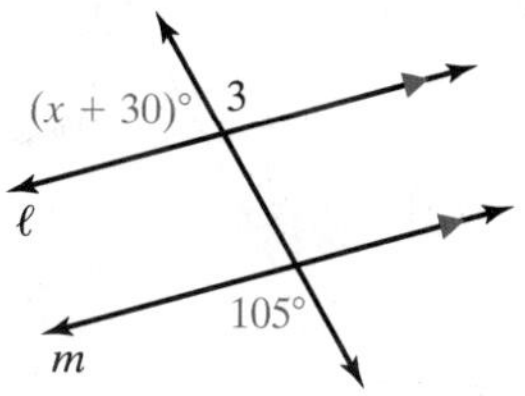

Solution

$m\angle 3 = 105°$	Alternate Exterior Angles Converse
$m\angle 3 + (x + 30)° = 180°$	Linear Pair Theorem
$105° + (x + 30)° = 180°$	Substitution
$x° + 135° = 180°$	Angle Addition Postulate
$x° = 45°$	Subtraction
or $x = 45$	

The value of x is 45. □

PRACTICE 4 Given $n \parallel q$, find the value of y.

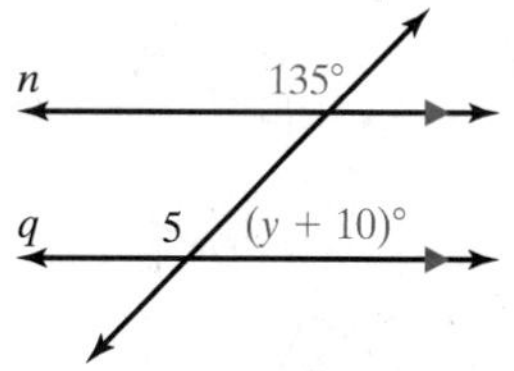

VOCABULARY & READINESS CHECK

Fill in the blank.

1. Given line ℓ and point P, how many lines can be drawn through P parallel to line ℓ? ____________
2. The postulate that gives us the answer to Exercise 1 above is called the ____________ Postulate.

3.3 EXERCISE SET MyMathLab®

Find the measure of $\angle 1$ and $\angle 2$. Justify each answer. See Examples 1 and 2.

1.

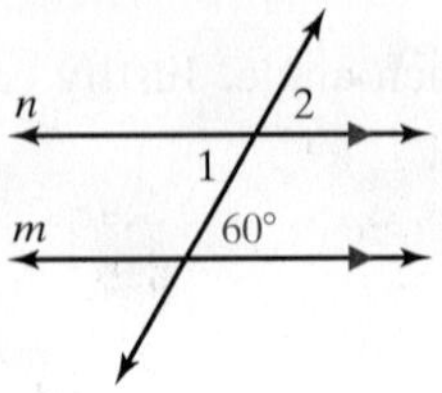

2.

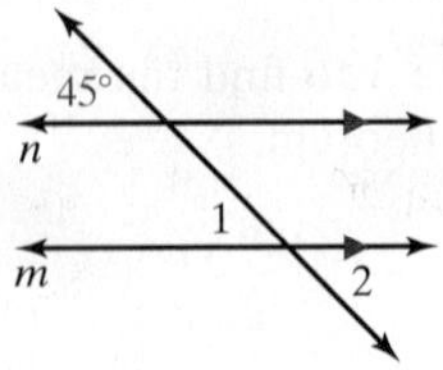

3.

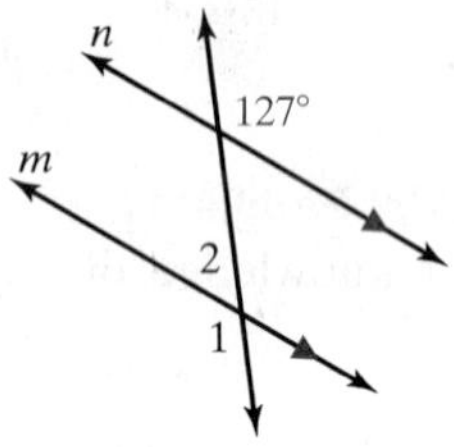

4.

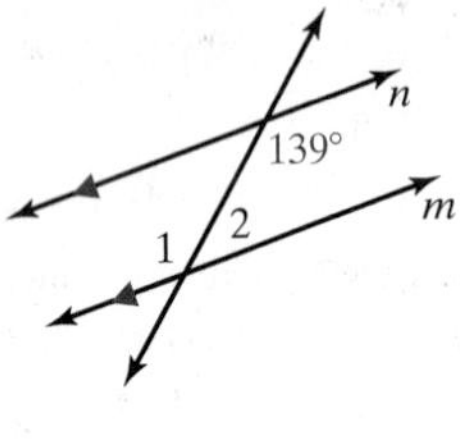

5.

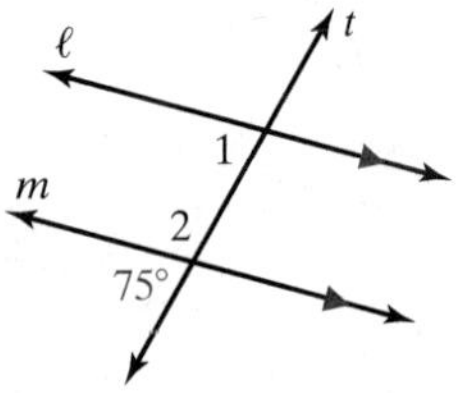

6.

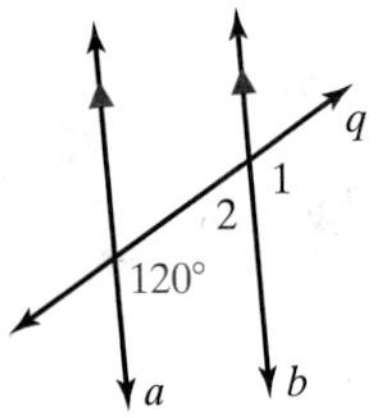

Identify all the numbered angles that are congruent to the angle whose measure is given. Justify your answers. See Examples 1 and 2.

7.

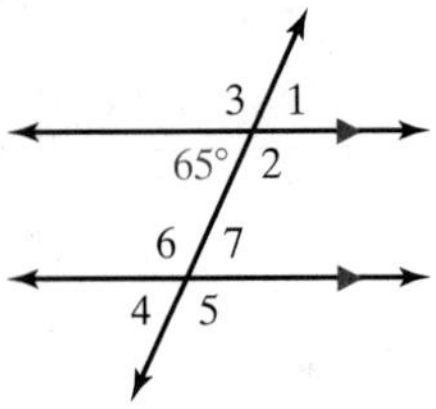

8.

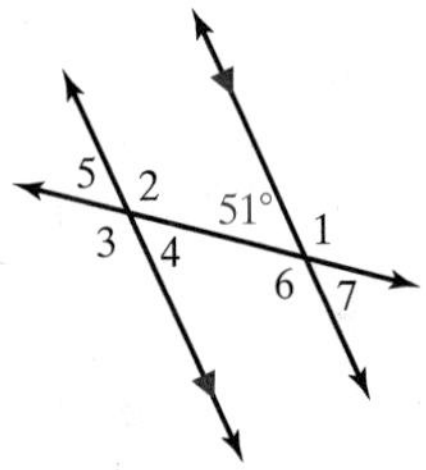

9.

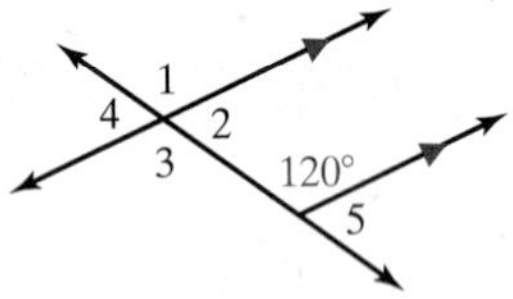

10.

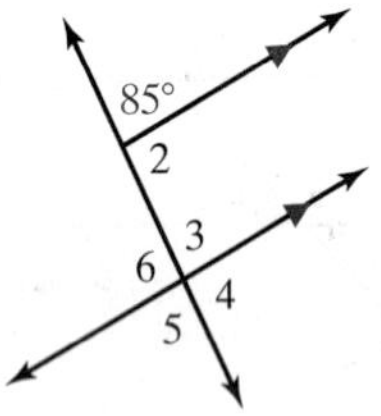

Use the given angle measure and the figure shown to find the measures of the other seven numbered angles. (Note that the values change in each new exercise.) See Example 2.

11. Suppose that $m\angle 3 = 73°$.

12. Suppose that $m\angle 1 = 68°$.

13. Suppose that $m\angle 6 = 103°$.

14. Suppose that $m\angle 8 = 111°$.

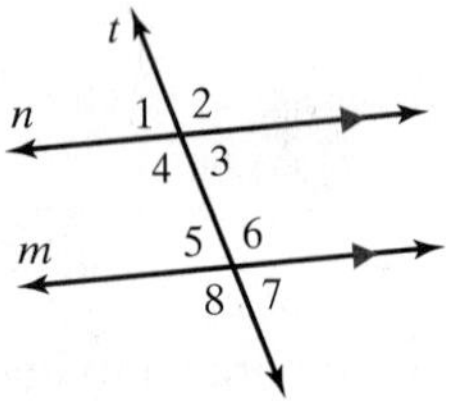

Use the given angle measure and the figure shown to find the measures of the other seven numbered angles. (Note that the values change in each new exercise.) See Example 2.

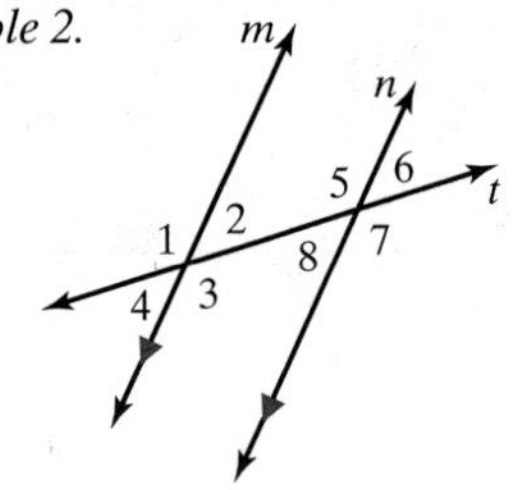

15. Suppose that $m\angle 2 = 28°$.

16. Suppose that $m\angle 4 = 34°$.

17. Suppose that $m\angle 5 = 149°$.

18. Suppose that $m\angle 7 = 154°$.

Find the values of x and y. See Examples 1 and 2.

19.

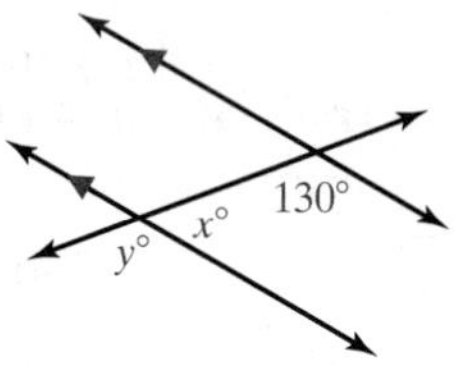

20.

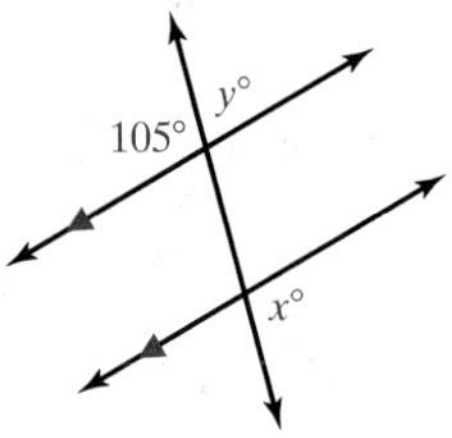

21.

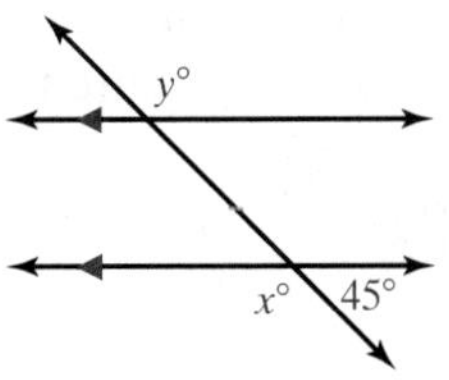

22.

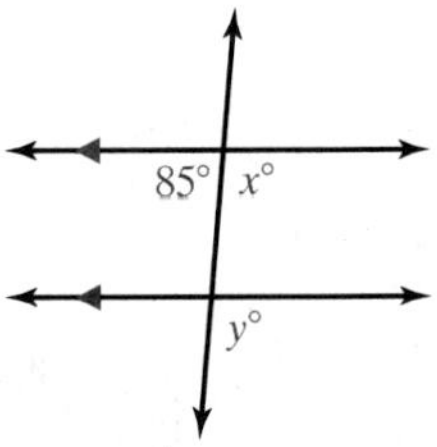

Find the measures of the numbered angles. Name the theorem that justifies each answer. See Example 3.

23.

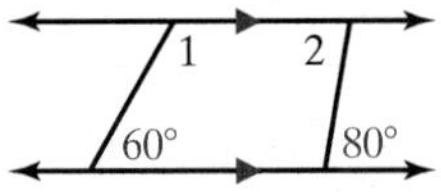

24.

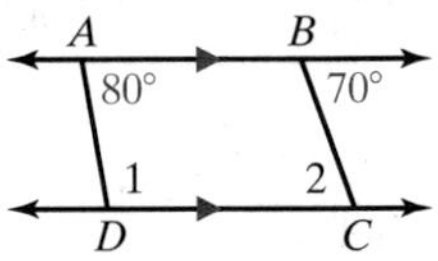

25.

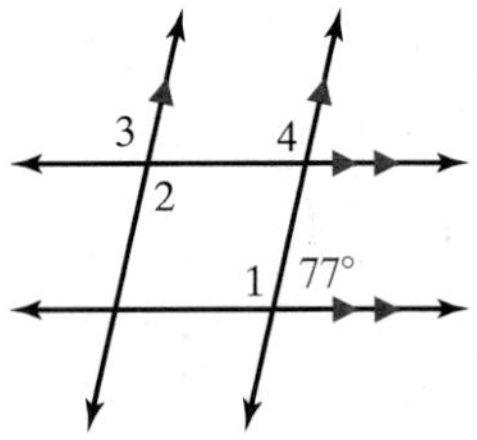

26.

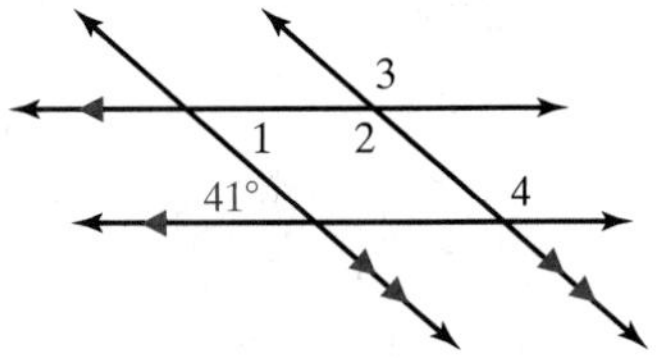

27.

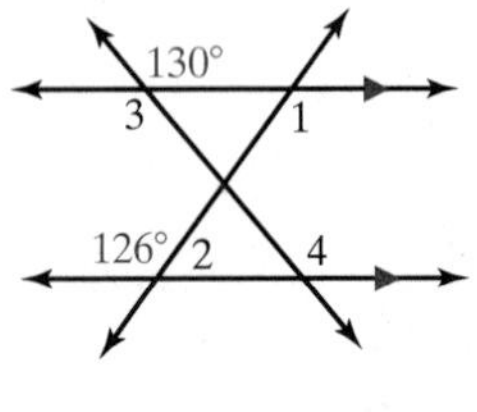

28.

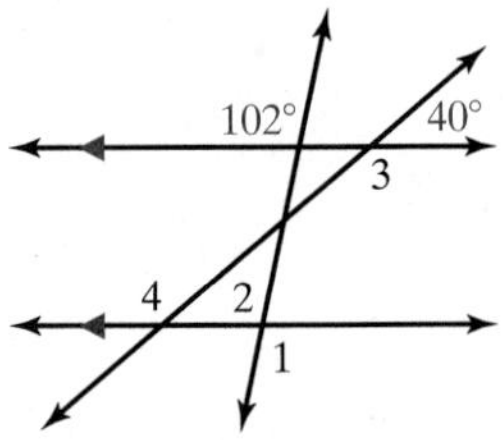

29.

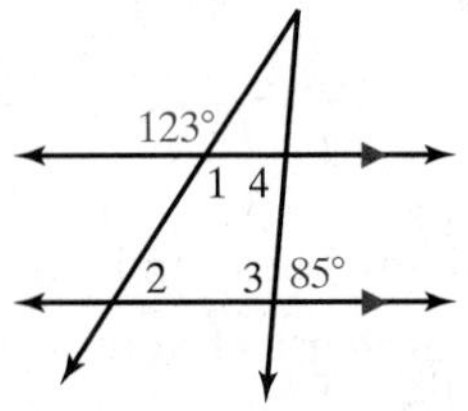

30.

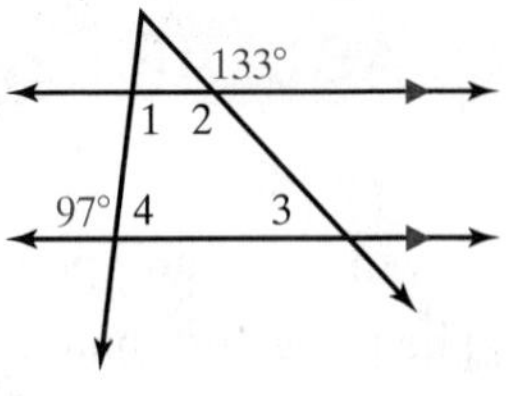

Find the measure of ∠1. See Examples 1–3.

31.

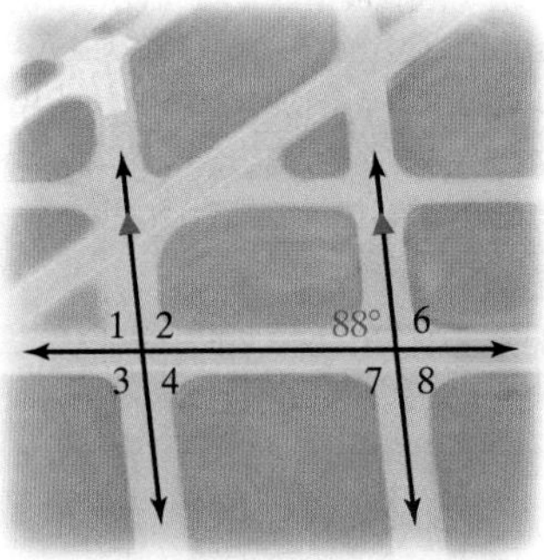

32.

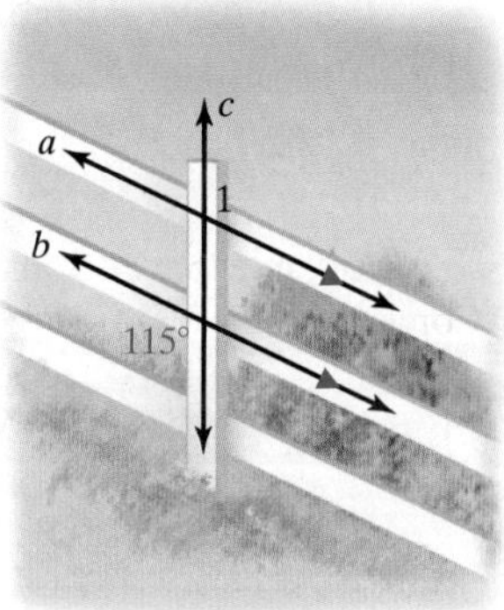

33.

34.

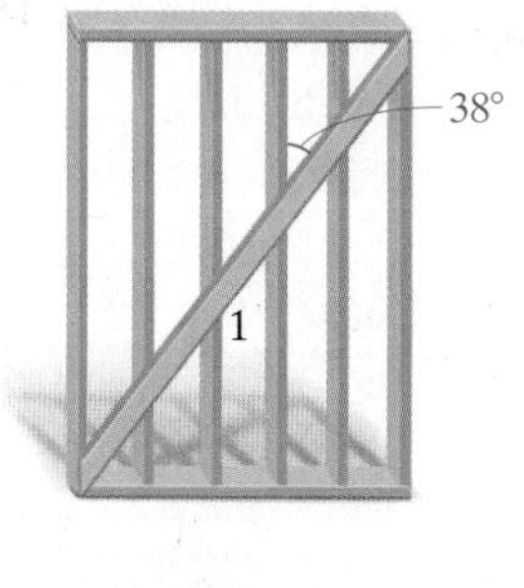

Find the value of x. See Example 4.

35.

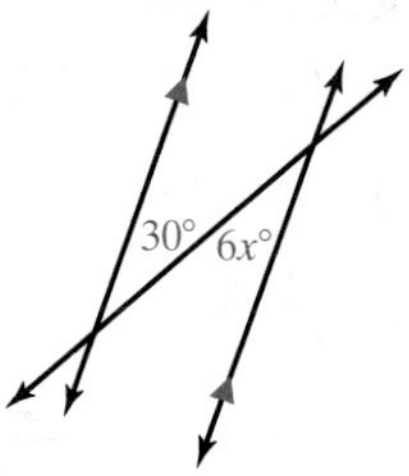

36.

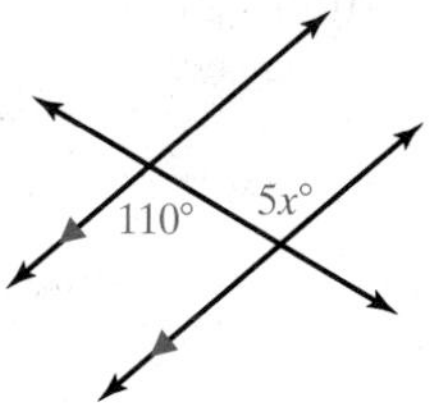

37.

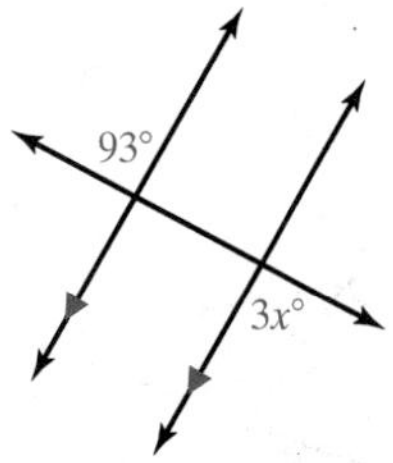

38.

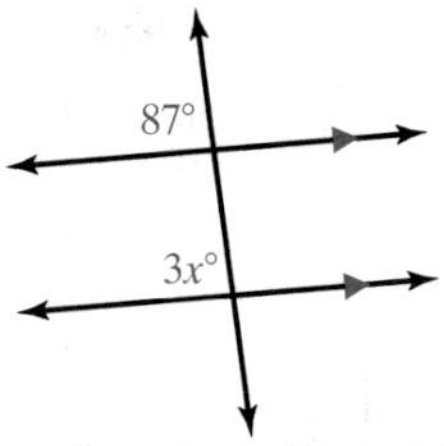

39.

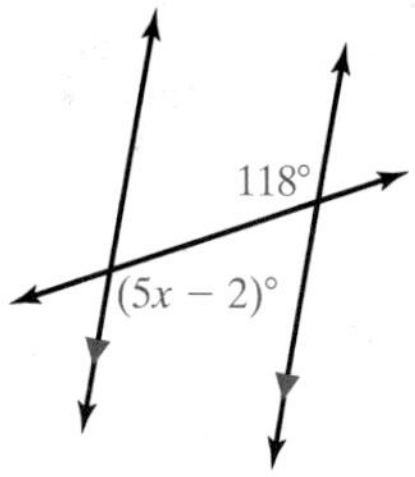

40.

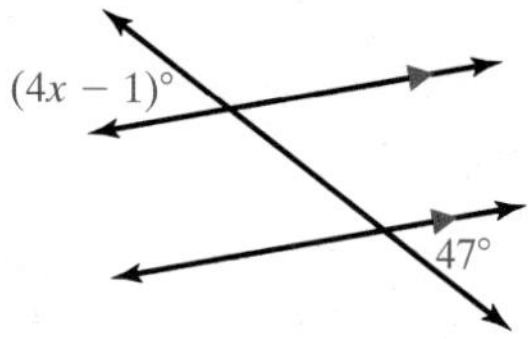

41.

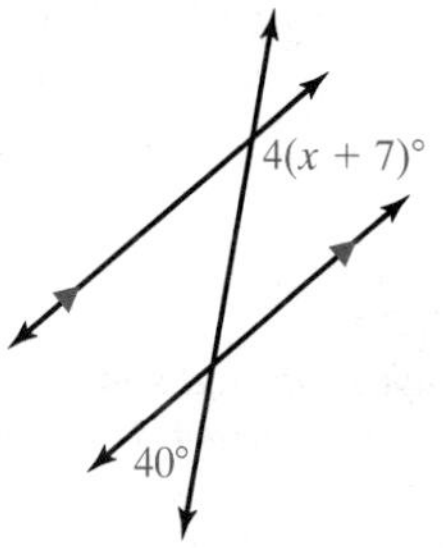

42.

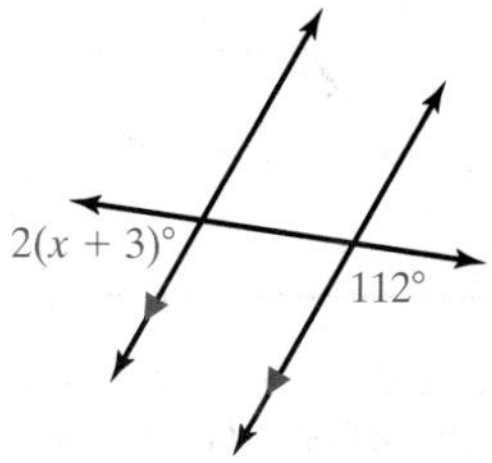

Find the value of x. (Hint: If needed, extend the parallel lines for each figure so that you can see two parallel lines and transversals.) Then find the measure of each labeled angle.

43.

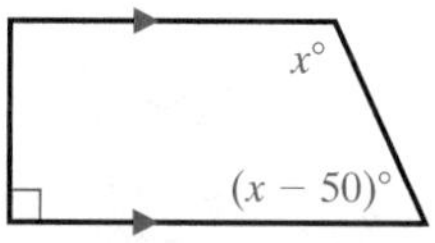

44.

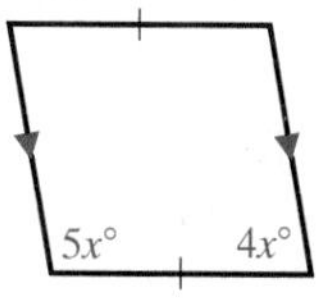

45.

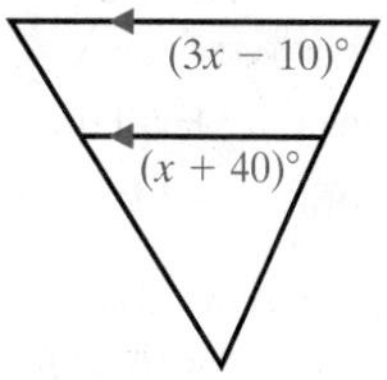

46.

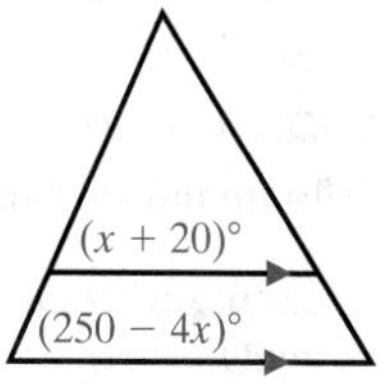

Proof *Supply the missing statements and reasons in the two-column proof.*

Given: $a \parallel b$, $c \parallel d$

Prove: $\angle 1 \cong \angle 3$

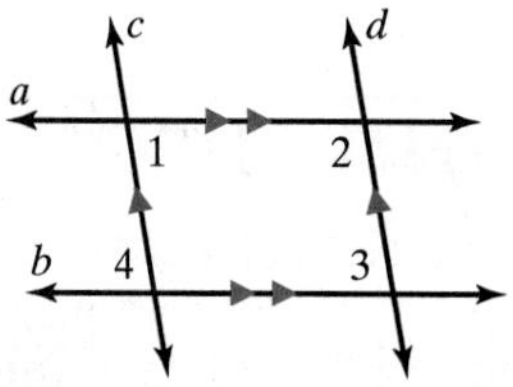

Statements	Reasons
1. $a \parallel b$	**1.** Given
2. ∠3 and ∠2 are supplementary.	**2.** __?__ (Exercise 47.)
3. __?__ (Exercise 48.)	**3.** Given
4. ∠1 and ∠2 are supplementary.	**4.** __?__ (Exercise 49.)
5. ∠1 ≅ ∠3	**5.** __?__ (Exercise 50.)

CONCEPT EXTENSIONS

Find the values of the variables.

51.

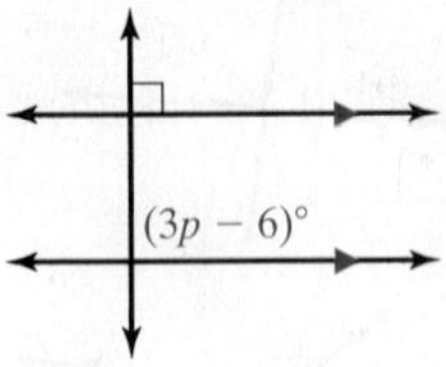

52.

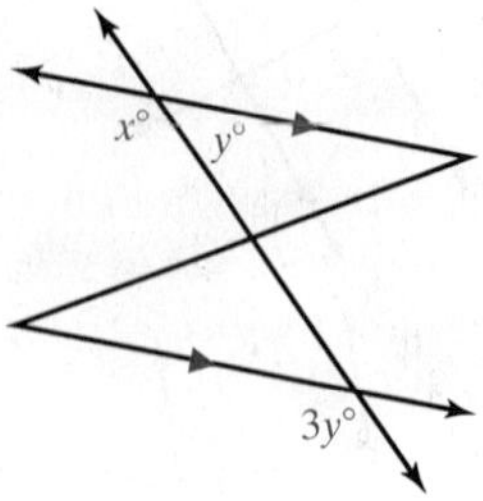

53.

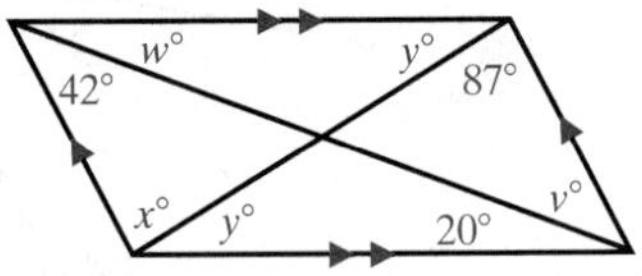

54.

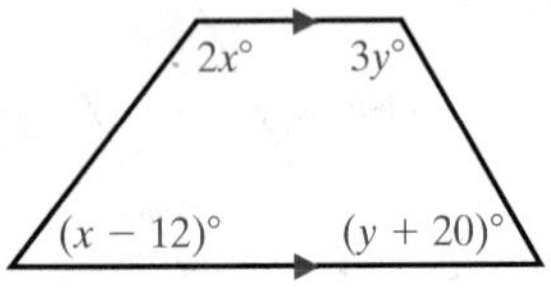

Find the Error *Determine whether each proposed solution for the value of x in the given figure is correct or incorrect. Explain.*

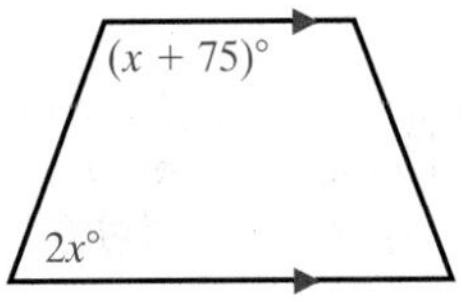

55.

2x = x + 75
x = 75

56.

2x + (x + 75) = 180
3x + 75 = 180
3x = 105
x = 35

57. Are same-side interior angles ever congruent? Explain.

58. Are alternate interior angles ever not congruent? Explain.

Proof *For Exercises 59 and 60, prove the theorems. Feel free to use the Alternate Interior Angles Converse and the Corresponding Angles Converse.*

59. Prove the Same-Side Interior Angles Converse (3.3-3).

60. Prove the Alternate Exterior Angles Theorem (3.3-4).

61. Campers often use a "bear bag" at night to avoid attracting animals to their food supply. In the bear bag system below, a camper pulls one end of the rope to raise and lower the food bag.

a. If the rope is taut between the two parallel trees, as shown, what is $m\angle 1$?

b. Are $\angle 1$ and the given 63° angle called alternate interior angles, same-side interior angles, or corresponding angles?

62. How are the Alternate Interior Angles Converse and the Alternate Exterior Angles Converse alike? How are they different?

REVIEW AND PREVIEW

Find $m\angle 1$ and $m\angle 2$. Justify each answer. See this section.

63.

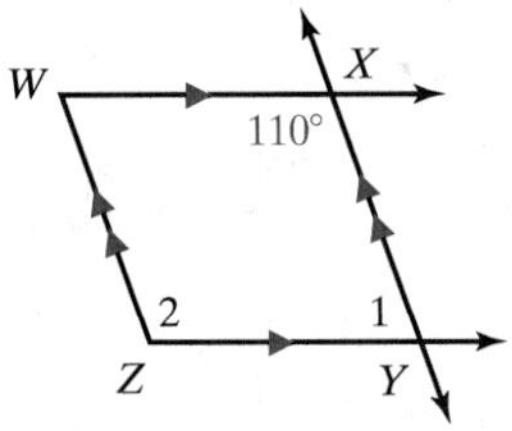

64.

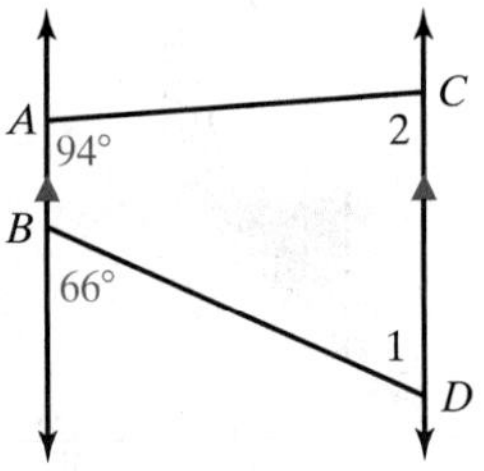

Determine whether each statement is always, sometimes, *or* never *true. See Sections 1.6 and 3.1.*

65. Perpendicular lines meet at right angles.

66. Two lines in intersecting planes are perpendicular.

67. Two lines in the same plane are parallel.

68. Two lines in parallel planes are perpendicular.

3.4 PROVING THEOREMS ABOUT PARALLEL AND PERPENDICULAR LINES

OBJECTIVES

1. Use and Prove Theorems About Parallel and Perpendicular Lines.
2. Use Algebra to Find Measures of Angles Related to Perpendicular Lines.

OBJECTIVE 1 ▶ Using and Proving Parallel and Perpendicular Theorems. In Section 3.2, we introduced the most important postulate in this text—the Parallel Postulate.

Postulate 3.2-1 Parallel Postulate

Through a point not on a line, there is one and only one line parallel to the given line.

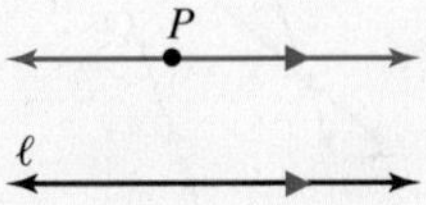

In this same section, we also introduced an important postulate about perpendicular lines, reviewed below.

Postulate 3.2-2 Perpendicular Postulate

Through a point not on a line, there is one and only one line perpendicular to the given line.

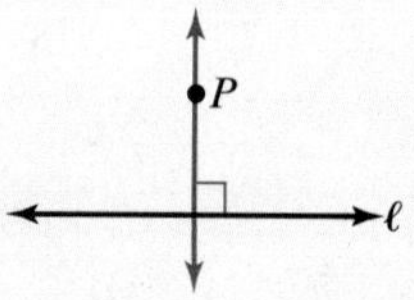

In this section, we introduce some additional theorems about perpendicular lines and parallel lines.

Theorem 3.4-1 Perpendicular Transversal Theorem

Theorem

In a plane, let two parallel lines be cut by a transversal. If the transversal is perpendicular to one of the parallel lines, then it is perpendicular to the other parallel line.

If . . .

notation for "is perpendicular to"

$\ell \parallel m$ and $t \perp \ell$

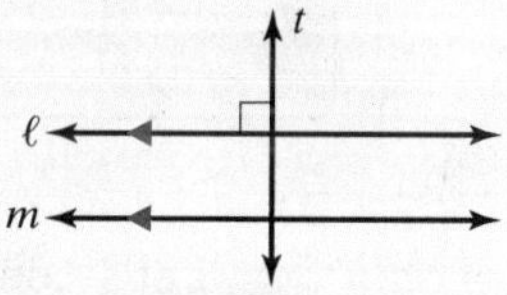

Then . . .

$t \perp m$ also.

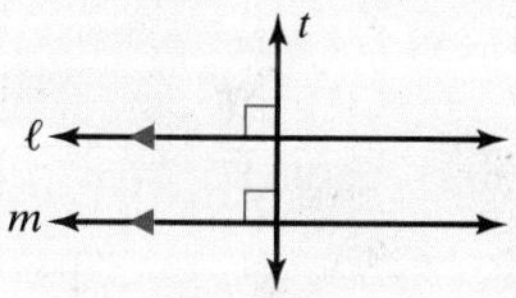

Theorem 3.4-2 Two Lines Parallel to a Third Line

Theorem

If two lines are parallel to the same line, then all three lines are parallel to each other.

If . . .

$a \parallel b$ and $b \parallel c$

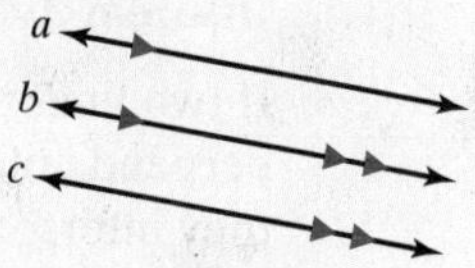

Then . . .

$a \parallel c$

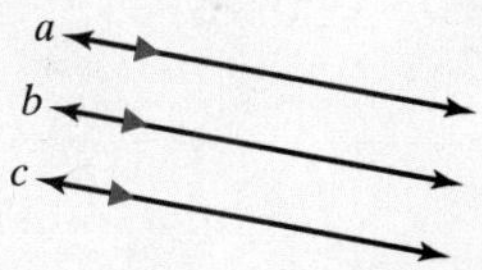

Helpful Hint

Remember, we may use any theorem that we proved earlier to help us prove these new theorems.

Recall that although we accept postulates as true, we must prove theorems; so let's work on some proofs of the theorems above.

EXAMPLE 1 Proving Perpendicular Transversal Theorem

Prove Theorem 3.4-1.

Given: In a plane, $\ell \parallel m$ and $t \perp \ell$

Prove: $t \perp m$

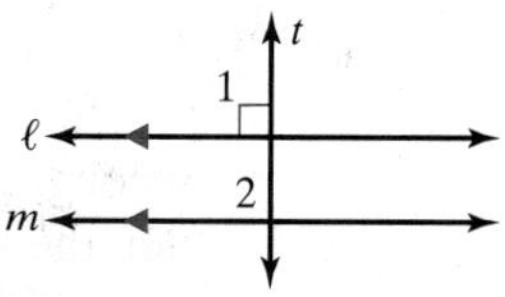

Solution

Statements	Reasons
1. $\ell \parallel m$ and $t \perp \ell$	**1.** Given
2. $\angle 1 \cong \angle 2$	**2.** Corresponding Angles Converse
3. $\angle 2$ is a right $\angle$	**3.** definition of right $\angle$s (Step 2)
4. $t \perp m$	**4.** definition of $\perp$ lines (Step 3)

□

PRACTICE 1 Fill in the blanks to complete the proof of Theorem 3.4-2.

Given: $a \parallel b$ and $b \parallel c$

Prove: $a \parallel c$

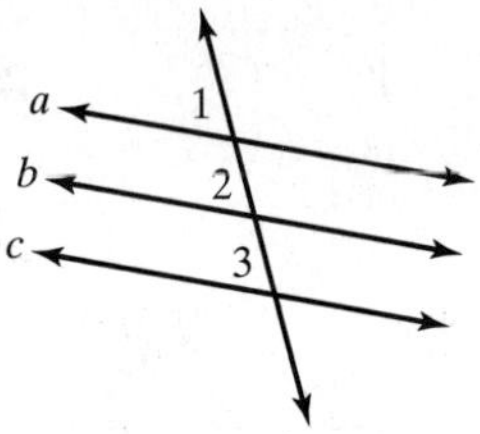

Statements	Reasons
1. $a \parallel b$	1. ________
2. $\angle 1 \cong \angle 2$	2. ________
3. $b \parallel c$	3. Given
4. $\angle 2 \cong \angle 3$	4. ________
5. ________	5. Substitution or Transitive Property
6. ________	6. Corresponding Angles Theorem

Helpful Hint

In Example 1, the Perpendicular Transversal Theorem (3.4-1) states that the lines must be *in a plane.* The diagram at the right shows why. In the rectangular solid, $\overleftrightarrow{AC}$ and $\overleftrightarrow{BD}$ are parallel. $\overleftrightarrow{EC}$ is perpendicular to $\overleftrightarrow{AC}$, but it is not perpendicular to $\overleftrightarrow{BD}$. In fact, $\overleftrightarrow{EC}$ and $\overleftrightarrow{BD}$ are skew because they are not in the same plane.

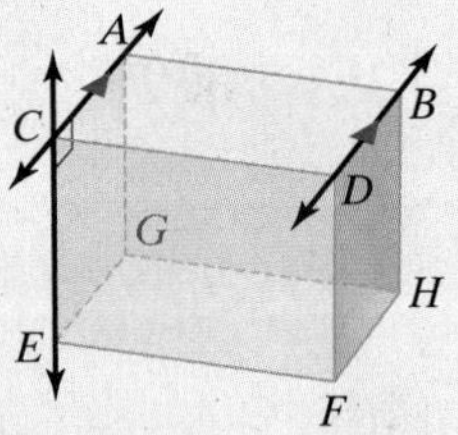

Let's continue to grow our mathematical system by introducing a few more theorems about perpendicular lines.

Theorems About Perpendicular Lines

Theorem	If . . .	Then . . .
Theorem 3.4-3 If two lines are perpendicular, then they intersect to form four right angles.	$l \perp m$ 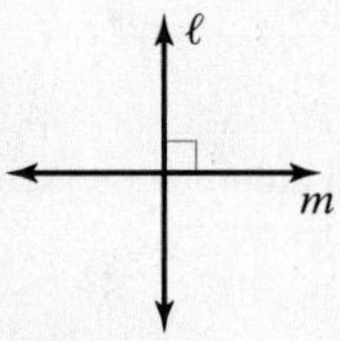	$m\angle 1 = m\angle 2 = m\angle 3 = m\angle 4 = 90°$ 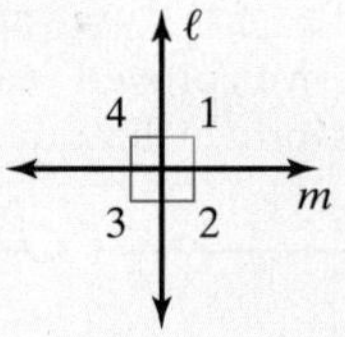
Theorem 3.4-4 If two lines intersect to form a linear pair of congruent angles, then the lines are perpendicular to each other.	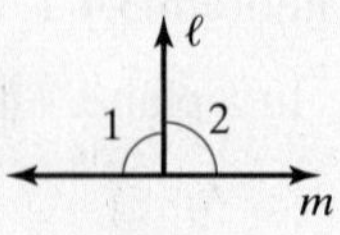 $\angle 1 \cong \angle 2$ and form a linear pair.	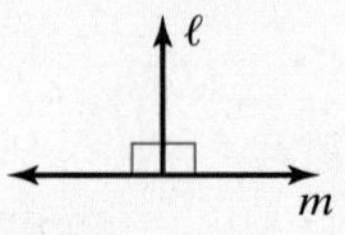 $l \perp m$

We prove these theorems in Exercise Set 3.4, Exercises 31 and 32.

OBJECTIVE 2 ▶ Using Algebra to Find Measures of Angles with Perpendicular Lines. Let's use algebra to find measures of angles related to perpendicular lines.

EXAMPLE 2 **Finding Measures of Angles**

Find the value of x.

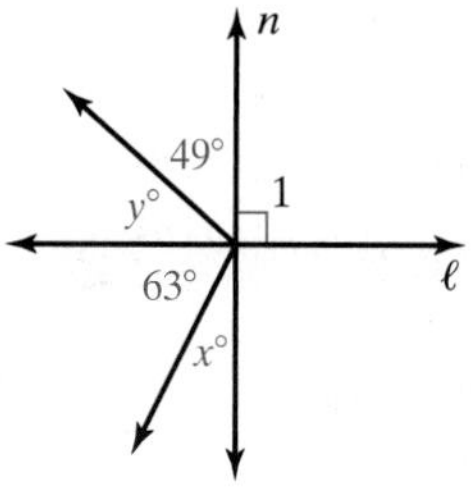

Solution Since $m\angle 1 = 90°$, then $n \perp \ell$.

If $n \perp \ell$, then all angles formed by these lines are right angles. Thus

$$63° + x° = 90°$$
$$x° = 27°$$

Thus, the value of x is 27. □

PRACTICE 2 Use the figure for Example 2 and find the value of y.

VOCABULARY & READINESS CHECK

Fill in the blank.

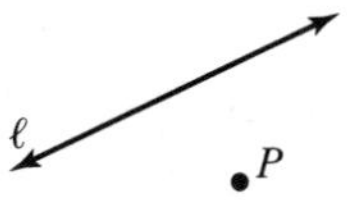

1. Given line ℓ and point P, how many lines can be drawn through P parallel to ℓ? ____________
2. Given line ℓ and point P, how many lines can be drawn through P perpendicular to ℓ? ____________
3. The postulate that gives us the answer to Exercise 1 above is called the ____________ Postulate.
4. The postulate that gives us the answer to Exercise 2 above is called the ____________ Postulate.

True or False.

5. Postulates are accepted as true and are not proved. ____________
6. Theorems are accepted as true and are not proved. ____________

3.4 EXERCISE SET MyMathLab®

For Exercises 1–8, use the theorems in this section to find the value of x. See Example 2.

1.

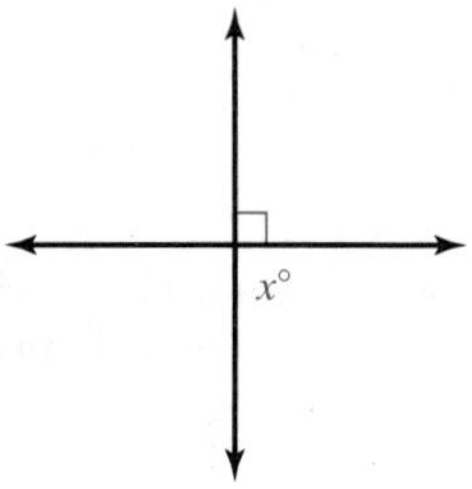

2.

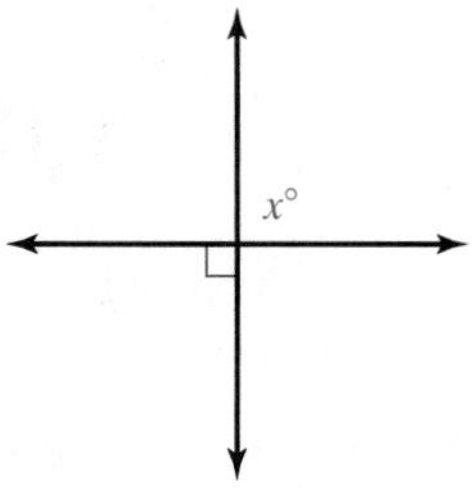

3.

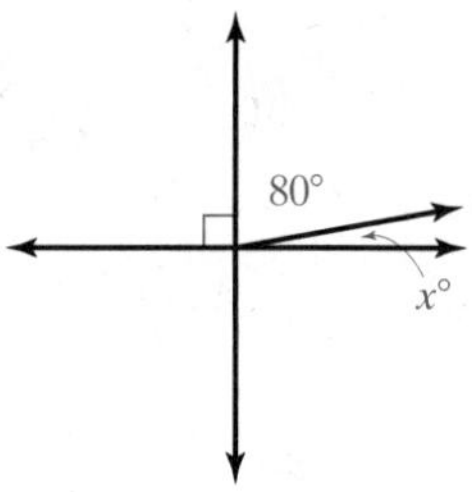

4.

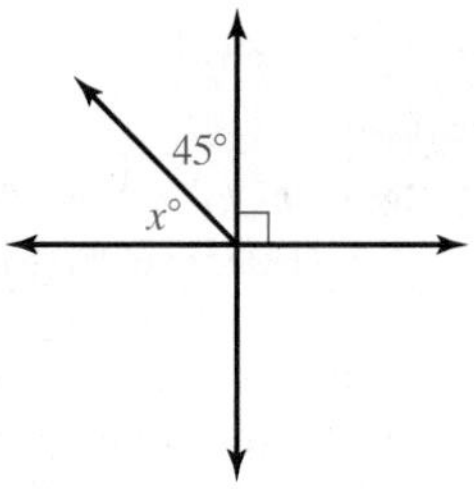

5.

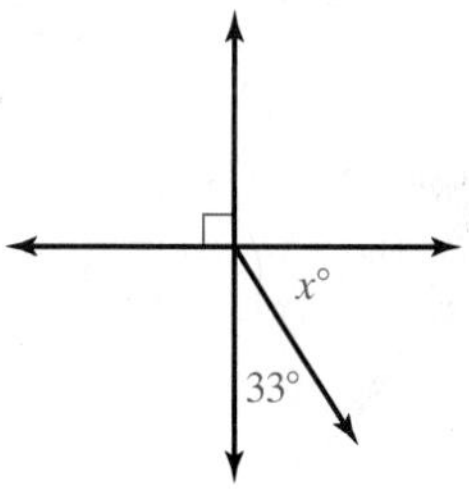

6.

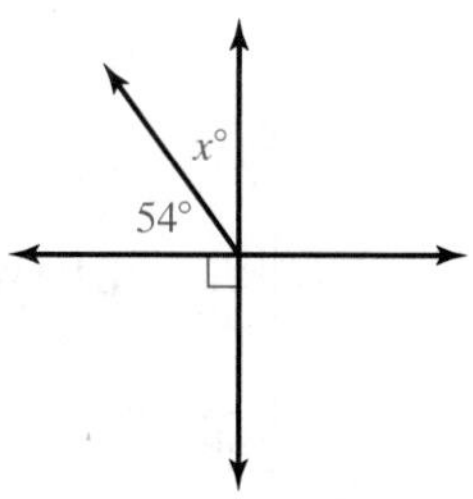

7.

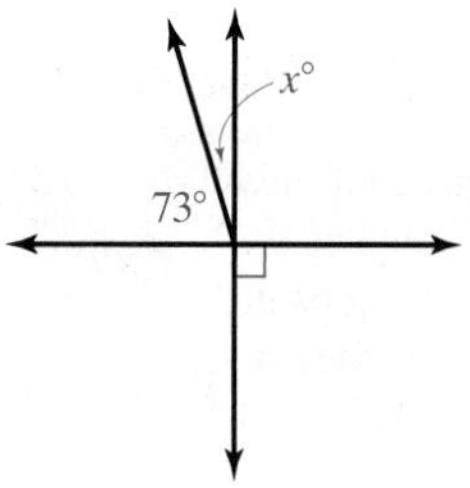

8.

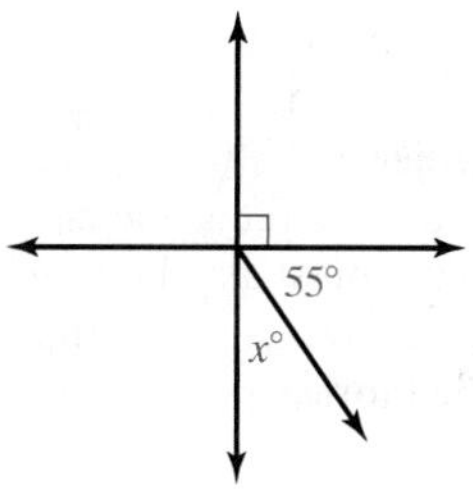

Fill in the Blank *Refer to the figure to answer Exercises 9–16. See Example 2.*

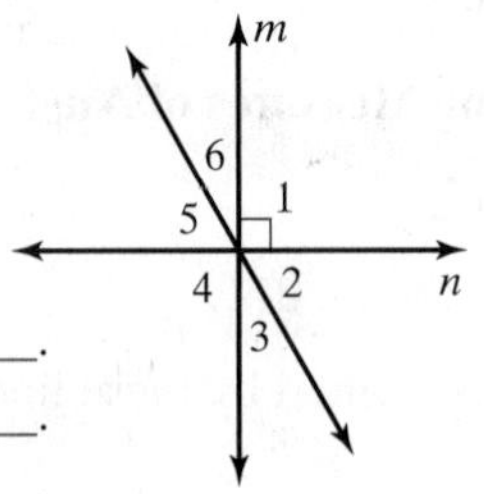

9. $m\angle 4 =$ ________

10. $m\angle 1 =$ ________

11. $m\angle 6 + m\angle 5 =$ ________

12. $m\angle 2 + m\angle 3 =$ ________

13. If $m\angle 5 = 61°$, then $m\angle 6 =$ ________.

14. If $m\angle 3 = 35°$, then $m\angle 2 =$ ________.

15. If $m\angle 6 = 30°$, then $m\angle 3 =$ ________.

16. If $m\angle 2 = 65°$, then $m\angle 5 =$ ________.

The following statements describe a ladder. Based only on the statement, make a conclusion about the rungs, one side, or both sides of the ladder. Explain.

17. The rungs are parallel and the top rung is perpendicular to one side.

18. The sides are parallel. The rungs are perpendicular to one side.

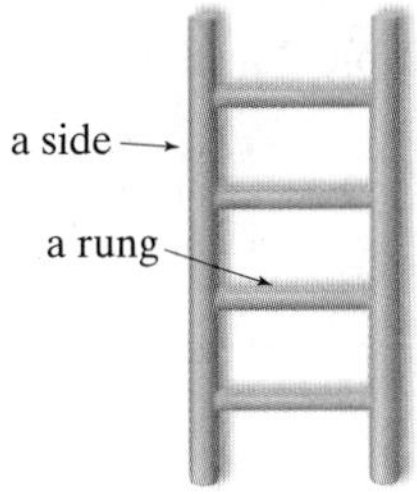

Ladder

For Exercises 19–24, a, b, c, and d are distinct lines in the same plane. For each combination of relationships, draw a diagram of the relationship. Then tell how a and d relate. Justify your answer.

19. $a \parallel b, b \parallel c, c \parallel d$

20. $a \parallel b, b \parallel c, c \perp d$

21. $a \parallel b, b \perp c, c \parallel d$

22. $a \perp b, b \parallel c, c \parallel d$

23. $a \parallel b, b \perp c, c \perp d$

24. $a \perp b, b \parallel c, c \perp d$

25. The map below is a section of a subway map. The yellow line is perpendicular to the brown line, the brown line is perpendicular to the blue line, and the blue line is perpendicular to the pink line. What conclusion can you make about the yellow line and the pink line? Explain.

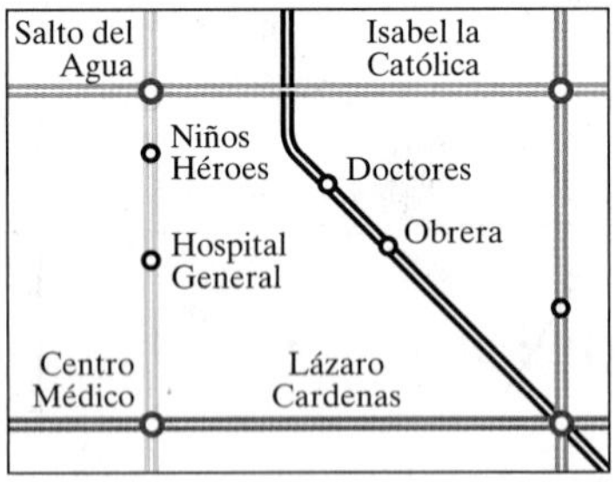

26. You plan to sew two triangles of fabric together to make a square for a quilting project. The triangles are both right triangles and have the same side and angle measures. What must also be true about the triangles in order to guarantee that the opposite sides of the fabric square are parallel? Explain.

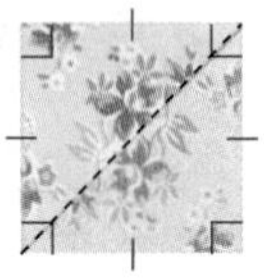

A carpenter is building a trellis for vines to grow on. The completed trellis will have two sets of diagonal pieces of wood that overlap each other.

27. If pieces A, B, and C must be parallel, what must be true of $\angle 1$, $\angle 2$, and $\angle 3$?

28. The carpenter attaches piece D so that it is perpendicular to piece A. If your answer to Exercise **27** is true, is piece D perpendicular to pieces B and C? Justify your answer.

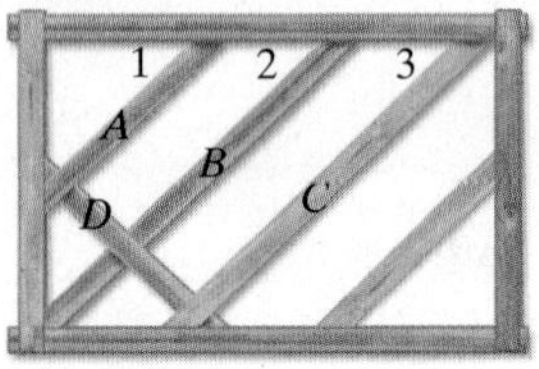

CONCEPT EXTENSIONS

Proof *For Exercises 29–32, prove each theorem using a two-column, statement/reason format. See Example 1.*

29. Write a proof.

Given: In a plane, $a \perp b$, $b \perp c$, and $c \parallel d$.

Prove: $a \parallel d$

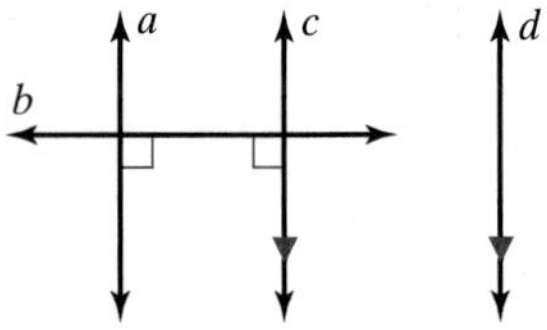

30. Write a proof.

Given: $q \parallel r$, $r \parallel s$, $b \perp q$, and $a \perp s$

Prove: $a \parallel b$

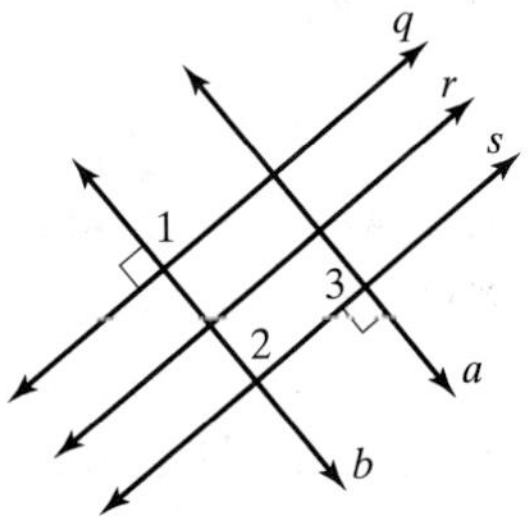

31. Prove Theorem 3.4-3.

32. Prove Theorem 3.4-4.

33. Find the Error A student sketched coplanar lines *m*, *n*, and *r* on his homework paper. He claims it shows that lines *m* and *n* are parallel. What other information do you need about line *r* in order for his claim to be true? Explain.

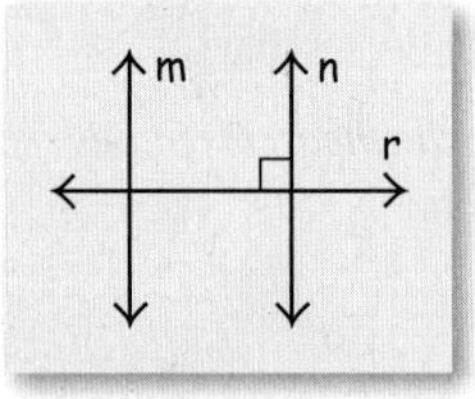

34. Main Street intersects Avenue A and Avenue B. Avenue A is parallel to Avenue B. Avenue A is also perpendicular to Main Street. How are Avenue B and Main Street related? Explain.

Proof *Copy and complete this paragraph proof of Theorem 3.4-2 for three coplanar lines.*

Given: $\ell \parallel k$ and $m \parallel k$

Prove: $\ell \parallel m$

Proof: Since $\ell \parallel k$, $\angle 2 \cong \angle 1$ by the _?_ (Exercise **35.**) Since $m \parallel k$, _?_ (Exercise **36.**) $\cong$ _?_ (Exercise **37.**) for the same reason. By the Transitive Property of Congruence, $\angle 2 \cong \angle 3$. By the _?_ (Exercise **38.**), $\ell \parallel m$.

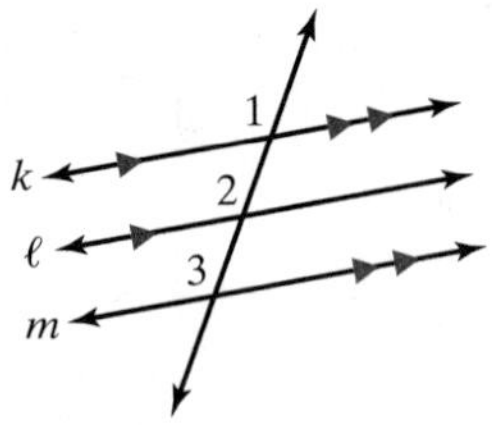

REVIEW AND PREVIEW

Determine the value of x that makes a || b. See Section 3.2.

39.

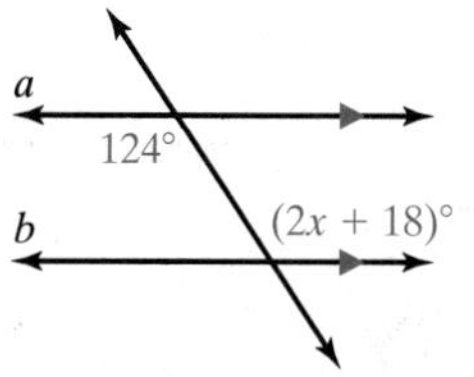

40.

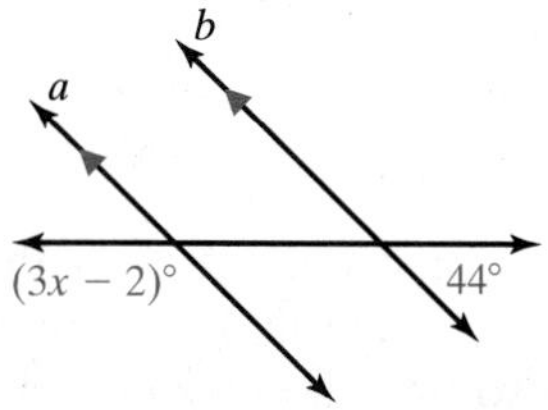

Go by sight and classify each angle as acute, right, or obtuse. See Section 1.5.

41.

42.

3.5 CONSTRUCTIONS—PARALLEL AND PERPENDICULAR LINES

OBJECTIVE

1 Construct Parallel and Perpendicular Lines.

OBJECTIVE 1 ▶ Constructing Parallel and Perpendicular Lines. We can use a straight edge and a compass to construct parallel and perpendicular lines.

In Section 3.2, we learned that through a point not on a line, there is one and only one line parallel to the given line. Example 1 shows the construction of this line.

EXAMPLE 1 **Constructing Parallel Lines**

Construct the line parallel to a given line and through a given point that is not on the line.

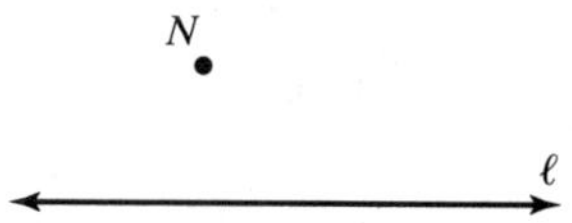

Given: Line ℓ and point N not on ℓ
Construct: Line m through N with $m \parallel \ell$

Solution

STEP 1. Label two points H and J on ℓ. Draw $\overleftrightarrow{HN}$.

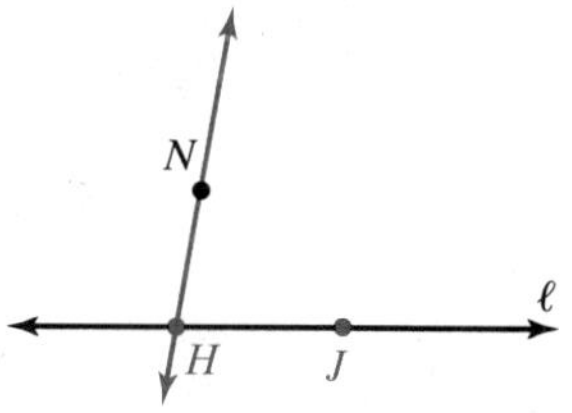

STEP 2. At N, construct $\angle 1$ congruent to $\angle NHJ$. Label the new line m.

$$m \parallel \ell$$

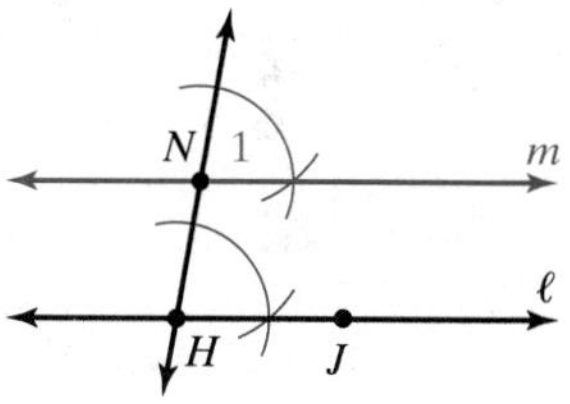

PRACTICE 1 Draw line n and point P not on line n. Construct line ℓ through P so that $\ell \parallel n$.

A quadrilateral is a four-sided figure. Examples of quadrilaterals include rectangles and squares. Let's see how we can construct a quadrilateral with one pair of parallel sides.

EXAMPLE 2 **Constructing a Special Quadrilateral**

Construct a quadrilateral with one pair of parallel sides of lengths b and a.

Given: Segments of lengths b and a
Construct: Quadrilateral $ABYZ$ with

$$AZ = a, BY = b, \text{and } \overleftrightarrow{AZ} \parallel \overleftrightarrow{BY}$$

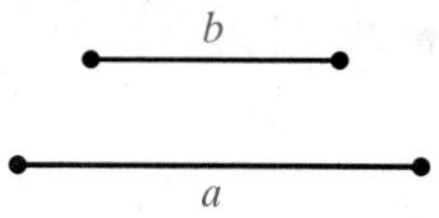

Helpful Hint
When constructing, try sketching the final figure first. This can help you visualize the construction steps you will need.

Solution

STEP 1. Draw a ray with endpoint A. Then draw $\overrightarrow{AB}$ such that point B is not on the first ray.

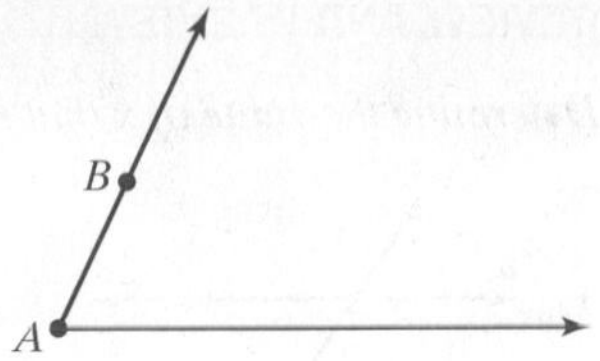

STEP 2. Through point B, draw a ray parallel to the first ray through A. Construct congruent corresponding angles so that the rays are parallel rays.

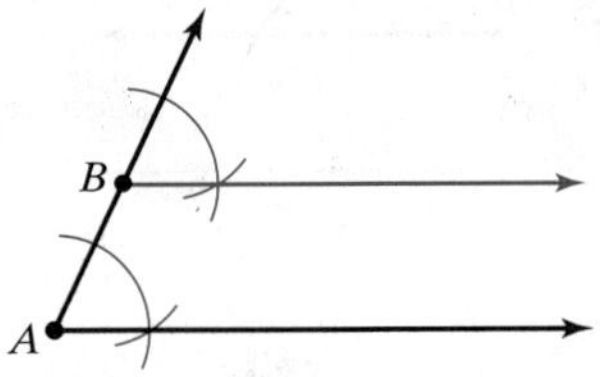

STEP 3. Draw sides of the given lengths b and a. Construct Y and Z so that $BY = b$ and $AZ = a$.

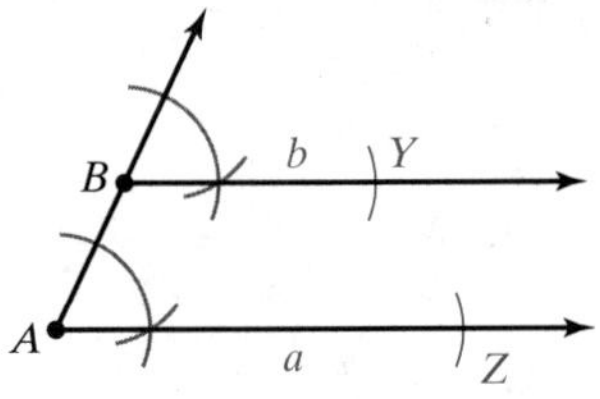

STEP 4. Draw $\overline{YZ}$ to complete the quadrilateral.

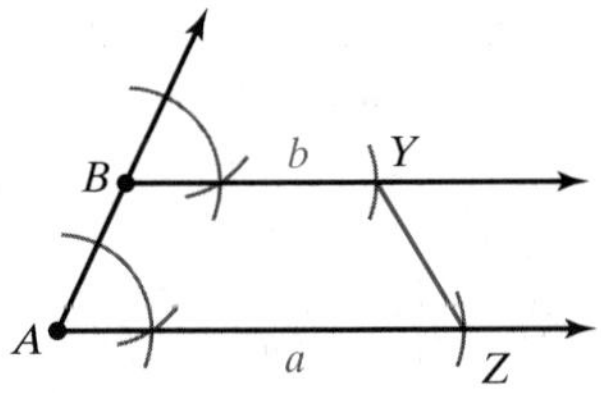

$ABYZ$ is a quadrilateral with parallel sides of lengths b and a. □

PRACTICE 2 Draw a segment. Label its length m. Construct quadrilateral $ABCD$ with $\overleftrightarrow{AB} \parallel \overleftrightarrow{CD}$ so that $AB = m$ and $CD = 2m$.

EXAMPLE 3 Constructing the Perpendicular at a Point on a Line

Construct the perpendicular to a given line at a given point on the line.

Given: Point P on line ℓ

Construct: $\overleftrightarrow{CP}$ with $\overleftrightarrow{CP} \perp \ell$

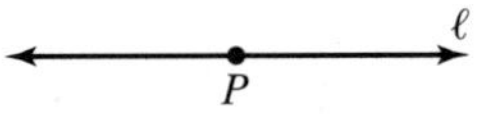

Solution

STEP 1. Construct two points on ℓ that are equidistant from P. Label the points A and B.

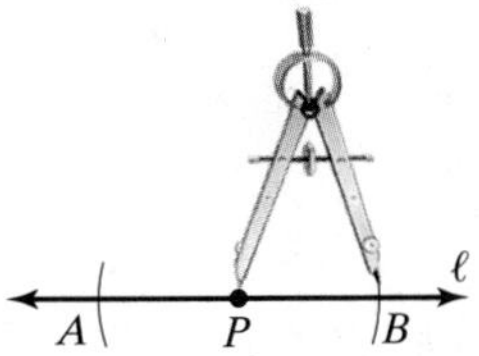

Helpful Hint
It is important to open our compass wider than $\frac{1}{2}AB$ for Example 3. If we don't, we won't be able to draw intersecting arcs above point P.

STEP 2. Open the compass wider so the opening is greater than $\frac{1}{2}AB$. With the compass tip on A, draw an arc above point P.

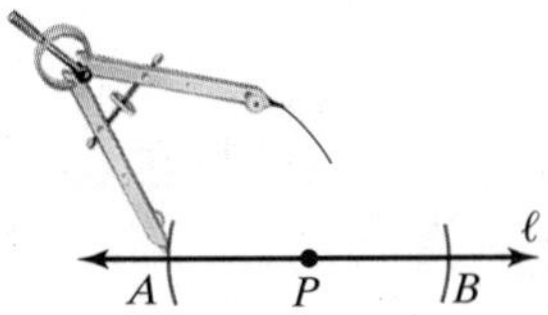

STEP 3. Without changing the compass setting, place the compass point on point B. Draw an arc that intersects the arc from Step 2. Label the point of intersection C.

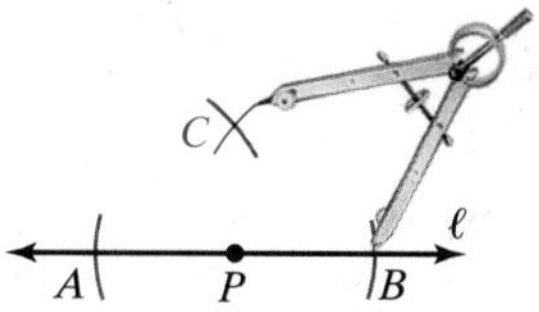

STEP 4. Draw $\overleftrightarrow{CP}$.

$\overleftrightarrow{CP} \perp \ell$

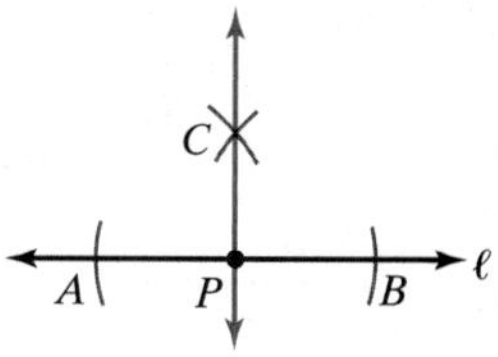

PRACTICE 3 Use a straight edge to draw $\overleftrightarrow{EF}$. Construct $\overleftrightarrow{FG}$ so that $\overleftrightarrow{FG} \perp \overleftrightarrow{EF}$ at point F.

We can also construct a perpendicular line from a point to a line. This perpendicular line is unique, according to the Perpendicular Postulate. We will prove in Chapter 5 that the shortest path from any point to a line is along this unique perpendicular line.

EXAMPLE 4 Constructing the Perpendicular from a Point to a Line

Construct the perpendicular to a given line through a given point not on the line.

Given: Line ℓ and point R not on ℓ

Construct: $\overleftrightarrow{RG}$ with $\overleftrightarrow{RG} \perp \ell$

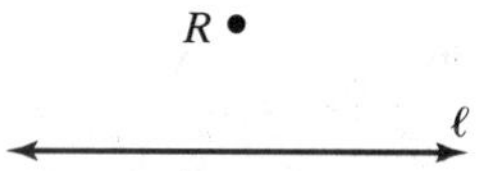

Solution

STEP 1. Open your compass to a size greater than the distance from R to ℓ. With the compass on point R, draw an arc that intersects ℓ at two points. Label the points E and F.

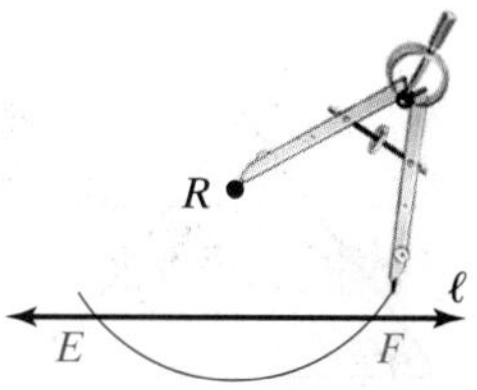

STEP 2. Place the compass point on E and make an arc.

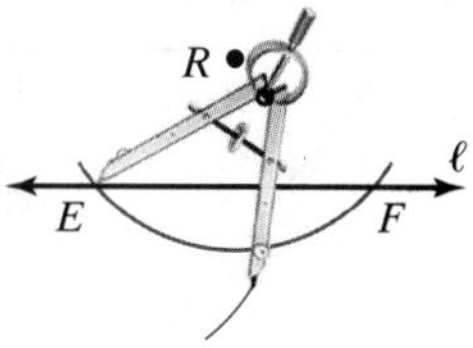

STEP 3. Keep the same compass setting. With the compass tip on F, draw an arc that intersects the arc from Step 2. Label the point of intersection G.

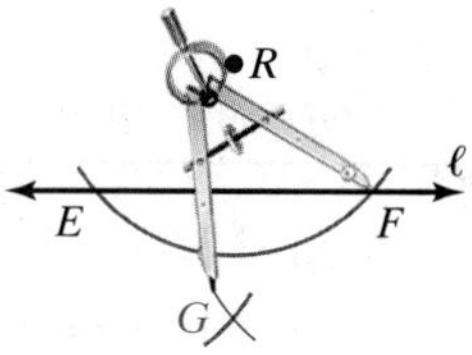

STEP 4. Draw $\overleftrightarrow{RG}$.

$\overleftrightarrow{RG} \perp \ell$

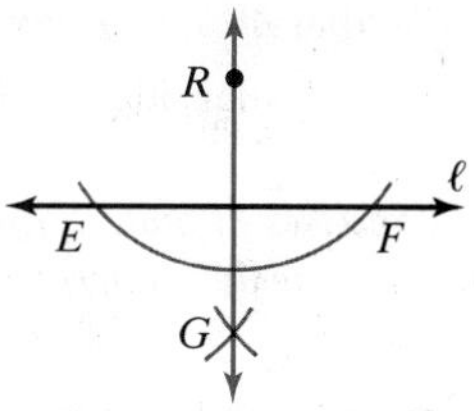

PRACTICE 4 Draw $\overleftrightarrow{CX}$ and a point Z not on $\overleftrightarrow{CX}$. Construct $\overleftrightarrow{ZB}$ so that $\overleftrightarrow{ZB} \perp \overleftrightarrow{CX}$.

3.5 EXERCISE SET MyMathLab®

Construction *For Exercises 1–6, draw a figure like the given one. Then construct the line through point J that is parallel to $\overleftrightarrow{AB}$. See Examples 1 and 2.*

1.

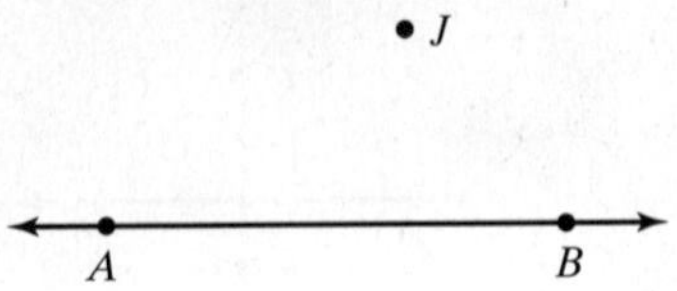

2.

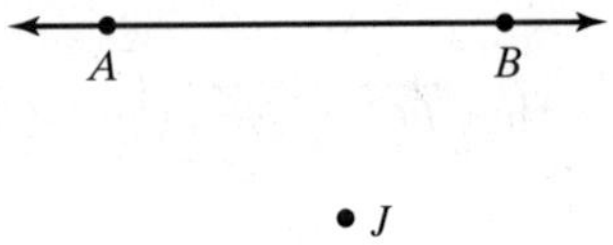

3.

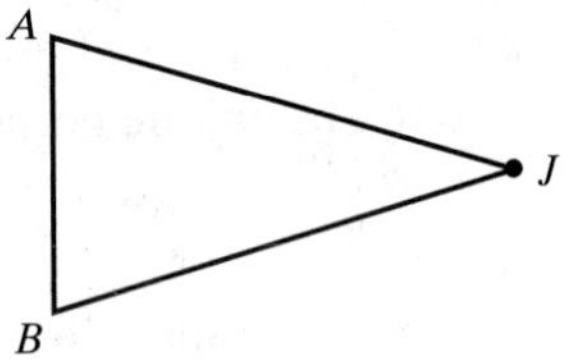

4.

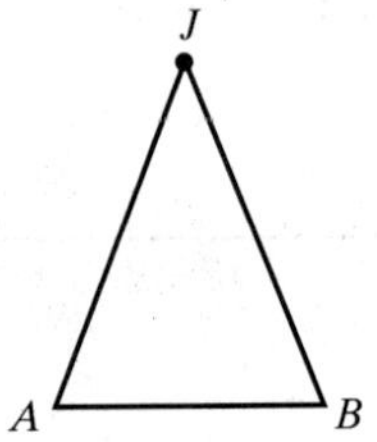

5.

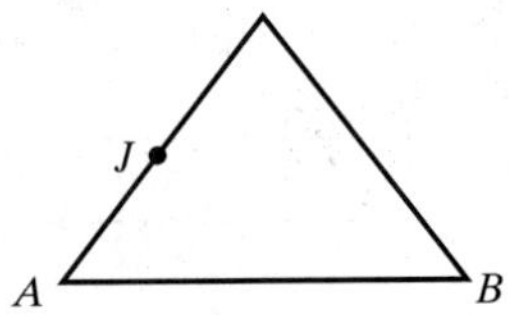

6.

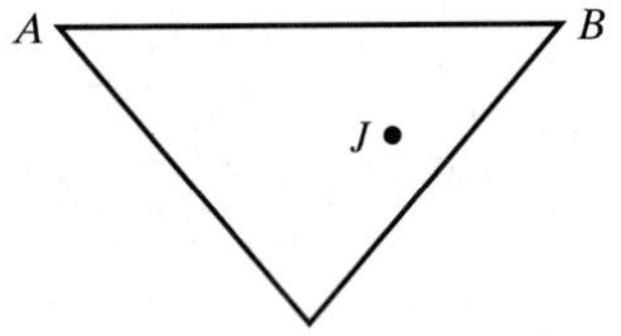

For Exercises 7–10, draw two segments. Label their lengths a and b. Construct a quadrilateral with one pair of parallel sides as described. See Example 2.

7. The parallel sides have lengths a and b.

8. The parallel sides have lengths $2a$ and b.

9. The parallel sides have lengths a and $2b$.

10. The parallel sides have lengths a and $\frac{1}{2}b$.

For Exercises 11 and 12, draw a figure like the given one. Then construct the line that is perpendicular to ℓ at point P. See Example 3.

11.

12.

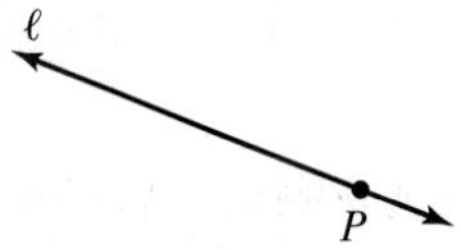

For Exercises 13–16, draw a figure like the given one. Then construct the line through point P that is perpendicular to $\overleftrightarrow{RS}$. See Example 4.

13.

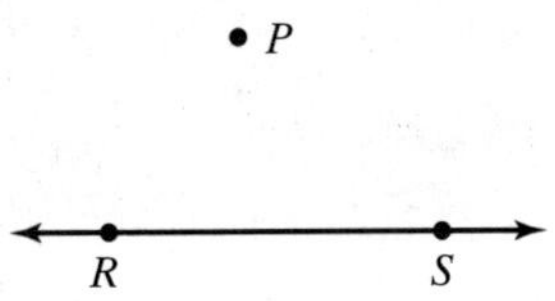

14.

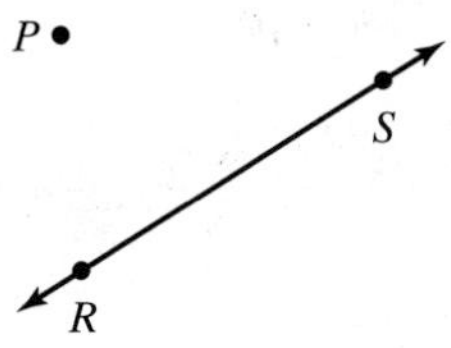

15.

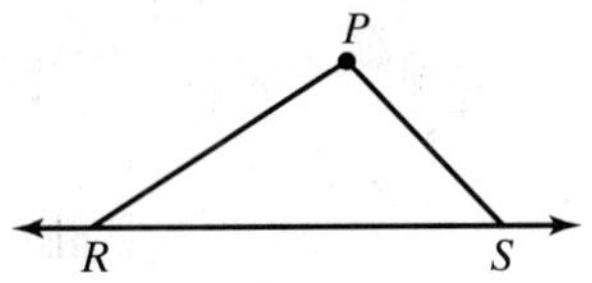

16.

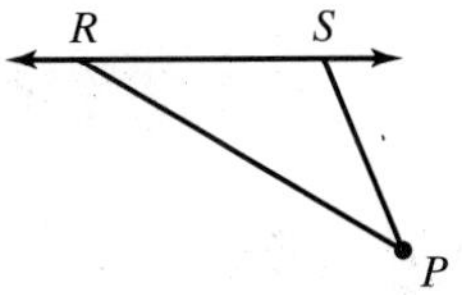

For Exercises 17–22, use the segments at the right.

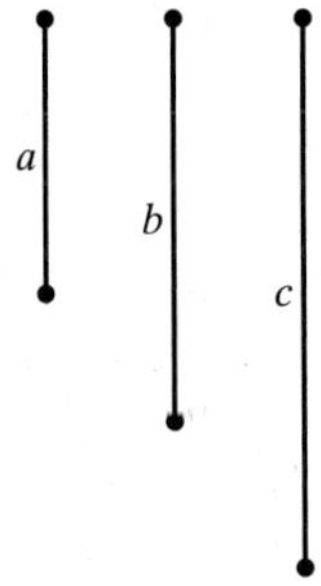

17. Construct a rectangle with width b and length c.

18. Construct a square with sides of length a.

19. Construct a rectangle with one side of length a and a diagonal of length b.

20. Construct a right triangle with legs of lengths a and b.

CONCEPT EXTENSIONS

Construction

21. Multiple Steps

a. Construct a triangle with sides of lengths a, b, and c.

b. Construct the midpoint of each side of the triangle.

c. Form a new triangle by connecting the midpoints.

d. How do the sides of the smaller triangle and the sides of the larger triangle appear to be related? Use a protractor, a ruler, or both to check the conjecture you made.

22. Multiple Steps

a. Construct a quadrilateral with a pair of parallel sides of length c.

b. What appears to be true about the other pair of sides in the quadrilateral you constructed?

c. Use a protractor, a ruler, or both to check the conjecture you made in part **b**.

The diagrams below show steps for a parallel line construction.

i.

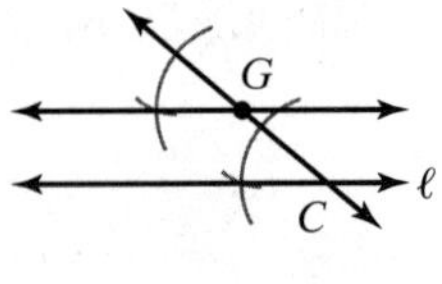

ii.

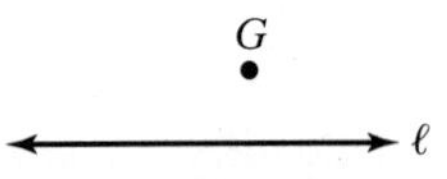

iii.

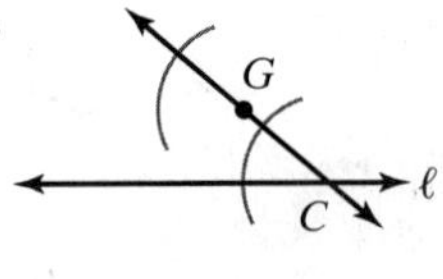

iv.

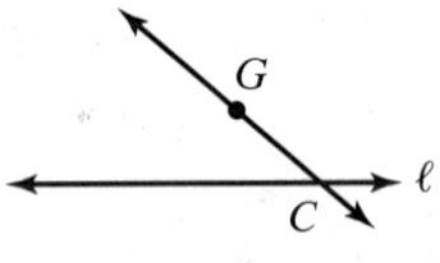

23. List the construction steps in the correct order.

24. For the steps that use a compass, describe the location(s) of the compass point.

Construction *Construct a rectangle with side lengths a and b that meets the given condition.*

25. $b = 2a$

26. $b = \frac{1}{2}a$

27. $b = \frac{1}{3}a$

28. $b = \frac{2}{3}a$

Construct a triangle with side lengths a, b, and c that meets the given conditions. If such a triangle is not possible, explain.

29. $a = b = c$

30. $a = b = 2c$

31. $a = 2b = 2c$

32. $a = b + c$

33. Draw $\overleftrightarrow{QR}$ and a point S on the line. Construct the line perpendicular to $\overleftrightarrow{QR}$ at point S.

34. Draw a line w and a point X not on the line. Construct the line perpendicular to line w through point X.

35. How are the constructions in Examples 3 and 4 similar? How are they different?

36. Suppose you use a wider compass setting in Step 1 of Example 4. Will you construct a different perpendicular line? Explain.

REVIEW AND PREVIEW

Simplify each ratio. See an appendix for review.

37. $\frac{2 - (-3)}{6 - (-4)}$

38. $\frac{1 - 4}{-2 - 1}$

39. $\frac{12 - 6}{2 - 5}$

40. $\frac{-7 - (-1)}{1 - (-1)}$

3.6 COORDINATE GEOMETRY—THE SLOPE OF A LINE

OBJECTIVES

1 Find the Slope of a Line.

2 Interpret the Slope-Intercept Form in an Application.

3 Compare the Slopes of Parallel and Perpendicular Lines.

VOCABULARY

- slope
- vertical change
- horizontal change
- rate of change
- y-intercept point
- y-intercept
- slope-intercept form
- perpendicular lines

OBJECTIVE 1 ▶ Finding Slope Given Two Points. You may have noticed by now that different lines often tilt differently. It is very important in many fields to be able to measure and compare the tilt, or **slope,** of lines. For example, a wheelchair ramp with a slope of $\frac{1}{12}$ means that the ramp rises 1 foot for every 12 horizontal feet. A road with a slope or grade of 11% $\left(\text{or } \frac{11}{100}\right)$ means that the road rises 11 feet for every 100 horizontal feet.

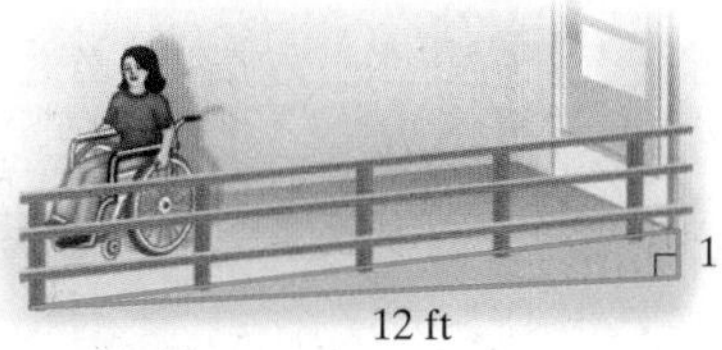

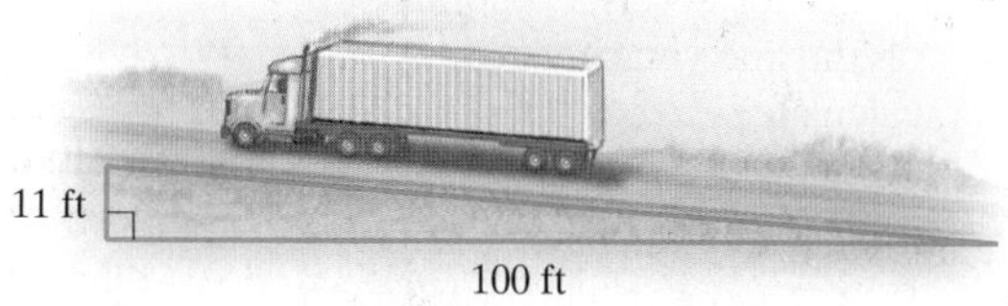

We measure the slope of a line as a ratio of **vertical change** to **horizontal change.** Slope is usually designated by the letter m.

Suppose that we want to measure the slope of the following line.

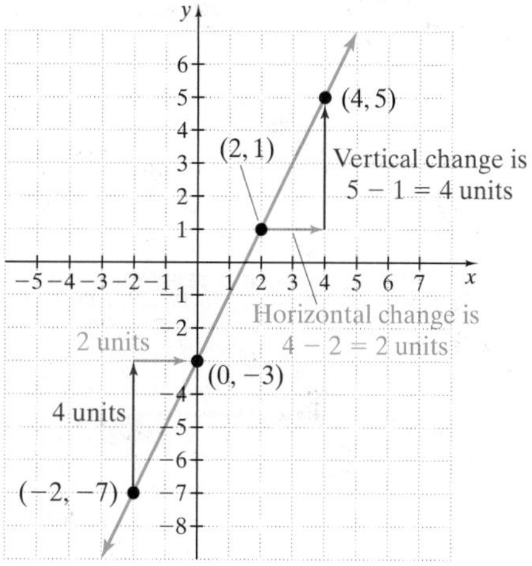

The vertical change between *both* pairs of points on the line is 4 units per horizontal change of 2 units. Then

$$\text{slope } m = \frac{\text{change in } y\text{ (vertical change)}}{\text{change in } x\text{ (horizontal change)}} = \frac{4}{2} = 2$$

We can also think of slope as a **rate of change** between points. A slope of 2 or $\frac{2}{1}$ means that between pairs of points on the line, the rate of change is a vertical change of 2 units per horizontal change of 1 unit.

The line in the box below passes through the points (x_1, y_1) and (x_2, y_2). (The notation x_1 is read "x-sub-one.") The vertical change, or *rise*, between these points is the difference of the y-coordinates: $y_2 - y_1$. The horizontal change, or *run*, between the points is the difference of the x-coordinates: $x_2 - x_1$.

Defined Term—Slope

Definition	*Formula*	*Diagram*
The **slope** m of a line is the ratio of the vertical change (rise) to the horizontal change (run) between any two points.	A line contains the points (x_1, y_1) and (x_2, y_2). $m = \frac{\text{rise}}{\text{run}} = \frac{y_2 - y_1}{x_2 - x_1}$	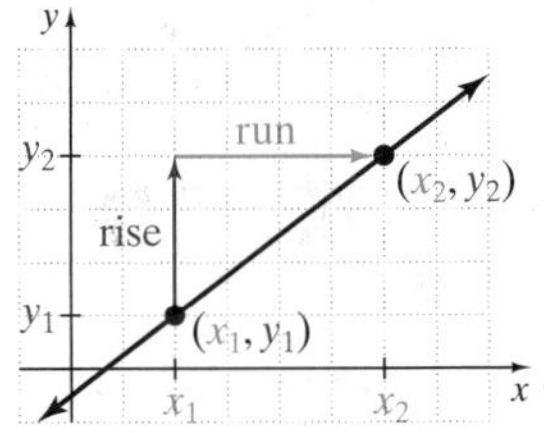

EXAMPLE 1 Finding Slopes of Lines

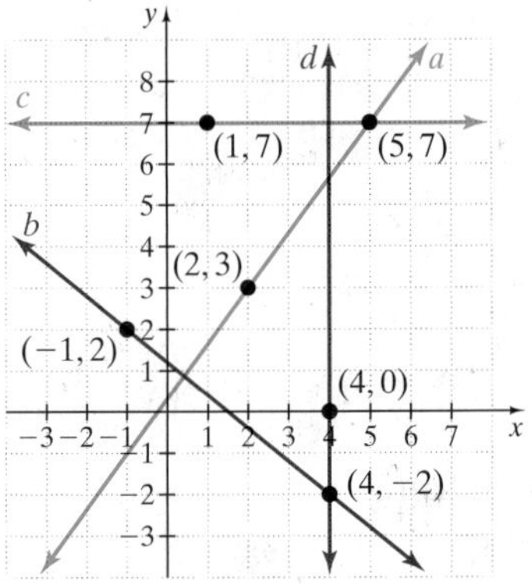

a. Find the slope of line b.

b. Find the slope of line d.

> **Helpful Hint**
> - The slope of a line is the same no matter which two points on the line you choose to calculate slope.
> - Also, the slope of a line is the same no matter which point is (x_1, y_1) and which is (x_2, y_2). Note: Once an x-value is called x_1, its y-value must be called y_1.

Solution

a. For line b, we use points $(-1, 2)$ and $(4, -2)$ and the slope formula.

$$m = \frac{2 - (-2)}{-1 - 4}$$
$$= \frac{4}{-5}$$
$$= -\frac{4}{5}$$

b. For line d, we use points $(4, 0)$ and $(4, -2)$ and the slope formula.

$$m = \frac{0 - (-2)}{4 - 4}$$
$$= \frac{2}{0} \quad \text{Undefined}$$

□

PRACTICE 1 Use the graph in Example 1.

a. What is the slope of line a?

b. What is the slope of line c?

As you saw in Example 1 and Practice 1, the slope of a line can be positive, negative, zero, or undefined. The sign of the slope tells us whether the line rises or falls as we follow it from left to right. A slope of zero tells us that the line is horizontal. An undefined slope tells us that the line is vertical.

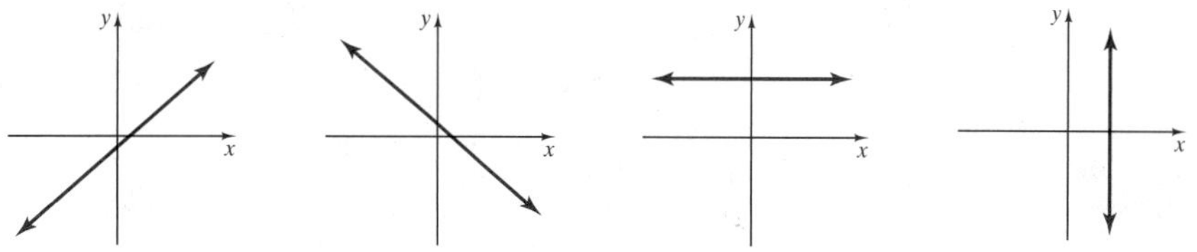

Positive slope | Negative slope | Zero slope | Undefined slope

Don't forget these important facts about slopes of vertical and horizontal lines.

> **Helpful Hint**
> Slope of 0 and undefined slope are not the same. Vertical lines have undefined slope, whereas horizontal lines have slope of 0.

The slope of any vertical line is undefined.

The slope of any horizontal line is 0.

We can also find the slope of a line when we know its equation. The equation of a line has different forms. Two forms are shown below. Recall that the **y-intercept point** of a line is the point where the line crosses the y-axis, so has the form $(0, y)$. Also, the **y-intercept** is the y coordinate only of this point, or simply y.

Forms of Linear Equations

Definition	*Symbols*
The **slope-intercept form** of an equation of a nonvertical line is $y = mx + b$, where m is the slope and b is the y-intercept.	$y = mx + b$ ↑ slope (under m) ↑ y-intercept; $(0, b)$ is the y-intercept point (under b)
The **point-slope form** of an equation of a nonvertical line is $y - y_1 = m(x - x_1)$, where m is the slope and (x_1, y_1) is a point on the line.	$y - y_1 = m(x - x_1)$ ↑ y-coordinate (under y_1) ↑ slope (under m) ↑ x-coordinate (under x_1)

Notice the slope-intercept form in the box above. When a linear equation is written in the form $y = mx + b$, m is the slope of the line and b is its y-intercept or $(0, b)$ is its y-intercept point. That's why the form $y = mx + b$ is appropriately called the **slope-intercept form.**

> **Helpful Hint**
> Remember that only when an equation is solved for y is the coefficient of x the slope.

EXAMPLE 2 Find the slope and the y-intercept point of the line $3x - 4y = 4$.

Solution We write the equation in slope-intercept form by solving for y.

$$3x - 4y = 4$$

$$-4y = -3x + 4 \quad \text{Subtract } 3x \text{ from both sides.}$$

$$\frac{-4y}{-4} = \frac{-3x}{-4} + \frac{4}{-4} \quad \text{Divide both sides by } -4.$$

$$y = \frac{3}{4}x - 1 \quad \text{Simplify.}$$

The coefficient of x, $\frac{3}{4}$, is the slope, and the y-intercept point is $(0, -1)$. □

PRACTICE 2 Find the slope and the y-intercept point of the line $2x - 3y = 9$.

The appearance of a line can give us further information about its slope.

The graphs of $y = \frac{1}{2}x + 1$ and $y = 5x + 1$ are shown at the right. Recall that the graph of $y = \frac{1}{2}x + 1$ has a slope of $\frac{1}{2}$ and that the graph of $y = 5x + 1$ has a slope of 5.

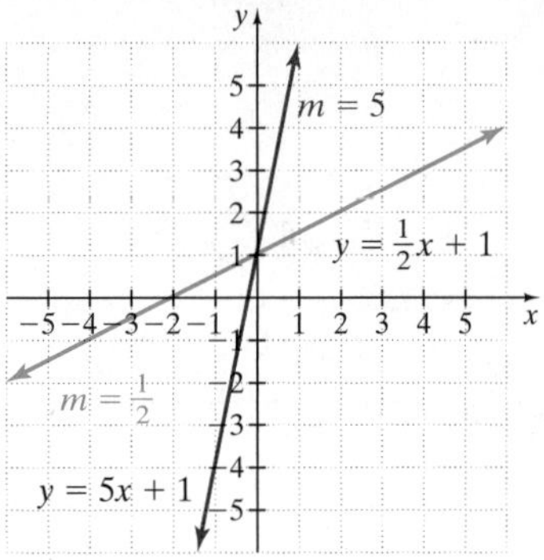

Notice that the line with the slope of 5 is steeper than the line with the slope of $\frac{1}{2}$. This is true in general for positive slopes.

For a line with positive slope m, as m increases, the line becomes steeper.

To see why this is so, compare the slopes from above.

$\frac{1}{2}$ means a vertical change of 1 unit per a horizontal change of 2 units

5 or $\frac{10}{2}$ means a vertical change of 10 units per a horizontal change of 2 units

For larger positive slopes, the vertical change is greater for the same horizontal change. Thus, larger positive slopes mean steeper lines.

Helpful Hint
The notation $0 \leftrightarrow 2000$ below the graph means that the number 0 corresponds to the year 2000, 1 corresponds to the year 2001, and so on.

OBJECTIVE 2 ▶ Interpreting Slope-Intercept Form. Below is a graph of one-day ticket prices at Disney World for the years shown.

Notice that the graph resembles the graph of a line. Recall that businesses often depend on equations that "closely fit" graphs like this one to model the data and to predict future trends. By a method called the **least squares** method, the linear equation $y = 3.2x + 48$ approximates the data shown, where x is the number of years since 2000 and y is the ticket price for that year.

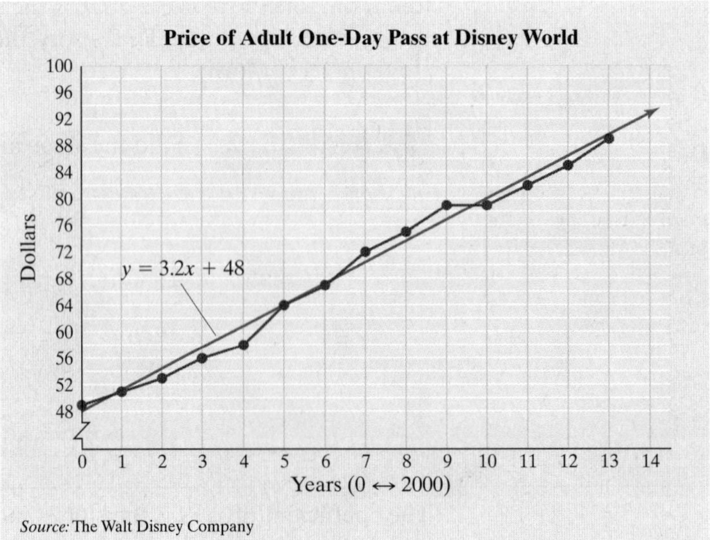

Source: The Walt Disney Company

EXAMPLE 3 **Predicting Future Prices**

The adult one-day pass price for Disney World is given by

$$y = 3.2x + 48$$

where x is the number of years since 2000.

a. Use this equation to predict the ticket price for the year 2020.

b. What does the slope of this equation mean?

c. What does the y-intercept point of this equation mean?

Solution

a. To predict the price of a pass in 2020, we need to find y when x is 20. (Since year 2000 corresponds to $x = 0$, year 2020 corresponds to $x = 20$.)

$$\begin{aligned} y &= 3.2x + 48 \\ &= 3.2(20) + 48 \quad \text{Let } x = 20. \\ &= 112 \end{aligned}$$

We predict that in the year 2020, the price of an adult one-day pass to Disney World will be about \$112.

b. The slope of $y = 3.2x + 48$ is 3.2. We can think of this number as $\frac{\text{rise}}{\text{run}}$ or $\frac{3.2}{1}$. This means that the ticket price increases on the average by \$3.20 every 1 year.

c. The y-intercept point of $y = 3.2x + 48$ is (0, 48).

$\uparrow$ year $\uparrow$ price

This means that at year 0, or 2000, the ticket price was about \$48. □

PRACTICE 3 Use the equation from Example 3 to predict the ticket price for the year 2025.

OBJECTIVE 3 ▶ Comparing Slopes of Parallel and Perpendicular Lines. Slopes of lines can help us determine whether lines are parallel. Parallel lines are distinct lines with the same steepness, so it follows that they have the same slope.

Parallel Lines

Two nonvertical lines are parallel if they have the same slope and different y-intercepts.

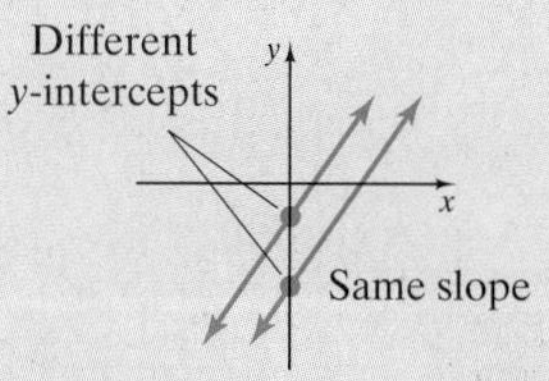

How do the slopes of perpendicular lines compare? (Two lines intersecting at right angles are called **perpendicular lines.**) Suppose that a line has a slope of $\frac{a}{b}$. If the line is rotated 90°, the rise and run are now switched, except that the run is now negative. This means that the new slope is $-\frac{b}{a}$. Notice that

$$\left(\frac{a}{b}\right) \cdot \left(-\frac{b}{a}\right) = -1$$

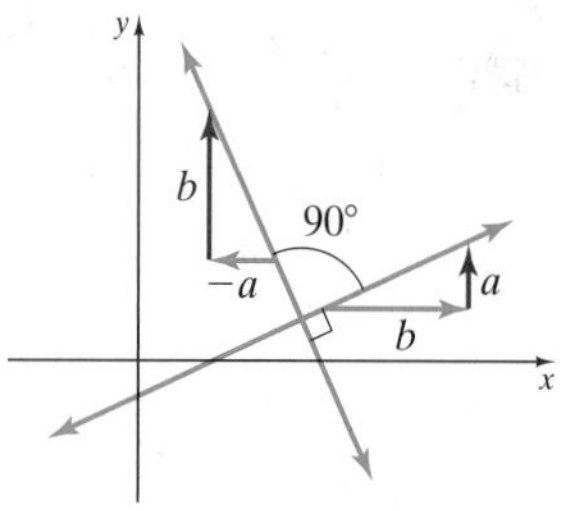

This is how we tell whether two lines are perpendicular.

Perpendicular Lines

Two nonvertical lines are perpendicular if the product of their slopes is -1.

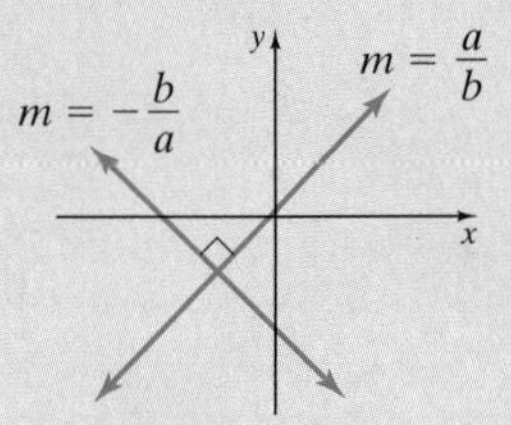

In other words, two nonvertical lines are perpendicular if the slope of one is the negative reciprocal of the slope of the other.

EXAMPLE 4 Are the following pairs of lines parallel, perpendicular, or neither?

a. $3x + 7y = 4$ $\quad$ $6x + 14y = 7$

b. $-x + 3y = 2$ $\quad$ $2x + 6y = 5$

Solution Find the slope of each line by solving each equation for y.

a.
$$3x + 7y = 4 \qquad 6x + 14y = 7$$
$$7y = -3x + 4 \qquad 14y = -6x + 7$$
$$\frac{7y}{7} = \frac{-3x}{7} + \frac{4}{7} \qquad \frac{14y}{14} = \frac{-6x}{14} + \frac{7}{14}$$
$$y = -\frac{3}{7}x + \frac{4}{7} \qquad y = -\frac{3}{7}x + \frac{1}{2}$$

↑ slope ↖ y-intercept $\qquad$ ↑ slope ↖ y-intercept

$\left(0, \frac{4}{7}\right)$ ← y-intercept points → $\left(0, \frac{1}{2}\right)$

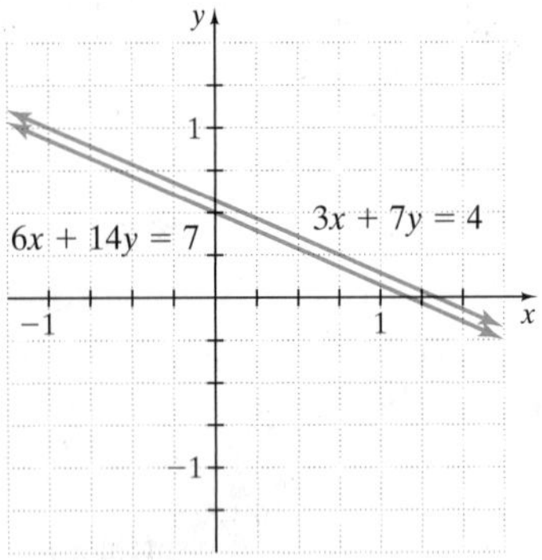

The slopes of both lines are $-\frac{3}{7}$.

The y-intercepts points are different, so the lines are not the same. Therefore, the lines are parallel.

b.
$$-x + 3y = 2 \qquad 2x + 6y = 5$$
$$3y = x + 2 \qquad 6y = -2x + 5$$
$$\frac{3y}{3} = \frac{x}{3} + \frac{2}{3} \qquad \frac{6y}{6} = \frac{-2x}{6} + \frac{5}{6}$$
$$y = \frac{1}{3}x + \frac{2}{3} \qquad y = -\frac{1}{3}x + \frac{5}{6}$$

↑ slope ↖ y-intercept $\qquad$ ↑ slope ↖ y-intercept

$\left(0, \frac{2}{3}\right)$ ← y-intercept points → $\left(0, \frac{5}{6}\right)$

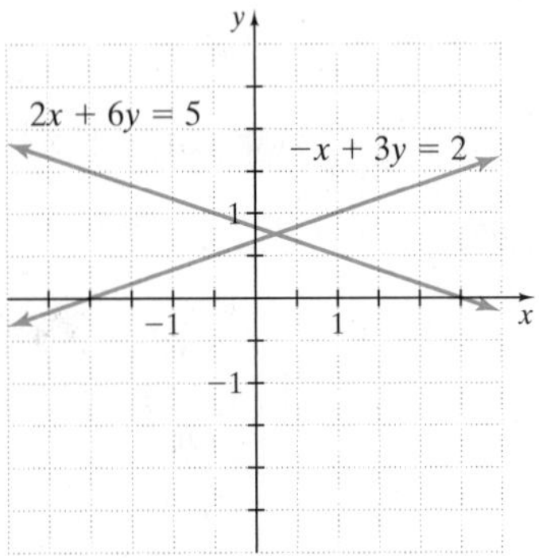

The slopes are not the same and their product is not $-1\left[\left(\frac{1}{3}\right)\cdot\left(-\frac{1}{3}\right) = -\frac{1}{9}\right]$. Therefore, the lines are neither parallel nor perpendicular. □

PRACTICE 4 Are the following pairs of lines parallel, perpendicular, or neither?

a. $x - 2y = 3$ $\quad$ $2x + y = 3$

b. $4x - 3y = 2$ $\quad$ $-8x + 6y = -6$

VOCABULARY & READINESS CHECK

Word Bank *Use the choices below to fill in each blank. Some choices may be used more than once and some not at all.*

horizontal	the same	-1	y-intercepts	$(0, b)$	slope	b
vertical	different	m	x-intercepts	$(b, 0)$	slope-intercept	

1. The measure of the steepness or tilt of a line is called ______________.
2. The slope of a line through two points is measured by the ratio of ______________ change to ______________ change.
3. If a linear equation is in the form $y = mx + b$, the slope of the line is ______________, the y-intercept point is ______________, and the y-intercept is ______________.
4. The form $y = mx + b$ is the ______________ form.
5. The slope of a(n) ______________ line is 0.
6. The slope of a(n) ______________ line is undefined.
7. Two nonvertical perpendicular lines have slopes whose product is ______________.
8. Two nonvertical lines are parallel if they have ______________ slope and different ______________.

Decision Making *Decide whether a line with the given slope slants upward or downward from left to right, or is horizontal or vertical.*

9. $m = \frac{7}{6}$ **10.** $m = -3$ **11.** $m = 0$ **12.** m is undefined.

3.6 EXERCISE SET MyMathLab®

Find the slope of the line that goes through the given points. See Example 1.

1. $(3, 2), (8, 11)$
2. $(1, 6), (7, 11)$
3. $(3, 1), (1, 8)$
4. $(2, 9), (6, 4)$
5. $(-2, 8), (4, 3)$
6. $(3, 7), (-2, 11)$
7. $(-2, -6), (4, -4)$
8. $(-3, -4), (-1, 6)$
9. $(-3, -1), (-12, 11)$
10. $(3, -1), (-6, 5)$
11. $(-2, 5), (3, 5)$
12. $(4, 2), (4, 0)$
13. $(-1, 1), (-1, -5)$
14. $(-2, -5), (3, -5)$
15. $(0, 6), (-3, 0)$
16. $(5, 2), (0, 5)$
17. $(-1, 2), (-3, 4)$
18. $(3, -2), (-1, -6)$

Decision Making *Two lines are graphed on each set of axes. Decide whether l_1 or l_2 has the greater slope. See the boxed material on page 134.*

19.

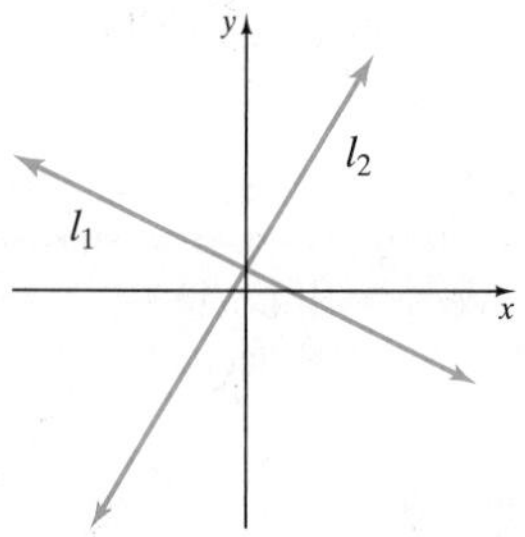

20.

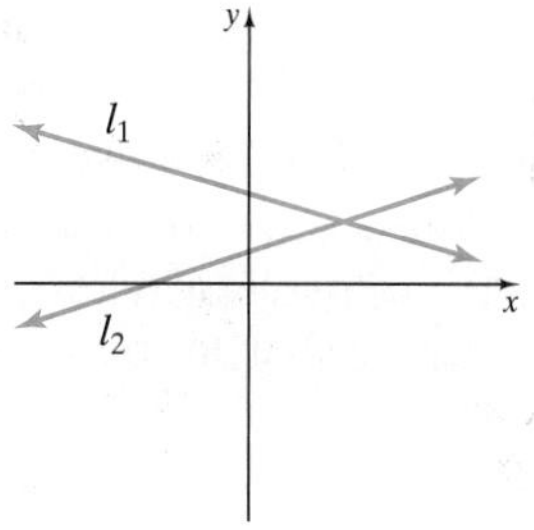

21.

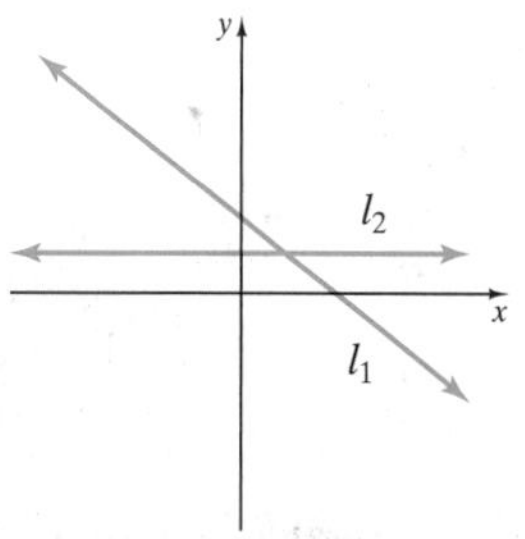

22.

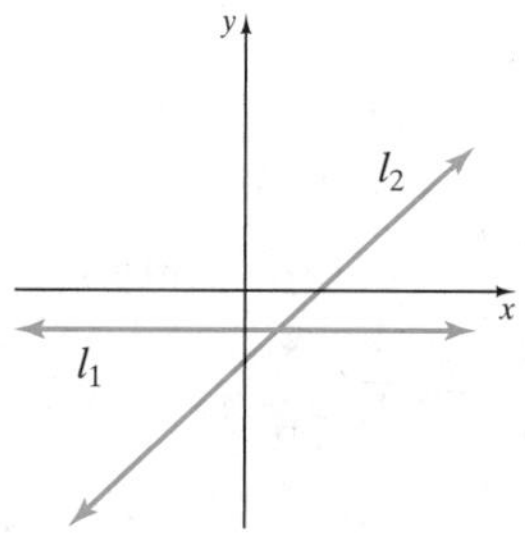

23.

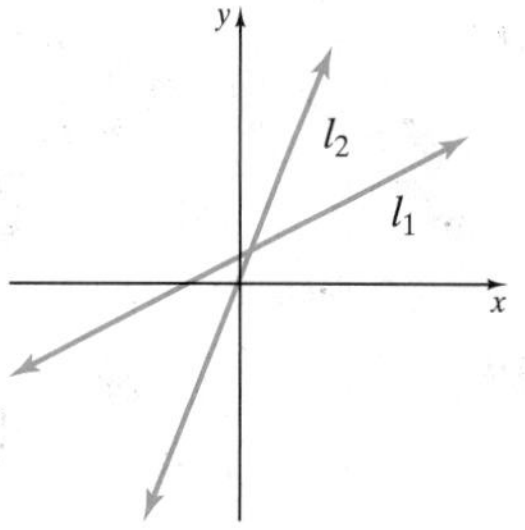

24.

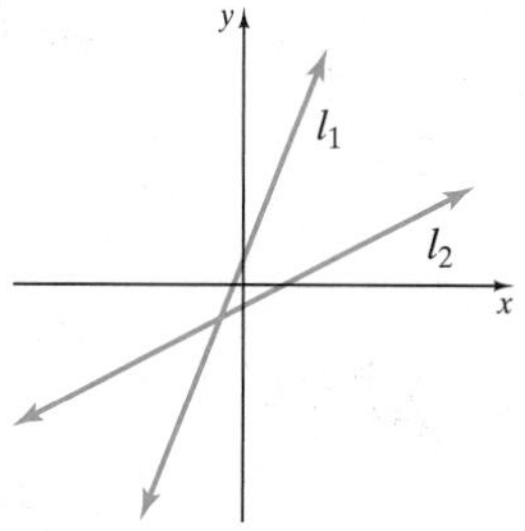

Find the slope and the y-intercept point of each line. See Example 2.

25. $y = 5x - 2$
26. $y = -2x + 6$
27. $2x + y = 7$
28. $-5x + y = 10$
29. $2x - 3y = 10$
30. $-3x - 4y = 6$
31. $y = \frac{1}{2}x$
32. $y = -\frac{1}{4}x$

Matching *Match each graph with its equation. See Examples 1 and 2.*

A.

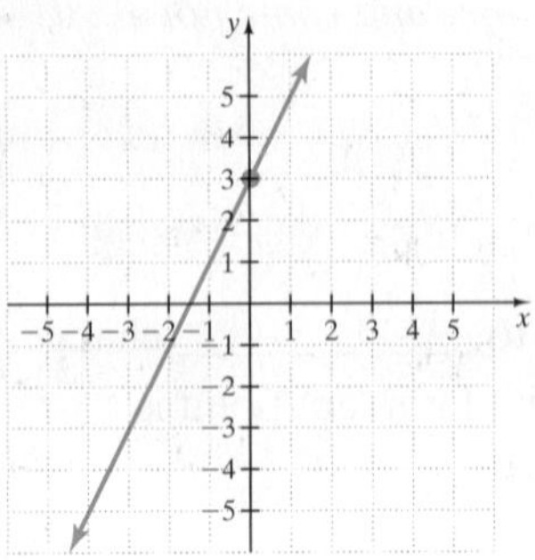

B.

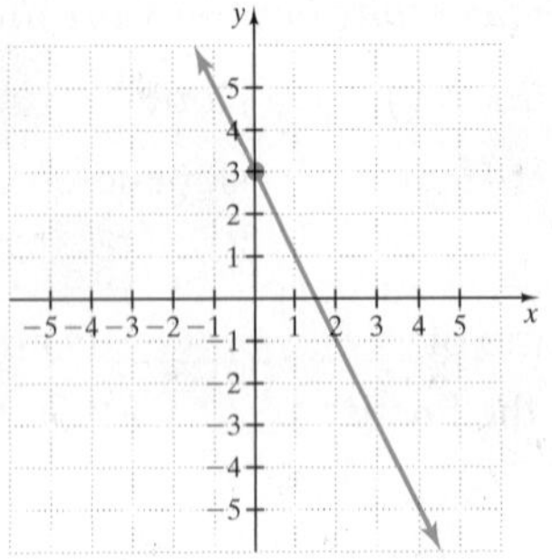

C.

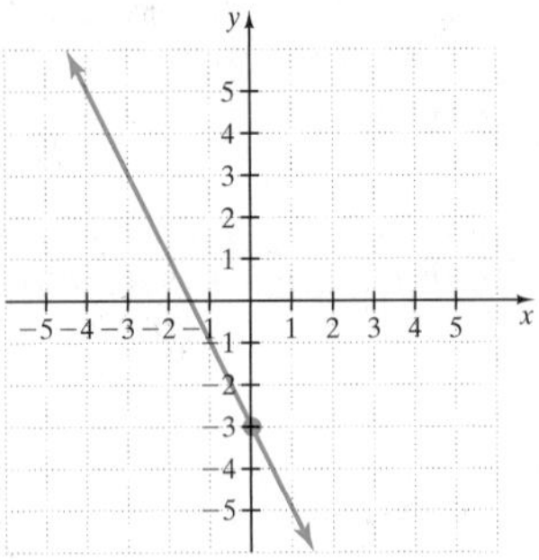

D.

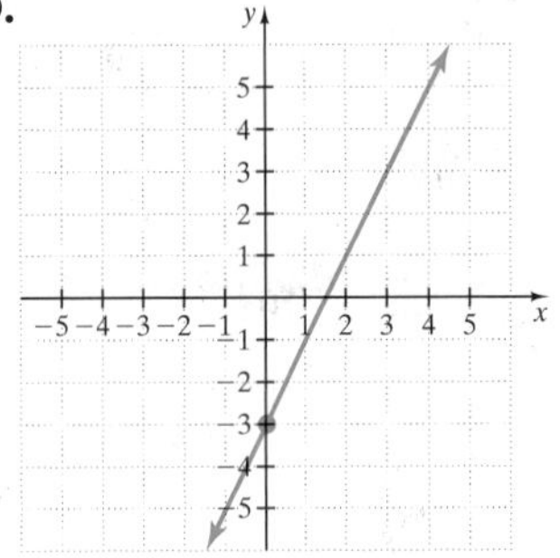

33. $y = 2x + 3$

34. $y = 2x - 3$

35. $y = -2x + 3$

36. $y = -2x - 3$

Find the slope of each line. Sample points are given. See Example 1.

37. $x = 1; (1, 5), (1, 0)$

38. $y = -2; (0, -2), (4, -2)$

39. $y = -3; (-2, -3), (-3, -3)$

40. $x = 4; (4, 1), (4, -1)$

41. $x + 2 = 0; (-2, 1), (-2, 0)$

42. $y - 7 = 0; (0, 7)(2, 7)$

Determine whether the lines are parallel, perpendicular, or neither. See Example 4.

43. $y = -3x + 6$
$y = 3x + 5$

44. $y = 5x - 6$
$y = 5x + 2$

45. $-4x + 2y = 5$
$2x - y = 7$

46. $2x - y = -10$
$2x + 4y = 2$

47. $-2x + 3y = 1$
$3x + 2y = 12$

48. $x + 4y = 7$
$2x - 5y = 0$

Use the points shown on the graphs to determine the slope of each line. See Example 1.

49.

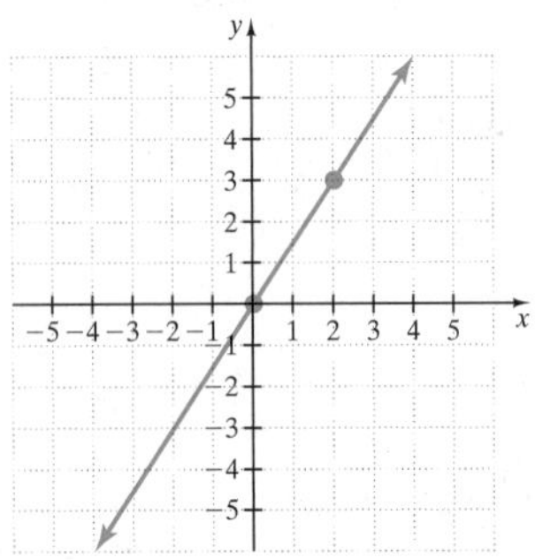

50.

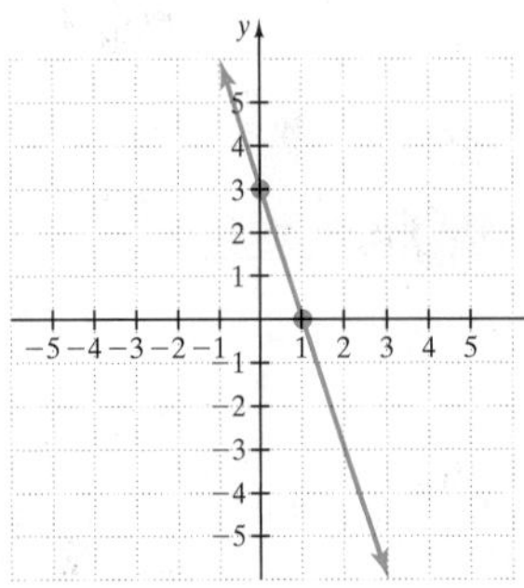

51.

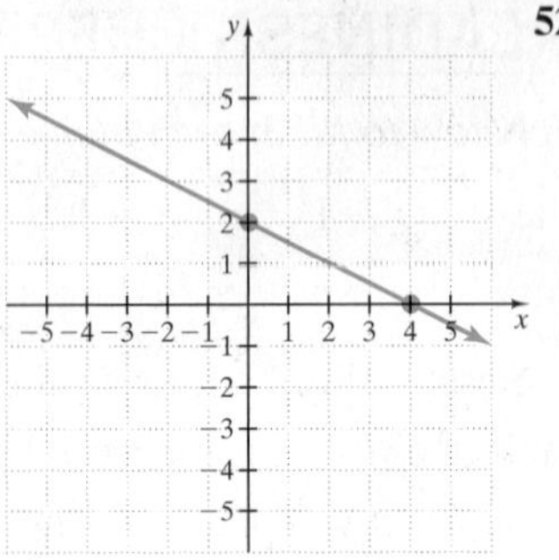

52.

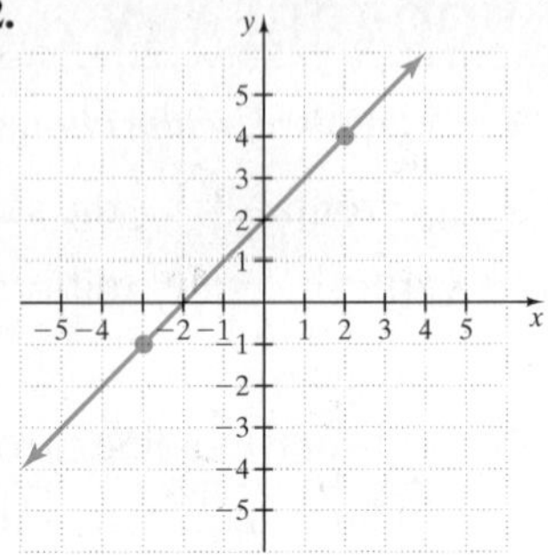

Find each slope. See Example 1.

53. Find the pitch, or slope, of the roof shown.

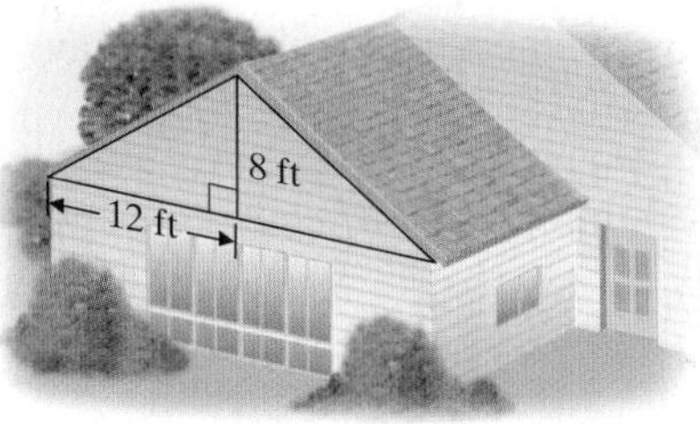

54. Upon takeoff, a Delta Airlines jet climbs to 3 miles as it passes over 25 miles of land below it. Find the slope of its climb.

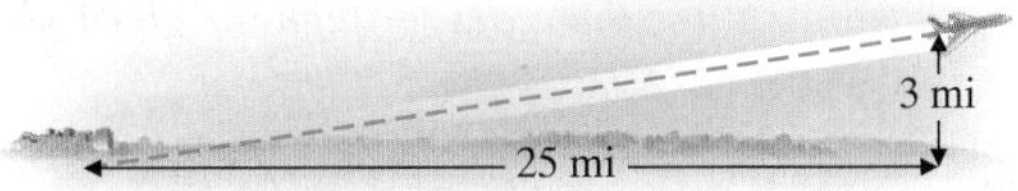

55. Driving down Bald Mountain in Wyoming, Bob Dean finds that he descends 1600 feet in elevation by the time he is 2.5 miles (horizontally) away from the high point on the mountain road. Find the slope of his descent, rounded to two decimal places (1 mile = 5280 feet).

56. Find the grade, or slope, of the road shown.

Multiple Steps *Solve. See Example 3.*

57. With wireless Internet (WiFi) gaining popularity, the number of public wireless Internet access points (in thousands) is projected to grow from 2011 to 2015 according to the equation

$$-1125x + y = 1300$$

where x is the number of years after 2011.

a. Find the slope and y-intercept of the linear equation.

b. What does the slope mean in this context?

c. What does the y-intercept point mean in this context?

58. One of the faster-growing occupations over the next few years is expected to be nursing. The number of people y in thousands employed in nursing in the United States can be estimated by the linear equation $-266x + 10y = 27{,}409$, where x is the number of years after 2000. (*Source:* Based on data from American Nurses Association)

a. Find the slope and y-intercept of the linear equation.

b. What does the slope mean in this context?

c. What does the y-intercept point mean in this context?

59. The yearly cost of tuition and required fees for attending a public four-year college full time can be estimated by the linear function

$$y = 291.5x + 2944.05$$

where x is the number of years after 2000 and y is the total cost. (*Source:* U.S. National Center for Education Statistics)

a. Find and interpret the slope of this equation.

b. Find and interpret the y-intercept point of this equation.

60. The yearly cost of tuition and required fees for attending a public two-year college full time can be estimated by the linear function

$$y = 107.3x + 1245.62$$

where x is the number of years after 2000 and y is the total cost. (*Source:* U.S. National Center for Education Statistics)

a. Find and interpret the slope of this equation.

b. Find and interpret the y-intercept point of this equation.

61. You want to construct a "funbox" at a local skate park. The skate park's safety regulations allow for the ramp on the funbox to have a maximum slope of $\frac{4}{11}$. If you use the funbox plan in the next column, can you build the ramp to meet the safety regulations? Explain.

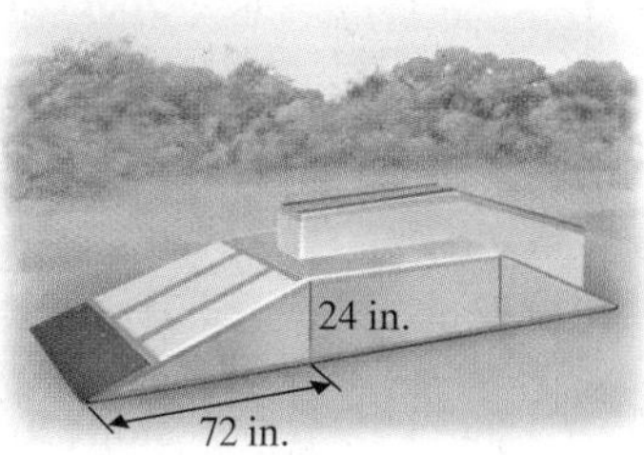

62. By law, the maximum slope of an access ramp in new construction is $\frac{1}{12}$. The plan for the new library shows a 3-foot height from the ground to the main entrance. The distance from the sidewalk to the building is 10 feet. If you assume the ramp does not have any turns, can you design a ramp that complies with the law? Explain.

CONCEPT EXTENSIONS

63. Find the slope of a line parallel to the line $y = -\frac{7}{2}x - 6$.

64. Find the slope of a line parallel to the line $y = x$.

65. Find the slope of a line perpendicular to the line

$$y = -\frac{7}{2}x - 6$$

66. Find the slope of a line perpendicular to the line $y = x$.

67. Find the slope of a line parallel to the line $5x - 2y = 6$.

68. Find the slope of a line parallel to the line $-3x + 4y = 10$.

REVIEW AND PREVIEW

Simplify and solve for y. See Section 2.6.

69. $y - 2 = 5(x + 6)$

70. $y - 0 = -3[x - (-10)]$

71. $y - (-1) = 2(x - 0)$

72. $y - 9 = -8[x - (-4)]$

3.7 COORDINATE GEOMETRY—EQUATIONS OF LINES

OBJECTIVES

1 Use the Slope-Intercept Form.

2 Use the Point-Slope Form.

3 Write Equations of Vertical and Horizontal Lines.

4 Find Equations of Parallel and Perpendicular Lines.

VOCABULARY

- point-slope form
- standard form
- vertical line
- horizontal line

Before this section begins, let's list the different forms of linear equations. These will be used throughout this section.

Forms of Linear Equations

$y = mx + b$	**Slope-intercept form** of a linear equation The slope is m, and the y-intercept point is $(0, b)$.
$y - y_1 = m(x - x_1)$	**Point-slope form** of a linear equation The slope is m, and (x_1, y_1) is a point on the line.
$Ax + By = C$	**Standard form** of a linear equation A and B are not both 0.
$y = c$	**Horizontal line** The slope is 0, and the y-intercept point is $(0, c)$.
$x = c$	**Vertical line** The slope is undefined, and the x-intercept point is $(c, 0)$.

OBJECTIVE 1 ▶ Using Slope-Intercept Form. In the last section, we learned that the slope-intercept form of a linear equation is $y = mx + b$. When a linear equation is written in this form, the slope of the line is the same as the coefficient m of x. Also, the y-intercept point of the line is $(0, b)$ or the y-intercept is simply b. For example, the slope of the line defined by $y = 2x + 3$ is 2, and its y-intercept point is $(0, 3)$ or its y-intercept is 3.

We may also use the slope-intercept form to write the equation of a line given its slope and y-intercept.

EXAMPLE 1 Write an equation of the line with y-intercept point $(0, -3)$ and slope of $\frac{1}{4}$.

Solution A y-intercept point of $(0, -3)$ means the y-intercept is -3. We want to write a linear equation in two variables that describes the line with y-intercept of -3 or y-intercept point $(0, -3)$ and has a slope of $\frac{1}{4}$. We are given the slope and the y-intercept. Let $m = \frac{1}{4}$ and $b = -3$, and write the equation in slope-intercept form, $y = mx + b$.

$$y = mx + b$$

$$y = \frac{1}{4}x + (-3) \quad \text{Let } m = \frac{1}{4} \text{ and } b = -3.$$

$$y = \frac{1}{4}x - 3 \quad \text{Simplify.}$$

□

PRACTICE 1 Write an equation of the line with y-intercept point $(0, 4)$ and slope of $-\frac{3}{4}$.

Given the slope and y-intercept (or y-intercept point) of a line, we may graph the line as well as write its equation. Let's graph the line from Example 1.

EXAMPLE 2 Graph $y = \frac{1}{4}x - 3$.

Solution The slope of the graph of $y = \frac{1}{4}x - 3$ is $\frac{1}{4}$ and the y-intercept point is $(0, -3)$. To graph the line, we first plot the y-intercept point $(0, -3)$. To find another point on the line, we recall that slope is $\frac{\text{rise}}{\text{run}} = \frac{1}{4}$. Another point may then be plotted by starting at $(0, -3)$, rising 1 unit up, and then running 4 units to the right. We are now at the point $(4, -2)$. The graph is the line through these two points.

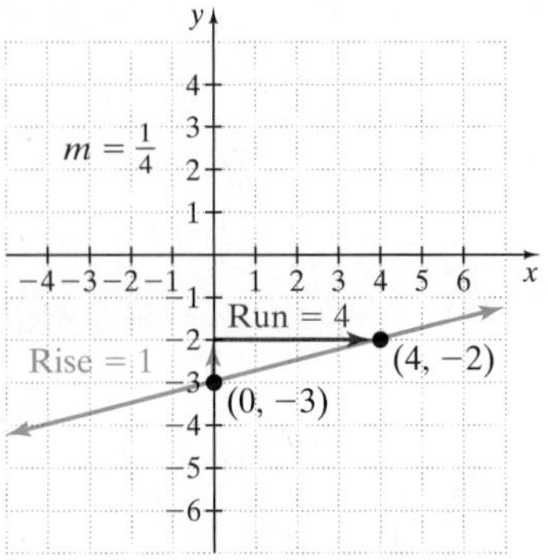

Notice that the line does have a y-intercept of -3 and a slope of $\frac{1}{4}$. □

PRACTICE 2 Graph $y = \frac{3}{4}x + 2$.

EXAMPLE 3 Graph $2x + 3y = 12$.

Solution First, we solve the equation for y to write it in slope-intercept form. In slope-intercept form, the equation is $y = -\frac{2}{3}x + 4$. Next we plot the y-intercept point $(0, 4)$. To find another point on the line, we use the slope $-\frac{2}{3}$, which can be written as $\frac{\text{rise}}{\text{run}} = \frac{-2}{3}$. We start at $(0, 4)$ and move down 2 units since the numerator of the slope is -2; then we move 3 units to the right since the denominator of the slope is 3. We arrive at the point $(3, 2)$. The line through these points is the graph shown below on the left.

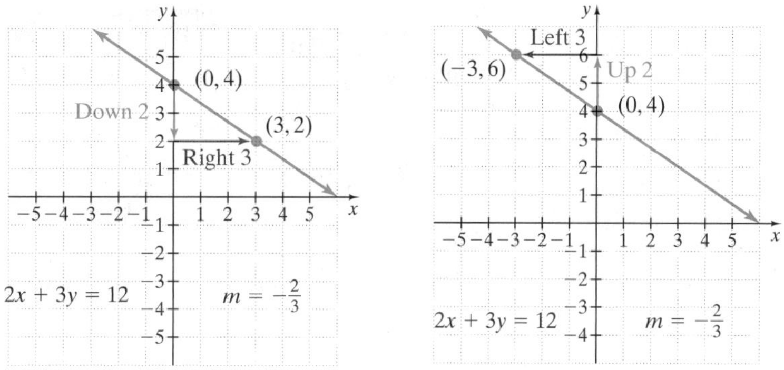

The slope $-\frac{2}{3}$ can also be written as $\frac{2}{-3}$, so to find another point we could start at $(0, 4)$ and move up 2 units and then 3 units to the left. We would arrive at the point $(-3, 6)$. The line through $(-3, 6)$ and $(0, 4)$ is the same line as shown previously through $(3, 2)$ and $(0, 4)$. See the graph above on the right. □

PRACTICE 3 Graph $x + 2y = 6$.

OBJECTIVE 2 ▶ Using Point-Slope Form. Recall the **point-slope form** of the equation of a line from Section 3.6.

Point-Slope Form of the Equation of a Line

The **point-slope form** of the equation of a line is

$$y - y_1 = m(x - x_1)$$

(slope: m; point: y-coordinate y_1, x-coordinate x_1)

where m is the slope of the line and (x_1, y_1) is a point on the line.

EXAMPLE 4 Find an equation of the line with slope -3 containing the point $(1, -5)$. Write the equation in slope-intercept form, $y = mx + b$.

Solution Because we know the slope and a point on the line, we use the point-slope form with $m = -3$ and $(x_1, y_1) = (1, -5)$.

$$y - y_1 = m(x - x_1) \quad \text{Point-slope form}$$
$$y - (-5) = -3(x - 1) \quad \text{Let } m = -3 \text{ and } (x_1, y_1) = (1, -5).$$
$$y + 5 = -3x + 3 \quad \text{Apply the Distributive Property.}$$
$$y = -3x - 2 \quad \text{Write in slope-intercept form.}$$

In slope-intercept form, the equation is $y = -3x - 2$. □

PRACTICE
4 Find an equation of the line with slope -4 containing the point $(-2, 5)$. Write the equation in slope-intercept form, $y = mx + b$.

Helpful Hint
Remember, "slope-intercept form" means the equation is "solved for y."

EXAMPLE 5 Find an equation of the line through points (4, 0) and $(-4, -5)$. Write the equation in slope-intercept form, $y = mx + b$.

Solution First, find the slope of the line.

$$m = \frac{-5 - 0}{-4 - 4} = \frac{-5}{-8} = \frac{5}{8}$$

Next, make use of the point-slope form. Replace (x_1, y_1) by either (4, 0) or $(-4, -5)$ in the point-slope form. We will choose the point (4, 0). The line through (4, 0) with slope $\frac{5}{8}$ is

$$y - y_1 = m(x - x_1) \quad \text{Point-slope form}$$
$$y - 0 = \frac{5}{8}(x - 4) \quad \text{Let } m = \frac{5}{8} \text{ and } (x_1, y_1) = (4, 0).$$
$$8y = 5(x - 4) \quad \text{Multiply both sides by 8.}$$
$$8y = 5x - 20 \quad \text{Apply the Distributive Property.}$$

To write the equation in slope-intercept form, we solve for y.

$$8y = 5x - 20$$
$$y = \frac{5}{8}x - \frac{20}{8} \quad \text{Divide both sides by 8.}$$
$$y = \frac{5}{8}x - \frac{5}{2} \quad \text{Simplify } \frac{20}{8}.$$
□

PRACTICE
5 Find an equation of the line through points $(-1, 2)$ and (2, 0). Write the equation in slope-intercept form, $y = mx + b$.

Helpful Hint
If two points of a line are given, either one may be used with the point-slope form to write an equation of the line.

EXAMPLE 6 Find an equation of the line graphed. Write the equation in **standard form**.

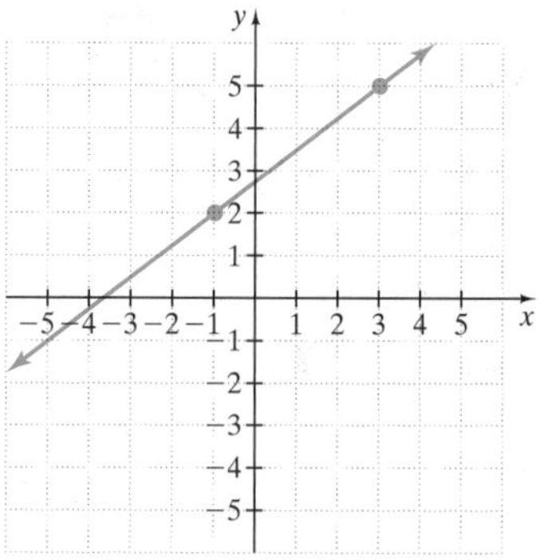

Solution First, find the slope of the line by identifying the coordinates of the noted points on the graph.

The points have coordinates $(-1, 2)$ and $(3, 5)$.

$$m = \frac{5 - 2}{3 - (-1)} = \frac{3}{4}$$

Next, use the point-slope form. We will choose (3, 5) for (x_1, y_1), although it makes no difference which point we choose. The line through (3, 5) with slope $\frac{3}{4}$ is

$$y - y_1 = m(x - x_1) \quad \text{Point-slope form}$$

$$y - 5 = \frac{3}{4}(x - 3) \quad \text{Let } m = \frac{3}{4} \text{ and } (x_1, y_1) = (3, 5).$$

$$4(y - 5) = 3(x - 3) \quad \text{Multiply both sides by 4.}$$

$$4y - 20 = 3x - 9 \quad \text{Apply the Distributive Property.}$$

> **Helpful Hint**
>
> Another standard-form solution to Example 6 is
>
> $$3x - 4y = -11.$$
>
> This equation is equivalent to $-3x + 4y = 11$ and can be obtained by multiplying the equation by -1 on both sides.

To write the equation in standard form, move x- and y-terms to one side of the equation and any numbers (constants) to the other side.

$$4y - 20 = 3x - 9$$

$$-3x + 4y = 11 \quad \text{Subtract } 3x \text{ from both sides and add 20 to both sides.}$$

The equation of the graphed line is $-3x + 4y = 11$. □

PRACTICE 6 Find an equation of the line graphed. Write the equation in standard form.

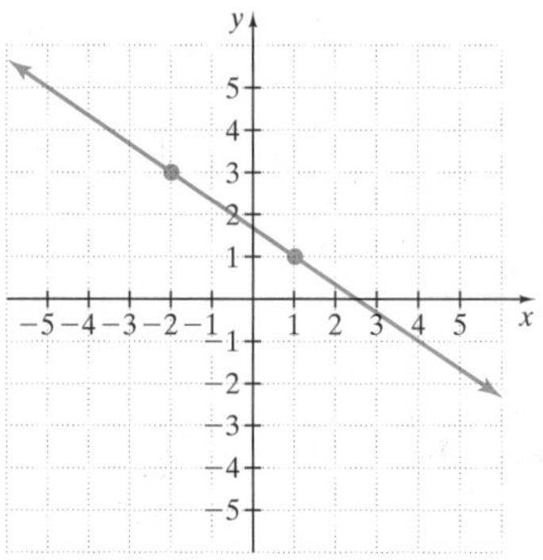

The point-slope form of an equation is very useful for solving real-world problems.

EXAMPLE 7 **Predicting Sales**

Southern Star Realty is an established real estate company that has enjoyed constant growth in sales since 2000. In 2002 the company sold 200 houses, and in 2007 the company sold 275 houses. Use these figures to predict the number of houses this company will sell in the year 2016.

Solution

1. UNDERSTAND. Read and reread the problem. Then let

$x =$ the number of years after 2000 and

$y =$ the number of houses sold in the year corresponding to x

The information provided then gives the ordered pairs (2, 200) and (7, 275). To better visualize the sales of Southern Star Realty, we graph the linear equation that passes through the points (2, 200) and (7, 275).

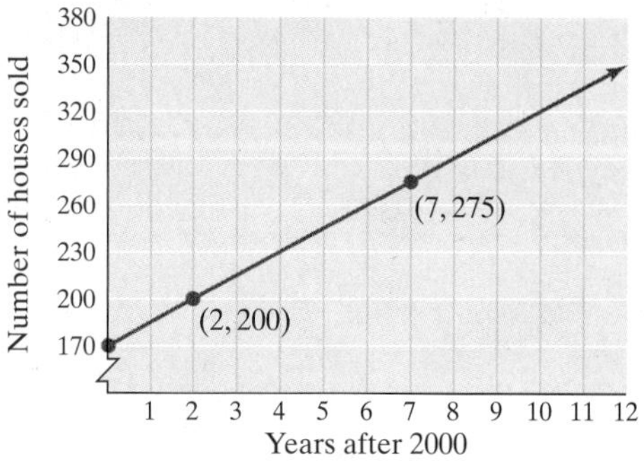

2. TRANSLATE. We write a linear equation that passes through the points (2, 200) and (7, 275). To do so, we first find the slope of the line.

$$m = \frac{275 - 200}{7 - 2} = \frac{75}{5} = 15$$

Then, using the point-slope form and the point (2, 200) to write the equation, we have

$$y - y_1 = m(x - x_1)$$
$$y - 200 = 15(x - 2) \quad \text{Let } m = 15 \text{ and } (x_1, y_1) = (2, 200).$$
$$y - 200 = 15x - 30 \quad \text{Multiply.}$$
$$y = 15x + 170 \quad \text{Add 200 to both sides.}$$

3. SOLVE. To predict the number of houses sold in the year 2016, we use $y = 15x + 170$ and complete the ordered pair (16,), since $2016 - 2000 = 16$.

$$y = 15(16) + 170 \quad \text{Let } x = 16.$$
$$y = 410$$

4. INTERPRET.

Check: Verify that the point (16, 410) is a point on the line graphed in Step 1.

State: Southern Star Realty should expect to sell 410 houses in the year 2016. □

PRACTICE 7 Southwest Florida, including Fort Myers and Cape Coral, has been a growing real estate market in past years. In 2002, there were 7513 house sales in the area, and in 2006, there were 9198 house sales. Use these figures to predict the number of house sales there will be in 2014.

OBJECTIVE 3 ▶ Writing Equations of Vertical and Horizontal Lines. A few special types of linear equations are linear equations whose graphs are **vertical** and **horizontal lines.**

EXAMPLE 8 Find an equation of the horizontal line containing the point (2, 3).

Solution A horizontal line has an equation of the form $y = b$. Since the line contains the point (2, 3), the equation is $y = 3$, as shown at the right.

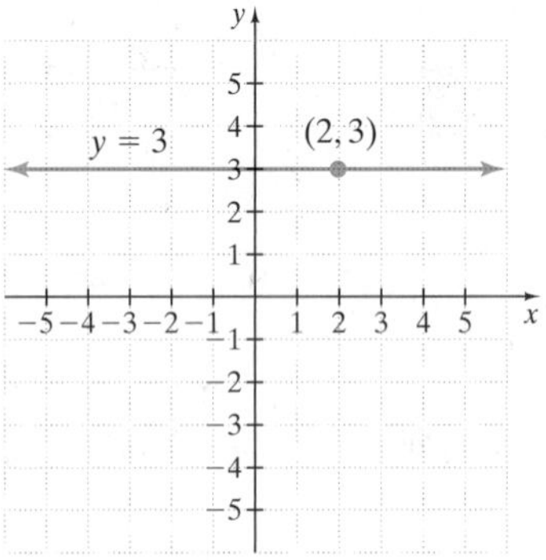

□

PRACTICE 8 Find the equation of the horizontal line containing the point $(6, -2)$.

EXAMPLE 9 Find an equation of the line containing the point (2, 3) with undefined slope.

Solution Since the line has undefined slope, the line must be vertical. A vertical line has an equation of the form $x = c$. Since the line contains the point (2, 3), the equation is $x = 2$, as shown at the right.

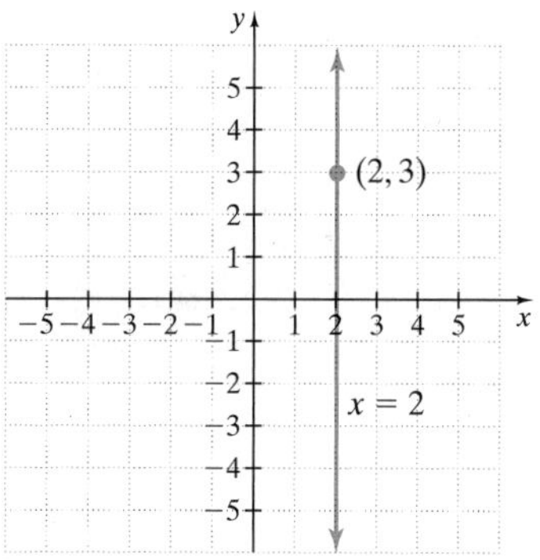

□

PRACTICE 9 Find an equation of the line containing the point $(6, -2)$ with undefined slope.

OBJECTIVE 4 ▶ Finding Equations of Parallel and Perpendicular Lines. Next, we find equations of parallel and perpendicular lines.

EXAMPLE 10 Find an equation of the line containing the point (4, 4) and parallel to the line $2x + 3y = -6$. Write the equation in standard form.

Solution Because the line we want to find is *parallel* to the line $2x + 3y = -6$, the two lines must have equal slopes. Find the slope of $2x + 3y = -6$ by writing it in the form $y = mx + b$. In other words, solve the equation for y.

$$2x + 3y = -6$$

$$3y = -2x - 6 \quad \text{Subtract } 2x \text{ from both sides.}$$

$$y = \frac{-2x}{3} - \frac{6}{3} \quad \text{Divide by 3.}$$

$$y = -\frac{2}{3}x - 2 \quad \text{Write in slope-intercept form.}$$

The slope of this line is $-\frac{2}{3}$. Thus, a line parallel to this line will also have a slope of $-\frac{2}{3}$. The equation we are asked to find describes a line containing the point (4, 4) with a slope of $-\frac{2}{3}$. We use the point-slope form.

$$y - y_1 = m(x - x_1)$$

$$y - 4 = -\frac{2}{3}(x - 4) \quad \text{Let } m = -\frac{2}{3}, x_1 = 4, \text{ and } y_1 = 4.$$

$$3(y - 4) = -2(x - 4) \quad \text{Multiply both sides by 3.}$$

$$3y - 12 = -2x + 8 \quad \text{Apply the Distributive Property.}$$

$$2x + 3y = 20 \quad \text{Write in standard form.}$$

□

▶ Helpful Hint
Multiply both sides of the equation $2x + 3y = 20$ by -1, and it becomes $-2x - 3y = -20$. Both equations are in standard form, and their graphs are the same line.

PRACTICE 10 Find an equation of the line containing the point $(8, -3)$ and parallel to the line $3x + 4y = 1$. Write the equation in standard form.

EXAMPLE 11 Write an equation that describes the line containing the point (4, 4) and perpendicular to the line $2x + 3y = -6$. Write the equation in slope-intercept form.

Solution In the previous example, we found that the slope of the line $2x + 3y = -6$ is $-\frac{2}{3}$. A line perpendicular to this line will have a slope that is the negative reciprocal of $-\frac{2}{3}$, or $\frac{3}{2}$. From the point-slope equation, we have

$$y - y_1 = m(x - x_1)$$

$$y - 4 = \frac{3}{2}(x - 4) \quad \text{Let } x_1 = 4,\ y_1 = 4, \text{ and } m = \frac{3}{2}.$$

$$2(y - 4) = 3(x - 4) \quad \text{Multiply both sides by 2.}$$

$$2y - 8 = 3x - 12 \quad \text{Apply the Distributive Property.}$$

$$2y = 3x - 4 \quad \text{Add 8 to both sides.}$$

$$y = \frac{3}{2}x - 2 \quad \text{Divide both sides by 2.}$$

□

Helpful Hint Parallel and Perpendicular Lines

Nonvertical parallel lines have the same slope.

The product of the slopes of two nonvertical perpendicular lines is −1.

PRACTICE 11 Write an equation that describes the line containing the point (8, −3) and is perpendicular to the line $3x + 4y = 1$. Write the equation in slope-intercept form.

EXAMPLE 12 **Writing Equations of Lines**

The baseball field below is on a coordinate grid with home plate at the origin. A batter hits a ground ball along the line shown. The player at (110, 70) runs along a path perpendicular to the path of the baseball. What is an equation of the line on which the player runs? Keep the equation in point-slope form.

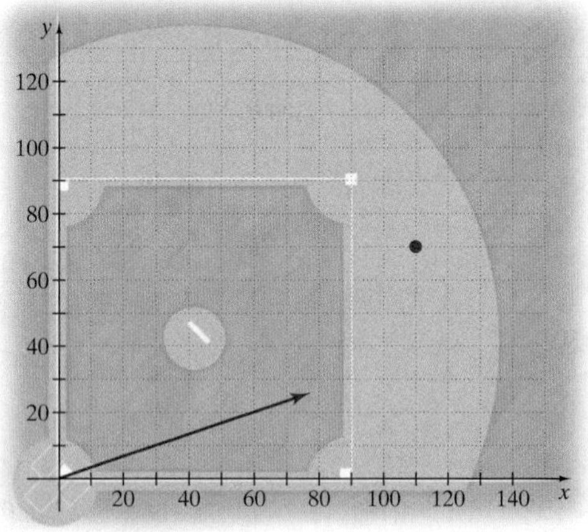

Solution

STEP 1. Find the slope of the baseball's path.

$$m_1 = \frac{y_2 - y_1}{x_2 - x_1} = \frac{20 - 10}{60 - 30} = \frac{10}{30} = \frac{1}{3} \quad \text{Points (30, 10) and (60, 20) are on the baseball's path.}$$

STEP 2. Find the slope of a line perpendicular to the baseball's path.

The perpendicular slope is the negative reciprocal of the slope $\frac{1}{3}$.

$$-\left(\frac{1}{\frac{1}{3}}\right) = -3$$

negative reciprocal

STEP 3. Write an equation of the line on which the player runs. The slope is -3 and a point on the line is (110, 70).

$y - y_1 = m(x - x_1)$ Point-slope form

$y - 70 = -3(x - 110)$ Substitute -3 for m and (110, 70) for (x_1, y_1). □

PRACTICE
12 Suppose a second player standing at (90, 40) misses the ball, turns around, and runs on a path parallel to the baseball's path. What is an equation of the line representing this player's path?

VOCABULARY & READINESS CHECK

State the slope and the y-intercept point of each line with the given equation.

1. $y = -4x + 12$

2. $y = \frac{2}{3}x - \frac{7}{2}$

3. $y = 5x$

4. $y = -x$

5. $y = \frac{1}{2}x + 6$

6. $y = -\frac{2}{3}x + 5$

Decision Making *Decide whether the lines are parallel, perpendicular, or neither.*

7. $y = 12x + 6$
$y = 12x - 2$

8. $y = -5x + 8$
$y = -5x - 8$

9. $y = -9x + 3$
$y = \frac{3}{2}x - 7$

10. $y = 2x - 12$
$y = \frac{1}{2}x - 6$

3.7 EXERCISE SET MyMathLab®

Use the slope-intercept form of the linear equation to write the equation of each line with the given slope and y-intercept point. See Example 1.

1. Slope -1; y-intercept point (0, 1)

2. Slope $\frac{1}{2}$; y-intercept point $(0, -6)$

3. Slope 2; y-intercept point $\left(0, \frac{3}{4}\right)$

4. Slope -3; y-intercept point $\left(0, -\frac{1}{5}\right)$

5. Slope $\frac{2}{7}$; y-intercept point (0, 0)

6. Slope $-\frac{4}{5}$; y-intercept point (0, 0)

Graph each linear equation. See Examples 2 and 3.

7. $y = 5x - 2$

8. $y = 2x + 1$

9. $4x + y = 7$

10. $3x + y = 9$

11. $-3x + 2y = 3$

12. $-2x + 5y = -16$

Find an equation of the line with the given slope and containing the given point. Write the equation in slope-intercept form. See Example 4.

13. Slope 3; through (1, 2)

14. Slope 4; through (5, 1)

15. Slope -2; through $(1, -3)$

16. Slope -4; through $(2, -4)$

17. Slope $\frac{1}{2}$; through $(-6, 2)$

18. Slope $\frac{2}{3}$; through $(-9, 4)$

19. Slope $-\frac{9}{10}$; through $(-3, 0)$

20. Slope $-\frac{1}{5}$; through $(4, -6)$

Find an equation of the line passing through the given points. Write the equation in slope-intercept form. See Example 5.

21. (2, 0), (4, 6)

22. (3, 0), (7, 8)

23. $(-2, 5), (-6, 13)$

24. $(7, -4), (2, 6)$

25. $(-2, -4), (-4, -3)$

26. $(-9, -2), (-3, 10)$

27. $(-3, -8), (-6, -9)$

28. $(8, -3), (4, -8)$

29. $\left(\frac{3}{5}, \frac{4}{10}\right)$ and $\left(-\frac{1}{5}, \frac{7}{10}\right)$

30. $\left(\frac{1}{2}, -\frac{1}{4}\right)$ and $\left(\frac{3}{2}, \frac{3}{4}\right)$

Find an equation of each line graphed. Write the equation in standard form. See Example 6.

31.

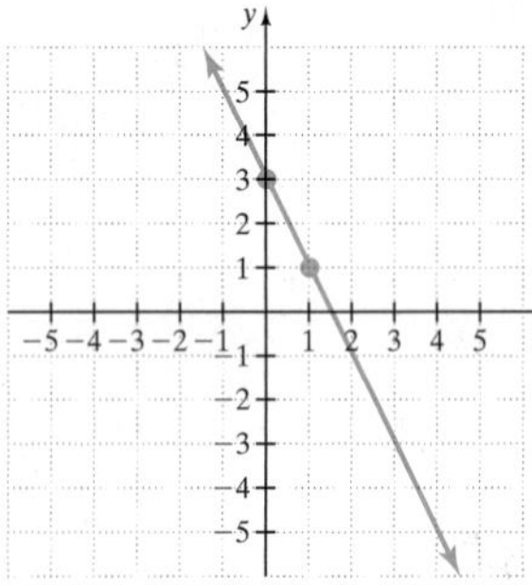

32.

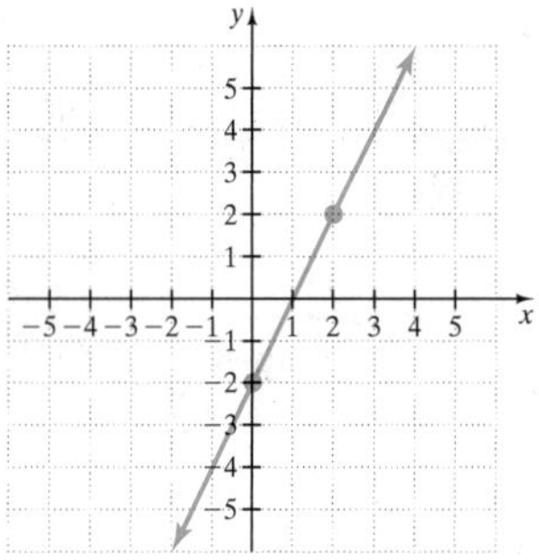

33.

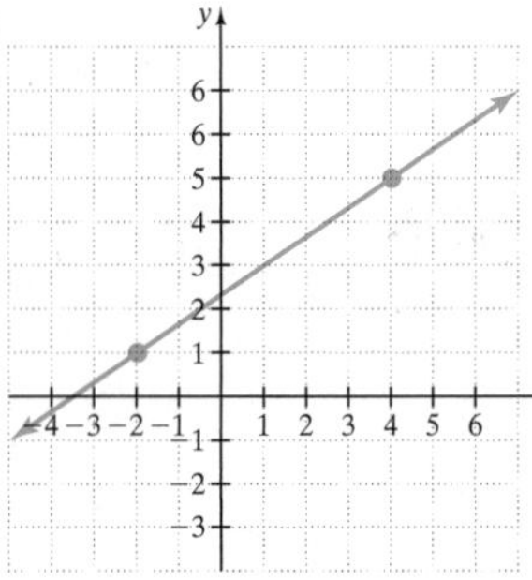

34.

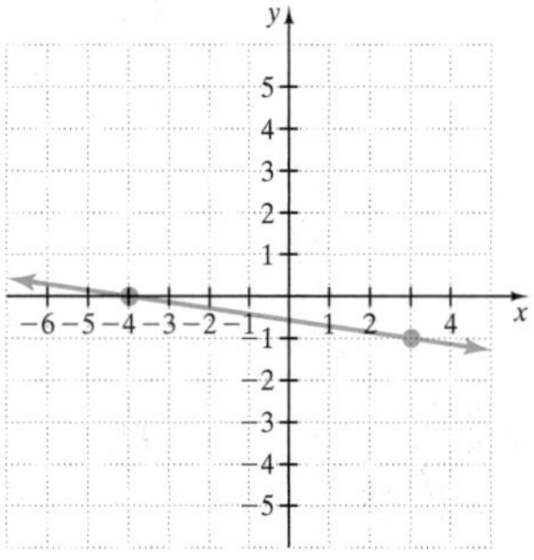

Write an equation of each line. See Examples 8 and 9.

35. Slope 0; through $(-2, -4)$

36. Horizontal; through $(-3, 1)$

37. Vertical; through $(4, 7)$

38. Vertical; through $(2, 6)$

39. Horizontal; through $(0, 5)$

40. Undefined slope; through $(0, 5)$

Find an equation of each line. Write the equation using slope-intercept form. See Examples 10 and 11.

41. Through $(3, 8)$; parallel to the line $y = 4x - 2$

42. Through $(1, 5)$; parallel to the line $y = 3x - 4$

43. Through $(2, -5)$; perpendicular to the line $y = \frac{1}{3}x - 2$

44. Through $(-4, 8)$; perpendicular to the line $y = \frac{2}{3}x - \frac{1}{3}$

45. Through $(-2, -3)$; parallel to the line $y = -\frac{3}{2}x + \frac{5}{2}$

46. Through $(-2, -3)$; perpendicular to the line $y = -\frac{3}{2}x + \frac{5}{2}$

MIXED PRACTICE

Find the equation of each line. Write the equation in standard form. See Examples 1, 4, 5, and 8 through 11.

47. Slope 2; through $(-2, 3)$

48. Slope 3; through $(-4, 2)$

49. Through $(1, 6)$ and $(5, 2)$

50. Through $(2, 9)$ and $(8, 6)$

51. With slope $-\frac{1}{2}$; y-intercept 11

52. With slope -4; y-intercept $\frac{2}{9}$

53. Vertical line; through $(-2, -10)$

54. Horizontal line; through $(1, 0)$

55. Through $(6, -2)$; parallel to the line $2x + 4y = 9$

56. Through $(8, -3)$; parallel to the line $6x + 2y = 5$

Multiple Steps *Solve. See Example 7.*

57. Del Monte Fruit Company recently released a new applesauce. By the end of its first year, profits on this product amounted to \$30,000. The anticipated profit for the end of the fourth year is \$66,000. The ratio of change in time to change in profit is constant. Let x be years and y be profit.

a. Write a linear equation y that expresses profit in terms of x.

b. Use this equation to predict the company's profit at the end of the seventh year.

c. Predict when the profit should reach \$126,000.

58. The value of a computer bought in 2009 depreciates, or decreases, as time passes. Two years after the computer was bought, it was worth \$2000; 4 years after it was bought, it was worth \$800.

a. If this relationship between number of years past 2009 and value of the computer is linear, write an equation describing this relationship. [Use ordered pairs of the form (years past 2009, value of computer).]

b. Use this equation to estimate the value of the computer in the year 2014.

59. The Pool Fun Company has learned that, by pricing a newly released Fun Noodle at \$3, sales will reach 10,000 Fun Noodles per day during the summer. Raising the price to \$5 will cause the sales to fall to 8000 Fun Noodles per day.

a. Assume that the relationship between sales price and number of Fun Noodles sold is linear and write an equation describing this relationship. Use ordered pairs of the form (sales price, number sold).

b. Predict the daily sales of Fun Noodles if the price is \$3.50.

60. The value of a building bought in 2000 appreciates, or increases, as time passes. Seven years after the building was bought, it was worth \$165,000; 12 years after it was bought, it was worth \$180,000.

a. If this relationship between number of years past 2000 and value of the building is linear, write an equation describing this relationship. [Use ordered pairs of the form (years past 2000, value of building).]

b. Use this equation to estimate the value of the building in the year 2020.

61. The number of people employed in the United States as medical assistants was 387 thousand in 2004. By the year 2014, this number is expected to rise to 589 thousand. Let y be the number of medical assistants (in thousands) employed in the United States in the year x, where $x = 0$ represents 2004. (*Source:* Bureau of Labor Statistics)

a. Write a linear equation that models the number of people (in thousands) employed as medical assistants in the year x.

b. Use this equation to estimate the number of people who will be employed as medical assistants in the year 2019.

62. The number of people employed in the United States as systems analysts was 487 thousand in 2004. By the year 2014, this number is expected to rise to 640 thousand. Let y be the number of systems analysts (in thousands) employed in the United States in the year x, where $x = 0$ represents 2004. (*Source:* Bureau of Labor Statistics)

a. Write a linear equation that models the number of people (in thousands) employed as systems analysts in the year x.

b. Use this equation to estimate the number of people who will be employed as systems analysts in the year 2018.

CONCEPT EXTENSIONS

True or False *Answer true or false.*

63. A vertical line is always perpendicular to a horizontal line.

64. A vertical line is always parallel to a vertical line.

Example:

Find an equation of the perpendicular bisector of the line segment whose endpoints are (2, 6) and (0, −2).

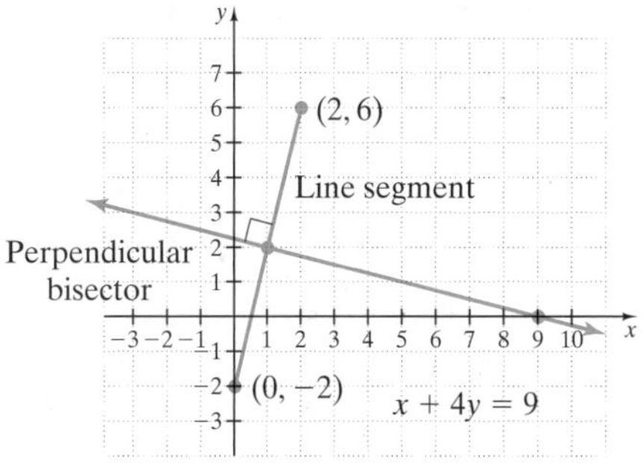

Solution:

A perpendicular bisector is a line that contains the midpoint of the given segment and is perpendicular to the segment.

STEP 1. The midpoint of the segment with endpoints (2, 6) and $(0, -2)$ is $(1, 2)$.

STEP 2. The slope of the segment containing points (2, 6) and $(0, -2)$ is 4.

STEP 3. A line perpendicular to this line segment will have slope of $-\frac{1}{4}$.

STEP 4. The equation of the line through the midpoint $(1, 2)$ with a slope of $-\frac{1}{4}$ will be the equation of the perpendicular bisector. This equation in standard form is $x + 4y = 9$.

Find an equation of the perpendicular bisector of the line segment whose endpoints are given. See the previous example.

65. $(3, -1); (-5, 1)$
66. $(-6, -3); (-8, -1)$
67. $(-2, 6); (-22, -4)$
68. $(5, 8); (7, 2)$

REVIEW AND PREVIEW

Find the length of each line segment whose endpoints were given in the previous exercises. See Section 1.7.

69. $(3, -1); (-5, 1)$
70. $(-6, -3); (-8, -1)$
71. $(-2, 6); (-22, -4)$
72. $(5, 8); (7, 2)$

Historical Note—The Parallel Postulate

For more than 2000 years, Euclidean Geometry, which was based on the Parallel Postulate (also called the Fifth Postulate), was the only recognized geometry.

Euclid's Fifth Postulate is actually stated something like the following. In two-dimensional geometry:

"If a line segment intersects two straight lines forming two interior angles on the same side that sum to less than two right angles, then the two lines, if extended indefinitely, meet on that side on which the angles sum to less than two right angles."

The Parallel Postulate or Euclid's Fifth Postulate that we state in this text sounds more like what is known as Playfair's Axiom:

"Given a line and a point not on it, at most one parallel to the given line can be drawn through the point."

This axiom is named after the Scottish mathematician John Playfair. Don't let the phrase in this axiom "at most" bother you because the rest of the postulates imply that there is exactly one parallel.

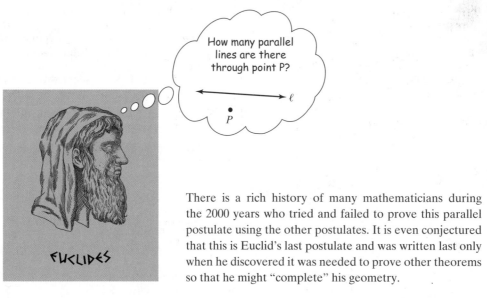

There is a rich history of many mathematicians during the 2000 years who tried and failed to prove this parallel postulate using the other postulates. It is even conjectured that this is Euclid's last postulate and was written last only when he discovered it was needed to prove other theorems so that he might "complete" his geometry.

Finally, during the early 1800s, two mathematicians, Bolyai (Hungarian) and Lobachevsky (Russian), at about the same time decided to prove that there was only one parallel through a point to a line by assuming that there was more than one. They hoped that this reasoning would lead to a contradiction. (See Indirect Proofs, Section 5.5) Amazingly, they did not arrive at a contradiction, but in fact, proved that if there are two parallels to a line through a single point, then there must be an infinite number of parallels. Both of these mathematicians are given credit for discovering a non-Euclidean geometry called Hyperbolic Geometry. The German mathematician Bernhard Riemann is also given much credit for discovering a non-Euclidean geometry called Elliptic Geometry (or Spherical Geometry.)

Final Note: In 2001, NASA launched the WMAP spacecraft. This spacecraft measured temperature differences in cosmic radiation, which in turn would be used to measure the geometry of the universe. As of 2010, WMAP is abandoned and in orbit around the sun. The data collected is still being researched and we still do not know the geometry of our universe.

CHAPTER 3 VOCABULARY CHECK

Word Bank *Use the choices below to fill in each blank. Some choices may be used more than once and some not at all.*

horizontal	the same	-1	y-intercepts	$(0, b)$	slope
vertical	different	m	x-intercepts	$(b, 0)$	slope-intercept
parallel lines	skew lines	b	alternate interior		
transversal	parallel planes		same-side interior		
interior	alternate exterior		corresponding		
exterior					

1. Planes that do not intersect are called ____________.

2. Coplanar lines that do not intersect are called ____________.

3. A line that intersects two or more coplanar lines at different points is called a(n) ____________.

4. ____________ are not coplanar; they are not parallel and do not intersect.

5. The measure of the steepness or tilt of a line is called ____________.

6. The slope of a line through two points is measured by the ratio of ____________ change to ____________ change.

7. If a linear equation is in the form $y = mx + b$, the slope of the line is ____________, the y-intercept point is ____________, and the y-intercept is ____________.

8. The form $y = mx + b$ is called the ____________ form.

9. The slope of a ____________ line is 0.

10. The slope of a ____________ line is undefined.

11. Two non-vertical perpendicular lines have slopes whose product is ____________.

12. Two non-vertical lines are parallel if they have ____________ slope and different ____________.

Use the given figures and the most descriptive choice in the Word Bank above to fill in each blank.

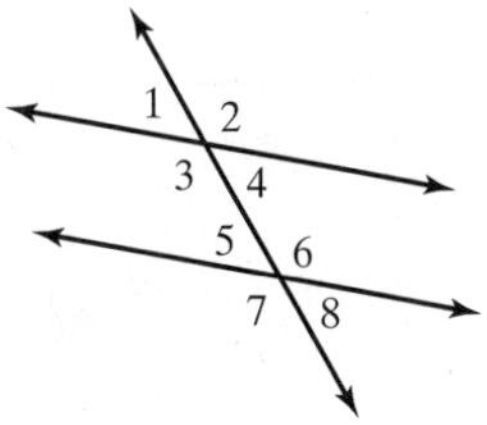

13. Angles 3, 4, 5, and 6 are called ____________ angles.

14. Angles 1, 2, 7, and 8 are called ____________ angles.

15. Angles 2 and 6 are called ____________ angles.

16. Angles 4 and 5 are called ____________ angles.

17. Angles 1 and 8 are called ____________ angles.

18. Angles 3 and 5 are called ____________ angles.

19. Angles 2 and 3 are called ____________ angles.

20. Angles 1 and 4 are called ____________ angles.

Fill in the Blank *These answers are not contained in the word bank.*

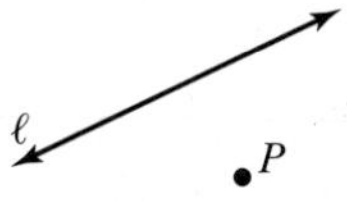

21. Given line ℓ and point P, how many lines can be drawn through P parallel to ℓ? ____________

22. Given line ℓ and point P, how many lines can be drawn through P perpendicular to ℓ? ____________

23. The postulate that gives us the answer to Exercise **21** above is called the ____________ Postulate.

24. The postulate that gives us the answer to Exercise **22** above is called the ____________ Postulate.

CHAPTER 3 REVIEW

(3.1) Identify all numbered angle pairs that form the given type of angle pair. Then name the two lines and transversal that form each pair.

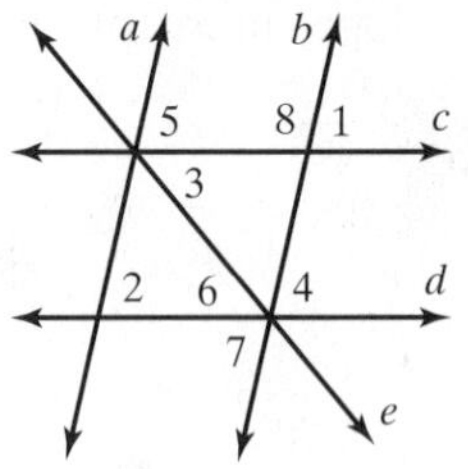

1. alternate interior angles
2. same-side interior angles
3. corresponding angles
4. alternate exterior angles

Classify the angle pair formed by ∠1 and ∠2.

5.

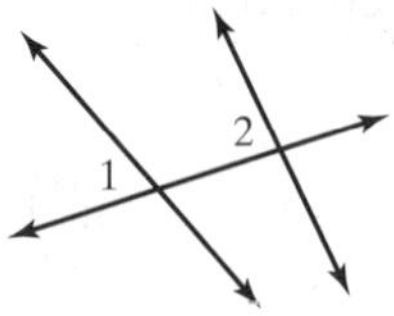

6.

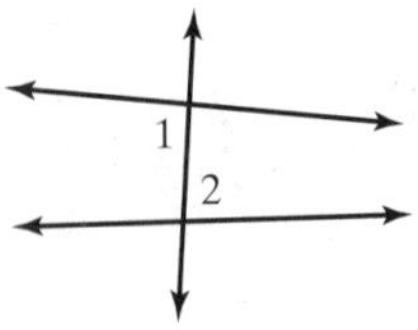

(3.2) Find the value of x that makes ℓ ∥ m.

7.

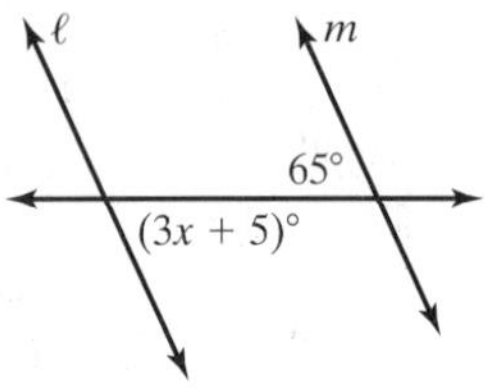

8.

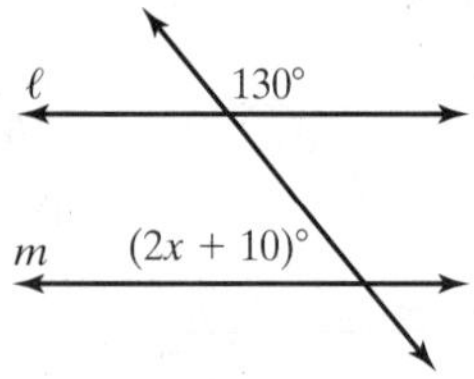

Use the given information to decide which lines, if any, are parallel. Justify your conclusion.

9. $\angle 1 \cong \angle 9$
10. $m\angle 3 + m\angle 6 = 180°$
11. $m\angle 2 + m\angle 3 = 180°$
12. $\angle 5 \cong \angle 11$

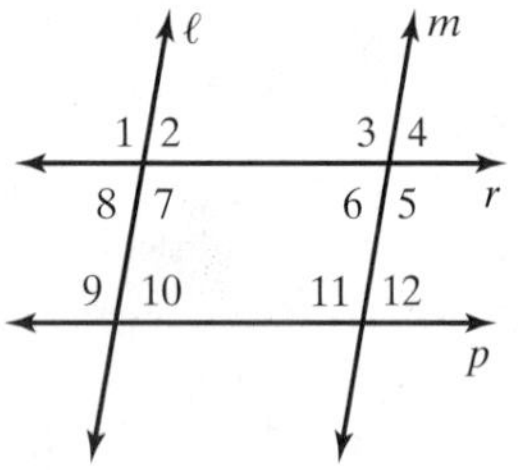

(3.3) Find m∠1 and m∠2. Justify your answers.

13.

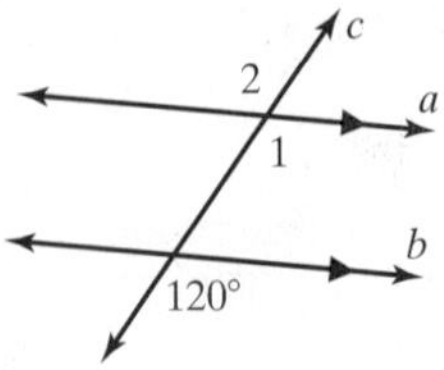

14.

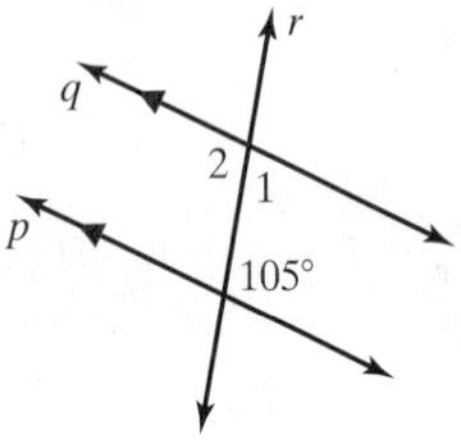

15. Find the value of x in the figure below.
16. Find the value of y in the figure below.

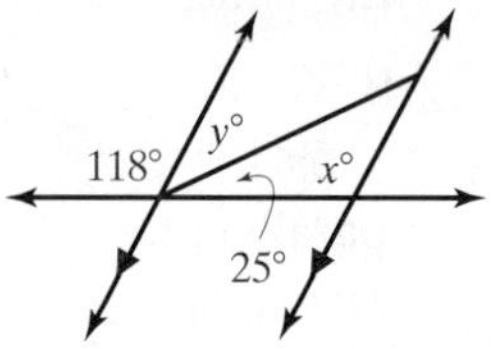

(3.4) Use the diagram at the right to complete each statement.

17. If $b \perp c$ and $b \perp d$, then c __?__ d.
18. If $c \parallel d$, then __?__ $\perp c$.

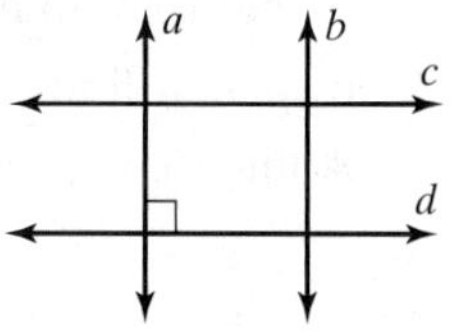

(3.5)

19. Draw a line m and point Q not on m. Construct a line perpendicular to m through Q.

Use the segments below.

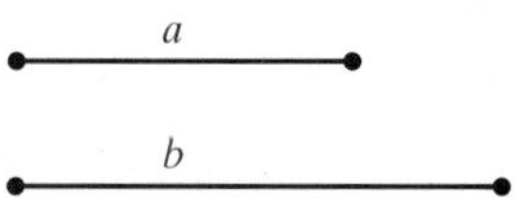

20. Construct a rectangle with side lengths a and b.
21. Construct a rectangle with side lengths a and $2b$.
22. Construct a quadrilateral with one pair of parallel opposite sides, each side of length $2a$.

(3.6) Find the slope of the line passing through the points.

23. $(6, -2), (1, 3)$
24. $(-7, 2), (-7, -5)$
25. Name the slope and y-intercept point of $y = 2x - 1$. Then graph the line.
26. Name the slope of and a point on $y - 3 = -2(x + 5)$. Then graph the line.

Write an equation of the line.

27. Slope $-\frac{1}{2}$; y-intercept 12
28. Passes through $(4, 2)$ and $(3, -2)$

(3.7) Determine whether $\overleftrightarrow{AB}$ and $\overleftrightarrow{CD}$ are parallel, perpendicular, or neither.

29. $A(-1, -4), B(2, 11), C(1, 1), D(4, 10)$
30. $A(2, 8), B(-1, -2), C(3, 7), D(0, -3)$
31. $A(-3, 3), B(0, 2), C(1, 3), D(-2, -6)$
32. $A(-1, 3), B(4, 8), C(-6, 0), D(2, 8)$
33. Write an equation of the line parallel to $y = 8x - 1$ that contains $(-6, 2)$.
34. Write an equation of the line perpendicular to $y = \frac{1}{6}x + 4$ that contains $(3, -3)$.

MIXED PRACTICE

35. Name two other pairs of corresponding angles in the figure.

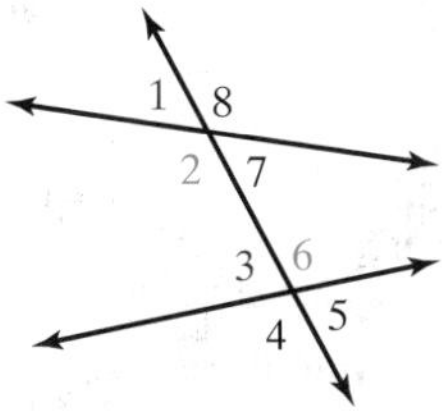

36. Which other angles measure 110°?

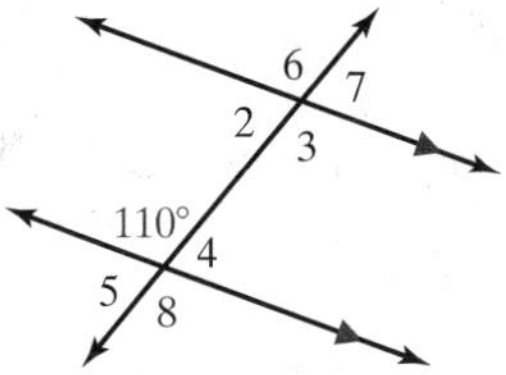

37. What is the value of x that makes $\ell \parallel m$?

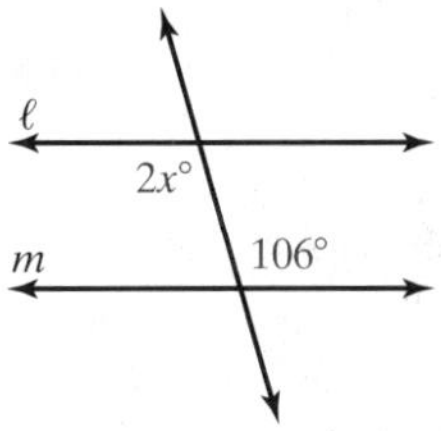

38. What are the pairs of parallel and perpendicular lines in the diagram?

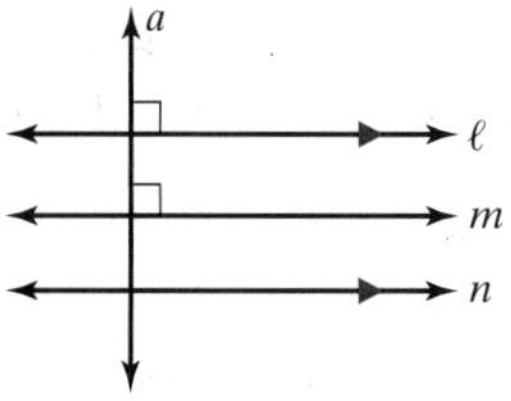

39. Morris Avenue intersects both 1st Street and 3rd Street at right angles. 3rd Street is parallel to 5th Street. How are 1st Street and 5th Street related? Explain.

40. What is an equation of the line with slope -5 and y-intercept point $(0, 6)$?

41. What is an equation of the line through $(-2, 8)$ with slope 3?

Write an equation of the line.

42. Slope 3; passes through $(1, -9)$

43. What is an equation of the line perpendicular to $y = 2x - 5$ that contains $(1, -3)$?

CHAPTER 3 TEST

TEST PREP VIDEO The fully worked-out solutions to any exercises you want to review are available in MyMathLab.

Use the figure to determine whether the pairs of lines are parallel lines, skew lines, or neither. (Assume lines and planes that appear to be parallel are parallel.)

1. $\overleftrightarrow{AB}$ and $\overleftrightarrow{DH}$

2. $\overleftrightarrow{BC}$ and $\overleftrightarrow{AD}$

3. $\overleftrightarrow{DA}$ and $\overleftrightarrow{HD}$

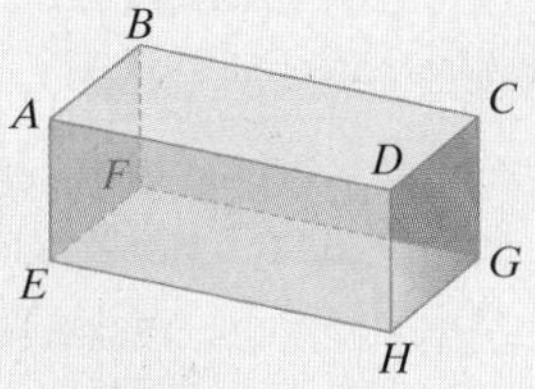

Find m∠1 and m∠2. Justify each answer.

4.

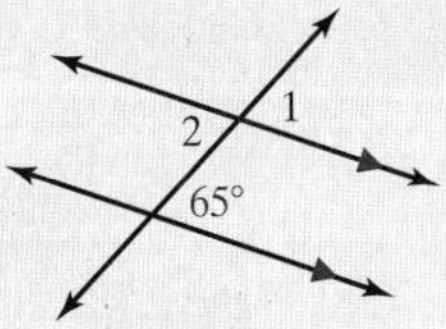

5.

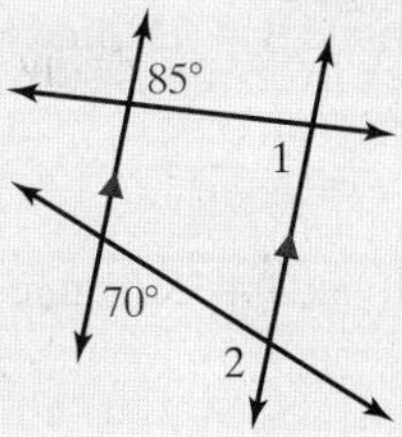

6.

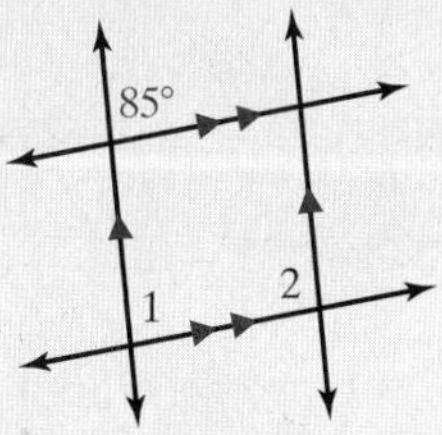

7.

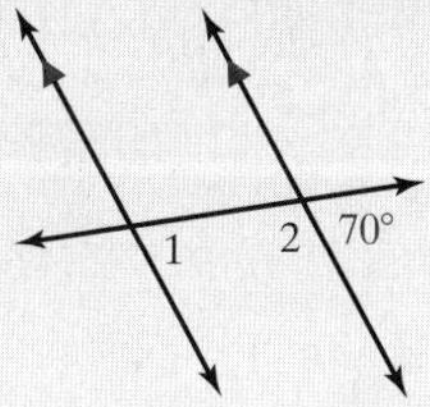

Find the value of x that makes $\ell \parallel m$.

8.

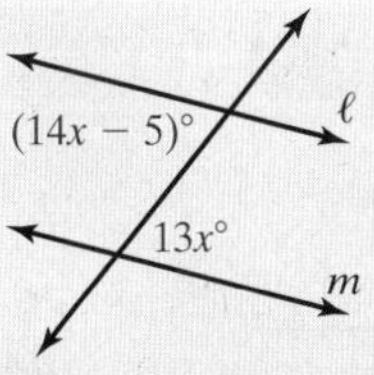

9.

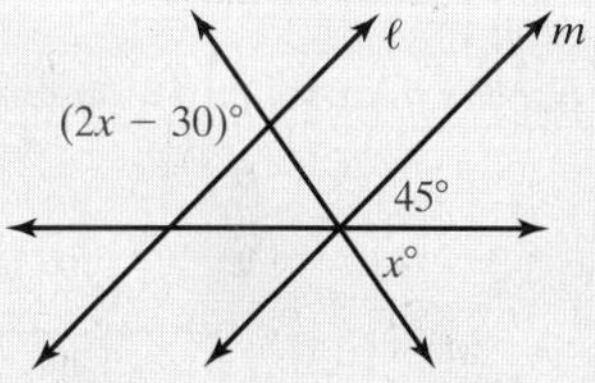

10. Find the value of x.

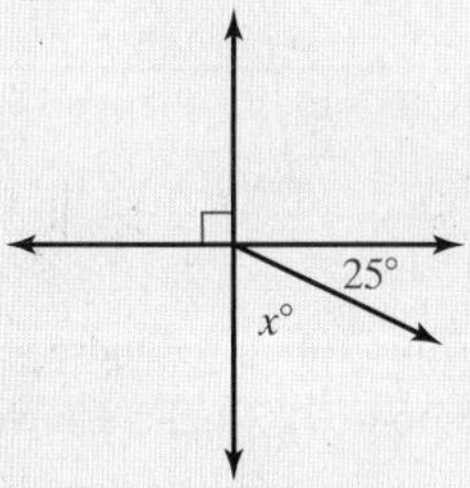

11. Proof Provide the reason for each step.

Given: $\ell \parallel m$, $\angle 2 \cong \angle 4$

Prove: $n \parallel p$

Statements	**Reasons**
1. $\ell \parallel m$	**1. a.** ?
2. $\angle 1 \cong \angle 2$	**2. b.** ?
3. $\angle 2 \cong \angle 4$	**3. c.** ?
4. $\angle 1 \cong \angle 4$	**4. d.** ?
5. $n \parallel p$	**5. e.** ?

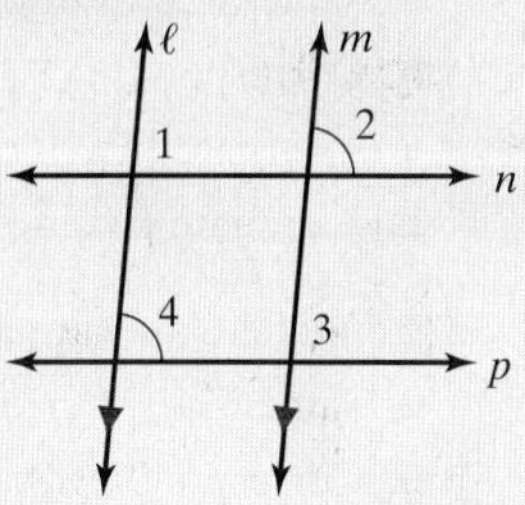

12. Draw a line m and a point T not on the line. Construct the line through T perpendicular to m.

13. Draw any $\angle ABC$. Then construct line m through A so that $m \parallel \overleftrightarrow{BC}$.

Graph each line.

14. $2x - 3y = -6$

15. $y = -3$

16. Find the slope of the line that passes through $(5, -8)$ and $(-7, 10)$.

17. Find the slope and the y-intercept point of the line $3x + 12y = 8$.

Find an equation of each line satisfying the given conditions. Write Exercises 18–21 in standard form. Write Exercise 22 using slope-intercept form.

18. Horizontal; through $(2, -8)$

19. Through $(4, -1)$; slope -3

20. Through $(4, -2)$ and $(6, -3)$

21. Through $(-1, 2)$; perpendicular to $y = 3x - 4$

22. Parallel to $y = -\frac{1}{2}x + \frac{3}{2}$; through $(3, -2)$

23. Line L_1 has the equation $2x - 5y = 8$. Line L_2 passes through the points $(1, 4)$ and $(-1, -1)$. Determine whether these lines are parallel lines, perpendicular lines, or neither.

CHAPTER 3 STANDARDIZED TEST

1. Coplanar lines that do not intersect are ____________.
   a. parallel planes
   b. parallel lines
   c. skew lines
   d. none of these

2. Noncoplanar lines that do not intersect are ____________.
   a. parallel planes
   b. parallel lines
   c. skew lines
   d. none of these

Use the figure to answer questions 3–10.

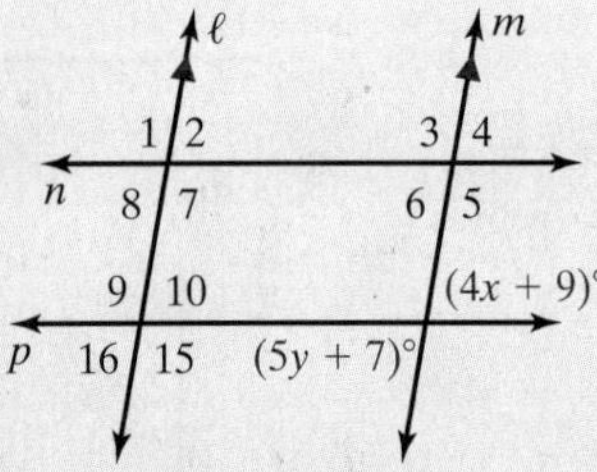

3. $\angle 1 \cong \angle 5$ by ____________.
   a. Same-Side Interior Angles Converse
   b. Corresponding Angles Converse
   c. Alternate Interior Angles Converse
   d. Alternate Exterior Angles Converse

4. $\angle 7 \cong \angle 5$ by ____________.
   a. Same-Side Interior Angles Converse
   b. Corresponding Angles Converse
   c. Alternate Interior Angles Converse
   d. Alternate Exterior Angles Converse

5. $\angle 7$ and $\angle 6$ are supplementary by ____________.
   a. Same-Side Interior Angles Converse
   b. Corresponding Angles Converse
   c. Alternate Interior Angles Converse
   d. Alternate Exterior Angles Converse

6. If $\angle 10 \cong \angle 8$, then $n \parallel p$ by ____________.
   a. Same-Side Interior Angles Theorem
   b. Corresponding Angles Theorem
   c. Alternate Interior Angles Theorem
   d. Alternate Exterior Angles Theorem

7. If $\angle 1 \cong \angle 9$, then $n \parallel p$ by ____________.
   a. Same-Side Interior Angles Theorem
   b. Corresponding Angles Theorem
   c. Alternate Interior Angles Theorem
   d. Alternate Exterior Angles Theorem

8. If $\angle 16 \cong \angle 2$, then $n \parallel p$ by ____________.
   a. Same-Side Interior Angles Theorem
   b. Corresponding Angles Theorem
   c. Alternate Interior Angles Theorem
   d. Alternate Exterior Angles Theorem

9. If $m\angle 3 = 75°$, find the value of x for which $n \parallel p$.
   a. 16.5 b. 19.6
   c. 24 d. 13.6

10. If $m\angle 4 = 110°$, find the value of y for which $n \parallel p$.
    a. 20.6 b. 25.25
    c. 12.6 d. 15.25

11. The slope of the line that passes through $(2, -7)$ and $(8, -2)$ is ____________.
    a. $-\frac{5}{6}$ b. $\frac{6}{5}$
    c. $\frac{5}{6}$ d. $-\frac{6}{5}$

12. Find the slope and y-intercept point of the line $4x - 2y = 7$.

a. slope: -2; y-intercept: $(0, -\frac{7}{2})$

b. slope: 2; y-intercept: $(0, -\frac{7}{2})$

c. slope: $-\frac{1}{2}$; y-intercept: $(0, \frac{7}{2})$

d. slope: $-\frac{1}{2}$; y-intercept: $(0, -\frac{7}{2})$

Find an equation of the line satisfying the given conditions.

13. Horizontal; through $(6, 5)$

a. $y = 5$ **b.** $y = 6$

c. $x + y = 11$ **d.** $x = 6$

14. Through $(4, -7)$; slope $\frac{3}{2}$

a. $3x - 2y = 26$ **b.** $3y + 2y = -2$

c. $3x - 2y = -2$ **d.** $3x + 2y = 26$

15. Through $(6, 3)$; perpendicular to $4x - 7y = 8$

a. $y = \frac{7}{4}x - \frac{15}{2}$ **b.** $y = \frac{4}{7}x - \frac{3}{7}$

c. $y = -\frac{4}{7}x + \frac{45}{7}$ **d.** $y = -\frac{7}{4}x + \frac{27}{2}$

16. Line l has equation $3x + 2y = 7$. Line m passes through the points $(6, -2)$ and $(0, 7)$. Lines l and m are ____________.

a. parallel **b.** perpendicular

c. neither

CHAPTER

4 Triangles and Congruence

4.1 Types of Triangles

4.2 Congruent Figures

4.3 Congruent Triangles by SSS and SAS

4.4 Congruent Triangles by ASA and AAS

4.5 Proofs Using Congruent Triangles

4.6 Isosceles, Equilateral, and Right Triangles

Real-Life Examples of Triangles

Many bridge and roof trusses (frameworks) are triangular.

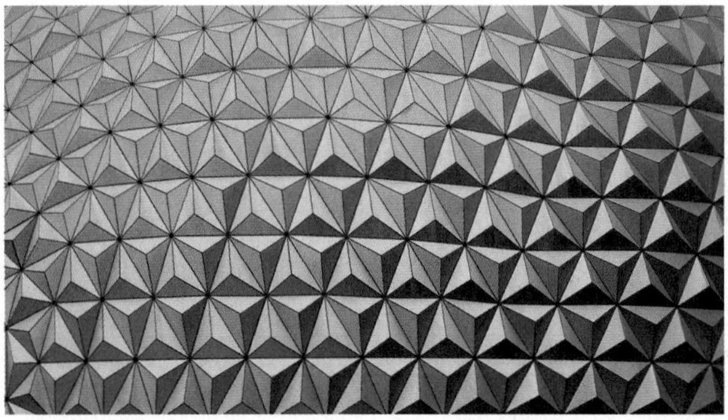

Close-up View of Epcot Ball—it is made of 11,324 isosceles triangles.

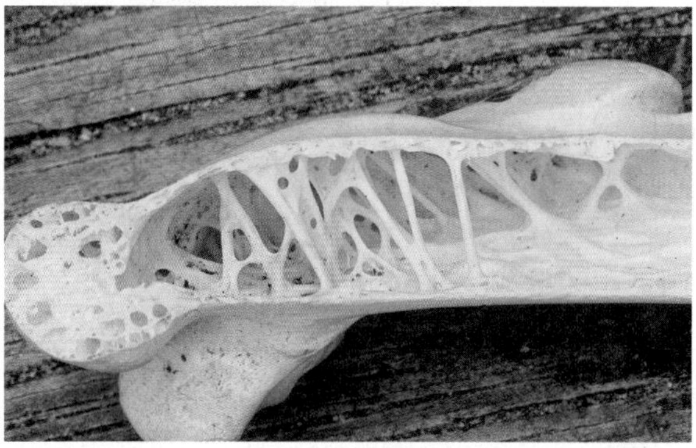

Drawing of a Slice of a Hollow Bird Bone—hollow bones of dinosours and modern birds are braced with a pattern of triangles.

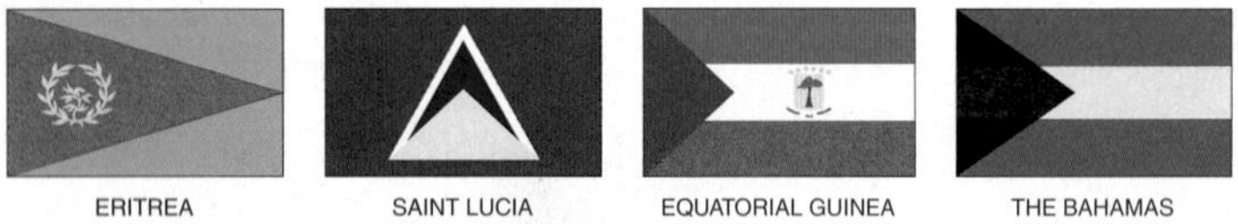

Examples of flags containing triangles.

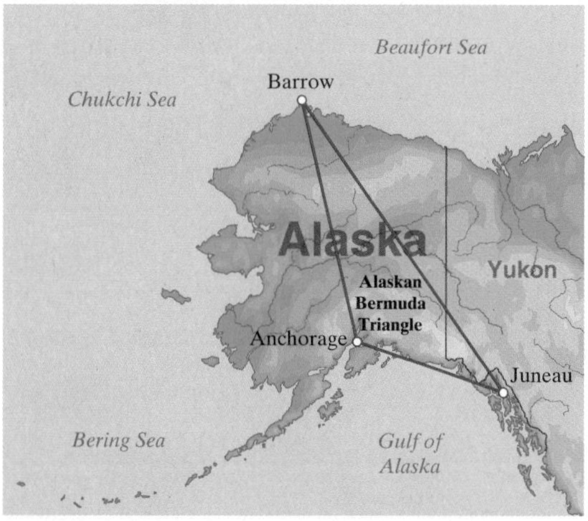

The Bermuda Triangle of Alaska—given that name because of the high rate of disappearances.

Many examples above show the beauty of triangles, but many also show their structural strength, which we review in Section 4.3.

4.1 TYPES OF TRIANGLES

OBJECTIVES

1 Learn the Vocabulary of Triangles.

2 Classify Triangles by Angles and Sides.

3 Find Angle Measures of Triangles.

VOCABULARY

- triangle
- vertex
- sides of a triangle
- adjacent sides
- opposite side and angle
- included side and angle
- acute triangle
- obtuse triangle
- equiangular triangle
- right triangle
- scalene triangle
- isosceles triangle
- equilateral triangle
- interior angle
- exterior angle
- corollary

OBJECTIVE 1 ▶ Learning the Vocabulary of Triangles. Let's use this chapter to expand our mathematical system called geometry. Here, we study a specific geometric figure called a triangle.

To begin, let's formally define a triangle and its parts. Let A, B, and C below be noncollinear points.

Defined Terms

Definitions	*Diagram*	*How to Name It*
A **triangle** is formed by 3 noncollinear points connected by segments.	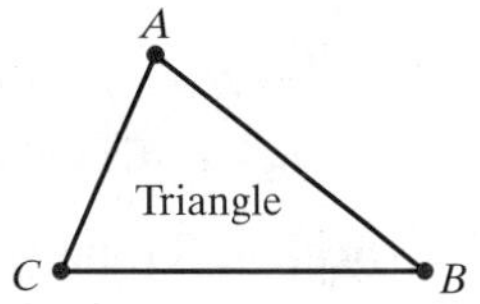	$\triangle ABC$ — Notation means "triangle."
The noncollinear points are called **vertices** (singular is **vertex**).		Points A, B, and C are vertices. Point A is a vertex. Point B is a vertex. Point C is a vertex.
The segments joining the points are called **sides.**		$\overline{AB}$, $\overline{BC}$, and $\overline{CA}$ are the sides.

There are many types of triangles. Before we classify them, let's define some commonly used words and phrases associated with triangles.

Adjacent sides are two sides that share a common vertex.

For example: $\overline{ED}$ and $\overline{DF}$ are adjacent sides.

Also, the side not adjacent to an angle is called the **side opposite the angle.**

Example: Side $\overline{EF}$ is opposite $\angle D$.

Sometimes, we also use the word **included.**

Examples: $\overline{DF}$ is the included side of $\angle D$ and $\angle F$.

$\angle E$ is the included angle of $\overline{ED}$ and $\overline{EF}$.

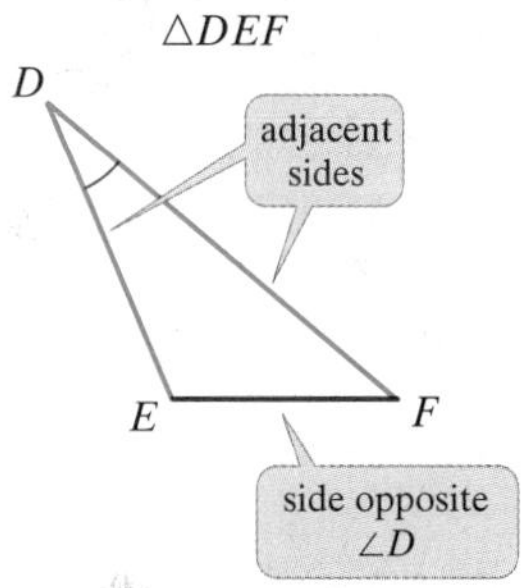

EXAMPLE 1 **Identifying Parts of a Triangle**

Given $\triangle PQR$:

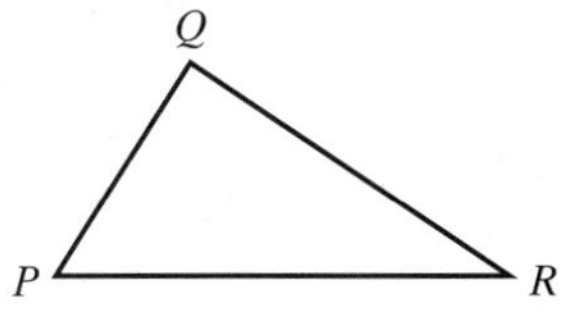

a. Which angle is opposite $\overline{PQ}$?

b. Which side is opposite $\angle Q$?

c. Which side is included between $\angle P$ and $\angle R$?

d. Which angle is included between $\overline{QR}$ and $\overline{PR}$?

Solution

a. $\angle R$ is opposite $\overline{PQ}$.

b. $\overline{PR}$ is opposite $\angle Q$.

c. $\overline{PR}$ is the included side of $\angle P$ and $\angle R$.

d. $\angle R$ is the included angle of $\overline{QR}$ and $\overline{PR}$. □

PRACTICE 1 Given $\triangle PQR$:

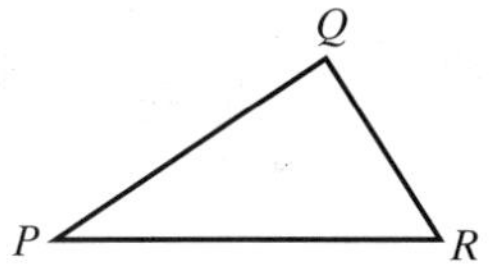

a. Which angle is opposite $\overline{PR}$?

b. Which side is opposite $\angle P$?

c. Which side is included between $\angle R$ and $\angle Q$?

d. Which angle is included between $\overline{PQ}$ and $\overline{PR}$?

OBJECTIVE 2 ▶ Classifying Triangles by Angles and Sides. We can classify triangles by their angles or their sides. In the charts below, equal angles mean that the angles are equal in measure, and equal sides mean that the sides are equal in measure.

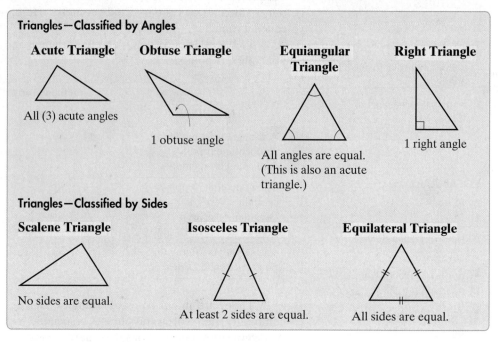

Triangles—Classified by Angles

Acute Triangle

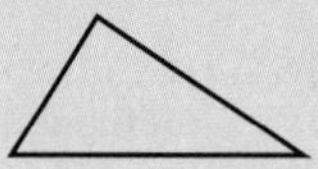

All (3) acute angles

Obtuse Triangle

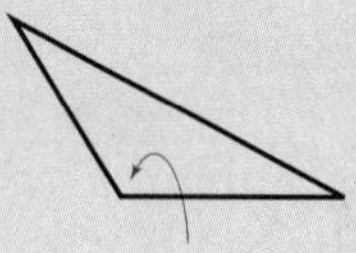

1 obtuse angle

Equiangular Triangle

All angles are equal. (This is also an acute triangle.)

Right Triangle

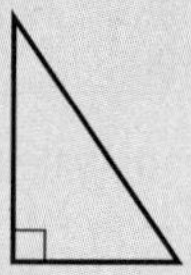

1 right angle

Triangles—Classified by Sides

Scalene Triangle

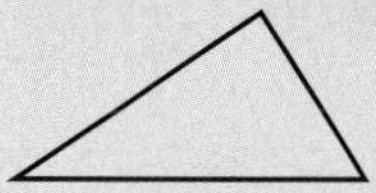

No sides are equal.

Isosceles Triangle

At least 2 sides are equal.

Equilateral Triangle

All sides are equal.

▶ Helpful Hint

Remember: The same number of tick marks per side of a figure means that the sides have the same measure.

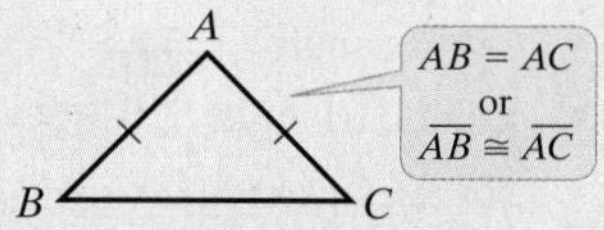

The same number of arcs per angle of a figure means that the angles have the same measure.

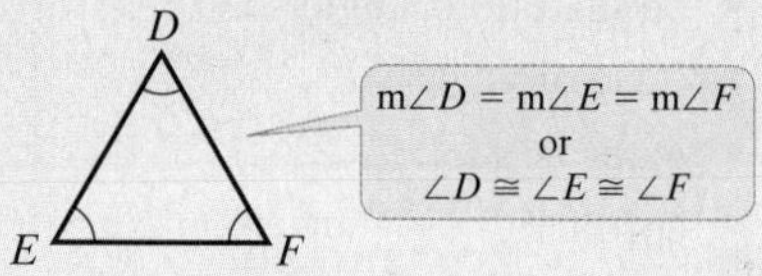

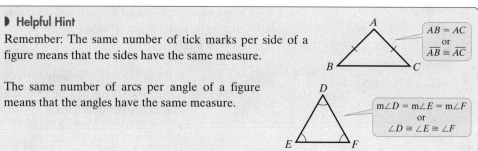

Reread the isosceles triangle and equilateral triangle definitions above. The conditional statement next is true:

If a triangle is equilateral, then it is isosceles. True. $(p \rightarrow q)$

Its converse is not true:

If a triangle is isosceles, then it is equilateral. False. (Converse, $q \rightarrow p$)

▶ Helpful Hint

- If a triangle is isosceles, it may or may not be equilateral.
- If a triangle is equilateral, then it is always isosceles.

Study the triangle chart above. Two types of triangles have special names for their sides.

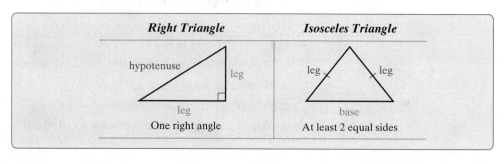

Right Triangle

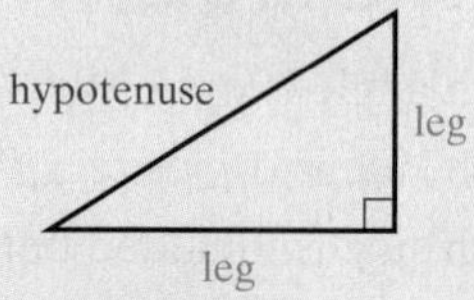

One right angle

Isosceles Triangle

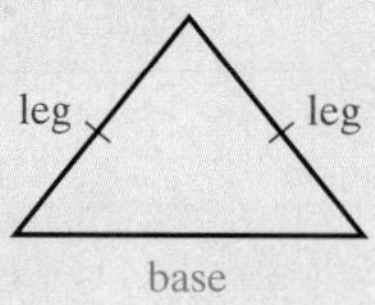

At least 2 equal sides

We will formally study these two special triangles and review their names in Section 4.6.

For now, let's practice classifying a triangle by its angles and sides.

EXAMPLE 2 Classifying Triangles

Classify each triangle by its angles and sides. Use the most specific name.

a.

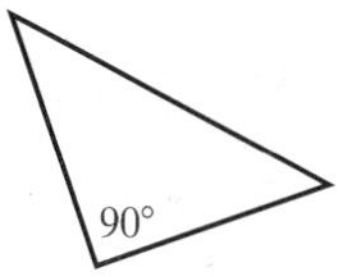

No sides are equal.

b.

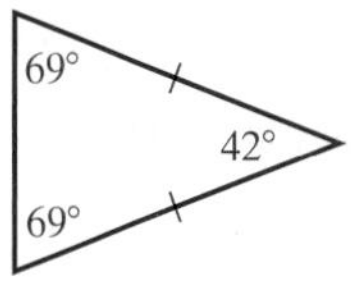

Solution

	Classified by Angles	*Classified by Sides*	*Complete Name of This Triangle*
a.	right triangle	scalene triangle	a right scalene triangle
b.	acute triangle	isosceles triangle	an acute isosceles triangle

□

PRACTICE 2 Classify each triangle by its angles and sides. Use the most specific name.

a.

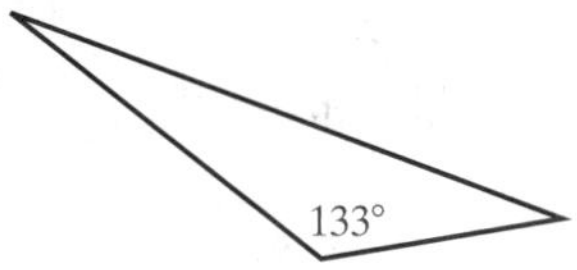

No sides are equal.

b.

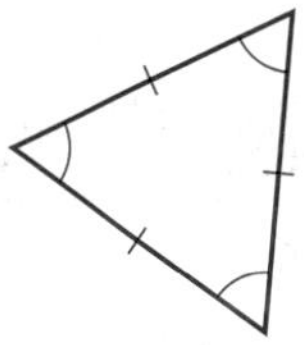

OBJECTIVE 3 ▶ Finding Angle Measures of Triangles. Now let's study the angles formed by extending the sides of a triangle. We call the 3 original angles of a triangle the **interior angles** of the triangle. The angles that are adjacent to the interior angles are the **exterior angles** of the triangle. There are two exterior angles associated with each interior angle, but since these two exterior angles are congruent vertical angles, we usually show only one exterior angle with each interior angle.

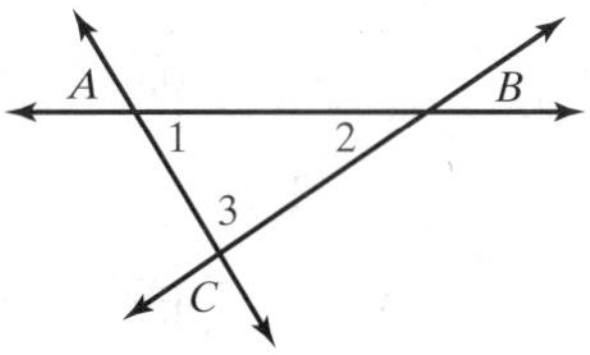

Interior Angles

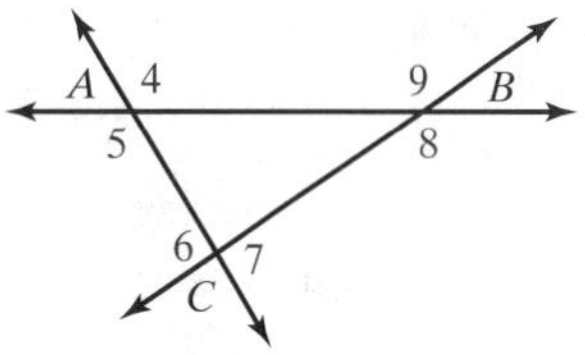

Exterior Angles

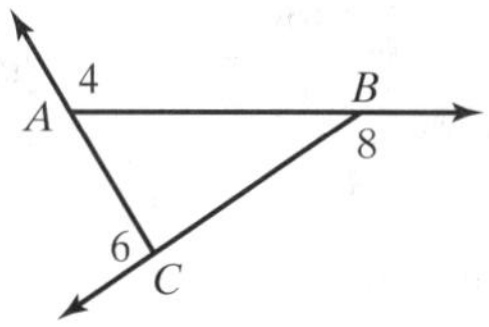

Exterior Angles Usually Shown

▶ Helpful Hint

To determine whether an angle is an exterior angle of a triangle, see whether the two sides of the angle are formed by

- an extension of a side of the triangle and
- a side of the triangle.

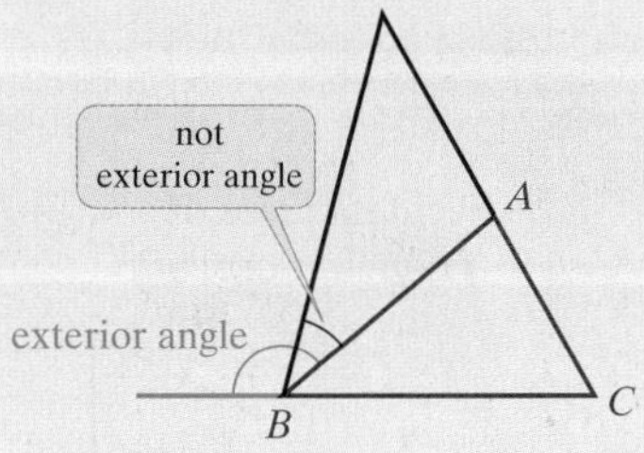

The following important theorem focuses on the sum of the measures of the interior angles of a triangle.

Theorem 4.1-1 Triangle Angle-Sum Theorem

The sum of the measures of the interior angles of a triangle is 180°.

$$m\angle 1 + m\angle 2 + m\angle 3 = 180°$$

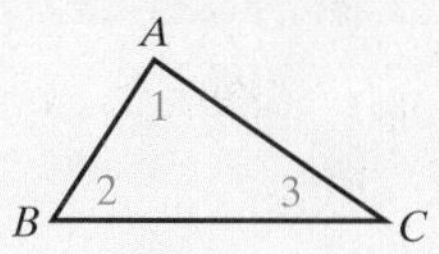

We prove this theorem next. To do so, we first perform a construction.

Proof of Theorem 4.1-1: Triangle Angle-Sum Theorem

Given: ΔABC

Prove: $m\angle 1 + m\angle 2 + m\angle 3 = 180°$

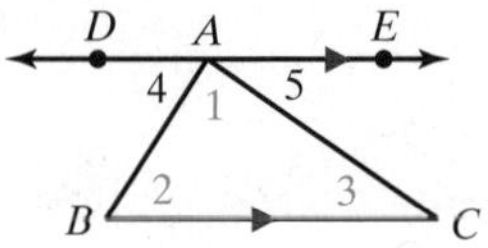

Construction: Use the Parallel Postulate.

Through point A not on line $\overleftrightarrow{BC}$, construct line $\overleftrightarrow{DE}$, the only line through point A parallel to $\overleftrightarrow{BC}$.

Statements	Reasons
1. $\overleftrightarrow{BC} \parallel \overleftrightarrow{DE}$ through A	**1.** Construction and Parallel Postulate
2. $\angle DAE$ is a straight angle, so $m\angle DAE = 180°$.	**2.** Definition of straight angle
3. $m\angle 4 + m\angle 1 + m\angle 5 = 180°$	**3.** Angle Addition Postulate
4. $m\angle 2 = m\angle 4$ and $m\angle 3 = m\angle 5$	**4.** Alternate Interior Angles Converse
5. $m\angle 2 + m\angle 1 + m\angle 3 = 180°$	**5.** Substitution—Steps 3 and 4

> **Helpful Hint**
> Don't forget that once a theorem is proved, we may use it as we want or need. We now know that the sum of the measures of angles of any triangle is 180°.

A **corollary** is a special name given to a theorem that is easy to prove as a direct result of another previously proved theorem.

> **Helpful Hint**
> A corollary is also a theorem.

Below are two corollaries to the Triangle Angle-Sum Theorem. (This simply means that these corollaries can easily be proved using the Triangle Angle-Sum Theorem.)

Corollary 4.1-2 Exterior Angle of a Triangle

The measure of each exterior angle of a triangle equals the sum of the measures of its two nonadjacent interior angles.

$$m\angle 1 = m\angle 2 + m\angle 3$$

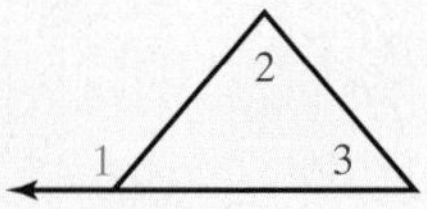

Corollary 4.1-3 Acute Angles of a Right Triangle

The two acute angles of a right triangle are complementary.

$$m\angle 1 + m\angle 2 = 90°$$

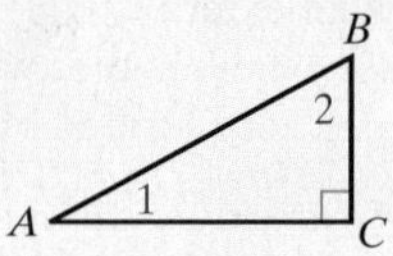

We prove these corollaries in Exercise Set 4.1, Exercises 67 and 68.

> **Helpful Hint**
> Examples of theorems/corollaries learned in this section:

Triangle Angle-Sum Theorem

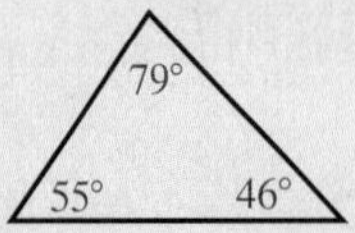

$55° + 79° + 46° = 180°$

Exterior Angle of a Triangle Corollary

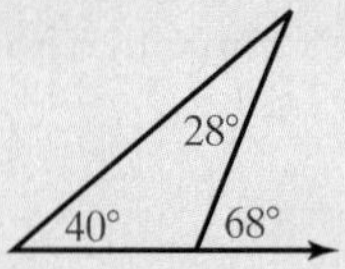

$40° + 28° = 68°$

Acute Angles of a Right Triangle Corollary

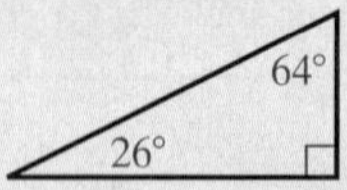

$26° + 64° = 90°$

EXAMPLE 3 Finding Angle Measures

Use the Triangle Angle-Sum Theorem to find the measure of each angle in the given triangle.

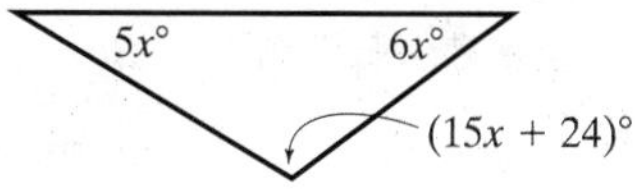

Solution Let's use the Triangle Angle-Sum Theorem to write an equation.

$5x + 6x + (15x + 24) = 180$ Sum of angles of a △ is 180° (Triangle Angle-Sum Theorem).

$26x + 24 = 180$ Combine like terms.

$26x = 156$ Subtraction Property of Equality

$x = 6$ Division Property of Equality

Now let's use the value of x and the given triangle to find the measure of each angle.

If $x = 6$, then $5x = 5 \cdot 6 = 30$,

$6x = 6 \cdot 6 = 36$, and

$15x + 24 = 15 \cdot 6 + 24 = 90 + 24 = 114.$

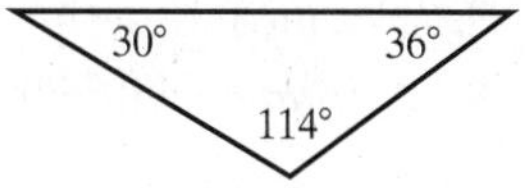

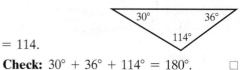

Check: $30° + 36° + 114° = 180°$. □

PRACTICE 3 Use the Triangle Angle-Sum Theorem to find the measure of each angle in the given triangle.

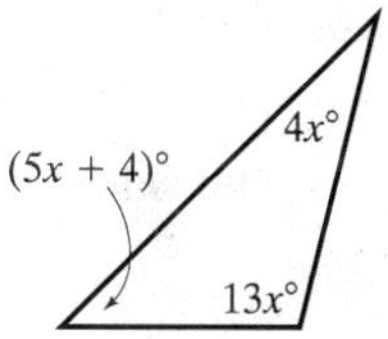

EXAMPLE 4 Finding Angle Measures

Use the Exterior Angle of a Triangle Corollary to find the measure of the exterior angle and the nonadjacent angle shown.

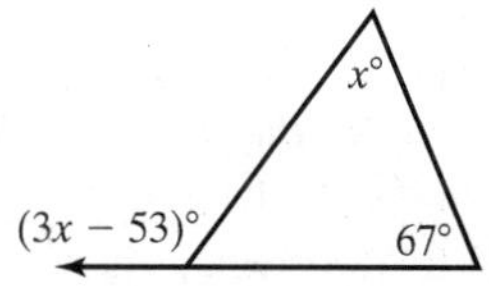

Solution Let's use the Exterior Angle of a Triangle Corollary to write an equation.

$3x - 53 = x + 67$ Measure of exterior angle equals sum of the two nonadjacent interior angles (the Exterior Angle of a Triangle Corollary).

$2x - 53 = 67$ Subtraction Property of Equality

$2x = 120$ Addition Property of Equality

$x = 60$ Division Property of Equality

Let's use the value of x and the given figure to find the measure of each angle.

Since $x = 60$, then

$3x - 53 = 3 \cdot 60 - 53 = 180 - 53 = 127.$

Also, $x° = 60°$.

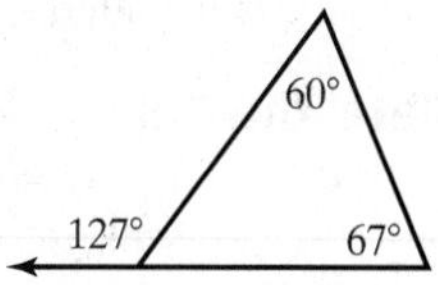

Check: $60° + 67° = 127°$. □

PRACTICE 4 Use the Exterior Angle of a Triangle Corollary to find the measure of the exterior angle and the nonadjacent angle shown.

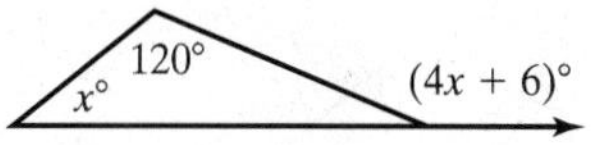

VOCABULARY & READINESS CHECK

Word Bank *Use the choices to fill in each blank.*

isosceles	equiangular	vertex	sides
scalene	equilateral	vertices	triangle

1. A(n) ____________ is formed by three noncollinear points connected by segments.
2. The segments joining the points in Exercise 1 above are called ____________.
3. The noncollinear points from Exercise 1 above are called ____________.
4. The singular of the word vertices is ____________.
5. A triangle with no equal sides is a(n) ____________ triangle.
6. A triangle with at least two equal sides is a(n) ____________ triangle.
7. The most descriptive name for a triangle with all sides equal is a(n) ____________ triangle.
8. A triangle with all angles equal is a(n) ____________ triangle.

Fill in the Blank

9. The interior angles are ____________.
10. The exterior angles are ____________.
11. The adjacent sides that share $\angle A$ are ____________.
12. The side opposite $\angle A$ is ____________.

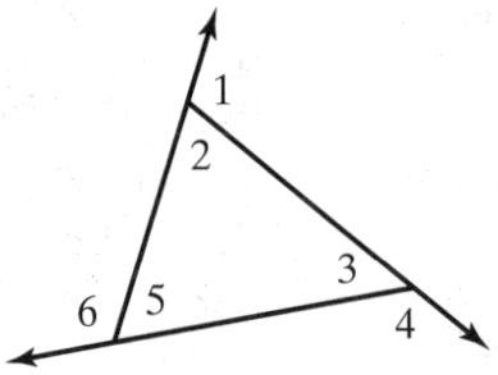

For Exercises 9 and 10.

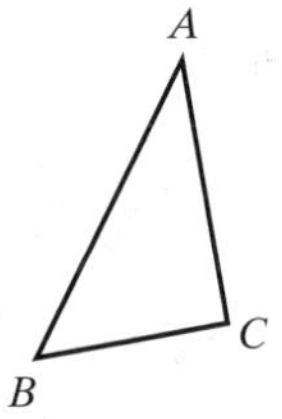

For Exercises 11 and 12.

4.1 EXERCISE SET MyMathLab®

Answer the questions for the given triangle ABC. See Example 1.

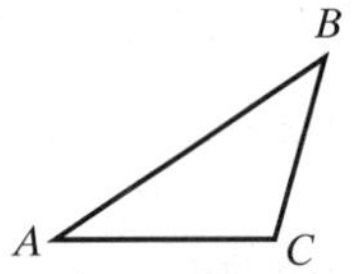

1. Which angle is opposite $\overline{BC}$?
2. Which angle is opposite $\overline{AB}$?
3. Which side is opposite $\angle B$?
4. Which side is opposite $\angle C$?
5. Which angle is included between $\overline{AC}$ and $\overline{BC}$?
6. Which angle is included between $\overline{AB}$ and $\overline{AC}$?
7. Which side is included between $\angle C$ and $\angle B$?
8. Which side is included between $\angle B$ and $\angle A$?

Complete the Table *Complete the table by classifying each triangle by angles (acute, obtuse, right, or equiangular) and by sides (scalene, isosceles, or equilateral). (Use the most specific name.) See Example 2.*

	Triangle	***Classify by Angles***	***Classify by Sides***
9.	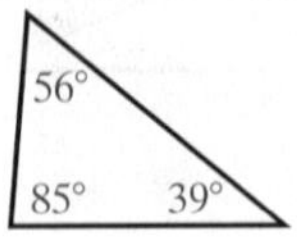 No equal sides	____________	____________
10.	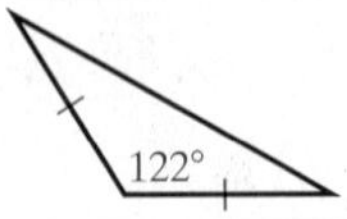	____________	____________
11.	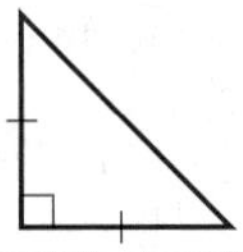	____________	____________
12.	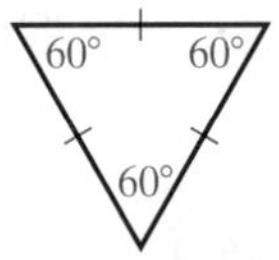	____________	____________

Continue the table for ΔABC in each Flag.

13.	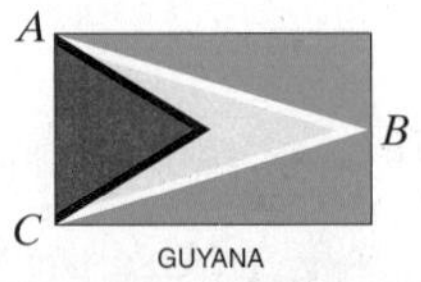	____________	____________
14.		____________	____________

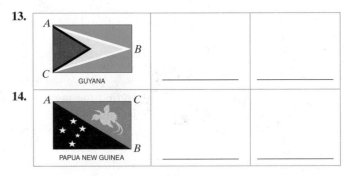

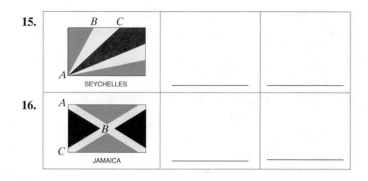

15.	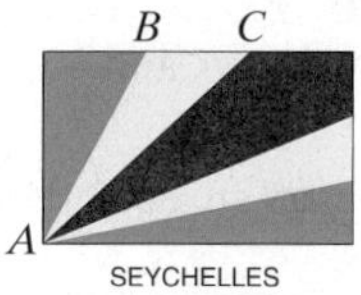	________	________
16.	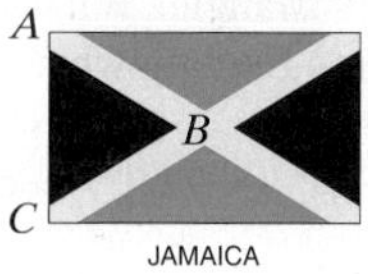	________	________

Matching *Match each triangle with the most specific name. See Example 2.*

17. Angles: 10°, 40°, 130° — A. Scalene

18. Angles: 40°, 50°, 90° — B. Isosceles

19. Angles: 60°, 60°, 60° — C. Equilateral

20. Sides: 7 mm, 7 mm, 10 mm — D. Right

21. Sides: 5 ft, 8 ft, 7 ft — E. Obtuse

22. Sides: 11 in., 11 in., 11 in. — F. Equiangular

Mixed Practice *Use the theorem and corollaries in this section to find the measures of the numbered angles. See Examples 3 and 4.*

23.
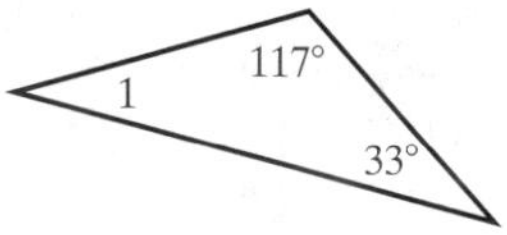

24.
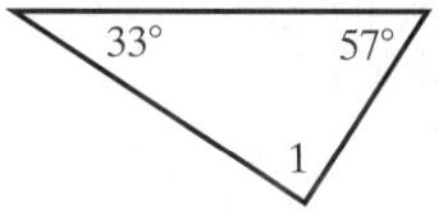

25.
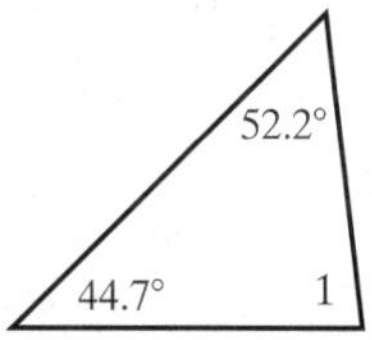

26.
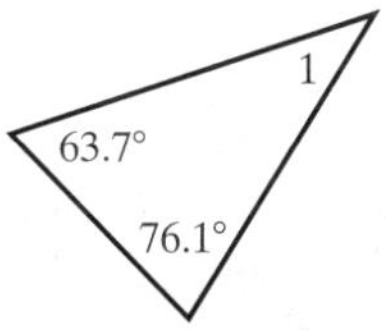

27.
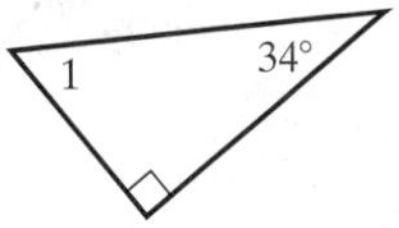

28.
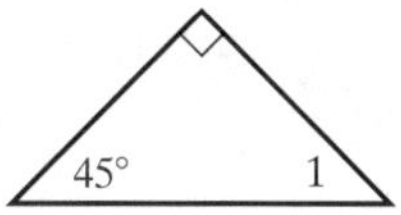

29.
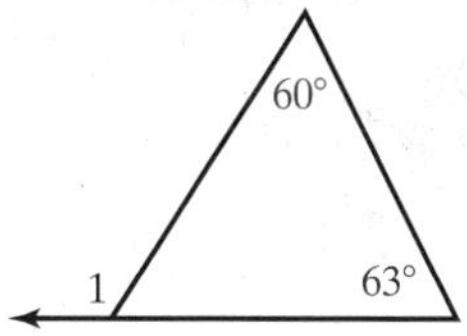

30.
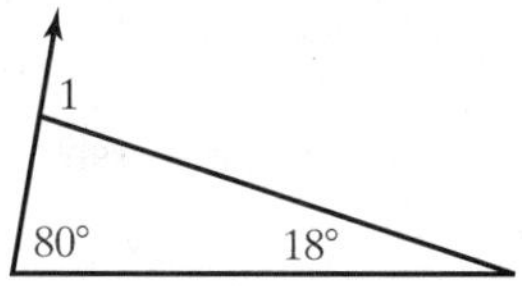

31.
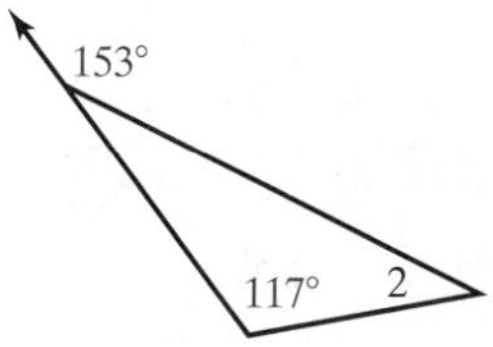

32.
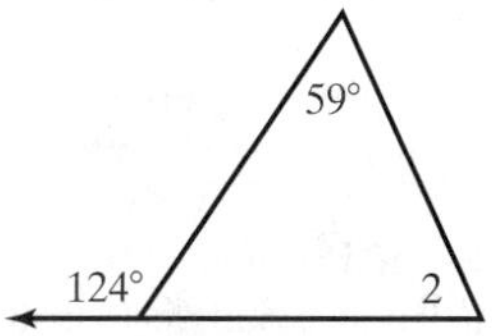

33.
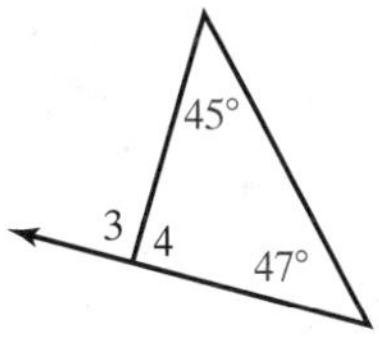

34.
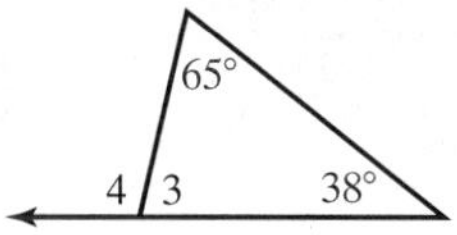

35.
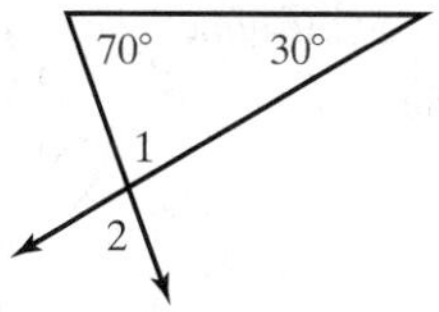

36.
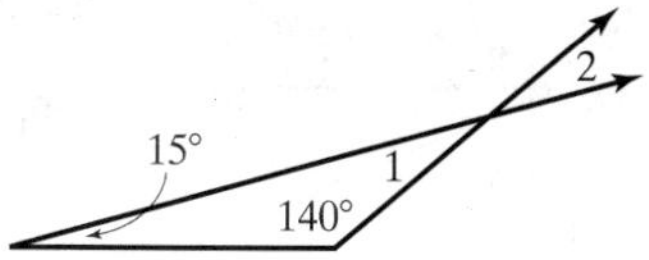

37.
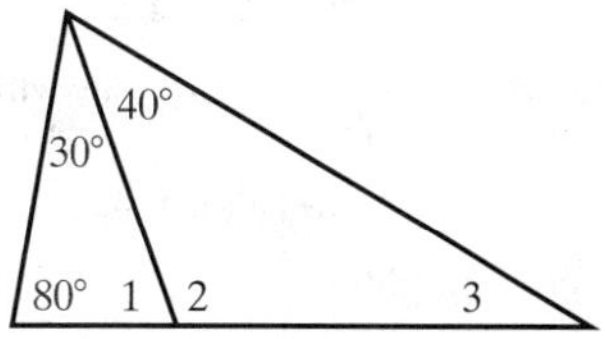

(*Hint:* Find $m\angle 1$, then $m\angle 2$, and then $m\angle 3$.)

38.
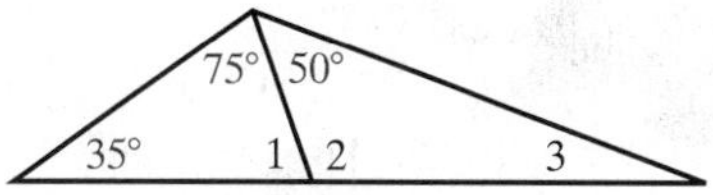

(Use the same hint as for Exercise 37.)

Mixed Practice *Find the values of the variables and the measures of the angles whose degree measure is noted on the figure. See Examples 3 and 4.*

39.
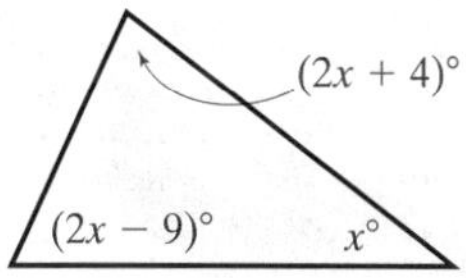

40.
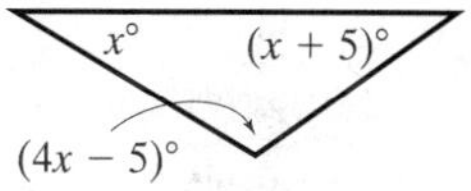

41.
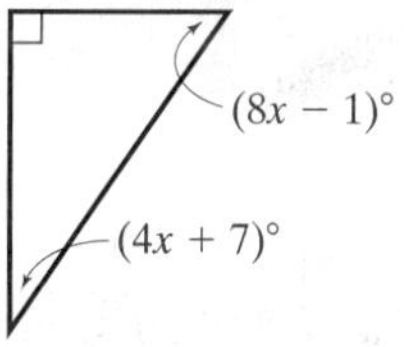

42.
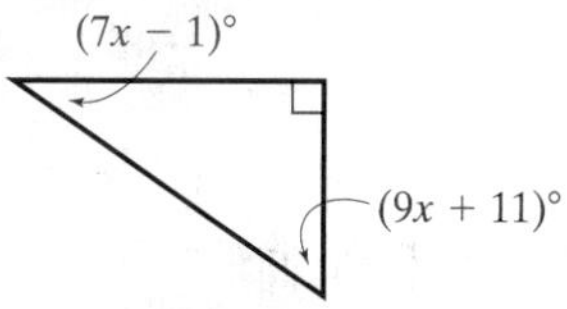

43.
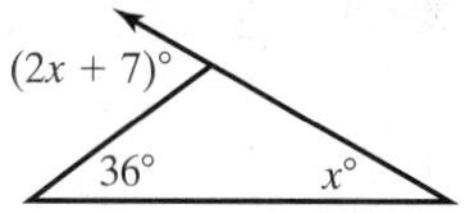

44.
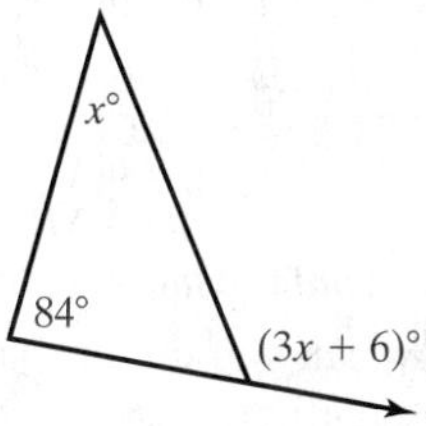

45.
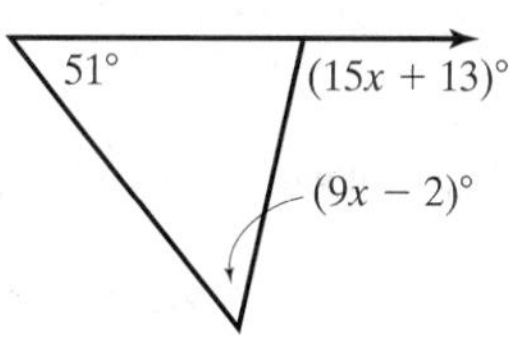

46.
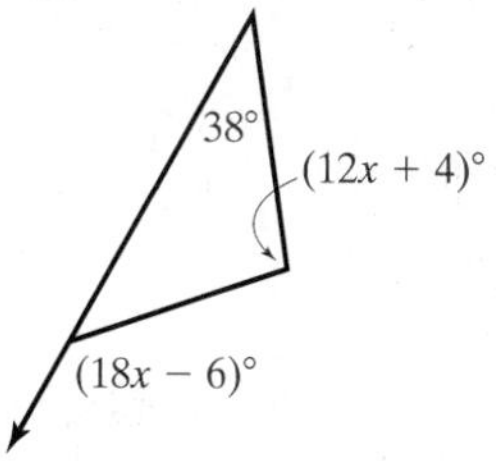

47.
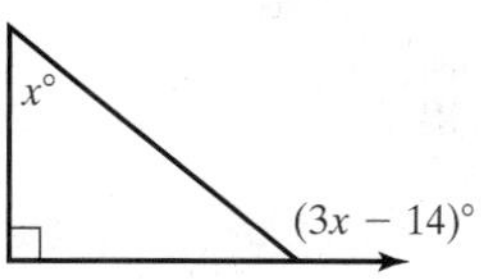

48.
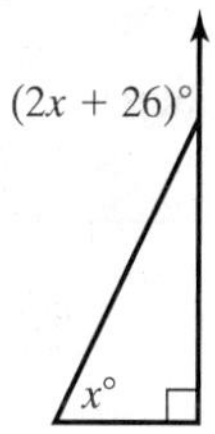

For Exercises 49 and 50, find the measures of the numbered angles.

49.

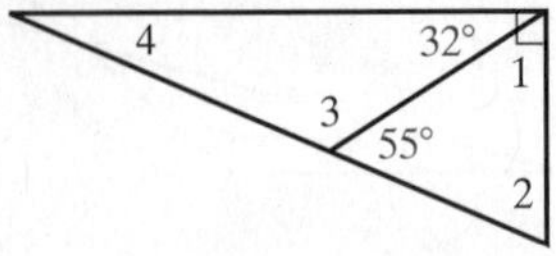

50.

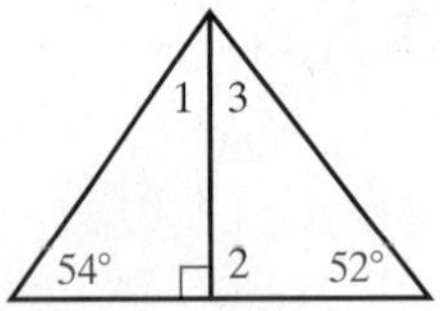

51. A ramp forms the angles shown below. What are the values of a and b?

52. A lounge chair has different settings that change the angles formed by its parts. Suppose $m\angle 2 = 71°$ and $m\angle 3 = 43°$. Find $m\angle 1$. (See the figure.)

CONCEPT EXTENSIONS

The angle measures of $\triangle RST$ are represented by $2x$, $x + 14$, and $x - 38$. What are the angle measures of $\triangle RST$?

53. How can you use the Triangle Angle-Sum Theorem to write an equation?

54. How can you check your answer?

Multiple Choice *For Exercises 55 and 56, which diagram below correctly represents each description? Explain your reasoning.*

i.

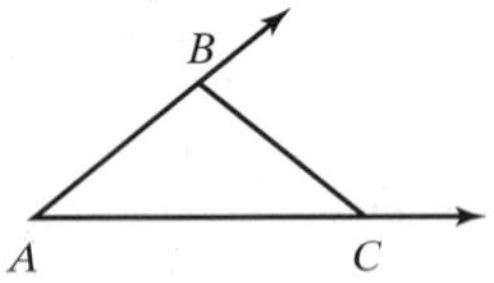

ii.

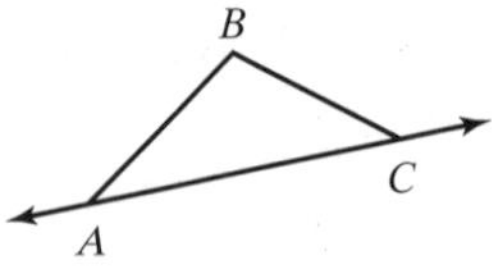

iii.

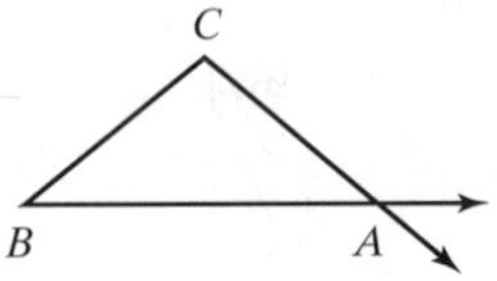

55. Draw any triangle. Label it $\triangle ABC$. Extend two sides of the triangle to form two exterior angles at vertex A.

56. Draw any triangle. Label it $\triangle ABC$. Extend two sides of the triangle to form an exterior angle at B and an exterior angle at C.

57. What is the measure of each angle of an equiangular triangle? Explain.

58. Suppose you have an isosceles right triangle. What are the measures of the angles of the triangle? Explain why.

Use the diagram at the right for Exercises 59 through 62.

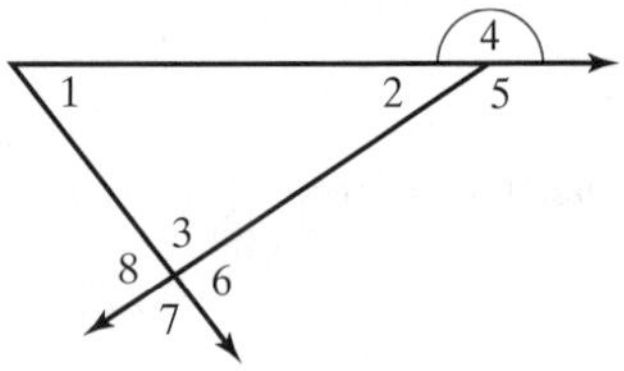

59. Which of the numbered angles are exterior angles?

60. Which of the numbered angles are interior angles?

61. How are angles 6 and 8 related?

62. How are angles 2 and 5 related?

63. In general, how many exterior angles are at each vertex of a triangle?

64. In general, how many exterior angles does a triangle have in all?

Find the value of x for Exercises 65 and 66.

65.

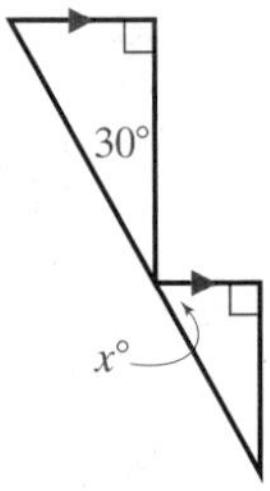

66.

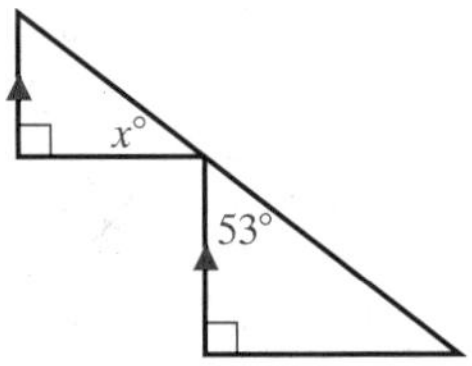

67. Proof Proof of the Exterior Angle of a Triangle Corollary (Corollary 4.1-2) Fill in the missing steps of the proof.

Given: $\triangle ABC$ with $\angle 1$ an exterior angle

Prove: $m\angle 1 = m\angle 2 + m\angle 3$

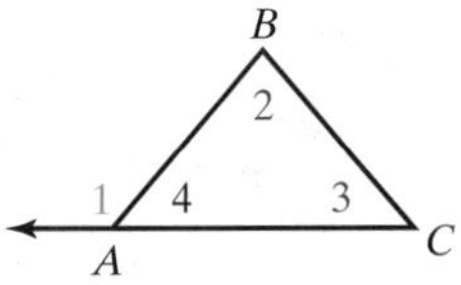

Statements	Reasons
1. $\angle 1$ is an exterior angle.	**1.** Given
2. $m\angle 1 + m\angle 4 = 180°$	**2.** ?(a)
3. ?(b)	**3.** Triangle Angle-Sum Theorem
4. $m\angle 1 + m\angle 4 = m\angle 2 + m\angle 3 + m\angle 4$	**4.** ?(c)
5. $m\angle 1 = m\angle 2 + m\angle 3$	**5.** ?(d)

68. Proof Proof of the Acute Angles of a Right Triangle Corollary (Corollary 4.1-3) Fill in the missing steps of the proof.

Given: Right triangle ABC

Prove: $m\angle 1 + m\angle 2 = 90°$

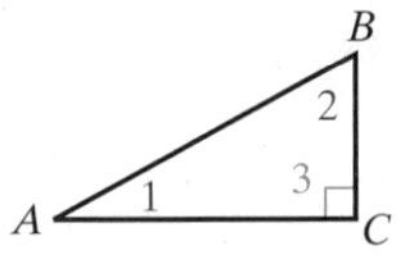

Statements	**Reasons**
1. Right ΔABC	**1.** Given
2. $m\angle 3 = 90°$	**2.** Definition of right Δ
3. ?(a)	**3.** Triangle Angle-Sum Theorem
4. $m\angle 1 + m\angle 2 + 90° = 180°$	**4.** ?(b)
5. $m\angle 1 + m\angle 2 = 90°$	**5.** ?(c)

REVIEW AND PREVIEW

Find the coordinates of the midpoint of $\overline{AB}$. *See Section 1.7.*

69. $A(-2, 3), B(4, 1)$

70. $A(7, 10), B(-5, -8)$

71. $A(0, 5), B(3, 6)$

72. $A(9, 0), B(4, 5)$

4.2 CONGRUENT FIGURES

OBJECTIVES

1 Identify Corresponding Parts in Congruent Figures.

2 Prove Triangles Are Congruent.

VOCABULARY

- congruent figures
- corresponding sides
- corresponding angles

Congruent figures have the exact same shape and size. All the figures below are congruent. As shown below, a flip, rotate (turn), or slide does not affect whether figures are congruent because they still have the same shape and size.

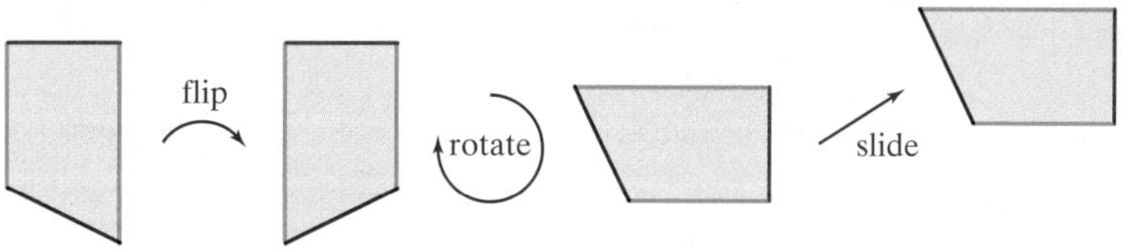

OBJECTIVE 1 ▶ Identifying Corresponding Parts in Congruent Figures. When figures are congruent, their **corresponding sides** are congruent, and their **corresponding angles** are congruent.

▶ Helpful Hint

When reading figures, the angles with the same number of arcs are congruent and the sides with the same number of tick marks are congruent.

$\overline{BC} \cong \overline{ST}$

$\angle A \cong \angle R$

$\Delta ABC \cong \Delta RST$

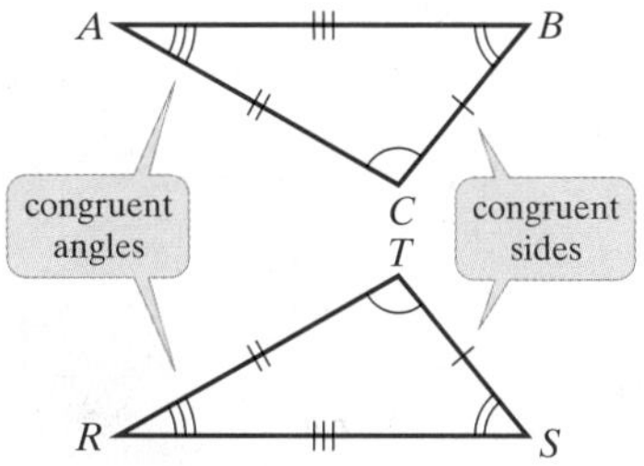

Corresponding Angles	Corresponding Sides
$\angle A \cong \angle R$	$\overline{AB} \cong \overline{RS}$
$\angle B \cong \angle S$	$\overline{BC} \cong \overline{ST}$
$\angle C \cong \angle T$	$\overline{CA} \cong \overline{TR}$

▶ Helpful Hint

When writing congruent figures, make sure that you list the congruent angles in the same order. Below are a few correct ways to list the congruent triangles above.

$\Delta ABC \cong \Delta RST \quad \Delta BCA \cong \Delta STR \quad \Delta CBA \cong \Delta TSR$

EXAMPLE 1 **Naming Congruent Parts**

For the two figures, we are given that $ABCD \cong TRQS$. Name the congruent corresponding angles and sides.

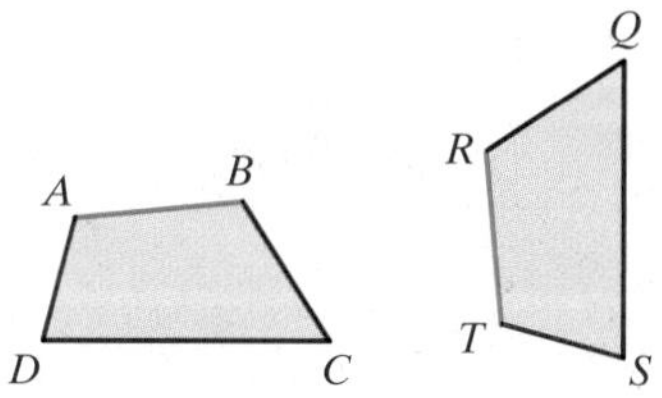

Solution The figures help, but having the congruent angles listed in the corresponding order is all that is needed.

$$ABCD \cong TRQS$$

Angles: $\angle A \cong \angle T, \angle B \cong \angle R, \angle C \cong \angle Q, \angle D \cong \angle S$

Sides: $\overline{AB} \cong \overline{TR}, \overline{BC} \cong \overline{RQ}, \overline{CD} \cong \overline{QS}, \overline{DA} \cong \overline{ST}$

□

PRACTICE **1** For the two figures, we are given that $ABCDE \cong MNPGH$. Name the congruent corresponding angles and sides.

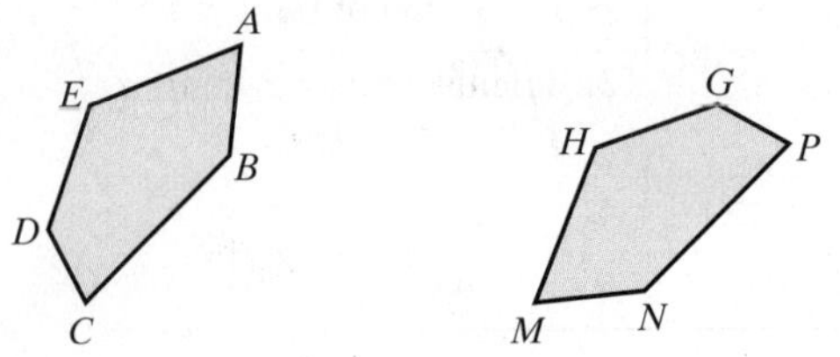

EXAMPLE 2 Using Congruent Triangles to Find Angle Measures

In the process of folding an origami cat, two congruent triangles are formed.

Given that $\Delta ABC \cong \Delta AEF$, find $m\angle F$.

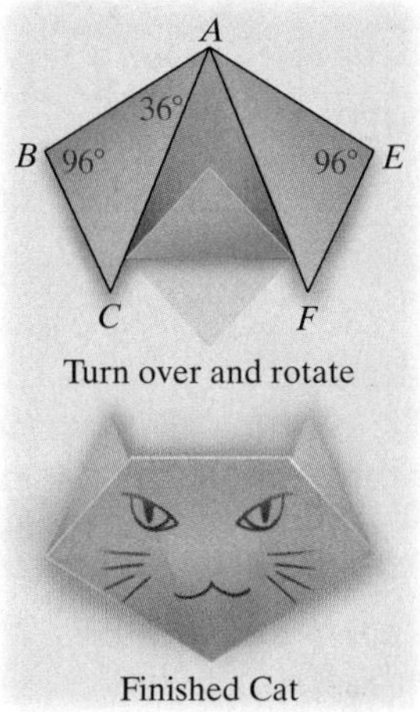

Solution First, let's find $m\angle C$ using ΔABC.

$m\angle C + 96° + 36° = 180°$ Triangle Angle-Sum Theorem

$m\angle C = 48°$ Subtraction Property of Equality

Since $\Delta ABC \cong \Delta AEF$,
$m\angle F = m\angle C = 48°$.

PRACTICE **2** The figure is an origami leaf that contains two congruent triangles. Given that $\Delta QRS \cong \Delta QTS$, find $m\angle T$.

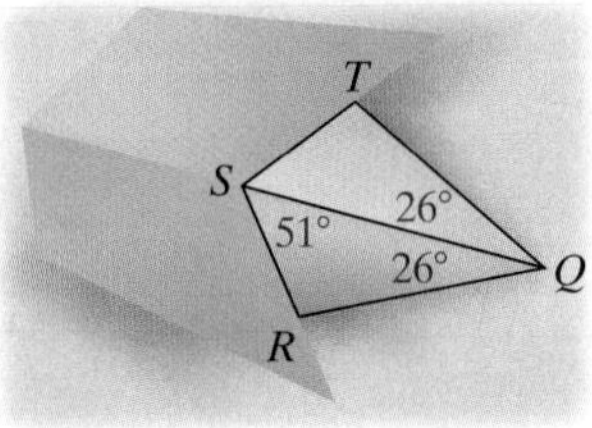

Recall that the sum of the measures of the angles in a triangle is 180°. The next theorem follows from that Triangle Angle-Sum Theorem.

Theorem 4.2-1 Third Angles Theorem

Theorem	If . . .	Then . . .
If two angles of one triangle are congruent to two angles of another triangle, then the third angles are congruent.	$\angle A \cong \angle D$ and $\angle B \cong \angle E$ 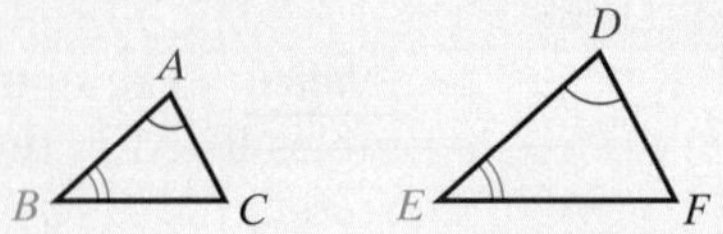	$\angle C \cong \angle F$

We prove this theorem in Exercise Set 4.2, Exercises 39–42.

EXAMPLE 3 **Using the Third Angles Theorem**

Find the value of x.

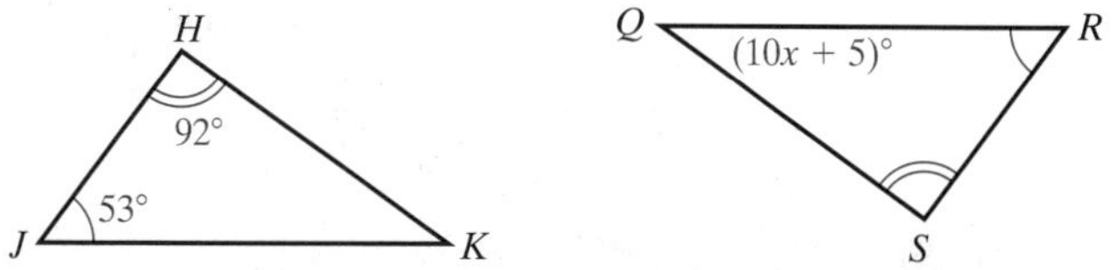

Solution From the figures, we have

$$\angle J \cong \angle R \quad \text{and} \quad \angle H \cong \angle S.$$

Thus, from the Third Angles Theorem,

$$\angle K \cong \angle Q \quad \text{or} \quad m\angle K = m\angle Q.$$

First, let's use ΔJHK to find $m\angle K$.

$$m\angle K = 180° - 53° - 92° = 35°$$

Now, $m\angle K = m\angle Q$ Third Angles Theorem

$35 = 10x + 5$ Substitution

$30 = 10x$ Subtraction Property of Equality

$3 = x$ Division Property of Equality □

The value of x is 3. To **check**, replace x with 3 and see that $10x + 5 = 35$. Then make sure that $53° + 92° + 35° = 180°$.

PRACTICE 3 Find the value of x.

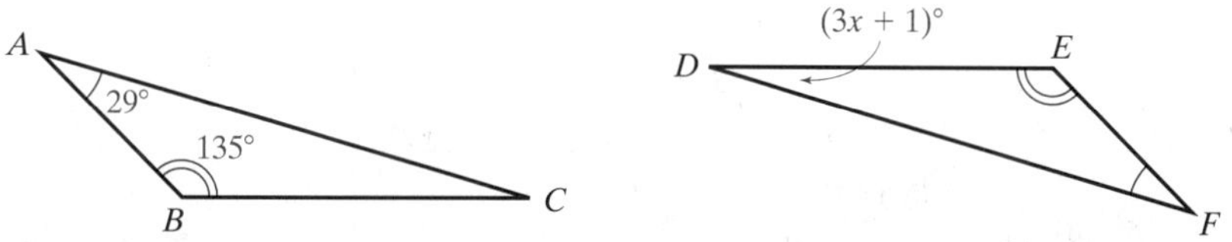

OBJECTIVE 2 ▶ Proving Triangles Are Congruent. By definition, we know that two triangles are congruent when they have the exact same shape and size. This means that all three corresponding sides are congruent and all three corresponding angles are congruent.

(In later sections, we will learn shortcuts for proving that triangles are congruent.)

EXAMPLE 4 **Proving Triangles Are Congruent**

Given: $\overline{LM} \cong \overline{LO}$, $\overline{MN} \cong \overline{ON}$,
$\angle M \cong \angle O$, $\angle MLN \cong \angle OLN$

Prove: $\Delta LMN \cong \Delta LON$

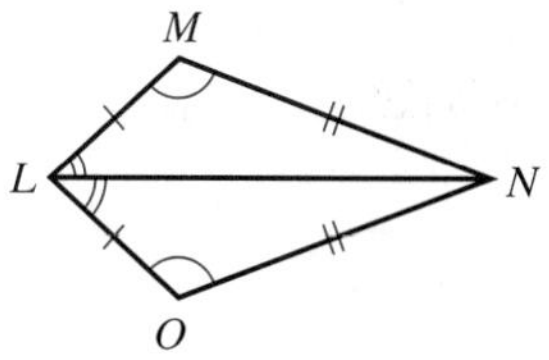

Solution

Statements	**Reasons**
1. $\overline{LM} \cong \overline{LO}$, $\overline{MN} \cong \overline{ON}$	**1.** Given
2. $\overline{LN} \cong \overline{LN}$	**2.** Reflexive Property of $\cong$
3. $\angle M \cong \angle O$, $\angle MLN \cong \angle OLN$	**3.** Given
4. $\angle MNL \cong \angle ONL$	**4.** Third Angles Theorem
5. $\Delta LMN \cong \Delta LON$	**5.** Definition of $\cong$ triangles (all corresponding sides and angles are $\cong$)

□

PRACTICE 4

Given: $\angle A \cong \angle D, \overline{AE} \cong \overline{DC}, \overline{EB} \cong \overline{CB}, \overline{BA} \cong \overline{BD}$

Prove: $\Delta AEB \cong \Delta DCB$

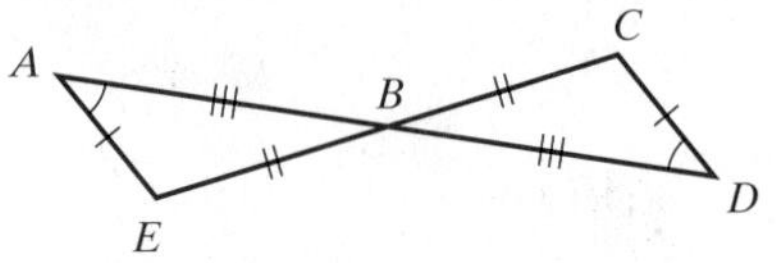

VOCABULARY & READINESS CHECK

Fill in each blank.

1. Figures that have the exact same shape and size are called ____________ figures.
2. When figures are congruent, their ____________ sides are congruent and their ____________ angles are congruent.

Given: $\Delta PQR \cong \Delta STV$. *For Exercises 3–8, use these triangles to fill in the blanks with corresponding parts.*

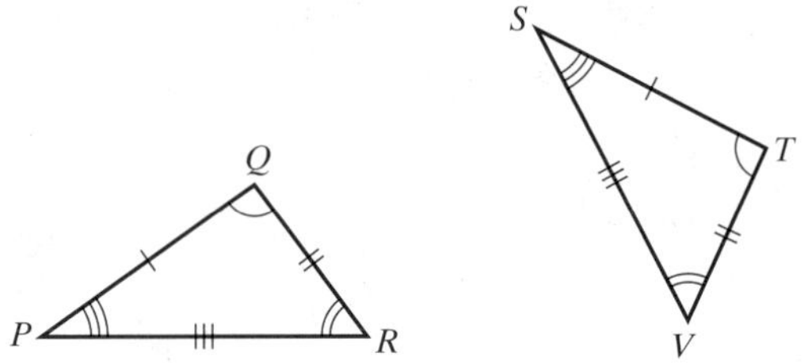

3. $\angle P \cong$ ____________
4. $\angle Q \cong$ ____________
5. $\angle R \cong$ ____________
6. $\overline{PQ} \cong$ ____________
7. $\overline{QR} \cong$ ____________
8. $\overline{RP} \cong$ ____________

4.2 EXERCISE SET MyMathLab®

$\Delta LMC \cong \Delta BJK$. Complete the congruence statements. See Example 1.

1. $\overline{LC} \cong$?
2. $\overline{KJ} \cong$?
3. $\overline{JB} \cong$?
4. $\angle L \cong$?
5. $\angle K \cong$?
6. $\angle M \cong$?
7. $\Delta CML \cong$?
8. $\Delta KBJ \cong$?
9. $\Delta MLC \cong$?
10. $\Delta JKB \cong$?

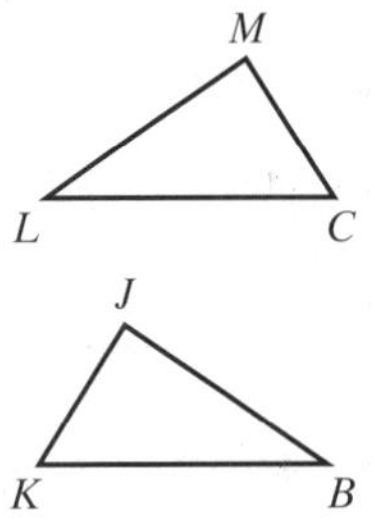

Given two congruent figures, MNEP ≅ SRTQ, list each of the following. See Example 1.

11. four pairs of congruent sides
12. four pairs of congruent angles

13. Builders use the king post truss (below left) for the top of a simple structure. In this truss, $\Delta ABC \cong \Delta ABD$. List the congruent corresponding parts. See Example 1.

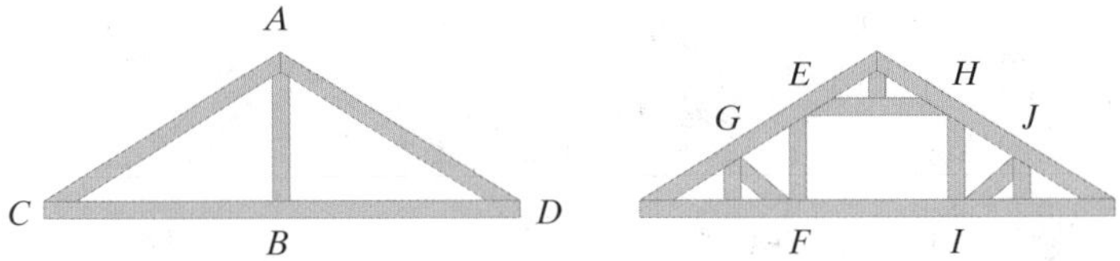

14. The attic frame truss (above right) provides open space in the center for storage. In this truss, $\Delta EFG \cong \Delta HIJ$. List the congruent corresponding parts. See Example 1.

At an archeological site, the remains of two ancient step pyramids are congruent. If ABCD ≅ EFGH, find each of the following. (Diagrams are not to scale.) See Example 1.

15. AD
16. GH
17. $m\angle GHE$
18. $m\angle BAD$
19. EF
20. BC
21. $m\angle DCB$
22. $m\angle EFG$

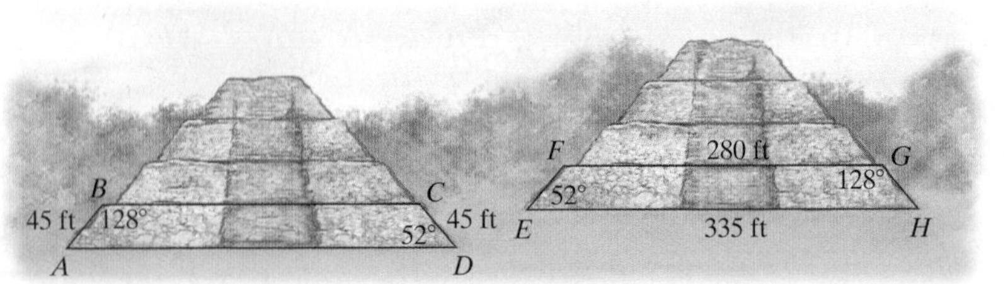

Mixed Practice *For Exercises 23 and 24, can you conclude that the triangles are congruent? Justify your answers. See Example 4.*

23. ΔTRK and ΔTUK

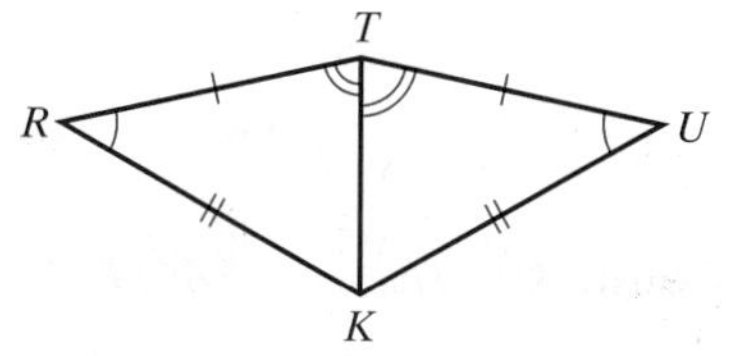

24. ΔSPQ and ΔTUV

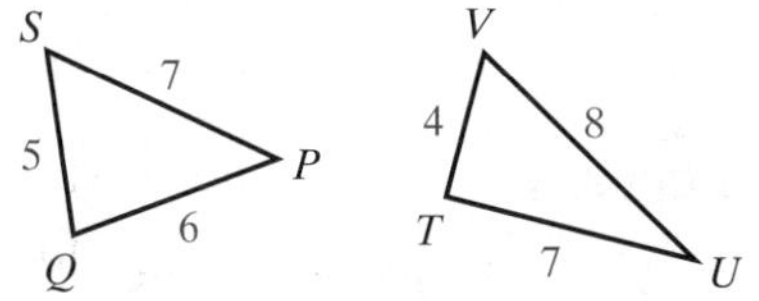

Multiple Choice *Use the choices below for Exercises 25 and 26. See Example 1.*

a. $\overline{DE} \cong \overline{MN}$ **b.** $\overline{FE} \cong \overline{NL}$

c. $\angle N \cong \angle F$ **d.** $\angle M \cong \angle F$

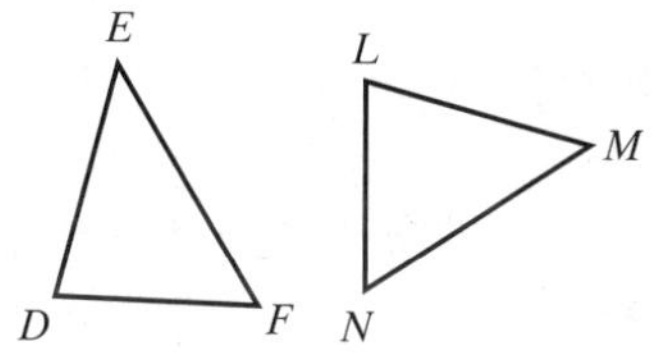

25. If $\Delta DEF \cong \Delta LMN$, choose the correct statement.

26. If $\Delta DEF \cong \Delta LNM$, choose the correct statement.

Complete each statement in two different ways.

27. $\Delta JLM \cong$ ___?___ **28.** $\Delta LJM \cong$ ___?___

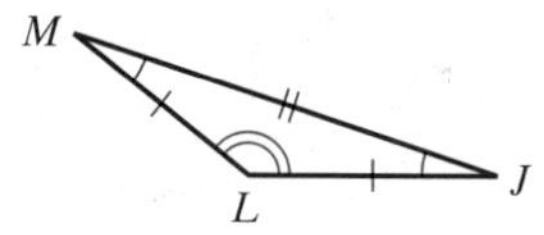

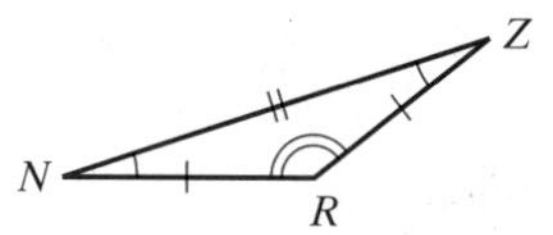

For each exercise, $\Delta ABC \cong \Delta DEF$. Find $m\angle F$. See Example 2.

29.

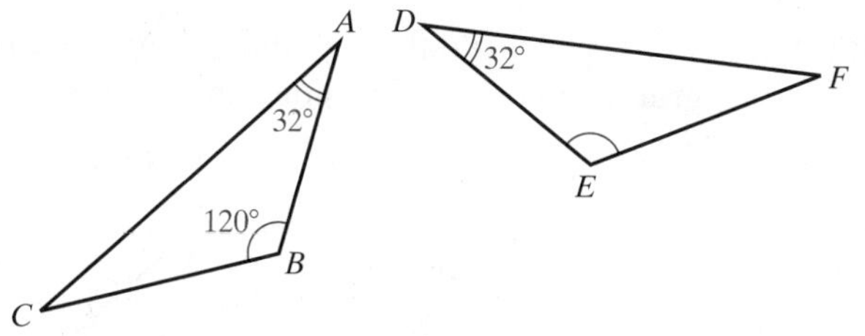

30.

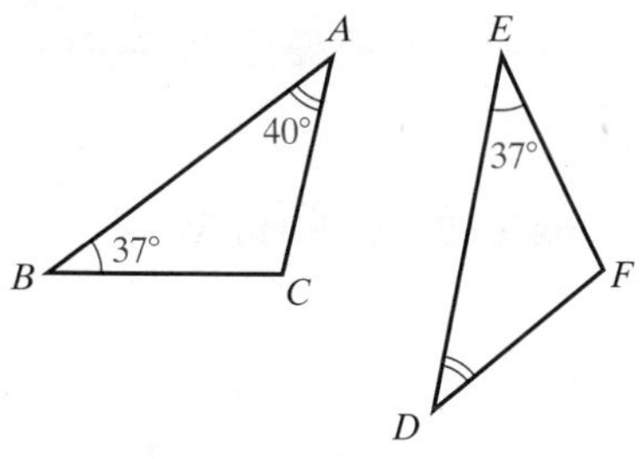

$\Delta ABC \cong \Delta DBE$. Use the given figure for Exercises 31 and 32. See Examples 2 and 3.

31. Find the value of y.

32. Find the value of x.

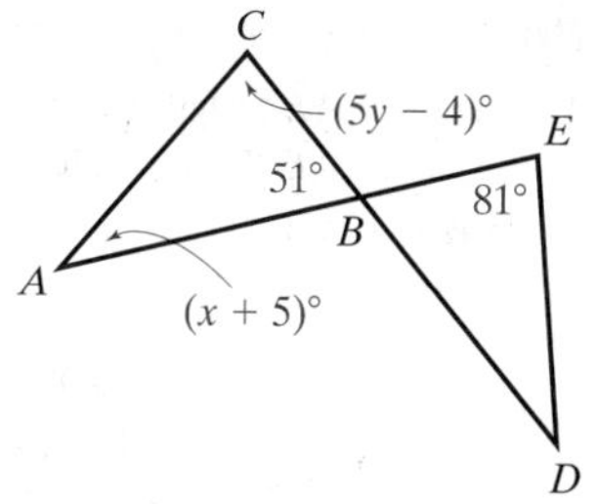

For Exercises 33 and 34, find the values of the variables. See Examples 2 and 3.

33.

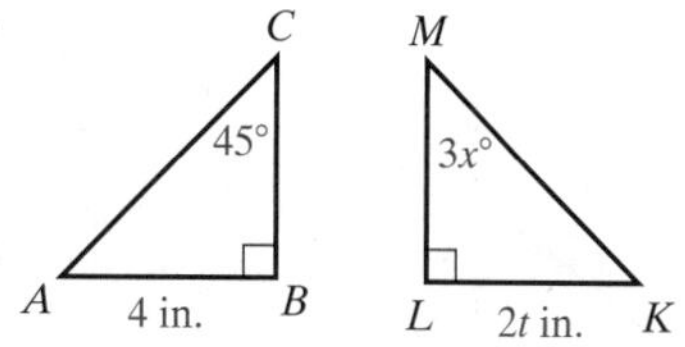

$\Delta ABC \cong \Delta KLM$

34.

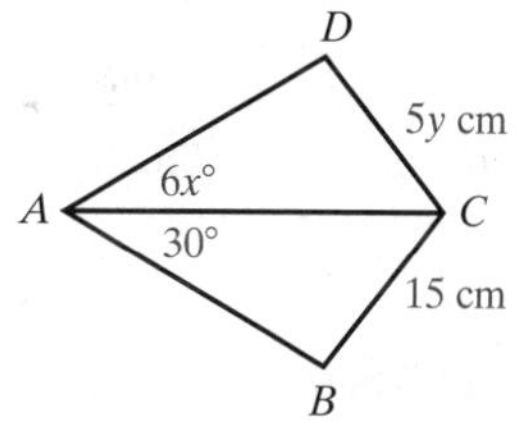

$\Delta ACD \cong \Delta ACB$

$\Delta ABC \cong \Delta DEF$. Find the measures of the given angles or the lengths of the given sides. See Example 3.

35. $m\angle A = (x + 10)°, m\angle D = 2x°$

36. $m\angle B = 3y°, m\angle E = (6y - 12)°$

37. $BC = 3z + 2, EF = z + 6$

38. $AC = 7a + 5, DF = 5a + 9$

Proof *In Exercises 39–42, complete the proof of the Third Angles Theorem, Theorem 4.2-1. See Example 4.*

Given: $\angle A \cong \angle D, \angle B \cong \angle E$

Prove: $\angle C \cong \angle F$

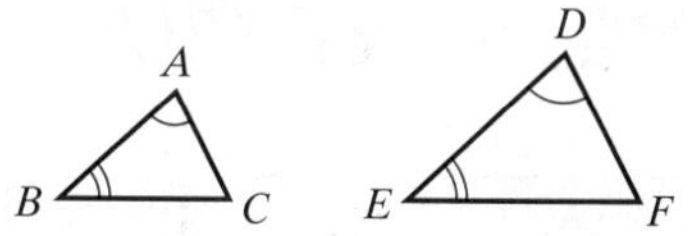

Statements	Reasons
1. $\angle A \cong \angle D, \angle B \cong \angle E$	**1.** Given
2. $m\angle A = m\angle D, m\angle B = \angle E$	**2.** Definition of $\cong$ angles
3. $m\angle A + m\angle B + m\angle C = 180$, $m\angle D + m\angle E + m\angle F = 180$	**3.** ? (Exercise 39)
4. $m\angle A + m\angle B + m\angle C = m\angle D + m\angle E + m\angle F$	**4.** ? (Exercise 40)
5. $m\angle D + m\angle E + m\angle C = m\angle D + m\angle E + m\angle F$	**5.** ? (Exercise 41)
6. $m\angle C = m\angle F$	**6.** ? (Exercise 42)
7. $\angle C \cong \angle F$	**7.** Definition of $\cong$ angles

Proof *For Exercises 43 and 44, see Example 4.*

43. Given: $\overline{AB} \parallel \overline{DC}$, $\angle B \cong \angle D$, $\overline{AB} \cong \overline{DC}$, $\overline{BC} \cong \overline{AD}$

Prove: $\triangle ABC \cong \triangle CDA$

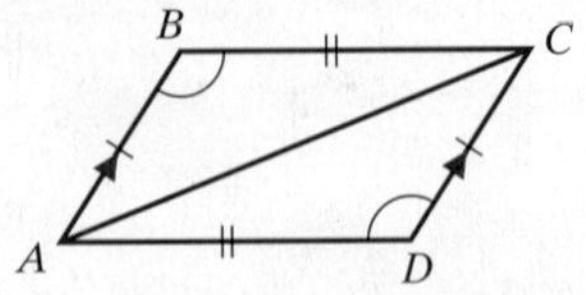

44. Given: $\overline{AB} \perp \overline{AD}$, $\overline{BC} \perp \overline{CD}$, $\overline{AB} \cong \overline{CD}$, $\overline{AD} \cong \overline{CB}$, $\overline{AB} \parallel \overline{CD}$

Prove: $\triangle ABD \cong \triangle CDB$

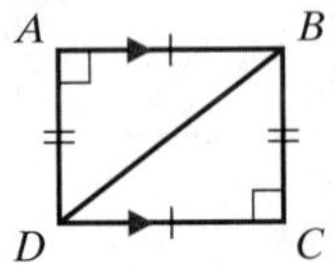

CONCEPT EXTENSIONS

45. Randall says he can use the information in the figure to prove $\triangle BCD \cong \triangle DAB$. Is he correct? Explain.

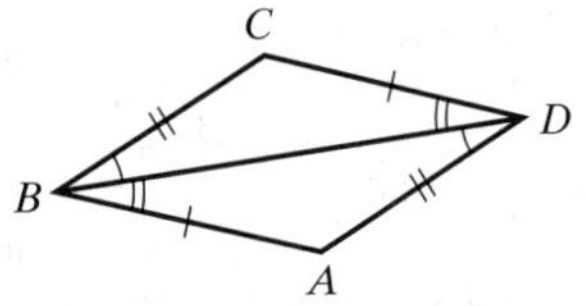

46. Write a congruence statement for two triangles. Then list the congruent sides and angles.

Proof *Given:* $\overline{PR} \parallel \overline{TQ}$, $\overline{PR} \cong \overline{TQ}$, $\overline{PS} \cong \overline{QS}$, $\overline{PQ}$ *bisects* $\overline{RT}$. *Use this information and the figure in the next column for Exercises 47 and 48.*

47. Use the parallel lines shown to prove congruent angles.

48. Prove: $\triangle PRS \cong \triangle QTS$

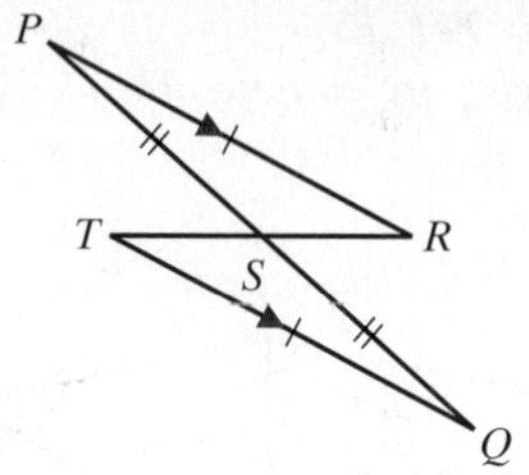

Coordinate Geometry *The vertices of* $\triangle GHJ$ *are* $G(-2, -1)$, $H(-2, 3)$, *and* $J(1, 3)$.

49. $\triangle KLM \cong \triangle GHJ$. Find KL, LM, and KM.

50. If L and M have coordinates $L(3, -3)$ and $M(6, -3)$, how many pairs of coordinates are possible for K? Find one such pair.

REVIEW AND PREVIEW

Write an equation for the line perpendicular to the given line that contains P. See Section 3.7.

51. $P(2, 7)$; $y = \frac{3}{2}x - 2$

52. $P(1, 1)$; $y = 4x + 3$

Find the distance between the points. If necessary, round to the nearest tenth. See Section 1.7.

53. $X(11, 24)$, $Y(-7, 24)$

54. $E(1, -12)$, $F(1, -2)$

What can you conclude from each diagram about possible congruent sides or angles, or possible parallel lines? See Sections 1.4 and 3.2.

55.

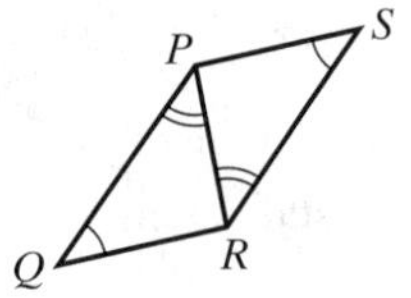

56.

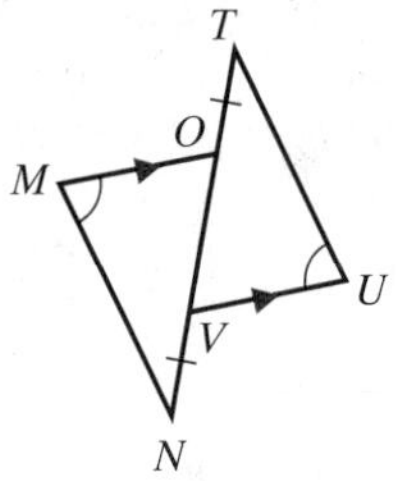

4.3 CONGRUENT TRIANGLES BY SSS AND SAS

OBJECTIVE

1 Prove Two Triangles Are Congruent Using the SSS and SAS Postulates.

In the last section, we learned that if two triangles have three pairs of congruent corresponding angles and three pairs of congruent corresponding sides, then the triangles are congruent.

If you know …

$\angle F \cong \angle J$ $\quad \overline{FG} \cong \overline{JK}$

$\angle G \cong \angle K$ $\quad \overline{GH} \cong \overline{KL}$

$\angle H \cong \angle L$ $\quad \overline{FH} \cong \overline{JL}$

…then you know $\triangle FGH \cong \triangle JKL$.

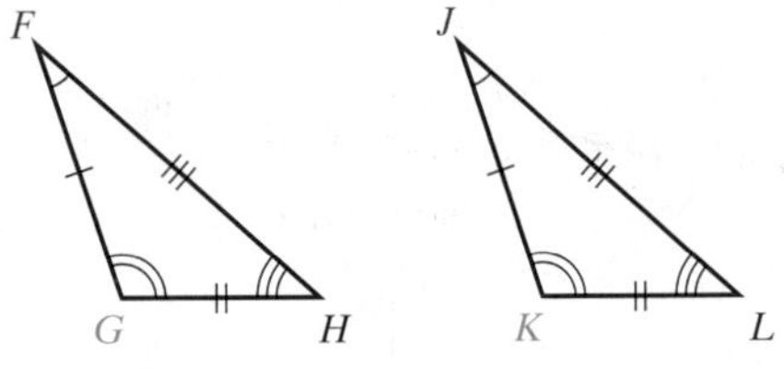

OBJECTIVE 1 ▶ Proving Triangles Are Congruent Using SSS and SAS. We can prove that two triangles are congruent without showing that *all* six corresponding parts are congruent. In this section, we introduce two postulates that can be used to prove that two triangles are congruent: the Side-Side-Side (SSS) Postulate and the Side-Angle-Side (SAS) Postulate.

Postulate 4.3-1 Side-Side-Side (SSS) Postulate

Postulate	If . . .	Then . . .
If the three sides of one triangle are congruent to the three sides of another triangle, then the two triangles are congruent.	$\overline{AB} \cong \overline{DE}, \overline{BC} \cong \overline{EF}, \overline{AC} \cong \overline{DF}$ Side - Side - Side	$\Delta ABC \cong \Delta DEF$

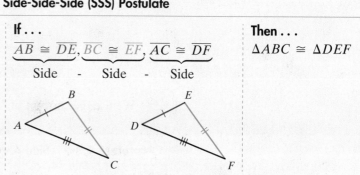

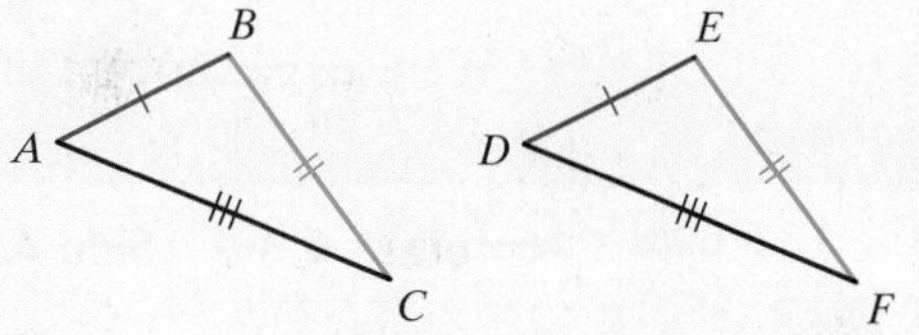

▶ **Helpful Hint**
Remember that a postulate is an accepted statement of fact, so we do not need to prove it.

In engineering, a diagonal brace (even a temporary one) is commonly used to reinforce a rectangular structure. Can you understand why? Much construction is in the form of rectangles, like a wall, a window, or a doorway. Let's see what can happen to a wooden rectangle.

A rectangle has a 90° angle in each corner.

With a little stress or weight, the best of constructions may shift.

Even with the four corners securely attached, imagine the movement possible.

A triangle whose three vertices are securely attached is a rigid shape. This brace transforms the rectangle into two triangles, and the construction is now rigid.

We explore this idea further in the following example.

EXAMPLE 1 **Proving Triangles are Congruent**

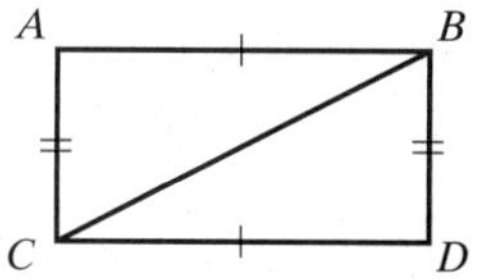

Given: $\overline{AB} \cong \overline{CD}, \overline{AC} \cong \overline{BD}$

Prove: $\Delta ABC \cong \Delta DCB$

Solution

Statements	Reasons
1. $\overline{AB} \cong \overline{CD}$, $\overline{AC} \cong \overline{BD}$	**1.** Given
2. $\overline{CB} \cong \overline{CB}$	**2.** Reflexive Property of $\cong$
3. $\Delta ABC \cong \Delta DCB$	**3.** SSS Postulate (Steps 1 and 2)

□

PRACTICE **1**

Given: $\overline{MQ} \cong \overline{MR}, \overline{QP} \cong \overline{RP}$

Prove: $\Delta MQP \cong \Delta MRP$

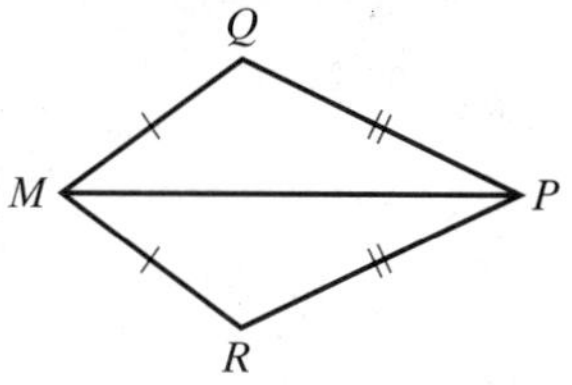

Let's review what we mean by the word **included**.

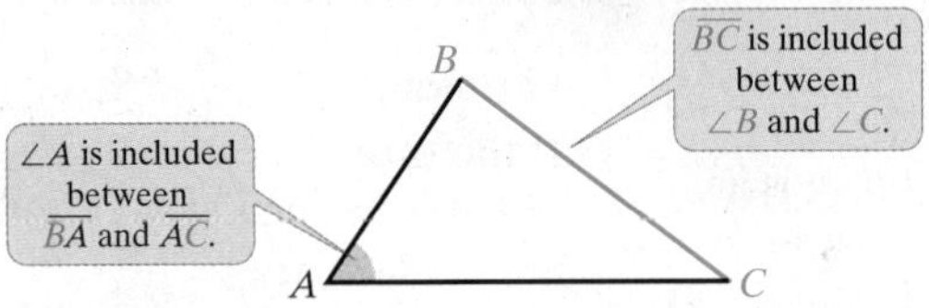

We use the concept of "included" in the next postulate.

Postulate 4.3-2 Side-Angle-Side (SAS) Postulate

Postulate	If . . .	Then . . .
If two sides and the included angle of one triangle are congruent to two sides and the included angle of another triangle, then the two triangles are congruent.	$\overline{AB} \cong \overline{DE}$, $\angle A \cong \angle D$, $\overline{AC} \cong \overline{DF}$ Side - Angle - Side 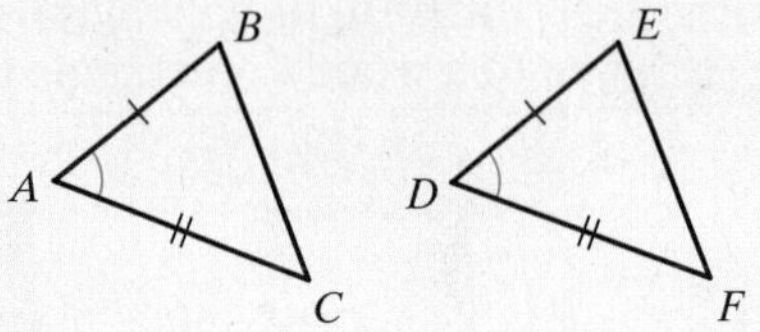	$\Delta ABC \cong \Delta DEF$

Helpful Hint

Notice that the SAS Postulate does not refer to any two sides and any one angle. Rather, SAS refers to two sides and the only *included angle* of the two sides.

EXAMPLE 2 Proving Triangles are Congruent

Given: The figure with congruent segments shown by equal number of tick marks

Prove: $\Delta ABE \cong \Delta CBD$

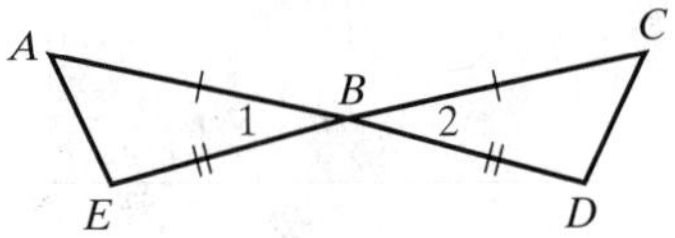

Solution

Statements	Reasons
1. $\overline{AB} \cong \overline{BC}$, $\overline{EB} \cong \overline{BD}$	**1.** Given
2. $\angle 1 \cong \angle 2$	**2.** Vertical Angles Theorem
3. $\Delta ABE \cong \Delta CBD$	**3.** SAS Postulate (Steps 1 and 2)

□

PRACTICE 2

Given: $\overline{AC}$ bisects $\angle BAF$, $\overline{AB} \cong \overline{AF}$

Prove: $\Delta ABC \cong \Delta AFC$

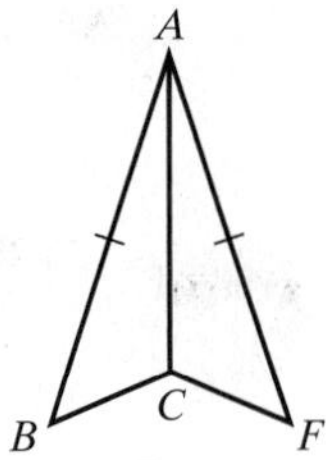

EXAMPLE 3 Identifying Congruent Triangles

Can we use SSS or SAS to prove the triangles on the next page are congruent? If there is not enough information to prove by SSS or SAS, then write "not enough information" and explain why.

a.

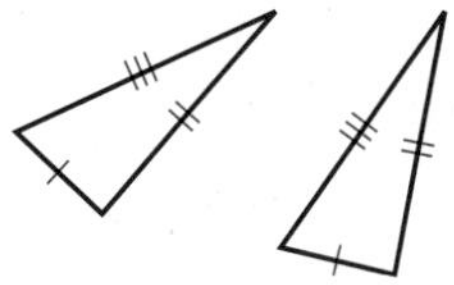

b.

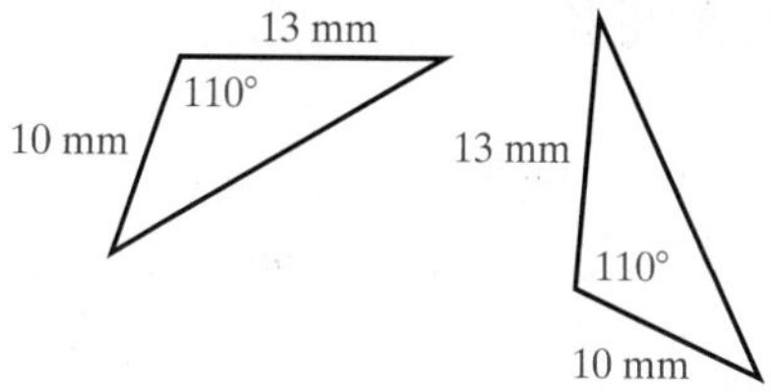

c.

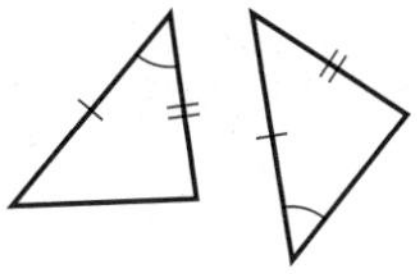

d.

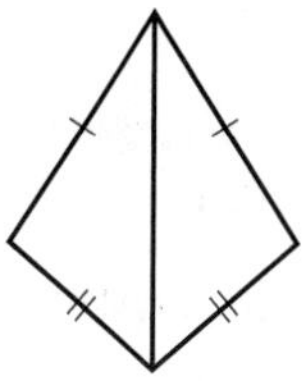

Solution

a. We can use SSS to prove the triangles are congruent. Here, we have all three pairs of corresponding sides congruent.

b. We can use SAS to prove the triangles are congruent. Here, we have two pairs of corresponding sides congruent, and their included angles congruent.

c. Not enough information. Two pairs of corresponding sides are congruent, but notice that we do not have that the included angles are congruent. (In the second triangle, the arc is not on the included angle.)

d. We can use SSS. Three pairs of corresponding sides are congruent because the triangles share a common side. □

PRACTICE 3 Use the same directions as for Example 3.

a.

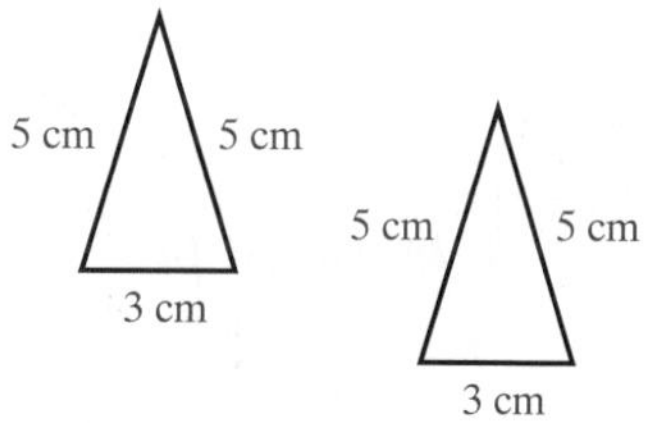

b.

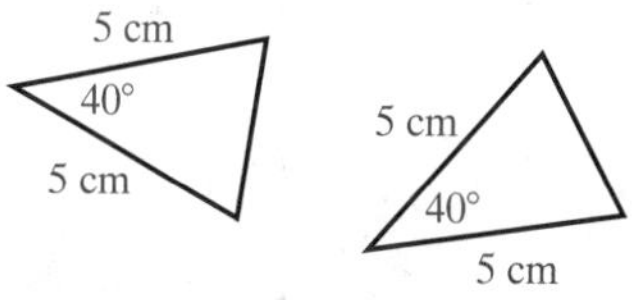

c.

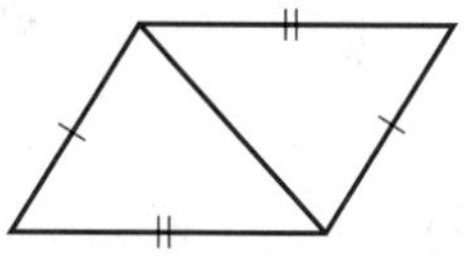

d.

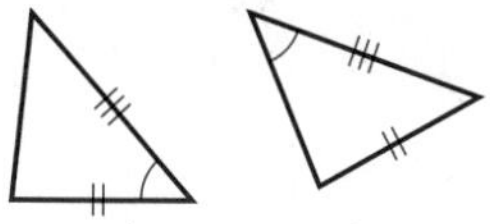

VOCABULARY & READINESS CHECK

Fill in the Blank *Name the postulate that proves each pair of triangles is congruent. List the vertices of the congruent triangle in correct corresponding order.*

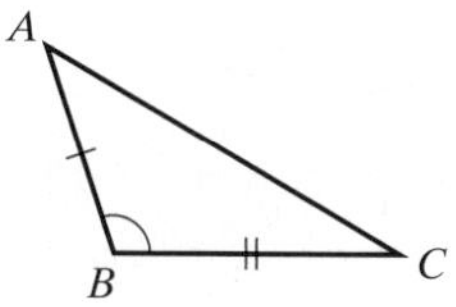

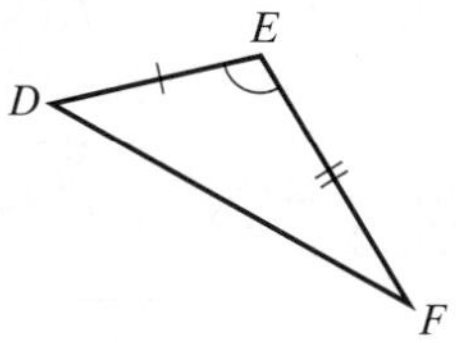

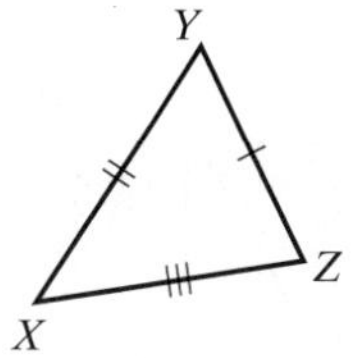

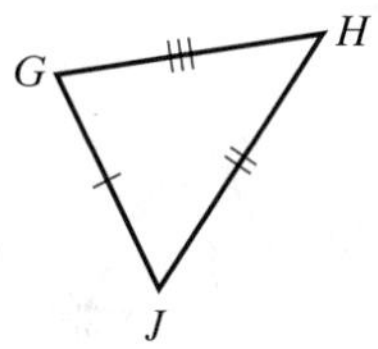

1. $\Delta ABC \cong \Delta$_____ by the _____________ Postulate.

2. $\Delta XYZ \cong \Delta$_____ by the _____________ Postulate.

4.3 EXERCISE SET MyMathLab®

Mixed Practice *See Examples 1–3.*

Draw a triangle of any kind and label it ΔRST. Use your drawing to name the angle included between the given sides.

1. $\overline{RT}$ and $\overline{ST}$ **2.** $\overline{ST}$ and $\overline{RS}$

Draw another triangle and label it ΔHAT. Use your drawing to name the sides between which the given angle is included.

3. $\angle A$ **4.** $\angle H$

Name the postulate you would use to prove the triangles congruent.

5.

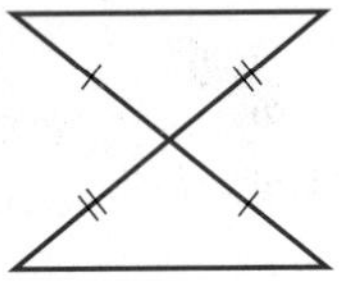

6.

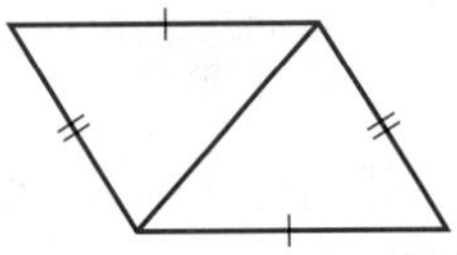

What other information, if any, do you need to prove the two triangles congruent by SAS? Explain.

7.

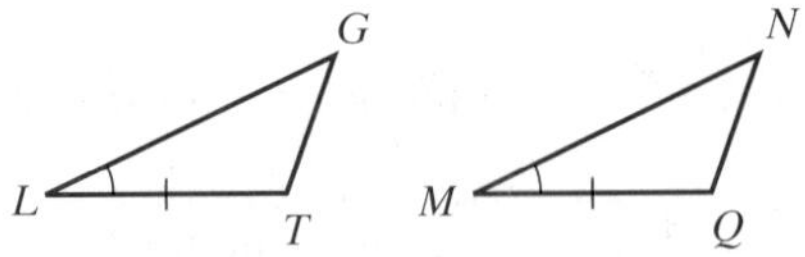

8.

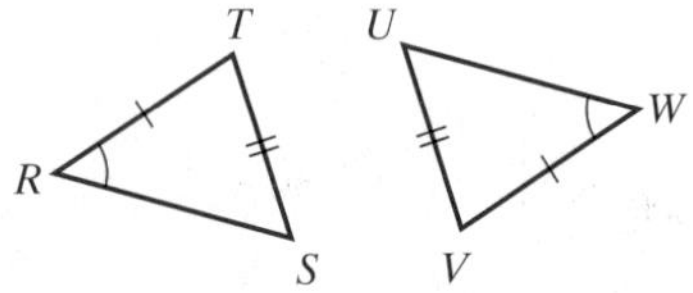

Would you use SSS or SAS to prove the triangles congruent? If there is not enough information to prove the triangles congruent by SSS or SAS, write "not enough information." Explain your answer. See Example 3.

9.

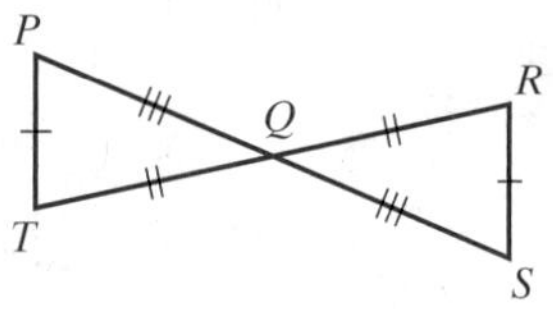

10.

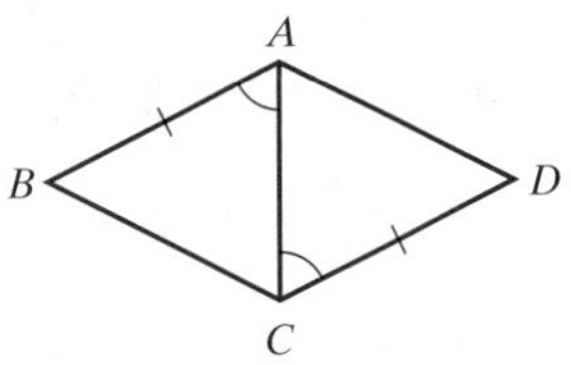

11.

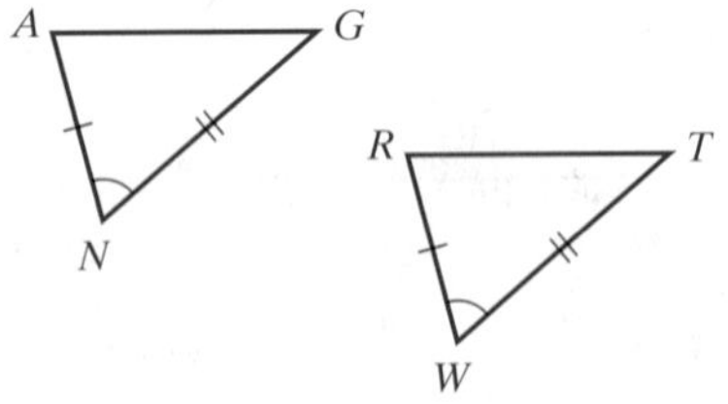

12.

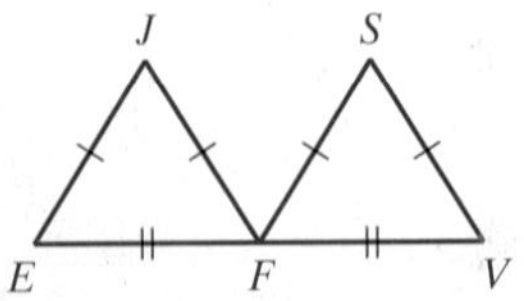

13.

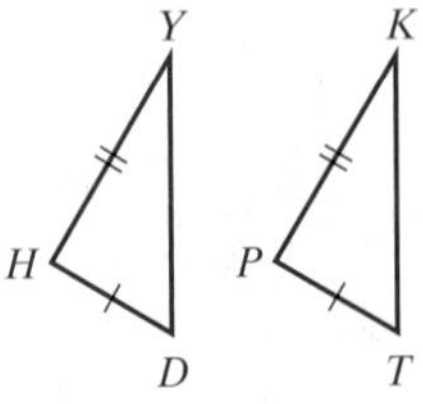

14.

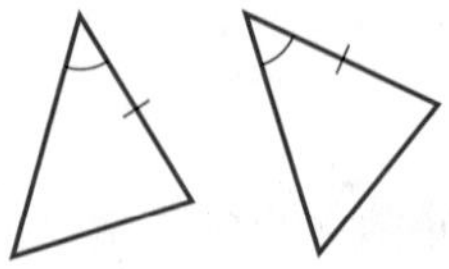

15.

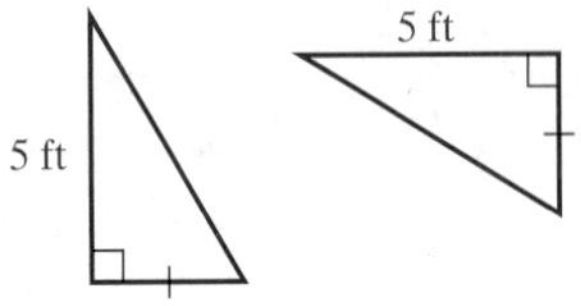

16.

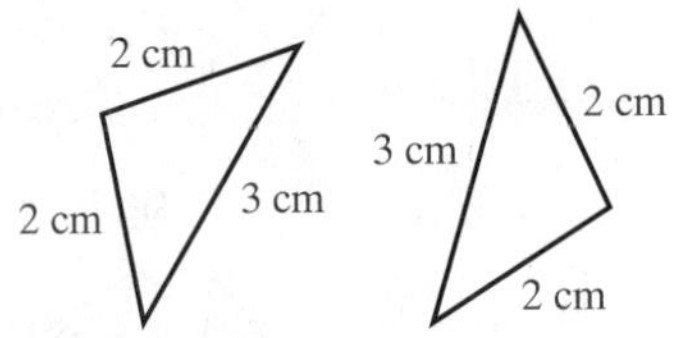

Fill in the Blank *For Exercises 17–22, fill in the blanks to complete the proof. See Example 2.*

Given: $\overline{AC} \perp \overline{DC}$,
$\overline{AC} \perp \overline{AB}$,
$\overline{AB} \cong \overline{DC}$

Prove: $\Delta DCA \cong \Delta BAC$

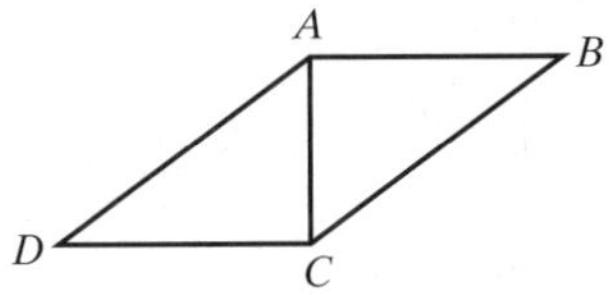

Statements	Reasons
1. $\overline{AC} \perp \overline{DC}$, $\overline{AC} \perp \overline{AB}$	1. ? (Exercise 17)
2. $\angle DCA$ and $\angle CAB$ are right angles.	2. ? (Exercise 18)
3. ? (Exercise 19)	3. Right angles are congruent.
4. ? (Exercise 20)	4. Reflexive Property
5. $\overline{AB} \cong \overline{DC}$	5. ? (Exercise 21)
6. $\Delta DCA \cong \Delta BAC$	6. ? (Exercise 22)

MIXED PRACTICE

Proof *For Exercises 23–30, complete each proof. See Examples 1–3.*

23. Given: $\overline{IE} \cong \overline{GH}$, $\overline{EF} \cong \overline{HF}$, F is the midpoint of $\overline{GI}$

Prove: $\Delta EFI \cong \Delta HFG$

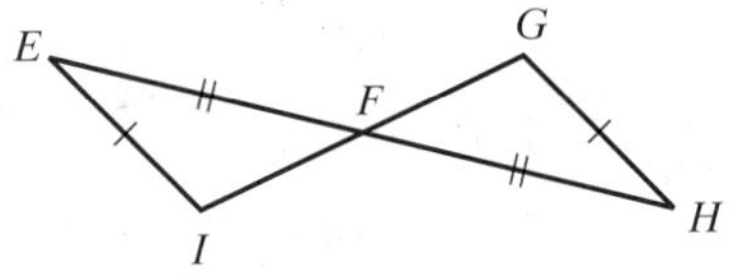

24. Given: $\overline{WZ} \cong \overline{ZS} \cong \overline{SD} \cong \overline{DW}$

Prove: $\Delta WZD \cong \Delta SDZ$

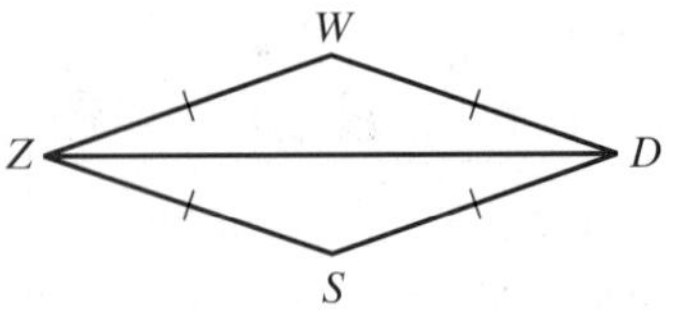

25. Given: $\overline{GK}$ bisects $\angle JGM$, $\overline{GJ} \cong \overline{GM}$

Prove: $\Delta GJK \cong \Delta GMK$

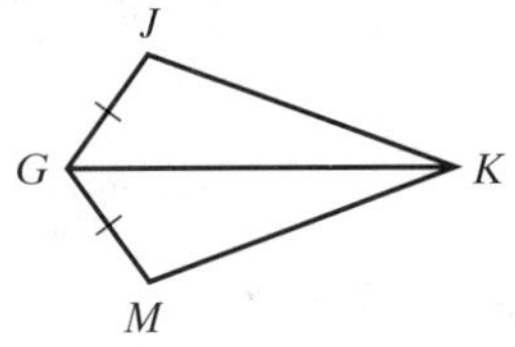

26. Given: $\overline{AE}$ and $\overline{BD}$ bisect each other
Prove: $\Delta ACB \cong \Delta ECD$

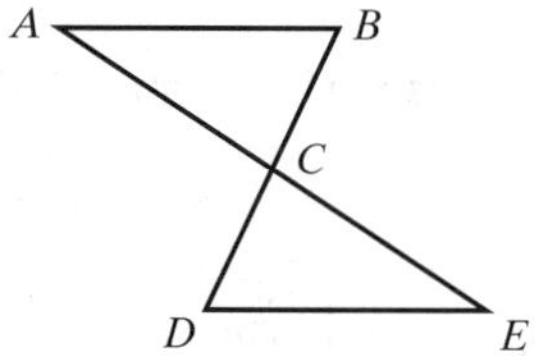

27. Given: $\overline{FG} \parallel \overline{KL}, \overline{FG} \cong \overline{KL}$
Prove: $\Delta FGK \cong \Delta KLF$

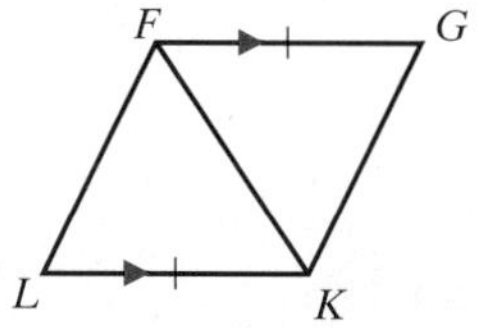

28. Given: $\overline{AB} \perp \overline{CM}$, $\overline{AB} \perp \overline{DB}$, $\overline{CM} \cong \overline{DB}$, M is the midpoint of $\overline{AB}$
Prove: $\Delta AMC \cong \Delta MBD$

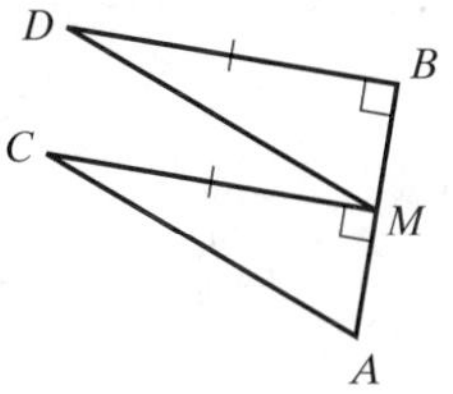

29. Given: $\overline{BC} \cong \overline{DA}, \angle CBD \cong \angle ADB$
Prove: $\Delta BCD \cong \Delta DAB$

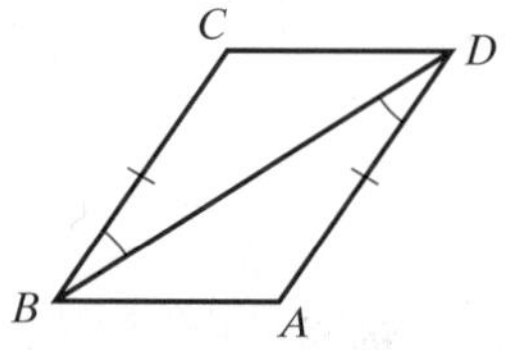

30. Given: X is the midpoint of $\overline{AG}$ and $\overline{NR}$
Prove: $\Delta ANX \cong \Delta GRX$

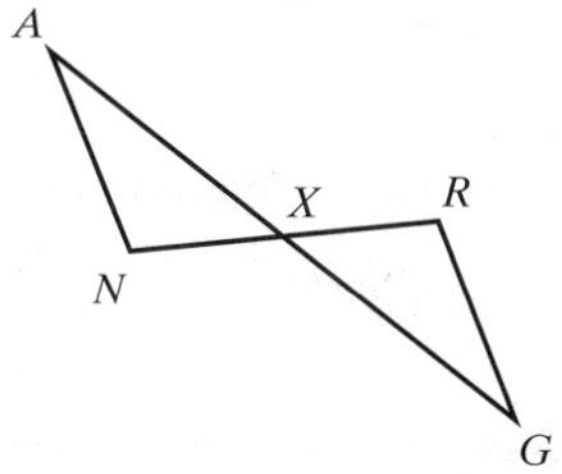

Coordinate Geometry *Use the Distance Formula to determine whether ΔABC and ΔDEF are congruent. Justify your answer.*

31. $A(1, 6), B(1, 1), C(5, 1), D(7, -5), E(2, -5), F(2, -1)$

32. $A(3, 1), B(3, 3), C(6, 1), D(-1, -3), E(-1, -1), F(-4, -3)$

33. $A(1, 4), B(5, 5), C(2, 2); D(-5, 1), E(-1, 0), F(-4, 3)$

34. $A(3, 8), B(8, 12), C(10, 5); D(3, -1), E(7, -7), F(12, -2)$

CONCEPT EXTENSIONS

35. How are the SSS Postulate and the SAS Postulate alike? How are they different?

36. Find the Error Your friend thinks that the triangles shown below are congruent by SAS. Is your friend correct? Explain.

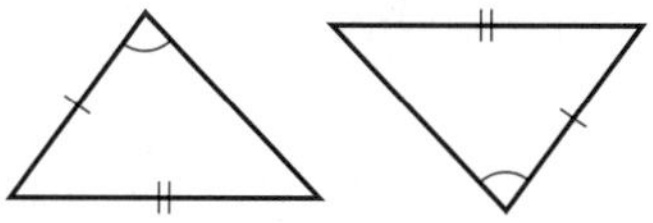

Construction *Use a straight edge to draw any triangle JKL. Then construct $\Delta MNP \cong \Delta JKL$ using the given postulate.*

37. SSS

38. SAS

39. Suppose $\overline{GH} \cong \overline{JK}, \overline{HI} \cong \overline{KL}$, and $\angle I \cong \angle L$. Is ΔGHI congruent to ΔJKL? Explain.

40. Sierpinski's Triangle Sierpinski's triangle is a famous geometric pattern. To draw Sierpinski's triangle, start with a single triangle and connect the midpoints of the sides to draw a smaller triangle. If you repeat this pattern over and over, you will form a figure like the one shown. This particular figure started with an isosceles triangle. Are the triangles outlined in red congruent? Explain.

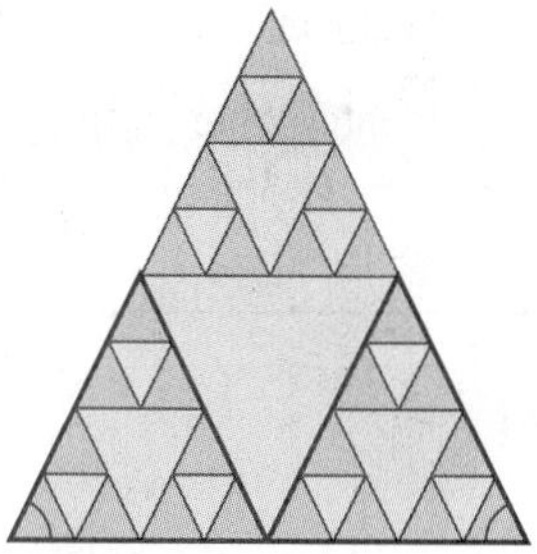

41. List three real-life uses of congruent triangles. For each real-life use, describe why you think congruence is necessary.

42. Four sides of polygon $ABCD$ are congruent, respectively, to the four sides of polygon $EFGH$. Are $ABCD$ and $EFGH$ congruent? Is a quadrilateral a rigid figure? If not, what could you add to make it a rigid figure? Explain.

Proof *For Exercises 43 and 44, write a two-column proof.*

43. Given: $\overline{HK} \cong \overline{LG}, \overline{HF} \cong \overline{LJ}, \overline{FG} \cong \overline{JK}$
Prove: $\Delta FGH \cong \Delta JKL$

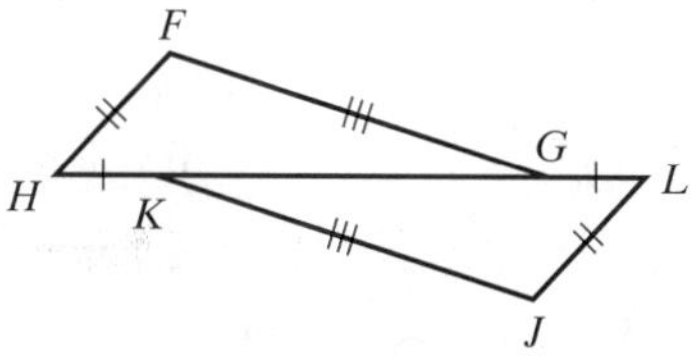

44. Given: $\angle N \cong \angle L, \overline{MN} \cong \overline{OL}, \overline{NO} \cong \overline{LM}$
Prove: $\overline{MN} \parallel \overline{OL}$

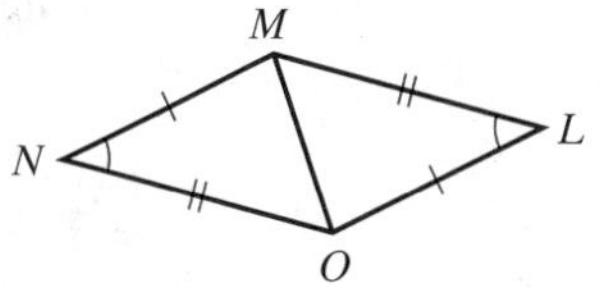

REVIEW AND PREVIEW

Given congruent figures, $ABCD \cong EFGH$, name the angle or side that corresponds to each part. See Section 4.1.

45. $\angle A$

46. $\overline{EF}$

47. $\overline{BC}$

48. $\angle G$

Write the converse of each statement. Determine whether the statement and its converse are true or false. See Section 2.2.

49. If $x = 3$, then $2x = 6$.

50. If $x = 3$, then $x^2 = 9$.

Solve. See Sections 4.1 and 4.3.

51. In ΔJHK, name the side that is included between $\angle J$ and $\angle H$.

52. In ΔNLM, name the angle that is included between $\overline{NM}$ and $\overline{LN}$.

4.4 CONGRUENT TRIANGLES BY ASA AND AAS

OBJECTIVES

1. Prove Two Triangles Are Congruent Using the ASA Postulate and the AAS Theorem.
2. Identify When to Use SSS, SAS, ASA, or AAS to Prove Triangles Congruent.

OBJECTIVE 1 ▸ Proving Two Triangles Are Congruent Using the ASA Postulate or the AAS Theorem. We already know that triangles are congruent if

- three pairs of sides are congruent (SSS) or if
- two pairs of sides and their included angles are congruent (SAS).

We can also prove triangles are congruent using other groupings of angles and sides.

Next, we will introduce a postulate that will allow us to prove triangles are congruent by using one pair of corresponding sides and two pairs of corresponding angles.

Postulate 4.4-1 Angle-Side-Angle (ASA) Postulate

Postulate	If . . .	Then . . .
If two angles and the included side of one triangle are congruent to two angles and the included side of another triangle, then the two triangles are congruent.	$\underbrace{\angle A \cong \angle D}_{\text{Angle}}, \underbrace{\overline{AC} \cong \overline{DF}}_{\text{Side}}, \underbrace{\angle C \cong \angle F}_{\text{Angle}}$ 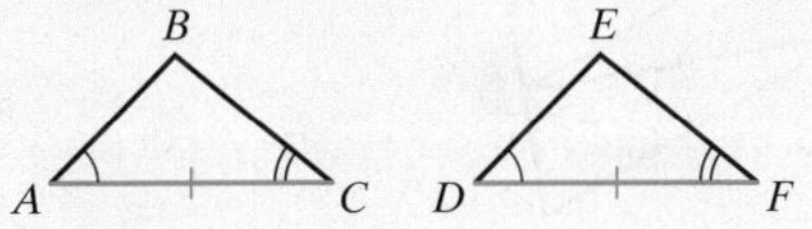	$\triangle ABC \cong \triangle DEF$

EXAMPLE 1 Identifying ASA

Multiple Choice Choose two triangles that are congruent by the ASA Postulate. Explain why.

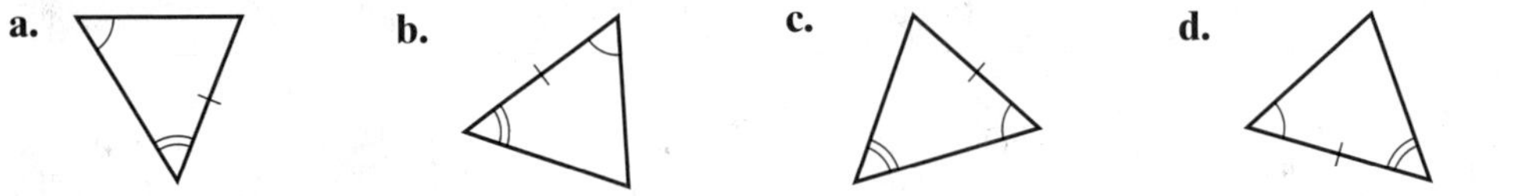

Solution Choices **b** and **d** are congruent by ASA because for these two triangles, the sides marked congruent are the included sides of the two congruent angles. □

PRACTICE 1 Multiple Choice

Choose two triangles that are congruent by the ASA Postulate. Explain why.

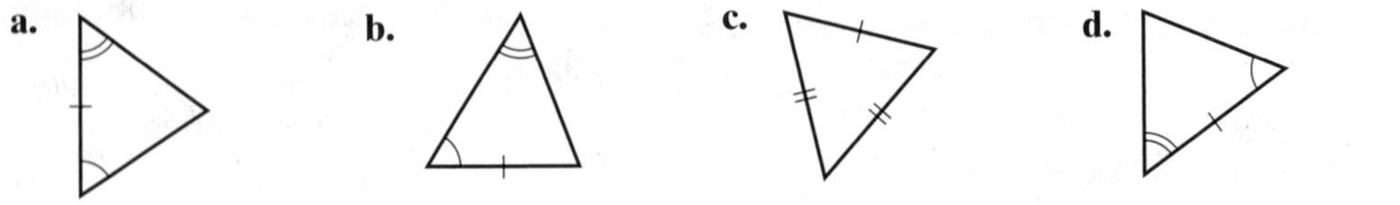

As usual, we can use our definitions, postulates, and other theorems to "grow" our geometry mathematical system. Here, we introduce a theorem that can be used to prove triangles are congruent.

> **Helpful Hint**
> Notice this is a theorem that we prove and not a postulate that we do not prove.

Theorem 4.4-2 Angle-Angle-Side (AAS) Theorem

Theorem	If . . .	Then . . .
If two angles and a nonincluded side of one triangle are congruent to two angles and the corresponding nonincluded side of another triangle, then the triangles are congruent.	$\underbrace{\angle A \cong \angle D}_{\text{Angle}}, \underbrace{\angle B \cong \angle E}_{\text{Angle}}, \underbrace{\overline{AC} \cong \overline{DF}}_{\text{Side}}$ 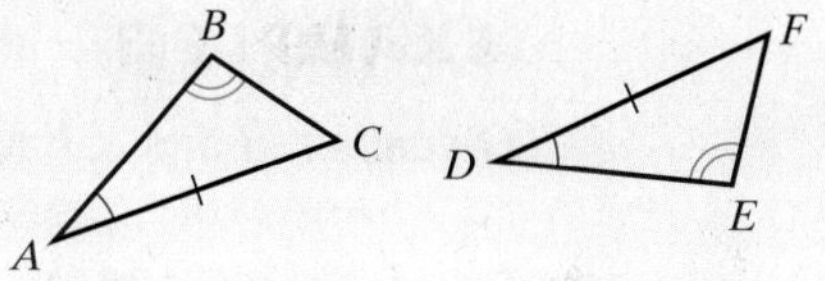	$\Delta ABC \cong \Delta DEF$

Proof of Theorem 4.4-2: Angle-Angle-Side Theorem

Given: $\angle A \cong \angle D, \angle B \cong \angle E, \overline{AC} \cong \overline{DF}$

Prove: $\Delta ABC \cong \Delta DEF$

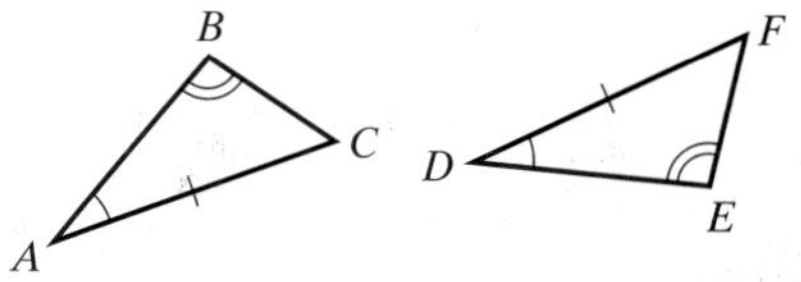

Statements	Reasons
1. $\angle A \cong \angle D, \angle B \cong \angle E$	**1.** Given
2. $\angle C \cong \angle F$	**2.** Third Angles Theorem
3. $\overline{AC} \cong \overline{DF}$	**3.** Given
4. $\Delta ABC \cong \Delta DEF$	**4.** ASA Postulate (Steps 1, 2, 3)

> **Helpful Hint**
> Think about the AAS Theorem for a moment. If two pairs of corresponding angles are congruent, then the third pair is congruent (Third Angles Theorem). After that, as long as a pair of corresponding sides is congruent, the ASA Postulate can be used.

Thus far, we have introduced three types of proofs—two-column (statement/reason), paragraph, and flow proof. Each method of organizing a proof is valid. In this text, we will usually present a proof in two-column (statement/reason) style for ease of reading the proof and following the logic.

EXAMPLE 2 **Proving Triangles are Congruent**

Given: $\overrightarrow{DB}$ bisects $\angle ABC, \angle 1 \cong \angle 2$

Prove: $\Delta DAB \cong \Delta DCB$

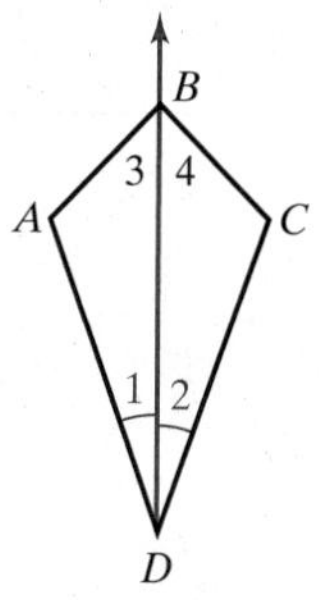

Solution

Statements	Reasons
1. $\overrightarrow{DB}$ bisects $\angle ABC$.	**1.** Given
2. $\angle 3 \cong \angle 4$	**2.** Definition of angle bisector
3. $\angle 1 \cong \angle 2$	**3.** Given
4. $\overline{DB} \cong \overline{DB}$	**4.** Reflexive Property
5. $\Delta DAB \cong \Delta DCB$	**5.** ASA Postulate (Steps 2, 3, 4)

□

PRACTICE
2 **Given:** D is the midpoint of $\overline{EB}$,
$\angle E \cong \angle B$
Prove: $\Delta EDA \cong \Delta BDC$

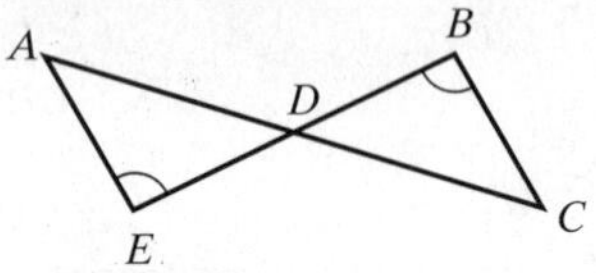

EXAMPLE 3 Proving Triangles are Congruent

Given: $\angle B$ and $\angle E$ are right angles, C is the midpoint of $\overline{AD}$

Prove: $\Delta ABC \cong \Delta DEC$

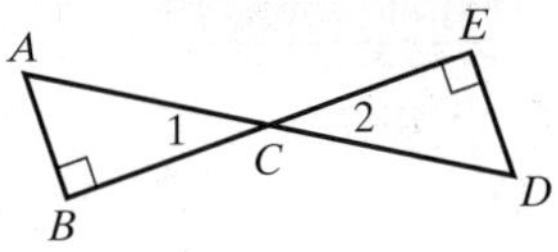

Solution

Statements	Reasons
1. $\angle B$ and $\angle E$ are right angles.	**1.** Given
2. $\angle B \cong \angle E$	**2.** Definition of right angles
3. C is the midpoint of $\overline{AD}$.	**3.** Given
4. $\overline{AC} \cong \overline{CD}$	**4.** Definition of midpoint
5. $\angle 1 \cong \angle 2$	**5.** Vertical Angles Theorem
6. $\Delta ABC \cong \Delta DEC$	**6.** AAS Theorem (Steps 2, 4, 5)

PRACTICE
3 **Given:** $\overline{DB} \perp \overline{AC}$,
$\angle A \cong \angle C$
Prove: $\Delta ADB \cong \Delta CDB$

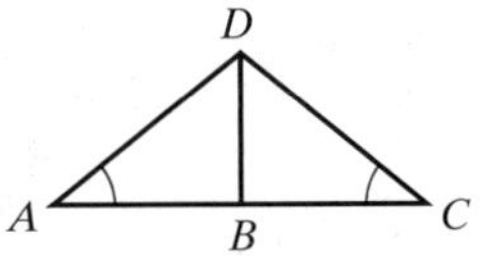

> **Helpful Hint**
> Remember: The methods we have presented for proving triangles congruent (S is for "side" and A is for "angle"): SSS, SAS, ASA, and AAS.

OBJECTIVE 2 ▶ Identifying When to Use SSS, SAS, ASA, or AAS. Let's now practice identifying all of our congruence postulates and theorem for triangles.

EXAMPLE 4 Identifying SSS, SAS, ASA, and AAS

Each pair of triangles is congruent by SSS, SAS, ASA, or AAS. Identify the postulate or theorem that immediately confirms their congruence.

a.
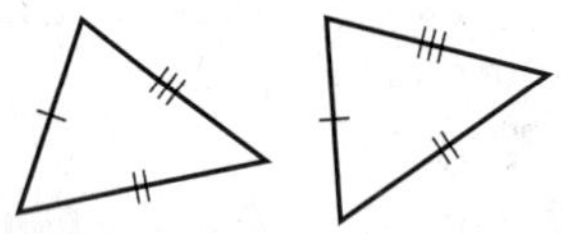

b.
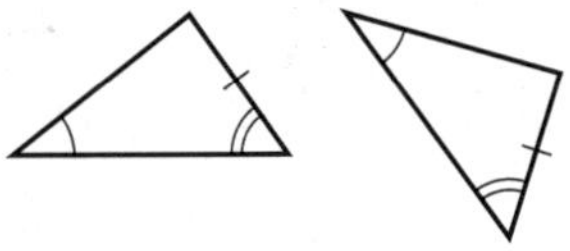

c.
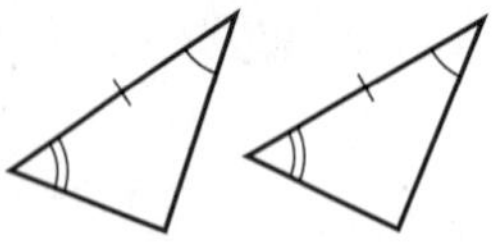

d.
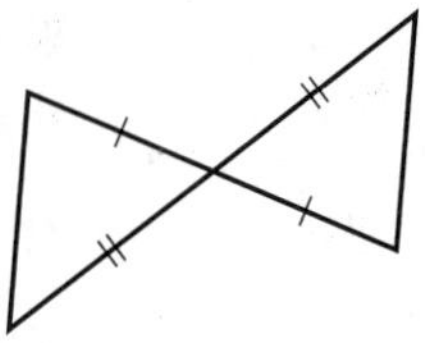

Solution **a.** SSS **b.** AAS **c.** ASA **d.** SAS

PRACTICE
4 Use SSS, SAS, ASA, or AAS to identify the postulate or theorem that immediately confirms the congruence for each pair of triangles.

a.
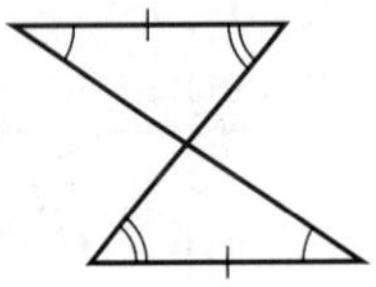

b.
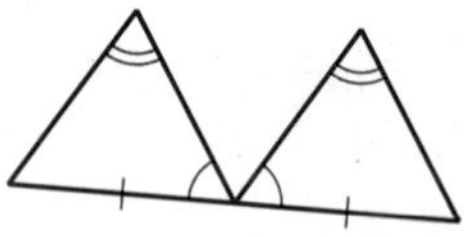

c.

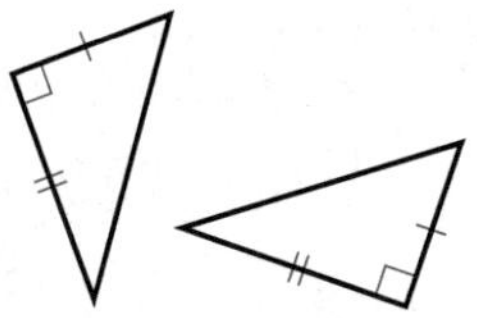

d.

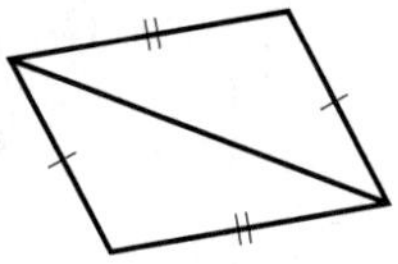

> **Helpful Hint**
> Important: Note that AAA is NOT a postulate nor a theorem for congruence. We will study AAA in a later chapter, but as it turns out, AAA guarantees the same shape but *not* the same size.

VOCABULARY & READINESS CHECK

Fill in the Blank *Name the postulate or theorem that proves each pair of triangles is congruent. List the vertices of the congruent triangles in correct corresponding order.*

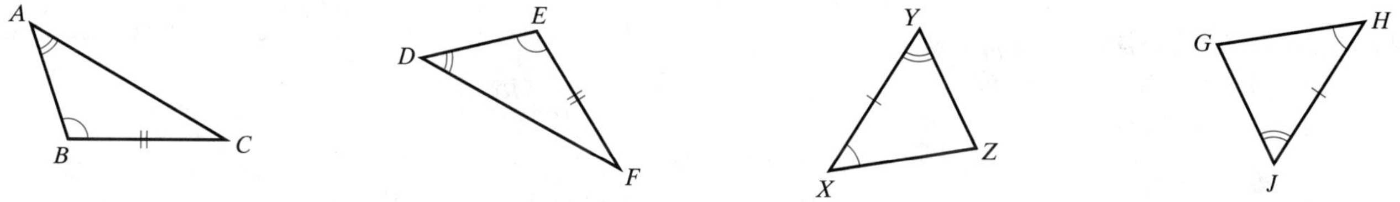

1. $\Delta ABC \cong \Delta$______ by the ______________ Postulate/Theorem.
2. $\Delta XYZ \cong \Delta$______ by the ______________ Postulate/Theorem.

4.4 EXERCISE SET MyMathLab®

MIXED PRACTICE

Which postulate or theorem immediately proves $\Delta ABC \cong \Delta DEF$? See Examples 1–3.

1.

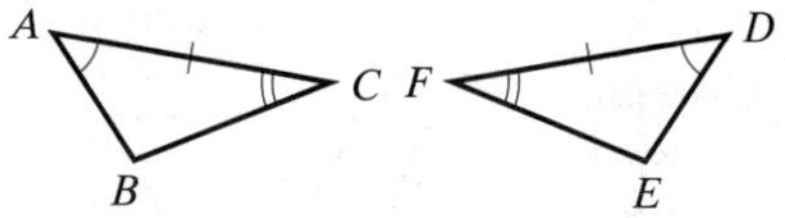

2.

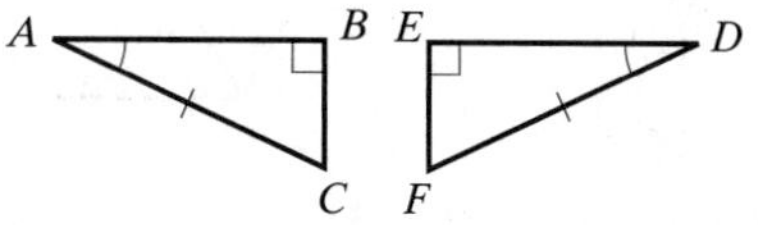

Name two triangles that are congruent by ASA. See Example 1.

3.

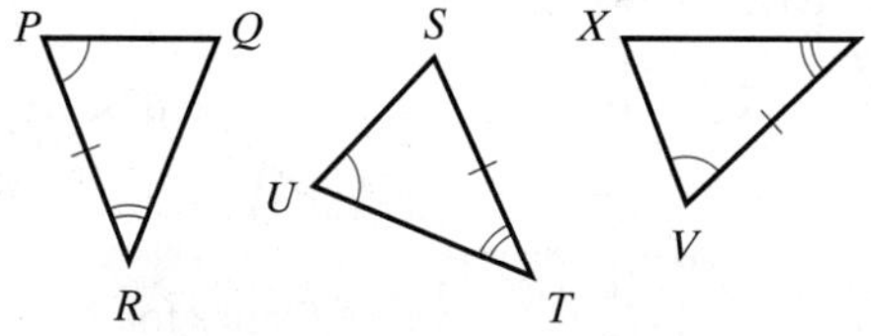

4.

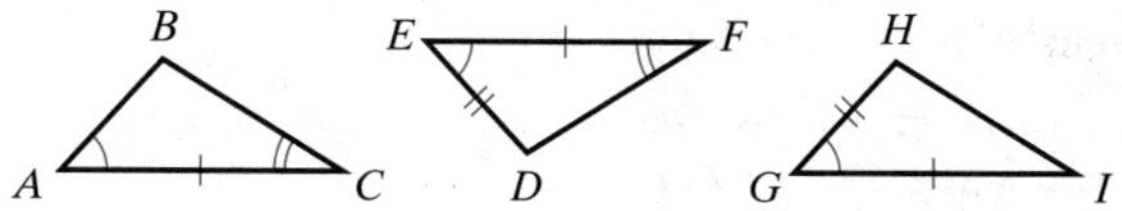

Determine whether the triangles are congruent. If so, name the postulate or theorem that justifies your answer. If not, explain. See Example 4.

5.

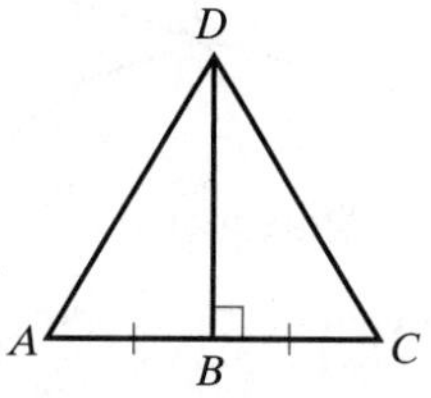

6.

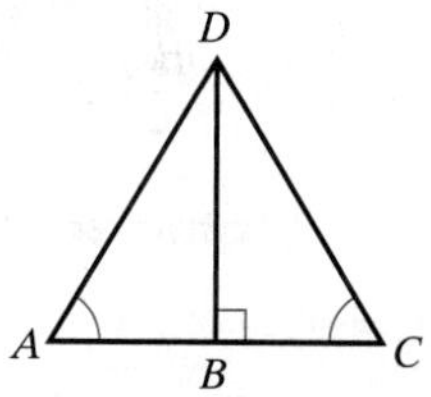

7.

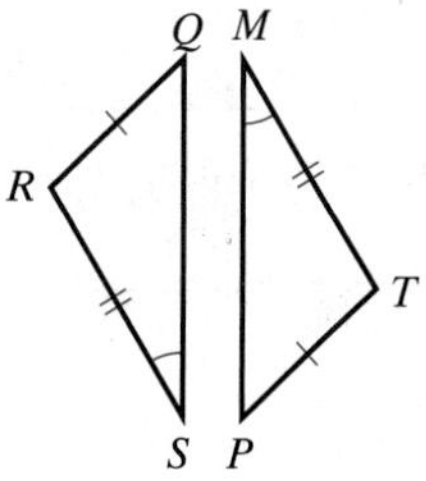

8.

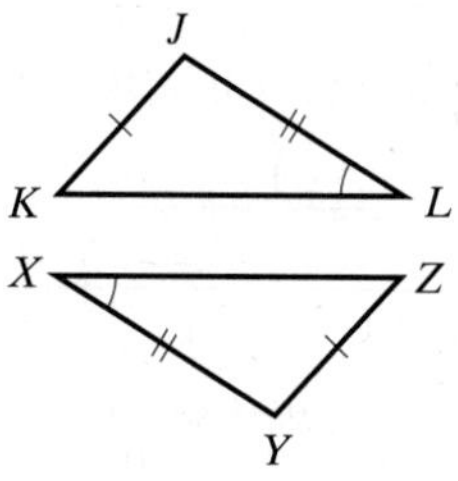

9.

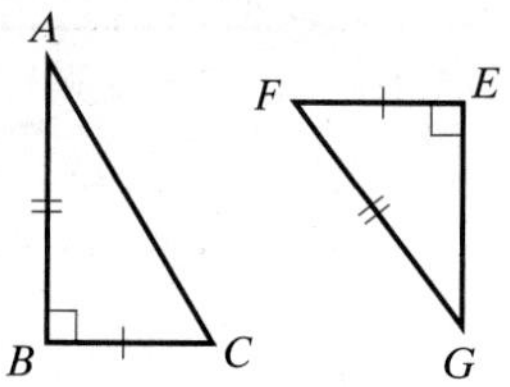

10.

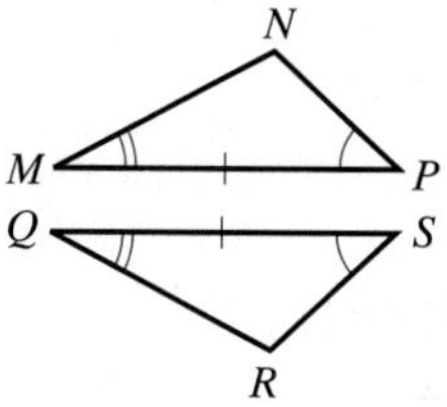

11.

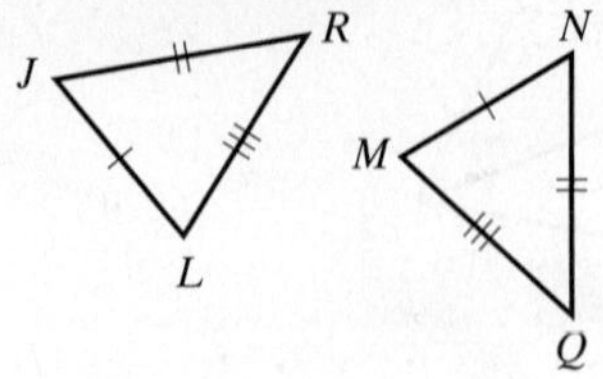

12.

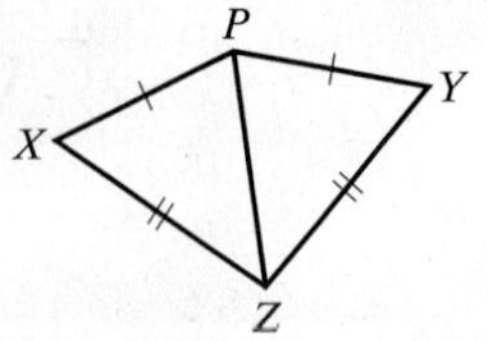

13.

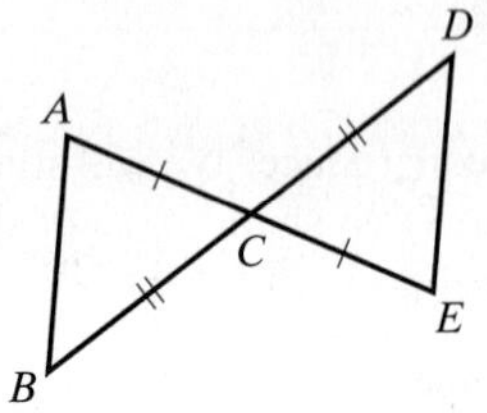

14.

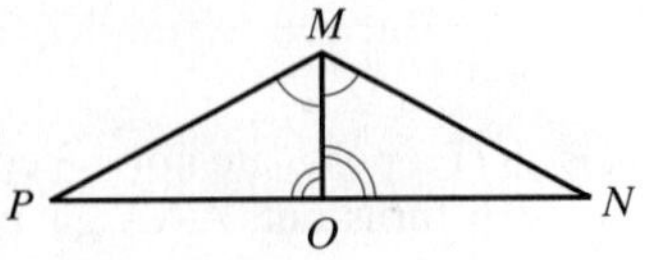

15.

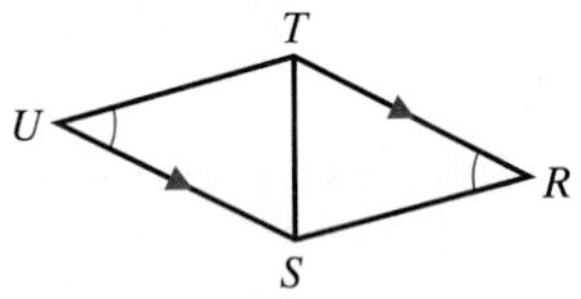

16.

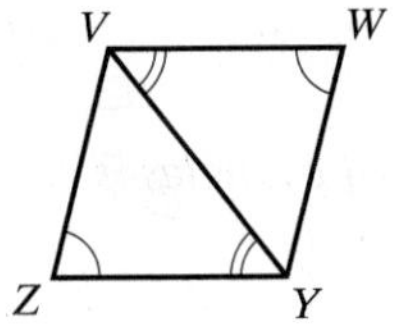

Proof *Complete the two-column proofs in Exercises 17 and 18 by filling in the blanks. See Examples 2 and 3.*

17. Given: $\angle LKM \cong \angle JKM$, $\angle LMK \cong \angle JMK$

Prove: $\Delta LKM \cong \Delta JKM$

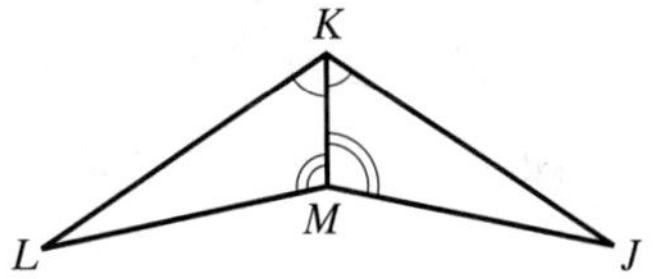

Statements	Reasons
1. $\angle LKM \cong \angle JKM$, $\angle LMK \cong \angle JMK$	1. Given
2. $\overline{KM} \cong \overline{KM}$	2. ? (a.)
3. $\Delta LKM \cong \Delta JKM$	3. ? (b.)

18. Given: $\angle N \cong \angle S$, line ℓ bisects $\overline{TR}$ at Q

Prove: $\Delta NQT \cong \Delta SQR$

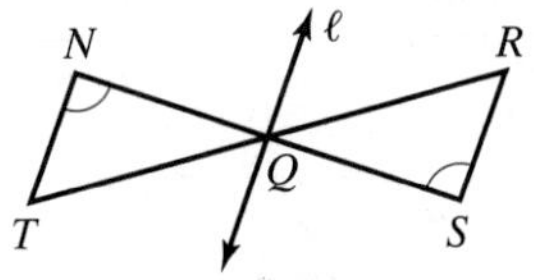

Statements	Reasons
1. $\angle N \cong \angle S$	1. Given
2. $\angle NQT \cong \angle SQR$	2. ? (a.)
3. Line ℓ bisects $\overline{TR}$ at Q.	3. ? (b.)
4. ? (c.)	4. Definition of bisector
5. $\Delta NQT \cong \Delta SQR$	5. ? (d.)

Proof *Complete each proof. See Example 4.*

19. Given: $\angle BAC \cong \angle DAC$, $\overline{AC} \perp \overline{BD}$

Prove: $\Delta ABC \cong \Delta ADC$

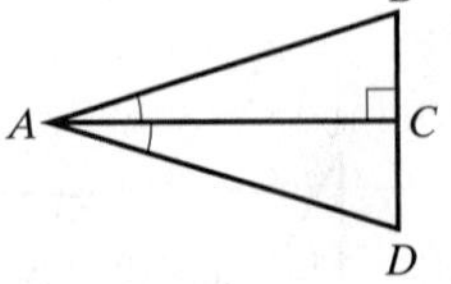

20. Given: $\overline{QR} \cong \overline{TS}$, $\overline{RQ} \parallel \overline{TS}$

Prove: $\Delta QRT \cong \Delta TSQ$

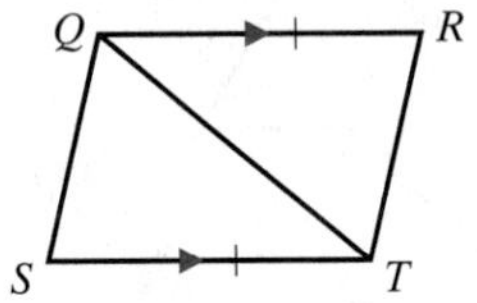

21. Given: $\angle V \cong \angle Y$, $\overline{WZ}$ bisects $\angle VWY$

Prove: $\Delta VWZ \cong \Delta YWZ$

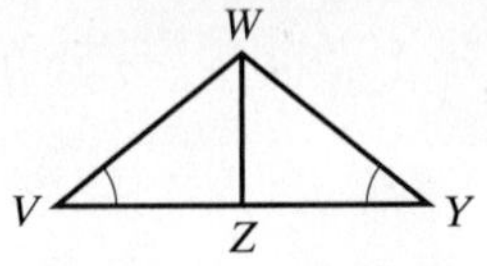

22. Given: $\overline{PQ} \perp \overline{QS}$, $\overline{RS} \perp \overline{SQ}$, T is the midpoint of $\overline{PR}$

Prove: $\Delta PQT \cong \Delta RST$

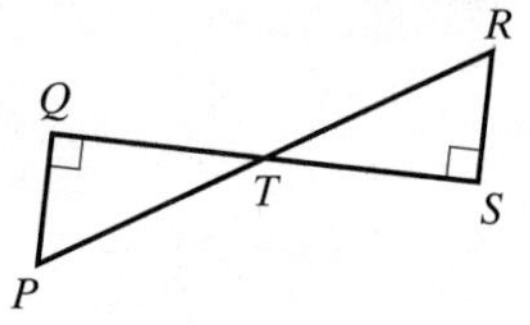

23. Given: $\angle N \cong \angle P$, $\overline{MO} \cong \overline{QO}$

Prove: $\Delta MON \cong \Delta QOP$

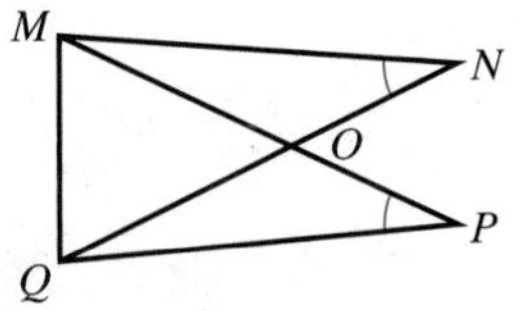

24. Given: $\angle FJG \cong \angle HGJ$, $\overline{FG} \parallel \overline{JH}$

Prove: $\Delta FGJ \cong \Delta HJG$

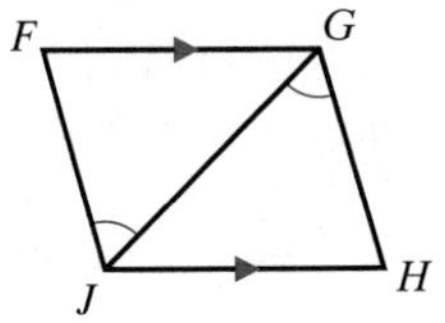

25. Given: $\overline{AE} \parallel \overline{BD}$, $\overline{AE} \cong \overline{BD}$, $\angle E \cong \angle D$

Prove: $\Delta AEB \cong \Delta BDC$

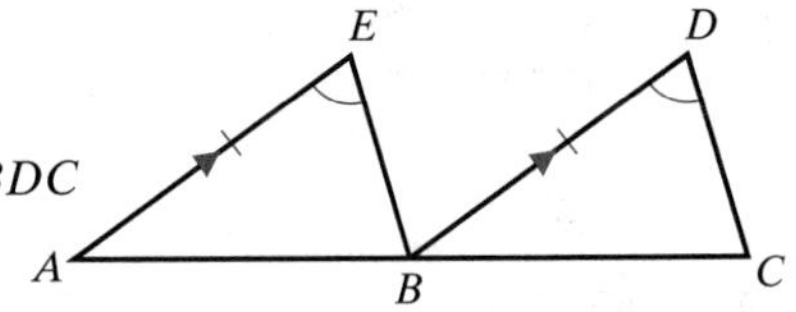

26. Given: $\angle 1 \cong \angle 2$, $\overline{DH}$ bisects $\angle BDF$

Prove: $\Delta BDH \cong \Delta FDH$

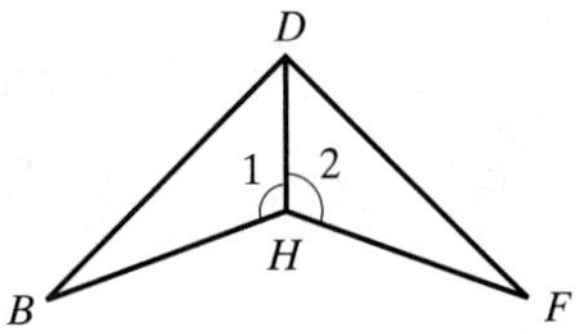

27. Given: $\overline{AB} \parallel \overline{DC}$, $\overline{AD} \parallel \overline{BC}$

Prove: $\Delta ABC \cong \Delta CDA$

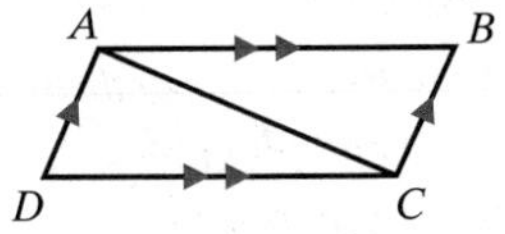

28. Given: ΔEFH is equilateral, ΔFGH is equilateral

Prove: $\Delta EFH \cong \Delta FGH$

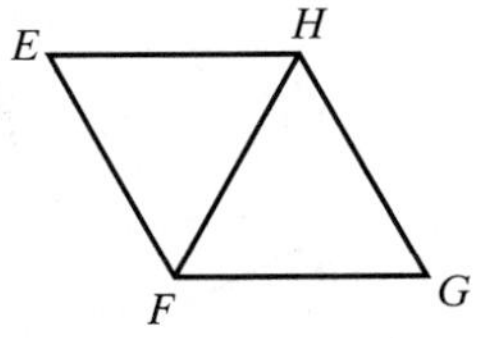

29. Given: $\overline{AB} \cong \overline{BC}$, $\angle 1 \cong \angle 2$

Prove: $\Delta ABE \cong \Delta CBD$

30. Given: $\overline{EB} \cong \overline{BD}$, $\angle 1 \cong \angle 2$

Prove: $\Delta ABE \cong \Delta CBD$

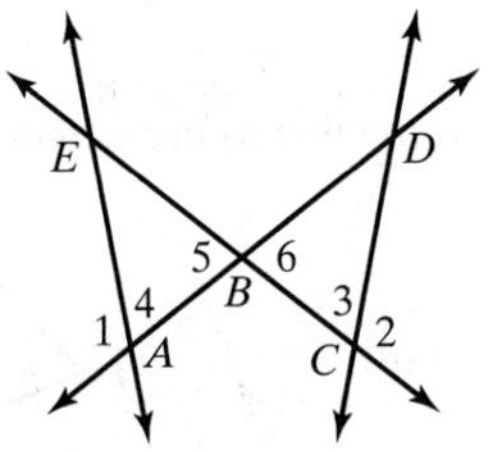

Figure for 29 and 30

31. Given: ΔACD is equilateral, $\angle 1 \cong \angle 2$, $\overline{BC} \cong \overline{DE}$

Prove: $\Delta ABC \cong \Delta AED$

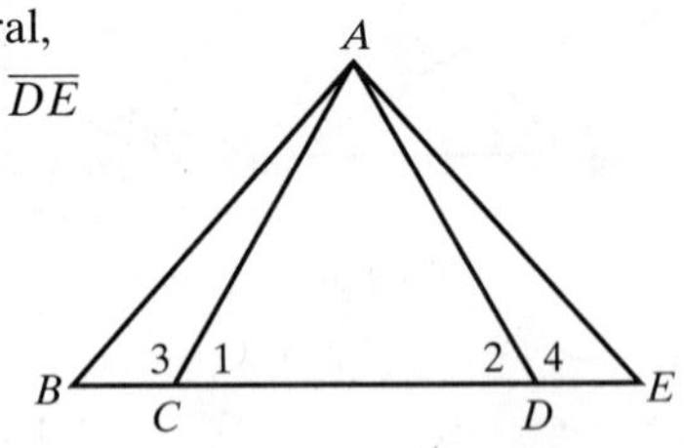

32. Given: ΔACD is isosceles with base $\overline{CD}$,
$\angle 1 \cong \angle 2$,
$\overline{BD} \cong \overline{CE}$
Prove: $\Delta ABC \cong \Delta AED$

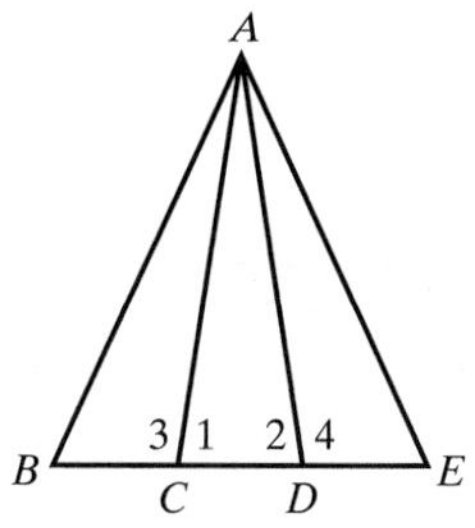

33. Given: $AB = CD, AC = BD$
Prove: $\Delta ABC \cong \Delta DCB$

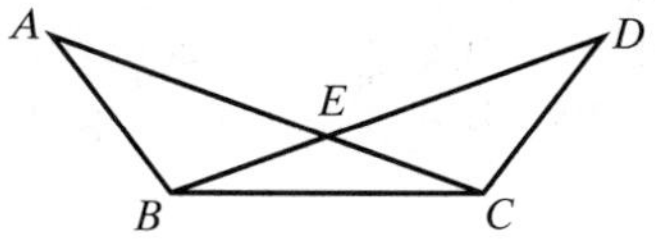

34. Given: $AD = BD, AE = BC$
Prove: $\Delta ACD \cong \Delta BED$

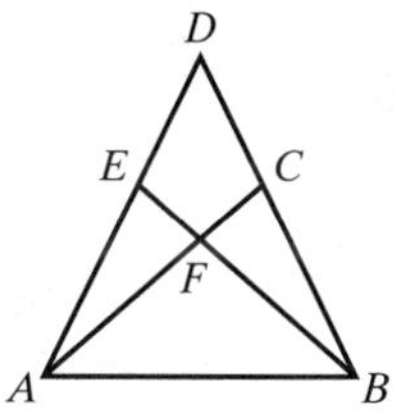

MIXED PRACTICE

$\Delta ABC \cong \Delta DEF$. Use this information for Exercises 35–40. See Sections 4.3 and 4.4.

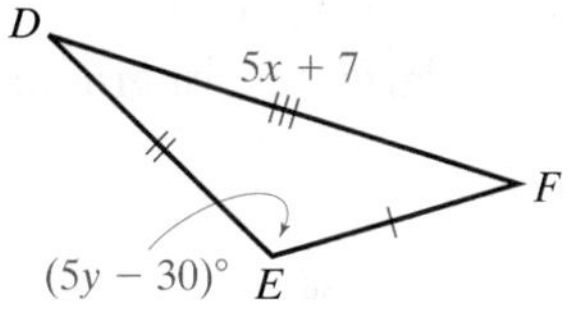

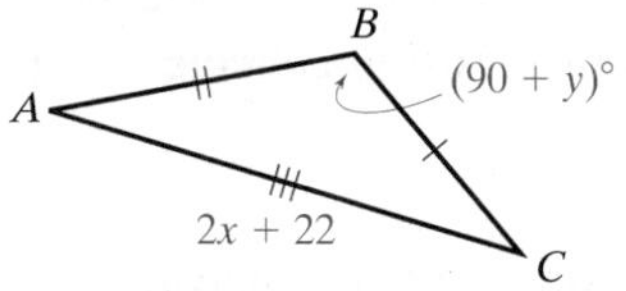

35. Find x.
36. Find y.
37. Find DF.
38. Find AC.
39. Find $m\angle E$.
40. Find $m\angle B$.

CONCEPT EXTENSIONS

41. How are the ASA Postulate and the SAS Postulate alike? How are they different?

42. Find the Error Your friend asks you for help on a geometry exercise. Below is your friend's paper. What error did your friend make? Explain.

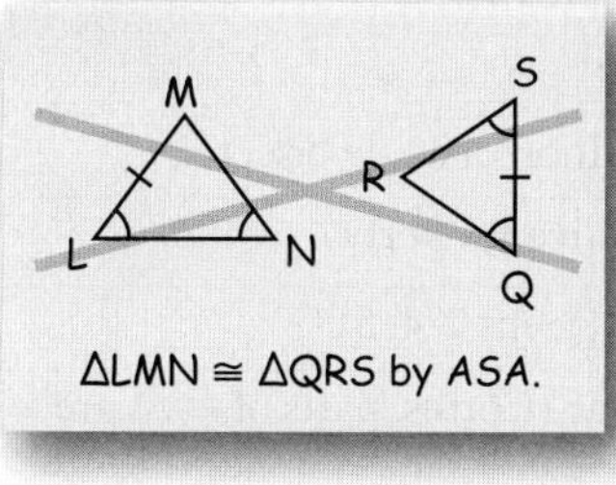

43. Given $\overline{AD} \parallel \overline{BC}$ and $\overline{AB} \parallel \overline{DC}$, name as many pairs of congruent triangles as you can.

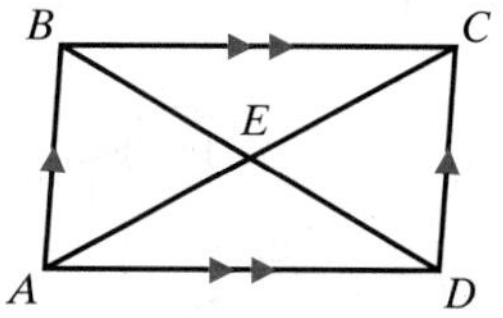

44. Sketch Draw two noncongruent triangles that have two pairs of congruent angles and one pair of congruent sides.

45. Can you prove that the triangles below are congruent? Justify your answer.

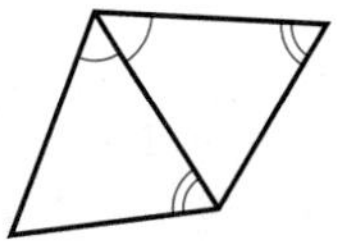

46. Anita says that you can rewrite any proof that uses the AAS Theorem as a proof that uses the ASA Postulate. Do you agree with Anita? Explain.

47. Construction In ΔRST below, $RS = 5$, $RT = 9$, and $m\angle T = 30°$. Show that there is no SSA congruence rule by constructing ΔUVW with $UV = RS$, $UW = RT$, and $m\angle W = m\angle T$, but with $\Delta UVW \not\cong \Delta RST$.

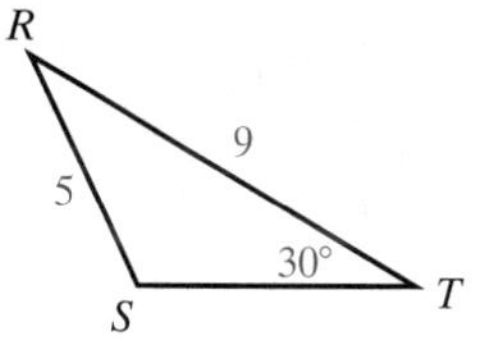

48. Construction Use a straight edge to draw a triangle. Label it ΔJKL. Construct $\Delta MNP \cong \Delta JKL$ so that the triangles are congruent by ASA.

While helping your family clean out the attic, you find the piece of paper shown below. The paper contains clues to locate a time capsule buried in your backyard. The maple tree is due east of the oak tree in your backyard.

> Mark a line on the ground from the oak tree to the maple tree. From the oak tree, walk along a path that forms a 70° angle with the marked line, keeping the maple tree to your right. From the maple tree, walk along a path that forms a 40° angle with the marked line, keeping the oak tree to your left. The time capsule is buried where the paths meet.

49. Will the clues always lead you to the same spot? Explain.

50. How can you use a diagram to help you?

51. What type of geometric figure do the paths and the marked line form?

52. How does the position of the marked line relate to the positions of the angles?

REVIEW AND PREVIEW

Would you use SSS or SAS to prove the triangles congruent? Explain. See Section 4.3.

53.

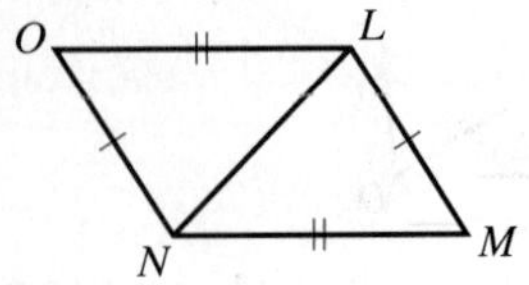

54.

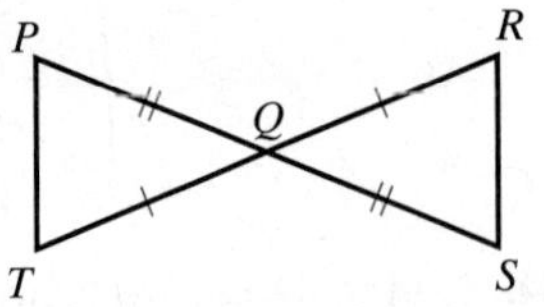

For ΔTIC ≅ ΔLOK, list the indicated parts. See Section 4.2.

55. congruent corresponding angles

56. congruent corresponding sides

4.5 PROOFS USING CONGRUENT TRIANGLES

OBJECTIVES

1. Use Triangle Congruence and Corresponding Parts of Congruent Triangles to Prove that Parts of Two Triangles Are Congruent.
2. Prove Two Triangles Are Congruent Using Other Congruent Triangles.

VOCABULARY

- cpoctac

OBJECTIVE 1 ▸ Using Congruent Parts of Triangles. With SSS, SAS, ASA, and AAS, we know how to use three congruent parts of two triangles to show that the two triangles are congruent. Once we know that two triangles are congruent, we know that the other corresponding parts of these congruent triangles are congruent.

For example, suppose we know that two triangles are congruent by the SSS postulate.

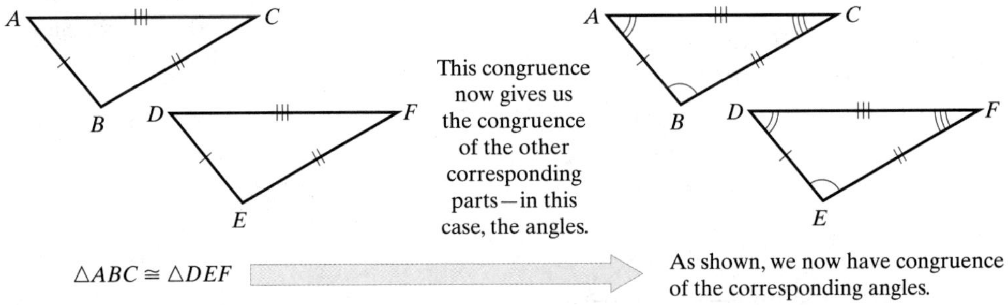

Thus far we have been proving that two triangles are congruent (or not).

Sometimes we need to show that specific parts of two triangles are congruent, such as two angles or two segments. In this section, we will do this by showing that these parts are

Corresponding Parts of Congruent Triangles (and thus Are Congruent).

We abbreviate this in a proof using **cpoctac** (Corresponding Parts Of Congruent Triangles Are Congruent).

EXAMPLE 1 **Using Congruent Triangles**

Given: $\overrightarrow{AB}$ bisects ∠*CAD*, $\overrightarrow{BA}$ bisects ∠*CBD*

Prove: ∠1 ≅ ∠2

Solution Mark our figure with the given information. See if we can prove the two triangles are congruent.

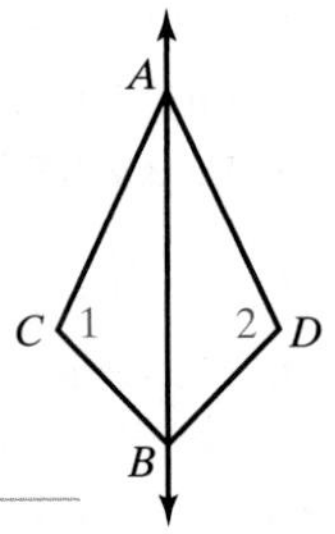

Statements	Reasons
1. $\overrightarrow{AB}$ bisects ∠*CAD*.	**1.** Given
2. ∠*CAB* ≅ ∠*BAD*	**2.** Definition of angle bisector
3. $\overrightarrow{BA}$ bisects ∠*CBD*.	**3.** Given
4. ∠*CBA* ≅ ∠*ABD*	**4.** Definition of angle bisector
5. $\overline{AB} \cong \overline{AB}$	**5.** Reflexive Property
6. Δ*ACB* ≅ Δ*ADB*	**6.** ASA (Steps 2, 4, 5)
7. ∠1 ≅ ∠2	**7.** cpoctac (Corres. parts of ≅ Δs are ≅)

Once we prove the two triangles are ≅ (Step 6), we may state any corresponding parts are ≅ as needed (Step 7).

□

PRACTICE 1

Given: $\overline{PR} \cong \overline{RT}, \overline{SR} \cong \overline{RQ}$

Prove: $\angle Q \cong \angle S$

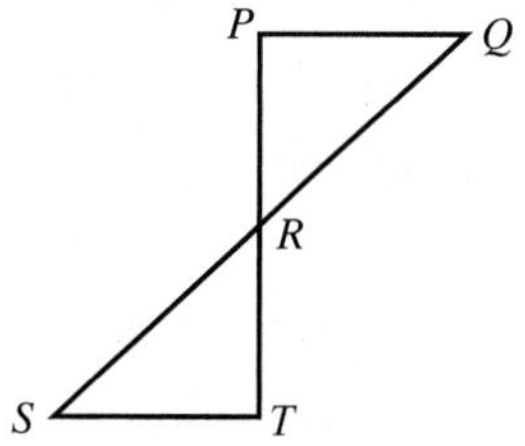

EXAMPLE 2 Proving Triangle Parts Congruent to Measure Distance

Thales, a Greek philosopher, is said to have developed a method to measure the distance to a ship at sea. He made a compass by nailing two sticks together. Standing on top of a tower, he would hold one stick vertical and tilt the other until he could see the ship S along the line of the tilted stick. With this compass setting, he would find a landmark L on the shore along the line of the tilted stick. How far would the ship be from the base of the tower?

Given: $\angle TRS$ and $\angle TRL$ are right angles, $\angle RTS \cong \angle RTL$

Prove: $\overline{RS} \cong \overline{RL}$

Solution Mark our figure with the given information. See if we can prove that the two triangles are congruent.

Statements	Reasons
1. $\angle RTS \cong \angle RTL$	**1.** Given
2. $\overline{TR} \cong \overline{TR}$	**2.** Reflexive Property of $\cong$
3. $\angle TRS$ and $\angle TRL$ are right angles.	**3.** Given
4. $\angle TRS \cong \angle TRL$	**4.** All right angles are $\cong$.
5. $\Delta TRS \cong \Delta TRL$	**5.** ASA Postulate (Steps 1, 2, 4)
6. $\overline{RS} \cong \overline{RL}$	**6.** cpoctac

The distance between the ship and the base of the tower would be the same as the distance between the base of the tower and the landmark. □

PRACTICE 2

Given: $\overline{AB} \cong \overline{AC}$, M is the midpoint of $\overline{BC}$

Prove: $\angle AMB \cong \angle AMC$

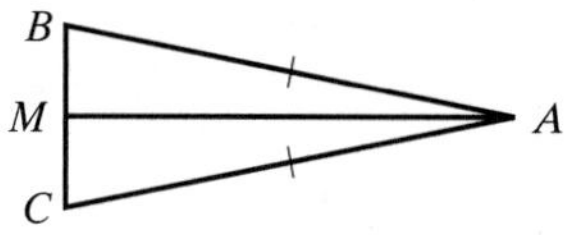

Now that we have the tools to prove that triangles are congruent, we are ready to prove a theorem from Chapter 3—the Alternate Interior Angles Theorem, stated below.

Recall Theorem 3.2-4, which states that if two lines and a transversal form alternate interior angles that are congruent, then the two lines are parallel.

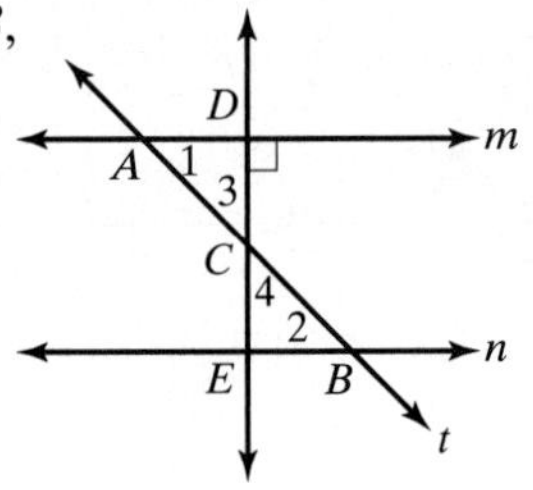

Given: Lines m and n are cut by transversal t at points A and B, $\angle 1 \cong \angle 2$

Prove: $m \parallel n$

Construction: Find the midpoint of $\overline{AB}$ and call it C.

Construct $\overleftrightarrow{CD} \perp m$ through at D.

Statements	Reasons
1. m and n are cut by transversal t.	**1.** Given
2. C is the midpoint of $\overline{AB}$.	**2.** Construction
3. $\overline{AC} \cong \overline{CB}$	**3.** Definition of midpoint
4. $CD \perp m$ at D	**4.** Construction
5. $\angle ADC$ is a right angle.	**5.** Definition of $\perp$ lines
6. $\angle 3 \cong \angle 4$	**6.** Vertical Angles Theorem
7. $\angle 1 \cong \angle 2$	**7.** Given
8. $\Delta ACD \cong \Delta BCE$	**8.** ASA Postulate (Steps 3, 6, 7)
9. $\angle ADC \cong \angle CEB$	**9.** cpoctac
10. $\angle CEB$ is a right angle.	**10.** Substitution (Steps 5, 9)
11. $\overleftrightarrow{CD} \perp n$	**11.** Step 10 and definition of a right angle
12. $m \parallel n$	**12.** Lines $\perp$ to the same line (in a plane) are $\parallel$.

OBJECTIVE 2 ▶ Proving Two Triangles Are Congruent Using Other Congruent Triangles. Until now we have examined only triangles that do not overlap. Let's now look at overlapping triangles, which may have a common side or angle. When doing so, it may be easier to work with these overlapping triangles by separating and redrawing the triangles.

EXAMPLE 3 **Identifying Common Parts of Overlapping Triangles**

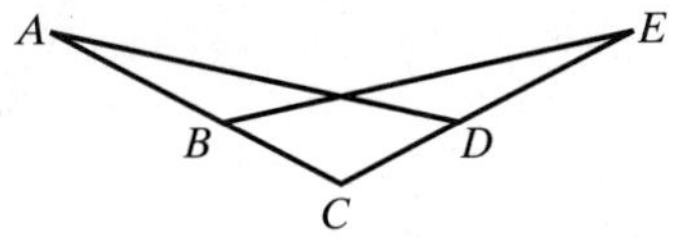

What common angle do ΔACD and ΔECB share?

Solution If needed, separate and redraw ΔACD and ΔECB.

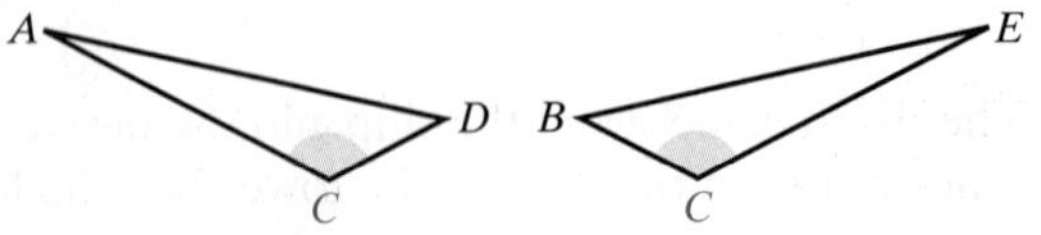

The common angle is $\angle C$. □

PRACTICE 3

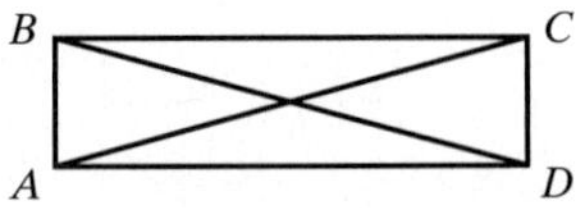

a. What is the common side in ΔABD and ΔDCA?

b. What is the common side in ΔABD and ΔBAC?

Next, let's practice working with overlapping triangles in a proof.

EXAMPLE 4 Using Common Parts of Overlapping Triangles

Given: $\angle ZXW \cong \angle YWX$, $\angle ZWX \cong \angle YXW$

Prove: $\overline{ZW} \cong \overline{YX}$

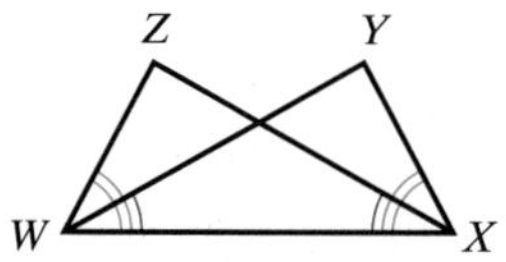

Solution Since $\overline{WX}$ is common to ΔZWX and ΔYXW, we can use that information to prove the overlapping triangles are congruent. From there, we try to prove $\overline{ZW} \cong \overline{YX}$.

Statements	Reasons
1. $\angle ZXW \cong \angle YWX$	**1.** Given
2. $\overline{WX} \cong \overline{WX}$	**2.** Reflexive Property of $\cong$
3. $\angle ZWX \cong \angle YXW$	**3.** Given
4. $\Delta ZWX \cong \Delta YXW$	**4.** ASA Postulate (Steps 1, 2, 3)
5. $\overline{ZW} \cong \overline{YX}$	**5.** cpoctac

□

PRACTICE 4

Given: $\Delta ACD \cong \Delta BDC$

Prove: $\overline{CE} \cong \overline{DE}$

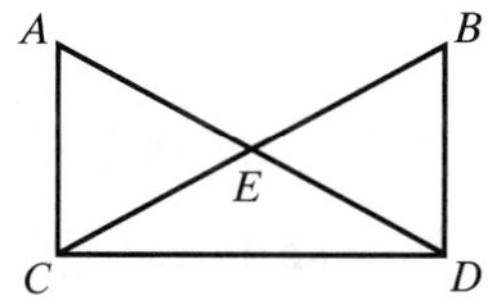

> **Helpful Hint**
> If you have trouble working with overlapping triangles, redraw the overlapping triangles so that you have separate triangles.

In some proofs, we must prove the congruence of two triangles before we can prove the congruence of any other triangles.

EXAMPLE 5 Using Two Pairs of Triangles

Given: In the origami design, E is the midpoint of $\overline{AC}$ and $\overline{DB}$

Prove: $\Delta GED \cong \Delta JEB$

Solution Take a moment to redraw and mark our figure with the given information. We will use the given information to first find that the two larger triangles are congruent, ΔAED and ΔCEB.

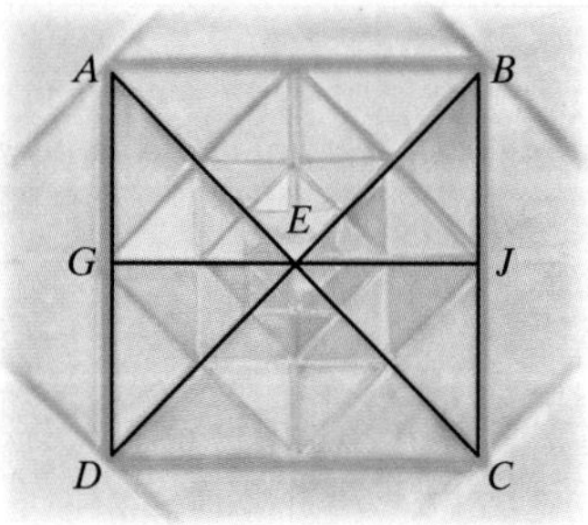

Statements	Reasons
1. E is the midpoint of $\overline{AC}$ and $\overline{DB}$.	**1.** Given
2. $\overline{AE} \cong \overline{CE}$	**2.** Definition of midpoint
3. $\overline{DE} \cong \overline{BE}$	**3.** Definition of midpoint
4. $\angle AED \cong \angle BEC$	**4.** Vertical Angles Theorem
5. $\triangle AED \cong \triangle CEB$	**5.** SAS Postulate (Steps 2, 3, 4)
6. $\angle D \cong \angle B$	**6.** cpoctac
7. $\angle GED \cong \angle JEB$	**7.** Vertical Angles Theorem
8. $\triangle GED \cong \triangle JEB$	**8.** ASA Postulate (Steps 3, 6, 7)

PRACTICE 5

Given: $\overline{PS} \cong \overline{RS}$, $\angle PSQ \cong \angle RSQ$

Prove: $\triangle QPT \cong \triangle QRT$

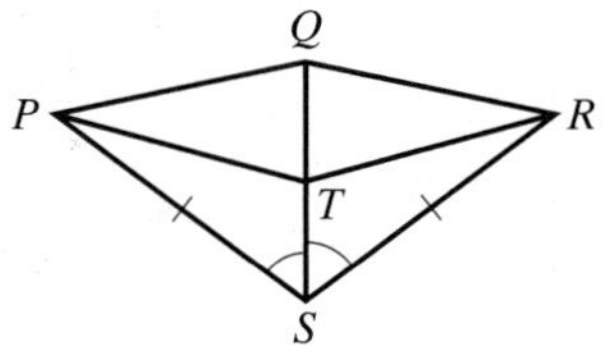

VOCABULARY & READINESS CHECK

1. What does cpoctac mean? ____________
2. This is an overlapping triangle. Use this diagram to draw two different pairs of separate overlapping triangles.

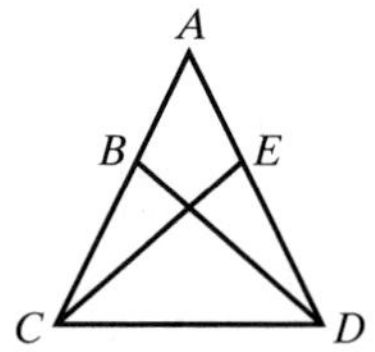

4.5 EXERCISE SET MyMathLab®

Identify any common angles or sides. See Example 3.

1. $\triangle MKJ$ and $\triangle LJK$

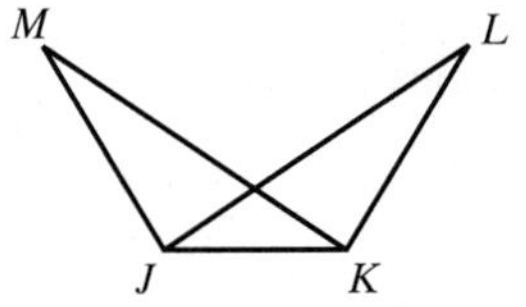

2. $\triangle DEH$ and $\triangle DFG$

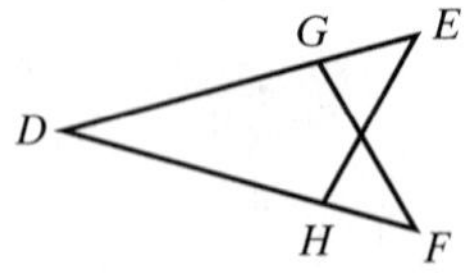

Separate and redraw the overlapping triangles. Label the vertices. See Example 3.

3.

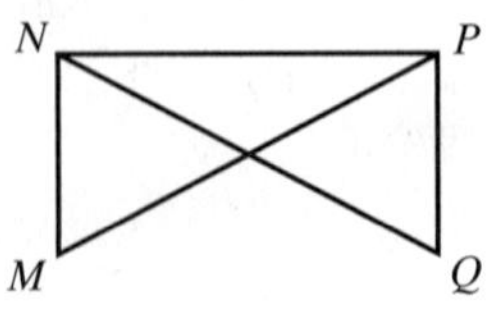

4.

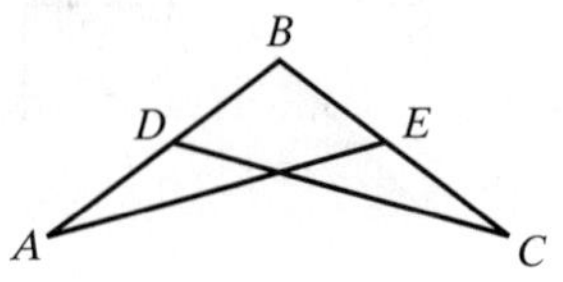

In each diagram, the red and blue triangles are congruent. Identify their common side or angle. See Example 3.

5.

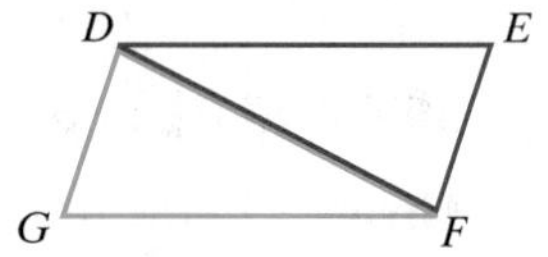

6.

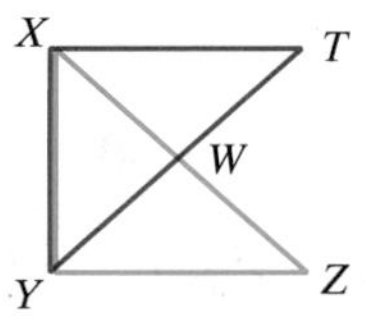

Separate and redraw the indicated triangles. Identify any common angles or sides. See Example 3.

7. $\triangle PQS$ and $\triangle QPR$

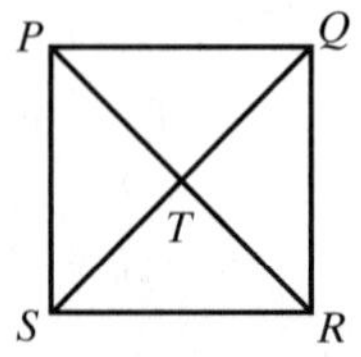

8. $\triangle ACB$ and $\triangle PRB$

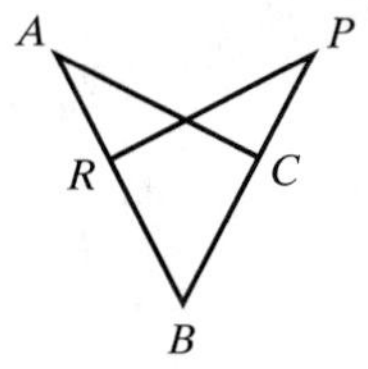

For Exercises 9 and 10, tell why the two triangles are congruent. Give the congruence statement. Then list all the other corresponding parts of the triangles that are congruent. See Examples 1 and 2.

9.

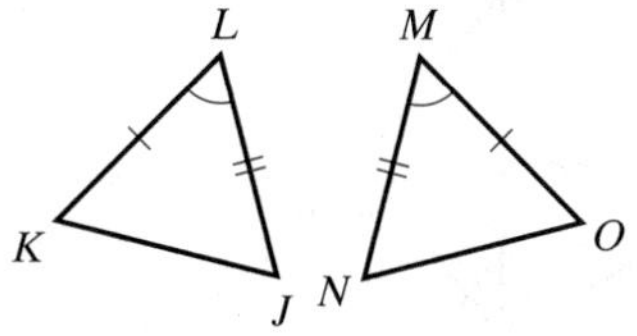

10.

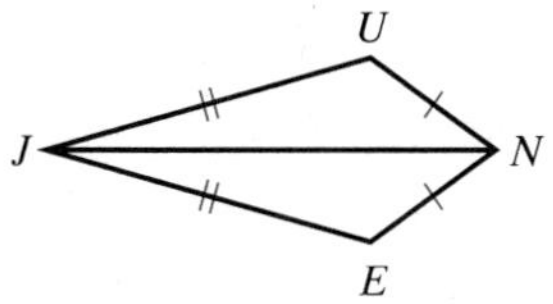

Proof *A balalaika is a stringed instrument. Prove that the bases of the balalaikas are congruent by filling in the blanks.*

Given: $\overline{RA} \cong \overline{NY}, \angle R \cong \angle N, \angle A \cong \angle Y$

Prove: $\overline{KA} \cong \overline{JY}$

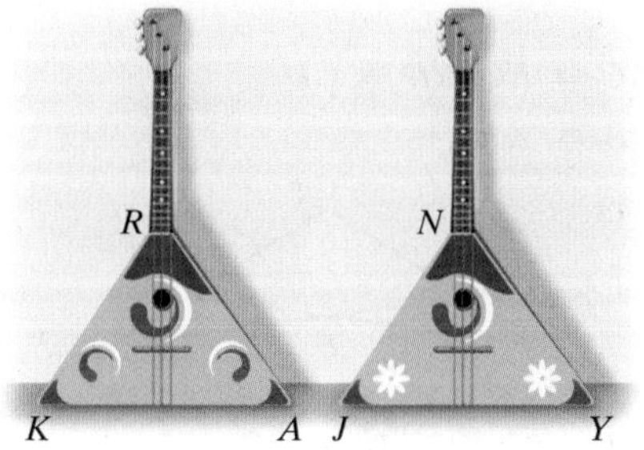

Statements	Reasons
1. $\overline{RA} \cong \overline{NY}, \angle R \cong \angle N, \angle A \cong \angle Y$	**1.** Given
2. $\Delta KRA \cong \Delta JNY$	**2.** ? (Exercise 11)
3. $\overline{KA} \cong \overline{JY}$	**3.** ? (Exercise 12)

Proof *For Exercises 13–16, complete each proof. See Examples 1 and 2.*

13. Given: $\overline{OM} \cong \overline{ER}, \overline{ME} \cong \overline{RO}$

Prove: $\angle M \cong \angle R$

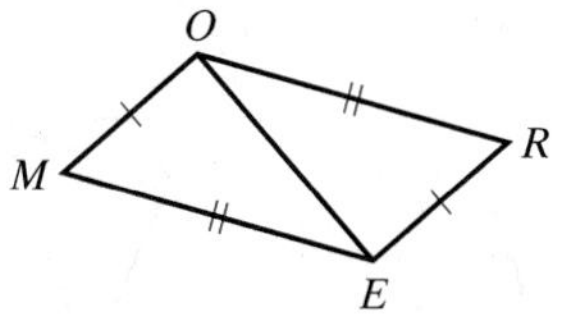

14. Given: $\angle ABD \cong \angle CBD, \angle BDA \cong \angle BDC$

Prove: $\overline{AB} \cong \overline{CB}$

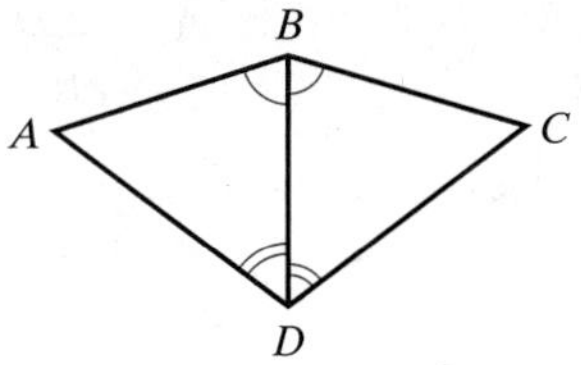

15. Given: $\angle SPT \cong \angle OPT, \overline{SP} \cong \overline{OP}$

Prove: $\angle S \cong \angle O$

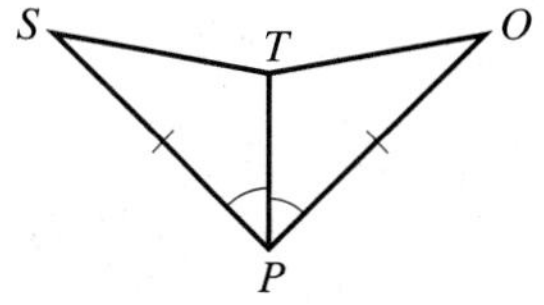

16. Given: $\overline{YT} \cong \overline{YP}, \angle C \cong \angle R, \angle T \cong \angle P$

Prove: $\overline{CT} \cong \overline{RP}$

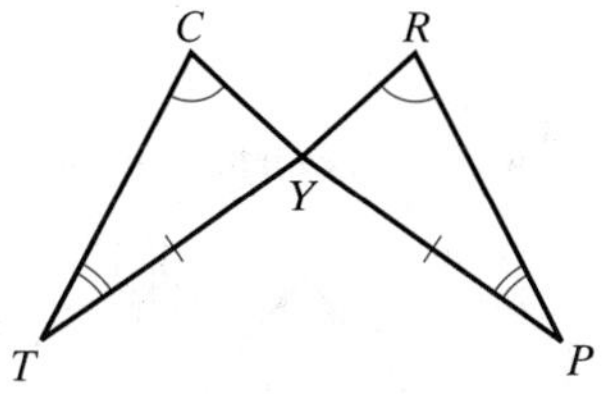

For Exercises 17 and 18, copy and mark the figure to show the given information. Explain how you would prove $\angle P \cong \angle Q$. (Hint: Work 17 and 18 separately.)

17. Given: $\overline{PK} \cong \overline{QK}, \overline{KL}$ bisects $\angle PKQ$

18. Given: $\overline{KL}$ is the perpendicular bisector of $\overline{PQ}$

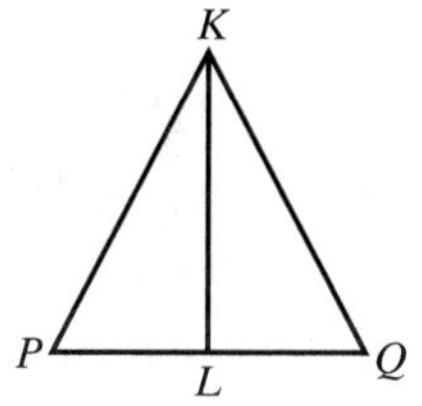

Proof *Complete each proof. See Examples 1 and 2.*

19. Given: $\overline{BA} \cong \overline{BC}, \overline{BD}$ bisects $\angle ABC$

Prove: $\overline{BD} \perp \overline{AC}, \overline{BD}$ bisects $\overline{AC}$

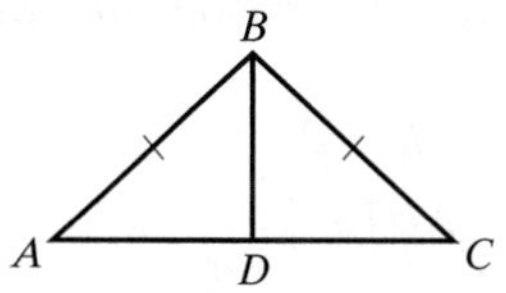

20. Given: $\ell \perp \overline{AB}$, ℓ bisects $\overline{AB}$ at C, P is on ℓ

Prove: $PA = PB$

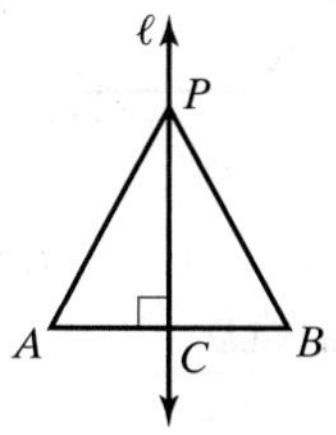

21. Given: $\overline{BE} \perp \overline{AC}, \overline{DF} \perp \overline{AC}, \overline{BE} \cong \overline{DF}, \overline{AF} \cong \overline{CE}$

Prove: $\overline{AB} \cong \overline{CD}$

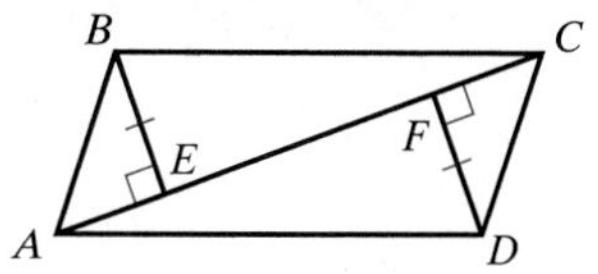

22. Given: $\overline{JK} \parallel \overline{QP}$, $\overline{JK} \cong \overline{PQ}$

Prove: $\overline{KQ}$ bisects $\overline{JP}$

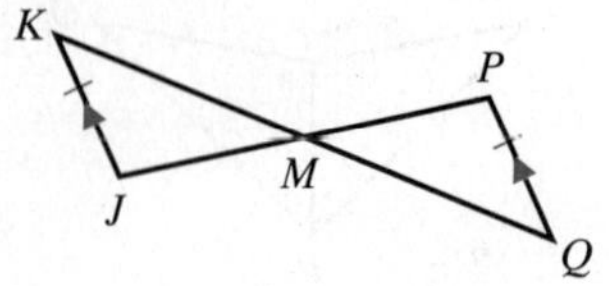

Proof *Complete the proof by filling in the blanks.*

Given: $\angle T \cong \angle R$, $\overline{PQ} \cong \overline{PV}$

Prove: $\angle PQT \cong \angle PVR$

(Hint: If needed, redraw the diagram to show two separate triangles.)

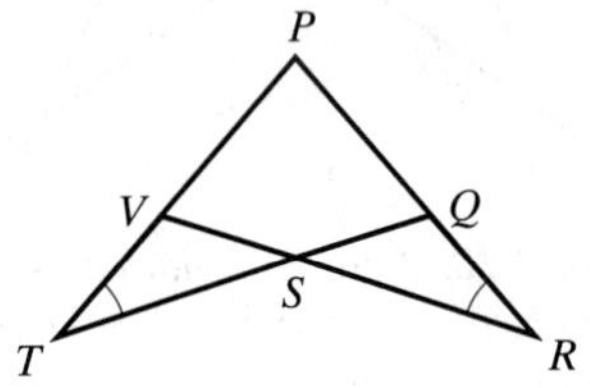

Statements	Reasons
1. $\angle T \cong \angle R$, $\overline{PQ} \cong \overline{PV}$	**1.** ? (Exercise 23)
2. $\angle TPQ \cong \angle RPV$	**2.** ? (Exercise 24)
3. $\Delta TPQ \cong \Delta RPV$	**3.** ? (Exercise 25)
4. $\angle PQT \cong \angle PVR$	**4.** ? (Exercise 26)

MIXED PRACTICE

Proof *If needed, redraw the diagram to show separate triangles. See Examples 4 and 5.*

27. Given: $\overline{RS} \cong \overline{UT}$, $\overline{RT} \cong \overline{US}$

Prove: $\Delta RST \cong \Delta UTS$

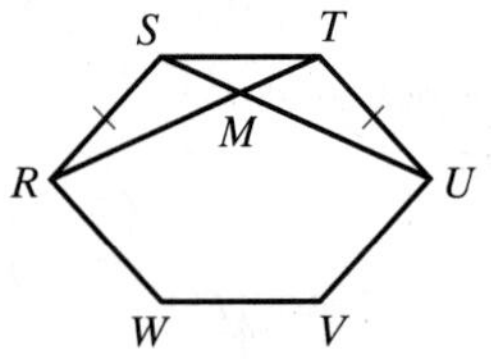

28. Given: $\overline{QD} \cong \overline{UA}$, $\angle QDA \cong \angle UAD$

Prove: $\Delta QDA \cong \Delta UAD$

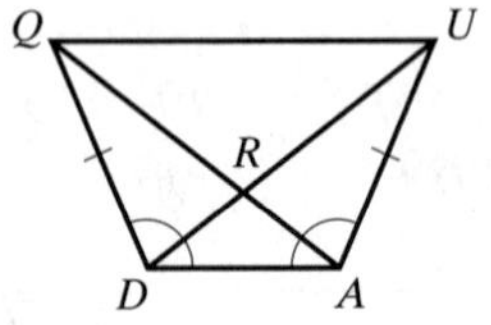

29. Given: $\angle 1 \cong \angle 2$, $\angle 3 \cong \angle 4$

Prove: $\Delta QET \cong \Delta QEU$

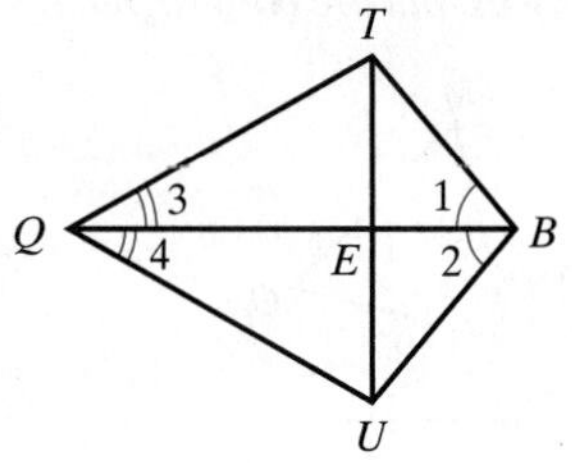

30. Given: $\overline{AD} \cong \overline{ED}$, D is the midpoint of $\overline{BF}$

Prove: $\Delta ADC \cong \Delta EDG$

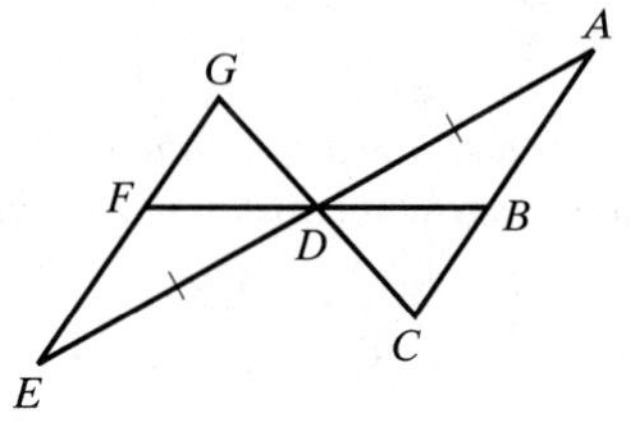

31. Given: $\overline{AC} \cong \overline{EC}$, $\overline{CB} \cong \overline{CD}$

Prove: $\angle A \cong \angle E$

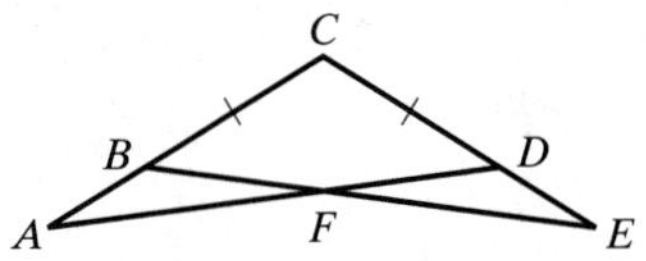

32. Given: $\overline{QT} \perp \overline{PR}$, $\overline{QT}$ bisects $\overline{PR}$, $\overline{QT}$ bisects $\angle VQS$

Prove: $\overline{VQ} \cong \overline{SQ}$

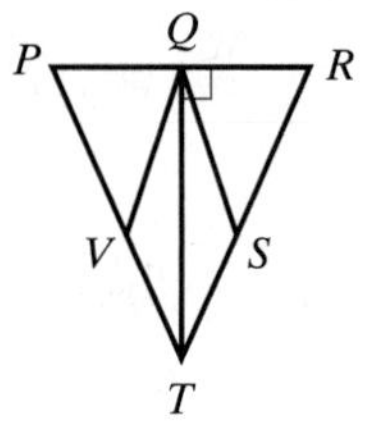

33. Given: $\overline{TE} \cong \overline{RI}$, $\overline{TI} \cong \overline{RE}$, $\angle TDI$ and $\angle ROE$ are right angles.

Prove: $\overline{TD} \cong \overline{RO}$

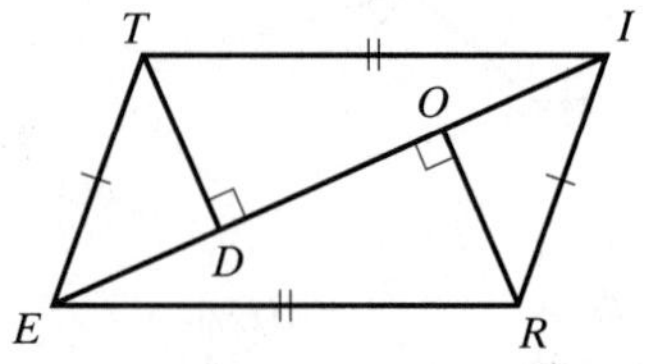

34. Given: $\overline{AB} \perp \overline{BC}$, $\overline{DC} \perp \overline{BC}$, $\angle A \cong \angle D$

Prove: $\overline{AE} \cong \overline{DE}$

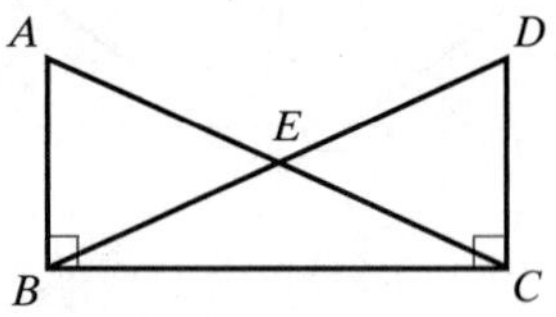

The figure below is part of a clothing design pattern, and it has the following relationships.

- $\overline{GC} \perp \overline{AC}$
- $\overline{AB} \perp \overline{BC}$
- $\overline{AB} \| \overline{DE} \| \overline{FG}$
- $m\angle A = 50°$
- ΔDEC is isosceles with base $\overline{DC}$.

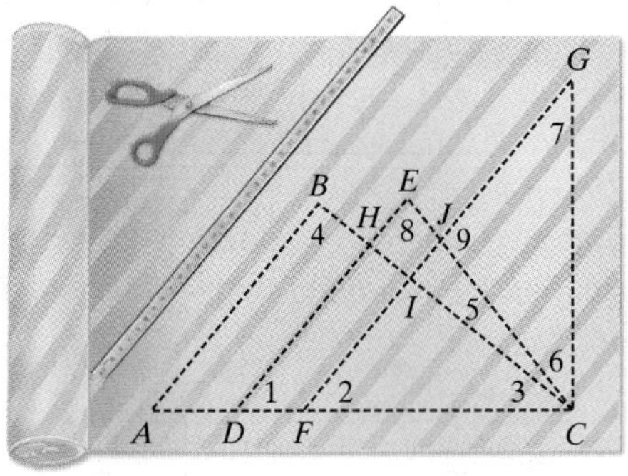

35. Find the measures of all the numbered angles in the figure.

36. Suppose $\overline{AB} \cong \overline{FC}$. Name two congruent triangles and explain how you can prove them congruent.

CONCEPT EXTENSIONS

37. Rangoli is a colorful design pattern drawn outside houses in India, especially during festivals. A student plans to use the pattern below as the base of her design. In this pattern, $\overline{RU}$, $\overline{SV}$, and $\overline{QT}$ bisect each other at O. $RS = 6$, $RU = 12$, $\overline{RU} \cong \overline{SV}$, $\overline{ST} \| \overline{RU}$, and $\overline{RS} \| \overline{QT}$. What is the perimeter of the hexagon? (Recall that perimeter means "distance around.")

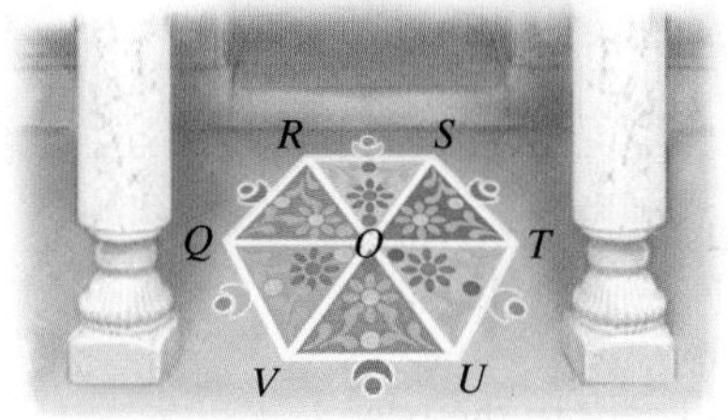

38. Proof Identify a pair of overlapping congruent triangles in the diagram. Then use the given information to write a proof to show that the triangles are congruent.

Given: $\overline{AC} \cong \overline{BC}$, $\angle A \cong \angle B$

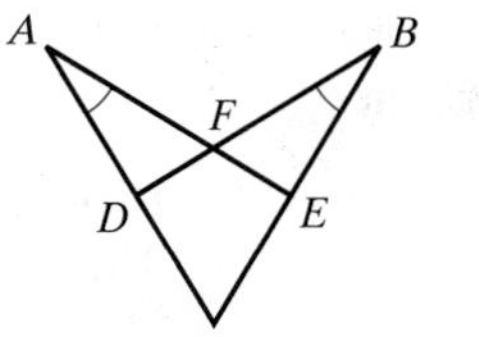

Proof *In the diagram that follows,* $\overline{BA} \cong \overline{KA}$ *and* $\overline{BE} \cong \overline{KE}$.

39. Prove: S is the midpoint of $\overline{BK}$

40. Prove: $\overline{BK} \perp \overline{AE}$

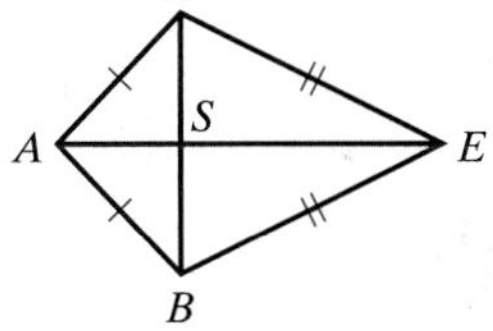

41. Find the Error In the diagram, $\Delta PSY \cong \Delta SPL$. Based on that fact, your friend claims that $\Delta PRL \not\cong \Delta SRY$. Explain why your friend is incorrect.

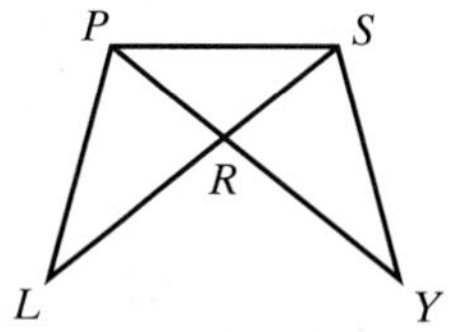

42. Find the Error Find and correct the error(s) in the proof.

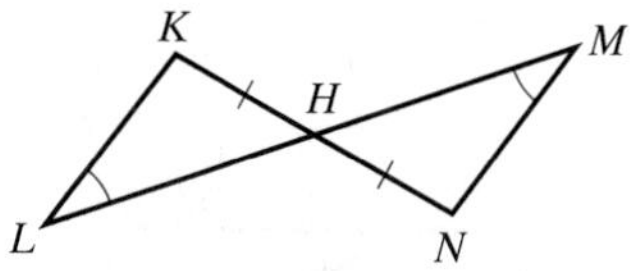

Given: $\overline{KH} \cong \overline{NH}$, $\angle L \cong \angle M$

Prove: H is the midpoint of $\overline{LM}$

Proof: $\overline{KH} \cong \overline{NH}$ because it is given. $\angle L \cong \angle M$ because it is given. $\angle KHL \cong \angle NHM$ because vertical angles are congruent. So, $\Delta KHL \cong \Delta MHN$ by ASA Postulate. Since corresponding parts of congruent triangles are congruent, $\overline{LH} \cong \overline{MH}$. By the definition of midpoint, H is the midpoint of $\overline{LM}$.

43. In the figure below, which pair of triangles could you prove congruent first in order to prove that $\Delta ACD \cong \Delta CAB$? Explain.

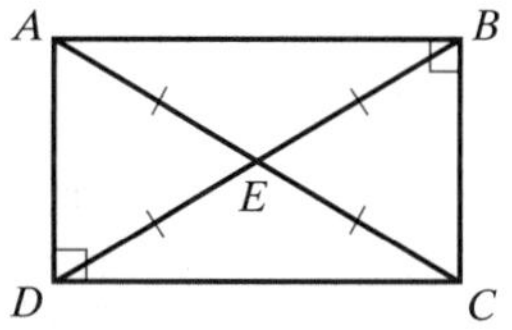

44. How does the fact that corresponding parts of congruent triangles are congruent relate to the definition of congruent triangles?

Sketch *For Exercises 45–46, draw the diagram described.*

45. Draw a vertical segment on your paper. On the right side of the segment draw two triangles that share the vertical segment as a common side.

46. Draw two triangles that have a common angle.

For Exercises 47 and 48, draw a quadrilateral ABCD with $\overline{AB} \| \overline{DC}$, $\overline{AD} \| \overline{BC}$, *and diagonals* $\overline{AC}$ *and* $\overline{DB}$ *intersecting at E. Label your diagram to indicate the parallel sides.*

47. List all the pairs of congruent segments in your diagram.

48. Explain how you know that the segments you listed in Exercise **47** are congruent.

49. The construction of a line perpendicular to line $\overleftrightarrow{AB}$ through point P on the line is shown. Explain why you can conclude that $\overleftrightarrow{CP}$ is perpendicular to $\overleftrightarrow{AB}$.

a. How can you use congruent triangles to justify the construction?

b. Which lengths or distances are equal by construction?

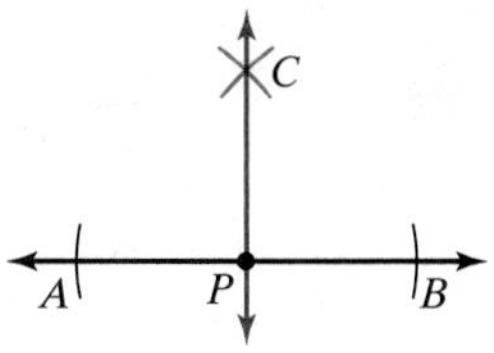

50. The construction of $\angle B$ congruent to given $\angle A$ is shown. $\overline{AD} \cong \overline{BF}$ because they are congruent radii. $\overline{DC} \cong \overline{FE}$ because both arcs have the same compass settings. Explain why you can conclude that $\angle A \cong \angle B$.

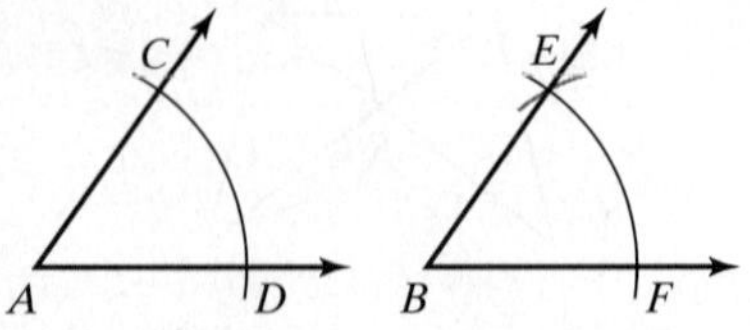

REVIEW AND PREVIEW

For Exercises 51 and 52, give the postulate or theorem that you can immediately use to prove the triangles congruent. See Sections 4.3 and 4.4.

51.

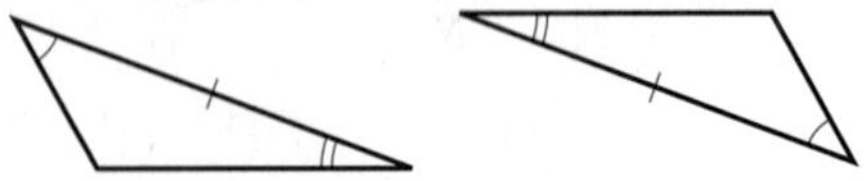

52.

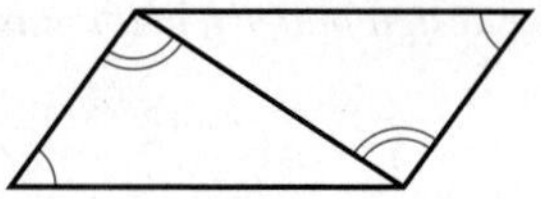

Solve. Use the figure below for Exercises 53–56. See Sections 4.1 and 4.3.

53. What is the side opposite $\angle ABC$?

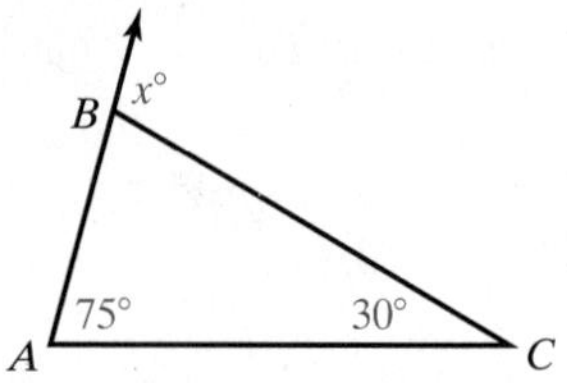

54. What is the angle opposite side $\overline{AB}$?

55. What is the angle opposite side $\overline{BC}$?

56. Find the value of x.

4.6 ISOSCELES, EQUILATERAL, AND RIGHT TRIANGLES

OBJECTIVES

1. Use Properties of Isosceles and Equilateral Triangles.
2. Use Properties of Right Triangles.

VOCABULARY

- legs of an isosceles triangle
- base of an isosceles triangle
- vertex angle of an isosceles triangle
- base angles of an isosceles triangle
- hypotenuse
- legs of a right triangle

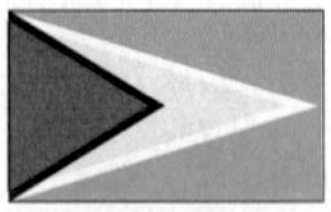

GUYANA

Flag of Guyana. See the red isosceles triangle.

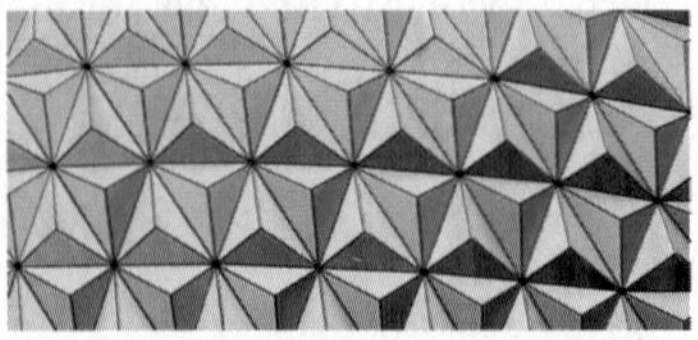

The "Epcot ball" contains 11,324 isosceles triangles. (Notice that 3 of the 3D isosceles triangles appear to form a 2D equilateral triangle.)

OBJECTIVE 1 ▸ Using Properties of Isosceles and Equilateral Triangles. In Section 4.1 when we classified triangles by sides, we learned that

- a triangle with at least two equal sides is an isosceles triangle, and
- a triangle with three equal sides is an equilateral triangle.

> **Helpful Hint**
> Recall that:
> - All equilateral triangles are also isosceles triangles.
> - Some isosceles triangles are equilateral triangles.

If an isosceles triangle has exactly two congruent (equal measure) sides, then these two sides are its **legs.** The third side is the **base.** The angle opposite the base is the **vertex angle.** The other two angles adjacent to the base are the **base angles.**

Let's now use what we know about isosceles triangles and proofs to learn more.

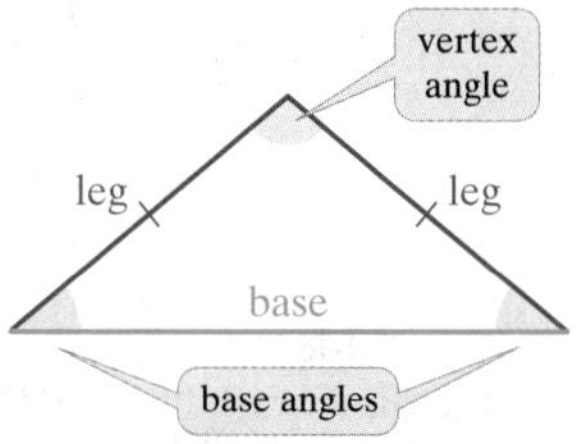

Isosceles triangle with 2 equal sides

Theorem 4.6-1 Isosceles Base Angles Theorem

Theorem	If . . .	Then . . .
If two sides of a triangle are congruent, then the angles opposite those sides are congruent.	$\overline{AB} \cong \overline{AC}$ 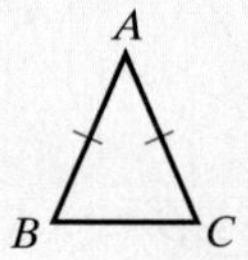	$\angle B \cong \angle C$ 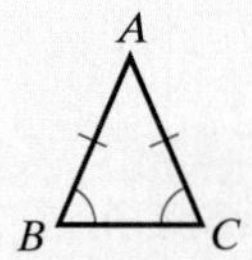

EXAMPLE 1 Prove the Isosceles Base Angles Theorem

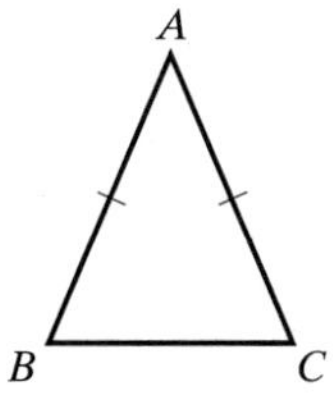

Given: $\Delta ABC, \overline{AB} \cong \overline{AC}$

Prove: $\angle B \cong \angle C$

Solution We first need a construction.

Construction: Draw $\overline{AD}$, the angle bisector of $\angle A$.

Proof:

Statements	Reasons
1. $\overline{AB} \cong \overline{AC}$	**1.** Given
2. $\angle 1 \cong \angle 2$	**2.** Construction
3. $\overline{AD} \cong \overline{AD}$	**3.** Reflexive Propert of ≅
4. $\Delta BAD \cong \Delta CAD$	**4.** SAS Postulate (Steps 1, 2, 3)
5. $\angle B \cong \angle C$	**5.** cpoctac

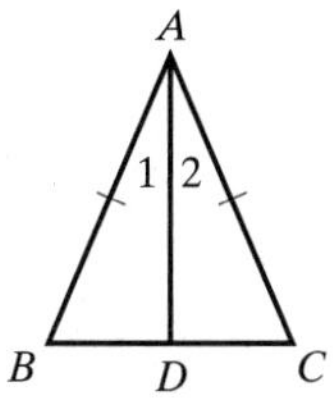

The converse of this theorem is also true.

Theorem 4.6-2 Converse of the Isosceles Base Angles Theorem

Theorem
If two angles of a triangle are congruent, then the sides opposite those angles are congruent.

If . . .
$\angle B \cong \angle C$

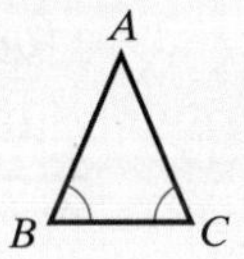

Then . . .
$\overline{AB} \cong \overline{AC}$

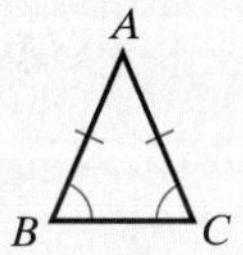

PRACTICE 1 Prove the theorem above. (*Hint:* No angle bisector is needed. Prove $\Delta ABC \cong \Delta ACB$. Think of ΔACB as the "flip" of ΔABC.)

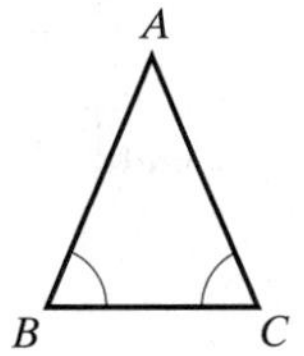

EXAMPLE 2 Using Isosceles Triangles

Use the figure and markings to find the values of x and y.

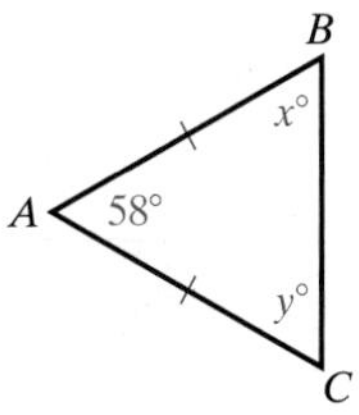

Solution ΔABC is an isosceles triangle, so the base angles are equal in measures ($m\angle C = m\angle B$, or $x = y$).

Using the sum of the measures of the angles of a triangle,

$$58 + x + y = 180$$
$$x + y = 122 \quad \text{Subtract.}$$
$$x + x = 122 \quad \text{Substitute } x \text{ for } y \text{ since } x = y.$$
$$2x = 122 \quad \text{Combine like terms.}$$
$$x = 61 \quad \text{Divide both sides by 2.}$$

Thus, $x = 61$. Since $x = y$, then $y = 61$ also.
To **check**, see that $58° + 61° + 61° = 180°$.

PRACTICE 2 Find the values of x and y.

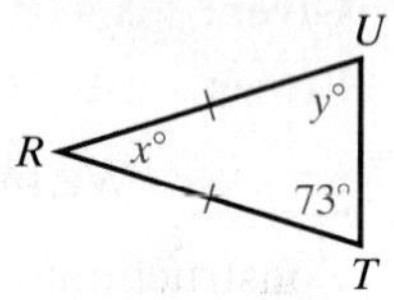

Theorem 4.6-3 Perpendicular Bisector of the Base of an Isosceles Triangle

Theorem	If . . .	Then . . .
If a line bisects the vertex angle of an isosceles triangle, then the line is also the perpendicular bisector of the base.	$\overline{AB} \cong \overline{AC}$ and $\angle 1 \cong \angle 2$ 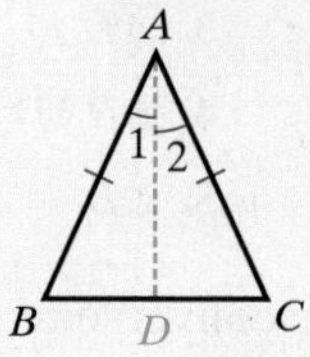	$\overline{AD} \perp \overline{BC}$ and $\overline{BD} \cong \overline{DC}$ 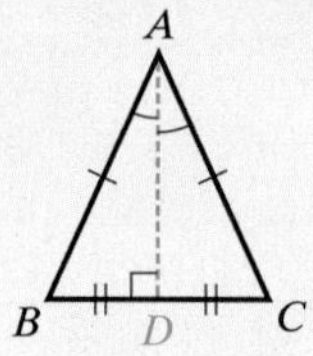

We prove this theorem in Exercise Set 4.6, Exercise 57.

EXAMPLE 3 Using Isosceles Triangles

Use the given figure to find the value of x.

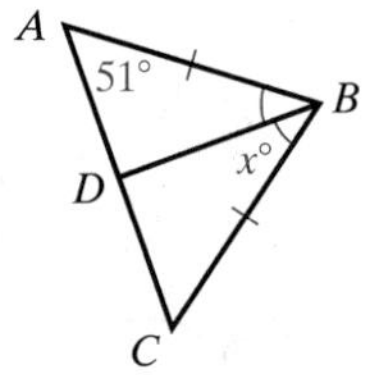

Solution Since $\overline{AB} \cong \overline{CB}$ by the Isosceles Base Angles Theorem, we have, $\angle A \cong \angle C$. Thus, $m\angle C = 51°$.

From the markings on the triangle, we see that $\overline{BD}$ bisects $\angle ABC$. Thus, we know by the theorem above that $\overline{BD} \perp \overline{AC}$. So $m\angle BDC = 90°$.

$m\angle C + m\angle BDC + m\angle DBC = 180$ Triangle Angle-Sum Theorem

$51 + 90 + x = 180$ Substitute.

$x = 39$ Subtract 141 $(51 + 90)$ from each side. □

The value of x is 39.

PRACTICE 3 Suppose $m\angle A = 27$. What is the value of x? (Use the same figure from Example 3.)

Recall that a **corollary** is a theorem that can be proved easily using another theorem.

> **Helpful Hint**
> Don't forget that
> - A corollary is also a theorem and must be proved.
> - The word corollary is simply used to help us know how to prove it.

Corollary 4.6-4 If Equilateral then Equiangular Triangle (Corollary to Theorem 4.6-1)

Corollary	If . . .	Then . . .
If a triangle is equilateral, then the triangle is equiangular.	$\overline{AB} \cong \overline{BC} \cong \overline{CA}$ 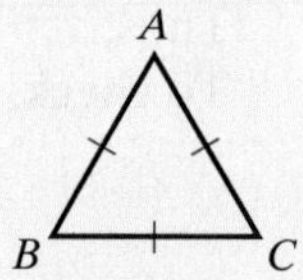	$\angle A \cong \angle B \cong \angle C$ 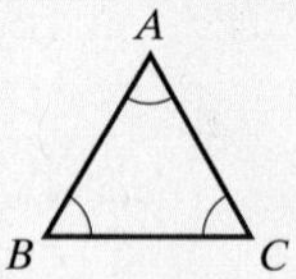

Corollary 4.6-5 If Equiangular then Equilateral Triangle (Corollary to Theorem 4.6-2)

Corollary	If . . .	Then . . .
If a triangle is equiangular, then the triangle is equilateral.	$\angle A \cong \angle B \cong \angle C$ 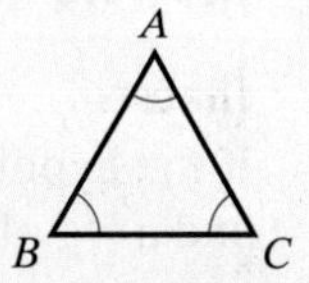	$\overline{AB} \cong \overline{BC} \cong \overline{CA}$ 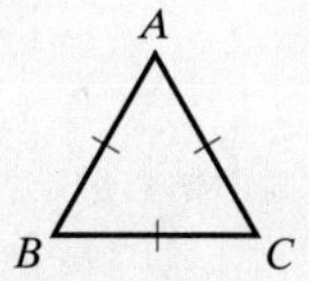

We prove these corollaries in Exercise Set 4.6, Exercises 59 and 60.

> **Helpful Hint**
> The two corollaries basically say that a triangle is equilateral if and only if it is equiangular.

EXAMPLE 4 **Using Equilateral Triangles**

Each triangle in the floorpattern shown is an equilateral triangle. Find the measure of each angle in ΔXYZ.

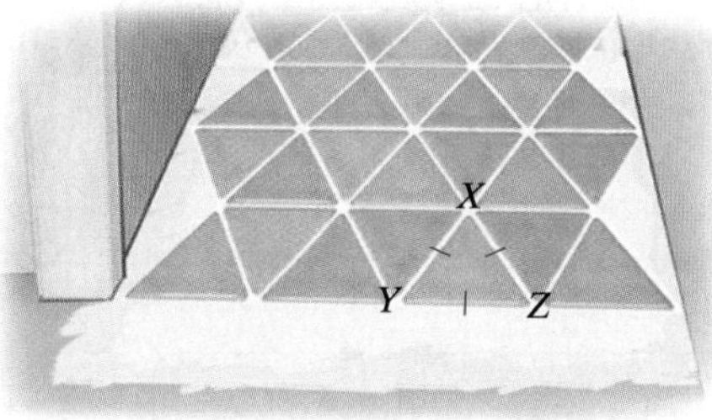

Solution Since ΔXYZ is equilateral, it is also equiangular. Thus, we have a triangle with 3 angles of equal measure.

$$m\angle X = m\angle Y = m\angle Z$$

Also, we know that the angle sum is 180°. Thus

$$m\angle X + m\angle Y + m\angle Z = 180° \quad \text{Triangle angle sum}$$
$$m\angle X + m\angle X + m\angle X = 180° \quad \text{Substitution, since } m\angle X = m\angle Y = m\angle Z$$
$$3(m\angle X) = 180° \quad \text{Combine like terms.}$$
$$m\angle X = 60° \quad \text{Divide both sides by 3.}$$

Thus, $m\angle X = m\angle Y = m\angle Z = 60°$. □

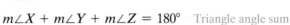

PRACTICE 4 Find the measure of each angle in ΔABC.

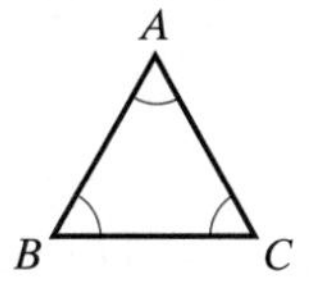

> **Helpful Hint**
> We now know that those facts listed below are all equivalent and that if you know one of them, you know them all.
> - equilateral Δ
> - equiangular Δ
> - a triangle with all angles of 60°

> **Helpful Hint**
> Thus far we have four ways to prove that triangles are congruent.
> - SSS (Side-Side-Side) Congruence Postulate
> - SAS (Side-Angle-Side) Congruence Postulate
> - ASA (Angle-Side-Angle) Congruence Postulate
> - AAS (Angle-Angle-Side) Congruence Theorem

OBJECTIVE 2 ▶ Using Properties of Right Triangles. Recall that in a right triangle, the side opposite the right angle is called the **hypotenuse.** It is the longest side in the triangle. The other two sides are called **legs.**

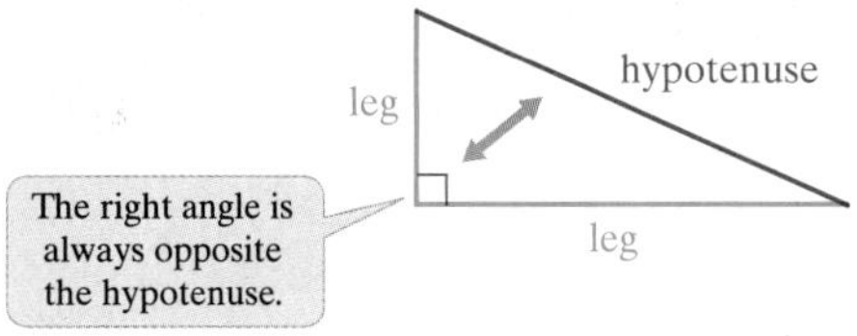

Now we study a special theorem that can be used to prove that two right triangles are congruent.

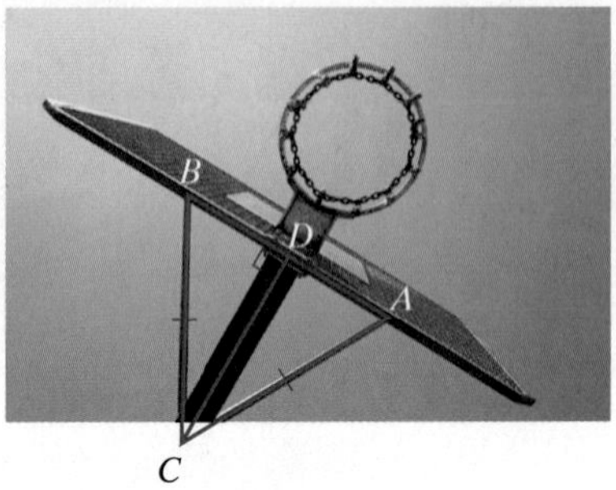

Using the H-L Theorem, the backboard brackets are two congruent right triangles.

Theorem 4.6-6 Hypotenuse-Leg (H-L) Theorem

Theorem	If . . .	Then . . .
If the hypotenuse and a leg of one right triangle are congruent to the hypotenuse and a leg of another right triangle, then the triangles are congruent.	ΔPQR and ΔXYZ are right triangles, $\overline{PR} \cong \overline{XZ}$, and $\overline{PQ} \cong \overline{XY}$ 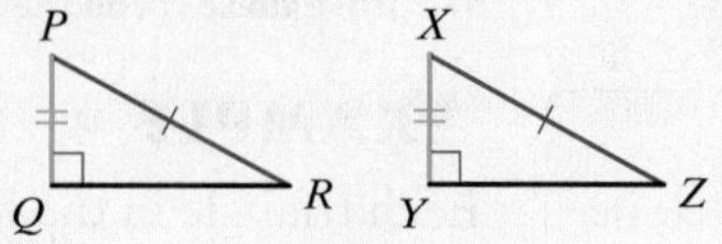	$\Delta PQR \cong \Delta XYZ$

The proof of this theorem is found in the back of this book.

Conditions for Using the H-L Theorem

To use the H-L Theorem, the triangles must meet three conditions.

- There are two right triangles.
- The triangles have congruent hypotenuses.
- There is at least one pair of congruent legs.

Helpful Hint
There is no SSA congruence (Side-Side-Angle) postulate or theorem for triangles in general.

EXAMPLE 5 Using the H-L Theorem

Given: $\overline{BE}$ bisects $\overline{AD}$ at C, $\overline{AB} \perp \overline{BC}$, $\overline{DE} \perp \overline{EC}$, $\overline{AB} \cong \overline{DE}$

Prove: $\Delta ABC \cong \Delta DEC$

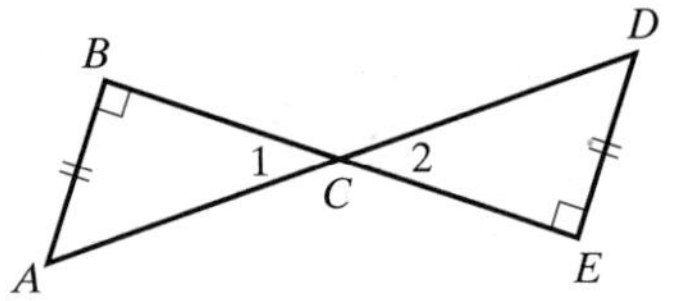

Solution To use the H-L Theorem, make sure you note that the two triangles are right triangles.

Statements	Reasons
1. $\overline{BE}$ bisects $\overline{AD}$ at C.	1. Given
2. $\overline{AC} \cong \overline{CD}$	2. Definition of bisector
3. $\overline{AB} \perp \overline{BC}$, $\overline{DE} \perp \overline{EC}$	3. Given
4. $\angle B$ and $\angle E$ are right angles.	4. Definition of $\perp$
5. ΔABC and ΔDEC are right triangles.	5. Definition of right triangles
6. $\overline{AB} \cong \overline{DE}$	6. Given
7. $\Delta ABC \cong \Delta DEC$	7. H-L Theorem (Steps 2, 5, 6)

Helpful Hint
There are other ways to prove $\Delta ABC \cong \Delta DEC$. Notice, for example, that $\angle 1$ and $\angle 2$ are vertical angles, so $\angle 1 \cong \angle 2$.

PRACTICE
5 **Given:** $\overline{CD} \cong \overline{EA}$, $\overline{AD}$ is the perpendicular bisector of $\overline{CE}$

Prove: $\Delta CBD \cong \Delta EBA$

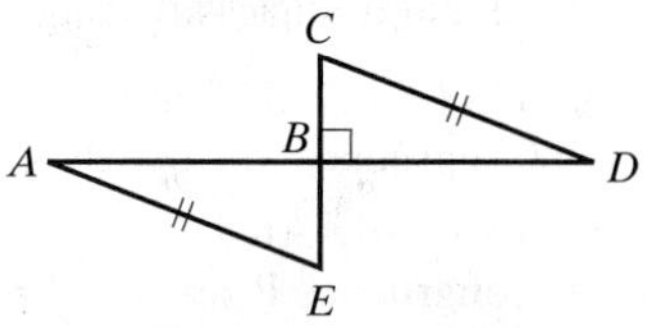

VOCABULARY & READINESS CHECK

Word Bank *Use the choices to fill in each blank. Some choices may be used more than once.*

leg	equilateral	vertex
base	hypotenuse	isosceles

1. A triangle with at least two equal sides is a(n) ____________ triangle.
2. The most specific name for a triangle with three equal sides is a(n) ____________ triangle.

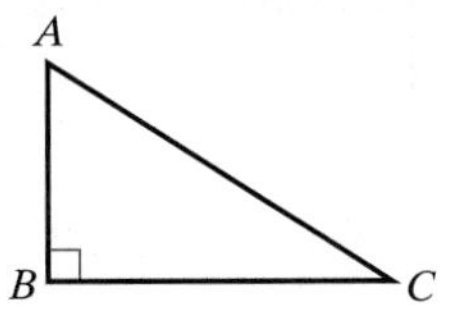

Right triangle

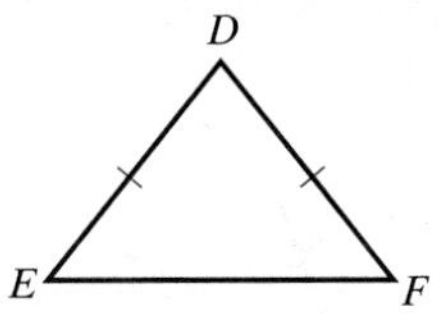

Isosceles triangle

3. In ΔABC, side $\overline{AC}$ is called the ____________.
4. In ΔABC, side $\overline{AB}$ and side $\overline{BC}$ are each called a(n) ____________.
5. In ΔDEF, side $\overline{EF}$ is called the ____________.
6. In ΔDEF, side $\overline{DE}$ and side $\overline{DF}$ are each called a(n) ____________.
7. In ΔDEF, $\angle D$ is called the ____________ angle.
8. In ΔDEF, $\angle E$ and $\angle F$ are each called a(n) ____________ angle.

Circle the correct word to make the statement true.

9. All or some isosceles triangles are equilateral triangles.
10. All or some equilateral triangles are isosceles triangles.
11. All or some equiangular triangles are equilateral triangles.
12. All or some equilateral triangles are equiangular triangles.

4.6 EXERCISE SET MyMathLab®

MIXED PRACTICE

Is there enough information to determine whether the two triangles are congruent? If so, write the congruence statement. See Example 1 and 5.

1.

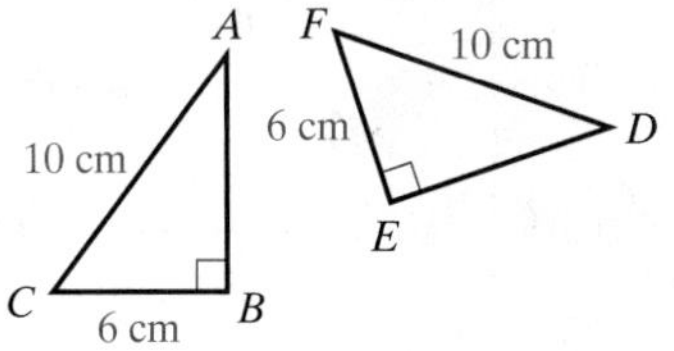

2.

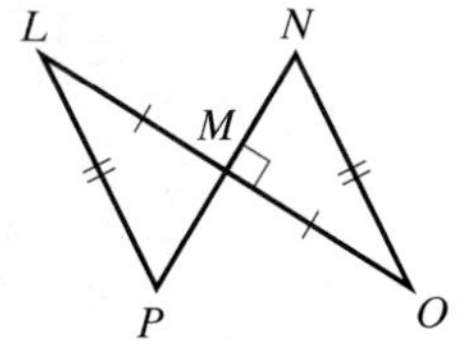

3.

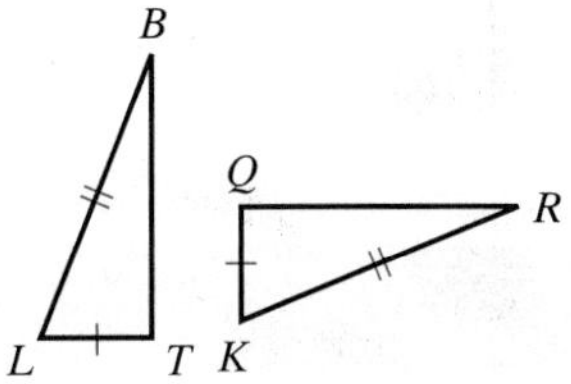

4.

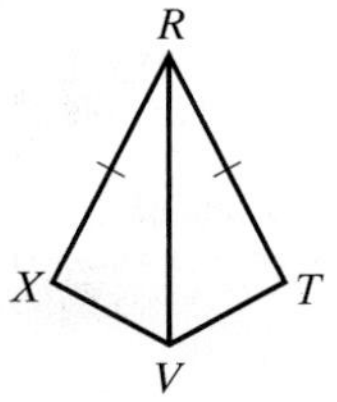

Find the unknown angle measures. See Examples 2–4.

5.

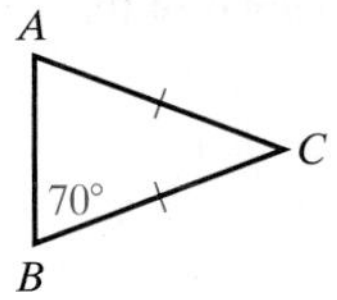

6.

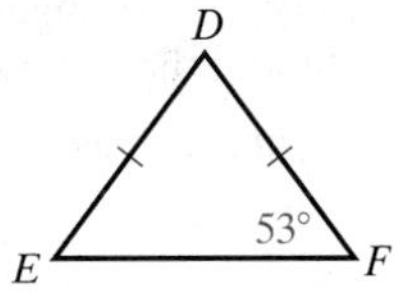

7.

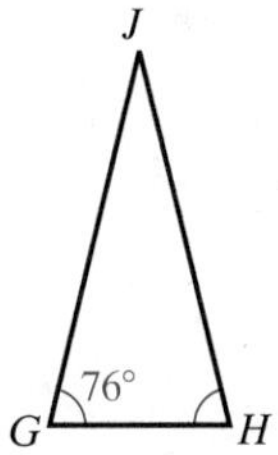

8.

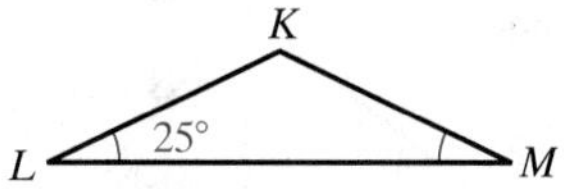

9.

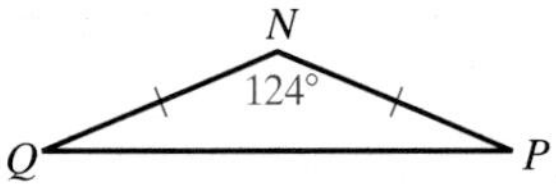

10.

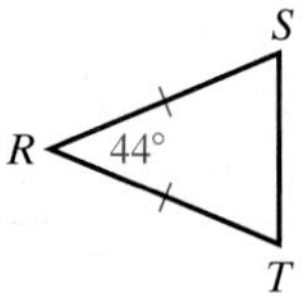

11.

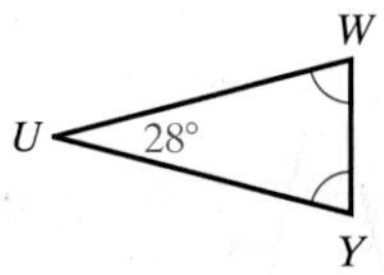

12.

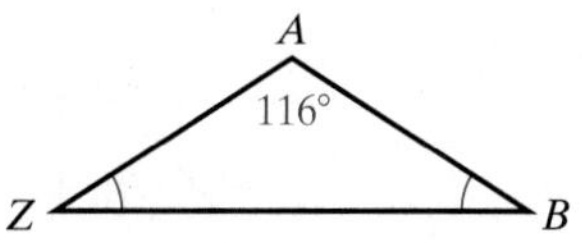

13.

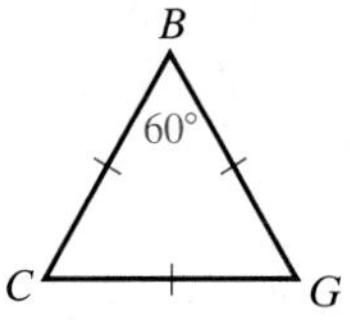

14.

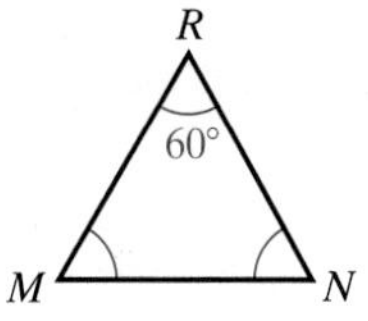

15. 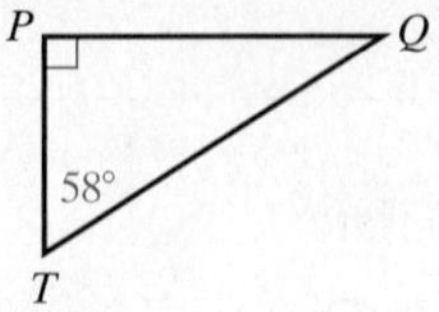

16. 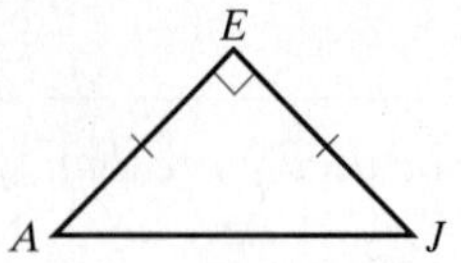

Fill in the Blank *Read Theorem 4.6-3 again and use it to fill in the blanks for Exercises 17–28.*

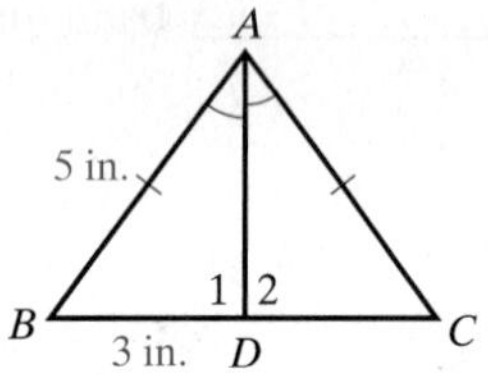

17. $DC =$ ______ in.
18. $BC =$ ______ in.
19. $m\angle 1 =$ ______°
20. $m\angle 2 =$ ______°
21. $\angle B \cong \angle$______
22. $\angle BAD \cong \angle$______

Given isosceles ΔJKL *with base* $\overline{JL}$, *find each value.*

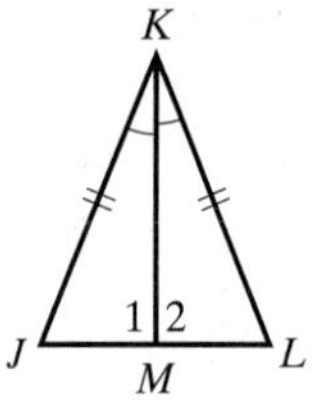

23. $m\angle 1 =$ ______
24. $m\angle 2 =$ ______
25. If $m\angle L = 58°$, then $m\angle LKJ =$ ___?___.
26. If $JL = 5$, then $ML =$ ___?___.
27. If $m\angle JKM = 48°$, then $m\angle J =$ ___?___.
28. If $m\angle J = 55°$, then $m\angle JKM =$ ___?___.

Mixed Practice *Complete each statement. Explain why it is true.*

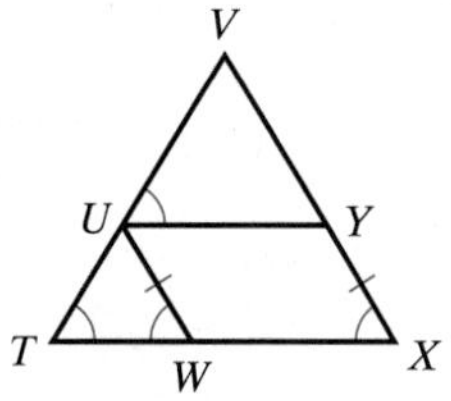

29. $\overline{VT} \cong$ ___?___
30. $\overline{UT} \cong$ ___?___ $\cong \overline{YX}$
31. $\overline{VU} \cong$ ___?___
32. $\angle VYU \cong$ ___?___

Find the values of x and y. See Examples 2–4.

33. 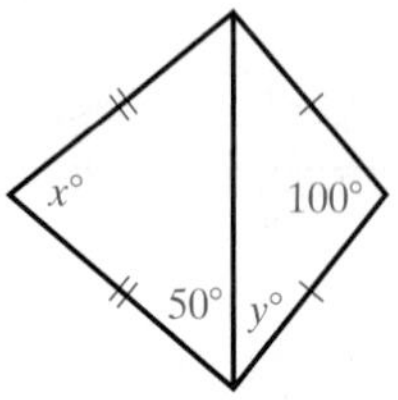

34. 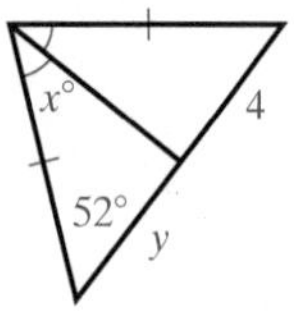

35. 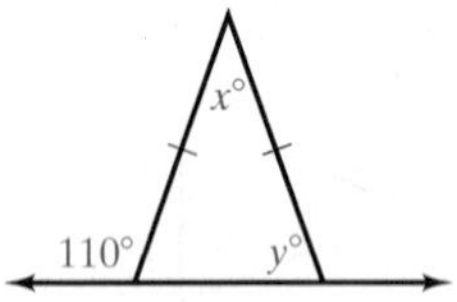

36. 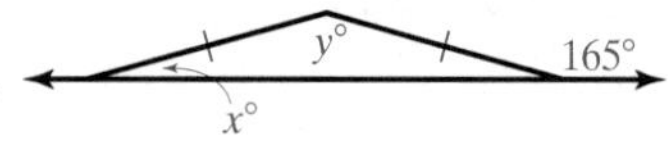

Find the values of x and y.

37. 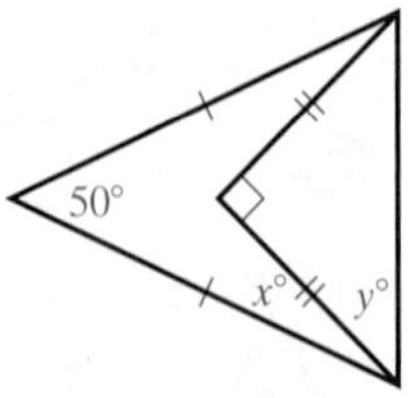

38. 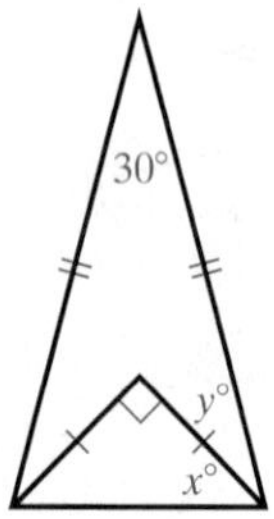

39. 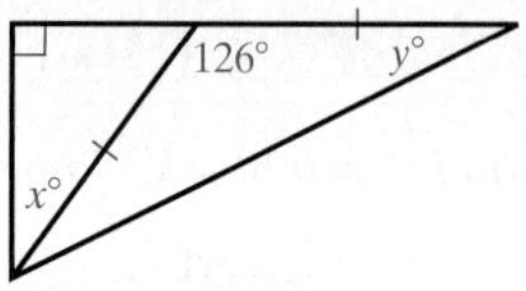

40. 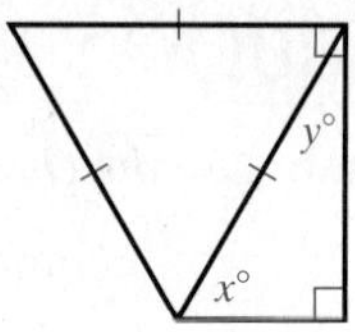

For what values of x and y are the triangles congruent by H-L? See Example 5.

41. 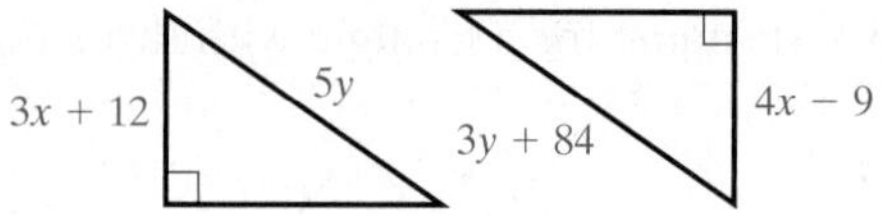

42. 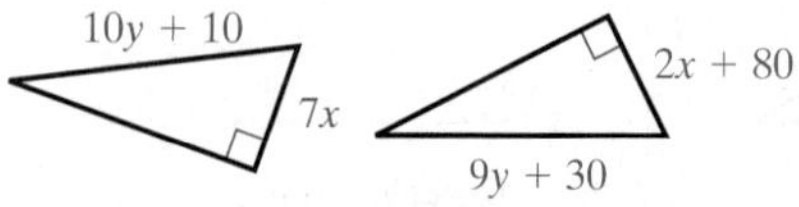

43. An equilateral triangle and an isosceles triangle share a common side. What is the measure of $\angle ABC$?

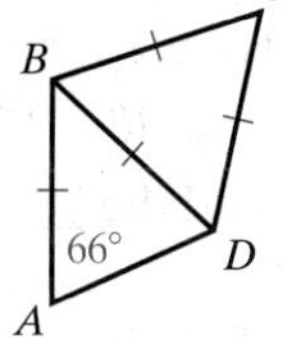

44. Each face of the Great Pyramid at Giza is an isosceles triangle with a 76° vertex angle. What are the measures of the base angles?

45. What are the measures of the base angles of a right isosceles triangle? Explain.

46. ΔABC and ΔPQR are right triangular sections of a fire escape, as shown. Is each story of the building the same height? Explain.

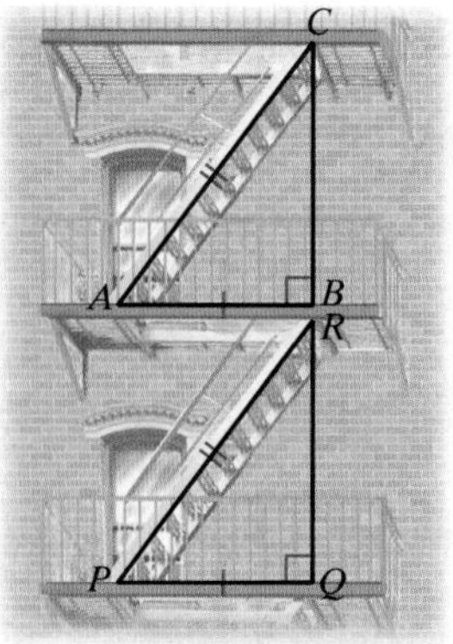

47. **Multiple Steps** In the diagram on the next page, assume that the cables of the same height have equal lengths.

a. What type of triangle is formed by the cables of the same height and the ground?

b. What are the two different total base lengths of the triangles?

c. How is the tower related to each of the triangles?

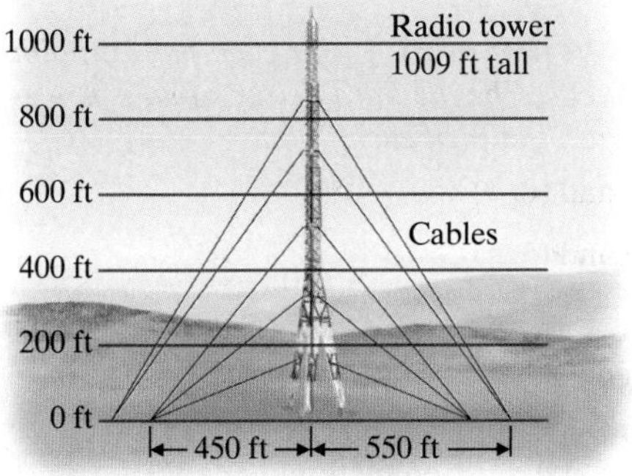

48. The length of the base of an isosceles triangle is x. The length of a leg is $2x - 5$. The perimeter of the triangle is 20. Find x.

CONCEPT EXTENSIONS

Proof *For Exercises 49 and 50, complete the proof. See Examples 1 and 5.*

49. Given: $\overline{PS} \cong \overline{PT}$, $\angle 1 \cong \angle 2$

Prove: $\Delta PRS \cong \Delta PRT$

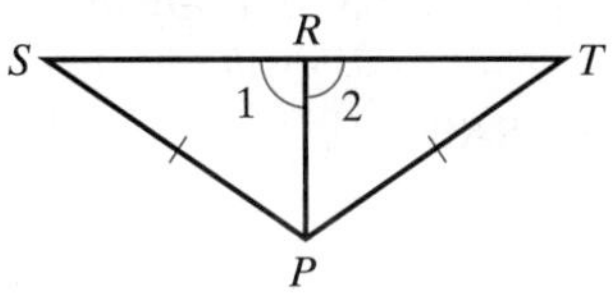

Statements	Reasons
1. $\angle 1 \cong \angle 2$	**1.** Given
2. $\angle 1$ and $\angle 2$ are supplementary.	**2.** Angles that form a linear pair are supplementary.
3. $\angle 1$ and $\angle 2$ are right angles.	**3.** ? (a)
4. ΔPRS and ΔPRT are right triangles.	**4.** ? (b)
5. $\overline{PS} \cong \overline{PT}$	**5.** ? (c)
6. $\overline{PR} \cong \overline{PR}$	**6.** ? (d)
7. $\Delta PRS \cong \Delta PRT$	**7.** ? (e)

50. Given: $\angle A$ and $\angle D$ are right angles, $\overline{AB} \cong \overline{DE}$

Prove: $\Delta ABE \cong \Delta DEB$

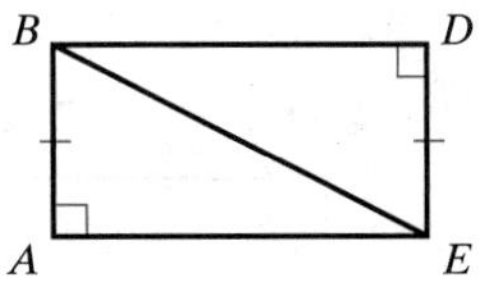

Statements	Reasons
1. $\angle A$ and $\angle D$ are right angles.	**1.** Given
2. $\angle A \cong \angle D$	**2.** ? (a)
3. $\overline{BE} \cong \overline{BE}$	**3.** ? (b)
4. $\overline{AB} \cong \overline{DE}$	**4.** ? (c)
5. $\Delta ABE \cong \Delta DEB$	**5.** ? (d)

Proof *For Exercises 51–60, complete each proof. See Examples 1 and 5.*

51. Given: $\overline{HV} \perp \overline{GT}$, $\overline{GH} \cong \overline{TV}$, I is the midpoint of $\overline{HV}$

Prove: $\Delta IGH \cong \Delta ITV$

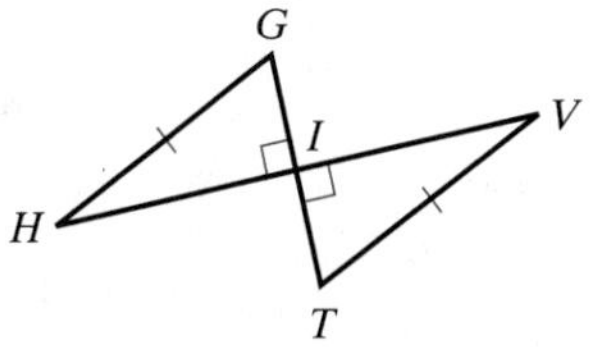

52. Given: $\overline{PM} \cong \overline{RJ}$, $\overline{PT} \perp \overline{TJ}$, $\overline{RM} \perp \overline{TJ}$, M is the midpoint of $\overline{TJ}$

Prove: $\Delta PTM \cong \Delta RMJ$

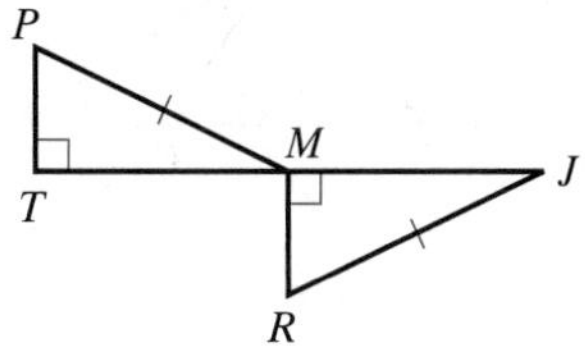

53. Given: $\overline{RS} \cong \overline{TU}$, $\overline{RS} \perp \overline{ST}$, $\overline{TU} \perp \overline{UV}$, T is the midpoint of $\overline{RV}$

Prove: $\Delta RST \cong \Delta TUV$

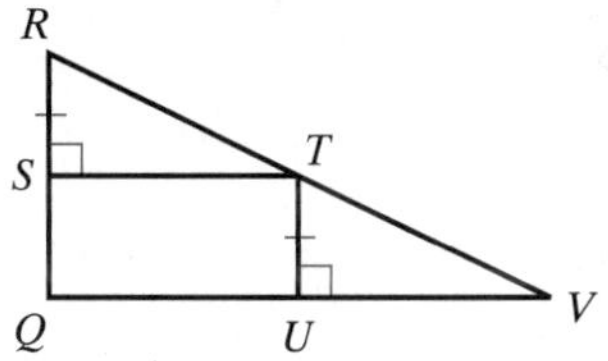

54. Given: ΔLNP is isosceles with base $\overline{NP}$, $\overline{MN} \perp \overline{NL}$, $\overline{QP} \perp \overline{PL}$, $\overline{ML} \cong \overline{QL}$

Prove: $\Delta MNL \cong \Delta QPL$

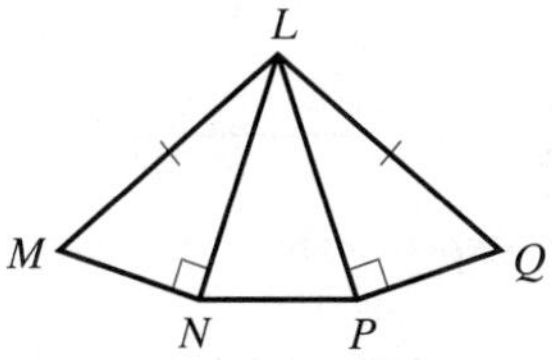

55. Given: ΔGKE is isosceles with base $\overline{GE}$, $\angle L$ and $\angle D$ are right angles, K is the midpoint of $\overline{LD}$

Prove: $\overline{LG} \cong \overline{DE}$

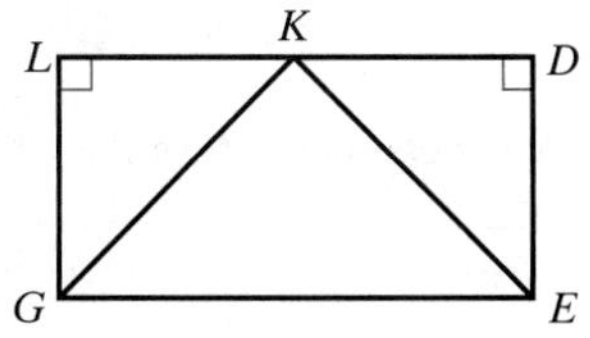

56. Given: $\overline{LO}$ bisects $\angle MLN$, $\overline{OM} \perp \overline{LM}$, $\overline{ON} \perp \overline{LN}$

Prove: $\Delta LMO \cong \Delta LNO$

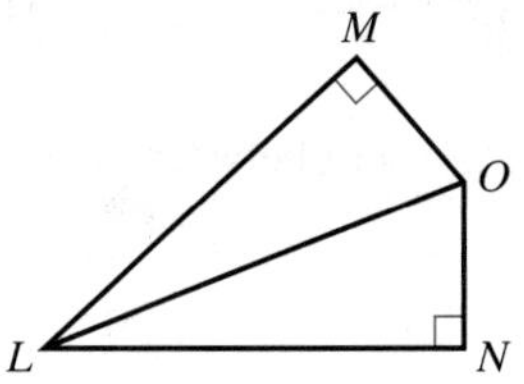

57. Prove Theorem 4.6-3. To do so, use the information below.

Given: $\overline{AB} \cong \overline{AC}$, $\angle 1 \cong \angle 2$

Prove: $\overline{AD} \perp \overline{BC}$ and $\overline{BD} \cong \overline{DC}$

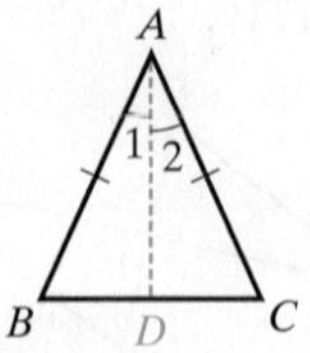

58. Given: $\overline{AE} \cong \overline{DE}$, $\overline{AB} \cong \overline{DC}$

Prove: $\Delta ABE \cong \Delta DCE$

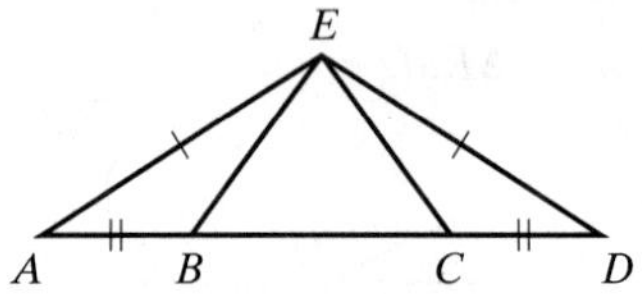

59. Prove Corollary 4.6-4 (to Theorem 4.6-1). To do so, use the information below.

Given: $\overline{AB} \cong \overline{BC} \cong \overline{CA}$

Prove: $\angle A \cong \angle B \cong \angle C$

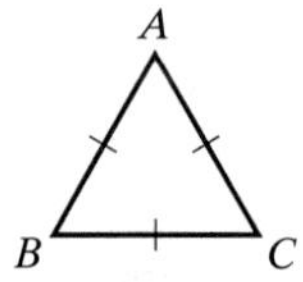

60. Prove Corollary 4.6-5 (to Theorem 4.6-2). To do so, use the information below.

Given: $\angle A \cong \angle B \cong \angle C$

Prove: $\overline{AB} \cong \overline{BC} \cong \overline{CA}$

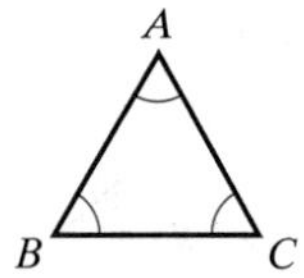

61. Find the Error Explain why the marking of the isosceles, but not equilateral, triangle is incorrect.

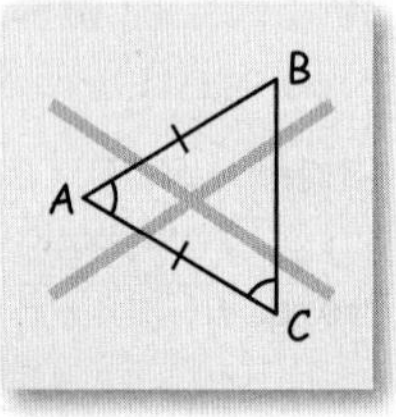

62. Find the Error Your classmate says that there is not enough information to determine whether the two triangles below are congruent. Is your classmate correct? Explain.

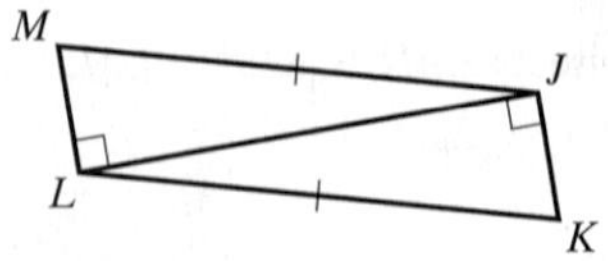

63. What is the relationship between sides and angles for each type of triangle?

a. isosceles

b. equilateral

64. A right triangle has side lengths of 5 cm, 12 cm, and 13 cm. Which length is the length of the hypotenuse? How do you know?

Coordinate Geometry *For each pair of points, there are six points that could be the third vertex of an isosceles right triangle. Find the coordinates of each point.*

65. (4, 0) and (0, 4)

66. (0, 0) and (5, 5)

Geometry in Three Dimensions *For Exercises 67 and 68, use the figure at the right.*

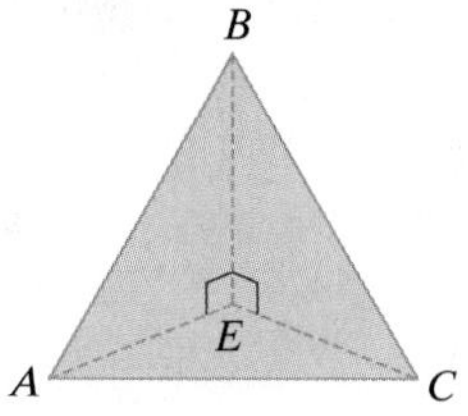

67. Given: $\overline{BE} \perp \overline{EA}$, $\overline{BE} \perp \overline{EC}$, ΔABC is equilateral

Prove: $\Delta AEB \cong \Delta CEB$

68. Given: $\Delta AEB \cong \Delta CEB$, $\overline{BE} \perp \overline{EA}$, $\overline{BE} \perp \overline{EC}$

Can you prove that ΔABC is equilateral? Explain.

MIXED PRACTICE (SECTIONS 4.3, 4.4, 4.6)

Construction *Copy the triangle and construct a triangle congruent to it using the given method.*

69. SAS

70. H − L

71. ASA

72. SSS

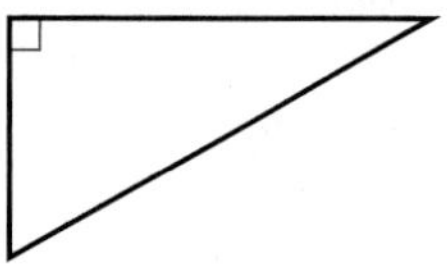

REVIEW AND PREVIEW

For Exercises 73 and 74, what type of triangle must ΔSTU be? Explain. See Section 4.1 and this section.

73. $\Delta STU \cong \Delta UTS$

74. $\Delta STU \cong \Delta UST$

Can you conclude that the triangles are congruent? Explain. See Sections 4.3, 4.4, and this section.

75. ΔABC and ΔLMN

76. ΔLMN and ΔHJK

77. ΔRST and ΔABC

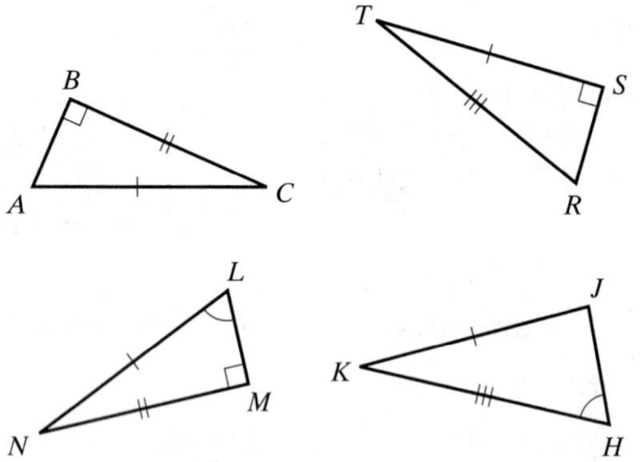

CHAPTER 4 VOCABULARY CHECK

Word Bank *Use the choices to fill in each blank. Some choices may be used more than once.*

leg	isosceles	equiangular	vertex	sides
base	scalene	equilateral	vertices	triangle
	congruent	corresponding	hypotenuse	

1. Figures that have the exact same shape and size are called ____________ figures.
2. When figures are congruent, their ____________ sides are congruent and their ____________ angles are congruent.
3. A(n) ____________ is formed by three noncollinear points connected by segments.
4. The segments joining the points in Exercise 3 above are called ____________.
5. The noncollinear points in Exercise 3 above are called ____________.
6. The singular of the word vertices is ____________.
7. A triangle with no equal sides is a(n) ____________ triangle.
8. A triangle with at least two equal sides is a(n) ____________ triangle.
9. The most specific name for a triangle with three equal sides is a(n) ____________ triangle.
10. A triangle that is equilateral is also ____________.

Fill in the Blank *Use the figures below for Exercises 11–16.*

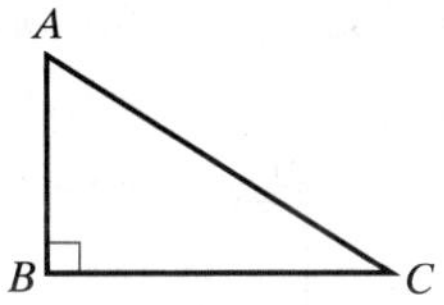

Right triangle

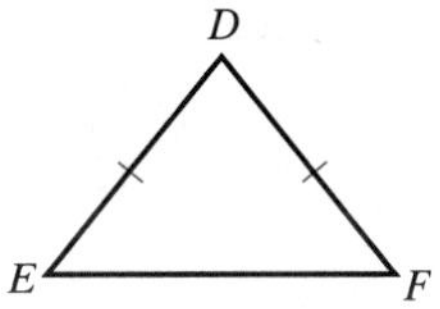

Isosceles triangle

11. In ΔABC, side $\overline{AC}$ is called the ____________.
12. In ΔABC, side $\overline{AB}$ and side $\overline{BC}$ are each called a(n) ____________.
13. In ΔDEF, side $\overline{EF}$ is called the ____________.
14. In ΔDEF, side $\overline{DE}$ and side $\overline{DF}$ are each called a(n) ____________.
15. In ΔDEF, $\angle D$ is called the ____________ angle.
16. In ΔDEF, $\angle E$ and $\angle F$ are each called a(n) ____________ angle.
17. Write the meaning of cpoctac. ____

CHAPTER 4 REVIEW

***(4.1)** Fill in the blank.*

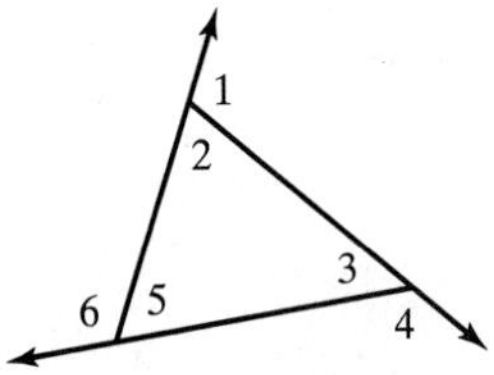

For Exercises 1 and 2

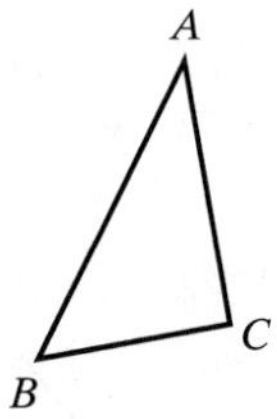

For Exercises 3 and 4

1. The interior angles are ____________.
2. The exterior angles are ____________.
3. The adjacent sides that share $\angle A$ are ____________.
4. The side opposite $\angle A$ is ____________.

Find m∠1.

5. 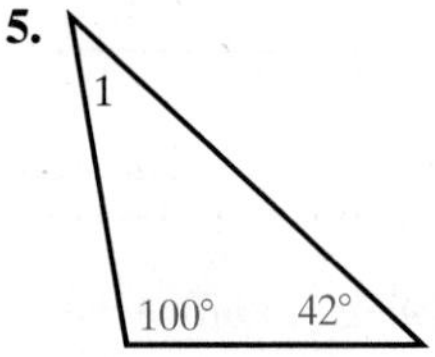

6. 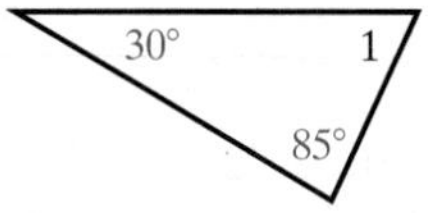

Find the values of the variables.

7. 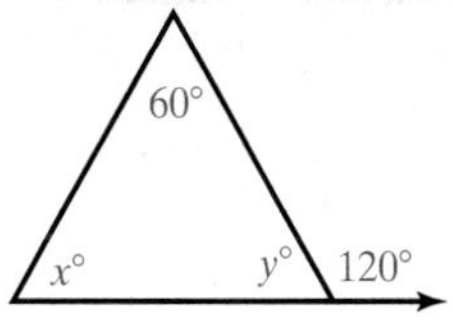

8. 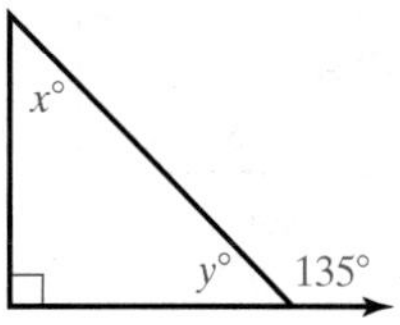

The measures of the three angles of a triangle are given. Find the value of x.

9. $x°, 2x°, 3x°$

10. $x° + 10°, x° - 20, x° + 25°$

(4.2) *RSTUV ≅ KLMNO. Complete the congruence statements.*

11. $\overline{TS} \cong$ ___?___ **12.** $\angle N \cong$ ___?___

13. $\overline{LM} \cong$ ___?___ **14.** $VUTSR \cong$ ___?___

WXYZ ≅ PQRS. Find each measure or length.

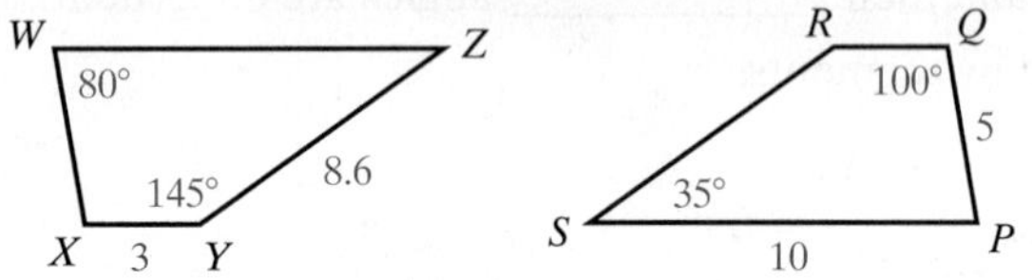

15. $m\angle P$ **16.** QR

17. WX **18.** $m\angle Z$

19. $m\angle X$ **20.** $m\angle R$

(4.3 and 4.4) *Fill in the blank. Name the postulate or theorem that immediately proves the triangles are congruent. List the vertices of the congruent triangle in correct corresponding order.*

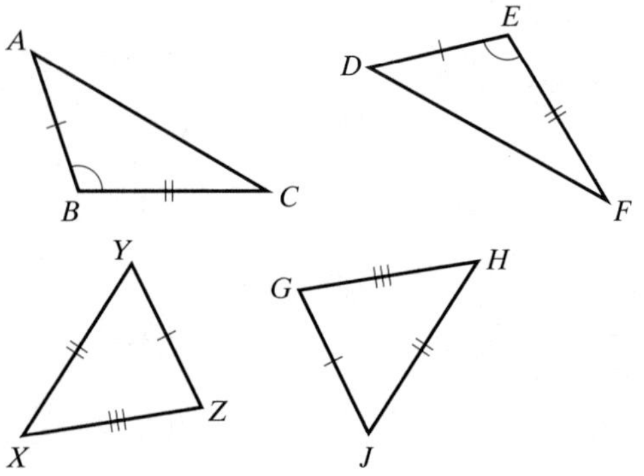

21. $\triangle ABC \cong \triangle$_____ by the __________ Postulate/Theorem.

22. $\triangle XYZ \cong \triangle$_____ by the __________ Postulate/Theorem.

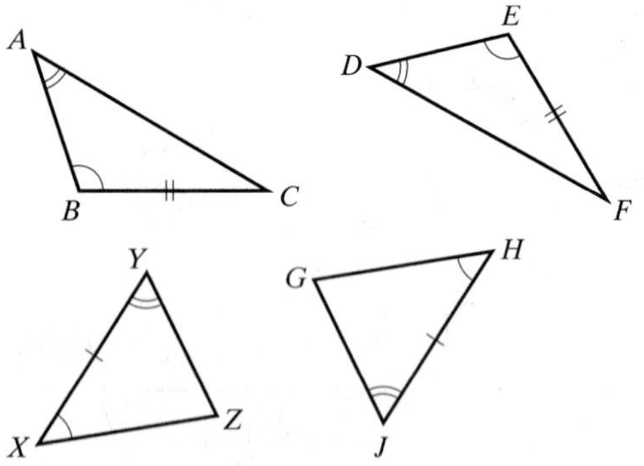

23. $\triangle ABC \cong \triangle$_____ by the __________ Postulate/Theorem.

24. $\triangle XYZ \cong \triangle$_____ by the __________ Postulate/Theorem.

Which postulate or theorem, if any, could you use to prove the two triangles are congruent? If there is not enough information to prove the triangles are congruent, write "not enough information."

25.

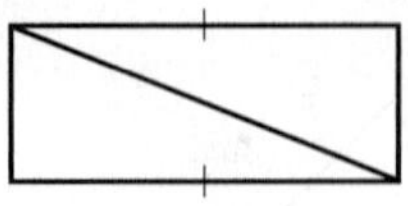

26.

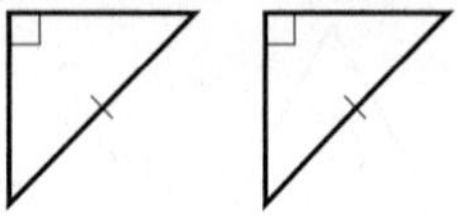

27.

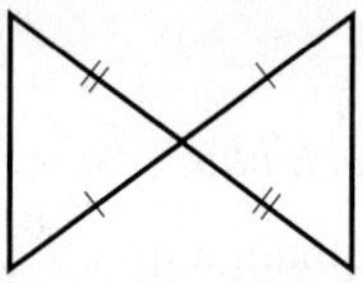

28.

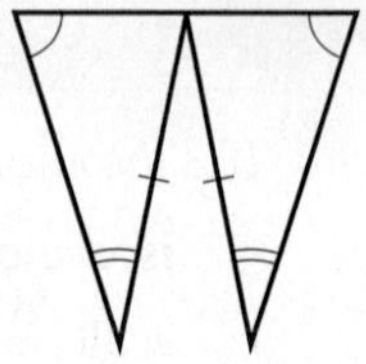

(4.5) *For each exercise, prove that the two triangles are congruent. Then use the congruent triangles to prove each statement is true.*

29. $\overline{TV} \cong \overline{YW}$

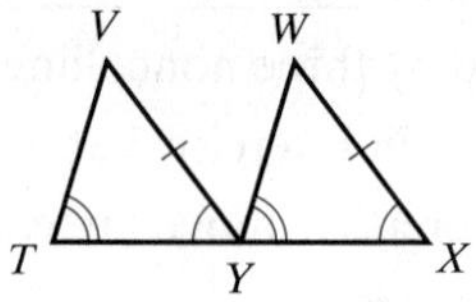

30. $\overline{BE} \cong \overline{DE}$

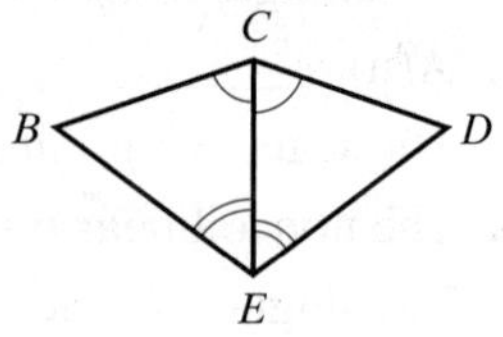

31. $\angle B \cong \angle D$

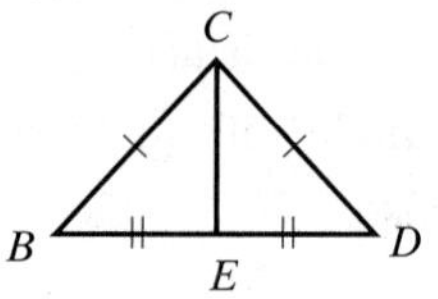

32. $\overline{KN} \cong \overline{ML}$

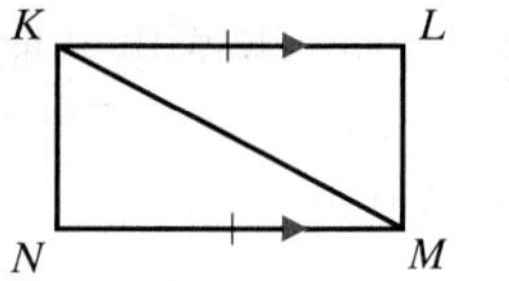

Name a pair of overlapping congruent triangles in each diagram. State whether the triangles are congruent by SSS, SAS, ASA, or AAS.

33.

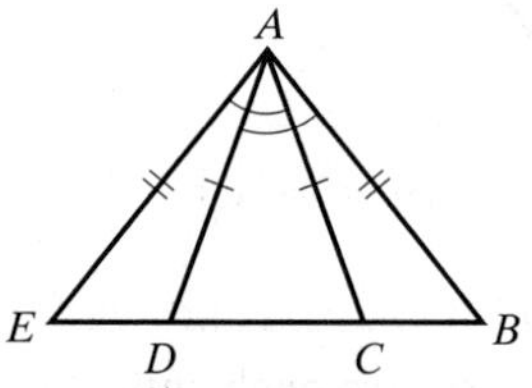

34.

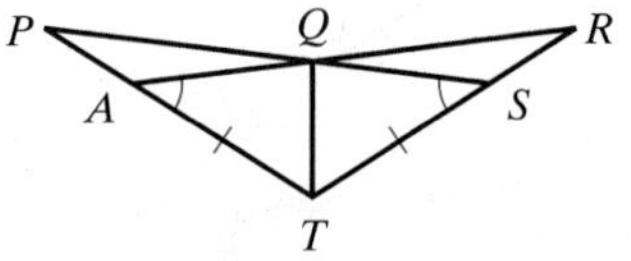

(4.6) *Find the values of x and y.*

35.

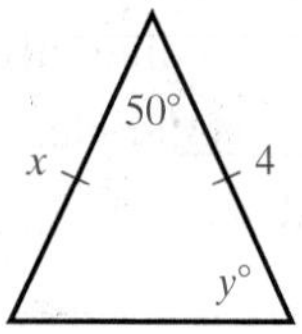

36.

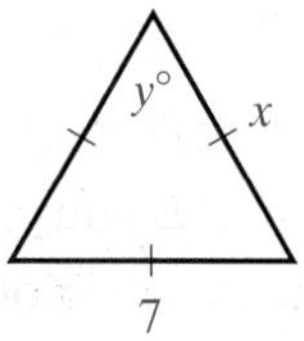

37.

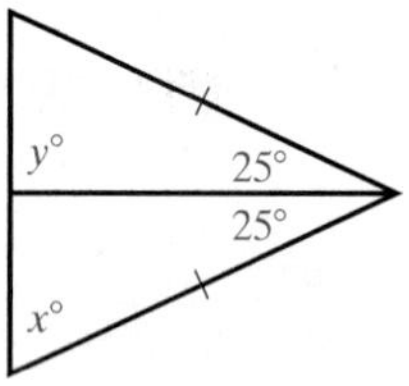

38.

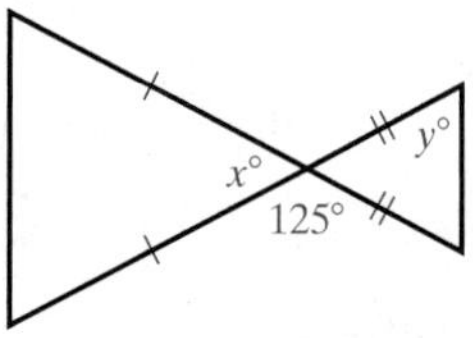

Proof *Write a proof for each of the following.*

39. Given: $\overline{LN} \perp \overline{KM}, \overline{KL} \cong \overline{ML}$

Prove: $\triangle KLN \cong \triangle MLN$

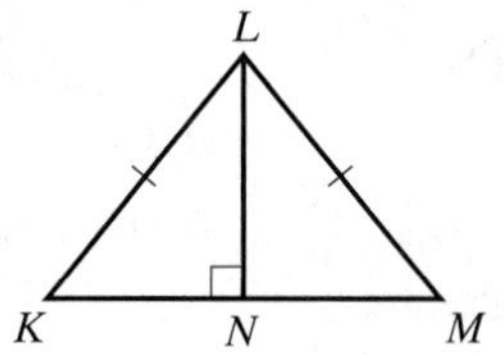

40. Given: $\overline{PS} \perp \overline{SQ}, \overline{RQ} \perp \overline{QS}$, $\overline{PQ} \cong \overline{RS}$

Prove: $\Delta PSQ \cong \Delta RQS$

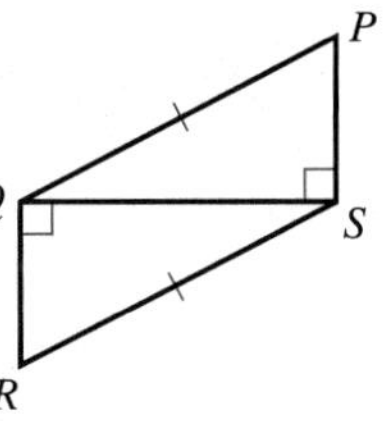

MIXED REVIEW

41. The measures of the three angles of a triangle are given. Find the value of x.

$$(20x + 10)°, (30x - 2)°, (7x + 1)°$$

42. What are the values of x and y?

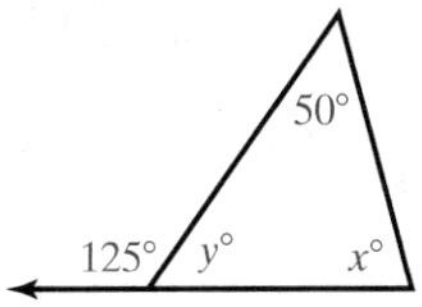

43. $HIJK \cong PQRS$. Write all possible congruence statements.

44. What postulate or theorem would you use to prove the triangles congruent?

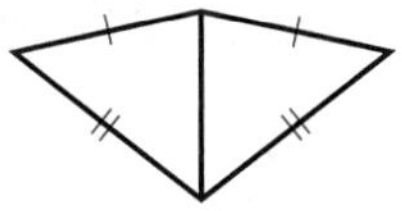

45. How can you use congruent triangles to prove $\angle Q \cong \angle D$?

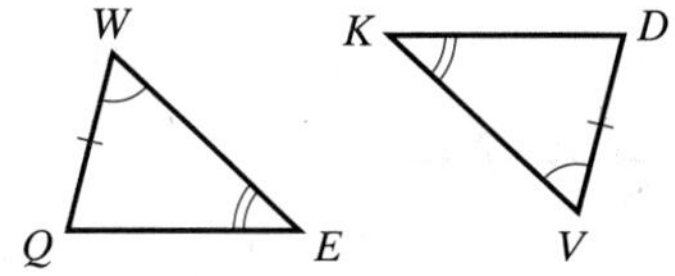

46. What is $m\angle G$?

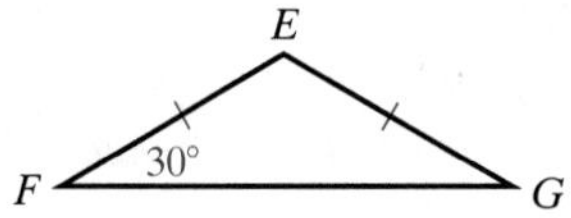

47. Which two triangles are congruent? Explain.

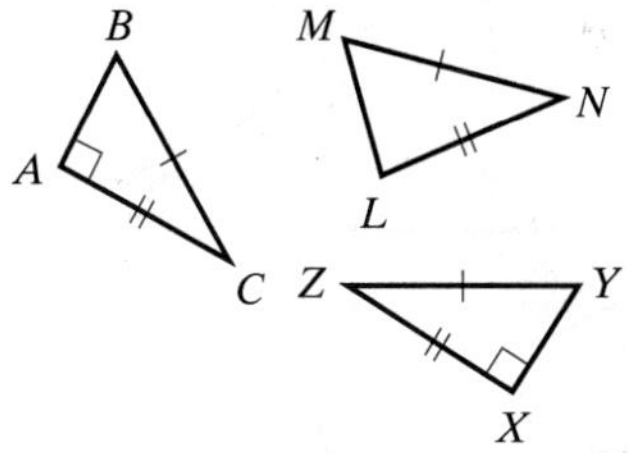

Name a pair of overlapping congruent triangles in the diagram. State whether the triangles are congruent by SSS, SAS, ASA, AAS, or H-L.

48.

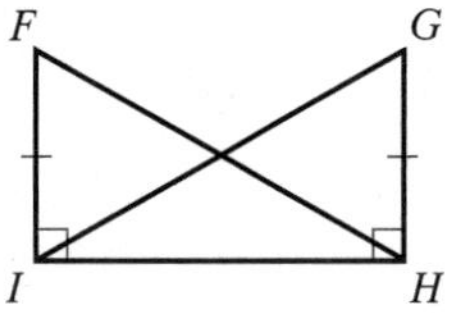

CHAPTER 4 TEST

TEST PREP VIDEO The fully worked-out solutions to any exercises you want to review are available in MyMathLab.

Complete a Table *Complete the table using the most specific name.*
Angles: acute, obtuse, right, equiangular
Sides: scalene, equilateral, isosceles

	Triangle	*Classify by Angles*	*Classify by Sides*
1.			
2.	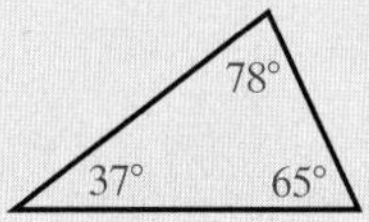		
3.	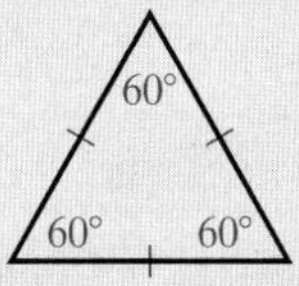		
4.	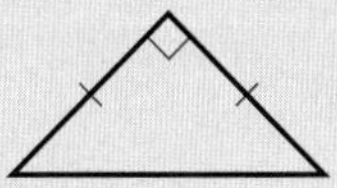		

Find the measures of the numbered angles.

5.

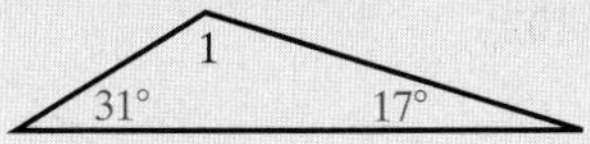

6.

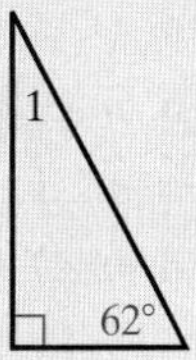

7.

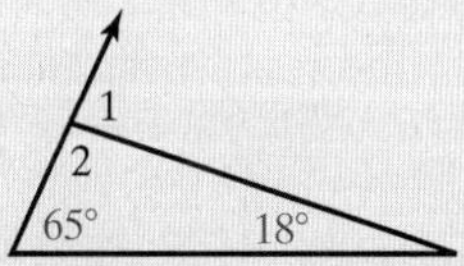

8.

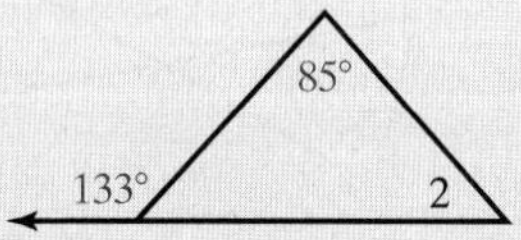

9.

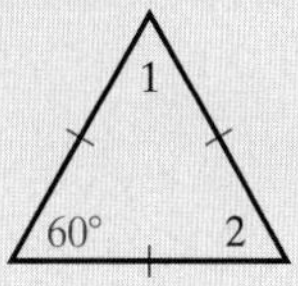

10.

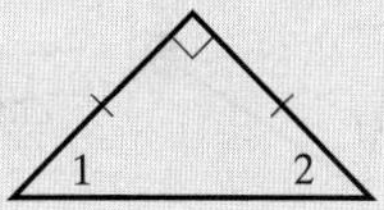

11.

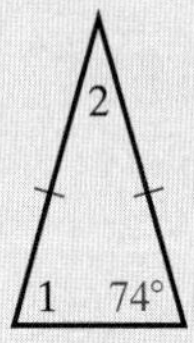

12.

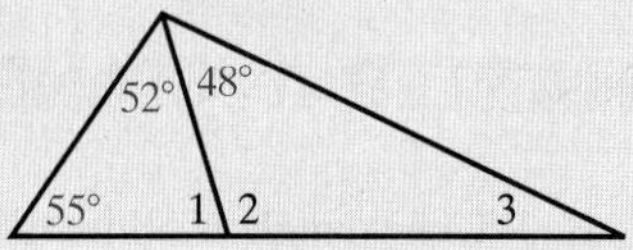

Fill in the Blank *Use ΔMQP for Exercises 13–18.*

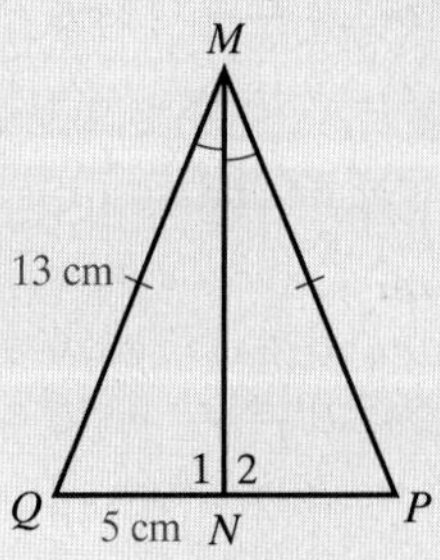

13. $NP =$ ______

14. $QP =$ ______

15. $m\angle 1 =$ ______

16. $m\angle 2 =$ ______

17. $\angle Q \cong \angle$ ______

18. $\angle QMN \cong \angle$ ______

Write a congruence statement for the pair of triangles. Name all of the pairs of corresponding congruent parts.

19.

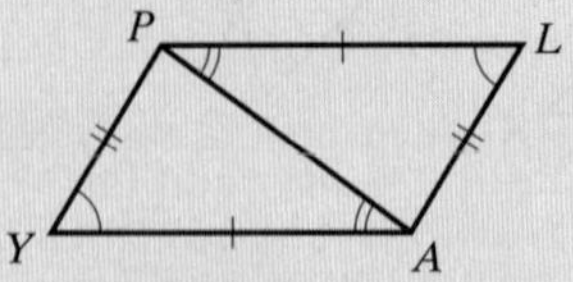

Which postulate or theorem, if any, could you use to prove the two triangles congruent? If not enough information is given, write "not enough information."

20.

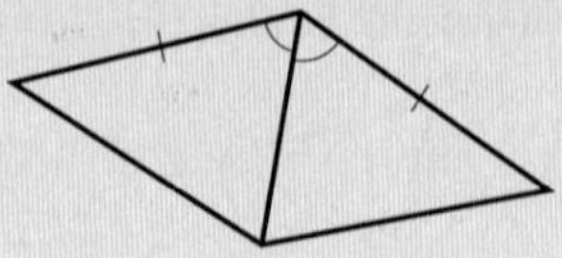

21.

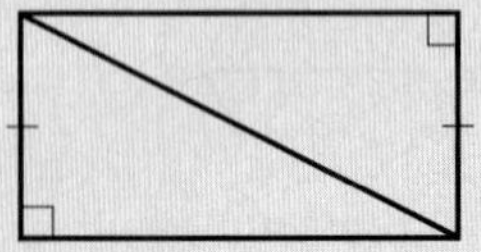

22.

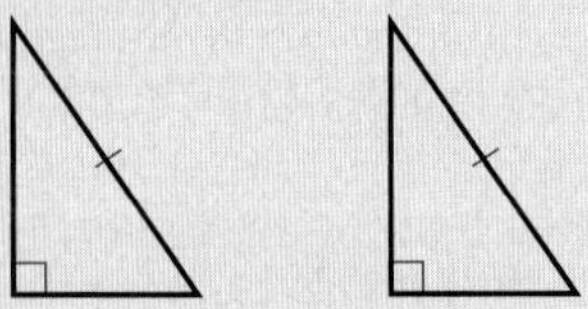

23.

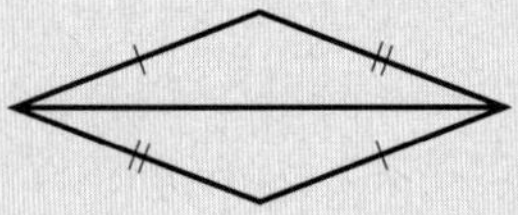

24.

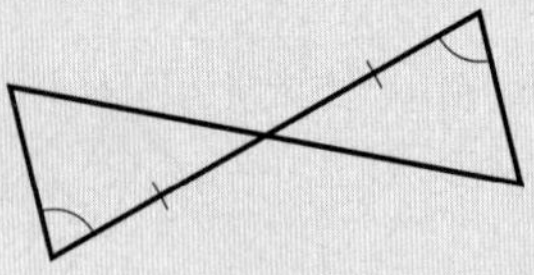

25.

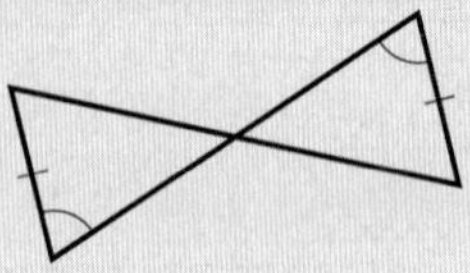

26. Find the value of x.

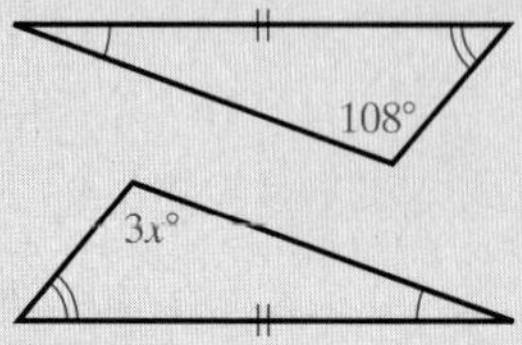

Name a pair of overlapping congruent triangles in each diagram. State whether the triangles are congruent by SSS, SAS, ASA, AAS, or H-L.

27. Given: $\overline{CE} \cong \overline{DF}$,
$\overline{CF} \cong \overline{DE}$

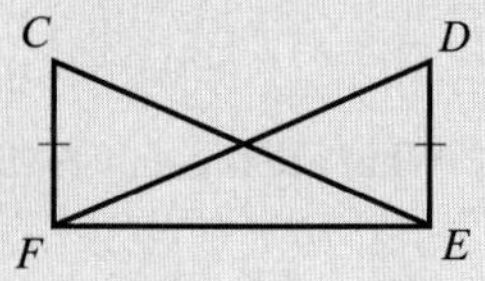

28. Given: $\overline{RT} \cong \overline{QT}$,
$\overline{AT} \cong \overline{ST}$

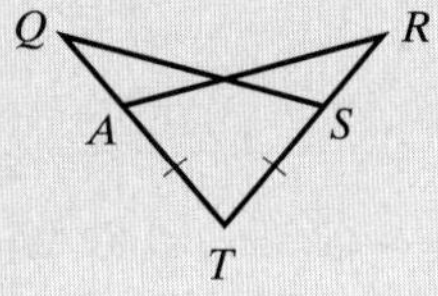

Proof *Write a proof for each of the following.*

29. Given: $\overline{AT} \cong \overline{GS}, \overline{AT} \parallel \overline{GS}$

Prove: $\Delta GAT \cong \Delta TSG$

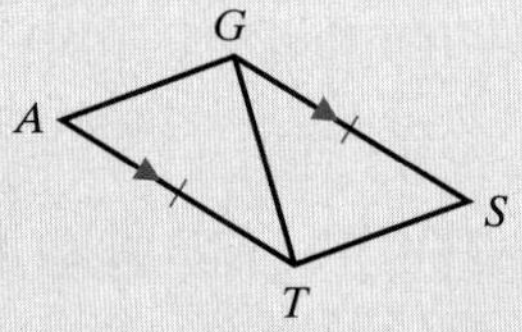

30. Given: $\overline{LN}$ bisects $\angle OLM$ and $\angle ONM$
Prove: $\overline{ON} \cong \overline{MN}$

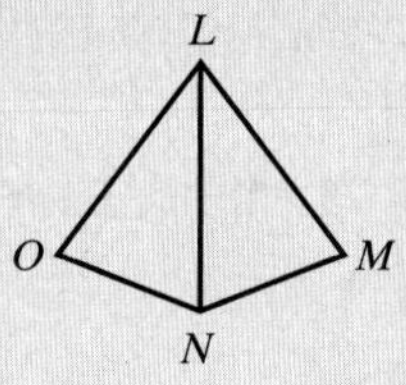

CHAPTER 4 STANDARDIZED TEST

1. Which triangle is obtuse and scalene?

a. $m\angle A = 130°, m\angle B = 25°, m\angle C = 25°$

b. $m\angle A = 92°, m\angle B = 45°, m\angle C = 43°$

c. $m\angle A = 88°, m\angle B = 46°, m\angle C = 46°$

d. $m\angle A = 88°, m\angle B = 40°, m\angle C = 52°$

2. Which triangle is acute and isosceles?

a. $m\angle A = 130°, m\angle B = 25°, m\angle C = 25°$

b. $m\angle A = 92°, m\angle B = 45°, m\angle C = 43°$

c. $m\angle A = 88°, m\angle B = 46°, m\angle C = 46°$

d. $m\angle A = 88°, m\angle B = 40°, m\angle C = 52°$

3. In $\Delta ABC, m\angle A = 46°$ and $m\angle C = 16°$. What is $m\angle B$?

a. 118° **b.** 44°

c. 74° **d.** 138°

4. ΔABC is isosceles with base $\overline{AB}$. If $m\angle C = 22°$, what is $m\angle B$?

a. 79° **b.** 68°

c. 22° **d.** 136°

5. ΔDEF is isosceles with base $\overline{DE}$. If $m\angle D = 22°$, what is $m\angle F$?

a. 79° **b.** 68°

c. 136° **d.** 22°

6. Choose the correct measures for the numbered angles.

a. $m\angle 1 = 35°, m\angle 2 = 107°$

b. $m\angle 1 = 38°, m\angle 2 = 104°$

c. $m\angle 1 = 55°, m\angle 2 = 87°$

d. $m\angle 1 = 35°, m\angle 2 = 127°$

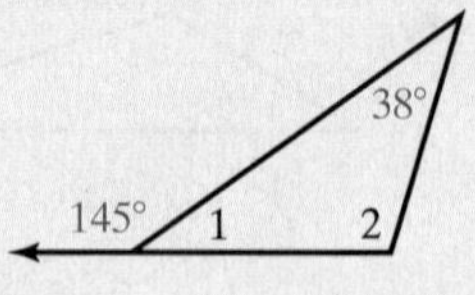

7. ΔDEF is a right triangle, with $\angle E$ a right angle. If $m\angle D = 27°$, what is $m\angle F$?

a. 27° **b.** 63°

c. 153° **d.** 53°

8. Choose the correct measures for the numbered angles.

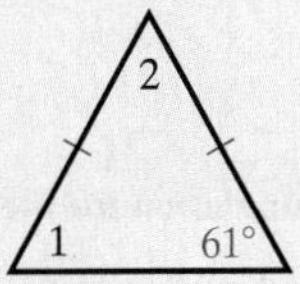

a. $m\angle 1 = 29°, m\angle 2 = 90°$

b. $m\angle 1 = 61°, m\angle 2 = 29°$

c. $m\angle 1 = 58°, m\angle 2 = 61°$

d. $m\angle 1 = 61°, m\angle 2 = 58°$

9. According to the diagram, which statement is true?

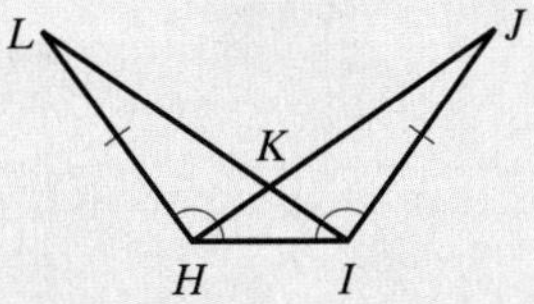

a. $\Delta LHI \cong \Delta JIH$ by SAS

b. $\Delta LHK \cong \Delta JIK$ by AAS

c. $\Delta LHI \cong \Delta JIH$ by SSS

d. $\Delta LHK \cong \Delta JIK$ by SAS

Use the diagram for Exercises 10–15. $\overline{PC} \perp \overline{AE}$ and $\overline{PC}$ bisects $\angle APE$ and $\angle BPD$.

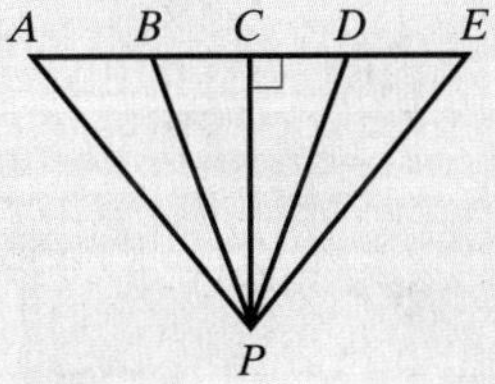

10. Suppose $AC = 25$ and $BC = 14$. What is the length of $\overline{DE}$?

a. 14 **b.** 11

c. 25 **d.** not enough information

11. Suppose $BC = 14$. What is the length of $\overline{BD}$?

a. 28 **b.** 14

c. not enough information **d.** 7

12. Which angle is congruent to $\angle EDP$?

a. $\angle CDP$ **b.** $\angle CBP$

c. $\angle ABP$ **d.** $\angle BAP$

13. Which angle is congruent to $\angle APB$?

a. $\angle EPD$ **b.** $\angle BPC$

c. $\angle CPD$ **d.** $\angle CPE$

14. Suppose $PB = 27$. What is the length of $\overline{PD}$?

a. not enough information

b. greater than 27

c. less than 27

d. 27

15. Which lengths could you use to find the length of $\overline{BC}$?

a. AB and DE **b.** AC and DE

c. AP and AB **d.** BE and BP

Use the diagram for Exercises 16–19.

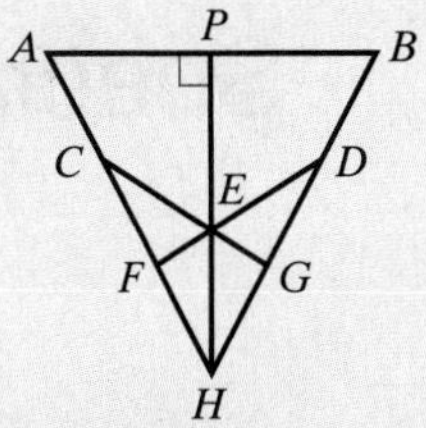

16. Given that $\overline{AH} \cong \overline{BH}$ and $\overline{HP} \perp \overline{AB}$, which statement is immediately true?

a. $\Delta APH \cong \Delta BPH$ by SSS

b. $\Delta APH \cong \Delta BPH$ by SAS

c. $\Delta APH \cong \Delta BPH$ by H-L

d. $\Delta APH \cong \Delta BPH$ by AAA

17. Given that $\overline{CH} \cong \overline{DH}$ and $\overline{FH} \cong \overline{GH}$, which statement is true?

a. $\Delta CHG \cong \Delta DHF$ by SSS

b. $\Delta CHG \cong \Delta DHF$ by SAS

c. $\Delta CFE \cong \Delta DGE$ by SAS

d. $\Delta CFE \cong \Delta DGE$ by SSS

18. Given that $\angle ECF \cong \angle EDG$ and $\overline{CF} \cong \overline{DG}$, which statement is true?

a. $\Delta CFE \cong \Delta DGE$ by SAS

b. $\Delta CGH \cong \Delta DFH$ by SAS

c. $\Delta CFE \cong \Delta DGE$ by AAS

d. $\Delta CGH \cong \Delta DFH$ by AAS

19. Given that $\overline{FE} \cong \overline{GE}$ and $\overline{FH} \cong \overline{GH}$, which statement is true?

a. $\Delta EFH \cong \Delta EGH$ by SSS

b. $\Delta EFH \cong \Delta EGH$ by SSA

c. $\Delta CFE \cong \Delta DGE$ by SAS

d. $\Delta EFH \cong \Delta EGH$ by SAS

For Exercises 20 and 21, suppose $\overline{BD} \cong \overline{HK}$ and $\angle B \cong \angle H$.

20. What additional information do you need to prove $\Delta BDF \cong \Delta HKS$ by ASA?

a. $\angle D \cong \angle K$ **b.** $\angle B \cong \angle S$

c. $\angle D \cong \angle S$ **d.** $\angle F \cong \angle K$

21. What additional information do you need to prove $\Delta BDF \cong \Delta HKS$ by SAS?

a. $\overline{FD} \cong \overline{SK}$ **b.** $\overline{BF} \cong \overline{SK}$

c. $\overline{FD} \cong \overline{HS}$ **d.** $\overline{BF} \cong \overline{HS}$

CHAPTER

5 Special Properties of Triangles

5.1 Perpendicular and Angle Bisectors

5.2 Bisectors of a Triangle

5.3 Medians and Altitudes of a Triangle

5.4 Midsegments of a Triangle

5.5 Indirect Proofs and Inequalities in One Triangle

5.6 Inequalities in Two Triangles

Mean Center of Population for the United States: 1790 to 2020 (projected)

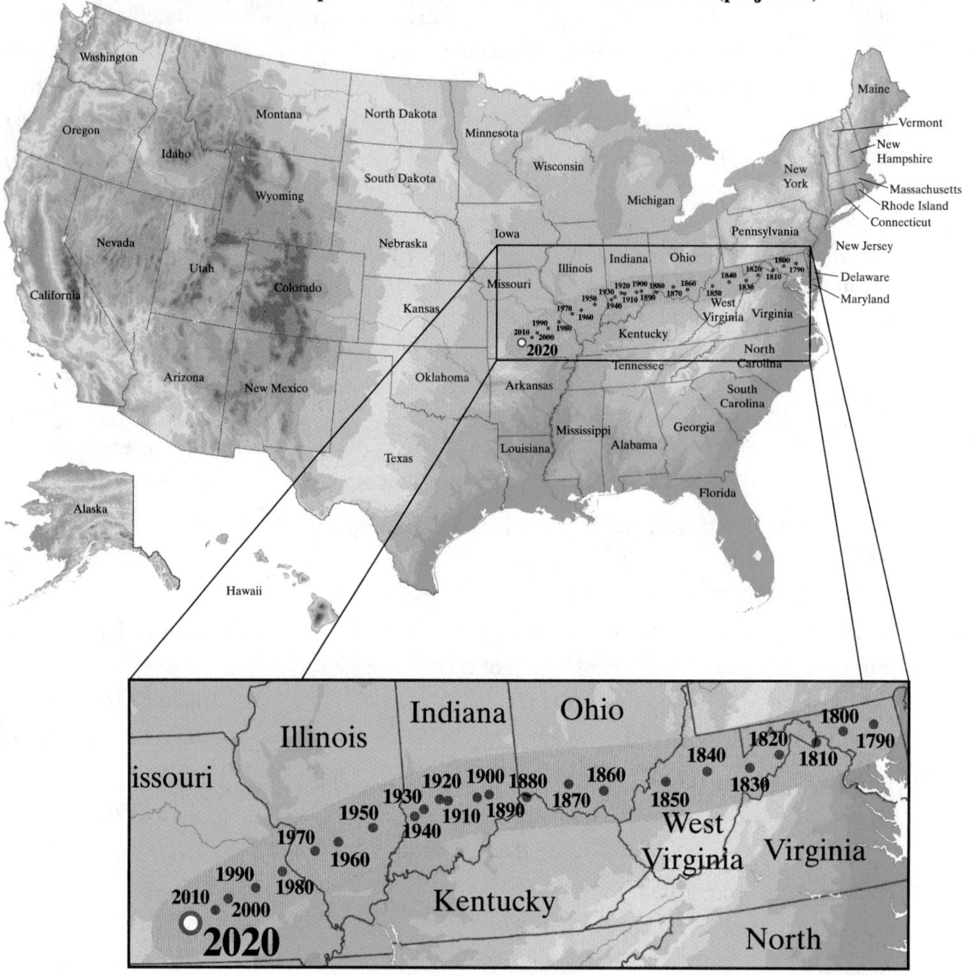

○ 2020 Projected Mean Center of Population located in Wright County, MO
● Mean Center of Population

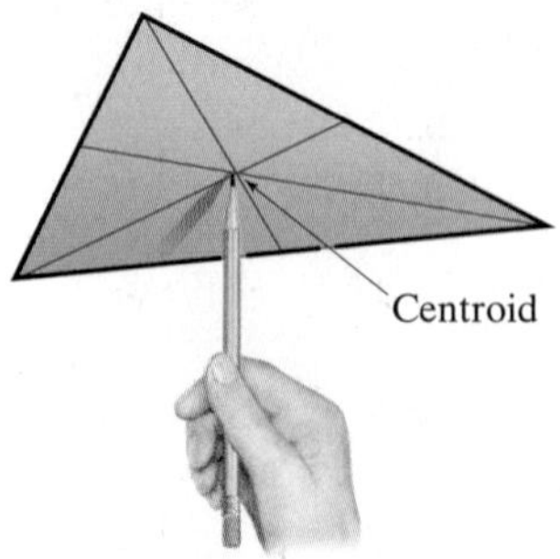

The mean center of U.S. population is the point at which an imaginary, flat, weightless, and rigid map of the United States would balance perfectly if weights of identical value were placed on it so that each weight represented the location of one person on the date of the census.

This is similar to a point called the centroid of a triangle, which we study in Section 5.3. It is the point where a triangular shape of uniform thickness will balance.

5.1 PERPENDICULAR AND ANGLE BISECTORS

OBJECTIVES

1. Use Perpendicular Bisectors to Solve Problems.
2. Use Angle Bisectors to Solve Problems.

VOCABULARY

- equidistant
- distance from a point to a line

OBJECTIVE 1 ▶ Using Perpendicular Bisectors. In Section 1.8, we learned that a **perpendicular bisector** of a segment is a line, a segment, a ray, or even a plane that is perpendicular to the segment at its midpoint.

▶ Helpful Hint

To remember the meaning of perpendicular bisector, think about what each word means.

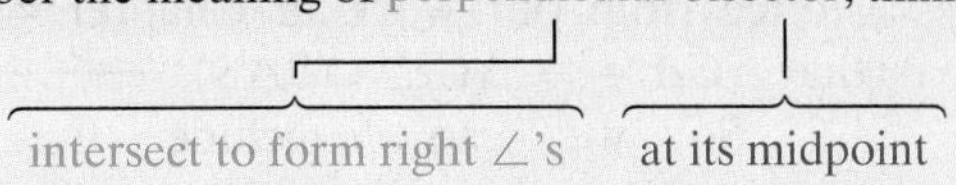

We also learned how to construct the perpendicular bisector of a segment in Section 1.8. Congruent triangles can be used to prove that our construction actually gives the perpendicular bisector.

Theorem 5.1-1 Construction of the Perpendicular Bisector

The construction shown in Section 1.8, Example 3, gives the perpendicular bisector of a segment.

We review the construction below and then prove Theorem 5.1-1.

Construction of Perpendicular Bisector

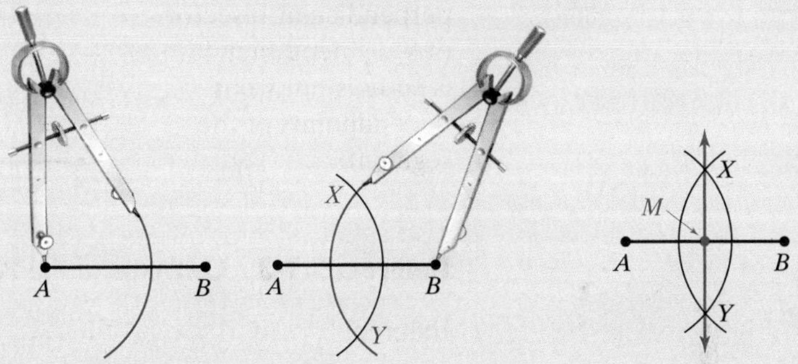

Proof of Construction of Perpendicular Bisector (Theorem 5.1-1)

Given: $\overline{AX} \cong \overline{AY} \cong \overline{BX} \cong \overline{BY}$ (by construction)

Prove: $\overleftrightarrow{XY}$ is the $\perp$ bisector of $\overline{AB}$

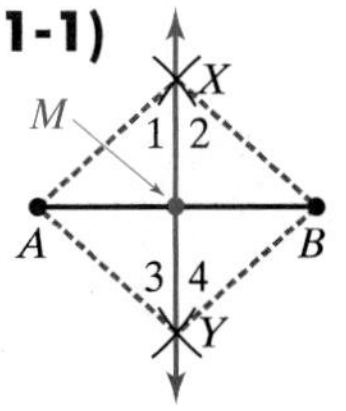

Figure for Proof

Statements	Reasons
1. $\overline{AX} \cong \overline{AY} \cong \overline{BX} \cong \overline{BY}$	**1.** Given by construction
2. $\overline{XY} \cong \overline{XY}$	**2.** Reflexive Property
3. $\Delta YAX \cong \Delta YBX$	**3.** SSS (Steps 1, 2)
4. $\angle 1 \cong \angle 2$	**4.** cpoctac
5. ΔAXB is an isosceles Δ.	**5.** $\overline{AX} \cong \overline{BX}$ (Step 1)
6. $\angle AXB$ is the vertex angle and $\overline{AB}$ is the base of the isosceles Δ.	**6.** Definitions of vertex angle and isosceles Δ
7. $\overleftrightarrow{XY}$ is the $\perp$ bisector of $\overline{AB}$.	**7.** Theorem 4.6-3 (If a line bisects the vertex angle of an isosceles Δ (Steps 4, 5), then the line is also the $\perp$ bisector of the base.)

Let's choose another point on the perpendicular bisector of $\overline{AB}$, say point C.

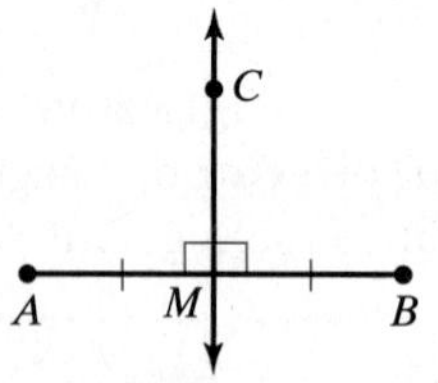

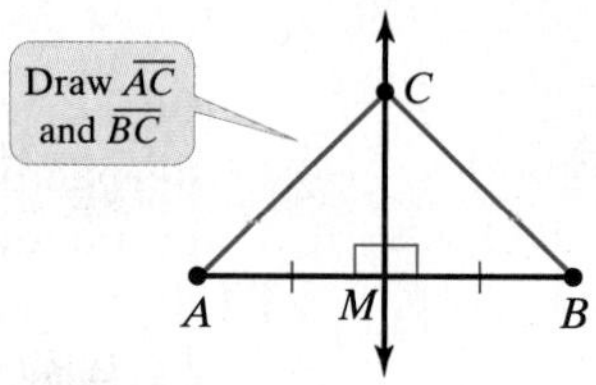

Notice from our work in Chapter 4 that $\Delta CAM \cong \Delta CBM$ (SAS).

So we can conclude that $\overline{CA} \cong \overline{CB}$, or that $CA = CB$. A point is **equidistant** from two objects if it is the same distance from the objects. So point C is equidistant from points A and B.

> **Helpful Hint**
>
> To prove that $\Delta CAM \cong \Delta CBM$,
>
> 1. $\overline{AM} \cong \overline{MB}$, $\angle AMC \cong \angle BMC$ — 1. Definition of perpendicular bisector
> 2. $\overline{CM} \cong \overline{CM}$ — 2. Reflexive property
> 3. $\Delta CAM \cong \Delta CBM$ — 3. SAS Postulate

This suggests a proof of Theorem 5.1-2, the Perpendicular Bisector Theorem. Its converse is also true and is stated as Theorem 5.1-3.

Theorem 5.1-2 Perpendicular Bisector Theorem

Theorem	If . . .	Then . . .
If a point is on the perpendicular bisector of a segment, then it is equidistant from the endpoints of the segment.	$\overleftrightarrow{PM} \perp \overline{AB}$ and $MA = MB$ 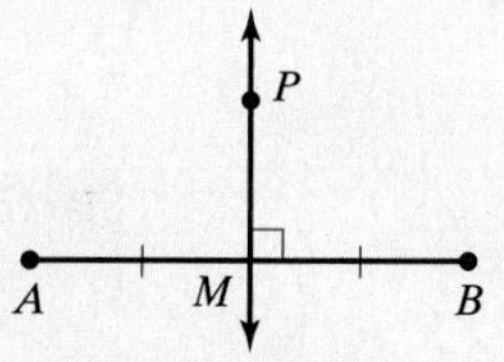	$PA = PB$ 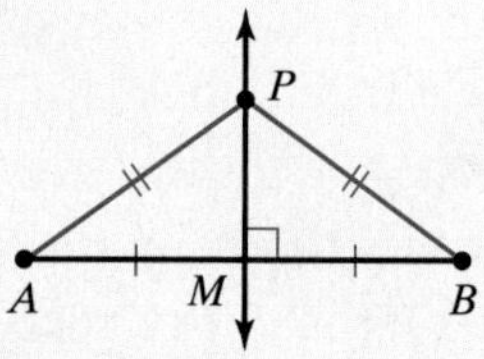

Theorem 5.1-3 Converse of the Perpendicular Bisector Theorem

Theorem	If . . .	Then . . .
If a point is equidistant from the endpoints of a segment, then it is on the perpendicular bisector of the segment.	$PA = PB$ 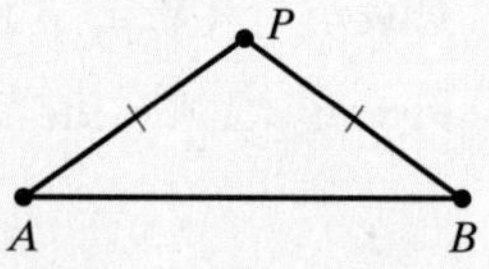	$\overleftrightarrow{PM} \perp \overline{AB}$ and $MA = MB$ 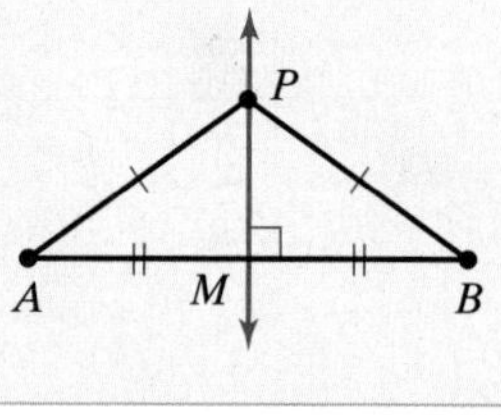

We prove these theorems in Exercise Set 5.1, Exercises 51–52.

EXAMPLE 1 Using the Perpendicular Bisector Theorem

Use the given figure to find the length of $\overline{AB}$.

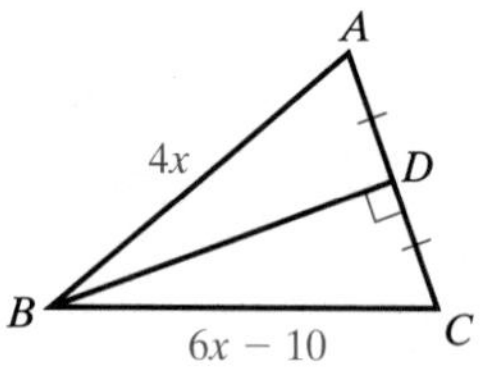

Solution $\overline{BD}$ is the perpendicular bisector of $\overline{AC}$, so B is equidistant from A and C.

$BA = BC$	Perpendicular Bisector Theorem
$4x = 6x - 10$	Substitute $4x$ for BA and $6x - 10$ for BC.
$-2x = -10$	Subtract $6x$ from each side.
$x = 5$	Divide each side by -2.

Now find AB. (Recall that AB is the same as BA.)

$AB = 4x$

$AB = 4(5) = 20$ Substitute 5 for x.

Check: See that $BA = BC$ by finding BC.

$BC = 6x - 10 = 6 \cdot 5 - 10 = 20$ Thus, $BA = BC = 20$ units. □

PRACTICE

1 Use the given figure to find the length of $\overline{QR}$.

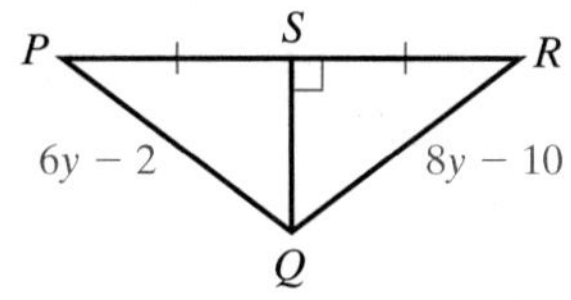

There are many real-world applications of the Perpendicular Bisector Theorem (Theorem 5.1-2). These applications exist in the fields of business, science, and geography, to name a few.

EXAMPLE 2 Using the Perpendicular Bisector Theorem to Solve an Application

A chef wants to open a new Japanese restaurant in the city of Metropolis. Currently, the city has two successful Japanese restaurants, so the chef is looking for a location within the city limits, but one that is equidistant, and as far away as possible, from the two current restaurants.

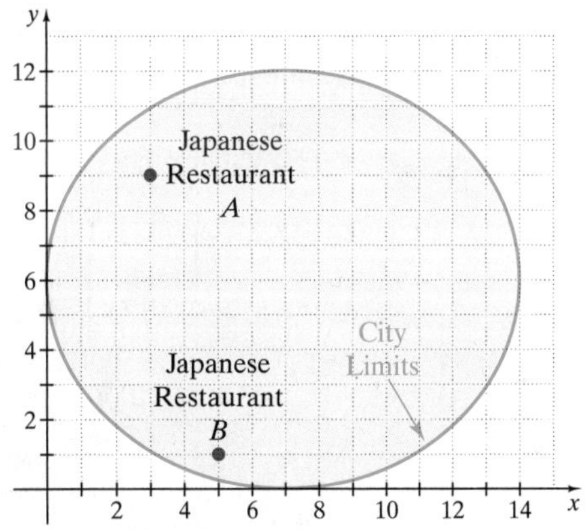

a. Explain the process for finding the location the chef is looking for.

b. Find the coordinates of the midpoint of $\overline{AB}$; call it point C.

c. Find the slope of $\overline{AB}$, and then the slope of a line perpendicular to $\overline{AB}$.

d. Locate point C and use the perpendicular slope from part **c** to draw the perpendicular bisector of $\overline{AB}$.

e. Approximate the coordinates of the chef's desired location for the new restaurant. (Use whole number coordinates within the city limits.)

Solution

a. "Equidistant" from the two restaurants tells us to draw the perpendicular bisector of $\overline{AB}$, say $\overleftrightarrow{XY}$. We then approximate the coordinates of the intersection of $\overleftrightarrow{XY}$ and the city limits furthest from the restaurants.

b. Point A has coordinates (3, 9). Point B has coordinates (5, 1).

$$\text{midpoint} = \left(\frac{x_1 + x_2}{2}, \frac{y_1 + y_2}{2}\right) = \left(\frac{3 + 5}{2}, \frac{9 + 1}{2}\right) \text{ or } (4, 5), \text{ point } C$$

c. $$\text{Slope of } \overline{AB} = \frac{y_2 - y_1}{x_2 - x_1} = \frac{1 - 9}{5 - 3} = \frac{-8}{2} = -4$$

The slope of the line $\perp$ to $\overline{AB}$ is the negative reciprocal of -4, or $-\frac{1}{(-4)} = \frac{1}{4}$.

d.

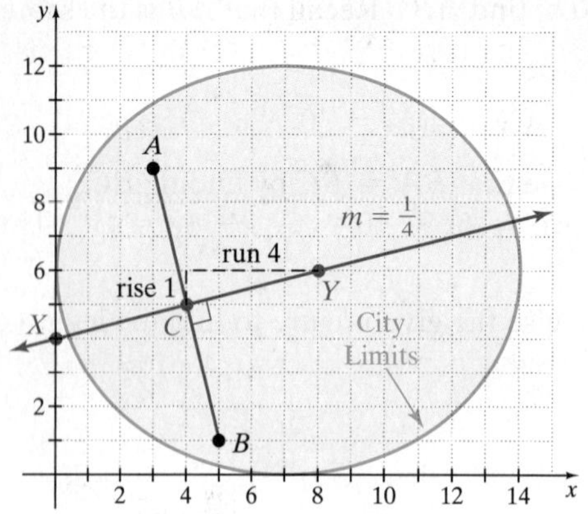

e. Using whole number coordinates within the city limits, the chef should look for property near the coordinates (13, 7) or (13, 8). □

PRACTICE 2 Use the same information and city limits graph as in Example 2, but change the current restaurants to two Mexican restaurants located at points D (6, 2) and E (8, 10). Answer parts b–e.

Other examples of applications of equidistances from two points are shown below.

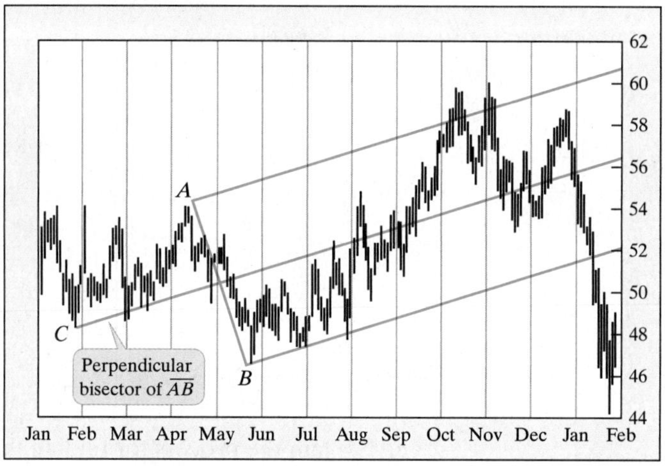

Developed by A. Andrews, Andrews' Pitchfork (in blue) is used to study stock market trends. In this diagram, the line through point C is sometimes the perpendicular bisector of the line segment through points A and B, as it is here.

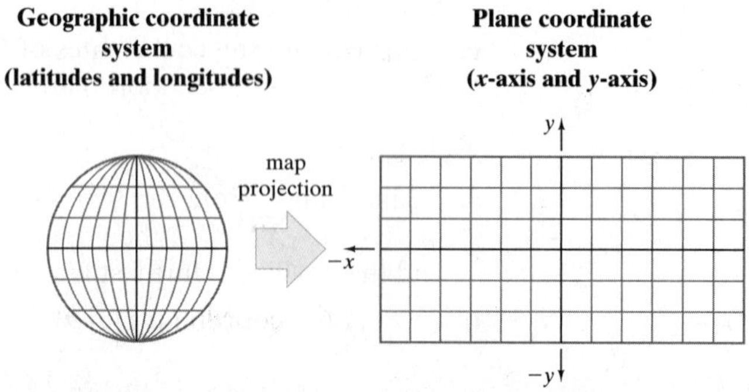

There are many ways to project maps of the earth's spherical surface onto a two-dimensional coordinate plane. One such map projection is called the Two-Point Equidistant projection.

For the Two-Point Equidistance projection, all distances are correct from two specified points (locations). This type of projection is often used to draw flat maps of larger continents, such as Asia. It is also used to determine the distance of a ship at sea from the start and end of a voyage as well as some diagrams of air routes for planes.

OBJECTIVE 2 ▶ Using Angle Bisectors. Again, in Section 1.8 we constructed the angle bisector of an angle. We now have the tools to prove that this construction is correct.

Theorem 5.1-4 Construction of the Angle Bisector

The construction shown in Section 1.8, Example 4, gives the angle bisector of an angle.

We review the construction below and then prove Theorem 5.1-4.

Construction of Angle Bisector

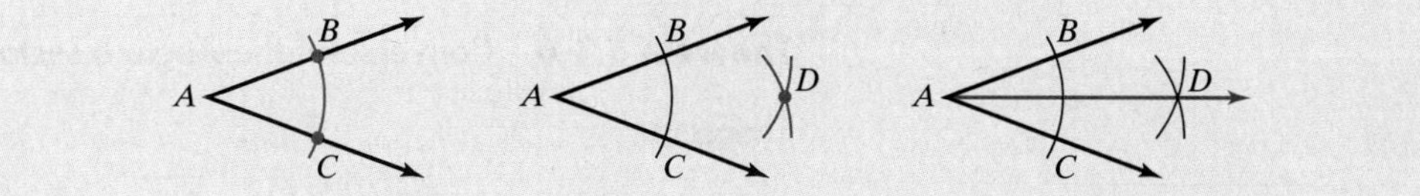

Proof of Angle Bisector (Theorem 5.1-4)

Given: $\overline{AB} \cong \overline{AC}, \overline{BD} \cong \overline{CD}$ (by construction)

Prove: $\angle 1 \cong \angle 2$

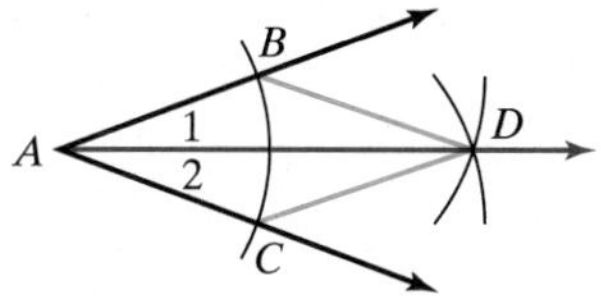

Figure for Proof

Statements	Reasons
1. $\overline{AB} \cong \overline{AC}, \overline{BD} \cong \overline{CD}$	**1.** Given by construction
2. $\overline{AD} \cong \overline{AD}$	**2.** Reflexive Property
3. $\Delta ABD \cong \Delta ACD$	**3.** SSS (Steps 1, 2)
4. $\angle 1 \cong \angle 2$	**4.** cpoctac

▶ Helpful Hint

When constructing the angle bisector, the compass setting for the arc giving points B and C may or may not be the same compass setting for the arcs giving point D. Either way, ΔABD and ΔACD are congruent by SSS.

There is a special relationship between the points on the bisector of an angle and the sides of the angle. Before we explain this relationship, let's define the distance from a point to a line.

The **distance from a point to a line** is the length of the perpendicular segment from the point to the line. In the figure at the right, the distance from A to ℓ is AM and the distance from B to ℓ is BN.

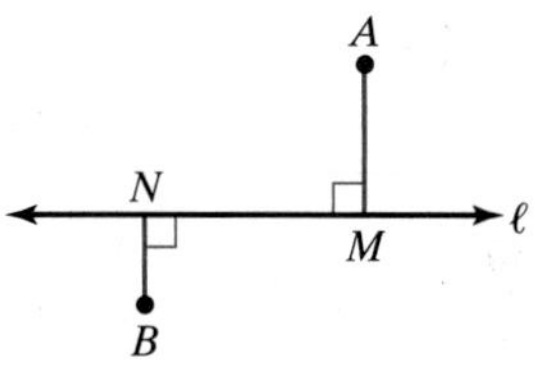

When a point is the same distance from two distinct lines, we say that the point is **equidistant from the two lines.**

The following theorems have to do with any point on the angle bisector and its distances from the sides of the angle.

Theorem 5.1-5 Angle Bisector Theorem

Theorem	If . . .	Then . . .
If a point is on the bisector of an angle, then the point is equidistant from the sides of the angle.	$\overrightarrow{QS}$ bisects $\angle PQR$, $\overline{SP} \perp \overrightarrow{QP}$, and $\overline{SR} \perp \overrightarrow{QR}$ 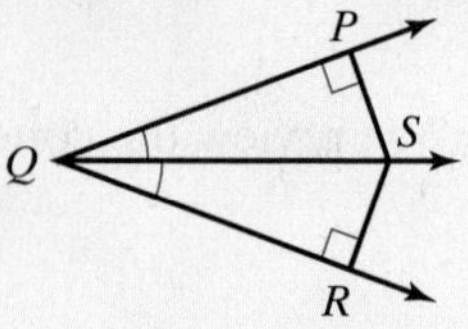	$SP = SR$ 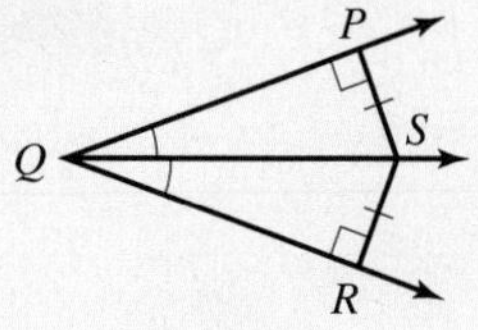

Theorem 5.1-6 Converse of the Angle Bisector Theorem

Theorem	If . . .	Then . . .
If a point in the interior of an angle is equidistant from the sides of the angle, then the point is on the angle bisector.	$\overline{SP} \perp \overrightarrow{QP}$, $\overline{SR} \perp \overrightarrow{QR}$, and $SP = SR$ 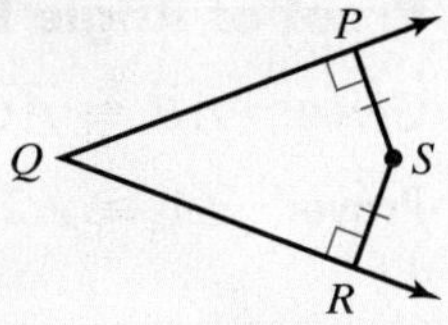	$\overrightarrow{QS}$ bisects $\angle PQR$ 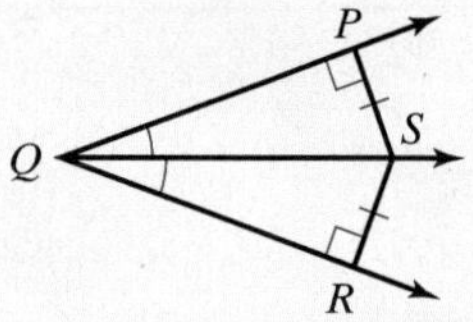

We prove these theorems in Exercise Set 5.1, Exercises 53 and 54.

EXAMPLE 3 **Using the Angle Bisector Theorem**

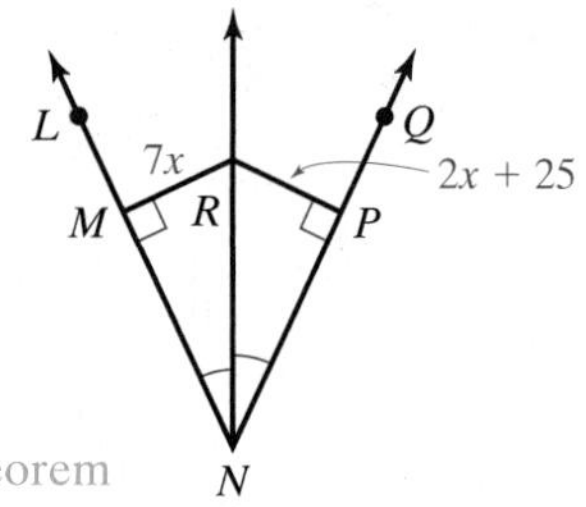

Use the given figure and find the length of $\overline{RM}$.

Solution From the marks on the figure, we are given $\overrightarrow{NR}$ bisects $\angle LNQ$, $\overline{RM} \perp \overrightarrow{NL}$, and $\overline{RP} \perp \overrightarrow{NQ}$. To find RM, we use the Angle Bisector Theorem.

$RM = RP$ The Angle Bisector Theorem

$7x = 2x + 25$ Substitute.

$5x = 25$ Subtract $2x$ from each side.

$x = 5$ Divide each side by 5.

Now find RM.

$RM = 7x$

$= 7(5) = 35$ Substitute 5 for x.

Check: See that $RM = RP$ by finding RP.

$RP = 2x + 25 = 2(5) + 25 = 35$ Thus, $RM = RP = 35$ units. □

PRACTICE 3 Use the given figure and find the length of $\overline{FB}$.

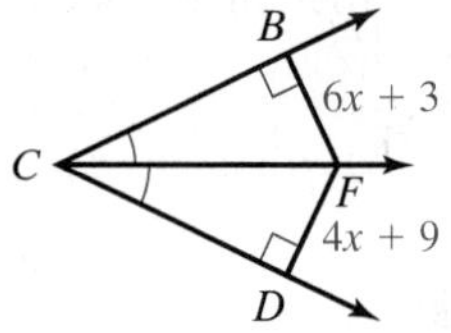

VOCABULARY & READINESS CHECK

Word Bank *Use the choices to fill in each blank. Some choices may be used more than once.*

midpoint	perpendicular bisector	congruent
equidistant	perpendicular	angle bisector

Use the figure to the right for Exercises 1–8. The numerical answers to Exercises 7 and 8 are not in contained in the Word Bank.

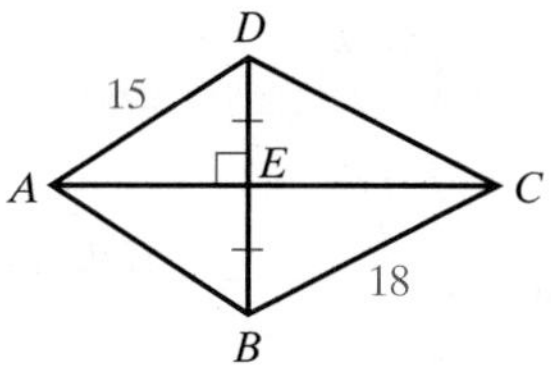

1. The point E is called the __________ of $\overline{BD}$.

2. $\overline{BD}$ is __________ to $\overline{AC}$.

3. The line containing $\overline{AC}$ is thus called the __________ of $\overline{BD}$.

4. From Exercise 3, all points on $\overline{AC}$ are __________ from points B and D.

5. The distance from a point to a line is the length of the __________ segment from the point to the line.

6. ΔAED is __________ to ΔAEB.

7. The length of $\overline{AB}$ is __________ units.

8. The length of $\overline{DC}$ is __________ units.

Use the figure at the right for Exercises 9–12. The answers to Exercises 10–12 are not in contained in the Word Bank.

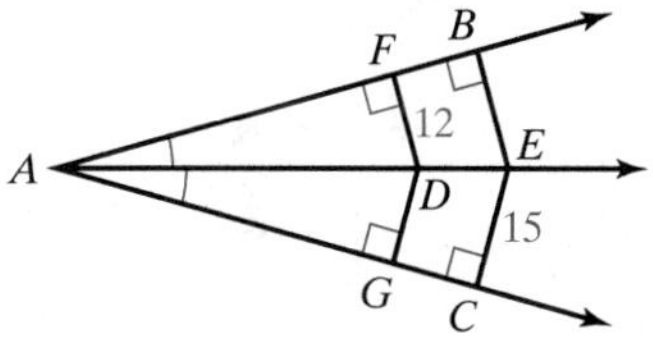

9. $\overrightarrow{AE}$ is the __________ of $\angle BAC$.

10. From Exercise 9, all points on $\overrightarrow{AE}$ are equidistant from the sides of __________.

11. The length of $\overline{DG}$ is __________ units.

12. The length of $\overline{EB}$ is __________ units.

5.1 EXERCISE SET MyMathLab®

Complete the Table *Use the two figures below and your knowledge of perpendicular bisectors to find each length. See Example 1.*

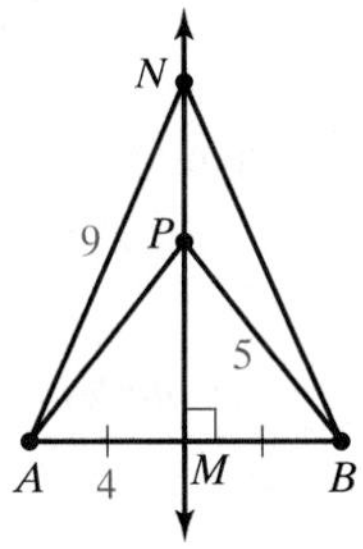

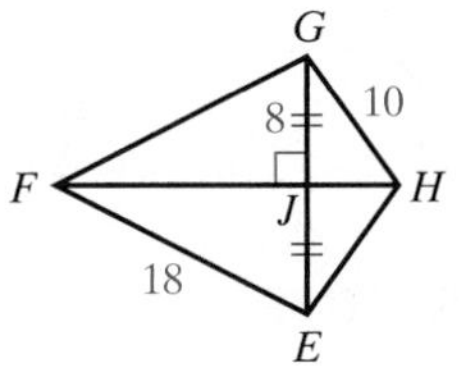

	Segment	*Length (in units)*
1.	$\overline{GF}$	
2.	$\overline{PA}$	
3.	$\overline{NB}$	
4.	$\overline{EH}$	
5.	$\overline{MB}$	
6.	$\overline{EJ}$	
7.	$\overline{AB}$	
8.	$\overline{EG}$	

Complete the Table *Use the two figures below and your knowledge of angle bisectors to find each length or measure. See Example 3.*

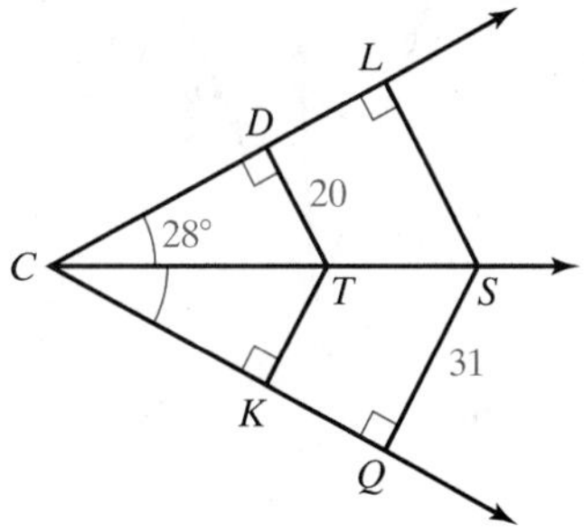

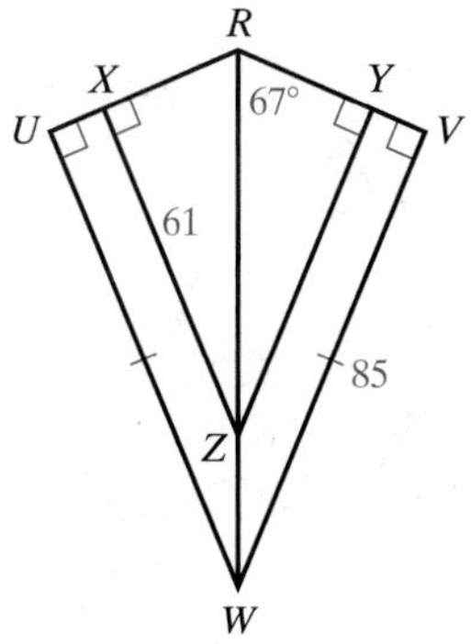

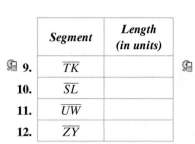

	Segment	*Length (in units)*
9.	$\overline{TK}$	
10.	$\overline{SL}$	
11.	$\overline{UW}$	
12.	$\overline{ZY}$	

	Angle	*Measure (in degrees)*
13.	∠QCS	
14.	∠XRZ	
15.	∠URV	
16.	∠KCD	

MIXED PRACTICE

Multiple Steps *For Exercises 17–20, each figure shows a perpendicular bisector or an angle bisector. See Examples 1 and 3.*

a. *Choose whether the figure shows a perpendicular bisector or an angle bisector.*

If a perpendicular bisector, fill in the blanks:
b. _______ *is the perpendicular bisector of* _______.
c. *Every point on* _______ *is equidistant from points* _______ *and* _______.

If an angle bisector, fill in the blanks:
b. _______ *is the angle bisector of* _______.
c. *Every point on* _______ *is equidistant from angle sides* _______ *and* _______.

17.

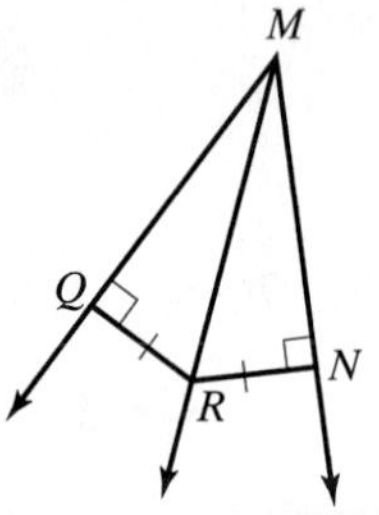

18.

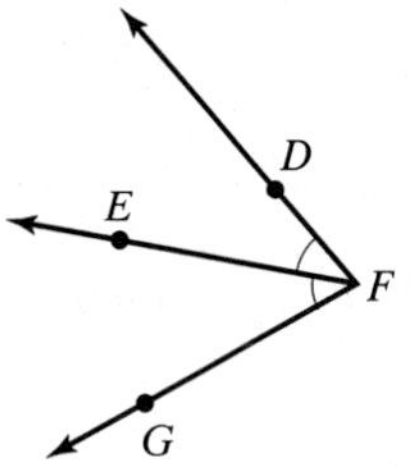

19.

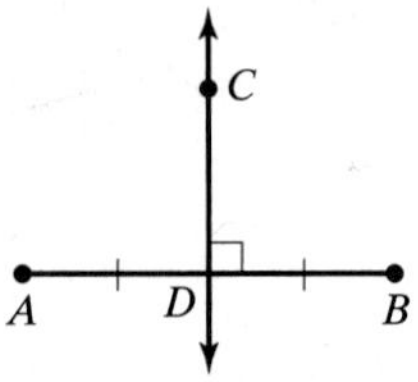

20.

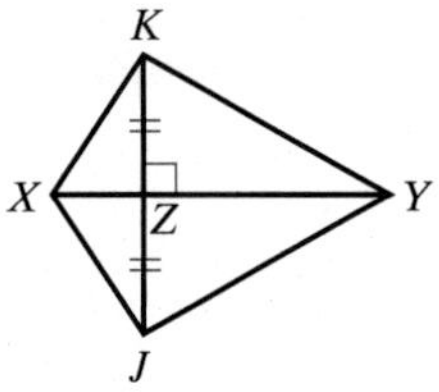

For Exercises 21–24, determine whether A must be on the bisector of ∠TXR. Explain.

21.

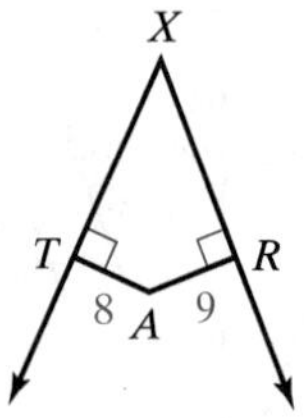

22.

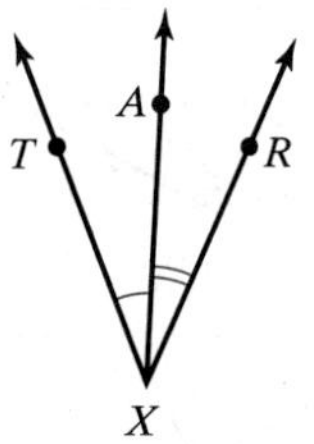

23.

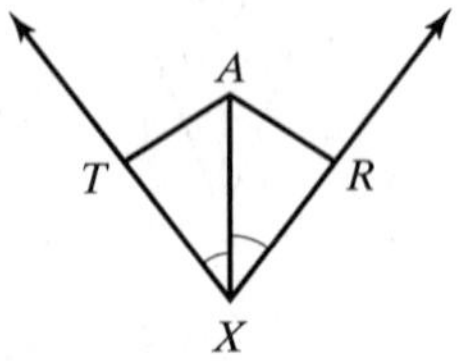

24.

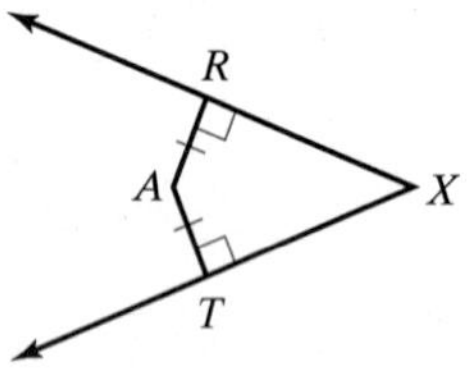

Use the figure below for Exercises 25–28. See Example 1.

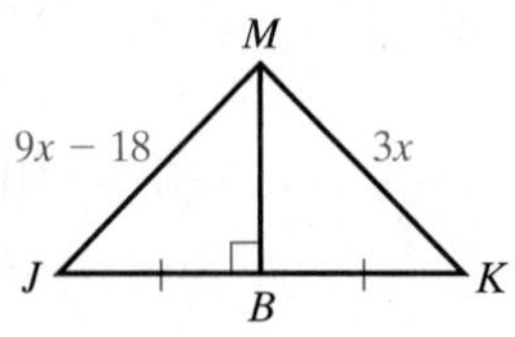

25. What is the relationship between $\overline{MB}$ and $\overline{JK}$?

26. What is the value of x?

27. Find JM.

28. Find MK.

Use the figure below for Exercises 29–32. See Example 3.

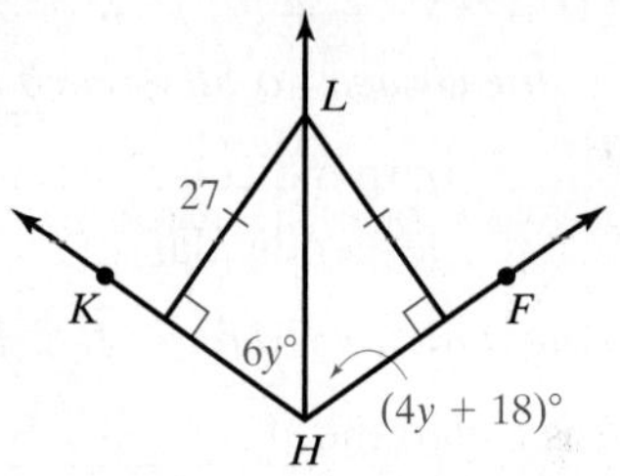

29. How far is L from $\overrightarrow{HF}$?

30. How is $\overrightarrow{HL}$ related to $\angle KHF$?

31. Find the value of y.

32. Find $m\angle KHL$ and $m\angle FHL$.

33. Find x, JK, and JM.

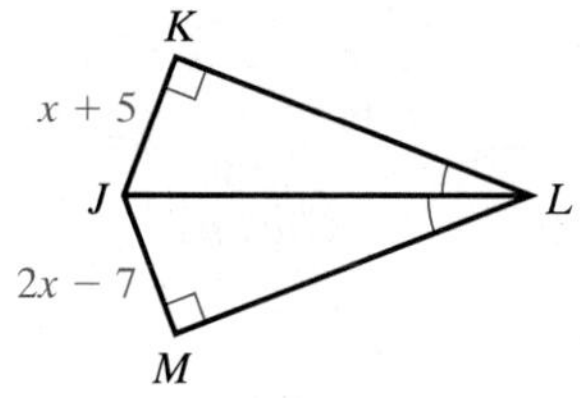

34. Find y, ST, and TU.

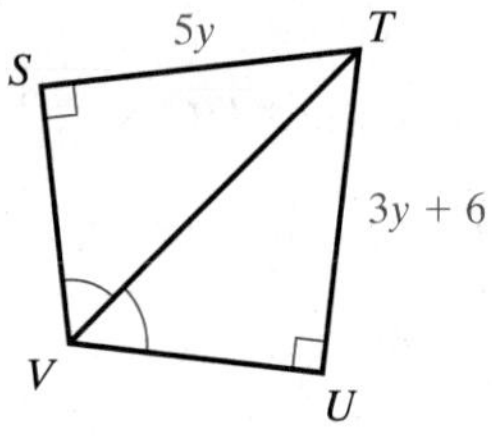

Use the figure below for Exercises 35–38.

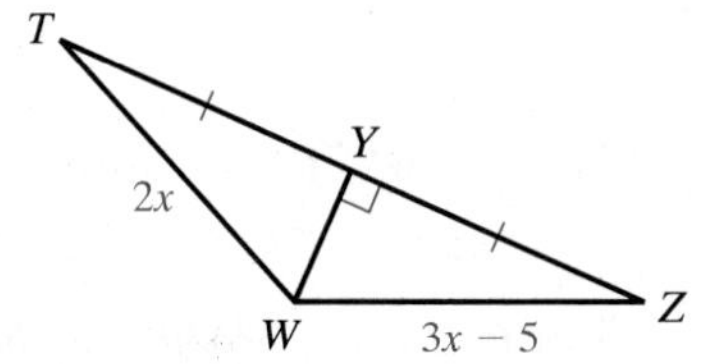

35. Find the value of x.

36. Find TW.

37. Find WZ.

38. What kind of triangle is ΔTWZ? Explain.

MIXED PRACTICE

Use the map to solve Exercises 39–42. See Examples 1 and 2.

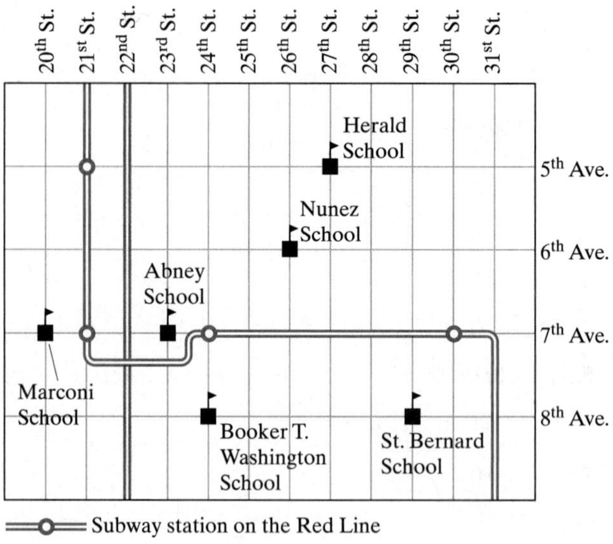

39. Any location equidistant from the two subway stations on 21st St. must lie on what street?

40. Any location equidistant from the subway stations on 24th St. and 30th St. must lie on what street?

41. Find the Red Line subway station at 24th St. and 7th Ave. and the station at 30th St. and 7th Ave. Which school is equidistant from these two stations?

42. Find the two Red Line subway stations on 21st St. Which school is equidistant from these two stations?

43. **Multiple Steps** A city planner for Smallville has been instructed to find a location for a new city park. This will be the third park in Smallville, and the mayor would like its location to be within city limits and equidistant from the other two parks, but as far away as possible from them. (See the diagram and Example 2.)

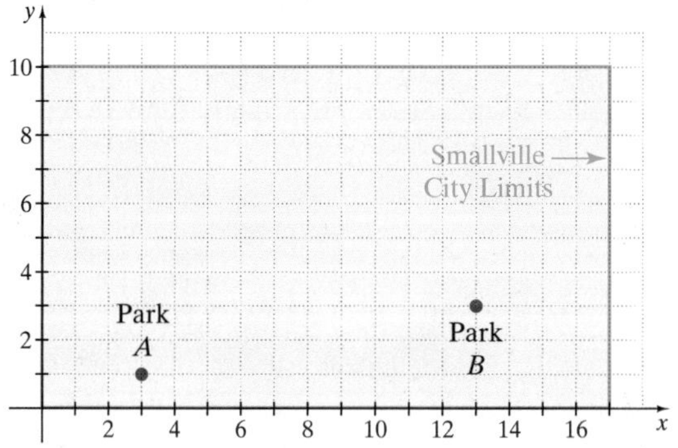

a. Explain the process for finding the approximate location of the new park.

b. Identify the coordinates for Park A and Park B.

c. Find the coordinates of the exact midpoint of $\overline{AB}$ and call it point D.

d. Find the slope of $\overline{AB}$. Use this slope to find the slope of a line perpendicular to $\overline{AB}$.

e. Draw a rectangular coordinate system and graph points A and B. Then draw $\overline{AB}$ and graph point D. Use the perpendicular slope (part **d**) to draw the perpendicular bisector of $\overline{AB}$.

f. Approximate the coordinates of the new park. (Use whole number coordinates within city limits.)

44. **Multiple Steps** Use the diagram for Exercise **43**. This time, position Park A at (3, 4) and Park B at (13, 2). Use these new coordinates to answer the questions in Exercise **43**, parts **c–f**.

CONCEPT EXTENSIONS

45. On a piece of paper, mark a point H for home and a point S for school. Describe how to find the set of points equidistant from H and S.

46. Point P is in the interior of $\angle LOX$. Describe how you can determine whether P is on the bisector of $\angle LOX$ without drawing the angle bisector.

47. a. **Construction** Draw a large triangle, ΔCDE. Construct the angle bisectors of each angle.

b. What appears to be true about the angle bisectors?

c. Test your conjecture (from part **b**) with another triangle.

48. a. **Construction** Draw a large acute scalene triangle, ΔPQR. Construct the perpendicular bisectors of each side.

b. What appears to be true about the perpendicular bisectors?

c. Test your conjecture (from part **b**) with another triangle.

49. Write Theorems 5.1-2 and 5.1-3 as a single biconditional statement.

50. Write Theorems 5.1-5 and 5.1-6 as a single biconditional statement.

PROOF

51. Prove the Perpendicular Bisector Theorem (5.1-2).

Given: $\overleftrightarrow{PM} \perp \overline{AB}$, $\overleftrightarrow{PM}$ bisects $\overline{AB}$

Prove: $AP = BP$

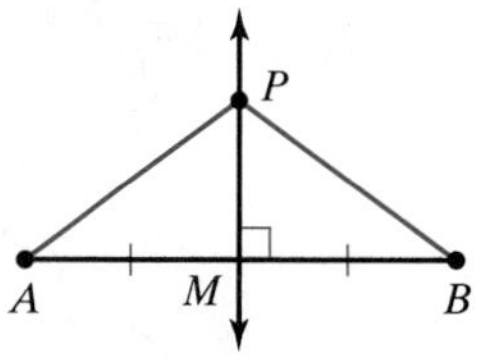

52. Prove the Converse of the Perpendicular Bisector Theorem (5.1-3).

Given: $PA = PB$ with $\overleftrightarrow{PM} \perp \overline{AB}$, at M.

Prove: P is on the perpendicular bisector of $\overline{AB}$.

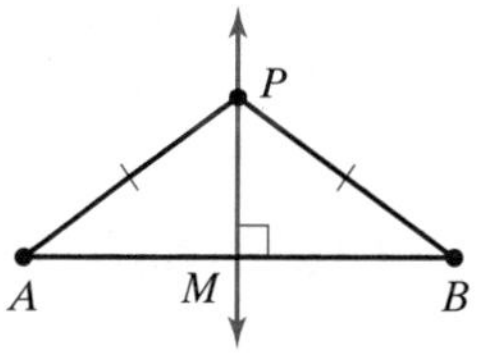

53. Prove the Angle Bisector Theorem (5.1-5).

Given: $\overrightarrow{QS}$ bisects $\angle PQR$, $\overline{SP} \perp \overrightarrow{QP}$, $\overline{SR} \perp \overrightarrow{QR}$

Prove: $SP = SR$

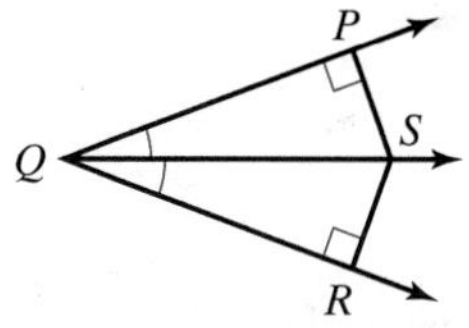

54. Prove the Converse of the Angle Bisector Theorem (5.1-6).

Given: $\overline{SP} \perp \overrightarrow{QP}$, $\overline{SR} \perp \overrightarrow{QR}$, $SP = SR$

Prove: $\overrightarrow{QS}$ bisects $\angle PQR$.

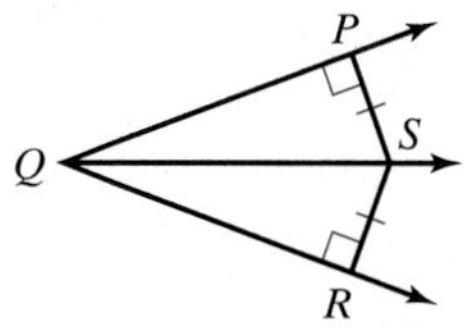

55. **Coordinate Geometry** Use points $A(6,8)$, $O(0,0)$, and $B(10,0)$.

a. Write equations of lines ℓ and m such that $\ell \perp \overleftrightarrow{OA}$ at A and $m \perp \overleftrightarrow{OB}$ at B.

b. Find the intersection C of lines ℓ and m.

c. Show that $CA = CB$.

d. Explain why C is on the bisector of $\angle AOB$.

56. Find the Error To prove that ΔPQR is isosceles, a student began by stating that since Q is on the segment perpendicular to $\overline{PR}$, Q is equidistant from the endpoints of $\overline{PR}$. What is the error in the student's reasoning?

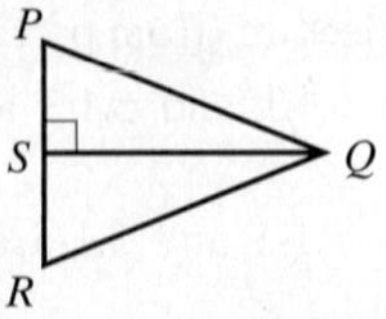

57. A, B, and C are three noncollinear points. Describe and sketch a line in plane ABC such that points A, B, and C are equidistant from the line. Justify your response.

58. M is the intersection of the perpendicular bisectors of two sides of ΔABC. Line ℓ is perpendicular to plane ABC at M. Explain why a point E on ℓ is equidistant from A, B, and C. (Hint: Show that $\Delta EAM \cong \Delta EBM \cong \Delta ECM$.)

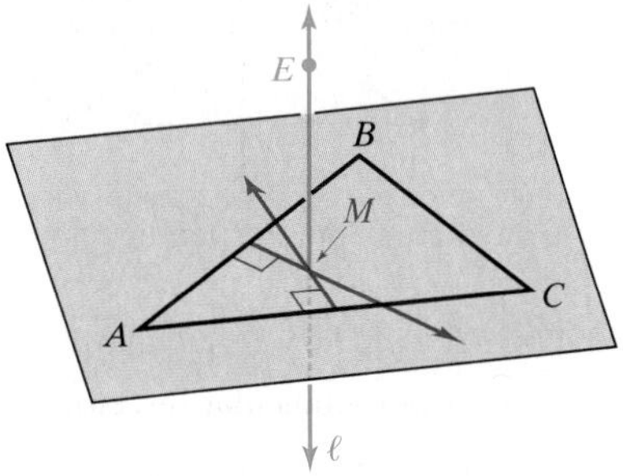

REVIEW AND PREVIEW

Solve. ∠1 and ∠2 are complementary and ∠1 and ∠3 are supplementary. See Section 1.6.

59. If $m\angle 2 = 30°$, find $m\angle 1$.

60. Use $m\angle 1$ from Exercise 59 to find $m\angle 3$.

Solve. See Sections 3.6 and 3.7.

61. What is the slope of a line that is parallel to the line $y = -3x + 4$?

62. What is the slope of a line that is perpendicular to the line $y = -3x + 4$?

63. Describe the line $x = 5$.

64. Describe the line $y = 2$.

5.2 BISECTORS OF A TRIANGLE

OBJECTIVES

1 Use Properties of the Perpendicular Bisectors of the Sides of a Triangle, Including the Circumcenter.

2 Use Properties of the Angle Bisectors of the Angles of a Triangle, Including the Incenter.

VOCABULARY

- concurrent
- point of concurrency
- circumcenter of a triangle
- circumscribed about
- incenter of a triangle
- inscribed in

OBJECTIVE 1 ▸ Using Properties of Perpendicular Bisectors of a Triangle, Including the Circumcenter. In the last section, we studied and proved properties of the perpendicular bisector of a line segment. Let's now discover what happens if we expand our work to triangles.

A **perpendicular bisector of a triangle** is a line (or a segment, a ray, or a plane) that is perpendicular to a side of the triangle at the side's midpoint.

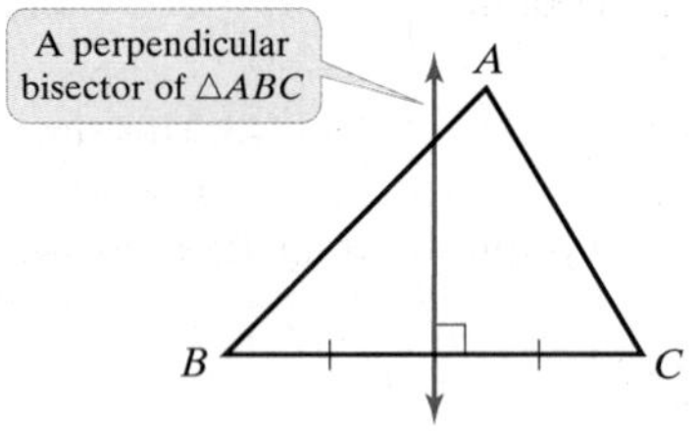

Given this definition, how many perpendicular bisectors does a triangle have? Since a triangle has 3 sides, it has 3 perpendicular bisectors.

TRY THE ACTIVITY BELOW

Paper-Folding Activity—Perpendicular Bisectors of a Triangle

In this activity, we use paper folding to investigate the perpendicular bisectors of the sides of a triangle.

STEP 1. Draw and cut out three different acute scalene triangles.

STEP 2. Choose a triangle. Use paper folding to make the perpendicular bisector of each side of your acute triangle. What do you notice about the perpendicular bisectors?

STEP 3. Repeat Step 2 with your other triangles. Does your discovery from Step 2 still hold true?

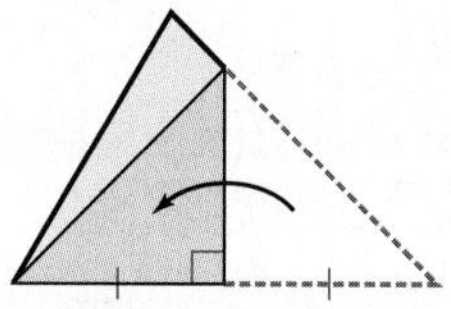

Folding a perpendicular bisector

When three or more lines intersect at one point, they are **concurrent.** The point at which they intersect is called the **point of concurrency.**

You may have conjectured from the activity above that the three perpendicular bisectors of a triangle are concurrent. This is true.

Theorem 5.2-1 Concurrency of Perpendicular Bisectors Theorem

Theorem

The perpendicular bisectors of the sides of a triangle are concurrent at a point equidistant from the vertices.

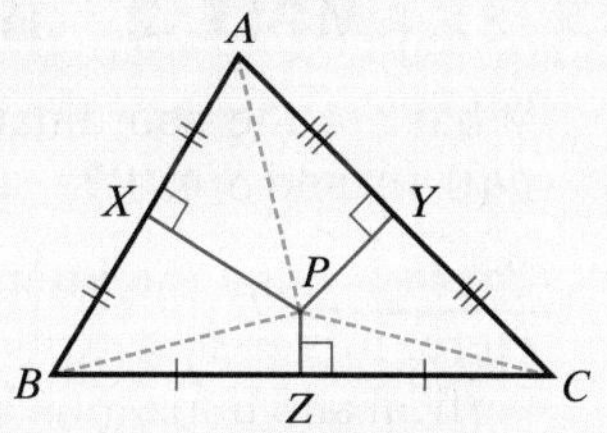

Perpendicular bisectors $\overline{PX}$, $\overline{PY}$, and $\overline{PZ}$ are concurrent at P. Also,

$PA = PB = PC$

(*Note:* A proof of Theorem 5.2-1 is in the back of this text.)

Since P lies on the perpendicular bisector of each side of the triangle, the distance from P to each vertex is the same, or

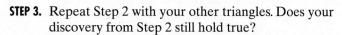

$\underbrace{PA}_{} = \underbrace{PB}_{} = \underbrace{PC}_{}$ We call P the point of concurrency.

→radius←

Using this distance as a radius and the point P as a center, we can draw a circle about ΔABC, as shown.

We say the circle is **circumscribed about** the triangle. P, the point of concurrency of the perpendicular bisectors of this triangle, is also called the **circumcenter of the triangle.**

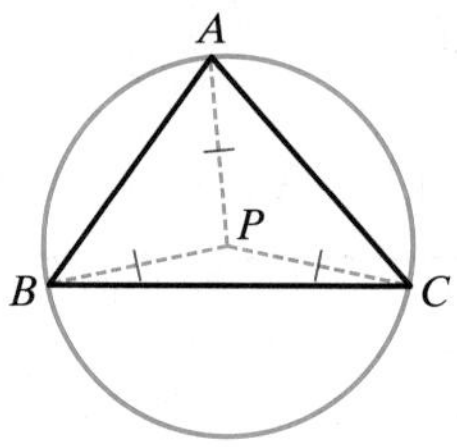

The circumcenter of a triangle can be inside, on, or outside a triangle. Notice the type of triangle below and each location of P, the circumcenter.

> **Helpful Hint**
> Remember:
> - The three perpendicular bisectors of a triangle are concurrent.
> - This point of concurrency is called the circumcenter.
> - The circumcenter can lie inside, on, or outside the triangle.

Acute triangle

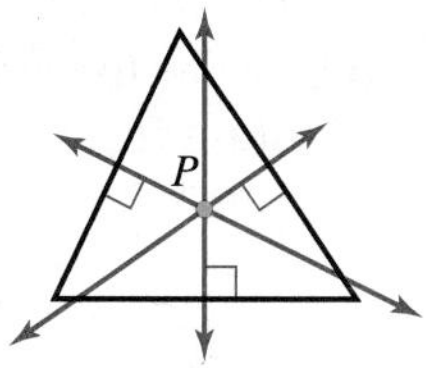

Right triangle

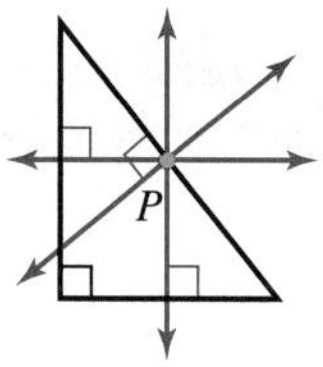

Obtuse triangle

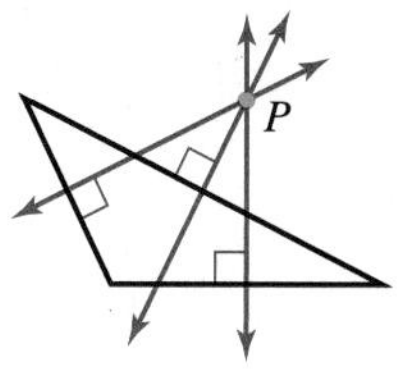

EXAMPLE 1 Using the Circumcenter of a Triangle

The circumcenter of ΔDEF is point R.

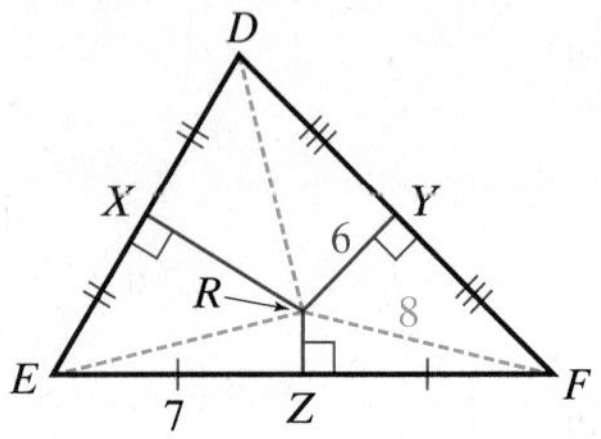

Fill in the blanks.

a. $RD =$ ________ $=$ ________ (Use segment distances here.)

b. $RD =$ ________ units

Solution Since R is the circumcenter, the distance from R to each vertex of ΔDEF is the same.

a. $RD = \underline{RE} = \underline{RF}$

b. $RF = 8$ units, so $RD = \underline{8}$ units. □

PRACTICE

1 The circumcenter of ΔABC is point Q.

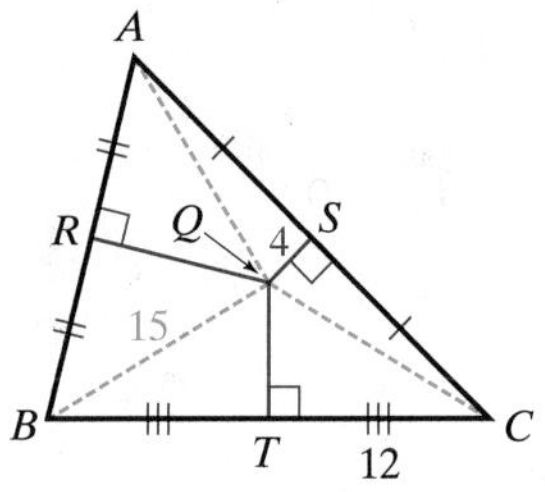

Fill in the blanks.

a. $QA =$ ____ $=$ ____ (Use segment distances here.)

b. $QA =$ ____ units

EXAMPLE 2 Finding the Circumcenter of a Triangle

What are the coordinates of the circumcenter of the triangle with vertices $P(0, 6)$, $O(0, 0)$, and $S(4, 0)$?

Solution Find the intersection point of two of the triangle's perpendicular bisectors. Here, it is easiest to find the perpendicular bisectors of $\overline{PO}$ and $\overline{OS}$ since they are vertical and horizontal lines.

Helpful Hint
ΔPOS is a right triangle, so its circumcenter lies on its hypotenuse.

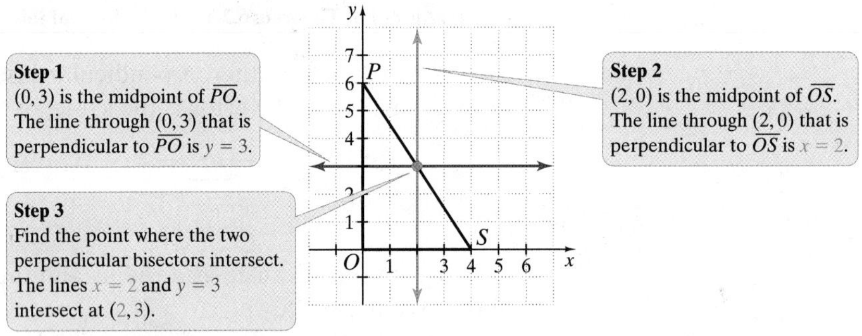
Step 1
(0, 3) is the midpoint of PO. The line through (0, 3) that is perpendicular to PO is y = 3.
Step 2
(2, 0) is the midpoint of OS. The line through (2, 0) that is perpendicular to OS is x = 2.
Step 3
Find the point where the two perpendicular bisectors intersect. The lines x = 2 and y = 3 intersect at (2, 3).
y
P
S
O
x

The coordinates of the circumcenter of the triangle are (2, 3). □

PRACTICE

2 What are the coordinates of the circumcenter of the triangle with vertices $A(2, 7)$, $B(10, 7)$, and $C(10, 3)$?

EXAMPLE 3 Using the Circumcenter in an Application

A recycling center is to be built to service three neighboring towns. To save fuel, the center is to be built equidistant from towns A, B, and C.

Where should the center be built?

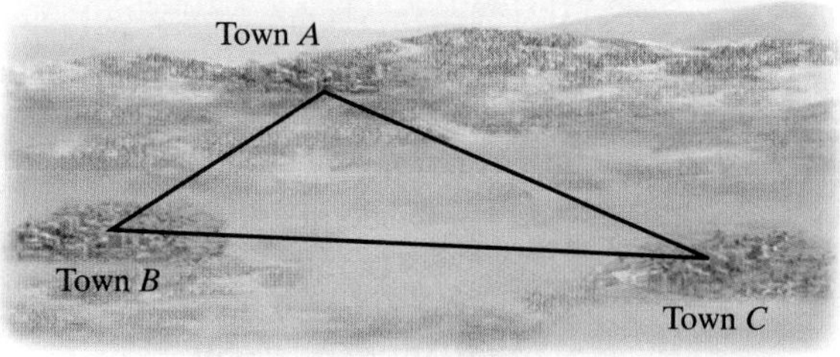

Solution To find a location equidistant from the towns, draw a triangle using towns A, B, and C as vertices. The location for the recycling center is point P, the circumcenter of ΔABC. This is the point of concurrency of the three perpendicular bisectors of ΔABC.

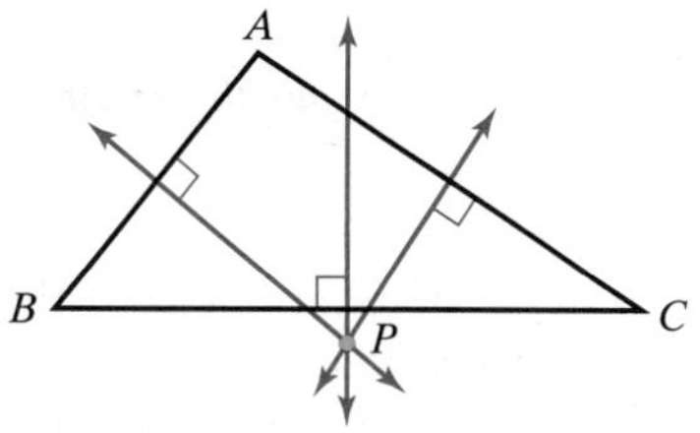

□

PRACTICE 3 A community college is planning to construct a computer center building equidistant from three other buildings on campus, buildings A, B, and C. Sketch the location of the computer center.

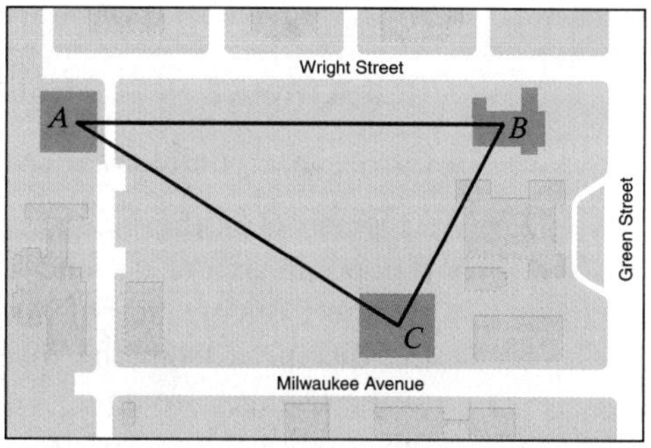

OBJECTIVE 2 ▶ Using Properties of Angle Bisectors of a Triangle, Including the Incenter. An **angle bisector of a triangle** is a bisector of an angle of the triangle.

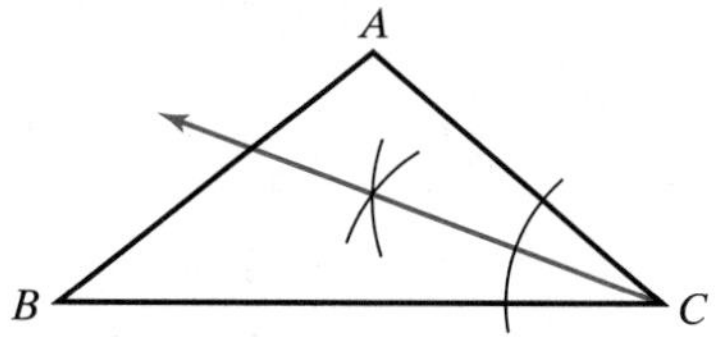

A bisector of $\angle C$ of $\triangle ABC$. (Each $\triangle$ has 3 angle bisectors.)

TRY THE ACTIVITY BELOW

Paper-Folding Activity—Bisectors of the Angles of a Triangle

In this activity, we will use paper folding to investigate the bisectors of the angles of a triangle.

STEP 1. Draw and cut out three different triangles.

STEP 2. Choose a triangle. Use paper folding to make the angle bisectors of each angle of your triangle. What do you notice about the angle bisectors?

STEP 3. Repeat Step 2 with your other triangles. Does your discovery from Step 2 still hold true?

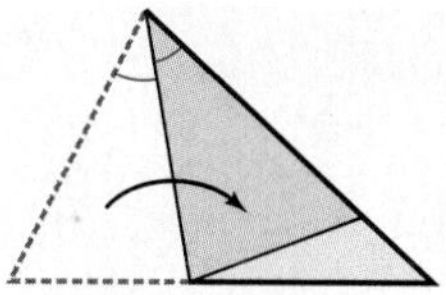

Folding an angle bisector

It appears that the three angle bisectors are concurrent. This is true.

Theorem 5.2-2 Concurrency of Angle Bisectors Theorem

Theorem
The bisectors of the angles of a triangle are concurrent at a point equidistant from the sides of the triangle.

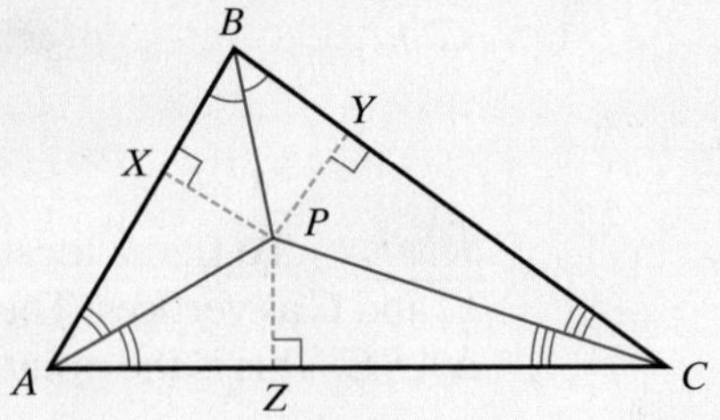

Angle bisectors $\overline{AP}$, $\overline{BP}$, and $\overline{CP}$ are concurrent at P. Also, $PX = PY = PZ$

We prove this theorem in Exercise set 5.2, Exercises 37–40.

Helpful Hint
Remember
- The three angle bisectors of a triangle are concurrent.
- This point of concurrency is called the incenter.
- The incenter always lies inside the triangle.

The point of concurrency of the angle bisectors of a triangle is called the **incenter of the triangle** and the incenter is always inside the triangle.

Here, the incenter P is the center of the circle that is **inscribed** in the triangle. Notice that the circle touches each side once, at points X, Y, and Z. The radius of the circle is the (perpendicular) distance from P to each side.

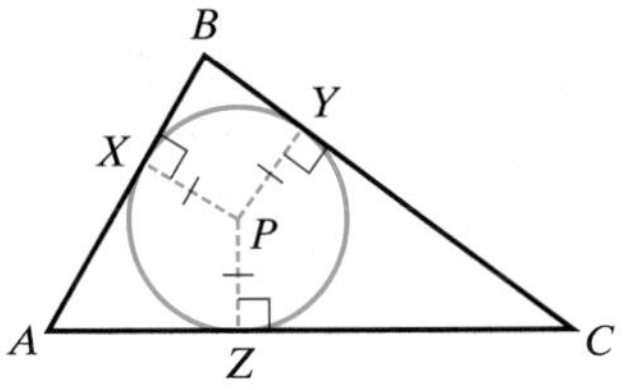

EXAMPLE 4 **Identifying and Using the Incenter of a Triangle**

$GE = 2x - 7$ and $GF = x + 4$. What is GH?

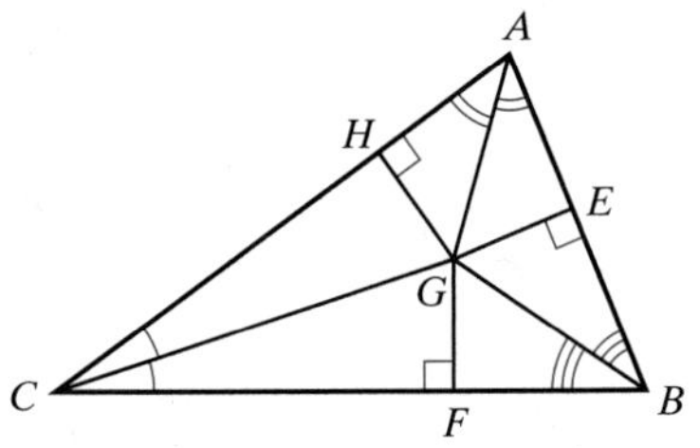

Solution G is the incenter of ΔABC because it is the point of concurrency of the angle bisectors. By the Concurrency of Angle Bisectors Theorem, the distances from the incenter to the three sides of the triangle are equal, so $GE = GF = GH$. Use this relationship to find x.

$2x - 7 = x + 4$ $GE = GF$

$2x = x + 11$ Add 7 to each side.

$x = 11$ Subtract x from each side.

Now find GF.

$GF = x + 4$

$= 11 + 4 = 15$ Substitute 11 for x.

Since $GE = GF = GH$, then $GH = 15$ units. □

PRACTICE 4 $QN = 5x + 36$ and $QM = 2x + 51$. What is QO?

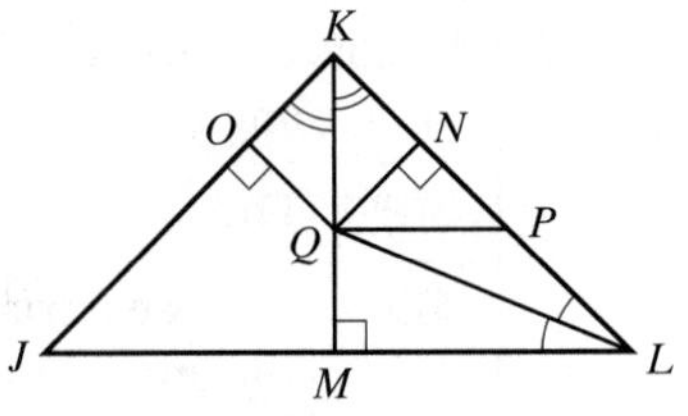

VOCABULARY & READINESS CHECK

Word Bank *Use the choices to fill in each blank.*

circumcenter	concurrent	inscribed in
incenter	point of concurrency	circumscribed about

1. When three or more lines intersect in one point, they are ___________.
2. The point of intersection from #1 above is called the ___________.
3. The three angle bisectors of a triangle intersect in one point called the ___________.
4. The three perpendicular bisectors of a triangle intersect in one point called the ___________.
5. The circumcenter of a triangle is the center of a circle ___________ the triangle.
6. The incenter of a triangle is the center of a circle ___________ the triangle.

Matching *For Exercises 7–10, choose the type of triangle(s) that matches each description.*

A. acute triangle *B. right triangle* *C. obtuse triangle*

7. A triangle's circumcenter is outside the triangle. ___________
8. A triangle's circumcenter is inside the triangle. ___________
9. A triangle's circumcenter is on the triangle. ___________
10. A triangle's incenter is inside the triangle. ___________

5.2 EXERCISE SET MyMathLab®

MIXED PRACTICE

Complete the Table *Mark the statements that are true for each point (circumcenter and incenter) by placing a checkmark, ✓, in the appropriate column. (You can check more than one column.) See Examples 1–4.*

		Circumcenter	*Incenter*
1.	Always lies inside the triangle		
2.	Can be inside, on, or outside the triangle		
3.	A point of concurrency of a triangle		
4.	Each triangle has only one of these points.		
5.	The point of concurrency of the perpendicular bisectors		
6.	The point of concurrency of the angle bisectors		
7.	Equidistant from the sides of a triangle		
8.	Equidistant from the vertices of a triangle		
9.	Center of circle that passes through the vertices of a triangle		
10.	Center of circle that touches each side of a triangle only once		

Fill in the blank *for Exercises 11–14. Complete parts* ***a*** *and* ***b****.*

a. Point P is the ___________.

b. $PA =$ ___________ units. See Examples 1–4.

11.

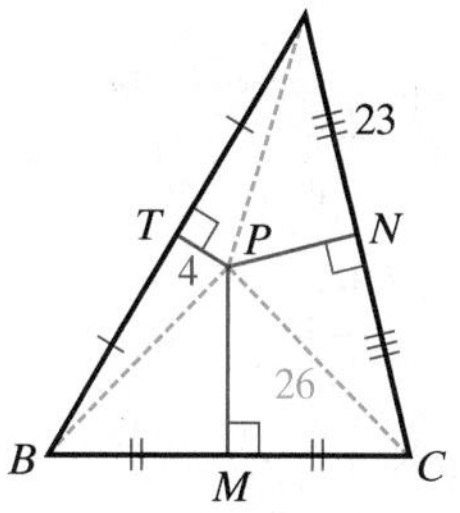

12.

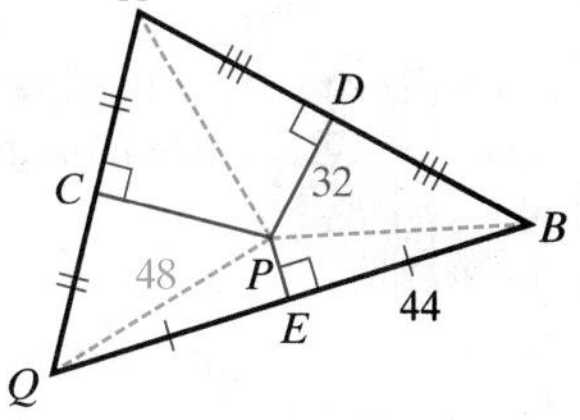

13.

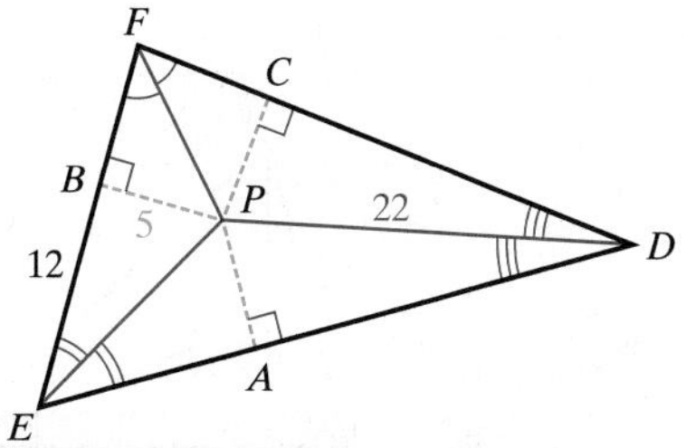

14.

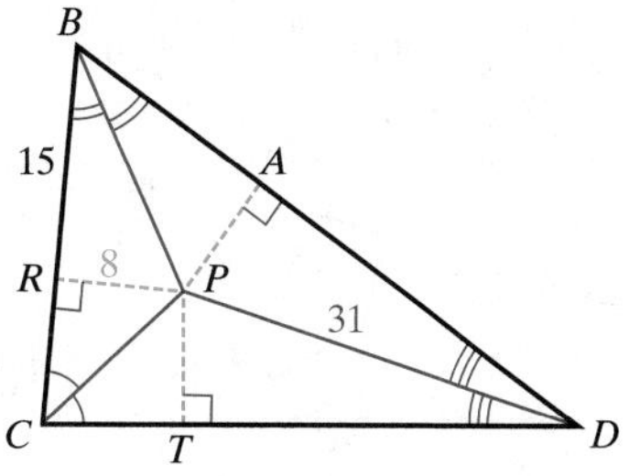

Coordinate Geometry *Find the coordinates of the circumcenter of each triangle. See Example 2.*

15.

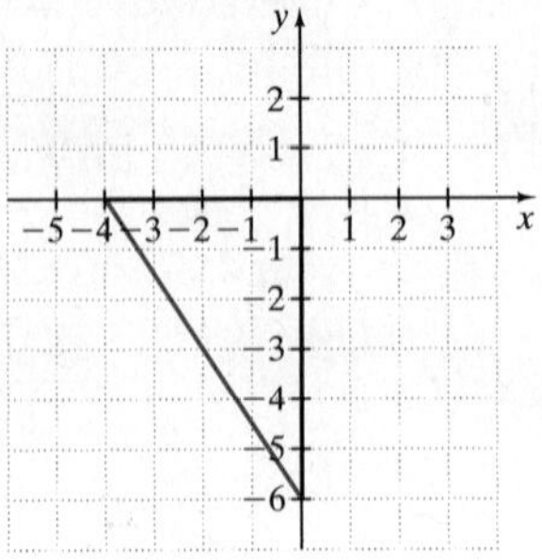

16.

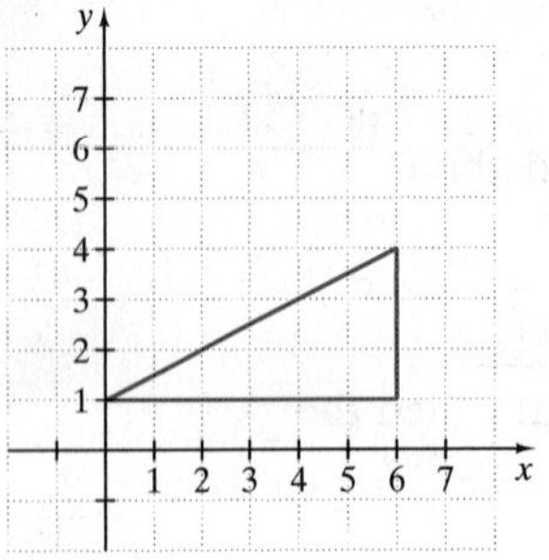

17.

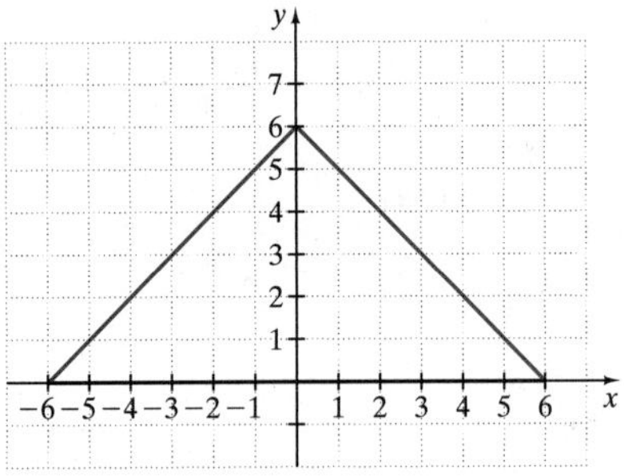

18.

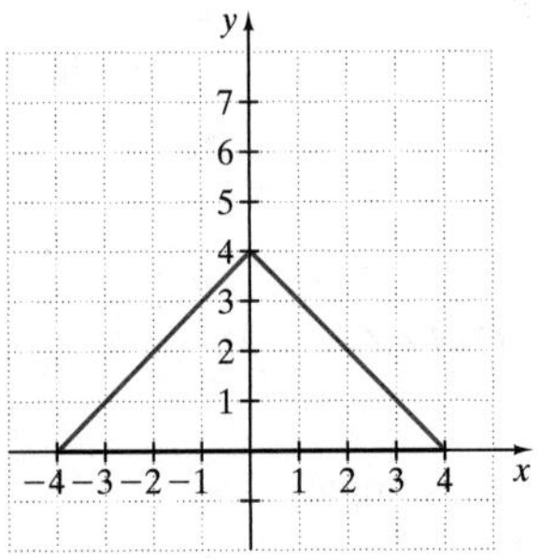

Coordinate Geometry *Find the coordinates of the circumcenter of ΔABC. See Example 2.*

19. $A(0,0)$
$B(3,0)$
$C(3,2)$

20. $A(0,0)$
$B(4,0)$
$C(4,-3)$

21. $A(-4,5)$
$B(-2,5)$
$C(-2,-2)$

22. $A(-1,-2)$
$B(-5,-2)$
$C(-1,-7)$

Name the point of concurrency of the angle bisectors. See Example 4.

23.

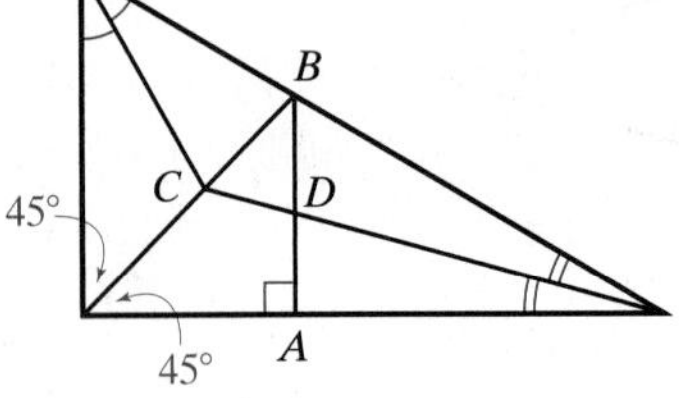

24.

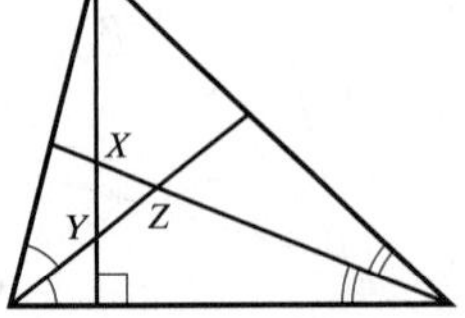

Find the value of x. See Example 4.

25.

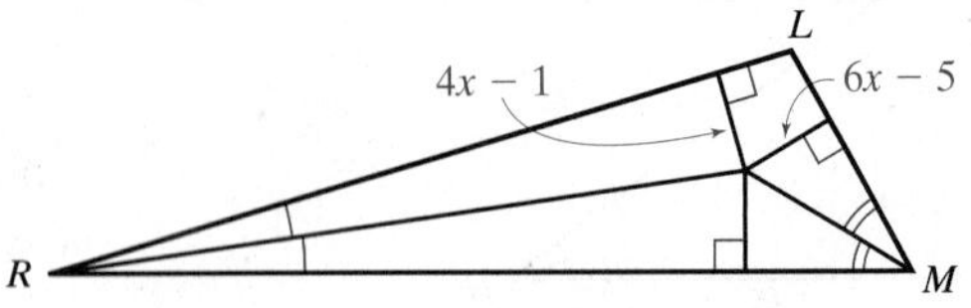

26.

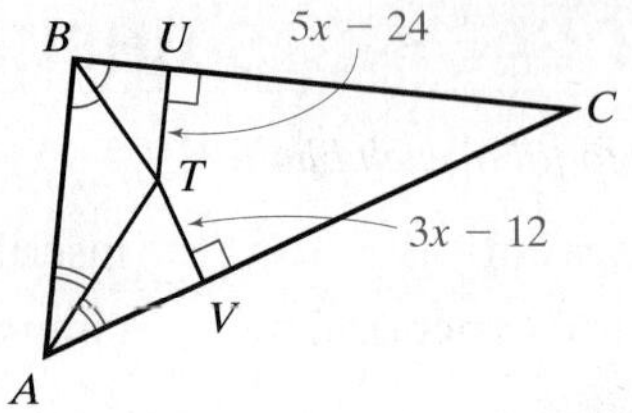

27. $RS = 4(x - 3) + 6$ and $RT = 5(2x - 6)$

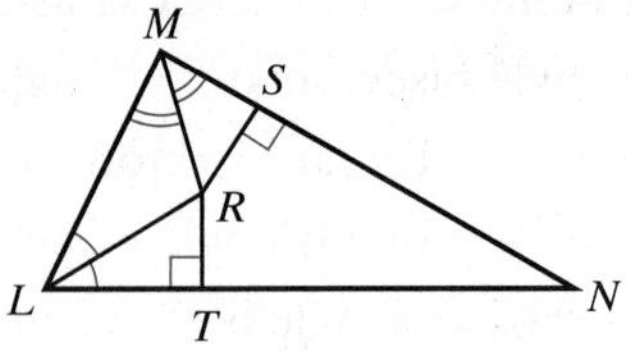

28. $DE = 12(x + 2) - 7$ and $DF = 7(3x - 4)$

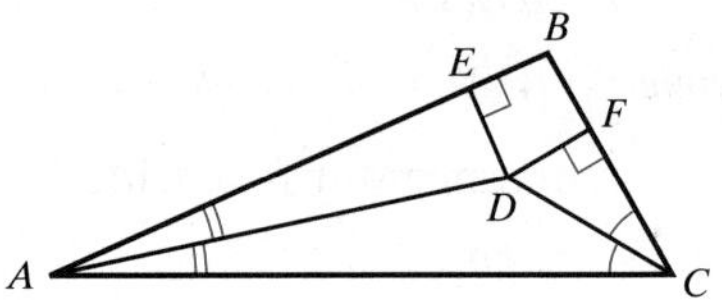

Find the Error *Explain why each statement is false.*

29.

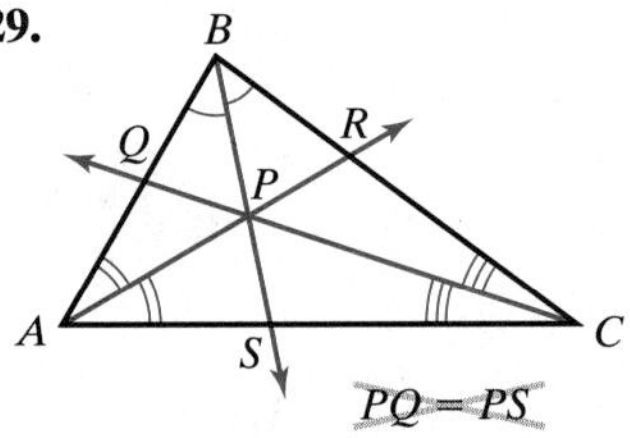

30.

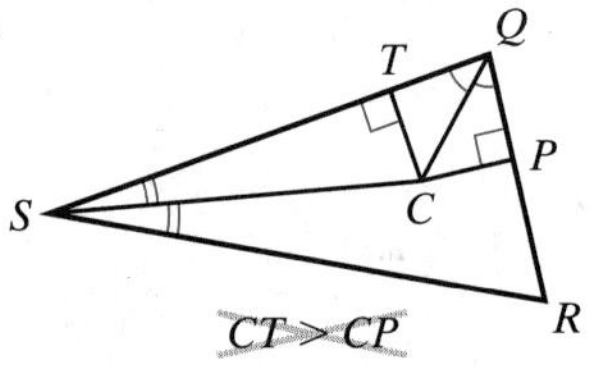

31. Copy the diagram of the beach. Show where town officials should place a recycling barrel so that it is equidistant from the lifeguard chair, the snack bar, and the volleyball court. Explain. See Example 3.

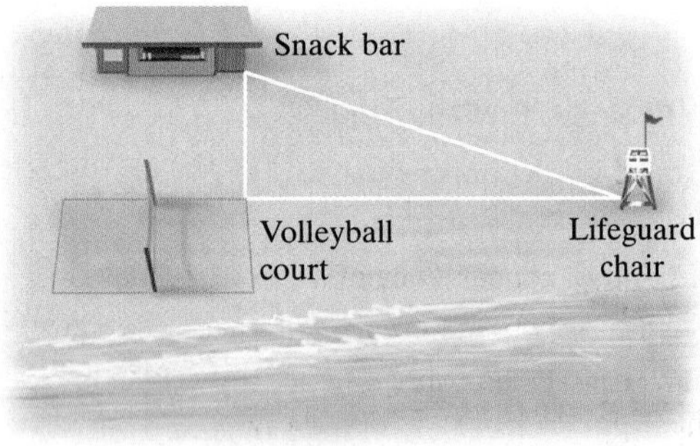

32. You are trying to talk to a friend on the phone in a busy bus station. The buses are so loud that you can hardly hear. Referring to the figure below, should you stand at P or C to be as far as possible from all the buses? Explain.

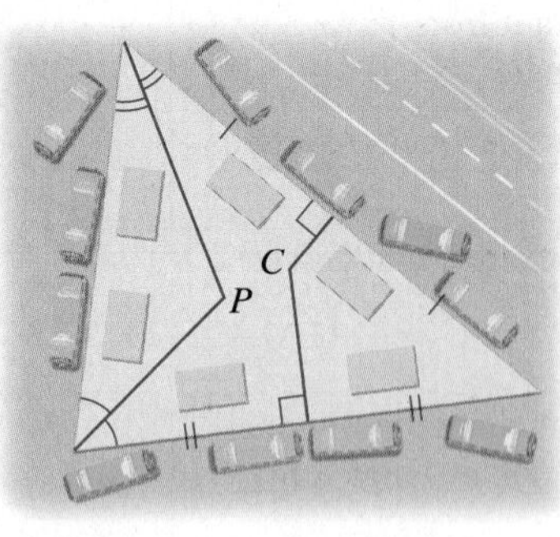

In the figure at the right, P is the incenter of ΔRST.

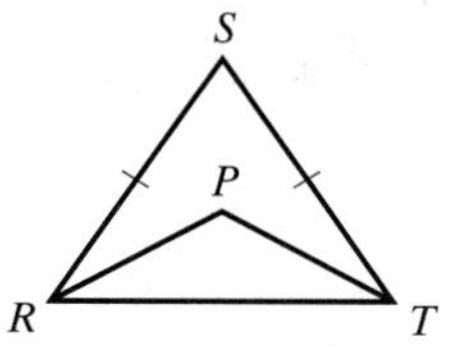

33. What type of triangle is ΔRST? Explain.

34. What do you know about the base angles of an isosceles triangle?

35. What segments determine the incenter of a triangle?

36. What type of triangle is ΔRPT? Explain.

Proof *For Exercises 37–40, use the diagram below to complete the proof of the Concurrency of Angle Bisectors Theorem.*

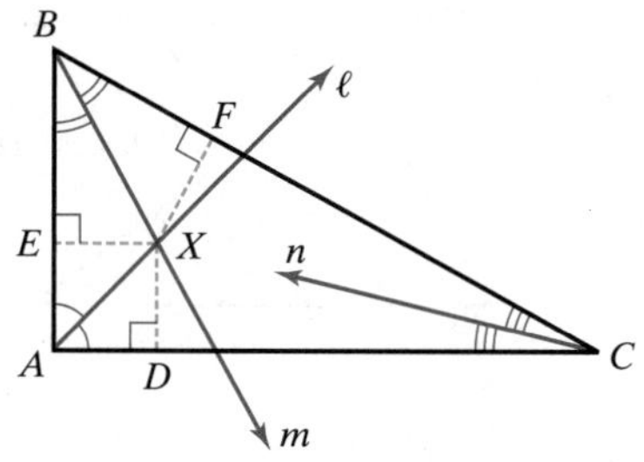

Given: Rays l, m, and n are bisectors of the angles of ΔABC. X is the intersection of rays l and m, $\overline{XD} \perp \overline{AC}$, $\overline{XE} \perp \overline{AB}$, and $\overline{XF} \perp \overline{BC}$.

Prove: Ray n contains point X, and $XD = XE = XF$.

Statements	**Reasons**
1. Rays l and m are bisectors of the angles of ΔABC; rays l and m intersect at point X; $\overline{XD} \perp \overline{AC}$, $\overline{XE} \perp \overline{AB}$, and $\overline{XF} \perp \overline{BC}$.	**1.** Given
2. $XE = XD$	**2.** Ray l bisects $\angle BAC$, so X is ? **(Exercise 37)** from the sides of $\angle BAC$.
3. $XE = XF$	**3.** Ray m bisects $\angle ABC$, so X is ? **(Exercise 38)** from the sides of $\angle ABC$.
4. $XD = XF$	**4.** ? **(Exercises 39)**
5. X is on the angle bisector of $\angle C$.	**5.** Converse of Angle Bisectors Theorem
6. $XE = XD = XF$	**6.** ? **(Exercise 40)**

Construction *Draw a triangle that fits the given description. Then construct the inscribed circle and the circumscribed circle. Describe your method.*

41. right triangle, ΔDEF

42. obtuse triangle, ΔSTU

Determine whether each statement is true or false. If the statement is false, give a counterexample.

43. The incenter of a triangle is equidistant from all three vertices.

44. The incenter of a triangle always lies inside the triangle.

45. You can circumscribe a circle about any three points in a plane.

46. If point C is the circumcenter of ΔPQR and the circumcenter of ΔPQS, then R and S must be the same point.

CONCEPT EXTENSIONS

In the diagram at the right, G is the incenter of ΔDEF, $m\angle DEF = 60°$, and $m\angle EFD = 2 \cdot m\angle EDF$. Find the following.

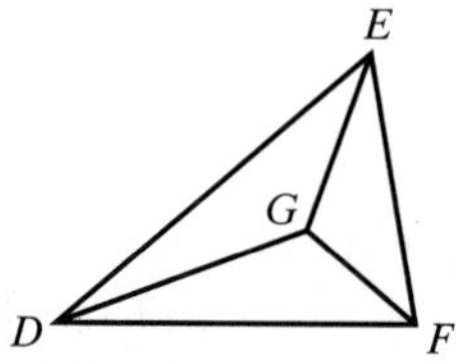

47. $m\angle DEG$

48. $m\angle GEF$

49. $m\angle EFD$ and $m\angle EDF$

50. $m\angle EFG$

51. $m\angle FDG$

52. $m\angle DGE$

53. $m\angle DGF$

54. $m\angle EGF$

55. Explain why the circumcenter of a right triangle is on one of the triangle's sides.

56. You want to find the circumcenter of a triangle. Why do you need to find the intersection of only two of the triangle's perpendicular bisectors, instead of all three?

Determine whether each statement is always, sometimes, or never true. Explain.

57. It is possible to find a point equidistant from three parallel lines in a plane.

58. The circles inscribed in and circumscribed about an isosceles triangle have the same center.

REVIEW AND PREVIEW

Use the figure below for Exercises 59 and 60. See Section 5.1.

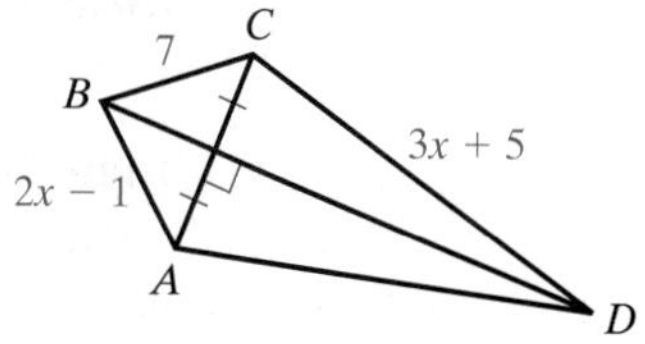

59. Find the value of x.

60. Find the length of $\overline{AD}$.

Find the coordinates of the midpoint of $\overline{AB}$ with the given endpoints. See Section 1.7.

61. $A(3, 0)$, $B(3, 16)$

62. $A(6, 8)$, $B(4, -1)$

5.3 MEDIANS AND ALTITUDES OF A TRIANGLE

OBJECTIVES

1 Use Properties of the Medians of a Triangle.

2 Use Properties of the Altitudes of a Triangle.

VOCABULARY

- median of a triangle
- centroid of a triangle
- altitude of a triangle
- orthocenter of a triangle

OBJECTIVE 1 ▶ Using Properties of Medians. In Section 5.2, we studied two points of concurrency of a triangle. Some of the properties of these points are reviewed below.

▶ Helpful Hint Review—Points of Concurrency

Circumcenter

- Intersection of the perpendicular bisectors of a triangle
- This point can be inside, outside, or on the triangle.
- 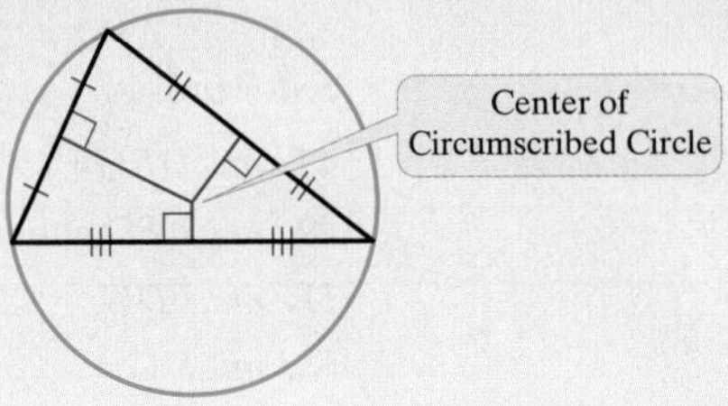

Incenter

- Intersection of the angle bisectors of the triangle
- This point always lies inside the triangle.
- 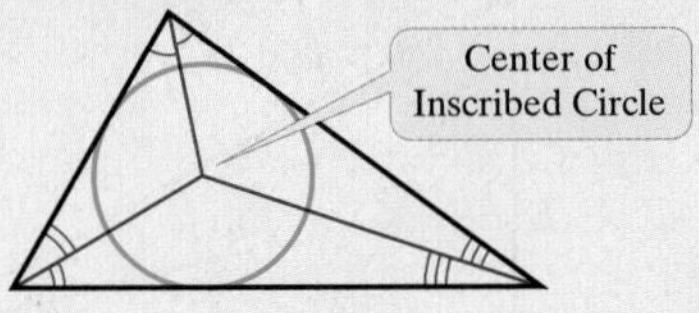

Now we study two other types of segments associated with a triangle and their corresponding points of concurrency.

A **median of a triangle** is a segment whose endpoints are a vertex and the midpoint of the opposite side. A triangle's three medians are always concurrent.

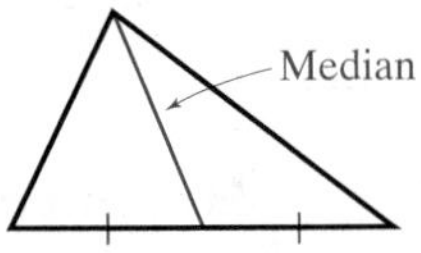

Theorem 5.3-1 Concurrency of Medians Theorem

The medians of a triangle are concurrent at a point that is two-thirds the distance from each vertex to the midpoint of the opposite side.

$DC = \frac{2}{3}DJ \qquad EC = \frac{2}{3}EG \qquad FC = \frac{2}{3}FH$

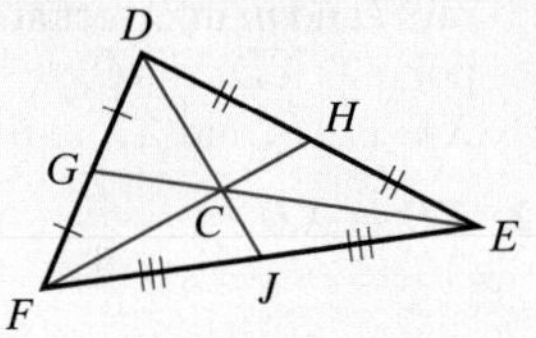

(*Note:* A proof of Theorem 5.3-1 is in the back of this text.) We demonstrate this theorem in Exercise 49.

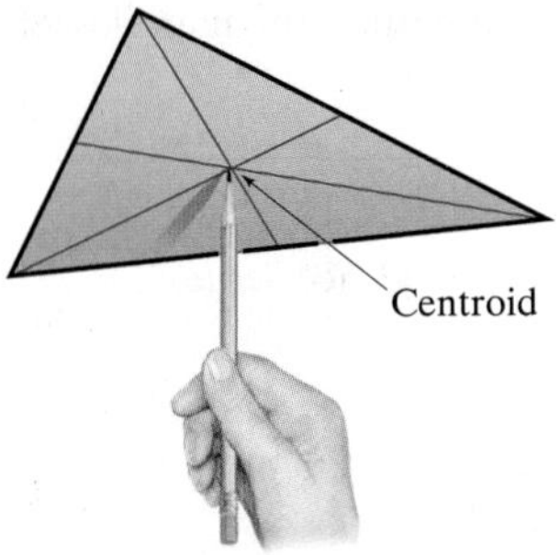

A triangle of uniform thickness balanced by a pencil at the centroid.

In a triangle, the point of concurrency of the medians is called the **centroid of the triangle**, also known as the *center of gravity* of a triangle because it is the point where a triangular shape of uniform thickness will balance. For any triangle, the centroid is always inside the triangle.

EXAMPLE 1 Finding the Length of a Median

In the diagram, $AC = 10$ units. Find AE.

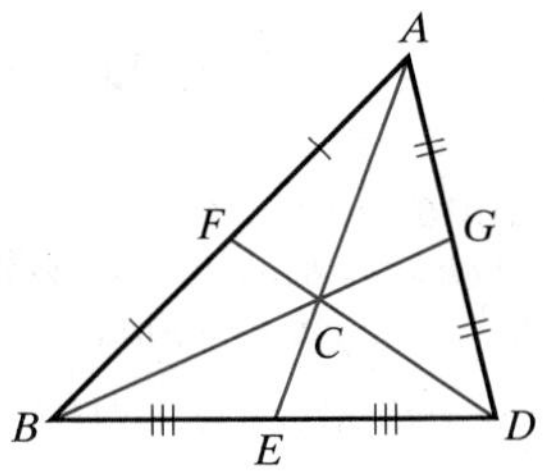

Solution Point C is the centroid of ΔABD because it is the point of concurrency of the medians. Thus:

$$AC = \frac{2}{3} \cdot AE \quad \text{Concurrency of Medians Theorem}$$

$$10 = \frac{2}{3} \cdot AE \quad \text{Given that } AC = 10 \text{ units}$$

$$\frac{3}{2} \cdot 10 = \frac{3}{2} \cdot \frac{2}{3} \cdot AE \quad \text{Multiply both sides by } \frac{3}{2}.$$

$$15 = AE \quad \text{Simplify.} \left(\frac{3}{2} \cdot \frac{2}{3} \text{ is 1.}\right)$$

Thus, $AE = 15$ units. □

PRACTICE 1 In the diagram for Example 1, suppose that $DC = 12$ units. If so, find DF.

OBJECTIVE 2 ▶ Using Properties of Altitudes. An **altitude of a triangle** is the perpendicular segment from a vertex of the triangle to the line containing the opposite side. An altitude of a triangle can be inside or outside the triangle, or it can be a side of the triangle.

> **Helpful Hint**
>
> Take care when constructing or identifying an altitude.
>
> Remember: An altitude is a segment from a vertex of the triangle perpendicular to the opposite side (or the opposite side extended).
>
> - Altitude is inside the Δ
>
> 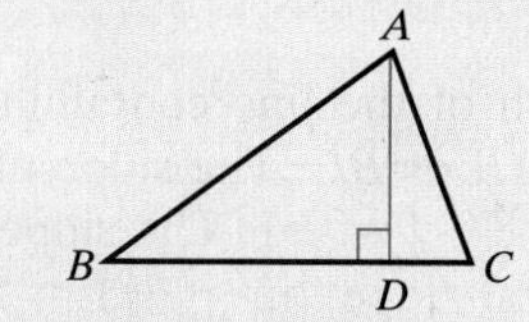
>
>
> - Altitude is outside the Δ
>
> 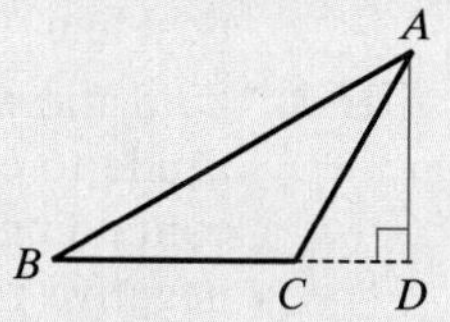
>
>
> - Altitude is a side of the Δ
>
> 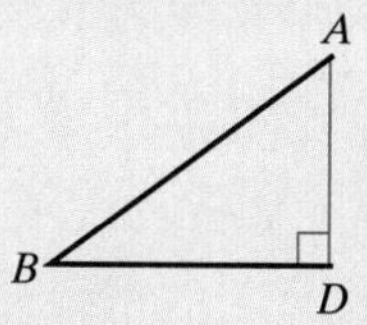
>
>
> In all 3 triangles, $\overline{AD}$ is an altitude of ΔABC.

Let's practice identifying medians and altitudes.

EXAMPLE 2 **Identifying Medians and Altitudes**

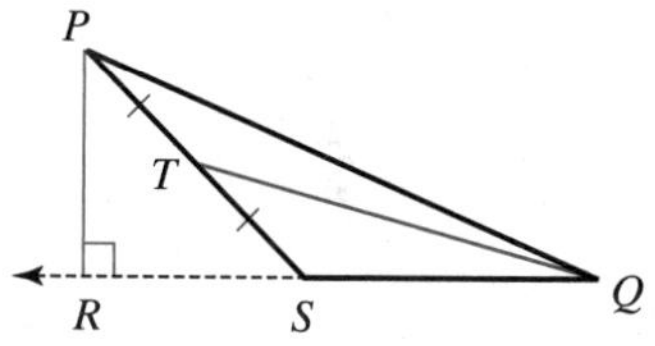

a. For ΔPQS, is $\overline{QT}$ a *median*, an *altitude*, or *neither*? Explain.

b. For ΔPQS, is $\overline{PR}$ a *median*, an *altitude*, or *neither*? Explain.

Solution

a. $\overline{QT}$ is a segment that extends from vertex Q to the side opposite Q. Since $\overline{PT} \cong \overline{TS}$, T is the midpoint of $\overline{PS}$. So $\overline{QT}$ is a median of ΔPQS.

b. $\overline{PR}$ is a segment that extends from vertex P to the line containing $\overline{SQ}$, the side opposite P. $\overline{PR} \perp \overrightarrow{QR}$, so $\overline{PR}$ is an altitude of ΔPQS. □

PRACTICE

2 For ΔABC, is each segment a *median*, an *altitude*, or *neither*? Explain.

a. $\overline{AD}$ **b.** $\overline{EG}$ **c.** $\overline{CF}$

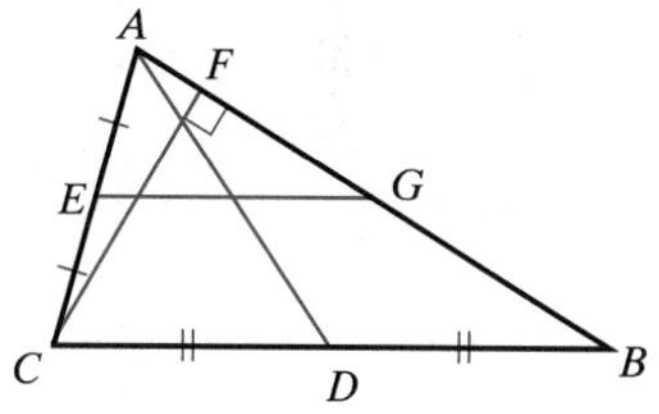

> **Theorem 5.3-2 Concurrency of Altitudes Theorem**
>
> The lines that contain the altitudes of a triangle are concurrent.

(*Note:* A proof of Theorem 5.3-2 is in the back of this text.) We demonstrate this theorem in Exercise 50.

The lines that contain the altitudes of a triangle are concurrent at the **orthocenter of the triangle.** The orthocenter of a triangle can be inside, on, or outside the triangle.

Acute triangle

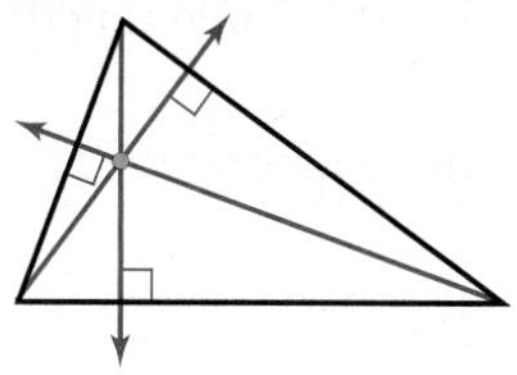

Right triangle

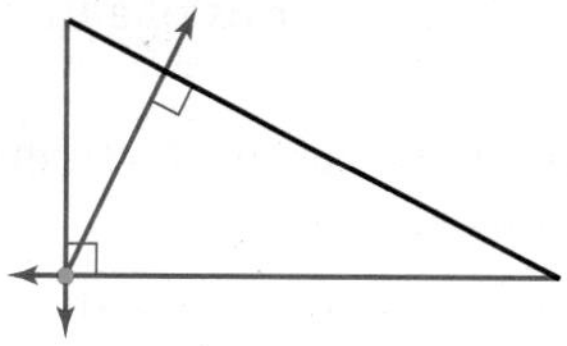

Obtuse triangle

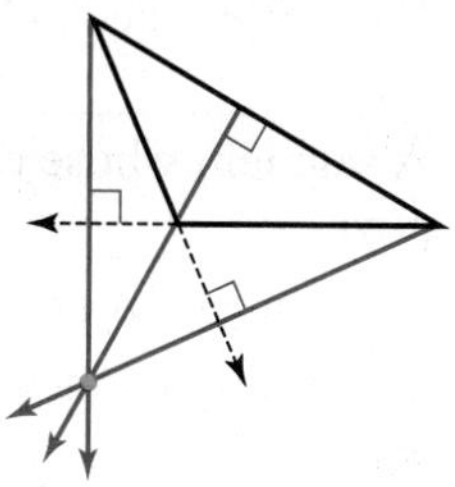

Each blue point is the orthocenter

> **Helpful Hint**
> Remember: To find a point of concurrency, we need to find the point of intersection of only two lines.

EXAMPLE 3 **Finding the Orthocenter of a Triangle**

ΔABC has vertices $A(1, 3)$, $B(2, 7)$, and $C(6, 3)$. What are the coordinates of the orthocenter of ΔABC?

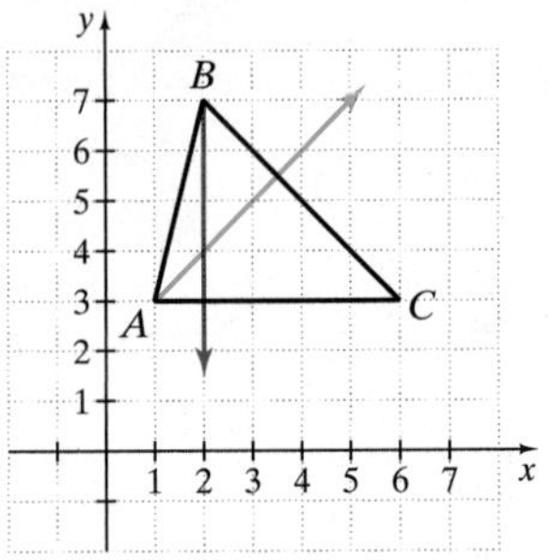

Solution

STEP 1. Use the given vertices to graph and study ΔABC.

STEP 2. Since $\overline{AC}$ is horizontal, the line containing the altitude to $\overline{AC}$ is vertical and passes through the vertex $B(2, 7)$. The equation of this line is the vertical line $x = 2$.

STEP 3. Find the equation of the line containing the altitude to either $\overline{AB}$ or $\overline{BC}$. The slope of the line containing $\overline{BC}$ is $\frac{3-7}{6-2} = -1$. The slope of a line perpendicular to $\overline{BC}$ has a negative reciprocal slope, or $-\left(\frac{1}{-1}\right) = 1$. The line containing the altitude to $\overline{BC}$ has slope 1.

The line passes through the vertex $A(1, 3)$. The equation of the line is $y - 3 = 1(x - 1)$, which simplifies to $y = x + 2$.

STEP 4. Find the orthocenter by solving this system of equations: $\begin{cases} x = 2 \\ y = x + 2 \end{cases}$

$y = x + 2$ Second equation in the system

$y = 2 + 2$ Substitute 2 for x in the second equation.

$y = 4$ Simplify.

The coordinates of the orthocenter are $(2, 4)$. □

PRACTICE

3 ΔDEF has vertices $D(1, 2)$, $E(1, 6)$, and $F(4, 2)$. What are the coordinates of the orthocenter of ΔDEF?

> **Helpful Hint** Review—Points of Concurrency
> Study the names and meanings of the points of concurrency of a triangle.

Perpendicular Bisectors	***Angle Bisectors***	***Medians***	***Altitudes***

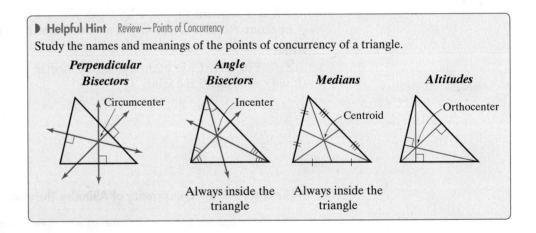

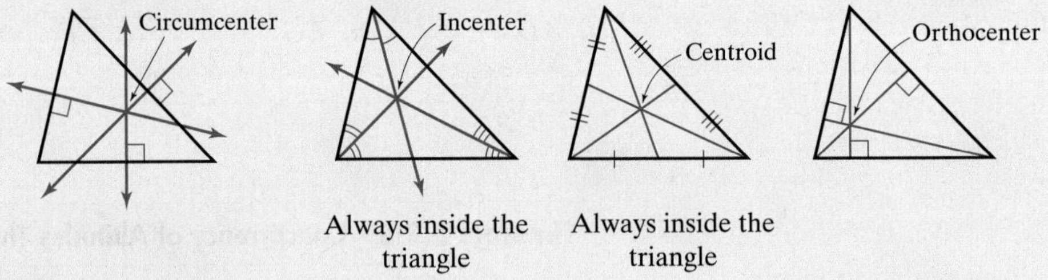

VOCABULARY & READINESS CHECK

Word Bank *Use the choices to fill in each blank. Some choices may be used more than once.*

centroid orthocenter median altitude

1. The perpendicular segment from a vertex of a triangle to the line containing the opposite side is called a(n) ____________ of the triangle.
2. A segment whose endpoints are a vertex of a triangle and the midpoint of the opposite side is called a(n) ____________ of the triangle.
3. The point of concurrency of the medians of a triangle is called the ____________.
4. The point of concurrency of the altitudes of a triangle is called the ____________.
5. The point of concurrency called the ____________ may lie inside, outside, or on a triangle.
6. The ____________ is also known as the center of gravity of a triangle.

5.3 EXERCISE SET MyMathLab®

In ΔTUV, Y is the centroid. Find the following. See Example 1.

1. If $YW = 9$, find TY and TW.
2. If $YU = 12$, find ZY and ZU.
3. If $VX = 33$, find VY and YX.
4. If $TW = 72$, find TY and YW.

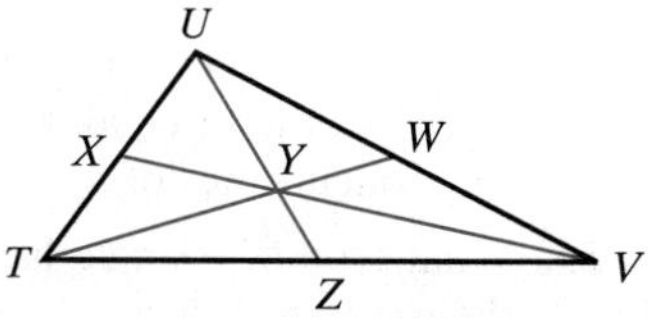

For ΔABC, is the red segment a median, an altitude, or neither? Explain. See Example 2.

5.

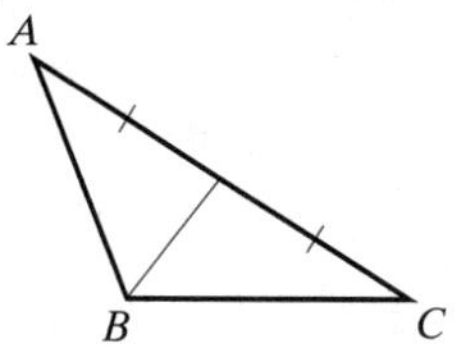

6.

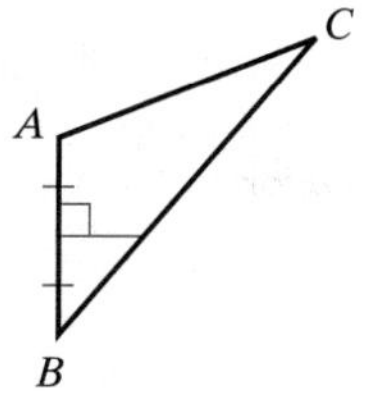

7.

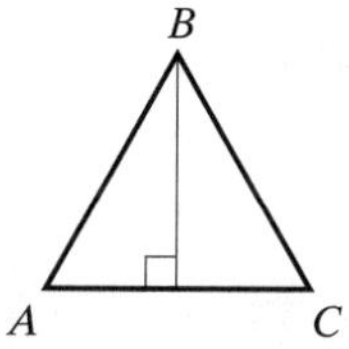

8.

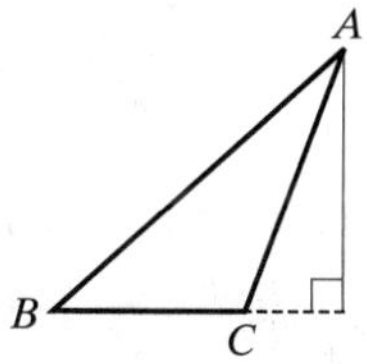

Use ΔABC for Exercises 9–14. See Example 2.

9. Is $\overline{AP}$ a median or an altitude?
10. Is $\overline{BQ}$ a median or an altitude?
11. If $AP = 18$, what is KP?
12. If $BK = 15$, what is KQ?
13. Is $\overline{BA}$ a median or an altitude?
14. Is $\overline{CA}$ a median or an altitude?

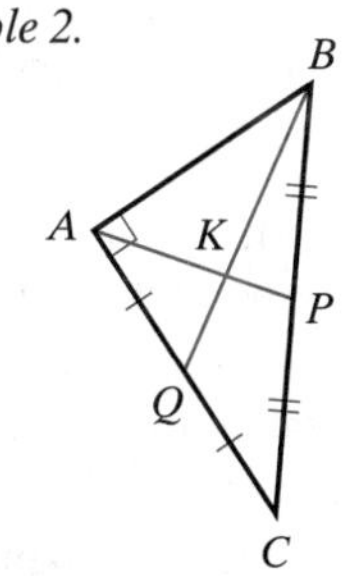

Use the information and diagram to answer each question. Point P is the centroid and $BD \perp AC$.

15. Find the length of $\overline{DC}$.
16. Find the length of $\overline{BE}$.
17. Find the length of $\overline{AC}$.
18. Find the length of $\overline{BC}$.
19. Find the length of $\overline{PD}$.
20. Find the length of $\overline{BD}$.
21. Find the perimeter of (distance around) ΔBDA.
22. Find the perimeter of (distance around) ΔABC.

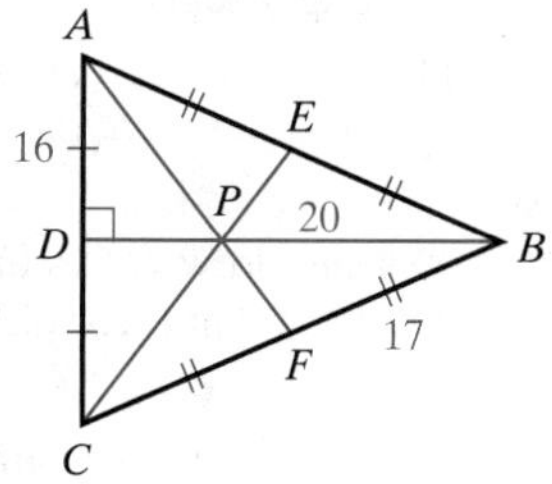

> **Helpful Hint**
> Note from ΔABC that if a median is also a perpendicular bisector, then the triangle is isosceles. (See Exercises 15–22.)

Name the centroid for each triangle.

23.

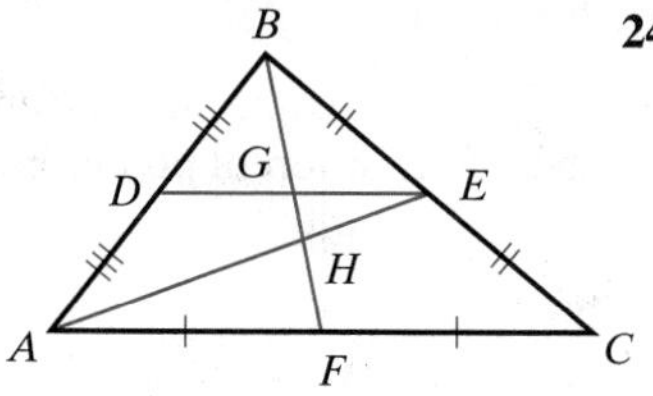

24.

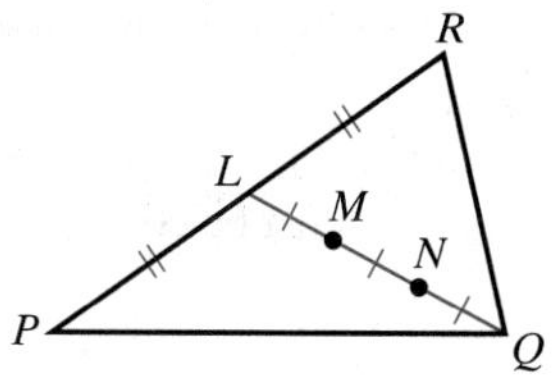

Name the orthocenter for each triangle.

25.

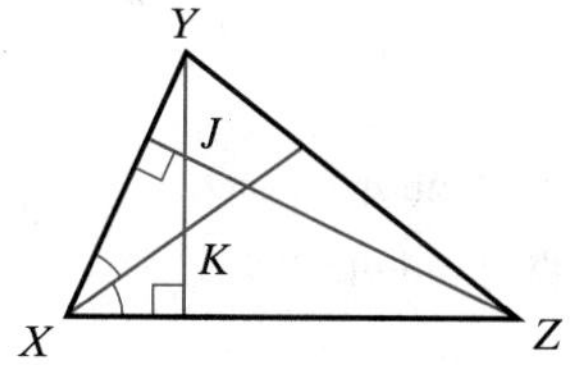

26.

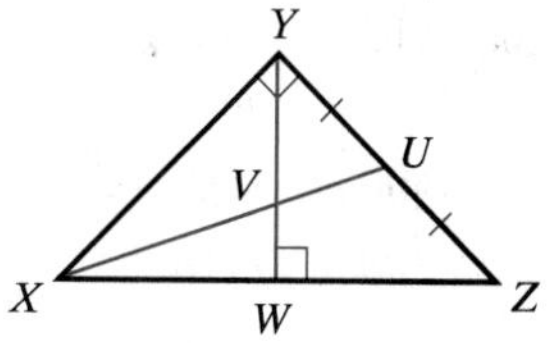

In Exercises 27–30, name each segment.

27. a median in ΔABC
28. an altitude in ΔABC
29. a median in ΔBDC
30. an altitude in ΔAOC

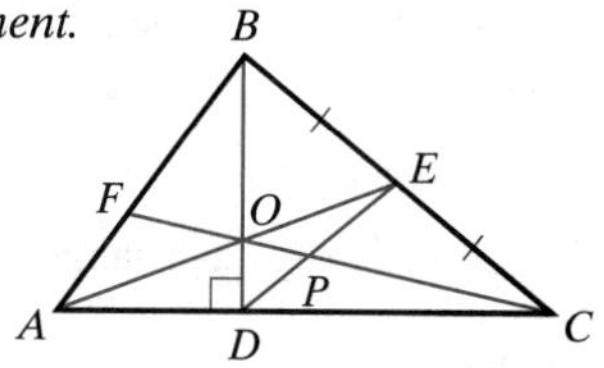

Coordinate Geometry *Find the coordinates of the orthocenter of ΔABC. See Example 3.*

31. $A(0, 0)$ $B(4, 0)$ $C(4, 2)$
32. $A(0, 0)$ $B(-2, 0)$ $C(0, 5)$
33. $A(2, 6)$ $B(8, 6)$ $C(6, 2)$
34. $A(0, -2)$ $B(4, -2)$ $C(-2, -8)$

Coordinate Geometry

For Exercises 35–36, find the following.

a. Find the midpoint of $\overline{BC}$. Call the midpoint M.
b. Find the length of median $\overline{AM}$.
c. Find the coordinates of the centroid P. (The coordinates will be $\frac{2}{3}$ of the distance from the vertex A.)

35.

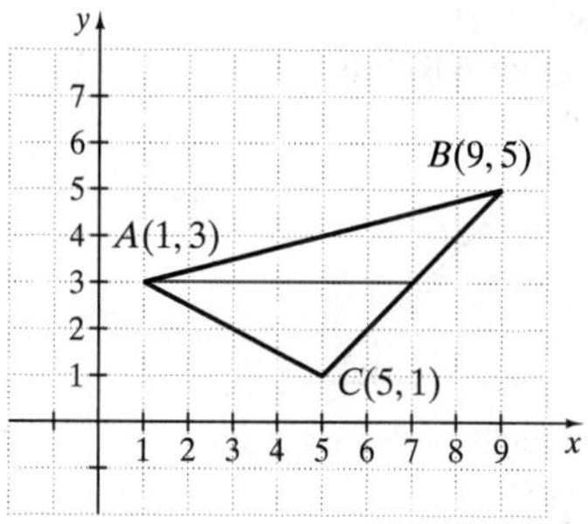

36.

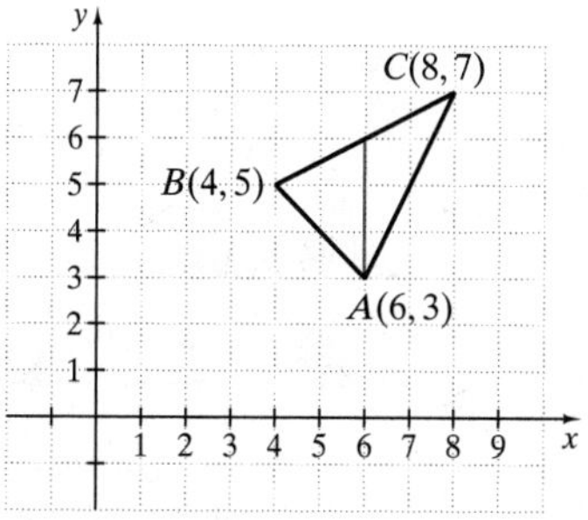

In the diagram at the right, $\overline{QS}$ and $\overline{PT}$ are altitudes and $m\angle R = 55°$.

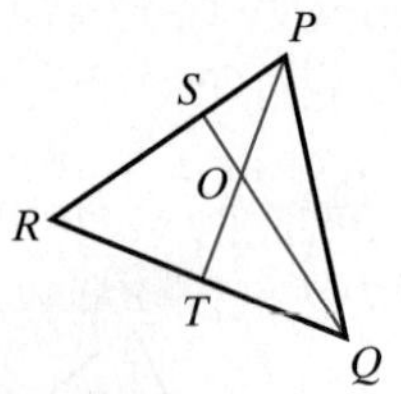

37. What does it mean for a segment to be an altitude?

38. What do you know about the sum of the angle measures in a triangle?

39. Find $m\angle SQR$.

40. Find $m\angle TOQ$.

41. Find $m\angle POQ$.

42. Find $m\angle POS$.

43. Find $m\angle OPS$.

44. Find $m\angle SOT$.

Construction *Draw a triangle that fits the given description. Then construct the centroid and the orthocenter.*

45. acute scalene triangle, ΔLMN

46. obtuse isosceles triangle, ΔRST

Multiple Choice *In the figure below, C is the centroid of ΔDEF.*

47. If $GF = 12x^2 + 6y$, which expression represents CF?

a. $6x^2 + 3y$
b. $4x^2 + 2y$
c. $8x^2 + 4y$
d. $8x^2 + 3y$

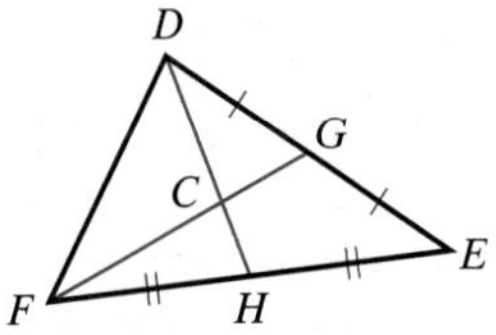

48. If $DH = 30x^2 + 18y$, which expression represents CH?

a. $20x^2 + 12y$
b. $10x^2 + 6y$
c. $15x^2 + 9y$
d. $15x^2 + 6y$

The figures below show how to construct altitudes and medians by paper folding. Refer to them for Exercises 49 and 50.

Folding an Altitude

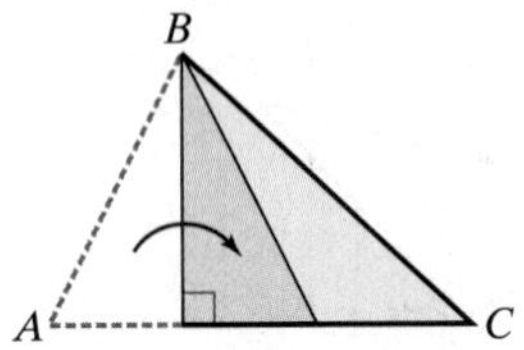

Fold the triangle so that side $\overline{AC}$ overlaps itself and the fold contains the opposite vertex B.

Folding a Median

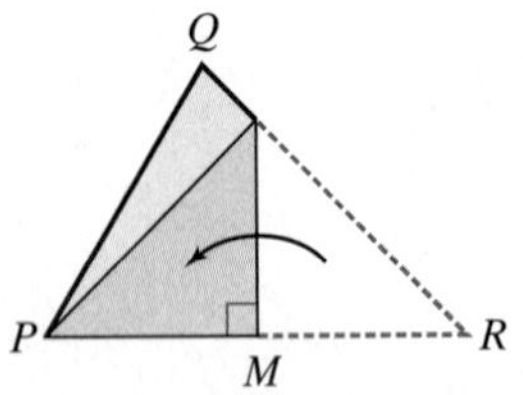

Fold one vertex R to another vertex P. This locates the midpoint M of the side.

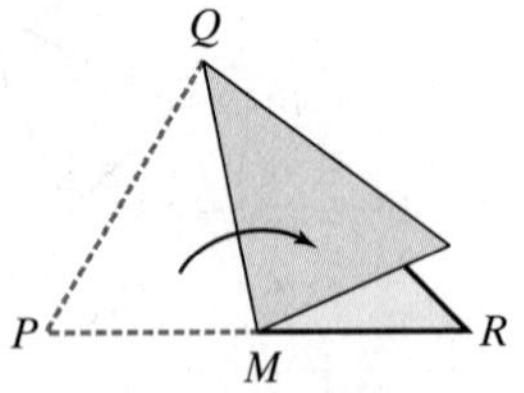

Unfold the triangle. Then fold it so that the fold contains the midpoint M and the opposite vertex Q.

49. Cut out a large triangle. Fold the paper carefully to construct the three medians of the triangle and demonstrate the Concurrency of Medians Theorem. Use a ruler to measure the length of each median and the distance of each vertex from the centroid.

50. Cut out a large acute triangle. Fold the paper carefully to construct the three altitudes of the triangle and demonstrate the Concurrency of Altitudes Theorem.

51. What type of triangle has its orthocenter on the exterior of the triangle? Draw a sketch to support your answer.

52. Explain why the median to the base of an isosceles triangle is also an altitude.

Coordinate Geometry *ΔABC has vertices $A(0, 0)$, $B(2, 6)$, and $C(8, 0)$. Complete the following steps to verify the Concurrency of Medians Theorem for ΔABC.*

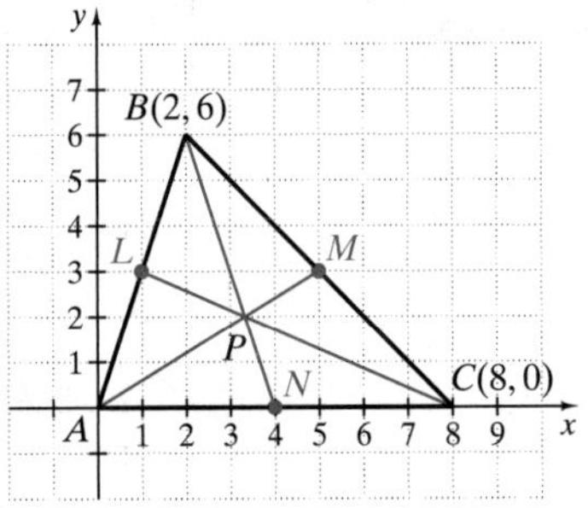

53. Find the coordinates of midpoints L, M, and N.

54. Find equations of $\overleftrightarrow{AM}$, $\overleftrightarrow{BN}$, and $\overleftrightarrow{CL}$.

55. Find the coordinates of P, the intersection of $\overleftrightarrow{AM}$ and $\overleftrightarrow{BN}$. This point is the centroid.

56. Use the Distance Formula to show that point P is $\frac{2}{3}$ of the distance from each vertex to the midpoint of the opposite side.

57. A centroid separates a median into two segments. What is the ratio of the length of the shorter segment to the length of the longer segment?

58. The orthocenter of ΔABC lies at vertex A. What can you conclude about $\overline{BA}$ and $\overline{AC}$? Explain.

59. **Find the Error** Your classmate says she drew $\overline{HJ}$ as an altitude of ΔABC. What error did she make?

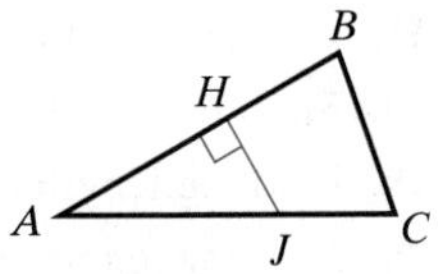

60. **Find the Error** Your classmate says he drew the centroid. What error did he make?

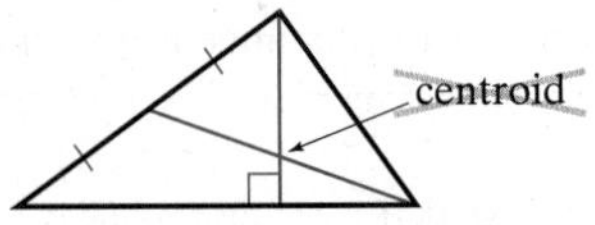

CONCEPT EXTENSIONS

61. **Construction** A, B, and O are three noncollinear points. Construct point C such that O is the orthocenter of ΔABC. Describe your method.

62. In an isosceles triangle, show that the circumcenter, incenter, centroid, and orthocenter can be four different points, but all four must be collinear.

A, B, C, and D are points of concurrency for the triangle. Determine whether each point is a circumcenter, incenter, centroid, or orthocenter. Explain.

63.

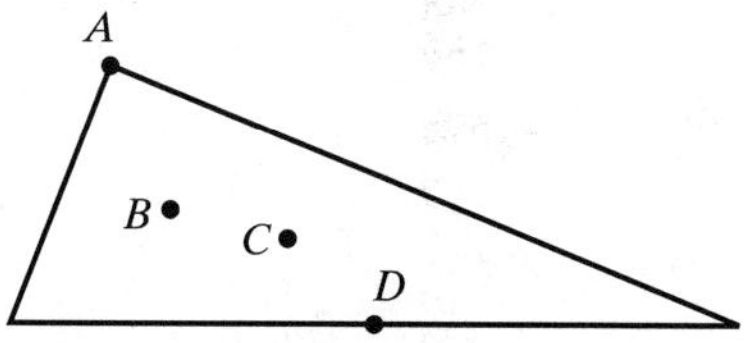

64.

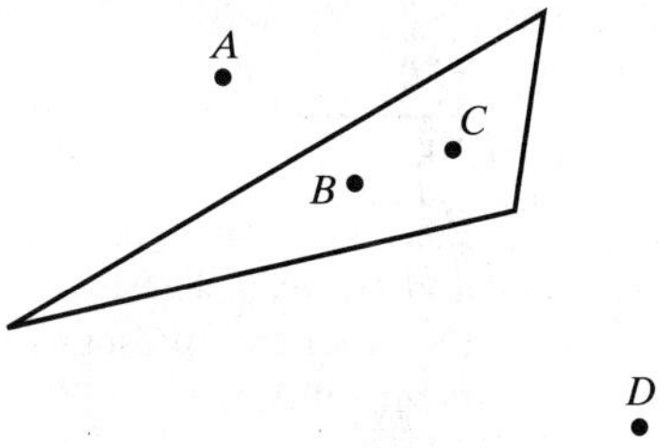

In 1765, Leonhard Euler proved that, for any triangle, three of the four points of concurrency are collinear. The line that contains these three points is known as Euler's Line. Use Exercises 63 and 64 to determine the following.

65. Which point of concurrency does not necessarily lie on Euler's Line?

66. Which three points of concurrency do you conjecture are always collinear?

REVIEW AND PREVIEW

Is $\overline{XY}$ a perpendicular bisector, an angle bisector, or neither? Explain. See Section 5.1.

67.

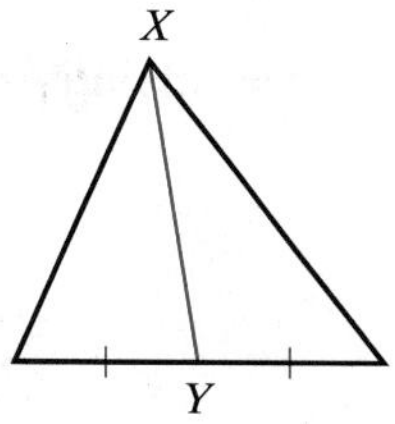

68.

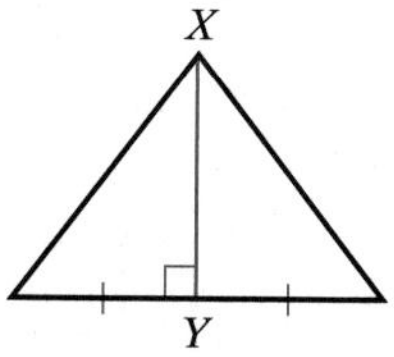

Write the negation of each statement. See Section 2.3.

69. You are not 16 years old.

70. Two angles are congruent.

71. $m\angle B \leq 90°$

72. $m\angle A < 90°$

5.4 MIDSEGMENTS OF TRIANGLES

OBJECTIVES

1. Use Properties of Midsegments of Triangles.
2. Use Coordinate Geometry with Midsegments.
3. Solve Applications of Midsegments.

VOCABULARY

- midsegment of a triangle

OBJECTIVE 1 ▶ Using Properties of Midsegments. In Sections 5.2 and 5.3, we studied four points of concurrency of a triangle. These points and some of their properties are reviewed below.

▶ Helpful Hint Review—Points of Concurrency

Point of Concurrency	*Intersection of What Lines/Segments*	*Point Lies Inside, On, or Outside Triangle*	*Special Feature(s)*
Circumcenter	Three perpendicular bisectors of the sides of a triangle	Can be inside, on, or outside the triangle	The center of the circle circumscribed about the triangle
Incenter	Three angle bisectors of a triangle	Always inside the triangle	The center of the circle inscribed in the triangle
Centroid	Three medians of a triangle	Always inside the triangle	• The point where a triangle of uniform thickness balances • Medians intersect $\frac{2}{3}$ the distance from vertex to midpoint of opposite side
Orthocenter	Three altitudes of a triangle	Can be inside, on, or outside the triangle	

Let's continue our study of triangles by introducing another special property—the midsegment of a triangle.

A **midsegment of a triangle** is a segment connecting the midpoints of two sides of the triangle.

TRY THE ACTIVITY BELOW

Paper-Folding Activity—Midsegments of a Triangle

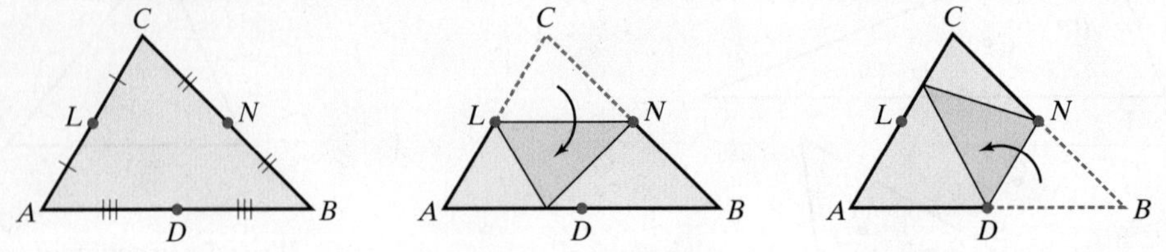

Cut out a triangle of any shape. Call it ΔABC. By folding, find the midpoint of each side of your triangle.

Fold the triangle on $\overline{LN}$ as shown. $\overline{LN}$ is called a midsegment.

Fold the triangle on $\overline{ND}$ and $\overline{DL}$ also. These are the other two midsegments of ΔABC.

There are two special relationships between a midsegment of a triangle and the third side of the triangle.

Theorem 5.4-1 Triangle Midsegment Theorem

Theorem	If...	Then...
If a segment joins the midpoints of two sides of a triangle, then the segment is parallel to the third side and is half as long.	D is the midpoint of $\overline{CA}$ and E is the midpoint of $\overline{CB}$	$\overline{DE} \parallel \overline{AB}$ and $DE = \frac{1}{2}AB$

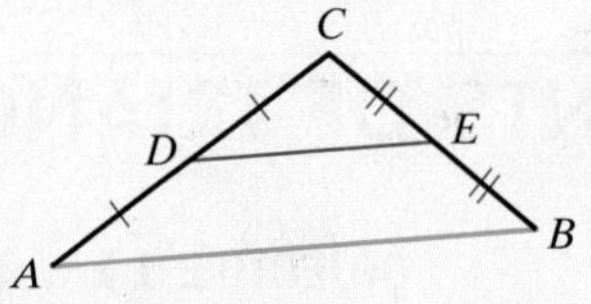

(*Note:* A proof of Theorem 5.4-1 is in the back of this text.) We verify this theorem in Example 4.

EXAMPLE 1 Identifying the Parallel Segments (Midsegments Theorem)

a. Name the midsegments and sides that are parallel in ΔDEF. Write your answer in the form of three parallel statements.

b. Name the midsegments that are half as long as the sides of ΔDEF. Write your answer in the form of three equations.

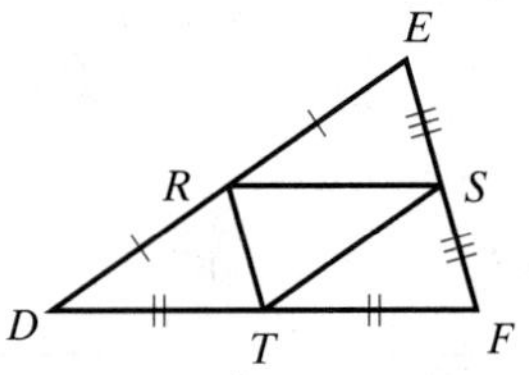

Solution

a. $\overline{RS}$, $\overline{ST}$, and $\overline{TR}$ are the midsegments of ΔDEF. By the Triangle Midsegment Theorem, $\overline{RS} \parallel \overline{DF}$, $\overline{ST} \parallel \overline{ED}$, and $\overline{TR} \parallel \overline{FE}$.

b. $TR = \frac{1}{2}(FE)$, $TS = \frac{1}{2}(ED)$, and $RS = \frac{1}{2}(DF)$ □

PRACTICE

1 Write your answers in the same forms as for Example 1.

a. Name the midsegments and sides that are parallel in ΔNQM.

b. Name the midsegments that are half as long as the sides of ΔNQM.

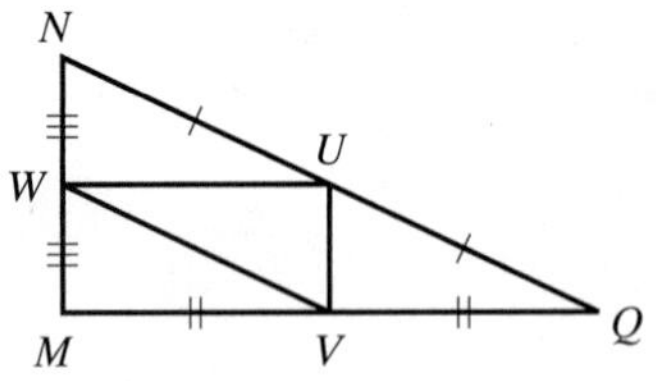

EXAMPLE 2 **Finding Lengths**

In ΔQRS, T, U, and B are midpoints. What are the lengths of $\overline{TU}$, $\overline{UB}$, and $\overline{QR}$?

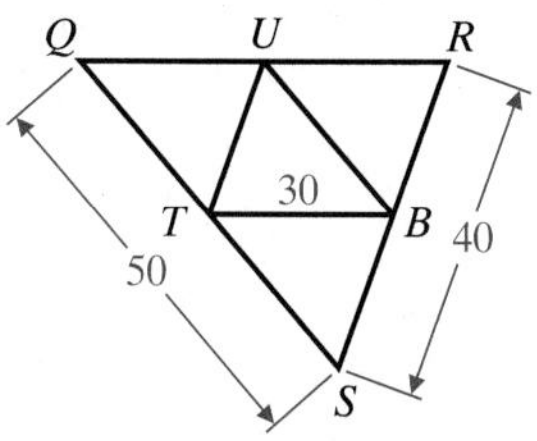

Solution

$TU = \frac{1}{2}SR$ | $UB = \frac{1}{2}QS$ | $TB = \frac{1}{2}QR$

$= \frac{1}{2}(40)$ | $= \frac{1}{2}(50)$ | $30 = \frac{1}{2}QR$

$= 20$ | $= 25$ | $60 = QR$

□

PRACTICE 2 In ΔABC, what are the lengths of $\overline{DC}$, $\overline{AC}$, $\overline{DF}$, and $\overline{DE}$?

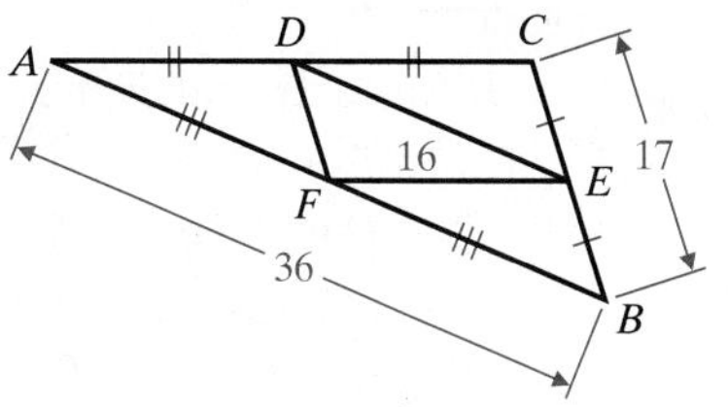

Let's look at the triangles in Example 1 and Practice 1 again. This time, we review our knowledge of parallel lines cut by a transversal and angle measures.

EXAMPLE 3 **Identifying Angle Measures**

Review the three pairs of midsegments and sides that are parallel in ΔDEF. (See Example 1.)

a. Find $m\angle EFD$. **b.** Find $m\angle FDE$.

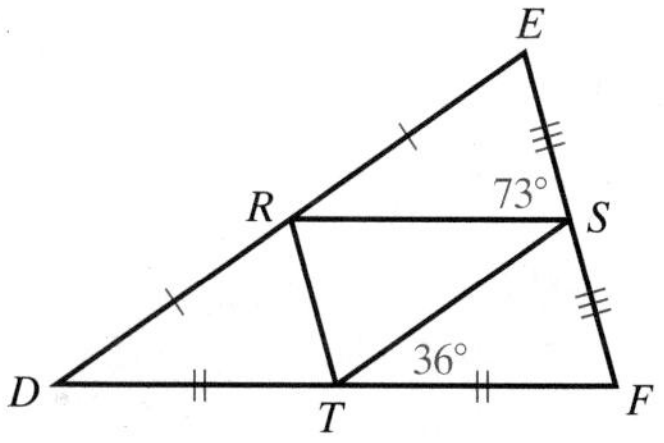

Solution

a. $RS \parallel DF$ with transversal $\overline{FE}$.
So, $m\angle EFD = m\angle ESR = 73°$.

b. $\overline{ST} \parallel \overline{ED}$ with transversal $\overline{DF}$. So, $m\angle FDE = m\angle FTS = 36°$. □

PRACTICE 3 Review the three pairs of midsegments and sides that are parallel in ΔNQM. (See Practice 1.)

a. Find $m\angle NUW$. **b.** Find $m\angle VUQ$.

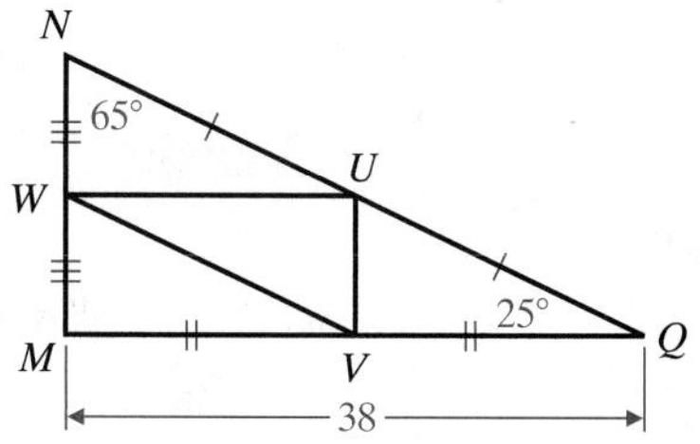

OBJECTIVE 2 ▶ Using Coordinate Geometry with Midsegments.

> **▶ Helpful Hint**
>
> For Example 4 and Practice 4 next, recall these formulas:
>
> Midpoint Formula: $M = \left(\frac{x_1 + x_2}{2}, \frac{y_1 + y_2}{2}\right)$
>
> Distance Formula: $d = \sqrt{(x_2 - x_1)^2 + (y_2 - y_1)^2}$
>
> Slope Formula: $m = \frac{y_2 - y_1}{x_2 - x_1}$
>
> For each formula, (x_1, y_1) and (x_2, y_2) are two points.

EXAMPLE 4 **Coordinate Geometry—Verifying the Triangle Midsegment Theorem**

Use the given triangle to verify the Triangle Midsegment Theorem. To do this, show that $\overline{DE} \parallel \overline{AB}$ and $DE = \frac{1}{2}AB$. (Points D and E are the midpoints of two sides of ΔABC.)

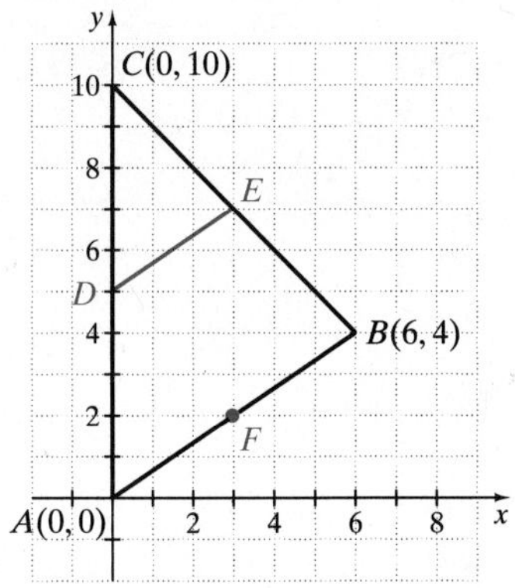

a. Find the coordinates of D and E.
b. Show that $\overline{DE} \parallel \overline{AB}$.
c. Show that $DE = \frac{1}{2}AB$.

Solution

a. Since D and E are midpoints, we use the Midpoint Formula.

D is the midpoint of $A(0, 0)$ and $C(0, 10)$, and
E is the midpoint of $B(6, 4)$ and $C(0, 10)$.

$$D = \left(\frac{0 + 0}{2}, \frac{0 + 10}{2}\right) = (0, 5)$$

$$E = \left(\frac{6 + 0}{2}, \frac{4 + 10}{2}\right) = (3, 7)$$

b. To show that $\overline{DE} \parallel \overline{AB}$, find the slope of each segment.

$$\text{slope of } \overline{DE} = \frac{7 - 5}{3 - 0} = \frac{2}{3} \qquad \text{slope of } \overline{AB} = \frac{4 - 0}{6 - 0} = \frac{4}{6} = \frac{2}{3}$$

Since the slopes are the same, $\overline{DE} \parallel \overline{AB}$.

c. To show $DE = \frac{1}{2}AB$, find each distance.

$$DE = \sqrt{(3 - 0)^2 + (7 - 5)^2} = \sqrt{9 + 4} = \sqrt{13}$$

$$AB = \sqrt{(6 - 0)^2 + (4 - 0)^2} = \sqrt{36 + 16} = \sqrt{52} = 2\sqrt{13}$$

It is true that $\sqrt{13} = \frac{1}{2}(2\sqrt{13})$. So, $DE = \frac{1}{2}(AB)$. □

PRACTICE 4 Use the same triangle for Example 4, except call point F the midpoint of $\overline{AB}$.

a. Find the coordinates of D and F.
b. Show that $\overline{DF} \parallel \overline{CB}$.
c. Show that $DF = \frac{1}{2}CB$.

OBJECTIVE 3 ▶ Solve Applications by Using Midsegments. We can use the Triangle Midsegment Theorem to find lengths of segments that might be difficult to measure directly.

EXAMPLE 5 **Using a Midsegment of a Triangle**

A geologist wants to determine the distance, AB, across a sinkhole. Choosing a point E outside the sinkhole, she finds the distances AE and BE. She locates the midpoints C and D of $\overline{AE}$ and $\overline{BE}$ and then measures $\overline{CD}$. What is the distance across the sinkhole?

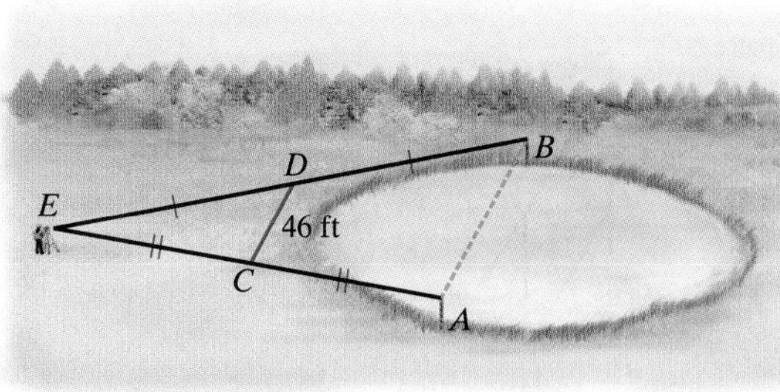

Solution CD is a midesegment of ΔAEB.

$$CD = \frac{1}{2}AB \quad \text{Δ Midsegment Theorem}$$
$$46 = \frac{1}{2}AB \quad \text{Substitute 46 for } CD.$$
$$92 = AB \quad \text{Multiply each side by 2.}$$

The distance across the sinkhole is 92 ft. □

PRACTICE

5 $\overline{CD}$ is a bridge being built over a lake, as shown in the figure at the right. What is the length of the bridge?

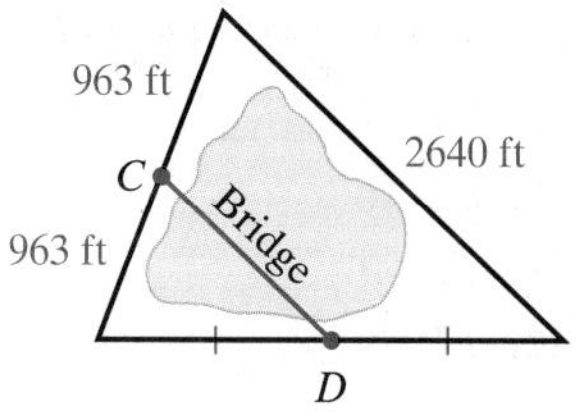

VOCABULARY & READINESS CHECK

Use the figure below for Exercises 1–4.

Multiple Choice

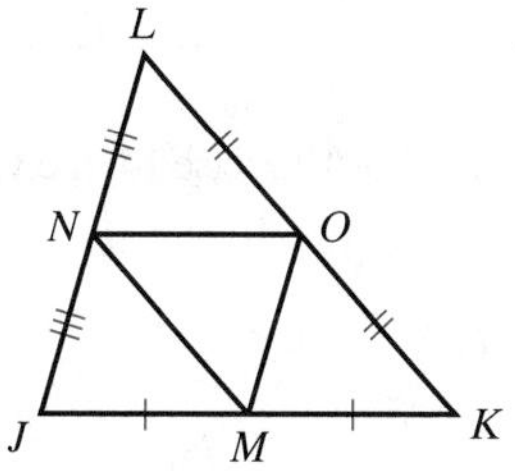

1. Choose a midsegment of ΔLJK.
   a. $\overline{LO}$ **b.** $\overline{MK}$ **c.** $\overline{OM}$ **d.** $\overline{LN}$
2. Midsegment $\overline{NO}$ has a length of
   a. $2(JK)$ **b.** $\frac{1}{2}(JK)$ **c.** OK **d.** $\frac{1}{2}(LK)$
3. Which segment is parallel to $\overline{JK}$?
4. If $LK = 46$, what is NM?

Matching

5. midsegment **A.** perpendicular segment from a vertex of a Δ to the line containing the opposite side
6. median **B.** point that bisects a segment into two congruent segments
7. altitude **C.** segment the midpoints of two sides of a Δ
8. midpoint **D.** segment whose endpoints are a vertex of a Δ and the midpoint of the opposite side

5.4 EXERCISE SET MyMathLab®

Multiple Steps

*For Exercises 1–4, answer parts **a** and **b**. See Example 1.*

a. *Name the midsegments and sides that are parallel in each triangle. Write your answer in the form of three parallel statements.*

b. *Name the midsegments that are half as long as each side of the triangle. Write your answer in the form of three equations.*

1. ΔABC

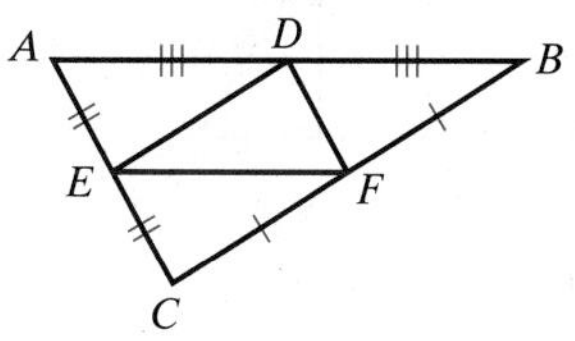

2. ΔTXV

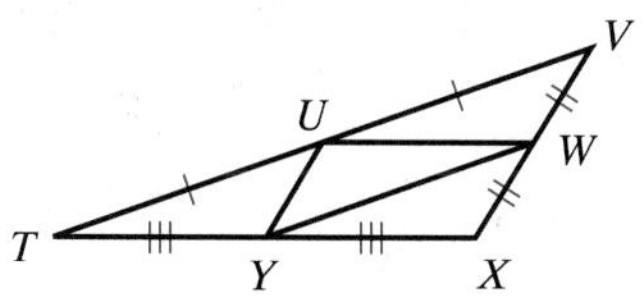

3. ΔFHK

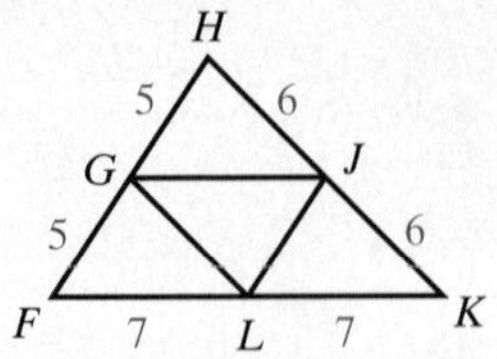

4. ΔMNP

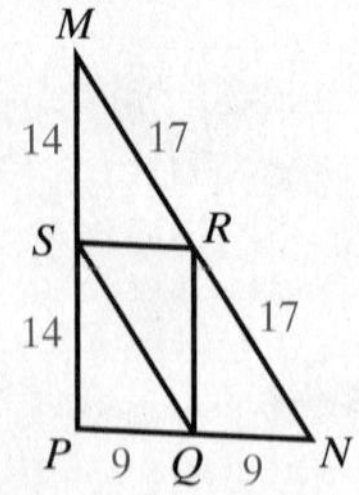

Fill in the blank *Use ΔABC and the Triangle Midsegment Theorem to name the segment that is parallel to the given segment. See Example 1.*

5. $\overline{AB} \parallel$ ________

6. $\overline{CA} \parallel$ ________

7. $\overline{GE} \parallel$ ________

8. $\overline{FG} \parallel$ ________

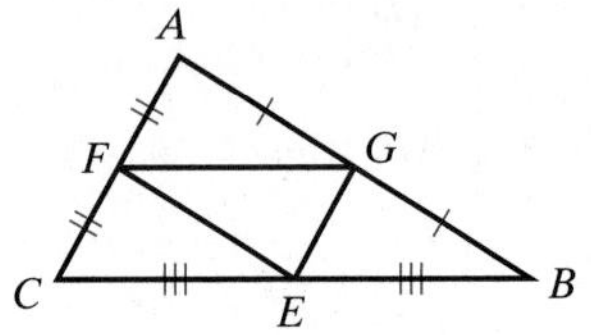

Fill in the blank *Use ΔTUV to determine each length. Given: $UV = 80$, $TV = 100$, and $HD = 80$. See Example 2.*

9. $HE =$ ________

10. $ED =$ ________

11. $TU =$ ________

12. $TE =$ ________

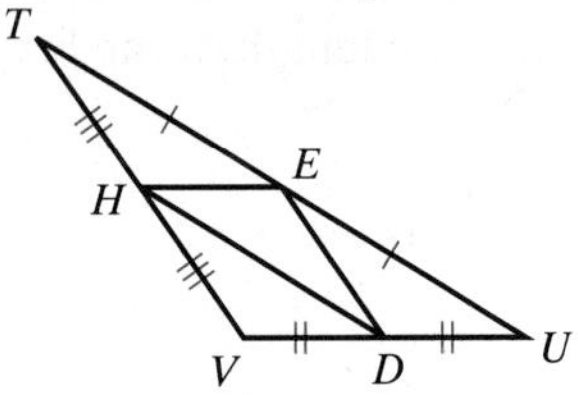

Multiple Choice

Find the value of x or y. See Example 2.

13.

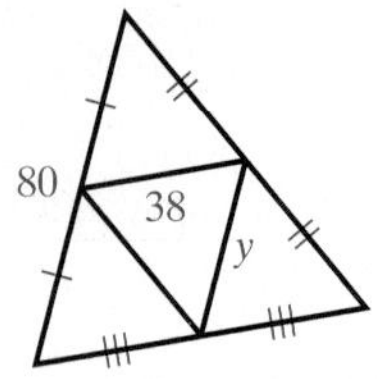

For Exercise **13**, the value of y is

a. 80 **b.** 38 **c.** 40 **d.** 76

14.

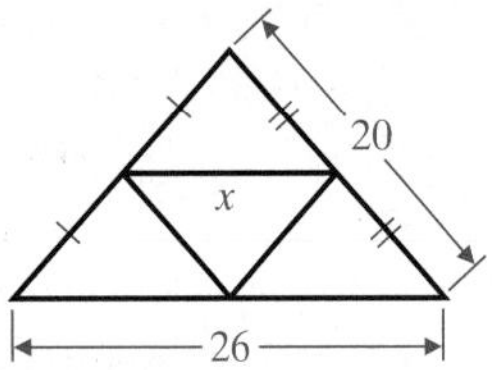

For Exercise **14**, the value of x is

a. 13 **b.** 26 **c.** 52 **d.** 10

Use ΔWUV for Exercises 15–18. See Example 3.

15. $m\angle V =$ ______

16. $m\angle UXY =$ ______

17. $m\angle UYX =$ ______

18. $m\angle TYW =$ ______

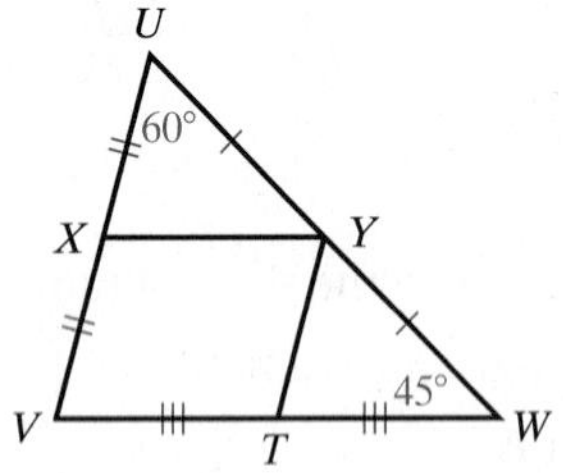

Find the value of x.

19.

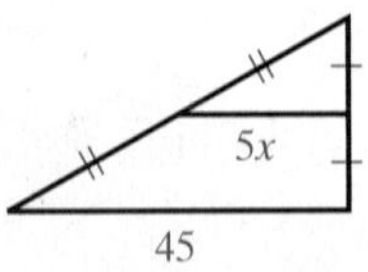

20.

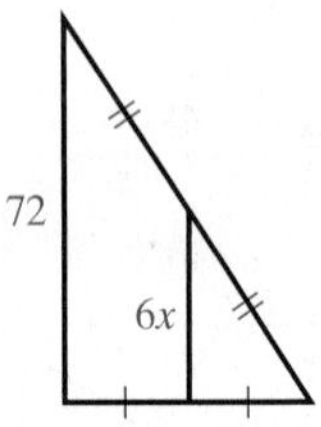

21.

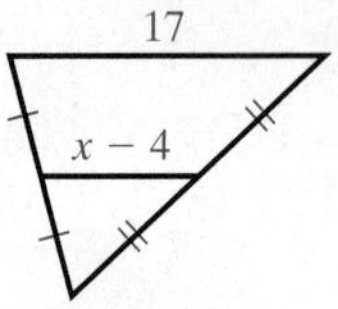

22.

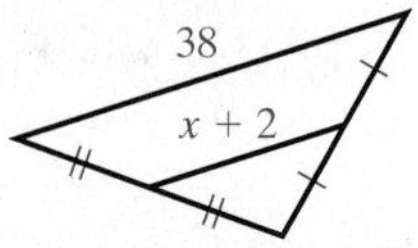

23.

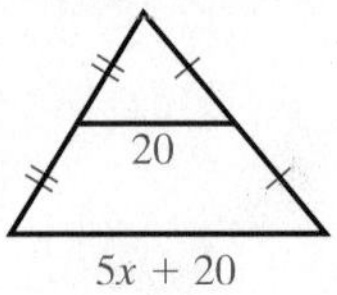

24.

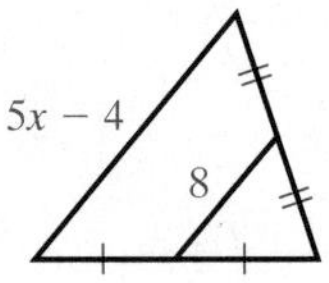

25.

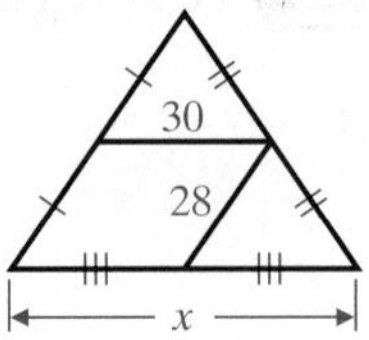

26.

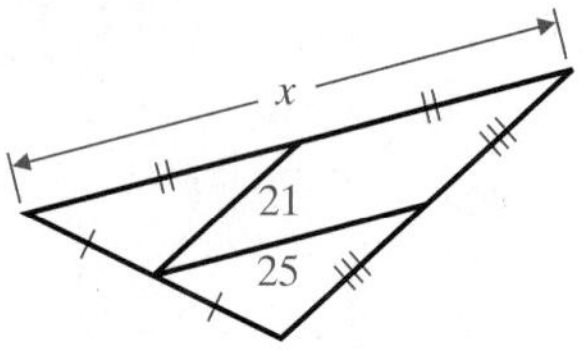

27.

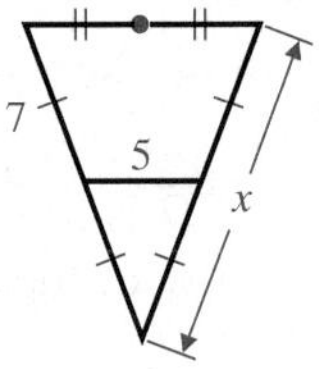

28.

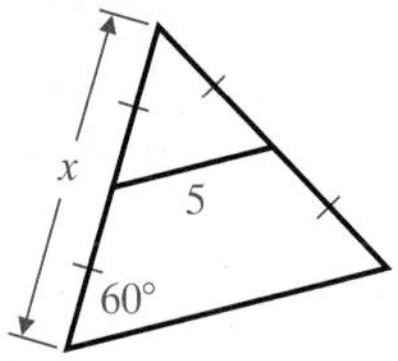

Use the following triangle for Exercises 29 and 30.

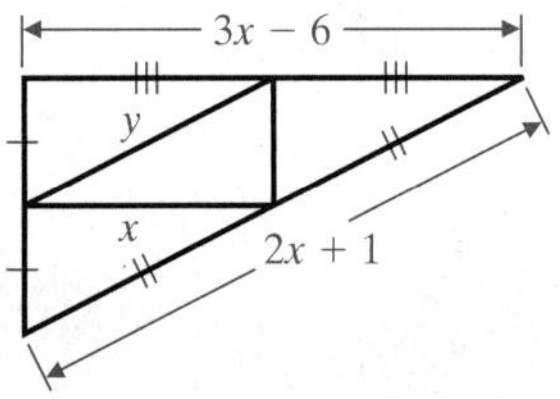

29. Find the value of x.

30. Find the value of y.

Coordinate Geometry *The coordinates of the vertices of a triangle are $E(1, 2)$, $F(5, 6)$, and $G(3, -2)$. Use this triangle for Exercises 31 and 32. See Example 4.*

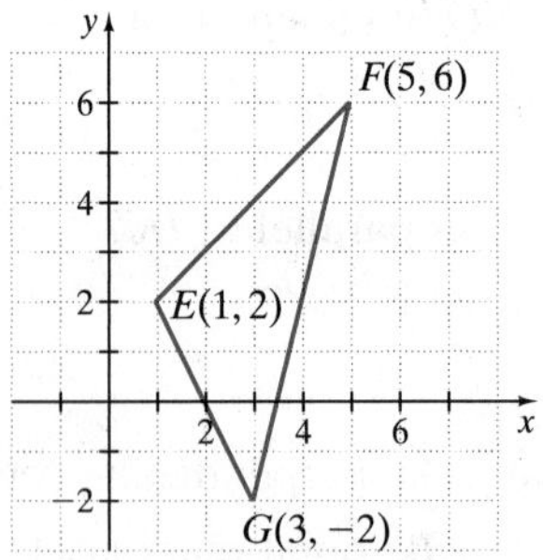

31. a. Find the coordinates of H, the midpoint of $\overline{EG}$, and the coordinates of J, the midpoint of $\overline{FG}$.

b. Show that $\overline{HJ} \parallel \overline{EF}$.

c. Show that $HJ = \frac{1}{2}EF$.

32. a. Find the coordinates of K, the midpoint of $\overline{EF}$, and the coordinates of H, the midpoint of $\overline{EG}$.

b. Show that $\overline{HK} \parallel \overline{FG}$.

c. Show that $HK = \frac{1}{2}FG$.

Solve. See Example 5.

33. A surveyor needs to measure the distance PQ across the lake. Beginning at point S, she locates the midpoints of $\overline{SQ}$ and $\overline{SP}$ at M and N. She then measures $\overline{NM}$ in yards.

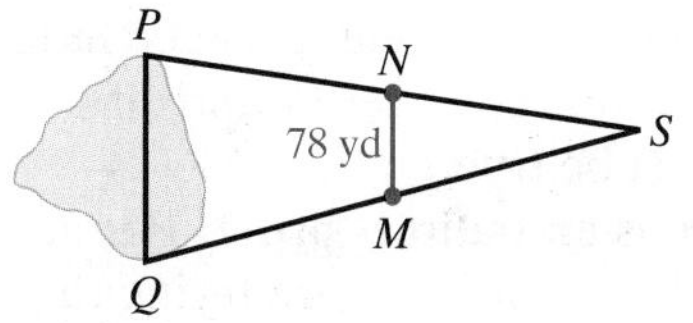

a. Find NM in feet. **b.** Find PQ in feet.

34. You want to paddle your kayak across a lake. To determine how far you must paddle, you pace out a triangle, counting the number of strides, as shown.

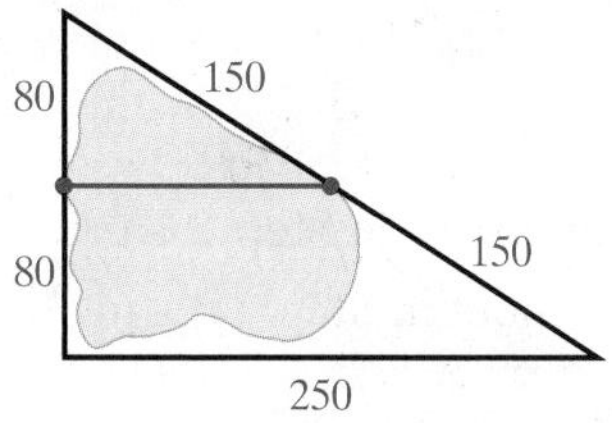

These units are in strides.
1 stride = 3.5 ft

a. If your strides average 3.5 ft, what is the length of the longest side of the triangle in feet?

b. What distance in feet must you paddle across the lake?

35. Multiple Choice You design a kite to look like the one below. Its diagonals measure 64 cm and 90 cm. You plan to use ribbon, represented by the purple rectangle, to connect the midpoints of its sides. How much ribbon do you need?

a. 77 cm

b. 122 cm

c. 154 cm

d. 308 cm

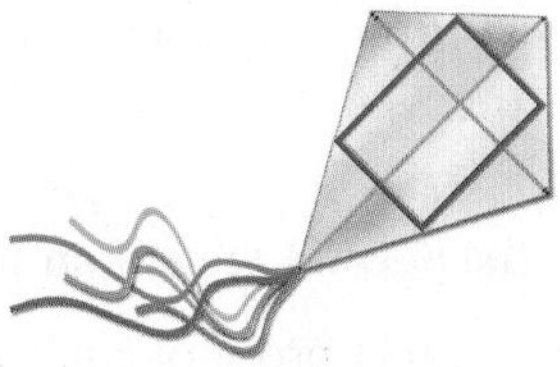

36. Multiple Choice Design a kite as in Exercise 35, except this time the diagonals measure 42 cm and 60 cm. How much ribbon is needed?

a. 102 cm **b.** 51 cm

c. 204 cm **d.** 36 cm

$\overline{IJ}$ is a midsegment of ΔFGH. $IJ = 7$, $FH = 10$, and $GH = 13$. Find the perimeter of each triangle.

37. ΔIJH

38. ΔFGH

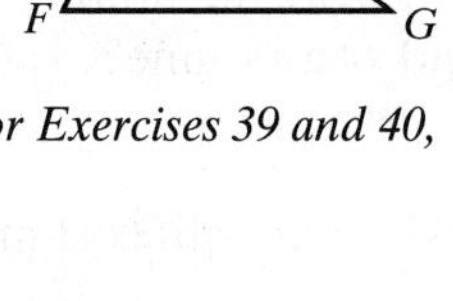

Use the figure below for Exercises 39–42. For Exercises 39 and 40, $DF = 24$, $BC = 6$, and $DB = 8$.

39. Find the perimeter of ΔADF.

40. Find the perimeter of ΔBCE.

41. If $BE = 2x + 6$ and $DF = 5x + 9$, find DF.

42. If $EC = 3x - 1$ and $AD = 5x + 7$, find EC.

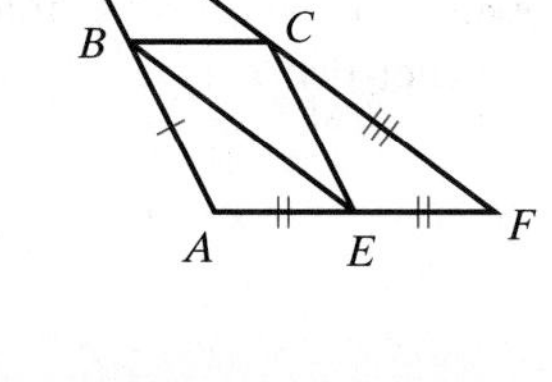

CONCEPT EXTENSIONS

43. In the diagram below, K, L, and M are the midpoints of the sides of ΔABC. The vertices of the three small purple triangles are the midpoints of the sides of ΔKBL, ΔAKM, and ΔMLC. The perimeter of ΔABC is 24 cm. What is the perimeter of the shaded region?

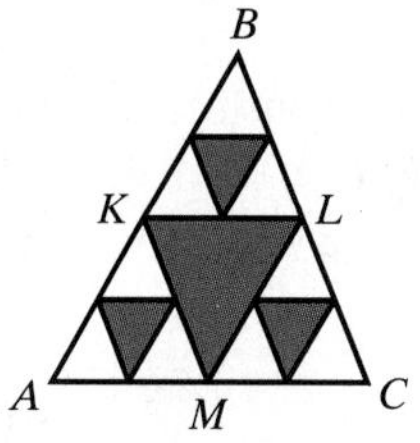

44. In the figure below, $m\angle QST = 40°$. What is $m\angle QPR$? Explain how you know.

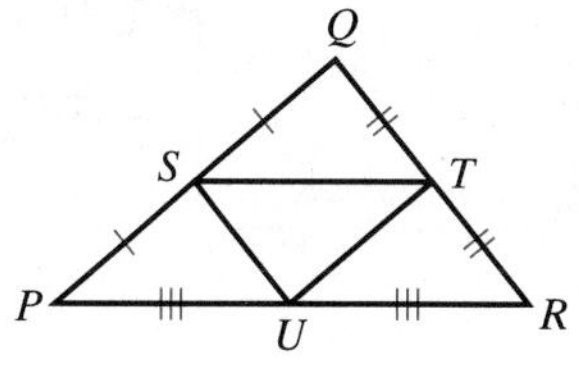

45. Find the Error A student sees this figure and concludes that $\overline{PL} \parallel \overline{NO}$. What is the error in the student's reasoning?

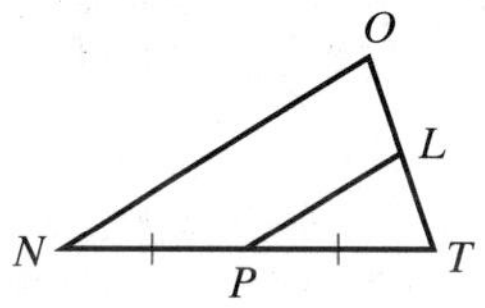

46. If two noncollinear segments in the coordinate plane have slope 3, what can you conclude?

47. Coordinate Geometry In ΔGHJ, $K(2, 3)$ is the midpoint of $\overline{GH}$, $L(4, 1)$ is the midpoint of $\overline{HJ}$, and $M(6, 2)$ is the midpoint of $\overline{GJ}$. Find the coordinates of G, H, and J.

48. Proof Complete the Prove statement and then write a proof.

Given: In ΔVYZ, S, T, and U are midpoints.

Prove: $\Delta YST \cong \Delta TUZ \cong \Delta SVU \cong$?

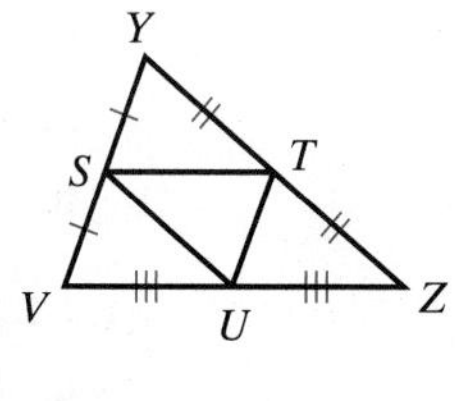

REVIEW AND PREVIEW

Use the figure below for Exercises 49 and 50. See Sections 4.5 and 4.6.

49. List all the pairs of congruent triangles that you can find in the figure.

50. Given: $\overline{FD} \cong \overline{FE}$, $\overline{BF} \cong \overline{CF}$, $\angle 1 \cong \angle 2$

Prove: $\overline{AB} \cong \overline{AC}$

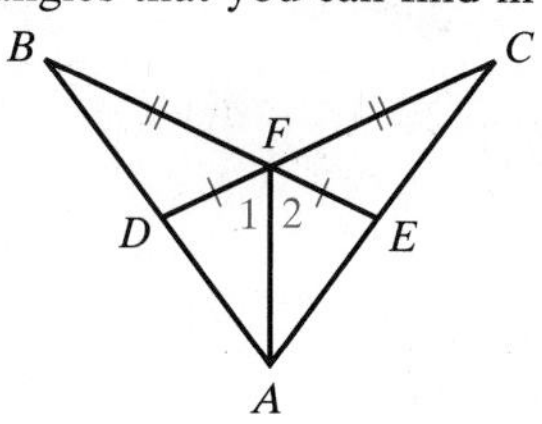

$\overline{TM}$ bisects $\angle STU$ so that $m\angle STM = (5x + 4)°$ and $m\angle MTU = (6x - 2)°$. Use this for Exercises 51 and 52. See Sections 1.5 and 1.6.

51. Find the value of x.

52. Find $m\angle STU$.

53. Construction Draw acute $\angle E$. Construct the bisector of $\angle E$.

5.5 INDIRECT PROOFS AND INEQUALITIES IN ONE TRIANGLE

OBJECTIVES

1. Use Indirect Reasoning to Write Proofs.
2. Learn the Triangle Relationship Between Length of a Side and Size of Its Opposite Angle.
3. Use the Triangle Inequality Theorem.

VOCABULARY

- indirect reasoning
- indirect proof

OBJECTIVE 1 ▶ Using Indirect Proofs. Thus far, all proofs in this text have used two laws of deductive reasoning—the Law of Detachment and the Law of Syllogism (see Section 2.5.) In this section, we learn a new type of reasoning called **indirect reasoning.**

In **indirect reasoning,** all possibilities are considered and then all but one are proved false. The remaining possibility must be true.

A proof involving indirect reasoning is an **indirect proof.** For an indirect proof, first we make sure we identify the original statement we are trying to prove. Next, we assume the opposite. Then we prove that the opposite is not possible, thus, the original statement (we were trying to prove) must be true. For this reason, indirect proof is sometimes called *proof by contradiction.*

How to Write an Indirect Proof

STEP 1. Identify the statement that we are trying to prove true.

STEP 2. Assume that the *opposite* of the statement in Step 1 is true.

STEP 3. Logically try to prove that the statement in Step 2 is true, but a contradiction should occur.

STEP 4. If we arrive at a contradiction, then the statement in Step 2 is false. Thus, the original statement that we wanted to prove (in Step 1) must be true.

▶ Helpful Hint

Make sure you can write the opposite or the contradiction of the statement that you wish to prove true (Step 2). This is a new process that needs practice.

In the second step of an indirect proof, we assume that the opposite of what we want to prove is true.

EXAMPLE 1 **Writing the Second Step of an Indirect Proof**

Suppose you want to write an indirect proof of each statement. As the second step of the proof, what would you assume?

a. An integer n is divisible by 5.

b. You do not have soccer practice today.

Solution

a. The opposite of "is divisible by" is "is not divisible by." Assume temporarily that n is not divisible by 5.

b. The opposite of "do not have" is "do have." Assume temporarily that you do have soccer practice today. □

PRACTICE 1 Suppose you want to write an indirect proof of each statement. As the second step of the proof, what would you assume?

a. ΔBOX is not acute.

b. At least one pair of shoes you bought cost more than $25.

To write an indirect proof, we also have to be able to identify a contradiction. (See Step 3 of How to Write an Indirect Proof.)

EXAMPLE 2 **Identifying Contradictions**

Which two statements contradict each other?

I. $\overline{FG} \parallel \overline{KL}$ **II.** $\overline{FG} \cong \overline{KL}$ **III.** $\overline{FG} \perp \overline{KL}$

Solution

Statements I and II: Segments can be parallel and congruent. Thus, statements I and II do not contradict each other.

Statements II and III: Segments can be congruent and perpendicular. Thus, statements II and III do not contradict each other.

Statements I and III: Parallel segments do not intersect, so they cannot be perpendicular. Thus, statements I and III DO contradict each other. □

PRACTICE 2 Which two statements contradict each other?

I. ΔXYZ is acute. **II.** ΔXYZ is scalene. **III.** ΔXYZ is equiangular.

OBJECTIVE 2 ▶ Learning Triangle Relationships. The angles and sides of a triangle have special relationships.

From algebra, we know the following inequality property.

Property 5.5-1 Comparison Property of Inequality

If $a = b + c$ and $c > 0$, then $a > b$.

We prove this property next.

Proof of Comparison Property of Inequality

Given: $a = b + c, c > 0$

Prove: $a > b$

Statements	Reasons
1. $c > 0$	**1.** Given
2. $b + c > b + 0$	**2.** Addition Property of Inequality
3. $b + c > b$	**3.** Identity Property of Addition
4. $a = b + c$	**4.** Given
5. $a > b$	**5.** Substitution

We can use this Comparison Property of Inequality to prove the following corollary to the Triangle Exterior Angle Theorem (Theorem 4.1-2).

Corollary 5.5-2 Corollary to the Triangle Exterior Angle Theorem

Corollary	If...	Then...
The measure of an exterior angle of a triangle is greater than the measure of either of its remote interior angles.	$\angle 1$ is an exterior angle 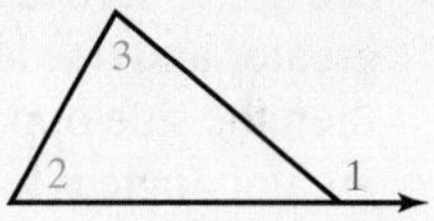	$m\angle 1 > m\angle 2$ and $m\angle 1 > m\angle 3$

We prove this corollary next.

Proof of Corollary 5.5-2

Given: $\angle 1$ is an exterior angle of the triangle.

Prove: $m\angle 1 > m\angle 2$ and $m\angle 1 > m\angle 3$.

Proof: By the Triangle Exterior Angle Theorem, recall that $m\angle 1 = m\angle 2 + m\angle 3$. Since $m\angle 2 > 0$ and $m\angle 3 > 0$, you can use the Comparison Property of Inequality and conclude that $m\angle 1 > m\angle 2$ and $m\angle 1 > m\angle 3$.

Helpful Hint

To determine whether an angle is an exterior angle of a triangle, see whether the two sides of the angle are formed by

- an extension of a side of the triangle and
- a side of the triangle.

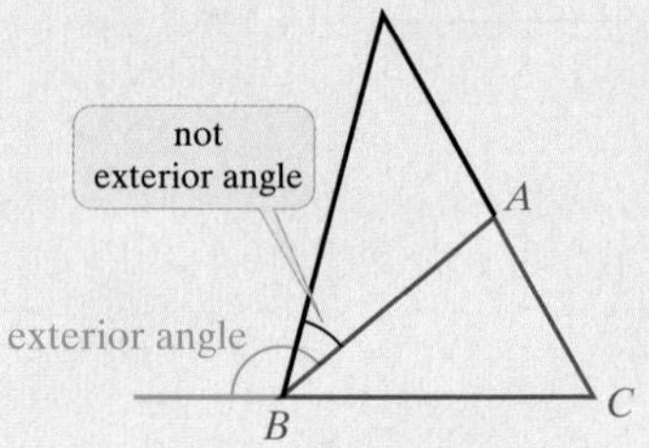

EXAMPLE 3 Applying the Corollary

Use the figure at the right. Why is $m\angle 2 > m\angle 3$?

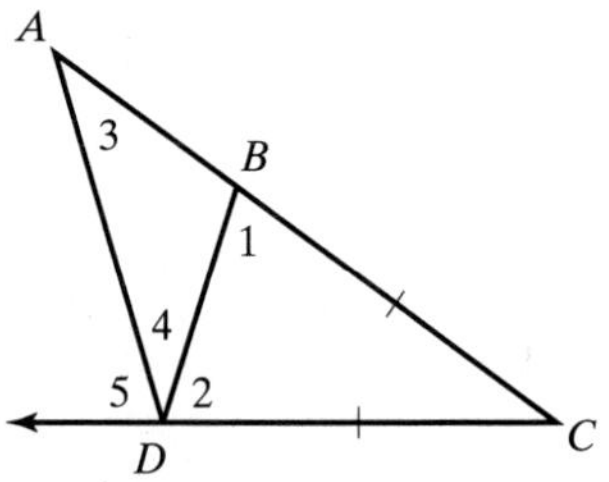

Solution In ΔBCD, $\overline{CB} \cong \overline{CD}$, so by the Isosceles Triangle Theorem, $m\angle 1 = m\angle 2$. $\angle 1$ is an exterior angle of ΔABD, so by the Corollary to the Triangle Exterior Angle Theorem, $m\angle 1 > m\angle 3$. Thus, $m\angle 2 > m\angle 3$ by substitution. □

PRACTICE 3 Why is $m\angle 5 > m\angle C$?

We can use Corollary 5.5-2 to prove the following theorem.

Theorem 5.5-3 Triangle Inequality Theorem

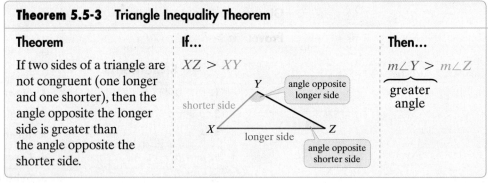

Theorem	If...	Then...
If two sides of a triangle are not congruent (one longer and one shorter), then the angle opposite the longer side is greater than the angle opposite the shorter side.	$XZ > XY$ 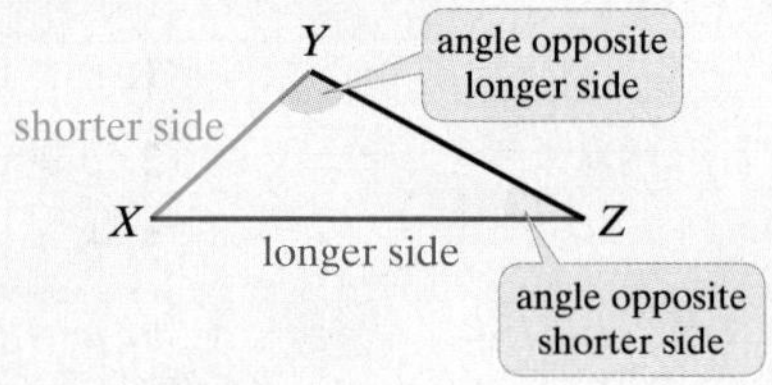	$m\angle Y > m\angle Z$ (greater angle)

We prove this theorem in Exercise 67, Exercise Set 5.5.

Theorem 5.5-4 below is the converse of Theorem 5.5-3. The proof of Theorem 5.5-4 relies on indirect reasoning.

Theorem 5.5-4 Converse of the Triangle Inequality Theorem

Theorem	If...	Then...
If two angles of a triangle are not congruent (one greater and one lesser), then the side opposite the greater angle is longer than the side opposite the lesser angle.	$m\angle A > m\angle B$ 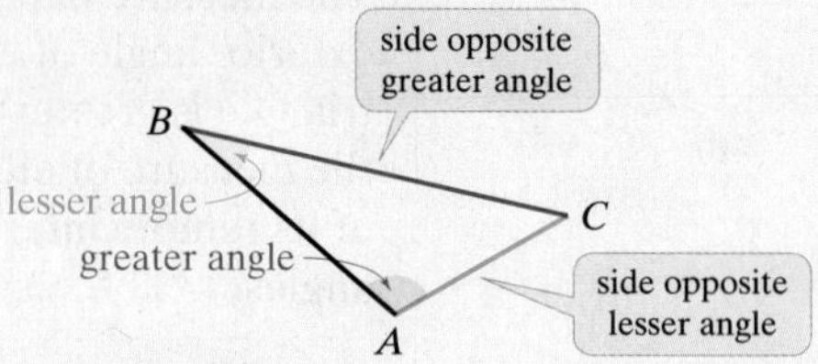	$BC > AC$ (longer side)

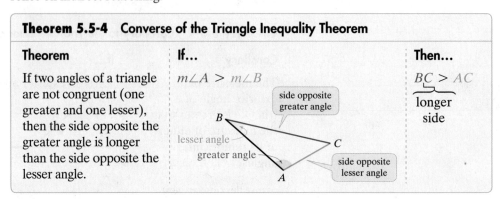

Let's prove this theorem next.

Indirect Proof of Theorem 5.5-4

Given: $m\angle A > m\angle B$

Prove: $BC > AC$

> **Helpful Hint**
> We use the steps for an indirect proof found earlier in this section.

STEP 1. We want to prove that $BC > AC$.

STEP 2. Assume temporarily that $BC \ngtr AC$. That is, assume temporarily that either $BC < AC$ or $BC = AC$.

STEP 3. If $BC < AC$, then $m\angle A < m\angle B$ (Theorem 5.5-3). This contradicts the given fact that $m\angle A > m\angle B$. Therefore, $BC < AC$ must be false.

If $BC = AC$, then $m\angle A = m\angle B$ (Isosceles Triangle Theorem). This also contradicts $m\angle A > m\angle B$. Therefore, $BC = AC$ must be false.

STEP 4. The temporary assumption $BC \ngtr AC$ is false, so $BC > AC$ is true.

> **Helpful Hint**
> - The angle of a triangle opposite a side is the only angle whose sides are not the side of the angle.
> - The side of a triangle opposite an angle is the only side that is not the sides of the angle.

EXAMPLE 4 **Triangle—Side Length vs. Angle Measure**

A town park is triangular. A landscape architect wants to place a bench at the corner with the largest angle. Which two streets form the corner with the largest angle?

Solution Hollingsworth Road is the longest street, so it is opposite the largest angle. MLK Boulevard and Valley Road form the largest angle. □

PRACTICE 4 Suppose the landscape architect wants to place a drinking fountain at the corner with the second-largest angle. Which two streets form the corner with the second-largest angle?

EXAMPLE 5 **Triangle—Angle Measure vs. Side Length**

Multiple Choice Which choice shows the sides of ΔTUV in order from shortest to longest?

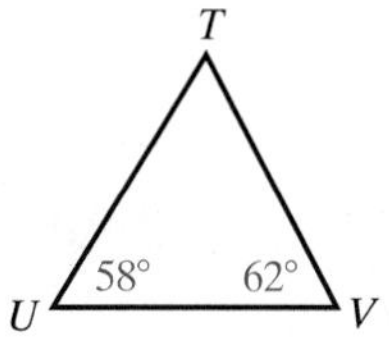

a. $\overline{TV}, \overline{UV}, \overline{UT}$ **b.** $\overline{UT}, \overline{UV}, \overline{TV}$

c. $\overline{UV}, \overline{UT}, \overline{TV}$ **d.** $\overline{TV}, \overline{UT}, \overline{UV}$

Solution By the Triangle Angle-Sum Theorem, $m\angle T = 60°$. $58° < 60° < 62°$, so $m\angle U < m\angle T < m\angle V$. By Theorem 5.5-4, $TV < UV < UT$. Choice **a** is correct. □

PRACTICE 5 In the figure at the right, $m\angle S = 24°$ and $m\angle O = 130°$. List the sides of ΔSOX from shortest to longest.

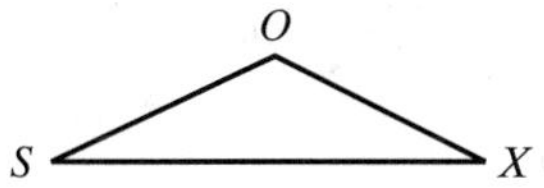

OBJECTIVE 3 ▶ Using the Triangle Inequality Theorem. For three segments to form a triangle, their lengths must be related in a certain way. Notice that only one of the sets of segments below can form a triangle. The sum of the smallest two lengths must be greater than the greatest length.

Theorem 5.5-5 Triangle Inequality Theorem for Sum of Lengths of Sides

The sum of the lengths of any two sides of a triangle is greater than the length of the third side.

$XY + YZ > XZ$ $YZ + XZ > XY$ $XZ + XY > YZ$

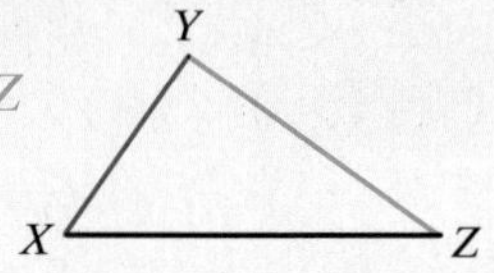

We prove this theorem in Exercise 76, Exercise Set 5.5.

▶ Helpful Hint

To check whether a triangle can be formed given three lengths, find the sum of pairs of lengths. Each of these sums must be greater than the third side.

EXAMPLE 6 Using the Triangle Inequality Theorem (Theorem 5.5-5)

Can a triangle have sides with the given lengths? Explain.

a. 3 ft, 7 ft, 8 ft

b. 5 ft, 10 ft, 15 ft

Solution

a. $3 + 7 > 8$ $7 + 8 > 3$ $8 + 3 > 7$

$10 > 8$ $15 > 3$ $11 > 7$

Yes. The sum of the lengths of any two sides is greater than the length of the third side.

Solution

b. $5 + 10 \not> 15$

$15 \not> 15$

No. The sum of 5 and 10 is not greater than 15. This contradicts Theorem 5.5-5. □

PRACTICE 6 Can a triangle have sides with the given lengths? Explain.

a. 2 m, 6 m, and 9 m

b. 4 yd, 6 yd, and 9 yd

EXAMPLE 7 Finding Possible Side Lengths

Two sides of a triangle are 5 ft and 8 ft long. What is the range of possible lengths for the third side?

Solution Let x represent the length of the third side. Use the Triangle Inequality Theorem for Sum of Lengths of Sides to write three inequalities. Then solve each inequality for x.

$x + 5 > 8$ $x + 8 > 5$ $5 + 8 > x$

$x > 3$ $x > -3$ $x < 13$

Numbers that satisfy $x > 3$ and $x > -3$ must be greater than 3. So, the third side must be greater than 3 ft and less than 13 ft long. □

PRACTICE 7 A triangle has side lengths of 4 in. and 7 in. What is the range of possible lengths for the third side?

Helpful Hint

From Example 7, let's check to see why 3 ft and 13 ft are not acceptable as third lengths. Remember, we are given that two sides of a triangle are 5 ft and 8 ft.

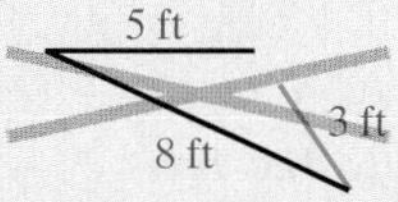

Here, $5 + 3 = 8$, so the Triangle Inequality is not true.

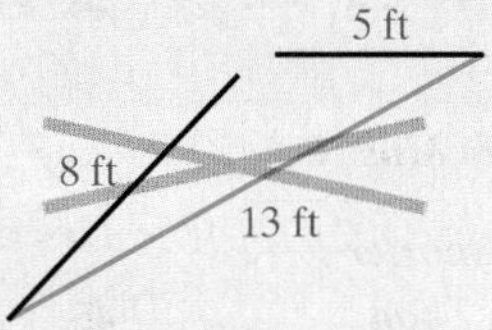

Here, $8 + 5 = 13$, so the Triangle Inequality is not true.

VOCABULARY & READINESS CHECK

Word Bank *Use the choices to fill in each blank. Some choices may be used more than once and some not used at all.*

proof by contradiction	indirect proof
Triangle Inequality Theorem for Sum of Lengths of Sides	deductive reasoning

1. An indirect proof is also called a ___________.
2. Law of Detachment and Law of Syllogism are two examples of ___________.
3. The ___________ says that the sum of the lengths of any two sides of a triangle must be greater than the length of the third side.
4. By the ___________, the following side lengths do/do not (circle one) form a triangle: 6, 8, and 14 units.

5.5 EXERCISE SET MyMathLab®

Explain why $m\angle 1 > m\angle 2$. See Example 3.

1.

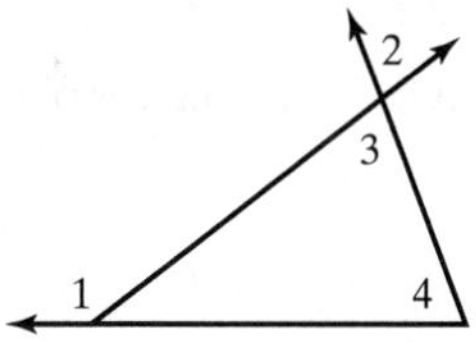

2.

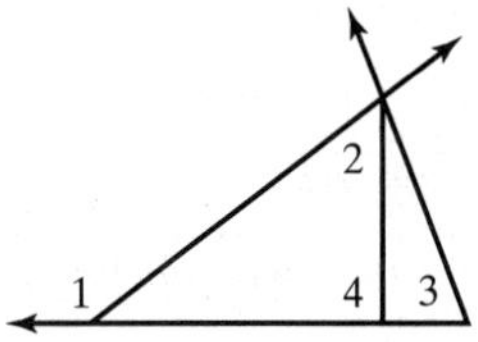

3.

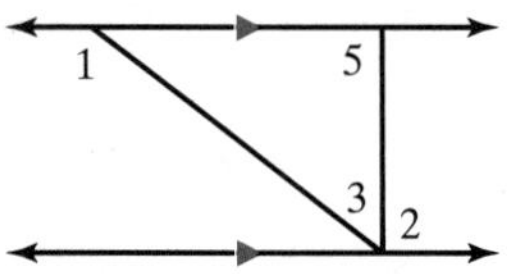

4.

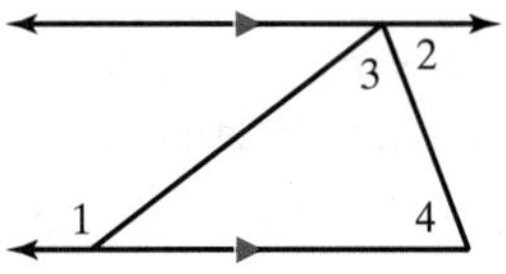

For Exercises 5–12, list the angles of each triangle in order from smallest to largest. See Example 4.

5.

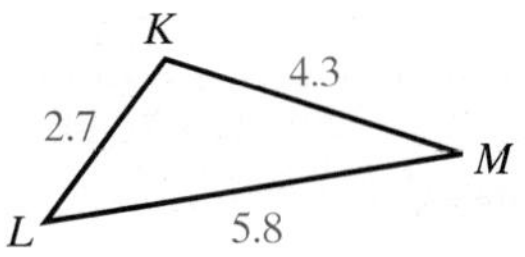

6.

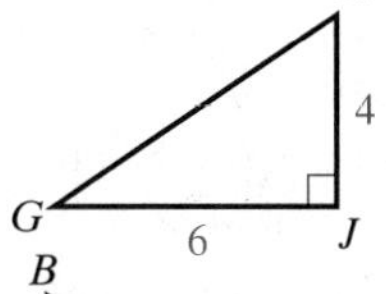

7.

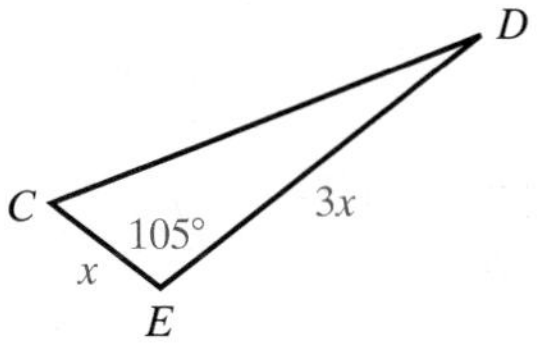

8.

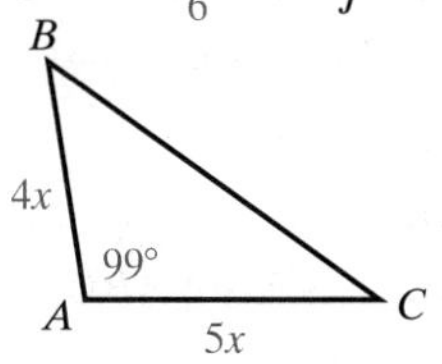

9. ΔABC, where $AB = 8$, $BC = 5$, and $CA = 7$

10. ΔDEF, where $DE = 15$, $EF = 18$, and $DF = 5$

11. ΔXYZ, where $XY = 12$, $YZ = 24$, and $ZX = 30$

12. ΔABC, where $AB = 14$, $BC = 18$, and $CA = 10$

For Exercise 13–20, list the sides of each triangle in order from shortest to longest. See Example 5.

13.

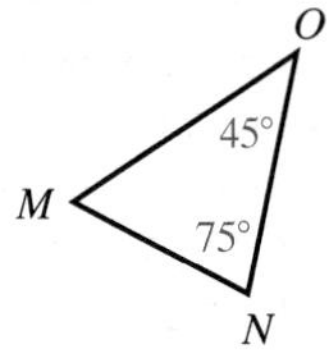

14.

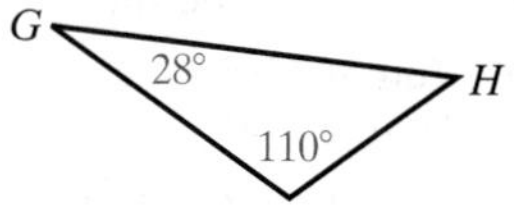

15.

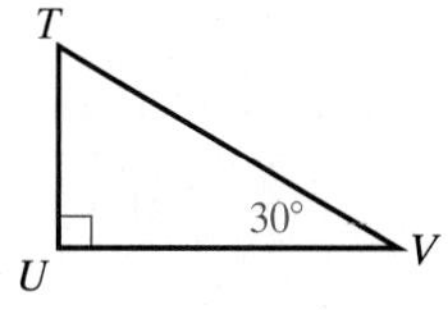

16.

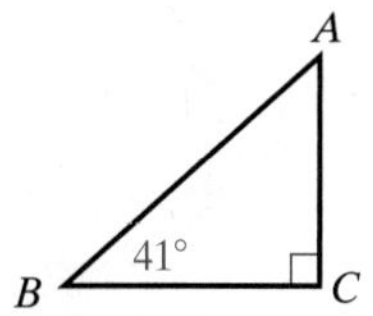

17. ΔGFH, with $m\angle G = 61°$, $m\angle F = 60°$, and $m\angle H = 59°$

18. ΔABC, with $m\angle A = 90°$, $m\angle B = 40°$, and $m\angle C = 50°$

19. ΔDEF, with $m\angle D = 20°$, $m\angle E = 120°$, and $m\angle F = 40°$

20. ΔXYZ, with $m\angle X = 51°$, $m\angle Y = 59°$, and $m\angle Z = 70°$

Can a triangle have sides with the given lengths? Explain. See Example 6.

21. 2 in., 3 in., 6 in.
22. 11 cm, 12 cm, 15 cm
23. 8 m, 10 m, 19 m
24. 1 cm, 15 cm, 15 cm
25. 2 yd, 9 yd, 10 yd
26. 4 m, 5 m, 9 m

Complete the Chart *Use the Triangle Inequality Theorem for Sum of Lengths of Sides to complete the table. (Assume that the units are the same.) See Example 7.*

	First Side Measure	Second Side Measure	Third Side Measure Must Be	
			Longer Than	Shorter Than
27.	8	12		
28.	16	5		
29.	9	9		
30.	21	21		
31.		19	6	32
32.		12.5	10	15
33.	0.3		1.2	1.8
34.	1.7		0.1	3.5

Find the longest side of ΔABC, with

35. $m\angle A = 70°$, $m\angle B = (2x - 10)°$, and $m\angle C = (3x + 20)°$.

36. $m\angle B = 32°$, $m\angle A = (7x - 1)°$, and $m\angle C = (14x + 2)°$.

You are setting up a study area where you will do your homework each evening. It is triangular with an entrance on one side. You want to put your computer in the corner with the largest angle and a bookshelf on the longest side.

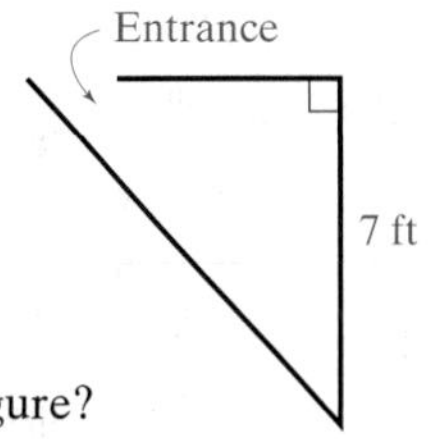

37. What type of triangle is shown in the figure?

38. Where should you place your computer?

39. Once you find the largest angle of a triangle, how do you find the longest side?

40. On which side should you place the bookshelf? Explain.

41. You and a friend compete in a scavenger hunt at a museum. The two of you walk from the Picasso exhibit to the Native American gallery along the dashed red line. When he sees that another team is ahead of you, your friend says, "They must have cut through the courtyard." Explain what your friend means.

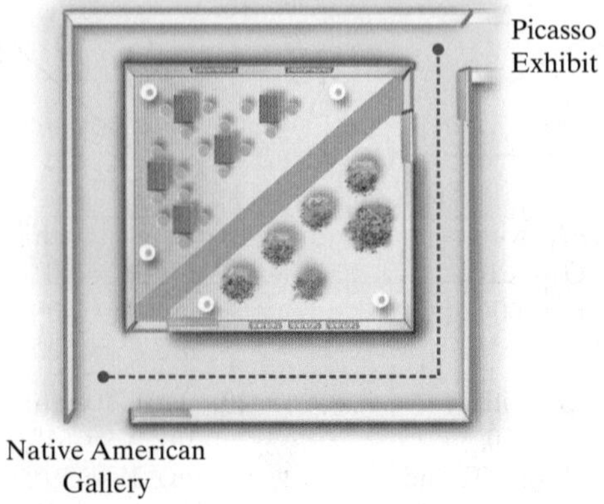

42. Is it possible to draw a right triangle with an exterior angle measuring 88? Explain your reasoning.

43. **Find the Error** Your family drives across Kansas on Interstate 70. A sign reads, "Wichita 90 mi, Topeka 110 mi." Your little brother says, "I didn't know that it was only 20 miles from Wichita to Topeka." Explain why the distance between the two cities does not have to be 20 mi.

44. **Find the Error** A friend tells you that she drew a triangle with perimeter 16 and one side of length 8. How do you know she made an error in her drawing?

Determine which segment is shortest in each diagram.

45.

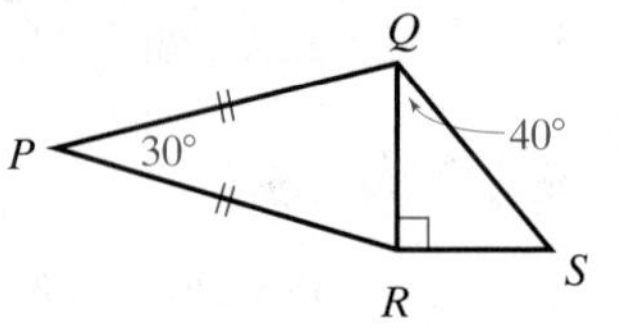

46.

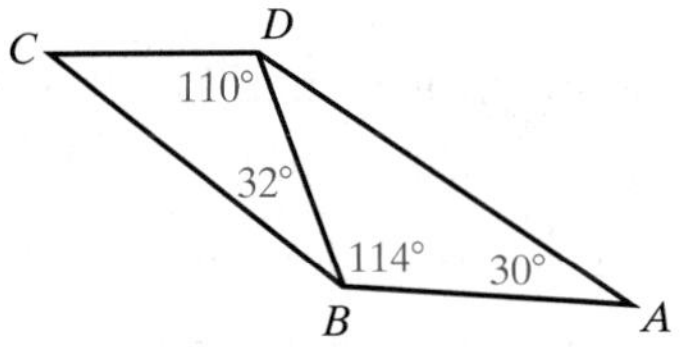

47.

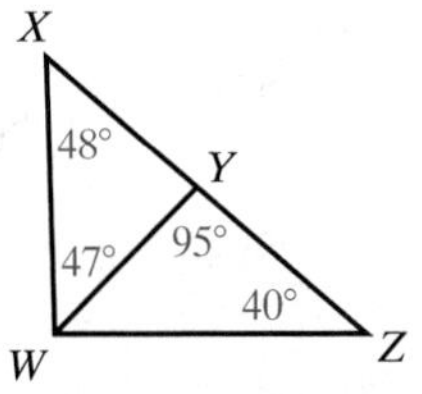

48.

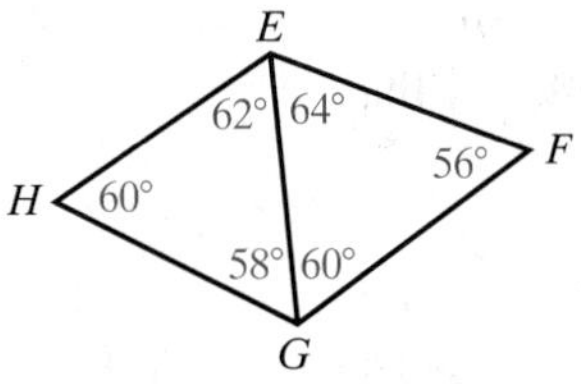

Write the second step of an indirect proof for the given statement. See Example 1.

49. It is raining outside.

50. $\angle J$ is not a right angle.

51. ΔPEN is isosceles.

52. At least one angle is obtuse.

53. $\overline{XY} \cong \overline{AB}$

54. $m\angle 2 > 90°$

Identify the two statements that contradict each other. See Example 2.

55. **I.** ΔPQR is equilateral.
II. ΔPQR is a right triangle.
III. ΔPQR is isosceles.

56. **I.** $\ell \parallel m$
II. ℓ and m do not intersect.
III. ℓ and m are skew.

57. **I.** Each of the two items that Val bought costs more than \$10.
II. Val spent \$34 for the two items.
III. Neither of the two items that Val bought costs more than \$15.

58. **I.** In right ΔABC, $m\angle A = 60°$.
II. In right ΔABC, $\angle A \cong \angle C$.
III. In right ΔABC, $m\angle B = 90°$.

For Exercises 59–60, begin to write an indirect proof.

Given: $\angle 1 \not\cong \angle 2$

Prove: $\ell \nparallel p$

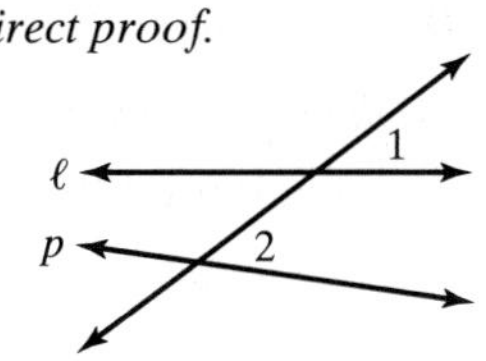

59. What assumption should be the second step of your proof?

60. In the figure, what type of angle pair do $\angle 1$ and $\angle 2$ form?

Write the second step of an indirect proof of the given statement. In other words, assume that the opposite of each given statement is true.

61. If a number n ends in 5, then it is not divisible by 2.

62. If point X is on the perpendicular bisector of $\overline{AB}$, then $\overline{XB} \cong \overline{XA}$.

63. Quadrilateral $ABCD$ has four right angles.

64. Write a statement that contradicts the following statement.

Lines a and b are parallel.

65. Find the Error A classmate began an indirect proof as shown below. Explain and correct your classmate's error.

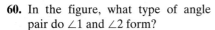

> ~~Given: $\triangle ABC$~~
> ~~Prove: $\angle A$ is obtuse.~~
> ~~Assume temporarily that $\angle A$ is acute.~~

66. Find the Error Your friend wants to prove indirectly that $\triangle ABC$ is equilateral. For a second step, he writes, "Assume temporarily that $\triangle ABC$ is scalene." What is wrong with your friend's statement? How can he correct himself?

CONCEPT EXTENSIONS

67. Proof Fill in the blanks for a proof of Theorem 5.5-3: If two sides of a triangle are not congruent, then the larger angle lies opposite the longer side.

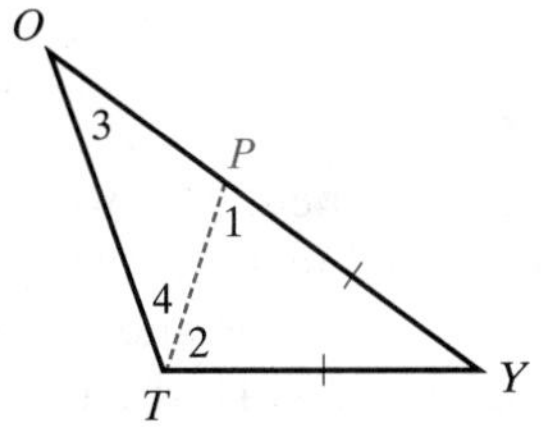

Given: $\triangle TOY$, with $YO > YT$

Prove: **a.** ? > **b.** ?

Mark P on $\overline{YO}$ so that $\overline{YP} \cong \overline{YT}$. Draw $\overline{TP}$.

Statements	Reasons
1. $\overline{YP} \cong \overline{YT}$	1. Ruler Postulate
2. $m\angle 1 = m\angle 2$	2. **c.** ?
3. $m\angle OTY = m\angle 4 + m\angle 2$	3. **d.** ?
4. $m\angle OTY > m\angle 2$	4. **e.** ?
5. $m\angle OTY > m\angle 1$	5. **f.** ?
6. $m\angle 1 > m\angle 3$	6. **g.** ?
7. $m\angle OTY > m\angle 3$	7. **h.** ?

68. Proof Prove this corollary to Theorem 5.5-4: The perpendicular segment from a point to a line is the shortest segment from the point to the line.

Given: $\overline{PT} \perp \overline{TA}$

Prove: $PA > PT$

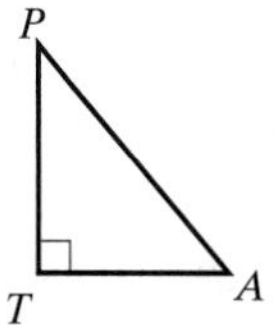

69. Proof Fill in the blanks to prove the following statement.
If the Yoga Club and Go Green Club together have fewer than 20 members and the Go Green Club has 10 members, then the Yoga Club has fewer than 10 members.

Given: The total membership of the Yoga Club and the Go Green Club is fewer than 20. The Go Green Club has 10 members.

Prove: The Yoga Club has fewer than 10 members.

Proof: Assume temporarily that the Yoga Club has 10 or more members.
This means that together the two clubs have **a.** ? members.
This contradicts the given information that **b.** ? .
The temporary assumption is false. Therefore, it is true that **c.** ? .

70. Proof Fill in the blanks to prove the following statement.
In a given triangle, $\triangle LMN$, there is at most one right angle.

Given: $\triangle LMN$

Prove: $\triangle LMN$ has at most one right angle.

Proof: Assume temporarily that $\triangle LMN$ has more than one **a.** ? . That is, assume that both $\angle M$ and $\angle N$ are **b.** ? . If $\angle M$ and $\angle N$ are both right angles, then $m\angle M = m\angle N =$ **c.** ? . By the Triangle Angle-Sum Theorem, $m\angle L + m\angle M + m\angle N =$ **d.** ? . Use substitution to write the equation $m\angle L +$ **e.** ? + **f.** ? $= 180$. When you solve for $m\angle L$, you find that $m\angle L =$ **g.** ? . This means that there is no $\triangle LMN$, which contradicts the given statement. So the temporary assumption that $\triangle LMN$ has **h.** ? must be false. Therefore, $\triangle LMN$ has **i.** ? .

71. A student has two straws. One is 6 cm long and the other is 9 cm long. She picks a third straw at random from a group of four straws whose lengths are 3 cm, 5 cm, 11 cm, and 15 cm. What is the probability that the straw she picks will allow her to form a triangle? Justify your answer.

For Exercises 72 and 73, x and y are integers such that $1 < x < 5$ and $2 < y < 9$.

72. What is the probability that you can draw an isosceles triangle that has sides 5 cm, x cm, and y cm, with x and y chosen at random?

73. The sides of a triangle are 5 cm, x cm, and y cm. List all possible (x, y) pairs.

Proof *For Exercises 74 and 75, write an indirect proof.*

74. Use the figure at the right.

Given: $\triangle ABC$ with $BC > AC$

Prove: $\angle A \not\cong \angle B$

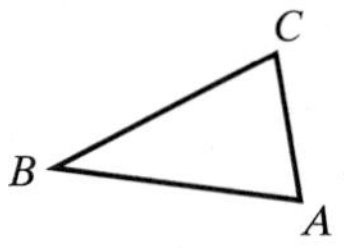

75. Given: ΔXYZ is isosceles.

Prove: Neither base angle is a right angle.

76. Proof Prove the Triangle Inequality Theorem for Sum of Lengths of Sides: The sum of the lengths of any two sides of a triangle is greater than the length of the third side.

Given: ΔABC

Prove: $AC + CB > AB$

(*Hint:* On $\overrightarrow{BC}$, mark a point D not on $\overline{BC}$ so that $DC = AC$. Draw $\overline{DA}$ and use Theorem 5.5-4 with ΔABD.)

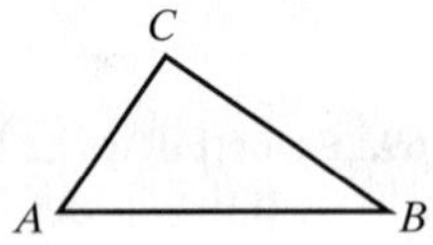

REVIEW AND PREVIEW

Use the figures below. See Section 4.1.

77. What is $m\angle D$?

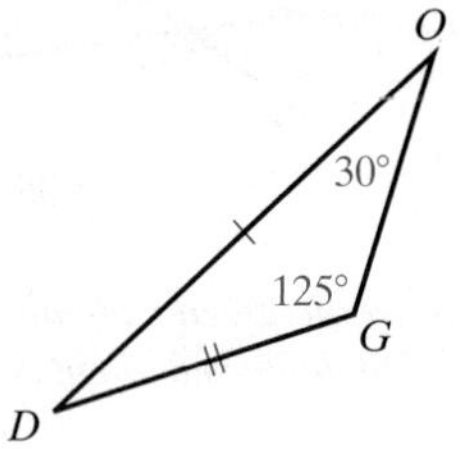

78. What is $m\angle P$?

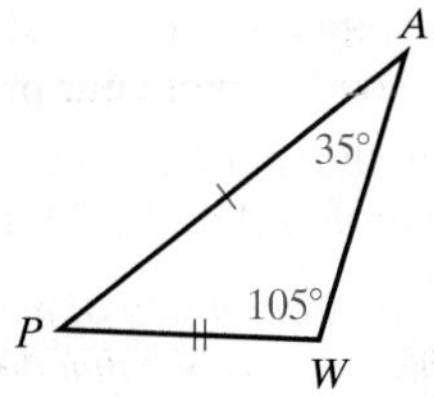

79. Which angle has greater measure, $\angle D$ or $\angle P$?

80. Is it possible for AW to equal OG?

5.6 INEQUALITIES IN TWO TRIANGLES

OBJECTIVE

1 Use the Hinge Theorem and Its Converse to Compare Measures of Sides and Angles of Two Triangles.

OBJECTIVE 1 ▶ Using the Hinge Theorem and Its Converse. In the last section, we compared measures of sides and angles within a single triangle. In this section, we extend this comparison to two triangles.

This comparison is sometimes called the **Hinge Theorem** because it can be described using the application of opening a door on hinges.

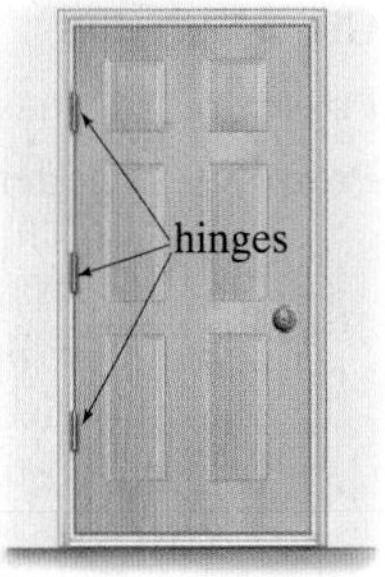

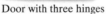

Door with three hinges

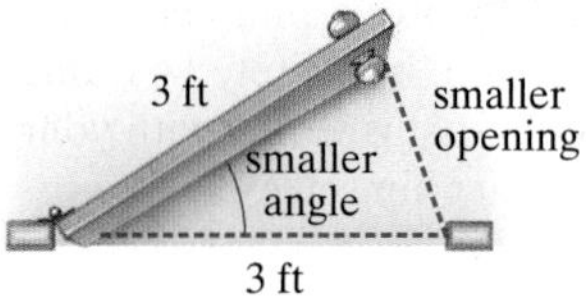

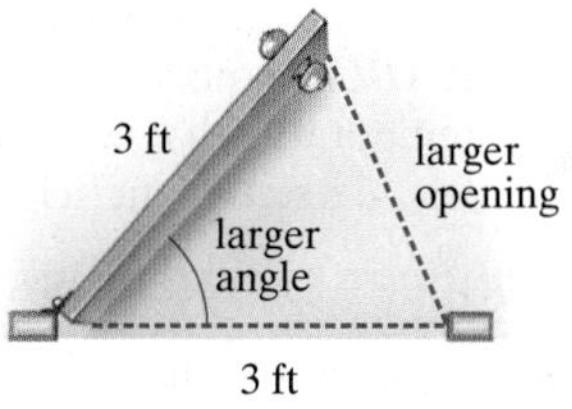

The figures show two top views of a door being opened. The width of the door and the door frame remain the same—in this case, both 3 feet. As the door is opened, the hinge angle increases and the opening (between the door and the frame) increases. If we think of the above diagram as two triangles, we have the Hinge Theorem, also called the SAS Inequality Theorem.

Theorem 5.6-1 The Hinge Theorem (SAS Inequality Theorem)

Theorem	If...	Then...
If two sides of one triangle are congruent to two sides of another triangle, and the included angles are not congruent, then the longer third side is opposite the larger included angle.	$m\angle A > m\angle E$ 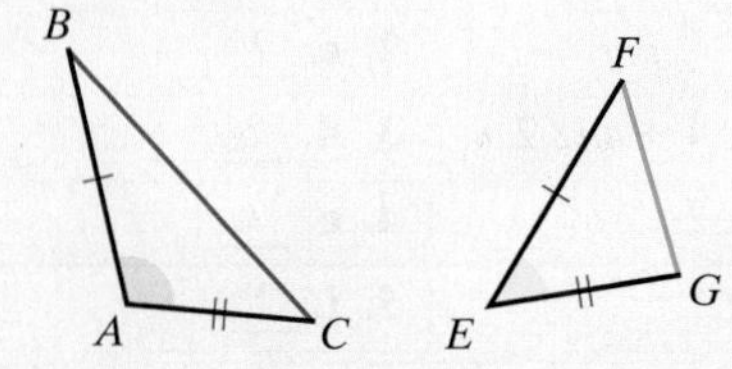	$BC > FG$

We prove this theorem in Practice 1 and Exercises 37 and 38.

The Converse of the Hinge Theorem is also true.

Theorem 5.6-2 Converse of the Hinge Theorem (SSS Inequality Theorem)

Theorem	If...	Then...
If two sides of one triangle are congruent to two sides of another triangle, and the third sides are not congruent, then the larger included angle is opposite the longer third side.	$BC > FG$ 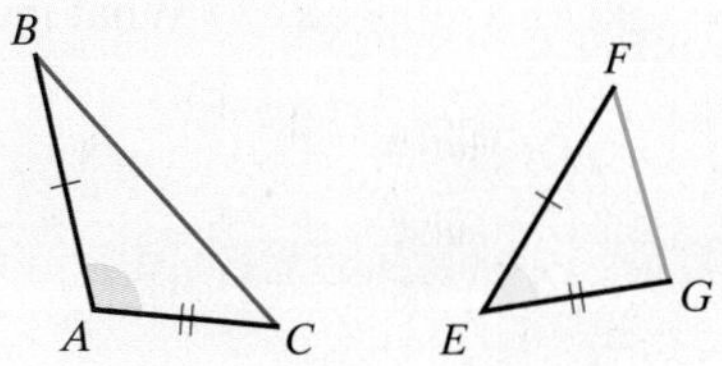	$m\angle A > m\angle E$

The proof of the converse is an indirect proof, shown in Example 1.

EXAMPLE 1 Indirect Proof of the Converse of the Hinge Theorem (SSS Inequality Theorem)

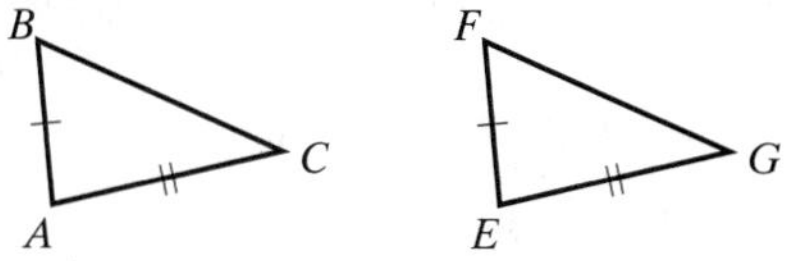

Given: $\overline{AB} \cong \overline{EF}$, $\overline{AC} \cong \overline{EG}$,
$BC > FG$

Prove: $m\angle A > m\angle E$

Solution

STEP 1. Prove $m\angle A > m\angle E$.

STEP 2. Assume $m\angle A \not> m\angle E$. If so, then either $m\angle A = m\angle E$ or $m\angle A < m\angle E$.

STEP 3. *Case 1.* If $m\angle A = m\angle E$, then $\angle A \cong \angle E$. Thus, $\Delta ABC \cong \Delta EFG$ by the SAS Congruence Postulate. If so, then $BC = FG$ since cpoctac. This is a contradiction to our given.

Case 2. If $m\angle A < m\angle E$, then $BC < FG$ by the Hinge Theorem. This is a contradiction to our given.

STEP 4. Since Step 3 led to a contradiction in both cases, our Step 2 assumption, $m\angle A \not> m\angle E$, must be incorrect. Thus, $m\angle A > m\angle E$ is true. □

PRACTICE

1 Prove a part of Theorem 5.6-1, the Hinge Theorem. We can prove the Hinge Theorem by considering three cases. Here, we prove one case. The other two cases are considered in Exercises **37** and **38**.

Given: $\overline{AB} \cong \overline{DE}$, $\overline{BC} \cong \overline{EF}$, $m\angle ABC > m\angle E$

Construction: Construct $\overrightarrow{BQ}$ so that $\angle QBC \cong \angle E$. Locate point P on $\overrightarrow{BQ}$ so that $\overline{BP} \cong \overline{ED}$.

Prove: $AC > DF$

There will be three cases depending on the location of point P. P lies on $\overrightarrow{BQ}$ for each case.

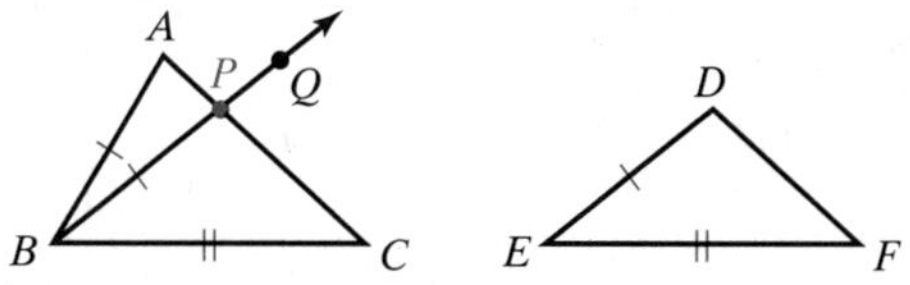

Case 1. Point P lies on $\overline{AC}$.

Case 1. Point P lies on $\overline{AC}$.

Case 2. Point P lies outside ΔABC. (See Exercise 37.)

Case 3. Point P lies inside ΔABC. (See Exercise 38.)

Prove Case 1 using a Statements/Reasons proof.

EXAMPLE 2 Using the Hinge Theorem

Multiple Choice Which of the following statements must be true?

a. $AS < YU$

b. $SK > YU$

c. $SK < YU$

d. $AK = YU$

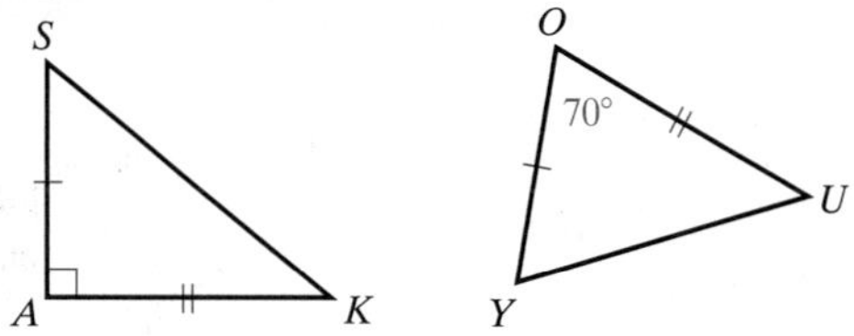

Solution $\overline{SA} \cong \overline{YO}$ and $\overline{AK} \cong \overline{OU}$, so the triangles have two pairs of congruent sides. The included angles, $\angle A$ and $\angle O$, are not congruent. Since $m\angle A > m\angle O$, $SK > YU$ by the Hinge Theorem. The correct answer is **b**. □

PRACTICE 2 What inequality relates LN and OQ in the figures below?

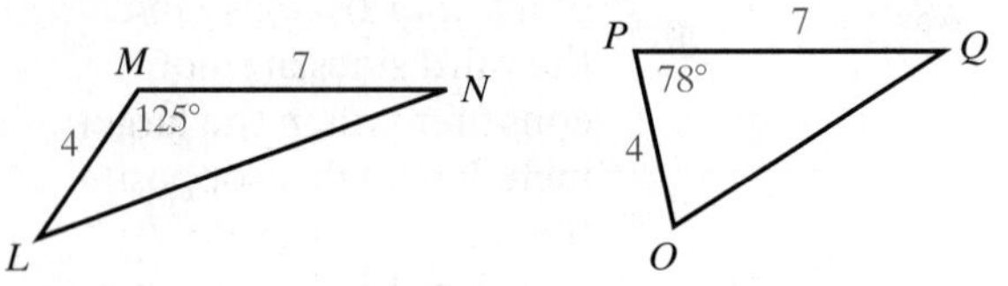

EXAMPLE 3 Using the Converse of the Hinge Theorem

Multiple Choice Choose the only possible measure for $\angle A$.

a. 35° **b.** 55° **c.** 45°

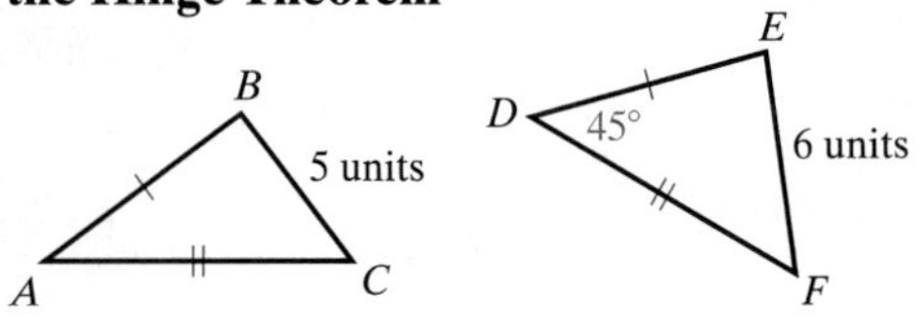

Solution Since $BC < EF$ (and the other pairs of sides are congruent), then $m\angle A < m\angle D$. The only choice less than 45° is choice **a**, 35°. □

PRACTICE 3 **Multiple Choice** Choose the only possible measure for $\angle C$.

a. 120° **b.** 118° **c.** 110°

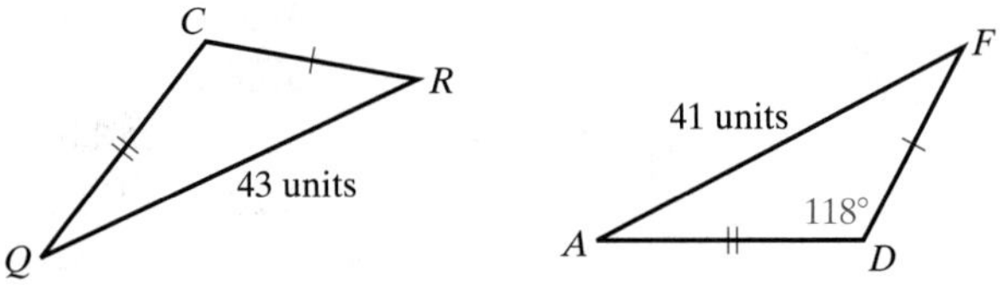

The Hinge Theorem is useful in applications.

EXAMPLE 4 An Application of the Hinge Theorem

Knott's Berry Farm has a swing ride called Screamin' Swing. It stands over 60 feet tall, and riders are air-launched. According to the diagram, which riders are farther from the base—riders on side $\overline{AD}$ or on side $\overline{DB}$?

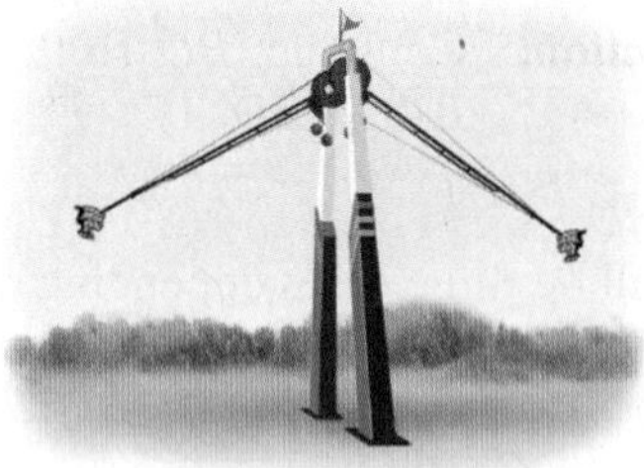

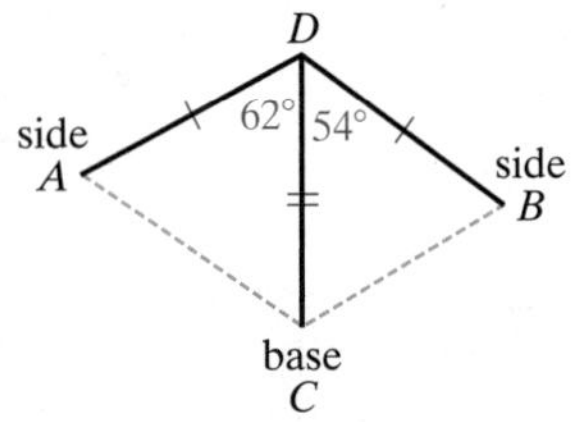

Solution Using the Hinge Theorem, since 62° > 54°, riders on side $\overline{AD}$ are farther from the base. □

PRACTICE 4 The diagram below shows a pair of scissors in two different positions. In which position is the distance between the tips of the two blades greater? Use the Hinge Theorem to justify your answer.

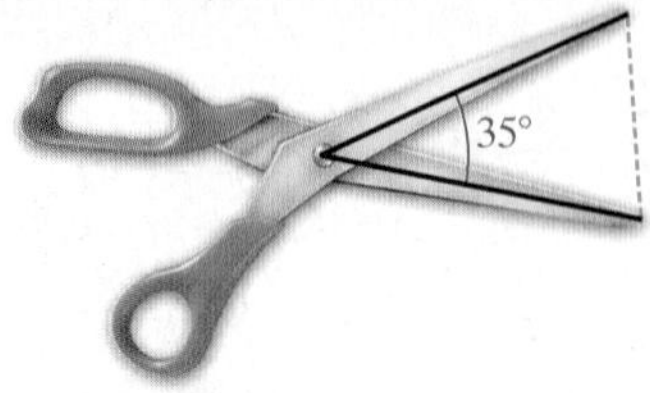

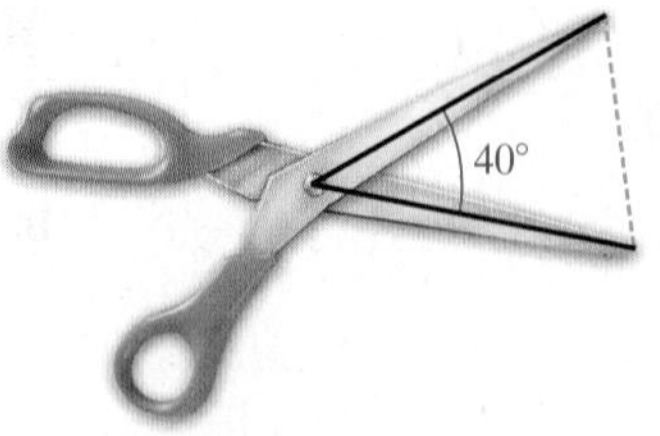

EXAMPLE 5 **Using the Converse of the Hinge Theorem**

What is the range of possible values for x?

Solution

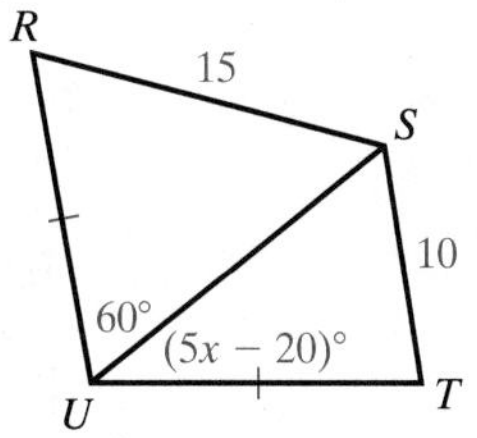

STEP 1. Find an upper limit for the value of x. $\overline{UT} \cong \overline{UR}$ and $\overline{US} \cong \overline{US}$, so ΔTUS and ΔRUS have two pairs of congruent sides and $RS > TS$, so you can use the Converse of the Hinge Theorem to write an inequality.

$m\angle RUS > m\angle TUS$ Converse of the Hinge Theorem

$60 > 5x - 20$ Substitute.

$80 > 5x$ Add 20 to each side.

$16 > x$ Divide each side by 5.

STEP 2. Find a lower limit for the value of x.

$m\angle TUS > 0$ The measure of an angle of a triangle is greater than 0.

$5x - 20 > 0$ Substitute.

$5x > 20$ Add 20 to each side.

$x > 4$ Divide each side by 5.

Rewrite $16 > x$ and $x > 4$ as $4 < x < 16$. □

PRACTICE 5 What is the range of possible values for x in the figure below?

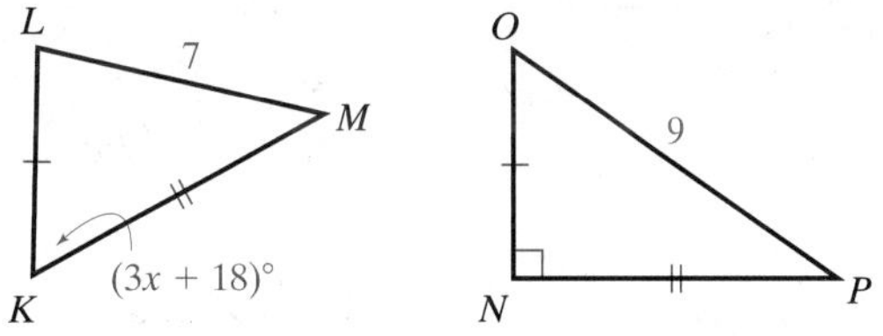

5.6 EXERCISE SET MyMathLab®

MIXED PRACTICE

Fill in the Blank *Use the Hinge Theorem, its converse, or a congruence postulate or theorem to fill in each blank with <, >, or =. See Examples 2–4.*

1. FD _______ BC

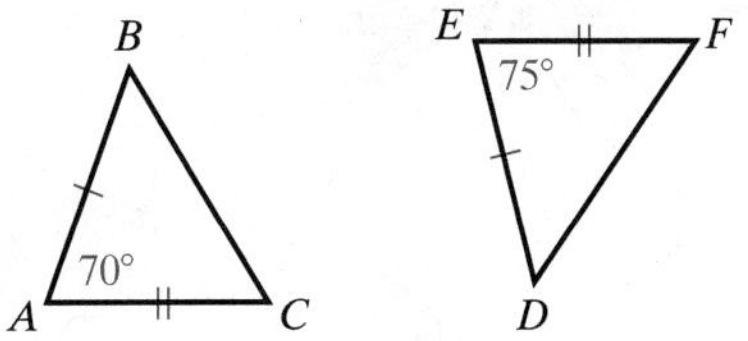

2. AC _______ DE

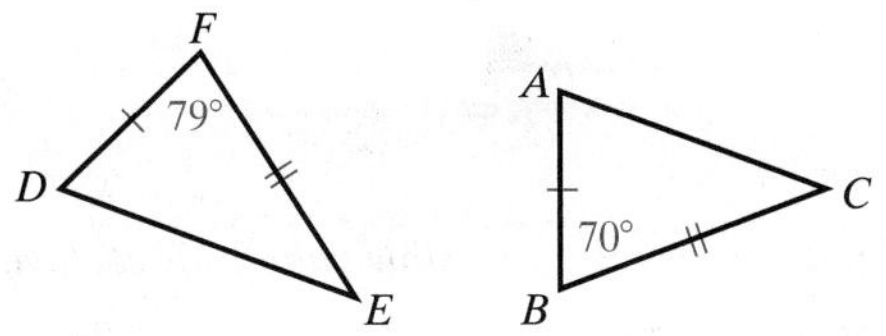

3. $m\angle 1$ _______ $m\angle 2$

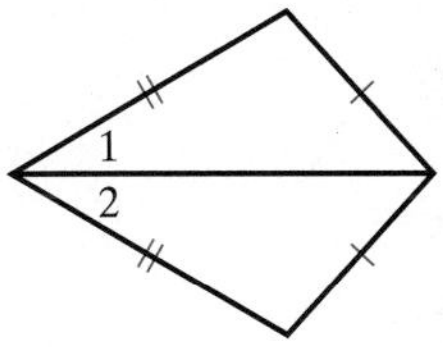

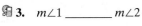

4. $m\angle 1$ _______ $m\angle 2$

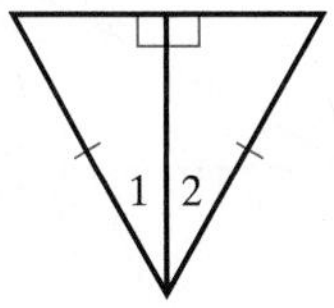

5. $m\angle 1$ _______ $m\angle 2$

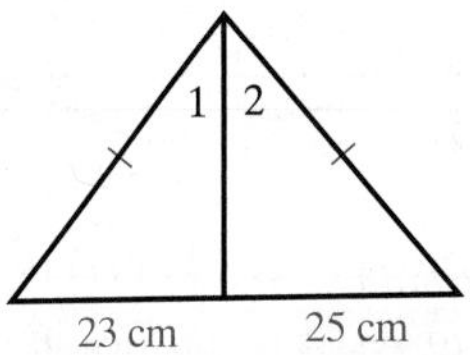

6. $m\angle 1$ _______ $m\angle 2$

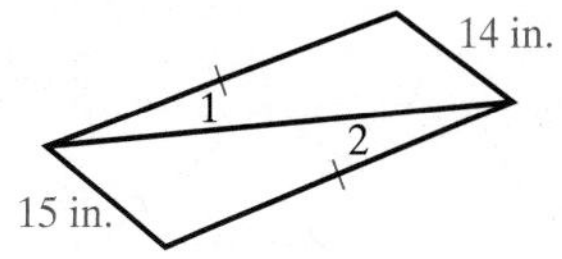

7. RT ________ VN

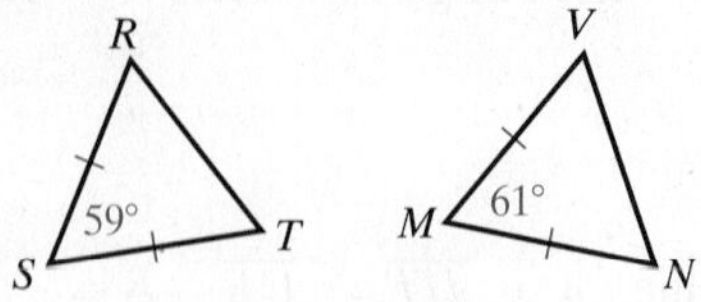

8. QL ________ AC

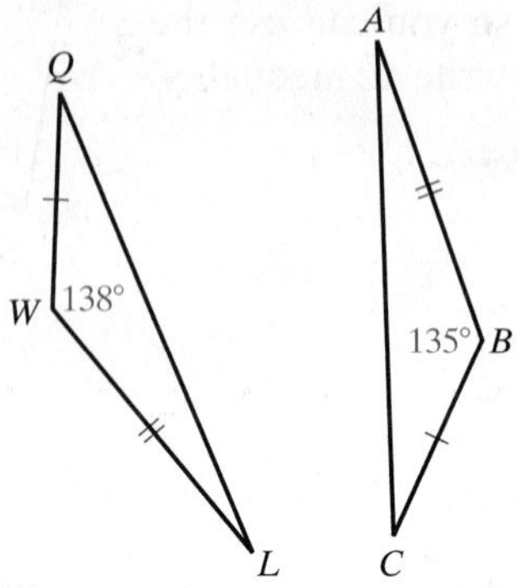

9. AE ________ ET

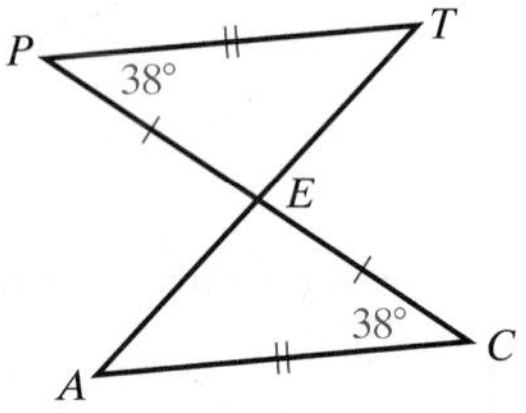

10. FD ________ HJ

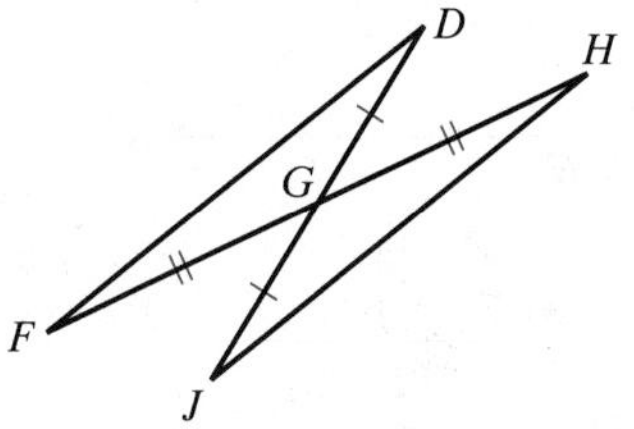

11. $m\angle 1$ ________ $m\angle 2$

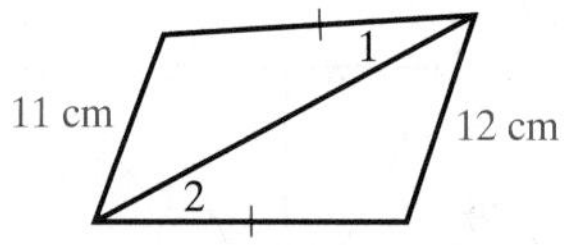

12. $m\angle 1$ ________ $m\angle 2$

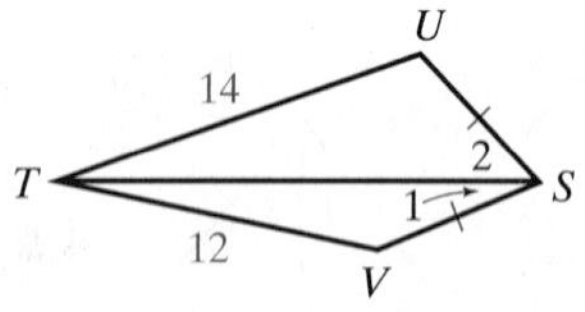

Write an inequality relating the given side lengths. If there is not enough information to reach a conclusion, write no conclusion. *See Examples 2–4.*

13. AB and AD

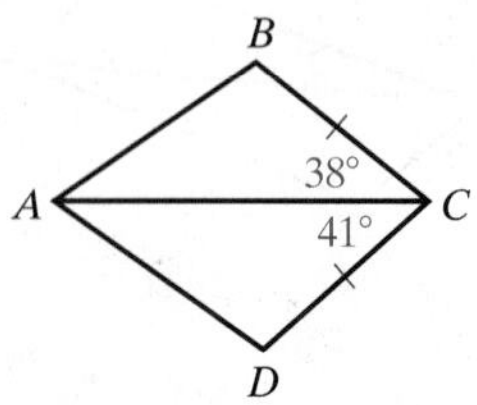

14. PR and RT

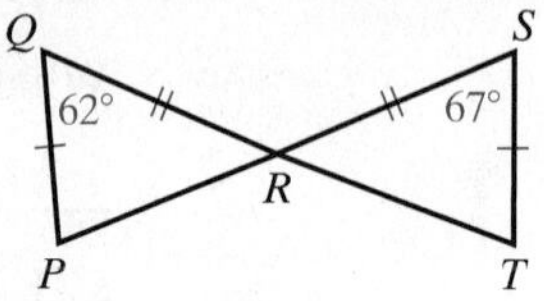

15. LM and KL

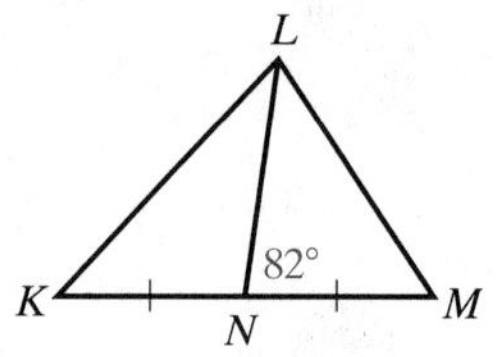

16. YZ and UV

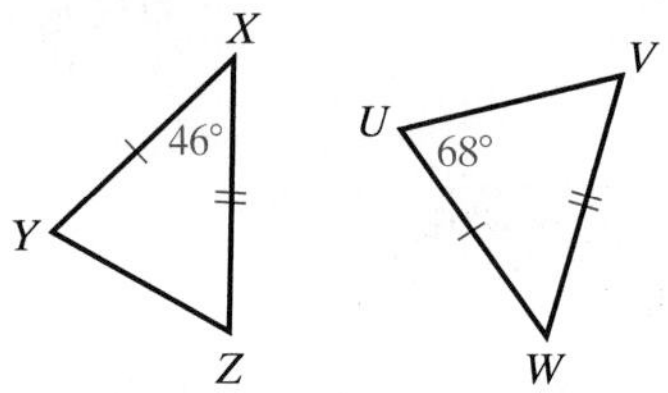

Solve. See Example 4.

17. The diagram below shows a robotic arm in two different positions. In which position (A or B) is the tip of the robotic arm closer to the base?

A.

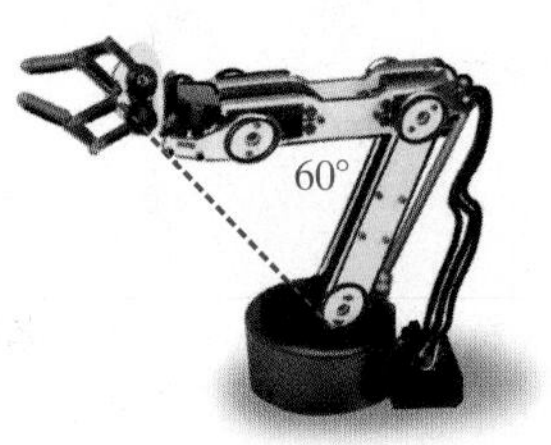

B.

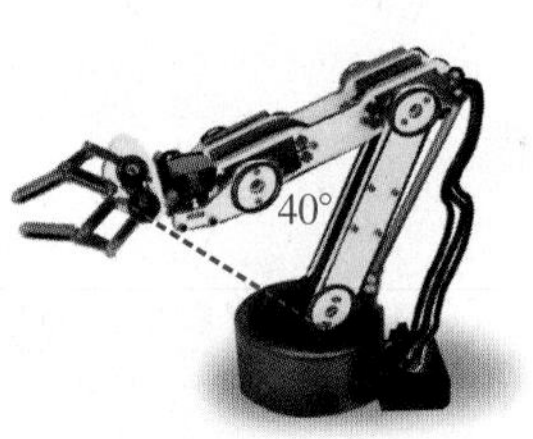

18. Which lamp bulb position is closer to the base—position A or position B?

A.

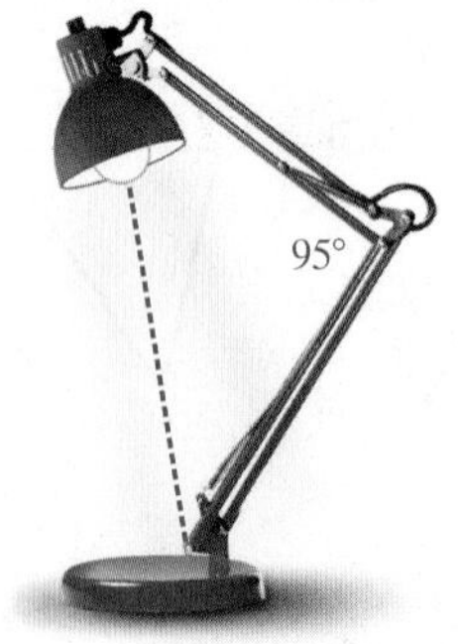

B.

Find the range of possible values for each variable. See Example 5.

19.

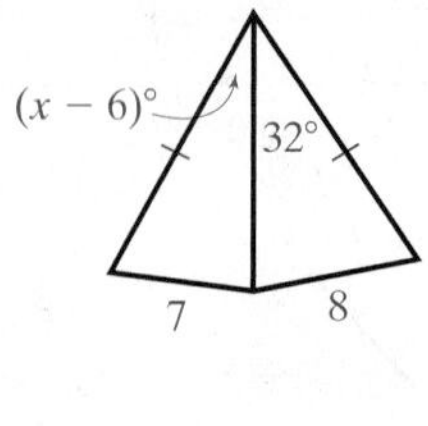

20.

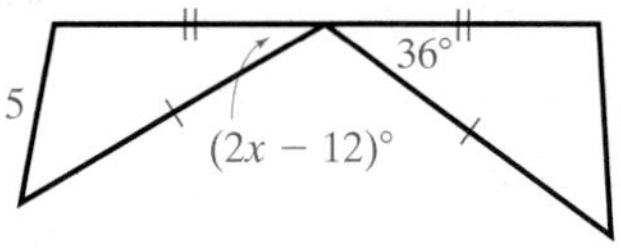

21.

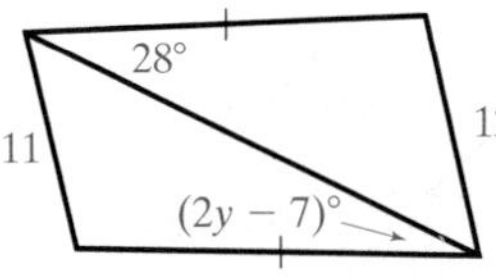

22.

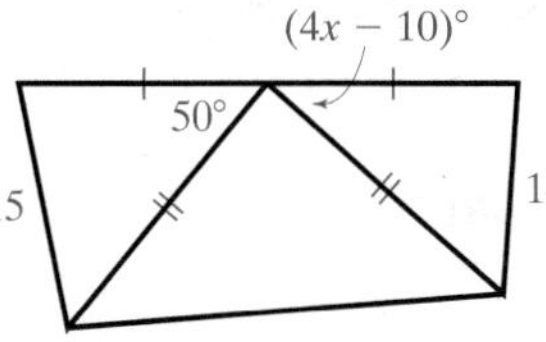

Complete with > or <. Explain your reasoning.

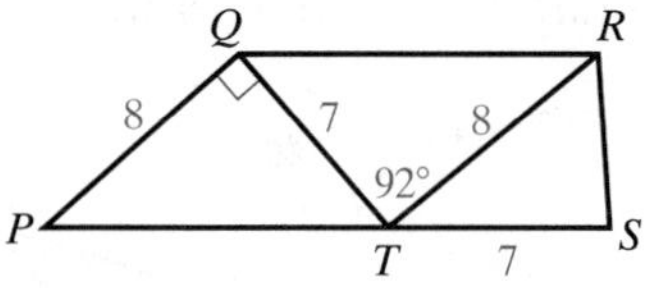

23. PT ________ QR

24. perimeter of ΔPQT ________ perimeter of ΔQTR

25. Find the Error Your classmate draws the figure below. Explain why the figure cannot have the labeled dimensions. Then describe a way you could change the dimensions to make the figure possible.

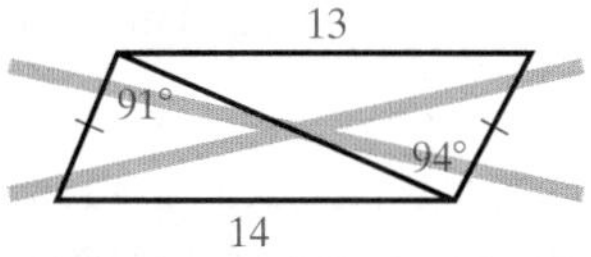

26. Find the Error From the figure below, your friend concludes that $m\angle BAD > m\angle BCD$. How would you correct your friend's mistake?

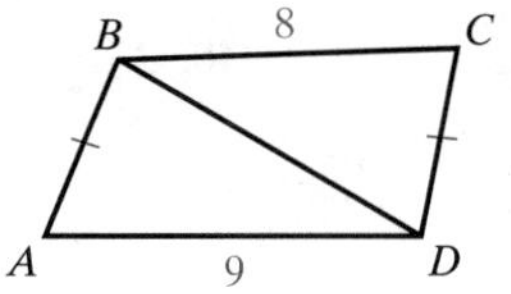

Proof *See Example 1.*

27. Complete the following proof.

Given: C is the midpoint of $\overline{BD}$,
$m\angle EAC = m\angle AEC$,
$m\angle BCA > m\angle DCE$

Prove: $AB > ED$

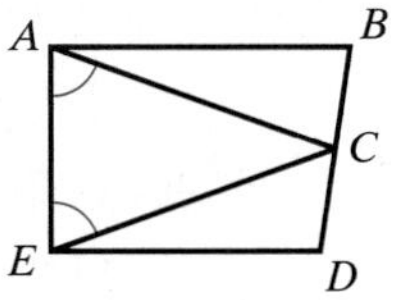

Statements	Reasons
1. $m\angle EAC = m\angle AEC$	**1.** Given
2. $AC = EC$	**2. a.** ?
3. C is the midpoint of $\overline{BD}$.	**3. b.** ?
4. $\overline{BC} \cong \overline{CD}$	**4. c.** ?
5. d. ?	**5.** ≅ segments have = length.
6. $m\angle BCA > m\angle DCE$	**6. e.** ?
7. $AB > ED$	**7. f.** ?

28. Complete the following proof.

Given: $BA = DE$, $BE > DA$

Prove: $m\angle BAE > m\angle BEA$

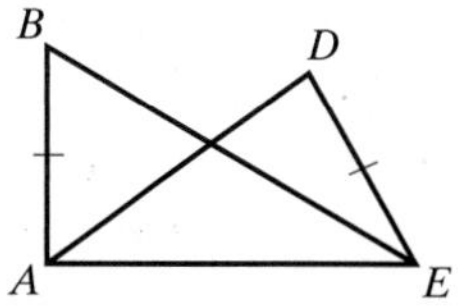

Statements	Reasons
1. $BA = DE$	**1.** Given
2. $AE = AE$	**2. a.** ?
3. $BE > DA$	**3.** Given
4. $m\angle BAE > m\angle DEA$	**4. b.** ?
5. $m\angle DEA = m\angle DEB + m\angle BEA$	**5. c.** ?
6. $m\angle DEA > m\angle BEA$	**6.** Comparison Property of Inequality
7. $m\angle BAE > m\angle BEA$	**7. d.** ?

29.

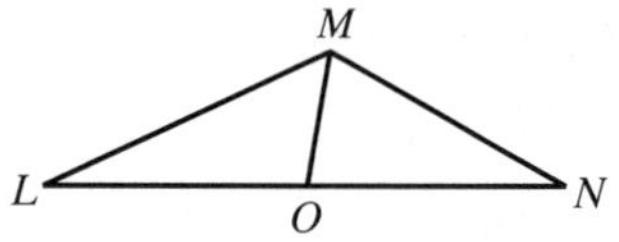

Given: $m\angle MON = 80°$, O is the midpoint of $\overline{LN}$

Prove: $LM > MN$

30. Use the figure below.

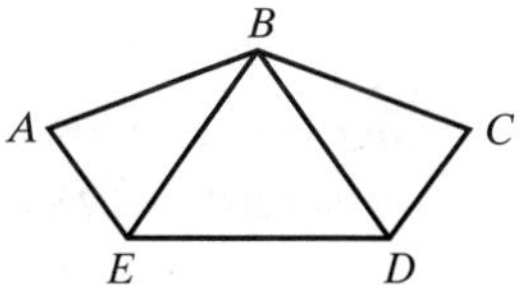

Given: ΔABE is isosceles with vertex $\angle B$,
$\Delta ABE \cong \Delta CBD$, $m\angle EBD > m\angle ABE$

Prove: $ED > AE$

The legs of a right isosceles triangle are congruent to the legs of an isosceles triangle with an 80° vertex angle.

31. Which triangle has a greater perimeter?

32. How do you know the answer to **31**?

33. Which of the following lists the segment lengths in order from least to greatest?

a. CD, AB, DE, BC, EF

b. EF, DE, AB, BC, CD

c. BC, DE, EF, AB, CD

d. EF, BC, DE, AB, CD

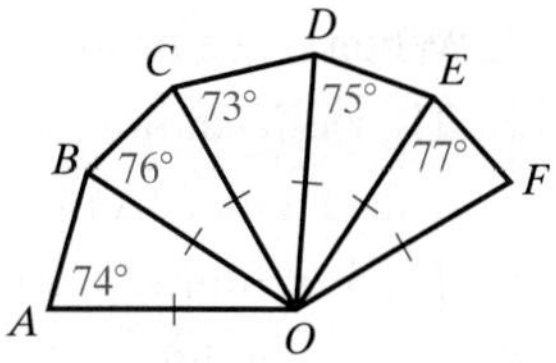

34. Ship A and Ship B leave from the same point in the ocean. Ship A travels 150 mi due west, turns 65° toward north, and then travels another 100 mi. Ship B travels 150 mi due east, turns 70° toward south, and then travels another 100 mi. Which ship is farther from the starting point? Explain. (See art on following page.)

- How can you use the given angle measures?
- How does the Hinge Theorem help you to solve this problem?

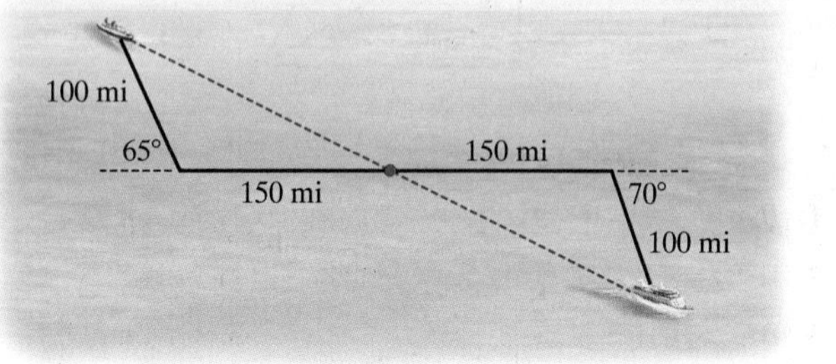

CONCEPT EXTENSIONS

35. Coordinate Geometry Triangles ΔAOB and ΔAOC have vertices $A(0, 7)$, $B(-1, -2)$, $C(2, -1)$, and $O(0, 0)$. Show that $m\angle AOB > m\angle AOC$.

36. The orthocenter of a triangle lies outside the triangle. Which of the following statements cannot be true?
a. The triangle has a 120° angle.
b. The triangle is isosceles.
c. The incenter is inside the triangle.
d. The triangle is acute.

Proof *In this section's Practice 1, we proved a part of the Hinge Theorem, Case 1. For Exercises 37 and 38, we ask you to prove the other two cases.*

Given: $\overline{AB} \cong \overline{DE}, \overline{BC} \cong \overline{EF}, m\angle ABC > m\angle E$

Construction: *Construct* $\overrightarrow{BQ}$ *so that* $\angle QBC \cong \angle E$. *Locate point* P *on* $\overrightarrow{BQ}$ *so that* $\overline{BP} \cong \overline{ED}$.

Prove: $AC > DF$

Recall that there are three cases depending on the location of point P. P lies on $\overrightarrow{BQ}$ *for each case.*

Case 1. Point P lies on $\overline{AC}$. (Practice 1)

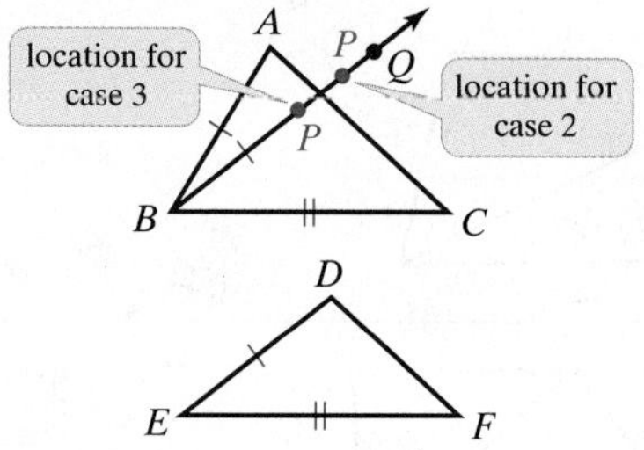

37. *Case 2.* Point P lies outside ΔABC.

38. *Case 3.* Point P lies inside ΔABC.

REVIEW AND PREVIEW

List the angles of each triangle in order from smallest to largest. See Section 5.5.

39.

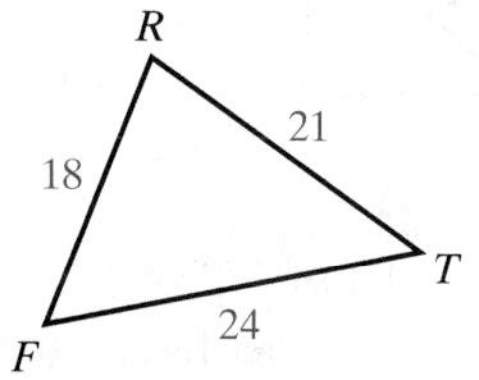

40.

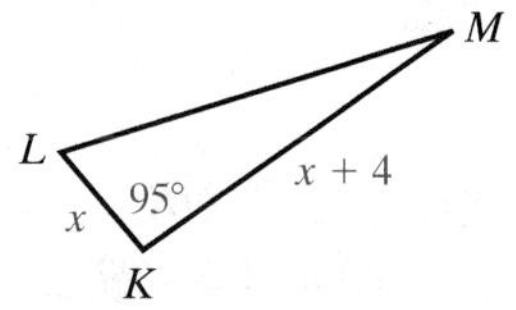

The lengths of two sides of a triangle are given. Find the range of possible lengths for the third side. See Section 5.5.

41. 15 cm, 19 cm
42. 3 in., 3 in.

Solve. See Section 4.1.

43. In ΔGHI, which side is included between $\angle G$ and $\angle H$?

44. In ΔGHI, which angle is included between $\overline{GH}$ and $\overline{IG}$?

Find the slope of the line through each pair of points. See Section 3.6.

45. $R(3, 8)$, $S(6, 0)$
46. $A(4, 3)$, $B(2, 1)$
47. $C(2, 5)$, $D(7, 8)$
48. $X(0, 6)$, $Y(4, 9)$

CHAPTER 5 VOCABULARY CHECK

Word Bank *Use the choices to fill in each blank. Some choices may be used more than once and some not at all.*

perpendicular	circumscribed about	point of concurrency	altitude
equidistant	circumcenter	indirect proof	median
proof by contradiction	incenter	deductive reasoning	centroid
inscribed in	concurrent	orthocenter	midsegment

1. When three or more lines intersect in one point, they are ____________________.
2. The point of intersection in number **1** above is called the ____________________.
3. The distance from a point to a line is the length of the ____________________ segment from the point to the line.
4. The incenter of a triangle is the center of a circle ____________________ the triangle.
5. When a point is the same distance from two distinct lines, we say that the point is ____________________ from the two lines.
6. A segment connecting the midpoints of two sides of a triangle is called a(n) ____________________.
7. The perpendicular segment from a vertex of a triangle to the line containing the opposite side is called a(n) ____________________ of the triangle.

8. An indirect proof is also called a(n) ____________________.

9. Law of Detachment and Law of Syllogism are two examples of ____________________.

10. A segment whose endpoints are a vertex of a triangle and the midpoint of the opposite side is called a(n) ____________________ of the triangle.

11. Two points of concurrency may lie inside, outside, or on a triangle. They are the ____________________ and the ____________________.

12. The point of concurrency of the medians of a triangle is called the ____________________.

13. The point of concurrency of the altitudes of a triangle is called the ____________________.

14. The ____________________ is also known as the center of gravity of a triangle.

15. The three perpendicular bisectors of a triangle intersect in one point called the ____________________.

16. The circumcenter of a triangle is the center of a circle ____________________ the triangle.

17. The three angle bisectors of a triangle intersect in one point called the ____________________.

CHAPTER 5 REVIEW

***(5.1)** In the figure, $QP = 4$ and $AB = 8$. Find each of the following.*

1. QR

2. CB

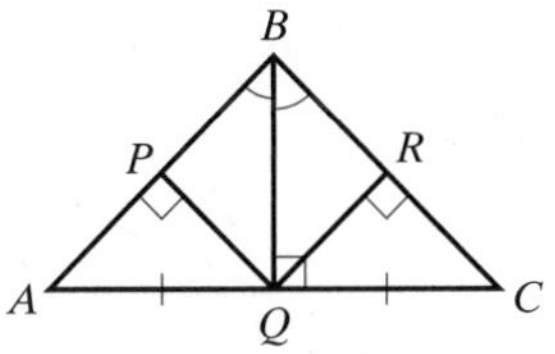

In the figure, $m\angle DBE = 50°$. Find each of the following.

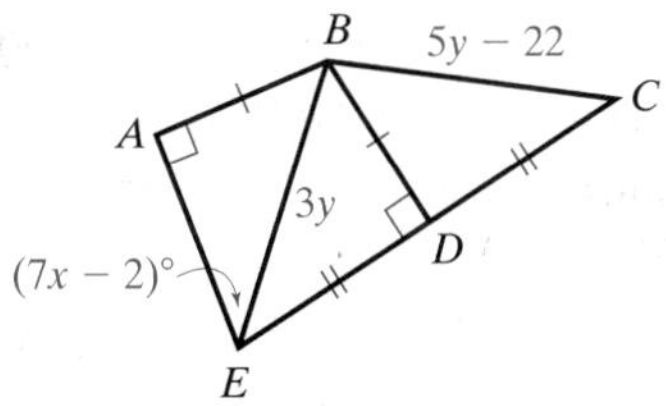

3. $m\angle BED$ **4.** $m\angle BEA$

5. x **6.** y

7. BE **8.** BC

(5.2)

9. Identify the incenter of ΔKLM.

10. Is X the circumcenter? Why or why not?

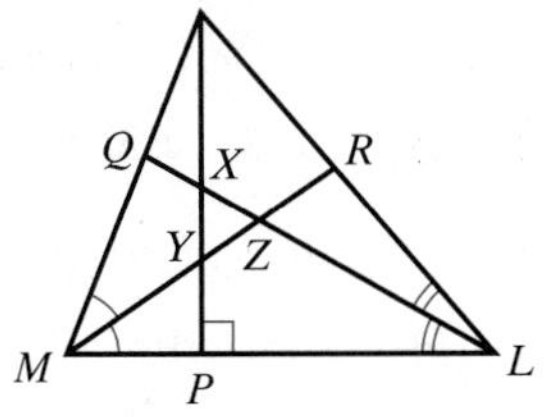

Find the coordinates of the circumcenter of ΔDEF.

11. $D(6,0), E(0,6), F(-6,0)$

12. $D(0,0), E(6,0), F(0,4)$

13. $D(5,-1), E(-1,3), F(3,-1)$

14. $D(2,3), E(8,3), F(8,-1)$

P is the incenter of ΔXYZ. Find the indicated angle measure.

15. $m\angle PXY$

16. $m\angle XYZ$

17. $m\angle PZX$

18. $m\angle XPY$

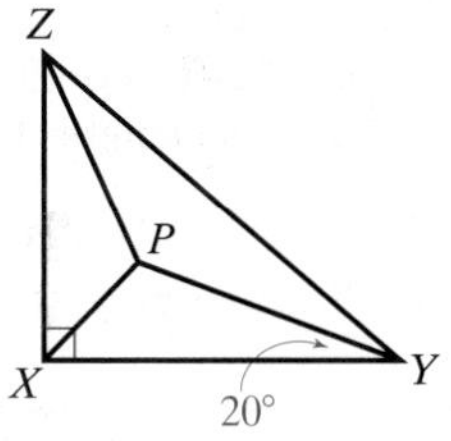

(5.3)

19. If $PB = 6$, what is SB?

20. If $CS = 6$, what is CR?

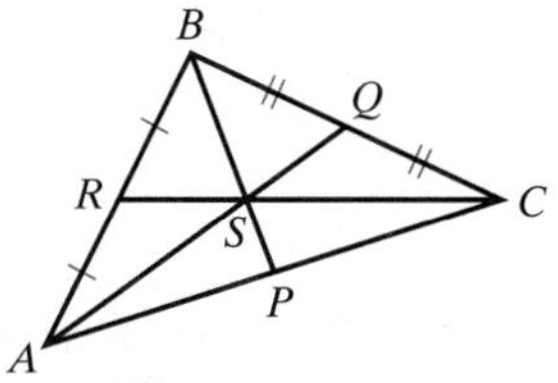

Determine whether $\overline{AB}$ is a median, an altitude, or neither. Explain.

21.

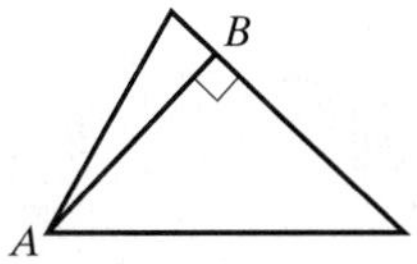

22.

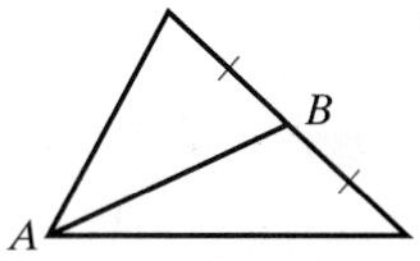

ΔABC has vertices $A(2, 3)$, $B(-4, -3)$, and $C(2, -3)$. Find the coordinates of each point of concurrency.

23. centroid **24.** orthocenter

***(5.4)** Use the given figure for Exercises 25–28.*

25. Find the value of x.

26. Find DE.

27. Find BC.

28. Explain why the lengths of $\overline{DE}$ and $\overline{BC}$ check.

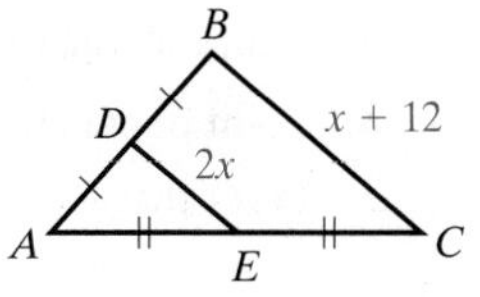

For Exercises 29 and 30, find the value of x.

29.

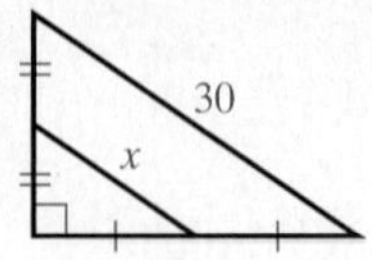

30.

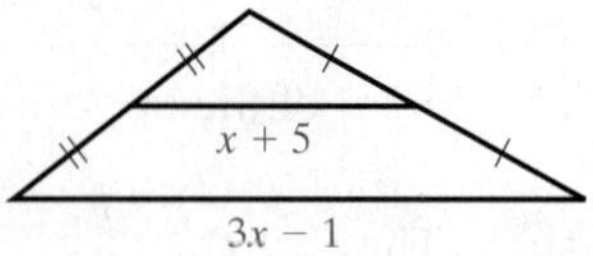

(5.5) *For Exercises 31 and 32, write a convincing argument that uses indirect reasoning.*

31. Show that a triangle can have at most one obtuse angle.

32. The product of two numbers is even. Show that at least one of the numbers must be even.

33. In $\triangle RST$, $m\angle R = 70°$ and $m\angle S = 80°$. List the sides of $\triangle RST$ in order from shortest to longest.

For Exercises 34 and 35, is it possible for a triangle to have sides with the given lengths? Explain.

34. 5 in., 8 in., 15 in.

35. 10 cm, 12 cm, 20 cm

36. The lengths of two sides of a triangle are 12 ft and 13 ft. Find the range of possible lengths for the third side.

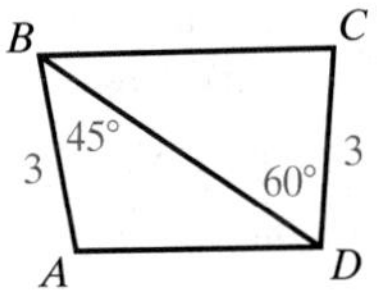

(5.6)

37. Which is greater, BC or AD?

Use the figure to the right to complete each statement with >, <, or =.

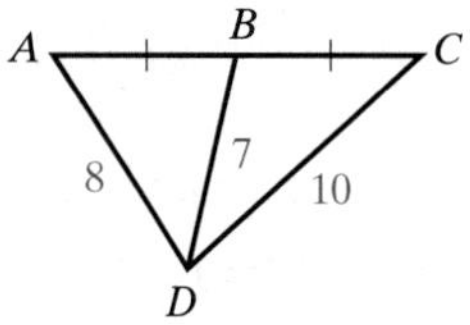

38. $m\angle BAD$ ____________ $m\angle ABD$

39. $m\angle CBD$ ____________ $m\angle BCD$

40. $m\angle ABD$ ____________ $m\angle CBD$

MIXED REVIEW

41. $\triangle PQR$ has medians $\overline{QM}$ and $\overline{PN}$ that intersect at Z. If $ZM = 4$, find QZ and QM.

42. $\triangle ABC$ has vertices $A(0, 0)$, $B(2, 2)$, and $C(5, -1)$. Find the coordinates of L, the midpoint of $\overline{AC}$, and the coordinates of M, the midpoint of $\overline{BC}$. Verify that $\overline{LM} \parallel \overline{AB}$ and $LM = \frac{1}{2}AB$.

43. Which two statements contradict each other?

I. The perimeter of $\triangle ABC$ is 14.

II. $\triangle ABC$ is isosceles.

III. The side lengths of $\triangle ABC$ are 3, 5, and 6.

Write a convincing argument that uses indirect reasoning.

44. Show that an equilateral triangle cannot have an obtuse angle.

45. The sum of three integers is greater than 9. Show that one of the integers must be greater than 3.

In the figure below, E is the incenter, J is the circumcenter, and G is the centroid of △ABC. Use this figure for Exercises 46–50.

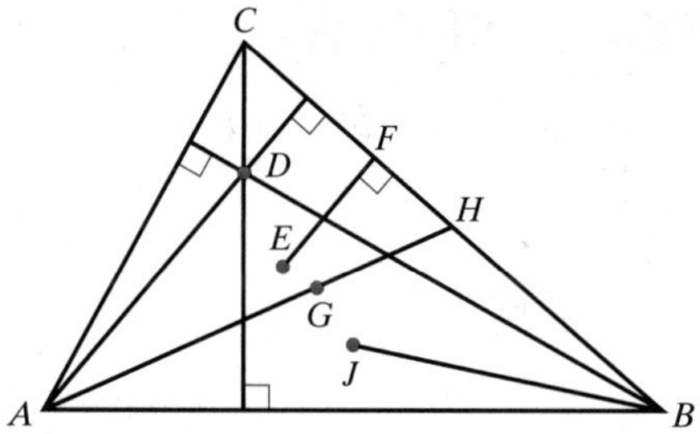

46. What is the radius of the inscribed circle?

47. What is the radius of the circumscribed circle?

48. If $GH = 3$ units, what is the measure of AG?

49. What is the orthocenter of $\triangle ABC$?

50. If $BH = 5$ units, what is BC?

CHAPTER 5 TEST

TEST PREP VIDEO The fully worked-out solutions to any exercises you want to review are available in MyMathLab.

Matching *Use these choices for Exercises 1–9.*

a. incenter **b.** circumcenter

c. orthocenter **d.** centroid

1. Point of concurrency of medians

2. Point of concurrency of angle bisectors

3. Point of concurrency of perpendicular bisectors

4. Point of concurrency of altitudes

5. Point that is equidistant from the sides of a triangle

6. Two points that always lie inside the triangle

7. Point where a triangle of uniform thickness will balance

8. Point that is the center of the circle inscribed inside the triangle

9. Point that is the center of the circle that circumscribes the triangle

Draw △ABC and find the coordinates of its circumcenter.

10. $A(0, 5), B(-4, 5), C(-4, -3)$

Draw △ABC and find the coordinates of its orthocenter.

11. $A(-1, -1), B(-1, 5), C(-4, -1)$

Identify the two statements that contradict each other.

12. **I.** $\triangle PQR$ is a right triangle.

II. $\triangle PQR$ is an obtuse triangle.

III. $\triangle PQR$ is scalene.

13. If $AB = 9$, $BC = 4\frac{1}{2}$, and $AC = 12$, list the angles of $\triangle ABC$ from smallest to largest.

14. Point P is inside $\triangle ABC$ and equidistant from all three sides. If $m\angle ABC = 60°$, what is $m\angle PBC$?

List the sides from shortest to longest.

15. 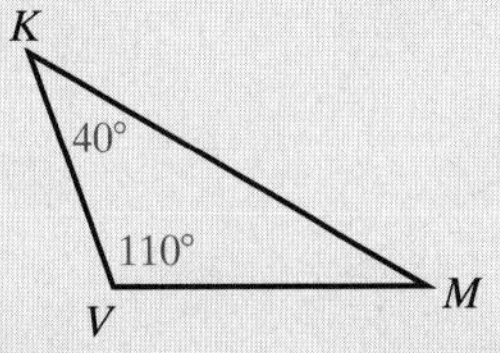

16. In ΔABC, $EP = 4$. What is PC?

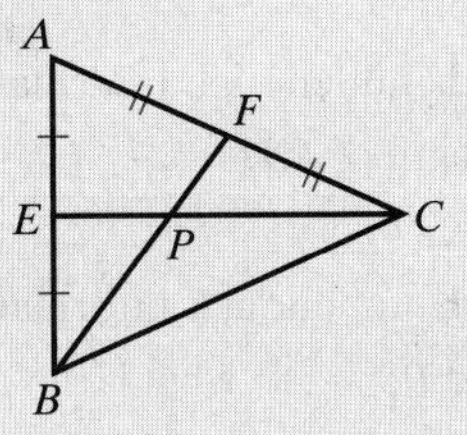

Use the figure below for Exercises 17–19.

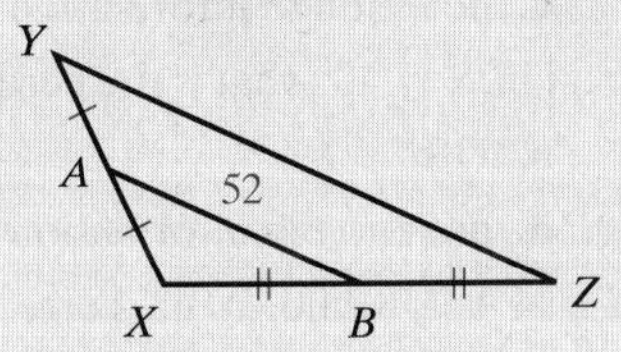

17. Find YZ.

18. $AX = 26$ and $BZ = 36$. Find the perimeter of ΔXYZ.

19. Which angle is congruent to $\angle XBA$? How do you know?

For the figure to the right, what can you conclude about each of the following? Explain.

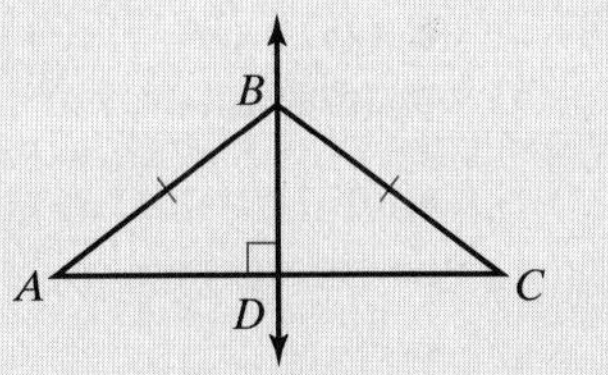

20. $\angle CDB$

21. ΔABD and ΔCBD

22. $\overline{AD}$ and $\overline{DC}$

23. Which is greater, AD or DC? Explain.

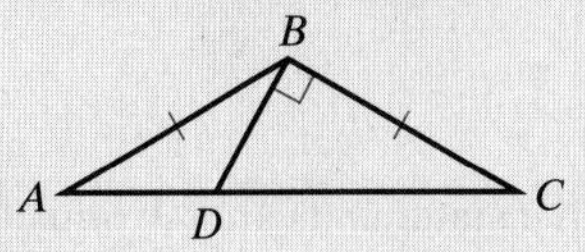

Find the value of x.

24. 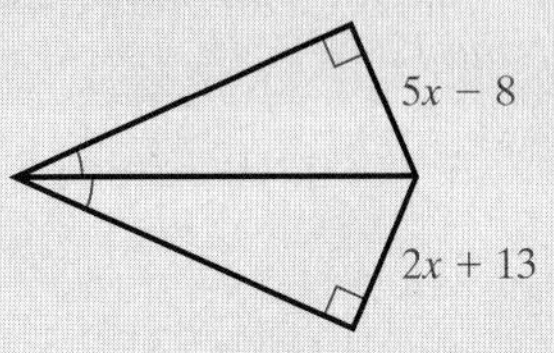

25. 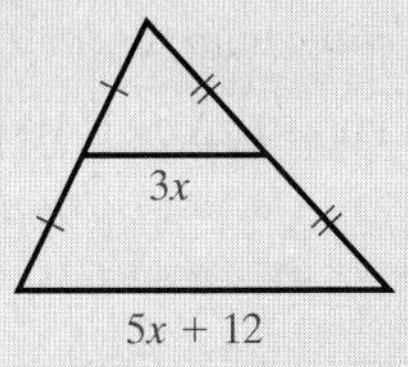

26. What can you conclude from the diagram at the right? Justify your answer.

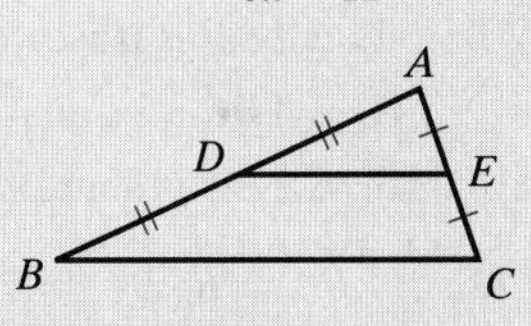

27. In the figure, $WK = KR$. What can you conclude about point A? Explain.

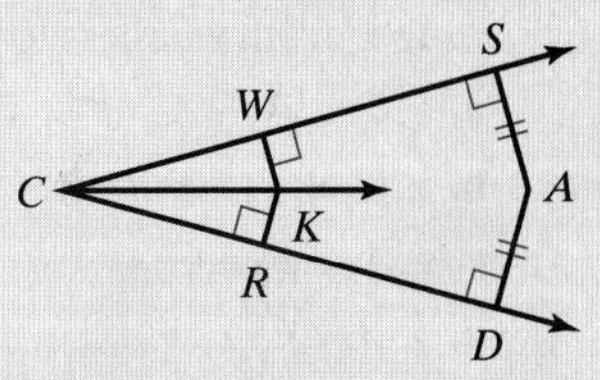

28. Use indirect reasoning to explain why the following statement is true: If an isosceles triangle is obtuse, then the obtuse angle is the vertex angle.

29. **Given:** $\overleftrightarrow{PQ}$ is the perpendicular bisector of $\overline{AB}$. $\overleftrightarrow{QT}$ is the perpendicular bisector of $\overline{AC}$.

Prove: $QC = QB$

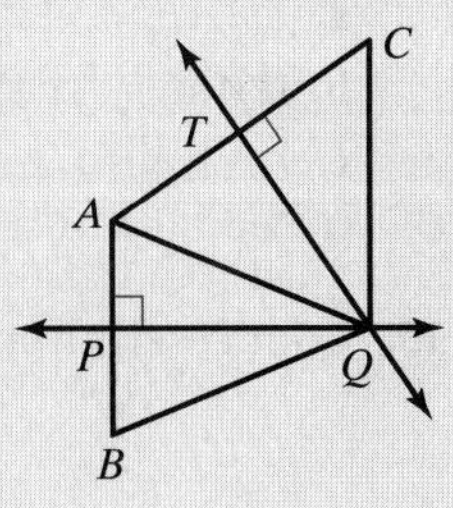

CHAPTER 5 STANDARDIZED TEST

Use these choices for Exercises 1–8.

a. incenter

b. centroid

c. circumcenter

d. orthocenter

1. Point where a triangle of uniform thickness would balance

2. Point that is the center of the circle inscribed inside the triangle

3. Point of concurrency of altitudes

4. Point that is equidistant from the sides of a triangle

5. Point of concurrency of perpendicular bisectors

6. Point of concurrency of angle bisectors

7. Point that is the center of the circle that circumscribes the triangle

8. Point of concurrency of medians

9. Find the coordinates of the orthocenter of ΔABC for $A(-6, 2)$, $B(-2, 2)$, and $C(-6, -4)$.

a. $(-6, 2)$ **b.** $(-2, 2)$
c. $(-6, 4)$ **d.** $(-4, -1)$

10. Find the coordinates of the circumcenter of ΔABC for $A(1, 4)$, $B(1, -2)$, and $C(-3, -2)$.

a. $(1, 4)$ **b.** $(1, -2)$
c. $(-3, -2)$ **d.** $(-1, 1)$

Use these choices for Exercises 11 and 12.

a. I and II
b. I and III
c. II and III
d. None of the statements are contradictory.

Identify the two statements that contradict each other. Use the choices above for Exercises 11 and 12.

11. **I.** ΔABC is scalene.
II. ΔABC is acute.
III. ΔABC is a right triangle.

12. **I.** ΔABC is isosceles.
II. ΔABC is equilateral.
III. ΔABC is acute.

13. If $AB = 6$, $BC = 7$, and $AC = 10$, list the angles of ΔABC from smallest to largest.

a. $\angle C, \angle B, \angle A$
b. $\angle C, \angle A, \angle B$
c. $\angle A, \angle B, \angle C$
d. $\angle B, \angle C, \angle A$

14. In ΔABC, $\overline{AM}$ bisects $\overline{BC}$, $\overline{BN}$ bisects $\overline{AC}$, and P is the point of intersection of $\overline{AM}$ and $\overline{BN}$. Which statement is true?

a. $\angle BAP \cong \angle PAC$
b. If $AM = 24$, then $AP = 16$.
c. $\angle AMB \cong \angle AMC$
d. $\overline{AP} \cong \overline{BP}$

15. List the sides from shortest to longest.

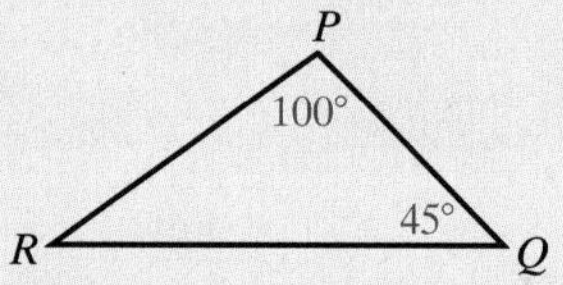

a. $\overline{PQ}, \overline{PR}, \overline{QR}$
b. $\overline{PR}, \overline{PQ}, \overline{QR}$
c. $\overline{QR}, \overline{PR}, \overline{PQ}$
d. $\overline{PQ}, \overline{QR}, \overline{PR}$

Use the figure below for Exercises 16–19.

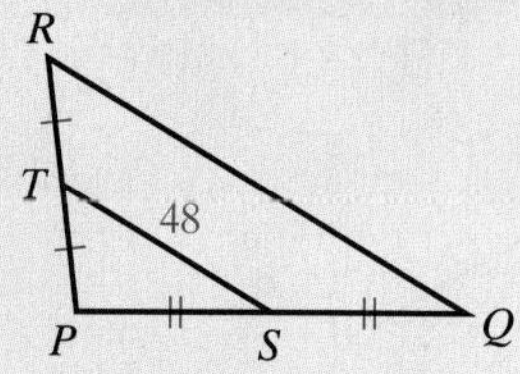

16. What is the length of $\overline{QR}$?

a. 24 **b.** 96
c. 72 **d.** not enough information

17. What is the length of $\overline{PS}$?

a. 24 **b.** 96
c. 72 **d.** not enough information

18. If $TR = 30$ and $PS = 40$, what is the perimeter of ΔPQR?

a. 148 **b.** 118
c. 236 **d.** not enough information

19. If $TR = 30$ and $PS = 40$, what is the perimeter of ΔPST?

a. 148 **b.** 118
c. 236 **d.** not enough information

20. Given the figure below, choose the correct statement.

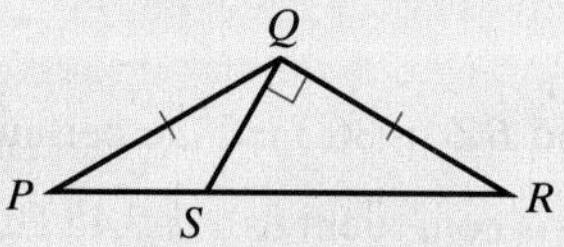

a. $\angle PQS \cong \angle QPS$ **b.** $\angle QSR \cong \angle QRS$
c. $QS = QR$ **d.** $SR > PS$

21. Find the value of x.

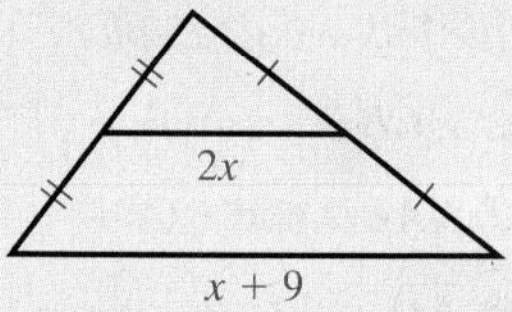

a. 6
b. 9
c. 3
d. not enough information

22. For an indirect proof that ΔABC is equilateral, choose the correct completion of the statement, "Assume temporarily that...."

a. ΔABC is scalene.
b. ΔABC has two sides that are not congruent.
c. ΔABC is isosceles.
d. ΔABC is obtuse.

CHAPTER

6 Quadrilaterals

6.1 Polygons

6.2 Parallelograms

6.3 Proving that a Quadrilateral Is a Parallelogram

6.4 Rhombuses, Rectangles, and Squares

6.5 Trapezoids and Kites

Ceiling fan blade

Quadrilateral-shaped

Ceiling fans can provide dramatic energy savings. By using a fan with your air conditioning during the summer, you can save approximately 4–8% for each degree you raise your thermostat.

A blade of this fan is approximately in the shape of a quadrilateral. Just as a triangle is a 3-sided polygon, a quadrilateral is a 4-sided polygon and special types of quadrilaterals are the main focus of this chapter.

6.1 POLYGONS

OBJECTIVES

1 Define and Name Polygons.

2 Find the Sum of the Measures of the Interior Angles of a Quadrilateral.

VOCABULARY

- polygon
- vertex
- *n*-gon
- concave polygon
- convex polygon
- quadrilateral
- regular polygon
- diagonal
- equilateral polygon
- equiangular polygon

OBJECTIVE 1 ▶ Defining and Naming Polygons. For the past two chapters, we have studied triangles. Triangles are part of a larger group of figures called *polygons*.

Polygon Definition

A figure is a **polygon** if it meets the following conditions:

1. It is a plane figure formed by three or more line segments called **sides.**
2. Sides that have a common endpoint are noncollinear.
3. Each side intersects exactly two other sides, but only at their endpoints.

The endpoints of the sides of a polygon are called the **vertices** (singular is **vertex**). Below are some examples of polygons. Each vertex of the middle polygon is labeled. A polygon can be named by listing its vertices consecutively in order, as shown.

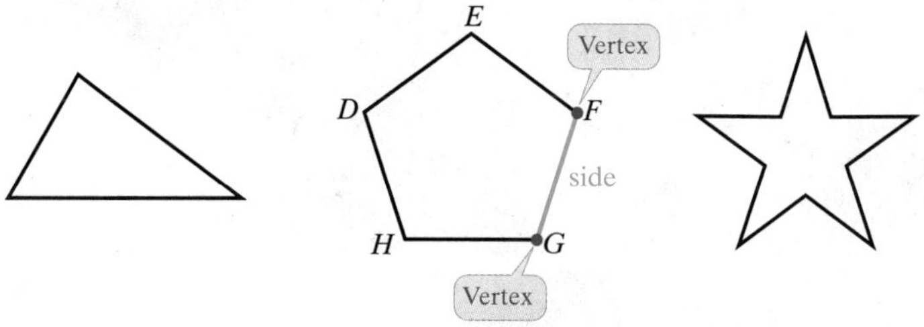

Polygon *DEFGH*
Polygon *EDHGF*
} two of many correct names

Let's see if we can correctly identify polygons.

EXAMPLE 1 **Identifying Polygons**

Identify the polygons. If not a polygon, state why.

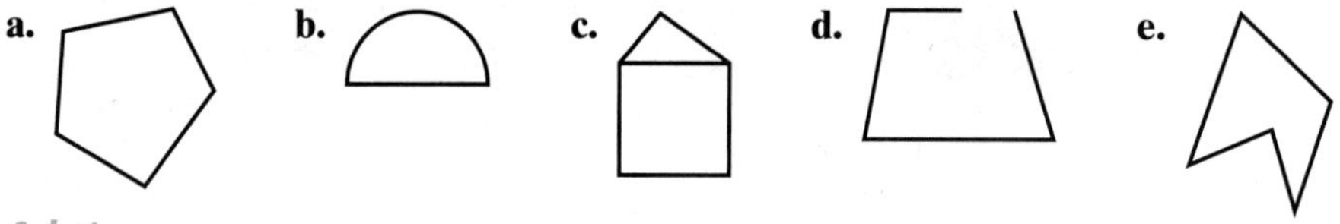

Solution

Figures *A* and *E* are polygons.

Figure *B* is not a polygon because there is a "side" that is a curve and not a line segment.

Figure *C* is not a polygon by our definition because there is a side that intersects more than two other sides.

Figure *D* is not a polygon because two sides intersect only one other side. □

PRACTICE 1 Identify the polygons. If not a polygon, state why.

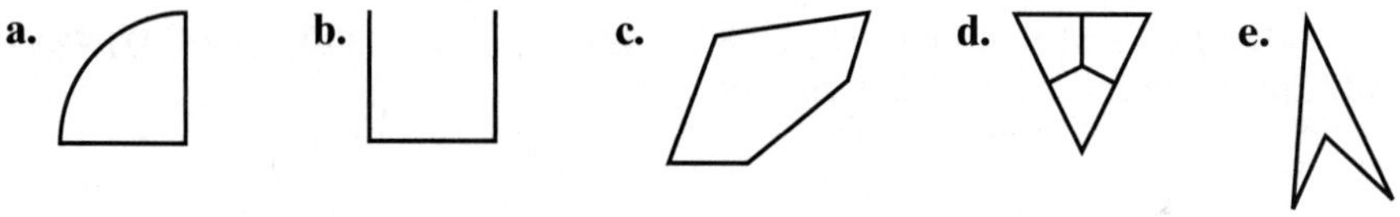

In this chapter, we concentrate on four-sided polygons called **quadrilaterals,** but let's learn the names of other polygons also. They are named according to the number of sides they have.

> **Helpful Hint**
> It is important to study and learn the names of these polygons.

Number of Sides	*Name of Polygon*
3	triangle
4	quadrilateral
5	pentagon
6	hexagon
7	heptagon
8	octagon
9	nonagon
10	decagon
12	dodecagon
n	n-gon

In general, a polygon with n sides is called an ***n*-gon.** For example, a polygon with 13 sides is called a 13-gon.

Another way to classify polygons is as convex or concave. A polygon is **convex** if no line containing a side contains a point within the interior of the polygon. A polygon is **concave** (or **nonconvex**) if it is not convex.

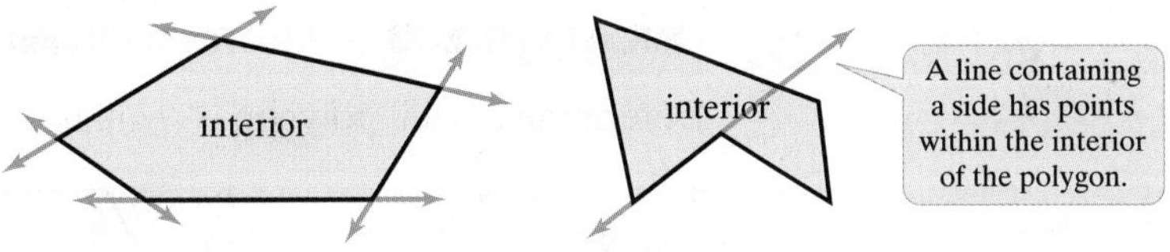

Convex Polygon

Concave or Nonconvex Polygon

EXAMPLE 2 **Identifying Convex and Concave Polygons**

Name the polygon according to its number of sides. Then identify whether it is convex or concave.

a. 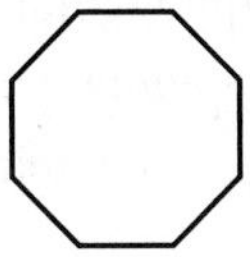

b. 

c. 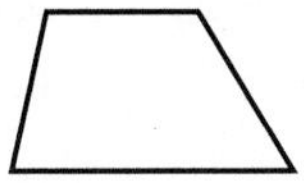

Solution

a. The polygon has 8 sides, so it is an octagon. None of the extended sides contain a point of the interior, so it is convex.

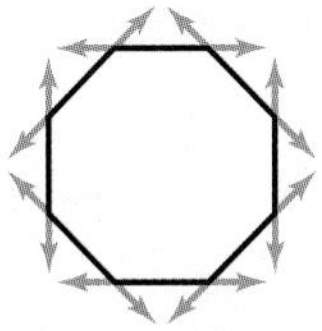

Convex octagon

b. The polygon has 6 sides, so it is a hexagon. Some of the extended sides contain a point of the interior, so it is concave (or nonconvex).

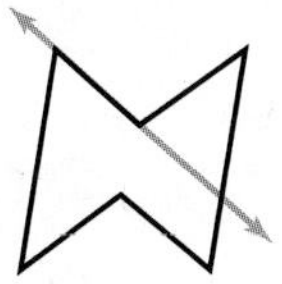

Concave hexagon

c. The polygon has 4 sides, so it is a quadrilateral. None of the extended sides contain a point within the interior, so it is convex.

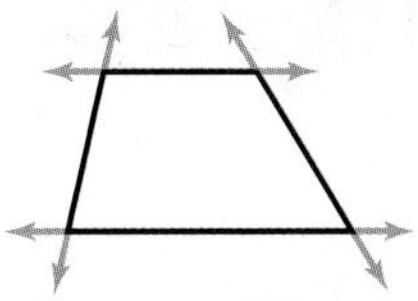

Convex quadrilateral

□

PRACTICE 2 Name the polygon according to the number of sides. Then identify whether it is convex or concave.

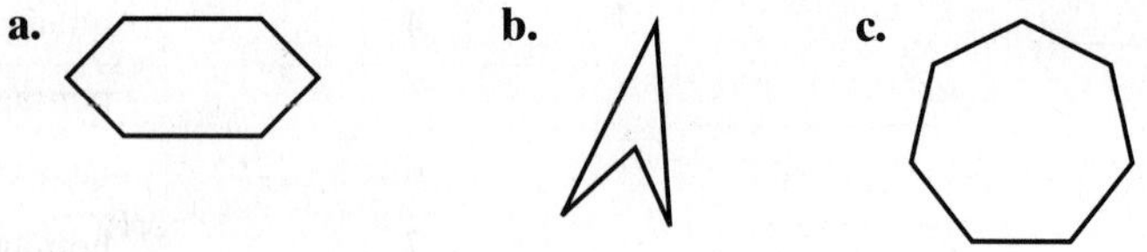

Let's learn another way to classify polygons.

An **equilateral polygon** is a polygon with all sides congruent.

An **equiangular polygon** is a polygon with all angles congruent.

A **regular polygon** is a polygon that is both equilateral and equiangular.

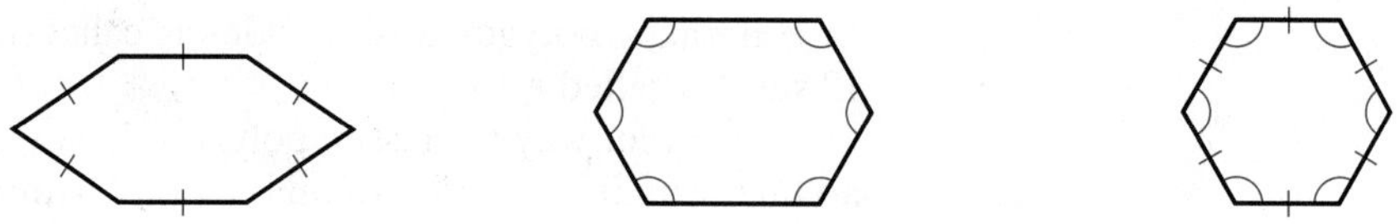

EXAMPLE 3 Identifying Regular Polygons

Determine if each polygon is regular or not. Explain your reasoning.

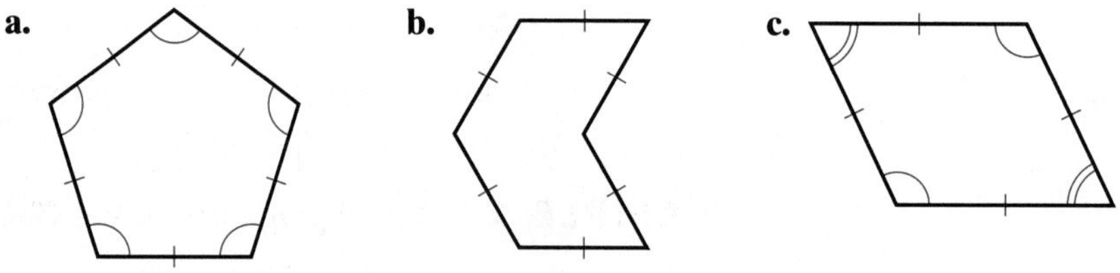

Solution

a. The pentagon is equilateral and equiangular, so it is a regular polygon.

b. The hexagon is equilateral, but not equiangular, so it is not a regular polygon.

c. The quadrilateral is equilateral, but not equiangular, so it is not a regular polygon. □

PRACTICE 3 Determine if each polygon is regular or not. Explain your reasoning.

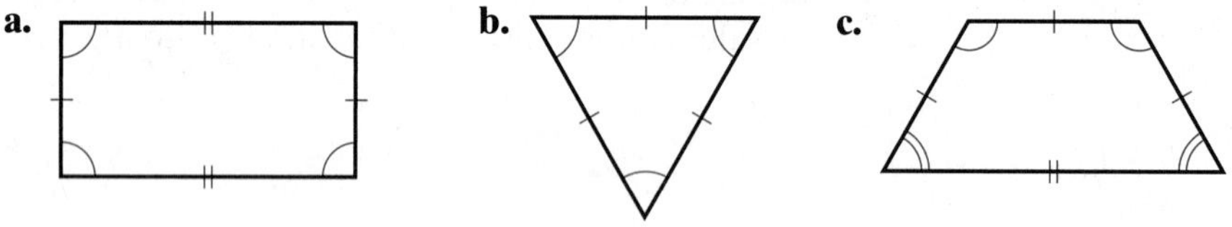

OBJECTIVE 2 ▶ Finding the Sum of the Measures of the Interior Angles of a Quadrilateral. Let's discover the sum of the measures of the interior angles of any quadrilateral. To do so, we need to know two things:

1. The sum of the measures of the interior angles of a triangle and
2. The definition of a diagonal of a polygon

We know from Section 4.1 that the answer to number **1** above is 180°, so let's continue to number **2** now and define a diagonal. A segment joining two nonconsecutive vertices of a convex polygon is called a **diagonal** of the polygon.

Examples of Diagonals in Red

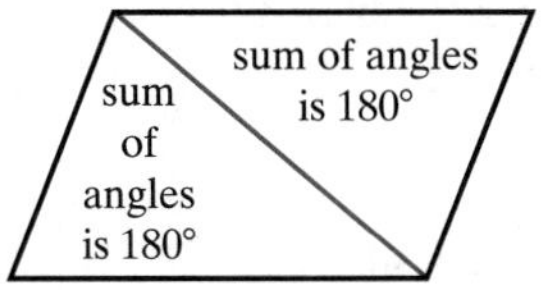

Quadrilateral (4 sides)
(*Note:* 1 diagonal forms 2 triangles)

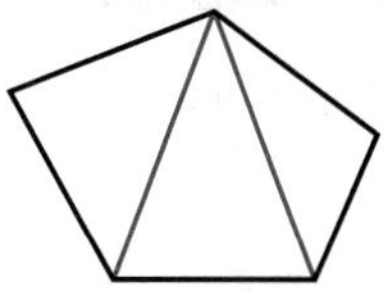

Pentagon (5 sides)
(*Note:* 2 diagonals form 3 triangles)

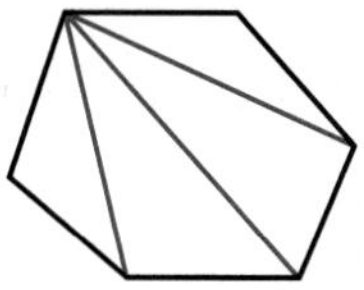

Hexagon (6 sides)
(*Note:* 3 diagonals form 4 triangles)

As shown in the first diagram above, a convex quadrilateral will always have one diagonal, which will always form 2 triangles. Thus, the sum of the measures of the interior angles of a convex quadrilateral is always $2 \cdot 180°$, or 360°.

Theorem 6.1-1 Interior Angle Sum of a Convex Quadrilateral

The sum of the measures of the interior angles of a convex quadrilateral is 360°.

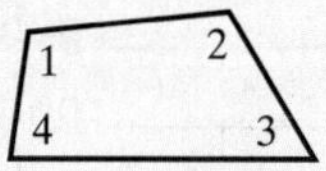

$$m\angle 1 + m\angle 2 + m\angle 3 + m\angle 4 = 360°$$

EXAMPLE 4 **Find x, and then $m\angle C$ and $m\angle D$.**

Solution The sum of these angle measures is 360°, thus:

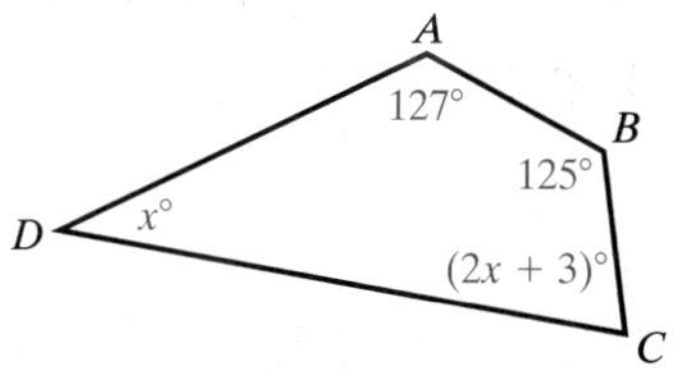

> **Helpful Hint**
> Notice that we leave degree units off for ease of solving equations.

$$x + (2x + 3) + 127 + 125 = 360 \quad \text{Sum of interior angles of quadrilateral is 360.}$$
$$3x + 255 = 360 \quad \text{Combine like terms.}$$
$$3x = 105 \quad \text{Subtract 255 from both sides.}$$
$$x = 35 \quad \text{Divide both sides by 3.}$$

Use $x = 35$ to find $m\angle C$ and $m\angle D$.

$$m\angle D = x° = 35°$$
$$m\angle C = (2x + 3)° = (2 \cdot 35 + 3)° = 73°$$

Check: To check, see that $m\angle A + m\angle B + m\angle C + m\angle D = 360°$, or

$$127° + 125° + 73° + 35° = 360°$$
$$360° = 360°$$

The solution checks. □

PRACTICE 4 Find x, and then $m\angle H$ and $m\angle E$.

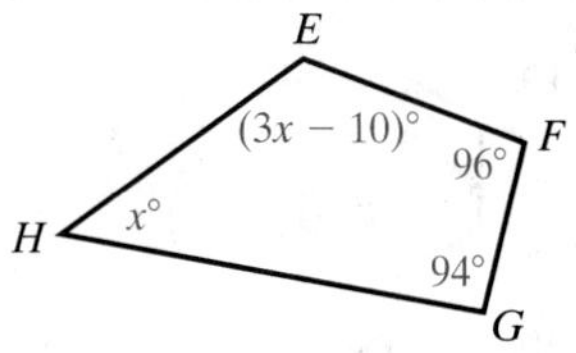

VOCABULARY & READINESS CHECK

Word Bank *Use the choices to fill in each blank.*

polygon	diagonal	convex
vertex	regular	equilateral
n-gon	concave	equiangular
quadrilateral		

1. A triangle and a quadrilateral are each special cases of a(n) ____________.
2. Each endpoint of a side of a polygon is called a(n) ____________.
3. A polygon with 4 sides is called a(n) ____________.
4. In general, a polygon with n sides is called a(n) ____________.
5. A polygon is ____________ if no line containing a side contains a point within the interior of the polygon.
6. A polygon is ____________, or nonconvex, if it is not convex.
7. A segment joining two nonconsecutive vertices of a convex polygon is called a(n) ____________ of the polygon.
8. A(n) ____________ polygon is a polygon with all sides congruent.
9. A(n) ____________ polygon is a polygon with all angles congruent.
10. A(n) ____________ polygon is a polygon that is equilateral and equiangular.

6.1 EXERCISE SET MyMathLab®

Identify each polygon. If not a polygon state why. See Example 1.

1. 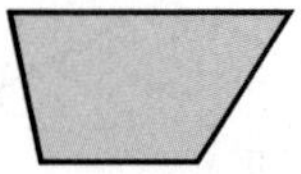

2. 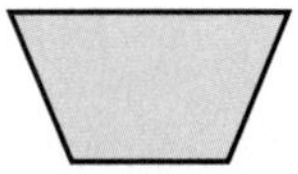

3. 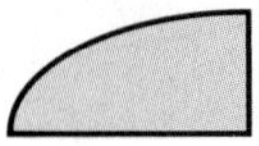

4. 

5. 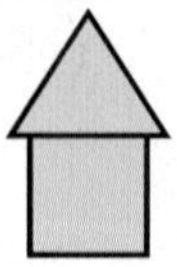

6. 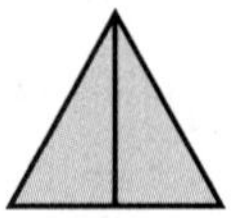

7. 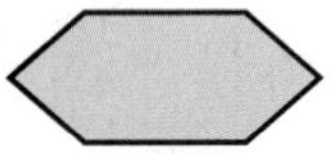

8. 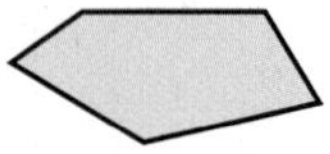

9. 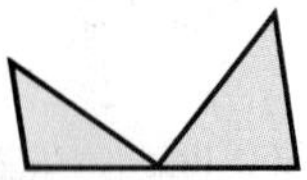

10. 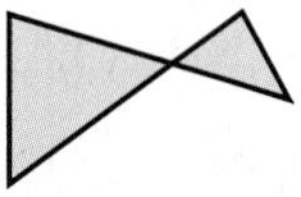

11. 

12. 

Name the polygon according to the number of sides. Then identify whether it is convex or concave. See Example 2.

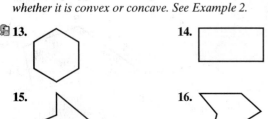

13. 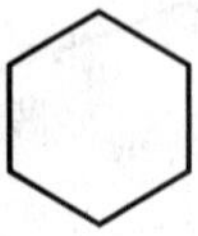

14. 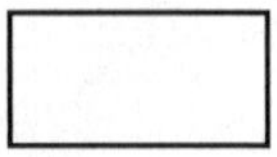

15. 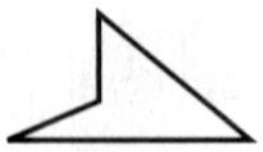

16. 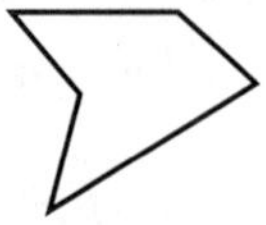

17. 

18. 

19. 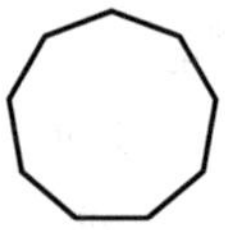

20. 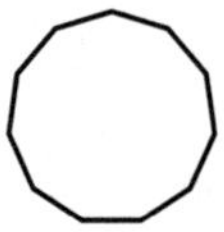

Multiple Steps *Answer the following for each traffic sign. (Note: When deciding whether a traffic sign is a polygon, disregard any slightly curved corners.) See Examples 1 and 3.*

a. Is the traffic sign a polygon? If yes, continue with parts **b–e**. If not a polygon, state why.
b. Name the polygon.
c. Is the polygon equiangular?
d. Is the polygon equilateral?
e. Is the polygon regular?

21. 

22. 

23. 

24. 

25. 

26. 

27. 

28. 

29. 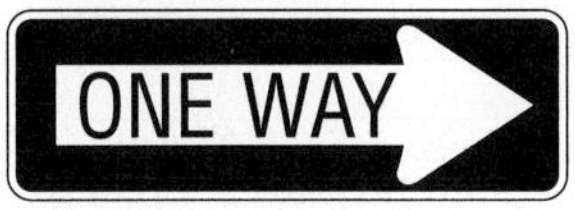

30. 

MIXED PRACTICE

Use polygon ABCDE to answer each of the following.

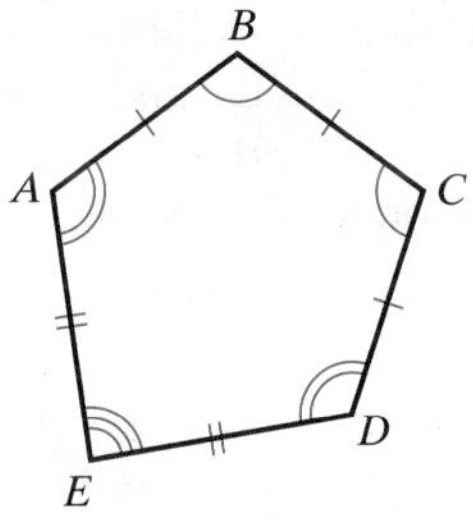

31. Name the vertices of the polygon.

32. Name the angles of the polygon.

33. Name the sides of the polygon.

34. Classify the polygon by the number of sides.

35. Classify the polygon as convex or concave.

36. Classify the polygon as regular or not regular.

37. Name the included side of $\angle C$ and $\angle D$.

38. Name the included angle of $\overline{BA}$ and $\overline{CB}$.

Find $m\angle C$ in each quadrilateral. See Example 4.

39. 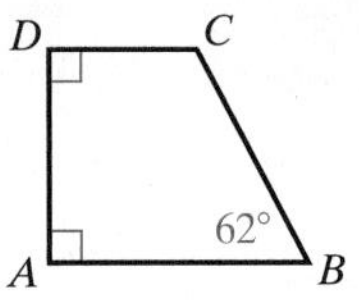

40. 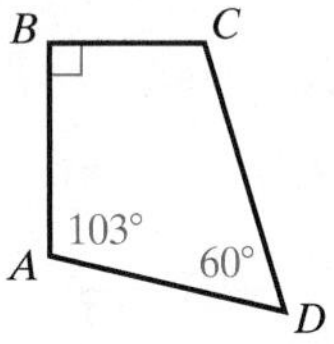

41. 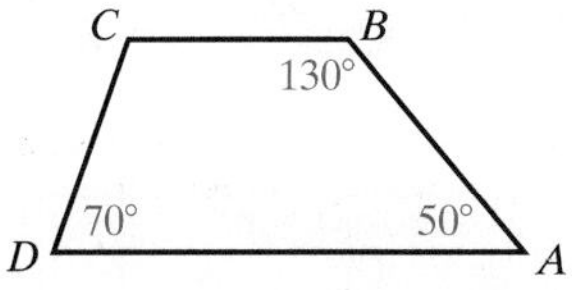

42. 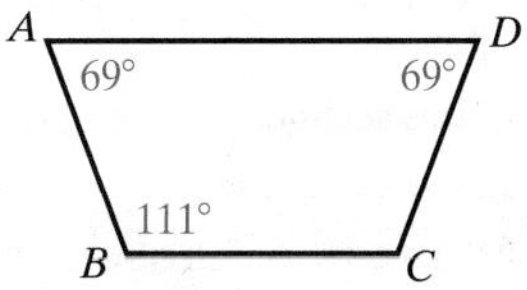

Find the value of each variable.

43. 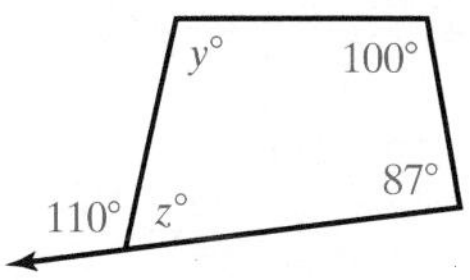

44. 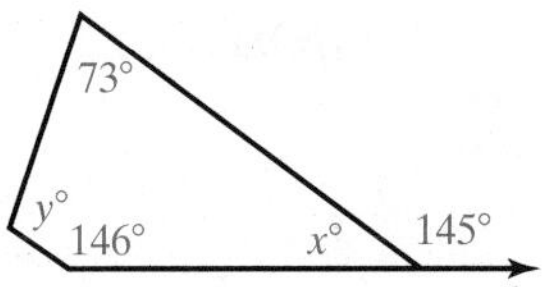

Find x, and then any unknown angles in each quadrilateral. See Example 4.

45. 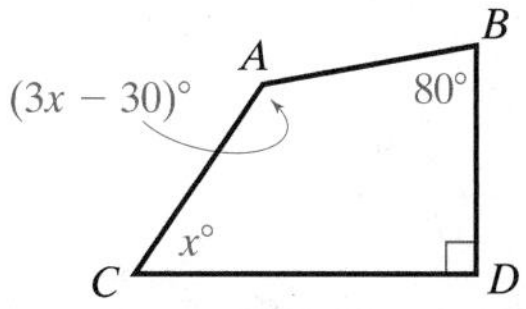

46. 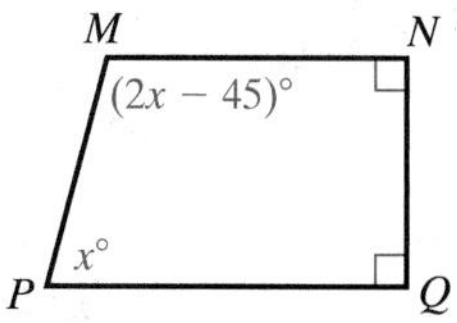

47. 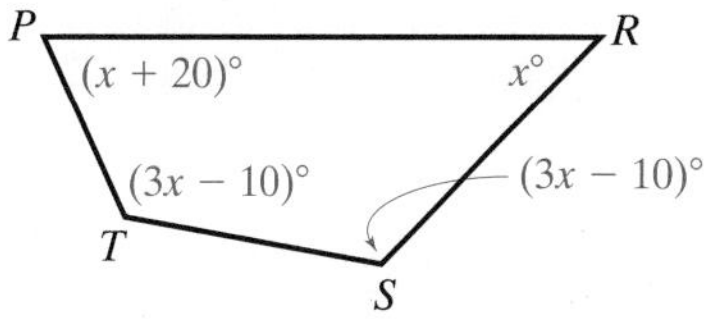

48. 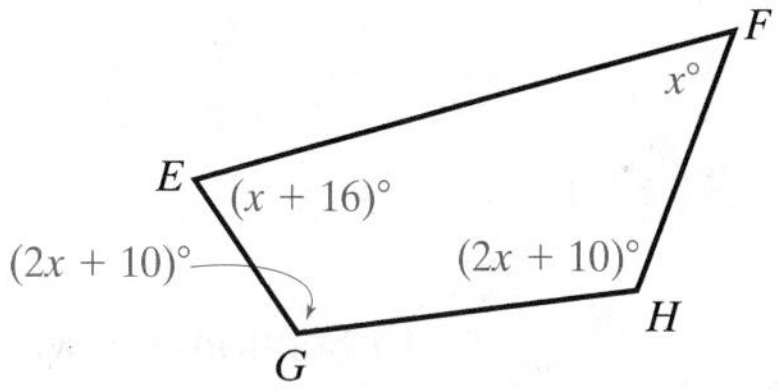

CONCEPT EXTENSIONS

The gift package at the right contains fruit and cheese. The fruit is in a container that has the shape of a regular octagon. The fruit container fits in a square box. A triangular cheese wedge fills each corner of the box.

49. Find the measure of each interior angle of a cheese wedge.

50. Show how to rearrange the four pieces of cheese to make a regular polygon.What is the measure of each interior angle of the polygon?

51. Your friend says she has another way to find the sum of the interior angle measures of a polygon. She picks a point inside the polygon, draws a segment to each vertex, and counts the number of triangles. She multiplies the total by 180, and then subtracts 360 from the product. Does her method work? Explain.

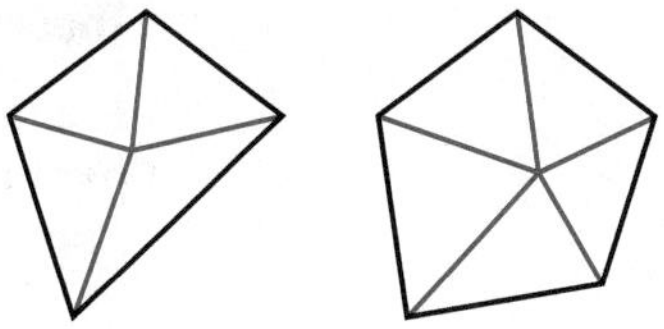

52. Sketch an equilateral polygon that is not equiangular.

REVIEW AND PREVIEW

53. If $\overline{AB} \cong \overline{CB}$ and $m\angle ABD < m\angle CBD$, which is longer, $\overline{AD}$ or $\overline{CD}$? Explain. See Sections 5.5 and 5.6.

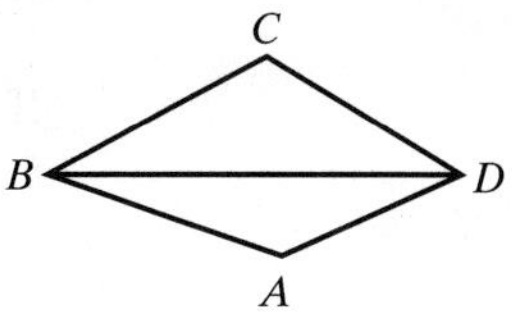

54. Fill in the Blank $m\angle 1 + m\angle 2$ ____ $m\angle 3$. See Section 4.1.

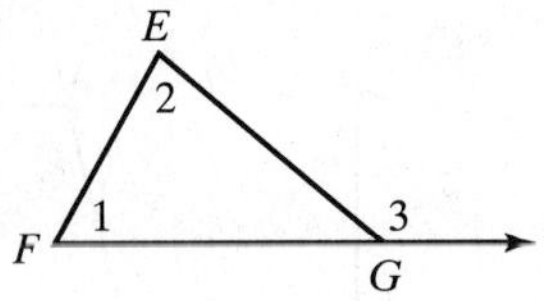

Name the property that justifies each statement. See Section 2.6.

55. $\overline{RS} \cong \overline{RS}$

56. If $\angle 1 \cong \angle 4$, then $\angle 4 \cong \angle 1$.

Use the figure below to answer Exercises 57 and 58. See Section 4.3.

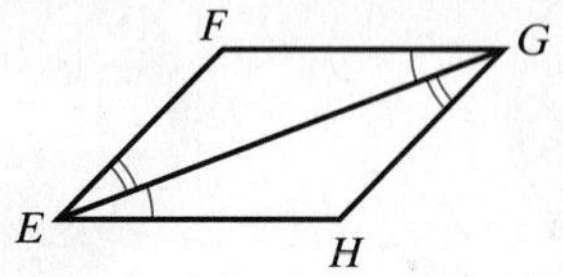

57. Name the postulate or theorem that justifies $\Delta EFG \cong \Delta GHE$.

58. Complete each statement.

a. $\angle FEG \cong$ ________ **b.** $\angle EFG \cong$ ________

c. $\angle FGE \cong$ ________ **d.** $\overline{EF} \cong$ ________

e. $\overline{FG} \cong$ ________ **f.** $\overline{GE} \cong$ ________

6.2 PARALLELOGRAMS

OBJECTIVES

1. Use Relationships Among Sides and Angles of Parallelograms.
2. Use Relationships Among Consecutive Angles and Diagonals of Parallelograms.

VOCABULARY

- parallelogram
- opposite sides
- opposite angles
- consecutive angles

In Section 6.1, we introduced polygons and concentrated on quadrilaterals, which are polygons with 4 sides. For the rest of this chapter, we study different types of quadrilaterals and their properties.

OBJECTIVE 1 ▶ Using Relationships Among Sides and Angles of Parallelograms. A **parallelogram** is a quadrilateral with both pairs of opposite sides parallel. Parallelograms have special properties regarding their sides, angles, and diagonals.

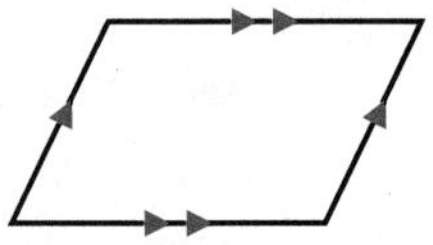

In a parallelogram (and in all other convex quadrilaterals), **opposite sides** do not share a vertex, and **opposite angles** do not share a side.

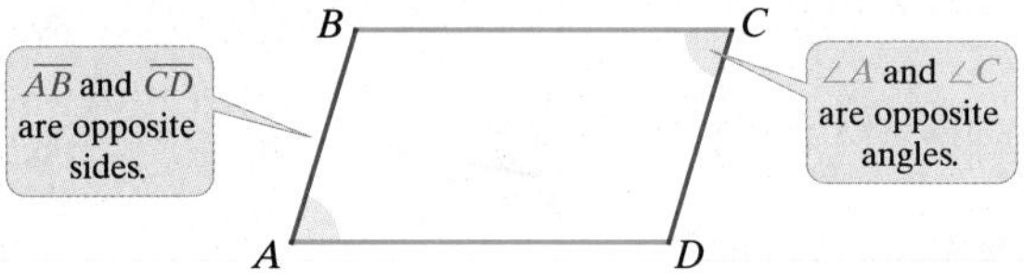

We can abbreviate *parallelogram* with the symbol ▱ and *parallelograms* with the symbol ▱s. In this section, we will use our knowledge of parallel lines and transversals to prove some basic theorems about parallelograms.

Theorem 6.2-1 Opposite Sides of a Parallelogram

Theorem	If . . .	Then . . .
If a quadrilateral is a parallelogram, then its opposite sides are congruent.	$ABCD$ is a ▱ 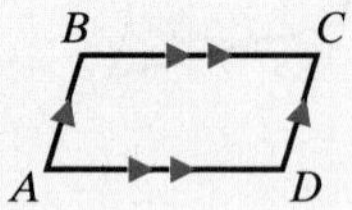	$\overline{AB} \cong \overline{CD}$ and $\overline{BC} \cong \overline{DA}$ 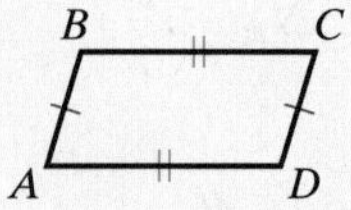

We prove this theorem in Example 1 of this section.

Theorem 6.2-2 Opposite Angles of a Parallelogram

Theorem	If . . .	Then . . .
If a quadrilateral is a parallelogram, then its opposite angles are congruent.	$ABCD$ is a ▱ 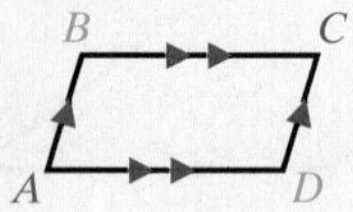	$\angle A \cong \angle C$ and $\angle B \cong \angle D$ 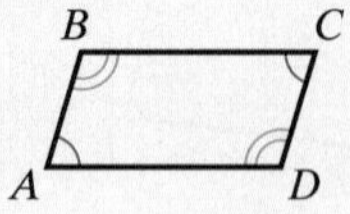

We prove this theorem in Practice 1 of this section.

EXAMPLE 1 **Proof of Theorem 6.2-1**

Use the given figure for a proof of Theorem 6.2-1.

Given: $\square ABCD$

Prove: $\overline{AB} \cong \overline{CD}$ and $\overline{BC} \cong \overline{DA}$

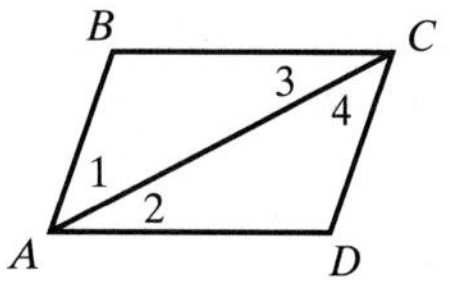

Solution

Statements	Reasons
1. $ABCD$ is a parallelogram.	**1.** Given
2. $\overline{AB} \parallel \overline{CD}$ and $\overline{BC} \parallel \overline{DA}$	**2.** Definition of parallelogram
3. $\angle 1 \cong \angle 4$ and $\angle 3 \cong \angle 2$	**3.** If lines are $\parallel$, then alternate interior $\angle$'s are $\cong$.
4. $\overline{AC} \cong \overline{AC}$	**4.** Reflexive Property of $\cong$
5. $\triangle ABC \cong \triangle CDA$	**5.** ASA
6. $\overline{AB} \cong \overline{CD}$ and $\overline{BC} \cong \overline{DA}$	**6.** Corresponding, parts of $\cong$ $\triangle$s are $\cong$. (cpoctac)

□

PRACTICE 1 Use the diagram to prove Theorem 6.2-2.

Given: $\square ABCD$ ($ABCD$ is a parallelogram)

Prove: $\angle A \cong \angle C$ and $\angle B \cong \angle D$

Construction: Draw $\overline{AC}$.

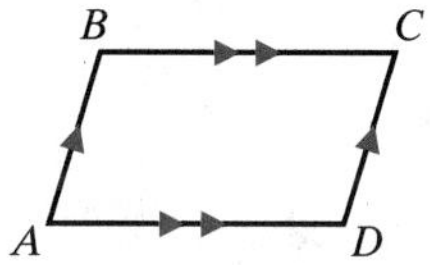

Hint: Use transversal $\overline{AC}$ and your knowledge of alternate interior angles to prove $\angle A \cong \angle C$.

Next, we define consecutive angles of a polygon. Once defined, we have another Hint (method to prove) for Practice 1 that we have written in the margin.

> **Helpful Hint**
> Another Hint for Practice 1:
> Consecutive angles of a parallelogram are the same-side interior angles of parallel lines and are thus supplementary. Then recall that supplements of the same angle are congruent.

OBJECTIVE 2 ▶ Using Relationships Among Consecutive Angles and Diagonals of Parallelograms. Angles of a polygon that share a side are **consecutive angles.** In the diagram, $\angle A$ and $\angle B$ are consecutive angles because they share side $\overline{AB}$.

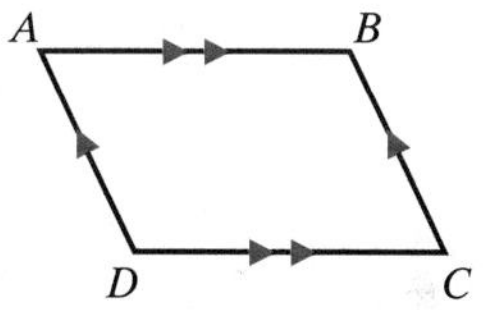

$\angle B$ and $\angle C$ are also consecutive angles.

Theorem 6.2-3 below uses the fact that consecutive angles of a parallelogram are same-side interior angles of parallel lines.

Also, the diagonals of parallelograms have a special property, as stated in Theorem 6.2-4.

Theorem 6.2-3 Consecutive Angles of a Parallelogram

Theorem	If . . .	Then . . .
If a quadrilateral is a parallelogram, then its consecutive angles are supplementary.	$ABCD$ is a $\square$ 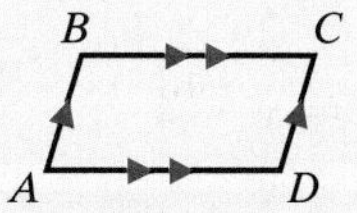	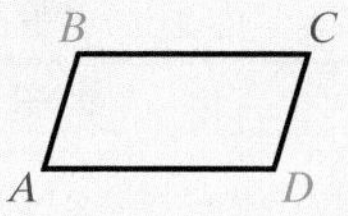 $m\angle A + m\angle B = 180°$ $m\angle B + m\angle C = 180°$ $m\angle C + m\angle D = 180°$ $m\angle D + m\angle A = 180°$

Theorem 6.2-4 Diagonals of a Parallelogram

Theorem	If . . .	Then . . .
If a quadrilateral is a parallelogram, then its diagonals bisect each other.	$ABCD$ is a $\square$ 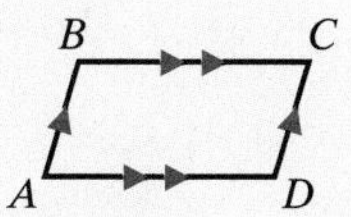	$\overline{AE} \cong \overline{CE}$ and $\overline{BE} \cong \overline{DE}$ 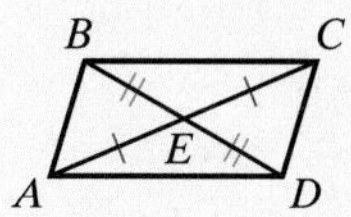

We prove these theorems in Exercises 52–53, Exercise Set 6.2.

Helpful Hint
We know that consecutive angles of a parallelogram are supplementary. We can use this to write an equation and solve for $m\angle P$.

EXAMPLE 2 Using Consecutive Angles

Multiple Choice What is $m\angle P$ in $\square PQRS$?

a. 26° **b.** 64° **c.** 116° **d.** 126°

Solution $\square PQRS$ means parallelogram $PQRS$.

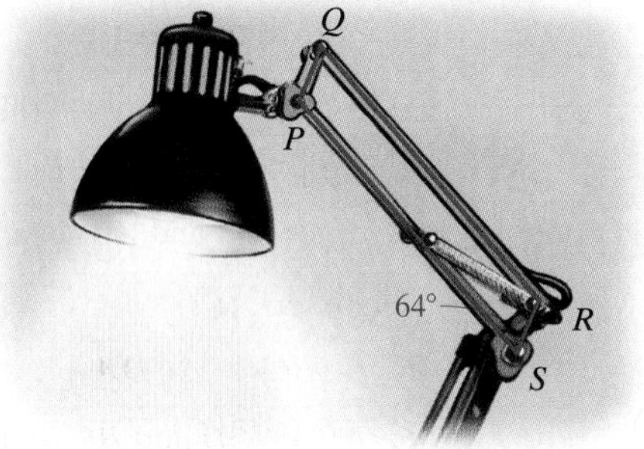

$m\angle P + m\angle S = 180°$ Consecutive angles of a $\square$ are supplementary.

$m\angle P + 64° = 180°$ Substitute.

$m\angle P = 116°$ Subtract 64° from each side.

The correct answer is **c.** □

PRACTICE 2 Suppose you adjust the lamp so that $m\angle S = 86°$. What is $m\angle R$ in $\square PQRS$?

We can use Theorem 6.2-4 to find unknown lengths in parallelograms.

EXAMPLE 3 Using Algebra to Find Lengths

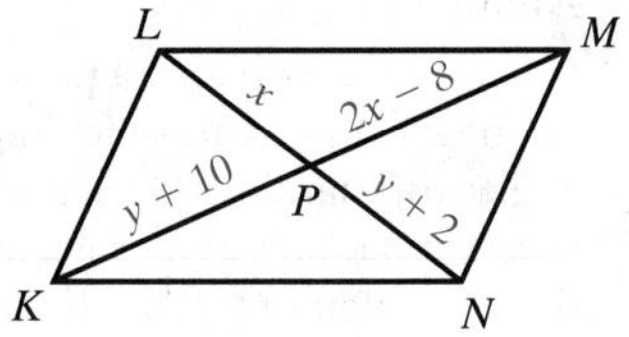

Solve a system of linear equations to find the values of x and y in $\square KLMN$. What are KM and LN?

Solution To find these lengths, we set up a system of two equations and solve by substitution.

Equation ① $\begin{cases} KP = MP \\ LP = NP \end{cases}$ Equation ② — The diagonals of a parallelogram bisect each other (Theorem 6.2-4).

① $y + 10 = 2x - 8$ Substitute the given expressions for KP and MP.

② $x = y + 2$ Substitute the given expressions for LP and NP.

① $y + 10 = 2x - 8$ Write Equation ①.

$y + 10 = 2(y + 2) - 8$ Substitute $y + 2$ for x from Equation ②.

$y + 10 = 2y + 4 - 8$ Distributive property

$y + 10 = 2y - 4$ Simplify.

$10 = y - 4$ Subtract y from both sides.

$14 = y$ Add 4 to both sides.

To find the value of x, choose Equation ① or Equation ② and substitute 14 for y. We choose Equation ②.

$x = y + 2$ Write Equation ②.

$x = 14 + 2$ Substitute 14 for y.

$= 16$

Now, let's find KM and LN. By looking at the parallelogram, we see that $KM = 2(KP)$ and $LN = 2(LP)$.

$$\begin{aligned} KM &= 2(KP) \\ &= 2(y + 10) \\ &= 2(14 + 10) \\ &= 48 \end{aligned} \qquad \begin{aligned} LN &= 2(LP) \\ &= 2(x) \\ &= 2(16) \\ &= 32 \end{aligned}$$

Thus, $x = 16$ and $y = 14$. Also, $KM = 48$ units and $LN = 32$ units. □

PRACTICE
3 Find the values of x and y in ▱$PQRS$ at the right. What are PR and SQ?

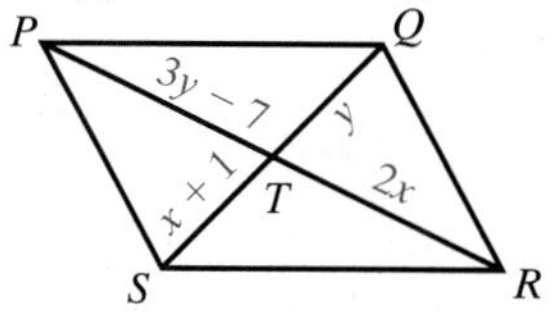

We can use parallelograms to prove the following theorem.

Theorem 6.2-5 Multiple Parallel Lines and Transversals

Theorem	If . . .	Then . . .
If three (or more) parallel lines cut off congruent segments on one transversal, then they cut off congruent segments on every transversal.	$\overleftrightarrow{AB} \parallel \overleftrightarrow{CD} \parallel \overleftrightarrow{EF}$ and $\overline{AC} \cong \overline{CE}$ 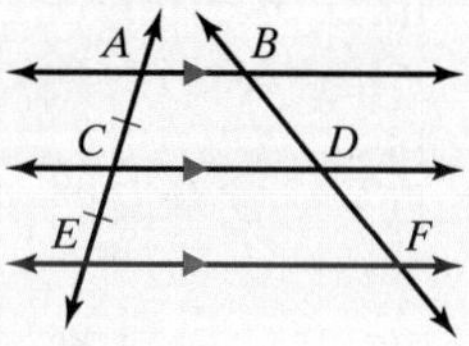	$\overline{BD} \cong \overline{DF}$ 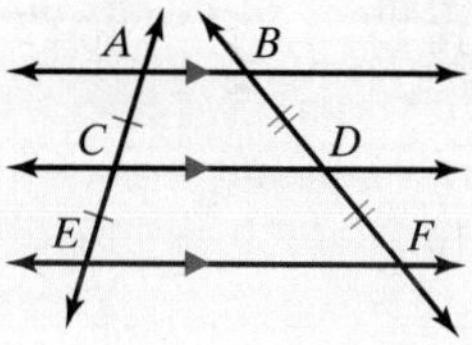

We prove this theorem in Exercise 61, Exercise Set 6.2.

EXAMPLE 4 **Using Parallel Lines and Transversals**

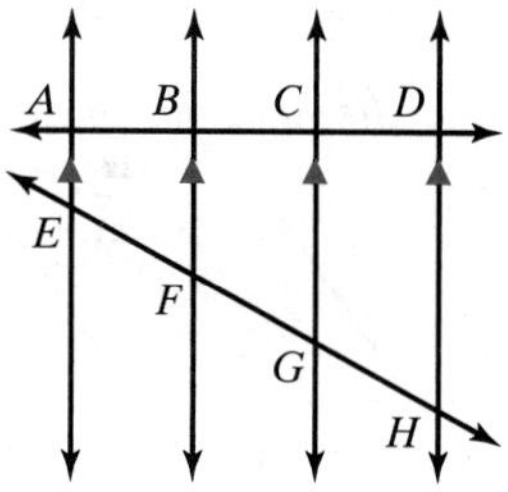

In the figure, $\overleftrightarrow{AE} \parallel \overleftrightarrow{BF} \parallel \overleftrightarrow{CG} \parallel \overleftrightarrow{DH}$, $AB = BC = CD = 2$, and $EF = 2.25$. What is EH?

Solution We are given the length of $\overline{EF}$. Thus, to find EH, we need the length of $\overline{FG}$ and $\overline{GH}$.

$EF = FG = GH$	Since ∥ lines divide $\overline{AD}$ into equal parts, they also divide $\overline{EH}$ into equal parts.
$2.25 = FG = GH$	Substitute 2.25 for EF.
$EH = EF + FG + GH$	Segment Addition Postulate
$EH = 2.25 + 2.25 + 2.25 = 6.75$	Substitute.

Thus, $EH = 6.75$ units. □

PRACTICE
4 Use the figure in Example 4. If $EF = FG = GH = 6$ and $AD = 15$, what is CD?

VOCABULARY & READINESS CHECK

Word Bank *Use the choices to fill in each blank.*

opposite sides   opposite angles   consecutive angles   parallelogram

1. In a quadrilateral, ___________ do not share a side.
2. In a quadrilateral, ___________ do not share a vertex.
3. Angles of a polygon that share a side are called ___________.
4. A quadrilateral with both pairs of opposite sides parallel is called a(n) ___________.

6.2 EXERCISE SET MyMathLab®

Multiple Choice *Use ▱MNQP and ▱EFGH to find each value. See Examples 1 and 2.*

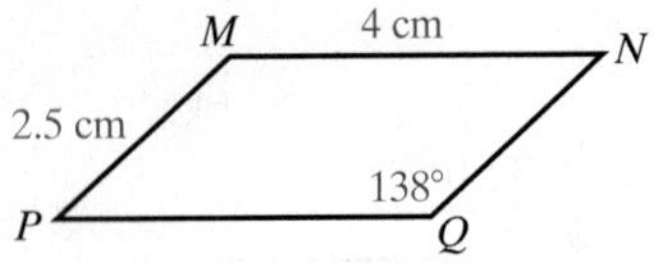

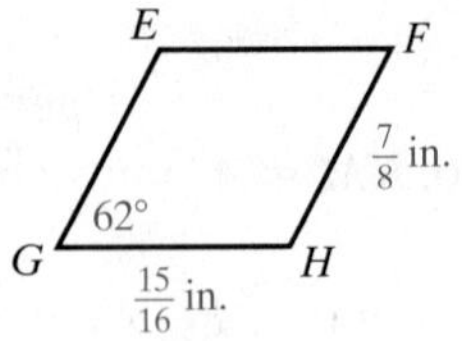

1. $m\angle M =$
   a. 138° b. 42° c. 222° d. 48°
2. $m\angle P =$
   a. 138° b. 42° c. 222° d. 48°
3. $NQ =$
   a. 6.5 cm b. 1.5 cm c. 4 cm d. 2.5 cm
4. $PQ =$
   a. 6.5 cm b. 1.5 cm c. 4 cm d. 2.5 cm
5. $EF =$
   a. $\frac{1}{16}$ in. b. $1\frac{13}{16}$ in. c. $\frac{15}{16}$ in. d. $\frac{7}{8}$ in.
6. $GE =$
   a. $\frac{1}{16}$ in. b. $1\frac{13}{16}$ in. c. $\frac{15}{16}$ in. d. $\frac{7}{8}$ in.
7. $m\angle H =$
   a. 28° b. 298° c. 62° d. 118°
8. $m\angle F =$
   a. 28° b. 298° c. 62° d. 118°

Find the value of x in each parallelogram. See Example 2.

9. 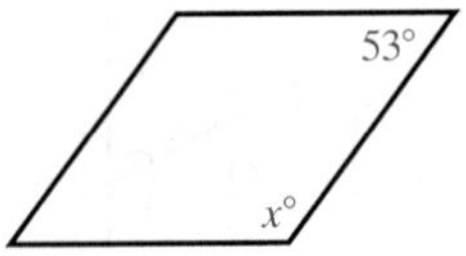

10. 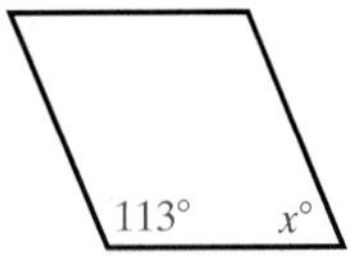

11. 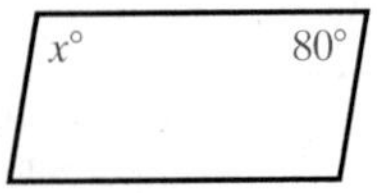

12. 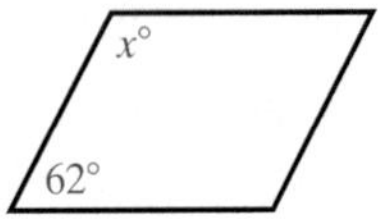

Find the value of the variable in each parallelogram. See Example 2.

13. 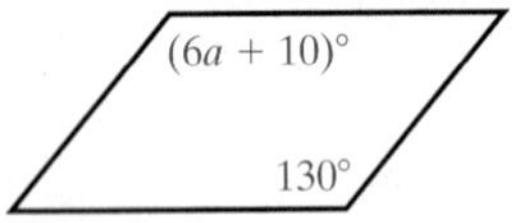

14. 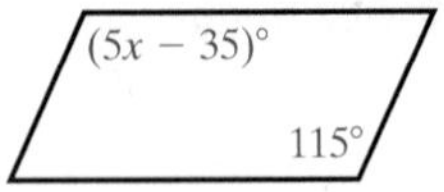

15. 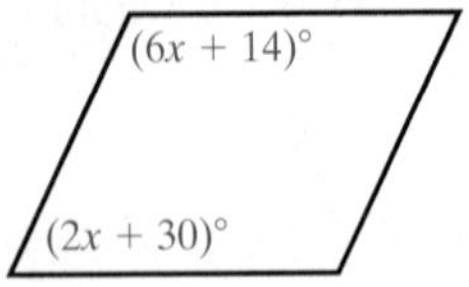

16. 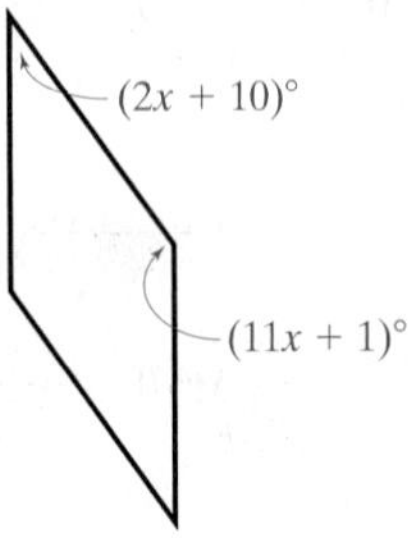

Use the diagram of ▱ABCD to find each value. See Examples 1 and 2.

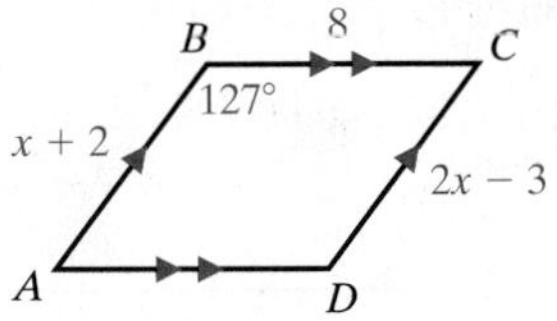

17. $m\angle A$
18. $m\angle D$
19. x
20. AB
21. AD
22. CD

Find the measures of the numbered angles for each parallelogram. See Examples 1–3.

23. 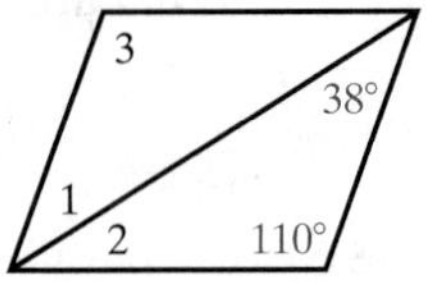

24. 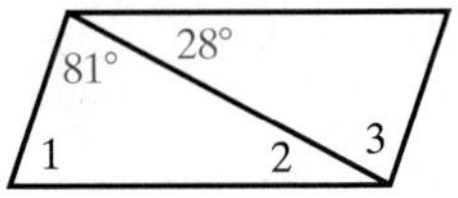

Find each measure in ▱ABDC. See Examples 1–3.

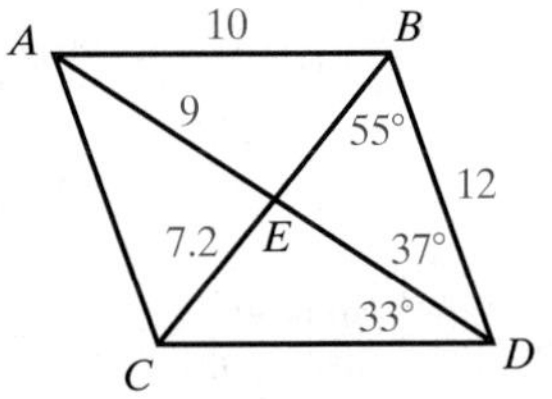

25. AD
26. EB
27. CD
28. AC
29. $m\angle EAB$
30. $m\angle BCA$
31. $m\angle BAC$
32. $m\angle ABD$
33. $m\angle BED$
34. $m\angle AEC$

In the figure, $PQ = QR = RS$. Find each length. See Example 4.

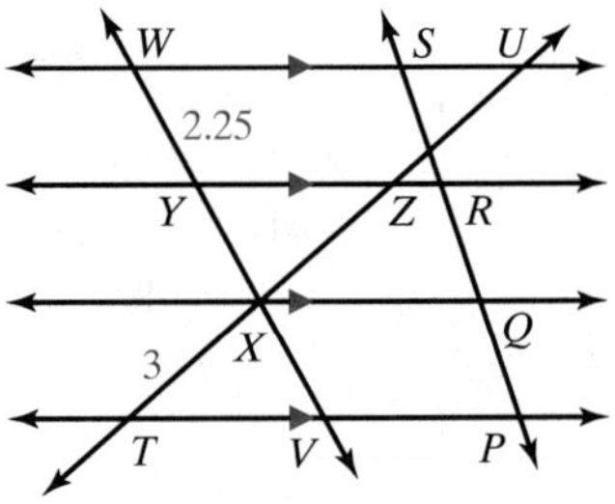

35. ZU
36. XZ
37. TU
38. XV
39. YX
40. YV
41. WX
42. WV

Find the values of x and y in ▱PQRS. See Example 3.

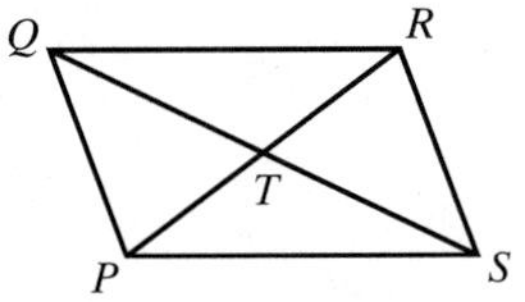

43. $PT = 2x$, $TR = y + 4$, $QT = x + 2$, $TS = y$
44. $PT = x + 2$, $TR = y$, $QT = 2x$, $TS = y + 3$

Find the value of a. Then find each side length or angle measure.

45. 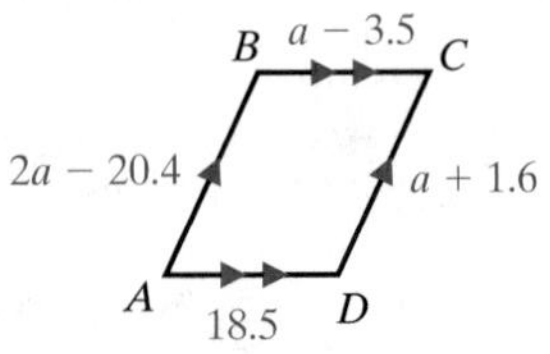

46. 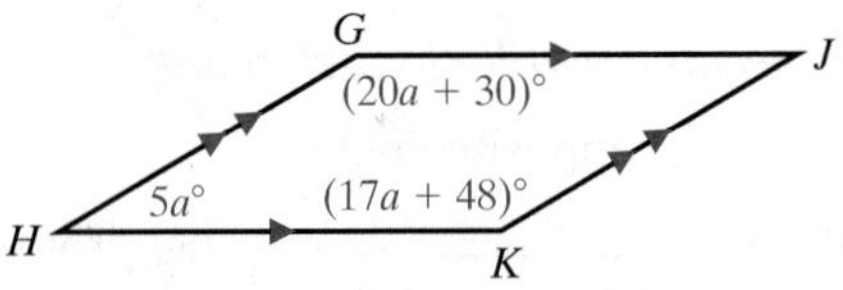

47. A pantograph is the expandable device shown below. Pantographs are used in the television industry in positioning lighting and other equipment. In the photo, points D, E, F, and G are the vertices of a parallelogram. ▱$DEFG$ is one of many parallelograms that change shape as the pantograph extends and retracts.

a. If $DE = 2.5$ ft, what is FG?

b. If $m\angle E = 129°$, what is $m\angle G$?

c. What happens to $m\angle D$ as $m\angle E$ increases or decreases? Explain.

48. Use the diagram given.

a. What are the values of x and y in the parallelogram?

b. How are the angles related?

c. Which variable should you solve for first?

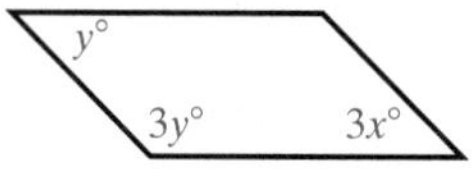

CONCEPT EXTENSIONS

49. If you know one angle measure of a parallelogram, how do you find the other three angle measures? Explain.

50. What is similar and what is different between a quadrilateral and a parallelogram?

51. Find the Error Your classmate says that $QV = 10$. Explain why the statement may not be correct.

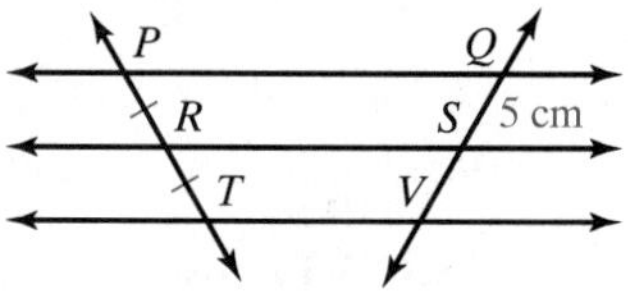

52. Proof Complete this two-column proof of Theorem 6.2-4.

Given: ▱$ABCD$

Prove: $\overline{AC}$ and $\overline{BD}$ bisect each other at E.

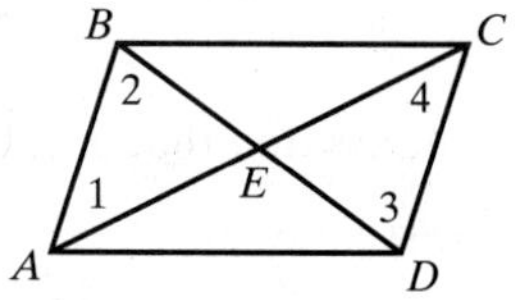

Statements	Reasons
1. $ABCD$ is a parallelogram.	1. Given
2. $\overline{AB} \parallel \overline{DC}$	2. a. ___?___
3. $\angle 1 \cong \angle 4$; $\angle 2 \cong \angle 3$	3. b. ___?___
4. $\overline{AB} \cong \overline{DC}$	4. c. ___?___
5. d. ___?___	5. ASA
6. $\overline{AE} \cong \overline{CE}$; $\overline{BE} \cong \overline{DE}$	6. e. ___?___
7. f. ___?___	7. Definition of bisector

53. Proof Prove Theorem 6.2-3.

Given: ▱$ABCD$

Prove: $\angle A$ is supplementary to $\angle B$. $\angle A$ is supplementary to $\angle D$.

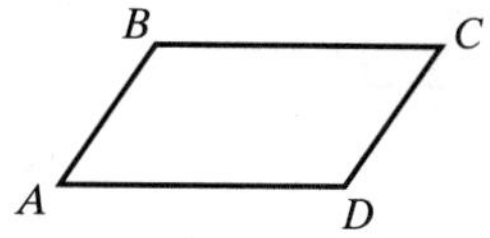

Proof *Use the diagram at the right for each proof.*

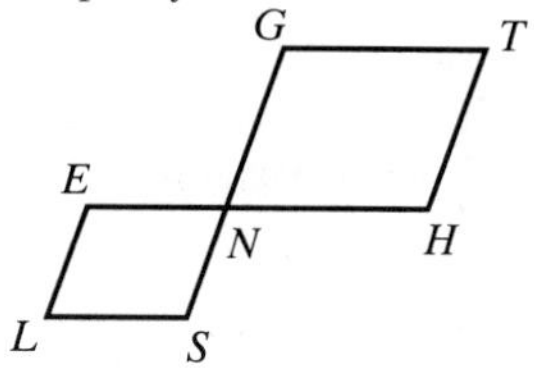

54. Given: ▱$LENS$ and ▱$NGTH$

Prove: $\angle L \cong \angle T$

55. Given: ▱$LENS$ and ▱$NGTH$

Prove: $\overline{LS} \parallel \overline{GT}$

56. Given: ▱$LENS$ and ▱$NGTH$

Prove: $\angle E$ is supplementary to $\angle T$.

Proof *Use the diagram at the right for each proof.*

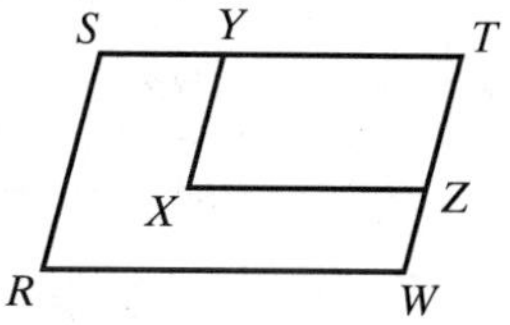

57. Given: ▱$RSTW$ and ▱$XYTZ$

Prove: $\angle R \cong \angle X$

58. Given: ▱$RSTW$ and ▱$XYTZ$

Prove: $\overline{XY} \parallel \overline{RS}$

Solve.

59. The perimeter of ▱$ABCD$ is 92 cm. AD is 7 cm more than twice AB. Find the lengths of all four sides of ▱$ABCD$.

60. Is there an SSSS congruence theorem for parallelograms? Explain.

61. Proof Prove Theorem 6.2-5. Use the diagram below.

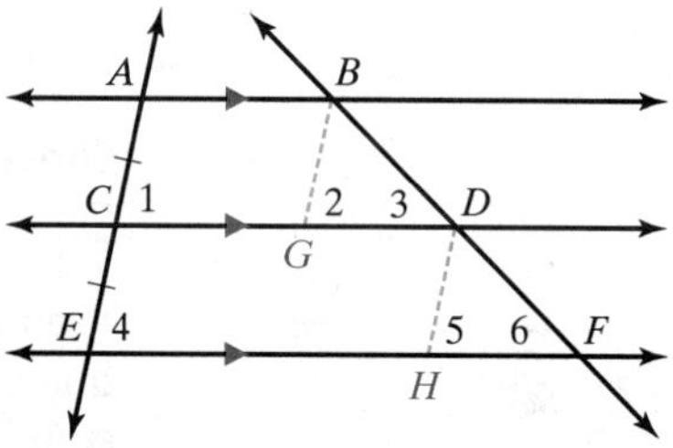

Given: $\overleftrightarrow{AB} \parallel \overleftrightarrow{CD} \parallel \overleftrightarrow{EF}$, $\overline{AC} \cong \overline{CE}$

Prove: $\overline{BD} \cong \overline{DF}$

(*Hint:* Draw lines through B and D parallel to $\overleftrightarrow{AE}$ and intersecting $\overleftrightarrow{CD}$ at G and $\overleftrightarrow{EF}$ at H.)

62. Explain how to separate a blank card into three strips that are the same height by using lined paper, a straight edge, and Theorem 6.2-5.

REVIEW AND PREVIEW

63. What additional information do you need to prove $\triangle ADC \cong \triangle ABC$ by the H-L Theorem? See Section 4.6.

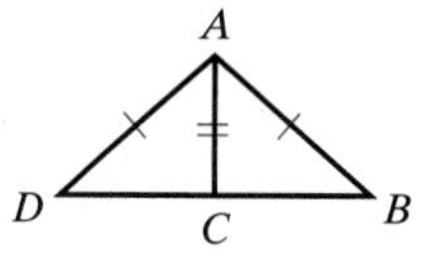

64. Two consecutive angles in a parallelogram have measures $x + 5$ and $4x - 10$. Find the measure of the smaller angle. See Section 6.2.

6.3 PROVING THAT A QUADRILATERAL IS A PARALLELOGRAM

OBJECTIVES

1. Determine Whether a Quadrilateral Is a Parallelogram.
2. Use Coordinate Geometry with Parallelograms.

OBJECTIVE 1 ▶ Determining Whether a Quadrilateral Is Also a Parallelogram. In Section 6-2, we learned theorems about the properties of parallelograms. In this section, we learn that the converses of those theorems are also true. That is, if a quadrilateral has certain properties, then it must be a parallelogram.

Theorems 6.3-1, 6.3-2, and 6.3-3 are the converses of Theorems 6.2-1, 6.2-2, and 6.2-3 respectively.

Recall Theorem 6.2-1 states that if a quadrilateral is a parallelogram, then its opposite sides are congruent.

▶ Helpful Hint
Don't forget that the converse of $p \to q$ is $q \to p$.

The converse of Theorem 6.2-1

Theorem 6.3-1 Converse of Opposite Sides Theorem

Theorem
If **both pairs of opposite sides** of a quadrilateral are **congruent,** then the quadrilateral is a parallelogram.

If . . .

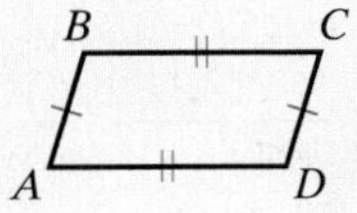

$\overline{AB} \cong \overline{CD}$
$\overline{BC} \cong \overline{DA}$

Then . . .
$ABCD$ is a ▱

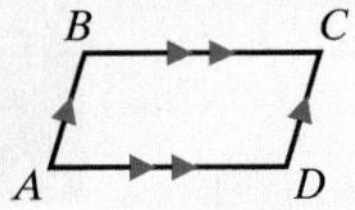

The converse of Theorem 6.2-2

Theorem 6.3-2 Converse of Opposite Angles Theorem

Theorem
If **both pairs of opposite angles** of a quadrilateral are **congruent,** then the quadrilateral is a parallelogram.

If . . .

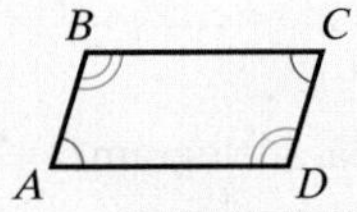

$\angle A \cong \angle C$
$\angle B \cong \angle D$

Then . . .
$ABCD$ is a ▱

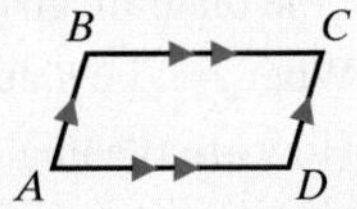

The converse of Theorem 6.2-3

Theorem 6.3-3 Converse of Consecutive Angles Theorem

Theorem
If an **angle** of a quadrilateral **is supplementary** to both of its consecutive angles, then the quadrilateral is a parallelogram.

If . . .

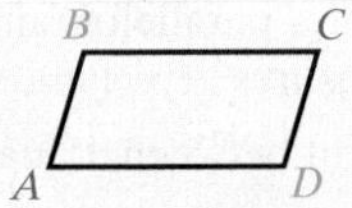

$m\angle A + m\angle B = 180$
$m\angle A + m\angle D = 180$

Then . . .
$ABCD$ is a ▱

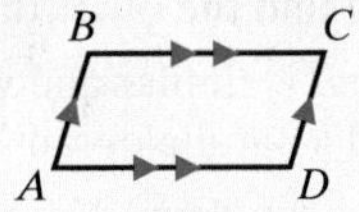

We prove these three theorems in Exercise Set 6.3, Exercises 35, 33, and 36, respectively.

We can use algebra together with Theorems 6.3-1, 6.3-2, and 6.3-3 to find segment lengths and angle measures that assume that a quadrilateral is a parallelogram.

EXAMPLE 1 **Finding Values of Variables in Parallelograms**

For what value of y must $PQRS$ be a parallelogram?

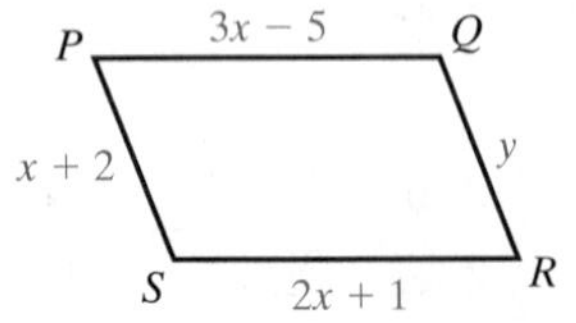

Solution Since we are given side values, we look for a value of y that makes both pairs of opposite sides congruent.

Since $PQ = SR$, if this is a parallelogram, and both these opposite sides have expressions in x, we start here.

STEP 1. Find x.

$3x - 5 = 2x + 1$	If opposite sides are $\cong$, then the quadrilateral is a ▱.
$x - 5 = 1$	Subtract $2x$ from each side.
$x = 6$	Add 5 to each side.

STEP 2. Find y. To do so, we let $QR = PS$.

$y = x + 2$	If opposite sides are $\cong$, then the quadrilateral is a ▱.
$= 6 + 2$	Substitute 6 for x.
$= 8$	Simplify.

For $PQRS$ to be a parallelogram, the value of y must be 8. □

PRACTICE

1 Use the diagram below. For what values of x and y must $EFGH$ be a parallelogram? (*Hint:* Use Theorem 6.3-3 and make sure you check to see that opposite angles are congruent.)

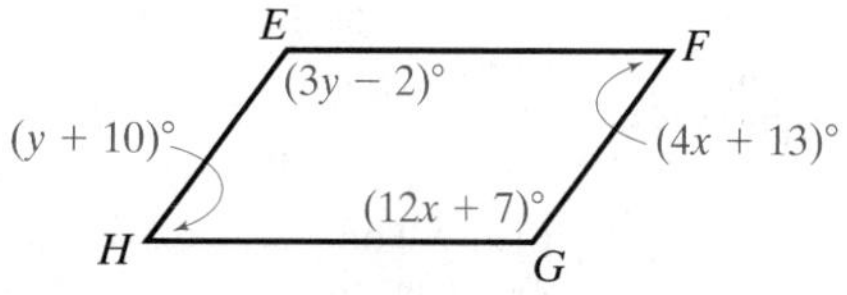

The converse of Theorem 6.2-4, from Section 6.2, is also true.

The converse of Theorem 6.2-4

Theorem 6.3-4 Converse of the Diagonals Theorem

Theorem	If . . .		Then . . .
If the **diagonals of a quadrilateral bisect each other,** then the quadrilateral is a parallelogram.	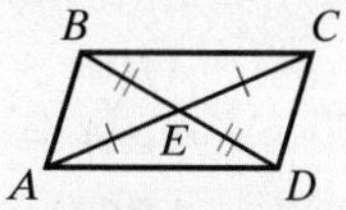	$\overline{AE} \cong \overline{CE}$ $\overline{BE} \cong \overline{DE}$	$ABCD$ is a ▱ 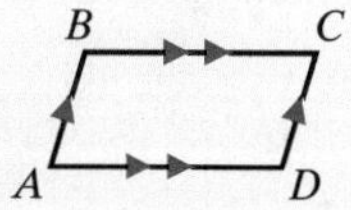

The proof of this theorem is shown below using a flow-proof format.

Proof of Theorem 6.3-4

Given: $\overline{AC}$ and $\overline{BD}$ bisect each other at E.

Prove: $ABCD$ is a parallelogram.

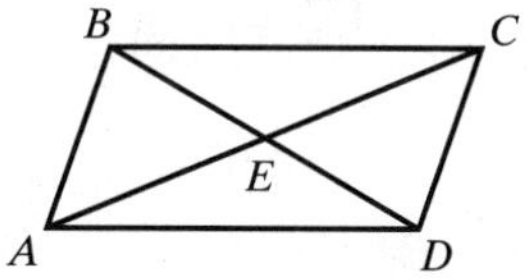

$\overline{AC}$ and $\overline{BD}$ bisect each other at E.
Given

$\angle AEB \cong \angle CED$
Vertical ∠s are $\cong$.

$\overline{AE} \cong \overline{CE}$, $\overline{BE} \cong \overline{DE}$
Definition of segment bisector

$\angle BEC \cong \angle DEA$
Vertical ∠s are $\cong$.

$\triangle AEB \cong \triangle CED$
SAS

$\triangle BEC \cong \triangle DEA$
SAS

$\angle BAE \cong \angle DCE$
Corresponding parts of $\cong$ △s are $\cong$.

$\angle ECB \cong \angle EAD$
Corresponding parts of $\cong$ △s are $\cong$.

$\overline{AB} \parallel \overline{CD}$
If alternate interior ∠s $\cong$, then lines are $\parallel$.

$\overline{BC} \parallel \overline{AD}$
If alternate interior ∠s $\cong$, then lines are $\parallel$.

$ABCD$ is a parallelogram.
Definition of parallelogram

The next theorem, Theorem 6.3-5, suggests that if you keep two objects of the same length parallel, such as cross-country skis, then the quadrilateral formed by connecting their endpoints is always a parallelogram.

Theorem 6.3-5 Quadrilateral as a Parallelogram

Theorem	If . . .		Then . . .
If **one pair of opposite sides** of a quadrilateral is both **congruent and parallel,** then the quadrilateral is a parallelogram.	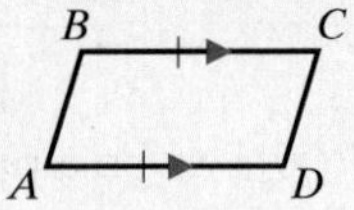	$\overline{BC} \cong \overline{AD}$ $\overline{BC} \parallel \overline{AD}$	*ABCD* is a ▱ 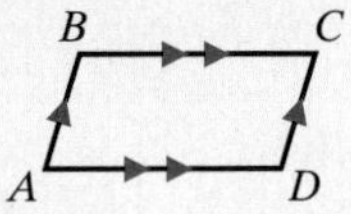

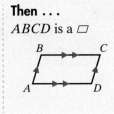

We prove this theorem in Exercise 34, Section 6.3.

Helpful Hint

Study Theorems 6.3-1 through 6.3-5 in this section so that you will know when you have enough information to show that a quadrilateral is also a parallelogram.

EXAMPLE 2 **Deciding Whether a Quadrilateral Is a Parallelogram**

Can you prove that the quadrilateral is a parallelogram based on the given information? Explain why or why not.

a. Given: $AB = 5$, $CD = 5$, $m\angle A = 50°$, $m\angle D = 130°$

Prove: *ABCD* is a parallelogram.

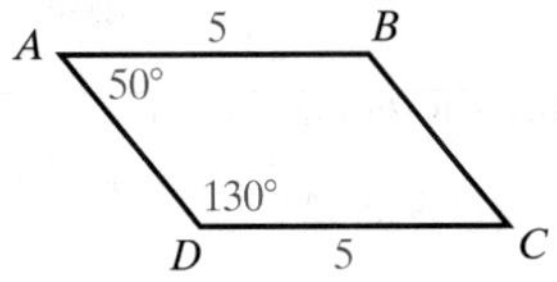

b. Given: $\overline{HI} \cong \overline{HK}$, $\overline{JI} \cong \overline{JK}$

Prove: *HIJK* is a parallelogram.

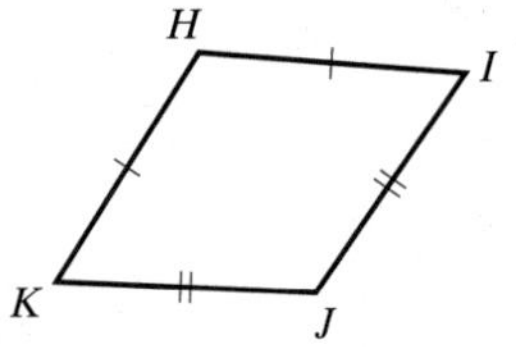

Solution

a. Yes. Same-side interior angles *A* and *D* are supplementary, so $\overline{AB} \parallel \overline{CD}$. Since $\overline{AB} \cong \overline{CD}$, *ABCD* is a parallelogram by Theorem 6.3-5.

Solution

b. No. By Theorem 6.3-1 you need to show that both pairs of *opposite* sides are congruent, not consecutive sides. □

PRACTICE 2 Can you prove that the quadrilateral is a parallelogram based on the given information? Explain why or why not.

a. Given: $\overline{EF} \cong \overline{DG}$, $\overline{DE} \parallel \overline{GF}$

Prove: *DEFG* is a parallelogram.

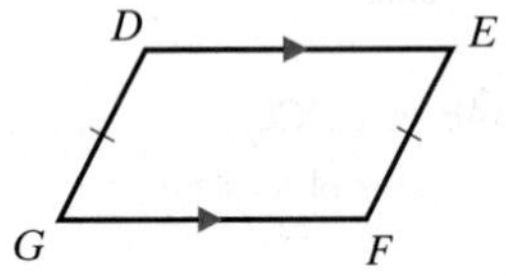

b. Given: $\angle ALN \cong \angle DNL$, $\angle ANL \cong \angle DLN$

Prove: *LAND* is a parallelogram.

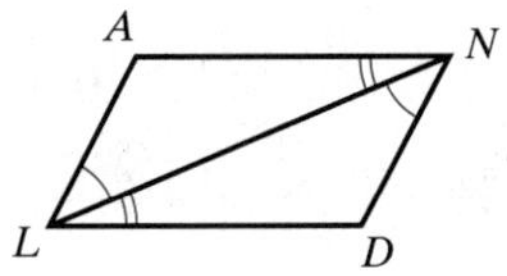

EXAMPLE 3 Identifying Parallelograms

A truck sits on the platform of a vehicle lift. Two moving arms raise the platform until a mechanic can fit underneath. Why will the truck always remain parallel to the ground as it is lifted? Explain.

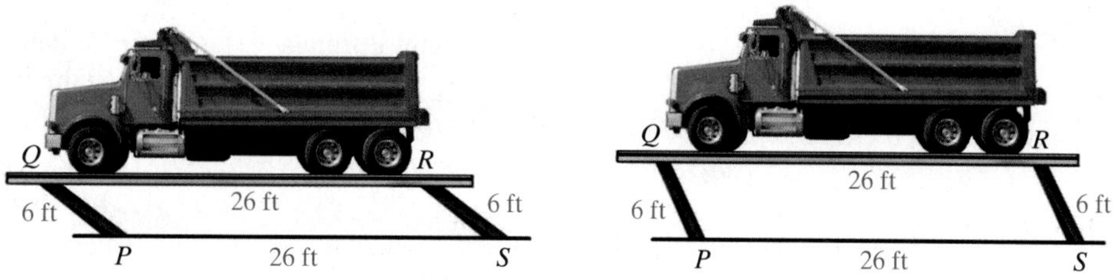

Solution The angles of $PQRS$ change as platform $\overline{QR}$ rises, but its side lengths remain the same. Both pairs of opposite sides are congruent, so $PQRS$ is a parallelogram by Theorem 6.3-1. Then, by the definition of a parallelogram, $\overline{PS} \parallel \overline{QR}$. Since the base of the lift $\overline{PS}$ lies along the ground, platform $\overline{QR}$, and therefore the truck, will always be parallel to the ground. □

PRACTICE

3 What is the maximum height that the vehicle lift can elevate the truck? Explain.

In the Helpful Hint below, we review the contents of the theorems in this section.

Helpful Hint Review—Proving that a Quadrilateral Is a Parallelogram Summary

	Diagram
• Prove that both pairs of opposite sides are parallel (definition of parallelogram).	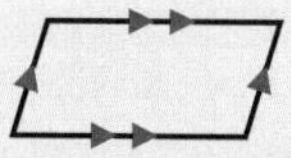
• Prove that both pairs of opposite sides are congruent.	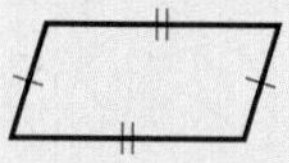
• Prove that an angle is supplementary to both of its consecutive angles.	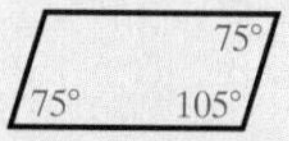
• Prove that both pairs of opposite angles are congruent.	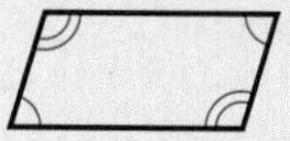
• Prove that the diagonals bisect each other.	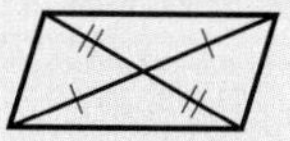
• Prove that one pair of opposite sides is congruent and parallel.	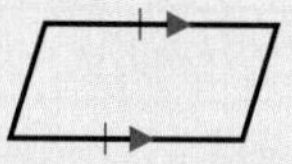

OBJECTIVE 2 ▶ Using the Coordinate Plane with Parallelograms. Let's begin by reviewing some important formulas in the coordinate plane that we may need.

Helpful Hint Review—Important Formulas in the Coordinate Plane

Formula	*When to Use It*
Distance Formula $d = \sqrt{(x_2 - x_1)^2 + (y_2 - y_1)^2}$	• To determine the distance between two points, (x_1, y_1) and (x_2, y_2)
Midpoint Formula $M = \left(\frac{x_1 + x_2}{2}, \frac{y_1 + y_2}{2}\right)$	• To determine the midpoint of a segment with endpoints (x_1, y_1) and (x_2, y_2)
Slope Formula $m = \frac{y_2 - y_1}{x_2 - x_1}$	• To determine whether lines are parallel or perpendicular using two points, (x_1, y_1) and (x_2, y_2)

EXAMPLE 4 **Showing that a Quadrilateral is a Parallelogram**

Show that a quadrilateral with vertices $A(2, 6)$, $B(9, 4)$, $C(10, -1)$, and $D(3, 1)$ is a parallelogram.

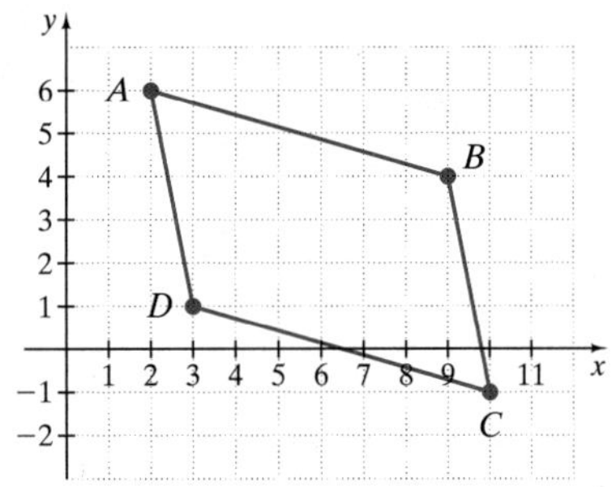

Solution There are many ways to do this. We show three ways below.

1. Show that pairs of opposite sides have the same length using the distance formula.

$$AB = \sqrt{(2-9)^2 + (6-4)^2} = \sqrt{(-7)^2 + 2^2} = \sqrt{49 + 4} = \sqrt{53}$$
$$DC = \sqrt{(3-10)^2 + (-1-1)^2} = \sqrt{(-7)^2 + (-2)^2} = \sqrt{49 + 4} = \sqrt{53}$$
$$BC = \sqrt{(9-10)^2 + [4-(-1)]^2} = \sqrt{(-1)^2 + 5^2} = \sqrt{1 + 25} = \sqrt{26}$$
$$AD = \sqrt{(2-3)^2 + (6-1)^2} = \sqrt{(-1)^2 + 5^2} = \sqrt{1 + 25} = \sqrt{26}$$

Opposite pairs have the same length, $AB = DC$ and $BC = AD$, so $ABCD$ is a parallelogram, by Theorem 6.3-1.

2. Show that opposite sides have the same slope using the slope formula. Then the opposite sides are parallel.

$$\text{slope of } \overline{AB} = \frac{6-4}{2-9} = \frac{2}{-7} \quad \text{or} \quad -\frac{2}{7}$$
$$\text{slope of } \overline{DC} = \frac{-1-1}{10-3} = \frac{-2}{7} \quad \text{or} \quad -\frac{2}{7}$$
$$\text{slope of } \overline{BC} = \frac{4-(-1)}{9-10} = \frac{5}{-1} \quad \text{or} \quad -5$$
$$\text{slope of } \overline{AD} = \frac{6-1}{2-3} = \frac{5}{-1} \quad \text{or} \quad -5$$

Opposite sides have the same slope, so $ABCD$ is a parallelogram, by the definition of a parallelogram.

3. Show that one pair of opposite sides is congruent and parallel.
Look at the slopes and sides of $\overline{AB}$ and $\overline{DC}$ above.

$$\text{slope of } \overline{AB} = -\frac{2}{7}; \quad \text{slope of } \overline{DC} = -\frac{2}{7}$$
$$AB = \sqrt{53} \quad \text{and} \quad DC = \sqrt{53}$$

Thus $ABCD$ is a parallelogram, by Theorem 6.3-5. □

PRACTICE 4 Show that a quadrilateral with vertices $M(2, 2)$, $N(7, 4)$, $Q(6, 0)$, and $P(1, -2)$ is a parallelogram.

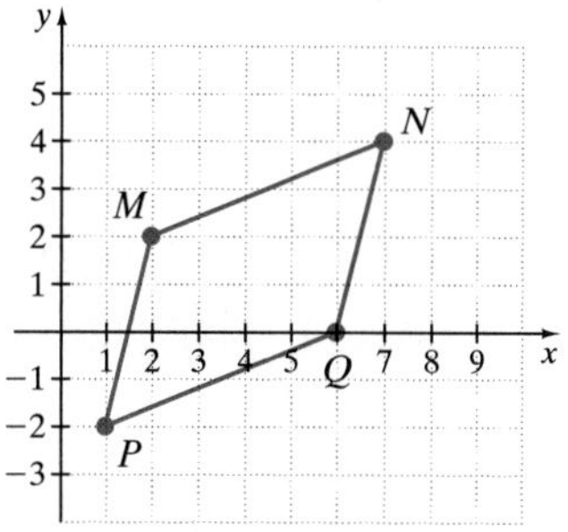

6.3 EXERCISE SET MyMathLab®

Find the values of the variables so that each figure is a parallelogram. See Example 1.

1.
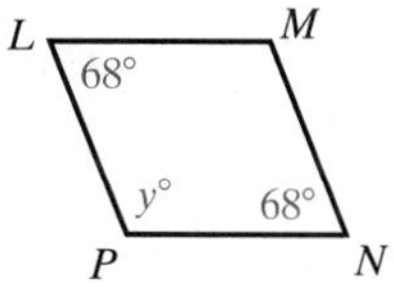

2.
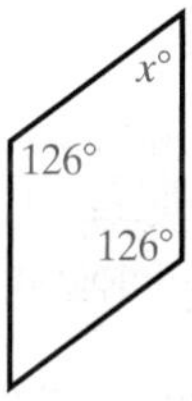

3.
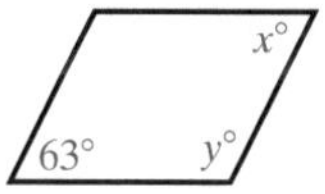

4.
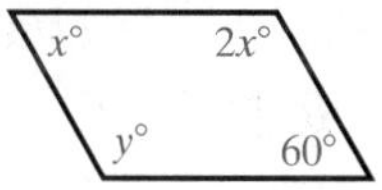

Is there enough information to prove that the quadrilateral is also a parallelogram? Explain why or why not. See Example 2.

5.
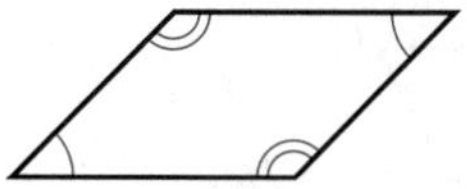

6.
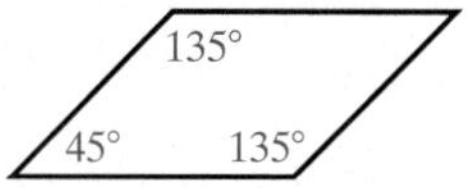

7.
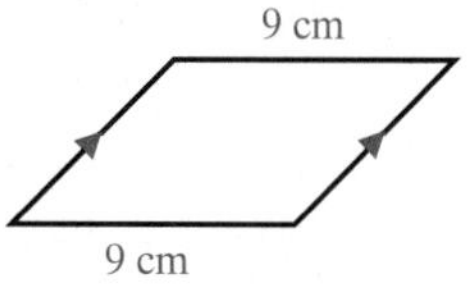

8.
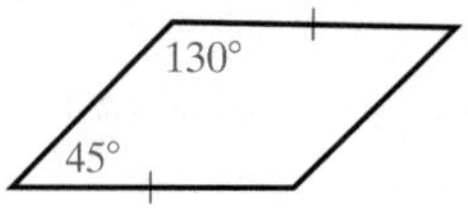

9.
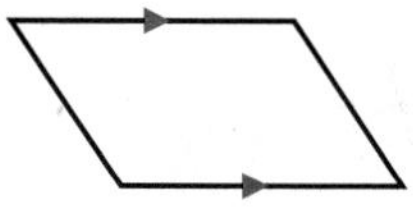

10.
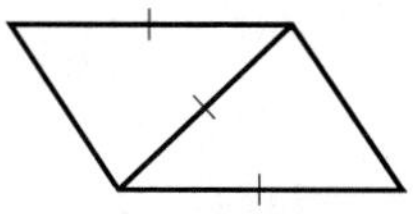

11.
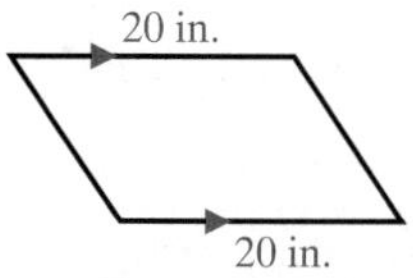

12.
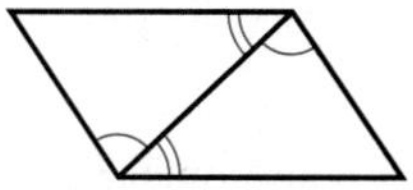

13.
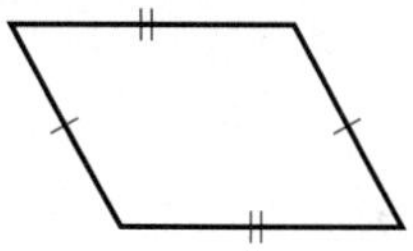

14.
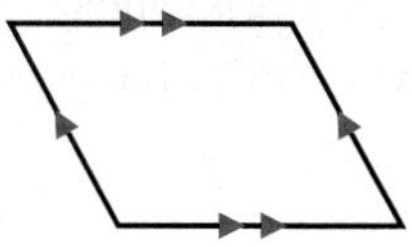

15.
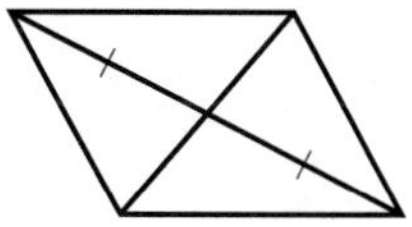

16.
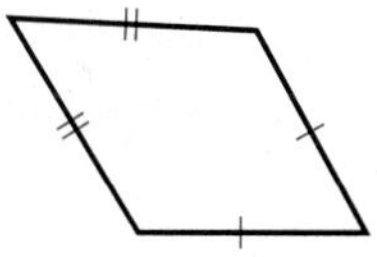

MIXED PRACTICE

For what values of the variables must ABCD be a parallelogram? See Examples 1–3.

17.
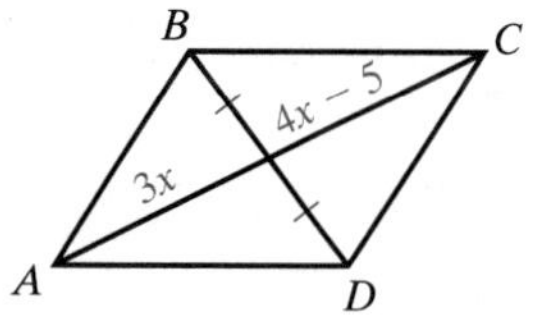

18.
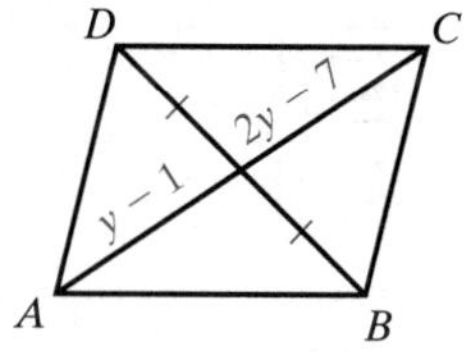

19.
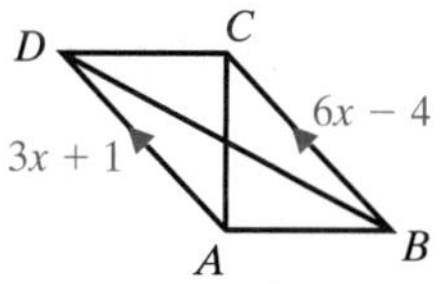

20.
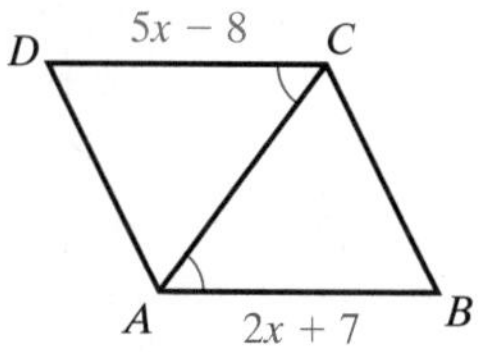

21.
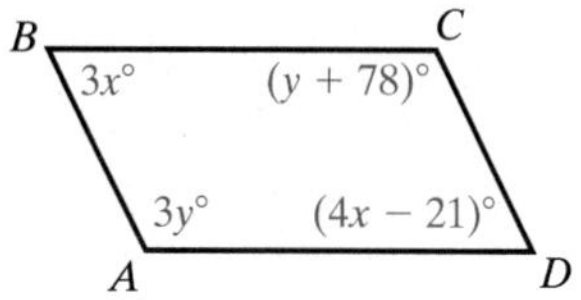

22.
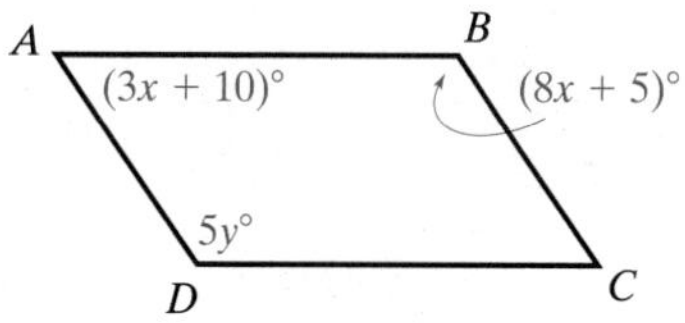

23.
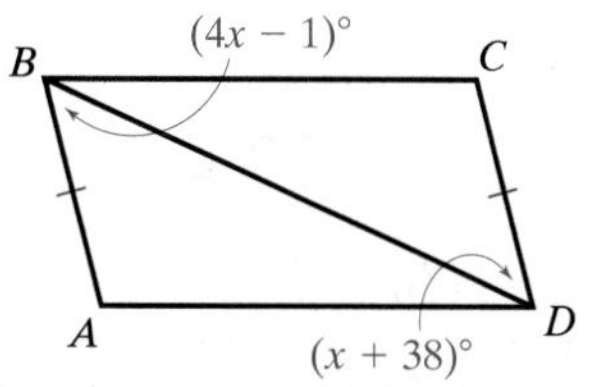

24.
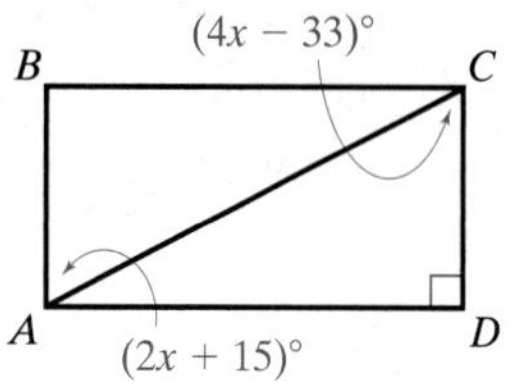

25.

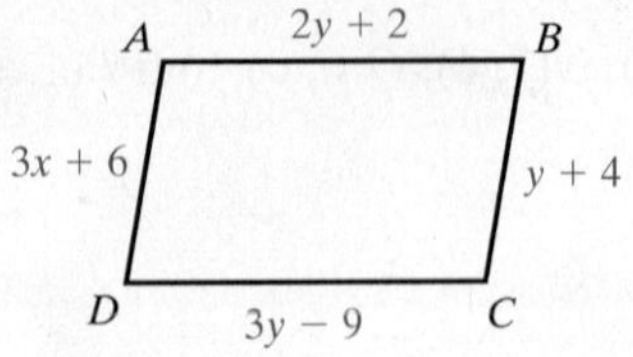

26.

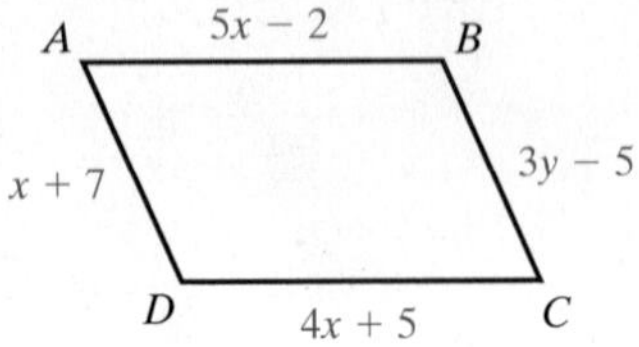

Show that a quadrilateral with the given vertices is a parallelogram. See Example 4.

27.

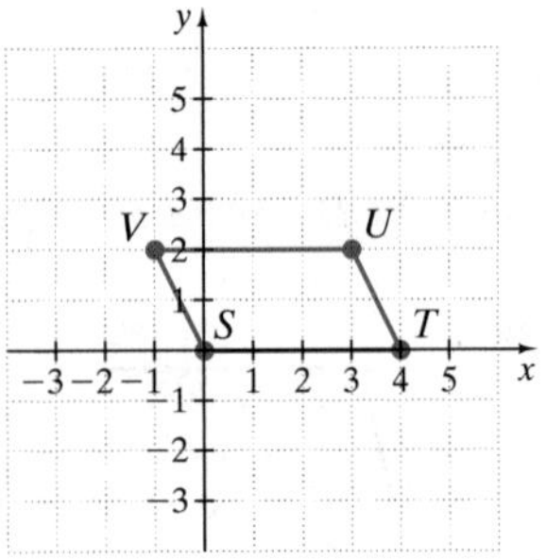

28.

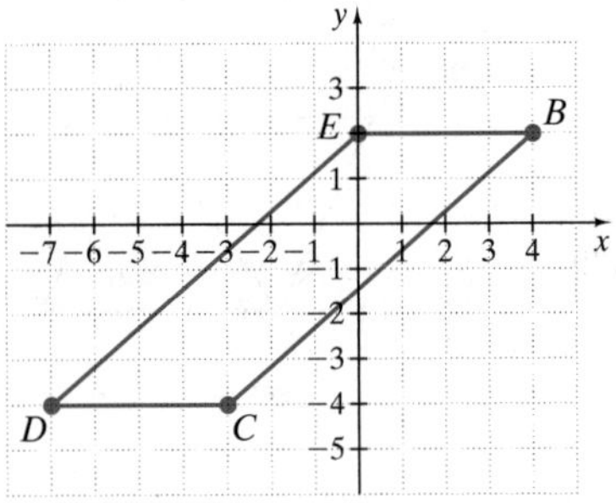

29.

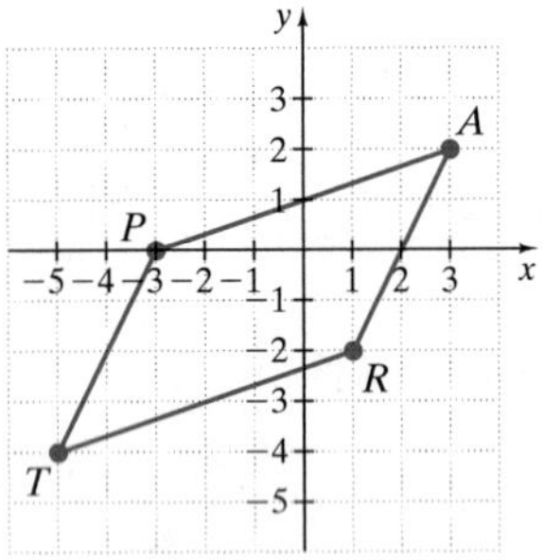

30.

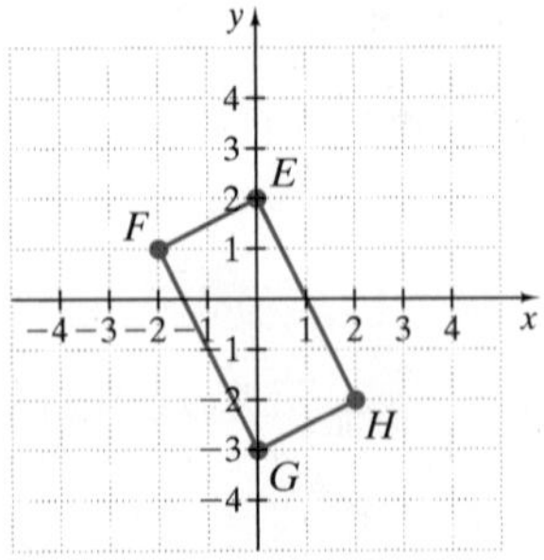

31. $A(-2, 0), B(0, 4), C(4, 5), D(2, 1)$

32. $N(-6, 4), P(-3, 1), Q(0, 2), R(-3, 5)$

CONCEPT EXTENSIONS

33. Proof Complete this two-column proof of Theorem 6.3-2.

Given: $\angle A \cong \angle C, \angle B \cong \angle D$

Prove: $ABCD$ is a parallelogram.

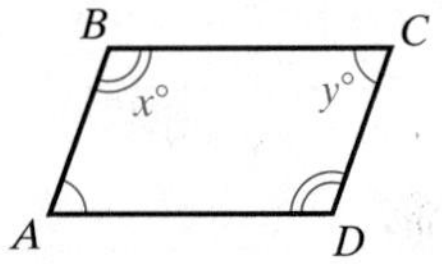

Statements	Reasons
1. $x + y + x + y = 360$	**1.** The sum of the measures of the angles of a quadrilateral is 360.
2. $2(x + y) = 360$	**2. a.** ___?___
3. $x + y = 180$	**3. b.** ___?___
4. $\angle A$ and $\angle B$ are supplementary $\angle A$ and $\angle D$ are supplementary	**4.** Definition of supplementary angles
5. c. ___?___ ∥ ___?___ , ___?___ ∥ ___?___	**5. d.** ___?___
6. $ABCD$ is a parallelogram	**6. e.** ___?___

Proof *Prove each theorem or statement.*

34. Prove Theorem 6.3-5.

Given: $\overline{BC} \parallel \overline{DA}, \overline{BC} \cong \overline{DA}$

Prove: $ABCD$ is a parallelogram.

- How can drawing diagonals help you?
- How can you use triangles in this proof?

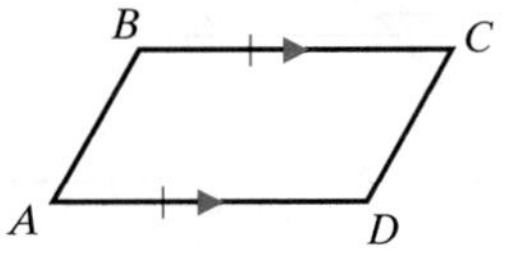

35. Prove Theorem 6.3-1.

Given: $\overline{AB} \cong \overline{CD}, \overline{BC} \cong \overline{DA}$

Prove: $ABCD$ is a parallelogram.

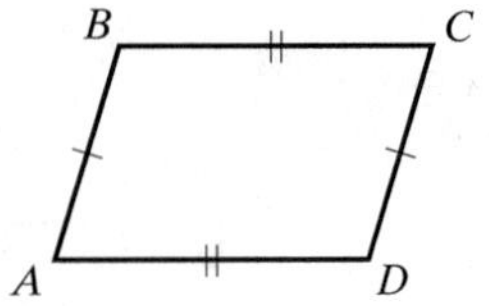

36. Prove Theorem 6.3-3.

Given: $\angle A$ is supplementary to $\angle B$.
$\angle A$ is supplementary to $\angle D$.

Prove: $ABCD$ is a parallelogram.

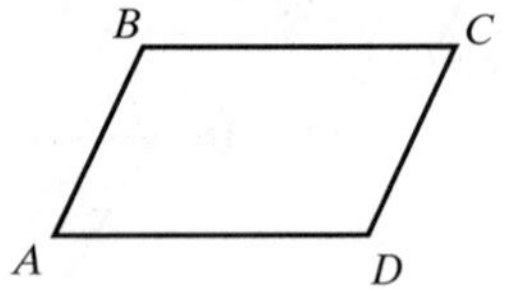

37. Given: $\Delta TRS \cong \Delta RTW$

Prove: *RSTW* is a parallelogram.

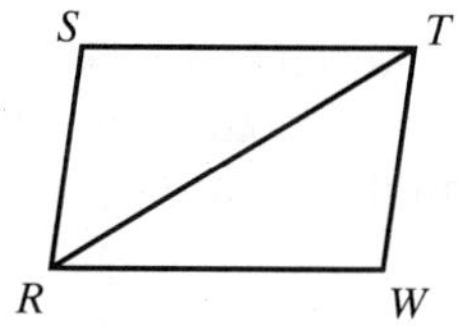

38. Use the diagram below.

Given: $\square ABCD, \overline{AK} \cong \overline{MK}$

Prove: $\angle BCD \cong \angle CMD$

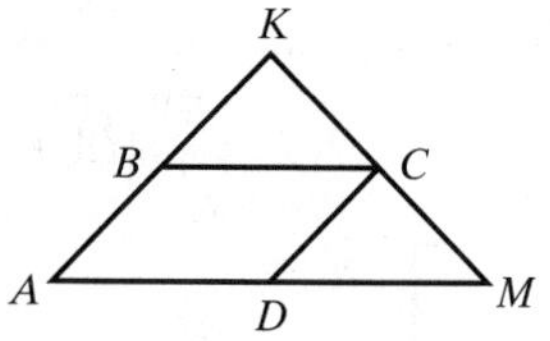

39. Quadrilaterals are formed on the side of this fishing tackle box by the adjustable shelves and connecting pieces. Explain why the shelves are always parallel to each other no matter what their position is.

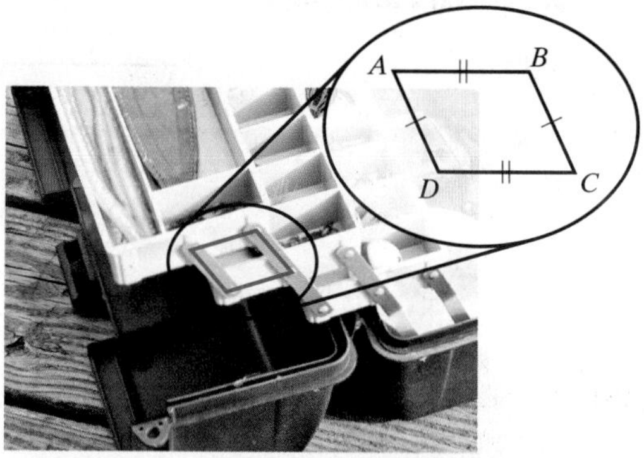

40. Sketch two noncongruent parallelograms *ABCD* and *EFGH* such that $\overline{AC} \cong \overline{EG}$ and $\overline{BD} \cong \overline{FH}$.

41. Explain why you can now write a biconditional statement regarding opposite sides of a parallelogram.

42. Combine each of Theorems 6.2-1, 6.2-2, 6.2-3, and 6.2-4 with its converse from this section into biconditional statements.

43. How is Theorem 6.3-4 in this section different from Theorem 6.2-4 in the previous section? In what situations should you use each theorem? Explain.

44. Find the Error Your friend says, "If a quadrilateral has a pair of opposite sides that are congruent and a pair of opposite sides that are parallel, then it is a parallelogram." What is your friend's error? Explain.

45. Construction In the figure at the right, point *D* is constructed by drawing two arcs. One has center *C* and radius *AB*. The other has center *B* and radius *AC*. Prove that $\overline{AM}$ is a median of ΔABC.

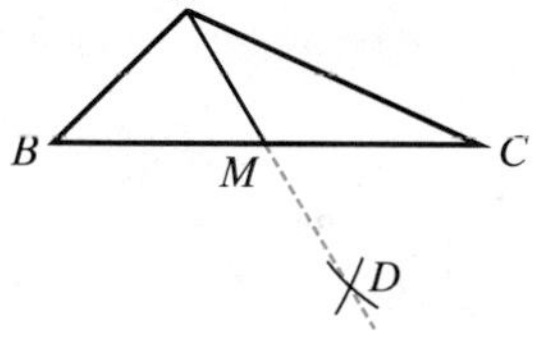

46. If two opposite angles of a quadrilateral measure 120 and the measures of the other angles are multiples of 10, what is the probability that the quadrilateral is a parallelogram?

REVIEW AND PREVIEW

Find the value of each variable in each parallelogram. See Section 6.2.

47.

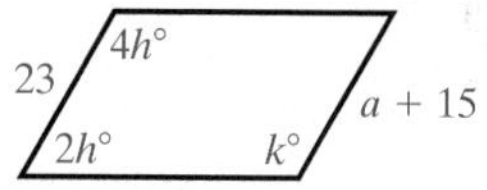

48.

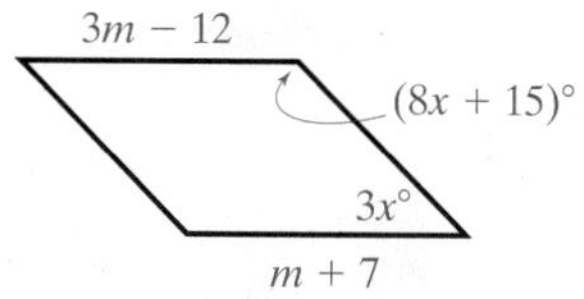

49.

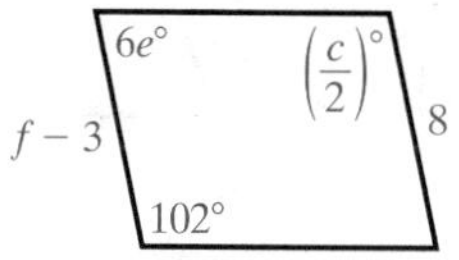

50. Explain how you can use overlapping congruent triangles to prove $\overline{AC} \cong \overline{BD}$. See Sections 4.4 and 4.5.

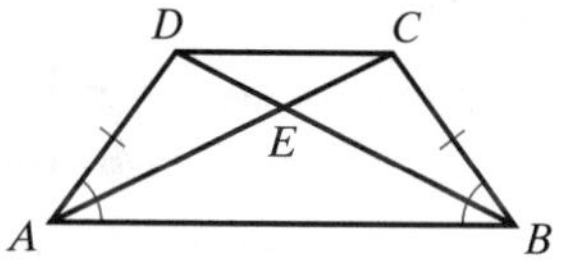

PACE is a parallelogram and $m\angle PAC = 124°$. Complete the following. If necessary, round answers to 2 decimal places. See Sections 5.1 and 6.2.

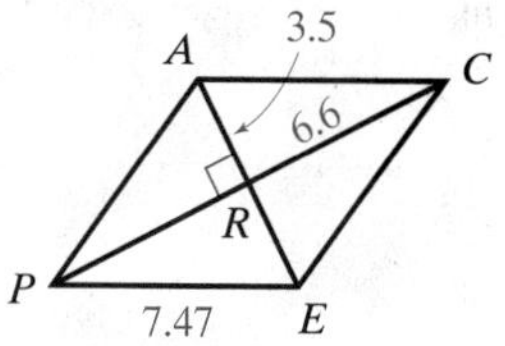

51. $AC =$ ____________

52. $CE =$ ____________

53. $PA =$ ____________

54. $RE =$ ____________

55. $CP =$ ____________

56. $m\angle CEP =$ ____________

57. $m\angle EPA =$ ____________

58. $m\angle ECA =$ ____________

59. $m\angle ACR =$ ____________

60. From which set of information can you conclude that ***RSTW*** is a parallelogram? (The diagonals of *RSTW* intersect at *Z*.)

a. $\overline{RS} \parallel \overline{WT}, \overline{RS} \cong \overline{ST}$

b. $\overline{RS} \parallel \overline{WT}, \overline{ST} \cong \overline{RW}$

c. $\overline{RS} \cong \overline{ST}, \overline{RW} \cong \overline{WT}$

d. $\overline{RZ} \cong \overline{TZ}, \overline{SZ} \cong \overline{WZ}$

6.4 RHOMBUSES, RECTANGLES, AND SQUARES

OBJECTIVES

1 Define and Classify Special Types of Parallelograms.

2 Use Properties of Diagonals of Rhombuses, Rectangles, and Squares.

3 Use Properties of Diagonals to Form Rhombuses, Rectangles, and Squares.

VOCABULARY

- rhombus
- rectangle
- square

Thus far, we know that one special type of quadrilateral is a parallelogram. There are also special types of parallelograms. In this section, we study three special parallelograms—the rhombus, the rectangle, and the square.

OBJECTIVE 1 ▶ Defining Special Types of Parallelograms.

Special Parallelograms

A **rhombus** is a parallelogram with four congruent sides.

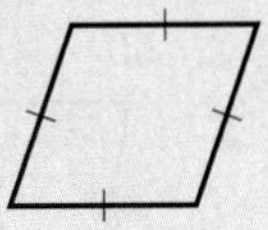

A **rectangle** is a parallelogram with four right angles.

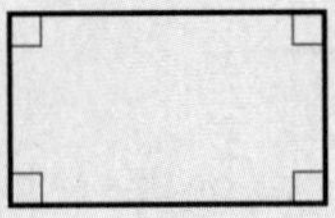

A **square** is a parallelogram with four congruent sides and four right angles.

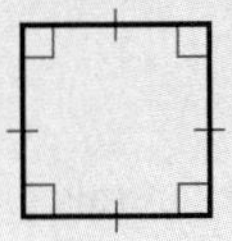

EXAMPLE 1 Using Properties of Special Parallelograms

The figure shown is a rhombus.

a. Find the value of x.

b. Find the measure of each side.

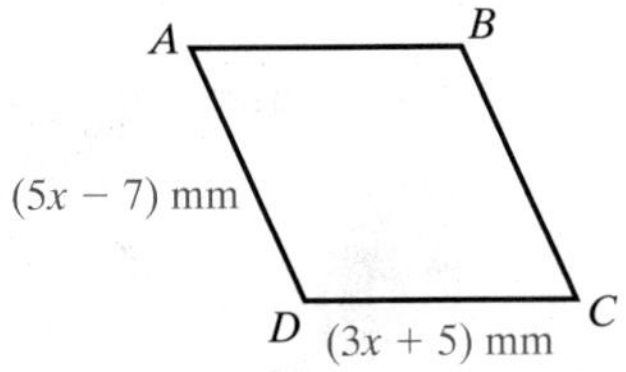

Solution

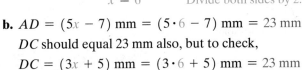

a. The sides of a rhombus are congruent, thus $AD = DC$.

$$AD = DC$$

$5x - 7 = 3x + 5$ Substitute $5x - 7$ for AD and $3x + 5$ for DC.

$2x = 12$ Subtract $3x$ and add 7 to both sides.

$x = 6$ Divide both sides by 2.

b. $AD = (5x - 7)\text{ mm} = (5 \cdot 6 - 7)\text{ mm} = 23\text{ mm}$

DC should equal 23 mm also, but to check,

$DC = (3x + 5)\text{ mm} = (3 \cdot 6 + 5)\text{ mm} = 23\text{ mm}$

Thus, $AD = DC = CB = BA = 23\text{ mm}$ □

Notice the polygons throughout this pattern.

PRACTICE

1 The figure shown is a rhombus.

a. Find the value of x.

b. Find the measure of each side.

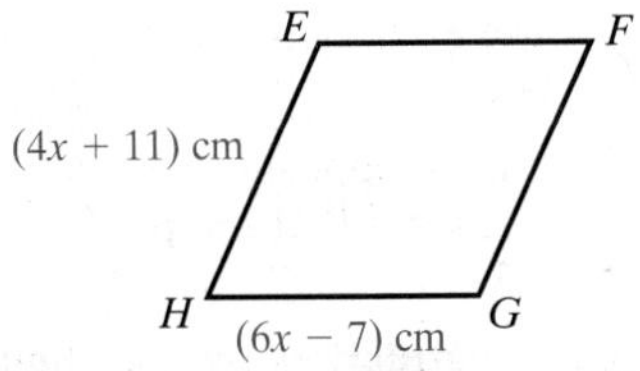

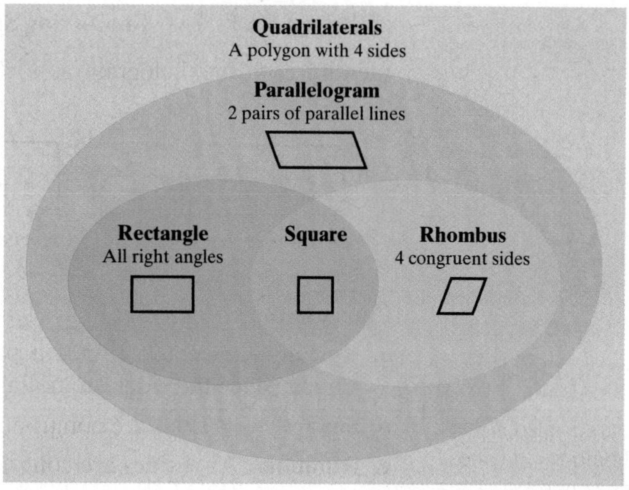

The Venn diagram above shows the special quadrilaterals we have learned thus far. Notice that the square lies in the intersection of rectangles and rhombuses. This means that a square is also a rectangle, a rhombus, a parallelogram, and a quadrilateral.

Helpful Hint Review—Relationships Among Some Quadrilaterals

This is another diagram reviewing the relationships among the quadrilaterals we have seen thus far.

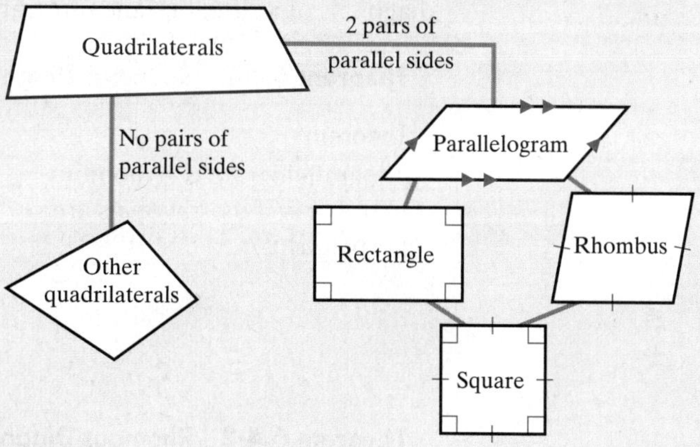

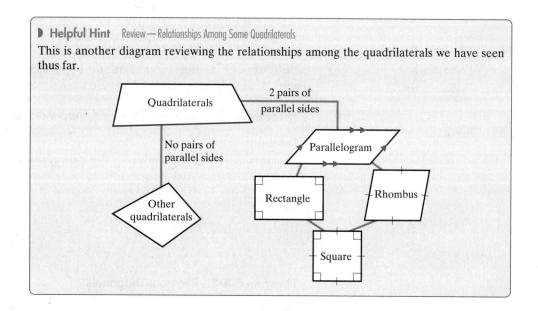

Since rhombuses, rectangles, and squares are also parallelograms, these special figures also have all the properties of parallelograms. Let's recall what we learned about parallelograms in Sections 6.2 and 6.3.

First, remember that a parallelogram is a quadrilateral with both pairs of opposite sides parallel. Also,

A quadrilateral is a parallelogram if and only if

- opposite sides are congruent.
- opposite angles are congruent.
- consecutive angles are supplementary.
- its diagonals bisect each other.

EXAMPLE 2 Identifying Special Parallelograms

Identify each parallelogram as a rhombus, a rectangle, or a square. Be as precise as possible.

a. 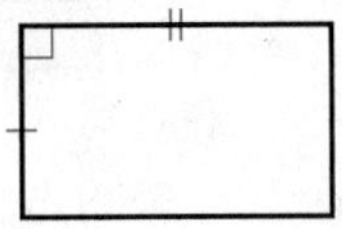

b. 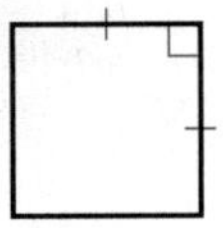

c. 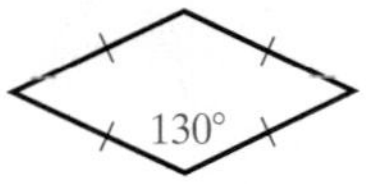

Solution

Let's use the properties of parallelograms to help us identify each.

a. Rectangle. Because this is a parallelogram, we know that all 4 angles are right angles. Since the sides are not all congruent, this is not a square nor a rhombus.

b. Square. All 4 sides are congruent and all 4 angles are right angles.

c. Rhombus. All 4 sides are congruent and the angles are not given as right angles. □

> **Helpful Hint**
> Although a square is also a rectangle and a rhombus, the most precise name is a square.

PRACTICE 2 Identify each parallelogram as a rhombus, a rectangle, or a square. Be as precise as possible.

a. 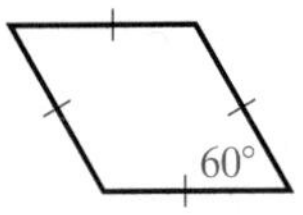

b. 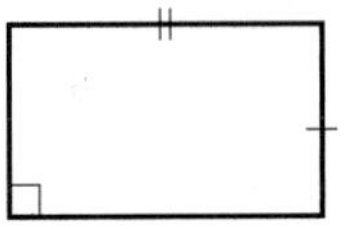

c. 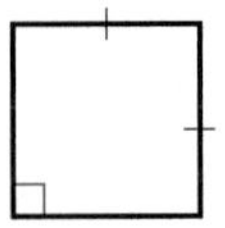

OBJECTIVE 2 ▶ Using Properties of Diagonals of Special Parallelograms. The diagonals of a rhombus and a rectangle have special properties.

Theorem 6.4-1 Rhombus Diagonal/Perpendicular Theorem

Theorem

A parallelogram is a rhombus if and only if its diagonals are perpendicular.

▱$ABCD$ is a rhombus if and only if $\overline{AC} \perp \overline{BD}$

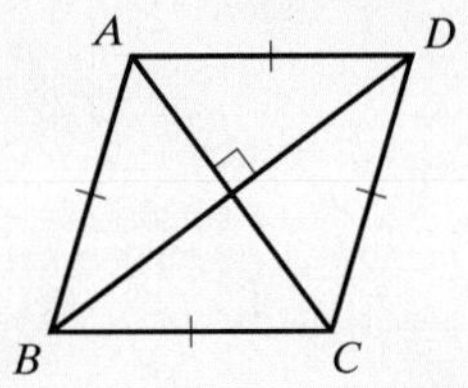

Theorem 6.4-2 Rhombus Diagonals

Theorem

A parallelogram is a rhombus if and only if each diagonal bisects a pair of opposite angles.

▱$ABCD$ is a rhombus if and only if

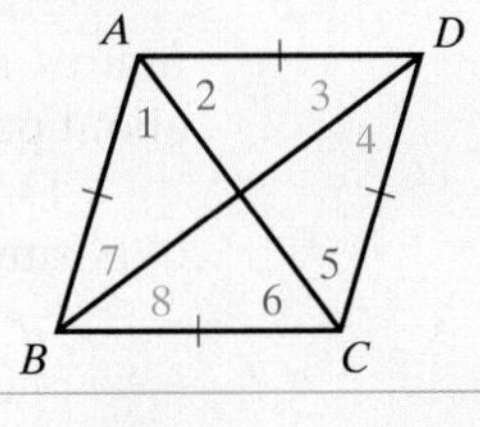

$\angle 1 \cong \angle 2$
$\angle 3 \cong \angle 4$
$\angle 5 \cong \angle 6$
$\angle 7 \cong \angle 8$

> **Helpful Hint**
> Remember that a square is also a rhombus, and a square is also a rectangle. Thus, these theorems all apply to squares, also.

Theorem 6.4-3 Rectangle Diagonals

Theorem

A parallelogram is a rectangle if and only if its diagonals are congruent.

▱$ABCD$ is a rectangle if and only if $\overline{AC} \cong \overline{BD}$

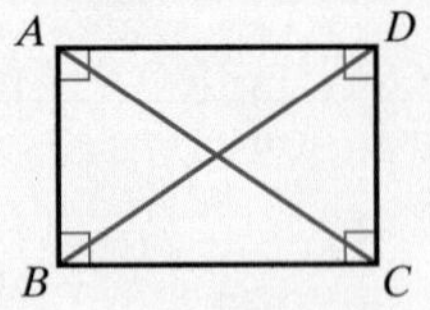

We prove Theorems 6.4-2 and 6.4-3 in Exercise Set 6.4, Exercises 59–62. We prove Theorem 6.4-1 in Example 3 and Practice 3.

The three theorems are written as "if and only if" statements. To prove each theorem, we can write each as a conditional and its converse. Then we prove the statements separately.

For example, Theorem 6.4-1 Rhombus Diagonals/Perpendicular Theorem.

Conditional Statement: If the diagonals of a parallelogram are perpendicular, then the parallelogram is a rhombus.

Converse: If the parallelogram is a rhombus, then its diagonals are perpendicular.

Let's now prove both statements.

EXAMPLE 3 Proving Part of Theorem 6.4-1

Given: $\square ABDC$, $\overline{AD} \perp \overline{BC}$

Prove: $\square ABDC$ is a rhombus.

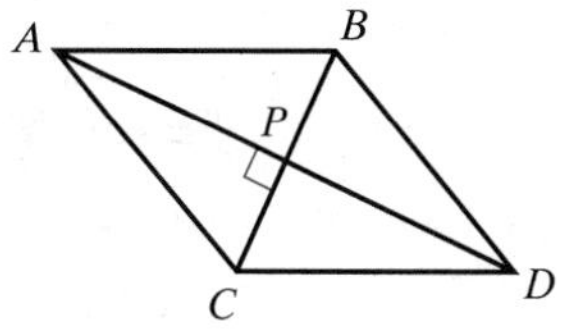

Solution To prove $\square ABDC$ is a rhombus, we prove $\overline{AB} \cong \overline{BD} \cong \overline{DC} \cong \overline{CA}$.

Statements	Reasons
1. $\square ABDC$ with $AD \perp BC$	**1.** Given
2. $m\angle APC = m\angle CPD = m\angle DPB = m\angle BPA = 90^\circ$	**2.** Definition of $\overline{AD} \perp \overline{BC}$
3. $\overline{CP} \cong \overline{PB}$; $\overline{AP} \cong \overline{PD}$	**3.** Diagonals of a parallelogram bisect each other.
4. $\Delta CPD \cong \Delta BPD \cong \Delta BPA \cong CPA$	**4.** SAS; Steps 2, 3
5. $\overline{CD} \cong \overline{BD} \cong \overline{AB} \cong \overline{AC}$	**5.** cpoctac
6. $\square ABDC$ is a rhombus.	**6.** $\square ABDC$ has 4 congruent sides; Step 5. □

PRACTICE 3 Let's finish proving Theorem 6.4-1 by proving the converse, stated above Example 3.

Given: $\square ABDC$ is a rhombus.

Prove: The diagonals are perpendicular, or $\overline{AD} \perp \overline{CB}$ at P

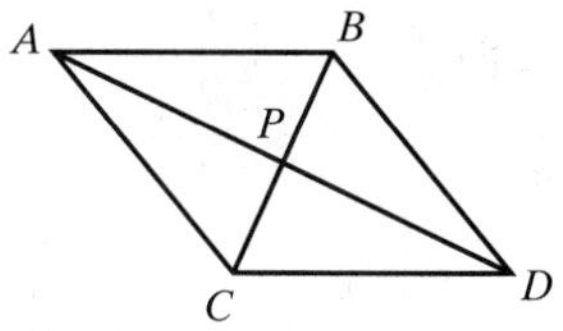

EXAMPLE 4 Finding Diagonal Length

Multiple Choice In rectangle $RSBF$, $SF = 2x + 15$ and $RB = 5x - 12$. What is the length of a diagonal?

a. 1 **b.** 9 **c.** 18 **d.** 33

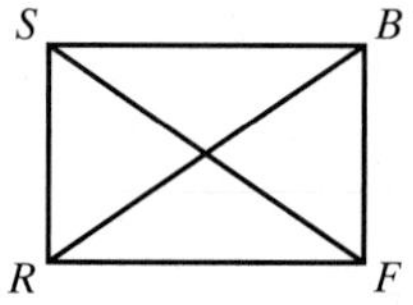

Solution We know that the diagonals of a rectangle are congruent, so their lengths are equal.

$SF = RB$	Set the lengths of the diagonals equal to each other.
$2x + 15 = 5x - 12$	Substitute the given expressions for SF and RB.
$15 = 3x - 12$	Subtract $2x$ from both sides.
$27 = 3x$	Add 12 to both sides.
$9 = x$	Divide both sides by 3.

To find the length of a diagonal, choose one diagonal expression and substitute 9 for x.

Continued on next page

$$RB = 5x - 12$$
$$= 5(9) - 12 \quad \text{Substitute 9 for } x \text{ in the expression.}$$
$$= 33 \quad \text{Simplify.}$$

The correct answer is **d.** □

PRACTICE
4 **a.** If $LN = 4x - 17$ and $MO = 2x + 13$, what are the lengths of the diagonals of rectangle $LMNO$?

b. What type of triangle is ΔPMN? Explain.

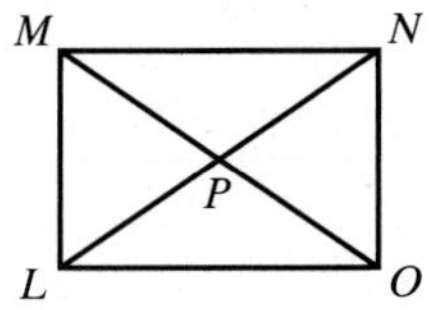

EXAMPLE 5 Finding Angle Measures

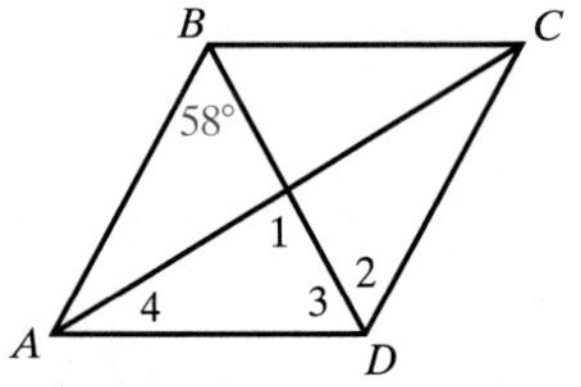

What are the measures of the numbered angles in rhombus $ABCD$?

Solution

To solve, remember everything we know about rhombuses, parallelograms, and any other polygon we need.

> **Helpful Hint**
> Rhombus $ABCD$ is also a ▱, thus $\overline{BC} \parallel \overline{AD}$ (and $\overline{AB} \parallel \overline{DC}$.)

> **Helpful Hint**
> The sum of the measures of the angles of a Δ is 180°.

$m\angle 1 = 90°$ The diagonals of a rhombus are ⊥.

$m\angle 2 = 58°$ Alternate Interior Angles Theorem

$m\angle 3 = 58°$ Each diagonal of a rhombus bisects a pair of opposite angles.

$m\angle 1 + m\angle 3 + m\angle 4 = 180°$ Triangle Angle-Sum Theorem

$90° + 58° + m\angle 4 = 180°$ Substitute.

$148° + m\angle 4 = 180°$ Simplify.

$m\angle 4 = 32°$ Subtract 148° from each side.

Thus, $m\angle 4 = 32°$, $m\angle 2 = m\angle 3 = 58°$, and $m\angle 1 = 90°$. □

PRACTICE
5 What are the measures of the numbered angles in rhombus $PQRS$?

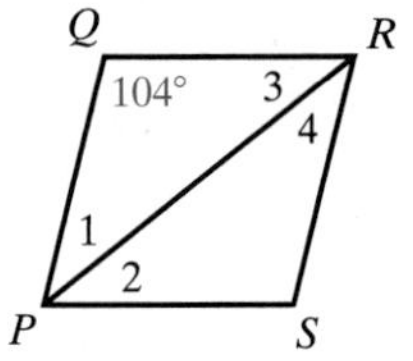

OBJECTIVE 3 ▶ Using Properties of Diagonals to Form Rhombuses, Rectangles, and Squares.

EXAMPLE 6 Using Properties of Special Parallelograms

For what value of x is ▱$ABCD$ a rhombus?

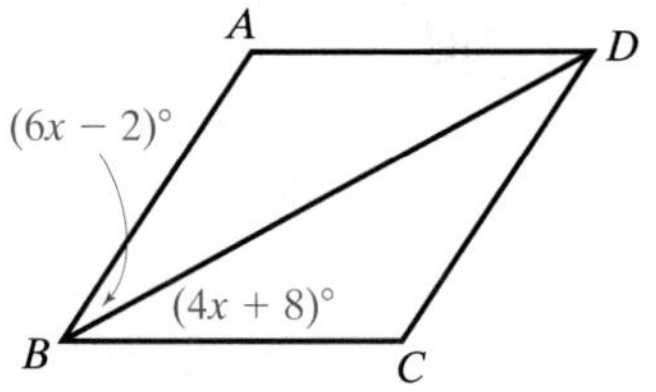

Solution For ▱$ABCD$ to be a rhombus, its diagonals must bisect a pair of opposite angles. Thus,

$m\angle ABD = m\angle CBD$

$6x - 2 = 4x + 8$ Substitute the given angle expressions.

$2x - 2 = 8$ Subtract $4x$ from both sides.

$2x = 10$ Add 2 to both sides.

$x = 5$ Divide both sides by 2.

If $x = 5$, then ▱$ABCD$ is a rhombus. □

PRACTICE 6 For what value of y is $\square DEFG$ a rectangle?

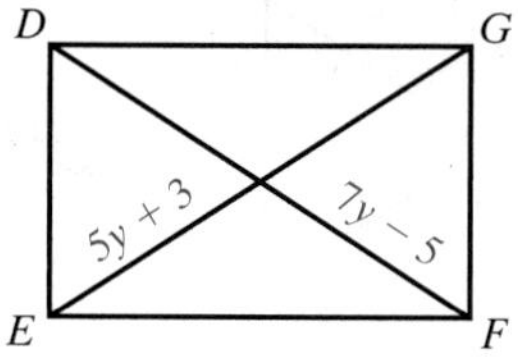

VOCABULARY & READINESS CHECK

Word Bank *Use the choices to fill in each blank. Use each choice only once.*

rectangle   rhombus   square   parallelogram

1. A rhombus, a rectangle, and a square are all examples of a ____________.
2. A ____________ is a parallelogram with four congruent sides and four right angles.
3. A ____________ is a parallelogram with four congruent sides.
4. A ____________ is a parallelogram with four right angles.

6.4 EXERCISE SET MyMathLab®

Find the values of the variables. Then find the side lengths. See Example 1.

1. rhombus

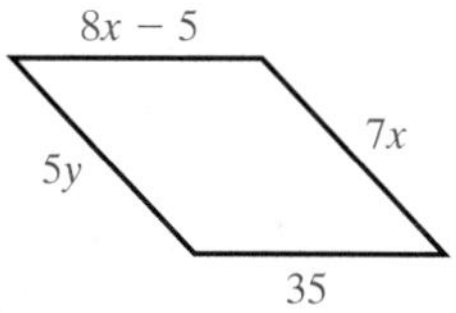

2. rhombus

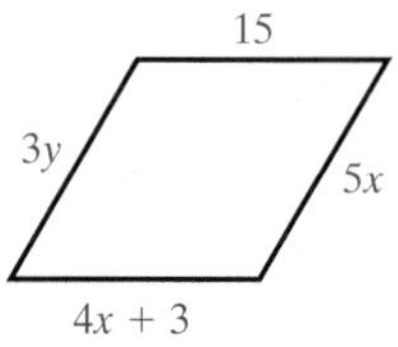

3. square

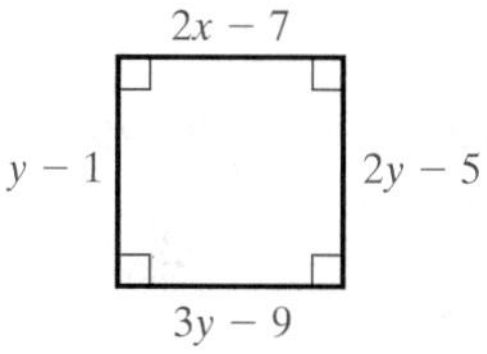

4. square

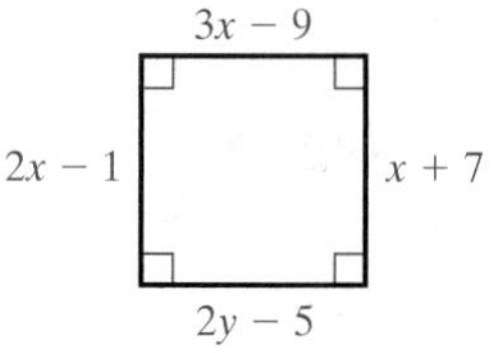

Determine the most precise name for each quadrilateral. See Example 2.

5.

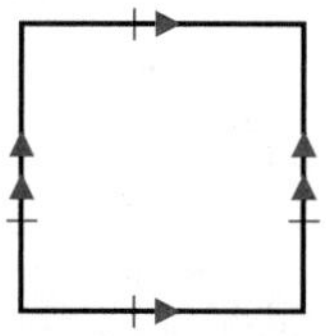

6.

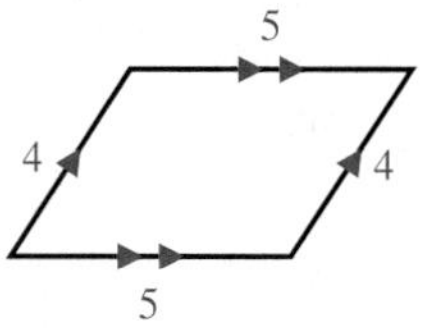

7.

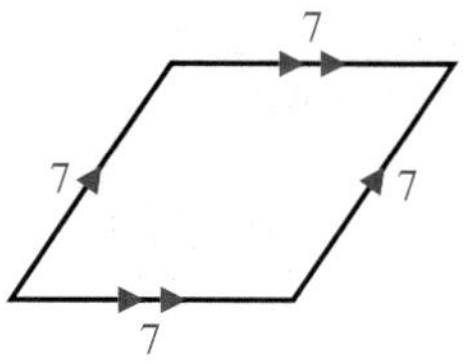

8.

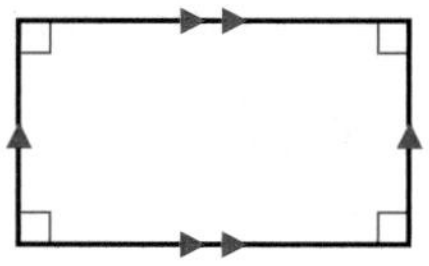

Decide whether the parallelogram is a rhombus, a rectangle, or a square. Explain. See Example 2.

9.

10.

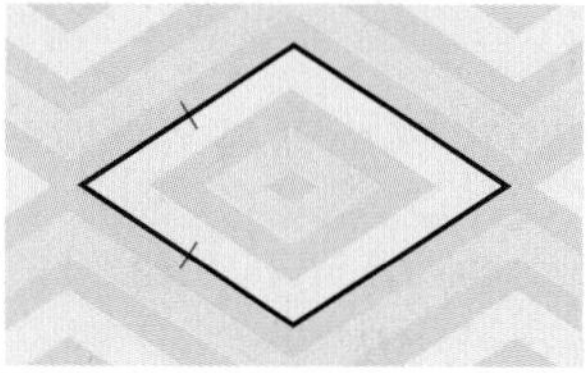

Answer for the parallelogram outlined in black.

Find the measures of the numbered angles in each rhombus. See Example 5.

11.

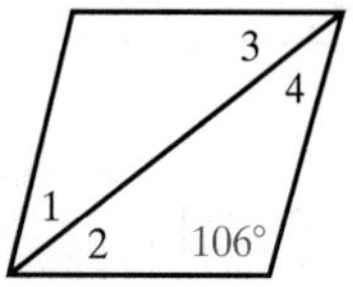

12.

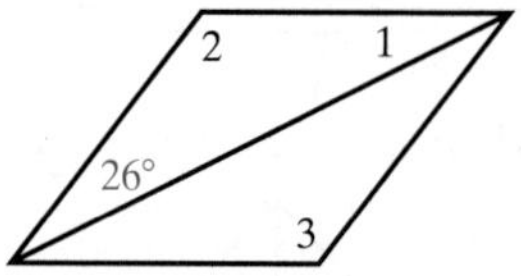

13.

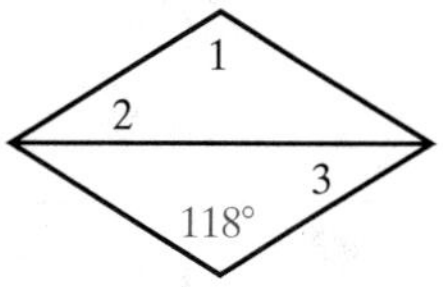

14.

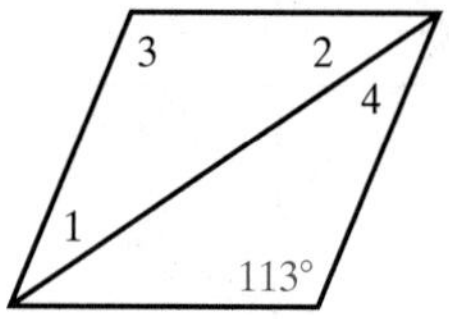

15.

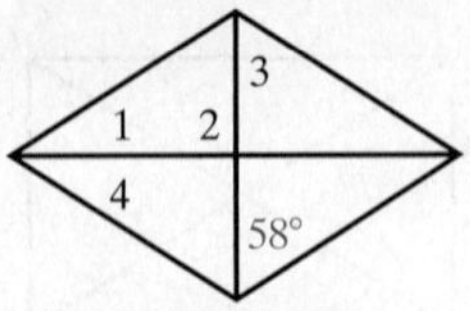

16.

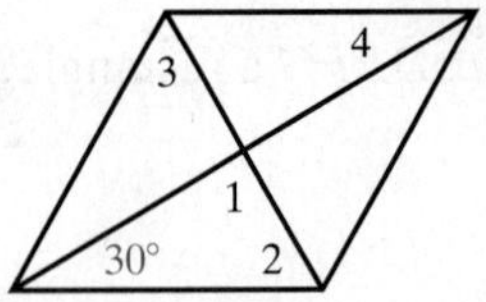

17.

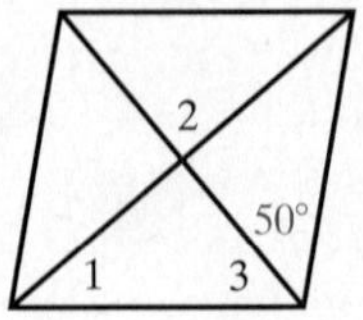

18.

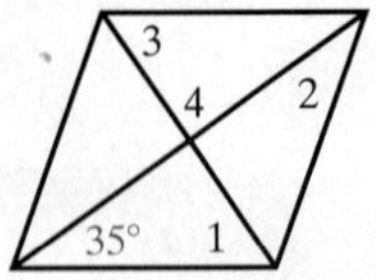

19.

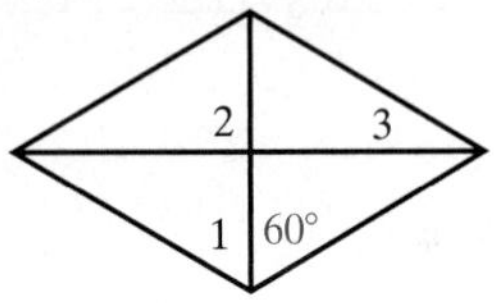

20.

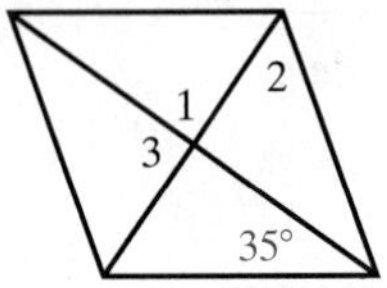

LMNP is a rectangle. Find the value of x and the length of each diagonal. See Example 4.

21. $LN = x$ and $MP = 2x - 4$

22. $LN = 5x - 8$ and $MP = 2x + 1$

23. $LN = 3x + 1$ and $MP = 8x - 4$

24. $LN = 9x - 14$ and $MP = 7x + 4$

25. $LN = 7x - 2$ and $MP = 4x + 3$

26. $LN = 3x + 5$ and $MP = 9x - 10$

Complete the Table *Fill in the table by indicating whether each quadrilateral has the property. Write yes or no. See Examples 1–6.*

	Property	*Parallelogram*	*Rectangle*	*Rhombus*	*Square*
27.	Opposite sides are ∥.				
28.	Opposite sides are ≅.				
29.	Opposite ∠s are ≅.				
30.	All sides are ≅.				
31.	It is equiangular.				
32.	It is equilateral.				
33.	It is equilateral and equiangular.				
34.	All ∠s are right ∠s.				
35.	Diagonals bisect each other.				
36.	Diagonals are ≅.				
37.	Each diagonal bisects opposite ∠s.				
38.	Diagonals are ⊥.				

Find the value(s) of the variable(s) for each parallelogram. See Examples 1, 4, and 5.

39. $RZ = 2x + 5$
$SW = 5x - 20$

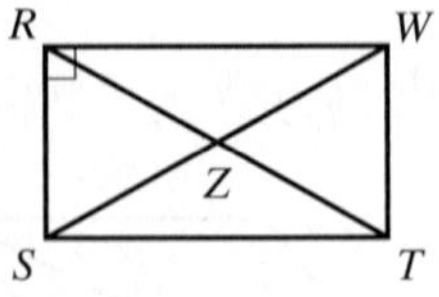

40. $JL = 4x - 12$
$MK = x$

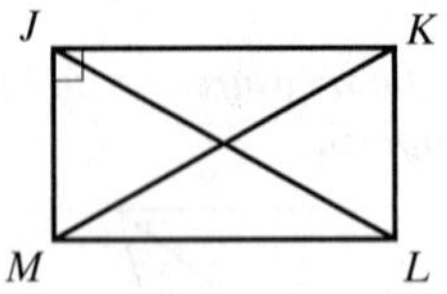

41. $m\angle 1 = 3y - 6$

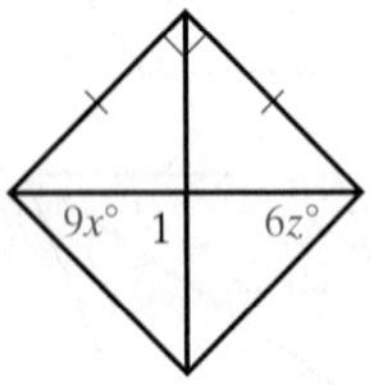

42. $ED = 4x - 16$
$BE = 5y - 1$

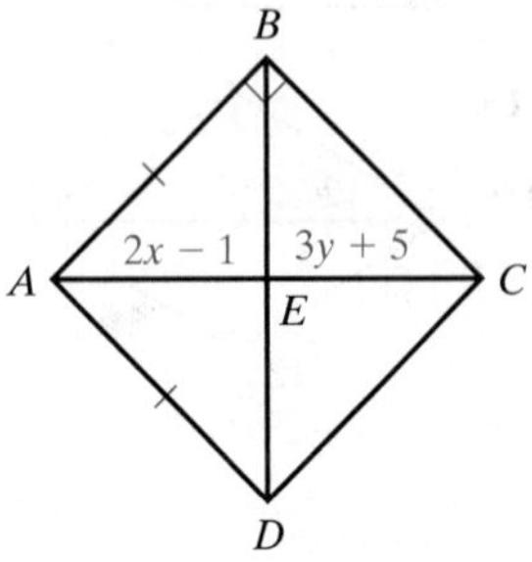

ABCD is a rectangle. Find the length of each diagonal. See Example 4.

43. $AC = 2(x - 3)$ and $BD = x + 5$

44. $AC = 2(5a + 1)$ and $BD = 2(a + 1)$

45. $AC = \frac{3y}{5}$ and $BD = 3y - 4$

46. $AC = \frac{3c}{9}$ and $BD = 4 - c$

Can you conclude that the parallelogram is a rhombus, a rectangle, or a square? Explain and be as precise as possible. See Examples 2 and 6.

47.

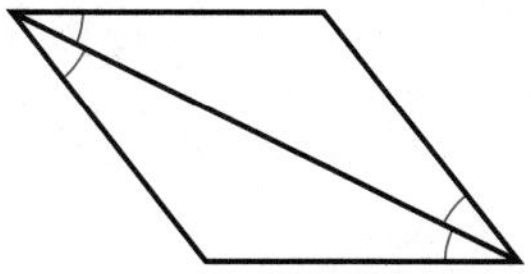

48.

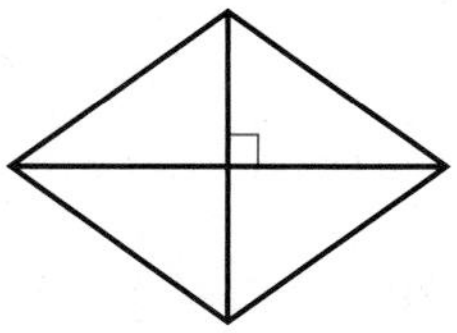

49.

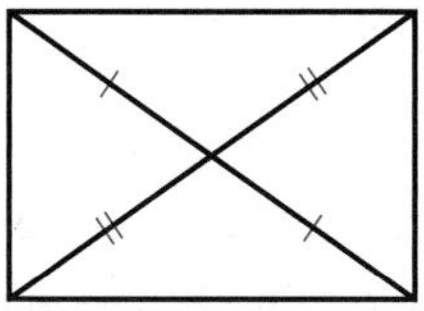

50.

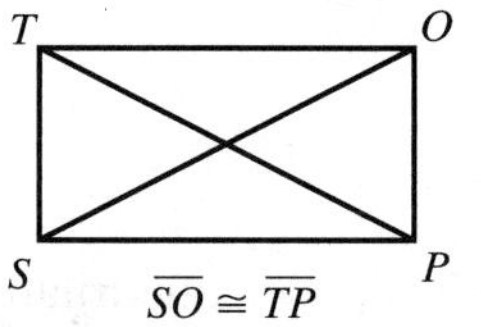

$\overline{SO} \cong \overline{TP}$

For what value of x is the figure the given special parallelogram? See Example 6.

51. rhombus

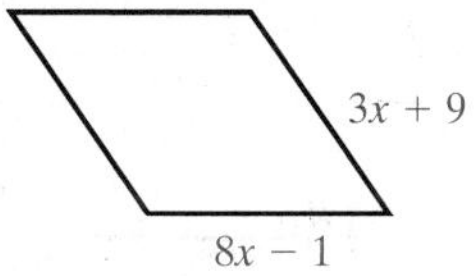

52. rhombus

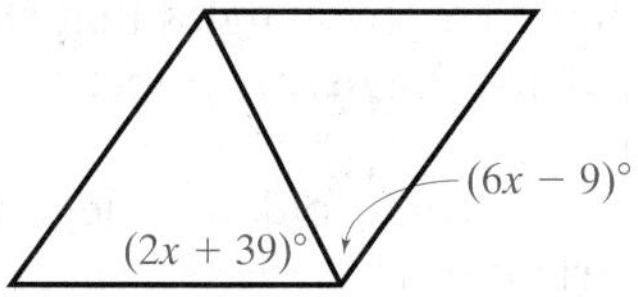

53. rectangle

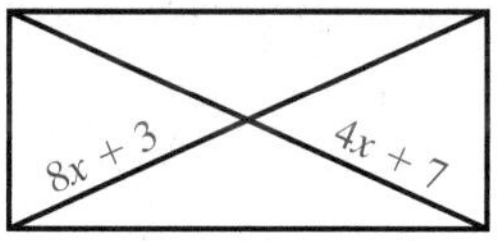

54. rectangle

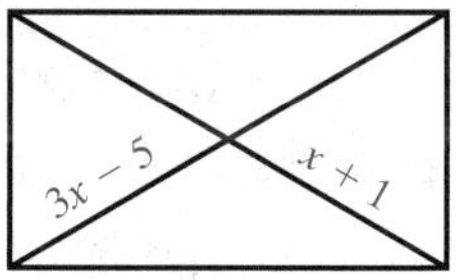

55. rectangle

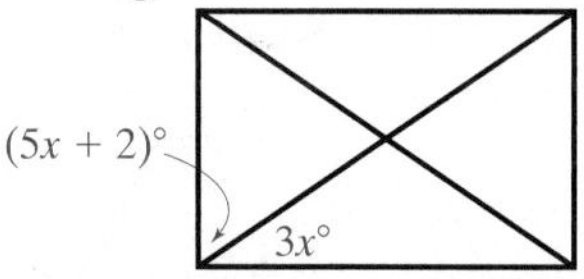

56. rectangle

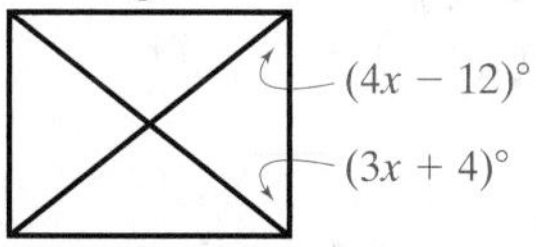

57. rhombus

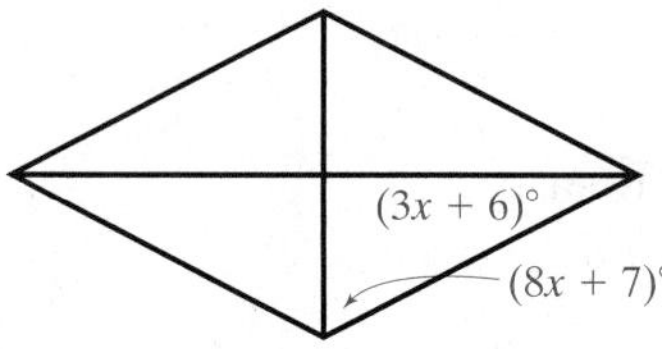

58. rhombus

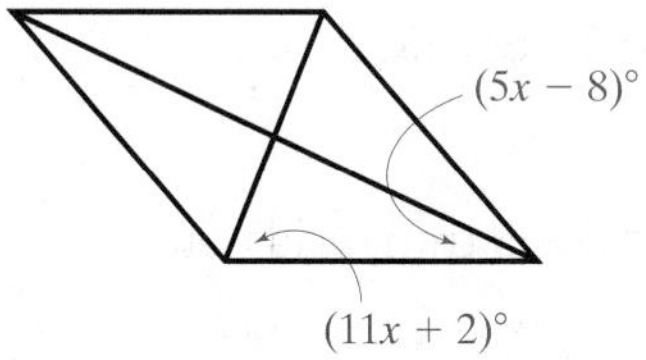

Proof *To prove Theorem 6.4-2, complete Exercises 59 and 60—the conditional and the converse of Theorem 6.4-2. See Example 3.*

59. Prove part of Theorem 6.4-2.
Given: *ABCD* is a rhombus.
Prove: $\overline{AC}$ bisects $\angle BAD$ and $\angle BCD$.

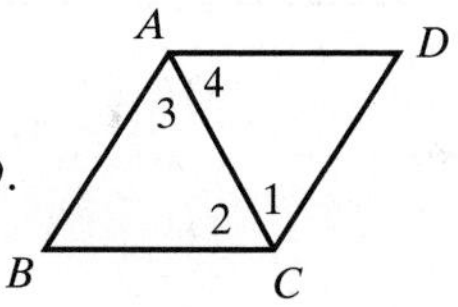

60. Prove part of Theorem 6.4-2.
Given: *ABCD* is a parallelogram.
$\overline{AC}$ bisects $\angle BAD$ and $\angle BCD$.
Prove: *ABCD* is a rhombus.

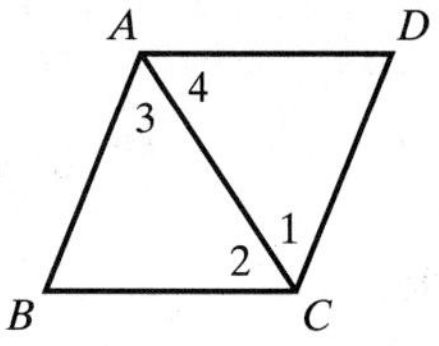

Proof *To prove Theorem 6.4-3, complete Exercises 61 and 62—the conditional and the converse of Theorem 6.4-3. See Example 3.*

61. Prove part of Theorem 6.4-3.
Given: *ABCD* is a rectangle.
Prove: $\overline{AC} \cong \overline{BD}$

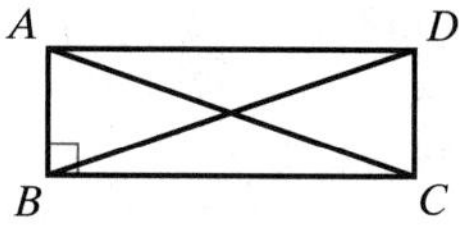

62. Prove part of Theorem 6.4-3.
Given: $\square ABCD$, $\overline{AC} \cong \overline{BD}$
Prove: *ABCD* is a rectangle.

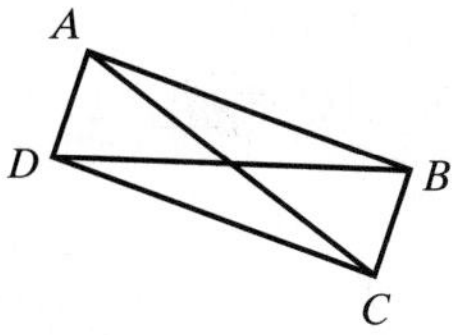

CONCEPT EXTENSIONS

63. Find the Error Your class needs to find the value of *x* for which $\square DEFG$ is a rectangle. A classmate's work is shown below to the right. What is the error? Explain.

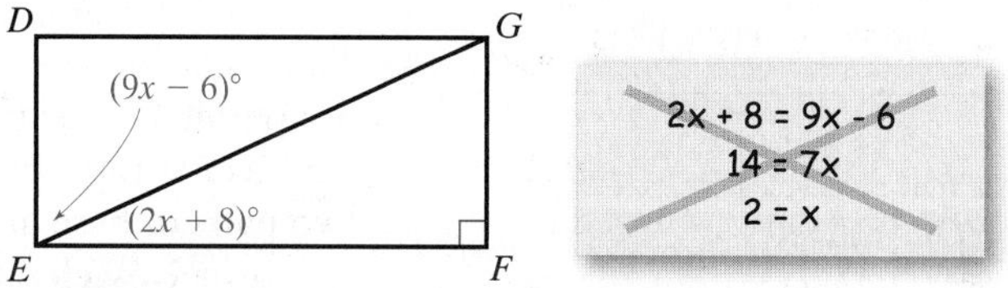

64. Find the Error Your friend says, "A parallelogram with perpendicular diagonals is a rectangle." What is your friend's error? Explain.

Given two segments with lengths a and b ($a \neq b$), what special parallelograms meet the given conditions? Show each sketch.

65. Both diagonals have length *a*.

66. The two diagonals have lengths *a* and *b*.

Find the value of x in the rhombus.

67.

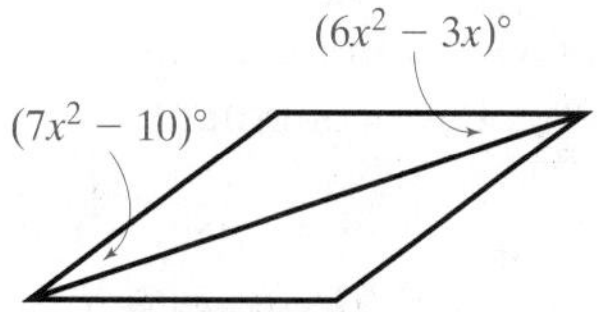

68.

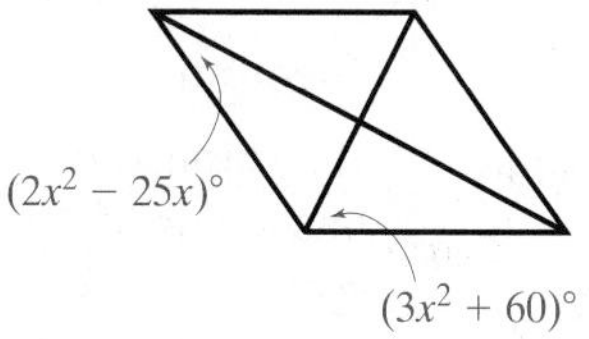

REVIEW AND PREVIEW

Can you conclude that the quadrilateral is a parallelogram? Explain. See Section 6.3.

69.

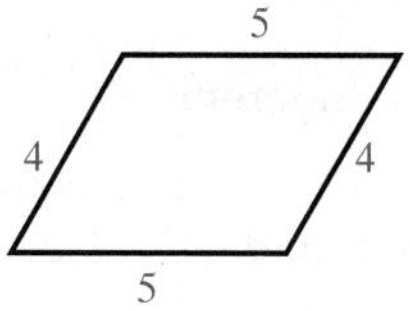

70.

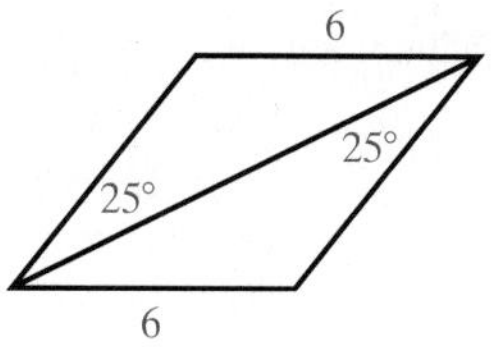

71.

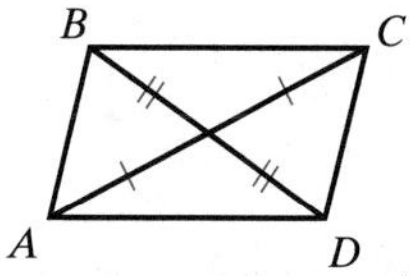

72.

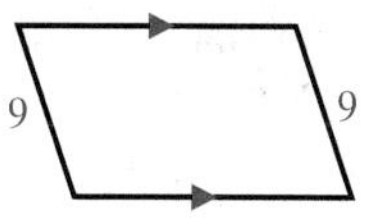

In △PQR, points S, T, and U are midpoints. Complete each statement. See Section 5.4.

73. $PQ = \underline{\ ?\ }$

74. $TU = \underline{\ ?\ }$

75. $\overline{SU} \parallel \underline{\ ?\ }$

76. $\overline{TU} \parallel \underline{\ ?\ }$

77. $\overline{PQ} \parallel \underline{\ ?\ }$

78. $TQ = \underline{\ ?\ }$

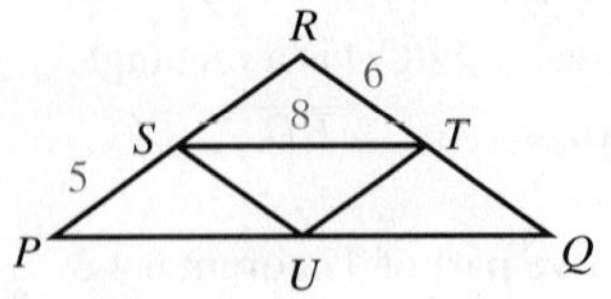

Sketch the following. See this section.

79. Draw a rectangle that is not a square.

80. Draw a rhombus that is not a square.

6.5 TRAPEZOIDS AND KITES

OBJECTIVES

1 Use Properties of Trapezoids.

2 Use Properties of Kites.

VOCABULARY

- trapezoid
- base
- leg
- base angle
- isosceles trapezoid
- midsegment of a trapezoid
- kite

Helpful Hint

Remember—a trapezoid is *not* a parallelogram.

In this section, we will learn about special quadrilaterals that are *not* parallelograms.

OBJECTIVE 1 ▶ Using Properties of Trapezoids. A **trapezoid** is a quadrilateral with exactly one pair of parallel sides. The parallel sides of a trapezoid are called **bases.** The nonparallel sides are called **legs.** The two angles that share a base of a trapezoid are called **base angles.** A trapezoid has two pairs of base angles.

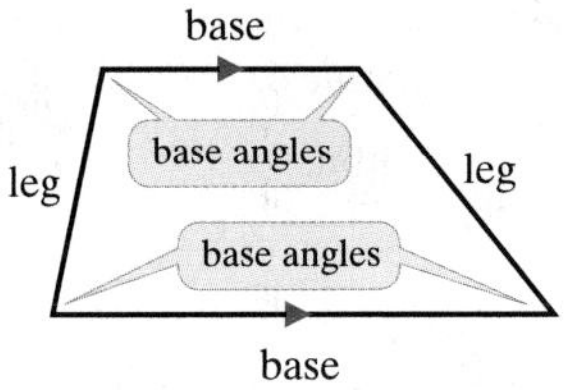

An **isosceles trapezoid** is a trapezoid with legs that are congruent. *ABCD* at the right is an isosceles trapezoid. The angles of an isosceles trapezoid have some unique properties.

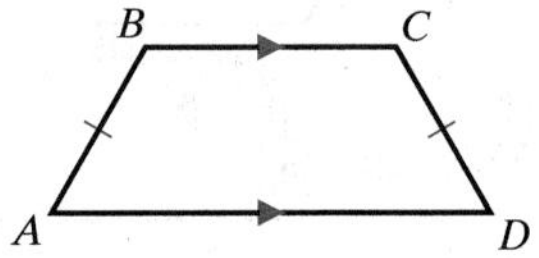

Theorem 6.5-1 Base Angles of an Isosceles Trapezoid

Theorem

If a trapezoid is isosceles, then each pair of base angles is congruent.

If . . .

TRAP is an isosceles trapezoid with bases $\overline{RA}$ and $\overline{TP}$

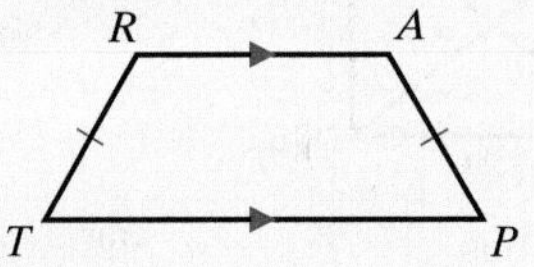

Then . . .

$\angle T \cong \angle P$, $\angle R \cong \angle A$

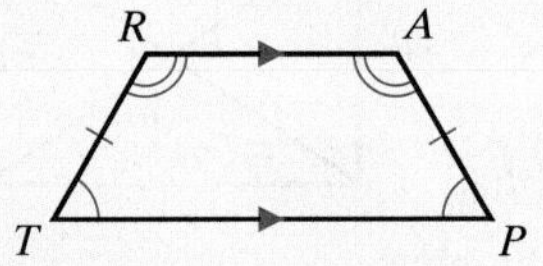

Theorem 6.5-2 Isosceles Trapezoid

Theorem

If a trapezoid has a pair of congruent base angles, then the trapezoid is isosceles.

If . . .

TRAP is a trapezoid with $\angle T \cong \angle P$

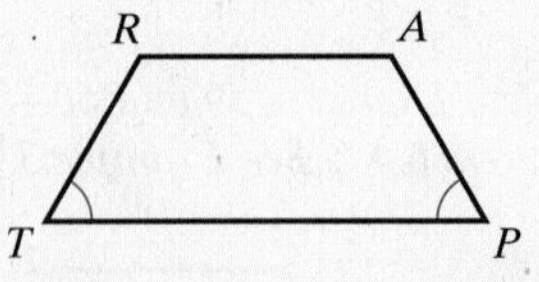

Then . . .

$TR \cong PA$

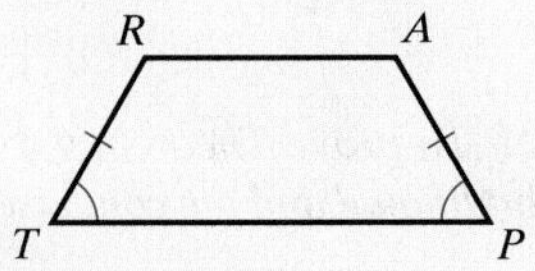

Theorem 6.5-3 Diagonals of an Isosceles Trapezoid

Theorem

A trapezoid is isosceles, if and only if its diagonals are congruent.

ABCD is an isosceles trapezoid if and only if $\overline{AC} \cong \overline{BD}$

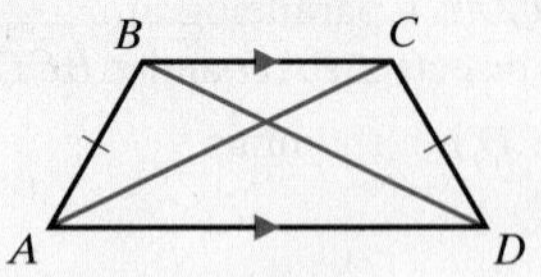

We prove these theorems in Exercises 39, 40, 47, and 48, Exercise Set 6.5.

EXAMPLE 1 Finding Angle Measures in Trapezoids

CDEF is an isosceles trapezoid. Calculate $m\angle D$, $m\angle E$, and $m\angle F$.

Solution $\overline{DE} \parallel \overline{CF}$, so $\angle D$ and $\angle C$ are same-side interior angles and are thus supplementary.

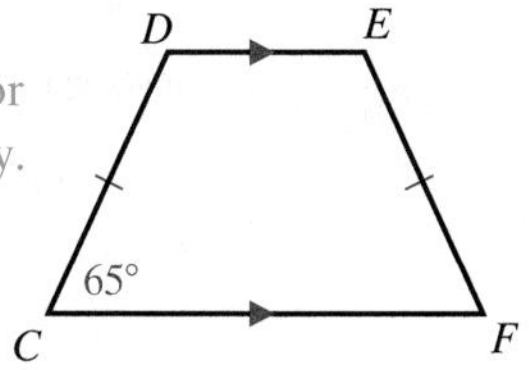

$m\angle C + m\angle D = 180°$	Two angles that form same-side interior angles along one leg are supplementary.
$65° + m\angle D = 180°$	Substitute.
$m\angle D = 115°$	Subtract 65° from each side.

Since each pair of base angles of an isosceles trapezoid is congruent, $m\angle C = m\angle F = 65°$ and $m\angle D = m\angle E = 115°$. □

PRACTICE

1 In the diagram, *PQRS* is an isosceles trapezoid. Calculate $m\angle P$, $m\angle Q$, and $m\angle S$?

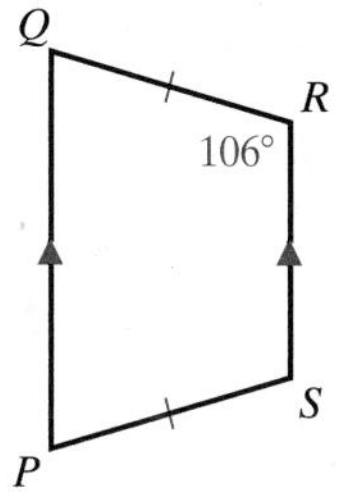

In Section 5.4, we learned about midsegments of triangles. Trapezoids also have midsegments. The **midsegment of a trapezoid** is the segment that joins the midpoints of its legs. The midsegment has two unique properties.

Theorem 6.5-4 Trapezoid Midsegment Theorem

Theorem

If a quadrilateral is a trapezoid, then
(1) the midsegment is parallel to the bases, and
(2) the length of the midsegment is half the sum of the lengths of the bases.

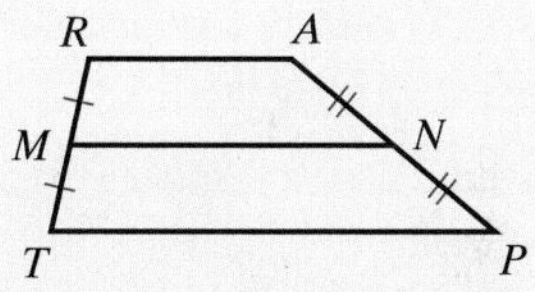

(1) $\overline{MN} \parallel \overline{TP}$, $\overline{MN} \parallel \overline{RA}$, and
(2) $MN = \frac{1}{2}(TP + RA)$

A proof of Theorem 6.5-4 is in the back of this text.

EXAMPLE 2 Using the Midsegment of a Trapezoid

$\overline{QR}$ is the midsegment of trapezoid *LMNP*. What is *x*?

Solution Use the formula for the length of the midsegment.

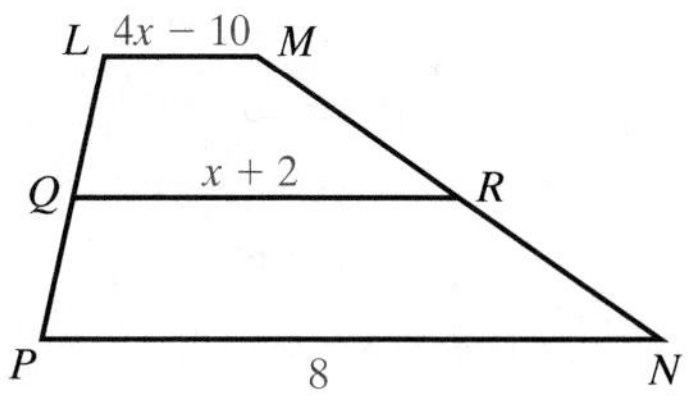

$QR = \frac{1}{2}(LM + PN)$	Trapezoid Midsegment Theorem
$x + 2 = \frac{1}{2}[(4x - 10) + 8]$	Substitute.
$x + 2 = \frac{1}{2}(4x - 2)$	Simplify.
$x + 2 = 2x - 1$	Distributive Property
$3 = x$	Subtract *x* and add 1 to each side.

The value of *x* is 3. To check, find *LM* and *QR* and see that *QR* is the average of *LM* and *PN*. □

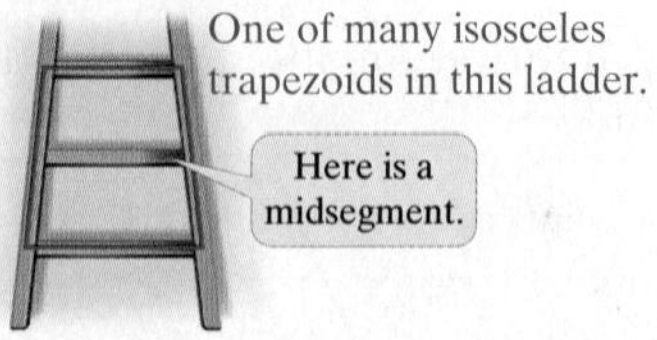

PRACTICE

2 $\overline{MN}$ is the midsegment of trapezoid $PQRS$. What is x? What is MN?

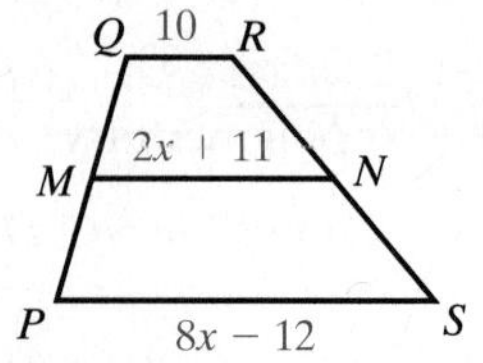

> **Helpful Hint**
> A kite is *not* a parallelogram.

OBJECTIVE 2 ▶ Using Properties of Kites. A **kite** is a quadrilateral with two pairs of consecutive sides congruent and no opposite sides congruent.

The angles, sides, and diagonals of a kite have certain properties.

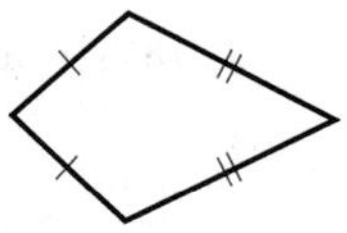

Theorem 6.5-5 Diagonals of a Kite

Theorem	If . . .	Then . . .
If a quadrilateral is a kite, then its diagonals are perpendicular.	$ABCD$ is a kite 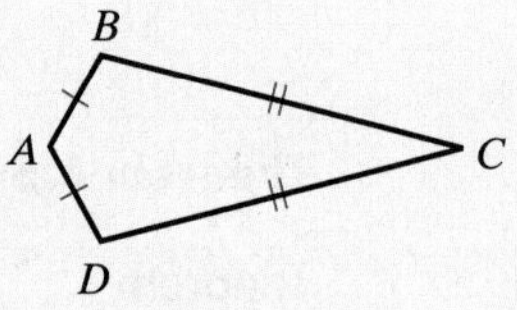	$\overline{AC} \perp \overline{BD}$ 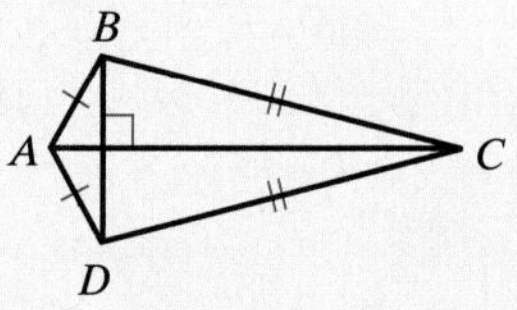

Theorem 6.5-6 Opposite Angles of a Kite

Theorem	If . . .	Then . . .
If a quadrilateral is a kite, then exactly one pair of opposite angles is congruent.	$ABCD$ is a kite 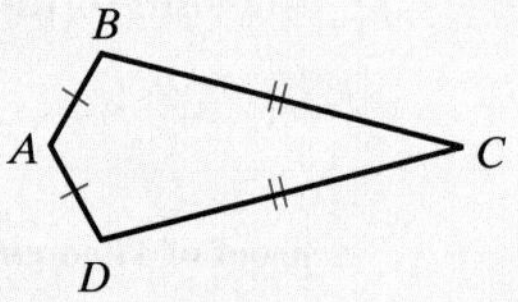	$\angle B \cong \angle D$ 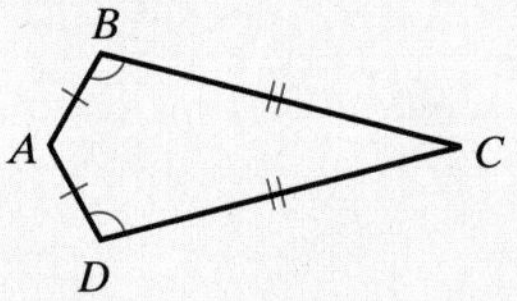

We prove these theorems in Exercises 51 and 52, Exercise Set 6.5.

EXAMPLE 3 Finding Angle Measures in Kites

Quadrilateral $DEFG$ is a kite. What are $m\angle 1$, $m\angle 2$, and $m\angle 3$?

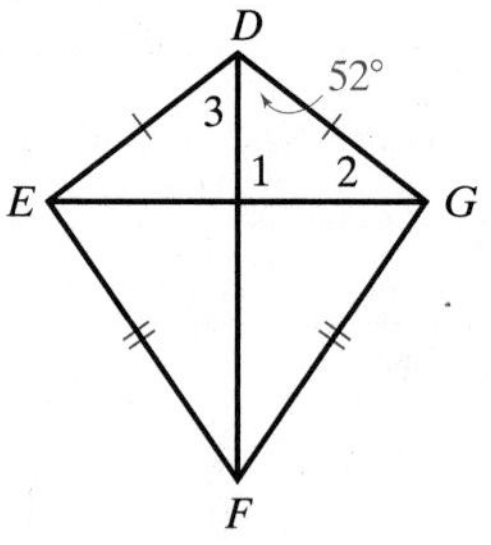

Solution

$m\angle 1 = 90°$	Diagonals of a kite are $\perp$.
$90° + m\angle 2 + 52° = 180°$	Triangle Angle-Sum Theorem
$142° + m\angle 2 = 180°$	Simplify.
$m\angle 2 = 38°$	Subtract 142° from each side.

$\Delta DEF \cong \Delta DGF$ by SSS. Since corresponding parts of congruent triangles are congruent, $m\angle 3 = m\angle GDF = 52°$.

Thus, $m\angle 1 = 90°$, $m\angle 2 = 38°$, and $m\angle 3 = 52°$. □

PRACTICE

3 Quadrilateral $KLMN$ is a kite. What are $m\angle 1$, $m\angle 2$, and $m\angle 3$?

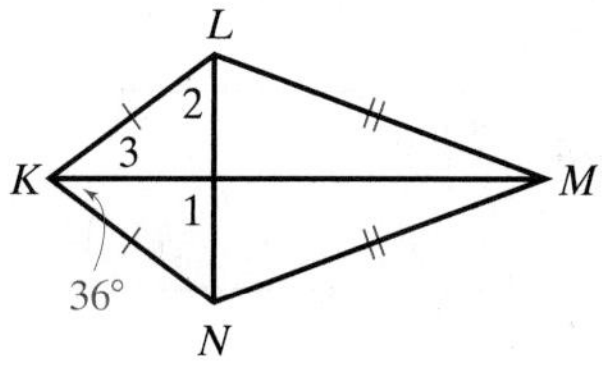

Helpful Hint Review—Summary of Relationships Among Quadrilaterals

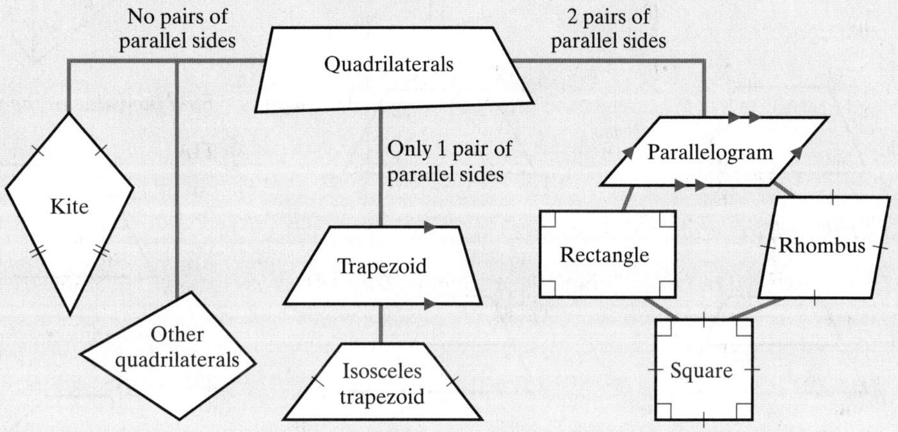

VOCABULARY & READINESS CHECK

Word Bank *Use the choices to fill in each blank. Some choices may be used more than once.*

trapezoid	kite	isosceles	leg
base angle(s)	base	midsegment	

1. A quadrilateral with two pairs of consecutive sides congruent and no opposite sides congruent is called a(n) __________.
2. A quadrilateral with exactly one pair of parallel sides is called a(n) __________.
3. The nonparallel sides of a trapezoid are each called a(n) __________.
4. The parallel sides of a trapezoid are each called a(n) __________.
5. The two angles that share a base of a trapezoid are called __________.
6. A trapezoid has two pairs of __________.
7. The segment that joins the midpoints of the legs of a trapezoid is called a(n) __________.
8. A trapezoid with legs that are congruent is called a(n) __________ trapezoid.

6.5 EXERCISE SET MyMathLab®

Find the measures of the numbered angles in each isosceles trapezoid. See Example 1.

1.

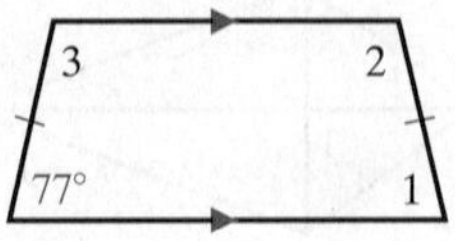

2.

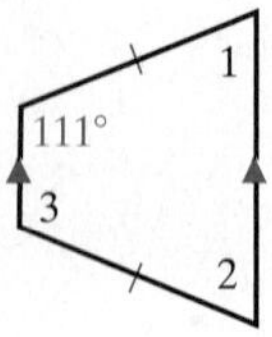

3.

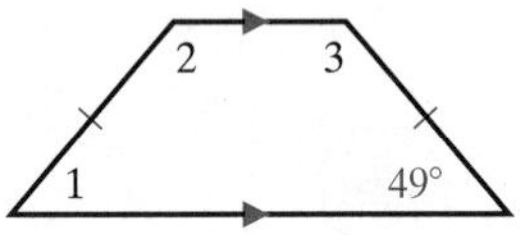

4.

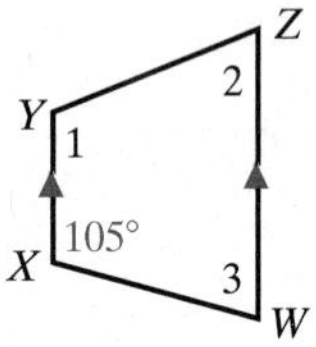

5.

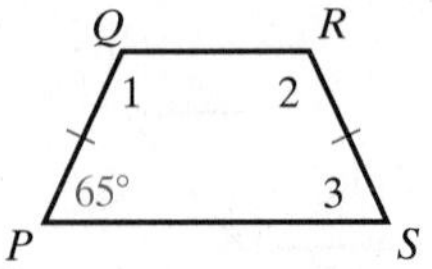

6.

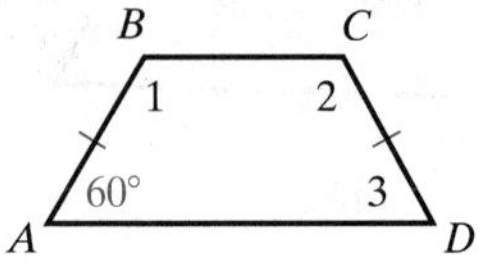

Find EF in each trapezoid. See Example 2.

7.

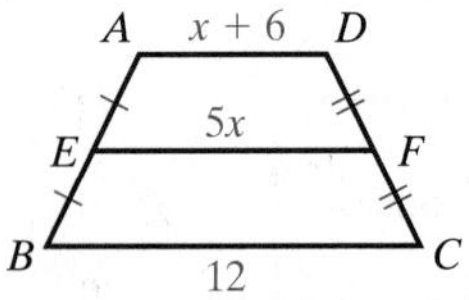

8.

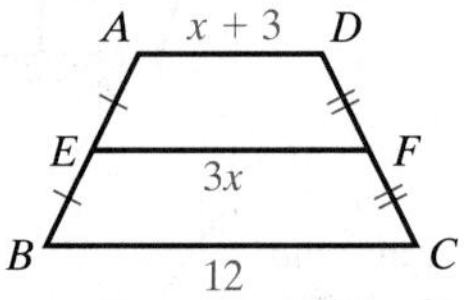

9.

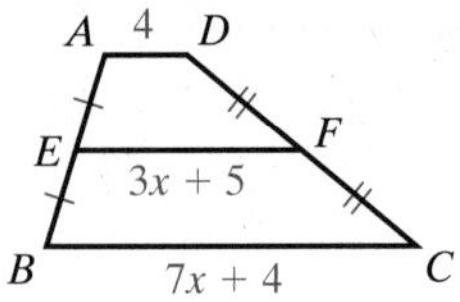

10.

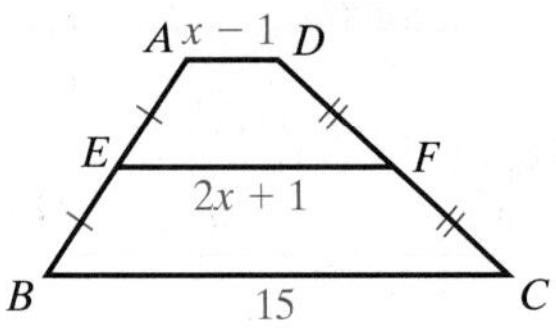

Find the measures of the numbered angles in each kite. See Example 3.

11.

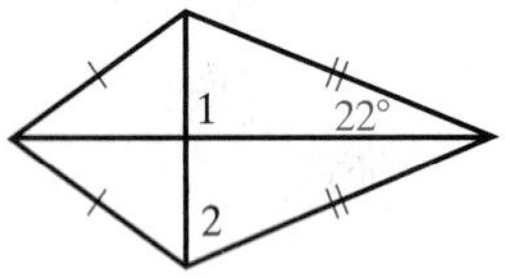

12.

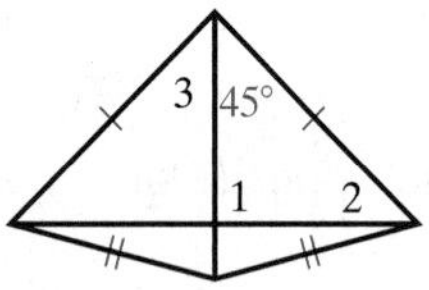

13.

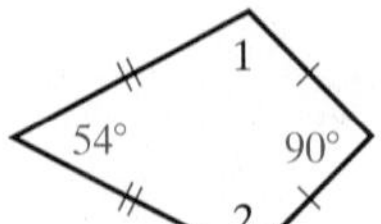

14.

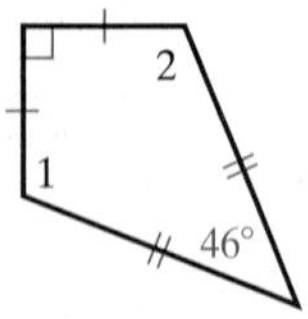

15.

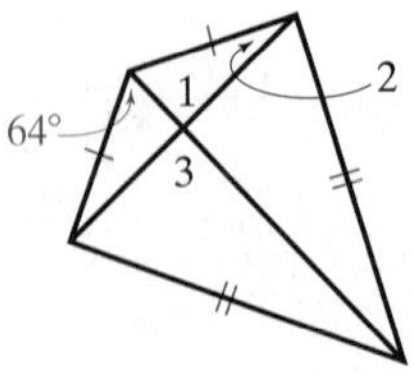

16.

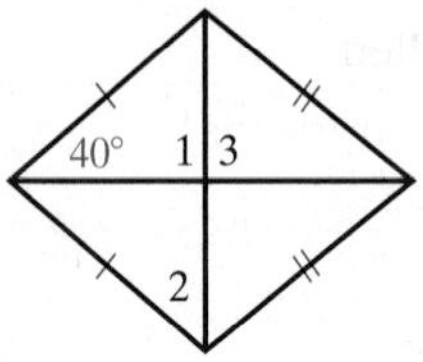

17.

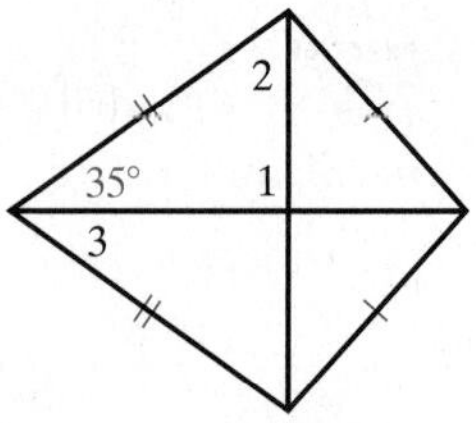

18.

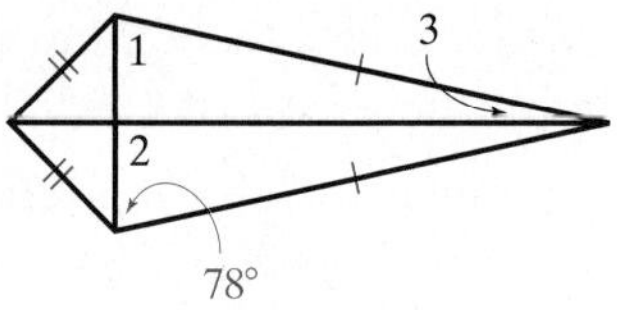

19.

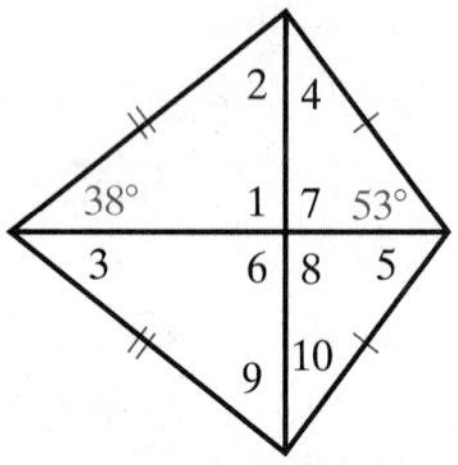

20.

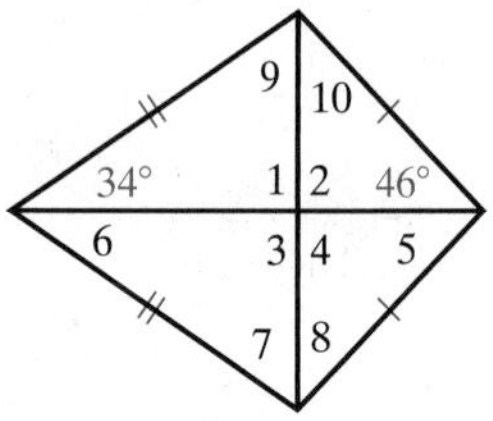

Find the value of the variable in each isosceles trapezoid.

21.

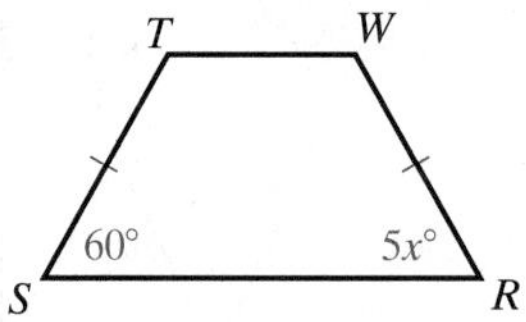

22.

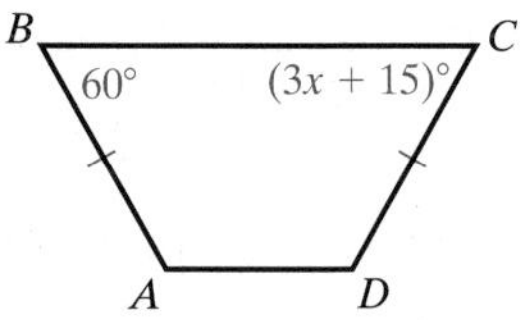

Find the lengths of the segments with variable expressions.

23.

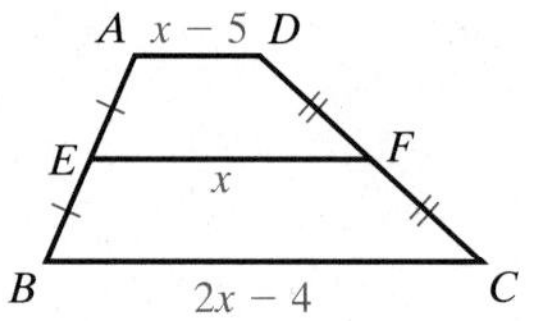

24.

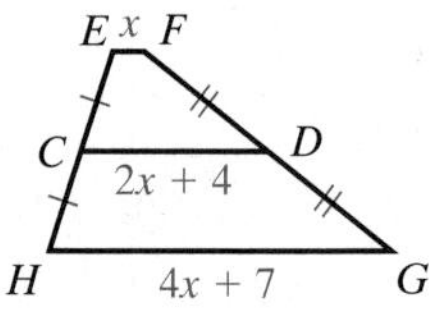

Find the value(s) of the variable(s) in each kite.

25.

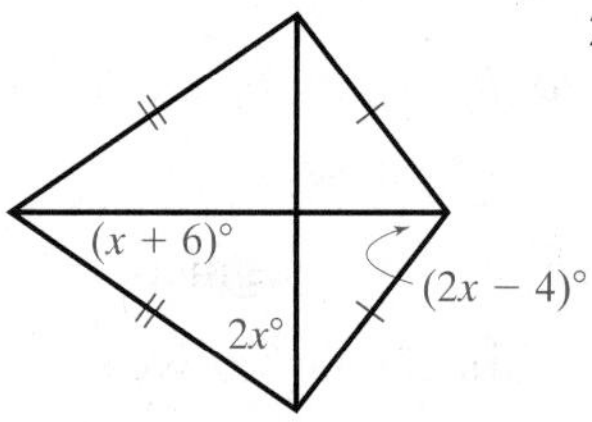

26.

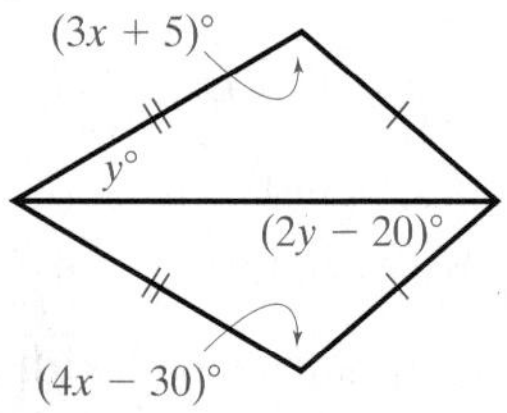

The beams of the bridge shown here form quadrilateral ABCD. $\triangle AED \cong \triangle CDE \cong \triangle BEC$ and $m\angle DCB = 120°$.

27. Classify the quadrilateral. Explain your reasoning.

28. Find the measures of the other interior angles of the quadrilateral.

For Exercises 29–34, can two angles of a kite be as follows? Explain.

29. opposite and acute

30. consecutive and obtuse

31. opposite and supplementary

32. consecutive and supplementary

33. opposite and complementary

34. consecutive and complementary

CONCEPT EXTENSIONS

35. If $KLMN$ is an isosceles trapezoid, is it possible for $\overline{KM}$ to bisect $\angle LMN$ and $\angle LKN$? Explain.

36. Find the Error Since a parallelogram has two pairs of parallel sides, it certainly has one pair of parallel sides. Therefore, a parallelogram must also be a trapezoid. What is the error in this reasoning? Explain.

37. How is a kite similar to a rhombus? How is it different? Explain.

38. Is a kite a parallelogram? Explain.

39. Proof The plan below suggests a proof of Theorem 6.5-1. Write a proof that follows the plan.

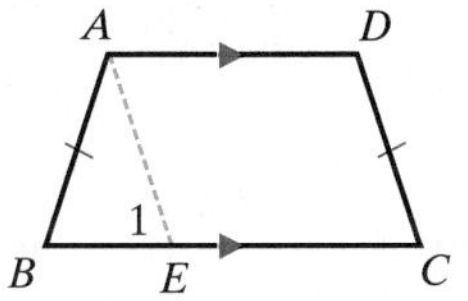

Given: Isosceles trapezoid $ABCD$ with $\overline{AB} \cong \overline{DC}$

Prove: $\angle B \cong \angle C$ and $\angle BAD \cong \angle D$

Plan: Begin by drawing $\overline{AE} \parallel \overline{DC}$ to form parallelogram $AECD$ so that $\overline{AE} \cong \overline{DC} \cong \overline{AB}$. $\angle B \cong \angle C$ because $\angle B \cong \angle 1$ and $\angle 1 \cong \angle C$. Also, $\angle BAD \cong \angle D$ because they are supplements of the congruent angles, $\angle B$ and $\angle C$.

40. Proof Prove Theorem 6.5-2: If a trapezoid has a pair of congruent base angles, then the trapezoid is isosceles.

For Exercises 41–46, name each type of special quadrilateral that can meet the given condition. Make sketches to support your answers.

41. exactly one pair of congruent sides

42. two pairs of parallel sides

43. four right angles

44. adjacent sides that are congruent

45. perpendicular diagonals

46. congruent diagonals

Proof

47. Prove the conditional part of Theorem 6.5-3.

Given: Isosceles trapezoid $ABCD$ with $\overline{AB} \cong \overline{DC}$

Prove: $\overline{AC} \cong \overline{DB}$

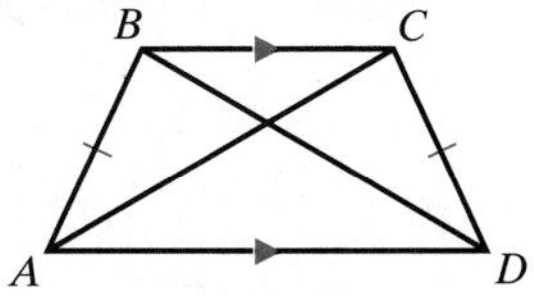

48. Prove the converse part of Theorem 6.5-3: If the diagonals of a trapezoid are congruent, then the trapezoid is isosceles.

49. Given: Isosceles trapezoid $TRAP$ with $\overline{TR} \cong \overline{PA}$

Prove: $\angle RTA \cong \angle APR$

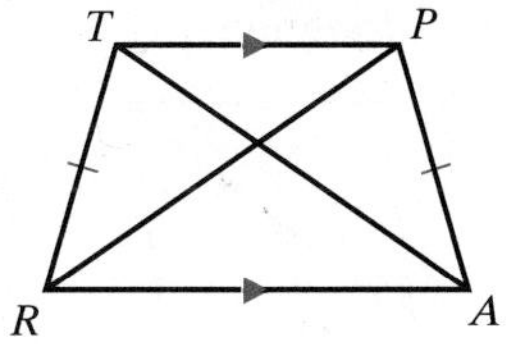

50. Prove that the angles formed by the noncongruent sides of a kite are congruent. (*Hint:* Draw a diagonal of the kite.)

51. Prove Theorem 6.5-5.

Given: $ABCD$ is a kite

Prove: $AC \perp BD$

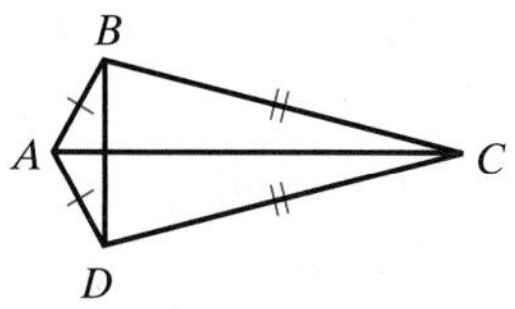

52. Prove Theorem 6.5-6.

Given: $ABCD$ is a kite

Prove: $\angle B \cong \angle D$

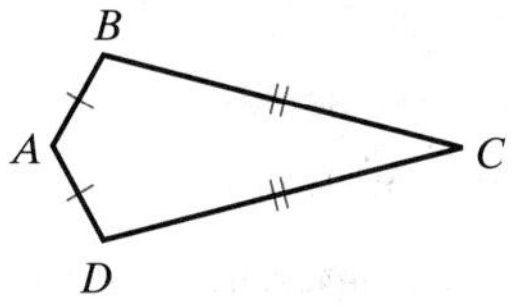

Determine whether each statement is true or false. Justify your response.

53. All squares are rectangles.

54. A trapezoid is a parallelogram.

55. A rhombus can be a kite.

56. Some parallelograms are squares.

57. Every quadrilateral is a parallelogram.

58. All rhombuses are squares.

For a trapezoid, consider the segment joining the midpoints of the two given segments below. How are its length and the lengths of the two parallel sides of the trapezoid related? Justify your answer.

59. the two nonparallel sides

60. the diagonals

REVIEW AND PREVIEW

Find the value of x for which the figure is the given special parallelogram. See Section 6.4.

61. Rhombus

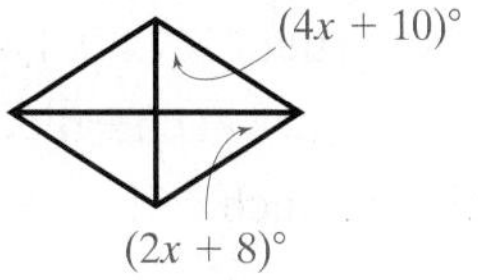

62. Rhombus

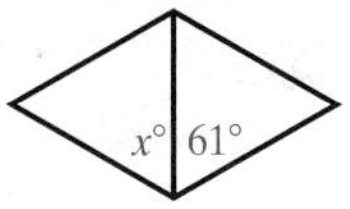

63. Rectangle

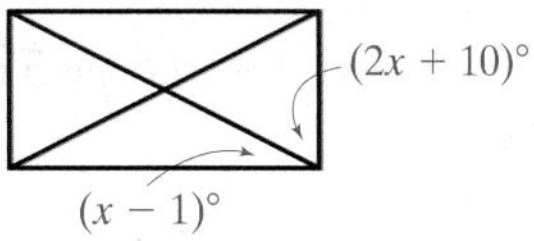

64. Find the value of c. See Section 4.1.

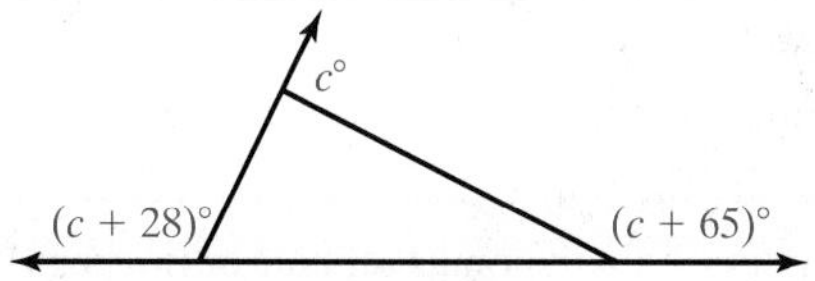

Solve. See Sections 1.7 and 3.6.

65. Find the slope of the line containing the points $C(-1, 5)$ and $D(7, 3)$.

66. Given $A(6, -2)$ and $B(-4, 8)$, find the midpoint and length of $\overline{AB}$.

CHAPTER 6 VOCABULARY CHECK

Word Bank *Use the choices to fill in each blank. Some choices may be used more than once. Always fill in the blank with the most precise polygon.*

polygon	quadrilateral	concave	kite
vertex	diagonal	convex	base
n-gon	regular	equilateral	leg
isosceles	rectangle	equiangular	
midsegment	square	rhombus	
trapezoid	base angle(s)	parallelogram	

1. A quadrilateral with both pairs of opposite sides parallel is called a(n) ____________.
2. A(n) ____________ is a parallelogram with four congruent sides and four right angles.
3. A(n) ____________ is a parallelogram with four congruent sides.
4. A(n) ____________ is a parallelogram with four right angles.
5. Each end point of a side of a polygon is called a(n) ____________.
6. In general, a polygon with *n* sides is called a(n) ____________.
7. A polygon with 4 sides is called a(n) ____________.
8. A polygon is ____________ if no line containing a side contains a point within the interior of the polygon.
9. A polygon is ____________, or nonconvex, if it is not convex.
10. A segment joining two nonconsecutive vertices of a convex polygon is called a(n) ____________ of the polygon.
11. A(n) ____________ polygon is a polygon with all sides congruent.
12. A(n) ____________ polygon is a polygon with all angles congruent.
13. A(n) ____________ polygon is a polygon that is equilateral and equiangular.
14. A quadrilateral with two pairs of consecutive sides congruent and no opposite sides congruent is called a(n) ____________.
15. A quadrilateral with exactly one pair of parallel sides is called a(n) ____________.
16. The segment that joins the midpoints of the legs of a trapezoid is called a(n) ____________.
17. A trapezoid with legs that are congruent is called a(n) ____________ trapezoid.
18. The nonparallel sides of a trapezoid are each called a(n) ____________.
19. The parallel sides of a trapezoid are each called a(n) ____________.
20. The two angles that share a base of a trapezoid are called ____________.
21. A trapezoid has two pairs of ____________.
22. A(n) ____________ is a plane figure formed by 3 or more line segments called sides. Each side intersects two other sides only at their endpoints.

CHAPTER 6 REVIEW

(6.1) *For each polygon,*

a. Name the polygon according to the number of sides.
b. Classify the polygon as convex or concave.
c. Classify the polygon as regular or not regular.

1.

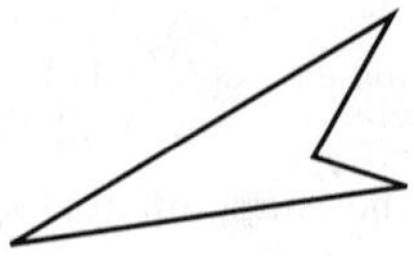

2.

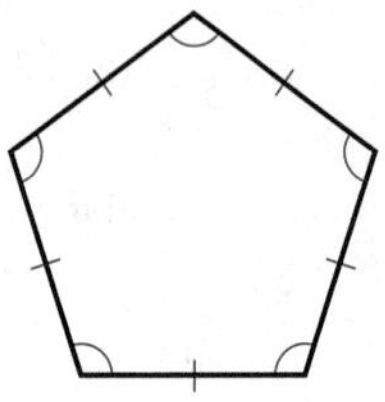

Find the measure of each missing angle.

3.

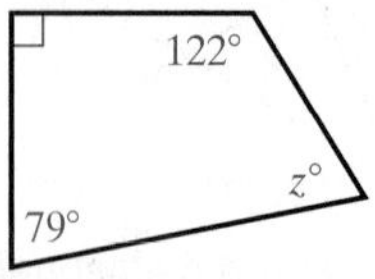

4.

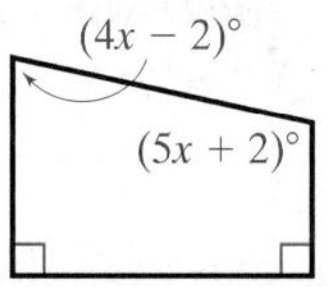

(6.2) *Find the measures of the numbered angles for each parallelogram.*

5.

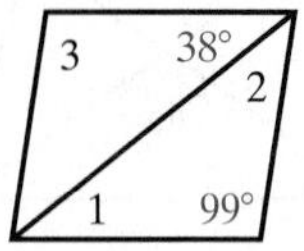

6.

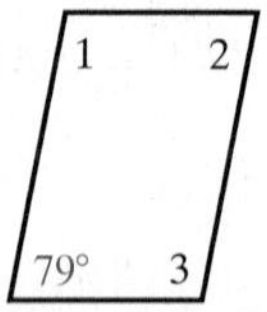

7.
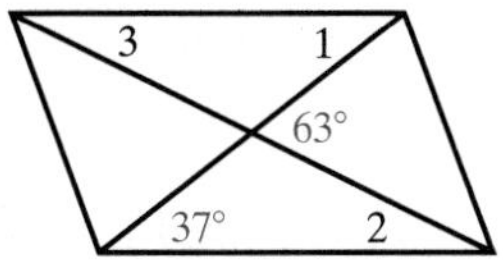

8.

Find the values of x and y in ▱ABCD.

9. $AB = 2y, BC = y + 3, CD = 5x - 1, DA = 2x + 4$

10. $AB = 2y + 1, BC = y + 1, CD = 7x - 3, DA = 3x$

(6.3) *Determine whether the quadrilateral must be a parallelogram.*

11.
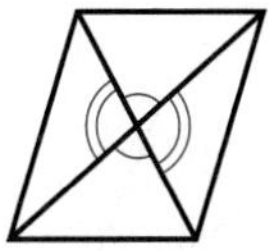

12.
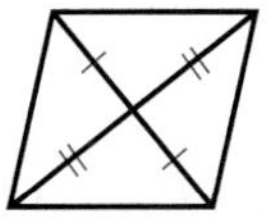

Find the values of the variables for which ABCD must be a parallelogram.

13.
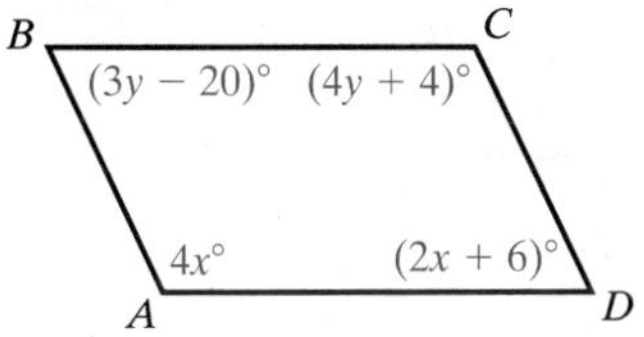

14.
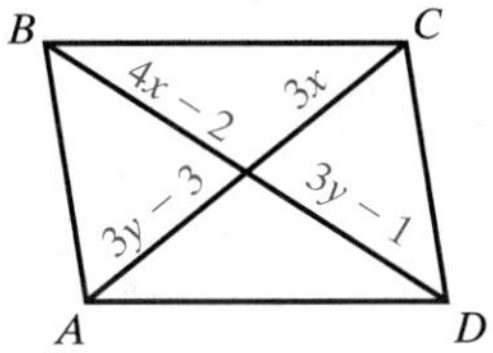

(6.4) *Find the measures of the numbered angles in each special parallelogram.*

15.
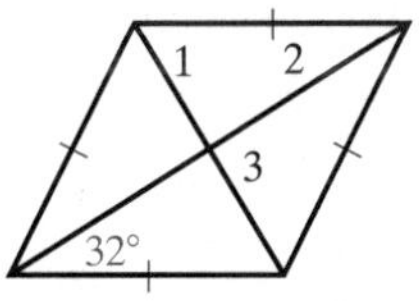

16.
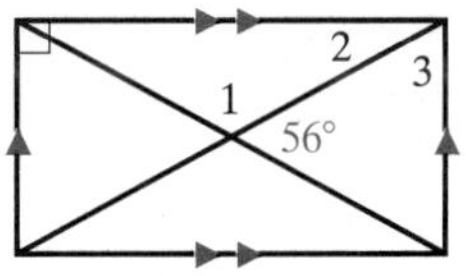

Determine whether each statement is always, sometimes, or never true.

17. A rhombus is a square.

18. A square is a rectangle.

19. A rhombus is a rectangle.

20. The diagonals of a parallelogram are perpendicular.

21. The diagonals of a parallelogram are congruent.

22. Opposite angles of a parallelogram are congruent.

Can you conclude that the parallelogram is a rhombus, rectangle, or square? Give the most precise answer and explain your answer.

23.
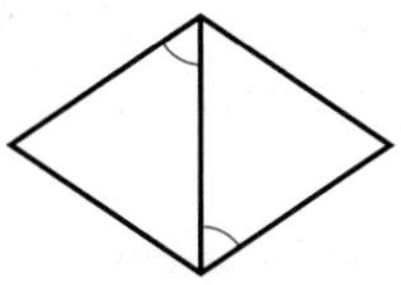

24.
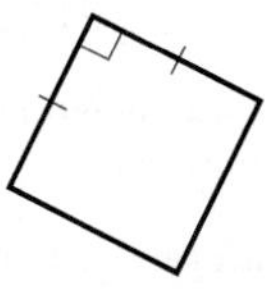

For what value of x is the figure the given parallelogram? Justify your answer.

25. Rhombus
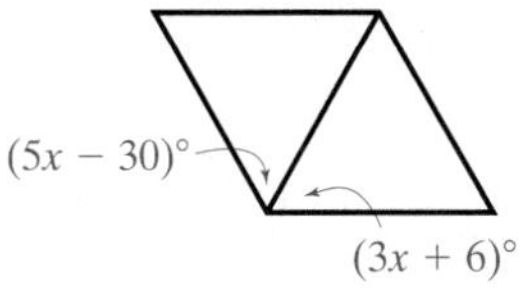

26. Rectangle
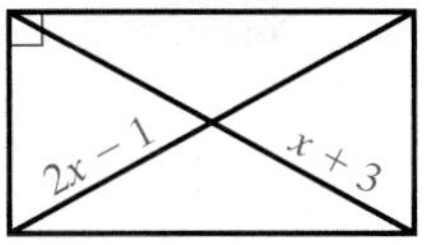

(6.5) *Find the measures of the numbered angles in each isosceles trapezoid.*

27.
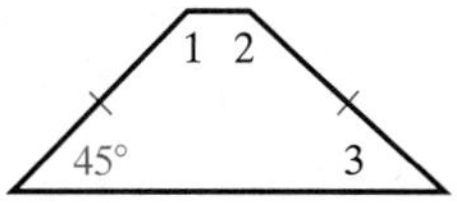

28.
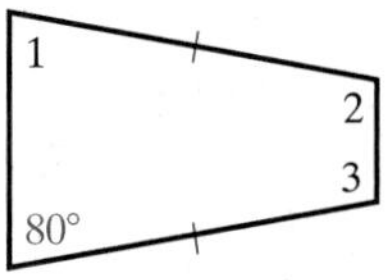

Find the measures of the numbered angles in each kite.

29.
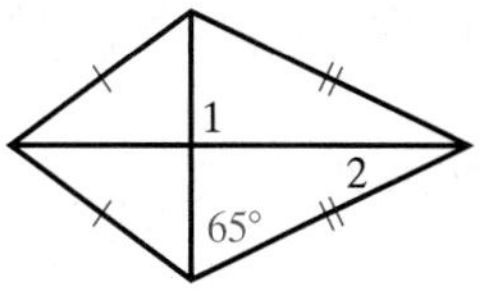

30.
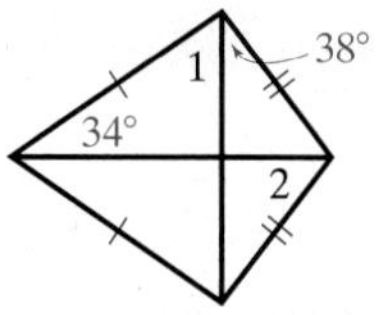

31. A trapezoid has base lengths of $(6x - 1)$ units and 3 units. Its midsegment has a length of $(5x - 3)$ units. What is the value of x?

32. Find the value of x, then find HG, CD, and EF.
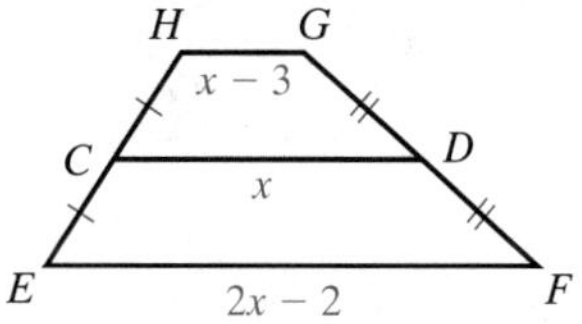

MIXED REVIEW

33. Find the measures of the numbered angles in the parallelogram.
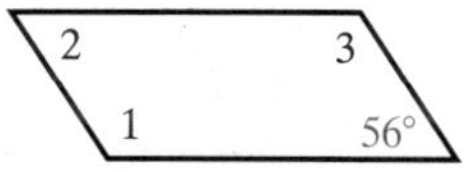

34. What are the measures of the numbered angles in the rhombus?
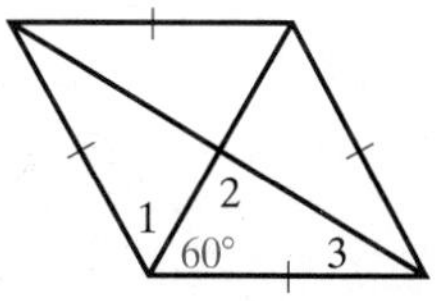

35. Can you conclude that the parallelogram is a rhombus, rectangle, or square? Give the most precise polygon.
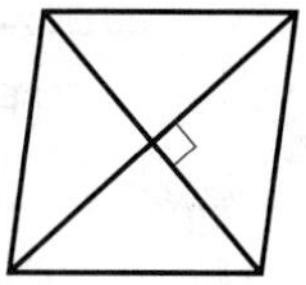

36. $ABCD$ is an isosceles trapezoid. What is $m\angle C$?
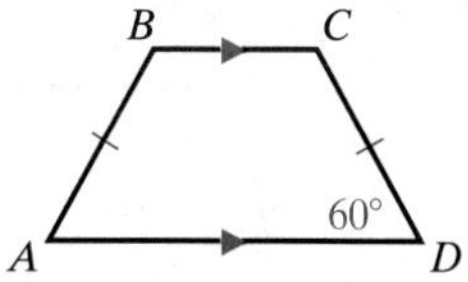

Find the values of the variables for each quadrilateral.

37.
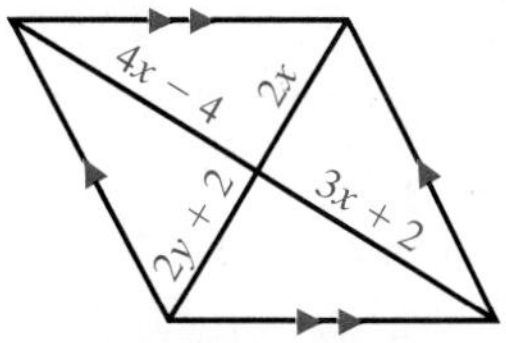

38.
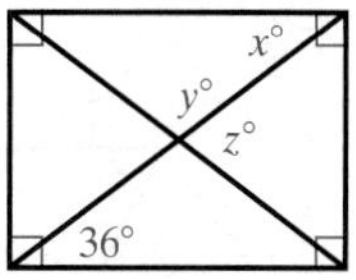

39.
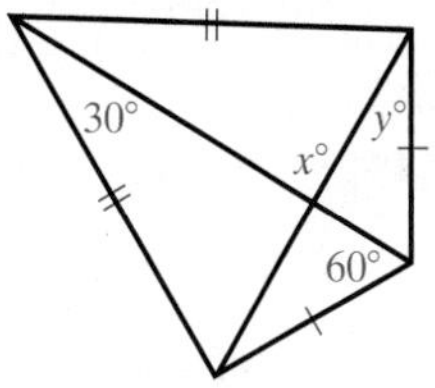

40.
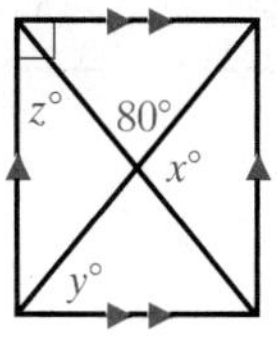

Show that a quadrilateral with the given vertices is a parallelogram.

41. Coordinate Geometry $G(2,5), R(5,8), A(-2,12), D(-5,9)$

CHAPTER 6 TEST

TEST PREP VIDEO The fully worked-out solutions to any exercises you want to review are available in MyMathLab.

Name the polygon according to the number of sides. Then identify whether it is convex or concave.

1. 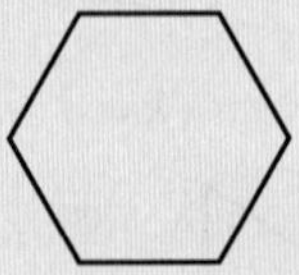

2. 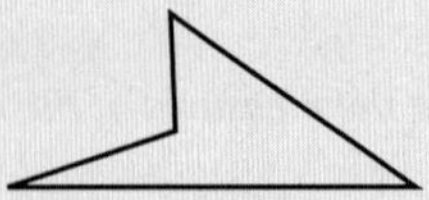

Classify the quadrilateral as precisely as possible. Explain your reasoning.

3. 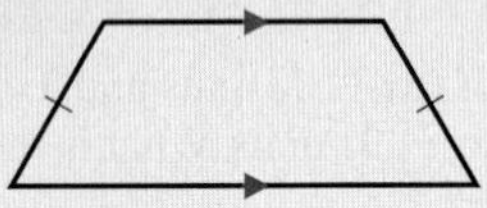

4. 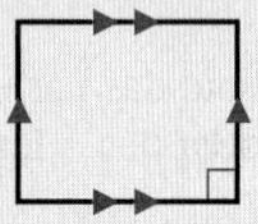

5. 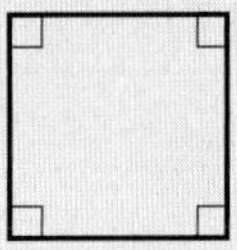

6. 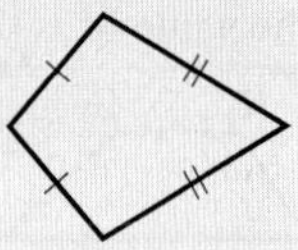

Find the value of each variable.

7. 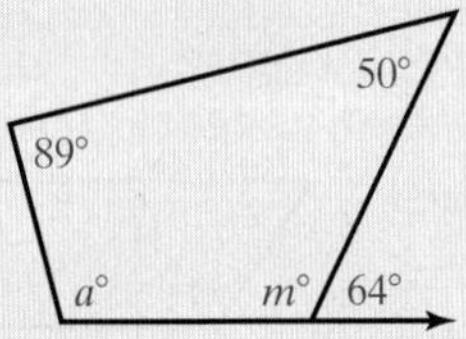

Find the values of the variables for each quadrilateral.

8. 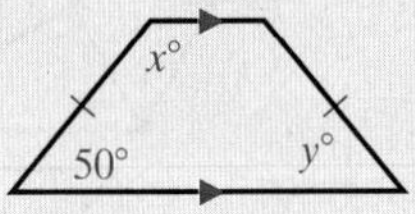

9. 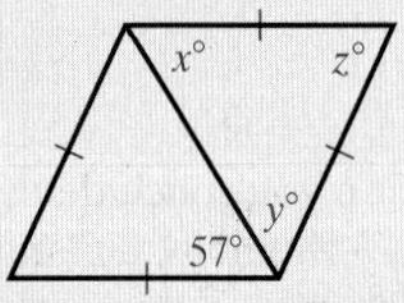

Find the values of the variables for which ABCD is a parallelogram.

10. 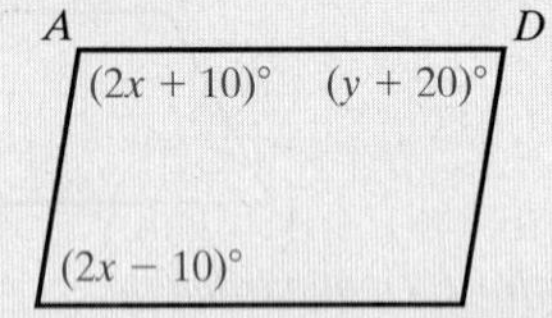

11. 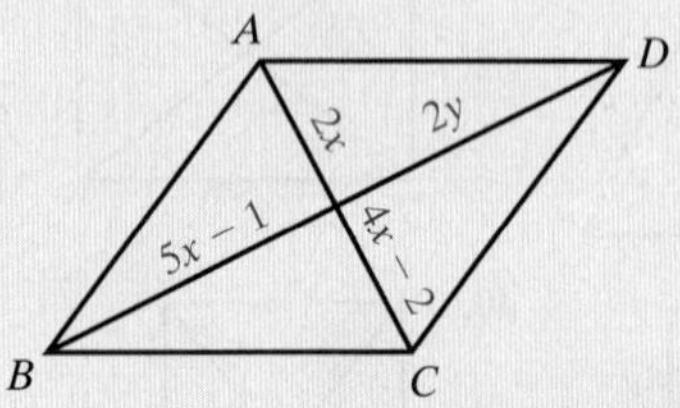

Classify the quadrilateral or precisely as possible. Then find the value(s) of the variable(s).

12. 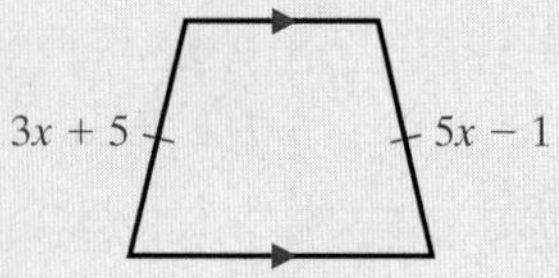

13. 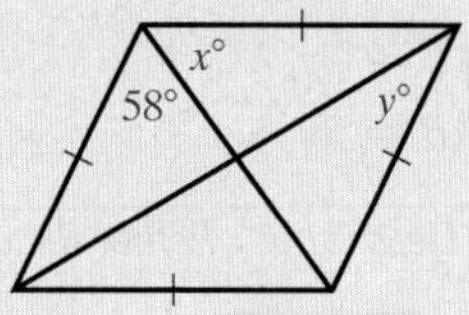

14. 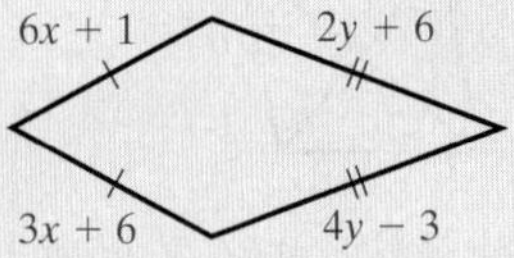

15. 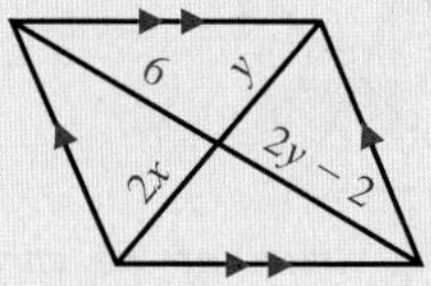

16. Find AB, CD, and EF.

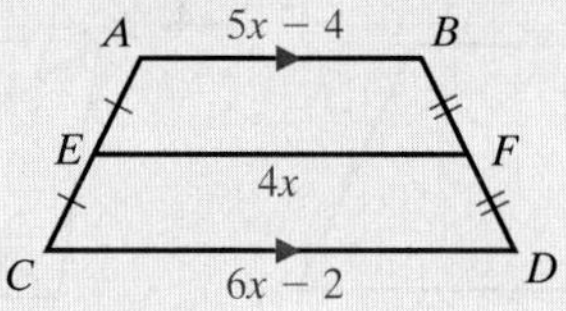

17. In the figure at the right, $\overleftrightarrow{AB} \parallel \overleftrightarrow{CD} \parallel \overleftrightarrow{EF}$. Find AE.

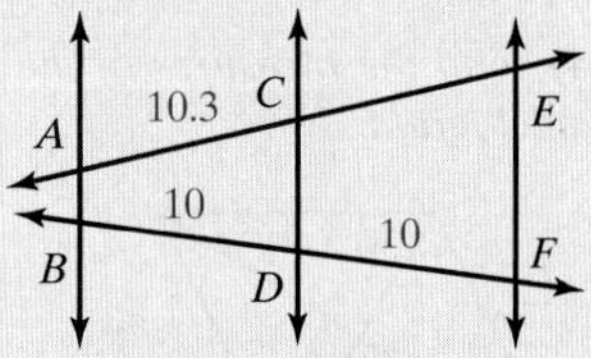

18. Proof

Given: $\square ABCD$, $\overline{AC}$ bisects $\angle DAB$.

Prove: $\overline{AC}$ bisects $\angle DCB$.

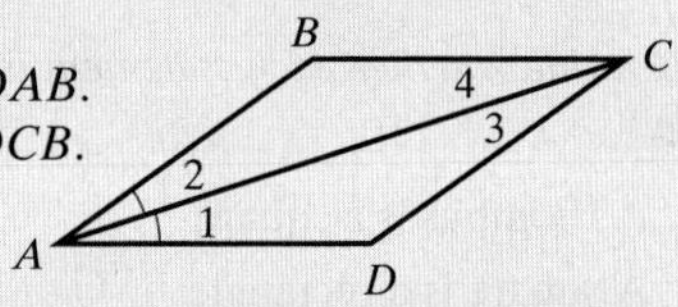

Coordinate Geometry *Graph each quadrilateral ABCD. Then determine the most precise name for it.*

19. $A(1, 2), B(11, 2), C(7, 5), D(4, 5)$

20. $A(3, -2), B(5, 4), C(3, 6), D(1, 4)$

Coordinate Geometry *Show that a quadrilateral with the given vertices is a parallelogram.*

21. $A(1, -4), B(1, 1), C(-2, 2), D(-2, -3)$

Decide whether the statement is true or false. If true, explain why. If false, show a counterexample.

22. A quadrilateral with congruent diagonals is either an isosceles trapezoid or a rectangle.

23. A quadrilateral with congruent and perpendicular diagonals must be a kite.

24. Each diagonal of a kite bisects two angles of the kite.

CHAPTER 6 STANDARDIZED TEST

Choose the correct name and classification of each polygon.

1.

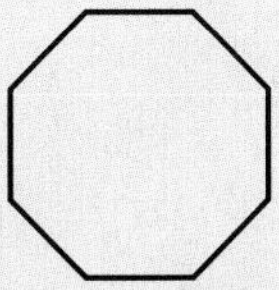

a. hexagon; concave **b.** octagon; convex
c. heptagon; convex **d.** octagon; concave

2.

a. quadrilateral; concave **b.** pentagon; convex
c. pentagon; concave **d.** none of these

Classify each quadrilateral as precisely as possible.

3.

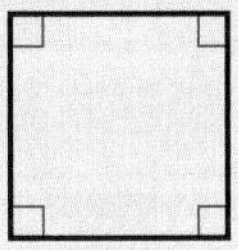

a. square **b.** parallelogram
c. rectangle **d.** none of these

4.

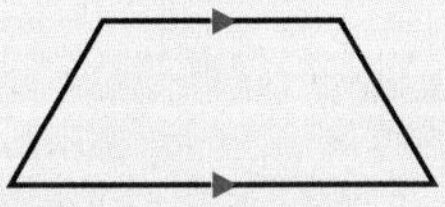

a. trapezoid **b.** isosceles trapezoid
c. parallelogram **d.** none of these

5.

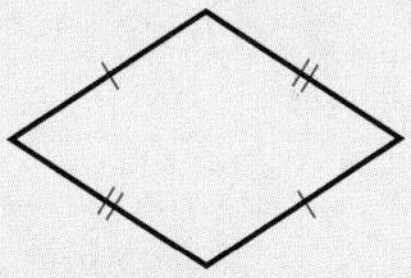

a. parallelogram **b.** rhombus
c. kite **d.** none of these

Choose the correct values for the variables.

6.

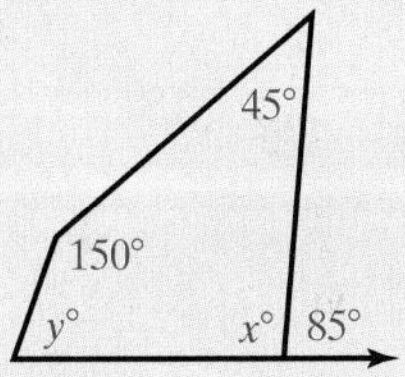

a. $x = 95, y = 85$ **b.** $x = 85, y = 80$
c. $x = 95, y = 70$ **d.** none of these

7.

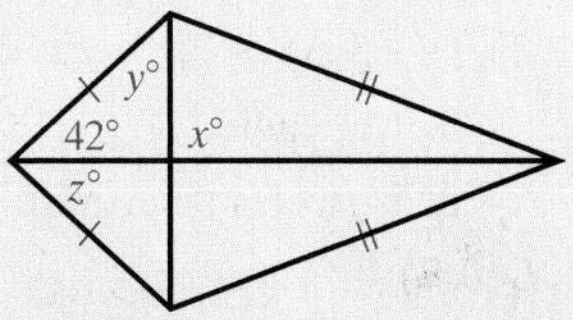

a. $x = 90, y = 42, z = 42$ **b.** $x = 90, y = 42, z = 48$
c. $x = 90, y = 48, z = 42$ **d.** none of these

8. Find the values of the variables for which the figure is a parallelogram.

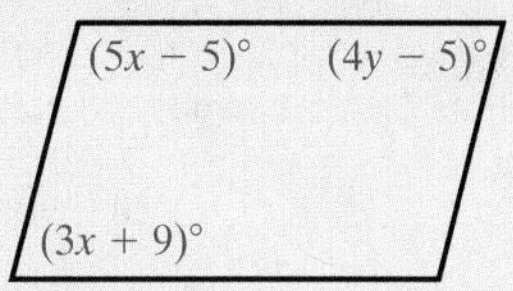

a. $x = 7, y = 8.75$ **b.** $x = 22, y = 20$
c. $x = 22, y = 27.5$ **d.** none of these

Classify the quadrilateral precisely and choose the correct values for the variables.

9.

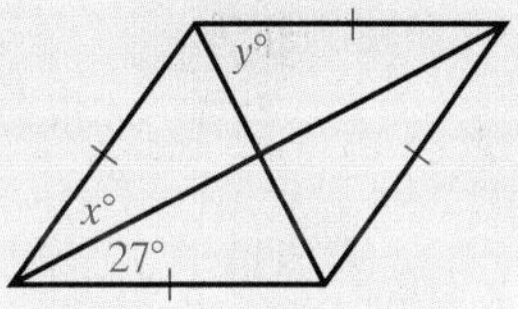

a. kite; $x = 27, y = 63$
b. rhombus; $x = 27, y = 27$
c. parallelogram; $x = 27, y = 63$
d. rhombus; $x = 27, y = 63$

10.

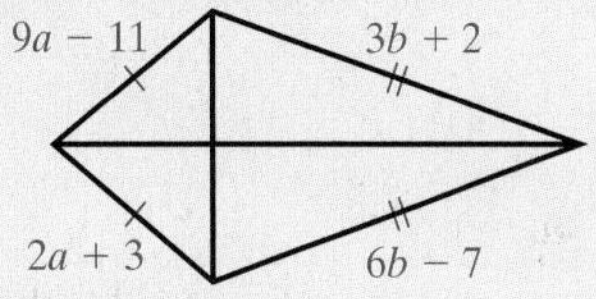

a. rhombus; $a = 2, b = 3$
b. kite; $a = 2, b = 3$
c. quadrilateral; $a = 3, b = 2$
d. parallelogram; $a = 3, b = 2$

11. Find AB, CD, and EF.

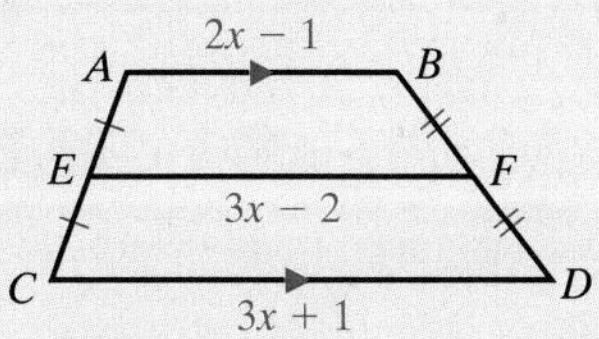

a. $AB = 3, CD = 6, EF = 3$
b. $AB = 7, CD = 13, EF = 10$
c. $AB = 6, CD = 12, EF = 9$
d. $AB = 9, CD = 17, EF = 13$

Coordinate Geometry *Give the most precise name for quadrilateral ABCD.*

12. $A(-5,-3), B(1,-1), C(1,3), D(-5,5)$

a. quadrilateral **b.** trapezoid
c. parallelogram **d.** isosceles trapezoid

13. $A(-1,3), B(3,4), C(4,8), D(0,7)$

a. rhombus **b.** kite
c. parallelogram **d.** square

14. $A(-3,2), B(-2,-5), C(5,-4), D(4,3)$

a. rhombus **b.** kite
c. parallelogram **d.** square

Decide whether each statement is true or false.

15. All equilateral polygons are regular.

a. True **b.** False

16. All equiangular polygons are regular.

a. True **b.** False

17. Every square is a rhombus.

a. True **b.** False

CHAPTER

7 Similarity

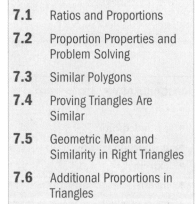

7.1 Ratios and Proportions

7.2 Proportion Properties and Problem Solving

7.3 Similar Polygons

7.4 Proving Triangles Are Similar

7.5 Geometric Mean and Similarity in Right Triangles

7.6 Additional Proportions in Triangles

History of Tallest Structures in Seattle

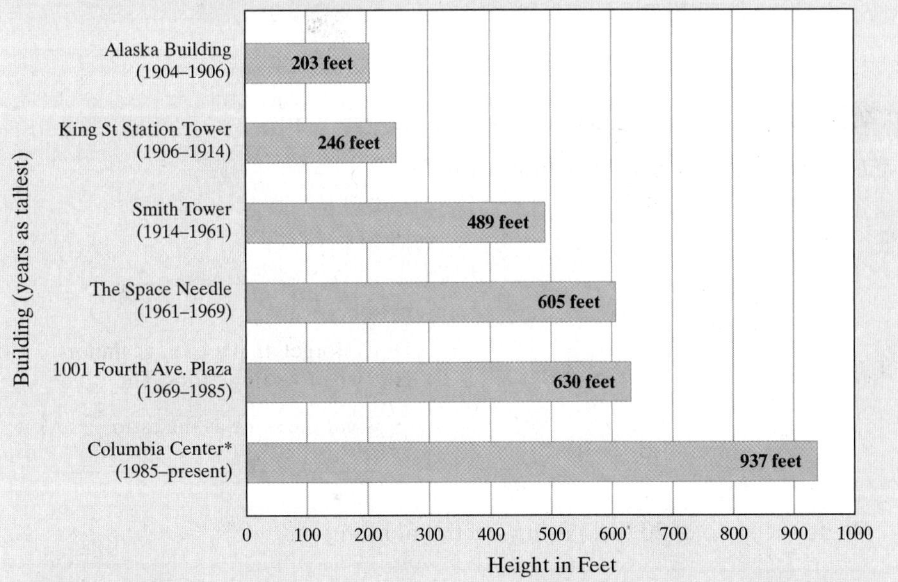

*Emerald City Tower I is being proposed for construction in Seattle with a height of 1000 feet.

Source: Wikipedia

The bar graph above shows the tallest structure history of Seattle, Washington. We look at this city because of a unique structure that once held the title of tallest in Seattle, as well as tallest structure west of the Mississippi. This unique structure is pictured above. It is called the "Space Needle" and is now the official landmark of Seattle. This flying-saucer-looking tower was built for the 1962 World's Fair, hosted by Seattle.

In Section 7.2, Exercise 43, and Section 7.4, Exercise 13, we will explore some interesting facts about the Seattle Space Needle. (*Source:* SpaceNeedle.com and other Internet research)

7.1 RATIOS AND PROPORTIONS

OBJECTIVES

1. Write Ratios as Fractions.
2. Write Ratios in Simplest Form.
3. Understand and Work with Extended Ratios.
4. Solve Proportions.

VOCABULARY

- ratio
- extended ratio
- proportion
- extremes
- means
- cross products

OBJECTIVE 1 ▶ Writing Ratios as Fractions. A **ratio** is the quotient of two quantities. In fact, ratios are no different from fractions, except that ratios are sometimes written using a notation other than fractional notation. For example, the ratio of 1 to 2 can be written as

$$1 \text{ to } 2 \quad \text{or} \quad \frac{1}{2} \quad \text{or} \quad 1{:}2$$

$\frac{1}{2}$ ↗ fractional notation; 1:2 ↑ colon notation

These ratios are all read as "the ratio of 1 to 2."

In this section, we write ratios using fractional notation. If the fraction happens to be an improper fraction, we do not write the fraction as a mixed number. Why? The mixed number form is not a quotient of two quantities, so it is not written in the form of a ratio.

Writing a Ratio as a Fraction

The order of the quantities is important when writing ratios. To write a ratio as a fraction, write the *first number* of the ratio as the *numerator* of the fraction and the *second number* as the *denominator.*

Helpful Hint

The ratio of 6 to 11 is $\frac{6}{11}$, *not* $\frac{11}{6}$.

EXAMPLE 1 **Writing a Ratio as a Fraction**

Write the ratio of 12 to 17 using fractional notation.

Solution The ratio is $\frac{12}{17}$.

Helpful Hint

Don't forget that order is important when writing ratios. The ratio $\frac{17}{12}$ is *not* the same as the ratio $\frac{12}{17}$.

PRACTICE 1 Write the ratio of 20 to 23 using fractional notation.

OBJECTIVE 2 ▶ Writing Ratios in Simplest Form. To simplify a ratio, we just write the fraction in simplest form. Common factors as well as common units can be divided out. If the ratio has unlike units, convert to like units before simplifying.

EXAMPLE 2 **Writing a Ratio as a Fraction in Simplest Form**

Write each ratio as a fraction in simplest form.

a. \$15 to \$10 **b.** 4 ft to 24 in. **c.** $\frac{500 \text{ cm}}{7 \text{ m}}$

Solution For parts **a** and **b**, we first write the ratio as a fraction. For parts b and c, we convert unlike units to like units so that the units divide out.

a. $\frac{\$15}{\$10} = \frac{15}{10} = \frac{3 \cdot \cancel{5}^1}{2 \cdot \cancel{5}_1} = \frac{3}{2}$ Dollars divide out.

b. $\frac{4\text{ ft}}{24\text{ in.}} = \frac{4 \cdot 12\text{ in.}}{24\text{ in.}} = \frac{4 \cdot \overset{1}{\cancel{12}}}{\underset{2}{\cancel{24}}} = \frac{4}{2} = \frac{2}{1}$ Inches divide out.

c. $\frac{500\text{ cm}}{7\text{ m}} = \frac{500\text{ cm}}{7 \cdot 100\text{ cm}} = \frac{500}{700} = \frac{5}{7}$ Centimeters divide out. □

PRACTICE 2 Write each ratio as a fraction in simplest form.

a. \$8 to \$6 **b.** $\frac{2\text{ yd}}{2\text{ ft}}$ **c.** 9 m to 1100 cm

Helpful Hint

In Example 2 above:

a. Although $\frac{3}{2} = 1\frac{1}{2}$, a ratio is a quotient of *two* quantities. For that reason, ratios are not written as mixed numbers.

b. Although $\frac{2}{1} = 2$, a ratio is a quotient of *two* quantities, so we write the ratio as $\frac{2}{1}$ and not 2.

EXAMPLE 3 Using Ratios in Geometry

Given the rectangle shown:

a. Find the ratio of its width to its length.

b. Find the ratio of its length to its perimeter.

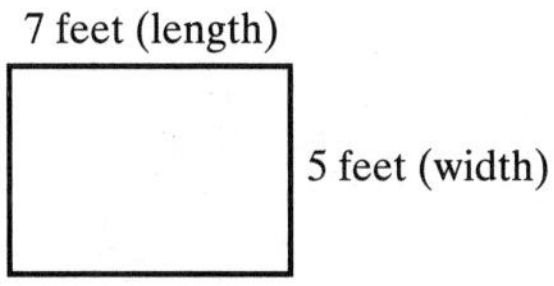

Solution

a. The ratio of its width to its length is

$$\frac{\text{width}}{\text{length}} = \frac{5\ \cancel{\text{feet}}}{7\ \cancel{\text{feet}}} = \frac{5}{7}$$

b. Recall that the perimeter of a rectangle is the distance around the rectangle: $7 + 5 + 7 + 5 = 24$ feet. The length is given as 7 feet. The ratio of its length to its perimeter is

$$\frac{\text{length}}{\text{perimeter}} = \frac{7\ \cancel{\text{feet}}}{24\ \cancel{\text{feet}}} = \frac{7}{24}$$ □

PRACTICE 3 Given the triangle shown:

a. Find the ratio of the length of the shortest side to the length of the longest side.

b. Find the ratio of the length of the longest side to the perimeter of the triangle.

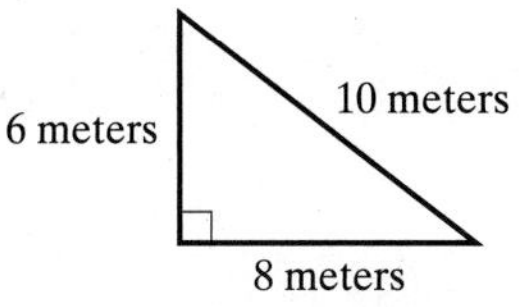

OBJECTIVE 3 ▶ Understanding Extended Ratios. An **extended ratio** compares three (or more) numbers. In the extended ratio $a:b:c$, the ratio of the first two numbers is $a:b$, the ratio of the last two numbers is $b:c$, and the ratio of the first and last numbers is $a:c$.

An extended ratio of three numbers is often used to provide information about triangles.

EXAMPLE 4 Using an Extended Ratio

The lengths of the sides of a triangle are in the extended ratio 3:5:6. The perimeter of the triangle is 98 units. What is the length of each side?

Solution Sketch the triangle. Use the given extended ratio to label the sides with expressions for their lengths. The perimeter will help us find exact side lengths.

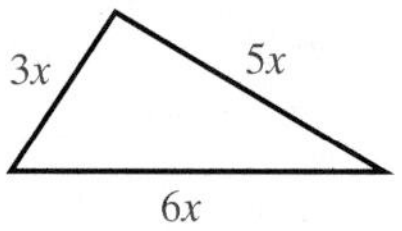

$$3x + 5x + 6x = 98 \quad \text{The perimeter is 98 units.}$$
$$14x = 98 \quad \text{Combine like terms.}$$
$$x = 7 \quad \text{Divide both sides by 14.}$$

Thus, the side lengths are

$$3x = 3(7) = 21 \text{ units}, 5x = 5(7) = 35 \text{ units, and } 6x = 6(7) = 42 \text{ units.}$$ □

To **check**, see that the perimeter 21 units + 35 units + 42 units is the given 98 units. It is.

> **▶ Helpful Hint**
> Remember: The extended ratio of 3:5:6 does *not* mean the lengths of the sides of the triangle are 3, 5, and 6 units.

PRACTICE 4 The lengths of the sides of a triangle are in the extended ratio 4:7:9. The perimeter is 60 cm. What are the lengths of the sides?

OBJECTIVE 4 ▶ Solving Proportions. An equation stating that two ratios are equal is called a **proportion.** The first and last numbers in a proportion are the **extremes.** The middle two numbers are the **means.**

extremes → $\frac{2}{3} = \frac{4}{6}$ ← means (extremes: 2 and 6; means: 3 and 4) or $2:3 = 4:6$ (extremes: 2 and 6; means: 3 and 4)

Cross products are the product of the means and also the product of the extremes.

Cross Products Property

Words	Symbols	Example
In a true proportion, the product of the extremes equals the product of the means.	If $\frac{a}{b} = \frac{c}{d}$, where $b \neq 0$ and $d \neq 0$, then $ad = bc$.	$\frac{2}{3} = \frac{4}{6}$ $2 \cdot 6 = 3 \cdot 4$ $12 = 12$

To see why the above is true, begin with $\frac{a}{b} = \frac{c}{d}$, where $b \neq 0$ and $d \neq 0$, and multiply through by bd.

> **▶ Helpful Hint**
> Don't forget that order does not matter when multiplying (Commutative Property). This means that $da = cb$ means the same as $ad = bc$.

$$bd \cdot \frac{a}{b} = \frac{c}{d} \cdot bd \quad \text{Multiply each side of the proportion by } bd.$$
$$\frac{\cancel{b}d}{1} \cdot \frac{a}{\cancel{b}} = \frac{c}{\cancel{d}} \cdot \frac{b\cancel{d}}{1} \quad \text{Divide out the common factors.}$$
$$ad = bc \quad \text{Simplify.}$$

EXAMPLE 5 **Solving a Proportion**

Solve each proportion for the variable.

a. $\frac{6}{x} = \frac{5}{4}$ **b.** $\frac{y+4}{9} = \frac{y}{3}$

Solution

a.
$$\frac{6}{x} = \frac{5}{4}$$
$$6(4) = 5x \quad \text{Cross Products Property}$$
$$24 = 5x \quad \text{Simplify.}$$
$$x = \frac{24}{5} \quad \text{Solve for the variable.}$$

The solution is $\frac{24}{5}$ or 4.8.

b.
$$\frac{y+4}{9} = \frac{y}{3}$$
$$3(y+4) = 9y$$
$$3y + 12 = 9y$$
$$12 = 6y$$
$$2 = y \quad \text{or} \quad y = 2$$

The solution is 2.

Check: To check, replace the variable with the proposed solution and see that a true statement results.

a.
$$\frac{6}{x} = \frac{5}{4}$$
$$\frac{6}{\frac{24}{5}} \stackrel{?}{=} \frac{5}{4}$$
$$6 \cdot 4 \stackrel{?}{=} \frac{24}{5} \cdot 5 \quad \text{Cross Products Property}$$
$$24 = 24 \checkmark \quad \text{True}$$

b.
$$\frac{y+4}{9} = \frac{y}{3}$$
$$\frac{2+4}{9} \stackrel{?}{=} \frac{2}{3}$$
$$\frac{6}{9} \stackrel{?}{=} \frac{2}{3}$$
$$6 \cdot 3 \stackrel{?}{=} 9 \cdot 2$$
$$18 = 18 \checkmark$$

PRACTICE 5 Solve each proportion for the variable.

a. $\frac{9}{2} = \frac{a}{14}$ **b.** $\frac{15}{m+1} = \frac{3}{m}$

VOCABULARY & READINESS CHECK

Word Bank *Use the words and phrases below to fill in each blank.*

means	cross products	true	ratio
extremes	proportion	false	

1. $\frac{4.2}{8.4} = \frac{x}{2}$ is called a(n) ____________ while $\frac{7}{8}$ is called a(n) ____________.

2. In $\frac{a}{b} = \frac{c}{d}$, $a \cdot d$ and $b \cdot c$ are called ____________.

3. In a proportion, if cross products are equal, the proportion is a(n) ____________ proportion.

4. In a proportion, if cross products are not equal, the proportion is a(n) ____________ proportion.

5. In $\frac{a}{b} = \frac{c}{d}$, the variables b and c are called the ____________.

6. In $\frac{a}{b} = \frac{c}{d}$, the variables a and d are called the ____________.

Answer each statement true or false.

7. The quotient of two quantities is called a ratio. ____________

8. The ratio $\frac{7}{5}$ means the same as the ratio $\frac{5}{7}$. ____________

9. The ratio 2 to 5 equals $\frac{5}{2}$ in fractional notation. ____________

10. The ratio 30:41 equals $\frac{30}{41}$ in fractional notation. ____________

7.1 EXERCISE SET MyMathLab®

Write each ratio using fractional notation. Do not simplify. See Example 1.

1. 11 to 14

2. 23 to 10

3. 2.8 to 7.6

4. 3.9 to 4.2

Mixed Practice *Write each ratio as a ratio of whole numbers using fractional notation. Write the fraction in simplest form. See Examples 1 through 3.*

5. 16 to 24

6. 25 to 150

7. \$32 to \$100

8. \$46 to \$102

9. 9 inches to 12 inches

10. 8 inches to 20 inches

11. 10 hours to 1 day

12. 14 hours to 2 days

13. 3 ft to 12 in.

14. 36 ft to 1 yd

15. 6 m to 900 cm

16. 800 cm to 10 m

Write the ratio described in each exercise as a fraction in simplest form. See Examples 2 and 3.

17. **Average Weight of Mature Whales**

Use the information above to find the ratio of the average weight of a mature Fin Whale to the average weight of a mature Blue Whale.

18. **Countries with Small Land Areas**

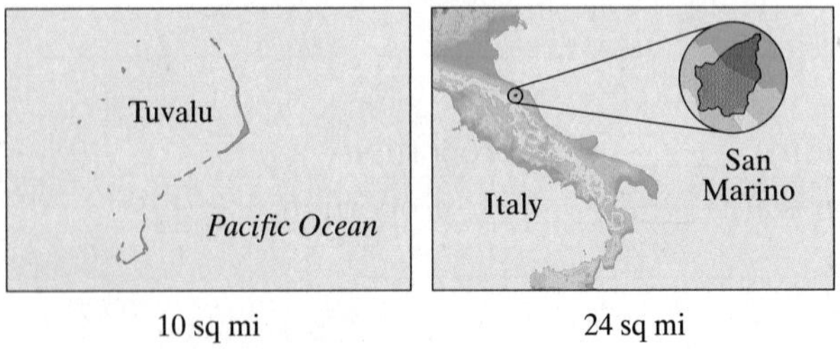

Use the information above to find the ratio of the land area of Tuvalu to the land area of San Marino. (*Source:* World Almanac)

19. For a regulation size basketball court, find the ratio of the width to the perimeter.

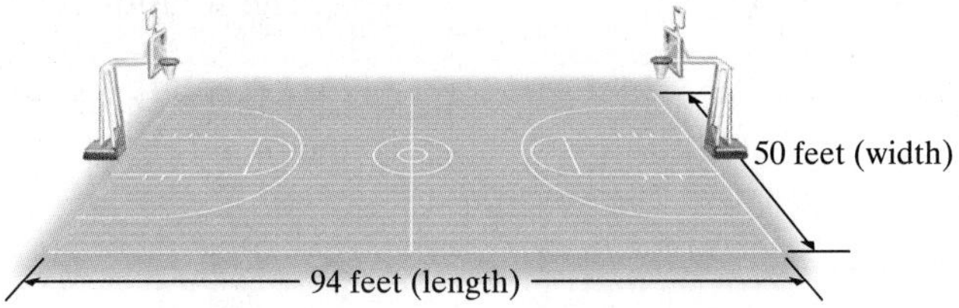

20. For the swimming pool shown, find the ratio of the width to the perimeter.

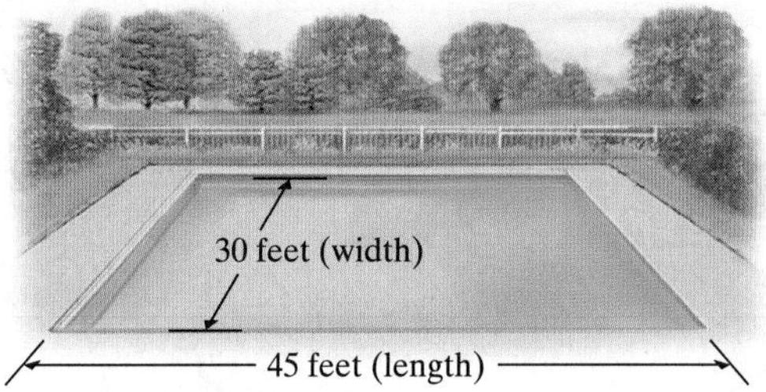

21. Find the ratio of the width (shorter side) to the length (longer side) of the sign below.

22. The circle graph below shows the 2012 Box Office ratings by MPAA. Find the ratio of G films to PG films.

2012 Box Office Films

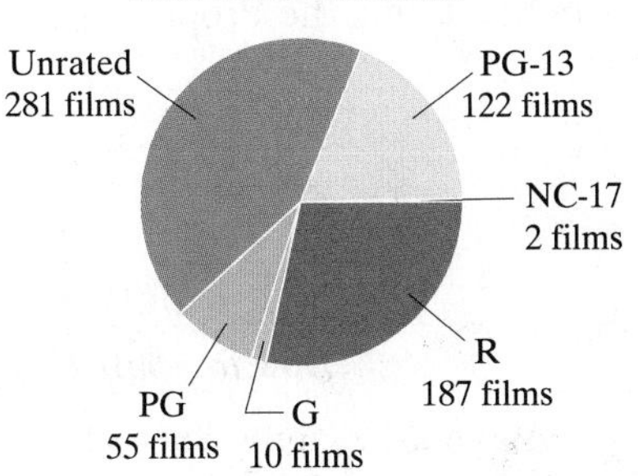

Source: MPAA (Motion Picture Association of America)

Solve. See Example 4.

23. The measures of three angles whose sum is 90° are in the extended ratio 2:3:5. Find the exact angle measures.

24. The measures of three angles whose sum is 70° are in the same extended ratio as in Exercise **23**; that is, 2:3:5. Find the exact angle measures.

25. The lengths of the sides of a triangle are in the extended ratio 6:7:9. The perimeter of the triangle is 88 cm. Find the lengths of the sides.

26. The lengths of the sides of a triangle are in the extended ratio 5:8:10. The perimeter of the triangle is 115 in. Find the lengths of the sides.

27. The measures of the angles of a triangle are in the extended ratio 4:3:2. Find the measures of the angles.

28. The measures of the angles of a triangle are in the extended ratio 7:3:2. Find the measures of the angles.

Solve each proportion for the variable. See Example 5.

29. $\frac{1}{3} = \frac{x}{12}$

30. $\frac{9}{5} = \frac{3}{x}$

31. $\frac{4}{x} = \frac{5}{9}$

32. $\frac{y}{10} = \frac{15}{25}$

33. $\frac{9}{24} = \frac{12}{n}$

34. $\frac{11}{14} = \frac{b}{21}$

35. $\frac{3}{5} = \frac{6}{x + 3}$

36. $\frac{y + 7}{9} = \frac{8}{5}$

37. $\frac{5}{x - 3} = \frac{10}{x}$

38. $\frac{n - 4}{6} = \frac{n}{8}$

39. $\frac{1}{7y - 5} = \frac{2}{9y}$

40. $\frac{4a + 1}{7} = \frac{2a}{3}$

41. $\frac{5}{x + 2} = \frac{3}{x + 1}$

42. $\frac{2b - 1}{4} = \frac{b - 2}{12}$

Coordinate Geometry *Use the graph. Write each ratio as a fraction in simplest form.*

43. $\frac{AC}{BD}$

44. $\frac{BC}{AD}$

45. slope of $\overline{EB}$

46. slope of $\overline{ED}$

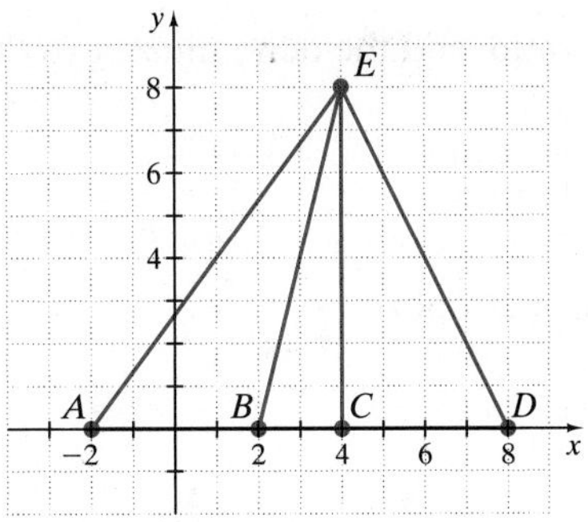

CONCEPT EXTENSIONS

47. The means of a proportion are 4 and 15. List all possible pairs of positive integers that could be the extremes of a true proportion.

48. Describe how to use the Cross Products Property to determine whether $\frac{10}{26} = \frac{16}{42}$ is a true proportion.

49. Write a proportion that has means 6 and 18 and extremes 9 and 12.

50. What is the difference between a ratio and a proportion?

51. **Find the Error** What is the error in the solution of the proportion shown at the right?

52. **Find the Error** What is the error in the solution of the proportion shown at the right?

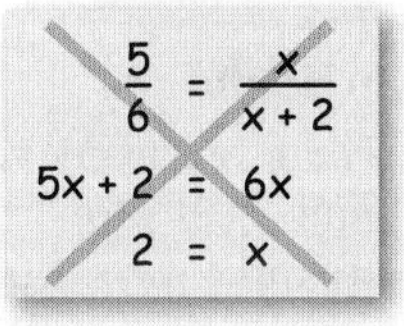

For Exercises 53–54, draw a triangle that satisfies this condition: The measures of the angles are in each extended ratio.

53. 4:4:1

54. 1:1:1

Complete each statement. Justify your answer.

55. If $4m = 9n$, then $\frac{m}{n} = \frac{\blacksquare}{\blacksquare}$.

56. If $30r = 18t$, then $\frac{t}{r} = \frac{\blacksquare}{\blacksquare}$.

Solve each proportion for the variable.

57. $\frac{x - 3}{3} = \frac{2}{x + 2}$

58. $\frac{3 - 4x}{1 + 5x} = \frac{1}{2 + 3x}$

Use the numbers in each proportion to write two other true proportions.

59. $\frac{9}{15} = \frac{3}{5}$

60. $\frac{6}{18} = \frac{1}{3}$

REVIEW AND PREVIEW

Insert $<$ or $>$ to form a true statement. See the Concept Review section at the end of this text.

61. 8.01 8.1

62. 7.26 7.026

63. $2\frac{1}{2}$ $2\frac{1}{3}$

64. $9\frac{1}{5}$ $9\frac{1}{4}$

Simplify each fraction. See the Concept Review section at the end of this text.

65. $\frac{75}{125}$

66. $\frac{11y}{99y}$

67. $\frac{12x}{42}$

68. $\frac{28y^2}{42y^3}$

7.2 PROPORTION PROPERTIES AND PROBLEM SOLVING

OBJECTIVES

1 Use Properties of Proportions to Write Equivalent Proportions.

2 Solve Problems by Writing Proportions.

OBJECTIVE 1 ▶ Using Properties of Proportions. Using the Properties of Equality, we can rewrite proportions in equivalent forms.

Properties of Proportions

a, b, c, and d do not equal zero.

Property	*How to Apply It*
(1) $\frac{a}{b} = \frac{c}{d}$ is equivalent to $\frac{b}{a} = \frac{d}{c}$.	Write the reciprocal of each ratio. $\frac{2}{3} = \frac{4}{6}$ becomes $\frac{3}{2} = \frac{6}{4}$.
(2) $\frac{a}{b} = \frac{c}{d}$ is equivalent to $\frac{a}{c} = \frac{b}{d}$.	Switch the means. $\frac{2}{3} = \frac{4}{6}$ becomes $\frac{2}{4} = \frac{3}{6}$.
(3) $\frac{a}{b} = \frac{c}{d}$ is equivalent to $\frac{a+b}{b} = \frac{c+d}{d}$.	In each ratio, add the denominator to the numerator. $\frac{2}{3} = \frac{4}{6}$ becomes $\frac{2+3}{3} = \frac{4+6}{6}$.

▶ Helpful Hint

There are many ways of writing equivalent proportions, but here are three ways.

We justify Property (1) next. We justify the other two properties in Exercises 57 and 58 of Exercise Set 7.2.

To see why Property (1) is true, use cross products.

$\frac{a}{b} = \frac{c}{d}$ is equivalent to $\frac{b}{a} = \frac{d}{c}$

$a \cdot d = b \cdot c$ Cross products form equivalent equations. $b \cdot c = a \cdot d$

EXAMPLE 1 Using Properties of Proportions

Use the Properties of Proportions to write three proportions equivalent to $\frac{3}{x} = \frac{4}{y}$.

Solution

Property (1)—Write the reciprocal of each ratio:

$$\frac{3}{x} = \frac{4}{y} \text{ is equivalent to } \frac{x}{3} = \frac{y}{4}.$$

Property (2)—Switch the means:

$$\frac{3}{x} = \frac{4}{y} \text{ is equivalent to } \frac{3}{4} = \frac{x}{y}.$$

Property (3)—In each ratio, add the denominator to the numerator:

$$\frac{3}{x} = \frac{4}{y} \text{ is equivalent to } \frac{3+x}{x} = \frac{4+y}{y}.$$

□

PRACTICE 1 Use the Properties of Proportions to write three proportions equivalent to $\frac{z}{2} = \frac{x}{7}$.

EXAMPLE 2 Writing Equivalent Proportions

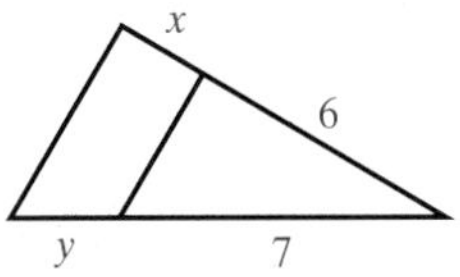

In the diagram, $\frac{x}{6} = \frac{y}{7}$. What ratio completes the equivalent proportion $\frac{x}{y} = \frac{\square}{\square}$? Justify your answer.

Solution Notice the positions of x and y in the first proportion. Now see where x and y appear in the second proportion.

A method using Property (2)

$\frac{x}{6} = \frac{y}{7}$

$\frac{x}{y} = \frac{6}{7}$ Property of Proportions (2)

A method using cross products

$\frac{x}{6} = \frac{y}{7}$

$7x = 6y$ Cross Products Property

$\frac{7x}{7y} = \frac{6y}{7y}$ To solve for $\frac{x}{y}$, divide each side by $7y$.

$\frac{x}{y} = \frac{6}{7}$ Simplify.

The ratio that completes the proportion is $\frac{6}{7}$. □

> **Helpful Hint**
> Look at how the positions of the known parts of the incomplete proportion relate to their positions in the original proportion.
> This information will help you choose which property to use.

PRACTICE
2 For parts **a** and **b**, use the proportion $\frac{x}{6} = \frac{y}{7}$. What ratio completes the equivalent proportion? Justify your answer.

a. $\frac{6}{x} = \frac{\square}{\square}$ **b.** $\frac{\square}{\square} = \frac{y + 7}{7}$

OBJECTIVE 2 ▶ Solving Problems by Writing Proportions. Writing proportions is a powerful tool for solving problems in almost every field, including business, chemistry, biology, health sciences, and engineering, as well as in daily life. Given a specified ratio (or rate) of two quantities, a proportion can be used to determine an unknown quantity.

In this section, we label the problem-solving steps that we will use.

EXAMPLE 3 Determining Distances from a Map

On a chamber of commerce map of Abita Springs, 5 miles corresponds to 2 inches. How many miles correspond to 7 inches?

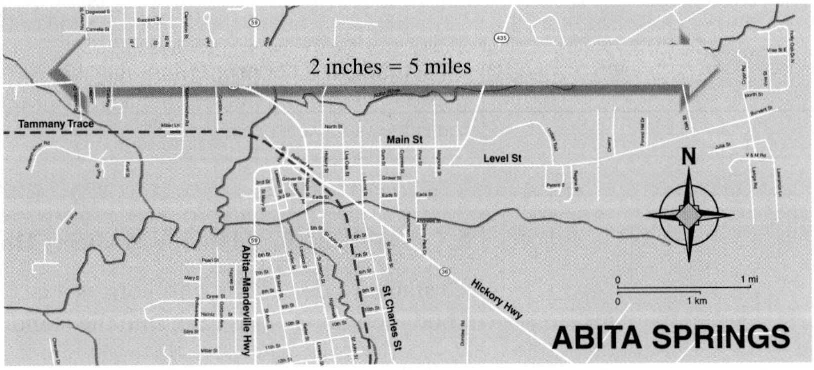

Solution

1. UNDERSTAND. Read and reread the problem. You may want to draw a diagram.

15 miles ?

5 miles	5 miles	5 miles	5 miles	Total is between 15 and 20 miles
2 inches	2 inches	2 inches	2 inches	Total is 7 inches

6 inches 1 inch

From the diagram we can see that a reasonable solution is between 15 and 20 miles.

2. TRANSLATE. We will let x be our unknown number. Since 5 miles corresponds to 2 inches as x miles corresponds to 7 inches, we have the proportion

$$\begin{array}{lcl} \text{miles} \rightarrow & \frac{5}{2} = \frac{x}{7} & \leftarrow \text{miles} \\ \text{inches} \rightarrow & & \leftarrow \text{inches} \end{array}$$

3. SOLVE: Now we solve the proportion.

$$\frac{5}{2} = \frac{x}{7}$$

$$5 \cdot 7 = 2 \cdot x \quad \text{Set the cross products equal to each other.}$$

$$35 = 2x \quad \text{Multiply.}$$

$$\frac{35}{2} = \frac{2x}{2} \quad \text{Divide both sides by 2.}$$

$$17\frac{1}{2} = x \text{ or } x = 17.5 \quad \text{Simplify.}$$

Helpful Hint

Since the answer is not a ratio, it can be a mixed number or a decimal.

4. INTERPRET. *Check* your work. This result is reasonable and fits our estimate since it is between 15 and 20 miles. *State* your conclusion: 7 inches corresponds to 17.5 miles. □

PRACTICE

3 On an architect's blueprint, 1 inch corresponds to 4 feet. How long is a wall represented by a $4\frac{1}{4}$-inch line on the blueprint?

Helpful Hint

We can also solve Example 3 by writing the proportion

$$\frac{2 \text{ inches}}{5 \text{ miles}} = \frac{7 \text{ inches}}{x \text{ miles}}$$

Although other proportions may be used to solve Example 3, we solve by writing proportions so that the numerators have the same unit measures and the denominators have the same unit measures.

EXAMPLE 4 Finding Medicine Dosage

The standard dose of an antibiotic is 4 cc (cubic centimeters) for every 25 pounds (lb) of body weight. At this rate, find the standard dose for a 140-lb woman.

Solution

1. UNDERSTAND. Read and reread the problem. You may want to draw a diagram to estimate a reasonable solution.

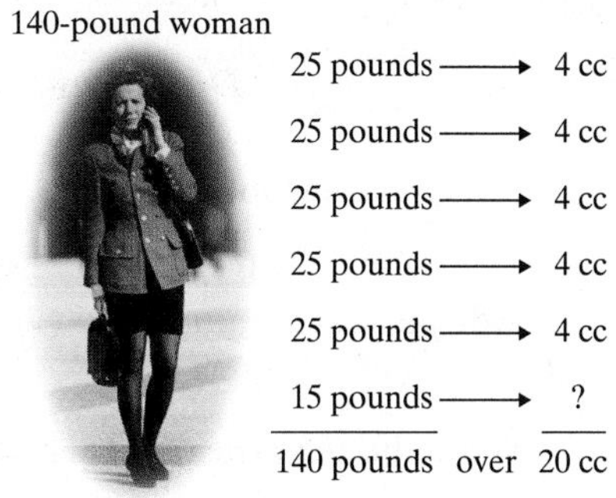

From the diagram, we can see that a reasonable solution is a little over 20 cc.

2. TRANSLATE. We will let x be the unknown number. From the written example, we know that 4 cc is to 25 pounds as x cc is to 140 pounds, or

$$\begin{array}{rcl} \text{cubic centimeters} \rightarrow & \dfrac{4}{25} = \dfrac{x}{140} & \leftarrow \text{cubic centimeters} \\ \text{pounds} \rightarrow & & \leftarrow \text{pounds} \end{array}$$

3. SOLVE:

$$\frac{4}{25} = \frac{x}{140}$$

$$4 \cdot 140 = 25 \cdot x \quad \text{Set the cross products equal to each other.}$$

$$560 = 25x \quad \text{Multiply.}$$

$$\frac{560}{25} = \frac{25x}{25} \quad \text{Divide both sides by 25.}$$

$$22\frac{2}{5} = x \text{ or } x = 22.4 \quad \text{Simplify.}$$

4. INTERPRET. *Check* your work. This result is reasonable since it is a little over 20 cc. *State* your conclusion: The standard dose for a 140-lb woman is 22.4 cc. □

PRACTICE 4 An auto mechanic recommends that 5 ounces of isopropyl alcohol be mixed with a tankful of gas (16 gallons) to increase the octane of the gasoline for better engine performance. At this rate, how many gallons of gas can be treated with an 8-ounce bottle of alcohol?

VOCABULARY & READINESS CHECK

Fill in the chart with T for true or F for false

	Original Proportion	*Directions*	*The Directions Have Been Followed*	*True/False*
Example:	$\frac{x}{y} = \frac{9}{z}$	Write the reciprocal of each ratio.	$\frac{y}{x} = \frac{z}{9}$	T
1.	$\frac{x}{y} = \frac{9}{z}$	Switch the means.	$\frac{x}{z} \stackrel{?}{=} \frac{9}{y}$	
2.	$\frac{x}{y} = \frac{9}{z}$	Add each denominator to each numerator.	$\frac{x+y}{y} \stackrel{?}{=} \frac{9}{z}$	
3.	$\frac{a}{11} = \frac{b}{8}$	Write the reciprocal of each ratio.	$\frac{a}{11} \stackrel{?}{=} \frac{8}{b}$	

	Original Proportion	Directions	The Directions Have Been Followed	True/False
4.	$\frac{a}{11} = \frac{b}{8}$	Switch the means.	$\frac{a}{b} \stackrel{?}{=} \frac{11}{8}$	
5.	$\frac{2}{5} = \frac{x}{y}$	Add each denominator to each numerator.	$\frac{2+5}{5} \stackrel{?}{=} \frac{x+y}{y}$	
6.	$\frac{2}{5} = \frac{x}{y}$	Write the reciprocal of each ratio.	$\frac{5}{2} \stackrel{?}{=} \frac{y}{x}$	

7.2 EXERCISE SET MyMathLab®

Fill in the Chart *Let's practice writing equivalent proportions using the properties in this section. See Examples 1 and 2.*

	Original Proportion	Directions	Equivalent Proportion (by following directions)
1.	$\frac{a}{2} = \frac{e}{5}$	Switch the means.	
2.	$\frac{17}{x} = \frac{3}{y}$	Switch the means.	
3.	$\frac{a}{2} = \frac{e}{5}$	Write the reciprocal of each ratio.	
4.	$\frac{17}{x} = \frac{3}{y}$	Write the reciprocal of each ratio.	
5.	$\frac{a}{2} = \frac{e}{5}$	Add each denominator to each numerator.	
6.	$\frac{17}{x} = \frac{3}{y}$	Add each denominator to each numerator.	

For $\frac{a}{7} = \frac{13}{b}$ complete each equivalent proportion. See Examples 1 and 2.

7. $\frac{a}{\_\_} = \frac{7}{\_\_}$

8. $\frac{a}{13} = \frac{\_\_}{\_\_}$

9. $\frac{a+7}{7} = \frac{\_\_}{\_\_}$

10. $\frac{\_\_}{\_\_} = \frac{13+b}{b}$

11. $\frac{7}{a} = \frac{\_\_}{\_\_}$

12. $\frac{\_\_}{\_\_} = \frac{b}{13}$

In the diagram, $\frac{a}{b} = \frac{3}{4}$. Complete each statement. Justify your answer. See Examples 1 and 2.

13. $\frac{b}{a} = \frac{\_\_}{\_\_}$

14. $\frac{b}{\_\_} = \frac{4}{\_\_}$

15. $4a = \_\_$

16. $\_\_ = 3b$

17. $\frac{\_\_}{\_\_} = \frac{7}{4}$

18. $\frac{a+b}{b} = \frac{\_\_}{\_\_}$

19. $\frac{\_\_}{\_\_} = \frac{b}{4}$

20. $\frac{a}{3} = \frac{\_\_}{\_\_}$

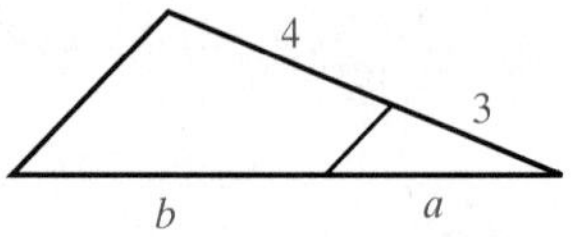

Solve. For Exercises 21 and 22, the solutions have been started for you. See Examples 3 and 4.

An NBA basketball player averages 45 baskets for every 100 attempts.

21. If he attempted 800 baskets, how many baskets did he make?

Start the solution:

1. UNDERSTAND the problem. Reread it as many times as needed. Let

   x = how many baskets he made

2. TRANSLATE into an equation.

   $$\begin{array}{lcccl} \text{baskets} & \rightarrow & \frac{45}{100} = \frac{x}{800} & \leftarrow & \text{baskets} \\ \text{attempts} & \rightarrow & & \leftarrow & \text{attempts} \end{array}$$

3. SOLVE the equation. Set cross products equal to each other and solve.

   $$\frac{45}{100} \times \frac{x}{800}$$

   Finish by SOLVING and **4.** INTERPRET.

22. If he made 225 baskets, how many did he attempt?

Start the solution:

1. UNDERSTAND the problem. Reread it as many times as needed. Let

   x = how many baskets attempted

2. TRANSLATE into an equation.

   $$\begin{array}{lcccl} \text{baskets} & \rightarrow & \frac{45}{100} = \frac{225}{x} & \leftarrow & \text{baskets} \\ \text{attempts} & \rightarrow & & \leftarrow & \text{attempts} \end{array}$$

3. SOLVE the equation. Set cross products equal to each other and solve.

   $$\frac{45}{100} \times \frac{225}{x}$$

   Finish by SOLVING and **4.** INTERPRET.

For Exercises 23–42, solve the problem using the given situation. See Examples 3 and 4.

On average, it takes a particular gem cutter 90 minutes to map the cutting of 4 gems.

23. Find how long it takes her to map the cutting of 22 gems.

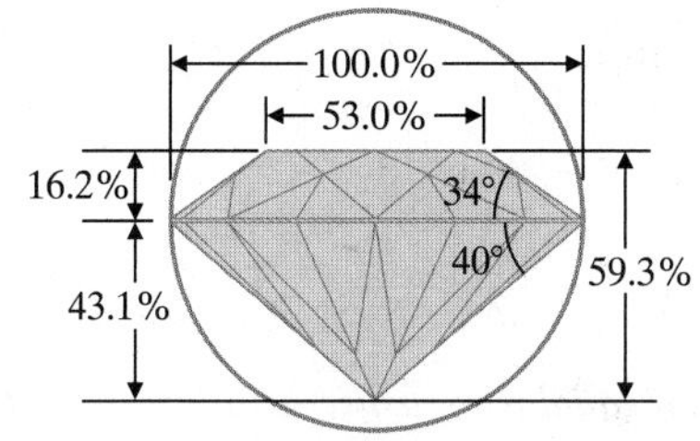

24. Find how many gems she can draw maps for in 4.5 hours.

University Law School accepts 2 out of every 7 applicants.

25. If the school accepted 180 students, find how many applications they received.

26. If the school accepted 150 students, find how many applications they received.

On an architect's blueprint, 1 inch corresponds to 8 feet.

27. Find the length of a wall that is represented by a line $2\frac{7}{8}$ inches long on the blueprint.

28. Find the length of a wall that is represented by a line $5\frac{1}{4}$ inches long on the blueprint.

A human-factors expert recommends that there be at least 9 square feet of floor space in a college classroom for every student in the class.

29. Find the minimum floor space that 30 students require.

30. Due to a lack of space, a university converts a 21-by-15-foot conference room into a classroom. Find the maximum number of students the room can accommodate.

A hybrid car averages 627 miles on a 12.3-gallon tank of gas.

31. Manuel Lopez is planning a 1250-mile vacation trip in his hybrid. Find how many gallons of gas he can expect to burn. Round to the nearest gallon.

32. Ramona Hatch has enough money to put 6.9 gallons of gas in her hybrid. She is planning on driving her brother home from college for the weekend. If their home is 290 miles away, should she be able to make it home before she runs out of gas?

The scale on a map of Italy states that 1 centimeter corresponds to 30 kilometers.

33. Find how far apart Milan and Rome are if their corresponding points on the map are 15 centimeters apart.

34. On the map, a small Italian village is located 0.4 centimeter from the Mediterranean Sea. Find the actual distance.

A bag of lawn fertilizer covers 3000 square feet of lawn.

35. Find how many bags of fertilizer should be purchased to cover a rectangular lawn 260 feet by 180 feet.

36. Find how many bags of fertilizer should be purchased to cover a square lawn measuring 160 feet on each side.

A self-tanning lotion advertises that a 3-oz bottle will provide four applications.

37. Jen Haddad found a great deal on a 14-oz bottle of the self-tanning lotion. Based on the advertising claims, how many applications of the self-tanner should Jen expect? Round down to the nearest whole number.

38. The Community College theater group needs fake tans for a play they are doing. If the play has a cast of 35, how many ounces of self-tanning lotion should the cast purchase? Round up to the nearest whole ounce.

The school's computer lab goes through 5 reams of printer paper every 3 weeks.

39. Find out how long a case of printer paper is likely to last (a case of paper holds 8 reams of paper). Round to the nearest week.

40. How many cases of printer paper should be purchased to last the entire semester of 15 weeks? Round up to the next case.

A recipe for pancakes calls for 2 cups flour and $1\frac{1}{2}$ cups milk to make a serving for four people.

41. Ming has plenty of flour, but only 4 cups milk. How many servings can he make?

42. The swim team has a weekly breakfast after early practice. How much flour will it take to make pancakes for 18 swimmers?

Solve.

43. In the Seattle Space Needle, the elevators whisk you to the revolving restaurant at a speed of 800 feet in 60 seconds. If the revolving restaurant is 500 feet up, how long will it take you to reach the restaurant by elevator? (*Source:* Seattle Space Needle)

44. A 16-oz grande Tazo Black Iced Tea at Starbucks has 80 calories. How many calories are in a 24-oz venti Tazo Black Iced Tea? (*Source:* Starbucks Coffee Company)

45. To eliminate mosquito larvae, a certain granular substance can be applied to standing water in a ratio of 1 tsp per 25 sq ft of standing water.

a. At this rate, find how many teaspoons of granules must be used for 450 square feet.

b. If 3 tsp = 1 tbsp, how many tablespoons of granules must be used?

46. Another type of mosquito control is liquid, where 3 oz of pesticide is mixed with 100 oz of water. This mixture is sprayed on roadsides to control mosquito breeding grounds hidden by tall grass.

a. If one mixture of water with this pesticide can treat 150 feet of roadway, how many ounces of pesticide are needed to treat one mile? (*Hint:* 1 mile = 5280 feet)

b. If 8 liquid ounces equals one cup, write your answer to part **a** in cups. Round to the nearest cup.

47. A student would like to estimate the height of the Statue of Liberty in New York City's harbor. The length of the Statue of Liberty's right arm is 42 feet. The student's right arm is 2 feet long and her height is $5\frac{1}{3}$ feet. Use this information to estimate the height of the Statue of Liberty. How close is your estimate to the statue's actual height of 111 feet, 1 inch from heel to top of head? (*Source:* National Park Service)

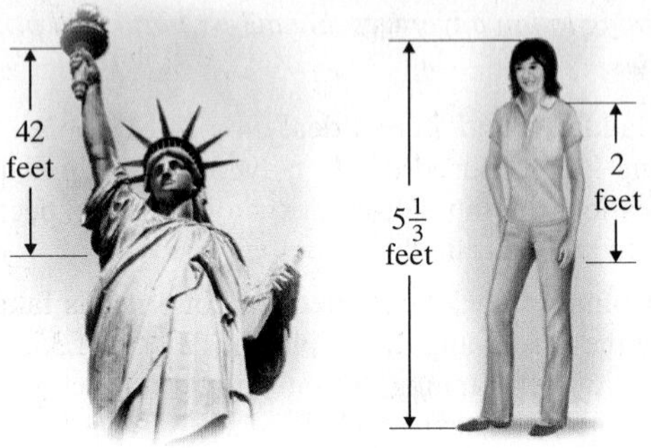

48. The length of the Statue of Liberty's index finger is 8 feet while the height to the top of the head is about 111 feet. Suppose your measurements are proportionally the same as this statue's and your height is 5 feet.

a. Use this information to find the proposed length of your index finger. Give an exact measurement and then a decimal rounded to the nearest hundredth.

b. Measure your index finger and write it as a decimal in feet rounded to the nearest hundredth. How close is the length of your index finger to the answer to part **a**? Explain why.

49. Trump World Tower in New York City is 881 feet tall and contains 72 stories. The Empire State Building contains 102 stories. If the Empire State Building has the same number of feet per floor as the Trump World Tower, approximate its height rounded to the nearest foot. (*Source:* skyscrapers.com)

50. Medication is prescribed in 7 out of every 10 hospital emergency room visits that involve an injury. If a large urban hospital had 620 emergency room visits involving an injury in the past month, how many of these visits would you expect to have included a prescription for medication? (*Source:* National Center for Health Statistics)

51. The gas/oil ratio for a certain chainsaw is 50 to 1.

a. How much oil (in gallons) should be mixed with 5 gallons of gasoline?

b. If 1 gallon equals 128 fluid ounces, write the answer to part **a** in fluid ounces. Round to the nearest whole ounce.

52. The gas/oil ratio for a certain tractor mower is 20 to 1.

a. How much oil (in gallons) should be mixed with 10 gallons of gas?

b. If 1 gallon equals 4 quarts, write the answer to part **a** in quarts.

CONCEPT EXTENSIONS

Complete each statement. Justify your answer.

53. If $\frac{4}{9} = \frac{n}{m}$, then $\frac{m}{n} = \frac{}{—}$.

54. If $\frac{30}{t} = \frac{18}{r}$, then $\frac{t}{r} = \frac{}{—}$.

55. If $\frac{a+5}{5} = \frac{b+2}{2}$, then $\frac{a}{5} = \frac{}{—}$.

56. If $\frac{a}{b} = \frac{c}{d}$, then $\frac{a+b}{c+d} = \frac{}{—}$.

Use properties of equality to justify each Property of Proportions.

57. $\frac{a}{b} = \frac{c}{d}$ is equivalent to $\frac{a}{c} = \frac{b}{d}$. Property (2)

58. $\frac{a}{b} = \frac{c}{d}$ is equivalent to $\frac{a+b}{b} = \frac{c+d}{d}$. Property (3)

Exercises 59–62 involve liquid drug preparations, where the mass of the drug is contained in a volume of solution. The following description of mg and ml will help.

mg means milligrams (A paper clip is about a gram. A milligram is about the mass of $\frac{1}{1000}$ of a paper clip.)

ml means milliliter (A liter is about a quart. A milliliter is about the amount of liquid in $\frac{1}{1000}$ of a quart.)

(Hint: Use the proportion $\frac{mg}{ml} = \frac{mg}{ml}$ if you'd like.)

A solution strength of 15 mg of medicine in 1 ml of solution is available.

59. If a patient needs 12 mg of medicine, how many ml do you administer?

60. If a patient needs 33 mg of medicine, how many ml do you administer?

A solution strength of 8 mg of medicine in 1 ml of solution is available.

61. If a patient needs 10 mg of medicine, how many ml do you administer?

62. If a patient needs 6 mg of medicine, how many ml do you administer?

63. Explain how to use two different Properties of Proportions to change the proportion $\frac{3}{4} = \frac{12}{16}$ into the proportion $\frac{12}{3} = \frac{16}{4}$.

64. Thus far, we know that $\frac{x}{2} = \frac{y}{3}$ is equivalent to $\frac{x+2}{2} = \frac{y+3}{3}$. Is it also true that $\frac{x}{2} = \frac{y}{3}$ is equivalent to $\frac{x-2}{2} = \frac{y-3}{3}$? Explain why or why not.

REVIEW AND PREVIEW

Given: $\Delta ABC \cong \Delta HIJ$

65. Name three pairs of congruent angles.

66. Name three pairs of congruent sides.

7.3 SIMILAR POLYGONS

OBJECTIVES

1. Identify Similar Polygons.
2. Use Similar Polygons to Solve Applications.

VOCABULARY

- similar figures
- similar polygons
- extended proportion
- scale factor

OBJECTIVE 1 ▶ Identifying Similar Polygons. **Similar figures** have the same shape but not necessarily the same size. We will abbreviate "is similar to" with the symbol ~.

We can use ratios and proportions to decide whether two polygons are similar and to find unknown side lengths of similar figures.

Similar Polygons

<table>
<tr><th>Define</th><th>Diagram</th><th>Symbols</th></tr>
<tr><td>Two polygons are similar polygons if corresponding angles are congruent and if the lengths of corresponding sides are proportional.</td><td>$ABCD \sim GHIJ$
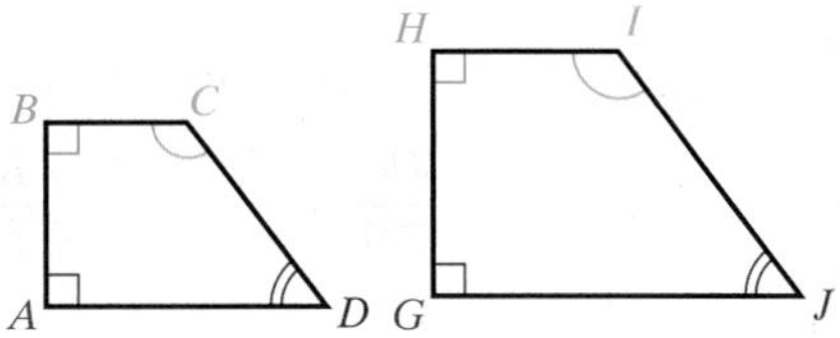
</td><td>$\angle A \cong \angle G$
$\angle B \cong \angle H$
$\angle C \cong \angle I$
$\angle D \cong \angle J$
$\frac{AB}{GH} = \frac{BC}{HI} = \frac{CD}{IJ} = \frac{AD}{GJ}$</td></tr>
</table>

We write a similarity statement with corresponding vertices in order, the same way we write a congruence statement. When three or more ratios are equal, we can write an **extended proportion.** The proportion $\frac{AB}{GH} = \frac{BC}{HI} = \frac{CD}{IJ} = \frac{AD}{GJ}$ is an extended proportion.

EXAMPLE 1 **Understanding Similarity and Using Extended Proportions**

$\Delta MNP \sim \Delta SRT$

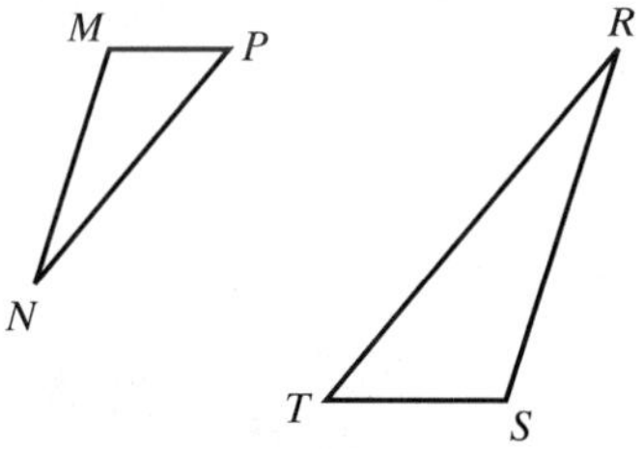

a. What are the pairs of congruent angles?

b. What is the extended proportion for the ratios of corresponding sides?

Solution

a. Use the order of the vertices in the similarity statement $\Delta MNP \sim \Delta SRT$ to write pairs of congruent angles.

$$\angle M \cong \angle S, \angle N \cong \angle R, \text{ and } \angle P \cong \angle T$$

b. Since $\Delta MNP \sim \Delta SRT$, we know that $\overline{MN}$ corresponds to $\overline{SR}$, so $\frac{MN}{SR}$ is a ratio of corresponding sides. The same is true for the other two pairs of corresponding sides. Thus,

$$\frac{MN}{SR} = \frac{NP}{RT} = \frac{MP}{ST}$$

□

PRACTICE 1 $DEFG \sim HJKL$

a. What are the pairs of congruent angles?

b. What is the extended proportion for the ratios of the lengths of corresponding sides?

A **scale factor** is the ratio of corresponding linear measurements of two similar figures. The ratio of the lengths of corresponding sides $\overline{BC}$ and $\overline{YZ}$, or more simply stated, the ratio of corresponding sides, is $\frac{BC}{YZ} = \frac{20}{8} = \frac{5}{2}$. So, the scale factor of ΔABC to ΔXYZ is $\frac{5}{2}$ or 5:2.

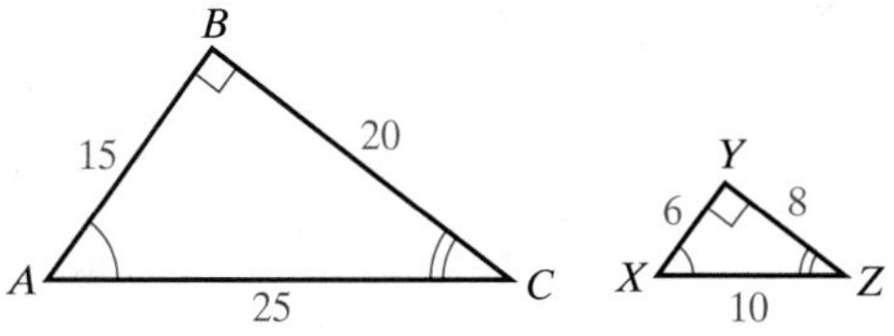

$\Delta ABC \sim \Delta XYZ$

EXAMPLE 2 Determining Similarity Using Scale Factors

For the figures in **a** and then in **b**, answer Part 1 and Part 2.

PART 1 Are the polygons similar?

PART 2 If they are, write a similarity statement and give the scale factor.

a. *JKLM* and *TUVW*

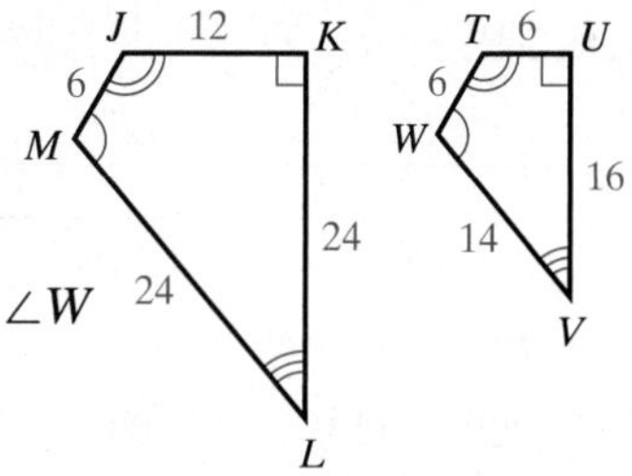

Solution

PART 1 Identify pairs of congruent angles.

$\angle J \cong \angle T, \angle K \cong \angle U, \angle L \cong \angle V$, and $\angle M \cong \angle W$

Then compare the ratios of corresponding sides.

$\frac{JK}{TU} = \frac{12}{6} = \frac{2}{1}$ $\quad \frac{KL}{UV} = \frac{24}{16} = \frac{3}{2}$

$\frac{LM}{VW} = \frac{24}{14} = \frac{12}{7}$ $\quad \frac{JM}{TW} = \frac{6}{6} = \frac{1}{1}$

PART 2 Corresponding sides are not proportional since the scale factors are not the same, so the polygons are not similar.

b. ΔABC and ΔDEF

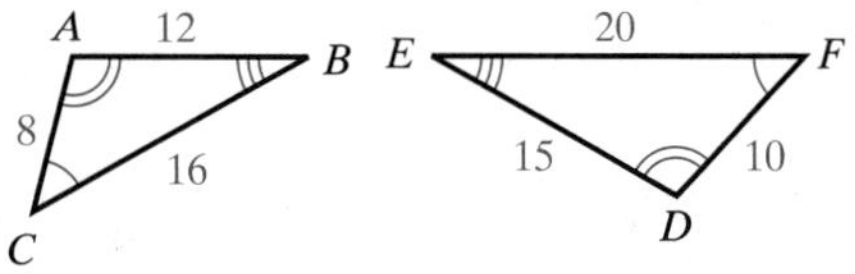

Solution

PART 1 Identify pairs of congruent angles.

$\angle A \cong \angle D, \angle B \cong \angle E$, and $\angle C \cong \angle F$

Then compare the ratios of corresponding sides.

$\frac{AB}{DE} = \frac{12}{15} = \frac{4}{5}$ $\quad \frac{BC}{EF} = \frac{16}{20} = \frac{4}{5}$ $\quad \frac{AC}{DF} = \frac{8}{10} = \frac{4}{5}$

PART 2 Yes; $\Delta ABC \sim \Delta DEF$ and the scale factor is $\frac{4}{5}$ or 4:5. □

▶ Helpful Hint
If the scale factor of two similar polygons is a proper fraction $\left(\frac{\text{smaller number}}{\text{larger number}}\right)$, then the first polygon is smaller in area than the second polygon.

PRACTICE

2 Are the polygons similar? If they are, write a similarity statement and give the scale factor.

a.

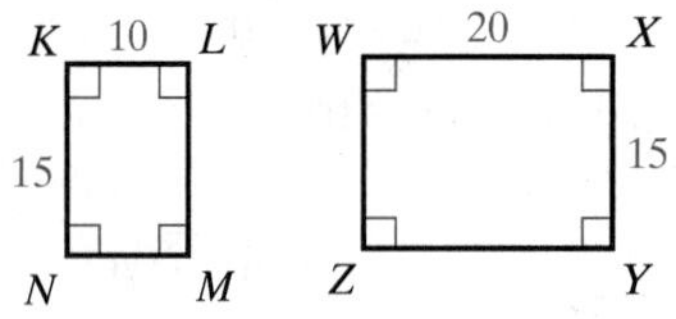

b.

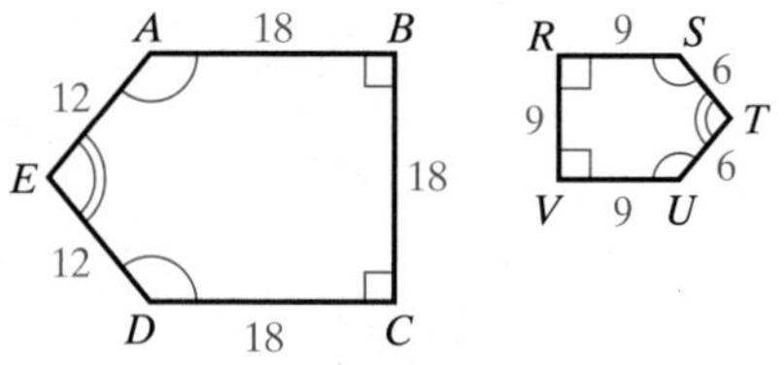

We can use what we've learned so far about similarity and algebra to find unknown values in similar polygons.

EXAMPLE 3 Using Similar Polygons to Find Unknown Values

ABCD ~ *EFGD*. What is the value of *x*?

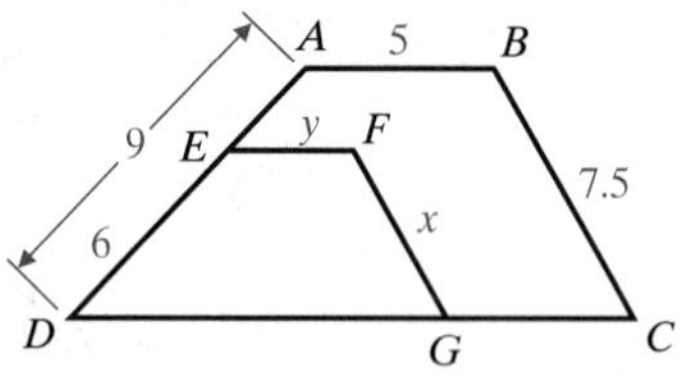

a. 4.5

b. 5

c. 7.2

d. 11.25

Solution Use the similarity statement to write a proportion.

$$\frac{FG}{BC} = \frac{ED}{AD}$$ Corresponding sides of similar polygons are proportional.

$$\frac{x}{7.5} = \frac{6}{9}$$ Substitute side lengths.

$$9x = 45$$ Use the Cross Products Property.

$$x = 5$$ Divide both sides by 9.

The value of x is 5. The correct answer is **b**. □

PRACTICE 3 Use the diagram in Example 3. What is the value of y?

OBJECTIVE 2 ▶ Using Similar Polygons in Applications.

EXAMPLE 4 Using Similarity

Your class is making a rectangular poster for a rally. The poster's design is 6 in. in width by 10 in. in length. The space allowed for the poster is 4 ft in width by 8 ft in length. What are the dimensions of the largest poster that will still fit in the space?

10 in.

6 in.

Poster Design
(ratio must remain the same)

8 ft

4 ft

Space Allowed (not to scale)

Solution

> ▶ **Helpful Hint**
> Why can't the poster be enlarged 9.6 times? If done, the width will be 6 in. · 9.6 or 57.6 in., which is greater than the 4 ft allowed.

STEP 1. Determine whether the width or the length will fill the space first.

Width:	4 ft = 48 in.	Length:	8 ft = 96 in.
How many 6 in. will fit in 4 ft?	48 in. ÷ 6 in. = 8	How many 10 in. will fit in 8 ft?	96 in. ÷ 10 in. = 9.6

The design can be enlarged at most 8 times (the smaller of the two quotients).

STEP 2. The greatest width is 48 in., so find the length.

$$\frac{6}{48} = \frac{10}{x}$$ Corresponding sides of similar polygons are proportional.

$$6x = 480$$ Cross Products Property

$$x = 80$$ Divide each side by 6.

The largest poster that will fit is 48 in. by 80 in. or 4 ft by $\frac{20}{3}$ ft or 4 ft by $6\frac{2}{3}$ ft. □

PRACTICE 4 Use the same poster design in Example 4. What are the dimensions of the largest complete poster that will fit in a space 3 ft in width by 4 ft in length?

VOCABULARY & READINESS CHECK

Word Bank *Use the words and phrases below to fill in each blank.*

congruent	scale factor	extended
similar	proportional	

1. Figures that have the same shape but not necessarily the same size are called ____________ figures.
2. Two polygons are similar polygons if corresponding angles are ____________ and corresponding sides are ____________.
3. The proportion $\frac{AB}{XY} = \frac{BC}{YZ} = \frac{CA}{ZX}$ is called a(n) ____________ proportion.
4. The ratio of corresponding linear measurements of two similar figures is called the ____________.

7.3 EXERCISE SET MyMathLab®

List the pairs of congruent angles and the extended proportion that relates the corresponding sides for the similar polygons. See Example 1.

1. $RSTV \sim DEFG$

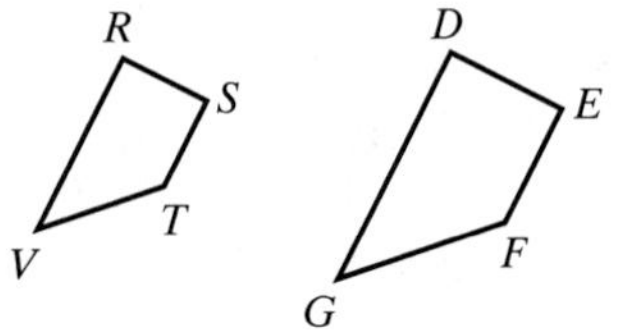

2. $\Delta CAB \sim \Delta WVT$

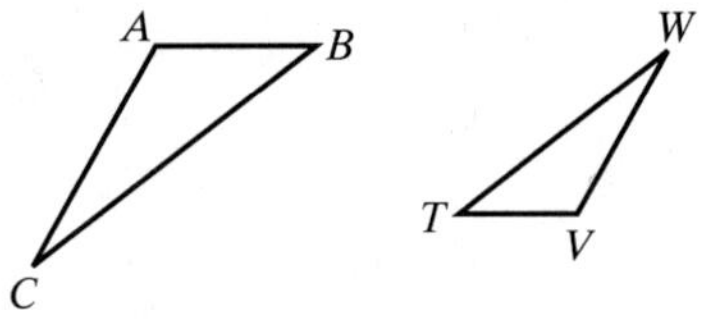

Determine whether the polygons are similar. If they are, write a similarity statement and give the scale factor. If not, explain. See Example 2.

3.

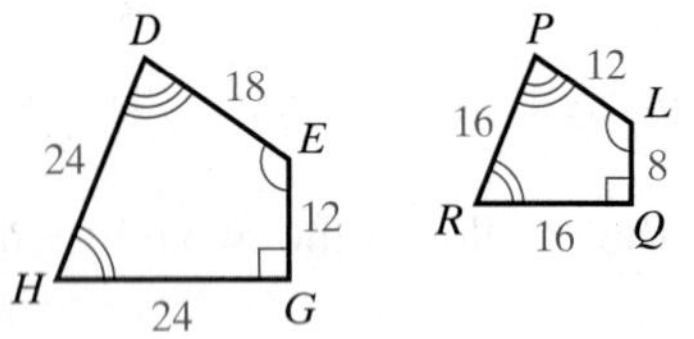

4.

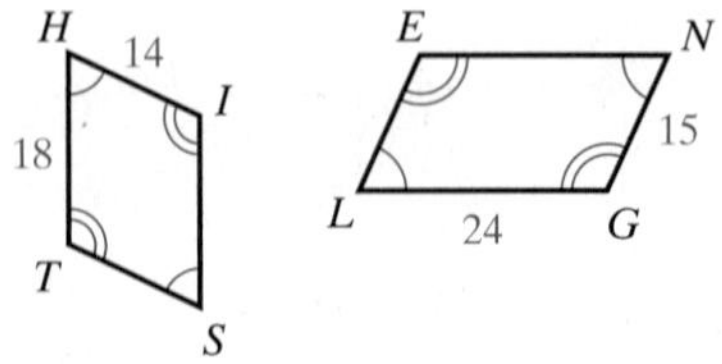

5.

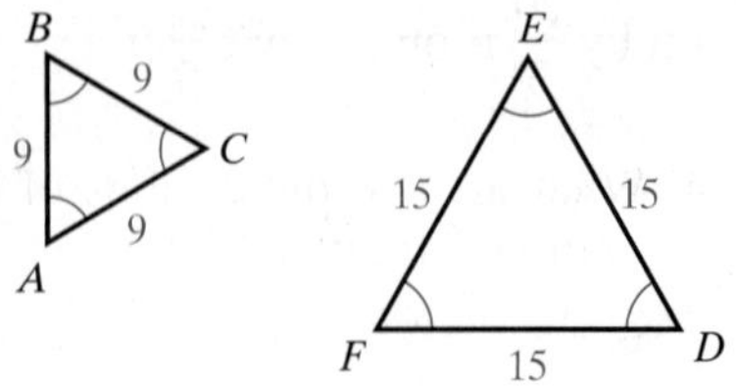

6.

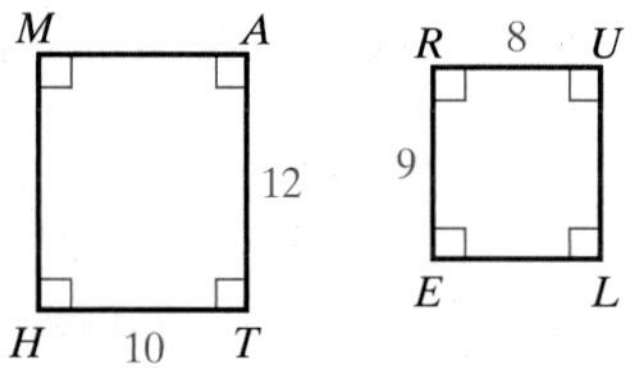

7.

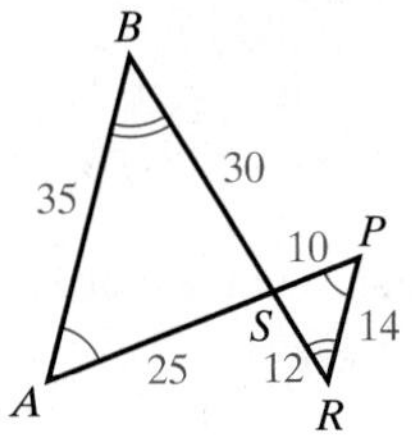

8.

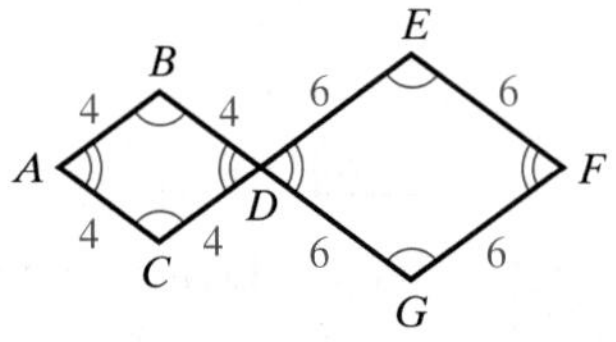

9.

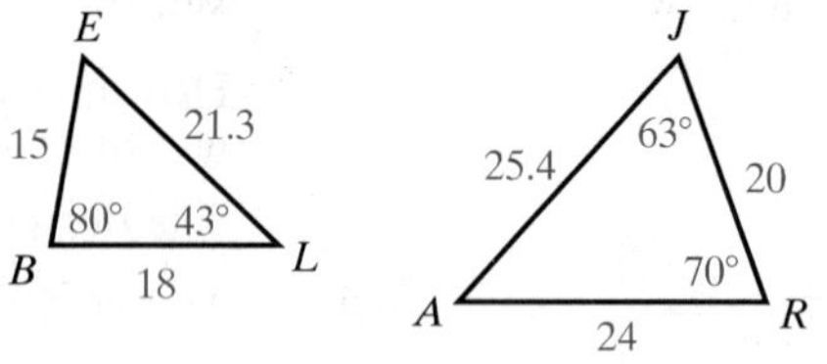

10.

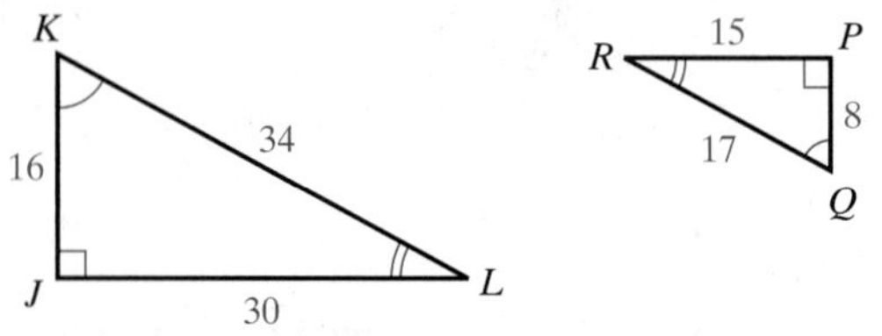

The polygons are similar. Find the value of each variable. See Example 3.

11.

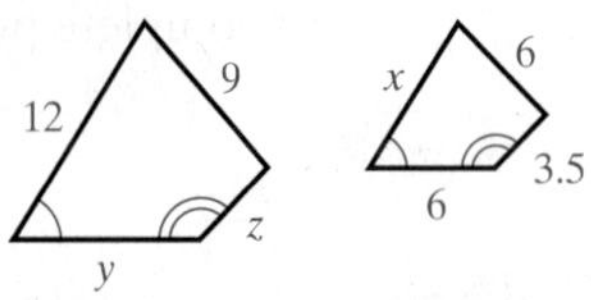

12.

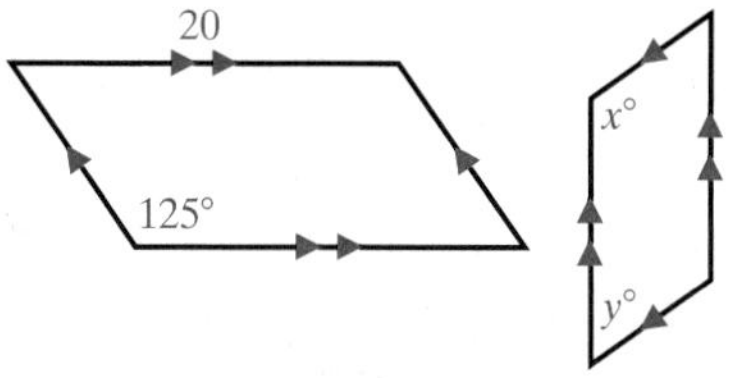

13.

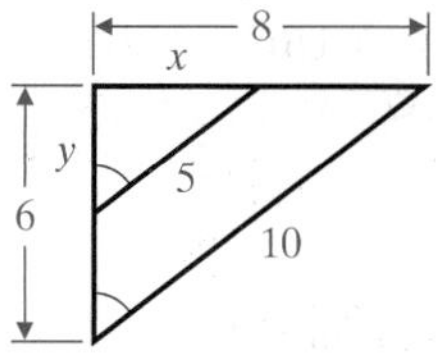

14.

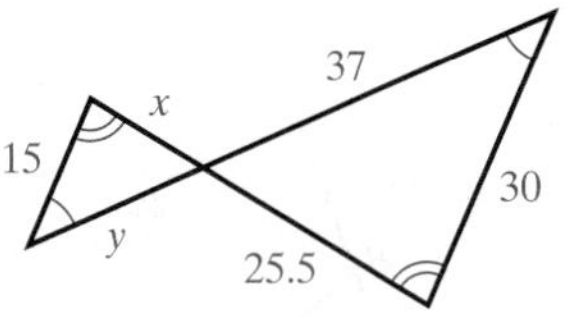

Mixed Practice *In the diagram below, $\Delta DFG \sim \Delta HKM$. Find each of the following.*

15. the scale factor of ΔHKM to ΔDFG

16. $\dfrac{GD}{MH}$

17. $m\angle K$

18. $m\angle M$

19. MK

20. GD

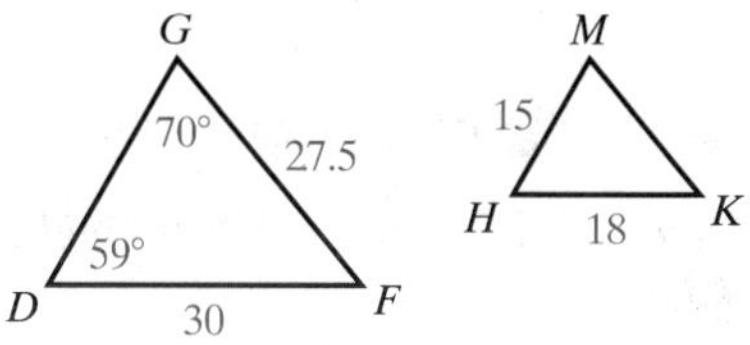

Solve. See Example 4.

21. You want to make a scale drawing of New York City's Empire State Building using the scale 1 in. = 250 ft. If the building is 1250 ft tall, how tall should you make the building in your scale drawing?

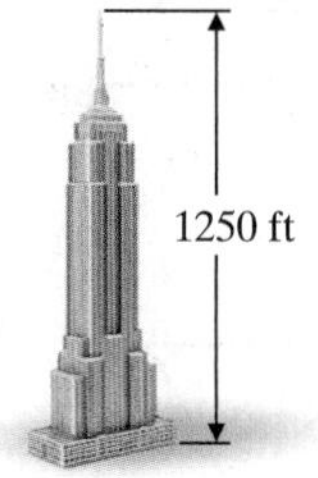

22. A cartographer is making a map of Pennsylvania. She uses the scale 1 in. = 10 mi. The actual distance between Harrisburg and Philadelphia is about 95 mi. How far apart should she place the two cities on the map?

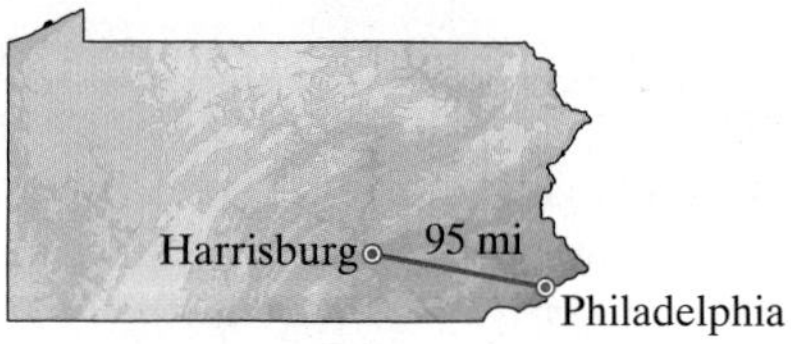

23. A company produces a standard-size U.S. flag that is 3 ft by 5 ft. The company also produces a giant-size flag that is similar to the standard-size flag. If the shorter side of the giant-size flag is 36 ft, what is the length of its longer side?

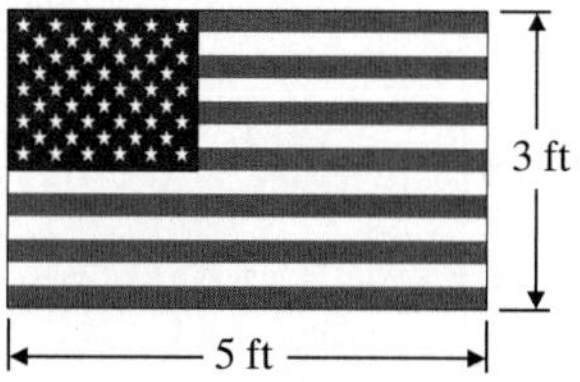

24. The scale drawing at the right is part of a floor plan for a home. The scale is 1 cm = 10 ft. What are the actual dimensions of the family room? (Each square on the drawing paper is 1 cm by 1 cm.)

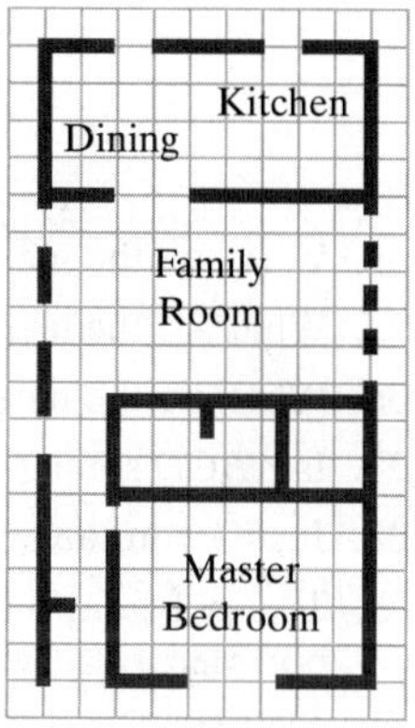

25. The space allowed for the mascot on a school's Web page is 120 pixels wide by 90 pixels high. Its digital image is 500 pixels wide by 375 pixels high. What is the largest image of the mascot that will fit on the Web page?

26. The design for a mural is 16 in. wide and 9 in. high. What are the dimensions of the largest possible complete mural that can be painted on a wall 24 ft wide by 14 ft high?

Determine whether each statement is always, sometimes, or never true.

27. Any two regular pentagons are similar.

28. A hexagon and a triangle are similar.

29. A square and a rhombus are similar.

30. Two similar rectangles are congruent.

Mixed Practice

The polygons are similar. Find the value of x. Give the scale factor of the left polygon to the right polygon.

31. $\Delta WLJ \sim \Delta QBV$

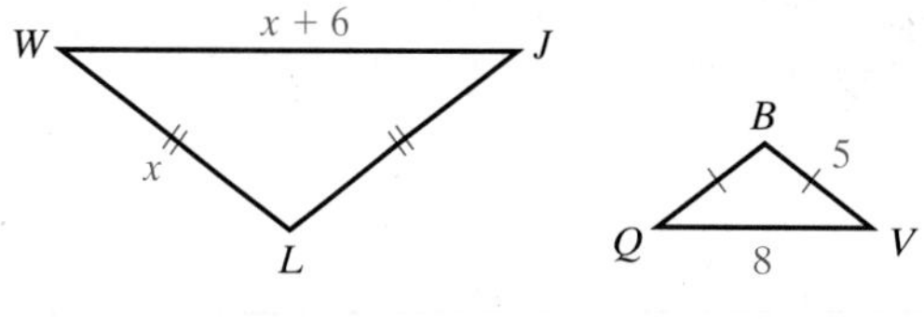

32. $GKNM \sim VRPT$

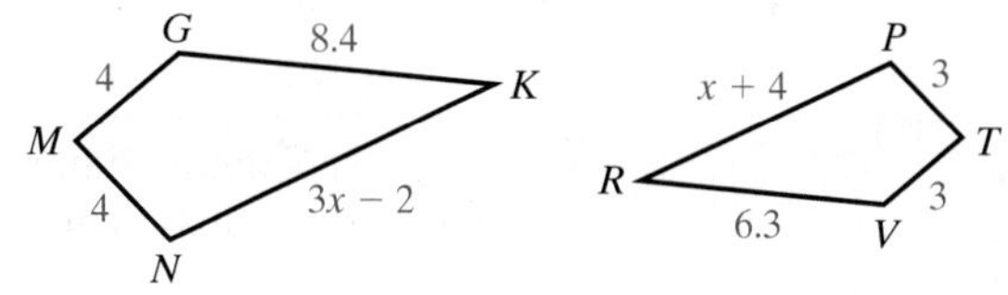

CONCEPT EXTENSIONS

33. What are the measures of $\angle A$, $\angle ABC$, $\angle BCD$, and $\angle CDA$? Explain.

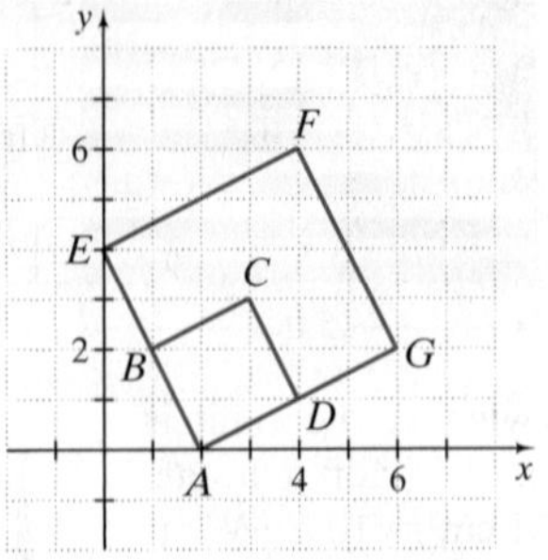

34. What are the measures of $\angle E$, $\angle F$, and $\angle G$? Explain.

35. What are the lengths of $\overline{AB}$, $\overline{BC}$, $\overline{CD}$, and $\overline{DA}$?

36. What are the measures of $\overline{AE}$, $\overline{EF}$, $\overline{FG}$, and $\overline{AG}$?

37. Is $ABCD$ similar to $AEFG$?

38. Justify your answer to Exercise **37**.

39. Two polygons have corresponding side lengths that are proportional. Can you conclude that the polygons are similar? Justify your reasoning.

40. Explain why two congruent figures must also be similar. Include scale factor in your explanation.

41. ΔJLK and ΔRTS are similar. The scale factor of ΔJLK to ΔRTS is 3:1. What is the scale factor of ΔRTS to ΔJLK?

42. Is similarity reflexive? Transitive? Symmetric? Justify your reasoning.

Find the Error *The polygons at the right are similar. Is each similarity statement correct or incorrect? Explain.*

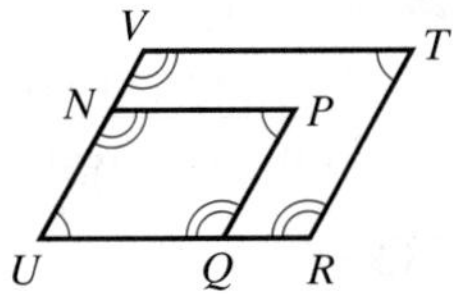

43. $TRUV \sim NPQU$

44. $RUVT \sim QUNP$

Choose a scale and make a scale drawing of each rectangular playing surface.

45. A soccer field is 110 yd by 60 yd.

46. A volleyball court is 60 ft by 30 ft.

47. A tennis court is 78 ft by 36 ft.

48. A football field is 360 ft by 160 ft.

PROOF

49. In rectangle $BCEG$, $BC:CE = 2:3$. In rectangle $LJAW$, $LJ:JA = 2:3$. Show that $BCEG \sim LJAW$.

50. Prove the following statement: If $\Delta ABC \sim \Delta DEF$ and $\Delta DEF \sim \Delta GHK$, then $\Delta ABC \sim \Delta GHK$.

REVIEW AND PREVIEW

Use the diagram for Exercises 51–54. See Section 4.6.

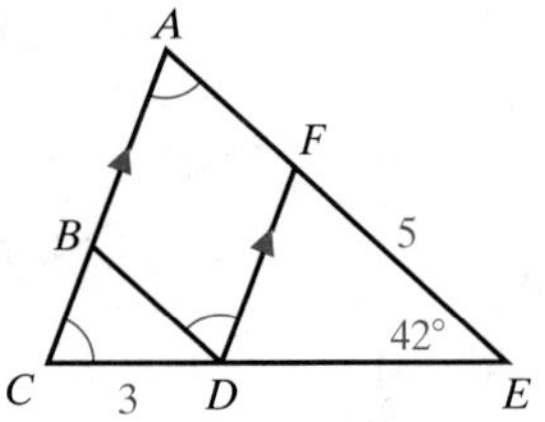

51. Name the isosceles triangles in the figure.

52. $\overline{CD} \cong$ __?__ $\cong$ __?__

53. $AE =$ __?__

54. $m\angle A =$ __?__

If $\frac{x}{7} = \frac{y}{9}$, *complete each statement using the Properties of Proportions. See Section 7.2.*

55. $9x =$ ___ **56.** $\frac{x}{y} =$ ___

How can you prove that the triangles are congruent? See Sections 4.3 and 4.4.

57.

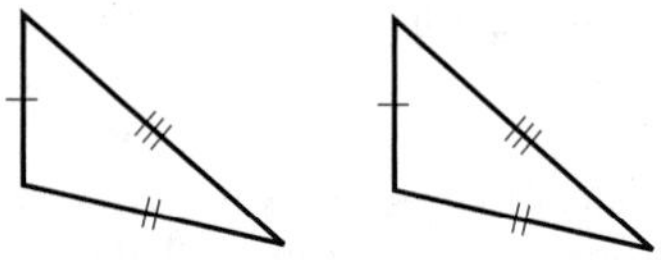

58.

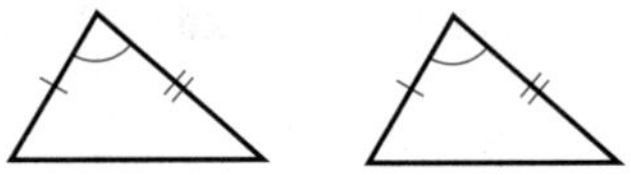

59.

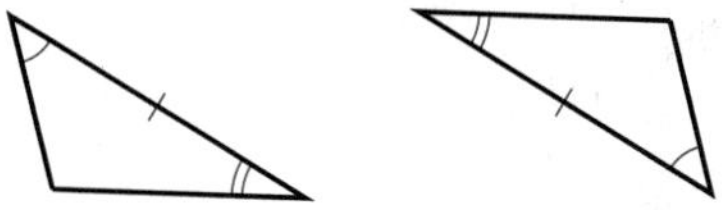

60.

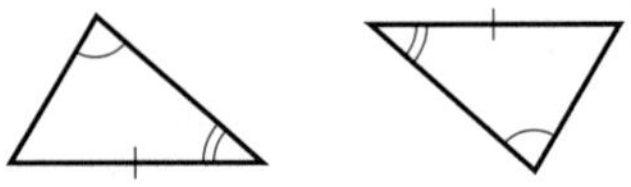

7.4 PROVING TRIANGLES ARE SIMILAR

OBJECTIVES

1 Use the AA ~ Postulate and the SAS ~ and SSS ~ Theorems.

2 Use Similarity to Find Indirect Measurements.

VOCABULARY

- indirect measurement

In the previous section, we studied similar polygons. In this section, let's concentrate on similar triangles.

Just as for congruent triangles (same exact shape and size), we can discover postulates and prove theorems that help us show that two triangles are similar (same exact shape but not necessarily same size).

OBJECTIVE 1 ▶ Using the AA Postulate and the SAS and SSS Similarity Theorems. We can show that two triangles are similar when we know the relationships between only two or three pairs of corresponding parts. The following postulate illustrates this concept.

Helpful Hint

This AA Postulate shows that two triangles are similar. There is *no* such postulate to show that two triangles are congruent.

Postulate 7.4-1 Angle-Angle Similarity (AA~) Postulate

Postulate	If ...	Then ...
If two angles of one triangle are congruent to two angles of another triangle, then the triangles are similar.	$\angle S \cong \angle M$ and $\angle R \cong \angle L$ 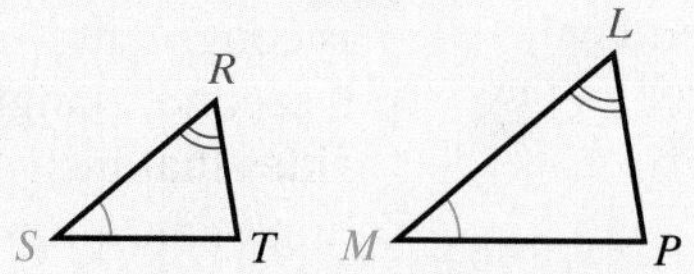	$\Delta SRT \sim \Delta MLP$

EXAMPLE 1 **Using the AA ~ Postulate**

Determine whether ΔRSW and ΔVSB are similar. Explain.

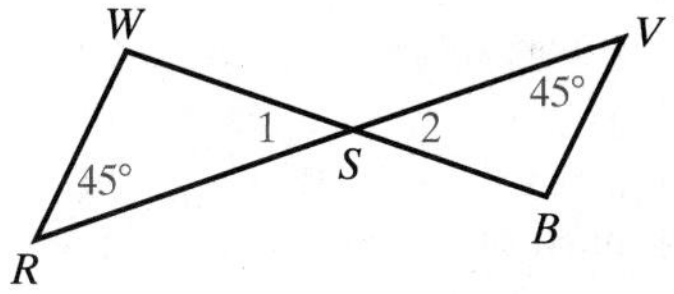

Solution By studying the diagram, we see that we can use the AA ~ Postulate. Let's show that two pairs of angles are congruent.

$\angle R \cong \angle V$ because both angles measure 45°.
$\angle 1 \cong \angle 2$ because vertical angles are congruent.
So, $\Delta RSW \sim \Delta VSB$ by the AA ~ Postulate. □

PRACTICE 1 Determine whether the two triangles are similar. Explain.

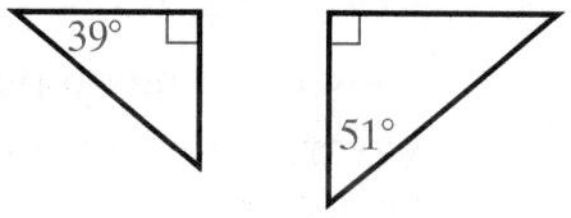

EXAMPLE 2 **Using the AA ~ Postulate**

Determine whether ΔJKL and ΔPQR are similar. Explain.

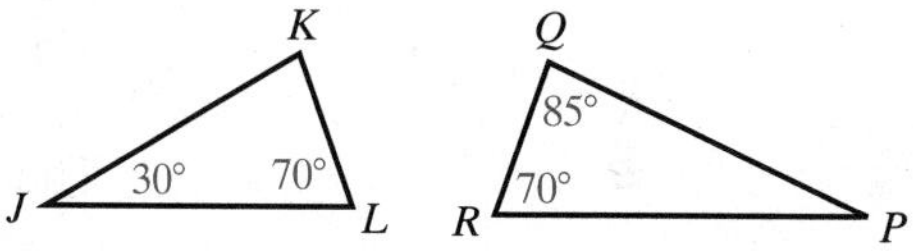

Solution By studying the diagram, we see that we can use the AA ~ Postulate. Let's show that two pairs of angles are congruent.

$\angle L \cong \angle R$ because both angles measure 70°.
By the Triangle Angle-Sum Theorem,
$m\angle K = 180° - 30° - 70° = 80°$ and $m\angle P = 180° - 85° - 70° = 25°$.
Only one pair of angles is congruent. So, ΔJKL and ΔPQR are *not* similar. □

Helpful Hint

Remember that the sum of the measures of the angles of any Δ is 180°.

PRACTICE 2 Determine whether the two triangles are similar. Explain.

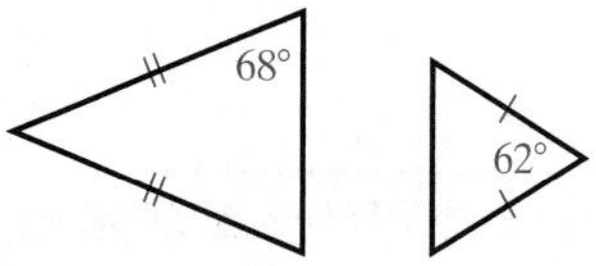

Now we will explore two other ways to determine whether two triangles are similar: the SAS and SSS theorems.

> **Helpful Hint**
> Don't forget. There is a SAS for ≅ and now a SAS for ~. For ≅, the sides must be ≅. For ~, the sides are in proportion.

Theorem 7.4-2 Side-Angle-Side Similarity (SAS ~) Theorem

Theorem	If ...	Then ...
If an angle of one triangle is congruent to an angle of a second triangle, and the sides that include the two angles are proportional, then the triangles are similar.	$\frac{AB}{QR} = \frac{AC}{QS}$ and $\angle A \cong \angle Q$	$\Delta ABC \sim \Delta QRS$

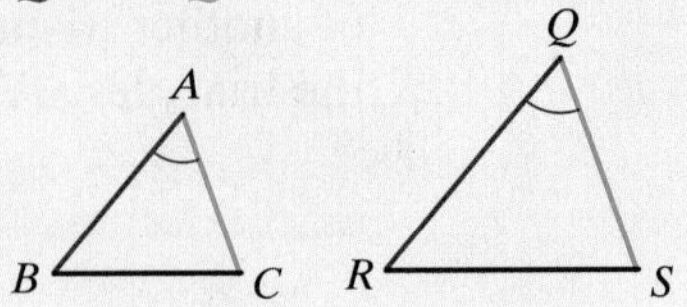

> **Helpful Hint**
> There is a SSS for ≅ and now a SAS for ~. For ≅, the sides must be ≅. For ~, the sides are in proportion.

Theorem 7.4-3 Side-Side-Side Similarity (SSS ~) Theorem

Theorem	If ...	Then ...
If the corresponding sides of two triangles are proportional, then the triangles are similar.	$\frac{AB}{QR} = \frac{AC}{QS} = \frac{BC}{RS}$	$\Delta ABC \sim \Delta QRS$

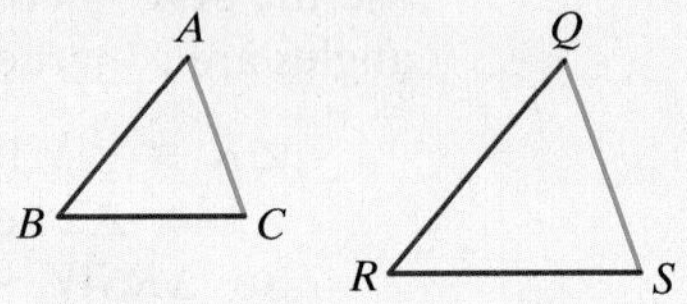

We prove these theorems in Exercises 41 and 42, Exercise Set 7.4.

We sometimes have to decide when to use each postulate or theorem to show that two triangles are similar. The next two examples offer some practice in making this decision.

EXAMPLE 3 Verifying Triangle Similarity

Determine whether the triangles are similar. If they are, write a similarity statement for the triangles. Use the AA ~ Postulate, SAS ~ Theorem, or SSS ~ Theorem.

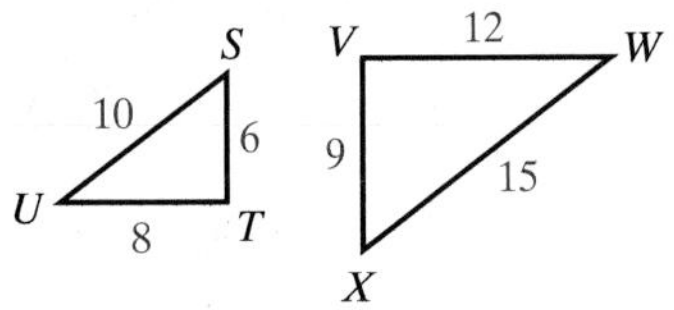

Solution Let's use the side lengths to identify corresponding sides. Then we'll set up ratios for each pair of corresponding sides.

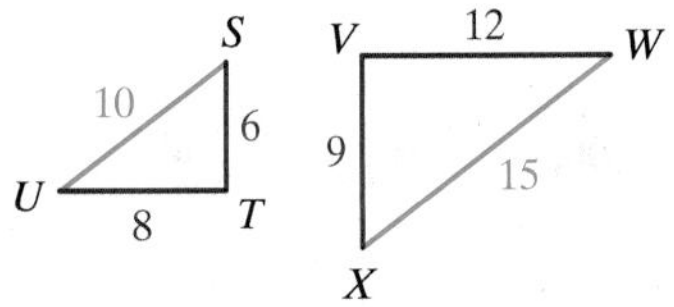

Shortest sides $\frac{ST}{XV} = \frac{6}{9} = \frac{2}{3}$

Longest sides $\frac{US}{WX} = \frac{10}{15} = \frac{2}{3}$

Remaining sides $\frac{TU}{VW} = \frac{8}{12} = \frac{2}{3}$

All three ratios are equal, so corresponding sides are proportional.
$\Delta STU \sim \Delta XVW$ by the SSS ~ Theorem. □

PRACTICE 3 Determine whether the triangles are similar. If so, write a similarity statement for the triangles and explain how you know the triangles are similar. Use the AA ~ Postulate, SAS ~ Theorem, or SSS ~ Theorem.

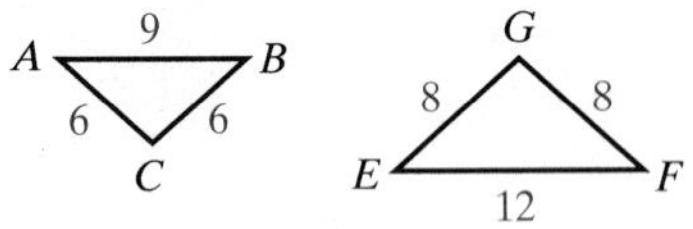

EXAMPLE 4 **Verifying Triangle Similarity**

Determine whether the triangles are similar. If they are, write a similarity statement for the triangles. Use the AA ~ Postulate, SAS ~ Theorem, or SSS ~ Theorem.

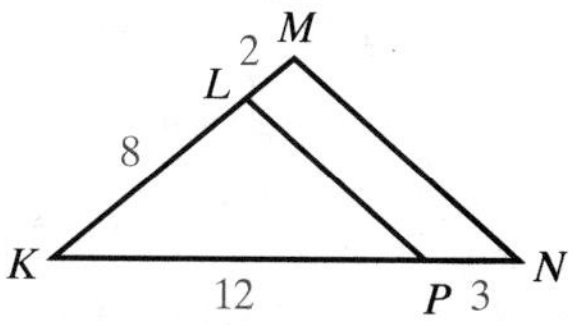

> **Helpful Hint**
> If needed, draw triangles separately, as shown here.

Solution Since we have an included angle, $\angle K$, let's try using the SAS ~ Theorem.

$\angle K \cong \angle K$ by the Reflexive Property of Congruence.

$$\frac{KL}{KM} = \frac{8}{10} = \frac{4}{5} \text{ and } \frac{KP}{KN} = \frac{12}{15} = \frac{4}{5}$$

So, $\Delta KLP \sim \Delta KMN$ by the SAS ~ Theorem.

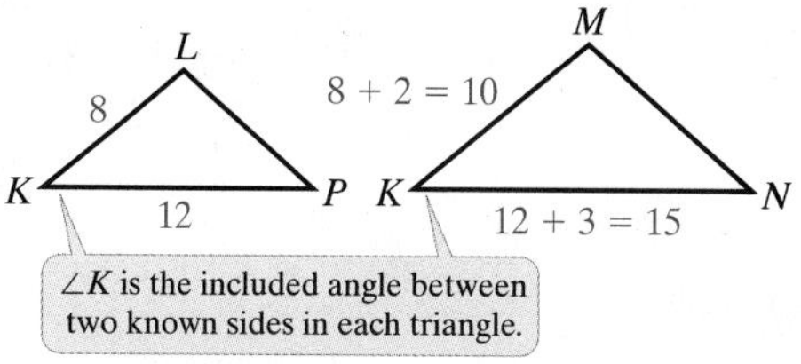

PRACTICE 4 Determine whether the triangles are similar. If so, write a similarity statement for the triangles and explain how you know the triangles are similar. Use the AA ~ Postulate, SAS ~ Theorem, or SSS ~ Theorem.

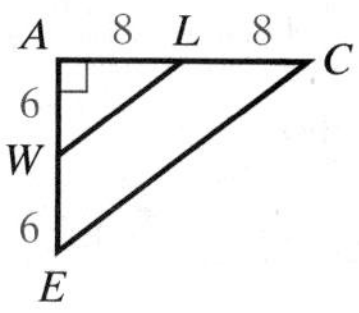

Let's try proving that two triangles are similar.

EXAMPLE 5 **Proving Triangles Similar**

Given: $\overline{FG} \cong \overline{GH}$,
$\overline{JK} \cong \overline{KL}$,
$\angle F \cong \angle J$

Prove: $\Delta FGH \sim \Delta JKL$

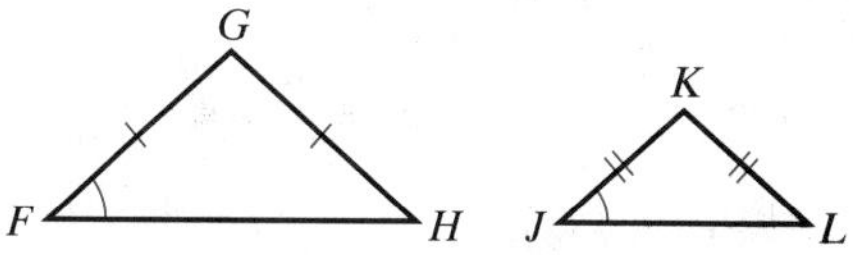

Solution Study the figures and notice that both triangles are isosceles. This should be helpful in our proof.

Statements	Reasons
1. $\overline{FG} \cong \overline{GH}, \overline{JK} \cong \overline{KL}$	1. Given
2. ΔFGH is isosceles. ΔJKL is isosceles.	2. Definition of an isosceles triangle
3. $\angle F \cong \angle H, \angle J \cong \angle L$	3. Base angles of an isosceles triangle are congruent.
4. $\angle F \cong \angle J$	4. Given
5. $\angle H \cong \angle J$	5. Transitive Property of ≅ or Substitution (Steps 3, 4)
6. $\angle H \cong \angle L$	6. Transitive Property of ≅ or Substitution (Steps 3, 4, 5)
7. $\Delta FGH \sim \Delta JKL$	7. AA ~ Postulate (Steps 4, 6)

PRACTICE 5 **Given:** $\overline{MP} \parallel \overline{AC}$

Prove: $\Delta ABC \sim \Delta PBM$

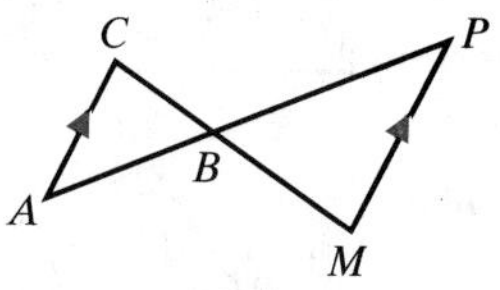

OBJECTIVE 2 ▶ Using Similar Triangles to Find Measurements. Sometimes we can use similar triangles to find lengths that cannot be measured easily.

We use **indirect measurement** to find such lengths.

Note: One method of indirect measurement uses the fact that light reflects off a mirror at the same angle at which it hits the mirror.

EXAMPLE 6 **Finding Lengths Using Similar Triangles**

Before rock climbing, a student wants to know how high he will climb. He places a mirror on the ground and walks backward until he can see the top of the cliff in the mirror. What is the height of the cliff?

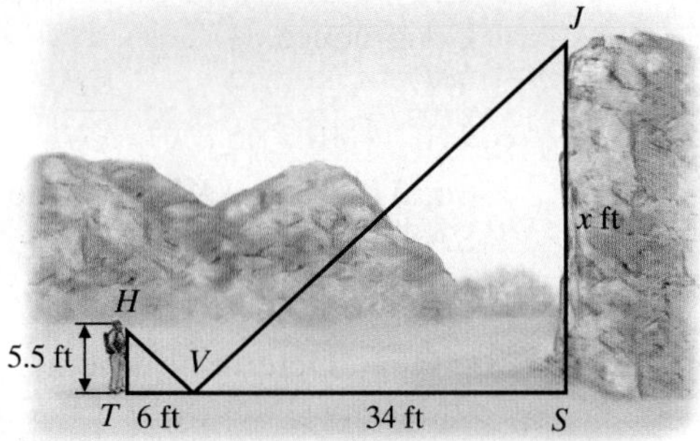

Solution We can use proportional sides if the triangles formed, ΔHTV and ΔJSV, are similar.

$\angle T \cong \angle S$ since both measure 90°. Also, $\angle HVT \cong \angle JVS$ since the light reflects on and off the mirror at the same angle.

Thus, $\Delta HTV \sim \Delta JSV$ by the AA $\sim$ Postulate.

Now let's set up a proportion to help us find x.

$$\frac{HT}{JS} = \frac{TV}{SV} \quad \text{Corresponding sides of} \sim \text{triangles are proportional.}$$

$$\frac{5.5}{x} = \frac{6}{34} \quad \text{Substitute the side lengths.}$$

$$187 = 6x \quad \text{Use the Cross Products Property.}$$

$$31.2 \approx x \quad \text{Solve for } x.$$

The cliff is a little over 31 ft high. □

PRACTICE

6 Use the same figure as for Example 6. Find the height of the cliff if the student's height at eye level is 5 feet, *not* 5.5 feet.

VOCABULARY & READINESS CHECK

For each exercise, the triangles are similar. Write a correct similarity statement. (There's more than one correct answer for each exercise.)

1.

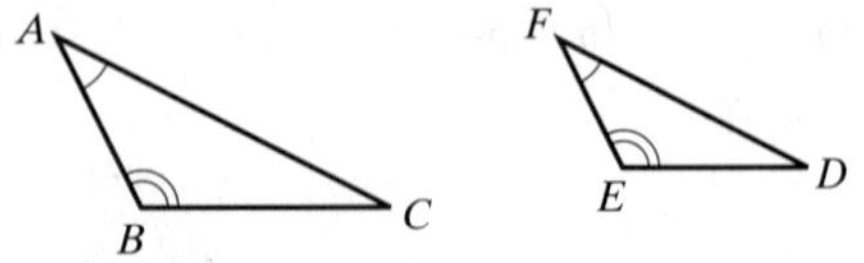

2.

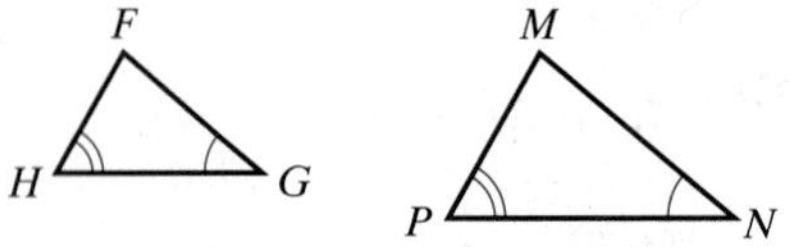

3.

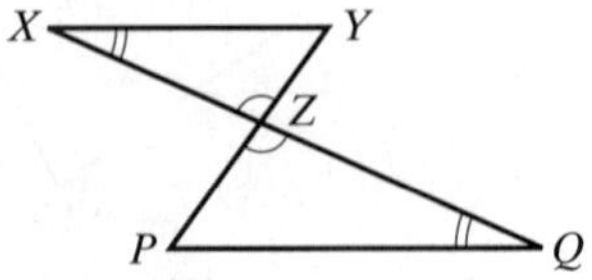

4.

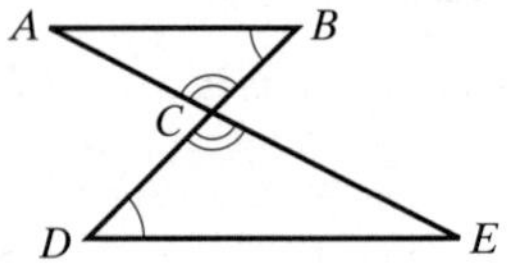

5.

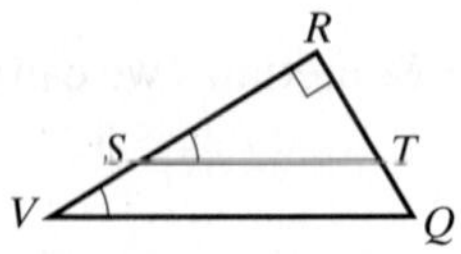

6.

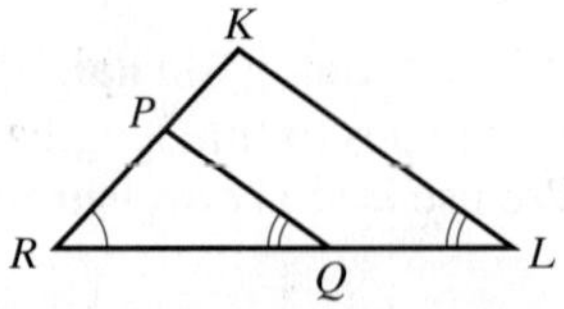

7.4 EXERCISE SET MyMathLab®

Mixed Practice

Determine whether the triangles are similar. If they are, write a similarity statement and name the postulate or theorem you used. If not, explain. See Examples 1–4.

1.

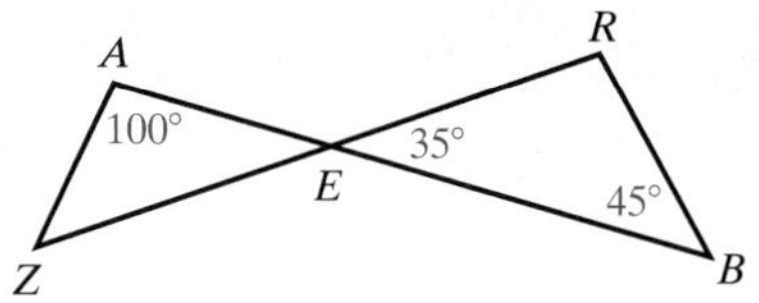

2.

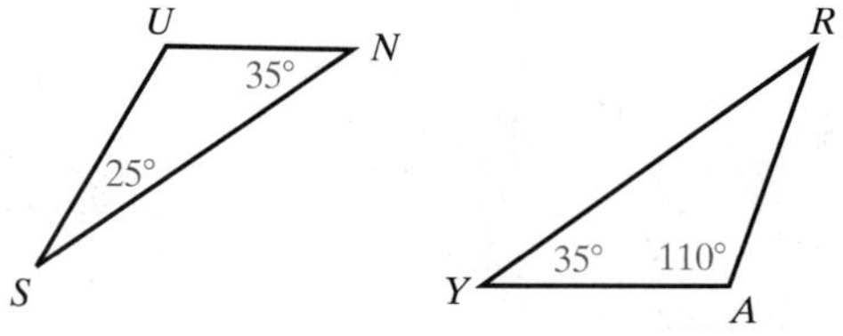

3.

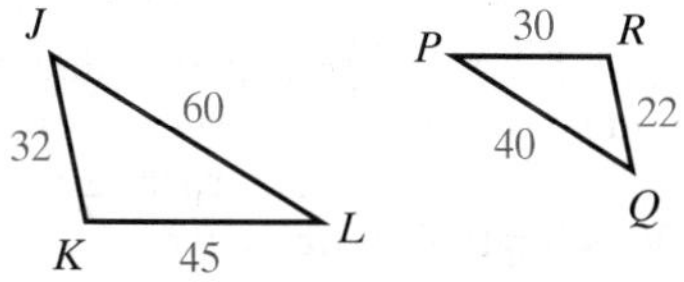

4.

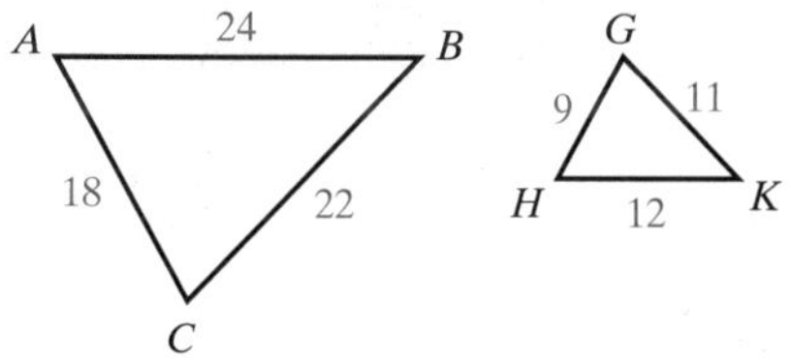

5.

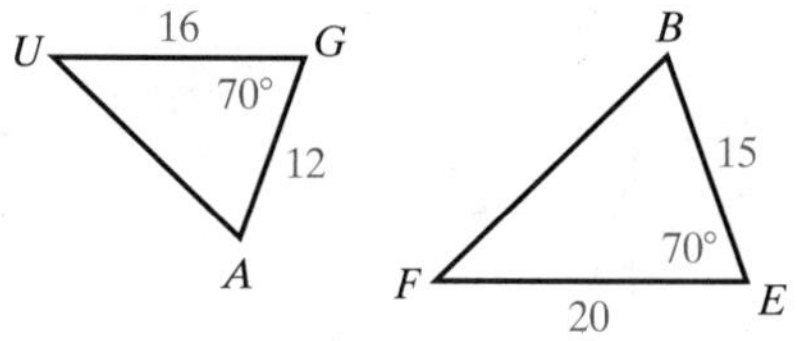

6.

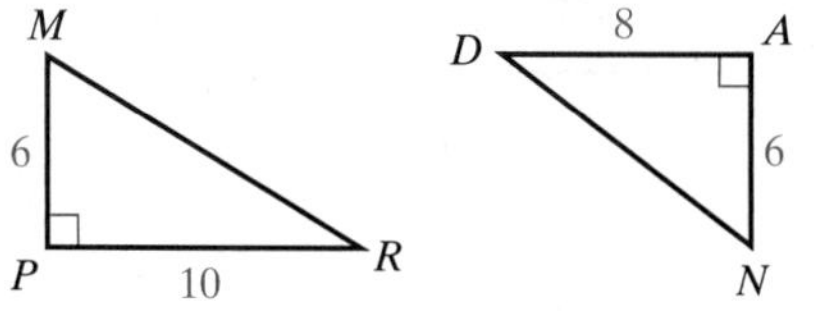

7.

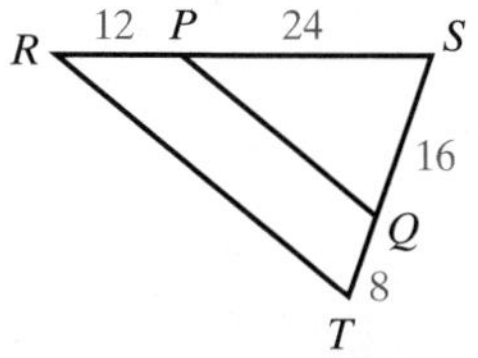

8.

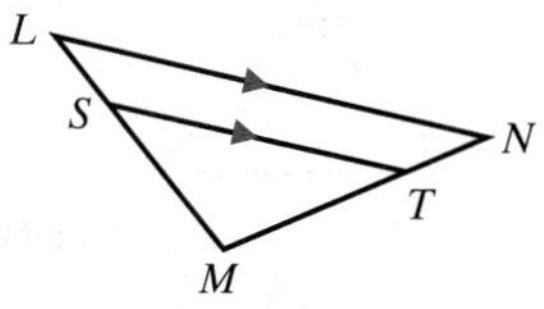

9.

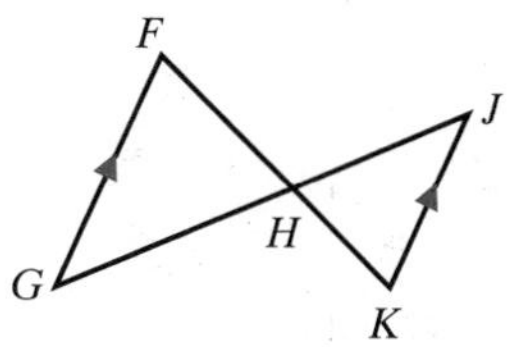

10.

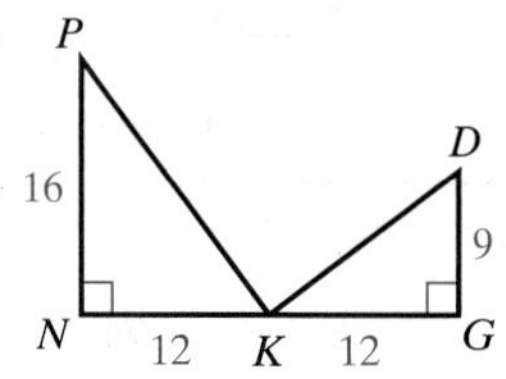

11.

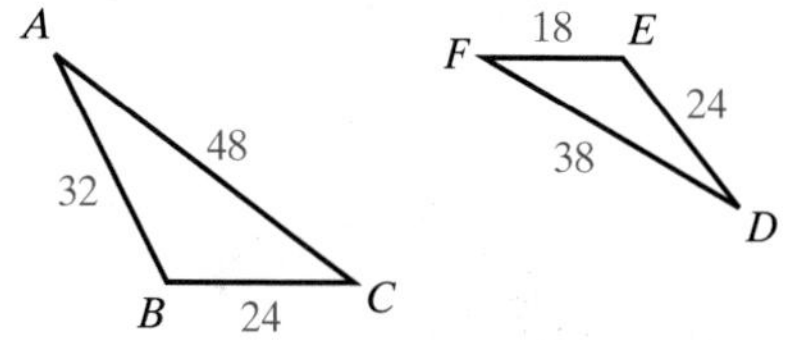

12.

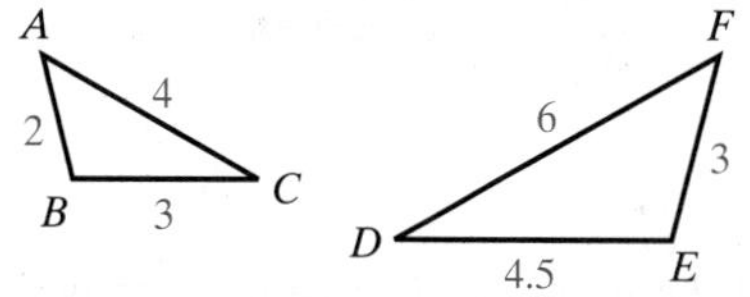

Solve. For Exercises 13 and 14, the solutions have been started for you. See Example 6.

13. Given the following diagram, approximate the height of the observation deck in the Seattle Space Needle in Seattle, Washington. (*Source:* Seattle Space Needle)

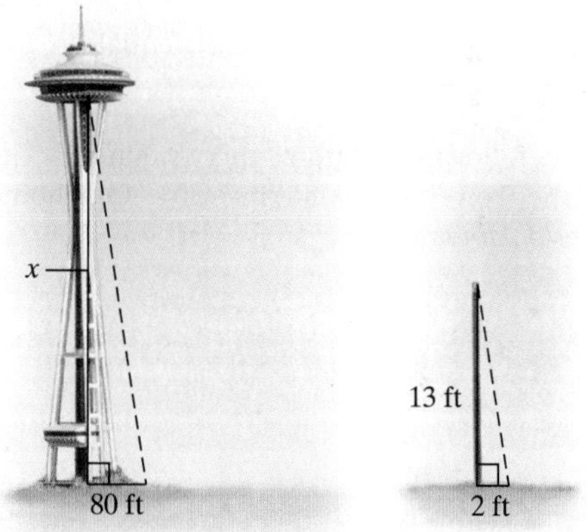

Start the solution:

1. UNDERSTAND the problem. Reread it as many times as needed.
2. TRANSLATE into a proportion using the similar triangles formed. (Fill in the blanks.)

height of observation deck	→	$\frac{x}{13} = \frac{\_\_}{\_\_}$	←	length of Space Needle shadow
height of pole	→		←	length of pole shadow

3. SOLVE by setting cross products equal.
4. INTERPRET.

14. Fountain Hills, Arizona, boasts the tallest fountain in the world. The fountain sits in a 28-acre lake and shoots up a column of water every hour. Based on the diagram on the following page, what is the approximate height of the fountain?

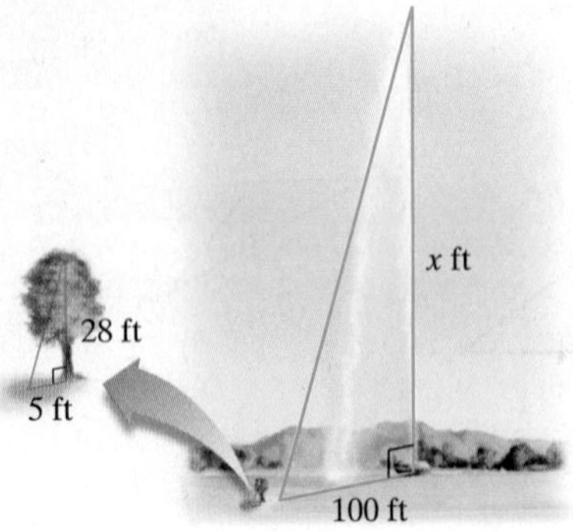

Start the solution:

1. UNDERSTAND the problem. Reread it as many times as needed.
2. TRANSLATE into a proportion using the similar triangles formed. (Fill in the blanks.)

height of tree	→	$\dfrac{28}{x} = \dfrac{\_\_}{\_\_}$	←	length of tree shadow
height of fountain	→		←	length of fountain shadow

3. SOLVE by setting cross products equal.
4. INTERPRET.

15. Given the following diagram, approximate the height of the Bank One Tower in Oklahoma City, Oklahoma. (*Source: The World Almanac*)

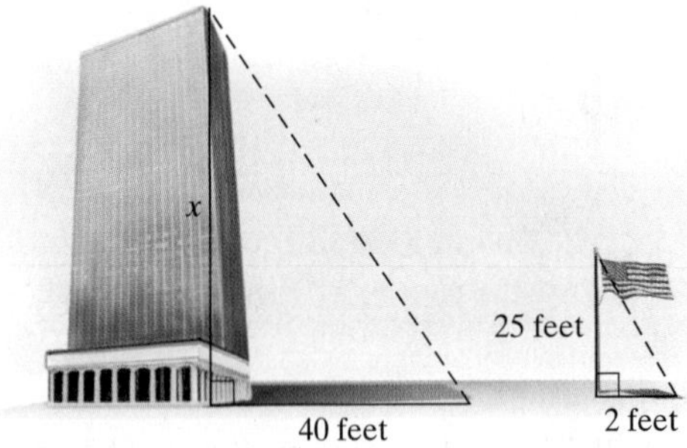

16. The tallest tree standing today is a redwood located in the Humboldt Redwoods State Park near Ukiah, California. Given the following diagram, approximate its height. (*Source: Guinness World Records*)

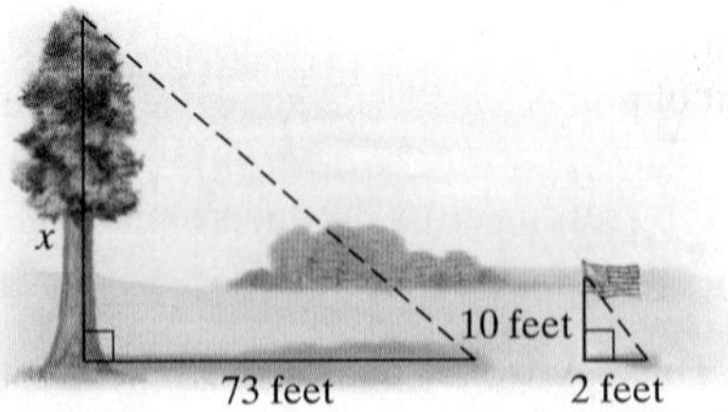

17. If a 30-foot tree casts an 18-foot shadow, find the length of the shadow cast by a 24-foot tree.

18. If a 24-foot flagpole casts a 32-foot shadow, find the length of the shadow cast by a 44-foot antenna. Round to the nearest tenth.

Explain why the triangles are similar. Then find the distance represented by x. See Example 6.

19.

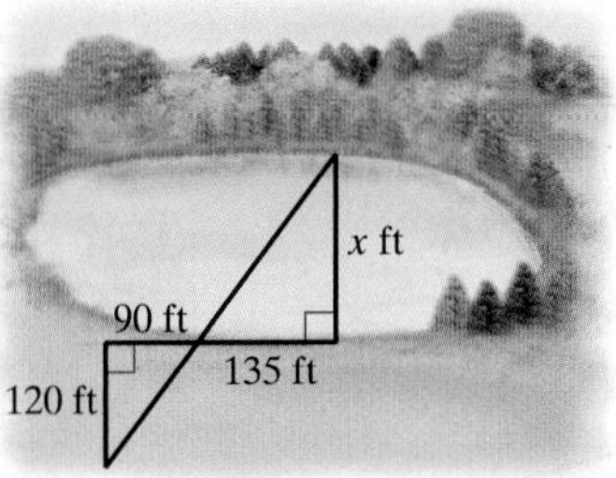

20.

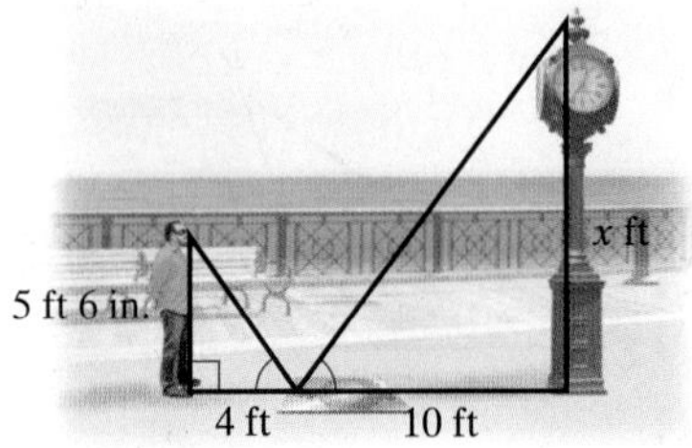

21. At a certain time of day, a 1.8-m-tall person standing next to the Washington Monument casts a 0.7-m shadow. At the same time, the Washington Monument casts a 65.8-m shadow. How tall is the Washington Monument?

22. A 2-ft vertical post casts a 16-in. shadow at the same time a nearby cell phone tower casts a 120-ft shadow. How tall is the cell phone tower?

Mixed Practice *For each pair of similar triangles, find the value of x.*

23.

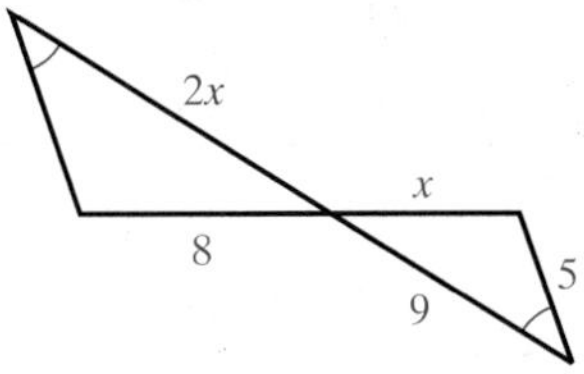

24.

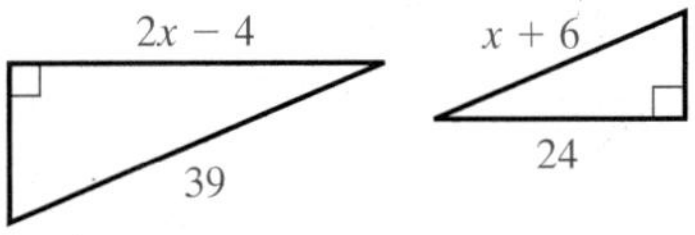

25.

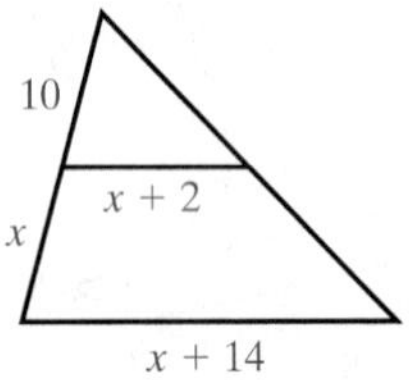

26.

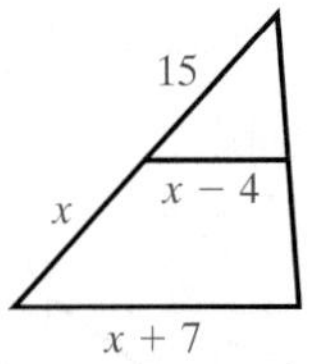

CONCEPT EXTENSIONS

Proof *See Example 5.*

27. Given: $\angle ABC \cong \angle ACD$

Prove: $\Delta ABC \sim \Delta ACD$

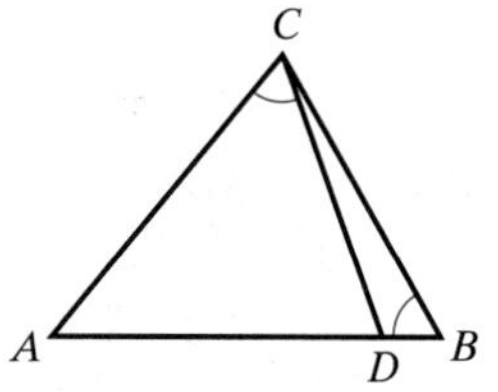

28. Given: $PR = 2NP$,
$PQ = 2MP$

Prove: $\Delta MNP \sim \Delta QRP$

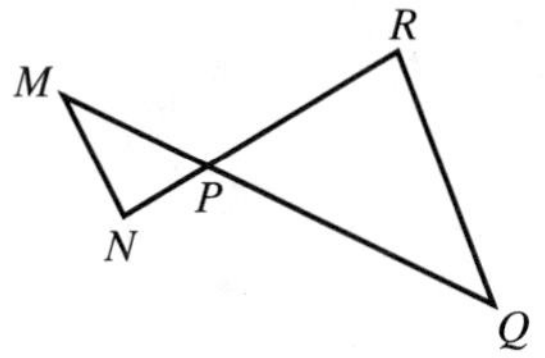

29. Given: $\overline{PQ} \perp \overline{QT}, \overline{ST} \perp \overline{TQ}, \dfrac{PQ}{ST} = \dfrac{RQ}{TV}$

Prove: ΔVKR is isosceles.

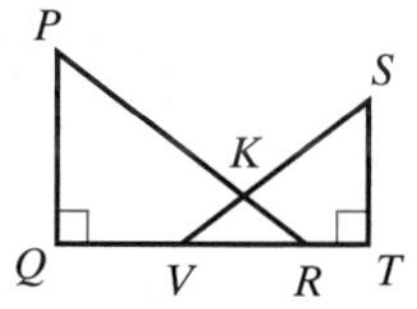

30. Given: $\overline{AB} \parallel \overline{CD}, \overline{BC} \parallel \overline{DG}$

Prove: $AB \cdot CG = CD \cdot AC$

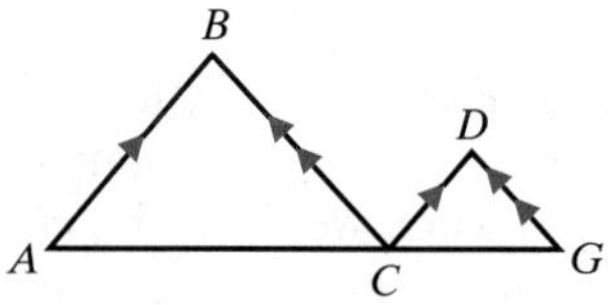

31. Are two isosceles triangles always similar? Explain.

32. Are two right isosceles triangles always similar? Explain.

33. Does any line that intersects two sides of a triangle and is parallel to the third side of the triangle form two similar triangles? Justify your reasoning.

34. In the diagram below, $\Delta PMN \sim \Delta SRW$. $\overline{MQ}$ and $\overline{RT}$ are altitudes. The scale factor of ΔPMN to ΔSRW is 4:3. What is the ratio of $\overline{MQ}$ to $\overline{RT}$? Explain how you know.

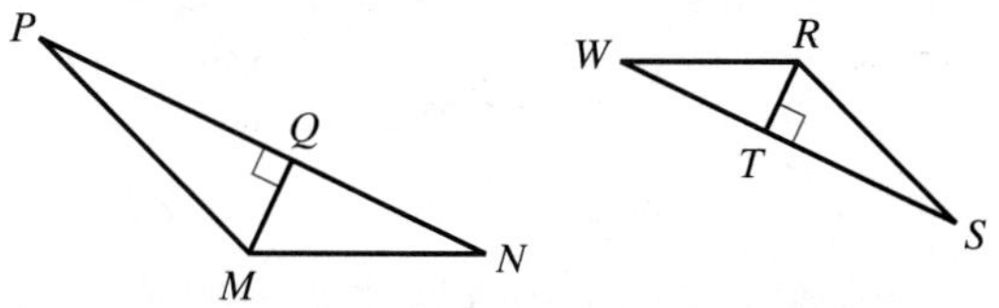

Find the Error *For Exercises 35 and 36, is each solution for the value of x in the figure below correct or incorrect? Explain.*

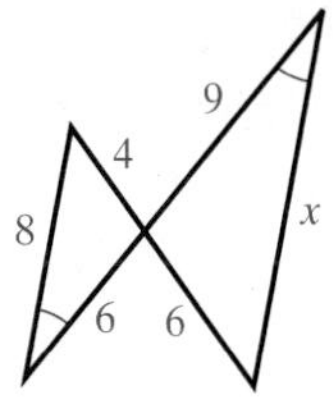

35.

$\dfrac{4}{8} = \dfrac{8}{x}$

$4x = 72$

$x = 18$

36.

$\dfrac{8}{x} = \dfrac{4}{6}$

$48 = 4x$

$12 = x$

37. How are the SAS Similarity Theorem and the SAS Congruence Postulate alike? How are they different?

38. How are the SSS Similarity Theorem and the SSS Congruence Postulate alike? How are they different?

Proof *See Example 5.*

39. Write a proof of the following: Any two nonvertical parallel lines have equal slopes.

Given: Nonvertical lines ℓ_1 and ℓ_2, $\ell_1 \parallel \ell_2$, $\overline{EF}$ and $\overline{BC}$ are $\perp$ to the x-axis

Prove: $\dfrac{BC}{AC} = \dfrac{EF}{DF}$

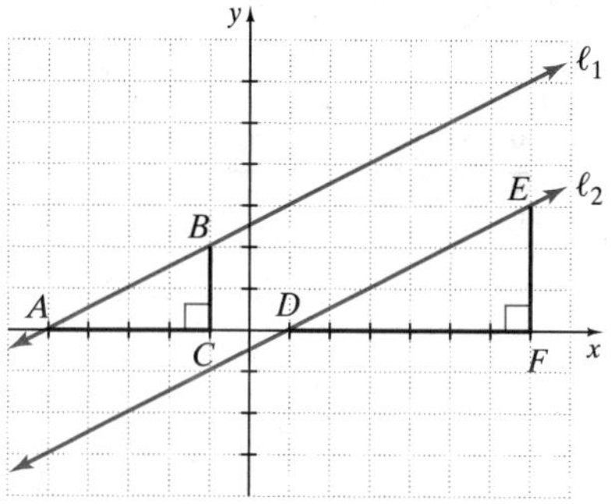

40. Use the diagram in Exercise **39**. Prove: Any two nonvertical lines with equal slopes are parallel.

41. Prove the Side-Angle-Side Similarity Theorem (Theorem 7.4-2).

Given: $\dfrac{AB}{QR} = \dfrac{AC}{QS}, \angle A \cong \angle Q$

Prove: $\Delta ABC \sim \Delta QRS$

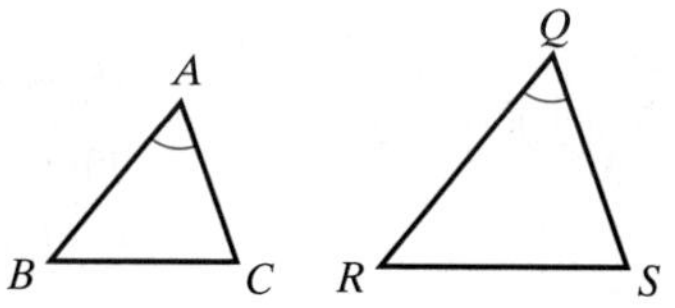

Plan for Proof: Choose X on $\overline{RQ}$ so that $QX = AB$. Draw $\overleftrightarrow{XY} \parallel \overline{RS}$ such that Y lies on QS. Show that $\Delta QXY \sim \Delta QRS$ by the AA ~ Postulate.

Then use the proportion $\dfrac{QX}{QR} = \dfrac{QY}{QS}$ and the given proportion

$\dfrac{AB}{QR} = \dfrac{AC}{QS}$ to show that $AC = QY$.

Then prove that $\Delta ABC \cong \Delta QXY$.

Finally, prove that $\Delta ABC \sim \Delta QRS$ by the AA ~ Postulate.

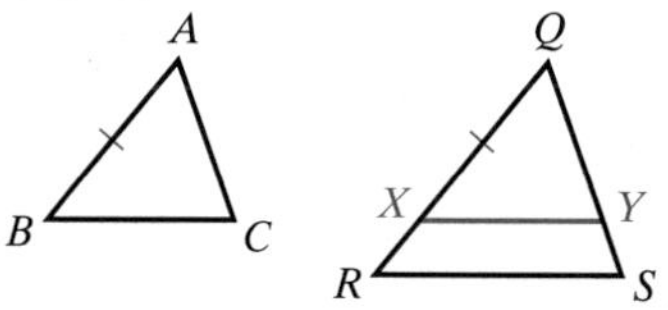

42. Prove the Side-Side-Side Similarity Theorem (Theorem 7.4-3).

Given: $\frac{AB}{QR} = \frac{AC}{QS} = \frac{BC}{RS}$

Prove: $\Delta ABC \sim \Delta QRS$

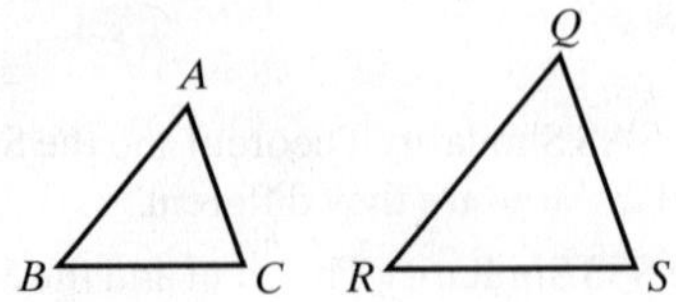

43. ΔABC has vertices $A(0, 0)$, $B(2, 4)$, and $C(4, 2)$. ΔRST has vertices $R(0, 3)$, $S(-1, 5)$, and $T(-2, 4)$. Prove that $\Delta ABC \sim \Delta RST$. (*Hint:* Graph ΔABC and ΔRST in the coordinate plane.)

REVIEW AND PREVIEW

TRAP ~ EZYD. Use the diagram below to find the following. See Section 7.3.

44. the scale factor of *TRAP* to *EZYD*

45. $m\angle R$

46. DY

47. $\frac{DE}{PT}$

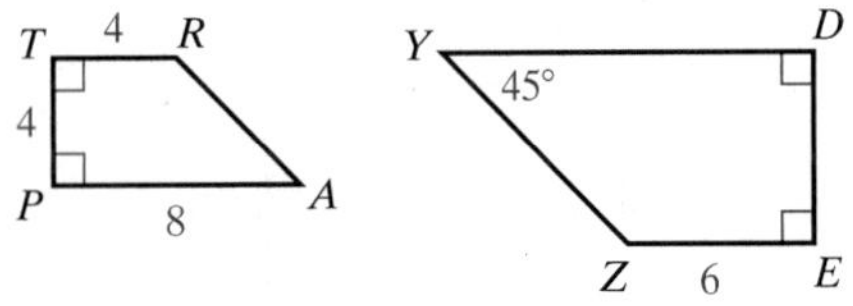

Use a protractor to find the measure of each angle. Classify the angle as "acute," "right," "obtuse," or "straight." See Section 1.5.

48.

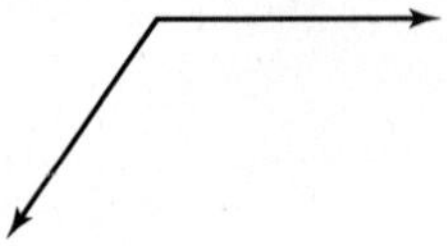

49.

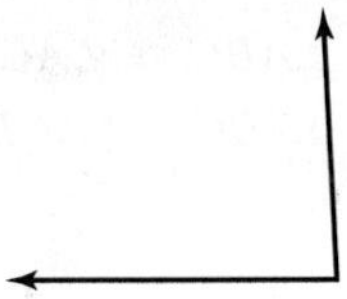

50.

51.

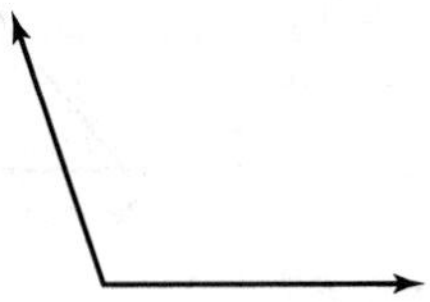

Identify the means and extremes of each proportion. Then solve for x. See Section 7.1.

52. $\frac{x}{8} = \frac{18}{24}$

53. $\frac{12}{m} = \frac{18}{20}$

54. $\frac{15}{x + 2} = \frac{9}{x}$

55. $\frac{x - 3}{x + 4} = \frac{5}{9}$

7.5 GEOMETRIC MEAN AND SIMILARITY IN RIGHT TRIANGLES

OBJECTIVES

1. Use Altitudes of Right Triangles to Prove Similarity.
2. Find the Geometric Mean of the Lengths of Segments in a Right Triangle.
3. Solve Applications Involving Right Triangles.

VOCABULARY

- geometric mean

OBJECTIVE 1 ▶ Using Altitudes of Right Triangles. In this section, we learn new ways to think about the proportions that come from similar right triangles.

Theorem 7.5-1 shows that when we draw the *altitude to the hypotenuse* of a right triangle, three pairs of similar right triangles are formed.

Theorem 7.5-1 Altitude of a Right Triangle

Theorem

The altitude to the hypotenuse of a right triangle divides the triangle into two triangles that are similar to the original triangle and to each other.

If ...

ΔABC is a right triangle with right $\angle ACB$, and $\overline{CD}$ is the altitude to the hypotenuse

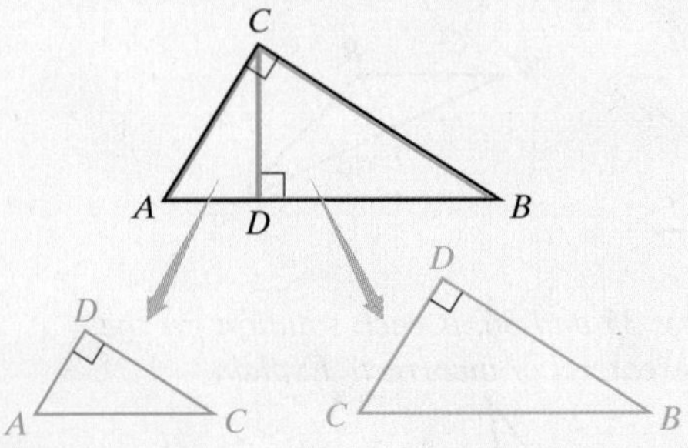

Then ...

$\Delta ABC \sim \Delta ACD$
$\Delta ABC \sim \Delta CBD$
$\Delta ACD \sim \Delta CBD$

Let's prove this theorem.

Proof of Theorem 7.5-1

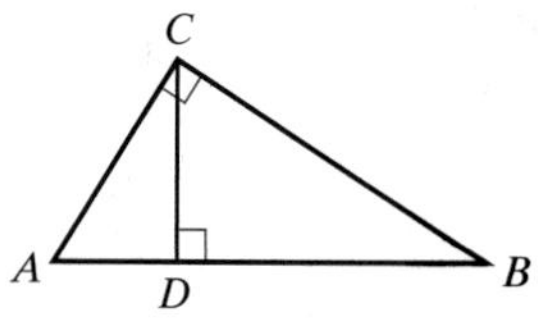

Given: Right ΔABC with right $\angle ACB$ and altitude $\overline{CD}$

Prove: $\Delta ACD \sim \Delta ABC$, $\Delta CBD \sim \Delta ABC$, $\Delta ACD \sim \Delta CBD$

Statements	Reasons
1. $\angle ACB$ is a right angle.	**1.** Given
2. $\overline{CD}$ is an altitude.	**2.** Given
3. $\overline{CD} \perp \overline{AB}$	**3.** Definition of altitude
4. $\angle ADC$ and $\angle CDB$ are right angles.	**4.** Definition of perpendicularity
5. $\angle ADC \cong \angle ACB$, $\angle CDB \cong \angle ACB$	**5.** All right angles are congruent.
6. $\angle A \cong \angle A$, $\angle B \cong \angle B$	**6.** Reflexive Property of Congruence
7. $\Delta ACD \sim \Delta ABC$, $\Delta CBD \sim \Delta ABC$	**7.** AA ~ Postulate (Steps 5, 6)
8. $\angle ACD \cong \angle B$	**8.** Corresponding angles of ~ triangles are congruent.
9. $\angle ADC \cong \angle CDB$	**9.** All right angles are congruent.
10. $\Delta ACD \sim \Delta CBD$	**10.** AA ~ Postulate (Steps 8, 9)

Now we can use this theorem to practice writing similarity statements.

EXAMPLE 1 **Writing Similarity Statements**

What similarity statement can you write relating the three triangles in the diagram?

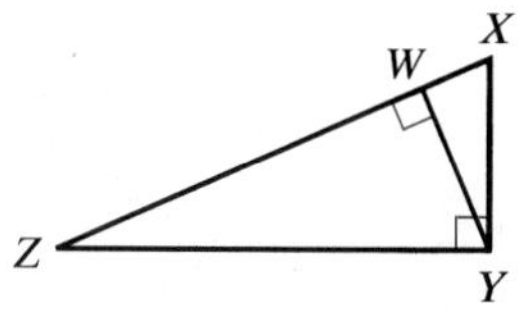

Solution To help, let's sketch the triangles separately in the same orientation. $\overline{YW}$ is the altitude to the hypotenuse of right ΔXYZ, so we can use Theorem 7.5-1. There are three similar triangles.

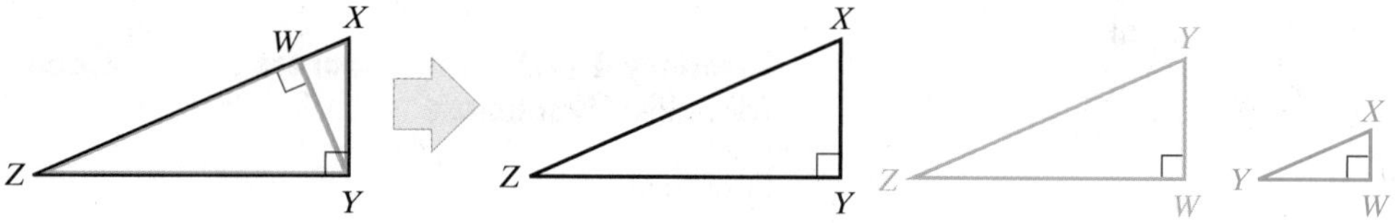

$\Delta XYZ \sim \Delta YWZ \sim \Delta XWY$ □

PRACTICE
1 What similarity statement can you write relating the three triangles in the diagram?

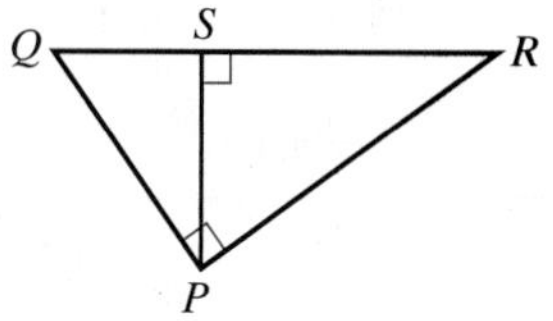

OBJECTIVE 2 ▶ Finding Geometric Means. Proportions in which the means are equal occur frequently in geometry.

For any two positive numbers a and b, the **geometric mean** of a and b is the positive number x such that

$$\frac{a}{x} = \frac{x}{b}.$$

EXAMPLE 2 Finding the Geometric Mean

What is the geometric mean of 6 and 15?

a. 90 **b.** $3\sqrt{10}$ **c.** $9\sqrt{10}$ **d.** 30

Solution Let x be the geometric mean. Use the definition of geometric mean above to set up a proportion. Then solve for x.

$$\frac{6}{x} = \frac{x}{15} \quad \text{Definition of geometric mean}$$

$$x^2 = 90 \quad \text{Cross Products Property}$$

$$x = \sqrt{90} \quad \text{Take the positive square root of each side.}$$

$$x = 3\sqrt{10} \quad \text{Write in simplest radical form.}$$

The geometric mean of 6 and 15 is $3\sqrt{10}$. The correct answer is **b**. □

PRACTICE

2 What is the geometric mean of 4 and 18?

In Practice 1, we used the figure below and wrote a similarity statement. Let's revisit the similarity and concentrate on two similar triangles $\Delta SQP \sim \Delta SPR$ to write a proportion with a geometric mean.

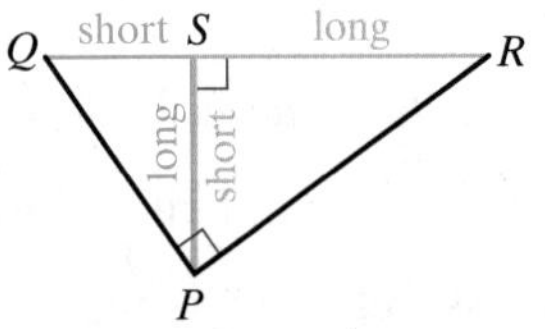

$\Delta SQP \sim \Delta SPR$

$$\frac{\text{short leg}}{\text{short leg}} = \frac{\text{long leg}}{\text{long leg}}$$

$$\frac{SQ}{SP} = \frac{SP}{SR}$$

SP is the geometric mean of *SQ* and *SR*.

This illustrates the first of two important corollaries of Theorem 7.5-1.

Corollary 1 (7.5-2) to Theorem 7.5-1 Geometric Mean in Similar Right Triangles: Hypotenuse

Corollary	If ...	Then ...
The length of the altitude to the hypotenuse of a right triangle is the geometric mean of the lengths of the segments of the hypotenuse.	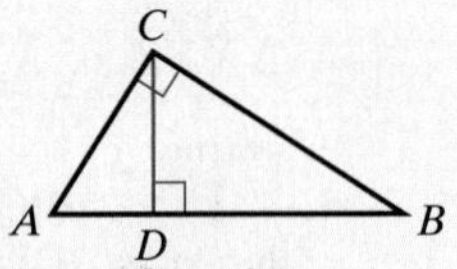	$\frac{AD}{CD} = \frac{CD}{DB}$

Example

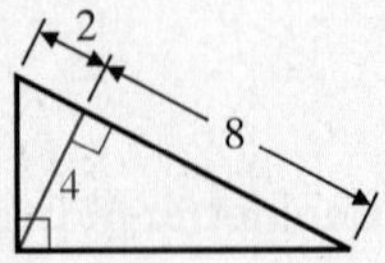

Segments of hypotenuse → $\frac{2}{4} = \frac{4}{8}$ ← Altitude of hypotenuse (geometric mean)

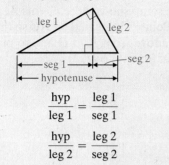

> **Helpful Hint**
>
> To remember this corollary, the following may help:
>
> 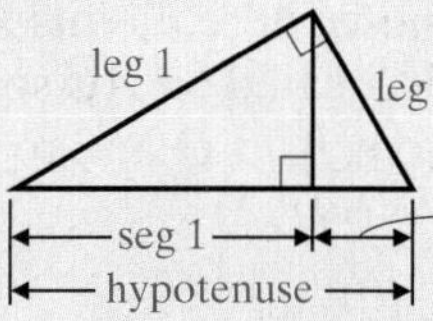
>
>
> $$\frac{\text{hyp}}{\text{leg 1}} = \frac{\text{leg 1}}{\text{seg 1}}$$
>
> $$\frac{\text{hyp}}{\text{leg 2}} = \frac{\text{leg 2}}{\text{seg 2}}$$

Corollary 2 (7.5-3) to Theorem 7.5-1 Geometric Mean in Similar Right Triangles: Legs

Corollary	If ...	Then ...
The altitude to the hypotenuse of a right triangle separates the hypotenuse so that the length of each leg of the triangle is the geometric mean of the length of the hypotenuse and the length of the segment of the hypotenuse adjacent to the leg.	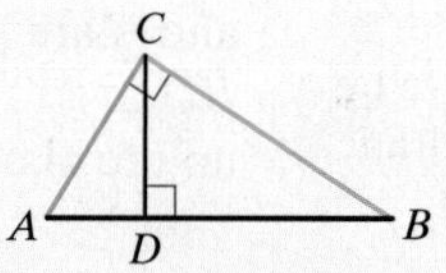	$\frac{AB}{AC} = \frac{AC}{AD}$ $\frac{AB}{CB} = \frac{CB}{DB}$

Example

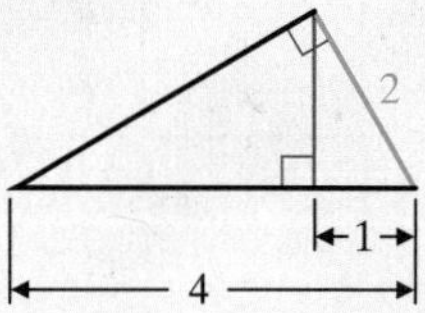

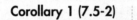

Hypotenuse → $\frac{4}{2} = \frac{2}{1}$ ← Leg; 1 ← Segment of hypotenuse adjacent to leg

We prove these corollaries in Exercises 35 and 36, Exercise Set 7.5.

The two corollaries to Theorem 7.5-1 give us ways to write proportions using lengths in similar right triangles. To help remember these corollaries, consider the following diagram and these properties.

Corollary 1 (7.5-2)	Corollary 2 (7.5-3)	
$\frac{s_1}{a} = \frac{a}{s_2}$	$\frac{h}{\ell_1} = \frac{\ell_1}{s_1}, \frac{h}{\ell_2} = \frac{\ell_2}{s_2}$	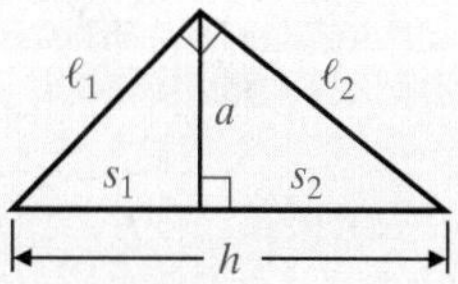

> **Helpful Hint**
>
> How might we decide which corollary to use?
>
> If you are using or finding an altitude, use Corollary 1. If you are using or finding a leg or hypotenuse, use Corollary 2.

EXAMPLE 3 Using the Corollaries and Algebra

What are the values of x and y?

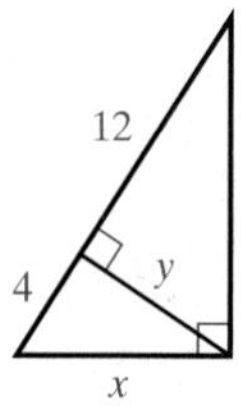

Solution To find the altitude y, we can use Corollary 1. After that, since x is a leg of the right triangle, let's try Corollary 2.

Use Corollary 1.

$$\frac{12}{y} = \frac{y}{4}$$
$$y^2 = 48$$
$$y = \sqrt{48}$$
$$y = 4\sqrt{3}$$

Use Corollary 2.

$$\frac{4 + 12}{x} = \frac{x}{4} \quad \text{Write a proportion.}$$
$$\frac{16}{x} = \frac{x}{4} \quad \text{Add: } 4 + 12.$$
$$x^2 = 64 \quad \text{Cross Products Property}$$
$$x = \sqrt{64} \quad \text{Take the positive square root.}$$
$$x = 8 \quad \text{Simplify.}$$

□

The value of x is 8 units and the value of y is $4\sqrt{3}$ units.

PRACTICE 3 What are the values of x and y?

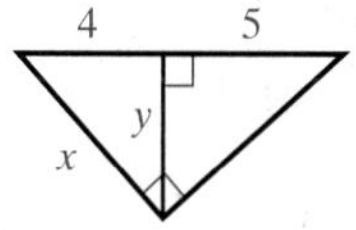

OBJECTIVE 3 ▶ Solving Applications Involving Right Triangles.

> **▶ Helpful Hint**
> To solve this quadratic equation, first write it in standard form, $ax^2 + bx + c = 0$. Then solve by factoring or using the quadratic formula.

EXAMPLE 4 Finding a Distance

You are preparing for a robotics competition using the setup shown here. Points A, B, and C are located so that $AB = 20$ in. and $\overline{AB} \perp \overline{BC}$. Point D is located on $\overline{AC}$ so that $\overline{BD} \perp \overline{AC}$ and $DC = 9$ in. You program the robot to move from A to D and to pick up the plastic bottle at D. How far does the robot travel from A to D?

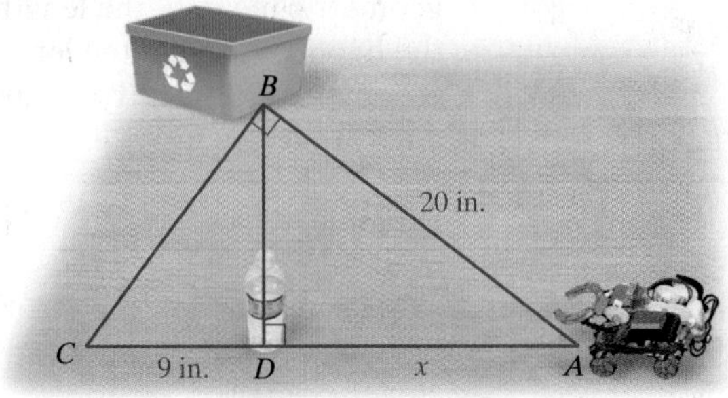

Solution Concentrate on right ΔABC with altitude $\overline{BD}$. We try Corollary 2 since we are *not* trying to find the altitude length BD.

> **▶ Helpful Hint**
> Unlike Example 3, these are unlike terms and may not be combined (added).

$\dfrac{x+9}{20} = \dfrac{20}{x}$	Corollary 2
$x(x+9) = 400$	Cross Products Property
$x^2 + 9x - 400 = 0$	Distributive Property and subtract 400 from each side.
$(x-16)(x+25) = 0$	Factor.
$x - 16 = 0$ or $(x+25) = 0$	Zero Product Property
$x = 16$ or $x = -25$	Solve for x.

Only the positive solution makes sense in this situation. The robot travels 16 in. □

PRACTICE 4 From point D, the robot must turn right and move to point B to put the bottle in the recycling bin. How far does the robot travel from D to B?

VOCABULARY & READINESS CHECK

Find the geometric mean of each pair of numbers.

1. 4 and 9 **2.** 4 and 12

Use the figure to complete each proportion.

3. $\dfrac{g}{e} = \dfrac{e}{\_\_}$ **4.** $\dfrac{j}{d} = \dfrac{d}{\_\_}$ **5.** $\dfrac{\_\_}{f} = \dfrac{f}{\_\_}$ **6.** $\dfrac{j}{\_\_} = \dfrac{\_\_}{g}$

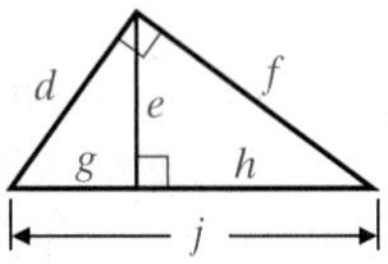

Identify the following in ΔRST.

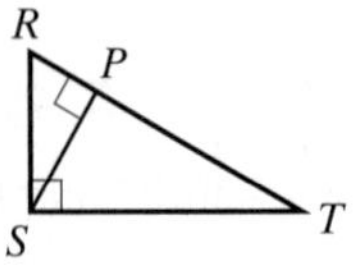

7. the hypotenuse

8. the altitude shown

9. the segments of the hypotenuse defined by the altitude

10. the segment of the hypotenuse adjacent to leg $\overline{ST}$

7.5 EXERCISE SET MyMathLab®

Write a similarity statement relating the three triangles in each diagram. See Example 1.

1. 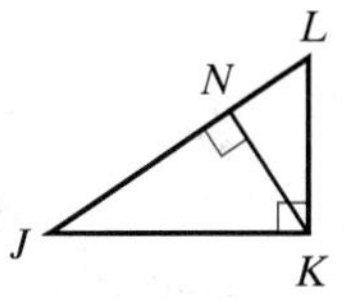

2. 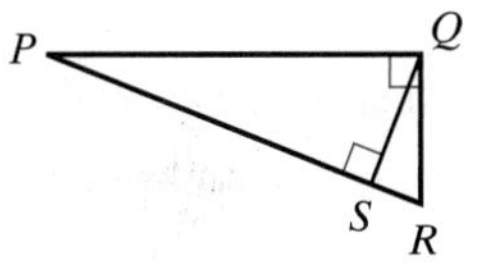

Find the geometric mean of each pair of numbers. See Example 2.

3. 4 and 10
4. 3 and 48
5. 5 and 125
6. 7 and 9
7. 3 and 16
8. 4 and 49
9. 5 and 1.25
10. $\frac{1}{2}$ and 2
11. $\sqrt{8}$ and $\sqrt{2}$
12. $\sqrt{28}$ and $\sqrt{7}$

Solve for x. See Example 3.

13. 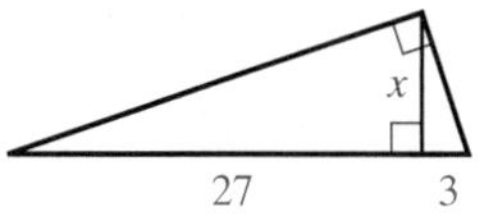

14. 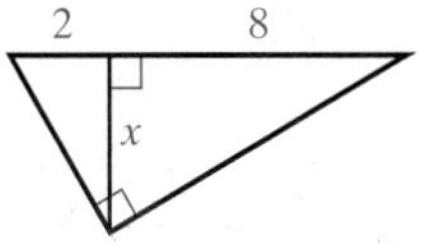

15. 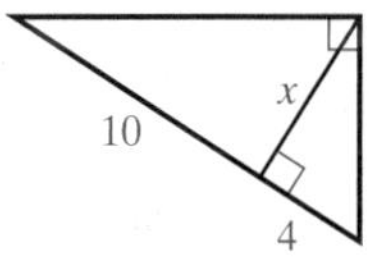

16. 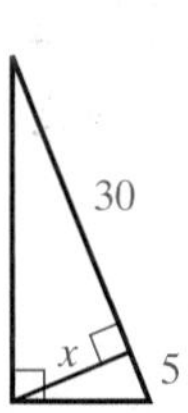

Solve for x and y. See Example 3.

17. 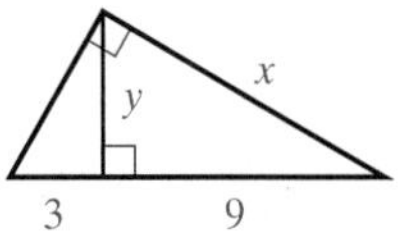

18. 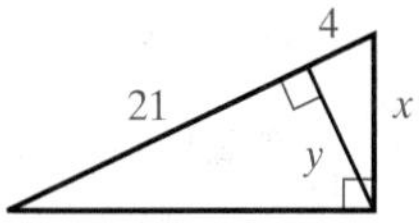

19. 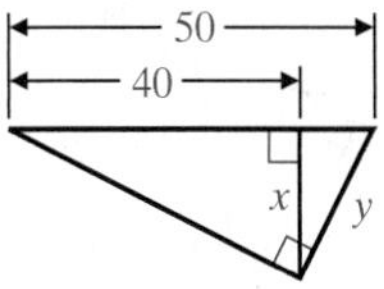

20. 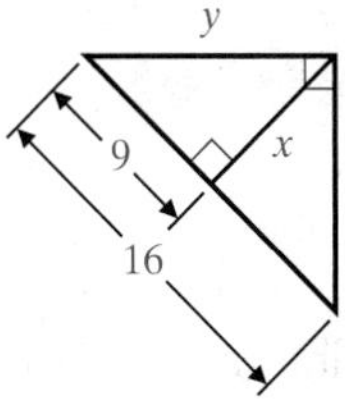

Find the value of x.

21. 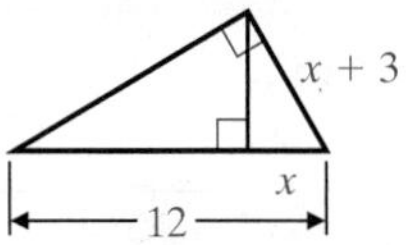

22. 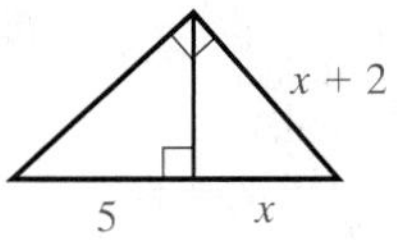

23. 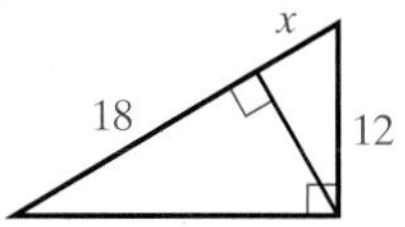

24. 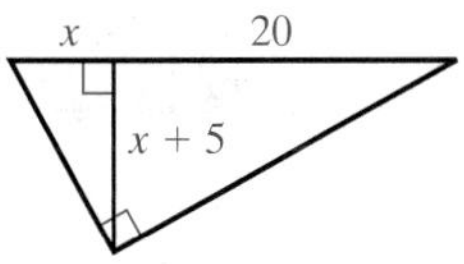

The diagram shows the parts of a right triangle with an altitude to the hypotenuse. For the two given measures, find the other four. Use simplest radical form.

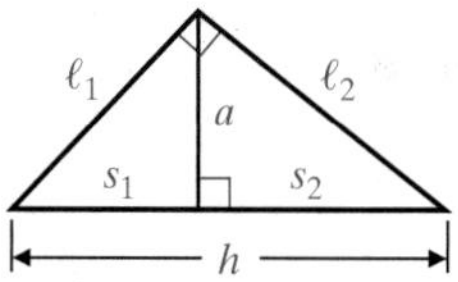

25. $h = 2, s_1 = 1$
26. $a = 6, s_1 = 6$
27. $\ell_1 = 2, s_2 = 3$
28. $s_1 = 3, \ell_2 = 6\sqrt{3}$

Solve. See Example 4.

29. The architect's side-view drawing of a saltbox-style house shows a post that supports the roof ridge. The support post is 10 ft tall. How far from the front of the house is the support post positioned?

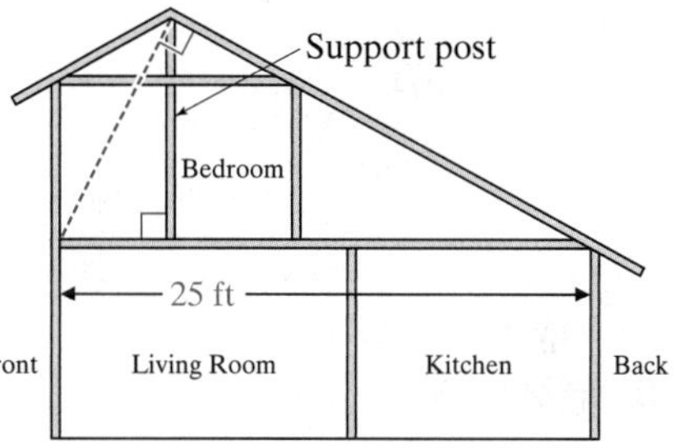

30. To estimate the height of a stone figure, Anya holds a small square up to her eyes and walks backward from the figure. She stops when the bottom of the figure aligns with the bottom edge of the square and the top of the figure aligns with the top edge of the square. Her eye level is 1.84 m from the ground. She is 3.50 m from the figure. What is the height of the figure to the nearest hundredth of a meter?

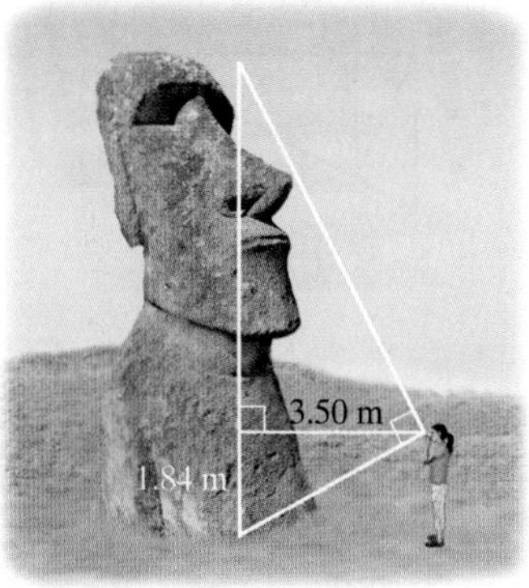

CONCEPT EXTENSIONS

31. A classmate says the following statement is true: The geometric mean of positive numbers a and b is $\sqrt{ab}$. Do you agree? Explain.

32. The altitude to the hypotenuse of a right triangle divides the hypotenuse into segments 2 cm and 8 cm long. Find the length of the altitude to the hypotenuse.

33. **Find the Error** A classmate wrote an incorrect proportion to find x. Explain and correct the error.

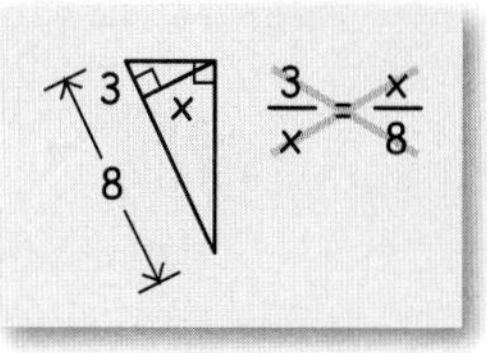

34. The altitude to the hypotenuse of a right triangle divides the hypotenuse into segments with lengths in the ratio 1:2. The length of the altitude is 8. How long is the hypotenuse?

Proof *Use the figure below for Exercises 35 and 36.*

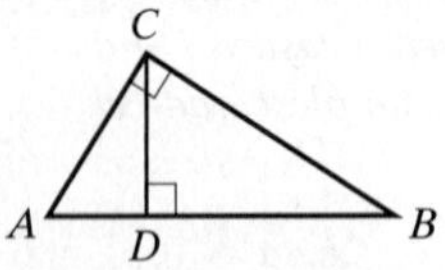

35. Prove Corollary 1 to Theorem 7.5-1.

Given: Right ΔABC with altitude to the hypotenuse $\overline{CD}$

Prove: $\frac{AD}{CD} = \frac{CD}{DB}$

36. Prove Corollary 2 to Theorem 7.5-1.

Given: Right ΔABC with altitude to the hypotenuse $\overline{CD}$

Prove: $\frac{AB}{AC} = \frac{AC}{AD}, \frac{AB}{BC} = \frac{BC}{DB}$

PROOF

37. Given: Right ΔABD with altitude to the hypotenuse $\overline{BE}$, and equilateral ΔABC

Prove: $BE = AE\sqrt{3}$

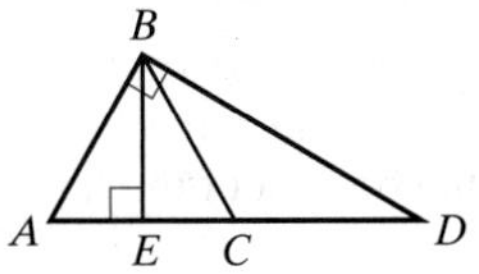

38. Given: In right ΔABC, $\overline{BD} \perp \overline{AC}$ and, $\overline{DE} \perp \overline{BC}$.

Prove: $\frac{AD}{DC} = \frac{BE}{EC}$

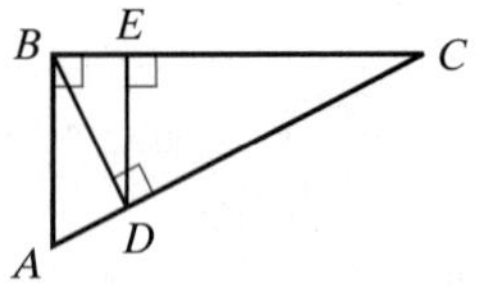

REVIEW AND PREVIEW

Use the figure below to answer Exercises 39 and 40. See Section 7.4.

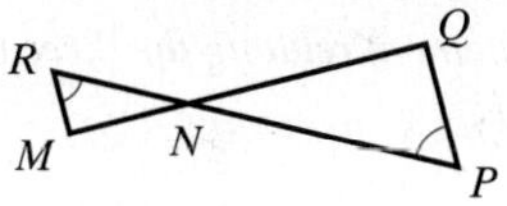

39. Write a similarity statement for the two triangles.

40. How do you know they are similar?

For Exercises 41 and 42, use the figure to find the values of x and y in ▱RSTV. See Section 6.2.

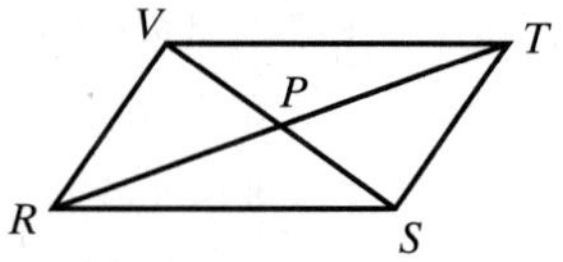

41. $RP = 2x, PT = y + 2, VP = y, PS = x + 3$

42. $RV = 2x + 3, VT = 5x, TS = y + 5, SR = 4y - 1$

The two triangles in each diagram are similar. Find the value of x in each. See Section 7.3.

43.

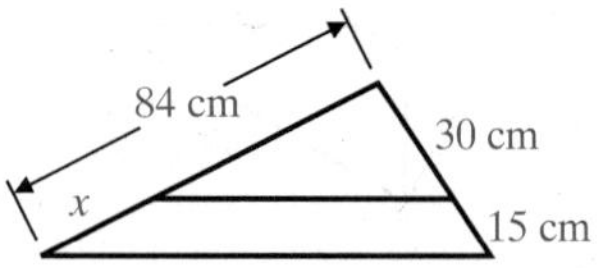

44.

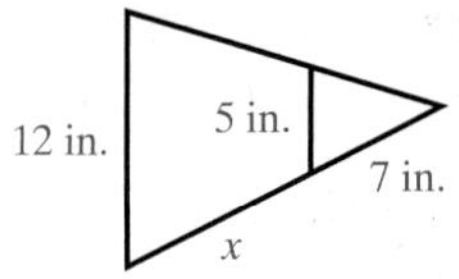

45.

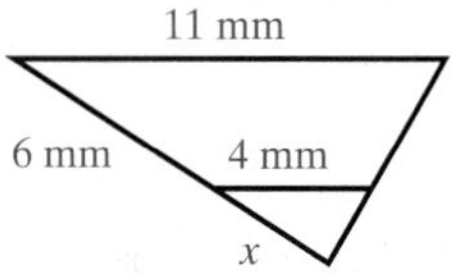

7.6 ADDITIONAL PROPORTIONS IN TRIANGLES

OBJECTIVES

1 Use the Side-Splitter Theorem.

2 Use the Triangle-Angle-Bisector Theorem.

OBJECTIVE 1 ▶ Using the Side-Splitter Theorem. In this section, we learn how to use proportions to find lengths of segments formed by parallel lines that intersect two or more transversals. This can be done using the following theorem.

Theorem 7.6-1 Side-Splitter Theorem

Theorem	If ...	Then ...
If a line is parallel to one side of a triangle and intersects the other two sides, then it divides those sides proportionally.	$\overleftrightarrow{RS} \parallel \overleftrightarrow{XY}$ 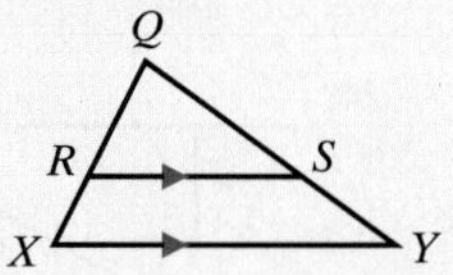	$\frac{XR}{RQ} = \frac{YS}{SQ}$

Next, we prove this theorem. Then we will use this theorem in an example.

Proof of Theorem 7.6-1: Side-Splitter Theorem

Given: ΔQXY with $\overleftrightarrow{RS} \parallel \overleftrightarrow{XY}$

Prove: $\dfrac{XR}{RQ} = \dfrac{YS}{SQ}$

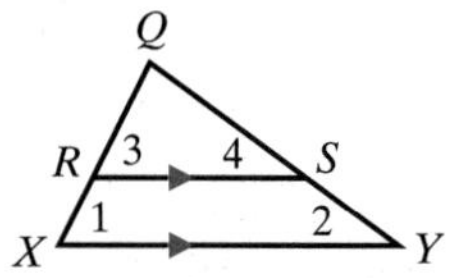

Statements	Reasons
1. $\overleftrightarrow{RS} \parallel \overleftrightarrow{XY}$	**1.** Given
2. $\angle 1 \cong \angle 3, \angle 2 \cong \angle 4$	**2.** If lines are parallel, then corresponding angles are congruent.
3. $\Delta QXY \sim \Delta QRS$	**3.** AA $\sim$ Postulate (Step 2)
4. $\dfrac{XQ}{RQ} = \dfrac{YQ}{SQ}$	**4.** Corresponding sides of similar triangles are proportional.
5. $XQ = XR + RQ$; $YQ = YS + SQ$	**5.** Segment Addition Postulate
6. $\dfrac{XR + RQ}{RQ} = \dfrac{YS + SQ}{SQ}$	**6.** Substitution Property (Steps 4, 5)
7. $\dfrac{XR}{RQ} = \dfrac{YS}{SQ}$	**7.** Property of Proportions (3) (Step 6)

EXAMPLE 1 **Using the Side-Splitter Theorem**

What is the value of x in the diagram at the right?

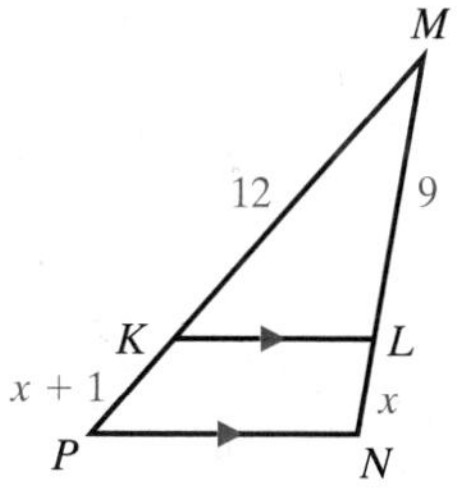

Solution $\overline{KL}$ is parallel to $\overline{PN}$, a side of ΔPMN. Let's set up a proportion using the Side-Splitter Theorem.

$$\frac{PK}{KM} = \frac{NL}{LM} \quad \text{Side-Splitter Theorem}$$

$$\frac{x+1}{12} = \frac{x}{9} \quad \text{Substitute.}$$

$$9x + 9 = 12x \quad \text{Cross Products Property}$$

$$9 = 3x \quad \text{Subtract } 9x \text{ from each side.}$$

$$3 = x \quad \text{Divide each side by 3.}$$

The value of x is 3 units. □

PRACTICE 1 What is the value of a in the diagram at the right?

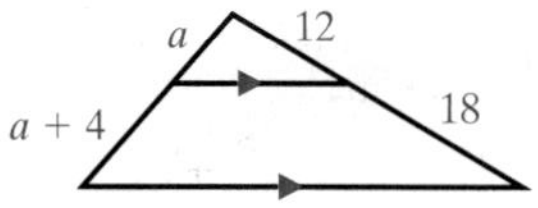

The Side-Splitter Theorem has the following corollary.

Corollary 7.6-2 Corollary to the Side-Splitter Theorem

Corollary	If ...	Then ...
If three parallel lines intersect two transversals, then the segments intercepted on the transversals are proportional.	$a \parallel b \parallel c$ 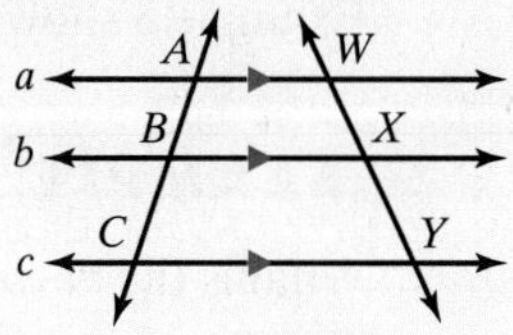	$\dfrac{AB}{BC} = \dfrac{WX}{XY}$

We prove this corollary in Exercise 41 of Exercise Set 7.6.

Let's use this corollary in the next example.

EXAMPLE 2 Finding a Length

Three campsites are shown in the diagram. What is the length of Site A along the river?

Solution Study the figure and note the parallel lines. Think of $\overline{MN}$ and $\overline{PQ}$ as transversals and use the Corollary to the Side-Splitter Theorem.

Let x be the length of Site A along the river.

$$\frac{x}{8} = \frac{9}{7.2} \quad \text{Corollary (7.6-2) to the Side-Splitter Theorem}$$

$$7.2x = 72 \quad \text{Cross Products Property}$$

$$x = 10 \quad \text{Divide each side by 7.2.}$$

The length of Site A along the river is 10 yd.

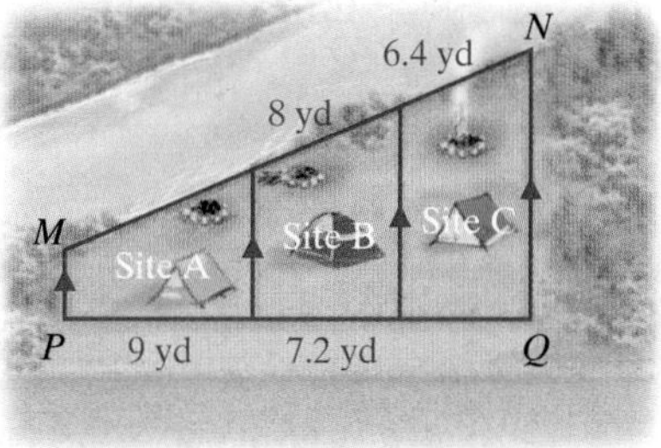

□

PRACTICE

2 Using the same diagram and situation in Example 2, find the length of Site C along the road.

OBJECTIVE 2 ▶ Using the Triangle-Angle-Bisector Theorem. The bisector of an angle of a triangle divides the opposite side into two segments with lengths proportional to the sides of the triangle that form the angle. This is known as the Triangle-Angle-Bisector Theorem, which we state formally next.

Theorem 7.6-3 Triangle-Angle-Bisector Theorem

Theorem	If ...	Then ...
If a ray bisects an angle of a triangle, then it divides the opposite side into two segments that are proportional to the other two sides of the triangle.	$\overrightarrow{AD}$ bisects $\angle CAB$ 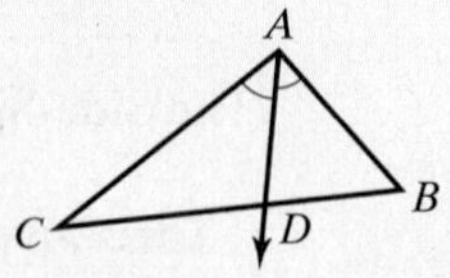	$\frac{CD}{DB} = \frac{CA}{BA}$

We prove this theorem in Exercise 42 of Exercise Set 7.6.

Example 3 shows us one way that this theorem can be used.

EXAMPLE 3 Using the Triangle-Angle-Bisector Theorem

What is the value of x in the diagram at the right?

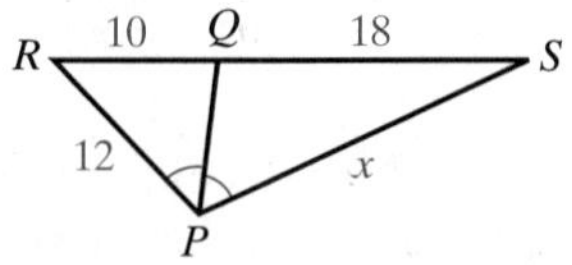

Solution $\overline{PQ}$ bisects $\angle RPS$, so use the Triangle-Angle-Bisector Theorem and write a proportion.

$$\frac{RQ}{QS} = \frac{PR}{PS}$$ Use the Triangle-Angle-Bisector Theorem.

$$\frac{10}{18} = \frac{12}{x}$$ Substitute the known values.

$$10x = 216$$ Set cross products equal.

$$x = 21.6$$ Divide both sides by 10. □

The value of x is 21.6 units.

PRACTICE 3 What is the value of y in the diagram at the right?

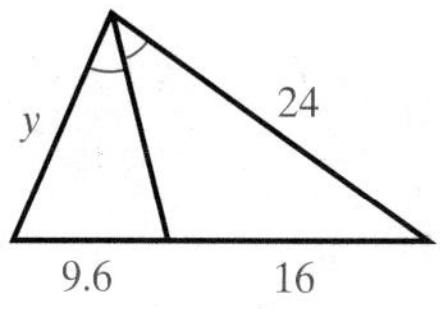

VOCABULARY & READINESS CHECK

Use the figure to complete each proportion.

1. $\frac{a}{b} = \frac{\underline{\quad}}{e}$

2. $\frac{b}{\underline{\quad}} = \frac{e}{f}$

3. $\frac{c}{a} = \frac{f}{\underline{\quad}}$

4. $\frac{\underline{\quad}}{c} = \frac{d}{f}$

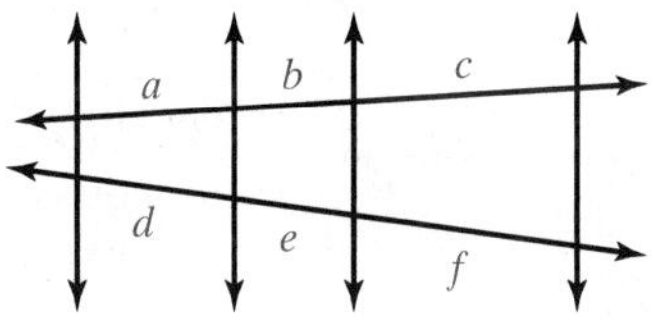

Write a proportion that can be used to find the value of x. Do not solve.

5.

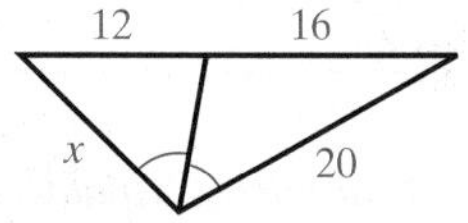

6.

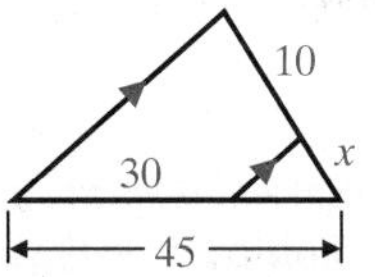

7.6 EXERCISE SET MyMathLab®

Solve for x. See Example 1.

1.

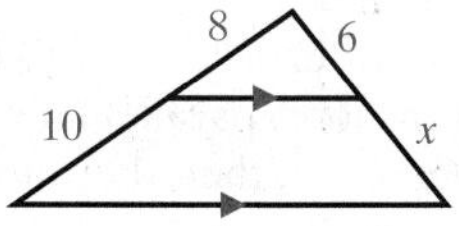

2.

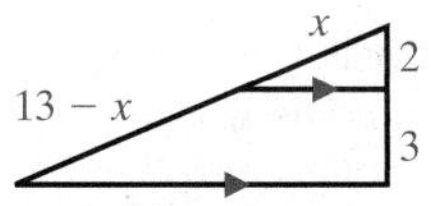

3.

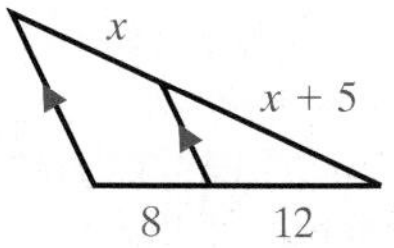

4.

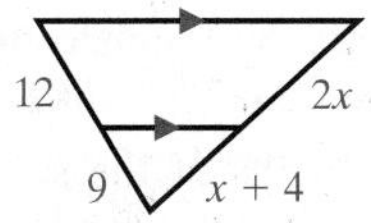

Use the information shown on the image of an auger shell. See Example 2.

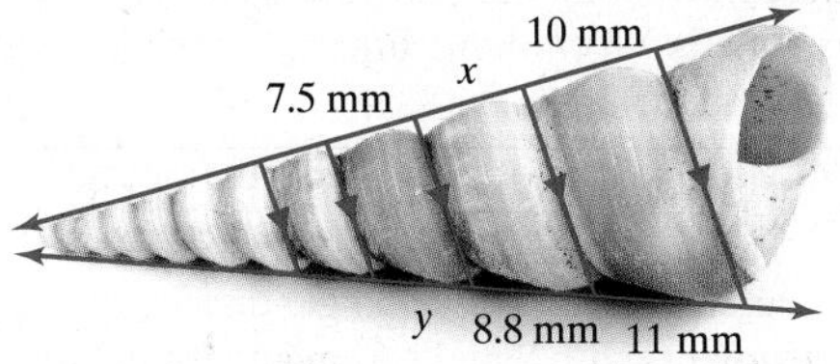

5. What is the value of x?

6. What is the value of y?

Solve for x. See Example 2.

7.

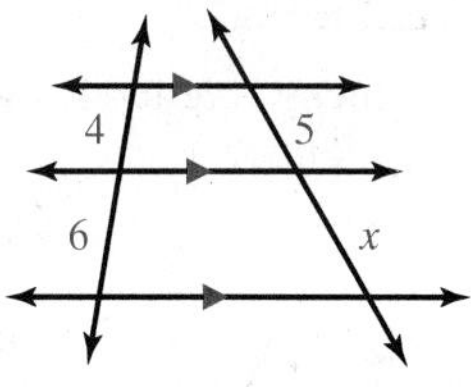

8.

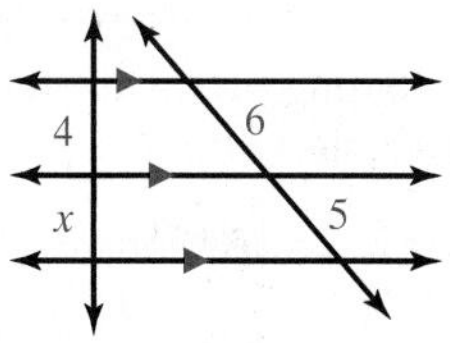

9.

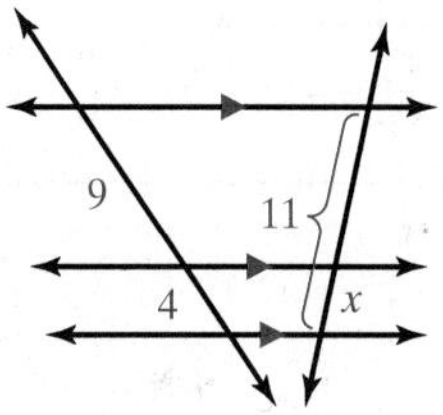

10.

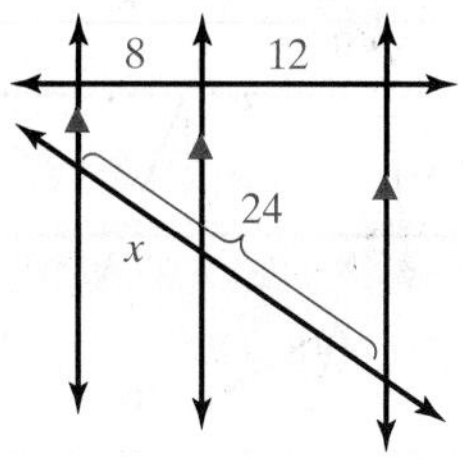

Solve for x. See Example 3.

11.

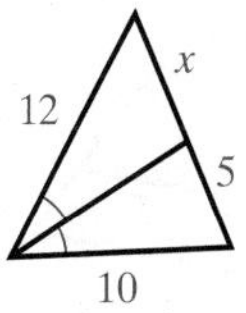

12.

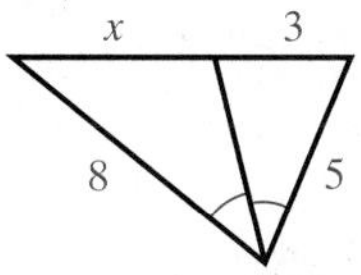

13.

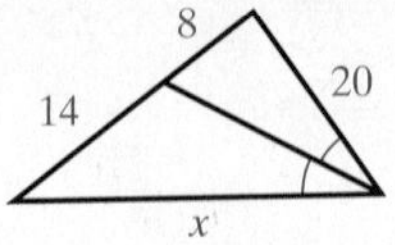

14.

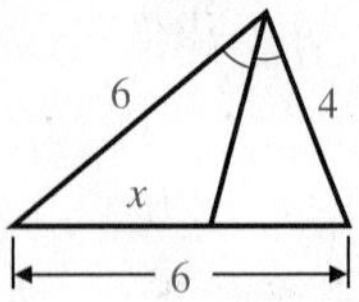

Mixed Practice

Use the figure below to complete each proportion. Justify your answer. See Examples 1–3.

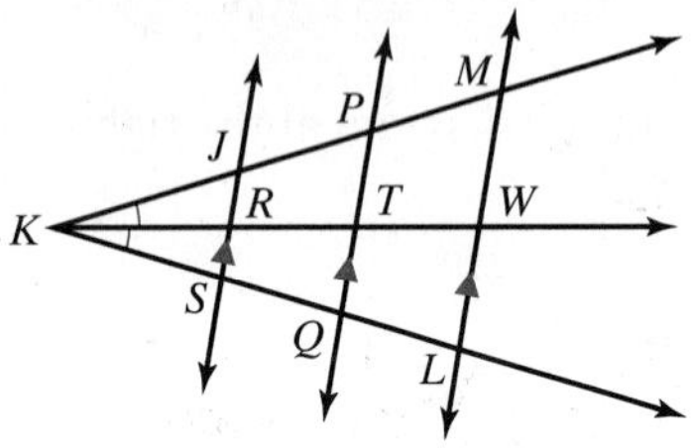

15. $\dfrac{RS}{—} = \dfrac{JR}{KJ}$

16. $\dfrac{KJ}{JP} = \dfrac{KS}{—}$

17. $\dfrac{QL}{PM} = \dfrac{SQ}{—}$

18. $\dfrac{PT}{—} = \dfrac{TQ}{KQ}$

19. $\dfrac{KL}{LW} = \dfrac{—}{MW}$

20. $\dfrac{—}{KP} = \dfrac{LQ}{KQ}$

In Washington, D.C., E. Capitol Street, Independence Avenue, C Street, and D Street are parallel streets that intersect Kentucky Avenue and 12th Street.

21. How long (to the nearest foot) is Kentucky Avenue between C Street and D Street?

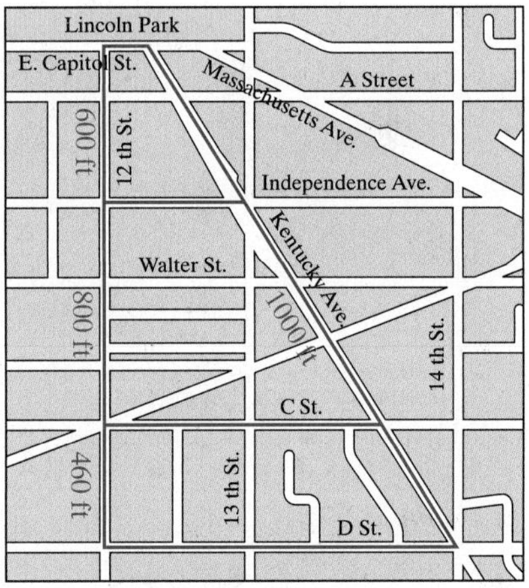

22. How long (to the nearest foot) is Kentucky Avenue between E. Capitol Street and Independence Avenue?

Mixed Practice *Solve for x.*

23.

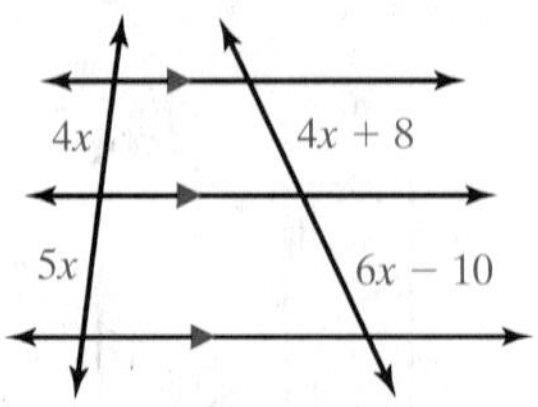

24.

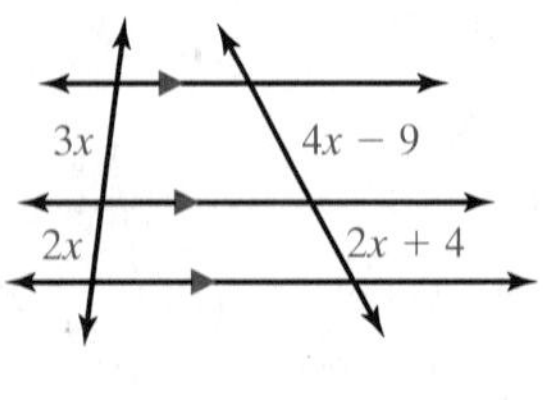

25.

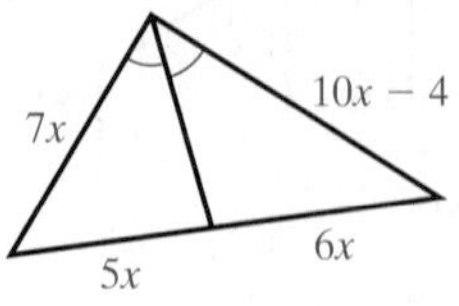

26.

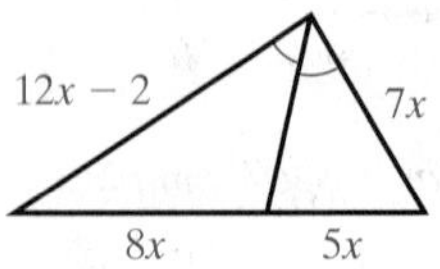

27.

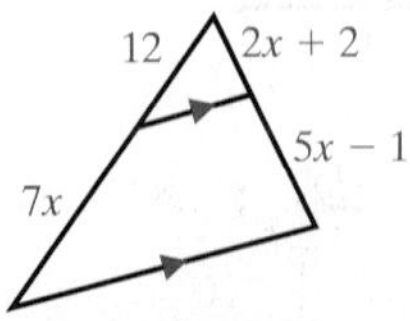

28.

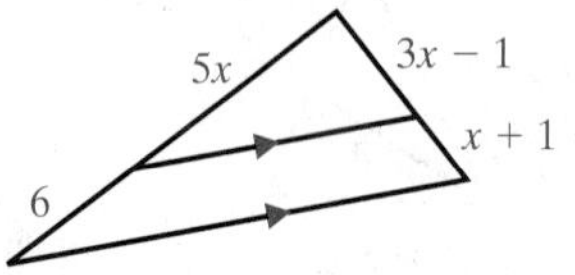

29.

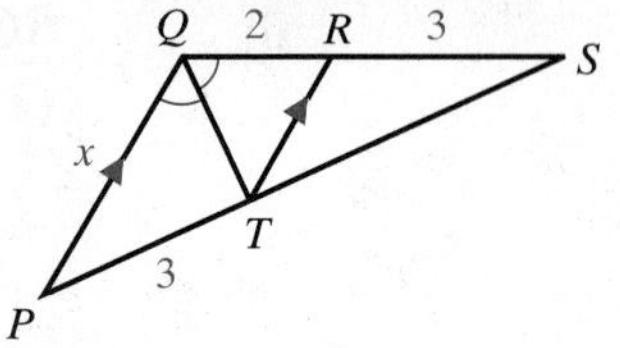

30.

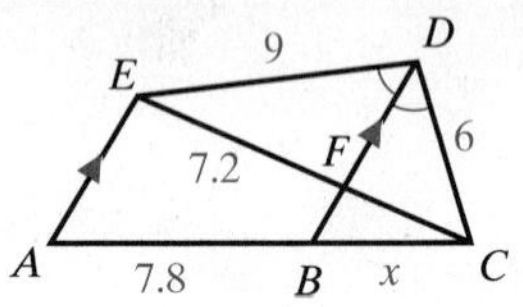

CONCEPT EXTENSIONS

31. Multiple Steps The perimeter of the triangular lot below is 50 m. The surveyor's tape bisects an angle.

a. Use the perimeter to write an equation in x and y.

b. Use the Triangle-Angle-Bisector Theorem to write an equation in x and y.

c. Find the lengths of x and y.

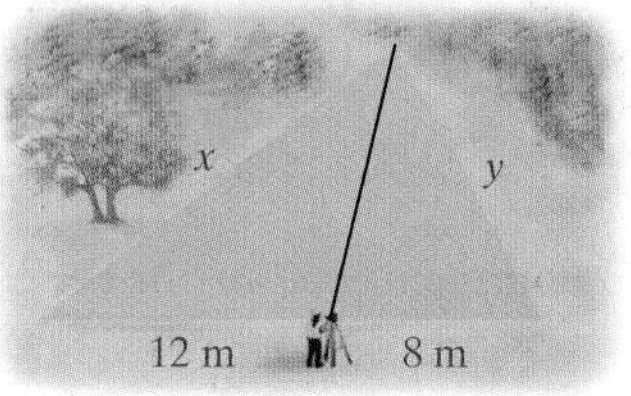

32. Prove the Converse of the Side-Splitter Theorem: If a line divides two sides of a triangle proportionally, then it is parallel to the third side.

Given: $\dfrac{XR}{RQ} = \dfrac{YS}{SQ}$

Prove: $\overline{RS} \parallel \overline{XY}$

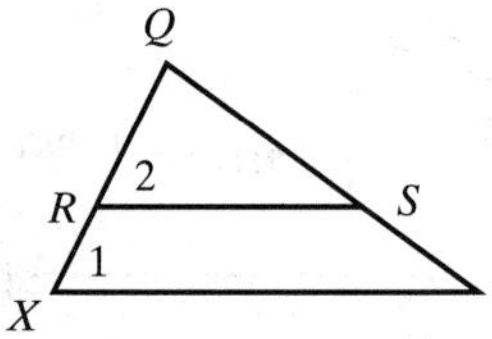

Determine whether the red segments are parallel. Explain each answer. You can use the theorem proved in Exercise ***32.***

33.

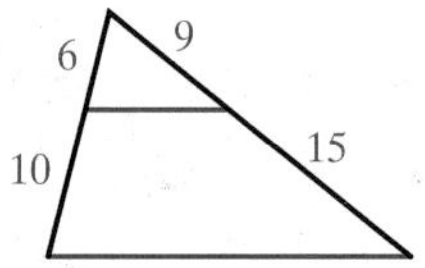

34.

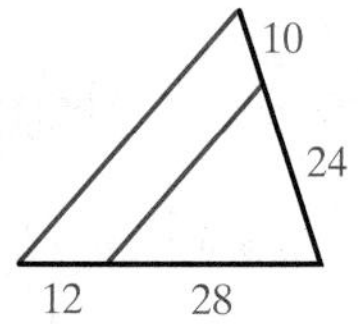

35. How is the Corollary to the Side-Splitter Theorem (7.6-2) related to Theorem 6.2-5: If three (or more) parallel lines cut off congruent segments on one transversal, then they cut off congruent segments on every transversal?

36. How are the Triangle-Angle-Bisector Theorem and Corollary 1 to Theorem 7.5-1 alike? How are they different?

37. The lengths of the sides of a triangle are 5 cm, 12 cm, and 13 cm. Find the lengths, to the nearest tenth, of the segments into which the bisector of each angle divides the opposite side.

38. An angle bisector of a triangle divides the opposite side of the triangle into segments 5 cm and 3 cm long. A second side of the triangle is 7.5 cm long. Find all possible lengths for the third side of the triangle.

39. In ΔABC, the bisector of $\angle C$ bisects the opposite side. What type of triangle is ΔABC? Explain your reasoning.

40. In a triangle, the bisector of an angle divides the opposite side into two segments with lengths 6 cm and 9 cm. How long could the other two sides of the triangle be? [*Hint:* Make sure the three sides satisfy the Triangle Inequality Theorem for Sum of Lengths of Sides (5.5-5).]

PROOF

41. Prove the Corollary to the Side-Splitter Theorem. In the diagram given with the corollary, draw the auxiliary line $\overleftrightarrow{CW}$ (in red) and label its intersection with line b as point P.

Given: $a \parallel b \parallel c$

Prove: $\frac{AB}{BC} = \frac{WX}{XY}$

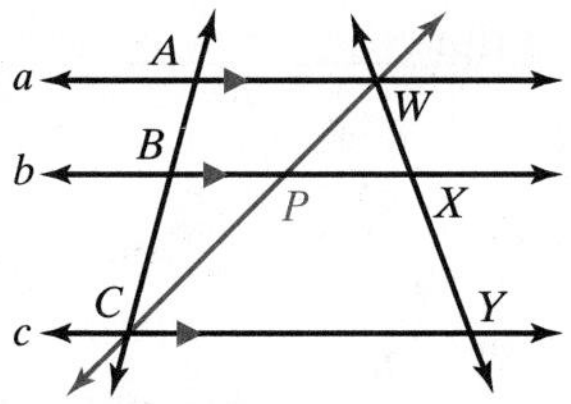

42. Prove the Triangle-Angle-Bisector Theorem. In the diagram given with the theorem, draw the auxiliary line $\overleftrightarrow{BE}$ (in red) so that $\overleftrightarrow{BE} \parallel \overline{DA}$. Extend $\overline{CA}$ to meet $\overleftrightarrow{BE}$ at point F.

Given: $\overrightarrow{AD}$ bisects $\angle CAB$.

Prove: $\frac{CD}{DB} = \frac{CA}{BA}$

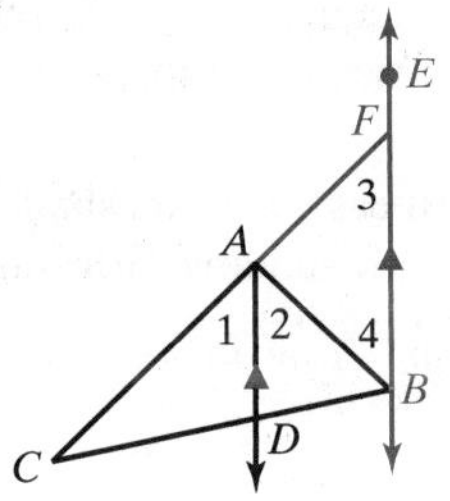

Solve.

43. Use the definition in part **a** to prove the statements in parts **b** and **c**.

a. Write a definition for a midsegment of a parallelogram.

b. A parallelogram midsegment is parallel to two sides of the parallelogram.

c. A parallelogram midsegment bisects the diagonals of a parallelogram.

44. State the converse of the Triangle-Angle-Bisector Theorem. Give a convincing argument that the converse is true or a counterexample to prove that it is false.

In $\triangle ABC$, the bisectors of $\angle A$, $\angle B$, and $\angle C$ cut the opposite sides into lengths a_1 and a_2, b_1 and b_2, and c_1 and c_2, respectively, labeled in order counterclockwise around $\triangle ABC$. Find the perimeter of $\triangle ABC$ for each set of values.

45. $b_1 = 16, b_2 = 20, c_1 = 18$

46. $a_1 = \frac{5}{3}, a_2 = \frac{10}{3}, b_1 = \frac{15}{4}$

REVIEW AND PREVIEW

Use the figure to complete each proportion. See Section 7.5

47. $\frac{n}{h} = \frac{h}{\underline{\quad}}$

48. $\frac{\underline{\quad}}{b} = \frac{b}{c}$

49. $\frac{n}{a} = \frac{a}{\underline{\quad}}$

50. $\frac{m}{h} = \frac{\underline{\quad}}{n}$

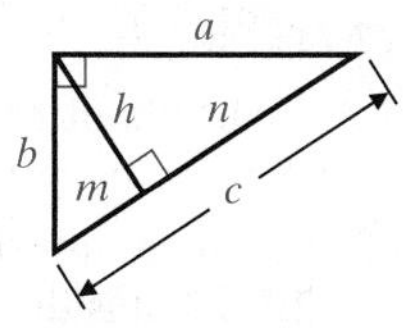

Find the center of the circle that you can circumscribe about each $\triangle ABC$. See Section 5.2

51. $A(0, 0)$
$B(6, 0)$
$C(0, -6)$

52. $A(2, 5)$
$B(-2, 5)$
$C(-2, -1)$

Square the lengths of the sides of each triangle. See the Concept Reviews at the end of this text.

53.

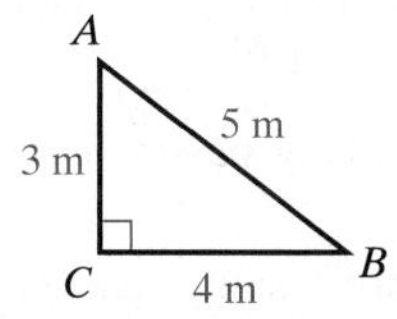

54.

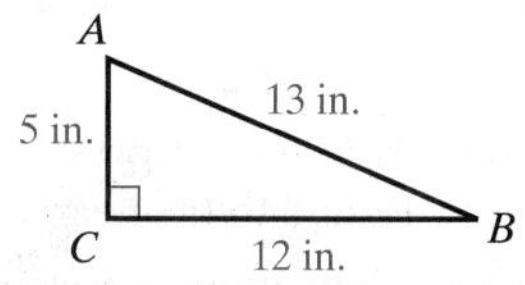

55.

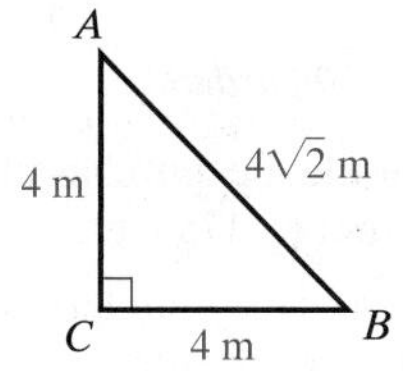

CHAPTER 7 VOCABULARY CHECK

Word Bank *Fill in each blank with one of the words or phrases listed below.*

means	cross products	extremes	geometric mean	similar	scale factor
ratio	proportion	proportional	extended	congruent	

1. A(n) ____________ is the quotient of two numbers. It can be written as a fraction, using a colon, or using the word *to*.

2. $\frac{x}{2} = \frac{7}{16}$ is an example of a(n) ____________.

3. The ____________ of positive numbers a and b is the positive number x such that $\frac{a}{x} = \frac{x}{b}$.

4. In $\frac{a}{b} = \frac{c}{d}$, the variables b and c are called the ____________.

5. In $\frac{a}{b} = \frac{c}{d}$, the variables a and d are called the ____________.

6. In the proportion $\frac{x}{2} = \frac{7}{16}$, $x \cdot 16$ and $2 \cdot 7$ are called ____________.

7. ____________ figures have exactly the same shape but not necessarily the same size.

8. Two polygons are similar polygons if corresponding angles are ____________ and corresponding sides are ____________.

9. The proportion $\frac{AB}{XY} = \frac{BC}{YZ} = \frac{CA}{ZX}$ is called a(n) ____________ proportion.

10. The ratio of corresponding linear measurements of two similar figures is called the ____________.

CHAPTER 7 REVIEW

(7.1)

1. A high school has 16 math teachers for 1856 math students. What is the ratio of math teachers to math students?

2. The measures of two complementary angles are in the ratio 2:3. What is the measure of the smaller angle?

Solve each proportion.

3. $\frac{x}{7} = \frac{18}{21}$

4. $\frac{6}{11} = \frac{15}{2x}$

5. $\frac{x}{3} = \frac{x+4}{5}$

6. $\frac{8}{x+9} = \frac{2}{x-3}$

(7.2) *Solve.*

The ratio of a quarterback's completed passes to attempted passes is 3 to 7.

7. If he attempted 32 passes, find how many passes he completed. Round to the nearest whole pass.

8. If he completed 15 passes, find how many passes he attempted.

One bag of pesticide covers 4000 square feet of garden.

9. Find how many bags of pesticide should be purchased to cover a rectangular garden that is 180 feet by 175 feet.

10. Find how many bags of pesticide should be purchased to cover a square garden that is 250 feet on each side.

On a road map of Texas, 0.75 inch represents 80 miles.

11. Find the distance from Houston to Corpus Christi if the distance on the map is about 2 inches.

12. The distance from El Paso to Dallas is 1025 miles. Find the distance between these cities on the map. Round to the nearest tenth of an inch.

(7.3) and (7.4)

The polygons are similar. Write a similarity statement and give the scale factor.

13.

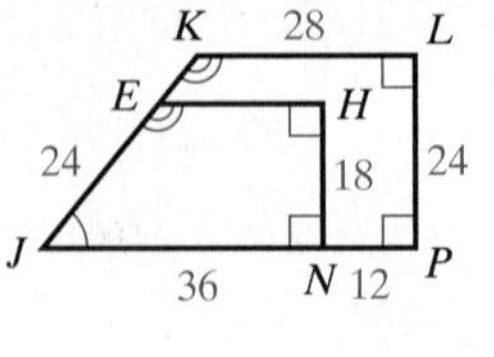

14.

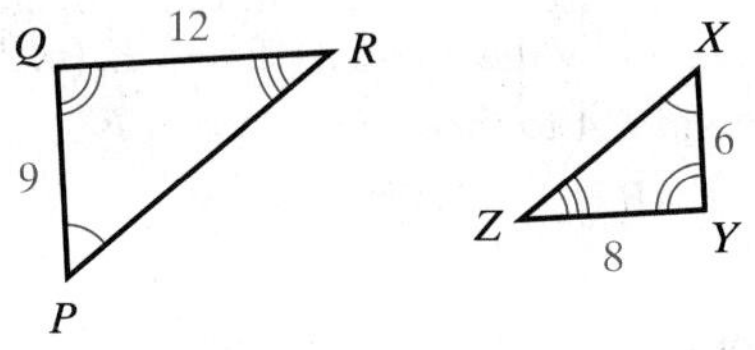

15. The length of a rectangular playground in a scale drawing is 12 in. If the scale is 1 in. = 10 ft, what is the actual length?

16. A 3-ft vertical post casts a 24-in. shadow at the same time a pine tree casts a 30-ft shadow. How tall is the pine tree?

Are the triangles similar? How do you know?

17.

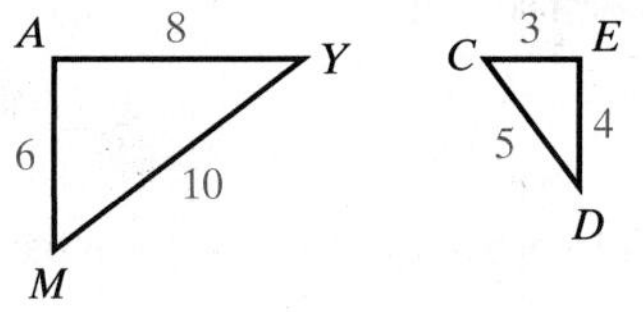

18.

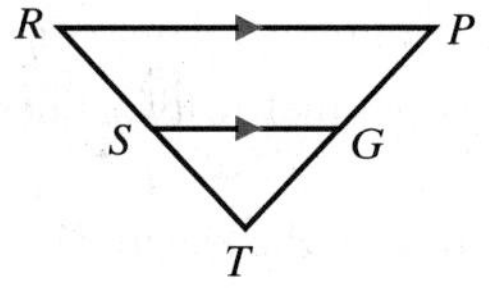

(7.5)

Find the geometric mean of each pair of numbers.

19. 9 and 16

20. 5 and 12

Find the value of each variable. Write your answer in simplest radical form.

21.

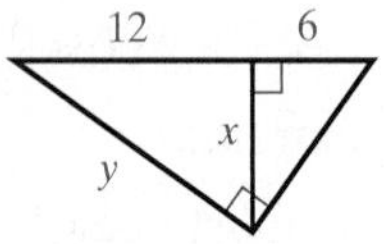

22.

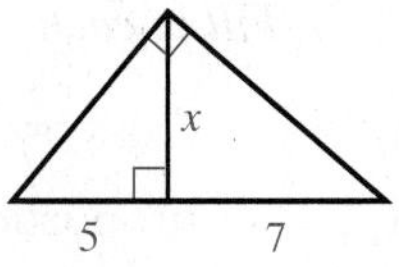

23.

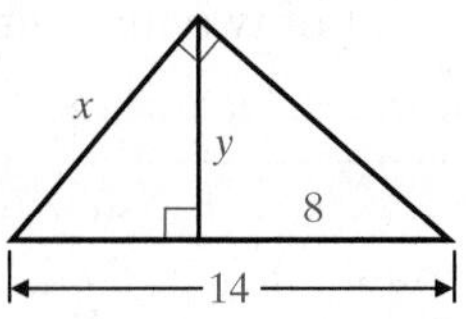

24.

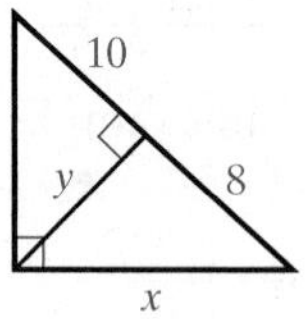

(7.6)

Find the value of x.

25.

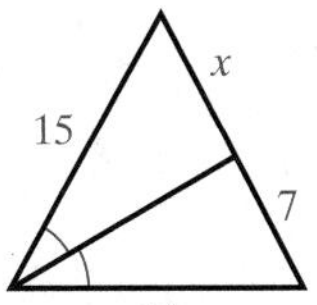

26.

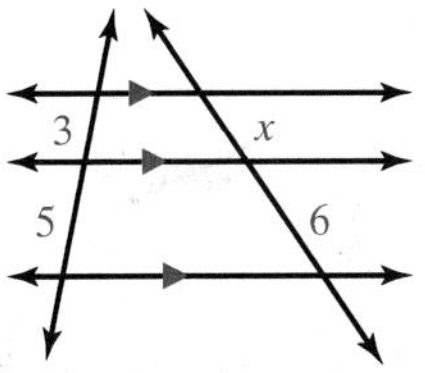

27.

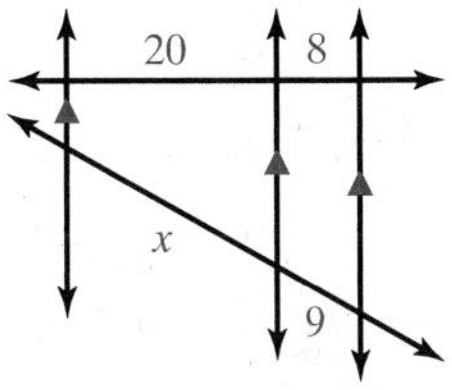

28.

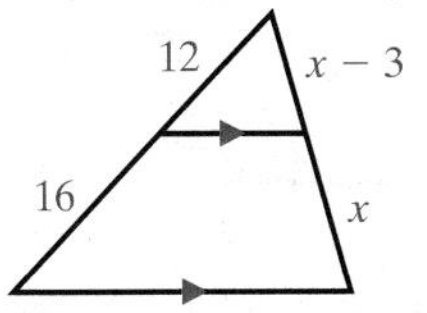

29.

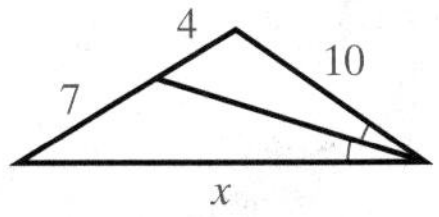

30.

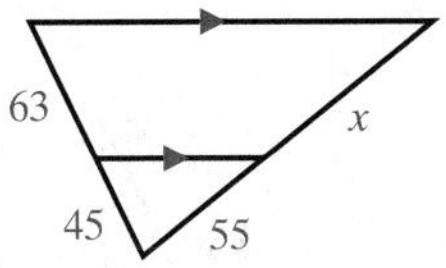

MIXED REVIEW

Write each ratio as a fraction in simplest form.

31. 15 to 25

32. 3 pints to 81 pints

33. 2 feet to 18 inches

34. 6 yards to 24 feet

Find the unknown number x in each proportion.

35. $\frac{3}{x} = \frac{15}{8}$

36. $\frac{5}{4} = \frac{x}{20}$

37. What is the solution of $\frac{x}{x + 3} = \frac{4}{6}$?

38. Is ΔABC similar to ΔRQP? How do you know?

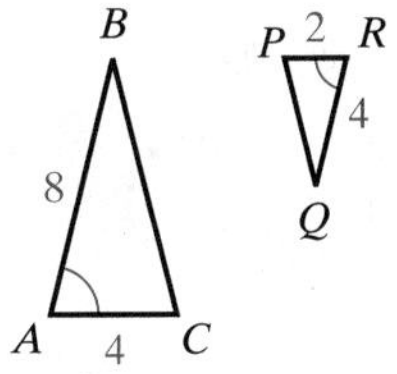

39. What is the value of x?

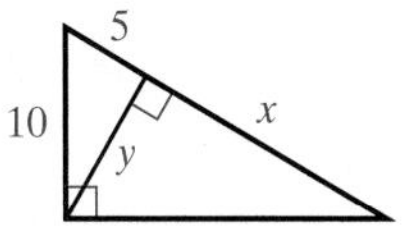

40. What is the value of x?

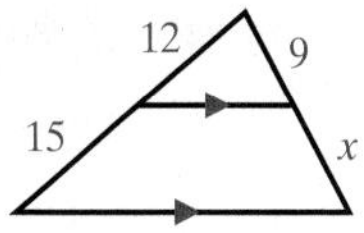

41. A meter stick perpendicular to the ground casts a 1.5-m shadow. At the same time, a telephone pole casts a shadow that is 9 m. How tall is the telephone pole?

42. A design company is making a triangular sail for a model sailboat. The model sail is to be the same shape as a life-size sailboat's sail. Use the following diagram to find the unknown lengths x and y.

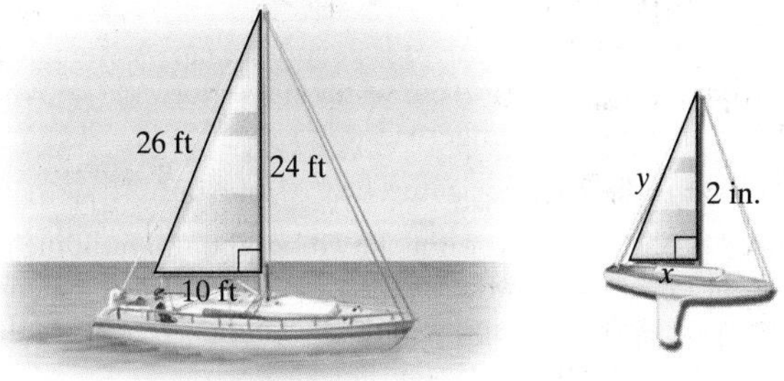

43. Explain why the triangles are similar, then find the length, x.

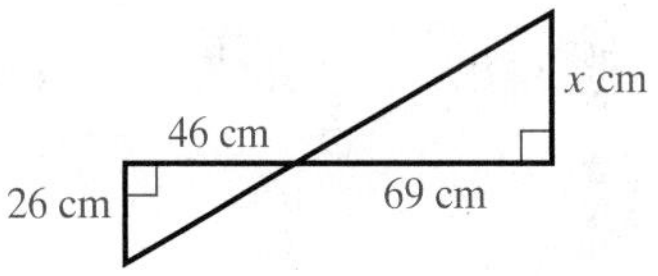

CHAPTER 7 TEST

TEST PREP VIDEO The fully worked-out solutions to any exercises you want to review are available in MyMathLab.

Solve each proportion.

1. $\frac{x}{3} = \frac{8}{12}$

2. $\frac{4}{x + 2} = \frac{16}{9}$

3. Are the polygons below similar? If they are, write a similarity statement and give the scale factor.

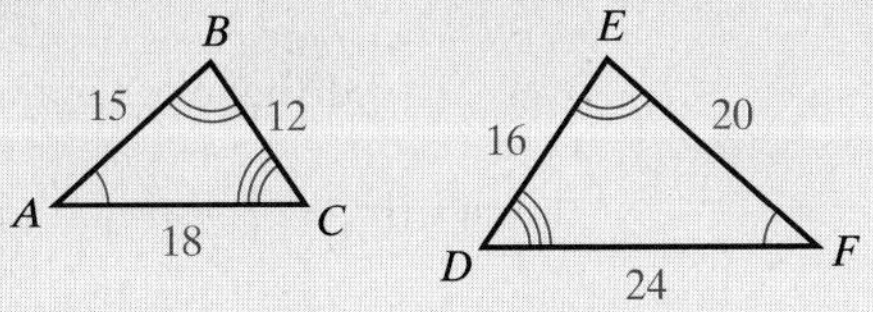

The figures in each pair are similar. Find the value of each variable.

4.

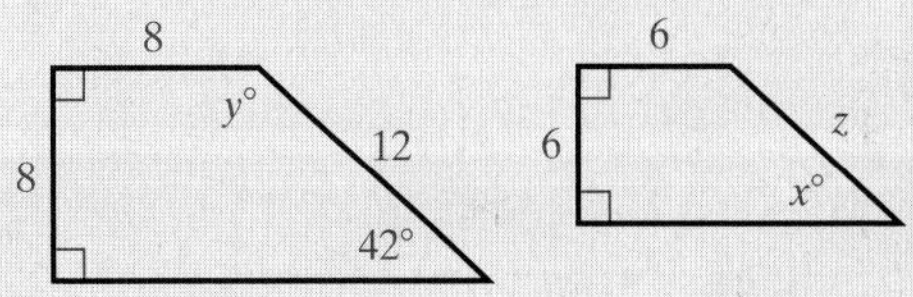

5.

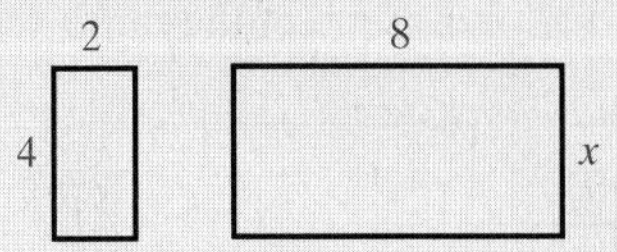

6. 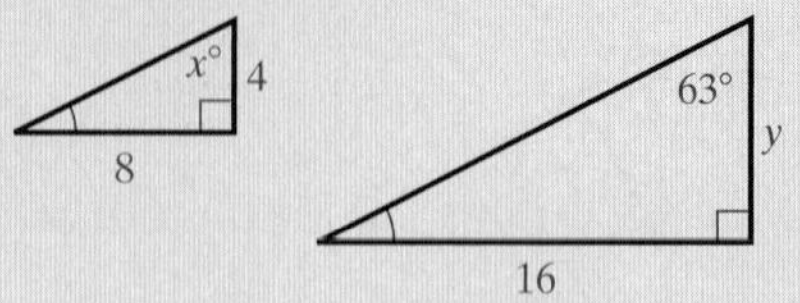

Determine whether the triangles are similar. If they are, write a similarity statement and name the postulate or the theorem you used. If not, explain.

7. 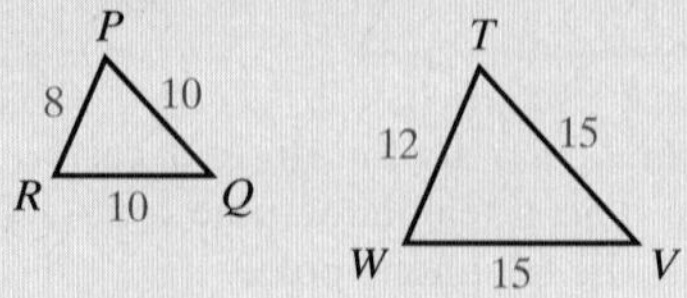

8. 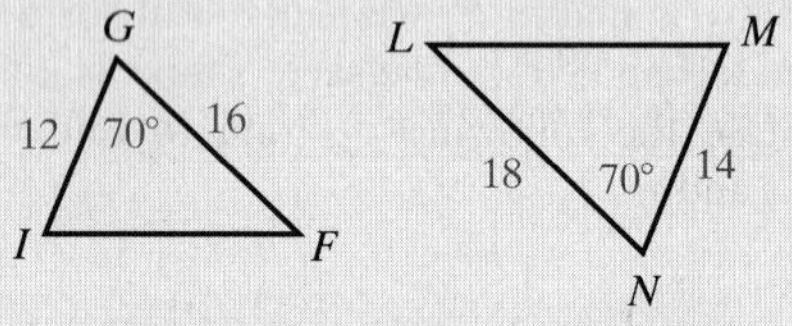

9. A photographic negative is 3 cm by 2 cm. A similar print from the negative is 9 cm wide on its shorter side. What is the length of the longer side?

10. What is the geometric mean of 10 and 15?

11. A 440-foot cell tower not shown behind the building casts a 449-foot shadow at the same time that this building casts a 90-foot shadow. Find the height of the building rounded to the nearest tenth.

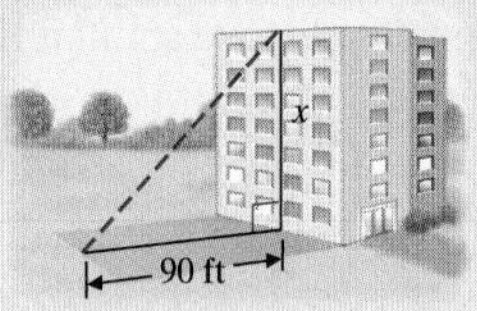

Find the value of x.

12. 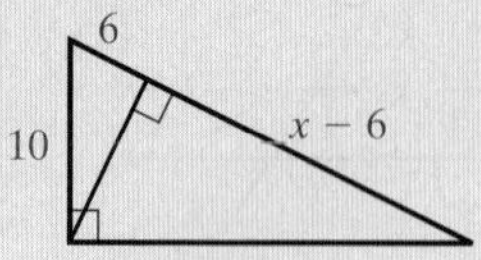

13. 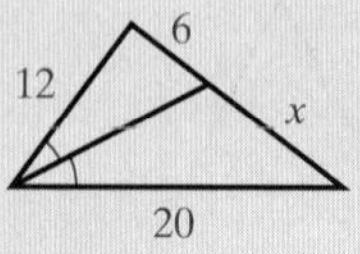

14. 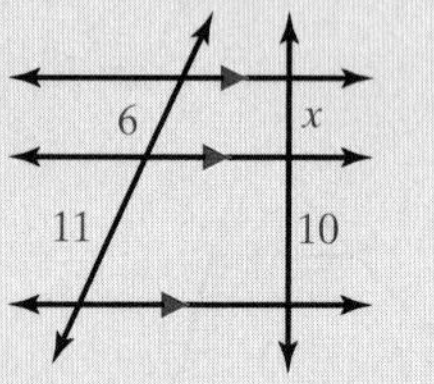

15. 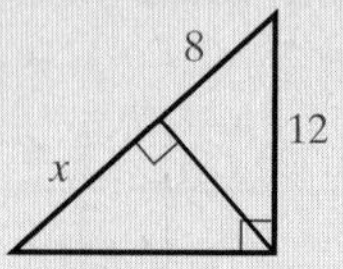

Explain why the triangles are similar. Then find the value of x.

16. 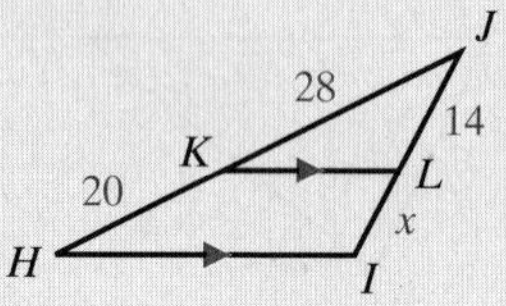

17. 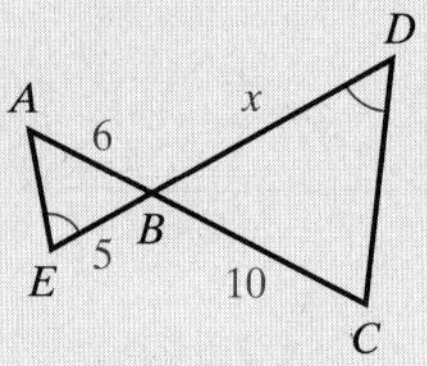

Determine whether each statement is always, sometimes, or never true.

18. A parallelogram is similar to a trapezoid.

19. Two rectangles are similar.

20. If the vertex angles of two isosceles triangles are congruent, then the triangles are similar.

CHAPTER 7 STANDARDIZED TEST

Solve each proportion.

1. $\frac{6}{x} = \frac{9}{12}$

   a. 18 b. $\frac{3}{4}$

   c. 8 d. none of the these

2. $\frac{5}{x-2} = \frac{30}{42}$

   a. $\frac{5}{7}$ b. 7

   c. 9 d. none of the these

Use the similar figures for Questions 3–6.

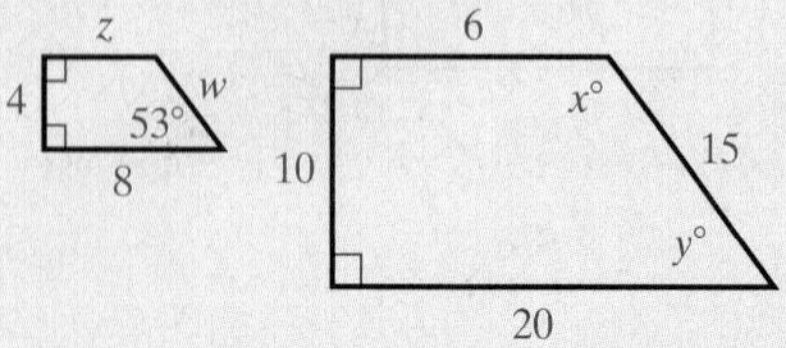

3. $x =$ ______

   a. 127 b. 53

   c. 37 d. none of the these

4. $y =$ ______

   a. 127 b. 53

   c. 37 d. none of the these

5. $z =$ ______

   a. $\frac{5}{3}$ b. $\frac{12}{5}$

   c. 3 d. none of the these

6. $w =$ ______

   a. 7.5 b. 6

   c. 8 d. none of the these

Use the figure for Questions 7–11.

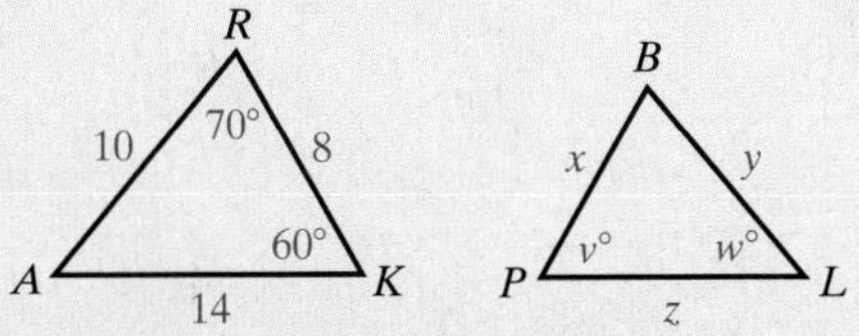

7. If $v = 60$ and $w = 50$, then

   a. $\Delta ARK \sim \Delta PBL$ by the AA ~ Postulate.

   b. $\Delta ARK \sim \Delta LBP$ by the AA ~ Postulate.

   c. There is not enough information to conclude that the triangles are similar.

   d. The triangles are not similar.

8. If $x = 4$, $y = 5$, and $z = 7$, then

   a. $\Delta ARK \sim \Delta LBP$ by the SSS ~ Theorem.

   b. $\Delta ARK \sim \Delta PBL$ by the SSS ~ Theorem.

   c. There is not enough information to conclude that the triangles are similar.

   d. The triangles are not similar.

9. If $x = 6$, $y = 7.5$, and $v = 60$, then

   a. $\Delta ARK \sim \Delta PBL$ by the SAS ~ Theorem.

   b. $\Delta ARK \sim \Delta LBP$ by the SAS ~ Theorem.

   c. There is not enough information to conclude that the triangles are similar.

   d. The triangles are not similar.

10. If $x = 5$, $z = 7$, and $v = 60$, then

    a. $\Delta ARK \sim \Delta PBL$ by the SSS ~ Theorem.

    b. $\Delta ARK \sim \Delta BLP$ by the SAS ~ Theorem.

    c. There is not enough information to conclude that the triangles are similar.

    d. The triangles are not similar.

11. If $x = 7.5$, $z = 10.5$, and $v = 50$, then

    a. $\Delta ARK \sim \Delta PBL$ by the SAS ~ Theorem.

    b. $\Delta ARK \sim \Delta BLP$ by the SAS ~ Theorem.

    c. There is not enough information to conclude that the triangles are similar.

    d. The triangles are not similar.

12. A yard stick perpendicular to the ground casts a 5-ft shadow. At the same time, a tree casts a shadow that is 30 ft long. How tall is the tree?

    a. 6 ft **b.** 25 ft

    c. 18 ft **d.** 18 yd

13. What is the geometric mean of 8 and 20?

    a. $\frac{5}{16}$ **b.** 160

    c. $\frac{16}{5}$ **d.** $4\sqrt{10}$

Use the figure for Questions 14 and 15.

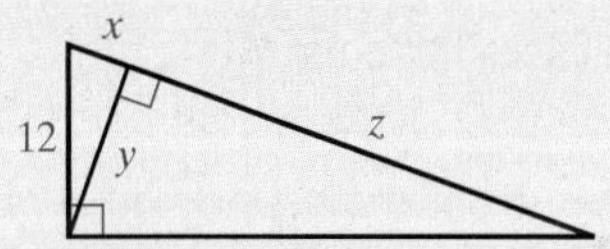

14. If $x = 4$, find the value of z.

    a. 36 **b.** 16

    c. 20 **d.** 32

15. If $y = 8$, what are the values of x and z?

    a. $x = 4$, $z = 16$

    b. $x = \frac{8}{3}\sqrt{3}$, $z = 8\sqrt{3}$

    c. $x = 4\sqrt{5}$, $z = \frac{16}{5}\sqrt{5}$

    d. $x = 4$, $z = 32$

16. Find the value of x.

    a. 6.75 **b.** 7

    c. 6 **d.** 8

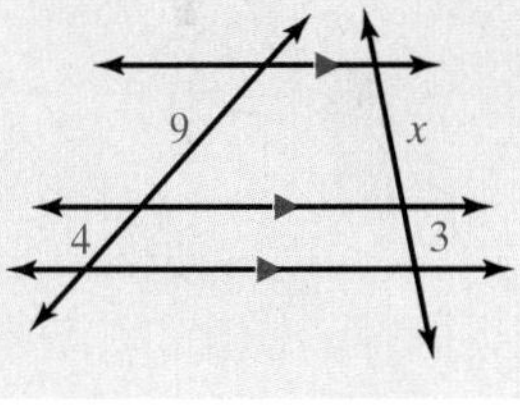

CHAPTER

8 Transformations

8.1 Rigid Transformations
8.2 Translations
8.3 Reflections
8.4 Rotations
8.5 Dilations
8.6 Compositions of Reflections
Extension—Frieze Patterns

Original Photo Photo Reflected

A larger background was needed for a film, so a single photo was reflected to create the larger background.

Larger Background Created

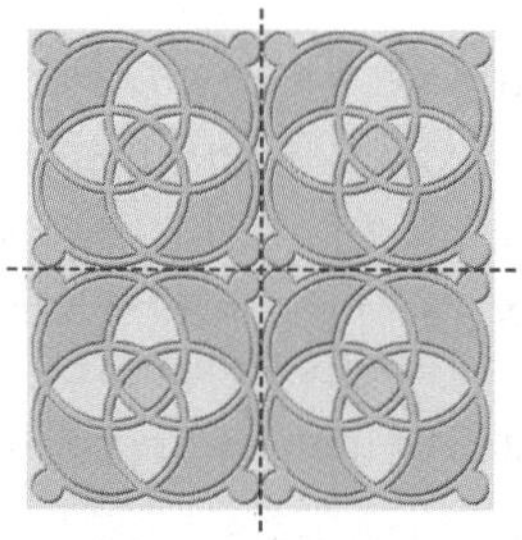

This tile design is made by transforming one quadrant of this end result.

The word *transform* means "to change." In geometry, a transformation of a figure is a change in the position, shape, or size of the figure. If the original shape and size of a figure stay the same, we have transformed the original figure to a congruent figure. If the original shape stays the same, but the size changes, we have transformed the original figure to a similar figure.

The real-life applications of transformations are abundant, especially in the areas of computer programming, stenciling, tile design, photography, architecture, cartooning, and gaming, just to name a few. You may have recently transformed a digital photograph by flipping, rotating, enlarging, and/or reducing it.

8.1 RIGID TRANSFORMATIONS

OBJECTIVES

1 Identify Rigid Transformations or Isometries.

2 Name Images and Corresponding Parts.

VOCABULARY

- transformation
- preimage
- image
- isometry
- rigid transformation

OBJECTIVE 1 ▶ Identifying Rigid Transformations or Isometries. In this section, we begin to understand how a figure in a plane can be translated (slid) or reflected or rotated to result in a new figure. These movements or transformations are more formally defined in the following sections.

Recall that the word *transform* means "to change." Thus, a **transformation** of a figure in a plane is a change in the position, shape, or size of the figure.

In a transformation, the original figure is the **preimage**. The resulting figure is the **image**. An **isometry** is a transformation in which the preimage and image are congruent. This means that angle measures and side lengths are the same in the preimage and the image. An isometry is also called a **rigid transformation** because the original shape and size of the figure do not change.

In this section, we study three basic isometries or rigid transformations. They are called **translations** (or slides), **reflections** (or flips), and **rotations** (or turns).

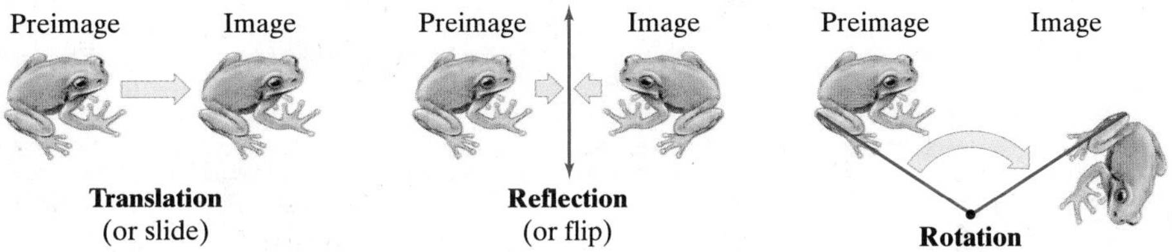

Translation (or slide)   **Reflection** (or flip)   **Rotation** (or turn)

▶ **Helpful Hint**
In an isometry, the image and the preimage must be congruent.

EXAMPLE 1 **Identifying an Isometry**

Does the transformation below appear to be an isometry? Explain.

Solution No, this transformation involves a change in size. Since the sides of the preimage square and the sides of its image are not congruent, this transformation is not an isometry.

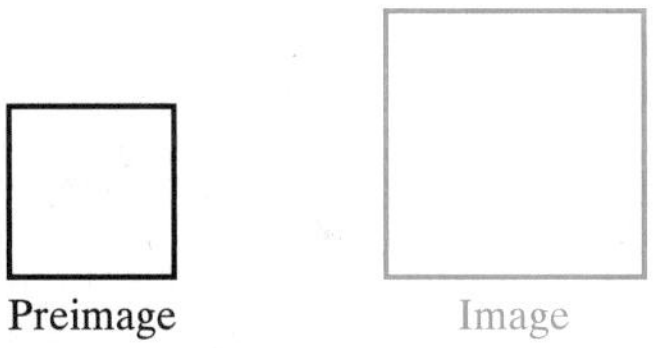

PRACTICE 1 Does each transformation appear to be an isometry? Explain.

a.

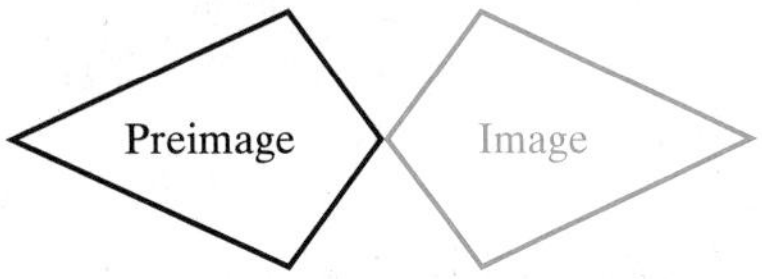

b.

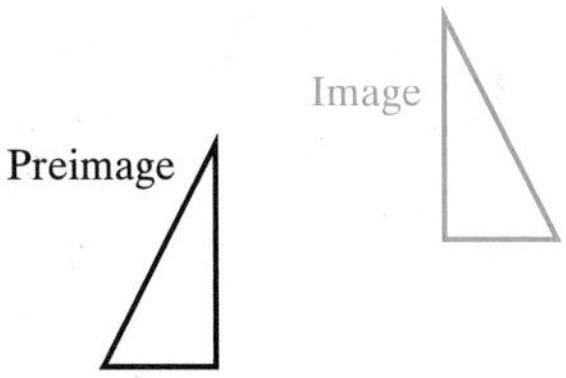

EXAMPLE 2 **Identifying Single Rigid Transformations**

Identify the single transformation from the preimage to each image.

a.

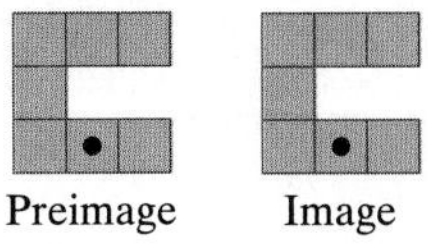

b.

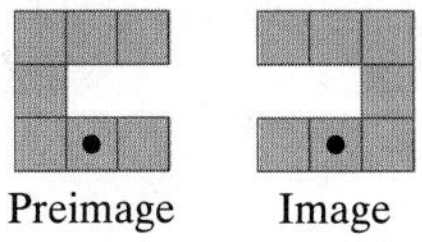

c.

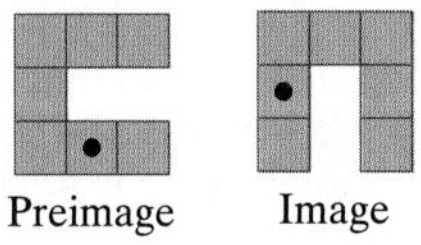

Solution

a. The transformation is a translation.
b. The transformation is a reflection.
c. The transformation is a rotation. □

PRACTICE 2 Identify the single transformation from the preimage to each individual image.

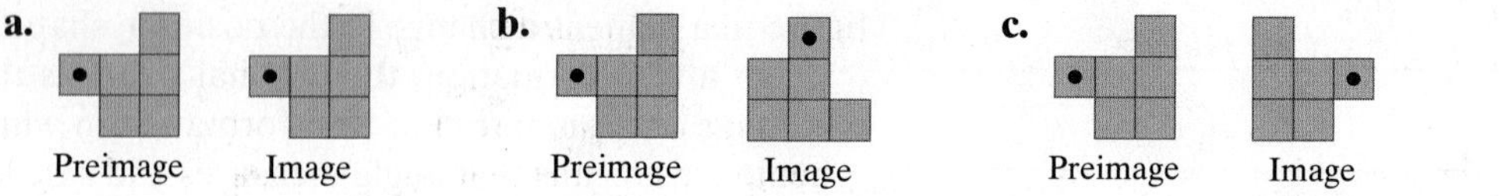

EXAMPLE 3 Identifying Single Rigid Transformations

Identify the single transformation from the preimage to each individual image.

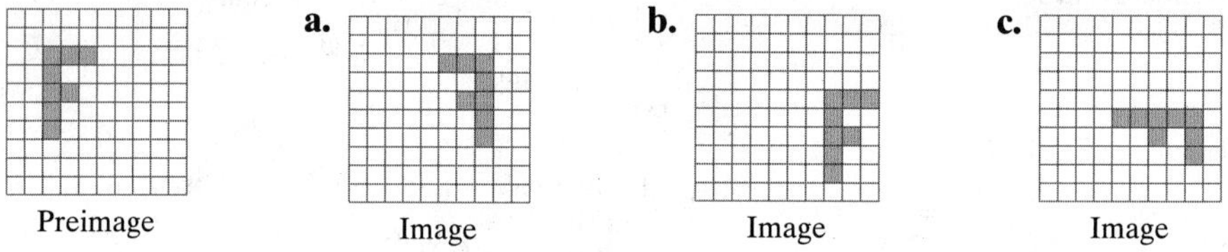

Solution

a. The transformation from the preimage to the image in part **a** above, is a reflection.
b. The transformation from the preimage to the image in part **b** above, is a translation.
c. The transformation from the preimage to the image in part **c** above, is a rotation. □

PRACTICE 3 Identify the single transformation from the preimage to each individual image.

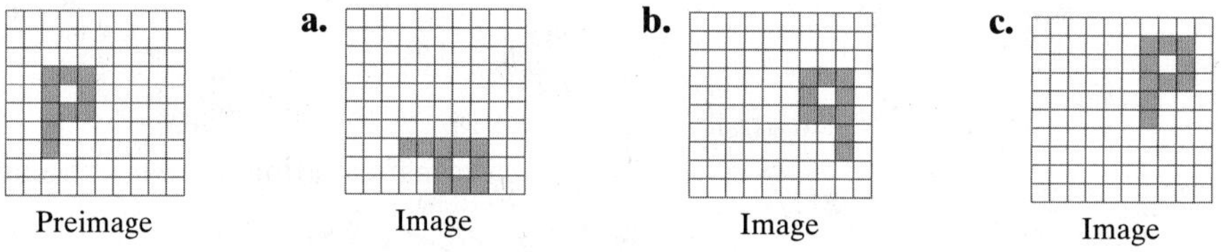

OBJECTIVE 2 ▶ Naming Images and Corresponding Parts. A transformation maps (or moves) a figure to its image and may be described with arrow notation (→). Prime notation (′) is sometimes used to identify image points. In the diagram below, K' is the image of K, or we may write $K \to K'$.

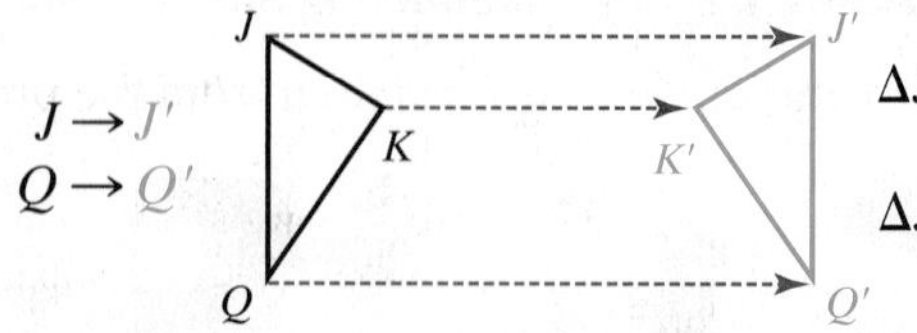

ΔJKQ maps to $\Delta J'K'Q'$
or
$\Delta JKQ \to \Delta J'K'Q'$.

Notice that we list corresponding points of the preimage and image in corresponding order, just as we do for corresponding points of congruent or similar figures.

EXAMPLE 4 **Naming Images and Corresponding Parts**

In the diagram, $EFGH$ maps to $E'F'G'H'$.

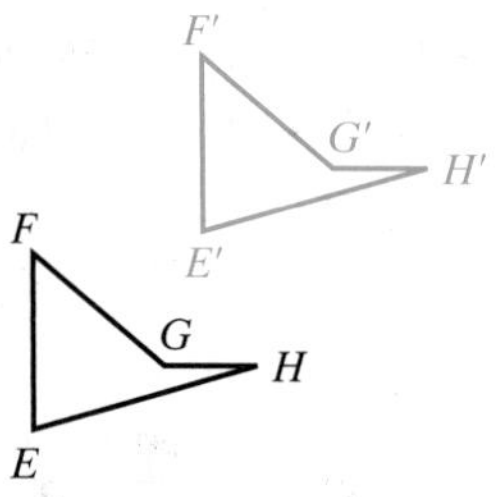

$EFGH \rightarrow E'F'G'H'$

a. What are the images of point F and point H?

b. What are the pairs of congruent corresponding sides?

Solution We are given that $EFGH \rightarrow E'F'G'H'$. Use this to help answer parts **a** and **b**.

a. $F \rightarrow F'$ and $H \rightarrow H'$. The image of F is F'. The image of H is H'.

b. The pairs of corresponding sides are:

$\overline{EF} \cong \overline{E'F'}$ $\quad \overline{FG} \cong \overline{F'G'}$

$\overline{EH} \cong \overline{E'H'}$ $\quad \overline{GH} \cong \overline{G'H'}$

□

PRACTICE 4 In the diagram, $\Delta NID \rightarrow \Delta SUP$.

a. What are the images of point I and point D?

b. What are the pairs of congruent corresponding sides?

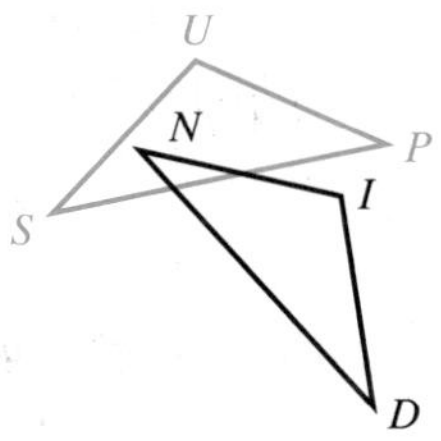

VOCABULARY & READINESS CHECK

Word Bank *Use the choices to fill in each blank. Some choices may be used more than once and some not at all.*

preimage	isometry	rigid transformation	flip	translation	to change
image	transformation	turn	slide	rotation	reflection

1. The word transform means "____________."
2. A(n) ____________ of a figure is a change in the position, shape, or size of the figure.
3. The resulting figure after transformation has occurred is called the ____________.
4. The original figure before transformation has occurred is called the ____________.
5. A(n) ____________ is a transformation in which the preimage and the image are congruent.
6. A(n) isometry is also called a ____________.
7. Three rigid transformations are called a(n) ____________, a(n) ____________, and a(n) ____________.
8. Another word for a translation is a(n) ____________.
9. Another word for a reflection is a(n) ____________.
10. Another word for a rotation is a(n) ____________.

8.1 EXERCISE SET MyMathLab®

Tell whether the transformation appears to be an isometry. Explain. See Example 1.

1.

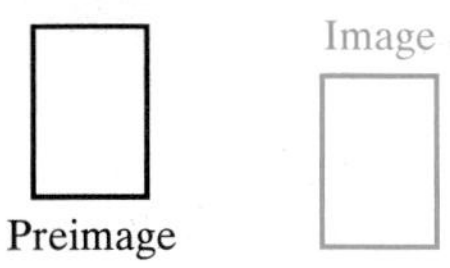

2.

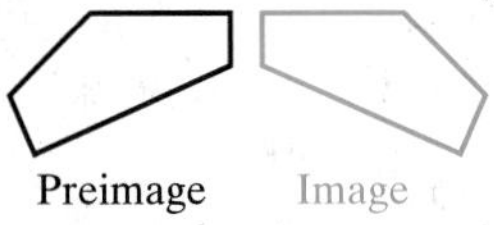

3.

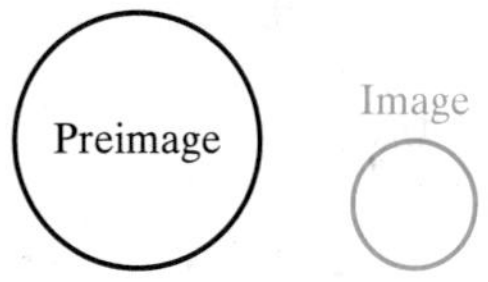

4.

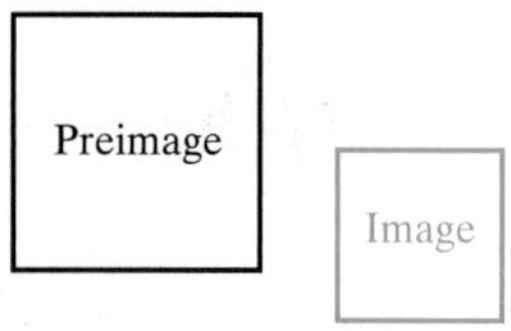

In each diagram, the red figure is an image of the black figure. For Exercises 5 and 6, prime notation is used. See Example 4.

a. *Choose a point from the preimage and name its image point.*

b. *List all pairs of congruent corresponding sides.*

5.

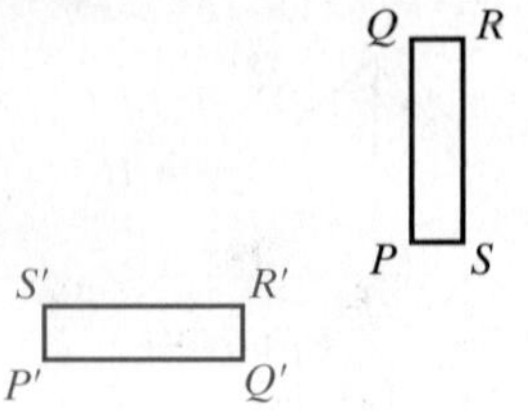

6.

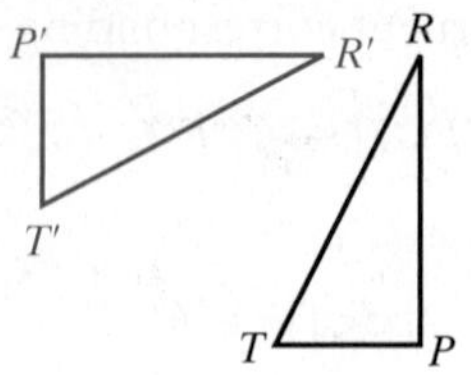

7.

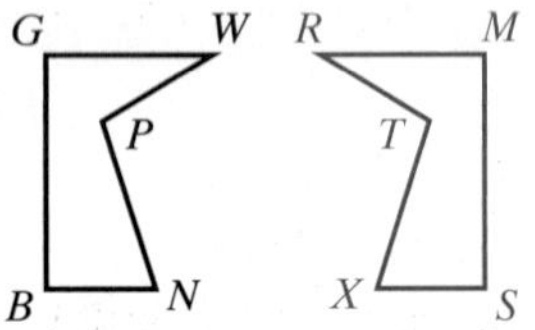

8.

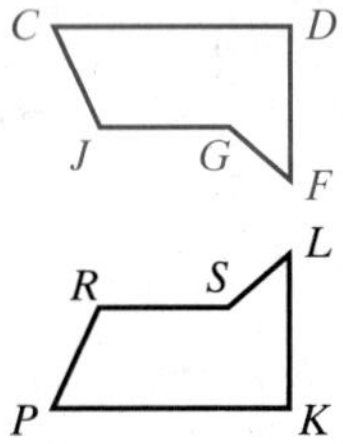

Identify the single transformation (translation, reflection, or rotation) from the preimage to each image. See Example 2.

9. a.

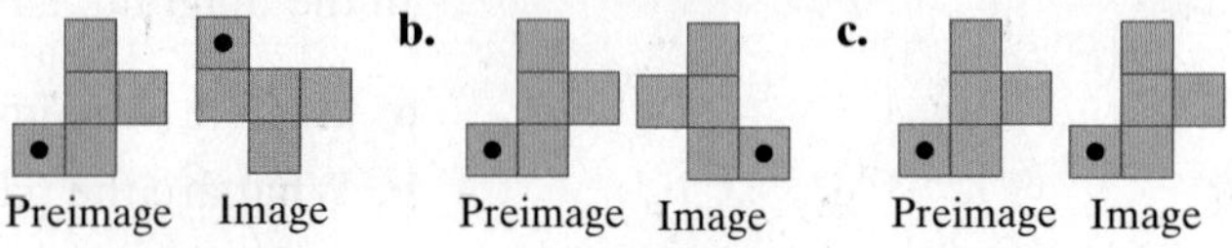

10. a.

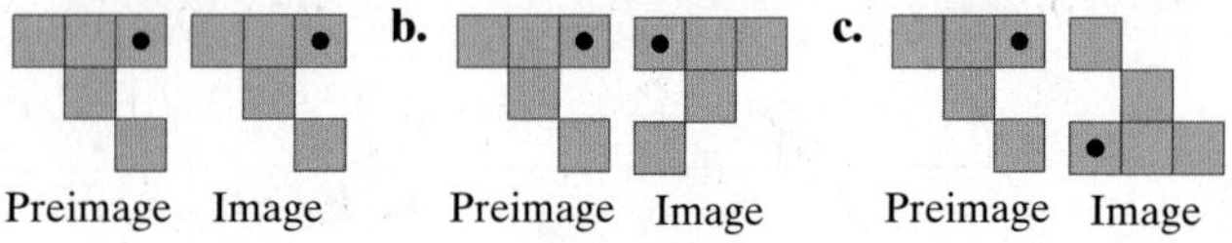

11. a.

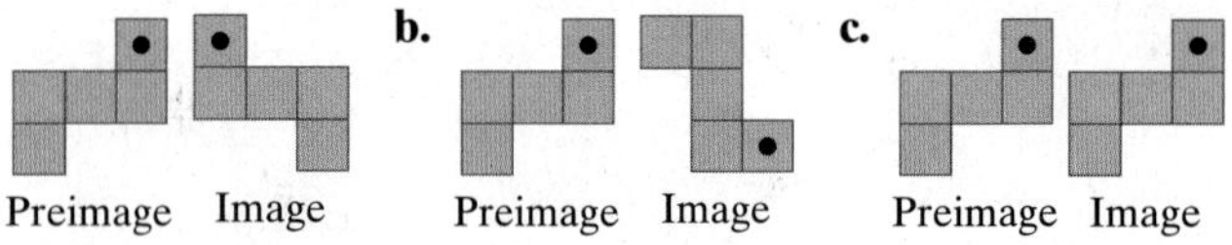

12. a.

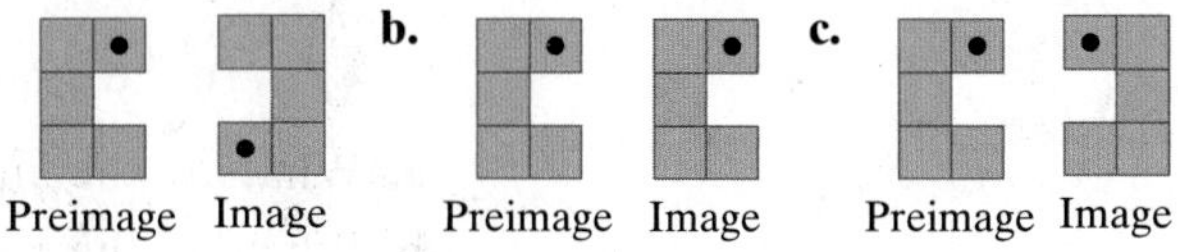

Identify the single transformation (translation, reflection, or rotation) from the preimage to each individual image. See Example 3.

13.

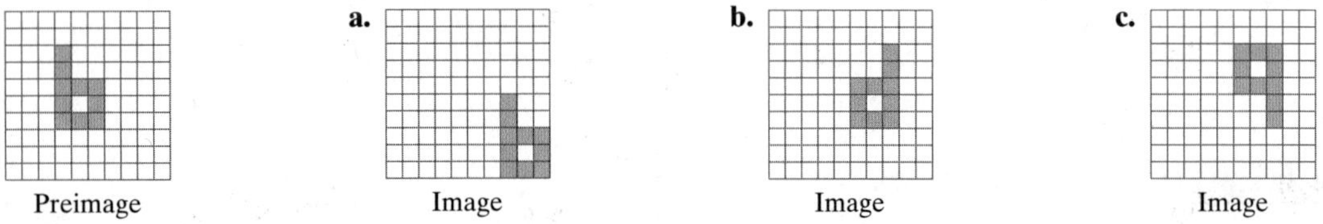

14.

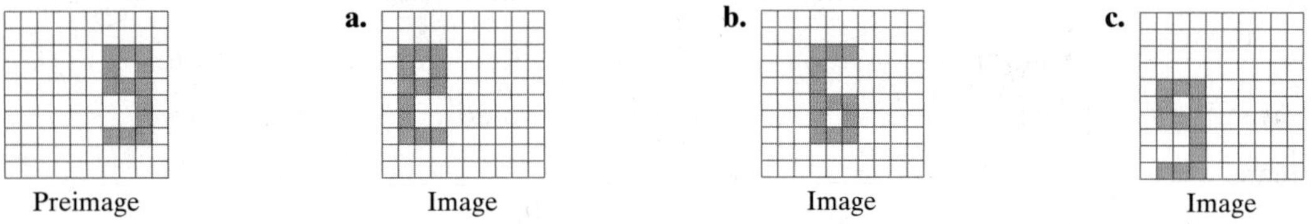

15.

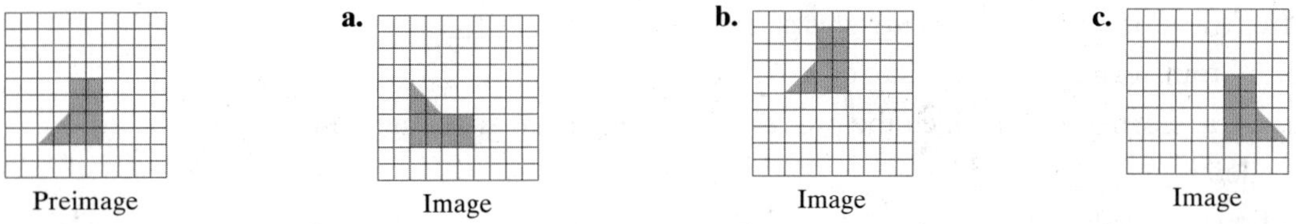

16.

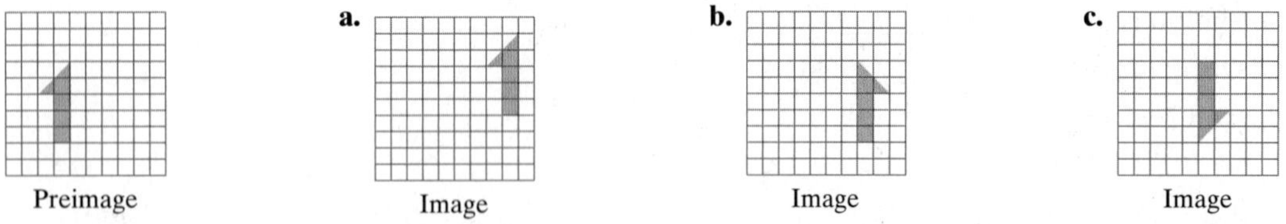

17.

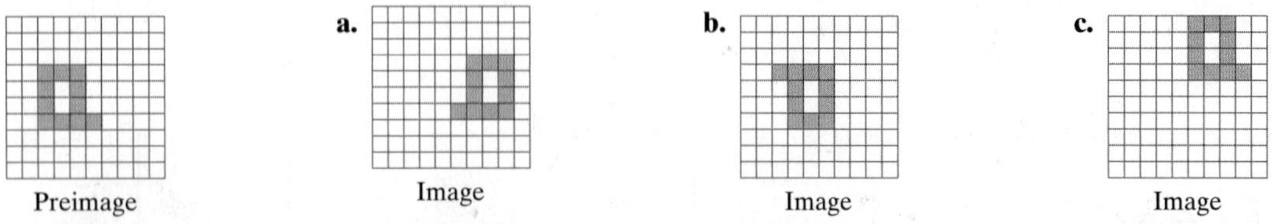

18.

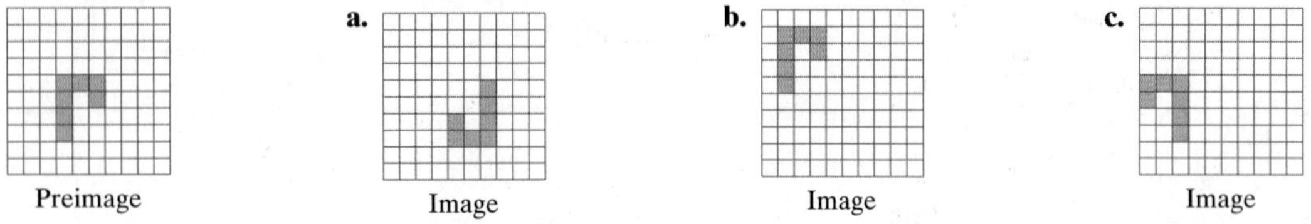

Use the diagrams to complete each mapping statement. These mappings are all isometries. See Example 4.

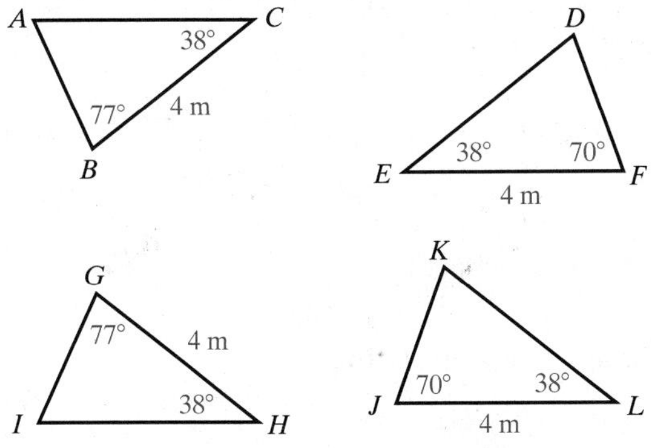

19. $\Delta EDF \rightarrow$ _______ **20.** $\Delta ABC \rightarrow$ _______

21. $\Delta HIG \rightarrow$ _______ **22.** $\Delta JLK \rightarrow$ _______

23. _______ $\rightarrow \Delta KJL$ **24.** _______ $\rightarrow \Delta GHI$

CONCEPT EXTENSIONS

If each transformation is an isometry, find the value of each variable.

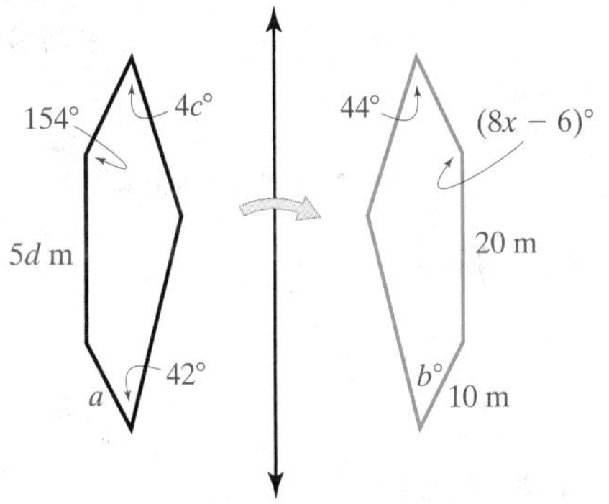

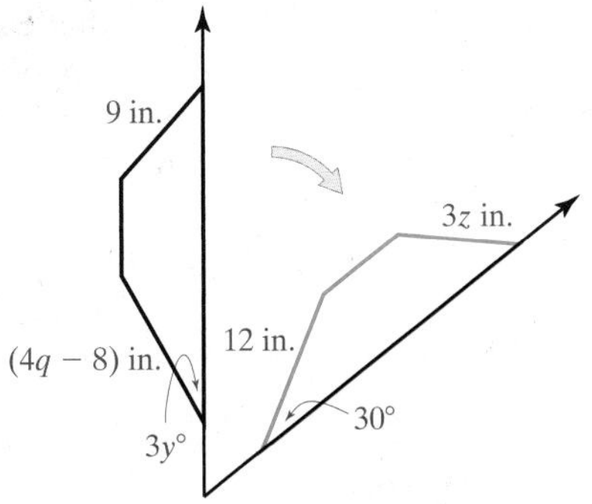

25. $a =$ ____ **26.** $b =$ ____ **27.** $c =$ ____

28. $d =$ ____ **29.** $y =$ ____ **30.** $z =$ ____

31. $q =$ ____ **32.** $x =$ ____

REVIEW AND PREVIEW

The lengths of two sides of a triangle are given. What are the possible lengths for the third side? See Section 5.5.

33. 16 in., 26 in. **34.** 19.5 ft, 20.5 ft

35. 9 m, 9 m **36.** $4\frac{1}{2}$yd, 8 yd

Find the area of each figure. See Section 2.1.

37. a square with 5-cm sides

38. a rectangle with base 4 in. and height 7 in.

39. a 4.6-m-by-2.5-m rectangle

40. a rectangle with length 3 ft and width $\frac{1}{2}$ ft

8.2 TRANSLATIONS

OBJECTIVE

1 Find Translation Images of Figures.

VOCABULARY

- translation
- composition of transformations

Helpful Hint

Because rigid transformations, such as translations, retain shape and size, it is true that $\Delta ABC \cong \Delta A'B'C'$.

OBJECTIVE 1 ▶ Finding Translation Images of Figures. Recall from Section 8.1 a translation is a rigid transformation, or an isometry. This means that the size, shape, and orientation of a figure in a plane stay the same when we "slide" it.

Translation

A **translation** is a transformation that maps all points of a figure the same distance in the same direction.
A translation is an isometry.

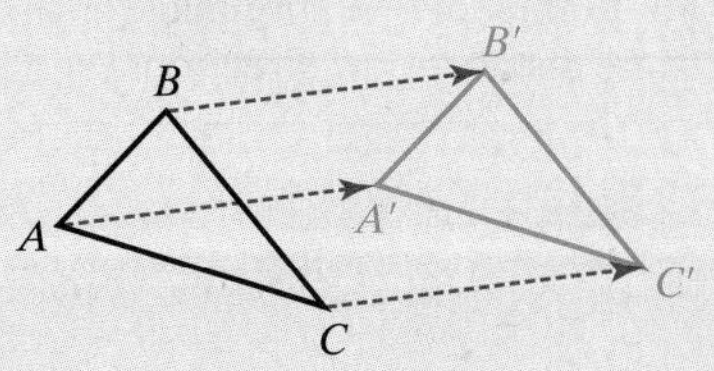

$AA' = BB' = CC'$

The diagram on the next page shows a translation in the coordinate plane. Each point of the black square moves 4 units right and 2 units down. Using ordered pair notation, we say that each (x, y) point in the original figure is mapped to (x', y'), where $x' = x + 4$ and $y' = y - 2$. We will use mapping notation to write this *translation rule* as:

$$(x, y) \rightarrow (x + 4, y - 2)$$

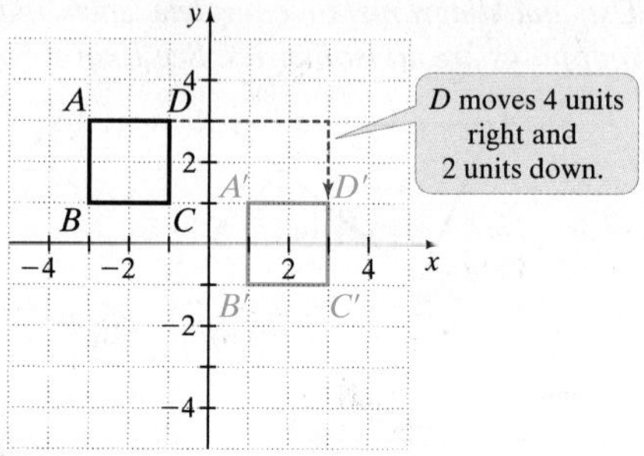
D moves 4 units right and 2 units down.

EXAMPLE 1 **Finding the Image of a Translation**

a. Find the image of each vertex of ΔPQR for the translation $(x, y) \rightarrow (x - 2, y - 5)$.

b. Graph the image of ΔPQR.

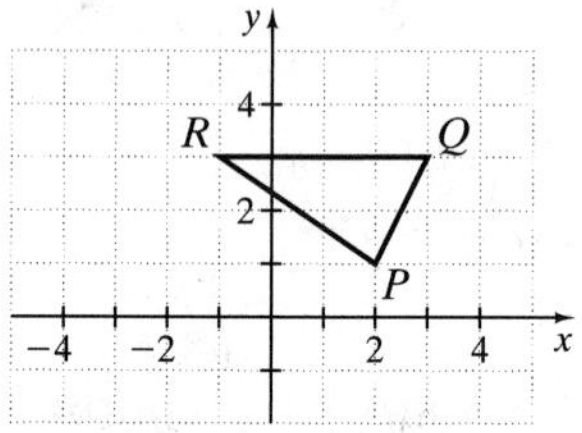

Solution The translation rule $(x,y) \rightarrow (x - 2, y - 5)$ tells us how to move each point of the preimage, ΔPQR.

a. Identify the coordinates of each vertex. Then use the translation rule to find the coordinates of each vertex of the image.

$$(x, y) \rightarrow (x - 2, y - 5)$$
$$P(2, 1) \rightarrow (2 - 2, 1 - 5), \text{ or } P'(0, -4)$$
$$Q(3, 3) \rightarrow (3 - 2, 3 - 5), \text{ or } Q'(1, -2)$$
$$R(-1, 3) \rightarrow (-1 - 2, 3 - 5), \text{ or } R'(-3, -2)$$

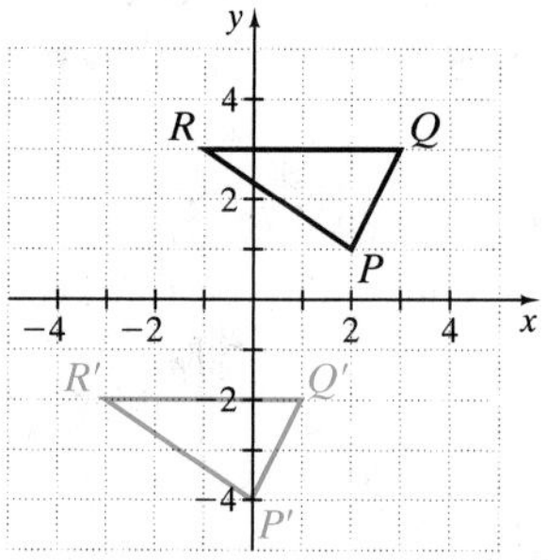

b. To graph the image of ΔPQR, first graph P', Q', and R'. Then draw $\overline{P'Q'}$, $\overline{Q'R'}$, and $\overline{R'P'}$.

Know that $\Delta RQP \cong R'Q'P'$.

PRACTICE 1

a. Find the image of each vertex of ΔABC for the translation $(x, y) \rightarrow (x + 1, y - 4)$.

b. Graph ΔABC and its image.

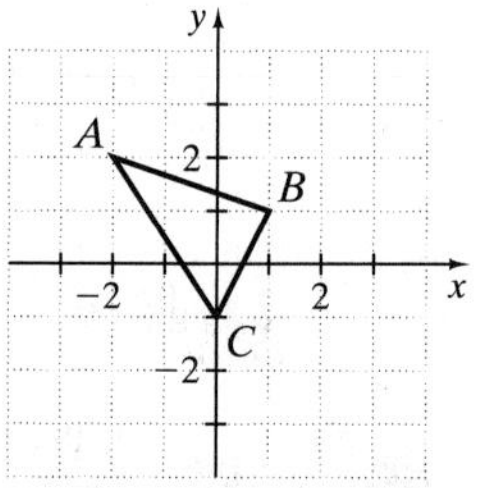

> **Helpful Hint**
> Recall that a translation moves all points of the preimage the same distance and direction.

EXAMPLE 2 **Writing a Rule to Describe a Translation**

What is a translation rule that describes the translation $PQRS \rightarrow P'Q'R'S'$?

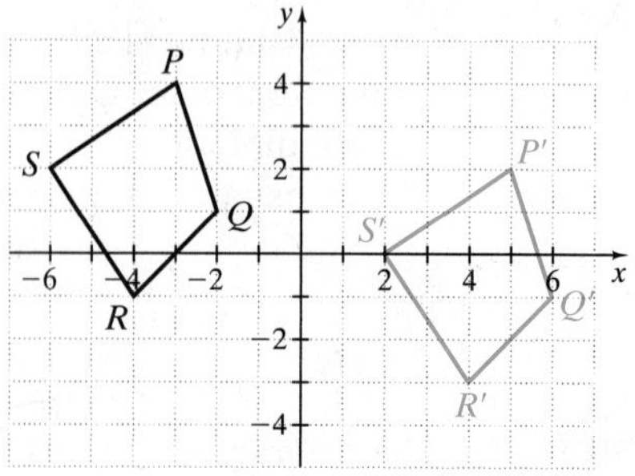

Solution Choose a preimage point and its corresponding image point, say, P and P'. Find the coordinates of both points. Then find the horizontal change and the vertical change needed to move from P to P'. This gives the translation rule.

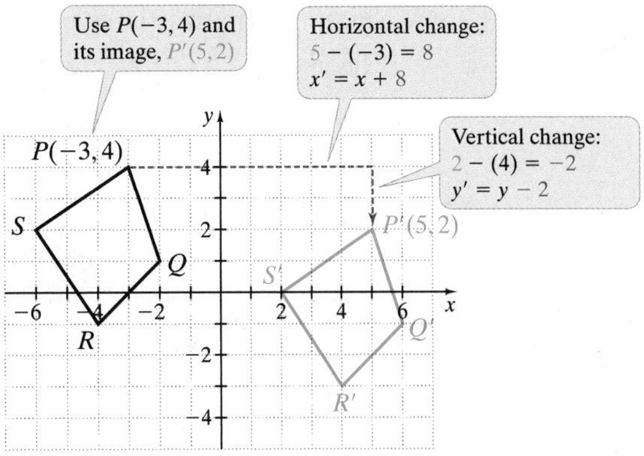

The translation rule is $(x, y) \rightarrow (x + 8, y - 2)$.

Check by applying this rule to a few preimage points.

PRACTICE

2 The translation image of ΔLMN is $\Delta L'M'N'$ with $L'(1, -2)$, $M'(3, -4)$, and $N'(6, -2)$. What is a translation rule that describes the translation?

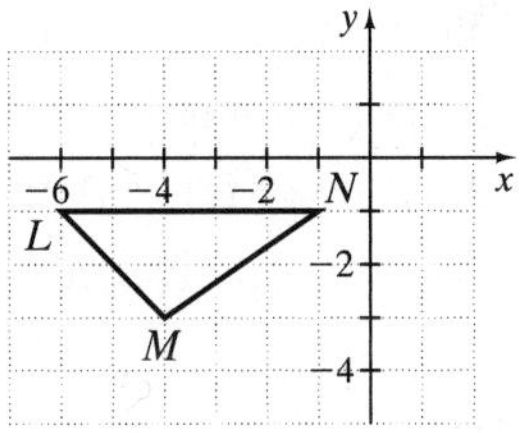

A **composition of transformations** is a combination of two or more transformations. In a composition, we perform each transformation on the image of the preceding transformation.

In this section, we study a composition of transformations involving translations only.

In the diagram at the right, the field hockey ball can move from Player 3 to Player 5 by a direct pass. This translation is represented by the blue arrow. The ball can also be passed from Player 3 to Player 9, and then from Player 9 to Player 5. The two red arrows represent this composition of translations. Notice that the preimage (ball with player 3) and the final image (ball with player 5) is the same for both.

In general, the composition of any two translations is another translation.

EXAMPLE 3 Composing Translations

The diagram at the right shows two moves of the black bishop in a chess game. Where is the bishop in relation to its original position?

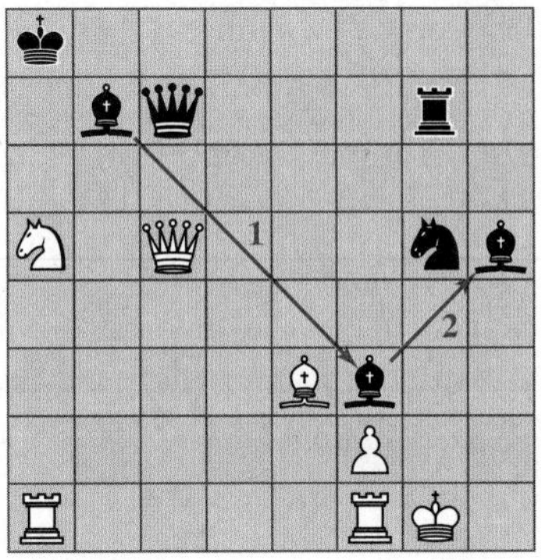

Solution Use $(0, 0)$ to represent the bishop's original position. Write translation rules to represent each move.

Translation 1: $(x, y) \rightarrow (x + 4, y - 4)$ followed by — The bishop first moves 4 squares right and 4 squares down.

Translation 2: $(x, y) \rightarrow (x + 2, y + 2)$ — The bishop next moves 2 squares right and 2 squares up.

The bishop's current position is the composition of the two translations.

First, $(0, 0)$ translates to $(0 + 4, 0 - 4)$, or $(4, -4)$.
Then, $(4, -4)$ translates to $(4 + 2, -4 + 2)$, or $(6, -2)$.

The bishop is 6 squares right and 2 squares down from its original position. □

PRACTICE 3 The bishop (after the moves in Example 3) next moves diagonally upward until the chess piece is located along the top row of the board game diagram for Example 3. Where is the bishop in relation to its original position at the beginning of Example 3?

VOCABULARY & READINESS CHECK

Word Bank *Use the choices to fill in each blank. Some choices may be used more than once and some not at all.*

slide	isometry	composition of transformations	flip
rotation	direction	distance	translation

1. A translation is a transformation that maps all points of a figure the same distance in the same ____________.
2. A combination of two or more transformations is called a(n) ____________.
3. True or false: A translation is an isometry. ____________
4. Fill in the Blank: On the rectangular coordinate system, the translation rule $(x, y) \rightarrow (x + 2, y - 5)$ slides a figure ____________ units left, right (circle one) and ____________ units up, down (circle one).

8.2 EXERCISE SET MyMathLab®

Suppose that $\Delta HZT \rightarrow \Delta H'Z'T'$. Find each image. See Example 1.

1. The image of Z is ____________.
2. The image of T is ____________.
3. The image of $\overline{HZ}$ is ____________.
4. The image of $\overline{TH}$ is ____________.

Write an ordered pair translation rule that describes each translation. See Examples 1 and 2.

5. Point $A(x, y)$ moves 5 units right and 2 units up.
6. Point $B(x, y)$ moves 10 units right and 7 units up.
7. Point $C(x, y)$ moves 9 units right and 5 units down.
8. Point $D(x, y)$ moves 3 units right and 6 units down.
9. Point $E(x, y)$ moves 4 units left and 1 unit up.
10. Point $F(x, y)$ moves 12 units left and 8 units up.
11. Point $G(x, y)$ moves 20 units left and 10 units down.
12. Point $H(x, y)$ moves 15 units left and 9 units down.

Copy each graph.

a. Find the image of each vertex.

b. Graph the image of each figure under the given translation. See Example 1.

13. $(x, y) \rightarrow (x + 3, y + 2)$

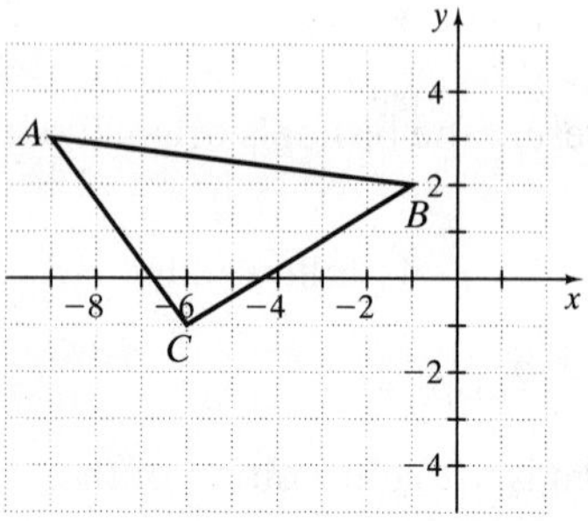

14. $(x, y) \rightarrow (x + 4, y - 2)$

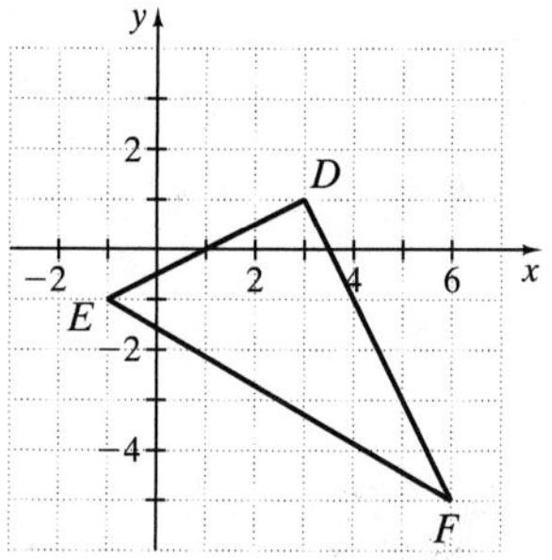

15. $(x, y) \rightarrow (x + 5, y - 1)$

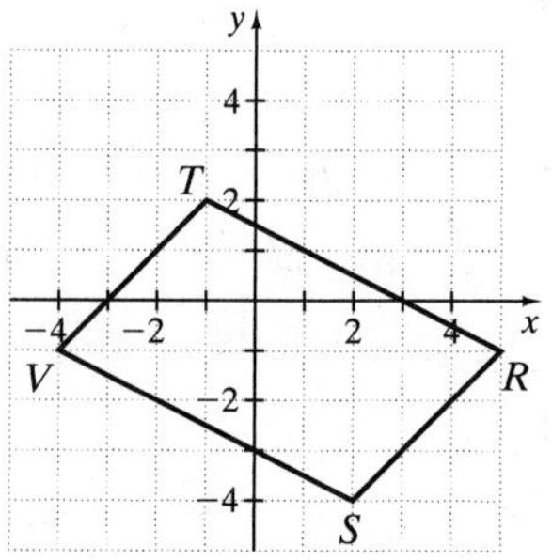

16. $(x, y) \rightarrow (x - 3, y - 4)$

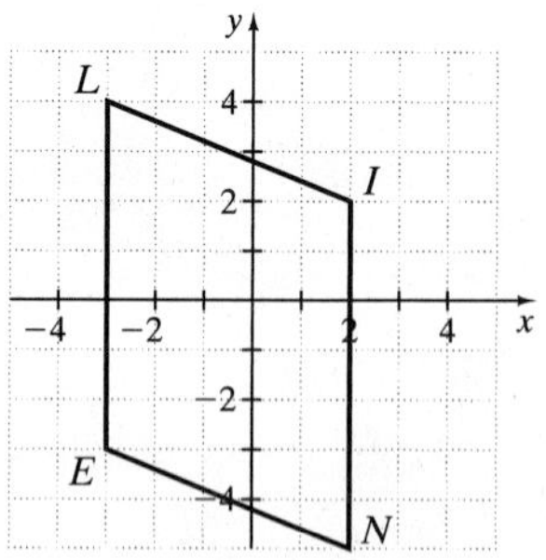

17. $(x, y) \rightarrow (x - 2, y + 5)$

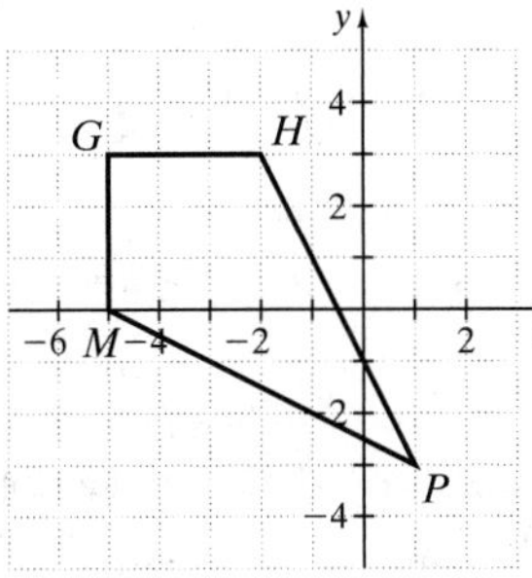

18. $(x, y) \rightarrow (x + 1, y - 3)$

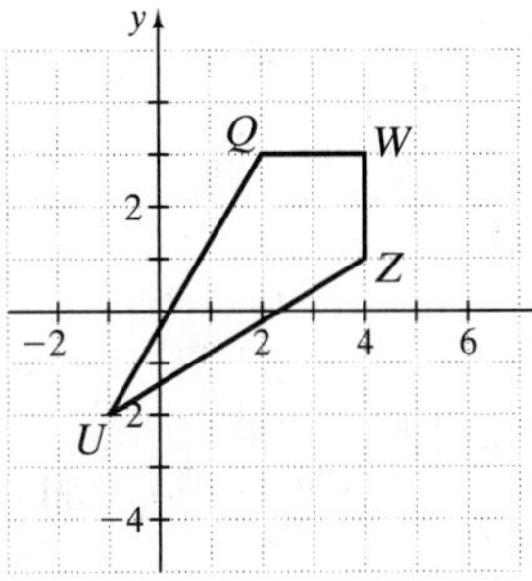

The blue figure is a translation image of the black figure. Write a rule to describe each translation. See Example 2.

19.

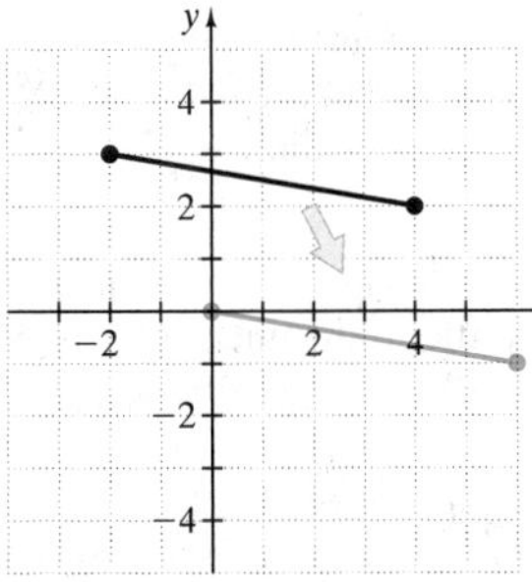

20.

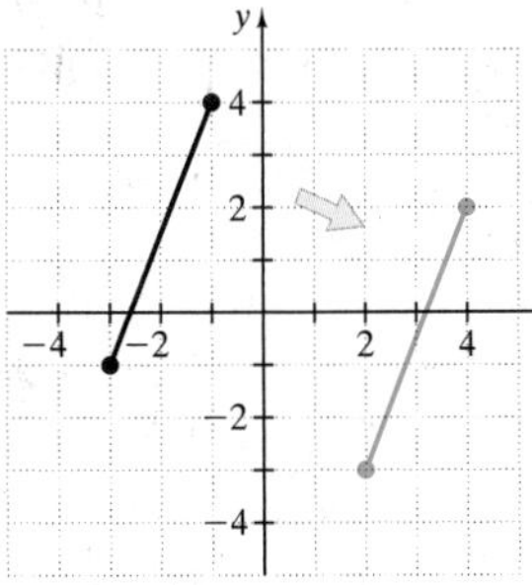

21.

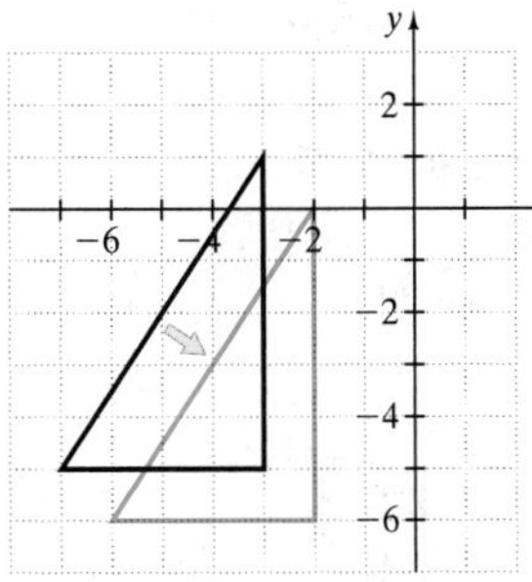

22.

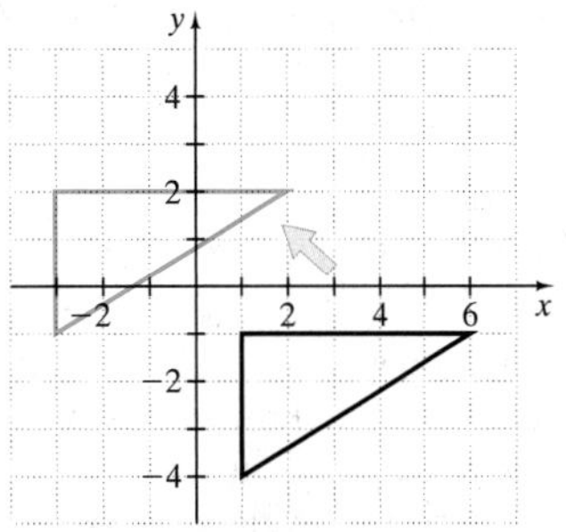

23.

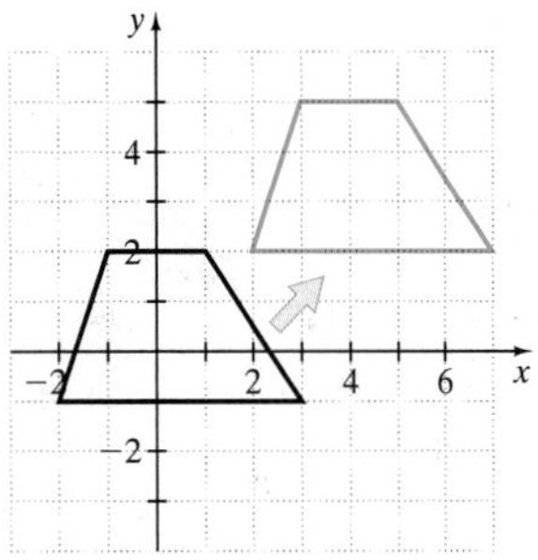

24.

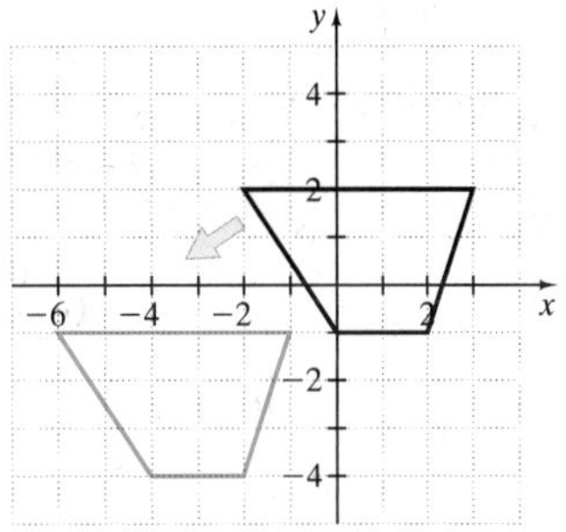

Use the ordered pair translation rule $(x, y) \rightarrow (x - 10, y + 7)$ for Exercises 25–32. See Examples 1 and 2.

25. Find the image of $(0, 0)$.

26. Find the image of $(9, 2)$.

27. Find the preimage of $(20, 20)$.

28. Find the preimage of $(0, 0)$.

29. Find the image of $(3, -8)$.

30. Find the image of $(-5, 11)$.

31. Find the preimage of $(-6, 4)$.

32. Find the preimage of $(1, -1)$.

Solve. See Example 3.

33. You are visiting San Francisco. From your hotel near Union Square, you walk 4 blocks east and 4 blocks north to the Wells Fargo History Museum. Then you walk 5 blocks west and 3 blocks north to the Cable Car Barn Museum. Where is the Cable Car Barn Museum in relation to your hotel?

34. Your friend and her parents are visiting colleges. They leave their home in Enid, Oklahoma, and drive to Tulsa, which is 107 mi east and 18 mi south of Enid. From Tulsa, they go to Norman, 83 mi west and 63 mi south of Tulsa. Where is Norman in relation to Enid?

Find a translation that has the same effect as each composition of translations.

35. $(x, y) \rightarrow (x + 2, y + 5)$ followed by $(x, y) \rightarrow (x - 4, y + 9)$

36. $(x, y) \rightarrow (x + 12, y + 0.5)$ followed by $(x, y) \rightarrow (x + 1, y - 3)$

37. You write a computer animation program to help young children learn the alphabet. The program draws a letter, erases the letter, and makes it reappear in a new location two times. The program uses the following composition of translations to move the letter.

$$(x, y) \rightarrow (x + 5, y + 7) \text{ followed by}$$
$$(x, y) \rightarrow (x - 9, y - 2)$$

Suppose the program makes the letter W by connecting the points $(1, 2)$, $(2, 0)$, $(3, 2)$, $(4, 0)$, and $(5, 2)$. What points does the program connect to make the last W?

38. You write a computer animation program similar to the one in Exercise **37**. This time, the program uses the following composition of translations:

$$(x, y) \rightarrow (x + 2, y + 5) \text{ followed by}$$
$$(x, y) \rightarrow (x - 10, x + 1)$$

Suppose the program makes the letter N by connecting the points $(1, 0)$, $(1, 2)$, $(3, 0)$, and $(3, 2)$. What points does the program connect to make the last N?

39. The translation $(x, y) \rightarrow (x + 5, y + 7)$ maps ΔMNO to $\Delta M'N'O'$. What translation rule maps $\Delta M'N'O'$ to ΔMNO?

40. In the diagram below, the yellow figure is a translation image of the red figure. Write a rule that describes the translation.

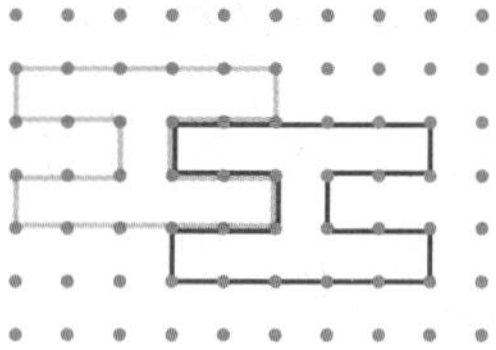

41. ΔMUG has coordinates $M(2, -4)$, $U(6, 6)$, and $G(7, 2)$. A translation maps point M to $M'(-3, 6)$.

a. Find a rule that describes the translation.

b. What are the coordinates of U' and G' for this translation?

42. Use the graph below. Write three different translation rules for which the image of ΔJKL has a vertex at the origin.

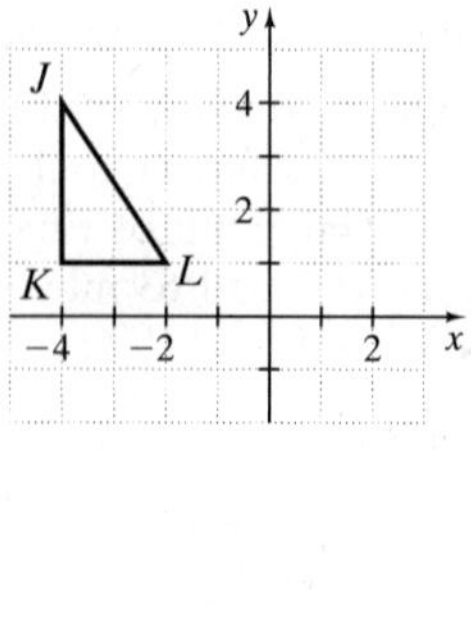

CONCEPT EXTENSIONS

43. Find the Error Your friend says the transformation $\Delta ABC \rightarrow \Delta PQR$ is a translation. Explain and correct her error.

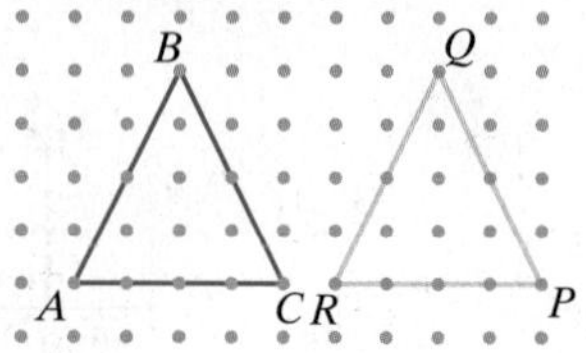

44. Write the translation $(x, y) \rightarrow (x + 1, y - 3)$ as a composition of a horizontal translation and a vertical translation.

45. The diagram below shows the site plan for a backyard storage shed. Local law, however, requires the shed to sit at least 15 ft from property lines. Describe how to move the shed to comply with the law.

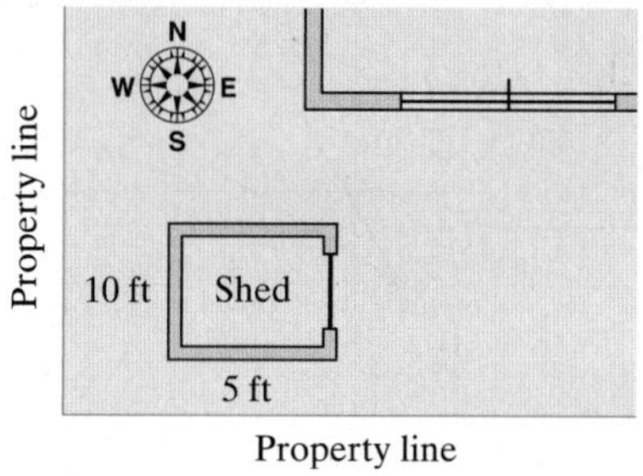

46. Explain how to use translations to draw a parallelogram.

47. Coordinate Geometry $PLAT$ has vertices $P(-2, 0)$, $L(-1, 1)$, $A(0, 1)$, and $T(-1, 0)$. $P'L'A'T'$ is the image of $PLAT$ for the translation $(x, y) \rightarrow (x + 2, y - 3)$. Show that $\overline{PP'}$, $\overline{LL'}$, $\overline{AA'}$, and $\overline{TT'}$ are all parallel.

48. Coordinate Geometry ΔABC has vertices $A(-2, 5)$, $B(-4, -1)$, and $C(2, -3)$. Show that the images of the midpoints of the sides of ΔABC are the midpoints of the sides of $\Delta A'B'C'$ for the translation $(x, y) \rightarrow (x + 4, y + 2)$.

Geometry in Three Dimensions *Follow the sample below. Use each figure, graph paper, and the given translation to draw a three-dimensional figure.*

Sample *Use the rectangle and* $(x, y) \rightarrow (x + 3, y + 1)$ *to draw a box.*

STEP 1.

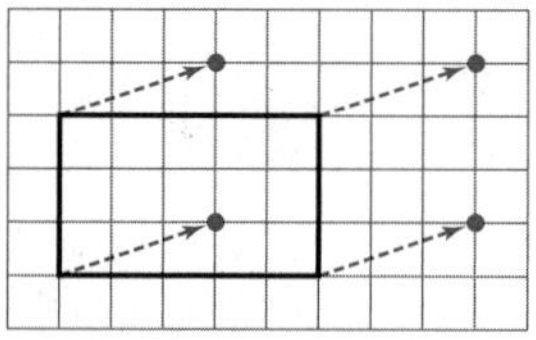

STEP 2.

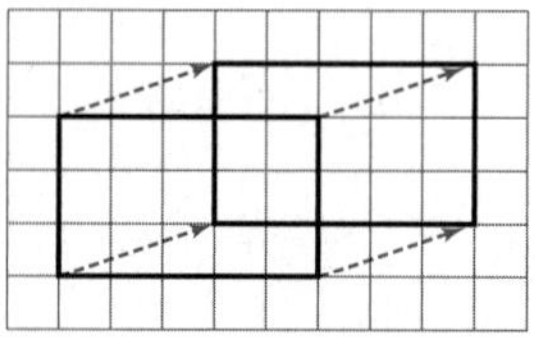

49. $(x, y) \rightarrow (x + 2, y - 1)$

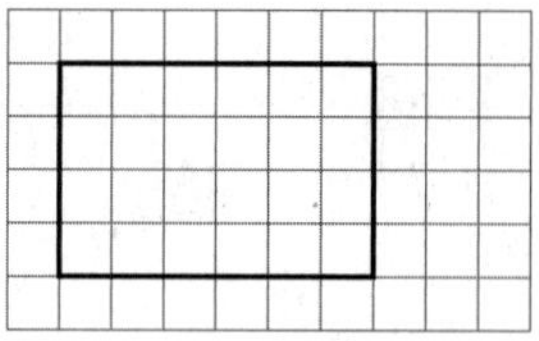

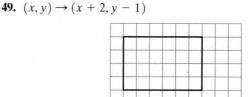

50. $(x, y) \rightarrow (x - 3, y - 1)$

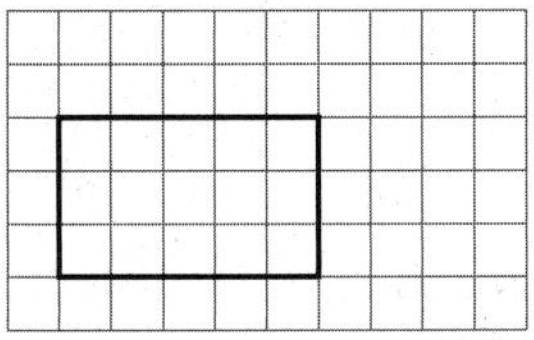

51. $(x, y) \rightarrow (x - 2, y + 2)$

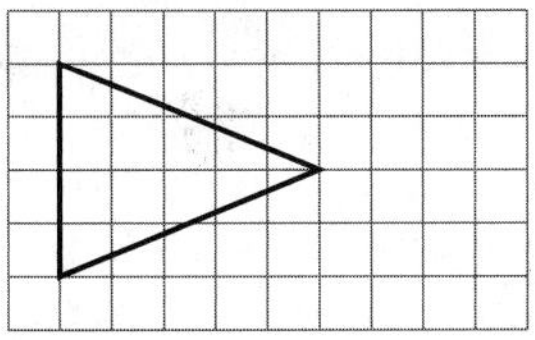

52. $(x, y) \rightarrow (x - 3, y - 5)$

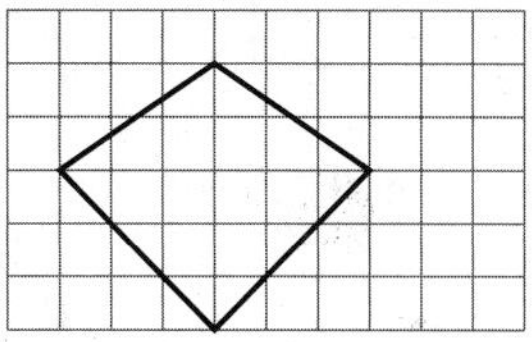

REVIEW AND PREVIEW

Write an equation for the line through point A and perpendicular to the given line. See Section 3.7.

53. $A(1, -2); x = -2$

54. $A(-5, 4); x = 3$

55. $A(-1, -1); y = 1$

56. $A(-1, 2); y = -3$

ACTIVITIES (USE WITH SECTION 8.3)

Paper Folding and Reflections

In Activity 1, we see how a figure and its *reflection* image are related. In Activity 2, we use these relationships to construct a reflection image.

Activity 1

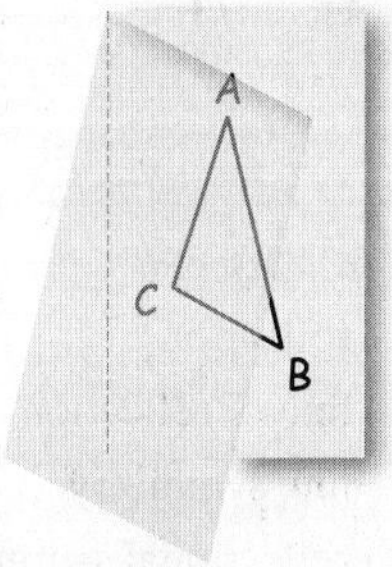

STEP 1. Use a piece of tracing paper and a straight edge. Using less than half the page, draw a large scalene triangle. Label its vertices A, B, and C.

STEP 2. Fold the paper so that your triangle is covered. Trace ΔABC using a straight edge.

STEP 3. Unfold the paper. Label the traced points corresponding to A, B, and C as A', B', and C', respectively. $\Delta A'B'C'$ is a reflection image of ΔABC. The fold is the reflection line.

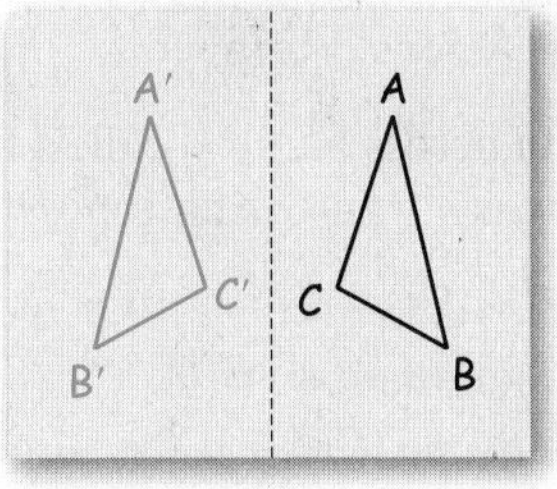

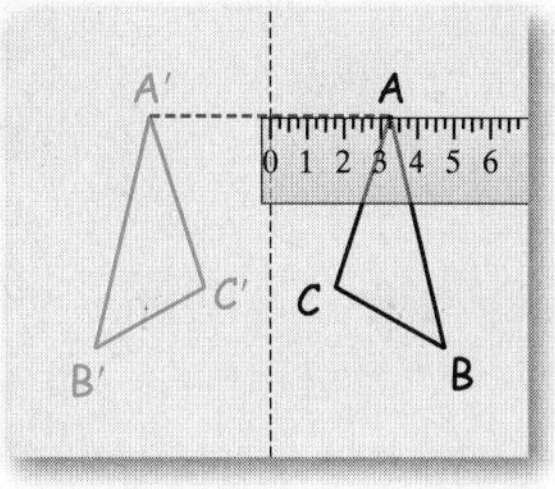

1. Use a ruler to draw $\overline{AA'}$. Measure the perpendicular distances from A to the fold and from A' to the fold. What do you notice?
2. Measure the angles formed by the fold and $\overline{AA'}$.
3. Repeat Questions 1 and 2 for B and B' and for C and C'. Then, make a conjecture: How is the reflection line related to the segment joining a point and its image?

Activity 2

STEP 1. On regular paper, draw a simple shape or design made of segments. Use less than half the page. Draw a reflection line near your figure.

(continued)

STEP 2. Use a compass and straight edge to construct a line perpendicular to the reflection line through one point of your drawing.

4. Explain how you can use a compass and the perpendicular line you drew to find the reflection image of the point you chose.

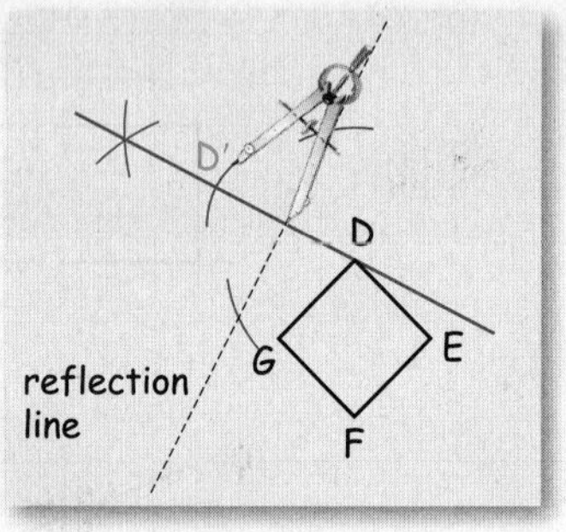

5. Connect the reflection images for several points of your shape and complete the image. Check the accuracy of the reflection image by folding the paper along the reflection line and holding it up to a light source.

8.3 REFLECTIONS

OBJECTIVES

1 Find Reflection Images of Figures.

2 Identify Line Symmetry.

VOCABULARY

- reflection
- line of reflection
- line symmetry
- reflectional symmetry
- line of symmetry

OBJECTIVE 1 ▶ Finding Reflection Images of Figures. Notice below that when a figure flips across a line, the preimage and its image are congruent and have *opposite orientations*. In other words, choose a way to read the preimage triangle below, say ΔBUG. It maps to $\Delta B'U'G'$. Read each triangle in these order of vertices. If one reads clockwise, then the other reads counterclockwise or vice versa.

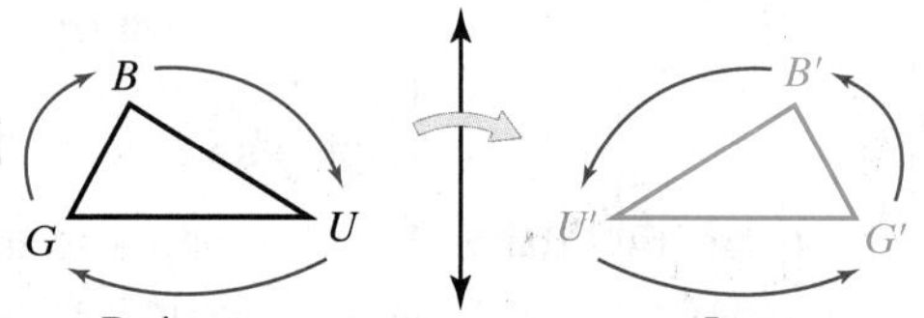

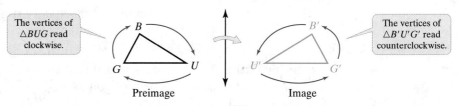

The size and shape of a geometric figure stay the same when we flip the figure across a line. Just remember that its orientation reverses.

Reflection Across a Line

Reflection across a line r, called the **line of reflection**, is a transformation with these two properties:

- If A is on r, then $A' = A$.
- If B is not on r, then r is the perpendicular bisector of $\overline{BB'}$.

A reflection across a line is an isometry.

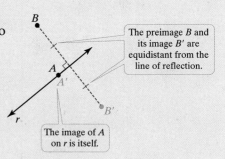

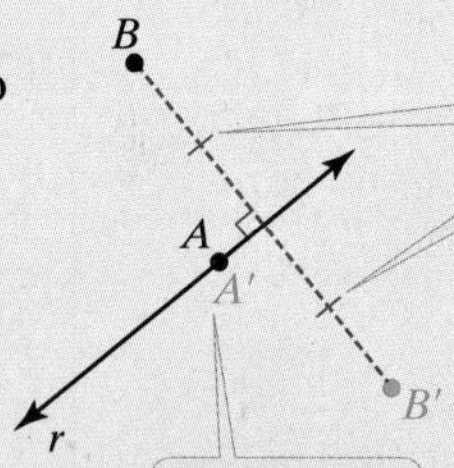

▶ Helpful Hint

The reflection of any figure across a line is best visualized when a graph is drawn.

EXAMPLE 1 **Reflecting a Point Across a Line**

Multiple Choice If point $P(3, 4)$ is reflected across the line $y = 1$, what are the coordinates of its reflection image?

a. $(3, -4)$ **b.** $(0, 4)$ **c.** $(3, -2)$ **d.** $(-3, -2)$

Solution Graph point P and the line of reflection $y = 1$. Point P and its reflection image across the line are equidistant from the line.

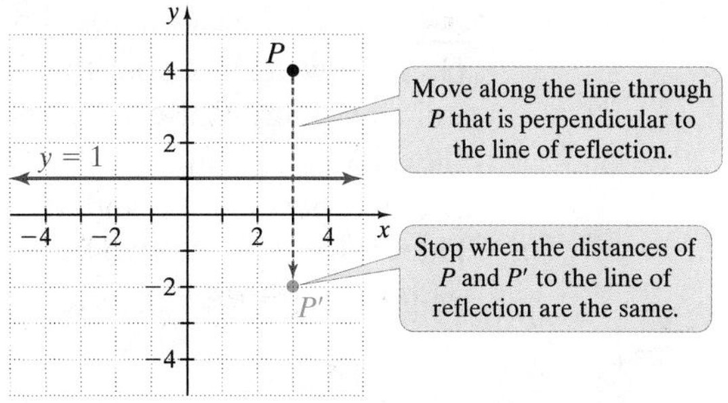

P is 3 units above the line $y = 1$, so P' must be 3 units below the line $y = 1$. The line $y = 1$ is the perpendicular bisector of $\overline{PP'}$ if P' is $(3, -2)$. The correct answer is **c**. □

PRACTICE
1 What is the image of $P(3, 4)$ reflected across the line $x = -1$?

EXAMPLE 2 Graphing a Reflection Image

Coordinate Geometry Graph points $A(-3, 4)$, $B(0, 1)$, and $C(4, 2)$. What is the image of ΔABC reflected across the y-axis?

Solution

> **Helpful Hint**
> Don't forget that the image of any point on the line of reflection is itself.

First:
Graph ΔABC. The y-axis is the line of reflection and is shown in red.

Next:
Find A', B', and C'. B' is the same position as B because B is on the line of reflection. Locate A' and C' so that the y-axis is the perpendicular bisector of $\overline{AA'}$ and $\overline{CC'}$.

Finally:
Draw $\Delta A'B'C'$.

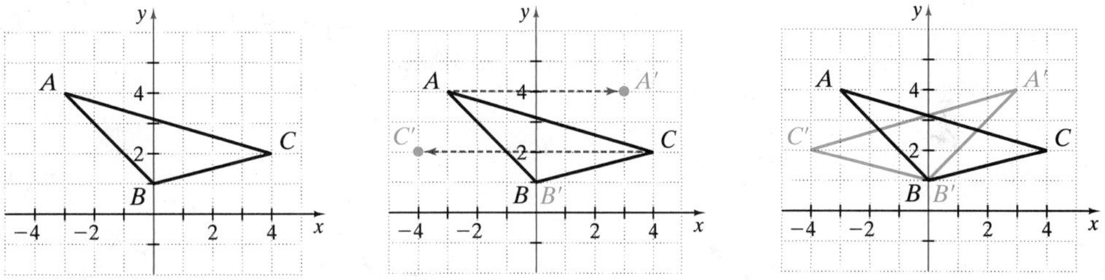

□

PRACTICE
2 Graph ΔABC from Example 2. What is the image of ΔABC reflected across the x-axis?

EXAMPLE 3 Minimizing a Distance

Beginning from a single point on Summit Trail, a hiking club will build two trails–one trail to the Overlook and one trail to Balance Rock. Working under a tight budget, the club members want to minimize the total length of the two trails. How can we find the point on Summit Trail where the two new trails should start? Assume the trails will be straight and that they will cover similar terrain.

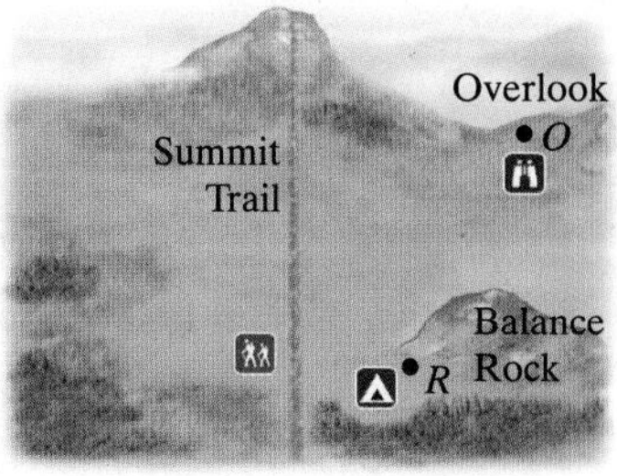

Solution Reflect O across line t and obtain O'. Then draw $\overline{RO'}$. Label the intersection of $\overline{RO'}$ and t as point P. Because the shortest distance between two points is a line, $\overline{RO'}$ is the shortest distance between R and O'. Also $PO = PO'$. Thus, if we start the two trails at point P, the two trails $\overline{PO}$ and $\overline{PR}$ will have a minimum distance.

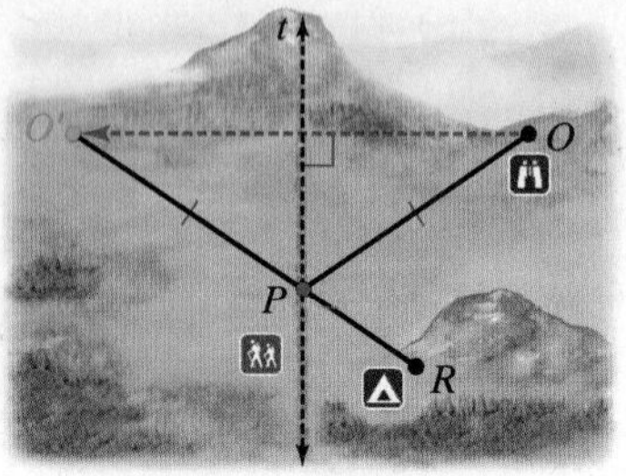

□

PRACTICE
3 Your classmate began to solve the problem above by reflecting point R in line t. Will this method work? Explain.

OBJECTIVE 2 ▶ Identifying Line Symmetry. Some figures appear unchanged after a reflection across a line. These figures are said to have line *symmetry*.

Line Symmetry

A plane figure has **line symmetry** or **reflectional symmetry** if the figure on one side of the line is the reflection of the figure on the other side of the line. The line of reflection is called a **line of symmetry**. It divides the plane figure into congruent halves.

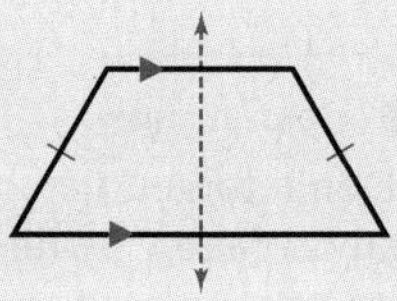

EXAMPLE 4 Identifying Lines of Symmetry

How many lines of symmetry does a regular hexagon have?

Solution

First, let's sketch a hexagon and then study our figure for ways it will reflect across a line.

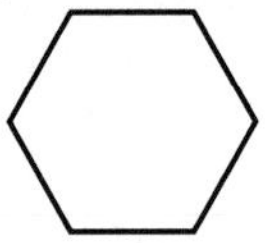

Sketch

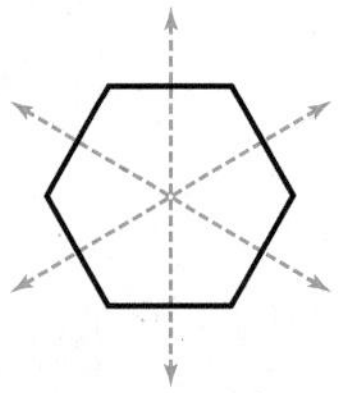

The hexagon reflects onto itself across each line that passes through the midpoints of a pair of parallel sides.

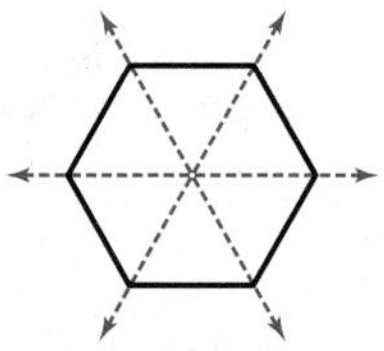

The hexagon also reflects onto itself across each diagonal that passes through the center of the hexagon.

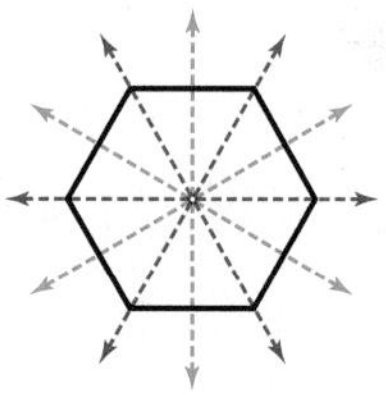

A regular hexagon has six lines of symmetry.

□

PRACTICE
4 Draw a rectangle that is not a square. How many lines of symmetry does your rectangle have?

VOCABULARY & READINESS CHECK

Word Bank *Use the choices to fill in each blank. Some choices may be used more than once and some not at all.*

slide	isometry	composition of transformations	flip	translation
rotation	reflectional	line of reflection	reflection	itself

1. A reflection across line r is a transformation that maps any point on line r to ____________.

2. The line r in Exercise 1 above is called a(n) ____________.

3. Another word for reflection is a ____________.
4. Another term for line symmetry is called ____________ symmetry.
5. A line of symmetry is also a line of ____________.
6. True or false: A reflection is an isometry. ____________

8.3 EXERCISE SET MyMathLab®

Each point is reflected across the line indicated. Find the coordinates of each image. See Example 1.

1. Q across $x = 1$
2. V across $y = -1$
3. U across the y-axis
4. T across the x-axis
5. R across $y = 2$
6. S across $x = 3$

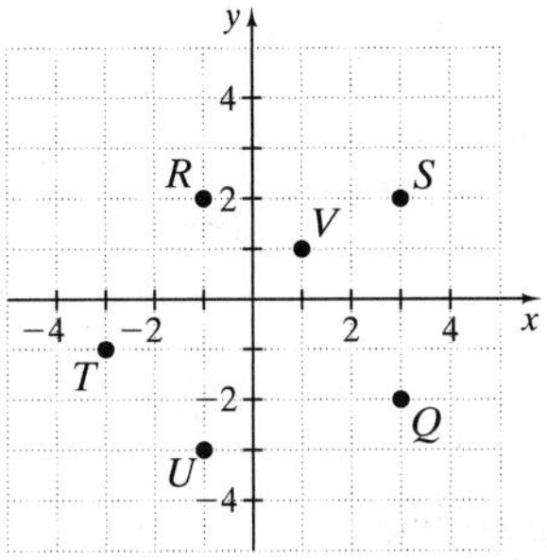

Coordinate Geometry *Given points $J(1, 4)$, $A(3, 5)$, and $R(2, 1)$, graph ΔJAR and its reflection image across each line. See Example 2.*

7. the x-axis
8. the y-axis
9. $y = 2$
10. $y = 5$
11. $x = -1$
12. $x = 2$

Copy each figure and line ℓ. Draw each figure's reflection image across line ℓ.

13. 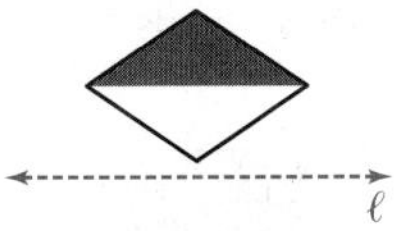

14. 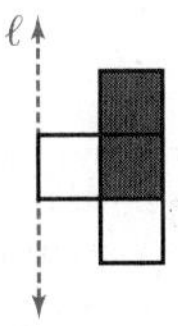

15. 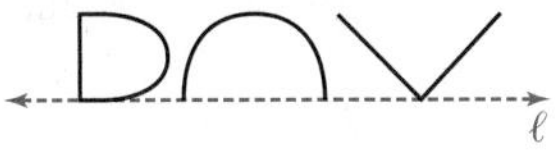

16. 

If the figure has line symmetry, sketch the figure and the line(s) of symmetry. If the figure has no line symmetry, write "no line symmetry." See Example 4.

17. 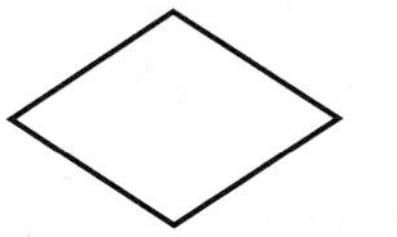

18. 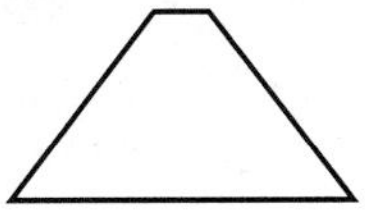

19. 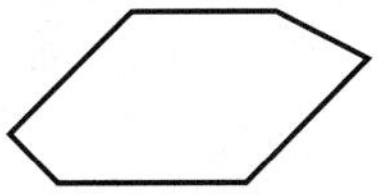

20. 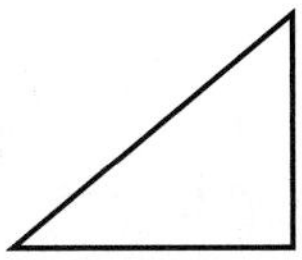

21. 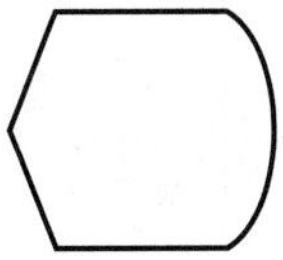

22. 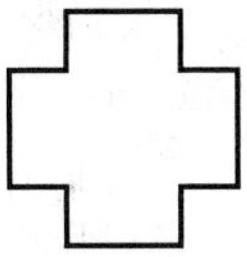

Determine how many lines of symmetry each type of quadrilateral has. Include a sketch to support your answer. See Example 4.

23. rhombus
24. kite
25. square
26. parallelogram

Use the letters of the alphabet below.

English: ABCDEFGHIJKLMNOPQRSTUVWXYZ

Classify the letters of the alphabet. You will list some letters in more than one category.

27. Horizontal Line Symmetry:
28. Vertical Line Symmetry:

Use the graph of ΔFGH.

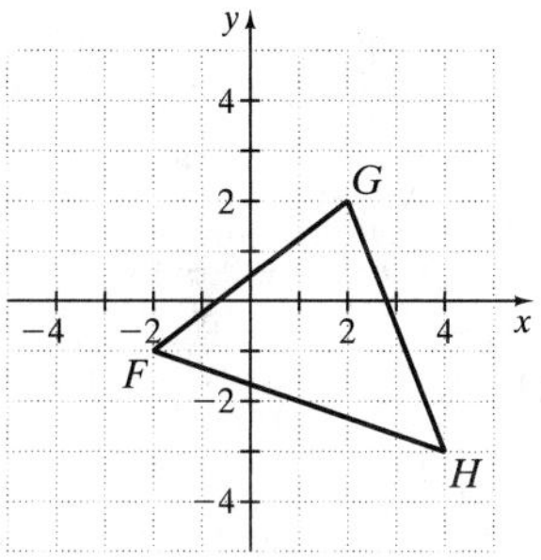

29. What are the coordinates of H reflected across the y-axis?
30. What are the coordinates of G reflected across the line $x = 3$?
31. Graph ΔFGH and its reflection image across the line $y = 4$.
32. Graph ΔFGH and its reflection image across the line $x = 2$.

Solve. See Example 3.

33. Town officials in Waterville and Drighton plan to construct a water pumping station along the Franklin Canal. The station will provide both towns with water. Where along the canal should the officials build the pumping station to minimize the total length of pipe needed?

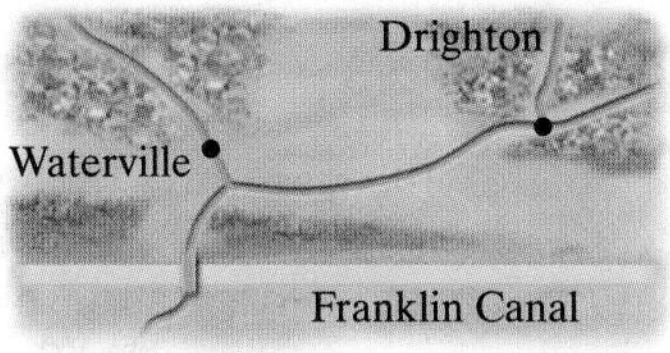

34. Cable is to be buried from point A to road r and from point B to road r. Find point C on road r so that $AC + BC$ is the minimum length.

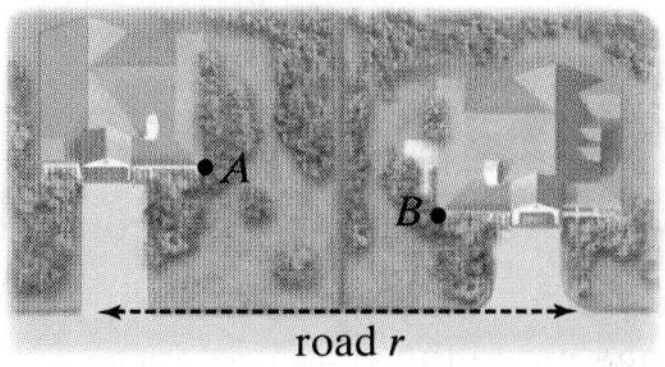

Copy each pair of figures. Then draw the line of reflection you can use to map one figure onto the other.

35. 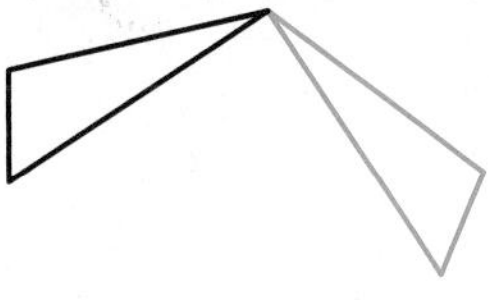

36. 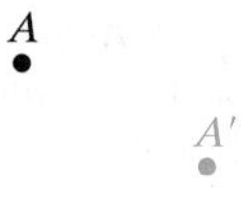

37.

38.

Find the image of $O(0, 0)$ after two reflections, first across line ℓ_1 and then across line ℓ_2.

39. $\ell_1: y = 3, \ell_2$: x-axis

40. $\ell_1: x = -2, \ell_2$: y-axis

41. $\ell_1: x = -2, \ell_2: y = 3$

42. $\ell_1: y = 3, \ell_2: x = -2$

43. Coordinate Geometry The following steps explain how to reflect point A across the line $y = x$.

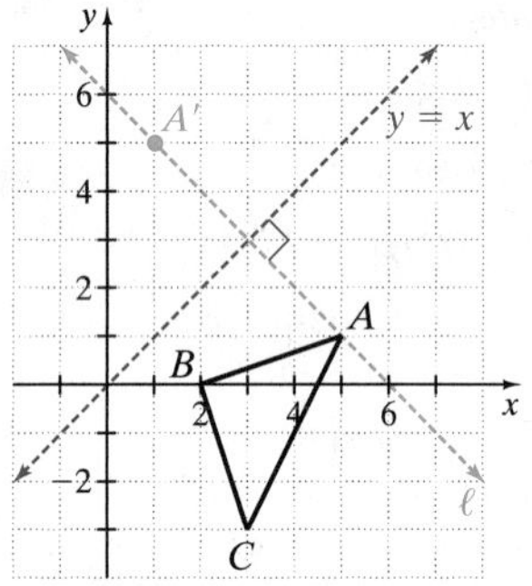

STEP 1 Draw line ℓ through $A(5, 1)$ perpendicular to the line $y = x$. The slope of $y = x$ is 1, so the slope of line ℓ is $1 \cdot (-1)$, or -1.

STEP 2 From A, move two units left and two units up to $y = x$. Then move two more units left and two more units up to find the location of A' on line ℓ. The coordinates of A' are $(1, 5)$.

a. Copy the diagram. Then draw the lines through B and C that are perpendicular to the line $y = x$. What is the slope of each line?

b. Reflect B and C across the line $y = x$. What are the coordinates of B' and C'?

c. Graph $\Delta A'B'C'$.

d. Compare the coordinates of the vertices of ΔABC and $\Delta A'B'C'$. Make a conjecture about the coordinates of the point $P(a, b)$ reflected across the line $y = x$.

44. Coordinate Geometry ΔABC has vertices $A(-3, 5)$, $B(-2, -1)$, and $C(0, 3)$. Graph ΔABC and $\Delta A'B'C'$, the reflection of ΔABC across the line $y = -x$. (*Hint*: See Exercise **43**.)

CONCEPT EXTENSIONS

45. What is the relationship between a line of reflection and a segment joining corresponding points of the preimage and image?

46. When you reflect a figure across a line, does every point on the preimage move the same distance? Explain.

47. Find the Error A classmate reflected point A across line r as shown in the diagram. Explain your classmate's error. Copy point A and line r and show the correct location of A'.

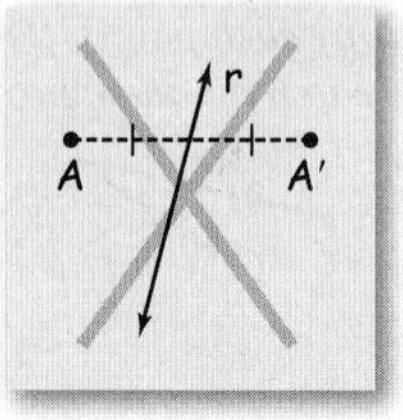

48. What are the coordinates of a point $P(x, y)$ reflected across the y-axis? Across the x-axis?

Each diagram shows a reflection across a line. Find the value of each variable.

49.

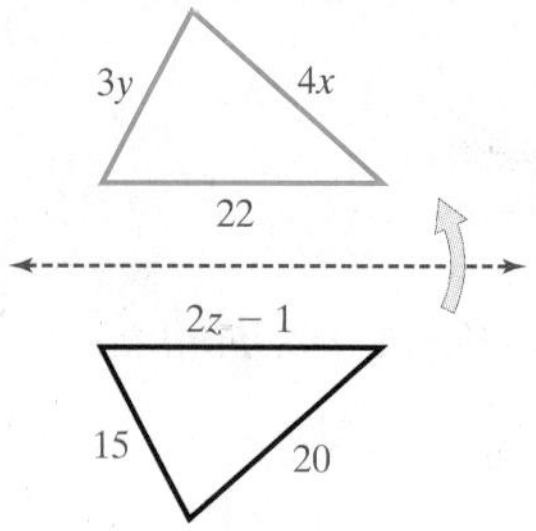

50.

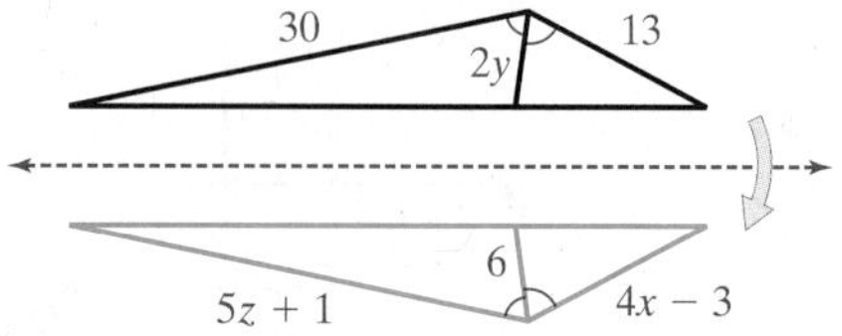

Find the image of $O(0, 0)$ after two reflections, first across line ℓ_1 and then across line ℓ_2. The letters a and b below are constants.

51. $\ell_1: x = -2, \ell_2: y = x$

52. $\ell_1: x = 4, \ell_2: y = x$

53. $\ell_1: x = a, \ell_2: y = x$

54. $\ell_1: y = b, \ell_2: y = x$

55. Is the line that contains the bisector of an angle also a line of symmetry of the angle? Explain.

56. Is the line that contains the bisector of an angle of a triangle also a line of symmetry of the triangle? Explain.

57. If you stack the letters of MATH vertically, you can find a vertical line of symmetry. Find two other words for which this is true.

58. The equation $\frac{10}{10} - 1 = 0 \div \frac{83}{83}$ is not only true, but also symmetrical (horizontally). Write four other equations or inequalities that are both true and symmetrical.

Coordinate Geometry *A figure that has a vertex at $(3, 4)$ has the given line of symmetry. Tell the coordinates of another vertex of the figure.*

59. the y-axis

60. the x-axis

Coordinate Geometry *Graph each equation and describe any line symmetry.*

61. $y = x^2$

62. $y = |x|$

63. $x^2 + y^2 = 9$

64. $y = (x + 2)^2$

Use the diagram below. Find the coordinates of the given point.

65. A', the reflection image of A across $y = x$

66. A'', the reflection image of A' across $y = -x$

67. A''', the reflection image of A'' across $y = x$

68. A'''', the reflection image of A''' across $y = -x$

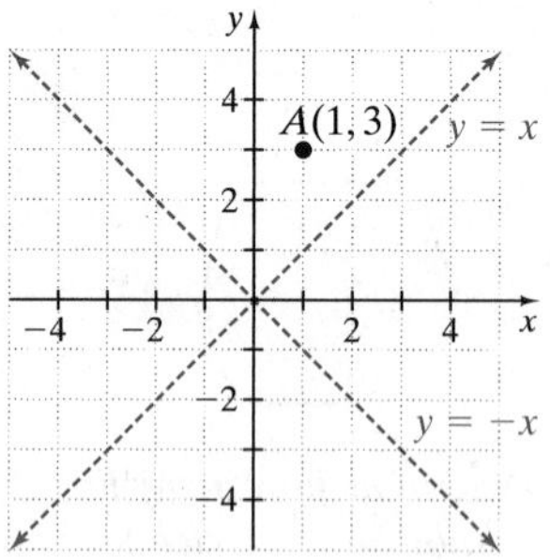

Can you form the given type of quadrilateral by drawing a triangle and then reflecting one or more times? Explain.

69. parallelogram
70. isosceles trapezoid
71. kite
72. rhombus
73. rectangle
74. square

REVIEW AND PREVIEW

For the given points, $\overline{A'B'}$ is a translation of $\overline{AB}$. Write a rule to describe each translation. See Section 8.2.

75. $A(-1, 5), B(2, 0), A'(3, 3), B'(6, -2)$

76. $A(-9, -4), B(-7, 1), A'(-4, -3), B'(-2, 2)$

Use a protractor to draw an angle with the given measure. See Section 1.5.

77. 120° **78.** 90° **79.** 45° **80.** 36°

8.4 ROTATIONS

OBJECTIVES

1. Draw and Identify Rotation Images of Figures.
2. Identify Rotational Symmetry.

VOCABULARY

- rotation
- center of rotation
- angle of rotation
- center of a regular polygon
- symmetry
- rotational symmetry

OBJECTIVE 1 ▶ Drawing and Identifying Rotation Images of Figures. In this section, we learn how to recognize and construct rotations of geometric figures.

The size, shape, and orientation of a geometric figure stay the same when we rotate or turn the figure about a point.

Rotation About a Point

A **rotation** of $x°$ about a point R, called the **center of rotation**, is a transformation with these two properties:

- The image of R is itself (that is, $R' = R$).
- For any other point V, $RV' = RV$ and $m\angle VRV' = x°$.

The positive number of degrees a figure rotates is the **angle of rotation**.

A rotation about a point is an isometry.

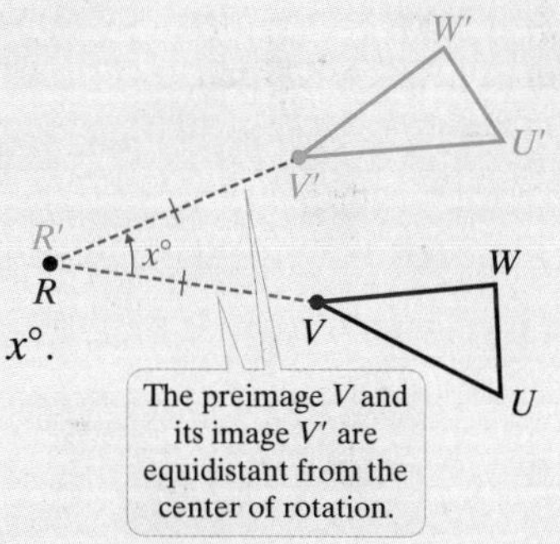

Note: Unless stated otherwise, rotations in this text are counterclockwise.

EXAMPLE 1 **Drawing a Rotation Image**

What is the image of ΔLOB for a 100° rotation about C?

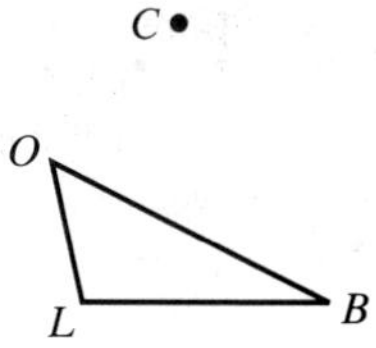

Solution Choose a vertex, say, O. We know $CO \cong CO'$ and $m\angle OCO' = 100°$. This is how we find O', then L', and then B':

STEP 1. Draw $\overline{CO}$. Use a protractor to draw a 100° angle with vertex C and side $\overline{CO}$.

STEP 2. Use a compass to construct $\overline{CO'} \cong \overline{CO}$.

STEP 3. Locate B' and L' in a similar manner.

STEP 4. Draw $\Delta L'O'B'$.

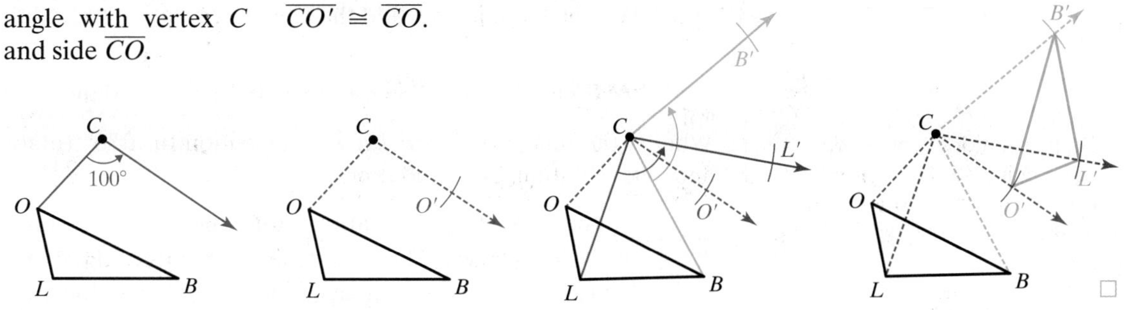

□

PRACTICE

1 Copy ΔLOB from Example 1. What is the image of ΔLOB for a 50° rotation about B?

■

> **Helpful Hint**
>
> The center of a regular n-gon is equidistant from the vertices. Also, all of the sides of a regular *n*-gon are congruent as shown below.
>
> This means that the *n* triangles formed are congruent by SSS.
>
> Example:

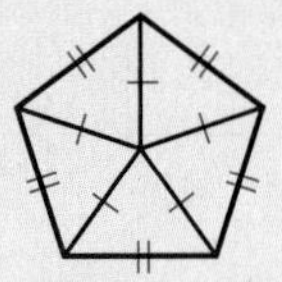

The **center of a regular polygon** is the point that is equidistant from its vertices. The center and the vertices of a regular *n*-gon determine *n* congruent triangles. We can use this fact to find rotation images of regular polygons.

EXAMPLE 2 Identifying a Rotation Image

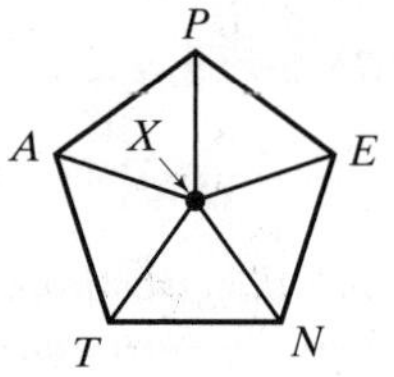

Point X is the center of regular pentagon $PENTA$. What is the image of the given point or segment for the given rotation?

a. 72° rotation of T about X

b. 216° rotation of $\overline{TN}$ about X

Solution

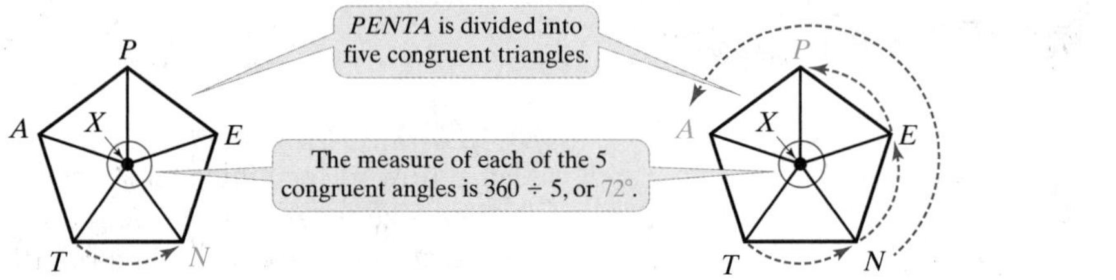

> **Helpful Hint**
>
> Since $360° \div 5 = 72°$,
> $m\angle AXT = 72°$,
> $m\angle TXN = 72°$,
> $m\angle NXE = 72°$,
> $m\angle EXP = 72°$, and
> $m\angle PXA = 72°$.

When T rotates 72° about X, it moves one vertex counterclockwise. The image of T is N, or $T' = N$. (Also, $N' = E$, and so on.)

$216° \div 72° = 3$, so a 216° rotation about X moves each vertex three vertices counterclockwise. The image of $\overline{TN}$ is $\overline{PA}$. □

PRACTICE 2 What is the image of E for a 144° rotation about X? Use the figure in Example 2.

EXAMPLE 3 Finding an Angle of Rotation

Hubcaps of car wheels often have interesting designs that involve rotation. What is the angle of rotation, in degrees, about C that maps Q to X?

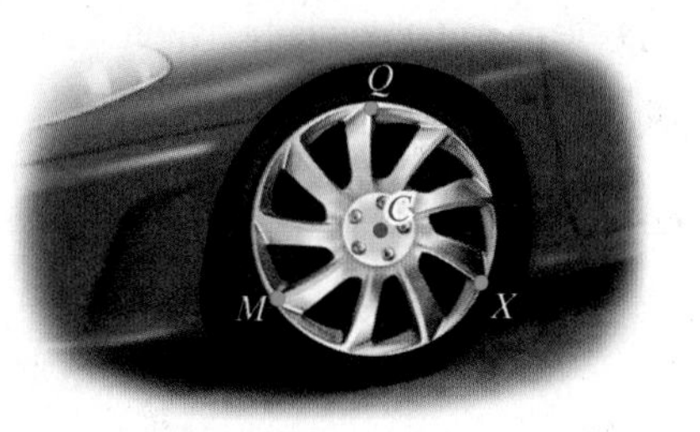

Solution The design consists of 9 spokes, so it divides the circle into 9 congruent parts. Since $360° \div 9 = 40°$, each part has a 40° angle at the center. Moving counterclockwise, Q touches 6 spokes as it rotates to point X. The angle of rotation that maps Q to X is $6 \cdot 40°$, or 240°. □

> **Helpful Hint**
>
> Rotations can be clockwise or counterclockwise. Don't forget that we will be using counterclockwise rotations.

PRACTICE 3 In Example 3, what is the angle of rotation about C that maps M to Q?

A composition of rotations about the same point is itself a rotation about that point. To sketch the image, add the angles of rotation to find the total rotation angle.

EXAMPLE 4 Finding a Composition of Rotations

> **Helpful Hint**
>
> Both rotations are about the same point, so the angle of rotation of the composition is equal to the sum of the two angles of rotation.

What is the image of $KITE$ for a composition of a 30° rotation and a 60° rotation, both about point K?

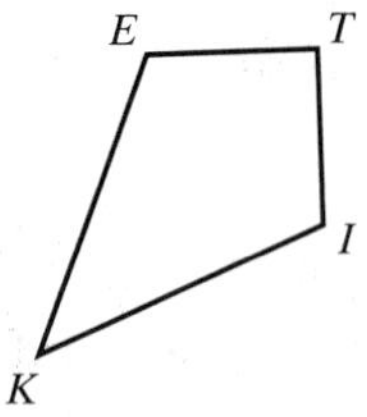

Solution The angle of rotation for the composition is $30° + 60° = 90°$. Draw $KITE$. Locate image points of the vertices for a 90° rotation. Use the image points to sketch the entire image.

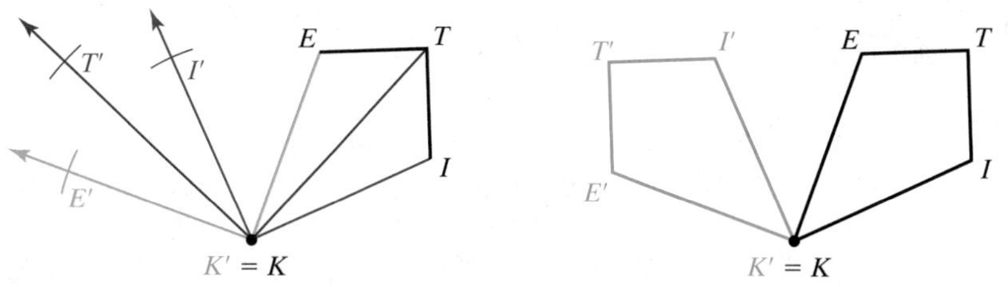

PRACTICE 4 What are the coordinates of the image of point $A(-2, 3)$ for a composition of two $90°$ rotations about the origin?

OBJECTIVE 2 ▶ Identifying Rotational Symmetry. In general, a figure has **symmetry** if there is an isometry that maps the figure onto itself.

Rotational

A figure has **rotational symmetry** if there is a rotation of $180°$ or less for which the figure is its own image. The angle of rotation for rotational symmetry is the smallest angle needed for the figure to rotate onto itself.

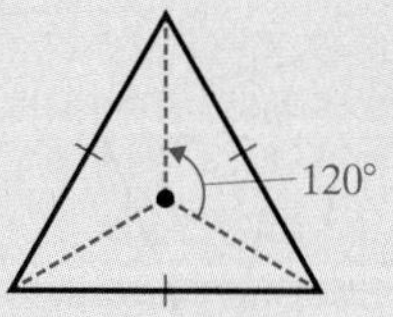

▶ **Helpful Hint**

To identify rotational symmetry, look for a possible center point. Think about the angles formed by joining preimage-image pairs to the center. All these angles must be congruent for the figure to have rotational symmetry.

EXAMPLE 5 Identifying Rotational Symmetry

Does the figure have rotational symmetry? If so, what is the angle of rotation?

a.

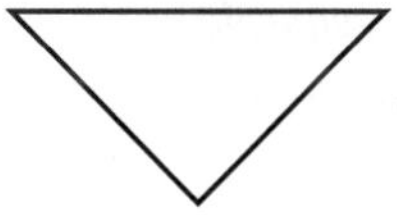

b.

Solution

a. There is no center point about which the triangle will rotate onto itself. This figure does not have rotational symmetry. (This triangle is not equilateral.)

Solution

b. The star has rotational symmetry. The angle of rotation is $72°$.

PRACTICE 5 Does the figure at the right have rotational symmetry? If so, what is the angle of rotation?

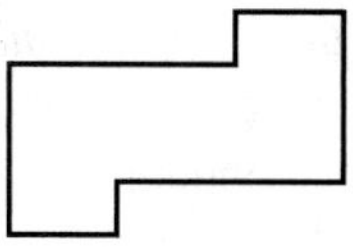

VOCABULARY & READINESS CHECK

Word Bank *Use the choices to fill in each blank. Some choices may be used more than once and some not at all.*

slide	turn	center of rotation	flip	translation	center
rotation	rotational symmetry	angle of rotation	itself	symmetry	

1. A rotation about a point R is a transformation that maps point R to ____________.
2. The point R in Exercise 1 is called a(n) ____________.
3. Another word for rotation is a(n) ____________.
4. The positive number of degrees a figure rotates is called the ____________.
5. A figure has ____________ if there is a rotation of $180°$ or less for which the figure is its own image.
6. True or false: A rotation about a point is an isometry ____________.
7. The ____________ is the point that is equidistant from the vertices of a regular polygon.

8.4 EXERCISE SET MyMathLab®

Copy each figure and point P. Draw the image of each figure for the given rotation about P. Use prime notation to label the vertices of the image. See Example 1.

1. 60°

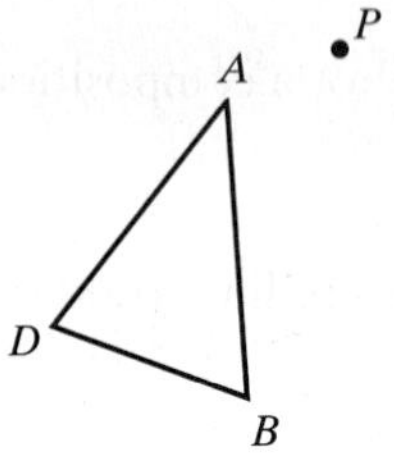

2. 70°

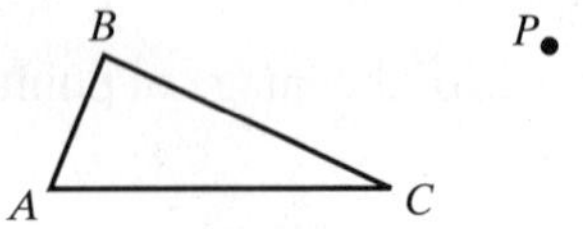

3. 180°

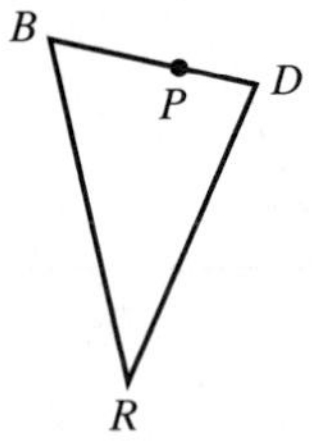

4. 90°

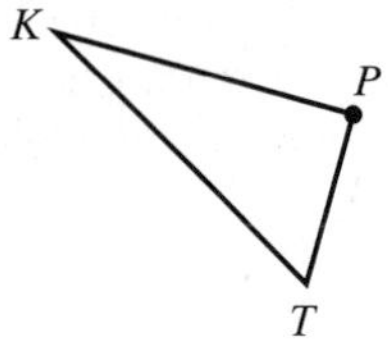

Copy each figure and point P. Then draw the image of $\overline{JK}$ for a 180° rotation about P. Use prime notation to label the vertices of the image.

5.

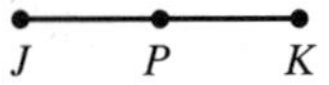

6.

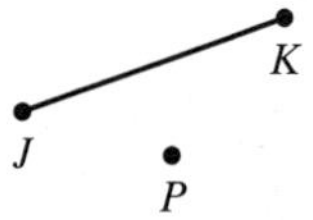

7.

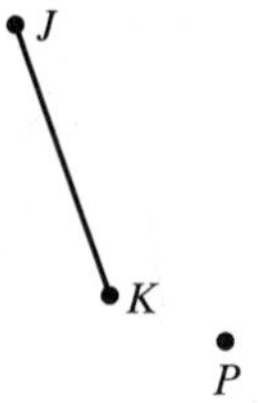

8.

In the figure at the right, point A is the center of square SQRE.

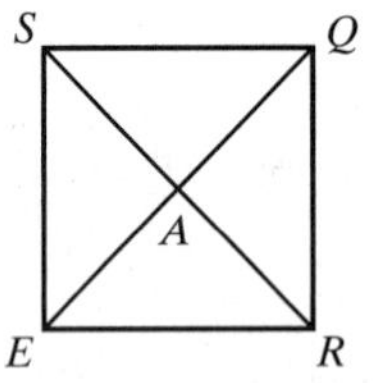

9. What is the image of *E* for a 90° rotation about point *A*?

10. What is the image of $\overline{RQ}$ for a 180° rotation about point *A*?

11. What is the image of *S* for a composition of a 30° rotation and a 240° rotation, both about point *A*?

12. What is the image of *Q* for a composition of a 30° rotation and a 330° rotation, both about point *A*?

Point O is the center of regular hexagon HEXAGN. Find the image of the given point or segment for the given rotation. See Example 2.

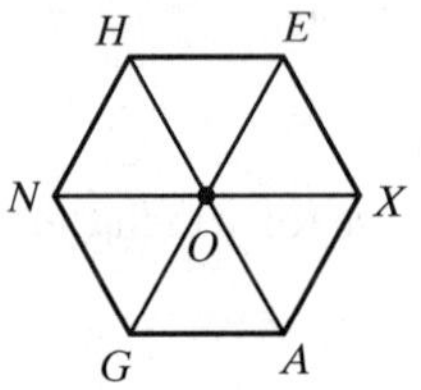

13. 60° rotation of *E* about *O*

14. 120° rotation of $\overline{AX}$ about *O*

15. 240° rotation of $\overline{NG}$ about *O*

16. 360° rotation of *H* about *O*

*Find the angle of rotation about C that **(a)** maps Q to X and **(b)** maps X to Q. Use counterclockwise rotation. See Example 3.*

17.

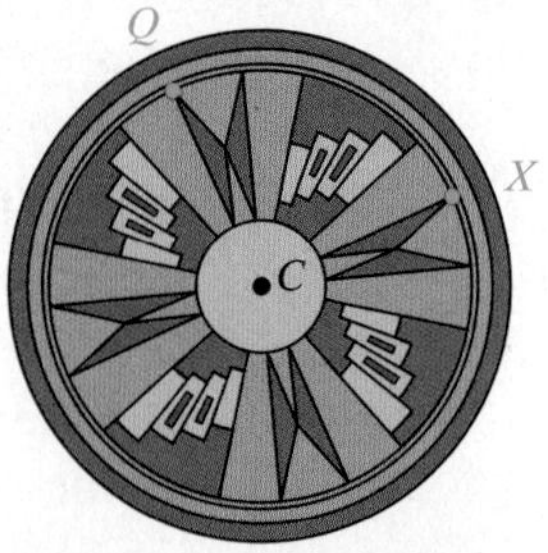

18.

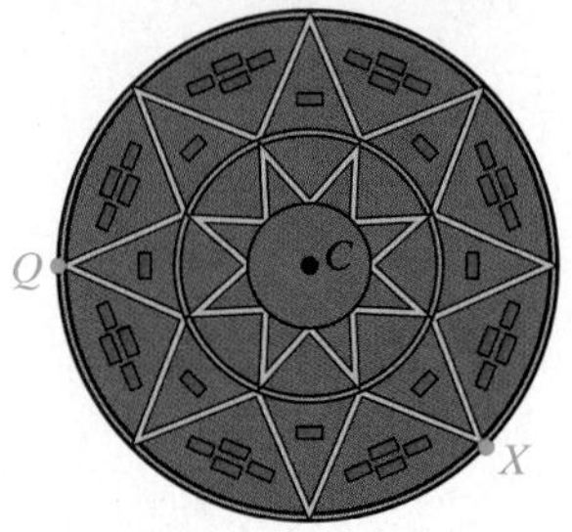

19.

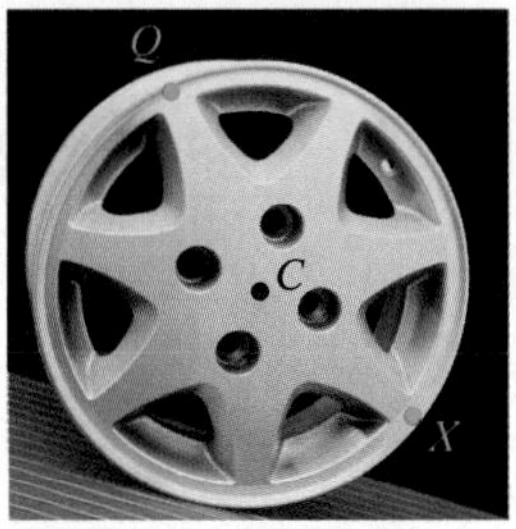

20.

For Exercises 21–26, copy ΔXYZ. Draw the image of ΔXYZ for the given composition of rotations about the given point. See Example 4.

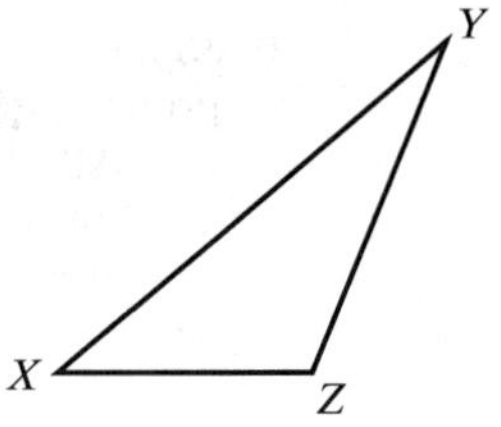

21. 45°, then 45°; *X*

22. 45°, then 45°; *Y*

23. 20°, then 160°; *Z*

24. 30°, then 30°; *Z*

25. 135°, then 135°; *Y*

26. 180°, then 180°; *X*

Study each figure. If it has rotational symmetry, tell the angle of rotation. See Example 5.

27.

28.

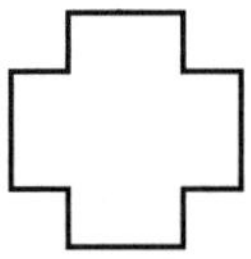

29.

30.

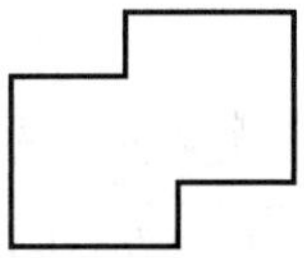

31.

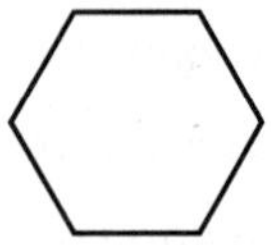

32.

33.

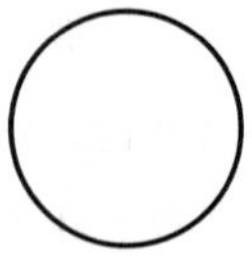

34.

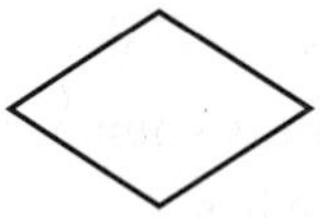

MIXED PRACTICE

Tell what type(s) of symmetry each figure has. For line symmetry, sketch the figure and the line(s) of symmetry. For rotational symmetry, tell the angle of rotation.

35.

36.

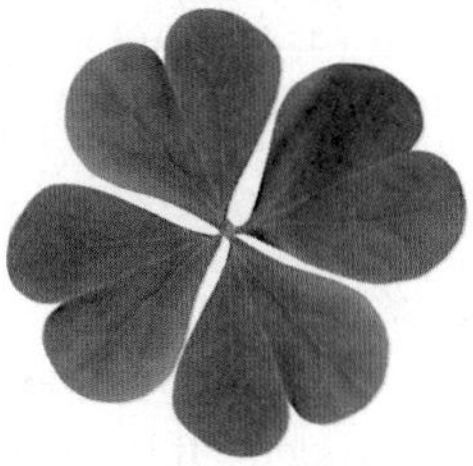

Describe the types of symmetry, if any, of each automobile logo.

37.

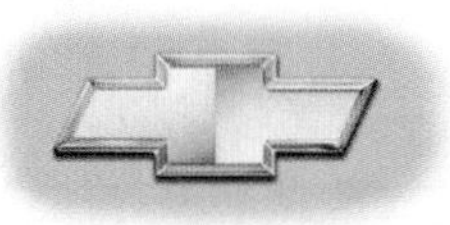

38.

39.

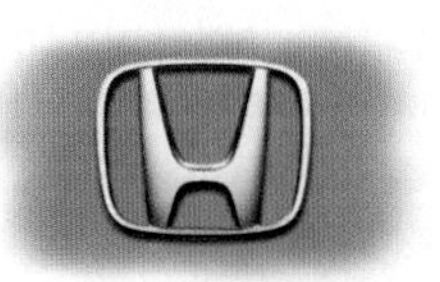

40.

41. In the diagram below, $\overline{M'N'}$ is the rotation image of $\overline{MN}$ about point E. Name all pairs of congruent angles and all pairs of congruent segments in the diagram.

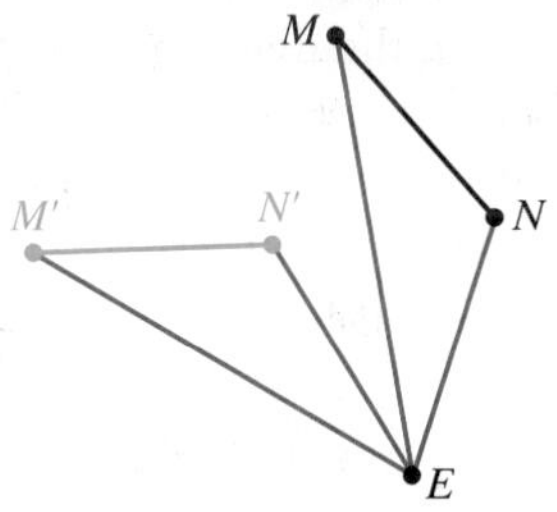

42. The symbol ə is called a *schwa*. It is used in dictionaries to represent neutral vowel sounds such as *a* in *ago*, *i* in *sanity*, and *u* in focus. What transformation maps a ə to a lowercase e?

Find the angle of rotation about C that maps the black figure to the blue figure.

43.

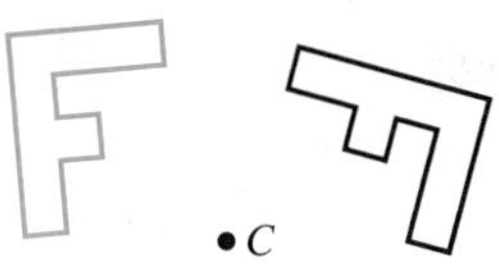

44.

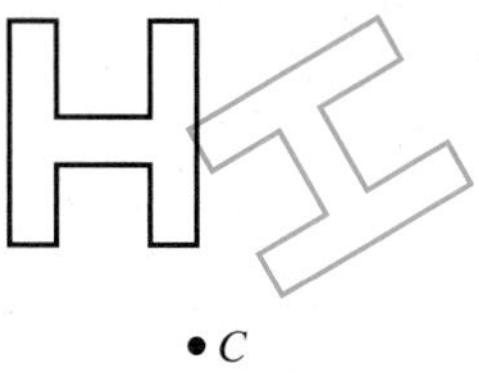

45.

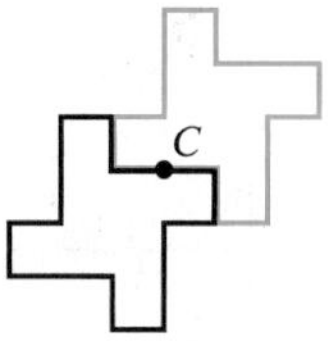

46.

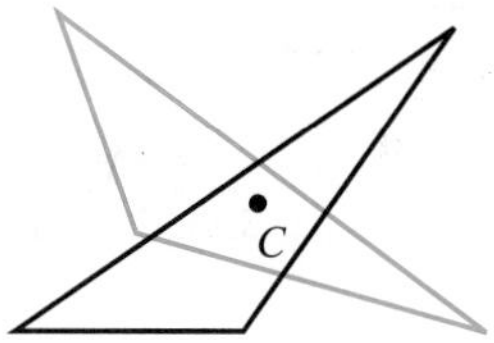

Use the figure to the right for Exercises 47 and 48.

47. Does the figure have line symmetry? If so, copy the figure and draw its line(s) of symmetry.

48. Does the figure have rotational symmetry? If so, what is the angle of rotation?

Point O is the center of the regular nonagon shown at the right.

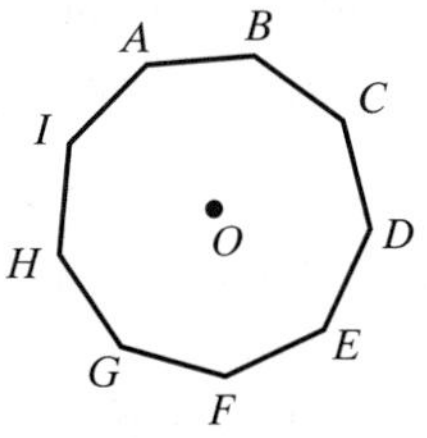

49. Find the angle of rotation that maps F to H.

50. Find the Error Your friend says that $\overline{AB}$ is the image of $\overline{ED}$ for a 120° counterclockwise rotation about O. What is wrong with your friend's statement?

51. The Millennium Wheel, also known as the London Eye, contains 32 observation cars, and 32 stopped positions as cars are emptied and filled.

a. Find the angle of rotation that a car travels when it moves one position counterclockwise.

b. How many positions does Car 3 move to be in the position of Car 18?

c. Determine the angle of rotation that will bring Car 3 to the position of Car 18.

52. Coordinate Geometry Graph $A(5, 2)$. Graph B, the image of A for a 90° rotation about the origin O. Graph C, the image of A for a 180° rotation about O. Graph D, the image of A for a 270° rotation about O. What type of quadrilateral is $ABCD$? Explain.

In the figure at the right, the large triangle, the quadrilateral, and the hexagon are regular. Find the image of each point or segment for the given rotation or composition of rotations. (Hint: Adjacent green segments form 30° angles.)

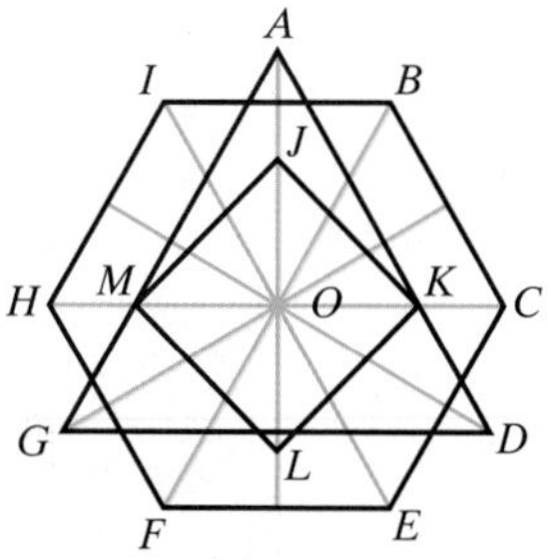

53. 120° rotation of B about O

54. 270° rotation of L about O

55. 300° rotation of $\overline{IB}$ about O

56. 60° rotation of E about O

57. 180° rotation of $\overline{JK}$ about O

58. 270° rotation of M about L

59. 120° rotation of F about H

60. 240° rotation of G about O

61. 60° rotation followed by 120° rotation of I about O

62. 45° rotation followed by 225° rotation of M about O

CONCEPT EXTENSIONS

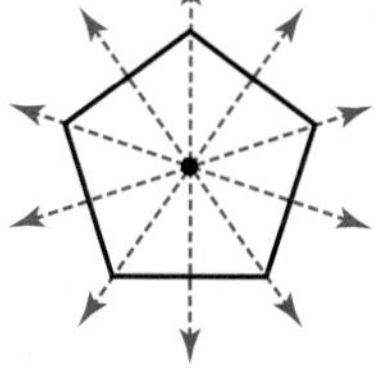

63. Find the Error Your friend thinks that the regular pentagon in the diagram at the right has 10 lines of symmetry. Explain and correct your friend's error.

64. The word CHECKBOOK has a horizontal line of symmetry. Find two other words for which this is true.

65. For center of rotation P, does an $x°$ rotation followed by a $y°$ rotation give the same image as a $y°$ rotation followed by an $x°$ rotation? Explain.

66. Describe compositions of rotations that have the same effect as a 360° rotation about a point X.

67. **Coordinate Geometry** Draw ΔLMN with vertices $L(2, -1)$, $M(6, -2)$, and $N(4, 2)$. Find the coordinates of the vertices after a 90° rotation about the origin and about each of the points L, M, and N.

68. **Reasoning** If you are given a figure and a rotation image of the figure, how can you find the center and angle of rotation?

REVIEW AND PREVIEW

69. Which capital letters of the alphabet are rotation images of themselves? Draw each letter and give an angle of rotation ($< 360°$). See Section 8.4.

70. Three vertices of an isosceles trapezoid are $(-2, 1)$, $(1, 4)$, and $(4, 4)$. Find all possible coordinates for the fourth vertex. See Section 6.5.

Determine the scale drawing dimensions of each room using a scale of $\frac{1}{4}$ *in.* $=$ *1 ft. See Sections 7.1 and 7.2.*

71. kitchen: 12 ft by 16 ft

72. laundry room: 6 ft by 9 ft

73. bedroom: 8 ft by 10 ft

74. bedroom: 11 ft by 12 ft

ACTIVITIES (USE WITH SECTION 8.4)

Tracing Paper Transformations

In Section 8.2, we learned how to describe a translation using variables. In these activities, we will use tracing paper to perform translations, rotations, and reflections. We will also describe certain rotations and reflections using variables.

Activity 1

We can use the arrow shown in the diagram to represent the translation $(x, y) \rightarrow (x + 4, y + 2)$. The translation shifts ΔABC with $A(-3, 3)$, $B(-1, 1)$, and $C(1, 4)$ to $\Delta A'B'C'$ with $A'(1, 5)$, $B'(3, 3)$, and $C'(5, 6)$. We can see this translation using tracing paper as follows:

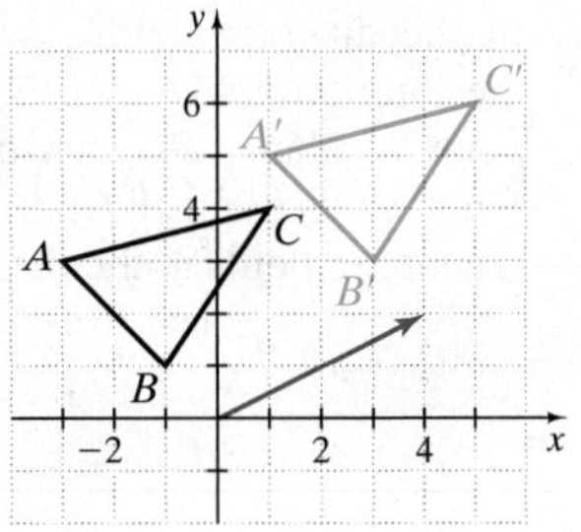

STEP 1. Draw ΔABC and the arrow on graph paper. Also, show the line containing the arrow.

STEP 2. Trace ΔABC and the arrow.

STEP 3. Move your tracing of the arrow along the line until the tail of the tracing is on the head of the original arrow. The vertices of your tracing of ΔABC should now be at $A'(1, 5)$, $B'(3, 3)$, and $C'(5, 6)$.

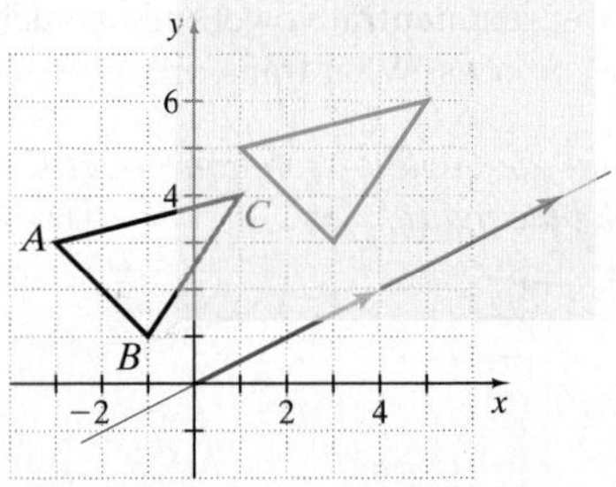

Use tracing paper. Find the translation image of each triangle for the given arrow.

1.
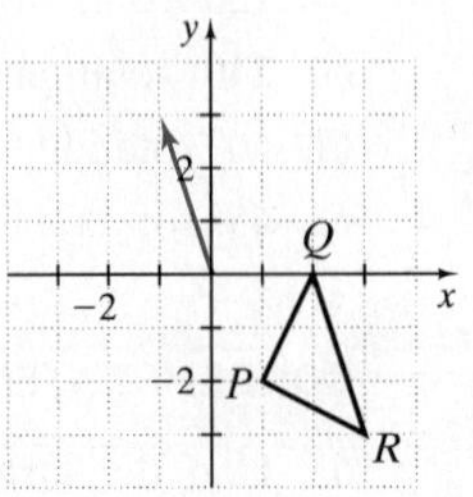

2.
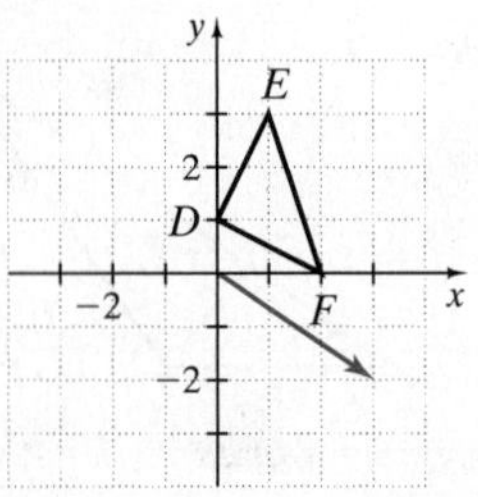

3.
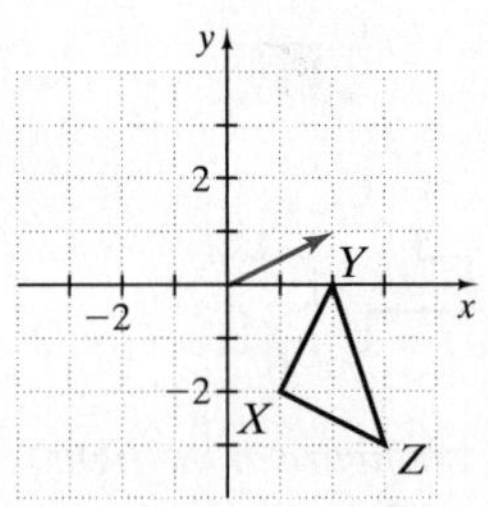

4. Show that the composition of the translation in Question 1 followed by the translation in Question 2 gives you the translation in Question 3.

Activity 2

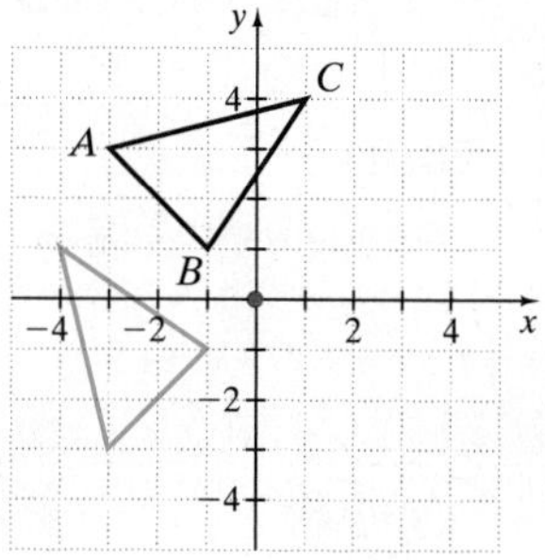

To rotate a figure 90° about the origin, trace the figure, one axis, and the origin. Then turn your tracing paper counterclockwise, keeping the origin in place and aligning the traced axis with the other original axis.

5. Use ΔABC from Activity 1 with $A(-3, 3)$, $B(-1, 1)$, and $C(1, 4)$. What is the image of ΔABC for a 90° rotation about the origin?

6. Copy and complete the table. Use tracing paper to find point P', the image of point P, for a rotation of 90° about the origin.

P	P'
$(3, 4)$	______
$(-3, 4)$	______
$(-3, -4)$	______
$(3, -4)$	______
$(3, 0)$	______
$(0, 4)$	______

7. Study the pattern in your table. Complete this rule for a rotation of 90° about the origin: $(x, y) \rightarrow (\_, \_)$.

8. Test your rule with tracing paper. For ΔTRN with $T(0, 2)$, $R(3, 0)$, and $N(4, 5)$, a 90° rotation about the origin should result in $\Delta T'R'N'$ with $T'(-2, 0)$, $R'(0, 3)$, and $N'(-5, 4)$. Does it?

9. In parts **a** and **b** below, what should be the result of each composition?

a. a 90° rotation about the origin followed by a 90° rotation about the origin

b. $(x, y) \rightarrow (-y, x)$ followed by $(x, y) \rightarrow (-y, x)$

c. Use tracing paper. Test your conjectures from parts **a** and **b** on ΔABC from Activity 1.

Activity 3

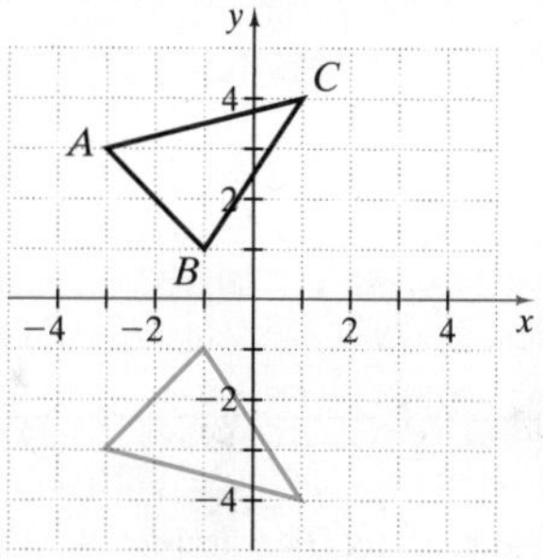

To reflect a figure across an axis using tracing paper, trace the figure, the axis, and the origin. Then turn over your tracing paper, keeping the origin in place and aligning the traced axis with the original axis.

10. What is the reflection image of ΔABC across the x-axis? Across the y-axis?

11. Copy and complete the table. Use tracing paper to find point P_x and P_y, the reflection images of point P across the x-axis and y-axis, respectively.

P	P_x	P_y
$(3, 4)$	______	______
$(-3, 4)$	______	______
$(-3, -4)$	______	______
$(3, -4)$	______	______
$(3, 0)$	______	______
$(0, 4)$	______	______

12. Study the table. Complete these rules for reflections across the axes.

a. x-axis: $(x, y) \rightarrow (\_, \_)$

b. y-axis: $(x, y) \rightarrow (\_, \_)$

13. Test your rules on ΔKLM with $K(-2, 3)$, $L(3, 1)$, and $M(1, -2)$.

a. Across the x-axis, ΔKLM should map to _?_. Does it?

b. Across the y-axis, ΔKLM should map to _?_. Does it?

14. In parts **a** and **b** below, what should be the result of each composition?

a. $(x, y) \rightarrow (x, -y)$ followed by $(x, y) \rightarrow (-x, y)$

b. a reflection across the x-axis followed by a reflection across the y-axis

c. Use tracing paper. Test your conjectures from parts **a** and **b** on ΔABC from Activity 1.

15. Compare the results of Questions 9 and 14. Make a conjecture about the compositions suggested by each.

16. Use tracing paper to find a rule for a reflection across the line $y = x$. Test your rule on ΔABC from Activity 1.

8.5 DILATIONS

OBJECTIVE

1 Understand Dilation Images of Figures.

VOCABULARY

- dilation
- center of dilation
- scale factor of a dilation
- enlargement
- reduction

The pupil is the opening in the iris that lets light into the eye. Depending on the amount of light available, the size of the pupil changes.

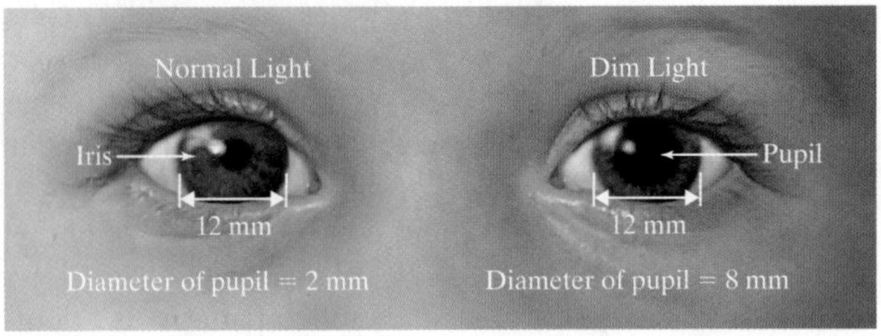

Suppose our definition of similar polygons we learned in Chapter 7 is extended to apply to curved figures. Now think about the following: Is the pupil in dim light similar to the pupil in normal light? Is the iris in dim light similar to the iris in normal light?

OBJECTIVE 1 ▶ Understanding Dilation Images of Figures. Above, we looked at how the pupil of an eye changes in size, or *dilates*. In this section, we learn how to dilate geometric figures.

We can use a scale factor to make a larger or smaller copy of a figure that is also similar to the original figure.

Dilation

A **dilation** with **center** C and **scale factor** n, $n > 0, n \neq 1$, is a transformation with these two properties:

- The image of C is itself (that is, $C' = C$).
- For any other point R, R' is on $\overrightarrow{CR}$ and $CR' = n \cdot CR$, or $n = \frac{CR'}{CR}$.

The image of a dilation is similar to its preimage.

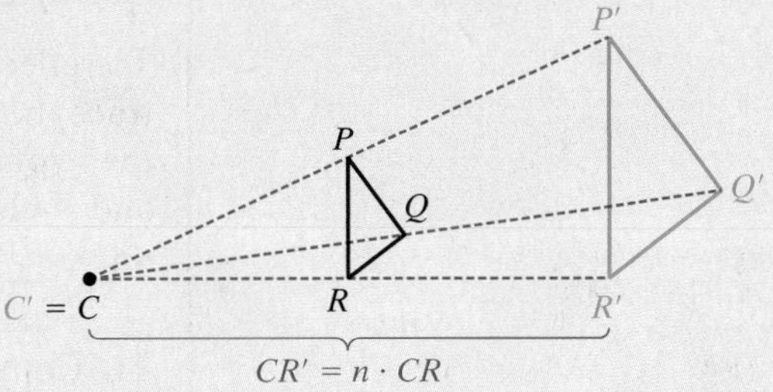

Helpful Hint
Take note: The image of a dilation is similar to its preimage.

We can show that the scale factor of a dilation has the same value as the scale factor of the similar figures (preimage and image), with the image length in the numerator. For the figure shown above, $n = \frac{CR'}{CR} = \frac{R'P'}{RP}$.

A dilation is an **enlargement** if the scale factor is greater than 1. The dilation is a **reduction** if the scale factor is between 0 and 1.

Helpful Hint
Enlargement: Scale factor > 1
Reduction: Scale factor between 0 and 1

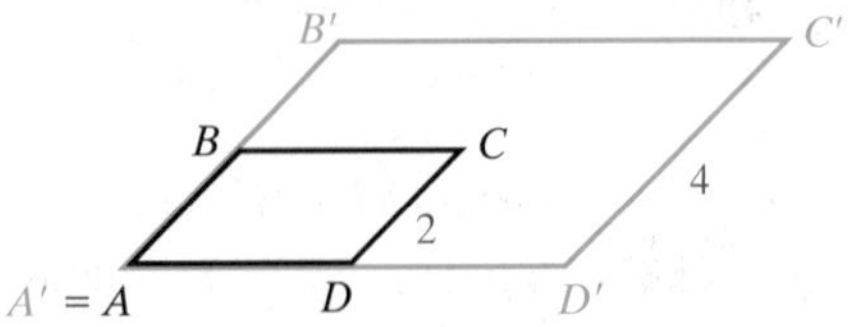

Enlargement center A, scale factor 2

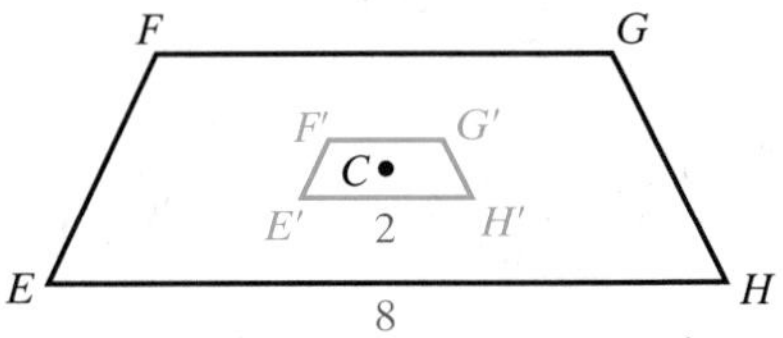

Reduction center C, scale factor $\frac{1}{4}$

Helpful Hint
When finding the scale factor, the numerator always has the image measurement.

EXAMPLE 1 **Finding a Scale Factor**

Multiple Choice $\Delta X'T'R'$ is a dilation image of ΔXTR. The center of dilation is X. Is the dilation an enlargement or a reduction? What is the scale factor of the dilation?

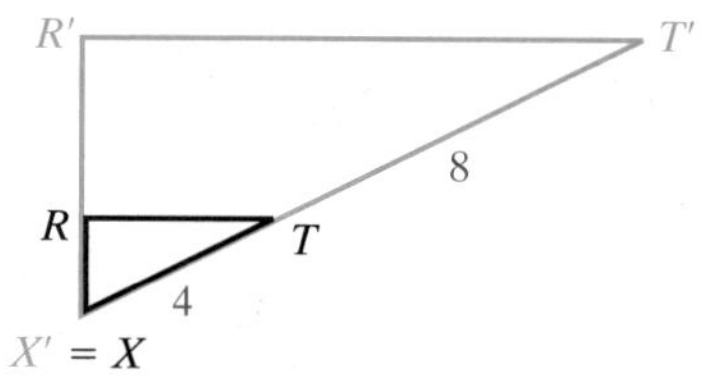

a. enlargement; scale factor 2

b. enlargement; scale factor 3

c. reduction; scale factor $\frac{1}{3}$

d. reduction; scale factor 3

Solution The image is larger than the preimage, so the dilation is an enlargement. Use the ratio of the lengths of corresponding sides to find the scale factor.

$$n = \frac{X'T'}{XT} = \frac{4+8}{4} = \frac{12}{4} = 3$$

$\Delta X'T'R'$ is an enlargement of ΔXTR, with a scale factor of 3. The correct answer is **b.**

PRACTICE

1 $J'K'L'M'$ is a dilation image of $JKLM$. The center of dilation is O. Is the dilation an enlargement or a reduction? What is the scale factor of the dilation?

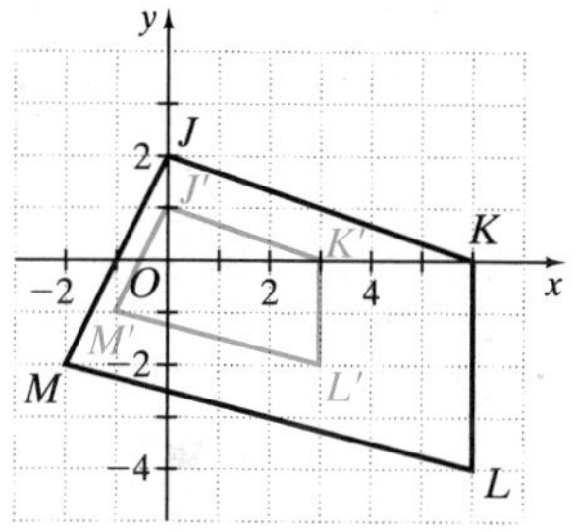

In Practice 1 above, we looked at a dilation of a figure drawn in the coordinate plane. In this text, all dilations of figures in the coordinate plane have the origin as the center of dilation. So we can find the dilation image of a point $P(x, y)$ by multiplying the coordinates of P by the scale factor n.

$$P(x, y) \rightarrow P'(nx, ny)$$

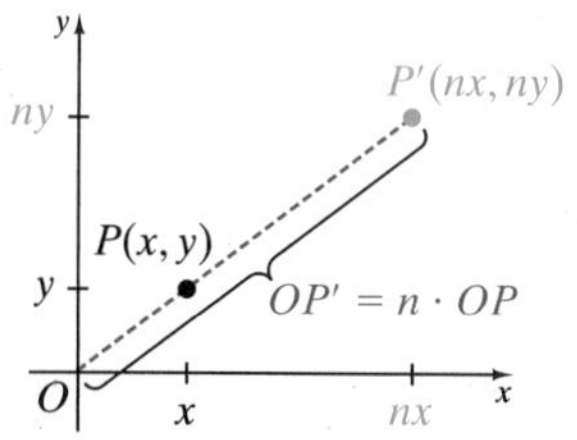

EXAMPLE 2 Finding a Dilation Image

What are the images of the vertices of ΔPZG for a dilation with center $(0, 0)$ and scale factor 2? Graph the image of ΔPZG.

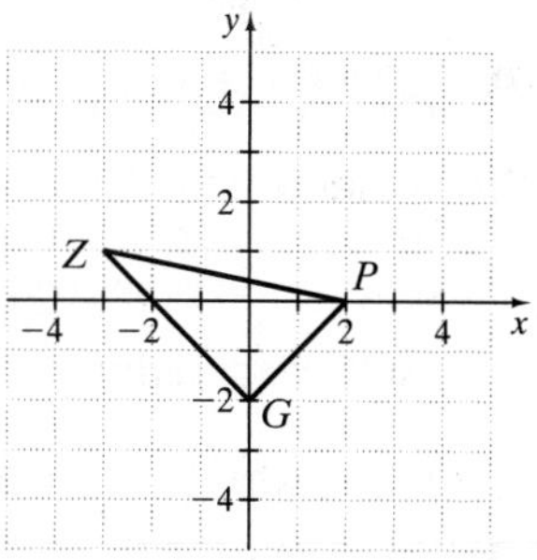

Solution First, let's identify the coordinates of each vertex. They are listed below. The center of dilation is the origin and the scale factor is 2, so use the dilation rule $(x, y) \rightarrow (2x, 2y)$.

$P(2, 0) \rightarrow (2 \cdot 2, 2 \cdot 0)$, or $P'(4, 0)$.

$Z(-3, 1) \rightarrow (2 \cdot (-3), 2 \cdot 1)$, or $Z'(-6, 2)$.

$G(0, -2) \rightarrow (2 \cdot 0, 2 \cdot (-2))$, or $G'(0, -4)$.

To graph the image of ΔPZG, graph P', Z', and G'. Then draw $\Delta P'Z'G'$, as shown.

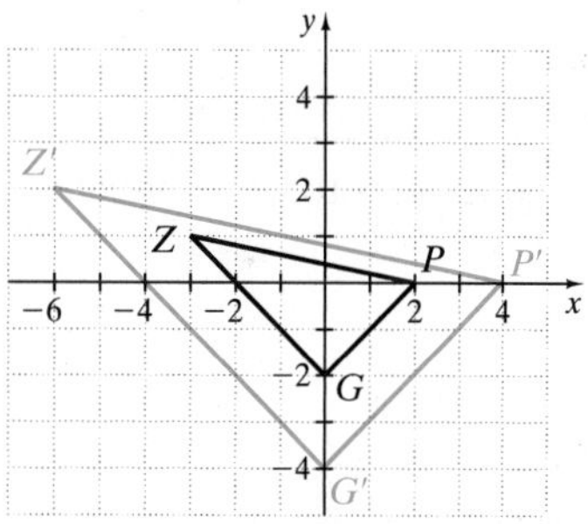

PRACTICE 2 Use the figures in Example 2.

a. What are the images of the vertices of ΔPZG for a dilation with center $(0, 0)$ and scale factor $\frac{1}{2}$?

b. In Example 2, how can you prove $\Delta P'Z'G' \sim \Delta PZG$ using a postulate or theorem from Chapter 7?

Dilations and scale factors help us understand real-world enlargements and reductions, such as images seen through a microscope or on a computer screen.

EXAMPLE 3 **Using a Scale Factor to Find a Length**

A magnifying glass shows you an image of an object that is 7 times the object's actual size. So the scale factor of the enlargement is 7. The photo shows an apple seed under this magnifying glass. What is the actual length of the apple seed?

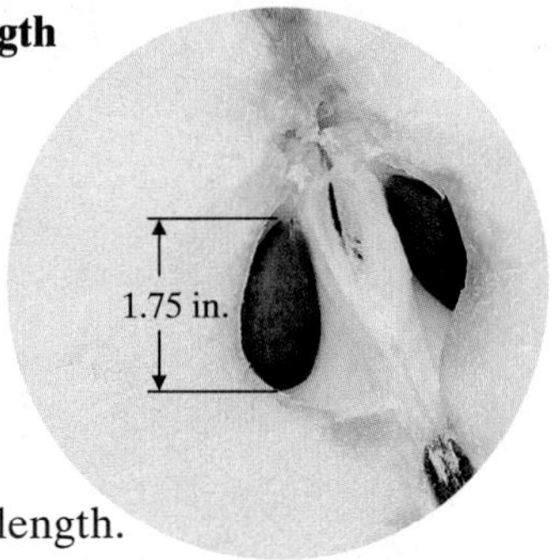

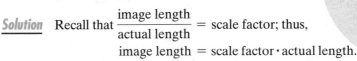

Solution Recall that $\frac{\text{image length}}{\text{actual length}} = \text{scale factor}$; thus,

$$\text{image length} = \text{scale factor} \cdot \text{actual length}.$$

Let's let p be the actual length and notice that 1.75 in. is the image length.

$$1.75 = 7 \cdot p \quad \text{image length} = \text{scale factor} \cdot \text{actual length}$$

$$0.25 = p \quad \text{Divide each side by 7.}$$

The actual length of the apple seed is 0.25 in.

PRACTICE 3 The height of a document on your computer screen is 20.4 cm. When you change the zoom setting on your screen from 100% to 25%, the new image of your document is a dilation of the previous image with scale factor 0.25. What is the height of the new image?

VOCABULARY & READINESS CHECK

Word Bank *Use the choices to fill in each blank. Some choices may be used more than once and some not at all.*

slide	isometry	enlargement	flip	translation	rotation
itself	reduction	symmetry	$\overleftrightarrow{CR}$	$\angle R$	

A dilation with center C and scale factor n is a transformation with these two properties:

1. The image of C is ______________.

2. For any other point R, the image of R is on ______________.

Fill in the Blank:

3. A dilation is a(n) ______________ if the scale factor is greater than 1.

4. A dilation is a(n) ______________ if the scale factor is between 0 and 1.

True or false:

5. A dilation is an isometry. ______________

6. The image of a dilation is congruent to its preimage. ______________

8.5 EXERCISE SET MyMathLab®

The blue figure is a dilation image of the black figure. The labeled point is the center of dilation. Tell whether the dilation is an enlargement or a reduction. Then find the scale factor of the dilation. See Example 1.

1.
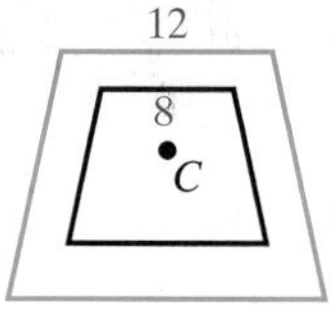

2.
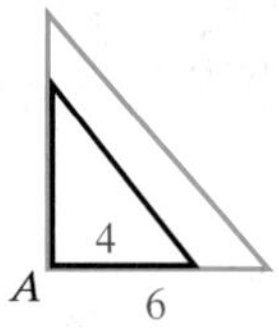

3.
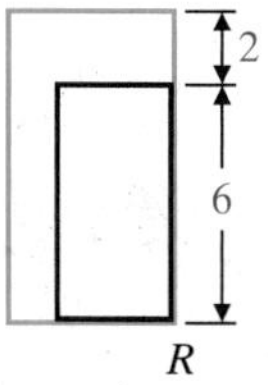

4.
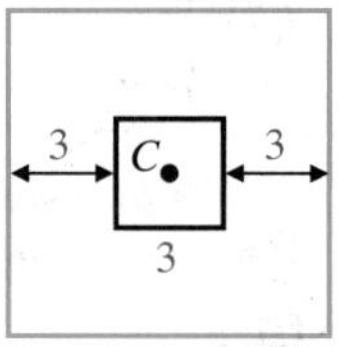

5.
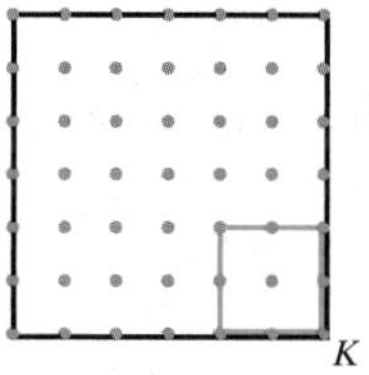

6.
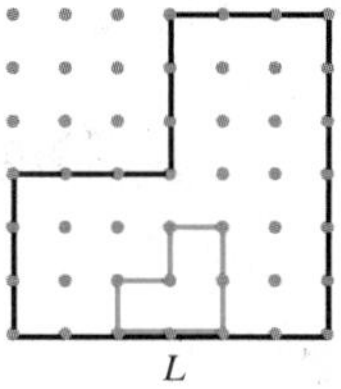

7.
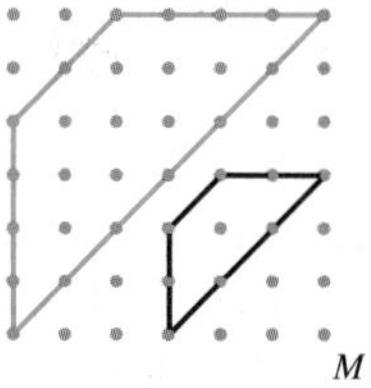

8.
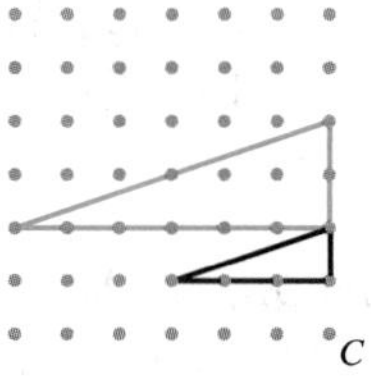

9.
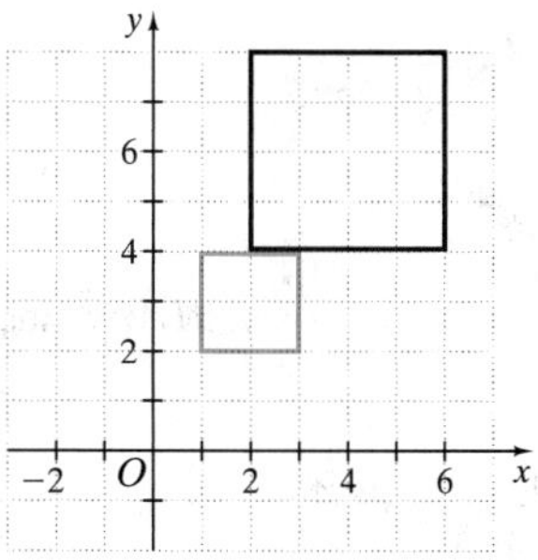

10.
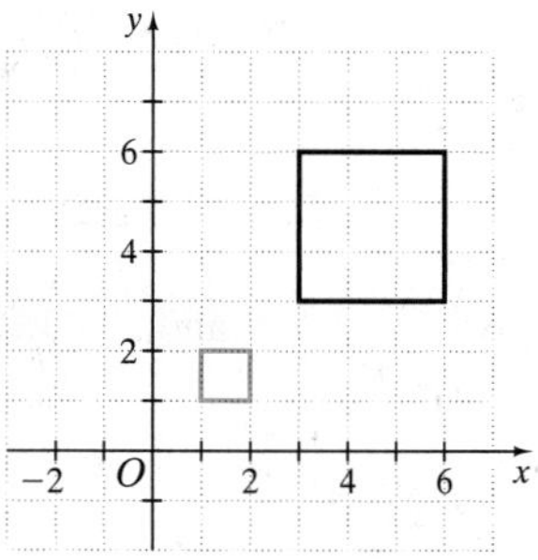

11.
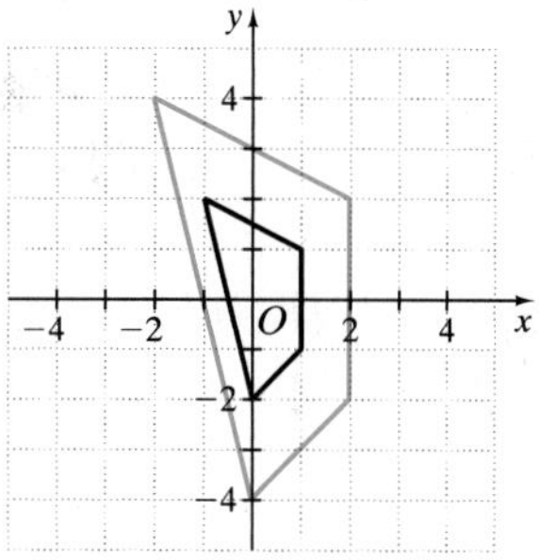

12.
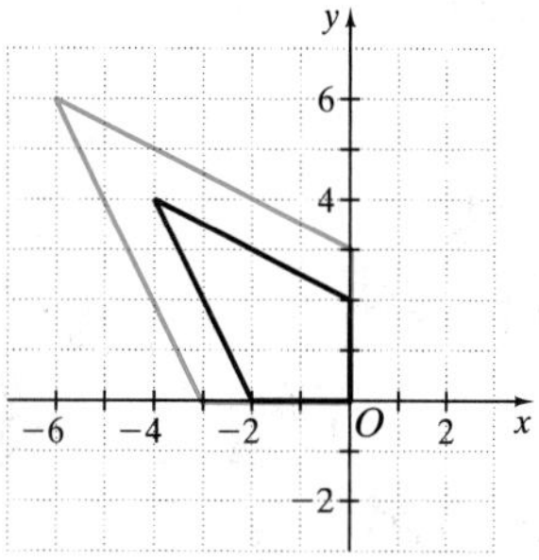

A dilation has center $(0, 0)$. *Find the image of each point for the given scale factor.*

13. $D(1, -5); 2$

14. $M(0, 0); 10$

15. $T(0, 6); \frac{1}{3}$

16. $D(8, -4); \frac{1}{2}$

Find the images of the vertices of ΔPQR *for a dilation with center* $(0, 0)$ *and the given scale factor. Graph the image. See Example 2.*

17. scale factor 10

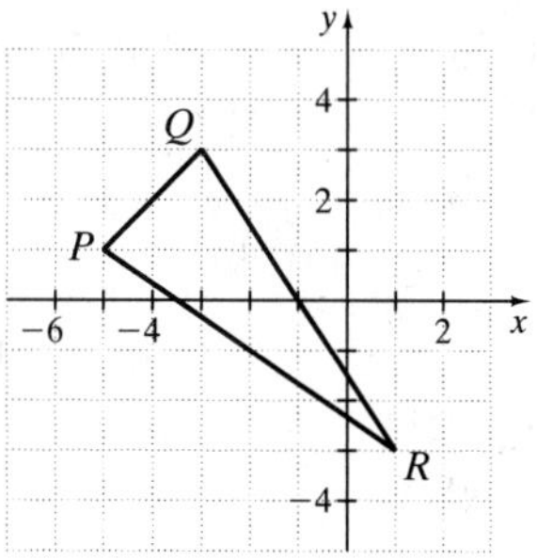

18. scale factor 3

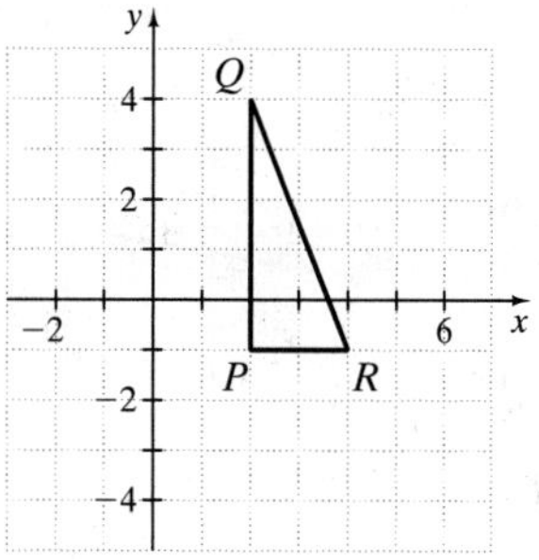

19. scale factor $\frac{3}{4}$

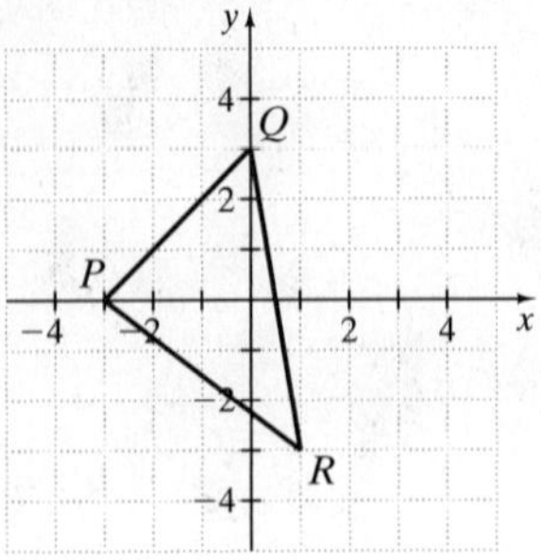

20. scale factor $\frac{2}{3}$

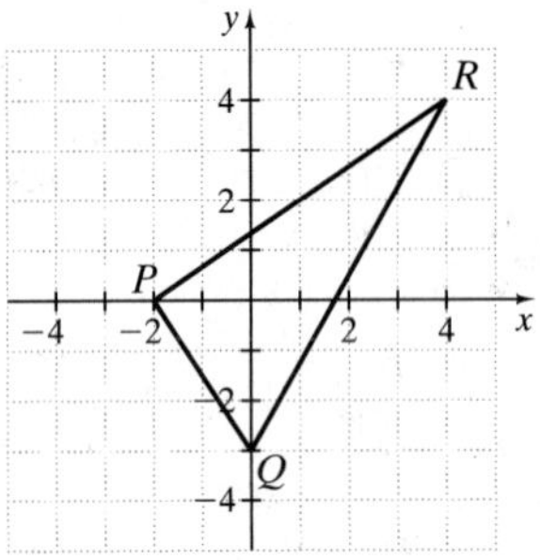

You look at each object described in Exercises 21–24 under a magnifying glass. Find the actual dimension of each object. See Example 3.

21. The image of a button is 5 times the button's actual size and has a diameter of 6 cm.

22. The image of a pinhead is 8 times the pinhead's actual size and has a width of 1.36 cm.

23. The image of an ant is 7 times the ant's actual size and has a length of 1.4 cm.

24. The image of a capital letter N is 6 times the letter's actual size and has a height of 1.68 cm.

A dilation has center (0, 0). Find the image of each point for the given scale factor.

25. $L(-3, 0); 5$

26. $N(-4, 7); 0.2$

27. $A(-6, 2); 1.5$

28. $F(3, -2); \frac{1}{3}$

29. $B\left(\frac{5}{4}, -\frac{3}{2}\right); \frac{1}{10}$

30. $Q\left(6, \frac{\sqrt{3}}{2}\right); \sqrt{6}$

Use the graph below. Find the vertices of the image of QRTW for a dilation with center (0, 0) and the given scale factor.

31. $\frac{1}{4}$

32. $\frac{1}{3}$

33. 0.6

34. 0.9

35. 10

36. 100

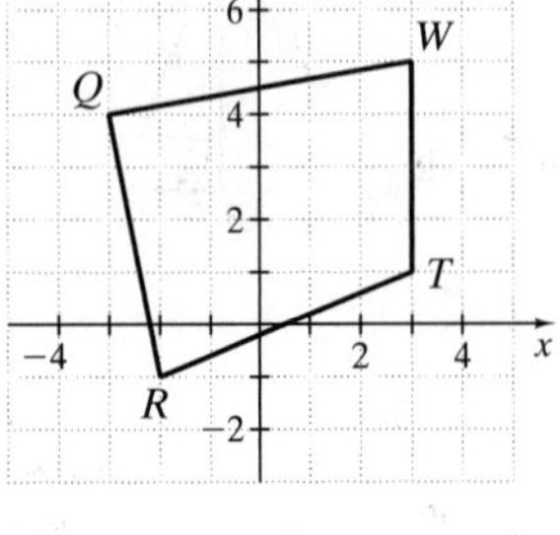

Coordinate Geometry *Graph MNPQ and its image M′N′P′Q′ for a dilation with center (0, 0) and the given scale factor.*

37. $M(1, 3), N(-3, 3), P(-5, -3), Q(-1, -3); 3$

38. $M(2, 6), N(-4, 10), P(-4, -8), Q(-2, -12); \frac{1}{4}$

A dilation maps ΔHIJ onto $\Delta H'I'J'$. Find the missing values.

39. $HI = 8$ in. $\quad H'I' = 16$ in.
$IJ = 5$ in. $\quad I'J' = \_$ in.
$HJ = 6$ in. $\quad H'J' = \_$ in.

40. $HI = \_$ ft $\quad H'I' = 8$ ft
$IJ = 30$ ft $\quad I'J' = \_$ ft
$HJ = 24$ ft $\quad H'J' = 6$ ft

Copy ΔTBA and point O for each of Exercises 41–44. Draw the dilation image $\Delta T'B'A'$ for the given center and scale factor.

41. center O, scale factor 2

42. center B, scale factor 3

43. center T, scale factor $\frac{1}{3}$

44. center O, scale factor $\frac{1}{2}$

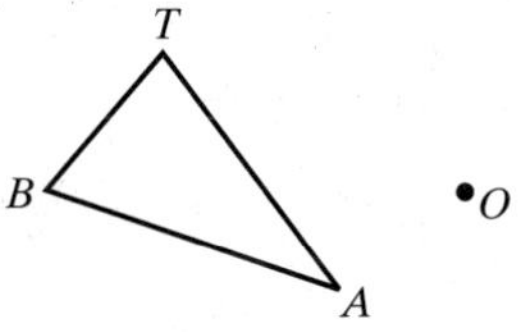

CONCEPT EXTENSIONS

Find the Error *The blue figure is a dilation image of the black figure for a dilation with center A.*

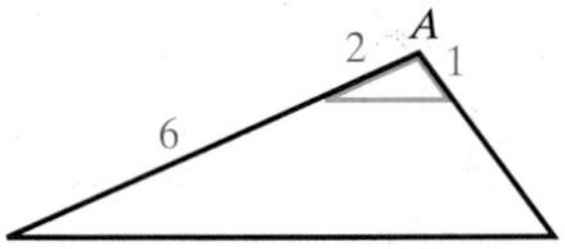

Two students made errors when asked to find the scale factor. Explain and correct their errors.

45.

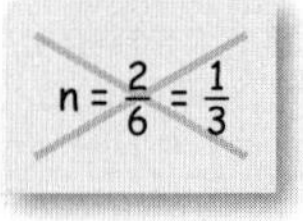

46.

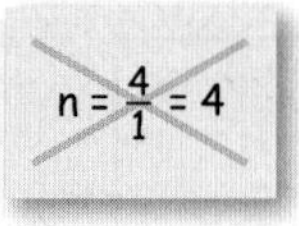

47. **Multiple Steps** The diagram below shows ΔLMN and its image $\Delta L'M'N'$ for a dilation with center P.

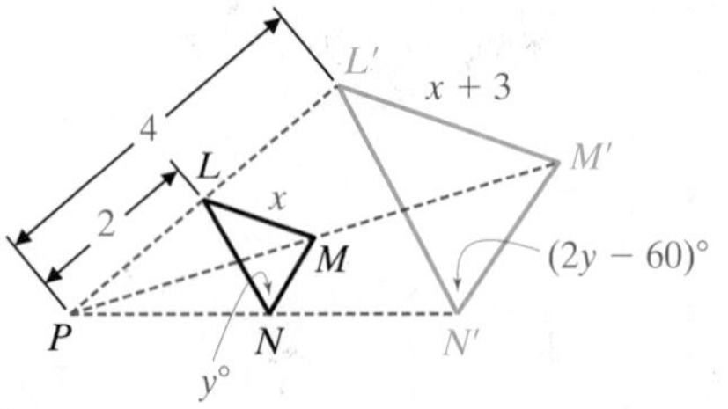

a. What is the relationship between ΔLMN and $\Delta L'M'N'$?

b. What is the scale factor of the dilation?

c. Find the values of x and y.

48. A flashlight projects an image of rectangle $ABCD$ on a wall so that each vertex of $ABCD$ is 3 ft away from the corresponding vertex of $A'B'C'D'$. The length of $\overline{AB}$ is 3 in. The length of $\overline{A'B'}$ is 1 ft. How far from each vertex of $ABCD$ is the light?

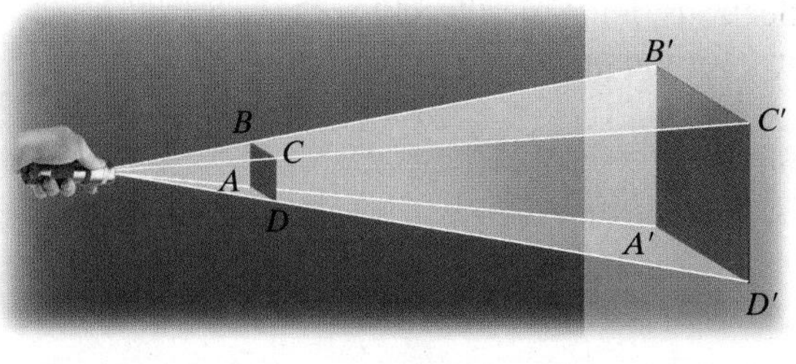

49. Compare the definition of scale factor of a dilation to the definition of scale factor of two similar polygons. How are they alike? How are they different?

50. An equilateral triangle has 4-in. sides. Describe its image for a dilation with center at one of the triangle's vertices and scale factor 2.5.

51. Your picture of your family crest is 4.5 in. wide. You need a reduced copy for the front page of the family newsletter. The copy must fit in a space 1.8 in. wide. What scale factor should you use on the copy machine to adjust the size of your picture of the crest?

52. You are given $\overline{AB}$ and its dilation image $\overline{A'B'}$ with A, B, A', and B' noncollinear. Explain how to find the center of dilation and scale factor.

Reasoning *Write "true" or "false" for Exercises 53–56. Explain your answers.*

53. A dilation is an isometry.

54. A dilation with a scale factor greater than 1 is a reduction.

55. For a dilation, corresponding angles of the image and preimage are congruent.

56. A dilation image cannot have any points in common with its preimage.

Coordinate Geometry *In the coordinate plane, you can extend dilations to include scale factors that are negative numbers. For Exercises 57 and 58, use ΔPQR with vertices $P(1, 2)$, $Q(3, 4)$, and $R(4, 1)$.*

57. Graph ΔPQR and its image for a dilation centered at $(0, 0)$ with scale factor -3.

58. Graph ΔPQR and its image for a dilation centered at $(0, 0)$ with scale factor -1.

REVIEW AND PREVIEW

Coordinate Geometry *A figure with a vertex at $(-2, 7)$ has the given type of symmetry. State the coordinates of another vertex of the figure. See Section 8.3.*

59. line symmetry across the y-axis

60. line symmetry across the x-axis

Given points $R(-1, 1)$, $S(-4, 3)$, and $T(-2, 5)$, graph ΔRST and its reflection image across each line. See Section 8.3.

61. the y-axis

62. the x-axis

63. $y = 1$

64. $x = 1$

8.6 COMPOSITIONS OF REFLECTIONS

OBJECTIVES

1 Find Compositions of Reflections, Including Glide Reflections.

2 Classify Isometries.

VOCABULARY

- glide reflection

OBJECTIVE 1 ▶ Finding Compositions of Reflections. In this section, we will see that any isometry can be expressed as a composition of reflections.

If two figures in a plane are congruent, we can map one onto the other using a composition of reflections.

The theorems in this section lead to the fact stated above. Complete proofs of these theorems are beyond the scope of this course, but Examples 1 and 2 suggest approaches to the proofs of Theorems 8.6-1 and 8.6-2. Theorem 8.6-2 is the converse of Theorem 8.6-1.

Theorem 8.6-1

A translation or rotation is a composition of two reflections.

Theorem 8.6-2

A composition of reflections across two parallel lines is a translation.

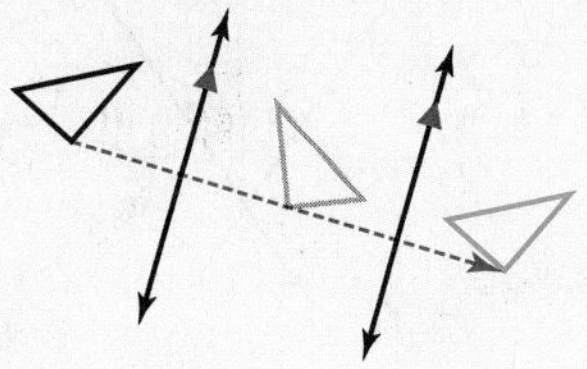

A composition of reflections across two intersecting lines is a rotation.

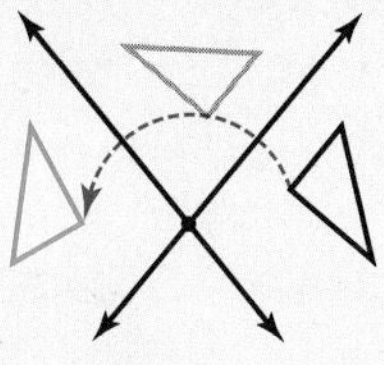

> **Helpful Hint**
> Recall what we know about reflections across a line (Section 8.3).
> - If P is on ℓ, then $P = P'$.
> - If P is not on ℓ, then ℓ is the perpendicular bisector of $\overline{PP'}$.

EXAMPLE 1 Composing Reflections Across Parallel Lines

What is the image of R reflected first across line ℓ and then across line m? What are the direction and distance of the resulting translation?

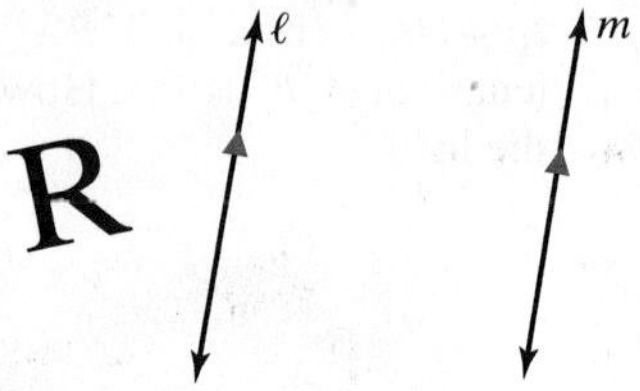

R
ℓ
m

Solution As we do the two reflections, let's keep track of the distance moved by a point P of the preimage.

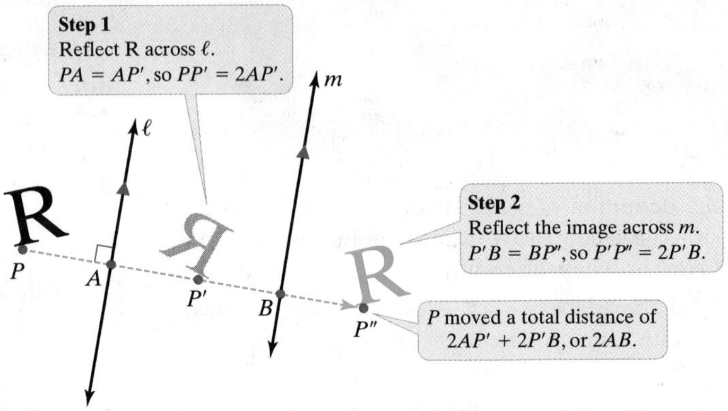

Step 1
Reflect R across ℓ.
PA = AP', so PP' = 2AP'.
Step 2
Reflect the image across m.
P'B = BP'', so P'P'' = 2P'B.
P moved a total distance of
2AP' + 2P'B, or 2AB.
ℓ
m
P
A
P'
B
P''

The blue arrow shows the direction of the translation, so the direction of the translation is determined by the line through P perpendicular to ℓ and m. Points A and B are the intersection of the perpendicular line with lines l and m respectively. The total distance P moved is $2 \cdot AB$. Because $\overleftrightarrow{AB} \perp \ell$, AB is the distance between ℓ and m. The distance of the translation is twice the distance between parallel lines ℓ and m. □

PRACTICE 1

a. Draw parallel lines ℓ and m as in Example 1. Draw capital letter R between ℓ and m. What is the image of R reflected first across line ℓ and then across line m? What are the direction and distance of the resulting translation?

b. Use the results of part **a** and Example 1. Make a conjecture about the direction and distance of any translation that is the result of a composition of reflections across two parallel lines.

EXAMPLE 2 Composing Reflections Across Intersecting Lines

Lines ℓ and m intersect at point C and form an acute angle that measures 70°. What is the image of R reflected first across line ℓ and then across line m? What are the center of rotation and the angle of rotation for the resulting rotation?

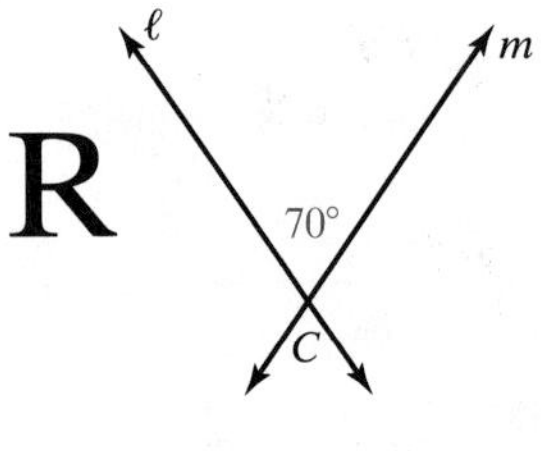

R
ℓ
m
70°
C

Solution After we do the two reflections, let's follow the path of a point P of the preimage.

> **Helpful Hint**
> How do we know that $m\angle 1 = m\angle 2$? By drawing $\overline{PP'}$; then $\Delta APC \cong \Delta AP'C$ by SSS.

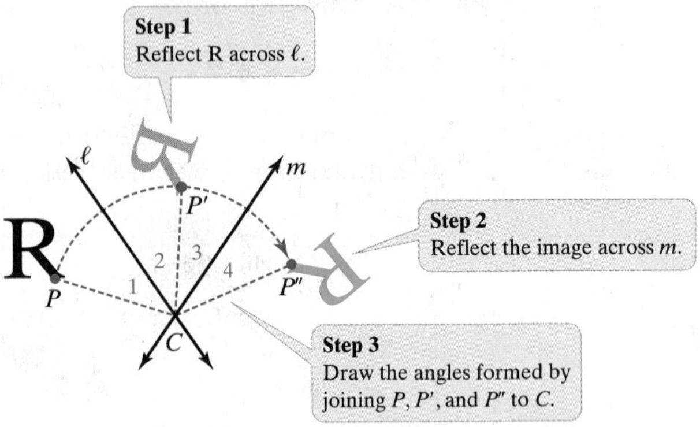

Step 1
Reflect R across ℓ.
Step 2
Reflect the image across m.
Step 3
Draw the angles formed by
joining P, P', and P'' to C.
ℓ
m
P
P'
P''
C
1
2
3
4

R is rotated clockwise about the intersection point of the lines. The center of rotation is C. We know that $m\angle 2 + m\angle 3 = 70°$. We can use the definition of reflection to show that $m\angle 1 = m\angle 2$ and $m\angle 3 = m\angle 4$. So, $m\angle 1 + m\angle 2 + m\angle 3 + m\angle 4 = 140°$. The angle of rotation is 140° clockwise. □

PRACTICE 2

a. Use the diagram at the right. What is the image of R reflected first across line a and then across line b? What are the center of rotation and the angle of rotation for the resulting rotation?

b. Use the results of part **a** and Example 2. Make a conjecture about the center of rotation and the angle of rotation for any rotation that is the result of a composition of reflections across two intersecting lines.

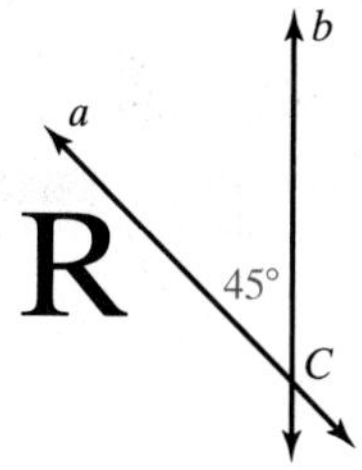

Helpful Hint

Make sure you are aware of the orientations of a preimage and image thus far:

Transformation	*Orientation*
Translation	same
Reflection	opposite
Rotation	same

In a plane, any two congruent figures with the same orientation are related by either a translation or a rotation, and therefore by two reflections.

Suppose two congruent plane figures A and B have opposite orientations. Reflect A and you get a figure A' that has the same orientation as B. Thus, B is a translation or rotation image of A'. By Theorem 8.6-1, two reflections map A' to B. The end result is that three reflections map A to B. This is summarized in what is sometimes called the Fundamental Theorem of Isometries.

Theorem 8.6-3 Fundamental Theorem of Isometries

In a plane, one of two congruent figures can be mapped to the other by a composition of at most three reflections.

If two figures are congruent and have opposite orientations (but are not simply reflections of each other), then there are a translation and a reflection that will map one onto the other. A **glide reflection** is the composition of a translation (a glide) and a reflection across a line parallel to the direction of translation. For example, we can map a left paw print onto a right paw print with a glide reflection.

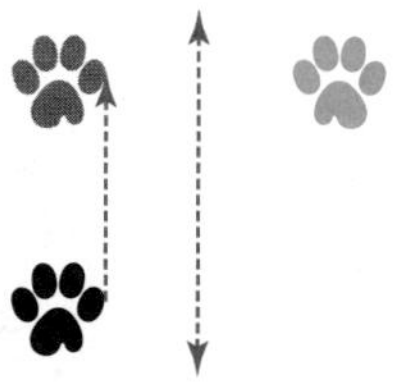

EXAMPLE 3 Finding a Glide Reflection Image

Coordinate Geometry What is the image of ΔTEX for a glide reflection where the translation is $(x, y) \rightarrow (x, y - 5)$ and the line of reflection is $x = 0$?

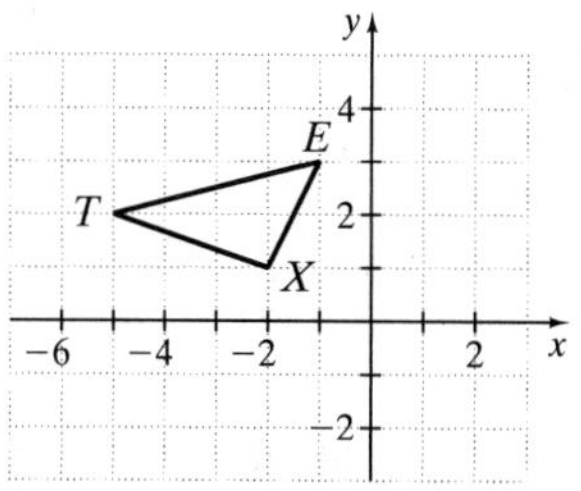

Solution Let's use the translation rule to translate ΔTEX. Then we can reflect this image across the line of reflection, $x = 0$, which is the y-axis.

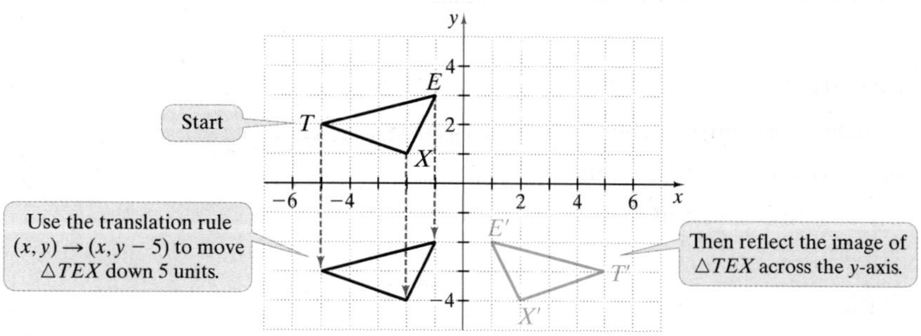

□

PRACTICE 3

Graph ΔTEX from Example 3. What is the image of ΔTEX for a glide reflection where the translation is $(x, y) \rightarrow (x + 1, y)$ and the line of reflection is $y = -2$?

OBJECTIVE 2 ▸ Classifying Isometries. We can map one of two congruent figures in a plane onto the other by a single reflection, translation, rotation, or glide reflection.

> **▸ Helpful Hint**
> Remember: An isometry is a rigid transformation that retains shape and size, or congruence of the preimage.

Theorem 8.6-4 Isometry Classification Theorem

There are only four isometries.

Translation Rotation

Orientations are the same.

Reflection Glide Reflection

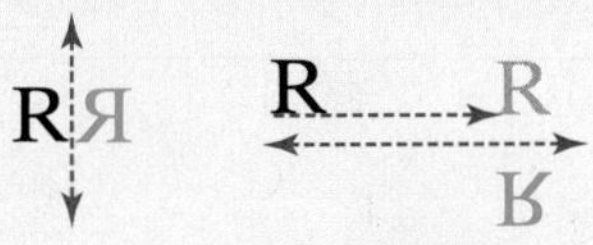

Orientations are opposite.

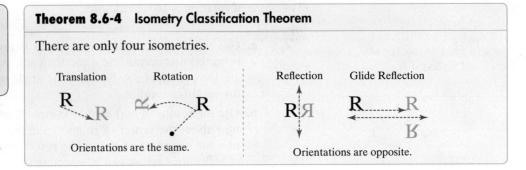

EXAMPLE 4 Classifying Isometries

Each transformation is an isometry. Are the orientations of the preimage and image the same or opposite? What type of isometry maps the preimage to the image?

a.

Preimage Image

b.

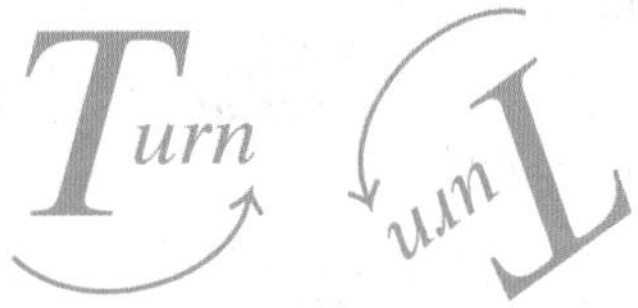

Preimage Image

Solution Compare the directions of the arrows. The orientations are opposite. The image is not a simple reflection of the preimage. The isometry that leads to opposite orientation but is not a reflection is a glide reflection.

Solution Compare the directions of the arrows. The orientations are the same. The image is not a translation of the preimage. The isometry that leads to the same orientation but is not a translation is a rotation. □

PRACTICE 4 Each transformation is an isometry. Are the orientations of the preimage and image the same or opposite? What type of isometry maps the preimage to the image?

a.

Preimage Image

b.

Preimage Image

c.

Preimage Image

VOCABULARY & READINESS CHECK

Answer each true or false.

1. A translation is a composition of two reflections. ____________
2. A rotation is a composition of two reflections. ____________
3. A reflection is not an isometry. ____________
4. A translation is an isometry. ____________
5. A reflection is an isometry. ____________
6. A dilation is an isometry. ____________
7. A dilation is a composition of two translations. ____________
8. A reflection is a composition of two translations. ____________

8.6 EXERCISE SET MyMathLab®

Find the image of each letter for a reflection first across line ℓ and then across line m. Is the resulting transformation a translation or a rotation? For a translation, describe the direction and distance. For a rotation, tell the center of rotation and the angle of rotation. See Examples 1 and 2.

1.

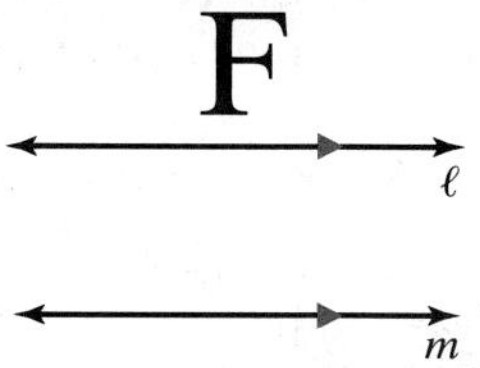

2.

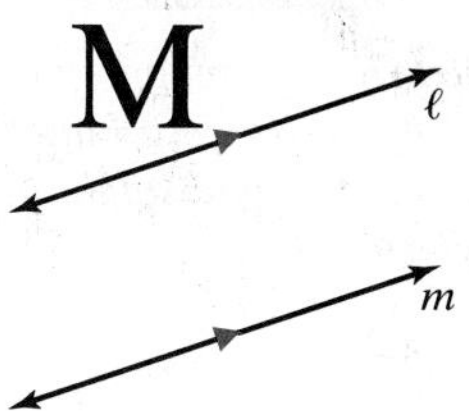

3.

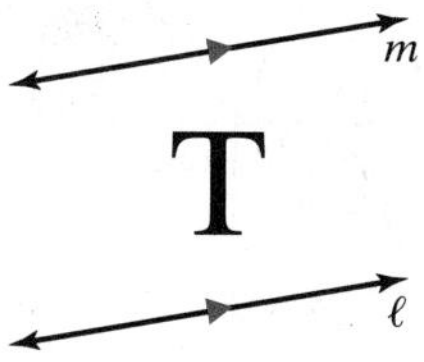

4.

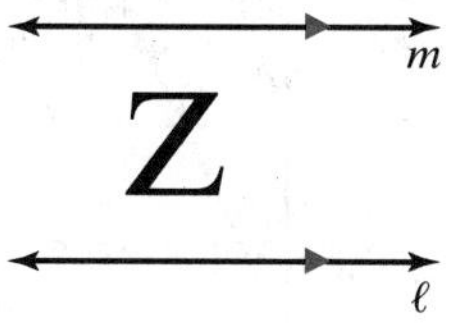

5.

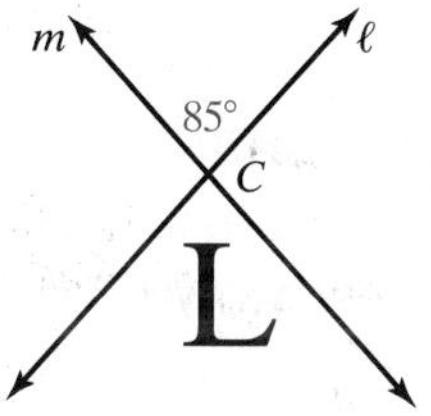

6.

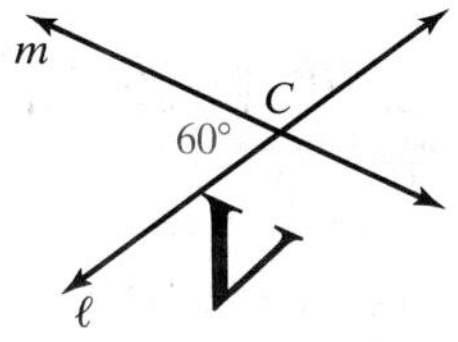

7.

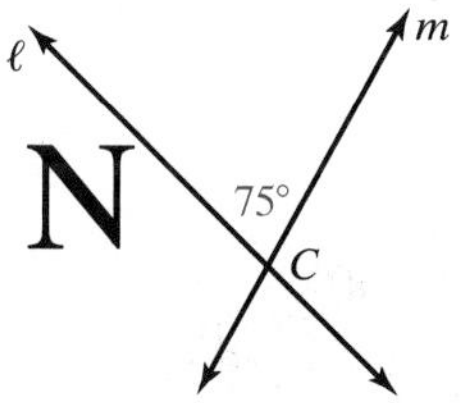

8.

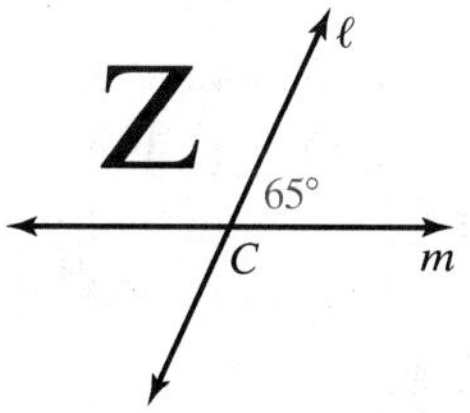

Graph ΔPNB and its glide reflection image for the given translation and reflection line. See Example 3.

9. $(x, y) \to (x + 2, y);\ y = 3$

10. $(x, y) \to (x, y - 3);\ x = 0$

11. $(x, y) \to (x + 2, y + 2);\ y = x$

12. $(x, y) \to (x - 1, y + 1);\ y = -x$

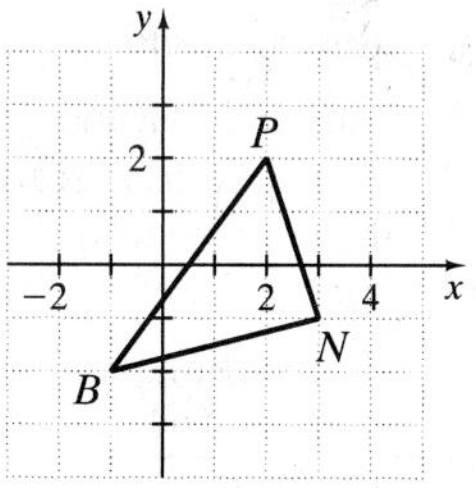

Each transformation is an isometry. Tell whether the two figures' orientations are the same or opposite. Then classify the isometry. See Example 4.

13.

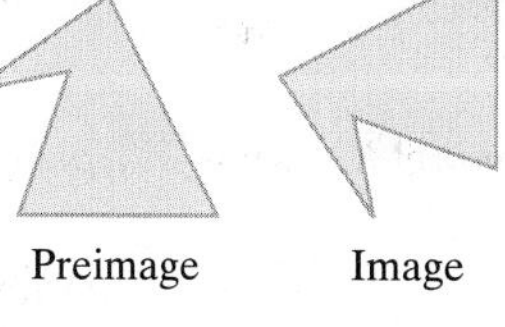

Preimage Image

14.

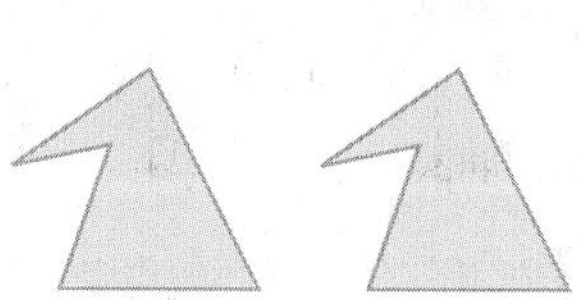

Preimage Image

15.

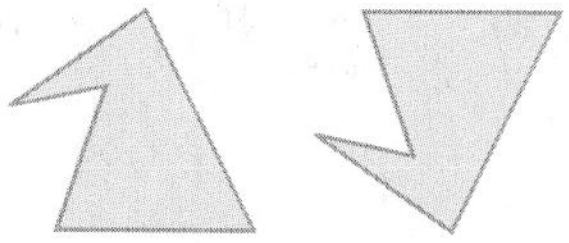

Preimage Image

16.

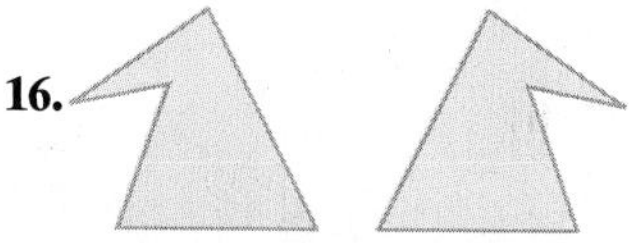

Preimage Image

Each transformation is an isometry. Tell whether the two figures' orientations are the same or opposite. Then classify the isometry.

17.

Preimage Image

18.

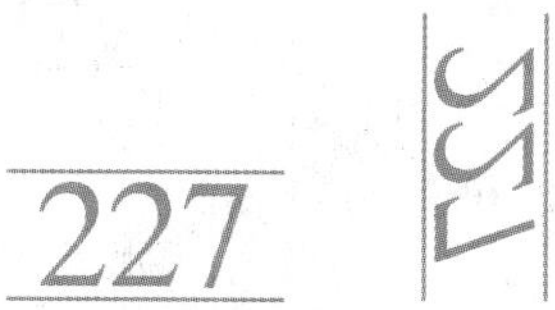

Preimage Image

19.

Preimage Image

20.

Preimage Image

Use the given points and lines. Graph $\overline{AB}$ and its image $\overline{A'B'}$ after a reflection first across ℓ_1 and then across ℓ_2. Is the resulting transformation a translation or a rotation? For a translation, describe the direction and distance. For a rotation, tell the center of rotation and the angle of rotation.

21. $A(1, 5)$ and $B(2, 1)$; $\ell_1: x = 3$; $\ell_2: x = 7$

22. $A(2, -5)$ and $B(-1, -3)$; $\ell_1: y = 0$; $\ell_2: y = 2$

23. $A(-4, -3)$ and $B(-4, 0)$; $\ell_1: y = x$; $\ell_2: y = -x$

24. $A(2, 4)$ and $B(3, 1)$; ℓ_1: x-axis; ℓ_2: y-axis

25. $A(6, -4)$ and $B(5, 0)$; $\ell_1: x = 6$; $\ell_2: x = 4$

26. $A(-1, 0)$ and $B(0, -2)$; $\ell_1: y = -1$; $\ell_2: y = 1$

27. **Multiple Steps** $T \to T'(1, 5)$ by a glide reflection where the translation is $(x, y) \to (x + 3, y)$ and the line of reflection is $y = 1$.

a. Should T be to left or to the right of T'?

b. Should T be above or below T'?

c. What are the coordinates of T?

28. ΔPQR has vertices $P(0, 5)$, $Q(5, 3)$, and $R(3, 1)$. What are the vertices of the image of ΔPQR for a glide reflection where the translation is $(x, y) \to (x + 4, y)$ and the reflection line is $y = -2$?

For Exercises 29 and 30, the two figures are congruent. Describe the single isometry by writing a formula that maps the black figure onto the blue figure.

29.

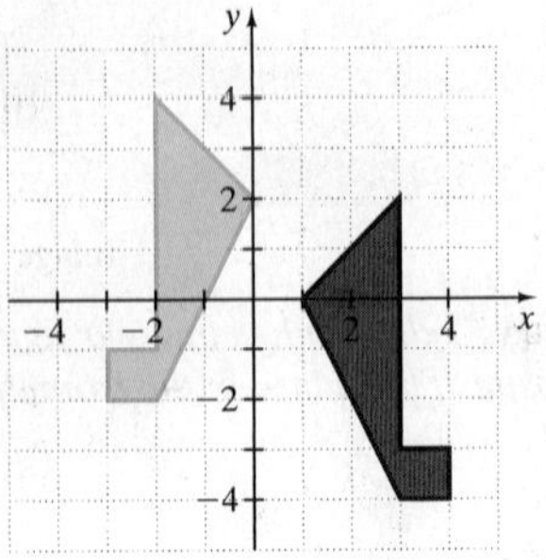

30.

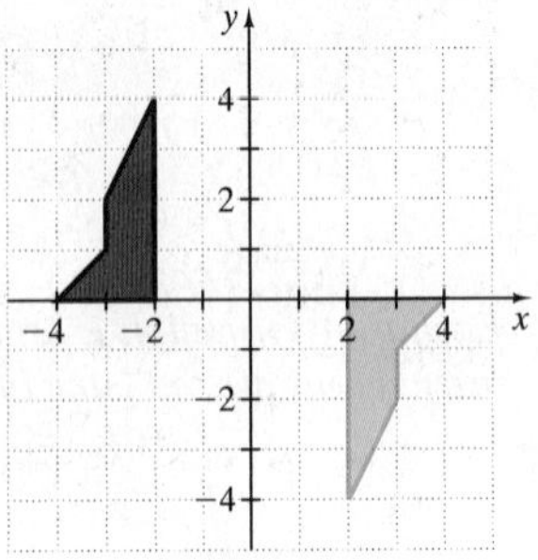

31. Multiple Choice Which transformation maps the black triangle onto the blue triangle?

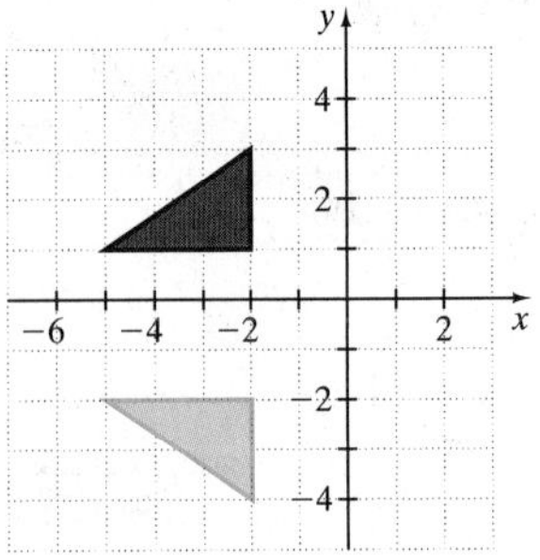

a. a glide reflection where the translation is $(x, y) \rightarrow (x, y - 3)$ and the line of reflection is $x = -2$

b. a 180° rotation about the origin

c. a reflection across the line $y = -\frac{1}{2}$

d. a reflection across the y-axis followed by a 180° rotation about the origin

32. Find the Error You reflect ΔDEF first across line m and then across line n. Your friend says you can get the same result by reflecting ΔDEF first across line n and then across line m. Explain your friend's error.

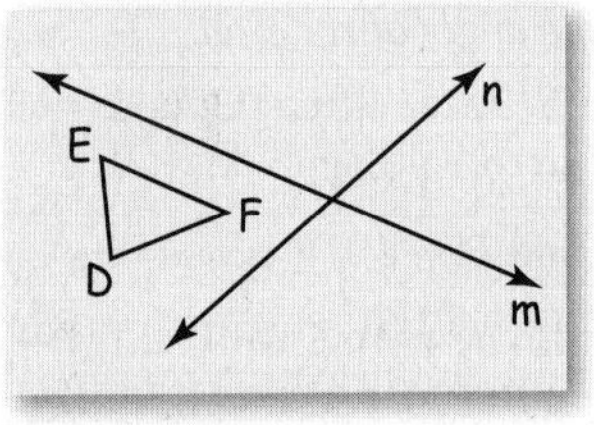

The images seen in a kaleidoscope are produced by compositions of reflections in intersecting mirrors.

How? The easiest way to understand this, and to calculate the angle, is to think of a single slice of pizza that we will place between two mirrors and duplicate to represent the whole pizza. For example, a 90° slice of pizza, *, will need to be duplicated only 4 times, or* $4(90°) = 360°$*, to form the whole pizza. Thus* $n(m\angle A) = 360°$*, where n is how many sectors there are. Determine the measure of the angle between the mirrors in each kaleidoscope image.*

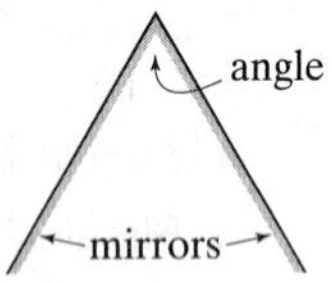

33.

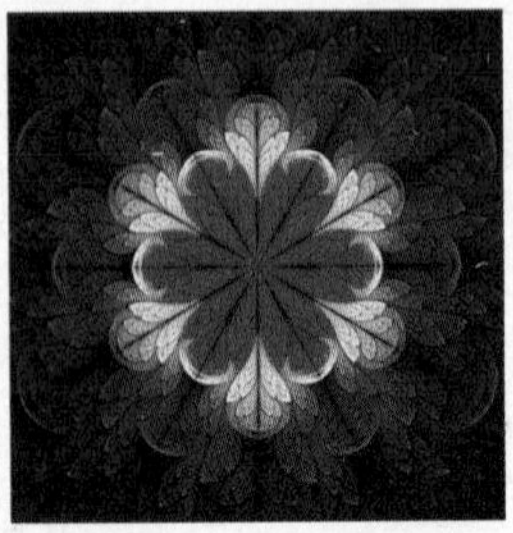

34.

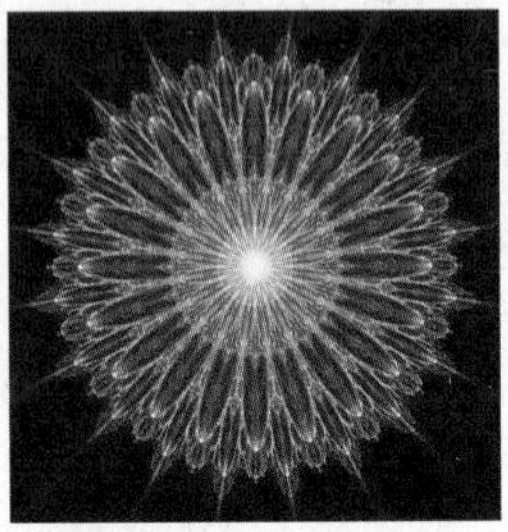

35.

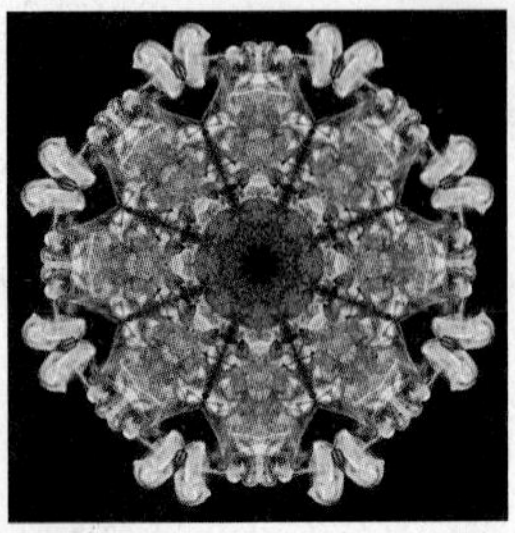

36.

Identify each mapping as a single translation, reflection, rotation, or glide reflection. Find the translation rule, reflection line, center of rotation and angle of rotation, or glide translation rule and reflection line.

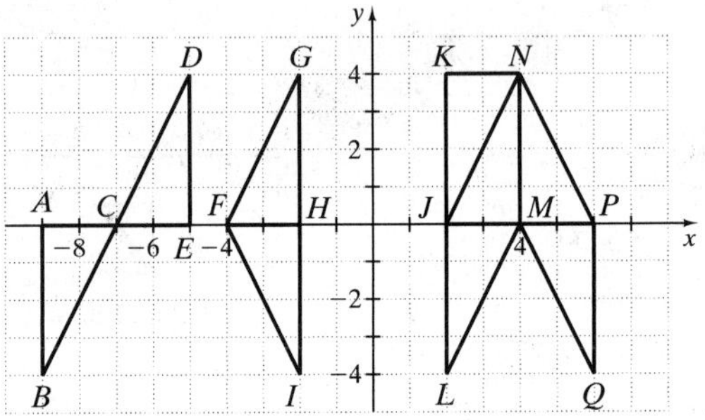

37. $\Delta EDC \rightarrow \Delta PQM$

38. $\Delta PQM \rightarrow \Delta KJN$

39. $\Delta MNJ \rightarrow \Delta EDC$

40. $\Delta KJN \rightarrow \Delta ABC$

41. $\Delta HIF \rightarrow \Delta HGF$

42. $\Delta PQM \rightarrow \Delta JLM$

43. $\Delta JLM \rightarrow \Delta MNJ$

44. $\Delta HGF \rightarrow \Delta KJN$

$P \rightarrow P'(3, -1)$ *for the given translation and reflection line. Find the coordinates of P.*

45. $(x, y) \rightarrow (x - 3, y); y = 2$

46. $(x, y) \rightarrow (x, y - 3); y = 2$

47. $(x, y) \rightarrow (x - 3, y - 3); y = x$

48. $(x, y) \rightarrow (x + 4, y - 4); y = -x$

CONCEPT EXTENSIONS

49. In a glide reflection, what is the relationship between the direction of the translation and the line of reflection?

50. Reflections and glide reflections are *odd isometries,* while translations and rotations are *even isometries*. Use what you have learned in this section to explain why these categories make sense.

51. The definition states that a glide reflection is the composition of a translation and a reflection. Explain why these can occur in either order.

52. Describe a glide reflection that maps the black R to the blue R.

For the given transformation mapping $\overline{XY}$ *to* $\overline{X'Y'}$*, prove that* $\overline{XY} \cong \overline{X'Y'}$*.*

53. a translation

54. a reflection

55. a rotation

56. Does an $x°$ rotation about a point P followed by a reflection across a line ℓ give the same image as a reflection across ℓ followed by an $x°$ rotation about P? Explain.

REVIEW AND PREVIEW

Coordinate Geometry *Find the image of ΔABC for a dilation with center $(0, 0)$ and the given scale factor. See Section 8.5.*

57. $A(0, 4), B(0, 0), C(-3, -1)$; scale factor 3

58. $A(2, 3), B(-4, -2), C(5, -3)$; scale factor 2

59. $A(4, 2), B(2, 8), C(8, 0)$; scale factor 1.5

60. $A(7, 8), B(5, 4), C(9, 6)$; scale factor 0.5

Identify the two statements that contradict each other. See Section 5.5.

61. **I.** ΔABC is a right triangle.
II. ΔABC is equiangular.
III. ΔABC is isosceles.

62. **I.** In right $\Delta ABC, m\angle B = 90°$.
II. In right $\Delta ABC, m\angle A = 80°$.
III. In right $\Delta ABC, m\angle C = 90°$.

Classify the polygon with the given number of sides. See Section 6.1.

63. five

64. eight

EXTENSION—FRIEZE PATTERNS

A frieze pattern is a design that repeats itself along a straight line. You can see frieze patterns in design trim. They often appear along the edges of buildings or as wallpaper borders of rooms. Every frieze pattern can be mapped onto itself by a translation.

Some frieze patterns can be mapped onto themselves by reflections.

a.

b.

Some frieze patterns have repeated rotational symmetry.

c.

d.
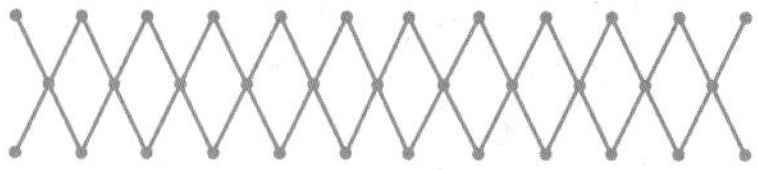

Some frieze patterns show glide reflections.

e.

f.

EXTENSION—FRIEZE PATTERNS MyMathLab®

For Exercises 1–4, refer to the frieze patterns above. You may find tracing paper helpful.

1. For each frieze pattern, find a portion (as small as possible) that you could translate repeatedly to form the entire pattern.

2. Which patterns show reflectional symmetry? Find their reflection lines.

3. Which patterns show rotational symmetry? Find their centers of rotation.

4. Which patterns show glide reflections? Find a glide translation and reflection line.

For each frieze pattern that follows, describe all transformations that map the pattern onto itself.

5.

6.
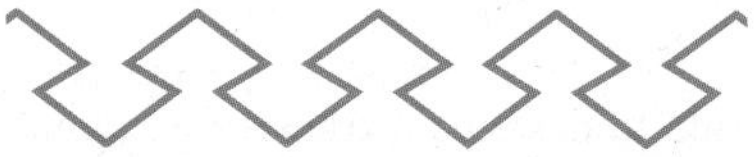

7.

8.

9.

10.

CHAPTER 8 VOCABULARY CHECK

Word Bank *Use the choices to fill in each blank. Some choices may be used more than once and some not at all.*

preimage	isometry	rigid transformation	translation	slide
image	transformation	composition of transformations	flip	turn
reflection	rotation	reduction	to change	enlargement

1. The word transform means "____________."
2. A(n) ____________ of a figure is a change in the position, shape, or size of the figure.
3. In a transformation, the resulting figure is called the ____________.
4. In a transformation, the original figure is called the ____________.
5. A(n) ____________ is a transformation in which the preimage and the image are congruent.
6. An isometry is also called a ____________.
7. Three rigid transformations are called a ____________, a ____________, and a ____________.
8. Another word for a translation is a ____________.
9. Another word for a reflection is a ____________.
10. Another word for a rotation is a ____________.
11. A combination of two or more transformations is called a(n) ____________.
12. A dilation is a(n) ____________ if the scale factor is greater than 1.
13. A dilation is a(n) ____________ if the scale factor is between 0 and 1.
14. True or false: The image of a dilation is congruent to its preimage. ____________

CHAPTER 8 REVIEW

(8.1 and 8.2)

1. **a.** A transformation maps $ZOWE$ onto $LFMA$. Does the transformation appear to be an isometry? Explain.

   b. What is the image of $\overline{ZE}$? What is the preimage of M?

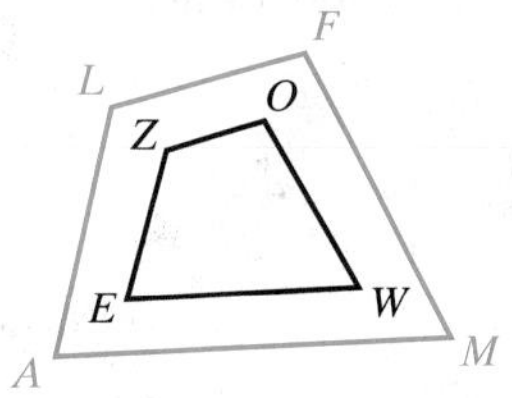

2. ΔRST has vertices $R(0, -4)$, $S(-2, -1)$, and $T(-6, 1)$. Graph the image of ΔRST for the translation $(x, y) \rightarrow (x - 4, y + 7)$.
3. Write a translation rule to describe a translation 5 units left and 10 units up.
4. Write a single translation rule that has the same effect as the following composition of translations.
   $(x, y) \rightarrow (x - 5, y + 7)$ followed by $(x, y) \rightarrow (x + 3, y)$

***(8.3 and 8.4)** Given points $A(6, 4), B(-2, 1)$, and $C(5, 0)$, graph ΔABC and its reflection image across each line.*

5. the x-axis
6. $x = 4$
7. Find the image of $P(-4, 1)$ for a $180°$ rotation about the origin.
8. Copy the diagram. Then draw the image of ΔZXY for a $90°$ rotation about P. Label the vertices of the image using prime notation.

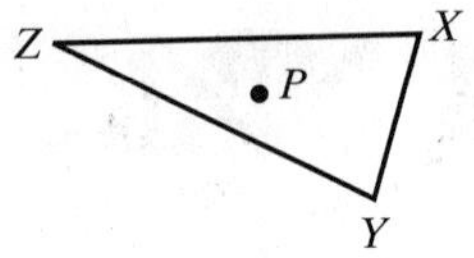

Point O is the center of regular pentagon NMPQR.

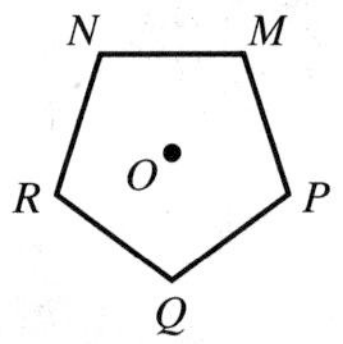

9. What is the image of point N for a composition of a $72°$ rotation and a $144°$ rotation about O?
10. What is the angle of rotation that maps point P to point Q?

(8.5)

11. The blue figure is a dilation image of the black figure. The center of dilation is O. Tell whether the dilation is an enlargement or a reduction. Then find the scale factor.

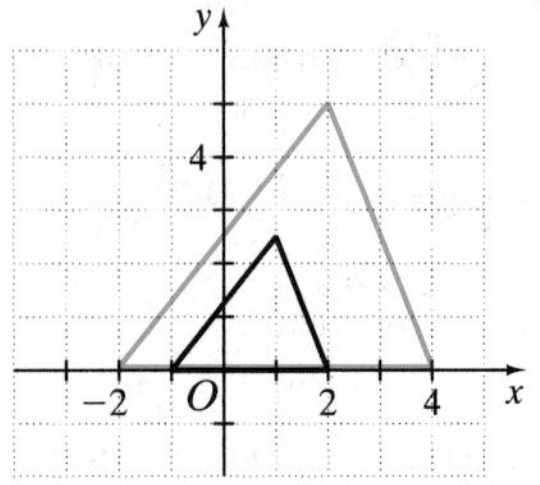

12. The blue figure is a dilation image of the black figure. The center of dilation is A. Is the dilation an enlargement or a reduction? What is the scale factor?

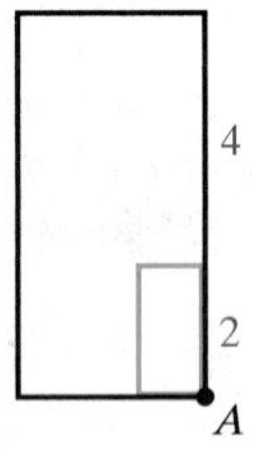

Graph the polygon with the given vertices. Then graph its image for a dilation with center $(0, 0)$ and the given scale factor.

13. $F(-4, 0), U(5, 0), N(-2, -5)$; scale factor $\frac{1}{2}$

14. $M(-3, 4), A(-6, -1), T(0, 0), H(3, 2)$; scale factor 5

(8.6)

15. Sketch and describe the result of reflecting E first across line ℓ and then across line m.

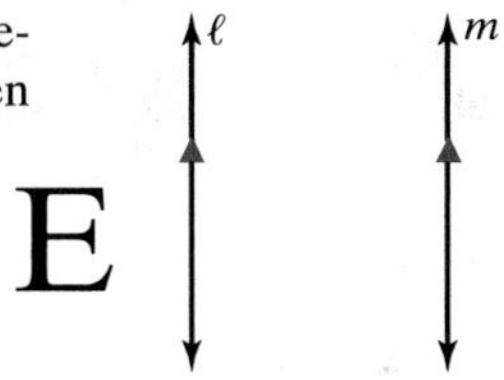

16. Describe the result of reflecting P first across line ℓ and then across line m.

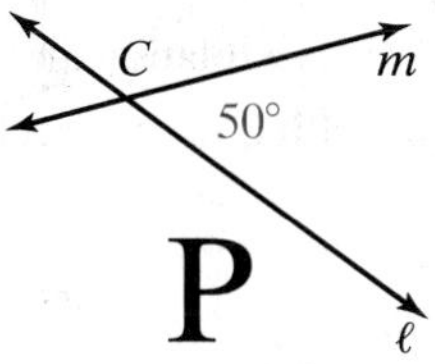

Each transformation is an isometry image. Tell whether their orientations are the same or opposite. Then classify the isometry.

17.

Preimage Image

18.

Preimage Image

19.

Preimage Image

20. ΔTAM has vertices $T(0, 5), A(4, 1)$, and $M(3, 6)$. Find the glide reflection images of ΔTAM for the translation $(x, y) \rightarrow (x - 4, y)$ followed by reflection across the line $y = -2$.

MIXED REVIEW

21. What are the coordinates of the image of $A(5, -9)$ for the translation $(x, y) \rightarrow (x - 2, y + 3)$?

22. Graph points $P(1, 0), Q(3, -2)$ and $R(4, 0)$ and draw ΔPQR. What is the image of ΔPQR reflected across the y-axis?

23. Given points $A(6, 4), B(-2, 1)$, and $C(5, 0)$, graph ΔABC and its reflection image across the line: $y = x$

Tell what type(s) of symmetry each figure has. If it has line symmetry, sketch the figure and the line(s) of symmetry. If it has rotational symmetry, state the angle of rotation.

24.

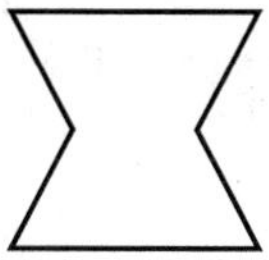

25.

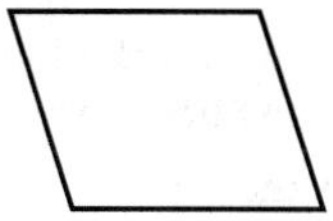

26.

27. How many lines of symmetry does an isosceles trapezoid have?

28. How many lines of symmetry does an equilateral triangle have?

29. What type(s) of symmetry does a square have?

30. Give an example of a two-dimensional object that has rotational symmetry about a line.

31. A dilation maps ΔLMN onto $\Delta L'M'N'$. $LM = 36$ ft, $LN = 26$ ft, $MN = 45$ ft, and $L'M' = 9$ ft. Find $L'N'$ and $M'N'$.

CHAPTER 8 TEST

TEST PREP VIDEO — The fully worked-out solutions to any exercises you want to review are available in MyMathLab.

Tell whether the transformation appears to be an isometry. Explain.

1.

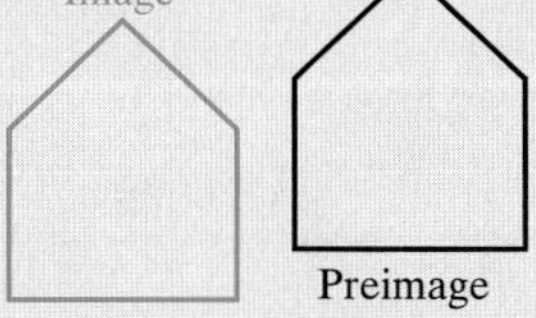

Image Preimage

2. What rule describes the translation 3 units right and 7 units down?

Find the reflection image of $(5, -3)$ across each line.

3. the x-axis

4. the y-axis

5. Write the translation rule that maps $P(-4, 2)$ onto $P'(-1, -1)$.

What type of transformation has the same effect as each composition of transformations?

6. translation $(x, y) \rightarrow (x, y - 5)$ followed by reflection across the line $x = 6$

7. translation $(x, y) \rightarrow (x - 3, y + 2)$ followed by translation $(x, y) \rightarrow (x + 8, y - 4)$

8. reflection across the line $x = -2$ and then across the line $x = 4$

9. reflection across the line $y = -x$ and then across the line $y = x$

For Exercises 10–16, find the coordinates of the vertices of the image of ABCD for each transformation.

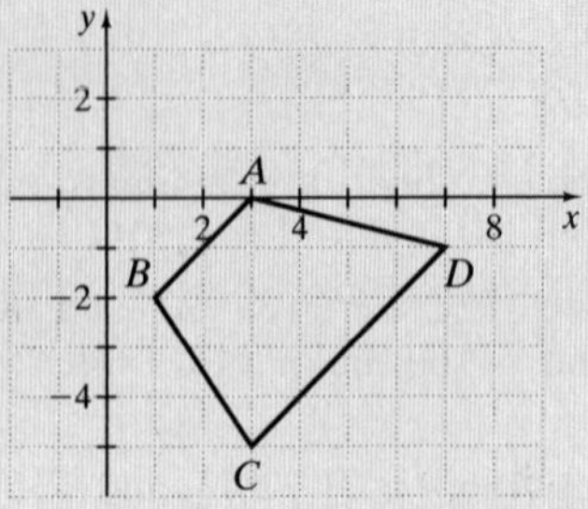

10. reflection across the line $x = -4$

11. translation $(x, y) \rightarrow (x - 6, y + 8)$

12. rotation of 90° about the point $(0, 0)$

13. dilation with center $(0, 0)$ and scale factor $\frac{2}{3}$

14. glide reflection with translation $(x, y) \rightarrow (x, y + 5)$ and reflection across the line $x = 0$

15. reflection across the line $y = x$

16. dilation with center $(0, 0)$ and scale factor 3

What type(s) of symmetry does each figure have?

17.

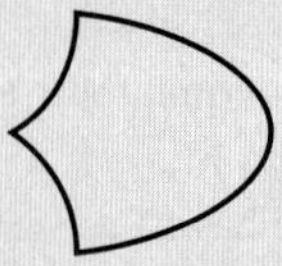

18.

19.

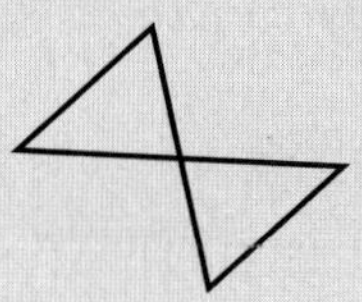

Identify the isometry that maps the black figure onto the blue figure.

20.

21.

22. Is a dilation an isometry? Explain.

23. Line m intersects $\overline{UH}$ at N, and $UN = NH$. Must H be the reflection image of U across line m? Explain your reasoning.

24. **Coordinate Geometry** A dilation with center $(0, 0)$ and scale factor 2.5 maps (a, b) onto $(10, -25)$. What are the values of a and b?

25. **Find the Error** A classmate says that a certain figure has 50° rotational symmetry. Explain your classmate's error.

CHAPTER 8 STANDARDIZED TEST

1. What translation is described by the rule $(x, y) \rightarrow (x - 21, y + 5)$?

a. 21 units right and 5 units up

b. 21 units left and 5 units up

c. 21 units right and 5 units down

d. none of these

2. What translation is described by the rule $(x, y) \rightarrow (x + 10, y + 7)$?

a. 10 units left and 7 units up

b. 10 units right and 7 units down

c. 10 units left and 7 units down

d. none of these

Use the figure for Exercises 3–12.

For Exercises 3–11, find the coordinates of the endpoints of the image of $\overline{AB}$ for each transformation.

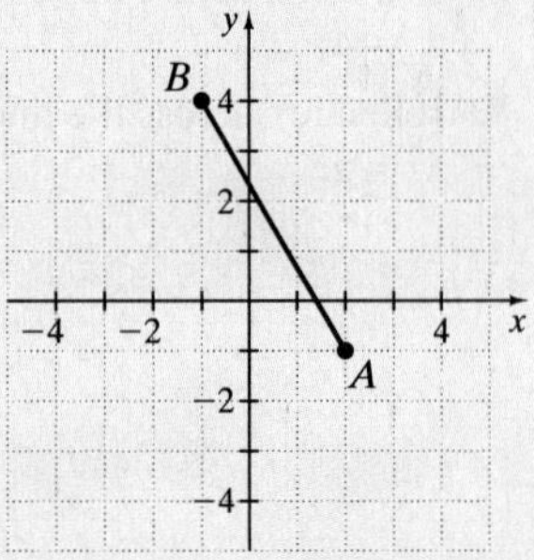

3. reflection across the line $y = 0$

a. $A'(-1, 2), B'(1, -4)$

b. $A'(2, 1), B'(-1, -4)$

c. $A'(1, -2), B'(-4, 1)$

d. $A'(-2, -1), B(1, 4)$

4. reflection across the line $x = 0$

a. $A'(-1, 2), B'(1, -4)$

b. $A'(2, 1), B'(-1, -4)$

c. $A'(1, -2), B'(-4, 1)$

d. $A'(-2, -1), B'(1, 4)$

5. reflection across the line $x = 1$

a. $A'(0, -1), B'(0, 4)$

b. $A'(3, -1), B'(0, 4)$

c. $A'(3, -1), B'(3, 4)$

d. $A'(0, -1), B'(3, 4)$

6. translation $(x, y) \rightarrow (x - 3, y + 3)$

a. $A'(-1, 2), B'(-4, 7)$

b. $A'(-1, 2), B'(4, -1)$

c. $A'(5, -4), B'(2, 1)$

d. none of these

7. rotation of 90° about the point $(0, 0)$

a. $A'(-1, -2), B'(4, 1)$

b. $A'(-2, 1), B'(1, -4)$

c. $A'(1, 2), B'(-4, -1)$

d. none of these

8. rotation of 180° about the point $(0, 0)$

a. $A'(-1, -2), B'(4, 1)$

b. $A'(-2, 1), B'(1, -4)$

c. $A'(1, 2), B'(-4, -1)$

d. none of these

9. a dilation with center $(0, 0)$ and scale factor $\frac{1}{2}$

a. $A'(4, -2), B'(-2, 8)$

b. $A'\left(-1, \frac{1}{2}\right), B'\left(\frac{1}{2}, -2\right)$

c. $A'\left(1, -\frac{1}{2}\right), B'\left(-\frac{1}{2}, 2\right)$

d. none of these

10. a glide reflection with translation $(x, y) \rightarrow (x + 2, y)$ and reflection across the line $y = 2$

a. $A'(4, 1), B'(1, -4)$

b. $A'(4, 5), B'(1, 0)$

c. $A'(4, 1), B'(1, 0)$

d. none of these

11. reflection across the line $y = -x$

a. $A'(-1, 2), B'(1, -4)$

b. $A'(2, 1), B'(-1, -4)$

c. $A'(1, -2), B'(-4, 1)$

d. $A'(-2, -1), B'(1, 4)$

12. The image of $\overline{AB}$ is shown (see previous page). What type of transformation maps $\overline{AB}$ to $\overline{A'B'}$?

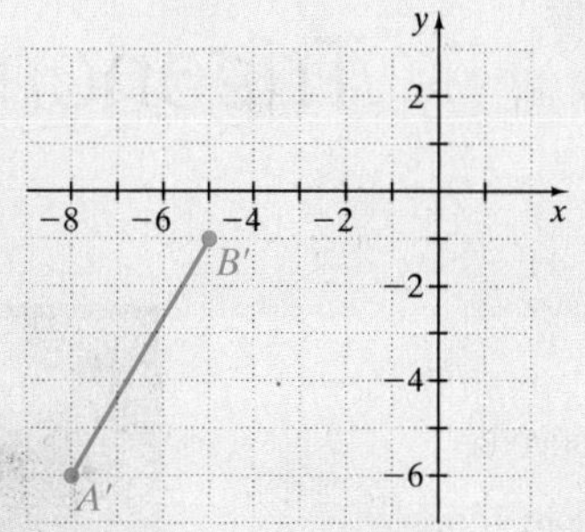

a. glide reflection

b. rotation

c. translation

d. dilation

Fill in the blank.

13. A dilation is ___________ an isometry.

a. always

b. sometimes

c. never

14. A rotation ___________ preserves orientation.

a. always

b. sometimes

c. never

15. A glide reflection ___________ preserves orientation.

a. always

b. sometimes

c. never

CHAPTER

9 Right Triangles and Trigonometry

9.1 The Pythagorean Theorem and Its Converse

9.2 Special Right Triangles

9.3 Trigonometric Ratios

9.4 Solving Right Triangles

9.5 Vectors

Extension—Law of Sines

Extension—Law of Cosines

Listen to the same note played on a piano and a violin. The notes have a different quality or "tone." Tone depends on the way an instrument vibrates.

When a note is played, it vibrates in a specific way. Two sounds from tuning forks can be modeled by the trigonometric ratio called sine. In this chapter, we will be introducing the trigonometric ratios of sine, cosine, and tangent and some applications of these ratios. Early applications of trigonometry were in the areas of surveying, architecture, and navigation, just to name a few.

9.1 THE PYTHAGOREAN THEOREM AND ITS CONVERSE

OBJECTIVES

1 Use the Pythagorean Theorem.

2 Use the Converse of the Pythagorean Theorem.

VOCABULARY

- Pythagorean triple

Helpful Hint

- Remember—the Pythagorean Theorem may be used for right triangles only.
- Also, it makes no difference which leg is called leg_1 and which is called leg_2.

OBJECTIVE 1 ▸ Using the Pythagorean Theorem. Recall that if we know the lengths of any two sides of a right triangle, we can find the length of the third side by using the Pythagorean Theorem. This theorem is named after the Greek mathematician Pythagoras, who lived about 500 BC. (See the Historical Note at the end of this chapter.)

Theorem 9.1-1 Pythagorean Theorem

Theorem	If . . .	Then . . .
If a triangle is a right triangle, then the sum of the squares of the lengths of the legs is equal to the square of the length of the hypotenuse.	ΔABC is a right triangle 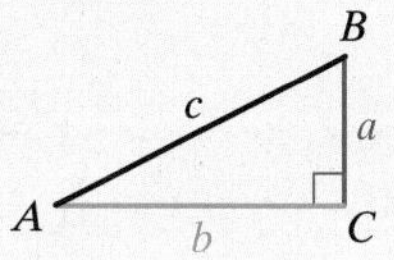	$(\text{leg}_1)^2 + (\text{leg}_2)^2 = (\text{hypotenuse})^2$ or $a^2 + b^2 = c^2$ (The hypotenuse, which is opposite the right angle)

We prove this theorem in Exercise Set 9.1, Exercise 47.

Helpful Hint

Recall that the set of positive integers is {1, 2, 3, 4, . . . }.

A **Pythagorean triple** is a set of positive integers *a*, *b*, and *c* that satisfy the equation $a^2 + b^2 = c^2$. Notice from this equation that *c* must be the greatest number. Below are some common Pythagorean triples.

3, 4, 5   5, 12, 13   8, 15, 17   7, 24, 25

For example, $3^2 + 4^2 = 5^2$, or
$9 + 16 = 25$, a true statement.

Of course, $a^2 + b^2 = c^2$ "sounds" familiar because it comes from the Pythagorean Theorem. Thus, if you have a right triangle, then the lengths of the sides will form a Pythagorean triple.

EXAMPLE 1 Finding the Length of the Hypotenuse

Find the length of the hypotenuse of ΔABC. Check to see that the side lengths of ΔABC form a Pythagorean triple. Explain why.

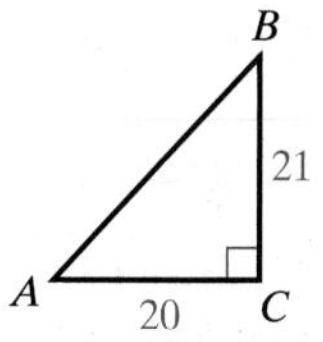

Solution Use either given form of the Pythagorean Theorem.

$$(\text{leg}_1)^2 + (\text{leg}_2)^2 = (\text{hypotenuse})^2 \quad \text{Pythagorean Theorem}$$
$$a^2 + b^2 = x^2 \quad \text{Also the Pythagorean Theorem}$$
$$21^2 + 20^2 = x^2 \quad \text{Substitute 21 for } a \text{ and 20 for } b.$$
$$441 + 400 = x^2 \quad \text{Square 21 and square 20.}$$
$$841 = x^2 \quad \text{Add.}$$
$$\sqrt{841} = x \quad \text{Take the positive square root.}$$
$$29 = x \quad \text{Simplify.}$$

The length of the hypotenuse is 29. The side lengths 20, 21, and 29 form a Pythagorean triple because they are positive integers that satisfy $a^2 + b^2 = c^2$, or $21^2 + 20^2 = 29^2$ or $441 + 400 = 841$, a true statement. □

PRACTICE 1

a. The legs of a right triangle have lengths 10 and 24. Find the length of the hypotenuse.

b. Check to see that the side lengths in part **a** form a Pythagorean triple. Explain why.

EXAMPLE 2 Finding the Length of a Leg

Find the value of x. Write the answer in simplest radical form.

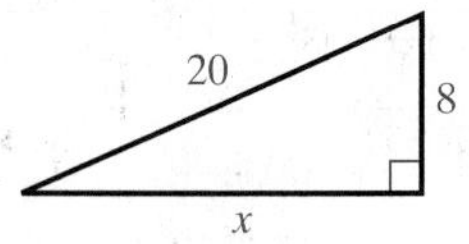

Solution The unknown length, x, is a leg of the right triangle, so for $a^2 + b^2 = c^2$, substitute x for a or b.

$a^2 + b^2 = c^2$ Pythagorean Theorem

$8^2 + x^2 = 20^2$ Substitute 8 for a, x for b, and 20 for c, the hypotenuse.

$64 + x^2 = 400$ Find 8^2 and 20^2.

$x^2 = 336$ Subtract 64 from both sides.

$x = \sqrt{336}$ Take the positive square root.

$x = \sqrt{16(21)}$ Factor out a perfect square factor of 16.

$x = 4\sqrt{21}$ Simplify.

The exact value of x is $4\sqrt{21}$ units.

PRACTICE 2

The hypotenuse of a right triangle has length 12. One leg has length 6. Find the length of the other leg. Write the answer in simplest radical form.

In real-life applications, often an approximate length is requested. In such cases, simply use a calculator to approximate. In Example 2, $x = 4 \cdot \sqrt{21} \approx 18.33$ rounded to two decimal places.

EXAMPLE 3 Calculating Placement of a Wire

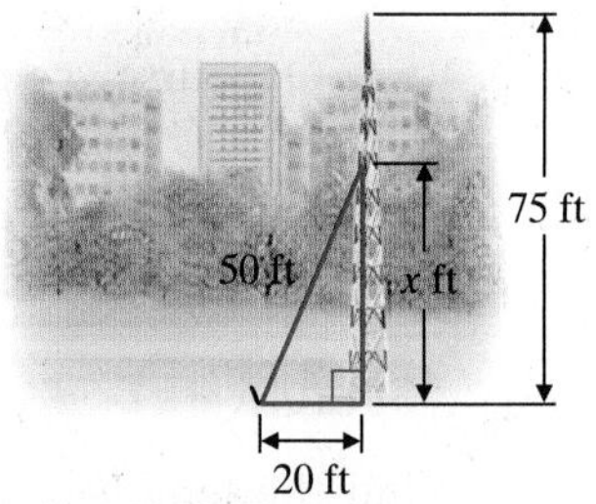

A 50-foot supporting wire is to be attached to a 75-foot antenna. Because of surrounding buildings, sidewalks, and roadways, the wire must be anchored exactly 20 feet from the base of the antenna. How high from the base of the antenna must the wire be attached? Give an exact answer and a one-decimal-place approximation.

Solution Since a right triangle is formed, we use the Pythagorean Theorem. Here, the unknown length, x, is a leg of the right triangle.

$a^2 + b^2 = c^2$ Pythagorean Theorem

$20^2 + x^2 = 50^2$ Let $a = 20$, $b = x$, and $c = 50$, since 50 is the hypotenuse.

$400 + x^2 = 2500$ Find 20^2 and 50^2.

$x^2 = 2100$ Subtract 400 from both sides.

$x = \sqrt{2100}$ Take the positive square root.

$x = \sqrt{100 \cdot 21}$ Factor out a perfect square factor of 100.

$= 10\sqrt{21}$ Simplify.

The wire is attached exactly $10\sqrt{21}$ feet from the base of the pole, or approximately 45.8 feet. □

PRACTICE

3 The size of a rectangular computer monitor is given as the measure of its diagonal. You want to buy a 19-inch monitor that has a width of 11 inches. What is the length of the monitor? Round to the nearest tenth of an inch.

OBJECTIVE 2 ▶ Using the Converse of the Pythagorean Theorem. The Converse of the Pythagorean Theorem is also true. We may use the Converse of the Pythagorean Theorem to determine whether a triangle is a right triangle.

Theorem 9.1-2 Converse of the Pythagorean Theorem

Theorem	If . . .	Then . . .
If the sum of the squares of the lengths of two sides of a triangle is equal to the square of the length of the third side, then the triangle is a right triangle.	$a^2 + b^2 = c^2$ 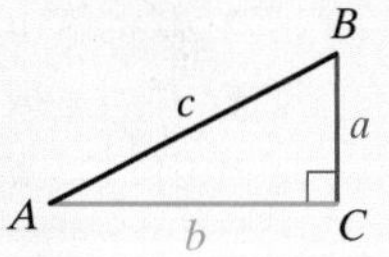	ΔABC is a right triangle

We prove this theorem in Exercise Set 9.1, Exercise 48.

EXAMPLE 4 Identifying a Right Triangle

A triangle has side lengths 85, 84, and 13. Is the triangle a right triangle? Explain.

Solution To determine whether we have a right triangle, we use the Converse of the Pythagorean Theorem and see whether the lengths form a Pythagorean triple.

> **Helpful Hint**
> If using $a^2 + b^2 = c^2$, don't forget to let c be the greatest number.

$a^2 + b^2 = c^2$ Pythagorean Theorem

$13^2 + 84^2 \stackrel{?}{=} 85^2$ Substitute 13 for a, 84 for b, and 85, the greatest number, for c.

$169 + 7056 \stackrel{?}{=} 7225$ Find the square of each number.

$7225 = 7225$ ✓ Add.

Since $7225 = 7225$ is a true statement, then yes, the triangle is a right triangle. □

PRACTICE

4 A triangle has side lengths 16, 48, and 50. Is the triangle a right triangle? Explain.

Sometimes, it is hard to determine by sight whether an unknown angle of a triangle is acute, obtuse, or right.

The theorems below allow us to determine this. These theorems relate to the Hinge Theorem, which states that the longer side is opposite the larger angle and the shorter side is opposite the smaller angle.

Theorem 9.1-3

Theorem	If . . .	Then . . .
If the square of the length of the longest side of a triangle is greater than the sum of the squares of the lengths of the other two sides, then the triangle is obtuse.	$c^2 > a^2 + b^2$ 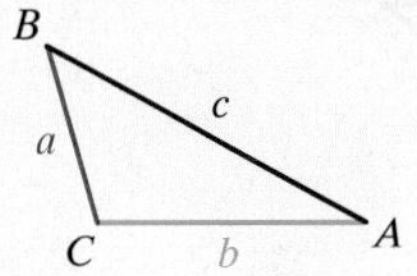	ΔABC is obtuse

Theorem 9.1-4

Theorem	If...	Then...
If the square of the length of the longest side of a triangle is less than the sum of the squares of the lengths of the other two sides, then the triangle is acute.	$c^2 < a^2 + b^2$	ΔABC is acute

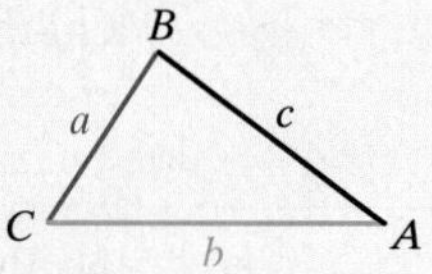

We prove these theorems in Exercise Set 9.1, Exercises 49 and 50.

Helpful Hint

Notice the patterns formed. Below, *a*, *b*, and *c* are the lengths of the sides of a Δ and *c* is the greatest number.

- If $c^2 < a^2 + b^2$, Δ is acute.
- If $c^2 = a^2 + b^2$, Δ is right.
- If $c^2 > a^2 + b^2$, Δ is obtuse.

EXAMPLE 5 **Classifying a Triangle**

A triangle has side lengths 6, 11, and 14. Is it acute, obtuse, or right?

Solution Let's compare c^2 and $a^2 + b^2$. (See the Helpful Hint above.)

c^2 ___ $a^2 + b^2$	Compare c^2, the greatest number, to $a^2 + b^2$.
14^2 ___ $6^2 + 11^2$	Substitute the greatest number for *c* and the other numbers for *a* and *b*.
196 ___ 36 + 121	Square each number.
196 ___ 157	Add.
196 > 157	Compare the numbers.

Since $c^2 > a^2 + b^2$, the triangle is obtuse. □

PRACTICE 5 Is a triangle with side lengths 7, 8, and 9 acute, obtuse, or right?

VOCABULARY & READINESS CHECK

Fill in the Blank

1. A set of three positive integers *a*, *b*, and *c* that satisfy the equation $a^2 + b^2 = c^2$ is called a(n) ____________.

True or False

2. The Pythagorean Theorem can be used for any triangle. ____________
3. If $c^2 > a^2 + b^2$, where *c* is the longest side of a triangle, then the triangle is acute. ____________
4. This triangle is a right triangle. ____________

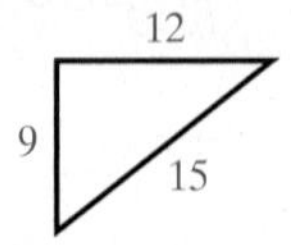

9.1 EXERCISE SET MyMathLab®

Find the value of x. If necessary, write your answer in simplest radical form. See Examples 1 and 2.

1.

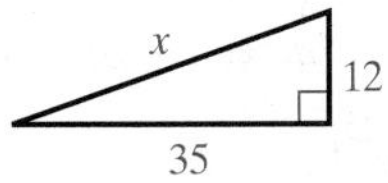

2.

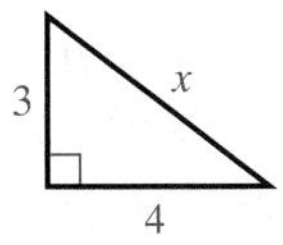

3.

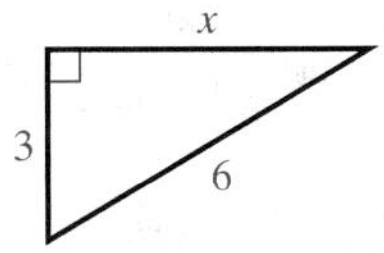

4.

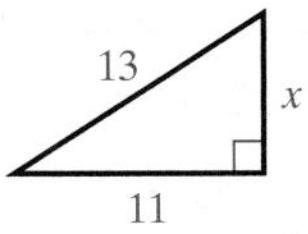

5.

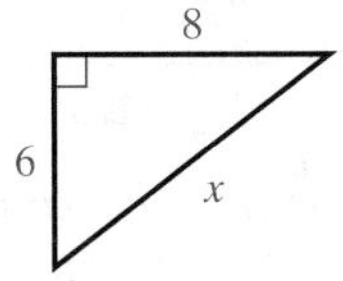

6.

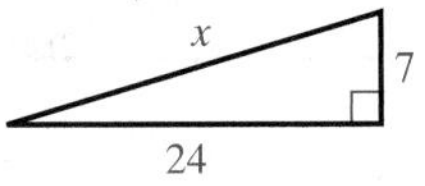

7.

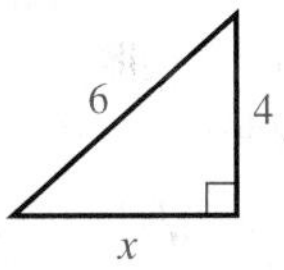

8.

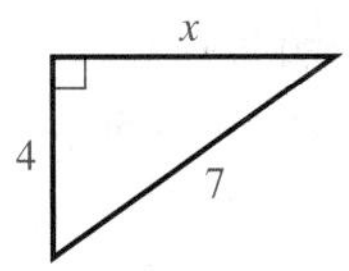

9.

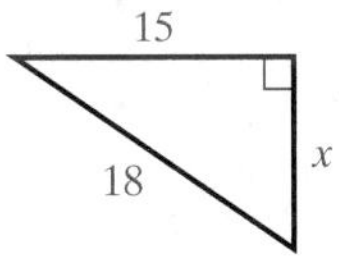

10.

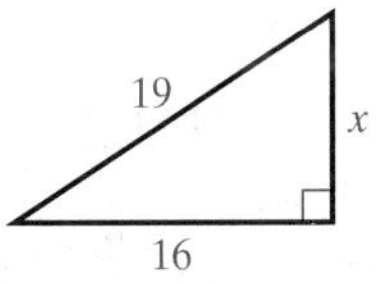

11.

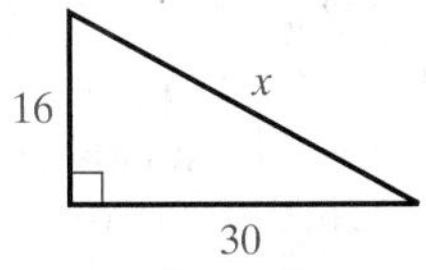

12.

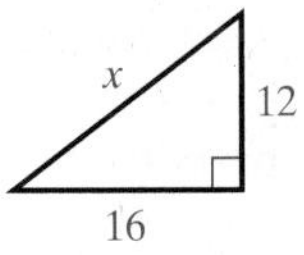

13.

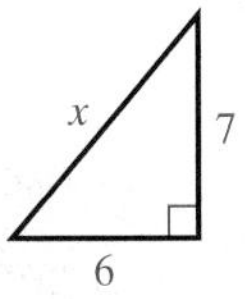

14.

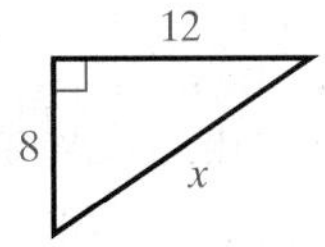

15.

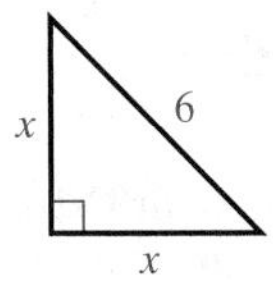

16.

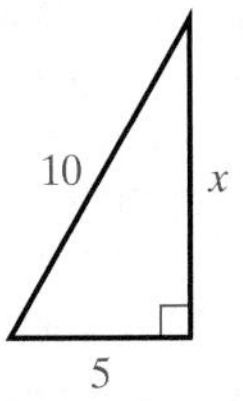

Does each set of numbers form a Pythagorean triple? Explain.

17. 4, 5, 6

18. 9, 11, 13

19. 10, 24, 26

20. 15, 20, 25

Solve. Give exact answers and two-decimal-place approximations where appropriate. See Example 3.

21. A wire is needed to support a vertical pole 15 feet tall. The cable will be anchored to a stake 8 feet from the base of the pole. How much cable is needed?

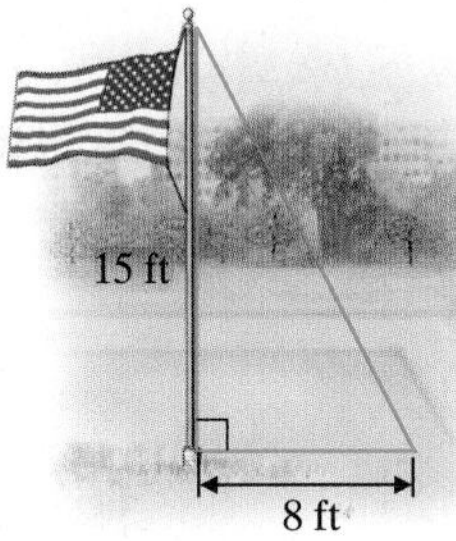

22. One of the tallest structures in the United States is a TV tower in Blanchard, North Dakota. Its height is 2063 feet. A 2382-foot length of wire is to be used as a guy wire attached to the top of the tower. Approximate to the nearest foot how far from the base of the tower the guy wire must be anchored. (*Source:* U.S. Geological Survey)

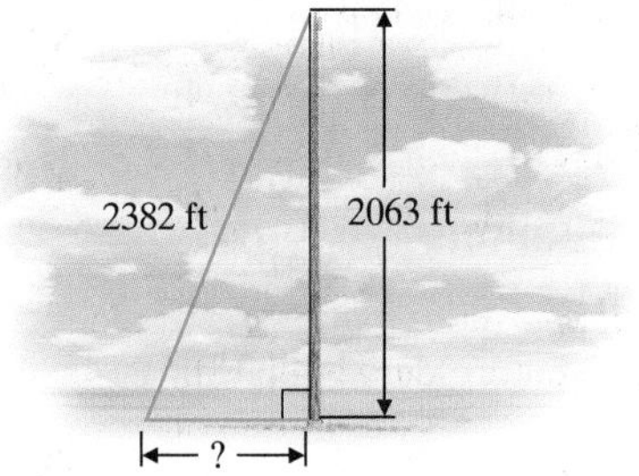

23. A spotlight is mounted on the eaves of a house, 12 feet above the ground. A flower bed runs between the house and the sidewalk, so the closest a ladder can be placed to the house is 5 feet. How long a ladder is needed so that an electrician can reach the place where the light is mounted?

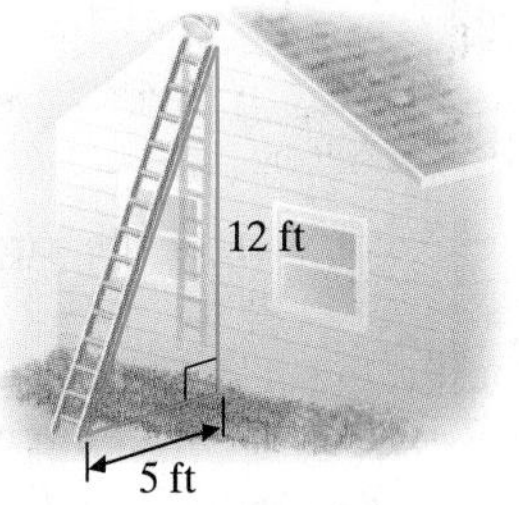

24. A wire is to be attached to support a telephone pole. Because of surrounding buildings, sidewalks, and roadways, the wire must be anchored exactly 15 feet from the base of the pole. Telephone company workers have only 30 feet of cable, and 2 feet of that must be used to attach the cable to the pole and to the stake on the ground. How high from the base of the pole can the wire be attached?

25. A walkway forms one diagonal of a square playground. The walkway is 24 m long. To the nearest meter, how long is a side of the playground?

26. A painter leans a 15-ft ladder against a house. The base of the ladder is 5 ft from the house. To the nearest tenth of a foot, how high on the house does the ladder reach?

Is each triangle a right triangle? Explain. See Example 4.

27.
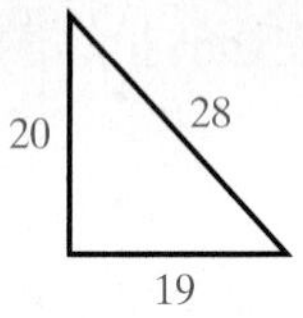

28.
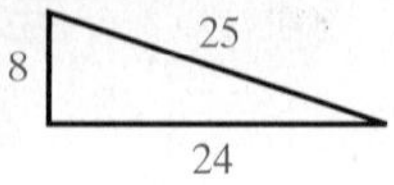

29.
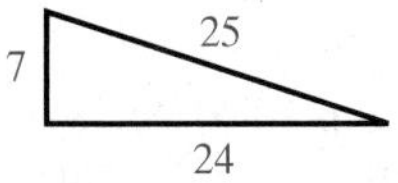

30.
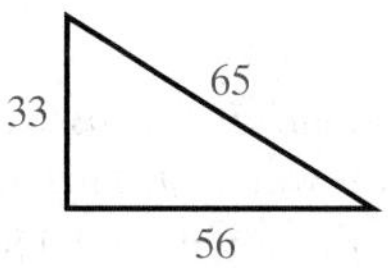

The lengths of the sides of a triangle are given. Classify each triangle as "acute," "right," or "obtuse." See Example 5.

31. 4, 5, 6 **32.** 11, 12, 15 **33.** 0.3, 0.4, 0.6

34. $\sqrt{3}$, 2, 3 **35.** $\sqrt{23}$, $\sqrt{7}$, 4 **36.** 30, 40, 50

CONCEPT EXTENSIONS

37. Multiple Steps You want to embroider a square design. You have an embroidery hoop with a 6-in. diameter.

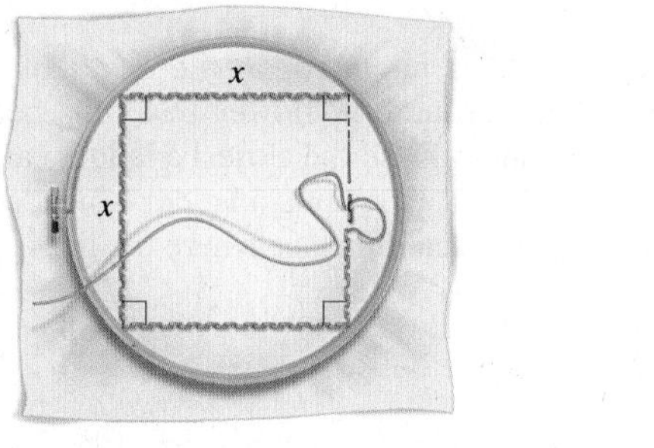

a. What does the diameter of the circle represent in the square?

b. Find the largest value of x so that the entire square will fit in the hoop. Round to the nearest tenth.

PROOF

38. Coordinate Geometry We can use the Pythagorean Theorem to prove the Distance Formula. Let points $P(x_1, y_1)$ and $Q(x_2, y_2)$ be the endpoints of the hypotenuse of a right triangle.

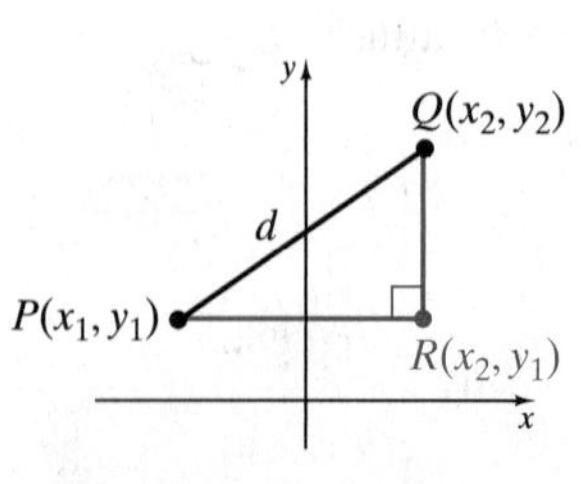

a. Write an algebraic expression to complete each of the following: PR = __?__ and QR = __?__.

b. By the Pythagorean Theorem, $d^2 = PR^2 + QR^2$. Rewrite this statement by substituting the algebraic expressions you found for PR and QR in part a.

c. Complete the proof by taking the square root of each side of the equation that you wrote in part b.

Find the value of x. If your answer is not an integer, express it in simplest radical form.

39.
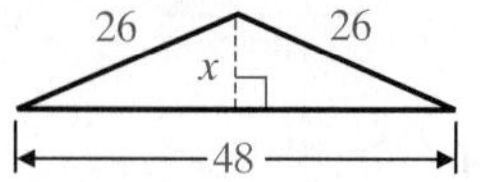

40.
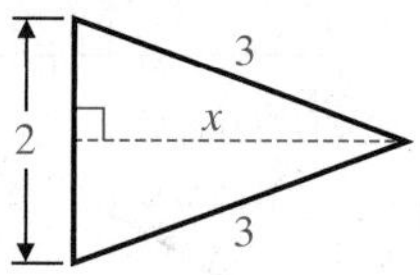

41.
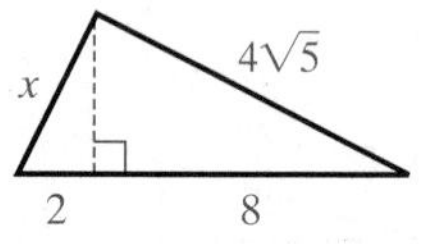

42.
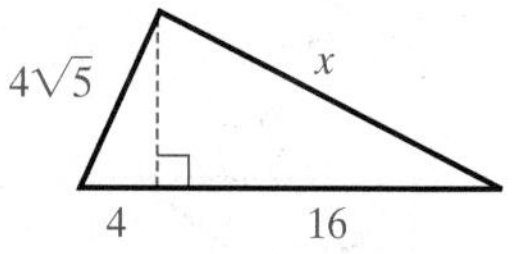

For each pair of numbers, find a third whole number such that the three numbers form a Pythagorean triple. Use the equation $a^2 + b^2 = c^2$.

43. $a = 20, b = 21$ **44.** $a = 14, b = 48$

45. $a = 13, c = 85$ **46.** $a = 12, C = 37$

PROOF

47. Prove the Pythagorean Theorem.

Given: ΔABC is a right triangle.

Prove: $a^2 + b^2 = c^2$

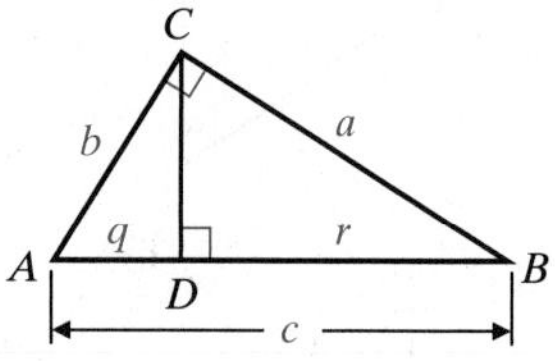

Plan: Write two proportions suggested by Corollary 7.5-3. Next, set the cross products equal in each proportion. Then add the left sides and the right sides of each cross product equation so that one side is $a^2 + b^2$. Study the given figure to see how the other side of the equation becomes c^2.

48. Use the plan and write a proof of Theorem 9.1-2 (Converse of the Pythagorean Theorem).

Given: ΔABC with sides of lengths a, b, and c, where $a^2 + b^2 = c^2$

Prove: ΔABC is a right triangle.

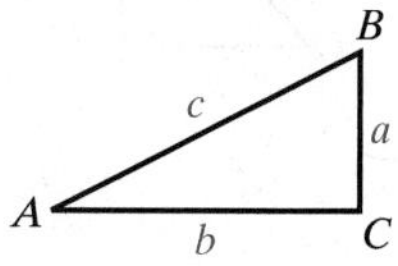

Plan: Draw a right triangle (not ΔABC) with legs of lengths a and b. Label the hypotenuse x. By the Pythagorean Theorem, $a^2 + b^2 = x^2$. Use substitution to compare the lengths of the sides of your triangle and ΔABC. Then prove that the triangles are congruent.

49. Use the plan and write a proof of Theorem 9.1-3.

Given: ΔABC with sides of lengths a, b, and c, where $c^2 > a^2 + b^2$

Prove: ΔABC is an obtuse triangle.

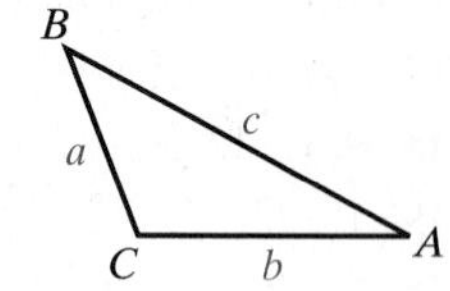

Plan: Draw a right triangle (not ΔABC) with legs of lengths a and b. Label the hypotenuse x. By the Pythagorean Theorem, $a^2 + b^2 = x^2$. Use substitution to compare lengths c and x. Then use the Converse of the Hinge Theorem to compare $\angle C$ to the right angle.

50. Prove Theorem 9.1-4.

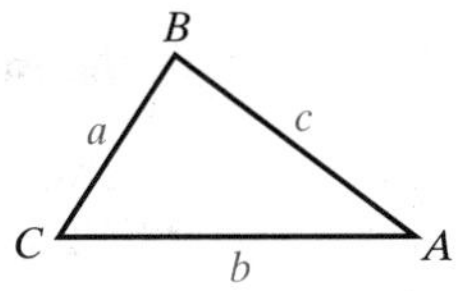

Given: ΔABC with sides of lengths a, b, and c, where $c^2 < a^2 + b^2$

Prove: ΔABC is an acute triangle.

51. Describe the conditions that a set of three numbers must meet in order to form a Pythagorean triple.

52. Find the Error A triangle has side lengths 16, 34, and 30. Your friend says it is not a right triangle. Look at your friend's work and describe the error.

$16^2 + 34^2 \stackrel{?}{=} 30^2$
$256 + 1156 \stackrel{?}{=} 900$
$1412 \neq 900$

Find integers j and k such that (a) the two given integers and j represent the side lengths of an acute triangle and (b) the two given integers and k represent the side lengths of an obtuse triangle.

53. 4, 5 **54.** 2, 4 **55.** 6, 9

56. 5, 10 **57.** 6, 7 **58.** 9, 12

59. The Hubble Space Telescope orbits 600 km above Earth's surface. Earth's radius is about 6370 km. Use the Pythagorean Theorem to find the distance x from the telescope to Earth's horizon. Round your answer to the nearest ten kilometers. (Diagram is not to scale.)

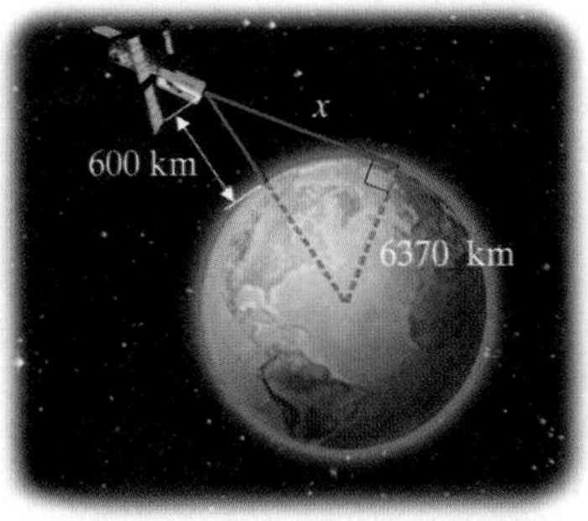

60. Proof Multiple Steps Prove that if the slopes of two lines have product −1, then the lines are perpendicular. Use parts a–c to write a coordinate proof.

a. First, argue that neither line can be horizontal or vertical.

b. Then, tell why the lines must intersect. (*Hint:* Use indirect reasoning.)

c. Use the lines in the given coordinate plane. Choose a point on ℓ_1 and find a related point on ℓ_2. To complete the proof, use $C(a, b)$ and $A(0, 0)$, and find the coordinates of B. Then find distances and show that $AC^2 + BA^2 = CB^2$ so that the angle at A is a right angle.

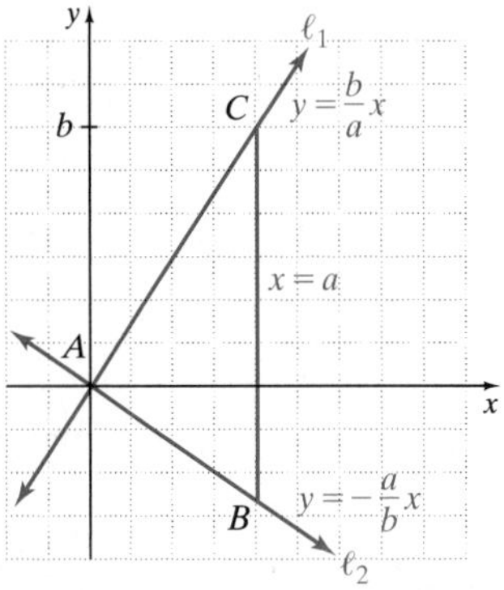

REVIEW AND PREVIEW

Solve. See Section 1.6.

61. Find the complement of a 39° angle.

62. Find the supplement of a 39° angle.

Simplify each expression. See the appendix.

63. $\sqrt{9} \div \sqrt{3}$ **64.** $\sqrt{30} \div \sqrt{2}$ **65.** $\frac{16}{\sqrt{16}}$

66. $\frac{25}{\sqrt{25}}$ **67.** $(\sqrt{7})^2$ **68.** $(\sqrt{11})^2$

Simplify and rationalize each denominator. See the appendix.

69. $\frac{16}{\sqrt{3}}$ **70.** $\frac{8}{\sqrt{5}}$ **71.** $\frac{1}{\sqrt{2}}$ **72.** $\frac{1}{\sqrt{3}}$

9.2 SPECIAL RIGHT TRIANGLES

OBJECTIVES

1 Use the Properties of 45°-45°-90° Triangles.

2 Use the Properties of 30°-60°-90° Triangles.

OBJECTIVE 1 ▶ Using Properties of 45°-45°-90° Triangles. Certain right triangles have properties that allow us to use shortcuts to determine side lengths without using the Pythagorean Theorem.

The acute angles of a right isosceles triangle are both 45° angles. Another name for an isosceles right triangle is a 45°-45°-90° triangle. If each leg has length x and the hypotenuse has length y, we can solve for y in terms of x.

$x^2 + x^2 = y^2$ Use the Pythagorean Theorem.

$2x^2 = y^2$ Combine like terms.

$\sqrt{2 \cdot x^2} = y$ Take the positive square root of each side.

$x\sqrt{2} = y$ Simplify.

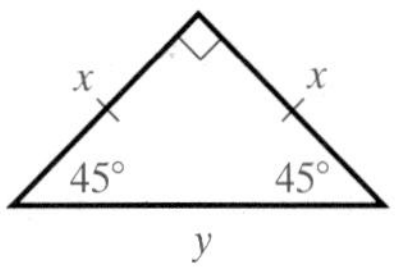

We just proved the following theorem.

> **Helpful Hint**
> For a 45°-45°-90° triangle, if hypotenuse $= \sqrt{2} \cdot$ leg, then $\dfrac{\text{hypotenuse}}{\sqrt{2}} =$ leg.

Theorem 9.2-1 45°-45°-90° Triangle Theorem

In a 45°-45°-90° triangle, both legs are congruent and the length of the hypotenuse is $\sqrt{2}$ times the length of a leg.

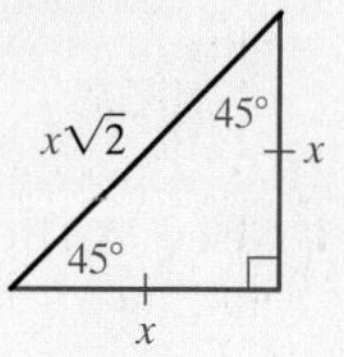

hypotenuse $= \sqrt{2} \cdot$ leg or from the triangle,
hypotenuse $= \sqrt{2} \cdot x$ or $x\sqrt{2}$

EXAMPLE 1 Finding the Length of the Hypotenuse

Find the value of each variable.

a.

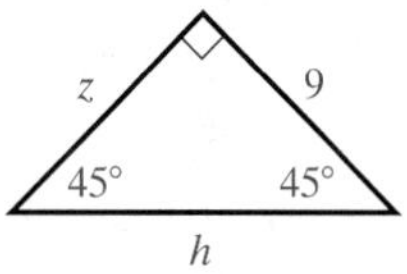

b.

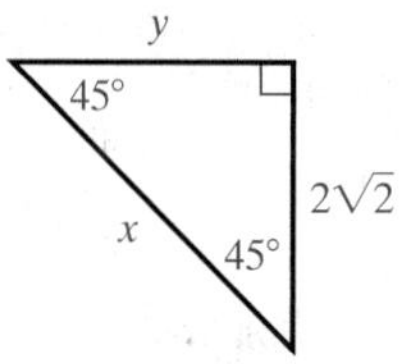

Solution Because both triangles are isosceles right triangles, we use the formula hypotenuse $= \sqrt{2} \cdot$ leg.

a. hypotenuse $= \sqrt{2} \cdot$ leg   45°-45°-90° Δ Theorem
$h = \sqrt{2} \cdot 9$ Substitute.
$h = 9\sqrt{2}$ Simplify.

Thus, $h = 9\sqrt{2}$ units. Also, $z = 9$ units since the triangle is isosceles.

b. hypotenuse $= \sqrt{2} \cdot$ leg   45°-45°-90° Δ Theorem
$x = \sqrt{2} \cdot 2\sqrt{2}$ Substitute.
$x = 4$ Simplify.

Thus, $x = 4$ units. Also, $y = 2\sqrt{2}$ units since the triangle is isosceles.

Parts **a** and **b** can be checked using the Pythagorean Theorem to see that $(\text{leg}_1)^2 + (\text{leg}_2)^2 = (\text{hypotenuse})^2$. □

> **Helpful Hint**
> If you forget the formula used for Example 1, you can always use the Pythagorean Theorem and/or your knowledge of isosceles triangles.

PRACTICE 1 Find the length of the hypotenuse of a 45°-45°-90° triangle with leg length $5\sqrt{3}$.

EXAMPLE 2 Finding the Length of a Leg

Multiple Choice Find the value of x.

a. 3   **b.** $3\sqrt{2}$ **c.** 6   **d.** $6\sqrt{2}$

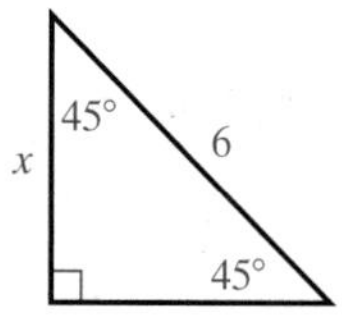

Solution We begin with our formula for a 45°-45°-90° triangle.

hypotenuse $= \sqrt{2} \cdot$ leg   45°-45°-90° Triangle Theorem
$6 = \sqrt{2} \cdot x$ Substitute.

Here, we are solving for a leg, so we need to divide both sides of the equation by $\sqrt{2}$.

$x = \dfrac{6}{\sqrt{2}}$ Divide both sides by $\sqrt{2}$.

$x = \dfrac{6}{\sqrt{2}} \cdot \dfrac{\sqrt{2}}{\sqrt{2}}$ Multiply by 1 in the form of $\dfrac{\sqrt{2}}{\sqrt{2}}$ to rationalize the denominator.

$x = \dfrac{6\sqrt{2}}{2}$ Simplify.

$x = 3\sqrt{2}$ Simplify.

The correct answer is **b**. □

PRACTICE 2 The length of the hypotenuse of a 45°-45°-90° triangle is 10. Find the length of one leg.

When we apply the 45°-45°-90° Triangle Theorem to a real-life example, we are sometimes asked to approximate the answer.

EXAMPLE 3 **Finding Distance**

A high school softball diamond is a square. The distance from base to base is 60 ft. To the nearest foot, how far does a catcher throw the ball from home plate to second base?

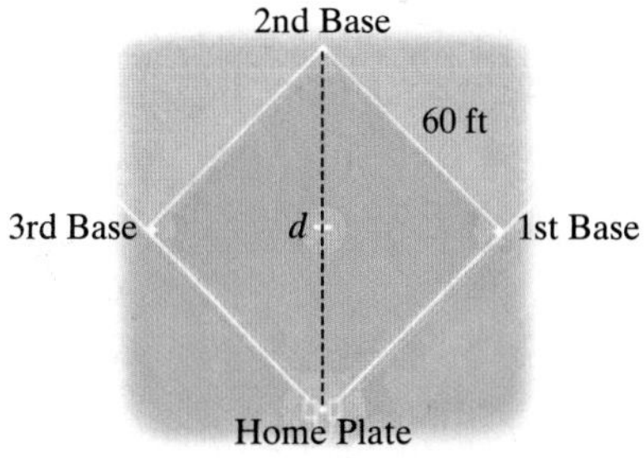

> **Helpful Hint**
> Don't forget that multiplication is commutative, so $\sqrt{2} \cdot \text{leg} = \text{leg} \cdot \sqrt{2}$. Also, $\sqrt{2} \cdot 60 = 60\sqrt{2}$.

Solution Because the distance from base to base along the square is the same, the distance d is the length of the hypotenuse of a 45°-45°-90° triangle.

$$d = 60\sqrt{2} \qquad \text{hypotenuse} = \sqrt{2} \cdot \text{leg} (\text{or leg} \cdot \sqrt{2})$$
$$d \approx 84.85281374 \qquad \text{Use a calculator.}$$

The catcher throws the ball about 85 ft from home plate to second base. □

PRACTICE
3 You plan to build a path along one diagonal of a 100-ft-by-100-ft square garden. To the nearest foot, how long will the path be?

OBJECTIVE 2 ▶ Using the Properties of 30°-60°-90° Triangles. Another type of special right triangle is a 30°-60°-90° triangle.

Theorem 9.2-2 30°-60°-90° Triangle Theorem

In a 30°-60°-90° triangle, the length of the hypotenuse is twice the length of the shorter leg. The length of the longer leg is $\sqrt{3}$ times the length of the shorter leg.

$$\text{hypotenuse} = 2 \cdot \text{shorter leg}$$
$$\text{longer leg} = \sqrt{3} \cdot \text{shorter leg}$$

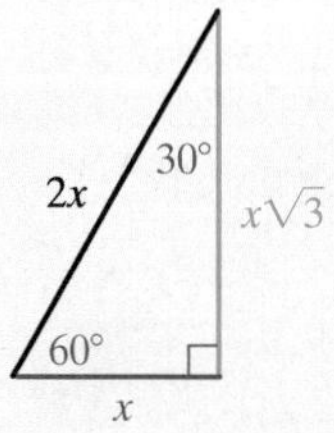

Proof of Theorem 9.2-2: 30°-60°-90° Triangle Theorem

Given: The figure at the right.

For equilateral ΔWXZ, altitude $\overline{WY}$ bisects $\angle W$ and is the perpendicular bisector of $\overline{XZ}$. So, $\overline{WY}$ divides ΔWXZ into two congruent 30°-60°-90° triangles.

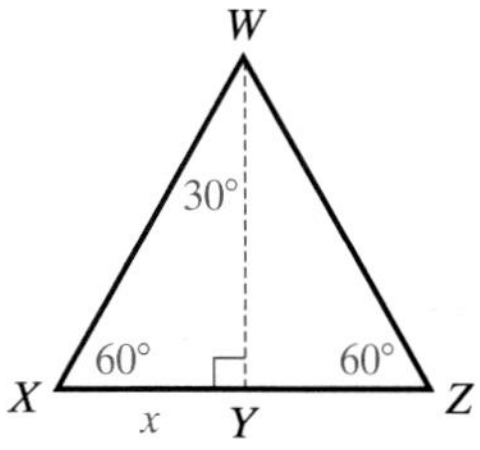

Thus, $XY = \frac{1}{2}XZ = \frac{1}{2}XW$, or $XW = 2XY = 2x$.

$XY^2 + YW^2 = XW^2$	Use the Pythagorean Theorem.
$x^2 + YW^2 = (2x)^2$	Substitute x for XY and $2x$ for XW.
$YW^2 = 4x^2 - x^2$	Subtract x^2 from both sides.
$YW^2 = 3x^2$	Combine like terms.
$YW = \sqrt{3x^2}$	Take the positive square root of each side.
$YW = x\sqrt{3}$	Simplify.

We can also use the 30°-60°-90° Triangle Theorem to find side lengths. With 30°-60°-90° triangles, we must be able to identify the longer leg and the shorter leg, in addition to the hypotenuse.

EXAMPLE 4 Finding the Length of the Shorter Leg

Find the value of d in simplest radical form.

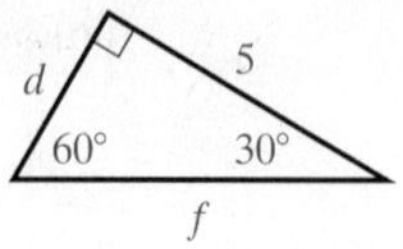

Solution We are given the length of the longer leg and are asked to find the length of the shorter leg, so that tells us the formula to start with.

$$\text{longer leg} = \sqrt{3}\cdot\text{shorter leg}$$

$$5 = d\sqrt{3} \quad \text{Substitute 5 for longer leg and } d \text{ for shorter leg.}$$

To solve for d, we divide both sides by $\sqrt{3}$.

$$\frac{5}{\sqrt{3}} = d \quad \text{Divide both sides by } \sqrt{3}.$$

$$\frac{5}{\sqrt{3}}\cdot\frac{\sqrt{3}}{\sqrt{3}} = d \quad \text{Rationalize the denominator by multiplying by 1 in the form } \frac{\sqrt{3}}{\sqrt{3}}.$$

$$\frac{5\sqrt{3}}{3} = d \quad \text{Simplify.}$$

Thus, the length of d is exactly $\frac{5\sqrt{3}}{3}$ units. □

PRACTICE 4 In Example 4, find the value of f in simplest radical form.

EXAMPLE 5 Applying the 30°-60°-90° Triangle Theorem

An artisan makes pendants in the shape of equilateral triangles. The height of each pendant is 18 mm. What is the length s of each side of a pendant to the nearest tenth of a millimeter?

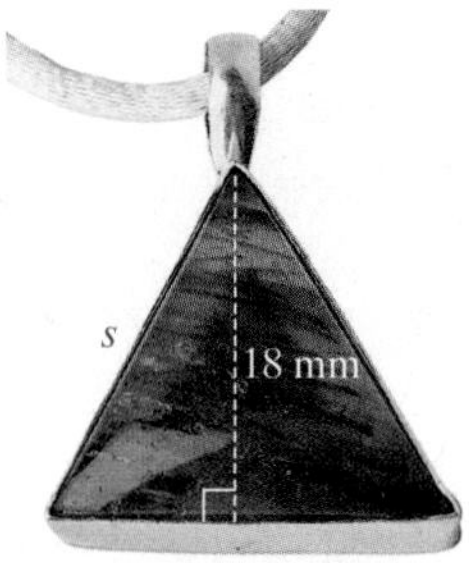

Solution The equilateral triangle can be divided into two 30°-60°-90° triangles, as shown. The hypotenuse of each 30°-60°-90° triangle is s. The shorter leg is $\frac{1}{2}s$. Now let's use the formula:

$$\text{longer leg} = \sqrt{3}\cdot\text{shorter leg}$$

$$18 = \sqrt{3}\left(\frac{1}{2}s\right) \quad \text{Substitute.}$$

$$18 = \frac{\sqrt{3}}{2}s \quad \text{Simplify.}$$

$$\frac{2}{\sqrt{3}}\cdot 18 = \frac{2}{\sqrt{3}}\cdot\frac{\sqrt{3}}{2}s \quad \text{Multiply each side by } \frac{2}{\sqrt{3}}.$$

$$\frac{36}{\sqrt{3}} = s \quad \text{Simplify.}$$

$$s \approx 20.78460969 \quad \text{Use a calculator to approximate.}$$

Each side of a pendant is about 20.8 mm long. □

PRACTICE 5 Suppose the sides of a pendant in the shape of an equilateral triangle are 18 mm long. What is the height of the pendant to the nearest tenth of a millimeter?

VOCABULARY & READINESS CHECK

Fill in the Blank *Fill in the blanks for these special triangles.*

1. $m\angle B =$ ____________

2. $m\angle F =$ ____________

3. $a =$ ____________

4. $c =$ ____________

5. $d =$ ____________

6. $f =$ ____________

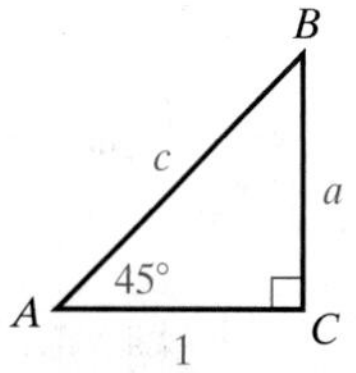

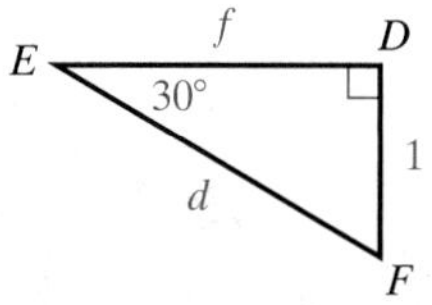

9.2 EXERCISE SET MyMathLab®

Find the value of each variable. If your answer is not an integer, write it in simplest radical form with the denominator rationalized. See Examples 1, 2, and 4.

1.

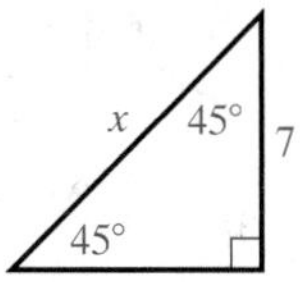

2.

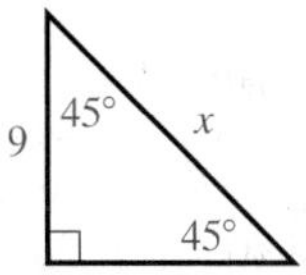

3.

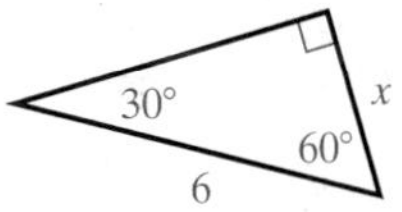

4.

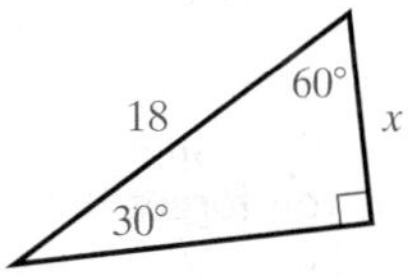

5.

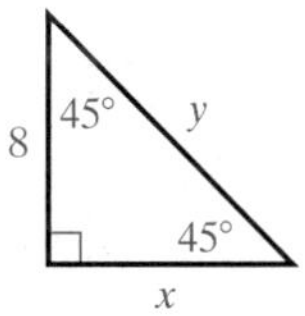

6.

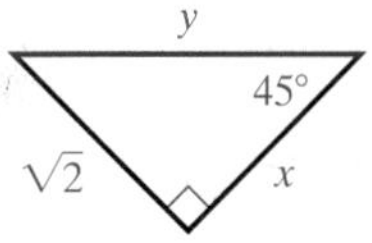

7.

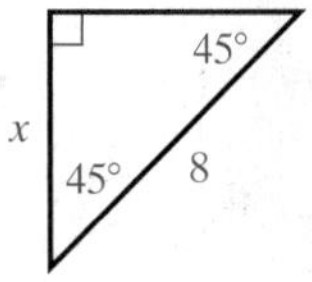

8.

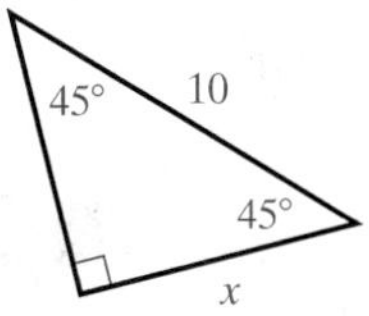

9.

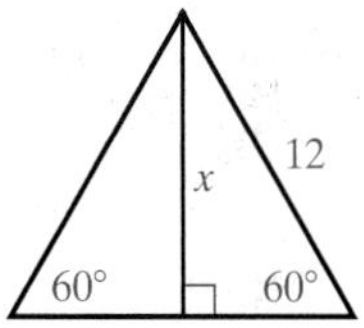

10.

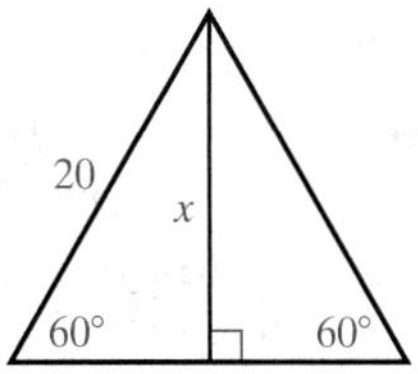

11.

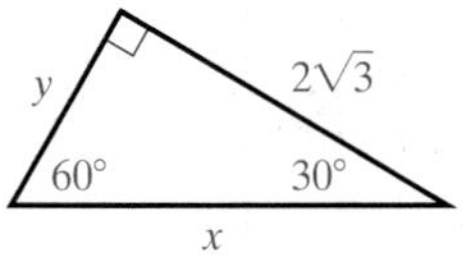

12.

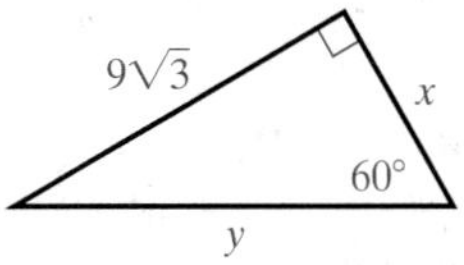

13.

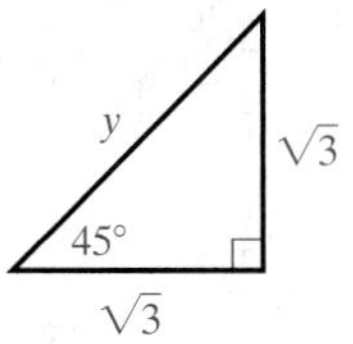

14.

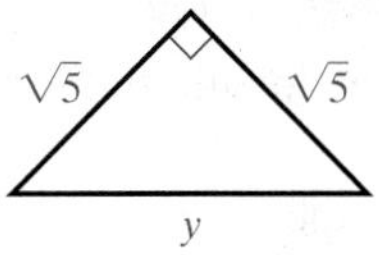

15.

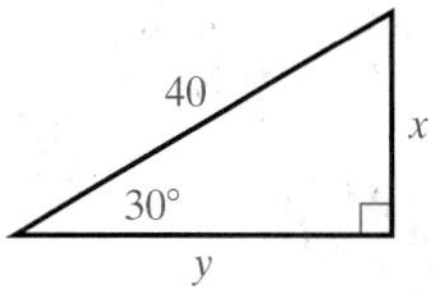

16.

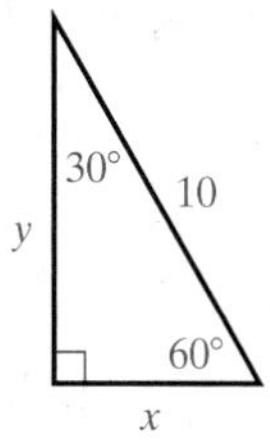

17.

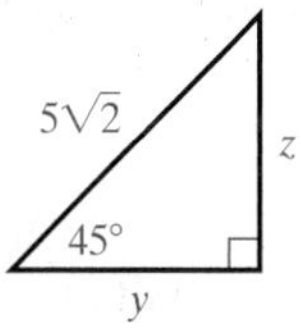

18.

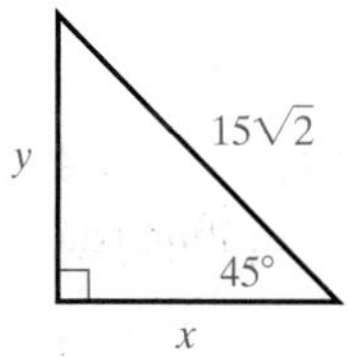

19.

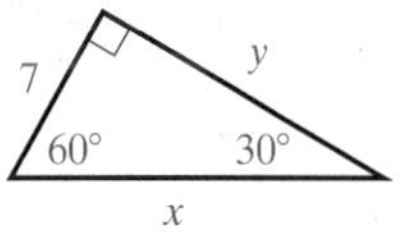

20.

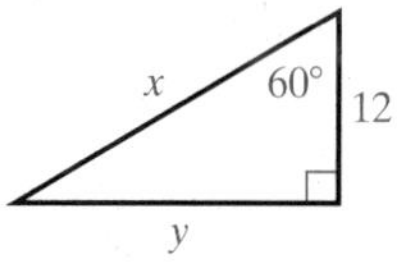

Solve. See Examples 3 and 5.

21. What is the side length of the smallest square plate on which a 20-cm chopstick can fit along a diagonal without any overhang? Round your answer to the nearest tenth of a centimeter.

22. The four blades of a helicopter meet at right angles and are all the same length. The distance between the tips of two adjacent blades is 36 ft. How long is each blade? Round your answer to the nearest tenth of a foot.

23. An escalator lifts people to the second floor of a building, 25 ft above the first floor. The escalator rises at a 30° angle. To the nearest foot, how far does a person travel from the bottom to the top of the escalator?

24. Jefferson Park sits on one square city block 300 ft on each side. Sidewalks across the park join opposite corners. To the nearest foot, how long is each diagonal sidewalk?

25. A bridge has 45°-45°-90° right triangular supports. If the hypotenuse of each support is 10 ft, find the length of each leg of a support.

26. A 12-ft ladder was placed on the side of a house so that it formed the hypotenuse of a 45°-45°-90° right triangle. Use this information to find how far up the side of the house the ladder lies.

27. A drafting triangle is in the shape of a 30°-60°-90° right triangle. If the manufacturer wants the hypotenuse of the outer triangle to be 1 ft, find the lengths of the legs of the triangle, in inches.

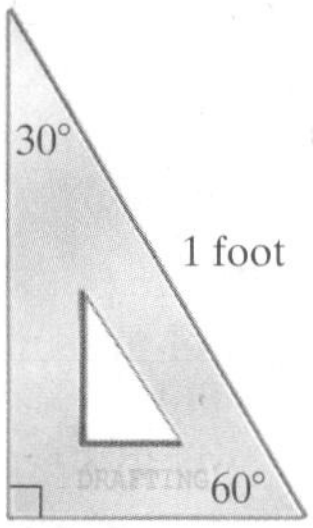

28. A home owner finds that one side of the upper part of his house forms a 30°-60°-90° right triangle. If the length of the longer leg is measured to be 20 ft, find the length of the hypotenuse, to the nearest foot.

CONCEPT EXTENSIONS

Mixed Practice *Find the value of each variable. If your answer is not an integer, express it in simplest radical form. See Examples 1–5.*

29.

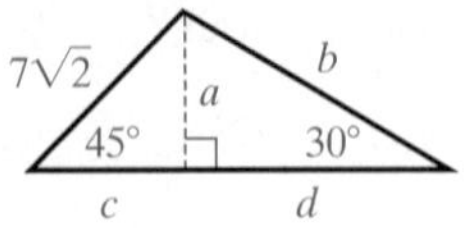

30.

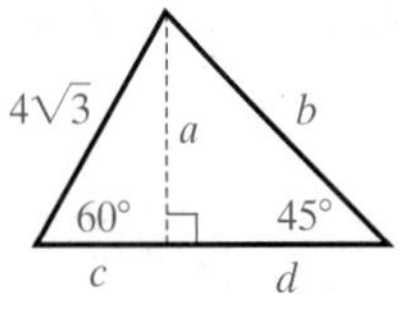

31.

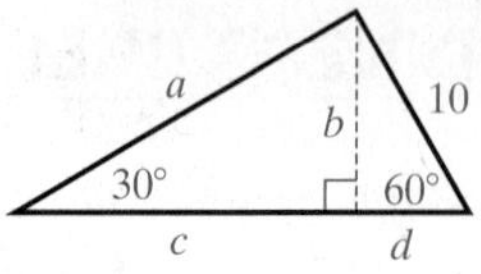

32.

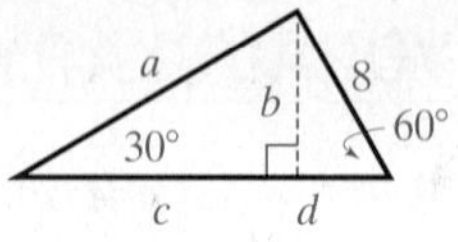

33.

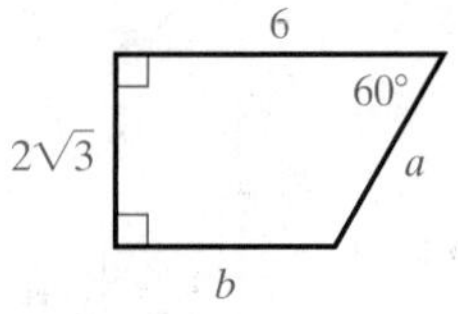

34.

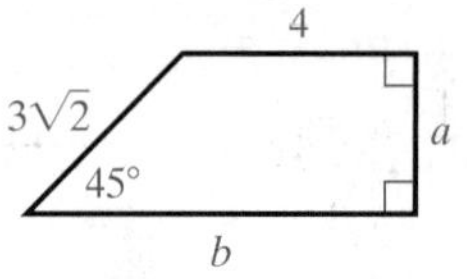

35.

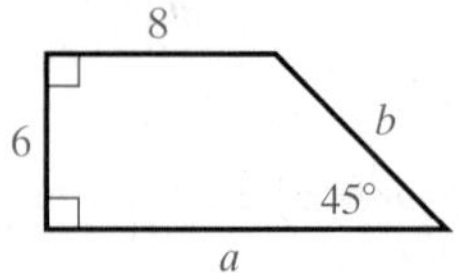

36.

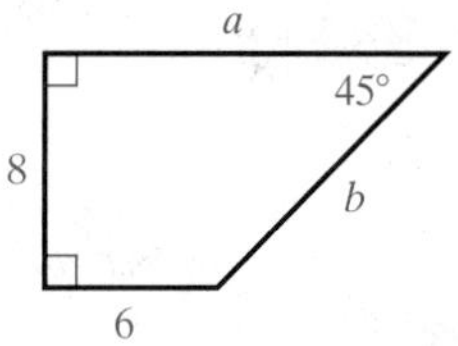

37. Find the Error Sandra drew the triangle shown here. Rika said that the labeled lengths are not possible. With which student do you agree? Explain.

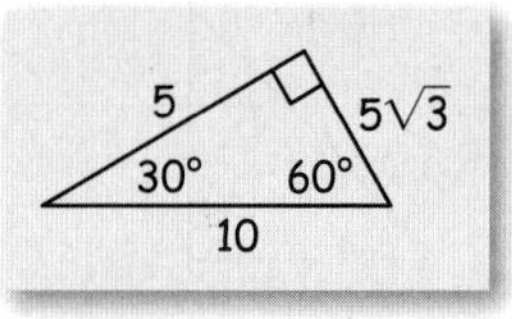

38. A test question asks you to find two side lengths of a 45°-45°-90° triangle. You know that the length of one leg is 6, but you forgot the special formula for 45°-45°-90° triangles. Explain how you can still determine the other side lengths. What are the other side lengths?

Multiple Steps *A farmer's conveyor belt carries bales of hay from the ground to the barn loft. The conveyor belt moves at 100 ft/min.*

39. Which part of a right triangle does the conveyor belt represent?

40. How are minutes and seconds related?

41. Find the length of the conveyor belt.

42. How many *seconds* does it take for a bale of hay to go from the ground to the barn loft?

Multiple Steps *After heavy winds damaged a house, workers placed a 6-m brace against its side at a 45° angle. Then, at the same spot on the ground, they placed a second, longer brace to make a 30° angle with the side of the house.*

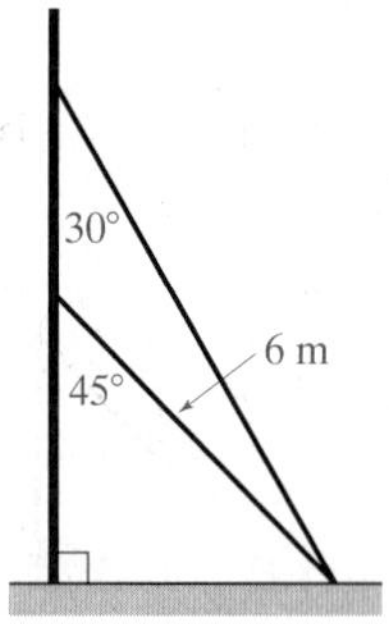

43. How long is the longer brace? Round to the nearest tenth of a meter.

44. To the nearest tenth of a meter, how much higher does the longer brace reach than the shorter brace?

45. Constructions Construct a 30°-60°-90° triangle using a segment that is the given side.

a. the shorter leg **b.** the hypotenuse **c.** the longer leg

46. **Constructions** Construct a 45°-45°-90° triangle using a segment that is the given side.

a. a leg **b.** the hypotenuse

Geometry in Three Dimensions *Find the length d, in simplest radical form, of the diagonal of a cube with edges of the given length.*

47. 1 unit

48. 2 units

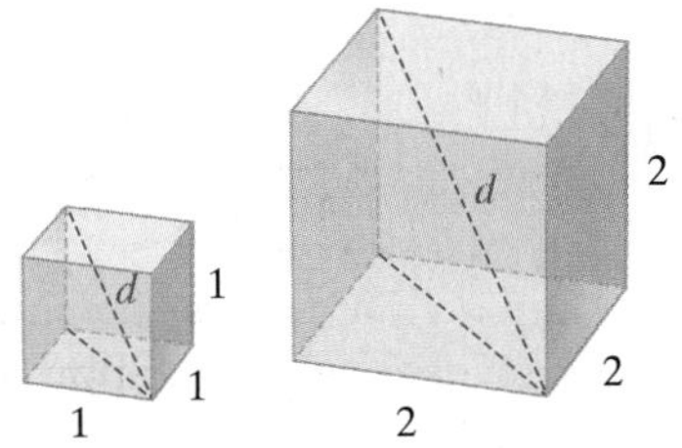

REVIEW AND PREVIEW

Solve. See Section 9.1

49. A right triangle has a 6-in. hypotenuse and a 5-in. leg. Find the length of the other leg in simplest radical form.

50. An isosceles triangle has 20-cm legs and a 16-cm base. Find the length of the altitude to the base in simplest radical form.

Solve each proportion. See Section 7.1.

51. $\frac{x}{3} = \frac{4}{7}$

52. $\frac{6}{11} = \frac{x}{9}$

53. $\frac{8}{15} = \frac{4}{x}$

54. $\frac{5}{x} = \frac{7}{12}$

9.3 TRIGONOMETRIC RATIOS

OBJECTIVES

1 Use the Sine, Cosine, and Tangent Ratios to Determine Side Lengths in Right Triangles.

2 Use the Sine, Cosine, and Tangent Ratios to Determine Angle Measures in Right Triangles.

VOCABULARY

- trigonometric ratios
- sine
- cosine
- tangent

If we know certain combinations of side lengths and angle measures of a right triangle, we can use trigonometric ratios to find other side lengths and angle measures.

OBJECTIVE 1 ▶ Using Sine, Cosine, and Tangent Ratios to Find Side Lengths. The word *trigonometry* is derived from ancient Greek and means "three-angle measurement" or "triangle measurement."

Recall that any two similar right triangles have equal angle measures and the ratios of their corresponding sides are equivalent. Because these similar right triangles have equal ratios, we call these ratios **trigonometric ratios** and we give them special names, as we see next.

Trigonometric Ratios

Let ΔABC be a right triangle with acute $\angle A$.

$$\textbf{sine of } \angle A = \frac{\text{length of leg opposite } \angle A}{\text{length of hypotenuse}} = \frac{a}{c}$$

$$\textbf{cosine of } \angle A = \frac{\text{length of leg adjacent to } \angle A}{\text{length of hypotenuse}} = \frac{b}{c}$$

$$\textbf{tangent of } \angle A = \frac{\text{length of leg opposite } \angle A}{\text{length of leg adjacent to } \angle A} = \frac{a}{b}$$

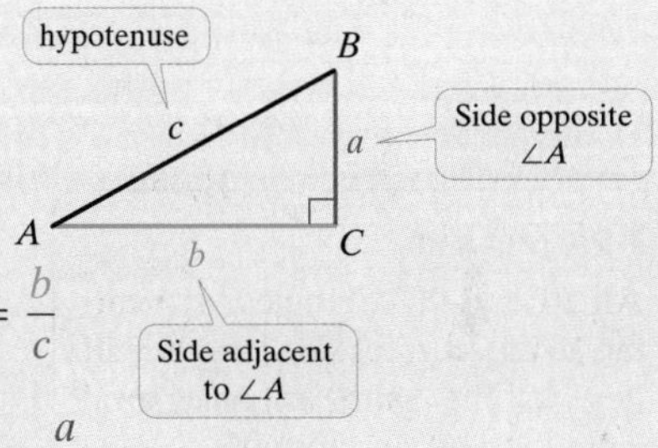

We can abbreviate the ratios as

$$\sin A = \frac{\text{opp.}}{\text{hyp.}}, \cos A = \frac{\text{adj.}}{\text{hyp.}}, \text{ and } \tan A = \frac{\text{opp.}}{\text{adj.}}$$

▶ Helpful Hint

Make sure you understand where the angle is located and where the ratio value is located in a trigonometric equation.

$$\sin A = \frac{a}{c}$$

a ratio or a value

an angle

EXAMPLE 1 **Writing Trigonometric Ratios**

What are the sine, cosine, and tangent ratios for $\angle T$?
(See the figure on the next page.)

Solution The opposite value and the adjacent value depend on the angle. We are interested in $\angle T$, so opposite $\angle T$ is 8 units and the leg adjacent to $\angle T$ is 15 units.

	Exact values	Approximations
$\sin T = \dfrac{\text{opp.}}{\text{hyp.}} =$	$\dfrac{8}{17}$	≈ 0.4706
$\cos T = \dfrac{\text{adj.}}{\text{hyp.}} =$	$\dfrac{15}{17}$	≈ 0.8824
$\tan T = \dfrac{\text{opp.}}{\text{adj.}} =$	$\dfrac{8}{15}$	≈ 0.5333

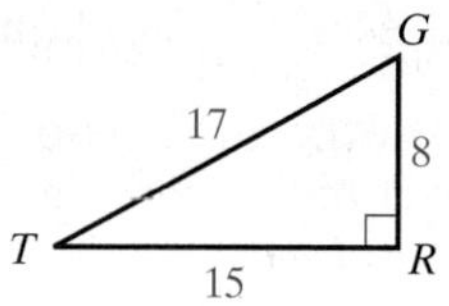

□

PRACTICE
1 Use the triangle in Example 1. What are the sine, cosine, and tangent ratios for $\angle G$? Give exact values and four-decimal-place approximations.

Let's now find the trigonometric values of the special right triangles we learned about in the last section.

EXAMPLE 2 Writing Trigonometric Ratios for 45°

Find the sine, the cosine, and the tangent of 45°. Give an exact value and a four-decimal-place approximation.

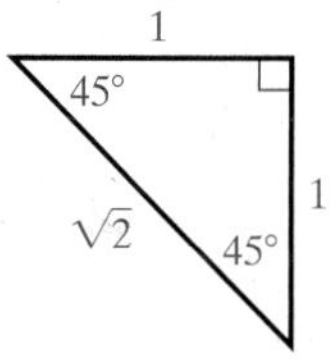

> **Helpful Hint**
> All 45°-45°-90° triangles are similar, so any correct side lengths will give us the same trigonometric values for 45°.

Solution Since the legs are the same length, choose one of the two 45° angles.

a form of 1

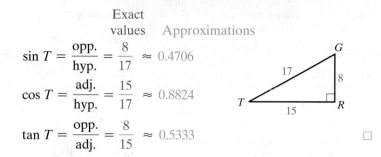

$$\sin 45° = \frac{\text{opp.}}{\text{hyp.}} = \frac{1}{\sqrt{2}} = \frac{1}{\sqrt{2}} \cdot \frac{\sqrt{2}}{\sqrt{2}} = \frac{\sqrt{2}}{\sqrt{2}} \approx 0.7071$$

$$\cos 45° = \frac{\text{adj.}}{\text{hyp.}} = \frac{1}{\sqrt{2}} = \frac{\sqrt{2}}{2} \approx 0.7071$$

$$\tan 45° = \frac{\text{opp.}}{\text{adj.}} = \frac{1}{1} = 1$$

□

> **Helpful Hint**
> All 30°-60°-90° triangles are similar, so any correct side lengths will give us the same trigonometric values for 30° and for 60°.

PRACTICE
2 Find the sine, the cosine, and the tangent of 30° and 60°. Give an exact value and a four-decimal-place approximation.

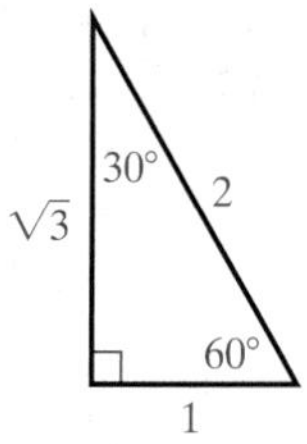

It is important to note that the trigonometric ratio of an angle depends on the angle and not on the size of the triangle. Why? Remember that similar triangles have equal corresponding angles and that their corresponding sides are in proportion, which means that their ratios of corresponding sides are the same.

To see this, let's find the sine, the cosine, and the tangent of 45° using a triangle similar to the triangle in Example 2, but larger.

$$\sin 45° = \frac{1.5}{1.5(\sqrt{2})} = \frac{1}{\sqrt{2}} = \frac{\sqrt{2}}{2} \approx 0.7071$$

$$\cos 45° = \frac{1.5}{1.5(\sqrt{2})} = \frac{1}{\sqrt{2}} = \frac{\sqrt{2}}{2} \approx 0.7071$$

$$\tan 45° = \frac{1.5}{1.5} = 1$$

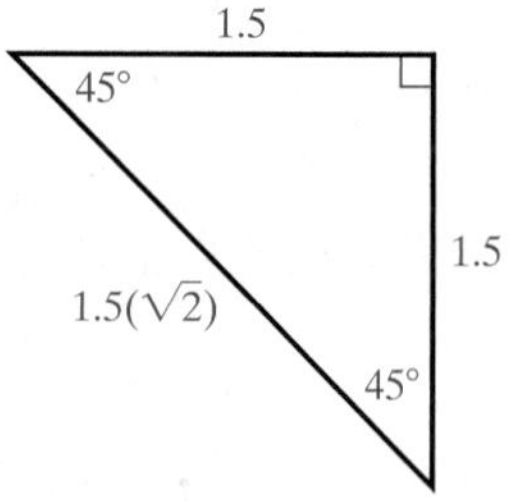

The trigonometric values are the same as in Example 2. Feel free to use your calculator to check the values above.

Important: Before you check any values, make sure your calculator is in degree mode.

Let's now use trigonometry to find an indirect measurement.

EXAMPLE 3 **Finding the Distance Across a Lake**

To find the distance across a lake, a surveyor took the measurements shown in the figure. Use these measurements to determine how far it is across the lake. Round to the nearest yard.

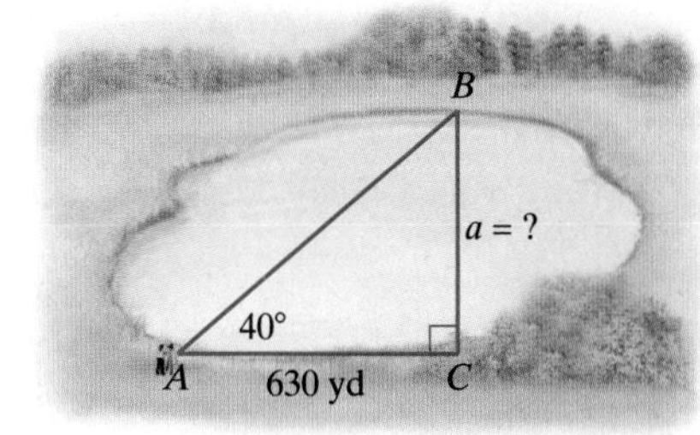

Solution Notice the right triangle formed. If we had the lengths of two sides, we could use the Pythagorean Theorem, but we do not.

We do have an angle of 40°. The unknown value is opposite the 40° angle, and 630 yd is adjacent to the 40° angle. The trigonometric function having to do with opposite and adjacent is tangent.

$$\tan 40^\circ = \frac{\text{opp.}}{\text{adj.}} \quad \text{The ratio for } \tan 40^\circ$$

$$\tan 40^\circ = \frac{a}{630} \quad \text{Substitute.}$$

$$630 \cdot \tan 40^\circ = 630 \cdot \frac{a}{630} \quad \text{Multiply both sides by 630.}$$

$$630 \tan 40^\circ = a \quad \text{Simplify.}$$

$$630(0.8391) \approx a \quad \text{Approximate } \tan 40^\circ.$$

$$529 \approx a \quad \text{Approximate the product.}$$

The distance across the lake is approximately 529 yards. □

PRACTICE

3 A plane rises from takeoff and flies at an angle of 10° with the horizontal runway. When it has gained 500 feet, find the distance, to the nearest foot, the plane has flown.

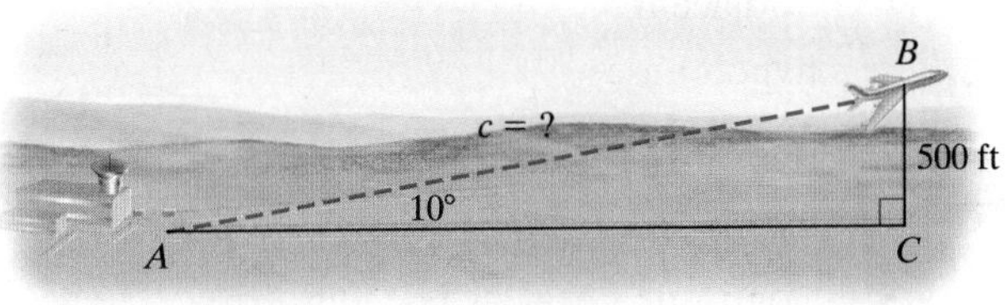

OBJECTIVE 2 ▶ Using Sine, Cosine, and Tangent Ratios to Find Angle Measures. If we know the sine, cosine, or tangent ratio for an angle, we can use an inverse ($\sin^{-1}$, $\cos^{-1}$, or $\tan^{-1}$) to find the measure of the angle.

EXAMPLE 4 **Using Inverses to Find Angle Measures**

What is $m\angle X$ to the nearest degree?

a.

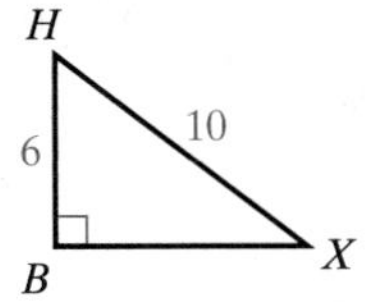

b.

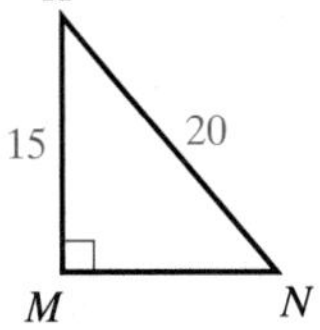

Solution

a. We know the lengths of the hypotenuse and the side opposite $\angle X$.

We use the sine ratio.

$\sin X = \frac{\text{opp.}}{\text{hyp.}}$ Write the ratio.

$\sin X = \frac{6}{10}$ Substitute.

$m\angle X = \sin^{-1}\left(\frac{6}{10}\right)$ Use the inverse to find the angle measure.

SIN⁻¹ 6 ÷ 10 ENTER

$m\angle X \approx 36.86989765°$ Use a calculator.

$\approx 37°$

Thus, $m\angle X \approx 37°$.

Solution

a. We know the lengths of the hypotenuse and the side adjacent to $\angle X$.

We use the cosine ratio.

$\cos X = \frac{\text{adj.}}{\text{hyp.}}$ Write the ratio.

$\cos X = \frac{15}{20}$ Substitute.

$m\angle X = \cos^{-1}\left(\frac{15}{20}\right)$ Use the inverse to find the angle measure.

COS⁻¹ 15 ÷ 20 ENTER

$m\angle X \approx 41.40962211°$ Use a calculator.

$\approx 41°$

Thus, $m\angle X \approx 41°$. □

Helpful Hint

Use the inverse key when you want to find the measure of an angle. Make sure your calculator is in degree mode.

PRACTICE 4

a. Use the figure at the right. What is $m\angle Y$ to the nearest degree?

b. Use the figure at the right. What is $m\angle P$ to the nearest degree?

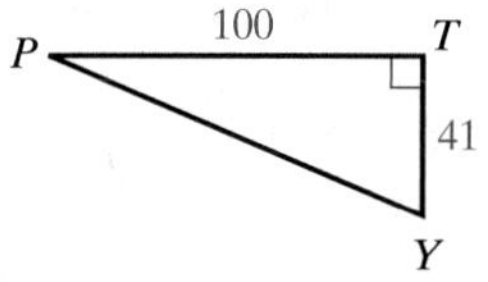

VOCABULARY & READINESS CHECK

Word Bank *Use the choices to fill in each blank.*

tangent cosine sine trigonometric

1. ____________ $A = \frac{\text{adjacent}}{\text{hypotenuse}}$

2. ____________ $A = \frac{\text{opposite}}{\text{adjacent}}$

3. ____________ $A = \frac{\text{opposite}}{\text{hypotenuse}}$

4. A ratio of the sides of a right triangle is called a ____________ ratio.

Fill in the Blank *Use the given triangles to fill in each blank.*

5. $\cos Y =$ _____

6. $\tan X =$ _____

7. $\sin B =$ _____

8. $\tan A =$ _____

9. _____ $X = \frac{12}{13}$

10. _____ $Y = \frac{5}{12}$

11. _____ $B = \frac{a}{c}$

12. _____ $A = \frac{b}{c}$

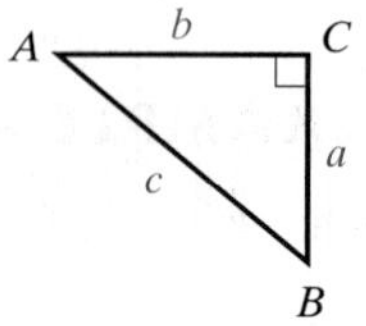

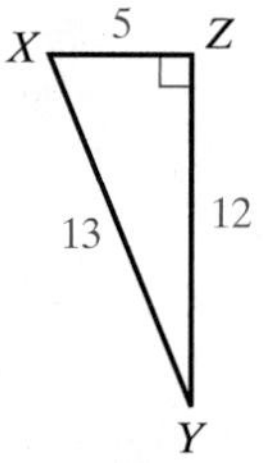

9.3 EXERCISE SET MyMathLab®

Write the ratios for sin M, cos M, and tan M. Give the exact value and a four-decimal-place approximation. See Example 1.

1.
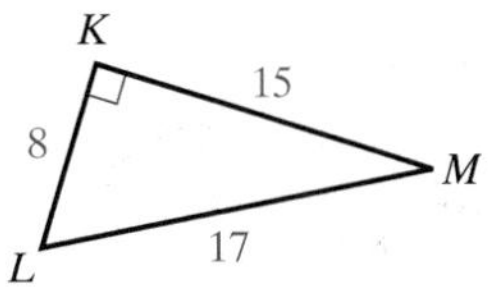

2.
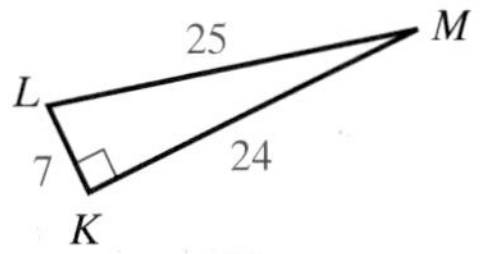

3.
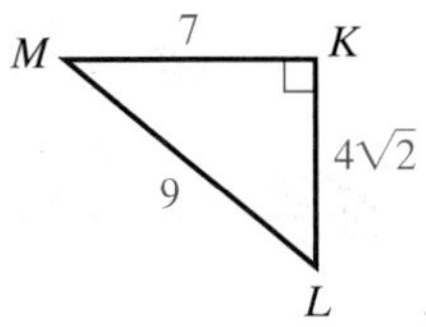

4.
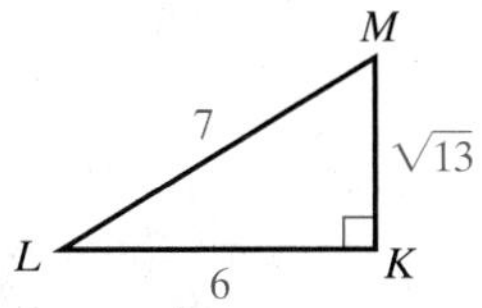

5.
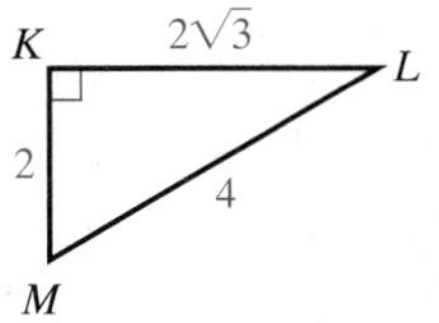

6.
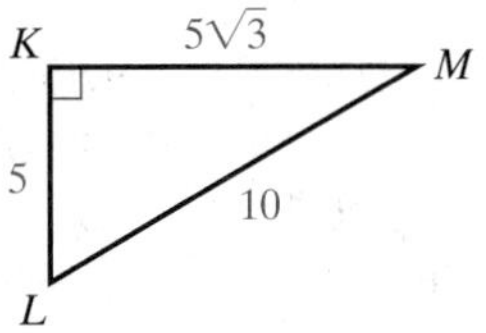

Use a calculator and write a four-decimal-place approximation for each value. See Example 2.

7. sin 35° **8.** cos 11° **9.** tan 45°
10. sin 82° **11.** cos 58° **12.** tan 72°
13. sin 5° **14.** cos 4° **15.** tan 9°
16. sin 21° **17.** cos 44° **18.** tan 63°

Find the value of x. Round to the nearest tenth. See Example 3.

19.
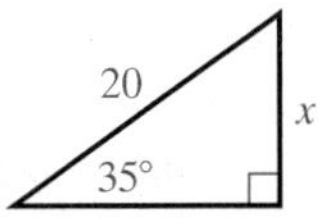

20.
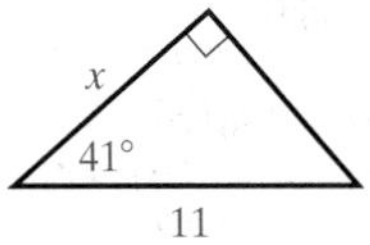

21.
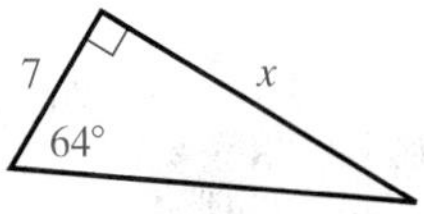

22.
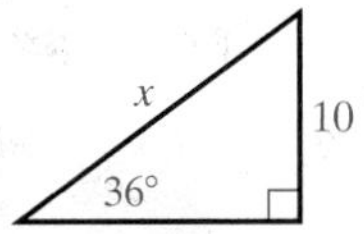

23.
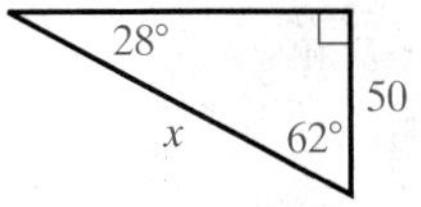

24.
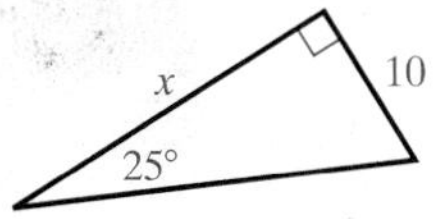

Solve. See Example 3.

25. A road is inclined at an angle of 5°. After driving 5000 feet along this road, find the driver's increase in altitude. Round to the nearest foot.

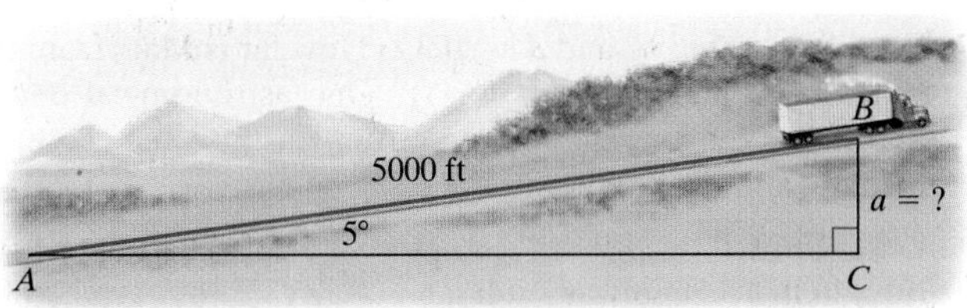

26. A guy wire 75 feet long is attached from the ground to the top of a pole. If the angle between the wire and the pole is 37°, find the height of the pole to the nearest whole foot.

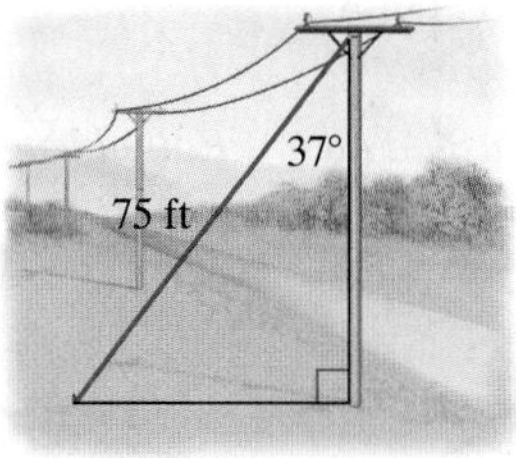

27. A skateboarding ramp is 12 in. high and rises at an angle of 17°. How long is the base of the ramp? Round to the nearest inch.

28. An escalator in the subway station has a vertical rise of 195 ft 9.5 in., and rises at an angle of 10.4°. How long is the escalator? Round to the nearest foot.

Use a calculator to approximate acute ∠A to the nearest whole degree. See Example 4.

29. $\sin A = 0.7$ **30.** $\sin A = 0.6$ **31.** $\cos A = 0.2$
32. $\cos A = 0.3$ **33.** $\tan A = 1.1$ **34.** $\tan A = 3.3$
35. $\sin A = 0.1736$ **36.** $\sin A = 0.4226$
37. $\cos A = 0.9455$ **38.** $\cos A = 0.8387$
39. $\tan A = 0.1944$ **40.** $\tan A = 11.43$

Find the value of x. Round to the nearest degree. See Example 4.

41.
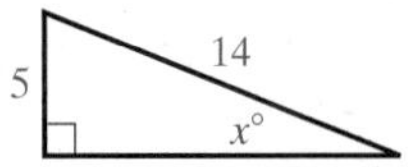

42.
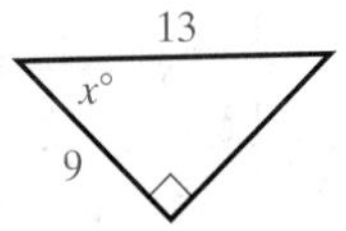

43.
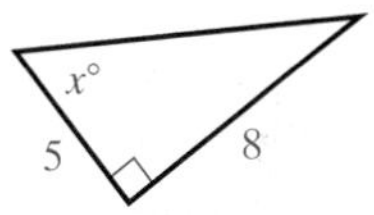

44.
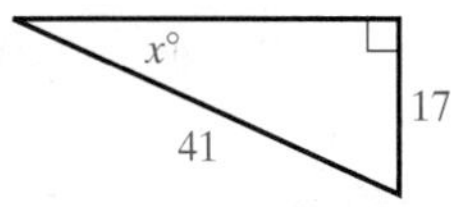

45.
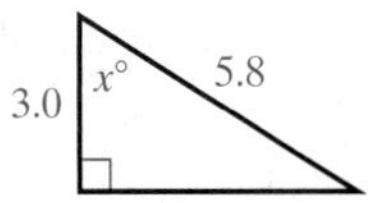

46.
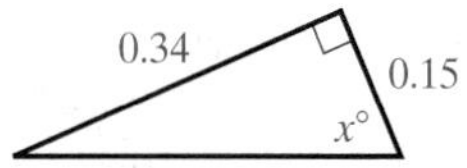

CONCEPT EXTENSIONS

47. The lengths of the diagonals of a rhombus are 2 in. and 5 in. Find the measures of the angles of the rhombus to the nearest degree.

48. Carlos plans to build a grain bin with a radius of 15 ft. The recommended slant of the roof is 25°. He wants the roof to overhang the edge of the bin by 1 ft. What should the length x be? Give your answer in feet and inches.

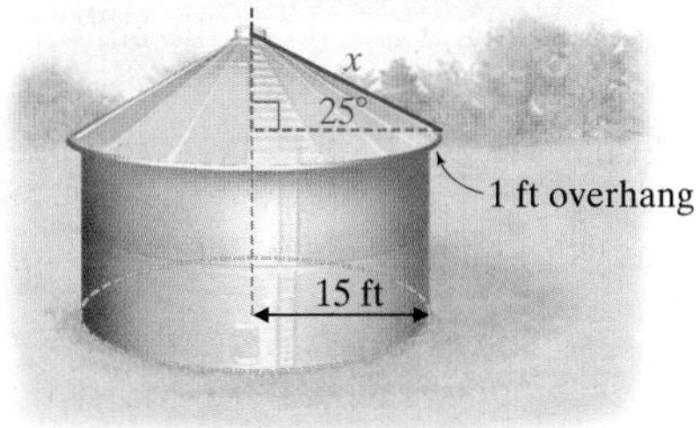

An **identity** *is an equation that is true for all the allowed values of the variable. Write the definition of each trigonometric ratio to show that each equation is an identity.*

49. $\tan X = \dfrac{\sin X}{\cos X}$

50. $\cos X = \dfrac{\sin X}{\tan X}$

51. $\sin X = \cos X \cdot \tan X$

52. $\tan X = \sin X \cdot \dfrac{1}{\cos X}$

Find the values of w and then x. Round lengths to the nearest tenth and angle measures to the nearest degree.

53.

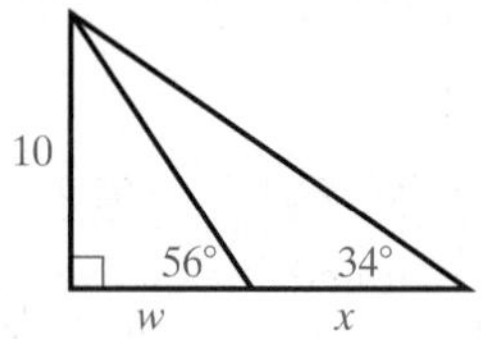

54.

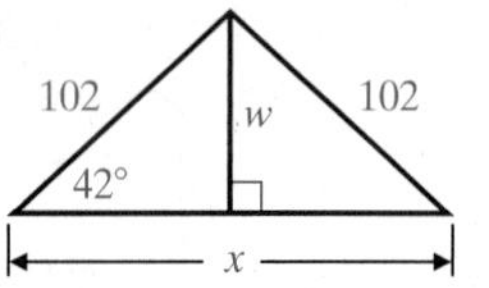

PROOF

For right △ABC with right ∠C, prove each of the following.

55. $\sin A < 1$

56. $\cos A < 1$

57. All but two of the pyramids built by the ancient Egyptians have faces inclined at 52° angles. Suppose an archaeologist discovers the ruins of a pyramid. Most of the pyramid has eroded, but the archaeologist is able to determine that the length of a side of the square base is 82 m. How tall was the pyramid, assuming its faces were inclined at 52°? Round your answer to the nearest meter.

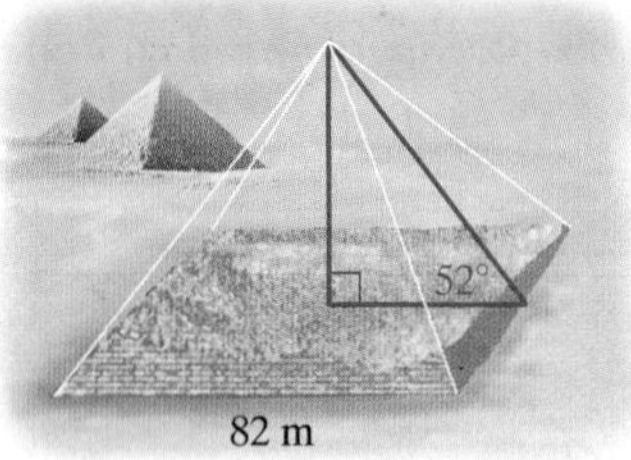

58. Multiple Steps Use the table feature of your graphing calculator to study sin X as X gets close to (but not equal to) 90. In the y = screen, enter Y1 = sin X.

a. Use the tblset feature so that X starts at 80 and changes by 1. Access the table. From the table, what is sin X for X = 89?

b. Perform a "numerical zoom-in." Use the tblset feature so that X starts with 89 and changes by 0.1. What is sin X for X = 89.9?

c. Continue to zoom in numerically on values close to 90. What is the greatest value you can get for sin X on your calculator? How close is X to 90? Does your result contradict what you are asked to prove in Exercise 55?

d. Use right triangles to explain the behavior of sin X found above.

59. Some people use SOH-CAH-TOA to remember the trigonometric ratios for sine, cosine, and tangent. Why do you think that word might help? (*Hint:* Think of the first letters of the ratios.)

60. Find the Error A student states that $\sin A > \sin X$ because the lengths of the sides of $\triangle ABC$ are greater than the lengths of the sides of $\triangle XYZ$. What is the student's error? Explain.

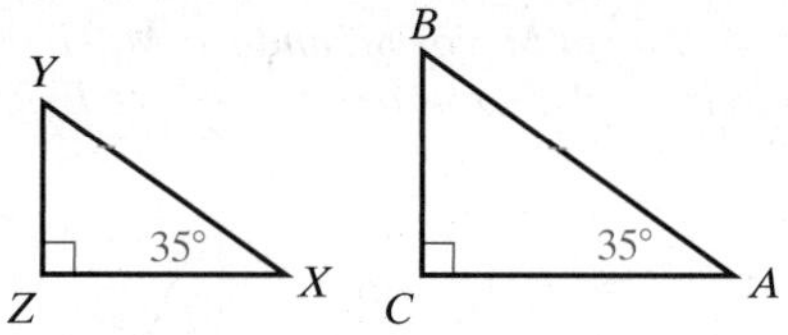

Multiple Steps

61. a. In $\triangle ABC$ at the right, how does sin A compare to cos B? Is this true for the acute angles of other right triangles?

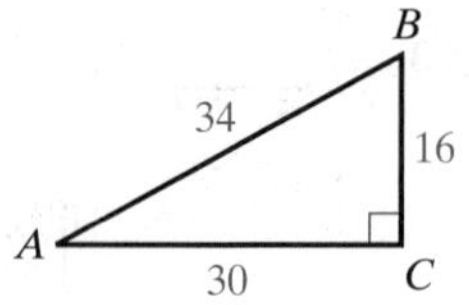

b. The word cosine is derived from the words *complement's sine*. Which angle in $\triangle ABC$ is the complement of $\angle A$? Of $\angle B$?

c. Explain why the derivation of the word cosine makes sense.

62. a. Explain why $\tan 60° = \sqrt{3}$. Include a diagram with your explanation.

b. How are the sine and cosine of a 60° angle related? Explain.

Verify that each equation is an identity. Use the given triangle to write the trigonometric ratio, and then show that each expression on the left simplifies to 1.

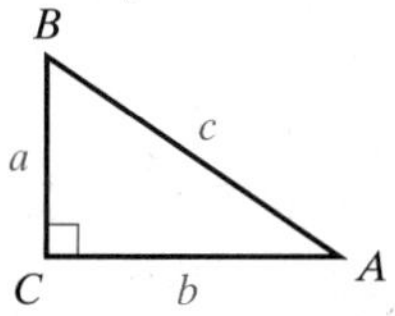

63. $(\sin A)^2 + (\cos A)^2 = 1$

64. $(\sin B)^2 + (\cos B)^2 = 1$

65. $\dfrac{1}{(\cos A)^2} - (\tan A)^2 = 1$

66. $\dfrac{1}{(\sin A)^2} - \dfrac{1}{(\tan A)^2} = 1$

The Polish astronomer Nicolaus Copernicus devised a method for determining the sizes of the orbits of planets farther from the sun than Earth. His method involved noting the number of days between the times that a planet was in the positions labeled A and B in the diagram. Using this time and the number of days in each planet's year, he calculated c and d.

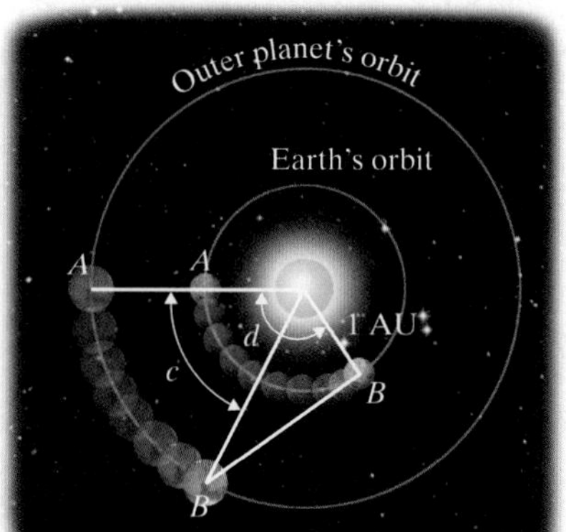

67. For Mars, $c = 55.2$ and $d = 103.8$. How far is Mars from the sun in astronomical units (AU)? One astronomical unit is defined as the average distance from Earth to the center of the sun, about 93 million miles.

68. For Jupiter, $c = 21.9$ and $d = 100.8$. How far is Jupiter from the sun in astronomical units?

The sine, cosine, and tangent ratios each have a reciprocal ratio. The reciprocal ratios are cosecant (csc), secant (sec), and cotangent (cot). Use ΔABC and the definitions below to write each ratio.

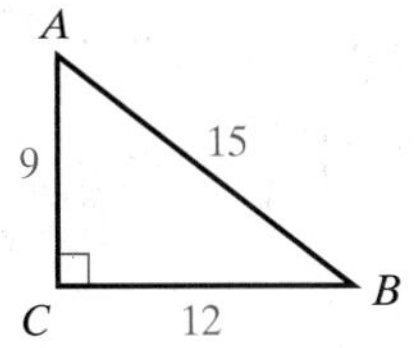

$$\csc X = \frac{1}{\sin X} \qquad \sec X = \frac{1}{\cos X} \qquad \cot X = \frac{1}{\tan X}$$

69. csc A **70.** sec A

71. cot A **72.** csc B

73. sec B **74.** cot B

REVIEW AND PREVIEW

Solve. See Section 9.2.

75. The length of the hypotenuse of a 30°-60°-90° triangle is 8. What are the lengths of the legs?

76. A diagonal of a square is 10 units. Find the length of a side of the square. Express your answer in simplest radical form.

Use rectangle ABCD to complete each statement. See Sections 3.3 and 6.4.

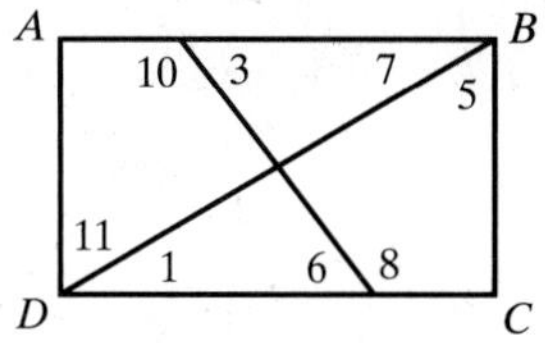

77. $\angle 1 \cong$ ___?___

78. $\angle 5 \cong$ ___?___

79. $\angle 3 \cong$ ___?___

80. $m\angle 1 + m\angle 5 =$ ___?___

9.4 SOLVING RIGHT TRIANGLES

OBJECTIVES

1 Solve Right Triangles.

2 Use Angles of Elevation and Depression to Solve Problems.

VOCABULARY

- solving the triangle
- angle of elevation
- angle of depression

OBJECTIVE 1 ▶ Solving Right Triangles. When we proved that two triangles were congruent, we talked about the three sides and three angles of each triangle. In this chapter, we are working with single right triangles. The process of determining the three angles and the lengths of the three sides of a triangle is called **solving the triangle**.

A right triangle can be solved if we know either

- the lengths of two sides or
- the length of one side and the measure of one acute angle

> **▶ Helpful Hint**
> Don't forget that we are solving right triangles. This means that
> - one angle is a 90° (right) angle
> - we can use the Pythagorean Theorem
> - we can use our trigonometric ratios

Let's solve a few triangles.

EXAMPLE 1 Solving a Right Triangle Given One Side and One Angle

Solve the right triangle. If needed, round any answers to one decimal place.

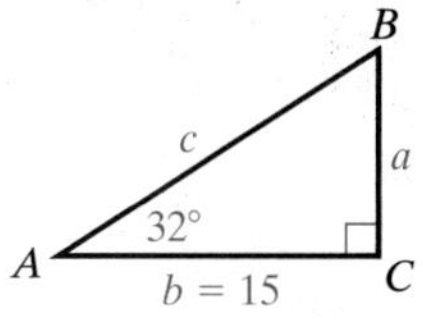

Solution $m\angle B$ can be found since $\angle A$ and $\angle B$ are complements. Thus

$$m\angle B = 90° - m\angle A = 90° - 32° = 58°$$

Now let's use trigonometric ratios. We can use either $\angle A$ or $\angle B$ now, but let's use the given $m\angle A$.

$\cos A = \frac{\text{adj.}}{\text{hyp.}}$	The ratio for cosine	$\tan A = \frac{\text{opp.}}{\text{adj.}}$	The ratio for tangent
$\cos 32° = \frac{15}{c}$	Substitute known values.	$\tan 32° = \frac{a}{15}$	Substitute known values.
$c \cdot \cos 32° = 15$	Multiply both sides by c.	$15 \tan 32° = a$	Multiply both sides by 15.
$c = \frac{15}{\cos 32°}$	Divide both sides by cos 32°.	$15(0.6249) \approx a$	Approximate tan 32°.
$c \approx \frac{15}{0.8480}$	Approximate cos 32°.	$9.4 \approx a$	Multiply.
$c \approx 17.7$	Divide.		

Thus, $m\angle B = 58°$, $a \approx 9.4$ units, and $c \approx 17.7$ units.

Check: To check, see that $m\angle A + m\angle B + m\angle C = 180°$. Also, see that $a^2 + b^2 \approx c^2$. We use an approximation sign here since two sides were rounded. □

PRACTICE 1 Solve the right triangle. If needed, round any answers to one decimal place.

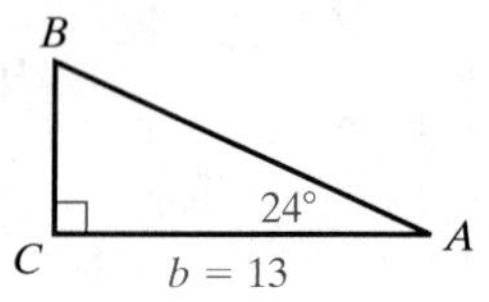

EXAMPLE 2 Solving a Right Triangle Given Two Sides

Solve the right triangle. If needed, round any answers to one decimal place.

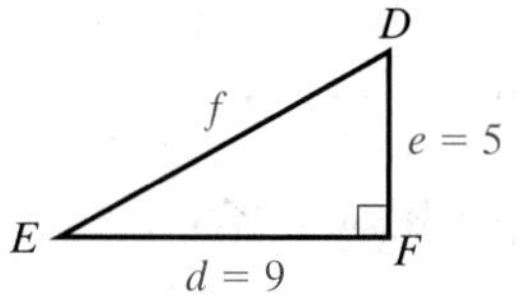

Solution To find f, we can use the Pythagorean Theorem.

$$(\text{leg})^2 + (\text{leg})^2 = (\text{hypotenuse})^2 \quad \text{Pythagorean Theorem}$$
$$9^2 + 5^2 = f^2 \quad \text{Substitute known values.}$$
$$81 + 25 = f^2 \quad \text{Find } 9^2 \text{ and } 5^2.$$
$$106 = f^2 \quad \text{Add.}$$
$$\sqrt{106} = f \quad \text{Take the positive square root of both sides.}$$
$$10.3 \approx f \quad \text{Approximate } \sqrt{106}.$$

Next, find $m\angle D$ or $m\angle E$. We'll choose $m\angle D$.

$$\tan D = \frac{\text{opp.}}{\text{adj.}}$$
$$\tan D = \frac{9}{5}$$
$$D = \tan^{-1}\left(\frac{9}{5}\right)$$

Using a calculator, we have $m\angle D \approx 60.9°$.
Since $\angle E$ and $\angle D$ are complements, we have

$$m\angle E = 90° - m\angle D \approx 90° - 60.9° = 29.1°$$

$\approx$ since $m\angle D \approx 60.9°$

Thus, $m\angle D \approx 60.9°$, $m\angle E \approx 29.1°$, and $f \approx 10.3$ units. □

PRACTICE 2 Solve the right triangle. If needed, round any answers to one decimal place.

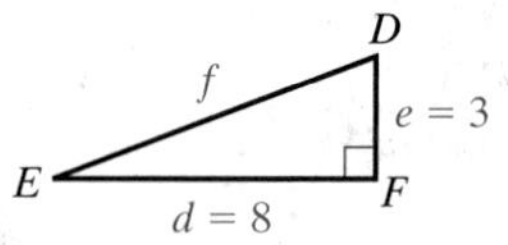

Helpful Hint

When solving right triangles, a triangle may not be completely marked.

- We often use capital letters for vertices and the same lower-case letters for the opposite side lengths.
- Also, a common triangle marking is shown:

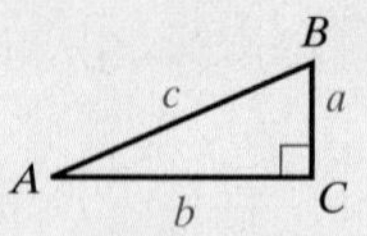

Transit

Sextant

OBJECTIVE 2 ▶ Using Angles of Elevation and Depression to Solve Problems. Many applications of right triangle trigonometry involve the angle made with an imaginary horizontal line. As shown in the figure below, an angle formed by a horizontal line and the line of sight to an object that is above the horizontal line is called the **angle of elevation**. The angle formed by a horizontal line and the line of sight to an object that is below the horizontal line is called the **angle of depression**. Transits and sextants are instruments used to measure such angles.

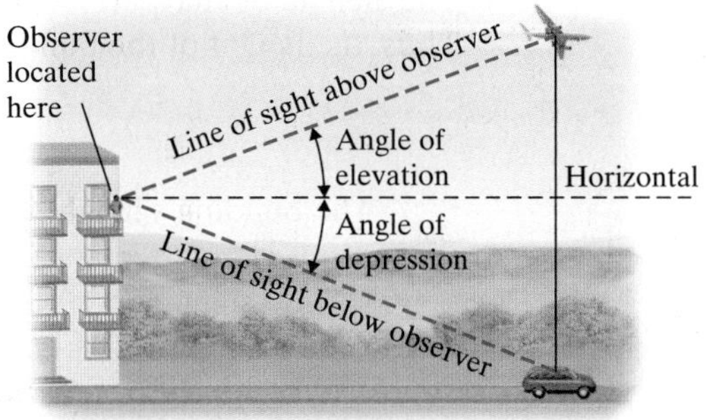

We can use the angles of elevation and depression as the acute angles of right triangles formed by a horizontal distance and a vertical height.

EXAMPLE 3 **Identifying Angles of Elevation and Depression**

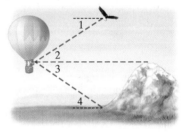

What is a description of the angle as it relates to the situation shown?

a. $\angle 1$ **b.** $\angle 4$

Solution

a. $\angle 1$ is the angle of depression from the bird to the person in the hot-air balloon.

b. $\angle 4$ is the angle of elevation from the base of the mountain to the person in the hot-air balloon. □

PRACTICE

3 Using the diagram in Example 3, what is a description of the angle as it relates to the situation shown?

a. $\angle 2$ **b.** $\angle 3$

EXAMPLE 4 **Problem Solving Using an Angle of Elevation**

Sighting the top of a building, a surveyor measured the angle of elevation to be 22°. The transit is 5 feet above the ground and 300 feet from the building. Find the building's height to the nearest whole foot.

Solution Study the figure below. Let a be the height of the part of the building that lies above the transit. The height of the building is the transit's height, 5 feet, plus a. The trigonometric ratio that will make it possible to find a is tangent. In terms of the 22° angle, we are looking for the side opposite the angle. Also, the side adjacent to the 22° angle is 300 feet, and tangent's ratio is $\frac{\text{opp.}}{\text{adj.}}$.

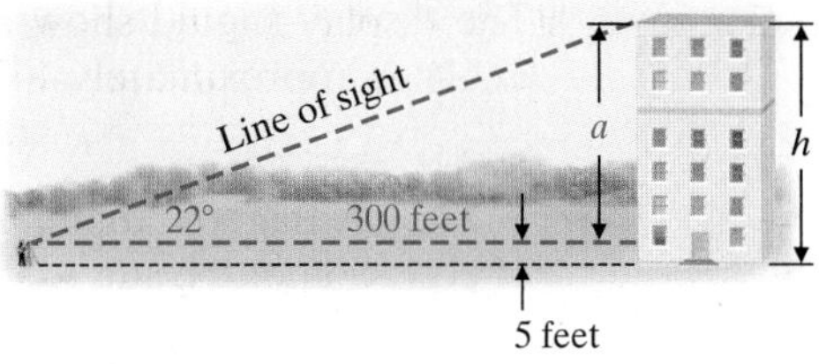

$$\tan 22° = \frac{\text{opp.}}{\text{adj.}} \quad \text{The ratio for tangent}$$

$$\tan 22° = \frac{a}{300}$$

$$a = 300 \tan 22° \quad \text{Multiply both sides by 300.}$$

$$a \approx 121 \quad \text{Use a calculator and round to the nearest whole.}$$

The height of the part of the building above the transit is approximately 121 feet. Thus, the height of the building is

$$h \approx \underset{\text{transit height}}{5} + 121 = 126$$

The building's height is approximately 126 feet. □

PRACTICE

4 You sight a rock climber on a cliff at a 32° angle of elevation. Your eye level is 6 ft above the ground and you are 1000 ft from the base of the cliff. What is the approximate height of the rock climber from the ground?

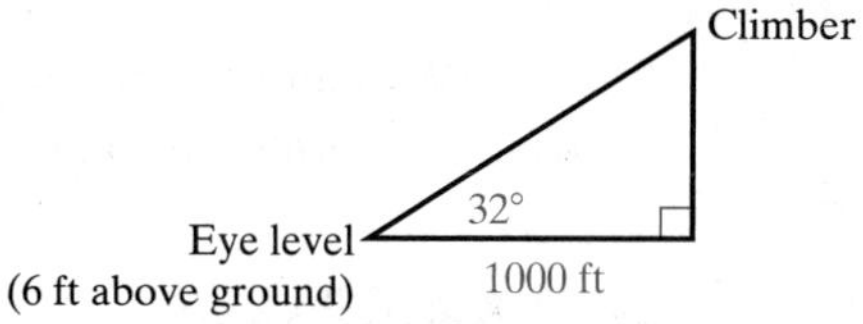

EXAMPLE 5 Determining the Angle of Elevation

A building that is 21 meters tall casts a shadow 25 meters long. Find the angle of elevation of the sun to the nearest degree.

Solution See the figure below. We are asked to find $m\angle A$.

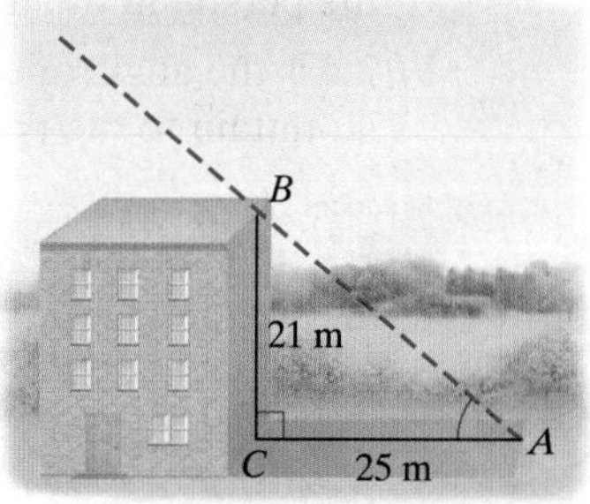

We begin with the tangent function.

> **Helpful Hint**
> Remember to use the inverse when solving for an angle measure.

$$\tan A = \frac{\text{opp.}}{\text{adj.}}$$

$$\tan A = \frac{21}{25}$$

$$A = \tan^{-1}\frac{21}{25}$$

We use a calculator in the degree mode to find $m\angle A$.

TAN⁻¹ 21 ÷ 25 ENTER

The display should show approximately 40. Thus, the angle of elevation of the sun, $m\angle A$, is approximately 40°. □

PRACTICE

5 A flagpole that is 14 meters tall casts a shadow 10 meters long. Find the angle of elevation of the sun to the nearest degree.

VOCABULARY & READINESS CHECK

Fill in the Blank *Fill in each blank so that the statement correctly describes each angle.*

1. ∠1 is an angle of $\frac{\text{elevation/depression}}{\text{(circle one)}}$ from point ______ to point ______.
2. ∠2 is an angle of $\frac{\text{elevation/depression}}{\text{(circle one)}}$ from point ______ to point ______.
3. ∠3 is an angle of $\frac{\text{elevation/depression}}{\text{(circle one)}}$ from point ______ to point ______.
4. ∠4 is an angle of $\frac{\text{elevation/depression}}{\text{(circle one)}}$ from point ______ to point ______.
5. ∠5 is an angle of $\frac{\text{elevation/depression}}{\text{(circle one)}}$ from point ______ to point ______.
6. Name two pairs of congruent angles in the diagram.

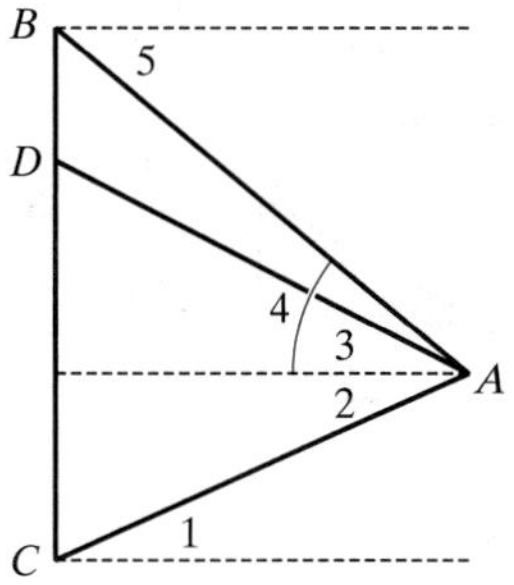

9.4 EXERCISE SET MyMathLab®

Solve the Right Triangle *If needed, round any answers to one decimal place. See Examples 1 and 2.*

1.
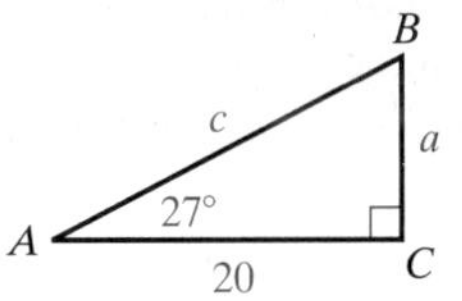

2.
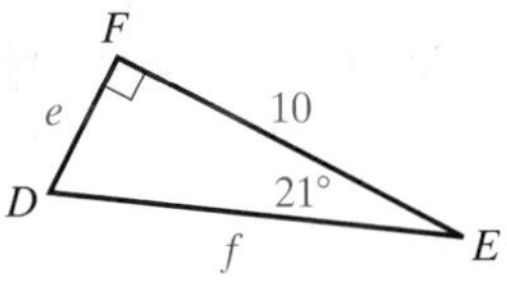

3.
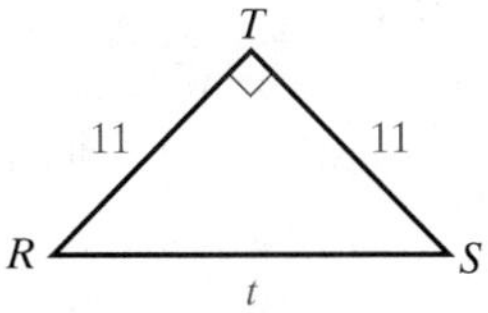

4.
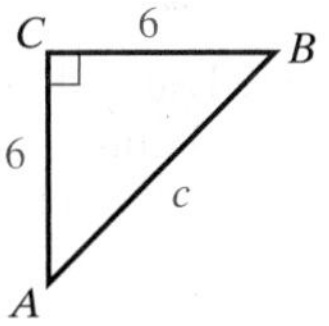

5.
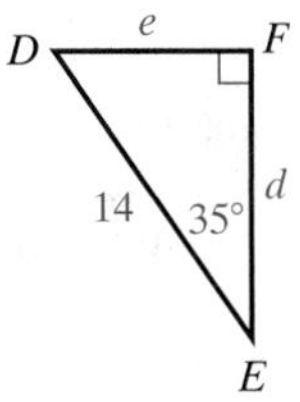

6.
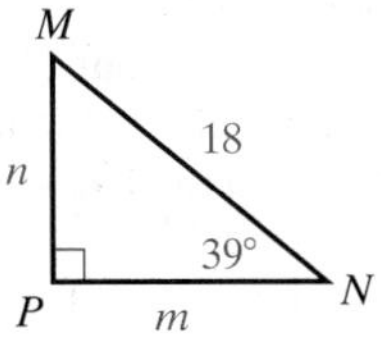

7.
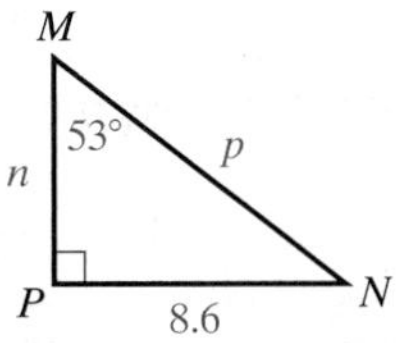

8.
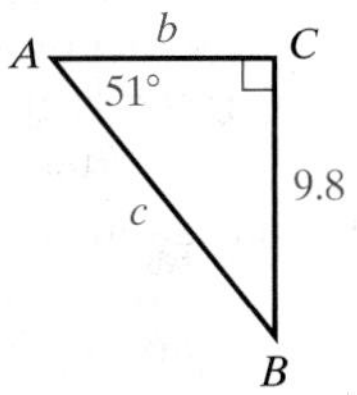

9.
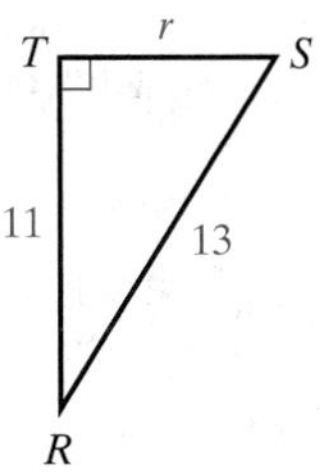

10.
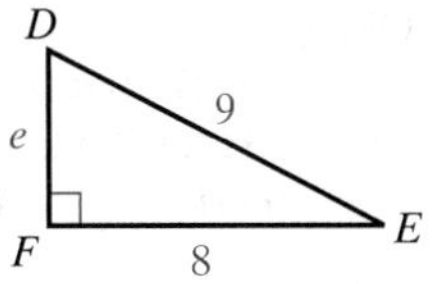

11.
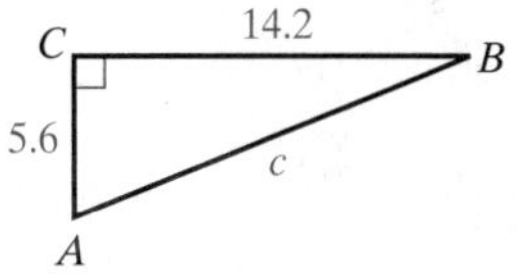

12.
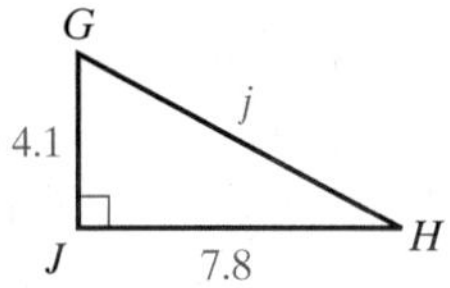

Describe each angle as it relates to the situation in the diagram. See Example 3.

13. ∠1 14. ∠2 15. ∠3 16. ∠4
17. ∠5 18. ∠6 19. ∠7 20. ∠8

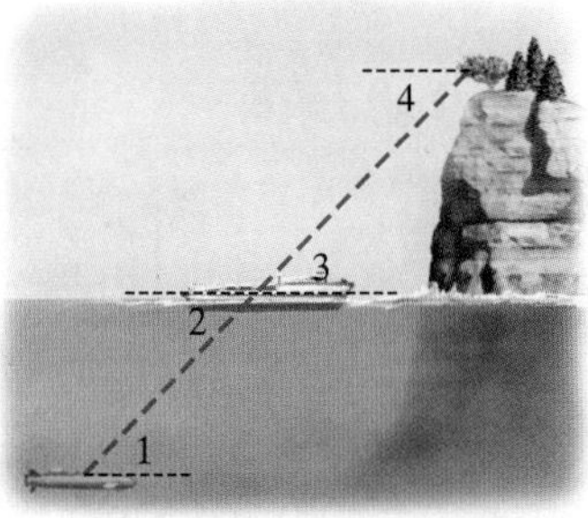

Mixed Practice *See Sections 9.3 and 9.4. Find the value of x. Round to the nearest tenth of a unit.*

21.
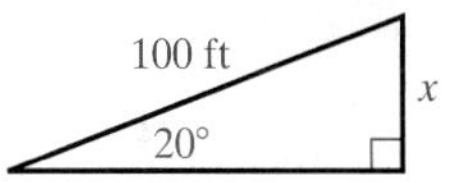

22.
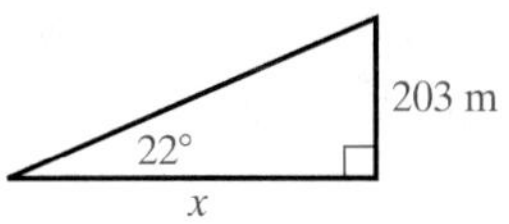

23.

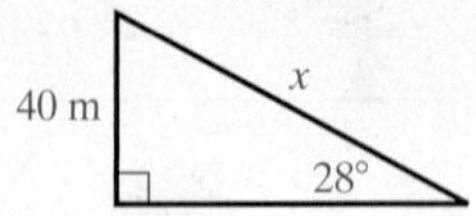

24.

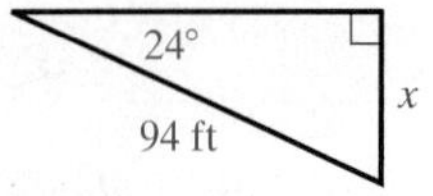

Find the value of x. Round to the nearest tenth of a unit. See Examples 4 and 5.

25.

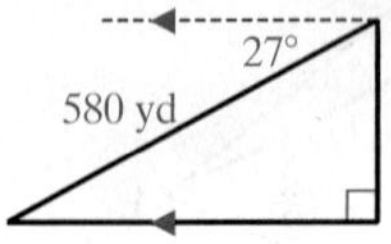

26.

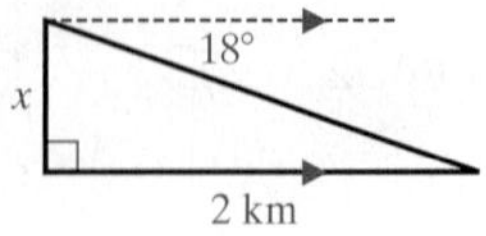

MIXED PRACTICE

Solve. See Sections 9.3 and 9.4.

27. At a certain time of day, the angle of elevation of the sun is 40°. To the nearest foot, find the height of a tree whose shadow is 35 feet long.

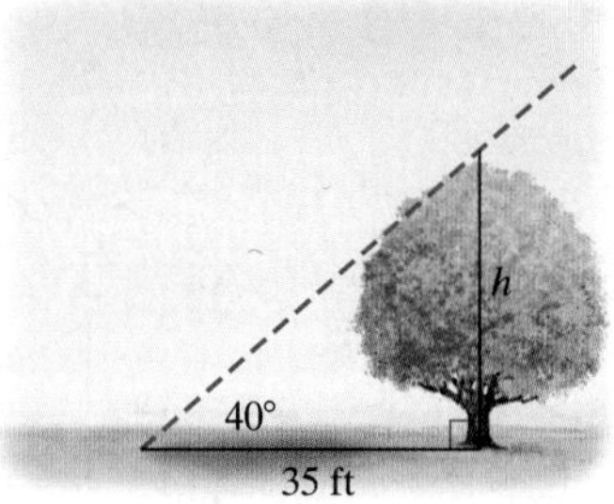

28. At a certain time of day, the angle of elevation of the sun is 50°. To the nearest foot, find the height of a building whose shadow is 90 feet long.

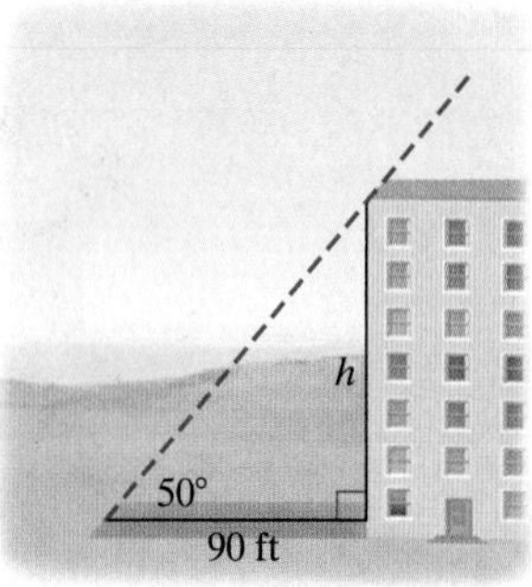

29. A tower that is 125 feet tall casts a shadow 172 feet long. Find the angle of elevation of the sun to the nearest degree.

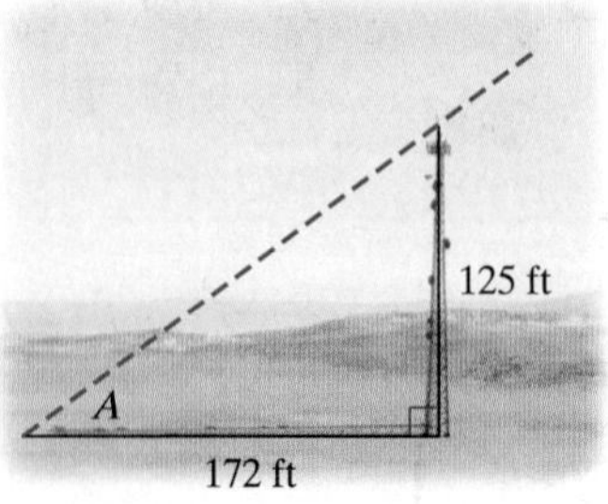

30. The Washington Monument is 555 feet high. If you are standing one-quarter of a mile, or 1320 feet, from the base of the monument and are looking to the top, find the angle of elevation to the nearest degree.

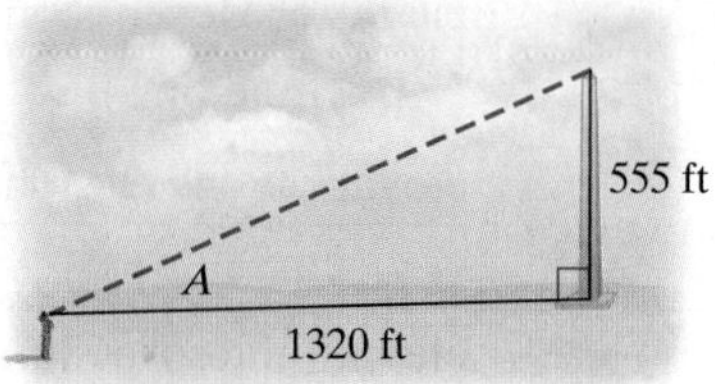

31. Smoke is sighted due north of Lookout Tower 1. From Lookout Tower 2, which is 8.2 miles west of Tower 1, a ranger reports that the smoke is 48.8° east of due north. How far is the smoke from Lookout Tower 1?

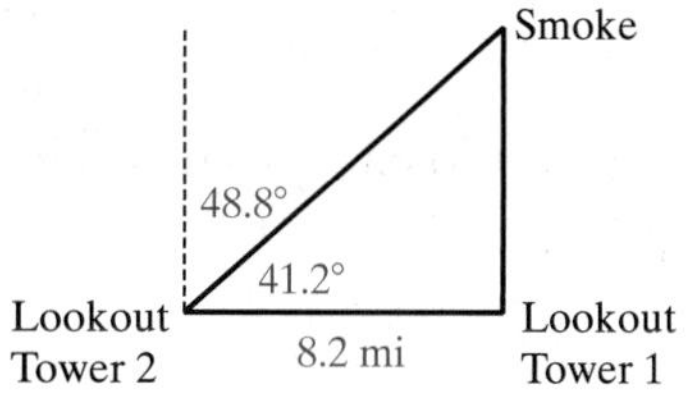

32. Use Exercise **31** and find the distance the smoke is from Lookout Tower 2.

33. A train track through a mountain is 2000 feet long and makes an angle of 1.5° with the horizontal. What is the change in elevation from one end of the tunnel to the other?

34. A meteorologist measures the angle of elevation of a weather balloon as 41°. A radio signal from the balloon indicates that it is 1503 m from his location. To the nearest meter, how high above the ground is the balloon?

35. There are over 3500 hot-air balloons in the United States. Suppose a hot-air balloon is flying at an altitude of 2100 feet. If the angle of depression from the pilot of the balloon to a house is 31°, how far is the house from the pilot?

36. A tourist looks out from the crown of the Statue of Liberty, approximately 250 ft above ground. The tourist sees a ship coming into the harbor and measures the angle of depression as 18°. Find the distance from the base of the statue to the ship to the nearest foot.

37. The world's tallest unsupported flagpole is a 282-ft-tall steel pole in Surrey, British Columbia. The shortest shadow cast by the pole during the year is 137 ft long. To the nearest degree, what is the angle of elevation of the sun when casting the flagpole's shortest shadow?

38. A communication tower is 956 feet in height. If it casts a shadow of 670 feet, find the angle of elevation of the sun, to the nearest whole degree.

39. To approach runway 17 of the Ponca City Municipal Airport in Oklahoma, the pilot must begin a 3° descent starting from a height of 2714 ft above sea level. The airport is 1007 ft above sea level. To the nearest tenth of a mile, how far from the runway is the airplane at the start of this approach?

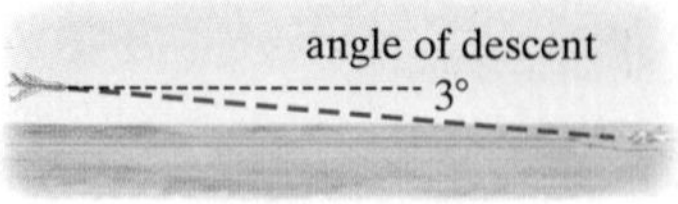

40. An airplane pilot sights a life raft at a 26° angle of depression. The airplane's altitude is 3 km. What is the airplane's horizontal distance d from the raft?

41. A blimp provides aerial television views of a football game. The television camera sights the stadium at a 7° angle of depression. The altitude of the blimp is 400 m. What is the line-of-sight distance from the television camera to the base of the stadium? Round to the nearest hundred meters.

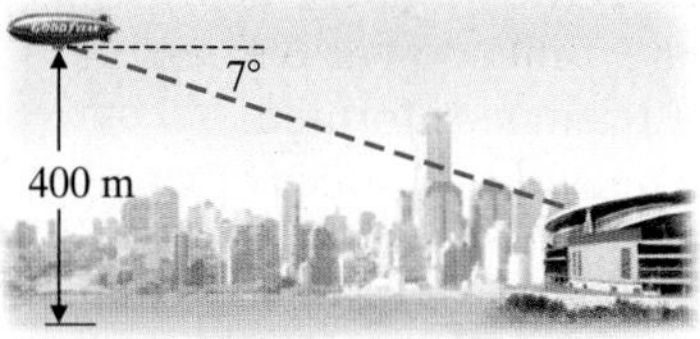

42. Two office buildings are 51 m apart. The height of the taller building is 207 m. The angle of depression from the top of the taller building to the top of the shorter building is 15°. Find the height of the shorter building to the nearest meter.

A communications tower is located on a plot of flat land. The tower is supported by several guy wires. Assume that you are able to measure distances along the ground as well as angles formed by the guy wires and the ground. Explain how you could estimate each of the following measurements.

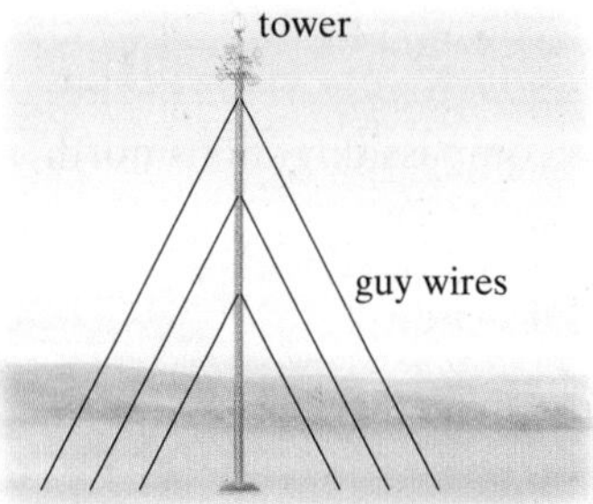

43. the length of any guy wire

44. how high on the tower each wire is attached

CONCEPT EXTENSIONS

45. How is an angle of elevation formed?

46. Find the Error A homework question says that the angle of depression from the bottom of a house window to a ball on the ground is 20°. The following is your friend's sketch of the situation. Describe your friend's error.

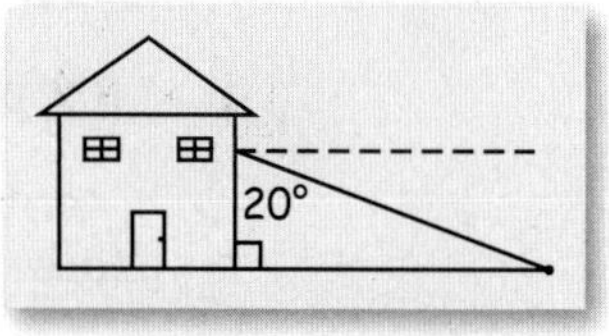

The angle of elevation e from A to B and the angle of depression d from B to A are given. Find the measure of each angle.

47. e: $(7x - 5)°$, d: $4(x + 7)°$ **48.** e: $(3x + 1)°$, d: $2(x + 8)°$

49. e: $(x + 21)°$, d: $3(x + 3)°$ **50.** e: $5(x - 2)°$, d: $(x + 14)°$

51. Solve A firefighter on the ground sees fire break through a window near the top of a building. The angle of elevation to the windowsill is 28°. The angle of elevation to the top of the building is 42°. The firefighter is 75 ft from the building and her eyes are 5 ft above the ground. What roof-to-windowsill distance can she report by radio to firefighters on the roof?

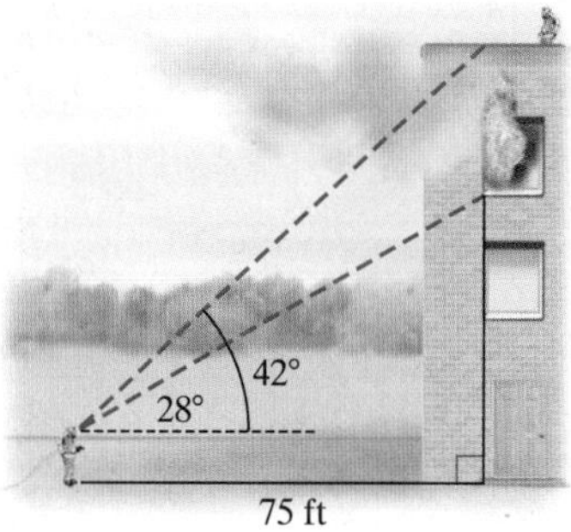

52. For locations in the United States, the relationship between the latitude ℓ and the greatest angle of elevation a of the sun at noon on the first day of summer is $a = 90° - \ell + 23.5°$. Find the latitude of your town. Then determine the greatest angle of elevation of the sun for your town on the first day of summer.

REVIEW AND PREVIEW

Find the distance between each pair of points. See Section 1.7.

53. (0, 0) and (8, 2) **54.** (0, 0) and (3, 10)

55. (−15, −2) and (0, 0) **56.** (−2, 12) and (0, 0)

9.5 VECTORS

OBJECTIVES

1 Describe Vectors.

2 Solve Problems Involving Vector Addition.

OBJECTIVE 1 ▶ Describing Vectors. A **vector** is any quantity with magnitude (size) and direction. For example, a car's speed and direction together represent a vector. There are many other examples that can be modelled by a vector.

We can use an arrow for a vector, as shown by the velocity vector $\overrightarrow{KW}$ in the photo. The **magnitude** corresponds to the distance from the **initial point** K to the **terminal point** W. The direction corresponds to the direction in which the arrow points.

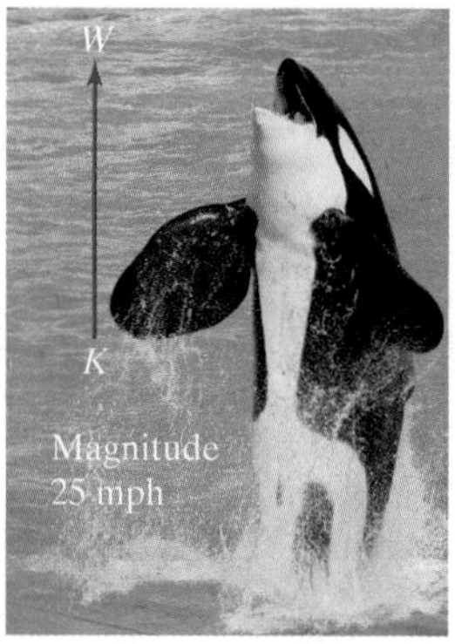

VOCABULARY
- vector
- magnitude
- initial point
- terminal point
- resultant
- component form

In the coordinate plane, we can also use the **component form** $\langle x, y\rangle$ to represent a vector. The magnitude and direction of the vector correspond to the distance and direction of $\langle x, y\rangle$ from the origin. (Note: (x, y) is a point and $\langle x, y\rangle$ is a vector.)

EXAMPLE 1 Describing a Vector

Coordinate Geometry What is the component form of $\overrightarrow{OL}$? Round the coordinates to the nearest tenth.

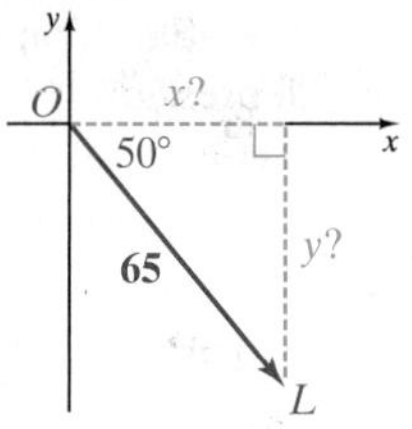

Solution Think of plotting the point L from the origin, 0. To do so, we move x units to the right, and then y units down. We can find the values of x and y because a right triangle is formed.

Let's use the sine and cosine ratios to find the values of x and y.

Helpful Hint
In the coordinate plane,
- the ordered pair notation, (x, y), represents a point and
- the component form notation $\langle x, y\rangle$ represents a vector (arrow).

$\cos 50° = \dfrac{x}{65}$	Write the ratios.	$\sin 50° = \dfrac{y}{65}$	Write the ratios.
$x = 65(\cos 50°)$	Multiply both sides by 65.	$y = 65(\sin 50°)$	Multiply both sides by 65.
≈ 41.78119463	Use a calculator.	≈ 49.7928888	Use a calculator.

L is in the fourth quadrant, so the y-coordinate is negative. $\overrightarrow{OL} \approx \langle 41.8, -49.8\rangle$. □

PRACTICE 1 What is the component form of the vector at the right? Round the coordinates to the nearest tenth.

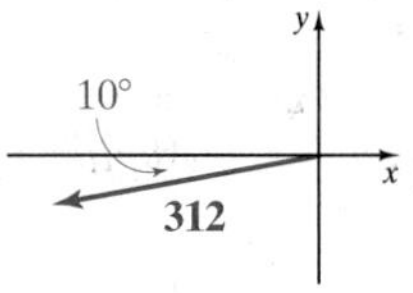

In many applications of vectors, we use the compass directions north, south, east, and west to describe the direction of a vector.

EXAMPLE 2 Describing a Vector Direction

What is the direction of each vector using compass directions?

a.

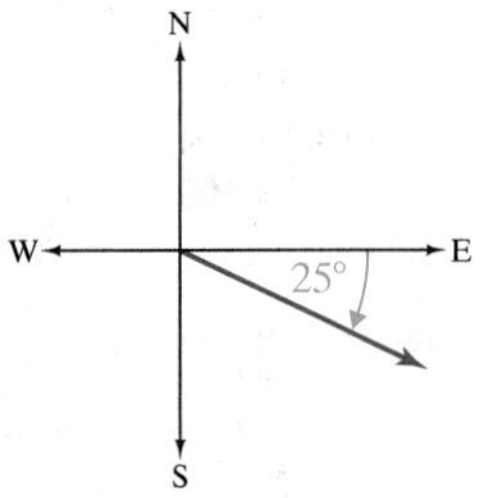

b.

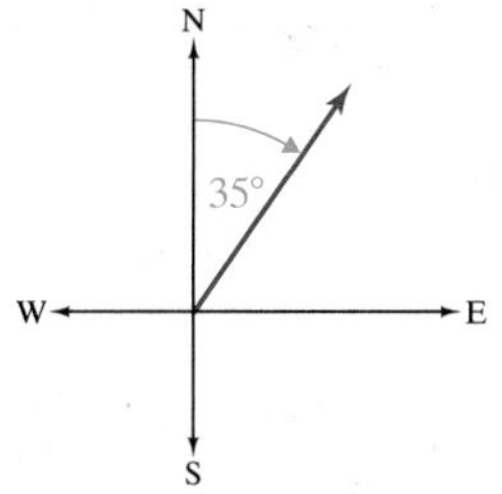

Solution

a. The angle is below (south) the west-east line, on the east side. The vector direction is 25° south of east.

b. The angle is to the right (east) of the north-south line, on the north side. The vector direction is 35° east of north. □

Helpful Hint
To use compass directions to describe a vector, look at the given angle in relation to the north-south line (vertical) or the west-east line (horizontal).

PRACTICE 2

a. What is the direction of the vector at the right?

b. Is there more than one way to describe the direction of this vector? Explain.

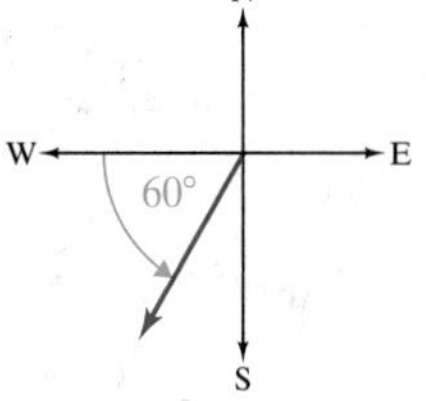

We can find the magnitude and direction of a vector when the vector is described as an ordered pair.

EXAMPLE 3 Finding the Magnitude and Direction of a Vector

Multiple Choice An airplane lands 40 km west and 25 km south from where it took off. This trip can be described by the vector $\langle -40, -25 \rangle$. What are the approximate magnitude and direction of its flight vector?

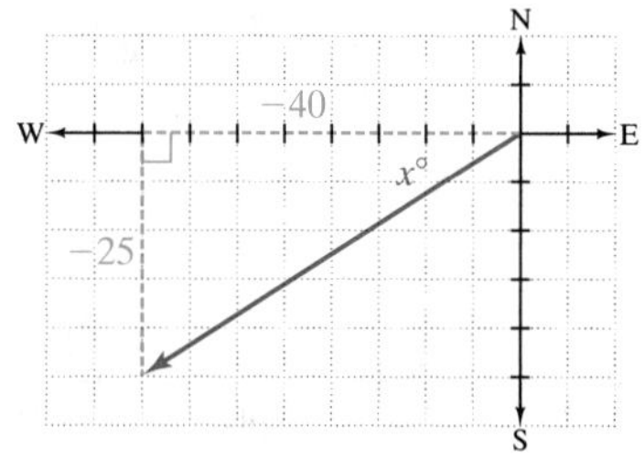

a. 47 km at 32° south of west

b. 47 km at 58° south of west

c. 2225 km at 32° west of south

d. 2225 km at 58° west of south

Solution Remember that the magnitude of a vector is the distance from the initial point to the terminal point (or length) of the arrow. The direction corresponds to the direction the vector points.

Use the Distance Formula to find the magnitude or distance between (0, 0) and (−40, −25).

$$\begin{aligned} d &= \sqrt{(-40 - 0)^2 + (-25 - 0)^2} \\ &= \sqrt{1600 + 625} \\ &= \sqrt{2225} \\ &\approx 47.16990566 \\ &\approx 47 \end{aligned}$$

Next, find the vector's direction. To do so, first find $x°$ using the formed right triangle.

$$\tan x° = \frac{\text{opp.}}{\text{adj.}}$$

$$\tan x° = \frac{25}{40}$$

$$\begin{aligned} x &= \tan^{-1}\left(\frac{25}{40}\right) \\ &\approx 32.00538321 \\ &\approx 32 \end{aligned}$$

The airplane flew about 47 km at 32° south of west. The correct answer is **a.** □

PRACTICE

3 An airplane lands 246 mi east and 76 mi north from where it took off. What are the approximate magnitude and direction of its flight vector?

Remember that a vector has magnitude (size) and direction. Thus far, we have learned two ways to uniquely describe a vector:

- In component form—the magnitude is the distance from the origin and the direction is described by using an angle (in degrees) along with compass directions.
- As an arrow on the coordinate plane—the magnitude (size) and angle direction is noted.

Our examples thus far have helped us understand how to convert from one way (to describe a vector) to the other way, described above.

Helpful Hint

From Example 3, here are two ways to describe the same vector.

$\langle -40, -25 \rangle$ — The Example 3 vector

A vector with magnitude 47 km and direction 32° south of east — An approximation of the same vector in Example 3

OBJECTIVE 2 ▶ Solving Problems Involving Vector Addition. We can use a single lowercase letter, such as $\vec{\mathbf{u}}$, to name a vector.

The map below shows vectors representing a flight from Albuquerque to Salt Lake City, with a stopover in Flagstaff. The vector from Albuquerque to Salt Lake City is the sum, or **resultant**, of the other two vectors. We write this as $\vec{\mathbf{w}} = \vec{\mathbf{u}} + \vec{\mathbf{v}}$.

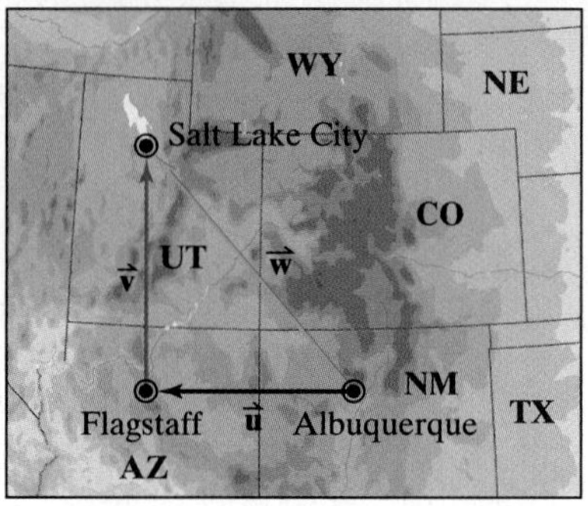

We can add two vectors by adding their coordinates. We can also show the sum geometrically.

Adding Vectors

For $\vec{\mathbf{a}} = \langle x_1, y_1 \rangle$ and $\vec{\mathbf{c}} = \langle x_2, y_2 \rangle$,

$$\vec{\mathbf{a}} + \vec{\mathbf{c}} = \langle x_1 + x_2, y_1 + y_2 \rangle$$

EXAMPLE 4 **Adding Vectors**

Vectors $\vec{\mathbf{a}}$ and $\vec{\mathbf{c}}$ are shown at the right. $\vec{\mathbf{a}} = \langle -4, -3 \rangle$ and $\vec{\mathbf{c}} = \langle 1, -2 \rangle$.

a. What is the resultant $\vec{\mathbf{e}}$ of the two vectors as an ordered pair?

b. Draw $\vec{\mathbf{e}}$.

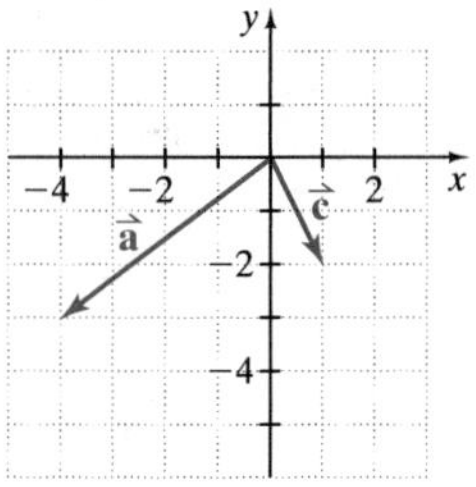

Solution

a. To add the vectors, add their x- and then their y-coordinates.

$$\begin{aligned} \vec{\mathbf{e}} &= \vec{\mathbf{a}} + \vec{\mathbf{c}} \\ &= \langle -4, -3 \rangle + \langle 1, -2 \rangle \\ &= \langle -4 + 1, -3 + (-2) \rangle \quad \text{Add the coordinates.} \\ &= \langle -3, -5 \rangle \quad \text{Simplify.} \end{aligned}$$

$\langle -3, -5 \rangle$ is the resultant.

b. Now that we have $\vec{\mathbf{e}} = \langle -3, -5 \rangle$, there are two ways to draw $\vec{\mathbf{e}}$. One way is to follow Steps 1–3.

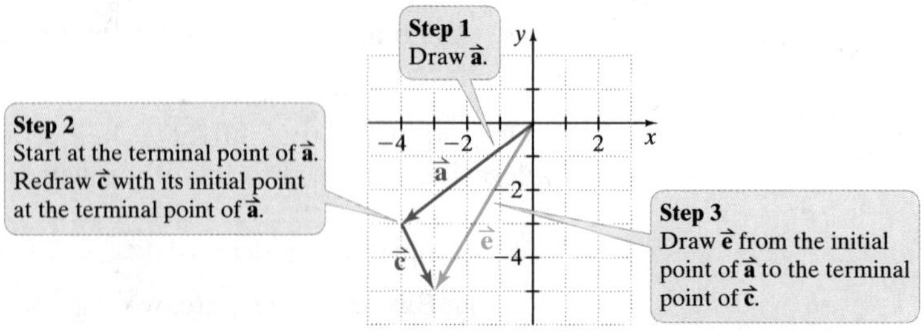

The other way is to simply draw an arrow starting at the origin and ending at point $(-3, -5)$. □

Helpful Hint

Remember, $\langle -3, -5 \rangle$ is not a point, but is a vector (arrow) that starts at the origin and ends at the point with coordinates $(-3, -5)$.

PRACTICE 4 What is the resultant of $\langle 2, 3 \rangle$ and $\langle -4, -2 \rangle$ as an ordered pair?

A vector sum can show what happens when vectors occur in sequence, as in the airplane flight described on the previous page.

A vector sum can also show what happens when vectors act at the same time. When you row a boat in water that has a current, you might paddle in a direction different from that of the current.

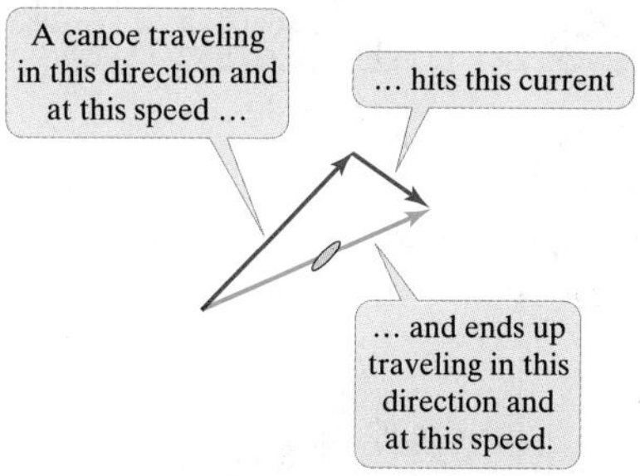

EXAMPLE 5 **Applying Vectors**

Your friend invites you to go boating on the river in a powerboat. The speed of the powerboat in still water is 35 mph. The river flows directly south at 8 mph. If the powerboat heads directly west, what are the boat's approximate resultant speed and direction?

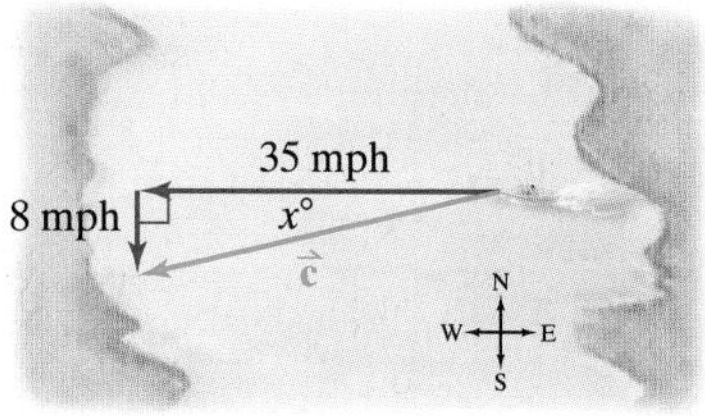

Solution The diagram shows the sum of the two vectors, $\vec{\mathbf{c}}$. Use the Pythagorean Theorem to find the powerboat's resultant speed, which is the magnitude or length of vector $\vec{\mathbf{c}}$.

$$\begin{aligned} c^2 &= 35^2 + 8^2 && \text{The lengths of the legs are 35 and 8.} \\ c^2 &= 1289 && \text{Simplify the right side.} \\ c &= \sqrt{1289} && \text{Take the positive square root.} \\ c &\approx 35.90264614 && \text{Use a calculator.} \\ c &\approx 36 && \text{Round to the nearest whole.} \end{aligned}$$

Use trigonometry to find $x°$ and then the powerboat's resultant direction.

$$\begin{aligned} \tan x° &= \frac{8}{35} && \text{Use the tangent ratio.} \\ x &= \tan^{-1}\left(\frac{8}{35}\right) && \text{Use the inverse of tangent.} \\ x &\approx 12.87500156 && \text{Use a calculator.} \\ x &\approx 13 && \text{Round to the nearest whole.} \end{aligned}$$

The powerboat's direction is about 13° south of west. Its speed is about 36 mph. □

PRACTICE
5 In Example 5, at what angle should the powerboat head upriver in order to travel directly west?

VOCABULARY & READINESS CHECK

Word Bank *Use the choices to fill in each blank.*

magnitude	vector	terminal point	arrow
direction	resultant	initial point	

1. A(n) ____________ is any quantity with magnitude and direction.
2. The ____________ of a vector is the distance from the initial point of the vector to the terminal point.
3. A vector can be represented by a(n) ____________.
4. The ____________ of a vector is the direction in which the vector points.
5. The starting point of a vector is called the ____________ and the ending point of the vector is called the ____________.
6. Another word for the sum of two vectors is the ____________ of the vectors.

9.5 EXERCISE SET MyMathLab®

Describe each vector in component form. Round the coordinates to the nearest tenth. See Example 1.

1.

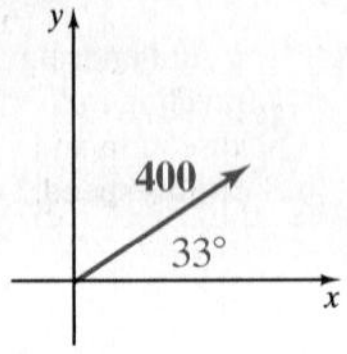

2.

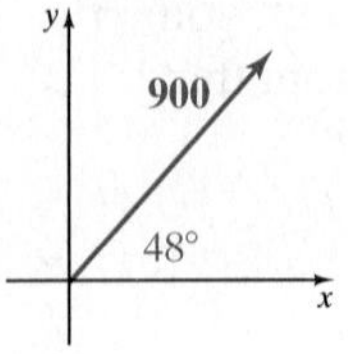

3.

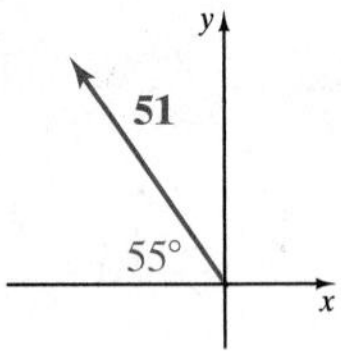

4.

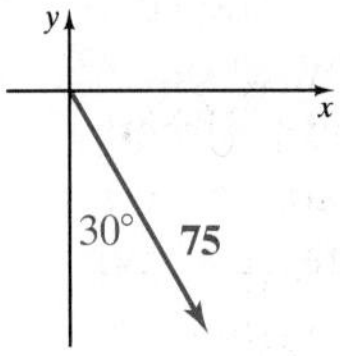

Use compass directions to describe the direction of each vector. See Example 2.

5.

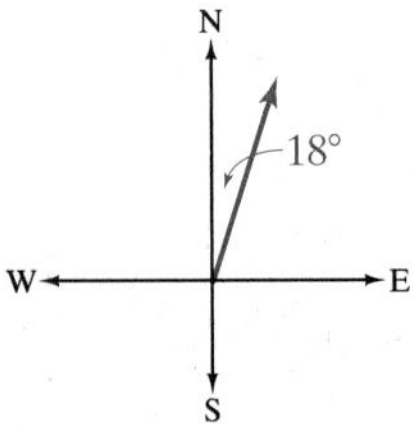

6.

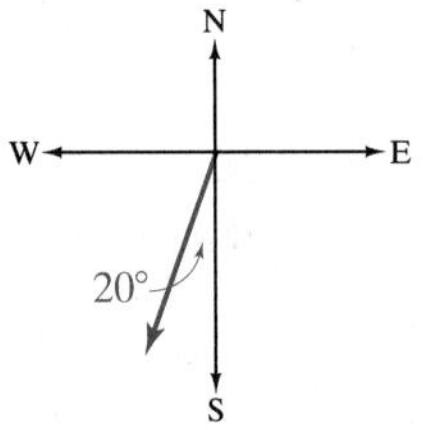

7.

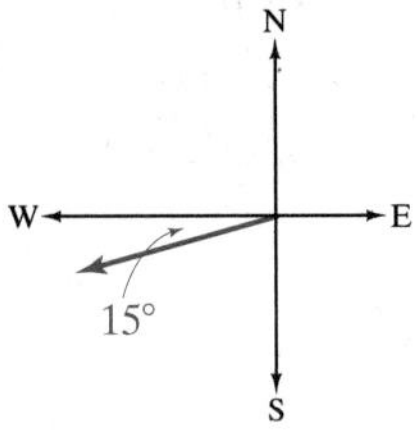

8.

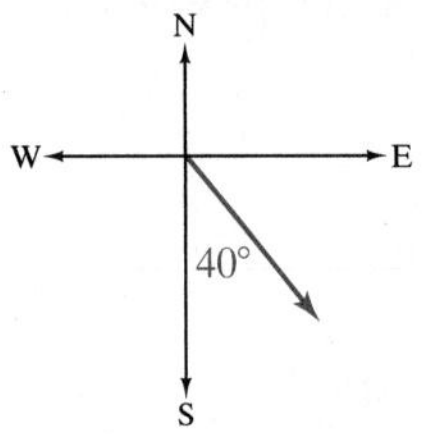

Sketch a vector that has the given direction. See Example 2.

9. 50° south of east
10. 20° north of west
11. 45° northeast
12. 70° west of north
13. 45° southwest
14. 10° east of south
15. 35° west of north
16. 12° south of east

Use $\vec{\mathbf{a}} = \langle 6, 1 \rangle$, $\vec{\mathbf{b}} = \langle 4, 4 \rangle$, $\vec{\mathbf{c}} = \langle 2, 5 \rangle$ *and* $\vec{\mathbf{d}} = \langle 3, 1 \rangle$ *to find each of the following. See Example 3.*

17. the magnitude of $\vec{\mathbf{a}}$
18. the magnitude of $\vec{\mathbf{b}}$
19. the magnitude of $\vec{\mathbf{c}}$
20. the magnitude of $\vec{\mathbf{d}}$

Find the magnitude and direction of each vector. Round each to the nearest whole. See Example 3.

21.

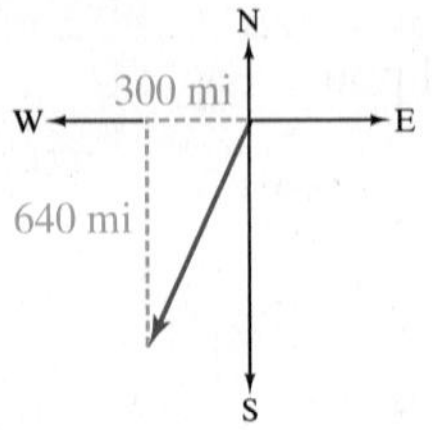

22.

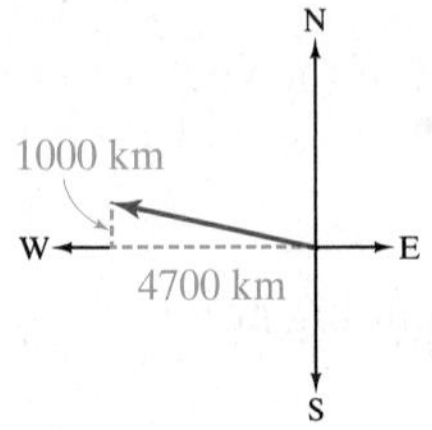

23.

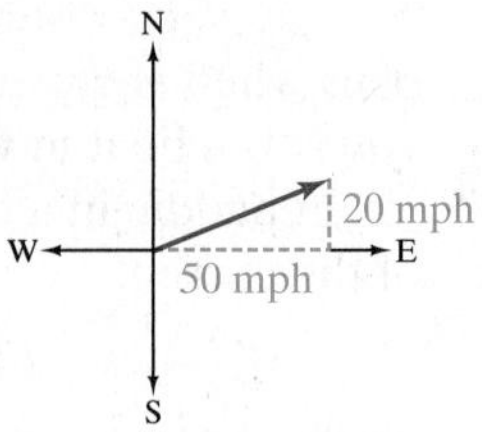

24.

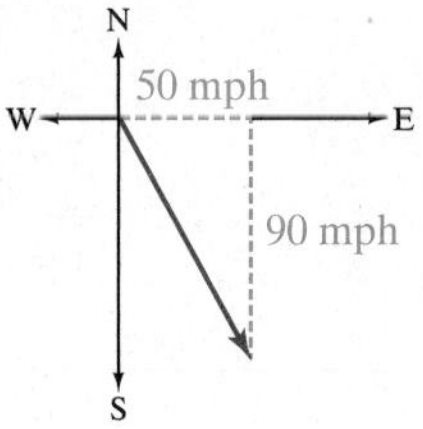

Use $\vec{\mathbf{a}} = \langle 6, 1 \rangle$, $\vec{\mathbf{b}} = \langle 4, 4 \rangle$, *and* $\vec{\mathbf{c}} = \langle 2, 5 \rangle$ *to find each of the following. See Example 4.*

25. $\vec{\mathbf{a}} + \vec{\mathbf{b}}$
26. $\vec{\mathbf{b}} + \vec{\mathbf{c}}$
27. $\vec{\mathbf{c}} + \vec{\mathbf{a}}$
28. $\vec{\mathbf{a}} + \vec{\mathbf{a}}$

Write the resultant of the two vectors in component form. See Example 4.

29. $\langle -5, -7 \rangle$ and $\langle 0, 0 \rangle$
30. $\langle 0, 0 \rangle$ and $\langle 4, -6 \rangle$
31. $\langle 2, 1 \rangle$ and $\langle -3, 2 \rangle$
32. $\langle -1, 1 \rangle$ and $\langle -1, 2 \rangle$

For Exercises 33–38, (a) write the resultant of the two vectors in component form and (b) draw the resultant.

33.

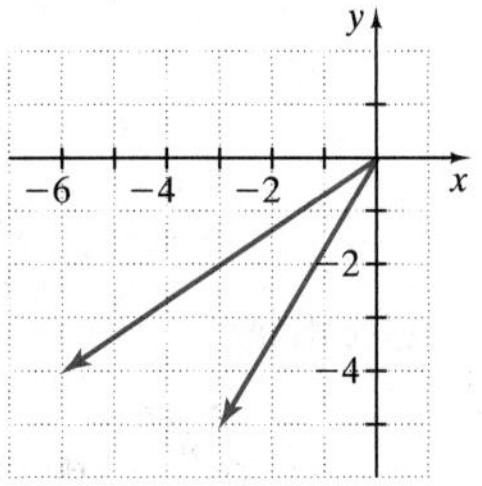

34.

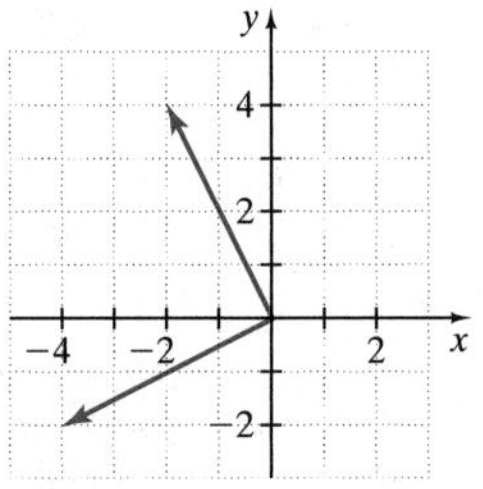

35.

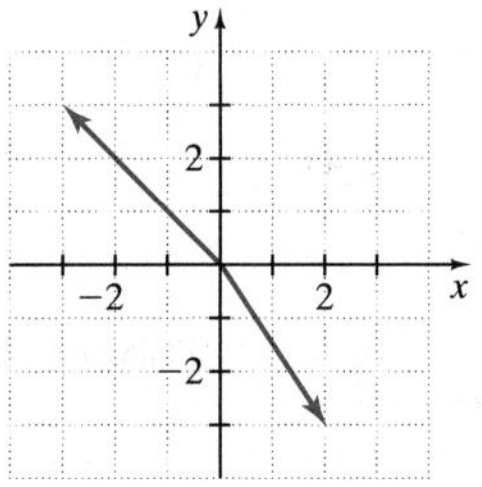

36.

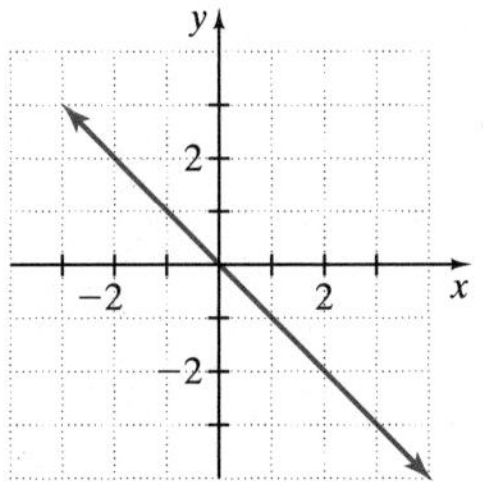

37.

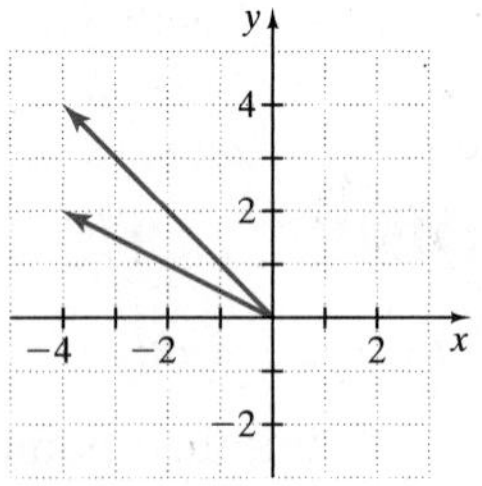

38.

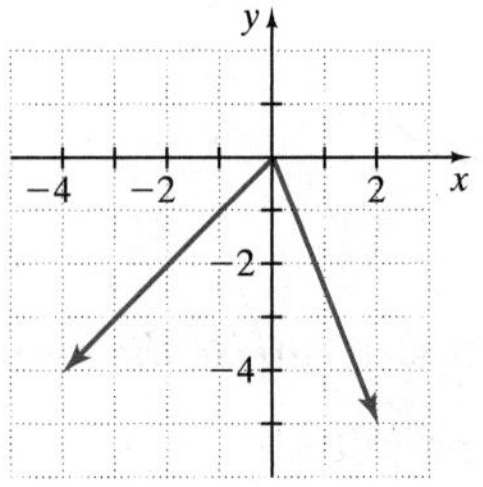

Solve. See Example 5.

39. Homing pigeons have the ability, or instinct, to find their way home when released hundreds of miles away from home. Homing pigeons carried news of Olympic victories to various cities in ancient Greece. Suppose one such pigeon took off from Athens and landed in Sparta, which is 73 mi west and 64 mi south of Athens. What are the magnitude and direction of its flight vector?

40. The sophomore class at your high school went on a backpacking trip. One morning, the group left the base camp to hike 11 km due north and 11 km due east. What are the distance and direction of the group's hike that day? Round each to the nearest tenth.

41. Navigation A ferry shuttles people from one side of a river to the other. The speed of the ferry in still water is 25 mph. The river flows directly south at 7 mph.

a. The ferry heads directly west across the river. What are the resulting speed and direction of the boat? Round to the nearest tenth.

b. At what angle should the ferry head upriver in order to travel directly west?

42. Aviation A twin-engine airplane has a speed of 300 mph in still air. Suppose this airplane heads directly south and encounters a 50 mph wind blowing due east. Find the resulting speed and direction of the plane. Round to the nearest unit.

43. Use the diagrams below to write a definition of *equal vectors*.

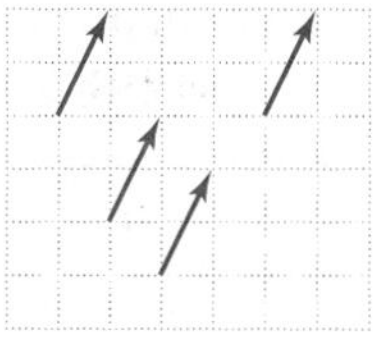

These vectors are equal.

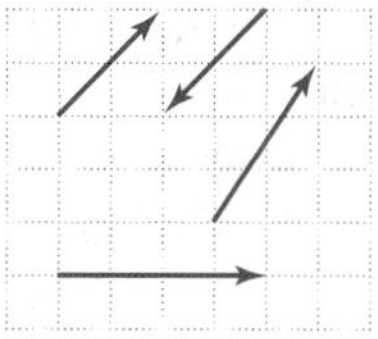

No two of these vectors are equal.

44. Use the diagrams below to write a definition of *parallel vectors*.

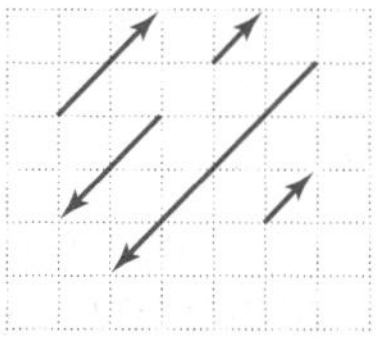

These vectors are parallel.

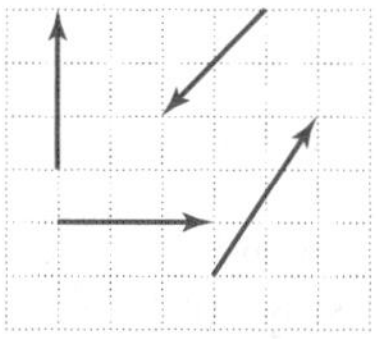

No two of these vectors are parallel.

CONCEPT EXTENSIONS

Use $\vec{u} = \langle 2, 4 \rangle$, $\vec{v} = \langle 3, 1 \rangle$, *and* $\vec{w} = \langle 5, -1 \rangle$.

45. Find $\vec{u} + \vec{v}$ and $\vec{v} + \vec{u}$.

46. Find $\vec{u} + (\vec{v} + \vec{w})$ and $(\vec{u} + \vec{v}) + \vec{w}$.

47. Which properties of real number addition seem to hold true for vector addition?

48. Prove your conjecture in Exercise **47** algebraically and geometrically.

Give the sum of $\vec{a}$ *and* $\vec{b}$. *Draw* $\vec{a}$ *and* $\vec{b}$ *and their sum in the coordinate plane.*

49. $\vec{a} = \langle -5, -2 \rangle, \vec{b} = \langle 2, -5 \rangle$

50. $\vec{a} = \langle 5, -2 \rangle, \vec{b} = \langle -5, -2 \rangle$

51. Matching An airplane takes off from a runway in the direction 10° east of south. When it reaches 5000 ft, it turns right 45°. It cruises at this altitude for 60 mi. Then it turns left 160°, descends, and lands. Match each vector with the appropriate portion of the flight.

A. The plane takes off.
B. The plane cruises.
C. The plane lands.

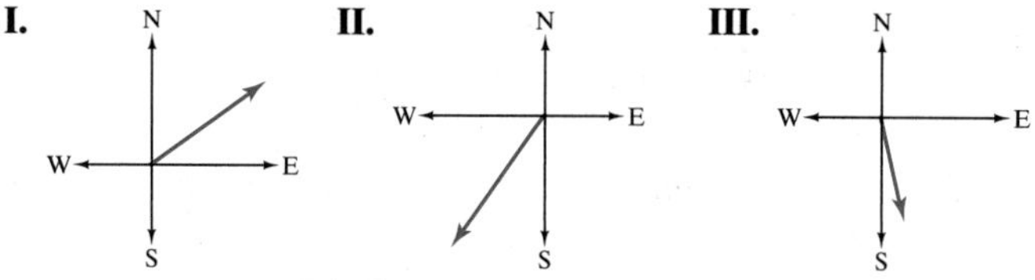

Multiple Steps for Exercises 52–54.

52. The cruising speed of one model of commercial jet in still air is 530 mph. Suppose that this jet is cruising directly east when it encounters an 80 mph wind blowing 40° south of west.

a. Sketch the vectors for the velocities of the airplane and the wind.

b. Express both vectors from part a in ordered pair notation.

c. Find the sum of the vectors from part b.

d. Find the magnitude and direction of the vector from part c.

53. A fishing boat leaves its home port and travels 150 mi directly east. It then changes course and travels 40 mi due north.

a. In what direction should the boat head to return to its home port?

b. If the boat averages 23 mph, how long will the return trip take?

54. A Red Cross helicopter takes off and flies to deliver some relief supplies. Then the helicopter flies to another location to pick up three nurses. The helicopter's flight path is shown.

a. What information does the diagram give you?

b. What is the helicopter's distance from its point of origin? Round to the nearest kilometer.

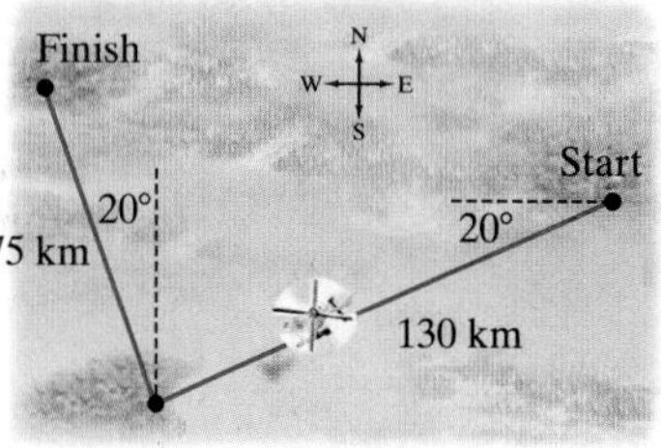

55. Name four other vectors with the same magnitude as $\langle -7, -24 \rangle$.

56. Name four other vectors with the same direction as $\langle -7, -24 \rangle$.

57. What are the similarities between rays and vectors? What are the differences?

58. One classmate describes the direction of a vector as 35° south of east. Another classmate describes it as 55° east of south. Could they be describing the same vector? Explain.

59. Find the Error Your friend says that the magnitude of vector $\langle 10, 7 \rangle$ is greater than that of vector $\langle -10, -7 \rangle$ because the coordinates of $\langle 10, 7 \rangle$ are positive and the coordinates of $\langle -10, -7 \rangle$ are negative. Explain why your friend's statement is incorrect.

60. Think of the number zero and its properties. Define a *zero vector* and justify your definition.

61. Multiple Steps Suppose you ride your bike 5 mi on a bike path. Your friend rides 10 mi on the same path in the same direction. If $\vec{w}$ represents your bike ride, then $2\vec{w}$ represents your friend's bike ride. This is an example of *scalar multiplication* of vectors. Suppose $\vec{\mathbf{w}} = \langle 2, 4 \rangle$.

a. Describe $2\vec{\mathbf{w}}$ as an ordered pair.

b. Find the magnitudes of $\vec{\mathbf{w}}$ and $2\vec{\mathbf{w}}$. What is their relationship?

c. Based on your results for parts **a** and **b**, what is the effect of multiplying a vector by a constant? Explain.

62. Geometry in Three Dimensions A hot-air ballon traveled 2000 ft north and 900 ft east, while rising 400 ft. This trip can be described with the three-coordinate vector $\langle 2000, 900, 400 \rangle$. What is the magnitude of the vector? What is the angle of elevation of the balloon from its starting point?

REVIEW AND PREVIEW

$\Delta ABC \cong \Delta EFG$. Complete the congruence statements. See Section 4.2.

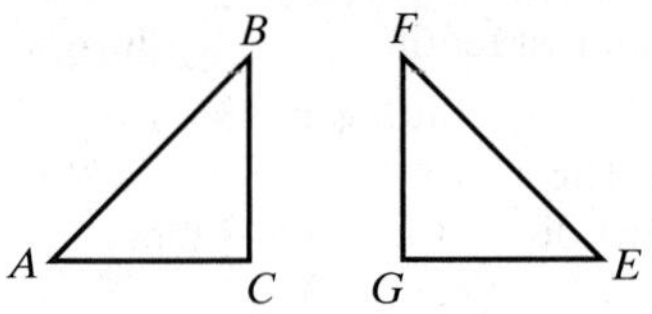

63. $\overline{AB} \cong$?

64. $\overline{EG} \cong$?

65. $\overline{FG} \cong$?

66. $\angle C \cong$?

67. $\angle E \cong$?

68. $\angle B \cong$?

EXTENSION—LAW OF SINES

OBJECTIVES

1 Use the Law of Sines to Solve Oblique Triangles.

2 Use the Law of Sines to Solve, if Possible, the Triangle or Triangles in the Ambiguous Case.

3 Find the Area of an Oblique Triangle Using the Sine Function.

VOCABULARY

- oblique triangle
- Law of Sines
- ambiguous case

Helpful Hint

Do not apply relationships that are valid for right triangles to oblique triangles. For example, the Pythagorean Theorem, $a^2 + b^2 = c^2$, applies only to right triangles.

Helpful Hint

The Law of Sines can also be expressed with the sines in the numerator:

$$\frac{\sin A}{a} = \frac{\sin B}{b} = \frac{\sin C}{c}$$

In this section, we consider triangles other than right triangles.

The Law of Sines and Its Proof An **oblique triangle** is a triangle that does not contain a right angle. The next figure shows that an oblique triangle has either three acute angles or two acute angles and one obtuse angle. Notice that the angles are labeled *A*, *B*, and *C*. The sides opposite each angle are labeled *a*, *b*, and *c*, respectively.

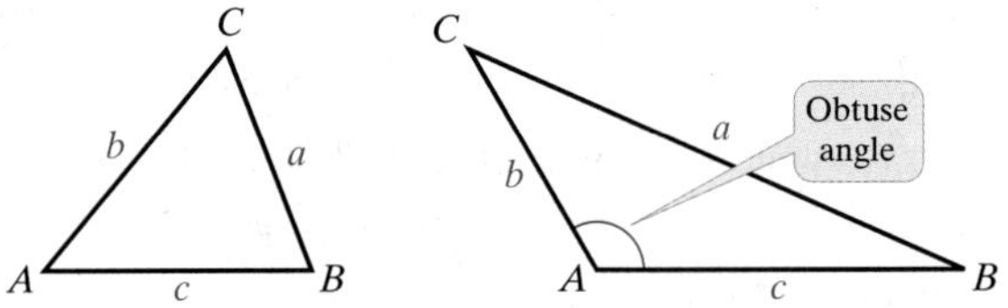

Oblique triangles

The trigonometric ratios we defined earlier apply only to right triangles and are not valid for oblique triangles. Thus, we must develop new relationships to work with oblique triangles.

One such relationship is called the **Law of Sines**.

The Law of Sines

If *A*, *B*, and *C* are the measures of the angles of a triangle, and *a*, *b*, and *c* are the lengths of the sides opposite these angles, then

$$\frac{a}{\sin A} = \frac{b}{\sin B} = \frac{c}{\sin C}$$

The ratio of the length of the side of any triangle to the sine of the angle opposite that side is the same for all three sides of the triangle.

To prove the Law of Sines, we draw an altitude of length *h* from one of the vertices of the triangle. In the oblique triangle in the next page margin, the altitude is drawn from vertex *C*. Two smaller triangles are formed, ΔACD and ΔBCD. Note that both are right triangles. Thus, we can use the definition of the sine of an angle of a right triangle.

$$\sin B = \frac{h}{a} \qquad \sin A = \frac{h}{b} \qquad \sin \theta = \frac{\text{opp.}}{\text{hyp.}}$$

$$h = a \sin B \qquad h = b \sin A \qquad \text{Solve each equation for } h.$$

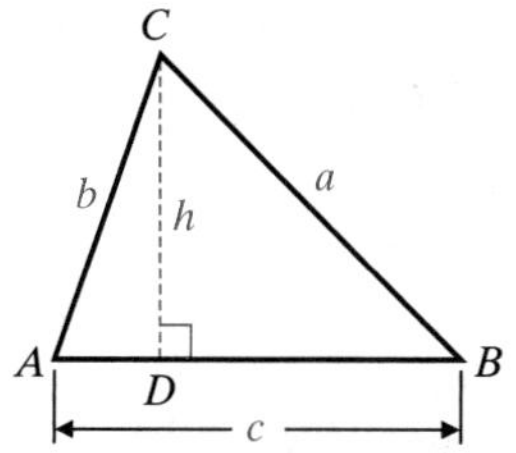

Drawing an altitude to prove the Law of Sines

Because we have found two expressions for h, we can set these expressions equal to each other.

$$a \sin B = b \sin A \quad \text{Equate the expressions for } h.$$

$$\frac{a \sin B}{\sin A \sin B} = \frac{b \sin A}{\sin A \sin B} \quad \text{Divide both sides by } \sin A \sin B.$$

$$\frac{a}{\sin A} = \frac{b}{\sin B} \quad \text{Simplify.}$$

This proves part of the Law of Sines. If we use the same process and draw an altitude of length h from vertex A, we obtain the following result:

$$\frac{b}{\sin B} = \frac{c}{\sin C}$$

When this equation is combined with the previous equation, we obtain the Law of Sines. (The Law of Sines is derived in a similar manner if the oblique triangle contains an obtuse angle.)

Helpful Hint

Use the Law of Sines when you know either

- SAA or
- ASA or
- SSA, called the ambiguous case

OBJECTIVE 1 ▶ Solving Oblique Triangles. Solving an oblique triangle means finding the lengths of its three sides and the measurements of its three angles. The Law of Sines can be used to solve a triangle in which one side and two angles are known. The three known measurements can be abbreviated as SAA (a side and two angles are known) or ASA (two angles and the side between them are known).

In this extension section and the next one, instead of writing $m\angle C = 63°$, we abbreviate and write $C = 63°$.

EXAMPLE 1 Solving an SAA Triangle Using the Law of Sines

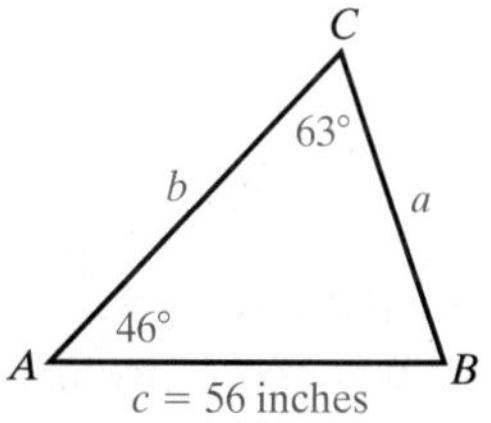

Solving an oblique SAA triangle

Solve the triangle shown in the margin with $A = 46°$, $C = 63°$, and $c = 56$ inches. Round lengths of sides to the nearest tenth.

Solution We begin by finding B, the third angle of the triangle. We do not need the Law of Sines to do this. Instead, we use the fact that the sum of the measures of the interior angles of a triangle is 180°

$$A + B + C = 180°$$

$$46° + B + 63° = 180° \quad \text{Substitute the given values: } A = 46° \text{ and } C = 63°.$$

$$109° + B = 180° \quad \text{Add.}$$

$$B = 71° \quad \text{Subtract } 109° \text{ from both sides.}$$

When we use the Law of Sines, we must be given one of the three ratios. In this example, we are given c and C: $c = 56$ and $C = 63°$. Thus, we use the ratio $\frac{c}{\sin C}$, or $\frac{56}{\sin 63°}$, to find the other two sides. Use the Law of Sines to find a.

$$\frac{a}{\sin A} = \frac{c}{\sin C} \quad \text{The ratio of any side to the sine of its opposite angle equals the ratio of any other side to the sine of its opposite angle.}$$

$$\frac{a}{\sin 46°} = \frac{56}{\sin 63°} \quad A = 46°, c = 56, \text{ and } C = 63°$$

$$a = \frac{56 \sin 46°}{\sin 63°} \quad \text{Multiply both sides by } \sin 46° \text{ and solve for } a.$$

$$a \approx 45.2 \text{ inches} \quad \text{Use a calculator.}$$

We use the Law of Sines again, this time to find b.

$$\frac{b}{\sin B} = \frac{c}{\sin C} \qquad \text{We use the given ratio, } \frac{c}{\sin C}\text{, to find } b.$$

$$\frac{b}{\sin 71^\circ} = \frac{56}{\sin 63^\circ} \qquad \text{We found that } B = 71^\circ\text{. We are given } c = 56 \text{ and } C = 63^\circ.$$

$$b = \frac{56 \sin 71^\circ}{\sin 63^\circ} \qquad \text{Multiply both sides by } \sin 71^\circ \text{ and solve for } b.$$

$$b \approx 59.4 \text{ inches} \qquad \text{Use a calculator.}$$

The solution is $B = 71^\circ$, $a \approx 45.2$ inches, and $b \approx 59.4$ inches. □

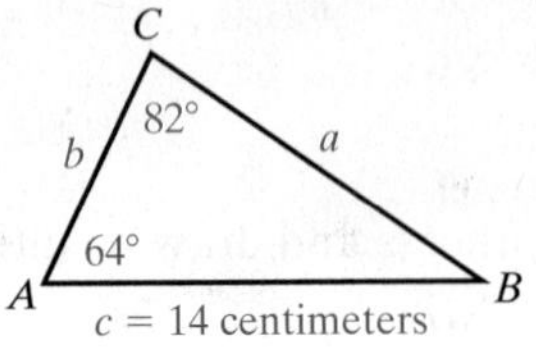

PRACTICE 1 Solve the triangle shown in the margin with $A = 64^\circ$, $C = 82^\circ$, and $c = 14$ centimeters. Round as in Example 1.

EXAMPLE 2 Solving an ASA Triangle Using the Law of Sines

Solve triangle ABC if $A = 50^\circ$, $C = 33.5^\circ$, and $b = 76$. Round measures to the nearest tenth.

Solution We begin by drawing a picture of triangle ABC and labeling it with the given information. The figure in the margin shows the triangle that we must solve. We begin by finding B.

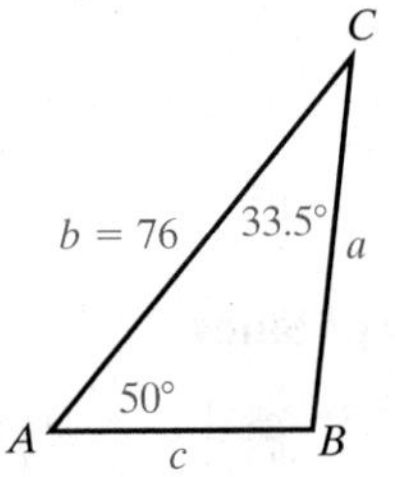

Solving an ASA triangle

$$A + B + C = 180^\circ \qquad \text{The sum of the measures of a triangle's interior angles is } 180^\circ.$$

$$50^\circ + B + 33.5^\circ = 180^\circ \qquad A = 50^\circ \text{ and } C = 33.5^\circ$$

$$83.5^\circ + B = 180^\circ \qquad \text{Add.}$$

$$B = 96.5^\circ \qquad \text{Subtract } 83.5^\circ \text{ from both sides.}$$

Keep in mind that we must be given one of the three ratios to apply the Law of Sines. In this example, we are given that $b = 76$ and we found that $B = 96.5^\circ$. Thus, we use the ratio $\frac{b}{\sin B}$, or $\frac{76}{\sin 96.5^\circ}$, to find the other two sides. Use the Law of Sines to find a and c.

This is the known ratio.

Find a:

$$\frac{a}{\sin A} = \frac{b}{\sin B}$$

$$\frac{a}{\sin 50^\circ} = \frac{76}{\sin 96.5^\circ}$$

$$a = \frac{76 \sin 50^\circ}{\sin 96.5^\circ} \approx 58.6$$

Find c:

$$\frac{c}{\sin C} = \frac{b}{\sin B}$$

$$\frac{c}{\sin 33.5^\circ} = \frac{76}{\sin 96.5^\circ}$$

$$c = \frac{76 \sin 33.5^\circ}{\sin 96.5^\circ} \approx 42.2$$

The solution is $B = 96.5^\circ$, $a \approx 58.6$, and $c \approx 42.2$. □

PRACTICE 2 Solve triangle ABC if $A = 40^\circ$, $C = 22.5^\circ$, and $b = 12$. Round as in Example 2.

> **Helpful Hint**
>
> Sometimes we recognize this ambiguous case by describing it as "two sides and the angle associated with one of the sides."
>
> Examples:
>
> a, b, A a, b, B
>
> b, c, C b, c, B and so on

OBJECTIVE 2 ▶ Solving the Ambiguous Case (SSA). If we are given two sides and an angle opposite one of them (SSA), does this determine a unique triangle? Can we solve this case using the Law of Sines? Such a case is called the **ambiguous case** because the given information may result in one triangle, two triangles, or no triangle at all. For example, in the figure in the margin, we are given a, b, and A. Because a is shorter than h, it is not long enough to form a triangle. The number of possible triangles, if any, that can be formed in the SSA case depends on h, the length of the altitude, where $h = b \sin A$.

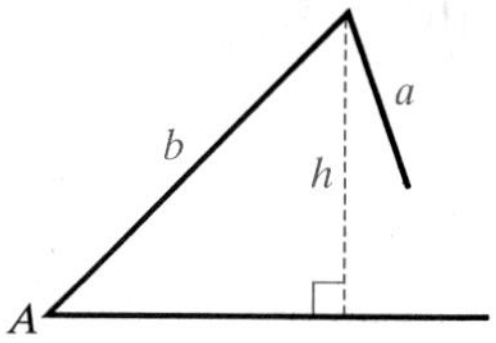

Given SSA, no triangle may result.

The Ambiguous Case (SSA)

Consider a triangle in which a, b, and A are given. This information may result in

One Triangle

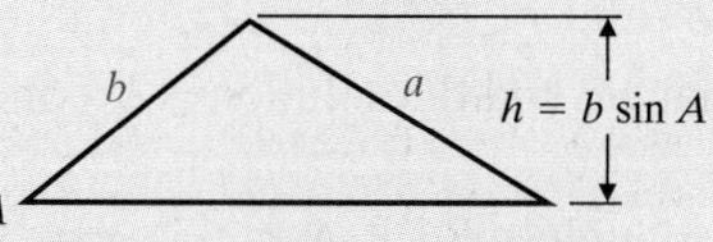

a is greater than h and a is greater than b. One triangle is formed.

One Right Triangle

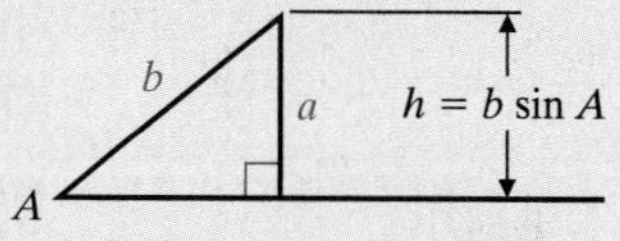

$a = h$ and is just the right length to form a right triangle.

No Triangle

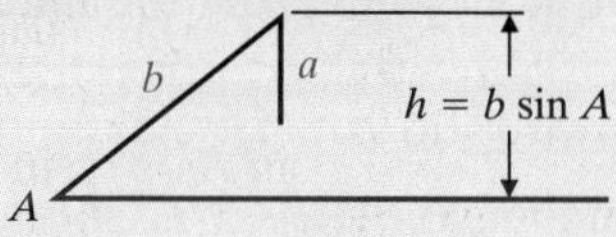

a is smaller than h and is not long enough to form a triangle.

Two Triangles

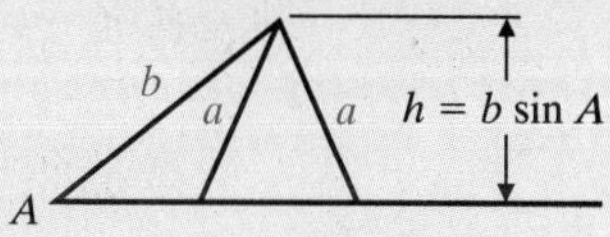

a is greater than h and a is less than b. Two distinct triangles are formed.

In an SSA situation, it is not necessary to draw an accurate sketch like those shown in the box. The Law of Sines determines the number of triangles, if any, and gives the solution for each triangle.

EXAMPLE 3 **Solving an SSA Triangle Using the Law of Sines (One Solution)**

Solve triangle ABC if $A = 43°$, $a = 81$, and $b = 62$. Round lengths of sides to the nearest tenth and angle measures to the nearest degree.

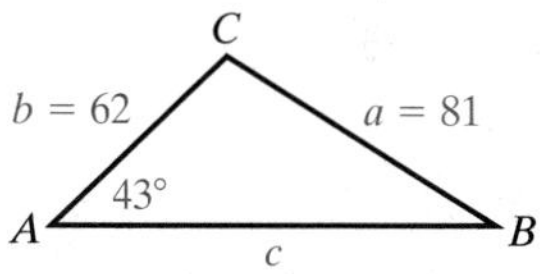

Solving an SSA triangle; the ambiguous case

Solution We begin with the sketch in the margin. The known ratio is $\frac{a}{\sin A}$, or $\frac{81}{\sin 43°}$. Because side b is given, we use the Law of Sines to find angle B.

$$\frac{a}{\sin A} = \frac{b}{\sin B} \quad \text{Apply the Law of Sines.}$$

$$\frac{81}{\sin 43°} = \frac{62}{\sin B} \quad a = 81, b = 62, \text{ and } A = 43°$$

$$81 \sin B = 62 \sin 43° \quad \text{Cross multiply: If } \frac{a}{b} = \frac{c}{d}, \text{ then } ad = bc.$$

$$\sin B = \frac{62 \sin 43°}{81} \quad \text{Divide both sides by 81 and solve for } \sin B.$$

$$\sin B \approx 0.5220 \quad \text{Use a calculator.}$$

There are two angles B between 0° and 180° for which $\sin B \approx 0.5220$.

$$B_1 \approx 31° \qquad B_2 \approx 180° - 31° = 149°$$

Obtain the acute angle with your calculator: $\sin^{-1} 0.5220$.

A second possibility is $B_2 = 180° - B_1$. Check with your calculator.

Look at the triangle in the previous page margin. Given that $A = 43°$, can you see that $B_2 \approx 149°$ is impossible? By adding 149° to the given angle, 43°, we exceed a 180° sum:

$$43° + 149° = 192°, \text{ which is } > 180°$$

Thus, the only possibility is that $B_1 \approx 31°$. We find C using this approximation for B_1 and the measure that was given for A: $A = 43°$.

$$C = 180° - B_1 - A \approx 180° - 31° - 43° = 106°$$

Side c, which lies opposite this 106° angle, can now be found using the Law of Sines.

$$\frac{c}{\sin C} = \frac{a}{\sin A} \quad \text{Apply the Law of Sines.}$$

$$\frac{c}{\sin 106°} = \frac{81}{\sin 43°} \quad a = 81, C \approx 106°, \text{ and } A = 43°$$

$$c = \frac{81 \sin 106°}{\sin 43°} \approx 114.2 \quad \text{Multiply both sides by } \sin 106° \text{ and solve for } c.$$

There is one triangle and the solution is B_1 (or B) $\approx 31°$, $C \approx 106°$, and $c \approx 114.2$. □

PRACTICE 3 Solve triangle ABC if $A = 57°$, $a = 33$, and $b = 26$. Round as in Example 3.

EXAMPLE 4 Solving an SSA Triangle Using the Law of Sines (No Solution)

Solve triangle ABC if $A = 75°$, $a = 51$, and $b = 71$.

Solution The known ratio is $\frac{a}{\sin A}$, or $\frac{51}{\sin 75°}$. Because side b is given, we use the Law of Sines to find angle B.

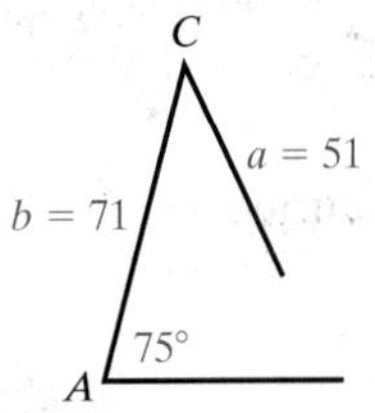

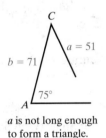

a is not long enough to form a triangle.

$$\frac{a}{\sin A} = \frac{b}{\sin B} \quad \text{Use the Law of Sines.}$$

$$\frac{51}{\sin 75°} = \frac{71}{\sin B} \quad \text{Substitute the given values.}$$

$$51 \sin B = 71 \sin 75° \quad \text{Cross multiply: If } \frac{a}{b} = \frac{c}{d}, \text{ then } ad = bc.$$

$$\sin B = \frac{71 \sin 75°}{51} \approx 1.34 \quad \text{Divide by 51 and solve for } \sin B.$$

You may try using your calculator to find B, but there is no angle B for which $\sin B \approx 1.34$. Thus, there is no triangle with the given measurements, as shown in the margin. □

PRACTICE 4 Solve triangle ABC if $A = 50°$, $a = 10$, and $b = 20$.

EXAMPLE 5 Solving an SSA Triangle Using the Law of Sines (Two Solutions)

Solve triangle ABC if $A = 40°$, $a = 54$, and $b = 62$. Round lengths of sides to the nearest tenth and angle measures to the nearest degree.

Solution The known ratio is $\frac{a}{\sin A}$, or $\frac{54}{\sin 40°}$. We use the Law of Sines to find angle B.

$$\frac{a}{\sin A} = \frac{b}{\sin B} \quad \text{Use the Law of Sines.}$$

$$\frac{54}{\sin 40°} = \frac{62}{\sin B} \quad \text{Substitute the given values.}$$

$$54 \sin B = 62 \sin 40° \quad \text{Cross multiply: If } \frac{a}{b} = \frac{c}{d}, \text{ then } ad = bc.$$

$$\sin B = \frac{62 \sin 40°}{54} \approx 0.7380 \quad \text{Divide by 54 and solve for } \sin B.$$

> **Helpful Hint**
> With the ambiguous case, SSA, try the first angle and its supplement with only the first angle you find.

There are two angles B between $0°$ and $180°$ for which $\sin B \approx 0.7380$.

$$B_1 \approx 48° \qquad B_2 \approx 180° - 48° = 132°$$

Use your calculator.

Try the second possibility: $B_2 = 180° - B_1$

If you add either angle to the given angle, $40°$, the sum does not exceed $180°$. Thus, there are two triangles with the given conditions, shown in the next figure, part a. The triangles, AB_1C_1 and AB_2C_2, are shown separately in parts.

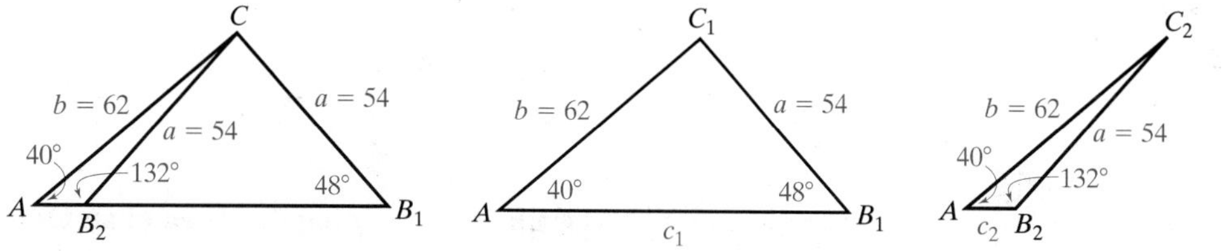

(a) Two triangles are possible with $A = 40°, a = 54$, and $b = 62$. **(b)** In one possible triangle, $B_1 = 48°$. **(c)** In the second possible triangle, $B_2 = 132°$.

We find angles C_1 and C_2 using a $180°$ angle sum in each of the two triangles.

$$C_1 = 180° - A - B_1 \approx 180° - 40° - 48° = 92° \qquad C_2 = 180° - A - B_2 \approx 180° - 40° - 132° = 8°$$

We use the Law of Sines to find c_1 and c_2.

$$\frac{c_1}{\sin C_1} = \frac{a}{\sin A} \qquad \frac{c_2}{\sin C_2} = \frac{a}{\sin A}$$

$$\frac{c_1}{\sin 92°} = \frac{54}{\sin 40°} \qquad \frac{c_2}{\sin 8°} = \frac{54}{\sin 40°}$$

$$c_1 = \frac{54 \sin 92°}{\sin 40°} \approx 84.0 \qquad c_2 = \frac{54 \sin 8°}{\sin 40°} \approx 11.7$$

> **Helpful Hint**
> If you keep track of the two triangles, one with the given information and $B_1 = 48°$, and the other with the given information and $B_2 = 132°$, you do not have to draw the figure to solve the triangles.

There are two triangles. In one triangle, the solution is $B_1 \approx 48°$, $C_1 \approx 92°$, and $c_1 \approx 84.0$. In the other triangle, $B_2 \approx 132°$, $C_2 \approx 8°$, and $c_2 \approx 11.7$. □

PRACTICE 5 Solve triangle ABC if $A = 35°$, $a = 12$, and $b = 16$. Round as in Example 5.

OBJECTIVE 3 ▶ Finding the Area of an Oblique Triangle. A formula for the area of an oblique triangle can be obtained using the procedure for proving the Law of Sines. We draw an altitude of length h from one of the vertices of the triangle, as shown in the triangle in the margin. We apply the definition of the sine of angle A, $\frac{\text{opposite}}{\text{hypotenuse}}$, in right triangle ACD:

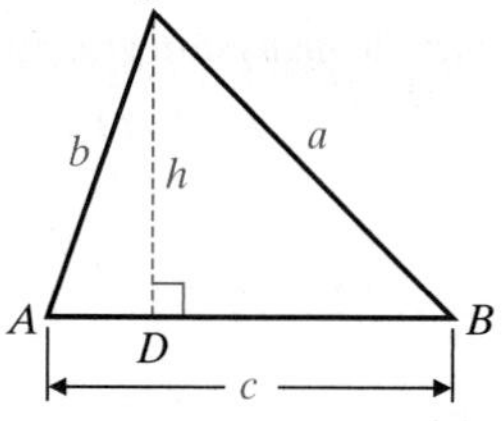

$$\sin A = \frac{h}{b}, \quad \text{so} \quad h = b \sin A$$

The area of a triangle is $\frac{1}{2}$ the product of any side and the altitude drawn to that side. Using the altitude h, we have

$$\text{Area} = \frac{1}{2}ch = \frac{1}{2}cb \sin A$$

Use the result from above: $h = b \sin A$.

This result, Area $= \frac{1}{2}cb \sin A$, or $\frac{1}{2}bc \sin A$, indicates that the area of the triangle is one-half the product of b and c times the sine of their included angle. If we draw altitudes from the other two vertices, we see that we can use any two sides to compute the area.

Area of an Oblique Triangle

The area of a triangle equals one-half the product of the lengths of two sides times the sine of their included angle. From our triangle, this wording can be expressed by the formulas

$$\text{Area} = \frac{1}{2}bc \sin A = \frac{1}{2}ab \sin C = \frac{1}{2}ac \sin B$$

EXAMPLE 6 **Finding the Area of an Oblique Triangle**

Find the area of a triangle having two sides of lengths 24 meters and 10 meters and an included angle of 62°. Round to the nearest square meter.

Solution The triangle is shown in the margin. Its area is half the product of the lengths of the two sides times the sine of the included angle.

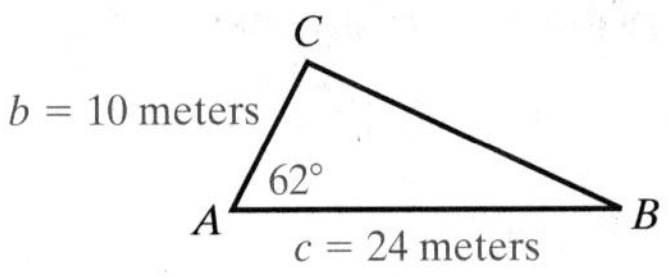

Finding the area of an SAS triangle

$$\text{Area} = \frac{1}{2}(24)(10)(\sin 62°) \approx 106$$

The area of the triangle is approximately 106 square meters. □

PRACTICE 6 Find the area of a triangle having two sides of lengths 8 meters and 12 meters and an included angle of 135°. Round to the nearest square meter.

EXTENSION—LAW OF SINES MyMathLab®

In Exercises 1–8, solve each triangle. Round lengths of sides to the nearest tenth and angle measures to the nearest degree.

1.

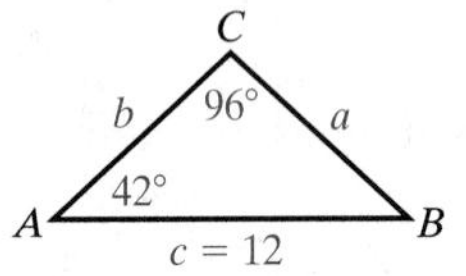

2.

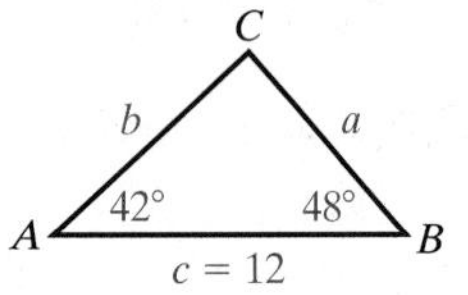

3.

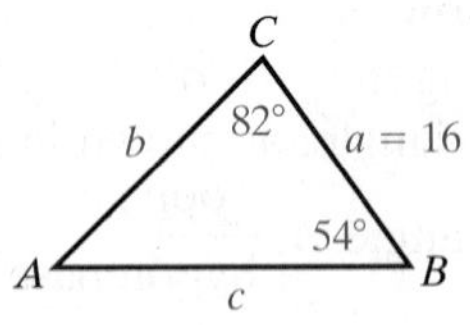

4.

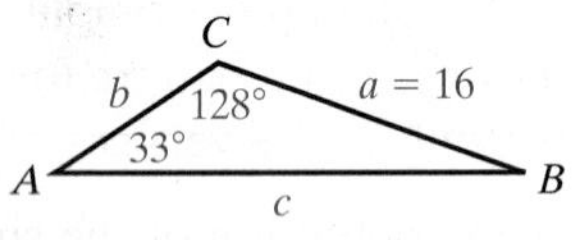

5.

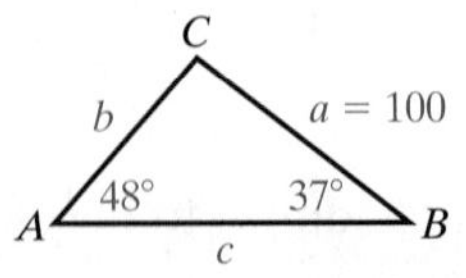

6.

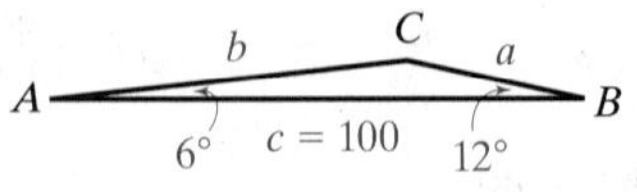

7.

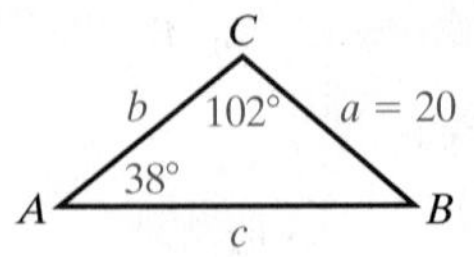

8.

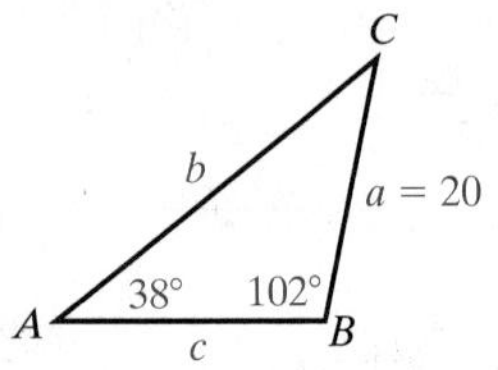

In Exercises 9–16, solve each triangle. Round lengths to the nearest tenth and angle measures to the nearest degree.

9. $A = 44°, B = 25°, a = 12$

10. $A = 56°, C = 24°, a = 22$

11. $B = 85°, C = 15°, b = 40$

12. $A = 85°, B = 35°, c = 30$

13. $A = 115°, C = 35°, c = 200$

14. $B = 5°, C = 125°, b = 200$

15. $A = 65°, B = 65°, c = 6$

16. $B = 80°, C = 10°, a = 8$

In Exercises 17–32, two sides and an angle (SSA) of a triangle are given. Determine whether the given measurements produce one triangle, two triangles, or no triangle at all. Solve each triangle that results. Round to the nearest tenth and the nearest degree for sides and angles, respectively.

17. $a = 20, b = 15, A = 40°$

18. $a = 30, b = 20, A = 50°$

19. $a = 10, c = 8.9, A = 63°$

20. $a = 57.5, c = 49.8, A = 136°$

21. $a = 42.1, c = 37, A = 112°$

22. $a = 6.1, b = 4, A = 162°$

23. $a = 10, b = 40, A = 30°$

24. $a = 10, b = 30, A = 150°$

25. $a = 16, b = 18, A = 60°$

26. $a = 30, b = 40, A = 20°$

27. $a = 12, b = 16.1, A = 37°$

28. $a = 7, b = 28, A = 12°$

29. $a = 22, c = 24.1, A = 58°$

30. $a = 95, c = 125, A = 49°$

31. $a = 9.3, b = 41, A = 18°$

32. $a = 1.4, b = 2.9, A = 142°$

In Exercises 33–38, find the area of the triangle having the given measurements. Round to the nearest square unit.

33. $A = 48°, b = 20$ feet, $c = 40$ feet

34. $A = 22°, b = 20$ feet, $c = 50$ feet

35. $B = 36°, a = 3$ yards, $c = 6$ yards

36. $B = 125°, a = 8$ yards, $c = 5$ yards

37. $C = 124°, a = 4$ meters, $b = 6$ meters

38. $C = 102°, a = 16$ meters, $b = 20$ meters

APPLICATION EXERCISES

39. Two fire-lookout stations are 10 miles apart, with Station B directly east of Station A. Both stations spot a fire. The bearing of the fire from Station A is N25°E and the bearing of the fire from Station B is N56°W. How far, to the nearest tenth of a mile, is the fire from each lookout station? (N25°E means 25° east of north; N56°W means 56° west of north.)

40. The Federal Communications Commission is attempting to locate an illegal radio station. It sets up two monitoring stations, A and B, with Station B 40 miles east of Station A. Station A measures the illegal signal from the radio station as coming from a direction of 48° east of north. Station B measures the signal as coming from a point 34° west of north. How far is the illegal radio station from monitoring stations A and B? Round to the nearest tenth of a mile.

41. The figure shows a 1200-yard-long sand beach and an oil platform in the ocean. The angle made with the platform from one end of the beach is 85° and from the other end is 76°. Find the distance of the oil platform, to the nearest tenth of a yard, from each end of the beach.

42. A surveyor needs to determine the distance between two points that lie on opposite banks of a river. The figure shows that 300 yards are measured along one bank. The angles from each end of this line segment to a point on the opposite bank are 62° and 53°. Find the distance between A and B to the nearest tenth of a yard.

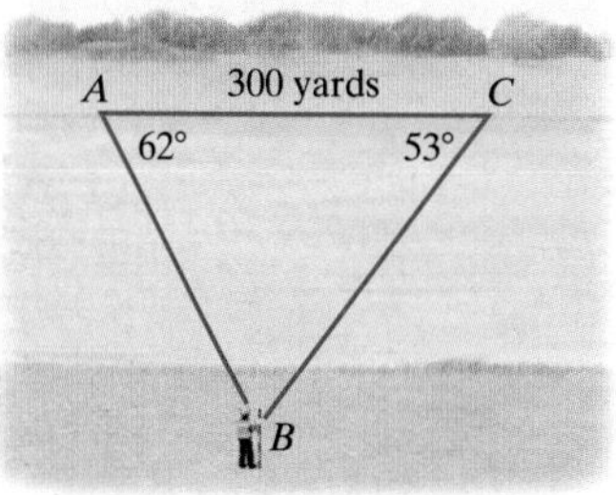

43. The Leaning Tower of Pisa in Italy leans at an angle of about 84.7°. The figure shows that 171 feet from the base of the tower, the angle of elevation to the top is 50°. Find the distance, to the nearest tenth of a foot, from the base to the top of the tower.

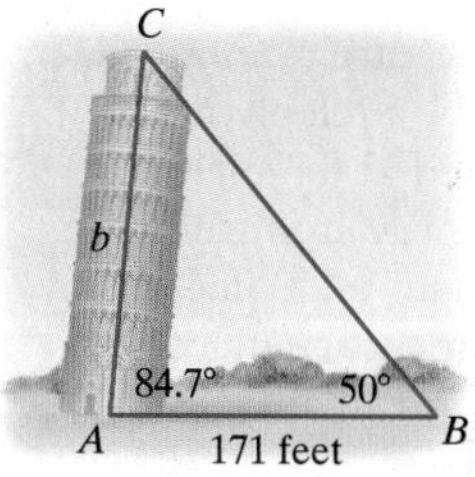

44. A pine tree growing on a hillside makes a 75° angle with the hill. From a point 80 feet up the hill, the angle of elevation to the top of the tree is 62° and the angle of depression to the bottom is 23°. Find, to the nearest tenth of a foot, the height of the tree.

45. The figure shows a shotput ring. The shot is tossed from A and lands at B. Using modern electronic equipment, the distance of the toss can be measured without the use of measuring tapes. When the shot lands at B, an electronic transmitter placed at B sends a signal to a device in the official's booth above the track. The device determines the angles at B and C. At a track meet, the distance from the official's booth to the shotput ring is 562 feet. If $B = 85.3°$ and $C = 5.7°$, determine the length of the toss to the nearest tenth of a foot.

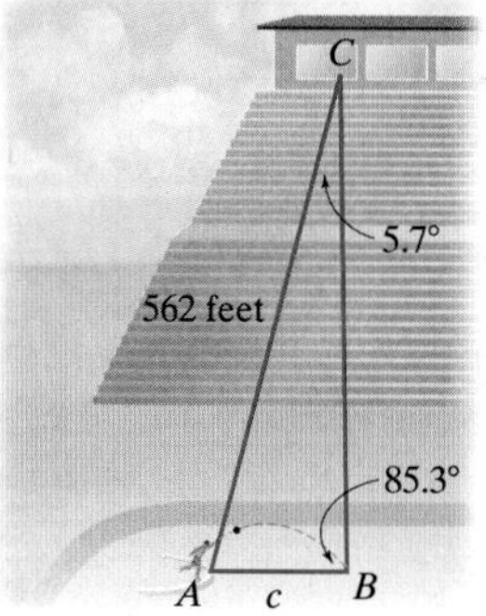

46. A pier forms an 85° angle with a straight shore. At a distance of 100 feet from the pier, the line of sight to the tip forms a 37° angle. Find the length of the pier to the nearest tenth of a foot.

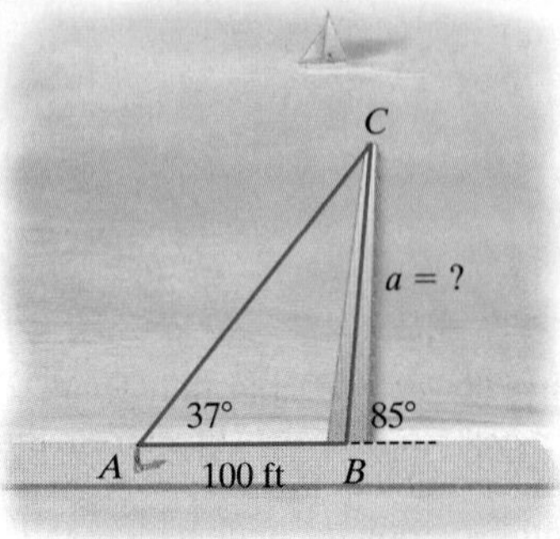

47. When the angle of elevation of the sun is 62°, a telephone pole that is tilted at an angle of 8° directly away from the sun casts a shadow 20 feet long. Determine the length of the pole to the nearest tenth of a foot.

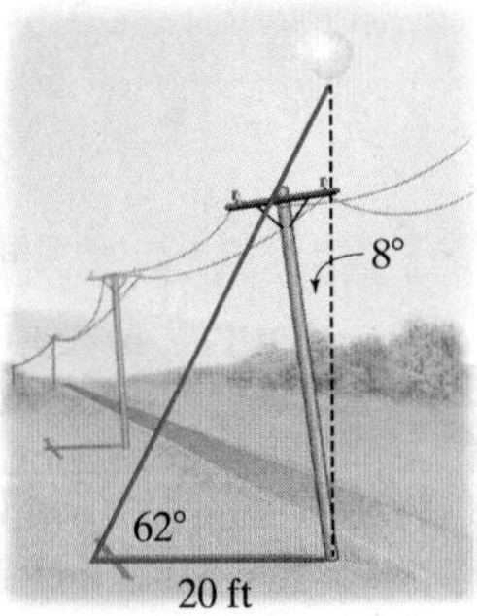

48. A leaning wall is inclined 6° from the vertical. At a distance of 40 feet from the wall, the angle of elevation to the top is 22°. Find the height of the wall to the nearest tenth of a foot.

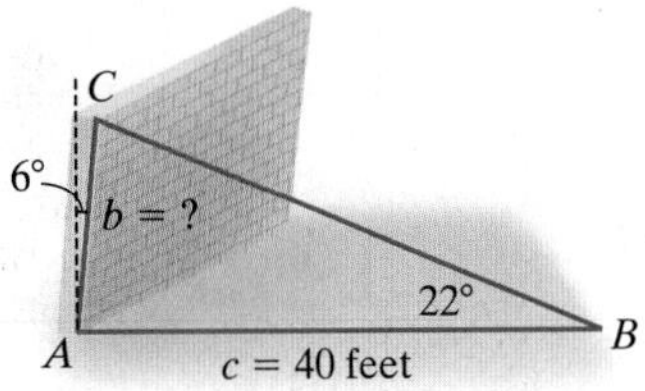

49. Redwood trees in California's Redwood National Park are hundreds of feet tall. The height of one of these trees is represented by h in the figure shown.

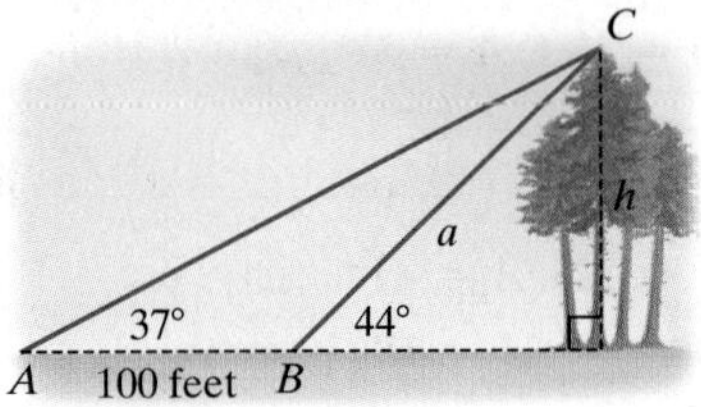

a. Use the measurements shown to find a, to the nearest tenth of a foot, in oblique triangle ABC.

b. Use the right triangle shown to find the height, to the nearest tenth of a foot, of a typical redwood tree in the park.

50. The figure shows a cable car that carries passengers from A to C. Point A is 1.6 miles from the base of the mountain. The angles of elevation from A and B to the mountain's peak are 22° and 66°, respectively.

a. Determine, to the nearest tenth of a foot, the distance covered by the cable car.

b. Find a, to the nearest tenth of a foot, in oblique triangle ABC.

c. Use the right triangle to find the height of the mountain to the nearest tenth of a foot.

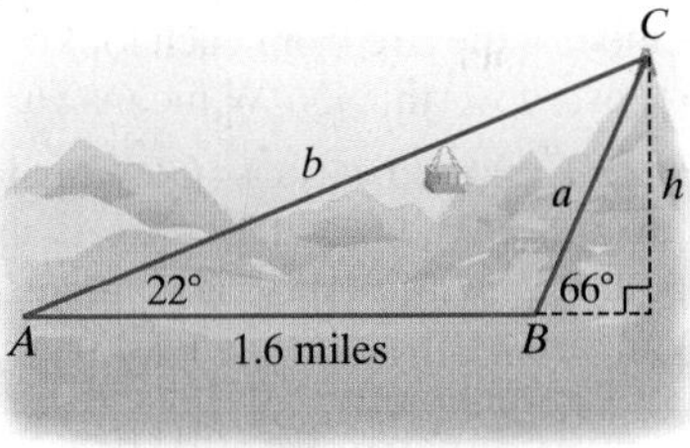

EXTENSION—LAW OF COSINES

OBJECTIVE

1 Use the Law of Cosines to Solve Oblique Triangles.

VOCABULARY

- Law of Cosines

Helpful Hint

Use the Law of Cosines when you know either

- SAS or
- SSS

Triangles in which we know the measures of three sides and need to find the measures of the missing angles cannot be solved by the Law of Sines. Thus, we need another law. In this extension, we study the Law of Cosines.

The Law of Cosines and Its Proof We now look at another relationship that exists among the sides and angles in an oblique triangle. The **Law of Cosines** is used to solve triangles in which two sides and the included angle (SAS) are known, or those in which three sides (SSS) are known.

The Law of Cosines

If A, B, and C are the measures of the angles of a triangle, and a, b, and c are the lengths of the sides opposite these angles, then

$$a^2 = b^2 + c^2 - 2bc \cos A$$

$$b^2 = a^2 + c^2 - 2ac \cos B$$

$$c^2 = a^2 + b^2 - 2ab \cos C$$

The square of a side of a triangle equals the sum of the squares of the other two sides minus twice their product times the cosine of their included angle.

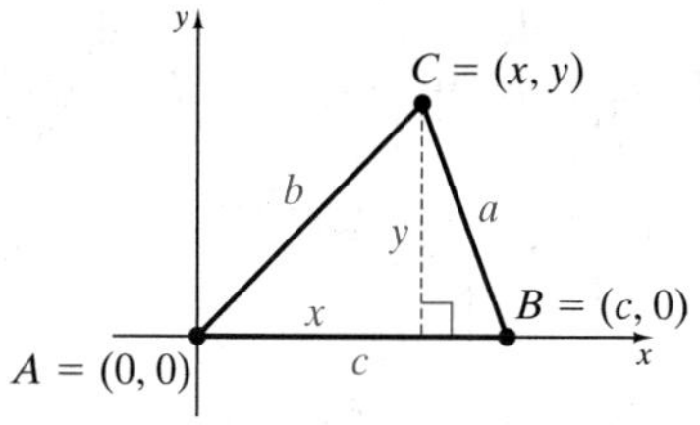

To prove the Law of Cosines, we place triangle ABC in a rectangular coordinate system. The figure in the margin shows a triangle with three acute angles. The vertex A is at the origin and side c lies along the positive x-axis. The coordinates of C are (x, y). Using the right triangle that contains angle A, we apply the definitions of the cosine and the sine.

$$\cos A = \frac{x}{b} \qquad \sin A = \frac{y}{b}$$

$$x = b\cos A \qquad y = b\sin A$$

Multiply both sides of each equation by b and solve for x and y, respectively.

Thus, the coordinates of C are $(x, y) = (b\cos A, b\sin A)$. Although triangle ABC shows angle A as an acute angle, if A were obtuse, the coordinates of C would still be $(b\cos A, b\sin A)$. This means that our proof applies to both kinds of oblique triangles.

We now apply the Distance Formula to the side of the triangle with length a. Notice that a is the distance from (x, y) to $(c, 0)$.

$$a = \sqrt{(x-c)^2 + (y-0)^2} \qquad \text{Use the Distance Formula.}$$

$$a^2 = (x-c)^2 + y^2 \qquad \text{Square both sides of the equation.}$$

$$a^2 = (b\cos A - c)^2 + (b\sin A)^2 \qquad x = b\cos A \text{ and } y = b\sin A$$

$$a^2 = b^2\cos^2 A - 2bc\cos A + c^2 + b^2\sin^2 A \qquad \text{Square the two expressions.}$$

$$a^2 = b^2\sin^2 A + b^2\cos^2 A + c^2 - 2bc\cos A \qquad \text{Rearrange terms.}$$

$$a^2 = b^2(\sin^2 A + \cos^2 A) + c^2 - 2bc\cos A \qquad \text{Factor } b^2 \text{ from the first two terms.}$$

$$a^2 = b^2 + c^2 - 2bc\cos A \qquad \sin^2 A + \cos^2 A = 1$$

The resulting equation is one of the three formulas for the Law of Cosines. The other two formulas are derived in a similar manner.

OBJECTIVE 1 ▶ Solving Oblique Triangles. If we are given two sides and an included angle (SAS) of an oblique triangle, none of the three ratios in the Law of Sines is known. This means that we do not begin solving the triangle using the Law of Sines. Instead, we apply the Law of Cosines and the following procedure:

Solving an SAS Triangle

STEP 1. Use the Law of Cosines to find the side opposite the given angle.

STEP 2. Use the Law of Sines to find the angle opposite the shorter of the two given sides. This angle is always acute.

STEP 3. Find the third angle by subtracting the measure of the given angle and the angle found in Step 2 from 180°.

EXAMPLE 1 **Solving an SAS Triangle**

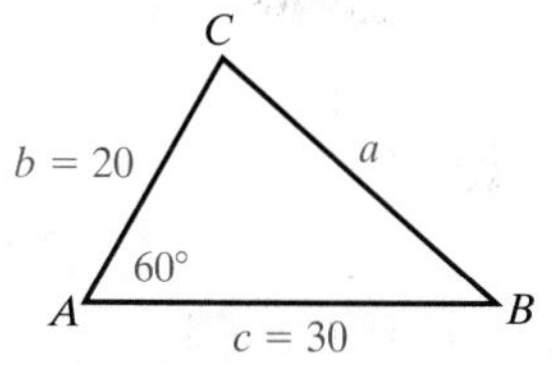

Solving an SAS triangle

Solve the triangle in the margin with $A = 60°$, $b = 20$, and $c = 30$. Round lengths of sides to the nearest tenth and angle measures to the nearest degree.

Solution We are given two sides and an included angle. Therefore, we apply the three-step procedure for solving an SAS triangle.

STEP 1. Use the Law of Cosines to find the side opposite the given angle. Thus, we will find a.

$$a^2 = b^2 + c^2 - 2bc\cos A \qquad \text{Apply the Law of Cosines to find } a.$$

$$a^2 = 20^2 + 30^2 - 2(20)(30)\cos 60° \qquad b = 20, c = 30, \text{ and } A = 60°$$

$$= 400 + 900 - 1200(0.5) \qquad \text{Perform the indicated operations.}$$

$$= 700$$

$$a = \sqrt{700} \approx 26.5 \qquad \text{Take the square root of both sides and solve for } a.$$

STEP 2. **Use the Law of Sines to find the angle opposite the shorter of the two given sides. This angle is always acute.** The shorter of the two given sides is $b = 20$. Thus, we will find acute angle B.

$$\frac{b}{\sin B} = \frac{a}{\sin A} \qquad \text{Apply the Law of Sines.}$$

$$\frac{20}{\sin B} = \frac{\sqrt{700}}{\sin 60°} \qquad \text{We are given } b = 20 \text{ and } A = 60°. \text{ Use the exact value of } a, \sqrt{700}, \text{ from Step 1.}$$

$$\sqrt{700} \sin B = 20 \sin 60° \qquad \text{Cross multiply: If } \frac{a}{b} = \frac{c}{d}, \text{ then } ad = bc.$$

$$\sin B = \frac{20 \sin 60°}{\sqrt{700}} \approx 0.6547 \qquad \text{Divide by } \sqrt{700} \text{ and solve for } \sin B.$$

$$B \approx 41° \qquad \text{Find } \sin^{-1} 0.6547 \text{ using a calculator.}$$

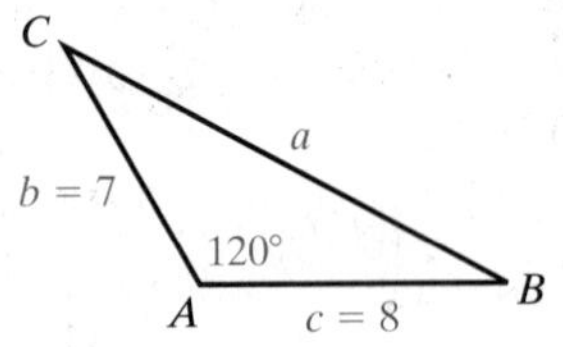

STEP 3. **Find the third angle. Subtract the measures of the given angle and the angle found in Step 2 from 180°.**

$$C = 180° - A - B \approx 180° - 60° - 41° = 79°$$

The solution is $a \approx 26.5$, $B \approx 41°$, and $C \approx 79°$. □

PRACTICE

1 Solve the triangle shown in the margin with $A = 120°$, $b = 7$, and $c = 8$. Round as in Example 1.

If we are given three sides of a triangle (SSS), solving the triangle involves finding the three angles. We use the following procedure:

Solving an SSS Triangle

STEP 1. Use the Law of Cosines to find the angle opposite the longest side.

STEP 2. Use the Law of Sines to find either of the two remaining acute angles.

STEP 3. Find the third angle by subtracting the measures of the angles found in Steps 1 and 2 from 180°.

EXAMPLE 2 Solving an SSS Triangle

Solve triangle ABC if $a = 6$, $b = 9$, and $c = 4$. Round angle measures to the nearest degree.

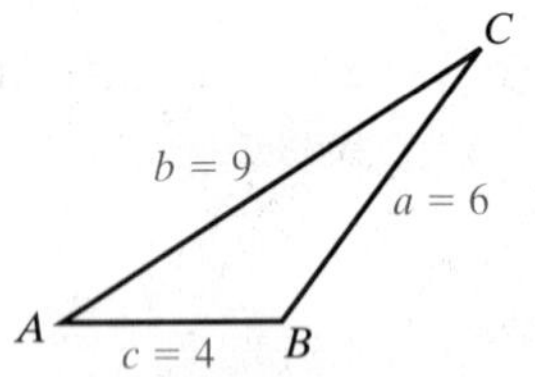

Solving an SSS triangle

Solution We are given three sides. Therefore, we apply the three-step procedure for solving an SSS triangle. The triangle is shown in the margin.

STEP 1. **Use the Law of Cosines to find the angle opposite the longest side.** The longest side is $b = 9$. Thus, we will find angle B.

$$b^2 = a^2 + c^2 - 2ac \cos B \qquad \text{Apply the Law of Cosines to find } B.$$

$$2ac \cos B = a^2 + c^2 - b^2 \qquad \text{Solve for } \cos B.$$

$$\cos B = \frac{a^2 + c^2 - b^2}{2ac}$$

$$\cos B = \frac{6^2 + 4^2 - 9^2}{2 \cdot 6 \cdot 4} = -\frac{29}{48} \qquad a = 6, b = 9, \text{ and } c = 4$$

Using a calculator, $\cos^{-1}\left(-\frac{29}{48}\right) \approx 127°$.

STEP 2. Use the Law of Sines to find either of the two remaining acute angles. We will find angle A.

$$\frac{a}{\sin A} = \frac{b}{\sin B}$$ Apply the Law of Sines.

$$\frac{6}{\sin A} = \frac{9}{\sin 127°}$$ We are given $a = 6$ and $b = 9$. We found that $B \approx 127°$.

$$9 \sin A = 6 \sin 127°$$ Cross multiply.

$$\sin A = \frac{6 \sin 127°}{9} \approx 0.5324$$ Divide by 9 and solve for $\sin A$.

$$A \approx 32°$$ Find $\sin^{-1} 0.5324$ using a calculator.

Helpful Hint

You can use the Law of Cosines in Step 2 to find either of the remaining angles. However, it is simpler to use the Law of Sines. Because the largest angle has been found, the remaining angles must be acute. Thus, there is no need to be concerned about two possible triangles or an ambiguous case.

STEP 3. Find the third angle. Subtract the measures of the angles found in Steps 1 and 2 from 180°.

$$C = 180° - B - A \approx 180° - 127° - 32° = 21°$$

The solution is $B \approx 127°$, $A \approx 32°$, and $C \approx 21°$. □

PRACTICE
2 Solve triangle ABC if $a = 8$, $b = 10$, and $c = 5$. Round angle measures to the nearest degree.

EXTENSION—LAW OF COSINES MyMathLab®

In Exercises 1–8, solve each triangle. Round lengths of sides to the nearest tenth and angle measures to the nearest degree.

1. 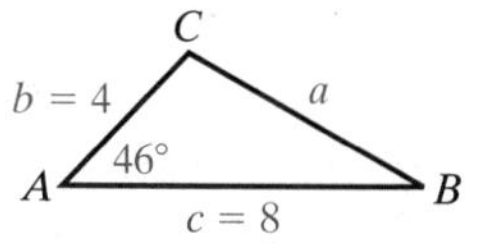

2. 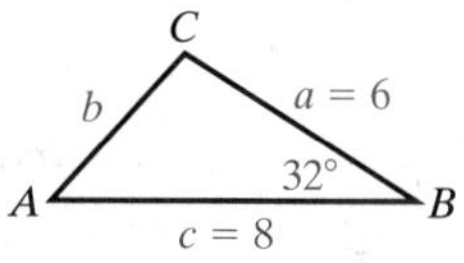

3. 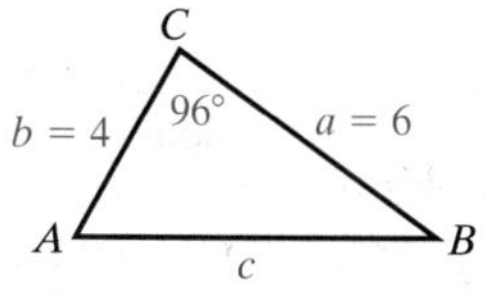

4. 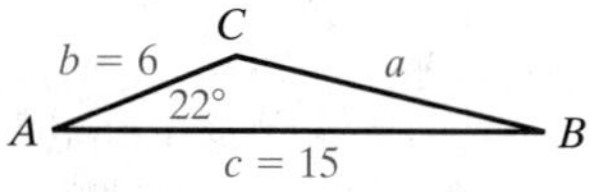

5. 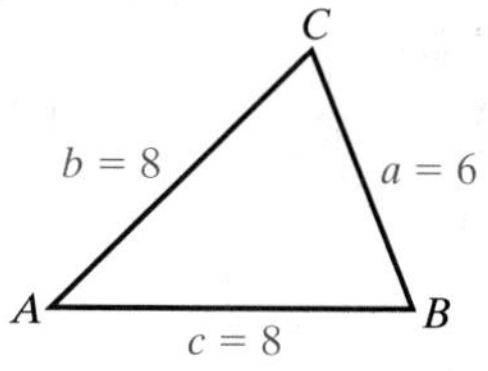

6. 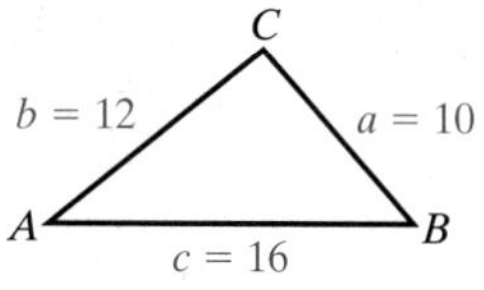

7. 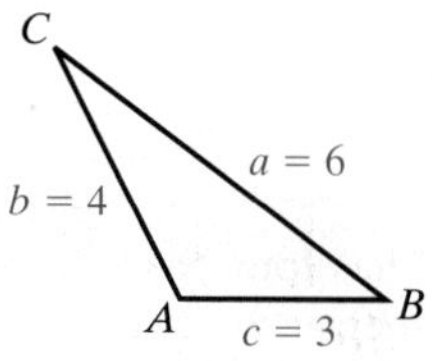

8. 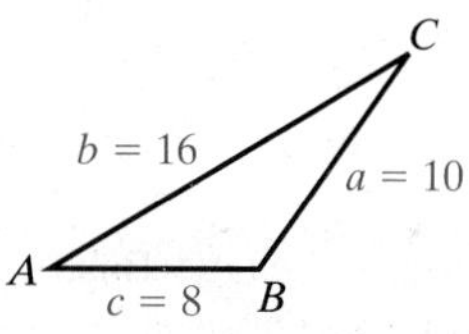

In Exercises 9–24, solve each triangle. Round lengths to the nearest tenth and angle measures to the nearest degree.

9. $a = 5, b = 7, C = 42°$
10. $a = 10, b = 3, C = 15°$
11. $b = 5, c = 3, A = 102°$
12. $b = 4, c = 1, A = 100°$
13. $a = 6, c = 5, B = 50°$
14. $a = 4, c = 7, B = 55°$
15. $a = 5, c = 2, B = 90°$
16. $a = 7, c = 3, B = 90°$
17. $a = 5, b = 7, c = 10$
18. $a = 4, b = 6, c = 9$
19. $a = 3, b = 9, c = 8$
20. $a = 4, b = 7, c = 6$
21. $a = 3, b = 3, c = 3$
22. $a = 5, b = 5, c = 5$
23. $a = 63, b = 22, c = 50$
24. $a = 66, b = 25, c = 45$

APPLICATION EXERCISES

25. Use the figure to find the pace angle, to the nearest degree, for the carnivore.

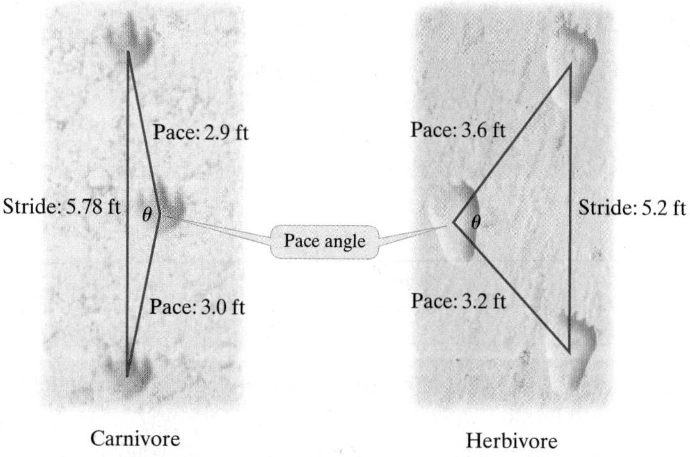

Dinosaur footprints

Source: Kuban, Glen, Dinosaur Footprints, adapted from *An Overview of Dinosaur Tracking*. Reprinted by permission of the author

26. Use the figure for Exercise **25** to find the pace angle, to the nearest degree, for the herbivore.

27. Two ships leave a harbor at the same time. One ship travels on a bearing of S12°W at 14 miles per hour. The other ship travels on a bearing of N75°E at 10 miles per hour. How far apart will the ships be after three hours? Round to the nearest tenth of a mile. (S12°W, for example, means 12° west of south.)

28. A plane leaves Airport A and travels 580 miles to Airport B on a bearing of N34°E. The plane later leaves Airport B and travels to Airport C 400 miles away on a bearing of S74°E. Find the distance from Airport A to Airport C to the nearest tenth of a mile.

29. Find the distance across the lake from A to C, to the nearest yard, using the measurements shown in the figure.

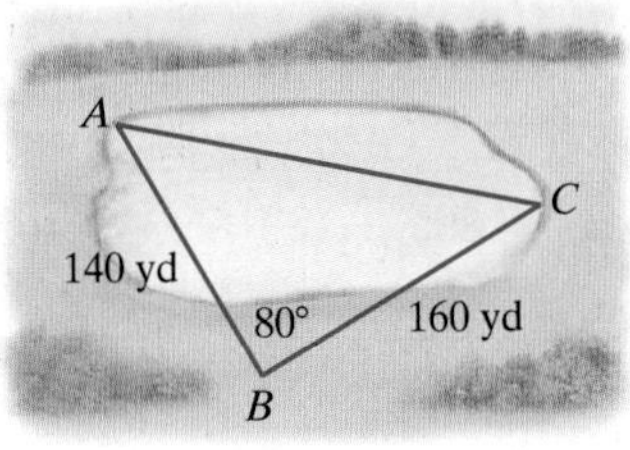

30. To find the distance across a protected cove at a lake, a surveyor makes the measurements shown in the figure. Use these measurements to find the distance from A to B to the nearest yard.

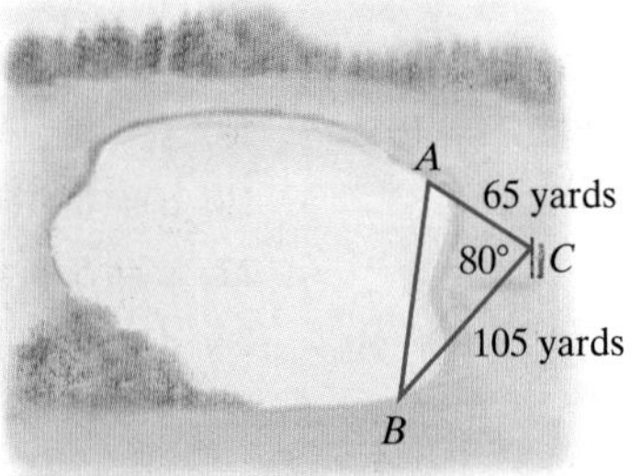

The diagram shows three islands in Florida Bay. You rent a boat and plan to visit each of these remote islands. Use the diagram to solve Exercises 31–32.

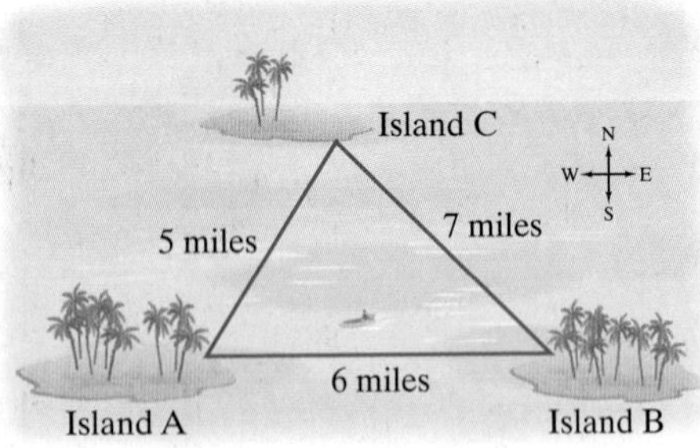

31. If you are on Island A, on what bearing should you navigate to go to Island C?

32. If you are on Island B, on what bearing should you navigate to go to Island C?

33. You are on a fishing boat that leaves its pier and heads east. After traveling for 25 miles, there is a report warning of rough seas directly south. The captain turns the boat and follows a bearing of S40°W for 13.5 miles. (See figure in the next column.)

a. At this time, how far are you from the boat's pier? Round to the nearest tenth of a mile.

b. What bearing could the boat have originally taken to arrive at this spot?

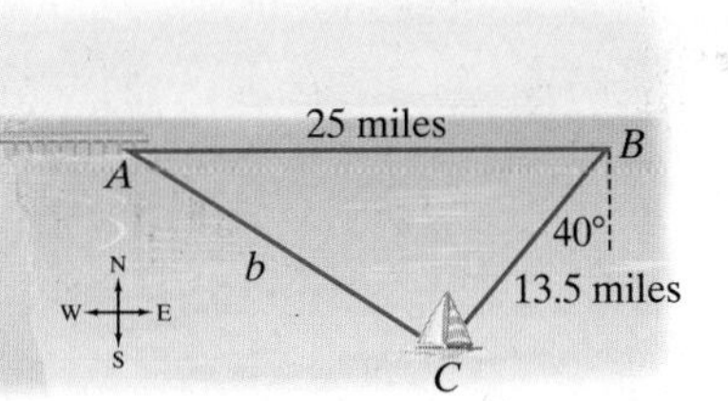

34. You are on a fishing boat that leaves its pier and heads east. After traveling for 30 miles, there is a report warning of rough seas directly south. The captain turns the boat and follows a bearing of S45°W for 12 miles.

a. At this time, how far are you from the boat's pier? Round to the nearest tenth of a mile.

b. What bearing could the boat have originally taken to arrive at this spot?

35. The figure shows a 400-foot tower on the side of a hill that forms a 7° angle with the horizontal. Find the length of each of the two guy wires that are anchored 80 feet uphill and downhill from the tower's base and extend to the top of the tower. Round to the nearest tenth of a foot.

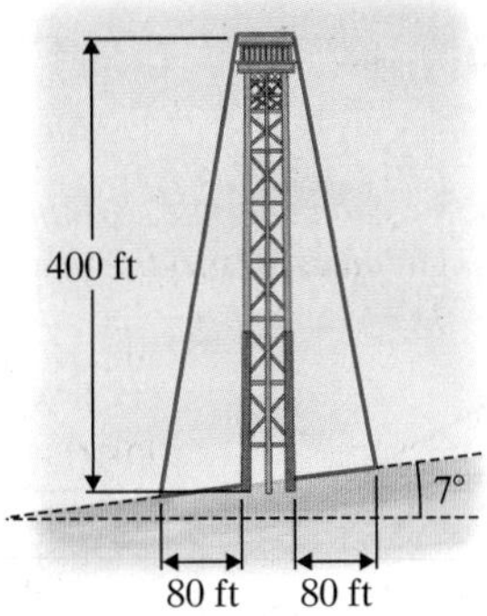

36. The figure shows a 200-foot tower on the side of a hill that forms a 5° angle with the horizontal. Find the length of each of the two guy wires that are anchored 150 feet uphill and downhill from the tower's base and extend to the top of the tower. Round to the nearest tenth of a foot.

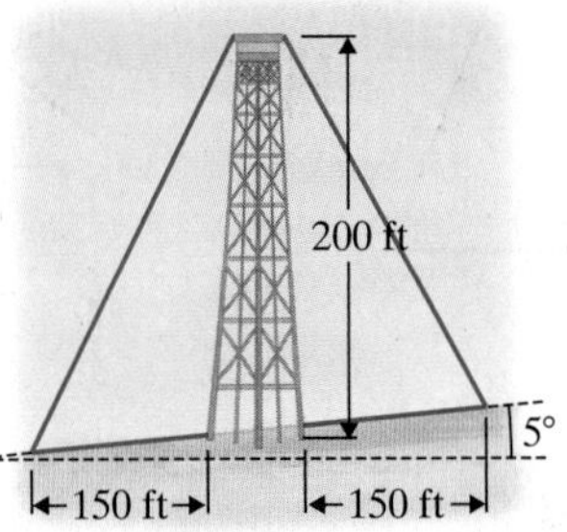

37. A Major League Baseball diamond has four bases forming a square whose sides measure 90 feet each. The pitcher's mound is 60.5 feet from home plate on a line joining home plate and second base. Find the distance from the pitcher's mound to first base. Round to the nearest tenth of a foot.

38. A Little League baseball diamond has four bases forming a square whose sides measure 60 feet each. The pitcher's mound is 46 feet from home plate on a line joining home plate and second base. Find the distance from the pitcher's mound to third base. Round to the nearest tenth of a foot.

Historical Note—The Life of Pythagoras

Although the Greek mathematician (~570–495 BC) is probably best known for his name on the Pythagorean Theorem, he was also a philosopher, musician, astronomer, and the founder of a religious movement called Pythagoreanism.

In music, he is given the credit for discovering that simple ratios in lengths of strings lead to different octaves. In mathematics, he is given credit for writing a proof of the Pythagorean Theorem, although there is evidence that this theorem was used by the Babylonians and possibly the Indians before Pythagoras' time.

Unfortunately, not much was written about Pythagoras until after his death. This led to some historians believing that many of his accomplishments may have actually been those of his students. It is believed that he thought of himself as a philosopher and he actually started a secret religious society, practicing rites that he developed. His society was a school, a brotherhood (there is some evidence that a few women were members), a secret club, and—most importantly—a way of life. His followers took an active role in politics and that may have led to the burning of his meeting places.

It is greatly believed that religion and science were closely connected in those days. Since much of religion and science back then had to do with astronomy and other strange phenomenon to men of those times, it is not hard to believe that this connection existed and was a strong one.

It is believed that Pythagoras and his Pythagoreans greatly influenced Plato (~424–348 BC), a well-known Greek philosopher and mathematician. Plato was a student of Socrates and a teacher of Aristotle.

We have only touched the surface of the life of Pythagoras. Feel free to read more about this man who—at the least—led an interesting life.

Question: Which U.S. president wrote a new proof of the Pythagorean Theorem?

CHAPTER 9 VOCABULARY CHECK

Word Bank *Use the choices to fill in each blank.*

angle of depression	Pythagorean triple	terminal point
angle of elevation	resultant	trigonometric
cosine	sine	vector
initial point	tangent	arrow
magnitude		

1. Any quantity that has magnitude and direction is called a(n) ____________.

2. The sum of two vectors is the ____________.

3. A(n) ____________ is formed by a horizontal line and the line of sight above that line.

4. A(n) ____________ is formed by a horizontal line and the line of sight below that line.

5. A set of three positive integers that satisfy $a^2 + b^2 = c^2$ form a(n) ____________.

6. A ratio of the sides of a right triangle is called a(n) ____________ ratio.

7. The ____________ of a vector is the distance from the initial point of the vector to the terminal point.

8. A vector can be represented by a(n) ____________.

9. The starting point of a vector is called the ____________ and the ending point of the vector is called the ____________.

For Exercises 10–12, let A be an acute angle of a right triangle.

10. ____________ $A = \dfrac{\text{opposite}}{\text{adjacent}}$

11. ____________ $A = \dfrac{\text{adjacent}}{\text{hypotenuse}}$

12. ____________ $A = \dfrac{\text{opposite}}{\text{hypotenuse}}$

CHAPTER 9 REVIEW

(9.1) *Find the value of x. If your answer is not an integer, express it in simplest radical form.*

1.

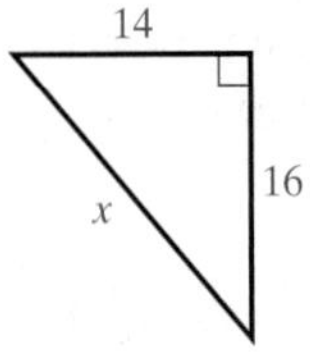

2.

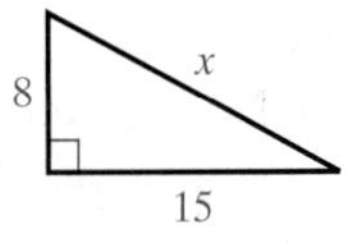

3.

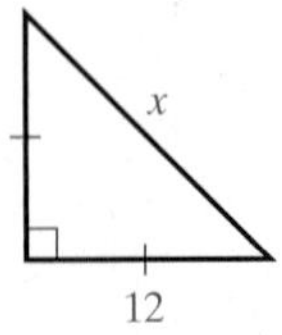

4.

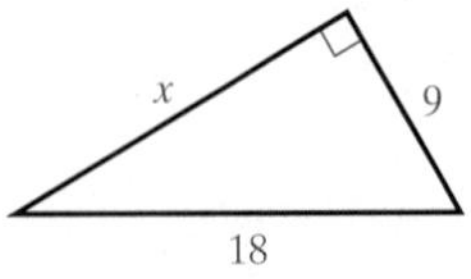

(9.2) *Find the value of each variable. If your answer is not an integer, express it in simplest radical form.*

5.

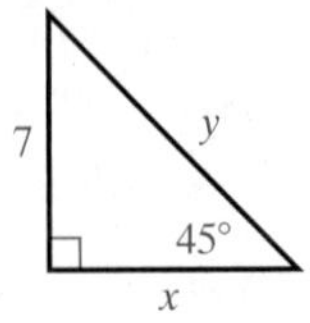

6.

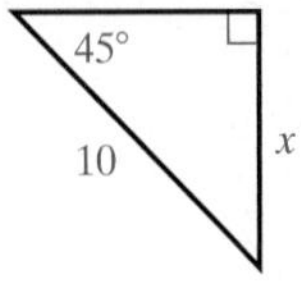

7.

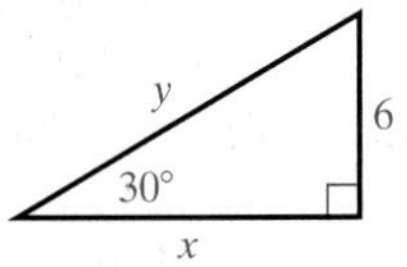

8.

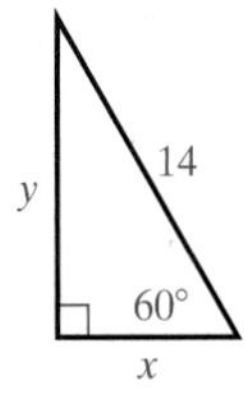

(9.3, 9.4) *Express sin A, cos A, and tan A as ratios.*

9.

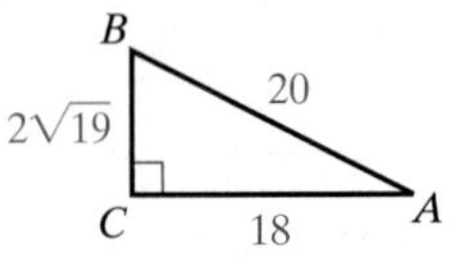

10.

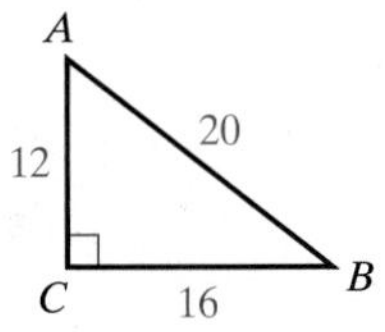

Find the value of x to the nearest tenth.

11.

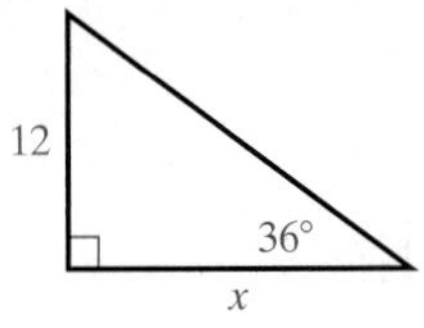

12.

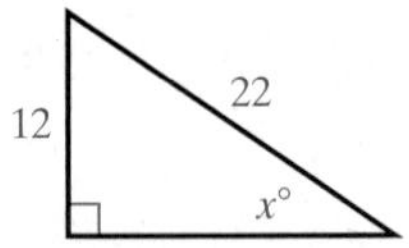

(9.5) Find the magnitude and direction of each vector. Round to the nearest tenth.

13.

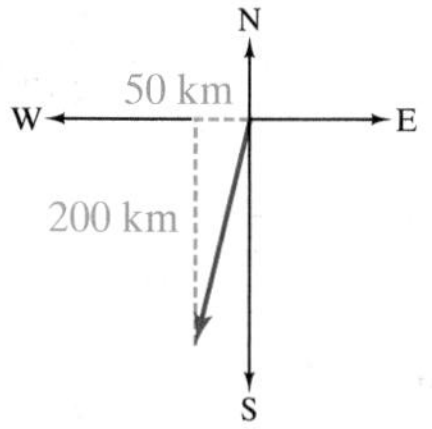

14.

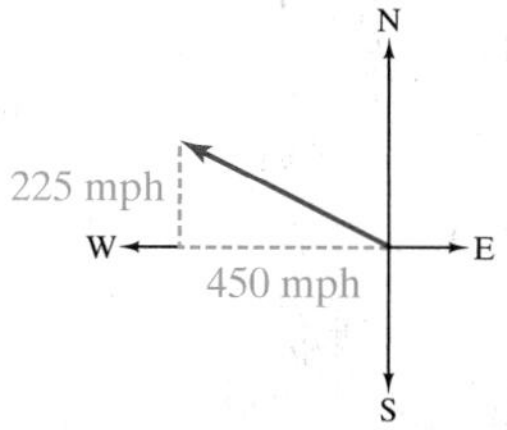

15. Find the sum of the vectors at the right. Express your answer as an ordered pair.

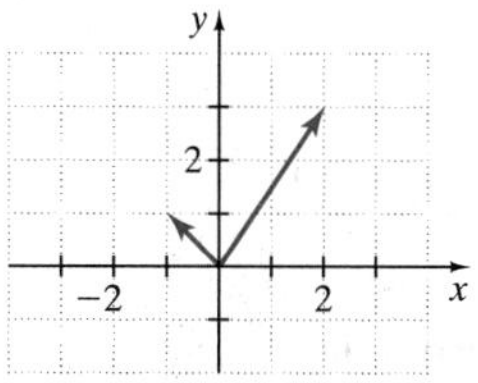

16. A whale-watching boat leaves port and travels 12 mi due north. Then the boat travels 5 mi due east. In what direction should the boat head to return directly to port?

MIXED REVIEW

17. What is the value of x?

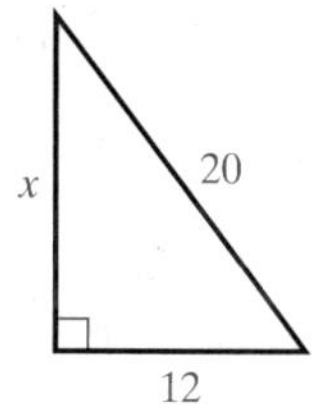

18. What is the value of x?

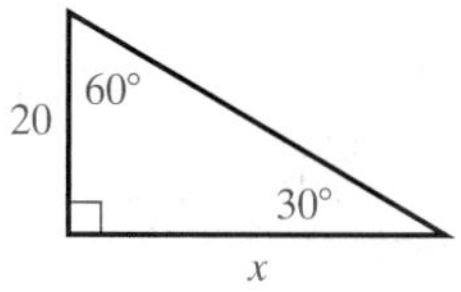

19. A square garden has sides 50 ft long. You stretch a hose from one corner of the garden to another corner along the garden's diagonal. To the nearest tenth, how long is the hose?

20. What is FE to the nearest tenth?

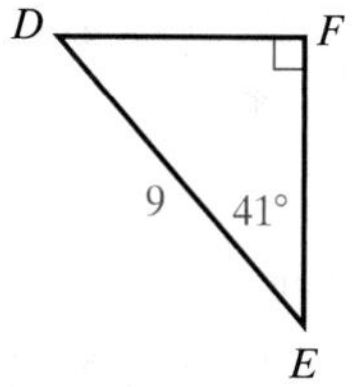

21. While flying a kite, Linda lets out 45 ft of string and anchors it to the ground. She determines that the angle of elevation of the kite is 58°. What is the height of the kite from the ground? Round to the nearest tenth.

22. What is the magnitude of the vector?

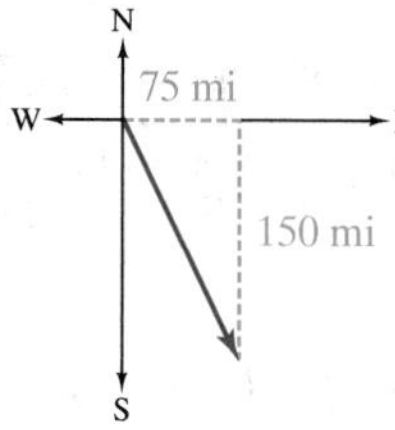

CHAPTER 9 TEST

TEST PREP VIDEO The fully worked-out solutions to any exercises you want to review are available in MyMathLab.

Find the value of each variable. Express your answer in simplest radical form.

1.

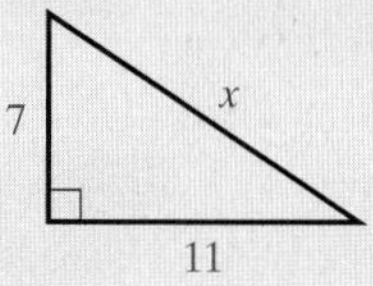

2.

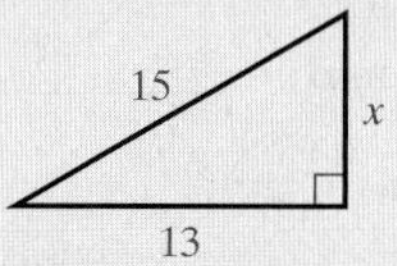

3.

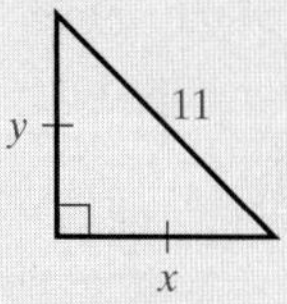

4.

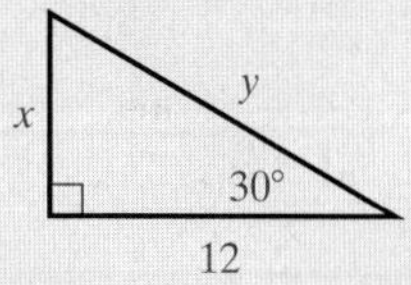

Given the following triangle side lengths, identify the triangle as "acute," "right," or "obtuse."

5. 9 cm, 10 cm, 12 cm

6. 8 m, 15 m, 17 m

7. 5 in., 6 in., 10 in.

Does any set of numbers form a Pythagorean triple? Explain.

8. 32, 60, 68

9. 1, 2, 3

10. 2.5, 6, 6.5

Express sin B, cos B, and tan B as ratios.

11.

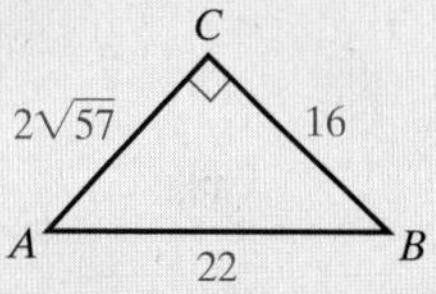

Find each missing value. Round angles to the nearest tenth. Round other answers to four decimal places.

12. tan ___° = 1.11

13. sin 34° = ___

14. cos ___° $= \frac{12}{15}$

15. A woman stands 15 ft from a statue. She looks up at an angle of 55° to see the top of the statue. Her eye level is 5 ft above the ground. How tall is the statue to the nearest foot?

16. **Landscaping** A landscaper uses a 13-ft wire to brace a tree. The wire is attached to a protective collar around the trunk of the tree. If the wire makes a 60° angle with the ground, how far up the tree is the protective collar located? Round to the nearest tenth of a foot.

Find the value of x. Round lengths to the nearest tenth and angle measures to the nearest degree.

17.

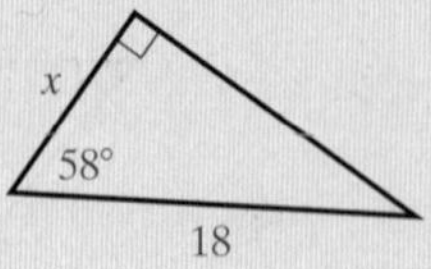

18.

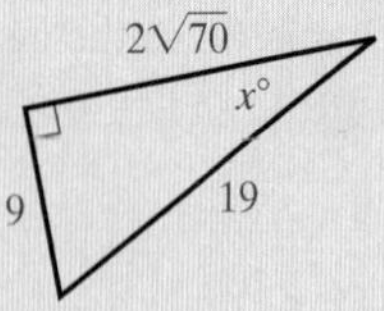

Solve the right triangles. If needed, round any answers to the nearest tenth.

19.

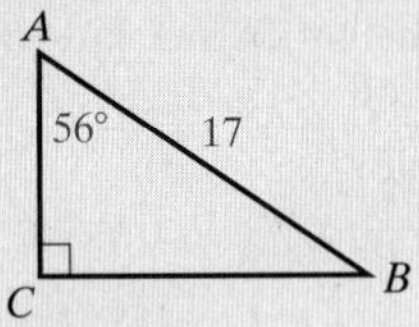

20.

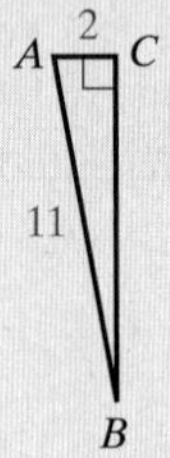

Find the magnitude and direction of each vector. Round to the nearest tenth.

21.

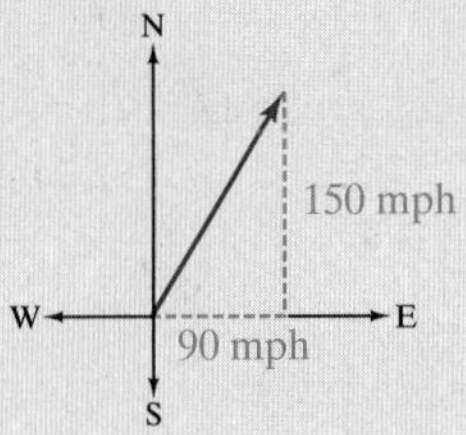

22.

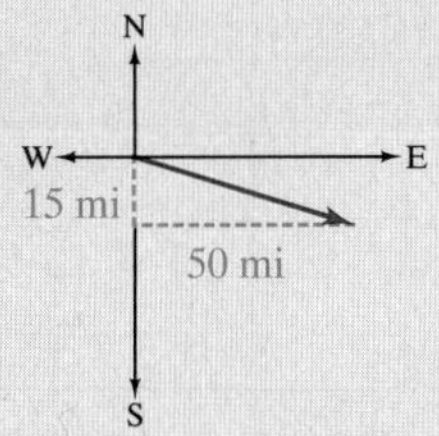

23. Describe the resultant of the vectors as an ordered pair.

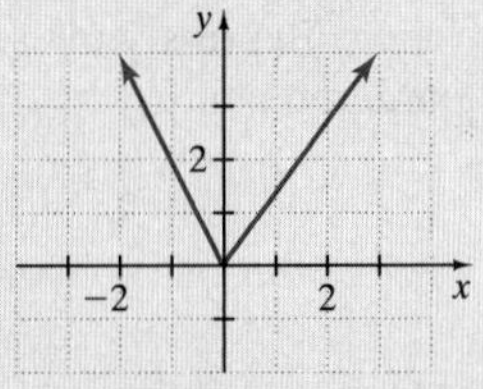

24. A canoe heading west is paddled at a rate of 7 mph. The current pushes the canoe south at a rate of 3 mph. Find the approximate resultant speed and direction of the canoe. Round the speed to the nearest tenth of a mph and the direction to the nearest whole degree.

CHAPTER 9 STANDARDIZED TEST

Use the right triangle to answer Exercises 1–4.

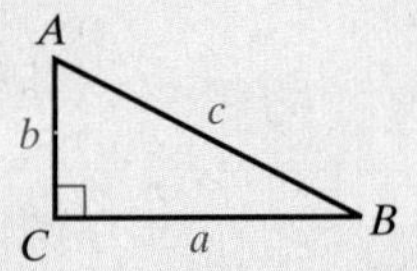

1. If $a = 8$ and $c = 14$, find the length b.

a. 260 **b.** $2\sqrt{65}$ **c.** 132 **d.** $2\sqrt{33}$

2. If $m\angle B = 30°$ and $c = 22$, find the lengths a and b.

a. $a = 11\sqrt{3}, b = 11\sqrt{2}$

b. $a = 11, b = 11\sqrt{3}$

c. $a = 11\sqrt{2}, b = 11$

d. $a = 11\sqrt{3}, b = 11$

3. If $m\angle A = 45°$ and $c = 14$, find the length b.

a. $14\sqrt{2}$ **b.** 28 **c.** $7\sqrt{2}$ **d.** 7

4. If $a = 9$ and $b = 24$, find the length c.

a. $\sqrt{33}$ **b.** $3\sqrt{73}$ **c.** 657 **d.** $3\sqrt{55}$

Given the triangle side lengths, identify the triangle as

a. acute

b. right

c. obtuse

5. 3.5 in., 12 in., 12.5 in.

6. 14 cm, 18 cm, 22 cm

7. 15 m, 32 m, 37 m

8. Which set of numbers is a Pythagorean triple?

a. 4.5, 6, 7 **b.** 3, 4, 6

c. 16, 30, 34 **d.** none of the above

Use the triangle for Exercises 9–11.

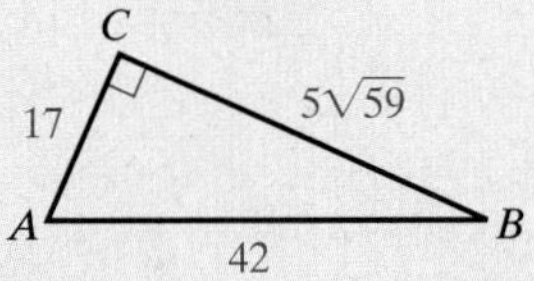

9. $\sin A =$ __________

a. $\frac{5\sqrt{59}}{42}$ **b.** $\frac{5\sqrt{59}}{17}$

c. $\frac{17}{42}$ **d.** none of the above

10. $\tan B =$ __________

a. $\frac{5\sqrt{59}}{42}$ **b.** $\frac{5\sqrt{59}}{17}$

c. $\frac{17}{42}$ **d.** none of the above

11. $\cos B =$ __________

a. $\frac{5\sqrt{59}}{42}$ **b.** $\frac{5\sqrt{59}}{17}$

c. $\frac{17}{42}$ **d.** none of the above

12. If $\cos D \approx 0.6624$, then

a. $D \approx 48.5°$ **b.** $D \approx 0.85°$

c. $D \approx 0.79$ **d.** none of the above

13. A woman on top of a zip-line tower notices that the angle of depression to the base of the next tower, 200 ft away, is 18°. How tall, to the nearest foot, is the tower the woman is on?

a. 62 ft **b.** 616 ft

c. 190 ft **d.** 65 ft

14. A facade in a stage set is supported by a 16-ft-long brace that makes an angle of 38° with the floor. How high up on the facade is the brace attached.

a. 9.9 ft **b.** 12.6 ft

c. 12.5 ft **d.** none of the above

15. Solve the right triangle.

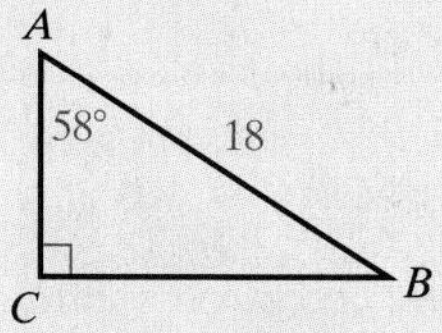

a. $B = 32°, a \approx 11.2, b \approx 14.1$

b. $B = 32°, a \approx 14.1, b \approx 11.2$

c. $B = 32°, a \approx 15.3, b \approx 9.5$

d. none of the above

16. Solve the right triangle.

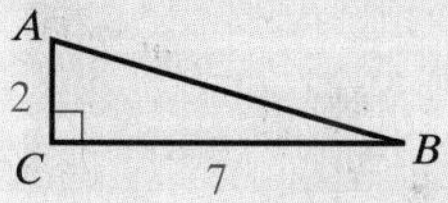

a. $c \approx 7.3, B \approx 16.6°, A \approx 73.4°$

b. $c \approx 7.3, B \approx 15.9°, A \approx 74.1°$

c. $c \approx 6.7, A \approx 73.2°, B \approx 16.8°$

d. none of the above

17. Find the magnitude and direction of the vector.

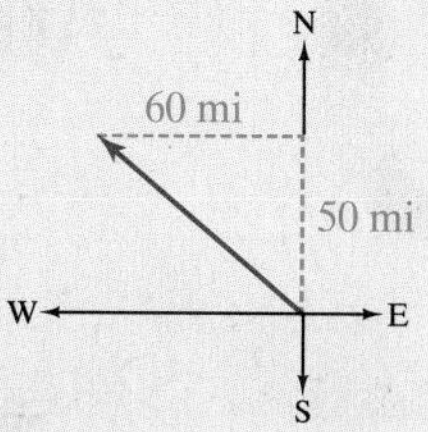

a. 6100 mi at 39.8° north of west

b. 78.1 mi at 50.2° west of north

c. 78.1 mi at 39.8° west of north

d. none of the above

18. What is the resultant of the vectors $\vec{\mathbf{u}} = \langle 4, -8 \rangle$ and $\vec{\mathbf{v}} = \langle -2, -3 \rangle$?

a. $\langle 6, -5 \rangle$ **b.** $\langle 6, -11 \rangle$

c. $\langle 2, -11 \rangle$ **d.** $\langle 2, -5 \rangle$

19. A boat heads due east across a river at 30 ft/min. The river flows south at 20 ft/min. To the nearest tenth what is the resultant speed and direction of the boat?

a. 36.1 ft/min, 33.7° south of east

b. 36.1 ft/min, 41.8° south of east

c. 22.4 ft/min, 41.8° north of east

d. none of the above

CHAPTER

10 Area

10.1 Angle Measures of Polygons and Regular Polygon Tessellations

10.2 Areas of Triangles and Quadrilaterals with a Review of Perimeter

10.3 Areas of Regular Polygons

10.4 Perimeters and Areas of Similar Figures

10.5 Arc Measures, Circumferences, and Arc Lengths of Circles

10.6 Areas of Circles and Sectors

10.7 Geometric Probability

You may have seen snowboarding half-pipe competitions during the Winter Olympics, but what is a half-pipe?

A half-pipe is a semicircular ditch with a flat bottom that is usually constructed on a downward slope. The depth of the ditch can be between 8 and 22 feet. As you see below, there are many measurements taken to ensure a safe and approved half-pipe. In addition, the snow must be carefully prepared and packed.

Throughout this chapter, we work with angles, perimeter, circumference, area, and other geometric concepts that give us an appreciation of the work that goes into constructing a half-pipe.

FIS Snowboard World Cup Half-Pipe Dimensions for an 18-Foot (≈ 5.4-m) Pipe

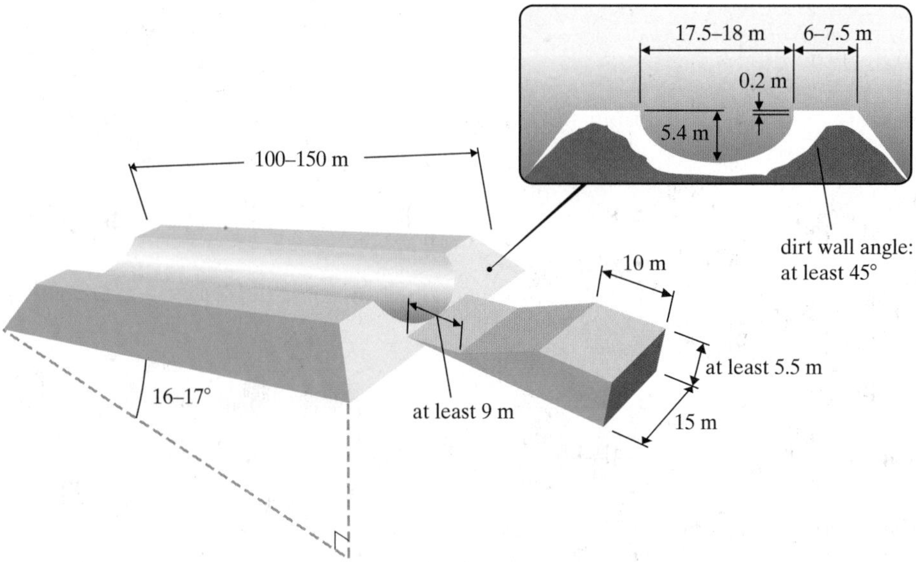

10.1 ANGLE MEASURES OF POLYGONS AND REGULAR POLYGON TESSELLATIONS

OBJECTIVES

1 Find the Measures of Interior Angles of Polygons.

2 Find the Measures of Exterior Angles of Polygons.

3 Determine Whether a Tessellation of Regular Polygons Is Formed.

VOCABULARY

- exterior angles of the polygon
- tessellation

OBJECTIVE 1 ▶ Finding the Measures of Interior Angles of Polygons. In Section 6.1, we learned that polygons are named according to the number of sides they have.

Number of Sides	*Name of Polygon*	*Number of Sides*	*Name of Polygon*
3	triangle	8	octagon
4	quadrilateral	9	nonagon
5	pentagon	10	decagon
6	hexagon	12	dodecagon
7	heptagon	n	n-gon

Let's see if we find a relationship in a polygon between

- number of sides
- number of diagonals
- number of triangles formed by the diagonals from one vertex
- sum of measures of interior angles

Quadrilateral (4 sides)

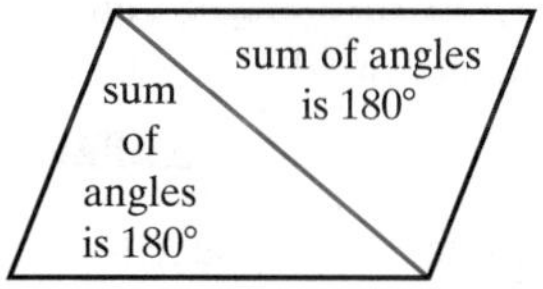

diagonals: 1
triangles formed: 2
sum of interior ∠s: 2(180°)

Pentagon (5 sides)

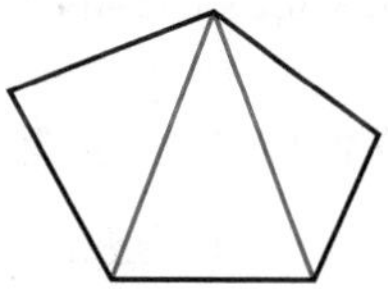

diagonals: 2
triangles formed: 3
sum of interior ∠s: 3(180°)

Hexagon (6 sides)

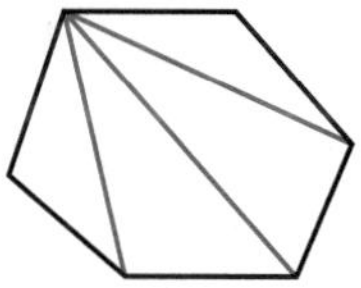

diagonals: 3
triangles formed: 4
sum of interior ∠s: 4(180°)

Let's continue this in a table.

Polygon	*Number of Sides*	*Number of Diagonals*	*Number of Triangles Formed*	*Sum of Measures of Interior Angles*
triangle	3	–	1	$1 \cdot 180° = 180°$
quadrilateral	4	1	2	$2 \cdot 180° = 360°$
pentagon	5	2	3	$3 \cdot 180° = 540°$
hexagon	6	3	4	$4 \cdot 180° = 720°$
⋮	↑ How are these numbers related?		↑	↑
n-gon	n ⟶		$(n - 2)$ ⟶	$(n - 2)180°$

Using the last row above, we now know how to find the sum of the measures of the interior angles of any n-gon.

Theorem 10.1-1 Polygon Interior Angle-Sum Theorem

The sum of the measures of the interior angles of a convex n-gon is

$$(n - 2) \cdot 180°$$

Corollary 10.1-2 Regular Polygon Interior Angle Corollary (Theorem to 10.1-1)

The measure of each interior angle of a regular n-gon is

$$\frac{1}{n}\cdot(n-2)\cdot 180°, \quad \text{or} \quad \frac{(n-2)\cdot 180°}{n}$$

We prove these theorems in Exercise Set 10.1, Exercises 61 and 62.

EXAMPLE 1 Finding the Sum of the Measures of the Angles of a Polygon

Find the sum of the measures of the interior angles of a convex octagon.

Solution An octagon has 8 sides. Thus

$$\begin{aligned}\text{sum of angles} &= (n-2)\cdot 180° \quad \text{Sum of angles formula}\\ &= (8-2)\cdot 180°\\ &= 6\cdot 180°\\ &= 1080°\end{aligned}$$

The interior angle sum of a convex octagon is 1080°. □

PRACTICE 1 Find the sum of the measures of the interior angles of a convex nonagon. ■

Now let's use algebra to help us find angle measures.

EXAMPLE 2 Finding the Measure of an Interior Angle

Find the value of x in the figure. Then use x to find $m\angle A$ and $m\angle B$.

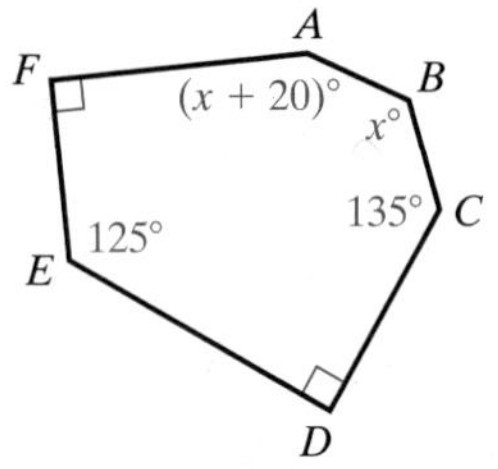

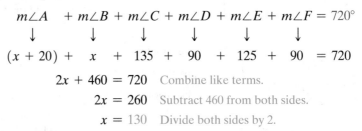

Solution This is a hexagon, which has 6 sides. The sum of the interior angles of any convex hexagon is:

$$\text{sum of angles} = (6-2)\cdot 180° = 4\cdot 180° = 720°$$

To find x, let's add the interior angle measures of the polygon and set the sum equal to 720°.

$$m\angle A + m\angle B + m\angle C + m\angle D + m\angle E + m\angle F = 720°$$

$$\downarrow \quad \downarrow \quad \downarrow \quad \downarrow \quad \downarrow \quad \downarrow$$

$$(x+20) + x + 135 + 90 + 125 + 90 = 720$$

$$\begin{aligned}2x + 460 &= 720 \quad \text{Combine like terms.}\\ 2x &= 260 \quad \text{Subtract 460 from both sides.}\\ x &= 130 \quad \text{Divide both sides by 2.}\end{aligned}$$

Since $x = 130$, then $m\angle B = x° = 130°$, and $m\angle A = (x+20)° = (130+20)° = 150°$. Thus, $x = 130$, $m\angle B = 130°$, and $m\angle A = 150°$.

Check to see that the angle sum is 720°. It will be. □

PRACTICE 2 Find the value of x in the figure. Then use x to find $m\angle M$ and $m\angle N$.

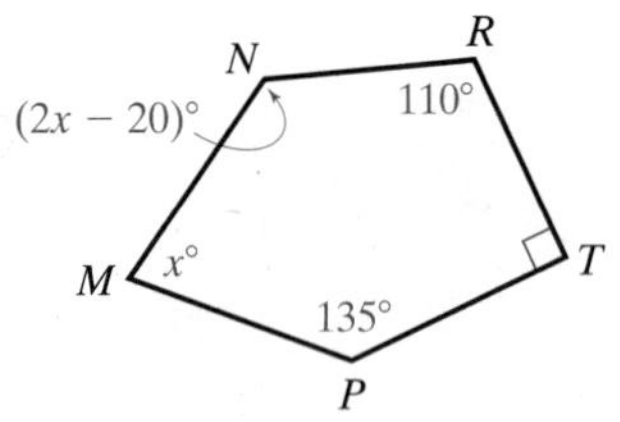

■

Recall that each interior angle of a regular polygon has the same measure.

EXAMPLE 3 **Using the Regular Polygon Interior-Angle Corollary**

The Sino-Steel Tower is a hexagonal, honey comb-looking "green" building, in Tianjin, China, designed by MAD Studios architects. Find the measure of each interior angle of one regular hexagon.

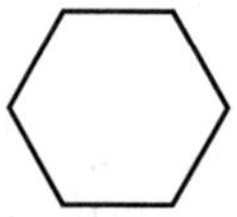

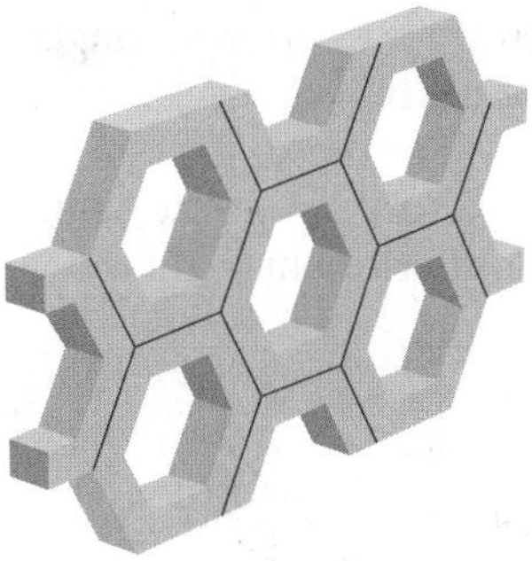

Solution Let's use Corollary 10.1-2, in which we basically find the angle sum of a hexagon, and then divide by the total number of angles, in this case 6. (Remember for a hexagon, $n = 6$.)

$$\text{measure of each angle} = \frac{(n-2)\cdot 180^\circ}{n} = \frac{(6-2)\cdot 180^\circ}{6} = \frac{4\cdot 180^\circ}{6} = 120^\circ$$

The measure of each interior angle of a regular hexagon is 120°. □

PRACTICE 3 Find the measure of each interior angle of a regular pentagon.

We can also use Corollary 10.1-2 to find the number of sides of a regular polygon.

EXAMPLE 4 **Finding the Number of Sides of a Regular Polygon**

The measure of an interior angle of a regular polygon is 144°. Find the number of sides of this polygon.

Solution Since this is a regular polygon, all angles have the same measure and we can use the formula in Corollary 10.1-2 and set it equal to 144 (for 144°).

$$\frac{(n-2)\cdot 180}{n} = 144 \quad \text{Set the formula equal to 144.}$$

$$n\cdot\frac{(n-2)\cdot 180}{n} = n\cdot 144 \quad \text{Multiply both sides by } n.$$

$$(n-2)\cdot 180 = 144n \quad \text{Simplify.}$$

$$180n - 360 = 144n \quad \text{Use the Distributive Property.}$$

$$36n = 360 \quad \text{Subtract } 144n\text{, and then add 360 to both sides.}$$

$$n = 10 \quad \text{Divide both sides by 36.}$$

The regular polygon has 10 sides. To **check**, let $n = 10$ in the Corollary 10.1-2 formula. The result will be 144 for 144°. □

PRACTICE 4 The measure of an interior angle of a regular polygon is 140°. Find the number of sides of this polygon.

OBJECTIVE 2 ▶ Finding the Measures of Exterior Angles of Polygons. We defined and studied exterior angles of a triangle in Sections 4.1 and 5.5. Let's extend our definition of an exterior angle to include any convex polygon. The angles that are adjacent to the interior angles of a convex polygon are the **exterior angles of the polygon**.

> **Helpful Hint**
> To determine whether an angle is an exterior angle of a triangle, see whether the two sides of the angle are formed by
> - an extension of a side of the triangle and
> - a side of the triangle.

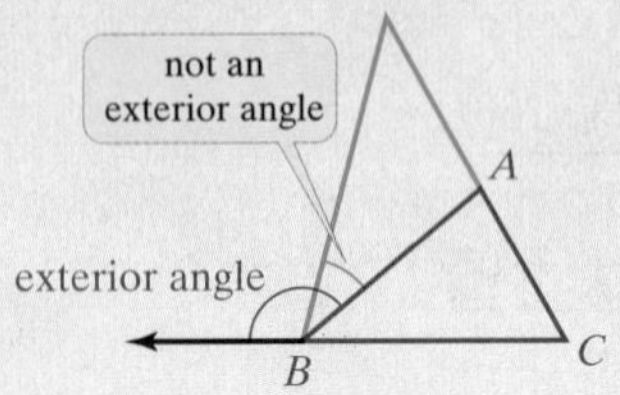

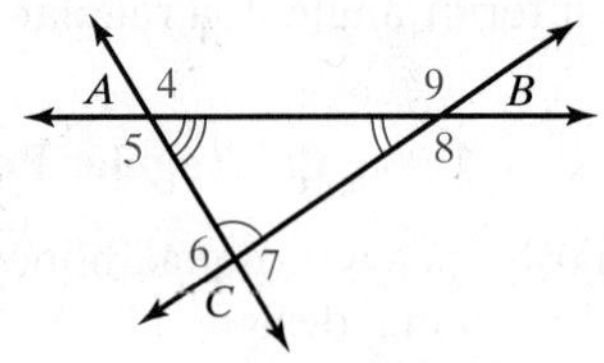

Exterior Angles (∠s 4, 5, 6, 7, 8, 9)

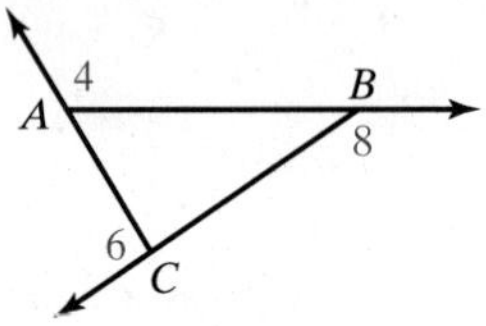

Exterior Angles Usually Shown

Each interior angle has two congruent exterior angles (for example, $\angle 6 \cong \angle 7$ since they are vertical angles).

We will study one exterior angle with each interior angle.

Let's use our reasoning skills and try to make a conjecture about the sum of the measures of the exterior angles of a convex polygon, one exterior angle at each vertex.

Triangle (3 sides)

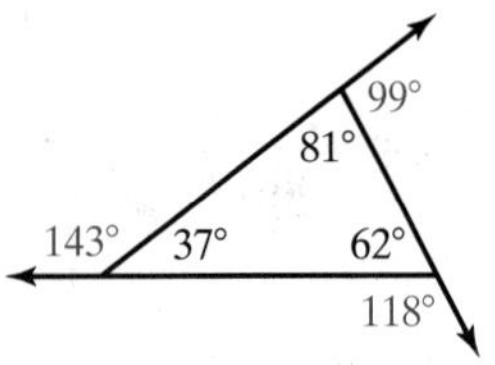

$143° + 99° + 118° = 360°$

Quadrilateral (4 sides)

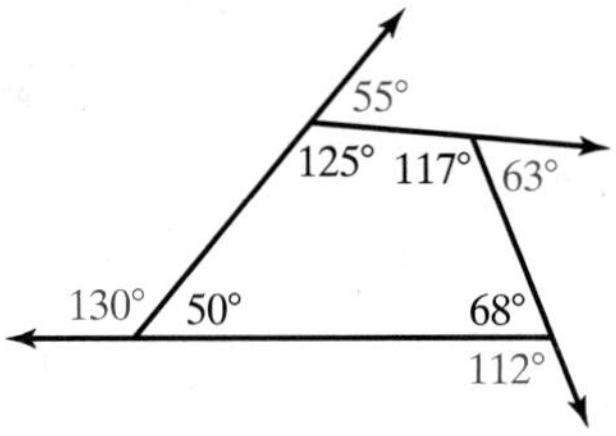

$130° + 55° + 63° + 112° = 360°$

Pentagon (5 sides)

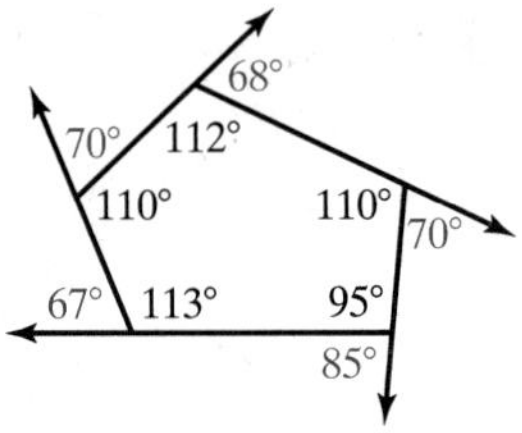

$67° + 70° + 68° + 70° + 85° = 360°$

Amazingly, the sum is 360° each time.

> **Helpful Hint**
> Notice that
> - The *sum of the interior ∠s* of a convex polygon depends on the number of sides of the polygon.
> - The *sum of the exterior ∠s* of a convex polygon, one at each vertex, is always 360°.

Theorem 10.1-3 Polygon Exterior Angle-Sum Theorem

The sum of the measures of the exterior angles of a convex polygon, one exterior angle at each vertex, is 360°.

For the hexagon (6 sides),

$$m\angle 1 + m\angle 2 + m\angle 3 + m\angle 4 + m\angle 5 + m\angle 6 = 360°$$

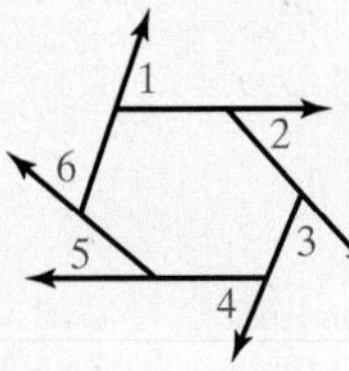

Corollary 10.1-4 Regular Polygon Exterior Angle Corollary (to Theorem 10.1-3)

The measure of each exterior angle of a regular n-gon is

$$\frac{1}{n} \cdot 360° \quad \text{or} \quad \frac{360°}{n}$$

For a regular hexagon (6 sides)

$$m\angle 1 = m\angle 2 = m\angle 3 = m\angle 4 = m\angle 5 = m\angle 6 = \frac{360°}{6} = 60°$$

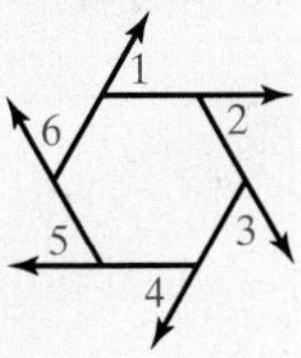

We prove these theorems in Exercise Set 10.1, Exercises 63 and 64.

> **Helpful Hint**
> For regular polygons,
> - Each interior ∠ has the same measure.
> - Each exterior ∠ has the same measure.

EXAMPLE 5 Finding the Exterior Angle Measure

Find the measure of each exterior angle of a regular pentagon.

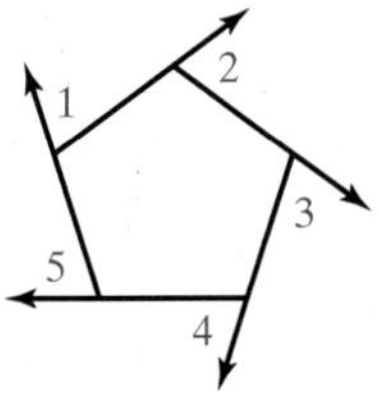

Solution Since this is a regular polygon, each exterior angle has the same measure. By Corollary 10.1-4,

$$m\angle 1 = m\angle 2 = m\angle 3 = m\angle 4 = m\angle 5 = \frac{360°}{5} = 72°$$ Since a pentagon has 5 sides

Each exterior angle measures 72°. □

PRACTICE 5 Find the measure of each exterior angle of a 15-gon.

EXAMPLE 6 Finding Exterior Angle Measures

Find the value of x. Then find each exterior angle measure.

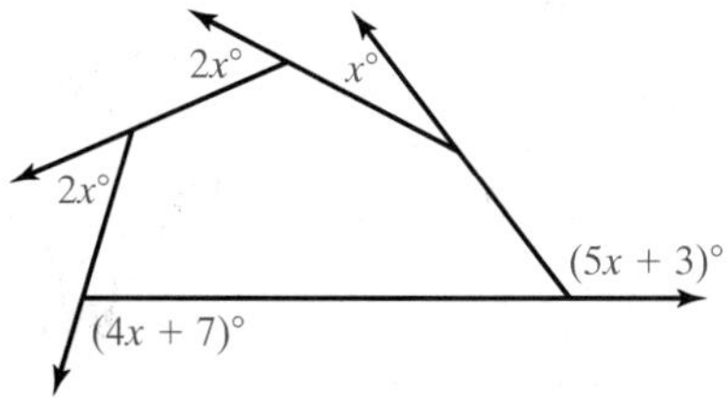

Solution The sum of the exterior angles is 360°; thus:

$$x + 2x + 2x + (4x + 7) + (5x + 3) = 360$$
$$14x + 10 = 360 \quad \text{Combine like terms.}$$
$$14x = 350 \quad \text{Subtract 10 from both sides.}$$
$$x = 25 \quad \text{Divide both sides by 14.}$$

The angle measures are:

$$x° = 25°;\ 2x° = 2 \cdot 25° = 50°;\ (4x + 7)° = (4 \cdot 25 + 7)° = 107°;$$
$$(5x + 3)° = (5 \cdot 25 + 3)° = 128°$$

To **check**, see that the exterior angle sum is 360°: 25° + 50° + 50° + 107° + 128° = 360°.

□

PRACTICE 6 Find the value of x. Then find each exterior angle measure.

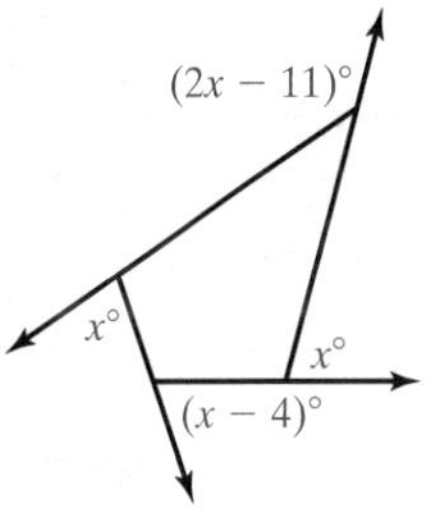

OBJECTIVE 3 ▶ Determining Whether a Tessellation of Regular Polygons Is Formed. To form a tiling pattern with no gaps or overlaps, we make sure that the sum of the angles where the polygons meet is 360°. Below are a few familiar patterns.

Square Pattern

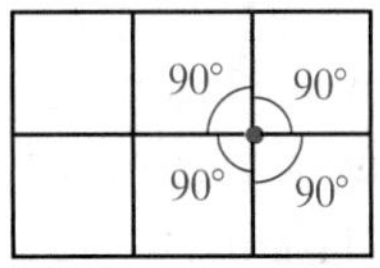

90° + 90° + 90° + 90° = 360°

Offset Rectangle Pattern

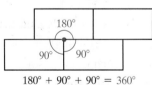

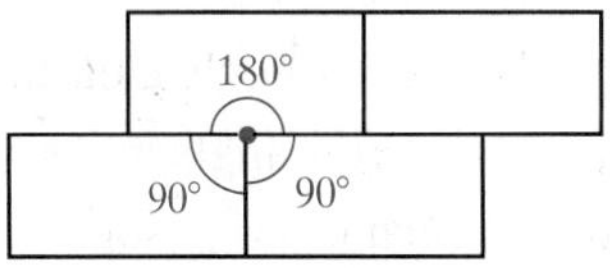

180° + 90° + 90° = 360°

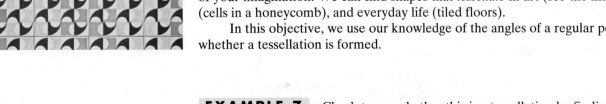

We call such tiling patterns tessellations.

A **tessellation**, or *tiling*, is a repeating pattern of figures that completely covers a plane, without gaps or overlaps. Tessellations can be made with polygons or figures of your imagination. We can find shapes that *tessellate* in art (see the margin), nature (cells in a honeycomb), and everyday life (tiled floors).

In this objective, we use our knowledge of the angles of a regular polygon to see whether a tessellation is formed.

EXAMPLE 7

Check to see whether this is a tessellation by finding the sum of the numbered angles shown in the equilateral triangles.

Solution Each angle of an equilateral triangle measures 60°, so we have

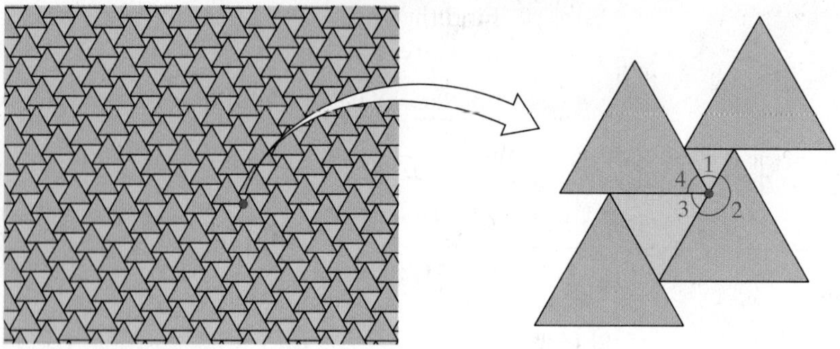

$$m\angle 1 + m\angle 2 + m\angle 3 + m\angle 4 = 60° + 180° + 60° + 60°$$
$$= 360°$$

Yes, this forms a tessellation. □

PRACTICE 7 Check to see whether this is a tessellation by finding the sum of the numbered angles of the regular polygons.

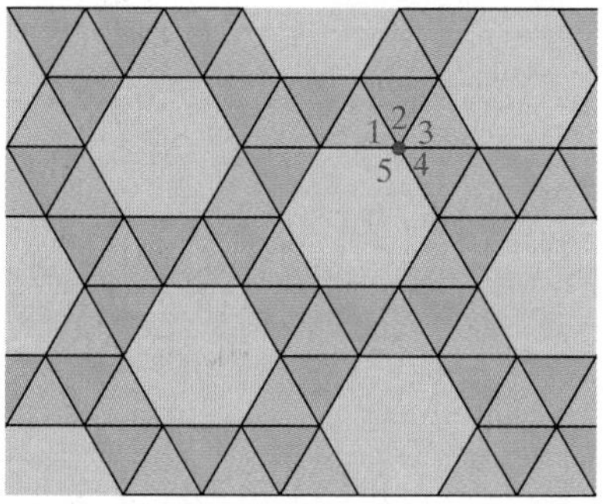

You may have noticed that a tesselation is formed by transformations (see Chapter 8) of a figure(s) so that the resulting repeating pattern fills a plane.

VOCABULARY & READINESS CHECK

Fill in the Blank *Use the choices below to fill in each blank. Some choices may not be used.*

$\frac{(11-2)\cdot 180°}{11}$ $\quad \frac{360°}{11}$ $\quad 360°$ $\quad (11-2)\cdot 180°$ $\quad 11\cdot 180°$ $\quad 180°$ $\quad n$

1. The sum of the interior angle measures of a convex 11-gon is ______________.
2. The sum of the exterior angle measures of a convex 11-gon is ______________.
3. The measure of each interior angle of a regular 11-gon is ______________.
4. The measure of each exterior angle of a regular 11-gon is ______________.
5. The measure of an interior angle and an exterior angle at the same vertex of a convex polygon is always ______________.
6. A convex n-gon has how many exterior angles (counting one at each vertex)? ______________

10.1 EXERCISE SET MyMathLab®

Find the sum of the interior angle measures of each convex polygon. See Example 1.

1.

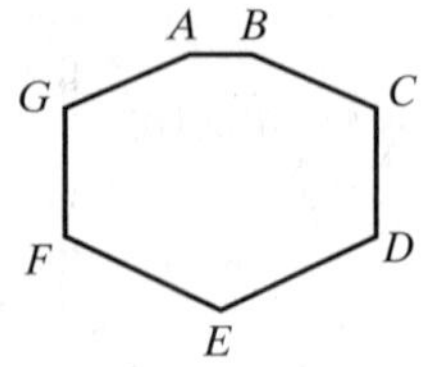

2.

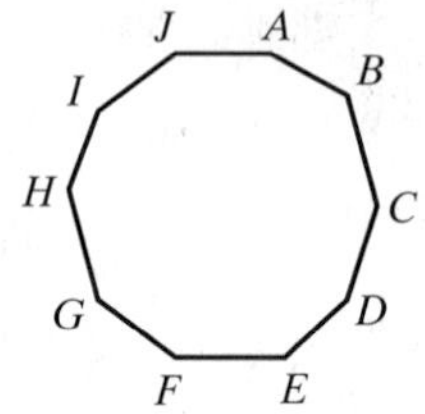

3. 13-gon

4. 15-gon

5. 20-gon

6. 30-gon

7. 202-gon

8. 1002-gon

Find the missing angle measures. See Examples 1 and 2.

9.

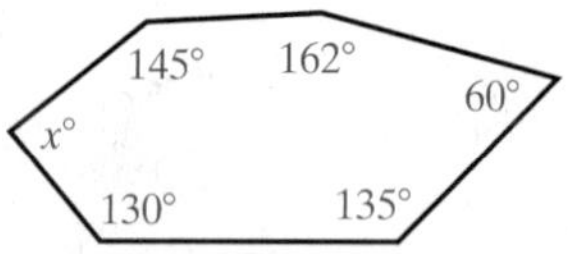

10.

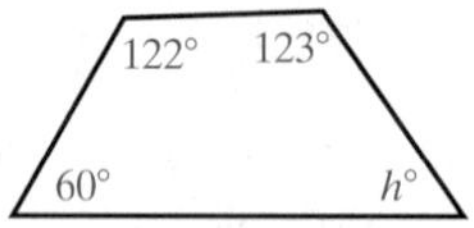

11. 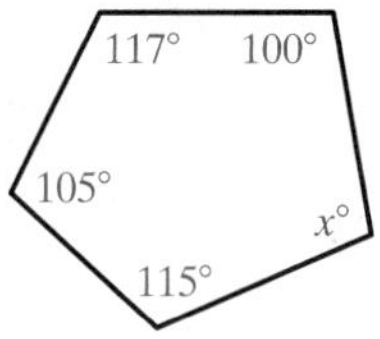

12. 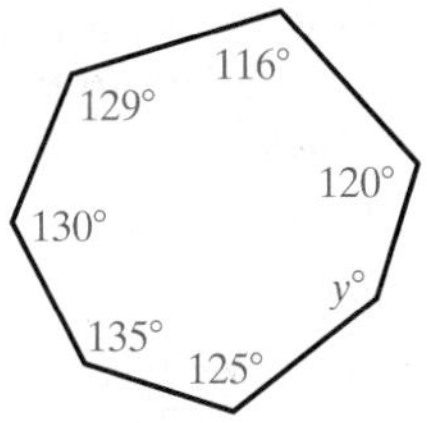

Find the measure of one interior angle in each regular polygon. See Example 3.

13. 

14. 

15. 

16. 

The sum of the interior angle measures of a polygon with n *sides is given. Find* n. *See Example 4.*

17. 180° **18.** 1080° **19.** 1980° **20.** 2880°

Find the measure of an exterior angle of each regular polygon. See Example 5.

21. nonagon **22.** 36-gon

23. 18-gon **24.** 100-gon

25. Find the measures of an interior angle and an exterior angle of a regular decagon.

26. What is the sum of the measures of an interior angle and a corresponding exterior angle?

Find the value of each variable. See Example 6.

27. 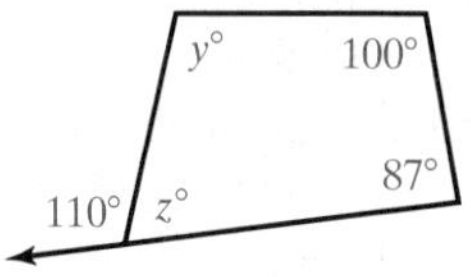

28. 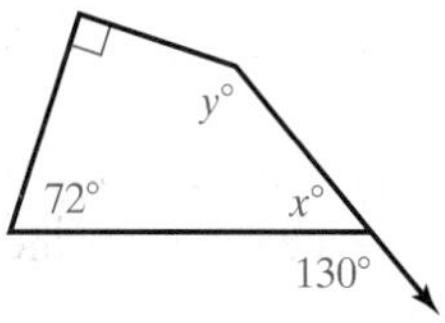

29. 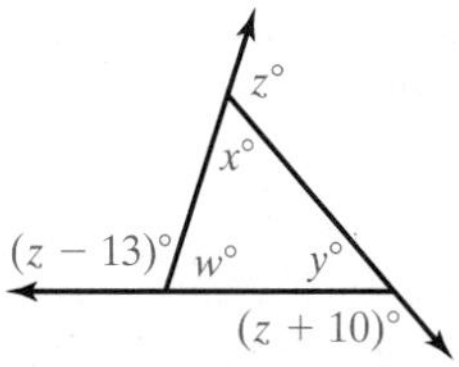

30. 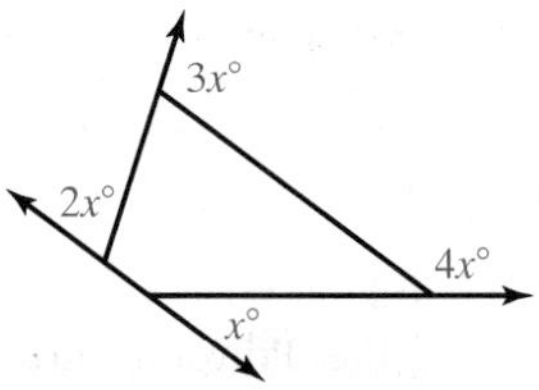

Multiple Steps *The measure of an exterior angle of a regular polygon is given. For each exercise,* **a.** *Find the measure of an interior angle.* **b.** *Then find the number of sides. See Examples 4 and 5.*

31. 72° **32.** 36° **33.** 18° **34.** 30°

Multiple Steps *Check the tilings below to see that there will be no gaps or overlaps by answering parts a–c. See Example 7.*

a. *Name and find the measure of the interior angle of each regular polygon. (Note: Exercise 44 does contain rectangles.)*

b. *Find the measures of the numbered angles.*

c. *See that the sum of the numbered angles is 360°.*

35. Hexagon Pattern

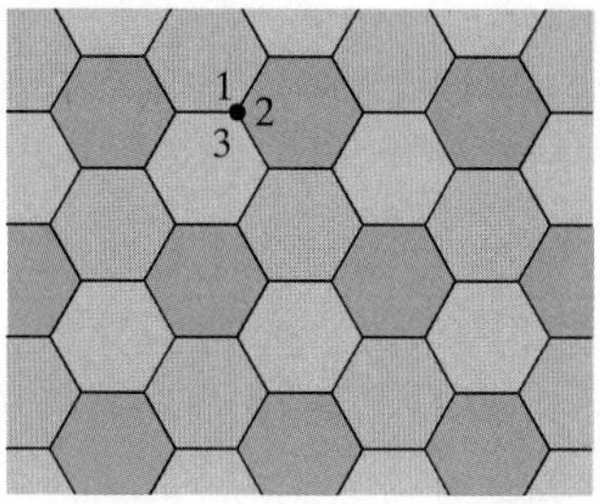

36. Equilateral Triangle Pattern

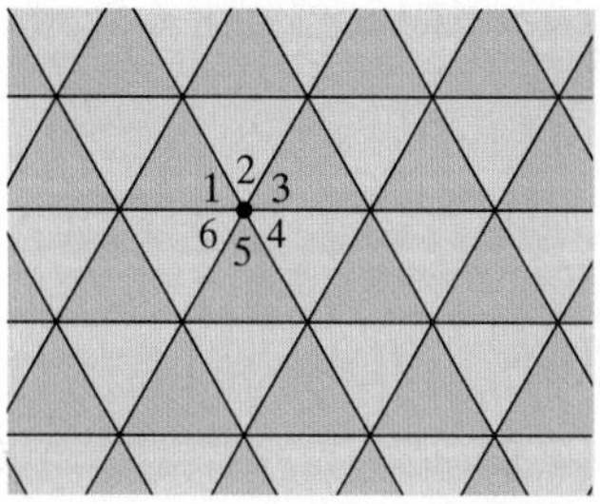

37. Square and Equilateral Triangle Pattern

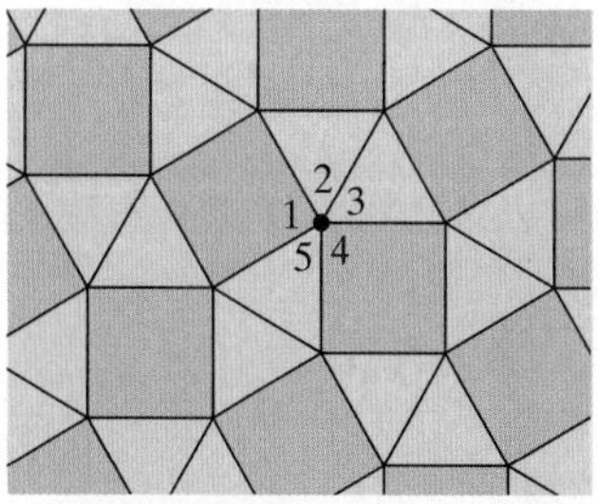

38. Square and Octagon Pattern

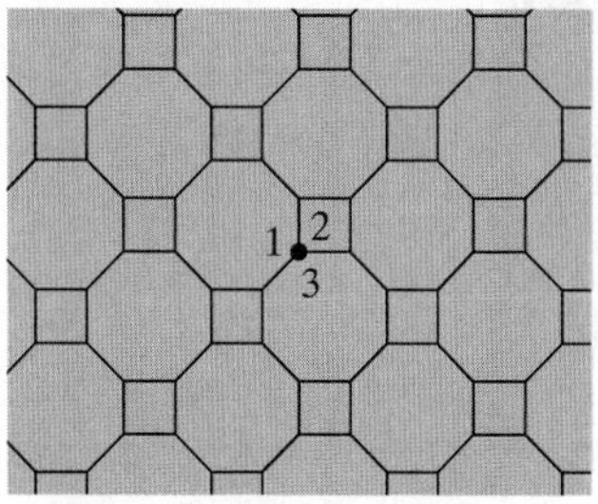

39. 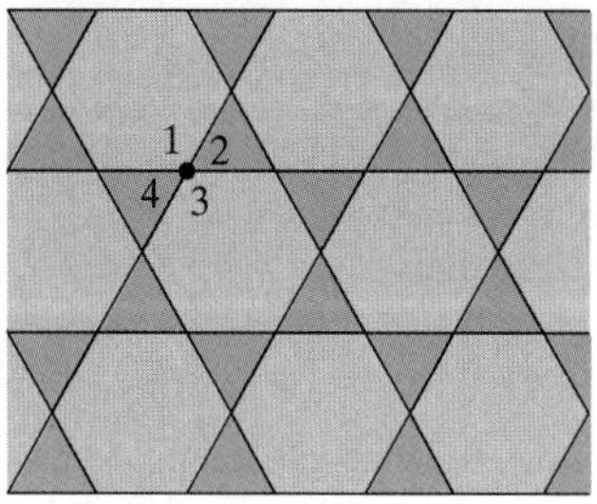

40.

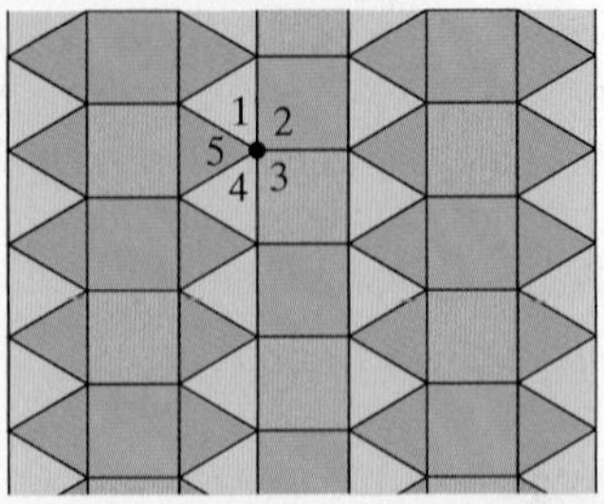

41.

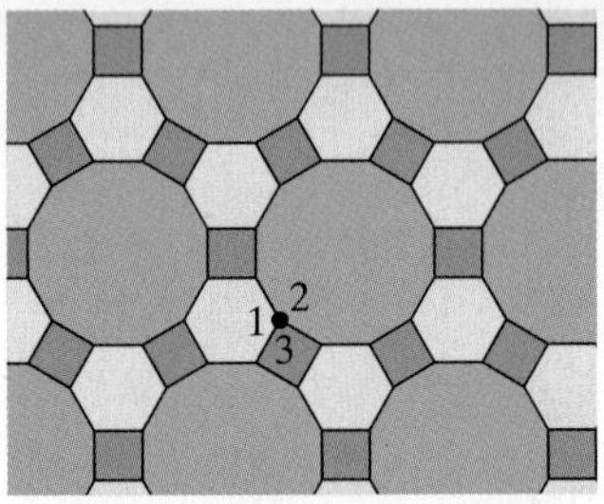

42.

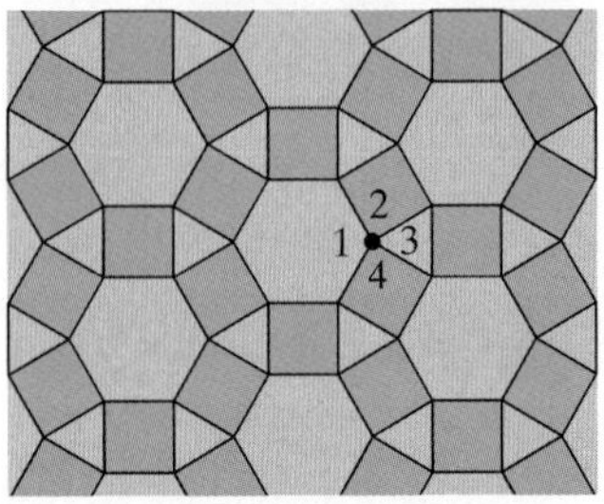

43.

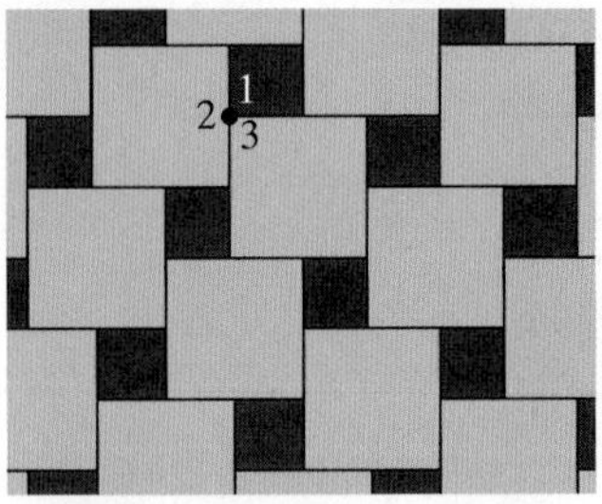

44.

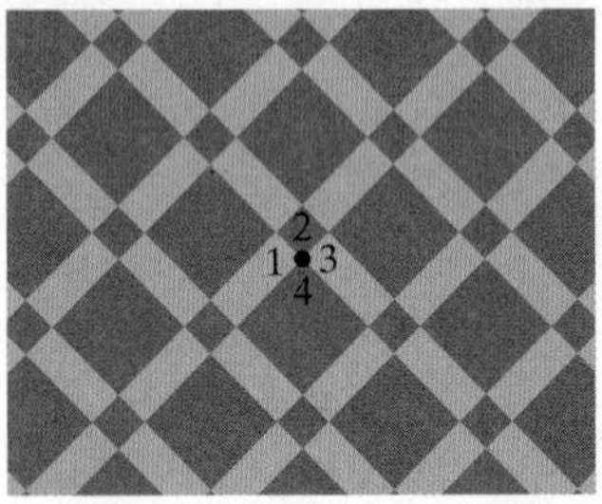

CONCEPT EXTENSIONS

45. Sketch an equilateral polygon that is not equiangular.

46. Can you draw an equiangular polygon that is not equilateral? Explain.

47. Which angles are the exterior angles for ∠1? What do you know about their measures? Explain.

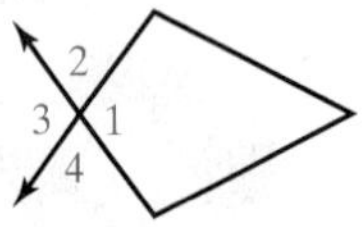

48. **Find the Error** Your friend says that she measured an interior angle of a regular polygon as 130°. Explain why this result is impossible.

Determine whether each regular polygon will tessellate a plane. Explain. (Hint: Find the measure of an interior angle of the regular polygon. Then see if the angle measure divides evenly into 360°.)

49. equilateral triangle | **50.** square

51. regular pentagon | **52.** regular heptagon

53. regular octagon | **54.** regular nonagon

55. A *regular tessellation* is a tessellation made up of congruent copies of one regular polygon. Explain why there are three, and only three, regular tessellations of the plane. (*Hint:* See Exercises **49–54**.)

56. The art of tessellation is used to form puzzle pieces. Which puzzle piece can tessellate a plane using *only* translation images of itself?

a. 

b. 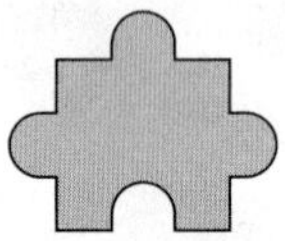

c. 

d. 

Packaging *The gift package at the below contains fruit and cheese. The fruit is in a container that has the shape of a regular octagon. The fruit container fits in a square box. A triangular cheese wedge fills each corner of the box.*

57. Find the measure of each interior angle of a cheese wedge.

58. Show how to rearrange the four pieces of cheese to make a regular polygon. Name the regular polygon and find the measure of each interior angle of the polygon.

59. A triangle has two congruent interior angles and an exterior angle that measures 100°. Find two possible sets of interior angle measures for the triangle.

60. The measure of an interior angle of a regular polygon is three times the measure of an exterior angle of the same polygon. What is the name of the polygon?

PROOF

61. **Fill in the Blank** Complete the proof of the Polygon Interior-Angle Sum Theorem (10.1-1) by filling in the blanks.

Given: A polygon with n sides.

Prove: The sum of the measures of the interior angles is $(n - 2)180°$.

Proof: Use the figure drawn, but let the number of sides be a number, n.

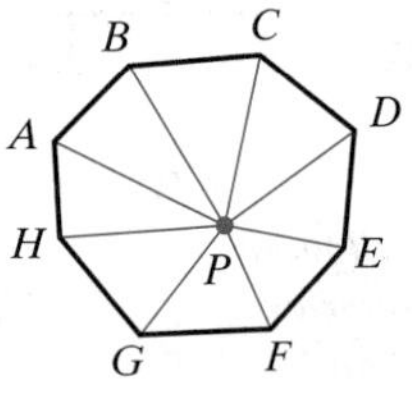

A polygon with n sides.

Choose a point P in the interior of the polygon and join P to each vertex of the polygon.

a. Since the polygon has n sides, how many triangles are formed? ____________

b. Each triangle formed has an angle sum of ____________.

c. Thus, the angle sum of all the triangles is the product ____________.

d. All the angles of the triangles formed at vertex P are not part of the polygon angle sum, so we must subtract them from our answer at part **c**.

Looking at the figure, we see that all these angles formed at vertex P have a sum of ____________.

e. From parts **c** and **d**, we have $n(180°) - 360°$. Let's rewrite this subtraction. $n(180°) - 360° = n(180°) - \_\_\_\ (180°) = (n - 2)180°$.

62. Use the Polygon Interior-Angle Sum Theorem (10.1-1) to prove Corollary 10.1-2, the Regular Interior Angle Corollary.

Given: A regular polygon with n sides

Prove: The measure of each interior angle is $\dfrac{(n-2)180°}{n}$.

PROOF

63. Fill in the Blank Complete the proof of the Polygon Exterior Angle-Sum Theorem (10.1-3).

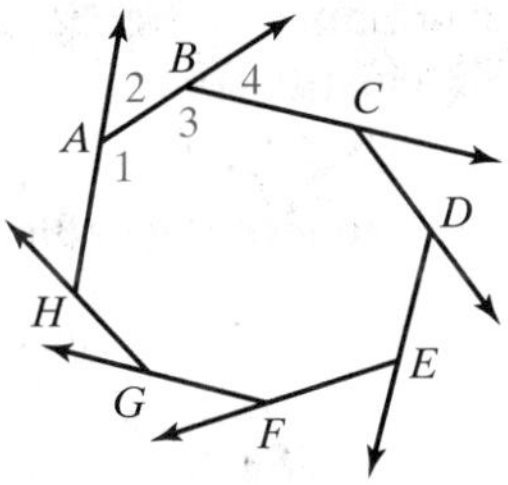

A polygon with n sides.

Given: A polygon with n sides.

Prove: The sum of the measures of the exterior angles (one at each vertex) is 360°.

Proof: Use the figure drawn, but let the number of sides be a number, n.

a. Let's study vertex A and angles 1 and 2. $\angle 1$ is an interior angle, $\angle 2$ is an exterior angle and $m\angle 1 + m\angle 2 =$ ____________.

b. The same thing happens at vertex B so that $m\angle 3 + m\angle 4 =$ ____________.

c. This means that at each of the n vertices, the sum of the interior angle of the polygon and the exterior angle is ____________.

d. The total measure of all these straight angles for the n vertices is ____________.

e. Recall from Theorem 10.1-1 that the sum of the measures of the interior angles of a polygon with n sides is ____________.

f. To find the exterior angle sum, we subtract the result of part **e** from the result of part **d**.

$$\begin{aligned}\text{exterior angle sum} &= (n)180° - (n-2)180° \\ &= (n)180° - (n)180° \underline{\qquad\quad} \\ &= 360°\end{aligned}$$

Use the distributive property.

64. Use the Exterior Angle-Sum Theorem (10.1-3) to prove Corollary 10.1-4, the Regular Polygon Exterior Angle Corollary.

Given: A regular polygon with n sides.

Prove: The measure of each exterior angle is $\dfrac{360°}{n}$.

65. Your friend says she has another way to find the sum of the interior angle measures of a polygon. She picks a point inside the polygon, draws a segment to each vertex, and counts the number of triangles. She multiplies the total by 180, and then subtracts 360 from the product. Does her method work? Explain.

66. Determine whether the statement is "always," "sometimes," or "never" true. Explain. A scalene triangle will tessellate.

A semiregular tessellation is made up of two or more different regular polygons, with the same arrangement of the polygons at each vertex.

67. Can you use regular dodecagons and equilateral triangles, all with side length 1 unit, to make a semiregular tessellation?

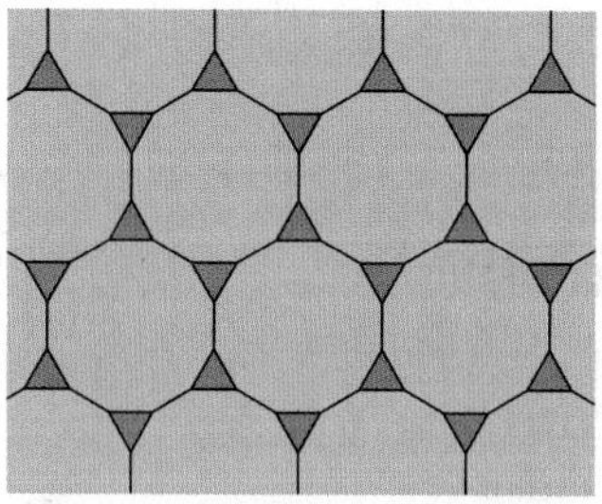

68. What is the sum of the measures of the angles at each vertex?

Can you make a semiregular tessellation using the given pair of regular polygons? If so, draw a sketch.

69.

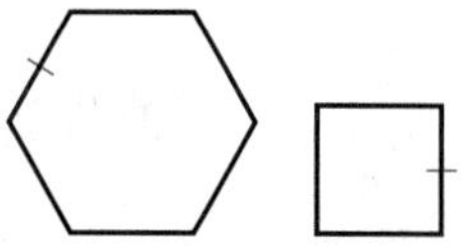

70.

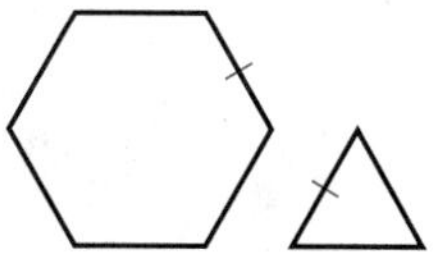

REVIEW AND PREVIEW

Construction *Use a compass and a straight edge only to construct each figure with the given side length. See Sections 1.8 and 3.5.*

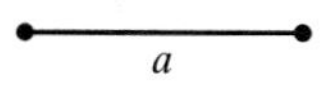

71. an equilateral triangle

72. a square

Three lengths are given. Is it possible to construct a triangle with these side lengths? See Section 5.5.

73. 16 in., 26 in., 42 in.

74. 9 m, 9 m, 18 m

75. 19.5 ft, 20.5 ft, 10 ft

76. $4\frac{1}{2}$ yd, 8 yd, 5 yd

Find the area of each figure. See Section 2.1.

77. a square with 5-cm sides

78. a square with 4-in. sides

79. a 4.6-m-by-2.5-m rectangle

80. a rectangle with length 3 ft and width $\frac{1}{2}$ ft

10.2 AREAS OF TRIANGLES AND QUADRILATERALS WITH A REVIEW OF PERIMETER

OBJECTIVES

1 Find Areas of Squares, Rectangles, Parallelograms, and Triangles.

2 Find Areas of Trapezoids, Rhombuses, and Kites.

Helpful Hint

- Perimeter is measured in units.
- Area is measured in square units.

VOCABULARY

- base of a parallelogram
- height of a parallelogram
- base of a triangle
- height of a triangle
- height of a trapezoid

In Section 2.1, we studied basic formulas for perimeter and area of squares, rectangles, and triangles. In this section, we review and justify these formulas and continue with areas of other quadrilaterals.

First, it is important to review the meaning of perimeter and area and the difference between these two concepts. Recall that perimeter is the distance around a figure. Thus, perimeter is given in units. We will not list perimeter formulas here, because if you know your polygon, you can find the distance around it. Area is the number of square units a geometric figure encloses.

Rectangular Room

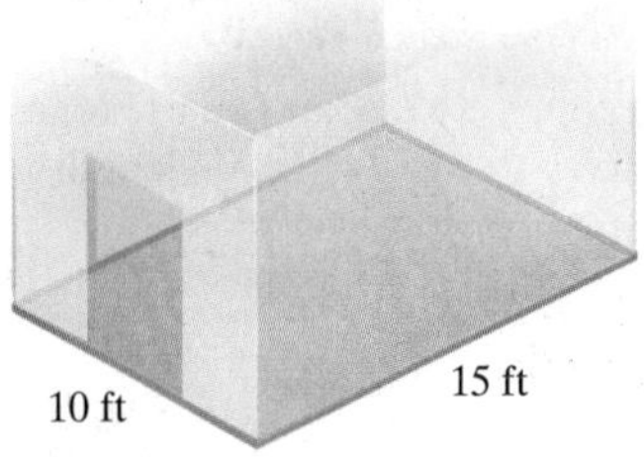

perimeter = 10 ft + 15 ft + 10 ft + 15 ft = 50 ft

area = 10 ft · 15 ft = 150 square ft

Example of perimeter: baseboard needed; 50 ft

Example of area: carpeting or tile needed; 150 sq ft

OBJECTIVE 1 ▶ Finding Areas of Squares, Rectangles, Parallelograms, and Triangles. The next postulates will help us justify some area formulas.

Postulate 10.2-1 Area Congruence Postulate

If two polygons are congruent, then their areas are the same.

Postulate 10.2-2 Area Addition Postulate

The area of a region is the sum of the areas of its nonoverlapping parts.

Now let's review some area formulas.

Postulate 10.2-3 Area of a Square

The area of a square is the square of its side length.

$$A = s^2$$

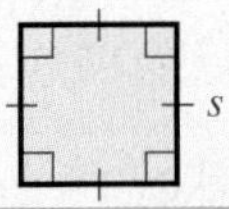

Theorem 10.2-4 Area of a Rectangle

The area of a rectangle is the product of its base and height.

$$A = bh$$

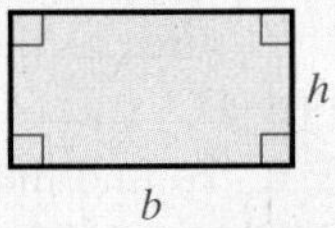

Theorem 10.2-5 Area of a Parallelogram

The area of a parallelogram is the product of a base and its corresponding height.

$$A = bh$$

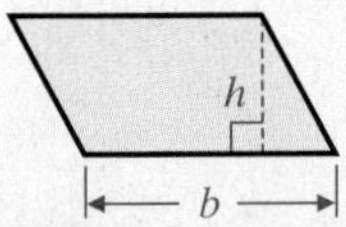

A **base of a parallelogram** can be any one of its sides. The corresponding **height** is the perpendicular length from the side opposite the chosen base to the line containing the base.

Theorem 10.2-6 Area of a Triangle

The area of a triangle is half the product of a base and its corresponding height.

$$A = \frac{1}{2}bh$$

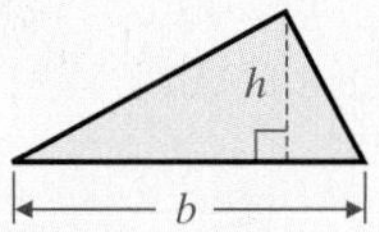

A **base of a triangle** can be any of its sides. The corresponding **height** is the length of the segment from the vertex (opposite the chosen base) perpendicular to the line containing that base.

- Notice that the area formulas for rectangle and parallelogram are the same. To see why, study the figures below.

Justify the Parallelogram Formula

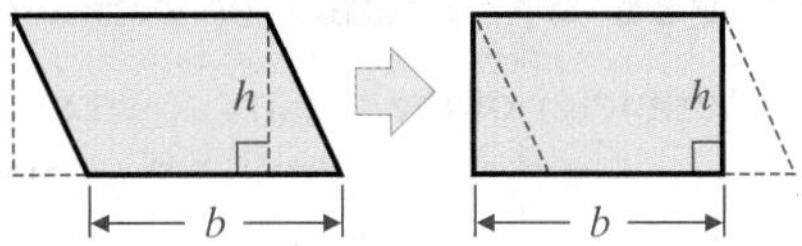

$A = bh$

- To understand the area formula for triangles, study the figure below.

Justify the Triangle Formula

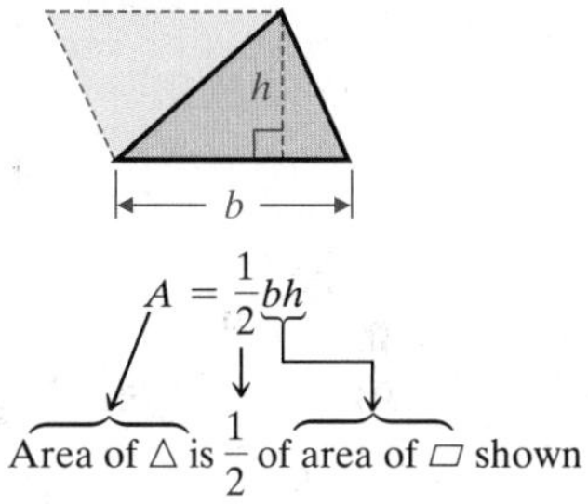

> **Helpful Hint**
> For the area formula of a parallelogram,
> - *b*, base, is the length of any side.
> - *h*, height, is the ⊥ distance from the base to the ∥ side.

EXAMPLE 1 Finding the Area of a Parallelogram

What is the area of each parallelogram?

a.

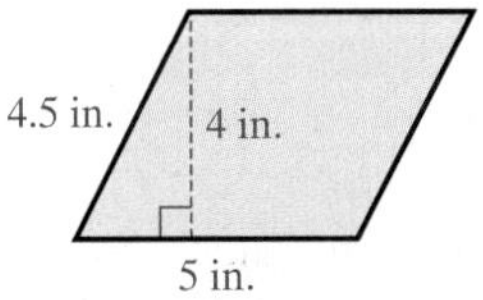

b.

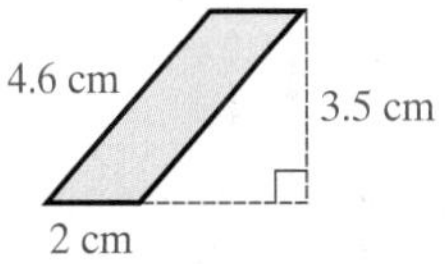

Solution We are given each height. Choose the perpendicular side to use as the base.

a. $A = bh$

$= 5(4) = 20$ Substitute for b and h.

The area is 20 sq in. or 20 in.^{2}.

b. $A = bh$

$= 2(3.5) = 7$ Substitute for b and h.

The area is 7 sq cm or 7 cm^2. □

PRACTICE 1 What is the area of a parallelogram with base length 12 m and height 9 m?

> **Helpful Hint**
> For the area formula of a triangle,
> - *b*, base, is the length of any side.
> - *h*, height, is the ⊥ distance to that base from the opposite vertex.

EXAMPLE 2 Finding the Area of a Triangle

To make two triangular sails like the ones shown, how many square feet of material are needed?

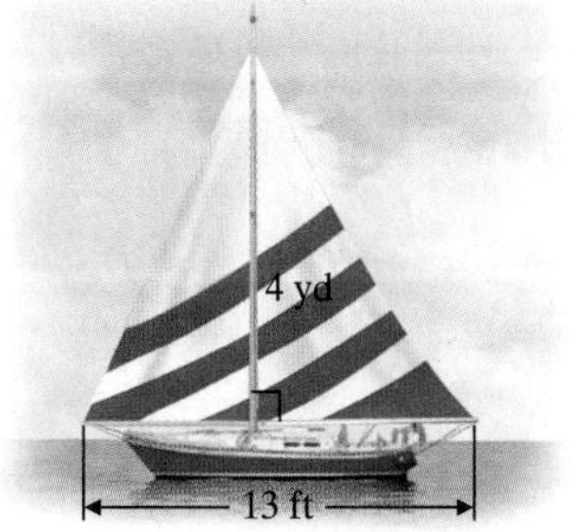

Solution To calculate material needed, let's treat the two adjacent sails as 1 triangle with length 13 ft and ⊥ height 4 yd. Since we are asked for the area in square feet, we first convert the height to feet.

$4 \text{ yd} = 4 \cdot 1 \text{ yd} = 4 \cdot 3 \text{ ft} = 12 \text{ ft}$ Recall that 1 yd = 3 ft.

Next, find the area of the triangle.

$$
\begin{aligned}
A &= \frac{1}{2}bh \\
&= \frac{1}{2}(13\text{ ft})(12\text{ ft}) \quad \text{Substitute 13 ft for } b \text{ and 12 ft for } h. \\
&= 78\text{ sq ft} \quad \text{Multiply.}
\end{aligned}
$$

We need 78 sq ft of material for the two sails. □

PRACTICE 2 What is the area of the triangle in square inches?

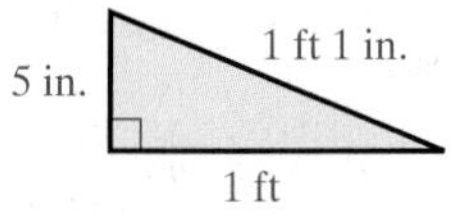

Before we find the area of an irregular room, let's practice finding the perimeter.

EXAMPLE 3 Finding the Perimeter of an Irregular Room

Find the perimeter of the room shown at the right.

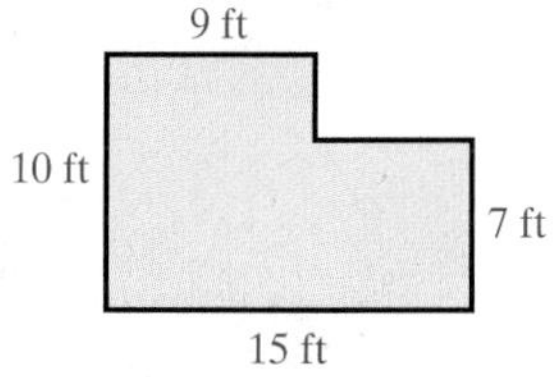

Solution To find the perimeter of the room, we first need to find the lengths of all sides of the room.

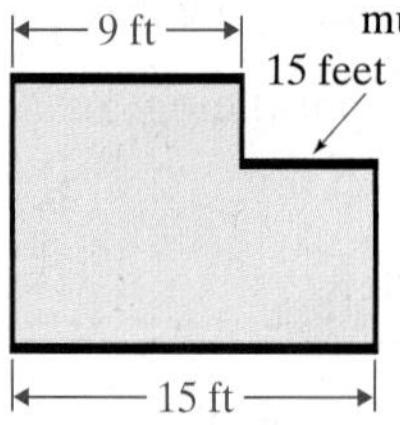

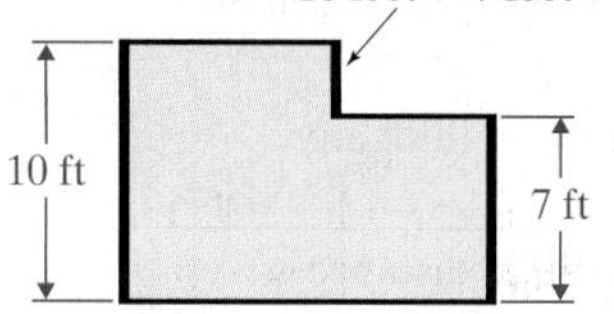

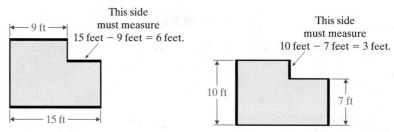

Now that we know the measures of all sides of the room, we can add the measures to find the perimeter.

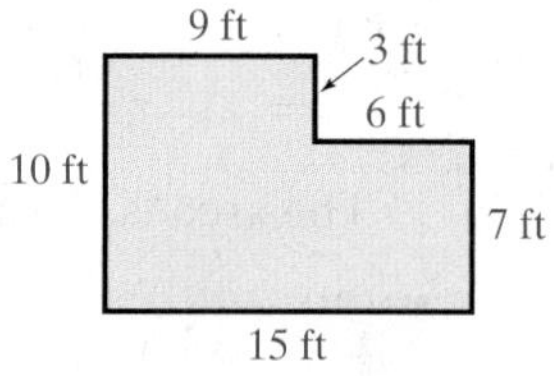

$$
\begin{aligned}
\text{perimeter} &= 10\text{ ft} + 9\text{ ft} + 3\text{ ft} + 6\text{ ft} + 7\text{ ft} + 15\text{ ft} \\
&= 50\text{ ft}
\end{aligned}
$$

The perimeter of the room is 50 feet. □

PRACTICE 3 Find the perimeter of the room shown.

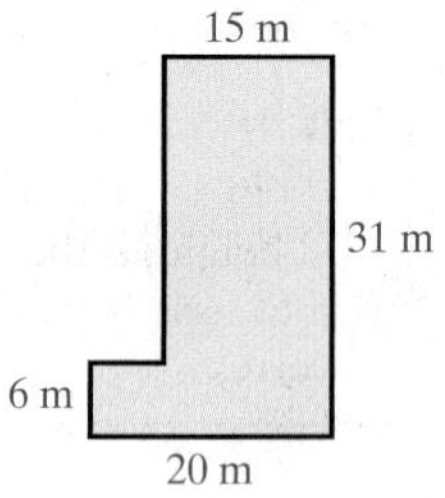

Next, we practice calculating area.

> **Helpful Hint**
> When finding the area of figures, check to make sure that all measurements are in the same units before calculations are made.

EXAMPLE 4 **Finding the Area of an Irregular Room**

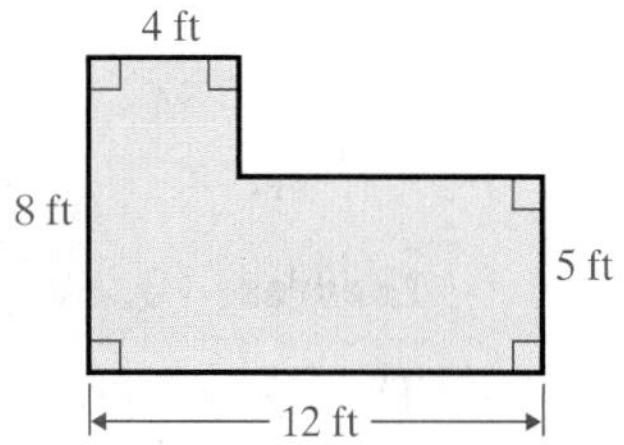

Solution Split the figure into two rectangles. To find the area of the figure, we find the sum of the areas of the two rectangles.

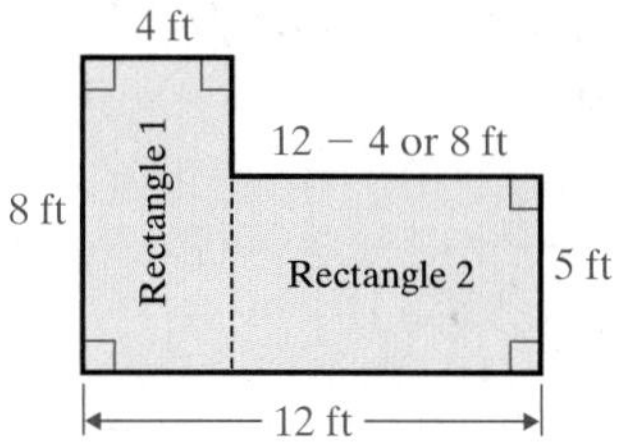

area of Rectangle 1 $= lw$
$= 8 \text{ feet} \cdot 4 \text{ feet}$
$= 32 \text{ square feet}$

Notice that the length of Rectangle 2 is 12 feet – 4 feet or 8 feet.

area of Rectangle 2 $= lw$
$= 8 \text{ feet} \cdot 5 \text{ feet}$
$= 40 \text{ square feet}$

area of the figure $=$ area of Rectangle 1 $+$ area of Rectangle 2
$= 32 \text{ square feet} + 40 \text{ square feet}$
$= 72 \text{ square feet}$ □

PRACTICE 4 Find the area of the figure.

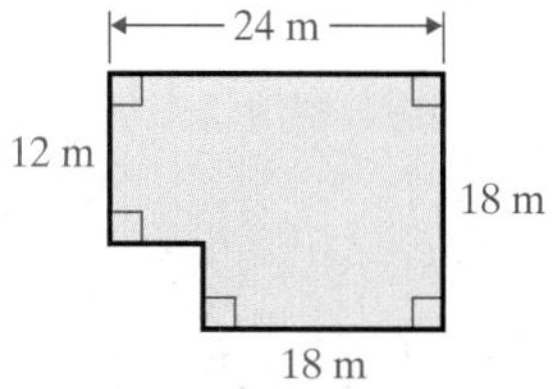

> **Helpful Hint**
> The figure in Example 4 could also be split into two rectangles as shown.

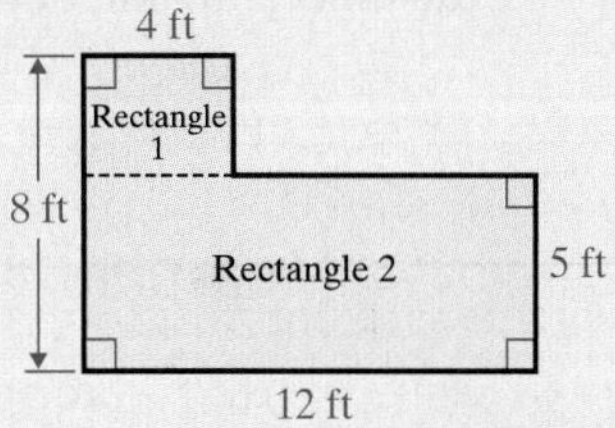

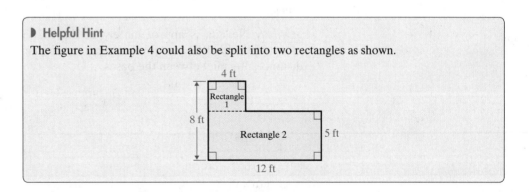

OBJECTIVE 2 ▶ Finding Areas of Trapezoids, Rhombuses, and Kites. The **height of a trapezoid** is the perpendicular distance between the bases. We use height in our formula for the area of a trapezoid.

We can find the area of a rhombus or a kite when we know the lengths of its diagonals.

Theorem 10.2-7 Area of a Trapezoid

The area of a trapezoid is half the product of the height and the sum of the bases.

$$A = \frac{1}{2}h(b_1 + b_2)$$

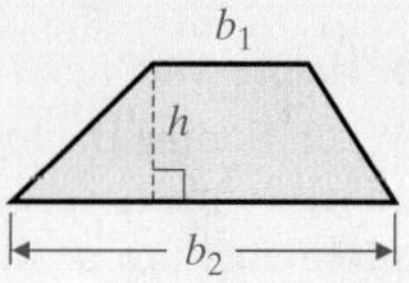

Theorem 10.2-8 Area of a Rhombus or a Kite

The area of a rhombus or a kite is half the product of the lengths of its diagonals.

$$A = \frac{1}{2}d_1d_2$$

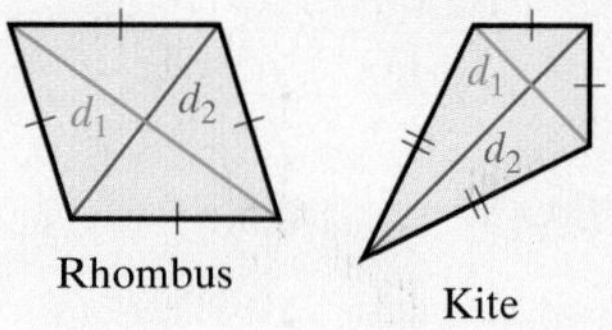

Notice how we can justify our formula for the area of a trapezoid by "cutting" the trapezoid along its midsegment and rotating the top half so that it becomes a parallelogram.

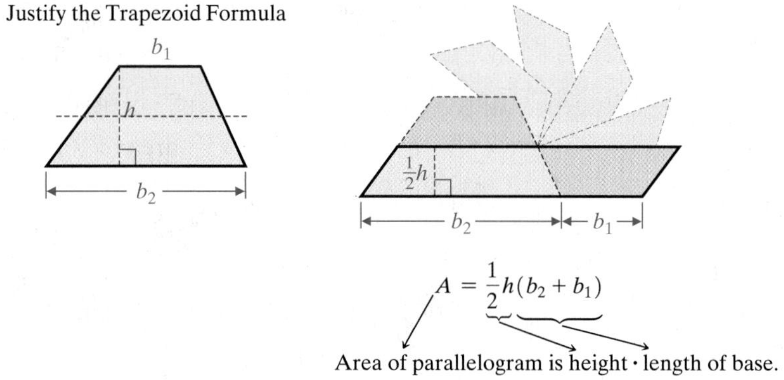

EXAMPLE 5 Area of a Trapezoid

What is the approximate area of the state of Nevada? Round to the nearest hundred square miles.

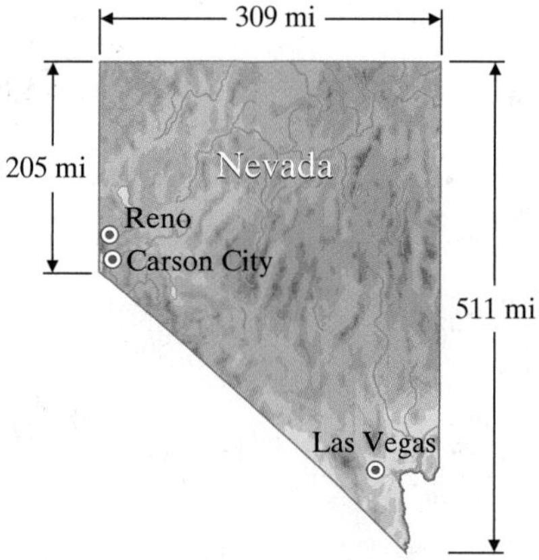

Solution Nevada is approximately in the shape of a trapezoid. We are given the perpendicular distance, 309 mi, between the bases.

$A = \frac{1}{2}h(b_1 + b_2)$	Use the formula for area of a trapezoid.
$= \frac{1}{2}(309)(205 + 511)$	Substitute 309 for h, 205 for b_1, and 511 for b_2.
$= 110{,}622$	Simplify.
$\approx 110{,}600$	Rounded to the nearest hundred

The area of Nevada is about 110,600 mi^2 or 110,600 sq mi. □

PRACTICE

5 What is the area of a trapezoid with height 7 cm and bases 12 cm and 15 cm?

> **Helpful Hint**
>
> Remember the special relationships in a 30°-60°-90°Δ and a 45°-45°-90°Δ.
>
> From the Pythagorean Theorem, we have:
>
> 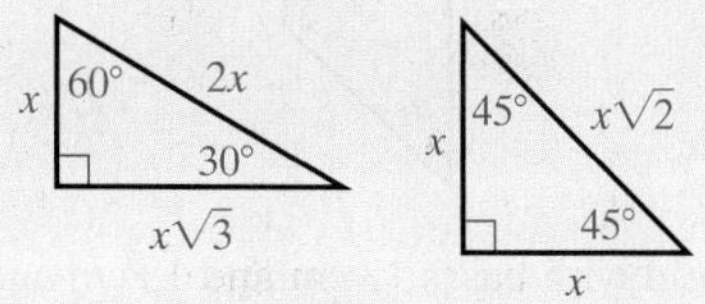
>

EXAMPLE 6 Finding Area Using a Right Triangle

Find the area of trapezoid *PQRS*.

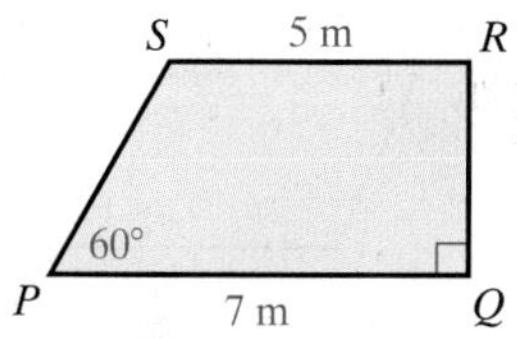

Solution We can draw an altitude that divides the trapezoid into a rectangle and a 30°-60°-90° triangle. Since the opposite sides of a rectangle are congruent, the longer base of the trapezoid is divided into segments of lengths 2 m and 5 m.

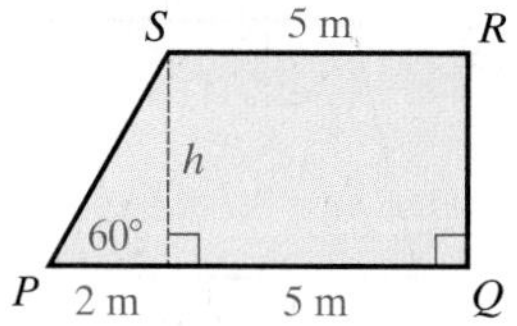

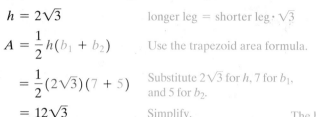

The height of Δ is $2 \cdot \sqrt{3}$ m. See the Helpful Hint in the margin.

$h = 2\sqrt{3}$ longer leg = shorter leg $\cdot \sqrt{3}$

$A = \frac{1}{2}h(b_1 + b_2)$ Use the trapezoid area formula.

$= \frac{1}{2}(2\sqrt{3})(7 + 5)$ Substitute $2\sqrt{3}$ for h, 7 for b_1, and 5 for b_2.

$= 12\sqrt{3}$ Simplify.

The area of trapezoid *PQRS* is $12\sqrt{3}$ sq m or $12\sqrt{3}$ m^2. □

PRACTICE 6 Find the area of trapezoid *PQRS*.

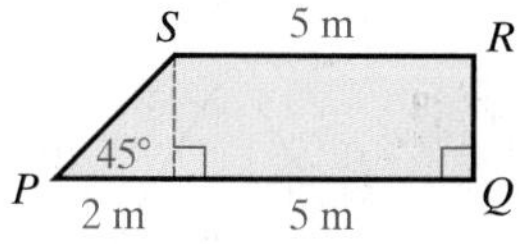

EXAMPLE 7 Finding the Area of a Kite

Find the area of kite *KLMN*.

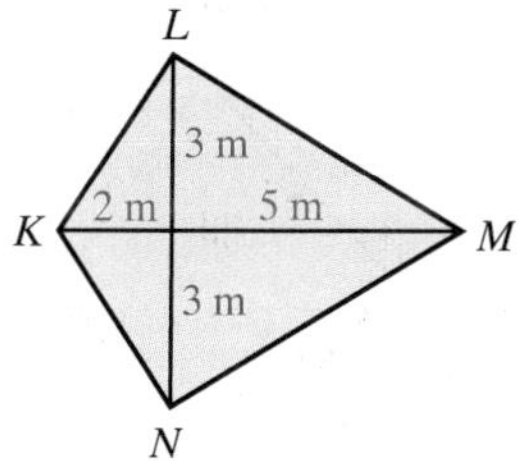

Solution To use our area formula, we first find the lengths of the two diagonals.

$KM = 2 + 5 = 7$m and $LN = 3 + 3 = 6$m

$A = \frac{1}{2}d_1d_2$ Use the formula for area of a kite.

$= \frac{1}{2}(7)(6)$ Substitute 7 for d_1 and 6 for d_2.

$= 21$ Simplify.

The area of kite *KLMN* is 21 m^2. □

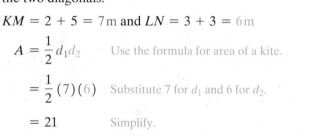

PRACTICE 7 Find the area of a kite with diagonals that are 12 in. and 9 in. long.

VOCABULARY & READINESS CHECK

Word Bank *Use the choices below to fill in each blank. Some choices may be used more than once and some not at all.*

area	height
base	perimeter

1. The ____________ of a polygon is the sum of the lengths of its sides.
2. The ____________ of a polygon or circle is the amount of space the figure encloses.
3. The height of a parallelogram is perpendicular to its ____________.
4. Any side of a triangle may be called its ____________.
5. The height of a triangle must be perpendicular to its ____________.
6. ____________ of a plane figure is measured in units.
7. ____________ of a plane figure is measured in square units.

10.2 EXERCISE SET MyMathLab®

Find the area of each parallelogram. See Example 1.

1.
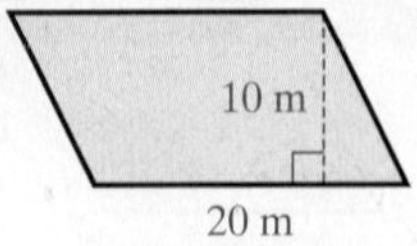

2.
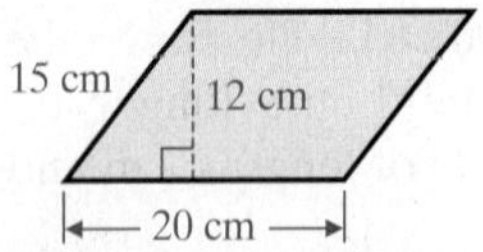

3.
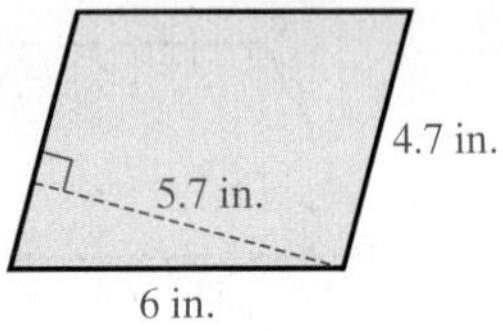

4.
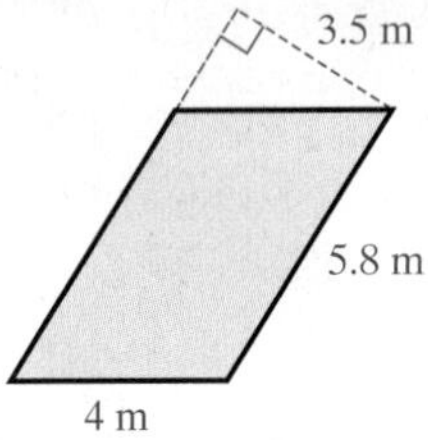

Find the area of each triangle. See Example 2.

5.
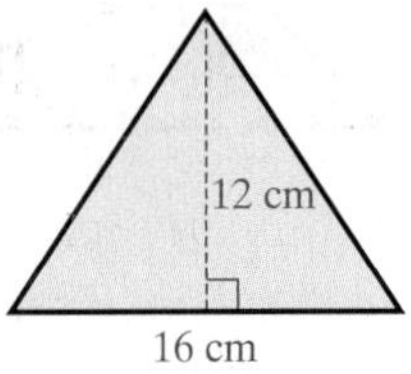

6.
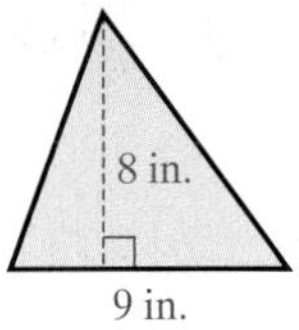

7.
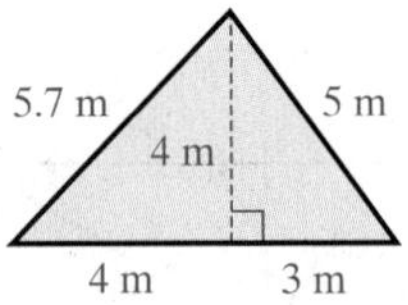

8.
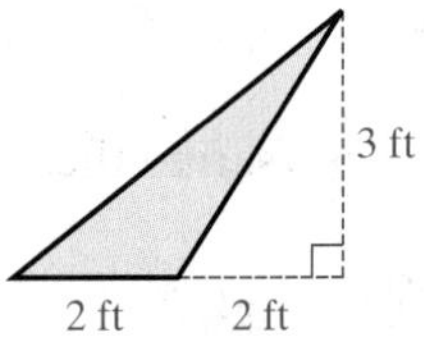

Find the perimeter and area of each figure. See Examples 3 and 4.

9.
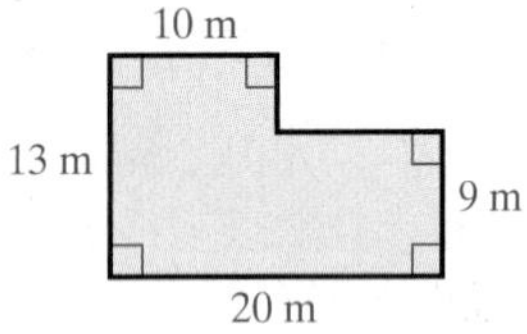

10.
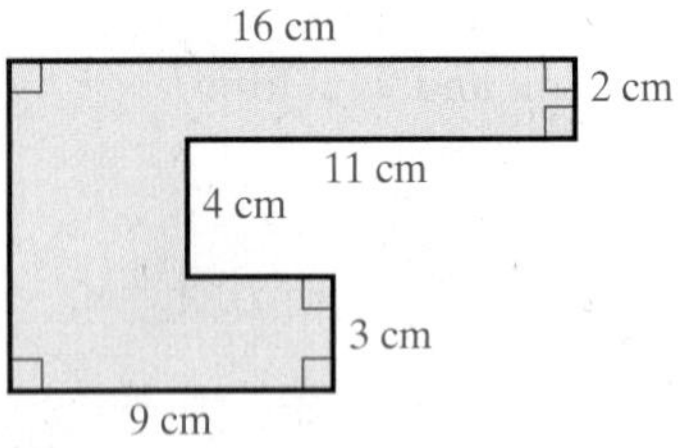

11.
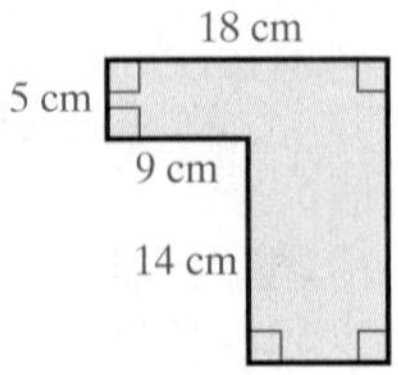

12.
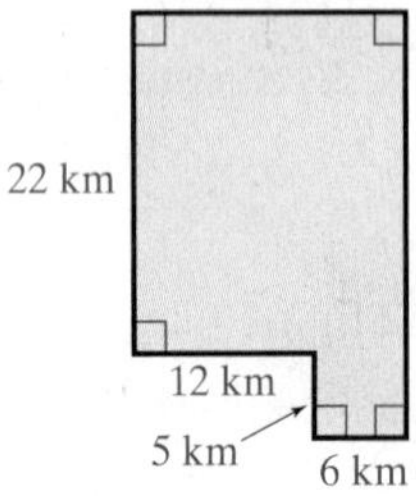

Find the area of each trapezoid. See Example 5.

13.
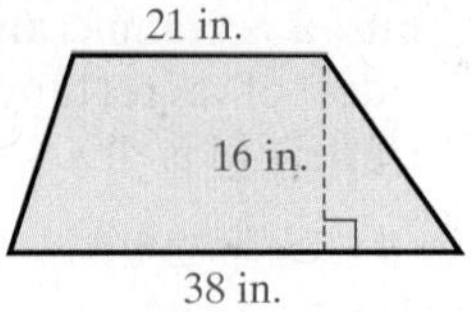

14.
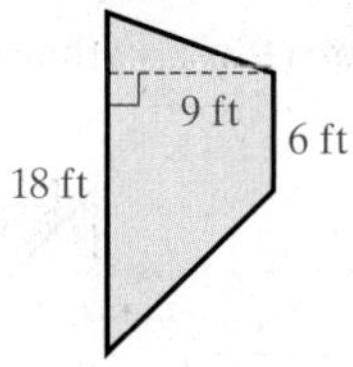

15. Find the area of a trapezoid with bases 12 cm and 18 cm and height 10 cm.

16. Find the area of a trapezoid with bases 2 ft and 3 ft and height $\frac{1}{3}$ ft.

Find the area of each trapezoid. If your answer is not an integer, leave it in simplest radical form. See Example 6.

17.
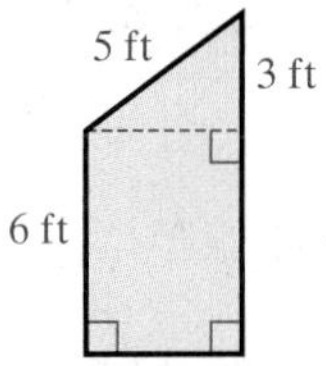

18.
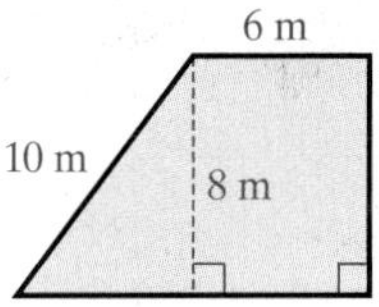

19.
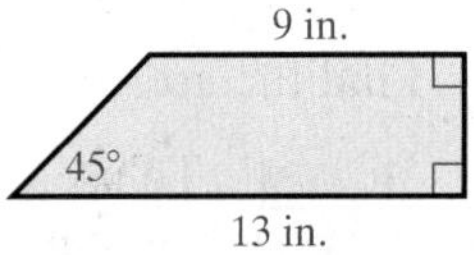

20.
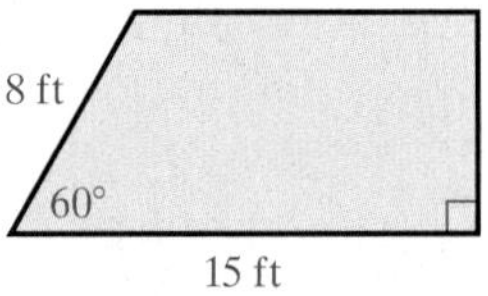

Find the area of each kite. See Example 7.

21.
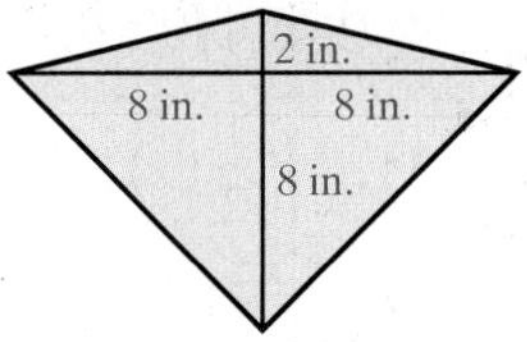

22.
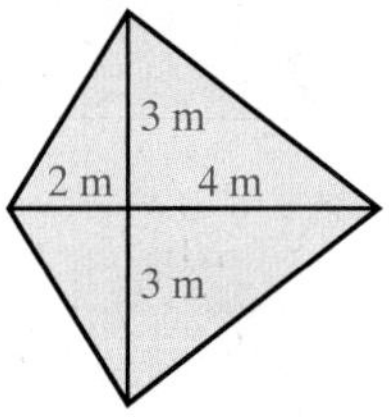

23.
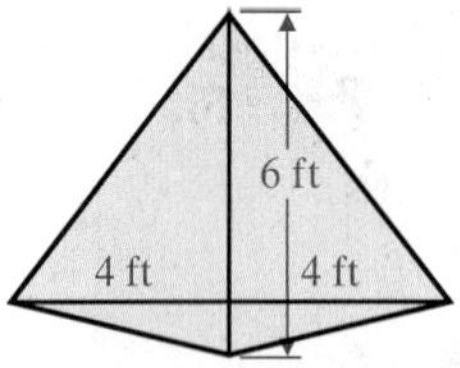

24.
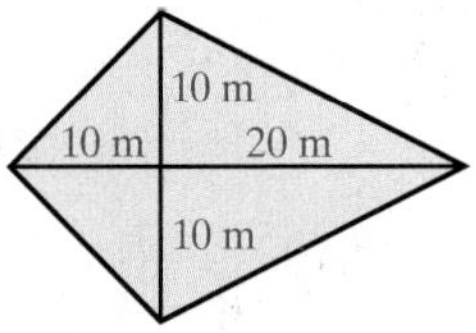

Find the area of each rhombus. See Example 7.

25.
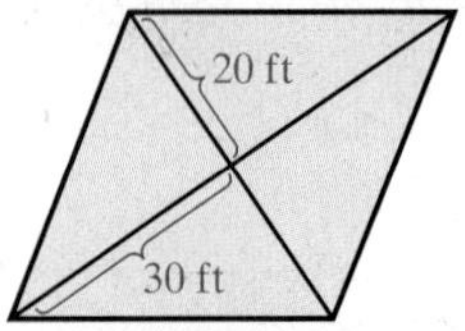

26.
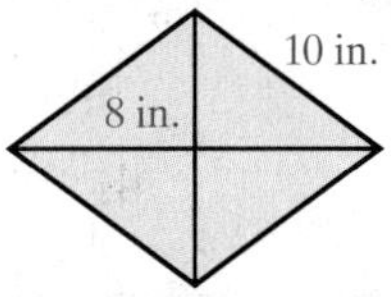

27.
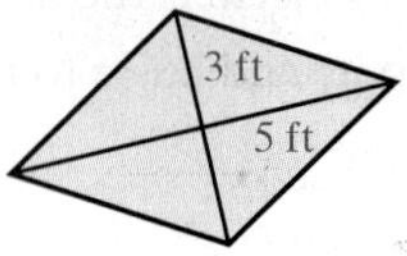

28.
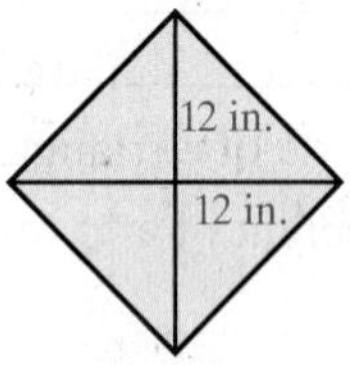

Find the area of each figure.

29.

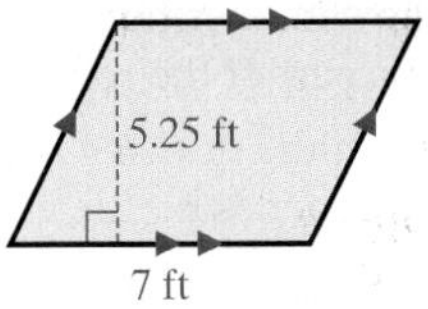

30.

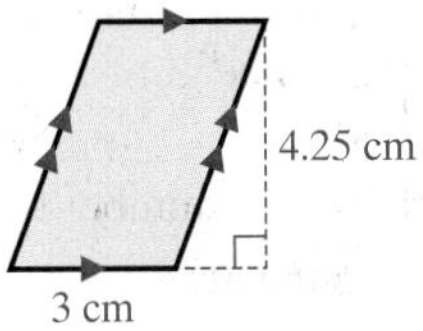

31.

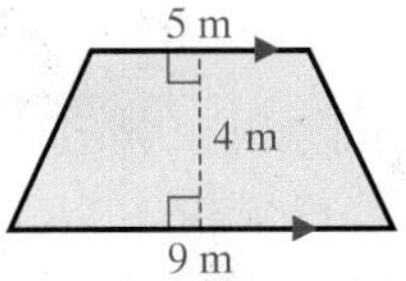

32.

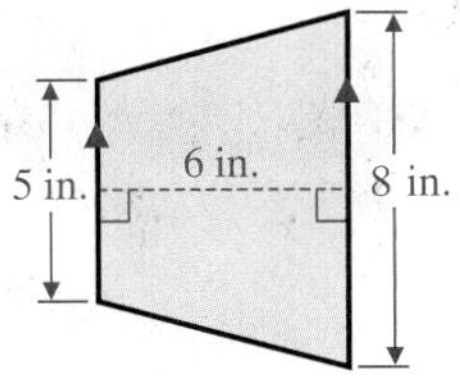

33.

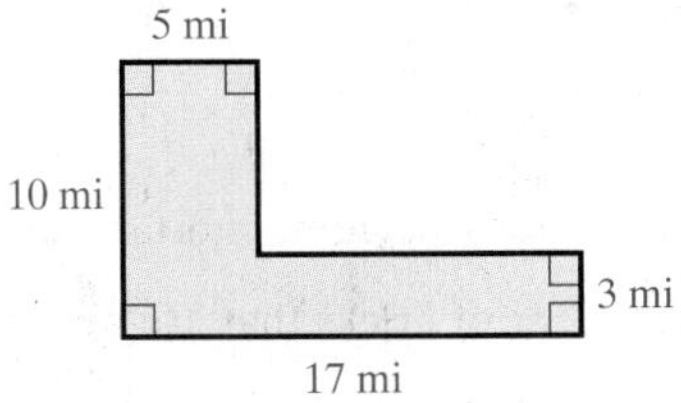

34.

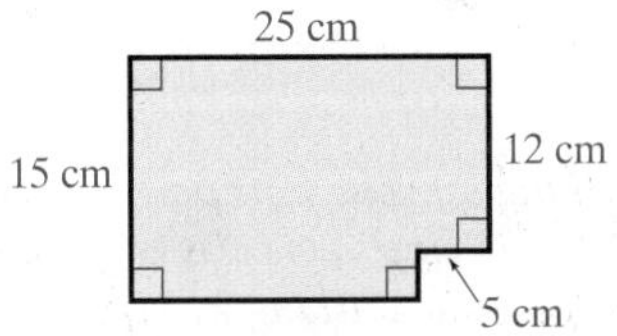

35.

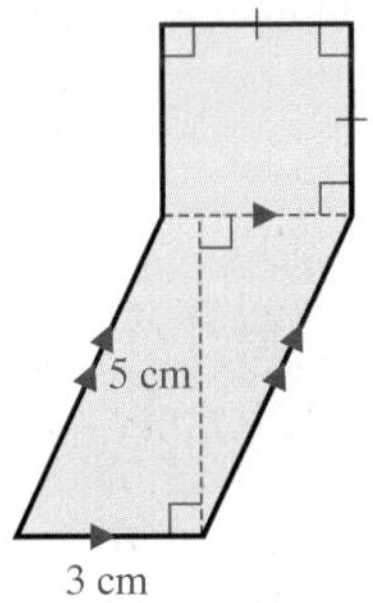

36.

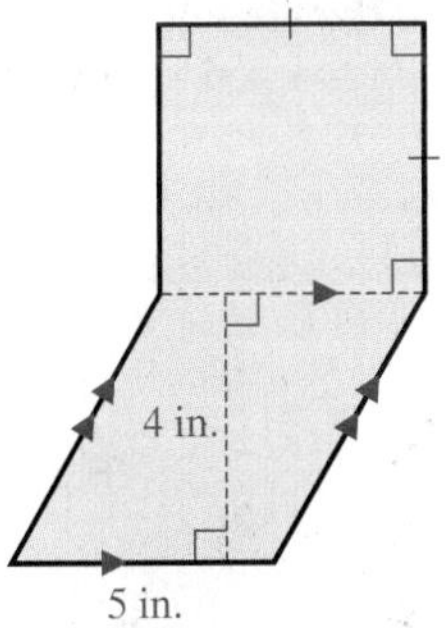

37.

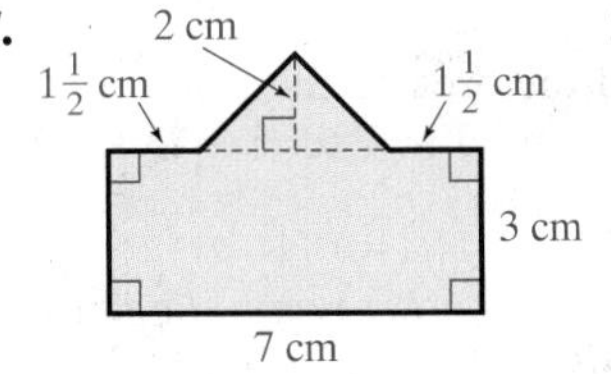

38.

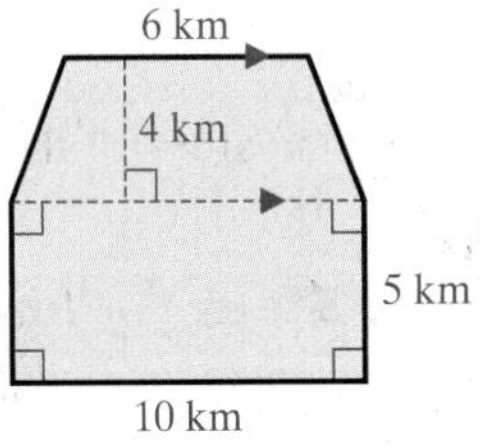

39. The diagram shows the approximate dimensions of the state of Utah. Find the area.

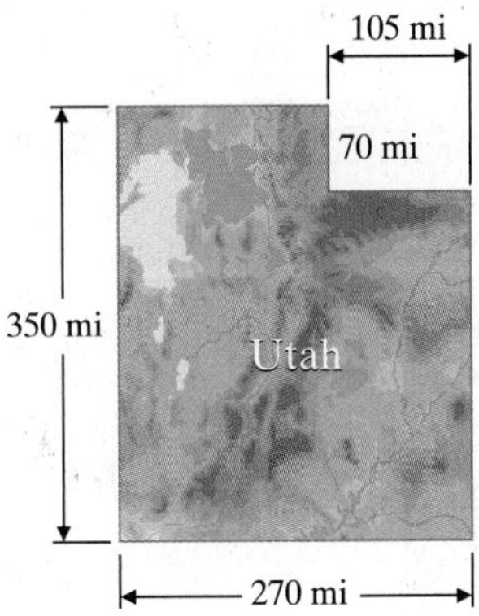

40. The border of Tennessee resembles a trapezoid with bases 340 mi and 440 mi and height 110 mi. Estimate the area of Tennessee by finding the area of the trapezoid.

41. **Multiple Choice** What is the area of the figure at the right?

a. 64 cm^2 **b.** 88 cm^2
c. 96 cm^2 **d.** 112 cm^2

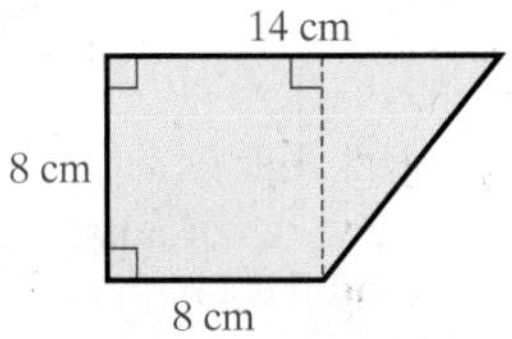

42. **Multiple Choice** What is the area of the kite at the right?

a. 90 m^2 **b.** 108 m^2
c. 135 m^2 **d.** 216 m^2

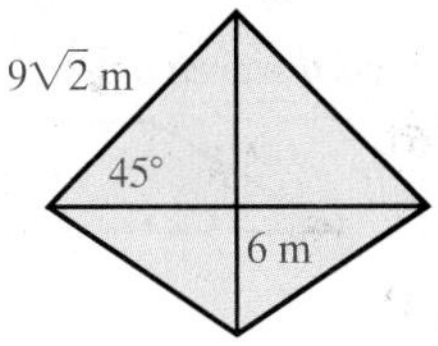

Find the area of each figure.

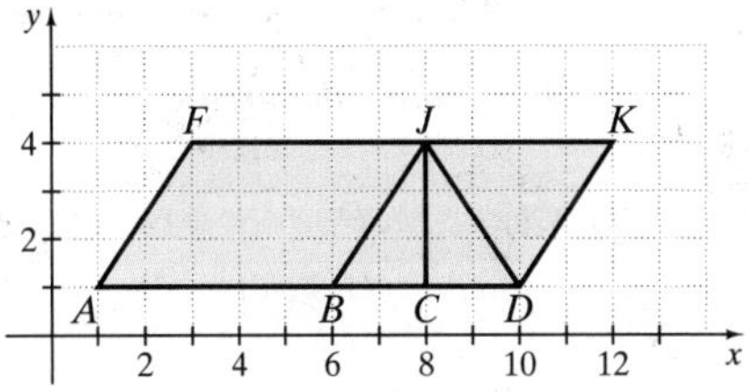

43. $\square ABJF$

44. ΔBDJ

45. ΔDKJ

46. $\square BDKJ$

47. $\square ADKF$

48. ΔBCJ

49. trapezoid $ADJF$

50. trapezoid $AFJC$

Coordinate Geometry *Find the area of quadrilateral QRST.*

51.

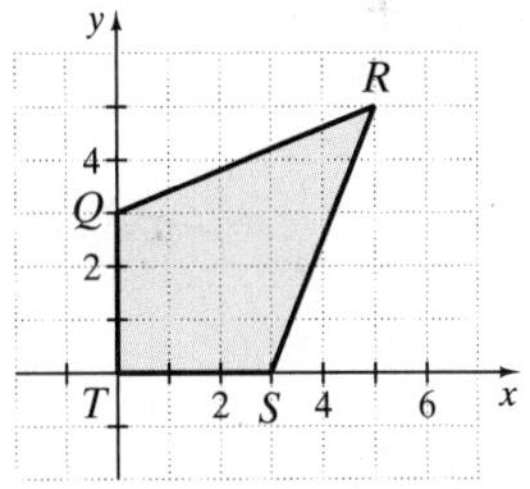

52.

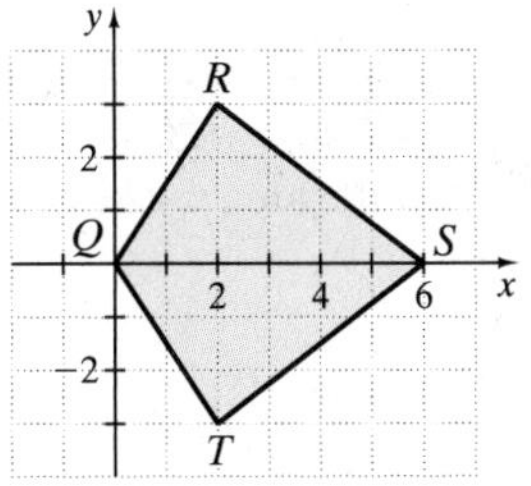

53.

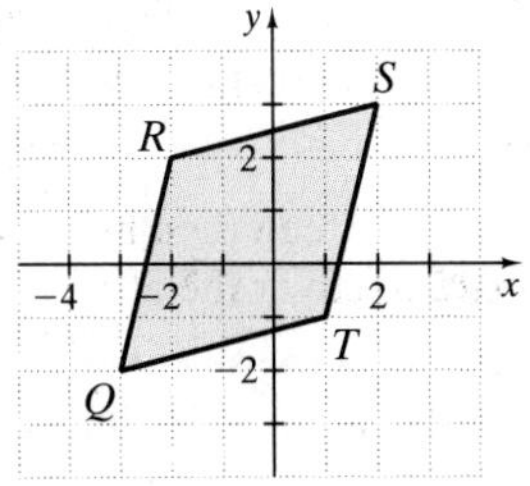

54.

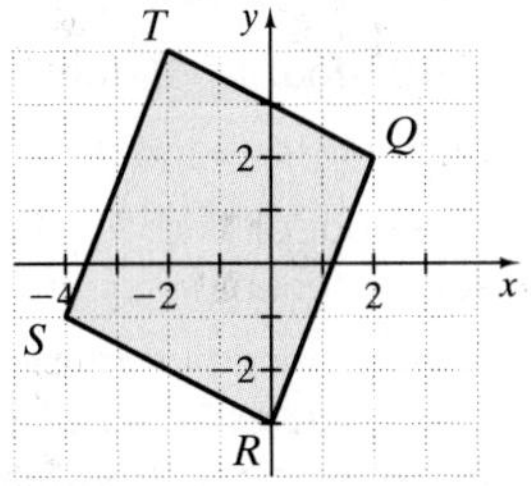

Find the area of each rhombus. Leave your answer in simplest radical form.

55.

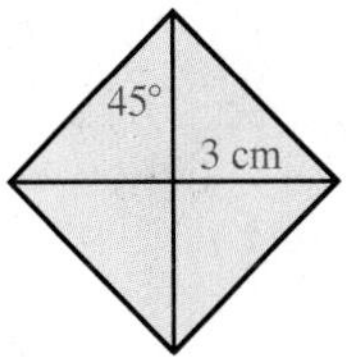

56.

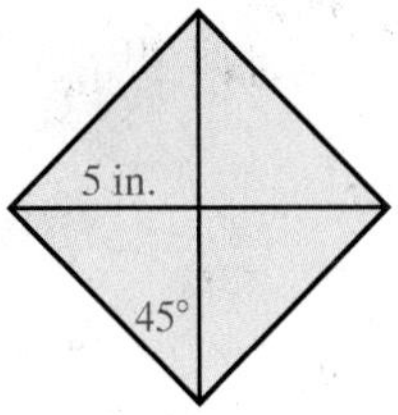

57.

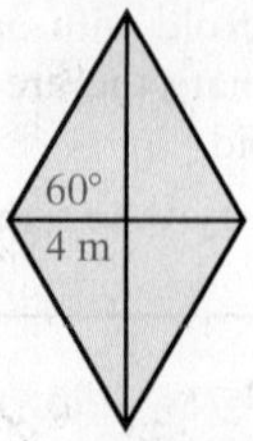

58.

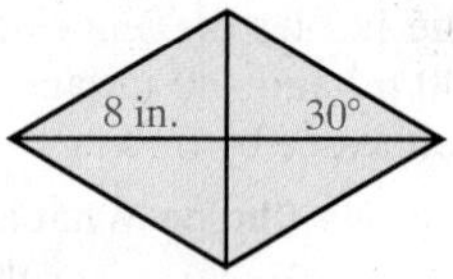

59. The area of a parallelogram is 24 $in.^2$ and the height is 6 in. Find the length of the corresponding base.

60. A right isosceles triangle has area 98 cm^2. Find the length of each leg.

Find the area of each trapezoid to the nearest tenth.

61.

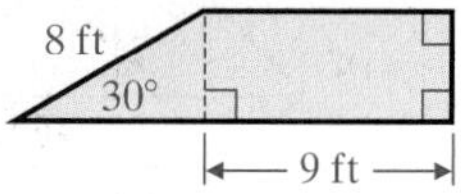

62.

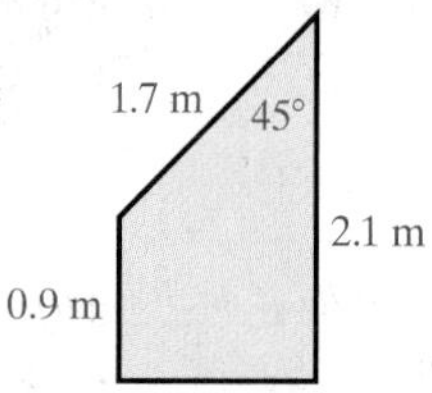

Coordinate Geometry *Find the area of a polygon with the given vertices.*

63. $A(3, 9), B(8, 9), C(2, -3), D(-3, -3)$

64. $E(1, 1), F(4, 5), G(11, 5), H(8, 1)$

65. $D(0, 0), E(2, 4), F(6, 4), G(6, 0)$

66. $K(-7, -2), L(-7, 6), M(1, 6), N(7, -2)$

67. The end of a gold bar has the shape of a trapezoid with the measurements shown. Find the area of the end.

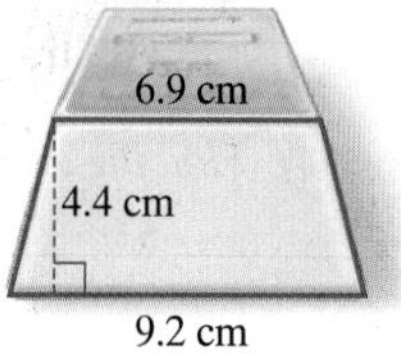

68. Find the area to the nearest tenth.

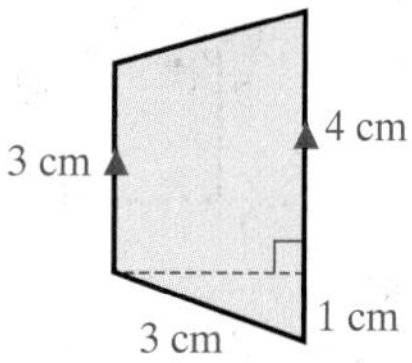

69. A $10\frac{1}{2}$-foot-by-16-foot concrete wall is to be built using concrete blocks. Find the area of the wall.

70. The floor of Terry's attic is 24 feet by 35 feet. Find how many square feet of insulation are needed to cover the attic floor.

71. The outlined part of the roof shown is in the shape of a trapezoid and needs to be shingled. The number of shingles to buy depends on the area.

a. Use the dimensions given to find the area of the outlined part of the roof to the nearest whole square foot.

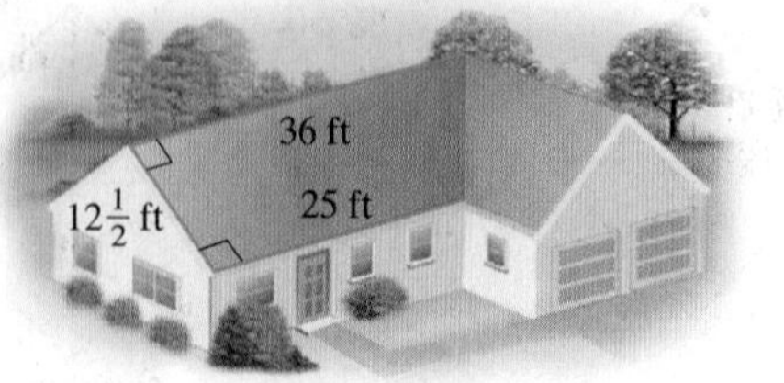

b. Shingles are packaged in a unit called a "square." If a "square" covers 100 square feet, how many whole squares need to be purchased to shingle this part of the roof?

72. The entire side of the building shaded in the drawing is to be bricked. The number of bricks to buy depends on the area.

a. Find the area.

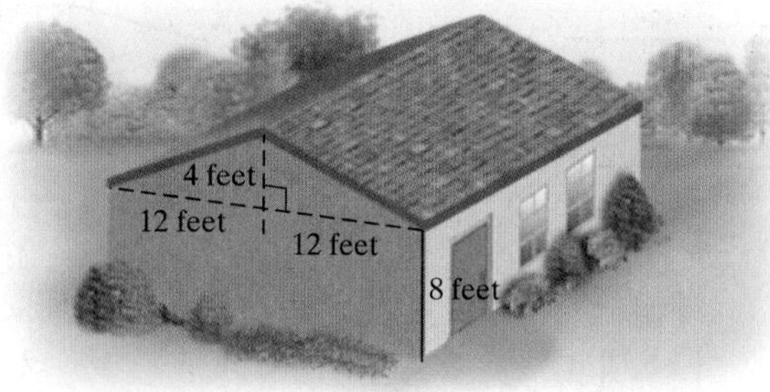

b. If the side area of each brick (including mortar room) is $\frac{1}{6}$ square ft, find the number of bricks that are needed to brick the end of the building.

CONCEPT EXTENSIONS

A bakery has a 50-ft-by-31-ft parking lot. The four parking spaces are congruent parallelograms, the driving region is a rectangle, and the two areas for flowers are congruent triangles.

73. Find the area of the paved surface by adding the areas of the driving region and the four parking spaces.

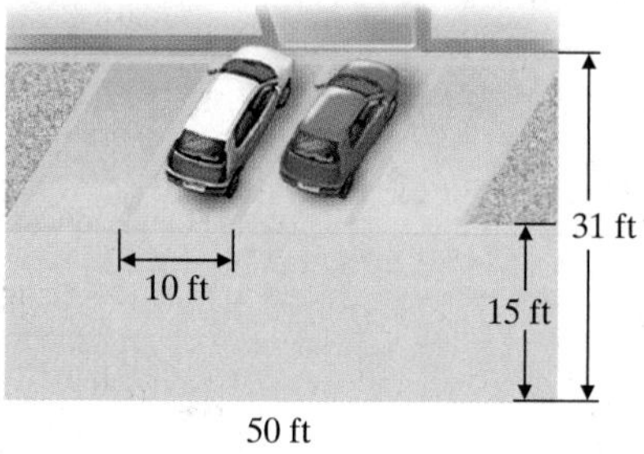

74. Use another method to find the area of the paved surface. Then compare with the area found in Exercise **73** to check your work.

Multiple Steps *For Exercises 75–78, (a) graph the lines and (b) find the area of the triangle enclosed by the lines.*

75. $y = x, x = 0, y = 7$

76. $y = x + 2, y = 2, x = 6$

77. $y = -\frac{1}{2}x + 3, y = 0, x = -2$

78. $y = \frac{3}{4}x + 2, y = -2, x = 4$

Find the area of each figure.

79.

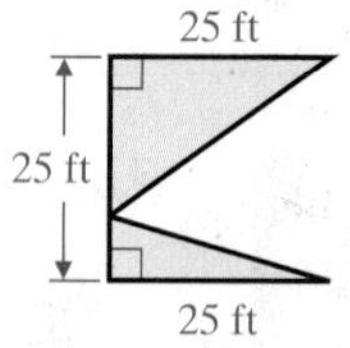

80.

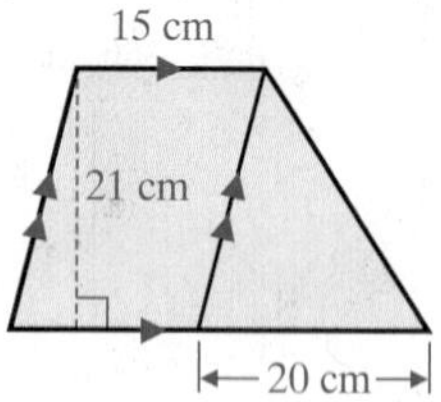

The Greek mathematician Heron is famous for this formula for the area of a triangle in terms of the lengths of its sides a, b, and c.

$$A = \sqrt{s(s-a)(s-b)(s-c)}, \text{ where } s = \frac{1}{2}(a+b+c)$$

Use Heron's Formula and a calculator to find the area of each triangle. Round your answer to the nearest whole number.

81. $a = 8$ in., $b = 9$ in., $c = 10$ in.

82. $a = 15$ m, $b = 17$ m, $c = 21$ m

83. Use Heron's Formula to find the area of this triangle.

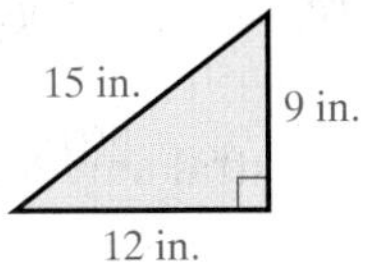

84. Verify your answer to Exercise **83** by using the formula $A = \frac{1}{2}bh$.

85. Suppose the height of a triangle is tripled. How does this affect the area of the triangle? Explain.

86. Draw a kite. Measure the lengths of its diagonals. Find its area.

87. The kite has diagonals d_1 and d_2 congruent to the sides of the rectangle. Explain why the area of the kite is $\frac{1}{2}d_1d_2$.

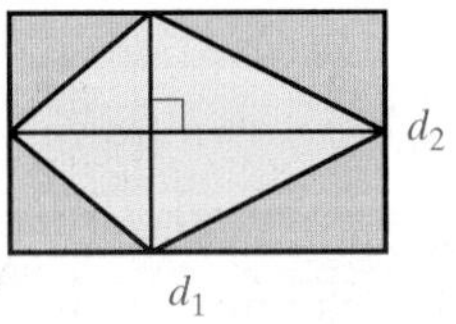

88. Draw a trapezoid. Label its bases b_1 and b_2 and its height h. Then draw a diagonal of the trapezoid.

a. Write equations for the area of each of the two triangles formed.

b. Explain how you can justify the trapezoid area formula using the areas of the two triangles.

89. Find how many square feet of grass are in the following plot.

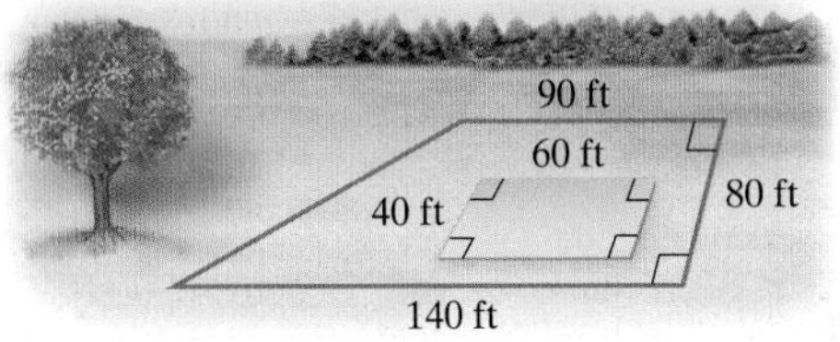

90. For Gerald Gomez to determine how much grass seed he needs to buy, he must know the size of his yard. Use the drawing to determine how many square feet are in his yard.

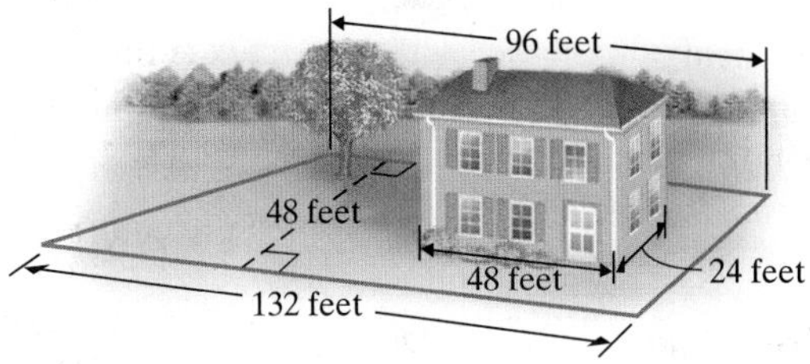

91. Does an altitude of a triangle have to lie inside the triangle? Explain.

92. How can you show that a parallelogram and a rectangle with the same bases and heights have equal areas?

93. Can a trapezoid and a parallelogram with the same bases and heights have the same area? Explain.

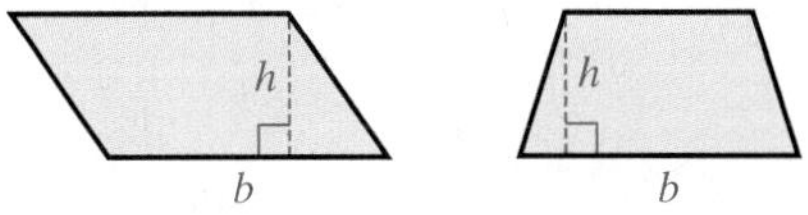

94. Do you need to know all the side lengths to find the area of a trapezoid?

95. Can you find the area of a rhombus if you know only the lengths of its sides? Explain.

96. Do you need to know the lengths of the sides to find the area of a kite? Explain.

REVIEW AND PREVIEW

Solve. See Sections 2.1 and 9.2.

97. Find the area of a right isosceles triangle that has one leg of length 12 cm.

98. A right isosceles triangle has area 112.5 ft^2. Find the length of each leg.

Find the area of each regular polygon. Leave radicals in simplest form.

99.

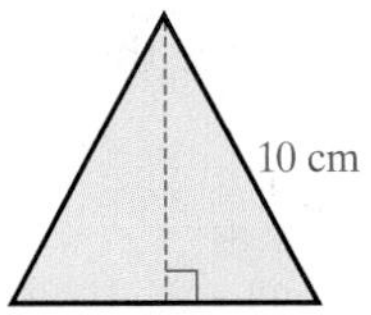

100.

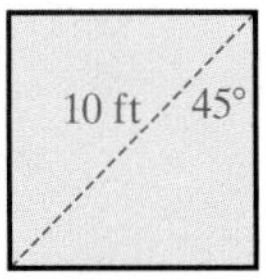

10.3 AREAS OF REGULAR POLYGONS

OBJECTIVES

1 Find the Area of a Regular Polygon.

2 Find the Area of a Regular Polygon Using Trigonometric Ratios.

OBJECTIVE 1 ▶ Finding Areas of Regular Polygons. The area of a regular polygon is related to the distance from the center to a side.

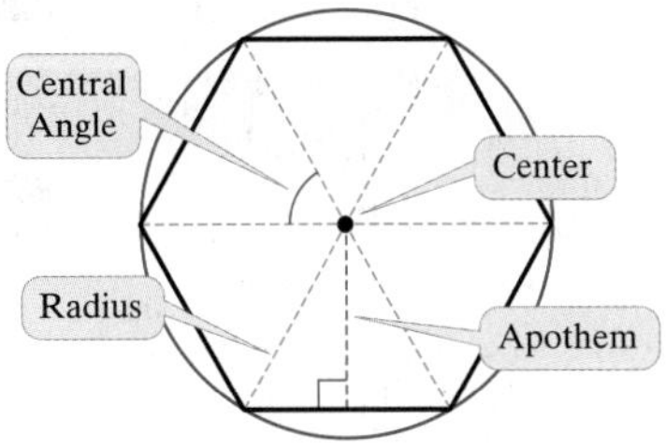

VOCABULARY

- center of a regular polygon
- radius of a regular polygon
- apothem
- central angle of a regular polygon

We can circumscribe a circle about any regular polygon. The **center of a regular polygon** is the center of the circumscribed circle. The **radius of a regular polygon** is the distance from the center to a vertex. The **apothem** is the perpendicular distance from the center to a side. Lastly, a **central angle of a regular polygon** is an angle whose vertex is the center of the polygon and whose sides contain two consecutive radii (*radii* is the plural of *radius*).

Helpful Hint

A regular polygon with n sides has n central angles.

The measure of each central angle is then

$$\frac{360^\circ}{n}$$

EXAMPLE 1 Finding Angle Measures

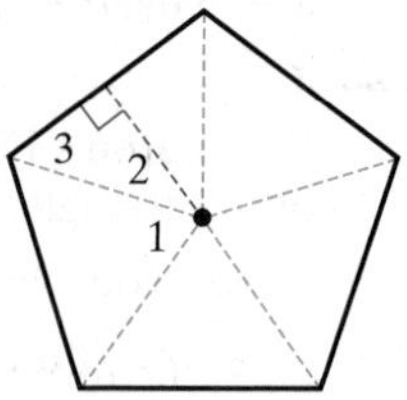

The figure at the right is a regular pentagon with radii and an apothem drawn. What is the measure of each numbered angle?

Solution Notice that $\angle 1$ is a central angle. Thus,

$$m\angle 1 = \frac{360^\circ}{5} = 72^\circ$$ Divide 360° by the number of sides.

$$m\angle 2 = \frac{1}{2}m\angle 1$$ The apothem bisects the vertex angle of the isosceles triangle formed by the radii.

$$= \frac{1}{2}(72^\circ) = 36^\circ$$

$$90^\circ + 36^\circ + m\angle 3 = 180^\circ$$ The sum of the measures of the angles of a triangle is 180°.

$$m\angle 3 = 54^\circ$$

Thus, $m\angle 1 = 72^\circ$, $m\angle 2 = 36^\circ$, and $m\angle 3 = 54^\circ$ □

PRACTICE

1 At the right, a portion of a regular octagon has radii and an apothem drawn. What is the measure of each numbered angle?

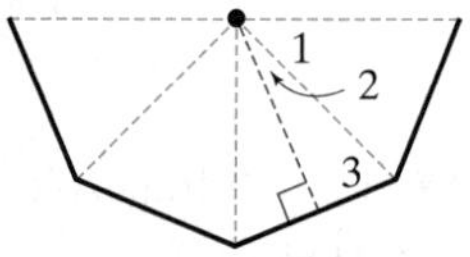

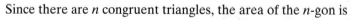

Suppose we have a regular n-gon with side s. The radii divide the figure into n congruent isosceles triangles. By Postulate 10.2-1, in the previous section, the areas of the isosceles triangles are equal. Each triangle has a height of a and a base of length s, so the area of each triangle is $\frac{1}{2}as$.

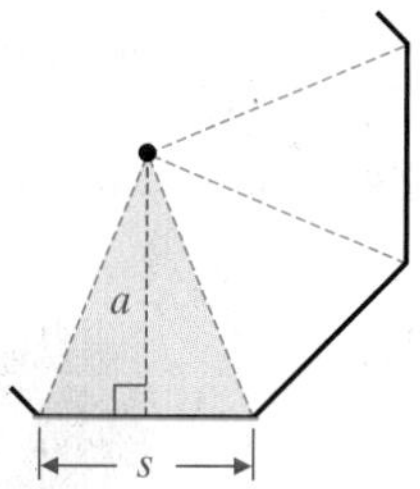

Since there are n congruent triangles, the area of the n-gon is

$$A = n \cdot \frac{1}{2}as$$

The perimeter P of the n-gon is the number of sides n times the length of a side s, or ns. By substitution, the area can be expressed as

$$A = \frac{1}{2}aP$$

Theorem 10.3-1 Area of a Regular Polygon

The area of a regular polygon is half the product of the apothem and the perimeter.

$$A = \frac{1}{2}aP$$

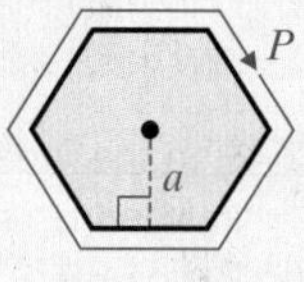

EXAMPLE 2 **Finding the Area of a Regular Polygon**

What is the area of the regular decagon at the right?

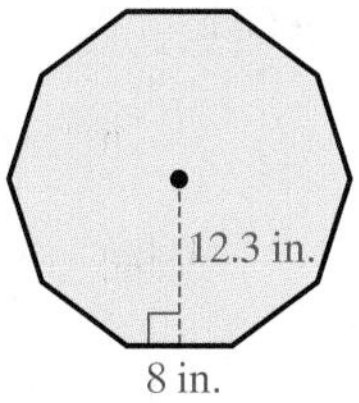

Solution To use the area formula, $A = \frac{1}{2}aP$, let's first find P, the perimeter.

A decagon has 10 sides and each side of this regular polygon is 8 in., so the perimeter is:

$$P = 10(8) = 80 \text{ in.}$$

Now let's find the area of the regular decagon.

$A = \frac{1}{2}aP$ Use the formula for the area of a regular polygon.

$= \frac{1}{2}(12.3)(80)$ Substitute 12.3 for a and 80 for P.

$= 492$

The regular decagon has an area of 492 sq in. or 492 in.^{2}. □

Helpful Hint

Remembering the side lengths of special right triangles will always save time.

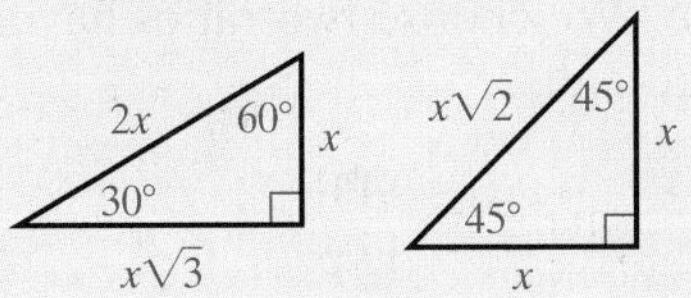

PRACTICE 2 What is the area of a regular pentagon with an 8-cm apothem and 11-cm sides?

EXAMPLE 3 **Using Special Triangles to Find the Area of a Regular Polygon**

A honeycomb is made up of regular hexagonal cells. The length of a side of a cell is 3 mm. What is the area of a cell? Round to the nearest square mm.

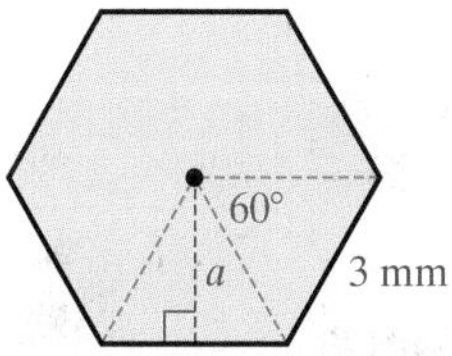

Solution To use the area formula, $A = \frac{1}{2}aP$, we need P, the perimeter, and a, the apothem.

A hexagon has 6 sides, and each side length is 3 mm; thus:

$$P = 6(3) = 18 \text{ mm}$$

To find the apothem, we see that the radii form six 60° angles at the center, so we can use a 30°-60°-90° triangle.

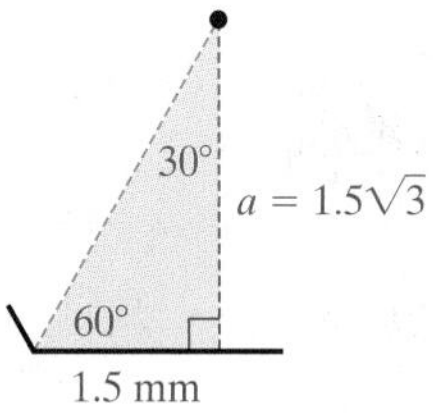

$a = 1.5\sqrt{3}$ longer leg (across 60°) = $\sqrt{3}$ × shorter leg (across 30°)

Now let's find the area.

$A = \frac{1}{2}aP$ Use the formula for the area of a regular polygon.

$= \frac{1}{2}(1.5\sqrt{3})(18)$ Substitute $1.5\sqrt{3}$ for a and 18 for P.

≈ 23.3826859 Use a calculator.

The area is about 23 mm^2. □

PRACTICE 3 The side of a regular hexagon is 16 ft. What is the area of the hexagon? Round the answer to the nearest square foot.

OBJECTIVE 2 ▶ Using Trigonometric Ratios to Find Area. Sometimes, we need to apply our knowledge of trigonometric ratios to be able to use our area formula.

EXAMPLE 4 Using Trigonometry to Find Area of a Regular Polygon

An architect in Europe is designing a regular dodecagon tile feature for a mall. If the radius is planned to be 10 meters, find the area of the regular dodecagon rounded to the nearest tenth of a square meter.

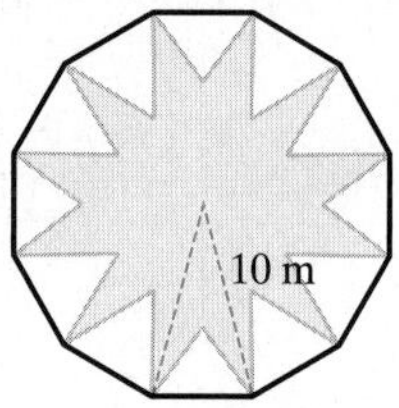

Solution To use our area formula, we need the perimeter and apothem of our regular dodecagon (12 sides).

Let's concentrate on ΔABC from the polygon.

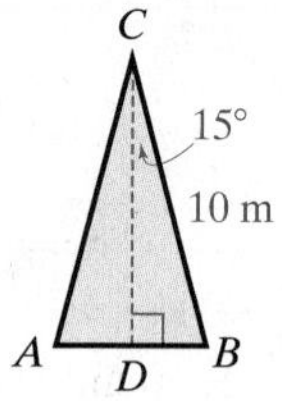

$$m\angle C = \frac{360°}{12} = 30°$$

For right ΔCDB, we now know that $CB = 10$ meters and $m\angle DCB = 15°$.

How do we find the apothem, CD, and the perimeter, starting with DB?

Let's use our trigonometric ratios with this right Δ. We choose two ratios having to do with 15° that contain our known value of the hypotenuse.

$$\cos 15° = \frac{\text{adj.}}{\text{hyp.}} \qquad\qquad \sin 15° = \frac{\text{opp.}}{\text{hyp.}}$$

$$\cos 15° = \frac{CD}{10} \qquad \text{and} \qquad \sin 15° = \frac{DB}{10}$$

$$10\cos 15° = CD \qquad\qquad 10\sin 15° = DB$$

$$9.7 \approx CD \qquad\qquad 2.6 \approx DB$$

Thus, the apothem is about 9.7 m and since half the side length is about 2.6 m, we have side length $\approx 2(2.6) = 5.2$ m. Thus

$$P \approx 12(5.2) = 62.4 \text{ m}$$

Now we find area.

$$A = \frac{1}{2}aP \approx \frac{1}{2}(9.7)(62.4) \approx 302.6 \text{ sq m}$$

The area is approximately 302.6 sq m. □

PRACTICE 4 Suppose the architect is also drawing a design with the same radius, but having to do with a regular heptagon. Find the area of the heptagon rounded to the nearest tenth of a square meter.

Now let's use trigonometry to find the area of a regular polygon when we know the length of a side.

EXAMPLE 5 Using Trigonometry to Find Area of a Regular Polygon

What is the area of the regular nonagon shown with 10-cm sides?

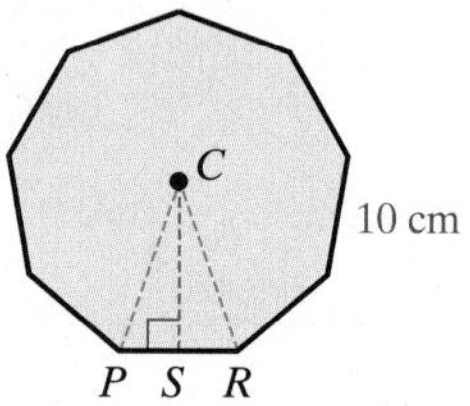

Solution Recall that a nonagon has 9 sides and that to find the area, we use the formula $A = \frac{1}{2}aP$.

perimeter, $P = 9 \cdot 10 = 90$ cm   The polygon has 9 sides, and each side length is 10 cm.

Also, $m\angle PCR = \frac{360°}{9} = 40°$.

Now let's concentrate on ΔPCS, with $\overline{CS}$ the apothem, a.

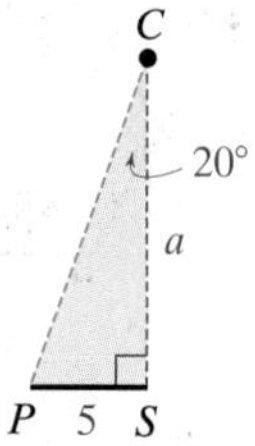

$$m\angle PCS = \frac{1}{2}m\angle PCR = 20° \text{ and } PS = \frac{1}{2}PR = 5 \text{ cm}$$

Let a represent CS. Find a and substitute into the area formula.

$$\tan 20° = \frac{5}{a} \qquad \text{Use the tangent ratio.}$$

$$a = \frac{5}{\tan 20°} \qquad \text{Solve for } a.$$

$$A = \frac{1}{2}aP$$

$$= \frac{1}{2} \cdot \frac{5}{\tan 20°} \cdot 90 \qquad \text{Substitute } \frac{5}{\tan 20°} \text{ for } a \text{ and 90 for } P.$$

$$\approx 618.1824194 \qquad \text{Use a calculator.}$$

The area of the regular nonagon is about 618 cm^2.

PRACTICE

5 What is the area of a regular pentagon with 4-in. sides? Round the answer to the nearest square inch.

VOCABULARY & READINESS CHECK

Word Bank *Use the choices below to fill in each blank. Some choices may be used more than once and some not at all.*

central apothem center radius

1. The ____________ of a regular polygon is the distance from the center to a vertex.

2. The ____________ of a regular polygon is the center of the circumscribed circle.

3. The ____________ of a regular polygon is the perpendicular distance from the center to a side.

4. A(n) ____________ angle of a regular polygon is an angle whose vertex is the center of the polygon and whose sides contain two consecutive radii.

10.3 EXERCISE SET MyMathLab®

Each regular polygon has radii and an apothem as shown. Find the measure of each numbered angle. See Example 1.

1.

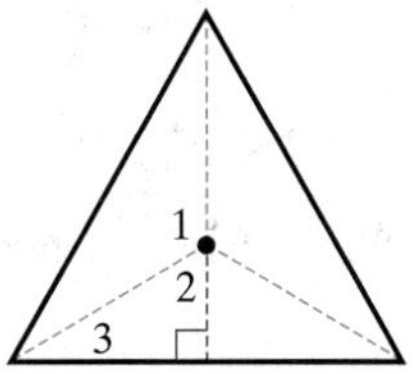

2.

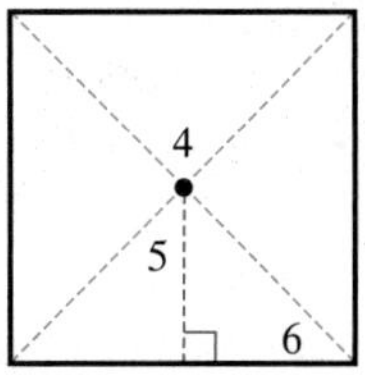

3.

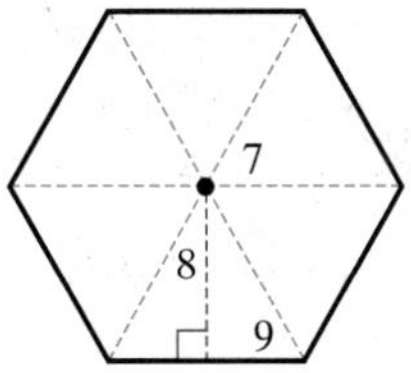

4.

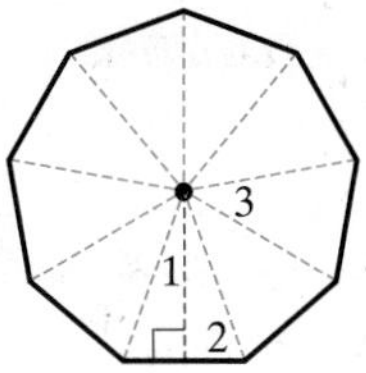

Find the perimeter and the area of each regular polygon with the given apothem a *and side length* s. *See Example 2.*

5. 7-gon, $a = 29$ ft, $s = 28$ ft

6. pentagon, $a = 24$ cm, $s = 35$ cm

7. nonagon, $a = 27.5$ in., $s = 20$ in.

8. octagon, $a = 60.4$ in., $s = 50$ in.

9. dodecagon, $a = 26.1$ cm, $s = 14$ cm

10. decagon, $a = 19$ m, $s = 12.3$ m

Find the perimeter and the area of each regular polygon. Round your answer to the nearest tenth. See Example 3.

11.

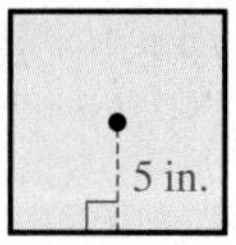

12.

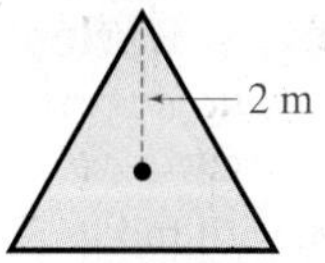

13.

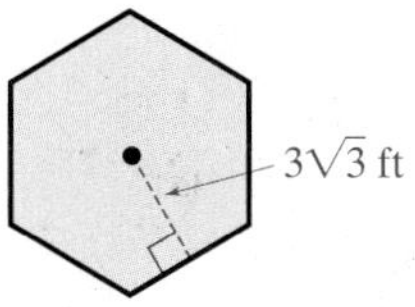

14.

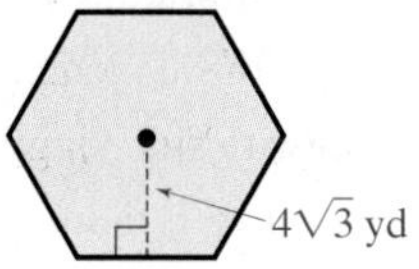

15.

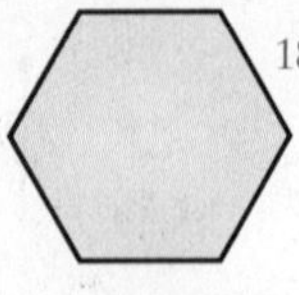

16.

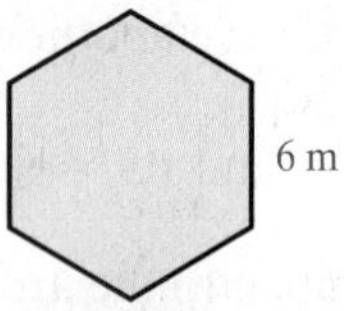

17.

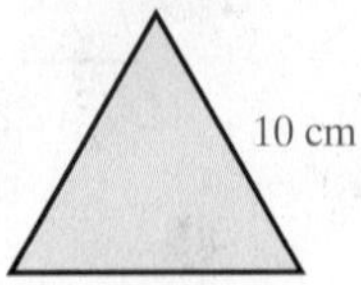

18.

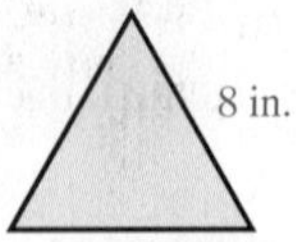

Find the area of each regular polygon with the given radius or apothem. If your answer is not an integer, leave it in simplest radical form. Then round it to the nearest tenth.

19.

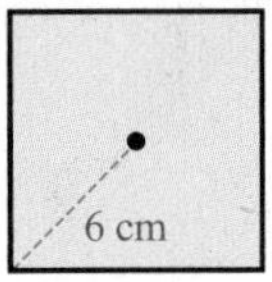

20.

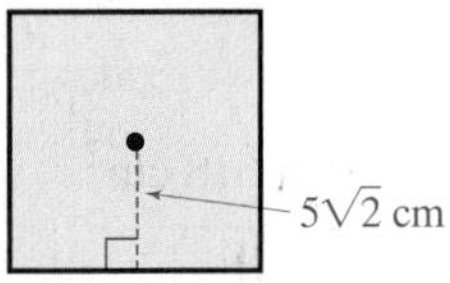

21.

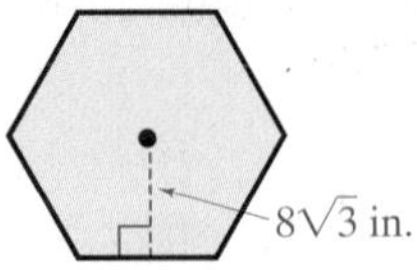

22.

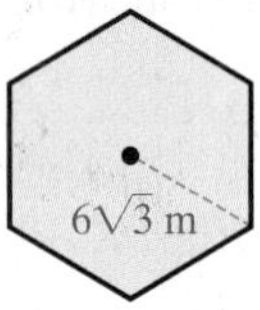

23.

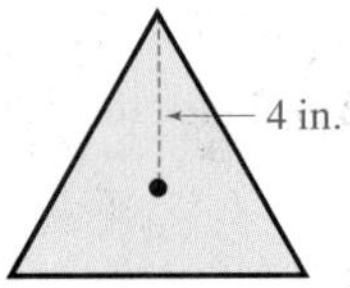

24.

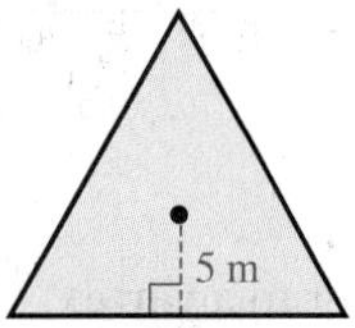

Find the area of each regular polygon. Show your answers in simplest radical form and rounded to the nearest tenth.

25.

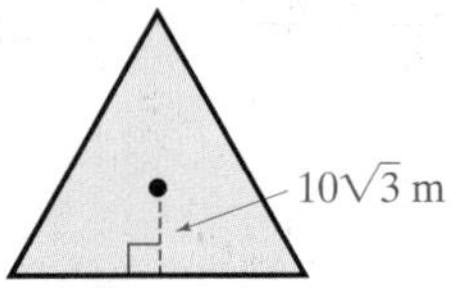

26.

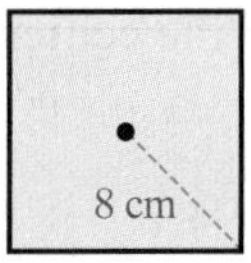

27.

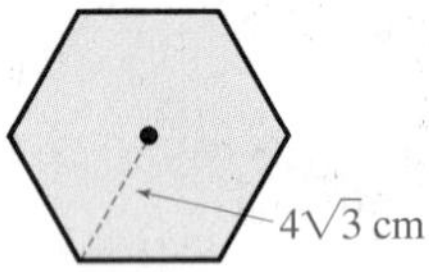

28.

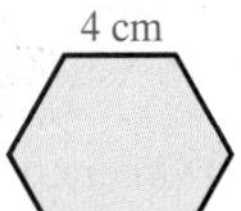

Multiple Steps *Find the measures of the angles formed by* **a.** *two consecutive radii and* **b.** *a radius and a side of the given regular polygon.*

29. pentagon

30. octagon

31. nonagon

32. dodecagon

Trigonometric Ratios *Find the area of each regular polygon. Round your answers to the nearest tenth. Trigonometric ratios may be needed. See Examples 4 and 5.*

33. pentagon with radius 3 ft

34. nonagon with radius 7 in.

35. octagon with side length 6 cm

36. decagon with side length 4 yd

37. dodecagon with radius 20 cm

38. 20-gon with radius 2 mm

39. 18-gon with perimeter 72 mm

40. 15-gon with perimeter 180 cm

NASA mission patches are often in the shape of a regular polygon. Find the area of each patch. Round to the nearest hundredth.

41. radius = 4 in.

NASA Manned Orbiting Lab Commemorative Patch

42. radius: use 2 inches

NASA Mission Patch for STS-73

A portion of a regular decagon has radii and an apothem drawn.

43. Find the measure of ∠1 and ∠2.

44. Find the measure of ∠3.

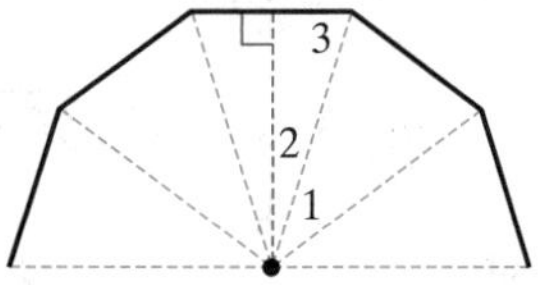

Trigonometric Ratios *Find the perimeter and area of each regular polygon to the nearest tenth.*

45.

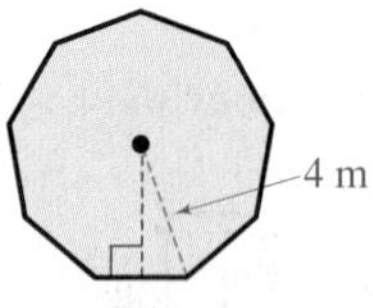

46.

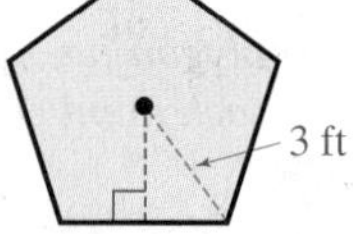

47.

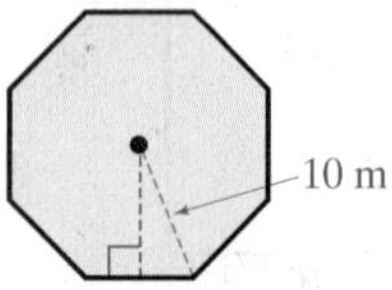

48.

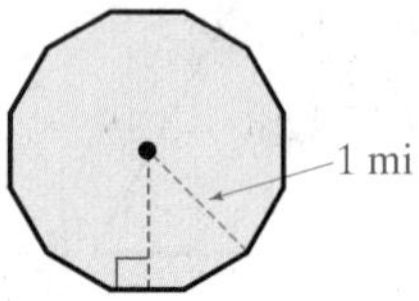

CONCEPT EXTENSIONS

49. Construction

a. Use a compass to construct a circle.

b. Construct two perpendicular diameters of the circle.

c. Construct diameters that bisect each of the four right angles.

d. Connect the consecutive points where the diameters intersect the circle. What regular polygon have you constructed?

50. Construction Complete parts **a**, **b**, and **d** of Exercise **49**.

51. The area of a regular polygon is 36 in.^{2}. Find the length of a side if the polygon has the given number of sides. Round your answer to the nearest tenth.

a. 3 **b.** 4 **c.** 6

d. **Estimation** Suppose the polygon is a pentagon. What would you expect the length of a side to be? Explain.

52. A regular hexagon has perimeter 120 m. Find its area.

53. What is the difference between a radius and an apothem?

54. Explain why the radius of a regular polygon is greater than the apothem.

Find the relationship between the side length and the apothem in each figure.

55. a square **56.** a regular hexagon

57. an equilateral triangle

58. **Find the Error** Your friend says you can use special triangles to find the apothem of any regular polygon. What is your friend's error? Explain.

59. One of the smallest space satellites ever developed has the shape of a pyramid. Each of the four faces of the pyramid is an equilateral triangle with sides about 13 cm long. What is the area of one equilateral triangular face of the satellite? Round your answer to the nearest whole number.

60. You are painting a mural of colored equilateral triangles. The radius of each triangle is 12.7 in. What is the area of each triangle to the nearest square inch?

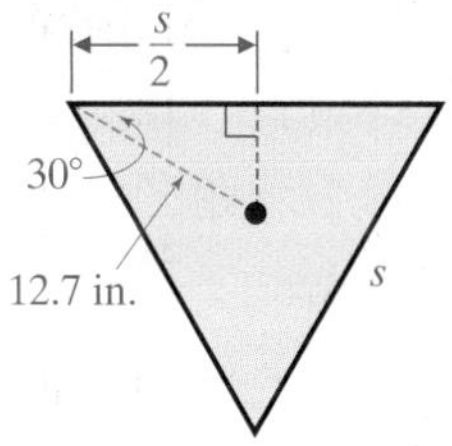

61. The gazebo in the photo is built in the shape of a regular octagon. Each side is 8 ft long, and the enclosed area is 310.4 ft^2. What is the length of the apothem?

62. Complete Exercise **61** again, but this time, the gazebo has the shape of a regular hexagon.

To find the area of an equilateral triangle, you can use the formula $A = \frac{1}{2}bh$ *or* $A = \frac{1}{2}aP$. *A third way to find the area of an equilateral triangle is to use the formula* $A = \frac{1}{4}s^2\sqrt{3}$. *Verify the formula* $A = \frac{1}{4}s^2\sqrt{3}$ *in two ways as follows:*

63. Find the area of Figure 1 using the formula $A = \frac{1}{2}bh$.

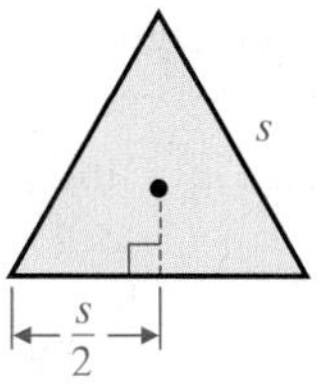

Figure 1

64. Find the area of Figure 2 using the formula $A = \frac{1}{2}aP$.

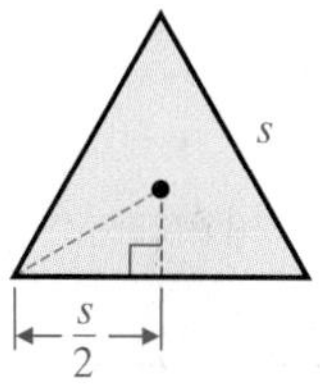

Figure 2

65. **Proof** For Example 1 on page 22, write a proof that the apothem bisects the vertex angle of an isosceles triangle formed by two radii.

66. **Proof** Prove that the bisectors of the angles of a regular polygon are concurrent and that they are, in fact, radii of the polygon. (*Hint:* For regular n-gon $ABCDE\ldots$, let P be the intersection of the bisectors of $\angle ABC$ and $\angle BCD$. Show that $\overline{DP}$ must be the bisector of $\angle CDE$.)

Coordinate Geometry *A regular octagon with center at the origin and radius 4 is graphed in the coordinate plane.*

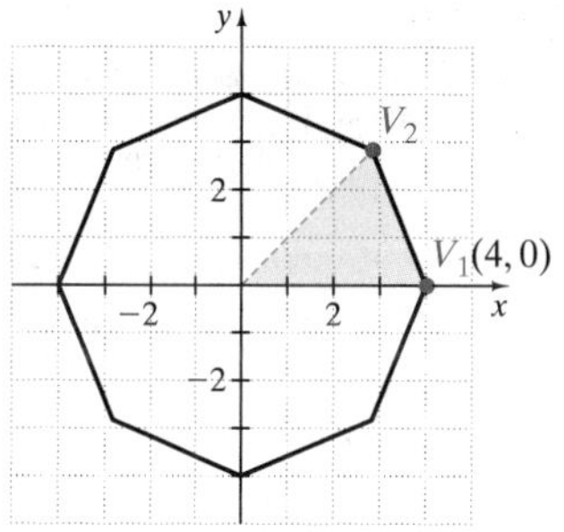

67. Since V_2 lies on the line $y = x$, its x- and y-coordinates are equal. Use the Distance Formula to find the coordinates of V_2 to the nearest tenth.

68. Use the coordinates of V_2 and the formula $A = \frac{1}{2}bh$ to find the area of ΔV_1OV_2 to the nearest tenth. Use your answer to find the area of the octagon to the nearest whole number.

REVIEW AND PREVIEW

Solve. See Section 10.2.

69. What is the area of a kite with diagonals 8 m and 11.5 m?

70. The area of a trapezoid is 42 m^2. The trapezoid has a height of 7 m and one base of 4 m. What is the length of the other base?

Find the perimeter and area of each figure. See Section 2.1.

71.

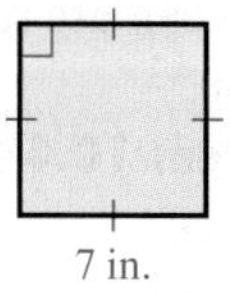

72.

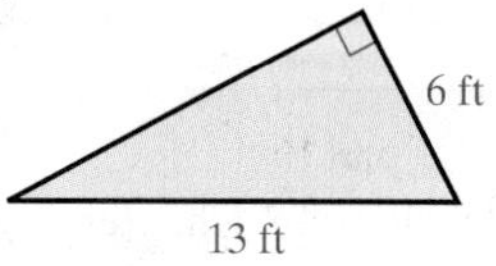

73.

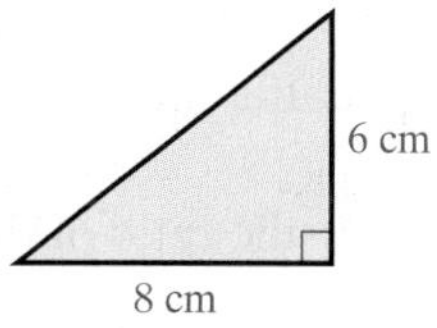

74.

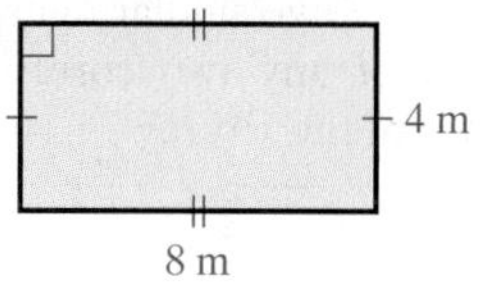

10.4 PERIMETERS AND AREAS OF SIMILAR FIGURES

OBJECTIVE

1 Find the Perimeters and Areas of Similar Figures.

VOCABULARY

- scale factor

OBJECTIVE 1 ▶ Finding Perimeters and Areas of Similar Figures. Today wide-screen, high-definition TVs and movie theaters have what we call a 16:9 aspect ratio, although many conventional TVs and some television shows still use the 4:3 aspect ratio (the ratio of length to width).

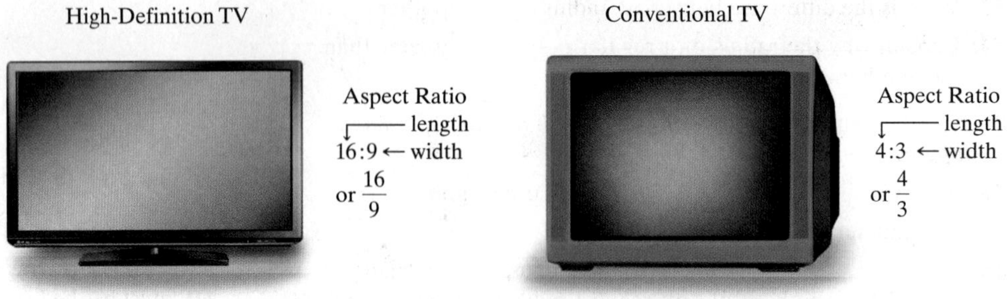

Let's compare similar rectangles whose length-to-width ratio is 4:3 or $\frac{4}{3}$, as in conventional TVs.

Helpful Hint
In each row of this table, we are concentrating on one rectangle.

Similar Rectangles (dimensions in inches)	*Simplified Ratio (length/width)*	*Perimeter (in.)*	*Area (sq in.)*
Rectangle 1 (16 by 12)	$\frac{16}{12}$ or $\frac{4}{3}$	56	192
Rectangle 2 (28 by 21)	$\frac{28}{21}$ or $\frac{4}{3}$	98	588
Rectangle 3 (40 by 30)	$\frac{40}{30}$ or $\frac{4}{3}$	140	1200

Helpful Hint
In each row of this table, we are comparing two rectangles.

Compare Similar Rectangles	*Scale Factor* $\left(\frac{\text{length}}{\text{length}} \text{ or } \frac{\text{width}}{\text{width}}\right)$	*Ratio of Perimeters*	*Ratio of Areas*
Rectangle 2 to Rectangle 1	$\left(\frac{28}{16}\right) = \frac{7}{4}$	$\left(\frac{98}{56}\right) = \frac{7}{4}$	$\frac{49}{16}$ or $\frac{7^2}{4^2}$
Rectangle 3 to Rectangle 2	$\left(\frac{40}{28}\right) = \frac{10}{7}$	$\left(\frac{140}{98}\right) = \frac{10}{7}$	$\frac{100}{49}$ or $\frac{10^2}{7^2}$
Rectangle 3 to Rectangle 1	$\left(\frac{40}{16}\right) = \frac{5}{2}$	$\left(\frac{140}{56}\right) = \frac{5}{2}$	$\frac{25}{4}$ or $\frac{5^2}{2^2}$

Recall from Section 7.3 that the **scale factor** of two similar figures is the ratio of any two corresponding sides of the figures.

In the second table above, we compared the perimeters and areas of similar rectangles.

We can use ratios to compare the perimeters and areas of similar figures, in general.

Helpful Hint
Recall from Chapter 7 and the second table above that the scale factor of two similar figures is the ratio of any two corresponding sides of the figures.

Theorem 10.4-1 Perimeters and Areas of Similar Figures

If the scale factor of two similar figures is $\frac{a}{b}$, then

1. the ratio of their perimeters is $\frac{a}{b}$ and
2. the ratio of their areas is $\frac{a^2}{b^2}$.

EXAMPLE 1 Finding Scale Factor, Side Lengths, and Ratios of Similar Polygons

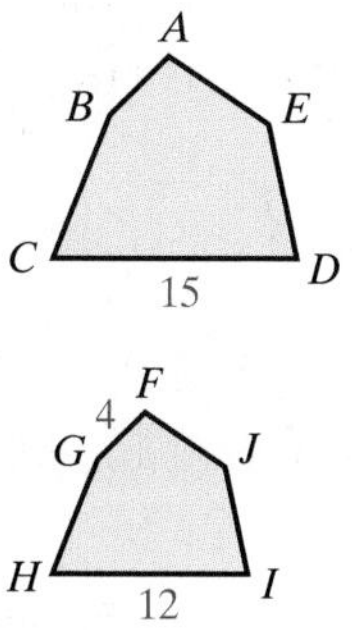

Figure $ABCDE \sim$ figure $FGHIJ$.

a. Find the scale factor of the larger figure to the smaller figure.

b. Given the scale factor, find AB.

c. Find the ratio of the perimeters of the larger figure to the smaller figure.

d. Find the ratio of the areas of the larger figure to the smaller figure.

Solution

a. To find the scale factor, identify two corresponding sides.

$$\text{scale factor} = \frac{CD}{HI} = \frac{15}{12} = \frac{5}{4}$$

b. To find AB, notice that $\frac{AB}{GF}$ must equal $\frac{5}{4}$, or $\frac{AB}{GF} = \frac{5}{4}$.

By substituting 4 for GF, we have $\frac{AB}{4} = \frac{5}{4}$. Thus $AB = 5$ units.

c. The ratio of the perimeters is the same as the scale factor, or $\frac{5}{4}$ or 5:4.

d. The ratio of the areas is $\frac{5^2}{4^2} = \frac{25}{16}$ or 25:16. □

PRACTICE 1 Figure $KLMN \sim$ figure $PQRS$.

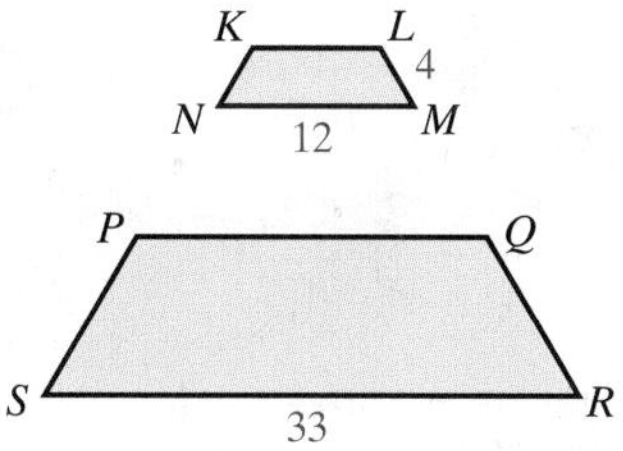

a. Find the scale factor of the smaller figure to the larger figure.

b. Given the scale factor, find QR.

c. Find the ratio of the perimeters of the smaller figure to the larger figure.

d. Find the ratio of the areas of the smaller figure to the larger figure.

When we know the area of one of two similar polygons, we can use a proportion to find the area of the other polygon.

EXAMPLE 2 Finding Areas Using Similar Figures

Multiple Choice The area of the smaller regular pentagon is about 27.5 cm^2. Choose the best approximation for the area of the larger regular pentagon.

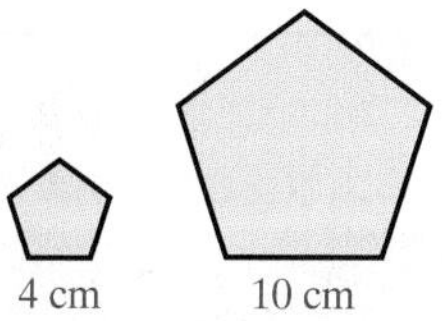

a. 11 cm^2 **b.** 69 cm^2 **c.** 172 cm^2 **d.** 275 cm^2

Solution Regular pentagons are similar because all angles measure 108° $\left[\frac{(5-2)\cdot 180°}{5}\right]$ and all sides in each pentagon are congruent. Here, the ratio of corresponding side lengths is $\frac{4}{10}$, or $\frac{2}{5}$. The ratio of the areas is $\frac{2^2}{5^2}$, or $\frac{4}{25}$.

To find the area of the smaller pentagon, write a proportion using the ratio of the areas.

$$\text{Ratio of areas}\begin{cases}\dfrac{4}{25} = \dfrac{27.5}{A} & \leftarrow \text{Area of smaller pentagon} \\ & \leftarrow \text{Area of larger pentagon}\end{cases}$$

$$4A = 687.5 \quad \text{Set cross products equal to each other.}$$

$$A = \frac{687.5}{4} \quad \text{Divide both sides by 4.}$$

$$A = 171.875 \quad \text{Simplify.}$$

The area of the larger pentagon is about 172 cm^2. The correct answer is **c**. □

PRACTICE 2 The scale factor of two similar parallelograms is $\frac{3}{4}$. The area of the larger parallelogram is 96 in.^{2}. What is the area of the smaller parallelogram?

EXAMPLE 3 Applying Area Ratios

During the summer, a group of high school students cultivated a plot of city land and harvested 13 bushels of vegetables, which they donated to a food pantry. Next summer, the city will let them use a larger plot of land that is similar to the previous plot of land. In the new plot, each dimension is 2.5 times the corresponding dimension of the original plot. How many bushels can the students expect to harvest next year?

Solution The ratio of the dimensions is $\frac{2.5}{1}$ or 2.5:1. Thus the ratio of the areas is $\frac{(2.5)^2}{1^2}$ or $(2.5)^2:1^2$, which is $\frac{6.25}{1}$ or 6.25:1. With 6.25 times as much land next year, the students can expect to harvest 6.25(13), or about 81, bushels. □

PRACTICE 3 The scale factor of the dimensions of two similar pieces of window glass is 3:5. The smaller piece costs \$2.50. How much should the larger piece cost?

If we know the ratio of the areas of two similar figures, we can work backward to find the ratio of their perimeters.

EXAMPLE 4 Finding Perimeter Ratios

The triangles at the right are similar. What is the scale factor? What is the ratio of their perimeters?

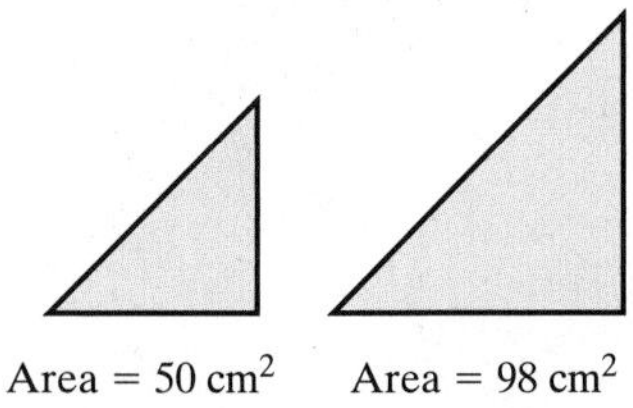

Solution Since we know the areas, we can write the area ratio. From there, we can find the perimeter ratio.

$$\frac{a^2}{b^2} = \frac{50}{98} \quad \text{Use } a^2:b^2 \text{ for the ratio of the areas.}$$

$$\frac{a^2}{b^2} = \frac{25}{49} \quad \text{Simplify the fraction on the right side.}$$

$$\frac{a}{b} = \frac{5}{7} \quad \text{Take the positive square root of each side of the proportion.}$$

Thus, the ratio of the perimeters, which equals the scale factor, is $\frac{5}{7}$ or 5:7. □

PRACTICE 4 The areas of two similar rectangles are 1875 ft^2 and 135 ft^2. What is the ratio of their perimeters?

VOCABULARY & READINESS CHECK

Fill in the Blank.

1. The ____________ of two similar figures is the ratio of any two corresponding sides of the figures.

If the scale factor of two similar figures is $\frac{x}{y}$, then

2. the ratio of their perimeters is ____________, and

3. the ratio of their areas is ____________.

10.4 EXERCISE SET MyMathLab®

Multiple Steps *The figures in each pair are similar. Compare the first figure to the second.*

a. *Find the scale factor.* **b.** *Give the ratio of the perimeters.*

c. *Give the ratio of the areas. See Example 1.*

1.

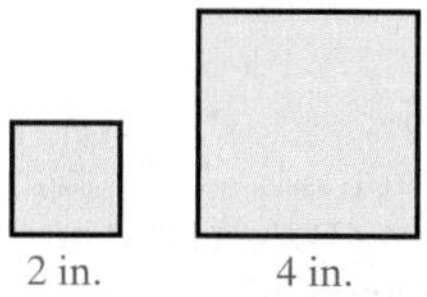

2.

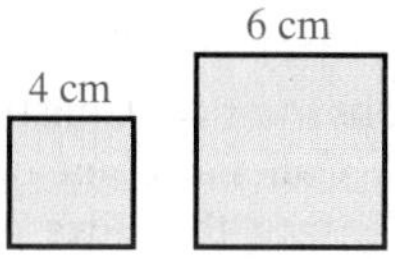

3.

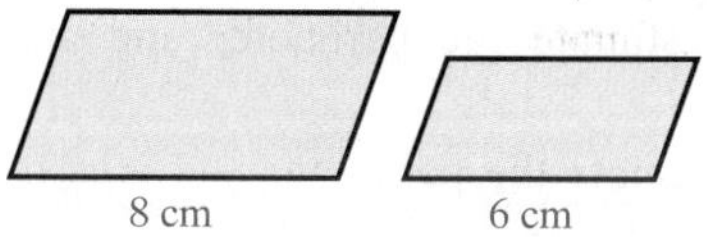

4.

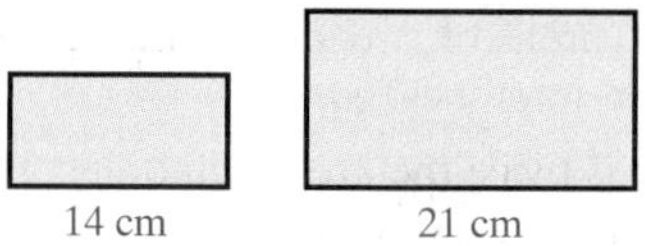

5.

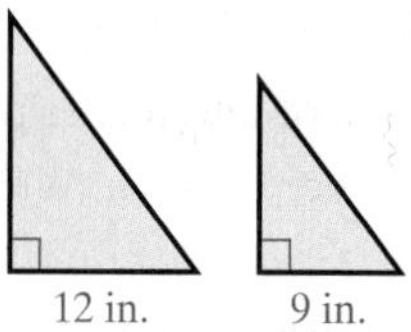

6.

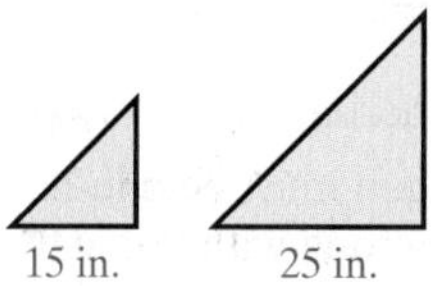

The figures in each pair are similar. The area of one figure is given. Find the area of the other figure to the nearest whole number. See Example 2.

7.

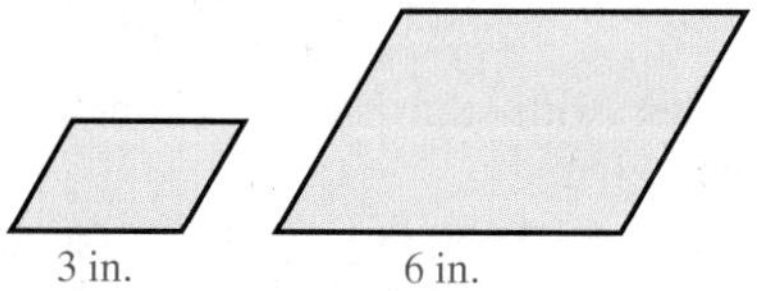

Area of smaller parallelogram = 6 in.^{2}

8.

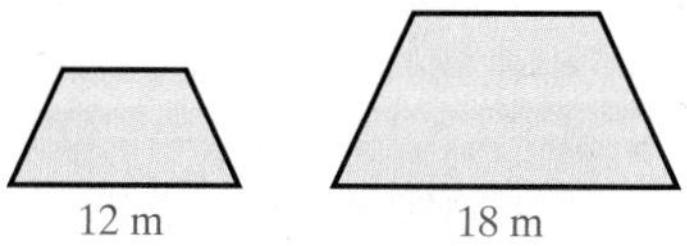

Area of larger trapezoid = 121 m^2

9.

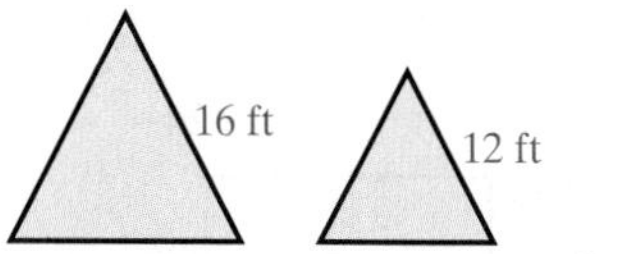

Area of larger triangle = 105 ft^2

10.

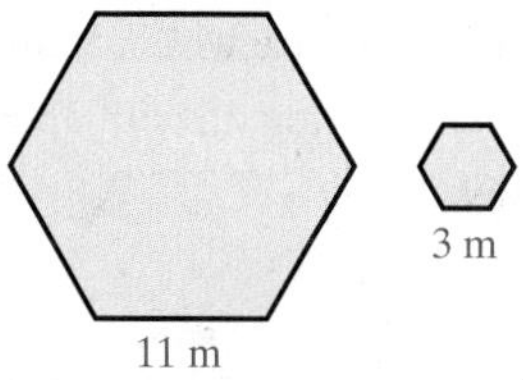

Area of smaller hexagon = 23 m^2

11. Given two similar triangles with scale factor $\frac{4}{3}$ or 4:3, if the area of the smaller triangle is about 39 ft^2, what is the area of the larger triangle to the nearest tenth?

12. Given two similar hexagons with scale factor $\frac{9}{5}$ or 9:5, if the area of the smaller hexagon is 47 sq m, what is the area of the larger hexagon to the nearest tenth?

Solve. See Example 3.

13. The scale factor of the dimensions of two similar wood floors is 4:3. It costs $216 to refinish the smaller wood floor. At that rate, how much would it cost to refinish the larger wood floor?

14. An empty rectangular lot of land costs $24,000. Another empty lot in the same neighborhood is three times as long and three times as wide. How much would you expect to pay for this lot of land?

15. An embroidered placemat costs $3.95. An embroidered tablecloth is similar to the placemat, but four times as long and four times as wide. How much would you expect to pay for the tablecloth?

16. The scale factor of the dimensions of two similar tile floors is 5:8. It costs $500 to clean the tile and grout of the larger floor. At this rate, how much would it cost to clean the smaller floor?

Find the scale factor and the ratio of perimeters for each pair of similar figures. See Example 4.

17. two regular octagons with areas 4 ft^2 and 16 ft^2

18. two triangles with areas 75 m^2 and 12 m^2

19. two trapezoids with areas 49 cm^2 and 9 cm^2

20. two parallelograms with areas 18 in.^{2} and 32 in.^{2}

21. two equilateral triangles with areas $16\sqrt{3}$ ft^2 and $\sqrt{3}$ ft^2

22. two circles with areas 2π cm^2 and 200π cm^2

The scale factor of two similar polygons is given. Find the ratio of their perimeters and the ratio of their areas.

23. 3:1 **24.** 2:5 **25.** $\frac{2}{3}$ **26.** $\frac{7}{4}$ **27.** 6:1 **28.** 9:1

29. The area of a regular decagon is 50 cm^2. Choose the area of a regular decagon with sides four times the sides of the smaller decagon.

a. 200 cm^2 **b.** 500 cm^2 **c.** 800 cm^2 **d.** 2000 cm^2

30. The area of a regular nonagon is 120 sq ft. Choose the area of a regular nonagon with sides five times the sides of the smaller nonagon.

a. 3000 sq ft **b.** 600 sq ft **c.** 24 sq ft **d.** 1080 sq ft

Find the values of x and y when the smaller triangle shown here has the given area. Leave your answer in the simplest radical form.

31. 3 cm^2 **32.** 6 cm^2
33. 12 cm^2 **34.** 16 cm^2
35. 24 cm^2 **36.** 48 cm^2

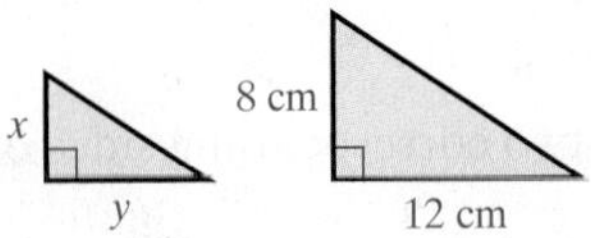

Multiple Steps *Compare the blue outline larger figure to the red outline smaller figure. Find the ratios of* **a.** *their perimeters (or circumferences) and* **b.** *their areas.*

37.

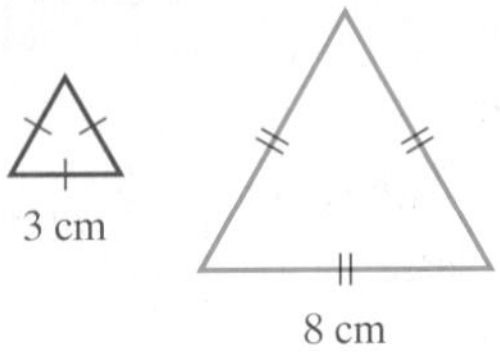

38.

39.

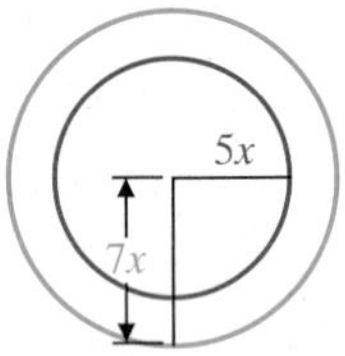

40.

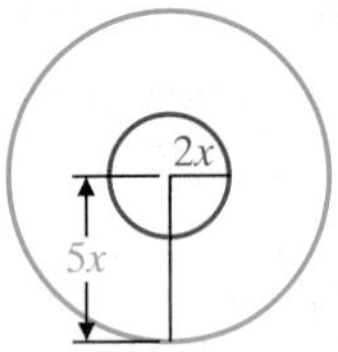

CONCEPT EXTENSIONS

Solve.

41. Find the area of a regular hexagon with sides 2 cm long. Leave your answer in simplest radical form.

42. Use your answer to Exercise **41** and Theorem 10.4-1 to find the areas of the regular hexagons shown at the right.

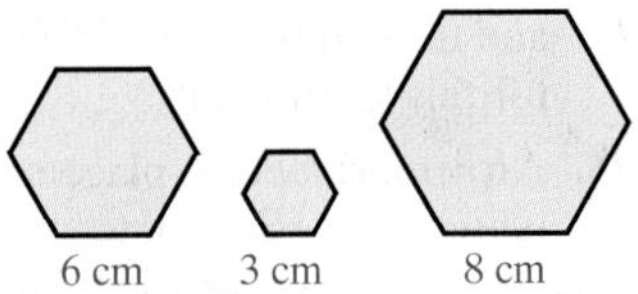

43. For some medical imaging, the scale of the image is 3:1. That means that if an image is 3 cm long, the corresponding length on the person's body is 1 cm. Find the actual area of a lesion if its image has area 2.7 cm^2.

44. The longer sides of a parallelogram are 5 m. The longer sides of a similar parallelogram are 15 m. The area of the smaller parallelogram is 28 m^2. What is the area of the larger parallelogram?

Two similar rectangles have areas 27 in.^{2} and 48 in.^{2}. The length of one side of the larger rectangle is 16 in.

45. What are the dimensions of the smaller rectangle?

46. What are the dimensions of the larger rectangle?

In ΔRST, $RS = 2$ cm, $ST = 5$ cm, and $RT = 4$ cm.

47. Construction Construct ΔRST using the exact measurements.

48. Construction Choose a convenient scale factor. Then use a straight edge and compass to draw $\Delta R'S'T' \sim \Delta RST$.

49. Construction Construct an altitude of $\Delta R'S'T'$ and measure its length. Find the area of $\Delta R'S'T'$.

50. Estimate the area of ΔRST.

The enrollment at an elementary school is going to increase from 200 students to 395 students. A parents' group is planning to increase the rectangular 100-ft-by-200-ft playground area to a larger, similar rectangular area that is 200 ft by 400 ft.

51. Find the scale factor of the two rectangles.

52. Find the perimeter of each rectangle.

53. Find the area of each rectangle.

54. What would you tell the parents' group when they ask your opinion about whether the new playground will be large enough?

A surveyor measured one side and two angles of a field, as shown in the diagram. Use the diagram for Exercises 55–58.

55. Construction Use a ruler and a protractor to construct a similar triangle, $\Delta A'B'C'$. (Use $\frac{1}{4}$ in. = 10 yd as your scale.)

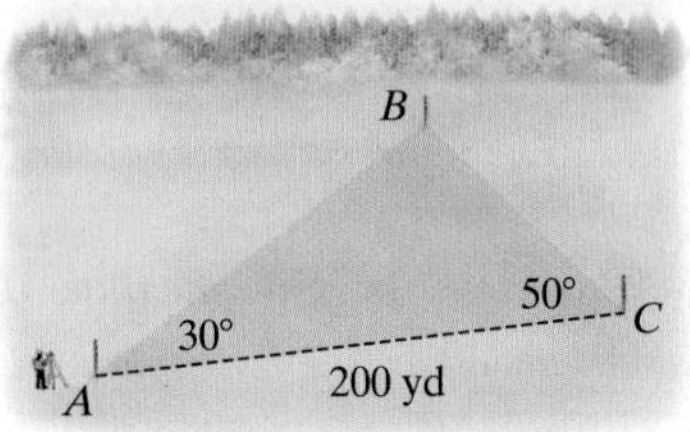

56. Construction Construct an altitude of $\Delta A'B'C'$ from B'. Call the point of intersection with side $A'C'$, point D'. Measure the sides and altitude of $\Delta A'B'C'$. Round each measurement to the nearest tenth of an inch.

57. Use Exercise **56** to estimate the perimeter and area of $\Delta A'B'C'$.

58. Use Exercise **57** to estimate the perimeter and area of the field, ΔABC.

59. How does the ratio of the areas of two similar figures compare to the ratio of their perimeters? Explain.

60. The area of one rectangle is twice the area of another. What is the ratio of their perimeters? How do you know?

61. Find the Error Your friend says that since the ratio of the perimeters of two polygons is $\frac{1}{2}$, the area of the smaller polygon must be one-half the area of the larger polygon. What is wrong with this statement? Explain.

62. How is the relationship between the areas of two congruent figures different from the relationship between the areas of two similar figures?

Complete each statement with "always," "sometimes," or "never." Justify your answers.

63. Two similar rectangles with the same perimeter are ___?___ congruent.

64. Two rectangles with the same area are ___?___ similar.

65. Two rectangles with the same area and different perimeters are ___?___ similar.

66. Similar figures ___?___ have the same area.

REVIEW AND PREVIEW

Find the area of each regular polygon. See Section 10.3.

67. a square with a 5-cm radius

68. a pentagon with apothem 13.8 and side length 20

69.

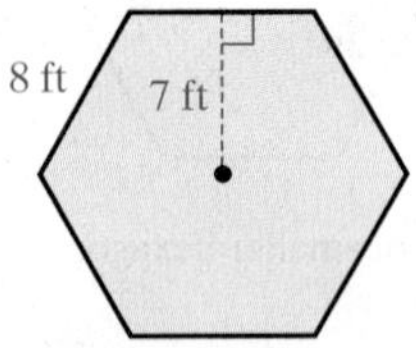

70.

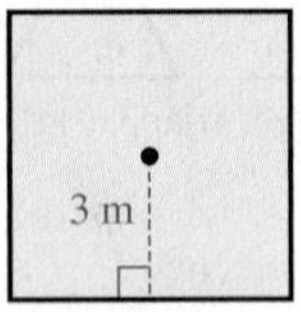

10.5 ARC MEASURES, CIRCUMFERENCES, AND ARC LENGTHS OF CIRCLES

OBJECTIVES

1 Find the Measures of Central Angles and Arcs.

2 Find the Circumference and Arc Length.

OBJECTIVE 1 ▶ Finding Measures of Central Angles and Arcs. In a plane, a **circle** is the set of all points equidistant from a given point called the **center**. We name a circle by its center. Circle $P(\odot P)$ is shown below.

A **diameter** is a segment that contains the center of a circle and has both endpoints on the circle. A **radius** is a segment that has one endpoint at the center and the other endpoint on the circle. **Congruent circles** have congruent radii. A **central angle** is an angle whose vertex is the center of the circle.

VOCABULARY

- circle
- center
- diameter
- radius
- congruent circles
- central angle
- semicircle
- minor arc
- major arc
- adjacent arcs
- circumference
- pi
- concentric circles
- arc length
- congruent arcs

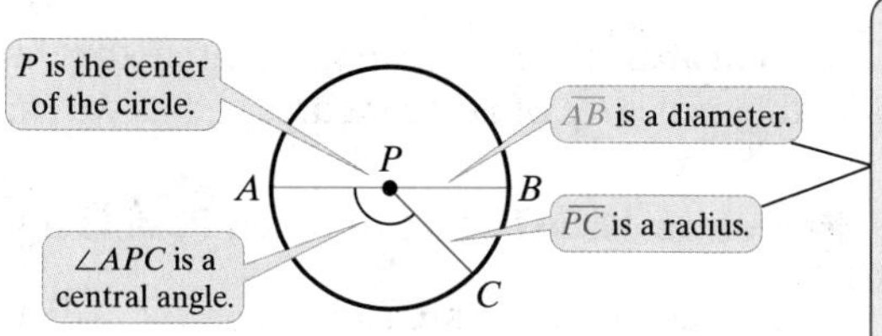

▶ Helpful Hint

Can you see that the diameter is twice the circle's radius?

$d = 2r$ or $\frac{d}{2} = r$

- A circle has many diameters and many radii when discussing segments.
- A circle has 1 diameter and 1 radius when discussing lengths.

We can find the length of part of a circle's circumference by relating it to an angle in the circle.

An arc is a part of a circle. One type of arc, a **semicircle**, is half of a circle. A **minor arc** is smaller than a semicircle. A **major arc** is larger than a semicircle. We name a minor arc by its endpoints and a major arc or a semicircle by its endpoints *and* another point on the arc.

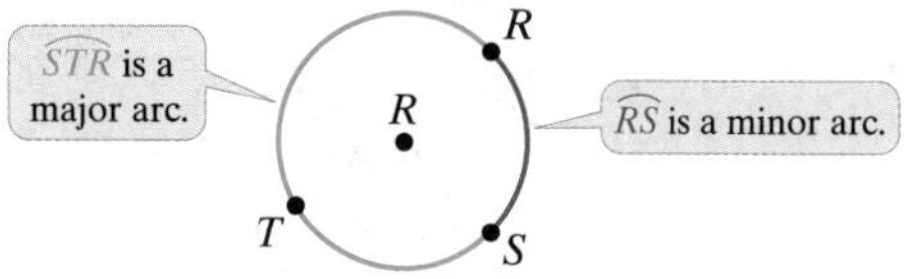

▶ Helpful Hint

Notice that diameter and radius each have two definitions.

- A diameter or a radius of a circle are each segments.
- The diameter or the radius are each lengths.

Let's practice naming arcs.

EXAMPLE 1 **Naming Arcs**

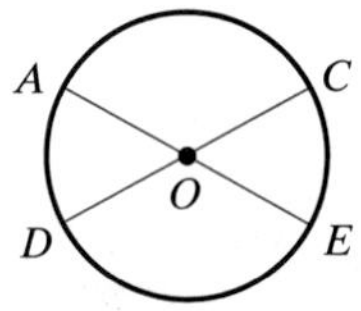

a. What are the minor arcs of $\odot O$?

b. What are the semicircles of $\odot O$?

c. What are the major arcs of $\odot O$ that contain point A?

Solution

a. The minor arcs are $\widehat{AD}$, $\widehat{CE}$, $\widehat{AC}$, and $\widehat{DE}$.

b. The semicircles are $\widehat{ACE}$, $\widehat{CED}$, $\widehat{EDA}$, and $\widehat{DAC}$.

c. The major arcs that contain point A are $\widehat{ACD}$, $\widehat{CEA}$, $\widehat{EDC}$, and $\widehat{DAE}$. □

▶ Helpful Hint

Make sure a major arc is named by its endpoints and a third point so that there is no confusion as to the arc named.

PRACTICE 1

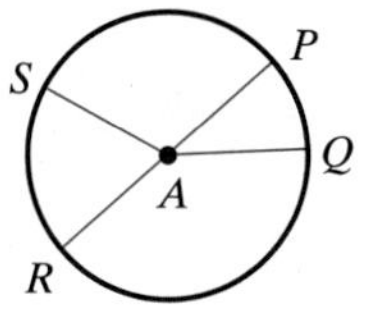

a. What are the minor arcs of $\odot A$?

b. What are the semicircles of $\odot A$?

c. What are the major arcs of $\odot A$ that contain point Q?

▶ Helpful Hint

Remember—

- A minor arc is smaller than a semicircle.
- A major arc is larger than a semicircle.

Now let's see how we measure arcs.

Arc Measure

Arc Measure	Example
• minor arc—The measure of a minor arc is equal to the measure of its corresponding central angle. • major arc—The measure of a major arc is the measure of the related minor arc subtracted from 360°. • semicircle—The measure of a semicircle is 180°.	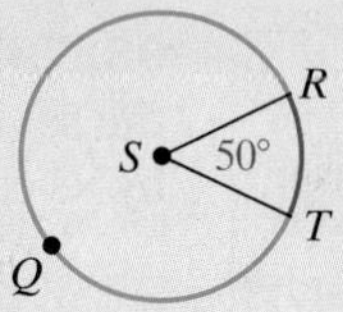 $m\overset{\frown}{RT} = m\angle RST = 50°$ $m\overset{\frown}{TQR} = 360° - m\overset{\frown}{RT}$ $= 310°$

Adjacent arcs are arcs of the same circle that have exactly one point in common. We can add the measures of adjacent arcs just as we can add the measures of adjacent angles.

Postulate 10.5-1 Arc Addition Postulate

The measure of the arc formed by two adjacent arcs is the sum of the measures of the two arcs.

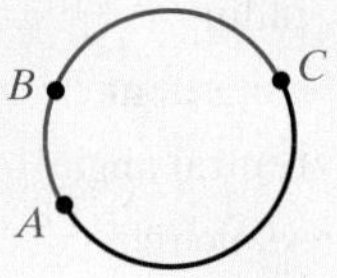

$$m\overset{\frown}{ABC} = m\overset{\frown}{AB} + m\overset{\frown}{BC}$$

EXAMPLE 2 **Finding the Measures of Arcs**

What is the measure of each arc in $\odot O$?

a. $\overset{\frown}{BC}$ **b.** $\overset{\frown}{BD}$ **c.** $\overset{\frown}{ABC}$ **d.** $\overset{\frown}{AB}$

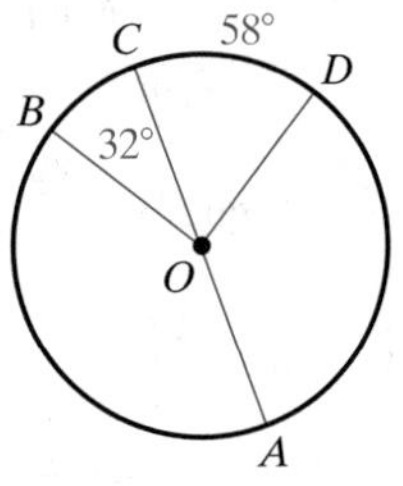

Solution

a. $m\overset{\frown}{BC} = m\angle BOC = 32°$

b. $m\overset{\frown}{BD} = m\overset{\frown}{BC} + m\overset{\frown}{CD}$

$m\overset{\frown}{BD} = 32° + 58° = 90°$

c. $\overset{\frown}{ABC}$ is a semicircle.

$m\overset{\frown}{ABC} = 180°$

d. $m\overset{\frown}{AB} = 180° - 32° = 148°$

PRACTICE

2 What is the measure of each arc in $\odot C$?

a. $m\overset{\frown}{PR}$ **b.** $m\overset{\frown}{RS}$

c. $m\overset{\frown}{PRQ}$ **d.** $m\overset{\frown}{PQR}$

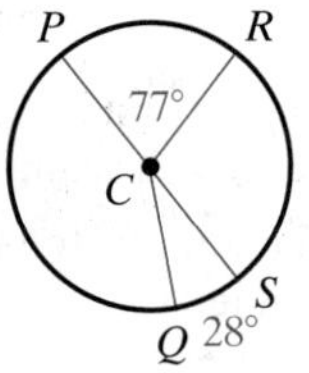

OBJECTIVE 2 ▶ Finding Circumferences and Arc Lengths. Recall from Section 2.1 that the **circumference** of a circle is the distance around the circle. Also, the number **pi** (π) is the circumference of any circle divided by its diameter.

If you did not do so in Chapter 2, try the following experiment. Take any can and measure its circumference and its diameter.

▶ Helpful Hint

Another way to describe π is as a ratio (or fraction).

$$\pi = \frac{\text{circumference}}{\text{diameter}}$$

The can in the previous figure has a circumference of about 23.5 centimeters and a diameter of about 7.5 centimeters. Now divide the circumference by the diameter.

$$\frac{\text{circumference}}{\text{diameter}} = \frac{23.5 \text{ cm}}{7.5 \text{ cm}} \approx 3.13$$

Try this with other sizes of cylinders and circles—you should always get a number close to 3.1. The exact ratio of circumference to diameter is the number π.

Theorem 10.5-2 Circumference of a Circle

The circumference of a circle is π times the diameter.

$$C = \pi d \text{ or } C = 2\pi r$$

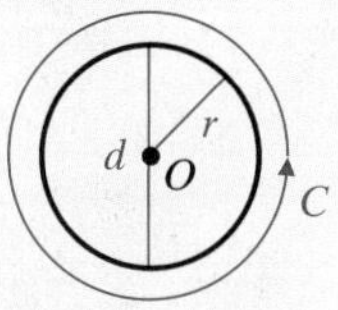

The number π is irrational, so we cannot write it as a terminating or repeating decimal. To approximate π, we can use 3.14, $\frac{22}{7}$, or the π key on a calculator. Thus,

$$\pi \approx 3.14 \text{ or } \pi \approx \frac{22}{7} \text{ or } \pi \approx \pi \text{ key on a calculator}$$

Coplanar circles that have the same center are **concentric circles**.

Let's try an application of concentric circles.

Diamond is the hardest material known to man, so it is probably no surprise that diamond drill bits are used in the fields of interstate construction, space construction, and tile and glass drilling, just to name a few.

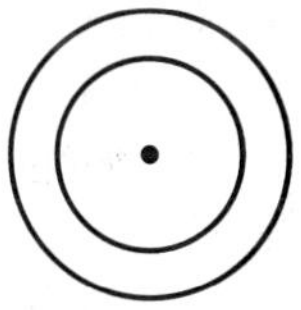

Concentric circles

EXAMPLE 3 Finding Difference in Circumferences

A 6-inch-diameter drill blade has a $5\frac{1}{4}$-inch-diameter hole. Approximate the difference in the outer and inner circumferences.

Solution To find the difference in the circumferences in the blade, let's find each circumference, and then subtract. If the outer diameter is 6 in., then the outer radius is 3 in.

$$\begin{aligned} \text{outer blade circumference} &= 2\pi r \\ &= 2\pi \cdot 3 \text{ in.} \\ &= 6\pi \text{ in.} \\ &\approx 18.850 \text{ in.} \end{aligned}$$

If the inner diameter is $5\frac{1}{4}$ in., or $\frac{21}{4}$ in., then the inner radius is $\frac{1}{2} \cdot \frac{21}{4} = \frac{21}{8}$ in.

$$\begin{aligned} \text{inner blade circumference} &= 2\pi r \\ &= 2\pi \cdot \frac{21}{8} \text{ in.} \\ &\approx 16.493 \text{ in.} \end{aligned}$$

The difference in circumferences is $18.850 - 16.493 \approx 2.4$ in. □

PRACTICE 3 A 5-inch-diameter drill blade has a $4\frac{1}{4}$-inch-diameter hole. Approximate the difference in the circumferences.

The measure of an arc is in degrees while the **arc length** is a fraction of a circle's circumference. An arc of 60° represents $\frac{60°}{360°}$ or $\frac{1}{6}$ of the circle. Its arc length is $\frac{1}{6}$ the circumference of the circle. This observation suggests the following theorem.

Theorem 10.5-3 Arc Length

The length of an arc of a circle is the product of the ratio $\frac{\text{measure of the arc}}{360°}$ and the circumference of the circle.

$$\text{length of } \widehat{AB} = \frac{m\widehat{AB}}{360°} \cdot 2\pi r \quad \text{or}$$

$$= \frac{m\widehat{AB}}{360°} \cdot \pi d$$

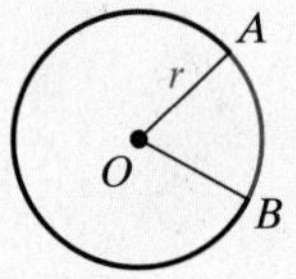

Helpful Hint
- arc—measured in degrees (same as measure of corresponding central angle)
- arc length—measured in given units

EXAMPLE 4 Finding Arc Length

What is the length of each arc shown in red? Leave your answer in terms of π.

a.

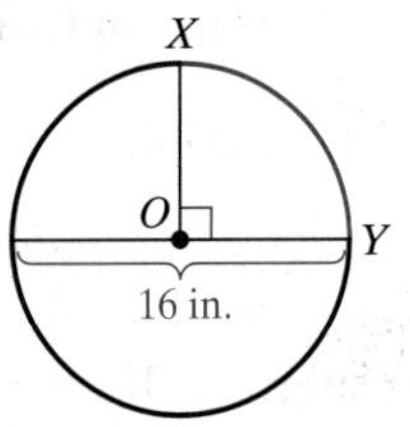

b.

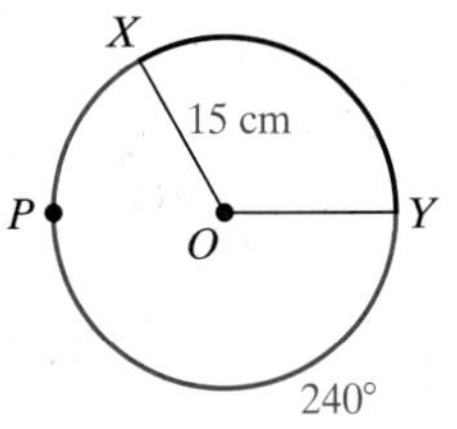

Solution

a. $\text{length of } \widehat{XY} = \frac{m\widehat{XY}}{360°} \cdot \pi d$ Use a formula for arc length.

$= \frac{90°}{360°} \cdot \pi(16)$ Substitute.

$= 4\pi$ in. Simplify.

b. $\text{length of } \widehat{XPY} = \frac{m\widehat{XPY}}{360°} \cdot 2\pi r$ Use a formula for arc length.

$= \frac{240°}{360°} \cdot 2\pi(15)$ Substitute.

$= 20\pi$ cm Simplify. □

PRACTICE 4 What is the length of a semicircle with radius 1.3 m? Leave your answer in terms of π.

It is possible for two arcs of different circles to have the same measure but different lengths. It is also possible for two arcs of different circles to have the same length but different measures. **Congruent arcs** are arcs that have the same measure *and* are in the same circle or in congruent circles.

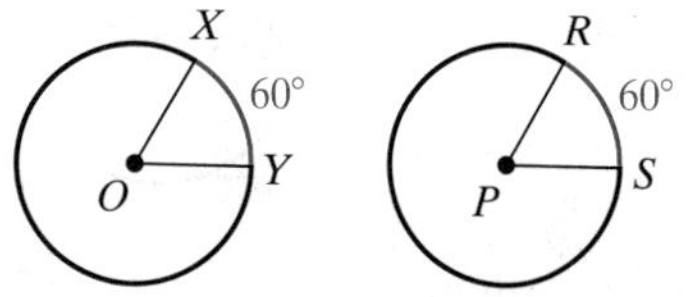

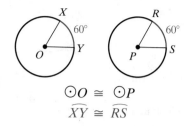

$\odot O \cong \odot P$
$\overset{\frown}{XY} \cong \overset{\frown}{RS}$

VOCABULARY & READINESS CHECK

Word Bank *Use the choices below to fill in each blank. Some choices may be used more than once and some not at all.*

congruent	minor	circle	radius	concentric
central	major	center	circumference	arc length
semicircle	adjacent	diameter	pi	degrees

Exercises 1–6 have to do with points in a single plane.

1. A(n) ____________ is a segment that contains the center of a circle and has both endpoints on the circle.
2. A(n) ____________ is a segment that has one endpoint at the center of a circle and the other endpoint on the circle.
3. A(n) ____________ is the set of all points equidistant from a point called the ____________.
4. Circles with congruent radii or congruent diameters are called ____________ circles.
5. A(n) ____________ angle of a circle is one whose vertex is the center.
6. Coplanar circles that have the same center are ____________ circles.

Exercises 7–12 have to do with arcs.

7. We measure an arc in ____________ while ____________ is a fraction of the circle's circumference.
8. A(n) ____________ is half of a circle.
9. A(n) ____________ arc is smaller than its corresponding semicircle.
10. A(n) ____________ arc is larger than its corresponding semicircle.
11. ____________ arcs are arcs from the same circle and have exactly one point in common.
12. ____________ arcs are arcs that have the same measure and are in the same circle or congruent circles.

10.5 EXERCISE SET MyMathLab®

Name the following in ⊙O. See Example 1.

1. the semicircles with endpoint F
2. the semicircles with endpoint B
3. the minor arcs contained within $\overset{\frown}{FEC}$
4. the minor arcs contained within $\overset{\frown}{FBC}$
5. the major arcs containing point B
6. the major arcs containing point E

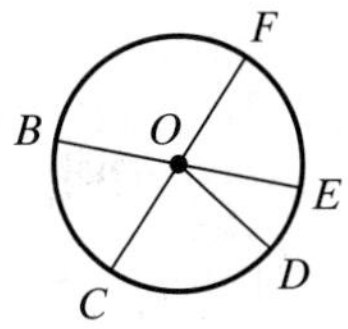

Find the measure of each arc in ⊙P. See Example 2.

7. $\overset{\frown}{TC}$ **8.** $\overset{\frown}{TBD}$ **9.** $\overset{\frown}{BTC}$ **10.** $\overset{\frown}{TCB}$
11. $\overset{\frown}{CD}$ **12.** $\overset{\frown}{DB}$ **13.** $\overset{\frown}{TCD}$ **14.** $\overset{\frown}{CBD}$
15. $\overset{\frown}{TDC}$ **16.** $\overset{\frown}{BCD}$ **17.** $\overset{\frown}{TB}$ **18.** $\overset{\frown}{BC}$

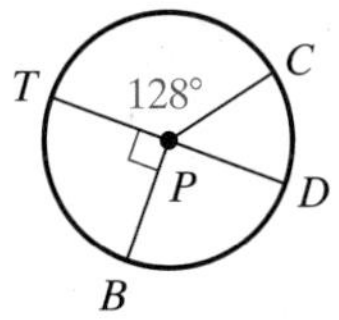

Find the circumference of each circle. Leave your answer in terms of π. See Example 3.

19.
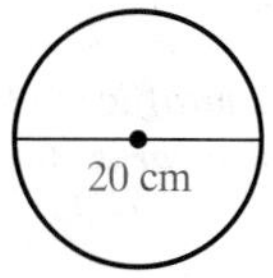

20.
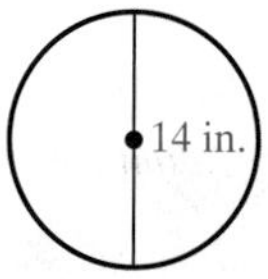

21.
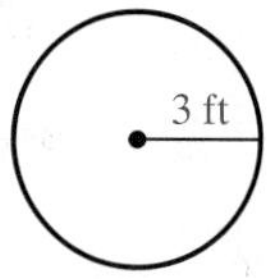

22.
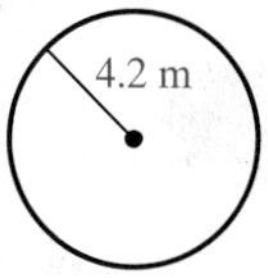

Mixed Practice *Use ⊙P at the right to answer each question. See Examples 1–3.*

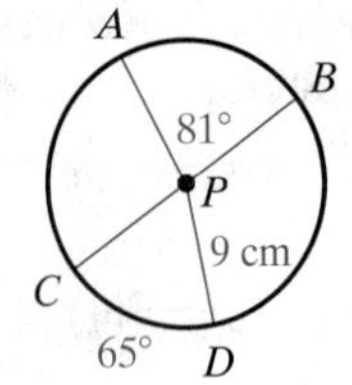

23. What is the name of a minor arc whose measure is less than 81°?

24. What is the name of a minor arc whose measure is greater than 81°?

25. What is the name of a major arc whose measure is 279°?

26. What is the name of a major arc whose measure is 261°?

27. What is the name of a semicircle that contains point D?

28. What is the name of a semicircle that contains point A?

29. What is $m\widehat{AB}$?

30. What is $m\widehat{CD}$?

31. What is the circumference of ⊙P in terms of π?

32. Approximate the circumference of ⊙P using the approximation 3.14 for π.

33. What is the length of $\widehat{BD}$?

34. What is the length of $\widehat{AC}$?

Find the length of each arc shown in red. Leave your answer in terms of π. See Example 4.

35.

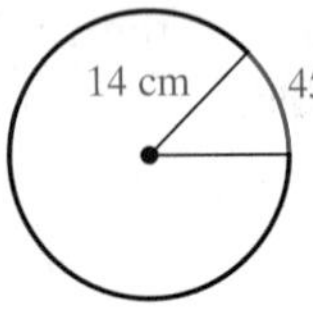

36.

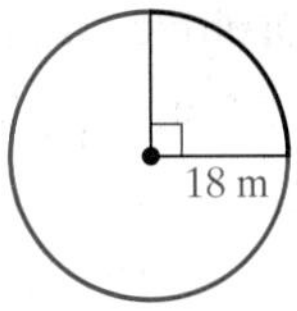

37.

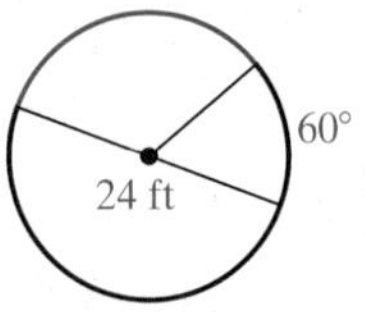

38.

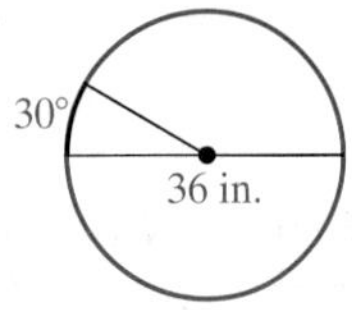

39.

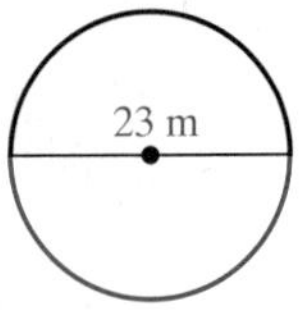

40.

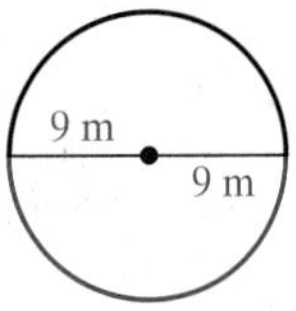

Find each indicated measure for ⊙O.

41. $m\angle EOF$

42. $m\widehat{EJH}$

43. $m\widehat{FH}$

44. $m\angle FOG$

45. $m\widehat{JEG}$

46. $m\widehat{HFJ}$

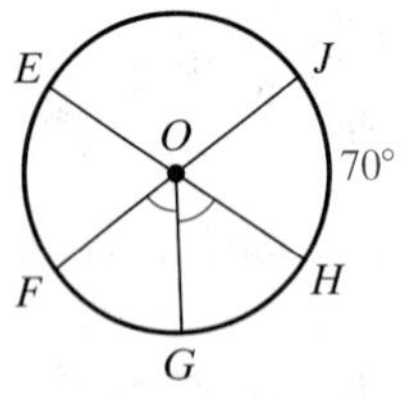

Solve. See Example 3. Five streets come together at a traffic circle, as shown below. The diameter of the circle traveled by a car is 200 ft.

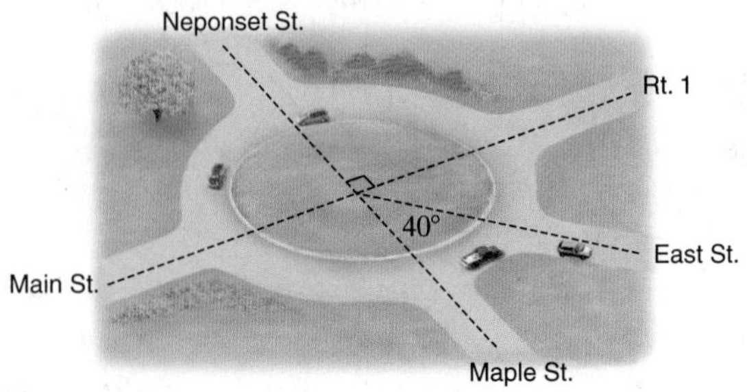

47. If traffic travels counterclockwise, what is the approximate distance from East St. to Neponset St.?

a. 227 ft **b.** 244 ft **c.** 401 ft **d.** 384 ft

48. If traffic travels clockwise, what is the approximate distance from East St. to Main St.?

a. 227 ft **b.** 244 ft **c.** 401 ft **d.** 384 ft

49. The wheel of a compact car has a 25-in. diameter. The wheel of a pickup truck has a 31-in. diameter. To the nearest inch, how much farther does the pickup truck wheel travel in one revolution than the compact car wheel?

50. A hamster wheel has a 7-in. diameter. How many feet will a hamster travel in 100 revolutions of the wheel?

The World Archery Federation, also known as FITA (the Fédération Internationale de Tir à l'Arc), is the governing body of archery sports. FITA recognizes two diameter sizes of outdoor target faces. Find the circumference of each.

51. diameter: 122 cm

52. diameter: 80 cm

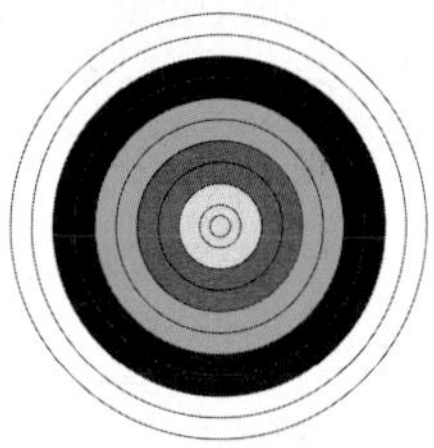

Find the value of each variable.

53.

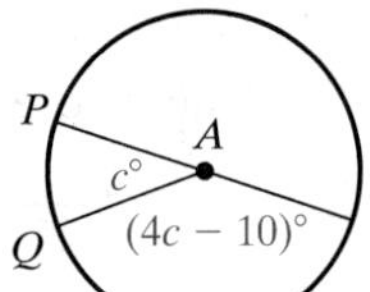

54.

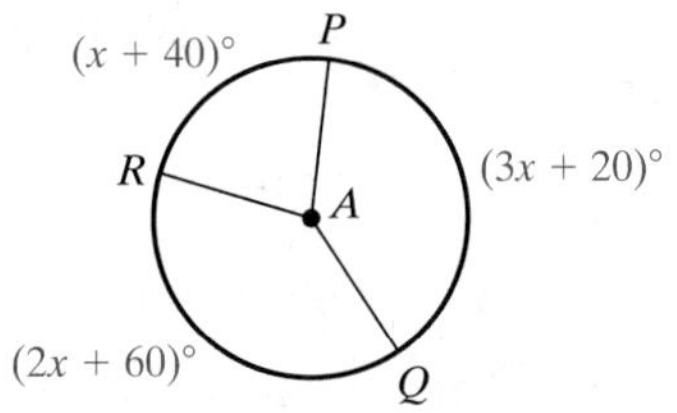

Hands of a clock suggest an angle whose measure is continually changing. How many degrees does a minute hand move through during each time interval?

55. 1 min

56. 5 min

57. 20 min

58. 50 min

59. A landscape architect is constructing a curved path through a rectangular yard. The curved path consists of two 90° arcs. He plans to edge the two sides of the path with plastic edging. What is the total length of plastic edging he will need? Round your answer to the nearest meter.

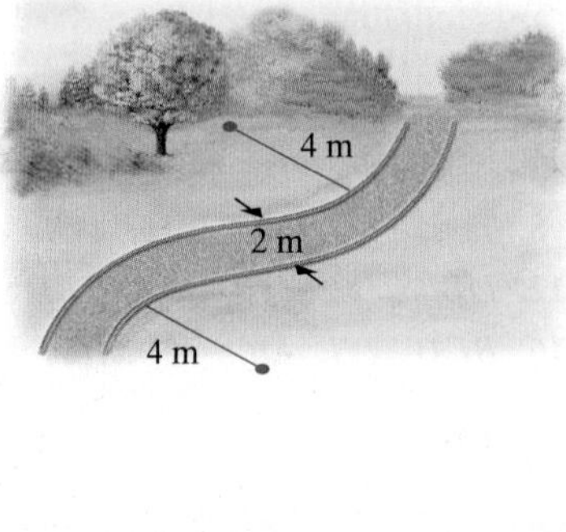

60. Nina designed a semicircular arch made of wrought iron for the top of a mall entrance. The nine segments between the two concentric semicircles are each 3 ft long. What is the total length of wrought iron used to make this structure? Round your answer to the nearest foot.

Find the length of each arc shown in red. Leave your answer in terms of π.

61.

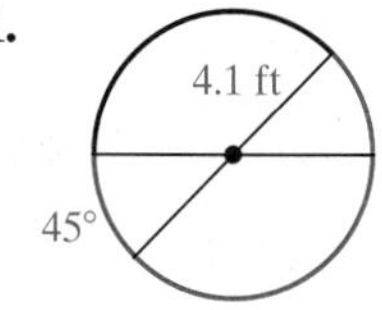

62.

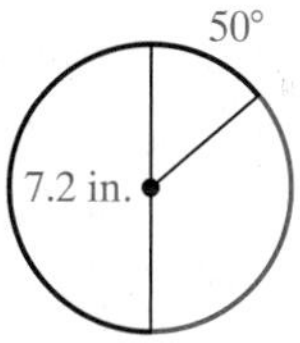

63.

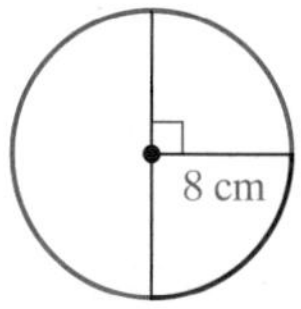

64.

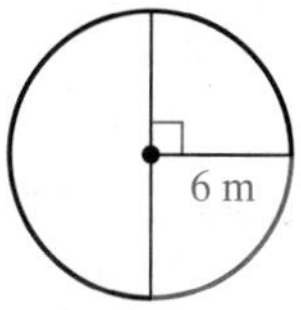

CONCEPT EXTENSIONS

65. Coordinate Geometry Find the length of a semicircle with endpoints (1, 3) and (4, 7). Give an exact answer and then an approximation rounded to the nearest tenth.

66. Coordinate Geometry Find the length of a semicircle with endpoints (2, 1) and (7, 13). Give an exact answer and then an approximation rounded to the nearest tenth.

67. A 60° arc of $\odot A$ has the same length as a 45° arc of $\odot B$. What is the ratio of the radius of $\odot A$ to the radius of $\odot B$?

68. In $\odot O$, the length of $\overset{\frown}{AB}$ is 6π cm and $m\overset{\frown}{AB}$ is 120°. What is the diameter of $\odot O$?

69. Suppose the radius of a circle is doubled. How does this affect the circumference of the circle? Explain.

70. Describe two ways to find the arc length of a major arc if you are given the measure of the corresponding minor arc and the radius of the circle.

71. What is the difference between the measure of an arc and arc length? Explain.

72. Find the Error Your class must find the length of $\overset{\frown}{AB}$. A classmate submits the following solution. What is the error?

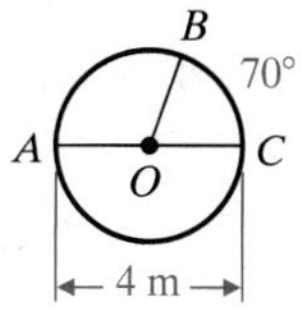

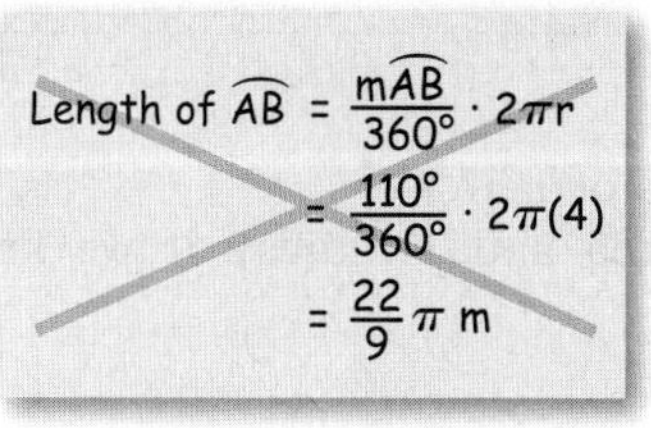

An athletic field is a 100-yd-by-40-yd rectangle, with a semicircle at each of the short sides. A running track 10 yd wide surrounds the field.

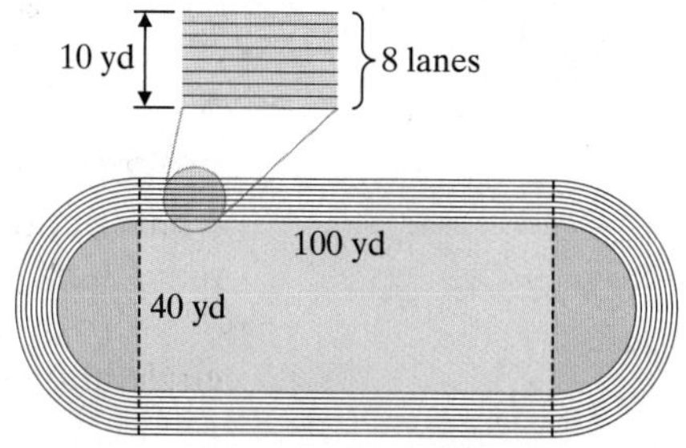

73. What is the distance around the track along the inside edge of the first lane?

74. What is the distance around the track along the outside edge of the outer lane?

75. What is the distance around the track along the inside edge of each of the first four lanes?

76. What is the distance around the track along the inside edge of each of the last four lanes?

77. The diagram below shows two concentric circles, and $\overline{AR} \cong \overline{RW}$. Show that the length of $\overset{\frown}{ST}$ is equal to the length of $\overset{\frown}{QR}$.

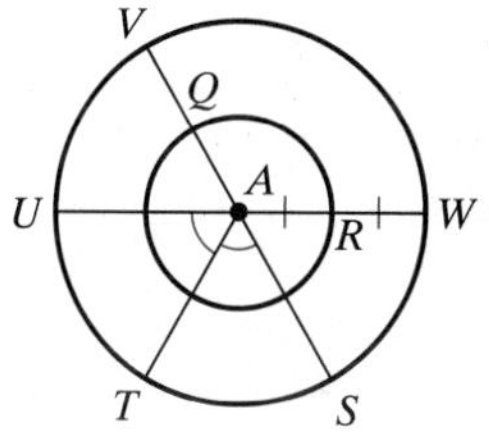

78. Given: $\odot P$ with $\overline{AB} \parallel \overline{PC}$

Prove: $m\overset{\frown}{BC} = m\overset{\frown}{CD}$

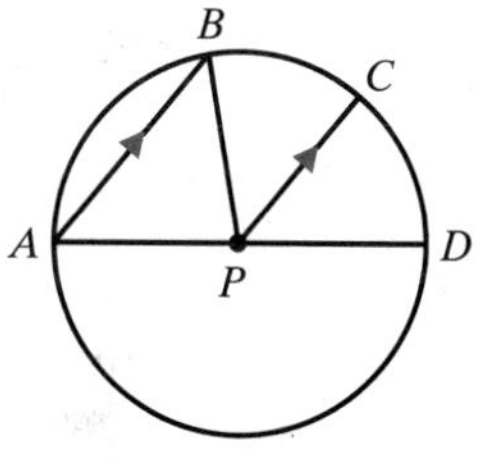

REVIEW AND PREVIEW

Part of a regular dodecagon is shown at the right. See Section 10.3.

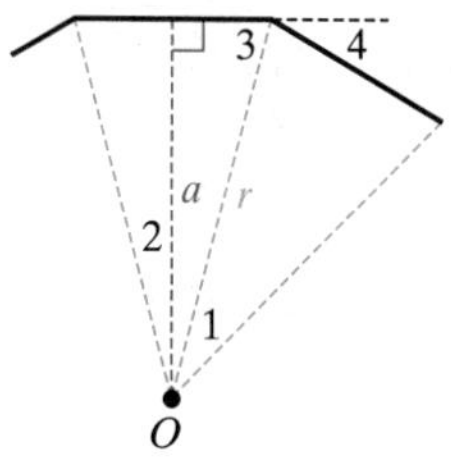

79. What is the measure of each numbered angle?

80. The radius is 19.3 mm. What is the apothem?

81. What is the perimeter of the dodecagon to the nearest millimeter?

82. What is the area of the dodecagon to the nearest square millimeter?

Can you conclude that the figure is a parallelogram? Explain. See Section 6.3.

83.

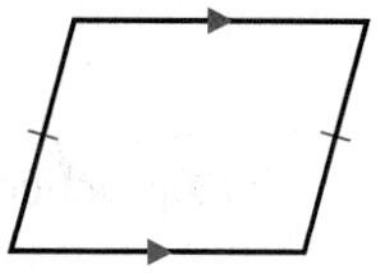

84.

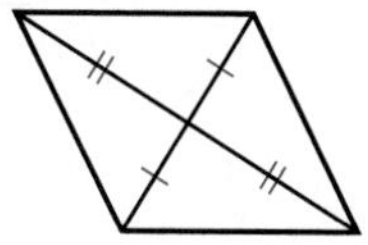

Solve. See Section 2.1.

85. What is the exact circumference of a circle with diameter 16 in.?

86. What is the exact area of a circle with diameter 16 in.?

10.6 AREAS OF CIRCLES AND SECTORS

OBJECTIVE

1 Find the Areas of Circles, Sectors, and Segments of Circles.

VOCABULARY

- sector of a circle
- segment of a circle

OBJECTIVE 1 ▶ Finding the Areas of Circles, Sectors, and Segments of Circles. Recall from Section 2.1 that we can find the area of a circle when we know its radius (or its diameter).

To better understand the formula for the area of a circle, try the following. Cut a circle into many pieces as shown:

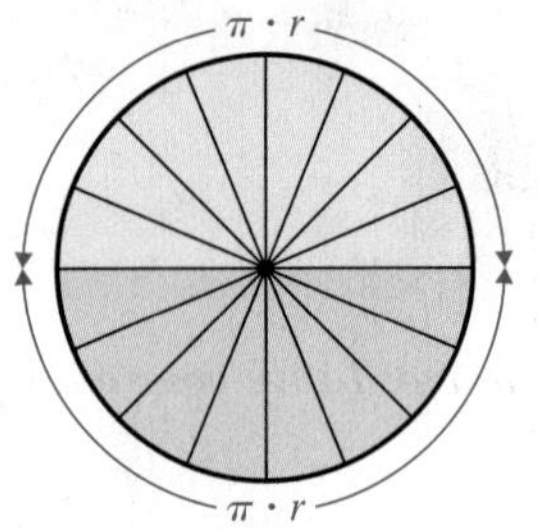

Recall that the circumference of a circle is $2 \cdot \pi \cdot r$. This means that the circumference of half a circle is half of $2 \cdot \pi \cdot r$, or $\pi \cdot r$.

Then unfold the two halves of the circle and place them together as shown:

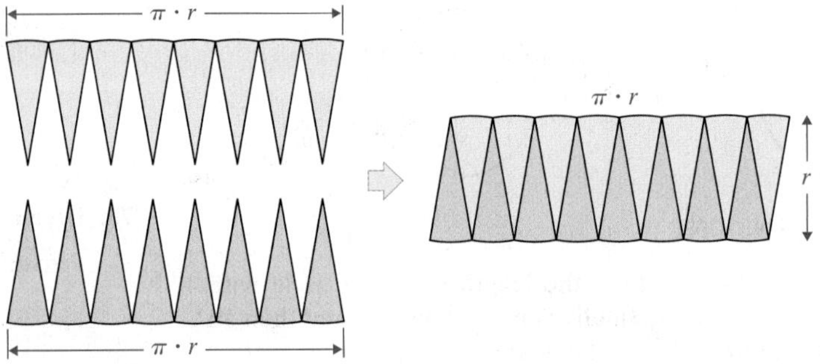

The figure on the right is almost a parallelogram with a base of $\pi \cdot r$ and a height of r. The area is

$$\begin{aligned} A &= \text{base} \cdot \text{height} \\ &= (\pi \cdot r) \cdot r \\ &= \pi \cdot r^2 \end{aligned}$$

This is the formula for the area of a circle.

▶ Helpful Hint

Recall that

$$\pi \approx 3.14 \text{ or } \pi \approx \frac{22}{7}$$

The exact value of π is π.

Theorem 10.6-1 Area of a Circle

The area of a circle is the product of π and the square of the radius.

$$A = \pi r^2$$

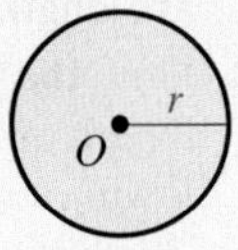

Let's practice using the formula for the area of a circle.

Since $A = \pi r^2$ has two unknowns (A area and r radius), if we know one unknown, we can find the other.

EXAMPLE 1 **Using the Area of a Circle Formula**

a. Find the exact area, and then a two-decimal-place approximation.

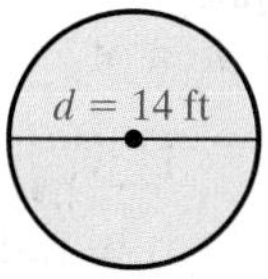

b. Find the exact radius, and then a two-decimal-place approximation.

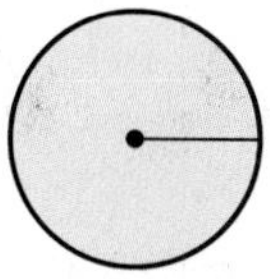

Area = 83 sq m

Solution

a. Since diameter = 14 ft,

$$\text{radius} = \frac{d}{2} = \frac{14}{2} = 7 \text{ feet}$$

$$A = \pi r^2$$

$$= \pi \cdot 7^2$$

$$= 49\pi \text{ sq ft}$$

$$\approx 153.94 \text{ sq ft}$$

The exact area is 49 π sq ft.
The approximate area is 153.94 sq ft.

b. Since area, $A = 83$ sq m,

$83 = \pi r^2$ Let $A = 83$ sq m.

$\frac{83}{\pi} = r^2$ Divide both sides by π.

$\sqrt{\frac{83}{\pi}} = r$ Take the positive square root of both sides.

$5.14 \approx r$

The exact radius is $\sqrt{\frac{83}{\pi}}$ m.
The approximate radius is 5.14 m. □

PRACTICE 1

a. Find the exact area, and then a two-decimal-place approximation.

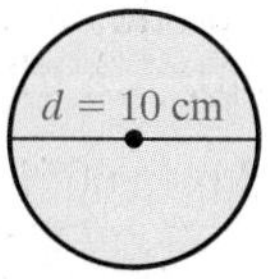

b. Find the exact radius, and then a two-decimal-place approximation.

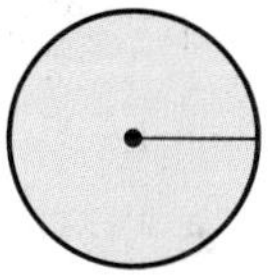

Area = 117 sq yd

Now that we are once again familiar with the formulas for area and circumference of a circle, let's discuss an application.

How to Measure a Tree

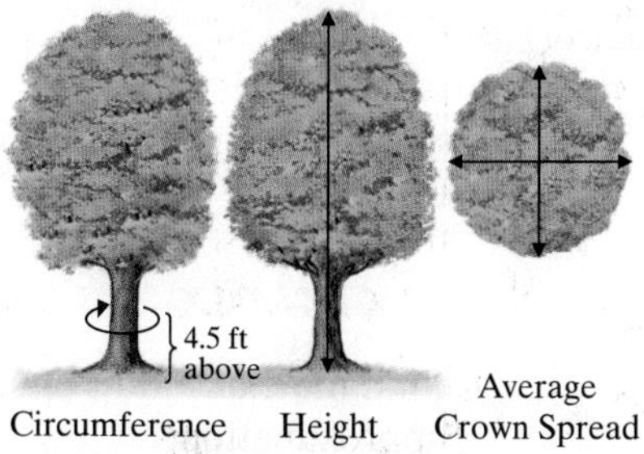

In Maryland, about 1925, the Big Tree Contest was started, and a formula was decided upon to award points for trees. The formula is:

$$\text{total points} = \text{circumference (in inches)} + \text{height (in feet)} + \frac{1}{4} \text{ of average crown spread}$$

Can you imagine all the mathematics and geometry needed to use this formula? Let's now see how we can use the circumference of a circle to calculate its area.

EXAMPLE 2 Going from Circumference of a Circle to Its Area

An artist has been commissioned to construct a circular table from a "slice" of a tree. The customer would like the table surface to be about 400 square inches in area. The artist found a tree with a nice circular girth (circumference) of 72 inches. Before the tree is cut down, let's find the area of a slice of this tree, rounded to two decimal places.

Solution We know $C = 72$ in.

Let's use this to find the radius, r.

$C = 2\pi r$ Circumference formula

$72 = 2\pi r$ Let $C = 72$.

$36 = \pi r$ Divide both sides by 2.

$\frac{36}{\pi} = r$ Divide both sides by π.

$11.46 \approx r$

The radius is about 11.46 in.

Now find the area using the approximate radius.

$A = \pi r^2$ Area formula

$A = \pi(11.46)^2$ Let $r = 11.46$.

$A \approx 412.59$ sq in.

The area of a slice of the tree is approximately 412.59 sq in.

This tree will make a table the size the customer wants. □

PRACTICE

2 A white oak tree in the U.S. has a girth (circumference) of 150 in. Find the area of a slice of the tree.

Note: There are formulas for approximating the age of trees without cutting them down and counting the annual rings. The white oak tree in Practice 2 is about 120 years old. This type of tree can live more than 300 years.

Now let's continue our discussion of areas of circles by first defining a sector.

A **sector of a circle** is a region bounded by an arc of the circle and the two radii to the arc's endpoints. We name a sector using one arc endpoint, the center of the circle, and the other arc endpoint.

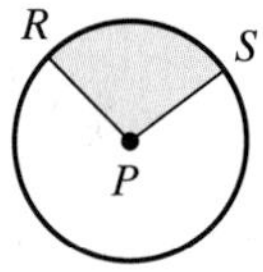

Sector RPS

The area of a sector is a fractional part of the area of a circle. The area of a sector formed by a $60°$ arc is $\frac{60°}{360°}$, or $\frac{1}{6}$, of the area of the circle, πr^2.

Helpful Hint

Notice that our formula for the area of a sector contains πr^2, our formula for the area of a circle. That is because

[area of sector] = [fractional part] of [area of a circle]

Theorem 10.6-2 Area of a Sector of a Circle

The area of a sector of a circle is the product of the ratio $\frac{\text{measure of the arc}}{360°}$ and the area of the circle.

$$\text{area of sector } AOB = \frac{m\widehat{AB}}{360°} \cdot \pi r^2$$

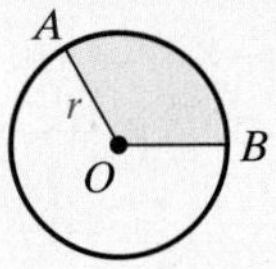

EXAMPLE 3 Finding the Area of a Sector of a Circle

Find the exact area of sector GPH. Then give a two-decimal-place approximation.

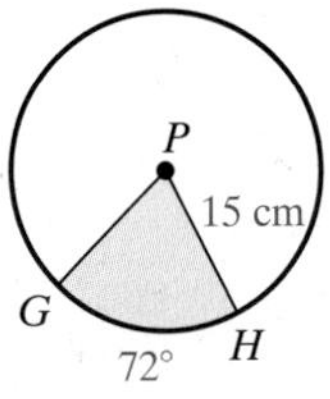

Solution Let's use our formula for the area of a sector.

$\text{area of sector } GPH = \frac{m\widehat{GH}}{360°} \cdot \pi r^2$

$= \frac{72°}{360°} \cdot \pi(15)^2$ Substitute $72°$ for $m\widehat{GH}$ and 15 for r.

$= 45\pi$ Simplify.

≈ 141.37

The area of sector GPH is exactly 45π sq cm or approximately 141.37 sq cm. □

PRACTICE
3 A circle has a radius of 4 in. Find the exact area of a sector bounded by a 45° minor arc. Then give a two-decimal-place approximation.

A part of a circle bounded by an arc and the segment joining its endpoints is a **segment of a circle**.

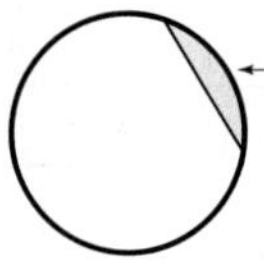

To find the area of a segment for a minor arc, draw radii to form a sector. The area of the segment equals the area of the sector minus the area of the triangle formed.

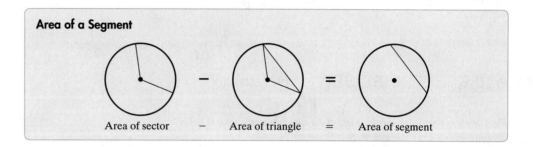

Area of a Segment

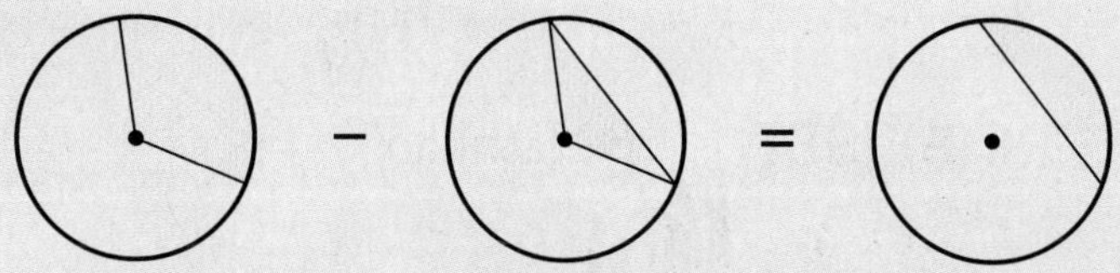

Helpful Hint
Once again, we need to remember our knowledge of special right triangles.

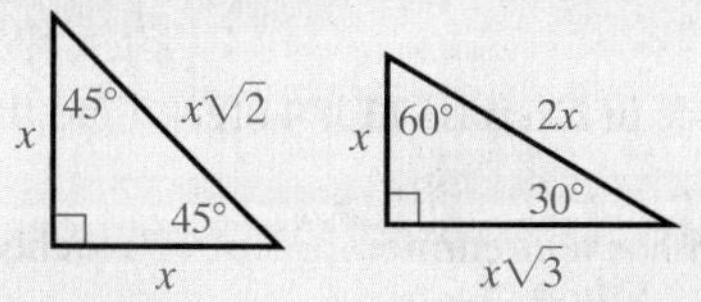

EXAMPLE 4 **Finding the Area of a Segment of a Circle**

Find the area of the shaded segment shown at the right. Round your answer to the nearest tenth.

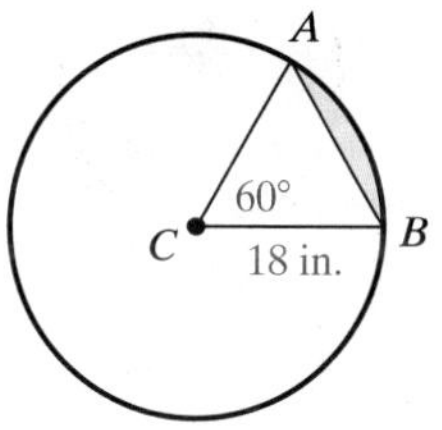

Solution Let's find the area of the sector, then the area of ΔACB, and then subtract these areas.

$$\text{area of sector } ACB = \frac{m\widehat{AB}}{360°}\cdot \pi r^2 \quad \text{Use the formula for the area of a sector.}$$

$$= \frac{60°}{360°}\cdot \pi(18)^2 \quad \text{Substitute } 60° \text{ for } \widehat{AB} \text{ and 18 for } r.$$

$$= 54\pi \quad \text{Simplify.}$$

Next, notice that ΔACB is equilateral. The altitude forms a 30°-60°-90° triangle.

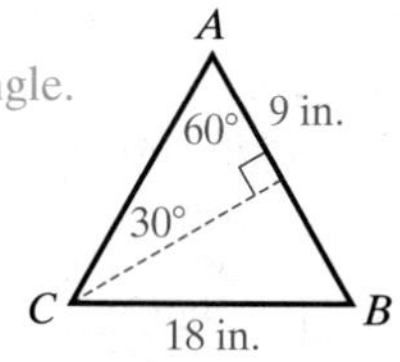

$$\text{area of } \Delta ACB = \frac{1}{2}bh \quad \text{Use the formula for the area of a triangle.}$$

$$= \frac{1}{2}(18)(9\sqrt{3}) \quad \text{Substitute 18 for } b \text{ and } 9\sqrt{3} \text{ for } h.$$

$$= 81\sqrt{3} \quad \text{Simplify.}$$

$$\text{area of shaded segment} = \text{area of sector } ACB - \text{area of } \Delta ACB$$

$$= 54\pi - 81\sqrt{3} \quad \text{Substitute.}$$

$$\approx 29.34988788 \quad \text{Use a calculator.}$$

The area of the shaded segment is about 29.3 in.^{2} or 29.3 sq in. □

PRACTICE
4 Find the area of the shaded segment shown at the right. Round your answer to the nearest tenth.

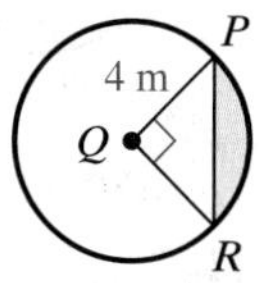

VOCABULARY & READINESS CHECK

Word Bank *Use the choices below to fill in each blank. Some choices may be used more than once and some not at all.*

segment	(area of sector) − (area of Δ)	area of the circle
sector	(area of Δ) − (area of sector)	perimeter of the circle

1. A(n) ___________ of a circle is part of the circle bounded by an arc and the segment joining its endpoints.
2. A(n) ___________ of a circle is a region bounded by an arc of the circle and the radii to the arc's endpoints.
3. Area of sector $= \dfrac{\text{measure of the arc}}{360°} \cdot$ ___________
4. Area of segment = ___________

10.6 EXERCISE SET MyMathLab®

Find the area of each circle. Find the exact area and also a two-decimal-place approximation. See Example 1.

1.

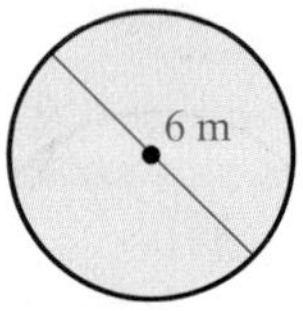

2.

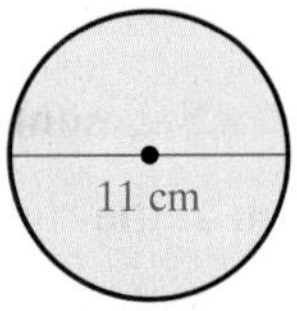

3.

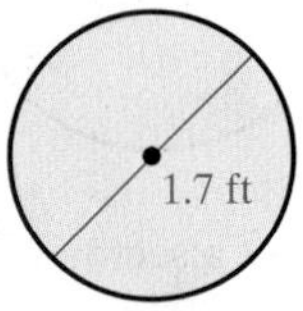

4.

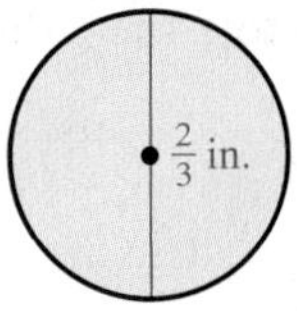

5. Some farmers use a circular irrigation method. An irrigation arm acts as the radius of an irrigation circle. How much land is covered with an irrigation arm of 300 ft?
6. You use an online store locator to search for a store within a 5-mi radius of your home. What is the area of your search region?

The area of each circle is given. Find a two-decimal-place approximation for its radius. See Example 1.

7.

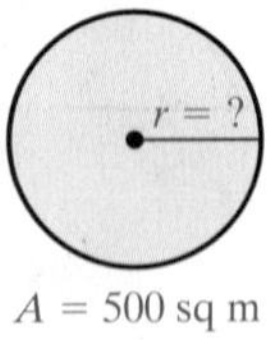

8.

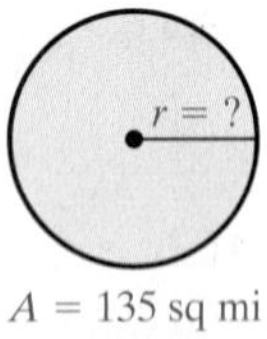

9.

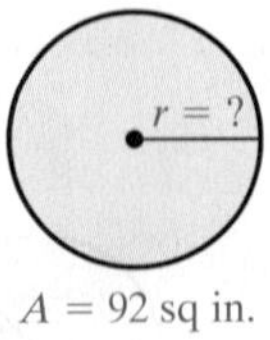

10.

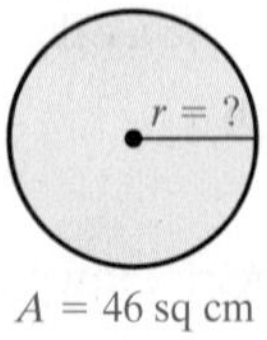

Solve. Round each answer to two decimal places. See Example 2.

11. The area of the surface of a circular pond is needed. A measure about the pond gives a circumference of about 38 feet. Find the area.

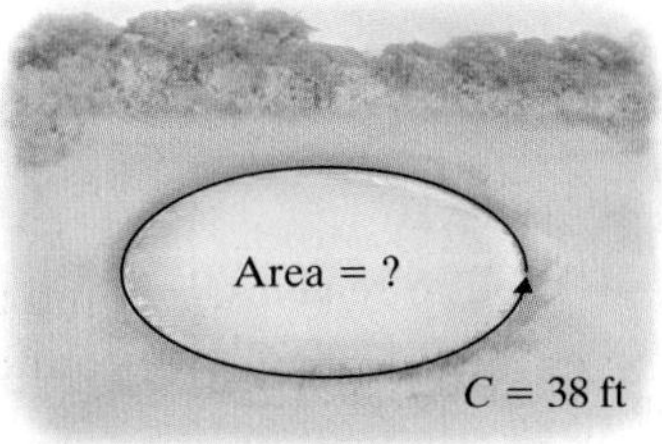

12. Suppose the circumference in Exercise **11** is 44 feet. Find the area.
13. A white oak in maryland has a circumference of 120 inches. Find the area of a circle with that circumference.

14. The volume of a column in the shape of a cylinder is needed. We will perform only the first step by measuring the circumference of the column and finding the area of a circle with that circumference. If the circumference is measured to be 19 feet, find the area.

Find the area of each shaded sector of a circle. Find the exact area and then a two-decimal-place approximation. See Example 3.

15.
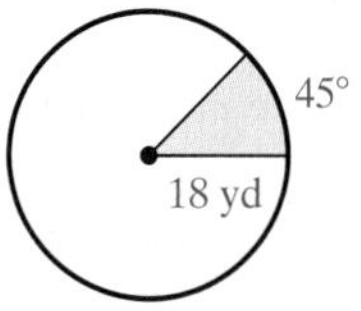

16.
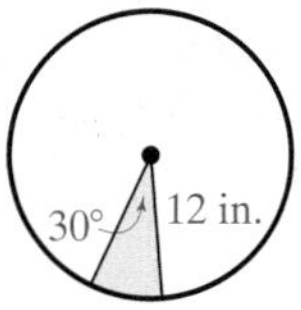

17.
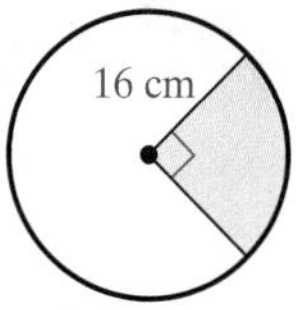

18.
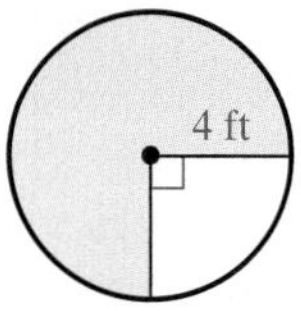

19.
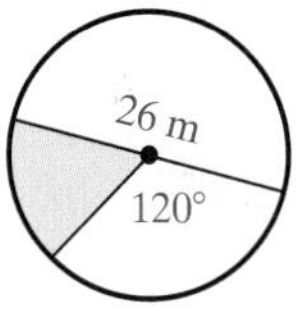

20.
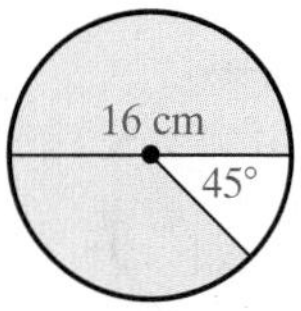

Find the area of sector TOP in ⊙O using the given information. Leave your answer in terms of π. See Example 3.

21. $r = 5$ m, $m\widehat{TP} = 90°$

22. $r = 6$ ft, $m\widehat{TP} = 15°$

23. $d = 16$ in., $m\widehat{PT} = 135°$

24. $d = 15$ cm, $m\widehat{POT} = 180°$

Find the area of each shaded segment. Round your answer to the nearest tenth. See Example 4.

25.
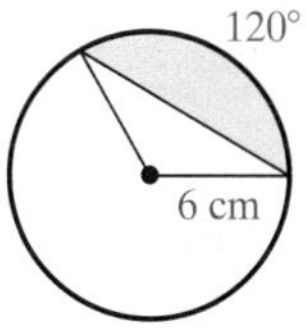

26.
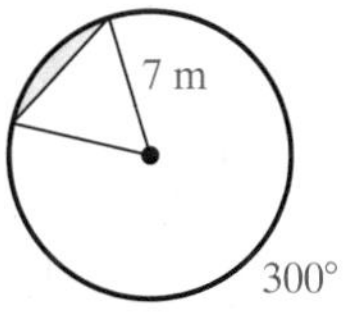

27.
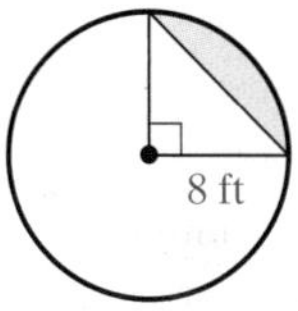

28.
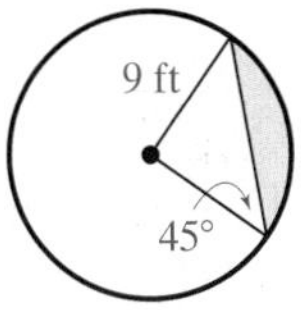

29.
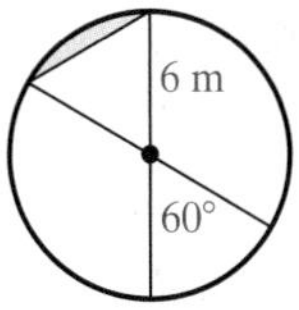

30.
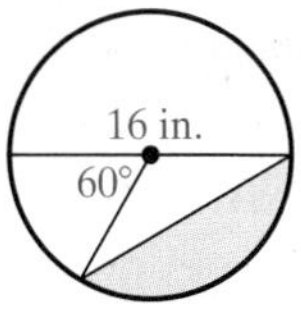

Find the area of the shaded region. Leave your answer in terms of π and in simplest radical form.

31.
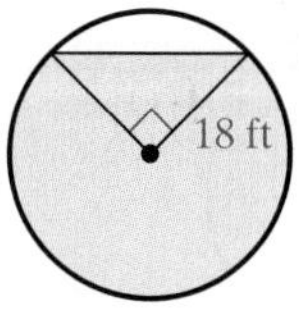

32.
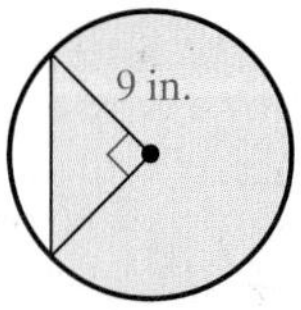

33.
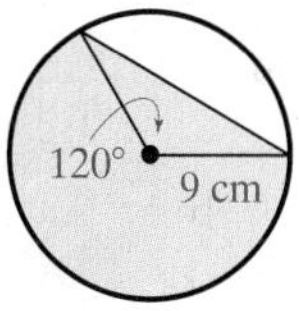

34.
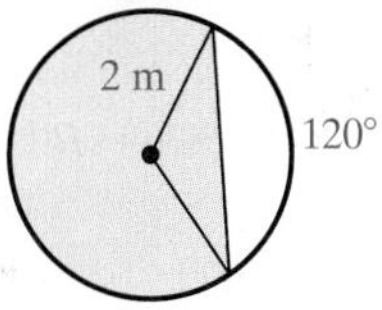

35.
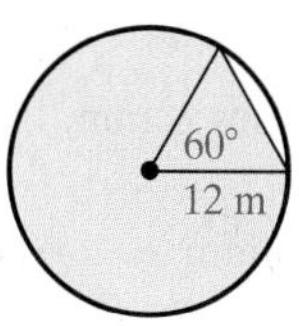

36.
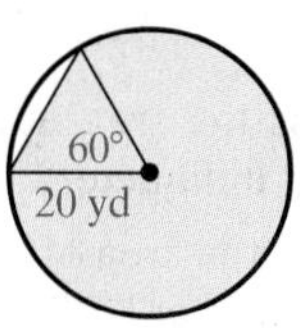

37.
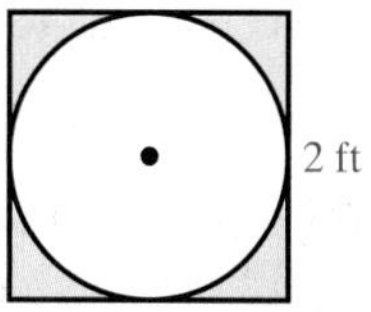

38.
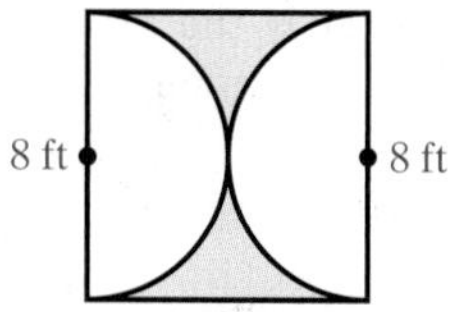

39.
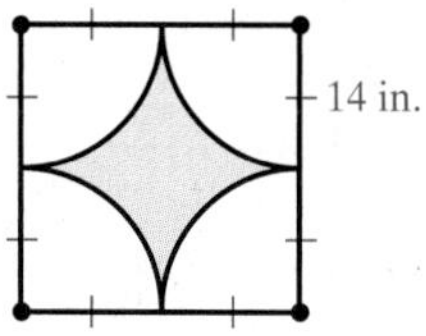

40.
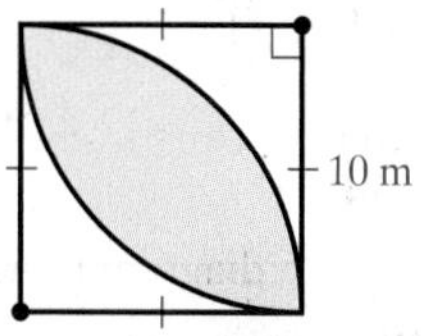

41. A town provides bus transportation to students living beyond 2 mi of the high school. What area does *not* have the bus service? Round to the nearest tenth.

42. A homeowner wants to build a circular patio. If the diameter of the patio is 20 ft, what is its area to the nearest whole number?

Multiple Steps *Solve Exercises 43–46. Round answers to two decimal places.*

43. A discus circle is shown. It has a diameter of 98.5 inches that includes a 2-inch-wide painted circle.

a. Find the area of the discus circle.

b. Find the area of the inner circle without the painted ring.

c. Use parts **a** and **b** to find the area of the painted ring.

44. A lamp is formed by two semicircles, as shown. To calculate cost, the area of the 3-inch-wide circle formed by the two semicircles is needed.

a. Draw the 3-inch-wide circle.

b. Find the area of the inner circle, or hole.

c. Find the area of the outer circle.

d. Use parts **b** and **c** to find the area of the 3-inch-wide circle.

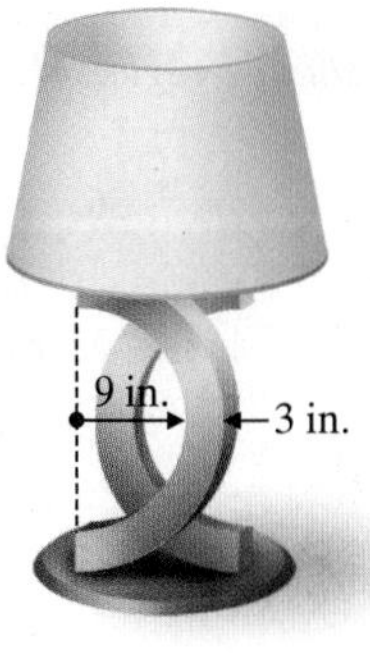

45. Refer to the diagram of the regular hexagonal nut.

a. What is the area of a hexagon with radius 8 mm, to the nearest millimeter?

b. What is the area of a circle with radius 4 mm, to the nearest millimeter?

c. What is the area of the hexagonal face shown, to the nearest millimeter?

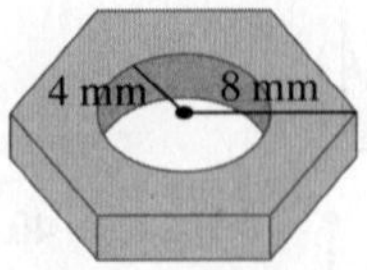

46. A circular mirror is 24 in. wide and has a 4-in. frame around it.

a. Draw a diagram to help solve the problem.

b. What is the area of the mirror?

c. What is the area of the mirror and the frame?

d. What is the area of the frame?

CONCEPT EXTENSIONS

47. What is the difference between a sector of a circle and a segment of a circle?

48. $\overline{AB}$ and $\overline{CD}$ are diameters of $\odot O$. Is the area of sector AOC equal to the area of sector BOD? Explain.

49. Find the Error Your class must find the area of a sector of a circle determined by a 150° arc. The radius of the circle is 6 cm. What is your classmate's error? Explain.

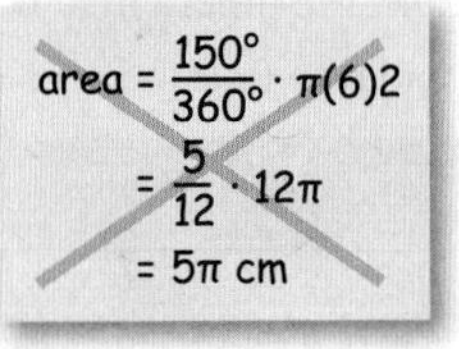

50. Suppose a sector of $\odot P$ has the same area as a sector of $\odot O$. Can you conclude that $\odot P$ and $\odot O$ have the same area? Explain.

51. A circle with radius 12 mm is divided into 20 sectors of equal area. What is the area of one sector to the nearest tenth?

52. Draw a circle and a sector so that the area of the sector is 16π cm^2. Give the radius of the circle and the measure of the sector's arc.

53. The circumference of a circle is 26π in. What is its area? Leave your answer in terms of π.

54. In a circle, a 90° sector has area 36π in.^{2}. What is the radius of the circle?

Find the area of the shaded region. Leave your answer in terms of π.

55.

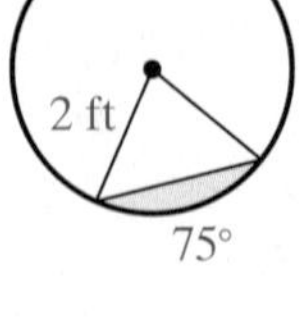

56.

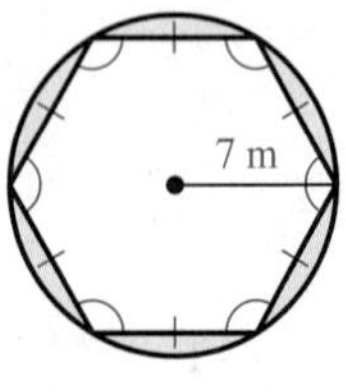

Mixed Practice (Sections 10.3 and 10.6)

Use the given figures and complete the table. Leave your answers in terms of pi and in simplest radical form. Each figure measurement is in units. (For Exercise 60, approximate answers to two decimal places.)

Equilateral Triangle

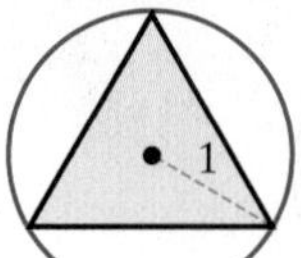

Square

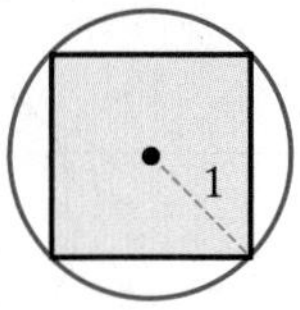

Regular Hexagon

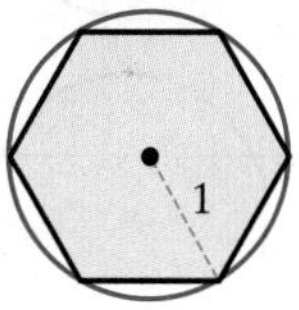

Regular Octagon

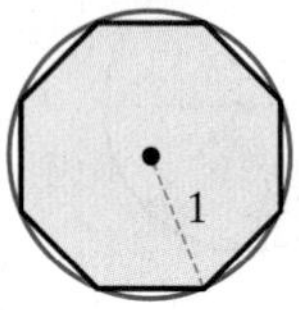

Circle

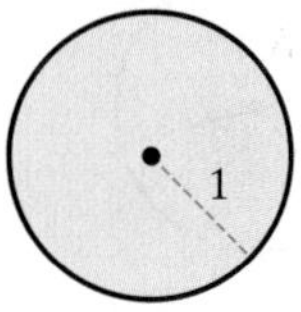

	Figure	*Number of Sides*	*Length of a Side, s*	*Apothem, a*	*Perimeter*	*Area*
57.	Equilateral Triangle					
58.	Square					
59.	Regular Hexagon					
60.	Regular Octagon					
					Circumference	*Area*
61.	Circle					

62. a. Compute and compare the ratios $\frac{\text{perimeter}}{\text{circumference}}$ and $\frac{\text{area of polygon}}{\text{area of circle}}$ as the number of sides of the regular polygon increases in your completed table.

b. What do you think happens as the number of sides of the regular polygon increases?

REVIEW AND PREVIEW

Find the length of $\widehat{AB}$ in each circle. Leave your answers in terms of π. See Section 10.5.

63.

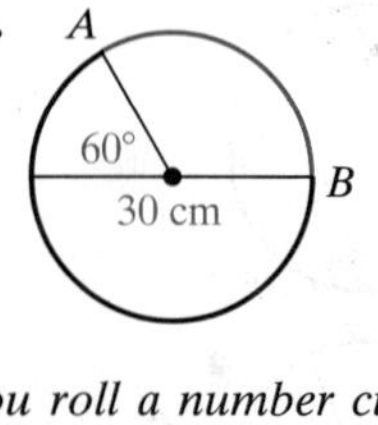

64.

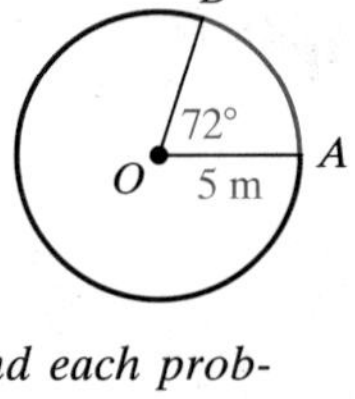

You roll a number cube or die. Find each probability. See the appendix on probability.

65. $P(4)$

66. $P(2 \text{ or } 5)$

67. $P(\text{odd number})$

68. $P(\text{prime number})$

Die

10.7 GEOMETRIC PROBABILITY

OBJECTIVE

1 Use Segment and Area Models to Find the Probabilities of Events.

VOCABULARY

- geometric probability

The probability of an event, written *P*(event), is the likelihood that the event will occur.

When the possible outcomes are equally likely, the theoretical probability of an event is the ratio of the number of favorable outcomes to the number of possible outcomes.

$$P(\text{event}) = \frac{\text{number of favorable outcomes}}{\text{number of possible outcomes}}$$

Recall that a probability can be expressed as a fraction, a decimal, or a percent.

For example, if a coin is tossed, what is the probability that heads occurs?

Heads

Tails

1 way the outcome is favorable

H, T

2 possible outcomes

$$\text{probability} = \frac{1}{2} \begin{array}{l}\text{Number of favorable outcomes}\\ \text{Number of possible outcomes}\end{array}$$

Since one of two equally likely possible outcomes is heads, the probability is $\frac{1}{2}$.

Note that the probability of an event is always between 0 and 1, inclusive (i.e., including 0 and 1). A probability of 0 means that an event won't occur, and a probability of 1 means that an event is certain to occur.

If a coin is tossed twice, how do we find the probability of tossing heads on the first toss and then heads again on the second toss (H, H)?

1 way the event can occur

H, T, H, H, T, H, T, T

4 possible outcomes

$$\text{probability} = \frac{1}{4} \begin{array}{l}\text{Number of favorable outcomes}\\ \text{Number of possible outcomes}\end{array}$$

The probability of tossing heads and then heads is $\frac{1}{4}$.

OBJECTIVE 1 ▶ Using Segment and Area Models to Find Probabilities of Events. We can use geometric models to solve certain types of probability problems.

In **geometric probability**, points on a segment or in a region of a plane represent outcomes. The geometric probability of an event is a ratio that involves geometric measures such as length or area.

▶ Helpful Hint

Notice that the denominator (AB) contains the length of $\overline{MN}$.

Probability and Length

Point S on $\overline{AB}$ is chosen at random. The probability that S is on $\overline{MN}$ is the ratio of the length of $\overline{MN}$ to the length of $\overline{AB}$.

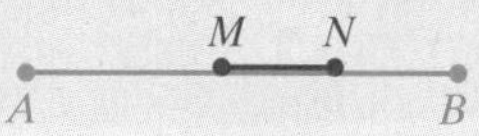

$$P(S \text{ on } \overline{MN}) = \frac{\text{length of } \overline{MN}}{\text{length of } \overline{AB}}$$

EXAMPLE 1 Using Segments to Find Probability

Point K on $\overline{ST}$ is chosen at random.
What is the probability that K lies on $\overline{QR}$?

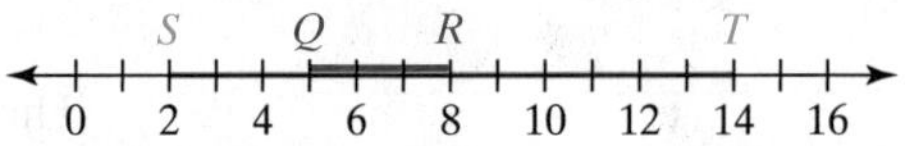

Solution

$$P(K \text{ on } \overline{QR}) = \frac{\text{length of } \overline{QR}}{\text{length of } \overline{ST}} = \frac{|5-8|}{|2-14|} = \frac{3}{12}, \text{ or } \frac{1}{4}$$

The probability that K is on $\overline{QR}$ is $\frac{1}{4}$, or 0.25, or 25%. □

PRACTICE 1 Use the diagram in Example 1. Point H on $\overline{ST}$ is selected at random. What is the probability that H lies on $\overline{SR}$?

EXAMPLE 2 Using Segments to Find Probability

Transportation A commuter train stops at Comet Station every 25 min. If a commuter arrives at the station at a random time, what is the probability that the commuter will wait at least 10 min for the train?

For example:

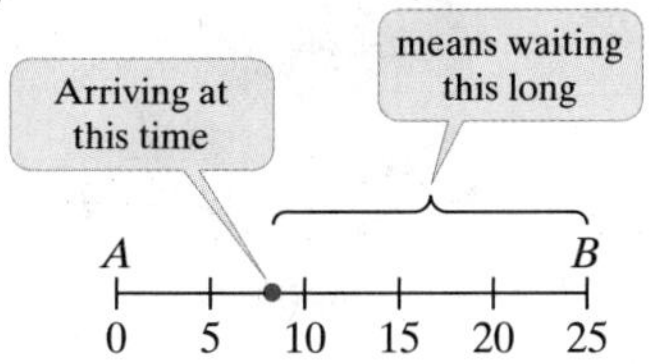

Solution Let's draw a line segment to model the situation. The length of the entire segment represents the amount of time between trains. A commuter will wait at least 10 min for the train if the commuter arrives at any time between 0 and 15 min.

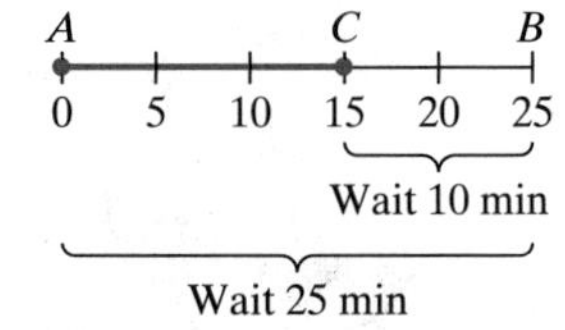

$$P(\text{waiting at least 10 min}) = \frac{\text{length of favorable segment}}{\text{length of entire segment}} = \frac{15}{25}, \text{ or } \frac{3}{5}$$

The probability that a commuter will wait at least 10 min for the train is $\frac{3}{5}$, or 0.6, or 60%. □

PRACTICE 2 What is the probability that a commuter will wait no more than 5 min for the train?

When the points of a region represent equally likely outcomes, we can find probabilities by comparing areas.

Helpful Hint
Area of region R does include the area of region N.

Probability and Area

Point S in region R is chosen at random. The probability that S is in region N is the ratio of the area of region N to the area of region R.

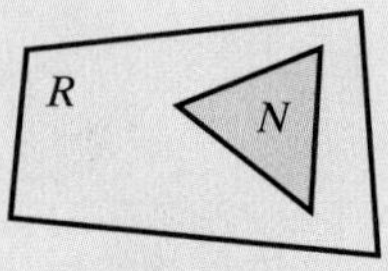

$$P(S \text{ in region } N) = \frac{\text{area of region } N \leftarrow \text{favorable region}}{\text{area of region } R \leftarrow \text{entire region}}$$

EXAMPLE 3 Using Area to Find Probability

A circle is inscribed in a square. Point Q in the square is chosen at random. What is the probability that Q lies in the shaded region?

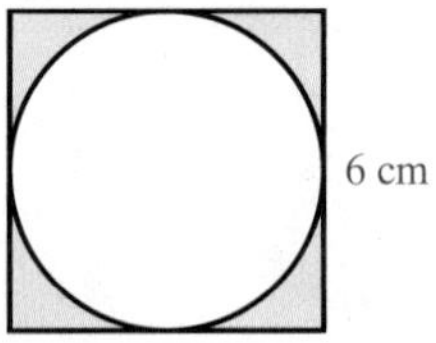

Solution The entire region is the square (goes in denominator), and the favorable region is the shaded region (goes in numerator).

Notice that a side of the square is 6 cm, so the radius of the circle is $\frac{6}{2}$ or 3 cm.

$$\begin{aligned}\text{area of square} &= (\text{side})^2\\ &= 6^2\\ &= 36 \text{ sq units}\\ \text{area of shaded region} &= \text{area of square} - \text{area of circle}\\ &= 36 - \pi(3)^2\\ &= 36 - 9\pi\end{aligned}$$

$$P(Q \text{ lies in shaded region}) = \frac{\text{area of shaded region}}{\text{area of square}} \begin{matrix}\leftarrow \text{favorable region}\\ \leftarrow \text{entire region}\end{matrix}$$

$$= \frac{36 - 9\pi}{36} \approx 0.215$$

The probability that Q lies in the shaded region is about 0.215, or 21.5%. □

PRACTICE

3 A triangle is inscribed in a square. Point T in the square is selected at random. What is the probability that T lies in the shaded region?

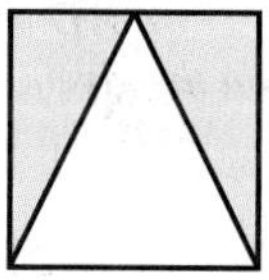

EXAMPLE 4 Using Area to Find Probability

The face of an archery target has 5 colored scoring zones formed by concentric circles. One official target's face diameter is 80 cm. The radius of the center yellow circle is 8 cm. The width of each of the other rings is also 8 cm. If an arrow hits the target face at a random point, what is the probability that it hits the blue ring?

Solution First, let's find the area of the entire region, which is a circle with a diameter of 80 cm, or a radius of 40 cm.

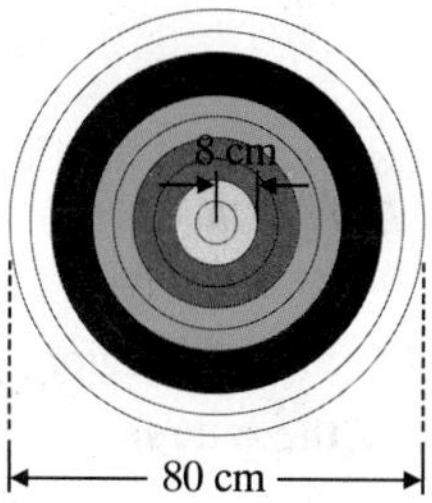

$$\begin{aligned}\text{area of entire target face} &= \pi r^2\\ &= \pi(40)^2\end{aligned}$$

To find the area of the favorable blue ring, we subtract areas of circles.

$$\begin{aligned}\text{area of favorable blue ring} &= \begin{matrix}\text{area of circle with}\\ \text{outer blue ring}\\ (r = 8 + 8 + 8)\end{matrix} - \begin{matrix}\text{area of circle with}\\ \text{outer red ring}\\ (r = 8 + 8)\end{matrix}\\ &= \pi(24)^2 - \pi(16)^2\end{aligned}$$

$$P(\text{arrow hits blue ring}) = \frac{\text{area of blue ring}}{\text{area of entire target}} = \frac{\pi(24)^2 - \pi(16)^2}{\pi(40)^2}$$

$$= 0.2$$

The probability of an arrow hitting the blue ring is 0.2, or 20%. □

PRACTICE

4 Another official target face has a diameter of 122 cm. The yellow center circle has a radius of 12.2 cm, and each of the other ring colors has a width of 12.2 cm. If an arrow randomly hits the target face, find the probability that it hits the red ring.

VOCABULARY & READINESS CHECK

Fill in the Blank.

1. The segment probability formula is $P(\text{a point on } \overline{QR}) =$ ____________.

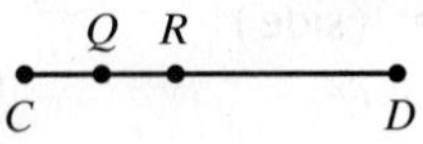

2. The area probability formula is $P(\text{a point in region } A) =$ ____________.

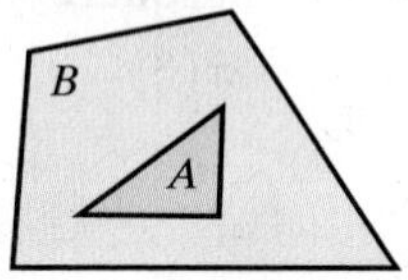

3. For ____________ probability, we use points on a segment or in a region to represent outcomes.

10.7 EXERCISE SET MyMathLab®

Point T on $\overline{AD}$ is chosen at random. What is the probability that T lies on the given segment? See Examples 1 and 2.

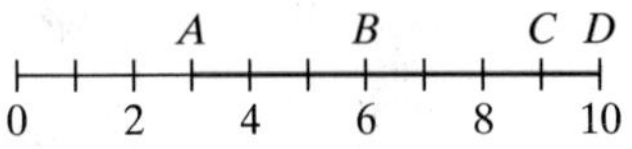

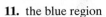

1. $\overline{AB}$
2. $\overline{AC}$
3. $\overline{BD}$
4. $\overline{BC}$

A point on $\overline{AK}$ is chosen at random. Find the probability that the point lies on the given segment. See Examples 1 and 2.

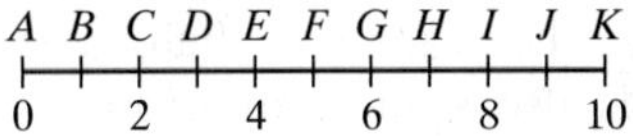

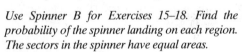

5. $\overline{CH}$
6. $\overline{FG}$
7. $\overline{DJ}$
8. $\overline{EI}$
9. $\overline{AK}$
10. $\overline{GK}$

Use Spinner A for Exercises 11–14. Find the probability of the spinner landing on each region. The sectors in the spinner have equal areas.

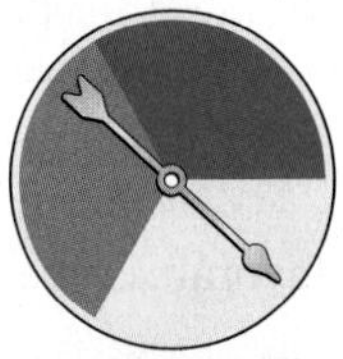

Spinner A

11. the blue region
12. the red region
13. the yellow or red region
14. the yellow, red, or blue region

Use Spinner B for Exercises 15–18. Find the probability of the spinner landing on each region. The sectors in the spinner have equal areas.

Spinner B

15. region 2
16. region 3
17. region 2 or 3
18. region 1, 2, or 4

A point in the figure is chosen at random. Find the probability that the point lies in the shaded region. See Examples 3 and 4.

19. 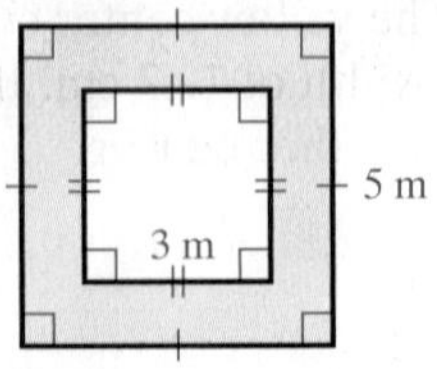

20. 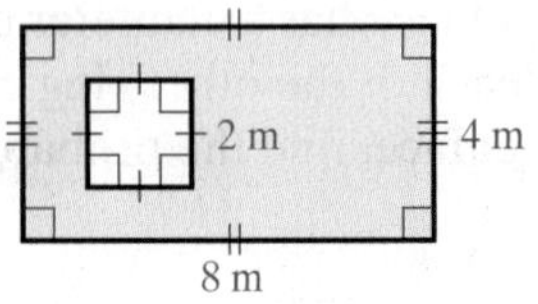

21. 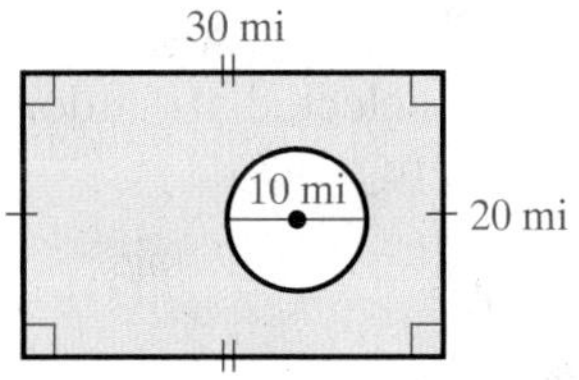

22. 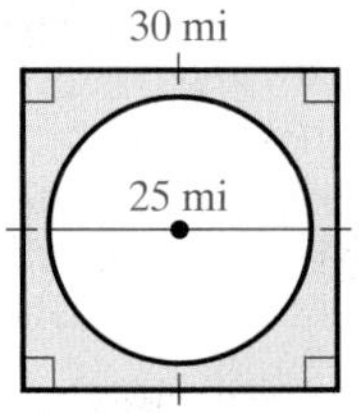

23. 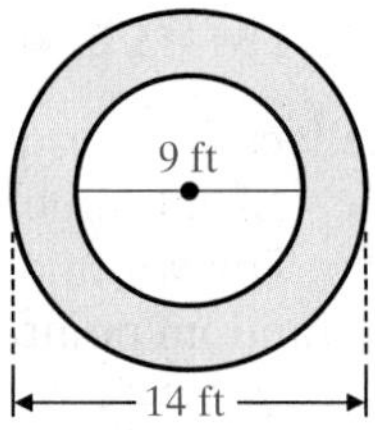

24. 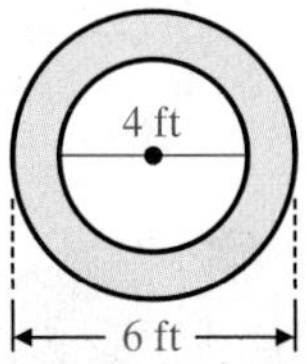

25. 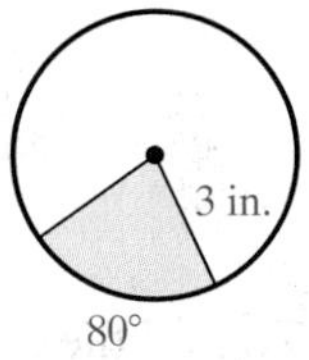

26. 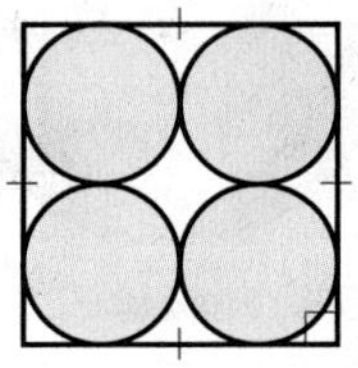

A target with a diameter of 14 cm has 4 scoring zones formed by concentric circles. The diameter of the center circle is 2 cm. The width of each ring is 2 cm. A dart hits the target at a random point. Find the probability that it will hit a point in the indicated region.

27. the center region
28. the blue region
29. the red region
30. the yellow region

Solve. See Example 3.

31. The cycle of the traffic light on Main Street at the intersection of Main Street and Commercial Street is 40 seconds green, 5 seconds yellow, and 30 seconds red. If you reach the intersection at a random time, what is the probability that the light is red?

32. Your friend is supposed to call you at a random time between 3 p.m. and 4 p.m. At 3:20 p.m., you realize that your cell phone is off and you immediately turn it on. What is the probability that you missed your friend's call?

33. At a given bus stop, a city bus stops every 16 min. If a student arrives at his bus stop at a random time, what is the probability that he will not have to wait more than 4 min for the bus?

34. Using Exercise **33**, find the probability that a student will not have to wait more than 10 minutes for the bus.

An archery target is sold with a face diameter of 30 inches. The radius of the center yellow ring is 3 in. The width of each of the other rings is 3 in.

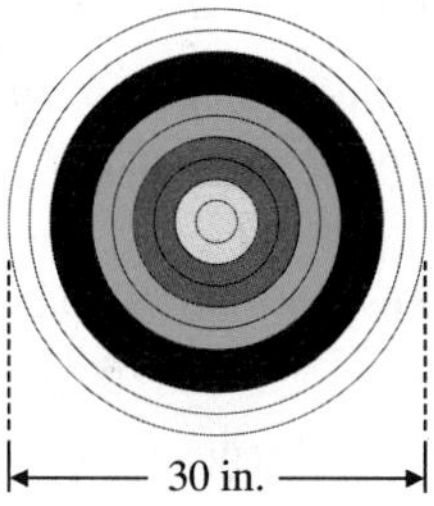

35. Find the radius of the yellow circle.

36. Find the radius of each colored ring.

Find the probability of an arrow randomly landing in each indicated region.

37. the yellow circle

38. the red ring

39. the black ring

40. the white ring

41. Points M and N are on $\overline{ZB}$ with M between Z and N. $ZM = 5$, $NB = 9$, and $ZB = 20$. A point on $\overline{ZB}$ is chosen at random. What is the probability that the point is on $\overline{MN}$?

42. $\overline{BZ}$ contains $\overline{MN}$ and $BZ = 20$. A point on $\overline{BZ}$ is chosen at random. The probability that the point is also on $\overline{MN}$ is 0.3, or 30%. Find MN.

CONCEPT EXTENSIONS

43. Suppose a point in the regular pentagon is chosen at random. What is the probability that the point is *not* in the shaded region? Explain.

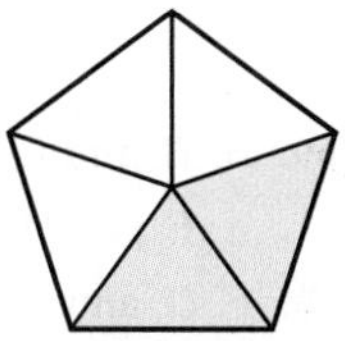

44. A point K in the regular hexagon is chosen at random. What is the probability that K lies in the region that is *not* shaded?

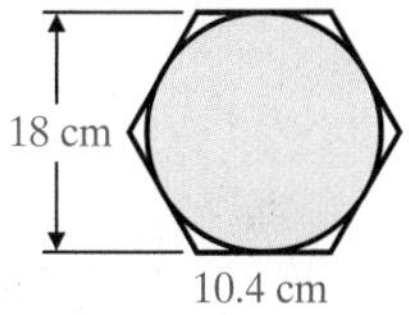

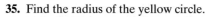

45. In the figure at the right, $\frac{SQ}{QT} = \frac{1}{2}$. What is the probability that a point on $\overline{ST}$ chosen at random will lie on $\overline{QT}$? Explain.

46. **Find the Error** Your class needs to find the probability that a point A in the square chosen at random lies in the shaded region. Your classmate's work is shown below. What is the error? Explain.

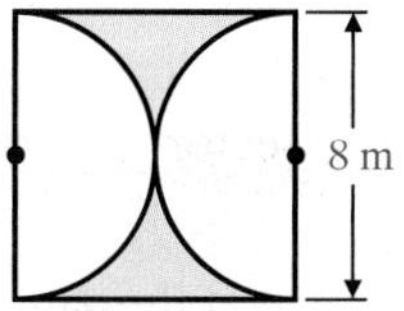

P (A in shaded region) = $\frac{\text{Area of semicircles}}{\text{Area of square}}$

$= \frac{16\pi}{64}$

≈ 0.785, or 79%

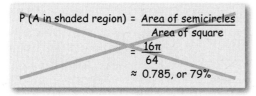

47. Meteorites (mostly dust-particle size) are continually bombarding Earth. The surface area of Earth is about 65.7 million mi^2. The area of the United States is about 3.7 million mi^2. What is the probability that a meteorite landing on Earth will land in the United States?

48. What is the probability that a point chosen at random on the circumference of $\odot C$ lies on $\widehat{AB}$? Explain how you know.

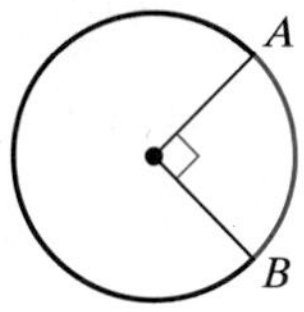

49. Every 20 min from 4:00 p.m. to 7:00 p.m., a commuter train crosses Boston Road. For 3 min, a gate stops cars from crossing over the tracks as the train goes by. What is the probability that a motorist randomly arriving at the train crossing during this time interval will have to stop for a train?

50. A bus arrives at a stop every 16 min and waits 3 min before leaving. What is the probability that a person arriving at the bus stop at a random time has to wait more than 10 min for a bus to arrive?

Find the probability that coordinate x of a point chosen at random on $\overline{\text{AK}}$ *satisfies the inequality.*

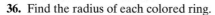

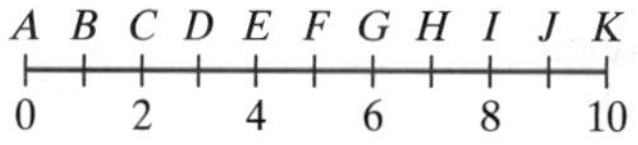

51. $2 \le x \le 8$

52. $2x \le 8$

53. $5 \le 11 - 6x$

54. $\frac{1}{2}x - 5 > 0$

55. $2 \le 4x \le 3$

56. $-7 \le 1 - 2x \le 1$

Assume that a dart you throw will land on the 12-in.-by-12-in. square dartboard and is equally likely to land at any point on the board. The diameter of the center circle is 2 in., and the width of each ring is 1 in.

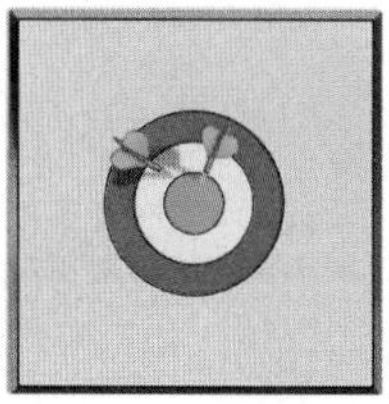

57. What is the probability of hitting either the blue or the yellow region?

58. What is the probability the dart will *not* hit the gray region?

59. **Multiple Steps** You have a 4-in. straw and a 6-in. straw. You want to cut the 6-in. straw into two pieces so that the three pieces form a triangle.

a. If you cut the straw to get two 3-in. pieces, can you form a triangle?

b. If the two pieces are 1 in. and 5 in., can you form a triangle?

c. If you cut the straw at a random point, what is the probability that you can form a triangle?

60. Multiple Steps To win a prize at a carnival game, you must toss a quarter so that it lands entirely within a circle, as shown below. Assume that the center of a tossed quarter is equally likely to land at any point within the 8-in. square.

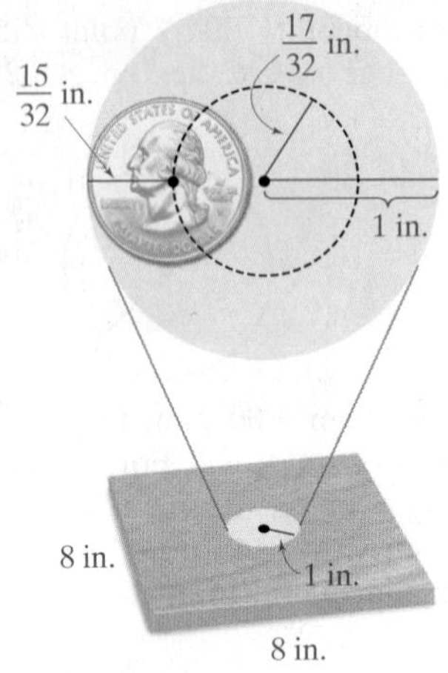

a. What is the probability that the quarter lands entirely in the circle in one toss?

b. On average, how many coins must you toss to win a prize? Explain.

61. One type of dartboard is a square of radius 10 in. You throw a dart and hit the target. Assume you hit the target at a random point. What is the probability that the dart lies within $\sqrt{10}$ in. of the square's center?

62. The traffic lights at Fourth and State Streets repeat themselves in 1-min cycles. A motorist will face a red light 60% of the time. Use this information to estimate how long the Fourth Street light is red during each 1-min cycle.

63. A circular dartboard has radius 1 m and a yellow circle in the center. Assume you hit the target at a random point. For what radius of the yellow center region would *P*(hitting yellow) equal each of the following? Use the table feature of a calculator to generate all six answers. Round to the nearest centimeter.

a. 0.2 **b.** 0.4 **c.** 0.5
d. 0.6 **e.** 0.8 **f.** 1.0

64. You and your friend agree to meet for lunch between 12 p.m and 1 p.m. Each of you agrees to wait 15 min for the other before giving up and eating lunch alone. If you arrive at 12:20, what is the probability you and your friend will eat lunch together?

REVIEW AND PREVIEW

Solve. See Section 10.6.

65. A circle has circumference 20π ft. What is its area in terms of π?

66. A circle has radius 12 cm. What is the area of a sector of the circle with a 30° central angle in terms of π?

Tell what type(s) of symmetry each figure has. See Sections 8.3 and 8.4.

67.

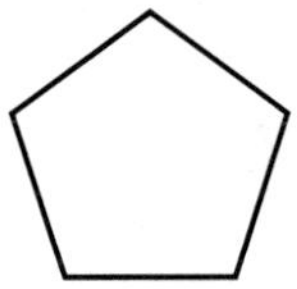

68.

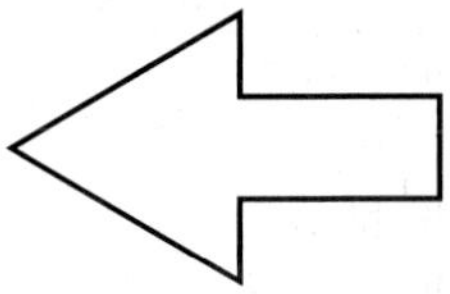

Name the plane (using four points) that contains the indicated points. See Section 1.3.

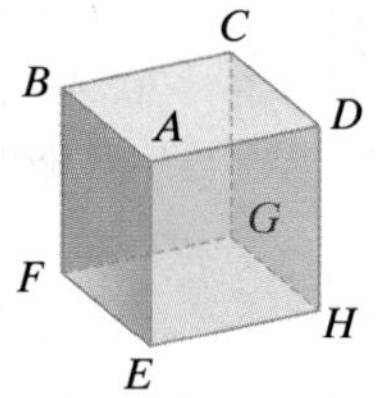

69. *A*, *B*, and *C*
70. *B*, *F*, and *G*
71. *E*, *F*, and *H*
72. *A*, *D*, and *G*
73. *F*, *D*, and *G*
74. *A*, *C*, and *G*

CHAPTER 10 VOCABULARY CHECK

Fill in the Blank *Use the choices below to fill in each blank. Some choices may not be used.*

$n \cdot 180°$	circle	semicircle	180°	n	segment	concentric
$(n-2) \cdot 180°$	center	minor arc	360°	$n+1$	congruent	
$(n-2) \cdot 360°$	3.14	central angle	$\frac{360°}{n}$	$\frac{22}{7}$	diameter	$\frac{(n-2) \cdot 180°}{n}$
adjacent	major arc	apothem	$\frac{360°}{n+1}$	sector	radius	

1. The sum of the measure of an interior angle and an exterior angle at the same vertex of a convex polygon is always ______________.

2. A convex n-gon has how many exterior angles (counting one at each vertex)? ______________

3. The sum of the exterior angle measures of a convex n-gon is ______________.

4. The measure of each exterior angle of a regular n-gon is ______________.

5. The measure of each interior angle of a regular n-gon is ______________.

6. The sum of the interior angle measures of a convex n-gon is ____________.
7. Two approximations for π are ____________ and ____________.
8. Circles with congruent radii or congruent diameters are called ____________ circles.
9. A(n) ____________ is the set of all points equidistant from a point called the ____________.

Use Figure 1 for Exercises 10–12.

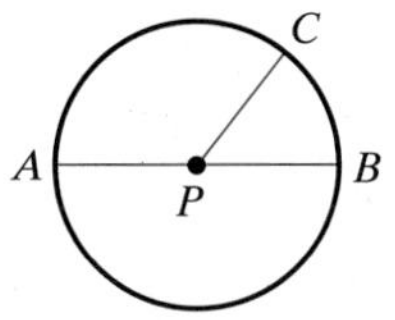

Figure 1

10. $\overline{PC}$ is a(n) ____________.
11. $\overline{AB}$ is a(n) ____________.
12. $\angle CPB$ is a(n) ____________.

Use Figure 2 for Exercises 13–16.

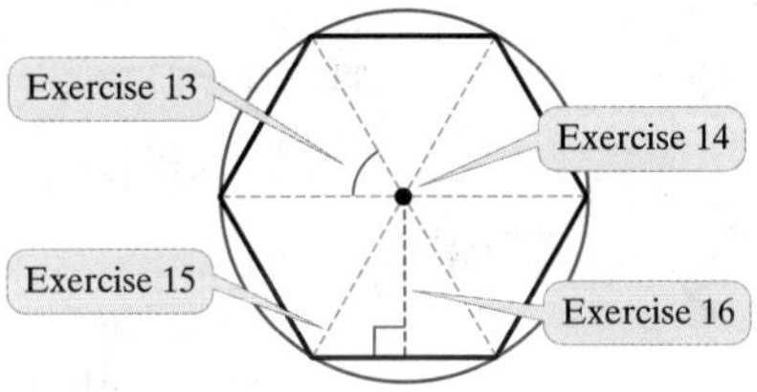

Figure 2

For Exercises 13–16, name each part.

13. ____________ 14. ____________
15. ____________ 16. ____________

Use Figure 3 for Exercise 17.

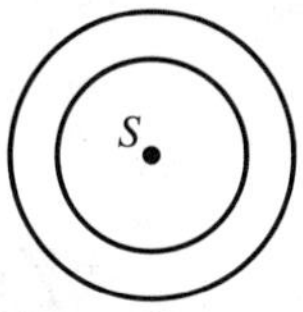

Figure 3

17. The circles in Figure 3 are called ____________ circles.

Use Figure 4 for Exercises 18–21.

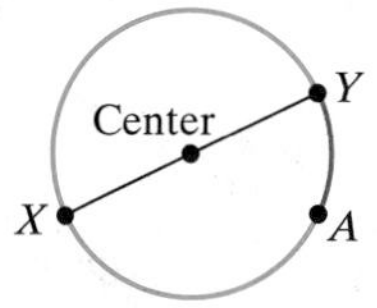

Figure 4

18. $\widehat{XY}$ is called a(n) ____________.
19. AXY is called a(n) ____________.
20. $\widehat{AY}$ is called a(n) ____________.
21. Arcs $\widehat{AY}$ and $\widehat{XA}$ are called ____________ arcs.

Use Figure 5 for Exercises 22–23.

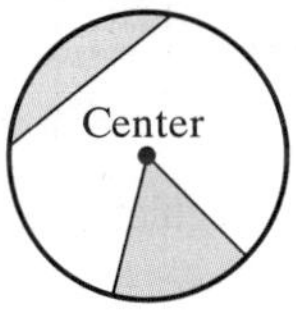

Figure 5

22. The yellow area is called a(n) ____________.
23. The blue area is called a(n) ____________.

CHAPTER 10 REVIEW

***(10.1)** Find the measure of an interior angle and an exterior angle of each regular polygon.*

1. hexagon
2. 16-gon
3. pentagon
4. What is the sum of the exterior angles for each polygon in Exercises 1–3?

Find the measure of the missing angle.

5.

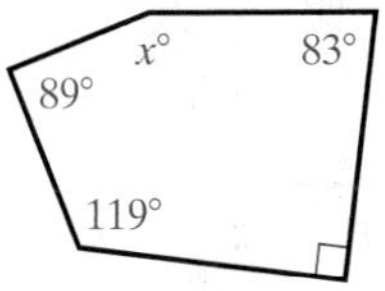

6.

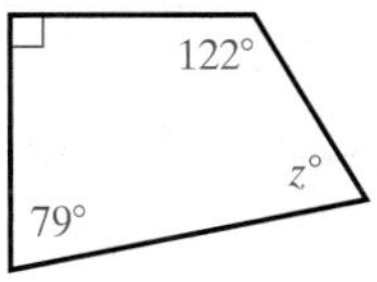

***(10.2)** Find the area of each figure.*

7.

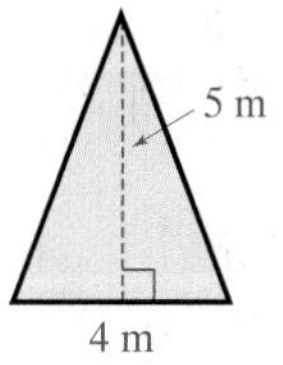

8.

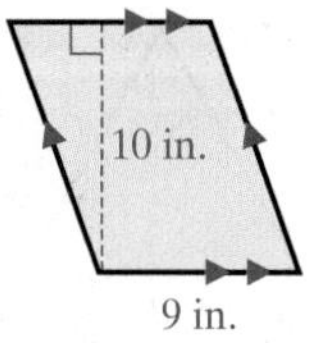

9.

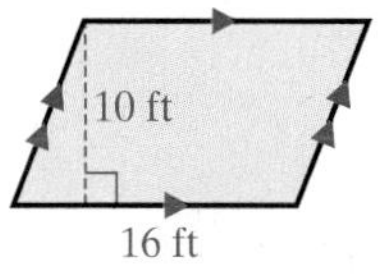

10.

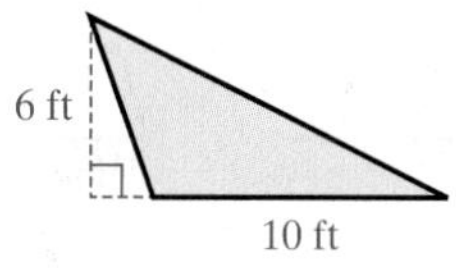

Find the area of each figure. If necessary, leave your answer in simplest radical form.

11.

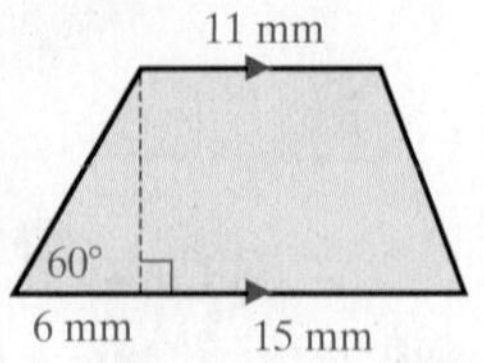

12.

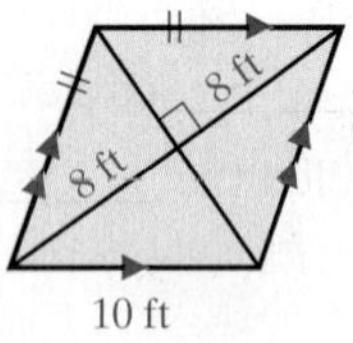

13.

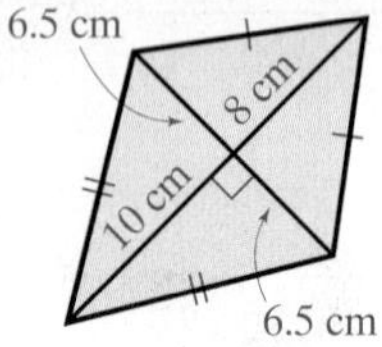

14.

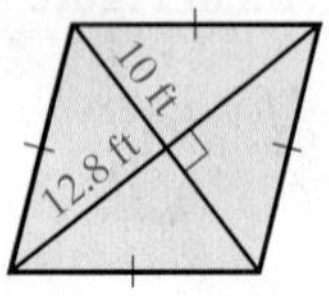

15. A right triangle has legs measuring 5 ft and 12 ft, and a hypotenuse measuring 13 ft. What is its area?

16. A trapezoid has a height of 6 m. The length of one base is three times the length of the other base. The sum of the base lengths is 18 m. What is the area of the trapezoid?

(10.3) *Find the area of each regular polygon. If your answer is not an integer, leave it in simplest radical form.*

17.

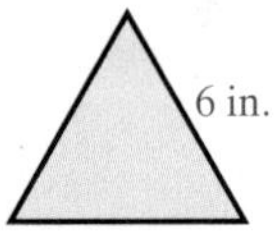

18.

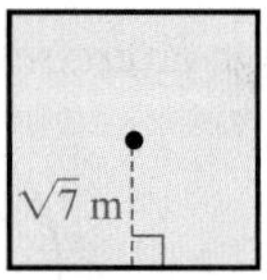

19. What is the area of a regular hexagon with a perimeter of 240 cm?

20. What is the area of a square with radius 7.5 m?

Sketch each regular polygon with the given radius. Then find its area to the nearest tenth.

21. hexagon; radius 7 cm

22. square; radius 8 mm

(10.4) *For each pair of similar figures, find the ratio of the area of the first figure to the area of the second.*

23.

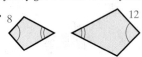

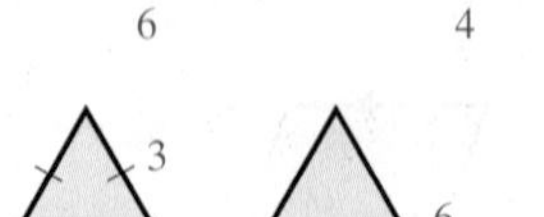

24.

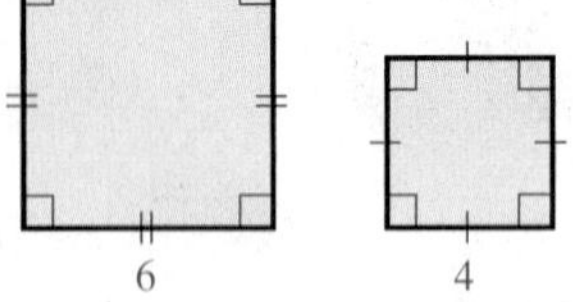

25.

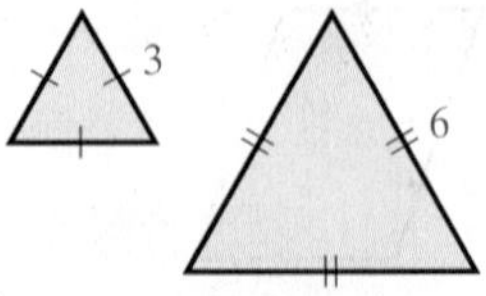

26.

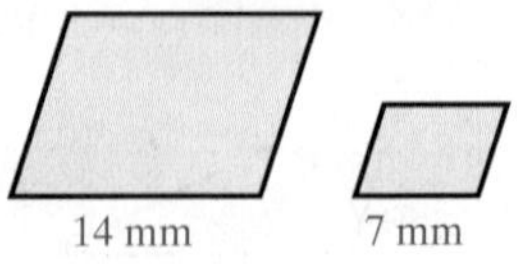

Find the area of each polygon. Round your answers to the nearest tenth. Trigonometric ratios may be needed.

27. regular decagon with radius 5 ft

28. regular pentagon with apothem 8 cm

29. regular hexagon with apothem 6 in.

30. regular quadrilateral with radius 2 m

(10.5) *Find each measure.*

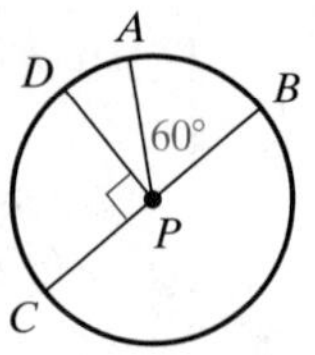

31. $m\angle APD$

32. $m\widehat{AC}$

33. $m\widehat{ABD}$

34. $m\angle CPA$

Find the length of each arc shown in red. Leave your answer in terms of π.

35.

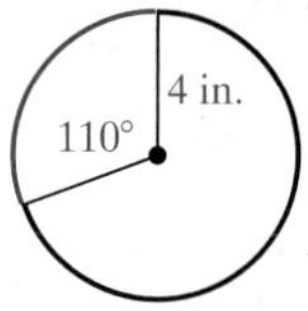

36.

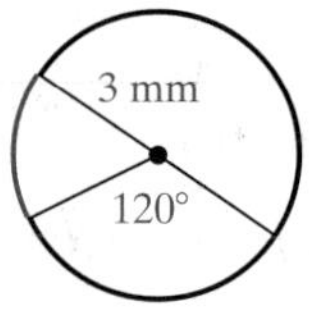

37.

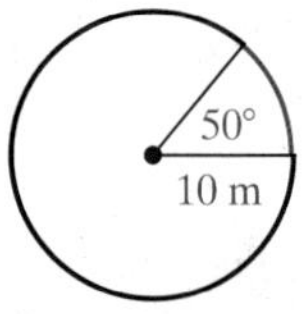

38.

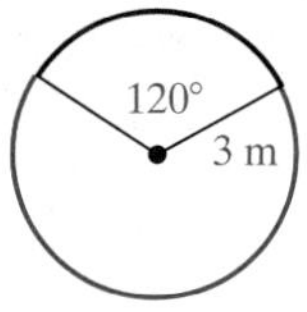

(10.6) *What is the area of each circle? Leave your answer in terms of π.*

39.

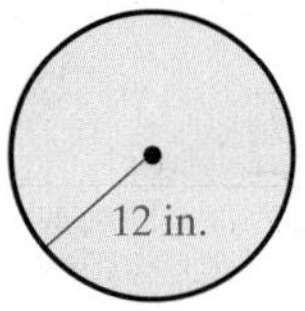

40.

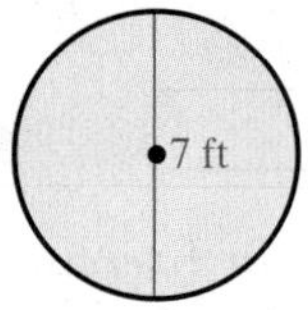

Find the area of each shaded region. Round your answer to the nearest tenth.

41.

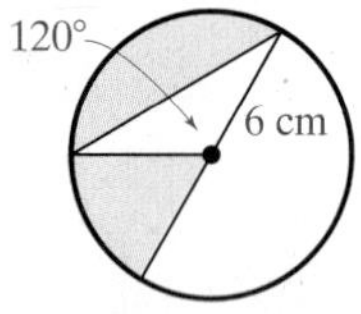

42.

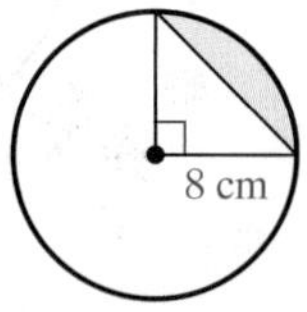

(10.7) *A dart hits each dartboard at a random point. Find the probability that it lands in the shaded region.*

43.

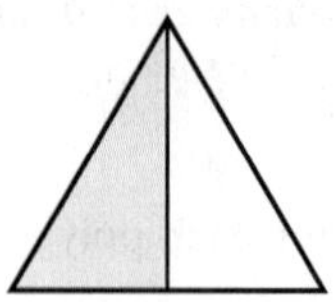

44.

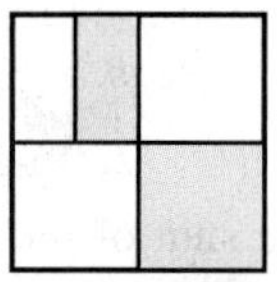

45.

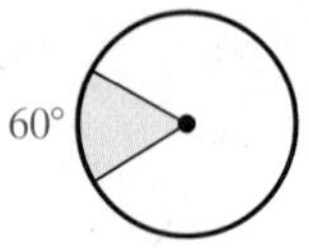

46.

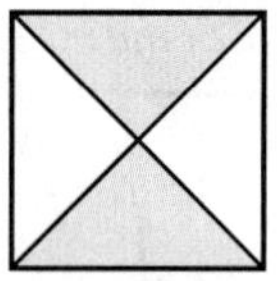

MIXED PRACTICE

47. Find the measure of an interior angle of a regular 20-gon.

48. What is the area of the parallelogram?

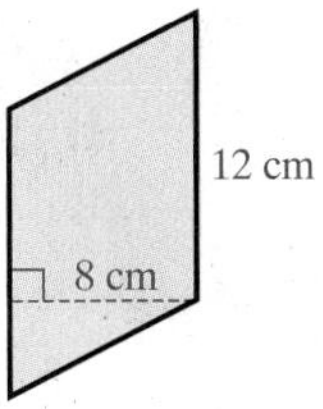

49. What is the area of the trapezoid?

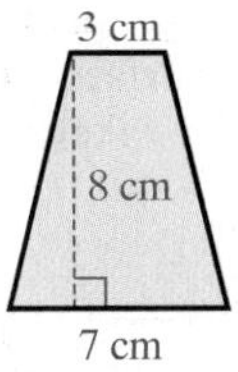

50. Sketch a regular triangle with a radius of 4 in. Then find its area to the nearest tenth.

51. What is the area of a hexagon with apothem 17.3 mm and perimeter 120 mm?

52. If the ratio of the areas of two similar hexagons is 8:25, what is the ratio of their perimeters?

53. If the ratio of the areas of two similar figures is $\frac{4}{9}$, what is the ratio of their perimeters?

Find the area of each polygon. Round your answers to the nearest tenth. Trigonometric ratios may be needed.

54. regular octagon with apothem 10 ft

55. regular heptagon with radius 3 ft

56. A circle has a radius of 5 cm. What is the length of an arc measuring 80°?

57. A circle has a radius of 20 cm. What is the area of the smaller segment of the circle formed by a 60° arc? Round to the nearest tenth.

58. What is the area of the shaded region?

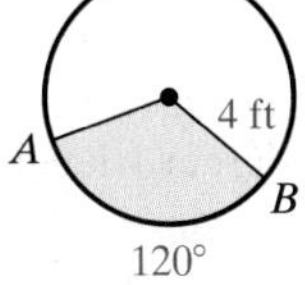

59. A dart hits a dartboard at random. Find the probability that it lands in the shaded region.

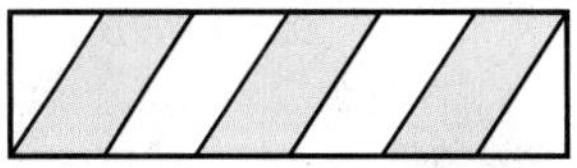

60. A ball hits the target at a random point. What is the probability that it lands in the shaded region?

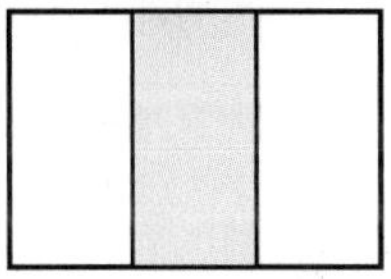

CHAPTER 10 TEST

TEST PREP VIDEO The fully worked-out solutions to any exercises you want to review are available in MyMathLab.

Find the value of each variable.

1.

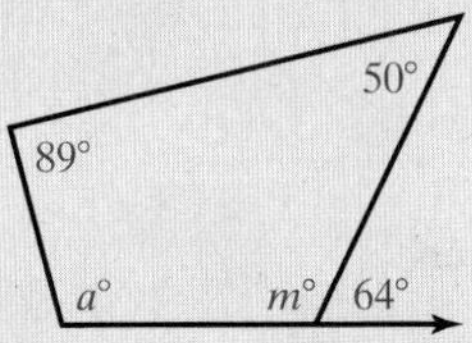

2.

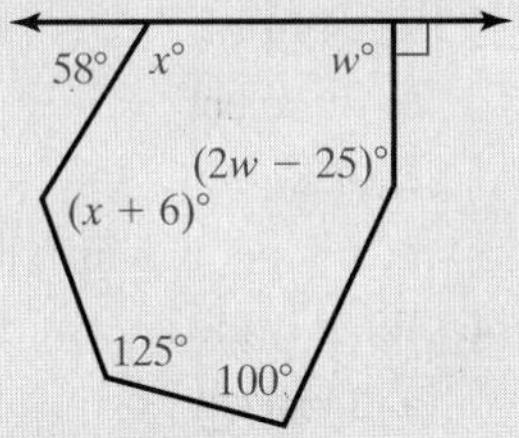

3. Find the measure of an interior and an exterior angle of a regular 36-gon.

Find the area of each figure.

4.

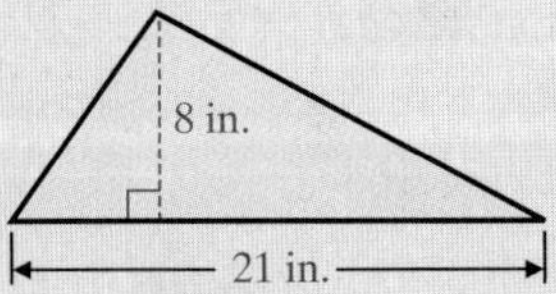

5.

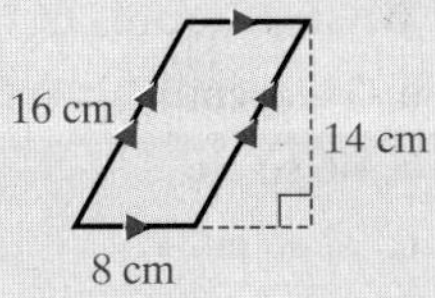

6. If the base of a triangle is 10 cm, and its area is 35 cm^2, what is the height of the triangle?

7. An equilateral triangle has a perimeter of 60 m and a height of 17.3 m. What is its area?

Find the area of each figure. If needed, round to the nearest tenth.

8.

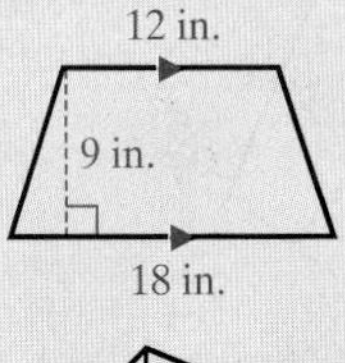

9.

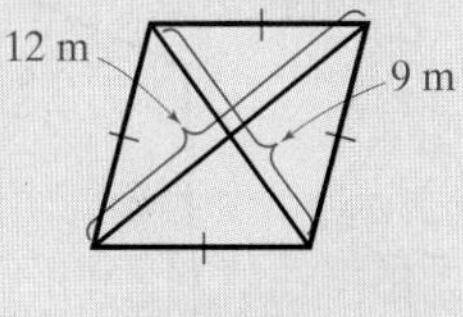

10.

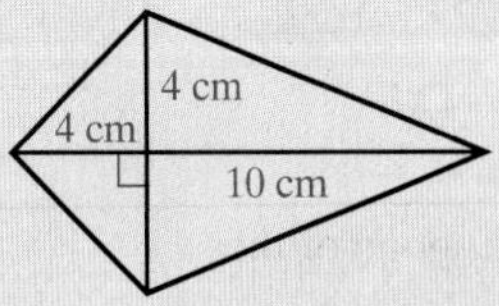

11.

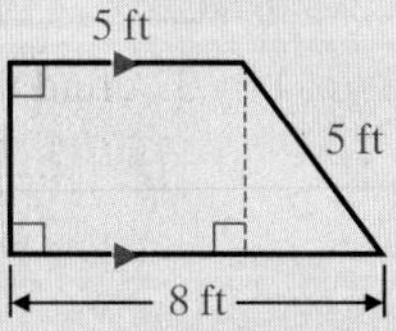

Find the area of each regular polygon. Round to the nearest tenth.

12.

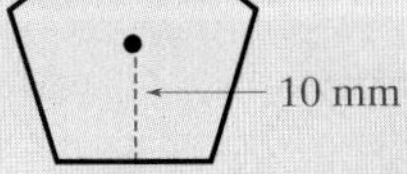

13.

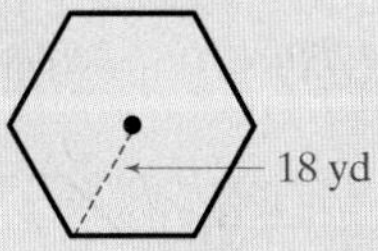

14. A regular octagon has sides 15 cm long. What is the area of the octagon?

15. For the similar trapezoids, find the ratio of the area of the first to that of the second.

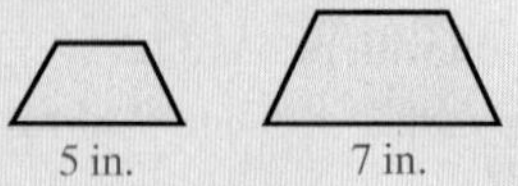

The scale factor of ΔABC to ΔDEF is 3:5. Fill in the missing information.

16. The perimeter of ΔABC is 36 in. The perimeter of ΔDEF is ______?______.

17. The area of ΔABC is ______?______. The area of ΔDEF is 125 in.^{2}.

18. The ratio of the perimeters of two similar triangles is 1:3. The area of the larger triangle is 27 ft^2. What is the area of the smaller triangle?

Find each measure for $\odot P$.

19. $m\angle BPC$ **20.** $m\widehat{AB}$

21. $m\widehat{ADC}$ **22.** $m\widehat{ADB}$

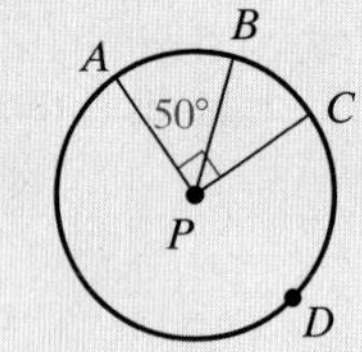

Find the length of each arc shown in red. Leave your answer in terms of π.

23.

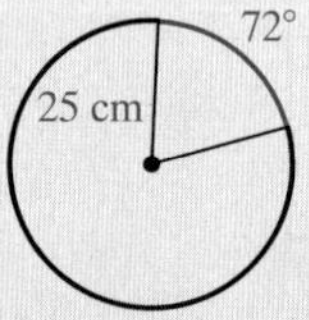

24.

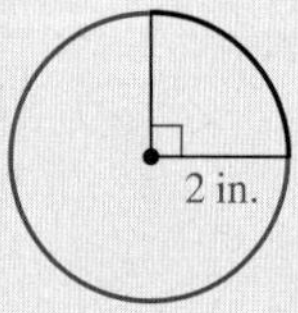

Find the area of each shaded region to the nearest hundredth.

25.

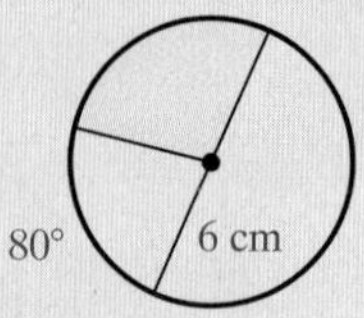

26.

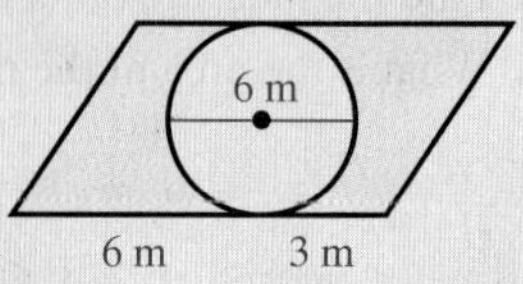

27.

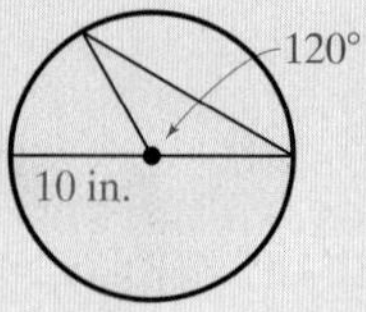

28. Fly A lands on the edge of the ruler at a random point. Fly B lands on the surface of the target at a random point. Which fly is more likely to land in a yellow region? Explain.

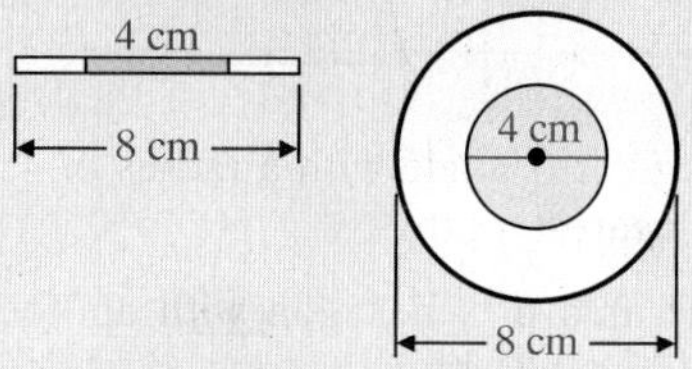

CHAPTER 10 STANDARDIZED TEST

Use the figure for Questions 1 and 2.

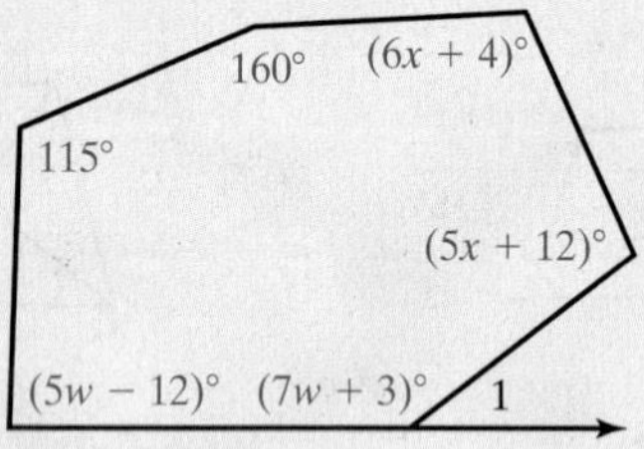

1. If $m\angle 1 = 37°$, find the value of x.

a. $x = 20$ **b.** $x = 18$

c. $x = 22$ **d.** none of these

2. If $x = 24$, find $m\angle 1$.

a. 42° **b.** 75.5°

c. 104.5° **d.** none of these

Use the figure for Exercises 3–8.

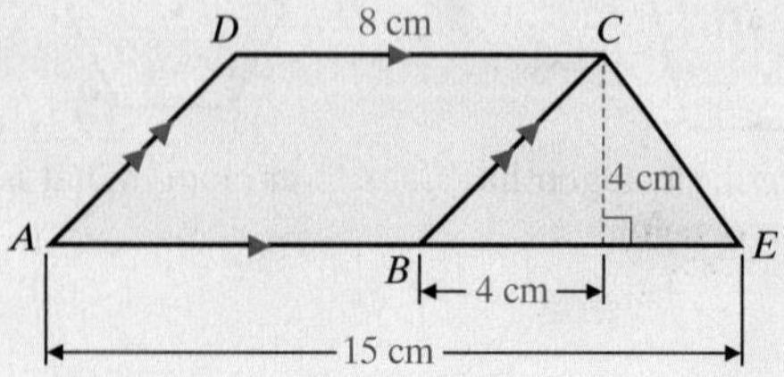

3. What is the area of $\square ABCD$?

a. 14 sq cm

b. 46 sq cm

c. 32 sq cm

d. not enough information given

4. What is the area of ΔBCE?

a. 14 sq cm

b. 46 sq cm

c. 32 sq cm

d. not enough information given

5. What is the area of trapezoid $AECD$?

a. 14 sq cm

b. 46 sq cm

c. 32 sq cm

d. not enough information given

6. What is the approximate perimeter of ΔBCE?

a. 13.7 cm

b. 16.7 cm

c. 17.7 cm

d. not enough information given

7. What is the approximate perimeter of $\square ABCD$?

a. 27 cm

b. 33 cm

c. 34 cm

d. not enough information given

8. What is the approximate perimeter of trapezoid $AECD$?

a. 27 cm

b. 33 cm

c. 34 cm

d. not enough information given

9. A rhombus has sides of length 6 in. and diagonals of length 4 in. and 11 in. What is the area of the rhombus?

a. 33 sq in. **b.** 18 sq in.

c. 44 sq in. **d.** 22 sq in.

10. A regular hexagon has radius 12 mm long. What is the area of the hexagon?

a. $108\sqrt{2}$ sq mm **b.** $432\sqrt{3}$ sq mm

c. $108\sqrt{3}$ sq mm **d.** $216\sqrt{3}$ sq mm

11. What is the approximate area of a regular pentagon with apothem of length 14 ft?

a. 712 sq ft

b. 3016 sq ft

c. 1349 sq ft

d. 1424 sq ft

12. The scale factor of ΔABC to ΔDEF is 5:7. If the perimeter of ΔDEF is 35 cm, what is the perimeter of ΔABC?

a. 25 cm **b.** 49 cm

c. $17\frac{6}{7}$ cm **d.** none of these

13. The scale factor of ΔABC to ΔDEF is 3:8. If the area of ΔDEF is 18 sq ft, what is the area of ΔABC?

a. 128 sq ft **b.** $6\frac{3}{4}$ sq ft

c. $2\frac{17}{32}$ sq ft **d.** none of these

Use $\odot P$ for Exercises 14–19.

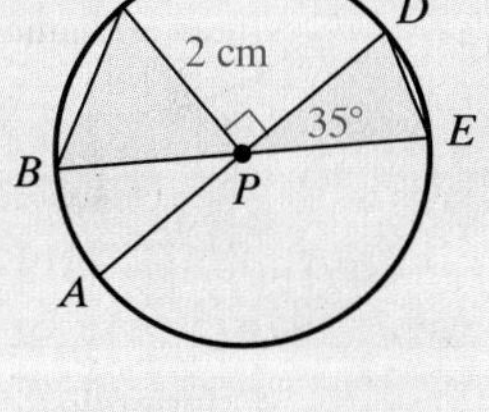

14. What is $m\overset{\frown}{BC}$?

a. 125° **b.** 55°

c. 35° **d.** none of these

15. What is $m\overset{\frown}{ACE}$?

a. 145° **b.** 215°

c. 235° **d.** none of these

16. What is the length of $\overset{\frown}{AB}$?

a. $\frac{7\pi}{36}$ cm **b.** $\frac{11\pi}{18}$ cm

c. $\frac{7\pi}{18}$ cm **d.** none of these

17. What is the length of $\overset{\frown}{AE}$?

a. $\frac{29\pi}{18}$ cm **b.** $\frac{43\pi}{18}$ cm

c. $\frac{29\pi}{36}$ cm **d.** none of these

18. What is the approximate area of the segment indicated between $\overline{PB}$ and $\overline{PC}$?

a. 0.3 sq cm **b.** 1.9 sq cm

c. 1.6 sq cm **d.** none of these

19. What is the approximate area of the sector indicated between $\overline{PD}$ and $\overline{PE}$?

a. 0.1 sq cm **b.** 1.2 sq cm

c. 1.1 sq cm **d.** none of these

A fly lands at random on $\overline{AM}$. Find the probability of the fly landing on each segment given in Exercises 20–22.

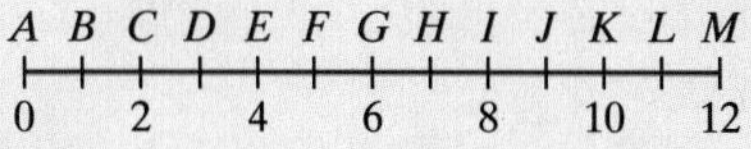

20. on $\overline{CG}$

a. $\frac{1}{6}$ **b.** $\frac{1}{4}$

c. $\frac{2}{5}$ **d.** $\frac{1}{3}$

21. on $\overline{BK}$

a. 67% **b.** 90%

c. 80% **d.** 75%

22. not on $\overline{GJ}$

a. 25% **b.** 70%

c. 80% **d.** 75%

CHAPTER

11 Surface Area and Volume

11.1 Solids and Cross Sections

11.2 Surface Areas of Prisms and Cylinders

11.3 Surface Areas of Pyramids and Cones

11.4 Volumes of Prisms and Cylinders and Cavalieri's Principle

11.5 Volumes of Pyramids and Cones

11.6 Surface Areas and Volumes of Spheres

11.7 Areas and Volumes of Similar Solids

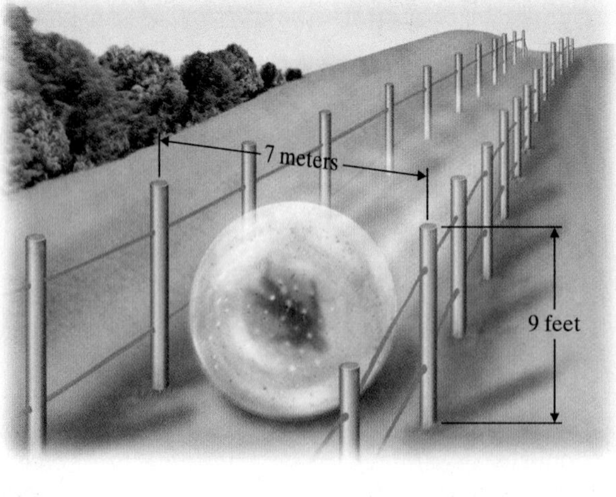

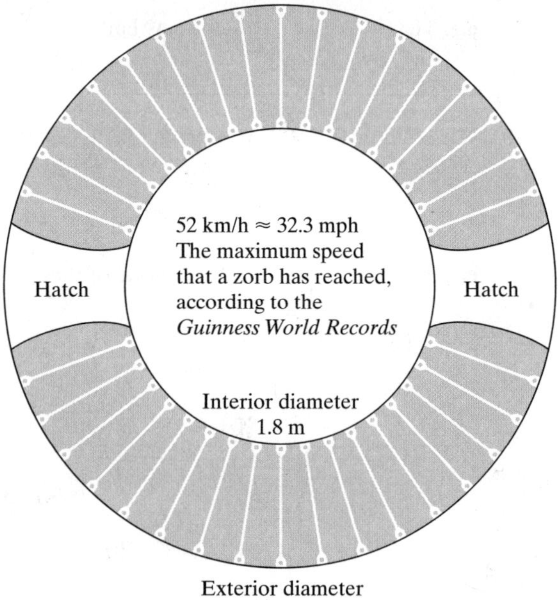

What is a zorb? Simply put, a zorb is a large inflated ball within a ball, and zorbing is a recreational activity that may involve rolling down a hill strapped in a zorb. Zorbing started in New Zealand (as did bungee jumping) and was invented by Andrew Akers and Dwane van der Sluis. The first site was set up in New Zealand's North Island. This downhill course has a length of about 490 feet and you can reach speeds of up to 20 mph.

11.1 SOLIDS AND CROSS SECTIONS

OBJECTIVES

1. Recognize Polyhedra and Their Parts.
2. Visualize Cross Sections of Solids.
3. Visualize Solids Formed by Revolving a Region About a Line.

VOCABULARY

- polyhedron
- face
- edge
- vertex
- polyhedra
- net
- cross section
- topographic map
- contour map

OBJECTIVE 1 ▶ Recognizing Polyhedra. A **polyhedron** is a solid, or three-dimensional figure, whose surface is made up of polygons. Each polygon is a **face** of the polyhedron. An **edge** is a segment that is formed by the intersection of two faces. A **vertex** is a point where three or more edges intersect. The plural of polyhedron is **polyhedra** or polyhedrons, and notice that polyhedra enclose regions of space.

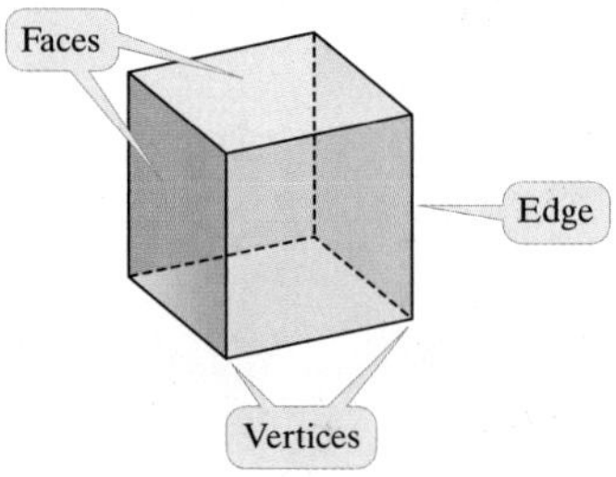

We can analyze a three-dimensional figure by using the relationships among its vertices, edges, and faces.

EXAMPLE 1 **Identifying Vertices, Edges, and Faces**

How many vertices, edges, and faces are in the polyhedron at the right? List them.

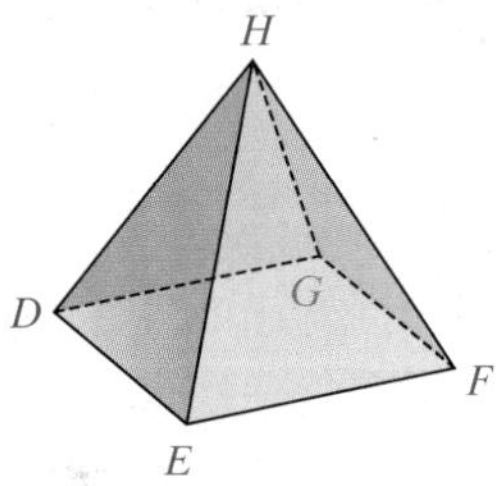

Solution Remember that these are three-dimensional figures drawn on a two-dimensional piece of paper. The dashed segments represent edges of the polyhedron in the back of the solid, which is not seen.

There are five vertices: D, E, F, G, and H.
There are eight edges: $\overline{DE}, \overline{EF}, \overline{FG}, \overline{GD}, \overline{DH}, \overline{EH}, \overline{FH}$, and $\overline{GH}$.
There are five faces: ΔDEH, ΔEFH, ΔFGH, ΔGDH, and quadrilateral $DEFG$. □

PRACTICE 1

a. How many vertices, edges, and faces are in the polyhedron at the right? List them.

b. Is $\overline{TV}$ an edge? Explain why or why not.

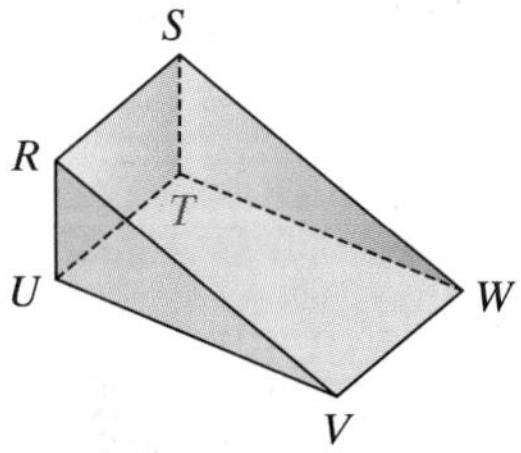

Another way to analyze a polyhedron is to draw a **net.** For our use, a net is a pattern that we can cut and fold to make a solid. We will study the example below in the next section when we introduce surface area.

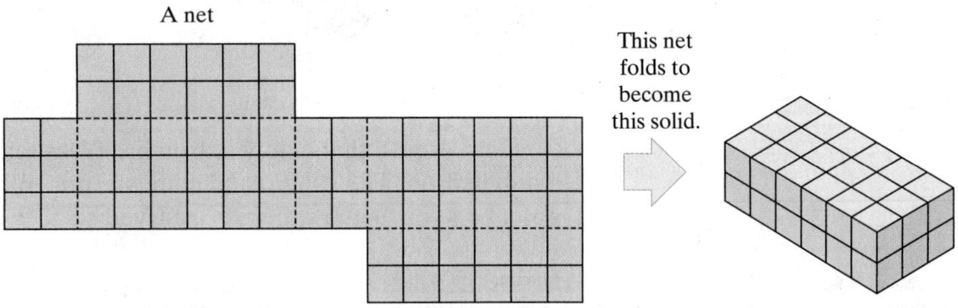

> **Helpful Hint**
> Don't forget that each face of a polyhedron is a polygon.

Leonhard Euler, a Swiss mathematician, discovered a relationship among the numbers of faces, vertices, and edges of any polyhedron. The result is known as Euler's Formula.

Euler's Formula

The sum of the number of faces (F) and vertices (V) of a polyhedron is two more than the number of its edges (E).

$$F + V = E + 2$$

In Example 1, we had

5 faces (F), 5 vertices (V), 8 edges (E), and

$$F + V = E + 2 \quad \text{or}$$
$$5 + 5 \stackrel{?}{=} 8 + 2 \quad \text{Let } F = 5, V = 5, \text{ and } E = 8.$$
$$10 = 10 \quad \text{True}$$

EXAMPLE 2 **Using Euler's Formula**

How many vertices, edges, and faces does the polyhedron at the right have? Use your results to verify Euler's Formula.

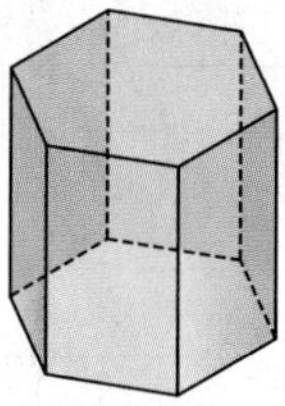

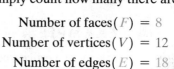

Solution In Example 1, we actually listed the vertices, edges, and faces. In this example, we simply count how many there are.

$$\text{Number of faces}(F) = 8$$
$$\text{Number of vertices}(V) = 12$$
$$\text{Number of edges}(E) = 18$$

Euler's Formula:

$$F + V = E + 2$$
$$8 + 12 \stackrel{?}{=} 18 + 2 \quad \text{Let } F = 8, V = 12, \text{ and } E = 18.$$
$$20 = 20 \checkmark \quad \text{True.}$$

□

PRACTICE 2 For each polyhedron, use Euler's Formula to find the missing number.

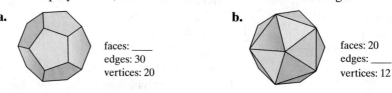

a.

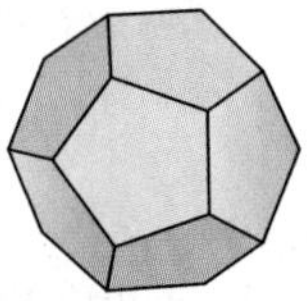

faces: ____
edges: 30
vertices: 20

b.

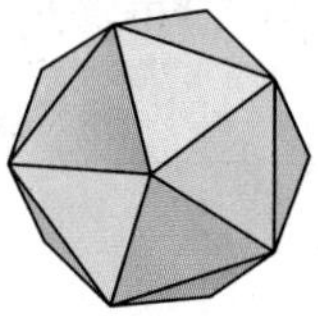

faces: 20
edges: ____
vertices: 12

After this section, we will concentrate on the five solids shown in Example 3 and Practice 3.

EXAMPLE 3 **Which of the three solids shown are polyhedra?**

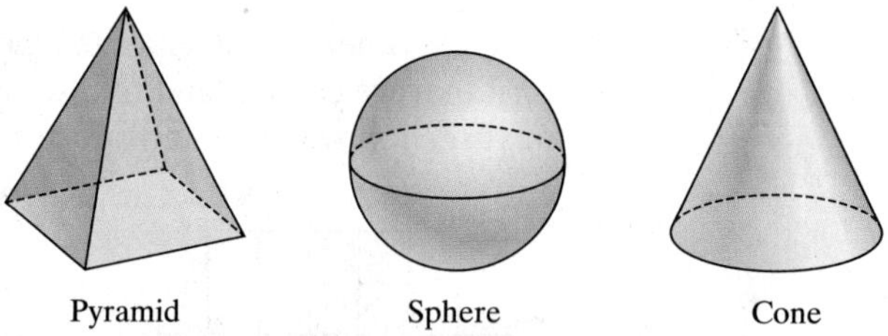

Solution Recall that a polyhedron has faces that are all polygons. Thus, the pyramid is a polyhedron. The sphere and cone are not. Why? The base of the cone is a circle and not a polygon, and a sphere is a curved surface and contains no polygons. □

PRACTICE 3 Which of the two solids shown is a polyhedron?

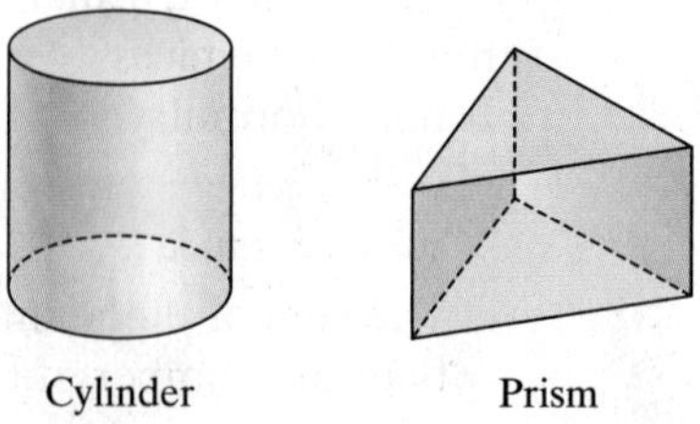

OBJECTIVE 2 ▶ Visualizing Cross Sections of Solids. A **cross section** is the intersection of a solid and a plane. We can think of a cross section as a very thin slice of the solid.

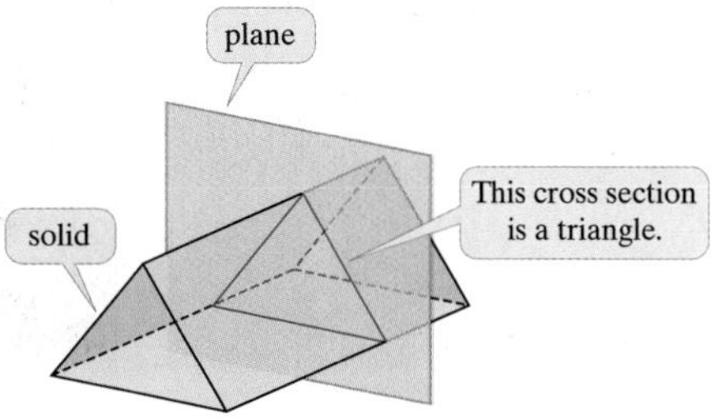

EXAMPLE 4 **Describing a Cross Section**

What is the cross section formed by the plane and the solid shown? (Assume that each plane is perpendicular or parallel to the circular bases of the solid.)

a.

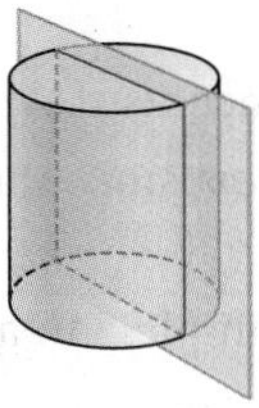

b.

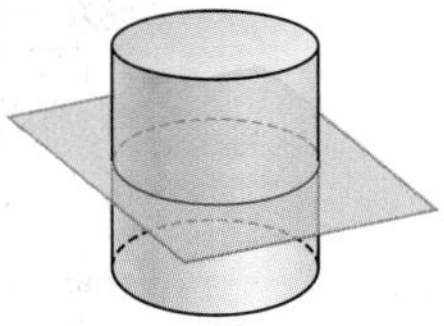

Solution

a. The cross section is a rectangle. **b.** The cross section is a circle. □

PRACTICE 4 What is the cross section formed by the planes and the solid shown?

a.

b.

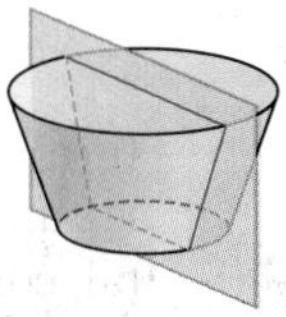

There are many cross-section applications. An important one is topographic maps. A **topographic map,** or **contour map,** shows elevation contour lines, in addition to detailed physical characteristics of the landscape.

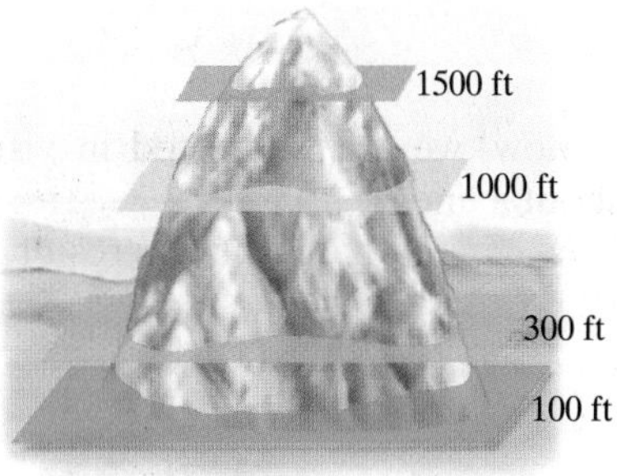

Elevation layers are measured using digital surveying methods.

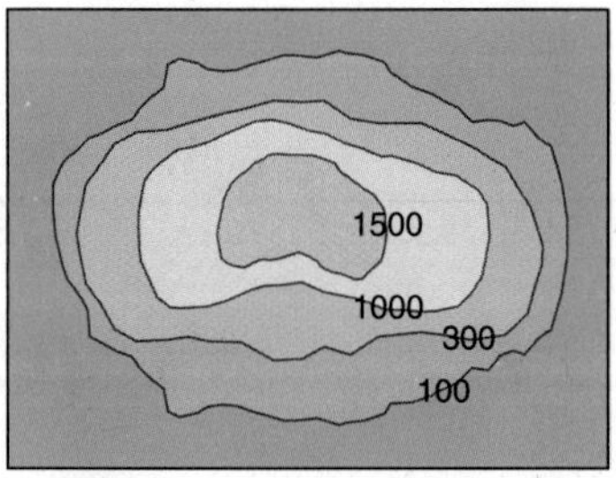

Topographic or Contour Map
The contour of the landscape is shown for each measured elevation.

EXAMPLE 5 **Reading a Contour Map**

Use the given contour map to answer each question.

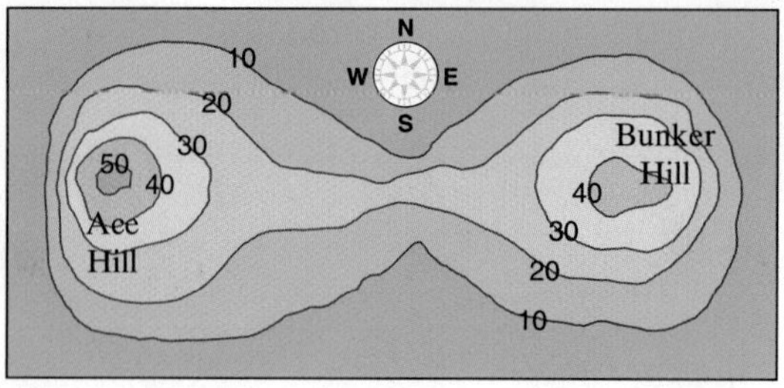

Elevation in meters.

a. Which hill is greater in elevation?

b. Which side of Ace Hill—north, south, east, or west—is steepest?

Solution

a. Ace Hill is greater in elevation since its elevation is 50 m.

b. The west side of Ace Hill is steepest since the contour lines are closest. □

PRACTICE 5 Use the contour map in Example 5 to answer each question.

a. What is the elevation of Bunker Hill?

b. Which side of Bunker Hill is steepest?

OBJECTIVE 3 ▶ Visualizing Solids of Revolution. You may have noticed stairs with wooden spindles and wondered how they're manufactured so that they are all the same size. This is done with a lathe.

A lathe is a machine for shaping an article of wood, metal, etc., by holding and turning it rapidly against the edge of a cutting tool. Two-person lathes date back to around 1300 B.C. in ancient Egypt. They continue to evolve and today's lathes are digital, contain computers, and have a multitude of applications.

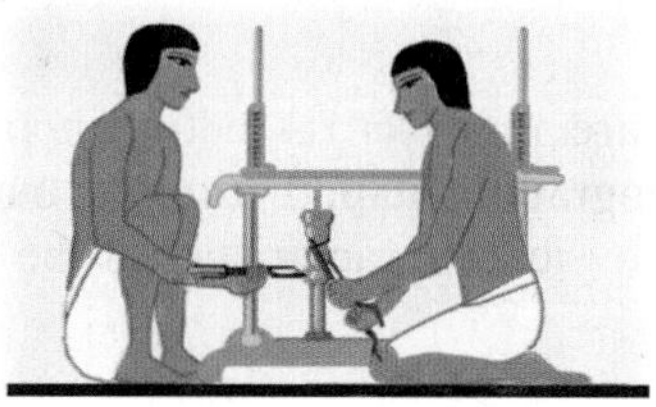

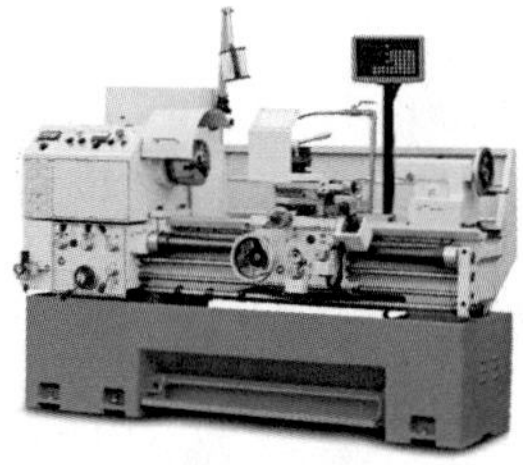

Computer Numerical Controlled (CNC) Lathe

For now, we are interested in visualizing the end result of revolving a plane region about a line ℓ.

For example, what is the end result of rotating a rectangular region about a line ℓ?

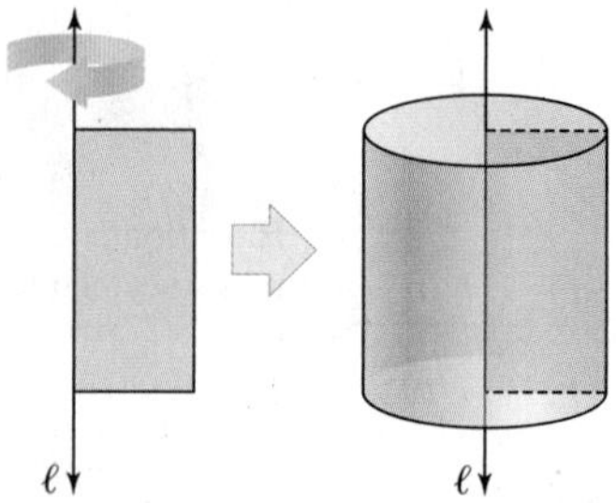

The end result is a cylinder. Thus, a cylinder is a solid of revolution. We can also say that a cylinder has rotational symmetry.

EXAMPLE 6 **Visualizing Solids of Revolution**

Describe the solid of revolution obtained by rotating the given plane region about line ℓ.

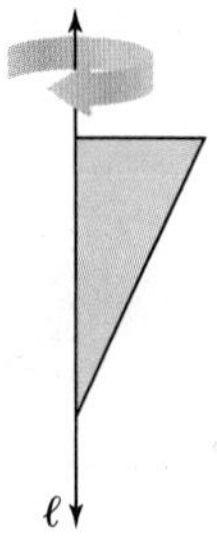

Solution The solid of revolution is a cone.

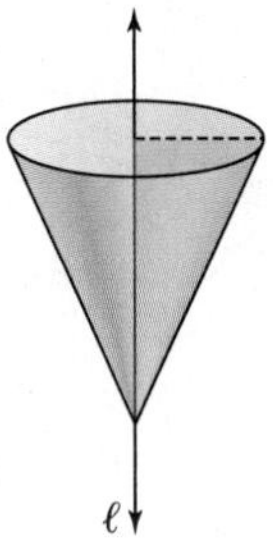

□

PRACTICE 6 Describe the solid of revolution obtained by rotating the given plane region about line ℓ.

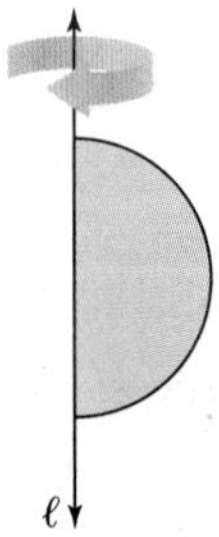

VOCABULARY & READINESS CHECK

Word Bank *Use the words and phrases to fill in each blank.*

net	edge	polyhedron	cross section
face	vertex	polyhedra	topographic map

1. A pattern that can be cut and folded to make a solid is called a(n) ____________.
2. A(n) ____________ is the intersection of a solid and a plane.
3. A(n) ____________ shows elevation contour lines as well as physical characteristics of the landscape.
4. A(n) ____________ is a solid, or three-dimensional figure, whose surfaces are polygons.
5. A plural of polyhedron is polyhedrons or ____________.
6. Each polygon of a polyhedron is called a(n) ____________.
7. A(n) ____________ of a polyhedron is a segment that is formed by the intersection of two faces.
8. A point of a polyhedron where three or more edges intersect is called a(n) ____________.

11.1 EXERCISE SET MyMathLab®

For each polyhedron, how many vertices, edges, and faces are there? List them. See Example 1.

1.

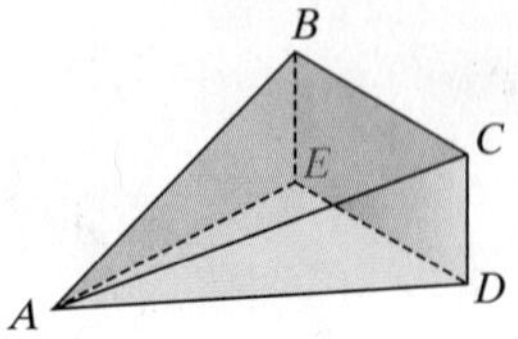

2.

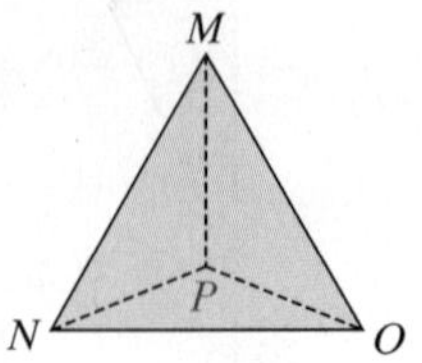

3.

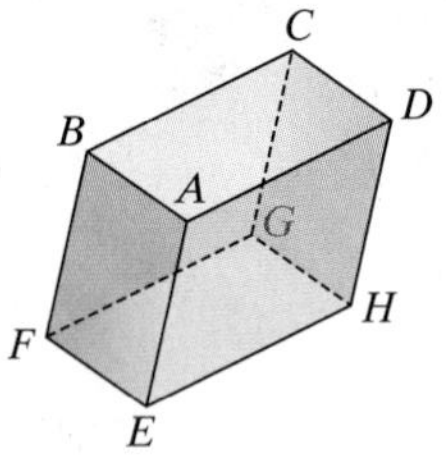

4.

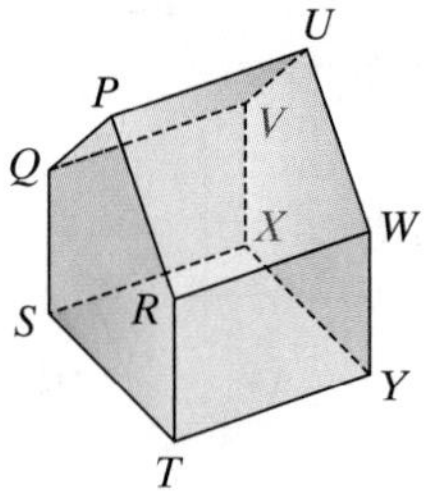

For each polyhedron, use Euler's Formula to find the missing number. See Example 2.

5. faces: ____________
edges: 15
vertices: 9

6. faces: ____________
edges: 12
vertices: 8

7. faces: 8
edges: ____________
vertices: 6

8. faces: 8
edges: ____________
vertices: 16

9. faces: 20
edges: 30
vertices: ____________

10. faces: 22
edges: 34
vertices: ____________

Tell whether the solid is a polyhedron. See Example 3.

11.

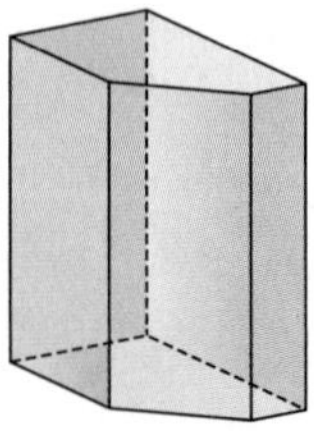

12.

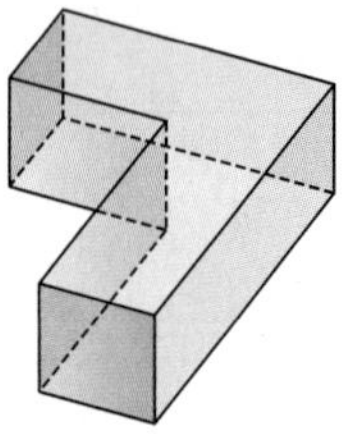

13.

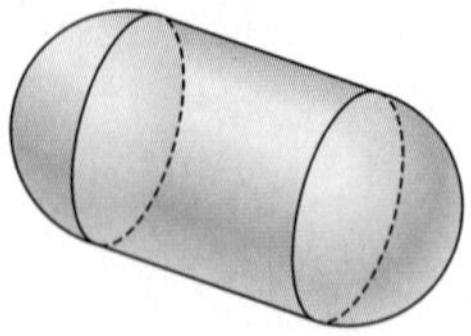

14.

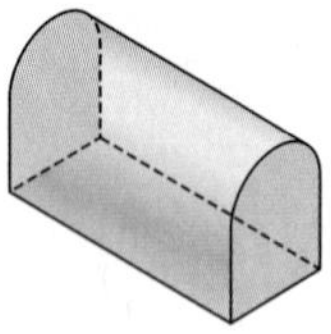

Describe each cross section. See Example 4. Use the same plane assumptions as for Example 4.

15.

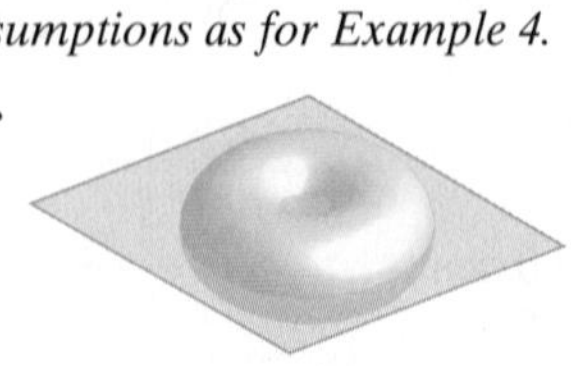

16.

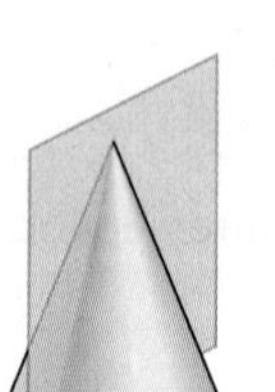

17.

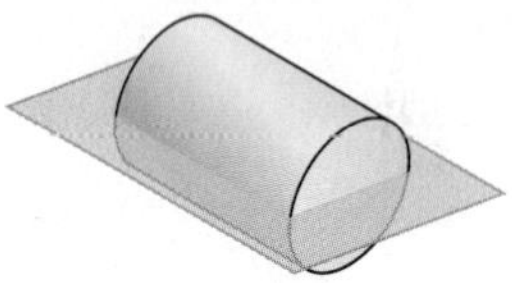

18.

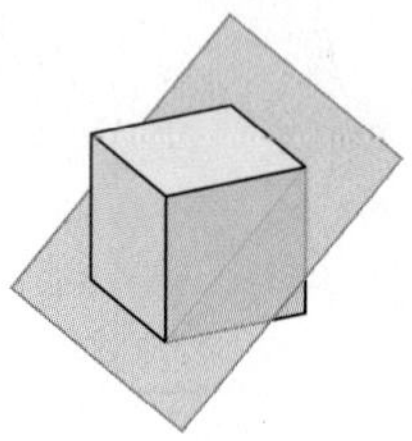

19.

20.

For Exercises 21–24, what is the shape formed by the cut? (Each answer is a two-dimension or plane figure.)

21.

22.

23.

24.

Use the contour map to answer each question. Elevations are in feet and elevation lines are in increments of 50 ft. See Example 5.

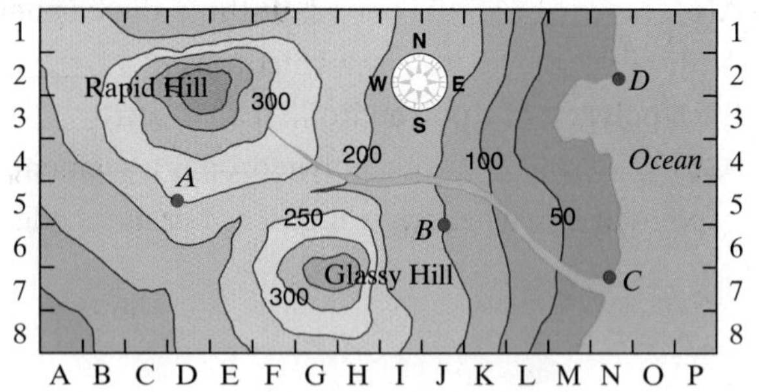

25. The elevation at point B (grid location J-6) is

a. 120 ft **b.** 150 ft **c.** 200 ft **d.** 100 ft

26. The elevation at point A (grid location D-5) is

a. 280 ft **b.** 150 ft **c.** 250 ft **d.** 200 ft

27. The elevation at grid location H-1 is

a. 200–250 ft **b.** 150–200 ft **c.** 50–100 ft **d.** 0–50 ft

28. The elevation at grid location M-4 is

a. 200–250 ft **b.** 150–200 ft **c.** 50–100 ft **d.** 0–50 ft

29. What side of Rapid Hill is the steepest?

a. north **b.** south **c.** east **d.** west

30. What side of Glassy Hill is the steepest?

a. north **b.** south **c.** east **d.** west

31. Which has the greatest elevation?

a. Rapid Hill **b.** Glassy Hill **c.** point A **d.** point C

32. Which has the lowest elevation?

a. Rapid Hill **b.** Glassy Hill **c.** point A **d.** point C

Draw the solid of revolution obtained by rotating the given plane region about line ℓ. See Example 6.

33.

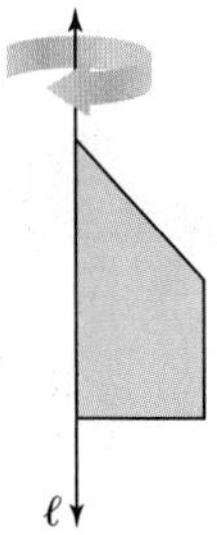

34.

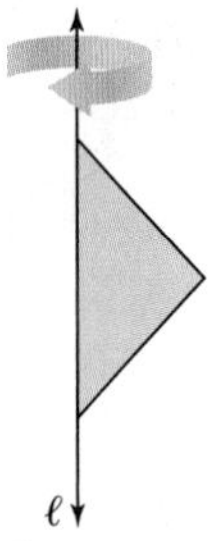

35.

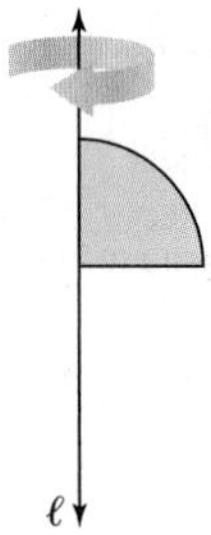

36.

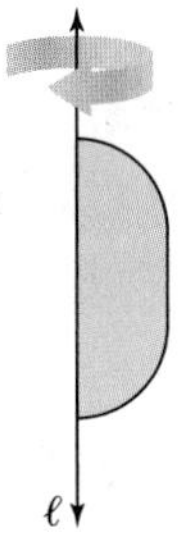

There are five regular polyhedrons. They are called regular because all their faces are congruent regular polygons, and the same number of faces meet at each vertex. They are also called Platonic solids after the Greek philosopher Plato, who first described them in his work Timaeus (about 350 B.C.).

Fill in the Table *For the last column, match each net below with a Platonic solid.*

a.

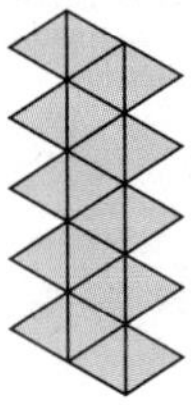

b.

c.

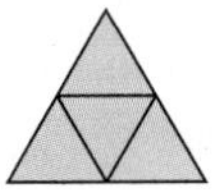

d.

e.

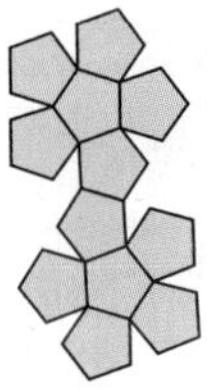

Platonic Solids

	Platonic Solid	*Number of Faces*	*Shape of Each Regular Polygon*	*Number of Vertices*	*Number of Edges*	*Matching: Unfolded Polyhedron, or Net (answer a, b, c, d, or e from above)*
37.	Tetrahedron 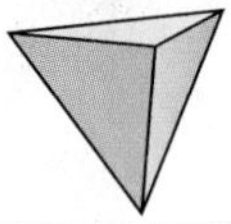					
38.	Cube 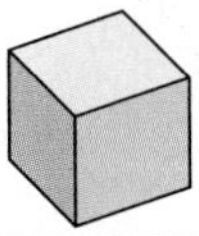					
39.	Octahedron 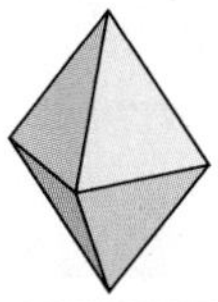					
40.	Dodecahedron 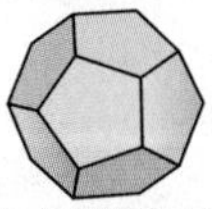			20	30	
41.	Icosahedron 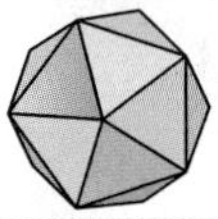			12	30	

42. Show that Euler's Formula is true for the first three Platonic solids in the table.

Complete the table for Exercises 43–48.

	Solid (3-D)	***Number of Faces***	***Number of Edges***	***Number of Vertices***	***(2-D) Parts of the Net***	***Net Drawing***
43.	Cube 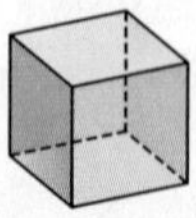				Number of: Squares____	
44.	Prism (square base) 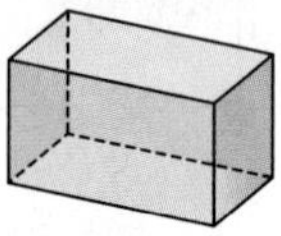				Number of: Squares____ Rectangles____	
45.	Cylinder 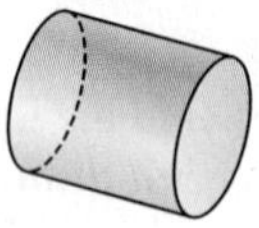				Number of: Circles____ Rectangles____	
46.	Pyramid (triangle base) 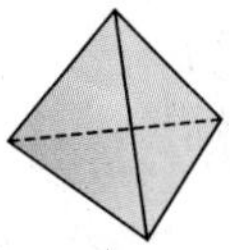				Number of: Triangles____	
47.	Pyramid (square base) 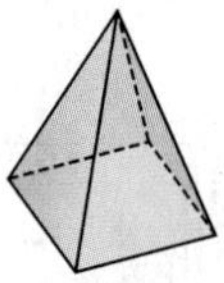				Number of: Squares____ Triangles____	
48.	Prism (pentagon base) 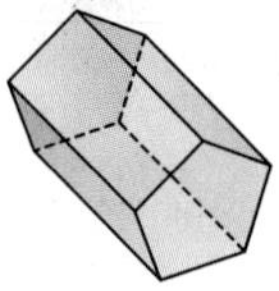				Number of: Rectangles____ Pentagons____	

Recall that three-dimensional objects can have rotational symmetry about a line.

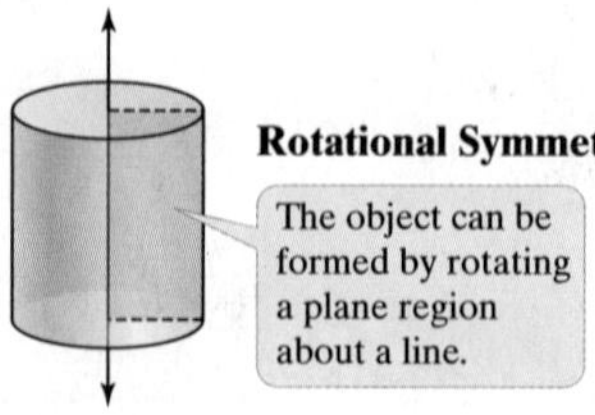

Does each object have rotational symmetry about a line? See Example 6.

49.

50. the base of the lamp

51.

52. the lampshade of the lamp in Exercise **50**

CONCEPT EXTENSIONS

53. **Find the Error** Your math class is drawing polyhedrons. Which figure does not belong in the diagram below? Explain.

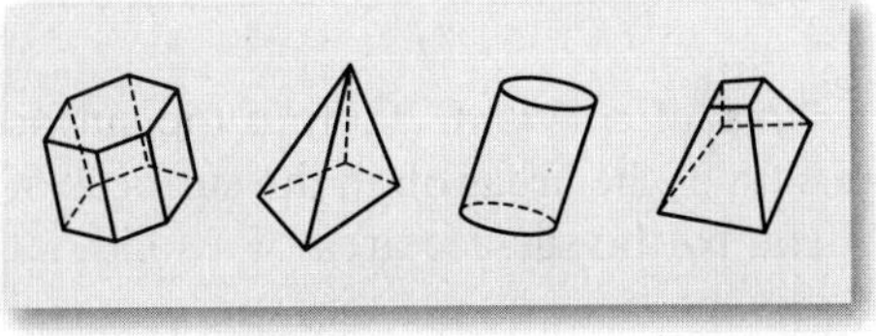

54. Suppose you build a polyhedron from two octagons and eight squares. Without using Euler's Formula, how many edges does the solid have? Explain.

55. A cube has a net with area 216 sq in. How long is an edge of the cube?

56. A cube has a net with area 150 sq cm. How long is an edge of the cube?

57. For the figure shown at the right, sketch each of following.

a. a horizontal cross section

b. a vertical cross section that contains the vertical line of symmetry

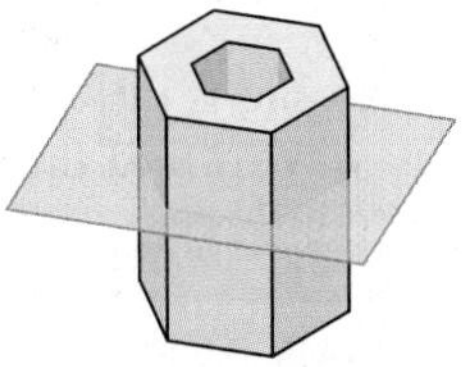

58. a. Sketch a polyhedron whose faces are all rectangles. Label the lengths of its edges.

b. Use graph paper to draw two different nets for the polyhedron.

59. Can you find a cross section of a cube that forms a triangle? Explain.

60. Suppose the number of faces in a certain polyhedron is equal to the number of vertices. Can the polyhedron have nine edges? Explain.

Draw and describe a cross section formed by a plane intersecting the cube as follows.

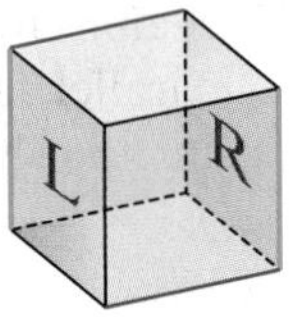

61. The plane contains the red edges of the cube.

62. The plane contains the blue edges of the cube.

63. The plane is tilted and intersects the left (L) and right (R) faces of the cube.

64. The plane cuts off a corner of the cube.

Draw the solid of revolution obtained by rotating the given plane region about line ℓ.

65.

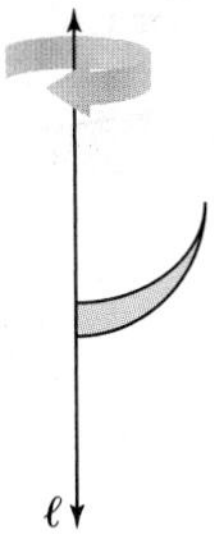

66.

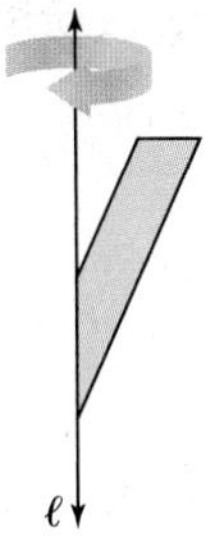

67.

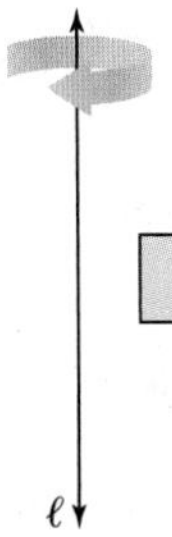

68.

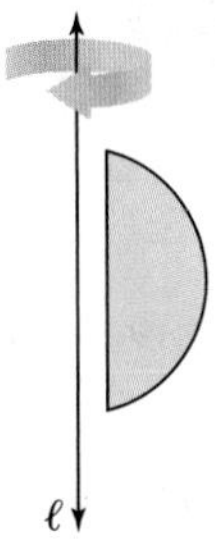

Some balls are made from panels that suggest polygons. A soccer ball suggests a polyhedron with 20 regular hexagons and 12 regular pentagons.

69. How can you determine the number of edges in a solid if you know the types of polygons that form the faces?

70. How many vertices does this polyhedron have?

71. Cross sections are used in medical training and research. Research and write a paragraph on how magnetic resonance imaging (MRI) is used to study cross sections of the brain.

72. Research topographic or contour maps. Although still incomplete, National Geographic's TOPO! offers topographic maps on CD-Rom by state. See if your state or area where you live has a map available.

Sketch *Draw a plane intersecting a cube to get the cross section indicated.*

73. scalene triangle

74. isosceles triangle

75. trapezoid

76. isosceles trapezoid

REVIEW AND PREVIEW

Find the value of x to the nearest tenth. See Section 9.3.

77.

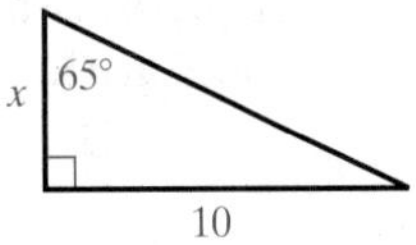

78.

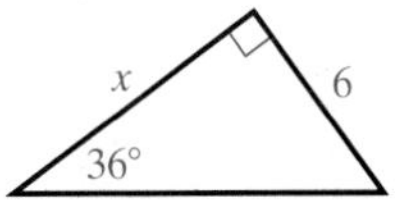

Find the area of each net. See Sections 2.1 and 10.3.

79.

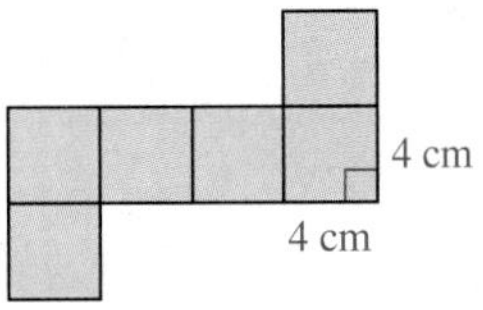

80.

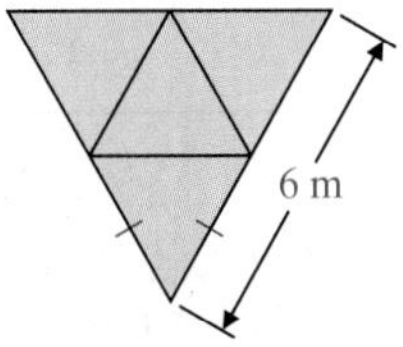

81.

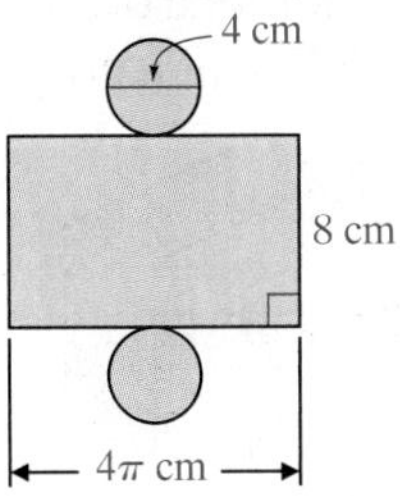

11.2 SURFACE AREAS OF PRISMS AND CYLINDERS

OBJECTIVES

1. Find the Surface Area of a Prism.
2. Find the Surface Area of a Cylinder.

VOCABULARY

- surface area
- prism (base, lateral face, altitude, height, lateral area)
- right prism
- oblique prism
- cylinder (base, altitude, height, lateral area, surface area)
- right cylinder
- oblique cylinder

Helpful Hint
Surface area of a solid is the area of the surface of the solid, so it is measured in square units.

In this section, we learn properties of three-dimensional figures by investigating their surfaces.

OBJECTIVE 1 ▶ Finding the Surface Area of a Prism. To find the surface area of a three-dimensional figure, we find the sum of the areas of all the surfaces of the figure.

To understand fully what we mean by the surface area of a solid, let's look at a net of a box.

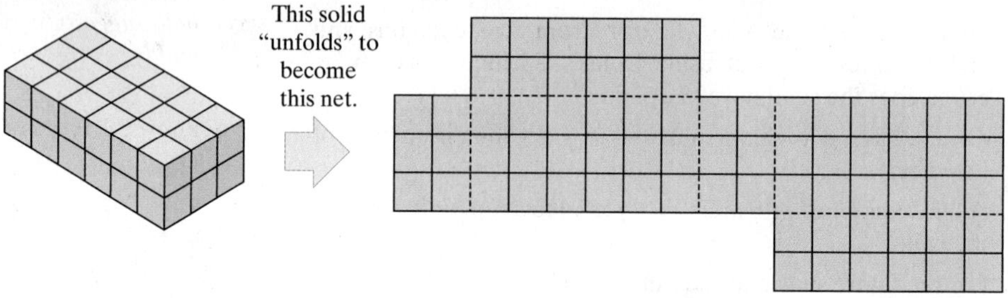

Solid | A Net of the solid

The **surface area** (SA) of a solid is just what the phrase says—it is the area of the surface of the solid. Thus, one way to find this surface area is to find the sum of the areas of the rectangles.

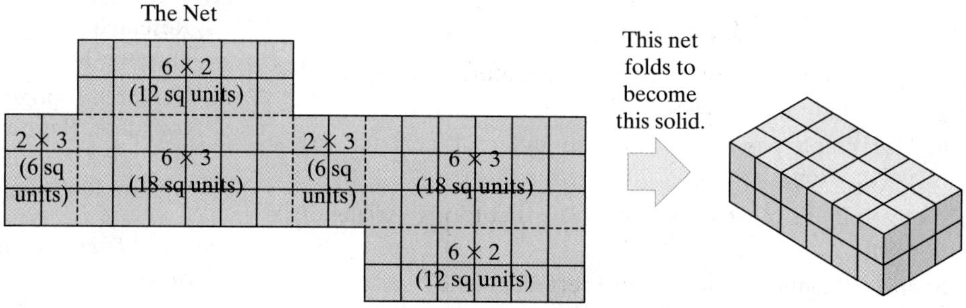

$$\text{Surface Area} = (6 + 12 + 18 + 6 + 18 + 12)\text{ sq units}$$
$$= 72 \text{ sq units}$$

Next, let's see if we can use our definition of surface area to discover a formula for the surface area of a prism. First, we introduce the names for the parts of a prism.

A **prism** is a polyhedron with two congruent, parallel faces, called **bases**. The other faces are **lateral faces**. You can name a prism using the shape of its bases.

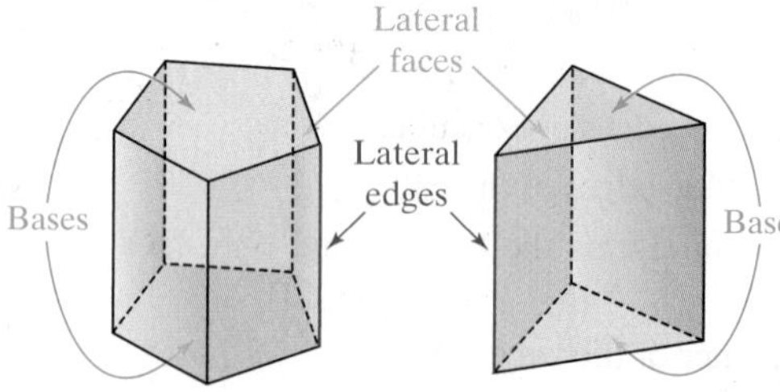

Pentagonal prism | Triangular prism

An **altitude** of a prism is a perpendicular segment that joins the planes of the bases. The **height** h of a prism is the length of an altitude. A prism may either be right or oblique.

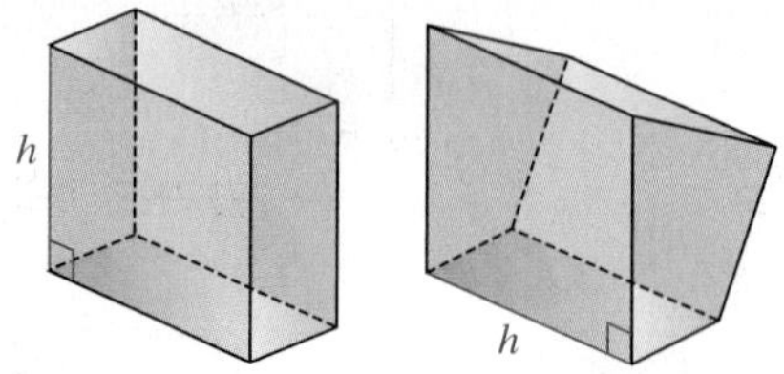

Right prisms

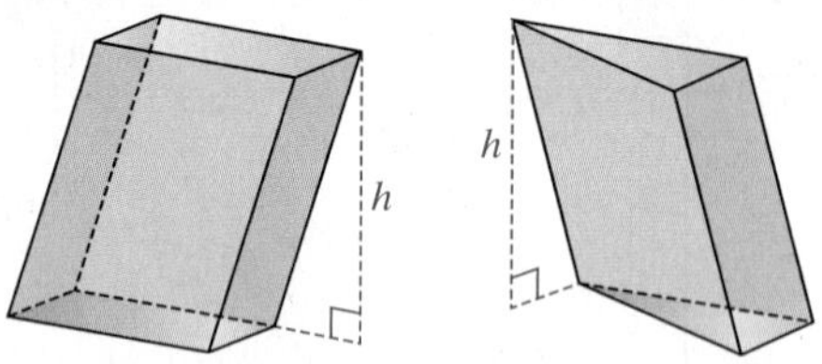

Oblique prisms

In a **right prism**, the lateral faces are rectangles and a lateral edge is an altitude. In an **oblique prism**, some or all of the lateral faces are nonrectangular. In this text, we will assume that a prism is a right prism unless stated or pictured otherwise.

The **lateral area** (LA) of a prism is the sum of the areas of the lateral faces. The **surface area** (SA) is the sum of the lateral area and the area of the two bases.

EXAMPLE 1 Using a Net to Find the Surface Area of a Prism

What is the surface area of the prism at the right? Use a net.

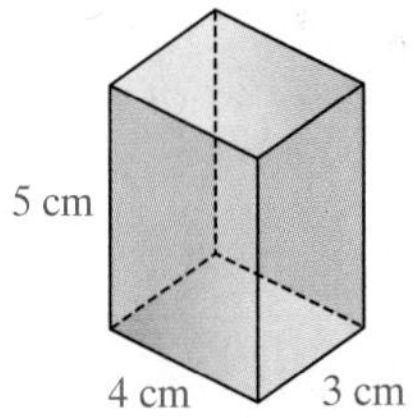

Solution First, we study the prism so that we may draw a net for the prism. Then we calculate the surface area by finding the sum of the areas of the surfaces of the faces. Each measure is in cm, so each area is in square cm.

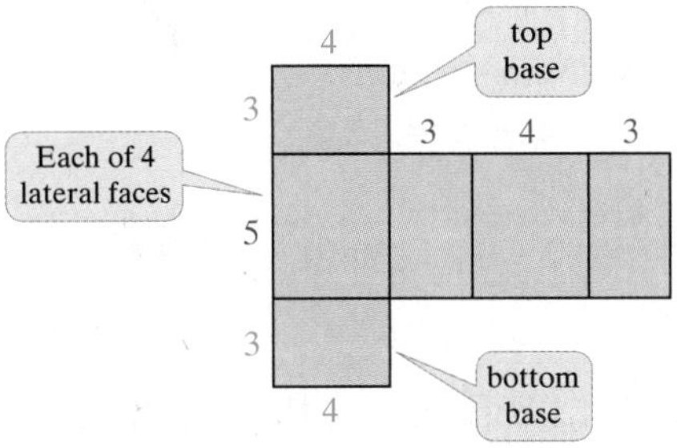

$$\begin{aligned} SA &= \text{sum of areas of all the faces} \\ &= 5 \cdot 4 + 5 \cdot 3 + 5 \cdot 4 + 5 \cdot 3 + 3 \cdot 4 + 3 \cdot 4 \\ &= 20 + 15 + 20 + 15 + 12 + 12 \\ &= 94 \end{aligned}$$

The surface area of the prism is 94 sq cm. □

PRACTICE 1 What is the surface area of the triangular prism? Use a net.

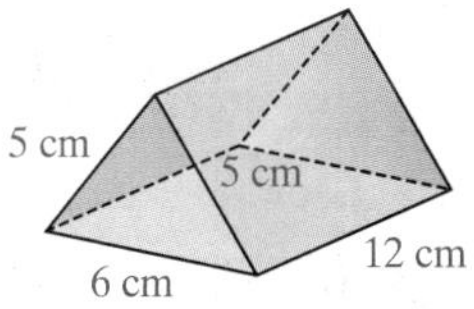

We can find formulas for the lateral area and the surface area of a prism by using a net.

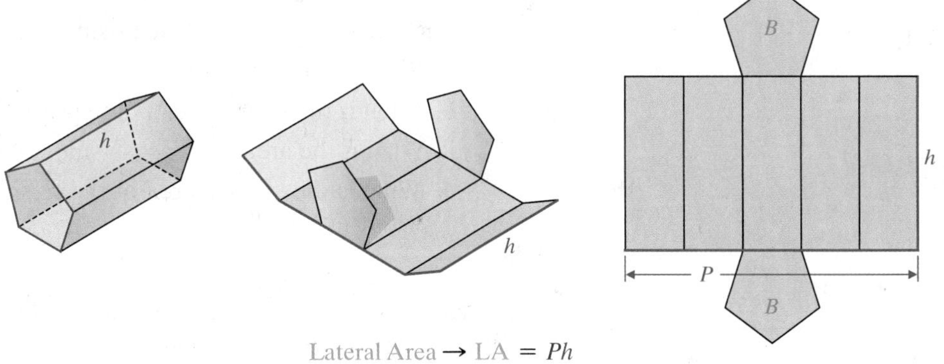

Lateral Area → LA $= Ph$

Surface Area → SA $= 2B + Ph$

We may use these formulas with any right prism.

Theorem 11.2-1 Surface Area of a Prism

The surface area SA of a right prism is

$$SA = 2B + Ph$$

where B is the area of a base, P is the perimeter of a base, and h is the height of the prism.

EXAMPLE 2 **Using Formulas to Find Surface Area of a Prism**

Find the surface area of the prism by answering parts **a–d**.

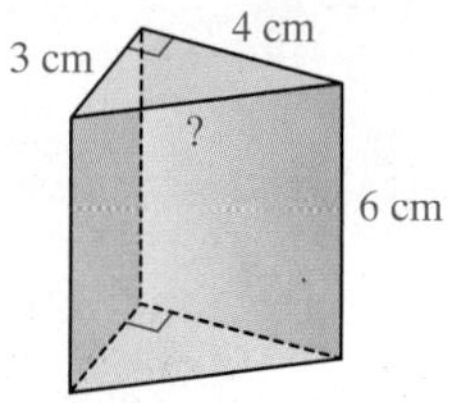

a. What is the perimeter of a base?

b. What is the lateral area of the prism?

c. What is the area of a base in simplest radical form?

d. What is the surface area of the prism rounded to a whole number?

Solution

a. Since the base is a right triangle, use the Pythagorean Theorem to find the length of the hypotenuse.

$$\text{hypotenuse} = \sqrt{3^2 + 4^2}\text{ cm} \quad \text{by the Pythagorean Theorem}$$
$$= \sqrt{25}\text{ cm} = 5\text{ cm}$$
$$P = 3 + 4 + 5 = 12$$

b. $\text{LA} = Ph$ Use the formula for lateral area.
$= 12 \cdot 6$ Substitute 12 for P and 6 for h.
$= 72$ Simplify.

c. To find the area of a base, notice that this base is a triangle.

$B = \frac{1}{2}bh$ Use the formula for the area of a triangle.
$= \frac{1}{2}(3 \cdot 4)$ Substitute 3 for b and 4 for h.
$= 6$

d. $\text{SA} = 2B + Ph$ Use the formula for surface area.
$= 2(6) + 72$ Substitute 6 for B and 72 for Ph.
$= 84$ Simplify.

The surface area of the prism is 84 sq cm.

PRACTICE 2

Find the surface area of the prism by answering parts **a–d**.

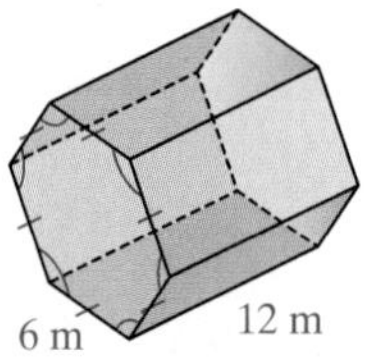

a. What is the perimeter of a base?

b. What is the lateral area of the prism?

c. What is the area of a base in simplest radical form?

d. What is the surface area of the prism rounded to a whole number?

OBJECTIVE 2 ▶ Finding the Surface Area of a Cylinder. A **cylinder** is a solid that has two congruent parallel **bases** that are circles. An **altitude** of a cylinder is a perpendicular segment that joins the planes of the bases. The **height** h of a cylinder is the length of an altitude.

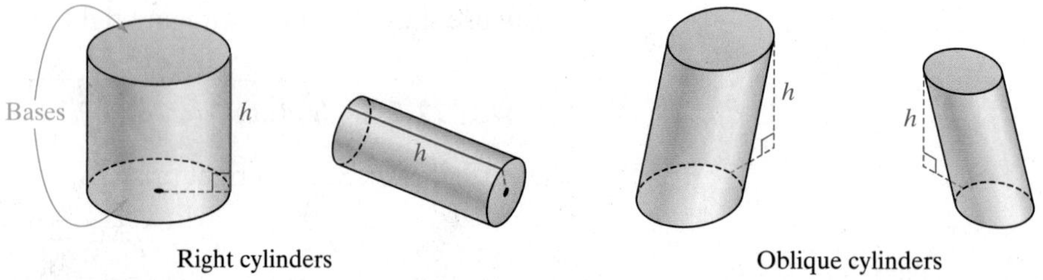

In a **right cylinder**, the segment joining the centers of the bases is an altitude. In an **oblique cylinder**, the segment joining the centers is not perpendicular to the planes containing the bases. In this text, we will assume that a cylinder is a right cylinder unless stated or pictured otherwise.

To find the area of the curved surface of a cylinder, visualize "unrolling" it. The area of the resulting rectangle is the **lateral area** of the cylinder. The **surface area** of a cylinder is the sum of the lateral area and the areas of the two circular bases. We can find formulas for these areas by looking at a net for a cylinder.

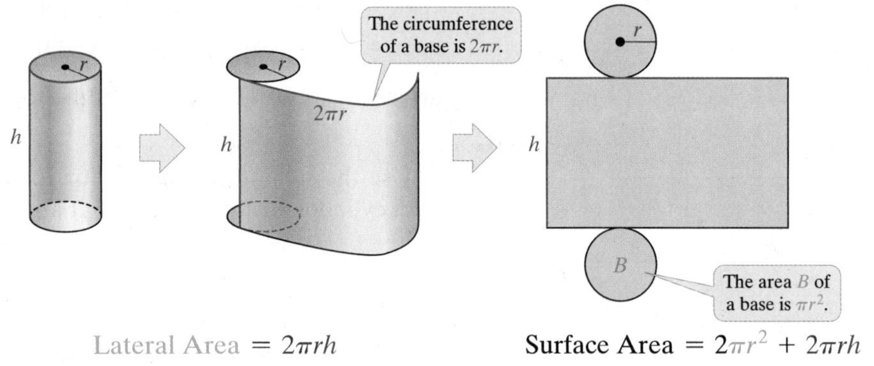
The circumference of a base is $2\pi r$.
r
h
$2\pi r$
B
The area B of a base is πr^2.
Lateral Area $= 2\pi rh$
Surface Area $= 2\pi r^2 + 2\pi rh$

Theorem 11.2-2 Surface Area of a Cylinder

The surface area SA of a right cylinder is

$$\text{SA} = 2B + 2\pi rh, \text{ or SA} = 2\pi r^2 + 2\pi rh$$

where B is the area of a base, r is the radius of a base, and h is the height of the cylinder.

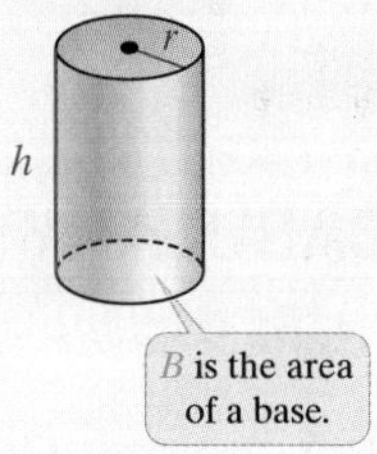
r
h
B is the area of a base.

EXAMPLE 3 **Finding Surface Area of a Cylinder**

Multiple Choice The radius of the base of a cylinder is 4 in. and its height is 6 in. What is the surface area of the cylinder in terms of π?

a. 32π sq in. **b.** 42π sq in. **c.** 80π sq in. **d.** 120π sq in.

Solution Use the formula for surface area of a cylinder.

$\text{SA} = 2\pi r^2 + 2\pi rh$	Use the formula for surface area of a cylinder.
$= 2\pi(4^2) + 2\pi(4)(6)$	Substitute 4 for r and 6 for h.
$= 32\pi + 48\pi$	Simplify.
$= 80\pi$	Add the like terms.

The surface area of the cylinder is 80π sq in. The correct choice is **c.** □

PRACTICE

3 A cylinder has a height of 9 cm and a radius of 10 cm. What is the surface area of the cylinder in terms of π?

EXAMPLE 4 **Finding Surface Area Without One Base**

Standing 82 ft tall with a radius of approximately 11 feet, the "AquaDom" is located in the Radisson Hotel in Berlin, Germany, and is titled the "World's Largest Cylindrical Fish Tank." Find the surface area of this tank without the area of the top base. Find the exact surface area and use $\pi \approx 3.14$ to approximate the surface area.

Solution We are given that $h = 82$ ft and $r = 11$ ft. Since we do not want to include the area of the top base, we use the formula

$$\begin{aligned} \text{SA without top base} &= 1(\pi r^2) + 2\pi rh && \text{SA of cylinder formula for 1 base (not 2)} \\ &= 1(\pi \cdot 11^2) + 2\pi \cdot 11 \cdot 82 && \text{Let } r = 11 \text{ and } h = 82. \\ &= 121\pi + 1804\pi && \text{Multiply.} \\ &= 1925\pi \text{ sq ft} && \text{Add like terms.} \\ &\approx 6044.5 \text{ sq ft} && \text{Let } \pi \approx 3.14. \end{aligned}$$

The surface area of the tank without the top base is 1925π sq ft ≈ 6044.5 sq ft. □

PRACTICE 4 Use the same directions as for Example 4. This is a jellyfish aquarium in the shape of a cylinder with a radius of 6 ft and a height of 7.5 ft.

VOCABULARY & READINESS CHECK

Word Bank *Use the choices to fill in each blank. Some choices may be used more than once and some not at all.*

prism	lateral area	bases	oblique	height
cylinder	surface area	right	lateral faces	altitude

1. A(n) ____________ is a solid that has two congruent parallel circles.
2. The circles in a cylinder are called ____________.
3. When the segment joining the centers of the bases of a cylinder is an altitude, it is called a(n) ____________ cylinder.
4. When the segment joining the centers is not perpendicular to the planes containing the bases of a cylinder, it is called a(n) ____________ cylinder.
5. The area of the curved surface of a cylinder is called the ____________.
6. The ____________ of a cylinder is the sum of the lateral area and the areas of the two circular bases.
7. A(n) ____________ is a polyhedron with two congruent, parallel faces.
8. The parallel faces from the exercise above are called ____________.
9. The faces of a prism that are not the bases are called the ____________.
10. A(n) ____________ of a prism or a cylinder is a perpendicular segment that joins the planes of the bases.
11. The ____________ of a prism or cylinder is the length of an altitude.
12. When the lateral faces are rectangles and a lateral edge is an altitude, this prism is called a(n) ____________ prism.
13. When some or all of the lateral faces are nonrectangular, this prism is called a(n) ____________ prism.
14. The ____________ of a prism is the sum of the areas of the lateral faces.
15. The ____________ of a prism is the sum of the lateral area and the area of the two bases.

11.2 EXERCISE SET MyMathLab®

For Exercises 1–10, give an exact answer and if necessary one rounded to the nearest tenth. Use a net to find the surface area of each prism. See Example 1.

1.

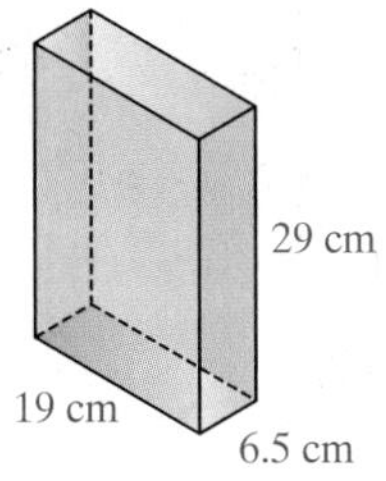

2.

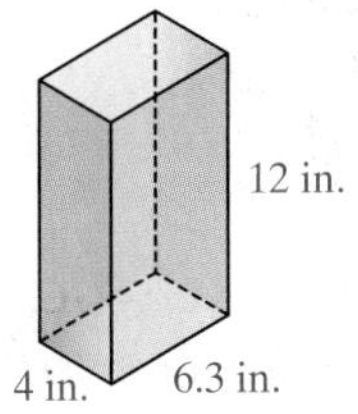

3.

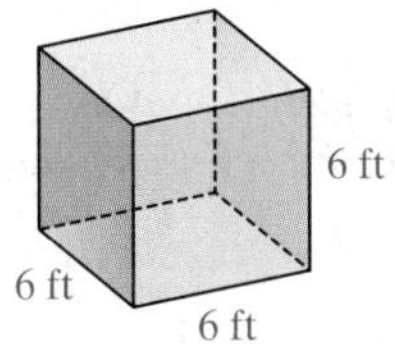

4.

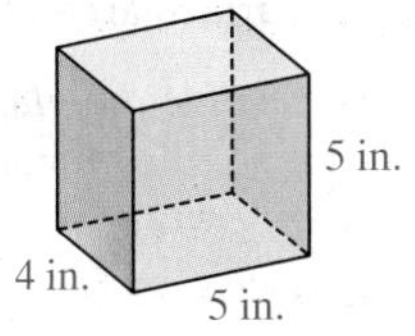

5.

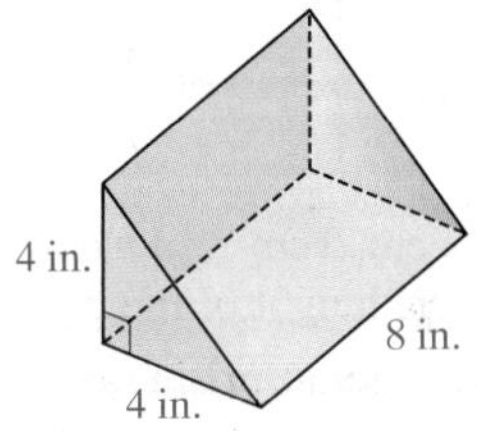

6.

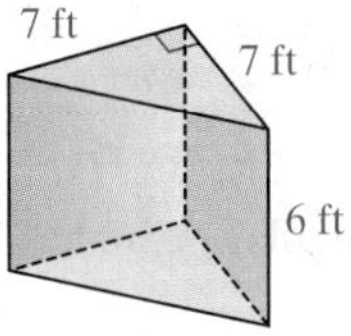

For Exercises 7–10, use the given prism to answer each question.

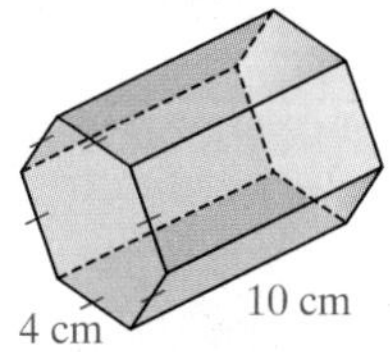

7. Classify the prism.

8. Find the lateral area of the prism.

9. The bases are regular hexagons. Find the sum of their areas.

10. Find the surface area of the prism.

Use formulas to find the surface area of each prism. Round your answer to the nearest whole number. See Example 2.

11.

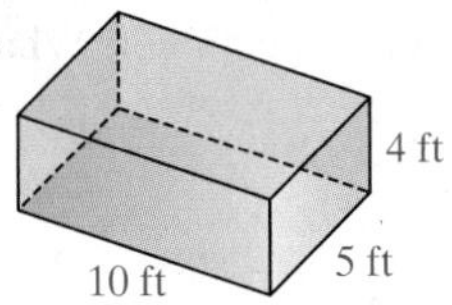

12.

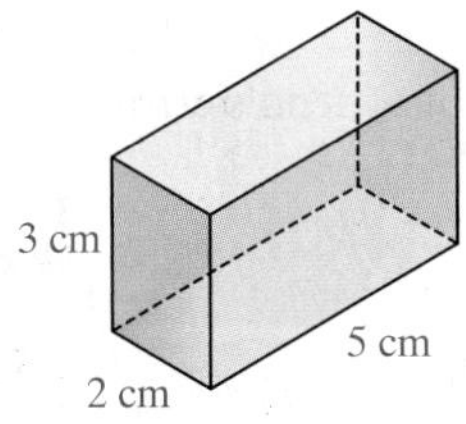

13.

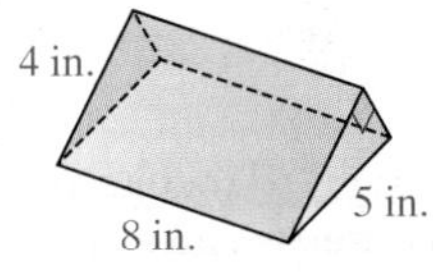

14.

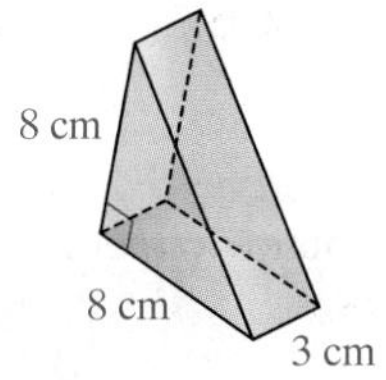

15. Regular octagon

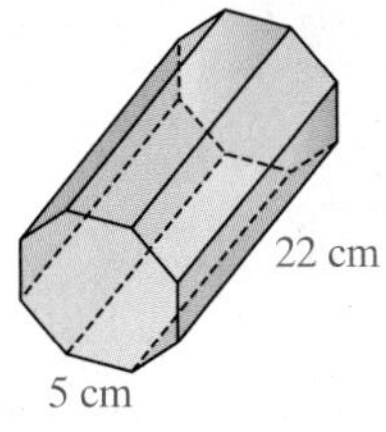

16. Regular hexagon

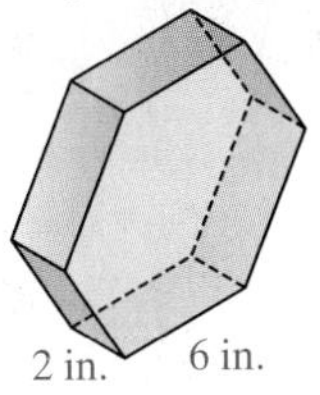

Find the lateral area of each cylinder to the nearest whole number. See Examples 3 and 4.

17.

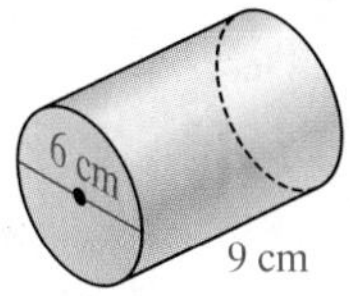

18.

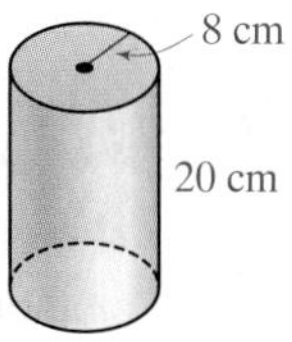

19.

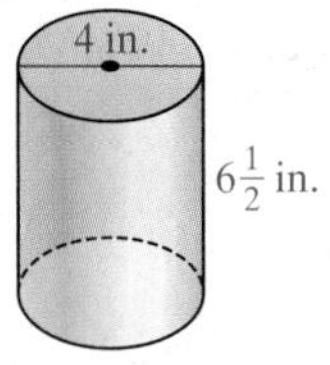

20.

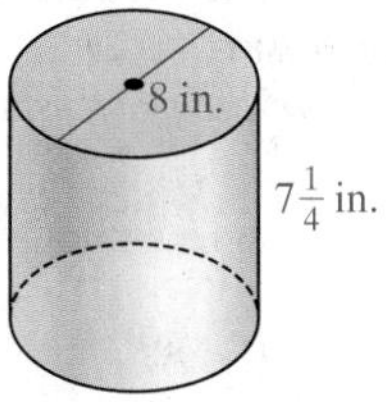

Find the surface area of each cylinder in terms of π. See Examples 3 and 4.

21.

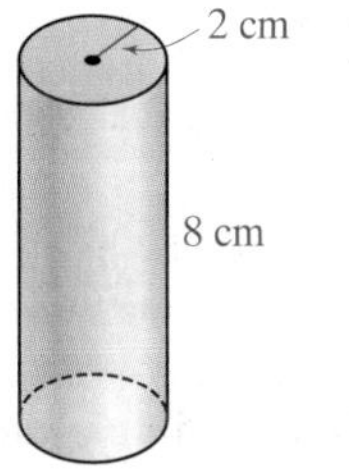

22.

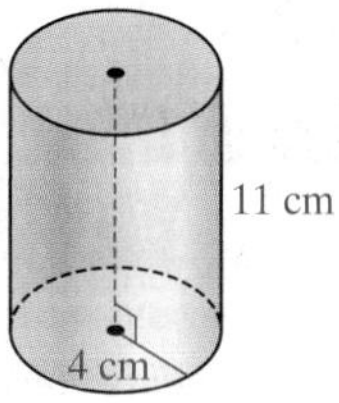

23.

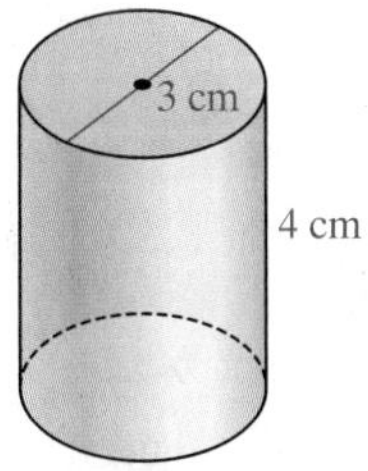

24.

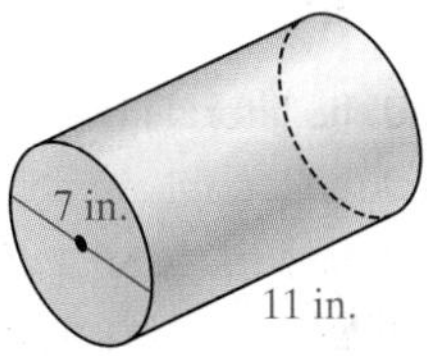

What is the surface area of each cylinder? Give an exact answer and one rounded to the nearest tenth.

25.

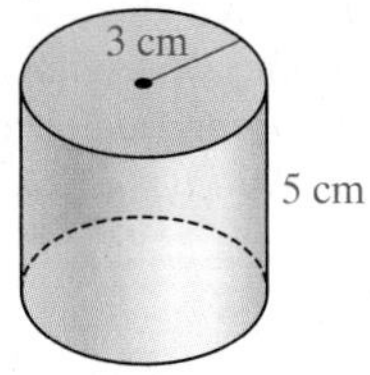

26.

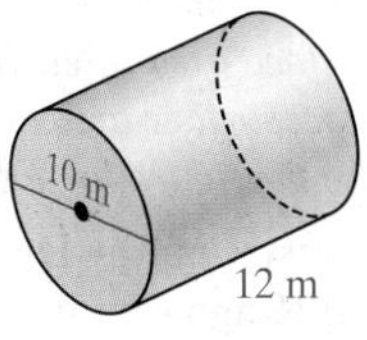

Solve. See Example 4.

27. A cylindrical carton of oatmeal with radius 3.5 in. is 9 in. tall. If all surfaces except the top are made of cardboard, how much cardboard is used to make the oatmeal carton? Assume no surfaces overlap. Round your answer to the nearest square inch.

28. A cylindrical can of cocoa has the dimensions shown below. What is the approximate surface area for the open can? Round to the nearest square inch.

A flour moth trap has the shape of a triangular prism that is open on both ends. An environmentally safe chemical draws the moth inside the prism, which is lined with an adhesive. What is the surface area of the prism-shaped trap?

29.

30.

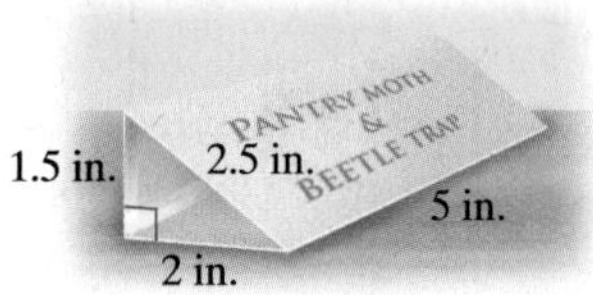

A new cereal is being developed, and two box sizes are being experimented with. A less expensive material is used for the lateral area of box packaging and a more expensive material is used for the bases.

31. Find the lateral area of Box A.
32. Find the lateral area of Box B.
33. Find the surface area of the bases of Box A.
34. Find the surface area of the bases of Box B.

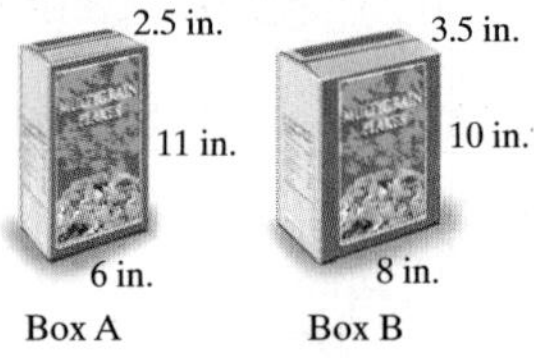

35. Find the total surface area of Box A.
36. Find the total surface area of Box B.

Find the surface area of each cylindrical aquarium without the top. Find the exact surface area and the surface area rounded to the nearest foot. Use the figure for Exercises 37–40. (Source: Living Color)

37. Rainforest Cafe at Disney's Animal Kingdom; diameter: 8 ft; height: 8 ft
38. Coco Beach Surf Company; diameter: 12 ft; height: 6 ft
39. The Orlando Airport custom aquarium; diameter: 8 ft; height: 7.5 ft (Note: This aquarium weighs more than a Learjet 35!)
40. Corporate headquarters; diameter: 8 ft; height: $2\frac{1}{6}$ ft

41. Find the surface area of a cube with edges 4.95 cm long.
42. Find the surface area of a cube with edges 3.25 in. long.

Coordinate Geometry *Use the diagram at the right.*

43. Find the three coordinates of each vertex A, B, C, and D of the rectangular prism.
44. Find AB.
45. Find BC.
46. Find CD.
47. Find the area of $AGCB$.
48. Find the area of $BCDE$.
49. Find the area of $ABEF$.
50. Find the surface area of the prism.

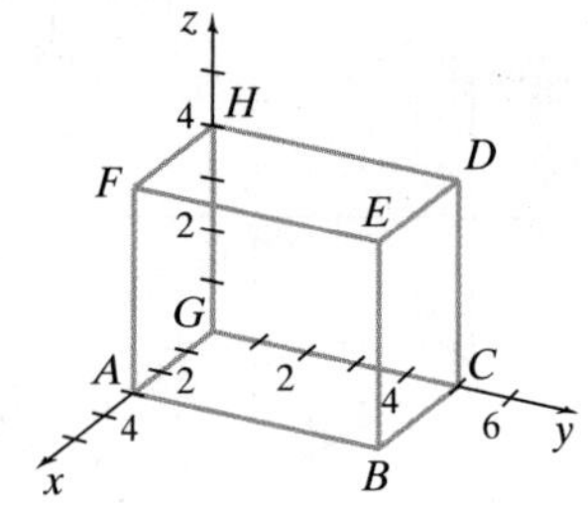

Suppose we revolve the plane region completely about the given line to form a solid of revolution. Describe the solid and find its surface area in terms of π.

51. the y-axis
52. the x-axis
53. the line $y = 2$
54. the line $x = 4$

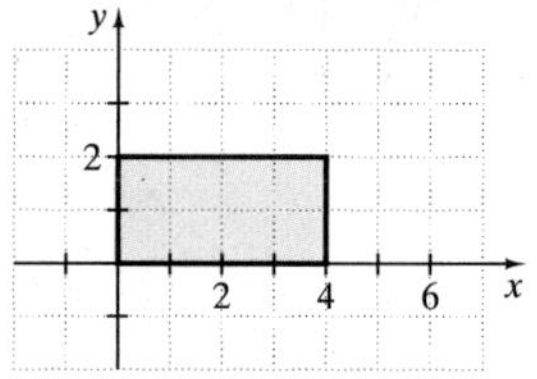

55. A triangular prism has base edges 4 cm, 5 cm, and 6 cm long. Its lateral area is 300 cm^2. What is the height of the prism?

56. A hexagonal pencil is a regular hexagonal prism, as shown below. A base edge of the pencil has a length of 4 mm. The pencil (without eraser) has a height of 170 mm. What is the area of the surface of the pencil that gets painted?

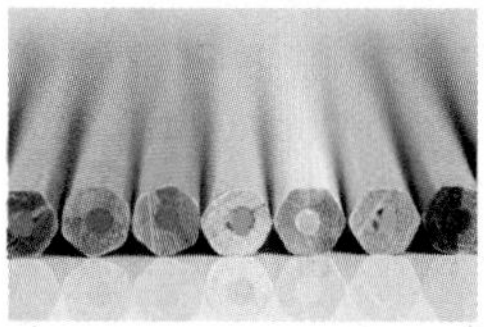

CONCEPT EXTENSIONS

57. **Find the Error** Your friend drew a net of a cylinder. What is your friend's error? Explain.

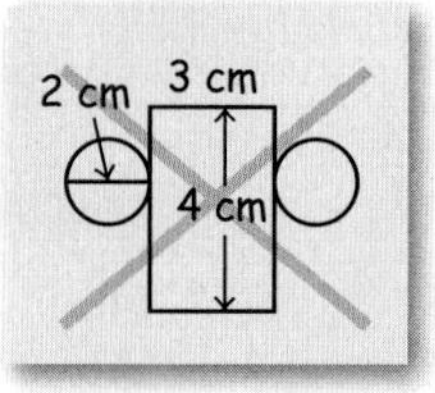

58. **Find the Error** Your class is drawing right triangular prisms. Your friend's paper is below. What is your friend's error?

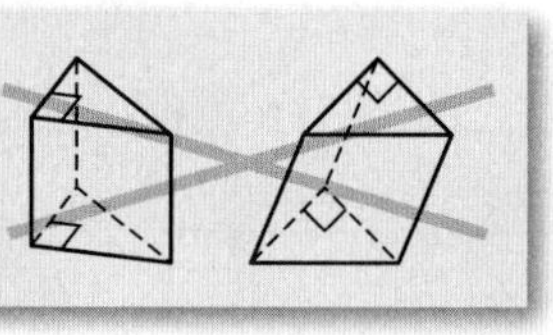

59. Name the lateral faces and the bases of the prism.

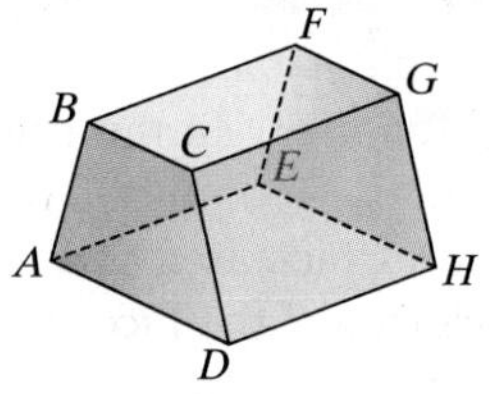

60. Explain how a cylinder and a prism are alike and how they are different.

Suppose that a cylinder has a radius of r units, and that the height of the cylinder is also r units. The lateral area of the cylinder is 98π square units.

61. Find the value of r.

62. Find the surface area of the cylinder.

63. Draw a net for a rectangular prism with a surface area of 220 sq cm.

64. Draw a net for a rectangular prism with a surface area of 400 sq in.

65. Consider a box with dimensions 3, 4, and 5.

a. Find its surface area.

b. Double each dimension and then find the new surface area.

c. Find the ratio of the new surface area to the original surface area.

66. Repeat parts **a–c** in Exercise **65** for a box with dimensions 6, 9, and 11.

d. How does doubling the dimensions of a rectangular prism affect the surface area?

67. Suppose you double the radius of a right cylinder.

a. How does that affect the lateral area?

b. How does that affect the surface area?

c. Use the formula for surface area of a right cylinder to explain why the surface area in part b was not doubled.

68. Some cylinders have wrappers with a spiral seam. Peeled off, the wrapper has the shape of a parallelogram. The wrapper for a biscuit container has base 7.5 in. and height 6 in.

a. Find the radius and height of the container.

b. Find the surface area of the container.

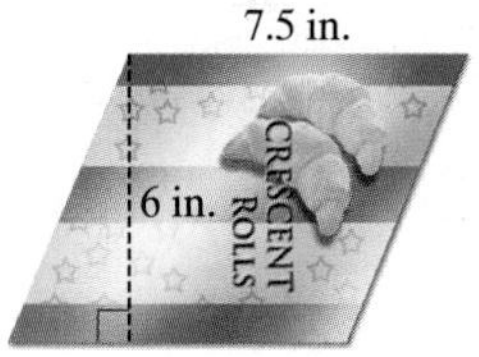

For Exercises 69–72, what is the surface area of each solid in terms of π?

69.

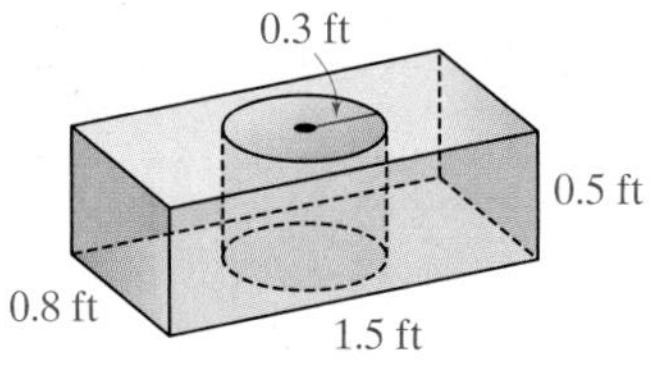

70.

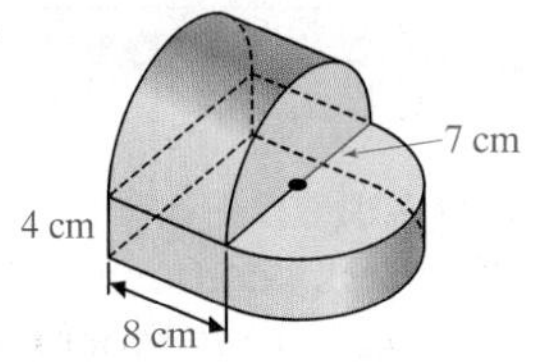

71.

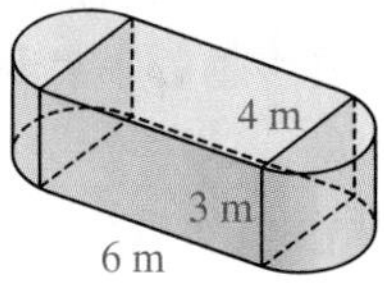

72.

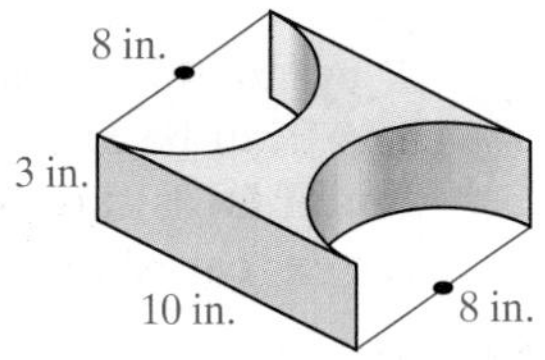

73. Each edge of the smaller cubes at the right is 12 inches long. The large cube is painted on the outside, and then cut into 27 smaller cubes as shown. Answer these questions about the 27 cubes.

a. How many are painted on 4, 3, 2, 1, and 0 faces?

b. What is the total surface area that is unpainted?

74. The sum of the height and radius of a cylinder is 9 m. The surface area of the cylinder is 54π sq m. Find the height and the radius.

REVIEW AND PREVIEW

Sketch each solid and then draw a net for it. Label the net with its dimensions. See Section 11.1.

75. a rectangular prism with height 5 cm and a base 3 cm by 4 cm

76. a cylinder with a 72π-in. circumference and a 22-in. height

Find the area of each part of the circle to the nearest tenth. See Section 10.6.

77. sector QOP

78. the segment of the circle bounded by $\overline{QP}$ and $\widehat{QP}$

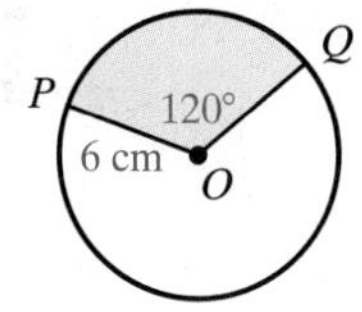

Find the length of the hypotenuse in simplest radical form. See Section 9.1.

79.

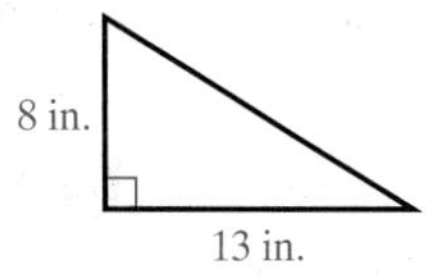

80.

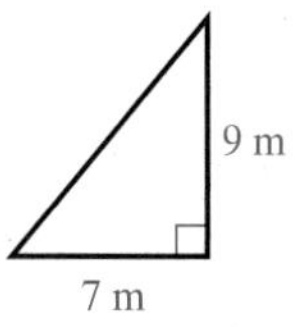

81.

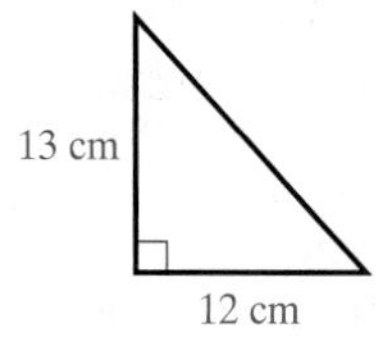

82.

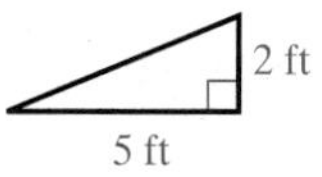

11.3 SURFACE AREAS OF PYRAMIDS AND CONES

OBJECTIVES

1 Find the Surface Area of a Pyramid.

2 Find the Surface Area of a Cone.

VOCABULARY

- pyramid (base, lateral face, vertex, altitude, height, slant height, lateral area, surface area)
- regular pyramid
- cone (base, altitude, vertex, height, slant height, lateral area, surface area)
- right cone

OBJECTIVE 1 ▶ Finding the Surface Area of a Pyramid. In this section, we will learn to name additional three-dimensional figures and to use formulas to find their surface areas.

Remember from Section 11.2, to find the surface area of a three-dimensional figure, we find the sum of the areas of all the surfaces of the faces of the figure.

A **pyramid** is a polyhedron in which one face (the **base**) can be any polygon and the other faces (the **lateral faces**) are triangles that meet at a common vertex (called the **vertex** of the pyramid).

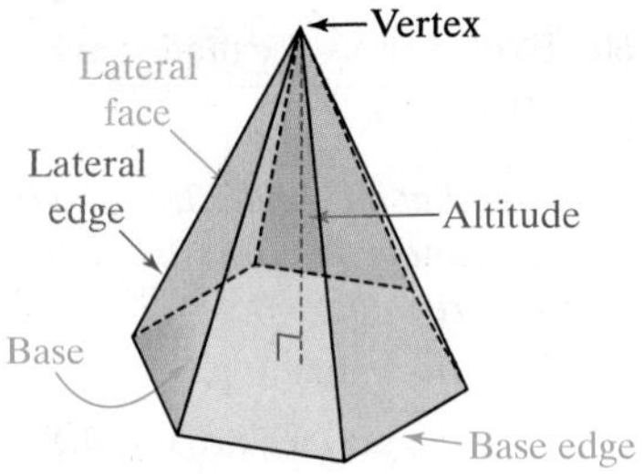

Hexagonal pyramid

We name a pyramid by the shape of its base. The **altitude** of a pyramid is the perpendicular segment from the vertex to the plane of the base. The length of the altitude is the **height** h of the pyramid.

A **regular pyramid** is a pyramid whose base is a regular polygon and whose lateral faces are congruent isosceles triangles. The **slant height** ℓ is the length of the altitude of a lateral face of the pyramid.

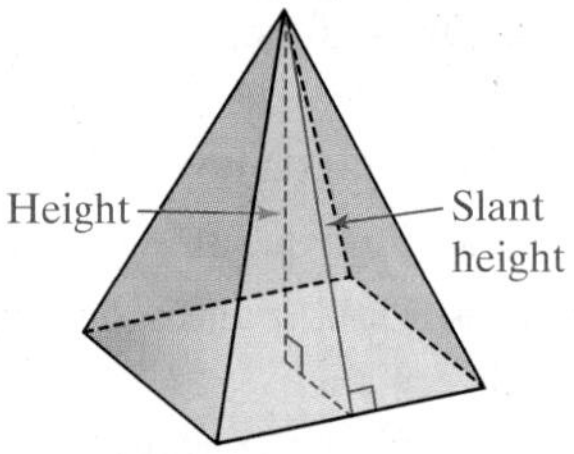

Square pyramid

In this text, we assume that a pyramid is regular unless stated otherwise.

Just as with prisms and cylinders, the **lateral area** of a pyramid is the sum of the areas of the lateral faces. For the pyramids in this section, the lateral faces are congruent. We can find a formula for the lateral area of a pyramid by looking at its net.

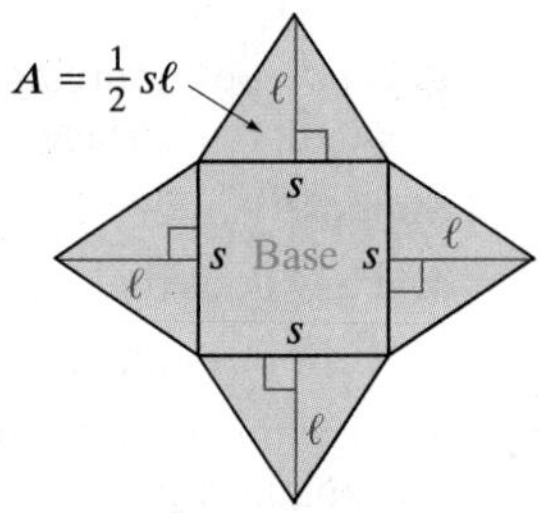

$$\text{LA} = 4\left(\frac{1}{2}s\ell\right) \quad \text{The area of each lateral face is } \frac{1}{2}s\ell.$$

$$= \frac{1}{2}(4s)\ell \quad \text{Commutative and Associative Properties of Multiplication}$$

$$= \frac{1}{2}P\ell \quad \text{The perimeter } P \text{ of the base is } 4s.$$

To find the **surface area** of a pyramid, we add the area of its base to its lateral area.

Theorem 11.3-1 Surface Area of a Pyramid

The surface area SA of a regular pyramid is

$$\text{SA} = B + \overbrace{\frac{1}{2}P\ell}^{\text{lateral area}}$$

where B is the area of the base, P is the perimeter of the base, and ℓ is the slant height.

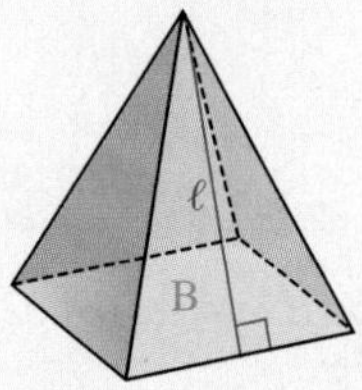

> **Helpful Hint**
>
> For Example 1:
>
> - The perimeter
>
> $P = 6(6) = 36\text{ in.}$
> - The slant height value ℓ is 9 in.
> - The area of the base
>
> 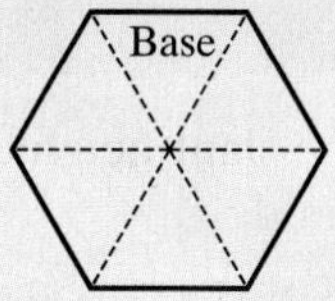
>
>
> $$B = \frac{1}{2}aP$$
> $$= \frac{1}{2}(3\sqrt{3})(36)$$

EXAMPLE 1 **Finding the Surface Area of a Pyramid**

What is the surface area of the hexagonal pyramid?

Solution

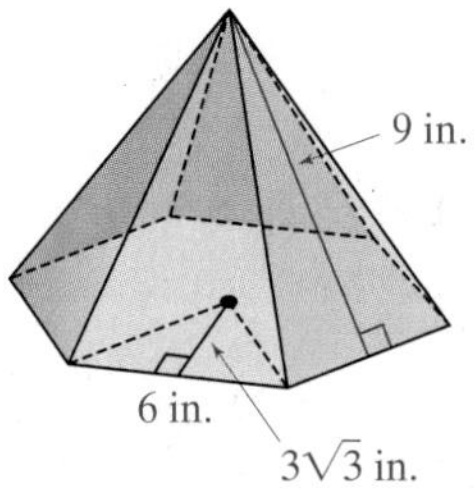

$SA = B + \frac{1}{2}P\ell$	Use the formula for surface area.
$= \frac{1}{2}aP + \frac{1}{2}P\ell$	Substitute the formula for B.
$= \frac{1}{2}(3\sqrt{3})(36) + \frac{1}{2}(36)(9)$	Substitute.
≈ 255.5307436	Use a calculator.

The surface area of the pyramid is about 256 sq in. □

PRACTICE

1 A square pyramid has base edges of 5 m and a slant height of 3 m. What is the surface area of the pyramid?

When the slant height of a pyramid is not given, we must calculate it before we can find the lateral area or surface area.

EXAMPLE 2 The Great Pyramid in Egypt is surrounded by a complex of buildings among which are smaller pyramids including the Pyramid of Menkaure. Use the dimensions to find the surface area of this pyramid.

Solution To find the slant height, use the Pythagorean Theorem.

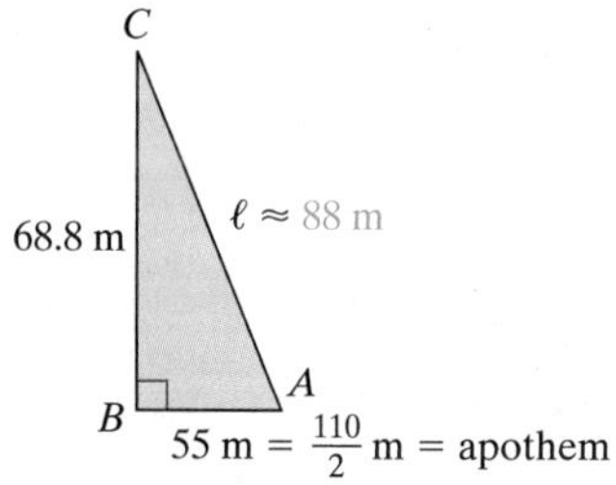

$\ell = \sqrt{AB^2 + BC^2}$	Use the Pythagorean Theorem.
$= \sqrt{55^2 + 68.8^2}$	Let $AB = 55\text{ m}$ and $BC = 68.8\text{ m}$
$= \sqrt{7758.44}$	Simplify.
$\approx 88\text{ m}$	Simplify.

The formula for the surface area of a pyramid is

$$SA = B + \frac{1}{2}P\ell$$

(perimeter of base; slant height; area of base)

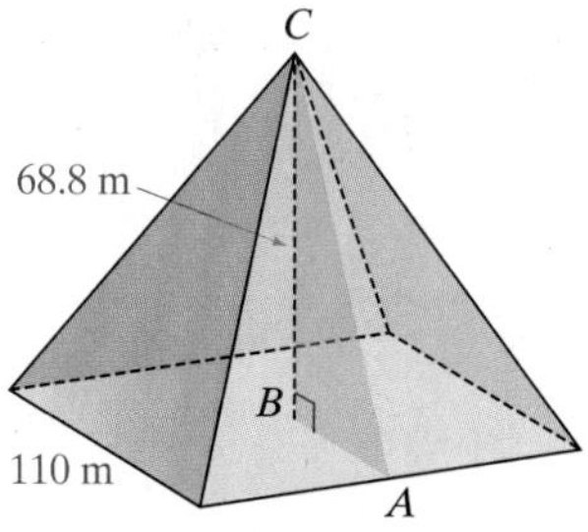

Pyramid of Menkaure in Egypt

$$SA \approx (110)^2 + 4(110)(88)$$
$$\approx 12{,}100 + 440(88)$$

(area of base; LA of Pyramid)

$\approx 50{,}820\text{ sq m}$ SA of pyramid □

PRACTICE

2 A meditation pyramid is built similar to the Great Pyramid. Find the surface area of the pyramid shown to the right rounded to the nearest whole foot. (The pyramid to the right has the same dimensions as the meditation pyriamid.)

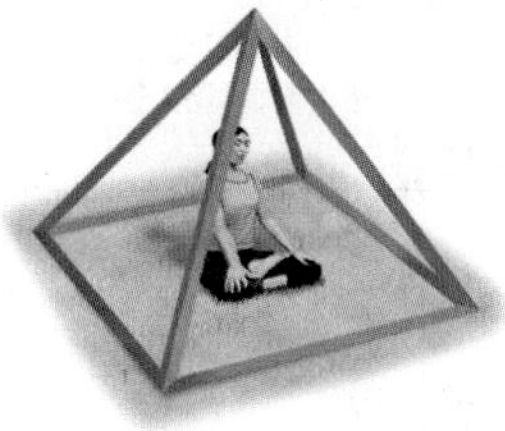

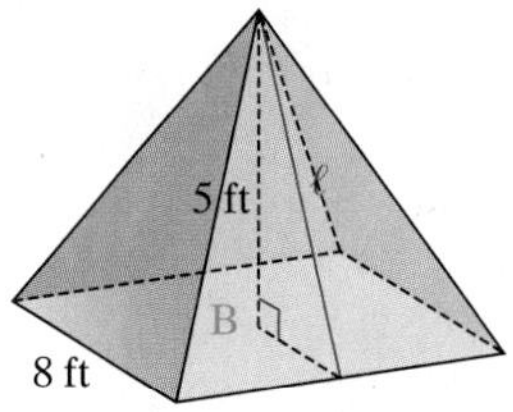

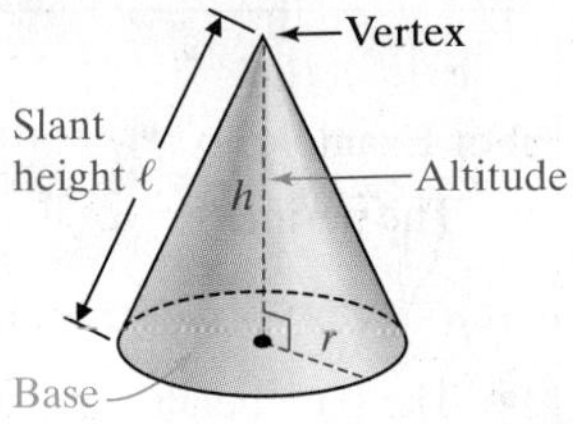

OBJECTIVE 2 ▶ Finding the Surface Area of a Cone. Like a pyramid, a **cone** is a solid that has one base and a vertex that is not in the same plane as the base. However, the **base** of a cone is a circle. In a **right cone**, the **altitude** is a perpendicular segment from the vertex to the center of the base. The **height** h is the length of the altitude. The **slant height** ℓ is the distance from the vertex to a point on the edge of the base. In this text, we will assume that a cone is a right cone unless stated or pictured otherwise.

The **lateral area** is half the circumference of the base times the slant height. The formula for the **surface area** of a cone is similar to that for a pyramid.

> **▶ Helpful Hint**
> Notice the similarity between the surface area formulas for pyramids and cones.
> - Both contain the area of the Base.
> - The lateral area part contains the Perimeter or the Circumference of the base.

Theorem 11.3-2 Surface Area of a Cone

The surface area of a cone is

$$\text{SA} = B + \overbrace{\pi r \ell}^{\text{lateral area}}$$

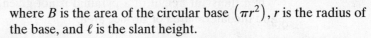

where B is the area of the circular base (πr^2), r is the radius of the base, and ℓ is the slant height.

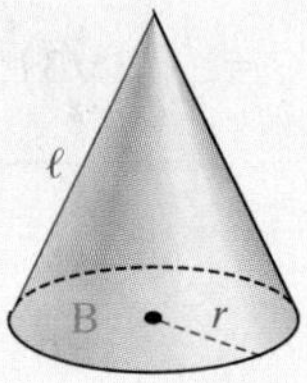

EXAMPLE 3 Finding the Surface Area of a Cone

What is the surface area of the cone in terms of π?

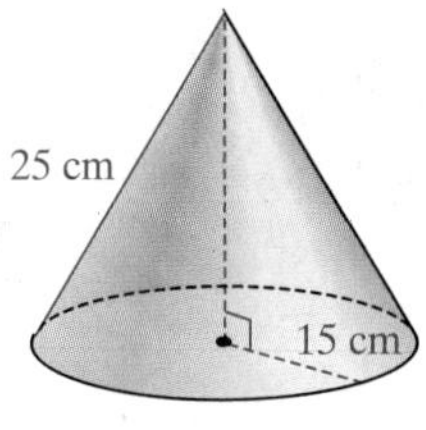

Solution Use the surface area formula for a cone.

$\text{SA} = B + \pi r \ell$	Use the formula for surface area.
$= \pi r^2 + \pi r \ell$	Substitute the formula for B.
$= \pi(15)^2 + \pi(15)(25)$	Substitute 15 for r and 25 for ℓ.
$= 225\pi + 375\pi$	Multiply.
$= 600\pi$	Combine like terms.

The surface area of the cone is 600π sq cm. □

PRACTICE

3 The radius of the base of a cone is 16 m. Its slant height is 28 m. What is the surface area in terms of π?

> **▶ Helpful Hint**
> By cutting a cone and laying it out flat, we can see how the formula for the lateral area of a cone $\left(\text{LA} = \frac{1}{2} \cdot C_{\text{base}} \cdot \ell\right)$ resembles that for the area of a triangle $\left(A = \frac{1}{2}bh\right)$.

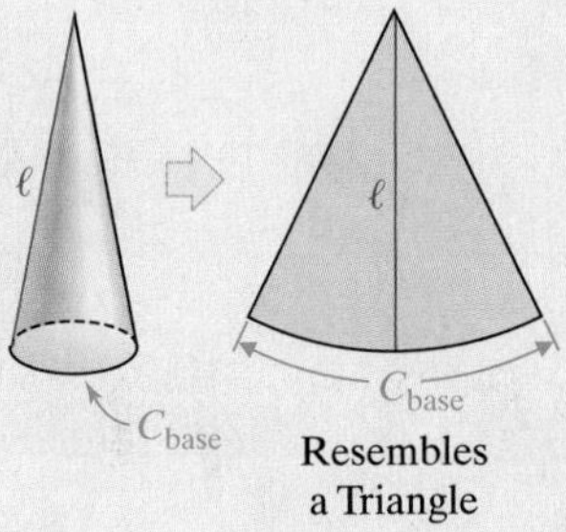

EXAMPLE 4 **Finding the Lateral Area of a Cone**

In a chemistry lab experiment, we use the conical filter funnel shown at the right. How much filter paper do we need to line the funnel?

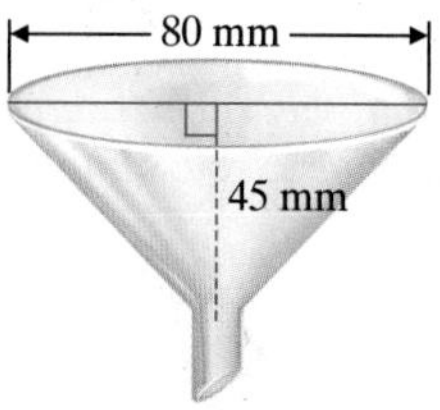

Solution To line the funnel, we do not line the base, or the top of it. Thus, we calculate the lateral area of a cone with a diameter of 80 mm and a height of 45 mm.

$LA = \frac{1}{2} \cdot 2\pi r \cdot \ell$ Think of the area of a Δ.

$LA = \pi r \ell$ Simplify the formula for lateral area of a cone.

$= \pi(40)\left(\sqrt{40^2 + 45^2}\right)$ Substitute $\frac{1}{2}$ of 80, or 40, for r. To find the slant height ℓ, use the Pythagorean Theorem.

≈ 7565.957013 Use a calculator to approximate.

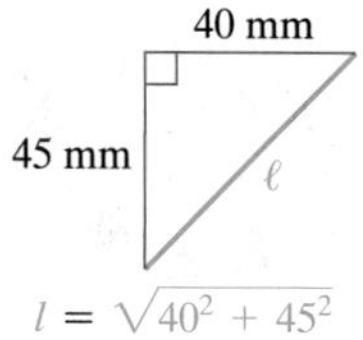

$l = \sqrt{40^2 + 45^2}$

We need about 7566 sq mm of filter paper to line the funnel. □

PRACTICE
4 What is the lateral area of a traffic cone with radius 10 in. and height 28 in.? Round to the nearest whole number.

VOCABULARY & READINESS CHECK

Word Bank *Use the choices to fill in each blank. Some choices may be used more than once and some not at all.*

pyramid	lateral area	base	oblique	height	slant height
cone	surface area	right	lateral faces	altitude	vertex

1. A(n) ____________ is a solid that has one base and a vertex that is not in the same plane as the base.
2. The ____________ of a cone is a circle.
3. In a right cone, the ____________ is a perpendicular segment from the vertex to the center of the base.
4. The ____________ of a cone is the distance from the vertex to a point on the edge of the base.
5. A(n) ____________ is a polyhedron in which one face can be any polygon and the other faces are triangles that meet at a point.
6. The ____________ of a pyramid can be any polygon.
7. The triangular faces of a pyramid are called the ____________.
8. The triangles of a pyramid meet at a common point called the ____________.
9. The ____________ of a pyramid is the perpendicular segment from the vertex to the plane of the base.
10. The ____________ of a pyramid or cone is the length of an altitude.
11. When the lateral faces are congruent isosceles triangles and the base is a regular polygon, this pyramid is called a(n) ____________ pyramid.
12. The ____________ of a pyramid is the length of the altitude of a lateral face.
13. The ____________ of a pyramid is the sum of the areas of the lateral faces.
14. The ____________ of a pyramid or a cone is the sum of the lateral area and the area of its base.

11.3 EXERCISE SET MyMathLab®

Find the surface area of each pyramid to the nearest whole number. See Example 1.

1.

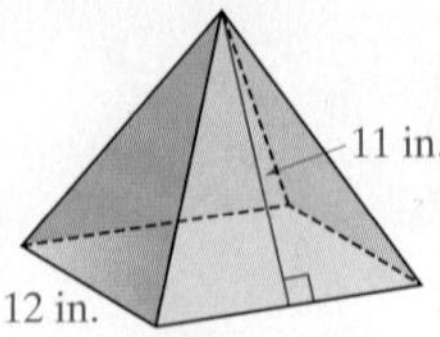

2.

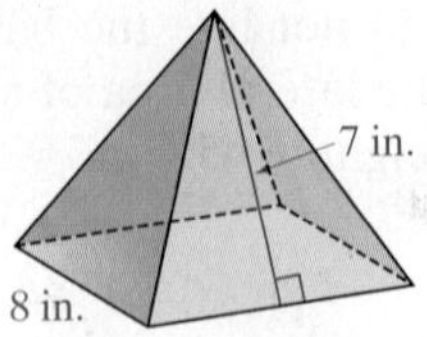

3.

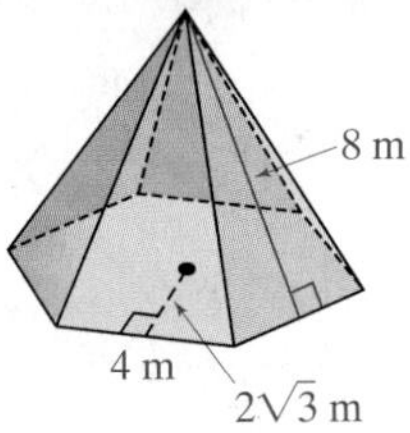

4.

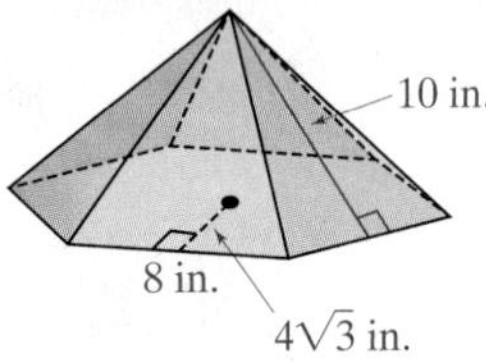

Find the lateral area of each pyramid to the nearest whole number. See Example 2.

5.

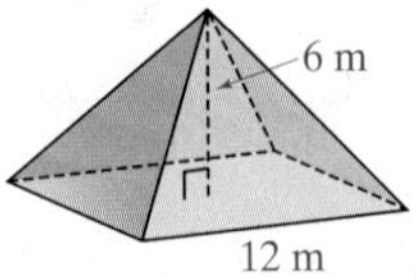

6.

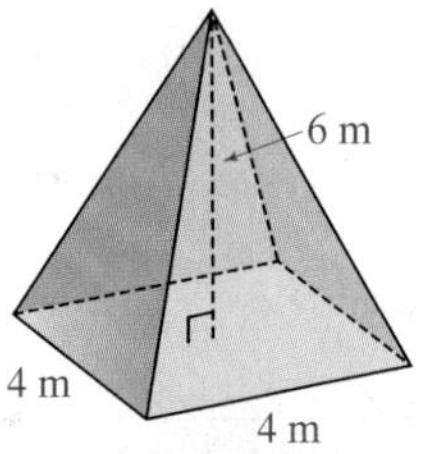

7.

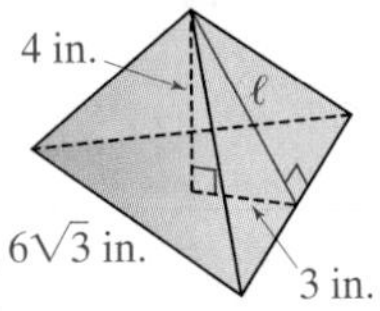

8.

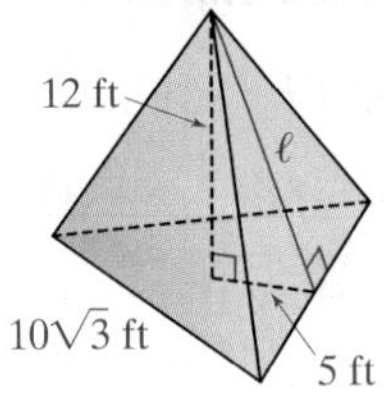

Find the lateral area of each cone to the nearest whole number. See Example 4.

9.

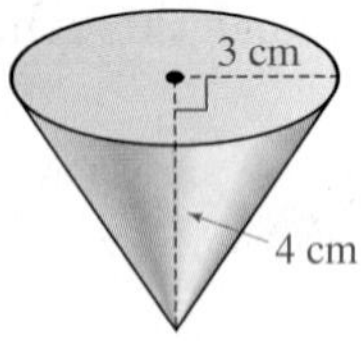

10.

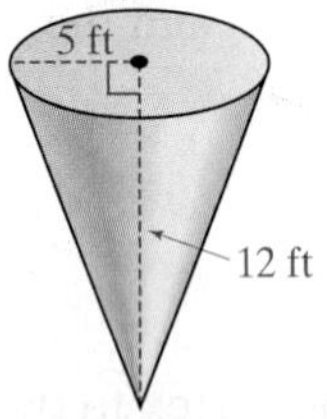

11.

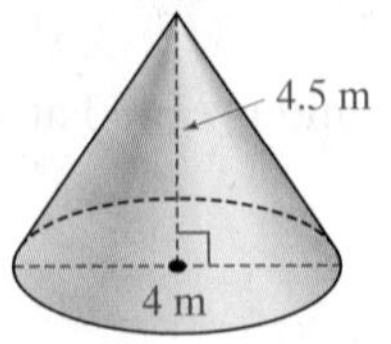

12.

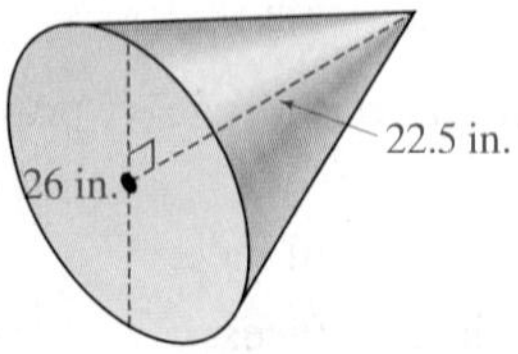

Find the surface area of each cone in terms of π. See Example 3.

13.

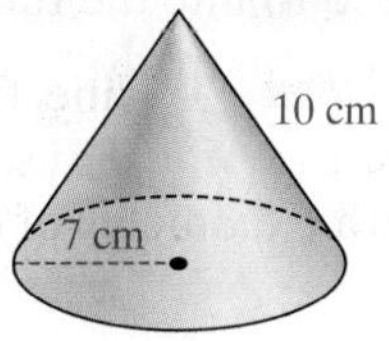

14.

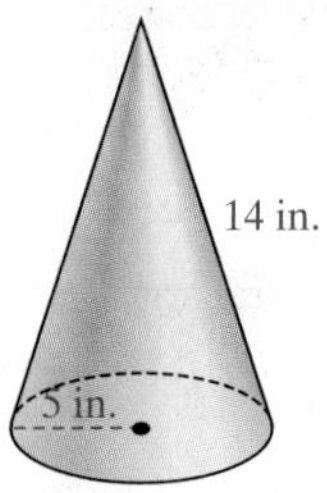

15.

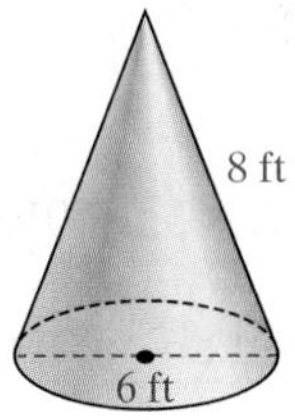

16.

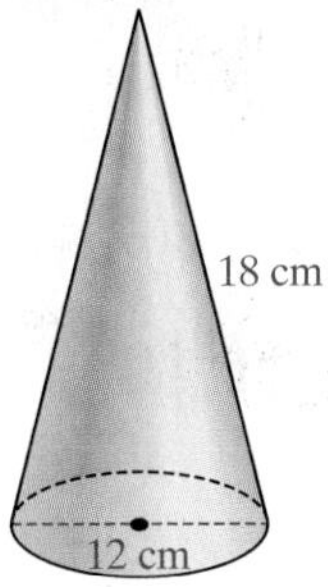

Use the diagram of the square pyramid at the right.

17. What is the lateral area of the pyramid?

18. What is the surface area of the pyramid?

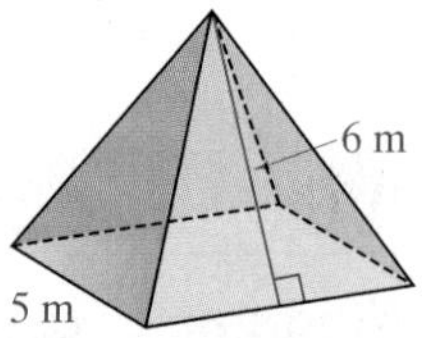

Use the diagram of the cone at the right. Round each answer to the nearest whole number.

19. What is the lateral area of the cone?

20. What is the surface area of the cone?

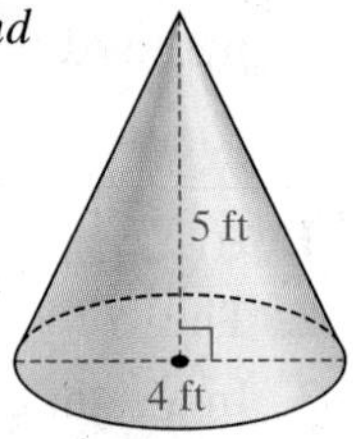

Suppose you revolve the plane region completely about the given line to form a solid of revolution. Describe the solid. Then find its surface area in terms of π.

21. the y-axis

22. the x-axis

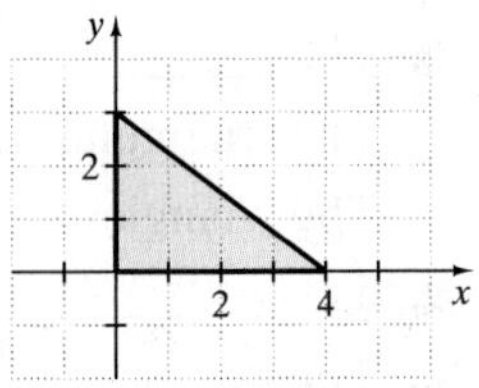

The surface area of figures has a lot to do with cost of materials, production cost, etc. Find each lateral area.

23. Find the lateral area of a waffle cone with a diameter of 2.6 in. and a slant height of 6.1 in. Give your answer in terms of π.

24. Find the lateral area of a sno-cone cup with a diameter of 3.2 in. and a slant height of 4.3 in. Give your answer in terms of π.

25. A traffic cone with a height of 28 in. is ideal for traffic control. Find the lateral area of a cone with a perpendicular height of 28 in. and a diameter of 10 in. Round your answer to the nearest whole number. (*Hint:* The height given is not the slant height.)

26. Safety cones are used in driver training. Find the lateral area of a cone with a perpendicular height of 6 in. and a perpendicular diameter of 11 in. Round your answer to the nearest whole number. (*Hint:* See the hint for Exercise **25**.)

"Pyramid of numbers is a graphic representation of various levels of individuals starting with producers forming the base and top carnivores at the tip." The shapes of the pyramids vary from ecosystem to ecosystem. A company manufactures and sells different-size pyramids as desktop giveaways for employees of companies.

27. Find the surface area of a square-base pyramid with slant height of 3.6 in. and base side length of 4 in.

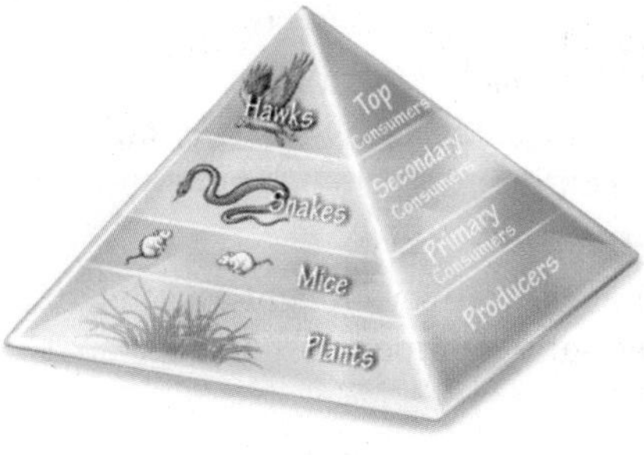

Example of a Pyramid of Numbers in a Grassland Ecosystem

28. Find the surface area of a square-base pyramid with slant height of 9 cm and base side length of 10 cm.

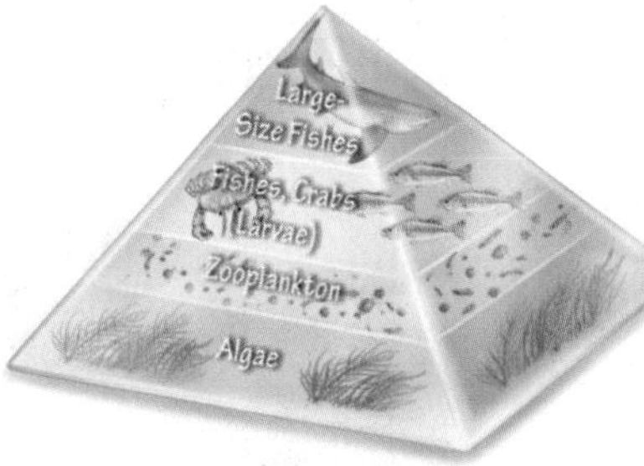

Example of a Pyramid of Numbers in an Aquatic Ecosystem

29. The original height of the Pyramid of Khafre, located next to the Great Pyramid in Egypt, was about 471 ft. Each side of its square base was about 708 ft. What is the lateral area, to the nearest square foot, of a pyramid with those dimensions?

30. The lateral area of a cone is 4.8π sq in. The radius is 1.2 in. Find the slant height.

Suppose you could climb to the top of the Great Pyramid.

31. Which route would be shorter, a route along a lateral edge or a route along the slant height of a side?

32. Which of these routes in Exercise **31** is steeper? Explain your answer.

Slove.

33. A pyramid with a square base of edge length 2 in. has surface area 32 sq in. What is its slant height?

34. A pyramid with a square base of edge length 5 cm has surface area of 90 sq cm. What is its slant height?

CONCEPT EXTENSIONS

35. How do the height and the slant height of a pyramid differ?

36. How many lateral faces does a pyramid have if its base is pentagonal? Hexagonal? n-sided?

37. **Find the Error** A cone has height 7 and radius 3. Your classmate calculates its lateral area. What is your classmate's error? Explain.

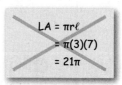

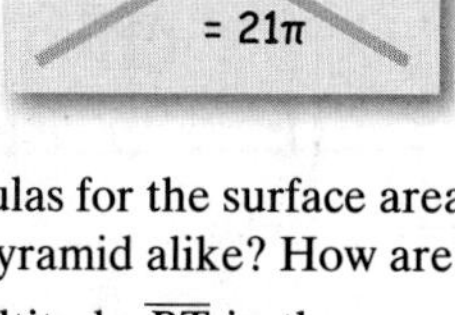

38. How are the formulas for the surface area of a prism and the surface area of a pyramid alike? How are they different?

39. Explain why the altitude $\overline{PT}$ in the pyramid below must be shorter than all of the lateral edges $\overline{PA}$, $\overline{PB}$, $\overline{PC}$, and $\overline{PD}$.

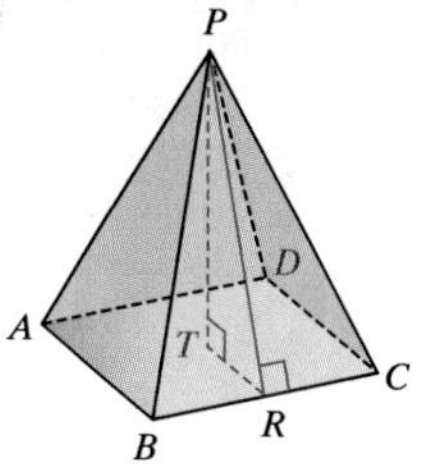

40. The figure below shows two glass cones inside a cylinder. Which has a greater surface area, the two cones or the cylinder? Explain.

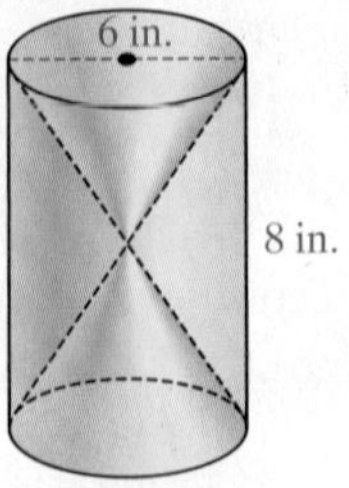

41. You can use the formula $SA = (\ell + r)r\pi$ to find the surface area of a cone. Explain why this formula works.

42. How are a right prism and a regular pyramid alike? How are they different?

Find the surface area to the nearest whole number.

43.

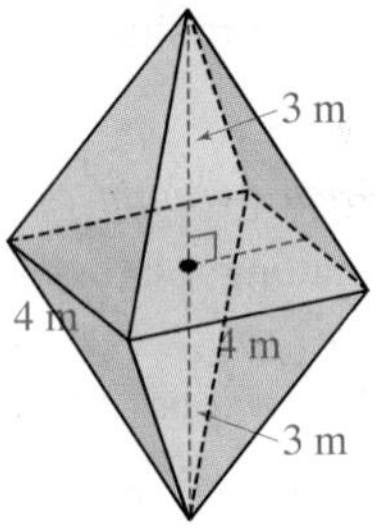

44.

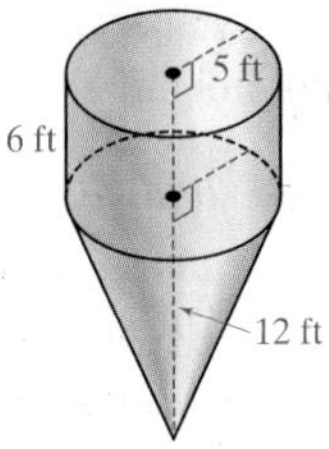

45.

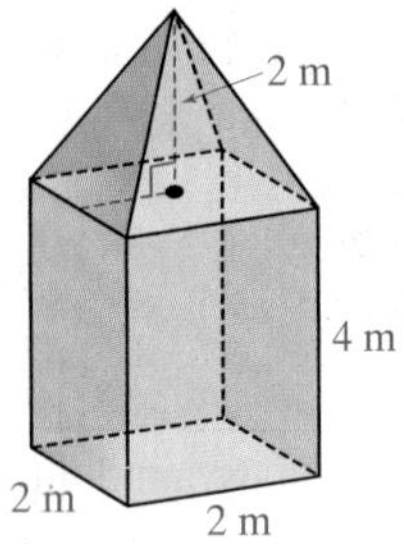

46.

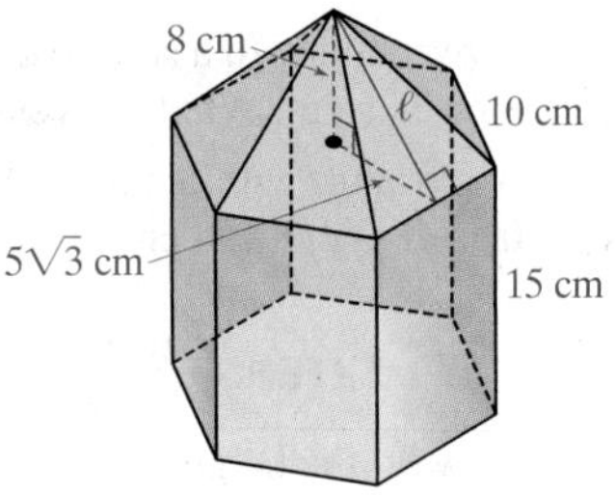

Find a formula for each of the following.

47. the slant height of a cone in terms of the surface area and radius

48. the radius of a cone in terms of the surface area and slant height

The length of a side (s) of the base, slant height (ℓ), height (h), lateral area (LA), and surface area (SA) are measurements of a square pyramid. Given two of the measurements, find the other three to the nearest tenth.

49. $s = 3$ in., SA $= 39$ sq in.

50. $h = 8$ m, $\ell = 10$ m

51. The lateral area of a pyramid with a square base is 240 ft^2. Its base edges are 12 ft long. Find the height of the pyramid.

52. Draw a square pyramid with a lateral area of 48 sq cm. Label its dimensions. Then find its surface area.

Suppose you revolve the plane region completely about the given line to form a solid of revolution. Describe the solid. Then find its surface area in terms of π.

53. the line $x = 4$

54. the line $y = 3$

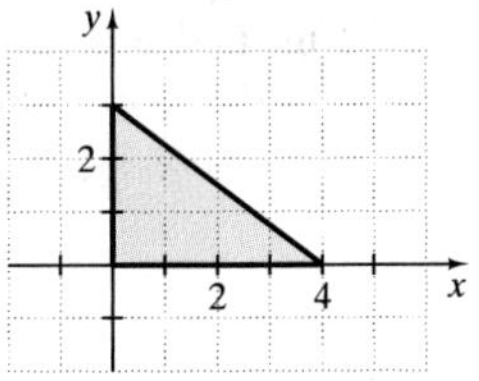

Each given figure fits inside a 10-cm cube. The figure's base is in one face of the cube and is as large as possible. The figure's vertex is in the opposite face of the cube. Draw a sketch and find the lateral and surface areas of the figure.

55. a square pyramid

56. a cone

REVIEW AND PREVIEW

Solve. See Section 11.2.

57. How much cardboard do you need to make a closed box 4 ft by 5 ft by 2 ft?

58. How much posterboard do you need to make a cylinder, open at each end, with height 9 in. and diameter $4\frac{1}{2}$ in.? Round your answer to the nearest square inch.

Find the area of each figure. If necessary, round to the nearest tenth. See Section 2.1.

59. a square with side length 2 cm

60. a circle with diameter 15 in.

11.4 VOLUMES OF PRISMS AND CYLINDERS AND CAVALIERI'S PRINCIPLE

OBJECTIVES

1. Find the Volume of a Prism.
2. Find the Volume of a Cylinder.
3. Find the Volume of Composite Solids.

OBJECTIVE 1 ▶ Finding the Volume of a Prism. **Volume** is the space that a figure occupies. It is measured in cubic units such as cubic inches ($in.^3$), cubic feet (ft^3), or cubic centimeters (cm^3). The volume V of a cube is the cube of the length of its edge e, or $V = e^3$.

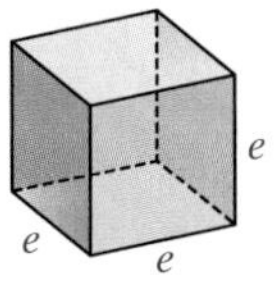

We can find the volume of a prism when we know its height and the area of its base. Both stacks of paper below contain the same number of sheets.

VOCABULARY
- volume
- composite solid

The first stack forms an oblique prism. The second forms a right prism. The stacks have the same height. The area of every cross section parallel to a base is the area of one sheet of paper. The stacks have the same volume. These stacks illustrate the following principle.

Theorem 11.4-1 Cavalieri's Principle

If two solids have the same height and the same cross-sectional area at every level, then they have the same volume.

The area of each shaded cross section below is 6 cm^2. Since the prisms have the same height and the same cross section area, their volumes must be the same by Cavalieri's Principle.

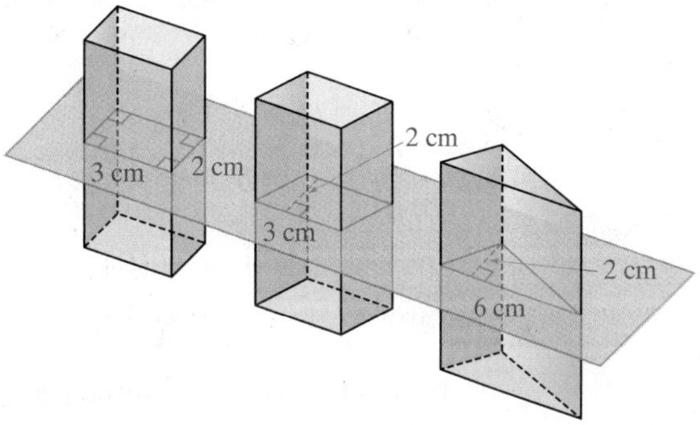

We can find the volume of a right prism by multiplying the area of the base by the height. Cavalieri's Principle lets us extend this idea to any prism.

Theorem 11.4-2 Volume of a Prism

The volume of a prism is

$$V = Bh$$

where B is the area of the base and h is the height of the prism.

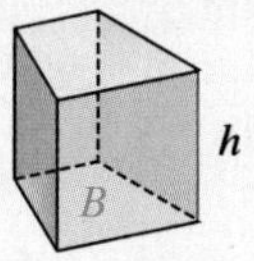

EXAMPLE 1 **Finding the Volume of a Rectangular Prism**

What is the volume of the rectangular prism at the right?

Solution Use the formula for volume of a prism.

$V = Bh$	Use the formula for the volume of a prism.
$= 480 \cdot 10$	The area of the base B is $24 \cdot 20$, or 480 sq cm, and the height is 10 cm.
$= 4800$	Multiply.

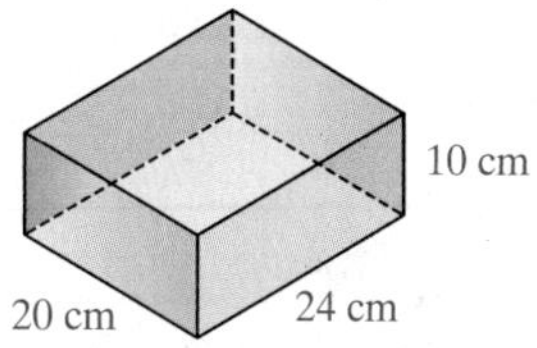

The volume of the rectangular prism is 4800 cubic cm. □

PRACTICE 1

a. What is the volume of the rectangular prism at the right?

b. Suppose the prism at the right is turned so that the base is 4 ft by 5 ft and the height is 3 ft. Does the volume change? Explain why or why not.

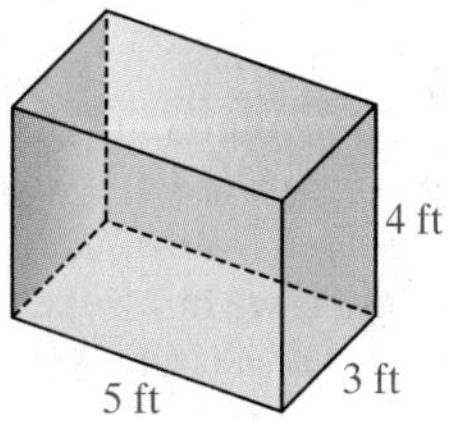

Helpful Hint

Don't forget:
- Perimeter, Circumference } measured in units
- Area, Surface area } measured in square units
- Volume—measured in cubic units

EXAMPLE 2 Finding the Volume of a Triangular Prism

Multiple Choice What is the approximate volume of the triangular prism?

a. 188 cu in. **b.** 277 cu in.

c. 295 cu in. **d.** 554 cu in.

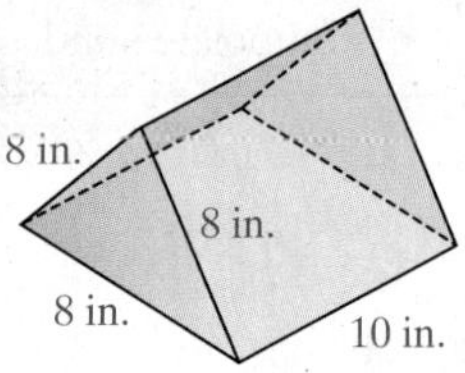

Solution To use the formula for volume of a prism, we first need to find the area of the base of the prism, which is a triangle.

> **Helpful Hint**
> Recall the special 30°-60°-90° triangle.
>
> 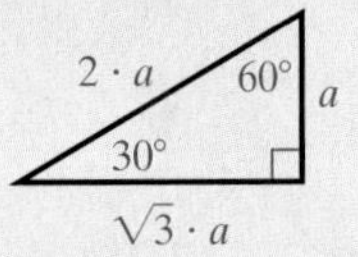
>

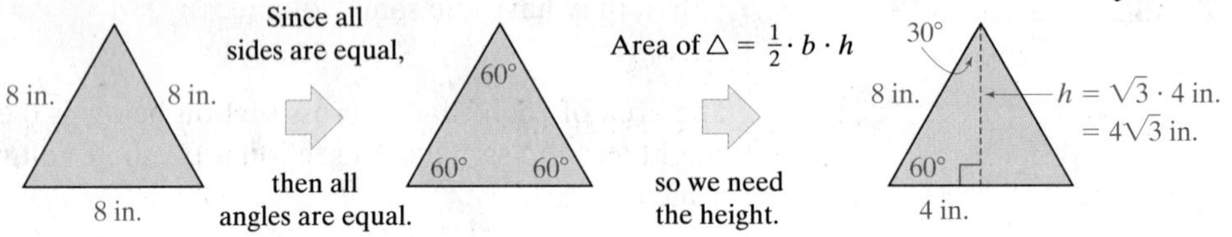

A base of the prism

To find h, either use the Pythagorean Theorem, or recall what we know about a 30°-60°-90° triangle. (See the Helpful Hint.)

$$B = \frac{1}{2}bh \qquad \text{Use the formula for the area of a triangle.}$$

$$= \frac{1}{2}(8)(4\sqrt{3}) \qquad \text{Substitute 8 for } b \text{ and } 4\sqrt{3} \text{ for } h.$$

$$= 16\sqrt{3} \qquad \text{Multiply.}$$

Now let's use our formula for the volume of the prism.

$$V = Bh \qquad \text{Use the formula for the volume of a prism.}$$

$$= 16\sqrt{3} \cdot 10 \qquad \text{Substitute } 16\sqrt{3} \text{ for } B \text{ and 10 for } h.$$

$$= 160\sqrt{3} \qquad \text{Multiply.}$$

$$\approx 277.1281292 \qquad \text{Use a calculator.}$$

The volume of the triangular prism is about 277 cu in. The correct answer is b. □

PRACTICE 2 What is the volume of the triangular prism shown here?

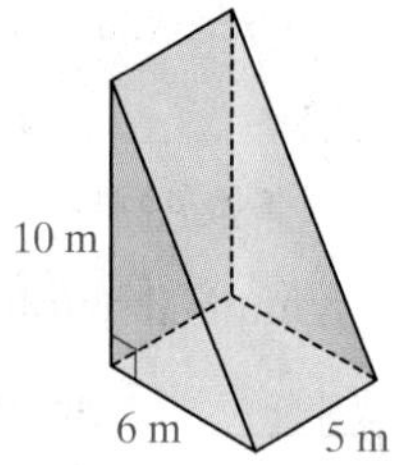

OBJECTIVE 2 ▶ Finding the Volume of a Cylinder. To find the volume of a cylinder, we use the same formula $V = Bh$ that we used to find the volume of a prism. Now, however, B is the area of the circle, so we use the formula $B = \pi r^2$ to find its value.

> **Helpful Hint**
> The volume of a cylinder and a prism are the same. They are both $V = Bh$, except with a cylinder, we know that $B = \pi r^2$, the area of a circle.

Theorem 11.4-3 Volume of a Cylinder

The volume of a cylinder is

$$V = Bh, \text{ or } V = \pi r^2 h$$

where B is the area of the circle base (so we know that $B = \pi r^2$) and h is the height of the cylinder.

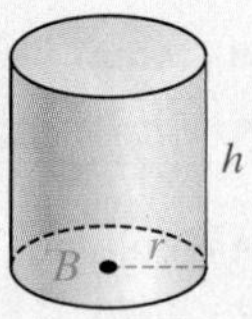

EXAMPLE 3 **Finding the Volume of a Cylinder**

What is the volume of the cylinder in terms of π?

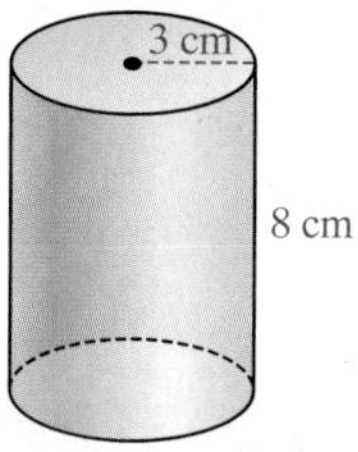

Solution Let's use our formula for the volume of a cylinder.

$V = \pi r^2 h$ Use the formula for the volume of a cylinder.
$= \pi(3)^2(8)$ Substitute 3 for r and 8 for h.
$= \pi(72)$ Simplify.

The volume of the cylinder is 72π cu cm. □

PRACTICE 3 What is the volume of the cylinder below in terms of π?

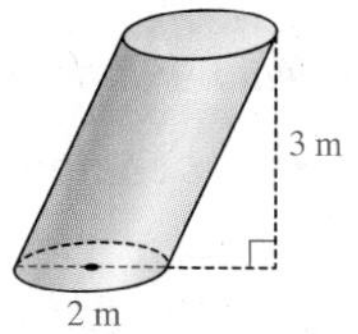

OBJECTIVE 3 ▶ Finding the Volume of Composite Solids. A **composite solid** is a three-dimensional figure that is the combination of two or more simpler figures. We can find the volume of a composite solid by adding the volumes of the figures that are combined.

EXAMPLE 4 **Finding Volume of a Composite Figure**

What is the approximate volume of the bullnose aquarium to the nearest cubic inch?

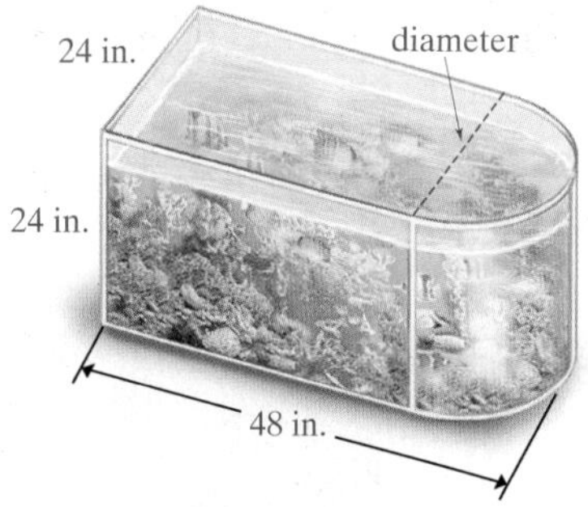

Solution This composite solid is formed by a prism and half of a cylinder. To find the total volume, find the volume of each piece to the nearest whole; then add.

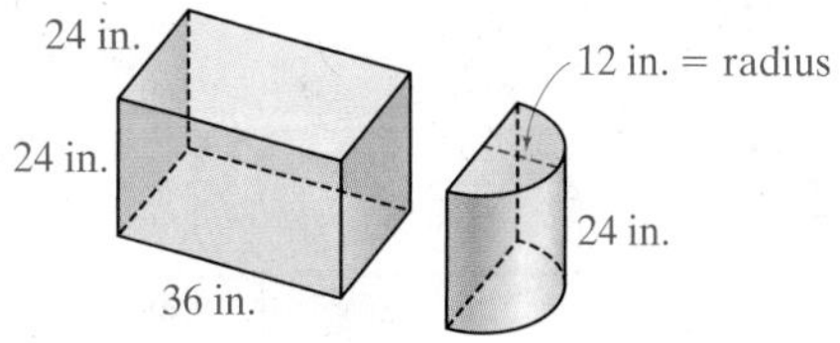

Prism

$V_{\text{of prism}} = Bh$
$= (24 \cdot 36)(24)$
$= 20{,}736$

Half a Cylinder

Since the diameter is 24 in., the radius is 12 in.

$V_{\text{of half the cylinder}} = \frac{1}{2}\pi r^2 h$
$= \frac{1}{2}\pi(12)^2(24)$
≈ 5429

Now, let's add the two volumes.

$$20{,}736 + 5429 = 26{,}165$$

The approximate volume of the aquarium is 26,165 cu in. □

PRACTICE 4 What is the approximate volume of the lunch box shown here? Round to the nearest cubic inch.

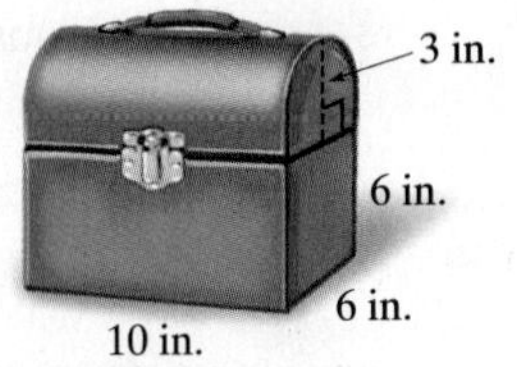

VOCABULARY & READINESS CHECK

Word Bank *Use the choices to fill in each blank. Some choices may be used more than once.*

units	cubic units	composite solid
square units	volume	

1. The measure of the space that a figure occupies is called ____________.
2. Perimeter is measured in ____________.
3. Area is measured in ____________.
4. Volume is measured in ____________.
5. Surface area is measured in ____________.
6. Circumference is measured in ____________.
7. If two solids have the same height and the same cross sectional area at all levels, then they have the same ____________.
8. A(n) ____________ is a three-dimensional figure that is the combination of two or more simpler figures.

11.4 EXERCISE SET MyMathLab®

Find the volume of each rectangular prism. See Example 1.

1.

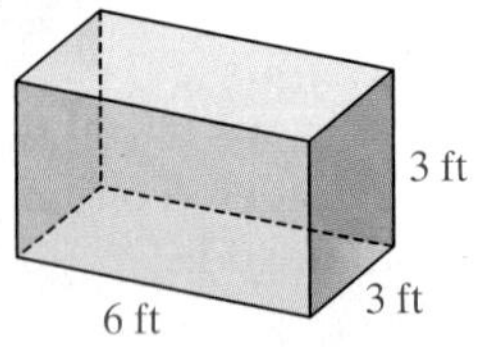

2.

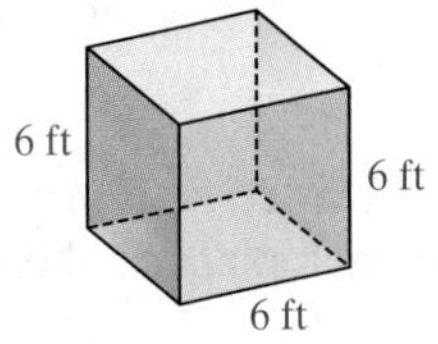

3.

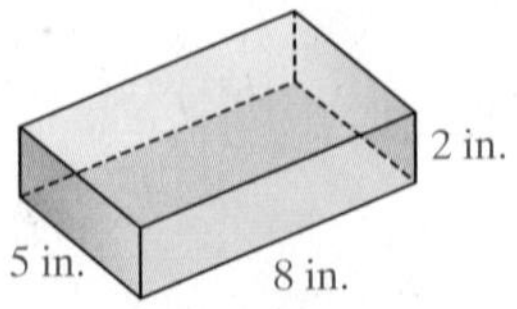

4.

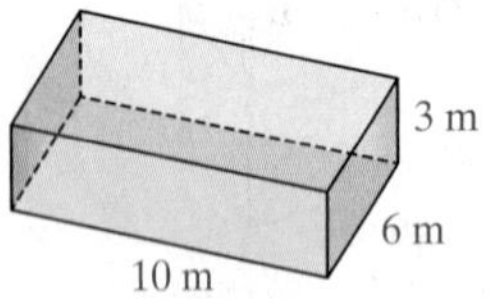

Find the volume of each triangular prism. See Example 2.

5.

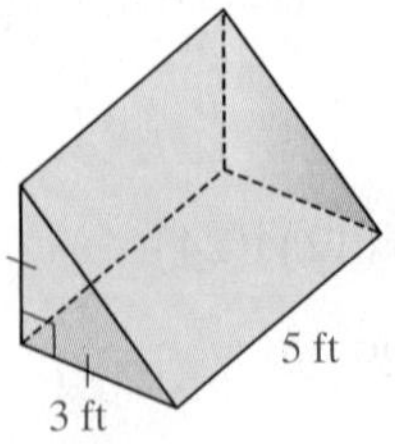

6.

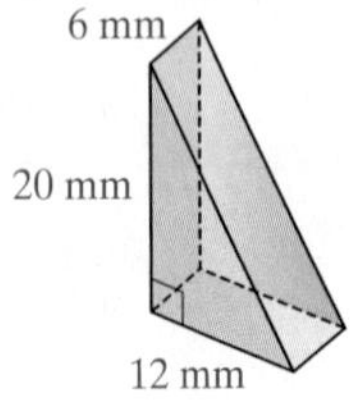

7.

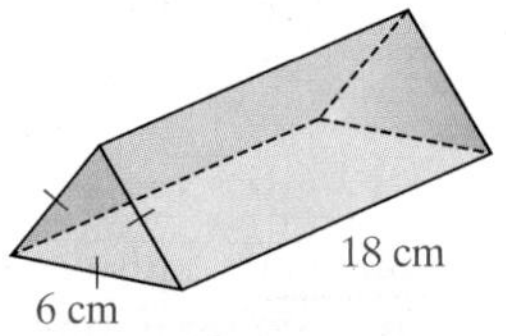

8.

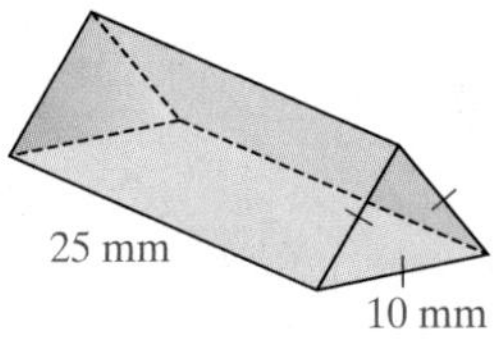

Find the volume of each cylinder in terms of π and to the nearest tenth. See Example 3.

9.

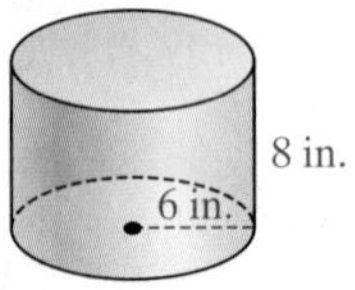

10.

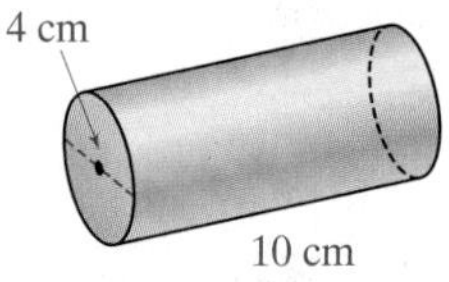

11.

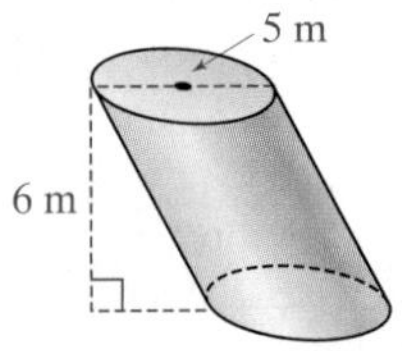

12.

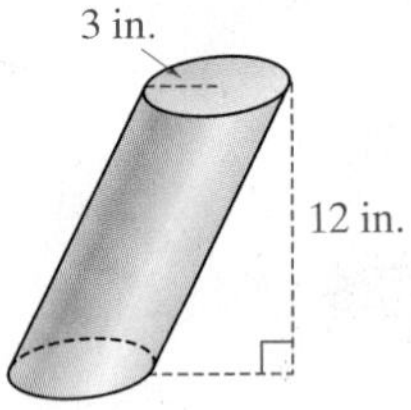

Solve. See Examples 1–3.

13. The Beijing National Aquatics Center was built for the swimming competitions of the 2008 Summer Olympics. Although it is called the Water Cube, the aquatic center is really a rectangular box (cuboid) with side lengths of 584 feet and a height of 102 feet. Find the volume of this building.

(Note: The outer wall is very interesting because it was based on the Weaire–Phelan structure, the cross section of the natural pattern of bubbles in soap lather.)

14. The Atlas Building in the Netherlands has a length and width of 44 meters and a height of 28 meters. Find the volume of this building.

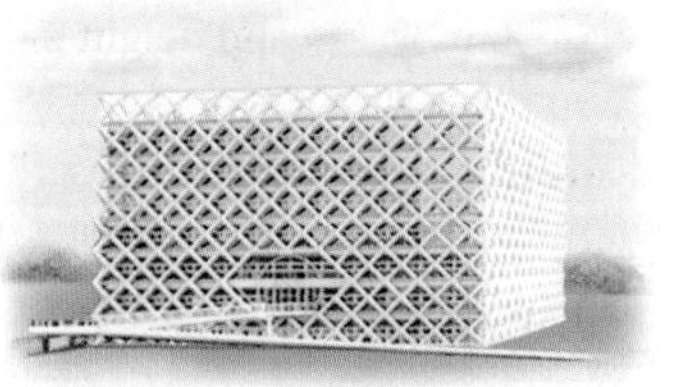

15. Find the volume and surface area of a rectangular box 2 ft by 1.4 ft by 3 ft.

16. Find the volume and surface area of a box in the shape of a cube that is 5 ft on each side.

17. Metal silos in the shape of cylinders are used in Kenya to store grain. These silos prevent the infestation of pests and avoid the development of toxins that can result if the grain is not fully dried. Find the volume of a metal silo that is 180 cm in height and 86 cm in diameter. Round to the nearest whole number.

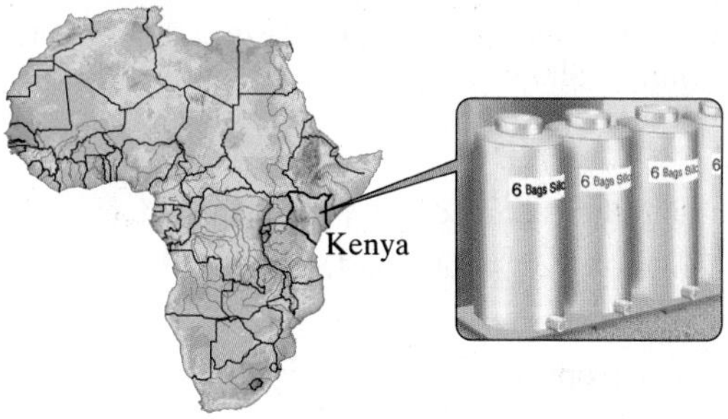

18. Find the volume of a stainless steel silo in the shape of a cylinder that is 12 feet in diameter and 60 feet in height. Round to the nearest whole number.

Find the volume of each cylinder rounded to the nearest whole number.

19.

20.

Use the diagram of the backpack at the right, formed by a rectangular prism and half of a cylinder. See Example 4.

21. Find the volume of the prism.

22. Find the volume of the half-cylinder in terms of π.

23. What is the volume of the backpack in terms of π?

24. What is the volume of the backpack to the nearest cubic inch?

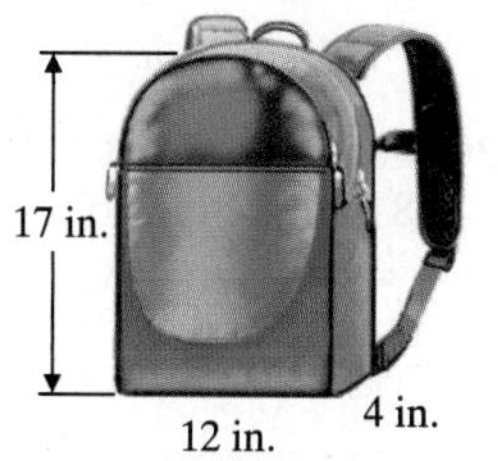

Find the volume of each composite solid to the nearest whole number.

25.

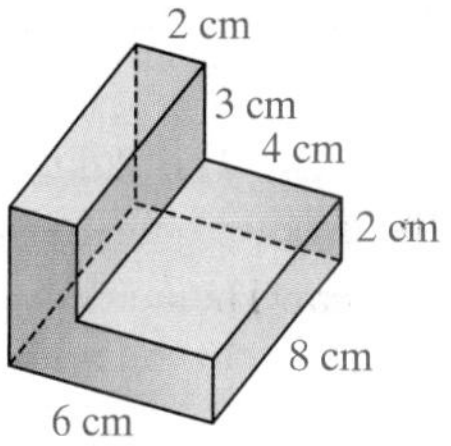

26.

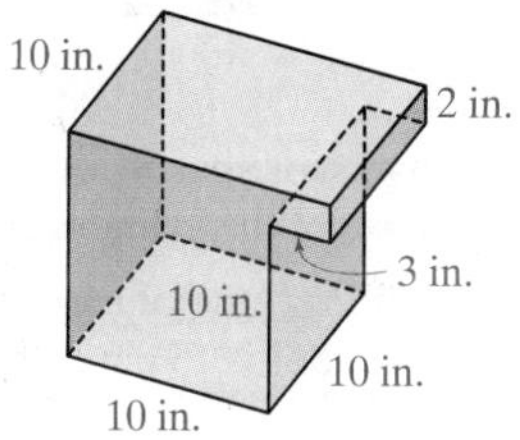

27.

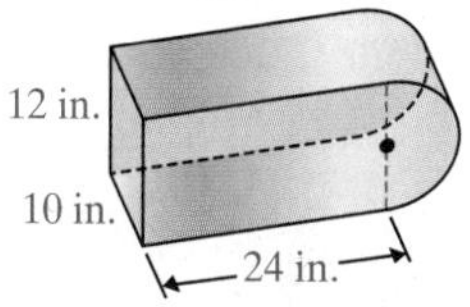

28.

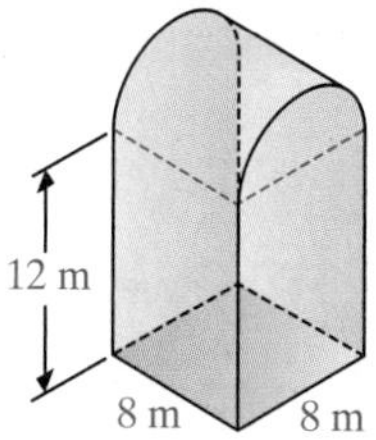

A full waterbed mattress is 7 ft by 4 ft by 1 ft. The mattress is to be filled with water that weighs 62.4 lb/ft^3.

29. Find the volume of the mattress.

30. Find the weight of the water in the mattress to the nearest pound.

Use the rectangular prisms below for Exercises 31–33.

31. Find the volume of each rectangular prism.

32. Find the surface area of each rectangular prism.

33. Do two rectangular solids with the same volume have the same surface area? To see, study your answers to Exercises **31** and **32**.

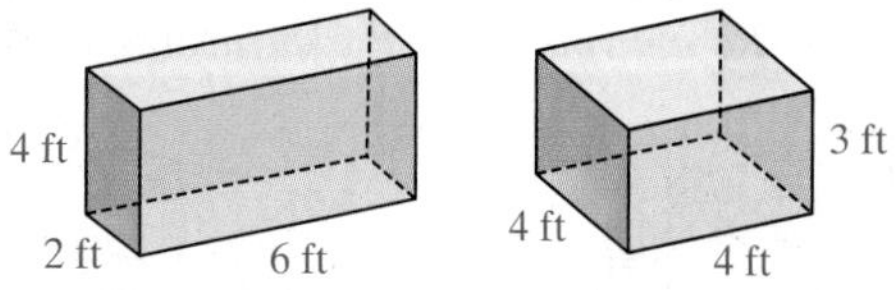

34. Give the dimensions of two rectangular prisms that have volumes of 80 cubic cm each but also have different surface areas.

Find the height of each figure with the given volume.

35.
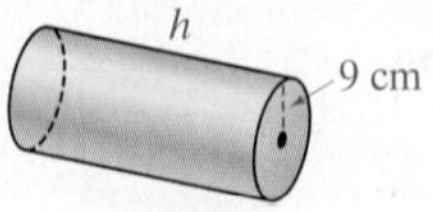

$V = 3240\pi$ cu cm

36.
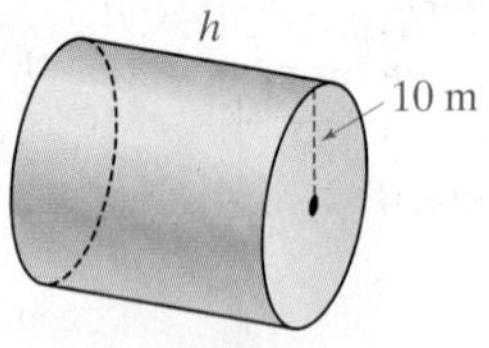

$V = 2000\pi$ cu m

37.
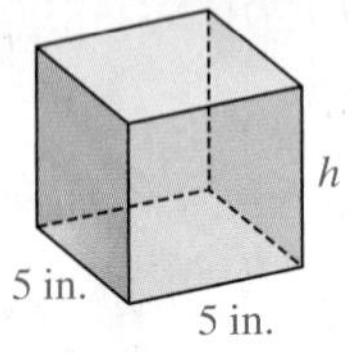

$V = 125$ cu in.

38.
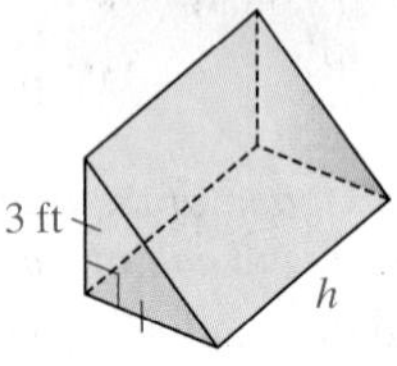

$V = 27$ cu ft

39. A can of tennis balls has a diameter of 3 in. and a height of 8 in. Find the volume of the can to the nearest cubic inch.

40. What is the volume of the oblique prism shown below?

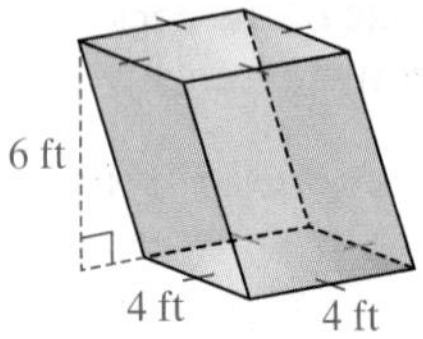

41. The volume of a cylinder is 600π cubic cm. The radius of a base of the cylinder is 5 cm. What is the height of the cylinder?

42. The volume of a cylinder is 135π cubic cm. The height of the cylinder is 15 cm. What is the radius of a base of the cylinder?

Coordinate Geometry *Find the volume of each rectangular prism below.*

43.
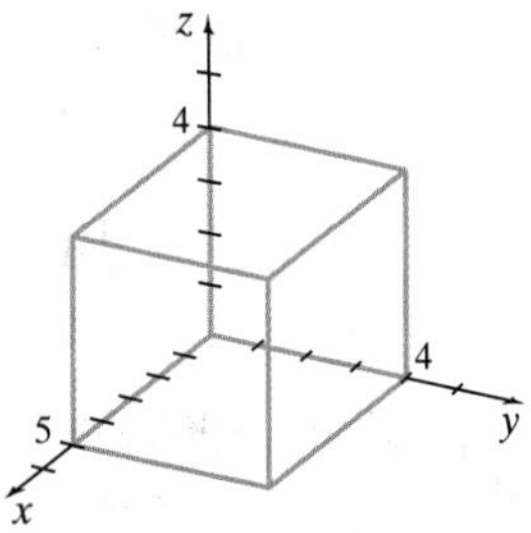

44.
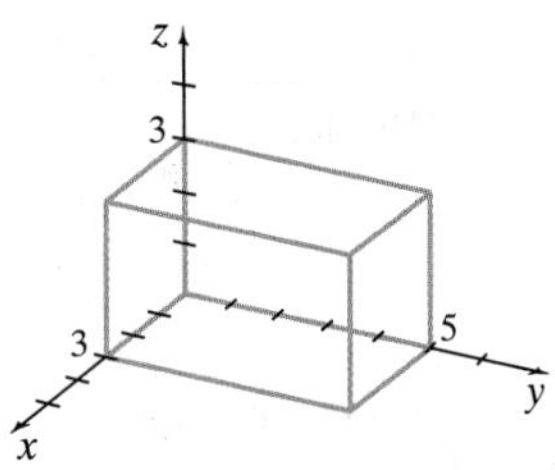

The closed box at the right is shaped like a regular pentagonal prism. The exterior of the box has base edge 10 cm and height 14 cm. The interior has base edge 7 cm and height 11 cm. Find each measurement.

45. the interior surface area

46. the exterior surface area

47. the volume of the interior of the box

48. the volume of the exterior of the box

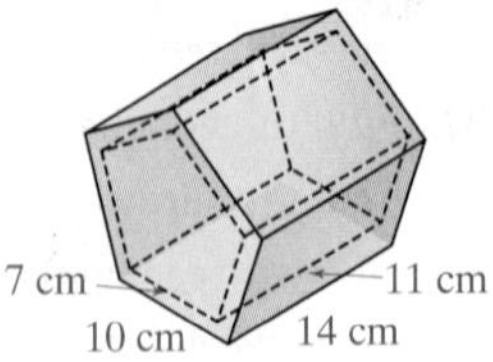

A cylinder has been cut out of each solid. Find the volume of the remaining solid. Round your answer to the nearest tenth.

49.
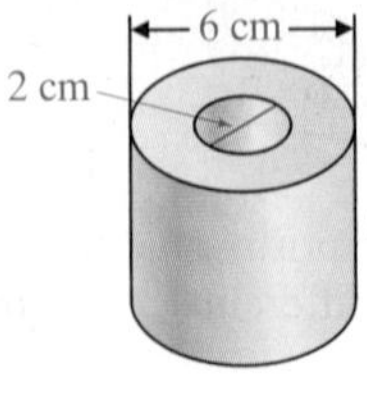

50.
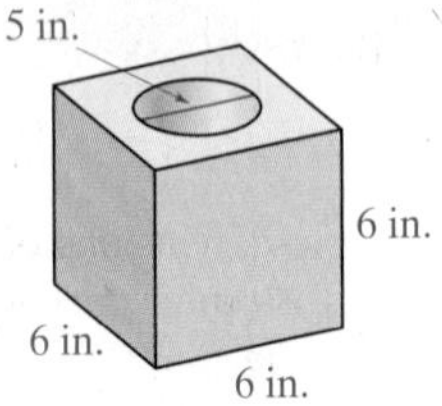

Suppose you revolve the plane region completely about the given line to form a solid of revolution. Describe the solid and find its volume in terms of π.

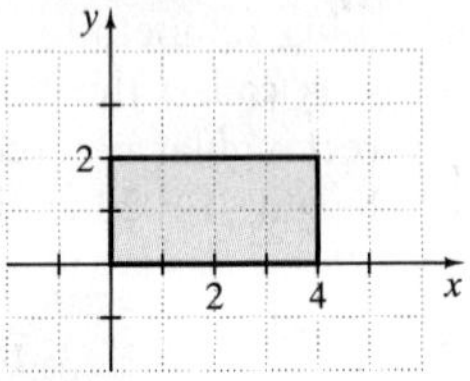

51. the x-axis

52. the y-axis

53. the line $y = 2$

54. the line $x = 5$

CONCEPT EXTENSIONS

55. How are the formulas for the volume of a prism and the volume of a cylinder alike? How are they different?

56. Is the figure at the right a composite space figure? Explain.

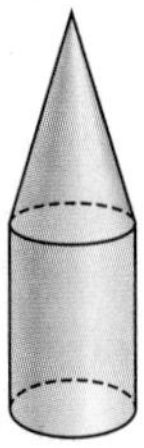

57. How is the volume of a rectangular prism with base 2 m by 3 m and height 4 m related to the volume of a rectangular prism with base 3 m by 4 m and height 2 m? Explain.

58. The figures below can be covered by equal numbers of straws that are the same length. Describe how Cavalieri's Principle could be adapted to compare the areas of these figures.

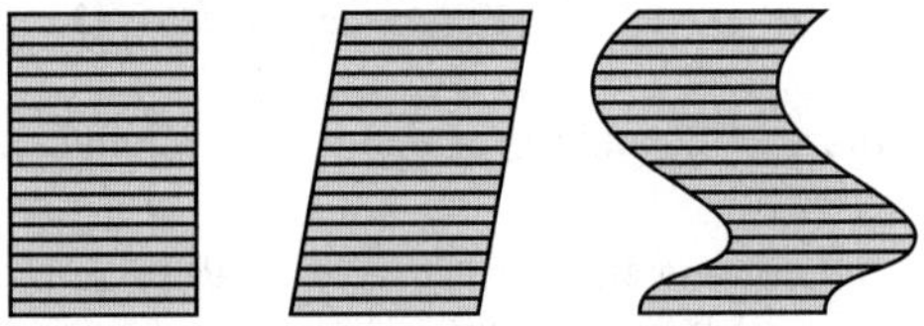

59. The approximate dimensions of an Olympic-size swimming pool are 164 ft by 82 ft by 6.6 ft.

a. Find the volume of the pool to the nearest cubic foot.

b. If 1 cu ft $\approx$ 7.48 gal, about how many gallons does the pool hold?

60. To landscape her 70-ft-by-60-ft rectangular backyard, your aunt is planning first to put down a 4-in. layer of topsoil. Option 1: She can buy bags of topsoil at $2.50 per 3-cu-ft bag, with free delivery. Option 2: Or, she can buy bulk topsoil for $22.00/cu yd, plus a $20 delivery fee.

a. Find the volume of soil needed in cu ft.

b. Find the cost of Option 1.

c. Find the cost of Option 2.

d. Which option is less expensive? Explain.

61. Suppose the dimensions of a rectangular prism are tripled. How does this affect its volume? Explain.

62. The radius of Cylinder B is twice the radius of Cylinder A. The height of Cylinder B is half the height of Cylinder A. Compare their volumes.

Any rectangular sheet of paper can be rolled into a right cylinder in two ways.

63. Use two 8.5 in. by 11 in. sheets of paper to model the two cylinders described in the directions. Compute the volume of each cylinder. How do they compare?

64. Of all sheets of paper with perimeter 39 in., which size can be rolled into a right cylinder with greatest volume? (*Hint:* Make a table and use whole numbers for lengths (longer side).)

REVIEW AND PREVIEW

Find the lateral area of each figure to the nearest tenth. See Section 11.3.

65. a right circular cone with height 12 mm and radius 5 mm

66. a regular hexagonal pyramid with base edges 9.2 ft long and slant height 17 ft

Solve. See Section 9.1.

67. Find h in the figure at the right.

68. A right triangle has hypotenuse 300 ft and leg 180 ft. What is the length of the other leg?

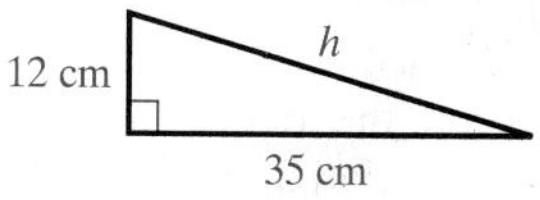

11.5 VOLUMES OF PYRAMIDS AND CONES

OBJECTIVES

1 Find the Volume of a Pyramid.

2 Find the Volume of a Cone.

OBJECTIVE 1 ▶ Finding the Volume of a Pyramid. The volume of a pyramid is related to the volume of a prism with the same base and height.

Theorem 11.5-1 Volume of a Pyramid

The volume of a pyramid is one-third the volume of a prism. It is

$$V = \frac{1}{3}Bh$$

where B is the area of the base and h is the height of the pyramid.

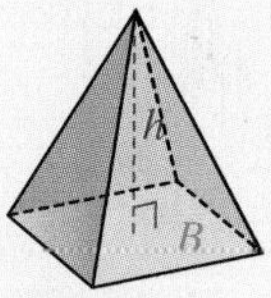

Because of Cavalieri's Principle, the volume formula is true for all pyramids. The height h of an oblique pyramid is the length of the perpendicular segment from its vertex to the plane of the base.

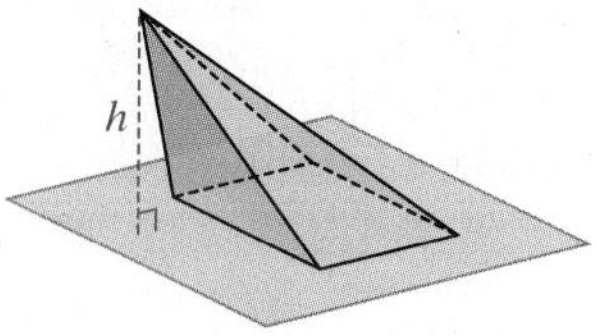

EXAMPLE 1 Finding Volume of a Pyramid

The entrance to the Louvre Museum in Paris, France, is a pyramid with a square base and a height of 21.64 m. What is the approximate volume of the Louvre Pyramid?

Solution The side length of the base is shown to be 35.4 m. B is the area of a square. We can calculate B now or simply insert side · side for B.

$$V = \frac{1}{3}Bh \quad \text{Use the formula for volume of a pyramid.}$$

$$= \frac{1}{3}(35.4 \cdot 35.4)(21.64) \quad \text{Substitute for } B \text{ and } h.$$

$$= 9039.4608 \quad \text{Multiply.}$$

The volume is about 9039 cubic m.

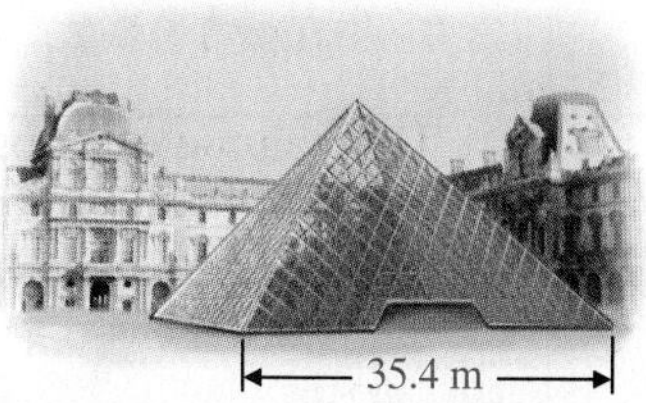

PRACTICE 1 A sports arena shaped like a pyramid has a base area of about 300,000 sq ft and a height of 321 ft. What is the approximate volume of the arena?

> **Helpful Hint**
>
> Notice how the slant height ℓ and the perpendicular height h are related. They form a leg and a hypotenuse of a right triangle.
>
> 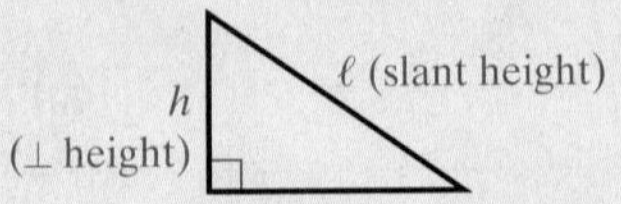
>

EXAMPLE 2 Finding the Volume of a Pyramid

What is the volume in cubic feet of a square pyramid with base edges 40 ft and slant height 25 ft?

Solution The slant height of a pyramid is important when finding its surface area, but for volume, we use a perpendicular height, h.

First, let's find the height of the pyramid.

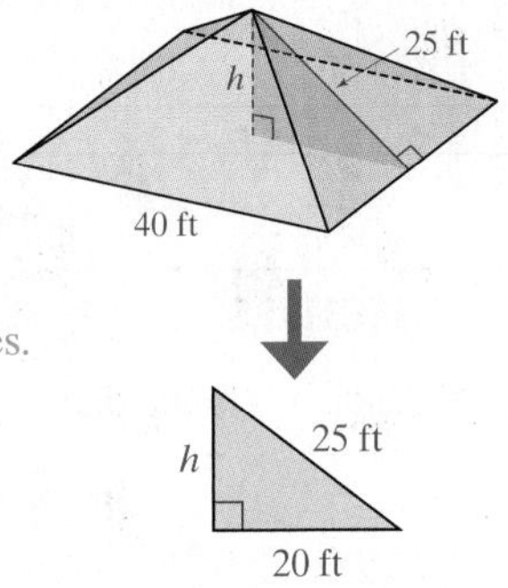

Use the Pythagorean Theorem to find h.

$25^2 = h^2 + 20^2$ Use the Pythagorean Theorem.

$625 = h^2 + 400$ Find 25^2 and 20^2.

$h^2 = 225$ Solve for h^2.

$h = \sqrt{225}$ Take the positive square root of both sides.

$h = 15$ Simplify.

Now we can find the volume of the pyramid.

$V = \frac{1}{3}Bh$ Use the formula for volume of a pyramid.

$= \frac{1}{3}(40 \cdot 40)(15)$ Substitute $40 \cdot 40$ for B and 15 for h.

$= 8000$ Simplify.

The volume of the pyramid is 8000 cubic ft. □

PRACTICE 2 What is the volume of a square pyramid with base edges 24 m and slant height 13 m?

OBJECTIVE 2 ▶ Finding the Volume of a Cone. The volume of a cone is related to the volume of a cylinder with the same base and height.

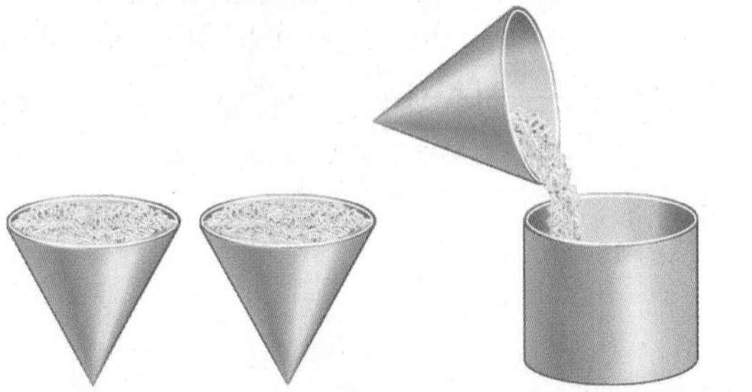

The cones and the cylinder have the same base and height.
It takes three cones full of rice to fill the cylinder.

> **Helpful Hint**
>
> The volume of a pyramid and a cone are both
>
> $$V = \frac{1}{3}Bh$$
>
> With a cone, we know that B is the area of a circle, so
>
> $$B = \pi r^2$$

> **Theorem 11.5-2 Volume of a Cone**
>
> The volume of a cone is one-third the volume of a cylinder. It is
>
> $$V = \frac{1}{3}Bh, \text{ or } V = \frac{1}{3}\pi r^2 h$$
>
> where B is the area of the circle base (so we know that $B = \pi r^2$) and h is the height of the cone.
>
> 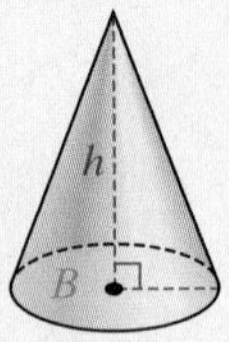
>

A cone-shaped structure can be particularly strong, as downward forces at the vertex are distributed to all points in its circular base.

EXAMPLE 3 **Finding the Volume of a Cone**

The covering on a tepee rests on poles that come together like concurrent lines. The resulting structure approximates a cone. If the tepee pictured is 12 ft high with a base diameter of 14 ft, what is its approximate volume?

Solution We use the formula for volume of a cone.

$V = \frac{1}{3}\pi r^2 h$ Use the formula for the volume of a cone.

$= \frac{1}{3}\pi(7)^2(12)$ Substitute $\frac{14}{2}$, or 7, for r and 12 for h.

≈ 615.7521601 Use a calculator to approximate.

The volume of the tepee is approximately 616 cubic ft. □

PRACTICE

3 The height and radius of a child's tepee are half those of the tepee in Example 3. What is the volume of the child's tepee to the nearest cubic foot?

This volume formula applies to all cones, including oblique cones.

EXAMPLE 4 **Finding the Volume of an Oblique Cone**

What is the volume of the oblique cone below at the right? Give your answer in terms of π and also rounded to the nearest cubic foot.

Solution For volume of an oblique cone, we still need radius r and the perpendicular height, h.

$V = \frac{1}{3}\pi r^2 h$ Use the formula for volume of a cone.

$= \frac{1}{3}\pi(15)^2(25)$ Substitute 15 for r and 25 for h.

$= 1875\pi$ Multiply.

≈ 5890.486225 Use a calculator to approximate.

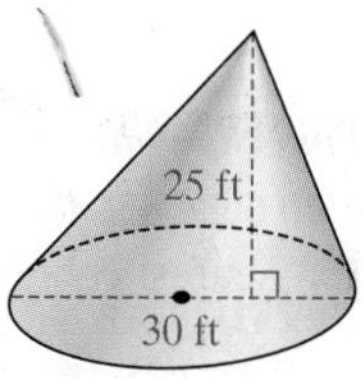

$d = 30$ ft or $r = 15$ ft

The volume of the cone is 1875π cubic ft, or about 5890 cubic ft. □

PRACTICE

4 What is the volume of the oblique cone at the right in terms of π and rounded to the nearest cubic meter?

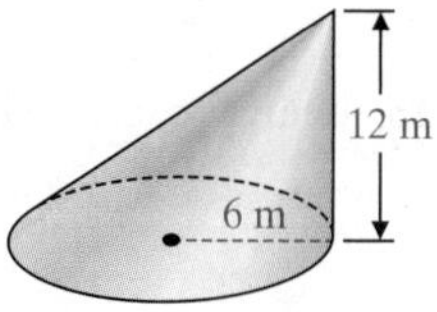

11.5 EXERCISE SET MyMathLab®

Find the volume of each square pyramid. If necessary, round to the nearest tenth. See Example 1.

1.

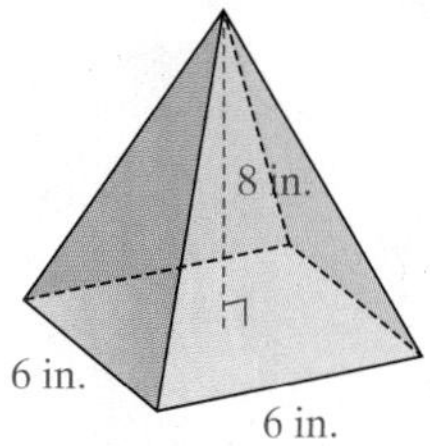

2.

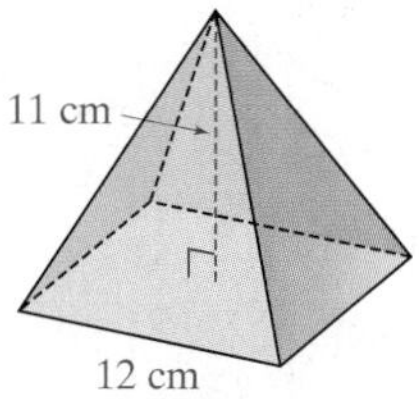

3.

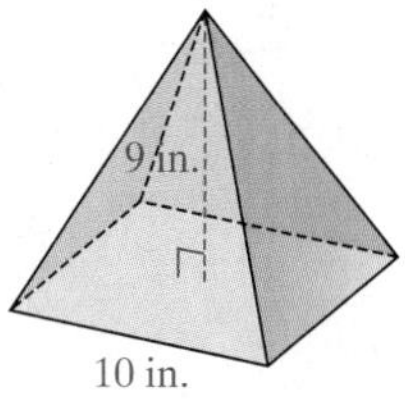

4.

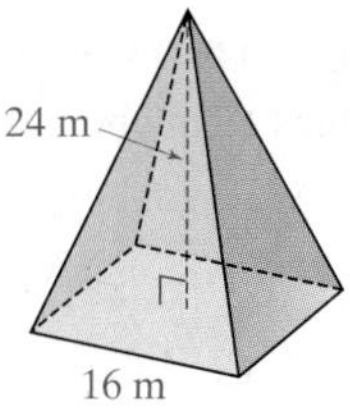

Find the volume of each square pyramid, given its slant height. Round to the nearest tenth. See Example 2.

5.

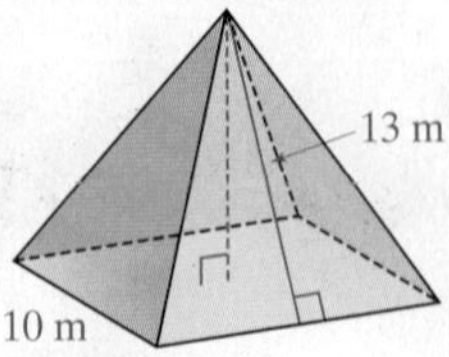

6.

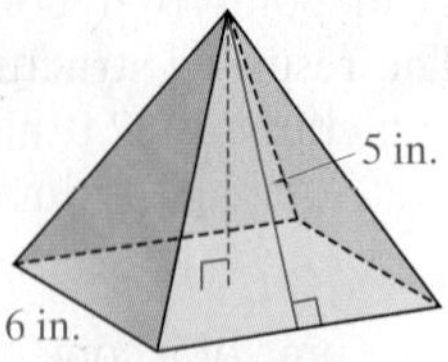

7.

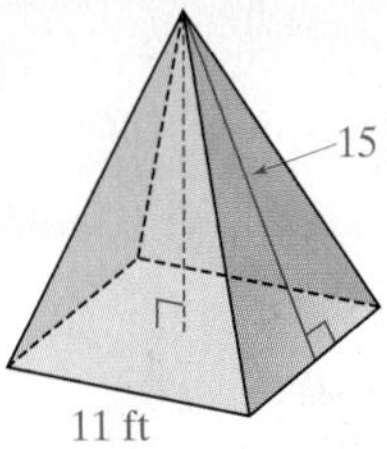

8.

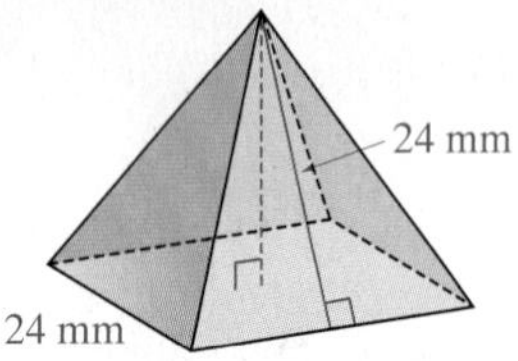

Find the volume of each cone. Round each volume to the nearest whole number. See Example 3.

9.

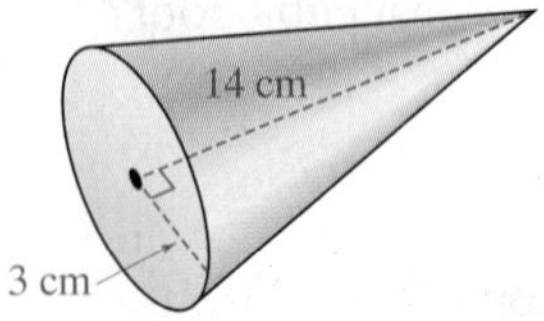

10.

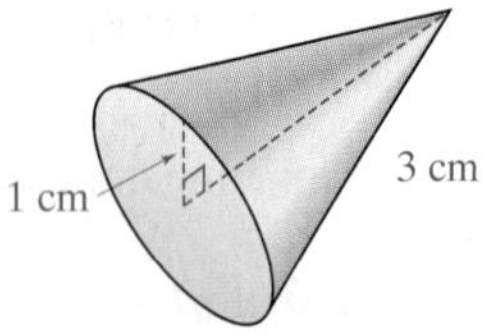

11.

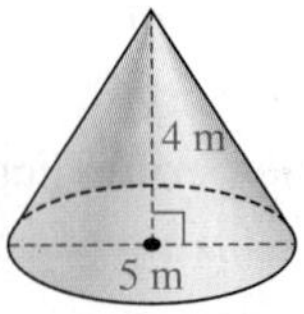

12.

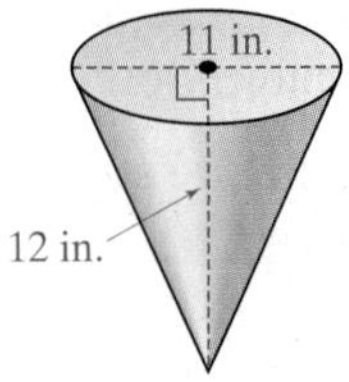

Find the volume of each cone in terms of π and also rounded as indicated. See Example 4.

13. nearest cubic foot

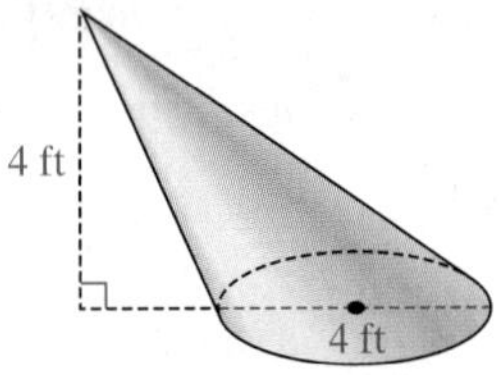

14. nearest cubic mile

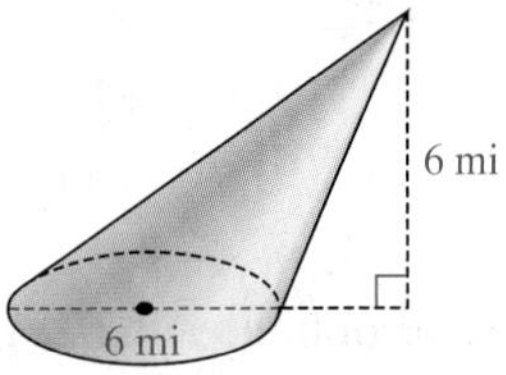

15. nearest cubic inch

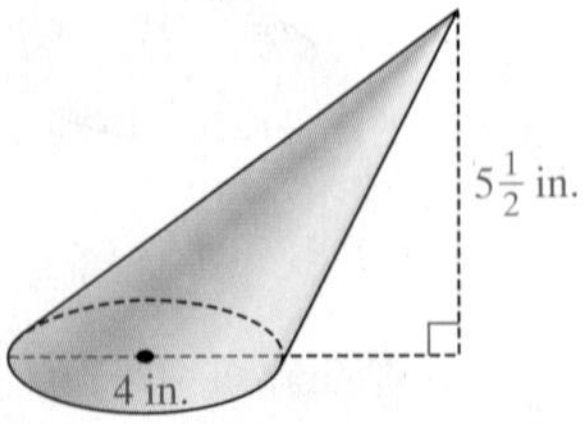

16. nearest cubic meter

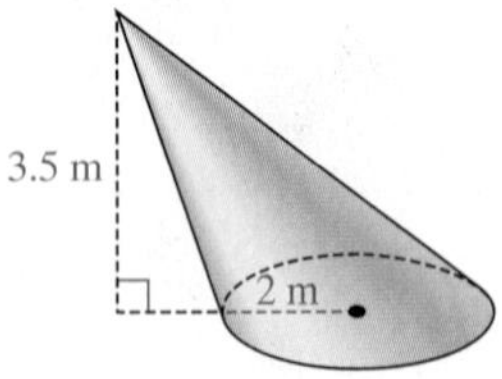

Find the volume of each square pyramid.

17. base edges 10 cm, height 6 cm

18. base edges 5 m, height 6 m

19. base edges 18 in., height 12 in.

20. base edges 20 yd, height 5 yd

Mixed Practice *Solve. If necessary, round to the nearest whole number.*

21. The Transamerica Pyramid Building in San Francisco is 853 ft tall with a square base that is 149 ft on each side. To the nearest thousand cubic feet, what is the volume of the Transamerica Pyramid?

22. A paperweight is in the shape of a square-based pyramid 20 centimeters tall. If an edge of the base is 12 centimeters, find the volume of the paperweight.

For Exercises 23 and 24, the volume of figures has a lot to do with the cost of the product sold, profit, etc. Find each volume. We found the surface area of these two products in Section 11.3.

23. Find the volume of a waffle cone with a diameter of 2.6 in. and an approximate height of 6 in.

24. Find the volume of a sno-cone cup with a diameter of 3.2 in. and an approximate height of 4 in.

25. In a chemistry lab you use a filter paper cone to filter a liquid. The diameter of the cone is 6.5 cm and its height is 6 cm. How much liquid will the cone hold when it is full?

26. This cone has a filter that was being used to remove impurities from a solution but became clogged and stopped draining. The remaining solution is represented by the shaded region. How many cubic centimeters of the solution remain in the cone?

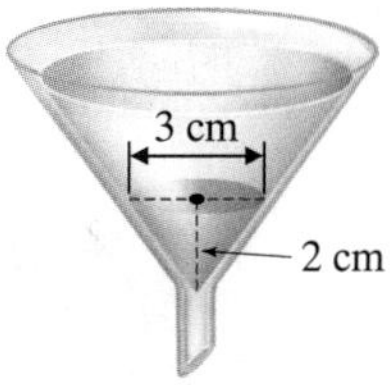

27. Mount Fuji, in Japan, is considered the most beautiful composite volcano in the world. The mountain is in the shape of a cone whose height is about 3.5 kilometers and whose base radius is about 3 kilometers. Approximate the volume of Mt. Fuji in cubic kilometers.

28. Find the exact volume of a waffle ice cream cone with a 3-in. diameter and a height of 7 inches.

Note: Exercises 29 and 30 contain more dimensions than needed. Find the volume of each pyramid rounded to the nearest whole number.

29.

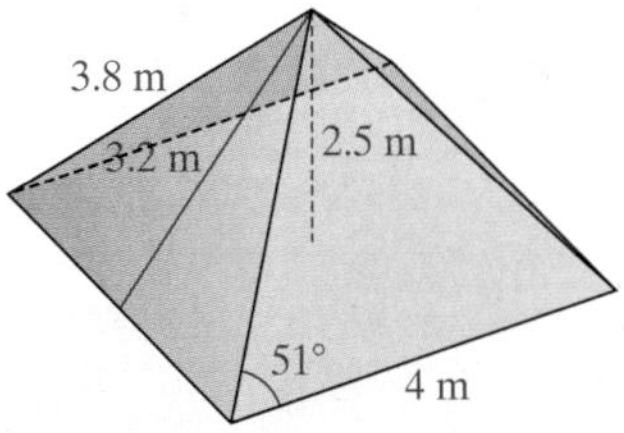

30. These are the dimensions of the Great Pyramid in royal cubits, an early Egyptian unit of measure. Find the volume in cubic royal cubits.

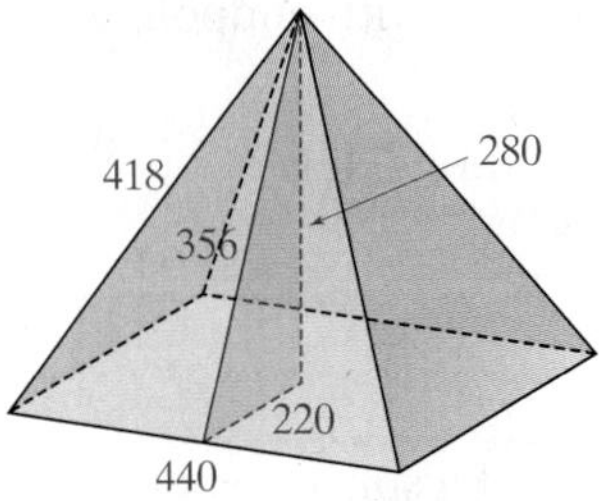

Note: 1 royal cubit = 7 palms = 28 fingers
Today: 1 royal cubit ≈ 52.5 cm

Suppose you revolve the plane region completely about the given line to form a solid of revolution. Describe the solid. Then find its volume in terms of π.

31. the y-axis

32. the x-axis

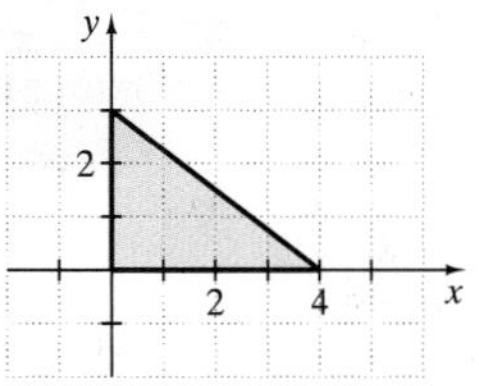

Find the volume to the nearest whole number.

33.

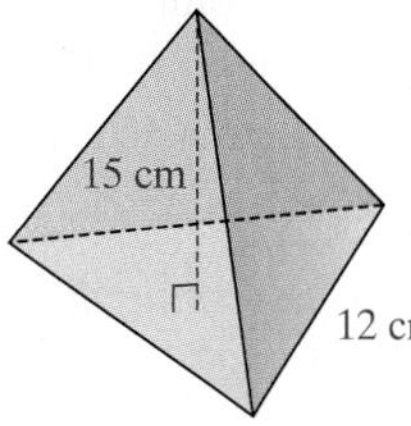

Equilateral base

34.

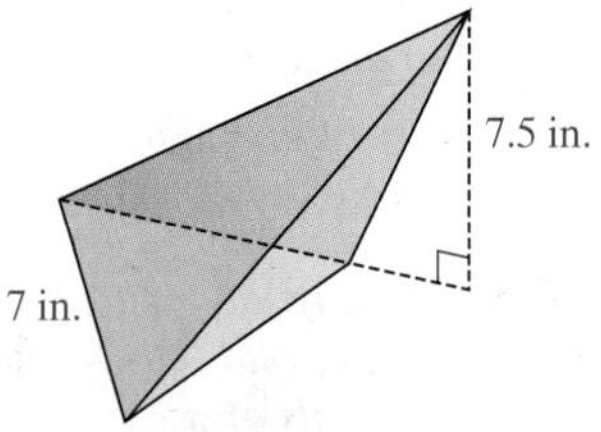

Equilateral base

35.

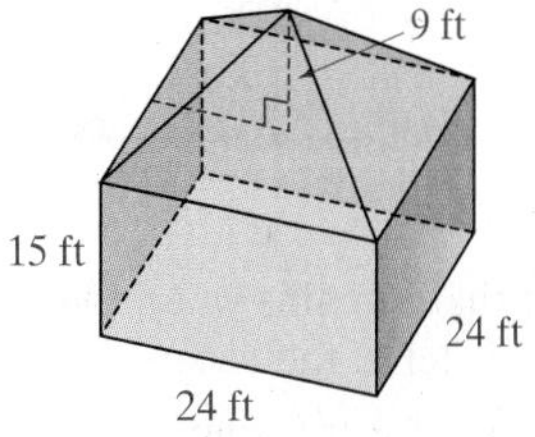

Square base

36.

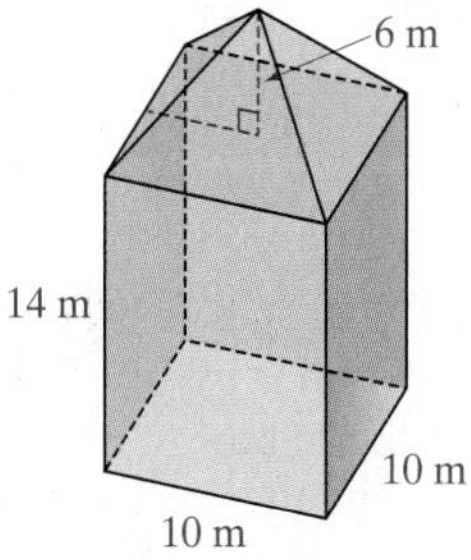

Square base

Find the value of the variable in each figure. Leave answers in simplest radical form. The diagrams are not to scale.

37.

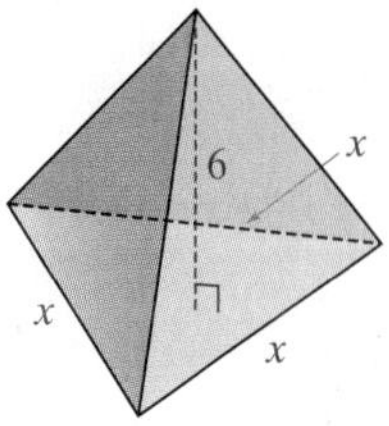

Volume = $18\sqrt{3}$ cu. units

38.

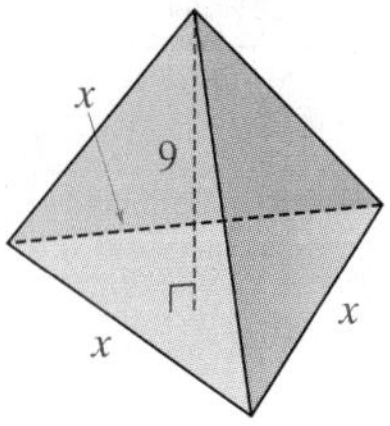

Volume = $75\sqrt{3}$ cu. units

39.

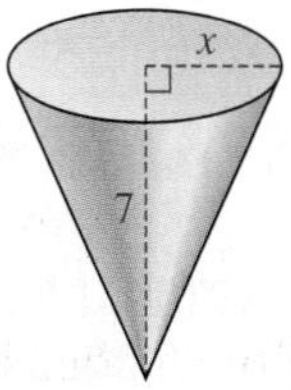

Volume = 21π cu. units

40.

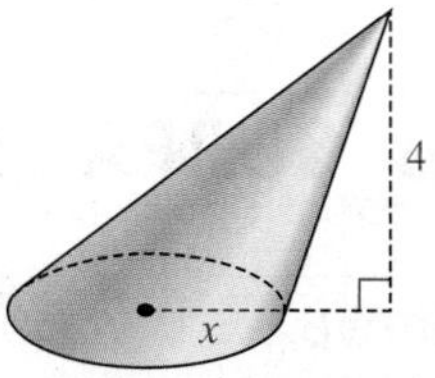

Volume = 24π cu. units

A cone with radius 3 ft and height 10 ft has a volume of 30π cubic ft. What is the volume of the cone formed when the following happens to the original cone?

41. The radius is doubled.

42. The height is doubled.

43. The radius and the height are both doubled.

44. The radius and the height are both tripled.

CONCEPT EXTENSIONS

45. Find the Error A square pyramid has base edges 13 ft and height 10 ft. A cone has diameter 13 ft and height 10 ft. Your friend claims the figures have the same volume because the volume formulas for a pyramid and a cone are the same: $V = \frac{1}{3}Bh$. What is the error?

46. How are the formulas for the volume of a pyramid and the volume of a cone alike? How are they different?

47. Without doing any calculations, explain how the volume of a cylinder with $B = 5\pi$ cm and $h = 20$ cm compares to the volume of a cone with the same base area and height.

48. Suppose the height of a pyramid is halved. How does this affect its volume? Explain.

Suppose you revolve the plane region completely about the given line to form a solid of revolution. Describe the solid. Then find its volume in terms of π.

49. the line $x = 4$

50. the line $y = -1$

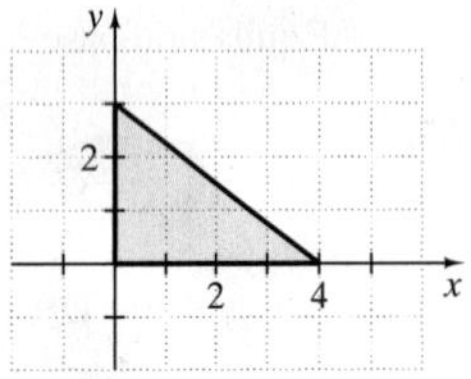

y
2
2
4
x

51. A *frustum* of a cone is the part that remains when the top of the cone is cut off by a plane parallel to the base.

a. Explain how to use the formula for the volume of a cone to find the volume of a frustum of a cone.

b. A popcorn container 9 in. tall is the frustum of a cone. Its small radius is 4.5 in. and its large radius is 6 in. What is its volume? Note: In the diagram below, $H = 36$ in.

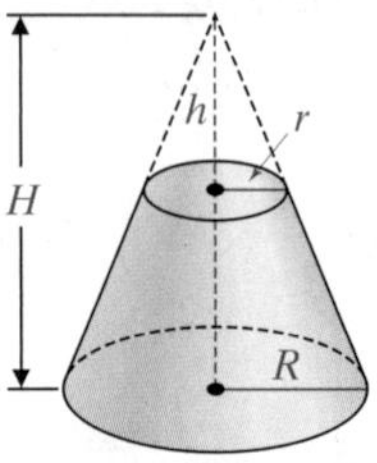

H
h
r
R

Frustum of cone

52. A disk has radius 10 m. A 90° sector is cut away, and a cone is formed.

a. What is the circumference of the base of the cone?

b. What is the area of the base of the cone?

c. What is the volume of the cone? (*Hint:* Use the slant height and the radius of the base to find the height.)

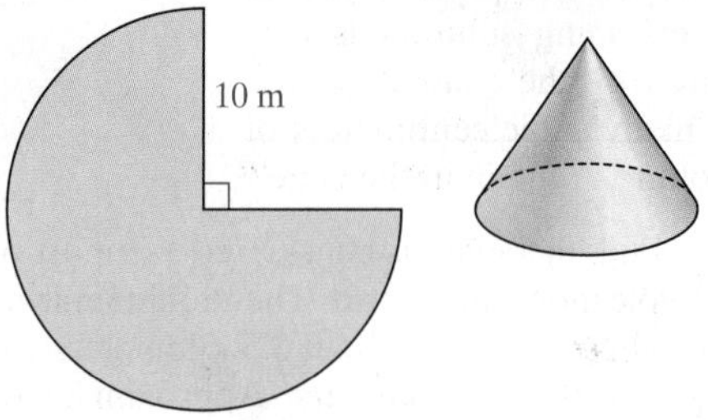

10 m

REVIEW AND PREVIEW

53. A triangular prism has height 30 cm. Its base is a right triangle with legs 10 cm and 24 cm. What is the volume of the prism? See Section 11.4.

54. Given ΔJAC and ΔKIN, you know $\overline{JA} \cong \overline{KI}$, $\overline{AC} \cong \overline{IN}$, and $m\angle A > m\angle I$. What can you conclude about JC and KN? See Section 5.6.

Solve. See Section 2.1.

55. Find the area of a circle with diameter 3 in. to the nearest tenth.

56. Find the circumference of a circle with radius 2 cm to the nearest centimeter.

11.6 SURFACE AREAS AND VOLUMES OF SPHERES

OBJECTIVE

1 Find the Surface Area and Volume of a Sphere.

VOCABULARY

- sphere
- center of a sphere
- radius of a sphere
- diameter of a sphere
- circumference of a sphere
- great circle
- hemisphere

OBJECTIVE 1 ▶ Finding the Surface Area and the Volume of a Sphere. A **sphere** is the set of all points in space equidistant from a given point called the **center.** A **radius** is a segment that has one endpoint at the center and the other endpoint on the sphere. A **diameter** is a segment passing through the center with endpoints on the sphere.

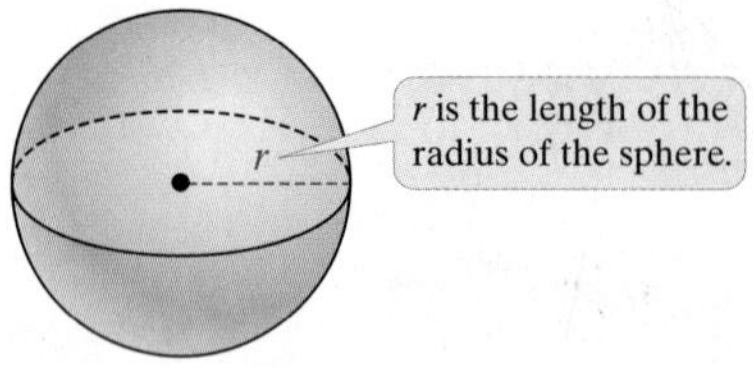

r
r is the length of the radius of the sphere.

We can find the surface area and the volume of a sphere when we know its radius.

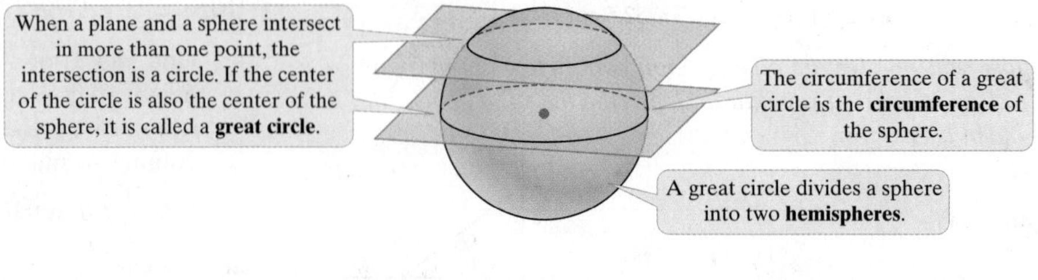

When a plane and a sphere intersect in more than one point, the intersection is a circle. If the center of the circle is also the center of the sphere, it is called a great circle.
The circumference of a great circle is the circumference of the sphere.
A great circle divides a sphere into two hemispheres.

> **Helpful Hint**
> Remember: Area and surface area—measured in square units
> Volume—measured in cubic units

Theorem 11.6-1 Surface Area of a Sphere

The surface area of a sphere is

$$SA = 4\pi r^2$$

where r is the radius of the sphere.

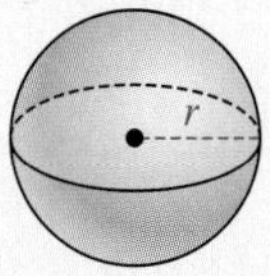

EXAMPLE 1 **Finding the Surface Area of a Sphere**

What is the surface area of the sphere in terms of π?

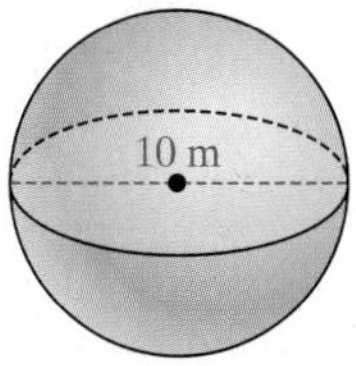

Solution To use our formula, we only need the radius of the sphere. The diameter is 10 m, so the radius is $\frac{10}{2}$ m, or 5 m.

$$\begin{aligned} SA &= 4\pi r^2 && \text{Use the formula for surface area of a sphere.} \\ &= 4\pi(5)^2 && \text{Substitute 5 for } r. \\ &= 100\pi && \text{Multiply.} \end{aligned}$$

The surface area is 100π sq m. □

PRACTICE 1 What is the surface area of a sphere with a diameter of 14 in.? Give your answer in terms of π and rounded to the nearest square inch.

We can use spheres to approximate the surface areas of real-world objects.

EXAMPLE 2 **Finding Surface Area from Circumference**

Earth's equator is about 24,902 mi long. What is the approximate surface area of Earth? Round to the nearest thousand square miles.

Solution

STEP 1. Find the radius of Earth.

$$\begin{aligned} C &= 2\pi r && \text{Use the formula for circumference.} \\ 24{,}902 &= 2\pi r && \text{Substitute 24,902 for } C. \\ \frac{24{,}902}{2\pi} &= r && \text{Divide each side by } 2\pi. \\ r &\approx 3963.276393 && \text{Use a calculator.} \end{aligned}$$

STEP 2. Use the radius to find the surface area of Earth.

$$\begin{aligned} SA &= 4\pi r^2 && \text{Use the formula for surface area.} \\ &= 4\pi\,\text{ANS}\ \boxed{x^2}\ \boxed{\text{enter}} && \text{Use a calculator. ANS uses the value of } r \text{ from Step 1.} \\ &\approx 197387017.5 \end{aligned}$$

The surface area of Earth is about 197,387,000 sq mi. □

PRACTICE 2 What is the surface area of a melon with circumference 18 in.? Round your answer to the nearest 10 square inches.

The following model suggests a formula for the volume of a sphere.

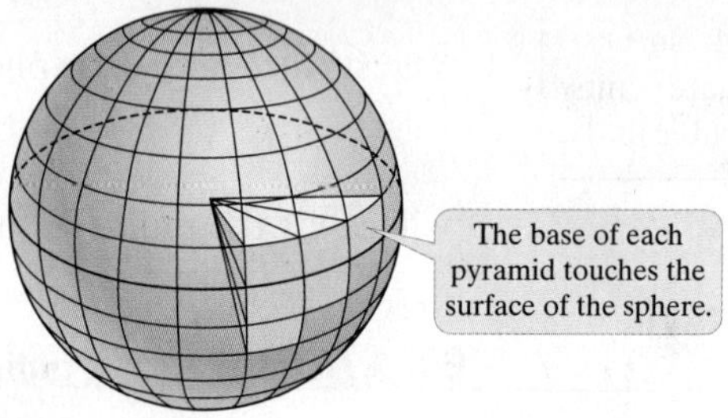
The base of each pyramid touches the surface of the sphere.

Fill a sphere with a large number n of small pyramids. The vertex of each pyramid is the center of the sphere. The height of each pyramid is approximately the radius r of the sphere. The sum of the areas of the bases of the n pyramids approximates the surface area of the sphere. The sum of the volumes of the n pyramids should approximate the volume of the sphere.

Volume of each pyramid $= \frac{1}{3}Bh$

Sum of the volumes of n pyramids $\approx n \cdot \frac{1}{3}Br$ — Substitute r for h.

$= \frac{1}{3} \cdot (nB)r$ — Commutative and Associative Properties of Multiplication

$\approx \frac{1}{3} \cdot (4\pi r^2)r$ — Replace nB with the surface area of a sphere.

$= \frac{4}{3}\pi r^3$ — Simplify.

It is reasonable to conjecture that the volume of a sphere with radius r is $\frac{4}{3}\pi r^3$.

Theorem 11.6-2 Volume of a Sphere

The volume of a sphere is

$$V = \frac{4}{3}\pi r^3$$

where r is the radius of the sphere.

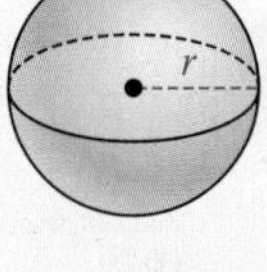
r

EXAMPLE 3 Finding the Volume of a Sphere

What is the volume of the sphere in terms of π?

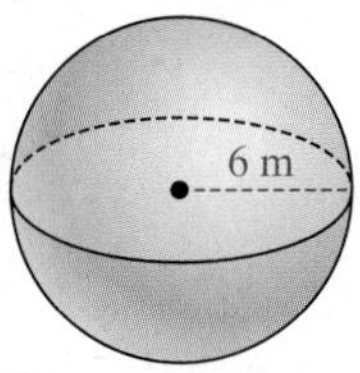
6 m

Solution We use the volume of a sphere formula.

$V = \frac{4}{3}\pi r^3$ — Use the formula for volume of a sphere.

$= \frac{4}{3}\pi(6)^3$ — Substitute.

$= 288\pi$ — Multiply.

The volume of the sphere is 288π cubic m. □

Helpful Hint

To calculate volume and surface area of a sphere, we only need its radius.

PRACTICE 3 A sphere has a diameter of 60 in. What is its volume to the nearest cubic inch?

If we know the volume of a sphere, we can solve and find its radius. Once the radius is known, we can find its surface area.

EXAMPLE 4 **Using Volume to Find Surface Area**

The volume of a sphere is 5000 cubic m. What is its surface area to the nearest square meter?

Solution Use the volume of a sphere formula to find the approximate radius.

Find the radius of the sphere.

$$V = \frac{4}{3}\pi r^3 \quad \text{Use the formula for volume of a sphere.}$$

$$5000 = \frac{4}{3}\pi r^3 \quad \text{Substitute 5000 for } V.$$

$$\frac{3}{4}\cdot 5000 = \frac{3}{4}\cdot\frac{4}{3}\pi r^3 \quad \text{Multiply both sides by } \frac{3}{4}\text{, the reciprocal of } \frac{4}{3}.$$

$$3750 = \pi r^3 \quad \text{Simplify.}$$

$$\frac{3750}{\pi} = r^3 \quad \text{Divide both sides by } \pi.$$

$$\sqrt[3]{\frac{3750}{\pi}} = r \quad \text{Take the cube root of each side.}$$

$$r \approx 10.60784418 \quad \text{Use a calculator to approximate.}$$

$$r \approx 10.6078$$

Now use the approximation of r to find the surface area of the sphere.

$$SA = 4\pi r^2 \quad \text{Use the formula for surface area of a sphere.}$$

$$= 4\pi\, 10.6078^2 \quad \text{Substitute 10.6078 for } r.$$

$$\approx 1414.036142 \quad \text{Use a calculator to approximate.}$$

The surface area of the sphere is about 1414 sq m. □

PRACTICE 4 The volume of a sphere is 4200 cubic ft. What is its surface area to the nearest tenth?

VOCABULARY & READINESS CHECK

Word Bank *Use the choices to fill in each blank. Some choices may be used more than once and some not at all.*

hemispheres	sphere	diameter	circumference	half
great circle	radius	center	twice	

1. The intersection of a sphere and a plane passing through the center of the sphere is called a(n) ____________.
2. The ____________ of a sphere is measured by the circumference of a great circle of the sphere.
3. A great circle divides a sphere into two ____________.
4. The set of all points in space equidistant from a given point is called a(n) ____________.
5. The given point in Exercise 4 above is called the ____________.
6. A(n) ____________ of a sphere is a segment that has one endpoint at the center and the other endpoint on the sphere.
7. A(n) ____________ of a sphere is a segment passing through the center of a sphere with endpoints on the sphere.
8. The length of a diameter of a sphere is ____________ the length of a radius of the same sphere.

11.6 EXERCISE SET MyMathLab®

Find the surface area of the sphere with the given diameter or radius. Leave your answer in terms of π. See Example 1.

1. $r = 10$ in. **2.** $r = 100$ yd

3. $d = 30$ m **4.** $d = 32$ mm

Find the surface area of each ball. Leave each answer in terms of π.

5.

$d = 68$ mm

6.

$d = 21$ cm

7.

$d = 2\frac{1}{16}$ in.

8.

$d = 1\frac{7}{16}$ ft

Use the given circumference to find the surface area of each spherical object. Round your answer to the nearest whole number. See Example 2.

9. a grapefruit with $C = 14$ cm

10. a bowling ball with $C = 27$ in.

11. a pincushion with $C = 8$ cm

12. a head of lettuce with $C = 22$ in.

Find the volume of each sphere. Give each answer in terms of π and rounded to the nearest cubic unit. See Example 3.

13.

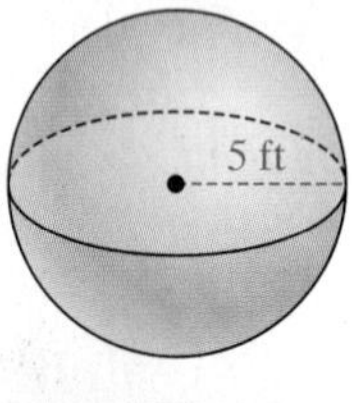

14.

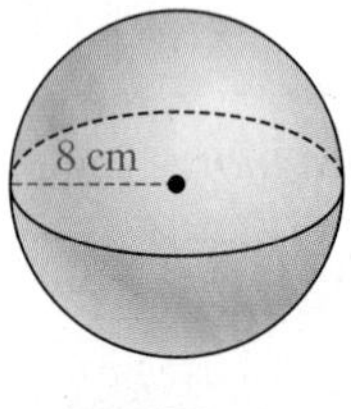

15.

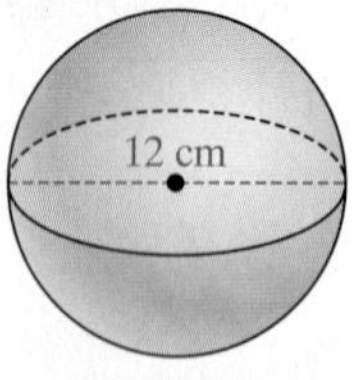

16.

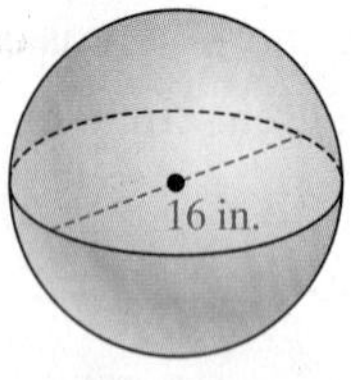

17.

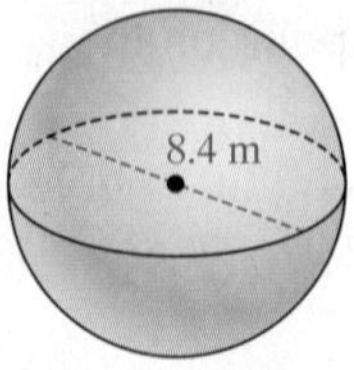

18.

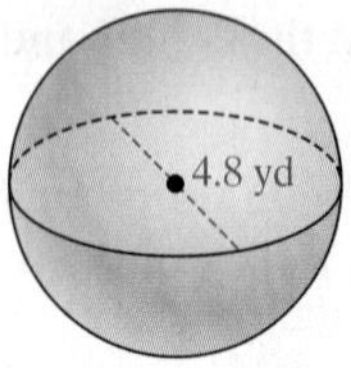

A sphere has the volume given. Find its surface area to the nearest whole number. See Example 4.

19. $V = 900$ cu in. **20.** $V = 3000$ cu m

21. $V = 140$ cu cm **22.** $V = 280$ cu mi

Nestled deep in the west coast rainforest in Qualicum Beach on Vancouver Island, these wooden spheres, called "Free Spirit Spheres," are hung from the trees by small networks of ropes. Directions for building your own sphere are available for purchase, but these are mainly used for rentals.

The exterior of a sphere has a diameter of about 10 feet 6 inches. The interior of a sphere has a diameter of about 10 feet 4 inches. Find each volume to the nearest whole cubic inch.

23. the volume of the exterior of a sphere

24. the volume of the interior of a sphere

Light spheres are designed to measure, in lumens, visible light or the flux of different lights, lamps, etc. The National Institute of Standards and Technology (NIST) establishes standards for how these commercial spheres are calibrated.

These spheres are sold in different diameters that determine the lamp length that can be tested. Find the surface area of the light spheres below, rounded to the nearest square unit.

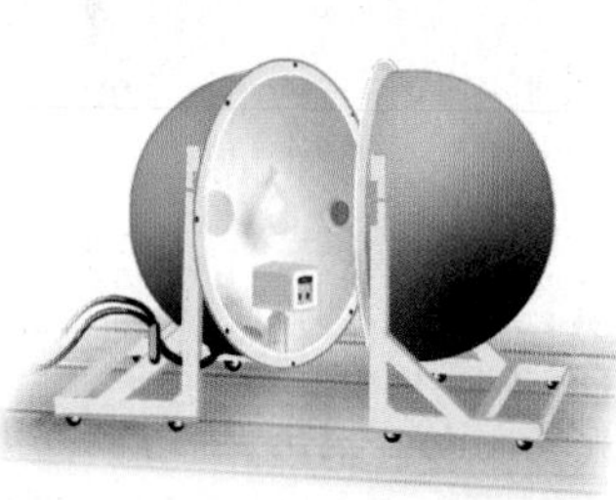

25. the surface area of a 40-inch light-measurement sphere

26. the surface area of a 65-inch light-measurement sphere

27. As mentioned at the beginning of this chapter, zorbing is an extreme sport invented by two New Zealanders. You are strapped in and sent down a zorbing hill. A standard zorb is 3 m in diameter. Find the exact volume of a zorb, and approximate the volume to the nearest tenth of a cubic meter.

28. A snow globe has a diameter of 6 inches. Find its exact volume. Then approximate its volume to the nearest tenth of an inch.

29. A water storage tank is in the shape of a hemisphere (half a sphere). If the radius is 14 ft, approximate the volume of the tank in cubic feet.

30. A birdbath is made in the shape of a hemisphere (half a sphere). If its radius is 10 inches, approximate its volume in cubic inches.

31. The Adler Museum in Chicago recently added a new planetarium, its StarRider Theater, which has a diameter of about 56 feet. Find the volume of its hemispheric (half a sphere) dome rounded to the nearest tenth. (*Source:* The Adler Museum)

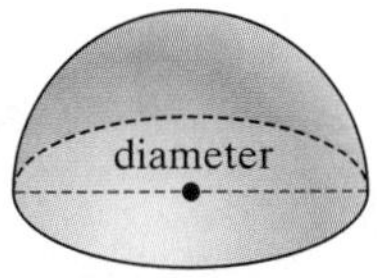

hemisphere

32. The Hayden Planetarium, at the Museum of Natural History in New York City, boasts a dome that has a diameter of 20 m. The dome is a hemisphere, or half a sphere. What is the volume enclosed by the dome at the Hayden Planetarium? Round to the nearest hundredth. (*Source:* Hayden Planetarium)

The diameter of a sphere is 12 ft.

33. What is its surface area in terms of π?

34. What is its volume to the nearest tenth?

Find the surface area and the volume of each sphere. Round each to the nearest whole number.

35.

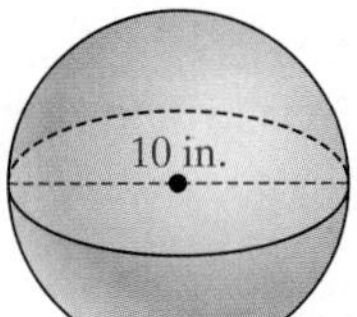

36.

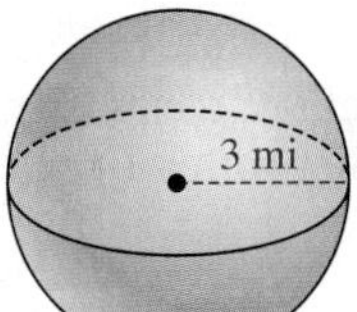

37. Virtusphere has developed the locomotion platform illustrated. It consists of a 10-foot hollow sphere that is placed on a special platform that allows the sphere to rotate freely in any direction according to the user's steps. Find the surface area and the volume of the sphere. Round each to the nearest whole number. (See the diagram at the top of the next column.)

38. Science on a Sphere (SOS) is a giant, 68-inch globe suspended from the ceiling. It was developed by NOAA (National Oceanic and Atmospheric Administration) and uses computers and video projectors to illustrate various elements of Earth science. Thus far, there are more than 75 SOS exhibits around the world. Find the surface area and the volume of the globe (sphere). Round each to the nearest whole number.

The region enclosed by the semicircle is revolved completely about the x-axis.

39. Describe the solid of revolution that is formed and find its volume in terms of π.

40. Find its surface area in terms of π.

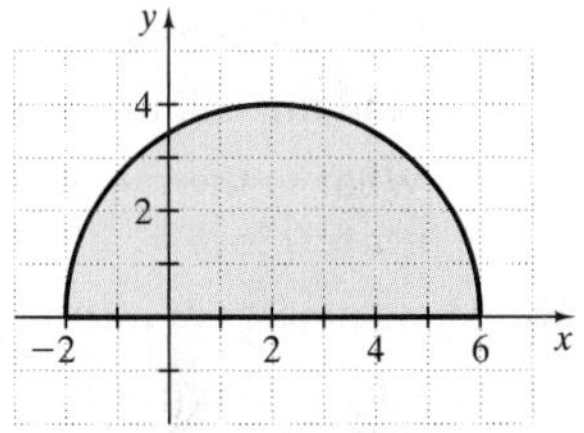

41. The sphere at the right fits snugly inside a cube with 6-in. edges. What is the approximate volume of the space between the sphere and cube?

a. 28.3 cu in.

b. 76.5 cu in.

c. 102.9 cu in.

d. 113.1 cu in.

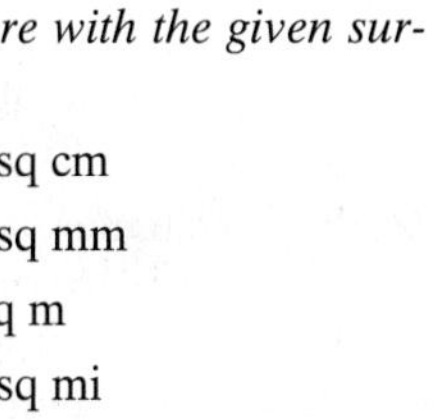

42. The volume of a sphere is 80π cu cm. What is its surface area to the nearest whole number?

Find the volume in terms of π of each sphere with the given surface area.

43. 36π sq in.

44. 144π sq cm

45. 4π sq m

46. 100π sq mm

47. 25π sq yd

48. 49π sq m

49. 9π sq ft

50. 225π sq mi

Coordinate Geometry *A sphere has center (0, 0, 0) and radius 5.*

51. Name the coordinates of six points on the sphere.

Tell whether each of the following points is inside, outside, or on the sphere in Exercise **51**.

52. $A(0, -3, 4)$ **53.** $B(1, -1, -1)$ **54.** $C(4, -6, -10)$

Find the surface area and volume of each figure.

55.

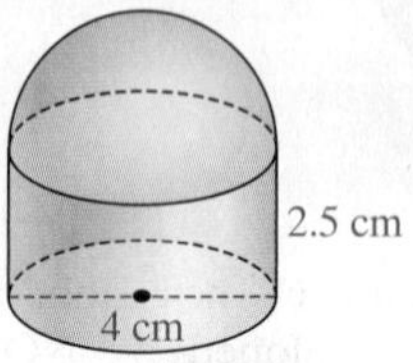

56.

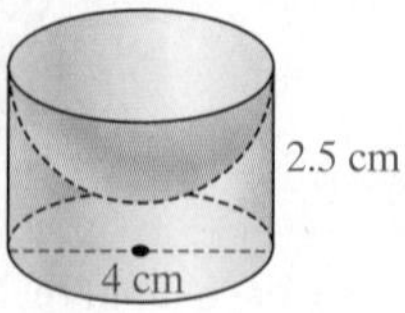

57.

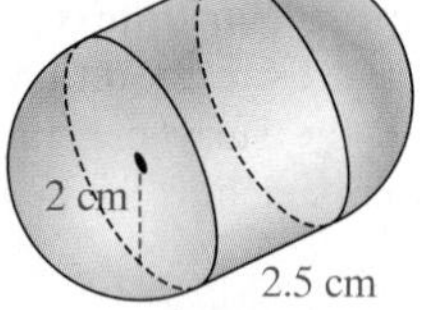

58.

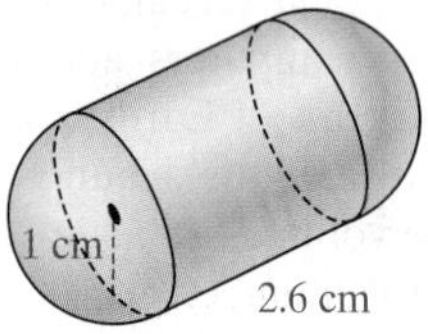

CONCEPT EXTENSIONS

59. What is the ratio of the area of a great circle to the surface area of the sphere?

60. Find the Error Your classmate claims that if you double the radius of a sphere, its surface area and volume will quadruple. What is your classmate's error? Explain.

61. A spherical balloon has a 14-in. diameter when it is fully inflated. Half of the air is let out of the balloon. Assume that the balloon remains a sphere.

a. Find the volume of the fully inflated balloon in terms of π.

b. Find the volume of the half-inflated balloon in terms of π.

c. What is the diameter of the half-inflated balloon to the nearest inch?

62. The diameter of a golf ball is 1.68 in.

a. Approximate the surface area of the golf ball.

Do you think that the value you found in part a is greater than or less than the actual surface area of the golf ball? Explain.

63. Which is greater, the total volume of three spheres, each of which has diameter 3 in., or the volume of one sphere that has diameter 8 in.?

64. How many great circles does a sphere have? Explain.

65. The diameter of Earth is about 7926 mi. The diameter of the moon is about 27% of the diameter of Earth. What percent of the volume of Earth is the volume of the moon? Round your answer to the nearest whole percent.

66. The circumference of Earth at the equator is approximately 40,075 km. About 71% of Earth is covered by oceans and other bodies of water. To the nearest thousand square kilometers, how much of Earth's surface is land?

A cube with edges 6 in. long fits snugly inside a sphere as shown at the right. The diagonal of the cube is the diameter of the sphere.

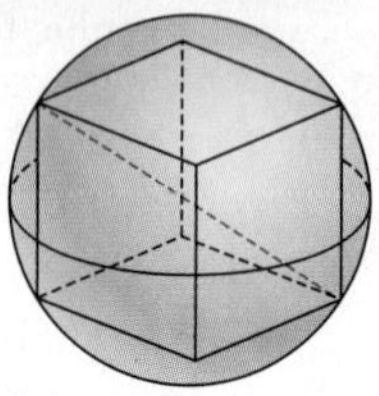

67. Find the length of the diagonal and the radius of the sphere. Leave your answer in simplest radical form.

68. What is the volume of the space between the sphere and the cube to the nearest tenth?

Find the radius of a sphere with the given property.

69. The number of square meters of surface area equals the number of cubic meters of volume.

70. The ratio of surface area in square meters to volume in cubic meters is 1:5.

71. Suppose a cube and a sphere have the same volume.

a. Which has the greater surface area? Explain.

b. Explain why spheres are rarely used for packaging.

72. At the right, the sphere fits snugly inside the cylinder. Archimedes (about 287–212 B.C.) requested that such a figure be put on his gravestone along with the ratio of their volumes, a finding that he regarded as his greatest discovery. What is that ratio?

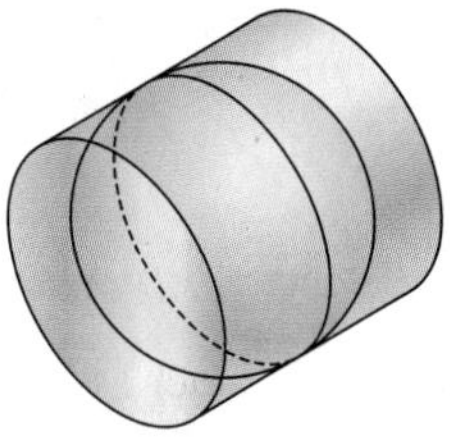

REVIEW AND PREVIEW

Find the volume of each figure to the nearest cubic unit. See Section 11.5.

73.

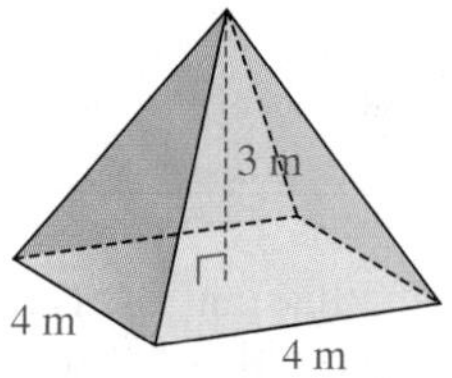

74.

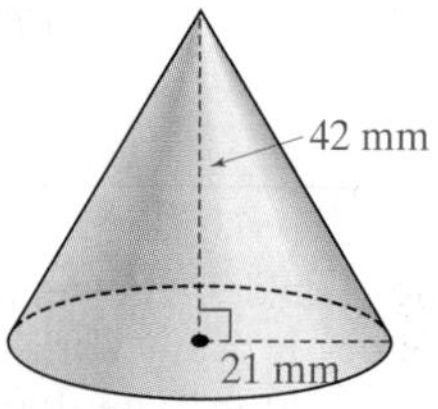

Solve. See Sections 6.4 and 9.3.

75. A leg of a right triangle has a length of 4 cm and the hypotenuse has a length of 7 cm. Find the measure of each acute angle of the triangle to the nearest degree.

76. The length of each side of a rhombus is 16. The longer diagonal has length 26. Find the measures of the angles of the rhombus to the nearest degree.

Are the figures similar? If so, give the scale factor. See Section 7.3.

77. two squares, one with 3-in. sides and the other with 1-in. sides

78. two right isosceles triangles, one with a 3-cm hypotenuse and the other with a 1-cm leg

11.7 AREAS AND VOLUMES OF SIMILAR SOLIDS

OBJECTIVE

1 Compare and Find the Areas and Volumes of Similar Solids.

VOCABULARY

- similar solids

Helpful Hint

Notice that cubes and spheres always have the same shape. Thus, any two cubes are similar, as are any two spheres.

OBJECTIVE 1 ▶ Comparing and Finding the Areas and Volumes of Similar Solids. We can use ratios to compare the areas and volumes of similar solids.

Similar solids have the same shape, and all their corresponding dimensions are proportional.

Recall that the ratio of corresponding linear dimensions of two similar solids is the scale factor.

EXAMPLE 1 **Identifying Similar Solids**

Are the two rectangular prisms similar? If so, what is the scale factor of the first figure to the second figure?

a.

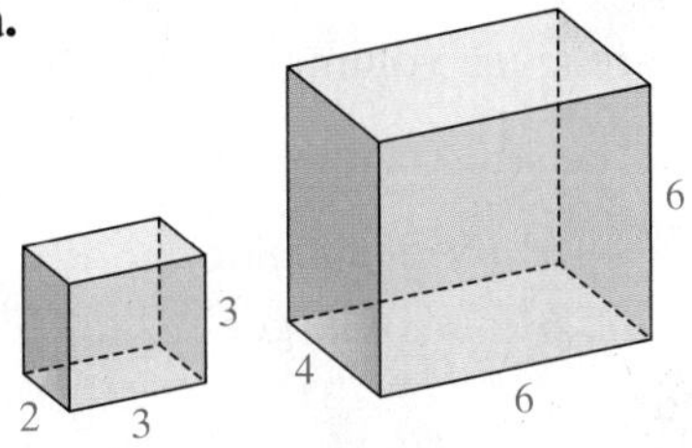

b.

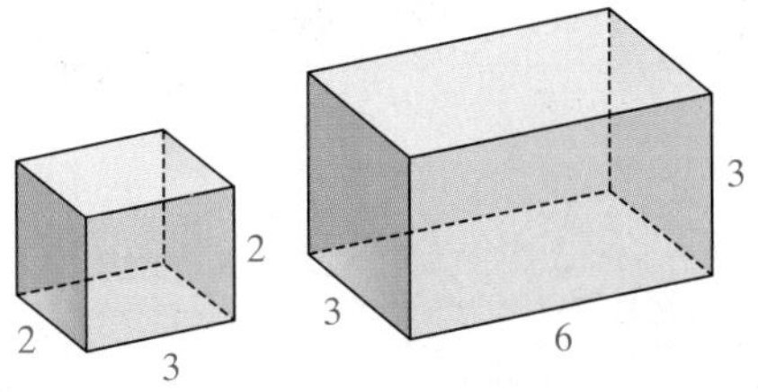

Solution For parts **a** and **b**, we check to see that the ratios of corresponding dimensions are the same.

a. $\frac{3}{6} = \frac{2}{4} = \frac{3}{6}$

The prisms are similar because the corresponding linear dimensions are proportional.

The scale factor is $\frac{1}{2}$.

b. $\frac{2}{3} = \frac{2}{3} \neq \frac{3}{6}$

The prisms are not similar because the corresponding linear dimensions are not proportional.

□

PRACTICE

1 Are the two cylinders similar? If so, what is the scale factor of the first figure to the second figure?

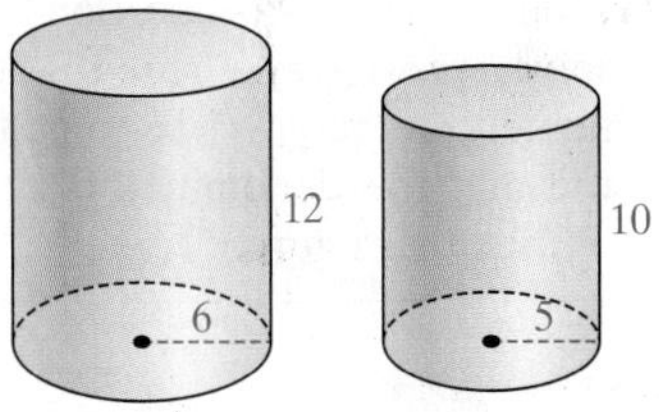

The two similar prisms shown here suggest two important relationships for similar solids.

The ratio of the side lengths is 1:2 or $\frac{1}{2}$.

The ratio of the surface areas is 22:88, or $\frac{22}{88}$ or $\frac{1}{4}$.

The ratio of the volumes is 6:48, or $\frac{6}{48}$ or $\frac{1}{8}$.

Notice that: $\left(\frac{1}{2}\right)^2 = \frac{1}{4}$

$(\text{scale factor})^2 = \text{ratio of surface areas}$

Also $\left(\frac{1}{2}\right)^3 = \frac{1}{8}$

$(\text{scale factor})^3 = \text{ratio of volumes}$

These two facts apply to all similar solids.

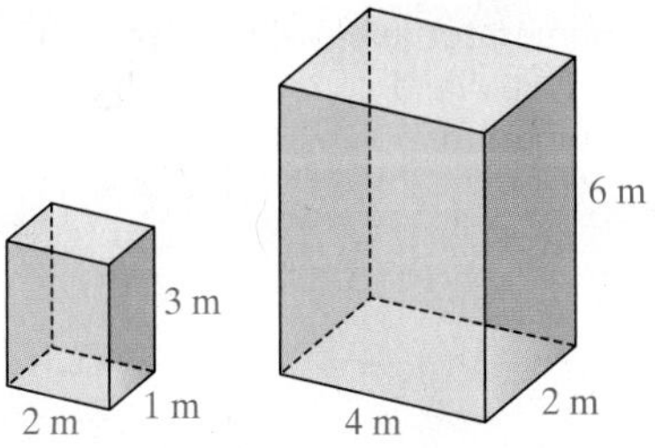

SA = 22 sq m
V = 6 cu m

SA = 88 sq m
V = 48 cu m

Theorem 11.7-1 Areas and Volumes of Similar Solids

If the scale factor of two similar solids is $a{:}b$ or $\frac{a}{b}$, then

- the ratio of their corresponding areas is $a^2{:}b^2$ or $\frac{a^2}{b^2}$
- the ratio of their volumes is $a^3{:}b^3$ or $\frac{a^3}{b^3}$

EXAMPLE 2 Finding the Scale Factor

The square prisms below at the right are similar. What is the scale factor of the smaller prism to the larger prism?

Solution Since we are given volumes, we use the volume ratio of $\frac{a^3}{b^3}$. Then we find $\frac{a}{b}$.

Since we want the scale factor of the smaller prism to the larger prism, we use

$$\frac{a^3 \ (\text{V of smaller})}{b^3 \ (\text{V of larger})}$$

$$\frac{a^3}{b^3} = \frac{729}{1331} \quad \text{The ratio of the volumes is } a^3{:}b^3.$$

$$\frac{a}{b} = \frac{9}{11} \quad \text{Take the cube root of each side.}$$

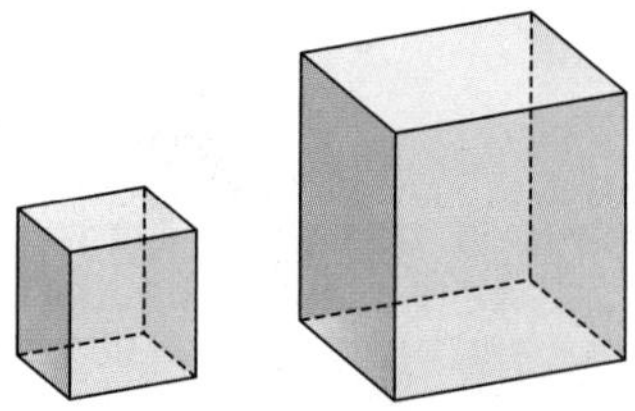

$V = 729$ cu cm $\quad V = 1331$ cu cm

The scale factor is $\frac{9}{11}$ or 9:11. □

PRACTICE

2 What is the scale factor of two similar prisms with surface areas 144 sq m and 324 sq m?

> **Helpful Hint**
> When writing proportions to solve, we will make sure that the numerators refer to the same figure and the denominators refer to the other figure.

EXAMPLE 3 Using a Scale Factor

The lateral areas of two similar paint cans are 425 sq cm and 1019 sq cm. The volume of the smaller can is 1157 cubic cm. What is the volume of the larger can?

Solution Since we have the lateral areas of both cans, let's use those to find the scale factor, $\frac{a}{b}$. We can then use the scale factor to find the unknown volume. Let's first find the scale factor $\frac{a}{b}$.

$$\frac{a^2}{b^2} = \frac{425}{1019} \quad \text{The ratio of the surface areas is } a^2{:}b^2.$$

$$\frac{a}{b} = \frac{\sqrt{425}}{\sqrt{1019}} \quad \text{Take the positive square root of each side.}$$

$$\frac{a}{b} \approx \frac{1}{1.548} \quad \text{Divide numerator and denominatory by } \sqrt{425}.$$

> **Helpful Hint**
> To simplify to a fraction with a numerator of 1, divide the numerator and denominator by the numerator.
>
> $$\frac{a}{b} = \frac{\sqrt{425}}{\sqrt{1019}} = \frac{\sqrt{425} \div \sqrt{425}}{\sqrt{1019} \div \sqrt{425}} \approx \frac{1}{1.548}$$

Next, use the scale factor to find the volume.

$$\frac{V_{\text{small}}}{V_{\text{large}}} \approx \frac{1^3}{(1.548)^3} \quad \text{The ratio of the volumes is } a^3{:}b^3.$$

$$\frac{1157}{V_{\text{large}}} \approx \frac{1}{(1.548)^3} \quad \text{Substitute 1157 for } V_{\text{small}}. \text{ Also, } 1^3 = 1.$$

$$1 \cdot V_{\text{large}} \approx 1157 \cdot (1.548)^3 \quad \text{Solve for } V_{\text{large}} \text{ by cross multiplying.}$$

$$V_{\text{large}} \approx 4295.475437 \quad \text{Use a calculator.}$$

The volume of the larger paint can is about 4295 cubic cm. □

PRACTICE
3 The volumes of two similar solids are 128 cu m and 250 cu m. The surface area of the larger solid is 250 sq m. What is the surface area of the smaller solid?

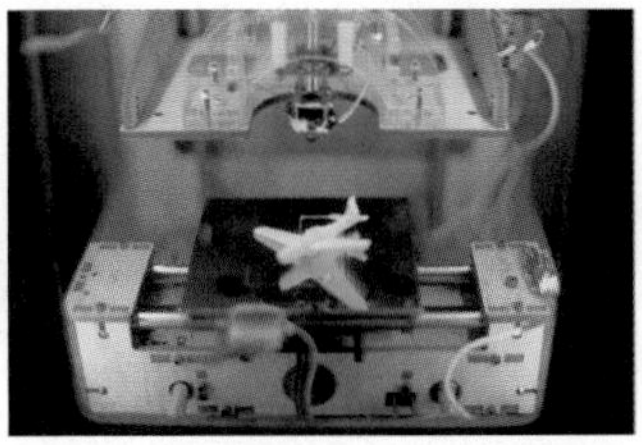

We can compare the capacities and weights of similar objects. The capacity of an object is the amount of fluid the object can hold. The capacities and weights of similar objects made of the same material are proportional to their volumes.

The use of scale factors for areas and volumes is widely used by architects, engineers, and artists, just to name a few occupations. Incredibly, there are now 3D printers. In other words, 3D printing produces a three-dimensional object of any shape from a digital model. This is achieved by adding successive layers of materials as needed to match the model.

EXAMPLE 4 Using a Scale Factor to Find Volume

The Ford Mustang has a volume of about 83 cubic feet of interior passenger room for many years. Find the interior room of a toy-size model car made in the likeness of a Ford Mustang. Use the scale factor of 1:64. Find this volume in cubic inches and round to the nearest hundredth.

Solution: Since the scale factor is 1:64, the ratio of these volumes is $1^3:64^3$ or 1:262,144.

Let x = volume of interior of toy-size model car in cubic inches.

Before we write a proportion to solve, lets convert 83 cu ft to cubic inches. How many cubic inches are in 1 cubic foot? Study the figure to see that

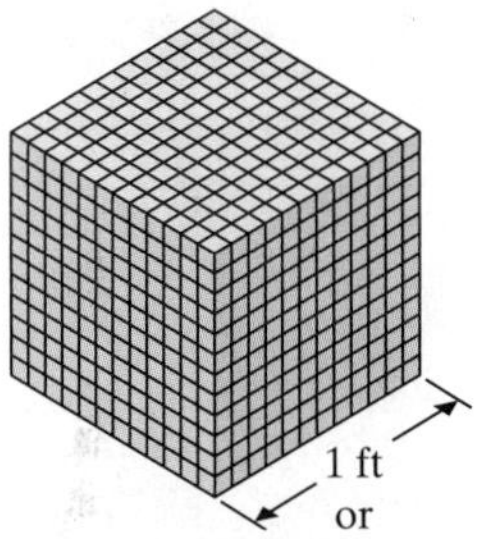

$$1 \text{ cu ft} = 12 \text{ in.} \cdot 12 \text{ in.} \cdot 12 \text{ in.} = 1728 \text{ cu in. Thus,}$$

$$83 \text{ cu ft} = 83 \cdot 1728 \text{ cu in.}$$
$$= 143{,}424 \text{ cu in.}$$

Now let's write our proportion:

$$\frac{1}{262{,}144} = \frac{x}{143{,}424}$$

$$x \approx 0.55 \text{ cu in.}$$

The interior room in a toy-size Ford Mustang is about 0.55 cu in.

PRACTICE
4 A marble paperweight shaped like a pyramid weighs 0.15 lb. How much does a similarly shaped marble paperweight weigh if each dimension is three times as large?

VOCABULARY & READINESS CHECK

Word Bank *Use the choices to fill in each blank. Some choices may be used more than once and some not at all.*

Scale factor   Similar solids

1. ____________ have the same shape, and all their corresponding dimensions are proportional.

2. The ratio of corresponding linear dimensions of two similar solids is called the ____________.

Fill in the Blank *Name two solid shapes that are always similar*

3. ____________   **4.** ____________

11.7 EXERCISE SET MyMathLab®

For Exercises 1–6, are the two figures similar? If so, give the scale factor of the first figure to the second figure. See Example 1.

1.

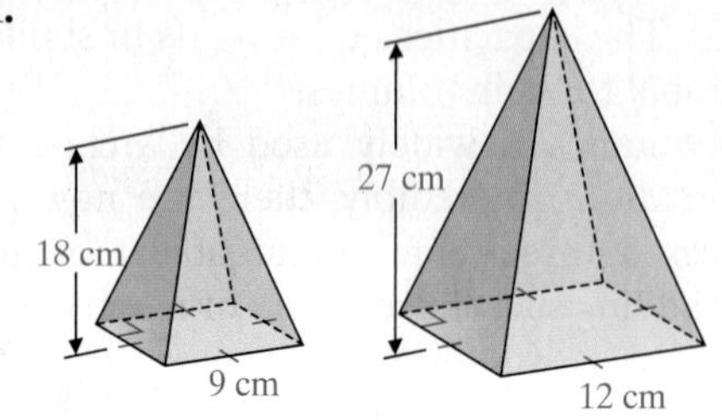

2.

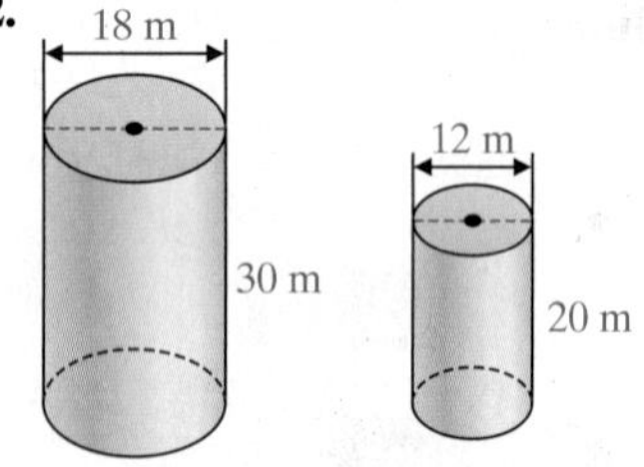

3.

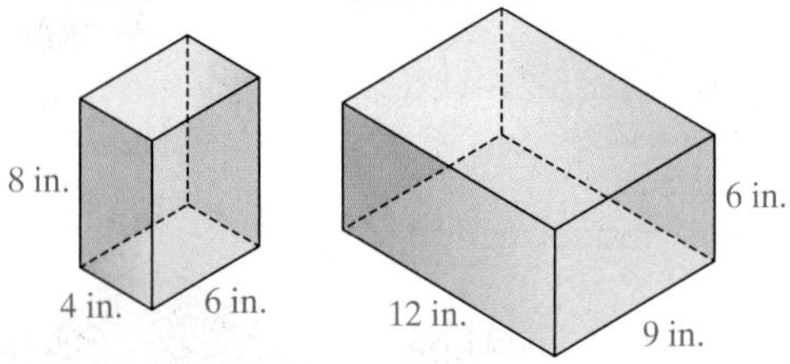

4.

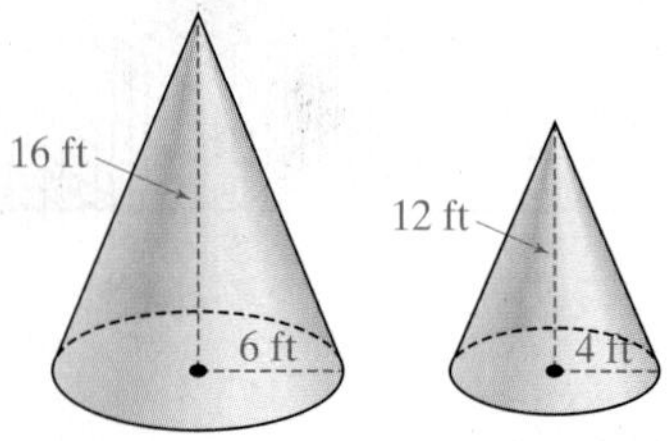

5. two cubes, one with 3-cm edges, the other with 4.5-cm edges

6. a cylinder and a square prism both with 3-in. radius and 1-in. height

Each pair of figures is similar. Use the given information to find the scale factor of the smaller figure to the larger figure. See Example 2.

7.

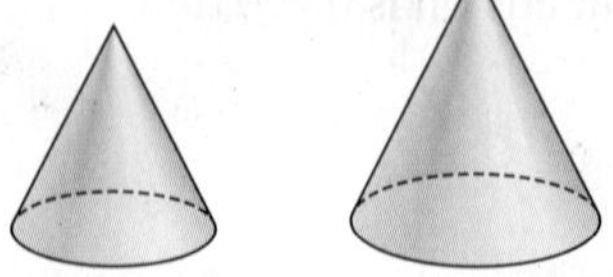

$V = 250\pi$ cu ft $V = 432\pi$ cu ft

8.

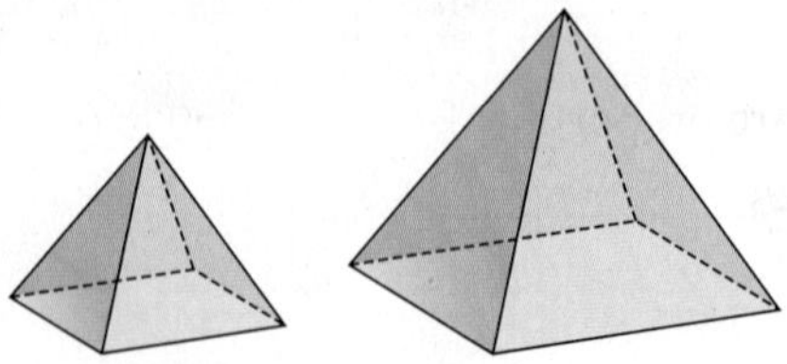

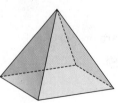

$V = 216$ cu in. $V = 343$ cu in.

9.

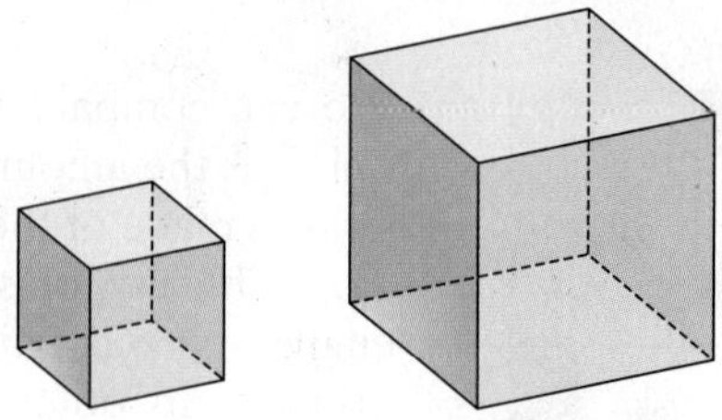

SA = 18 sq m SA = 32 sq m

10.

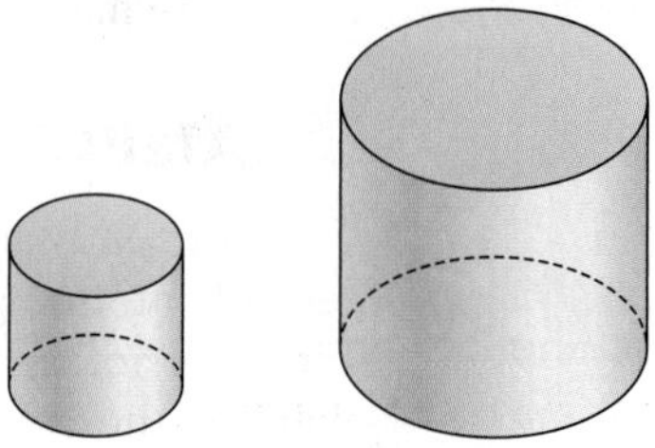

SA = 20π sq yd SA = 125π sq yd

The surface areas of two similar figures are given. The volume of the larger figure is given. Find the volume of the smaller figure. See Example 3.

11. SA = 248 sq in.
SA = 558 sq in.
V = 810 cu in.

12. SA = 68 sq mi
SA = 425 sq mi
V = 1375 cu mi

13. SA = 192 sq m
SA = 1728 sq m
V = 4860 cu m

14. SA = 52 sq ft
SA = 208 sq ft
V = 192 cu ft

A solid is similar to a larger solid. Using the given scale factor, find the surface area, SA, and the volume, V, of the larger solid.

15. Scale factor: $\frac{1}{5}$

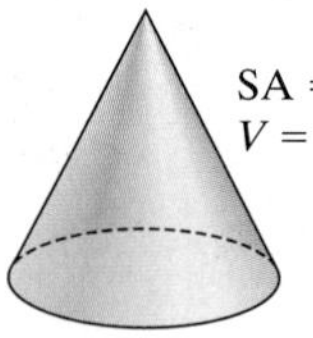

SA = 20π sq m
$V = 15\pi$ cu m

16. Scale factor: $\frac{1}{3}$

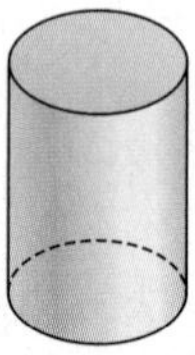

SA = 30π sq cm
$V = 24\pi$ cu cm

17. Scale factor: $\frac{3}{4}$

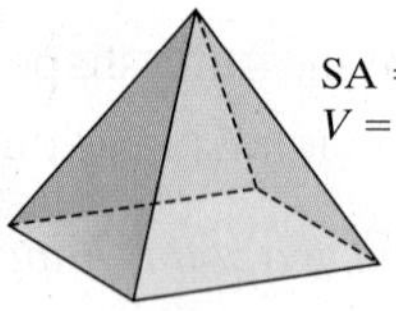

SA = 280 sq in.
V = 300 cu in.

18. Scale factor: $\frac{4}{5}$

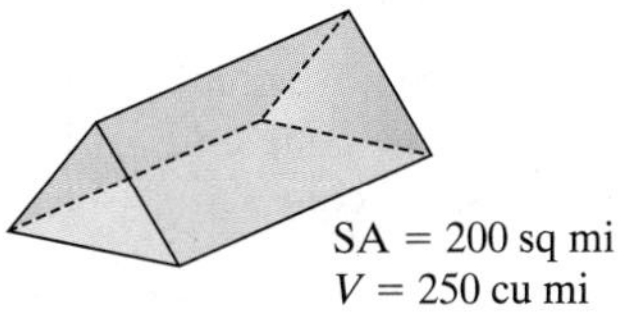

Fill in the Blank *Fill in the blank with "always," "sometimes," or "never."*

19. Two spheres are ____________ similar.

20. Two cubes are ____________ similar.

21. Two prisms are ____________ similar.

22. Two pyramids are ____________ similar.

23. A cylinder and a sphere are ____________ similar.

24. A cube and a sphere are ____________ similar.

Which two of the following three figures are similar? What is their scale factor?

25.

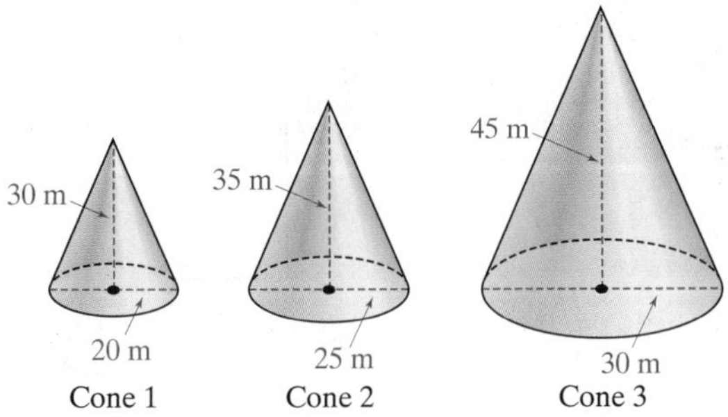

26.

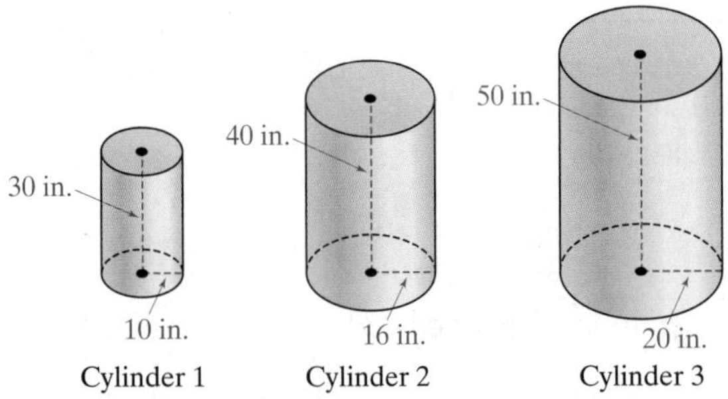

For Exercises 27–30, solve parts a–c.

a. What is their scale factor?

b. What is the ratio of their surface areas?

c. What is the ratio of their volumes?

27. two similar prisms with heights 4 cm and 10 cm

28. two similar pyramids with heights 9 in. and 12 in.

29. two similar spheres with radii 6 ft and 14 ft

30. two similar cubes with side lengths 20 m and 25 m

The volumes of two similar figures are given. The surface area of the smaller figure is given. Find the surface area of the larger figure.

31. $V = 27$ cu in.
$V = 125$ cu in.
$SA = 63$ sq in.

32. $V = 27$ cu m
$V = 64$ cu m
$SA = 63$ sq m

33. $V = 2$ cu yd
$V = 250$ cu yd
$SA = 13$ sq yd

34. $V = 5$ cu cm
$V = 135$ cu cm
$SA = 13$ sq cm

Solve. See Example 4.

35. There are 750 toothpicks in a regular-sized box. If a jumbo box is made by doubling all the dimensions of the regular-sized box, how many toothpicks will the jumbo box hold?

36. A cylinder with a 4-in. diameter and a 6-in. height holds 1 lb of oatmeal. To the nearest ounce, how much oatmeal will a similar, 10-in.-high cylinder hold? (*Hint:* 1 lb = 16 oz)

A regular pentagonal prism has 9-cm base edges. A larger, similar prism of the same material has 36-cm base edges. How does each indicated measurement for the larger prism compare to the same measurement for the smaller prism?

37. the volume

38. the weight

Copy and complete the table for the similar solids.

	Similarity Ratio	*Ratio of Surface Areas*	*Ratio of Volumes*
39.	1:2	__:__	__:__
40.	3:5	__:__	__:__
41.	__:__	49:81	__:__
42.	__:__	__:__	125:512

43. Two similar pyramids have lateral areas 20 sq ft and 45 sq ft. The volume of the smaller pyramid is 8 cu ft. Find the volume of the larger pyramid.

44. The volumes of two similar containers are 115 cu in. and 67 cu in. The surface area of the smaller container is 108 sq in. What is the surface area of the larger container?

CONCEPT EXTENSIONS

45. How are similar solids different from similar polygons? Explain.

46. **Find the Error** Two cubes have surface areas 49 sq cm and 64 sq cm. Your classmate tried to find the scale factor of the larger cube to the smaller cube. Explain and correct your classmate's error.

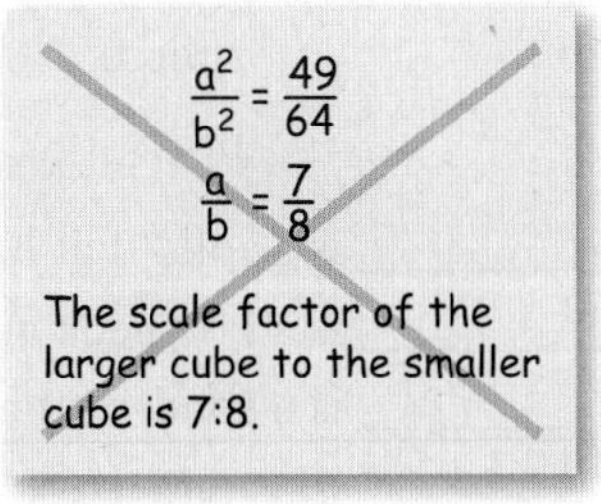

47. Is there a value of x for which the rectangular prisms are similar? Explain.

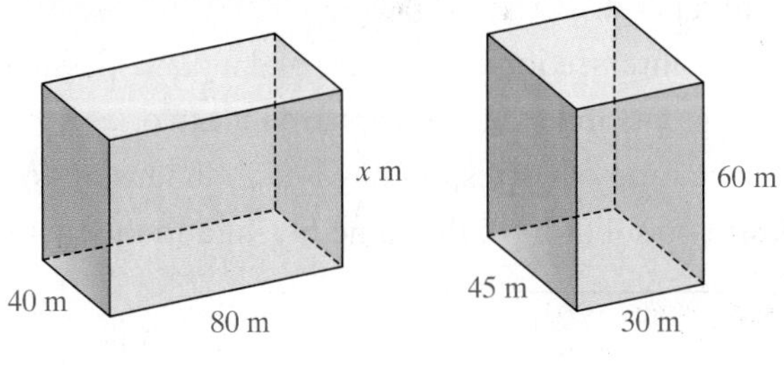

48. A carpenter is making a blanket chest based on an antique chest. Both chests have the shape of a rectangular prism. The length, width, and height of the new chest will all be 4 in. greater than the respective dimensions of the antique chest. Will the chests be similar? Explain.

49. The volume of a spherical balloon with radius 3.1 cm is about 125 cu cm. Estimate the volume of a similar balloon with radius 6.2 cm.

50. A clown's face on a balloon is 4 in. tall when the balloon holds 108 cu in. of air. How much air must the balloon hold for the face to be 8 in. tall?

51. The volumes of two spheres are 729 cu in. and 27 cu in.

a. Find the ratio of their radii.

b. Find the ratio of their surface areas.

52. The volumes of two similar pyramids are 1331 cu cm and 2744 cu cm.

a. Find the ratio of their heights.

b. Find the ratio of their surface areas.

53. Square pyramids A and B are similar. In pyramid A, each base edge is 12 cm. In pyramid B, each base edge is 3 cm and the volume is 6 cu cm.

a. Find the volume of pyramid A.

b. Find the ratio of the surface area of A to the surface area of B.

c. Find the surface area of each pyramid.

54. A cone is cut by a plane parallel to its base. The small cone on top is similar to the large cone. The ratio of the slant heights of the cones is 1:2. Find each ratio.

a. the surface area of the large cone to the surface area of the small cone

b. the volume of the large cone to the volume of the small cone

c. the lateral area of the frustum to the lateral area of the large cone and to the lateral area of the small cone

d. the volume of the frustum to the volume of the large cone and to the volume of the small cone

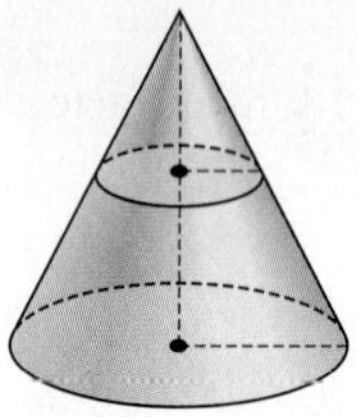

REVIEW AND PREVIEW

The circumference of a regulation basketball is between 75 cm and 78 cm. See Section 11.6.

55. What is the smallest surface area that a basketball can have? Give your answer to the nearest whole unit.

56. What is the largest surface area that a basketball can have? Give your answer to the nearest whole unit.

Find the volume of each sphere to the nearest tenth. See Section 11.6.

57. diameter = 6 in.

58. radius = 6 in.

Find the value of x. See Section 9.1.

59.

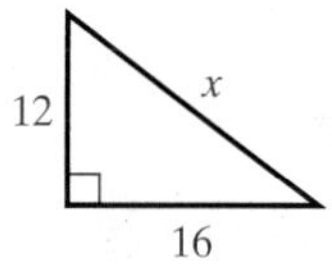

60.

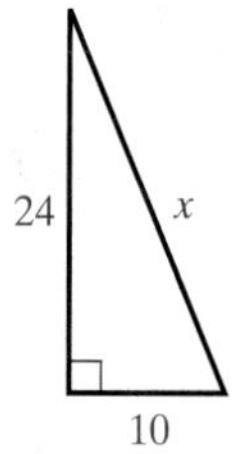

61.

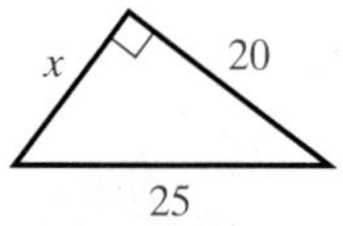

62.

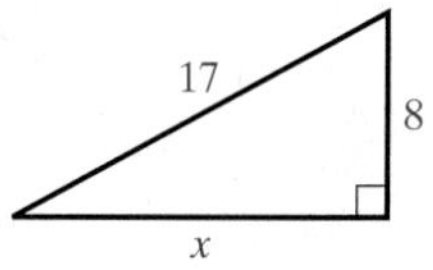

CHAPTER 11 VOCABULARY CHECK

Word Bank *Use the choices to fill in each blank. Some choices may be used more than once and some not at all.*

polyhedron	sphere	cone	scale factor	hemisphere
pyramid	prism	cylinder	volume	great circle

1. A(n) ____________ is a solid that has two congruent parallel circles.

2. A(n) ____________ is a polyhedron with two congruent, parallel faces.

3. A(n) ____________ is a solid, or three-dimensional figure, whose surfaces are polygons.

4. A(n) ____________ is a solid that has one base and a vertex that is not in the same plane as the base.

5. The set of all points in space equidistant from a given point is called a(n) ____________.

6. A(n) ____________ is a polyhedron in which one face can be any polygon and the other faces are triangles that meet at a point.

7. The intersection of a sphere and a plane passing through the center of the sphere is called a(n) ____________.

8. A great circle divides a sphere into two ____________.

9. The ratio of corresponding linear dimensions of two similar solids is called the ____________.

10. If two solids have the same height and the same cross sectional area at all levels, then they have the same ____________.

Fill in the Blank *Fill in each blank with units, square units, or cubic units.*

11. Volume is measured in ____________.

12. Surface area is measured in ____________.

13. Perimeter and circumference are both measured in ____________.

14. Area is measured in ____________.

CHAPTER 11 REVIEW

(11.1) *Draw a net for each three-dimensional figure.*

1.

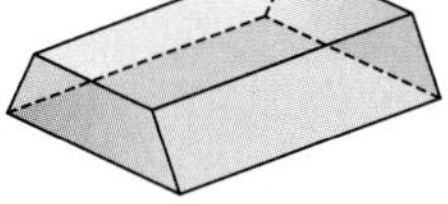

2.

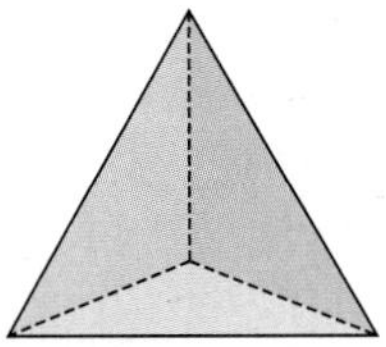

Use Euler's Formula to find the missing number.

3. $F = 5, V = 5, E =$ ________

4. $F = 6, V =$ ________, $E = 12$

5. How many vertices are there in a solid with 4 triangular faces and 1 square base?

6. Describe the cross section in the figure at the right.

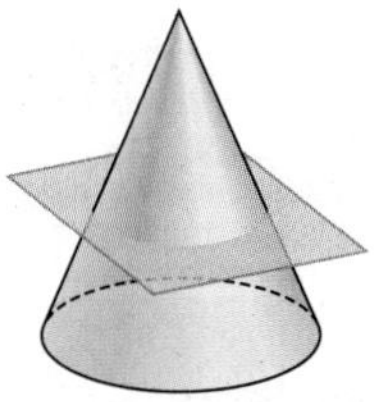

(11.2) *Find the surface area of each figure. Leave your answers in terms of* π *where applicable.*

7.

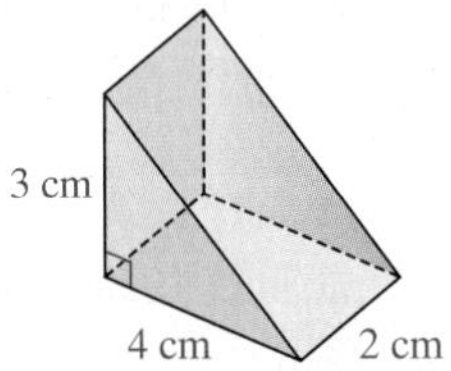

8.

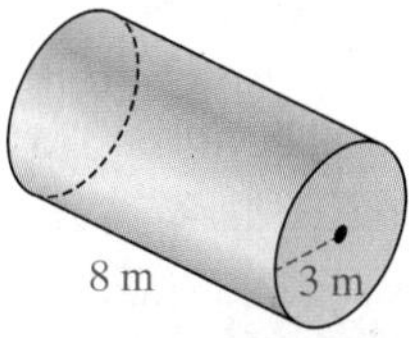

9.

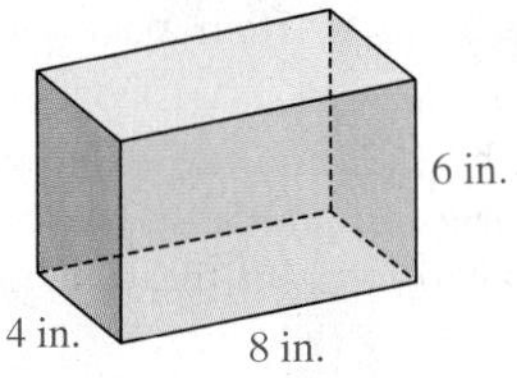

10.

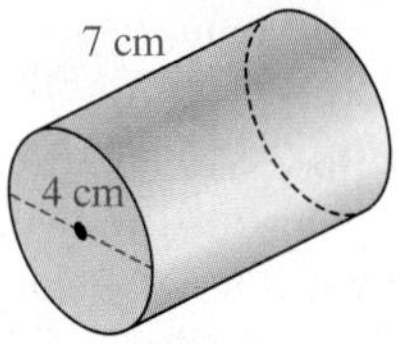

(11.3) *Find the surface area of each figure. Round your answers to the nearest tenth.*

11.

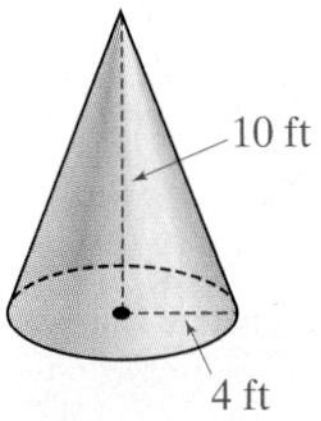

12.

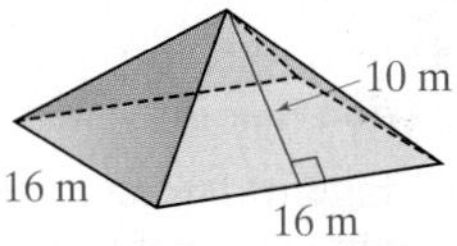

13.

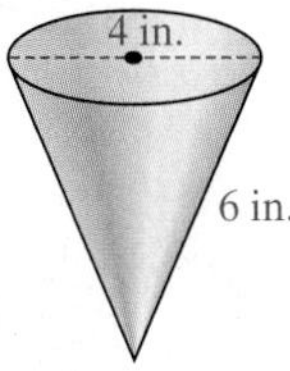

14.

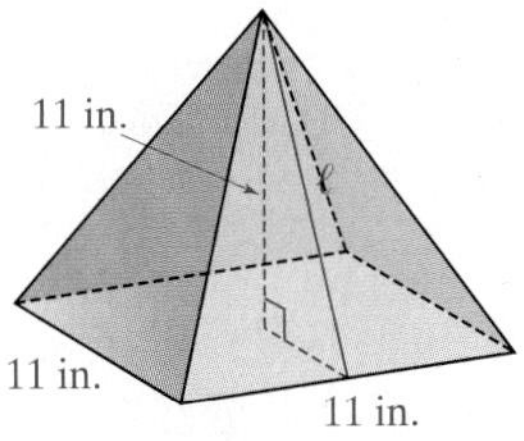

(11.4 and 11.5) *Find the volume of each figure. If necessary, round to the nearest tenth.*

15.

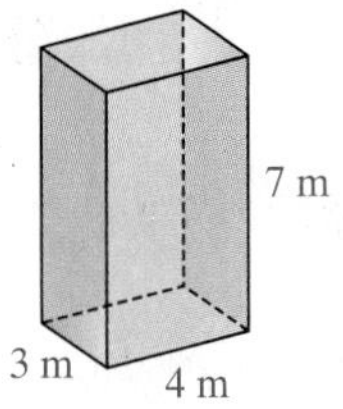

16.

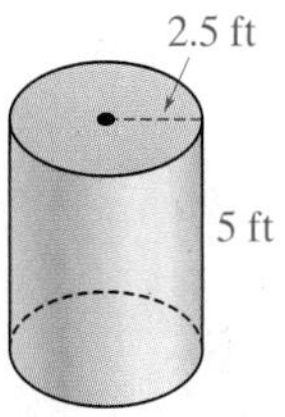

17.

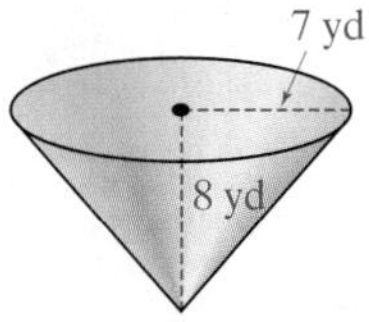

18.

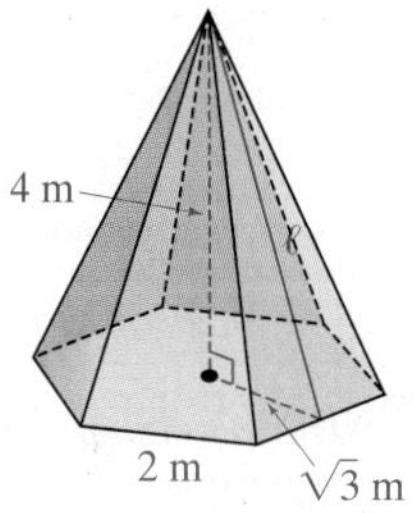

(11.6) *Find the surface area and volume of a sphere with the given radius or diameter. Round your answers to the nearest tenth.*

19. $r = 5$ in.

20. $d = 7$ cm

21. $d = 4$ ft

22. $r = 0.8$ ft

23. What is the volume of a sphere with a surface area of 452.39 sq cm? Round your answer to the nearest hundredth.

24. What is the surface area of a sphere with a volume of 523.6 cu m? Round your answer to the nearest square meter.

(11.7) *For each pair of similar solids, find the ratio of the volume of the first figure to the volume of the second.*

25.

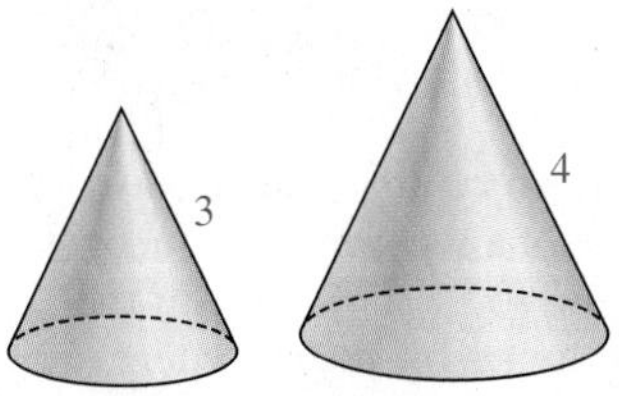

26.

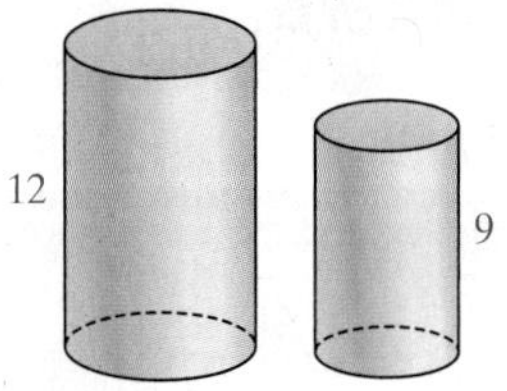

MIXED PRACTICE

27. How many faces and edges does the polyhedron have?

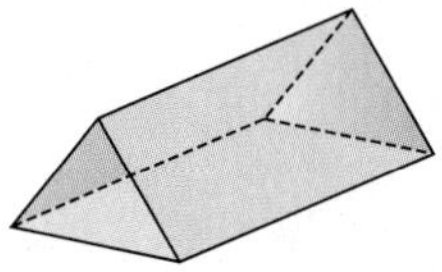

28. What is the surface area of a cylinder with radius 3 m and height 6 m? Leave your answer in terms of π.

29. A cylinder has radius 2.5 cm and lateral area 20π sq cm. What is the surface area of the cylinder in terms of π?

30. What is the surface area of a cone with radius 3 in. and slant height 10 in.? Leave your answer in terms of π.

31. What is the volume of a rectangular prism with base 3 cm by 4 cm and height 8 cm?

32. Find the formula for the base area of a prism in terms of surface area and lateral area.

33. What is the surface area of a sphere with radius 7 ft? Round your answer to the nearest tenth.

34. The circumference of a lacrosse ball is 8 in. Find its volume to the nearest tenth of a cubic inch.

35. Is a cylinder with radius 4 in. and height 12 in. similar to a cylinder with radius 14 in. and height 35 in.? If so, give the scale factor.

36. There are 12 pencils in a regular-sized box. If a jumbo box is made by tripling all the dimensions of the regular-sized box, how many pencils will the jumbo box hold?

CHAPTER 11 TEST

TEST PREP VIDEO The fully worked-out solutions to any exercises you want to review are available in MyMathLab.

Draw a net for each figure. Label the net with appropriate dimensions.

1.

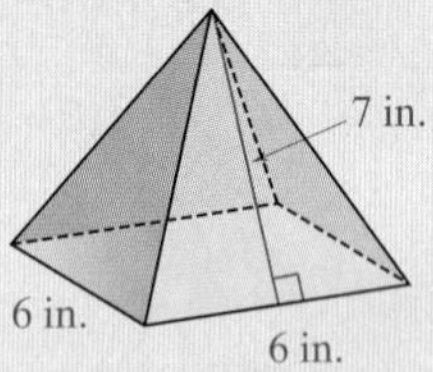

2.

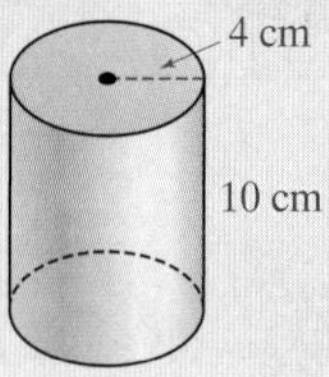

Use the polyhedron at the right for Exercises 3 and 4.

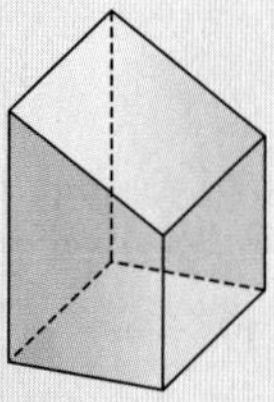

3. Verify Euler's Formula for the polyhedron.

4. Draw a net for the polyhedron. Verify $F + V = E + 1$ for the net.

5. What is the number of edges in a pyramid with seven faces?

Describe the cross section formed in each diagram.

6.

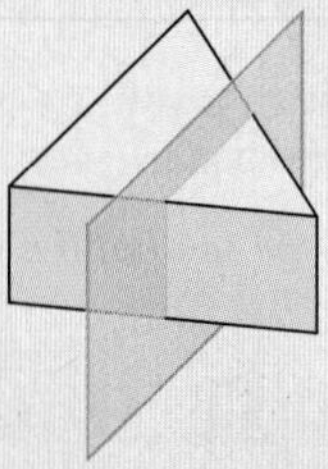

7.

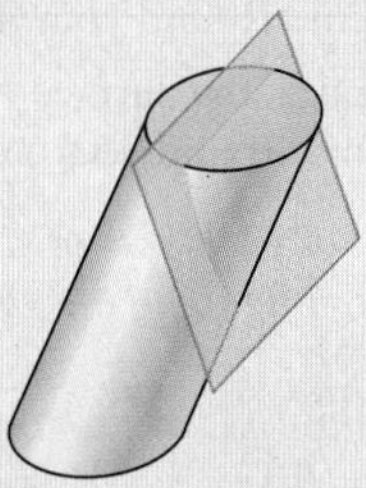

8. The flight data recorders on commercial airlines are rectangular prisms. The base of a recorder is 15 in. by 8 in. Its height ranges from 15 in. to 22 in. What are the largest and smallest possible volumes for the recorder?

Find the volume and surface area of each figure to the nearest tenth.

9.

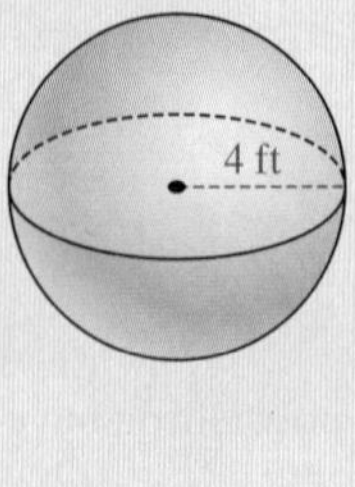

10.

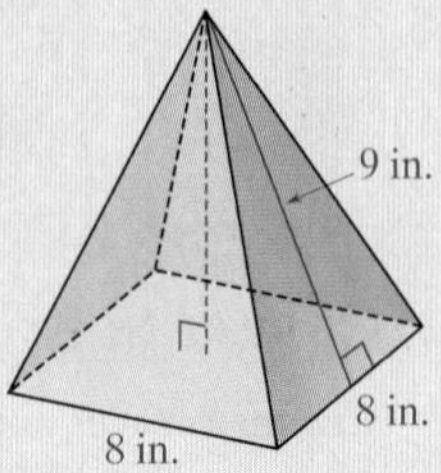

11.

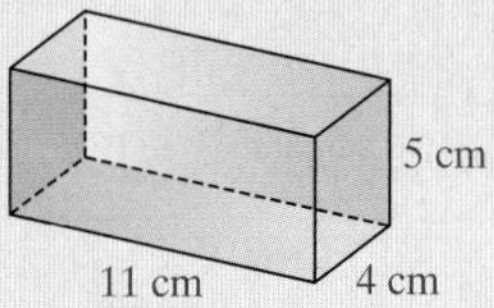

12.

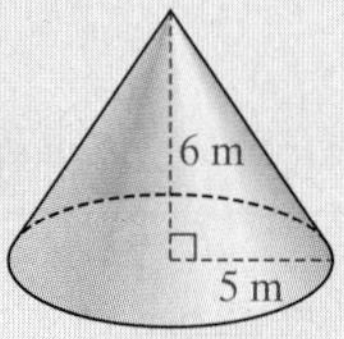

13.

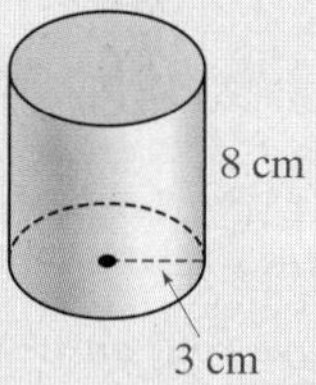

14.

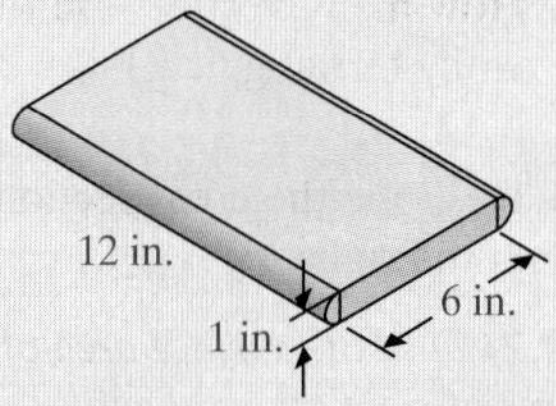

15. List these solids in order from the one with the least volume to the one with the greatest volume.

a. a cube with edge 5 cm

b. a cylinder with radius 4 cm and height 4 cm

c. a square pyramid with base edges 6 cm and height 6 cm

d. a cone with radius 4 cm and height 9 cm

e. a rectangular prism with a 5-cm-by-5-cm base and height 6 cm

16. The floor of a bedroom is 12 ft by 15 ft and the walls are 7 ft high. One gallon of paint covers about 450 sq ft. How many gallons of paint do you need to paint the walls of the bedroom?

17. What solid has a cross section that could either be a circle or a rectangle?

18. The triangle is revolved completely about the y-axis.

a. Describe the solid of revolution that is formed.

b. Find its lateral area and volume in terms of π.

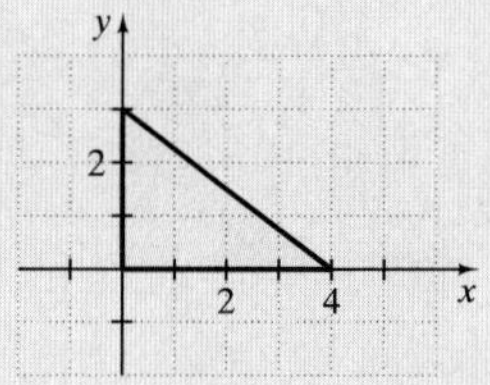

CHAPTER 11 STANDARDIZED TEST

Use the figure for Questions 1 and 2.

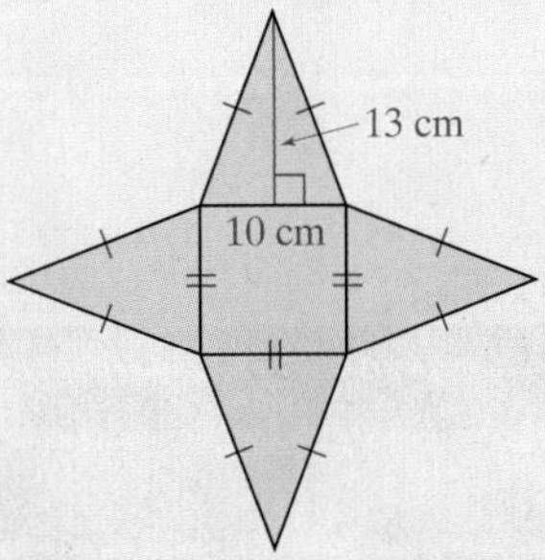

1. Identify the figure corresponding to the net.
   a. square pyramid with base edges 10 cm and height 13 cm
   b. square pyramid with base edges 10 cm and height $\sqrt{69}$ cm
   c. square pyramid with base edges 10 cm and height 12 cm
   d. none of these

2. What is the volume of the figure corresponding to the net?
   a. $\approx$433 cu cm b. $\approx$277 cu cm
   c. 1200 cu cm d. 400 cu cm

3. Which of the following are possible for a polyhedron?
   I. 7 faces, 9 vertices, 14 edges
   II. 10 faces, 12 vertices, 20 edges
   a. I only b. II only
   c. both I and II d. neither I nor II

4. What is the number of edges in a prism with seven faces?
   a. 21 edges b. 15 edges
   c. 12 edges d. none of these

5. Which figure is not a possible cross section when a cylinder is intersected by a plane?
   a. circle b. square
   c. rectangle d. none of these

6. What are the surface area and volume of a sphere with radius 3 feet?
   a. $SA = 36\pi$ sq ft.; $V = 36\pi$ cu ft
   b. $SA = 36\,\pi$ sq ft; $V = 108\,\pi$ cu ft
   c. $SA = 12\pi$ sq ft; $V = 36\pi$ cu ft
   d. $SA = 12\pi$ sq ft; $V = 108\pi$ cu ft

7. What are the surface area and volume of a square pyramid with base edges 8 in. and slant height 5 in.?
   a. $SA = 144$ sq in.; $V = 64$ cu in.
   b. $SA = 144$ sq in.; $V \approx 133$ cu in.
   c. $SA = 80$ sq in.; $V = 64$ cu in.
   d. $SA = 80$ sq in.; $V \approx 133$ cu in.

8. What are the surface area and volume of a cone with radius 6 in. and slant height 10 in.?
   a. $SA \approx 302$ sq in.; $V \approx 377$ cu in.
   b. $SA \approx 364$ sq in.; $V \approx 377$ cu in.
   c. $SA \approx 302$ sq in.; $V \approx 302$ cu in.
   d. $SA \approx 364$ sq in.; $V \approx 302$ cu in.

9. What are the surface area and volume of a cylinder with diameter 4 m and height 8 m?
   a. $SA \approx 201$ sq m; $V \approx 402$ cu m
   b. $SA \approx 126$ sq m; $V \approx 101$ cu m
   c. $SA \approx 113$ sq m; $V \approx 101$ cu m
   d. $SA \approx 126$ sq m; $V \approx 402$ cu m

10. A storage tank is in the shape of a cylinder topped by a hemisphere. The radius of the tank is 10 ft and the height of the cylinder part of the tank is 30 ft. The top and lateral surface are to be painted with two coats of paint. If one gallon of paint covers 350 sq ft, how many gallons of paint must be purchased?
    a. 8 gallons b. 18 gallons
    c. 15 gallons d. none of these

11. Which solid has greater volume, a sphere with radius 3 feet or a cylinder with radius 3 feet and height 4 feet?
    a. the sphere
    b. the cylinder
    c. the volumes are equal

12. If two similar solids have volumes of 648 cu cm and 2187 cu cm, what is the scale factor of the smaller solid to the larger solid?
    a. $\frac{8}{27}$ b. $\frac{2}{3}$
    c. $\frac{4}{9}$ d. none of these

CHAPTER

12 Circles and Other Conic Sections

12.1 Circle Review and Tangent Lines

12.2 Chords and Arcs

12.3 Inscribed Angles

12.4 Additional Angle Measures and Segment Lengths

12.5 Coordinate Plane—Circles

12.6 Locus

Extension—Parabolas

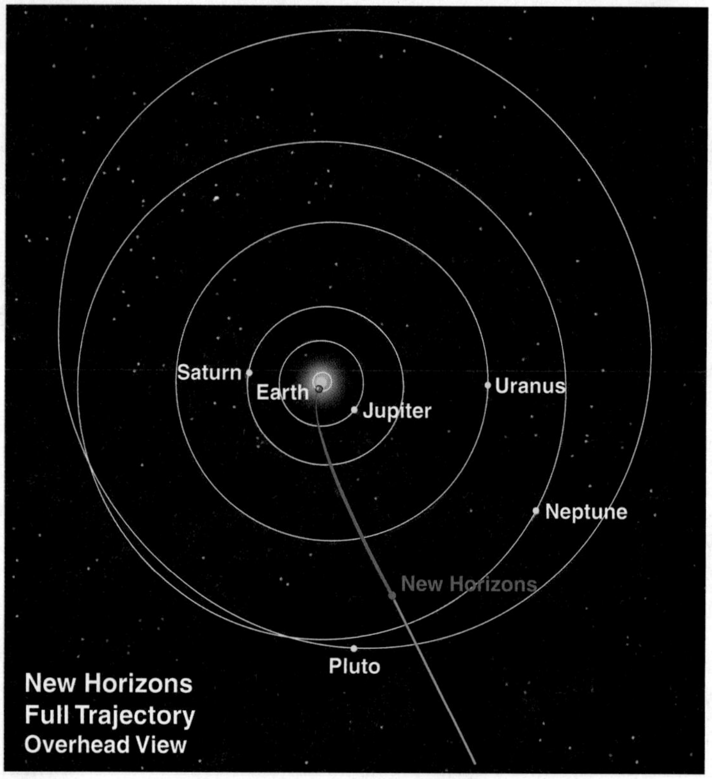

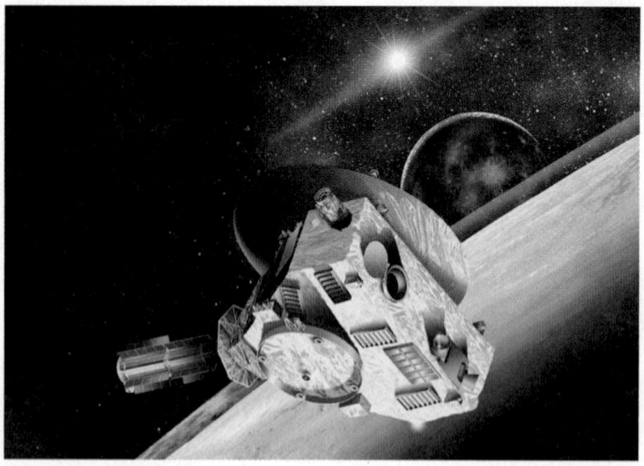

- February 28, 2007—Jupiter flyby.
- March 5, 2007—End of Jupiter encounter phase.
- June 8, 2008—The probe passed Saturn's orbit.
- March 18, 2011—The probe passed Uranus' orbit.
- August 24, 2014—The probe will pass Neptune's orbit.
- July 14, 2015—Flyby of Pluto.
- July 14, 2015—Flyby of Charon, one of Pluto's moons.
- 2016–2020—Possible flyby of Kuiper Belt objects. The Kuiper Belt is a massive region of the Solar System extending from the orbit of Neptune.
- 2029—The probe will leave the Solar System.

New Horizons is NASA's robotic spacecraft mission. This spacecraft is about the size of a grand piano, with a satellite dish attached. It was launched in January 2006, and above are previous and future key dates. As you can see, the path of New Horizons is different from the orbits of the planets. Historically, the orbits of planets were thought to be circular, until they were later discovered to be elliptical; and the Sun is located not at the center of the orbits, but rather at one focus, with the speed of each planet not constant. Simply put, elliptical and circular orbits are "captive" orbits, and parabolic and hyperbolic orbits are "escape" orbits. (Note: A study of ellipses and hyperbolas is included in the back of this text.)

12.1 CIRCLE REVIEW AND TANGENT LINES

OBJECTIVES

1. Review Circles and Arcs.
2. Use Properties of a Tangent Line to a Circle.

VOCABULARY

- tangent to a circle
- point of tangency
- tangent ray
- tangent segment
- common tangent
- line of centers
- tangent circles

In Chapter 10, we studied area and circumference of a circle. In Sections 10.5 and 10.6, we included lengths of arcs and areas of sectors. Let's review some definitions and concepts from those sections.

OBJECTIVE 1 ▶ Reviewing Circles and Arcs.

- In a plane, a **circle** is the set of all points equidistant from a given point called the **center**.
- A **diameter** is a segment that contains the center of a circle and has both endpoints on the circle. The length of this segment is also called the diameter.
- A **radius** is a segment that has one endpoint at the center and the other endpoint on the circle. The length of this segment is also called the radius.
- **Congruent circles** have congruent radii.
- A **central angle** is an angle whose vertex is the center of the circle.

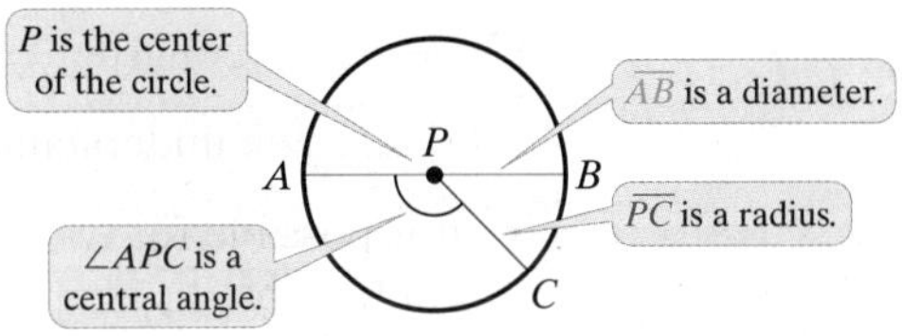

An arc is a part of a circle.

- A **semicircle** is half of a circle.
- A **minor arc** is smaller than a semicircle.
- A **major arc** is larger than a semicircle.

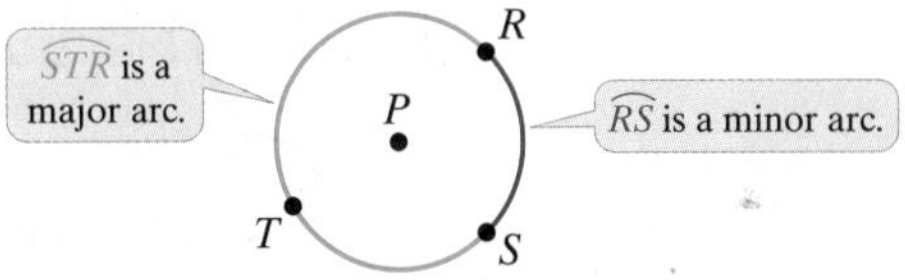

Next, let's recall how we measure these arcs.

Arc Measure

Arc Measure	***Example***
• minor arc—The measure of a minor arc is equal to the measure of its corresponding central angle. • major arc—The measure of a major arc is the measure of the related minor arc subtracted from 360°. • semicircle—The measure of a semicircle is 180°.	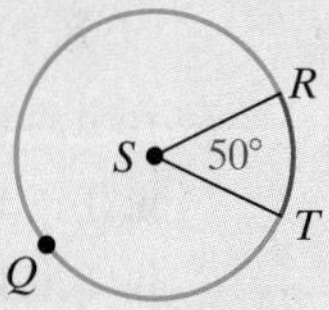 $m\widehat{RT} = m\angle RST = 50°$ $m\widehat{TQR} = 360° - m\widehat{RT}$ $= 360° - 50°$ $= 310°$

Finally, let's review definitions of concentric circles and congruent arcs, and how we measure arc length.

- Coplanar circles that have the same center are **concentric circles**.
- The measure of an arc is in degrees while the **arc length** is a fraction of a circle's circumference.

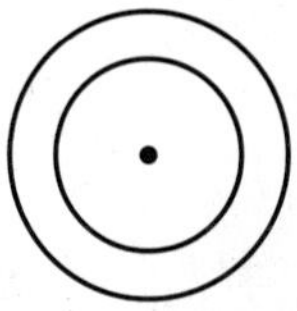

Concentric circles

Recall:

Theorem 10.5-3 Arc Length

The length of an arc of a circle is the product of the ratio $\frac{\text{measure of the arc}}{360°}$ and the circumference of the circle.

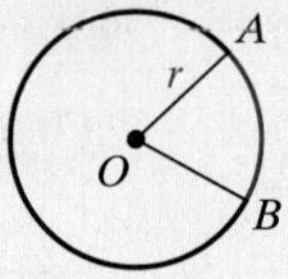

$$\text{length of } \widehat{AB} = \frac{m\widehat{AB}}{360°} \cdot 2\pi r \quad \text{or}$$
$$= \frac{m\widehat{AB}}{360°} \cdot \pi d$$

Congruent arcs are arcs that have the same measure *and* are in the same circle or in congruent circles.

$\odot O \cong \odot P$ Congruent circles and same measure (60°), thus

$\widehat{XY} \cong \widehat{RS}$ Congruent arcs

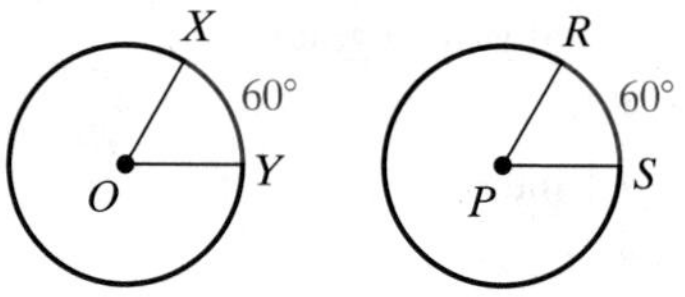

Make sure you understand:

- It is possible for two arcs of different circles to have the same measure but different lengths.
- It is also possible for two arcs of different circles to have the same length but different measures.

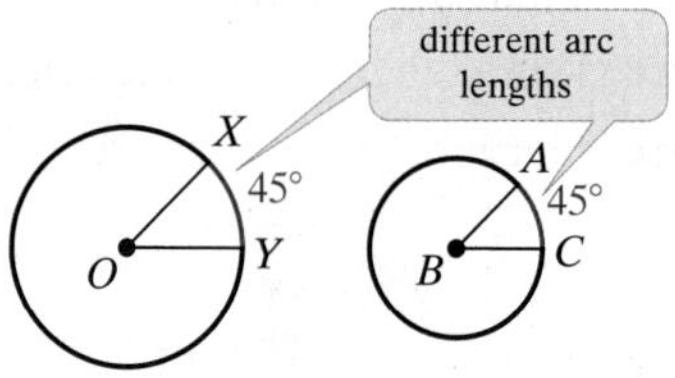

$\widehat{XY} \ncong \widehat{AC}$

Arcs not congruent:

- same measure
- *not* same or congruent circles, so different lengths

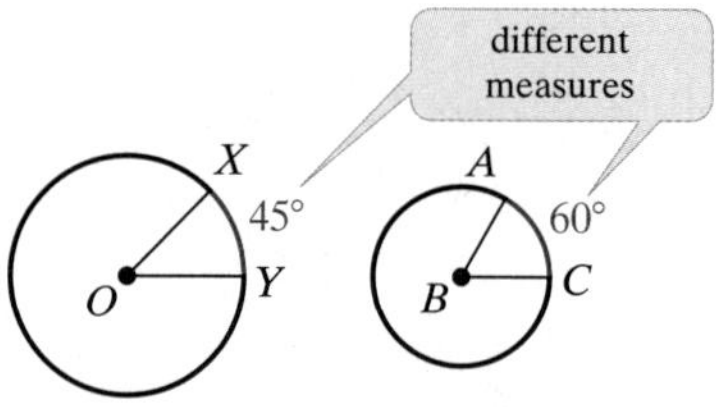

$\widehat{XY} \ncong \widehat{AC}$

Arcs not congruent:

- same lengths
- *not* same or congruent circles, so different measures

EXAMPLE 1 Answer parts **a** through **c** about $\odot O$. The radius of $\odot O$ is 18 m.

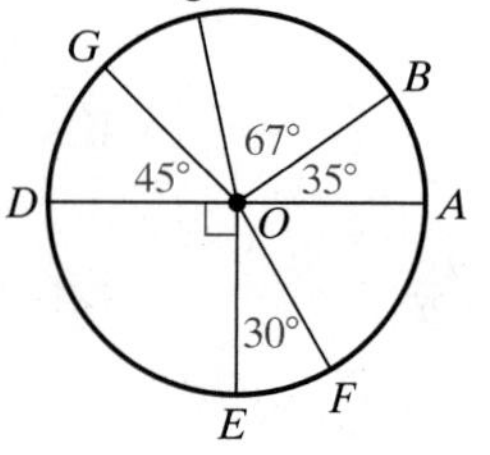

a. Find $m\widehat{BE}$.

b. Find $m\widehat{AED}$.

c. Find the length of $\widehat{DEF}$.

Solution

a. $m\widehat{BE} = m\widehat{BA} + m\widehat{AE}$
$= 35° + 90°$
$= 125°$

b. $m\widehat{AED} = 180°$ since $\widehat{AED}$ is a semicircle.

c. $\text{length of } \widehat{DEF} = \frac{m\widehat{DEF}}{360°} \cdot 2\pi r = \frac{120°}{360°} \cdot 2\pi(18 \text{ m})$
$= 12\pi \text{ m}$

□

PRACTICE 1 Use the same circle as in Example 1, except that the diameter of $\odot O$ is 40 ft.

a. Find $m\widehat{GE}$. **b.** Find $m\widehat{BED}$. **c.** Find the length of $\widehat{EDG}$.

> **Helpful Hint**
> Remember—a tangent to a circle is a line. It intersects the circle in one point.

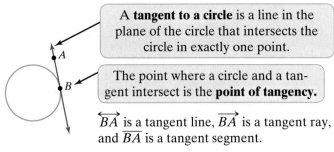

OBJECTIVE 2 ▶ Using Properties of Tangents.

A **tangent to a circle** is a line in the plane of the circle that intersects the circle in exactly one point.

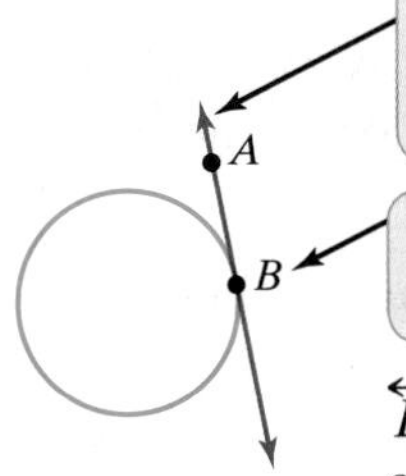

The point where a circle and a tangent intersect is the **point of tangency.**

$\overleftrightarrow{BA}$ is a tangent line, $\overrightarrow{BA}$ is a tangent ray, and $\overline{BA}$ is a tangent segment.

A radius of a circle and the tangent that intersects the endpoint of the radius on the circle have a special relationship.

Theorem 12.1-1 Tangent-Radius Theorem

Theorem	If . . .	Then . . .
If a line is tangent to a circle, then the line is perpendicular to the radius at the point of tangency.	$\overleftrightarrow{AB}$ is tangent to $\odot O$ at P 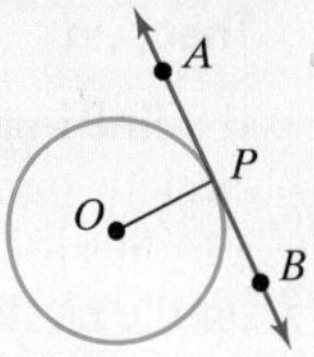	$\overleftrightarrow{AB} \perp \overline{OP}$ 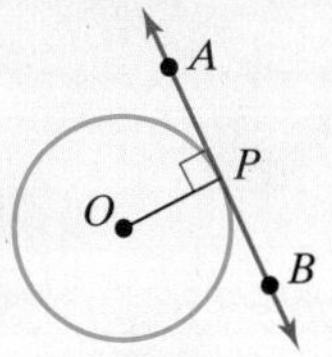

We prove this theorem next.

Indirect Proof of Theorem 12.1-1

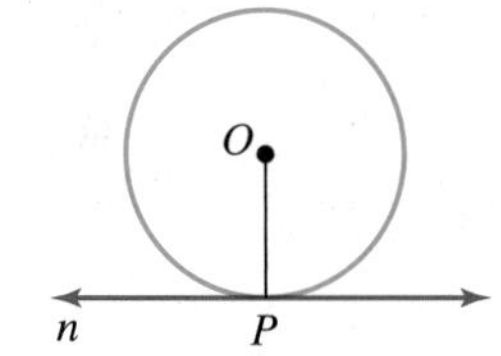

Given: n is tangent to $\odot O$ at P.

Prove: $n \perp \overline{OP}$

STEP 1. Assume that n is not perpendicular to $\overline{OP}$.

STEP 2. If line n is not perpendicular to $\overline{OP}$, then, for some other point L on n, $\overline{OL}$ must be perpendicular to n. Also, there is a point K on n such that $\overline{LK} \cong \overline{LP}$. $\angle OLK \cong \angle OLP$ because perpendicular lines form congruent adjacent angles. $\overline{OL} \cong \overline{OL}$. So, $\triangle OLK \cong \triangle OLP$ by SAS.

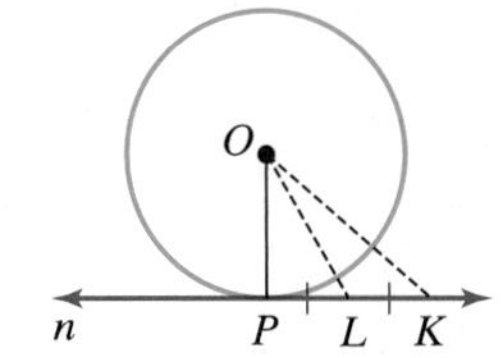

Since corresponding parts of congruent triangles are congruent, $\overline{OK} \cong \overline{OP}$. So K and P are both on $\odot O$ by the definition of a circle. For two points on n to also be on $\odot O$ contradicts the given fact that n is tangent to $\odot O$ at P. So the assumption that n is not perpendicular to $\overline{OP}$ must be false.

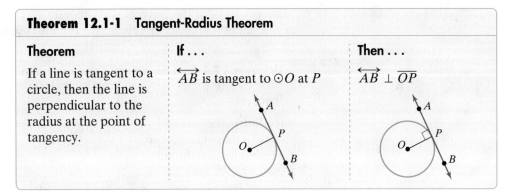

> **Helpful Hint**
> Remember—if line n is tangent to $\odot O$, then n intersects the circle in exactly one point.

STEP 3. Therefore, $n \perp \overline{OP}$ must be true.

EXAMPLE 2 Finding Angle Measures

Multiple Choice $\overline{ML}$ and $\overline{MN}$ are tangent to $\odot O$. What is the value of x?

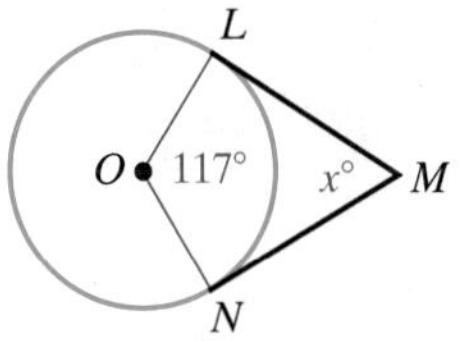

a. 58 **b.** 63

c. 90 **d.** 117

Solution Recall that a tangent line is perpendicular to the radius at the point of tangency. Since $\overline{ML}$ and $\overline{MN}$ are tangent to $\odot O$, $\angle L$ and $\angle N$ are right angles. Next, $LMNO$ is a quadrilateral. So the sum of the angle measures is 360°.

$$m\angle L + m\angle M + m\angle N + m\angle O = 360^\circ$$

$$90^\circ + m\angle M + 90^\circ + 117^\circ = 360^\circ \quad \text{Substitute.}$$

$$297^\circ + m\angle M = 360^\circ \quad \text{Combine like terms.}$$

$$m\angle M = 63^\circ \quad \text{Subtract } 297^\circ \text{ from both sides.}$$

Since $m\angle M = 63^\circ$, or x°, then $x = 63$. The correct answer is b. □

PRACTICE 2 $\overline{ED}$ is tangent to $\odot O$. What is the value of x?

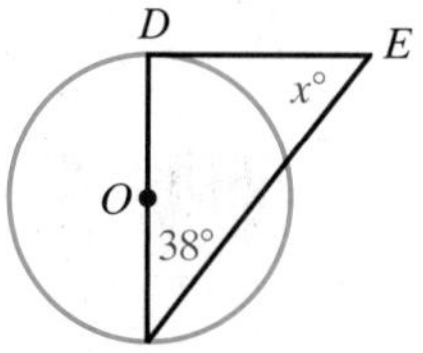

Theorem 12.1-2 is the converse of Theorem 12.1-1. We can use it to prove that a line or a segment is tangent to a circle. We can also use it to construct a tangent to a circle.

Theorem 12.1-2 Converse of Tangent-Radius Theorem

Theorem	If . . .	Then . . .
If a line in the plane of a circle is perpendicular to a radius at its endpoint on the circle, then the line is tangent to the circle.	$\overleftrightarrow{AB} \perp \overline{OP}$ at P 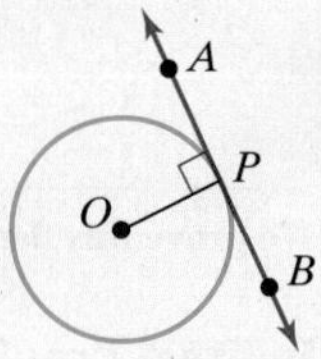	$\overleftrightarrow{AB}$ is tangent to $\odot O$

We prove this theorem in Exercise Set 12.1, Exercise 82.

EXAMPLE 3 Finding a Radius

What is the radius of $\odot C$?

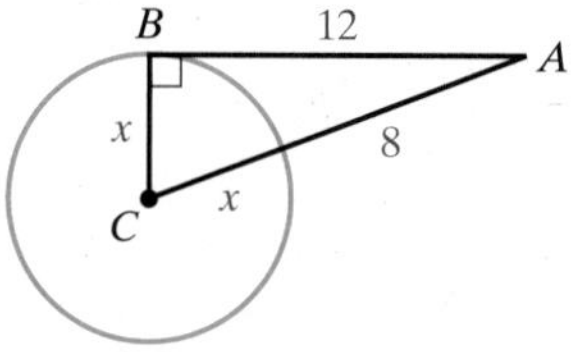

Solution From the diagram, to find the radius x, we use the Pythagorean Theorem since ΔABC is a right triangle.

$$AC^2 = AB^2 + BC^2 \quad \text{Pythagorean Theorem}$$

$$(x + 8)^2 = 12^2 + x^2 \quad \text{Substitute.}$$

$$x^2 + 16x + 64 = 144 + x^2 \quad \text{Replace } (x + 8)^2 \text{ with } x^2 + 16x + 64.$$

$$16x = 80 \quad \text{Subtract } x^2 \text{ and 64 from both sides.}$$

$$x = 5 \quad \text{Divide both sides by 16.}$$

The radius is 5. □

PRACTICE 3 What is the radius of $\odot O$?

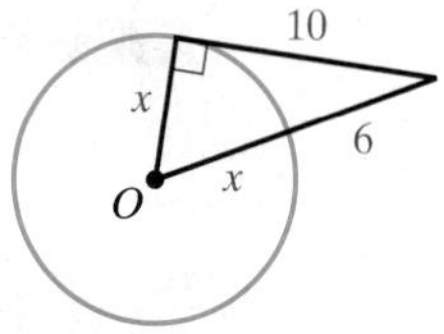

EXAMPLE 4 **Identifying a Tangent**

Is $\overline{ML}$ tangent to $\odot N$ at L? Explain.

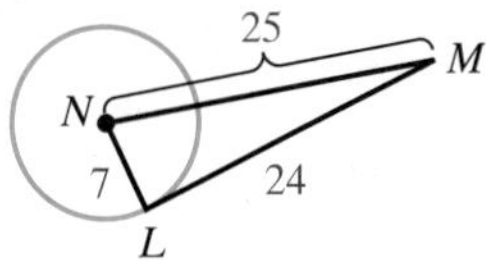

Solution

If $\overline{ML}$ is tangent to $\odot N$ at L, then $\overline{ML}$ must be perpendicular to $\overline{LN}$. To see if this is true, we see if ΔMLN is a right triangle. To do so, let's use the Converse of the Pythagorean Theorem.

$NL^2 + ML^2 \stackrel{?}{=} NM^2$	If a right Δ, then $\overline{NM}$ must be the hypotenuse since it is the longest side.
$7^2 + 24^2 \stackrel{?}{=} 25^2$	Substitute.
$49 + 576 \stackrel{?}{=} 625$	Square each number.
$625 = 625$	Simplify. Result is a true statement.

By the Converse of the Pythagorean Theorem, ΔLMN is a right triangle with $\overline{ML} \perp \overline{NL}$. So $\overline{ML}$ is tangent to $\odot N$ at L because it is perpendicular to the radius at the point of tangency (Theorem 12.1-2). □

Real-Life Example of a Tangent

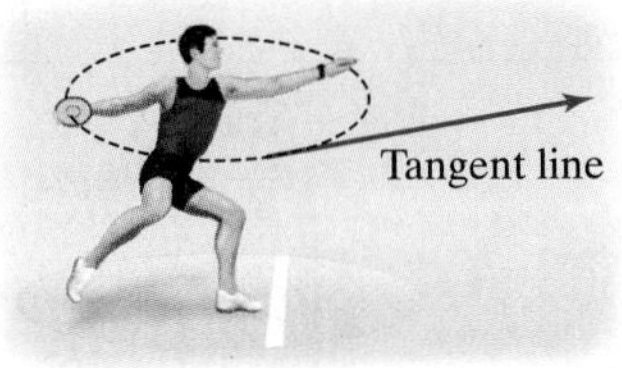

Spinning in a circular orbit and releasing an object will result in the object traveling in a path tangent to the circular orbit.

PRACTICE 4 Use the diagram in Example 4. If $NL = 4$, $ML = 7$, and $NM = 8$, is $\overline{ML}$ tangent to $\odot N$ at L? Explain.

Two circles in a plane can share a tangent line. **Common tangents** are lines or segments or rays that are tangent to more than one circle. Let's sketch the possible positions of two distinct circles in a plane. These two circles can intersect in two points, in one point, or not at all, as shown below.

Possible Intersections of Two Circles

Two Points of Intersection

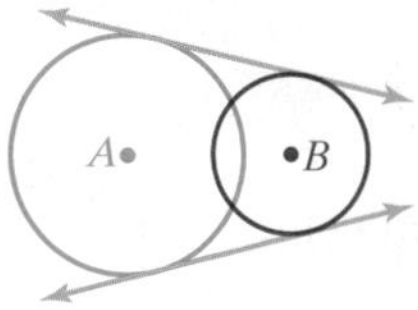

2 Common Tangents

2 external tangents (green)

0 internal tangents

One Point of Intersection

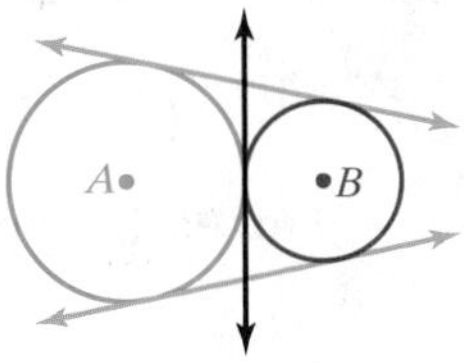

3 Common Tangents

(2 externally tangent circles)

2 external tangents (green)

1 internal tangent (black)

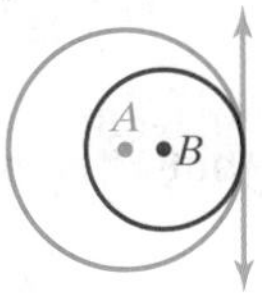

1 Common Tangent

(2 internally tangent circles)

1 external tangent (green)

0 internal tangents

No Points of Intersection

0 Common Tangents

(2 concentric circles)

0 external tangents

0 internal tangents

(one circle floating inside the other, without touching)

0 external tangents

0 internal tangents

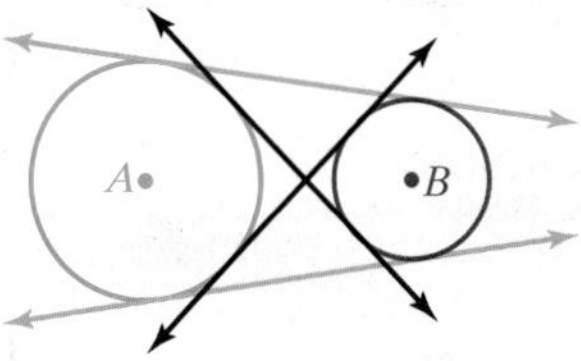

4 Common Tangents

2 external tangents (green)

2 internal tangents (black)

The line that passes through the centers of two circles is called their **line of centers**. If two circles are tangent, internally or externally (see middle examples above), then their common point of tangency is on their line of centers, and these circles are called **tangent circles**.

Theorem 12.1-3 Congruent Tangent Segments

Theorem	If . . .	Then . . .
If two tangent segments to a circle share a common endpoint outside the circle, then the two segments are congruent.	$\overrightarrow{BA}$ and $\overrightarrow{BC}$ are tangent to $\odot O$ 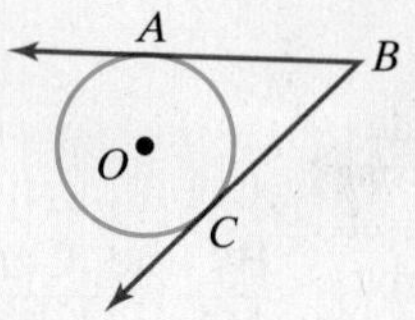 Note: ($\overline{BA}$ and $\overline{BC}$ are tangent segments.)	$\overline{BA} \cong \overline{BC}$

We prove this in Exercise Set 12.1, Exercise 81.

EXAMPLE 5 Finding Tangent Segment Lengths

If $\overleftrightarrow{AC}$ and $\overleftrightarrow{AB}$ are tangents to $\odot D$, find the value of x.

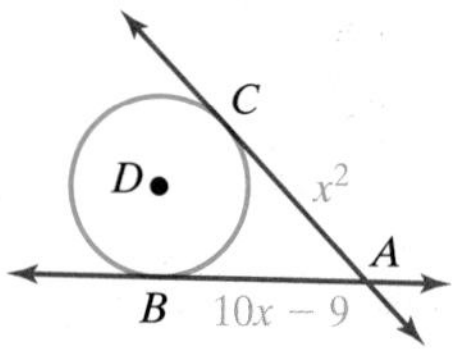

Solution From Theorem 12.1-3, we know that $AC \cong AB$. Thus,

$$AC = AB \quad \text{Lengths of tangent segments from exterior points are equal.}$$
$$x^2 = 10x - 9 \quad \text{Substitution}$$
$$x^2 - 10x + 9 = 0 \quad \text{Quadratic equation, so set equal to 0.}$$
$$(x - 9)(x - 1) = 0 \quad \text{Factor.}$$
$$x - 9 = 0 \text{ or } x - 1 = 0 \quad \text{Set each factor equal to 0.}$$
$$x = 9 \text{ or } x = 1 \quad \text{Solve.}$$

The value of x is 9 or 1.

To **check**, see that x^2 is the same as $10x - 9$ for each proposed value for x. Let $x = 9$.

$$x^2 = 9^2 = 81 \text{ units}$$
$$10x - 9 = 10 \cdot 9 - 9 = 81 \text{ units} \quad \text{The distances are equal.}$$

Let $x = 1$.

$$x^2 = 1^2 = 1 \text{ unit}$$
$$10x - 9 = 10 \cdot 1 - 9 = 1 \text{ unit} \quad \text{The distances are equal.}$$

□

PRACTICE 5 Use the same $\odot D$ as for Example 5, except that $AC = x^2$ units and $AB = 8x - 12$ units. Find the value of x.

In the figure below, the sides of the triangle are tangent to the circle. The circle is *inscribed in* the triangle. The triangle is *circumscribed about* the circle.

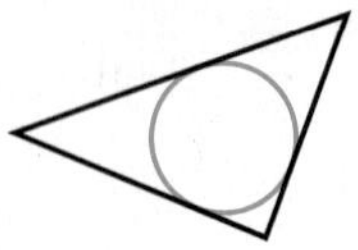

EXAMPLE 6 **Circles Inscribed in Polygons**

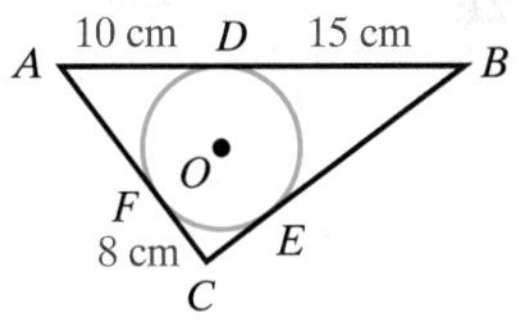

$\odot O$ is inscribed in ΔABC. What is the perimeter of ΔABC?

Solution From Theorem 12.1-3, we have congruent tangent segments.

$$AD = AF = 10 \text{ cm}$$
$$BD = BE = 15 \text{ cm}$$
$$CF = CE = 8 \text{ cm}$$

Two segments tangent to a circle from a point outside the circle are congruent, so they have the same length.

$P = AB + BC + CA$ — Definition of perimeter P

$= AD + DB + BE + EC + CF + FA$ — Segment Addition Postulate

$= 10 + 15 + 15 + 8 + 8 + 10$ — Substitute.

$= 66$ — Add.

The perimeter is 66 cm. □

PRACTICE

6 $\odot O$ is inscribed in ΔPQR, which has a perimeter of 88 cm. What is the length of $\overline{QY}$?

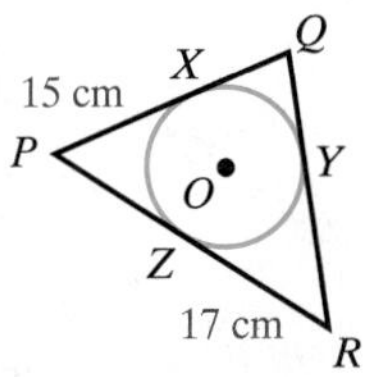

VOCABULARY & READINESS CHECK

Word Bank *Use the choices to fill in each blank. Some choices may be used more than once and some not at all.*

tangent ray	tangent to a circle	point of tangency
tangent segment	line of centers	
tangent circles	common tangent	

1. A line in the plane of a circle that intersects the circle in one point is called a ____________.
2. The point where a circle and a tangent intersect is the ____________.
3. If $\overleftrightarrow{AB}$ is a tangent to a circle at point B, then $\overrightarrow{BA}$ is called a ____________ and $\overline{BA}$ is called a ____________.
4. The line that passes through the center of two circles is called a ____________.
5. If two circles are tangent, they are called ____________.
6. A line or segment or ray that is tangent to more than one circle is called a ____________.

12.1 EXERCISE SET MyMathLab®

Use $\odot A$ to answer Exercises 1–10, given that $\overline{BF}$ is a diameter of $\odot A$. See Example 1.

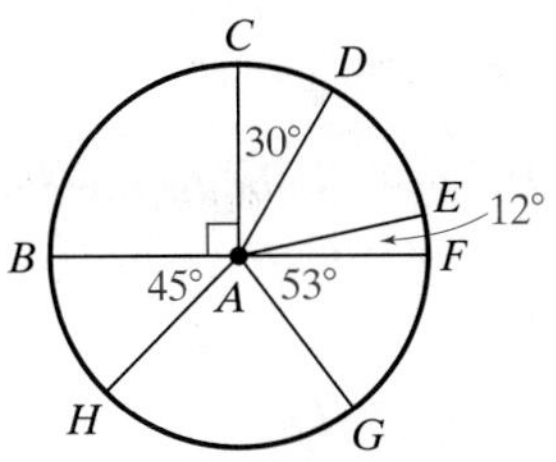

1. $m\widehat{EG}$ = ______
2. $m\widehat{BD}$ = ______
3. $m\widehat{BGF}$ = ______
4. $m\widehat{BHF}$ = ______
5. $m\widehat{DE}$ = ______
6. $m\widehat{HG}$ = ______
7. $m\widehat{FBC}$ = ______
8. $m\widehat{BDG}$ = ______
9. $m\widehat{FHG}$ = ______
10. $m\widehat{GBF}$ = ______

Use $\odot A$ to find each length in Exercises 11–18. The radius of $\odot A$ is 60 cm. Leave your answers in terms of π. See Example 1.

11. $\widehat{BC}$
12. $\widehat{BGF}$
13. $\widehat{CH}$
14. $\widehat{CD}$
15. $\widehat{CGD}$
16. $\widehat{BDH}$
17. $\widehat{EBF}$
18. $\widehat{EHD}$

Lines that appear to be tangent are tangent. O is the center of each circle. What is the value of x? See Example 2.

19.

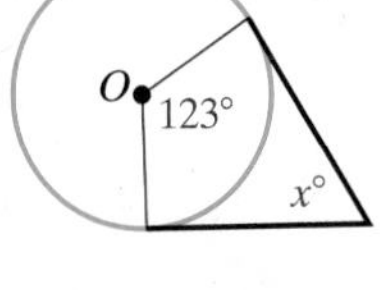

20.

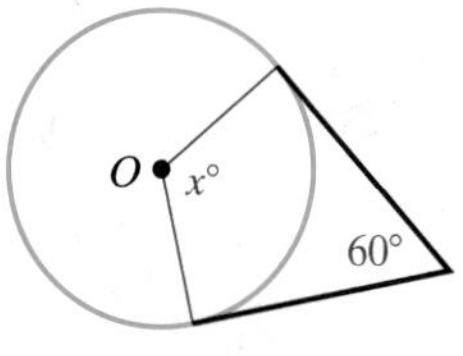

21.

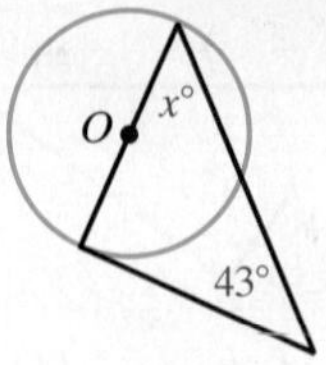

22.

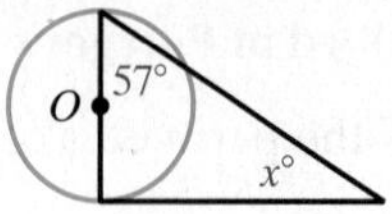

23.

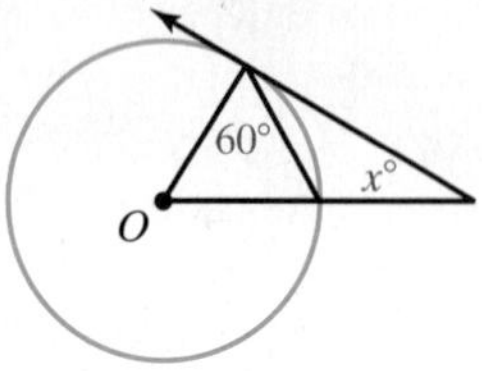

24.

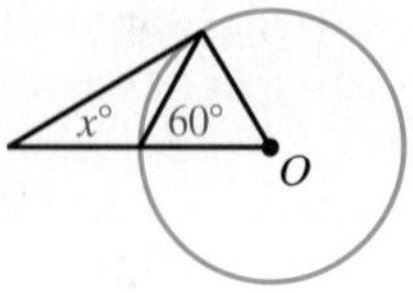

In each circle, what is the value of x? If necessary, round to the nearest tenth. See Example 3.

25.

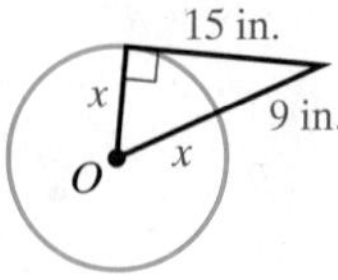

26.

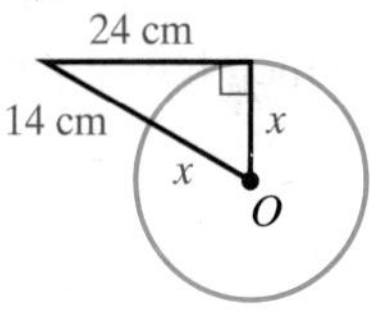

27.

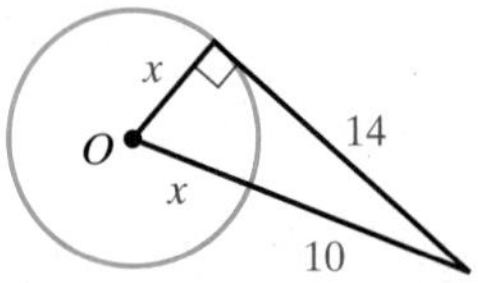

28.

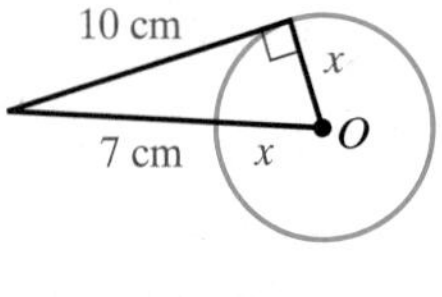

Mixed Practice *Use the figure to solve Exercises 29–34. See Examples 2 and 3.*

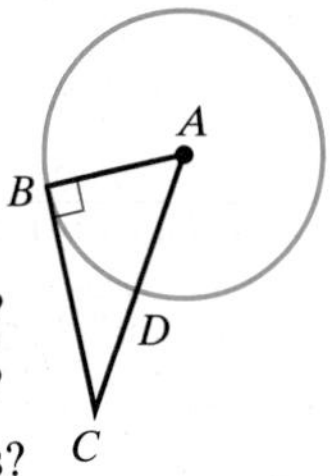

29. If $m\angle A = 58°$, what is $m\angle ACB$?

30. If $m\angle C = 28°$, what is $m\angle BAC$?

31. If $BC = 8$ and $DC = 4$, what is the radius?

32. If $BC = 12$ and $DC = 7$, what is the radius?

33. If $AC = 12$ and $BC = 9$, what is the radius?

34. If $AC = 13$ and $BC = 10$, what is the radius?

Determine whether a tangent is shown in each diagram. Explain. See Example 4.

35.

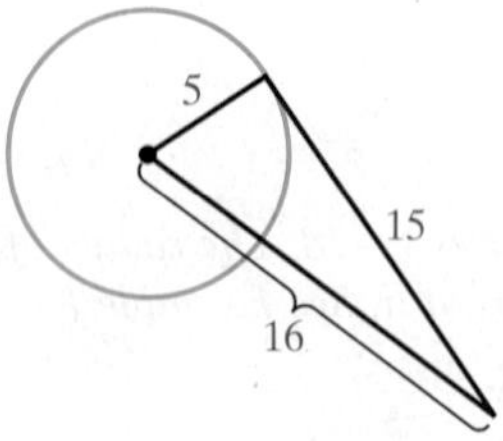

36.

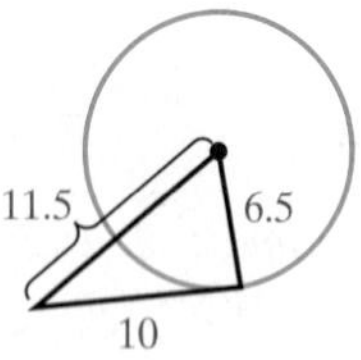

37.

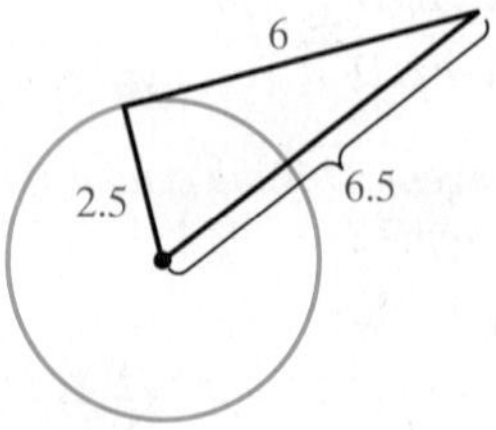

38.

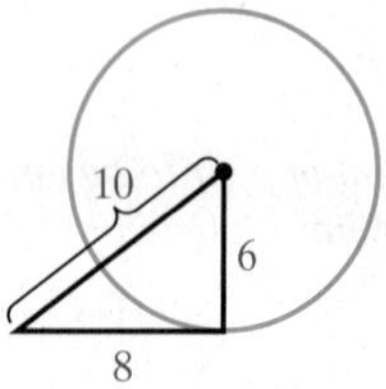

If $\overline{PQ}$ and $\overline{PR}$ are tangent to $\odot E$, find the value of x. See Example 5.

39.

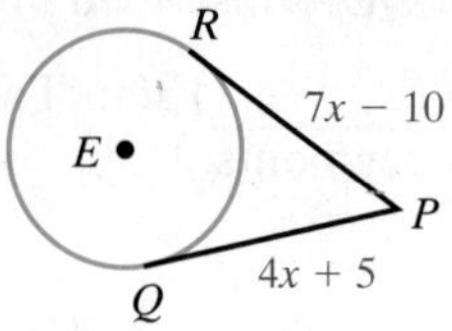

40.

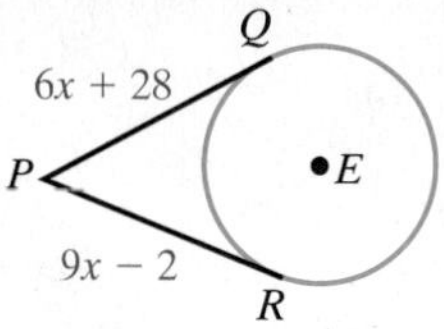

41.

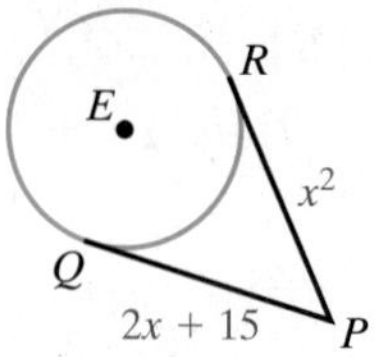

42.

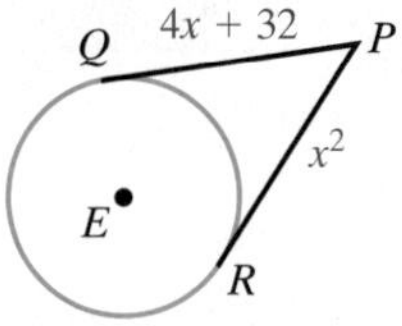

Each polygon circumscribes a circle. What is the perimeter of each polygon? See Example 6.

43.

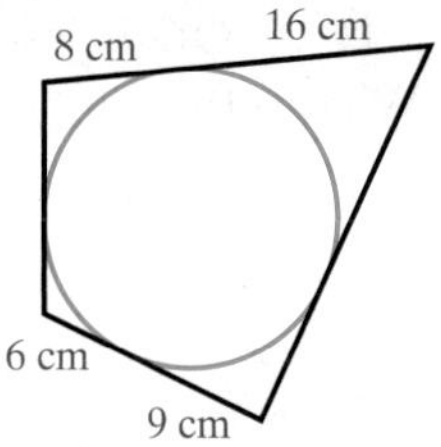

44.

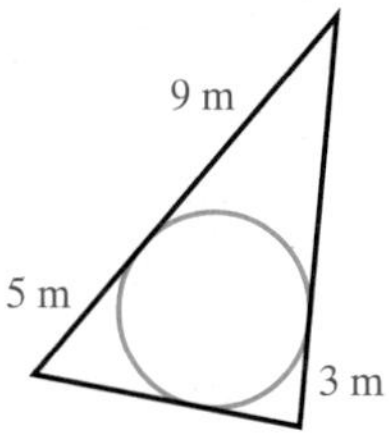

45.

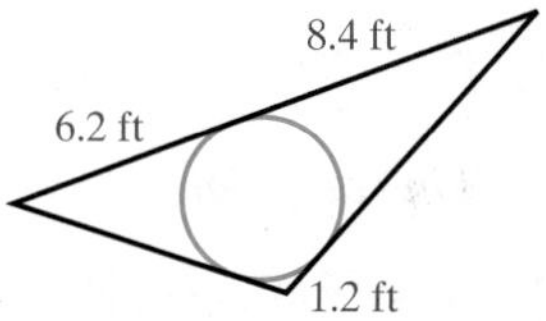

46.

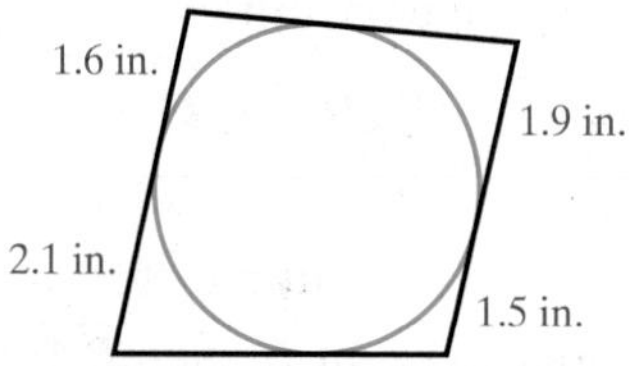

Tell whether the common tangents are internal or external. (Study the figures after Practice 4.)

47.

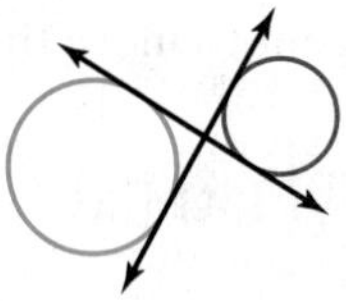

48.

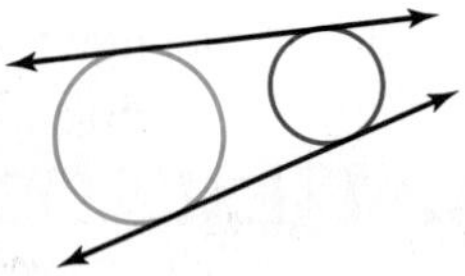

49.

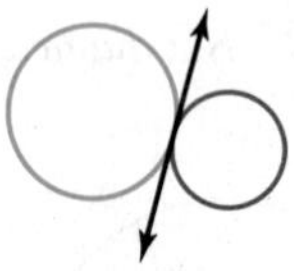

50.

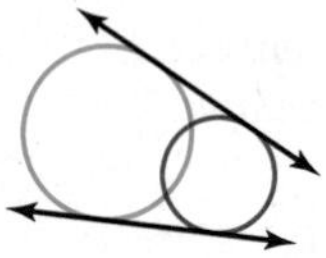

Draw circles in the same relationships as shown. Then sketch the common tangents and tell how many there are.

51.

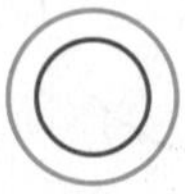

52.

53.

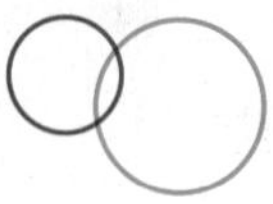

54.

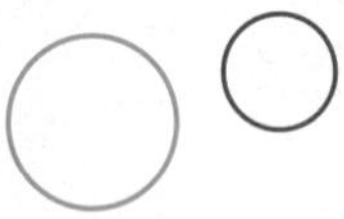

The circle at the right represents Earth. The radius of Earth is about 6400 km or 4000 mi. Find the distance d to the horizon that a person can see on a clear day from each of the following heights h above Earth. Round your answer to the nearest tenth of a kilometer or tenth of a mile.

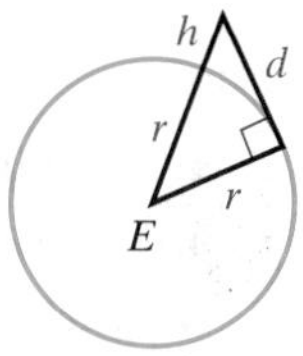

55. 5 km (~height of mount Blackburn in Alaska)

Mount Blackburn in Alaska

56. 1600 m (a mile~1600 m)

57. 1454 ft (~height of tip of Empire State Building in NYC) (Note: Watch your units of measure.)

Empire State Building

58. 12,338 ft (height of Mount Fuji)

Mount Fuji in Japan

Use the figure at the right for Exercises 59–68. $\overline{BD}$ and $\overline{CK}$ at the right are diameters of ⊙A. $\overline{BP}$ and $\overline{QP}$ are tangents to ⊙A. Find the measure of each angle. (Hint: First notice what is true about ΔABP and ΔAQP.)

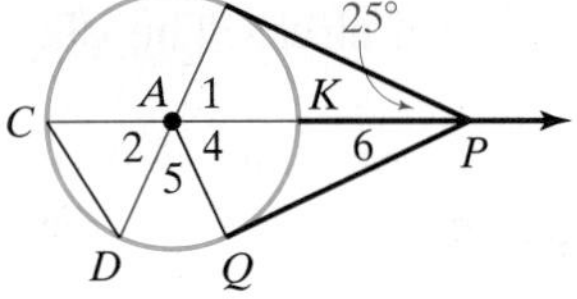

59. $m\angle 1$ **60.** $m\angle 4$

61. $m\angle 2$ **62.** $m\angle BAC$

63. $m\angle 5$ **64.** $m\angle 6$

65. What type of triangle is ΔACD?

66. What type of triangle is ΔADQ?

67. Find $m\angle ADC$.   **68.** Find $m\angle ADQ$.

An important real-life application is the belt or pulley problem, used in the design of bicycle gears, car engines, and other machinery. Knowledge of tangent lines to circles is needed to calculate the length of belts needed to connect these pulleys.

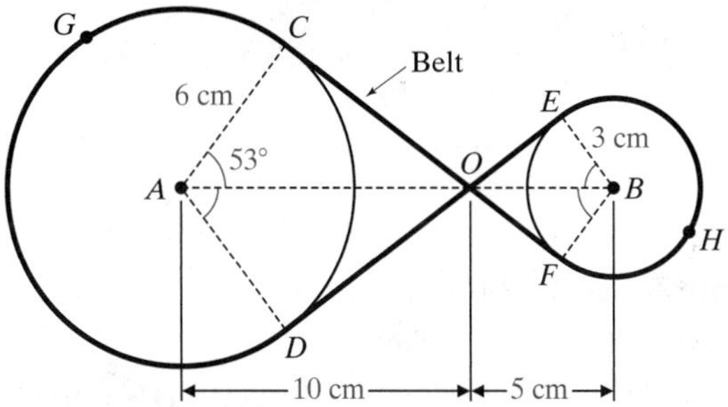

Two circular pullies are to be connected by a belt. Answer 69–74 using the figure above the photo. Note: $\overline{CF}$ and $\overline{DE}$ are tangent to both circles at their endpoints. Also, all angles with one arc mark measure 53°.

69. Use ΔACO and find the length of $\overline{CO}$ and $\overline{DO}$.

70. Use ΔOEB and find the length of $\overline{EO}$ and $\overline{FO}$.

71. Find $m\widehat{CGD}$. Then use the radius to calculate the length of this arc, rounded to the nearest tenth of a cm.

72. Find $m\widehat{EHF}$. Then use the radius to calculate the length of this arc, rounded to the nearest tenth of a cm.

73. Use the answers to Exercises **69–72** to find the approximate length of the belt.

74. The length of arc $\widehat{CGD}$ is twice the length of $\widehat{EHF}$. Tell why.

Common tangents to two circles may be internal or external. If you draw a segment joining the centers of the circles, a common internal tangent will intersect the segment. A common external tangent will not. For this cross-sectional diagram of the sun, moon, and Earth during a solar eclipse, use the terms above to describe the types of tangents of each color.

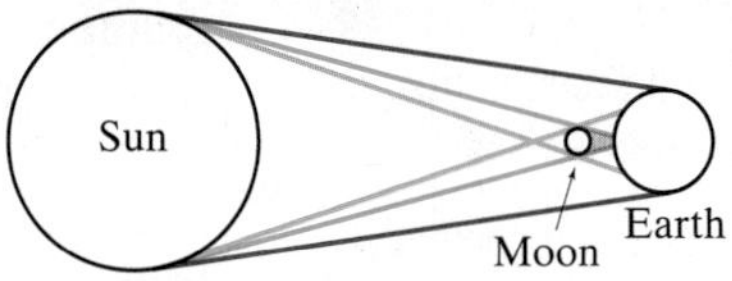

75. red   **76.** blue   **77.** green

78. Which tangents show the extent on Earth's surface of total eclipse? Of partial eclipse?

CONCEPT EXTENSIONS

79. Find the Error A classmate insists that $\overline{DF}$ is a tangent to $\odot E$. Explain how to show that your classmate is wrong.

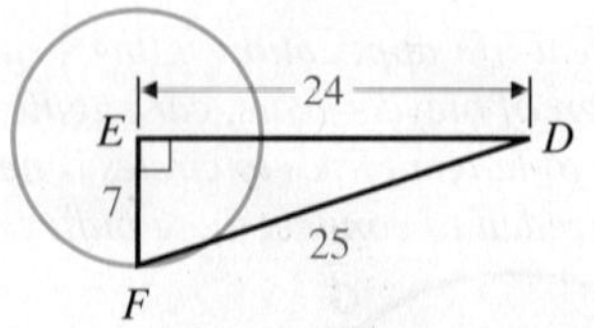

80. A classmate insists that $\overline{EF}$ is a tangent to $\odot E$ since it only touches the circle once. Explain the error in your classmate's thinking.

Proof

81. Prove Theorem 12.1-3.

Given: $\overline{BA}$ and $\overline{BC}$ are tangent to $\odot O$ at A and C, respectively.

Prove: $\overline{BA} \cong \overline{BC}$

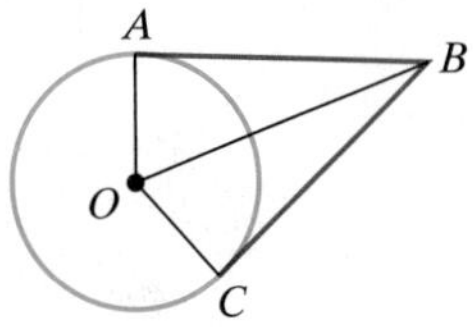

82. Write an indirect proof of Theorem 12.1-2.

Given: $\overline{AB} \perp \overline{OP}$ at P.

Prove: $\overline{AB}$ is tangent to $\odot O$.

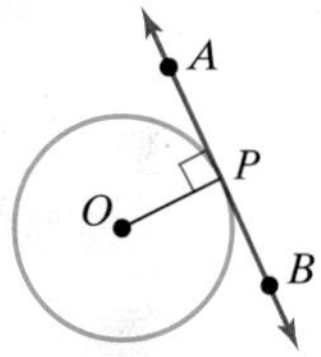

83. Given: $\overline{BC}$ is tangent to $\odot A$ at D.

$\overline{DB} \cong \overline{DC}$

Prove: $\overline{AB} \cong \overline{AC}$

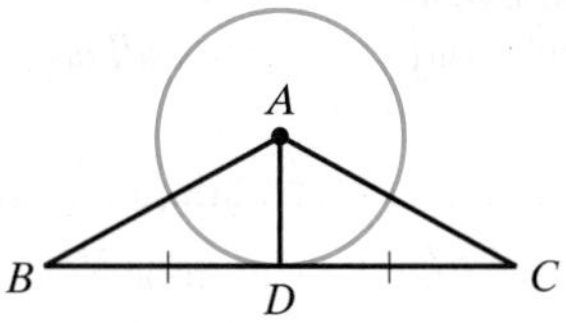

84. Given: $\odot A$ and $\odot B$ with common tangents $\overline{DF}$ and $\overline{CE}$

Prove: $\Delta GDC \sim \Delta GFE$

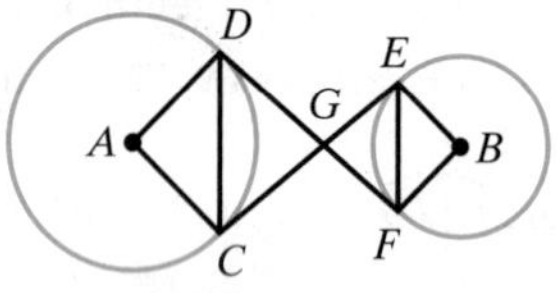

A belt fits snugly around the two circular pulleys. $\overline{CE}$ is an auxiliary line from E to $\overline{BD}$. $\overline{CE} \parallel \overline{BA}$.

85. a. What type of quadrilateral is $ABCE$? Explain.

b. What is the length of $\overline{CE}$?

c. What is the length of $\overline{BC}$?

d. What is the distance between the centers of the pulleys to the nearest tenth?

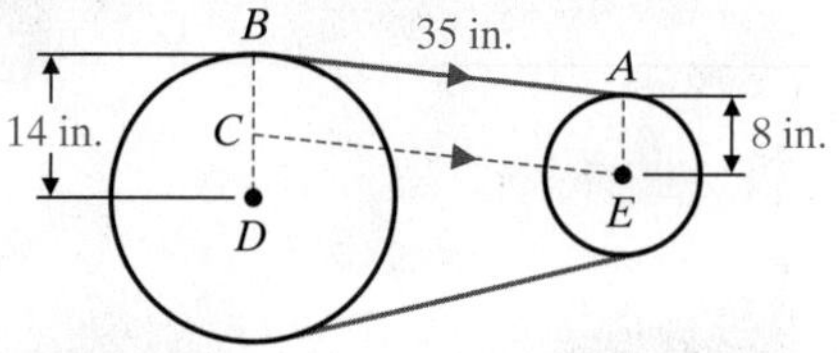

86. Construction Draw a circle. Label the center T. Locate a point on the circle and label it R. Construct a tangent to $\odot T$ at R.

REVIEW AND PREVIEW

Two cubes have sides of 6 in. and 8 in. Find each ratio. Use the order of smaller cube to larger cube. See Sections 10.4 and 11.7.

87. scale factor

88. ratio of the areas of one side

89. ratio of surface areas

90. ratio of volumes

Find the measure of $\angle x$. Round answers to the nearest tenth. See Section 9.3.

91.

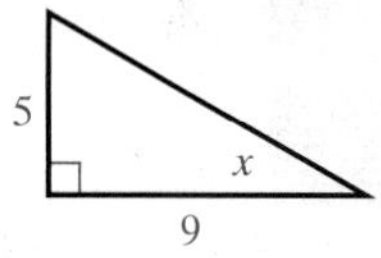

92.

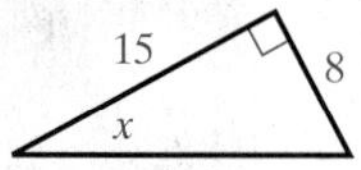

Find the value of each variable. Leave your answer in simplest radical form. See Section 9.2.

93.

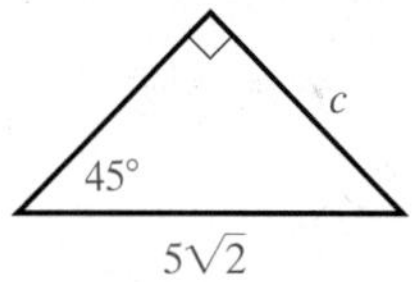

94.

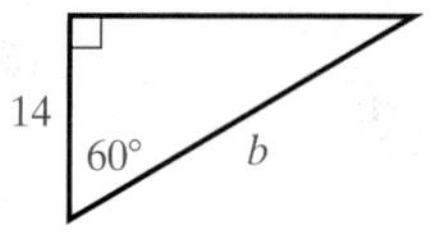

12.2 CHORDS AND ARCS

OBJECTIVES

1. Use Congruent Chords, Arcs, and Central Angles.
2. Use Perpendicular Bisectors to Chords.

VOCABULARY

- chord

OBJECTIVE 1 ▶ Using Congruent Chords, Arcs, and Central Angles. A **chord** is a segment whose endpoints are on a circle. The diagram shows the chord $\overline{PQ}$ and its related arc, $\overset{\frown}{PQ}$.

We can use information about congruent parts of a circle (or congruent circles) to find information about other parts of the circle (or circles).

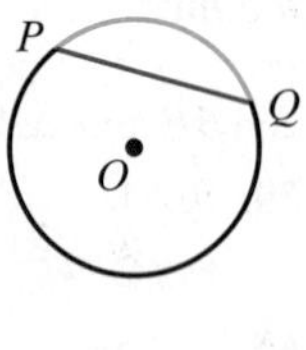

The following theorems and their converses confirm that if we know that chords, arcs, or central angles in a circle are congruent, then we know the other two parts are congruent.

> **Helpful Hint**
> In a circle, or in congruent circles, if we know one of the below, then we know all of the below.
> - Congruent central angles
> - Congruent arcs
> - Congruent chords

Theorem 12.2-1 and Its Converse (12.2-2) Congruent Central Angles and Arcs

Theorem 12.2-1

Within a circle or in congruent circles, congruent central angles have congruent arcs.

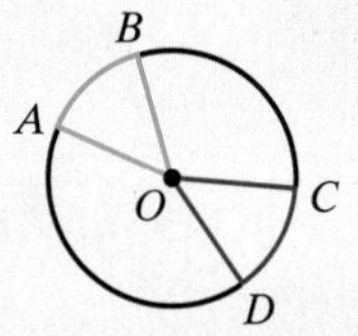

Converse (Theorem 12.2-2)

Within a circle or in congruent circles, congruent arcs have congruent central angles.

If $\angle AOB \cong \angle COD$, then $\overset{\frown}{AB} \cong \overset{\frown}{CD}$.

If $\overset{\frown}{AB} \cong \overset{\frown}{CD}$, then $\angle AOB \cong \angle COD$.

Theorem 12.2-3 and Its Converse (12.2-4) Congruent Central Angles and Chords

Theorem 12.2-3

Within a circle or in congruent circles, congruent central angles have congruent chords.

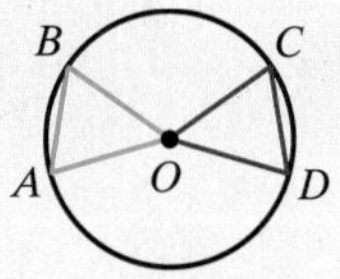

Converse (Theorem 12.2-4)

Within a circle or in congruent circles, congruent chords have congruent central angles.

If $\angle AOB \cong \angle COD$, then $\overline{AB} \cong \overline{CD}$.

If $\overline{AB} \cong \overline{CD}$, then $\angle AOB \cong \angle COD$.

Theorem 12.2-5 and Its Converse (12.2-6) Congruent Chords and Arcs

Theorem 12.2-5

Within a circle or in congruent circles, congruent chords have congruent arcs.

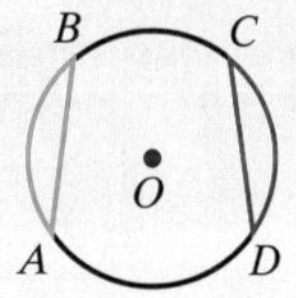

Converse (Theorem 12.2-6)

Within a circle or in congruent circles, congruent arcs have congruent chords.

If $\overline{AB} \cong \overline{CD}$, then $\overset{\frown}{AB} \cong \overset{\frown}{CD}$.

If $\overset{\frown}{AB} \cong \overset{\frown}{CD}$, then $\overline{AB} \cong \overline{CD}$.

We prove Theorem 12.2-1 and its converse in Exercises 35 and 43.
We prove Theorem 12.2-3 and its converse in Exercises 36 and 44.
We prove Theorem 12.2-5 and its converse in Exercises 37 and 45.

> **Helpful Hint**
> Remember: Two circles may have central angles with congruent chords, but the central angles will not be congruent unless the circles are congruent.

EXAMPLE 1 **Using Congruent Chords**

In the diagram, $\odot O \cong \odot P$. Given that $\overline{BC} \cong \overline{DF}$, what can you conclude?

Solution $\angle O \cong \angle P$ because, within congruent circles, congruent chords have congruent central angles (Converse of Theorem 12.2-3). $\overset{\frown}{BC} \cong \overset{\frown}{DF}$ because, within congruent circles, congruent chords have congruent arcs (Theorem 12.2-5).

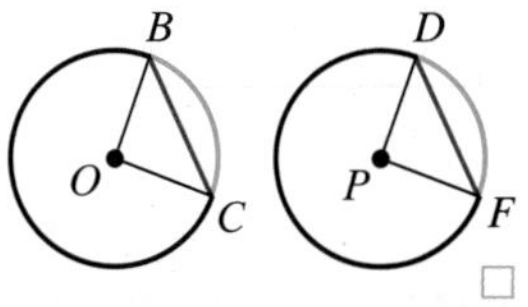

□

PRACTICE

1 Use the diagram in Example 1. Suppose you are given $\odot O \cong \odot P$ and $\angle OBC \cong \angle PDF$. How can you show $\angle O \cong \angle P$? From this, what else can you conclude?

Next, we study congruent chords and their distances from the center of their circles.

Theorem 12.2-7 and Its Converse (12.2-8) Chords Equidistant from the Center Are Congruent

Theorem 12.2–7

Within a circle or in congruent circles, chords equidistant from the center or centers are congruent.

Converse (Theorem 12.2-8)

Within a circle or in congruent circles, congruent chords are equidistant from the center (or centers).

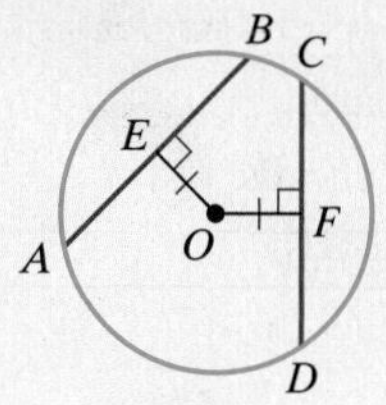

If $OE = OF$, then $\overline{AB} \cong \overline{CD}$.

If $\overline{AB} \cong \overline{CD}$, then $OE = OF$.

We prove the converse, Theorem 12.2-8, in Exercise 46.

Proof of Theorem 12.2-7

Given: $\odot O$, $\overline{OE} \cong \overline{OF}$, $\overline{OE} \perp \overline{AB}$, $\overline{OF} \perp \overline{CD}$

Prove: $\overline{AB} \cong \overline{CD}$

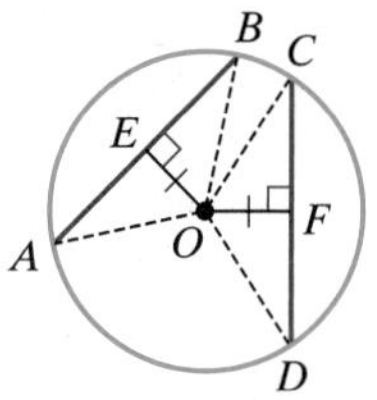

Statements	Reasons
1. $\overline{OA} \cong \overline{OB} \cong \overline{OC} \cong \overline{OD}$	1. Radii of a circle are congruent.
2. $\overline{OE} \cong \overline{OF}$, $\overline{OE} \perp \overline{AB}$, $\overline{OF} \perp \overline{CD}$	2. Given
3. $\angle AEO$ and $\angle CFO$ are right angles.	3. Definition of perpendicular segments
4. $\Delta AEO \cong \Delta CFO$	4. H-L Theorem
5. $\angle A \cong \angle C$	5. Corresponding parts of $\cong$ Δ are $\cong$ (cpoctac.)
6. $\angle B \cong \angle A$, $\angle C \cong \angle D$	6. Base ⦣ of an isosceles Δ are $\cong$.
7. $\angle B \cong \angle D$	7. Transitive Property of Congruence (Steps 5, 6)
8. $\angle AOB \cong \angle COD$	8. If two ⦣ of a Δ are $\cong$ to two ⦣ of another Δ, then the third ⦣ are $\cong$. (Third Angles Theorem)
9. $\overline{AB} \cong \overline{CD}$	9. $\cong$ central angles have $\cong$ chords.

EXAMPLE 2 Finding the Length of a Chord

What is the length of $\overline{RS}$ in $\odot O$?

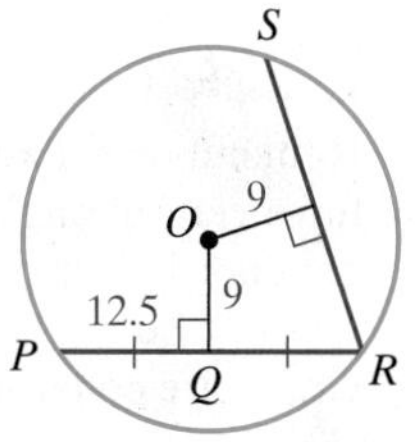

Solution Recall that chords equidistant from the center of a circle are congruent. Thus, $PR = RS$. Let's find PR, and we'll have RS.

$PQ = QR = 12.5$ Given in the diagram

$PQ + QR = PR$ Segment Addition Postulate

$12.5 + 12.5 = PR$ Substitute.

$25 = PR$ Add.

$RS = PR$ Chords equidistant from the center of a circle are congruent.

$RS = 25$ Substitute.

Thus, the length of $\overline{RS}$ is 25 units. □

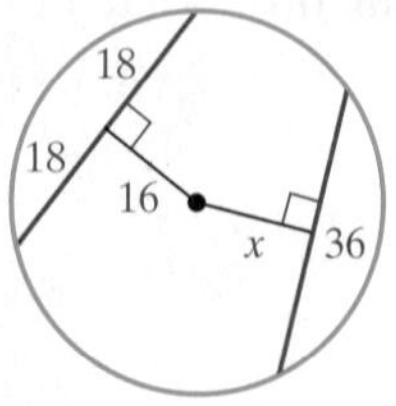

PRACTICE 2 Study the figure in the margin. What is the value of x? Justify your answer.

OBJECTIVE 2 ▶ Using Perpendicular Bisectors to Chords of Circles. The Converse of the Perpendicular Bisector Theorem from Section 5.1 has special applications to a circle and its diameters, chords, and arcs.

THEOREMS ABOUT CHORDS OF CIRCLES

Theorem 12.2-9

Theorem

In a circle, if a diameter is perpendicular to a chord, then it bisects the chord and its arc.

If . . .

$\overline{AB}$ is a diameter and $\overline{AB} \perp \overline{CD}$

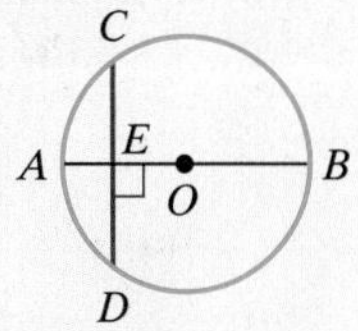

Then . . .

$\overline{CE} \cong \overline{ED}$ and $\overset{\frown}{CA} \cong \overset{\frown}{AD}$

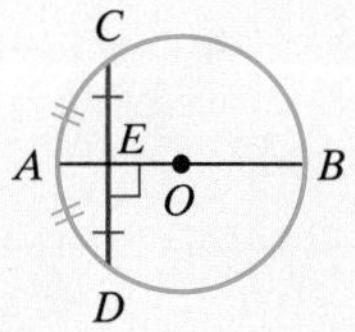

Theorem 12.2-10

Theorem

In a circle, if a diameter bisects a chord (that is not a diameter), then it is perpendicular to the chord.

If . . .

$\overline{AB}$ is a diameter and $\overline{CE} \cong \overline{ED}$

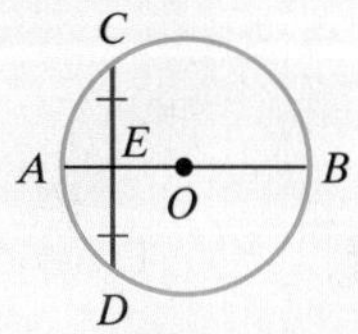

Then . . .

$\overline{AB} \perp \overline{CD}$

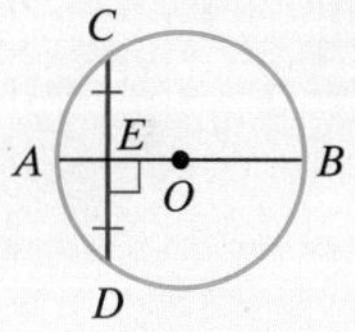

Theorem 12.2-11

Theorem

In a circle, the perpendicular bisector of a chord contains the center of the circle.

If . . .

$\overline{AB}$ is the perpendicular bisector of chord $\overline{CD}$

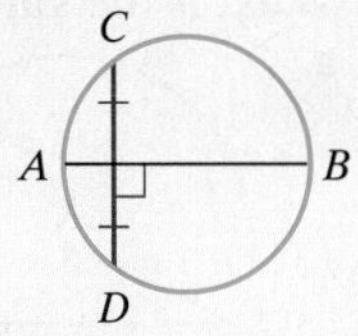

Then . . .

$\overline{AB}$ contains the center of $\odot O$

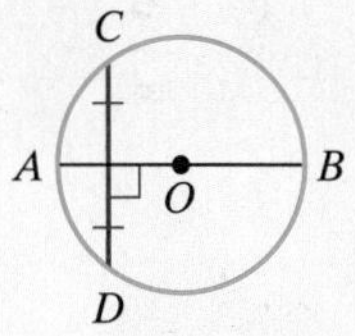

We prove Theorem 12.2-9 in Exercise 38. We prove Theorem 12.2-10 next.
We prove Theorem 12.2-11 in Exercise 41.

Proof of Theorem 12.2-10

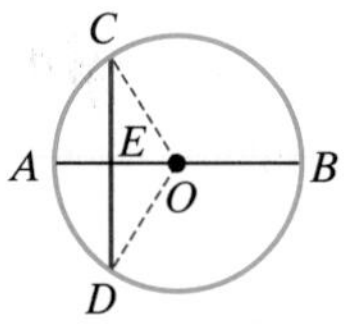

Given: $\odot O$ with diameter $\overline{AB}$ bisecting $\overline{CD}$ at E

Prove: $\overline{AB} \perp \overline{CD}$

Proof: $OC = OD$ because the radii of a circle are congruent. $CE = ED$ by the definition of bisect. Thus, O and E are both equidistant from C and D. By the Converse of the Perpendicular Bisector Theorem, both O and E are on the perpendicular bisector of $\overline{CD}$. Two points determine one line or segment, so $\overline{OE}$ is the perpendicular bisector of $\overline{CD}$. Since $\overline{OE}$ is part of $\overline{AB}$, $\overline{AB} \perp \overline{CD}$.

EXAMPLE 3 Finding the Center of a Given Circle

Given a circle, describe how to find the center of the circle and the length of its raduis.

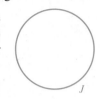

Solution Without the center of circle J given, we cannot find the length of its radius. However, we can use an earlier theorem (12.2-11) to find the center. This theorem says that the perpendicular bisector of a chord contains the center of the circle.

Let's use this theorem to find the center of a circle. Follow the steps below.

STEP 1. Draw two chords (not parallel to each other).

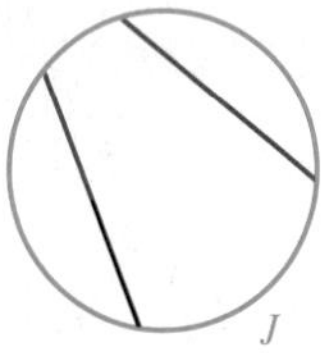

STEP 2. Construct the perpendicular bisector of each chord.

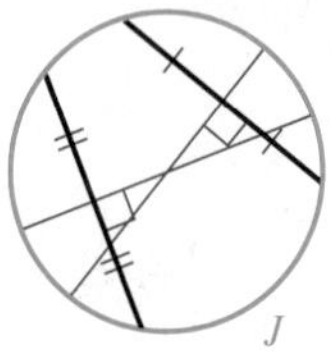

STEP 3. The center is located at the intersection of the perpendicular bisectors.

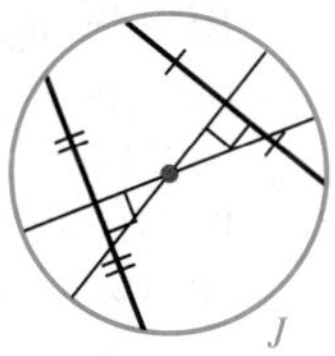

From there, the radius is the length from the center to any point of the circle. □

PRACTICE 3 Trace a coin. Find its center and radius.

EXAMPLE 4 Finding Measures in a Circle

What is the value of each variable to the nearest tenth?

a.

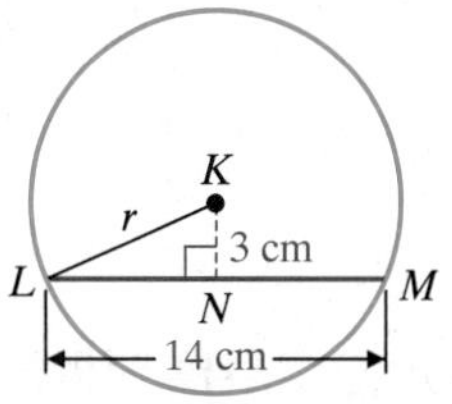

b.

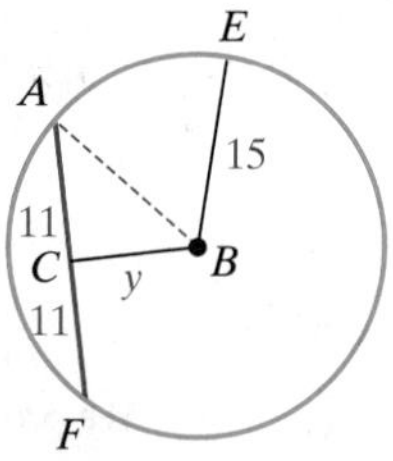

Solution Study each figure. Let's see if we can locate a right triangle, and then use the Pythagorean Theorem.

a. $LN = \frac{1}{2}(14) = 7$ A diameter ⊥ to a chord bisects the chord.

$r^2 = 3^2 + 7^2$ Use the Pythagorean Theorem.

$r^2 = 58$ $3^2 = 9, 7^2 = 49$, and their sum is 58.

$r \approx 7.6$ Find the positive square root of each side, rounded to the nearest tenth.

b. $\overline{BC} \perp \overline{AF}$ A diameter that bisects a chord that is not a diameter is ⊥ to the chord.

$BA = BE = 15$ Draw an auxiliary $\overline{BA}$. The auxiliary $\overline{BA} \cong \overline{BE}$ because they are radii of the same circle.

$y^2 + 11^2 = 15^2$ Use the Pythagorean Theorem.

$y^2 = 104$ Solve for y^2.

$y \approx 10.2$ Find the positive square root of each side, rounded to the nearest tenth. □

PRACTICE 4 Find the value of r to the nearest tenth.

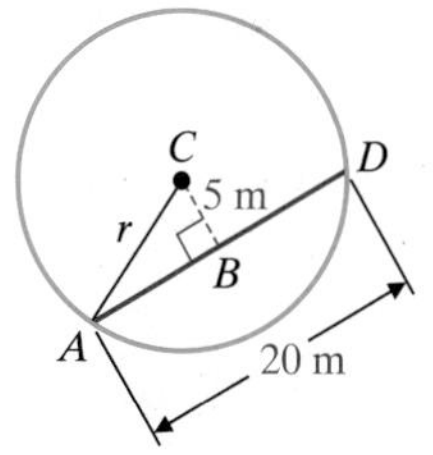

VOCABULARY & READINESS CHECK

Fill in the Blank.

1. A segment whose endpoints are on a circle is called a(n) ____________.
2. A segment whose endpoints are on a circle and contains the center of a circle is called a(n) ____________.

Answer the following true or false.

3. All diameters are chords.
4. All chords are diameters.
5. All congruent chords have congruent central angles.
6. All congruent central angles in the same circle have congruent chords.

12.2 EXERCISE SET MyMathLab®

In Exercises 1 and 2, the circles are congruent. What can you conclude? See Example 1.

1.

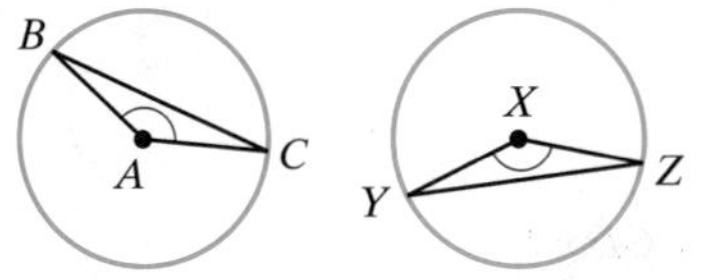

2.

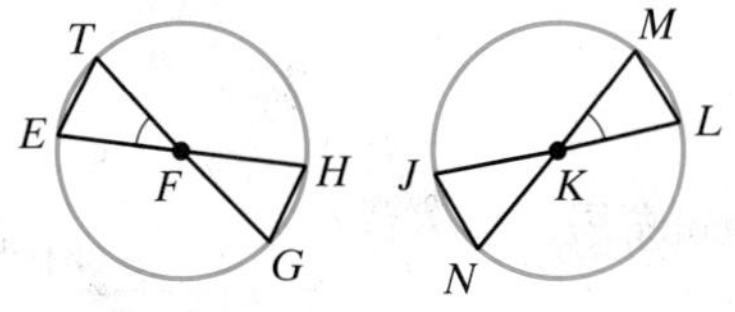

Find the value of x. See Example 2.

3.

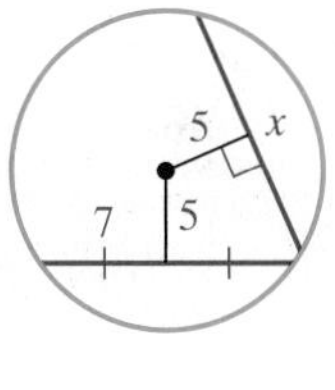

4.

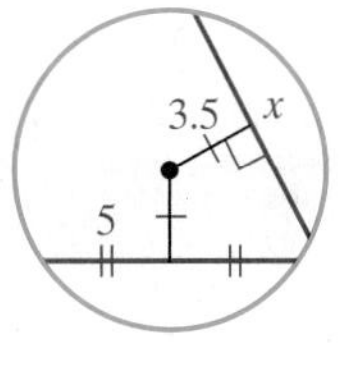

5.

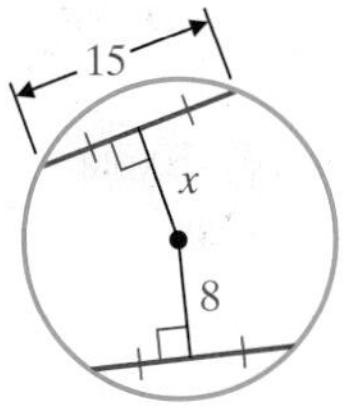

6.

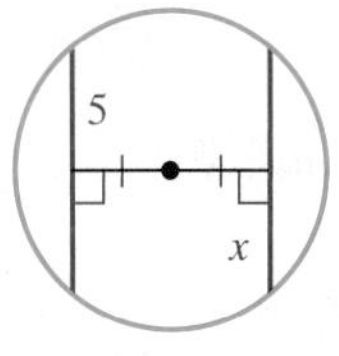

MIXED PRACTICE

In $\odot O$, $m\overset{\frown}{CD} = 50°$ *and* $\overline{CA} \cong \overline{BD}$. *Also, the center of the circle, point O, is the intersection of* $\overline{CB}$ and $\overline{AD}$.

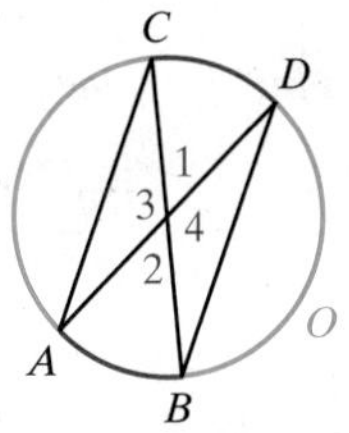

7. Find $m\angle 1$.

8. Find $m\angle 2$.

9. What is $m\overset{\frown}{AB}$?

10. What is true of $\overset{\frown}{CA}$ and $\overset{\frown}{BD}$? Why?

11. Find $m\angle 3$.

12. Find $m\angle 4$.

13. Find $m\overset{\frown}{CA}$.

14. Since $CA = BD$, what do you know about the distance of $\overline{CA}$ and $\overline{BD}$ from the center of $\odot O$?

Solve. See Examples 3 and 4.

15. In the diagram below, $\overline{GH}$ and $\overline{KM}$ are perpendicular bisectors of the chords they intersect. What can you conclude about the center of the circle? Justify your answer.

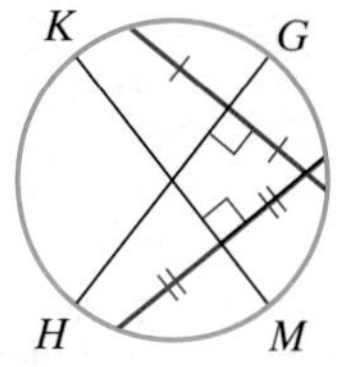

16. In $\odot O$, $\overline{AB}$ is a diameter of the circle and $\overline{AB} \perp \overline{CD}$. What conclusions can you make?

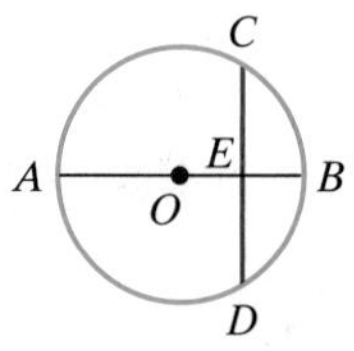

Find the value of x to the nearest tenth.

17.

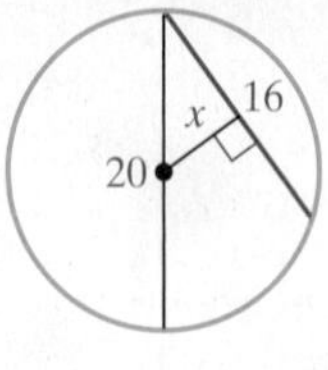

18.

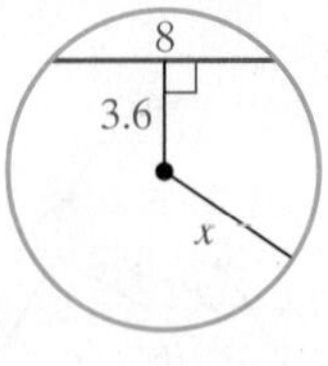

19.

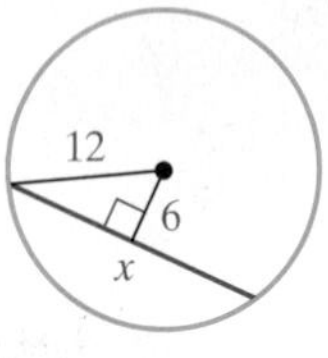

20.

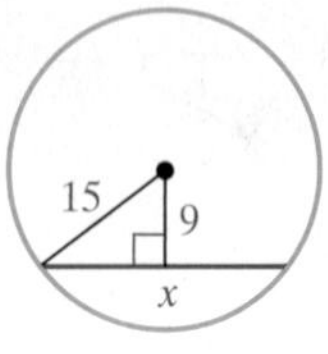

⊙A and ⊙B are congruent. $\overline{CD}$ is a chord of both circles. Use this figure for Exercises 21–24.

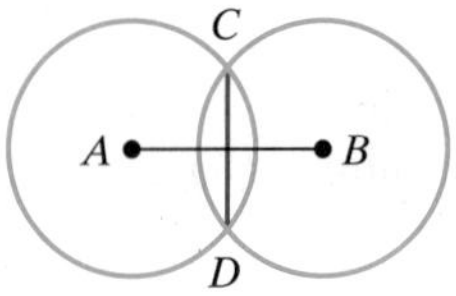

21. If $AB = 8$ in. and $CD = 6$ in., how long is a radius?

22. If $AB = 24$ cm and a radius $= 13$ cm, how long is $\overline{CD}$?

23. If a radius $= 13$ ft and $CD = 24$ ft, how long is $\overline{AB}$?

24. If a radius $= 17$ m and $CD = 16$ m, how long is $\overline{AB}$?

For Exercises 25–28, find $m\widehat{AB}$.

25.

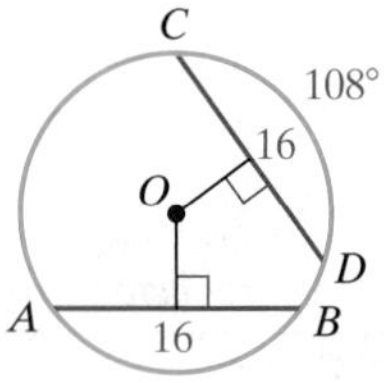

26.

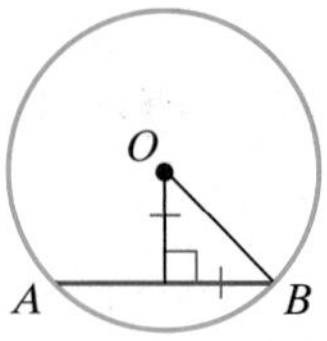

Find the value of x in ⊙O.

27.

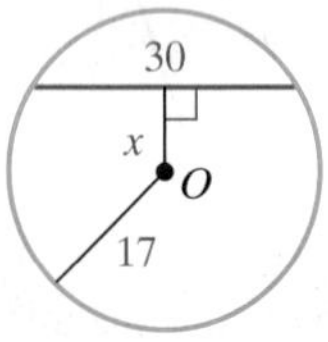

28.

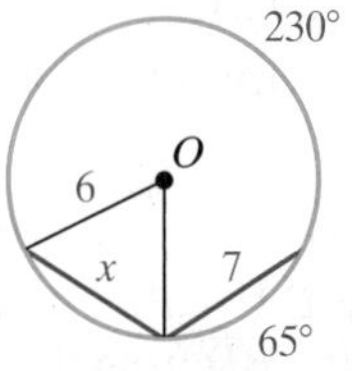

29. In the diagram below, the endpoints of the chord are the points where the line $x = 2$ intersects the circle $x^2 + y^2 = 25$. What is the length of the chord? Round your answer to the nearest tenth.

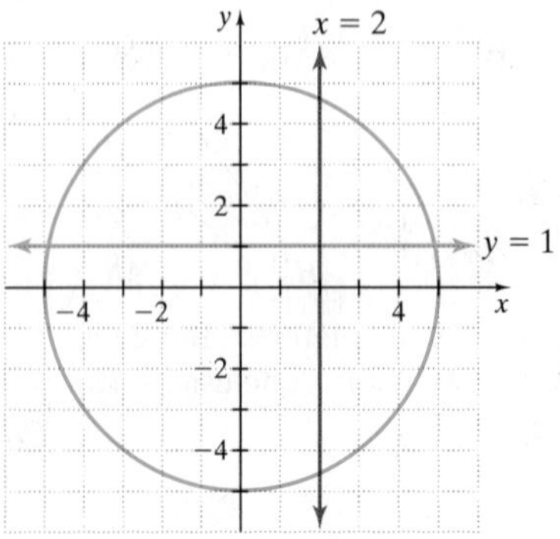

30. Use the same directions as for Exercise **29**, except use the chord found on the line $y = 1$.

31. In the figure below, sphere O with radius 13 cm is intersected by a plane 5 cm from center O. Find the radius of the cross section $\odot A$.

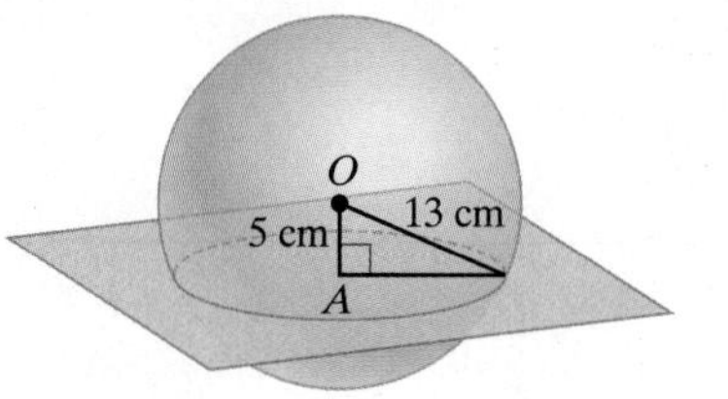

32. A plane intersects a sphere that has radius 10 in., forming the cross section $\odot B$ with radius 8 in. How far is the plane from the center of the sphere?

CONCEPT EXTENSIONS

33. Construction Use a circular object such as a can or a saucer to draw a circle. Construct the center of the circle.

34. Construction Use Theorem 12.2-3 to construct a regular octagon.

PROOF

35. Prove Theorem 12.2-1.

Given: $\odot O$ with $\angle AOB \cong \angle COD$

Prove: $\widehat{AB} \cong \widehat{CD}$

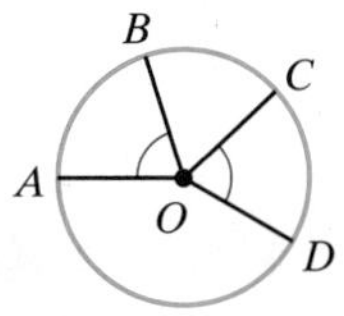

36. Prove Theorem 12.2-3.

Given: $\odot O$ with $\angle AOB \cong \angle COD$

Prove: $\overline{AB} \cong \overline{CD}$

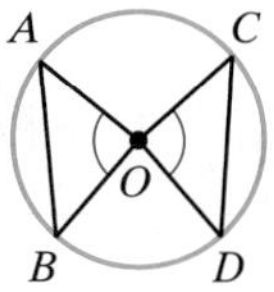

37. Prove Theorem 12.2-5.

Given: $\odot O$ with $\overline{AB} \cong \overline{CD}$

Prove: $\widehat{AB} \cong \widehat{CD}$

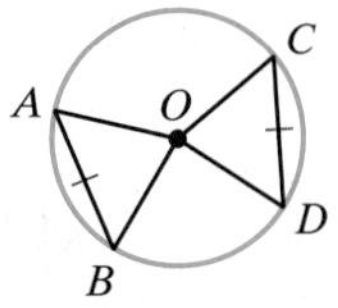

38. Prove Theorem 12.2-9.

Given: $\odot O$ with diameter $\overline{ED} \perp \overline{AB}$ at C

Prove: $\overline{AC} \cong \overline{BC}$, $\widehat{AD} \cong \widehat{BD}$

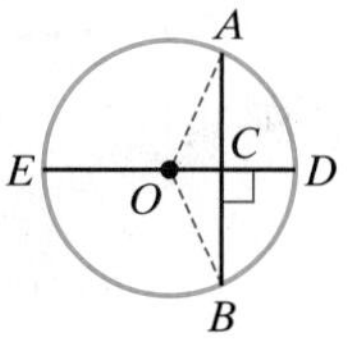

39. Find the Error What is the error in the diagram?

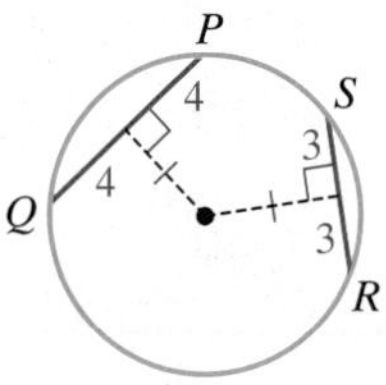

40. Is a radius a chord? Is a diameter a chord? Explain your answers.

Proof

41. Prove Theorem 12.2-11.

Given: $\odot X$ with ℓ the $\perp$ bisector of $\overline{WY}$.

Prove: ℓ contains the center of $\odot X$.

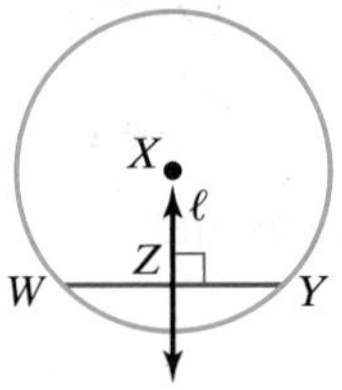

42. Given: $\odot A$ with $\overline{CE} \perp \overline{BD}$

Prove: $\widehat{BC} \cong \widehat{DC}$

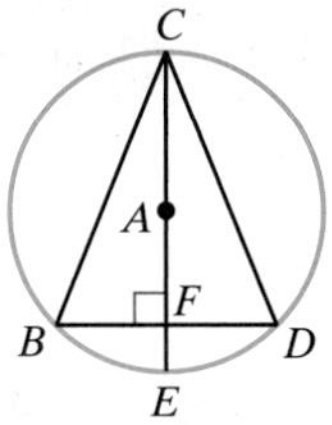

Proof

43. Prove Theorem 12.2-2, the Converse of Theorem 12.2-1: Within a circle or in congruent circles, congruent arcs have congruent central angles.

44. Prove Theorem 12.2-4, the Converse of Theorem 12.2-3: Within a circle or in congruent circles, congruent chords have congruent central angles.

45. Prove Theorem 12.2-6, the Converse of Theorem 12.2-5: Within a circle or in congruent circles, congruent arcs have congruent chords.

46. Prove Theorem 12.2-8, the Converse of Theorem 12.2-7: Within a circle or congruent circles, congruent chords are equidistant from the center (or centers).

REVIEW AND PREVIEW

Assume that the lines that appear to be tangent are tangent. O is the center of each circle. Find the value of x to the nearest tenth. See Section 12.1.

47.

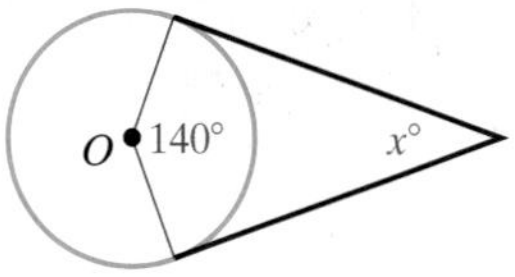

48.

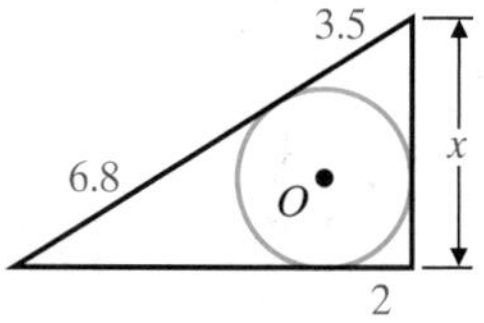

Identify the following in $\odot P$ at the right. See Section 12.1.

49. a semicircle

50. a minor arc that measures 86°

51. a minor arc that measures 145°

52. a major arc

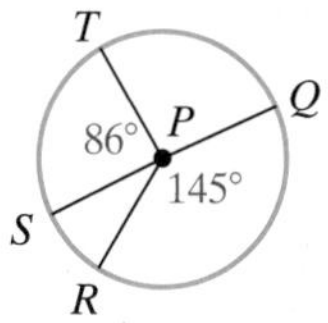

Find the measure of each arc in $\odot P$. See Section 10.5.

53. $\widehat{ST}$ **54.** $\widehat{STQ}$ **55.** $\widehat{RT}$ **56.** $\widehat{RTQ}$

12.3 INSCRIBED ANGLES

OBJECTIVES

1. Find the Measure of an Inscribed Angle.
2. Find the Measure of an Angle Formed by a Tangent and a Chord.

VOCABULARY

- inscribed angle
- intercepted arc

OBJECTIVE 1 ▶ Finding the Measure of an Inscribed Angle. An angle whose vertex is on the circle and whose sides are chords of the circle is an **inscribed angle**. An arc with endpoints on the sides of an inscribed angle and whose other points are in the interior of the angle is an **intercepted arc**. In the diagram, inscribed $\angle C$ intercepts $\widehat{AB}$.

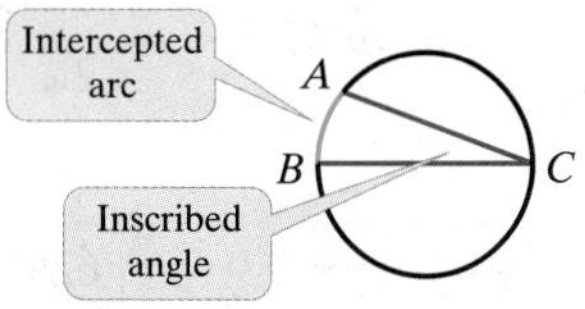

Angles formed by intersecting lines have a special relationship to the arcs the intersecting lines intercept. In this section, we study arcs formed by inscribed angles.

Theorem 12.3-1 Inscribed Angle Theorem

The measure of an inscribed angle is half the measure of its intercepted arc.

$$m\angle B = \frac{1}{2}m\widehat{AC}$$

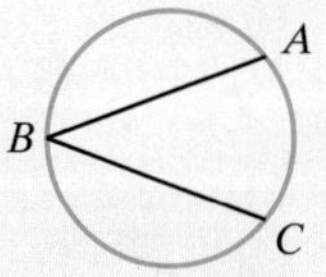

To prove Theorem 12.3-1, there are three cases to consider.

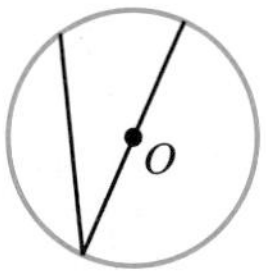

I: The center is on a side of the angle.

II. The center is inside the angle.

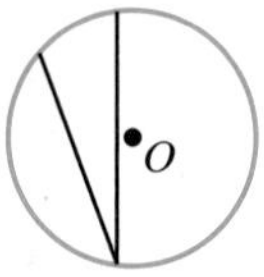

III: The center is outside the angle.

Below is a proof of Case I. We prove Case II and Case III in Exercises 45 and 46.

Proof of Theorem 12.3-1, Case I

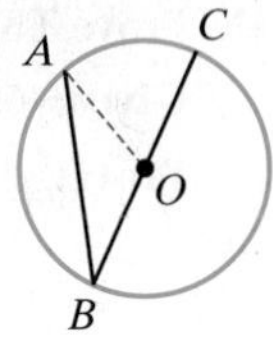

Given: $\odot O$ with inscribed $\angle B$ and diameter $\overline{BC}$

Prove: $m\angle B = \frac{1}{2}m\widehat{AC}$

Construction: Draw radius $\overline{OA}$.

Proof:

Statements	Reasons
1. ΔAOB is isosceles.	**1.** $OA = OB$ since both are radii.
2. $m\angle A = m\angle B$	**2.** Base angles of an isosceles Δ are $\cong$.
3. $m\angle AOC = m\angle A + m\angle B$	**3.** Triangle Exterior Angle Theorem
4. $m\widehat{AC} = m\angle AOC$	**4.** Definition of measure of an arc
5. $m\widehat{AC} = m\angle A + m\angle B$	**5.** Substitute. (Steps 3, 4)
6. $m\widehat{AC} = 2m\angle B$	**6.** Substitute and simplify. (Steps 2, 5)
7. $\frac{1}{2}m\widehat{AC} = m\angle B$	**7.** Divide each side by 2.

EXAMPLE 1 **Finding Measures of Inscribed Angles**

a. Find $m\angle ABD$.

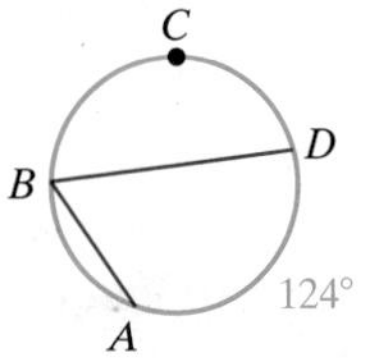

b. Find $m\widehat{EFG}$.

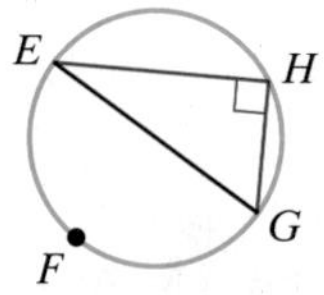

c. Find $m\angle T$ and $m\angle R$.

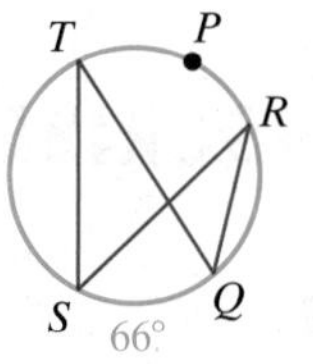

Solution We use the Inscribed Angle Theorem for parts **a**, **b**, and **c**.

a.
$$m\angle ABD = \frac{1}{2}m\widehat{AD}$$
$$= \frac{1}{2}(124°)$$
$$= 62°$$

b. We are given that $m\angle H = 90°$. Thus,
$$m\angle H = \frac{1}{2}m\widehat{EFG}$$
$$90° = \frac{1}{2}m\widehat{EFG}$$
$$180° = m\widehat{EFG}$$ Multiply both sides by 2.

c. $\angle T$ and $\angle R$ have the same intercepted arc. Thus
$$m\angle T = m\angle R = \frac{1}{2}m\widehat{SQ}$$
$$= \frac{1}{2}(66°)$$
$$= 33°$$

□

PRACTICE 1

a. Find $m\angle E$.

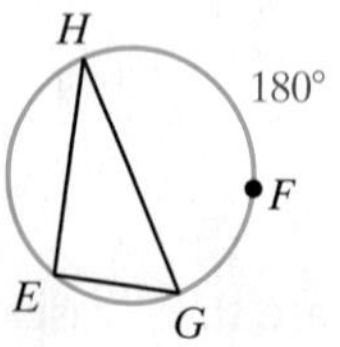

Given: $m\widehat{GFH} = 180°$

b. Find $m\widehat{WZ}$.

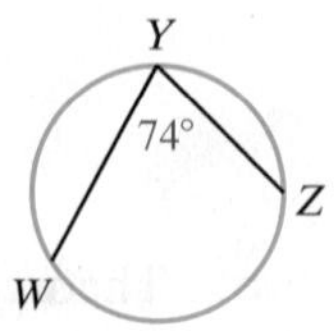

c. Find $m\angle A$ and $m\angle B$.

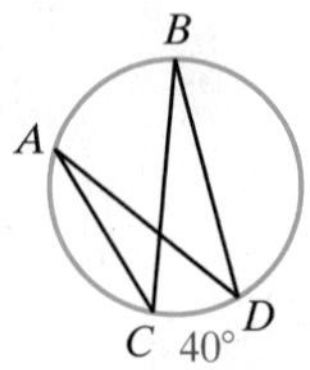

EXAMPLE 2 Using the Inscribed Angle Theorem

What are the values of a and b?

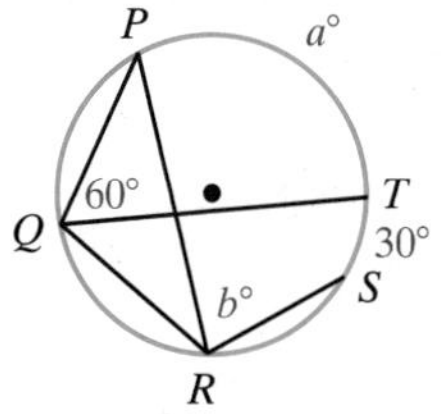

Solution Study the figure. Since the measure of inscribed $\angle PQT$ is 60°, we can find the value of a.

$$m\angle PQT = \frac{1}{2}m\widehat{PT} \quad \text{Inscribed Angle Theorem}$$

$$60° = \frac{1}{2}a° \quad \text{Substitute.}$$

$$120° = a° \quad \text{Multiply each side by 2.}$$

Now that we know the value of a, we can use this to find the value of b. Let's start with inscribed $\angle PRS$.

$$m\angle PRS = \frac{1}{2}m\widehat{PS} \quad \text{Inscribed Angle Theorem}$$

$$m\angle PRS = \frac{1}{2}(m\widehat{PT} + m\widehat{TS}) \quad \text{Arc Addition Postulate}$$

$$b° = \frac{1}{2}(120° + 30°) \quad \text{Substitute.}$$

$$b° = 75° \quad \text{Simplify.}$$

Thus, $b° = 75°$ or $b = 75$ and $a° = 120°$ or $a = 120$. □

PRACTICE 2

a. In $\odot O$, what is $m\angle A$?

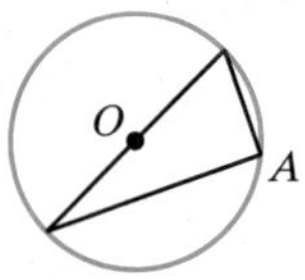

b. What are $m\angle A$, $m\angle B$, $m\angle C$, and $m\angle D$?

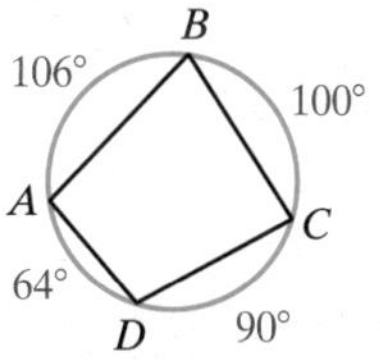

c. What do you notice about the sums of the measures of the opposite angles in the quadrilateral in part **b**?

We will use three corollaries to the Inscribed Angle Theorem to find measures of angles in circles.

Corollaries to Theorem 12.3-1: The Inscribed Angle Theorem

Corollary 1 (12.3-2)
Two inscribed angles that intercept the same arc are congruent.

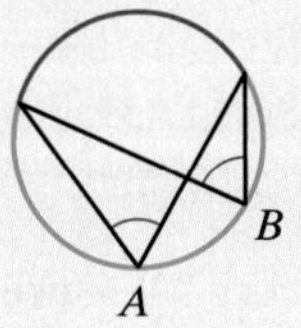

Corollary 2 (12.3-3)
An angle inscribed in a semicircle is a right angle.

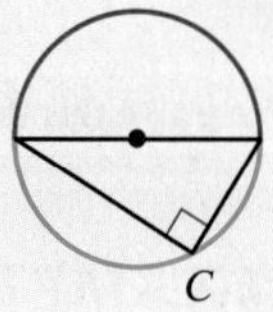

Corollary 3 (12.3-4)
The opposite angles of a quadrilateral inscribed in a circle are supplementary.

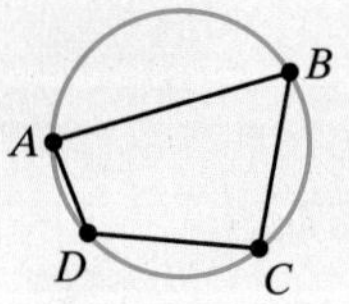

We prove these corollaries in Exercise Set 12.3, Exercises 51–53.

EXAMPLE 3 Using Corollaries to Find Angle Measures

What is the measure of each numbered angle?

a.

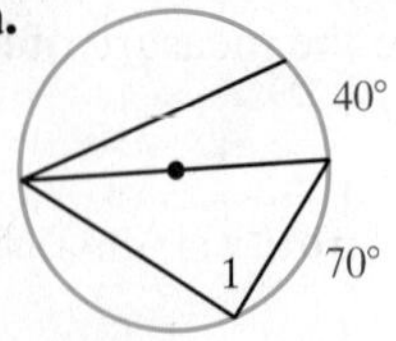

b.

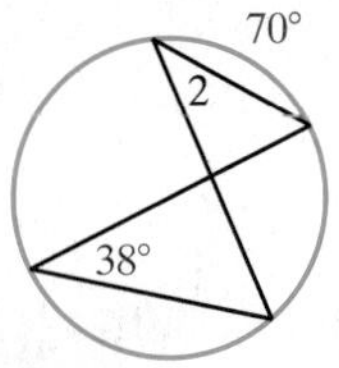

Solution Use only the information given in each diagram.

a. $\angle 1$ is inscribed in a semicircle. By Corollary 2, $\angle 1$ is a right angle, so $m\angle 1 = 90°$.

b. $\angle 2$ and the 38° angle intercept the same arc. By Corollary 1, the angles are congruent, so $m\angle 2 = 38°$. □

PRACTICE

3 In the diagram at the right, what is the measure of each numbered angle?

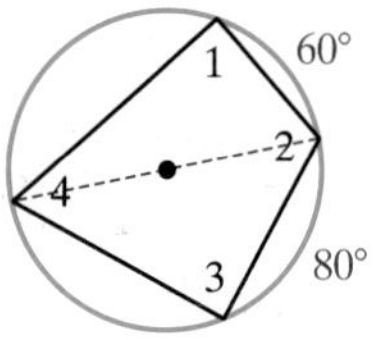

OBJECTIVE 2 ▶ Finding the Measure of an Angle Formed by a Tangent and a Chord. The following diagram shows point A moving along the circle until a tangent is formed. From the Inscribed Angle Theorem, you know that in the first three diagrams, $m\angle A$ is $\frac{1}{2}m\widehat{BC}$. As the last diagram suggests, this is also true when A and C coincide.

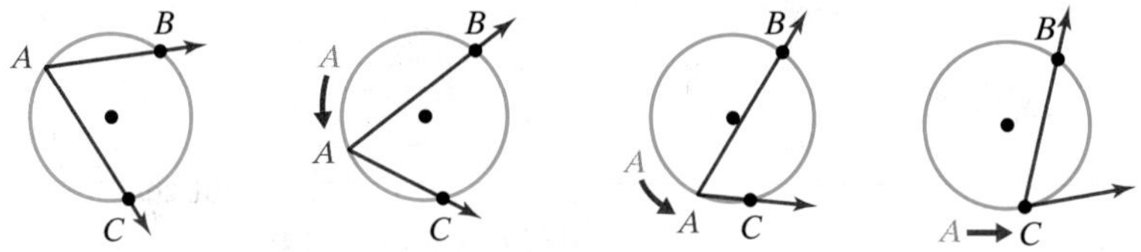

Theorem 12.3-5

The measure of an angle formed by a tangent and a chord is half the measure of the intercepted arc.

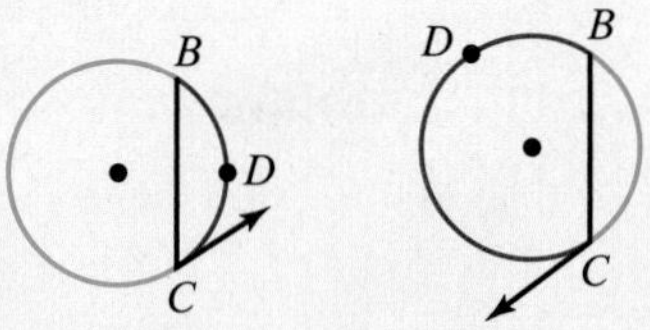

$$m\angle C = \frac{1}{2}m\widehat{BDC}$$

We prove this theorem in Exercise Set 12.3, Exercise 54.

EXAMPLE 4 Using Arc Measure

In the diagram, $\overleftrightarrow{SR}$ is a tangent to the circle at Q. If $m\widehat{PMQ} = 212°$, what is $m\angle 1$?

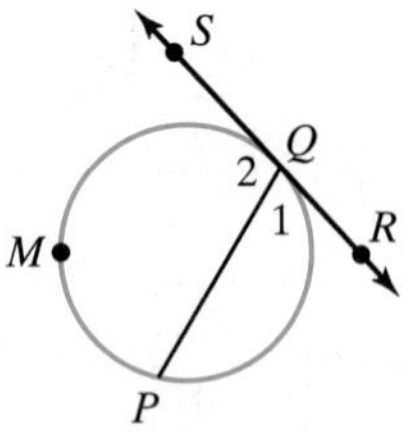

Solution First, let's find $m\angle 2$ since we are given the measure of its intercepted arc, $\widehat{PMQ}$.

$m\angle 2 = \frac{1}{2} m\widehat{PMQ}$ The measure of an $\angle$ formed by a tangent and a chord is $\frac{1}{2}$ the measure of the intercepted arc.

$m\angle 2 = \frac{1}{2}(212°)$ Substitute.

$m\angle 2 = 106°$ Simplify.

Next, recall that angles that form a linear pair are supplementary. Thus, $\angle 1$ and $\angle 2$ are supplementary.

$m\angle 1 + m\angle 2 = 180°$ Linear pair angels are supplementary.

$m\angle 1 + 106° = 180°$ Substitute.

$m\angle 1 = 74°$ Simplify.

Thus, $m\angle 1 = 74°$. □

PRACTICE
4 In the diagram at the right, $\overline{KJ}$ is tangent to $\odot O$. What are the values of x and y?
(*Hint:* The inscribed angle $(\angle Q)$ and the angle formed by a tangent and chord $(\angle KJL)$ intercept the same arc.)

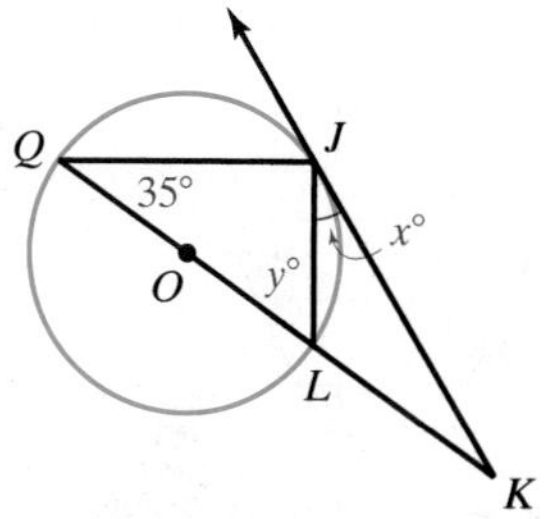

VOCABULARY & READINESS CHECK

Fill in the Blank *Use the diagrams at the right as indicated. Answers to Exercises 1 and 2 are in degrees. Answers to Exercises 3–8 are in arc measures.*

1. $m\angle D =$ ____________
2. $m\angle E =$ ____________
3. $m\angle 1 = \frac{1}{2} \cdot$ ____________
4. $m\angle 2 = \frac{1}{2} \cdot$ ____________
5. $m\angle 3 = \frac{1}{2} \cdot$ ____________
6. $m\angle 4 = \frac{1}{2} \cdot$ ____________
7. $m\angle 5 = \frac{1}{2} \cdot$ ____________
8. $m\angle 3 + m\angle 4 =$ ____________

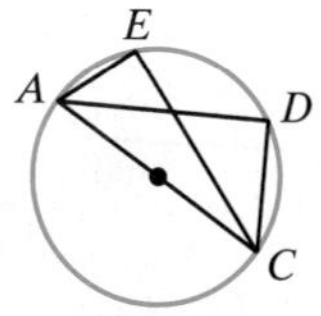

Exercises 1, 2

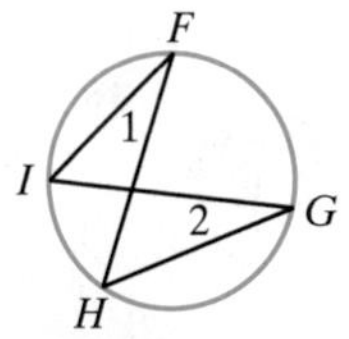

Exercises 3, 4

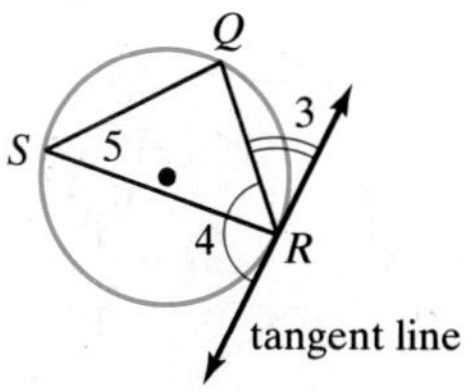

Exercises 5–8

Use the diagram to the right for Exercises 9–14.

9. Which arc does $\angle A$ intercept?
10. Which arc does $\angle B$ intercept?
11. Which angle intercepts $\widehat{ABC}$?
12. Which angle intercepts $\widehat{DAB}$?
13. Which angle of quadrilateral $ABCD$ is supplementary to $\angle B$?
14. Which angle is supplementary to $\angle C$?

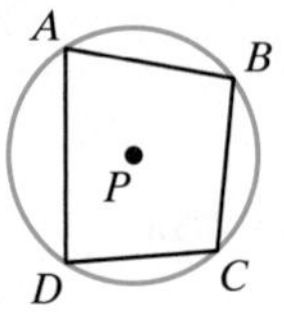

12.3 EXERCISE SET MyMathLab®

Mixed Practice *Find the value of each variable. For each circle, the dot (point) represents the center. See Examples 1–3.*

1.
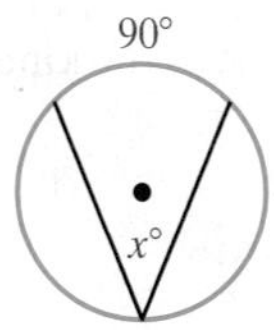

2.
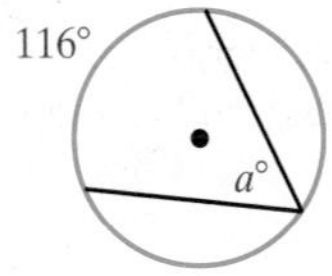

3.
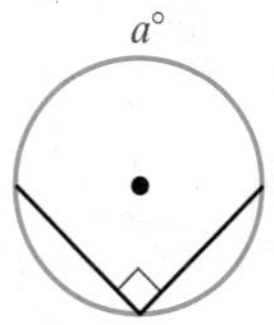

4.
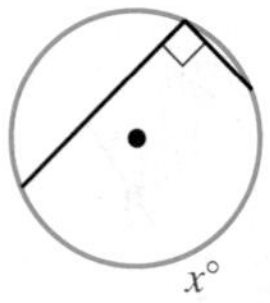

5.
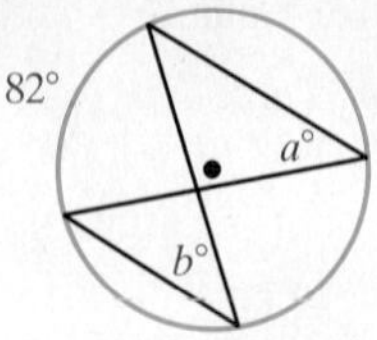

6.
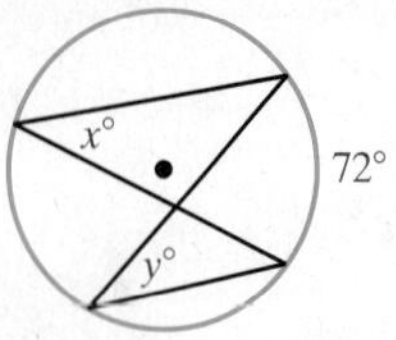

7.
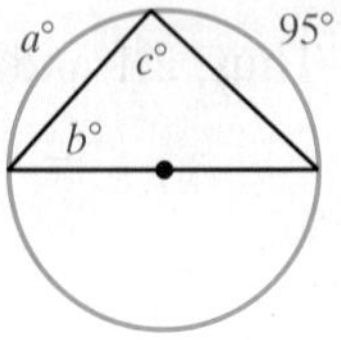

8.
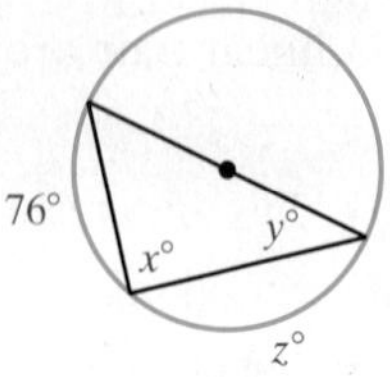

9.
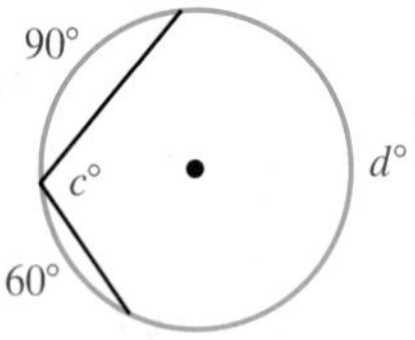

10.
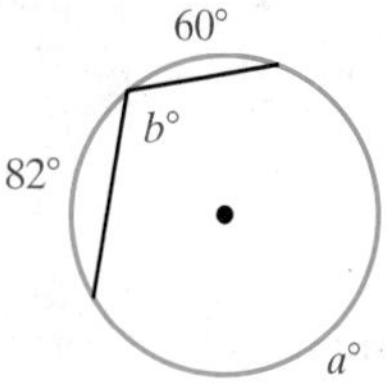

11.
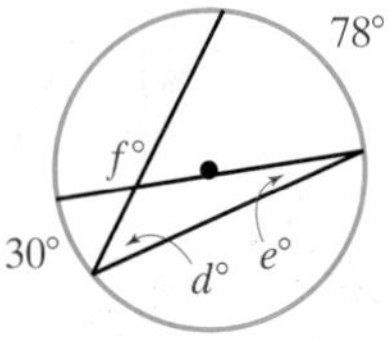

12.
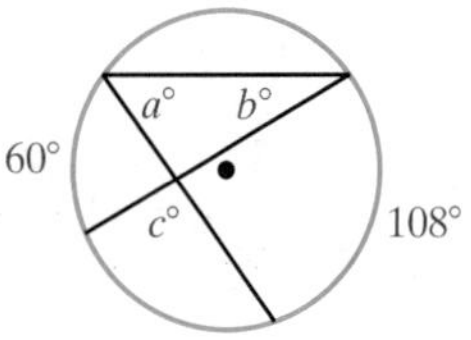

13.
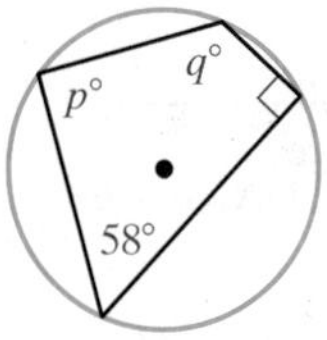

14.
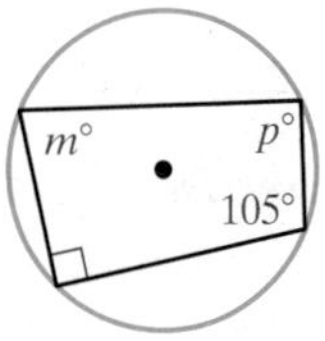

15.
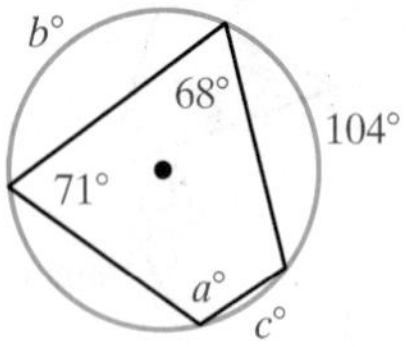

16.
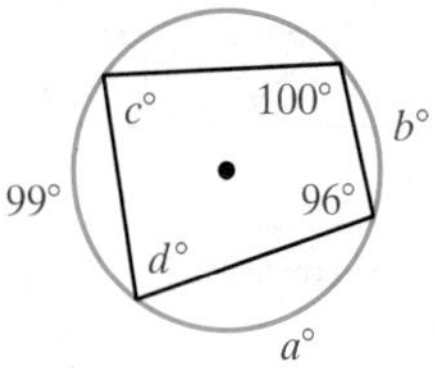

17.
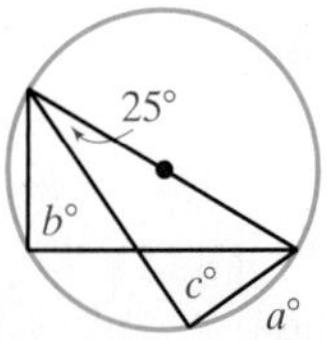

18.
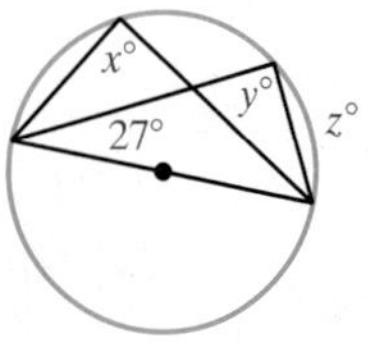

Find the indicated measure. Lines that appear to be tangent are tangent. See Example 4.

19. $m\angle 3$

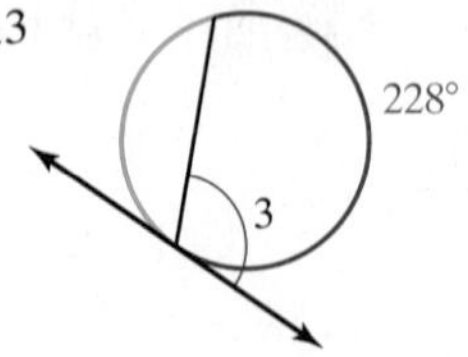

20. $m\angle 2$

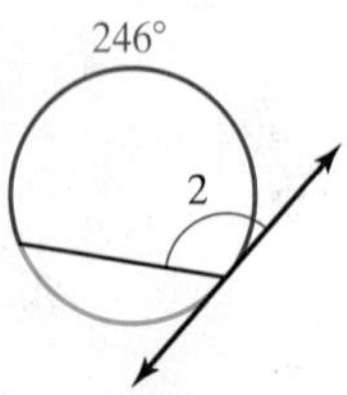

21. $m\angle 1$

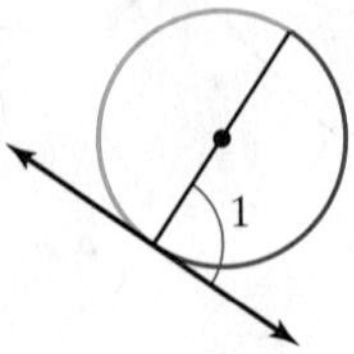

22. $m\widehat{ABC}$

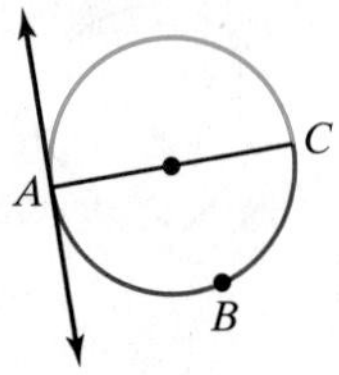

23. $m\angle 1$; $m\widehat{DF}$

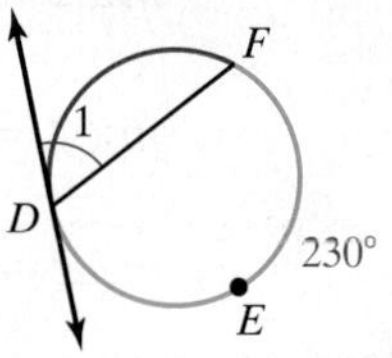

24. $m\angle 2$; $m\widehat{KJ}$

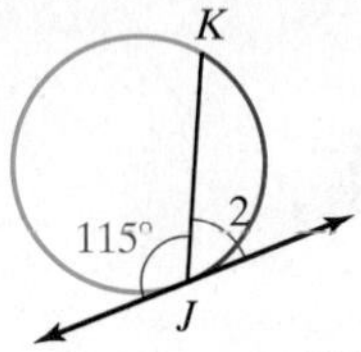

25. $m\angle 1$; $m\angle 2$; $m\widehat{QRP}$

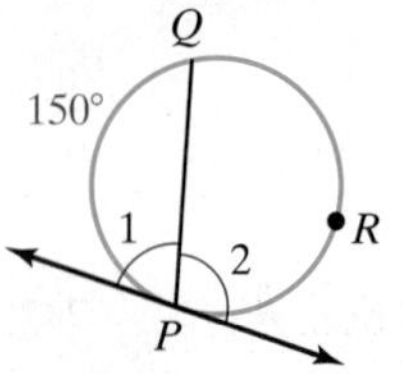

26. $m\angle 3$; $m\angle 4$; $m\widehat{TVS}$

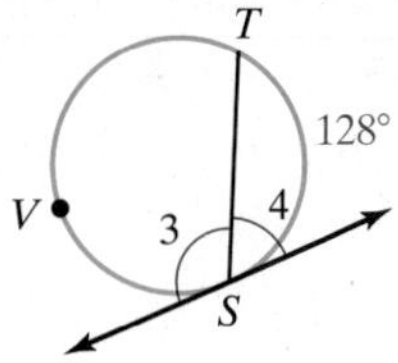

Find each indicated measure for $\odot O$.

27. $m\widehat{BC}$

28. $m\angle B$

29. $m\angle C$

30. $m\widehat{AB}$

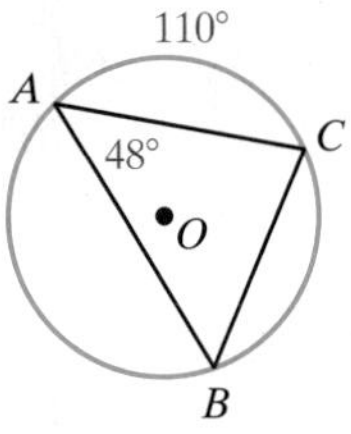

Find each indicated measure for $\odot Q$.

31. $m\angle A$

32. $m\widehat{CE}$

33. $m\angle C$

34. $m\angle D$

35. $m\angle ABE$

36. $m\widehat{ACE}$

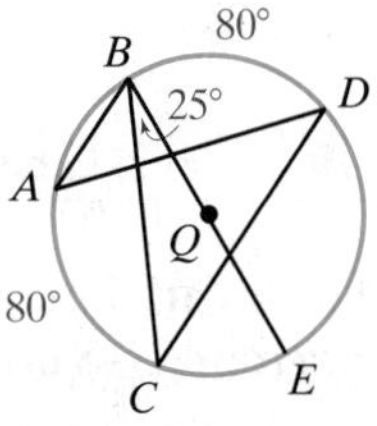

Find the value of each variable. For each circle, the dot represents the center, and lines that appear tangent are tangent.

37.
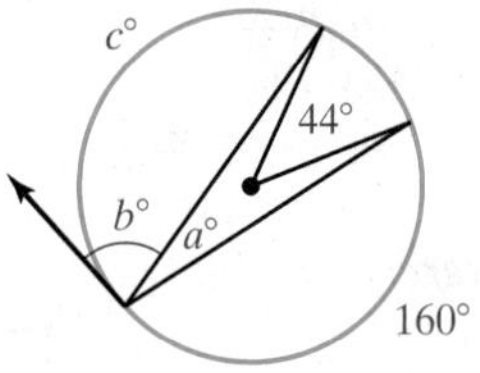

38.
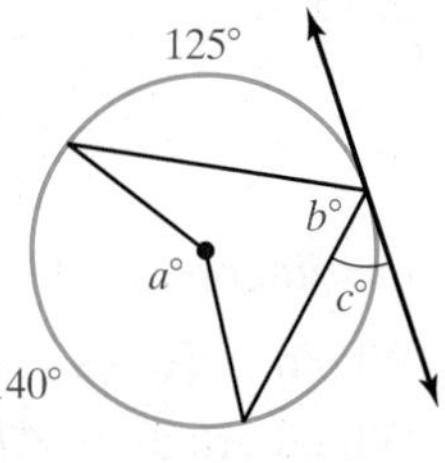

39.
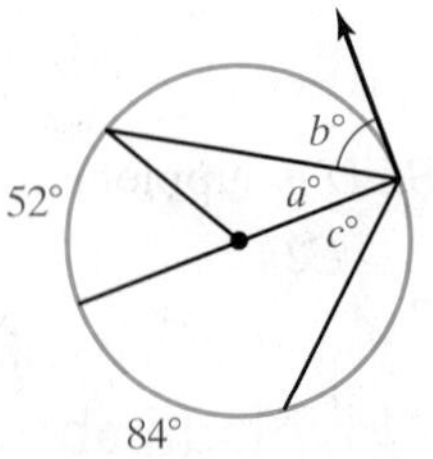

40.
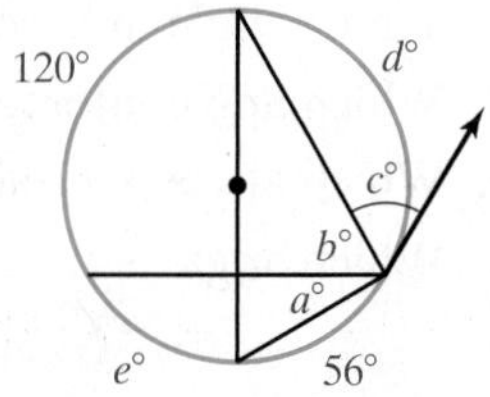

CONCEPT EXTENSIONS

41. What kind of trapezoid can be inscribed in a circle? Justify your response.

42. A parallelogram inscribed in a circle must be what kind of parallelogram? Explain.

Use the figure to the right for Exercises 43 and 44.

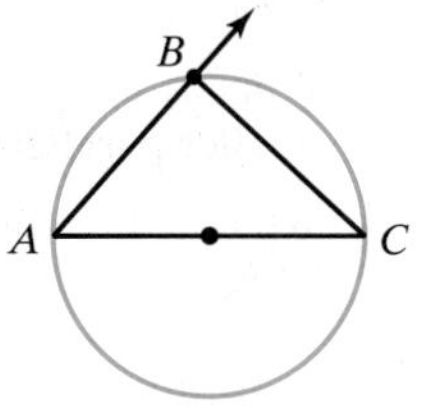

43. Find the Error A classmate says that $m\angle A = 90°$. What is your classmate's error?

44. Find the Error Another classmate says that $m\angle C = 90°$. What is this classmate's error?

Proof *Write a proof for Exercises 45 and 46.*

45. Inscribed Angle Theorem, Case II

Given: $\odot O$ with inscribed $\angle ABC$

Prove: $m\angle ABC = \frac{1}{2}m\widehat{AC}$

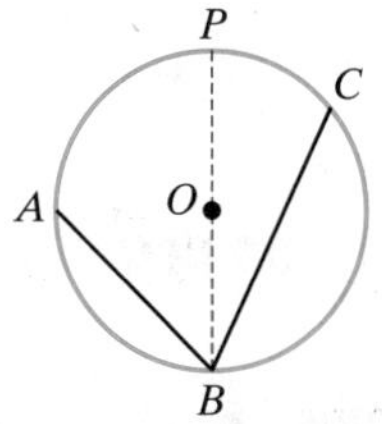

(*Hint:* Use the Inscribed Angle Theorem, Case I.)

46. Inscribed Angle Theorem, Case III

Given: $\odot S$ with inscribed $\angle PQR$

Prove: $m\angle PQR = \frac{1}{2}m\widehat{PR}$

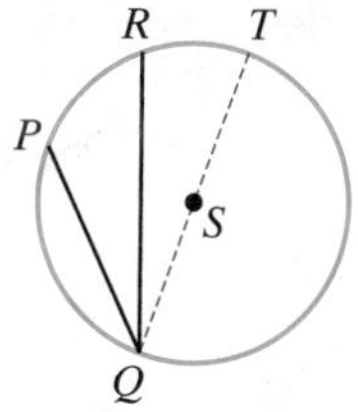

(*Hint:* Use the Inscribed Angle Theorem, Case I.)

47. Can a square be inscribed in a circle? Justify your answer.

48. Can a rhombus that is not a square be inscribed in a circle? Justify your answer.

49. Construction The diagrams below show the construction of a tangent to a circle from a point outside the circle. Explain why $\overleftrightarrow{BC}$ must be tangent to $\odot A$. (*Hint:* Copy the third diagram and draw $\overline{AC}$.)

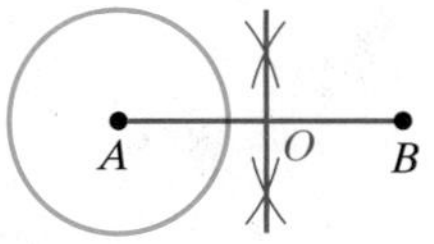

Given: $\odot A$ and point B
Construct the midpoint of $\overline{AB}$. Label the point O.

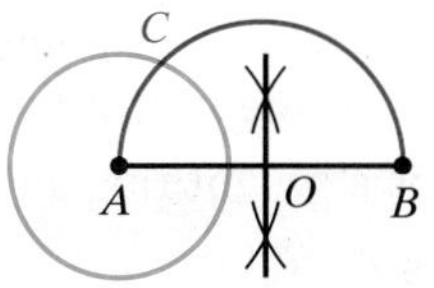

Construct a semicircle with radius OA and center O.
Label its intersection with $\odot A$ as C.

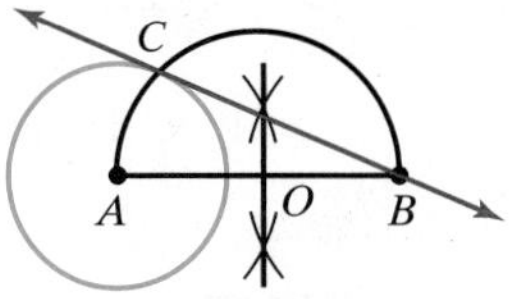

Draw $\overleftrightarrow{BC}$.

50. Construction Construct a circle A and place point P outside the circle. Now construct the two tangents to $\odot A$ through point P.

Proof *Write a proof for Exercises 51–54.*

51. Inscribed Angle Theorem, Corollary 1 (12.3-2)

Given: $\odot O$, $\angle A$ intercepts $\widehat{BC}$, $\angle D$ intercepts $\widehat{BC}$.

Prove: $\angle A \cong \angle D$

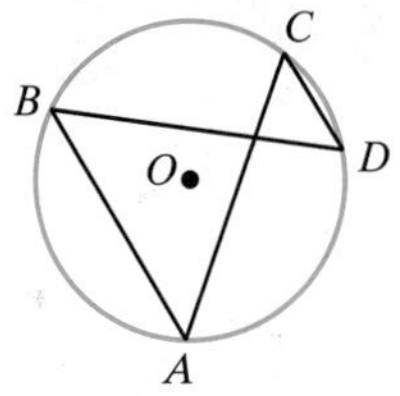

52. Inscribed Angle Theorem, Corollary 2 (12.3-3)

Given: $\odot O$ with $\angle CAB$ inscribed in a semicircle

Prove: $\angle CAB$ is a right angle.

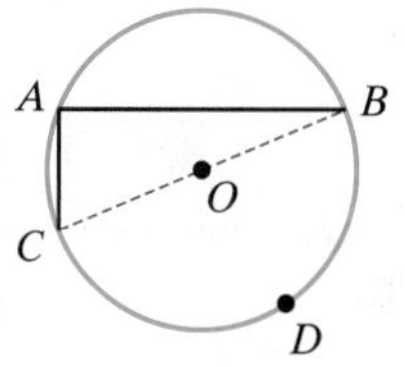

53. Inscribed Angle Theorem, Corollary 3 (12.3-4)

Given: Quadrilateral $ABCD$ inscribed in $\odot O$

Prove: $\angle A$ and $\angle C$ are supplementary.
$\angle B$ and $\angle D$ are supplementary.

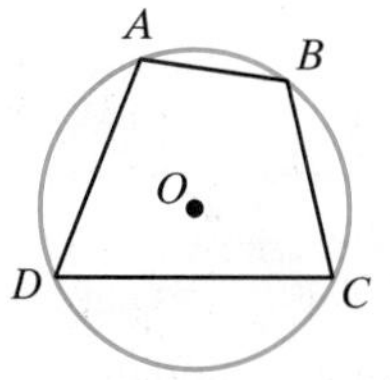

54. Theorem 12.3-5

Given: $\overline{GH}$ and tangent ℓ intersecting $\odot E$ at H

Prove: $m\angle GHI = \frac{1}{2}m\widehat{GFH}$

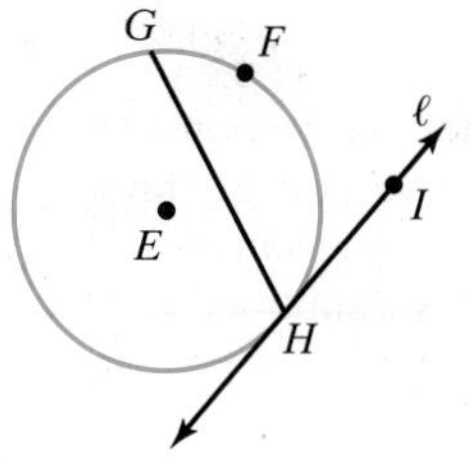

Is the statement true or false? If it is true, give a convincing argument. If it is false, give a counterexample.

55. If two angles inscribed in a circle are congruent, then they intercept the same arc.

56. If an inscribed angle is a right angle, then it is inscribed in a semicircle.

57. A circle can always be circumscribed about a quadrilateral whose opposite angles are supplementary.

58. A circle can always be circumscribed about any triangle.

59. Proof Prove that if two arcs of a circle are included between parallel chords, then the arcs are congruent.

60. Construction Draw two segments. Label their lengths x and y. Construct the geometric mean of x and y. (*Hint:* Recall a theorem about a geometric mean.)

REVIEW AND PREVIEW

Find the value of x in ⊙O, to the nearest tenth. See Section 12.2.

61.

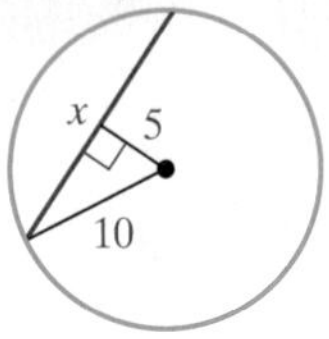

62.

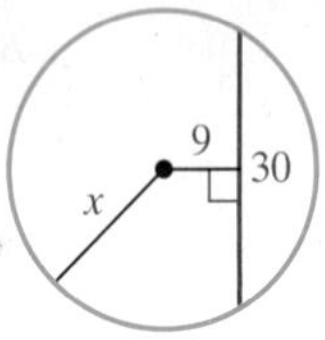

For Exercises 63 and 64, the areas of two similar parallelograms are 20 cm² and 3.2 cm². See Section 10.4.

63. What is the scale factor of the larger parallelogram to the smaller parallelogram?

64. What is the scale factor of the perimeter of the larger parallelogram to the perimeter of the smaller parallelogram?

In the diagram at the right, $\overrightarrow{FE}$ and $\overrightarrow{FD}$ are tangents to ⊙C at E and D, respectively. See Section 12.1.

65. Find $m\widehat{DE}$.

66. Find $m\angle AEC$.

67. Find CE.

68. Find CA.

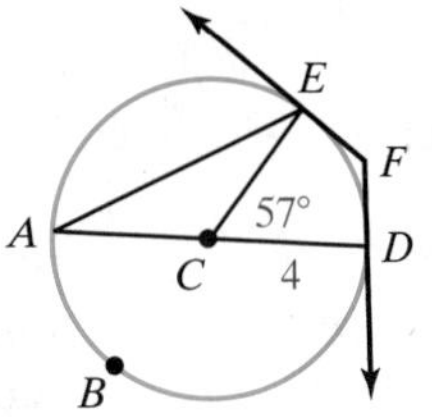

12.4 ADDITIONAL ANGLE MEASURES AND SEGMENT LENGTHS

OBJECTIVES

1. Find Measures of Angles Formed by Chords, Secants, and Tangents.
2. Find the Lengths of Segments Associated with Circles.

VOCABULARY

- secant

Helpful Hint

In this section, we are finding ways to calculate angle measure and—in Objective 2—ways to calculate segment length.

Helpful Hint

Notice that if the intersection is inside the circle, we add in the formula. If the intersection is outside the circle, we subtract.

OBJECTIVE 1 ▶ Finding Measures of Angles Formed by Chords, Secants, and Tangents. Angles formed by intersecting lines have a special relationship to the related arcs formed when the lines intersect a circle. In this section, we will study angles and arcs formed by lines intersecting either within a circle or outside a circle.

Theorem 12.4-1 Angle Measure—Lines Intersecting Inside a Circle

The measure of an angle formed by two lines that intersect inside a circle is half the sum of the measures of the intercepted arcs.

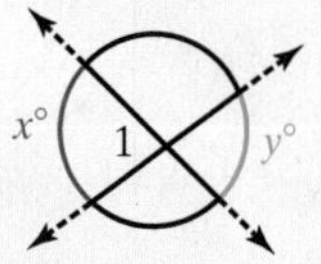

$$m\angle 1 = \frac{1}{2}(x° + y°)$$

Theorem 12.4-2 Angle Measure—Lines Intersecting Outside a Circle

The measure of an angle formed by two lines that intersect outside a circle is half the difference of the measures of the intercepted arcs.

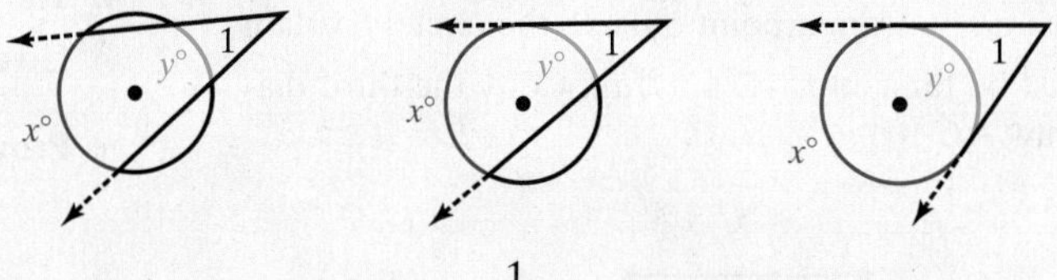

$$m\angle 1 = \frac{1}{2}(x° - y°)$$

We prove Theorem 12.4-1 next and Theorem 12.4-2 in Exercises 49 and 50.

In Theorem 12.4-1, the lines from a point outside the circle going through the circle are called secants. A **secant** is a line that intersects a circle at two points. $\overleftrightarrow{AB}$ is a secant, $\overrightarrow{AB}$ is a secant ray, and $\overline{AB}$ is a secant segment. A chord is part of a secant.

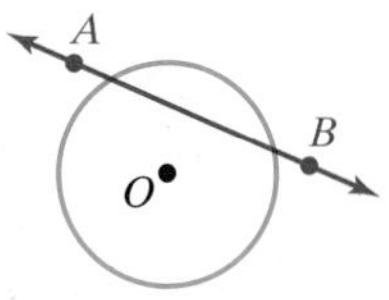

Proof of Theorem 12.4-1

Given: ⊙O with intersecting chords $\overline{AC}$ and $\overline{BD}$

Prove: $m\angle 1 = \frac{1}{2}(m\widehat{AB} + m\widehat{CD})$

Construction: Draw auxiliary $\overline{AD}$ as shown in the diagram.

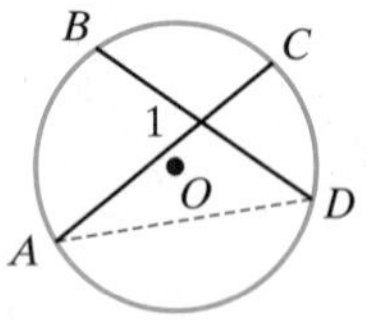

$m\angle BDA = \frac{1}{2}m\widehat{AB}$, and $m\angle CAD = \frac{1}{2}m\widehat{CD}$
Inscribed Angle Theorem

$m\angle 1 = m\angle BDA + m\angle CAD$
Δ Exterior Angle Theorem

$m\angle 1 = \frac{1}{2}m\widehat{AB} + \frac{1}{2}m\widehat{CD}$
Substitute.

$m\angle 1 = \frac{1}{2}(m\widehat{AB} + m\widehat{CD})$
Distributive Property

EXAMPLE 1 Finding Angle Measures

What is the value of each variable?

a.

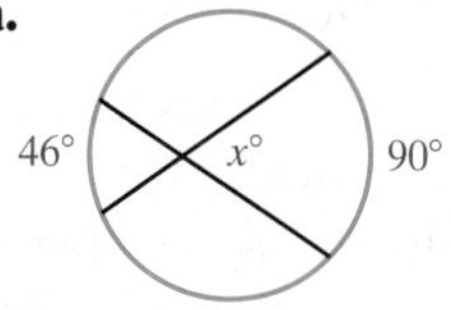

b.

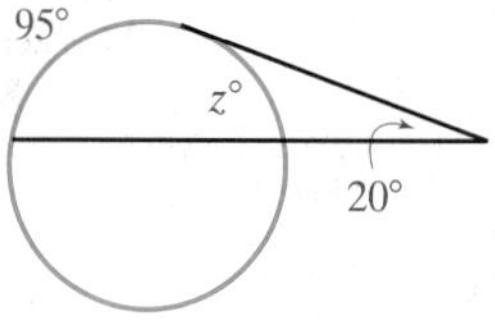

Solution

a. The lines intersect inside the circle, so we find $\frac{1}{2}$ (addition of arcs).

$x° = \frac{1}{2}(46° + 90°)$ Theorem 12.4-1

$x° = \frac{1}{2}(136°)$ Add.

$x° = 68°$ Multiply.

Thus, $x = 68$.

b. Here, the lines intersect outside the circle, so we find $\frac{1}{2}$ (difference of arcs).

$20° = \frac{1}{2}(95° - z°)$ Theorem 12.4-1

$40° = 95° - z°$ Multiply both sides by 2.

$z° = 55°$ Solve for z.

Thus, $z = 55$. □

> **Helpful Hint**
> Remember to add arc measures for arcs intercepted by lines that intersect inside a circle and subtract arc measures for arcs intercepted by lines that intersect outside a circle.

PRACTICE 1 What is the value of each variable?

a.

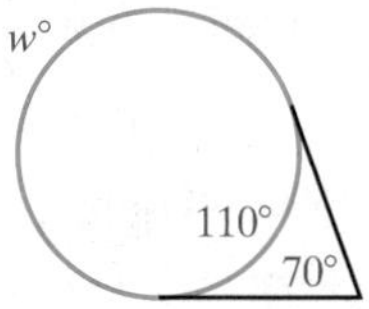

b.

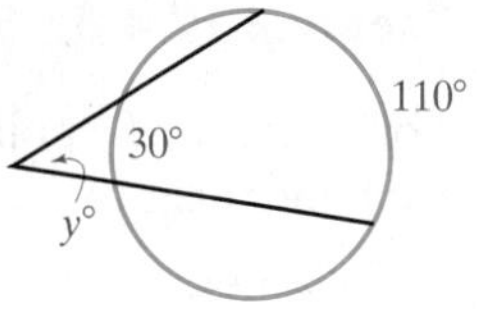

c.

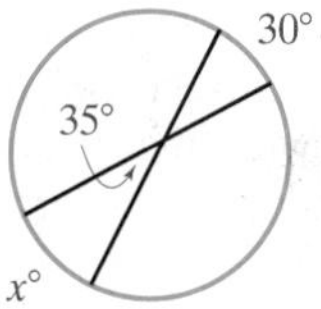

EXAMPLE 2 Finding an Arc Measure

A satellite in a geostationary orbit above Earth's equator has a viewing angle of Earth formed by the two tangents to the equator. The viewing angle is about 17.5°. What is the measure of the arc of Earth that is viewed from the satellite?

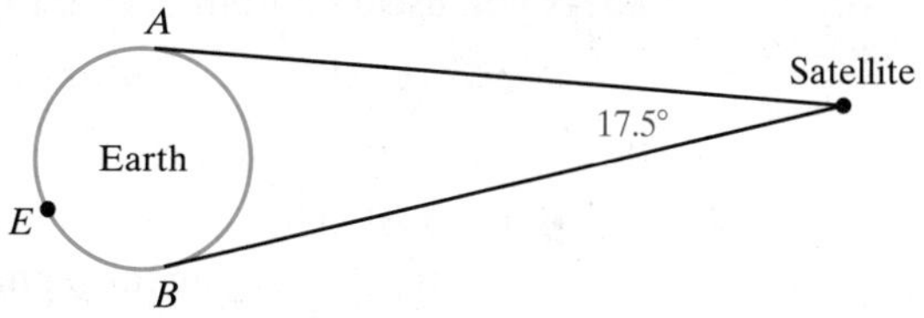

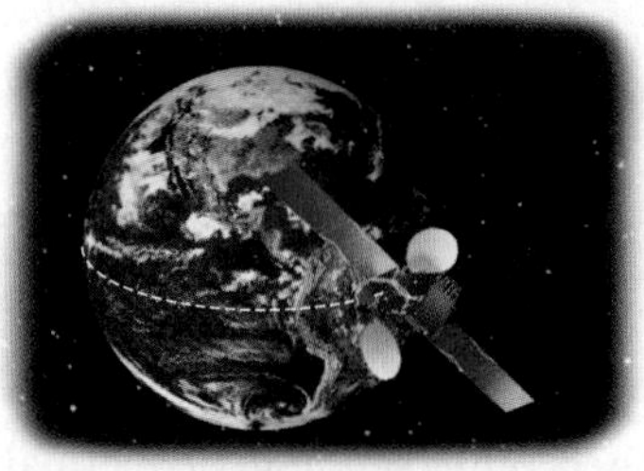

A geostationary orbit lies above the equator and follows the rotation of the Earth. Thus, from the ground looking up into the sky, the satellite appears motionless.

Solution Recall that the sum of the measures of the arcs of a circle is 360°. If we let $m\widehat{AB} = x°$,

then $m\widehat{AEB} = 360° - x°$

Our figure has 2 tangents intersecting outside the circle (Earth), thus,

$$17.5° = \frac{1}{2}(m\widehat{AEB} - m\widehat{AB}) \quad \text{Theorem 12.4-2}$$

$$17.5° = \frac{1}{2}[(360° - x°) - x°] \quad \text{Substitute.}$$

$$17.5° = \frac{1}{2}(360° - 2x°) \quad \text{Combine like terms.}$$

$$17.5° = 180° - x° \quad \text{Use the Distributive Property.}$$

$$x° = 162.5° \quad \text{Solve for } x°.$$

A 162.5° arc can be viewed from the satellite. □

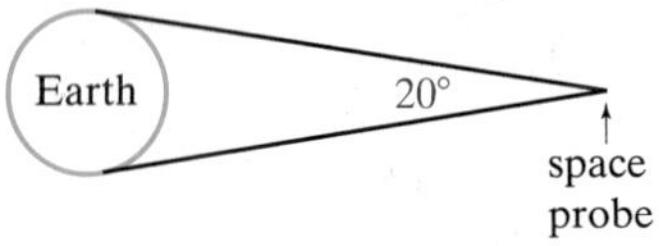

PRACTICE 2

a. A departing space probe sends back a picture of Earth as it crosses Earth's equator. The angle formed by the two tangents to the equator is 20°. What is the measure of the arc of the equator that is visible to the space probe?

b. Is the probe or the geostationary satellite in Example 2 closer to Earth? Explain.

OBJECTIVE 2 ▶ Finding Segment Lengths Associated with Circles. There is a special relationship between two intersecting chords, two intersecting secants, or a secant that intersects a tangent. This relationship allows us to find the lengths of unknown segments.

Through a given point P, we can draw an infinite number of chords or lines so that two segments to the circle are formed. For example, $\overline{PA_1}$ and $\overline{PB_1}$ lie along one such line. Theorem 12.4-3 states the surprising result that no matter which line we use, the product of the lengths $PA \cdot PB$ remains constant.

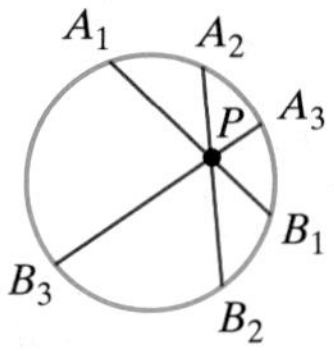

▶ Helpful Hint

To find the correct segment lengths, start with the point of intersection and go to the circle.

Theorem 12.4-3 Segment Products—Inside or Outside a Circle

For a circle and a point not on the circle, the product of the lengths of the two segments from the point to the circle is constant along any line through the point and circle.

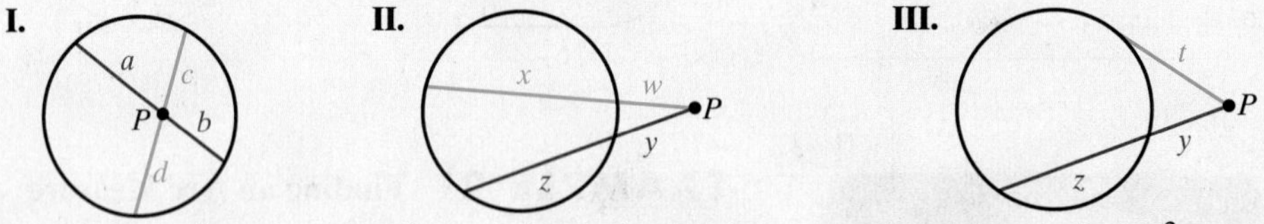

$a \cdot b = c \cdot d$ $\qquad$ $w(w + x) = y(y + z)$ $\qquad$ $y(y + z) = t^2$

As you use Theorem 12.4-3, remember the following.

- **Case I:** The products of the chord segments are equal.
- **Case II:** The products of the secant segments and their outer segments are equal.
- **Case III:** The product of a secant segment and its outer segment equals the square of the tangent segment.

Here is a proof for Case I. We prove Case II and Case III in Exercises 51 and 52.

Proof of Theorem 12.4-3, Case I

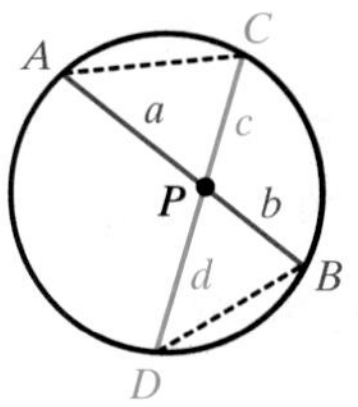

Given: A circle with chords $\overline{AB}$ and $\overline{CD}$ intersecting at P

Prove: $a \cdot b = c \cdot d$

Construction: Draw $\overline{AC}$ and $\overline{BD}$.

Proof: $\angle A \cong \angle D$ and $\angle C \cong \angle B$ because each pair intercepts the same arc, and angles that intercept the same arc are congruent. $\Delta APC \sim \Delta DPB$ by the Angle-Angle Similarity Postulate. The lengths of corresponding sides of similar triangles are proportional, so $\frac{a}{d} = \frac{c}{b}$. Therefore, $a \cdot b = c \cdot d$.

Helpful Hint

Study the movements below to see that for each case, the factors on one side of the equation are one line's segment lengths from P to the circle.

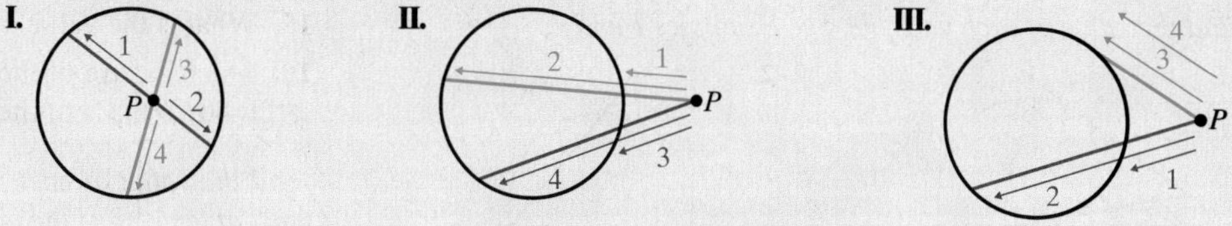

Each time: length 1 · length 2 = length 3 · length 4
factor · factor = factor · factor

EXAMPLE 3 Finding Segment Lengths

Find the value of the variable in $\odot N$.

a.

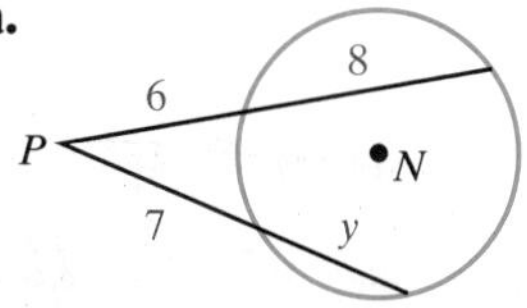

b.

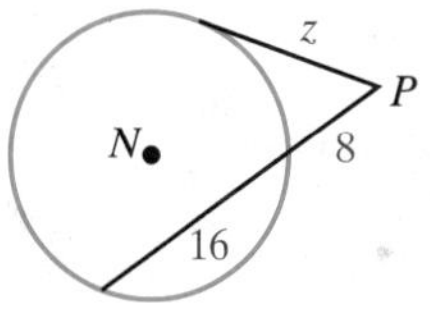

Solution

a. $6(6 + 8) = 7(7 + y)$ Theorem 12.4-3, Case II
$6 \cdot 14 = 7(7 + y)$ Add.
$84 = 49 + 7y$ Distributive Property
$35 = 7y$ Subtract 35 from both sides.
$5 = y$ Solve for y by dividing both sides by 7.

b. $8(8 + 16) = z^2$ Theorem 12.4-3, Case III
$8 \cdot 24 = z^2$ Add.
$192 = z^2$ Multiply.
$13.9 \approx z$ Approximate the positive square root of z. □

PRACTICE 3 What is the value of the variable to the nearest tenth?

a.

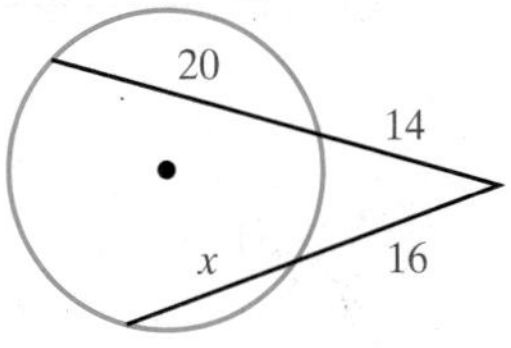

b.

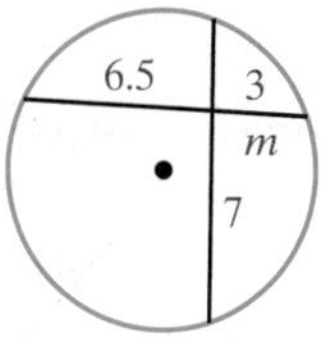

Helpful Hint

Remember: By definition,

- a chord is a segment whose endpoints are on the circle.
- a tangent is a line that intersects a circle in one point.
- a secant is a line that intersects a circle at two points.

VOCABULARY & READINESS CHECK

Matching *Use the word that best describes each object in the diagram. Use each choice only once.*

1. $\overleftrightarrow{EH}$	a. chord
2. $\overline{AD}$	b. diameter
3. $\overleftrightarrow{ID}$	c. radius
4. $\overline{ID}$	d. secant
5. $\overline{FC}$	e. secant segment
6. $\overline{EG}$	f. tangent line
7. $\overline{HD}$	g. tangent segment

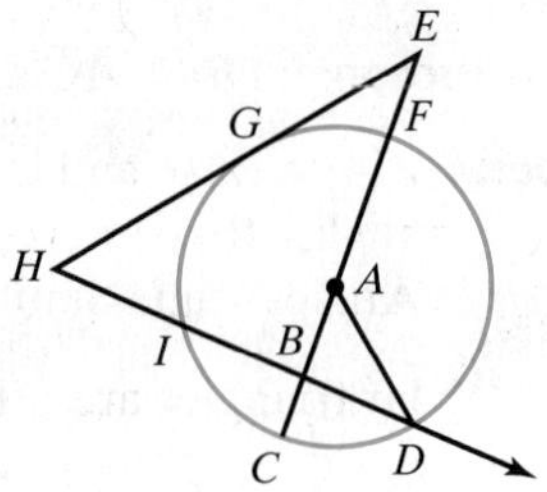

12.4 EXERCISE SET MyMathLab®

Find the value of each variable. See Examples 1 and 2.

1. 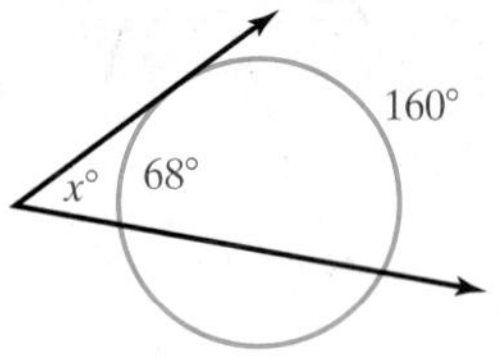

2. 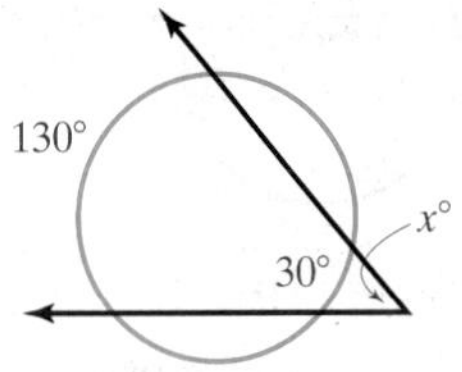

3. 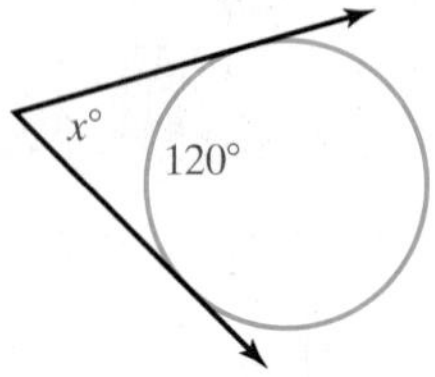

4. 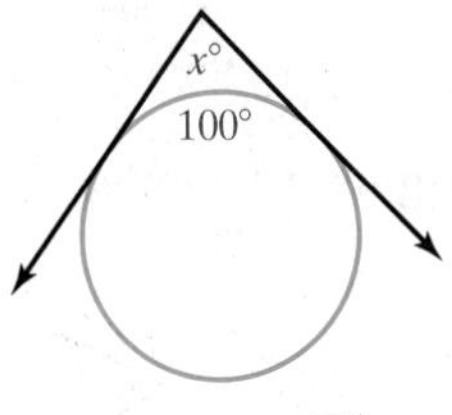

5. 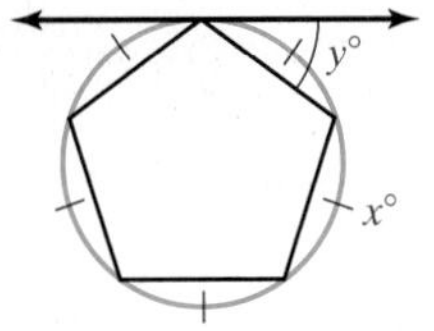

6. 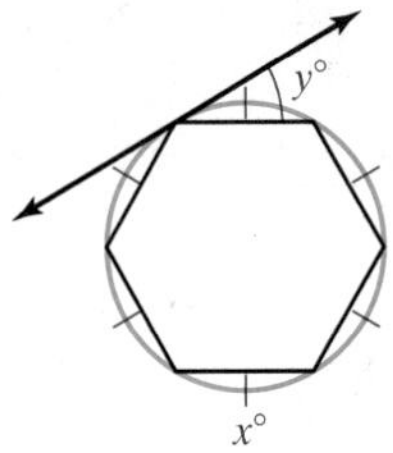

7. 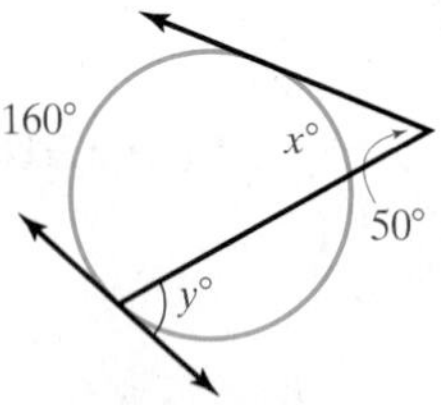

8. 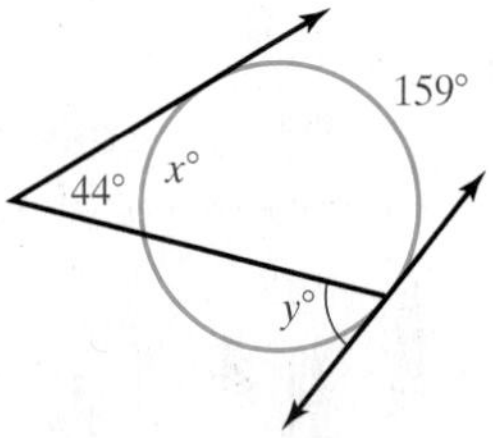

9. 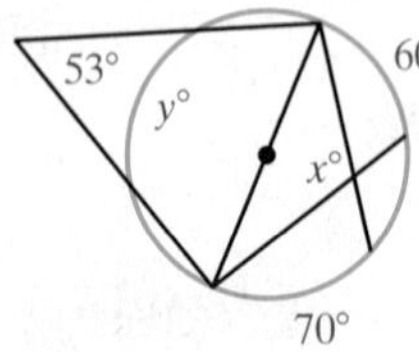

10. 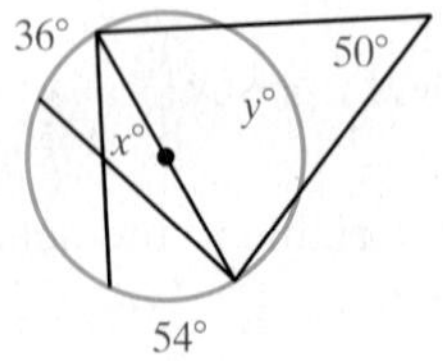

Mixed Practice *Use the diagram for Exercises 11–15.*

11. What is the value of x?
12. What is the value of y?
13. What is the value of a?
14. What is the value of b?

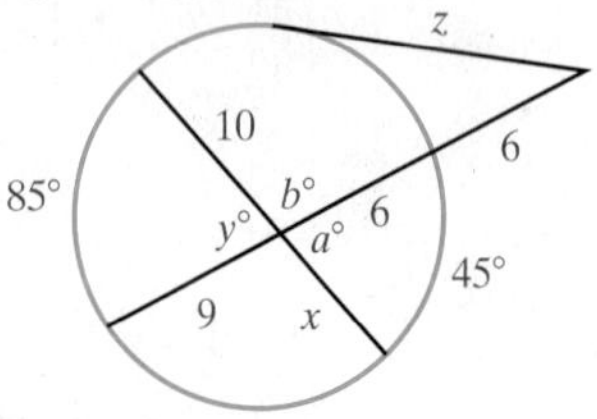

15. What is the value of z, to the nearest tenth?
16. The measure of the angle formed by two tangents to a circle is 80°. What are the measures of the intercepted arcs?

Find the value of each variable using the given chord, secant, or tangent lengths. If the answer is not a whole number, round to the nearest tenth. See Example 3.

17. 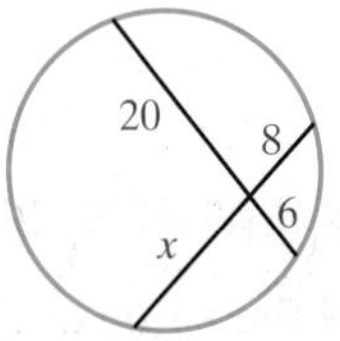

18. 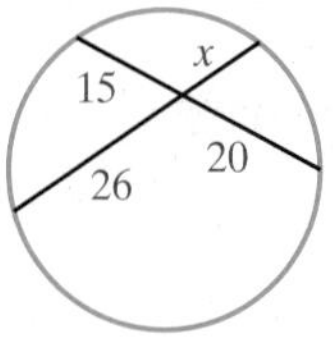

19. 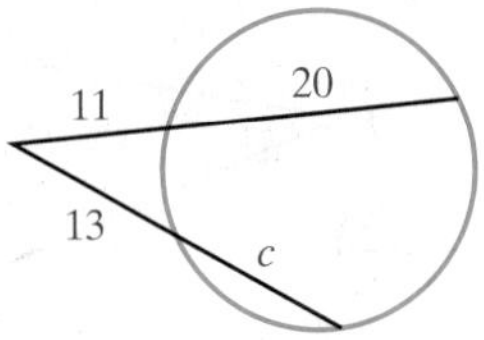

20. 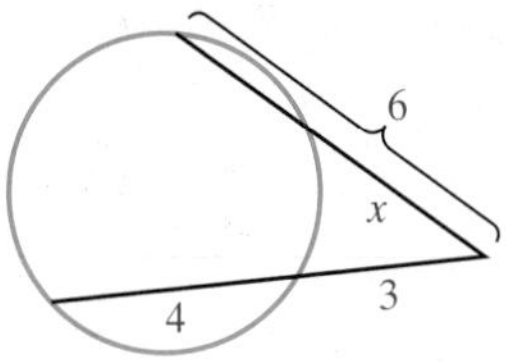

21. 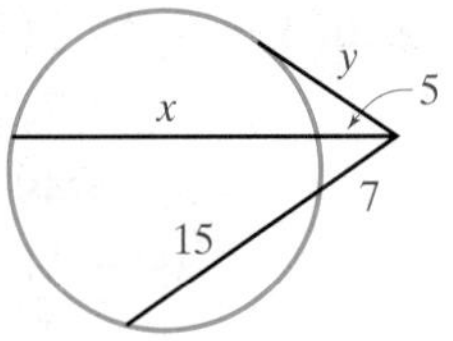

22. 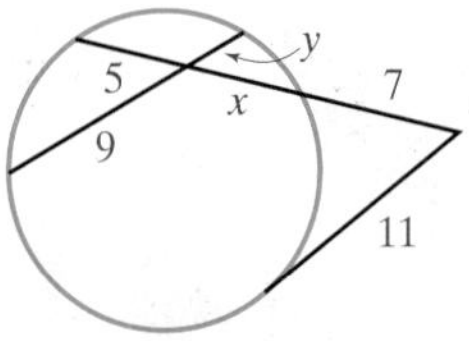

Find the diameter of $\odot O$. A line that appears to be tangent is tangent. If your answer is not a whole number, round it to the nearest tenth. See Example 3.

23. 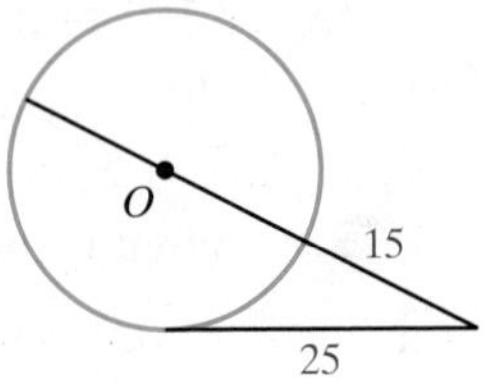

24. 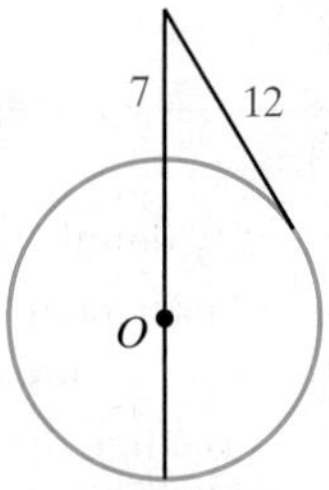

25. 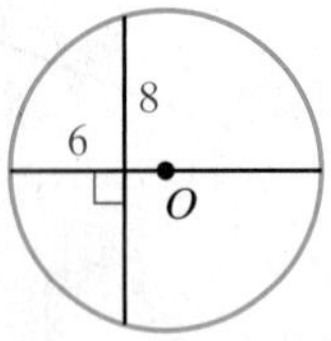

26. 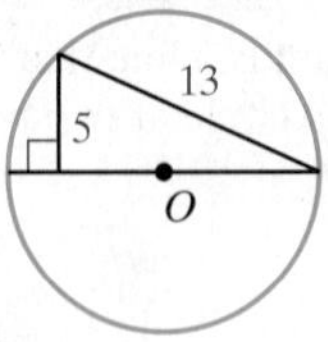

27. A circle is inscribed in a quadrilateral whose four angles have measures 85°, 76°, 94°, and 105°. Find the measures of the four arcs between consecutive points of tangency.

28. ΔPQR is inscribed in a circle with $m\angle P = 70°$, $m\angle Q = 50°$, and $m\angle R = 60°$. What are the measures of $\widehat{PQ}$, $\widehat{QR}$, and $\widehat{PR}$?

Find the values of x and y using the given chord, secant, and tangent lengths. If your answer is not a whole number, round it to the nearest tenth.

29.

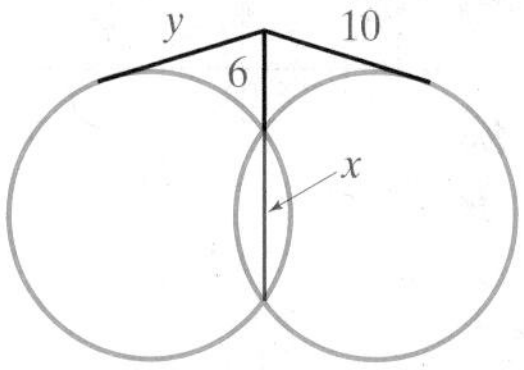

30.

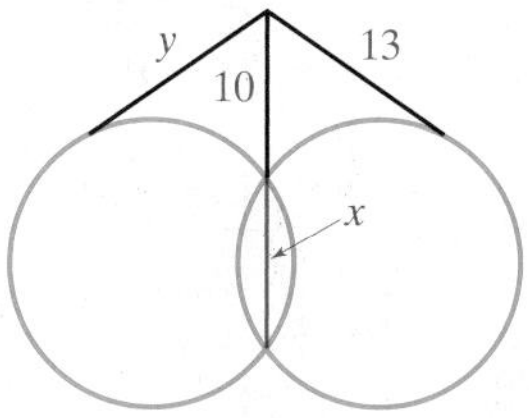

31.

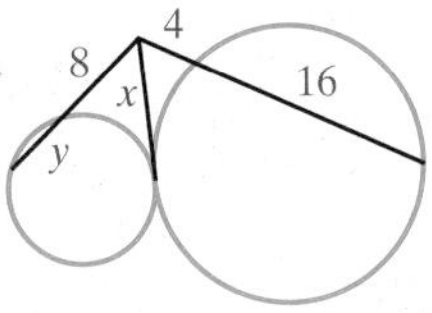

32.

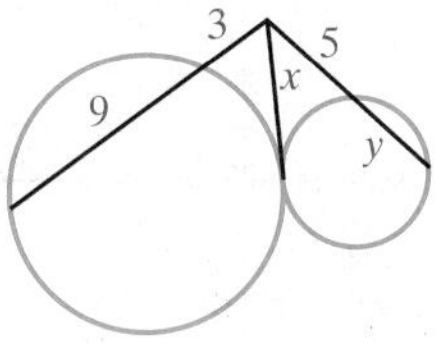

33.

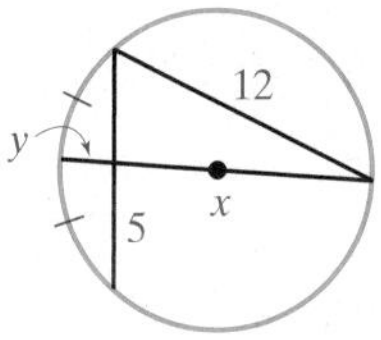

34.

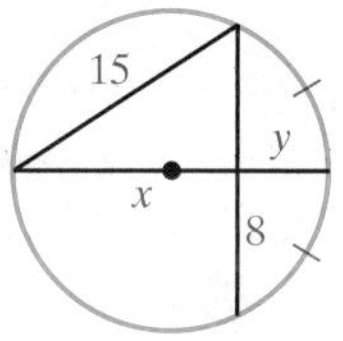

Solve. See Example 2.

35. A photo is being taken of a cylindrical silo. The camera is positioned at the vertex of a 35° angle formed by tangents to the silo. Find the measure of the arc along the silo that will be in the photograph.

36. A photo is being taken of a cylindrical basket in order to sell it on a website. The camera is positioned at the vertex of a 45° angle formed by tangents to the basket. Find the measure of the arc along the basket that will be in the photograph.

37. A departing space probe is sending back pictures of the Earth as it crosses the Earth's equator. The angle formed by the two tangents to the equator is 22.5°. Find the measure of the arc of the equator that is visible to the satellite.

38. The departing space probe from Exercise **37** continues to send back pictures of the Earth as it once again crosses the Earth's equator. The angle formed by the two tangents to the equator is now 12.5°. Find the measure of the arc of the equator that is now visible to the probe.

CONCEPT EXTENSIONS

39. Find the Error To find the value of x, a student wrote the equation $6(7.5) = x^2$. What error did the student make?

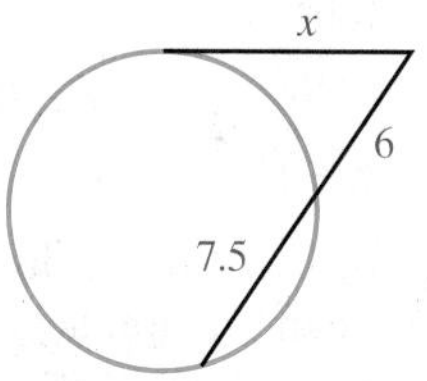

40. Find the Error To find the value of x, a student wrote: $8 \cdot 7 = 6 \cdot x$. Find the error and correct it.

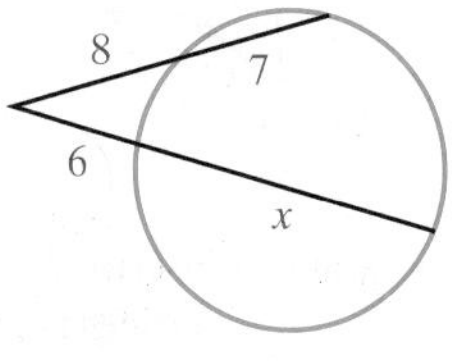

41. Describe the difference between a *secant* and a *tangent*.

42. Describe the difference between a *chord* and a *secant*.

$\overline{CA}$ and $\overline{CB}$ are tangents to $\odot O$. Write an expression for each arc or angle in terms of the given variable.

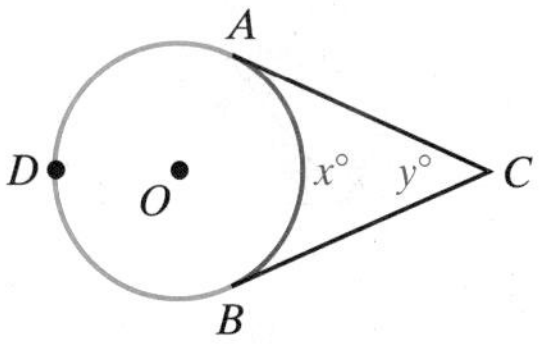

43. $m\widehat{ADB}$ using x

44. $m\angle C$ using x

45. $m\widehat{AB}$ using y (*Hint:* Use Exercise **44** and solve the answer for x.)

46. $m\widehat{ADB}$ using y

47. In the diagram, the circles are concentric. What is a formula you could use to find the value of c in terms of a and b? Explain.

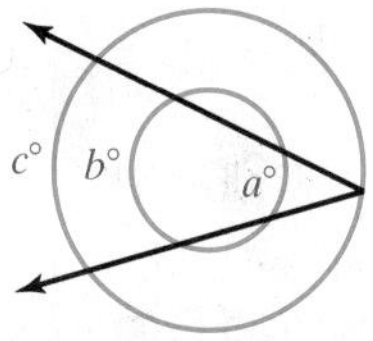

48. Use the diagram below. If you know the values of x and y, how can you find the measure of each numbered angle?

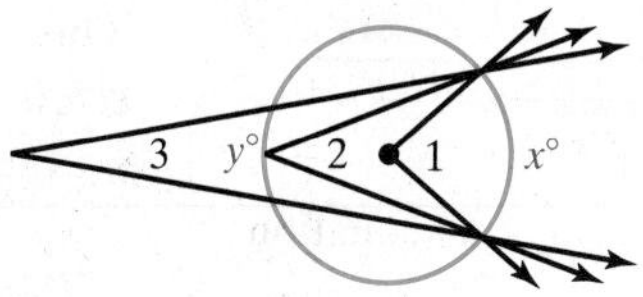

Proof

49. Prove Theorem 12.4-2 as it applies to two secant segments that intersect outside a circle.

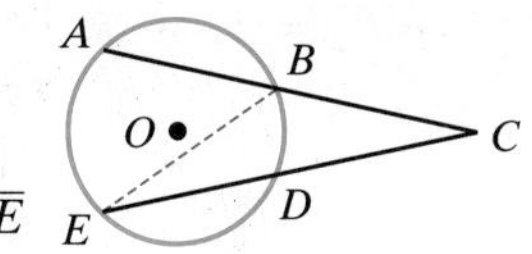

Given: $\odot O$ with secants $\overline{CA}$ and $\overline{CE}$

Prove: $m\angle ACE = \frac{1}{2}(m\widehat{AE} - m\widehat{BD})$

50. Prove the other two cases of Theorem 12.4-2. (See Exercise **49**.)

For Exercises 51 and 52, write proofs that use similar triangles.

51. Prove Theorem 12.4-3, Case II.

52. Prove Theorem 12.4-3, Case III.

Proof *For Exercises 53 and 54, use the diagram at the right. Prove each statement.*

53. $m\angle 1 + m\widehat{PQ} = 180°$

54. $m\angle 1 + m\angle 2 = m\widehat{QR}$

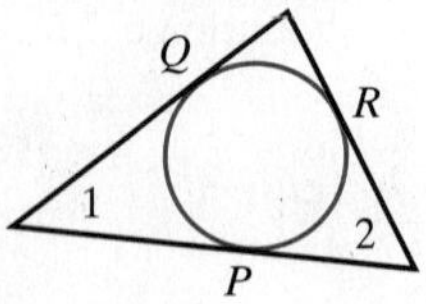

55. Use the diagram below and the theorems of this section to prove the Pythagorean Theorem.

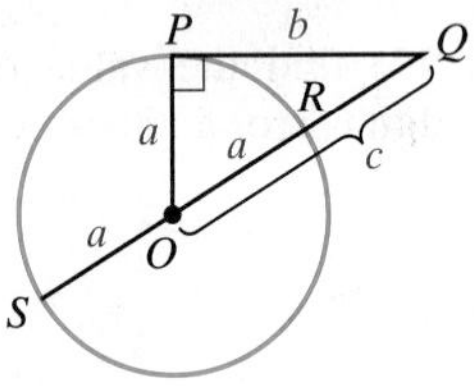

56. If an equilateral triangle is inscribed in a circle, prove that the tangents to the circle at the vertices form an equilateral triangle.

REVIEW AND PREVIEW

Find the value of each variable. See Section 12.3.

57.

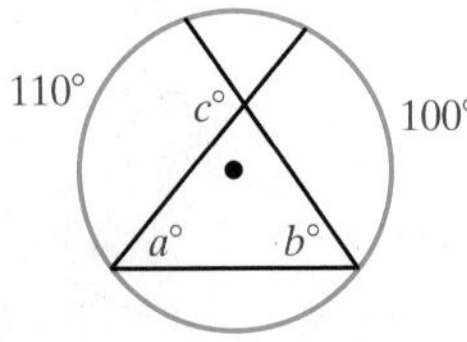

58.

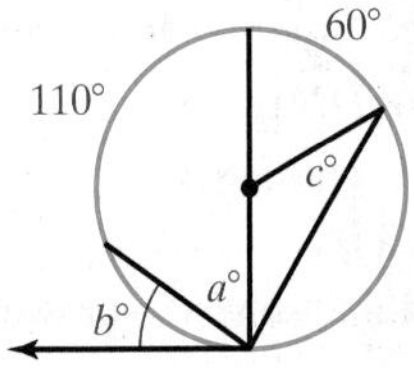

Find the value of x to the nearest whole number. See Section 9.3.

59.

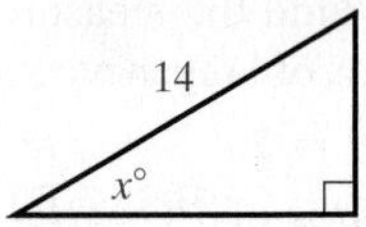

60.

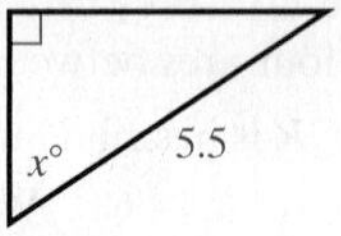

Find the length of each segment to the nearest tenth. See Section 1.7.

61.

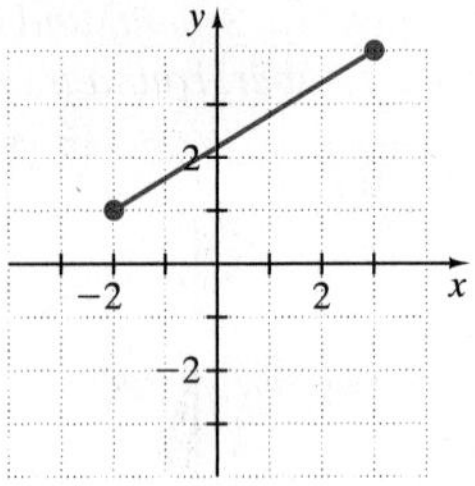

62.

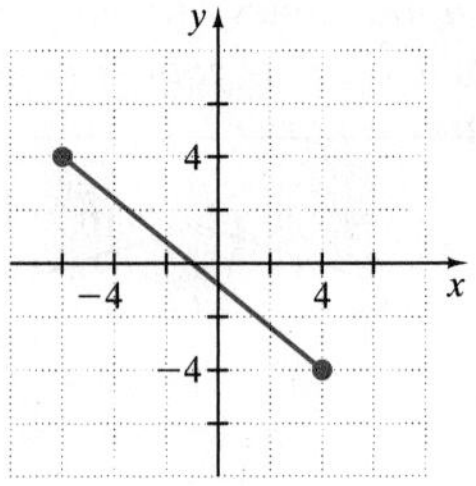

63.

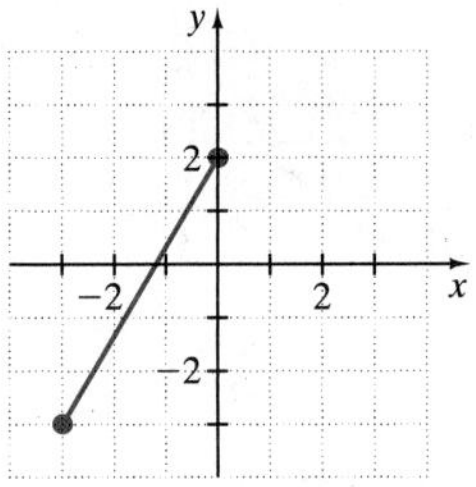

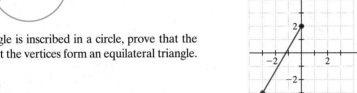

12.5 COORDINATE PLANE—CIRCLES

OBJECTIVES

1 Write an Equation of a Circle.

2 Find the Center and Radius of a Circle Written in Standard Form.

3 Complete the Square to Find the Center and Radius of a Circle.

VOCABULARY

- standard form of an equation of a circle
- standard equation of a circle

OBJECTIVE 1 ▶ Writing an Equation of a Circle. We can use the Distance Formula to find an equation of a circle with center (h, k) and radius r. Let (x, y) be any point on the circle. Then the radius r is the distance from (h, k) to (x, y).

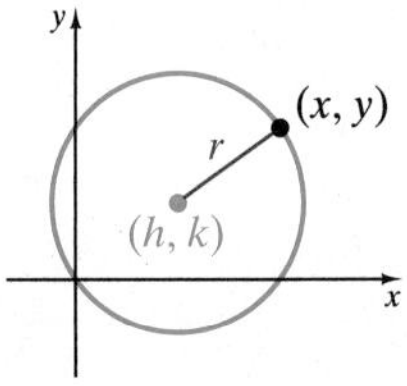

$$r = \sqrt{(x - h)^2 + (y - k)^2} \quad \text{Distance Formula}$$
$$r^2 = (x - h)^2 + (y - k)^2 \quad \text{Square both sides.}$$

The equation $(x - h)^2 + (y - k)^2 = r^2$ is the **standard form of an equation of a circle**. You may also call it the **standard equation of a circle**.

Let's write our results as a theorem.

Theorem 12.5-1 Equation of a Circle

An equation of a circle with center (h, k) and radius r is $(x - h)^2 + (y - k)^2 = r^2$.

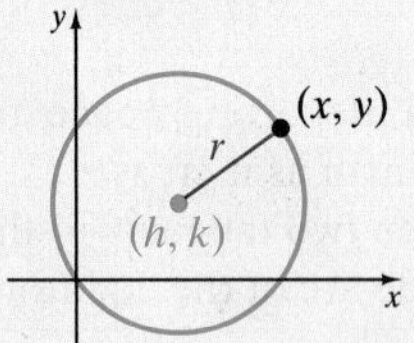

This information in the equation of a circle helps us graph a circle. Also, we can write the equation of a circle if we know its center and radius.

EXAMPLE 1 Writing the Equation of a Circle

What is the standard equation of the circle with center (5, −2) and radius 7?

Solution Let's use the given values $h = 5, k = -2, r = 7$, and the standard form of a circle.

$$(x - h)^2 + (y - k)^2 = r^2 \quad \text{Use the standard form of an equation of a circle.}$$
$$(x - 5)^2 + [y - (-2)]^2 = 7^2 \quad \text{Substitute } (5, -2) \text{ for } (h, k) \text{ and 7 for } r.$$
$$(x - 5)^2 + (y + 2)^2 = 49 \quad \text{Simplify.}$$

□

PRACTICE 1 What is the standard equation of each circle?

a. center (3, 5); radius 6 **b.** center (−2, −1); radius $\sqrt{2}$

EXAMPLE 2 Using the Center and a Point on a Circle

What is the standard equation of the circle with center (1, −3) that passes through the point (2, 2)?

Solution To use the standard form, we need the center and the radius. We have the center.

Let's find the radius. The radius is the distance between the center (1, −3) and the given point (2, 2).

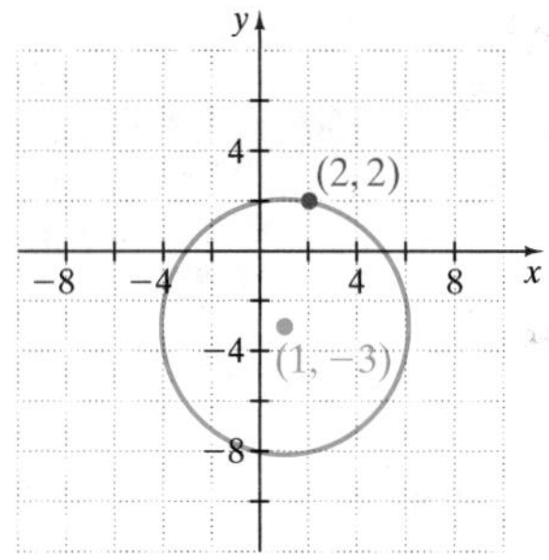

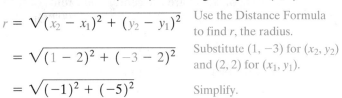

$$r = \sqrt{(x_2 - x_1)^2 + (y_2 - y_1)^2} \quad \text{Use the Distance Formula to find } r \text{, the radius.}$$
$$= \sqrt{(1 - 2)^2 + (-3 - 2)^2} \quad \text{Substitute } (1, -3) \text{ for } (x_2, y_2) \text{ and } (2, 2) \text{ for } (x_1, y_1).$$
$$= \sqrt{(-1)^2 + (-5)^2} \quad \text{Simplify.}$$
$$= \sqrt{26} \quad \text{Simplify.}$$

Next, we use the radius and the center to write an equation.

$$(x - h)^2 + (y - k)^2 = r^2 \quad \text{Use the standard form of an equation of a circle.}$$
$$(x - 1)^2 + [y - (-3)]^2 = \left(\sqrt{26}\right)^2 \quad \text{Substitute } (1, -3) \text{ for } (h, k) \text{ and } \sqrt{26} \text{ for } r.$$
$$(x - 1)^2 + (y + 3)^2 = 26 \quad \text{Simplify.}$$

To **check** and see that the circle passes through (2, 2), let $x = 2, y = 2$, and see that a true statement results. □

PRACTICE 2 What is the standard equation of the circle with center (4, 3) that passes through the point (−1, 1)?

> **Helpful Hint**
> How do we know that $(x - 1)^2 + (y + 3)^2 = 26$ passes through (2, 2)? Let $x = 2, y = 2$ and see that a true statement results.
>
> $$(2 - 1)^2 + (2 + 3)^2 \stackrel{?}{=} 26$$
> $$1^2 + 5^2 \stackrel{?}{=} 26$$
> $$26 = 26 \quad \text{True}$$

OBJECTIVE 2 ▶ Finding the Center and Radius of a Circle in Standard Form. If we know the standard equation of a circle, we can describe the circle by naming its center and radius. Then we use this information to graph the circle.

EXAMPLE 3 **Graphing a Circle Given Its Equation**

When we make a call on a cell phone, a tower receives and transmits the call. A way to monitor the range of a cell tower system is to use equations of circles. Suppose the equation $(x - 7)^2 + (y + 2)^2 = 64$ represents the position and the transmission range of a cell tower. What is the graph that shows the position and range of the tower?

Solution To graph this circle, we need the center and radius only.

Let's write the standard form of the equation $(x - 7)^2 + (y + 2)^2 = 64$ so we can identify (h, k), and radius r.

> **Helpful Hint**
> If a compass is not handy, we can sketch the circle by graphing a few points 8 units from the center. It is easiest to count vertically and horizontally, as shown by the red dots in the graph to the right.

$$(x - 7)^2 + (y + 2)^2 = 64 \quad \text{Use the standard equation of a circle.}$$

$$(x - \underset{h}{7})^2 + [y - (\underset{k}{-2})] = \underset{r}{8}^2 \quad \text{Rewrite to find } h, k, \text{ and } r.$$

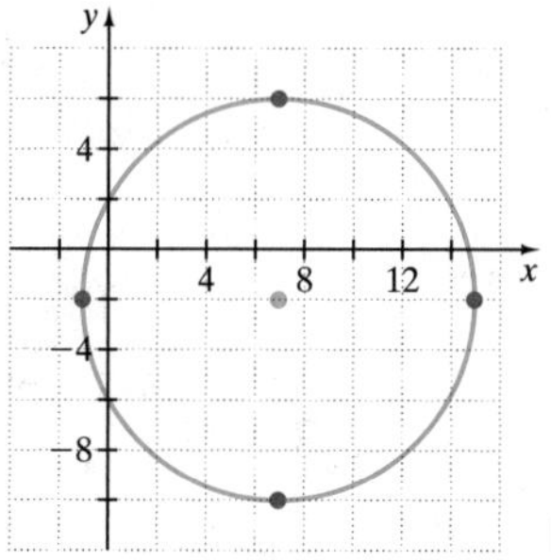

The center is $(7, -2)$ and the radius is 8.

To graph the circle, place the compass point at the center $(7, -2)$ and draw a circle with radius 8. □

PRACTICE 3

a. In Example 3, what does the center of the circle represent? What does the radius represent?

b. What are the center and radius of the circle with equation $(x - 2)^2 + (y - 3)^2 = 100$? Graph the circle.

EXAMPLE 4 Graph $x^2 + y^2 = 4$.

Solution The equation can be written in standard form as

$$(x - 0)^2 + (y - 0)^2 = 2^2$$

The center of the circle is $(0, 0)$, and the radius is 2. Its graph is shown.

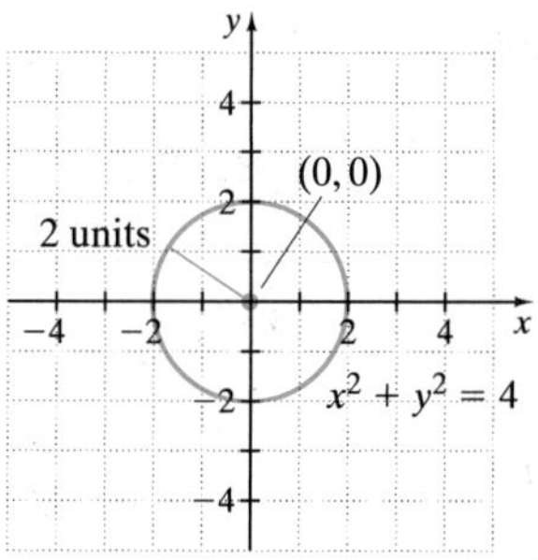

□

PRACTICE 4 Graph $x^2 + y^2 = 25$.

EXAMPLE 5 Graph $(x + 1)^2 + y^2 = 8$.

Solution The equation can be written as $(x + 1)^2 + (y - 0)^2 = 8$ with $h = -1$, $k = 0$, and $r = \sqrt{8}$. The center is $(-1, 0)$, and the radius is $\sqrt{8} = 2\sqrt{2} \approx 2.8$.

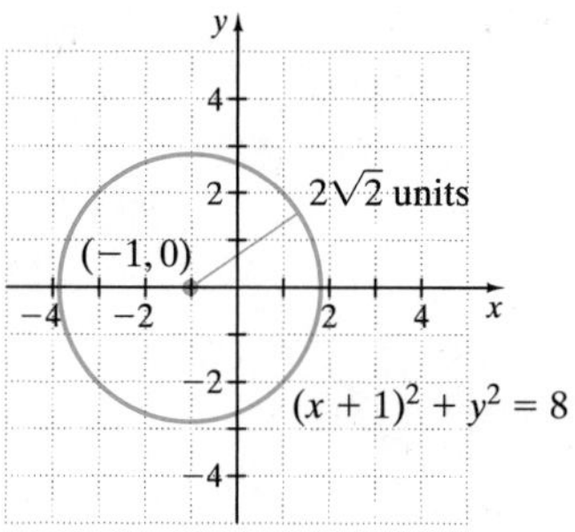

□

PRACTICE 5 Graph $(x - 3)^2 + (y + 2)^2 = 4$.

OBJECTIVE 3 ▶ Finding the Center and the Radius of a Circle by Completing the Square. To find the center and the radius of a circle from its equation, we use standard form. To write the equation of a circle in standard form, we complete the square on both x and y.

Helpful Hint

Later, we learn the following:

	Greatest Variable Power in Equation
Circles	x^2, y^2, same coefficients on same side of equation
Parabolas	x^2 or y^2, but not both
Ellipses	x^2, y^2, different coefficients, same sign on same side of equation
Hyperbolas	x^2, y^2, coefficients have opposite signs on same side of equation

EXAMPLE 6 Find the center and the radius; then graph $x^2 + y^2 + 4x - 8y = 16$.

Solution Since this equation contains x^2- and y^2-terms on the same side of the equation with equal coefficients, its graph is a circle. To write the equation in standard form, we group the terms involving x and the terms involving y, and then we complete the square on each variable.

$$(x^2 + 4x) + (y^2 - 8y) = 16$$

To complete the square on x, find the coefficient of x, 4, take half of 4, and then square the result. To complete the square on y, find the coefficient of y, -8, take half of -8, and then square the result.

Thus, $\frac{1}{2}(4) = 2$ and $2^2 = 4$. Also, $\frac{1}{2}(-8) = -4$ and $(-4)^2 = 16$.

Add 4 and then 16 to both sides.

$$(x^2 + 4x + 4) + (y^2 - 8y + 16) = 16 + 4 + 16$$

$$(x + 2)^2 + (y - 4)^2 = 36 \quad \text{Factor.}$$

This circle has the center $(-2, 4)$ and radius $\sqrt{36}$, or 6, as shown.

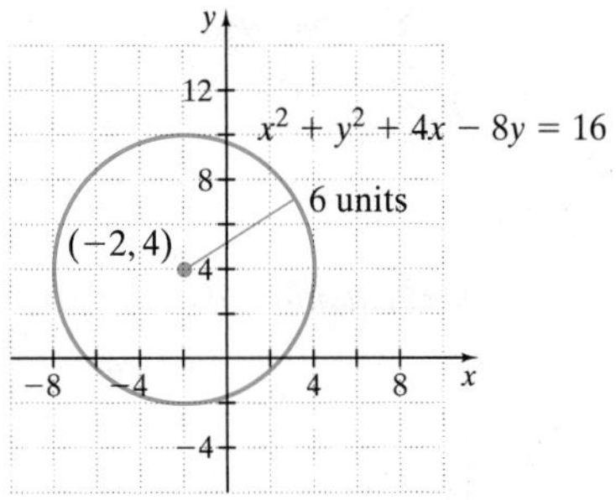

□

PRACTICE 6 Find the center and the radius; then graph $x^2 + y^2 + 6x - 2y = 6$.

VOCABULARY & READINESS CHECK

Word Bank *Use the choices below to fill in each blank. Some choices may be used more than once.*

radius diameter center circle

1. A ____________ is the set of all points in a plane that are the same distance from a fixed point. The fixed point is called the ____________.
2. The midpoint of a diameter of a circle is the ____________.
3. The distance from the center of a circle to any point on the circle is called the ____________.
4. Twice a circle's radius is its ____________.

12.5 EXERCISE SET MyMathLab®

Write the standard equation of each circle. See Example 1.

1. center $(2, -8)$; $r = 9$
2. center $(5, -1)$; $r = 12$
3. center $(-9, -4)$; $r = \sqrt{5}$
4. center $(-1, -1)$; $r = \sqrt{11}$
5. center $(-6, 3)$; $r = 8$
6. center $(-1, 6)$; $r = 5$
7. center $(0, 3)$; $r = 7$
8. center $(-4, 0)$; $r = 3$
9. center $(0, 0)$; $r = 4$
10. center $(0, 0)$; $r = 6$

Write a standard equation for each circle in the diagram at the right. See Example 2.

11. $\odot P$
12. $\odot Q$

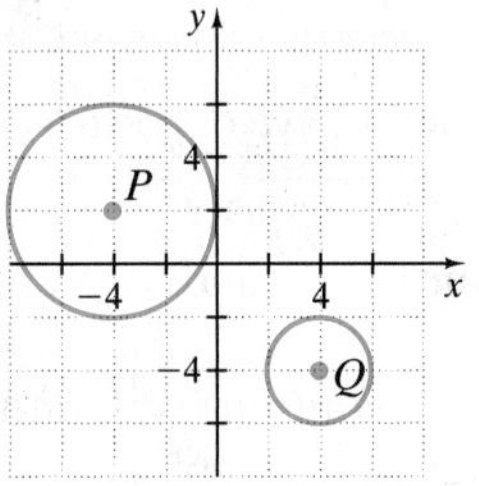

Write the standard equation of the circle with the given center that passes through the given point.

13. center $(-2, 6)$; point $(-2, 10)$
14. center $(7, -2)$; point $(1, -6)$
15. center $(6, 5)$; point $(0, 0)$
16. center $(1, 2)$; point $(0, 6)$
17. center $(-10, -5)$; point $(-5, 5)$
18. center $(-1, -4)$; point $(-4, 0)$

Find the center and radius of each circle. Then graph the circle. See Examples 3–5.

19. $(x + 7)^2 + (y - 5)^2 = 16$
20. $(x + 4)^2 + (y - 1)^2 = 25$
21. $(x - 3)^2 + (y + 8)^2 = 100$
22. $(x - 2)^2 + (y + 3)^2 = 1$
23. $x^2 + y^2 = 36$
24. $x^2 + y^2 = 4$
25. $(x + 8)^2 + y^2 = 7$
26. $x^2 + (y + 6)^2 = 10$

Write the standard equation of each circle.

27.

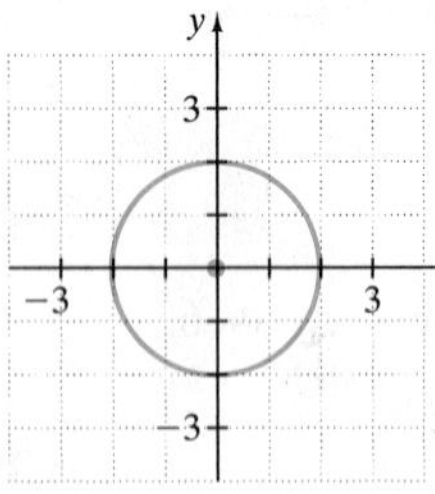

28.

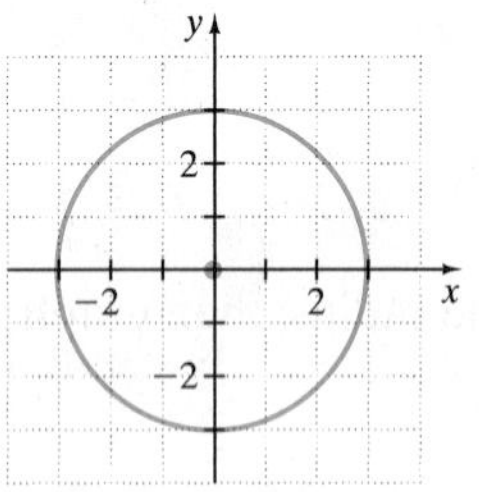

29.

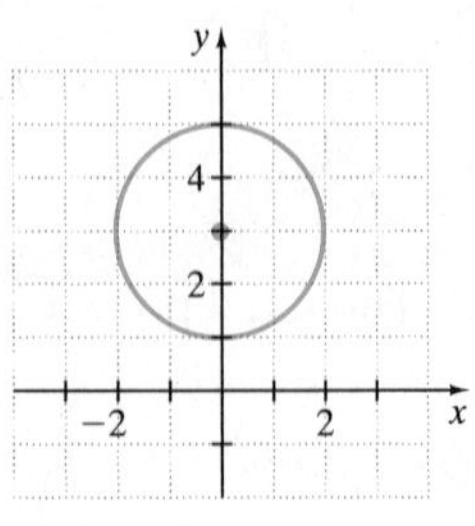

30.

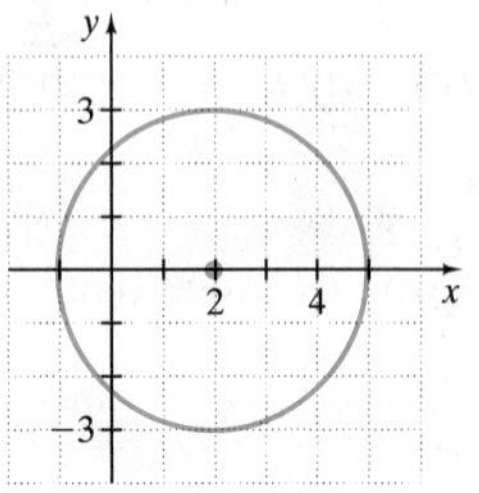

31.

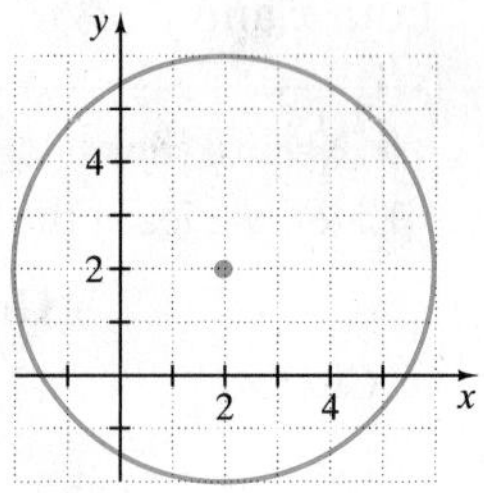

32.

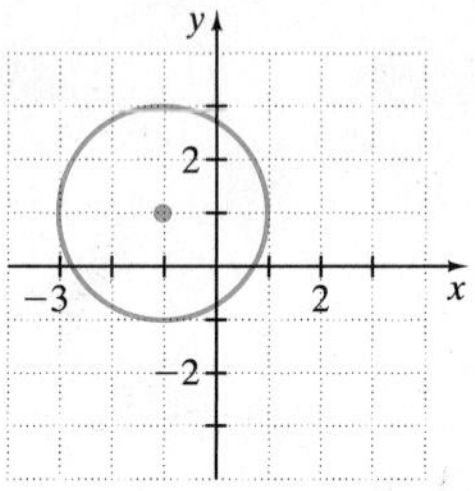

The graph of each equation is a circle. Find the center and the radius, and then sketch the circle. See Example 6.

33. $x^2 + y^2 + 6y = 0$
34. $x^2 + 10x + y^2 = 0$
35. $x^2 + y^2 + 2x - 4y = 4$
36. $x^2 + 6x - 4y + y^2 = 3$
37. $x^2 + y^2 - 4x - 8y - 2 = 0$
38. $x^2 + y^2 - 2x - 6y - 5 = 0$
39. $x^2 + y^2 + 6x + 10y - 2 = 0$
40. $x^2 + y^2 + 2x + 12y - 12 = 0$

41. As of this writing, the world's largest-diameter Ferris wheel currently in operation is the Star of Nanchung in Jiangxi Province, China. It has 60 compartments, each of which carries eight people. It is 160 meters tall, and the diameter of the wheel is 153 meters. (*Source:* China News Agency)

a. What is the radius of this Ferris wheel?
b. How close is the wheel to the ground?
c. How high is the center of the wheel from the ground?
d. Using the axes in the drawing, what are the coordinates of the center of the wheel?
e. Use parts **a** and **d** to write the equation of the wheel.

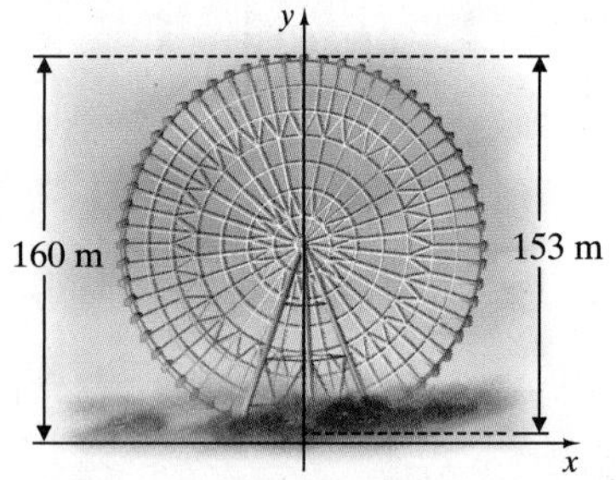

42. In 1893, Pittsburgh bridge builder George Ferris designed and built a gigantic revolving steel wheel whose height was 264 feet and whose diameter was 250 feet. This Ferris wheel opened at the 1893 exposition in Chicago. It had 36 wooden cars, each capable of holding 60 passengers. (*Source: The Handy Science Answer Book*)

a. What was the radius of this Ferris wheel?
b. How close was the wheel to the ground?
c. How high was the center of the wheel from the ground?
d. Using the axes in the drawing, what were the coordinates of the center of the wheel?
e. Use parts **a** and **d** to write the equation of the wheel.

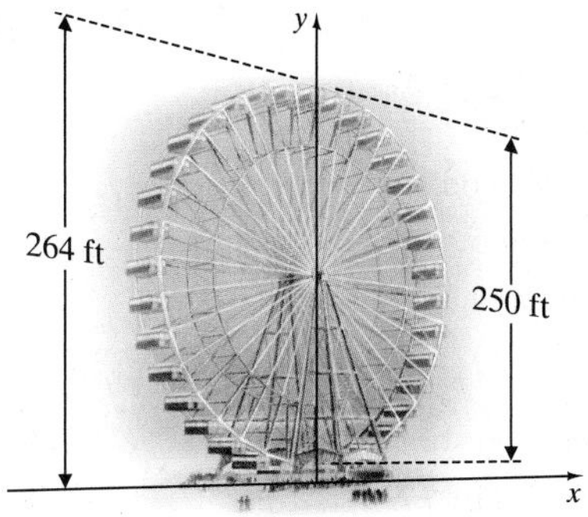

43. Opened in 2000 to celebrate the millennium, the British Airways London Eye is the world's biggest observation wheel. Each of the 32 enclosed capsules, which each hold 25 passengers, completes a full rotation every 30 minutes. Its diameter is 135 meters, and it is constructed on London's South Bank to allow passengers to enter the Eye at ground level. (*Source: Guinness Book of World Records*)

a. What is the radius of the London Eye?

b. How close is the wheel to the ground?

c. How high is the center of the wheel from the ground?

d. Using the axes in the drawing, what are the coordinates of the center of the wheel?

e. Use parts **a** and **d** to write the equation of the Eye.

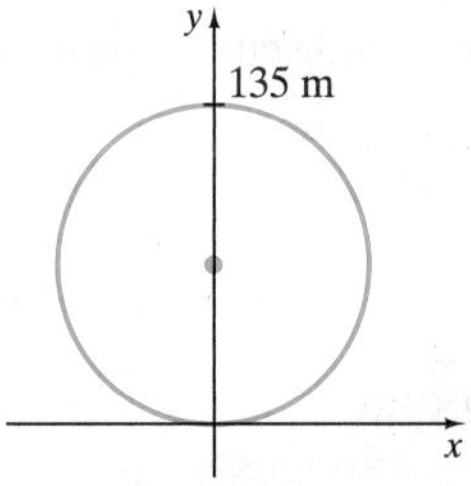

44. **The Sarsen Circle** The first image that comes to mind when one thinks of Stonehenge is the very large sandstone blocks with sandstone lintels across the top. The Sarsen Circle of Stonehenge is the outer circle of the sandstone blocks, each of which weighs up to 50 tons. There were originally 30 of these monolithic blocks, but only 17 remain upright today. The "altar stone" lies at the center of this circle, which has a diameter of 33 meters.

a. What is the radius of the Sarsen Circle?

b. What is the circumference of the Sarsen Circle? Round your result to two decimal places.

c. Since there were originally 30 Sarsen stones located on the circumference, how far apart would the centers of the stones have been? Round to the nearest tenth of a meter.

d. Using the axes in the drawing, what are the coordinates of the center of the circle?

e. Use parts **a** and **d** to write the equation of the Sarsen Circle.

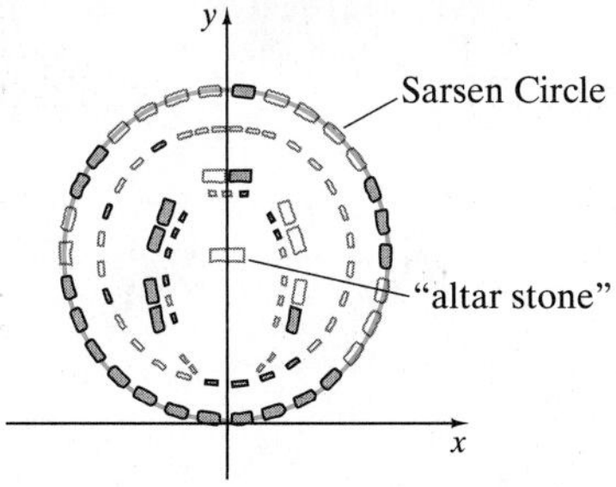

Write an equation of a circle with diameter $\overline{AB}$.

45. $A(0, 0), B(8, 6)$

46. $A(0, 0), B(5, 12)$

47. $A(3, 0), B(7, 6)$

48. $A(1, 1), B(5, 5)$

CONCEPT EXTENSIONS

Determine whether each equation is the equation of a circle. Justify your answer.

49. $(x - 1)^2 + (y + 2)^2 = 9$

50. $x + (y - 3)^2 = 9$

51. $x + y = 9$

52. $(x + 1)^2 + y = 3$

53. **Find the Error** A student says that the center of a circle with equation $(x - 2)^2 + (y + 3)^2 = 16$ is $(-2, 3)$. What is the student's error?

54. **Find the Error** A student says that the radius of a circle with equation $(x - 2)^2 + (y + 3)^2 = 16$ is 8. What is the student's error?

55. What is the least amount of information that you need to graph a circle? To write the equation of a circle?

56. Suppose you know the center of a circle and a point on the circle. How do you determine the equation of the circle?

A circle has the equation $(x - 9)^2 + (y - 3)^2 = 64$. Calculate each piece of information, and leave your answers in terms of π.

57. the circumference

58. the area

59. Write an equation of a circle with area 36π and center $(4, 7)$.

60. What are the x- and y-intercepts of the line tangent to the circle $(x - 2)^2 + (y - 2)^2 = 5^2$ at the point $(5, 6)$?

Sketch the graphs of each equation. Graph both equations on a single rectangle coordinate system. Find all points of intersection of each pair of graphs.

61. $\begin{cases} x^2 + y^2 = 13 \\ y = -x + 5 \end{cases}$

62. $\begin{cases} x^2 + y^2 = 17 \\ y = -\frac{1}{4}x \end{cases}$

63. $\begin{cases} x^2 + y^2 = 8 \\ y = 2 \end{cases}$

64. $\begin{cases} x^2 + y^2 = 20 \\ y = -\frac{1}{2}x + 5 \end{cases}$

65. $\begin{cases} (x + 1)^2 + (y - 1)^2 = 18 \\ y = x + 8 \end{cases}$

66. $\begin{cases} (x - 2)^2 + (y - 2)^2 = 10 \\ y = -\frac{1}{3}x + 6 \end{cases}$

67. The concentric circles $(x - 3)^2 + (y - 5)^2 = 64$ and $(x - 3)^2 + (y - 5)^2 = 25$ form a ring. The lines $y = \frac{2}{3}x + 3$ and $y = 5$ intersect the ring, making four sections. Find the area of each section. Round your answers to the nearest tenth of a square unit.

68. A close estimate of the radius of Earth's equator is 3960 mi.

a. Write the equation of the equator with the center of Earth as the origin.

b. Find the length of a 1° arc on the equator to the nearest tenth of a mile.

REVIEW AND PREVIEW

Find the value of each variable. See Section 12.4.

69.

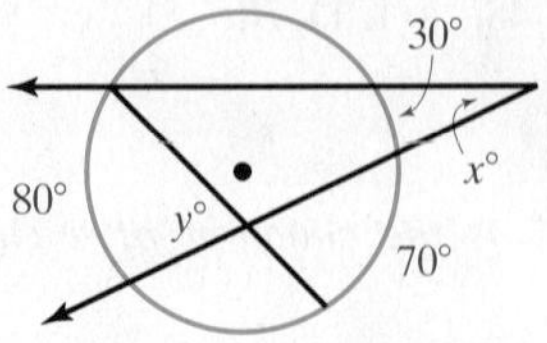

70.

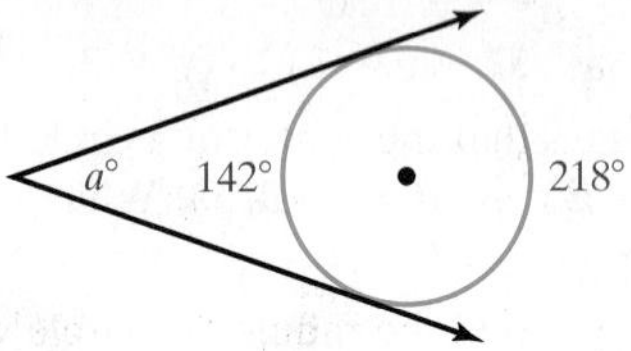

For the given vectors $\vec{a}$ and $\vec{c}$, write the sum $\vec{a} + \vec{c}$ as an ordered pair. See Section 9.5.

71. $\vec{a} = \langle -2, 5 \rangle$ and $\vec{c} = \langle 8, 7 \rangle$

72. $\vec{a} = \langle -3, -4 \rangle$ and $\vec{c} = \langle -2, 6 \rangle$

73. $\vec{a} = \langle 3, 1 \rangle$ and $\vec{c} = \langle 1, 3 \rangle$

74. $\vec{a} = \langle 9, -6 \rangle$ and $\vec{c} = \langle 2, -1 \rangle$

Construct each of the following. See Section 1.8.

75. the perpendicular bisector of $\overline{BC}$

76. $\angle EFG$ bisected by $\overrightarrow{FH}$

12.6 LOCUS

OBJECTIVE

1 Draw and Describe a Locus.

VOCABULARY

- locus

OBJECTIVE 1 ▶ Drawing and Describing a Locus of Points. A **locus** is a set of points, all of which meet a stated condition. *Loci* is the plural of locus. We can use the description of a locus to sketch a geometric relationship.

EXAMPLE 1 Describing a Locus in a Plane

What is a sketch and description for each locus of points in a plane?

a. the points in a plane 1 in. from a given point C

b. the points in a plane 1 cm from $\overline{AB}$

Solution

a. Draw a point C. Sketch several points 1 in. from C. Keep doing so until you see a pattern. Draw the figure the pattern suggests.

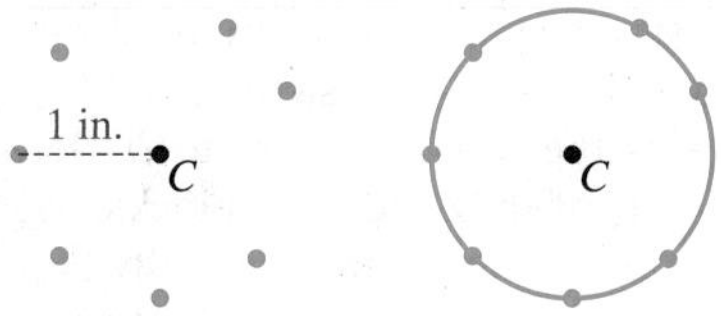

The locus is a circle with center C and radius 1 in.

b. Draw $\overline{AB}$. Sketch several points on either side of $\overline{AB}$. Also sketch points 1 cm from point A and point B. Keep doing so until you see a pattern. Draw the figure the pattern suggests.

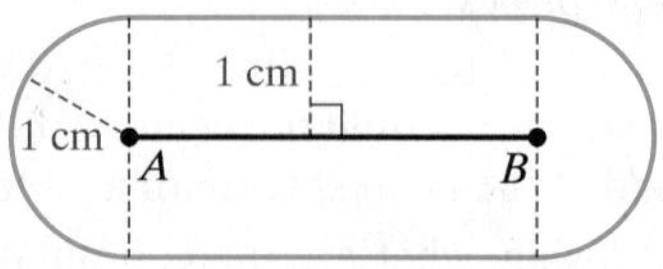

The locus is a pair of parallel segments, each 1 cm from $\overline{AB}$, and two semicircles with centers at A and B. □

PRACTICE 1 What is a sketch and a description of all points in a plane equidistant from two parallel lines?

We can use locus descriptions for geometric terms.

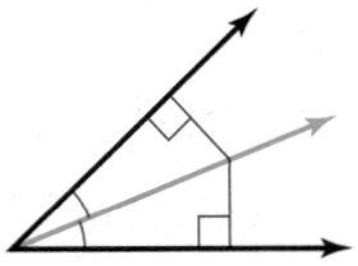

The locus of points in the interior of an angle that are equidistant from the sides of the angle is an angle bisector.

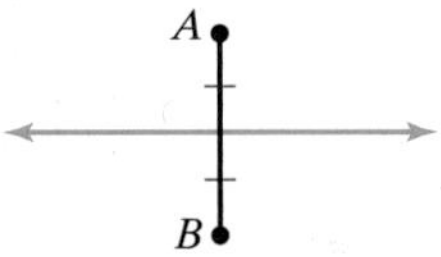

In a plane, the locus of points that are equidistant from a segment's endpoints is the perpendicular bisector of the segment.

Sometimes a locus is described by two conditions. We can draw the locus by first drawing the points that satisfy each condition. Then we find their intersection.

EXAMPLE 2 Drawing a Locus for Two Conditions

What is a sketch of the locus of points in a plane that satisfy these conditions?

- the points equidistant from intersecting lines k and m
- the points 5 cm from the point where k and m intersect

▶ **Helpful Hint**
Recall that the locus of points equidistant from a single point forms a circle.

Solution We start with intersecting lines k and m. Make a sketch to satisfy the first condition. Then sketch the second condition. Look for the points in common.

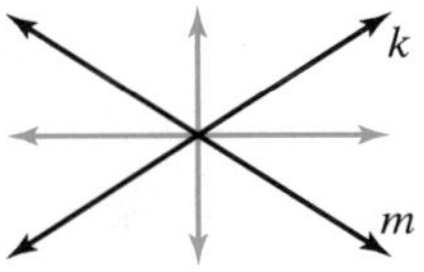

Sketch the points in a plane equidistant from lines k and m. These points form two lines that bisect the vertical angles formed by k and m.

Separately sketch the points in a plane 5 cm from the point where k and m intersect. (Recall that this is a circle.)

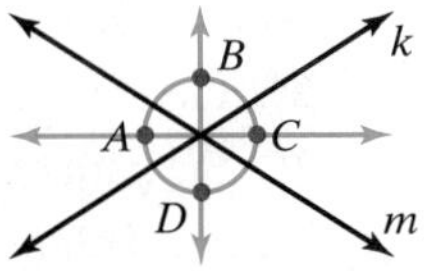

To locate the point, or set of points, that satisfies both conditions, we look for the intersection of our two sketches. This set of points is A, B, C, and D. □

PRACTICE 2 What is a sketch of the locus of points in a plane that satisfy these conditions?

- the points equidistant from two points X and Y
- the points 2 cm from the midpoint of $\overline{XY}$

EXAMPLE 3 Describing a Locus in Space

a. What is the locus of points in space that are c units from a point D?
b. What is the locus of points in space that are 3 cm from segment $\overline{AB}$?

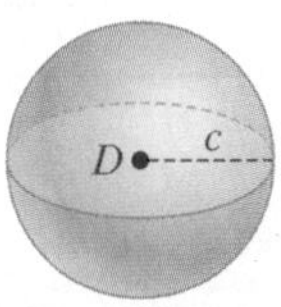

Solution

a. The locus is a sphere with center at point D and radius c units.
b. The locus is a cylinder with radius 3 cm and centerline $\overline{AB}$.

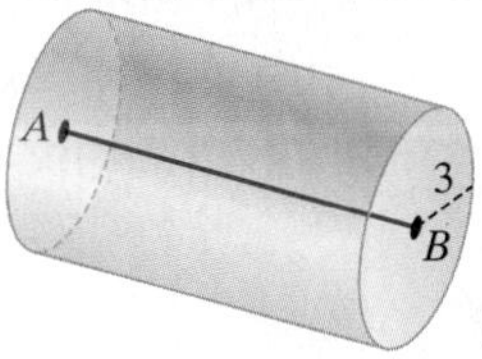

□

PRACTICE 3 What is each locus of points?

a. in a plane, the points that are equidistant from two parallel lines
b. in space, the points that are equidistant from two parallel planes

12.6 EXERCISE SET MyMathLab®

Mixed Practice *Sketch and describe each locus of points in a plane. See Examples 1 and 2.*

1. points 4 cm from a point X
2. points 2 in. from $\overline{UV}$
3. points 3 mm from $\overleftrightarrow{LM}$
4. points 1 in. from a circle with radius 3 in.
5. points equidistant from the endpoints of $\overline{PQ}$
6. points in the interior of $\angle ABC$ and equidistant from the sides of $\angle ABC$
7. points equidistant from two perpendicular lines
8. midpoints of radii of a circle with radius 2 cm
9. all points equidistant from two concentric circles with center, O, and radii 4 ft and 6ft
10. all points equidistant from two concentric circles with radii 2 in. and 2 ft
11. all points 7 units from a circle with radius 5 units
12. all points 5 units from a circle with radius 7 units
13. all points equidistant from parallel lines that are 6 units apart
14. all points equidistant from parallel lines that are 5 units apart

For Exercises 15–18, sketch the locus of points in a plane that satisfy the given conditions. See Example 2.

15. equidistant from points M and N and on a circle with center M and radius $= \frac{1}{2}MN$
16. 3 cm from $\overline{GH}$ and 5 cm from G, where $GH = 4.5\,\text{cm}$
17. equidistant from the sides of $\angle PQR$ and on a circle with center P and radius PQ
18. equidistant from points A and B and on a circle whose center is the midpoint of $\overline{AB}$ and diameter $= AB$.
19. Points equidistant from both points A and B and points C and D

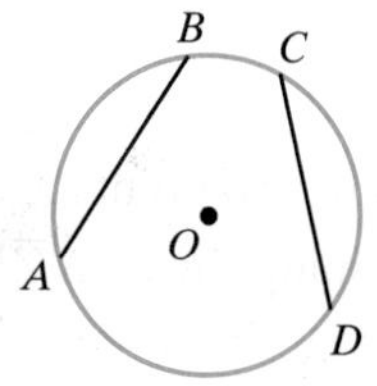

20. Points equidistant from the sides of $\angle JKL$ and on $\odot C$

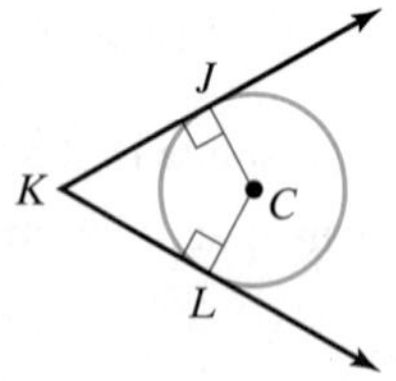

Use the graph to write the equation for each locus of points.

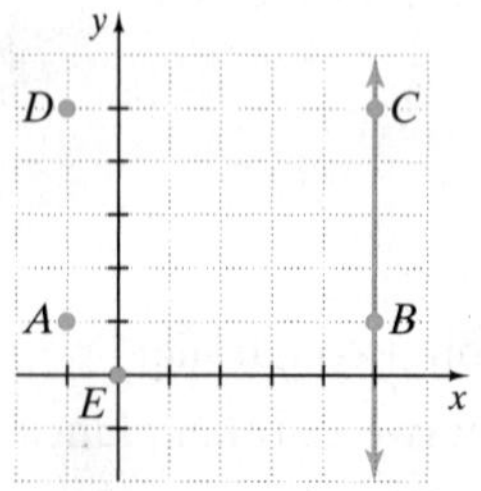

21. points equidistant from point A and point D
22. points equidistant from point D and point C
23. points 2 units from point E
24. points 3 units from point A
25. points 1 unit from $\overleftrightarrow{CB}$
26. points 2 units from $\overleftrightarrow{CB}$

Describe each locus of points in space. See Example 3.

27. points 3 cm from a point F
28. points 4 cm from $\overleftrightarrow{DE}$
29. points 1 in. from plane M
30. points 5 mm from $\overrightarrow{PQ}$

Describe the locus that each blue figure represents.

31.

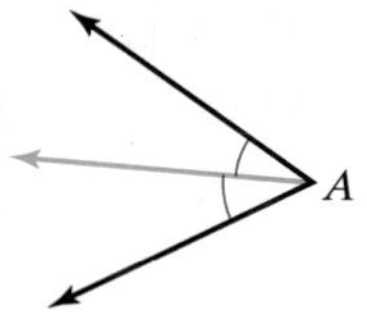

32.

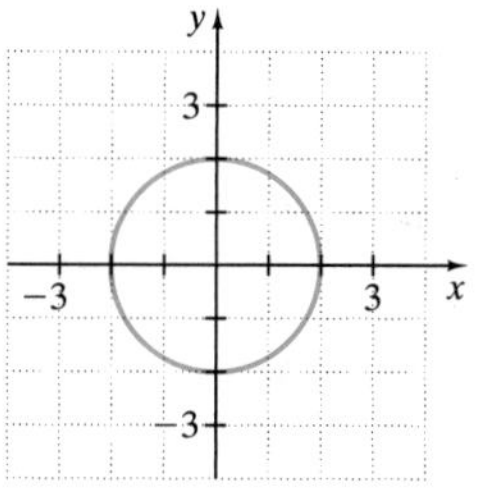

Coordinate Geometry *Write an equation for the locus of points in a plane equidistant from the two given points.*

33. $A(0, 2)$ and $B(2, 0)$
34. $P(1, 3)$ and $Q(5, 1)$

Make a drawing of each locus.

35. the path of a car as it turns to the right
36. the path of a doorknob as a door opens
37. the path of a knot in the middle of a jump rope as it is being used
38. the path of the tip of your nose as you turn your head
39. the path of a fast-pitched softball
40. the path of one foot when walking

The school plans to construct a fountain in front of the school. Use the diagram for Exercises 41 and 42.

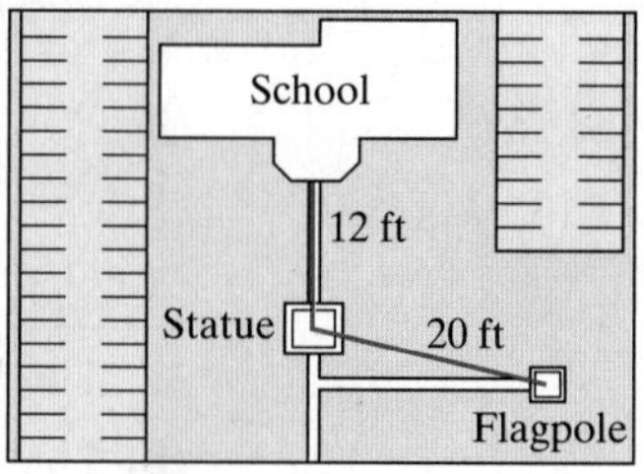

41. What are all the possible locations for a fountain if the fountain will be 8 ft from the statue and 16 ft from the flagpole?
42. Find the locations for a fountain if the fountain 16 ft from the statue and 8 ft from the flagpole.

Coordinate Geometry *Cell sites (usually a cellular phone tower) are commonly placed 1–2 miles apart (although closer in dense urban areas). Depending on terrain, the range of these sites can be from 2 to 50 miles. The technology of locating a cell phone is based on measuring signal strength and tower locations. For now, let's draw the locus of three cell sites with their range radii and find their intersection.*

43. At point $A(1, 2)$, the cell site has a dependable radius of 4 miles.

At point $B(-2, 1)$, the cell site radius is 3 miles.

At point $C(2, -3)$, the cell site radius is 7 miles.

Draw, then shade their intersection.

44. At point $A(0, 0)$ the cell site has a dependable radius of 40 miles.

At point $B(-10, 10)$, the cell site radius is 20 miles.

At point $C(-5, 20)$, the cell site radius is 10 miles.

Draw, then shade their intersection.

Coordinate Geometry *Draw each locus on the coordinate plane.*

45. all points 3 units from the origin

46. all points 2 units from $(-1, 3)$

47. all points 4 units from the y-axis

48. all points 5 units from $x = 2$

49. all points equidistant from $y = 3$ and $y = -1$

50. all points equidistant from $x = 4$ and $x = 5$

51. all points equidistant from the x- and y-axes

52. all points equidistant from $x = 3$ and $y = 2$

CONCEPT EXTENSIONS

Use the following two descriptions for Exercises 53 and 54.

- *in a plane, the points equidistant from points J and K*
- *in space, the points equidistant from points J and K*

53. How are the descriptions of the locus of points for each situation alike?

54. How are they different?

55. Give two examples of loci from everyday life, one in a plane and one in space.

56. A classmate says that it is impossible to find a point equidistant from three collinear points. Is the classmate correct? Explain.

Write a locus description of the points highlighted in blue on the coordinate plane. For Exercises 57 and 58, name each of the two conditions.

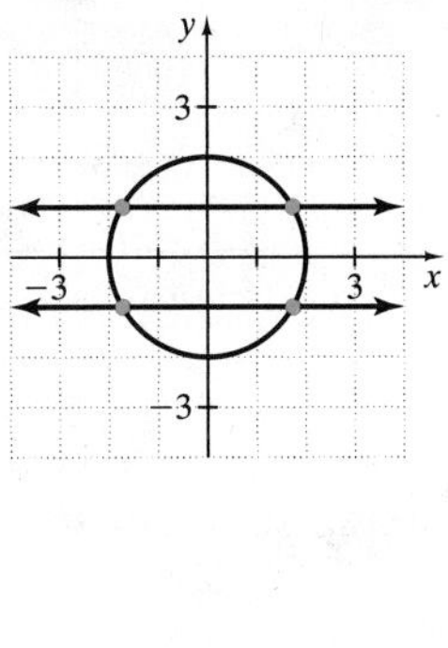

57. What is the condition with respect to the origin?

58. What are the conditions with respect to the x- and y-axes?

Solve.

59. Points A and B are 5 cm apart. Do the following loci in a plane have any points in common?

- the points 3 cm from A
- the points 4 cm from $\overline{AB}$

Illustrate your answer with a sketch.

60. a. Draw a segment to represent the base of an isosceles triangle. Locate three points that could be the vertex of the isosceles triangle.

b. Describe the locus of possible vertices for the isosceles triangle.

c. Explain why points in the locus you described are the only possibilities for the vertex of the isosceles triangle.

61. Describe the locus of points in a plane 3 cm from the points on a circle with radius 8 cm.

62. Describe the locus of points in a plane 8 cm from the points on a circle with radius 3 cm.

63. An interesting example of a locus is a cycloid. A cycloid is the path of a fixed point on a circle as the circle rolls along a straight line. Use a coin, a straight edge, and a pencil to draw a cycloid.

64. Sketch the locus of points for the air value on the tire of a bicycle as the bicycle moves down a straight path.

Think about the path of a child on each piece of playground equipment. Draw the path from (a) a top view, (b) a front view, and (c) a side view.

65. a swing

66. a straight slide

67. a firefighter's pole

68. a merry-go-round

REVIEW AND PREVIEW

Write an equation of the circle with center C and radius r. See Section 12.5.

69. $C(6, -10), r = 5$

70. $C(1, 7), r = 6$

Find the surface area of each figure to the nearest tenth. See Section 11.2.

71.

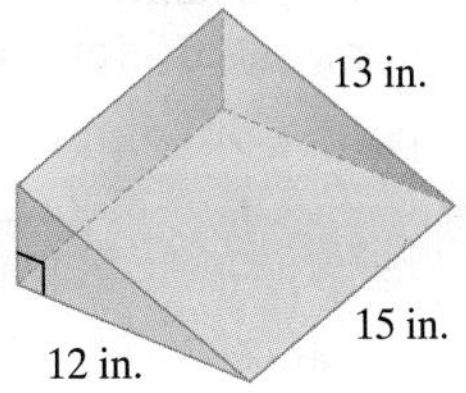

72.

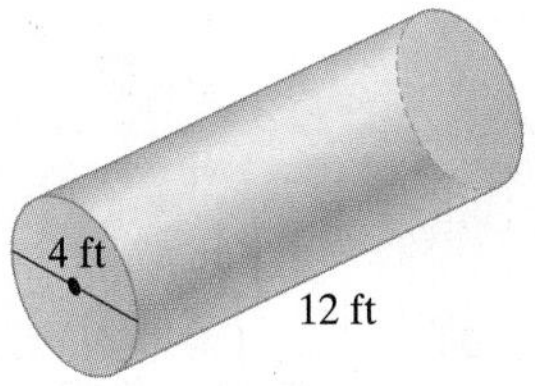

In ⊙O, find the area of sector AOB. Leave your answer in terms of π. See Section 10.6.

73. $OA = 4, m\widehat{AB} = 90°$

74. $OA = 8, m\widehat{AB} = 72°$

EXTENSION—PARABOLAS

OBJECTIVE

1 Find an Equation of a Parabola Given the Focus and Directrix.

VOCABULARY

- conic section
- parabola
- focus
- directrix

OBJECTIVE 1 ▶ Finding an Equation of a Parabola Given the Focus and Directrix. Parabolas, circles, ellipses, and hyperbolas are together called conic sections. A **conic section** is a curve generated by the intersection of a cone and a plane.

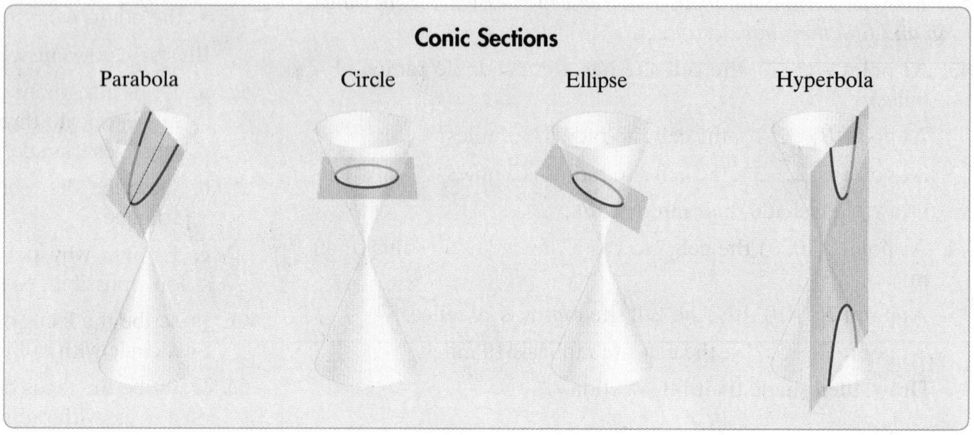

A **parabola** is the set of all points in a plane (the locus of points) such that the distance between a given point, called the focus, and a given line, called the directrix is equal. (Again, the given point is called the **focus** and the given line is called the **directrix**.)

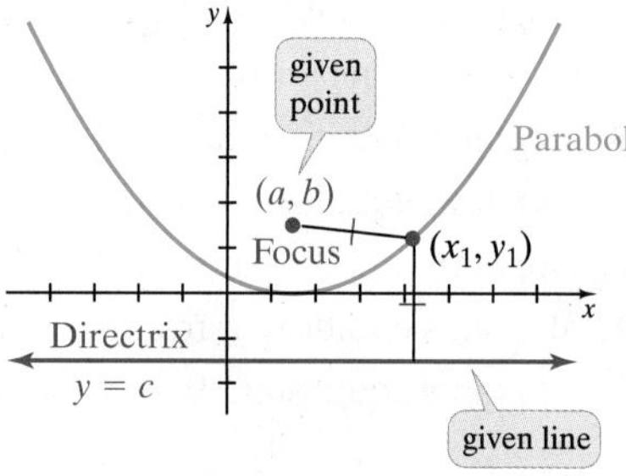

Our purpose for studying parabolas is to connect the above geometry definition of a parabola with the rectangular coordinate system. In other words, let's use the focus/directrix definition of a parabola and derive an algebraic equation that we will call the standard form equation of a parabola.

Here we go—we will use a horizontal line as our given line.

- Let (a, b) be the focus (given point) and
- let $y = c$ be the directrix (given line). Also,
- let (x_1, y_1) be a specific point on the parabola.

Remember our two distances that are equal:

Distance 1—the distance between the focus (a, b) and a point on the parabola, (x_1, y_1)

Distance 2—the distance between the directrix, $y = c$, and the same point, (x_1, y_1)

Let's calculate Distance 1 (d_1) and Distance 2 (d_2), and then set them equal to form our standard form equation.

Distance 1. Let's use the Distance Formula.

$$d_1 = \sqrt{(x_1 - a)^2 + (y_1 - b)^2}$$

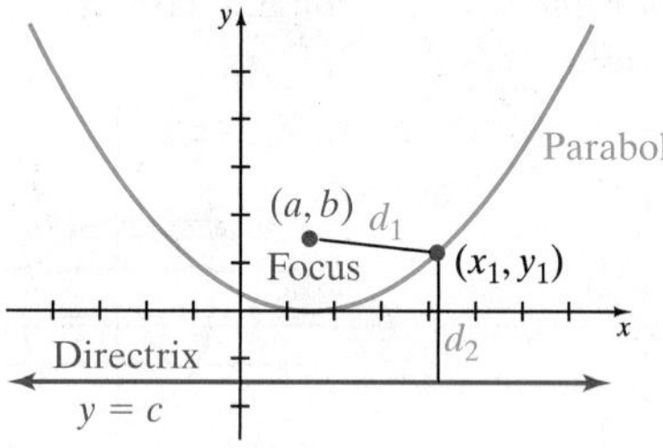

Distance 2. To calculate the vertical distance between the line $y = c$ and the point (x_1, y_1), just remember that a vertical distance is the distance between y-values, or

$$d_2 = |y_1 - c|$$

Since, by definition, these distances are the same, we set them equal and perform any simplifying necessary.

$$d_1 = d_2$$
$$\sqrt{(x_1 - a)^2 + (y_1 - b)^2} = |y_1 - c|$$
$$(x_1 - a)^2 + (y_1 - b)^2 = (y_1 - c)^2 \quad \text{Square both sides.}$$

Both sides will contain a $(y_1)^2$-term that will subtract out, so let's square any binomials that contain y_1.

$$(x_1 - a)^2 + (y_1)^2 - 2by_1 + b^2 = y_1^2 - 2cy_1 + c^2 \quad \text{Square } (y_1 - b)^2 \text{ and } (y_1 - c)^2.$$
$$(x_1 - a)^2 + b^2 - c^2 = 2by_1 - 2cy_1 \quad \text{Subtract } (y_1)^2 \text{ from both sides and move all } y_1\text{-terms to right side.}$$
$$(x_1 - a)^2 + b^2 - c^2 = 2(b - c)y_1 \quad \text{Factor out a 2 and a } y_1 \text{ from the terms on the right side.}$$

Finally, since this equation is true for all points (x, y) on the parabola, we use (x, y) instead of (x_1, y_1) for the standard form.

Helpful Hint

Notice that the equation of a parabola is in two variables, x and y, and that only one variable is squared.

Focus/Directrix Standard Form of a Parabola

A parabola with focus (a, b) and directrix $y = c$ has a standard form:

$$(x - a)^2 + b^2 - c^2 = 2(b - c)y$$

EXAMPLE 1 Find the equation of the parabola with focus (a, b) or $(4, 1)$ and directrix $y = 5$. Write the equation so it is solved for y.

Solution Let $a = 4$, $b = 1$, and $c = 5$.

$$(x - a)^2 + b^2 - c^2 = 2(b - c)y \quad \text{Focus/Directrix Standard Form of a Parabola}$$
$$(x - 4)^2 + 1^2 - 5^2 = 2(1 - 5)y \quad \text{Substitute } a = 4, b = 1, \text{ and } c = 5.$$
$$(x^2 - 8x + 16) + 1 - 25 = 2(-4)y \quad \text{Find } (x - 4)^2 \text{ and simplify.}$$
$$x^2 - 8x - 8 = -8y \quad \text{Combine like terms.}$$
$$\frac{x^2}{-8} - \frac{8x}{-8} - \frac{8}{-8} = \frac{-8y}{-8} \quad \text{Divide both sides by } -8 \text{ to solve for y.}$$
$$-\frac{1}{8}x^2 + x + 1 = y \quad \text{Simplify.}$$

Thus, the equation of the parabola is $y = -\frac{1}{8}x^2 + x + 1$. □

PRACTICE 1 Find the equation of the parabola with focus (2, 1) and directrix $y = 2$. Write the equation so it is solved for y.

EXTENSION—PARABOLAS MyMathLab®

Find the equation of each parabola with the given focus and directrix. Write the equation so it is solved for y. See Example 1.

1. Focus: (1, 5); Directrix: $y = 6$
2. Focus: (2, 7); Directrix: $y = 1$
3. Focus: (4, 3); Directrix: $y = -1$
4. Focus: (6, 2); Directrix: $y = -3$
5. Focus: $(5, -1)$; Directrix: $y = 4$
6. Focus: $(3, -3)$; Directrix: $y = 7$
7. Focus: (0, 4); Directrix: $y = -5$
8. Focus: (0, 8); Directrix: $y = -4$
9. Focus: (5, 0); Directrix: $y = 9$
10. Focus: (8, 0); Directrix: $y = 8$
11. Focus: $(-4, 2)$; Directrix: $y = -2$
12. Focus: $(-7, 6)$; Directrix: $y = -6$
13. Focus: $(-10, -3)$; Directrix: $y = 0$
14. Focus: $(-9, -5)$; Directrix: $y = 0$

CHAPTER 12 VOCABULARY CHECK

Fill in the Blank *Use the choices below to fill in each blank. Some choices may not be used.*

locus	tangent	inscribed angle	secant	radius
chord	center	intercepted arc	circle	

1. A segment whose endpoints are on a circle is called a(n) ____________.
2. A line that intersects a circle at two points is called a(n) ____________.
3. A(n) ____________ is a set of points, all of which meet a stated condition.
4. An angle whose vertex is on the circle and whose sides are chords of the circle is a(n) ____________.
5. A(n) ____________ to a circle is a line in the plane of the circle that intersects the circle in exactly one point.
6. An arc with endpoints on the sides of an inscribed angle and whose other points are in the interior of the angle is a(n) ____________.
7. The graph of the equation $(x - h)^2 + (y - k)^2 = r^2$ is a(n) ____________.
8. Using the form in Exercise 7, r is the ____________ and (h, k) is the ____________.

CHAPTER 12 REVIEW

(12.1) *Use* $\odot O$ *for Exercises 1–3. Use* $\odot R$ *for Exercise 4.*

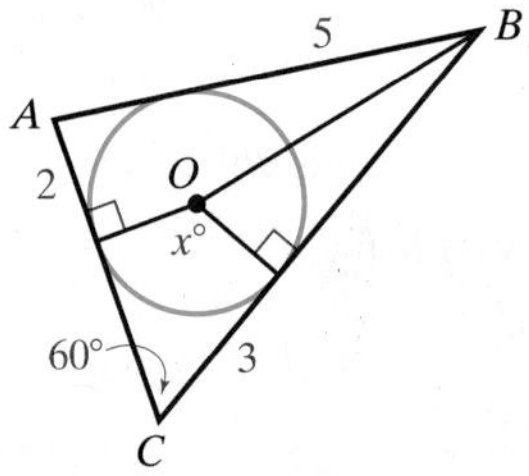

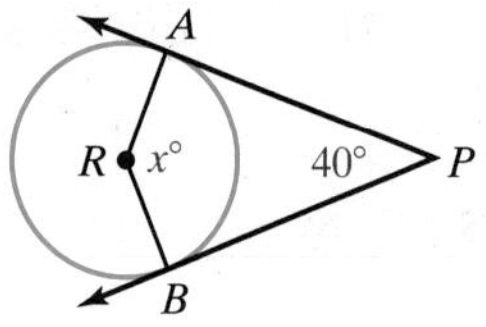

1. What is the perimeter of ΔABC?
2. $OB = \sqrt{28}$. What is the radius?
3. What is the value of x?
4. $\overrightarrow{PA}$ and $\overrightarrow{PB}$ are tangents. Find x.

(12.2) *Use the figure below for Exercises 5–7.*

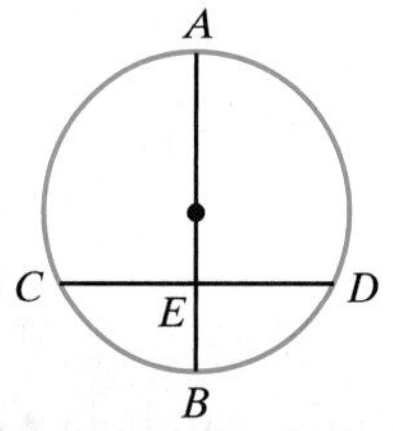

5. If $\overline{AB}$ is a diameter and $CE = ED$, then $m\angle AEC =$ ___?___.
6. If $\overline{AB}$ is a diameter and is at right angles to $\overline{CD}$, what is the ratio of CD to DE?
7. If $CE = \frac{1}{2}CD$ and $m\angle DEB = 90°$, what is true of $\overline{AB}$?

Use the circles below for Exercises 8–10. Give an exact answer. If needed, round to one decimal place.

8. What is the value of x?
9. What is the value of y?
10. What is the value of d?

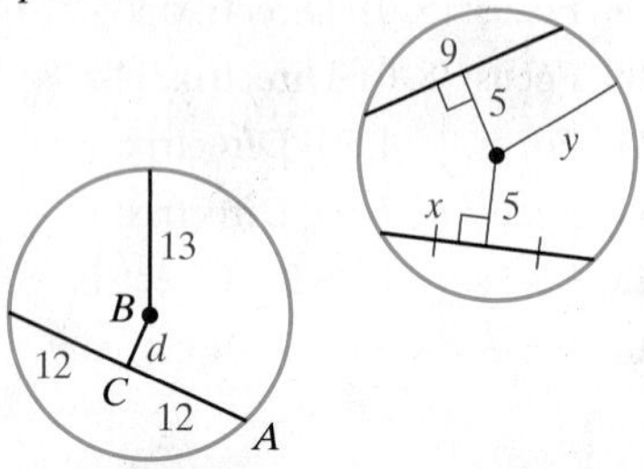

(12.3) *Find the value of each variable. Line* ℓ *is a tangent.*

11.

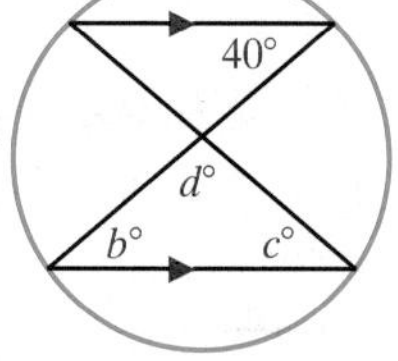

12.

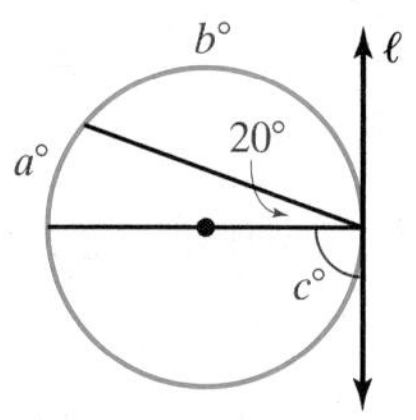

13.

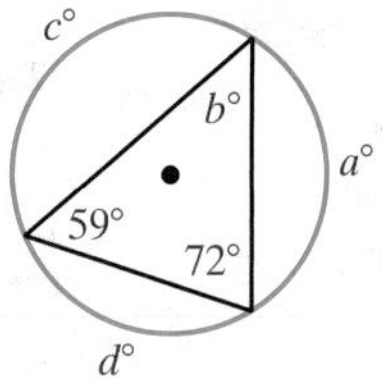

14.

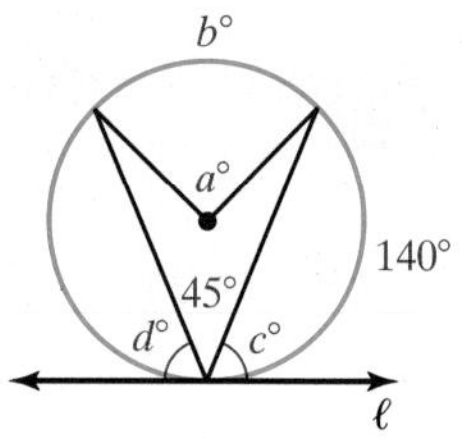

(12.4) *Find the value of each variable.*

15.

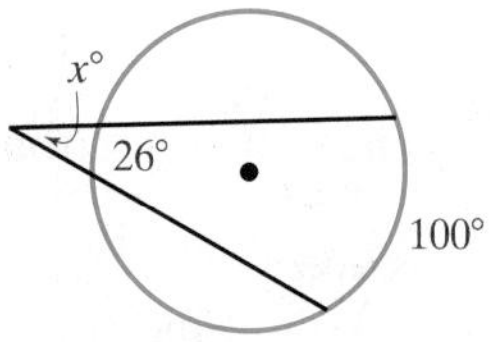

16.

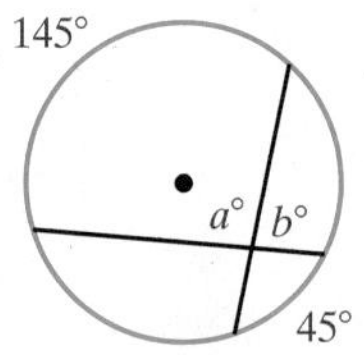

17.

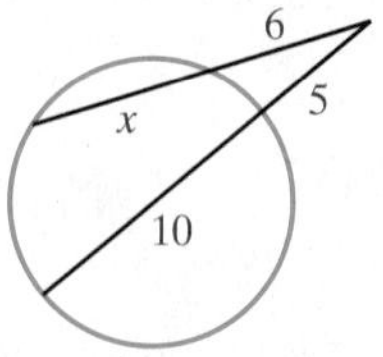

18.

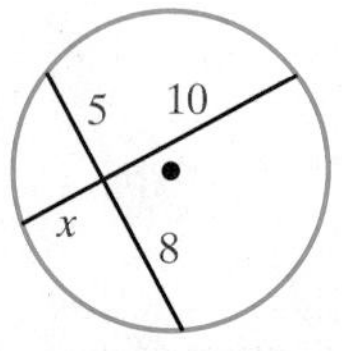

(12.5) *Write the standard equation of each circle below.*

19.

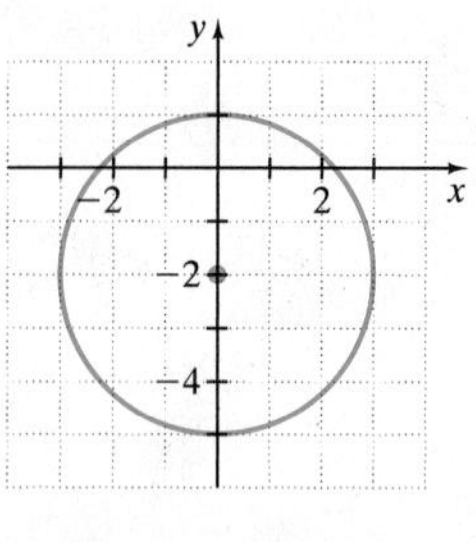

20.

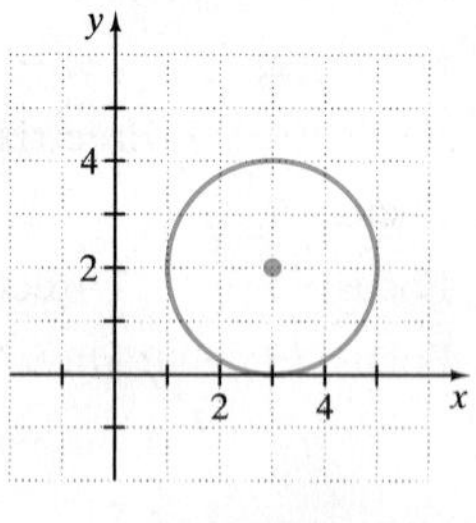

21. What is the standard equation of the circle with radius 5 and center $(-3, -4)$?

22. What is the standard equation of the circle with center $(1, 4)$ that passes through $(-2, 4)$?

23. What are the center and radius of the circle with equation $(x - 7)^2 + (y + 5)^2 = 36$?

24. Find the center and radius of the circle with equation $x^2 + y^2 + 4x + 2y = 1$.

(12.6) *Describe each locus of points.*

25. The set of all points in a plane that are in the interior of an angle and equidistant from the sides of the angle

26. The set of all points in a plane that are 5 cm from a circle with radius 2 cm

27. The set of all points in a plane at a distance 8 in. from a given line

28. The set of all points in space that are a distance 6 in. from $\overline{AB}$

MIXED PRACTICE

The polygon below circumscribes the circle. Find the perimeter of the polygon.

29.

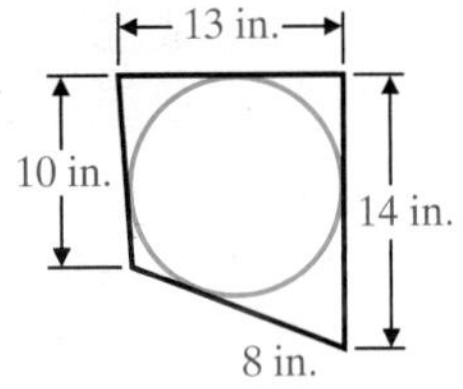

Find the value of each variable. Lines that appear to be tangent are tangent, and the dot represents the center.

30.

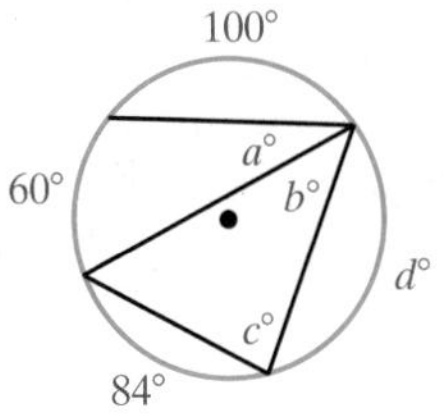

Find $m\widehat{AB}$.

31.

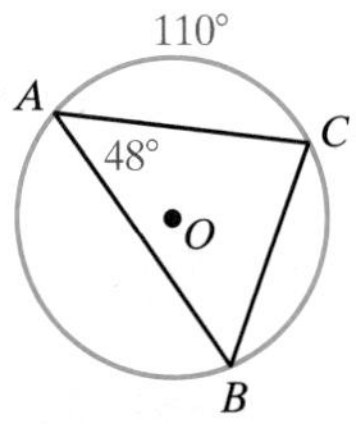

32.

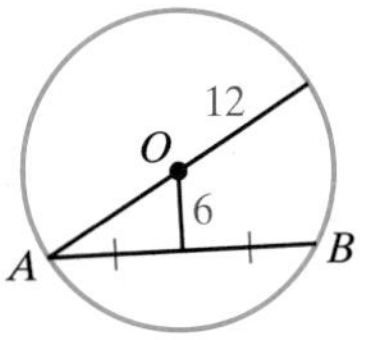

33. What is $m\widehat{PS}$? What is $m\angle R$?

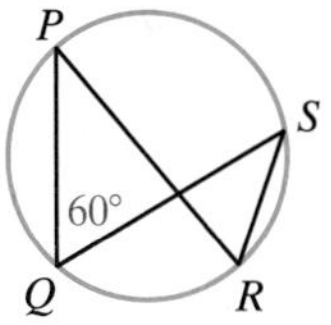

34. What is the value of x?

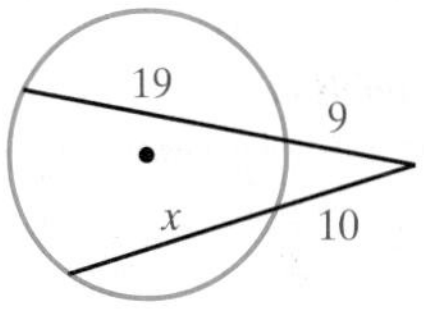

35. Write the standard equation of the circle shown.

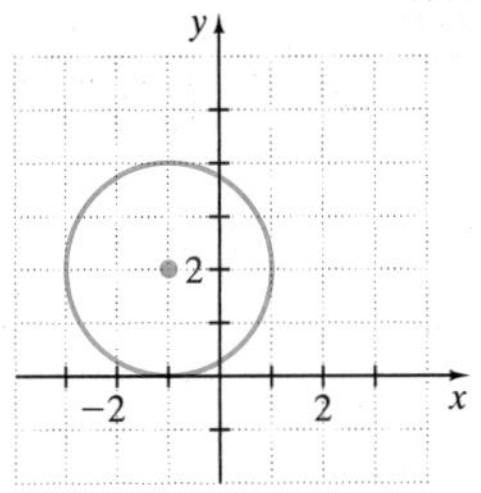

36. Sketch and describe the locus of points in a plane equidistant from points A and B.

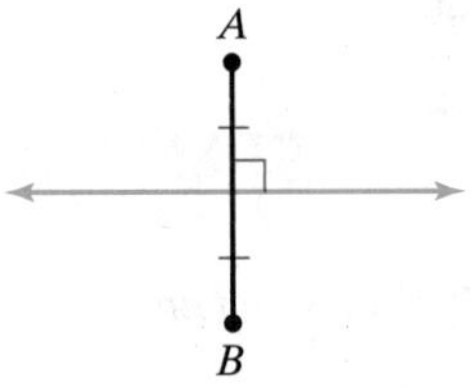

CHAPTER 12 TEST

TEST PREP VIDEO The fully worked-out solutions to any exercises you want to review are available in MyMathLab.

For Exercises 1–8, lines that appear tangent are tangent. Find the value of each variable. Round decimals to the nearest tenth.

1.

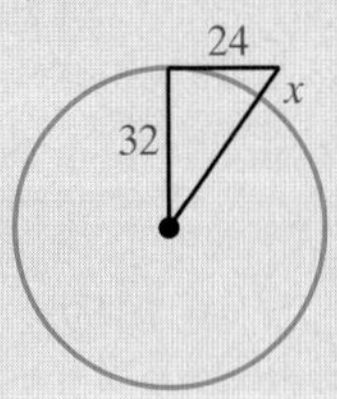

2.

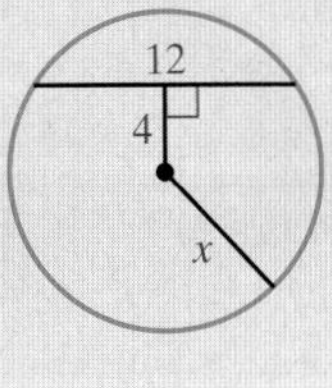

3.

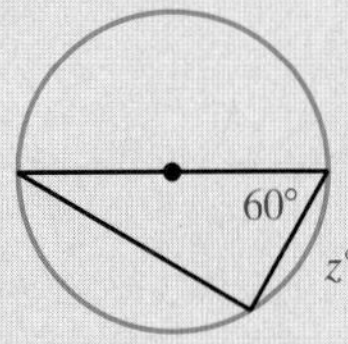

4.

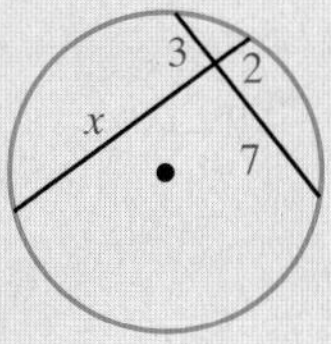

5.

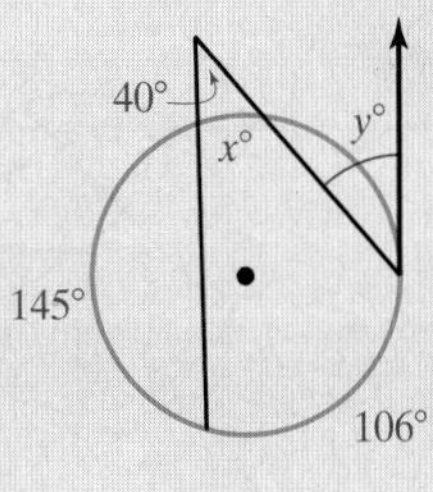

6.

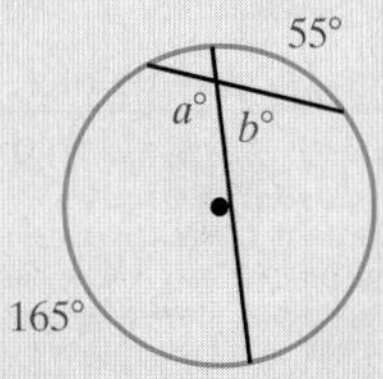

7.

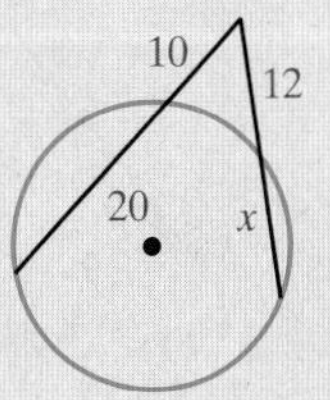

8.

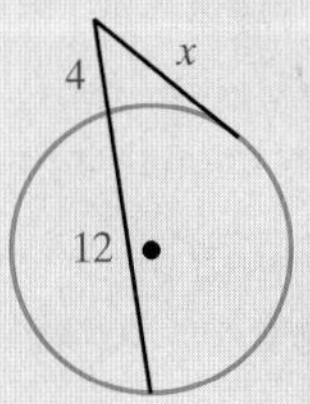

Find $m\widehat{AB}$.

9.

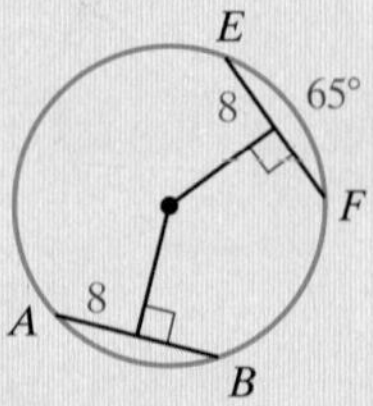

10.

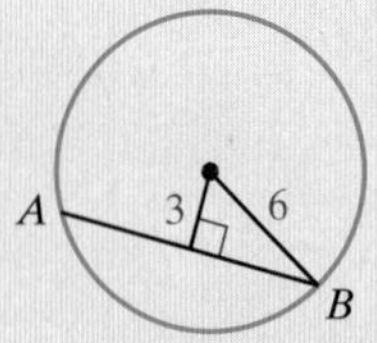

The triangle circumscribes the circle. Find the perimeter of the triangle.

11.

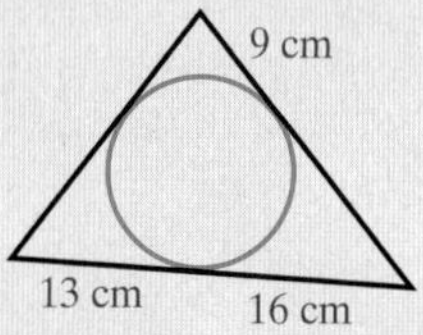

Find the value of each variable. Lines that appear to be tangent are tangent, and the dot represents the center.

12.

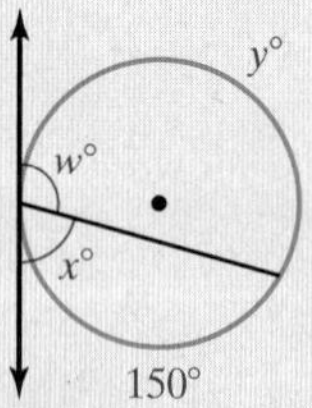

13.

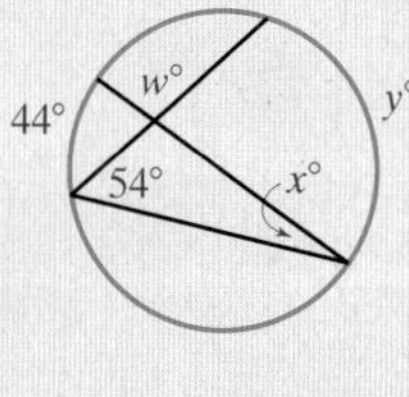

14. Graph $(x + 3)^2 + (y - 2)^2 = 9$. Then label the center and radius.

15. Write an equation of the circle with center $(3, 0)$ that passes through point $(-2, -4)$.

16. Graph $x^2 + y^2 = 10$. Label the center and radius.

17. Write an equation for the locus of points in the coordinate plane that are 4 units from $(-5, 2)$.

Write the standard equation of each circle.

18.

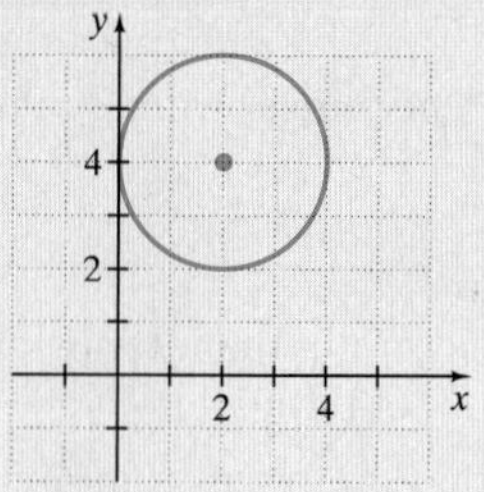

19.

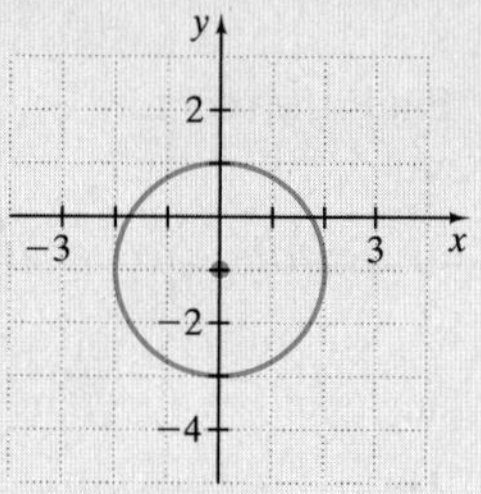

20. Find the center and radius of the circle with equation $x^2 + y^2 - 6x + 10y = -18$.

Sketch each locus on a coordinate plane.

21. the set of all points 3 units from the line $y = -2$

22. the set of all points equidistant from the axes

Write a two-column proof.

23. **Given:** $\odot A$ with $\overline{BC} \cong \overline{DE}$,
$\overline{AF} \perp \overline{BC}$,
$\overline{AG} \perp \overline{DE}$

Prove: $\angle AFG \cong \angle AGF$

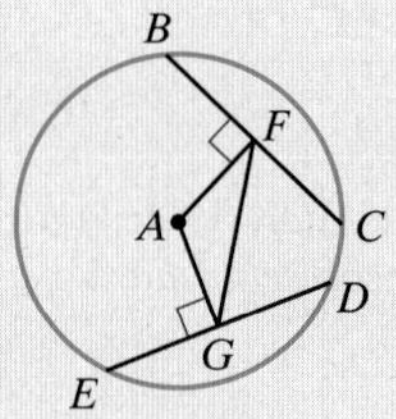

CHAPTER 12 STANDARDIZED TEST

For Questions 1–2, lines that appear tangent are tangent.

1. Find the value of x.

a. 50 **b.** 36 **c.** 2
d. not enough information given

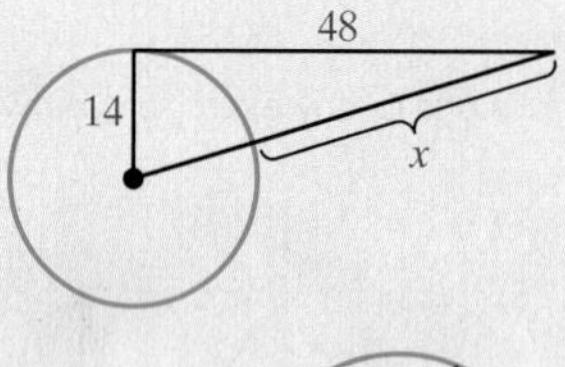

2. Find the value of x.

a. 12 **b.** ≈10.9 **c.** 13
d. not enough information given

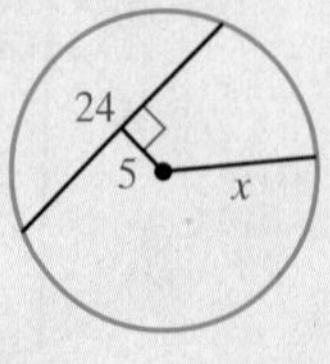

Use the figure for Questions 3–5.

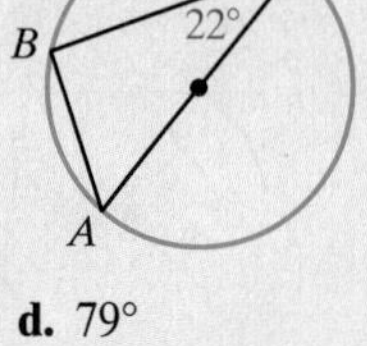

3. Find $m\angle A$.

a. 79° **b.** 68° **c.** 34°
d. not enough information given

4. Find $m\widehat{AB}$.

a. 22° **b.** 44° **c.** 11° **d.** 79°

5. Find $m\widehat{BC}$.

a. 34° **b.** 68° **c.** 136° **d.** 79°

Use the figure for Questions 6–8.

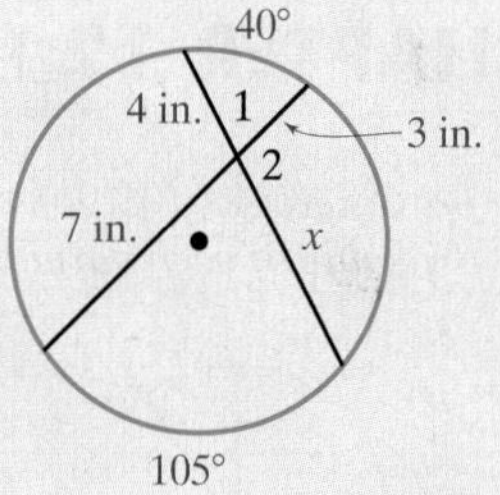

6. Find $m\angle 1$.

a. 90°
b. 32.5°
c. 20°
d. 72.5°

7. Find $m\angle 2$.

a. 90° **b.** 147.5° **c.** 168° **d.** 107.5°

8. Find the length labeled x.

a. 13.5 in. **b.** 6 in.
c. 5.25 in. **d.** 8 in.

Use the figure for Questions 9 and 10.

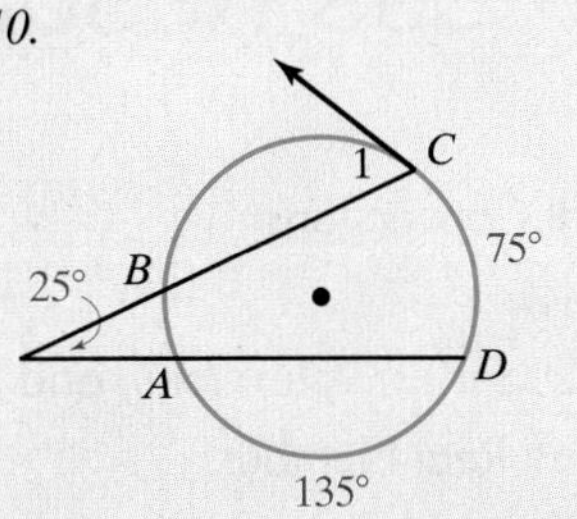

9. Find $m\overset{\frown}{AB}$.

a. 45°

b. 50°

c. 25°

d. not enough information given

10. Find $m\angle 1$.

a. 75° **b.** 100°

c. 62.5° **d.** not enough information given

Use the figure for Questions 11 and 12.

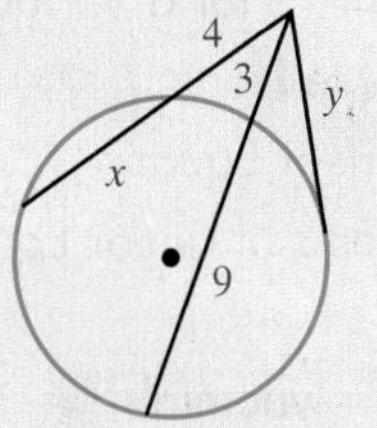

11. Find the value of x.

a. 5

b. 6.75

c. 8

d. 2

12. Find the value of y.

a. 6 **b.** ≈5.2

c. 3 **d.** not enough information given

13. Find the perimeter of the pentagon circumscribing the circle. Use the red points of tangency to help calculate the missing length.

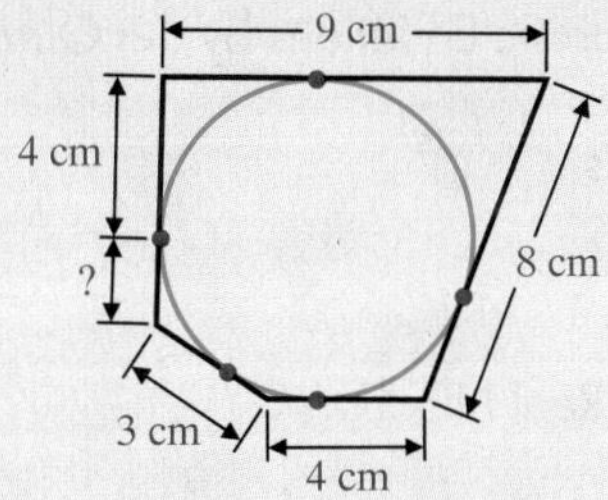

a. 28 cm

b. 31 cm

c. 32 cm

d. 30 cm

14. Which figure shows the graph of $(x + 1)^2 + (y - 1)^2 = 4$?

a.

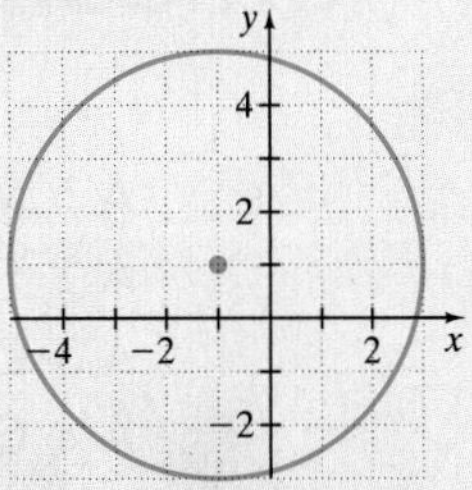

b.

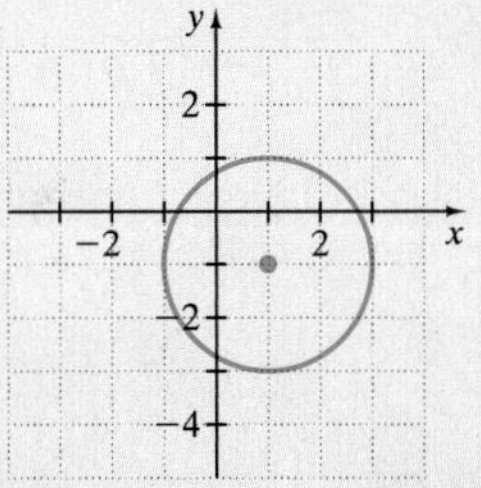

c.

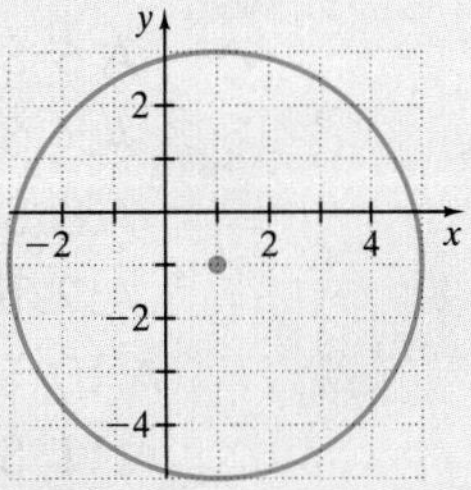

d.

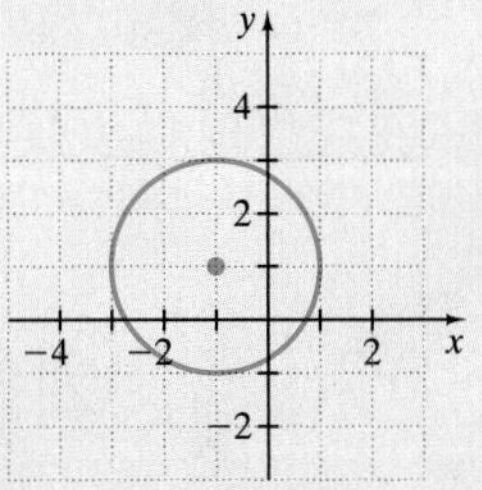

15. Which is the standard equation of the circle with center (8, −2) and radius 3?

a. $(x + 8)^2 + (y - 2)^2 = 3$

b. $(x - 8)^2 + (y + 2)^2 = 9$

c. $(x + 8)^2 + (y - 2)^2 = 9$

d. $(x - 8)^2 + (y + 2)^2 = 3$

16. Find the center and radius of the circle with equation $x^2 + y^2 - 4x + 6y = 23$.

a. center (−2, 3); radius 6

b. center (−2, 3); radius 36

c. center (2, −3); radius 36

d. center (2, −3); radius 6

17. Which is an equation for the locus of points in a plane equidistant from A (1, 1) and B (−1, −1).

a. $y = -x$ **b.** $y = 0$

c. $x = 0$ **d.** $y = x$

Student Success Resource Section

A Review of Basic Concepts

A.1 Measurement Conversions
A.2 Probability
A.3 Exponents, Order of Operations, and Variable Expressions
A.4 Operations on Real Numbers
A.5 Simplifying Expressions
A.6 Solving Linear Equations
A.7 Solving Linear Inequalities
A.8 Solving Formulas for a Variable
A.9 The Coordinate Plane
A.10 Graphing Linear Equations
A.11 Solving Systems of Linear Equations in Two Variables
A.12 Exponents
A.13 Multiplying Polynomials
A.14 Simplifying Radical Expressions
A.15 Solving Quadratic Equations by Factoring
A.16 Solving Quadratic Equations by the Square Root Property
A.17 Solving Quadratic Equations by the Quadratic Formula

Tables

1. Math Symbols
2. Formulas
3. Measures
4. Properties of Real Numbers

Postulates, Theorems, and Additional Proofs

B Additional Lessons

B.1 Ellipses and Hyperbolas
B.2 Measurement, Rounding Error, and Reasonableness
B.3 The Effect of Measurement Errors on Calculations

A REVIEW OF BASIC CONCEPTS

A.1 Measurement Conversions

To convert from one unit of measure to another, we multiply by a conversion factor in the form of a fraction. The numerator and denominator are in different units, but they represent the same amount. So we can think of this as multiplying by 1.

An example of a conversion factor is $\frac{1 \text{ ft}}{12 \text{ in.}}$. You can create other conversion factors using the measures table in the Tables section of these Student Success Resources.

EXAMPLE

1. Complete each statement.

a. 88 in. = ____ ft

$$88 \text{ in.} \cdot \frac{1 \text{ ft}}{12 \text{ in.}} = \frac{88}{12} \text{ ft} = 7\frac{1}{3} \text{ ft}$$

b. 5.3 m = ____ cm

$$5.3 \text{ m} \cdot \frac{100 \text{ cm}}{1 \text{ m}} = 5.3(100) \text{ cm} = 530 \text{ cm}$$

Perimeter and circumference are measured in units, area and surface area are measured in square units, and volume is measured in cubic units.

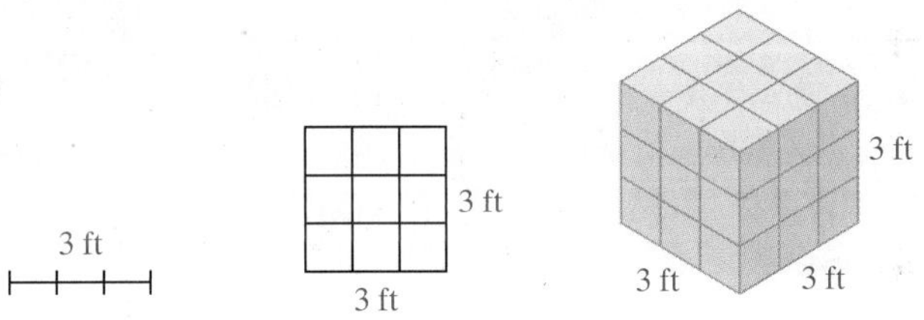

1 yd = 3 ft 1 sq yd = 9 sq ft 1 cu yd = 27 cu ft

EXAMPLE

1. Complete each statement.

a. 300 in.^{2} = _____ ft^2

1 ft = 12 in., so 1 ft^2 = (12 in.)^2 = 144 in.^{2}.

$$300 \text{ in.}^2 \cdot \frac{1 \text{ ft}^2}{144 \text{ in.}^2} = 2\frac{1}{12} \text{ ft}^2$$

b. 200,000 cm^3 = ____ m^3

1 m = 100 cm, so 1 m^3 = (100 cm)^3 = 1,000,000 cm^3

$$200{,}000 \text{ cm}^3 \cdot \frac{1 \text{ m}^3}{1{,}000{,}000 \text{ cm}^3} = 0.2 \text{ m}^3$$

A.1 EXERCISES MyMathLab®

Complete each statement.

1. 40 cm = ____ m
2. 1.5 kg = ____ g
3. 60 cm = ____ mm
4. 200 in. = ____ ft
5. 28 yd = ____ in.
6. 1.5 mi = ____ ft
7. 15 g = ____ mg
8. 430 mg = ____ g
9. 34 L = ____ mL
10. 1.2 m = ____ cm
11. 43 mm = ____ cm
12. 3600 sec = ____ min
13. 14 gal = ____ qt
14. 4500 lb = ____ t
15. 234 min = ____ hr
16. 3 ft^2 = ____ in.^{2}
17. 108 m^2 = ____ cm^2
18. 21 cm^2 = ____ mm^2
19. 1.4 yd^2 = ____ ft^2
20. 0.45 km^2 = ____ m^2
21. 1300 ft^2 = ____ yd^2
22. 1030 in.^{2} = ____ ft^2
23. 20,000,000 ft^2 = ____ mi^2
24. 1000 cm^3 = ____ m^3

A.2 Probability

Probability is a measure of the likelihood of an event occurring. All probabilities range from 0 to 1, where 0 is the probability of an event that cannot happen and 1 is the probability of an event that is certain to happen. An event with probability 0.5, or 50%, has an equal chance of happening or not happening.

The formula $P(E) = \dfrac{\text{number of favorable outcomes}}{\text{number of possible outcomes}}$ is used to calculate the probability of event E.

EXAMPLES

1. The numbers 2 through 21 are written on pieces of paper and placed in a hat. One piece of paper is drawn at random. Determine the probability of selecting a perfect square.

The total number of outcomes, 2, 3, 4, . . . , 21, for this event is 20.

There are 3 favorable outcomes: 4, 9, and 16.

$$P(\text{selecting a perfect square}) = \frac{3}{20}$$

2. Draw a tree diagram for tossing a coin and then choosing a number from 1 to 4.

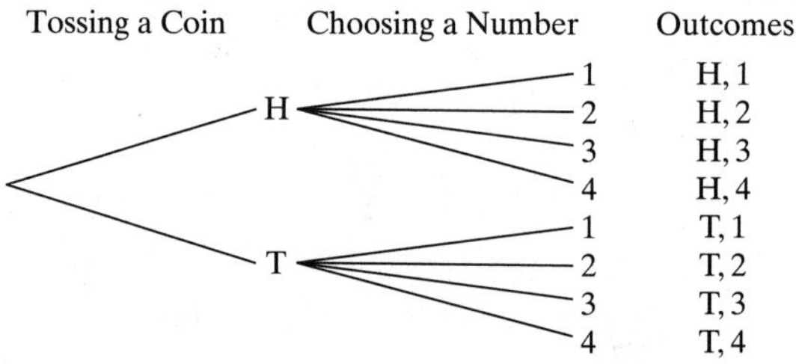

A.2 EXERCISES MyMathLab®

Draw a tree diagram for each experiment. Then use the diagram to find the number of possible outcomes.

1. Choosing a letter in the word MATH, then a number (1, 2, or 3)

2. Choosing a number (1 or 2) and then a vowel (a, e, i, o, u)

If a single die is tossed once, find the probability of each event.

3. A 5

4. A 9

5. A 1 or a 6

6. A 2 or a 3

7. An even number

8. An odd number

9. A number greater than 2

10. A number less than 6

Suppose the spinner shown is spun once. Find the probability of each event.

11. The result of the spin is 2.

12. The result of the spin is 3.

13. The result of the spin is 1, 2, or 3.

14. The result of the spin is not 3.

If a single choice is made from the bag of marbles shown, find the probability of each event.

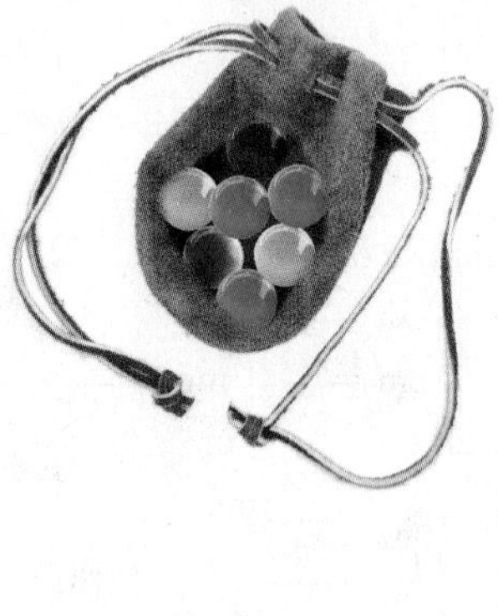

15. A red marble is chosen.

16. A blue marble is chosen.

17. A yellow marble is chosen.

18. A green marble is chosen.

A.3 Exponents, Order of Operations, and Variable Expressions

The expression a^n is an **exponential expression**. The number a is called the **base**; it is the repeated factor. The number n is called the **exponent**; it is the number of times that the base is a factor.

Order of Operations

1. Perform all operations within grouping symbols first, starting with the innermost set.
2. Evaluate exponential expressions.
3. Multiply or divide in order from left to right.
4. Add or subtract in order from left to right.

To evaluate an algebraic expression containing a variable, substitute a given number for the variable. Then simplify.

EXAMPLES **Evaluate**

1. $4^3 = 4 \cdot 4 \cdot 4 = 64$
2. $7^2 = 7 \cdot 7 = 49$
3. $$\frac{8^2 + 5(7-3)}{3 \cdot 7} = \frac{8^2 + 5(4)}{21} = \frac{64 + 5(4)}{21} = \frac{64 + 20}{21} = \frac{84}{21} = 4$$
4. Evaluate $x^2 - y^2$ when $x = 5$ and $y = 3$.

$$x^2 - y^2 = (5)^2 - 3^2 = 25 - 9 = 16$$

A.3 EXERCISES MyMathLab®

Evaluate.

1. 3^5
2. 5^4
3. 3^3
4. 4^4
5. $\left(\frac{2}{3}\right)^4$
6. $\left(\frac{6}{11}\right)^2$
7. $(1.2)^2$
8. $(0.4)^3$

Simplify each expression.

9. $5 + 6 \cdot 2$
10. $8 + 5 \cdot 3$
11. $4 \cdot 8 - 6 \cdot 2$
12. $12 \cdot 5 - 3 \cdot 6$
13. $18 \div 3 \cdot 2$
14. $48 \div 6 \cdot 2$
15. $2[5 + 2(8 - 3)]$
16. $3[4 + 3(6 - 4)]$
17. $\frac{|6 - 2| + 3}{8 + 2 \cdot 5}$
18. $\frac{15 - |3 - 1|}{12 - 3 \cdot 2}$
19. $2 + 3[10(4 \cdot 5 - 16) - 30]$
20. $3 + 4[8(5 \cdot 5 - 20) - 41]$

Evaluate each expression when $x = 1$, $y = 3$, and $z = 5$.

21. $\frac{z}{5x}$
22. $\frac{y}{2z}$
23. $|2x + 3y|$
24. $|5z - 2y|$

Evaluate each expression when $x = 12$, $y = 8$, and $z = 4$.

25. $x^2 - 3y + x$
26. $y^2 - 3x + y$
27. $\frac{x^2 + z}{y^2 + 2z}$
28. $\frac{y^2 + x}{x^2 + 3y}$

A.4 Operations on Real Numbers

Absolute Value

$$|a| = \begin{cases} a \text{ if } a \text{ is 0 or a positive number} \\ -a \text{ if } a \text{ is a negative number} \end{cases}$$

Adding Real Numbers

1. To add two numbers with the same sign, add their absolute values and attach their common sign.
2. To add two numbers with different signs, subtract the smaller absolute value from the larger absolute value and attach the sign of the number with the larger absolute value.

Subtracting Real Numbers

$$a - b = a + (-b)$$

Multiplying and Dividing Real Numbers

The product or quotient of two numbers with the same sign is positive.

The product or quotient of two numbers with different signs is negative.

A natural number **exponent** is a shorthand notation for repeated multiplication of the same factor.

The notation $\sqrt{a}$ is used to denote the **positive**, or **principal square root** of a nonnegative number a.

$$\sqrt{a} = b \text{ if } b^2 = a \text{ and } b \text{ is positive}$$

Also,

$$\sqrt[3]{a} = b \text{ if } b^3 = a$$

$$\sqrt[4]{a} = b \text{ if } b^4 = a \text{ and } b \text{ is positive.}$$

EXAMPLES **Find each absolute value.**

1. $|3| = 3$ **2.** $|0| = 0$ **3.** $|-7.2| = 7.2$

Perform the indicated operations.

4. $\frac{2}{7} + \frac{1}{7} = \frac{3}{7}$ **5.** $-5 + (-2.6) = -7.6$ **6.** $-18 + 6 = -12$

7. $20.8 + (-10.2) = 10.6$ **8.** $18 - 21 = 18 + (-21) = -3$

9. $(-8)(-4) = 32$ **10.** $8 \cdot 4 = 32$ **11.** $\frac{-8}{-4} = 2$ **12.** $\frac{8}{4} = 2$

13. $-17 \cdot 2 = -34$ **14.** $4(-1.6) = -6.4$ **15.** $\frac{-14}{2} = -7$ **16.** $\frac{22}{-2} = -11$

Evaluate.

17. $3^4 = 3 \cdot 3 \cdot 3 \cdot 3 = 81$ **18.** $\sqrt{49} = 7$ **19.** $\sqrt[3]{64} = 4$ **20.** $\sqrt[4]{16} = 2$

A.4 EXERCISES MyMathLab®

Find each absolute value.

1. $|-4|$
2. $|-6|$
3. $-|3|$
4. $-|11|$
5. $-\left|-\frac{2}{9}\right|$
6. $-\left|-\frac{5}{13}\right|$

Perform the indicated operations.

7. $-3 + 8$
8. $12 + (-7)$
9. $-14 + (-10)$
10. $-5 + (-9)$
11. $-4.3 - 6.7$
12. $-8.2 - (-6.6)$
13. $15 - (18)$
14. $13 - 17$
15. $-5 \cdot 12$
16. $-3 \cdot 8$
17. $-8(-10)$
18. $-4(-11)$
19. $0(-1)$
20. $0(-34)$
21. $\frac{-9}{3}$
22. $\frac{-20}{5}$
23. $\frac{-12}{-4}$
24. $\frac{-36}{-6}$
25. $\frac{0}{-2}$
26. $\frac{0}{-11}$
27. $\frac{-2}{0}$
28. $\frac{-22}{0}$

Evaluate or find the value of each expression.

29. -7^2
30. $(-7)^2$
31. $(-2)^3$
32. -2^3
33. $\left(-\frac{1}{3}\right)^3$
34. $\left(-\frac{1}{2}\right)^4$

Evaluate or find each root.

35. $\sqrt{49}$
36. $\sqrt{81}$
37. $\sqrt[3]{64}$
38. $\sqrt[4]{81}$
39. $\sqrt{\frac{4}{25}}$
40. $\sqrt{\frac{4}{81}}$

A.5 Simplifying Expressions

To combine like terms, add the numerical coefficients and multiply the result by the common variable factor.

To remove parentheses, apply the Distributive Property.

EXAMPLES **Simplify by combining like terms**

1. $9y + 3y$
$= 12y$

2. $-4z^2 + 5z^2 - 6z^2$
$= -5z^2$

Simplify the expression.

3. $-4(x + 7) + 10(3x - 1)$
$= -4x - 28 + 30x - 10$
$= 26x - 38$

A.5 EXERCISES MyMathLab®

Simplify each expression by combining any like terms.

1. $7y + 8y$

2. $3x + 2x$

3. $m - 4m + 2m - 6$

4. $a + 3a - 2 - 7a$

5. $7x^2 + 8x^2 - 10x^2$

6. $8x^3 + x^3 - 11x^3$

7. $6x + 0.5 - 4.3x - 0.4x + 3$

8. $0.4y - 6.7 + y - 0.3 - 2.6y$

Remove parentheses and simplify each expression.

9. $7(d - 3) + 10$

10. $9(z + 7) - 15$

11. $3(2x - 5) - 5(x - 4)$

12. $2(6x - 1) - (x - 7)$

13. $5(x + 2) - (3x - 4)$

14. $4(2x - 3) - (x + 1)$

15. $2 + 4(6x - 6)$

16. $8 + 4(3x - 4)$

17. $\frac{1}{2}(12x - 4) - (x + 5)$

18. $\frac{1}{3}(9x - 6) - (x - 2)$

Perform each indicated operation. Don't forget to simplify if possible.

19. Add $6x + 7$ to $4x - 10$.

20. Add $3y - 5$ to $y + 16$.

21. Subtract $5m - 6$ from $m - 9$.

22. Subtract $m - 3$ from $2m - 6$.

A.6 Solving Linear Equations

To Solve Linear Equations

1. Clear the equation of fractions.
2. Remove any grouping symbols such as parentheses.
3. Simplify each side by combining like terms.
4. Get all variable terms on one side and all numbers on the other side by using the Addition Property of Equality.
5. Get the variable alone by using the Multiplication Property of Equality.
6. Check the solution by substituting it into the original equation.

EXAMPLE

1. Solve: $\dfrac{5(-2x+9)}{6} + 3 = \dfrac{1}{2}$

$$6 \cdot \frac{5(-2x+9)}{6} + 6 \cdot 3 = 6 \cdot \frac{1}{2}$$

$$5(-2x+9) + 18 = 3 \quad \text{Apply the Distributive Property.}$$

$$-10x + 45 + 18 = 3$$

$$-10x + 63 = 3 \quad \text{Combine like terms.}$$

$$-10x + 63 - 63 = 3 - 63 \quad \text{Subtract 63.}$$

$$-10x = -60$$

$$\frac{-10x}{-10} = \frac{-60}{-10} \quad \text{Divide by } -10.$$

$$x = 6$$

A.6 EXERCISES MyMathLab®

Solve each equation.

1. $-2(3x-4) = 2x$

2. $-(5x-10) = 5x$

3. $5(2x-1) - 2(3x) = 1$

4. $3(2-5x) + 4(6x) = 12$

5. $-2y - 10 = 5y + 18$

6. $-7n + 5 = 8n - 10$

7. $0.50x + 0.15(70) = 35.5$

8. $0.40x + 0.06(30) = 9.8$

9. $5x - 5 = 2(x+1) + 3x - 7$

10. $3(2x-1) + 5 = 6x + 2$

11. $\dfrac{x}{2} - 1 = \dfrac{x}{5} + 2$

12. $\dfrac{x}{5} - 7 = \dfrac{x}{3} - 5$

13. $2(x+3) - 5 = 5x - 3(1+x)$

14. $4(2+x) + 1 = 7x - 3(x-2)$

A.7 Solving Linear Inequalities

Properties of inequalities are similar to properties of equations. However, if you multiply or divide both sides of an inequality by the same *negative* number, you must reverse the direction of the inequality symbol.

To Solve Linear Inequalities

1. Clear the inequality of fractions.
2. Remove grouping symbols.
3. Simplify each side by combining like terms.
4. Write all variable terms on one side and all numbers on the other side using the Addition Property of Inequality.
5. Get the variable alone by using the Multiplication Property of Inequality.

EXAMPLES **Solve each inequality**

1. $-2x \le 4$

$\frac{-2x}{-2} \ge \frac{4}{-2}$ Divide by -2; reverse the inequality symbol.

$x \ge -2$

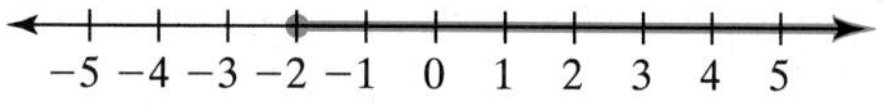

2. Solve: $3(x + 2) \le -2 + 8$

$3(x + 2) \le -2 + 8$ No fractions to clear

$3x + 6 \le -2 + 8$ Apply the Distributive Property.

$3x + 6 \le 6$ Combine like terms.

$3x + 6 - 6 \le 6 - 6$ Subtract 6.

$3x \le 0$

$\frac{3x}{3} \le \frac{0}{3}$ Divide by 3.

$x \le 0$

The solution set is $\{x \mid x \le 0\}$.

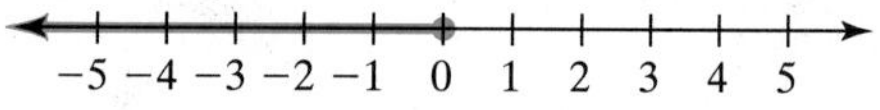

A.7 EXERCISES MyMathLab®

Solve each inequality. Graph the solution set. Write each answer using solution set notation.

1. $x - 2 \ge -7$

2. $x + 4 \le 1$

3. $3x - 5 > 2x - 8$

4. $3 - 7x \ge 10 - 8x$

5. $-8x \le 16$

6. $-5x < 20$

Solve each inequality. Write each answer using solution set notation.

7. $7(x + 1) - 6x \ge -4$

8. $10(x + 2) - 9x \le -1$

9. $3x - 7 < 6x + 2$

10. $2x - 1 \ge 4x - 5$

11. $-6x + 2 \ge 2(5 - x)$

12. $-7x + 4 > 3(4 - x)$

13. $3(x + 2) - 6 > -2(x - 3) + 14$

14. $7(x - 2) + x \le -4(5 - x) - 12$

15. $\frac{1}{4}(x + 4) < \frac{1}{5}(2x + 3)$

16. $\frac{1}{2}(x - 5) < \frac{1}{3}(2x - 1)$

A.8 Solving Formulas for a Variable

An equation that describes a known relationship among quantities is called a **formula**.

To solve a formula for a specified variable, use the same steps as for solving a linear equation. Treat the specified variable as the only variable of the equation.

EXAMPLE

1. Solve: $P = 2l + 2w$ for l.

$$P = 2l + 2w$$

$$P - 2w = 2l + 2w - 2w \quad \text{Subtract } 2w.$$

$$P - 2w = 2l$$

$$\frac{P - 2w}{2} = \frac{2l}{2} \quad \text{Divide by 2.}$$

$$\frac{P - 2w}{2} = l$$

A.8 EXERCISES MyMathLab®

Substitute the given values into each given formula and solve for the unknown variable.

1. $A = bh$; $A = 45$, $b = 15$ (Area of a parallelogram)

2. $d = rt$; $d = 195$, $t = 3$ (Distance Formula)

3. $A = \frac{1}{2}h(B + b)$; $A = 180$, $B = 11$, $b = 7$ (Area of a trapezoid)

4. $A = \frac{1}{2}h(B + b)$; $A = 60$, $B = 7$, $b = 3$ (Area of a trapezoid)

5. $C = 2\pi r$; $C = 15.7$ (Circumference of a circle) (Use the approximation 3.14 for π.)

6. $A = \pi r^2$; $r = 4$ (Area of a circle) (Use the approximation 3.14 for π.)

Solve each formula for the specified variable.

7. $f = 5gh$ for h

8. $x = 4\pi y$ for y

9. $V = lwh$ for w

10. $T = mnr$ for n

11. $3x + y = 7$ for y

12. $-x + y = 13$ for y

13. $P = a + b + c$ for a

14. $PR = x + y + z + w$ for z

15. $S = 2\pi rh + 2\pi r^2$ for h

16. $S = 4lw + 2wh$ for h

A.9 The Coordinate Plane

Two number lines that intersect at right angles form a coordinate plane. The horizontal axis is the x-axis and the vertical axis is the y-axis. The axes intersect at the origin and divide the coordinate plane into four sections called quadrants.

An ordered pair of numbers names the location of a point in the plane. These numbers are the coordinates of the point. Point B has coordinates $(-3, 4)$.

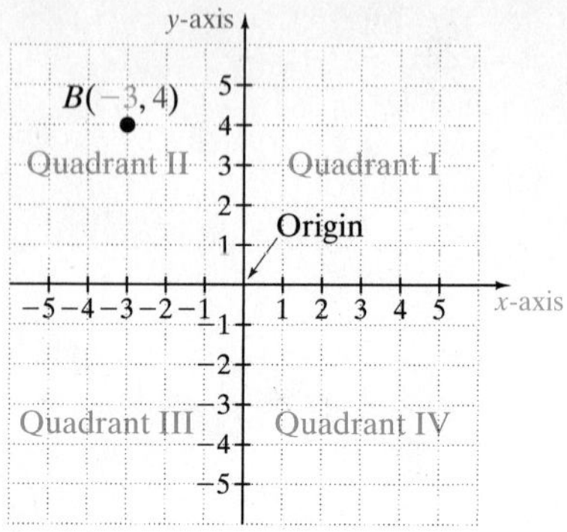

The first coordinate is the x-coordinate. $(-3, 4)$ The second coordinate is the y-coordinate.

We use the x-coordinate to tell how far to move right (positive) or left (negative) from the origin. We then use the y-coordinate to tell how far to move up (positive) or down (negative) to reach the point (x, y).

EXAMPLE

1. Graph each point in the coordinate plane. In which quadrant or on which axis would you find each point?

   a. Graph point $A(-2, 3)$ in the coordinate plane. To graph $A(-2, 3)$, move 2 units to the left of the origin. Then move 3 units up. Since the x-coordinate is negative and the y-coordinate is positive, point A is in Quadrant II.

   b. Graph point $B(2, 0)$ in the coordinate plane. To graph $B(2, 0)$, move 2 units to the right of the origin. Since the y-coordinate is 0, point B is on the x-axis.

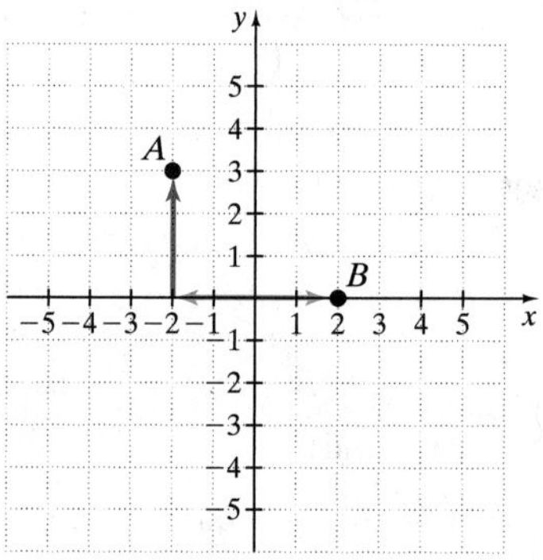

A.9 EXERCISES MyMathLab®

Plot each ordered pair. Use one rectangular coordinate system for the eight points in Exercise 1 and the same for Exercise 2. State in which quadrant or on which axis each point lies.

1. a. $(1, 5)$ **b.** $(-5, -2)$ **c.** $(-3, 0)$ **d.** $(0, -1)$

e. $(2, -4)$ **f.** $\left(-1, 4\frac{1}{2}\right)$ **g.** $(3.7, 2.2)$ **h.** $\left(\frac{1}{2}, -3\right)$

2. a. $(2, 4)$ **b.** $(0, 2)$ **c.** $(-2, 1)$ **d.** $(-3, -3)$

e. $\left(3\frac{3}{4}, 0\right)$ **f.** $(5, -4)$ **g.** $(-3.4, 4.8)$ **h.** $\left(\frac{1}{3}, -5\right)$

Find the x- and y-coordinates of each labeled point.

3. A **4.** B

5. C **6.** D

7. E **8.** F

9. G

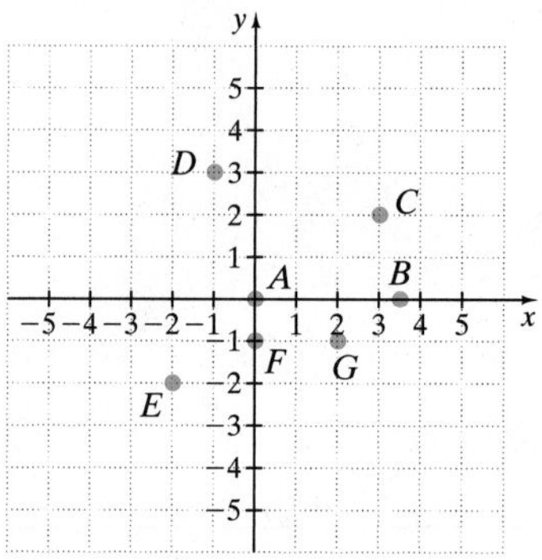

A.10 Graphing Linear Equations

A **linear equation in two variables** is an equation that can be written in the form $Ax + By = C$, where A and B are not both 0.

$$3x + 2y = -6 \qquad x = -5$$
$$y = 3 \qquad y = -x + 10$$
} Linear Equations in Two Variables

The form $Ax + By = C$ is called **standard form**.

$x + y = 10$ is in standard form.

To graph a linear equation in two variables, find three ordered pair solutions. Plot the solution points and draw the line connecting the points.

EXAMPLE

1. Graph: $x - 2y = 5$

x	y
5	0
1	−2
−1	−3

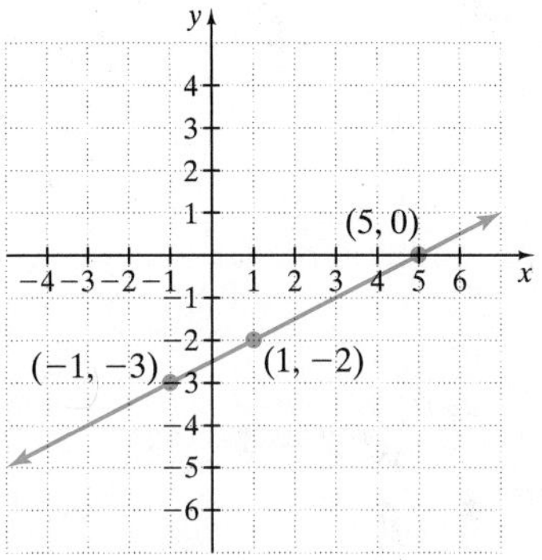

A.10 EXERCISES MyMathLab®

For each equation, find three ordered pair solutions by completing the table. Then use the ordered pairs to graph the equation.

1. $y = -4x$

x	y
1	
0	
−1	

2. $y = -5x$

x	y
1	
0	
−1	

3. $y = -4x + 3$

x	y
0	
1	
2	

4. $y = -5x + 2$

x	y
0	
1	
2	

Graph each linear equation.

5. $x - y = -2$

6. $-x + y = 6$

7. $x - 2y = 6$

8. $-x + 5y = 5$

9. $x = -4$

10. $x = -1$

11. $y = 5$

12. $y = 3$

13. $x = -3y$

14. $x = 4y$

15. $y = \frac{1}{2}x + 2$

16. $y = \frac{1}{4}x + 3$

A.11 Solving Systems of Linear Equations in Two Variables

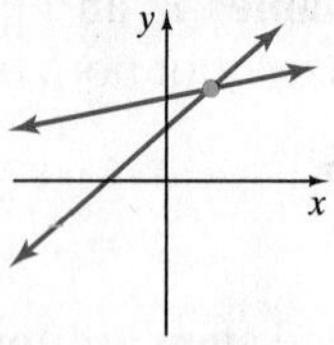

One solution:
Independent equations
Consistent system

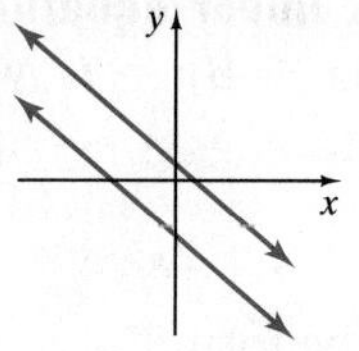

No solution:
Independent equations
Inconsistent system

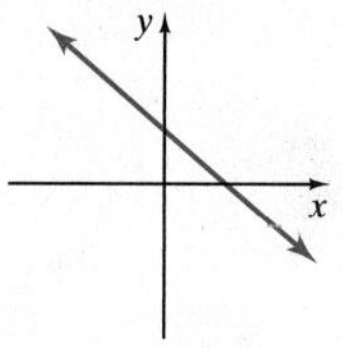

Infinite number of solutions:
Dependent equations
Consistent system

EXAMPLES

1. Solve by substitution:

$$\begin{cases} y = x + 2 \\ 3x - 2y = -5 \end{cases}$$

Since the first equation is solved for y, substitute $x + 2$ for y in the second equation.

$$\begin{aligned} 3x - 2y &= -5 && \text{Second equation} \\ 3x - 2(x + 2) &= -5 && \text{Let } y = x + 2. \\ 3x - 2x - 4 &= -5 \\ x - 4 &= -5 && \text{Simplify.} \\ x &= -1 && \text{Add 4.} \end{aligned}$$

To find y, let $x = -1$ in $y = x + 2$, so $y = -1 + 2 = 1$. The solution $(-1, 1)$ checks in both original equations.

2. Solve by elimination:

$$\begin{cases} x - 3y = -3 \\ -2x + y = 6 \end{cases}$$

Multiply both sides of the first equation by 2.

$$\begin{aligned} 2x - 6y &= -6 \\ -2x + y &= 6 \\ \hline -5y &= 0 && \text{Add.} \\ y &= 0 && \text{Divide by } -5. \end{aligned}$$

To find x, let $y = 0$ in an original equation.

$$\begin{aligned} x - 3y &= -3 \\ x - 3 \cdot 0 &= -3 \\ x &= -3 \end{aligned}$$

The solution $(-3, 0)$ checks in both original equations.

A.11 EXERCISES MyMathLab®

Use the substitution method to solve each system of equations.

1. $\begin{cases} x + y = 10 \\ y = 4x \end{cases}$

2. $\begin{cases} 5x + 2y = -17 \\ x = 3y \end{cases}$

3. $\begin{cases} 4x - y = 9 \\ 2x + 3y = -27 \end{cases}$

4. $\begin{cases} 3x - y = 6 \\ -4x + 2y = -8 \end{cases}$

Use the elimination method to solve each system of equations.

5. $\begin{cases} 5x + 2y = 1 \\ x - 3y = 7 \end{cases}$

6. $\begin{cases} 6x - y = -5 \\ 4x - 2y = 6 \end{cases}$

7. $\begin{cases} 5x - 2y = 27 \\ -3x + 5y = 18 \end{cases}$

8. $\begin{cases} 3x + 4y = 2 \\ 2x + 5y = -1 \end{cases}$

Solve each system of equations by substitution or elimination.

9. $\begin{cases} \frac{1}{3}x + y = \frac{4}{3} \\ -\frac{1}{4}x - \frac{1}{2}y = -\frac{1}{4} \end{cases}$

10. $\begin{cases} \frac{3}{4}x - \frac{1}{2}y = -\frac{1}{2} \\ x + y = -\frac{3}{2} \end{cases}$

11. $\begin{cases} 4x + 2y = 5 \\ 2x + y = -1 \end{cases}$

12. $\begin{cases} 3x + 6y = 15 \\ 2x + 4y = 3 \end{cases}$

13. $\begin{cases} 2x = 6 \\ y = 5 - x \end{cases}$

14. $\begin{cases} x = 3y + 4 \\ -y = 5 \end{cases}$

15. $\begin{cases} x = 3y + 2 \\ 5x - 15y = 10 \end{cases}$

16. $\begin{cases} x = 7y - 21 \\ 2x - 14y = -42 \end{cases}$

A.12 Exponents

a^n means the product of n factors, each of which is a.

Let m and n be integers and no denominators be 0.

Product Rule: $a^m \cdot a^n = a^{m+n}$

Power Rule: $(a^m)^n = a^{mn}$

Power of a Product Rule: $(ab)^n = a^n b^n$

Power of a Quotient Rule: $\left(\frac{a}{b}\right)^n = \frac{a^n}{b^n}$

Quotient Rule: $\frac{a^m}{a^n} = a^{m-n}$

Zero Exponent: $a^0 = 1, a \neq 0$

EXAMPLES **Evaluate**

1. $(-5)^3 = (-5)(-5)(-5) = -125$

2. $\left(\frac{1}{2}\right)^4 = \frac{1}{2}\cdot\frac{1}{2}\cdot\frac{1}{2}\cdot\frac{1}{2} = \frac{1}{16}$

Simplify.

3. $x^2 \cdot x^7 = x^{2+7} = x^9$

4. $(5^3)^8 = 5^{3\cdot 8} = 5^{24}$

5. $(7y)^4 = 7^4 y^4$

6. $\left(\frac{x}{8}\right)^3 = \frac{x^3}{8^3}$

7. $\frac{x^9}{x^4} = x^{9-4} = x^5$

8. $5^0 = 1$

A.12 EXERCISES MyMathLab®

Evaluate each expression.

1. 2^3 **2.** 3^1 **3.** $(-4)^2$

4. -4^2 **5.** $\left(\frac{1}{2}\right)^4$ **6.** $4 \cdot 3^2$

Evaluate each expression with the given replacement values.

7. $\frac{2z^4}{5}$ when $z = -2$

8. $\frac{10}{3y^3}$ when $y = -3$

Simplify each expression. Variables in these exercises are not equal to 0.

9. $(5y^4)(3y)$

10. $(-2z^3)(-2z^2)$

11. $(x^9y)(x^{10}y^5)$

12. $(a^2b)(a^{13}b^{17})$

13. $(x^9)^4$

14. $(y^7)^5$

15. $(pq)^8$

16. $(ab)^6$

17. $(x^2y^3)^5$

18. $(a^4b)^7$

19. $\left(\frac{r}{s}\right)^9$

20. $\left(\frac{q}{t}\right)^{11}$

21. $\left(\frac{-2xz}{y^5}\right)^2$

22. $\left(\frac{xy^4}{-3z^3}\right)^3$

23. $\frac{(-4)^6}{(-4)^3}$

24. $\frac{(-6)^{13}}{(-6)^{11}}$

25. $\frac{7x^2y^6}{14x^2y^3}$

26. $\frac{9a^4b^7}{27ab^2}$

27. $(2x)^0$

28. $(4y)^0$

29. $-7x^0$

30. $-2x^0$

31. $5^0 + y^0$

32. $-3^0 + 4^0$

33. $(2x^3)(-8x^4)$

34. $(3y^4)(-5y)$

35. $\left(\frac{3y^5}{6x^4}\right)^3$

36. $\left(\frac{2ab}{6yz}\right)^4$

A.13 Multiplying Polynomials

To multiply two polynomials, multiply each term of one polynomial by each term of the other polynomial, and then combine like terms.

The **FOIL method** may be used when multiplying two binomials.

Squaring a Binomial

$$(a + b)^2 = a^2 + 2ab + b^2$$
$$(a - b)^2 = a^2 - 2ab + b^2$$

Multiplying the Sum and Difference of Two Terms

$$(a + b)(a - b) = a^2 - b^2$$

EXAMPLES

1. Multiply.

$(2x + 1)(5x^2 - 6x + 2)$

$$= 2x(5x^2 - 6x + 2) + 1(5x^2 - 6x + 2)$$
$$= 10x^3 - 12x^2 + 4x + 5x^2 - 6x + 2$$
$$= 10x^3 - 7x^2 - 2x + 2$$

2. Multiply: $(5x - 3)(2x + 3)$

First Last

$(5x - 3)(2x + 3)$

Inner

Outer

F O I L

$$= (5x)(2x) + (5x)(3) + (-3)(2x) + (-3)(3)$$
$$= 10x^2 + 15x - 6x - 9$$
$$= 10x^2 + 9x - 9$$

3. Square the binomial.

$$(3x - 2y)^2 = (3x)^2 - 2(3x)(2y) + (2y)^2$$
$$= 9x^2 - 12xy + 4y^2$$

4. Multiply.

$$(6y + 5)(6y - 5) = (6y)^2 - 5^2$$
$$= 36y^2 - 25$$

A.13 EXERCISES MyMathLab®

Multiply.

1. $(a + 7)(a - 2)$ **2.** $(y - 10)(y + 11)$

3. $(2a - 3)(5a^2 - 6a + 4)$ **4.** $(3 + b)(2 - 5b - 3b^2)$

5. $(7xy - y)^2$ **6.** $(x^2 - 4)^2$

Multiply using the FOIL method.

7. $(x + 3)(x + 4)$ **8.** $(x + 5)(x + 1)$

9. $(x - 5)(x + 10)$ **10.** $(y - 12)(y + 4)$

11. $(y - 6)(4y - 1)$ **12.** $(2x - 9)(x - 11)$

Multiply.

13. $(2x - 1)^2$ **14.** $(5b - 4)^2$

15. $(5x + 9)^2$ **16.** $(6s + 2)^2$

17. $(a - 7)(a + 7)$ **18.** $(b + 3)(b - 3)$

19. $(9x + y)(9x - y)$ **20.** $(2x - y)(2x + y)$

21. $\left(\frac{1}{3}a^2 - 7\right)\left(\frac{1}{3}a^2 + 7\right)$ **22.** $\left(\frac{a}{2} + 4y\right)\left(\frac{a}{2} - 4y\right)$

23. $(3b + 7)(2b - 5)$ **24.** $(3y - 13)(y - 3)$

25. $(4x + 5)(4x - 5)$ **26.** $(3x + 5)(3x - 5)$

27. $(5x - 6y)^2$ **28.** $(4x - 9y)^2$

A.14 Simplifying Radical Expressions

The **positive or principal square root** of a positive number a is written as $\sqrt{a}$. The **negative square root** of a is written as $-\sqrt{a}$. $\sqrt{a} = b$ only if $b^2 = a$ and $b > 0$.

A square root of a negative number is not a real number.

Product Rule for Radicals

If $\sqrt{a}$ and $\sqrt{b}$ are real numbers, then

$$\sqrt{a \cdot b} = \sqrt{a} \cdot \sqrt{b}$$

A square root is in **simplified form** if the radicand contains no perfect square factors other than 1. To simplify a square root, factor the radicand so that one of its factors is a perfect square factor and then apply the product rule.

Quotient Rule for Radicals

If $\sqrt{a}$ and $\sqrt{b}$ are real numbers and $b \neq 0$, then

$$\sqrt{\frac{a}{b}} = \frac{\sqrt{a}}{\sqrt{b}}$$

EXAMPLES **Simplify**

1. $\sqrt{25} = 5$

2. $\sqrt{100} = 10$

3. $-\sqrt{9} = -3$

4. $\sqrt{\frac{4}{49}} = \frac{2}{7}$

5. $\sqrt{-4}$ is not a real number.

6. $\sqrt{45} = \sqrt{9 \cdot 5} = \sqrt{9} \cdot \sqrt{5} = 3\sqrt{5}$

7. $\sqrt{\frac{18}{25}} = \frac{\sqrt{18}}{\sqrt{25}} = \frac{\sqrt{9 \cdot 2}}{5} = \frac{\sqrt{9} \cdot \sqrt{2}}{5} = \frac{3\sqrt{2}}{5}$

A.14 EXERCISES MyMathLab®

Find each square root.

1. $\sqrt{36}$

2. $\sqrt{64}$

3. $\sqrt{\frac{1}{25}}$

4. $\sqrt{\frac{1}{64}}$

5. $-\sqrt{100}$

6. $-\sqrt{36}$

7. $\sqrt{-4}$

8. $\sqrt{-25}$

9. $-\sqrt{121}$

10. $-\sqrt{49}$

Approximate each square root to three decimal places.

11. $\sqrt{7}$

12. $\sqrt{10}$

13. $\sqrt{136}$

14. $\sqrt{8}$

Simplify each radical.

15. $\sqrt{20}$

16. $\sqrt{44}$

17. $\sqrt{33}$

18. $\sqrt{21}$

19. $\sqrt{180}$

20. $\sqrt{150}$

21. $-5\sqrt{27}$

22. $-6\sqrt{75}$

23. $\sqrt{\frac{36}{121}}$

24. $\sqrt{\frac{25}{144}}$

25. $\sqrt{\frac{27}{121}}$

26. $\sqrt{\frac{24}{169}}$

27. $\sqrt{\frac{11}{36}}$

28. $\sqrt{\frac{30}{49}}$

29. $-\sqrt{\frac{27}{144}}$

30. $-\sqrt{\frac{84}{121}}$

A.15 Solving Quadratic Equations by Factoring

To Solve Quadratic Equations by Factoring

STEP 1. Write the equation in standard form so that one side of the equation is 0.

STEP 2. Factor completely.

STEP 3. Set each factor containing a variable equal to 0.

STEP 4. Solve the resulting equations.

STEP 5. Check solutions in the original equation.

EXAMPLE

1. Solve: $3x^2 = 13x - 4$

STEP 1. $3x^2 - 13x + 4 = 0$

STEP 2. $(3x - 1)(x - 4) = 0$

STEP 3. $3x - 1 = 0$ or $x - 4 = 0$

STEP 4. $3x = 1$ $\quad x = 4$

$x = \frac{1}{3}$

STEP 5. Check both $\frac{1}{3}$ and 4 in the original equation.

A.15 EXERCISES MyMathLab®

Solve each equation.

1. $(2x + 3)(4x - 5) = 0$

2. $(3x - 2)(5x + 1) = 0$

3. $x^2 + 2x - 8 = 0$

4. $x^2 - 5x + 6 = 0$

5. $x^2 - 7x = 0$

6. $x^2 - 3x = 0$

7. $x^2 = 16$

8. $x^2 = 9$

9. $x(3x - 1) = 14$

10. $x(4x - 11) = 3$

11. $(2x + 3)(2x^2 - 5x - 3) = 0$

12. $(2x - 9)(x^2 + 5x - 36) = 0$

A.16 Solving Quadratic Equations by the Square Root Property

Square Root Property

If $x^2 = a$ for $a \geq 0$, then $x = \sqrt{a}$ or $x = -\sqrt{a}$.

EXAMPLE

1. Solve the equation.

$$(x - 1)^2 = 15$$

$$(x - 1) = \sqrt{15} \quad \text{or} \quad x - 1 = -\sqrt{15}$$

$$x = 1 + \sqrt{15} \qquad\qquad x = 1 - \sqrt{15}$$

A.16 EXERCISES MyMathLab®

Use the square root property to solve each quadratic equation.

1. $x^2 = 64$

2. $x^2 = 121$

3. $x^2 = 21$

4. $x^2 = 22$

5. $x^2 = \frac{1}{25}$

6. $x^2 = \frac{1}{16}$

7. $x^2 = -4$

8. $x^2 = -25$

9. $3x^2 = 13$

10. $5x^2 = 2$

11. $x^2 - 2 = 0$

12. $x^2 - 15 = 0$

13. $(x - 5)^2 = 49$

14. $(x + 2)^2 = 25$

15. $(p + 2)^2 = 10$

16. $(p - 7)^2 = 13$

17. $(3x - 7)^2 = 32$

18. $(5x - 11)^2 = 54$

A.17 Solving Quadratic Equations by the Quadratic Formula

Quadratic Formula

If a, b, and c are real numbers and $a \neq 0$, the quadratic equation $ax^2 + bx + c = 0$ has solutions

$$x = \frac{-b \pm \sqrt{b^2 - 4ac}}{2a}$$

To Solve a Quadratic Equation by the Quadratic Formula

STEP 1. Write the equation in standard form: $ax^2 + bx + c = 0$

STEP 2. If necessary, clear the equation of fractions.

STEP 3. Identify a, b, and c.

STEP 4. Replace a, b, and c in the quadratic formula with the identified values, and simplify.

EXAMPLE

1. Solve $3x^2 - 2x - 2 = 0$.

In this equation, $a = 3$, $b = -2$, and $c = -2$.

$$x = \frac{-(-2) \pm \sqrt{(-2)^2 - 4(3)(-2)}}{2 \cdot 3}$$

$$= \frac{2 \pm \sqrt{4 - (-24)}}{6}$$

$$= \frac{2 \pm \sqrt{28}}{6} = \frac{2 \pm \sqrt{4 \cdot 7}}{6} = \frac{2 \pm 2\sqrt{7}}{6}$$

$$= \frac{2(1 \pm \sqrt{7})}{2 \cdot 3} = \frac{1 \pm \sqrt{7}}{3}$$

A.17 EXERCISES MyMathLab®

Use the quadratic formula to solve each quadratic equation.

1. $x^2 - 3x + 2 = 0$

2. $x^2 - 5x - 6 = 0$

3. $3k^2 + 7k + 1 = 0$

4. $7k^2 + 3k - 1 = 0$

5. $m^2 - 12 = m$

6. $m^2 - 14 = 5m$

7. $3 - x^2 = 4x$

8. $10 - x^2 = 2x$

9. $5z^2 - 2z = \frac{1}{5}$

10. $9z^2 + 12z = -1$

11. $\frac{m^2}{2} = m + \frac{1}{2}$

12. $\frac{m^2}{2} = 3m - 1$

13. $6x^2 + 9x = 2$

14. $3x^2 - 9x = 8$

15. $7p^2 + 2 = 8p$

16. $11p^2 + 2 = 10p$

17. $x^2 - 6x + 2 = 0$

18. $x^2 - 10x + 19 = 0$

19. $2x^2 - 6x + 3 = 0$

20. $5x^2 - 8x + 2 = 0$

21. $3x^2 = 1 - 2x$

22. $5y^2 = 4 - y$

23. $4y^2 = 6y + 1$

24. $6z^2 = 2 - 3z$

25. $20y^2 = 3 - 11y$

26. $2z^2 = z + 3$

27. $x^2 + x + 2 = 0$

28. $k^2 + 2k + 5 = 0$

TABLES

Table 1 Math Symbols

Symbol	Words
$\ldots$	and so on
$=$	is equal to, equality
$\approx$	is approximately equal to
$\neq$	is not equal to
$>$	is greater than
$<$	is less than
$\geq$	is greater than or equal to
$\leq$	is less than or equal to
$\ngtr$	is not greater than
$\nless$	is not less than
$+$	plus (addition)
$-$	minus (subtraction)
$\cdot, \times$	times (multiplication)
n^2	square of n
$\sqrt{x}$	nonnegative square root of x
$\pm$	plus or minus
$\%$	percent
$\lvert a \rvert$	absolute value of a
$(\ ), [\ ]$	parentheses and brackets for grouping
$p \rightarrow q$	if p, then q
$p \leftrightarrow q$	p if and only if q
$\sim p$	not p
$\rightarrow$	maps to
$-a$	opposite of a
d	distance
M	midpoint
$^\circ$	degree(s)
$\overleftrightarrow{AB}$	line through points A and B
$\overline{AB}$	segment with endpoints A and B
$\overrightarrow{AB}$	ray with endpoint A and through point B
AB	length of $\overline{AB}$
$\angle A$	angle with vertex A
$\angle ABC$	angle with sides $\overrightarrow{BA}$ and $\overrightarrow{BC}$
$m\angle A$	measure of angle A
$\angle$s	angles
ΔABC	triangle with vertices A, B, and C
⌐	right angle symbol
$\triangle$s	traingles
$\cong$	is congruent to
$\ncong$	is not congruent to
$\sim$	is similar to
$\stackrel{?}{=}$	is this statement true?
▱$ABCD$	parallelogram with vertices A, B, C, and D
▱s	parallelograms
A'	image of A, A prime
A	area
s	length of a side
b	base length
h	height, length of an altitude
d	diameter
r	radius
P	perimeter
π	pi, ratio of the circumference of a circle to its diameter
C	circumference
b_1, b_2	bases of a trapezoid
d_1, d_2	lengths of diagonals
a	apothem
B	area of a base
LA	lateral area
SA	surface area
ℓ	slant height
V	volume
n-gon	polygon with n sides
$\odot A$	circle with center A
$\overset{\frown}{AB}$	arc with endpoints A and B
$\overset{\frown}{ABC}$	arc with endpoints A and C and containing B
$m\overset{\frown}{AB}$	measure of $\overset{\frown}{AB}$
$\parallel$	is parallel to
$\perp$	is perpendicular to
m	slope of a linear function
b	y-intercept of a linear function
$a{:}b, \frac{a}{b}$	ratio of a to b
$\tan A$	tangent of $\angle A$
$\sin A$	sine of $\angle A$
$\cos A$	cosine of $\angle A$
(a, b)	ordered pair with x-coordinate a and y-coordinate b
$\overrightarrow{\mathbf{AB}}$	vector with initial point A and terminal point B
$\langle x, y \rangle$	component form of a vector
$\vec{\mathbf{v}}$	vector **v**
$P(\text{event})$	probability of an event
$\begin{bmatrix} 1 & 2 \\ 3 & 4 \end{bmatrix}$	matrix

Table 2 Formulas

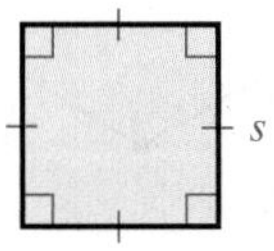

$P = 4s$

$A = s^2$

Square

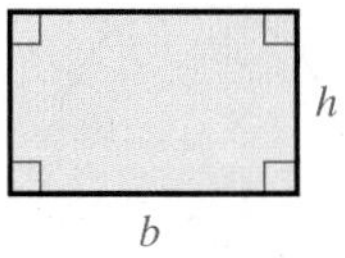

$P = 2b + 2h$

$A = bh$

Rectangle

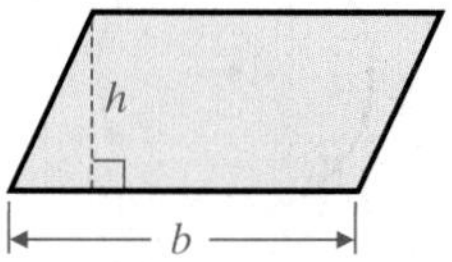

$A = bh$

Parallelogram

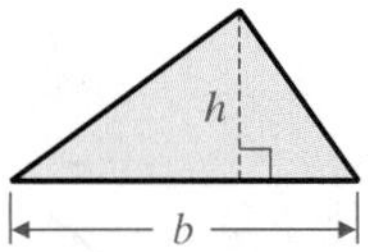

$A = \frac{1}{2}bh$

Triangle

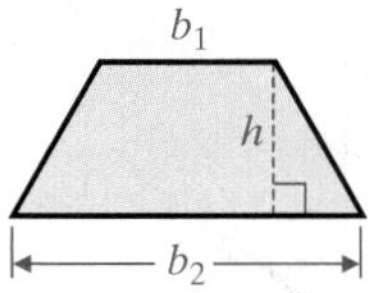

$A = \frac{1}{2}h(b_1 + b_2)$

Trapezoid

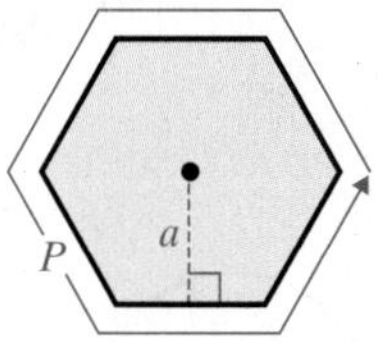

$A = \frac{1}{2}aP$

Regular Polygon

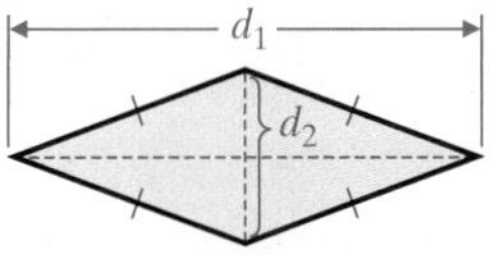

$A = \frac{1}{2}d_1d_2$

Rhombus (or Kite)

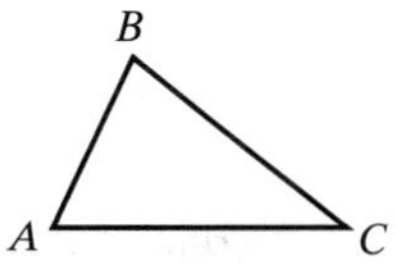

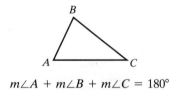

$m\angle A + m\angle B + m\angle C = 180°$

Triangle Angle Sum

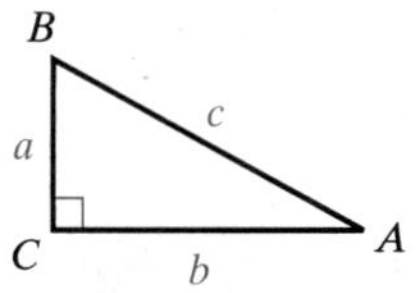

$a^2 + b^2 = c^2$

Pythagorean Theorem

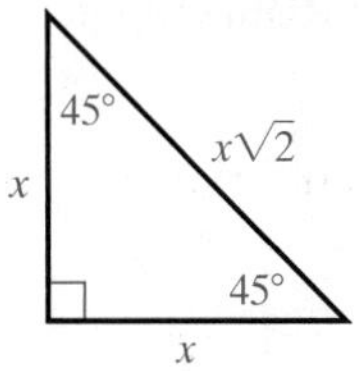

Ratio of sides $= 1:1:\sqrt{2}$

45°-45°-90° Triangle

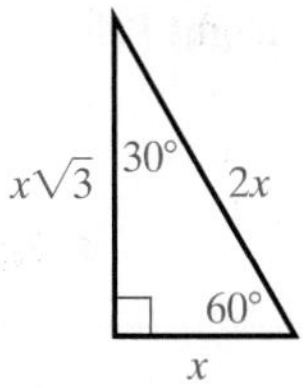

Ratio of sides $= 1:\sqrt{3}:2$

30°-60°-90° Triangle

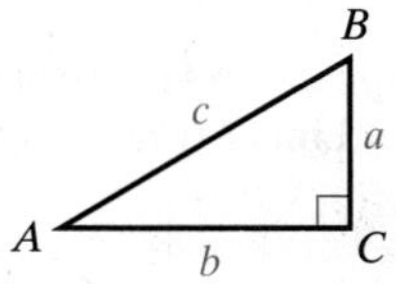

$\tan A = \frac{a}{b}$

$\sin A = \frac{a}{c}$; $\cos A = \frac{b}{c}$

Trigonometric Ratios

(Continued)

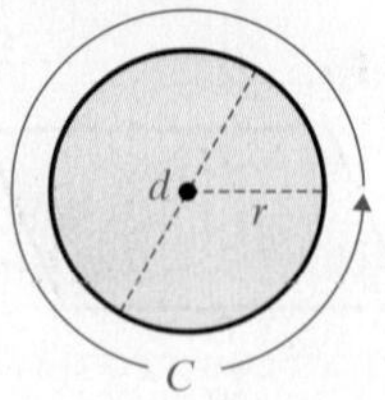

$C = \pi d$ or $C = 2\pi r$

$A = \pi r^2$

Circle

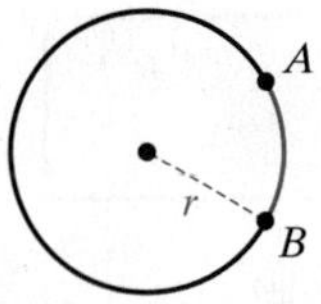

Length of $\widehat{AB} = \dfrac{m\widehat{AB}}{360°} \cdot 2\pi r$

Arc

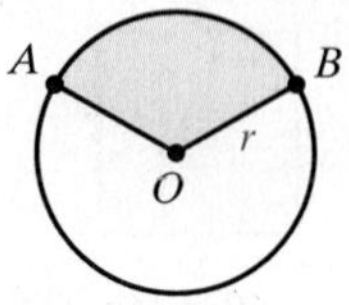

Area of sector $AOB = \dfrac{m\widehat{AB}}{360°} \cdot \pi r^2$

Sector of a Circle

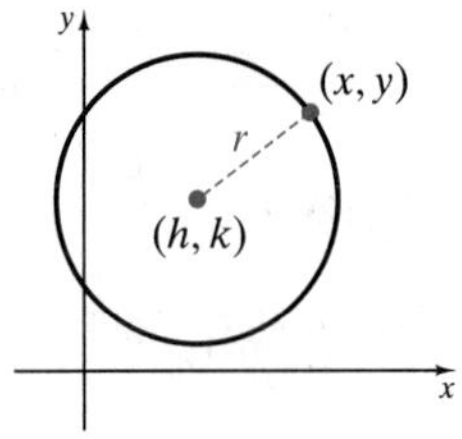

$(x - h)^2 + (y - k)^2 = r^2$

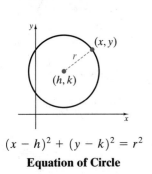

Equation of Circle

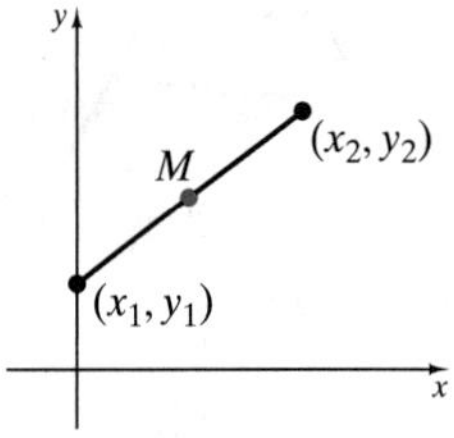

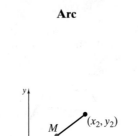

$d = \sqrt{(x_2 - x_1)^2 + (y_2 - y_1)^2}$

$M = \left(\dfrac{x_1 + x_2}{2}, \dfrac{y_1 + y_2}{2}\right)$

Distance and Midpoint

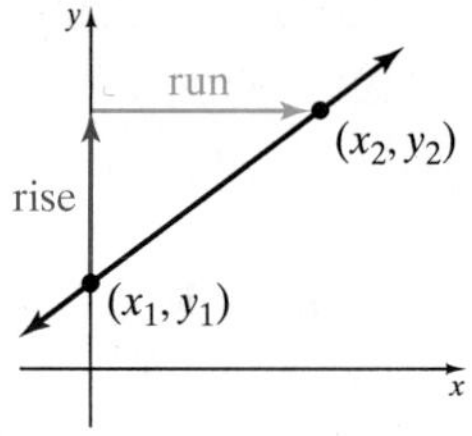

$m = \dfrac{\text{rise}}{\text{run}} = \dfrac{y_2 - y_1}{x_2 - x_1}$

Slope

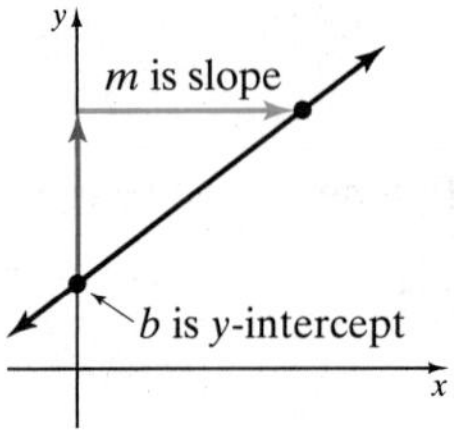

$y = mx + b$

Slope-Intercept Form of a Linear Equation

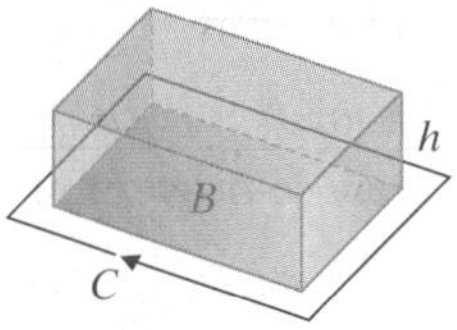

$LA = Ph$

$SA = LA + 2B$

$V = Bh$

Right Prism

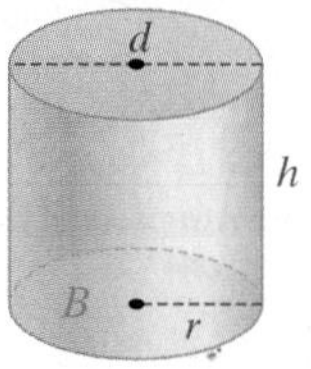

$LA = 2\pi rh$ or $LA = \pi dh$

$SA = LA + 2B$

$V = Bh$ or $V = \pi r^2 h$

Right Cylinder

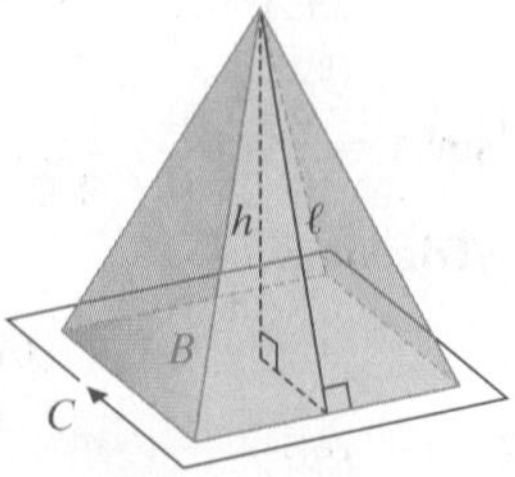

$LA = \dfrac{1}{2}P\ell$

$SA = LA + 2B$

$V = \dfrac{1}{3}Bh$

Regular Pyramid

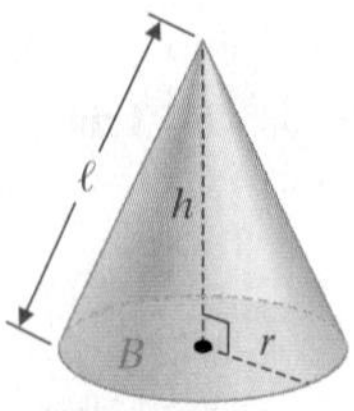

$LA = \pi r\ell$

$SA = LA + B$

$V = \dfrac{1}{3}Bh$ or $V = \dfrac{1}{3}\pi r^2 h$

Right Cone

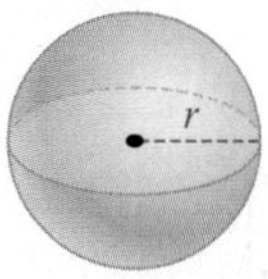

$SA = 4\pi r^2$

$V = \dfrac{4}{3}\pi r^3$

Sphere

Table 3 Measures

	United States Customary	Metric
Length	12 inches (in.) = 1 foot (ft) 36 in. = 1 yard (yd) 3 ft = 1 yard 5280 ft = 1 mile (mi) 1760 yd = 1 mile	10 millimeters (mm) = 1 centimeter (cm) 100 cm = 1 meter (m) 1000 mm = 1 meter 1000 m = 1 kilometer (km)
Area	144 square inches (in.^{2}) = 1 square foot (ft^2) 9 ft^2 = 1 square yard (yd^2) 43,560 ft^2 = 1 acre (a) 4840 yd^2 = 1 acre	100 square millimeters (mm^2) = 1 square centimeter (cm^2) 10,000 cm^2 = 1 square meter (m^2) 10,000 m^2 = 1 hectare (ha)
Volume	1728 cubic inches (in.^{3}) = 1 cubic foot (ft^3) 27 ft^3 = 1 cubic yard (yd^3)	1000 cubic millimeters (mm^3) = 1 cubic centimeter (cm^3) 1,000,000 cm^3 = 1 cubic meter (m^3)
Liquid Capacity	8 fluid ounces (fl oz) = 1 cup (c) 2 c = 1 pint (pt) 2 pt = 1 quart (qt) 4 qt = 1 gallon (gal)	1000 milliliters (mL) = 1 liter (L) 1000 L = 1 kiloliter (kL)
Weight or Mass	16 ounces (oz) = 1 pound (lb) 2000 pounds = 1 ton (t)	1000 milligrams (mg) = 1 gram (g) 1000 g = 1 kilogram (kg) 1000 kg = 1 metric ton
Temperature	32°F = freezing point of water 98.6°F = normal human body temperature 212°F = boiling point of water	0°C = freezing point of water 37°C = normal human body temperature 100°C = boiling point of water

	U.S. Customary Units and Metric Units
Length	1 in. = 2.54 cm 1 mi ≈ 1.61 km 1 ft ≈ 0.305 m
Capacity	1 qt ≈ 0.946 L
Weight and Mass	1 oz ≈ 28.4 g 1 lb ≈ 0.454 kg

Time		
60 seconds (sec) = 1 mintue (min)	4 weeks (approx.) = 1 month (mo)	12 months = 1 year
60 minutes = 1 hour (hr)	365 days = 1 year (yr)	10 years = 1 decade
24 hours = 1 day (d)	52 weeks (approx.) = 1 year	100 years = 1 century
7 days = 1 week (wk)		

(*Note:* Square units may also be written as: sq units or units^2; Cubic units may also be written as: cu units or units^3)

Table 4 Properties of Real Numbers

Unless otherwise stated, *a*, *b*, *c*, and *d* represent real numbers.

Identity Properties

Addition $a + 0 = a$ and $0 + a = a$

Multiplication $a \cdot 1 = a$ and $1 \cdot a = a$

Commutative Properties

Addition $a + b = b + a$

Multiplication $a \cdot b = b \cdot a$

Associative Properties

Addition $(a + b) + c = a + (b + c)$

Multiplication $(a \cdot b) \cdot c = a \cdot (b \cdot c)$

Inverse Properties

Addition The sum of a number and its *opposite*, or *additive inverse*, is zero.

$a + (-a) = 0$ and $-a + a = 0$

Multiplication The *reciprocal*, or *multiplicative inverse*, of a rational number $\frac{a}{b}$ is $\frac{b}{a}(a, b \neq 0)$.

$a \cdot \frac{1}{a} = 1$ and $\frac{1}{a} \cdot a = 1 (a \neq 0)$

Distributive Properties

$a(b + c) = ab + ac$ $(b + c)a = ba + ca$

$a(b - c) = ab - ac$ $(b - c)a = ba - ca$

Properties of Equality

Addition If $a = b$, then $a + c = b + c$.

Subtraction If $a = b$, then $a - c = b - c$.

Multiplication If $a = b$, then $a \cdot c = b \cdot c$.

Division If $a = b$ and $c \neq 0$, then $\frac{a}{c} = \frac{b}{c}$.

Substitution If $a = b$, then b can replace a in any expression.

Reflexive $a = a$

Symmetric If $a = b$, then $b = a$.

Transitive If $a = b$ and $b = c$, then $a = c$.

Properties of Proportions

$\frac{a}{b} = \frac{c}{d}$ $(a, b, c, d \neq 0)$ is equivalent to

1. $ad = bc$ **2.** $\frac{b}{a} = \frac{d}{c}$

3. $\frac{a}{c} = \frac{b}{d}$ **4.** $\frac{a + b}{b} = \frac{c + d}{d}$

Zero Factor Property

If $ab = 0$, then $a = 0$ or $b = 0$.

Properties of Inequality

Addition If $a > b$ and $c \geq d$, then $a + c > b + d$.

Multiplication If $a > b$ and $c > 0$, then $ac > bc$.
If $a > b$ and $c < 0$, then $ac < bc$.

Transitive If $a > b$ and $b > c$, then $a > c$.

Comparison If $a = b + c$, and $c > 0$, then $a > b$.

Properties of Exponents

For any nonzero numbers *a* and *b*, and any integers *m* and *n*,

Zero Exponent $a^0 = 1$

Negative Exponent $a^{-n} = \frac{1}{a^n}$

Product Rule $a^m \cdot a^n = a^{m+n}$

Quotient Rule $\frac{a^m}{a^n} = a^{m-n}$

Power Rule $(a^m)^n = a^{mn}$

Power of a Product Rule $(ab)^n = a^n b^n$

Power of a Quotient Rule $\left(\frac{a}{b}\right)^n = \frac{a^n}{b^n}$

Properties of Square Roots

For any nonnegative numbers *a* and *b*, and any positive number *c*,

Product of Square Roots $\sqrt{a} \cdot \sqrt{b} = \sqrt{ab}$

Quotient of Square Roots $\frac{\sqrt{a}}{\sqrt{c}} = \sqrt{\frac{a}{c}}$

POSTULATES, THEOREMS, AND ADDITIONAL PROOFS

Chapter 1 The Beginning of Geometry

Postulate 1.3-1
Through any two points there is exactly one line.

Postulate 1.3-2
If two distinct lines intersect, then they intersect in exactly one point.

Postulate 1.3-3
If two distinct planes intersect, then they intersect in exactly one line.

Postulate 1.3-4
Through any three noncollinear points there is exactly one plane.

Postulate 1.4-1 Ruler Postulate
The points on a line can be paired with real numbers. This makes a one-to-one correspondence between the points on the line and the real numbers.

Postulate 1.4-2 Segment Addition Postulate
If three points A, B, and C are collinear and B is between A and C, then $AB + BC = AC$.

Postulate 1.5-1 Protractor Postulate
Consider $\overrightarrow{AB}$ and a point C on one side of $\overrightarrow{AB}$. Every ray of the form $\overrightarrow{AC}$ can be paired one-to-one with a real number from 0 to 180.

Postulate 1.5-2 Angle Addition Postulate
If point P is in the interior of $\angle ABC$, then $m\angle ABP + m\angle PBC = m\angle ABC$.

The Midpoint Formulas

On a Number Line (Sec. 1.7)
The coordinate of the midpoint M of $\overline{AB}$ is $\frac{a+b}{2}$.

In the Coordinate Plane (Sec. 1.7)
Given $\overline{AB}$ where $A(x_1, y_1)$ and $B(x_2, y_2)$, the coordinates of the midpoint of $\overline{AB}$ are $M\left(\frac{x_1 + x_2}{2}, \frac{y_1 + y_2}{2}\right)$.

The Distance Formula (Sec. 1.7)
The distance between two points $A(x_1, y_1)$ and $B(x_2, y_2)$ is $d = \sqrt{(x_2 - x_1)^2 + (y_2 - y_1)^2}$.

Chapter 2 Introduction to Reasoning and Proofs

Law of Detachment (Sec. 2.5)
If the hypothesis of a true conditional is true, then the conclusion is true. In symbolic form:
If $p \rightarrow q$ is true and p is true, then q is true.

Law of Syllogism (Sec. 2.5)
If $p \rightarrow q$ is true and $q \rightarrow r$ is true, then $p \rightarrow r$ is true.

Properties of Congruence (sec. 2.6)

Reflexive Property
$\overline{AB} \cong \overline{AB}$ and $\angle A \cong \angle A$

Symmetric Property
If $\overline{AB} \cong \overline{CD}$, then $\overline{CD} \cong \overline{AB}$.
If $\angle A \cong \angle B$, then $\angle B \cong \angle A$.

Transitive Property
If $\overline{AB} \cong \overline{CD}$, and $\overline{CD} \cong \overline{EF}$, then $\overline{AB} \cong \overline{EF}$.
If $\angle A \cong \angle B$, and $\angle B \cong \angle C$, then $\angle A \cong \angle C$.
If $\angle B \cong \angle A$, and $\angle B \cong \angle C$, then $\angle A \cong \angle C$.

Theorem 2.7-1 Equal Complements Theorem
Complements of the same angle (or of congruent angles) are equal in measure.

Theorem 2.7-2 Equal Supplements Theorem
Supplements of the same angle (or of congruent angles) are equal in measure.

Theorem 2.7-3 Linear Pair Theorem
If two angles form a linear pair, then the angles are supplementary.

Theorem 2.7-4 Vertical Angles Theorem
Vertical angles are congruent.

Theorem 2.7-5
All right angles are congruent.

Theorem 2.7-6
If two angles are congruent and supplementary, then each is a right angle.

Chapter 3 Parallel and Perpendicular Lines

Postulate 3.2-1 Parallel Postulate
Through a point not on a line, there is one and only one line parallel to the given line.

Postulate 3.2-2 Perpendicular Postulate
Through a point not on a line, there is one and only one line perpendicular to the given line.

Theorem 3.2-3
In a plane, if two lines are perpendicular to the same line, then they are parallel to each other.

Theorem 3.2-4 Alternative Interior Angles Theorem
If two lines and a transversal form alternate interior angles that are congruent, then the two lines are parallel.

Theorem 3.2-5 Corresponding Angles Theorem
If two lines and a transversal form corresponding angles that are congruent, then the lines are parallel.

Theorem 3.2-6 Same-Side Interior Angles Theorem
If two lines and a transversal form same-side interior angles that are supplementary, then the two lines are parallel.

Theorem 3.2-7 Alternate Exterior Angles Theorem
If two lines and a transversal form alternate exterior angles that are congruent, then the two lines are parallel.

Theorem 3.3-1 Converse of the Alternate Interior Angles Theorem
If a transversal intersects two parallel lines, then alternate interior angles are congruent.

Theorem 3.3-2 Converse of the Corresponding Angles Theorem
If a transversal intersects two parallel lines, then corresponding angles are congruent.

Theorem 3.3-3 Converse of the Same-Side Interior Angles Theorem
If a transversal intersects two parallel lines, then same-side interior angles are supplementary.

Theorem 3.3-4 Converse of the Alternate Exterior Angles Theorem
If a transversal intersects two parallel lines, then alternate exterior angles are congruent.

Theorem 3.4-1 Perpendicular Transversal Theorem
In a plane, if a line is perpendicular to one of two parallel lines, then it is perpendicular to the other.

Theorem 3.4-2
If two lines are parallel to the same line, then they are parallel to each other.

Theorem 3.4-3
If two lines are perpendicular, then they intersect to form four right angles.

Theorem 3.4-4
If two lines intersect to form a linear pair of congruent angles then the lines are perpendicular to each other.

Slopes of Parallel Lines (Sec. 3.6)
If two nonvertical lines are parallel, then their slopes are equal. If the slopes of two distinct nonvertical lines are equal, then the lines are parallel. Any two vertical lines or horizontal lines are parallel.

Slopes of Perpendicular Lines (Sec. 3.6)
If two nonvertical lines are perpendicular, then the product of their slopes is -1. If the slopes of two lines have a product of -1, then the lines are perpendicular. Any horizontal line and vertical line are perpendicular.

Chapter 4 Triangles and Congruence

Theorem 4.1-1 Triangle Angle-Sum Theorem
The sum of the measures of the angles of a triangle is 180°.

Corollary 4.1-2 Triangle Exterior Angle Corollary
The measure of each exterior angle of a triangle equals the sum of the measures of its two remote interior angles.

Theorem 4.2-1 Third Angles Theroem
If the two angles of one triangle are congruent to two angles of another triangle, then the third angles are congruent.

Postulate 4.3-1 Side-Side-Side (SSS) Postulate
If the three sides of one triangle are congruent to the three sides of another triangle, then the two triangles are congruent.

Postulate 4.3-2 Side-Angle-Side (SAS) Postulate
If two sides and the included angle of one triangle are congruent to two sides and the included angle of another triangle, then the two triangles are congruent.

Postulate 4.4-1 Angle-Side-Angle (ASA) Postulate
If two angles and the included side of one triangle are congruent to two angles and the included side of another triangle, then the two triangles are congruent.

Theorem 4.4-2 Angle-Angle-Side (AAS) Theorem
If two angles and a nonincluded side of one triangle are congruent to two angles and the corresponding nonincluded side of another triangle, then the triangles are congruent.

Theorem 4.6-1 Isosceles Base Angles Theorem
If two sides of a triangle are congruent, then the angles opposite those sides are congruent.

Theorem 4.6-2 Converse of the Isosceles Base Angles Theorem
If two angles of a triangle are congruent, then the sides opposite the angles are congruent.

Theorem 4.6-3
If a line bisects the vertex angle of an isosceles triangle, then the line is also the perpendicular bisector of the base.

Corollary 4.6-4
If a triangle is equilateral, then the triangle is equiangular.

Corollary 4.6-5
If a triangle is equiangular, then the triangle is equilateral.

Theorem 4.6-6 Hypotenuse-Leg (H-L) Theorem
If the hypotenuse and a leg of one right triangle are congruent to the hypotenuse and a leg of another right triangle, then the triangles are congruent.

Chapter 5 Special Properties of Triangles

Theorem 5.1-1
Proof of the construction of the perpendicular bisector.

Theorem 5.1-2 Perpendicular Bisector Theorem
If a point is on the perpendicular bisector of a segment, then it is equidistant from the endpoints of the segment.

Theorem 5.1-3 Converse of the Perpendicular Bisector Theorem
If a point is equidistant from the endpoints of a segment, then it is on the perpendicular bisector of the segment.

Theorem 5.1-4
Proof of the construction of the angle bisector.

Theorem 5.1-5 Angle Bisector Theorem
If a point is on the bisector of an angle, then the point is equidistant from the sides of the angle.

Theorem 5.1-6 Converse of the Angle Bisector Theorem
If a point in the interior of an angle is equidistant from the sides of the angle, then the point is on the angle bisector.

Theorem 5.2-1 Concurrency of Perpendicular Bisectors Theorem
The perpendicular bisectors of the sides of a triangle are concurrent at a point equidistant from the vertices.

Theorem 5.2-2 Concurrency of Angle Bisectors Theorem
The bisectors of the angles of a triangle are concurrent at a point equidistant from the sides of the triangle.

Theorem 5.3-1 Concurrency of Medians Theorem
The medians of a triangle are concurrent at a point that is two-thirds the distance from each vertex to the midpoint of the opposite side.

Theorem 5.3-2 Concurrency of Altitudes Theorem
The lines that contain the altitudes of a triangle are concurrent.

Theorem 5.4-1 Triangle Midsegment Theorem
If a segment joins the midpoints of two sides of a triangle, then the segment is parallel to the third side and is half as long.

Theorem 5.5-1 Comparison Property of Inequality
If $a = b + c$ and $c > 0$, then $a > b$.

Corollary 5.5-2 Corollary to the Triangle Exterior Angle Theorem
The measure of an exterior angle of a triangle is greater than the measure of either of its remote interior angles.

Theorem 5.5-3
If two sides of a triangle are not congruent, then the larger angle lies opposite the longer side.

Theorem 5.5-4
If two angles of a triangle are not congruent, then the longer side lies opposite the larger angle.

Theorem 5.5-5 Triangle Inequality Theorem
The sum of the lengths of any two sides of a triangle is greater than the length of the third side.

Theorem 5.6-1 The Hinge Theorem (SAS Inequality Theorem)
If two sides of one triangle are congruent to two sides of another triangle and the included angles are not congruent, then the longer third side is opposite the larger included angle.

Theorem 5.6-2 Converse of the Hinge Theorem (SSS Inequality)
If two sides of one triangle are congruent to two sides of another triangle and the third sides are not congruent, then the larger included angle is opposite the longer third side.

Chapter 6 Quadrilaterals

Theorem 6.1-1 Interior Angle Sum of a Convex Quadrilateral
The sum of the measures of the interior angles of a convex quadrilateral is 360°.

Theorem 6.2-1
If a quadrilateral is a parallelogram, then its opposite sides are congruent.

Theorem 6.2-2
If a quadrilateral is a parallelogram, then its opposite angles are congruent.

Theorem 6.2-3
If a quadrilateral is a parallelogram, then its consecutive angles are supplementary.

Theorem 6.2-4
If a quadrilateral is a parallelogram, then its diagonals bisect each other.

Theorem 6.2-5
If three (or more) parallel lines cut off congruent segments on one transversal, then they cut off congruent segments on every transversal.

Theorem 6.3-1
If both pairs of opposite sides of a quadrilateral are congruent, then the quadrilateral is a parallelogram.

Theorem 6.3-2
If both pairs of opposite angles of a quadrilateral are congruent, then the quadrilateral is a parallelogram.

Theorem 6.3-3
If an angle of a quadrilateral is supplementary to both of its consecutive angles, then the quadrilateral is a parallelogram.

Theorem 6.3-4
If the diagonals of a quadrilateral bisect each other, then the quadrilateral is a parallelogram.

Theorem 6.3-5
If one pair of opposite sides of a quadrilateral is both congruent and parallel, then the quadrilateral is a parallelogram.

Theorem 6.4-1
A parallelogram is a rhombus if and only if its diagonals are perpendicular.

Theorem 6.4-2
A parallelogram is a rhombus if and only if each diagonal bisects a pair of opposite angles.

Theorem 6.4-3
A parallelogram is a rectangle if and only if its diagonals are congruent.

Theorem 6.5-1
If a trapezoid is isosceles, then each pair of base angles is congruent.

Theorem 6.5-2
If a trapezoid has a pair of congruent base angles, then the trapezoid is isosceles.

Theorem 6.5-3
A trapezoid is isosceles if and only if its diagonals are congruent.

Theorem 6.5-4 Trapezoid Midsegment Theorem
If a quadrilateral is a trapezoid, then

1. the midsegment is parallel to the bases, and
2. the length of the midsegment is half the sum of the lengths of the bases.

Theorem 6.5-5
If a quadrilateral is a kite, then its diagonals are perpendicular.

Theorem 6.5-6
If a quadrilateral is a kite, then exactly one pair of opposite angles is congruent.

Chapter 7 Similarity

Postulate 7.4-1 Angle-Angle Similarity (AA ~) Postulate
If two angles of one triangle are congruent to two angles of another triangle, then the triangles are similar.

Theorem 7.4-2 Side-Angle-Side Similarity (SAS ~) Theorem
If an angle of one triangle is congruent to an angle of a second triangle, and the sides that include the two angles are proportional, then the triangles are similar.

Theorem 7.4-3 Side-Side-Side Similarity (SSS ~) Theorem
If the corresponding sides of two triangles are proportional, then the triangles are similar.

Theorem 7.5-1
The altitude to the hypotenuse of a right triangle divides the triangle into two triangles that are similar to the original triangle and to each other.

Corollary 7.5-2
The length of the altitude to the hypotenuse of a right triangle is the geometric mean of the lengths of the segments of the hypotenuse.

Corollary 7.5-3
The altitude to the hypotenuse of a right triangle separates the hypotenuse so that the length of each leg of the triangle is the geometric mean of the length of the hypotenuse and the length of the segment of the hypotenuse adjacent to the leg.

Theorem 7.6-1 Side-Splitter Theorem
If a line is parallel to one side of a triangle and intersects the other two sides, then it divides those sides proportionally.

Corollary 7.6-2
If three parallel lines intersect two transversals, then the segments intercepted on the transversals are proportional.

Theorem 7.6-3 Triangle-Angle-Bisector Theorem
If a ray bisects an angle of a triangle, then it divides the opposite side into two segments that are proportional to the other two sides of the triangle.

Chapter 8 Transformations

Theorem 8.6-1
A translation or rotation is a composition of two reflections.

Theorem 8.6-2
A composition of reflections across two parallel lines is a translation. A composition of reflections across two intersecting lines is a rotation.

Theorem 8.6-3 Fundamental Theorem of Isometries
In a plane, one of two congruent figures can be mapped onto the other by a composition of at most three reflections.

Theorem 8.6-4 Isometry Classification Theorem
There are only four isometrie. They are translation, rotation, reflection, and glide reflection.

Chapter 9 Right Triangles and Trigonometry

Theorem 9.1-1 Pythagorean Theorem
If a triangle is a right triangle, then the sum of the squares of the lengths of the legs is equal to the square of the length of the hypotenuse.

$$a^2 + b^2 = c^2$$

Theorem 9.1-2 Converse of the Pythagorean Theorem
If the sum of the squares of the lengths of two sides of a triangle is equal to the square of the length of the third side, then the triangle is a right triangle.

Theorem 9.1-3
If the square of the length of the longest side of a triangle is greater than the sum of the squares of the lengths of the other two sides, then the triangle is obtuse.

Theorem 9.1-4
If the square of the length of the longest side of a triangle is less than the sum of the squares of the lengths of the other two sides, then the triangle is acute.

Theorem 9.2-1 45°-45°-90° Triangle Theorem
In a 45°-45°-90° triangle, both legs are congruent and the length of the hypotenuse is $\sqrt{2}$ times the length of a leg.
hypotenuse $= \sqrt{2} \cdot$ leg.

Theorem 9.2-2 30°-60°-90° Triangle Theorem
In a 30°-60°-90° triangle, the length of the hypotenuse is twice the length of the shorter leg. The length of the longer leg is $\sqrt{3}$ times the length of the shorter leg.

$$\text{hypotenuse} = 2 \cdot \text{shorter leg}$$

$$\text{longer leg} = \sqrt{3} \cdot \text{shorter leg}$$

Law of Sines

$$\frac{\sin A}{a} = \frac{\sin B}{b} = \frac{\sin C}{c}$$

Law of Cosines

$$a^2 = b^2 + c^2 - 2bc \cos A$$
$$b^2 = a^2 + c^2 - 2ac \cos B$$
$$c^2 = a^2 + b^2 - 2ab \cos C$$

Chapter 10 Area

Theorem 10.1-1 Polygon Interior Angle-Sum Theorem
The sum of the measures of the interior angles of an n-gon is $(n - 2)180°$.

Corollary 10.1-2
The measure of each interior angle of a regular n-gon is $\frac{(n-2)180°}{n}$.

Theorem 10.1-3 Polygon Exterior Angle-Sum Theorem
The sum of the measures of the exterior angles of a polygon, one at each vertex, is 360°.

Corollary 10.1-4
The measure of each exterior angle of a regular n-gon is $\frac{1}{n} \cdot 360°$ or $\frac{360°}{n}$.

Postulate 10.2-1
If two figures are congruent, then their areas are equal.

Postulate 10.2-2 Area Addition Postulate
The area of a region is the sum of the areas of its nonoverlapping parts.

Postulate 10.2-3 Area of a Square
The area of a square is the square of its side legnth.

$$A = s^2$$

Theorem 10.2-4 Area of a Rectangle
The area of a rectangle is the product of its base and height.

$$A = bh$$

Theorem 10.2-5 Area of a Parallelogram
The area of a parallelogram is the product of a base and the corresponding height.

$$A = bh$$

Theorem 10.2-6 Area of a Triangle
The area of a triangle is half the product of a base and the corresponding height.

$$A = \frac{1}{2}bh$$

Theorem 10.2-7 Area of a Trapezoid
The area of a trapezoid is half the product of the height and the sum of the bases.

$$A = \frac{1}{2}h(b_1 + b_2)$$

Theorem 10.2-8 Area of a Rhombus or a Kite
The area of a rhombus or a kite is half the product of the lengths of its diagonals.

$$A = \frac{1}{2}d_1 d_2$$

Theorem 10.3-1 Area of a Regular Polygon
The area of a regular polygon is half the product of the apothem and the perimeter.

$$A = \frac{1}{2}aP$$

Theorem 10.4-1 Perimeters and Areas of Similar Figures

If the scale factor of two similar figures is $\frac{a}{b}$, then

1. the ratio of their perimeters is $\frac{a}{b}$ and

2. the ratio of their areas is $\frac{a^2}{b^2}$.

Postulate 10.5-1 Arc Addition Postulate
The measure of the arc formed by two adjacent arcs is the sum of the measures of the two arcs.

$$m\widehat{ABC} = m\widehat{AB} + m\widehat{BC}$$

Theorem 10.5-2 Circumference of a Circle
The circumference of a circle is π times the diameter.
$C = \pi d$ or $C = 2\pi r$

Theorem 10.5-3 Arc Length
The length of an arc of a circle is the product of the ratio $\frac{\text{measure of the arc}}{360°}$ and the circumference of the circle.

$$\text{length of } \widehat{AB} = \frac{m\widehat{AB}}{360°} \cdot 2\pi r \quad \text{or}$$

$$\text{length of } \widehat{AB} = \frac{m\widehat{AB}}{360°} \cdot \pi d$$

Theorem 10.6-1 Area of a Circle
The area of a circle is the product of π and the square of the radius.

$$A = \pi r^2$$

Theorem 10.6-2 Area of a Sector of a Circle
The area of a sector of a circle is the product of the ratio $\frac{\text{measure of the arc}}{360^\circ}$ and the area of the circle.

$$\text{Area of sector } AOB = \frac{m\widehat{AB}}{360^\circ} \cdot \pi r^2$$

Chapter 11 Surface Area and Volume

Theorem 11.2-1 Lateral and Surface Areas of a Prism
The lateral area of a right prism is the product of the perimeter of the base and the height of the prism.

$$\text{LA} = Ph$$

The surface area of a right prism is the sum of the lateral area and the areas of the two bases.

$$\text{SA} = \text{LA} + 2B$$

Theorem 11.2-2 Lateral and Surface Areas of a Cylinder
The lateral area of a right cylinder is the product of the circumference of the base and the height of the cylinder.

$$\text{LA} = 2\pi rh, \text{ or LA} = \pi dh$$

The surface area of a right cylinder is the sum of the lateral area and areas of the two bases.

$$\text{SA} = \text{LA} + 2B, \text{ or SA} = 2\pi rh + 2\pi r^2$$

Theorem 11.3-1 Lateral and Surface Areas of a Pyramid
The lateral area of a regular pyramid is half the product of the perimeter P of the base and the slant height ℓ of the pyramid.

$$\text{LA} = \frac{1}{2}P\ell$$

The surface area of a regular pyramid is the sum of the lateral area and the area B of the base.

$$\text{SA} = \text{LA} + B$$

Theorem 11.3-2 Lateral and Surface Areas of a Cone
The lateral area of a right cone is half the product of the circumference of the base and the slant height of the cone.

$$\text{LA} = \frac{1}{2}\cdot 2\pi r\ell, \text{ or LA} = \pi r\ell$$

The surface area of a right cone is the sum of the lateral area and the area of the base.

$$\text{SA} = \text{LA} + B$$

Theorem 11.4-1 Cavalieri's Principle
If two space figures have the same height and the same cross-sectional area at every level, then they have the same volume.

Theorem 11.4-2 Volume of a Prism
The volume of a prism is the product of the area of the base and the height of the prism.

$$V = Bh$$

Theorem 11.4-3 Volume of a Cylinder
The volume of a cylinder is the product of the area of the base and the height of the cylinder.

$$V = Bh, \text{ or } V = \pi r^2 h$$

Theorem 11.5-1 Volume of a Pyramid
The volume of a pyramid is one-third the product of the area of the base and the height of the pyramid.

$$V = \frac{1}{3}Bh$$

Theorem 11.5-2 Volume of a Cone
The volume of a cone is one-third the product of the area of the base and the height of the cone.

$$V = \frac{1}{3}Bh, \text{ or } V = \frac{1}{3}\pi r^2 h$$

Theorem 11.6-1 Surface Area of a Sphere
The surface area of a sphere is four times the product of π and the square of the radius of the sphere.

$$\text{SA} = 4\pi r^2$$

Theorem 11.6-2 Volume of a Sphere
The volume of a sphere is four-thirds the product of π and the cube of the radius of the sphere.

$$V = \frac{4}{3}\pi r^3$$

Theorem 11.7-1 Areas and Volumes of Similar Solids
If the scale factor of two similar solids is $a{:}b$, then

- the ratio of their corresponding areas is $a^2{:}b^2$, and
- the ratio of their volumes is $a^3{:}b^3$.

Chapter 12 Circles and Other Conic Sections

Theorem 12.1-1
If a line is tangent to a circle, then the line is perpendicular to the radius at the point of tangency.

Theorem 12.1-2
If a line in the plane of a circle is perpendicular to a radius at its endpoint on the circle, then the line is tangent to the circle.

Theorem 12.1-3
If two segments are tangent to a circle from a point outside the circle, then the two segments are congruent.

Theorem 12.2-1
Within a circle or in congruent circles, congruent central angles have congruent arcs.

Converse 12.2-2
Within a circle or in congruent circles, congruent arcs have congruent central angles.

Theorem 12.2-3
Within a circle or in congruent circles, congruent central angles have congruent chords.

Converse 12.2-4
Within a circle or in congruent circles, congruent chords have congruent central angles.

Theorem 12.2-5
Within a circle or in congruent circles, congruent chords have congruent arcs.

Converse 12.2-6
Within a circle or in congruent circles, congruent arcs have congruent chords.

Theorem 12.2-7
Within a circle or in congruent circles, chords equidistant from the center (or centers) are congruent.

Converse 12.2-8
Within a circle or in congruent circles, congruent chords are equidistant from the center (or centers).

Theorem 12.2-9
In a circle, if a diameter is perpendicular to a chord, it bisects the chord and its arc.

Theorem 12.2-10
In a circle, if a diameter bisects a chord (that is not a diameter), it is perpendicular to the chord.

Theorem 12.2-11
In a circle, the perpendicular bisector of a chord contains the center of the circle.

Theorem 12.3-1 Inscribed Angle Theorem
The measure of an inscribed angle is half the measure of its intercepted arc.

Corollary 12.3-2
Two inscribed angles that intercept the same arc are congruent.

Corollary 12.3-3
An angle inscribed in a semicircle is a right angle.

Corollary 12.3-4
The opposite angles of a quadrilateral inscribed in a circle are supplementary.

Theorem 12.3-5
The measure of an angle formed by a tangent and a chord is half the measure of the intercepted arc.

Theorem 12.4-1
The measure of an angle formed by two lines that intersect inside a circle is half the sum of the measures of the intercepted arcs.

Theorem 12.4-2
The measure of an angle formed by two lines that intersect outside a circle is half the difference of the measures of the intercepted arcs.

Theorem 12.4-3
For a given point and circle, the product of the lengths of the two segments from the point to the circle is constant along any line through the point and circle.

Theorem 12.5-1
An equation of a circle with center (h, k) and radius r is $(x - h)^2 + (y - k)^2 = r^2$.

Additional Proofs Now, we prove the theorems left for the back of the text.

Theorem 3.2-3 Two Lines Perpendicular to a Third Line

Theorem	If ...	Then ...
In a plane, if two lines are perpendicular to the same line, then they are parallel to each other.	$m \perp t$ and $n \perp t$ 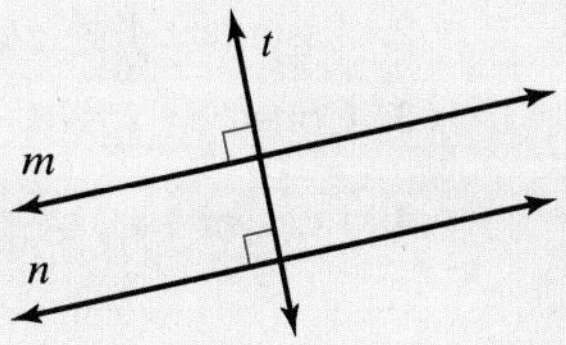	$m \parallel n$ 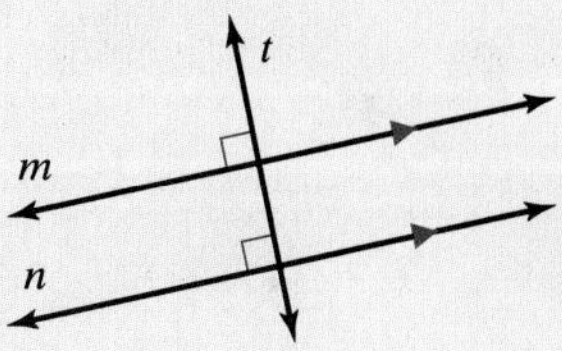

Let's use the concept of indirect proofs (section 5.5) to prove this theorem.

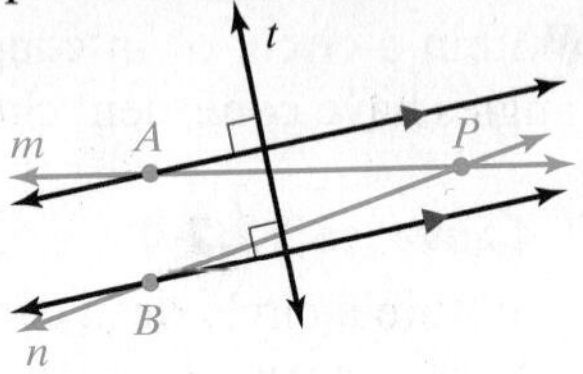

Proof

Given: Lines m, n, and t with $m \perp t$ and $n \perp t$

Prove: $m \parallel n$

Statements	Reasons
1. lines m, n, t with $m \perp t$ and $n \perp t$	**1.** Given
2. Assume $m \nparallel n$	**2.** Assumption
3. m and n intersect at some point, P	**3.** Definition of lines that are not parallel

We now have that both $\overleftrightarrow{PA} \perp t$ and $\overleftrightarrow{PB} \perp t$ through point P(Step 1), or $m \perp t$ and $n \perp t$ through point P(Step 1).

This means that through point P not on line t, we have two lines m and n perpendicular to the given line t. This is a contradiction to the Perpendicular Postulate (Postulate 3.2-2). Thus Step 2, the assumption $m \nparallel n$ is a contradiction and we must conclude that $m \parallel n$, which is what we wanted to prove.

Theorem 3.3-1 Alternate Interior Angles Converse (Converse of Theorem 3.2-4)

Theorem	If ...	Then ...
If two parallel lines are cut by a transversal, then alternate interior angles are congruent.	$\ell \parallel m$ 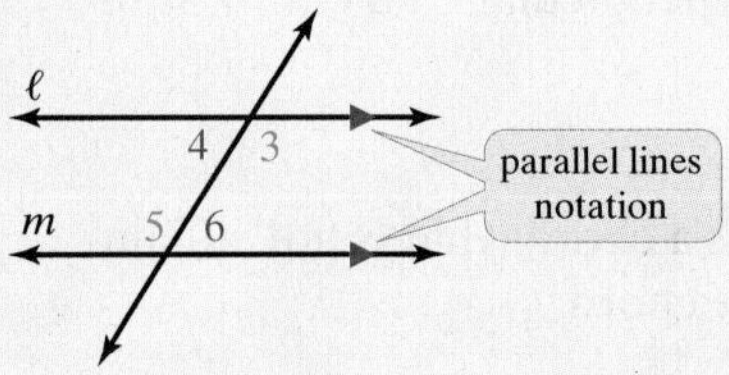	$\angle 4 \cong \angle 6$ $\angle 3 \cong \angle 5$

As stated in Section 3.3, we only needed the concept of indirect proofs (Section 5.5) to prove this theorem. Let's now use this concept and the figure below to prove Theorem 3.3-1.

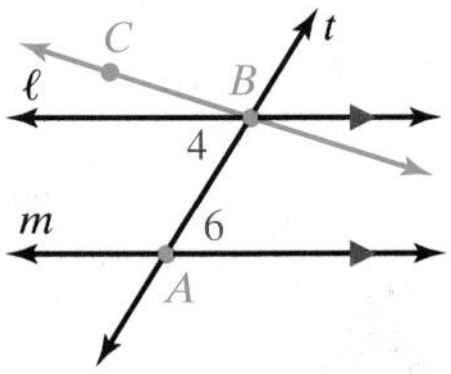

Proof

Given: $\ell \parallel m$ and both are intersected by transversal t

Prove: $\angle 4 \cong \angle 6$

Statements	Reasons
1. $\ell \parallel m$ and transversal t	**1.** Given
2. Assume $\angle 4 \ncong \angle 6$	**2.** Assumption
3. Construct $\angle ABC$ so that $\angle ABC \cong \angle 6$	**3.** Construction
4. $\overleftrightarrow{CB} \parallel m$	**4.** alternate interior angles are $\cong$ (Theorem 3.2-4)

We now have that $\overleftrightarrow{CB} \parallel m$ (Step 4) and $\ell \parallel m$ (Step 1). Since line l contains point B, this means that we now have two distinct lines ($\overleftrightarrow{CB}$ and line l) through point B parallel to line m. This contradicts the Parallel Postulate (Postulate 3.2-1). Thus Step 2, the assumption $\angle 4 \ncong \angle 6$ is a contradiction and we must conclude that $\angle 4 \cong 6$, which is what we wanted to prove. (A similar proof can be used to show that the other pair of alternate interior angles are congruent.)

Theorem 4.6-6 Hypotenuse-Leg (H-L) Theorem

Theorem

If the hypotenuse and a leg of one right triangle are congruent to the hypotenuse and a leg of another right triangle, then the triangles are congruent.

If ...

ΔPQR and ΔXYZ are right triangles, $\overline{PR} \cong \overline{XY}$, and $\overline{PQ} \cong \overline{XY}$

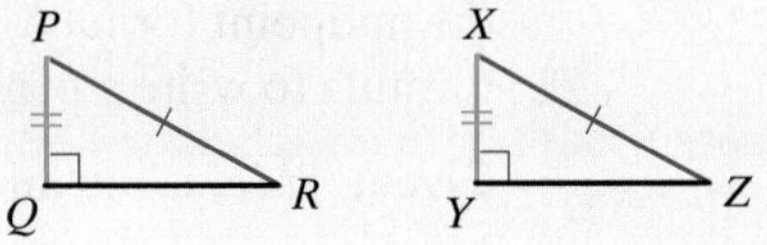

Then ...

$\Delta PQR \cong \Delta XYZ$

To prove this theorem, we need to draw auxiliary lines to form a third triangle.

Proof:

Given: ΔPQR and ΔXYZ are right triangles, with right angles Q and Y. $\overline{PR} \cong \overline{XZ}$ and $\overline{PQ} \cong \overline{XY}$.

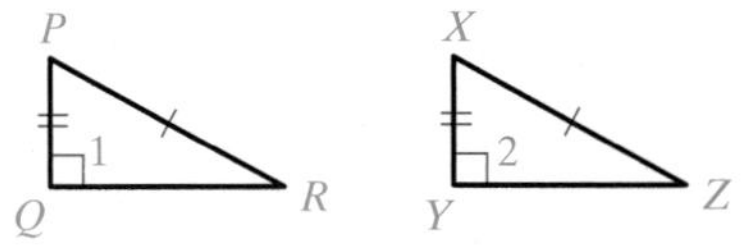

Prove: $\Delta PQR \cong \Delta XYZ$

Construction: On ΔXYZ, draw $\overrightarrow{ZY}$.

Mark point S so that $YS = QR$.

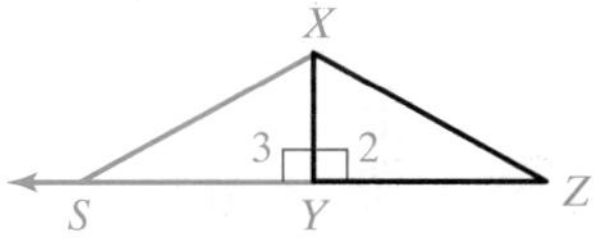

Statements	Reasons
1. ΔPQR and ΔXYZ are right Δ's, $\overline{PR} \cong \overline{XZ}, \overline{PQ} \cong \overline{XY}, \angle 1 = \angle 2 = 90°$	**1.** Given
2. On ΔXYZ, draw $\overrightarrow{ZY}$ mark point S so that $YS = QR$	**2.** Construction
3. $\Delta PQR \cong \Delta XYS$	**3.** SAS
4. $PR \cong XS$	**4.** cpoctac
5. $PR \cong XZ$	**5.** Given
6. $XS \cong XZ$	**6.** Substitution or Transitive Property (Steps 4, 5)
7. $\angle S \cong \angle Z$	**7.** Base $\angle s$ of isosceles Δ^s are $\cong$
8. $\Delta XYS \cong \Delta XYZ$	**8.** AAS (steps 7, 3, 6)
9. $\Delta PQR \cong \Delta XYZ$	**9.** Substitution or Transitive Property (Steps 3, 8)

Notice the Step 9 Statement. We proved the H-L Theorem.

Theorem 6.5-4 Trapezoid Midsegment Theorem

Theorem

If a quadrilateral is a trapezoid, then

(1) the midsegment is parallel to the bases, and

(2) the length of the midsegment is half the sum of the lengths of the bases.

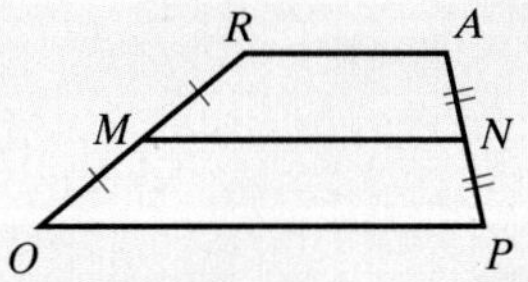

(1) $\overline{MN} \parallel \overline{OP}$, $\overline{MN} \parallel \overline{RA}$, and

(2) $MN = \frac{1}{2}(OP + RA)$

For this theorem, we will write a rectangular coordinate proof.

Proof

First, let's draw and label a figure. Midpoints will be involved, so use multiples of 2 to name coordinates.

Now we use our figure and our knowledge of the midpoint formula, slope formula, and the distance formula to write a proof.

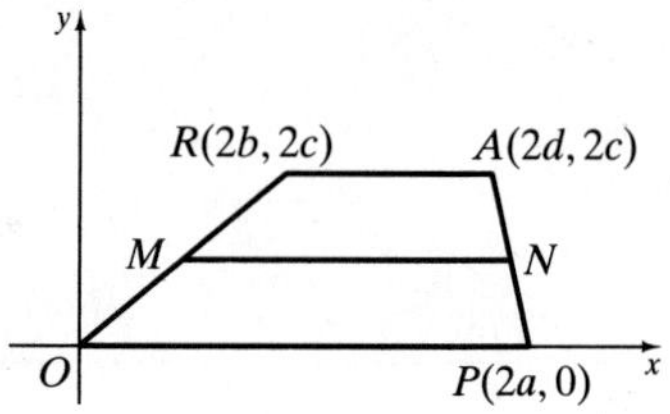

Given: $\overline{MN}$ is the midsegment of trapezoid $ORAP$.

Prove: $\overline{MN} \| \overline{OP}, \overline{MN} \| \overline{RA}, MN = \frac{1}{2}(OP + RA)$

First, let's prove that

(1) $\overline{MN} \| \overline{OP}$ and $\overline{MN} \| RA$.

Use the Midpoint Formula to find the coordinates of M and N.

$$M = \left(\frac{2b + 0}{2}, \frac{2c + 0}{2}\right) = (b, c)$$

$$N = \left(\frac{2a + 2d}{2}, \frac{0 + 2c}{2}\right) = (a + d, c)$$

Use the Slope Formula to determine whether $\overline{MN}$ is parallel to $\overline{OP}$ and $\overline{RA}$.

$$\text{slope of } \overline{MN} = \frac{c - c}{(a + d) - b} = 0$$

$$\text{slope of } \overline{RA} = \frac{2c - 2c}{2d - 2b} = 0$$

$$\text{slope of } \overline{OP} = \frac{0 - 0}{2a - 0} = 0$$

The three slopes are equal, so

(1) $\overline{MN} \| \overline{OP}$ and $\overline{MN} \| \overline{RA}$.

Now, let's prove that

(2) $MN \frac{1}{2}(OP + RA)$

Use the Distance Formula to find and compare MN, OP, and RA.

$$MN = \sqrt{[(a + d) - b)]^2 + (c - c)^2} = a + d - b$$

$$OP = \sqrt{(2a - 0)^2 + (0 - 0)^2} = 2a$$

$$RA = \sqrt{(2d - 2b)^2 + (2c - 2c)^2} = 2d - 2b$$

Now, check that $MN = \frac{1}{2}(OP + RA)$ is true.

$$MN \stackrel{?}{=} \frac{1}{2} OP + RA$$

$$a + d - b \stackrel{?}{=} \frac{1}{2}[2a + (2d - 2b)] \quad \text{Substitute.}$$

$$a + d - b = a + d - b \quad \text{Simplify.}$$

Statements (2) is true. Thus,

1. the midsegment of a trapezoid is parallel to its bases, and
2. the length of the midsegment of a trapezoid is half the sum of the lengths of the bases.

B ADDITIONAL LESSONS

OBJECTIVES

1. Find the Equation of an Ellipse Given Its Foci and Vertices.
2. Find the Equation of a Hyperbola Given Its Foci and Vertices.

B.1 Ellipses and Hyperbolas. Recall from the Chapter 12 Extension on Parabolas that parabolas, circles, ellipses, and hyperbolas are together called conic sections. Also, a **conic section** is a curve generated by the intersection of a cone and a plane.

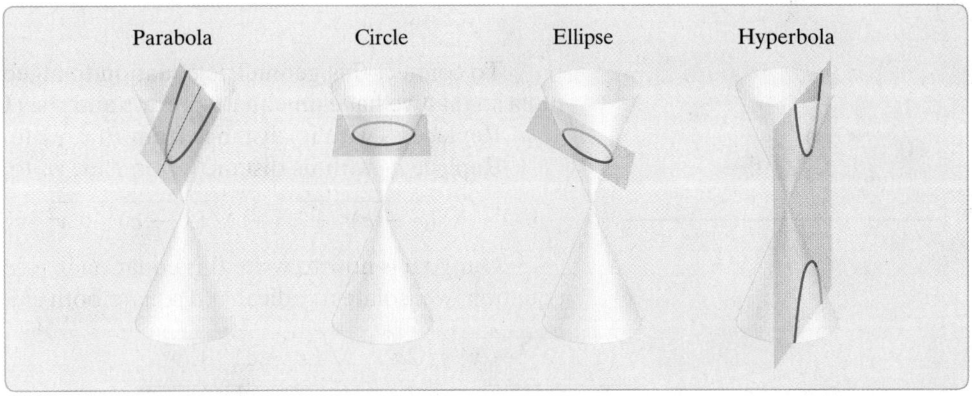

In this section, we concentrate on connecting the geometric definition of ellipses and hyperbolas with the rectangular coordinate system, or with an algebraic equation.

Objective 1 ▶ Finding the Equation of an Ellipse Given Its Foci and Vertices. Let's begin with a definition of an ellipse. An **ellipse** is the set of all points in the plane, the sum of whose distances from two fixed points is constant. The fixed points are called the **foci** (plural of **focus**).

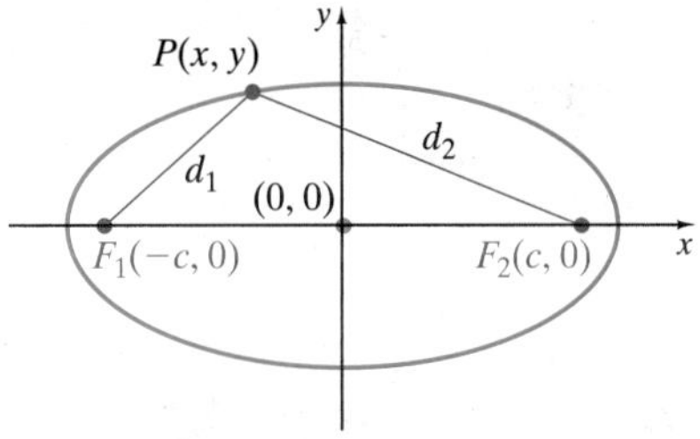

F_1 and F_2 are each a Focus.

An ellipse may remind you of a circle with opposite sides that have been stretched. In this section, we will study ellipses that have been stretched vertically or horizontally, as shown below.

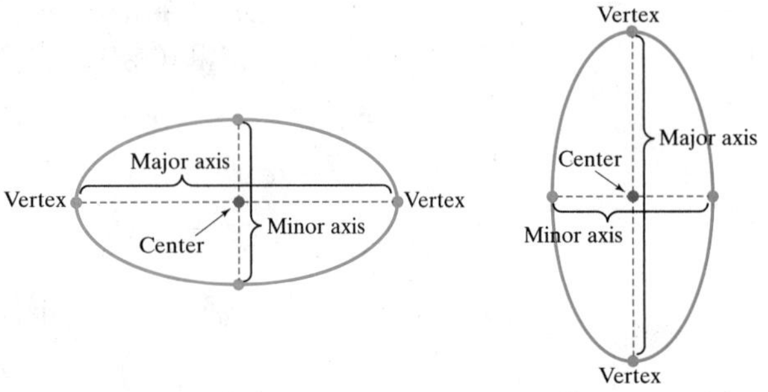

Ellipse stretched Horizontally   Vertex stretched Vertically

Make sure you understand the parts of an ellipse:

- **major axis**—line segment that passes through the foci, with endpoints on the ellipse. These endpoints are called **vertices** (singular is **vertex**)
- **center**—the midpoint of the major axis.
- **minor axis**—line segment perpendicular to the major axis at the center with endpoints on the ellipse

Now let's connect the geometric definition of an ellipse with an algebraic definition. For this exercise, let's use our horizontally stretched ellipse on the next page.

Our ellipse has foci $(-c, 0)$ and $(c, 0)$, on the x-axis, thus the center is the origin, $(0, 0)$. If $P(x, y)$ is any point on the ellipse, the sum of the distances to the two foci, $d_1 + d_2$, must be a constant number, say 2a. Thus, for any point $P(x, y)$ on this ellipse, we have

$$d_1 + d_2 = 2a$$

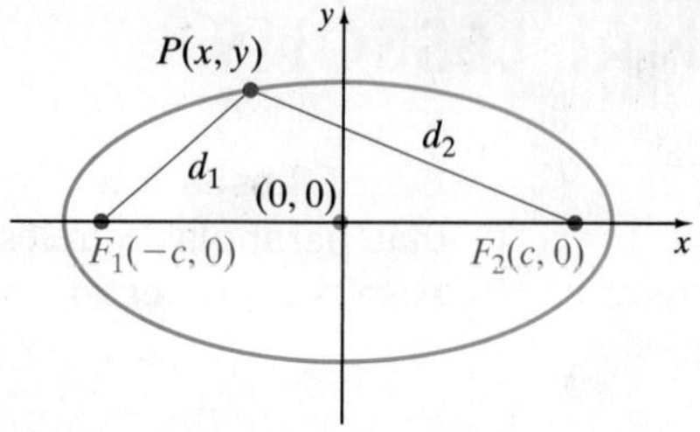

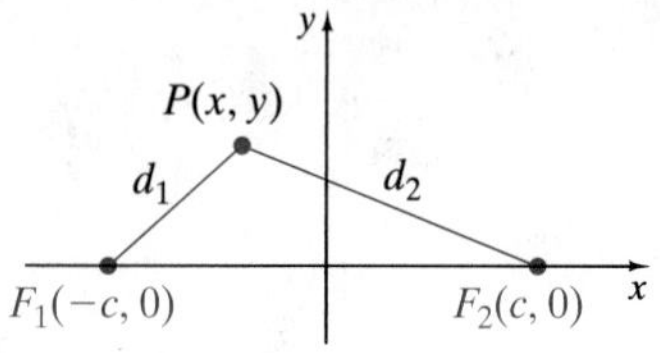

To connect this geometric equation to algebra, we will first use the distance formula so that we have an equation in x's and y's. (Use the figure in the margin.)

Replace d_1 with its distance from $P(x, y)$ to $F_1(-c, 0)$.

Replace d_2 with its distance from $P(x, y)$ to $F_2(c, 0)$.

$$\sqrt{(x+c)^2+y^2}+\sqrt{(x-c)^2+y^2}=2a \quad \text{Use the distance formula.}$$

Our goal is now to write this equation in a convenient form. Since this is a radical equation, we isolate a radical and square both sides.

$$\sqrt{(x+c)^2+y^2}=2a-\sqrt{(x-c)^2+y^2} \quad \text{Isolate a radical.}$$

$$\left(\sqrt{(x+c)^2+y^2}\right)^2=\left(2a-\sqrt{(x-c)^2+y^2}\right)^2 \quad \text{Square both sides.}$$

$$(x+c)^2+y^2=4a^2-4a\sqrt{(x-c)^2+y^2}+(x-c)^2+y^2 \quad \text{Simplify. (Recall how to square the right side.)}$$

$$4cx-4a^2=-4a\sqrt{(x-c)^2+y^2} \quad \text{Simplify further by finding } (x+c)^2 \text{ and } (x-c)^2\text{. Then isolate the radical.}$$

Let's continue by dividing each term by 4.

$$cx-a^2=-a\sqrt{(x-c)^2+y^2} \quad \text{Divide both sides by 4.}$$

$$(cx-a^2)^2=\left(-a\sqrt{(x-c)^2+y^2}\right)^2 \quad \text{Square both sides.}$$

$$c^2x^2-2a^2cx+a^4=a^2\left[(x-c)^2+y^2\right] \quad \text{Simplify. (Find } (cx-a)^2 \text{ on left side.)}$$

$$c^2x^2-2a^2cx+a^4=a^2(x^2-2cx+c^2+y^2) \quad \text{Find } (x-c)^2 \text{ on right side.}$$

$$c^2x^2-2a^2cx+a^4=a^2x^2-2a^2cx+a^2c^2+a^2y^2 \quad \text{Use the distributive property.}$$

$$c^2x^2+a^4=a^2x^2+a^2c^2+a^2y^2 \quad \text{Add } 2a^2cx \text{ to both sides.}$$

$$-a^2x^2+c^2x^2-a^2y^2=-a^4+a^2c^2 \quad \text{Rearrange terms.}$$

$$a^2x^2-c^2x^2+a^2y^2=a^4-a^2c^2 \quad \text{Multiply both sides by } -1.$$

$$(a^2-c^2)x^2+a^2y^2=a^2(a^2-c^2) \quad \text{Factor out } x^2 \text{ on left side and } a^2 \text{ on right side.}$$

$$\frac{(a^2-c^2)x^2}{a^2(a^2-c^2)}+\frac{a^2y^2}{a^2(a^2-c^2)}=\frac{a^2(a^2-c^2)}{a^2(a^2-c^2)} \quad \text{Divide both sides by } a^2(a^2-c^2).$$

$$\frac{x^2}{a^2}+\frac{y^2}{a^2-c^2}=1 \quad \text{Simplify.}$$

Let $b^2=a^2-c^2$, and we have

$$\frac{x^2}{a^2}+\frac{y^2}{b^2}=1 \quad \text{Substitute } b^2 \text{ for } a^2-c^2.$$

This is the **standard form of an ellipse** centered at $(0, 0)$, stretched horizontally.

Note 1: Let $y=0$ and $x=\pm a$. Thus, the x-intercept points are $(a, 0)$ and $(-a, 0)$.
Let $x=0$ and $y=\pm b$. Thus, the y-intercept points are $(0, b)$ and $(0, -b)$.

Note 2: Study our last triangular figure in the margin. The distance from F_1 to F_2 is 2c. From our Triangle Inequality, we know that

$$d_1+d_2>2c, \quad \text{or} \quad 2a>2c \quad \text{Since } d_1+d_2=2a.$$

$$\text{or} \quad a>c \quad \text{Divide both sides by 2.}$$

Also, we can show that $a>b$.

> **Helpful Hint**
> Notice that a^2, the larger denominator, tells us whether the major axis is horizontal or vertical. If the numerator of a^2 is x^2, the major axis is horizontal. If it is y^2, the major axis is vertical.

Ellipses—Standard-Form Equations with Center (0, 0) and Major Axis Length 2a

Major Axis	*Standard Form*	*Foci*
Horizontal	$\frac{x^2}{a^2} + \frac{y^2}{b^2} = 1$	$(-c, 0), (c, 0)$
Vertical	$\frac{y^2}{a^2} + \frac{x^2}{b^2} = 1$	$(0, -c), (0, c)$
where $c^2 = a^2 - b^2$		with $a > b > 0$

To sketch the graph of an ellipse, plot the center, the intercepts $\pm a$ on the major axis and $\pm b$ on the minor axis.

EXAMPLE 1 Graphing an Ellipse

Graph $\frac{y^2}{16} + \frac{x^2}{9} = 1$. Label the intercepts and foci.

Solution The equation is of the form $\frac{y^2}{a^2} + \frac{x^2}{b^2} = 1$, with $a^2 = 16$ and $b^2 = 9$, so its graph is an ellipse with center (0, 0), y-intercepts (0, 4) and $(0, -4)$, and x-intercepts (3, 0) and $(-3, 0)$.

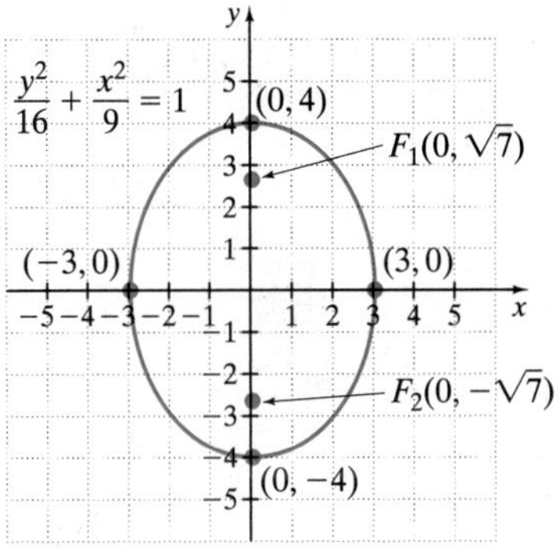

Find the foci by calculating c.

$$c^2 = a^2 - b^2 = 16 - 9 = 7$$

Thus, the foci are $(0, -\sqrt{7})$ and $(0, \sqrt{7})$. □

PRACTICE 1 Graph $\frac{x^2}{36} + \frac{y^2}{9} = 1$. Label the intercepts and foci.

EXAMPLE 2 Graphing an Ellipse

Graph $4x^2 + 16y^2 = 64$. Label the intercepts and foci.

Solution Since the standard form of the equation of an ellipse has 1 on one side, divide both sides of this equation by 64.

$$4x^2 + 16y^2 = 64$$

$$\frac{4x^2}{64} + \frac{16y^2}{64} = \frac{64}{64} \quad \text{Divide both sides by 64.}$$

$$\frac{x^2}{16} + \frac{y^2}{4} = 1 \quad \text{Simplify.}$$

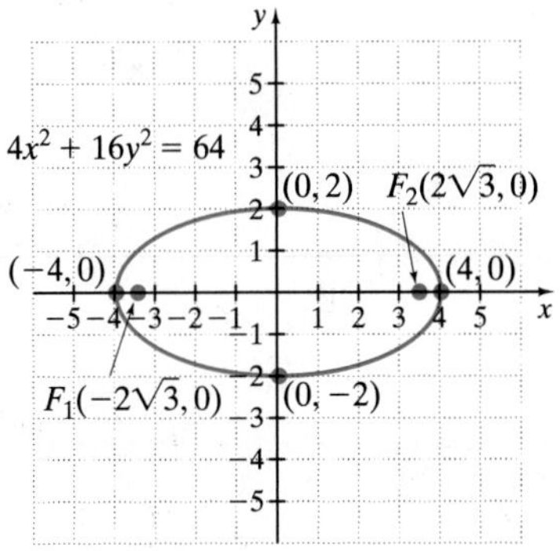

We now recognize the equation of an ellipse with $a^2 = 16$ and $b^2 = 4$. This ellipse has center (0, 0), x-intercepts (4, 0) and $(-4, 0)$, and y-intercepts (0, 2) and $(0, -2)$.

To find the foci, $c^2 = a^2 - b^2 = 16 - 4 = 12$. The foci are $(\sqrt{12}, 0)$ or $(2\sqrt{3}, 0)$ and also $(-\sqrt{12}, 0)$ or $(-2\sqrt{3}, 0)$. □

PRACTICE 2 Graph $25x^2 + 4y^2 = 100$. Label the intercepts and foci.

In Examples 1 and 2, we were given an equation of an ellipse and asked to find the intercepts and foci. For Example 3, we do just the opposite.

EXAMPLE 3 **Writing the Standard-Form Equation of an Ellipse**

Write the standard-form equation of the ellipse with foci at $(-2, 0)$ and $(2, 0)$ and vertices at $(-5, 0)$ and $(5, 0)$.

Solution The foci, $(-2, 0)$ and $(2, 0)$ are located on the x-axis, so the major axis is horizontal. The center of the ellipse is the midpoint of the major axis, so it is $(0, 0)$. Thus, the equation has standard form

$$\frac{x^2}{a^2} + \frac{y^2}{b^2} = 1$$

Let's now find the values for a^2 and b^2. The distance from the center $(0, 0)$ to either vertex, $(-5, 0)$ or $(5, 0)$, is 5 units. This means that

$$a = 5 \quad \text{or} \quad a^2 = 5^2 = 25.$$

To find b^2, we will use $c^2 = a^2 - b^2$. The distance from the center $(0, 0)$ to either given foci, $(-2, 0)$ or $(2, 0)$ is 2 units. Thus,

$$c = 2 \quad \text{or} \quad c^2 = 2^2 = 4.$$

Then since

$$c^2 = a^2 - b^2$$

$$4 = 25 - b^2 \quad \text{Replace } c^2 \text{ with 4 and } a^2 \text{ with 25.}$$

$$b^2 = 25 - 4 \quad \text{Add } b^2 \text{ and subtract 4 from both sides.}$$

$$b^2 = 21 \quad \text{Subtract.}$$

The standard form equation is then

$$\frac{x^2}{25} + \frac{y^2}{21} = 1$$

□

PRACTICE

3 Write the standard-form equation of the ellipse with foci at $(-3, 0)$ and $(3, 0)$ and vertices at $(-4, 0)$ and $(4, 0)$.

OBJECTIVE 2 ▶ Finding the Equation of a Hyperbola Given its Foci and Vertices. The final conic section to study is the hyperbola. A **hyperbola** is the set of points in a plane such that the absolute value of the difference of the distances from two fixed points, called foci, is constant.

The first figure to the right shows the two branches of a hyperbola. Some defined parts of a hyperbola are:

- **vertices**—the intersection of the line through the foci and the hyperbola
- **transverse axis**—the line segment that joins the vertices
- **center**—the midpoint of the transverse axis

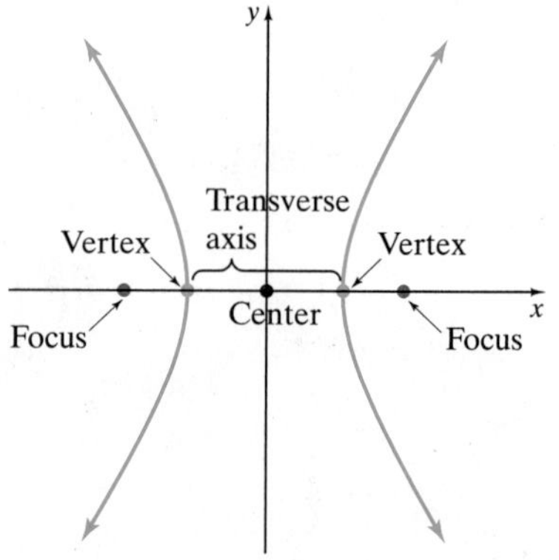

Hyperbola

Let's use the rectangular coordinate system to connect the geometric definition of a hyperbola to an algebraic one. In the second figure on the next page to the right, we have the foci of this hyperbola on the x-axis at $(-c, 0)$ and $(c, 0)$.

Thus, the center is $(0, 0)$ and we will let $P(x, y)$ be any point of the hyperbola.

Recall that for any point of the hyperbola the absolute value of the distances from the two foci, $|d_2 - d_1|$, must be a constant number, say 2a. Thus for any point $P(x, y)$ on this hyperbola, we have

$$|d_2 - d_1| = 2a$$

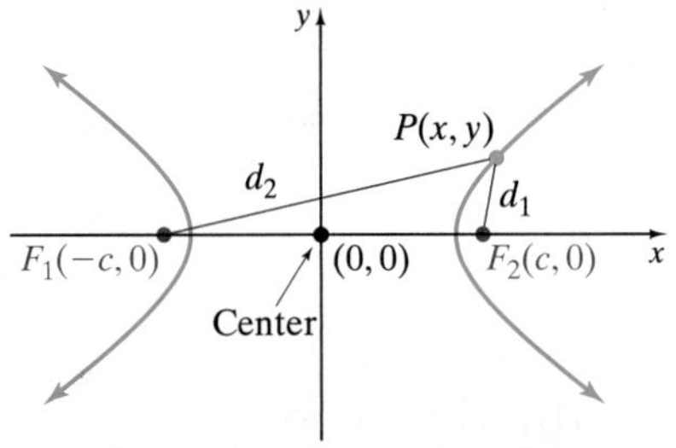

A hyperbola has two branches.

To connect this geometric equation to algebra, we will use the distance formula so that we have an equation in x's and y's. Thus

$$|\sqrt{(x + c)^2 + y^2} - \sqrt{(x - c)^2 + y^2}| = 2a \quad \text{Use the distance formula.}$$

Eliminate radicals and simplify, similar to our work with an ellipse, and we have

$$(c^2 - a^2)x^2 - a^2y^2 = a^2(c^2 - a^2)$$

$$\frac{(c^2 - a^2)x^2}{a^2(c^2 - a^2)} - \frac{a^2y^2}{a^2(c^2 - a^2)} = \frac{a^2(c^2 - a^2)}{a^2(c^2 - a)^2} \quad \text{Divide both sides by } a^2(c^2 - a^2).$$

$$\frac{x^2}{a^2} - \frac{y^2}{c^2 - a^2} = 1 \quad \text{Simplify.}$$

Let $b^2 = c^2 - a^2$ and we have

$$\frac{x^2}{a^2} - \frac{y^2}{b^2} = 1 \quad \text{Substitute } b^2 \text{ for } c^2 - a^2 = 0.$$

This is the **standard form of a hyperbola** with center $(0, 0)$ and foci and vertices on the x-axis.

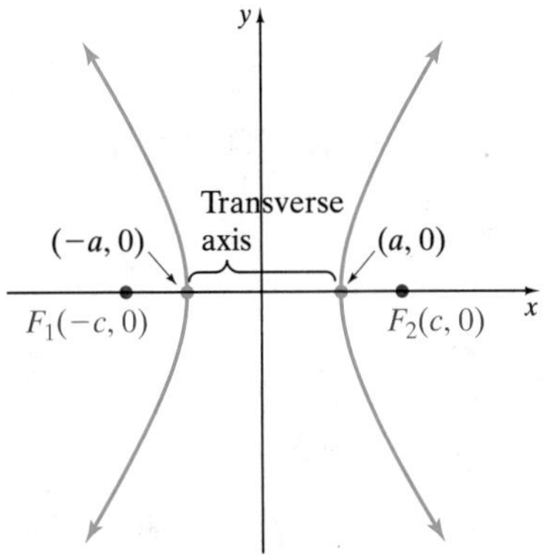

$$\frac{x^2}{a^2} - \frac{y^2}{b^2} = 1$$

(Horizontal Transverse Axis)

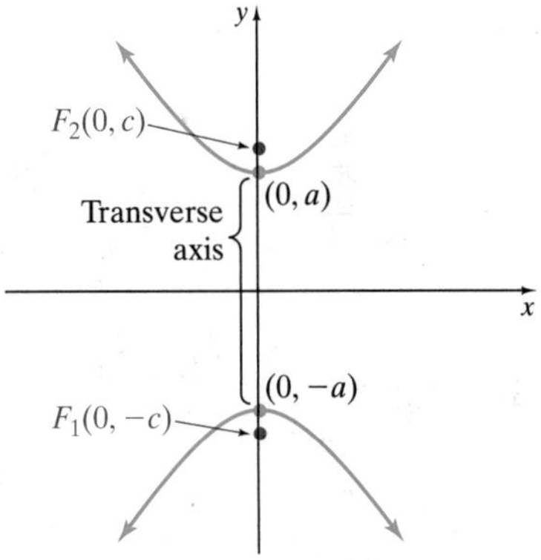

$$\frac{y^2}{a^2} - \frac{x^2}{b^2} = 1$$

(Vertical Transverse Axis)

Helpful Hint

Note: If the x^2 term is positive, the vertices and the foci are along the x-axis. If the y^2 term is positive, the vertices and the foci are along the y-axis.

Hyperbolas–Standard-Form Equations with Center (0, 0) and Transverse Axis Length 2a

Transverse Axis	*Standard Form*	*Foci*
Horizontal	$\frac{x^2}{a^2} - \frac{y^2}{b^2} = 1$	$(-c, 0), (c, 0)$
Vertical	$\frac{y^2}{a^2} - \frac{x^2}{b^2} = 1$	$(0, -c), (0, c)$
where $c^2 = a^2 + b^2$		

> **Helpful Hint**
>
> Make sure you notice the differences for finding c^2 (which lead to foci) for ellipses and hyperbolas.
>
> Ellipses: $c^2 = a^2 - b^2$
>
> Hyperbolas: $c^2 = a^2 + b^2$

Graphing a hyperbola such as $\frac{y^2}{b^2} - \frac{x^2}{a^2} = 1$ is made easier by recognizing one of its important characteristics. Examining the figure to the right, notice how the sides of the branches of the hyperbola extend indefinitely and seem to approach the dashed lines in the figure. These dashed lines are called the **asymptotes** of the hyperbola.

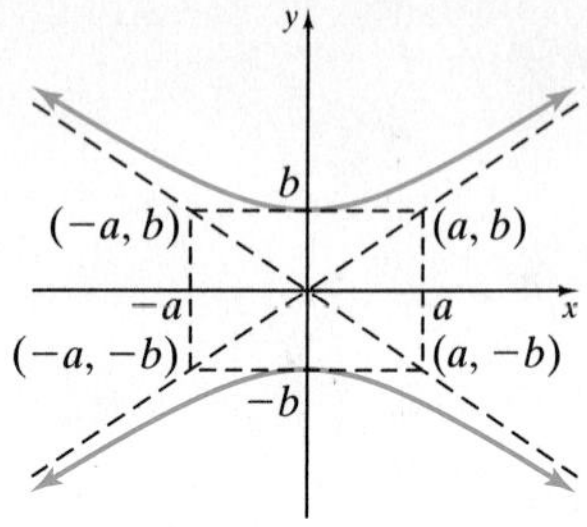

To sketch these lines, or asymptotes, draw a rectangle with vertices (a,b), $(-a, b)$, and $(a, -b)$, $(-a, -b)$. The asymptotes of the hyperbola are the extended diagonals of this rectangle.

EXAMPLE 4 **Graphing a Hyperbola**

Graph $\frac{x^2}{16} - \frac{y^2}{25} = 1$. Label the intercepts and foci.

Solution This equation has the form $\frac{x^2}{a^2} - \frac{y^2}{b^2} = 1$, with $a^2 = 16$ and $b^2 = 25$. Thus, its graph is a hyperbola that opens to the left and right. It has center $(0, 0)$ and x-intercepts $(4, 0)$ and $(-4, 0)$. To aid in graphing the hyperbola, we first sketch its asymptotes. The extended diagonals of the rectangle with corners $(4, 5)$, $(4, -5)$, $(-4, 5)$, and $(-4, -5)$ are the asymptotes of the hyperbola. Then we use the asymptotes to aid in sketching the hyperbola.

To find the foci, we use $c^2 = a^2 + b^2 = 16 + 25 = 41$. The foci are located at $(-\sqrt{41}, 0)$ and $(\sqrt{41}, 0)$.

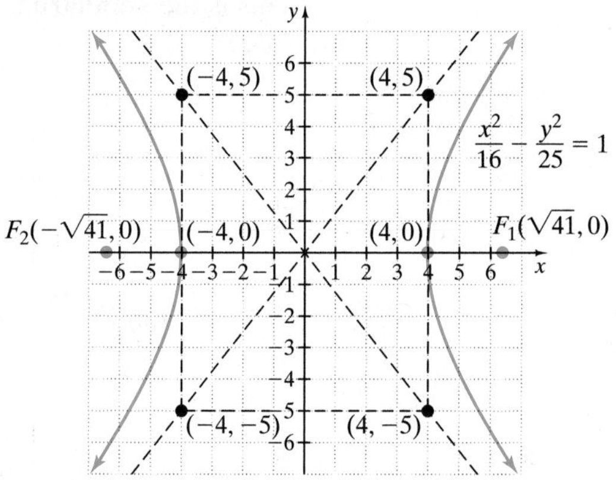

□

PRACTICE 4 Graph $\frac{x^2}{9} - \frac{y^2}{16} = 1$. Label the intercepts and foci.

EXAMPLE 5 **Graphing a Hyperbola**

Graph $4y^2 - 9x^2 = 36$. Label the intercepts and foci.

Solution Since this is a difference of squared terms in x and y on the same side of the equation, its graph is a hyperbola as opposed to an ellipse or a circle. The standard form of the equation of a hyperbola has a 1 on one side, so divide both sides of the equation by 36.

$$4y^2 - 9x^2 = 36$$

$$\frac{4y^2}{36} - \frac{9x^2}{36} = \frac{36}{36} \quad \text{Divide both sides by 36.}$$

$$\frac{y^2}{9} - \frac{x^2}{4} = 1 \quad \text{Simplify.}$$

The equation is of the form $\frac{y^2}{b^2} - \frac{x^2}{a^2} = 1$, with $a^2 = 4$ and $b^2 = 9$, so the hyperbola is centered at $(0, 0)$ with y-intercepts $(0, 3)$ and $(0, -3)$. The sketch of the hyperbola is shown.

To find the foci, we use $c^2 = a^2 + b^2 = 4 + 9 = 13$. The foci are located at $\left(0, -\sqrt{13}\right)$ and $\left(0, \sqrt{13}\right)$.

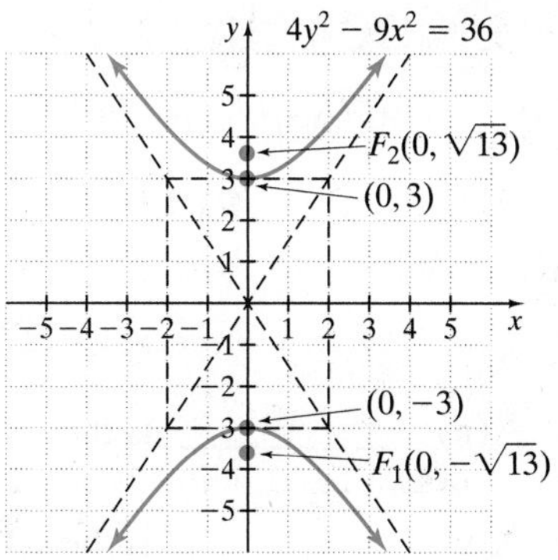

□

PRACTICE 5 Graph $9y^2 - 25x^2 = 225$. Label the intercepts and foci.

In Examples 4 and 5, we were given an equation of a hyperbola and asked to find the intercepts and foci. For Example 6, we do just the opposite.

EXAMPLE 6 Writing the Standard-Form Equation of a Hyperbola

Write the standard-form equation of the hyperbola with foci at $(0, -4)$ and $(0, 4)$ and vertices at $(0, -1)$ and $(0, 1)$.

Solution The foci, $(0, -4)$ and $(0, 4)$ are located on the y-axis, so the transverse axis is horizontal. The center of the hyperbola is the midpoint of the transverse axis, or midway between the foci, so it is $(0, 0)$. Thus, the equation has standard form

$$\frac{y^2}{a^2} - \frac{x^2}{b^2} = 1$$

Let's now find the values for a^2 and b^2. The distance from the center $(0, 0)$ to either vertex, $(0, -1)$ or $(0, 1)$, is 1 unit. This means that

$$a = 1 \text{ or } a^2 = 1^2 = 1.$$

To find b^2, we will use $c^2 = a^2 + b^2$. The distance from the center $(0, 0)$ to either given foci, $(0, -4)$ or $(0, 4)$ is 4 units. Thus,

$$c = 4 \text{ or } c^2 = 4^2 = 16.$$

Then since

$$c^2 = a^2 + b^2$$
$$16 = 1 + b^2 \quad \text{Replace } c^2 \text{ with 16 and } a^2 \text{ with 1.}$$
$$16 - 1 = b^2 \quad \text{Subtract 1 from both sides.}$$
$$15 = b^2 \quad \text{Simplify.}$$

The standard-form equation is then $\frac{y^2}{1} - \frac{x^2}{15} = 1$ □

PRACTICE 6 Write the standard-form equation of the hyperbola with foci at $(0, -5)$ and $(0, 5)$ and vertices at $(0, -2)$ and $(0, 2)$.

B.1 VOCABULARY & READINESS CHECK

Word Bank *Use the choices below to fill in each blank. Some choices will be used more than once and some not at all.*

ellipse	$(0, 0)$	focus	$(a, 0)$ and $(-a, 0)$	$(0, a)$ and $(0, -a)$
hyperbola	center	x	$(b, 0)$ and $(-b, 0)$	$(0, b)$ and $(0, -b)$
		y		

1. A(n) ____________ is the set of points in a plane such that the absolute value of the differences of their distances from two fixed points is constant.
2. A(n) ____________ is the set of points in a plane such that the sum of their distances from two fixed points is constant.

For Exercises 1 and 2 above,

3. The two fixed points are each called a(n) ____________.
4. The point midway between the foci is called the ____________.
5. The graph of $\frac{x^2}{a^2} - \frac{y^2}{b^2} = 1$ is a(n) __________ with center __________ and __________-intercepts of __________.
6. The graph of $\frac{x^2}{b^2} + \frac{y^2}{a^2} = 1$ is a(n) __________ with center __________ and x-intercepts of __________.

Decision Making *Identify the graph of each equation as an ellipse or a hyperbola.*

7. $\frac{x^2}{16} + \frac{y^2}{4} = 1$
8. $\frac{x^2}{16} - \frac{y^2}{4} = 1$
9. $x^2 - 5y^2 = 3$
10. $-x^2 + 5y^2 = 3$
11. $-\frac{y^2}{25} + \frac{x^2}{36} = 1$
12. $\frac{y^2}{25} + \frac{x^2}{36} = 1$

B.1 EXERCISE SET MyMathLab®

Ellipses and Hyperbolas

Sketch the graph of each equation. Label the intercepts and foci. See Examples 1 and 2.

1. $\frac{x^2}{4} + \frac{y^2}{25} = 1$
2. $\frac{x^2}{16} + \frac{y^2}{9} = 1$
3. $\frac{x^2}{9} + y^2 = 1$
4. $x^2 + \frac{y^2}{4} = 1$
5. $9x^2 + y^2 = 36$
6. $x^2 + 4y^2 = 16$
7. $4x^2 + 25y^2 = 100$
8. $36x^2 + y^2 = 36$

Sketch the graph of each equation. Label the intercepts and foci. See Examples 4 and 5.

9. $\frac{x^2}{4} - \frac{y^2}{9} = 1$
10. $\frac{x^2}{36} - \frac{y^2}{36} = 1$
11. $\frac{y^2}{25} - \frac{x^2}{16} = 1$
12. $\frac{y^2}{25} - \frac{x^2}{49} = 1$
13. $x^2 - 4y^2 = 16$
14. $4x^2 - y^2 = 36$
15. $16y^2 - x^2 = 16$
16. $4y^2 - 25x^2 = 100$

In Exercises 17–22, write the standard form of the equation of each ellipse and give the location of its foci.

17.

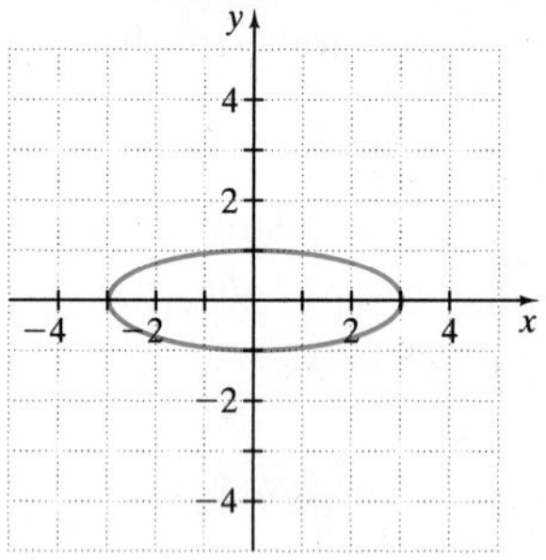

18.

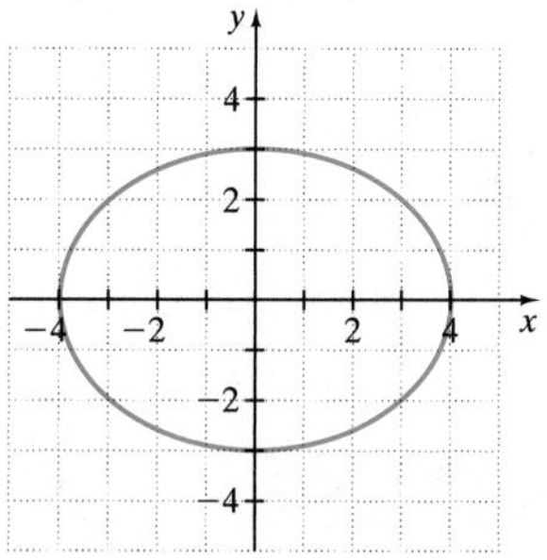

19.

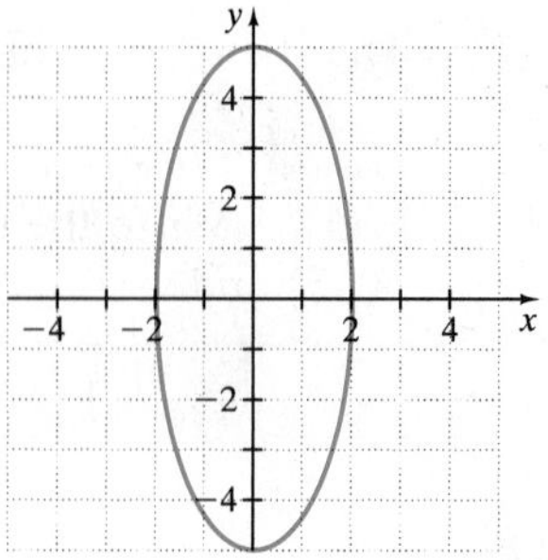

20.

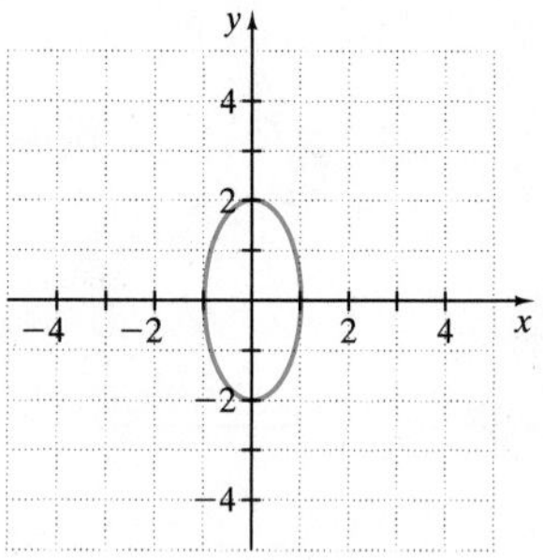

21.

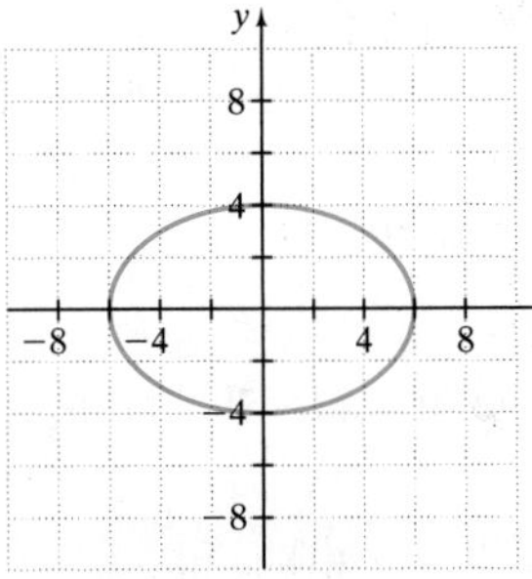

22.

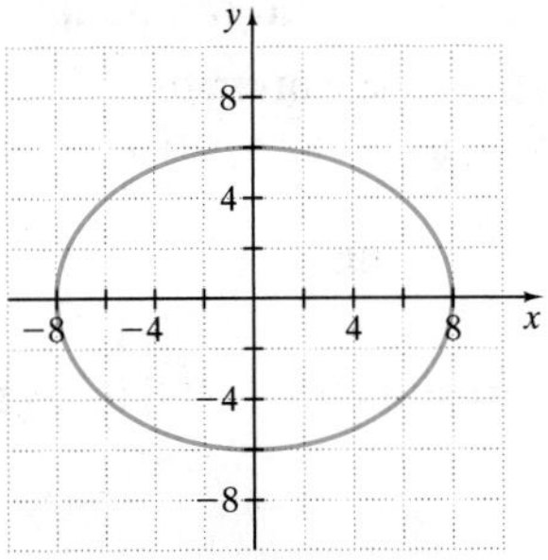

For Exercises 23–26, write the standard form of the equation of each ellipse. See Example 3.

23. Foci: $(-5, 0)$, $(5, 0)$ Vertices: $(-6, 0)$, $(6, 0)$

24. Foci: $(-4, 0)$, $(4, 0)$ Vertices: $(-7, 0)$, $(7, 0)$

25. Foci: $(0, -1)$, $(0, 1)$ Vertices: $(0, -4)$, $(0, 4)$

26. Foci: $(0, -3)$, $(0, 3)$ Vertices: $(0, -8)$, $(0, 8)$

Matching *For Exercises 27–30, find the vertices and locate the foci of each hyperbola with the given equation. Then match each equation to its graph, marked a, b, c, or d. See Examples 4–6.*

27. $\frac{y^2}{9} - \frac{x^2}{4} = 1$

28. $\frac{y^2}{4} - \frac{x^2}{9} = 1$

29. $\frac{x^2}{9} - \frac{y^2}{4} = 1$

30. $\frac{x^2}{4} - \frac{y^2}{9} = 1$

(a)

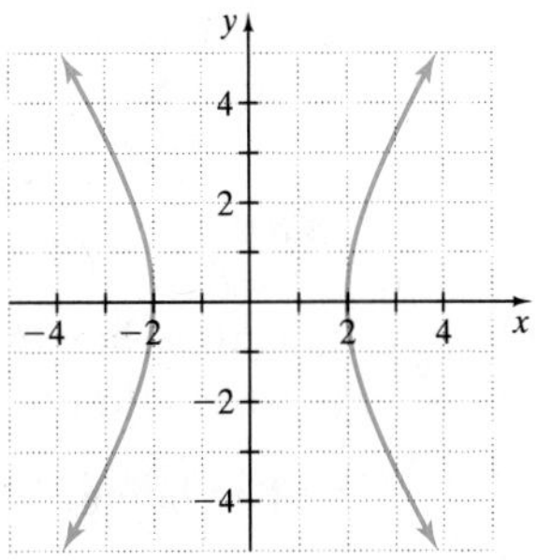

(b)

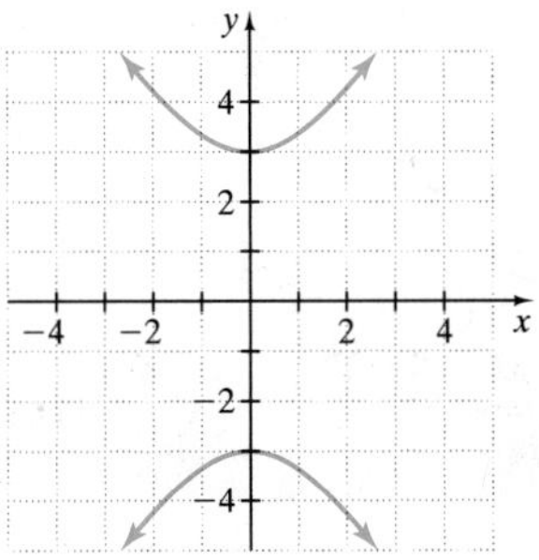

(c)

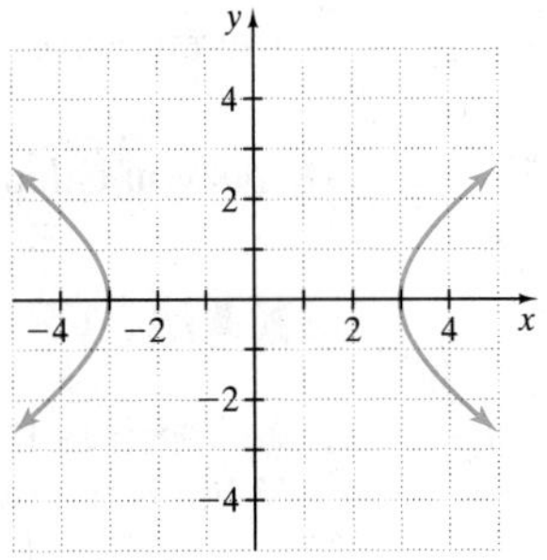

(d)

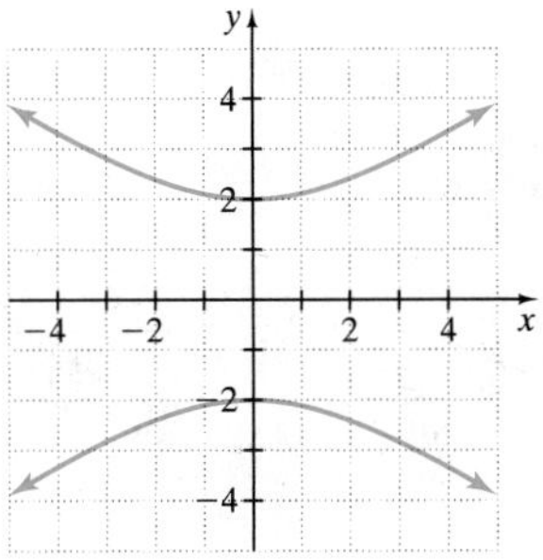

For Exercises 31–34, write the standard form of the equation of each hyperbola. See Example 6.

31. Foci: $(0, -4)$, $(0, 4)$ Vertices: $(0, -1)$, $(0, 1)$

32. Foci: $(0, -6)$, $(0, 6)$ Vertices: $(0, -4)$, $(0, 4)$

33. Foci: $(-3, 0)$, $(3, 0)$ Vertices: $(-2, 0)$, $(2, 0)$

34. Foci: $(-8, 0)$, $(8, 0)$ Vertices: $(-5, 0)$, $(5, 0)$

B.2 Measurement, Rounding Error, and Reasonableness

There is no such thing as an *exact* measurement. Measurements are always approximate. No matter how precise it is, a measurement actually represents a range of values.

EXAMPLE

1. Chris's height, to the nearest inch, is 5 ft 8 in. What range of values does this measurement represent?

The height is given to the nearest inch, so the error is $\frac{1}{2}$ in. Chris's height, then, is between 5 ft $7\frac{1}{2}$ in. and 5 ft $8\frac{1}{2}$ in., or 5 ft 8 in. $\pm$ $\frac{1}{2}$ in. Within this range are all the measures that, when rounded to the nearest inch, equal 5 ft 8 in.

As you calculate with measurements, errors can accumulate.

EXAMPLE

1. Jean drives 18 km to work each day. This distance is given to the nearest kilometer. What is the range of values for the round-trip distance?

The driving distance is between 17.5 and 18.5 km, or 18 $\pm$ 0.5 km. Double the lower limit, 17.5, and the upper limit, 18.5. Thus, the round trip can be anywhere between 35 and 37 km, or 36 $\pm$ 1 km. Notice that the error for the round trip is double the error for a single leg of the trip.

So that your answers will be reasonable, keep precision and error in mind as you calculate. For example, in finding AB, the length of the hypotenuse of ΔABC, it would be inappropriate to give the answer as 8.6533 if the sides are given to the nearest tenth. Round your answer to 8.7.

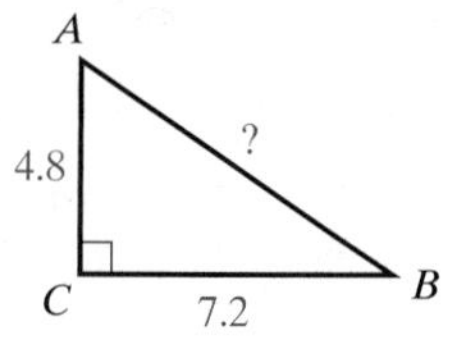

B.2 EXERCISES MyMathLab®

Each measurement is followed by its unit of greatest precision. Find the range of values that each measurement represents.

1. 24 ft (ft)
2. 124 cm (cm)
3. 340 mL (mL)
4. $5\frac{1}{2}$ mi $\left(\frac{1}{2}\text{ mi}\right)$
5. 73.2 mm (0.1 mm)
6. 34 yd^2 (yd^2)
7. The lengths of the sides of $TJCM$ are given to the nearest tenth of a centimeter. What is the range of values for the figure's perimeter?

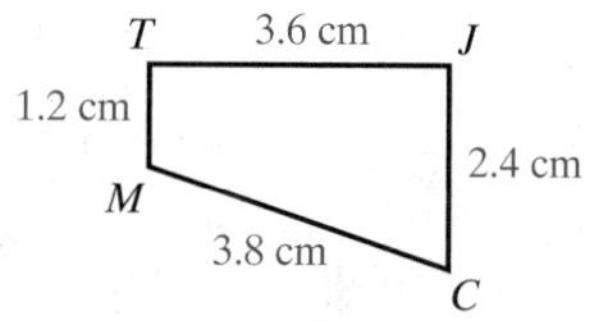

8. To the nearest degree, two angles of a triangle are 49° and 73°. What is the range of values for the measures of the third angle?
9. The lengths of the legs of a right triangle are measured as 131 m and 162 m. You use a calculator to find the length of the hypotenuse. The calculator display reads 208.33867. What should your answer be?

B.3 The Effect of Measurement Errors on Calculations

Measurements are always approximate, and calculations with these measurements produce error. Percent error is a measure of accuracy of a measurement or calculation. It is the ratio of the greatest possible error to the measurement.

$$\text{percent error} = \frac{\text{greatest possible error}}{\text{measurement}}$$

EXAMPLE

The dimensions of a box are measured as 18 in., 12 in., and 9 in. What is the percent error in calculating the box's volume?

The measurements are to the nearest inch, so the greatest possible error is 0.5 in.

Volume:

as measured	maximum value	minimum value
$V = \ell \cdot w \cdot h$	$V = \ell \cdot w \cdot h$	$V = \ell \cdot w \cdot h$
$= 18 \cdot 12 \cdot 9$	$= 18.5 \cdot 12.5 \cdot 9.5$	$= 17.5 \cdot 11.5 \cdot 8.5$
$= 1944 \text{ in.}^3$	$\approx 2196.9 \text{ in.}^3$	$\approx 1710.6 \text{ in.}^3$

Possible Error:

maximum value − measured	measured − minimum value
$2196.9 - 1944 = 252.9$	$1944 - 1710.6 = 233.4$

$$\begin{aligned}\text{percent error} &= \frac{\text{greatest possible error}}{\text{measurement}} \\ &= \frac{252.9}{1944} \\ &\approx 0.1300926\end{aligned}$$

The percent error is about 13%.

B.3 EXERCISES MyMathLab®

Find the percent error in calculating the volume of each box given its dimensions. Round to the nearest percent.

1. 10 cm by 5 cm by 20 cm
2. 1.2 mm by 5.7 mm by 2.0 mm
3. 1.24 cm by 4.45 cm by 5.58 cm
4. $8\frac{1}{4}$ in. by $17\frac{1}{2}$ in. by 5 in.

Find the percent error in calculating the perimeter of each figure.

5. 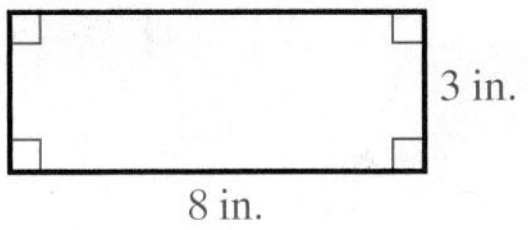

6. 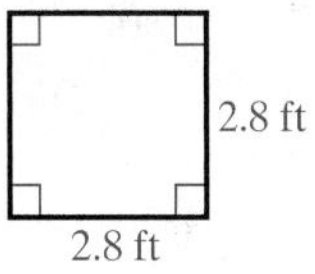

7. 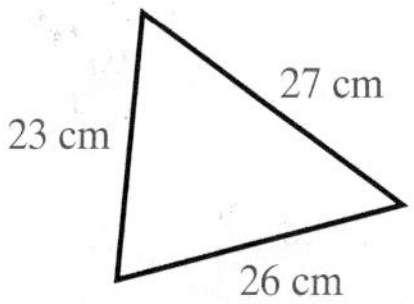

Answers to Selected Exercises

CHAPTER 1 THE BEGINNING OF GEOMETRY

Section 1.1

Answers may vary

Section 1.2

Practice

1. 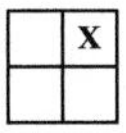

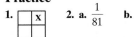

2. a. $\frac{1}{81}$ **b.** 3

3. No. None of the angles are marked, so we may not conclude that they have the same measure. **7.** 

B = Blue **9.** 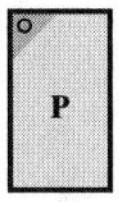

P = Purple

Vocabulary & Readiness Check

1. defined term **3.** theorem

Exercise Set 1.2

1.  **3.** 

5. 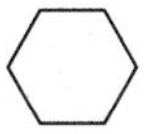 **7.**  **9.** 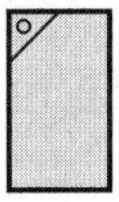

11. 30, 40 **13.** $\frac{1}{16}, \frac{1}{32}$ **15.** 256, 1024 **17.** $\frac{1}{1250}, \frac{1}{6250}$ **19.** 0, 7
21. no **23.** yes **25.** no **27.** yes **29.** yes **31.** no
33. $\frac{1}{16}$ Double the denominator to get the next denominator.
35.  The point is moved clockwise with no section skipped. **37.** Answers may vary. Sample: A theorem is proved and a postulate is not.

Section 1.3

Practice

1. a. $\overleftrightarrow{HE}$, line n **b.** Plane EFG, plane DEG, plane DFG **c.** Points D, E, and F **d.** Points D, E, F, and G **e.** Points H, E, and F (many other correct answers are possible) **2. a.** $\overrightarrow{GJ}$ or $\overrightarrow{JG}$, $\overrightarrow{JQ}$ or $\overrightarrow{QJ}$, $\overrightarrow{GQ}$ or $\overrightarrow{QG}$ **b.** $\overrightarrow{GQ}$ or $\overrightarrow{GJ}$, $\overrightarrow{JG}$, $\overrightarrow{JQ}$, $\overrightarrow{QJ}$ or $\overrightarrow{QG}$ **c.** $\overrightarrow{JQ}$ and $\overrightarrow{JG}$ **3. a.** 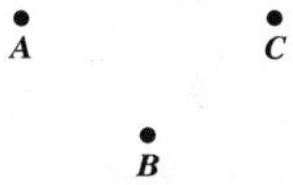

b., c., d. 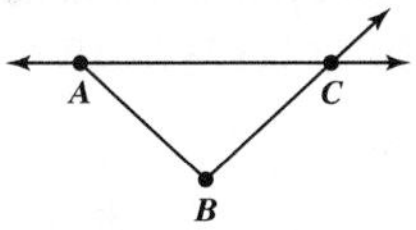

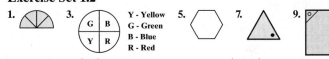

4. $\overrightarrow{AZ}$ and $\overrightarrow{AW}$; $\overrightarrow{AX}$ and $\overrightarrow{AY}$ **5.** $\overleftrightarrow{EH}$ **6.** The plane that passes at a slant through the figure contains points J, M, and Q. 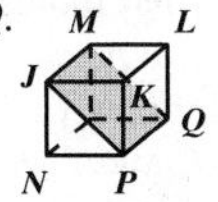

Vocabulary & Readiness Check 1.3

1. C **3.** D **5.** line **7.** ray **9.** Coplanar **11.** geometric figure **13.** point

Exercise Set 1.3

1. Answers may vary. Sample: $\overleftrightarrow{EB}$, $\overleftrightarrow{FB}$, line n **3.** Answers may vary. Sample: plane EBG, plane BFG **5.** E, B, F **7.** E, B, F, G
9. $\overline{RS}$ or $\overline{SR}$, $\overline{ST}$ or $\overline{TS}$, $\overline{TW}$ or $\overline{WT}$, $\overline{RT}$ or $\overline{TR}$, $\overline{RW}$ or $\overline{WR}$, $\overline{SW}$ or $\overline{WS}$
11. $\overrightarrow{RS}, \overrightarrow{SR}, \overrightarrow{ST}, \overrightarrow{TS}, \overrightarrow{TW}, \overrightarrow{WT}, \overrightarrow{TR}, \overrightarrow{RT}, \overrightarrow{WR}, \overrightarrow{RW}, \overrightarrow{WS}, \overrightarrow{SW}$ **13.** $\overrightarrow{TW}$ and $\overrightarrow{TS}$ or $\overrightarrow{TR}$ **15.** A, B, C **17.** $\overline{AB}$ or $\overline{BA}$, $\overline{BC}$ or $\overline{CB}$, $\overline{AC}$ or $\overline{CA}$
19. false **21.** true **23.** Answers may vary. Sample: N, R, Q
25. Answers may vary. Sample: R, T, V **27.** false **29.** true **31.** true
33. false **35.** $\overrightarrow{RS}$ **37.** $\overrightarrow{UV}$ **39.** plane QUX, plane QUV

41. plane XTQ, plane XTS **43.** T **45.** H **47.** E **49.** G
51. coplanar **53.** noncoplanar **55.** noncoplanar
57. 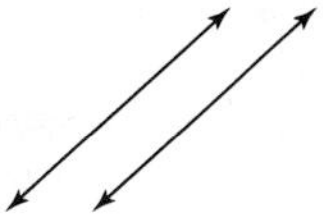 **59.** 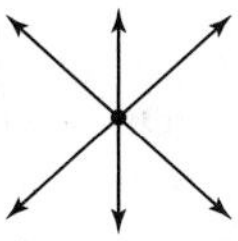 **61.** 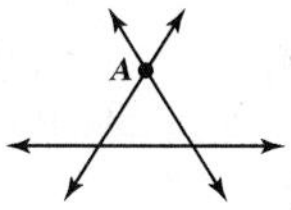

63. always **65.** always **67.** always
69. 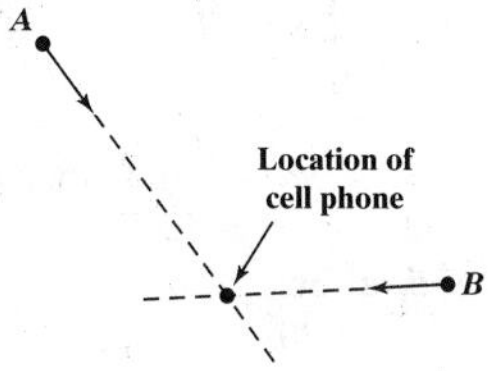

By Postulate 1.3-1, the location of the cell phone and point A determine a line, and the location of the cell phone and point B determine a line. By Postulate 1.3-2, the two lines intersect at, or share, only one point. Since the cell phone signal is on both lines, its location must be the intersection of the lines.

71. 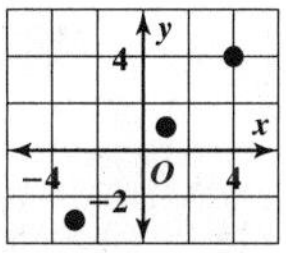

yes **73.** 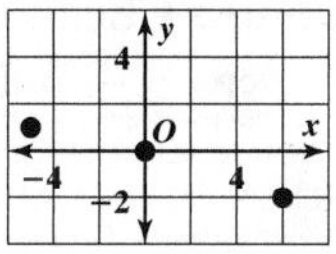

no

75. 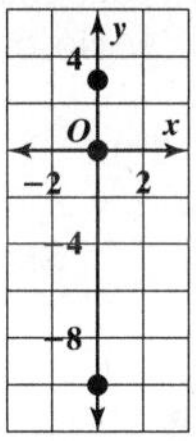

yes **77.** Infinitely many; answers may vary. Sample: The three collinear points are contained in one line. There are infinitely many planes that can intersect in that line. **79.** Answers may vary. Sample: Since the plane is flat, the line would have to curve in order to contain the two points and not lie in the plane, but lines are straight, so the line must also be in plane P.

Section 1.4

Practice

1. 15 **2. a.** $\overline{AB} \cong \overline{DE}$ and $AB = DC$; $\overline{BC} \cong \overline{EF}$ and $BC = EF$; $\overline{CA} \cong \overline{FD}$ and $CA = FD$ **b.** $DF = 10$ cm **c.** $FE = 25$ cm
3. yes **4.** $QR = 45$; $RS = 34$ **5.** $QP = PR = 69$; $QR = 138$

Vocabulary & Readiness Check 1.4

1. congruent **3.** between **5.** equal **7.** segment bisector

Exercise Set 1.4

1. yes **3.** no **5.** no **7.** no **9.** yes **11.** 9
13. 6 **15.** 12 m **17.** 5 in. **19.** 33 ft **21.** no
23. yes **25.** 24 **27. a.** 7 **b.** $RS = 60$; $ST = 36$
29. a. 3 **b.** $XA = AY = 9$; $XY = 18$ **31.** 33 **33.** 34
35. $XY = 4$; $ZW = 4$; congruent **37.** -2.5 or 2.5 **39.** -2 or 8
41. $y = 15$; $AC = 24$; $DC = 12$ **43.** about 1 h, 21 min **45.** The distance is $|80 - 65|$, or 15 miles. The driver added the values instead of subtracting them. **47.** 30 **49.** No. The points, P, Q, and R must be collinear to use the Segment Addition Postulate.

Section 1.5

Practice

1. a. $\angle NPQ$ or $\angle QPN$ **b.** $\angle QPM$ **2.** $m\angle TQN = 15°$, acute; $m\angle TQM = 135°$, obtuse; $m\angle TQR = 180°$, straight **3. a.** $\angle A \cong \angle D$, $m\angle A = m\angle D$, $\angle C \cong \angle F$, $m\angle C = m\angle F$ **b.** 25° **c.** 120°
4. $x = 60°$ **5.** $m\angle PQR = 38°$; $m\angle RQS = 142°$

Vocabulary & Readiness Check 1.5

1. protractor **3.** angle **5.** vertex **7.** B **9.** A

Exercise Set 1.5

1. $\angle XYZ, \angle ZYX, \angle Y$ **3.** $\angle JKM, \angle MKJ, \angle 2$ **5.** 70°, acute **7.** 110°, obtuse **9.** 85°, acute **11.**

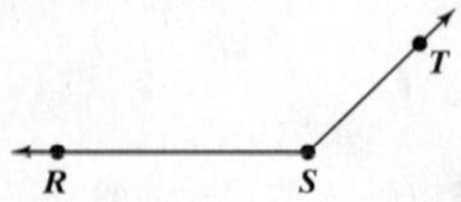

13.

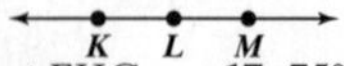

15. $\angle FHG$ **17.** 75° **19.** 90° **21.** 60° **23.** 36° **25.** 22.5° **27.** $m\angle ABC = 45°, m\angle DBC = 34°$ **29.** $m\angle RQS = 43°$, $m\angle TQS = 137°$ **31.** $m\angle RST = 39°$, $m\angle TSU = 51°$ **33.** about 60°, acute **35.** about 135°, obtuse **37.** $x = 8$; $m\angle AOB = 30°, m\angle BOC = 50°, m\angle COD = 30°$ **39.** 80° **41.** 75° **43.** No, the diagram is not marked with congruent angles or perpendicular lines.

Section 1.6

Practice

1. a. True **b.** True **c.** False **d.** False **2.** $\angle 5$ and $\angle 6$, $\angle 6$ and $\angle 7$, $\angle 7$ and $\angle 8$, $\angle 5$ and $\angle 8$ **3.** $\angle 5$ and $\angle 6$, $\angle 6$ and $\angle 7$, $\angle 7$ and $\angle 8$, $\angle 5$ and $\angle 8$ **4. a.** 164° **b.** 74° **5. a.** 36° **b.** 36° **6.** $x = 23$; $m\angle KHL = m\angle LHJ = 69°$ **7.** $x = 20, y = 15$; $m\angle DEA = m\angle CEB = 94°$; $m\angle AEB = m\angle DEC = 86°$

Vocabulary & Readiness Check 1.6

1. complementary **3.** vertical **5.** linear pair

Exercise Set 1.6

1. $\angle 4$ **3.** True **5.** False **7.** Yes, the angles share a common side and vertex, and have no interior points in common. **9.** No, they are supplementary **11.** $\angle DOC, \angle AOB$ **13.** $\angle EOC$ **15.** $\angle COD$ **17.** False **19.** True **21.** True **23.** False **25.** True **27.** Yes, they are marked as congruent. **29.** Yes, $\angle FAJ$ is a straight angle. **31.** Yes, they are marked as congruent. **33.** No, they are adjacent angles. **35.** 90° **37.** 155° **39.** 170°; 80° **41.** 92°; 2° **43.** 57°; There is no complement since $123° > 90°$. **45.** 29° **47.** 72° **49. a.** 11 **b.** 30° **c.** 30° **d.** 60° **51.** $x = 5$; $m\angle ABC = 50°$ **53.** $x = 11$; $m\angle ABC = 56°$ **55. a.** 19.5 **b.** $m\angle RQS = 43°, m\angle TQS = 137°$ **c.** Answers may vary. Sample: $43 + 137 = 180$ **57.** $x = 10$; $y = 40$; $m\angle ABD = m\angle EBC = 112°$; $m\angle ABE = m\angle DBC = 68°$ **59.** $x = 44$; $y = 23$; $m\angle MNQ = m\angle RNP = 83°$; $m\angle MNR = m\angle QNP = 97°$ **61.** 120°, 60° **63.** No, they do not have a common vertex **65. a.** $\angle CBD$, 41° **b.** 82° **c.** $m\angle ABE = 49°, m\angle DBF = 49°$ **67.** $\angle KML$ **69.** $\angle PMR, \angle KML, \angle KMQ, \angle MQP$

Section 1.7

Practice

1. -2.5 **2.** $(6.5, -4)$ **3.** $(11, -13)$ **4.** $\sqrt{17} \approx 4.1$

Vocabulary & Readiness Check 1.7

1. midpoint, point **3.** midpoint

Exercise Set 1.7

1. 3 **3.** -1.5 **5.** $(4, -2)$ **7.** $(-5, 2.5)$ **9.** $(3, 0)$ **11.** $(-0.5, 0.5)$ **13.** $(10, -20)$ **15.** $(0, -34)$ **17.** 6 **19.** 8 **21.** 10 **23.** $2\sqrt{37} \approx 12.2$ **25.** 8.2 **27.** 8.5 **29.** 33.5 m **31. a.** 5 **b.** $(4.5, 4)$ **33. a.** 5.8 **b.** $(1.5, 0.5)$ **35. a.** 5.4 **b.** $(3, 0.5)$ **37. a.** 10.8 **b.** $(3, -4)$ **39.** 6.7 mi **41.** 8.9 mi **43.** 3.2 mi **45. a.** Answers may vary. Sample: $(0, 2)$ and $(4, 2)$; $(2, 0)$ and $(2, 4)$; $(0, 4)$ and $(4, 0)$; $(0, 0)$ and $(4, 4)$ **b.** Infinitely many; draw a circle with center $(2, 2)$ and radius 4. Any diameter of that circle has length 8 and midpoint $(2, 2)$. **47.** S is in Quadrant IV; since $(0, 0)$ is the midpoint of $\overline{TS}$, the coordinates of S are the opposites of the coordinates of T, so S must be in Quadrant IV. **49. a.** Answers may vary. Sample: Distance Formula (find KP, then divide by 2). **b.** Answers may vary. Sample: Distance Formula (if M is the given midpoint, find KM, and then multiply it by 2).

Section 1.8

Practice

1.

2.

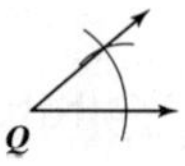

3.

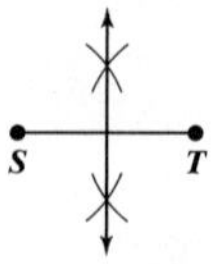

4.

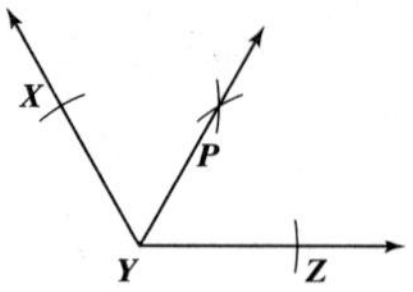

Exercise Set 1.8

1.

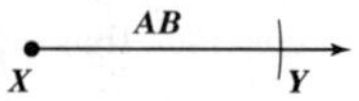

3.

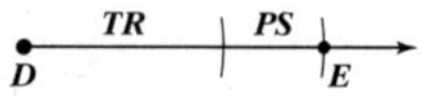

5.

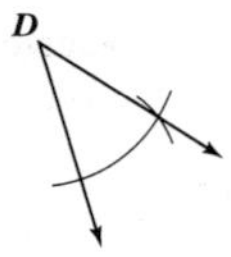

7.

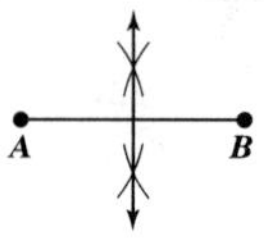

9.

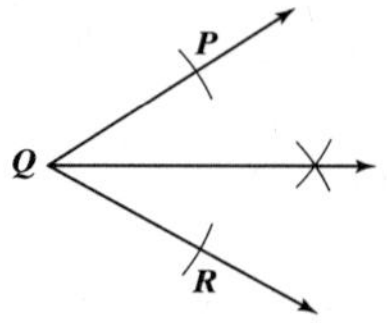

11. Answers may vary. Sample:

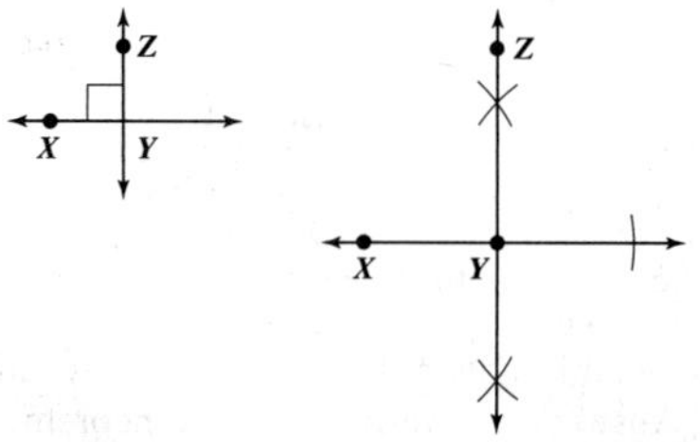

13. Answers may vary. Sample: Both constructions involve drawing arcs with the same radius from two different points, and using the point(s) of intersection of those arcs. Arcs must intersect at two points for the perpendicular bisector, but only one point for the angle bisector. **15.** A segment has exactly one midpoint; using the Ruler Postulate, each point corresponds with exactly one number, and exactly one number represents half the length of a segment. **17.** In the plane with the segment, there is one perpendicular bisector because only one line in that plane can be drawn through the midpoint so that it forms a right angle with the given segment. **19.**

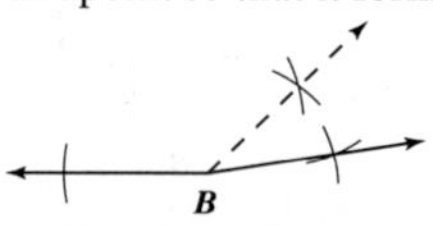

21. With P as center, draw an arc with radius slightly more than $\frac{1}{2}PQ$. Keeping that radius, draw an arc with Q as center. Those two arcs meet at two points; the line through those two points intersects $\overline{PQ}$ at its midpoint. Call the midpoint C. Repeat this process for segments $\overline{PC}$ and $\overline{CQ}$.

23. possible

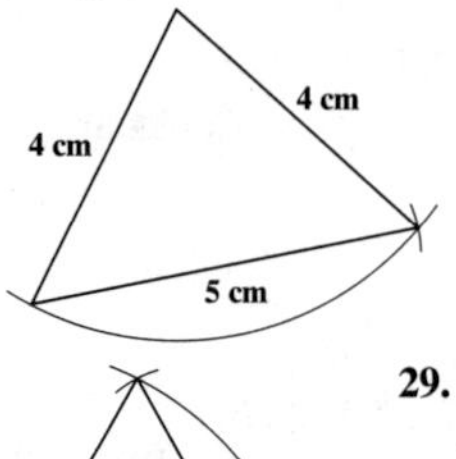

25. Not possible; the two 2-cm sides do not meet.

27. possible

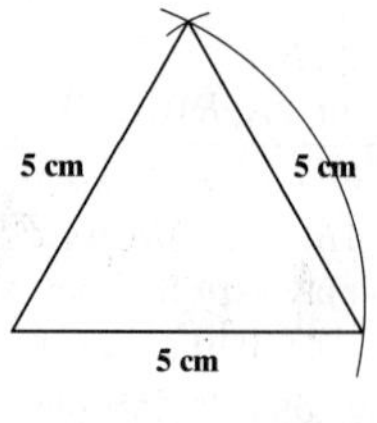

29. Draw a line segment. Construct the perpendicular bisector to get two 90° angles. Then construct the angle bisector of one of the 90° angles to get two 45° angles.

Chapter 1 Review

1. 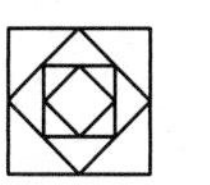 **3.** $\frac{1}{48}, \frac{1}{96}$ **5.** Answers may vary. Sample: $\overleftrightarrow{QA}$ and $\overleftrightarrow{AB}$ **7.** Answers may vary. Sample: A, B, C

9. True; Postulate 1.3-1 states, "Through any two points, there is exactly one line." **11.** $-7, 3$ **13.** 15 **15.** acute **17.** 36° **19.** $\angle ADB$ and $\angle BDC$ **21.** Answers may vary. Sample: $\angle ADC$ and $\angle EDF$ **23.** 31 **25.** 1.4 units **27.** $(0, 0)$ **29.** $(6, -2)$

31. 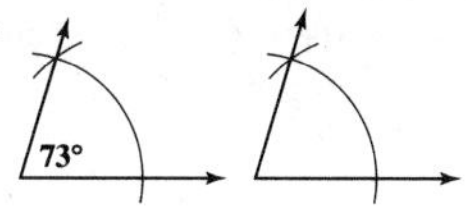

33. 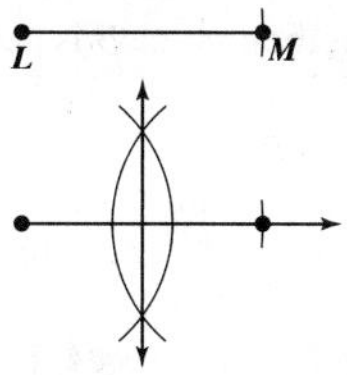

35. Yes; all four points are located in the plane AEF. **37.** No; sample: Points H, G, and F are in plane $\overleftrightarrow{HGF}$, and point B is not in that plane. **39. a.** $\overleftrightarrow{BF}$ **b.** $\overleftrightarrow{EH}$ **41.** $\overrightarrow{PQ}, \overrightarrow{PB}$ **43.** Answers may vary. Sample: D, P, C **45.** Answer may vary. Sample: $\overline{AP}, \overline{PR}, \overline{RQ}$ **47.** Answers may vary. Sample: $\angle APD$ and $\angle CPR$ **49.** 3 **51.** 3 **53.** 34 **55.** 80° **57.** No; C, F, and G are collinear, so $\overleftrightarrow{AB}$ must intersect $\overleftrightarrow{GF}$ at C in plane M.

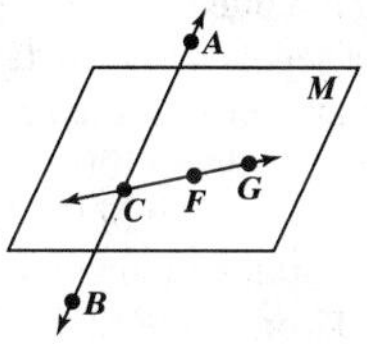

Chapter 1 Test

1. 31, 37 **2.**  **3.** B **4.** C **5.** C **6.** C **7.** A

8. $\angle AFC$ or $\angle 2$ **9.** $\angle 3$ or $\angle 1$ **10.** $\angle DEF$ and $\angle FEA$

11. True **12.** False **13.** True **14.** True **15.** False **16.** right **17.** obtuse **18.** acute **19.** straight **20.** 10 **21.** Not congruent; $AC = 6$ but $BD = 7$ **22. a.** 8 **b.** $XM = MY = 44; XY = 88$ **23.**

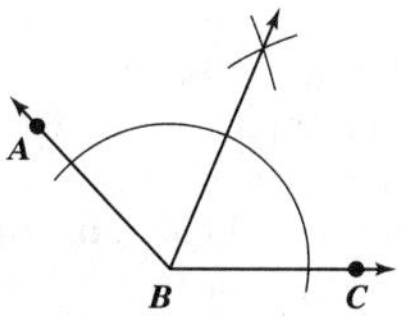

24. A, B, C **25.** point B **26. a.** 1 plane **b.** 1 plane **27.** Answers may vary. Sample: A, B, C, E **28.** 10 **29.** perpendicular bisector **30.** 7 **31.** $E, \overline{AY}$ **32.** 9 **33.** $m\angle RPT = 20°$

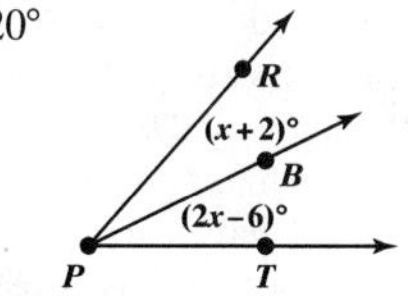

34. $(0.5, 5.5)$ **35.** $(7, -8)$ **36.** $\sqrt{73} \approx 8.5$

Chapter 1 Standardized Test

1. a **2.** a **3.** c **4.** c **5.** b **6.** b **7.** d **8.** d **9.** c **10.** a **11.** b **12.** b **13.** d **14.** a **15.** c **16.** a **17.** a **18.** c **19.** c **20.** d

CHAPTER 2 INTRODUCTION TO REASONING AND PROOFS

Section 2.1

Practice Exercises

1. a. 24 in. **b.** 32 in. **2. a.** 48π m **b.** 9.4 m **3.**

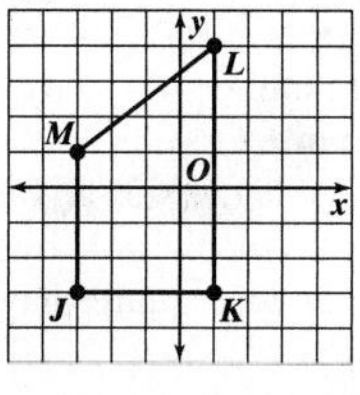

20 units **4.** 72 sq. ft **5. a.** 49π sq. ft **b.** 153.86 sq. ft

Vocabulary & Readiness Check 2.1

1. Perimeter **3.** π **5.** $\frac{22}{7}$ (or 3.14); 3.14 (or $\frac{22}{7}$)

Exercise Set 2.1

1. 64 ft **3.** 21 in. **5.** 32 in. **7.** 42 in. **9.** 155 cm **11.** 50π ft; 157 ft **13.** 16π mi; 50.24 mi **15.** 17π cm; 53.38 cm **17.** 7 sq. m **19.** $9\frac{3}{4}$ sq. yd **21.** 15 sq. yd **23.** 9π sq. in. $\approx$ 28.26 sq. in. **25.** 49 sq. cm **27.** 36π sq. in. $\approx$ 113.04 sq. in. **29.** $11 + \sqrt{13}$ units

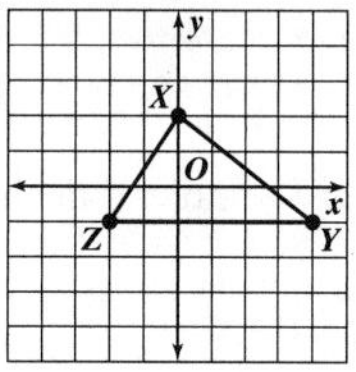

31. 16 units

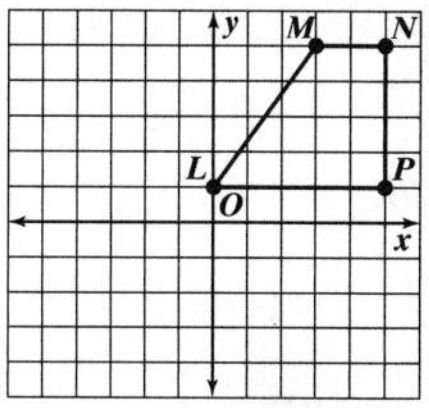

33. 6000 sq. ft **35.** 113, 625 sq. ft **37.** 128 ft **39.** 15π ft; 47.1 ft **41.** area **43.** area **45.** perimeter **47.** perimeter **49.** 27 sq. in. **51.** P = 10 units; A = 4 sq. units **53.** 6.25π sq. units **55.** d. **57.** Yes; A sample reason: because a square is a rectangle **59.** Answers may very; a sample: The classmate forgot to insert π when using the formula for finding the area. **61.** 25 **63.** 9 **65.** 5 **67.** 20

Section 2.2

Practice Exercises

1. a. 3456; multiply by 2, then 3, then 4, and repeat **b.** 7; add 8 twice, then subtract 14, and repeat **2. a.** 76; each number is the sum of the two previous numbers **b.** 257; the numbers being added are multiplied by 2 each step

3. 

4. a. We conjecture that the 11th circle is red. **b.** We conjecture that the 11th circle has 2^{11} or 2048 regions. **c.** The 11th circle is red with 2048 regions. **d.** The 30th circle is green with 2^{30} regions.

5. 74 hybrid cars **6. a.** Answers may vary. Sample: $\frac{1}{2} \cdot 8 = 4$ **b.** Answers may vary. Sample: A square with side length of 3 units has an area A of: $A = 3^2 = 9$ square units and 9 is not an even number.

Vocabulary & Readiness Check 2.2

1. conjecture **3.** inductive reasoning

Exercise Set 2.2

1. 80 **3.** 8 **5.** -4 **7.** 36 **9.** 720 **11.** 15 **13.** 864 **15.** 0.0001 **17.** $\frac{31}{32}$ **19.** 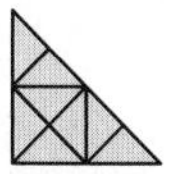 **21.** 

23. green **25.** star **27. a.** red **b.** circle **c.** A red circle **29.** Answers may vary. Sample: The sum of two odd numbers is even. **31.** Answers may vary. Sample: The product of two even numbers is even. **33.** Answers may vary. Sample: The sum of a number and its opposite is 0. **35. a.** red **b.** 26 regions **c.** a red oval with 26 regions **d.** a green oval with 58 regions **37. a.** purple, red, blue, and green going clockwise from the upper left square **b.** a square subdivided into 4 smaller squares **c.**

G	P
B	R

d.

B	G
R	P

39. 1 mi **41.** two right angles **43.** Answers may vary. Sample: -2 and -3

45. August begins with the letter A. **47.** 123454321 or 123,454,321 **49.** 102 cm **51.** (12, 3) does not fit the pattern. **53.** Answers may vary. Sample: 1, 2, 3, 4, 5,... and 1, 2, 4, 8, 16,... **55.**

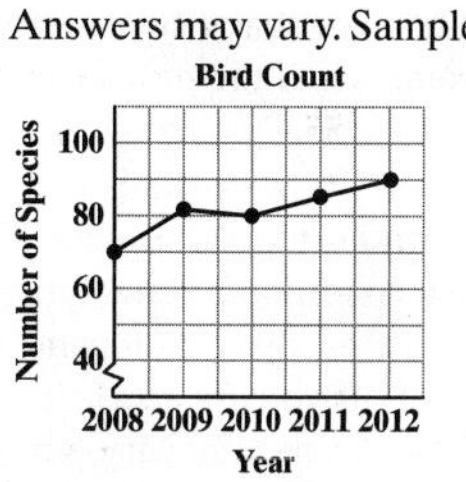

57. His conjecture is probably false because most people's growth slows down by age 18 until they stop growing some where between ages 18 and 22.

59. a. 1, 3, 6, 10, 15, 21
b.

n	1	2	3	4	5	6
$\frac{n^2 + n}{2}$	1	3	6	10	15	21

The values are the same.
c. $\frac{n^2 + n}{2} = \frac{n(n + 1)}{2}$; since the diagram represents $n(n + 1)$, half the diagram represents $\frac{n^2 + n}{2}$. **d.**

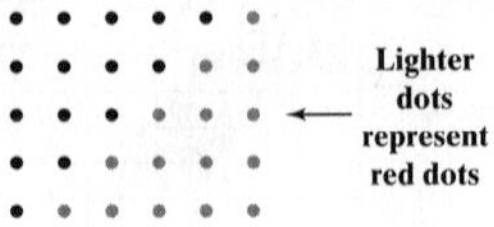
Lighter dots represent red dots

61. 2 **63.** True; explanations may vary. Sample: If the two even numbers are $2a$ and $2b$, the sum is $2a + 2b = 2(a + b)$, which is an even number.

Section 2.3

Practice Exercises

1. a. Hypothesis (p): an animal is a pig; Conclusion (q): the animal has 44 teeth **b.** Hypothesis (p): $x = 7$; Conclusion (q): $x^2 = 49$ **2. a.** If an angle measures 36°, then the angle is an acute angle. **b.** If you live in Texas, then you live in the continental U.S. **3.** If a number is a fraction, then it is a real number. **4. a.** False; there are nonadjacent complementary angles. **b.** True; the months that begin with the letter *M*, March and May, have 31 days. **5. a.** If the figure has 3 sides, then the figure is a triangle. **b.** If the figure is not a triangle, then it does not have 3 sides. **c.** If the figure does not have 3 sides, then it is not a triangle.

Vocabulary & Readiness Check 2.3

1. $p \rightarrow q$ **3.** conclusion **5.** inverse **7.** converse **9.** equivalent

Exercise Set 2.3

1. Hypothesis (p): you are an American citizen; Conclusion (q): you have the right to vote **3.** Hypothesis (p): a figure is a rectangle; Conclusion (q): it has four sides **5.** If the animal is an alligator, then it is a reptile. **7.** If $x = 9$, then $x^2 = 81$. **9.** If an angle is a straight angle, then it measures 180°. **11.** If a person is a pianist, then he or she is a musician. **13.** If $3x - 7 = 14$, then $3x = 21$. **15.** If the measure of a line segment is 12 inches, then the segment measures 1 foot. **17.** false; Mexico also borders U. S. **19.** true **21.** d **23.** b **25.** Converse: If you play football, then you are a quarterback. Inverse: If you are not a quarterback, then you do not play football. Contrapositive: If you do not play football, then you are not a quarterback. **27.** Converse: If $x = 5$, then $4x + 8 = 28$. Inverse: If $4x + 8 \neq 28$, then $x \neq 5$. Contrapositive: If $x \neq 5$, then $4x + 8 \neq 28$. **29.** contrapositive **31.** inverse **33.** none of these **35.** converse **37.** If a number is divisible by 2, then it is not odd. Explanations may vary. **39.** If I do not live in North America, then I do not live in Canada. Explanations may vary. **41.** If $\angle C$ is not obtuse, then $m\angle C \neq 105°$. Explanations may vary. **43.** If two distinct lines intersect, then they meet in exactly one point. **45.** Conditional: If a number is an odd natural number less than 8, then the number is prime. Converse: If a number is prime, then it is an odd natural number less than 8. Inverse: If a number is not an odd natural number less than 8, then the number is not prime. Contrapositive: If a number is not prime, then it is not an odd natural number less than 8. All four statements are false; counterexamples are 1 and 11. **47.** If $|x| = 6$, then $x = -6$; false; $x = 6$ is a counterexample. **49.** If $x^3 < 0$, then $x < 0$; true **51.** If an event has a probability of 1, then that event is certain to occur. **53.** If a group is half the people, then that group should make up half the Congress. **55.** Natalie is correct because a conditional statement and its contrapositive are equivalent statements. **57.** He is correct. The given conditional statement is true. Thus, the contrapositive is true because a conditional statement and its contrapositive are equivalent statements. **59.** If you come to IHOP hungry, then you will leave happy. **61.** If you are a kid, then Trix are for you. **63.** 6 statements **65.** Answers may vary. Sample: 4 collinear points **67.** 36 in. (or 3 ft or 1 yd) **69.** 162 in. (or $13\frac{1}{2}$ ft or $4\frac{1}{2}$ yd)

Section 2.4

Practice Exercises

1. Conditional statement: If two angles are complementary, then the sum of their measures is 90°. Converse: If the sum of the measures of two angles is 90°, then the angles are complementary. **2. a.** If $AB + BC = AC$, then point B is between points A and C. **b.** The converse statement is true. **c.** Point B is between points A and C if and only if $AB + BC = AC$. **3. a.** If $|x| = 7$, then $x = 7$. **b.** The converse statement is false. **c.** $x = -7$ is a counterexample. **4.** c is a good definition. **5.** The definition is a good one.

Vocabulary & Readiness Check 2.4

1. biconditional, or if and only if **3.** $q \rightarrow p$ **5.** biconditional or if and only if

Exercise Set 2.4

1. If a line bisects a segment, then it intersects the segment only at its midpoint. If a line intersects a segment only at its midpoint, then the line bisects the segment. **3.** If an integer is divisible by 100, then its last two digits are zeros. If the last two digits of an integer are zeros, then the integer is divisible by 100. **5.** If you live in Washington, D.C., then you live in the capital of the United States. If you live in the capital of the United States, then you live in Washington, D.C. **7.** If $x^2 = 144$, then $x = 12$ or $x = -12$. If $x = 12$ or $x = -12$, then $x^2 = 144$. **9. a.** If two segments are congruent, then they have the same length. **b.** true **c.** Two segments have the same length if and only if they are congruent. **11. a.** If $2x - 5 = 19$, then $x = 12$. **b.** true **c.** $x = 12$ if and only if $2x - 5 = 19$. **13. a.** If a number is an even number, then it is divisible by 20. **b.** false **c.** Counterexamples may vary. Sample: 4 **15. a.** If it is Independence Day in the United States, then it is July 4. **b.** true **c.** In the United States, it is July 4 if and only if it is Independence Day. **17. a.** If $|x| = 3$, then $x = 3$. **b.** false **c.** $x = -3$ **19. a.** If two angles form a linear pair, then they are adjacent. **b.** true **c.** If two angles are adjacent, then they form a linear pair. **d.** false **e.** Counterexamples may vary. Sample: adjacent complementary angles **21. a.** If angle A is acute, then $0° < m\angle A < 90°$. **b.** true **c.** If $0° < m\angle A < 90°$, then angle A is acute. **d.** true **e.** An angle A is acute if and only if $0° < m\angle A < 90°$. **23.** No, it is not reversible; some animals with whiskers are not cats. **25.** No, it is not precise; a point or a ray can be part of a line. **27.** Yes; points are collinear if and only if they lie on the same line. **29.** d **31.** Points are in Quadrant III if and only if they have two negative coordinates. **33.** A figure is a hexagon if and only if it is a six-sided polygon. **35.** The sum of the digits of an integer is divisible by 9 if and only if the integer is divisible by 9. **37. a.** yes **b.** yes; thus, the statement is a good definition **39.** As a biconditional, the statement is, "an angle is a right angle if and only if it has a greater measure than an acute angle." Counterexamples to that statement are obtuse angles and straight angles. **41.** The prefix *bi-* means "two." **43.** If $\angle A$ and $\angle B$ are a linear pair, then $\angle A$ and $\angle B$ are supplementary angles. **45.** If $\angle A$ and $\angle B$ are a linear pair, then $\angle A$ and $\angle B$ are adjacent and supplementary angles. **47.** good definition **49.** good definition **51.** If your grades suffer, then you do not sleep enough. **53.** true; **a** **55.** 60, 50 **57.** 4, −2

Section 2.5

Practice Exercises

1. a. If a number is a real number, then it is an integer; $q \rightarrow p$ **b.** If a number is not an integer, then it is not a real number; $\sim p \rightarrow \sim q$ **c.** If a number is not a real number, then it is not an integer; $\sim q \rightarrow \sim p$ **2.** You have a license plate on the front and back of your car. **3. a.** valid **b.** not valid **4.** If a natural number ends in 0, then it is divisible by 5. **5. a.** The figure has 3 sides; Law of Detachment **b.** All squares have diagonals that bisect each other; Law of Syllogism

Vocabulary & Readiness Check 2.5

1. Law of Detachment **3.** Deductive

Exercise Set 2.5

1. $m\angle A \neq m\angle B$ **3.** If $m\angle A = m\angle B$, then $\angle A$ is congruent to $\angle B$. **5.** If $m\angle A \neq m\angle B$, then $\angle A$ is congruent to $\angle B$. **7.** If $m\angle A \neq m\angle B$, then $\angle A$ is not congruent to $\angle B$. **9.** conditional statement **11.** inverse statement **13.** $\sim q \rightarrow \sim p$ **15.** If a figure is a heptagon, then the figure has seven sides. **17.** If a figure does not have seven sides, then the figure is not a heptagon. **19.** Dr. Ngemba should take an X-ray. **21.** No conclusion is possible; the hypothesis has not been satisfied. **23.** Points X, Y, and Z are collinear. **25.** If an animal is a Florida panther, then it is endangered. **27.** No conclusion is possible; the same statement does not appear as the conclusion of one conditional and as the hypothesis of the other conditional.
29. a. Tracy lives in New York. **b.** Law of Detachment **31. a.** If you are studying botany, then you are studying a science. **b.** Law of Syllogism

33. a.

7	11	100
42	66	600
50	74	608
25	37	304
21	33	300

b. n

$6n$

$6n + 8$

$\dfrac{6n + 8}{2} = 3n + 4$

$3n + 4 - 4 = 3n$

The result is 3 times the original number.

35. If a place is a national park, then it is an interesting place. Mammoth Cave is an interesting place. **37.** If you are in Key West, Florida, then the temperature is always above 32°F. No conclusion is possible because the hypothesis is not satisfied. **39.** Must be true; since **e** is true, then it is breakfast time by **a**, so Julie is drinking juice by **d**. **41.** May be true; by **e** and **a**, it is breakfast time. You don't know what Kira drinks at breakfast. **43.** May be true; by **e**, Maria is drinking juice, but you don't know that she isn't also drinking water. **45.** Answers may vary. **47.** $\angle AOB$, $\angle BOA$ **49.** $\overrightarrow{OB}$

Section 2.6

Practice Exercises

1.

Statements	Reasons
$9x - 45 = 45$	Given
$9x = 90$	Addition Property of Equality
$x = 10$	Division Property of Equality

2.

Statements	Reasons
$20x - 8(3 + 2x) = 28x$	Given
$20x - 24 - 16x = 28x$	Distributive Property
$4x - 24 = 28x$	Simplify
$-24 = 24x$	Subtraction Property of Equality
$-1 = x$	Division Property of Equality

3. a. Symmetric Property of Equality **b.** Transitive Property of Equality **c.** Substitution

4.

Statements	Reasons
1. $AB = CD$	**1.** Given
2. $BC = BC$	**2.** Reflexive Property of Equality
3. $AB + BC = BC + CD$	**3.** Addition Property of Equality
4. $AB + BC = AC$	**4.** Segment Addition Postulate
5. $BC + CD = BD$	**5.** Segment Addition Postulate
6. $AC = BD$	**6.** Substitution Property

5.

Statements	Reasons
1. $m\angle A = 32°$	**1.** Given
2. $m\angle B = 32°$	**2.** Given
3. $32° = m\angle B$	**3.** Symmetric Property of Equality
4. $m\angle A = m\angle B$	**4.** Transitive Property of Equality or Substitution Property
5. $\angle A \cong \angle B$	**5.** Definition of congruent angles

Vocabulary & Readiness Check 2.6

1. Symmetric **3.** Transitive

Exercise Set 2.6

1. a. Addition Property of Equality **b.** Division Property of Equality **3. a.** Subtraction Property of Equality **b.** Addition Property of Equality **c.** Division Property of Equality **5. a.** Multiplication Property of Equality **b.** Distributive Property **c.** Addition Property of Equality **7. a.** Distributive Property **b.** Simplification **c.** Subtraction Property of Equality **d.** Division Property of Equality **9. a.** Definition of supplementary angles **b.** Substitution Property **c.** Simplification **d.** Subtraction Property of Equality **e.** Division Property of Equality **11.** Subtraction Property of Equality **13.** Symmetric Property of Congruence **15.** F **17.** A **19.** B **21.** $\angle K$ **23.** 3 **25.** $\angle XYZ \cong \angle WYT$ **27. a.** Given **b.** A midpoint divides a segment into two congruent segments. **c.** Substitution **d.** $2x = 12$ **e.** Division Property of Equality

29.

Statements	Reasons
1. $m\angle GFI = 128°$	**1.** Given
2. $m\angle GFE + m\angle EFI = m\angle GFI$	**2.** Angle Addition Postulate
3. $m\angle GFE + m\angle EFI = 128°$	**3.** Substitution Property, (Steps 1, 2)
4. $m\angle GFE = (9x - 2)°$; $m\angle EFI = 4x°$	**4.** Given
5. $9x - 2 + 4x = 128$	**5.** Substitution Property
6. $13x - 2 = 128$	**6.** Simplify
7. $13x = 130$	**7.** Addition Property of Equality
8. $x = 10$	**8.** Division Property of Equality
9. $m\angle EFI = 40°$	**9.** Substitution Property, (Steps 4, 8)

31. Since $\overline{LR}$ and $\overline{RL}$ are two ways to name the same segment and $\angle CBA$ and $\angle ABC$ are two ways to name the same angle, then both statements are examples of saying that some thing is congruent to itself. **33.** $DB = BE$ **35.** reflexive, symmetric, and transitive; because "has the same birthday as" satisfies all three properties **37.** transitive only; A cannot be taller than A; if A is taller than B, then B is not taller than A. **39.** The error is in the 5th step when both sides are divided by $b - a$, which is 0, and division by 0 is not defined. **41.** 80° **43.** 125° **45.** 50

Section 2.7

Practice Exercises

1. If two angles are supplementary to the same angle (or to equal angles), then they are equal in measure. **2.** Linear pair angles are supplementary. **3.** If two angles are vertical angles, then they are congruent.

4.

Statements	Reasons
1. $\angle 1$ and $\angle 2$ are right angles.	1. Given
2. $m\angle 1 = 90°, m\angle 2 = 90°$	2. Definition of right angles
3. $m\angle 1 = m\angle 2$	3. Transitive Property
4. $\angle 1 \cong \angle 2$	4. Definition of congruent angles

5. 5.5

Vocabulary & Readiness Check 2.7

1. paragraph

Exercise Set 2.7

1. 65° 3. 65° 5. 90° 7. $m\angle 1 = 90°, m\angle 2 = 50°, m\angle 3 = 40°$ 9. 20 11. $x = 38, y = 104$ 13. $m\angle AEB = 120°, m\angle BEC = 60°, m\angle CED = 120°, m\angle DEA = 60°$ 15. $m\angle JKF = 104°, m\angle FKG = 76°, m\angle GKH = 104°, m\angle HKJ = 76°$ 17. a. Vertical Angles Theorem b. $\angle 1 \cong \angle 6$ c. Vertical Angles Theorem d. Transitive Property of Congruence 19. $\angle 1 \cong \angle 3$ and $\angle 2 \cong \angle 4$ by the Vertical Angles Theorem. 21. $\angle EIG \cong \angle FIH$ by the Right Angles Congruent Theorem. $\angle EIF \cong \angle GIH$ by the Equal Complements Theorem. 23. Equal Supplements Theorem

25.

Statements	Reasons
1. $\angle A$ and $\angle B$ are supplements; $\angle C$ and $\angle D$ are supplements	1. Given
2. $m\angle A = m\angle C$	2. Given
3. $m\angle A + m\angle B = 180°$; $m\angle C + m\angle D = 180°$	3. Definition of supplementary angles
4. $m\angle A + m\angle B = m\angle C + m\angle D$	4. Transitive Property
5. $m\angle A + m\angle B = m\angle A + m\angle D$	5. Substitution (Steps 2 and 4)
6. $m\angle B = m\angle D$	6. Subtraction Property of Equality

27. a. 35° b. 55° 29. $x = 15, y = 30; m\angle BMQ = 25°, m\angle BMD = 155°, m\angle DME = 25°, m\angle EMQ = 155°$ 31. $m\angle A = 60°, m\angle B = 30°$ 33. $m\angle A = 120°, m\angle B = 60°$ 35. $x = 30, y = 90$; 60°, 120°, 60°, 120° 37. $x = 35, y = 70$; 110°, 70°, 110°, 70° 39. Answers may vary. Sample: scissors 41. B can be any point on the positive y-axis. Sample: (0, 5) 43. Subtraction Property of Equality 45. Transitive Property 47. F, I, H, B 49. Yes 51. any three of $\overleftrightarrow{FI}$ (or $\overleftrightarrow{IF}$), $\overleftrightarrow{FH}$, $\overleftrightarrow{FB}$, $\overleftrightarrow{IH}$, $\overleftrightarrow{IB}$, $\overleftrightarrow{HB}$

Chapter 2 Vocabulary Check

1. circumference 2. counterexample 3. inductive 4. perimeter 5. area 6. Symmetric 7. Reflexive 8. Transitive 9. conclusion 10. hypothesis 11. equivalent 12. $\sim p$ 13. $p \rightarrow q$ 14. $q \rightarrow p$ 15. $\leftrightarrow$ 16. biconditional 17. if and only if 18. biconditional 19. $p \rightarrow q; q \rightarrow p$ 20. Law of Detachment 21. Law of Syllogism 22. Deductive 23. inductive 24. Conditional 25. inverse 26. contrapositive 27. converse 28. negation

Chapter 2 Review

1. $P = 32$ cm; $A = 64$ sq. cm 3. $C = 6\pi$ in.; $A = 9\pi$ sq. in. 5. Divide the previous term by 10; $1, \frac{1}{10}$ 7. Subtract 7 from the previous term; 6, −1 9. Answers may vary. Sample: $-1 \cdot 2 = -2$ and −2 is not greater than 2 11. If a person is a motorcyclist, then that person wears a helmet. 13. If two angles form a linear pair, then the angles are supplementary. 15. Converse: If the measure of an angle is greater than 90° and less than 180°, then the angle is obtuse. Inverse: If an angle is not obtuse, then it is not true that its measure is greater than 90° and less than 180°. Contrapositive: If it is not true that the measure of an angle is greater than 90° and less than 180°, then the angle is not obtuse. All four statements are true. 17. Converse: If you play an instrument, then you play the tuba. Inverse: If you do not play the tuba, then you do not play an instrument. Contrapositive: If you do not play an instrument, then you do not play the tuba. The converse and inverse are false and the conditional and contrapositive are true. 19. a good definition 21. a good definition 23. A phrase is an oxymoron if and only if it contains contradictory terms. 25. Colin will become a better player. 27. If two angles are vertical, then their measures are equal. 29. Given 31. Substitution 33. Subtraction Property of Equality or Addition Property of Equality 35. BY 37. 18 39. 74°

41.

Statements	Reasons
1. $\angle 1$ and $\angle 2$ are complementary; $\angle 3$ and $\angle 4$ are complementary	1. Given
2. $\angle 2 \cong \angle 4$	2. Given
3. $m\angle 1 + m\angle 2 = 90°$, $m\angle 3 + m\angle 4 = 90°$;	3. Definition of complementary angles
4. $m\angle 2 = m\angle 4$	4. Definition of congruent angles
5. $m\angle 1 + m\angle 2 = m\angle 3 + m\angle 4$	5. Transitive Property
6. $m\angle 1 + m\angle 2 = m\angle 3 + m\angle 2$	6. Substitution
7. $m\angle 1 = m\angle 3$	7. Subtraction Property of Equality
8. $\angle 1 \cong \angle 3$	8. Definition of congruent angles

43. If you are younger than 20, then you are a teenager. The converse is false.

45.

Statements	Reasons
1. $\angle 1 \cong \angle 4$	1. Given
2. $\angle 1$ and $\angle 2$ are vertical angles; $\angle 3$ and $\angle 4$ are vertical angles	2. Definition of vertical angles
3. $\angle 1 \cong \angle 2$; $\angle 3 \cong \angle 4$	3. Vertical Angles Theorem
4. $\angle 1 \cong \angle 3$	4. Transitive Property of Congruence
5. $\angle 2 \cong \angle 3$	5. Substitution (Steps 3 and 4)

47. If you play hockey, then you are a varsity athlete.

Chapter 2 Test

1. $P = 36$ cm; $A = 81$ sq. cm 2. 38 ft of fencing 3. $C = 9\pi$ cm; $A = 20.25\pi$ sq. cm 4. $C = 10\pi$ ft ≈ 31.4 ft; $A = 25\pi$ sq. ft 5. Divide the previous term by −2 (or multiply by $-\frac{1}{2}$); $-1, \frac{1}{2}$ 6. Find the square of 1, 2, 3, 4, …; 36, 49 7. Answers may vary. Sample: a garter snake 8. two 45° angles 9. Hypothesis: $x + 9 = 11$; Conclusion: $x = 2$ 10. If a figure is a quadrilateral, then the figure has four sides. 11. c 12. Converse: If a figure has at least two right angles, then it is a square. Inverse: If a figure is not a square, then it does not have at least two right angles. Contrapositive: If a figure does

not have at least two right angles, then it is not a square. **13.** Converse: If a square's perimeter is 12 meters, then the square has side length 3 meters. Inverse: If a square does not have side length 3 meter, then its perimeter is not 12 meters. Contrapositive: If a square's perimeter is not 12 meters, then it does not have side length 3 meters. **14.** If a fish is a bluegill, then it is a bluish, freshwater sunfish. If a fish is a bluish, freshwater sunfish, then it is a bluegill. **15. a.** If A, B, and C lie on the same line, then they are collinear. **b.** true **c.** A, B, and C are collinear if and only if they lie on the same line. **16.** d **17.** The statement, "If two angles are supplementary, then they form a straight line" is not true; a counterexample is any two angles of a rectangle. **18.** Transitive Property of Equality or Substitution Property of Equality **19.** Subtraction Property of Equality or Addition Property of Equality **20.** Reflexive Property of Congruence **21.** Symmetric Property of Congruence **22.** $\angle LNP \cong \angle VNM$ and $\angle LNV \cong \angle PNM$ by the Vertical Angles Theorem. **23.** $\angle BCE \cong \angle FCD$, given in the diagram; $\angle BCF \cong \angle DCE$ by the Equal Supplements Theorem. **24.** $x = 21$; $m\angle BZR = m\angle NZP = 42°$, $m\angle RZN = m\angle BZP = 138°$ **25.** not possible **26.** If the traffic light is red, then you must apply your brakes. **27. a.** B is the midpoint of $\overline{AC}$. **b.** Definition of midpoint **c.** Segment Addition Postulate **d.** Substitution **e.** $2AB = AC$ **f.** Division Property of Equality or Multiplication Property of Equality **28. a.** 90° **b.** Angle Addition Postulate **c.** Substitution or Transitive **d.** Definition of right angle

Chapter 2 Standardized Test

1. c **2.** d **3.** a **4.** b **5.** b **6.** b **7.** a **8.** b **9.** c **10.** d **11.** a **12.** b **13.** c **14.** c **15.** d **16.** a **17.** d **18.** c **19.** a **20.** a **21.** d

CHAPTER 3 PARALLEL AND PERPENDICULAR LINES

Section 3.1

Practice Exercises

1. a. $\overleftrightarrow{BC}, \overleftrightarrow{FG}, \overleftrightarrow{EH}$ **b.** $\overleftrightarrow{BF}, \overleftrightarrow{EF}$ **c.** $\overleftrightarrow{AB}, \overleftrightarrow{DC}, \overleftrightarrow{AE}, \overleftrightarrow{DH}$ **d.** plane $EFGH$ **2. a.** $\angle 7$ and $\angle 6$; $\angle 1$ and $\angle 4$ **b.** $\angle 8$ and $\angle 3$; $\angle 2$ and $\angle 5$ **3. a.** alternate interior **b.** corresponding

Vocabulary & Readiness Check 3.1

1. parallel planes **3.** transversal **5.** interior **7.** corresponding **9.** alternate exterior

Exercise Set 3.1

1. perpendicular **3.** skew **5.** parallel **7.** parallel **9.** plane JCL **11.** $\overleftrightarrow{FG}$ **13.** Answers may vary. Sample: $\overleftrightarrow{AB}$ **15.** alternate interior angles; corresponding angles **17.** alternate interior angles; corresponding angles **19.** corresponding angles **21.** $\angle 8$ and $\angle 9$; $\angle 7$ and $\angle 10$ **23.** $\angle 9$ and $\angle 11$ **25.** $\angle 1$ and $\angle 14$ **27.** $\angle 3$ and $\angle 4$ **29.** alternate interior **31.** corresponding **33.** corresponding **35.** alternate interior **37.** alternate exterior **39.** true **41.** true **43.** true **45.** 2 pairs **47.** 2 pairs **49.** always **51.** always **53.** sometimes **55.** Skew; answers may vary. Sample: Since the paths are not coplanar, they are skew. **57. a.** Lines may be intersecting, parallel, or skew. **b.** Answers may vary. Sample: In a classroom, two adjacent edges of the floor are intersecting, two opposite edges of the floor are parallel, and one edge of the floor is skew to each of the vertical edges of the opposite wall. **59.** Answers may vary. Sample: E illustrates corresponding angles; F illustrates same-side interior angles. **61.** No; if two planes intersect, their intersection is a single line, and the intersection of planes A and B is $\overleftrightarrow{CD}$. **63.** 121° **65.** 29.5° **67.** 150.5° **69.** corresponding angles **71.** alternate exterior angles

Section 3.2

Practice Exercises

1. $l \parallel m$; $\angle 6$ and $\angle 2$ are alternate interior angles formed by lines l and m with transversal b.

2. $\angle 3$ and $\angle 6$ are supplementary
Given

$\angle 5$ and $\angle 6$ are supplementary
Liner pairs are Supplementary

↘ ↙

$m\angle 3 + m\angle 6 = 180°$; $m\angle 5 + m\angle 6 = 180°$
Definition of supplementary angles

↓

$m\angle 3 + m\angle 6 = m\angle 5 + m\angle 6$
Substitution

↓

$m\angle 3 = m\angle 5$
Subtraction Property of Equality

↓

$\angle 3 \cong \angle 5$
Definition of congruent angles

↓

$l \parallel m$
Alternate Interior Angles Theorem

3. $\angle 2 \cong \angle 3$ (vertical angles are congruent), so $\angle 1 \cong \angle 3$ (Transitive Property of Congruence). Then $r \parallel s$ by the Corresponding Angles Theorem. **4.** 19

Vocabulary & Readiness Check 3.2

1. $\angle 7$ **3.** $\angle 10$ **5.** $\angle 3$ **7.** $\angle 15$ **9.** $\angle 6$

Exercise Set 3.2

1. Yes; Alternate Interior Angles Theorem **3.** No **5.** Yes; Alternate Exterior Angles Theorem **7.** No **9.** $\overleftrightarrow{BE} \parallel \overleftrightarrow{CG}$ by the Corresponding Angles Theorem **11.** $\overleftrightarrow{GH} \parallel \overleftrightarrow{EF}$ by the Alternate Exterior Angles Theorem **13.** $\overline{CA} \parallel \overline{HR}$ by the Corresponding Angles Theorem **15.** $\overline{HE} \parallel \overline{AD}$ by the Alternate Interior Angles Theorem **17.** $p \parallel q$ by the Alternate Exterior Angles Theorem **19.** $m \parallel n$ by the Corresponding Angles Theorem **21.** $q \parallel m$ by the Corresponding Angles Theorem $(18° + 98° = 19° + 97° = 116°)$ **23.** No **25.** 30 **27.** 59 **29.** Yes; $\angle 1$ and $\angle 2$ are alternate exterior angles, and if alternate exterior angles are congruent, then the lines are parallel **31.** $x = 10$; $m\angle 1 = m\angle 2 = 70°$ **33.** $x = 2.5$; $m\angle 1 = m\angle 2 = 30°$ **35.** The Alternate Interior Angles Theorem **37.** Converse of the Corresponding Angles Theorem

$\angle 12 \cong \angle 8$,
Given

$\angle 8 \cong \angle 4$
Given

↘ ↙

$\angle 12 \cong \angle 4$
Substitution or Transitivity of $\cong$

↓

$j \parallel k$
Corresponding $\angle$ Theorem

$l \parallel n$
Corresponding $\angle$ Theorem

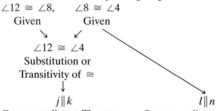

39. $a \parallel b$; Same-Side Interior Angles Theorem **41.** $a \parallel b$; Alternate Exterior Angles Theorem **43.** $a \parallel b$; Corresponding Angles Theorem **45.** $a \parallel b$; Corresponding Angles Theorem **47.** None **49.** 5 **51.** $\overleftrightarrow{DC}$ is the transversal, so by the Same-Side Interior Angles Theorem, $\overleftrightarrow{AD} \parallel \overleftrightarrow{BC}$. **53.** Never **55.** Never

Section 3.3

Practice Exercises

1. a. 48° **b.** 132° **c.** 132° **d.** 132° **2.** $m\angle 1 = m\angle 3 = m\angle 7 = m\angle 5 = 97°$; $m\angle 2 = m\angle 4 = m\angle 6 = m\angle 8 = 83°$

3. a. 75°; Alternate Interior Angles Converse **b.** 75°; Corresponding Angles Converse **c.** 105°; Vertical Angle Theorem **d.** 105°; Linear Pair Theorem **e.** 105°; Vertical Angle Theorem **f.** 105°; Corresponding Angles Converse or Same-Side Interior Angles Converse
4. 35

Vocabulary & Readiness Check 3.3

1. one

Exercise Set 3.3

1. $m\angle 1 = 60°$, Alternate Interior Angles Converse; $m\angle 2 = 60°$, Corresponding Angles Converse
3. $m\angle 1 = 127°$, Alternate Exterior Angles Converse; $m\angle 2 = 53°$, Linear Pair Theorem
5. $m\angle 1 = 75°$, Corresponding Angles Converse; $m\angle 2 = 105°$, Same-Side Interior Angles Converse
7. $\angle 1$, Vertical Angles; $\angle 7$, Alternate Interior Angles; $\angle 4$, Corresponding Angles
9. $\angle 1$, Corresponding Angles; $\angle 3$, Alternate Interior Angles
11. $m\angle 1 = m\angle 5 = m\angle 7 = 73°$; $m\angle 2 = m\angle 4 = m\angle 6 = m\angle 8 = 107°$
13. $m\angle 1 = m\angle 3 = m\angle 5 = m\angle 7 = 77°$; $m\angle 2 = m\angle 4 = m\angle 8 = 103°$
15. $m\angle 1 = m\angle 3 = m\angle 5 = m\angle 7 = 152°$; $m\angle 4 = m\angle 6 = m\angle 8 = 28°$
17. $m\angle 1 = m\angle 3 = m\angle 7 = 149°$; $m\angle 2 = m\angle 4 = m\angle 6 = m\angle 8 = 31°$
19. $x = 50$; $y = 130$ **21.** $x = 135$; $y = 135$ **23.** $m\angle 1 = 120°$, Same-Side Interior Angles Converse; $m\angle 2 = 80°$, Alternate Interior Angles Converse **25.** $m\angle 1 = 103°$, Linear Pair Theorem; $m\angle 4 = 103°$, Corresponding Angles Converse ($\angle 1$); $m\angle 3 = 103°$, Corresponding Angles Converse ($\angle 4$); $m\angle 2 = 103°$, Vertical Angles Theorem ($\angle 3$) **27.** $m\angle 2 = 54°$, Linear Pair Theorem; $m\angle 1 = 126°$, Same-Side Interior Angles Converse; $m\angle 3 = 130°$, Vertical Angles Theorem; $m\angle 4 = 130°$, Alternate Interior Angles Converse **29.** $m\angle 1 = 123°$, Vertical Angles Theorem; $m\angle 2 = 57°$, Same-Side Interior Angles Converse; $m\angle 3 = 95°$, Linear Pair Theorem; $m\angle 4 = 85°$, Same-Side Interior Angles Converse or Alternate Interior Angles Converse **31.** 88° **33.** 111° **35.** 5 **37.** 31 **39.** 24 **41.** 28
43. $x = 115$; $x - 50 = 65$ **45.** $x = 25$; $3x - 10 = x + 40 = 65$
47. Same-Side Interior Angles Converse **49.** Same-Side Interior Angles Converse **51.** 32 **53.** $x = 87, y = 31, v = 42, w = 20$
55. Incorrect; the marked angles are same-side interior angles, so they are supplementary, not congruent. **57.** Yes; same-side interior angles are congruent if they are both right angles, because two right angles are supplementary.
59.

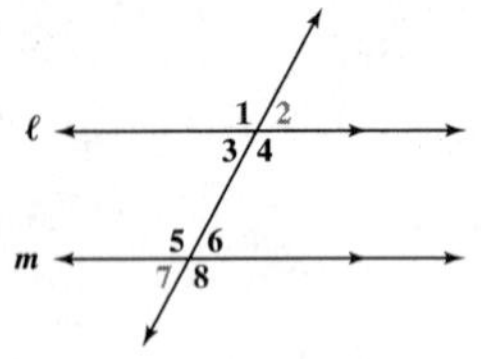

Statements	Reasons
1. $l \parallel m$	1. Given
2. $\angle 3 \cong \angle 6$; $\angle 4 \cong \angle 5$	2. Alternate Interior Angles Converse
3. $m\angle 3 + m\angle 4 = 180°$ $m\angle 5 + m\angle 6 = 180°$	3. Linear pair Theorem
4. $m\angle 6 + m\angle 4 = 180°$ $m\angle 5 + m\angle 3 = 180°$	4. Substitution (Steps 2, 3)
5. $\angle 6$ and $\angle 4$ are supplementary $\angle 5$ and $\angle 3$ are supplementary	5. Definition of supplementary (Step 4)

61. a. 117° **b.** same-side interior angles **63.** $m\angle 1 = 70°$, Same-Side Interior Angles Converse; $m\angle 2 = 110°$, Same-Side Interior Angles Converse **65.** Always **67.** Sometimes

Section 3.4

Practice Exercises

1. a. Given **b.** Corresponding Angles Converse **c.** Corresponding Angles Converse **d.** $\angle 1 \cong \angle 3$ **e.** $a \parallel c$ **2.** 41

Vocabulary & Readiness Check 3.4

1. one **3.** Parallel **5.** True

Exercise Set 3.4

1. 90 **3.** 10 **5.** 57 **7.** 17 **9.** 90° **11.** 90° **13.** 29°
15. 30° **17.** All of the rungs are ⊥ to one side. The side is ⊥ to the top rung, and since all the rungs are ∥ to each other, the side is ⊥ to all of the rungs. **19.** $a \parallel d$, Two Lines Parallel to a Third Line Theorem **21.** $a \perp d$, Perpendicular Transversal Theorem **23.** $a \parallel d$, Two Lines Perpendicular to a Third Line Theorem (gives $b \parallel d$) and Two Lines Parallel to a Third Line Theorem (gives $a \parallel d$) **25.** They are ⊥; answers may vary. Sample: The yellow and blue lines are ∥ because they are both ⊥ to the brown line. Since the pink line is ⊥ to one of those two ∥ lines, it must be ⊥ to the other. **27.** $\angle 1 \cong \angle 2 \cong \angle 3$

29.

Statements	Reasons
1. $a \perp b$, $b \perp c$	1. Given
2. $a \parallel c$	2. Two Lines Perpendicular to a Third Line Theorem
3. $c \parallel d$	3. Given
4. $a \parallel d$	4. Two Lines Parallel to a Third Line Theorem

31.

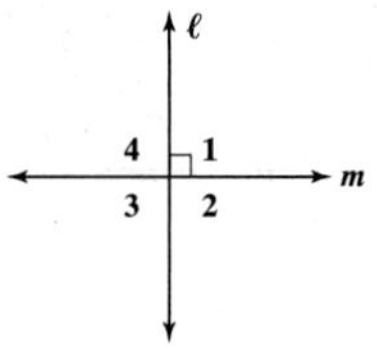

Given: $l \perp m$
To Prove: $m\angle 1 = m\angle 2 = m\angle 3 = m\angle 4 = 90°$

Statements	Reasons
1. $l \perp m$	1. Given
2. $\angle 1$ is a right angle	2. Definition of ⊥ Lines
3. $m\angle 1 = 90°$	3. Definition of Right Angle
4. $m\angle 4 = 180° - m\angle 1$	4. Linear Pair Theorem
5. $m\angle 4 = 180° - 90°$ $= 90°$	5. Substitution (Steps 3, 4)
6. $m\angle 1 = m\angle 3$, $m\angle 4 = m\angle 2$	6. Vertical Angles Theorem
7. $m\angle 3 = 90°$, $m\angle 2 = 90°$	7. Substitution (Steps 3, 5, 6)

33. The diagram should show that m and r are ⊥. **35.** Corresponding Angles Converse **37.** $\angle 3$ **39.** 53 **41.** obtuse

Section 3.5

Practice Exercises

1.

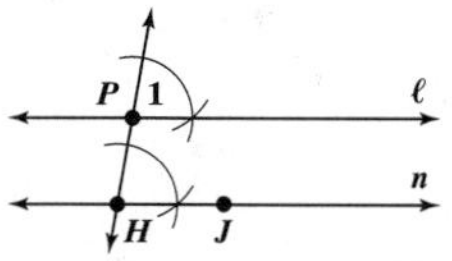

2. Answers may vary. Sample:

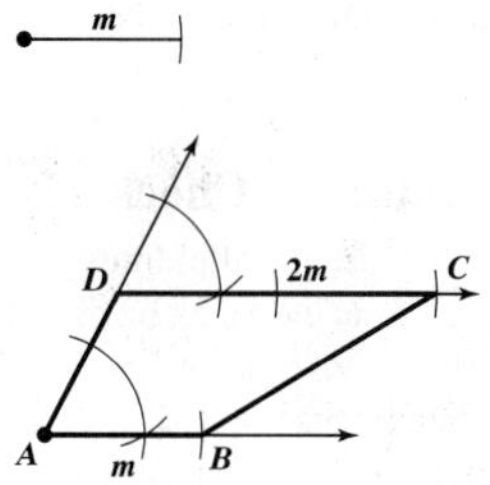

3.

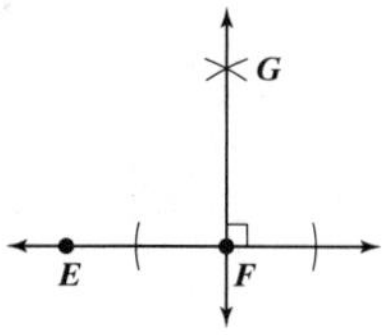

4.

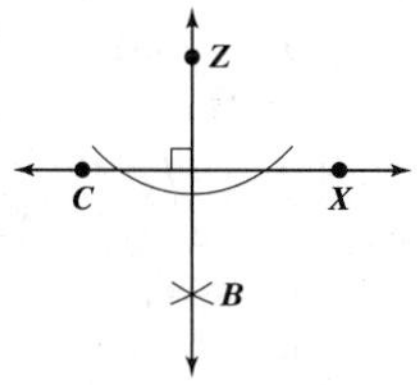

Exercise Set 3.5

1.

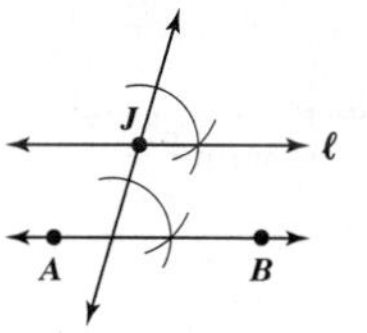

3.

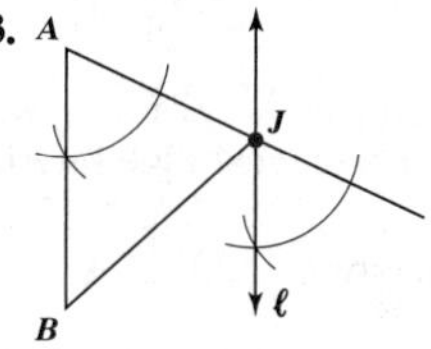

5.

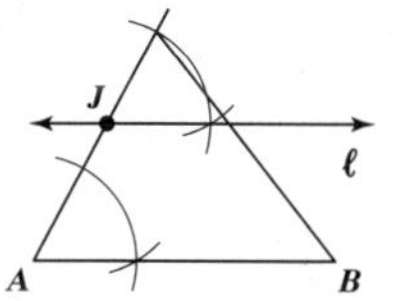

7., 9. Constructions may vary. Samples using the following segments are given.

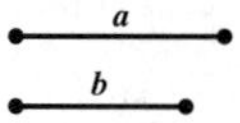

7. Answers may vary.

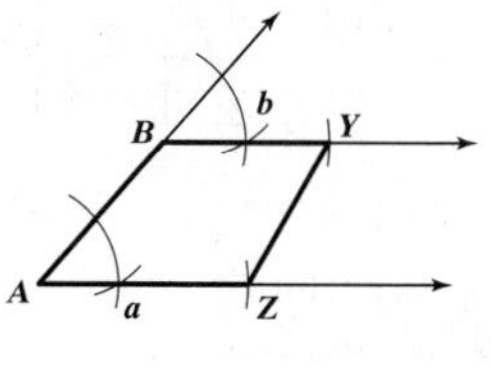

9. Answers may vary.

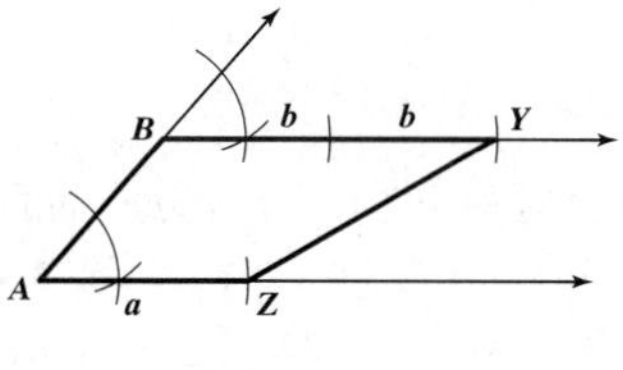

11.

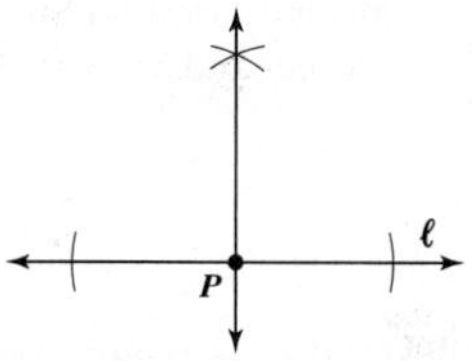

13.

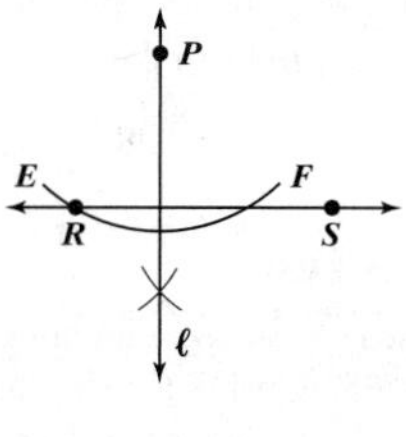

15.

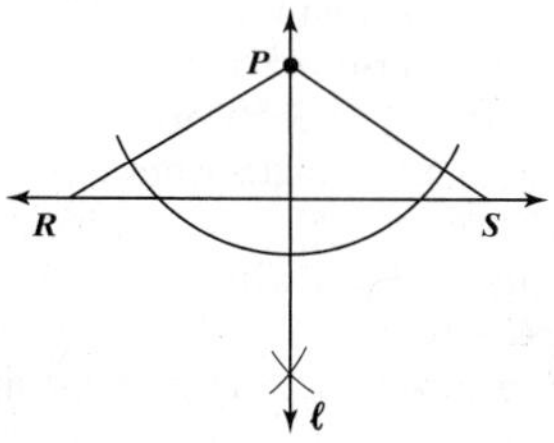

17.

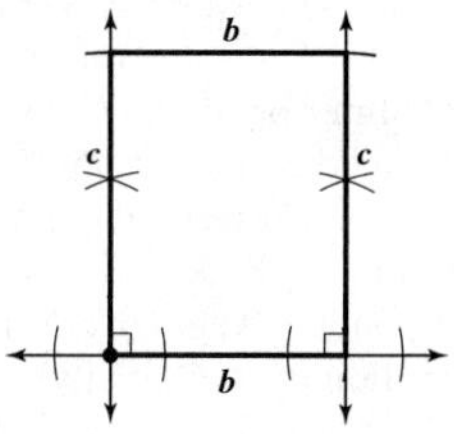

19.

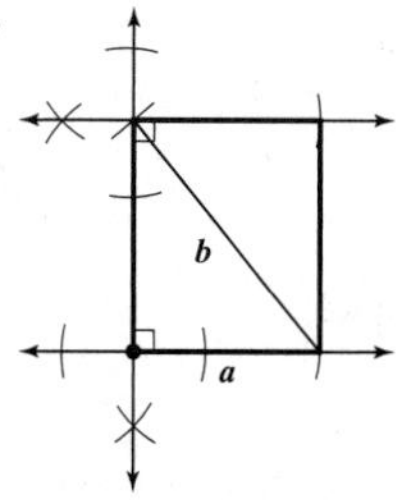

21. a.–c.

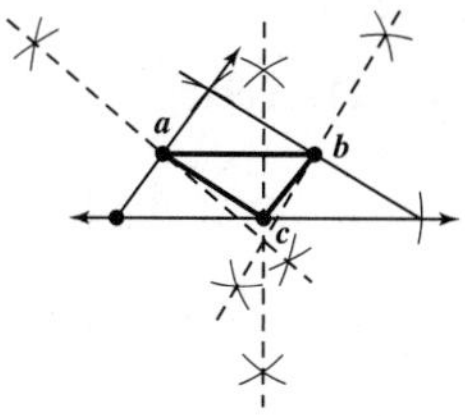

d. Sample: Each side of the smaller triangle is half the length and parallel to a side of the larger triangle. **23.** ii, iv, iii, i

25.

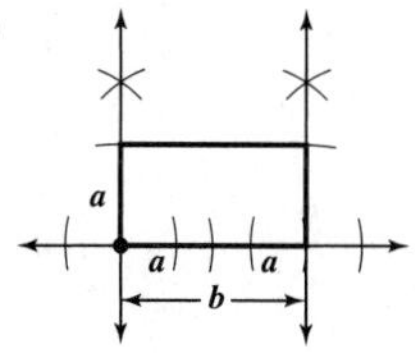

27.

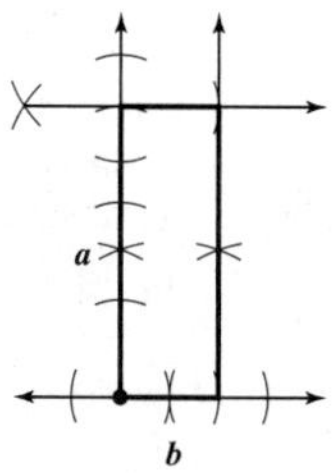

29.

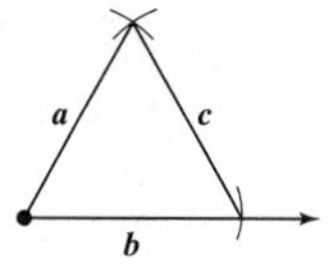

31. Not possible; if $a = 2b = 2c$, then $b = c$ and $a = b + c$. So, the shorter sides would meet at the midpoint of the longer side, forming a segment, not a triangle.

33.

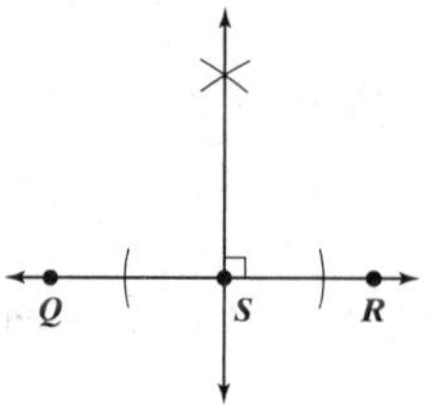

35. Similar: You are constructing a line ⊥ to a given line through a given point; different: The given point is on the given line in Example 3 and is not on the line in Example 4. **37.** $\frac{1}{2}$ **39.** -2

Section 3.6

Practice Exercises

1. a. $\frac{4}{3}$ **b.** 0 **2.** slope: $\frac{2}{3}$; y-intercept point: $(0, -3)$
3. \$128 **4. a.** perpendicular **b.** parallel

Vocabulary & Readiness Check 3.6

1. slope **3.** m; $(0, b)$; b **5.** horizontal **7.** -1 **9.** upward
11. horizontal

Exercise Set 3.6

1. $\frac{9}{5}$ **3.** $-\frac{7}{2}$ **5.** $-\frac{5}{6}$ **7.** $\frac{1}{3}$ **9.** $-\frac{4}{3}$ **11.** 0 **13.** undefined
15. 2 **17.** -1 **19.** l_2 **21.** l_2 **23.** l_2 **25.** slope: 5; y-intercept point: $(0, -2)$ **27.** slope: -2; y-intercept point: $(0, 7)$
29. slope: $\frac{2}{3}$; y-intercept point: $\left(0, -\frac{10}{3}\right)$ **31.** slope: $\frac{1}{2}$; y-intercept point: $(0, 0)$ **33.** A **35.** B **37.** undefined **39.** 0
41. undefined **43.** neither **45.** parallel **47.** perpendicular
49. $\frac{3}{2}$ **51.** $-\frac{1}{2}$ **53.** $\frac{2}{3}$ **55.** -0.12 **57. a.** slope: 1125; y-intercept point: $(0, 1300)$ **b.** The number of wireless Internet access points is expected to grow by 1125 thousand each year.
c. There were 1300 thousand wireless Internet access points in 2011.
59. a. slope: 291.5; the yearly cost is increasing by \$291.50 each year.
b. y-intercept point: $(0, 2944.05)$; the yearly cost was \$2944.05 in 2000.
61. Yes; if the ramp is 24 in. high and 72 in. long, the slope will be $\frac{24}{72} = 0.\overline{3}$, which is less than the maximum slope of $\frac{4}{11} = 0.\overline{36}$.
63. $-\frac{7}{2}$ **65.** $\frac{2}{7}$ **67.** $\frac{5}{2}$ **69.** $y = 5x + 32$ **71.** $y = 2x - 1$

Section 3.7

Practice Exercises

1. $y = -\frac{3}{4}x + 4$ **2.**

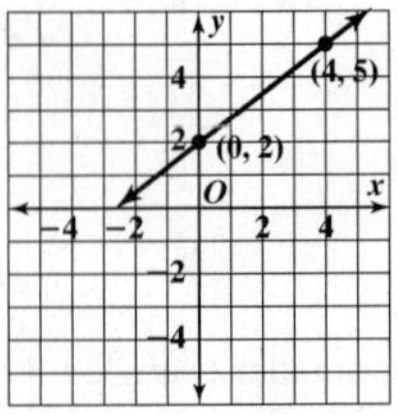

3.

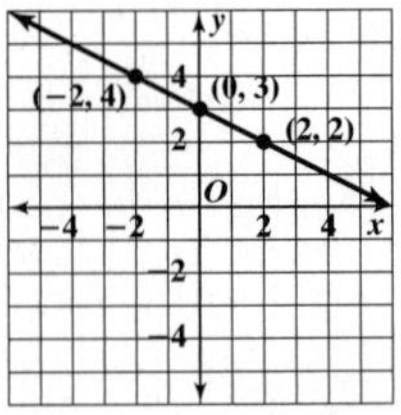

4. $y = -4x - 3$ **5.** $y = -\frac{2}{3}x + \frac{4}{3}$ **6.** $2x + 3y = 5$ **7.** House sales are predicted to be 12,568 in 2014. **8.** $y = -2$ **9.** $x = 6$ **10.** $3x + 4y = 12$ **11.** $y = \frac{4}{3}x - \frac{41}{3}$ **12.** $x - 3y = -30$

Vocabulary & Readiness Check 3.7

1. slope: −4; y-intercept point: (0, 12) **3.** slope: 5; y-intercept point: (0, 0) **5.** slope: $\frac{1}{2}$; y-intercept point: (0, 6) **7.** parallel **9.** neither

Exercise Set 3.7

1. $y = -x + 1$ **3.** $y = 2x + \frac{3}{4}$ **5.** $y = \frac{2}{7}x$

7.

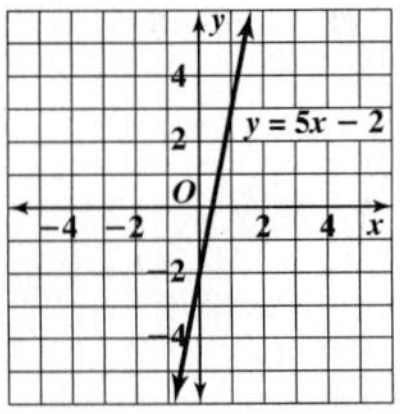

9.

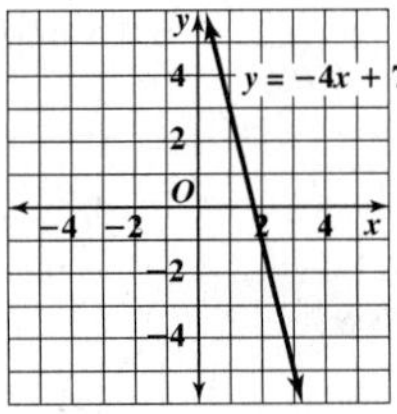

11.

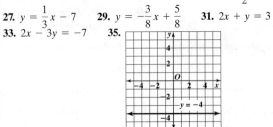

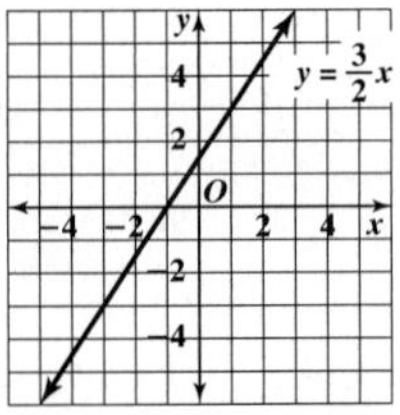

13. $y = 3x - 1$ **15.** $y = -2x - 1$ **17.** $y = \frac{1}{2}x + 5$ **19.** $y = -\frac{9}{10}x - \frac{27}{10}$

21. $y = 3x - 6$ **23.** $y = -2x + 1$ **25.** $y = -\frac{1}{2}x - 5$ **27.** $y = \frac{1}{3}x - 7$ **29.** $y = -\frac{3}{8}x + \frac{5}{8}$ **31.** $2x + y = 3$ **33.** $2x - 3y = -7$ **35.**

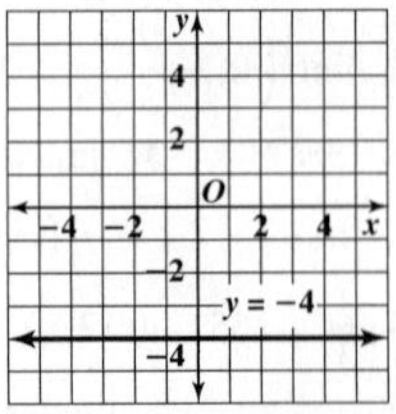

37.

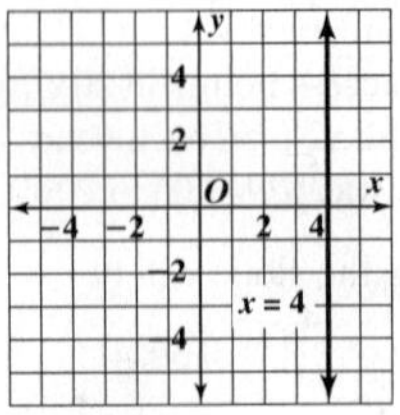

39.

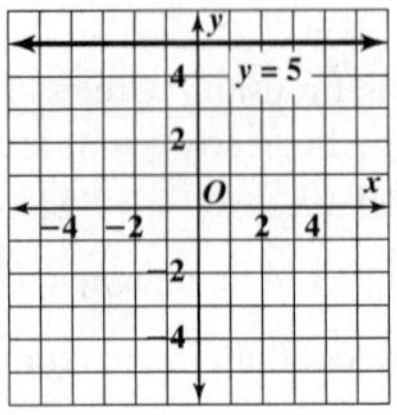

41. $y = 4x - 4$ **43.** $y = -3x + 1$ **45.** $y = -\frac{3}{2}x - 6$

47. $2x - y = -7$ **49.** $x + y = 7$ **51.** $x + 2y = 22$ **53.** $x = -2$ **55.** $2x + 4y = 4$ **57. a.** $y = 12{,}000x + 18{,}000$ **b.** \$102,000 **c.** 9 years **59. a.** $y = -1000x + 13{,}000$ **b.** 9500 Fun Noodles **61. a.** $y = 20.2x + 387$ **b.** 690 thousand **63.** true **65.** $4x - y = -4$ **67.** $2x + y = -23$ **69.** $2\sqrt{17}$ units **71.** $10\sqrt{5}$ units

Chapter 3 Vocabulary Check

1. parallel planes **2.** parallel lines **3.** transversal **4.** skew lines **5.** slope **6.** vertical; horizontal **7.** m; $(0, b)$; b **8.** slope-intercept **9.** horizontal **10.** vertical **11.** −1 **12.** the same; y-intercepts **13.** interior **14.** exterior **15.** corresponding **16.** alternate interior **17.** alternate exterior **18.** same-side interior **19.** vertical **20.** vertical **21.** one **22.** one **23.** Parallel **24.** Perpendicular

Chapter 3 Review

1. $\angle 2$ and $\angle 7$, a and b, transversal d; $\angle 3$ and $\angle 6$, c and d, transversal e; $\angle 3$ and $\angle 8$, b and e, transversal c
3. $\angle 1$ and $\angle 4$, c and d, transversal b; $\angle 2$ and $\angle 4$, a and b, transversal d; $\angle 2$ and $\angle 5$, c and d, transversal a; $\angle 1$ and $\angle 5$, a and b, transversal c
5. corresponding angles **7.** 20 **9.** $r \parallel p$, Corresponding Angles Theorem **11.** $\ell \parallel m$, Same-Side Interior Angles Theorem **13.** $m\angle 1 = 120°$, Corresponding Angles Converse; $m\angle 2 = 120°$, Vertical Angle Theorem ($\angle 1$) **15.** 118 **17.** $\parallel$

19.

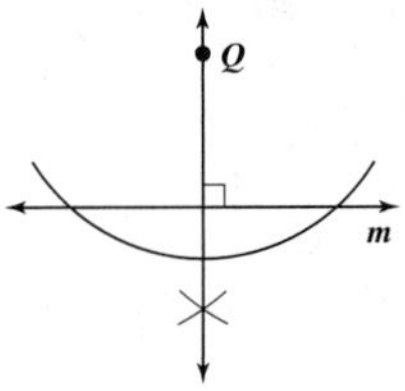

21.

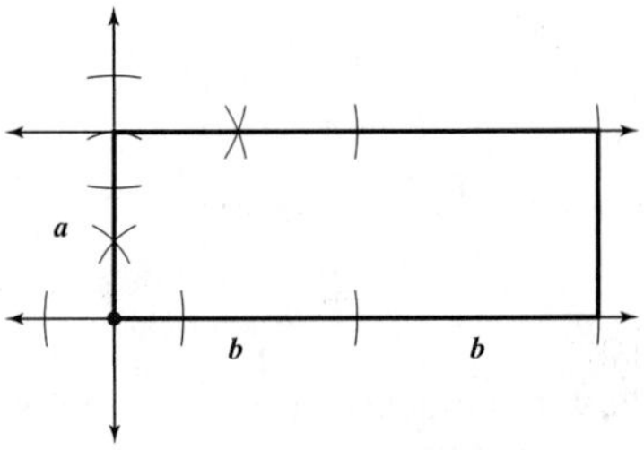

23. −1 **25.** slope: 2; y-intercept point: $(0, -1)$

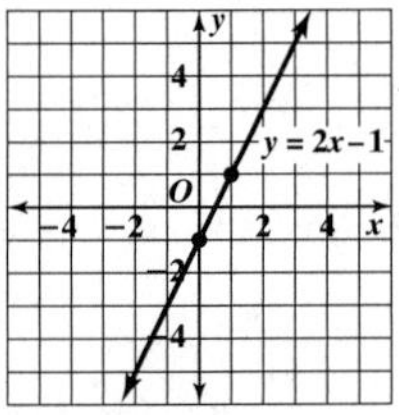

27. $y = -\frac{1}{2}x + 12$ **29.** neither **31.** perpendicular **33.** $y - 2 = 8(x + 6)$ **35.** Answers may vary. Sample: $\angle 1$ and $\angle 3$, $\angle 6$ and $\angle 8$ **37.** 53 **39.** 1st Street and 3rd Street are parallel because they are both perpendicular to Morris Avenue. Since 1st Street and 5th Street are both parallel to 3rd Street, 1st Street and 5th Street are parallel. **41.** $y - 8 = 3(x + 2)$ **43.** $y + 3 = -\frac{1}{2}(x - 1)$

Chapter 3 Test

1. skew lines **2.** parallel lines **3.** neither **4.** $m\angle 1 = 65°$, Corresponding Angles Converse; $m\angle 2 = 65°$, Alternate Interior Angles Converse or Vertical Angles Theorem ($\angle 1$) **5.** $m\angle 1 = 85°$, Alternate Interior Angles Converse; $m\angle 2 = 110°$, Same-Side Interior Angles Converse **6.** $m\angle 1 = 85°$, Corresponding Angles Converse; $m\angle 2 = 95°$, Same-Side Interior Angles Converse **7.** $m\angle 1 = 70°$, Corresponding Angles Converse; $m\angle 2 = 110°$, Linear Pair Theorem **8.** 5 **9.** 75 **10.** 65 **11. a.** Given **b.** Corresponding Angles Converse **c.** Given **d.** Substitution (or Transitivity) **e.** Corresponding Angles Theorem

12. 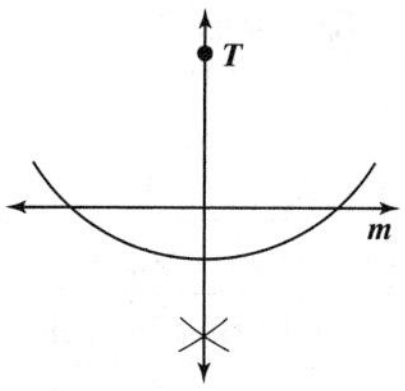

13. 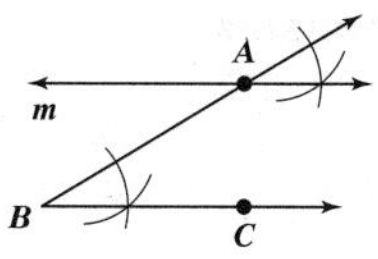

14. 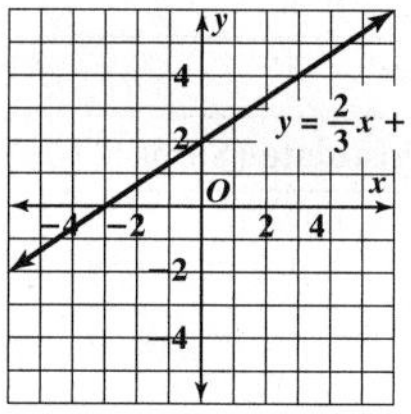

15. 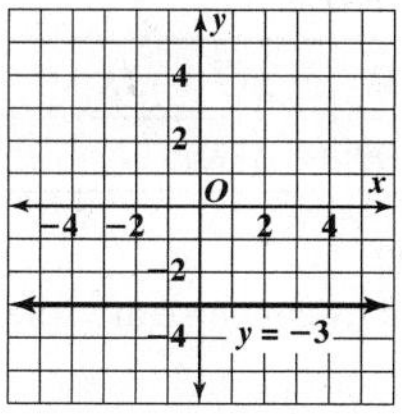

16. $-\frac{3}{2}$ **17.** slope: $-\frac{1}{4}$; y-intercept point: $\left(0, \frac{2}{3}\right)$ **18.** $y = -8$
19. $3x + y = 11$ **20.** $x + 2y = 0$ **21.** $x + 3y = 5$
22. $y = -\frac{1}{2}x - \frac{1}{2}$ **23.** neither

Chapter 3 Standardized Test

1. b **2.** c **3.** d **4.** b **5.** a **6.** c **7.** b **8.** d
9. c **10.** a **11.** c **12.** b **13.** a **14.** a **15.** d **16.** a

CHAPTER 4 TRIANGLES AND CONGRUENCE

Section 4.1

Practice Exercises

1. a. $\angle Q$ **b.** $\overline{QR}$ **c.** $\overline{QR}$ **d.** $\angle P$ **2. a.** obtuse triangle; scalene triangle; an obtuse scalene triangle **b.** equiangular triangle; equilateral triangle; an equiangular and equilateral triangle
3. $4x° = 32°$; $13x° = 104°$; $(5x + 4)° = 44°$ **4.** $x° = 38°$; $(4x + 6)° = 158°$

Vocabulary & Readiness Check 4.1

1. triangle **3.** vertices **5.** scalene **7.** equilateral (equiangular is also correct) **9.** $\angle 2, \angle 3, \angle 5$ **11.** $\overline{AB}$ and $\overline{AC}$

Exercise Set 4.1

1. $\angle A$ **3.** $\overline{AC}$ **5.** $\angle C$ **7.** $\overline{BC}$ **9.** acute; scalene
11. right; isosceles **13.** acute; isosceles **15.** obtuse; scalene
17. E **19.** F **21.** A **23.** $m\angle 1 = 30°$ **25.** $m\angle 1 = 83.1°$
27. $m\angle 1 = 56°$ **29.** $m\angle 1 = 123°$ **31.** $m\angle 2 = 36°$
33. $m\angle 3 = 92°$; $m\angle 4 = 88°$ **35.** $m\angle 1 = 80°$; $m\angle 2 = 80°$
37. $m\angle 1 = 70°$; $m\angle 2 = 110°$; $m\angle 3 = 30°$ **39.** $x = 37$; $x° = 37°$; $(2x + 4)° = 78°$; $(2x - 9)° = 65°$ **41.** $x = 7$; $(8x - 1)° = 55°$; $(4x + 7)° = 35°$ **43.** $x = 29$; $x° = 29°$; $(2x + 7)° = 65°$
45. $x = 6$; $(15x + 13)° = 103°$; $(9x - 2)° = 52°$ **47.** $x = 52$; $x° = 52°$; $(3x - 14)° = 142°$ **49.** $m\angle 1 = 58°$; $m\angle 2 = 67°$; $m\angle 3 = 125°$; $m\angle 4 = 23°$ **51.** $a = 162$; $b = 18$ **53.** answers may vary **55.** iii; answers may vary. Sample: This is the only Δ with 2 exterior angles at $\angle A$. **57.** 60°; answers may vary **59.** 5, 6, 8
61. They are vertical angles; they have the same measure. **63.** Two
65. 60 **67. a.** Definition of straight angle and Angle Addition Postulate
b. $m\angle 2 + m\angle 3 + m\angle 4 = 180°$ **c.** Substitution (Steps 2, 3)
d. Subtraction Property of = **69.** $(1, 2)$ **71.** $(1.5, 5.5)$

Section 4.2

Practice Exercises

1. Angles: $\angle A \cong \angle M, \angle B \cong \angle N, \angle C \cong \angle P, \angle D \cong \angle G, \angle E \cong \angle H$; sides: $\overline{AB} \cong \overline{MN}, \overline{BC} \cong \overline{NP}, \overline{CD} \cong \overline{PG}, \overline{DE} \cong \overline{GH}, \overline{EA} \cong \overline{HM}$ **2.** 103° **3.** $x = 5$

4.

Statements	Reasons
1. $\angle A \cong \angle D$	**1.** Given
2. $\angle ABE \cong \angle CBD$	**2.** Vertical Angles Theorem
3. $\angle E \cong \angle C$	**3.** Third Angles Theorem
4. $\overline{BA} \cong \overline{BD}, \overline{EB} \cong \overline{CB}, \overline{AE} \cong \overline{DC}$	**4.** Given
5. $\Delta AEB \cong \Delta DCB$	**5.** Definition of $\cong$ triangles

Vocabulary & Readiness Check 4.2

1. congruent **3.** $\angle S$ **5.** $\angle V$ **7.** $\overline{TV}$

Exercise Set 4.2

1. $\overline{BK}$ **3.** $\overline{ML}$ **5.** $\angle C$ **7.** ΔKJB **9.** ΔJBK **11.** $\overline{MN} \cong \overline{SR}, \overline{NE} \cong \overline{RT}, \overline{EP} \cong \overline{TQ}, \overline{PM} \cong \overline{QS}$ **13.** $\angle CAB \cong \angle DAB, \angle ABC \cong \angle ABD, \angle C \cong \angle D$; $\overline{AC} \cong \overline{AD}, \overline{CB} \cong \overline{DB}, \overline{BA} \cong \overline{BA}$
15. 335 ft **17.** 52° **19.** 45 ft **21.** 128° **23.** Yes; $\overline{TK} \cong \overline{TK}$ and $\angle RKT \cong \angle UKT$ **25.** c **27.** ΔZRN; ΔNRZ **29.** 28°
31. $y = 17$ **33.** $x = 15$; $t = 2$ **35.** $m\angle A = m\angle D = 20°$
37. $BC = EF = 8$
39. Δ Angle-Sum Theorem **41.** Substitution Property

43.

Statements	Reasons
1. $\overline{AB} \parallel \overline{DC}, \angle B \cong \angle D, \overline{AB} \cong \overline{DC}, \overline{BC} \cong \overline{AD}$	**1.** Given
2. $\overline{AC} \cong \overline{AC}$	**2.** Reflexive Property of $\cong$
3. $m\angle BAC = m\angle DCA$	**3.** Alternate Interior Angles Theorem
4. $m\angle ACB = m\angle CAD$	**4.** Third Angles Theorem
5. $\angle BAC \cong \angle DCA, \angle ACB \cong \angle CAD$	**5.** Definition of $\cong$ angles
6. $\Delta ABC \cong \Delta CDA$	**6.** Definition of $\cong$ triangles

45. Yes; answers may vary

47.

Statements	Reasons
1. $\overline{PR} \parallel \overline{TQ}$	**1.** Given
2. $m\angle RPS = m\angle TQS, m\angle PRS = m\angle QTS$	**2.** Alternate Interior Angles Theorem
3. $m\angle PSR = m\angle QST$	**3.** Third Angles Theorem
4. $\angle RPS \cong \angle TQS, \angle PRS \cong \angle QTS, \angle PSR \cong \angle QST$	**4.** Definition of $\cong$ angles

49. $KL = 4$; $LM = 3$; $KM = 5$ **51.** $y = -\frac{2}{3}x + \frac{25}{3}$ **53.** 18
55. $\angle QRP \cong \angle SPR$, thus $\overleftrightarrow{PS} \parallel \overleftrightarrow{QR}$; Also, since $\angle QPR \cong \angle SRP$, then $\overleftrightarrow{QP} \parallel \overleftrightarrow{RS}$

Section 4.3

Practice Exercises

1.

Statements	Reasons
1. $\overline{MQ} \cong \overline{MR}, \overline{QP} \cong \overline{RP}$	**1.** Given
2. $\overline{MP} \cong \overline{MP}$	**2.** Reflexive Property of $\cong$
3. $\Delta MQP \cong \Delta MRP$	**3.** SSS Postulate (Steps 1, 2)

2.

Statements	Reasons
1. $\angle BAC \cong \angle FAC, \overline{AB} \cong \overline{AF}$	1. Given
2. $\overline{AC} \cong \overline{AC}$	2. Reflexive Property of $\cong$
3. $\Delta ABC \cong \Delta AFC$	3. SAS Postulate (Steps 1, 2)

3. a. SSS **b.** SAS **c.** SSS **d.** not enough information; two pairs of corresponding sides are $\cong$, but the included $\angle$ in the first Δ is marked $\cong$ to an angle that is not the included $\angle$ in the second Δ.

Vocabulary & Readiness Check 4.3

1. *DEF*; SAS

Exercise Set 4.3

1. $\angle T$ **3.** $\overline{HA}$ and $\overline{TA}$ **5.** SAS **7.** $\overline{LG} \cong \overline{MN}$ **9.** SSS **11.** SAS **13.** not enough information; two pairs of sides are $\cong$, but we do not know if the included $\angle$s are $\cong$ **15.** SAS **17.** Given **19.** $\angle DCA \cong \angle CAB$ **21.** Given

23.

Statements	Reasons
1. $\overline{IE} \cong \overline{GH}, \overline{EF} \cong \overline{HF}$	1. Given
2. F is the midpoint of $\overline{GI}$.	2. Given
3. $\overline{IF} \cong \overline{GF}$	3. Definition of midpoint
4. $\Delta EFI \cong \Delta HFG$	4. SSS Postulate (Steps 1, 3)

25.

Statements	Reasons
1. $\overline{GK}$ bisects $\angle JGM$.	1. Given
2. $\angle JGK \cong \angle MGK$	2. Definition of bisector
3. $\overline{GJ} \cong \overline{GM}$	3. Given
4. $\overline{GK} \cong \overline{GK}$	4. Reflexive Property of $\cong$
5. $\Delta GJK \cong \Delta GMK$	5. SAS Postulate (Steps 2, 3, 4)

27.

Statements	Reasons
1. $\angle GFK \cong \angle LKF$	1. Alternate Interior Angles Theorem
2. $\overline{FG} \cong \overline{KL}$	2. Given
3. $\overline{FK} \cong \overline{FK}$	3. Reflexive Property of $\cong$
4. $\Delta FGK \cong \Delta KLF$	4. SAS Postulate (Steps 1, 2, 3)

29.

Statements	Reasons
1. $\overline{BC} \cong \overline{DA}, \angle CBD \cong \angle ADB$	1. Given
2. $\overline{BD} \cong \overline{BD}$	2. Reflexive Property of $\cong$
3. $\Delta BCD \cong \Delta DAB$	3. SAS Postulate (Steps 1, 2)

31. Congruent by SAS or SSS **33.** Congruent by SSS **35.** Answers may vary. Sample answer: Alike: Both use three pairs of $\cong$ parts to prove triangles $\cong$. Different: SSS uses three $\cong$ sides, while SAS uses two pairs of $\cong$ sides and their $\cong$ included angles.
37. Answers may vary. Sample:

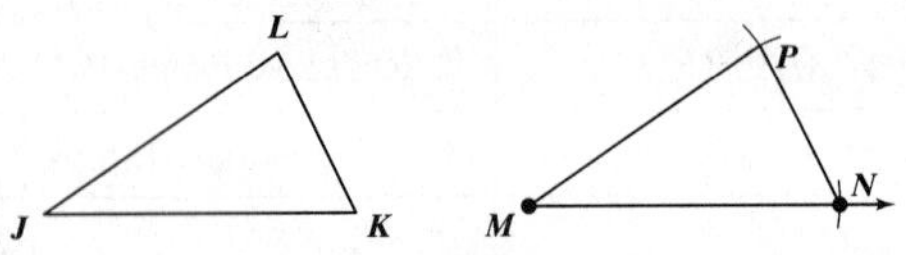

39. Not necessarily; the $\cong$ angles are not included between the pairs of $\cong$ sides. **41.** Answers may vary. Sample: roof trusses for a house, sections of a Ferris wheel, sawhorses used by a carpenter; explanations will vary.

43.

Statements	Reasons
1. $HK = LG$	1. Definition of $\cong$
2. $GK = GK$	2. Reflexive Property of $=$
3. $HK + GK = LG + GK$	3. Addition Property of $=$
4. $HG = LK$	4. Segment Addition Postulate
5. $\overline{HG} \cong \overline{LK}$	5. Definition of $\cong$
6. $\overline{HF} \cong \overline{LJ}, \overline{FG} \cong \overline{JK}$	6. Given
7. $\Delta FGH \cong \Delta JKL$	7. SSS Postulate (Steps 5, 6)

45. $\angle E$ **47.** $\overline{FG}$ **49.** Converse: If $2x = 6$, then $x = 3$. Both the statement and its converse are true. **51.** $\overline{JH}$

Section 4.4

Practice Exercises

1. a and d are congruent by ASA because the sides marked congruent in these triangles are the included sides of the two congruent angles.

2.

Statements	Reasons
1. $\angle E \cong \angle B$	1. Given
2. $\angle ADE \cong \angle CDB$	2. Vertical Angles Theorem
3. D is the midpoint of $\overline{EB}$.	3. Given
4. $\overline{ED} \cong \overline{BD}$	4. Definition of midpoint
5. $\Delta EDA \cong \Delta BDC$	5. ASA Postulate (Steps 1, 2, 4)

3.

Statements	Reasons
1. $\overline{DB} \perp \overline{AC}$	1. Given
2. $\angle DBA$ and $\angle DBC$ are right angles.	2. Definition of $\perp$
3. $\angle DBA \cong \angle DBC$	3. Definition of right angles
4. $\angle A \cong \angle C$	4. Given
5. $\overline{DB} \cong \overline{DB}$	5. Reflexive Property of $\cong$
6. $\Delta ADB \cong \Delta CDB$	6. AAS Theorem (Steps 3, 4, 5)

4. a. ASA **b.** AAS **c.** SAS **d.** SSS

Vocabulary & Readiness Check 4.4

1. *DEF*; AAS

Exercise Set 4.4

1. ASA **3.** ΔPRQ and ΔVWX **5.** Congruent by SAS **7.** Not congruent; the angles that are congruent are not the included angles of the congruent sides. **9.** Not congruent; the right angle is the included angle for the congruent sides in ΔABC but not in ΔEFG. **11.** Congruent by SSS **13.** Congruent by SAS **15.** Congruent by AAS **17. a.** Reflexive Property of $\cong$ **b.** ASA Postulate

19.

Statements	Reasons
1. $\overline{AC} \perp \overline{BD}$	1. Given
2. $\angle ACB$ and $\angle ACD$ are right angles.	2. Definition of $\perp$
3. $\angle ACB \cong \angle ACD$	3. Definition of right angles
4. $\overline{AC} \cong \overline{AC}$	4. Reflexive Property of $\cong$
5. $\angle BAC \cong \angle DAC$	5. Given
6. $\Delta ABC \cong \Delta ADC$	6. ASA Postulate (Steps 3, 4, 5)

21.

Statements	Reasons
1. $\overline{WZ}$ bisects $\angle VWY$.	**1.** Given
2. $\angle VWZ \cong \angle YWZ$	**2.** Definition of bisector
3. $\angle V \cong \angle Y$	**3.** Given
4. $\overline{WZ} \cong \overline{WZ}$	**4.** Reflexive Property of $\cong$
5. $\Delta VWZ \cong \Delta YWZ$	**5.** AAS Theorem (Steps 2, 3, 4)

23.

Statements	Reasons
1. $\angle N \cong \angle P$	**1.** Given
2. $\overline{MO} \cong \overline{QO}$	**2.** Given
3. $\angle QOP \cong \angle MON$	**3.** Vertical Angles Theorem
4. $\Delta MON \cong \Delta QOP$	**4.** AAS Theorem (Steps 1, 2, 3)

25.

Statements	Reasons
1. $\overline{AE} \parallel \overline{BD}$	**1.** Given
2. $\angle DBC \cong \angle A$	**2.** Corresponding Angles Theorem (transversal $\overline{AC}$)
3. $\overline{AE} \cong \overline{BD}, \angle E \cong \angle D$	**3.** Given
4. $\Delta AEB \cong \Delta BDC$	**4.** ASA Postulate (Steps 2, 3)

27.

Statements	Reasons
1. $\overline{AB} \parallel \overline{DC}$	**1.** Given
2. $\angle BAC \cong \angle DCA$	**2.** Alternate Interior Angles Theorem
3. $\overline{AD} \parallel \overline{BC}$	**3.** Given
4. $\angle ACB \cong \angle CAD$	**4.** Alternate Interior Angles Theorem
5. $\overline{AC} \cong \overline{CA}$	**5.** Reflexive Property of $\cong$
6. $\Delta ABC \cong \Delta CDA$	**6.** ASA Postulate (Steps 2, 4, 5)

29.

Statements	Reasons
1. $m\angle 1 + m\angle 4 = 180°$, $m\angle 2 + m\angle 3 = 180°$	**1.** Definition of straight angles
2. $m\angle 1 + m\angle 4 = m\angle 2 + m\angle 3$	**2.** Substitution
3. $\angle 1 \cong \angle 2$ or $m\angle 1 = m\angle 2$	**3.** Given and Definition of $\cong$ angles
4. $m\angle 2 + m\angle 4 = m\angle 2 + m\angle 3$	**4.** Substitution (Steps 2, 3)
5. $m\angle 4 = m\angle 3$	**5.** Subtraction Property of $=$
6. $\angle 4 \cong \angle 3$	**6.** Definition of $\cong$
7. $\angle 5 \cong \angle 6$	**7.** Vertical Angles Theorem
8. $\overline{AB} \cong \overline{BC}$	**8.** Given
9. $\Delta ABE \cong \Delta CBD$	**9.** ASA Postulate (Steps 6, 7, 8)

31.

Statements	Reasons
1. $m\angle 1 + m\angle 3 = 180°$, $m\angle 2 + m\angle 4 = 180°$	**1.** Definition of straight angles
2. $m\angle 1 + m\angle 3 = m\angle 2 + m\angle 4$	**2.** Substitution
3. $m\angle 1 = m\angle 2$	**3.** Given
4. $m\angle 3 = m\angle 4$	**4.** Substitution, then Subtraction Property of $=$
5. ΔACD is equilateral.	**5.** Given
6. $\overline{AC} \cong \overline{AD}$	**6.** Definition of equilateral Δ
7. $\overline{BC} \cong \overline{DE}$	**7.** Given
8. $\Delta ABC \cong \Delta AED$	**8.** SAS Postulate (Steps 4, 6, 7)

33.

Statements	Reasons
1. $AB = CD, AC = BD$	**1.** Given
2. $\overline{AB} \cong \overline{CD}, \overline{AC} \cong \overline{BD}$	**2.** Definition of $\cong$
3. $\overline{BC} \cong \overline{BC}$	**3.** Reflexive Property of $\cong$
4. $\Delta ABC \cong \Delta DCB$	**4.** SSS Postulate (Steps 2, 3)

35. 5 **37.** 32 **39.** 120° **41.** Answers may vary. Sample answer: Alike: Both postulates use three pairs of $\cong$ corresponding parts. Different: To use the ASA Postulate, the sides must be included between the pairs of corresponding angles, while to use the SAS Postulate, the angles must be included between the pairs of corresponding sides. **43.** ΔABC and ΔCDA; ΔABD and ΔCDB; ΔABE and ΔCDE; ΔADE and ΔCBE **45.** No; the common side is included between the two $\cong$ angles in one triangle, but it is not included between the $\cong$ angles in the other triangle.

47.

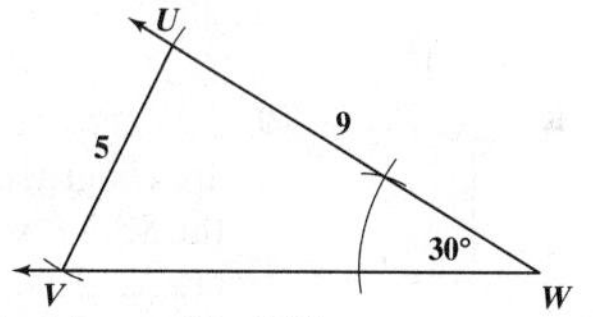

49. Answers may vary. Sample answer: Yes; ASA guarantees a unique triangle with vertices at the oak tree, the maple tree, and the time capsule.

51. a triangle **53.** SSS; answers may vary. Sample answer: We are given that 2 pairs of corresponding sides are $\cong$. Also, the two Δs share a common third side. **55.** $\angle T \cong \angle L, \angle I \cong \angle O, \angle C \cong \angle K$

Section 4.5

Practice Exercises

1.

Statements	Reasons
1. $\overline{PR} \cong \overline{RT}, \overline{SR} \cong \overline{RQ}$	**1.** Given
2. $\angle SRT \cong \angle QRP$	**2.** Vertical Angles Theorem
3. $\Delta SRT \cong \Delta QRP$	**3.** SAS Postulate (Steps 1 and 2)
4. $\angle Q \cong \angle S$	**4.** cpoctac

2.

Statements	Reasons
1. $\overline{AB} \cong \overline{AC}$	**1.** Given
2. M is the midpoint of $\overline{BC}$.	**2.** Given
3. $\overline{MB} \cong \overline{MC}$	**3.** Definition of midpoint
4. $\overline{AM} \cong \overline{AM}$	**4.** Reflexive Property of $\cong$
5. $\Delta AMB \cong \Delta AMC$	**5.** SSS Postulate (Steps 1, 3, 4)
6. $\angle AMB \cong \angle AMC$	**6.** cpoctac

3. a. $\overline{AD}$ b. $\overline{AB}$

4.

Statements	Reasons
1. $\Delta ACD \cong \Delta BDC$	1. Given
2. $\angle A \cong \angle B$	2. cpoctac
3. $\overline{AC} \cong \overline{BD}$	3. cpoctac
4. $\angle AEC \cong \angle BED$	4. Vertical Angles Theorem
5. $\Delta AEC \cong \Delta BED$	5. AAS Theorem (Steps 2, 3, 4)
6. $\overline{CE} \cong \overline{DE}$	6. cpoctac

5.

Statements	Reasons
1. $\overline{PS} \cong \overline{RS}$	1. Given
2. $\angle PSQ \cong \angle RSQ$	2. Given
3. $\overline{QS} \cong \overline{QS}$	3. Reflexive Property of $\cong$
4. $\Delta QPS \cong \Delta QRS$	4. SAS Postulate (Steps 1, 2, 3)
5. $\overline{QP} \cong \overline{QR}$	5. cpoctac
6. $\angle PQT \cong \angle RQT$	6. cpoctac
7. $\overline{QT} \cong \overline{QT}$	7. Reflexive Property of $\cong$
8. $\Delta QPT \cong \Delta QRT$	8. SAS Postulate (Steps 5, 6, 7)

Vocabulary & Readiness Check 4.5

1. Corresponding Parts of Congruent Triangles Are Congruent

Exercise Set 4.5

1. $\overline{JK}$ 3.

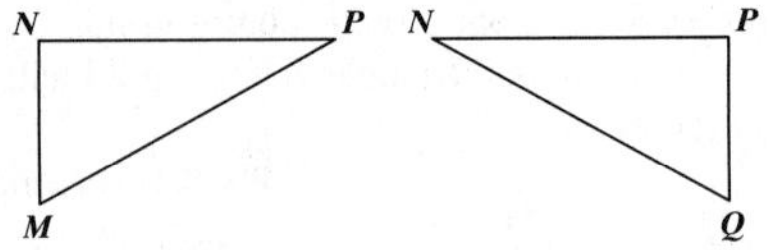

5. $\overline{DF}$ 7.

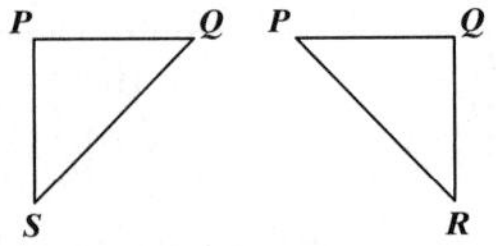

$\overline{PQ}$ 9. $\Delta JKL \cong \Delta NOM$ are congruent by the SAS Postulate. $\angle K \cong \angle O$, $\angle J \cong \angle N$, $\overline{JK} \cong \overline{ON}$

11. ASA Postulate

13.

Statements	Reasons
1. $\overline{OM} \cong \overline{ER}$, $\overline{ME} \cong \overline{RO}$	1. Given
2. $\overline{OE} \cong \overline{OE}$	2. Reflexive Property of $\cong$
3. $\Delta OME \cong \Delta ERO$	3. SSS Postulate (Steps 1, 2)
4. $\angle M \cong \angle R$	4. cpoctac

15.

Statements	Reasons
1. $\angle SPT \cong \angle OPT$, $\overline{SP} \cong \overline{OP}$	1. Given
2. $\overline{PT} \cong \overline{PT}$	2. Reflexive Property of $\cong$
3. $\Delta SPT \cong \Delta OPT$	3. SAS Postulate (Steps 1, 2)
4. $\angle S \cong \angle O$	4. cpoctac

17. $\overline{KL}$ bisects $\angle PKQ$, so $\angle PKL \cong \angle QKL$. $\overline{KL} \cong \overline{KL}$ by the Reflexive Property of $\cong$. $\Delta PKL \cong \Delta QKL$ by SAS, so $\angle P \cong \angle Q$ because cpoctac.

19.

Statements	Reasons
1. $\overline{BD}$ bisects $\angle ABC$.	1. Given
2. $\angle ABD \cong \angle CBD$	2. Definition of angle bisector
3. $\overline{BA} \cong \overline{BC}$	3. Given
4. $\overline{BD} \cong \overline{BD}$	4. Reflexive Property of $\cong$
5. $\Delta ABD \cong \Delta CBD$	5. SAS Postulate (Steps 2, 3, 4)
6. $m\angle ADB + m\angle CDB = 180°$	6. Definition of straight angles
7. $\angle ADB \cong \angle CDB$	7. cpoctac
8. $m\angle ADB = m\angle CDB$	8. Definition of $\cong$
9. $m\angle ADB + m\angle ADB = 180°$	9. Substitution (Steps 6, 8)
10. $2(m\angle ADB) = 180°$	10. Addition Property of =
11. $m\angle ADB = 90°$	11. Division Property of =
12. $m\angle ADB = m\angle CDB = 90°$	12. Substitution (Steps 8, 11)
13. $\overline{BD} \perp \overline{AC}$	13. Definition of $\perp$
14. $\overline{AD} \cong \overline{CD}$	14. cpoctac
15. $\overline{BD}$ bisects $\overline{AC}$.	15. Definition of bisector

21.

Statements	Reasons
1. $AE + EF = AF$, $CF + EF = CE$	1. Segment Addition Postulate
2. $\overline{AF} \cong \overline{CE}$	2. Given
3. $AF = CE$	3. Definition of $\cong$
4. $AE + EF = CF + EF$	4. Substitution (Steps 1, 3)
5. $AE = CF$	5. Subtraction Property of =
6. $\overline{AE} \cong \overline{CF}$	6. Definition of $\cong$
7. $\overline{BE} \perp \overline{AC}$, $\overline{DF} \perp \overline{AC}$	7. Given
8. $\angle AEB$ and $\angle CFD$ are right angles.	8. Definition of $\perp$
9. $\angle AEB \cong \angle CFD$	9. All right $\angle s$ are $\cong$
10. $\overline{BE} \cong \overline{DF}$	10. Given
11. $\Delta AEB \cong \Delta CFD$	11. SAS Postulate (Steps 6, 9, 10)
12. $\overline{AB} \cong \overline{CD}$	12. cpoctac

23. Given 25. AAS Theorem

27.

Statements	Reasons
1. $\overline{RS} \cong \overline{UT}$, $\overline{RT} \cong \overline{US}$	1. Given
2. $\overline{ST} \cong \overline{TS}$	2. Reflexive Property of $\cong$
3. $\Delta RST \cong \Delta UTS$	3. SSS Postulate (Steps 1, 2)

29.

Statements	Reasons
1. $\angle 1 \cong \angle 2, \angle 3 \cong \angle 4$	**1.** Given
2. $\overline{QB} \cong \overline{QB}$	**2.** Reflexive Property of $\cong$
3. $\Delta QTB \cong \Delta QUB$	**3.** ASA Postulate (Steps 1, 2)
4. $\overline{QT} \cong \overline{QU}$	**4.** cpoctac
5. $\overline{QE} \cong \overline{QE}$	**5.** Reflexive Property of $\cong$
6. $\Delta QET \cong \Delta QEU$	**6.** SAS Postulate (Steps 1, 4, 5)

31.

Statements	Reasons
1. $\overline{AC} \cong \overline{EC}, \overline{CB} \cong \overline{CD}$	**1.** Given
2. $\angle C \cong \angle C$	**2.** Reflexive Property of $\cong$
3. $\Delta ACD \cong \Delta ECB$	**3.** SAS Postulate (Steps 1, 2)
4. $\angle A \cong \angle E$	**4.** cpoctac

33.

Statements	Reasons
1. $\overline{TE} \cong \overline{RI}, \overline{TI} \cong \overline{RE}$	**1.** Given
2. $\overline{IE} \cong \overline{IE}$	**2.** Reflexive Property of $\cong$
3. $\Delta TEI \cong \Delta RIE$	**3.** SSS Postulate (Steps 1, 2)
4. $\angle TIE \cong \angle REI$	**4.** cpoctac
5. $\angle TDI \cong \angle ROE$	**5.** All right angles are $\cong$.
6. $\Delta TDI \cong \Delta ROE$	**6.** AAS Theorem (Steps 1, 4, 5)
7. $\overline{TD} \cong \overline{RO}$	**7.** cpoctac

35. $m\angle 1 = 50°; m\angle 2 = 50°; m\angle 3 = 40°; m\angle 4 = 90°; m\angle 5 = 10°; m\angle 6 = 40°; m\angle 7 = 40°; m\angle 8 = 80°; m\angle 9 = 100°$ **37.** 36

39.

Statements	Reasons
1. $\overline{BA} \cong \overline{KA}, \overline{BE} \cong \overline{KE}$	**1.** Given
2. $\overline{AE} \cong \overline{AE}$	**2.** Reflexive Property of $\cong$
3. $\Delta AKE \cong \Delta ABE$	**3.** SSS Postulate (Steps 1, 2)
4. $\angle KAS \cong \angle BAS$	**4.** cpoctac
5. $\overline{AS} \cong \overline{AS}$	**5.** Reflexive Property of $\cong$
6. $\Delta KAS \cong \Delta BAS$	**6.** SAS Postulate (Steps 1, 4, 5)
7. $\overline{KS} \cong \overline{BS}$	**7.** cpoctac
8. S is the midpoint of $\overline{BK}$.	**8.** Definition of midpoint

41. Answers may vary. Sample: Since $\Delta PSY \cong \Delta SPL, \overline{PL} \cong \overline{SY}$ and $\angle L \cong \angle Y$. Also, $\angle PRL \cong \angle SRY$ since they are vertical angles. Thus, $\Delta PRL \cong \Delta SRY$ by AAS Theorem. **43.** $\Delta DEC \cong \Delta BEA$ by SAS, which gives that $\angle ACD \cong \angle CAB$. Or $\Delta ADE \cong \Delta CBE$ by SAS, which gives that $\angle CAD \cong \angle ACB$. **45.** Answers may vary.
Sample answer:

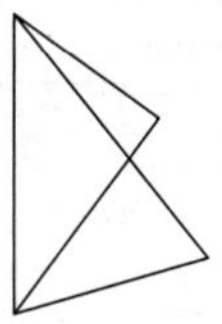

47. $\overline{AB} \cong \overline{DC}, \overline{AD} \cong \overline{BC}, \overline{BE} \cong \overline{ED}, \overline{AE} \cong \overline{EC}$
49. a. $\Delta ACP \cong \Delta BCP$ by SSS, which makes $\angle APC$ and $\angle BPC$ both right angles. **b.** $PA = PB$ and $AC = BC$ by construction.
51. ASA Postulate **53.** $\overline{AC}$ **55.** $\angle A$

Section 4.6
Practice Exercises

1.

Statements	Reasons
1. $\angle B \cong \angle C$	**1.** Given
2. $\angle A \cong \angle A$	**2.** Reflexive Property of $\cong$
3. $\overline{BC} \cong \overline{CB}$	**3.** Reflexive Property of $\cong$
4. $\Delta ABC \cong \Delta ACB$	**4.** AAS Theorem (Steps 1, 2, 3)
5. $\overline{AB} \cong \overline{AC}$	**5.** cpoctac

2. $x = 34; y = 73$ **3.** 63 **4.** $m\angle A = m\angle B = m\angle C = 60°$

5.

Statements	Reasons
1. $\overline{AD}$ is the perpendicular bisector of $\overline{CE}$.	**1.** Given
2. $\angle CBD$ and $\angle ABE$ are right angles.	**2.** Definition of $\perp$
3. $\overline{CB} \cong \overline{EB}$	**3.** Definition of bisector
4. $\overline{EA} \cong \overline{CD}$	**4.** Given
5. ΔCBD and ΔEBA are right triangles.	**5.** Definition of right triangles
6. $\Delta CBD \cong \Delta EBA$	**6.** H-L Theorem (Steps 3, 4, 5)

Vocabulary & Readiness Check 4.6

1. isosceles **3.** hypotenuse **5.** base **7.** vertex **9.** Some **11.** All

Exercise Set 4.6

1. $\Delta ABC \cong \Delta DEF$ **3.** There is not enough information. **5.** $m\angle A = 70°; m\angle C = 40°$ **7.** $m\angle H = 76°; m\angle J = 28°$ **9.** $m\angle Q = m\angle P = 28°$ **11.** $m\angle W = m\angle Y = 76°$ **13.** $m\angle C = m\angle G = 60°$ **15.** $m\angle Q = 32°$ **17.** 3 **19.** 90 **21.** C **23.** 90° **25.** 64° **27.** 42° **29.** $\overline{VX}$; answers may vary. Sample: ΔVTX is an isosceles triangle since $\angle T \cong \angle X$. **31.** $\overline{VY}$; answer may vary. Sample: $\overline{VT} \cong \overline{VX}$ and $\overline{YX} \cong \overline{UW} \cong \overline{UT}$. By subtraction, $VT - UT = VX - YX$ or $VU = VY$ or $\overline{VU} \cong \overline{VY}$. **33.** $x = 80; y = 40$ **35.** $x = 40; y = 70$ **37.** $x = 20; y = 45$ **39.** $x = 36; y = 27$ **41.** $x = 21; y = 42$ **43.** 108° **45.** 45° since the base angles have equal measures and their measures sum to 90°.
47. a. isosceles triangles **b.** 900 ft and 1100 ft **c.** It is the perpendicular bisector of the base of each triangle. **49. a.** Angles that are supplementary and congruent are right angles. **b.** Definition of right triangles **c.** Given **d.** Reflexive Property of $\cong$ **e.** H-L Theorem (Steps 4, 5, 6)

51.

Statements	Reasons
1. $\overline{HV} \perp \overline{GT}$	**1.** Given
2. $\angle HIG$ and $\angle VIT$ are right angles.	**2.** Definition of $\perp$
3. ΔIGH and ΔITV are right triangles.	**3.** Definition of right triangles
4. I is the midpoint of $\overline{HV}$.	**4.** Given
5. $\overline{HI} \cong \overline{VI}$	**5.** Definition of midpoint
6. $\overline{GH} \cong \overline{TV}$	**6.** Given
7. $\Delta IGH \cong \Delta ITV$	**7.** H-L Theorem (Steps 3, 5, 6)

53.

Statements	Reasons
1. $\overline{RS} \perp \overline{ST}$, $\overline{TU} \perp \overline{UV}$	**1.** Given
2. $\angle RST$ and $\angle TUV$ are right angles.	**2.** Definition of $\perp$
3. ΔRST and ΔTUV are right triangles.	**3.** Definition of right triangles
4. T is the midpoint of $\overline{RV}$.	**4.** Given
5. $\overline{RT} \cong \overline{TV}$	**5.** Definition of midpoint
6. $\overline{RS} \cong \overline{TU}$	**6.** Given
7. $\Delta RST \cong \Delta TUV$	**7.** H-L Theorem (Steps 3, 5, 6)

55.

Statements	Reasons
1. ΔGKE is isosceles with base $\overline{GE}$.	**1.** Given
2. $\overline{GK} \cong \overline{EK}$	**2.** Definition of isosceles triangle
3. $\angle L$ and $\angle D$ are right angles.	**3.** Given
4. ΔKLG and ΔKDE are right triangles.	**4.** Definition of right triangles
5. K is the midpoint of $\overline{LD}$.	**5.** Given
6. $\overline{KL} \cong \overline{KD}$	**6.** Definition of midpoint
7. $\Delta KLG \cong \Delta KDE$	**7.** H-L Theorem (Steps 2, 4, 6)
8. $\overline{LG} \cong \overline{DE}$	**8.** cpoctac

57.

Statements	Reasons
1. $\overline{AB} \cong \overline{AC}$, $\angle 1 \cong \angle 2$	**1.** Given
2. $\overline{AD} \cong \overline{AD}$	**2.** Reflexive Property of $\cong$
3. $\Delta ABD \cong \Delta ACD$	**3.** SAS Postulate (Steps 1, 2)
4. $\overline{BD} \cong \overline{DC}$	**4.** cpoctac
5. $\angle ADB \cong \angle ADC$	**5.** cpoctac
6. $\angle ADB$ and $\angle ADC$ are supplementary.	**6.** Angles that form a linear pair are supplementary.
7. $\angle ADB$ and $\angle ADC$ are right angles.	**7.** Angles that are supplementary and congruent are right angles.
8. $\overline{AD} \perp \overline{BC}$	**8.** Definition of $\perp$

59.

Statements	Reasons
1. $\overline{AB} \cong \overline{BC}$	**1.** Given
2. $\angle C \cong \angle A$	**2.** Isosceles Base Angles Theorem
3. $\overline{BC} \cong \overline{CA}$	**3.** Given
4. $\angle A \cong \angle B$	**4.** Isosceles Base Angles Theorem
5. $\angle A \cong \angle B \cong \angle C$	**5.** Transitive Property of $\cong$

61. If $\overline{AB} \cong \overline{AC}$, then $\angle C \cong \angle B$ (not $\angle A$ and $\angle C$). If $\angle A \cong \angle C$, then $\overline{AB} \cong \overline{BC}$ (not $\overline{AB}$ and $\overline{AC}$). **63. a.** Two sides have equal length and the angles opposite these two sides have equal measure. **b.** All three sides have the same length and all three angles have the same measure. **65.** (0, 0); (4, 4); (4, 8); (−4, 0); (0, −4); (8, 4)

67.

Statements	Reasons
1. $\overline{BE} \perp \overline{EA}$, $\overline{BE} \perp \overline{EC}$	**1.** Given
2. $\angle AEB$ and $\angle CEB$ are right angles.	**2.** Definition of $\perp$
3. ΔAEB and ΔCEB are right triangles.	**3.** Definition of right triangles
4. ΔABC is equilateral.	**4.** Given
5. $\overline{AB} \cong \overline{CB}$	**5.** Definition of equilateral triangle
6. $\overline{EB} \cong \overline{EB}$	**6.** Reflexive Property of $\cong$
7. $\angle AEB \cong \angle CEB$	**7.** H-L Theorem (Steps 3, 5, 6)

69.

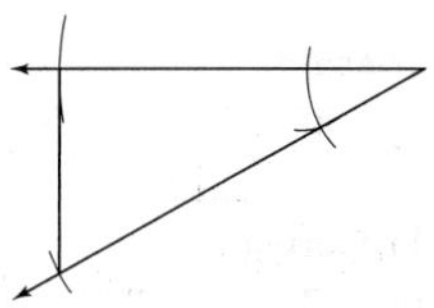

71.

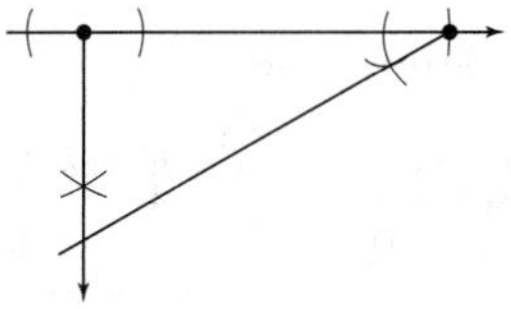

73. isosceles; answers may vary. Sample: Given the $\cong$ statement, $\overline{ST} \cong \overline{UT}$. Thus, the Δ is isosceles. **75.** Yes, by the H-L Theorem **77.** No, $\overline{AC} \cong \overline{ST}$ but $\overline{AC}$ is the hypotenuse of ΔABC while $\overline{ST}$ is a leg of ΔRST.

Chapter 4 Vocabulary Check

1. congruent **2.** corresponding; corresponding **3.** triangle **4.** sides **5.** vertices **6.** vertex **7.** scalene **8.** isosceles **9.** equilateral **10.** equiangular **11.** hypotenuse **12.** leg **13.** base **14.** leg **15.** vertex **16.** base **17.** Corresponding Parts Of Congruent Triangles Are Congruent

Chapter 4 Review

1. $\angle 2, \angle 3, \angle 5$ **3.** $\overline{AB}$ and $\overline{AC}$ **5.** 38° **7.** $x = 60; y = 60$ **9.** 30 **11.** $\overline{ML}$ **13.** $\overline{ST}$ **15.** 80° **17.** 5 **19.** 100° **21.** DEF; SAS **23.** DEF; AAS **25.** not enough information **27.** SAS

29.

Statements	Reasons
1. $\angle VTY \cong \angle WYX$, $\angle VYT \cong \angle WXY$, $\overline{VY} \cong \overline{WX}$	**1.** Given
2. $\Delta VTY \cong \Delta WYX$	**2.** AAS Theorem
3. $\overline{TV} \cong \overline{YW}$	**3.** cpoctac

31.

Statements	Reasons
1. $\overline{BC} \cong \overline{DC}$, $\overline{EB} \cong \overline{ED}$	**1.** Given
2. $\overline{CE} \cong \overline{CE}$	**2.** Reflexive Property of $\cong$
3. $\Delta BCE \cong \Delta DCE$	**3.** SSS (Steps 1, 2)
4. $\angle B \cong \angle D$	**4.** cpoctac

33. $\Delta AEC \cong \Delta ABD$ by SAS, ASA, or AAS **35.** $x = 4; y = 65$ **37.** $x = 65; y = 90$

39.

Statements	Reasons
1. $\overline{LN} \perp \overline{KM}$	**1.** Given
2. $\angle LNK$ and $\angle LNM$ are right angles.	**2.** Definition of $\perp$
3. ΔKLN and ΔMLN are right triangles.	**3.** Definition of right triangles
4. $\overline{KL} \cong \overline{ML}$	**4.** Given
5. $\overline{LN} \cong \overline{LN}$	**5.** Reflexive Property of $\cong$
6. $\Delta KLN \cong \Delta MLN$	**6.** H-L Theorem (Steps 3, 4, 5)

41. 3 **43.** Sides: $\overline{HI} \cong \overline{PQ}, \overline{IJ} \cong \overline{QR}, \overline{JK} \cong \overline{RS}, \overline{KH} \cong \overline{SP}$; angles: $\angle H \cong \angle P, \angle I \cong \angle Q, \angle J \cong \angle R, \angle K \cong \angle S$ **45.** Since $\Delta QWE \cong \Delta DVK$ by AAS, $\angle Q \cong \angle D$ because cpoctac. **47.** $\Delta ABC \cong \Delta XYZ$ by H-L

Chapter 4 Test

1. obtuse; isosceles **2.** acute; scalene **3.** equiangular; equilateral **4.** right; isosceles **5.** $m\angle 1 = 132°$ **6.** $m\angle 1 = 28°$ **7.** $m\angle 1 = 83°; m\angle 2 = 97°$ **8.** $m\angle 2 = 48°$ **9.** $m\angle 1 = m\angle 2 = 60°$ **10.** $m\angle 1 = m\angle 2 = 45°$ **11.** $m\angle 1 = 74°; m\angle 2 = 32°$ **12.** $m\angle 1 = 73°; m\angle 2 = 107°; m\angle 3 = 25°$ **13.** 5 cm **14.** 10 cm **15.** 90° **16.** 90° **17.** *P* **18.** *PMN* **19.** $\Delta AYP \cong \Delta PLA; \angle YAP \cong \angle LPA, \angle Y \cong \angle L, \angle APY \cong \angle PAL, \overline{PA} \cong \overline{PA}, \overline{AY} \cong \overline{PL}, \overline{YP} \cong \overline{LA}$ **20.** SAS Postulate **21.** H-L Theorem **22.** not enough information **23.** SSS Postulate **24.** ASA Postulate (or AAS Theorem) **25.** AAS Theorem (or ASA Postulate) **26.** $x = 36$ **27.** $\Delta CFE \cong \Delta DEF$ by SSS **28.** $\Delta QTS \cong \Delta RTA$ by SAS

29.

Statements	Reasons
1. $\overline{AT} \parallel \overline{GS}$	**1.** Given
2. $\angle ATG \cong \angle SGT$	**2.** Alternate Interior Angles Theorem
3. $\overline{AT} \cong \overline{GS}$	**3.** Given
4. $\overline{GT} \cong \overline{TG}$	**4.** Reflexive Property of $\cong$
5. $\Delta GAT \cong \Delta TSG$	**5.** SAS Postulate (Steps 2, 3, 4)

30.

Statements	Reasons
1. $\overline{LN}$ bisects $\angle OLM$.	**1.** Given
2. $\angle OLN \cong \angle MLN$	**2.** Definition of bisector
3. $\overline{LN}$ bisects $\angle ONM$.	**3.** Given
4. $\angle ONL \cong \angle MNL$	**4.** Definition of bisector
5. $\overline{LN} \cong \overline{LN}$	**5.** Reflexive Property of $\cong$
6. $\Delta LNO \cong \Delta LNM$	**6.** ASA Postulate (Steps 2, 4, 5)
7. $\overline{ON} \cong \overline{MN}$	**7.** cpoctac

Chapter 4 Standardized Test

1. b **2.** c **3.** a **4.** a **5.** c **6.** a **7.** b **8.** d **9.** a **10.** b **11.** a **12.** c **13.** a **14.** d **15.** b **16.** c **17.** b **18.** c **19.** a **20.** a **21.** d

CHAPTER 5 SPECIAL PROPERTIES OF TRIANGLES

Section 5.1

Practice Exercises

1. 22 units **2. b.** (7, 6) **c.** slope $\overline{DE} = 4$; slope of perpendicular $= -\frac{1}{4}$ **d.**

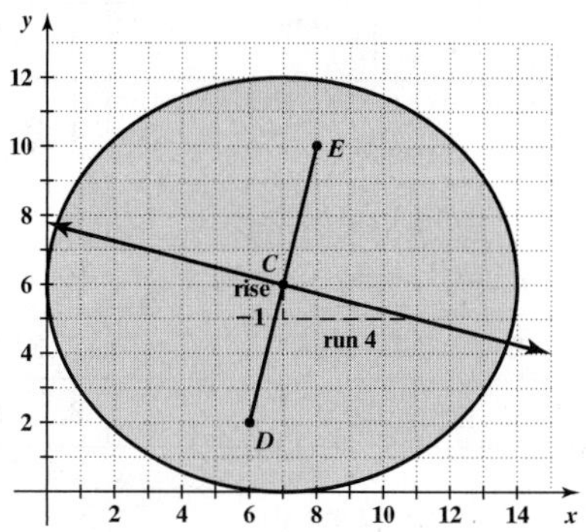

e. (1, 7), (1, 8), (13, 4), or (13, 5) **3.** 21 units

Vocabulary & Readiness Check 5.1

1. midpoint **3.** perpendicular bisector **5.** perpendicular **7.** 15 **9.** angle bisector **11.** 12

Exercise Set 5.1

1. 18 **3.** 9 **5.** 4 **7.** 8 **9.** 20 **11.** 85 **13.** 28° **15.** 134° **17. a.** angle bisector **b.** $\overrightarrow{MR}$; $\angle QMN$ **c.** $\overrightarrow{MR}$; $\overrightarrow{MQ}$; $\overrightarrow{MN}$ **19. a.** perpendicular bisector **b.** $\overleftrightarrow{CD}$; $\overline{AB}$ **c.** $\overleftrightarrow{CD}$; *A*; *B* **21.** No; *A* is not equidistant from the sides of $\angle TXR$. **23.** Yes; the markings show that $\angle TXA \cong \angle RXA$, so $\overrightarrow{XA}$ is the bisector of $\angle TXR$. **25.** $\overline{MB}$ is the $\perp$ bisector of $\overline{JK}$. **27.** 9 **29.** 27 units **31.** 9 **33.** $x = 12; JK = 17$ units; $JM = 17$ units **35.** 5 **37.** 10 units **39.** 6th Ave. **41.** Herald School **43. a.** Find the midpoint of the line segment between Park *A* and Park *B*. Construct the $\perp$ bisector of the line segment between Park *A* and *B*. Find the point on the $\perp$ bisector that is within the city limits and as far away as possible from both parks. **b.** Park *A*: (3, 1); Park *B*: (13, 3) **c.** *D* (8, 2) **d.** slope of $\overline{AB} = \frac{1}{5}$; slope of $\perp$ line: −5 **e.**

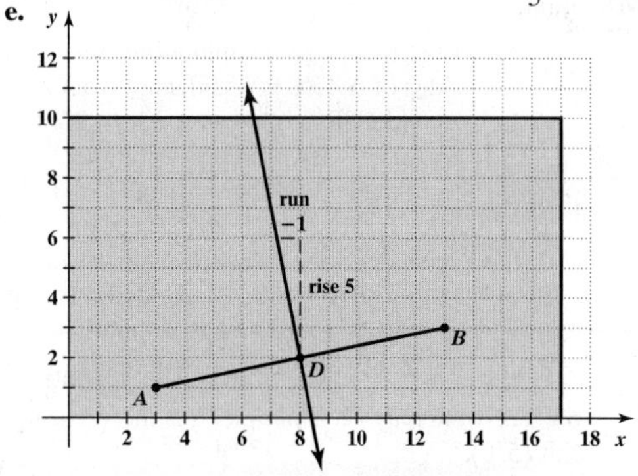

f. (6, 9) or (7, 9) **45.** Draw $\overline{HS}$ and find its midpoint *M*. Through *M*, construct the line $\perp$ to $\overline{HS}$. Any point on this line will be equidistant from *H* and *S*. **47. a.**

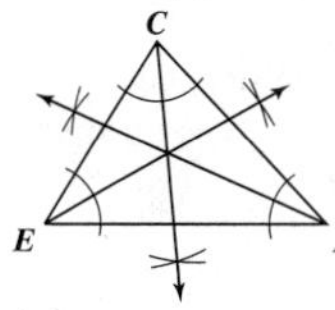

b. Answers may vary. Sample: They meet at a single point. **c.** Answers may vary **49.** A point is on the ⊥ bisector of a segment if and only if it is equidistant from the endpoints of the segment.

51.

Statements	Reasons
1. $\overleftrightarrow{PM}$ bisects $\overline{AB}$	**1.** Given
2. $\overline{AM} \cong \overline{BM}$	**2.** Definition of bisector
3. $\overleftrightarrow{PM} \perp \overline{AB}$	**3.** Given
4. $\angle PMA$ and $\angle PMB$ are right angles	**4.** Definition of ⊥
5. $\angle PMA \cong \angle PMB$	**5.** All right angles are ≅.
6. $\overline{PM} \cong \overline{PM}$	**6.** Reflexive Property
7. $\Delta PMA \cong \Delta PMB$	**7.** SAS (Steps 2, 5, 6)
8. $\overline{AP} \cong \overline{BP}$	**8.** cpoctac

53.

Statements	Reasons
1. $\overrightarrow{QS}$ bisects $\angle PQR$	**1.** Given
2. $\angle PQS \cong \angle RQS$	**2.** Definition of angle bisector
3. $\overline{SP} \perp \overrightarrow{QP}$, $\overline{SR} \perp \overrightarrow{QR}$	**3.** Given
4. $\angle QPS$ and $\angle QRS$ are right angles	**4.** Definition of ⊥
5. $\angle QPS \cong \angle QRS$	**5.** All right angles are ≅.
6. $\overline{QS} \cong \overline{QS}$	**6.** Reflexive Property
7. $\Delta QSP \cong \Delta QSR$	**7.** AAS (Steps 2, 5, 6)
8. $\overline{SP} \cong \overline{SR}$	**8.** cpoctac

55. a. l: $y = -\dfrac{3}{4}x + \dfrac{25}{2}$; m: $x = 10$ **b.** $C(10, 5)$
c. $CA = \sqrt{(10-6)^2 + (5-8)^2} = 5$;
$CB = \sqrt{(10-10)^2 + (5-0)^2} = 5$ **d.** C is equidistant from the sides of $\angle AOB$, so C is on the bisector of $\angle AOB$ by the Converse of the Angle Bisector Theorem.

57.

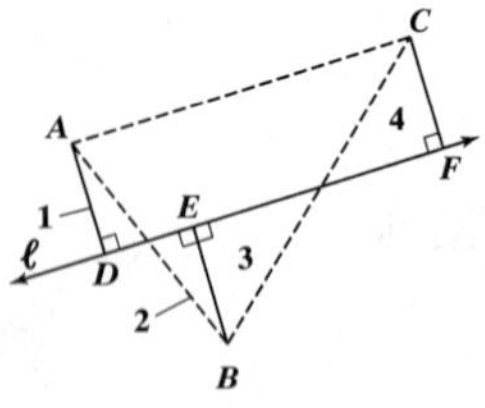

Line ℓ through the midpoints of two sides of ΔABC is equidistant from A, B, and C. This is because $\Delta 1 \cong \Delta 2$ and $\Delta 3 \cong \Delta 4$ by AAS (using the right angles and the ≅ vertical angles) $\overline{AD} \cong \overline{BE}$ and $\overline{BE} \cong \overline{CF}$ because cpoctac. By the Transitive Property of ≅, $\overline{AD} \cong \overline{BE} \cong \overline{CF}$. By the definition of ≅, $AD = BE = CF$, so points A, B, and C are equidistant from line ℓ.

59. 60° **61.** −3 **63.** Answers may vary. Sample: It is the vertical line that contains the point $(5, 0)$.

Section 5.2

Practice Exercises

1. a. QB; QC **b.** 15 **2.** $(6, 5)$
3.

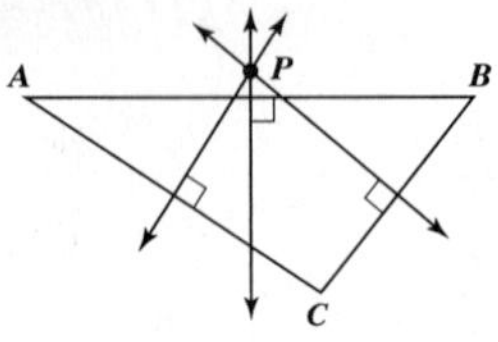

4. 61 units

Vocabulary & Readiness Check 5.2

1. concurrent **3.** incenter **5.** circumscribed about **7.** C **9.** B

Exercise Set 5.2

	Circumcenter	*Incenter*
1. Always lies inside the triangle		✓
3. A point of concurrency of a triangle	✓	✓
5. The point of concurrency of the perpendicular bisectors	✓	
7. Equidistant from the sides of a triangle		✓
9. Center of circle that passes through the vertices of a triangle	✓	

11. a. circumcenter **b.** 26 **13. a.** incenter **b.** 5 **15.** $(-2, -3)$ **17.** $(0, 0)$ **19.** $(1.5, 1)$ **21.** $(-3, 1.5)$ **23.** C **25.** 2 **27.** 4 **29.** Answers may vary. Sample: The distance from P to each side of the triangle is the same, but since $\overline{PQ}$ is not ⊥ to $\overline{AB}$ and $\overline{PS}$ is not ⊥ to $\overline{AC}$, $PQ \neq PS$. **31.** The circumcenter of the Δ formed by the lifeguard chair, snack bar, and volleyball court is equidistant from the vertices of the Δ. The recycling barrel should be placed at the intersection of two of the triangle's ⊥ bisectors.

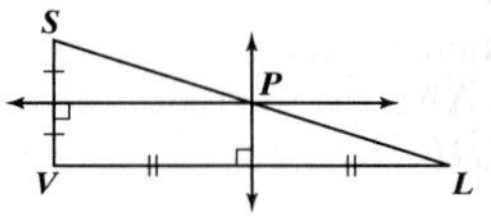

33. Since $RS = ST$, then ΔRST is an isosceles Δ with base $\overline{RT}$. **35.** Angle bisectors **37.** equidistant **39.** Transitive Property

41.

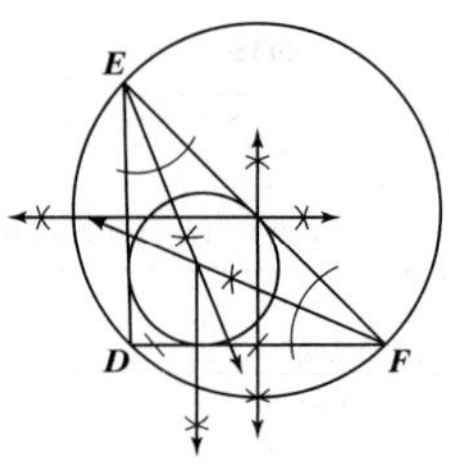

Inscribed circle: Construct two angle bisectors. Their intersection is the center of the circle. Construct a ⊥ line to any side from the center. The distance to the side is the radius. Construct a circle with this center and radius.

Circumscribed circle: Construct two ⊥ bisectors. Their intersection is the center of the circle. Draw a circle from that center through the vertices.

43. False; the circumcenter is equidistant from the vertices, and the incenter and circumcenter are different (unless the triangle is equilateral). **45.** False; if the points are collinear, then the ⊥ bisectors of the segments determined by the points are parallel. Since the ⊥ bisectors are parallel, they will not intersect, so there is no point that is equidistant from all three points. **47.** 30° **49.** $m\angle EFD = 80°$; $m\angle EDF = 40°$ **51.** 20° **53.** 120° **55.** Answers may vary. Sample: The two midsegments that extend from the midpoint of the hypotenuse are parallel to the two legs by the Δ Midsegment Theorem. Since the two legs form a right angle, the midsegments form right angles with the legs (Corresponding Angles Postulate). Hence, they are the ⊥ bisectors of the legs (Definition of ⊥ Bisector). Their meeting point—the midpoint of the hypotenuse—is the circumcenter, by definition. **57.** Never; if you have three parallel lines, l, m, and n, with m between l and n, then a point as equidistant from l as m would be (midway) between them. A point equidistant from m and n would be (midway) between those two lines. A point equidistant from all three lines would therefore have to be on both sides of m, which is impossible. **59.** 4 **61.** $(3, 8)$

Section 5.3

Practice Exercises

1. 18 units **2. a.** A median; it connects a vertex of ΔABC and the midpoint of the opposite side. **b.** Neither; E is a midpoint of ΔABC, but G is not a vertex of ΔABC. **c.** An altitude; it extends from a vertex of ΔABC and is ⊥ to the opposite side. **3.** $(1, 2)$

Vocabulary & Readiness Check 5.3

1. altitude **3.** centroid **5.** orthocenter

Exercise Set 5.3

1. $TY = 18$; $TW = 27$ **3.** $VY = 22$; $YX = 11$ **5.** Median; it connects a vertex of ΔABC and the midpoint of the opposite side. **7.** Altitude; it extends from a vertex of ΔABC and is $\perp$ to the opposite side. **9.** median **11.** 6 **13.** altitude **15.** 16 **17.** 32 **19.** 10 **21.** 80 **23.** H **25.** J **27.** $\overline{AE}$ **29.** $\overline{DE}$ **31.** $(4, 0)$ **33.** $(6, 4)$ **35. a.** $(7, 3)$ **b.** $AM = 6$ units **c.** $(5, 3)$ **37.** The segment has one endpoint at a vertex of the triangle and is $\perp$ to the opposite side. **39.** 35° **41.** 125° **43.** 35°

45.

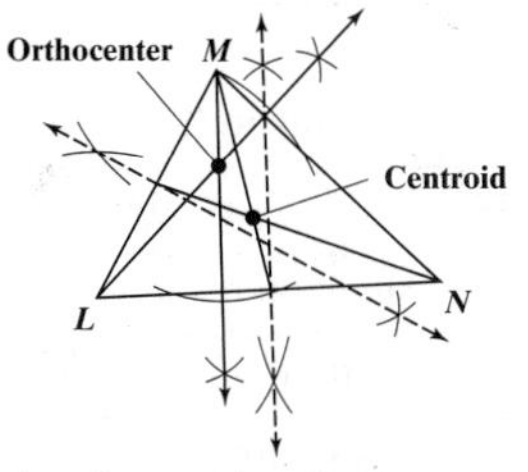

47. c, $8x^2 + 4y$ **49.** The folds should show the $\perp$ bisectors of the sides to identify the midpoint of each side, and also show the fold through each vertex and the midpoint of the opposite side.

51. An obtuse triangle

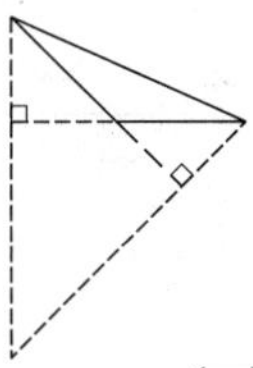

53. $L(1, 3)$; $M(5, 3)$; $N(4, 0)$ **55.** $\left(\frac{10}{3}, 2\right)$ **57.** 1:2 **59.** $\overline{HJ}$ does not contain a vertex of ΔABC, so it is not an altitude of ΔABC.

61.

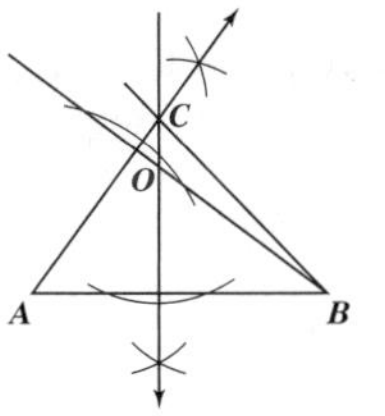

Draw $\overleftrightarrow{AB}$. Construct the $\perp$ to $\overleftrightarrow{AB}$ through O. Draw $\overleftrightarrow{BO}$. Construct the $\perp$ to $\overleftrightarrow{BO}$ through A. The two perpendiculars intersect at C. Draw $\overline{BC}$.

63. A is the intersection of the altitudes, so it is the orthocenter; B is the intersection of the angle bisectors, so it is the incenter; C is the intersection of the medians, so it is the centroid; D is the intersection of the $\perp$ bisectors of the sides, so it is the circumcenter. **65.** incenter **67.** Neither; $\overline{XY}$ connects vertex X and the midpoint of the opposite side, so $\overline{XY}$ is a median. **69.** You are 16 years old. **71.** $m\angle B > 90°$

Section 5.4

Practice Exercises

1. a. $\overline{NM} \parallel \overline{UV}$; $\overline{MQ} \parallel \overline{WU}$; $\overline{NQ} \parallel \overline{VW}$ **b.** $UV = \frac{1}{2}(NM)$, $WU = \frac{1}{2}(MQ)$, and $VW = \frac{1}{2}(NQ)$
2. $DC = 16$; $AC = 32$; $DF = 8.5$; $DE = 18$
3. $\overline{NM} \parallel \overline{UV}$; $\overline{MQ} \parallel \overline{DU}$; $\overline{QN} \parallel \overline{VD}$ **a.** 25° **b.** 65°
4. a. $D(0, 5)$; $F(3, 2)$ **b.** slope of $\overline{DF} = -1$; slope of $\overline{CB} = -1$ **c.** $CB = 6\sqrt{2}$; $DF = 3\sqrt{2} = \frac{1}{2}CB$ **5.** 1320 ft

Vocabulary & Readiness Check 5.4

1. c, $\overline{OM}$ **3.** $\overline{NO}$ **5.** C **7.** A

Exercise Set 5.4

1. $\overline{AB} \parallel \overline{EF}$; $\overline{BC} \parallel \overline{DE}$; $\overline{CA} \parallel \overline{FD}$; EF is $\frac{1}{2}(AB)$; DE is $\frac{1}{2}(BC)$; FD is $\frac{1}{2}(CA)$ **3.** $\overline{FH} \parallel \overline{LJ}$; $\overline{HK} \parallel \overline{GL}$; $\overline{KF} \parallel \overline{JG}$; LJ is $\frac{1}{2}(FH)$; GL is $\frac{1}{2}(HK)$; JG is $\frac{1}{2}(KF)$ **5.** $\overline{FE}$ **7.** $\overline{CA}$ (or $\overline{AF}$ or $\overline{FC}$) **9.** 40 **11.** 160 **13.** c **15.** 75° **17.** 45° **19.** 4.5 **21.** 12.5 **23.** 4 **25.** 60 **27.** 14 **29.** 6 **31. a.** $H(2, 0)$; $J(4, 2)$ **b.** Slope of $\overline{HJ} = 1$; slope of $\overline{EF} = 1$ **c.** $EF = 4\sqrt{2}$; $HJ = 2\sqrt{2} = \frac{1}{2}(EF)$ **33. a.** 234 ft **b.** 468 ft **35.** c **37.** 18.5 units **39.** 52 units **41.** 24 units **43.** 30 cm **45.** The student is assuming that L is the midpoint of $\overline{OT}$, which is not given. **47.** $G(4, 4)$; $H(0, 2)$; $J(8, 0)$ **49.** $\Delta FBD \cong \Delta FCE$; $\Delta BAE \cong \Delta CAD$; $\Delta DAF \cong \Delta EAF$; $\Delta ABF \cong \Delta ACF$ **51.** 6 **53.**

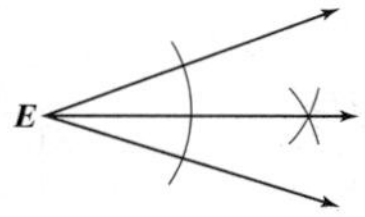

Section 5.5

Practice Exercises

1. a. Assume temporarily that ΔBOX is acute. **b.** Assume temporarily that no pairs of shoes you bought cost more than \$25. **2.** II and III **3.** $\angle 5$ is an exterior angle of ΔACD, so by the Corollary to the Δ Exterior Angle Theorem, $m\angle 5 > m\angle C$. **4.** Hollingsworth Road and MLK Boulevard **5.** $\overline{OX}, \overline{OS}, \overline{SX}$ **6. a.** No; $2 + 6 = 8 < 9$ **b.** Yes; the sum of the lengths of any two sides is greater than the length of the third side. **7.** Greater than 3 in. and less than 11 in.

Vocabulary & Readiness Check 5.5

1. proof by contradiction **3.** Triangle Inequality Theorem for Sum of Lengths of Sides

Exercise Set 5.5

1. $m\angle 1 > m\angle 3$ by the Corollary to the Δ Exterior Angle Theorem. $m\angle 3 = m\angle 2$ because they are vertical angles. Thus, $m\angle 1 > m\angle 2$ by substitution. **3.** $m\angle 1 > m\angle 5$ by the Corollary to the Δ Exterior Angle Theorem. $m\angle 5 = m\angle 2$ because parallel lines form alternate interior angles that are $\cong$. So, $m\angle 1 > m\angle 2$ by substitution. **5.** $\angle M, \angle L, \angle K$ **7.** $\angle D, \angle C, \angle E$ **9.** $\angle A, \angle B, \angle C$ **11.** $\angle Z, \angle X, \angle Y$ **13.** $\overline{MN}, \overline{ON}, \overline{MO}$ **15.** $\overline{TU}, \overline{UV}, \overline{TV}$ **17.** $\overline{GF}, \overline{GH}, \overline{FH}$ **19.** $\overline{EF}, \overline{DE}, \overline{DF}$ **21.** No; $2 + 3 \not> 6$ **23.** No; $8 + 10 \not> 19$ **25.** Yes; $2 + 9 > 10$, $9 + 10 > 2$, and $2 + 10 > 9$ **27.** 4; 20 **29.** 0; 18 **31.** 13 **33.** 1.5 **35.** $\overline{AB}$ **37.** right triangle **39.** The longest side is the side across the largest angle, the 90° angle. **41.** The dashed red line and the courtyard walkway determine three sides of a Δ, so by the Δ Inequality Theorem, the path that follows the dashed red line is longer than the courtyard walkway. **43.** The sign, Topeka, and Wichita are either collinear or they determine the vertices of a triangle. If D is the distance between Topeka and Wichita in miles, then $20 \le D \le 200$. **45.** $\overline{RS}$ **47.** $\overline{XY}$ **49.** Assume temporarily that it is not raining outside. **51.** Assume temporarily that ΔPEN is not isosceles. **53.** Assume temporarily that $\overline{XY}$ and $\overline{AB}$ are not $\cong$. **55.** I and II **57.** II and III **59.** Assume temporarily that lines l and p are parallel. **61.** Assume temporarily that the number is divisible by 2. **63.** Assume temporarily that at least one angle in quadrilateral $ABCD$ is not a right angle. **65.** The negation of "$\angle A$ is obtuse" is "$\angle A$ is not obtuse." **67. a.** $m\angle OTY$ **b.** $m\angle 3$ **c.** Isosceles Δ Theorem **d.** Angle Addition Postulate **e.** Comparison Property of Inequality **f.** Substitution **g.** Corollary to Δ Exterior Angle Theorem **h.** Transitive Property of Inequality **69. a.** 20 or more **b.** the total membership is fewer than 20 **c.** the Yoga Club has fewer than 10 members **71.** 0.5; the lengths of two sides of the Δ are 6 cm and 9 cm, so the length of the third side of the Δ must be greater than 3 cm and less than 15 cm, by the Δ Inequality Theorem. Since 2 of the 4 straws satisfy that condition, the probability that she can form a Δ is $\frac{2}{4}$ or $\frac{1}{2}$. **73.** (2, 4), (2, 5), (2, 6), (3, 3), (3, 4), (3, 5), (3, 6), (3, 7), (4, 3), (4, 4), (4, 5), (4, 6), (4, 7), (4, 8) **75.** We want to prove that neither base angle is a right angle. Assume temporarily that at least one base angle is a right angle. Then both base angles must be right angles, by the Isosceles Δ Theorem. But this contradicts the fact that a Δ is formed, because in a plane, two lines $\perp$ to the same line are parallel. Therefore, the temporary assumption that at least one base angle is a right angle is false, and we can conclude that neither base angle is a right angle. **77.** 25° **79.** $\angle P$

Section 5.6

Practice Exercises

1.

Statements	Reasons
1. $\overline{BC} \cong \overline{EF}$	**1.** Given
2. $\angle PBC \cong \angle DEF$	**2.** By construction
3. $\overline{BP} \cong \overline{ED}$	**3.** By construction
4. $\triangle PBC \cong \triangle DEF$	**4.** SAS Postulate (Steps 1, 2, 3)
5. $\overline{PC} \cong \overline{DF}$	**5.** cpoctac
6. Point P lies on AC.	**6.** Given (for this case)
7. $AC = AP + PC$	**7.** Segment Addition Postulate
8. $AC = AP + DF$	**8.** Substitution
9. $AC > DF$	**9.** Comparison Property of Inequality

2. $LN > OQ$ **3.** a **4.** The 40° opening; the lengths of the blades do not change as the scissors open. The included angle between the blades of the 40° opening is greater than the included angle of the 35° opening, so by the Hinge Theorem, the distance between the blades is greater for the 40° opening. **5.** $-6 < x < 24$

Exercise Set 5.6

1. $>$ **3.** $=$ **5.** $<$ **7.** $<$ **9.** $=$ **11.** $<$
13. $AB < AD$ **15.** $LM < KL$ **17.** B **19.** $6 < x < 38$
21. $3.5 < y < 17.5$ **23.** $PT < QR$; $QP = TR$, $QT = TQ$, and $m\angle PQT < m\angle RTQ$, so $PT < QR$ by the Hinge Theorem.
25. The two labeled angles are formed by $\cong$ corresponding sides of the two triangles, so the side opposite the 94° angle should be longer than the side opposite the 91° angle, by the Hinge Theorem. Thus, the side labeled "13" must be longer than the side labeled "14." Either switch the labels on the angles or switch the labels on the sides.
27. a. Converse of Isosceles $\triangle$ Theorem **b.** Given **c.** Definition of midpoint **d.** $BC = CD$ **e.** Given **f.** Hinge Theorem

29.

Statements	Reasons
1. O is the midpoint of $\overline{LN}$.	**1.** Given
2. $LO = ON$	**2.** Definition of midpoint
3. $\angle MOL$ and $\angle MON$ are supplementary angles.	**3.** Angle Addition Postulate
4. $m\angle MON = 80°$	**4.** Given
5. $m\angle MOL = 100°$	**5.** Supplementary angles
6. $\overline{MO} \cong \overline{MO}$	**6.** Reflexive Property
7. $m\angle MOL > m\angle MON$	**7.** Steps 4, 5
8. $LM > MN$	**8.** Hinge Theorem

31. The right isosceles triangle **33.** d **35.** $AO = 7$, $OB = OC = \sqrt{5}$, $AB = \sqrt{82}$, and $AC = \sqrt{68}$. Since $AB > AC$, then $m\angle AOB > m\angle AOC$ by the Converse of the Hinge Theorem.

37.

Statements	Reasons
1. $\overline{BC} \cong \overline{EF}$	**1.** Given
2 $\angle PBC \cong \angle DEF$, $\overline{BP} \cong \overline{ED}$	**2.** By construction
3. $\triangle PBC \cong \triangle DEF$	**3.** SAS Postulate (Steps 1, 2)
4. Locate R on $\overline{AC}$ so that $\overrightarrow{BR}$ bisects $\angle ABP$; then $\angle ABR \cong \angle RBP$	**4.** By construction; definition of angle bisector
5. $\overline{ED} \cong \overline{BA}$	**5.** Given
6. $\overline{BA} \cong \overline{BP}$	**6.** Transitive Property
7. $\overline{BR} \cong \overline{BR}$	**7.** Reflexive Property
8. $\triangle ABR \cong \triangle PBR$	**8.** SAS Postulate (Steps 4, 6, 7)
9. $\overline{AR} \cong \overline{PR}$	**9.** cpoctac
10. $AC = AR + RC$	**10.** Segment Addition Postulate
11. $AR = PR$	**11.** Definition of $\cong$
12. $AC = PR + RC$	**12.** Substitution
13. In $\triangle RPC$, $PC < PR + RC$	**13.** $\triangle$ Inequality Theorem
14. $\overline{PC} \cong \overline{DF}$	**14.** cpoctac (Step 3)
15. $PC = DF$	**15.** Definition of $\cong$
16. $DF < PR + RC$	**16.** Substitution
17. $DF < AR + RC$	**17.** Substitution
18. $DF < AC$	**18.** Substitution

39. $\angle T, \angle F, \angle R$ **41.** $4 \text{ cm} < x < 34 \text{ cm}$ **43.** $\overline{GH}$
45. $-\frac{8}{3}$ **47.** $\frac{3}{5}$

Chapter 5 Vocabulary Check

1. concurrent **2.** point of concurrency **3.** perpendicular **4.** inscribed in **5.** equidistant **6.** midsegment **7.** altitude **8.** proof by contradiction **9.** deductive reasoning **10.** median **11.** orthocenter, circumcenter **12.** centroid **13.** orthocenter **14.** centroid **15.** circumcenter **16.** circumscribed about **17.** incenter

Chapter 5 Review

1. 4 **3.** 40° **5.** 6 **7.** 33 **9.** Z **11.** (0, 0) **13.** (4, 4) **15.** 45° **17.** 25° **19.** 4 **21.** $\overline{AB}$ is an altitude; it is a segment from a vertex that is $\perp$ to the opposite side. **23.** (0, −1) **25.** 4 **27.** 16 **29.** 15 **31.** We want to prove that a triangle can have at most one obtuse angle. Assume temporarily that there is a triangle with two obtuse angles. Then the sum of the measures of the two angles is greater than 180°, which contradicts the $\triangle$ Angle-Sum Theorem. Therefore, the temporary assumption is false and a triangle can have at most one obtuse angle. **33.** $\overline{RS}, \overline{ST}, \overline{RT}$ **35.** Yes; $10 + 12 > 20$, $10 + 20 > 12$, and $12 + 20 > 10$ **37.** $\overline{BC}$ **39.** $>$ **41.** $QZ = 8$; $QM = 12$ **43.** II and III **45.** We want to prove that if three integers have a sum greater than 9, one of the integers is greater than 3. Assume temporarily that each of the three integers is less than or equal to 3. Then the sum of the three integers must be less than or equal to $3 \cdot 3$, which is 9. This contradicts the given statement that the sum of the three integers is greater than 9. Therefore, the temporary assumption is false, and you can conclude that one of the integers is greater than 3. **47.** JB **49.** D

Chapter 5 Test

1. D **2.** A **3.** B **4.** C **5.** A **6.** A, D **7.** D **8.** A **9.** B **10.** (−2, 1) **11.** (−1, −1) **12.** I and II **13.** $\angle A, \angle C, \angle B$ **14.** 30° **15.** $\overline{KV}, \overline{VM}, \overline{KM}$ **16.** 8 **17.** 104 **18.** 228 **19.** $\angle Z$; $\overline{AB}$ is a midsegment of $\triangle XYZ$, so by the $\triangle$ Midsegment Theorem, $\overline{AB}$ is parallel to $\overline{YZ}$. Then $\angle Z \cong \angle XBA$ because corresponding angles of parallel lines are $\cong$. **20.** $\angle CDB$ is a right angle; it is supplementary to $\angle ADB$. **21.** They are $\cong$ by H-L. **22.** They are $\cong$; they are corresponding parts in congruent triangles. **23.** $\overline{DC}$; since $m\angle ABC < 180°$, $m\angle ABC = m\angle ABD + m\angle DBC$, and $m\angle DBC = 90°$, we have $m\angle ABD + 90° < 180°$ by substitution. So $m\angle ABD < 90°$. Also, $\overline{AB} \cong \overline{CB}$ and $\overline{BD} \cong \overline{BD}$, so the Hinge Theorem applies. Hence, $AD < DC$. **24.** 7 **25.** 12 **26.** D

and E are midpoints of $\overline{AB}$ and $\overline{AC}$, so $\overline{DE}$ is parallel to $\overline{BC}$ and $DE = \frac{1}{2}BC$ by the Δ Midsegment Theorem. (Also $\angle B \cong \angle ADE$ and $\angle C \cong \angle AED$.) **27.** A is on $\overrightarrow{CK}$; points K and A are equidistant from the sides of $\angle SCD$, so points K and A are on the bisector of $\angle SCD$ by the Converse of the Δ Angle Bisector Theorem. Since $\overrightarrow{CK}$ is the bisector of $\angle SCD$, A must be on $\overrightarrow{CK}$. **28.** To prove: that an obtuse angle in an isosceles triangle must be the vertex angle. Assume temporarily that the obtuse angle is not the vertex angle. Then the obtuse angle must be a base angle. In an isosceles Δ, the base angles are $\cong$, so each base angle would have measure greater than 90° and their sum would be greater than 180°. This contradicts the Δ Angle-Sum Theorem. Therefore, the temporary assumption is false—if an isosceles Δ is obtuse, then the obtuse angle is the vertex angle.

29.

Statements	Reasons
1. $\overleftrightarrow{PQ}$ is the $\perp$ bisector of $\overline{AB}$; $\overleftrightarrow{QT}$ is the $\perp$ bisector of $\overline{AC}$	1. Given
2. $\overline{QB} \cong \overline{QA}$; $\overline{QA} \cong \overline{QC}$	2. $\perp$ Bisector Theorem
3. $\overline{QB} \cong \overline{QC}$	3. Transitive Property
4. $QB = QC$	4. Definition of $\cong$

Chapter 5 Standardized Test

1. b **2.** a **3.** d **4.** a **5.** c **6.** a **7.** c **8.** b **9.** a **10.** d **11.** c **12.** d **13.** b **14.** b **15.** a **16.** b **17.** d **18.** c **19.** b **20.** d **21.** c **22.** b

CHAPTER 6 QUADRILATERALS

Section 6.1

Practice Exercises

1. Figures C and E are polygons. Figure A is not a polygon because there is a "side" that is a curve and not a line segment. Figure B is not a polygon because two sides intersect only one other side. Figure D is not a polygon because there are sides that intersect more than two other sides. **2. a.** convex hexagon **b.** concave quadrilateral **c.** convex heptagon **3. a.** The quadrilateral is equiangular, but not equilateral, so it is not a regular polygon. **b.** The triangle is equilateral and equiangular, so it is a regular polygon. **c.** The quadrilateral is neither equilateral nor equiangular, so it is not a regular polygon. **4.** $x = 45$, $m\angle H = 45°$, $m\angle E = 125°$

Vocabulary & Readiness Check 6.1

1. polygon **3.** quadrilateral **5.** convex **7.** diagonal **9.** equiangular

Exercise Set 6.1

1. quadrilateral **3.** Not a polygon because there is a "side" that is a curve. **5.** Not a polygon because there is a side that intersects more than two other sides. **7.** hexagon **9.** Not a polygon because there is a side that intersects more than two other sides. **11.** quadrilateral **13.** convex hexagon **15.** concave quadrilateral **17.** concave dodecagon **19.** convex nonagon **21. a.** yes **b.** quadrilateral **c.** yes **d.** yes **e.** yes **23. a.** yes **b.** triangle **c.** no **d.** no **e.** no **25. a.** no **27. a.** yes **b.** octagon **c.** yes **d.** yes **e.** yes **29. a.** yes **b.** quadrilateral **c.** yes **d.** no **e.** no **31.** A, B, C, D, E **33.** $\overline{AB}, \overline{BC}, \overline{CD}, \overline{DE}, \overline{EA}$ **35.** convex **37.** $\overline{CD}$ **39.** 118° **41.** 110° **43.** $y = 103, z = 70$ **45.** $x = 55, m\angle C = 55°, m\angle A = 135°$ **47.** $x = 45, m\angle R = 45°, m\angle P = 65°, m\angle S = m\angle T = 125°$ **49.** 45°, 45°, 90° **51.** Yes; answers may vary. Sample: The sum of the measures of the interior angles is $(n - 2) \cdot 180°$. **53.** $\overline{CD}$; the longer side is opposite the larger angle. **55.** Reflexive Property of $\cong$ **57.** ASA Postulate

Section 6.2

Practice Exercises

1.

Statements	Reasons
1. $ABCD$ is a parallelogram	1. Given
2. $\overline{AB} \parallel \overline{CD}$; $\overline{BC} \parallel \overline{DA}$	2. Definition of parallelogram
3. $\angle A$ and $\angle B$ are same-side interior angles of parallel lines $\overleftrightarrow{AD}$ and $\overleftrightarrow{BC}$; $\angle B$ and $\angle C$ are same-side interior angles of parallel lines $\overleftrightarrow{AB}$ and $\overleftrightarrow{CD}$; $\angle C$ and $\angle D$ are same-side interior angles of parallel lines $\overleftrightarrow{AD}$ and $\overleftrightarrow{BC}$	3. Definition of same-side interior angles
4. $\angle A$ and $\angle B$ are supplementary; $\angle B$ and $\angle C$ are supplementary; $\angle C$ and $\angle D$ are supplementary	4. Same-side interior angles of parallel lines are supplementary.
5. $\angle A \cong \angle C$; $\angle B \cong \angle D$	5. Supplements of the same angle are $\cong$.

2. 94° **3.** $x = 4, y = 5, PR = 16, SQ = 10$ **4.** 5

Vocabulary & Readiness Check 6.2

1. opposite angles **3.** consecutive angles

Exercise Set 6.2

1. a **3.** d **5.** c **7.** d **9.** 127 **11.** 100 **13.** 20 **15.** 17 **17.** 53° **19.** 5 units **21.** 8 units **23.** $m\angle 1 = 38°$, $m\angle 2 = 32°, m\angle 3 = 110°$ **25.** 18 **27.** 10 **29.** 33° **31.** 70° **33.** 88° **35.** 3 **37.** 9 **39.** 2.25 **41.** 4.5 **43.** $x = 6, y = 8$ **45.** $a = 22, AB = CD = 23.6, BC = AD = 18.5$ **47. a.** 2.5 ft **b.** 129° **c.** Answers may vary. Sample: As $m\angle E$ increases, $m\angle D$ decreases. $\angle E$ and $\angle D$ are supplementary. **49.** Answers may vary. Sample: The angle opposite the given angle is congruent to it. The other two angles and the given angle are consecutive angles, so they are supplements of the given angle. **51.** It is not given that $\overleftrightarrow{PQ}$, $\overleftrightarrow{RS}$, and $\overleftrightarrow{TV}$ are parallel.

53.

Statements	Reasons
1. $ABCD$ is a parallelogram	1. Given
2. $\overline{AB} \parallel \overline{CD}$; $\overline{BC} \parallel \overline{DA}$	2. Definition of parallelogram
3. $\angle A$ and $\angle B$ are same-side interior angles of parallel lines $\overleftrightarrow{AD}$ and $\overleftrightarrow{BC}$; $\angle A$ and $\angle D$ are same-side interior angles of parallel lines $\overleftrightarrow{AB}$ and $\overleftrightarrow{CD}$.	3. Definition of same-side interior angles
4. $\angle A$ and $\angle B$ are supplementary; $\angle A$ and $\angle D$ are supplementary	4. Same-side interior angles of parallel lines are supplementary.

55.

Statements	Reasons
1. $LENS$ and $NGTH$ are parallelograms	1. Given
2. $\overline{LS} \parallel \overline{EN}$ and $\overline{NH} \parallel \overline{GT}$	2. Definition of parallelogram
3. $\overline{LS} \parallel \overline{GT}$	3. If two lines are parallel to the same line, they are parallel to each other.

57.

Statements	Reasons
1. $RSTW$ and $XYTZ$ are parallelograms	1. Given
2. $\angle R \cong \angle T$; $\angle X \cong \angle T$	2. Opposite angles of a parallelogram are $\cong$.
3. $\angle R \cong \angle X$	3. Transitive Property of $\cong$

59. $AB = CD = 13$ cm, $BC = AD = 33$ cm

61.

Statements	Reasons
1. $\overline{AB} \parallel \overline{CD}, \overline{CD} \parallel \overline{EF}$	**1.** Given
2. $\overline{BG} \parallel \overline{AC}, \overline{DH} \parallel \overline{CE}$	**2.** Construction
3. $ABGC$ and $CDHE$ are parallelograms	**3.** Definition of parallelogram
4. $\overline{AC} \cong \overline{BG}, \overline{CE} \cong \overline{DH}$	**4.** Opposite sides of a parallelogram are $\cong$.
5. $\overline{AC} \cong \overline{CE}$	**5.** Given
6. $\overline{BG} \cong \overline{DH}$	**6.** Transitive Property of $\cong$
7. $\overline{BG} \parallel \overline{DH}$	**7.** If two lines are parallel to the same line, then they are parallel to each other.
8. $\angle 3 \cong \angle 6$ and $\angle GBD \cong \angle HDF$	**8.** If lines are parallel, then corresponding angles are $\cong$.
9. $\Delta GBD \cong \Delta HDF$	**9.** AAS Theorem
10. $\overline{BD} \cong \overline{DF}$	**10.** cpoctac

63. $\overline{AC} \perp \overline{DB}$ or $\angle ACD$ and $\angle ACB$ are right angles.

Section 6.3

Practice Exercises

1. $x = 10, y = 43$ **2. a.** No; $DEFG$ could be an isosceles trapezoid. For $DEFG$ to be a parallelogram, one pair of sides must be both parallel and congruent. **b.** Yes; $\overline{AN} \parallel \overline{LD}$ and $\overline{AL} \parallel \overline{ND}$ because the pairs of alternate interior angles are $\cong$. **3.** 6 ft; explanations may vary. Sample: The maximum height occurs when $\overline{QP}$ is vertical. **4.** *Method 1*: $MN = \sqrt{29}$, $QP = \sqrt{29}$; $MP = \sqrt{17}$, $NQ = \sqrt{17}$ *Method 2*: slope of $\overline{MN} = \frac{2}{5}$, slope of $\overline{QP} = \frac{2}{5}$; slope of $\overline{MP} = 4$, slope of $\overline{NQ} = 4$ *Method 3*: slope of $\overline{MN} =$ slope of $\overline{QP} = \frac{2}{5}$ and $MN = QP = \sqrt{29}$

Exercise Set 6.3

1. 112 **3.** $x = 63, y = 117$ **5.** Yes; both pairs of opposite angles are $\cong$. **7.** No; it could be an isosceles trapezoid since the $\cong$ sides are not the parallel sides. **9.** No; it is not given that the parallel sides are $\cong$. **11.** Yes; one pair of opposite sides is both congruent and parallel. **13.** Yes; both pairs of opposite sides are congruent. **15.** No; one diagonal is bisected, but it is not given that the other diagonal is bisected. **17.** 5 **19.** $\frac{5}{3}$ **21.** $x = 21, y = 39$ **23.** 13 **25.** $x = 3, y = 11$ **27.** $ST = VU = 4$, $TU = SV = \sqrt{5}$; slope of $\overline{ST} =$ slope of $\overline{VU} = 0$, slope of $\overline{TU} =$ slope of $\overline{SV} = -2$ **29.** $PA = TR = 2\sqrt{10}$, $TP = RA = 2\sqrt{5}$; slope of $\overline{PA} =$ slope of $\overline{TR} = \frac{1}{3}$; slope of $\overline{TP} =$ slope of $\overline{RA} = 2$ **31.** $BC = AD = \sqrt{17}$, $AB = DC = 2\sqrt{5}$; slope of $\overline{BC} =$ slope of $\overline{AD} = \frac{1}{4}$; slope of $\overline{AB} =$ slope of $\overline{DC} = 2$ **33. a.** Distributive Property **b.** Division Property of Equality **c.** $\overline{AD} \parallel \overline{BC}, \overline{AB} \parallel \overline{DC}$ **d.** If same-side interior angles are supplementary, then the lines are parallel. **e.** Definition of parallelogram

35.

Statements	Reasons
1. Draw $\overline{BD}$	**1.** Construction
2. $\overline{AB} \cong \overline{CD}$ and $\overline{BC} \cong \overline{DA}$	**2.** Given
3. $\overline{BD} \cong \overline{BD}$	**3.** Reflexive Property of $\cong$
4. $\Delta ABD \cong \Delta CDB$	**4.** SSS Postulate
5. $\angle ADB \cong \angle CBD$ and $\angle CDB \cong \angle ABD$	**5.** cpoctac
6. $\overline{AB} \parallel \overline{DC}$ and $\overline{BC} \parallel \overline{AD}$	**6.** Converse of Alternative Interior Angles Theorem
7. $ABCD$ is a parallelogram	**7.** Definition of parallelogram

37.

Statements	Reasons
1. $\Delta TRS \cong \Delta RTW$	**1.** Given
2. $\overline{SR} \cong \overline{WT}$ and $\overline{ST} \cong \overline{WR}$	**2.** cpoctac
3. $RSTW$ is a parallelogram	**3.** Converse of Opposite Sides Theorem

39. The connecting pieces $\overline{AD}$ and $\overline{BC}$ are $\cong$ and the distances AB and CD between where the two pieces attach are $\cong$. The side lengths of $ABCD$ do not change as the tackle box opens and closes. Since both pairs of opposite sides are $\cong$, $ABCD$ is always a parallelogram. By the definition of a parallelogram, $\overline{AB} \parallel \overline{CD}$, so the shelves are always parallel to each other. **41.** Because the Opposite Sides of a Parallelogram Theorem and its converse are both true **43.** The theorems are converses of each other. Use Theorem 6.3-4 if you need to show that a figure is a parallelogram. Use Theorem 6.2-4 if it is given that the figure is a parallelogram.

45.

Statements	Reasons
1. $\overline{AB} \cong \overline{CD}, \overline{AB} \cong \overline{BD}$	**1.** Construction
2. $ABCD$ is a parallelogram	**2.** Converse of Opposite Sides Theorem
3. M is the midpoint of $\overline{BC}$	**3.** Diagonals of a Parallelogram Theorem
4. $\overline{AM}$ is a median of ΔABC	**4.** Definition of median

47. $a = 8, h = 30, k = 120$ **49.** $c = 204, e = 13, f = 11$ **51.** 7.47 **53.** 7.47 **55.** 13.2 **57.** 56° **59.** 28°

Section 6.4

Practice Exercises

1. a. 9 **b.** $EF = FG = GH = HE = 47$ cm **2. a.** rhombus **b.** rectangle **c.** square

3.

Statements	Reasons
1. A and D are equidistant from B and C; B and D are equidistant from A and C	**1.** All sides of a rhombus are $\cong$.
2. A and D are on the $\perp$ bisector of $\overline{BC}$; B and C are on the $\perp$ bisector of $\overline{AD}$	**2.** Converse of the Perpendicular Bisector Theorem
3. $\overline{AD} \perp \overline{BC}$	**3.** Through two points, there is one unique line (perpendicular to a given line).

4. a. 43 **b.** Isosceles because diagonals of a rectangle are $\cong$ and bisect each other. **5.** $m\angle 1 = m\angle 2 = m\angle 3 = m\angle 4 = 38°$ **6.** 4

Vocabulary & Readiness Check 6.4

1. parallelogram **3.** rhombus

Exercise Set 6.4

1. $x = 5, y = 7$, all sides are 35 **3.** $x = 5, y = 4$, all sides are 3 **5.** rhombus **7.** rhombus **9.** Rectangle; it is a parallelogram with 4 right angles and does not have 4 $\cong$ sides. **11.** $m\angle 1 = m\angle 2 = m\angle 3 = m\angle 4 = 37°$ **13.** $m\angle 1 = 118°, m\angle 2 = m\angle 3 = 31°$ **15.** $m\angle 1 = m\angle 4 = 32°, m\angle 2 = 90°, m\angle 3 = 58°$ **17.** $m\angle 1 = 40°, m\angle 2 = 90°, m\angle 3 = 50°$ **19.** $m\angle 1 = 60°, m\angle 2 = 90°, m\angle 3 = 30°$ **21.** $x = 4, LN = MP = 4$ **23.** $x = 1; LN = MP = 4$ **25.** $x = \frac{5}{3}; LN = MP = \frac{29}{3}$ **27.** parallelogram, rectangle, square, rhombus **29.** parallelogram, rectangle, square, rhombus **31.** rectangle, square **33.** square **35.** parallelogram, rectangle, square, rhombus **37.** rhombus, square **39.** 30 **41.** $x = 5, y = 32, z = 7.5$ **43.** $AC = BD = 16$ **45.** $AC = BD = 1$ **47.** Rhombus; one diagonal bisects a pair of opposite angles. **49.** No; you know only that the diagonals bisect each other, which is true of all parallelograms. **51.** 2 **53.** 1 **55.** 11 **57.** 7

59.

Statements	Reasons
1. $ABCD$ is a rhombus	1. Given
2. $\overline{AB} \cong \overline{AD}, \overline{CB} \cong \overline{CD}$	2. Definition of rhombus
3. $\overline{AC} \cong \overline{AC}$	3. Reflexive Property of $\cong$
4. $\Delta ABC \cong \Delta ADC$	4. SSS Postulate
5. $\angle 3 \cong \angle 4, \angle 1 \cong \angle 2$	5. cpoctac
6. $\overline{AC}$ bisects $\angle BAD$ and $\angle BCD$	6. Definition of angle bisector

61.

Statements	Reasons
1. $ABCD$ is a rectangle	1. Given
2. $ABCD$ is a parallelogram	2. Definition of rectangle
3. $\overline{AB} \cong \overline{DC}$	3. Opposite sides of a parallelogram are $\cong$.
4. $\angle ABC$ and $\angle DCB$ are right angles	4. Definition of rectangle
5. $\angle ABC \cong \angle DCB$	5. All right angles are $\cong$.
6. $\overline{BC} \cong \overline{BC}$	6. Reflexive Property of $\cong$
7. $\Delta ABC \cong \Delta DCB$	7. SAS Postulate (Steps 3, 5, 6)
8. $\overline{AC} \cong \overline{DB}$	8. cpoctac

63. The first step should be $2x + 8 + 9x - 6 = 90$. **65.** Rectangle, square; answers may vary. Sample:

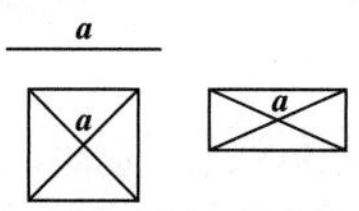

67. 2

69. Yes; both pairs of opposite sides of the quadrilateral are $\cong$. **71.** Yes; diagonals of the quadrilateral bisect each other. **73.** 16 **75.** $\overline{RQ}$ **77.** $\overline{ST}$ **79.** Answers may vary. Sample:

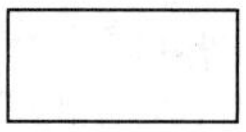

Section 6.5

Practice Exercises

1. $m\angle P = m\angle Q = 74°, m\angle S = 106°$ **2.** $x = 6, MN = 23$ **3.** $m\angle 1 = 90°, m\angle 2 = 54°, m\angle 3 = 36°$

Vocabulary & Readiness Check 6.5

1. kite **3.** leg **5.** base angles **7.** midsegment

Exercise Set 6.5

1. $m\angle 1 = 77°, m\angle 2 = 103°, m\angle 3 = 103°$ **3.** $m\angle 1 = 49°, m\angle 2 = 131°, m\angle 3 = 131°$ **5.** $m\angle 1 = 115°, m\angle 2 = 115°, m\angle 3 = 65°$ **7.** 10 **9.** 11 **11.** $m\angle 1 = 90°, m\angle 2 = 68°$ **13.** $m\angle 1 = 108°, m\angle 2 = 108°$ **15.** $m\angle 1 = 90°, m\angle 2 = 26°, m\angle 3 = 90°$ **17.** $m\angle 1 = 90°, m\angle 2 = 55°, m\angle 3 = 35°$ **19.** $m\angle 1 = 90°, m\angle 2 = 52°, m\angle 3 = 38°, m\angle 4 = 37°, m\angle 5 = 53°, m\angle 6 = 90°, m\angle 7 = 90°, m\angle 8 = 90°, m\angle 9 = 52°, m\angle 10 = 37°$ **21.** 12 **23.** $AD = 4, EF = 9, BC = 14$ **25.** 28 **27.** Isosceles trapezoid; $\overline{AB} \| \overline{DC}$ since the alternate interior angles are $\cong$, and $\overline{AD} \cong \overline{BC}$ since they are corresponding parts of $\cong$ triangles. **29.** Yes; the $\cong$ angles can be obtuse. **31.** Yes; if the two $\cong$ angles are both right angles, then they are supplementary. The other two angles will also be supplementary. **33.** Yes; if the two opposite $\cong$ angles each measure 45°, they are complementary. **35.** No; explanations may vary. Sample: Assume $\overline{KM}$ bisects both angles. Then $\angle MKL \cong \angle MKN \cong \angle KML \cong \angle KMN$. Both pairs of sides of $KLMN$ would be parallel, and $KLMN$ would be a parallelogram. It is impossible for an isosceles trapezoid to also be a parallelogram, so $\overline{KM}$ cannot bisect both $\angle LMN$ and $\angle LKN$. **37.** Answers may vary. Sample: Similarities: diagonals are $\perp$; consecutive sides are $\cong$. Differences: one diagonal of a kite bisects opposite angles, but the other does not; all sides of a rhombus are $\cong$.

39.

Statements	Reasons
1. Draw $\overline{AE} \| \overline{DC}$	1. Construction
2. $AEDC$ is a parallelogram	2. Definition of parallelogram
3. $\overline{AE} \cong \overline{DC}$	3. Opposite sides of a parallelogram are $\cong$.
4. $\angle 1 \cong \angle C$	4. If lines are parallel, corresponding angles are $\cong$.
5. $\angle B \cong \angle 1$	5. Isosceles Triangle Theorem
6. $\angle B \cong \angle C$	6. Transitive Property of $\cong$
7. $\angle D$ and $\angle C$ are supplementary; $\angle B$ and $\angle BAD$ are supplementary	7. If lines are parallel, same-side interior angles are supplementary.
8. $\angle D \cong \angle BAD$	8. Angles supplementary to $\cong$ angles are $\cong$.

41. Isosceles trapezoid

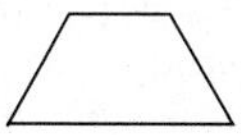

43. Rectangle, square

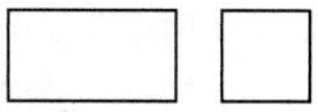

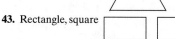

45. Kite, rhombus, square

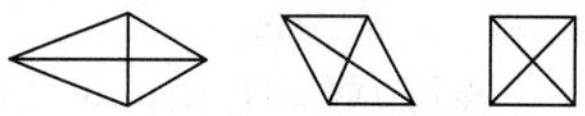

47.

Statements	Reasons
1. $\overline{AB} \cong \overline{DC}$	1. Given
2. $\angle BAD \cong \angle CDA$	2. Base angles of an isosceles triangle
3. $\overline{AD} \cong \overline{AD}$	3. Reflexive Property of $\cong$
4. $\Delta BAD \cong \Delta CDH$	4. SAS Postulate
5. $\overline{AC} \cong \overline{DB}$	5. cpoctac

49.

Statements	Reasons
1. Draw $\overline{TA}$ and $\overline{PR}$	1. Construction
2. $\overline{TR} \cong \overline{PA}$	2. Given
3. $\angle TRA \cong \angle PAR$	3. Base angles of an isosceles trapezoid are ≅.
4. $\overline{RA} \cong \overline{RA}$	4. Reflexive Property of ≅
5. $\Delta TRA \cong \Delta PAR$	5. SAS Postulate
6. $\angle RTA \cong \angle APR$	6. cpoctac

51.

Statements	Reasons
1. $ABCD$ is a kite with $\overline{AB} \cong \overline{AD}$ and $\overline{CB} \cong \overline{CD}$	1. Given
2. A and C lie on the ⊥ bisector of $\overline{BD}$	2. Converse of Perpendicular Bisector Theorem
3. $\overline{AC}$ is the ⊥ bisector of $\overline{BD}$	3. Two points determine a line.
4. $\overline{AC} \perp \overline{BD}$	4. Definition of perpendicular bisector

53. True; a square is a parallelogram with 4 right angles. **55.** False; a rhombus has 4 ≅ sides, and a kite does not. **57.** False; counterexample: Kites and trapezoids are not parallelograms. **59.** half the sum of the bases; Trapezoid Midsegment Theorem **61.** 12 **63.** 27 **65.** $-\frac{1}{4}$

Chapter 6 Vocabulary Check

1. parallelogram **2.** square **3.** rhombus **4.** rectangle **5.** vertex **6.** n-gon **7.** quadrilateral **8.** convex **9.** concave **10.** diagonal **11.** equilateral **12.** equiangular **13.** regular **14.** kite **15.** trapezoid **16.** midsegment **17.** isosceles **18.** leg **19.** base **20.** base angles **21.** base angles **22.** polygon

Chapter 6 Review

1. a. quadrilateral **b.** concave **c.** not regular **3.** 69° **5.** $m\angle 1 = 38°, m\angle 2 = 43°, m\angle 3 = 99°$ **7.** $m\angle 1 = 37°, m\angle 2 = 26°, m\angle 3 = 26°$ **9.** $x = 3, y = 7$ **11.** no **13.** $x = 29, y = 28$ **15.** $m\angle 1 = 58°, m\angle 2 = 32°, m\angle 3 = 90°$ **17.** sometimes **19.** sometimes **21.** sometimes **23.** No; two sides are parallel in all parallelograms. **25.** $x = 18$; a diagonal bisects a pair of angles in a rhombus **27.** $m\angle 1 = 135°, m\angle 2 = 135°, m\angle 3 = 45°$ **29.** $m\angle 1 = 90°, m\angle 2 = 25°$ **31.** 2 **33.** $m\angle 1 = 124°, m\angle 2 = 56°, m\angle 3 = 124°$ **35.** Yes; the diagonals are perpendicular, so the parallelogram is a rhombus. **37.** $x = 6, y = 5$ **39.** $x = 90, y = 30$ **41.** $GR = DA = 3\sqrt{2}, RA = DG = \sqrt{65}$; slope of $\overline{GR}$ = slope of $\overline{DA} = 1$, slope of $\overline{RA}$ = slope of $\overline{DG} = -\frac{4}{7}$

Chapter 6 Test

1. hexagon; convex **2.** quadrilateral; concave **3.** Isosceles trapezoid; the quadrilateral has one pair of parallel sides and another pair of congruent sides. **4.** Rectangle; the quadrilateral is a parallelogram with a right angle. **5.** Rectangle; the quadrilateral has 4 right angles. **6.** Kite; the quadrilateral has two pairs of consecutive sides ≅, and no opposite sides ≅. **7.** $a = 105, m = 116$ **8.** $x = 130, y = 50$ **9.** $x = 57, y = 57, z = 66$ **10.** $x = 45, y = 60$ **11.** $x = 1, y = 2$ **12.** isosceles trapezoid; $x = 3$ **13.** rhombus; $x = 58, y = 32$ **14.** kite; $x = \frac{5}{3}, y = \frac{9}{2}$ **15.** parallelogram; $x = 2, y = 4$ **16.** $AB = 6, CD = 10, EF = 8$ **17.** 20.6

18.

Statements	Reasons
1. $ABCD$ is a parallelogram	1. Given
2. $\angle 1 \cong \angle 4$ and $\angle 2 \cong \angle 3$	2. If lines are parallel, then alternate interior angles are ≅.
3. $\overline{AC}$ bisects $\angle DAB$	3. Given
4. $\angle 1 \cong \angle 2$	4. Definition of angle bisector
5. $\angle 3 \cong \angle 4$	5. Transitive property of ≅
6. $\overline{AC}$ bisects $\angle DCB$	6. Definition of angle bisector

19. trapezoid

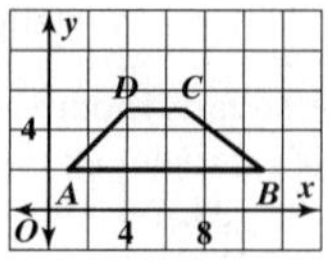

20. kite

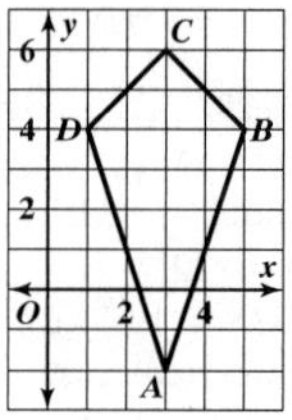

21. $AB = CD = 5, BC = AD = \sqrt{10}$; $\overline{AB}$ and $\overline{CD}$ both have undefined slope, slope of $\overline{BC}$ = slope of $\overline{AD} = -\frac{1}{3}$

22. False; counterexamples may vary. Sample:

23. False; counterexamples may vary. Sample:

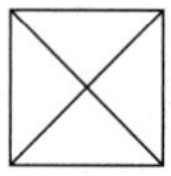

24. False; counterexamples may vary. Sample:

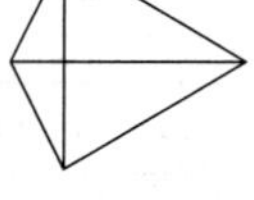

Chapter 6 Standardized Test

1. b **2.** c **3.** c **4.** a **5.** a **6.** c **7.** c **8.** b **9.** d **10.** b **11.** b **12.** d **13.** a **14.** d **15.** b **16.** b **17.** a

CHAPTER 7 SIMILARITY

Section 7.1

Practice Exercises

1. $\frac{20}{23}$ **2. a.** $\frac{4}{3}$ **b.** $\frac{3}{1}$ **c.** $\frac{9}{11}$ **3. a.** $\frac{3}{5}$ **b.** $\frac{5}{12}$ **4.** 12 cm, 21 cm, 27 cm **5. a.** 63 **b.** $\frac{1}{4}$

Vocabulary & Readiness Check 7.1

1. proportion; ratio **3.** true **5.** means **7.** true **9.** false

Exercise Set 7.1

1. $\frac{11}{14}$ **3.** $\frac{2.8}{7.6}$ **5.** $\frac{2}{3}$ **7.** $\frac{8}{25}$ **9.** $\frac{3}{4}$ **11.** $\frac{5}{12}$ **13.** $\frac{3}{1}$ **15.** $\frac{2}{3}$ **17.** $\frac{10}{29}$ **19.** $\frac{25}{144}$ **21.** $\frac{2}{3}$ **23.** 18°, 27°, 45° **25.** 24 cm, 28 cm, 36 cm **27.** 80°, 60°, 40° **29.** 4 **31.** $\frac{36}{5}$ **33.** 32 **35.** 7 **37.** 6 **39.** 2 **41.** $\frac{1}{2}$ **43.** $\frac{1}{1}$ **45.** $\frac{4}{1}$ **47.** 1 and 60, 2 and 30, 3 and 20, 4 and 15, 5 and 12, 6 and 10 **49.** $\frac{9}{6} = \frac{18}{12}, \frac{9}{18} = \frac{6}{12}, \frac{12}{6} = \frac{18}{9}$, or $\frac{12}{18} = \frac{6}{9}$ **51.** The second line should equate the product of the means and the product of the extremes: $7x = 12$. Then the third line would be $x = \frac{12}{7}$.

53. 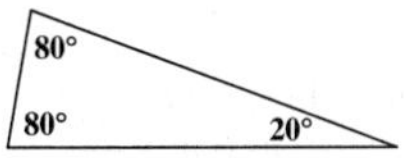

55. $\frac{9}{4}$; Divide each side by $4n$. **57.** -3 or 4 **59.** $\frac{9}{3} = \frac{15}{5}, \frac{5}{15} = \frac{3}{9}$, or $\frac{15}{9} = \frac{5}{3}$ **61.** < **63.** > **65.** $\frac{3}{5}$ **67.** $\frac{2x}{7}$

Section 7.2

Practice Exercises

1. Property (1): $\frac{2}{z} = \frac{7}{x}$; Property (2): $\frac{z}{x} = \frac{2}{7}$; Property (3): $\frac{z+2}{2} = \frac{x+7}{7}$
2. a. $\frac{7}{y}$; Property (1) **b.** $\frac{x+6}{6}$; Property (3) **3.** 17 ft
4. $25\frac{3}{5}$ or 25.6 gal

Vocabulary & Readiness Check 7.2

1. F **3.** F **5.** T

Exercise Set 7.2

1. $\frac{a}{e} = \frac{2}{5}$ **3.** $\frac{2}{a} = \frac{5}{e}$ **5.** $\frac{a+2}{2} = \frac{e+5}{5}$ **7.** $\frac{a}{13} = \frac{7}{b}$
9. $\frac{a+7}{7} = \frac{13+b}{b}$ **11.** $\frac{7}{a} = \frac{b}{13}$ **13.** $\frac{4}{3}$; Property (1)
15. $3b$; Cross Products **17.** $\frac{a+b}{b}$; Property (3) **19.** $\frac{a}{3}$; Property (2)
21. 360 baskets **23.** 495 min **25.** 630 applications **27.** 23 ft
29. 270 sq ft **31.** 25 gal **33.** 450 km **35.** 16 bags
37. 18 applications **39.** 5 weeks **41.** $10\frac{2}{3}$ servings **43.** 37.5 seconds
45. a. 18 tsp **b.** 6 tbsp **47.** 112 ft; 11-in. difference **49.** 1248 ft
51. a. 0.1 gal **b.** 13 fl oz **53.** $\frac{9}{4}$; Property (1) **55.** $\frac{b}{2}$; Property (3)
57. $\frac{a}{b} \times \frac{c}{d}$ is equivalent to $\frac{a}{c} \times \frac{b}{d}$
$a \cdot d = b \cdot c$ Cross products $a \cdot d = c \cdot d$ form equivalent equations.
59. 0.8 ml **61.** 1.25 ml **63.** $\frac{3}{4} = \frac{12}{16}$ is equivalent to $\frac{3}{12} = \frac{4}{16}$ by Property (2). $\frac{3}{12} = \frac{4}{16}$ is equivalent to $\frac{12}{3} = \frac{16}{4}$ by Property (1).
65. $\angle A \cong \angle H, \angle B \cong \angle I, \angle C \cong \angle J$

Section 7.3

Practice Exercises

1. a. $\angle D \cong \angle H, \angle E \cong \angle J, \angle F \cong \angle K, \angle G \cong \angle L$
b. $\frac{DE}{HJ} = \frac{EF}{JK} = \frac{FG}{KL} = \frac{GD}{LH}$ **2. a.** not similar
b. $ABCDE \sim SRVUT$ or $ABCDE \sim UVRST$; $\frac{2}{1}$ or 2:1 **3.** $\frac{10}{3}$
4. 28.8 in. high by 48 in. wide

Vocabulary & Readiness Check 7.3

1. similar **3.** extended

Exercise Set 7.3

1. $\angle R \cong \angle D, \angle S \cong \angle E, \angle T \cong \angle F, \angle V \cong \angle G$;
$\frac{RS}{DE} = \frac{ST}{EF} = \frac{TV}{FG} = \frac{VR}{GD}$ **3.** $DEGH \sim PLQR$; $\frac{3}{2}$ or 3:2
5. $\Delta ABC \sim \Delta DEF$ (in any order); $\frac{3}{5}$ or 3:5
7. $\Delta ABS \sim \Delta PRS$; $\frac{5}{2}$ or 5:2 **9.** Not similar; Sample explanation: The angle measures are not the same. **11.** $x = 8, y = 9, z = 5.25$
13. $x = 4, y = 3$ **15.** 3:5 **17.** 51° **19.** 16.5 **21.** 5 in.
23. 60 ft **25.** 120 pixels wide by 90 pixels high **27.** always
29. sometimes **31.** $x = 10$; 2:1 **33.** All four angles are right angles because the slopes of $\overline{AB}$ and $\overline{DC}$ are -2 while the slopes of $\overline{BC}$ and $\overline{AD}$ are $\frac{1}{2}$ (negative reciprocals). **35.** $AB = BC = CD = DA = \sqrt{5}$
37. yes **39.** No; for polygons with more than 3 sides, you also need to know that the angles are congruent in order to know that the figures are similar. **41.** 1:3 **43.** Incorrect; Sample explanation: In the diagram, $\angle T$ corresponds to $\angle P$ or to $\angle U$, but in the similarity statement $TRUV \sim NPQU$, $\angle T$ corresponds to $\angle N$. **45.** Answers may vary. Scale of rectangle should be $\frac{11}{6}$ or 11:6. **47.** Answers may vary. Scale of rectangle should be $\frac{13}{6}$ or 13:6. **49.** All angles in any rectangle are right angles, so all corresponding angles are congruent. The ratio of two pairs of consecutive sides for each rectangle is the same. Since opposite sides of a parallelogram are equal, the other two pairs of sides will also have the same ratio. So corresponding sides form equal ratios and are proportional. So $BCEG \sim LJAW$. **51.** $\Delta BDC, \Delta AEC, \Delta FED$
53. 8 **55.** 7y **57.** SSS **59.** ASA

Section 7.4

Practice Exercises

1. The measures of the two acute angles in each triangle are 39° and 51°, so the triangles are similar by the AA ~ Postulate. **2.** Each of the base angles in the triangle on the left measures 68°, while each of the base angles in the triangle on the right measures $\frac{1}{2}(180° - 62°) = 59°$, so the triangles are not similar. **3.** $\Delta ABC \sim \Delta EFG$ (or ΔFEG) by the SSS ~ Theorem **4.** $\Delta ALW \sim \Delta ACE$ by the SAS ~ Theorem

5.

Statements	Reasons
1. $\overline{MP} \parallel \overline{AC}$	1. Given
2. $\angle A \cong \angle P$; $\angle C \cong \angle M$	2. Alternate Interior Angles Theorem
3. $\Delta ABC \sim \Delta PBM$	5. AA ~ Postulate

6. The cliff is about 28.3 ft high.

Vocabulary & Readiness Check 7.4

1. $\Delta ABC \sim \Delta FED$ is one answer. **3.** $\Delta XYZ \sim \Delta QPZ$ is one answer.
5. $\Delta RST \sim \Delta RVQ$ is one answer.

Exercise Set 7.4

1. $\Delta AEZ \sim \Delta REB$ by the AA ~ Postulate **3.** Not similar; $\frac{JL}{PQ} = \frac{KL}{RP} = \frac{3}{2}$, but $\frac{JK}{QR} = \frac{16}{11}$ **5.** $\Delta AGU \sim \Delta BEF$ by the SAS ~ Theorem **7.** $\Delta PSQ \sim \Delta RST$ by the SAS ~ Theorem
9. $\Delta GHF \sim \Delta JHK$ by the AA ~ Postulate **11.** Not similar; $\frac{AB}{DE} = \frac{BC}{EF} = \frac{4}{3}$. but $\frac{AC}{DF} = \frac{24}{19}$. **13.** 520 ft **15.** 500 ft **17.** 14.4 ft
19. There is a pair of congruent vertical angles and a pair of congruent right angles, so the triangles are similar by the AA ~ Postulate; 180 ft
21. about 169.2 m **23.** 6 **25.** 10

27.

Statements	Reasons
1. $\angle ABC \cong \angle ACD$	1. Given
2. $\angle A \cong \angle A$	2. Reflexive Property of $\cong$
3. $\Delta ABC \sim \Delta ACD$	3. AA $\sim$ Postulate

29.

Statements	Reasons
1. $\overline{PQ} \perp \overline{QT}$; $\overline{ST} \perp \overline{TQ}$	1. Given
2. $\angle PQR$ and $\angle STV$ are right angles	2. Definition of $\perp$ lines
3. $\angle PQR \cong \angle STV$	3. All right angles are congruent
4. $\frac{PQ}{ST} = \frac{QR}{TV}$	4. Given
5. $\Delta PQR \sim \Delta STV$	5. SAS $\sim$ Theorem
6. $\angle KRV \cong \angle KVR$	6. Definition of $\sim$ Triangles
7. ΔVKR is isosceles	7. Definition of isosceles triangle

31. No; the ratios of the sides that form the vertex angles are the same, but the vertex angles may not be congruent. **33.** Yes; the two parallel lines and the two sides determine two pairs of congruent corresponding angles, so two triangles are similar by the AA $\sim$ Postulate.

35. Incorrect; the ratio $\frac{4}{8}$ does not use corresponding sides of two triangles.

37. Answers may vary. Sample: Both use two pairs of corresponding sides and the angles included by those sides, but the SAS $\sim$ Theorem uses pairs of equal ratios, while the SAS $\cong$ Theorem uses pairs of $\cong$ sides.

39.

Statements	Reasons
1. $\ell_1 \parallel \ell_2$	1. Given
2. $\angle BAC \cong \angle EDF$	2. Corresponding Angles Theorem
3. $\overline{BC} \perp \overline{AC}$; $\overline{EF} \perp \overline{DF}$	3. Given
4. $\angle BCA$ and $\angle EFD$ are right angles	4. Definition of perpendicular lines
5. $\angle BCA \cong \angle EFD$	5. All right angles are congruent
6. $\Delta ABC \sim \Delta DEF$	6. AA $\sim$ Postulate
7. $\frac{BC}{AC} = \frac{EF}{DF}$	7. Definition of similar triangles

41.

Statements	Reasons
1. Choose point X on $\overline{QR}$ so that $AB = QX$.	1. Construction
2. Draw $\overleftrightarrow{XY} \parallel \overline{RS}$ such that Y lies on $\overline{QS}$.	2. Parallel Postulate
3. $\angle A \cong \angle Q$	3. Given
4. $\angle QXY \cong \angle QRS$	4. Corresponding Angles Theorem
5. $\Delta QXY \sim \Delta QRS$	5. AA $\sim$ Postulate
6. $\frac{QX}{QR} = \frac{QY}{QS} = \frac{XY}{RS}$	6. Definition of similar triangles
7. $\frac{AB}{QR} = \frac{AC}{QS}$	7. Given
8. $\frac{QX}{QR} = \frac{AC}{QS}$	8. Substitution (steps 1 and 7)
9. $\frac{QX}{QR} = \frac{QY}{QS} = \frac{AC}{QS}$	9. Transitive Property of Equality (steps 6 and 8)
10. $QY = AC$	10. Multiplication Property of Equality
11. $\Delta ABC \cong \Delta QXY$	11. SAS $\cong$ Theorem
12. $\angle B \cong \angle QXY$	12. CPOCTAC
13. $\angle B \cong \angle R$	13. Transitive Property of $\cong$
14. $\Delta ABC \sim \Delta QRS$	14. AA $\sim$ Postulate

43. By the Distance Formula, $AB = AC = 2\sqrt{5}, BC = 2\sqrt{2}$, $RS = RT = \sqrt{5}$, and $ST = \sqrt{2}$. $\Delta ABC \sim \Delta RST$ by the SSS $\sim$ Theorem since $\frac{AB}{RS} = \frac{AC}{RT} = \frac{BC}{ST} = 2.$ **45.** 135° **47.** $\frac{3}{2}$ **49.** 88°, acute **51.** 110°, obtuse **53.** means: m, 18; extremes: 12, 20; $m = \frac{40}{3}$ **55.** means: $x + 4$, 5; extremes: $x - 3$, 9; $x = \frac{47}{4}$

Section 7.5

Practice Exercises

1. $\Delta PRQ \sim \Delta SRP \sim \Delta SPQ$ **2.** $6\sqrt{2}$ **3.** $x = 6, y = 2\sqrt{5}$ **4.** 12 in.

Vocabulary & Readiness Check 7.5

1. 6 **3.** h **5.** j, h or h, j **7.** $\overline{RT}$ **9.** $\overline{PR}$ and $\overline{PT}$

Exercise Set 7.5

1. Answers may vary. Sample: $\Delta JKL \sim \Delta JNK \sim \Delta KNL$ **3.** $2\sqrt{10}$ **5.** 25 **7.** $4\sqrt{3}$ **9.** 2.5 **11.** 2 **13.** 9 **15.** $2\sqrt{10}$ **17.** $x = 6\sqrt{3}, y = 3\sqrt{3}$ **19.** $x = 20, y = 10\sqrt{5}$ **21.** 3 **23.** 6 **25.** $l_1 = \sqrt{2}, l_2 = \sqrt{2}, a = 1, s_2 = 1$ **27.** $l_2 = 2\sqrt{3}, h = 4, a = \sqrt{3}$, $s_1 = 1$ **29.** 5 ft **31.** Yes; the proportion $\frac{a}{\sqrt{ab}} = \frac{\sqrt{ab}}{b}$ is true by the Cross Products Property and satisfies the definition of the geometric mean. **33.** the altitude to the hypotenuse is the geometric mean of the segments of the hypotenuse; $\frac{3}{x} = \frac{x}{(8-3)}$

35.

Statements	Reasons
1. Right ΔABC with altitude to the hypotenuse $\overline{CD}$	1. Given
2. $\Delta ACD \sim \Delta CBD$	2. Altitude of a Right Triangle Theorem
3. $\frac{AD}{CD} = \frac{CD}{BD}$	3. Definition of similar triangles

37.

Statements	Reasons
1. ΔABC is equilateral	1. Given
2. $\overline{AB} \cong \overline{CB}$	2. Definition of equilateral triangle
3. Right ΔABD with altitude to the hypotenuse $\overline{BE}$	3. Given
4. $\overline{BE} \perp \overline{AD}$	4. Definition of altitude
5. $\angle AEB$ and $\angle CEB$ are right angles	5. $\perp$ lines form right angles
6. ΔAEB and ΔCEB are right triangles	6. Definition of right triangle
7. $\Delta AEB \cong \Delta CEB$	7. Hypotenuse-Leg Theorem

Statements	Reasons
8. $\overline{AE} \cong \overline{CE}$	**8.** cpoctac
9. Let $x = AE = CE$; Then $2x = AC = AB = BC$	**9.** Segment addition and definition of equilateral Δ
10. $\frac{AD}{AB} = \frac{AB}{AE}$	**10.** Corollary 2
11. $\frac{AD}{2x} = \frac{2x}{x}$	**11.** Substitution Property of Equality
12. $x \cdot AD = 4x^2$	**12.** Cross Products Property
13. $AD = 4x$	**13.** Division Property of Equality
14. $AD = AE + ED$	**14.** Segment Addition Postulate
15. $4x = x + ED$	**15.** Substitution Property of Equality
16. $ED = 3x$	**16.** Subtraction Property of Equality
17. $\frac{AE}{BE} = \frac{BE}{ED}$	**17.** Corollary 1
18. $\frac{x}{BE} = \frac{BE}{3x}$	**18.** Substitution Property of Equality
19. $BE^2 = 3x^2$	**19.** Cross Products Property
20. $BE = x\sqrt{3}$	**20.** Square root of each side
21. $BE = AE\sqrt{3}$	**21.** Substitution Property of Equality

39. $\angle R \cong \angle P$ is given; $\angle RNM \cong \angle PNQ$ since they are vertical angles; $\Delta RNM \sim \Delta PNQ$ by the AA $\sim$ Postulate **41.** $x = 5, y = 8$ **43.** 28 cm **45.** $\frac{24}{7}$ mm

Section 7.6

Practice Exercises

1. 8 **2.** 5.76 yd **3.** 14.4

Vocabulary & Readiness Check 7.6

1. *d* **3.** *d* **5.** There are many. For example: $\frac{12}{16} = \frac{x}{20}$

Exercise Set 7.6

1. 7.5 **3.** 10 **5.** 8 mm **7.** 7.5 **9.** $\frac{44}{13}$ **11.** 6 **13.** 35 **15.** *KS* by the Triangle-Angle-Bisector Theorem **17.** *JP* by the Corollary to the Side-Splitter Theorem **19.** *KM* by the Triangle-Angle-Bisector Theorem **21.** 575 ft **23.** 20 **25.** 2.5 **27.** $\frac{2}{7}$ or 3 **29.** $\frac{10}{3}$ **31. a.** $x + y + 20 = 50$ **b.** $\frac{12}{8} = \frac{x}{y}$ **c.** $x = 18$ m, $y = 12$ m **33.** yes; $\frac{6}{10} = \frac{9}{15}$; Converse of the Side-Splitter Theorem **35.** Answers may vary. Sample: The Corollary to the Side-Splitter Theorem takes the same three (or more) parallel lines as in Theorem 6.2-5, but instead of cutting off congruent segments, it allows the segments to be proportional. **37.** 5-cm side: 2.4 cm, 2.6 cm; 12-cm side: about 3.3 cm, about 8.7 cm; 13-cm side: about 9.2 cm, about 3.8 cm **39.** Isosceles; *AC*:*BC* is 1:1 by the Triangle-Angle-Bisector Theorem.

41.

Statements	Reasons
1. $a \parallel b \parallel c$	**1.** Given
2. Line $\overleftrightarrow{CW}$ that intersects b at P	**2.** Construction
3. $\frac{AB}{BC} = \frac{WP}{PC}$; $\frac{WP}{PC} = \frac{WX}{XY}$	**3.** Side-Splitter Theorem
4. $\frac{AB}{BC} = \frac{WX}{XY}$	**4.** Transitive Property of Equality

43. a. Answers may vary. Sample: A midsegment of a parallelogram connects the midpoints of two opposite sides of the parallelogram.

b.

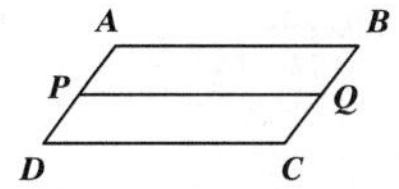

Given: $\overline{PQ}$ is a midsegment of parallelogram *ABCD*.
Prove: $\overline{PQ} \parallel \overline{AB}$, $\overline{PQ} \parallel \overline{DC}$

Statements	Reasons
1. *ABCD* is a parallelogram.	**1.** Given
2. $\overline{AD} \cong \overline{BC}$ and $\overline{AD} \parallel \overline{BC}$	**2.** Definition of parallelogram
3. $PD = \frac{1}{2}(AD) = \frac{1}{2}(BC) = QC$; $PA = \frac{1}{2}(AD) = \frac{1}{2}(BC) = BQ$	**3.** Definition of midpoint
4. *ABQP* and *PQCD* are parallelograms	**4.** A quadrilateral with a pair of opposite sides that are parallel and congruent is a parallelogram.
6. $\overline{PQ} \parallel \overline{AB}$; $\overline{PQ} \parallel \overline{DC}$	**6.** Opposite sides of a parallelogram are parallel.

c.

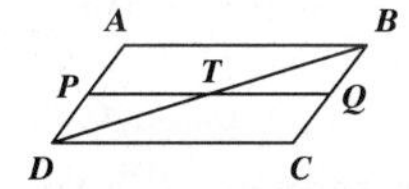

Given: Parallelogram *ABCD* and midsegment $\overline{PQ}$
Prove: $\overline{PQ}$ bisects $\overline{BD}$
From part (b), $\overline{AB} \parallel \overline{PQ} \parallel \overline{DC}$.
Since $\overline{AP} \cong \overline{PD}$ by the definition of midsegment, $\overline{DT} \cong \overline{TB}$ because if parallel lines cut off congruent segments on one transversal, they cut off congruent segments on every transversal. Since $\overline{PQ}$ contains the midpoint of $\overline{BD}$, then $\overline{PQ}$ bisects $\overline{BD}$ by the definition of bisection. Point *T* is the midpoint of both diagonals because the diagonals of a parallelogram have the same midpoint. So $\overline{PQ}$ bisects both diagonals of the parallelogram. **45.** 90 units **47.** *m* **49.** *c* **51.** $(3, -3)$ **53.** $(3\text{ m})^2 = 9\text{ m}^2$, $(4\text{ m})^2 = 16\text{ m}^2$, $(5\text{ m})^2 = 25\text{ m}^2$ **55.** $(4\text{ m})^2 = 16\text{ m}^2$, $(4\text{ m})^2 = 16\text{ m}^2$, $(4\sqrt{2}\text{ m})^2 = 32\text{ m}^2$

Chapter 7 Vocabulary Check

1. ratio **2.** proportion **3.** geometric mean **4.** means **5.** extremes **6.** cross products **7.** Similar **8.** congruent; proportional **9.** extended **10.** scale factor

Chapter 7 Review

1. $\frac{1}{116}$ or 1:116 **3.** 6 **5.** 6 **7.** 14 passes **9.** 8 bags **11.** $213\frac{1}{3}$ mi **13.** $JEHN \sim JKLP$; 3:4 **15.** 120 ft **17.** The ratio of each pair of corresponding sides is 2:1, so $\Delta AMY \sim \Delta ECD$ by the SSS $\sim$ Theorem. **19.** 12

21. $x = 6\sqrt{2}, y = 6\sqrt{6}$ **23.** $x = 2\sqrt{21}, y = 4\sqrt{3}$ **25.** 7.5 **27.** 22.5 **29.** 17.5 **31.** $\frac{3}{5}$ **33.** $\frac{4}{3}$ **35.** 1.6 **37.** 6 **39.** 15 **41.** 6 m **43.** 39 cm

Chapter 7 Test

1. 2 **2.** $\frac{1}{4}$ **3.** $\Delta ABC \sim \Delta FED$; 3:4 or $\frac{3}{4}$ **4.** $x = 42, y = 138, z = 9$ **5.** $x = 4$ **6.** $x = 63, y = 8$ **7.** $\Delta QRP \sim \Delta VWT$ by the SSS ~ Theorem **8.** No; the corresponding sides are not proportional. **9.** 13.5 cm **10.** $5\sqrt{6}$ **11.** 88.2 ft **12.** $\frac{50}{3}$ **13.** 10 **14.** $\frac{60}{11}$ **15.** 10 **16.** AA ~ Theorem (if lines are parallel, corresponding angles are congruent); $x = 10$ **17.** AA ~ Theorem ($\angle E \cong \angle D$ is given, vertical angles are congruent); $x = \frac{25}{3}$ **18.** never **19.** sometimes **20.** always

Chapter 7 Standardized Test

1. c **2.** c **3.** a **4.** b **5.** b **6.** b **7.** b **8.** a **9.** c **10.** d **11.** a **12.** c **13.** d **14.** d **15.** c **16.** a

CHAPTER 8 TRANSFORMATIONS

Section 8.1

Practice Exercises

1. a. Yes, the preimage and the image appear to be congruent. **b.** Yes, the preimage and the image appear to be congruent. **2. a.** translation **b.** rotation **c.** reflection **3. a.** rotation **b.** reflection **c.** translation **4. a.** point U; point P **b.** $\overline{NI} \cong \overline{SU}$; $\overline{ID} \cong \overline{UP}$; $\overline{ND} \cong \overline{SP}$

Vocabulary & Readiness Check 8.1

1. to change **3.** image **5.** isometry or rigid transformation **7.** translation; reflection; rotaion **9.** flip

Exercise Set 8.1

1. Yes, the preimage and the image appear to be congruent. **3.** No, the figures are not the same size. **5. a.** Answers may vary. Sample: $Q \to Q'$ **b.** $\overline{QR} \cong \overline{Q'R'}$; $\overline{RS} \cong \overline{R'S'}$; $\overline{SP} \cong \overline{S'P'}$; $\overline{PQ} \cong \overline{P'Q'}$ **7. a.** Answers may vary. Sample: $W \to R$ **b.** $\overline{GW} \cong \overline{MR}$; $\overline{WP} \cong \overline{RT}$; $\overline{PN} \cong \overline{TX}$; $\overline{NB} \cong \overline{XS}$; $\overline{BG} \cong \overline{SM}$ **9. a.** rotation **b.** reflection **c.** translation **11. a.** reflection **b.** rotation **c.** translation **13. a.** translation **b.** reflection **c.** rotation **15. a.** rotation **b.** translation **c.** reflection **17. a.** reflection **b.** rotation **c.** translation **19.** ΔLKJ **21.** ΔCAB **23.** ΔDFE **25.** 10 m **27.** 11 **29.** 10 **31.** 5 in. **33.** 10 in. $< x <$ 42 in. **35.** 0 m $< x <$ 18 m **37.** 25 sq cm **39.** 11.5 sq m

Section 8.2

Practice Exercises

1. a. $A'(-1, 2), B'(2, -3), C'(1, -5)$

b.

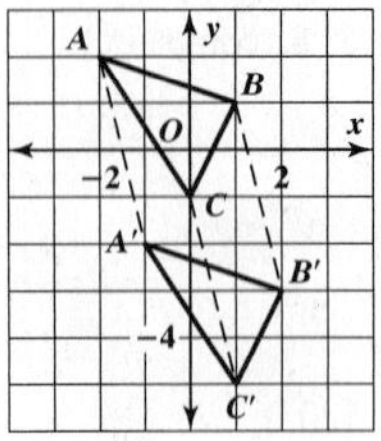

2. $(x, y) \to (x + 7, y - 1)$ **3.** $(3, 1)$ or the bishop is 3 squares right and 1 square up from its original position.

Vocabulary & Readiness Check 8.2

1. direction **3.** true

Exercise Set 8.2

1. Z' **3.** $\overline{H'Z'}$ **5.** $(x, y) \to (x + 5, y + 2)$ **7.** $(x, y) \to (x + 9, y - 5)$ **9.** $(x, y) \to (x - 4, y + 1)$ **11.** $(x, y) \to (x - 20, y - 10)$

13.

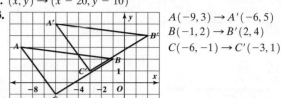

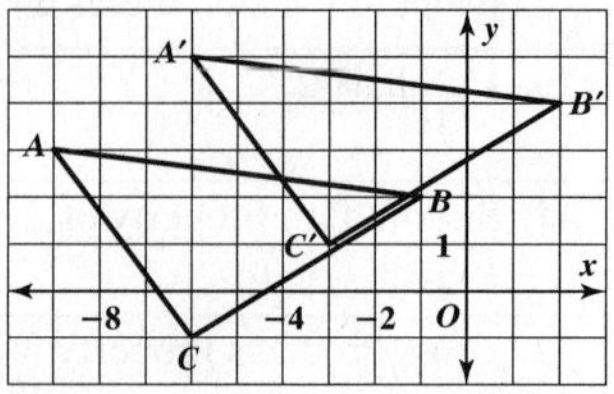

$A(-9, 3) \to A'(-6, 5)$
$B(-1, 2) \to B'(2, 4)$
$C(-6, -1) \to C'(-3, 1)$

15.

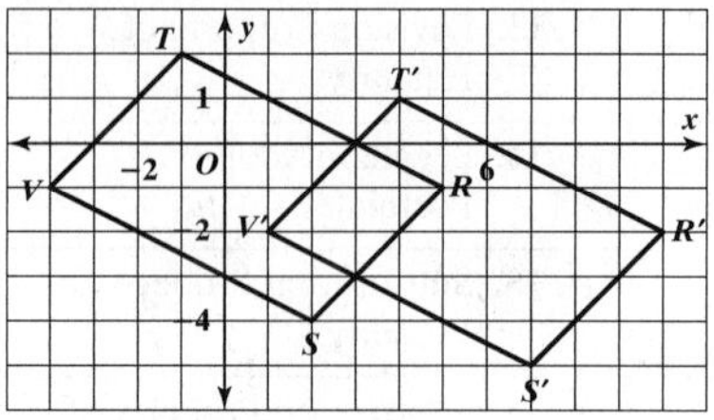

$T(-1, 2) \to T'(4, 1)$
$R(5, -1) \to R'(10, -2)$
$S(2, -4) \to S'(7, -5)$
$V(-4, -1) \to V'(1, -2)$

17.

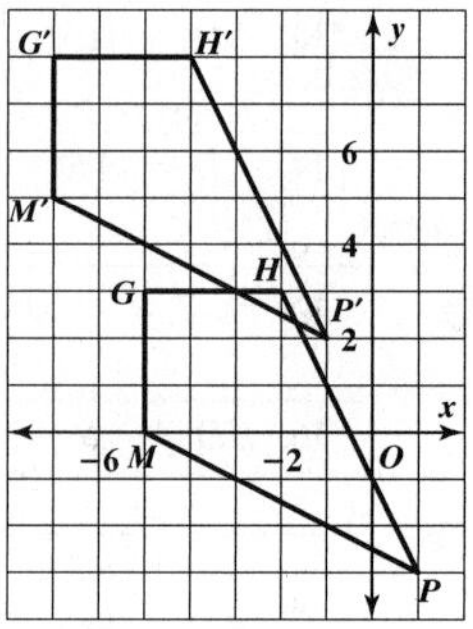

$G(-5, 3) \to G'(-7, 8)$
$H(-2, 3) \to H'(-4, 8)$
$P(1, -3) \to P'(-1, 2)$
$M(-5, 0) \to M'(-7, 5)$

19. $(x, y) \to (x + 2, y - 3)$ **21.** $(x, y) \to (x + 1, y - 1)$ **23.** $(x, y) \to (x + 4, y + 3)$ **25.** $(-10, 7)$ **27.** $(30, 13)$ **29.** $(-7, -1)$ **31.** $(4, -3)$ **33.** 1 block west and 7 blocks north **35.** $(x, y) \to (x - 2, y + 14)$ **37.** $(-3, 7), (-2, 5), (-1, 7), (0, 5)$, and $(1, 7)$ **39.** $(x, y) \to (x - 5, y - 7)$ **41. a.** $(x, y) \to (x - 5, y + 10)$ **b.** $U'(1, 16), G'(2, 12)$ **43.** The transformation $\Delta ABC \to \Delta PQR$ maps A to P and C to R, so it is a reflection, not a translation. $\Delta ABC \to \Delta RQP$ is a translation. **45.** at least 5 feet east and 10 feet north

47. The vertices of $P'L'A'T'$ are $P'(0, -3)$, $L'(1, -2), A'(2, -2)$, and $T'(1, -3)$.

slope of $\overline{PP'}$ = slope of $\overline{LL'}$
= slope of $\overline{AA'}$
= slope of $\overline{TT'}$
$= -\frac{3}{2}$

Thus, $\overline{PP'}, \overline{LL'}, \overline{AA'}$, and $\overline{TT'}$ are all parallel.

49.

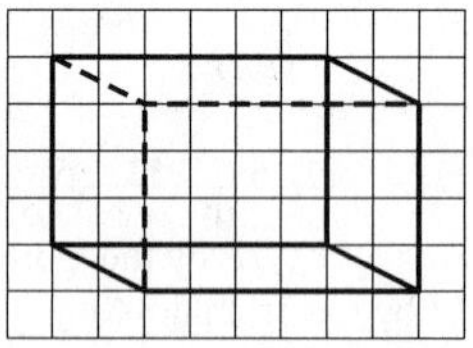

51.

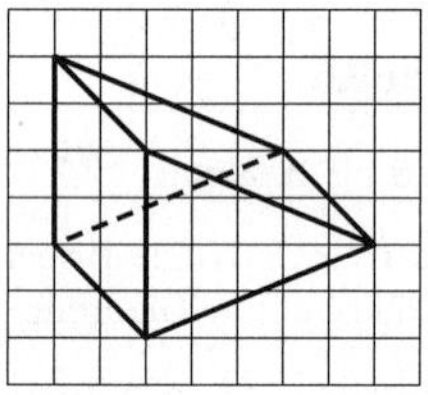

53. $y = -2$ **55.** $x = -1$

Section 8.3

Practice Exercises

1. $(-5, 4)$ **2.**

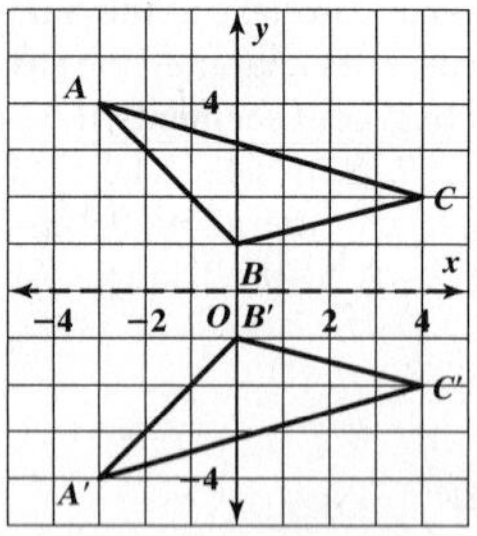

3. Yes, the intersection of $\overline{R'O}$ and line t will be the same point P. **4.** two

Vocabulary & Readiness Check 8.3

1. itself **3.** flip **5.** reflection

Exercise Set 8.3

1. $(-1, -2)$ **3.** $(1, -3)$ **5.** $(-1, 2)$ **7.**

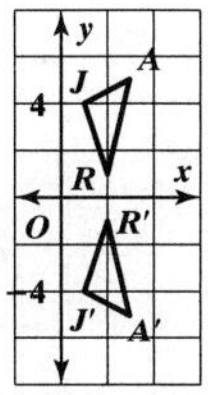

9.

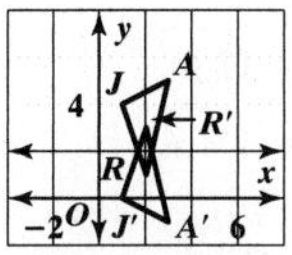

11.

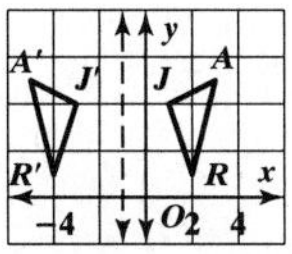

13.

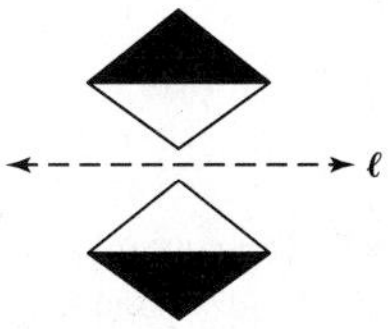

15.

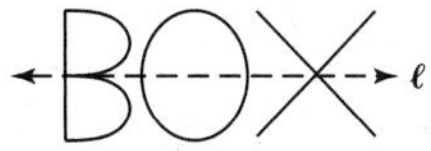

17.

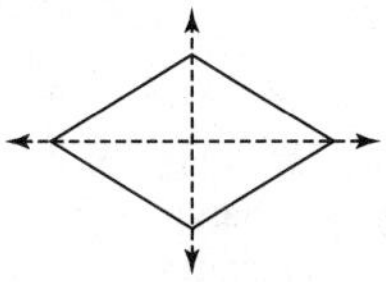

19. no line symmetry **21.**

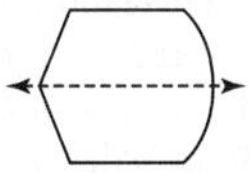

23. two

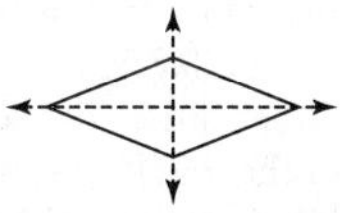

25. four

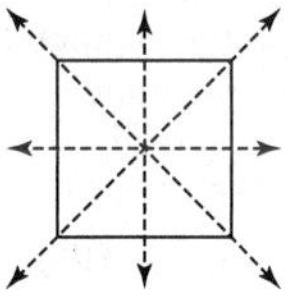

27. B, C, D, E, H, I, K, O, X **29.** $(-4, -3)$

31.

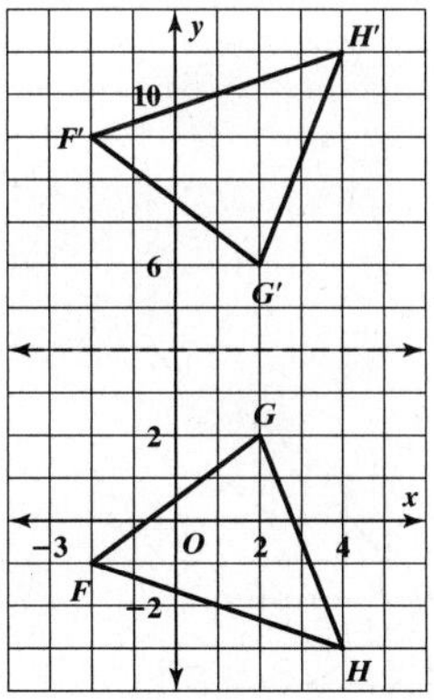

33. Reflect W over the canal to W'. Draw $\overline{DW'}$. The officials should build the pumping station at the point P where $\overline{DW'}$ intersects the canal.

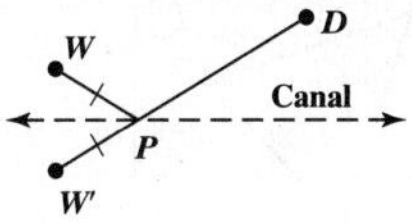

35.

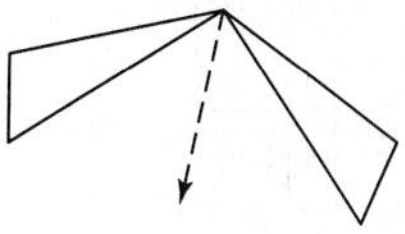

37.

39. $(0, -6)$ **41.** $(-4, 6)$ **43. a.** -1 **b.** $B'(0, 2)$, $C'(-3, 3)$

c.

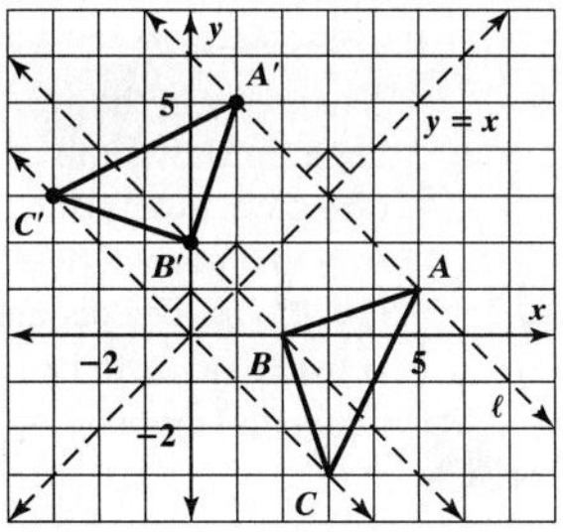

d. The coordinates of P' will be (b, a); the x- and y-coordinates will switch. **45.** The line of reflection is the perpendicular bisector of any segment whose endpoints are corresponding points of the preimage and image. **47.** $\overline{AA'}$ should be perpendicular to line r.

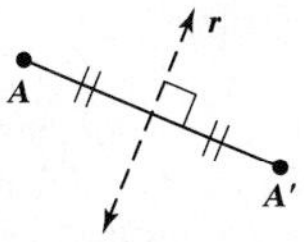

49. $x = 5, y = 5, Z = 11.5$ **51.** $(0, -4)$ **53.** $(0, 2a)$

55. Yes; explanations may vary. Sample: The angle bisector divides the angle into two congruent angles. By the definition of a line of symmetry, the angle bisector is a line of symmetry. **57.** Answers may vary. Sample: TOMATO, HOAX **59.** $(-3, 4)$

61. line symmetry across the y-axis

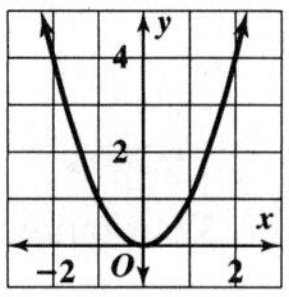

63. line symmetry across any line through the origin (the center of the circle).

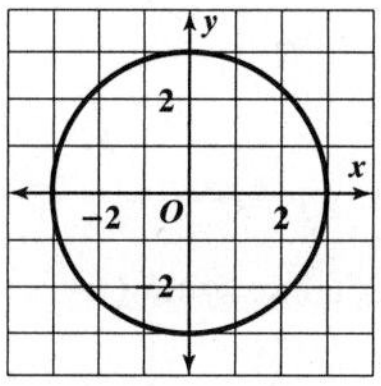

65. $(3, 1)$ **67.** $(-3, -1)$ **69.** Yes. **71.** Yes. **73.** Yes.

75. $(x, y) \rightarrow (x + 4, y - 2)$ **77.**

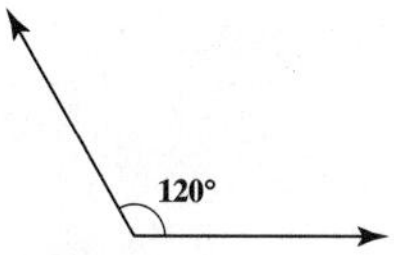

79.

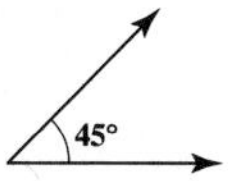

Section 8.4

Practice Exercises

1.

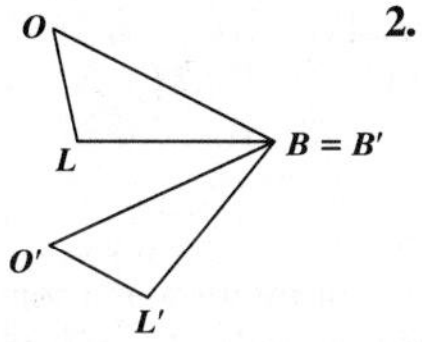

2. A **3.** 240° **4.** $(2, -3)$ **5.** Yes; 180°

Vocabulary & Readiness Check 8.4

1. itself **3.** turn **5.** rotational symmetry

Exercise Set 8.4

1.

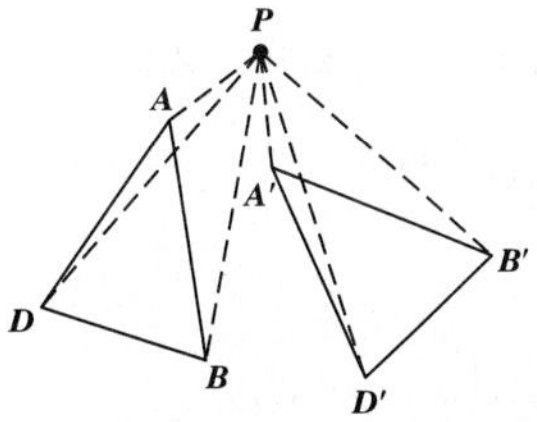

3.

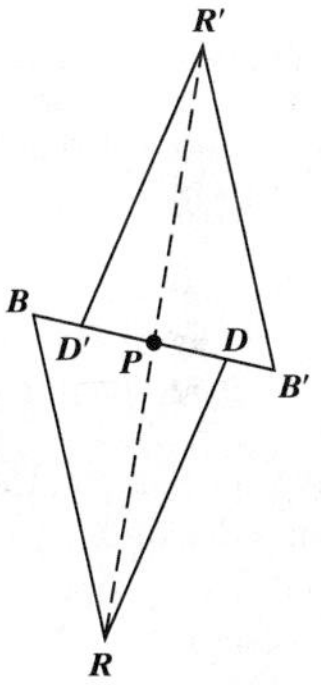

5.

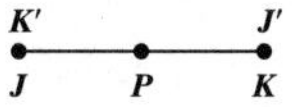

7.

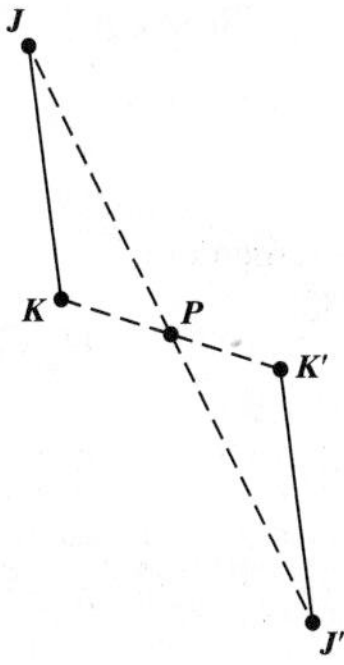

9. R **11.** Q **13.** H **15.** $\overline{EH}$ **17. a.** 270° **b.** 90° **19. a.** $205\frac{5}{7}°$ **b.** $154\frac{2}{7}°$

21.

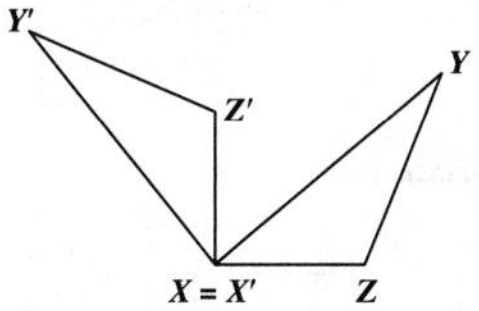

23.

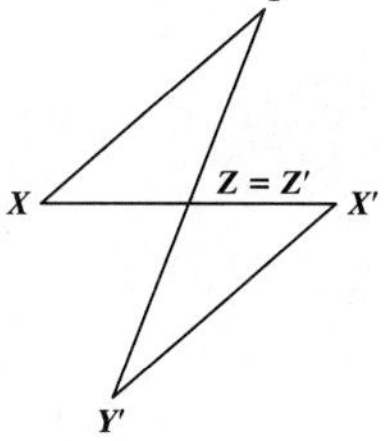

25. 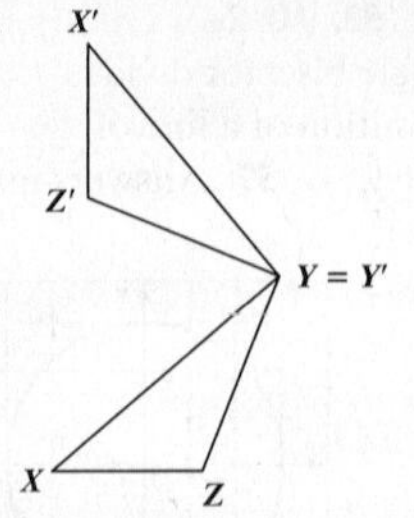

27. 90° **29.** 180° **31.** 60° **33.** any angle (or no angle if one considers the definition of rotational symmetry)

35. line symmetry

37. rotational **39.** line **41.** $\angle M \cong \angle M'$, $\angle N \cong \angle N'$, $\angle MEN \cong \angle M'EN'$; $\overline{MN} \cong \overline{M'N'}$, $\overline{NE} \cong \overline{N'E}$, $\overline{ME} \cong \overline{M'E}$ **43.** 110° **45.** 180°

47. yes

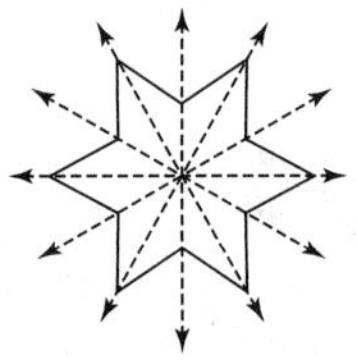

49. 280° **51. a.** 11.25° **b.** 15 **c.** 168.75° **53.** *H* **55.** $\overline{BC}$ **57.** $\overline{LM}$ **59.** *I* **61.** *E* **63.** Your friend counted the arrowheads instead of the lines; there are 5 lines of symmetry. **65.** Yes; the angle of rotation of a composition of two rotations is the sum of the two angles of the rotation. Since $x + y = y + x$, the two compositions give the same image.

67. about the origin: $L'(1, 2), M'(2, 6), N'(-2, 4)$
about L: $L'(2, -1), M'(3, 3), N'(-1, 1)$
about M: $L'(5, -6), M'(6, -2), N'(2, -4)$
about N: $L'(7, 0), M'(8, 4), N'(4, 2)$

69. H, 180°; I, 180°; N, 180°; O, any rotation; S, 180°; X, 180°; Z, 180°

71. 3 in. by 4 in. **73.** 2 in. by $2\frac{1}{2}$ in.

Section 8.5

Practice Exercises

1. reduction; $\frac{1}{2}$ **2. a.** $P'(1, 0), Z'\left(-\frac{3}{2}, \frac{1}{2}\right), G'(0, -1)$

b. Answers may vary. Sample: Use the Distance Formula to find the lengths of the sides of $\Delta P'Z'G'$ and ΔPZG. Then show that the corresponding sides are proportional, so the triangles are similar by the SSS Similarity Theorem. **3.** 5.1 cm

Vocabulary & Readiness Check 8.5

1. itself **3.** enlargement **5.** False

Exercise Set 8.5

1. enlargement; 1.5 **3.** enlargement; $\frac{4}{3}$ **5.** reduction; $\frac{1}{3}$

7. enlargement; 2 **9.** reduction; $\frac{1}{2}$ **11.** enlargement; 2

13. $D'(2, -10)$ **15.** $T'(0, 2)$

17. $P'(-50, 10), Q'(-30, 30), R'(10, -30)$

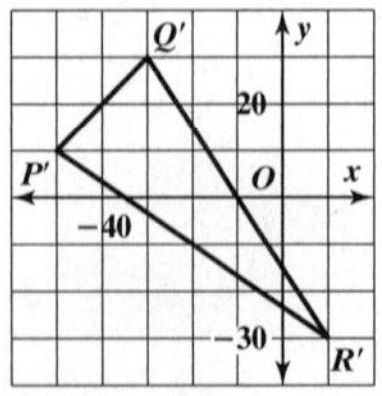

19. $P'\left(-\frac{9}{4}, 0\right), Q'\left(0, \frac{9}{4}\right), R'\left(\frac{3}{4}, -\frac{9}{4}\right)$

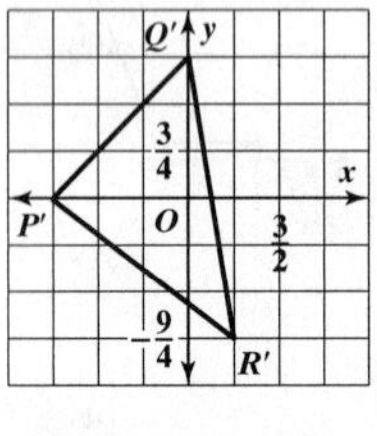

21. 1.2 cm **23.** 0.2 cm **25.** $L'(-15, 0)$ **27.** $A'(-9, 3)$

29. $B'\left(\frac{1}{8}, -\frac{3}{20}\right)$ **31.** $Q'\left(-\frac{3}{4}, 1\right), R'\left(-\frac{1}{2}, -\frac{1}{4}\right), T'\left(\frac{3}{4}, \frac{1}{4}\right)$, $W'\left(\frac{3}{4}, \frac{5}{4}\right)$ **33.** $Q'(-1.8, 2.4), R'(-1.2, -0.6), T'(1.8, 0.6)$, $W'(1.8, 3)$ **35.** $Q'(-30, 40), R'(-20, -10), T'(30, 10), W'(30, 50)$

37. 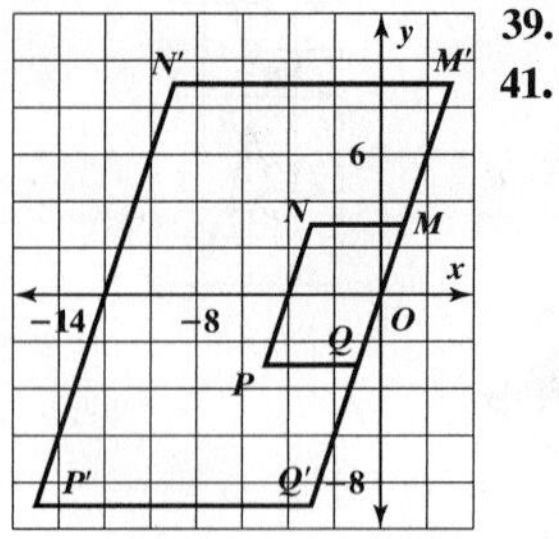

39. $I'J' = 10$ in, $H'J' = 12$ in.

41. 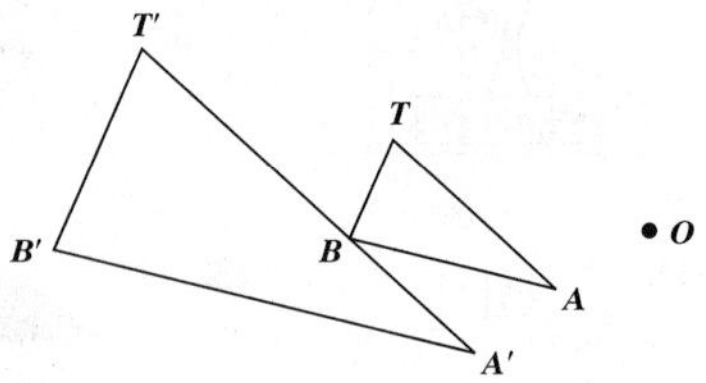

43. 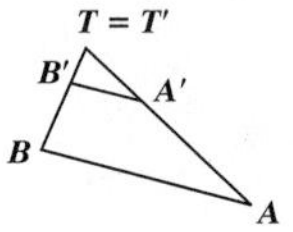

45. The student used 6, instead of $2 + 6 = 8$ as the preimage length in the denominator; the correct scale factor is $n = \frac{2}{2+6} = \frac{1}{4}$.

47. a. $\Delta LMN \sim \Delta L'M'N'$ **b.** 2 **c.** $x = 3, y = 60$

49. Answers may vary. Sample: Each type of scale factor is a constant ratio of corresponding lengths. For a dilation, the scale factor is always the ratio of an image length to a corresponding preimage length, while for similar figures, the scale factor ratio can relate the two figures in either order. The scale factor of two similar figures is always the ratio of the lengths of two corresponding sides, while the scale factor of a dilation is also the ratio of the distances of corresponding points from the center of dilation. If the center is not on the preimage, then these distances are not lengths of corresponding sides of the image and preimage. **51.** 0.4 **53.** False; a dilation does not map a segment to a congruent segment unless the scale factor is 1. **55.** True; the image and preimage are similar, so the corresponding angles are congruent.

57. 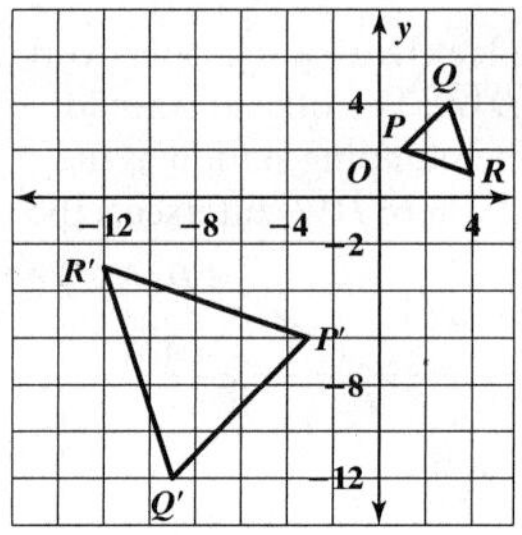

59. $(2, 7)$

61. 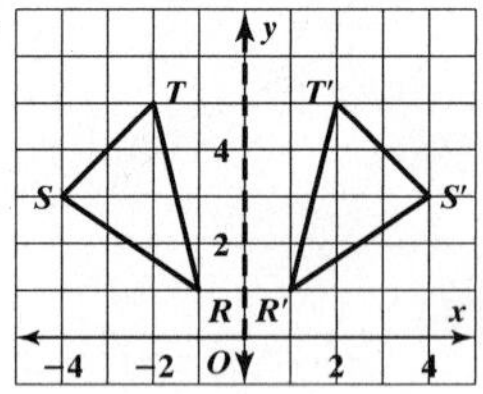

63. 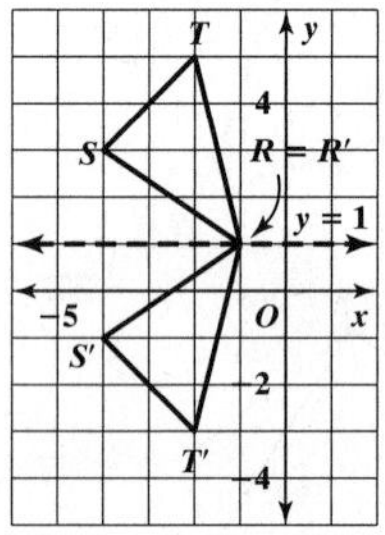

Section 8.6

Practice Exercises

1. a.

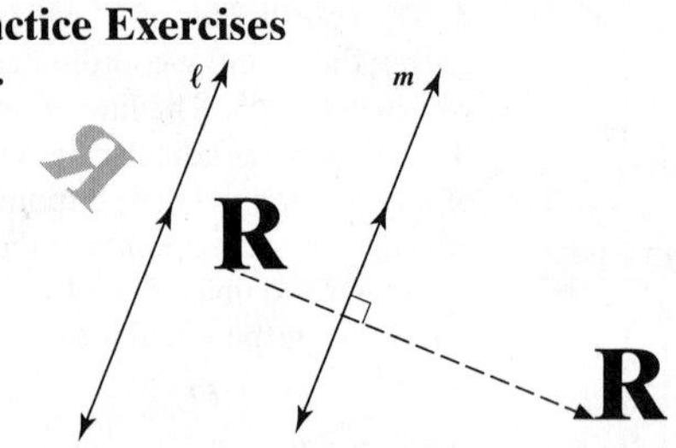

The arrow in the diagram shows the direction, determined by a line perpendicular to *l* and *m*. The distance is twice the distance between *l* and *m*.

b. The direction is from the first line of reflection toward the second line and is determined by a line perpendicular to the lines of reflection; the distance is two times the distance between the lines of reflection.

2. a.

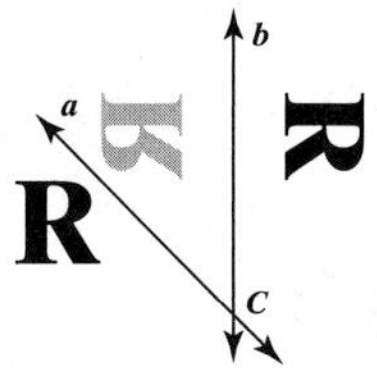

The center of rotation is C. The angle is 90° clockwise. **b.** The center of rotation is the intersection of the lines of reflection; the angle of rotation is two times the measure of the acute or right angle formed by the lines of reflection.

3.

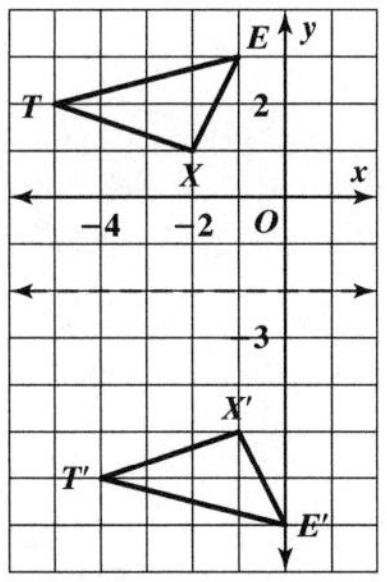

4. a. same; rotation **b.** same; translation **c.** opposite; glide reflection

Vocabulary & Readiness Check 8.6

1. true **3.** fasle **5.** true **7.** False

Exercise Set 8.6

1. A translation; the arrow in the answer diagram shows the direction, determined by a line perpendicular to l and m. The distance is twice the distance between l and m.

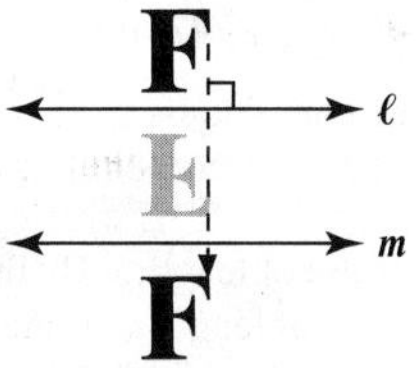

3. A translation; the arrow in the answer diagram shows the direction, determined by a line perpendicular to l and m. The distance is twice the distance between l and m.

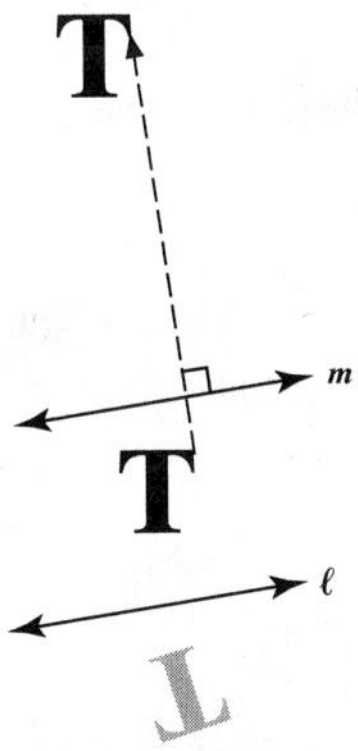

5. A rotation; the center of rotation is C. The angle of rotation is 190° clockwise.

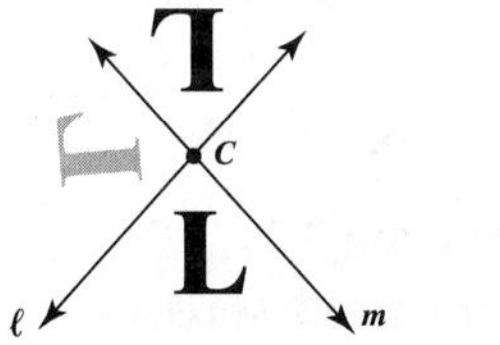

7. A rotation; the center of rotation is C. The angle of rotation is 150° clockwise.

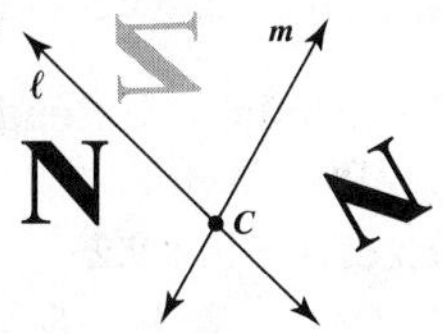

9.

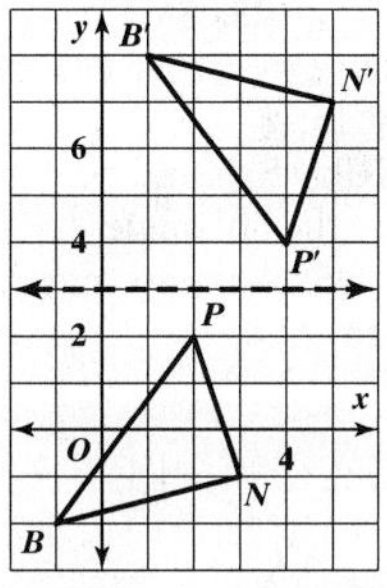

11.

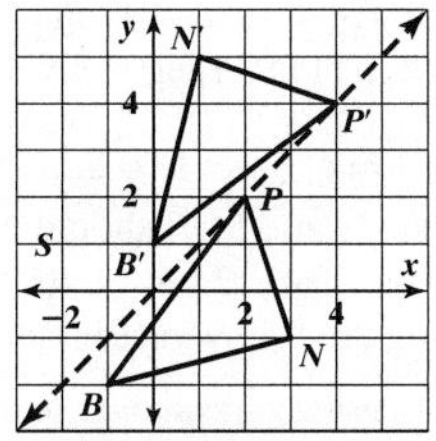

13. same; rotation **15.** opposite; glide reflection **17.** opposite; reflection **19.** same; translation

21. A translation 8 units to the right.

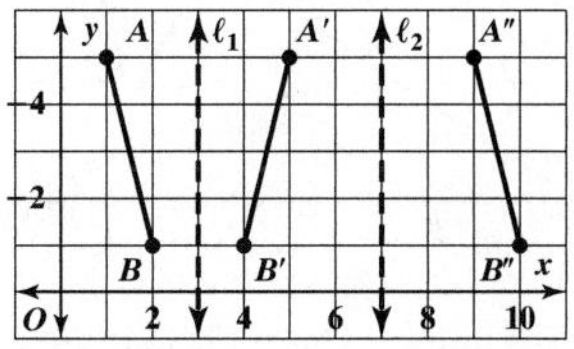

23. A 180° rotation about $(0, 0)$.

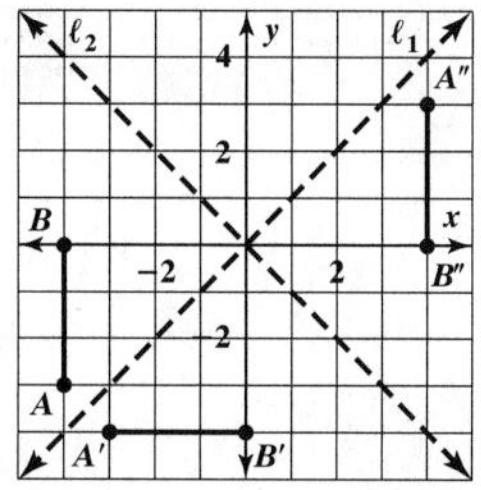

25. A translation 4 units left.

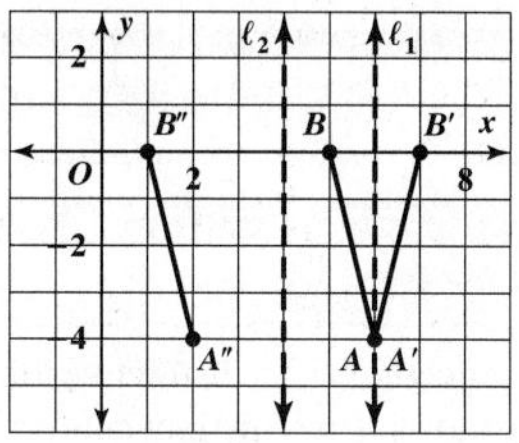

27. a. to the left **b.** below **c.** $(-2, -3)$ **29.** a glide reflection with translation $(x, y) \rightarrow (x, y + 2)$ and line of reflection $x = \frac{1}{2}$.

31. c **33.** 60° **35.** 45° **37.** glide reflection; $(x, y) \rightarrow (x + 11, y), y = 0$ **39.** translation; $(x, y) \rightarrow (x - 9, y)$ **41.** reflection; $y = 0$ **43.** rotation; center $(3, 0)$, angle of rotation 180° **45.** $(6, 5)$ **47.** $(2, 6)$ **49.** parallel **51.** Answers may vary. Sample: Since a reflection moves a point in the direction perpendicular to the translation, the order does not matter. **53.** $\overline{XX'} \parallel \overline{YY'}$ and $\overline{XX'} \cong \overline{YY'}$ so $XX'Y'Y$ is a parallelogram. Therefore, $\overline{XY} \cong \overline{X'Y'}$. **55.** If $\overline{XY}$ is rotated $x°$ about a point R, then $\overline{RX} \cong \overline{RX'}$ and $\overline{RY} \cong \overline{RY'}$. Also, $m\angle XRY + m\angle YRX' = m\angle YRX' + m\angle X'RY' = x°$, so $\angle XRY \cong \angle X'RY'$. So, $\Delta XRY \cong \Delta X'RY'$ by SAS and $\overline{XY} \cong \overline{X'Y'}$ because corresponding parts of congruent triangles are congruent. **57.** $A'(0, 12), B'(0, 0), C'(-9, -3)$ **59.** $A'(6, 3), B'(3, 12), C'(12, 0)$ **61.** I and II **63.** pentagon

Chapter Extension Exercise Set

1. Answers may vary. Samples are shown.

a.

b.

c.

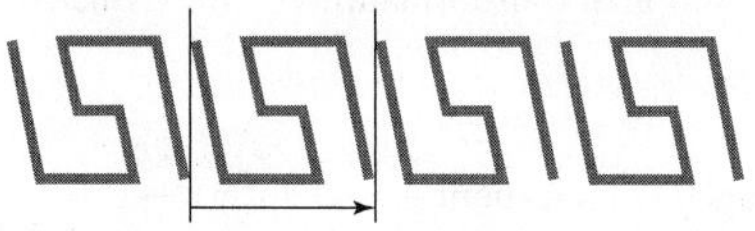

d.

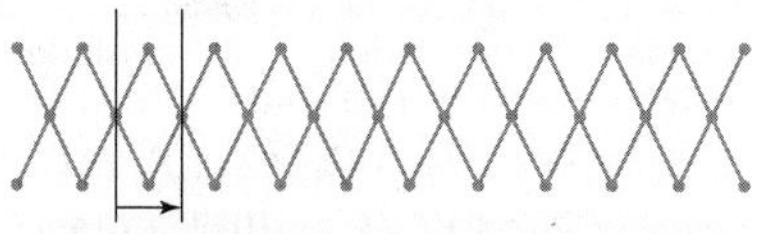

e.

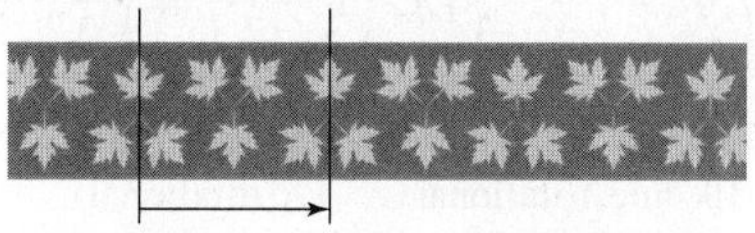

f.

3. Patterns with rotational symmetry have an infinite number of centers of rotation. The diagrams show the centers of rotation in one portion of the repeating pattern.

b. 

c. 

d. 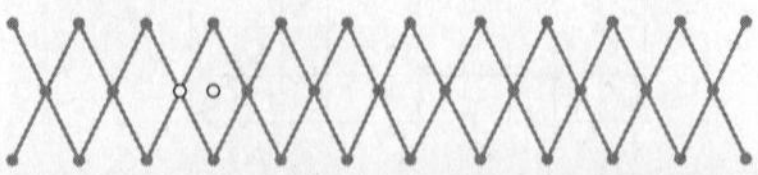

e. 

f. 

5. translation; reflection (two distinct vertical reflection lines) **7.** translation; rotation (two distinct centers) **9.** translation; reflection (one horizontal reflection line), glide reflection

Chapter 8 Review

1. a. No; the image and preimage are not ≅. **b.** $\overline{LA}$, W **3.** $(x, y) \rightarrow (x - 5, y + 10)$ **5.** $A'(6, -4), B'(-2, -1), C'(5, 0)$

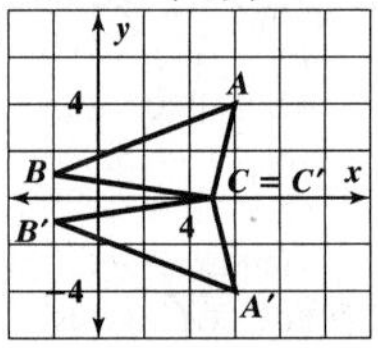

7. $P'(4, -1)$ **9.** P **11.** enlargement; 2

13.

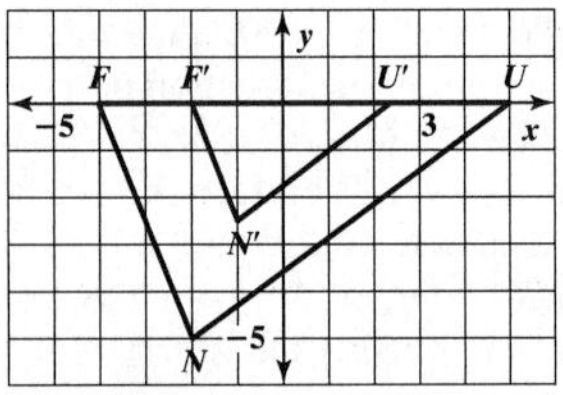

15.

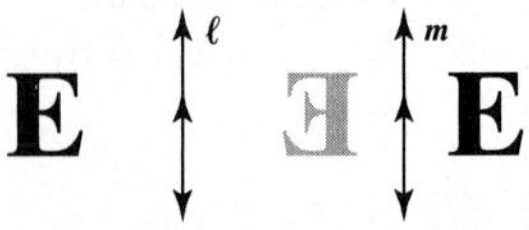

E is translated right, twice the distance between l and m.

17. same; rotation **19.** opposite; glide reflection **21.** $(3, -6)$ **23.** $A'(4, 6), B'(1, -2), C'(0, 5)$

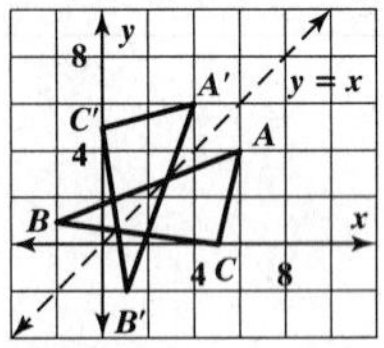

25. rotational: 180° **27.** one **29.** line, rotational **31.** $L'N' = 6.5$ ft, $M'N' = 11.25$ ft

Chapter 8 Vocabulary Check

1. to change **3.** image **5.** isometry **7.** translation; reflection; rotaion **9.** flip **11.** composition of transformations **13.** reduction

Chapter 8 Test

1. Yes, the preimage and image are congruent. **2.** $(x, y) \rightarrow (x + 3, y - 7)$ **3.** $(5, 3)$ **4.** $(-5, -3)$ **5.** $(x, y) \rightarrow (x + 3, y - 3)$ **6.** glide reflection **7.** translation **8.** translation **9.** rotation **10.** $A'(-11, 0), B'(-9, -2), C'(-11, -5), D'(-15, -1)$ **11.** $A'(-3, 8), B'(-5, 6), C'(-3, 3), D'(1, 7)$ **12.** $A'(0, 3), B'(2, 1), C'(5, 3), D'(1, 7)$ **13.** $A'(2, 0), B'\left(\frac{2}{3}, -\frac{4}{3}\right), C'\left(2, -\frac{10}{3}\right), D'\left(\frac{14}{3}, -\frac{2}{3}\right)$ **14.** $A'(-3, 5), B'(-1, 3), C'(-3, 0), D'(-7, 4)$ **15.** $A'(0, 3), B'(-2, 1), C'(-5, 3), D'(-1, 7)$ **16.** $A'(9, 0), B'(3, -6), C'(9, -15), D'(21, -3)$ **17.** line **18.** line, rotational **19.** rotational **20.** reflection **21.** rotation **22.** A dilation withscale factor $n \neq 1$ changes the size of the preimage, so it is not an isometry. **23.** No, line m does bisect $\overline{UH}$, but H is only the reflection image of U across m if m is also perpendicular to $\overline{UH}$. **24.** $a = 4, b = -10$ **25.** Answers may vary. Sample: For a figure to have $x°$ rotational symmetry, x must divide into 360 with no remainder. Since 50 leaves a remainder when divided into 360, a figure cannot have 50° rotational symmetry.

Chapter 8 Standardized Test

1. b **2.** d **3.** b **4.** d **5.** d **6.** a **7.** c **8.** b **9.** c **10.** b **11.** c **12.** a **13.** c **14.** a **15.** c

CHAPTER 9 RIGHT TRIANGLES AND TRIGONOMETRY

Section 9.1

Practice Exercises

1. a. 26 **b.** $10^2 + 24^2 = 100 + 576 = 676 = 26^2$ **2.** $6\sqrt{3}$ **3.** 15.5 in. **4.** No; $16^2 + 48^2 \neq 50^2$ **5.** acute

Vocabulary & Readiness Check 9.1

1. Pythagorean triple **3.** False

Exercise Set 9.1

1. 37 **3.** $3\sqrt{3}$ **5.** 10 **7.** $2\sqrt{5}$ **9.** $3\sqrt{11}$ **11.** 34 **13.** $\sqrt{85}$ **15.** $3\sqrt{2}$ **17.** No; $4^2 + 5^2 \neq 6^2$ **19.** Yes; $10^2 + 24^2 = 26^2$ **21.** 17 ft **23.** 13 ft **25.** 17 m **27.** No; $19^2 + 20^2 \neq 28^2$ **29.** Yes; $7^2 + 24^2 = 25^2$ **31.** acute **33.** obtuse **35.** right **37. a.** the diagonal **b.** 4.2 in. **39.** 10 **41.** $2\sqrt{5}$ **43.** 29 **45.** 84 **47.** From Corollary 7.5–3, $\frac{q}{b} = \frac{b}{c}$ and $\frac{r}{a} = \frac{a}{c}$. Setting cross products equal gives $b^2 = cq$ and $a^2 = cr$. Thus, $a^2 + b^2 = cq + cr = c(q + r)$. Substituting c for $q + r$ gives $a^2 + b^2 = c^2$. **49.** Draw right triangle ΔFDE with legs $\overline{DE}$ of length a and $\overline{EF}$ of length b, and hypotenuse of length x. By the Pythagorean Theorem, $a^2 + b^2 = x^2$. ΔABC has sides of length a, b, and c, where $c^2 > a^2 + b^2$. Thus, $c^2 > x^2$. If $c^2 > x^2$ then $c > x$ since these are positive values and by the Property of Inequalities. Since $c > x$, then $m\angle C > m\angle E$ by the Converse of the Hinge Theorem. An angle with measure greater than 90° is obtuse, so ΔABC is an obtuse triangle. **51.** The three numbers a, b, and c must be whole numbers that satisfy $a^2 + b^2 = c^2$. **53–57.** Answers may vary. Samples are given. **53. a.** 6 **b.** 7 **55. a.** 8 **b.** 11 **57. a.** 8 **b.** 10 **59.** 2830 km **61.** 51° **63.** $\sqrt{3}$ **65.** 4 **67.** 7 **69.** $\frac{16\sqrt{3}}{3}$ **71.** $\frac{\sqrt{2}}{2}$

Section 9.2

Practice Exercises

1. $5\sqrt{6}$ **2.** $5\sqrt{2}$ **3.** 141 ft **4.** $\frac{10\sqrt{3}}{3}$ **5.** 15.6 mm

Vocabulary & Readiness Check 9.2

1. 45° **3.** 1 **5.** 2

Exercise Set 9.2

1. $7\sqrt{2}$ **3.** 3 **5.** $x = 8, y = 8\sqrt{2}$ **7.** $4\sqrt{2}$ **9.** $6\sqrt{3}$ **11.** $x = 4, y = 2$ **13.** $\sqrt{6}$ **15.** $x = 20, y = 20\sqrt{3}$ **17.** $y = z = 5$ **19.** $x = 14, y = 7\sqrt{3}$ **21.** 14.1 cm **23.** 50 ft **25.** $5\sqrt{2}$ ft **27.** shorter leg = 6 in.; longer leg = $6\sqrt{3}$ in. **29.** $a = 7, b = 14, c = 7, d = 7\sqrt{3}$ **31.** $a = 10\sqrt{3}, b = 5\sqrt{3}, c = 15, d = 5$ **33.** $a = 4, b = 4$ **35.** $a = 14, b = 6\sqrt{2}$ **37.** Rika; 5 should be opposite the 30° angle and $5\sqrt{3}$ should be opposite the 60° angle. **39.** the hypotenuse **41.** 24 ft **43.** 8.5 m **45.** Answers may vary. Samples using the following segment are given

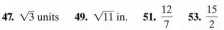

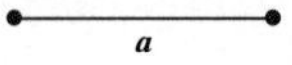

a.

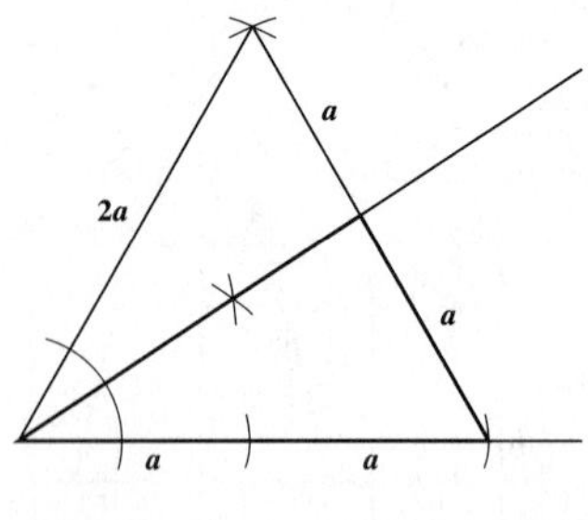

b.

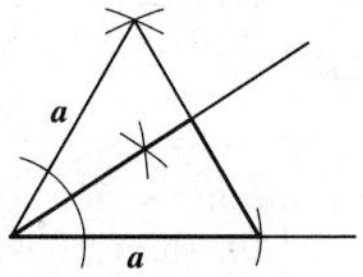

c.

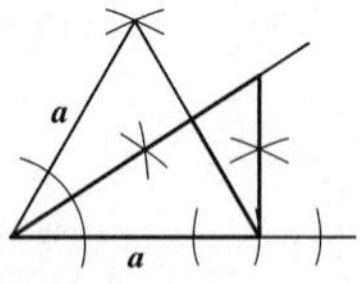

47. $\sqrt{3}$ units **49.** $\sqrt{11}$ in. **51.** $\frac{12}{7}$ **53.** $\frac{15}{2}$

Section 9.3

Practice Exercises

1. $\sin G = \frac{15}{17} \approx 0.8824$; $\cos G = \frac{8}{17} \approx 0.4706$; $\tan G = \frac{15}{8} = 1.875$ **2.** $\sin 30° = \frac{1}{2} = 0.5$; $\cos 30° = \frac{\sqrt{3}}{2} \approx 0.8660$; $\tan 30° = \frac{\sqrt{3}}{3} \approx 0.3774$; $\sin 60° = \frac{\sqrt{3}}{2} \approx 0.8660$; $\cos 60° = \frac{1}{2} = 0.5$; $\tan 60° = \sqrt{3} \approx 1.7321$

3. 2879 ft **4. a.** 68° **b.** 22°

Vocabulary & Readiness Check 9.3

1. cosine **3.** sine **5.** $\frac{12}{13}$ **7.** $\frac{b}{c}$ **9.** sin **11.** cos

Exercise Set 9.3

1. $\sin M = \frac{8}{17} \approx 0.4706$; $\cos M = \frac{15}{17} \approx 0.8824$; $\tan M = \frac{8}{15} \approx 0.5333$ **3.** $\sin M = \frac{4\sqrt{2}}{9} \approx 0.6285$; $\cos M = \frac{7}{9} \approx 0.7778$; $\tan M = \frac{4\sqrt{2}}{7} \approx 0.8081$ **5.** $\sin M = \frac{\sqrt{3}}{2} \approx 0.8660$; $\cos M = \frac{1}{2} = 0.5$; $\tan M = \sqrt{3} \approx 1.7321$

7. 0.5736 **9.** 1 **11.** 0.5299 **13.** 0.0872 **15.** 0.1584 **17.** 0.7193 **19.** 11.5 **21.** 14.4 **23.** 106.5 **25.** 436 ft **27.** 39 in. **29.** 44° **31.** 78° **33.** 48° **35.** 10° **37.** 19° **39.** 11° **41.** 21° **43.** 58° **45.** 59° **47.** 44° and 136°

49.
$$\frac{\sin X}{\cos X} = \sin X \cdot \frac{1}{\cos X} = \frac{\text{opposite}}{\text{hypotenuse}} \cdot \frac{\text{hypotenuse}}{\text{adjacent}} = \frac{\text{opposite}}{\text{adjacent}} = \tan X$$

51.
$$\cos X \cdot \tan X = \frac{\text{adjacent}}{\text{hypotenuse}} \cdot \frac{\text{opposite}}{\text{adjacent}} = \frac{\text{opposite}}{\text{hypotenuse}} = \sin X$$

53. $w \approx 6.7, x \approx 8.1$ **55.** Answers may vary. Sample: $\sin A = \frac{\text{opposite}}{\text{hypotenuse}}$, and the hypotenuse of a right triangle is always the longest side, so sin *A* is a proper fraction. Thus, $\sin A < 1$. **57.** 52 m **59.** The word is made up of the first letters of each ratio: $S = \frac{O}{H}$, $C = \frac{A}{H}$, and $T = \frac{O}{A}$. **61. a.** They are equal; yes, sine and cosine of complementary angles are equal. **b.** $\angle B$; $\angle A$ **c.** Sample: The cosine is the complement's sine.

63.
$$(\sin B)^2 + (\cos A)^2 = \left(\frac{a}{c}\right)^2 + \left(\frac{b}{c}\right)^2 = \frac{a^2}{c^2} + \frac{b^2}{c^2} = \frac{a^2 + b^2}{c^2} = \frac{c^2}{c^2} = 1$$

65.
$$\frac{1}{(\cos A)^2} - (\tan A)^2 = \left(1 \div \frac{b}{c}\right)^2 - \left(\frac{a}{b}\right)^2 = \frac{c^2}{b^2} - \frac{a^2}{b^2} = \frac{c^2 - a^2}{b^2} = \frac{b^2}{b^2} = 1$$

67. 1.5 AU **69.** $\frac{15}{12} = \frac{5}{4}$ **71.** $\frac{9}{12} = \frac{3}{4}$ **73.** $\frac{15}{12} = \frac{5}{4}$ **75.** $4, 4\sqrt{3}$ **77.** $\angle 7$ **79.** $\angle 6$

Section 9.4

Practice Exercises

1. $m\angle B = 66°, a \approx 5.8, c \approx 14.2$ **2.** $f \approx 8.5, m\angle D \approx 69.4°, m\angle E \approx 20.6$ **3. a.** the angle of elevation from the person in the hot-air balloon to the bird **b.** the angle of depression from the person in the hot-air balloon to the base of the mountain **4.** about 631 ft **5.** 54°

Vocabulary & Readiness Check 9.4

1. elevation; *C*; *A* **3.** elevation; *A*; *D* **5.** depression; *B*; *A*

Exercise Set 9.4

1. $m\angle B = 63°; a \approx 10.2; c \approx 22.4$ **3.** $t \approx 15.6; m\angle R = m\angle S = 45°$ **5.** $m\angle D = 55°; e \approx 8.0, d \approx 11.5$ **7.** $m\angle N = 37°; n \approx 6.5; p \approx 10.8$ **9.** $r \approx 6.9; m\angle R \approx 32.2°; m\angle S \approx 57.8°$ **11.** $c = 15.3; m\angle A \approx 68.5°; m\angle B \approx 21.5°$ **13.** angle of elevation from sub to boat **15.** angle of elevation from boat to tree **17.** angle of elevation from Max to top of waterfall **19.** angle of depression from top of waterfall to Max **21.** 34.2 ft **23.** 85.2 m **25.** 263.3 yd **27.** 29 ft **29.** 36° **31.** 7.2 mi **33.** 52.4 ft **35.** 4077.4 ft **37.** 64° **39.** about 6.2 mi **41.** 3300 m **43.** length of any guy wire = distance on the ground from the tower to the guy wire divided by the cosine of the angle formed by the guy wire and the ground. **45.** Answers may vary. Sample: An angle of elevation is formed by two rays with a common endpoint when one ray is horizontal and the other ray is above the horizontal ray. **47.** 72°, 72° **49.** 27°, 27° **51.** about 27.7 ft **53.** $2\sqrt{17} \approx 8.2$ **55.** $\sqrt{229} \approx 15.1$

Section 9.5

Practice Exercises

1. $\langle -307.3, -54.2 \rangle$ **2. a.** 60° south of west **b.** Yes; it can also be described as 30° west of south. **3.** about 257.5 mi at 17.2° north of east **4.** $\langle -2, 1 \rangle$ **5.** about 13.2° north of west

Vocabulary & Readiness Check 9.5

1. vector **3.** arrow **5.** initial point; terminal point

Exercise Set 9.5

1. $\langle 335.5, 217.9 \rangle$ **3.** $\langle -29.3, 41.8 \rangle$ **5.** 18° east of north (or 72° north of east) **7.** 15° south of west (or 75° west of south)

9.

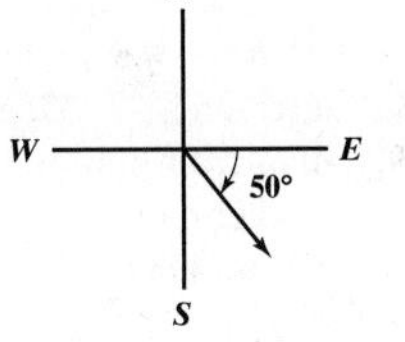

11.

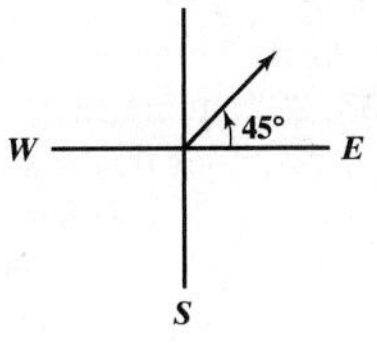

13.

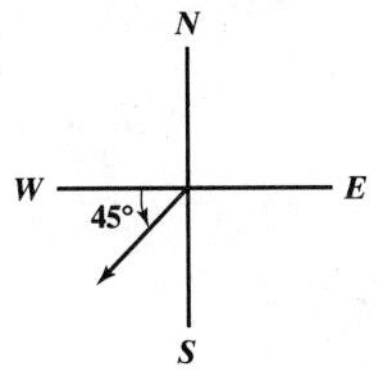

15.

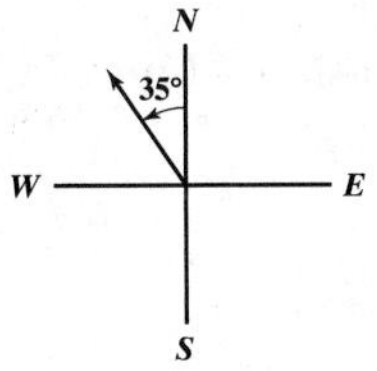

17. about 6.1 **19.** about 5.4 **21.** about 707 mi at 65° south of west **23.** about 54 mph at 22° north of east **25.** $\langle 10, 5\rangle$ **27.** $\langle 8, 6\rangle$ **29.** $\langle -5, -7\rangle$ **31.** $\langle -1, 3\rangle$ **33. a.** $\langle -9, -9\rangle$
b.

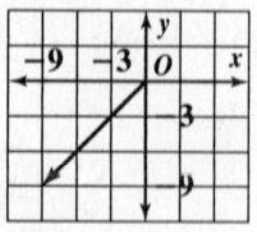

35. a. $\langle -1, 0\rangle$
b.

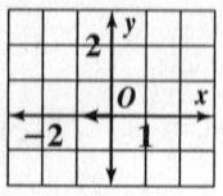

37. a. $\langle -8, 6\rangle$
b.

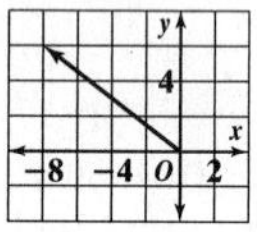

39. about 97 mi at 41° south of west **41. a.** 26.0 mi/h at 15.6° south of west **b.** 16.3° north of west **43.** Answers may vary. Sample: Two vectors are equal if and only if they have the same magnitude and the same direction. **45.** $\langle 5, 5\rangle, \langle 5, 5\rangle$ **47.** Commutative and Associative Properties
49. $\langle -3, -7\rangle$

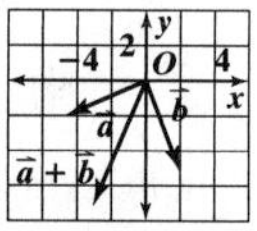

51. A: III; B: II; C: I **53. a.** 75.1° west of south or 14.9° south of west **b.** about 6.7 h **55.** Answers may vary. Sample: $\langle 7, 24\rangle$, $\langle 7, -24\rangle$, $\langle -7, 24\rangle$, $\langle 0, 25\rangle$ **57.** Answers may vary. Sample: Both have an endpoint. A ray extends indefinitely in a direction, while a vector has a terminal point and a magnitude. **59.** The magnitude of each vector is $\sqrt{419}$. **61. a.** $\langle 4, 8\rangle$ **b.** about 4.47; about 8.94; the magnitude of $2\vec{w}$ is two times the magnitude of $\vec{w}$. **c.** If $\vec{v} = \langle v_1, v_2\rangle$ and k is a constant, then $k\vec{v} = \langle kv_1, kv_2\rangle$. The magnitude of $k\vec{v}$ $= k$(magnitude of $\vec{v}$). **63.** $\overline{EF}$ **65.** $\overline{BC}$ **67.** $\angle A$

Chapter 9 Extension—Law of Sines

Practice Exercises

1. $B = 34°, a \approx 12.7$ cm, $b \approx 7.9$ cm **2.** $B = 117.5°, a \approx 8.7, c \approx 5.2$ **3.** $B \approx 41°, C \approx 82°, c \approx 39.0$ **4.** no triangle **5.** $B_1 \approx 50°, C_1 \approx 95°, c_1 \approx 20.8$ and $B_2 \approx 130°, C_2 \approx 15°, c_2 \approx 5.4$ **6.** 34 square meters

Extension—Law of Sines Exercise Set

1. $B = 42°, a = b \approx 8.1$ **3.** $A = 44°, b \approx 18.6, c \approx 22.8$ **5.** $C = 95°, b \approx 81.0, c \approx 134.1$ **7.** $B = 40°, b \approx 20.9, c \approx 31.8$ **9.** $C = 111°, b \approx 7.3, c \approx 16.1$ **11.** $A = 80°, a \approx 39.5, c \approx 10.4$ **13.** $B = 30°, a \approx 316.0, b \approx 174.3$ **15.** $C = 50°, a = b \approx 7.1$ **17.** one triangle: $B \approx 29°, C \approx 111°, c \approx 29.0$ **19.** one triangle: $C \approx 52°, B \approx 65°, b \approx 10.2$ **21.** one triangle: $C \approx 55°, B \approx 13°, b \approx 10.2$ **23.** no triangle **25.** two triangles: $B_1 \approx 77°, C_1 \approx 43°, c_1 \approx 12.6; B_2 \approx 103°, C_2 \approx 17°, c_2 \approx 5.4$ **27.** two triangles: $B_1 \approx 54°, C_1 \approx 89°, c_1 \approx 19.9; B_2 \approx 126°, C_2 \approx 17°, c_2 \approx 5.8$ **29.** two triangles: $C_1 \approx 68°, B_1 \approx 54°, b_1 \approx 21.0; C_2 \approx 112°, B_2 \approx 10°, b_2 \approx 4.5$ **31.** no triangle **33.** 297 square feet **35.** 5 square yards **37.** 10 square meters **39.** 5.7 miles from station A, 9.2 miles from station B **41.** 3576.4 yards from one end, 3671.8 yards from the other end **43.** 184.3 feet **45.** 56.0 feet **47.** 30.0 feet **49. a.** 493.8 feet **b.** 343.0 feet

Chapter 9 Extension—Law of Cosines

Practice Exercises

1. $a = 13, B \approx 28°, C \approx 32°$ **2.** $B = 98°, A \approx 52°, C \approx 30°$

Extension—Law of Cosines Exercise Set

1. $a \approx 6.0, B \approx 29°, C \approx 105°$ **3.** $c \approx 7.6, B \approx 32°, A \approx 52°$ **5.** $B \approx 68°, C \approx 68°, A \approx 44°$ **7.** $A \approx 117°, B \approx 36°, C \approx 27°$ **9.** $c \approx 4.7, A \approx 45°, B \approx 93°$ **11.** $a \approx 6.3, C \approx 28°, B \approx 50°$ **13.** $b \approx 4.7, C \approx 55°, A \approx 75°$ **15.** $b \approx 5.4, C \approx 22°, A \approx 68°$ **17.** $C \approx 112°, A \approx 28°, B \approx 40°$ **19.** $B \approx 100°, A \approx 19°, C \approx 61°$ **21.** $A = B = C = 60°$ **23.** $A \approx 117°, B \approx 18°, C \approx 45°$ **25.** 157° **27.** 61.7 miles **29.** 193 yards **31.** 78.5° north of east **33. a.** 19.3 mi **b.** 32.4° south of east **35.** uphill: 398.2 ft; downhill: 417.4 ft **37.** 63.7 ft

Chapter 9 Vocabulary Check

1. vector **2.** resultant **3.** angle of elevation **4.** angle of depression **5.** Pythagorean triple **6.** trigonometric **7.** magnitude **8.** arrow **9.** initial point; terminal point **10.** tangent **11.** cosine **12.** sine

Chapter 9 Review

1. $2\sqrt{113}$ **3.** $12\sqrt{2}$ **5.** $x = 7, y = 7\sqrt{2}$ **7.** $x = 6\sqrt{3}, y = 12$
9. $\sin A = \frac{2\sqrt{19}}{20} = \frac{\sqrt{19}}{10}; \cos A = \frac{18}{20} = \frac{9}{10}; \tan A = \frac{2\sqrt{19}}{18} = \frac{\sqrt{19}}{9}$
11. 16.5 **13.** 206.2 km; 76.0° south of west **15.** $\langle 1, 4\rangle$ **17.** 16 **19.** 70.7 ft **21.** 38.2 ft

Chapter 9 Test

1. $\sqrt{170}$ **2.** $2\sqrt{14}$ **3.** $\frac{11\sqrt{2}}{2}$ **4.** $x = 4\sqrt{3}, y = 8\sqrt{3}$ **5.** acute **6.** right **7.** obtuse **8.** yes; $32^2 + 60^2 = 68^2$ **9.** no; $1^2 + 2^2 \neq 3^2$ **10.** no; $2.5^2 + 6^2 = 6.5^2$, but the numbers have to be integers to be a Pythagorean triple. **11.** $\sin B = \frac{2\sqrt{57}}{22} = \frac{\sqrt{57}}{11}; \cos B = \frac{16}{22} = \frac{8}{11};$ $\tan B = \frac{2\sqrt{57}}{16} = \frac{\sqrt{57}}{8}$ **12.** 48.0° **13.** 0.5592 **14.** 36.9° **15.** 26 ft **16.** 11.3 ft **17.** 9.5 **18.** 28.3° **19.** $B = 34°, a \approx 14.1, b \approx 9.5$ **20.** $a = 10.8, B \approx 10.5°, A \approx 79.5°$ **21.** 174.9 m/h at 59.0° north of east **22.** 52.2 mi at 16.7° south of east **23.** $\langle 1, 8\rangle$ **24.** 7.6 mph in the direction 23° south of west (or 67° west of south)

Chapter 9 Standardized Test

1. d **2.** d **3.** c **4.** b **5.** b **6.** a **7.** c **8.** c **9.** a **10.** d **11.** a **12.** a **13.** d **14.** a **15.** c **16.** b **17.** b **18.** c **19.** a

CHAPTER 10 AREA

Section 10.1

Practice Exercises

1. 1260° **2.** $x = 75, m\angle M = 75°, m\angle N = 130°$ **3.** 108° **4.** 9 sides **5.** 24° **6.** $x = 75$; $x° = 75°$; $(x - 4)° = 71°$, $(2x - 11)° = 139°$
7. $m\angle 1 = m\angle 2 = m\angle 3 = m\angle 4 = 60°, m\angle 5 = 120°$; $m\angle 1 + m\angle 2 + m\angle 3 + m\angle 4 + m\angle 5 = 360°$; yes, this is a tessellation.

Vocabulary & Readiness Check 10.1

1. $(11 - 2)\cdot 180°$ **3.** $\frac{(11 - 2)\cdot 180°}{11}$ **5.** 180°

Exercise Set 10.1

1. 900° **3.** 1980° **5.** 3240° **7.** 36,000° **9.** 88° **11.** 103° **13.** 108° **15.** 150° **17.** 3 **19.** 13 **21.** 40° **23.** 20° **25.** interior angle: 144°; exterior angle: 36° **27.** $y = 103, z = 70$ **29.** $w = 72, x = 59, y = 49, z = 121$ **31.** interior angle: 108°; sides: 5 **33.** interior angle: 162°; sides: 20 **35. a.** regular hexagon: interior angles = 120° **b.** $m\angle 1 = m\angle 2 = m\angle 3 = 120°$ **c.** $m\angle 1 + m\angle 2 + m\angle 3 = 120° + 120° + 120° = 360°$ **37. a.** square:

interior angles $= 90°$; equilateral triangle: interior angles $= 60°$ **b.** $m\angle 1 = m\angle 4 = 90°; m\angle 2 = m\angle 3 = m\angle 5 = 60°$ **c.** $m\angle 1 + m\angle 2 + m\angle 3 + m\angle 4 + m\angle 5 = 90° + 60° + 60° + 90° + 60° = 360°$ **39. a.** regular hexagon: interior angles $= 120°$; equilateral triangle: interior angles $= 60°$ **b.** $m\angle 1 = m\angle 3 = 120°; m\angle 2 = m\angle 4 = 60°$ **c.** $m\angle 1 + m\angle 2 + m\angle 3 + m\angle 4 = 120° + 60° + 120° + 60° = 360°$ **41. a.** square: interior angles $= 90°$; regular hexagon: interior angles $= 120°$; regular 12-gon: interior angles $= 150°$ **b.** $m\angle 1 = 120°; m\angle 2 = 150°; m\angle 3 = 90°$ **c.** $m\angle 1 + m\angle 2 + m\angle 3 = 120° + 150° + 90° = 360°$ **43. a.** square: interior angles $= 90°$ **b.** $m\angle 1 = m\angle 3 = 90°; m\angle 2 = 180°$ **c.** $m\angle 1 + m\angle 2 + m\angle 3 = 90° + 180° + 90° = 360°$
45. Answers may vary. Sample:

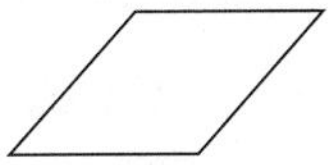

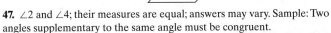

47. $\angle 2$ and $\angle 4$; their measures are equal; answers may vary. Sample: Two angles supplementary to the same angle must be congruent.
49. Yes; the measure of each angle is 60°, and 60 is a factor of 360.
51. No; the measure of each angle is 108°, and 108 is not a factor of 360.
53. No; the measure of each angle is 135°, and 135 is not a factor of 360.
55. A regular polygon with more than 6 sides must have angle measures greater than 120° and at least 3 polygons must meet at each vertex. The sum of 3 or more angles with measures greater than 120° is greater than 360°. So the only 3 regular polygons that tessellate are 3-, 4-, and 6-sided, since their angle measures are factors of 360. **57.** 45°, 45°, 90° **59.** 20°, 80°, 80° or 80°, 50°, 50° **61. a.** n triangles **b.** 180° **c.** $n \cdot 180°$ **d.** 360° **e.** 2 **63. a.** 180° **b.** 180° **c.** 180° **d.** $(n)180°$ **e.** $(n - 2)180°$ **f.** + 360° **65.** Yes; answers may vary. Sample: Multiplying the number of triangles by 180° gives the sum of the interior angles of the polygon, plus all the angles that meet at the point inside, which sum to 360°. **67.** Yes **69.** Not possible
71.

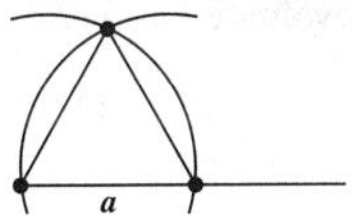

73. No; $16 + 26 = 42$ which is not greater than 42. **75.** Yes; $19.5 + 10 > 20.5$ **77.** 25 cm^2 **79.** 11.5 m^2

Section 10.2

Practice Exercises

1. 108 sq m **2.** 30 sq in. **3.** 102 m **4.** 396 sq m **5.** 94.5 sq cm **6.** 12 sq m **7.** 54 sq in.

Vocabulary & Readiness Check 10.2

1. perimeter **3.** base **5.** base **7.** area

Exercise Set 10.2

1. 200 sq m **3.** 26.79 sq in. **5.** 96 sq cm **7.** 14 sq m **9.** 66 m; 220 sq m **11.** 74 cm; 216 sq cm **13.** 472 sq in. **15.** 150 sq cm **17.** 30 sq ft **19.** 44 sq in. **21.** 80 sq in. **23.** 24 sq ft **25.** 1200 sq ft **27.** 30 sq ft **29.** 36.75 sq ft **31.** 28 sq m **33.** 86 sq mi **35.** 24 sq cm **37.** 25 sq cm **39.** 87,150 sq mi **41.** B **43.** 15 sq units **45.** 6 sq units **47.** 27 sq units **49.** 21 sq units **51.** 15 sq units **53.** 15 sq units **55.** 18 sq cm **57.** $32\sqrt{3}$ sq m **59.** 4 in. **61.** 49.9 sq ft **63.** 60 sq units **65.** 20 sq units **67.** 35.42 sq cm **69.** 168 sq ft **71. a.** 381 sq ft **b.** 4 squares **73.** 1390 sq ft **75. a.**

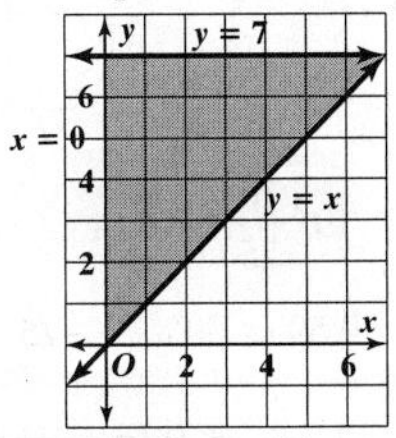

b. 24.5 sq units **77. a.**

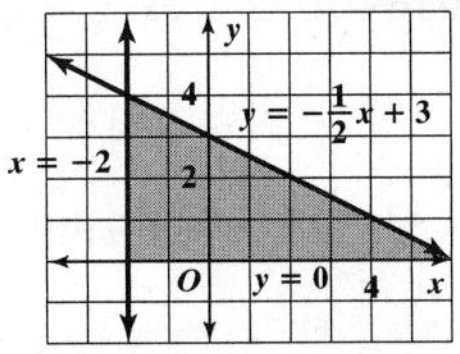

b. 16 sq units

79. 312.5 sq ft **81.** 34 sq in. **83.** 54 sq in.
85. The area is tripled; explanations may vary. Sample: If $A = \frac{1}{2}b \cdot h$, then $\frac{1}{2}(b \cdot 3h) = 3 \cdot \frac{1}{2}(b \cdot h) = 3A$. **87.** Explanations may vary. Sample: Each kite section is half of a corresponding rectangle section.
89. 6800 sq ft **91.** No; explanations may vary. Sample: Two altitudes of an obtuse triangle lie outside the triangle. The legs of a right triangle are two altitudes of the triangle. **93.** No; in the formula for the area of a trapezoid, half the sum of the bases would have to equal the base of the parallelogram in order for the areas to be the same. This is not possible since the other base of the trapezoid will be longer or shorter than the given base. **95.** No; unless the rhombus is a square, you cannot calculate the area without knowing the lengths of the diagonals.
97. 72 sq cm **99.** $25\sqrt{3}$ sq cm

Section 10.3

Practice Exercises

1. $m\angle 1 = 45°, m\angle 2 = 22.5°, m\angle 3 = 67.5°$ **2.** 220 sq cm **3.** 665 sq ft **4.** 270.9 sq m **5.** 28 sq in.

Vocabulary & Readiness Check 10.3

1. radius **3.** apothem

Exercise Set 10.3

1. $m\angle 1 = 120°, m\angle 2 = 60°, m\angle 3 = 30°$
3. $m\angle 7 = 60°, m\angle 8 = 30°, m\angle 9 = 60°$
5. 196 ft, 2842 sq ft **7.** 180 in., 2475 in. **9.** 168 cm, 2192.4 sq cm **11.** 40 in., 100 sq in. **13.** 36 ft, 93.5 sq ft **15.** 108 ft, 841.8 sq ft **17.** 30 cm, 43.3 sq cm **19.** 72 sq cm **21.** $384\sqrt{3}$ sq in. $\approx$ 665.1 sq in. **23.** $12\sqrt{3}$ sq in. $\approx$ 20.8 sq in. **25.** $900\sqrt{3}$ sq m $\approx$ 1558.8 sq m **27.** $72\sqrt{3}$ sq cm $\approx$ 62.4 sq cm **29. a.** 72° **b.** 54° **31. a.** 40° **b.** 70° **33.** 21.4 sq ft **35.** 173.8 sq cm **37.** 1200 sq cm **39.** 408.3 sq mm **41.** 38.04 sq in.
43. $m\angle 1 = 36°, m\angle 2 = 18°$
45. 24.6 m, 46.3 sq m **47.** 61.2 m, 282.8 sq m
49. a–d.

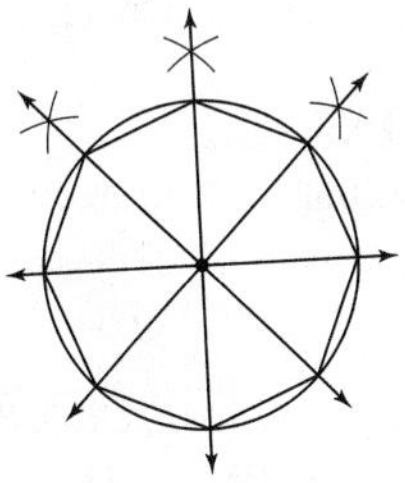

regular octagon

51. a. 9.1 in. **b.** 6 in. **c.** 3.7 in. **d.** Answers may vary. Sample: About 4.5 in.; the length of a side of a pentagon should be between 3.7 in. and 6 in. **53.** A radius is the distance from the center to a vertex, while the apothem is the perpendicular distance from the center to a side.

55. $s = 2a$ **57.** $s = 2\sqrt{3} \cdot a$ **59.** 73 sq cm **61.** 9.7 ft
63. $b = s, h = \frac{\sqrt{3}}{2}s; A = \frac{1}{2}bh = \frac{1}{2}s \cdot \frac{\sqrt{3}}{2}s = \frac{s^2\sqrt{3}}{4}$
65. The apothem is perpendicular to a side of the pentagon. Two right triangles are formed with the radii of the pentagon. The triangles are congruent by Hypotenuse-Leg. So the angles formed by the apothem and radii are congruent because corresponding parts of congruent triangles are congruent. Therefore, the apothem bisects the vertex angle.
67. (2.8, 2.8) **69.** 46 sq m **71.** 28 in., 49 sq in. **73.** 24 cm, 24 sq cm

Section 10.4

Practice Exercises

1. a. $\frac{4}{11}$ **b.** 11 **c.** $\frac{4}{11}$ **d.** $\frac{16}{121}$ **2.** 54 sq in.
3. \$6.94 **4.** $5\sqrt{5}:3$

Vocabulary & Readiness Check 10.4

1. scale factor **3.** $\frac{x^2}{y^2}$

Exercise Set 10.4

1. a. $\frac{1}{2}$ **b.** $\frac{1}{2}$ **c.** $\frac{1}{4}$ **3. a.** $\frac{4}{3}$ **b.** $\frac{4}{3}$ **c.** $\frac{16}{9}$ **5. a.** $\frac{4}{3}$ **b.** $\frac{4}{3}$ **c.** $\frac{16}{9}$ **7.** 24 sq in. **9.** 59 sq ft **11.** 69.3 sq ft **13.** \$384 **15.** \$63.20 **17.** $\frac{1}{2}; \frac{1}{2}$ **19.** $\frac{7}{3}; \frac{7}{3}$ **21.** $\frac{4}{1}; \frac{4}{1}$ **23.** $\frac{3}{1}; \frac{9}{1}$ **25.** $\frac{2}{3}; \frac{4}{9}$ **27.** $\frac{6}{1}; \frac{36}{1}$ **29.** C **31.** $x = 2$ cm; $y = 3$ cm **33.** $x = 4$ cm, $y = 6$ cm **35.** $x = 4\sqrt{2}$ cm, $y = 6\sqrt{2}$ cm **37. a.** $\frac{8}{3}$ **b.** $\frac{64}{9}$ **39. a.** $\frac{5}{7}$ **b.** $\frac{25}{49}$ **41.** $6\sqrt{3}$ sq cm **43.** 0.3 sq cm **45.** $2\frac{1}{4}$ in. by 12 in.

47.

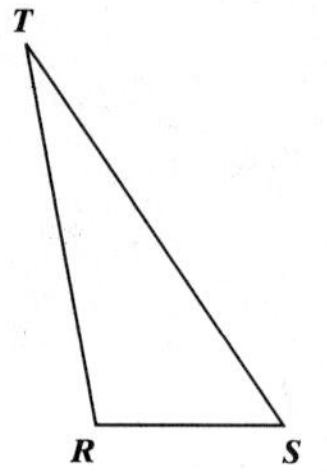

49. Answers will vary. **51.** $\frac{1}{2}$ **53.** 20,000 sq ft; 80,000 sq ft **55.** $\Delta A'B'C'$ is similar to ΔABC. For $\Delta A'B'C'$, the length $A'C'$ is 5 in., using the given scale. **57.** $P \approx 11.4$ in.; $A \approx 4.75$ sq in. **59.** For two similar figures, the ratio of their areas is the square of the ratio of the perimeters. **61.** Answers may vary. Sample: The ratios of perimeters and areas of similar figures are not equal unless the figures are congruent, in which case each ratio is 1. **63.** Always; similar rectangles with equal perimeters have a scale factor of 1:1, so they are congruent. **65.** Never; if the rectangles were similar, then the areas being the same would mean that the scale factor was 1:1, and the perimeters would have to be the same. **67.** 50 sq cm **69.** 168 sq ft

Section 10.5

Practice Exercises

1. a. $\widehat{SP}, \widehat{SQ}, \widehat{PQ}, \widehat{QR}, \widehat{RS}$ **b.** $\widehat{RSP}, \widehat{RQP}$ **c.** $\widehat{PQS}, \widehat{PSQ}, \widehat{SPR}, \widehat{QRS}, \widehat{RSQ}$ **2. a.** 77° **b.** 103° **c.** 208° **d.** 283° **3.** 2.356 in. **4.** 1.3π m

Vocabulary & Readiness Check 10.5

1. diameter **3.** circle; center **5.** central **7.** degrees; arc length **9.** minor **11.** adjacent

Exercise Set 10.5

1. $\widehat{CDF}, \widehat{CBF}$ **3.** $\widehat{FE}, \widehat{FD}, \widehat{ED}, \widehat{EC}, \widehat{DC}$ **5.** $\widehat{CBE}, \widehat{CBD}, \widehat{BED}, \widehat{BEC}, \widehat{DBF}, \widehat{EBF}, \widehat{BDF}, \widehat{DBE}$ **7.** 128° **9.** 218° **11.** 52° **13.** 180° **15.** 232° **17.** 90° **19.** 20π cm **21.** 6π ft **23.** $\widehat{CD}$ **25.** $\widehat{ACB}$ (or $\widehat{ADB}$) **27.** $\widehat{BDC}$ **29.** 81° **31.** 18π cm **33.** $\frac{23\pi}{4}$ cm **35.** $\frac{7\pi}{2}$ cm **37.** 8π ft **39.** $\frac{23\pi}{2}$ m **41.** 70° **43.** 110° **45.** 235° **47.** b **49.** 19 in. **51.** 122π cm ≈ 383.27 cm **53.** 38 **55.** 6° **57.** 120° **59.** 31 m **61.** 5.125π ft **63.** 12π cm **65.** 2.5π units ≈ 7.9 units **67.** 3:4 **69.** The circumference is doubled; explanations may vary. Sample: Since $C = 2\pi r$, doubling the radius results in $2\pi(2r) = 2(2\pi r) = 2C$. **71.** The measure of an arc corresponds to the measure of a central angle; an arc length is a fraction of the circle's circumference. **73.** 325.7 yd **75.** 325.7 yd, 333.5 yd, 341.4 yd, 349.2 yd **77.** Since $\overline{AR} \cong \overline{RW}$ and $AR + RW = AW$ by the Segment Addition Postulate, $AW = 2 \cdot AR$. So the radius of the outer circle is twice the radius of the inner circle. Because $\angle QAR$ and $\angle SAU$ are vertical angles and $m\angle SAT = \frac{1}{2} m\angle SAU$, $m\angle QAR = 2 \cdot m\angle SAT$. The length of $\widehat{ST}$ is then $\frac{m\angle SAT}{360} \cdot 2\pi(2 \cdot AR) = \frac{m\angle SAT}{90} \cdot \pi(AR)$ and the length of $\widehat{QR}$ is $\frac{m\angle QAR}{360} \cdot 2\pi(AR) = \frac{2 \cdot m\angle SAT}{360} \cdot 2\pi(AR) = \frac{m\angle SAT}{90} \cdot \pi(AR)$. Therefore, the length of $\widehat{ST}$ equals the length of $\widehat{QR}$ by the Transitive Property of Equality. **79.** $m\angle 1 = 30°$, $m\angle 2 = 15°$, $m\angle 3 = 75°$, $m\angle 4 = 30°$ **81.** 120 mm **83.** No, it could be an isosceles trapezoid **85.** 16π in.

Section 10.6

Practice Exercises

1. a. 25π sq cm; 78.54 sq cm **b.** $\sqrt{\frac{117}{\pi}}$ yd; 6.10 yd **2.** about 1790.49 sq in. **3.** 2π sq in.; 6.28 sq in. **4.** 4.6 sq m

Vocabulary & Readiness Check 10.6

1. segment **3.** area of the circle

Exercise Set 10.6

1. 9π sq m; 28.27 sq m **3.** 0.7225π sq ft; 2.27 sq ft **5.** $90{,}000\pi$ sq ft; 282,743.34 sq ft **7.** 12.62 m **9.** 5.41 in. **11.** 114.91 sq ft **13.** 1145.92 sq in. **15.** 40.5π sq yd; 127.23 sq yd **17.** 64π sq cm; 201.06 sq cm **19.** $\frac{169\pi}{6}$ sq m; 88.49 sq m **21.** $\frac{25\pi}{4}$ sq m **23.** 24π sq in. **25.** 22.1 sq cm **27.** 18.3 sq ft **29.** 3.3 sq m **31.** $(243\pi + 162)$ sq ft **33.** $(54\pi + 20.25\sqrt{3})$ sq ft **35.** $(120\pi + 36\sqrt{3})$ sq m **37.** $(4 - \pi)$ sq ft **39.** $(784 - 196\pi)$ sq in. **41.** 12.6 sq mi **43. a.** 7620.13 sq in. **b.** 7313.82 sq in. **c.** 306.31 sq in. **45. a.** 166 sq mm **b.** 50 sq mm **c.** 116 sq mm **47.** A sector of a circle is a region bounded by an arc and the two radii to the endpoints of the arc. A segment is a part of a circle bounded by an arc and the segment joining the arc's endpoints. **49.** 6^2 was incorrectly evaluated as $6 \cdot 2$. **51.** 22.6 sq mm **53.** 169π sq in. **55.** $\left(\frac{5\pi}{6} - 2\sin 75°\right)$ sq ft or $\left[\frac{5\pi}{6} - 4(\sin 37.5°)(\cos 37.5°)\right]$ sq ft **57.** 3; $\sqrt{3}$ units; $\frac{1}{2}$ units; $3\sqrt{3}$ units; $\frac{3\sqrt{3}}{4}$ sq units **59.** 6; 1 unit; $\frac{\sqrt{3}}{2}$ units; 6 units; $\frac{3\sqrt{3}}{2}$ sq units **61.** 2π units; π sq units **63.** 10π cm **65.** $\frac{1}{6}$ **67.** $\frac{1}{2}$

Section 10.7

Practice Exercises

1. $\frac{1}{2}$, 0.5, or 50% **2.** $\frac{1}{5}$, 0.2, or 20% **3.** $\frac{1}{2}$, 0.5, or 50% **4.** 0.12 or 12%

Vocabulary & Readiness Check 10.7

1. $\frac{\text{length of } \overline{QR}}{\text{length of } \overline{CD}}$ **3.** geometric

Exercise Set 10.7

1. $\frac{3}{7}$ **3.** $\frac{4}{7}$ **5.** $\frac{1}{2}$, 0.5, or 50% **7.** $\frac{3}{5}$, 0.6, or 60% **9.** 1 or 100% **11.** $\frac{1}{3}$ **13.** $\frac{2}{3}$ **15.** $\frac{1}{4}$, 0.25, or 25% **17.** $\frac{1}{2}$, 0.5, or 50% **19.** $\frac{16}{25}$, 0.64, or 64% **21.** $\frac{600 - 25\pi}{600} \approx 0.87$ or 87% **23.** $\frac{115}{196} \approx 0.59$ or 59% **25.** $\frac{2}{9} \approx 0.22$ or 22% **27.** $\frac{1}{49} \approx 0.02$ or 2% **29.** $\frac{8}{49} \approx 0.16$ or 16% **31.** $\frac{2}{5}$, 0.4, or 40% **33.** $\frac{1}{4}$, 0.25, or 25% **35.** 3 in. **37.** $\frac{1}{25}$, 0.04, or 4% **39.** $\frac{7}{25}$, 0.28, or 28% **41.** $\frac{3}{10}$, 0.3, or 30% **43.** $\frac{3}{5}$ or 60%; the probability that the point is in any one of the 5 sectors is $\frac{1}{5}$, and 3 sectors are not shaded.

45. $\frac{2}{3}$ or $66\frac{2}{3}\%$; since $\frac{SQ}{QT} = \frac{1}{2}$, cross multiply and $2(SQ) = QT$.
The probability that the point will lie on $\overline{QT} = \frac{\text{favorable length}}{\text{entire length}}$
$= \frac{QT}{ST} = \frac{2(SQ)}{3(SQ)} = \frac{2}{3}$. **47.** about 0.056 or 5.6% **49.** $\frac{3}{20}$, 0.15, or 15% **51.** $\frac{3}{5}$, 0.6, or 60% **53.** $\frac{1}{10}$, 0.1, or 10% **55.** $\frac{1}{40}$, 0.025, or 2.5% **57.** about 8.7% **59. a.** yes **b.** no **c.** $\frac{2}{3}$

61. $\frac{\pi}{20}$ or about 16% **63. a.** 45 cm **b.** 63 cm **c.** 71 cm **d.** 77 cm **e.** 89 cm **f.** 100 cm **65.** 100π sq ft
67. rotational, line or reflectional

69.

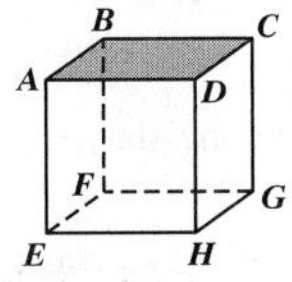

Plane ABCD

71.

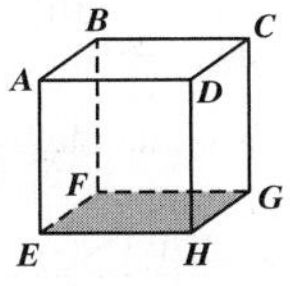

Plane EFGH

73.

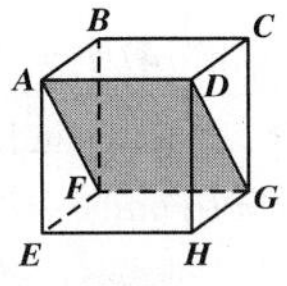

Plane ADGF

Chapter 10 Vocabulary Check

1. 180° **2.** n **3.** 360° **4.** $\frac{360°}{n}$ **5.** $\frac{(n-2)\cdot 180°}{n}$
6. $(n-2)\cdot 180°$ **7.** 3.14; $\frac{22}{7}$ **8.** congruent **9.** circle; center
10. radius **11.** diameter **12.** central angle **13.** central angle
14. center **15.** radius **16.** apothem **17.** concentric
18. semicircle **19.** major arc **20.** minor arc **21.** adjacent arcs
22. segment **23.** sector

Chapter 10 Review

1. interior: 120°; exterior: 60° **3.** interior: 108°; exterior: 72°
5. 159° **7.** 10 sq m **9.** 160 sq ft **11.** $96\sqrt{3}$ sq mm
13. 117 sq cm **15.** 30 sq ft **17.** $9\sqrt{3}$ sq in. **19.** $2400\sqrt{3}$ sq cm

21.

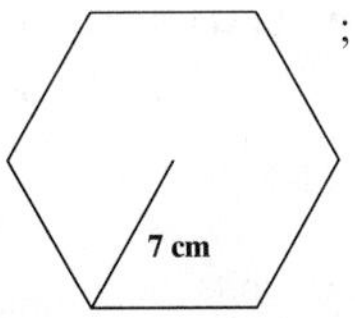

; 127.3 sq cm **23.** $\frac{4}{9}$ **25.** $\frac{1}{4}$ **27.** 73.5 sq ft
29. 124.7 sq in. **31.** 30°
35. $\frac{22\pi}{9}$ in. **37.** $\frac{25\pi}{9}$ m
39. 144π sq in. **41.** 41.0 sq cm

43. $\frac{1}{2}$ or 0.5 or 50% **45.** $\frac{1}{6}$ or about 16.7% **47.** 162°
49. 40 sq cm **51.** 1038 sq mm **53.** $\frac{2}{3}$ **55.** 24.6 sq ft
57. 36.2 sq cm **59.** $\frac{1}{2}$ or 0.5 or 50%

Chapter 10 Test

1. $a = 105, m = 116$ **2.** $x = 122, w = 90$ **3.** interior: 170°; exterior: 10° **4.** 84 sq in. **5.** 112 sq in. **6.** 7 cm **7.** 173 sq m
8. 135 sq in. **9.** 54 sq m **10.** 56 sq cm **11.** 26 sq ft
12. 363.3 sq mm **13.** 841.8 sq yd **14.** 1086.4 sq cm **15.** $\frac{25}{49}$
16. 60 in. **17.** 45 sq in. **18.** 3 sq ft **19.** 40° **20.** 50°
21. 270° **22.** 310° **23.** 10π cm **24.** 3π in. **25.** 31.42 sq cm
26. 25.73 sq m **27.** 252.74 sq in. **28.** Fly A; the probability of landing on the yellow region of the ruler is $\frac{1}{2}$ and the probability of landing on the yellow region of the target is $\frac{1}{4}$.

Chapter 10 Standardized Test

1. b **2.** b **3.** c **4.** a **5.** b **6.** c **7.** a **8.** c
9. d **10.** d **11.** a **12.** a **13.** c **14.** b **15.** b
16. c **17.** a **18.** a **19.** b **20.** d **21.** d **22.** d

CHAPTER 11 SURFACE AREA AND VOLUME

Section 11.1

Practice Exercises

1. a. Six vertices: R, S, T, U, V, W; nine edges: $\overline{RS}, \overline{RU}, \overline{RV}, \overline{ST}, \overline{SW}, \overline{TU}, \overline{TW}, \overline{UV}, \overline{VW}$; five faces: ΔRUV, ΔSTW, quadrilateral $RSTU$, quadrilateral $RSWV$, quadrilateral $TUVW$ **b.** No, an edge is a segment formed by the intersection of two faces. $\overline{TV}$ is a segment that is contained in only one face, so it is not an edge. **2. a.** 12 **b.** 30 **3.** the prism **4. a.** circle **b.** trapezoid **5. a.** 40 m **b.** east **6.** sphere

Vocabulary & Readiness Check 11.1

1. net **3.** topographic map **5.** polyhedra **7.** edge

Exercise Set 11.1

1. Five vertices: A, B, C, D, E; eight edges: $\overline{AB}, \overline{AC}, \overline{AD}, \overline{AE}, \overline{BC}, \overline{BE}, \overline{CD}, \overline{DE}$; five faces: ΔABC, ΔACD, ΔADE, ΔABE, quadrilateral $BCDE$ **3.** Eight vertices: A, B, C, D, E, F, G, H; twelve edges: $\overline{AB}, \overline{BC}, \overline{CD}, \overline{DA}, \overline{EF}, \overline{FG}, \overline{GH}, \overline{HE}, \overline{AE}, \overline{BF}, \overline{CG}, \overline{DH}$; six faces: quadrilaterals $ABCD$, $EFGH$, $ABFE$, $BFGC$, $CGHD$, $DHEA$
5. 8 **7.** 12 **9.** 12 **11.** yes **13.** no **15.** the area between two concentric circles **17.** rectangle **19.** triangle **21.** rectangle **23.** circle **25.** 150 ft; b **27.** 200–250 ft; a **29.** north; a
31. Rapid Hill; a

33.

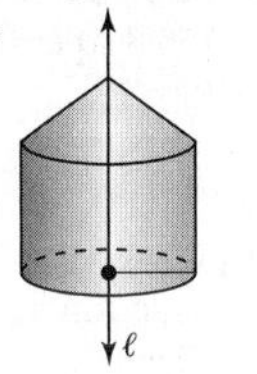

35.

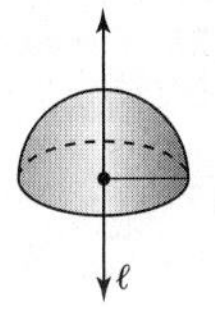

37. 4; equilateral triangle; 4; 6; c
39. 8; equilateral triangle; 6; 12; b
41. 20; equilateral triangle; a
43. 6; 12; 8; 6 squares;

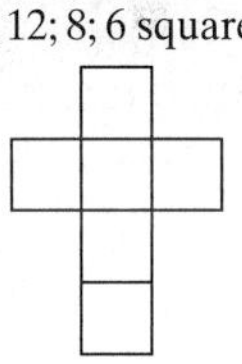

45. 2 circles, 1 rectangle;

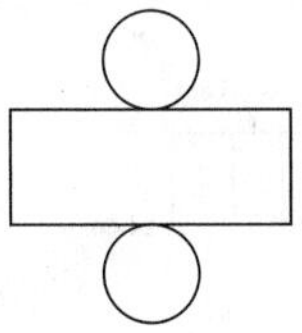

47. 5; 8; 5; 1 square, 4 triangles;

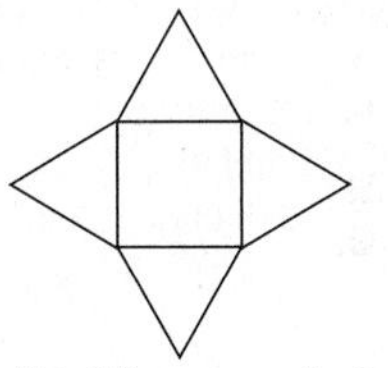

49. yes **51.** no **53.** The third figure; a cylinder is not a polyhedron because its faces are not polygons. **55.** 6 in.
57. a.

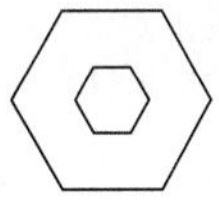

b.

59. Yes; sample: the plane intersects a corner of the cube.

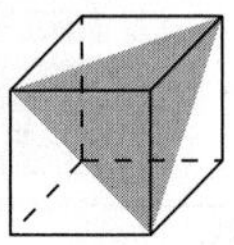

61.

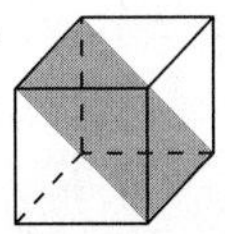

rectangle **63.**

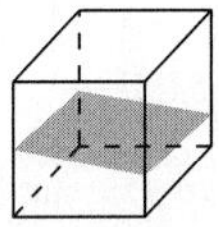

rectangle

65.

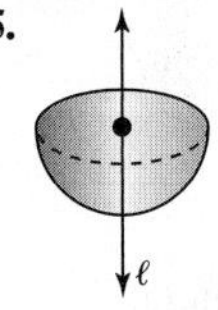

67.

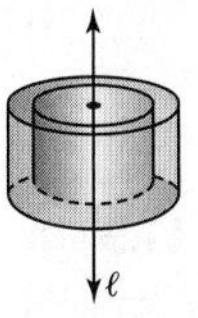

69. Each side of a polygon forming a face is an edge shared by two faces of the polyhedron, so the total number of edges is one-half of the total number of sides of the polygons. **71.** answers may vary **77.** 4.7 **79.** 96 sq cm **81.** 40π sq cm

73.

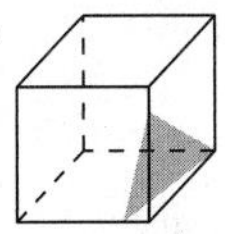

75.

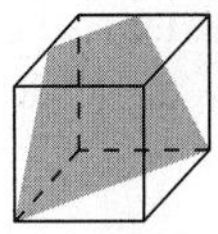

Section 11.2

Practice Exercises

1. 216 sq cm **2. a.** 36 m **b.** 432 sq m **c.** $54\sqrt{3}$ sq m **d.** 619 sq m **3.** 380π sq cm **4.** 126π sq ft $\approx$ 395.64 sq ft

Vocabulary & Readiness Check 11.2

1. cylinder **3.** right **5.** lateral area **7.** prism **9.** lateral faces **11.** height **13.** oblique **15.** surface area

Exercise Set 11.2

1. 1726 sq cm

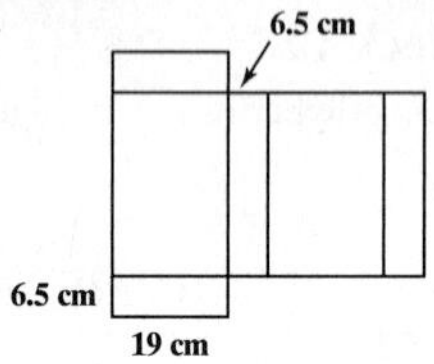

3. 216 sq ft

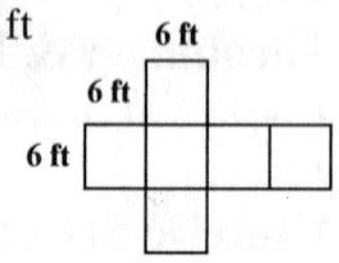

5. $(80 + 32\sqrt{2})$ sq in. $\approx$ 125.3 sq in.

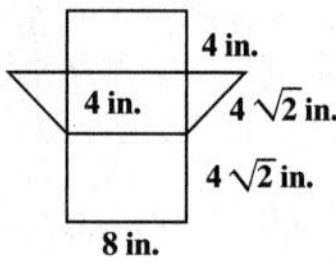

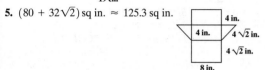

7. right hexagonal prism **9.** $48\sqrt{3}$ sq cm $\approx$ 83.1 sq cm **11.** 220 sq ft **13.** 108 sq in. **15.** 1121 sq cm **17.** 170 sq m **19.** 82 sq in. **21.** 40π sq cm **23.** 16.5π sq cm **25.** 48π sq cm $\approx$ 150.8 sq cm **27.** 236 sq in. **29.** 270 sq in. **31.** 187 sq in. **33.** 30 sq in. **35.** 217 sq in. **37.** 80π sq ft $\approx$ 251 sq ft **39.** 76π sq ft $\approx$ 239 sq ft **41.** 147.015 sq cm **43.** $A(3,0,0)$, $B(3,5,0)$, $C(0,5,0)$, $D(0,5,4)$ **45.** 3 units **47.** 15 sq units **49.** 20 sq units **51.** cylinder of radius 4 and height 2; 48π sq units **53.** cylinder of radius 2 and height 4; 24π sq units **55.** 20 cm **57.** The circumference of the circular bases does not match the length of the rectangle. If the diameter is 2 cm, then the length must be 2π cm, or if the length is 4 cm, then the diameter should be $\frac{4}{\pi}$ cm, or about 1.3 cm. **59.** Lateral faces: *BFGC, DCGH, ADHE, EFBA*; bases: *ABCD, EFGH* **61.** 7 units **63.** Answers may vary. Sample:

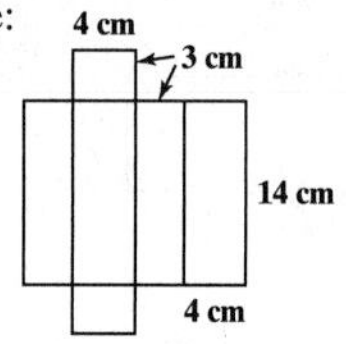

65. a. 94 sq units **b.** 376 units **c.** 4:1 **67. a.** The lateral area is doubled. **b.** The surface area is more than doubled. **c.** If r doubles,

$SA = 2\pi(2r)^2 + 2\pi(2r)h = 8\pi r^2 + 4\pi rh = 2(4\pi r^2 + 2\pi rh)$

So the surface area of $2\pi r^2 + 2\pi rh$ is more than doubled. **69.** $(4.7 + 0.12\pi)$ sq ft **71.** $(84 + 20\pi)$ sq m **73. a.** 4 faces: 0; 3 faces: 8; 2 faces: 12; 1 face: 6; 0 faces: 1 **b.** 15,552 sq in.

75.

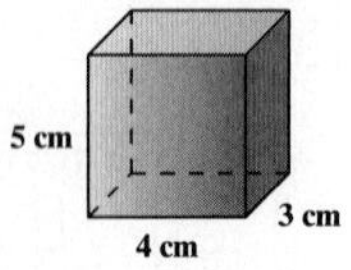

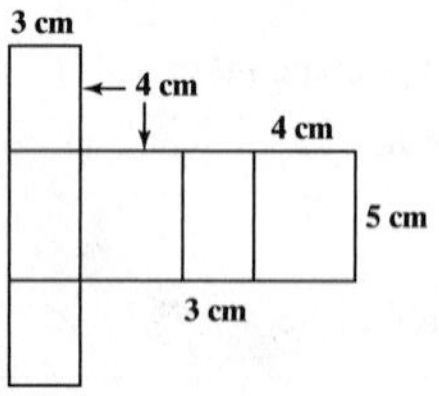

77. 37.7 sq cm **79.** $\sqrt{233}$ in. **81.** $\sqrt{313}$ cm

Section 11.3

Practice Exercises

1. 55 sq m **2.** 166 sq ft **3.** 704π sq m **4.** 934 sq in.

Vocabulary & Readiness Check 11.3

1. cone or pyramid **3.** altitude **5.** pyramid **7.** lateral faces **9.** altitude **11.** right **13.** lateral area

Exercise Set 11.3

1. 408 sq in. **3.** 138 sq m **5.** 204 sq m **7.** 78 sq in. **9.** 47 sq cm **11.** 31 sq m **13.** 119π sq cm **15.** 33π sq ft **17.** 60 sq m **19.** 34 sq ft **21.** cone with $r = 4$ and $h = 3$; 36π sq units **23.** 7.93π sq in. **25.** 447 sq in. **27.** 44.8 sq in. **29.** 834,308 sq ft **31.** slant height **33.** 7 in. **35.** The height is the distance from the vertex to the center of the base, while the slant height is the distance from the vertex to the midpoint of an edge of the base. **37.** The height 7 is not the slant height. The slant height is $\sqrt{7^2 + 3^2} = \sqrt{58}$, so the lateral area is $\pi rl = \pi(3)\sqrt{58} = 3\pi\sqrt{58}$ sq units. **39.** $\overline{PT}$ is a leg in each right triangle ΔPTA, ΔPTB, ΔPTC, and ΔPDT. Since $\overline{PA}$, $\overline{PB}$, $\overline{PC}$, and $\overline{PD}$ are each the hypotenuse in one of the right triangles, $\overline{PT}$ must be shorter than $\overline{PA}$, $\overline{PB}$, $\overline{PC}$, and $\overline{PD}$. **41.** $SA = B + \pi rl = \pi r^2 + \pi rl = (r + l)\pi r$ **43.** 58 sq m **45.** 45 sq m **47.** $l = \frac{SA}{\pi r} - r$ **49.** $l = 5$ in., $h = 4.8$ in., LA = 30 sq in. **51.** 8 ft **53.** cylinder of height 3 and radius 4 with a cone-shaped hole; 60π sq units **55.**

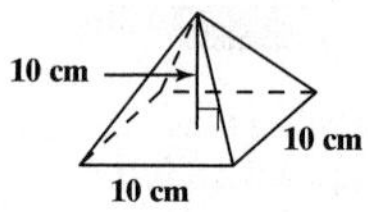

$LA = 100\sqrt{5}$ sq cm; $SA = (100\sqrt{5} + 100)$ sq cm

57. 76 sq ft **59.** 4 sq cm

Section 11.4

Practice Exercises

1. a. 60 cu ft **b.** No; explanations may vary. Sample: the volume is the product of the three dimensions, and multiplication is commutative. **2.** 150 cu m **3.** 3π cu m **4.** 501 cu in.

Vocabulary & Readiness Check 11.4

1. volume **3.** square units **5.** square units **7.** volume

Exercise Set 11.4

1. 54 cu ft **3.** 80 cu in. **5.** 22.5 cu ft **7.** about 280.6 cu cm **9.** 288π cu in. $\approx$ 904.8 cu in. **11.** 37.5π cu m $\approx$ 117.8 cu m **13.** 34,787,712 cu ft **15.** $V = 8.4$ cu ft; $SA = 26$ sq ft **17.** 1,045,585 cu cm **19.** 231 cu in. **21.** 528 cu in. **23.** $(528 + 72\pi)$ cu in. **25.** 144 cu cm **27.** 3445 cu in. **29.** 28 cu ft **31.** 48 cu ft; 48 cu ft **33.** no **35.** 40 cm **37.** 5 in. **39.** 57 cu in. **41.** 24 cm **43.** 80 cu units **45.** 533.6 sq cm **47.** 927.3 cu cm **49.** 125.7 cu cm **51.** cylinder with radius 2 and height 4; 16π cu units **53.** cylinder with radius 2 and height 4; 16π cu units **55.** Alike: Both are the product of the base area and the height. Different: For the prism, the base is a polygon. while for a cylinder, the base is a circle. **57.** The volumes are the same. 24 cu m, because multiplication is commutative. **59. a.** 88,757 cu ft **b.** 663,902 gal **61.** The volume is 27 times the original volume. Using $V = B \cdot h = l \cdot w \cdot h$ for a rectangular prism, $(3l) \cdot (3w) \cdot (3h) = 27 \cdot l \cdot w \cdot h = 27 \cdot V$. **63.** $C = 8.5$ in. and $h = 11$ in.: $V \approx 63.2$ cu in.; $C = 11$ in. and $h = 8.5$ in.: $V \approx 81.8$ cu in.; the cylinder with the greater circumference has the greater volume. **65.** 204.2 sq mm **67.** 37 cm

Section 11.5

Practice Exercises

1. 32,100,000 cu ft **2.** 960 cu m **3.** 77 cu ft **4.** 144π cu m $\approx$ 452 cu m

Exercise Set 11.5

1. 96 cu in. **3.** 300 cu in. **5.** 400 cu m **7.** 562.9 cu ft **9.** 132 cu cm **11.** 26 cu m **13.** $\frac{16}{3}\pi$ cu ft $\approx$ 17 cu ft **15.** $\frac{22}{3}\pi$ cu in. $\approx$ 23 cu in. **17.** 200 cu cm **19.** 1296 cu in. **21.** 6,312,000 cu ft **23.** about 10.6 cu in. **25.** about 66.4 cu cm **27.** about 33 cu km **29.** 13 cu m **31.** cone with radius 4 units and height 3 units; 16π cu units **33.** 312 cu cm **35.** 10,368 cu ft **37.** 6 **39.** 3 **41.** 120π cu ft **43.** 240π cu ft **45.** The areas of the bases are not equal; the area of the base of the pyramid is $13^2 = 169$ sq ft, but the area of the base of the cone is $\pi(6.5)^2 \approx 132.7$ sq ft. **47.** The volume of the cylinder is three times the volume of the cone, since the volume of the cylinder is Bh, while the volume of the cone is $\frac{1}{3}Bh$. **49.** cylinder with radius 4 units and height 3 units, with a cone of radius 4 units, height 3 units removed from it; 32π cu units **51. a.** The frustum has volume that is the difference between the volume of the entire cone and the volume of the small cone that was cut off. $V = \frac{1}{3}\pi R^2 H - \frac{1}{3}\pi r^2 h = \frac{1}{3}\pi(R^2 H - r^2 h)$ **b.** about 784.6 cu in. **53.** 3600 cu cm **55.** 7.1 sq in.

Section 11.6

Practice Exercises

1. 196π sq in. $\approx$ 616 sq in. **2.** 100 sq in. **3.** 113,097 cu in. **4.** 1258.9 sq in.

Vocabulary & Readiness Check 11.6

1. great circle **3.** hemispheres **5.** center **7.** diameter

Exercise Set 11.6

1. 400π sq in. **3.** 900π sq m **5.** 4624 sq mm **7.** $\frac{1089}{256}\pi$ sq in. **9.** 62 sq cm **11.** 20 sq cm **13.** $\frac{500}{3}\pi$ cu ft $\approx$ 524 cu ft **15.** 288π cu cm $\approx$ 905 cu cm **17.** $\frac{12{,}348}{125}\pi$ cu m $\approx$ 310 cu m **19.** 451 sq in. **21.** 130 sq cm **23.** 606 cu ft **25.** 5027 sq in. **27.** 4.5π cu m $\approx$ 14.1 cu m **29.** 5747 cu ft **31.** 45,976.2 cu ft **33.** 144π sq ft **35.** 314 sq in.; 524 cu in. **37.** 314 sq ft; 524 sq ft **39.** sphere with radius 4 units; $\frac{256}{3}\pi$ cu units **41.** C **43.** 36π cu in. **45.** $\frac{4}{3}\pi$ cu m **47.** $\frac{125}{6}\pi$ cu yd **49.** $\frac{9}{2}\pi$ cu ft **51.** Answers may vary. Sample: $(5, 0, 0)$, $(0, 5, 0)$, $(0, 0, 5)$, $(-5, 0, 0)$, $(0, -5, 0)$, $(0, 0, -5)$ **53.** inside **55.** 22π sq cm; $\frac{46}{3}\pi$ cu cm **57.** 26π sq cm; $\frac{62}{3}\pi$ cu cm **59.** 1:4 **61. a.** $457\frac{1}{3}\pi$ cu in. **b.** $228\frac{2}{3}\pi$ cu in. **c.** 11 in. **63.** The 8 in. diameter sphere; the volume of the three spheres is $3(4.5\pi)$ or 13.5π cu in., while the volume of the 8 in. sphere is $85\frac{1}{3}\pi$ cu in. **65.** 2% **67.** $6\sqrt{3}$ in.; $3\sqrt{3}$ in. **69.** 3 m **71. a.** Cube: explanations may vary. Sample: If $s^3 = \frac{4}{3}\pi r^3$, then $s = r \cdot \sqrt[3]{\frac{4\pi}{3}}$. So $6s^2 = 6\left(r \cdot \sqrt[3]{\frac{4\pi}{3}}\right)^2 \approx 15.6 \cdot r^2 > 4\pi r^2$, since $4\pi r^2 \approx 12.6r^2$. **b.** Answers may vary. Spheres are difficult to stack in a display or on a shelf. **73.** 16 cu m **75.** 35°; 55° **77.** yes; 3:1

Section 11.7

Practice Exercises

1. yes; 6:5 or $\frac{6}{5}$ **2.** $\frac{2}{3}$ or 2:3 **3.** 160 sq m **4.** 4.05 lb

Vocabulary & Readiness Check 11.7

1. similar solids **3.** cube (or sphere)

Exercise Set 11.7

1. no **3.** yes; $\frac{2}{3}$ **5.** yes; $\frac{2}{3}$ **7.** $\frac{5}{6}$ **9.** $\frac{3}{4}$ **11.** 240 cu in. **13.** 180 cu m **15.** $SA = 500$ sq m; $V = 1875$ cu m **17.** $SA = 497\frac{7}{9}$ sq in.; $V = 711\frac{1}{9}$ cu in. **19.** always **21.** sometimes **23.** never **25.** Cone 1 and cone 3 are similar; $\frac{2}{3}$ **27. a.** $\frac{2}{5}$ **b.** $\frac{4}{25}$ **c.** $\frac{8}{125}$ **29. a.** $\frac{3}{7}$ **b.** $\frac{9}{49}$ **c.** $\frac{27}{343}$ **31.** 175 sq in. **33.** 325 sq yd **35.** 6000 toothpicks **37.** It is 64 times the volume of the smaller prism. **39.** 1:4; 1:8 **41.** 7:9; 343:729 **43.** 27 cu ft **45.** Answers may vary. Sample: There are many relationships that must be true for the solids to be similar: all corresponding angles must be congruent, the corresponding faces must be similar, and all corresponding edges and heights must be proportional. **47.** yes, 60: $\frac{80}{60} = \frac{40}{30} = \frac{60}{45} = \frac{4}{3}$ **49.** about 1000 cu cm **51. a.** $\frac{3}{1}$ **b.** $\frac{9}{1}$ **53. a.** 384 cu cm **b.** $\frac{16}{1}$ **c.** A: 384 sq cm; B: 24 sq cm **55.** 1790 sq cm **57.** 113.1 cu in. **59.** 20 **61.** 15

Chapter 11 Vocabulary Check

1. cylinder **2.** prism **3.** polyhedron **4.** cone **5.** sphere **6.** pyramid **7.** great circle **8.** hemispheres **9.** scale factor **10.** volume **11.** cubic units **12.** square units **13.** units **14.** square units

Chapter 11 Review

1. Answers may vary. Sample:

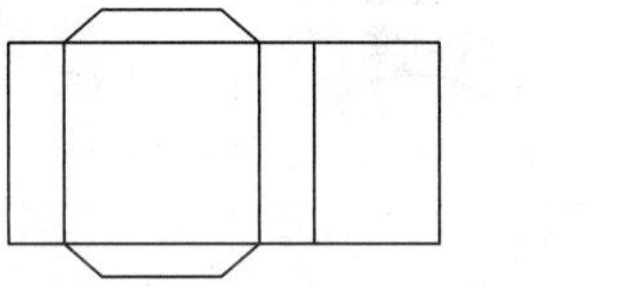

3. 8 **5.** 5 **7.** 36 sq cm **9.** 208 sq in. **11.** 185.6 sq ft **13.** 50.3 sq in. **15.** 84 cu m **17.** 410.5 cu yd **19.** $SA \approx 314.2$ sq in.; $V \approx 523.6$ cu in **21.** $SA \approx 50.3$ sq ft; $V \approx 33.5$ cu ft **23.** 904.78 cu cm **25.** $\frac{27}{64}$ **27.** five faces, nine edges **29.** 32.5π sq cm **31.** 96 cu cm **33.** 615.8 sq ft **35.** The cylinders are not similar.

Chapter 11 Test

1.

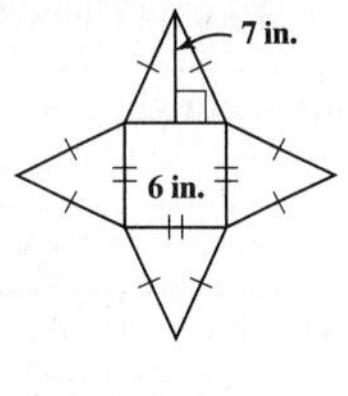

2.

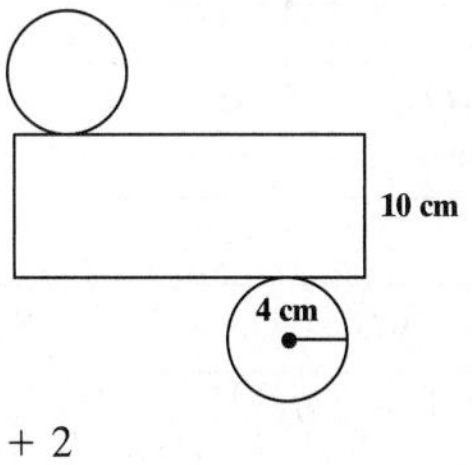

3. $F + V = E + 2$: $6 + 8 = 12 + 2$

4.

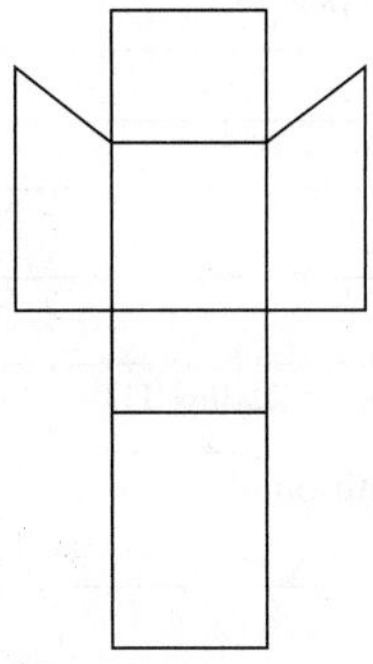

$6 + 14 = 19 + 1$

5. 12 edges **6.** rectangle **7.** an arc of an ellipse with a straight segment joining the two ends **8.** 2640 cu in.; 1800 cu in. **9.** 268.1 cu ft; 201.1 sq ft **10.** 172.0 cu in.; 208 sq in. **11.** 220 cu cm; 238 sq cm **12.** 157.1 cu m; 201.2 sq m **13.** 226.2 cu cm; 207.3 sq cm **14.** 81.4 cu in.; 195.3 sq in. **15. c, a, e, d, b** **16.** 1 gal **17.** cylinder **18. a.** cone with radius 4 units and height 3 units **b.** 20π sq units; 16π cu units

Chapter 11 Standardized Test

1. c **2.** d **3.** c **4.** b **5.** d **6.** a **7.** a **8.** c **9.** b **10.** c **11.** c **12.** b

CHAPTER 12 CIRCLES AND OTHER CONIC SECTIONS

Section 12.1

Practice Exercises

1. a. 135° **b.** 215° **c.** 15π ft **2.** 52° **3.** $5\frac{1}{3}$
4. no; $4^2 + 7^2 = 65 \neq 8^2$ **5.** $x = 2$ units or $x = 6$ units **6.** 12 cm

Vocabulary & Readiness Check 12.1

1. tangent to a circle **3.** tangent ray; tangent segment **5.** tangent circles

Exercise Set 12.1

1. 65° **3.** 180° **5.** 48° **7.** 270° **9.** 307° **11.** 30π cm **13.** 45π cm **15.** 110π cm **17.** 116π cm **19.** 57 **21.** 47 **23.** 30 **25.** 8 in. **27.** 4.8 **29.** 32° **31.** 6 **33.** $3\sqrt{7} \approx 7.9$ **35.** no; $5^2 + 15^2 = 250 \neq 16^2$ **37.** yes; $2.5^2 + 6^2 = 42.25 = 6.5^2$ **39.** $x = 5$ **41.** $x = -3$ or $x = 5$ **43.** 78 cm **45.** 31.6 ft **47.** internal **49.** internal **51.**

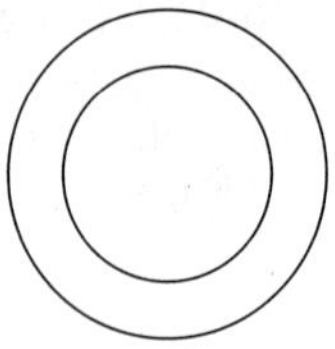

no common tangents

53.

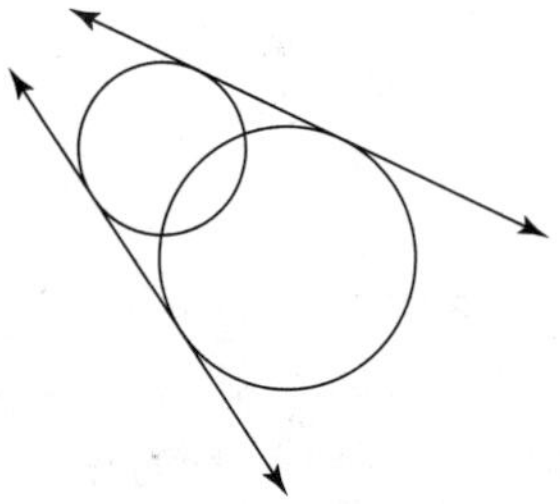

two common tangents

55. 253.0 km **57.** 46.9 mi **59.** 65° **61.** 65° **63.** 50° **65.** isosceles **67.** 57.5° **69.** 8 cm **71.** 254°; 26.6 cm **73.** 63.9 cm **75.** external **77.** internal **79.** If $\overline{DF}$ is tangent to $\odot E$, then $\overline{DF} \perp \overline{EF}$. That would mean that ΔDEF contains two right angles, which is impossible. So $\overline{DF}$ is not tangent to $\odot E$.

81.

Statements	Reasons
1. $\overline{BA}, \overline{BC}$ tangent to $\odot O$ at A and C.	1. Given
2. $\overline{AB} \perp \overline{OA}, \overline{BC} \perp \overline{OC}$	2. Tangent-Radius Theorem
3. ΔBAO and ΔBCO are right triangles	3. Definition of right triangle
4. $\overline{AO} \cong \overline{OC}$	4. Radii of a circle are $\cong$
5. $\overline{BO} \cong \overline{BO}$	5. Reflexive Property of $\cong$
6. $\Delta BAO \cong \Delta BCO$	6. Hypotenuse-Leg
7. $\overline{BA} \cong \overline{BC}$	7. cpoctac

83.

Statements	Reasons
1. $\overline{BC}$ tangent to $\odot A$ at D	1. Given
2. $\overline{DB} \cong \overline{DC}$	2. Given
3. $\overline{AD} \perp \overline{BC}$	3. Tangent-Radius Theorem
4. $\angle ADB$ and $\angle ADC$ are right angles	4. Definition of $\perp$
5. $\angle ADB \cong \angle ADC$	5. Right angles are $\cong$
6. $\overline{AD} \cong \overline{AD}$	6. Reflexive Property of $\cong$
7. $\Delta ADB \cong \Delta ADC$	7. SAS
8. $\overline{AB} \cong \overline{AC}$	8. cpoctac

85. a. $ABCE$ is a rectangle. $\overline{AB}$ is tangent to both $\odot D$ and $\odot E$, so $\overline{DB} \perp \overline{AB}$ and $\overline{AE} \perp \overline{AB}$ by the Tangent-Radius Theorem. This makes $\overline{BC} \parallel \overline{AE}$ since they are coplanar and both $\perp$ to $\overline{AB}$. So $ABCE$ is a parallelogram with two right angles, which is a rectangle. **b.** In a rectangle, $\overline{BA} \cong \overline{CE}$, so $CE = BA = 35$ in. **c.** In a rectangle, $\overline{BC} \cong \overline{AE}$, so $BC = AE = 8$ in. **d.** The horizontal distance between the centers of the pulleys is 35 inches, and the vertical distance is $14 - 8 = 6$ in. Use the Pythagorean Theorem to find the distance. $d^2 = (14 - 8)^2 + 35^2$; $d^2 = 36 + 1225$; $d^2 = 1261$; $d^2 = \sqrt{1261}$; $d \approx 35.5$in. **87.** 3:4 or $\frac{3}{4}$ **89.** 9:16 or $\frac{9}{16}$ **91.** 29.1° **93.** 5

Section 12.2

Practice Exercises

1. Since the circles are $\cong$, their radii are $\cong$, so ΔBOC and ΔDPF are isosceles, $\overline{OB} \cong \overline{OC} \cong \overline{PD} \cong \overline{PF}$. Since $\angle B \cong \angle D$ and the triangles are isosceles, $\angle B \cong \angle C \cong \angle D \cong \angle F$. So $\Delta BOC \cong \Delta DPF$ by AAS, which makes $\angle O \cong \angle P$ (cpoctac). Then $\overline{BC} \cong \overline{DF}$ (cpoctac or Congruent Central Angles and Chords) and $\overset{\frown}{BC} \cong \overset{\frown}{DF}$ (Congruent Central Angles and Arcs). **2.** 16; congruent chords are equidistant from the center. **3.** answers may vary **4.** $r \approx 11.2$m

Vocabulary & Readiness Check 12.2

1. chord **3.** true **5.** false

Exercise Set 12.2

1. $\overset{\frown}{BC} \cong \overset{\frown}{YZ}, \overline{BC} \cong \overline{YZ}$ **3.** 14 **5.** 8 **7.** 50° **9.** 50°; $\angle 1 \cong \angle 2$ because they are vertical angles, so $\overline{CD} \cong \overline{AB}$ because $\cong$ central angles have $\cong$ arcs. Therefore, $m\overset{\frown}{CD} = m\overset{\frown}{AB}$. **11.** 130° **13.** 130° **15.** The center is at the intersection of $\overline{GH}$ and $\overline{KM}$, because if a chord is the $\perp$ bisector of another chord, then the first chord is a diameter. Two diameters intersect at the center of a circle. **17.** 6 **19.** 20.8 **21.** 5 in. **23.** 10 ft **25.** 108° **27.** 8 **29.** 9.2 units **31.** 12 cm **33.** answers may vary **35.** Since $\angle AOB \cong \angle COD$, it fol lows that $m\angle AOB = m\angle COD$. By the definition of arc measure, $m\angle AOB = m\overset{\frown}{AB}$ and $m\angle COD = m\overset{\frown}{CD}$. Thus, $m\overset{\frown}{AB} = m\overset{\frown}{CD}$ by substitution, and $\overset{\frown}{AB} \cong \overset{\frown}{CD}$ by the definition of congruent arcs.

37.

Statements	Reasons
1. $\odot O$ with $\overline{AB} \cong \overline{CD}$	1. Given
2. $\overline{AO} \cong \overline{BO} \cong \overline{CO} \cong \overline{DO}$	2. All radii of a circle are congruent
3. $\Delta AOB \cong \Delta COD$	3. SSS
4. $\angle AOB \cong \angle COD$	4. cpoctac
5. $\overset{\frown}{AB} \cong \overset{\frown}{CD}$	5. Congruent Central Angles and Arcs Theorem

39. Chords $\overline{SR}$ and $\overline{QP}$ are equidistant from the center, so their lengths must be equal.

41.

Statements	Reasons
1. $\overline{XW} \cong \overline{XY}$	1. All radii of a circle are $\cong$
2. X is on the $\perp$ bisector of $\overline{WY}$	2. Converse of the $\perp$ Bisector Theorem
3. X is on l	3. Substitution
4. l contains the center of $\odot X$	4. Definition of the center of $\odot X$

43.

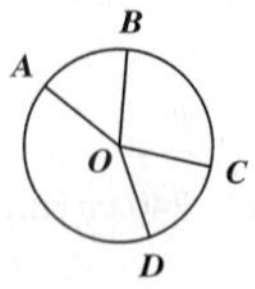

Given: $\odot O$ with $\overset{\frown}{AB} \cong \overset{\frown}{CD}$
Prove: $\angle AOB \cong \angle COD$

Statements	**Reasons**
1. $m\angle AOB = m\widehat{AB}$, $m\angle COD = m\widehat{CD}$	**1.** Definition of arc measure
2. $\widehat{AB} \cong \widehat{CD}$	**2.** Given
3. $m\widehat{AB} = m\widehat{CD}$	**3.** Definition of $\cong$ arcs
4. $m\angle AOB = m\angle COD$	**4.** Substitution
5. $\angle AOB \cong \angle COD$	**5.** Definition of $\cong$ angles

45.

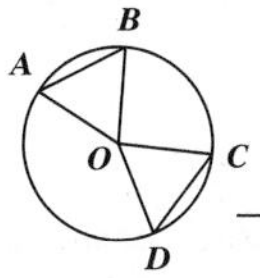

Given: $\odot O$ with $\widehat{AB} \cong \widehat{CD}$
Prove: $\overline{AB} \cong \overline{CD}$

Statements	**Reasons**
1. $\widehat{AB} \cong \widehat{CD}$	**1.** Given
2. $\angle AOB \cong \angle COD$	**2.** Congruent Central Angles and Arcs Theorem
3. $\overline{AO} \cong \overline{BO} \cong \overline{CO} \cong \overline{DO}$	**3.** All radii of a circle are $\cong$
4. $\Delta AOB \cong \Delta COD$	**4.** SAS
5. $\overline{AB} \cong \overline{CD}$	**5.** cpoctac

47. 40 **49.** $\widehat{STQ}$ or $\widehat{SRQ}$ **51.** $\widehat{QR}$ **53.** 86° **55.** 121°

Section 12.3

Practice Exercises

1. a. 90° **b.** 148° **c.** $m\angle A = m\angle B = 20°$ **2. a.** 90° **b.** $m\angle A = 95°, m\angle B = 77°, m\angle C = 85°, m\angle D = 103°$ **c.** The sum of the measures of opposite angles is 180°. **3.** $m\angle 1 = 90°$, $m\angle 2 = 110°, m\angle 3 = 90°, m\angle 4 = 70°$ **4.** $x = 35, y = 55$

Vocabulary & Readiness Check 12.3

1. 90° **3.** $m\widehat{IH}$ **5.** $m\widehat{QR}$ **7.** $m\widehat{QR}$ **9.** $\widehat{BD}$ or $\widehat{BCD}$ **11.** $\angle D$ **13.** $\angle D$

Exercise Set 12.3

1. 45 **3.** 180 **5.** $a = 41, b = 41$ **7.** $a = 85, b = 47.5, c = 90$ **9.** $c = 105, d = 210$ **11.** $d = 39, e = 15, f = 126$ **13.** $p = 90$, $q = 122$ **15.** $a = 112, b = 120, c = 38$ **17.** $a = 50, b = 90$, $c = 90$ **19.** 114° **21.** 90° **23.** $m\angle 1 = 65°, m\widehat{DF} = 130°$ **25.** $m\angle 1 = 75°, m\angle 2 = 105°, m\widehat{QRP} = 210°$ **27.** 96° **29.** 77° **31.** 40° **33.** 40° **35.** 65° **37.** $a = 22, b = 78, c = 156$ **39.** $a = 26, b = 64, c = 42$ **41.** Isosceles trapezoid; answers may vary. Sample answer: For inscribed trapezoid $ABCD$, $\angle A$ must be supplementary to $\angle C$ by Corollary 3 to the Inscribed Angle Theorem. $\angle C$ must be supplementary to $\angle B$ because they are same-side interior angles of parallel lines. Thus, $\angle A \cong \angle B$ and the trapezoid is isosceles. **43.** Rectangle; opposite angles are $\cong$ (because the figure is a parallelogram) and supplementary (Corollary 3 to the Inscribed Angle Theorem). Congruent supplementary angles are right angles, so the inscribed parallelogram must be a rectangle.

45.

Statements	**Reasons**
1. $\odot O$ with inscribed angle $\angle ABC$	**1.** Given
2. $m\angle ABO = \frac{1}{2}m\widehat{AP}, m\angle OBC = \frac{1}{2}m\widehat{PC}$	**2.** Inscribed Angle Theorem, Case 1
3. $m\angle ABO + m\angle OBC = m\angle ABC$	**3.** Angle Addition Postulate
4. $\frac{1}{2}m\widehat{AP} + \frac{1}{2}m\widehat{PC} = m\angle ABC$	**4.** Substitution
5. $\frac{1}{2}(m\widehat{AP} + m\widehat{PC}) = m\angle ABC$	**5.** Distributive Property
6. $\frac{1}{2}m\widehat{AC} = m\angle ABC$	**6.** Arc Addition Postulate

47. Yes; since the opposite angles of a square are supplementary, a square can be inscribed in a circle. **49.** $\angle ACB$ is a right angle because it is inscribed in semicircle $\widehat{ACB}$, so $\overline{AC} \perp \overleftrightarrow{BC}$. If a line is $\perp$ to a radius at its endpoint, it is tangent to the circle.

51.

Statements	**Reasons**
1. $\odot O$, $\angle A$ intercepts $\widehat{BC}$, $\angle D$ intercepts $\widehat{BC}$	**1.** Given
2. $m\angle A = \frac{1}{2}m\widehat{BC}$, $m\angle D = \frac{1}{2}m\widehat{BC}$	**2.** Inscribed Angle Theorem
3. $m\angle A = m\angle D$	**3.** Substitution
4. $\angle A \cong \angle D$	**4.** Definition of $\cong$ angles.

53.

Statements	**Reasons**
1. Quadrilateral $ABCD$ inscribed in $\odot O$	**1.** Given
2. $m\angle A = \frac{1}{2}m\widehat{BCD}$, $m\angle C = \frac{1}{2}m\widehat{BAD}$	**2.** Inscribed Angle Theorem
3. $m\angle A + m\angle C = \frac{1}{2}m\widehat{BCD} + \frac{1}{2}m\widehat{BAD}$	**3.** Addition Property
4. $m\widehat{BCD} + m\widehat{BAD} = 360°$	**4.** Arc measure of circle is 360°
5. $\frac{1}{2}m\widehat{BCD} + \frac{1}{2}m\widehat{BAD} = 180°$	**5.** Multiplication Property
6. $m\angle A + m\angle C = 180°$	**6.** Substitution
7. $\angle A$ and $\angle C$ are supplementary	**7.** Definition of supplementary angles
8. $m\angle B = \frac{1}{2}m\widehat{ADC}$, $m\angle D = \frac{1}{2}m\widehat{ABC}$	**8.** Inscribed Angle Theorem
9. $m\angle B + m\angle D = \frac{1}{2}m\widehat{ADC} + \frac{1}{2}m\widehat{ABC}$	**9.** Addition Property
10. $m\widehat{ADC} + m\widehat{ABC} = 360°$	**10.** Arc measure of circle of 360°
11. $\frac{1}{2}m\widehat{ADC} + \frac{1}{2}m\widehat{ABC} = 180°$	**11.** Multiplication Property
12. $m\angle B = m\angle D = 180°$	**12.** Substitution
13. $\angle B$ and $\angle D$ are supplementary	**13.** Definition of supplementary angles

55. False

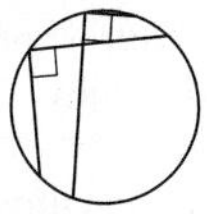

57. True; opposite angles in an inscribed quadrilateral intercept non-overlapping arcs totaling 360° and inscribed angles have half the measure of the intercepted arcs, so the opposite angles are supplementary.

59.

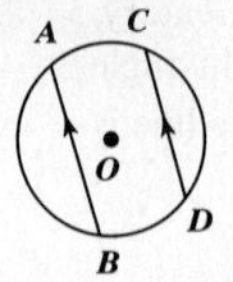

Given: $AB \parallel CD$
Prove: $m\overset{\frown}{AC} = m\overset{\frown}{BD}$

Statements	Reasons
1. $\odot O$ with $\overline{AB} \parallel \overline{CD}$	1. Given
2. Draw $\overline{AD}$	2. Construction: two points determine a line
3. $\angle CDA \cong \angle DAB$	3. Alternate Interior Angles of Parallel Lines are $\cong$
4. $m\overset{\frown}{AC} = m\overset{\frown}{BD}$	4. Congruent inscribed angles intercept congruent arcs

61. 17.3 **63.** 5:2 or $\frac{5}{2}$ **65.** 57° **67.** 4

Section 12.4

Practice Exercises

1. a. 250 **b.** 40 **c.** 40 **2. a.** 160° **b.** The probe is closer; as an observer moves away from Earth, the viewing angle decreases and the measure of the arc of Earth that is viewed gets larger and approaches 180°. **3. a.** 13.8 **b.** 3.2

Vocabulary & Readiness Check 12.4

1. f **3.** d **5.** b **7.** e

Exercise Set 12.4

1. 46 **3.** 60 **5.** $x = 72, y = 36$ **7.** $x = 60, y = 70$ **9.** $x = 115, y = 74$ **11.** 5.4 **13.** 65 **15.** 11.2 **17.** 15 **19.** 13.2 **21.** $x = 25.8, y = 12.4$ **23.** 26.7 **25.** 16.7 **27.** 95, 104, 86, 75 **29.** $x = 10.7, y = 10$ **31.** $x = 8.9, y = 2$ **33.** $x = 10.9, y = 2.3$ **35.** 145° **37.** 157.5° **39.** The student forgot to multiply by the length of the entire secant segment. The equation should be $(13.5)(6) = x^2$. **41.** A secant is a line that intersects a circle at two points. A tangent is a line that intersects a circle at one point. **43.** $360° - x$ **45.** $180° - y$ **47.** $c = b - a$; the measure of the angle is $\frac{1}{2}c$ when you consider the outer circle and $\frac{1}{2}(b - a)$ when you consider the inner circle.

49.

Statements	Reasons
1. $\odot O$ with secants $\overline{CA}$ and $\overline{CE}$	1. Given
2. Draw $\overline{BE}$	2. Construction: 2 points determine a line
3. $m\angle BEC = \frac{1}{2}m\overset{\frown}{BD}$, $m\angle ABE = \frac{1}{2}m\overset{\frown}{AE}$	3. The measure of an inscribed angle is half the measure of its intercepted arc.
4. $m\angle BEC + m\angle BCE = m\angle ABE$	4. Exterior Angle Theorem
5. $\frac{1}{2}m\overset{\frown}{BD} + m\angle BCE = \frac{1}{2}m\overset{\frown}{AE}$	5. Substitution
6. $m\angle BCE = \frac{1}{2}m\overset{\frown}{AE} - \frac{1}{2}m\overset{\frown}{BD}$	6. Subtraction Property of Equality
7. $m\angle BCE = \frac{1}{2}(m\overset{\frown}{AE} - m\overset{\frown}{BD})$	7. Distributive Property
8. $\angle BCE \cong \angle ACE$	8. Reflexive Property
9. $m\angle ACE = \frac{1}{2}(m\overset{\frown}{AE} - m\overset{\frown}{BD})$	9. Substitution

51.

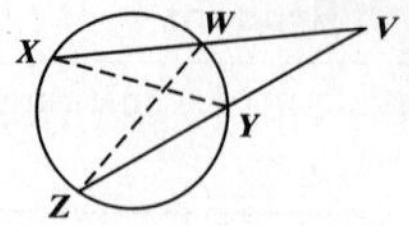

Given: A circle with secant segments $\overline{XV}$ and $\overline{ZV}$
Prove: $XV \cdot WV = ZV \cdot YV$

Statements	Reasons
1. Draw $\overline{XY}$ and $\overline{ZW}$	1. Construction: 2 points determine a line
2. $\angle XVY \cong \angle ZVW$	2. Reflexive Property
3. $\angle VXY \cong \angle WZV$	3. Inscribed angles that intercept the same arc are $\cong$.
4. $\Delta XVY \sim \Delta ZVW$	4. AA Similarity
5. $\frac{XV}{ZV} = \frac{YV}{WV}$	5. Corresponding sides of similar figures are proportional
6. $XV \cdot WV = ZV \cdot YV$	6. Property of Proportion

53. $m\angle 1 = \frac{1}{2}m\overset{\frown}{QRP} - \frac{1}{2}m\overset{\frown}{PQ}$ (vertex outside circle, angle measure is half the difference of intercepted arcs);
$m\angle 1 + m\overset{\frown}{PQ} = \frac{1}{2}m\overset{\frown}{QRP} + \frac{1}{2}m\overset{\frown}{PQ}$ (Addition Property of Equality);
$m\angle 1 + m\overset{\frown}{PQ} = \frac{1}{2}(m\overset{\frown}{QRP} + m\overset{\frown}{PQ})$ (Distributive Property);
$m\angle 1 = m\overset{\frown}{PQ} = \frac{1}{2}(360°)$ (arc measure of a circle is 360°);
$m\angle 1 + m\overset{\frown}{PQ} = 180°$ (simplification)

55.

Statements	Reasons
1. $(PQ)^2 = (QS)(QR)$	1. Square of the tangent equals the product of the secant times the outer segment
2. $b^2 = (c + a)(c - a)$	2. Substitution
3. $b^2 = c^2 - a^2$	3. Distributive Property
4. $a^2 + b^2 = c^2$	4. Addition Property of Equality

57. $a = 50, b = 55, c = 105$ **59.** 30 **61.** 5.8 units **63.** 5.8 units

Section 12.5

Practice Exercises

1. a. $(x - 3)^2 + (y - 5)^2 = 36$ **b.** $(x + 2)^2 + (y + 1)^2 = 2$ **2.** $(x - 4)^2 + (y - 3)^2 = 29$ **3. a.** The center of the circle represents the cell tower's position. The radius represents the cell tower's transmission range. **b.** center (2, 3); radius 10

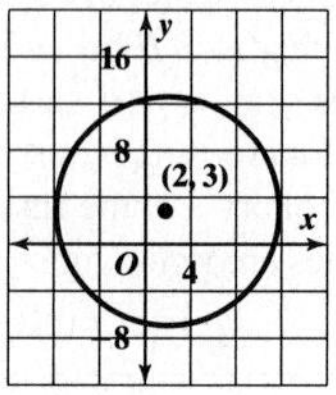

4.

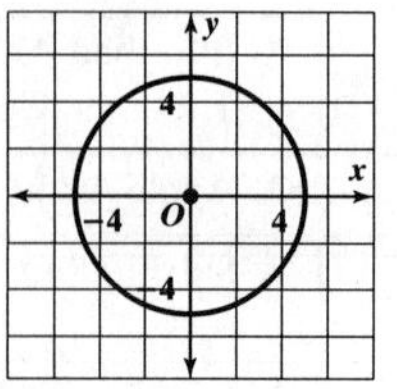

5.

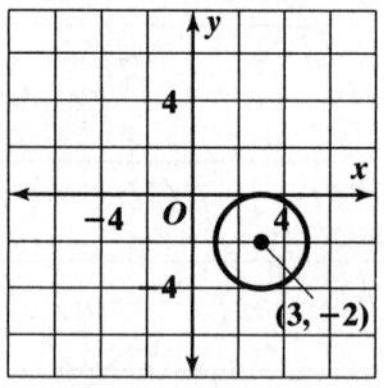

6. Center $(-3, 1)$, radius 4

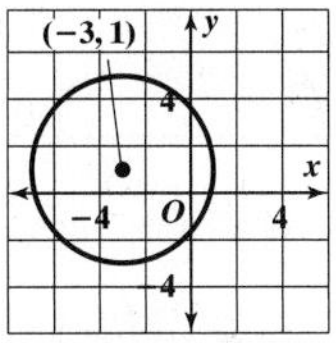

Vocabulary & Readiness Check 12.5

1. circle; center **3.** radius

Exercise Set 12.5

1. $(x - 2)^2 + (y + 8)^2 = 81$ **3.** $(x + 9)^2 + (y + 4)^2 = 5$ **5.** $(x + 6)^2 + (y - 3)^2 = 64$ **7.** $x^2 + (y - 3)^2 = 49$ **9.** $x^2 + y^2 = 16$ **11.** $(x + 4)^2 + (y - 2)^2 = 16$ **13.** $(x + 2)^2 + (y - 6)^2 = 16$ **15.** $(x - 6)^2 + (y - 5)^2 = 61$ **17.** $(x + 10)^2 + (y + 5)^2 = 125$

19. center $(-7, 5)$; radius 4

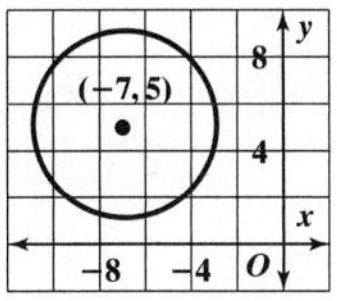

21. center $(3, -8)$; radius 10

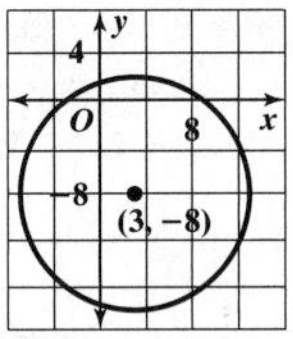

23. center $(0, 0)$; radius 6

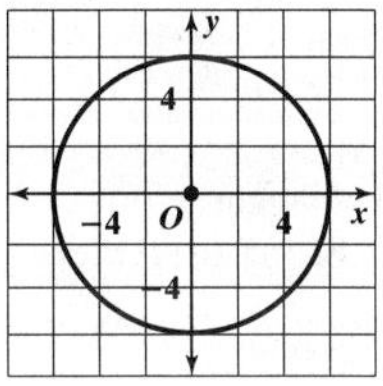

25. center $(-8, 0)$; radius $\sqrt{7}$

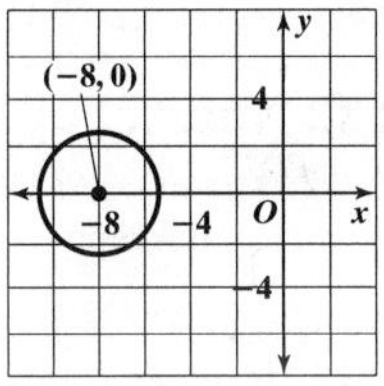

27. $x^2 + y^2 = 4$ **29.** $x^2 + (y - 3)^2 = 4$ **31.** $(x - 2)^2 + (y - 2)^2 = 16$

33. center $(0, -3)$; radius 3

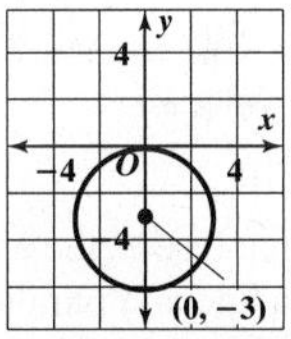

35. center $(-1, 2)$; radius 3

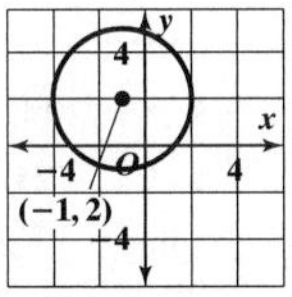

37. center $(2, 4)$; radius $\sqrt{22}$

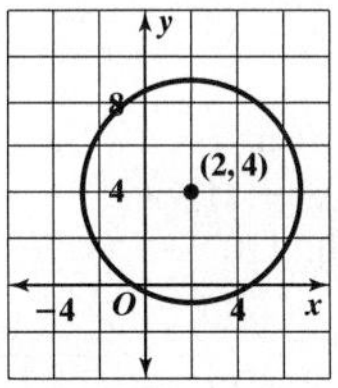

39. center $(-3, -5)$; radius 6

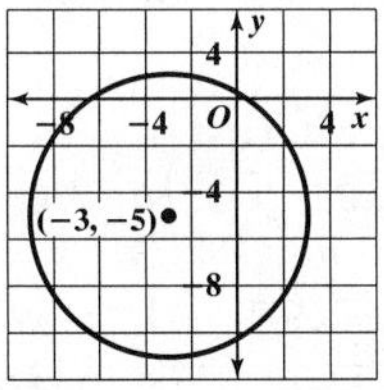

41. a. 76.5 meters **b.** 7 meters **c.** 83.5 meters **d.** $(0, 83.5)$ **e.** $x^2 + (y - 83.5)^2 = 76.5^2$ **43. a.** 67.5 meters **b.** 0 meters **c.** 67.5 meters **d.** $(0, 67.5)$ **e.** $x^2 + (y - 67.5)^2 = 67.5^2$ **45.** $(x - 4)^2 + (y - 3)^2 = 25$ **47.** $(x - 5)^2 + (y - 3)^2 = 13$ **49.** Yes; it is a circle with center $(1, -2)$ and radius 3. **51.** No; the x- and y-terms are not squared. **53.** Sample explanation: The student should have rewritten the equation as $(x - (2))^2 + (y - (-3))^2 = 16$ to realize that the center is $(2, -3)$. **55.** Its center and its radius; its center and its radius **57.** 16π units **59.** $(x - 4)^2 + (y - 7)^2 = 36$

61.

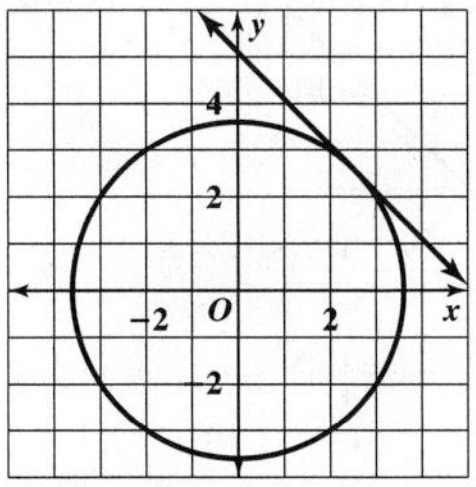

$(3, 2), (2, 3)$

63.

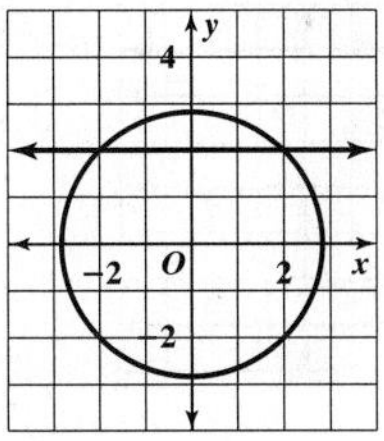

$(2, 2), (-2, 2)$

65.

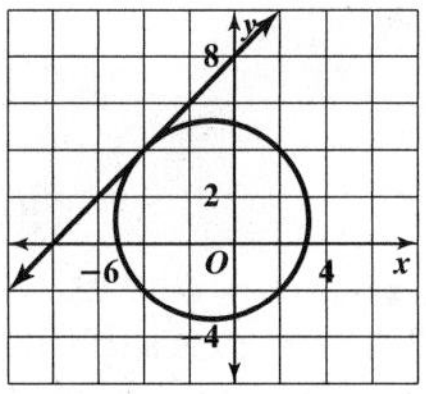

$(-4, 4)$

67. about 11.5, 11.5, 49.8, and 49.8 square units **69.** $x = 25, y = 75$ **71.** $\langle 6, 12\rangle$ **73.** $\langle 4, 4\rangle$ **75.**

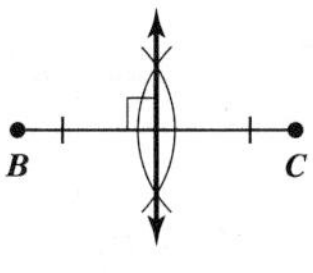

Section 12.6

Practice Exercises

1.

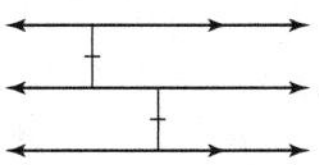

The line midway between the given parallel lines

2.

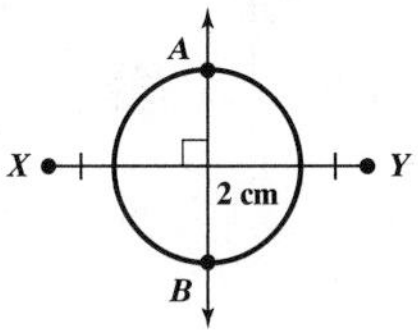

Points A and B satisfy both conditions. **3. a.** The locus is the line parallel to, and equidistant from, the given parallel lines (midway between them). **b.** The locus is the plane parallel to, and equidistant from, the given parallel planes (midway between them).

Exercise Set 12.6

1.

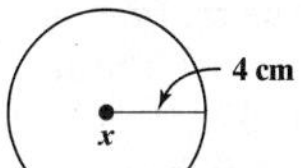

The locus is a circle with center X and radius 4 cm.

3.

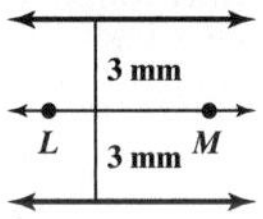

The locus is a pair of parallel lines, each 3 mm from $\overleftrightarrow{LM}$.

5.

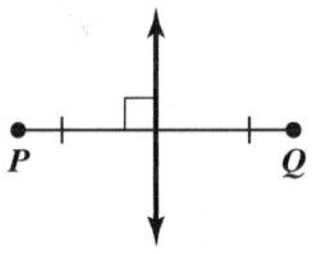

The locus is the $\perp$ bisector of $\overline{PQ}$.

7.

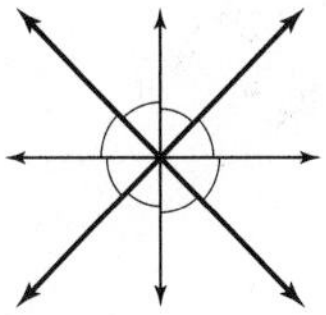

The locus is the two lines that bisect the right angles.

9.

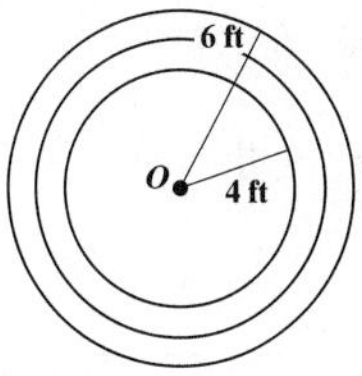

The locus is the circle with center O and radius 5 feet.

11.

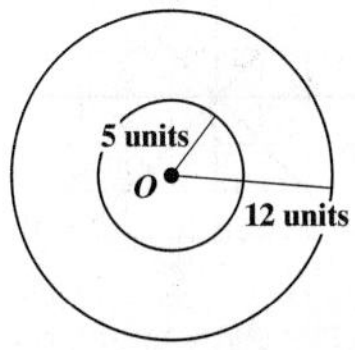

The locus is the circle concentric with the given circle of radius 12 units.

13.

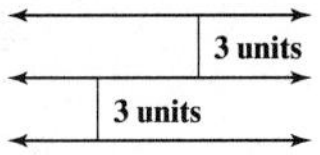

The locus is the line parallel to the given parallel lines, 3 units from each (midway between them).

15.

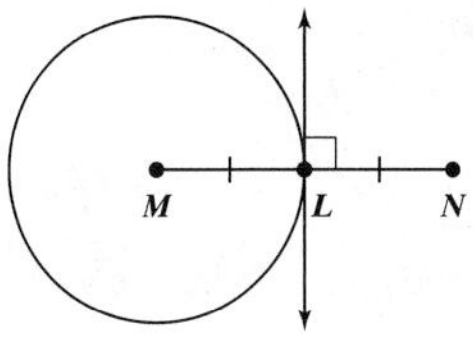

The locus is point L.

17. 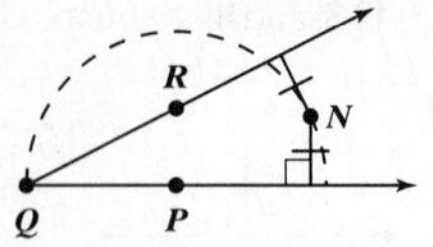

The locus is point N.

19. 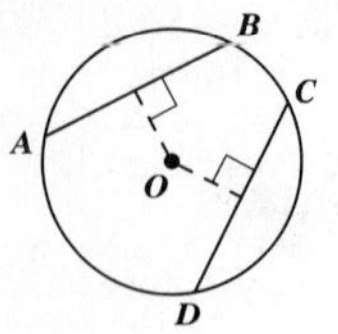

The locus is the center O.

21. $y = 3$ and $x = 6$ **23.** $x^2 + y^2 = 4$ **25.** $x = 4$ **27.** The locus is a sphere with center F and radius 3 cm. **29.** The locus is two planes, each parallel to plane M, and each 1 in. from plane M. **31.** The locus is the set of all points in the interior of $\angle A$ and equidistant from the sides of $\angle A$.

33. $y = x$ **35.** Answers may vary. Sample answer:

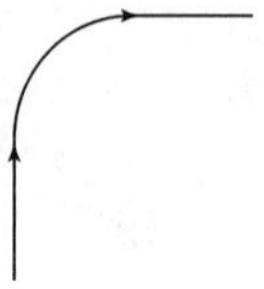

37. Answers may vary. Sample answer:

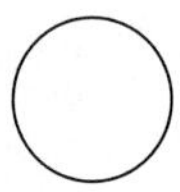

39. Answers may vary. Sample answer:

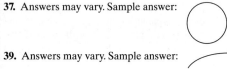

41. 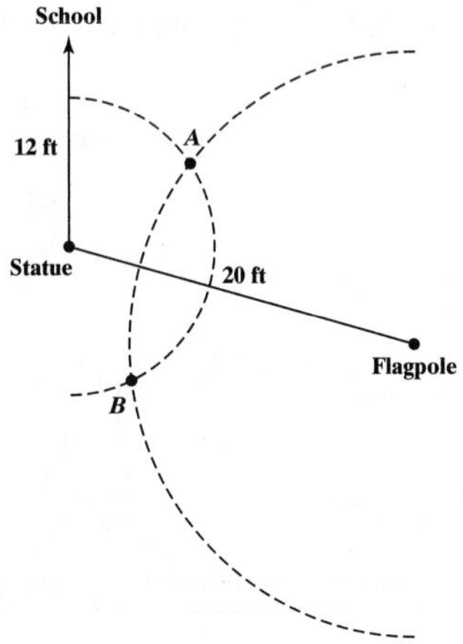

The radius of the arc from the statue represents 8 ft. The arc from the flagpole represents 16 ft. Points A and B are the two possible positions for the fountain.

43. 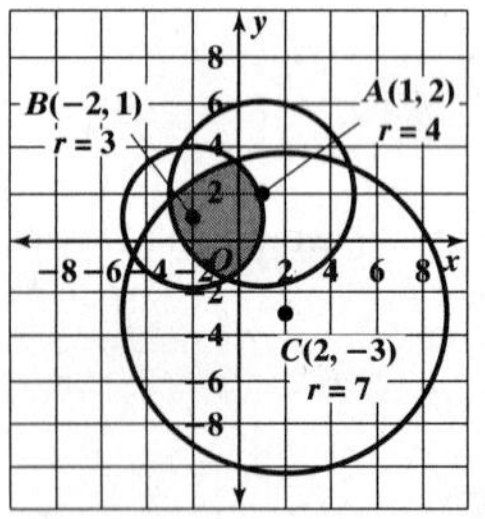

45. 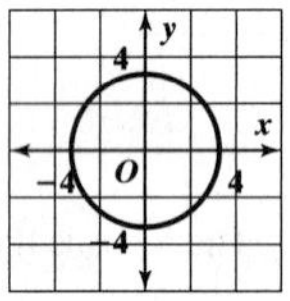

47. 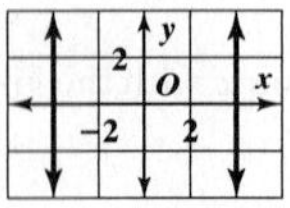

49. 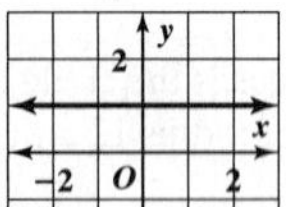

51. 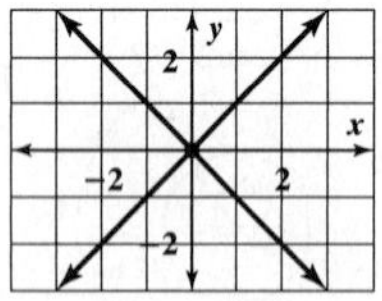

53. Both contain the midpoint of $\overline{JK}$ and are $\perp$ to $\overline{JK}$. **55.** Answers may vary. **57.** Points two units from the origin. **59.** No; the loci do not intersect.

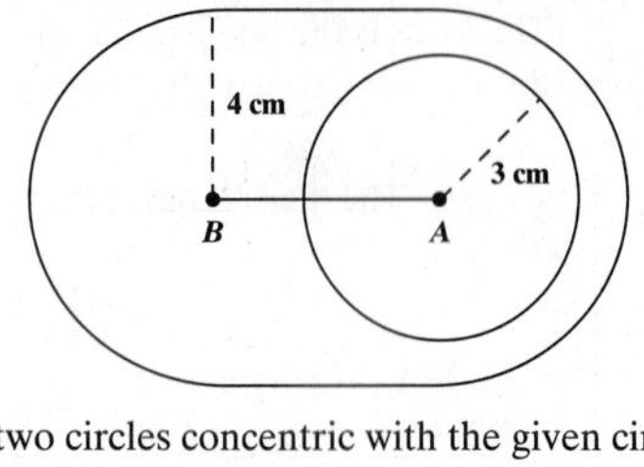

61. The locus is two circles concentric with the given circle, one with radius 5 cm and one with radius 11 cm.

63. 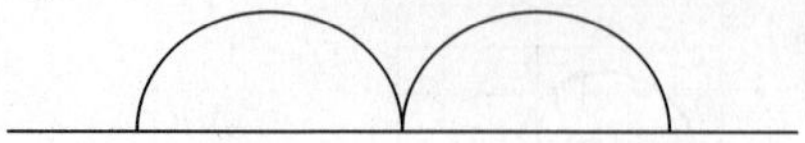

65. 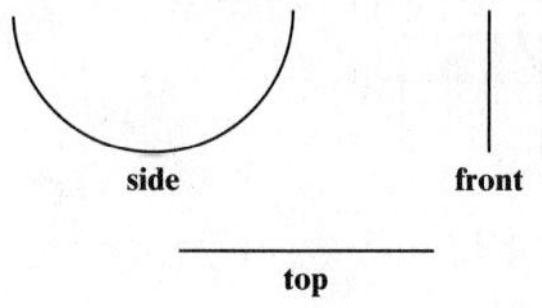

67. 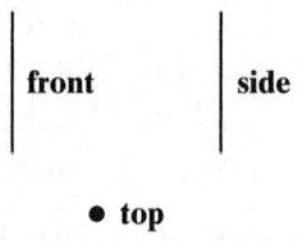

69. $(x - 6)^2 + (y + 10)^2 = 25$ **71.** 510 sq in. **73.** 4π sq units

Extension-Parabolas Practice Exercises

1. $y = -\frac{1}{2}x^2 + 2x - \frac{1}{2}$

Extension Exercise Set-Parabolas

1. $y = -\frac{1}{2}x^2 + x + 5$ **3.** $y = \frac{1}{8}x^2 - x + 3$
5. $y = -\frac{1}{10}x^2 + x - 1$ **7.** $y = \frac{1}{18}x^2 - \frac{1}{2}$
9. $y = -\frac{1}{18}x^2 + \frac{5}{9}x + \frac{28}{9}$ **11.** $y = \frac{1}{8}x^2 + x + 2$
13. $y = -\frac{1}{6}x^2 - \frac{10}{3}x - \frac{109}{6}$

Chapter 12 Vocabulary Check

1. chord **2.** secant **3.** locus **4.** inscribed angle
5. tangent **6.** intercepted arc **7.** circle **8.** radius; center

Chapter 12 Review

1. 20 units **3.** 120 **5.** 90° **7.** $\overline{AB}$ is a diameter of the circle.
9. $\frac{\sqrt{181}}{2} \approx 6.7$ **11.** $a = 80, b = 40, c = 40, d = 100$
13. $a = 118, b = 49, c = 144, d = 98$ **15.** 37 **17.** 6.5
19. $x^2 + (y + 2)^2 = 9$ **21.** $(x + 3)^2 + (y + 4)^2 = 25$
23. center $(7, -5)$; radius 6 **25.** The locus is the ray that bisects the angle. **27.** The locus is two lines, one on each side of the given line and parallel to the given line, each 8 in. from the given line. **29.** 48 in. **31.** 154°
33. $m\widehat{PS} = 120°, m\angle R = 60°$ **35.** $(x + 1)^2 + (y - 2)^2 = 4$

Chapter 12 Test

1. 8 **2.** 7.2 **3.** 60 **4.** 10.5 **5.** $x = 26, y = 41.5$
6. $a = 110, b = 70$ **7.** 13 **8.** 8 **9.** 65° **10.** 120° **11.** 76 cm
12. $w = 105, x = 75, y = 210$ **13.** $w = 104, x = 22, y = 108$

14. 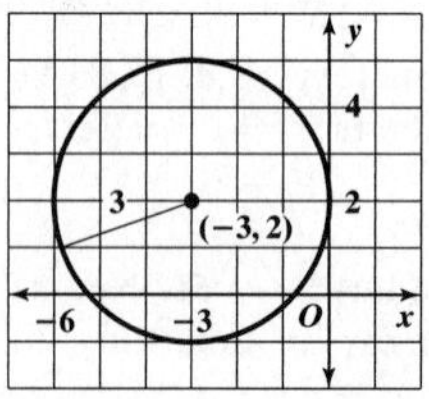

15. $(x - 3)^2 + y^2 = 41$

16. 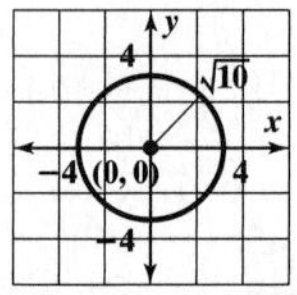

17. $(x + 5)^2 + (y - 2)^2 = 16$ **18.** $(x - 2)^2 + (y - 4)^2 = 4$
19. $x^2 + (y + 1)^2 = 4$ **20.** center $(3, -5)$; radius 4

21. 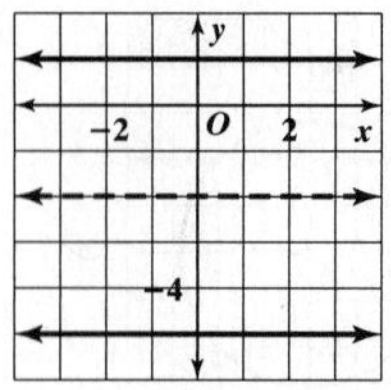

22. 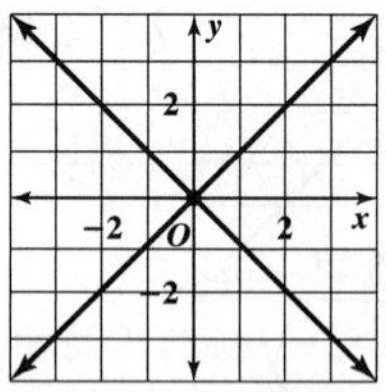

23.

Statements	Reasons
1. $\odot A$ with $\overline{BC} \cong \overline{DE}$, $\overline{AF} \perp \overline{BC}$, $\overline{AG} \perp \overline{DE}$	1. Given
2. $\overline{AF} \cong \overline{AG}$	2. Congruent chords in a circle are equidistant from the center
3. $\angle AFG \cong \angle AGF$	3. An isosceles triangle has $\cong$ base angles

Chapter 12 Standardized Test

1. b **2.** c **3.** b **4.** b **5.** c **6.** d **7.** d **8.** c **9.** c **10.** c **11.** a **12.** a **13.** d **14.** d **15.** b **16.** d **17.** a

APPENDIX STUDENT SUCCESS RESOURCE SECTION

A Review of Basic Concepts

A.1 Measurement Conversions

1. 0.4 m **3.** 600 mm **5.** 1008 in. **7.** 15,000 mg **9.** 34,000 mL **11.** 4.3 cm **13.** 56 qt **15.** 3.9 h **17.** 1,080,000 cm^2 **19.** 12.6 ft^2 **21.** $144\frac{4}{9}$ yd^2 **23.** $\frac{3125}{4356}$ $\text{mi}^2 \approx 0.717$ mi^2

A.2 Probability

1.

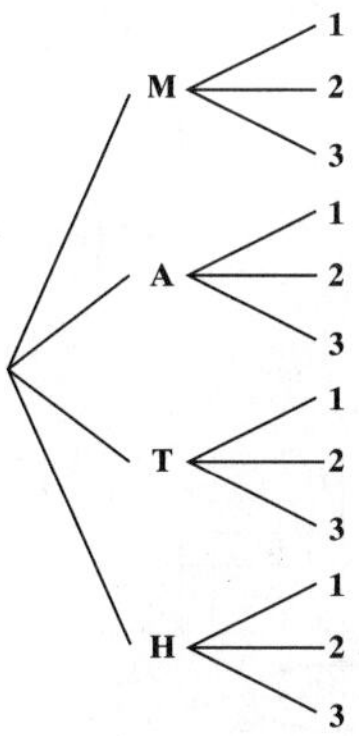

3. $\frac{1}{6}$ **5.** $\frac{1}{3}$ **7.** $\frac{1}{2}$ **9.** $\frac{2}{3}$ **11.** $\frac{1}{3}$ **13.** 1 **15.** $\frac{1}{7}$ **17.** $\frac{2}{7}$

A.3 Exponents, Order of Operations, and Variable Expressions

1. 243 **3.** 27 **5.** $\frac{16}{81}$ **7.** 1.44 **9.** 17 **11.** 20 **13.** 12 **15.** 30 **17.** $\frac{7}{18}$ **19.** 32 **21.** 1 **23.** 11 **25.** 132 **27.** $\frac{37}{18}$

A.4 Operations on Real Numbers

1. 4 **3.** -3 **5.** $-\frac{2}{9}$ **7.** 5 **9.** -24 **11.** -11 **13.** -3 **15.** -60 **17.** 80 **19.** 0 **21.** -3 **23.** 3 **25.** 0 **27.** undefined **29.** -49 **31.** -8 **33.** $-\frac{1}{27}$ **35.** 7 **37.** 4 **39.** $\frac{2}{5}$

A.5 Simplifying Expressions

1. $15y$ **3.** $-m - 6$ **5.** $5x^2$ **7.** $1.3x + 3.5$ **9.** $7d - 11$ **11.** $x + 5$ **13.** $2x + 14$ **15.** $-22 + 24x$ **17.** $5x - 7$ **19.** $10x - 3$ **21.** $-4m - 3$

A.6 Solving Linear Equations

1. 1 **3.** $\frac{3}{2}$ **5.** -4 **7.** 50 **9.** all real numbers **11.** 10 **13.** no solution

A.7 Solving Linear Inequalities

1. $\{x \mid x \geq -5\}$

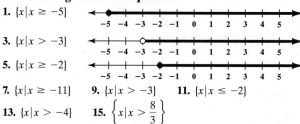

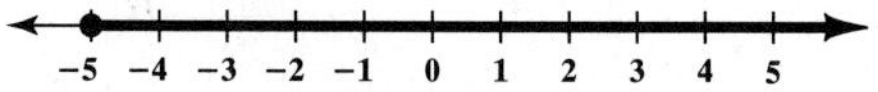

3. $\{x \mid x > -3\}$

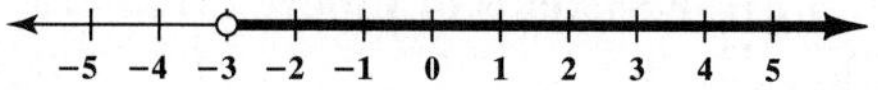

5. $\{x \mid x \geq -2\}$

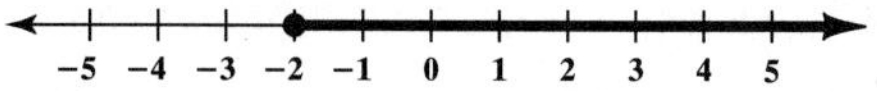

7. $\{x \mid x \geq -11\}$ **9.** $\{x \mid x > -3\}$ **11.** $\{x \mid x \leq -2\}$ **13.** $\{x \mid x > -4\}$ **15.** $\left\{x \mid x > \frac{8}{3}\right\}$

A.8 Solving Formulas

1. $h = 3$ **3.** $h = 20$ **5.** $r = 2.5$ **7.** $h = \frac{f}{5g}$ **9.** $w = \frac{V}{lh}$ **11.** $y = 7 - 3x$ **13.** $a = P - b - c$ **15.** $h = \frac{S - 2\pi r^2}{2\pi r}$

A.9 The Coordinate Plane

1.

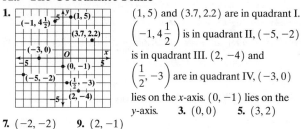

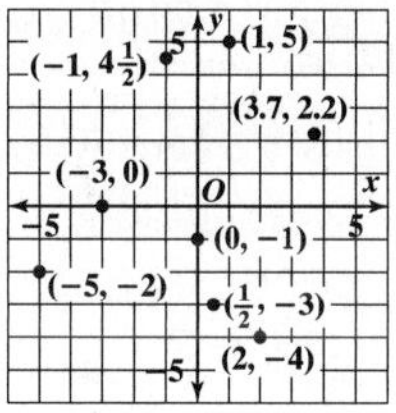

$(1, 5)$ and $(3.7, 2.2)$ are in quadrant I. $\left(-1, 4\frac{1}{2}\right)$ is in quadrant II, $(-5, -2)$ is in quadrant III. $(2, -4)$ and $\left(\frac{1}{2}, -3\right)$ are in quadrant IV, $(-3, 0)$ lies on the x-axis. $(0, -1)$ lies on the y-axis. **3.** $(0, 0)$ **5.** $(3, 2)$ **7.** $(-2, -2)$ **9.** $(2, -1)$

A.10 Graphing Linear Equations

1.

x	y
1	−4
0	0
−1	4

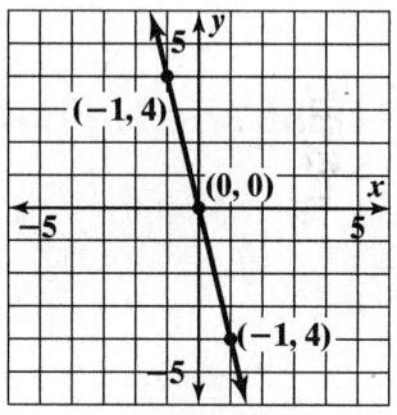

3.

x	y
0	3
1	−1
2	−5

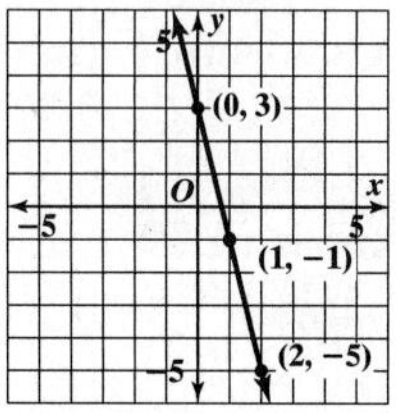

5.

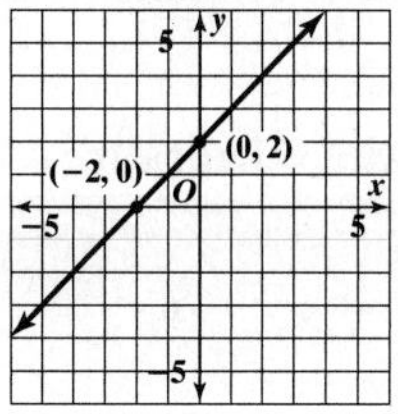

7.

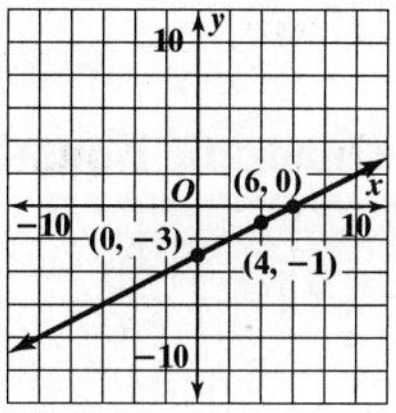

9.

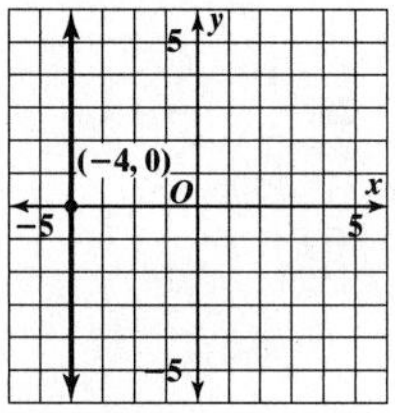

11.

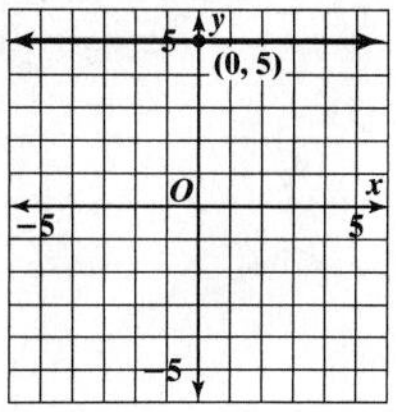

13. 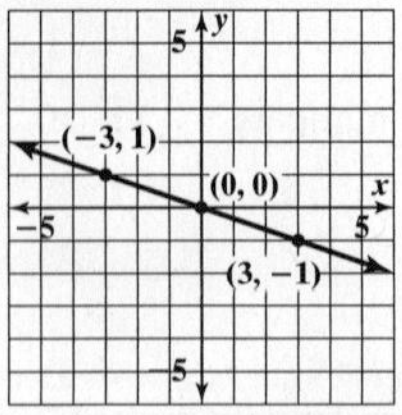

15. 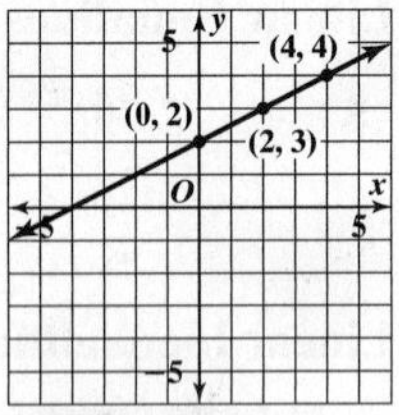

A.11 Solving Systems of Linear Equations

1. (2, 8) **3.** (0, −9) **5.** (1, −2) **7.** (9, 9) **9.** (−5, 3)
11. ∅ **13.** (3, 2) **15.** $\{(x, y)|x = 3y + 2\}$

A.12 Exponents

1. 8 **3.** 16 **5.** $\frac{1}{16}$ **7.** $\frac{32}{5}$ **9.** $15y^5$ **11.** $x^{19}y^6$ **13.** x^{36}
15. p^8q^8 **17.** $x^{10}y^{15}$ **19.** $\frac{r^9}{s^9}$ **21.** $\frac{4x^2z^2}{y^{10}}$ **23.** −64 **25.** $\frac{y^3}{2}$
27. 1 **29.** −7 **31.** 2 **33.** $-16x^7$ **35.** $\frac{y^{15}}{8x^{12}}$

A.13 Multiplying Polynomials

1. $a^2 + 5a - 14$ **3.** $10a^3 - 27a^2 + 26a - 12$ **5.** $49x^2y^2 - 14xy^2 + y^2$
7. $x^2 + 7x + 12$ **9.** $x^2 + 5x - 50$ **11.** $4y^2 - 25y + 6$
13. $4x^2 - 4x + 1$ **15.** $25x^2 + 90x + 81$ **17.** $a^2 - 49$
19. $81x^2 - y^2$ **21.** $\frac{1}{9}a^4 - 49$ **23.** $6b^2 - b - 35$ **25.** $16x^2 - 25$
27. $25x^2 - 60xy + 36y^2$

A.14 Simplifying Radical Expressions

1. 6 **3.** $\frac{1}{5}$ **5.** −10 **7.** not a real number **9.** −11 **11.** 2.646
13. 11.662 **15.** $2\sqrt{5}$ **17.** $\sqrt{33}$ **19.** $6\sqrt{5}$ **21.** $-15\sqrt{3}$
23. $\frac{6}{11}$ **25.** $\frac{3\sqrt{3}}{11}$ **27.** $\frac{\sqrt{11}}{6}$ **29.** $-\frac{\sqrt{3}}{4}$

A.15 Solving Quadratic Equations by Factoring

1. $-\frac{3}{2}, \frac{5}{4}$ **3.** −4, 2 **5.** 0, 7 **7.** 4, −4 **9.** $\frac{7}{3}, -2$
11. $-\frac{3}{2}, -\frac{1}{2}, 3$

A.16 Solving Quadratic Equations by the Square Root Method

1. ±8 **3.** $\pm\sqrt{21}$ **5.** $\pm\frac{1}{5}$ **7.** no real solution
9. $\pm\frac{\sqrt{39}}{3}$ **11.** $\pm\sqrt{2}$ **13.** 12, −2 **15.** $-2 \pm \sqrt{10}$
17. $\frac{7 \pm 4\sqrt{2}}{3}$

A.17 Solving Quadratic Equations by the Quadratic Formula

1. 2, 1 **3.** $\frac{-7 \pm \sqrt{37}}{6}$ **5.** −3, 4 **7.** $-2 \pm \sqrt{7}$ **9.** $\frac{1 \pm \sqrt{2}}{5}$
11. $1 \pm \sqrt{2}$ **13.** $\frac{-9 \pm \sqrt{129}}{12}$ **15.** $\frac{4 \pm \sqrt{2}}{7}$ **17.** $3 \pm \sqrt{7}$
19. $\frac{3 \pm \sqrt{3}}{2}$ **21.** $\frac{1}{3}, -1$ **23.** $\frac{3 \pm \sqrt{13}}{4}$ **25.** $\frac{1}{5}, -\frac{3}{4}$
27. no real solution

B Additional Lessons

B.1 Ellipses and Hyperbolas

Practice Exercises

1. 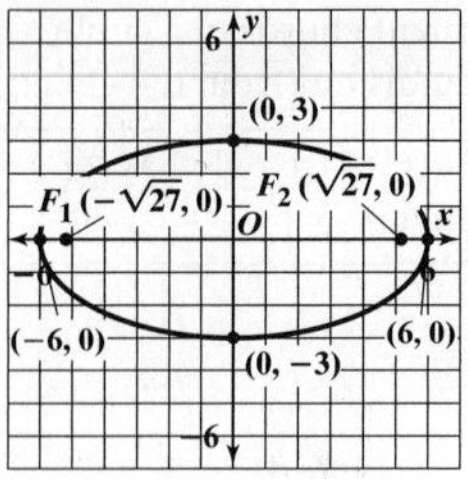

$\frac{x^2}{36} + \frac{y^2}{9} = 1$

2. 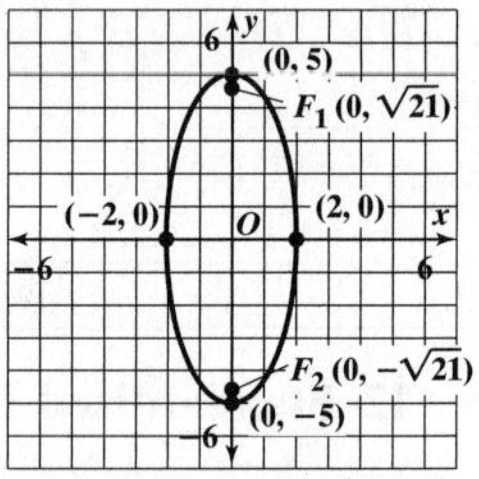

$\frac{x^2}{4} + \frac{y^2}{25} = 1$

3. $\frac{x^2}{16} + \frac{y^2}{7} = 1$

4. 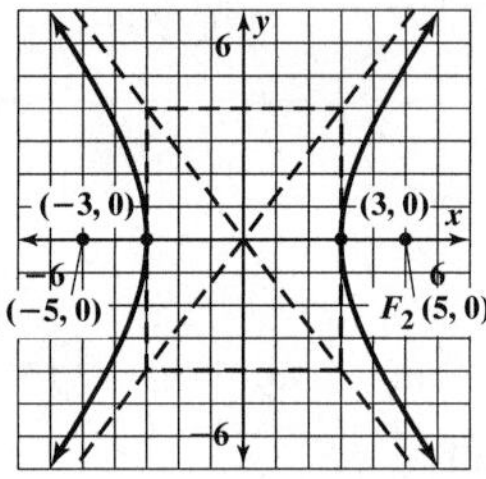

$\frac{x^2}{9} - \frac{y^2}{16} = 1$

5. 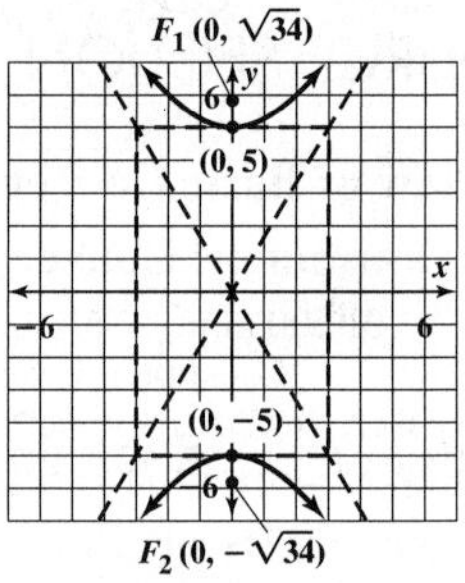

$\frac{y^2}{25} - \frac{x^2}{9} = 1$

6. $\frac{y^2}{4} - \frac{x^2}{21} = 1$

Vocabulary and Readiness Check

1. hyperbola **3.** focus **5.** hyperbola; (0, 0); x; $(a, 0)$ and $(-a, 0)$
7. ellipse **9.** hyperbola **11.** hyperbola

Ellipses and Hyperbolas Exercise Set

1. 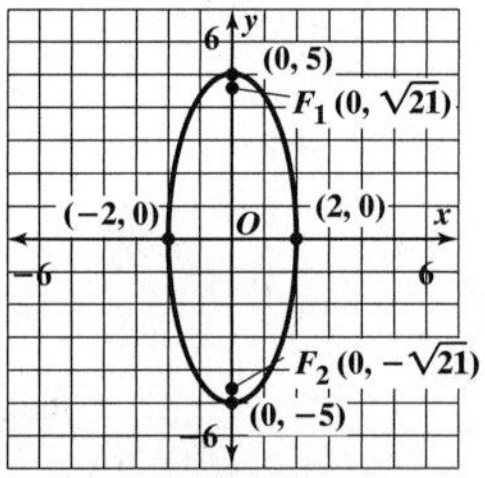

$\frac{x^2}{4} + \frac{y^2}{25} = 1$

3. 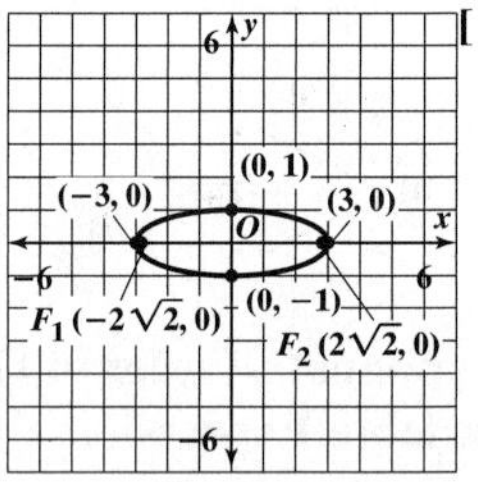

$\frac{x^2}{9} + y^2 = 1$

5. 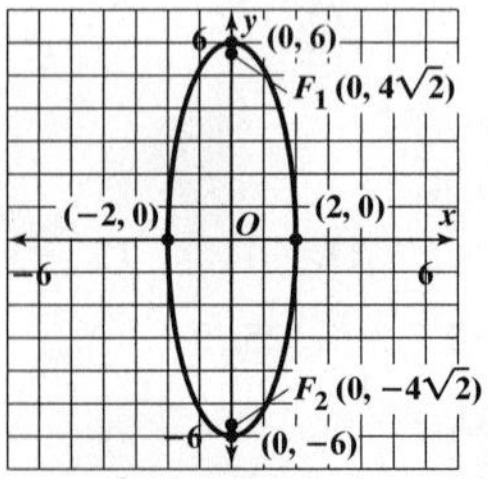

$\frac{x^2}{4} + \frac{y^2}{36} = 1$

7. 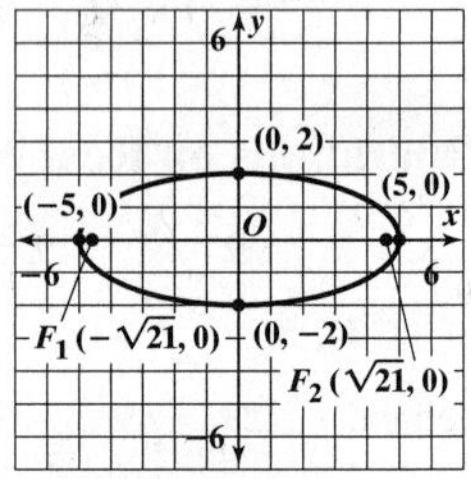

$\frac{x^2}{25} + \frac{y^2}{4} = 1$

9.

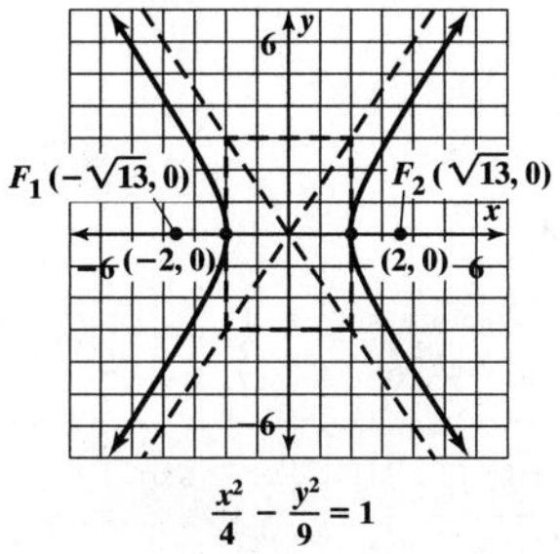

11.

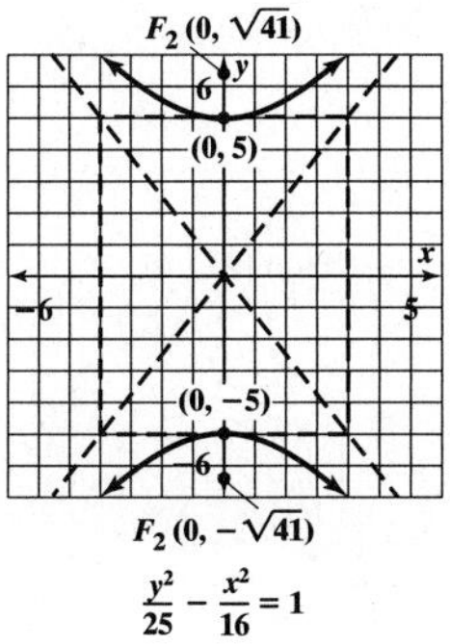

13.

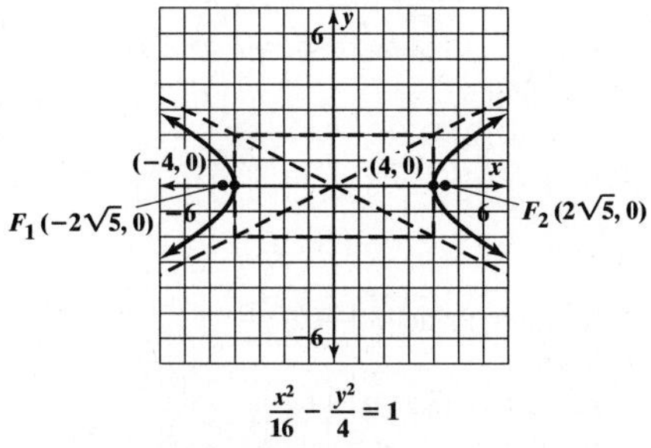

15.

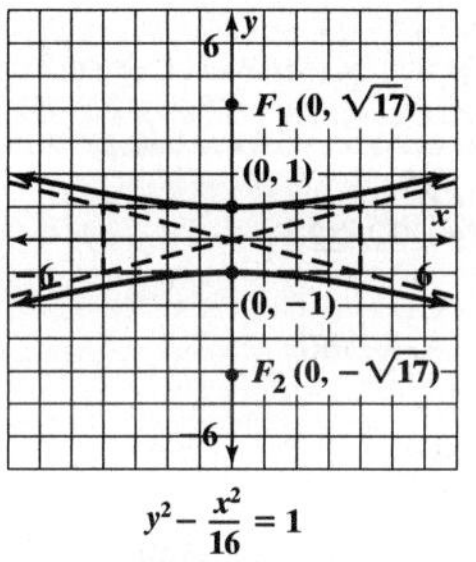

17. $\frac{x^2}{9} + \frac{y^2}{1} = 1$
Foci $(\sqrt{8}, 0)$ and $(-\sqrt{8}, 0)$ or $(2\sqrt{2}, 0)$ and $(-2\sqrt{2}, 0)$

19. $\frac{x^2}{4} + \frac{y^2}{25} = 1$
Foci $(0, \sqrt{21})$ and $(0, -\sqrt{21})$ **21.** $\frac{x^2}{36} + \frac{y^2}{16} = 1$
Foci $(\sqrt{20}, 0)$ and $(-\sqrt{20}, 0)$ or $(2\sqrt{5}, 0), (-2\sqrt{5}, 0)$

23. $\frac{x^2}{36} + \frac{y^2}{11} = 1$ **25.** $\frac{x^2}{15} + \frac{y^2}{16} = 1$ **27.** Vertices: $(0, 3)$ and $(0, -3)$; foci: $(0, \sqrt{13})$ and $(0, -\sqrt{13})$; b **29.** Vertices: $(3, 0)$ and $(-3, 0)$; foci $(\sqrt{13}, 0)$ and $(-\sqrt{13}, 0)$; c **31.** $\frac{y^2}{1} - \frac{x^2}{15} = 1$

33. $\frac{x^2}{4} - \frac{y^2}{5} = 1$

B.2 Measurement, Rounding Error, and Reasonableness

1. 23.5 ft to 24.5 ft or 24 ± 0.5 ft **3.** 339.5 mL to 341.5 mL or 340 ± 0.5 mL **5.** 73.15 mm to 73.25 mm or 73.2 ± 0.05 mm **7.** 10.8 cm to 11.2 cm or 11 ± 0.2 cm **9.** 208 m

B.3 The Effect of Measurement Errors on Calculations

1. 18% **3.** 1% **5.** 9% **7.** 2%

Index

A

Absolute value, 584
Acute angles, 26
Acute Angles of Right Triangle Corollary, 160
Acute triangles, 158
Addition
  of real numbers, 584
  vector, 404–405
Addition property of equality, 84–86, 587
Adjacent angles, 32
Adjacent sides, of triangles, 157
Algebraic properties of equality, 84–87
Alternate exterior angles, 104, 105
Alternate Exterior Angles Theorem, 111, 116
Alternate interior angles, 104, 105
Alternate Interior Angles Theorem
  converse of, 116–117, 612
  explanation of, 110
Altitude
  of cylinder, 494
  of prism, 492
  of right triangles, 320–321
  of triangles, 223–224
Altitude of Right Triangle Theorem, 320–321
Ambiguous case, 411–413
Angle Addition Postulate, 28, 29, 86
Angle-Angle-Side (AAS) Theorem, 177, 178
Angle-Angle Similarity (AA~) Postulate, 312–315
Angle bisectors
  construction of, 47
  explanation of, 34
  to find angle measures, 34–35
  of triangles, 217–218
  use of, 209–210
Angle Bisector Theorem, 210
Angle Measure—Lines Intersecting Inside Circle Theorem, 558
Angle Measure—Lines Intersecting Outside Circle Theorem, 558
Angle of depression, 397
Angle of elevation, 397–398
Angles
  acute, 26
  adjacent, 32
  alternate exterior, 104, 105
  alternate interior, 104, 105
  central, 457, 535
  classification of, 26–27
  complementary, 32, 33
  congruent, 27–28, 45–46, 165
  corresponding, 104, 105
  explanation of, 25
  exterior, 159
  exterior of, 25
  formed by chords, secants and tangents, 558–560
  formed by lines and transverse, 103–105, 115–119
  identifying and naming, 25
  inscribed, 551–555
  interior, 25, 159
  of isosceles triangle, 190
  measurement of, 26–29, 34–36, 116, 118–119
  obtuse, 26
  pairs of, 32–36
  of parallelograms, 260–262
  proving theorems about, 90–93
  right, 26, 27
  same-side interior, 104, 105
  sides of, 25
  straight, 26
  supplementary, 32–34
  vertex of, 25
  vertical, 32, 33
Angle-Side-Angle (ASA) Postulate, 176, 178
Apothem, 446
Arcs
  central, 457
  congruent, 461, 535, 536
  explanation of, 457, 535–537
  finding measures of, 458, 535, 559–560
  intercepted, 551
  length of, 460–461, 535–536
  major, 457, 535
  minor, 457, 535
Area. *See also* Surface area
  of circle, 54, 57, 464–467
  explanation of, 54
  to find probability, 472–473
  of kite, 441
  of parallelogram, 436, 437
  perimeter and, 436, 438
  of polygons, 427–431, 445–449
  of rectangle, 54, 57, 436
  of regular polygon tessellations, 431–432
  of rhombus, 439, 440
  of similar figures, 452–454
  of similar solids, 526–527
  of square, 54, 436
  of trapezoid, 439, 440
  of triangle, 54, 413, 414, 437–438, 441
  trigonometric ratios to find, 447–449
  use of right triangle to find, 441
Area Addition Postulate, 436
Area and Volumes of Similar Solids Theorem, 526
Area Congruence Postulate, 436
Area Formulas, 54
Area of Circle Formula, 364, 365
Area of Circle Theorem, 464
Area of Sector of Circle Theorem, 466
Asymptotes, 620
Axioms, 9, 14

B

Base
  of cylinder, 494
  of exponential expressions, 583
  of isosceles triangle, 190
  of parallelogram, 436
  of prism, 492
  of pyramid, 500
Base angle, of isosceles triangle, 190
Base Angles of Isosceles Trapezoid Theorem, 282
"between," 13, 21
Biconditional statements
  definitions and, 74–76
  explanation of, 73
  method to write, 73–74
  true or false, 73, 74
Bisect, 21
Bisectors
  angle, 34–35, 47, 209–210, 217–218
  perpendicular, 46, 205–208, 214–217
  segment, 21, 34
  triangle, 214–218, 224
Bolyai, János, 150

C

Cavalieri's Principle, 507
Center
  of circle, 457, 535, 548
  of ellipse, 615
  of hyperbola, 618
  of regular polygons, 354, 446
  of sphere, 518
Central angles, 457, 535
Centroid, of triangle, 222, 227
Chords
  angles formed by, 558–560
  congruent, 545–546
  explanation of, 544
  length of, 546
  perpendicular bisectors to, 547–549
  theorems about, 546, 547
Chords Equidistant from Center Theorem, 546
Circles
  area of, 54, 57, 464–467
  center of, 457, 535, 548, 565–567
  circumference of, 54–56, 458–460, 465, 466
  circumscribed about a triangle, 215
  concentric, 459, 535
  congruent, 535, 544–546
  equation of, 564–566
  explanation of, 457, 535–537
  inscribed in polygons, 541
  intersection of two, 539–540
  measures of angles in, 553–555
  radius of, 537, 565–567
  sector of, 466–467
  segment of, 467, 560–561
  symbol for, 55
  tangent to, 537, 540
Circumcenter, of triangle, 215–217, 227
Circumference
  explanation of, 54
  measurement of, 581
  method to find, 55–56, 458–460, 465, 466, 519
Circumference Formula, 54
Collinear points, 12, 13
Common tangents, 539–540
Comparison Property of Inequality, 235
Compass
  explanation of, 44
  geometric constructions with, 45–47
  steps to use, 44–45

Complementary angles
explanation of, 32, 33
measurement of, 34
Completing the square, 567
Composite solids, 509–510
Compositions, of reflections, 365–368
Concave polygons, 255
Concentric circles, 459
Conclusion, 67, 68
Concurrency
of altitudes, 223
of angle bisectors, 218
of medians, 222
points of, 227
Concurrency of Altitudes Theorem, 223
Concurrency of Angle Bisectors Theorem, 218
Concurrency of Medians Theorem, 222
Concurrency of Perpendicular Bisectors Theorem, 215
Concurrent lines, 215
Conditional statements
converse of, 70, 73, 74
explanation of, 67–68, 70
method to write, 68–70
related, 69–70, 78–79
true or false, 69, 73
Cones
explanation of, 502
oblique, 515
surface area of, 502–503
volume of, 514–515
Congruence
properties of, 86–87
symbol for, 20
Congruent angles
construction of, 45–46
explanation of, 27–28
Congruent arcs, 461, 535, 536
Congruent Central Angles and Arcs Theorem, 545
Congruent Central Angles and Chords Theorem, 545
Congruent chords, 545–546
Congruent Chords and Arcs Theorem, 545
Congruent circles, 535
Congruent figures
explanation of, 165
parts of, 165–167
Congruent segments
construction of, 45
explanation of, 20, 86
Congruent Tangent Segments, 540
Congruent triangles
by ASA and AAS, 176–179
explanation of, 166, 167
proofs using, 182–186
by SSS and SAS, 170–173
Conic section, 574, 615
Conjectures
explanation of, 61
formation of, 63–65
Consecutive Angles of Parallelogram Theorem, 261
Consecutive interior angles. *See* Same-side interior angles
Construction of Angle Bisector Theorem, 209
Construction of Perpendicular Bisector Theorem, 205
Constructions. *See also* Geometric constructions
explanation of, 44
Contour map. *See* Topographic map
Contradictions, identification of, 235
Contrapositive, of conditional statement, 70, 78, 79
Converse, of conditional statement, 70, 73, 74, 78, 79
Converse of Alternate Exterior Angles Theorem, 116
Converse of Alternate Interior Angles Theorem, 116–117
Converse of Angle Bisector Theorem, 210
Converse of Consecutive Angles Theorem, 266
Converse of Corresponding Angles Theorem, 117
Converse of Diagonals Theorem, 267
Converse of Hinge Theorem, 243
Converse of Isosceles Base Angles Theorem, 191
Converse of Opposite Angles Theorem, 266
Converse of Opposite Sides Theorem, 266
Converse of Perpendicular Bisector Theorem, 206
Converse of Pythagorean Theorem, 379
Converse of Same-Side Interior Angles Theorem, 117
Converse of Tangent-Radius Theorem, 538
Converse of Triangle Inequality Theorem, 236–237
Convex polygons, 255
Coordinate, of points, 19
Coordinate geometry
equations of lines and, 139–147
graph of reflection image and, 349
midsegments and, 229–230
slope of line and, 131–136
vectors and, 402
Coordinate plane
circles and, 564–567
explanation of, 590
midpoint on, 39–42
with parallelograms, 269–271
perimeter in, 56
Coplanar points, 12, 13
Corollaries, 160, 192
Corollaries to Inscribed Angle Theorem, 553, 554
Corollary to Side-Splitter Theorem, 327–328
Corollary to Triangle Exterior Angle Theorem, 235–236
Corresponding angles, in congruent figures, 165
Corresponding Angles Theorem
converse of, 117
explanation of, 110
proof of, 111
Corresponding sides, in congruent figures, 165
Cosine ratio
explanation of, 389
to find angle measures, 391–392
method to write, 390
Counterexamples
with conditional statements, 69
method to find, 65
Cross Product Property, 296
Cross products, 300
Cross sections, of solids, 485–486
Cubic units, 581
Cylinders
explanation of, 486, 494
lateral area of, 495
oblique, 494
right, 494
surface area of, 494–495
volume of, 508–509

D

Deductive reasoning
explanation of, 79
Law of Detachment and, 79–81
Law of Syllogism and, 80–81
Defined terms, 9
Definitions, 9, 74–76
Degrees, 26
Diagonals
of parallelograms, 261
of polygons, 256–257
of rectangles, 276–278
of special parallelograms, 276–279
Diagonals of Isosceles Trapezoid Theorem, 282
Diagonals of Kite Theorem, 284
Diagonals of Parallelogram Theorem, 261
Diameter, 535
Dilations
applications of, 362
explanation of, 360–362
Directrix, 574
Distance
between points, 19, 41–42
from point to line, 209
proofing triangle parts congruent to measure, 183–184
Distance Formula, 41, 270, 619
Distributive property, 84, 85
Division, of real numbers, 584
Division property of equality, 84, 85
Dodecagons, 448

E

Edge, of segment, 157
Ellipses
equation of, 615–618
explanation of, 615
graphs of, 617
parts of, 615
standard form of, 616
Elliptical geometry, 150
Elliptic geometry, 48
Endpoints, 40
Epoctac, 182
Equal Complements Theorem, 90, 91
Equality properties. *See* Properties of Equality
Equal Supplementary Angles Theorem, 93
Equal Supplements Theorem, 91
Equations
of circle, 565–566
of ellipse, 615–618
of horizontal lines, 145
of hyperbola, 618–621
of line, 133, 139–147
linear, 133, 139, 587, 591, 592
of parabola, 574–575
quadratic, 597–599
Equiangular polygons, 256
Equiangular triangles
explanation of, 158
properties of, 193–194
Equidistant, 206, 209
Equilateral polygons, 256

Equilateral triangles, 158
Equivalent statements, 70
Euclid, 150
Euclidean geometry, 48, 150
Euclid's Fifth Postulate, 150
Euler, Leonhard, 483
Euler's Formula, 483–484
Events, probabilities of, 471–473
Exams, 5
Exponential expressions, 583
Exponents, 583, 584, 594
Extended proportions, 307
Extended ratios, 296
Exterior, of angles, 25
Exterior Angle of Triangle Corollary, 160, 161
Exterior angles
  alternate, 104, 105, 111, 116
  explanation of, 103
  of polygons, 429–431
  of triangles, 159, 160
Extremes, in proportions, 296

F

Faces, of pyramid, 500
Factoring, 597
Fibonacci sequence, 62
Figures
  assumptions about unmarked, 9–10
  geometric, 13
Flow proof, 111–112
Focus, 574, 615
FOIL method, 595
Formulas, 589, 601. *See also* Specific formulas
45°-45°-90° Triangle Theorem, 383–384
Fractions, writting ratios as, 294
Fundamental Theorem of Isometries, 367

G

Geometric constructions, compass use for, 44–47
Geometric figures, 13
Geometric Mean in Similar Right Triangles: Hypothesis Corollary, 322, 323
Geometric Mean in Similar Right Triangles: Legs Corollary, 323
Geometric means
  explanation of, 321
  method to find, 322
  in similar right triangles, 322–323
Geometric probability
  explanation of, 471
  use of area to find, 472–475
  use of segment and area models to find, 471–472
Geometry. *See also* Coordinate geometry
  development of, 9–10
  Euclidean, 48
  explanation of, 7
  historical background of, 1, 48, 150
  learning terms and postulates of, 12–17
Glide reflection, 367

H

Hexagons, 257
Hinge Theorem
  converse of, 243, 245
  explanation of, 242, 379
  use of, 243–244
Horizontal change, 131, 132
Horizontal lines
  equations of, 145
  explanation of, 139
Hyperbolas
  equation of, 618–621
  explanation of, 618
  graphs of, 620
  standard form of, 619
Hyperbolic geometry, 48
  explanation of, 150
Hypotenuse
  explanation of, 193
  length of, 377–378, 384
Hypotenuse-Leg (H-L) Theorem, 194, 613
Hypothesis
  explanation of, 67
  identification of, 68

I

If Equiangular then Equilateral Triangle Corollary, 192
If Equilateral then Equiangular Triangle Corollary, 192
If-then form, 67
Image
  of rigid transformation, 337, 338
  of transformation, 337
  of translation, 342
Incenter, of triangles, 218, 227
Indirect measurement
  explanation of, 315–316
  use of trigonometry to find, 390–392
Indirect proofs, 234
Indirect reasoning, 234
Inductive reasoning
  explanation of, 61
  use of, 62, 63
Inequalities
  method solve linear, 588
  in triangles, 234–239
  in two triangles, 242–245
Inscribed angles
  explanation of, 551–552
  measures of, 552–555
Inscribed Angle Theorem, 551–553
Intercepted arcs, 551
Interior, of angles, 25
Interior angles
  explanation of, 103
  of polygons, 256–257, 427–429
  same-side, 104, 105
  of triangles, 159, 160
Interior Angle Sum of Convex Quadrilateral Theorem, 257
Intersecting lines, 366–367
Intersection
  of lines, 14–15
  of planes, 15–16
Inverse, of conditional statement, 70, 78, 79
Irrational numbers, π (pi) as, 55, 459
Isometries
  classification of, 368
  explanation of, 337
  Fundamental Theorem of, 367
  rotation about a point as, 353
  translation as, 341
Isometry Classification Theorem, 368
Isosceles Base Angles Theorem, 190, 191
Isosceles Trapezoid Theorem, 282
Isosceles triangles
  explanation of, 158
  properties of, 190–192, 383–384

K

Kites
  angle measures in, 284–285
  area of, 440, 441
  explanation of, 284
  properties of, 284–285

L

Lateral area
  of cone, 503
  of cylinder, 495
  of pyramid, 500
Lateral faces, 492, 500
Lathes, 486
Law of Cosines
  explanation of, 416–417
  oblique triangles solved with, 417–419
  proof of, 416, 417
Law of Detachment
  explanation of, 79–80
  use of, 80, 81
Law of Sines
  ambiguous case and, 411–413
  area of oblique triangle and, 413, 414
  explanation of, 408–409
  oblique triangles solved with, 409–410
  proof of, 408
Law of Syllogism, 80, 81
Least squares method, 134
Legs
  of isosceles triangle, 190
  of right triangles, 193, 378, 384, 386
Like terms, 586
Linear equations
  forms of, 133, 139
  method to graph, 591
  method to solve, 587
  systems of, 592
Linear inequalities, 588
Linear pairs, adjacent angles as, 32–33
Linear Pair Theorem, 91–92
Line of reflection, 348–349
Lines
  concurrent, 215
  equations of, 133, 139–147
  explanation of, 12–14
  horizontal, 139, 144
  intersection of, 14–15
  parallel, 102, 103, 108–109, 122, 127–128, 145
  perpendicular, 46, 123–125, 128–129, 146–147
  skew, 102, 103
  slope of, 131–136
  that do not intersect, 102–103
  transversal of, 103–105
  vertical, 139, 145
Line segments. *See* Segments
Line symmetry, 350
Lobachevsky, Nikolai, 150
Locus, 570–571
Logic
  explanation of, 8
  to recognize patterns, 8–9
  use of, 60–61

M

Magnitude, of vectors, 401–404
Major arcs, 457, 535
Major axis, of ellipse, 615
Mathematical system, 9–10
Mathematics, tips for success in, 2–6
Means
  geometric, 321–323
  in proportions, 296
Measurement
  of angles, 26–29, 34–36, 116, 118–119
  conversions in, 581
  errors in, 625
  indirect, 315–316, 390–392
  nature of, 624
  table of, 603
Medians, of triangles, 222–223
Metric units, 603
Midpoint
  on coordinate plane, 39–42
  method to find, 22–23
  of segment, 21, 39–42
Midpoint Formula, 39, 40, 270
Midsegments
  applications solved by using, 230–231
  of trapezoids, 283
  of triangles, 227–231
  use of coordinate geometry with, 229–230
Minor arcs, 457, 535
Minor axis, of ellipse, 615
Multiple Parallel Lines and Transversals Theorem, 263
Multiplication
  of polynomials, 595
  of real numbers, 584
Multiplication property of equality, 84, 587

N

Negation, 69
Negative square root, 594
Net, 483, 493
n-gon, 255, 446
Noncollinear points, planes and, 16–17
Non-Euclidean geometry, 48
Number line, midpoint on, 39–40
Numbers
  irrational, 55, 459
  real, 584, 604

O

Oblique cones, 515
Oblique cylinders, 494–495
Oblique triangles
  area of, 413–414
  explanation of, 408
  Law of Cosines to solve, 417–419
  Law of Sines to solve, 408–413
Obtuse angles, 26
Obtuse triangles, 158
Opposite Angles of Kite Theorem, 284
Opposite Angles of Parallelogram Theorem, 260, 261
Opposite rays, 13.15
Opposite Sides of Parallelogram Theorem, 260, 261
Orthocenter, of triangle, 223, 227
Overlapping triangles, 184, 185

P

Parabolas, 574–575
Paragraph proof, 92
Parallel lines
  composing reflections across, 366
  construction of, 127–128
  explanation of, 102, 103
  finding equations of, 145
  identification of, 110
  proving and using theorems about, 115–119
  proving that lines are, 108–113
  proving theorems about, 122
  slope of, 135–136
  theorems for, 108–109
Parallelograms
  area of, 436, 437
  base of, 436
  coordinate plane with, 269–271
  explanation of, 260, 275
  height of, 436
  identifying special, 276
  method to find values of variables in, 266–267
  method to identity, 269
  proof that quadrilaterals are, 266–271
  properties of diagonals of special, 276–279
  properties of special, 274–275
  quadrilaterals that are not, 282–285
  relationships among consecutive angles and diagonals of, 261–263
  relationships among sides and angles of, 260–261
  special types of, 274–279
Parallel planes, 102
Parallel Postulate, 109, 122, 150
Parallel segments
  explanation of, 103
  identification of, 228
Patterns
  inductive reasoning to study, 61–62
  logic to recognize, 8–9
  recognition of visual, 62–63
Pentagons
  diagonal of, 257
  explanation of, 427
Perimeter
  in coordinate plane, 56
  explanation of, 54
  of irregular room, 438–439
  measurement of, 581
  of rectangle, 54–55
  of similar figures, 452–454
  of square, 54
  of triangle, 54
Perimeter Formula, 54
Perpendicular bisectors
  construction of, 205
  explanation of, 46, 205–208
  of triangles, 214–217
  use of, 205–208
Perpendicular Bisector Theorem, 206
  use of, 206–208
Perpendicular lines
  construction of, 128–129
  explanation of, 46
  finding equations of, 146–147
  slope of, 135–136
  theorems about, 123–125
  using algebra to find measures of, 125
Perpendicular Postulate, 109, 123
Perpendicular Transverse Theorem
  explanation of, 123
  method to prove, 123–124
π (pi), 55, 458, 459
Planes
  explanation of, 12–13
  intersection of, 15–16
  locus in, 570–571
  parallel, 102
  that do not intersect, 102–103
Playfair, John, 150
Point of concurrency, 215
Points
  collinear, 12, 13
  coordinate of, 19
  coplanar, 12, 13
  distance between, 19
  explanation of, 12–13
  noncollinear, 16–17
Point-slope form
  explanation of, 133, 139, 141
  use of, 141–144
Polygon Exterior Angle-Sum Theorem, 430
Polygon Interior Angle-Sum Theorem, 427
Polygons
  area of, 427–431
  circles inscribed in, 541
  concave, 255
  convex, 255
  diagonal of, 256–257
  equiangular, 256
  equilateral, 256
  explanation of, 254
  identification of, 254–255
  interior angles of, 256–257, 427–429
  names for, 427
  n-gon, 255
  polyhedrons and, 483
  regular, 256, 354, 445–449
  similar, 307–309
Polyhedrons
  analysis of, 483–484
  explanation of, 483
  identification of, 484
  prisms as, 482–484
  pyramids as, 500–502
Polynomials, multiplication of, 595
Positive square roots, 584, 596
Postulates, 9, 14
Postulates and Theorems List, 605–611
Power of Product Rule, 594
Power of Quotient Rule, 594
Power Rule, 594
Preimage
  of rigid transformation, 338
  of transformation, 337
Principal square roots, 584, 596
Prisms
  explanation of, 492
  rectangular, 507
  right, 493, 507
  surface area of, 492–494
  triangular, 508
  volume of, 506–508
Probability
  explanation of, 471, 582
  geometric, 471–475
Problem-solving steps, 301–302
Product Rule, 594

Product Rule for Radicals, 596
Proofs
explanation of, 86
flow, 111–112
indirect, 234
method to write two-column, 86–89
Properties of Equality
explanation of, 84–87
to rewrite proportions in equivalent forms, 300, 303
Proportions. *See also* Ratios
equivalent, 300, 301
explanation of, 296
extended, 307
extremes in, 296
means in, 296
method to solve, 297
problem solving with, 301–303
properties of, 300–301
Protractor Postulate, 26
Protractors, 26
Pyramids
explanation of, 500
regular, 500
surface area of, 500–502
volume of, 513–514
Pythagorean Theorem
converse of, 379–380
Distance Formula and, 41
explanation of, 377
use of, 377–379, 494
Pythagorean triple, 377

Q

Quadratic equations, 597–599
Quadratic Formula, 599
Quadrilaterals. *See also* Parallelograms; Polygons; Rectangles; Rhombuses; Trapezoids
explanation of, 254, 427
interior angles of, 256–257
as parallelograms, 266–271, 275
relationships among, 275, 285
that are not parallelograms, 282–285
Quadrilaterals as Parallelogram Theorem, 268
Quotient Rule, 594
Quotient Rule for Radicals, 594

R

Radical expressions, 596
Radicals, 594
Radius
explanation of, 446, 518, 535
identification of, 539
method to find, 538
Rate of change, slope as, 132
Ratios
See also Proportions
area, 454
explanation of, 294
extended, 296
perimeter, 454
in simplest form, 294–295
tangent, 389–392
trigonometric, 389–392
written as fractions, 294
Rays
explanation of, 13
naming, 13–14
opposite, 13, 15
Real numbers
operations on, 584
properties of, 604
Reasoning
deductive, 78–81
indirect, 234
inductive, 61–63
Rectangle Diagonals Theorem, 276–277
Rectangles
area of, 54, 57, 436, 452
diagonals of, 276–278
explanation of, 274
perimeter of, 54–55, 452
properties of, 275
Rectangular coordinate system, 39
Rectangular prisms, 507
Reflectional symmetry, 350
Reflections
across lines, 348–350
compositions of, 365–368
glide, 367
image of figures, 347–350
Reflexive property of equality, 84–87
Regular Polygon Exterior Angle Corollary, 430
Regular Polygon Interior-Angle Corollary, 428, 429
Regular polygons
apothem of, 446
area of, 445–449
center of, 354, 446
central angle of, 446
explanation of, 256
radius of, 446
sides of, 429
Regular polygon tessellations, 431–432
Rhombus Diagonal/Perpendicular Theorem, 276–277
Rhombus Diagonals Theorem, 276–277
Rhombuses
angle measures of, 278
area of, 439, 440
diagonals of, 276–278
explanation of, 274
properties of, 275
Riemann, Bernhard, 150
Right angles, 26, 27
Right Angles Congruent Theorem, 93
Right cylinders, 494
Right prisms, 493, 507
Right triangles. *See also* Pythagorean Theorem
acute angles of, 160
altitudes of, 320–321
applications involving, 324
explanation of, 158
identification of, 379–380
method to solve, 395–398
30º-60º-90º, 385–386
45º-45º-90º, 383–384
properties of, 193–194
similar, 322–323
special, 383–386
Rigid transformations
explanation of, 337
identification of, 337–338
images and corresponding parts of, 338–339
Rotational symmetry, 355
Rotations
angle of, 353
center of, 353
composition of, 354–355
construction of, 353–354
explanation of, 353
identification of, 354
Rounding error, 624
Ruler Postulate, 29
Rulers, 44

S

SAA triangles, 409–410
Same-side interior angles, 104, 105
Same-Side Interior Angles Theorem, 110, 117
SAS Inequality Theorem. *See* Hinge Theorem
SAS triangles, 417–418
Scale factor
for dilations, 360–362
explanation of, 307–308, 452
method to find, 453, 526, 527
Scalene triangles, 158
Seattle Space Needle, 293
Secants
angles formed by, 558–560
explanation of, 558
Sector of circle, 466–467
Segment Addition Postulate, 21–22
Segment bisectors, 21, 34
Segment Products - Inside or Outside Circle Theorem, 560–561
Segments
angle bisectors of, 209–210
of circle, 467, 560–561
congruent, 20, 45, 86
edge of, 157
explanation of, 13
to find probability, 471–472
measures of, 19–23
midpoint of, 21, 39–42
naming, 13–14
parallel, 103
perpendicular bisectors of, 205–208
Semicircles, 457, 535
Side-Angle-Side (SAS) Postulate, 172, 173, 178
Side-Angle-Side Similarity (SAS~) Theorem, 314, 315
Side opposite the angle, of triangles, 157
Sides
of angles, 25
of parallelograms, 260
of polygons, 254, 429
of triangles, 157, 395–396
Side-Side-Side Similarity (SSS~) Theorem, 314, 315
Side-Side-Side (SSS) Postulate, 171–173, 178
Side-Splitter Theorem, 326–328
Similar figures, perimeter and area of, 452–454
Similarity
additional proportions in triangles and, 326–329
polygons and, 307–309
proportion properties and problem solving and, 300–303
ratios and proportions and, 294–297
in right triangles, 320–324
triangles and, 312–316
Similarity statements, 321
Similar polygons
in applicatrions, 309
explanation of, 307–309

Similar solids
areas and volumes of, 526–527
explanation of, 525
identification of, 525
Sine ratio
explanation of, 389
to find angle measures, 391–392
method to write, 390
Skew lines, 102, 103
Slant height, of pyramid, 500, 501
Slope
explanation of, 131
of horizontal line, 133
method to find, 131–134
of parallel lines, 135–136
of perpendicular lines, 135–136
of vertical line, 133
Slope Formula, 170
Slope-intercept form
explanation of, 133, 139, 140
method to interprete, 134–135
use of, 140–141
Solids
areas and volumes of similar, 525–527
polyhedrons and, 483–484
of revolution, 486–487
Solving the triangle
explanation of, 395–398
given one side and one angle, 395–396
given two sides, 396
use of angles of elevation to, 397–398
Space, 13
Special parallelograms. *See also* Parallelograms
diagonals of, 276–279
identification of, 276
properties of, 274–275
types of, 274
Spheres
explanation of, 518
surface area of, 518–519
volume of, 518, 520–521
Spherical geometry, 48
Square Root Property, 598
Square roots, 584, 596
Square root symbol, 584
Squares
area of, 54, 436
explanation of, 274
perimeter of, 54
properties of, 275
Square units, 581
SSA triangles, 411–413
SSS triangles, 418–419
Standard equation, of circle, 565–567
Standard form, of linear equations, 139, 143
Statements
biconditional, 73–76
conditional, 67–70, 73, 74, 78–79
giving reasons for, 84–85
Straight angles, 26
Straight edge, 44
Substitution property of equality, 84–86
Subtraction, of real numbers, 584
Subtraction property of equality, 84
Sudoku puzzle, 53
Supplementary angles
explanation of, 32–34
measurement of, 34
Surface area
of cone, 502–503
of cylinder, 494–495
explanation of, 492
method to find, 495–496
of prism, 492–494
of pyramid, 500–502
of sphere, 518–521
Surface Area of Cone Theorem, 502
Surface Area of Prism Theorem, 493
Surface Area of Pyramid Theorem, 500
Symbols
for circle, 55
for conditional statements, 68
congruent, 20
for parallelograms, 260
for right angles, 27
square root, 584
table of, 600
Symmetric property of equality, 84–87
Symmetry
identification of lines of, 350
rotational, 355
Systems of linear equations in two variables, 592

T

Tangent-Radius Theorem, 537
Tangent ratio
to find angle measures, 391–392
to find side length, 389, 390
Tangents
angles formed by, 554–555, 558–560
to circle, 537, 540
common, 539–540
properties of, 537–541
Terms
defined, 9
undefined, 9, 12–13
Tessellations, 431–432
Thales, 183
Theorems
about angles, 90–93
explanation of, 9
list of, 605–611
Third Angles Theorem, 166, 167
30°-60°-90° Triangle Theorem, 385–386
Time management, 6, 7
Time measures, 603
Topographic map, 485
Transformations
composition of, 343–344
composition of reflections and, 365–368
dilations and, 360–362
explanation of, 336, 337
image of, 337
isometry as, 337
preimage of, 337
reflections and, 348–350
rigid, 337–339
rotations and, 353–355
tracing paper, 358–359
translations and, 341–344
Transitive property of equality, 84–87
Translations
explanation of, 341
method to find image of, 341–344
Transversals
angle pairs formed by, 104–105
explanation of, 103
parallel lines and angles formed by, 115–119
of parallelograms, 260, 263
proving and using theorems about, 115–119
Transverse axis, of hyperbola, 618
Trapezoid, height of, 439
Trapezoid Midsegment Theorem, 283, 613–614
Trapezoids
area of, 439, 440
explanation of, 282
method to find angle measures in, 283
midsections of, 283–284
properties of, 282–283
Triangle-angle-Bisector Theorem, 328–329
Triangle Angle-Sum Theorem
explanation of, 159
proof of, 160
use of, 161
Triangle Inequality Theorem
explanation of, 236
use of, 238, 616
Triangle Inequality Theorem for Sum of Lengths of Sides, 238
Triangle Midsegment Theorem, 228, 230
Triangles. *See also* Right triangles
altitudes of, 223–224
angle bisectors of, 217–218, 224
angle measures of, 159–161, 166
area of, 54, 413–414, 437–438, 441
bisectors of, 214–218, 224
circumcenter of, 215–217
classification of, 158–159, 380
congruent, 166–168, 170–173, 176–179, 182–186
equilateral, 158, 193–194
explanation of, 157
exterior angles of, 159, 160
identifying parts of, 157, 184
incenter of, 218
indirect proofs and inequalities in, 234–239
inequalities in two, 242–245
interior angles of, 159, 160
isosceles, 158, 190–192
medians of, 222–224
midsegments of, 227–231
oblique, 408–414
overlapping, 184, 185
perimeter of, 54
perpendicular bisectors of, 214–217, 224
SAA, 409–410
SAS, 417–418
sides of, 157
Side-Splitter Theorem and, 326–328
similarity in, 312–316, 320–324, 326–329
SSA, 411–413
SSS, 418–419
Triangle-angle-Bisector Theorem and, 328–329
vertex of, 157
Triangular prisms, 508
Trigonometric ratios
explanation of, 389
to find area of regular polygons, 447–449
method to write, 389–390
use of, 390–392
Two Lines Perpendicular to Third Line Theorem, 109, 611–612

U

U. S. Customary Units, 603
Undefined terms
  examples of, 12–13
  explanation of, 9

V

Vector addition, 404–405
Vectors
  explanation of, 401–402
  magnitude and direction of, 401–404
Venn diagrams, of special quadrilaterals, 275
Vertex
  of angles, 25
  of ellipse, 615
  of hyperbola, 618
  of polyhedrons, 483
  of pyramid, 500
  of triangles, 157
Vertex angle, of isosceles triangle, 190
Vertical angles, 32, 33
Vertical Angle Theorem, 92, 93
Vertical change, 131–132
Vertical lines
  explanation of, 139
  writing equations of, 145
Volume
  of composite solid, 509–510
  of cone, 514–515
  of cylinder, 508–509
  explanation of, 506
  of prism, 506–508
  of pyramid, 513–514
  of similar solids, 526–527
  of sphere, 518, 520–521
Volume of Cone Theorem, 514
Volume of Pyramid Theorem, 513
Volume of Sphere Theorem, 520

X

x-axis, 590
x-coordinate, 590

Y

y-axis, 590
y-coordinate, 133
y-coordinate, 590
y-intercept, 133

Z

Zero Exponent, 594

Photo Credits

Chapter 1

CO (t) Alexey Pavluts/Fotolia; (b) Georgios Kollidas/Fotolia **p. 2** Monkey Business Images/Shutterstock **p. 8** (t) Joanne van Hoof/Shutterstock; (m) Raga Jose Fuste/Prisma Bildagentur AG/Alamy; (b) Nataliya Hora/Shutterstock **p. 16** Andrei/Shutterstock **p. 30** Foodlovers/Shutterstock **p. 31** (t) Radhoose/Shutterstock; (b) Aigarsr/Fotolia **p. 48** Library of Congress Division of Prints and Photographs [LC-USZC4-4940]

Chapter 2

p. 64 Michael Shake/Shutterstock **p. 65** Xiangdong Li/Fotolia **p. 68** (t) Cynoclub/Fotolia; (b) Anatolii/Fotolia **p. 71** (t) Kurapy/Fotolia; (b) Aleksey Stemmer/Fotolia **p. 72** (b) Sergejs Rahunoks/Fotolia; (t) Fotolia **p. 75** (l) Peter Waters/Fotolia; (r) Fovito/Fotolia **p. 77** Stephen Coburn/Fotolia **p. 79** Imaginis/Fotolia **p. 80** Ollyy/Shutterstock; (tl) Comugnero Silvana/Fotolia; (bl) Simone van den Berg/Fotolia **p. 82** (r) Bill Perry/Fotolia **p. 83** Thinkstock **p. 94** LesPalenik/Shutterstock **p. 150** (l) Alexey Pavluts/Fotolia

Chapter 3

CO (b) Senior Airman Asha Kin/U.S. Air Force; (t) Mass Communication Specialist 1st Class Johnie Hickmon/U.S. Navy **p. 102** Xuejun li/Fotolia **p. 105** Patrick Batchelder/Alamy **p. 107** (bl) Jpchret/Fotolia; (tl) Thomas Barrat/Shutterstock; (tr) Magnum Johansson/Shutterstock **p. 109** (b) Lee Barnwell/Fotolia; (m) Christian Colista/Shutterstock; (t) Shvak/Shutterstock **p. 112** Rikke/Fotolia **p. 143** Morgan Lane Photography/Shutterstock **p. 149** Val Lawless/Shutterstock

Chapter 4

p. 156 (l) Andy Mac/Fotolia; (r) Gilbert S. Grant/Science Source/Photo Researchers, Inc. (t) Gui Jun Peng/Shutterstock **p. 164** (b) Fotolia **p. 171** (b) Dirk Ott/Shutterstock; (t) Richard Thornton/Shutterstock **p. 194** Marcl Schauer/Shutterstock **p. 196** Brandelet Didier/Fotolia

Chapter 6

p. 253 Amy Walters/Shutterstock **p. 265** Cynthia Farmer/Shutterstock **p. 273** ArenaCreative/Fotolia **p. 274** Ratselmeister/Fotolia **p. 286** Shutterstock

Chapter 7

p. 293 Xuanlu Wang/Shutterstock **p. 325** Leonard Zhukovsky/Fotolia **p. 329** Keith Wilson/Fotolia

Chapter 8

CO (tr) Sally Wallis/Fotolia **p. 356** (bl) iStockphoto/Thinkstock/Getty Images; (br) Satawat Anukul/Fotolia **p. 357** (mr) Angelo Ferraris/Shutterstock; (tl) Boltenkoff/Fotolia; (tl) Le Do/Fotolia **p. 362** (tr) Reika/Fotolia **p. 370** (tr) Irantzuarb/Fotolia; (tr) Lprendy/Fotolia

Chapter 9

p. 377 (tm) Rossler/Fotolia **p. 386** (ml) Hemera Technologies/Thinkstock **p. 388** (tl) Rupert Robertson/Fotolia **p. 397** (ml) Vrihu/Fotolia; (tl) Apops/Fotolia **p. 400** (mr) Beachboyx10/Fotolia; (tm) Wildnerdpix/Fotolia **p. 401** (br) Jan Prchal/Fotolia **p. 421** (tm) Tupungato/Fotolia

Chapter 10

CO (tm) Mcoda/Fotolia **p. 431** (bl) Homydesign/Fotolia **p. 433** (m) Asaf Eliason/Shutterstock; (ml) Edsweb/Fotolia; (tm) Evlakhov Valeriy/Shutterstock; (tm) Shutterstock **p. 447** (ml) Dmytro Smaglov/Fotolia **p. 450** (l) NASA; (tr) NASA **p. 451** (bl) Fotolia **p. 454** (ml) Jim West/Alamy **p. 466** (tl) Leszek Glasner/Shutterstock **p. 468** (mr) Denphumi/Fotolia **p. 469** (mr) Nancy Hjxson/Fotolia **p. 471** (ml) Jackmicro/Fotolia; (tl) Jackmicro/Fotolia **p. 473** (ml) Matt Ragen/Shutterstock

Chapter 11

CO Jörg Hackemann/Fotolia **p. 486** (l) Steve Holderfield/Fotolia; (r) Mbongo/Fotolia **p. 491** Africa Studio/Fotolia **p. 495** Alamy **p. 498** Acik/Fotolia **p. 504** Anatoly Vartanov/Fotolia **p. 505** Miff32/Fotolia **p. 506** Pearson Education, Inc. **p. 511** (bl) ASL Creations/Fotolia; (tl) View Stock/Alamy **p. 515** Vojtech Vlk/Fotolia **p. 516** (waffle cone) Anatoly Vartanov/Fotolia; (tr) Rafael Ramirez/Fotolia **p. 517** Vacclav/Fotolia **p. 522** Jörg Hackemann/Fotolia **p. 524** Koufax73/Fotolia **p. 527** (tl) David Stock/Alamy

Chapter 12

CO (bl) NASA **p. 543** (bl) Fabfotos/Fotolia; (ml) Laurenthuet/Fotolia; (ml) Shutterstock; (tl) iStockphoto/Thinkstock **p. 563** (m) Josie/Fotolia **p. 569** (ml) Giemmephoto/Fotolia **p. 573** (tm) Vanich/Fotolia